Pro/ENGINEER 2000i²

Includes
Pro/NC and Pro/SHEETMETAL

Louis Gary Lamit
De Anza College

with technical assistance provided by
James Gee
De Anza College

BROOKS/COLE
THOMSON LEARNING™

Australia • Canada • Mexico • Singapore • Spain • United Kingdom • United States

Sponsoring Editor: *Bill Stenquist*
Marketing: *Chris Kelly*
Production Editor: *Janet Hill*
Cover Design: *Denise Davidson*
Print Buyer: *Kristine Waller*
Printing and Binding: *The Courier Company, Stoughton*
Cover Printing: *Phoenix Color Corporation*

For more information about this or any other Brooks/Cole product, contact:
BROOKS/COLE
511 Forest Lodge Road
Pacific Grove, CA 93950 USA
www.brookscole.com
1-800-423-0563 (Thomson Learning Academic Resource Center)

Parametric Technology Corporation
128 Technology Drive
Waltham, MA 02154
617-398-5000

Some material in this book was provided by CADTRAIN, Inc., and is based on the CBT product for Pro/ENGINEER®. We wish to thank CADTRAIN for the use of their product, COAch for Pro/ENGINEER®.
CADTRAIN, Inc.
2429 West Coast Highway
Newport Beach, CA 92663
800-631-5757

Printed in the United States of America

10 9 8 7 6 5 4 3 2 1

Library of Congress Cataloging-in-Publication Data
Lamit, Louis Gary
Pro/ENGINEER 2000i2 : includes Pro/NC and Pro/SHEETMETAL / Louis Gary Lamit ; with technical assistance provided by James McGee.
p. cm.
Includes index.
ISBN 0-534-37995-8
1. Pro/ENGINEER. 2. Computer-aided design. 3. Engineering design—Data processing. I. Title.
TA174 .L349 2001
620'. 0042'02855369—dc21 00-049841

About the Author

Louis Gary Lamit is the former head of the drafting department and CAD facility manager, and is currently an instructor, at De Anza College in Cupertino, California, where he teaches computer-aided drafting and design in the Manufacturing and Design Department..

Mr. Lamit has worked as a drafter, designer, numerical control (NC) programmer, technical illustrator, and engineer in the automotive, aircraft, and piping industries. A majority of his work experience is in the area of mechanical and piping design. Mr. Lamit started as a drafter in Detroit (as a job shopper) in the automobile industry, doing tooling, dies, jigs and fixture layout, and detailing at Koltanbar Engineering, Tool Engineering, Time Engineering, and Premier Engineering for Chrysler, Ford, AMC, and Fisher Body. Mr. Lamit has worked at Remington Arms and Pratt & Whitney Aircraft as a designer, and at Boeing Aircraft and Kollmorgan Optics as an NC programmer and aircraft engineer. Mr. Lamit also owns and operates his own consulting firm, and has been involved with advertising and patent illustrating.

Mr. Lamit received a BS degree from Western Michigan University in 1970 and did Masters' work at Wayne State University and Michigan State University. He has also done graduate work at the University of California at Berkeley and holds an NC programming certificate from Boeing Aircraft.

Since leaving industry, Mr. Lamit has taught at all levels (Melby Junior High School, Warren, Michigan; Carroll County Vocational Technical School, Carrollton, Georgia; Heald Engineering College, San Francisco, California; Cogswell Polytechnical College, San Francisco and Cupertino, California; Mission College, Santa Clara, California; Santa Rosa Junior College, Santa Rosa, California; Northern Kentucky University, Highland Heights, Kentucky; and De Anza College, Cupertino, California).

Mr. Lamit has written a number of textbooks, workbooks, tutorials, and articles, including:

Industrial Model Building, with Engineering Model Associates, Inc. (1981),
Piping Drafting and Design (1981), ***Piping Drafting and Design Workbook*** (1981),
Descriptive Geometry (1983), ***Descriptive Geometry Workbook*** (1983), and
Pipe Fitting and Piping Handbook (1984) were published by Prentice-Hall.

Drafting for Electronics, (3rd edition, 1998), and
CADD, (1987), were published by Charles Merrill (Macmillan-Prentice-Hall Publishing).

Technical Drawing and Design (1994), ***Technical Drawing and Design Worksheets and Problem Sheets*** (1994), ***Principles of Engineering Drawing***, (1994),
Fundamentals of Engineering Graphics and Design, (1997), and
Engineering Graphics and Design with Graphical Analysis (1997), ***Engineering Graphics and Design Worksheets and Problem Sheets*** (1997), were published by West Publishing (ITP/Delmar).

Basic Pro/ENGINEER in 20 Lessons (1998) (Pro/ENGINEER Rev. 18) and
Basic Pro/ENGINEER (with references to PT/Modeler) (1999) (Pro/ENGINEER Rev. 19 and PT/Modeler) were published by PWS Publishing (ITP).

Pro/ENGINEER 2000i (1999) (Pro/ENGINEER Rev. 2000i)
Pro/E 2000i^2 (includes Pro/NC and Pro/SHEETMETAL) (2000)
(Pro/ENGINEER Rev. 2000i^2) were published by Brooks/Cole Publishing (ITP).

Contents

Sections

Part Two Assemblies

Part Three Generating Drawings

Part Four Advanced Capabilities

Appendixes

Index

Preface

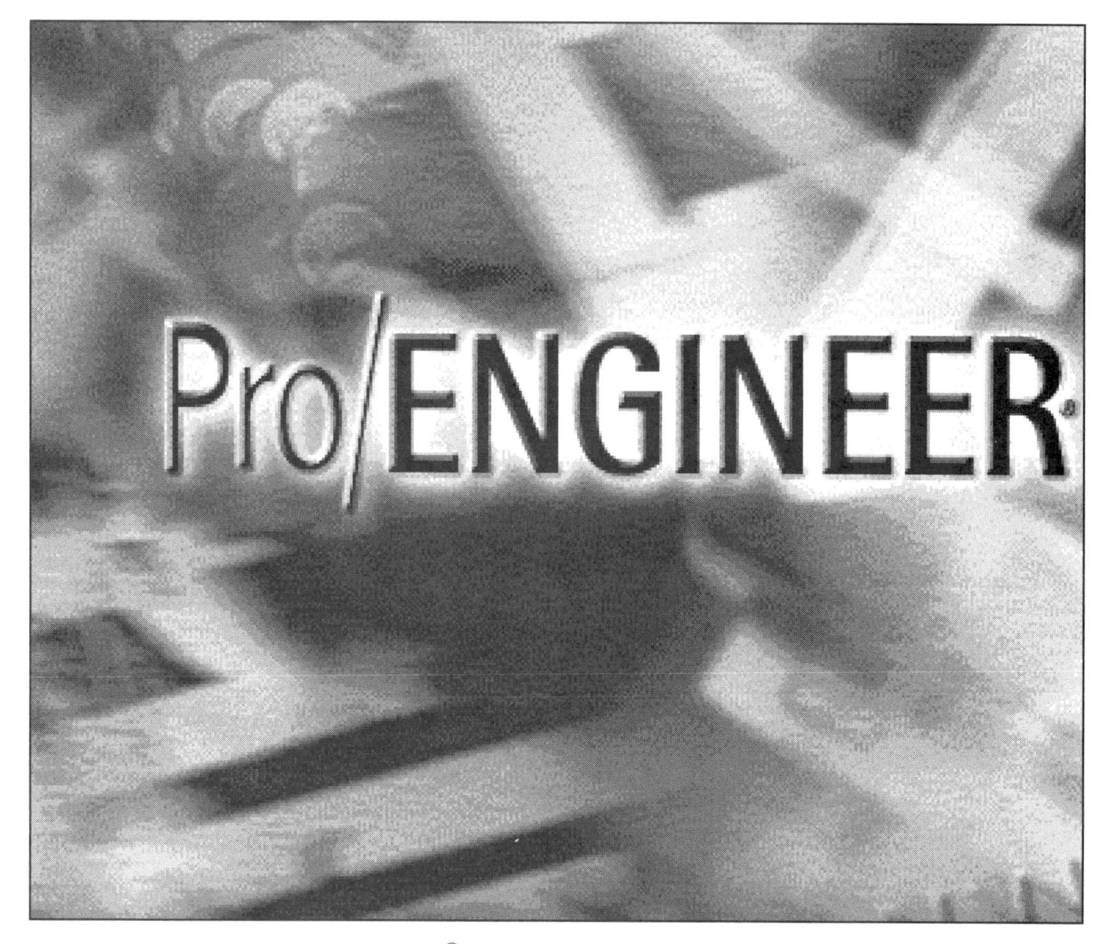

Pro/ENGINEER® is one of the most widely used CAD/CAM software programs in the world today. This book introduces you to the basics of the program and enables you to build on these basic commands to expand your knowledge beyond the scope of the book.

The book does not attempt to cover all of Pro/ENGINEER's features, but rather to provide an introduction to the software, make you reasonably proficient in its use, and establish a firm basis for exploring and growing with the program as you use it in your career or classroom.

The basic premise of this book is that the more parts, assemblies, and drawings you create using Pro/ENGINEER (Pro/E), the better you learn the software. With this in mind, each lesson introduces a new set of commands (with an incrementally harder set of parts), building on previous lessons. The parts created in the first thirteen lessons are used later in the text to create assemblies (Lessons 14-15), generate drawings (Lessons 16-20), and manufacture parts (Lesson 22). This procedure allows you to work with actual completed parts, assemblies, and drawings in a short-lesson format, instead of building large complex projects where basic commands may be overshadowed and lost in a complicated process.

The book is divided into one set of twelve *sections* and four *parts*: **Part One--Creating Parts**, **Part Two--Assemblies**, **Part Three--Generating Drawings,** and **Part Four--Advanced Capabilities**. Part Four contains three lessons. In this version, we have added a lesson covering **Pro/SHEETMETAL** (Lesson 23). Each **part** has a variety of individual **lessons.**

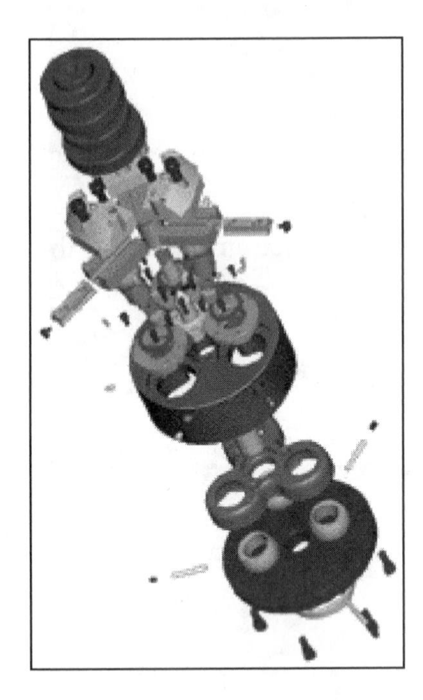

Every lesson introduces a new set of commands and concepts that are applied to a *part*, an *assembly*, or a *drawing*, depending on where in the book you are working.

Lessons involve creating a new part, an assembly, or a drawing, using a set of Pro/E commands that walk you through the process step by step. Each lesson starts with a list of objectives and ends with a **lesson project**. The lesson project consists of a part, assembly, or drawing that incorporates the lesson's new material and uses and expands on previously introduced material from other lessons. ***Projects require use of planning sheets from Appendix D as tools for establishing the design intent.***

The **Appendices** contain advanced projects, reference materials, a glossary, and design intent planning sheets.

An online demo tutorial from **CADTRAIN**, called **COAch for Pro/ENGINEER®**, has been referenced throughout the book. It is one of the best ways available for teaching and learning CAD/CAM software. **COAch** is used within lessons and as a reference. A sampler CD is included in the back of the text. This CD contains a variety of modules from various products offered in ***CADTRAIN's COAch for Pro/ENGINEER®***. You will need Pro/ENGINEER software (or the Student Edition) installed to run the tutorial and step-by-step functions of the CD. You can view the tutorial and step-by-step instructions without performing the procedures if you have **Netscape®**.

This book was written so that students and professionals will have an up-to-date Pro/E software manual. The book serves well as a home study guide for those wishing to expand their knowledge of Pro/E, as a training guide, or as a reference for the Pro/E user.

NOTE:

A CD with all book Pro/E files is available from the author for instructors who adopt the text:

lglamit@yahoo.com

or

www.netcom.com/~llamit/

If you wish to contact me concerning questions, changes, additions, suggestions, comments, and so on, please send email to one of the following:

lglamit@yahoo.com	Lamit and Associates
www.netcom.com/~llamit/	Web Site

Dedication

I dedicate this book to Dude.

Om Mani Padme Hum

Acknowledgments

The following people provided valuable assistance in preparing this version of the manuscript:

- **Ken Louie**
- **Thuy Dao Lamit**
- **Gary Mahany**
- **Victor Santini**
- **Max Jeffery Gilleland**
- **Christian Herrera**
- **Jovito Robledo**

I would like to thank the following for valuable comments during the review of the manuscript:

- **Jason Perry**, Design Engineer, PPC
- **Robert Conroy**, California Polytechnic State University, San Luis Obispo
- **Dale Carlile**, Rogue Community College

I also want to thank the following people and organizations for the support and materials they granted the author:

Parametric Technology Corporation for their support and timely software updates.

- **Larry Fire** *Parametric Technology Corporation*
- **Nick Maly** *Parametric Technology Corporation*

CADTRAIN, makers of COAch for Pro/ENGINEER®, for their Pro/ENGINEER tutorial software.

- **Dennis Stajic** CADTRAIN
- **Kathy Bennett** CADTRAIN
- **Ron Gates** CADTRAIN
- **Rick Navarre** CADTRAIN

Lastly, I would like to thank my editor Bill Stenquist.

Resources

A variety of information, books, online products, and job opportunities (**www.pejn.com**) are available. We have listed some of the more useful and important resources. You can also do a search on the Web with PTC, Proe, Pro/E, Pro/ENGINEER, and so on as keywords. Resources include Website addresses for a variety of companies, a list of books on Pro/ENGINEER, links to **InPart Design**, an Internet-based company that supplies 3D models of standard parts over the web, and **CADTRAIN**, the creator of **COAch for Pro/ENGINEER**.

Parametric Technology Corporation

Pro/ENGINEER 2000i^2 delivers the most comprehensive and tightly integrated applications available today to leverage the product model throughout the enterprise. Pro/ENGINEER 2000i^2 features more than 440 enhancements focused at helping customers' product development processes be more responsive, innovative, and connected to their customers and strategic suppliers.

Pro/ENGINEER 2000i^2 is a member of the PTC i-Series of flexible engineering solutions, which span the product development enterprise process, including conceptual design, styling, design and engineering, simulation, planning, manufacturing, and in-service support.

See
www.ptc.com
www.ptc.com/cs/index.htm

Access the PTC Website at **www.ptc.com**.

An email address and a valid software serial number are required for you to register. After registering, you will have your own login and password to enter the customer service site for technical support at **www.ptc.com/cs/index.htm**. Explore the different sections, including the Knowledge Base.

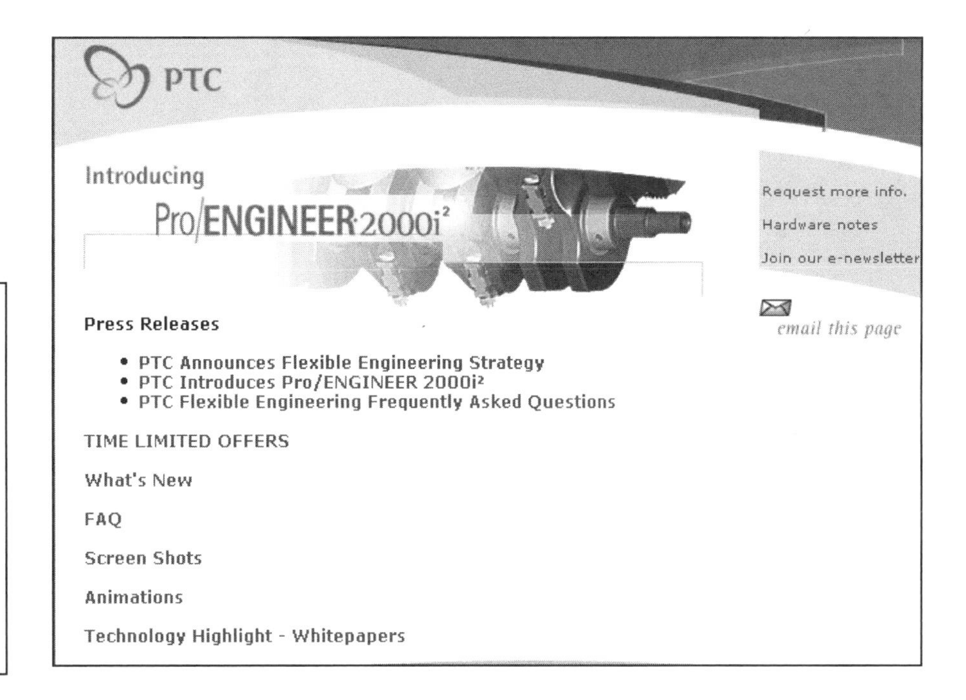

NOTE:

A CD with all book Pro/E files is available from the author for instructors who adopt the text:

lglamit@yahoo.com

or

www.netcom.com/~llamit/

InPart DESIGN

InPart, formerly called DesignSuite, represents the PTC strategy and technology for delivery of supplier components and information online, through your Web connection, for direct use in designs. InPart is an Internet-based portal of over 650,000 3D mechanical CAD models and over 13 million technical specifications. Customers subscribe on an annual basis to gain access to the data centers through an Internet connection. InPart delivers tools to quickly search and locate components that meet your design requirements. Models may be downloaded and assembled directly in your Pro/ENGINEER, CADDS 5, or Pro/DESKTOP designs. The InPart data center, created and maintained in partnership with leading component manufacturers, is expanding rapidly and continually updated with the most current supplier information. Leveraging the power of the Web and dynamic push technology, subscribers automatically receive updated part category indexes and related data upon each login at InPart.

See **www.ptc.com/products/inpart** or **www.inpart.com**

The company was founded in 1996 and launched its first product in February 1998. That product, DesignSuite (now called InPart), allows designers to search intuitively for and specify standard mechanical components from multiple manufacturers, including Aeroquip, Boston Gear, Parker Hannifin, The Torrington Company, Thomson Industries, and many more. A broad range of product categories are represented in the library, such as gears, actuators, automation equipment, hoses, clamps, tooling fixtures, motors, controls, connectors, pumps, fittings, and bearings.

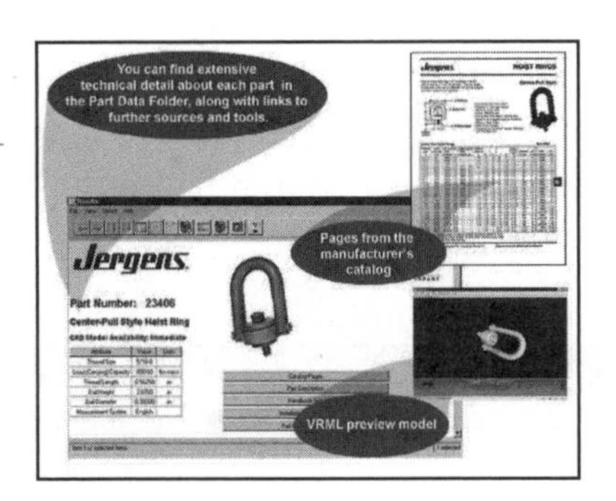

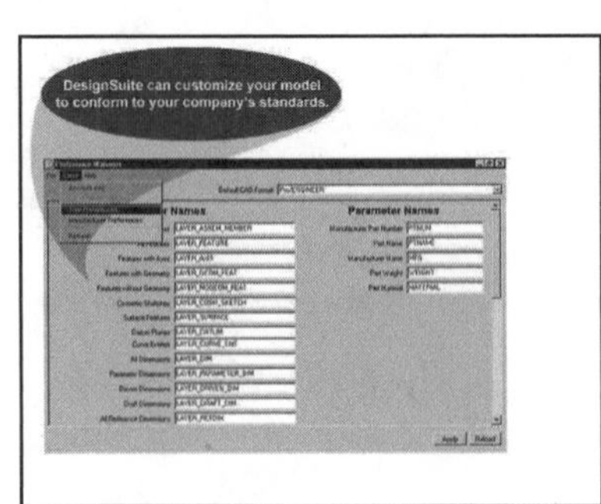

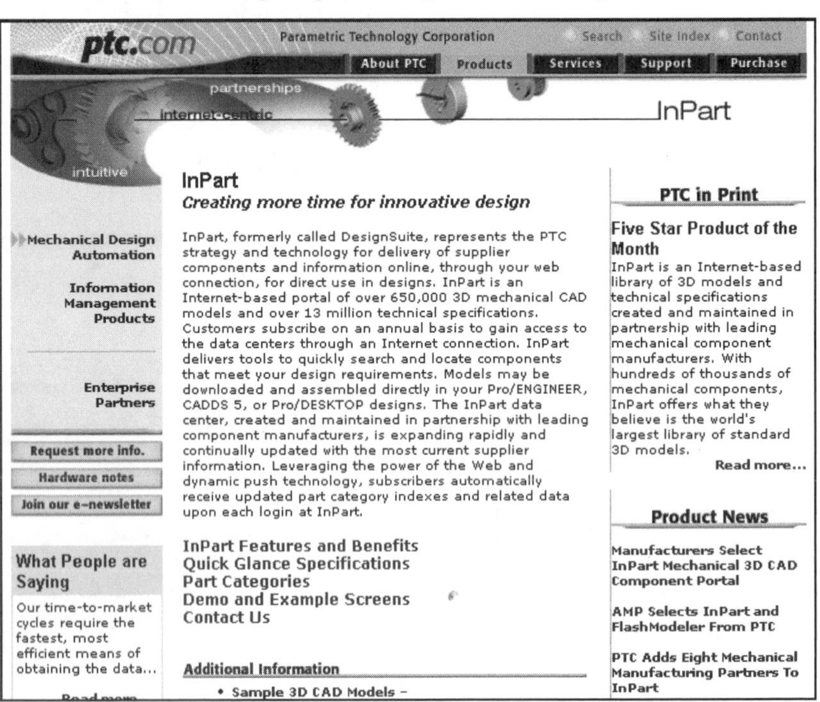

InPart's powerful search engine can locate and retrieve the right standard part for your Pro/E design. Efficient search algorithms give InPart an unprecedented high-performance query capability--results are returned to you within seconds.

See **www.pejn.com**

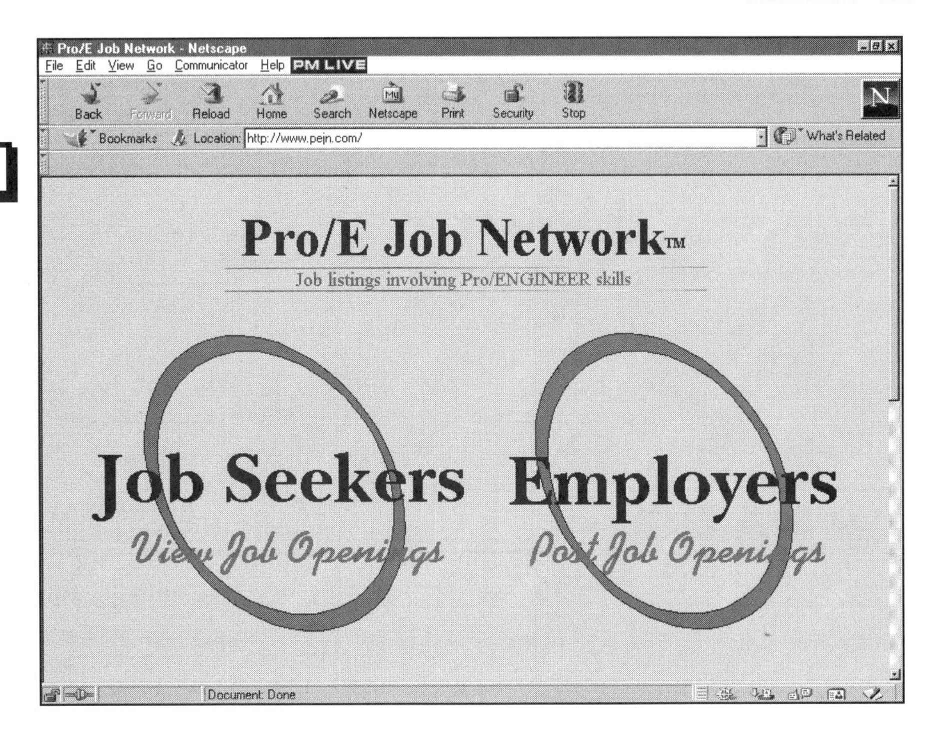

Pro/ENGINEER Job Network

A variety of services are available over the Internet:

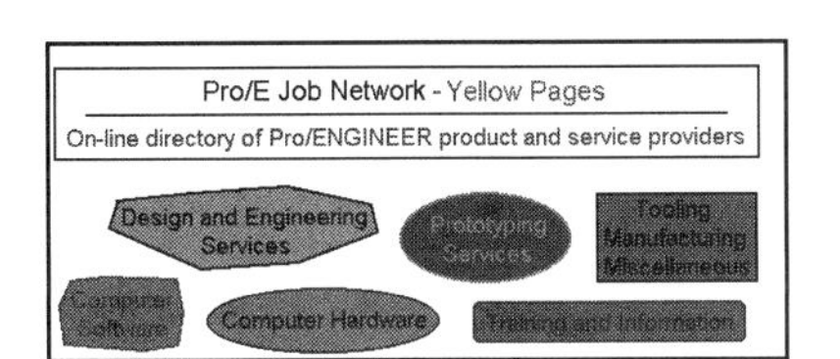

Employers: Announce your job openings on our Web site. Job seekers will fax, mail, or email their resumes directly to you. If you request it, your company name, address, and phone number will not be listed.

Job Seekers: Look over the job listings on our Web site, free of charge. Fax, mail, or email your resume directly to those employers whose jobs interest you.

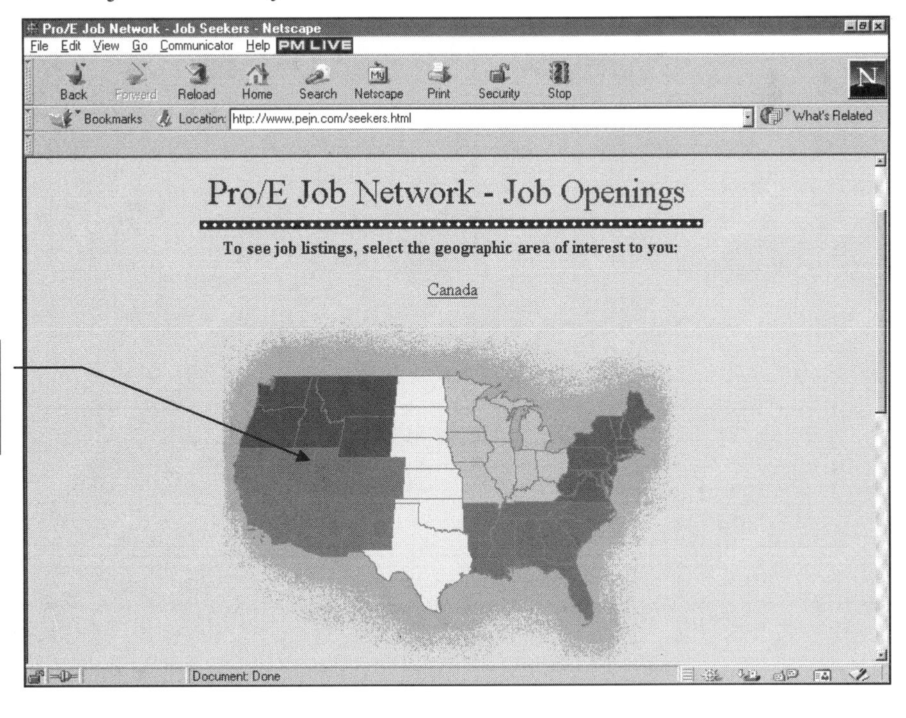

To see a list of jobs, pick on a section of the country where you are interested in working

See **www.cadtrain.com**

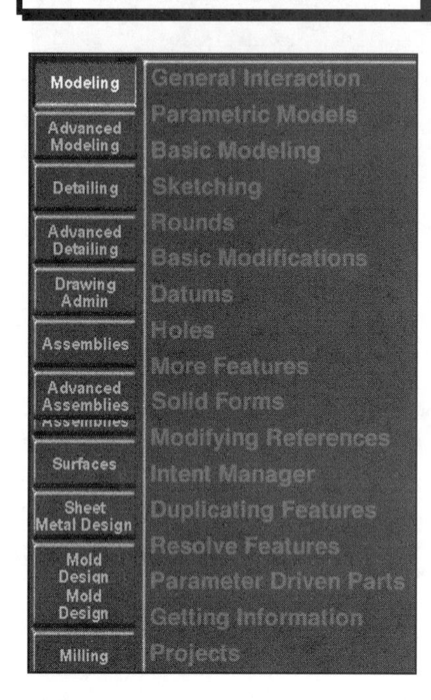

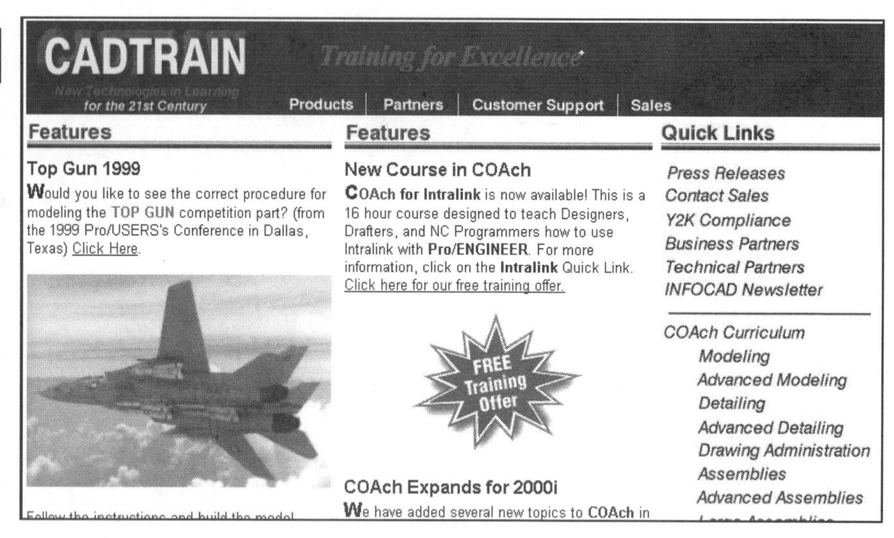

CADTRAIN's COAch for Pro/ENGINEER

COAch is a computer-based training (CBT) product designed to provide a comprehensive and affordable training program for Pro/ENGINEER users in their actual CAD environment. This self-paced onscreen training tool enables engineers, designers, drafters, and NC programmers to customize their training experience by following the learning sequence best suited to their individual needs. Working in an interactive environment, students are introduced to key concepts, are guided through demonstration exercises, answer test questions, and complete comprehensive modeling projects.

NOTE

The CD provided at the back of this text contains a sampler of CADTRAIN's COAch for Pro/ENGINEER and includes a variety of modules.

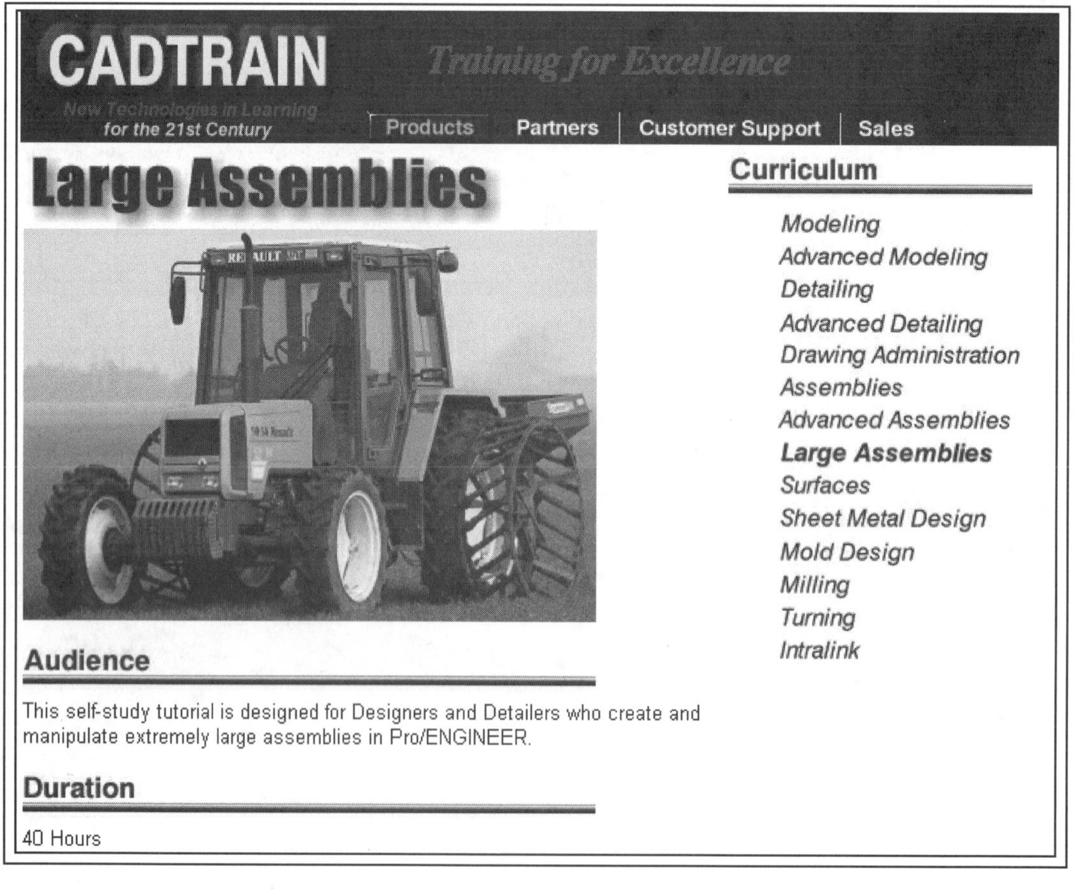

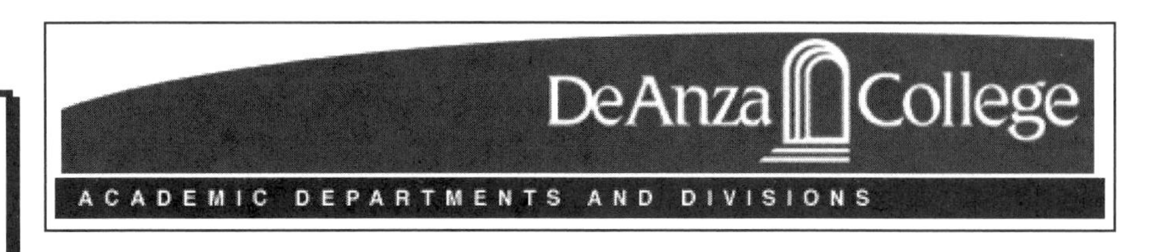

www.deanza.fhda.edu

The MFGD department Web page is listed under CAD/CAM as well as Manufacturing/Design

www.netcom.com/~llamit/
Website for Louis Gary Lamit and Lamit and Associates

De Anza College

The Manufacturing and Design Program at De Anza College offers a wide variety of degrees and certificates. In each degree or certificate option, students are provided with a broad yet in-depth curriculum that imparts a strong foundation for direct employment in local industries or transfer to a four-year college. Diversification is the hallmark of the program. The program's curriculum content is directed by four Industry Advisory Committees and teaches the same state-of-the-art theories, procedures, and computer programs utilized in Silicon Valley. It is the largest two-year program of its kind in the West.

De Anza College has a complete CAD drafting and design program that includes classes in Manufacturing, AutoCAD, SoildWorks, and **Pro/ENGINEER**.

Most of the Pro/ENGINEER classes can be taken off campus using the Internet, mail, and phone to interface with the instructor; email **lglamit@yahoo.com.**

You will be required to register at De Anza College for the class. De Anza is a California Community College. State, out-of state, and international fees are different. Credits are on quarter system units. Each class is 4 units.

You must have access to **Pro/ENGINEER** or **Pro/ENGINEER the student collection** (and a copy of this textbook).

* **MFGD 52A Pro/ENGINEER** part design **(Lessons 1-13)**
(can be taken off campus)

* **MFGD 52B Pro/ASSEMBLY and Pro/DETAIL**
assemblies and detail drawings **(Lessons 14-21)**
(can be taken off campus)

* **MFGD 52C Pro/MANUFACTURING** **(Lesson 22)**
(can be taken off campus with Pro/ENGINEER-
not the student collection)

* **MFGD 52D Pro/SURFACE product design**
(***cannot*** *be taken off campus*)

* **MFGD 51B Pro/SHEETMETAL** **(Lesson 23)**
(can be taken off campus)

Lesson Sequence

This text is written so that the student and instructor can work through the **lessons** in order; Lessons 1 through 23. But it is also possible to use different sequences depending on the class content and length. The text has been written for a two-term program using Lessons 1 through 21. Lesson 22 is for a class in Pro/NC, and Lesson 23 is for a class in Pro/SHEETMETAL.

*The **sections** at the front of the text are for study, not for completing commands.* They provide essential information for the mastery of Pro/ENGINEER and are to be read prior to completing the lessons.

An alternative approach to completing the work would be to choose corresponding lessons that cover modeling and drawing of the same part or assembly. An example of this approach would have you choosing *Lesson 1, Lessons 2 & 16, Lessons 3 & 17, Lessons 4 & 18, Lessons 14 & 19, and Lessons 15 & 20.* This approach would create parts used for detailing in later lessons and an assembly used for an assembly drawing. Lesson Projects in Lessons 5 through 8 would also be required in order to complete the assembly. Below is a listing that will give some guidance as to the possible sequences of sections and lessons:

Normal Sequence	Optional Sequence	Reading
Lesson 1		**Section 1-12** (overview)
Lesson 2	(then ***Lesson 16***)	**Section 12** (reread)
Lesson 3	(then ***Lesson 17***)	**Section 11** (reread)
Lesson 4	(then ***Lesson 18***)	**Section 7** (reread)
Lesson 5		**Section 5 & 8** (reread)
Lesson 6		**Section 9** (reread)
Lesson 7		**Section 10** (reread)
Lesson 8		**Section 6** (reread)
Lesson 9		
Lesson 10		
Lesson 11		
Lesson 12		
Lesson 13		
Lesson 14	(then ***Lesson 19***)	**Section 4** (reread)
Lesson 15		
Lesson 16		
Lesson 17		
Lesson 18		
Lesson 19		
Lesson 20		
Lesson 21		
Lesson 22		
Lesson 23		

One-Term or One-Semester Course

Lessons shown in *italics* constitute a one-term or one-semester course completed in order:

Lessons 1-8, 14, & 16-19

If you have sufficient time, the Pro/NC lesson (***Lesson 22***) uses the part in Lesson 8 and can also be completed during a one-semester term if manufacturing is desired in the course content.

Alternative (One-Term or One-Semester Course)
Lesson Course Sequence

Lessons shown in *italics* constitute a one-term or one-semester course (alternative sequence):

Lesson 1
Lesson 2
Lesson 16
Lesson 3
Lesson 17
Lesson 4
Lesson 18
Lesson 5
Lesson 6
Lesson 7
Lesson 8
Lesson 14
Lesson 19

Sections

Part Design

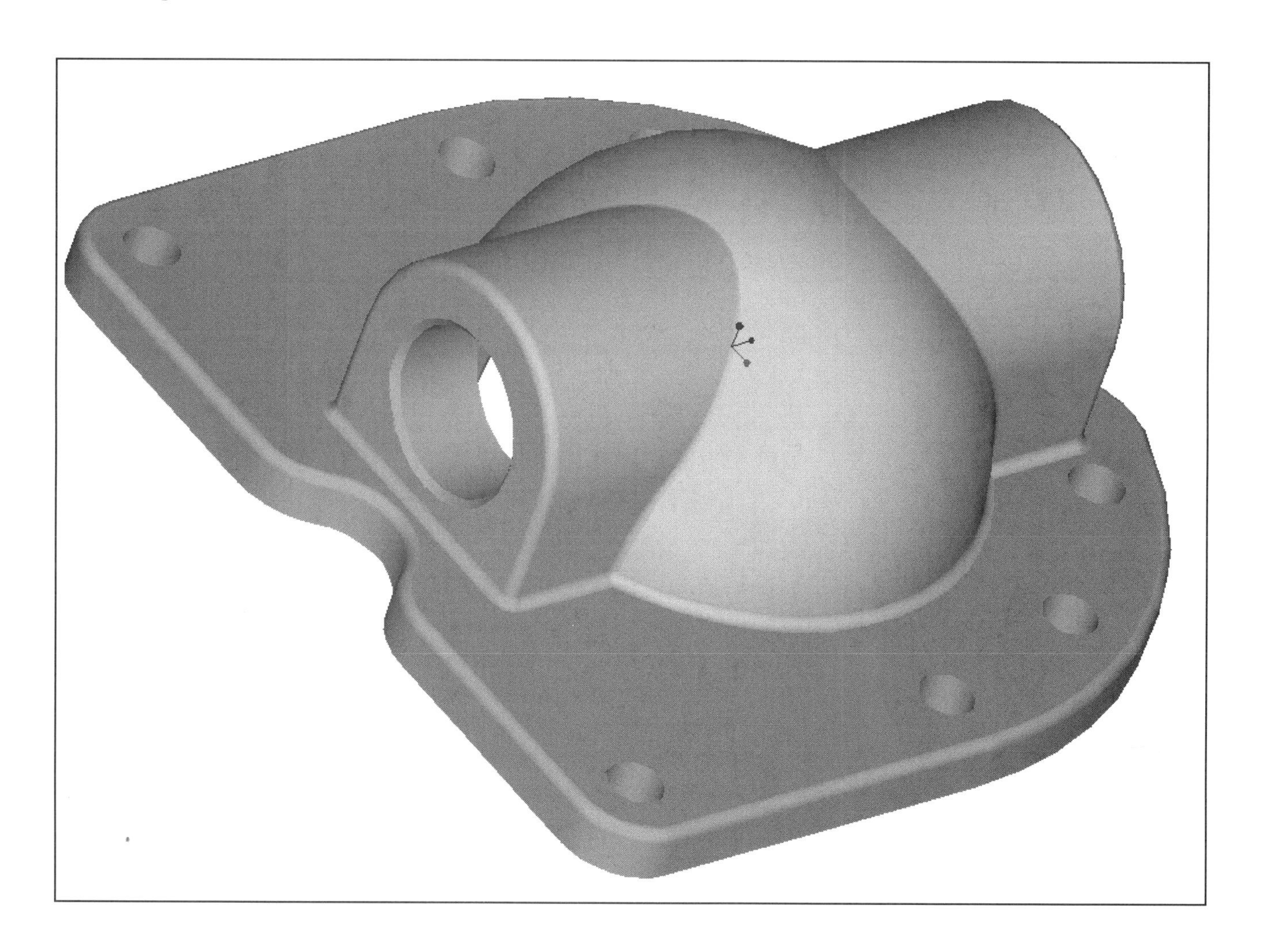

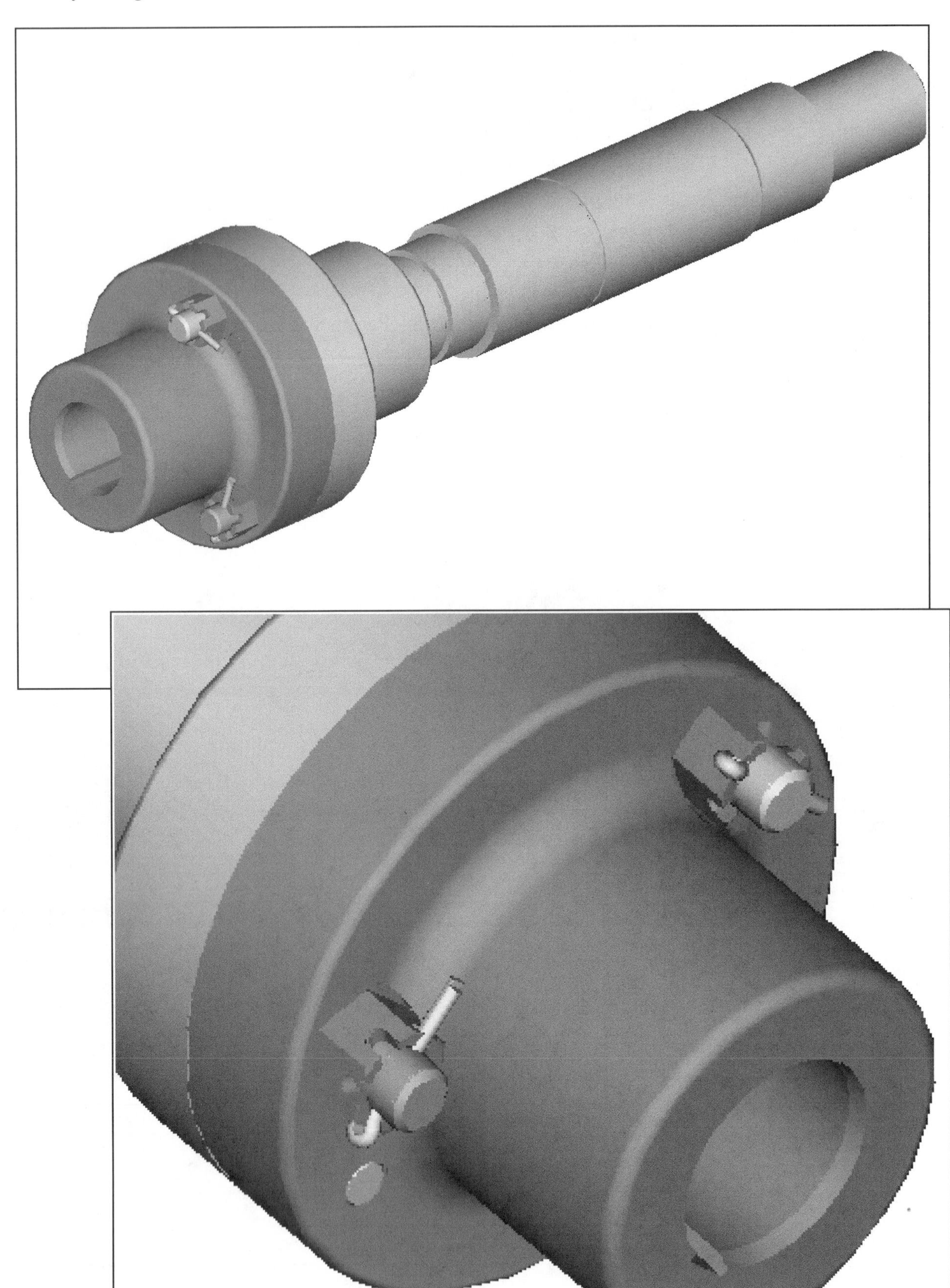

Section 1

Introduction

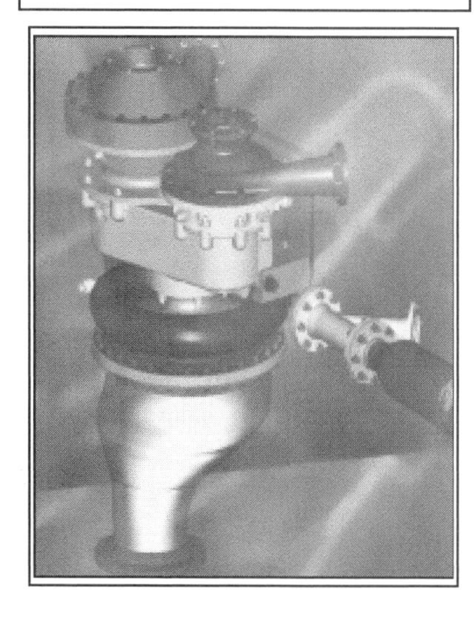

Figure 1.1
Part Design

This work text introduces the basic concepts of parametric design using Pro/ENGINEER (Pro/E) to create and document individual parts, assemblies, and drawings. **Parametric** can be defined as *any set of physical properties whose values determine the characteristics or behavior of an object*. **Parametric design** enables you to generate a variety of information about your design: its mass properties, a drawing, or a base model. To get this information, you must first model your part design (Fig. 1.1).

This section is intended to introduce you to parametric modeling philosophies used in Pro/E, including:

Feature-Based Modeling Parametric design represents solid models as combinations of engineering features (Fig. 1.2).

Creation of Assemblies Just as features are combined into parts, parts may be combined into assemblies, as shown in Figure 1.3.

Capturing Design Intent The ability to incorporate engineering knowledge successfully into the solid model is an essential aspect of parametric modeling (Fig. 1.4).

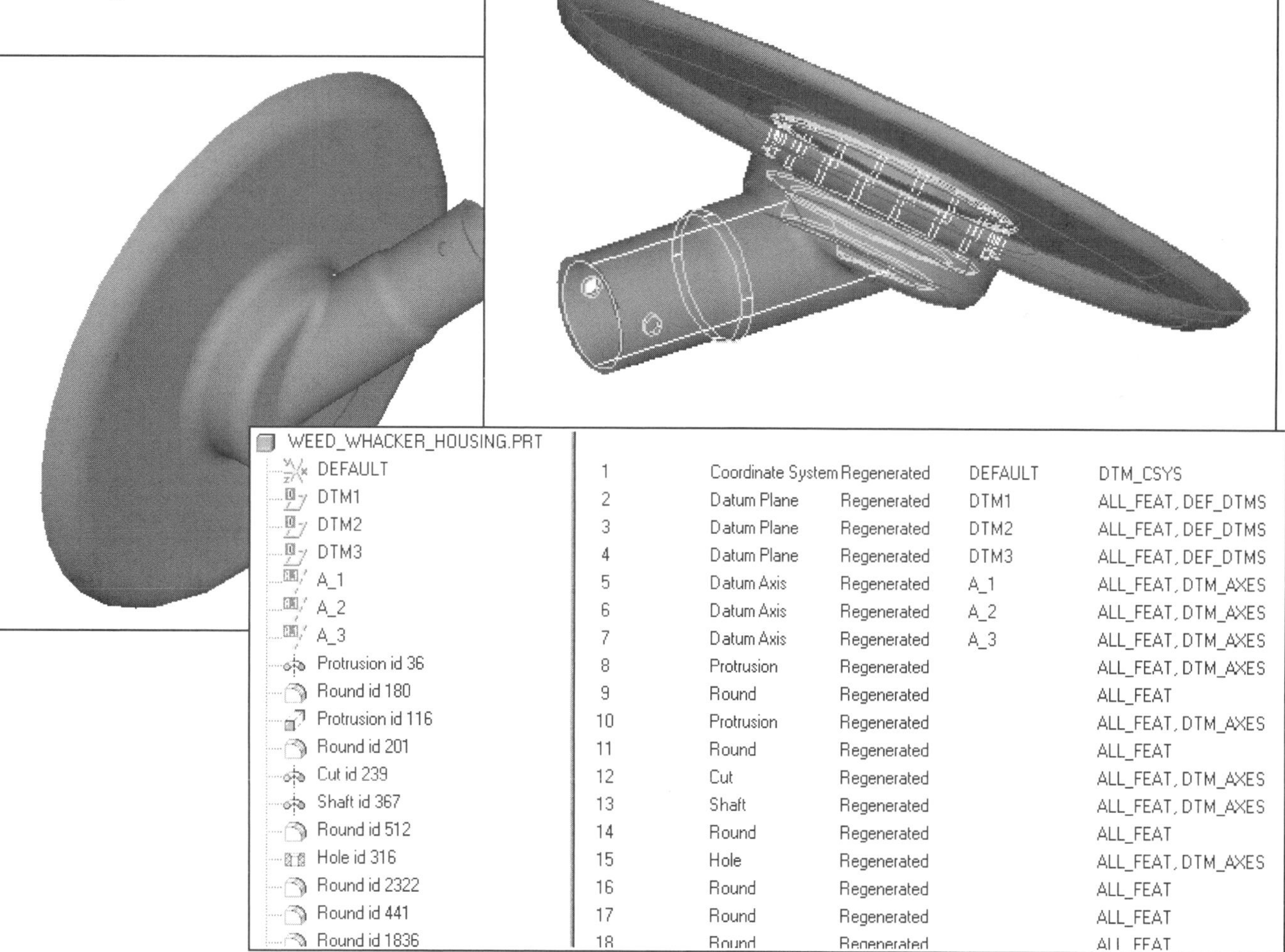

WEED_WHACKER_HOUSING.PRT					
DEFAULT	1	Coordinate System	Regenerated	DEFAULT	DTM_CSYS
DTM1	2	Datum Plane	Regenerated	DTM1	ALL_FEAT, DEF_DTMS
DTM2	3	Datum Plane	Regenerated	DTM2	ALL_FEAT, DEF_DTMS
DTM3	4	Datum Plane	Regenerated	DTM3	ALL_FEAT, DEF_DTMS
A_1	5	Datum Axis	Regenerated	A_1	ALL_FEAT, DTM_AXES
A_2	6	Datum Axis	Regenerated	A_2	ALL_FEAT, DTM_AXES
A_3	7	Datum Axis	Regenerated	A_3	ALL_FEAT, DTM_AXES
Protrusion id 36	8	Protrusion	Regenerated		ALL_FEAT, DTM_AXES
Round id 180	9	Round	Regenerated		ALL_FEAT
Protrusion id 116	10	Protrusion	Regenerated		ALL_FEAT, DTM_AXES
Round id 201	11	Round	Regenerated		ALL_FEAT
Cut id 239	12	Cut	Regenerated		ALL_FEAT, DTM_AXES
Shaft id 367	13	Shaft	Regenerated		ALL_FEAT, DTM_AXES
Round id 512	14	Round	Regenerated		ALL_FEAT
Hole id 316	15	Hole	Regenerated		ALL_FEAT, DTM_AXES
Round id 2322	16	Round	Regenerated		ALL_FEAT
Round id 441	17	Round	Regenerated		ALL_FEAT
Round id 1836	18	Round	Regenerated		ALL_FEAT

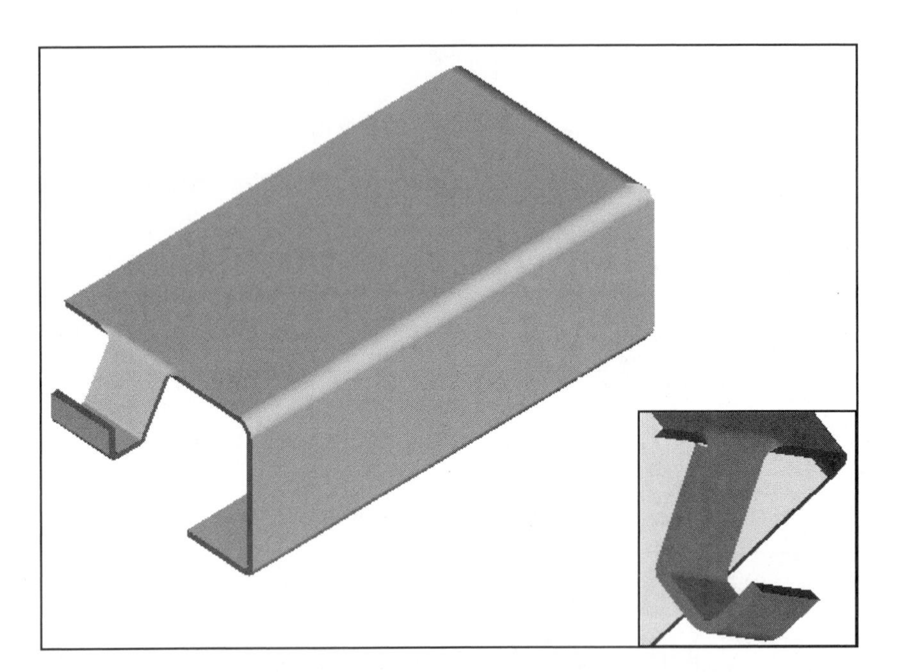

Figure 1.2
Feature-Based Modeling

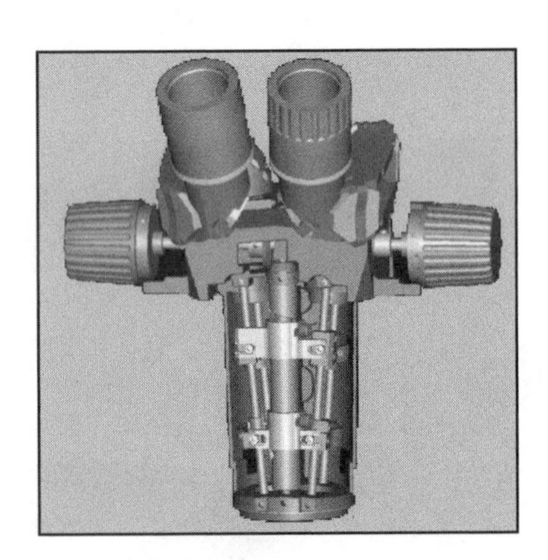

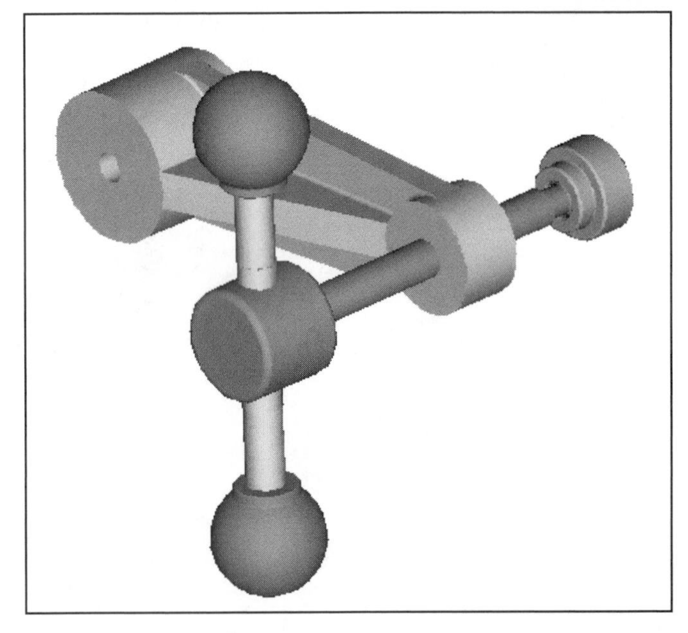

Figure 1.3
Creation of Assemblies

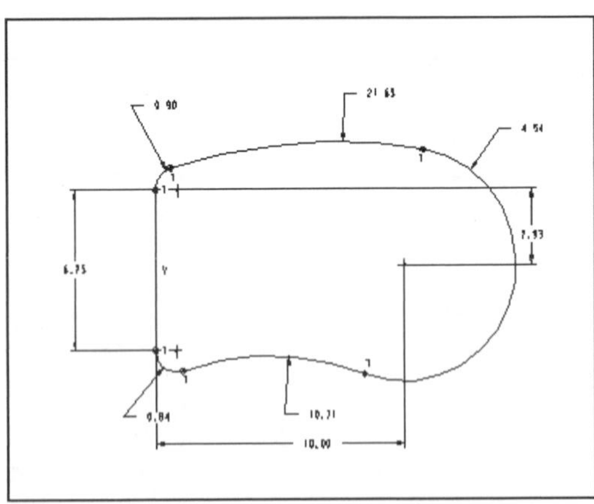

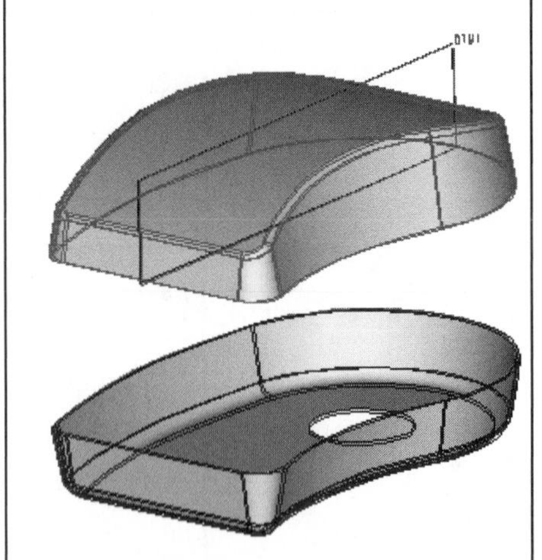

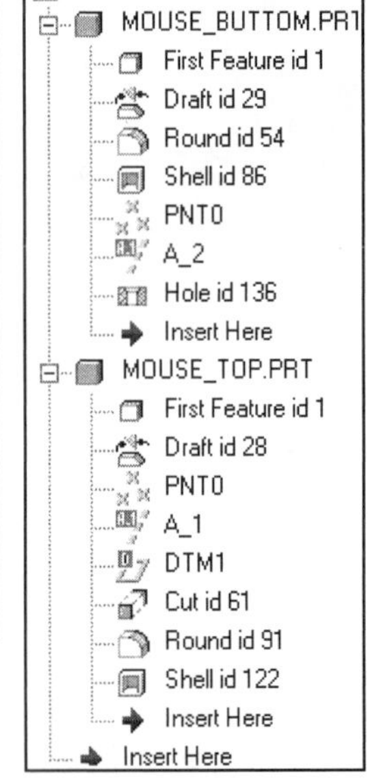

Figure 1.4
Capturing Design Intent

These methodologies are the principal aspects of successful parametric (solid) modeling. Figure 1.5 illustrates the role of each in the modeling process.

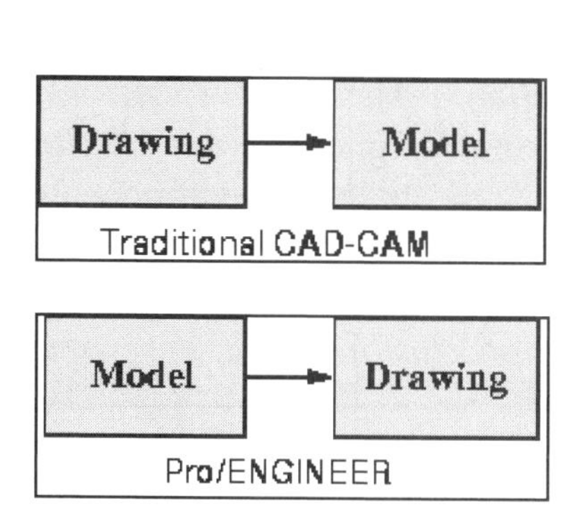

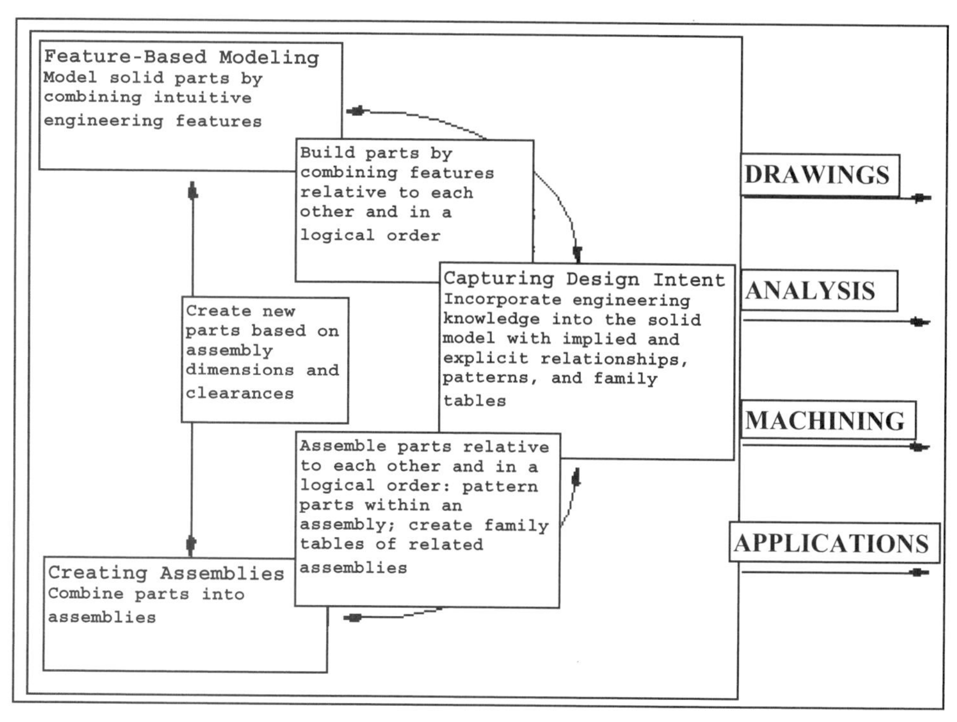

Figure 1.5
Parametric
Design Methodology

Modeling vs. Drafting

A primary and essential difference between parametric design and traditional computer-aided drafting packages is that parametric design models are three-dimensional. Increasingly, designs are represented in the form of solid models that capture design intent as well as design geometry. Engineering designs today are constructed as mathematical solid models instead of as 2D drawings. A solid model is one that represents a shape as a 3D object that has volume and mass properties.

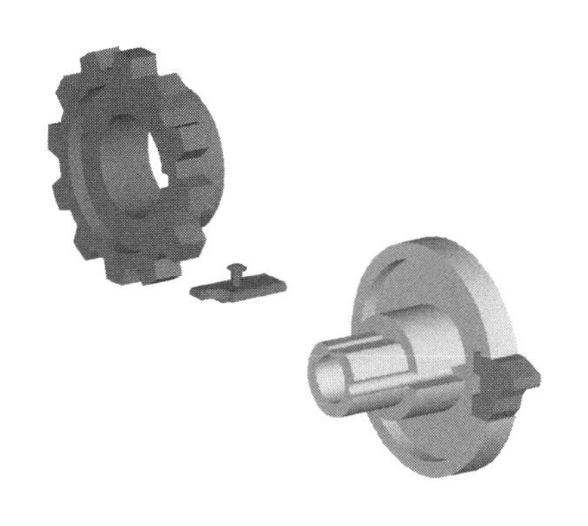

There are two main reasons for the move to solid models. First, solid modeling software packages can serve as an easy means of portraying parts for study by cross-functional concurrent-engineering teams. The solid model can be understood even by nontechnical members of the team, such as those from the marketing and sales departments.

Second, the capabilities of solid modelers have been upgraded so that the model can represent not only the geometry of the part being designed, but also the intent of the designer. This facility is most significant when the designer needs to make changes to the part geometry.

The designer must make far fewer changes in later-generation parametric solid models that capture design intent than in previous CAD/CAM modeling software. In parametric design, drawings are produced as views of the 3D model, rather than the other way around.

Parametric design models are not drawn so much as they are *sculpted* from solid volumes of materials.

Parametric Design Overview

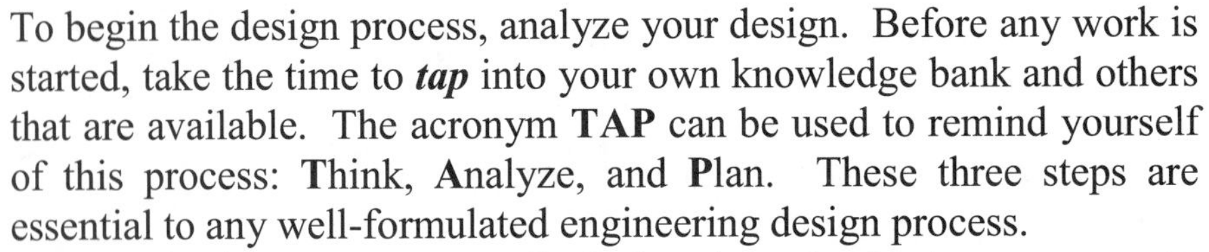

To begin the design process, analyze your design. Before any work is started, take the time to ***tap*** into your own knowledge bank and others that are available. The acronym **TAP** can be used to remind yourself of this process: **T**hink, **A**nalyze, and **P**lan. These three steps are essential to any well-formulated engineering design process.

Break down your overall design into its basic components, building blocks, or primary features. Identify the most fundamental feature of the part to sketch as the first, or base, feature. A variety of **base features** can be modeled using *protrusion-extrude, revolve, sweep,* and *blend* commands.

Sketched features (*cut, protrusion,* and *rib*) and pick-and-place features called **referenced features** (*holes, rounds,* and *chamfers*) are normally required to complete the design. With the **SKETCHER**, you use familiar 2D entities (points, lines, rectangles, circles, arcs, splines, and conics). There is no need to be concerned with the accuracy of the sketch. Lines can be at differing angles, arcs and circles can have unequal radii, and features can be sketched with no regard for the actual parts' dimensions. In fact, exaggerating the difference between entities that are similar but not exactly the same is actually a far better practice when using the SKETCHER.

The software enables you to apply logical geometric constraints to the sketch. **Constraints** clean up the sketch geometry according to the software assumptions. **Geometry assumptions** and constraints close ends of connected lines, align parallel lines, and snap sketched lines to horizontal and vertical orthogonal orientations. Additional constraints are added by means of **parametric dimensions** to control the size and shape of the sketch.

Feature-Based Modeling

Features are the basic building blocks you use to create a part. Features "understand" their fit and function as though "smarts" were built into the features themselves. For example, a hole or cut feature "knows" its shape and location and the fact that it has a negative volume. As you modify a feature, the entire part automatically updates after regeneration. The idea behind feature-based modeling is that the designer constructs a part so that it is composed of individual features that describe the way the geometry is supposed to behave if its dimensions change. This happens quite often in industry, as in the case of a design change.

Parametric modeling is the term used to describe the capturing of design operations as they take place, as well as future modifications and editing of the design (Fig. 1.6). The order of the design operations is significant. Suppose a designer specifies that two surfaces are parallel such that surface 2 is parallel to surface 1. Therefore, if surface 1 moves, surface 2 moves along with it to maintain the specified design relationship. Surface 2 is a **child** of surface 1 in this example. Parametric modelers allow the designer to **reorder** the steps in the part's creation.

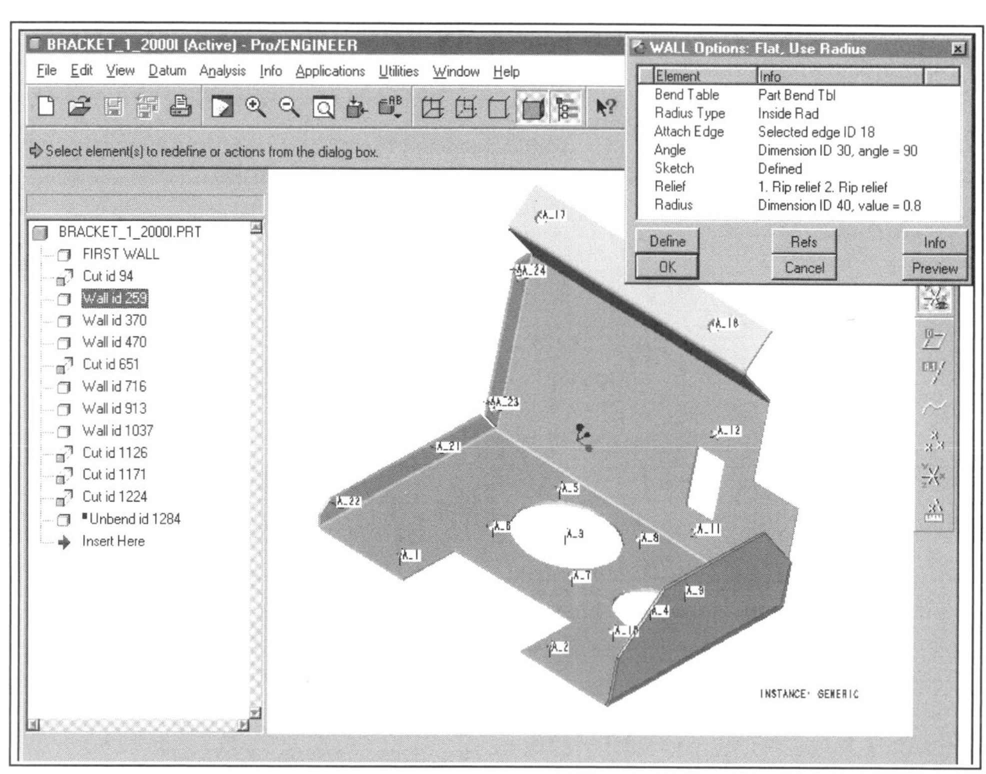

Figure 1.6
Editing a Design

The "chunks" of solid material from which parametric design models are constructed are called **features**. Features generally fall into one of the following categories:

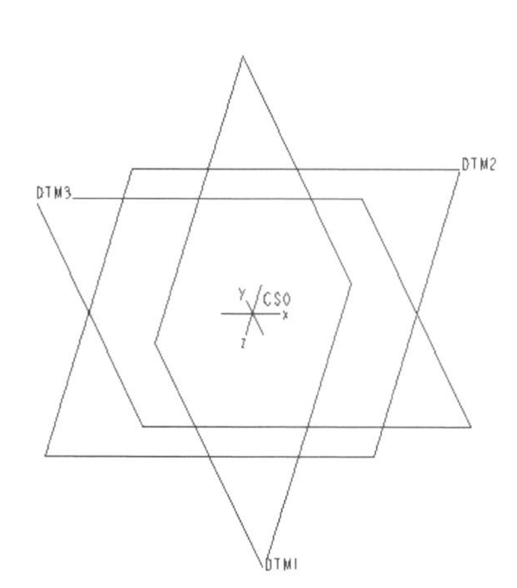

Base Feature The base feature may be either a sketched feature or datum plane(s) referencing the default coordinate system. The base feature is important because all future model geometry will reference this feature directly or indirectly; it becomes the root feature. Changes to the base feature will affect the geometry of the entire model (Fig. 1.7).

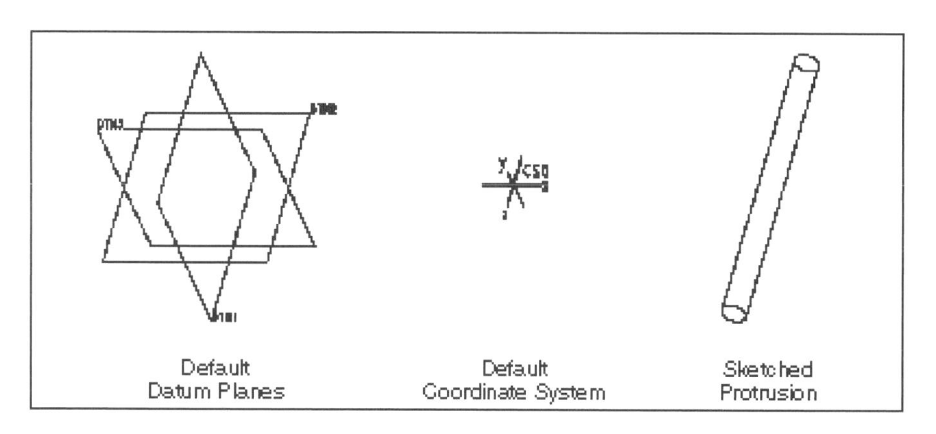

Figure 1.7
Base Features

Sketched Features Sketched features are created by extruding, revolving, blending, or sweeping a sketched cross section. Material may be added or removed by protruding or cutting the feature from the existing model (Fig. 1.8).

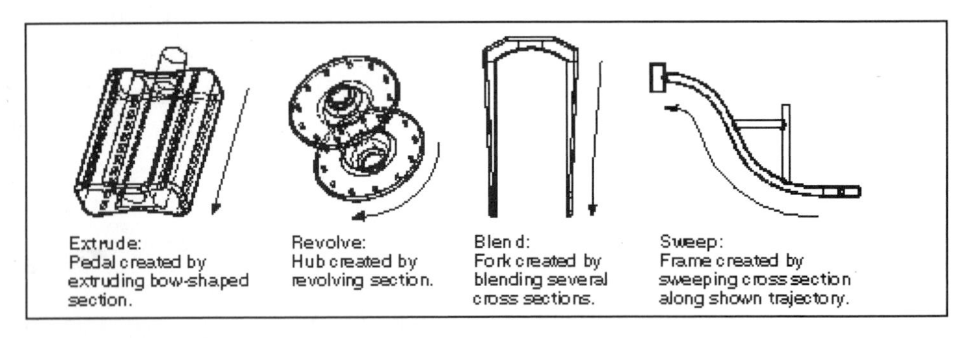

Figure 1.8
Sketched Features

Referenced Features Referenced features reference existing geometry and employ an inherent form; they do not need to be sketched. Some examples of referenced features are rounds, drilled holes, and shells (Fig. 1.9).

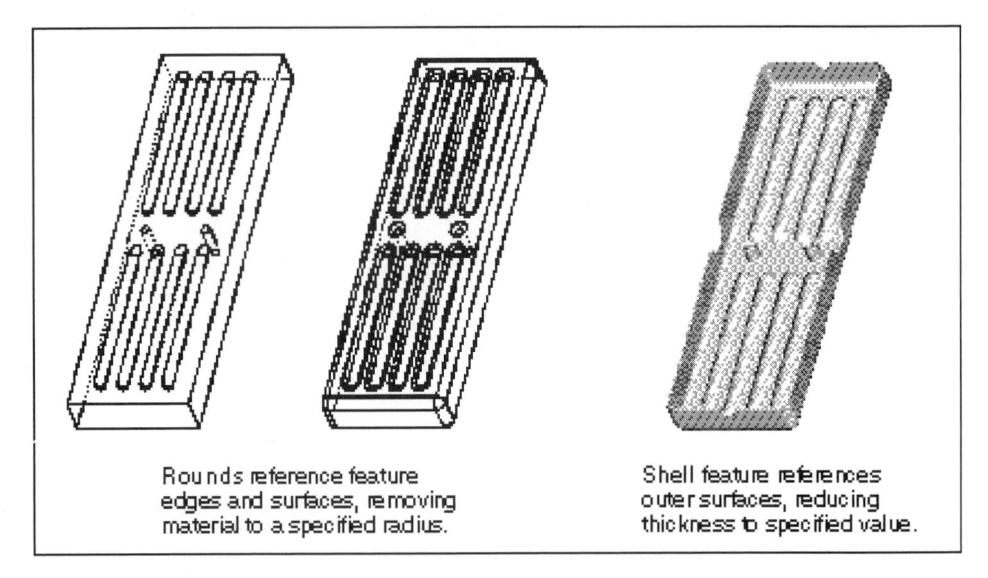

Figure 1.9
Referenced Features

Datum Features Datum features, such as planes, axes, curves, and points, are generally used to provide sketching planes and contour references for sketched and referenced features. Datum features do not have physical volume or mass and may be visually hidden without affecting solid geometry (Fig. 1.10).

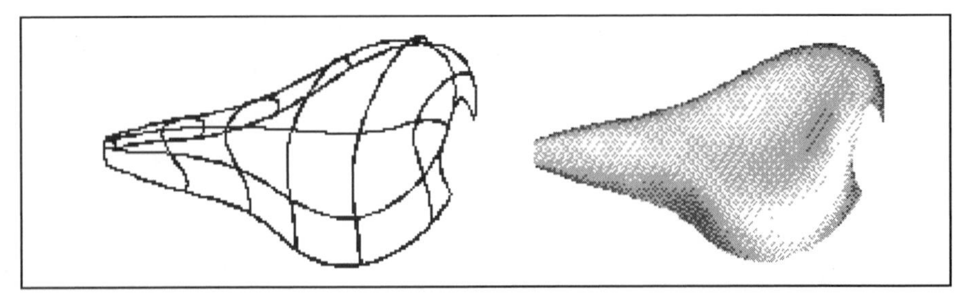

Figure 1.10
Datum Curves

The various types of features are used as building blocks in the progressive creation of solid parts. Figure 1.11 shows base features, datum features, sketched features, and referenced features.

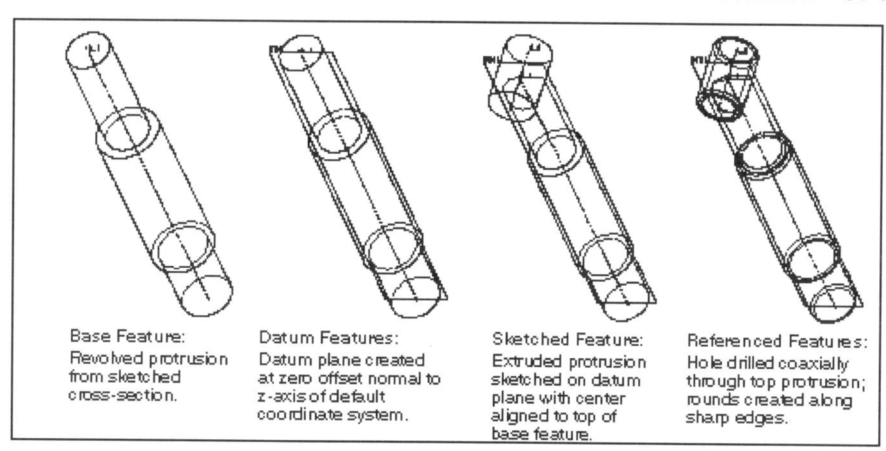

Figure 1.11
Feature Types

Conclusion

As you progress through this text's lessons, you will sometimes be prompted to return to a referenced section to review it. At this point in the process of learning Pro/E, you cannot expect to understand and apply every concept involved in this very complex software. The sections in the text are really references that are used throughout each lesson's step-by-step construction and each lesson's project. You will become more familiar with the concepts and capabilities of Pro/E as you complete each project. Reading the reference sections will help you improve this understanding, but nothing takes the place of actual practice. In reality, the lessons are presented in a building-block format that increases your knowledge incrementally and without unnecessary and complicated discussions. Return to the various sections to review material that you encounter throughout the lessons.

Besides consulting the sections, you should get comfortable with the online documentation. **Pro/HELP** can be accessed at any time during a working Pro/E session (Fig. 1.12, Fig. 1.13, and Fig. 1.14).

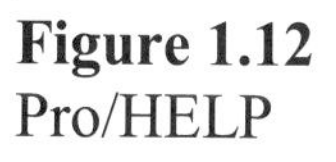

Figure 1.12
Pro/HELP

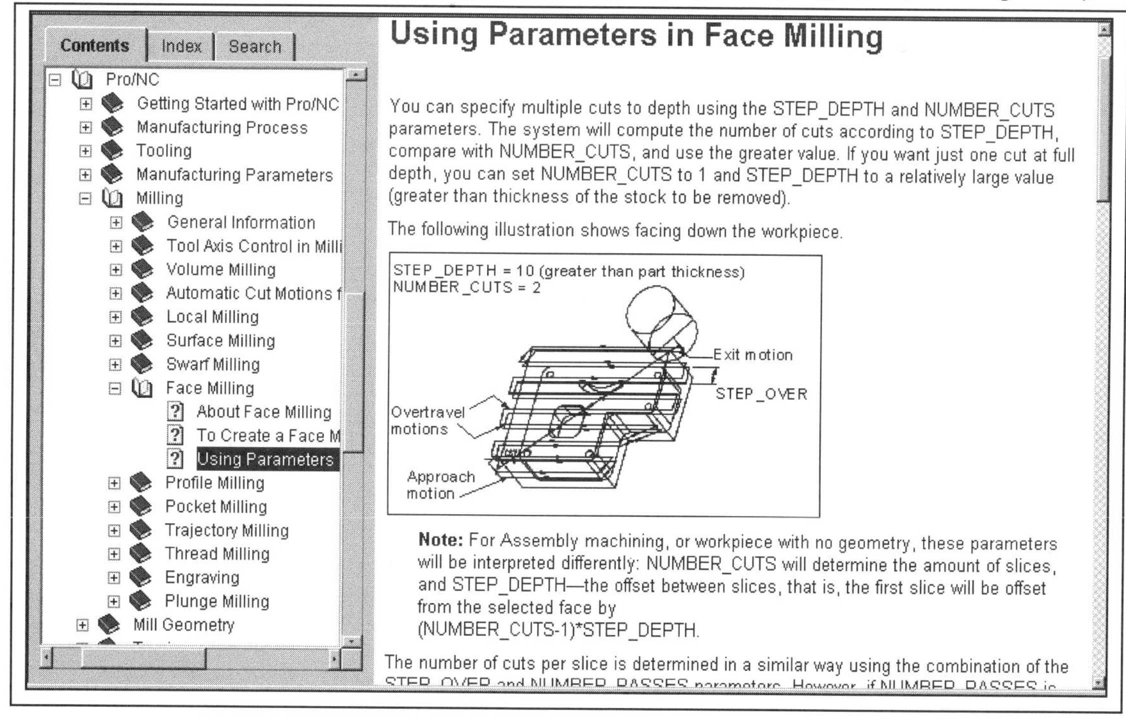

Figure 1.13
Pro/ENGINEER
Help System

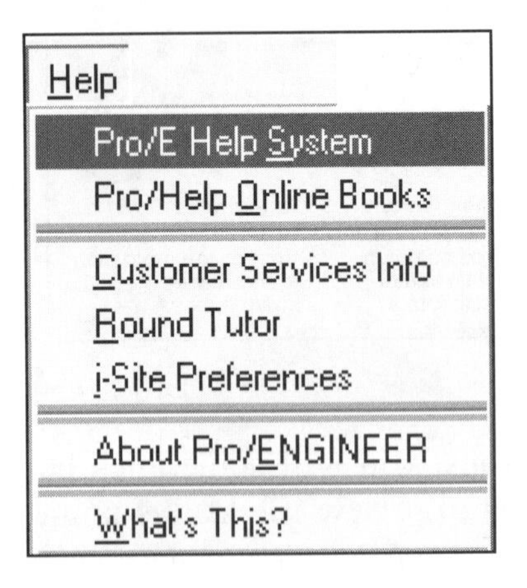

Welcome to PTC Help

Official PTC Website

PTC is continuing its commitment to provide integral internet/intranet enhancements through the i-series of software products, including the innovative changes in this release of PTC Help.

PTC Help is integrated with fully functional, highly flexible CAD/CAM and Data Management software available from PTC. It is easy to locate a desired topic. PTC Help can be installed on a Web server, allowing Web clients to access PTC Help without having direct mounts to file server machines, thus reducing network traffic and enabling a LAN/WAN configuration.

What's Available in PTC Help?

PTC Help offers:

⇒ A new help system with integrated table of contents, index, and search capability

⇒ Full certification of Internet Explorer 4.0 and Netscape 4.06

⇒ Web server installation capabilities

⇒ Full context-sensitive help, allowing access to PTC Help with a click of the mouse

⇒ Expanded context-sensitive help in dialog boxes

⇒ See Also hotspots for related topics

Where Are Particular Modules in PTC Help?

There are four main branches in the PTC Help table of contents: **Welcome, Pro/ENGINEER Foundation, Using Foundation Modules,** and **Using Additional Modules**. Consult the following list to find a particular module in the table of contents.

- Associative Topology Bus-CADDS 5 Using Foundation Modules

Figure 1.14
Online Documentation:
The Resolve Environment

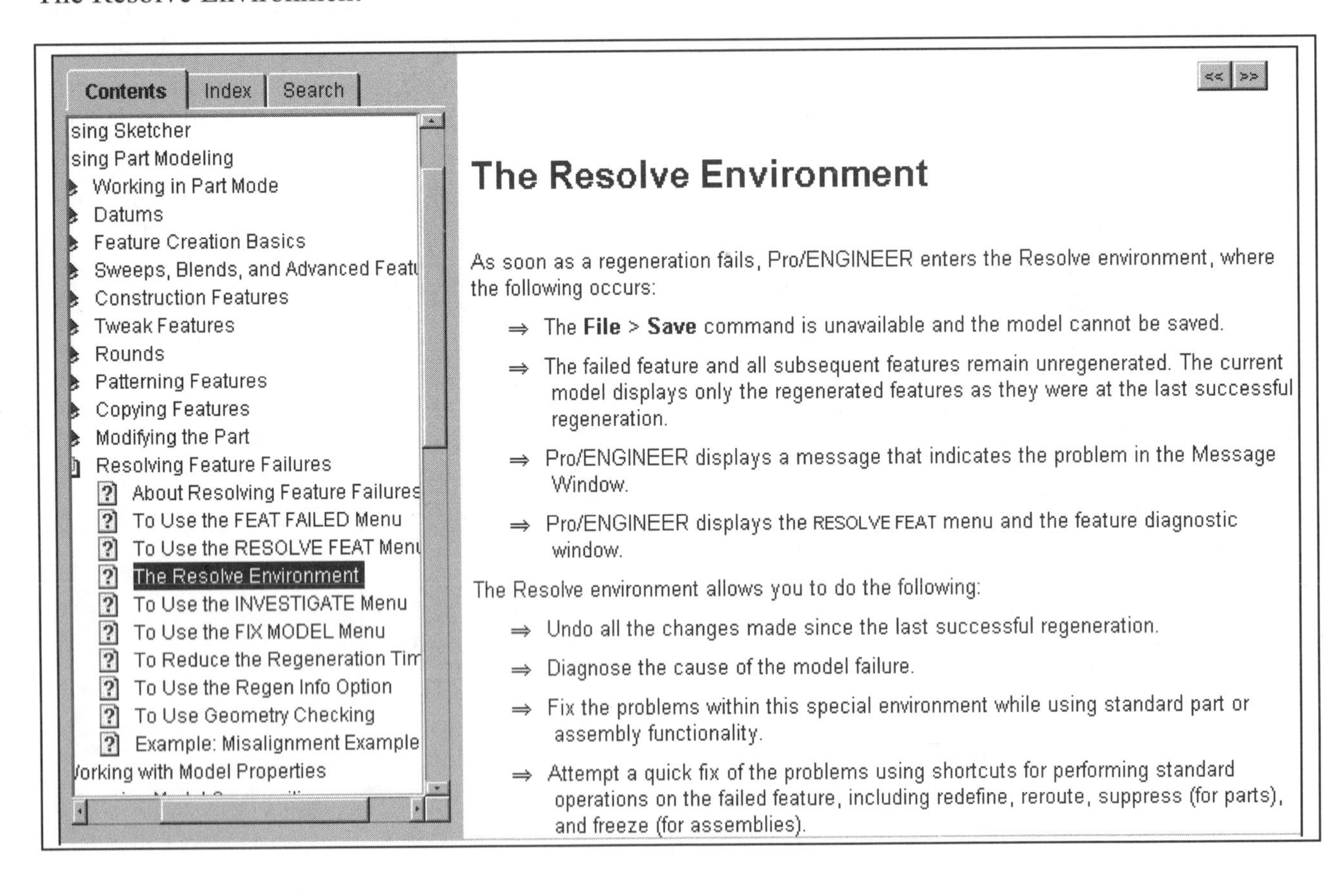

Section 2

Using the Text

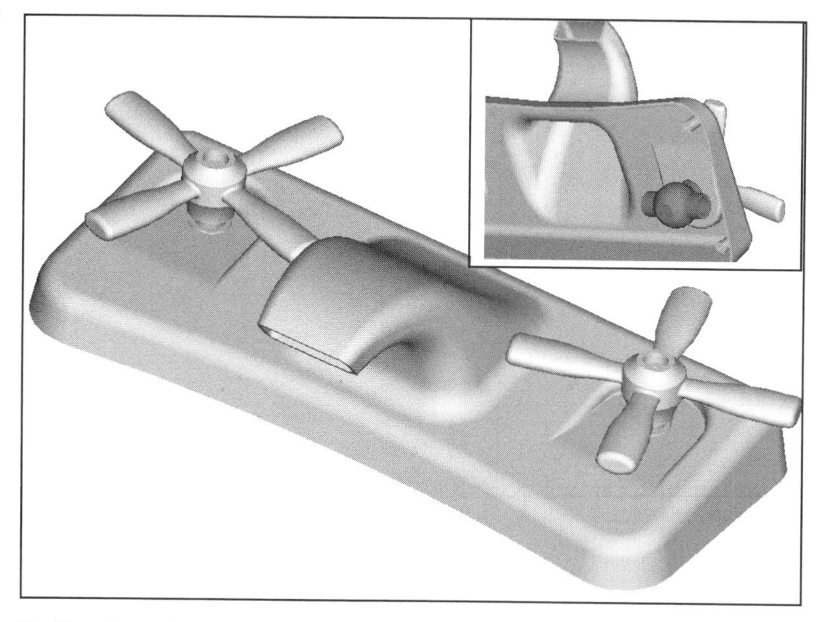

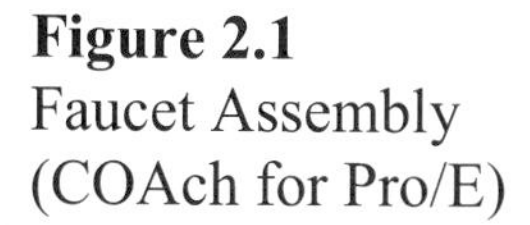

Figure 2.1
Faucet Assembly
(COAch for Pro/E)

Following the **sections** portion of the book introducing Pro/E, the main body of the text is divided into four parts: **Part One--Creating Parts**, **Part Two--Assemblies**, **Part Three--Generating Drawings**, and **Part Four--Advanced Capabilities.** Each **part** has a variety of individual **Lessons.** Every lesson introduces a new set of commands and concepts that are applied to a *part*, an *assembly*, or a *drawing*, depending on where in the text you are working.

Lessons involve creating a new part, an assembly, or a drawing using a set of commands that walk you through the process step by step. Each lesson starts with a set of objectives and ends with a new project that requires you to apply the material you have just learned in that lesson (also building on previous lessons).

Each lesson ends with a **Lesson Project**. The lesson project consists of a part, assembly (Fig. 2.1), or drawing that incorporates the lesson's new material and expands on and uses material introduced in earlier lessons. Student projects use planning sheets from Appendix D as tools for establishing the design intent.

Page S2-3 shows the typical lesson page layout. The left-hand column holds symbols that prompt you, at appropriate places, with a variety of messages, including when to save, ask for help, set up, configure, implement engineering change orders, incorporate hints, and apply notes. Sample menus will also be shown in this column.

Design Intent Planning Sheets

The *design intent* of a feature, a part, or an assembly is thought out and established *before* you rush into modeling. Appendix D provides a variety of sketching formats (Figs. 2.2 and 2.3) for planning a design. These sheets are referred to in the text as **DIPS** (**D**esign **I**ntent **P**lanning **S**heets). Design intent is included throughout this textbook.

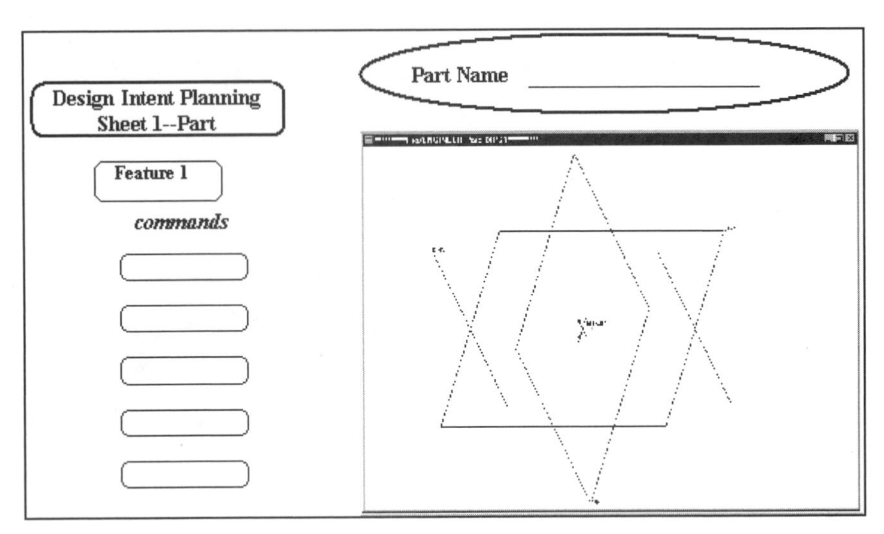

Figure 2.2
Design **I**ntent **P**lanning
Sheet 1--Part

We cannot emphasize enough the importance of manually sketching your project before starting the actual modeling. Planning your parent-child relationships, the features required for the part, etc. will in the end cut the actual time required to complete a project by 30%. We guarantee this! Most of the time you normally waste redoing, rerouting, modifying, and redefining your part will be reduced or eliminated!

Make sure you use these sheets or engineering grid paper for every lesson project (and if you're smart, every lesson).

The front inside cover has a check-off list that includes the category DIPS. In my class, I always say NO DIPS NO GRADE.

A few minutes of planning and sketching on paper will save countless hours of redoing your design on the computer system. Skipping this step in the design of a feature, part, or assembly is a recipe for disaster. In industry, there are thousands of stories of how a designer or engineer created a graphically correct part or assembly that looked "visually" correct. Upon closer examination, the model or assembly had too many or too few datum planes and a variety of parent-child relationships that were nothing but examples of the designer's incorrect use of Pro/E.

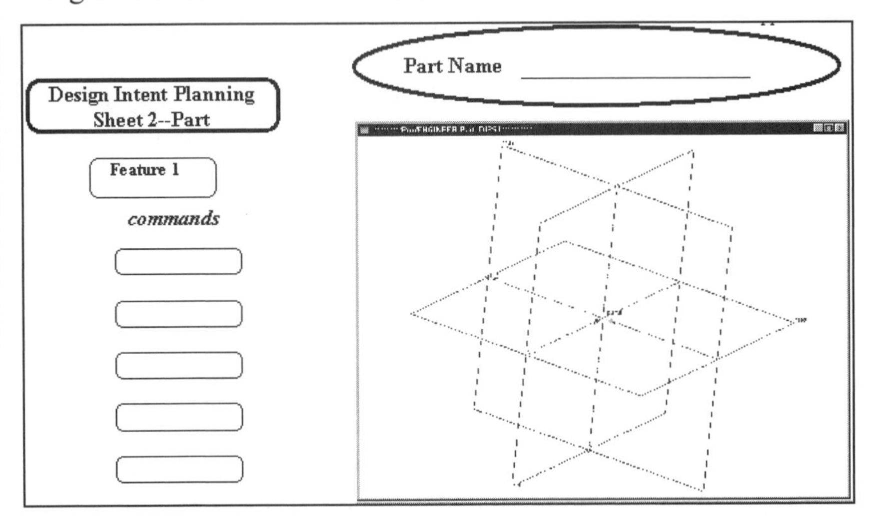

Figure 2.3
Design **I**ntent **P**lanning
Sheet 2--Part

A variety of other problems will occur downstream in the design and manufacturing process, including possible feature failures that result when minor engineering change orders are introduced. Without proper process planning, organization, and a well-defined design intent, the model is useless. In most cases, it would take more time to reorder, modify, redefine, and reroute the model to correct poor design habits than it would to remodel the part.

To get you used to the fact that changes are part of the design process, **engineering change orders** (ECOs) are introduced in lesson step-by-step assignments and in lesson projects. Figure 2.4 shows a sample page from a lesson project. See the ✔ ***EGD REFERENCE*** reference and the ✔ ***NOTE*** reference in the left-hand column of the page.

Symbols	Meanings
 **File ⇒ Save ⇒ ✔**	***File ⇒ Save ⇒ ✔*** The **File** symbol will remind you to save your object or design work as you complete a set of tasks or commands. **Save** your file every few minutes.
? Pro/HELP	***Help*** Move the cursor and **highlight** the command that you need online help with. Then press the right button on your mouse.
Feature ⇒ Create ⇒ Protrusion ⇒ Extrude ⇒ Solid	***Commands*** Commands are boldface and usually, *but not always*, enclosed in a box. The ⇒ symbol means to initiate the next command pick. When the command line has ⇒ **enter,** press the keyboard **enter** key or pick **✔.**
CONFIG.PRO ***def_layer_datum***	***Config.pro*** Configuration file settings are given to change default settings and help you become comfortable with customizing Pro/ENGINEER.
E C O	***ECO*** **Engineering Change Orders** are introduced at various times throughout the lessons. Changes may entail adding *parametric relations*, using *insert mode*, *redefining* the feature, *modifying* dimension values, *rerouting*, or *reordering* features.
COAch™ for Pro/ENGINEER	***COAch*** Refers the user to specific CADTRAIN tutorials.
✔ ***NOTE***	***Notes*** Notes are given to explain an aspect of design intent.
✔ ***EGD REFERENCE*** **Fundamentals of Engineering Graphics and Design** by L. Lamit and K. Kitto Read Chapter See page	***EGD References*** EGD references refer to chapter and page numbers in *Fundamentals of Engineering Graphics and Design,* by Lamit and Kitto (West/Delmar Publishing Co., 1997), that can be referenced for more information on a lesson part, project, or concept.
HINT **DATUM PLANES** will be the first features on all parts and assemblies.	***Hints*** Hints about commands or procedures are provided to assist you in completing each lesson.
 ***TOOLCHEST*** **Command ⇒ Command ⇒**	***Toolchest/Toolbar*** Create a mapkey for a newly introduced and a potential frequently used command. Add the mapkey to the Toolchest/Toolbar by using: **Utilities ⇒ Customize Screen ⇒ Commands ⇒ Mapkeys**

Figure 2.4 Sample Page

Lesson 14 Project

Coupling Assembly

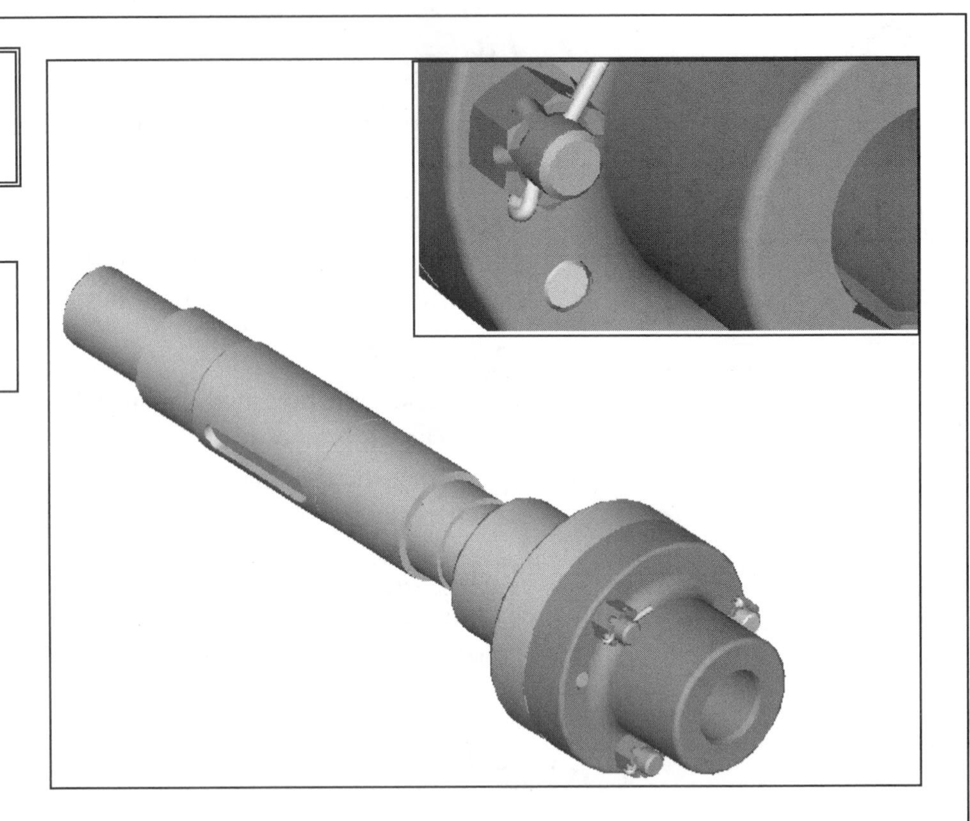

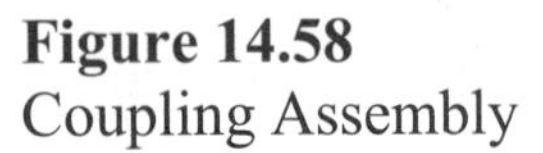

Figure 14.58
Coupling Assembly

Coupling Assembly

The fourteenth **lesson project** is an assembly that requires commands similar to those for the **Swing Clamp**. Model the parts and create the assembly shown in Figures 14.58 through 14.81. Analyze the assembly and plan out the steps required to assemble it. Use the **DIPS** in Appendix D to plan out the assembly component sequence and the parent-child relationships for the assembly.

You will use the **Coupling Shaft** from the Lesson 6 Project. The Coupling Shaft should be the first component assembled. The **Taper Coupling** from the Lesson 7 Project is also used in the assembly. The detail drawings for the second coupling are provided here, in this lesson project. Model this second **Coupling** *before* you start the assembly. Depending on the library parts available on your system, you may need to model the **Key**, the **Dowel**, and the **Washer**.

Since not all organizations purchase the libraries, details are provided for all the components required for the assembly, including the standard off-the-shelf parts available in Pro/E's library. Pro/LIBRARY commands to access the standard components are provided for those of you who have them loaded on your systems. The instance number is given for every standard component used in the assembly. The **Slotted Hex Nut**, **Socket Head Cap Screw**, **Hex Jam Nut**, and **Cotter Pin** are all standard parts from the library. The Cotter Pin is in *inch units,* and the remaining items are *metric.*

EGD REFERENCE

Fundamentals of Engineering Graphics and Design
by L. Lamit and K. Kitto
Read Chapter 23.
See pages 865-866.

NOTE

For this project, do not use the library parts directly in the assembly. Save each library part in your own directory with a new name, and then use the new part names in the assembly.

Section 3

Fundamentals

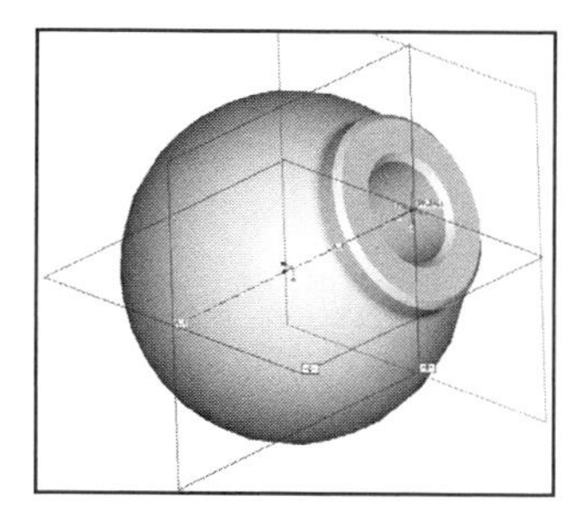

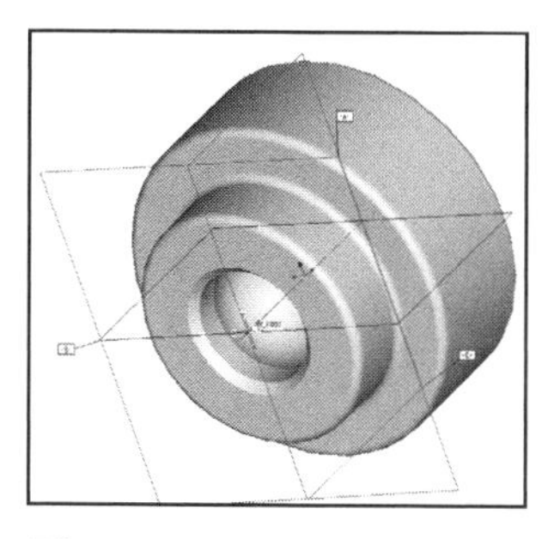

Figure 3.1
Assembly Drawing,
Assembly, and Part

The design of parts and assemblies, and the creation of related drawings, form the foundation of engineering graphics. When manufacturing with Pro/ENGINEER, many of the previous steps in the design process have been eliminated, streamlined, altered, refined, or expanded. The model you create as a part forms the basis for all engineering and design functions. The part model contains the geometric data describing the part's features, but it also includes nongraphical information embedded in the design itself. The part, its associated assembly, and the graphical documentation (drawings) are parametric. The physical properties described in the part drive (determine the characteristics and behavior of) the assembly and drawing. Any data established in the assembly mode, in turn, determine that aspect of the part and, subsequently, the drawings of the part and the assembly. In other words, all the information contained in the part, the assembly, and the drawing is interrelated, interconnected, and parametric (Fig. 3.1).

In many cases, the part will be the first component of this interconnected process. Therefore, in this text, the first set of lessons (1 through 13) covers part design.

Part Design

The *part* function in Pro/E is used to design components. Parts are started by sketching the basic form to produce the part's **base feature**.

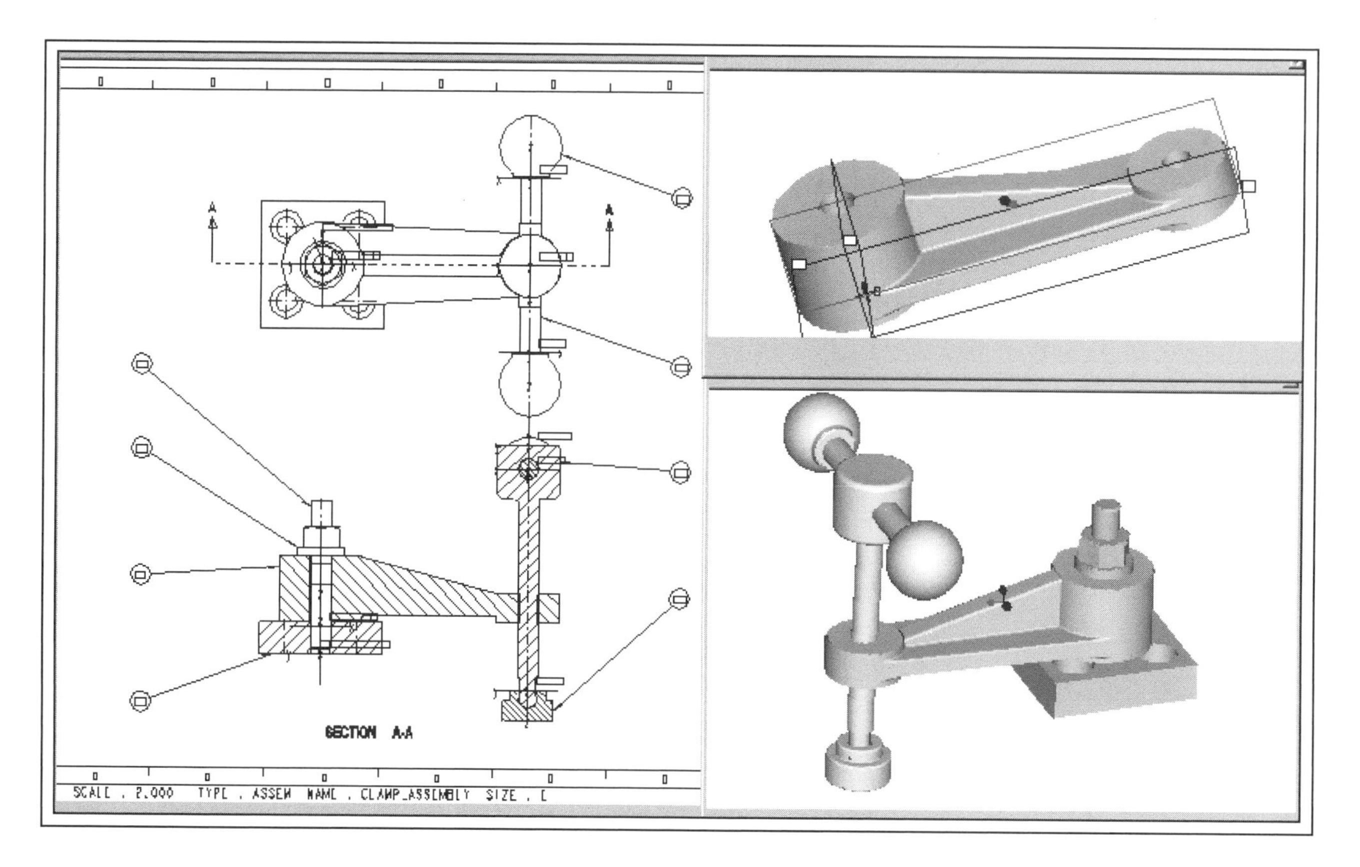

During part design (Fig. 3.2), you can accomplish the following:

* Define the base feature
* Define and redefine construction features to the base feature
* Modify the dimensional values of part features (Fig. 3.3)
* Embed design intent into the model using tolerance specifications and dimensioning schemes
* Create detail drawings of the part
* Create pictorial and shaded views of the component
* Create part families (family tables)
* Perform mass properties analysis and clearance checks
* List part, feature, layer, and other model information
* Measure and calculate model features

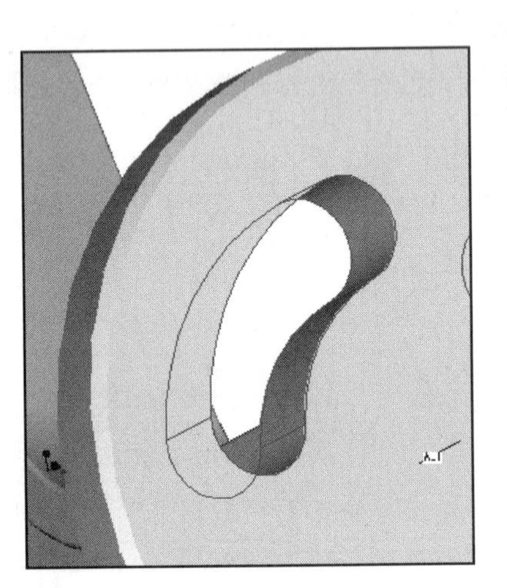

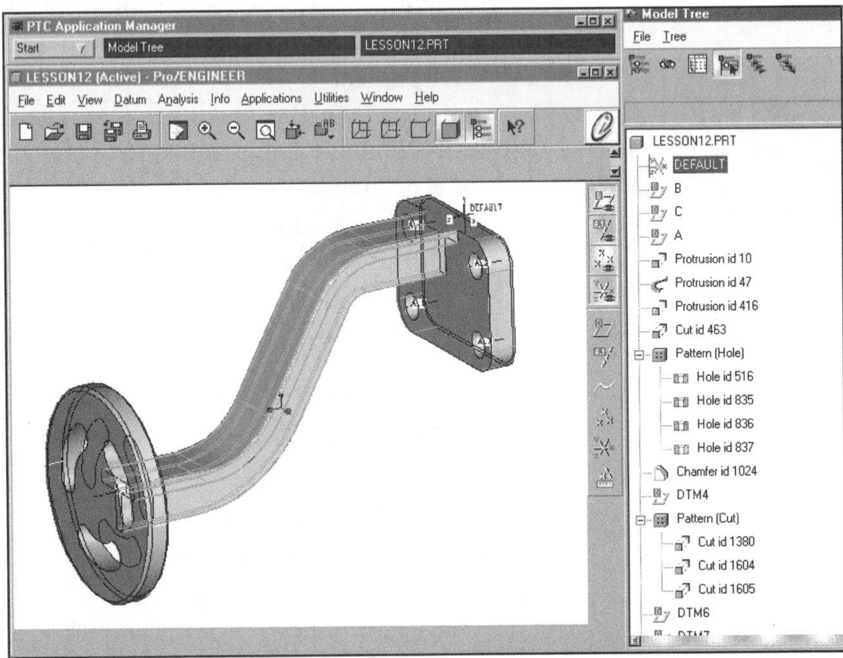

Figure 3.2
Part Design

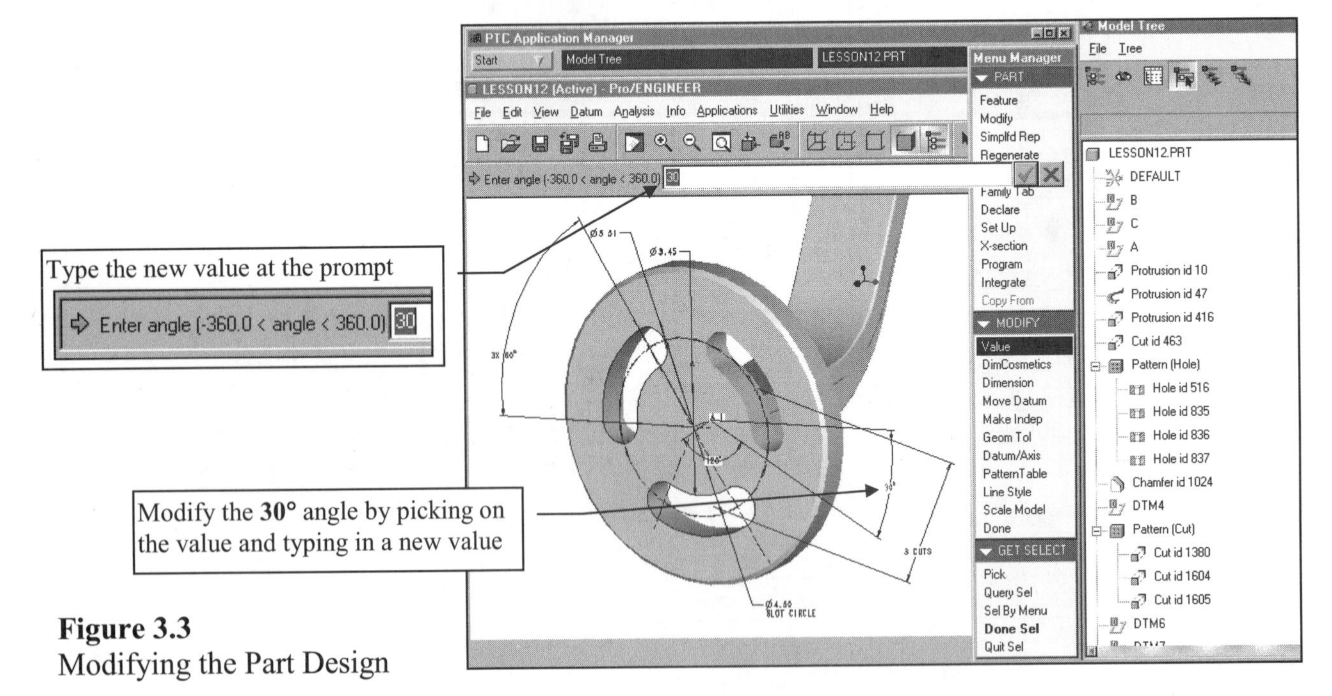

Figure 3.3
Modifying the Part Design

Establishing Part Features

The design of any part requires that the part be *confined*, *restricted*, *constrained*, and *referenced*. In parametric design, the easiest method to establish and control the geometry of your part design is to use three datum planes. Pro/E automatically creates the three **primary datums.** The default datum planes **(RIGHT**, **TOP**, and **FRONT)** constrain your design in all three directions. In Figure 3.4, three **default datum planes** and a **default coordinate system** were created when a NEW part was started. Note that in the **Model Tree** window they have become the first four features of the part, which means that they will be the **parents** of the features that follow.

The three *default datum planes* and the *coordinate system* appear in the Model Tree as the first four features of a new part **PRT0001.PRT**

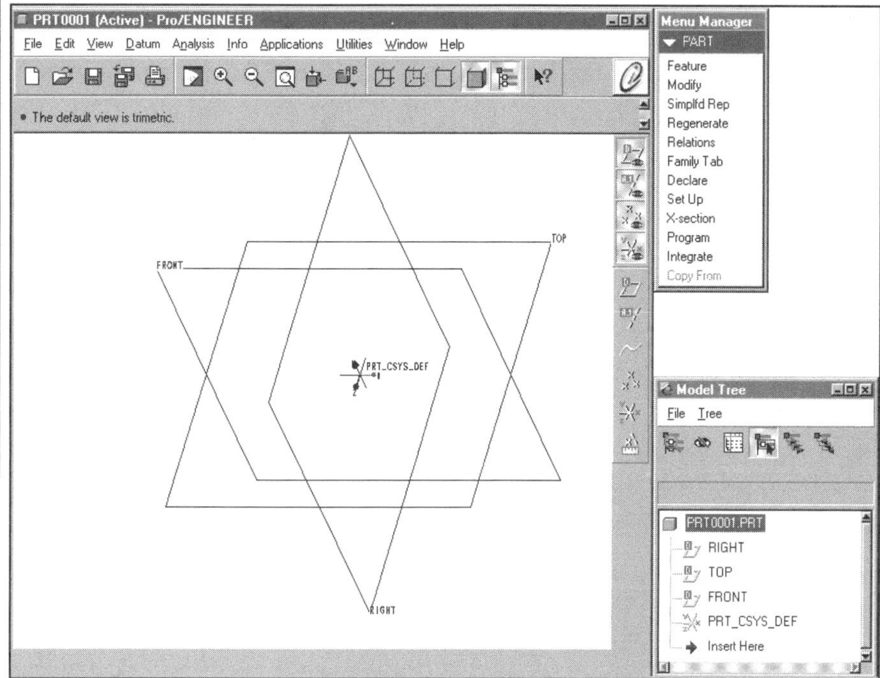

Figure 3.4
Default Datum Planes and Coordinate System

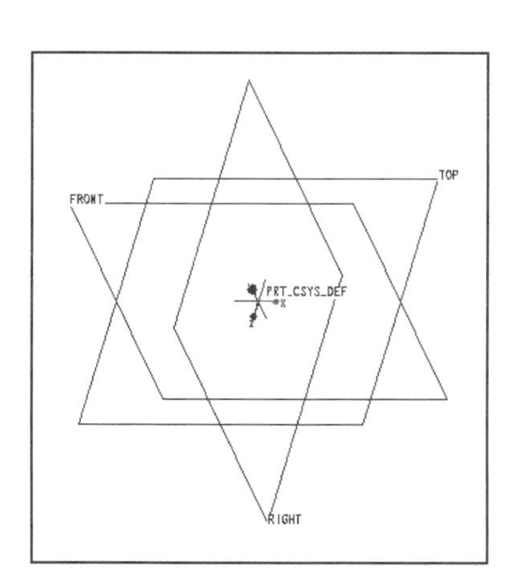

In order to see how datum planes work in the design of a part, try a simple exercise. Take a book or a box and put it on the floor of a room in your house or school. Choose the most important plane (flat side). You have now established **datum A**, the **primary datum** plane (**FRONT** in many, but not all, of the lessons and projects in this text). Choose the longest or second most important plane (flat side), and slide it up to and against a wall at the corner of the room. You have now established **datum B**, the **secondary datum** plane (**TOP** for many, but not all, of the lessons and projects in this text).

Finally, push the remaining side of the book or box against the other wall. You have now established **datum C**, the **tertiary datum** plane (**RIGHT** for many, but not all, of the lessons and projects in this text). The book/box is now constrained by these three planar surfaces. If the book/box were a real workpiece or stock material, you could secure the part to the floor with clamps (as though it were on a milling table) and machine it to the required design.

NOTE

All previous revisions of Pro/E *and* versions of this text used **DTM1** (now **RIGHT**), **DTM2** (now **TOP**), and **DTM3** (now **FRONT**) as the default names.

The name of a datum can be changed at any time using the command: **Set Up** ⇒ **Name** ⇒ (pick the name of the datum and type a new name) ⇒ ✔

Although this exercise and description are simplified and will not work for many parts, they do demonstrate the process of establishing your part in space using datums. In Pro/E, you can use any of the datums as sketching planes or, for that matter, any of the part planes for sketching geometry. Any number of other datums can be introduced into the part as required for feature creation, assembly operations, or manufacturing applications. The faucet in Figure 3.5 has multiple datum planes. Datum 13 is the 55th feature of this part. Figure 3.6 shows a part with information extracted on datum **A**, by displaying the features Children (many) and Parents (none).

DTM13 in Model Tree

Round id 1965	47
DTM6	48
DTM7	49
DTM8	50
DTM9	51
DTM10	52
DTM11	53
DTM12	54
DTM13	55
DTM14	56
Round id 2287	57

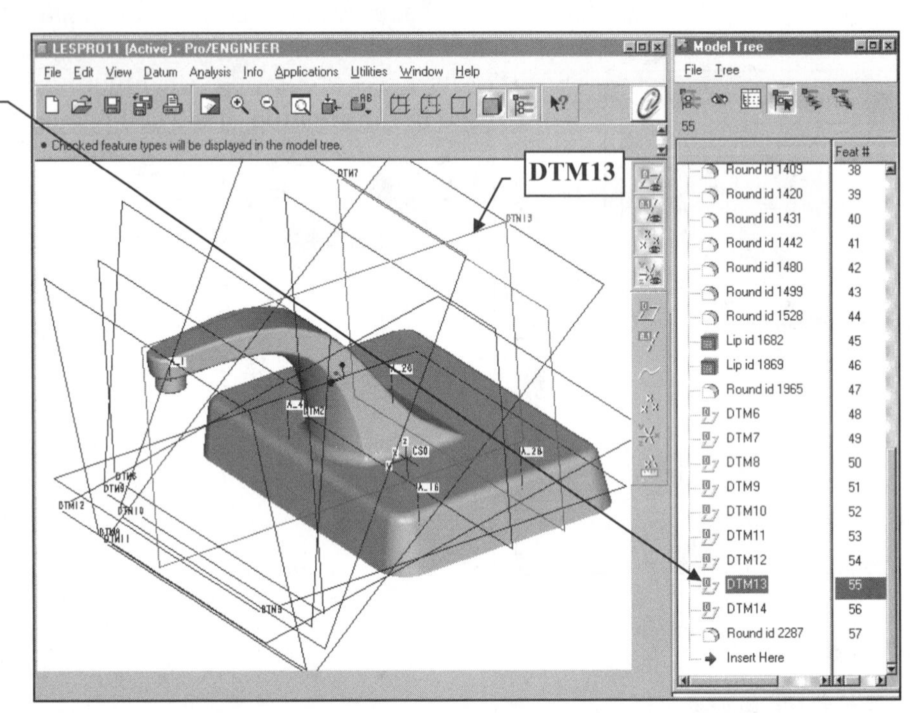

Figure 3.5
DTM13

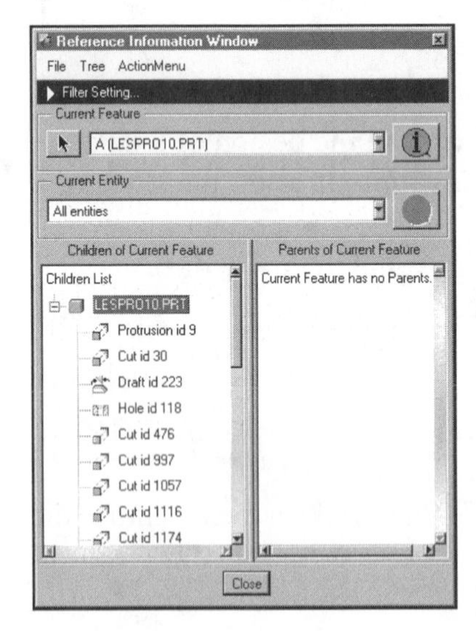

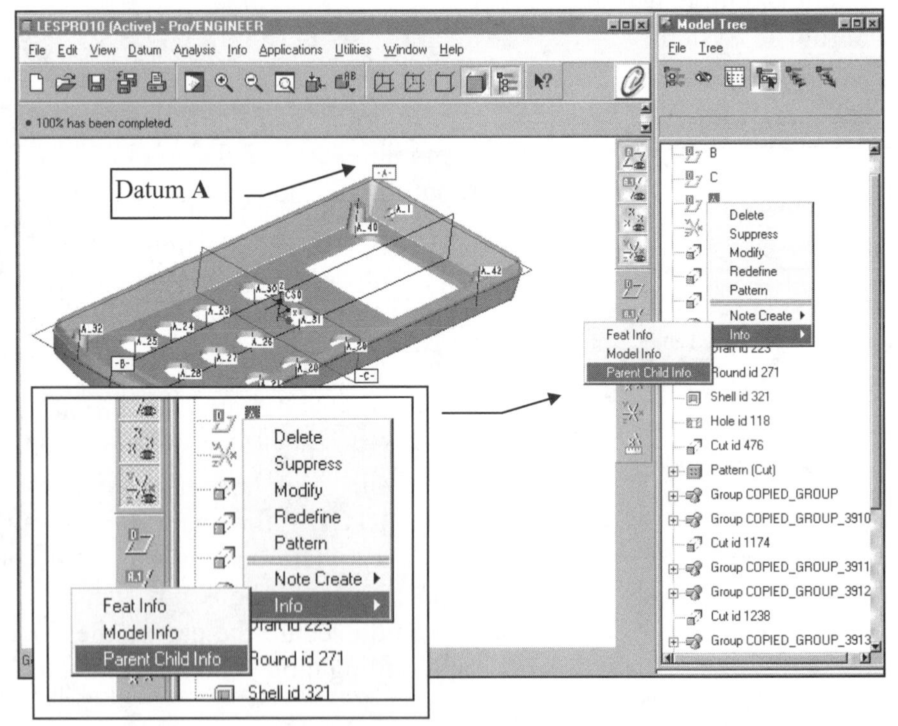

Figure 3.6
Datum **A** is the Parent of Most of the Part's other Features

Datum Features

Datum features are planes, axes, and points you use to place geometric features on the active part. We have discussed *default* datums. Datums other than defaults can be created at any time during the design process, as was done in the construction of the faucet shown in Figure 3.5, where a number of datums were introduced to locate sections of the swept blend.

There are three (primary) types of datum features: **datum planes, datum axes**, and **datum points** (there are also ***datum curves*** and ***datum coordinate systems*)**. You can display all types of datum features, but they do not define the surfaces or edges of the part or add to its mass properties. In Figure 3.7, a variety of datum planes are used in the creation of the cell phone.

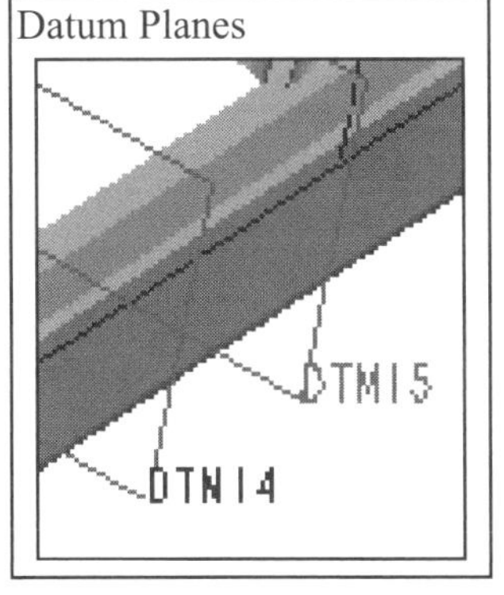

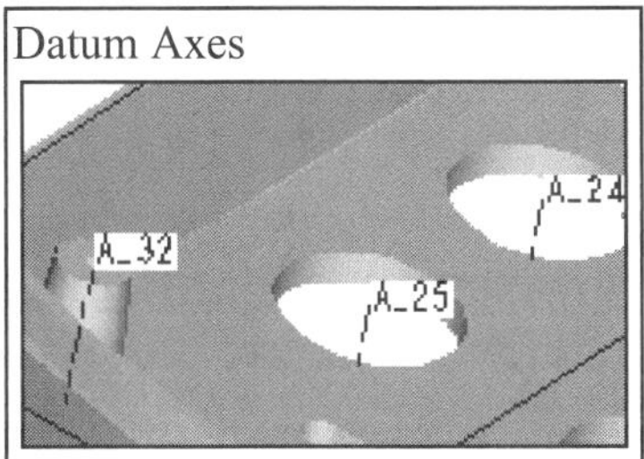

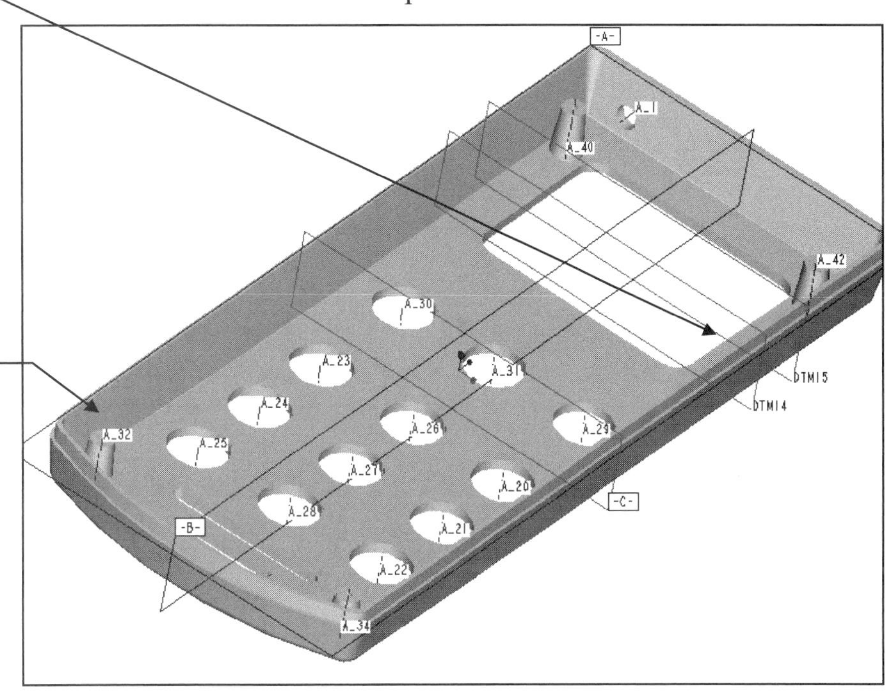

Figure 3.7
Datums in Part Design

Datum planes are infinite planes located in 3D model mode and associated with the object that was active at the time of their creation. To select a datum plane, you can pick on its name or anywhere on the perimeter edge. Datum planes are *parametric*--geometrically associated with the part. Parametric datum planes are associated with and dependent on the edges, surfaces, vertices, and axes of a part. For example, a datum plane placed parallel to a planar face and on the edge of a part moves whenever the edge moves and rotates about the edge if the face moves.

As you create parametric datum planes, relationships to the active part are determined by defining combinations of a placement option that link the datum plane to the part.

Datum planes are used to create a reference on a part that does not already exist. For example, you can sketch or place features on a datum plane when there is no appropriate planar surface. You can also dimension to a datum plane as though it were an edge.

A datum is created by specifying constraints that locate it with respect to existing geometry. For example, a datum plane might be made to pass through the axis of a hole and parallel to a planar surface. Chosen constraints must locate the datum plane relative to the model without ambiguity. You can also use and create datums in assembly mode.

Besides datum planes, datum axes and datum points can be created to assist in the design process. In Figure 3.8, a datum axis (**A_9**) passes through the center of a cylindrical feature.

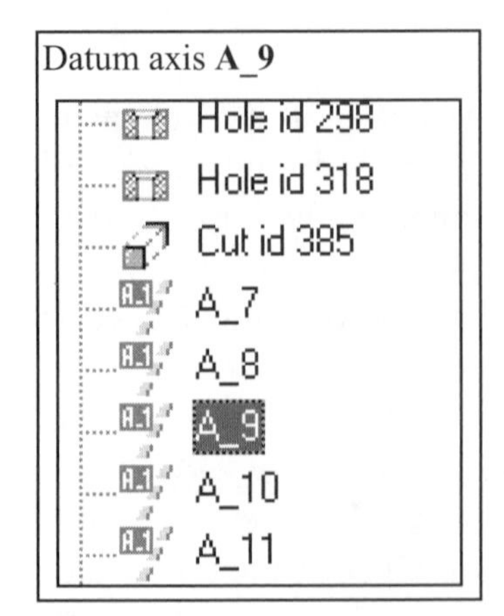

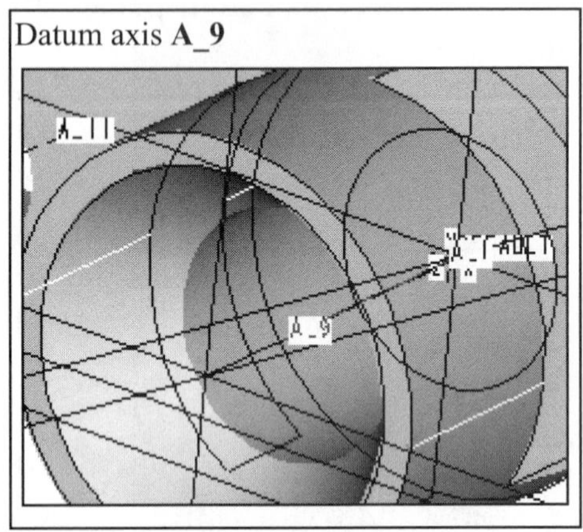

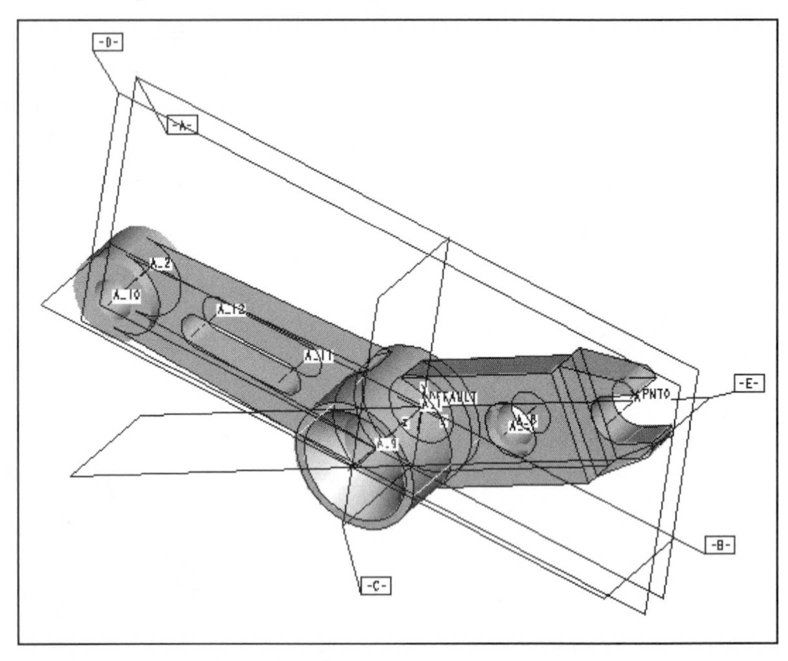

Figure 3.8
Created Datum Axes

You can also automatically create datum axes through cylindrical features such as holes and solid round features by setting this as a default in your Pro/E configuration file (the holes in the cell phone in Fig. 3.7 all have axes). The part in Figure 3.9 shows **A_1** through the hole. **A_1** is the default axis of the circular cut.

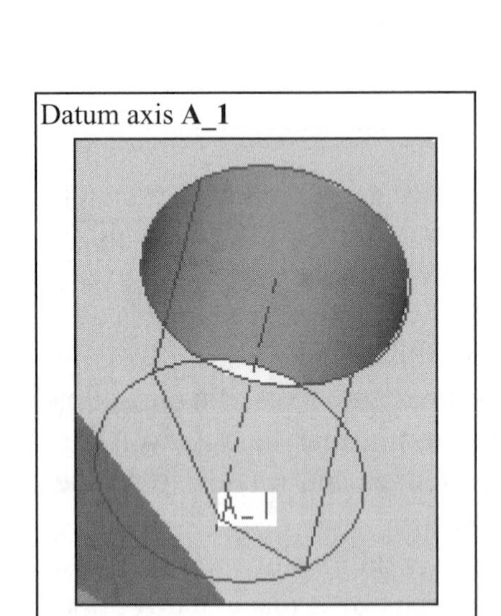

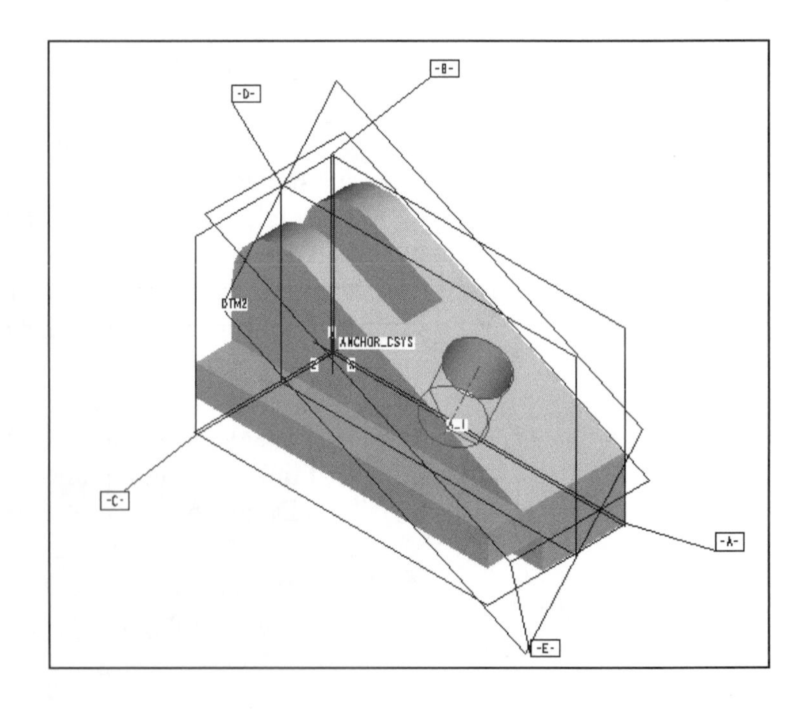

Figure 3.9
Feature Default Datum Axes

Parent-Child Relationships

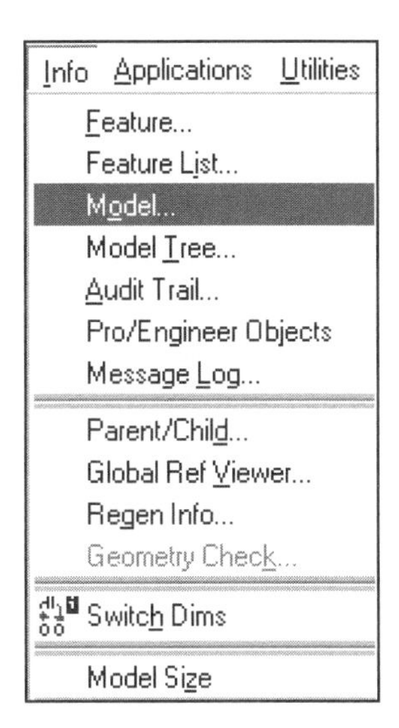

Because solid modeling is a cumulative process, certain features must, of necessity, precede others. Those that follow must rely on previously defined features for dimensional and geometric references. The relationships between features and those that reference them are termed ***parent-child relationships***. Because children reference parents, parent features can exist without children, but children cannot exist without their parents. Using **Info ⇒ Model**, Pro/E lists the models' information, including the parent-child relationships as shown in Figure 3.10.

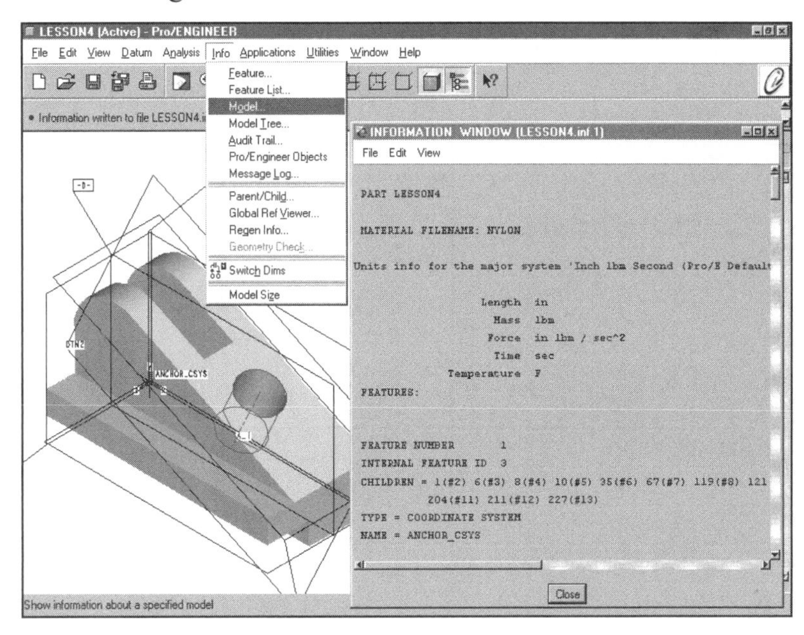

Figure 3.10
Parent-Child Relationships

To get the parent-child information for a feature, choose the following commands:

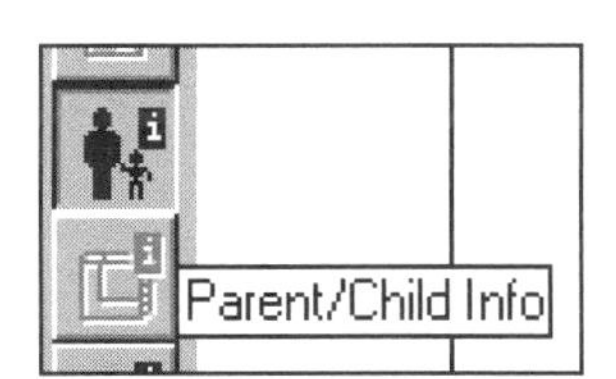

Info ⇒ Parent/Child (or pick the icon shown at left) ⇒ pick the feature in the Model Tree or Main Window (Fig. 3.11)

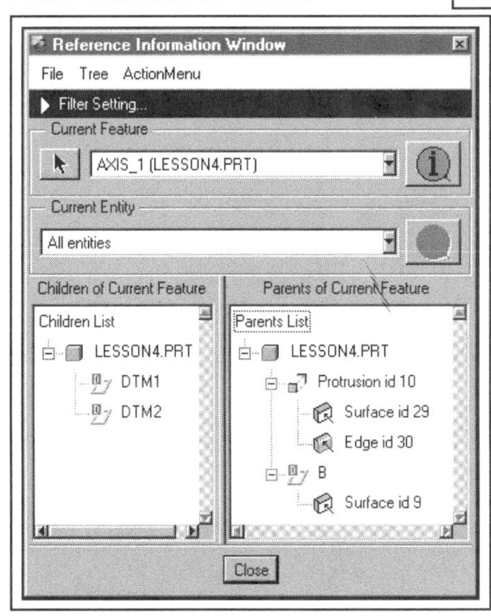

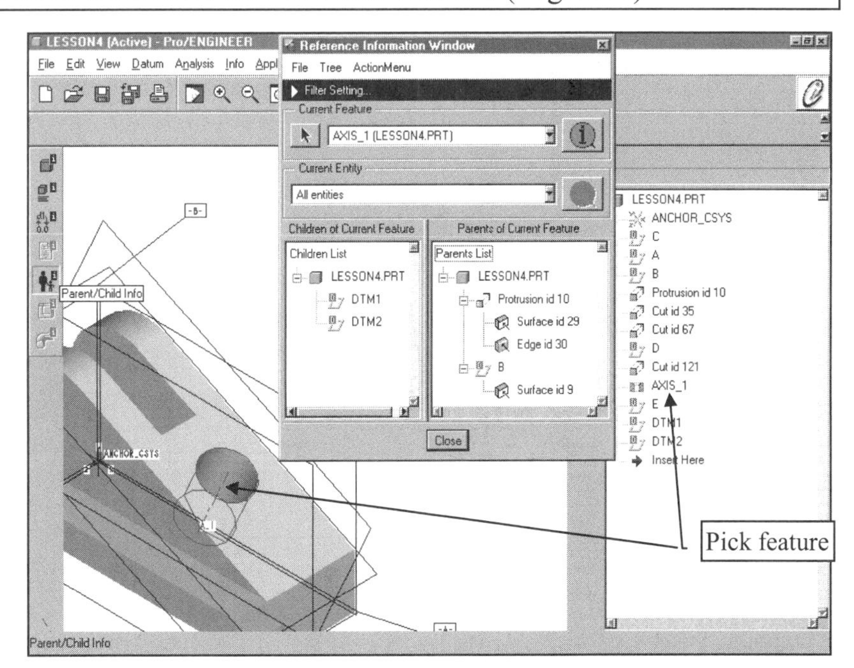

Figure 3.11
Parent-Child Information

Pick here for the feature's information

You can also extract the feature's information from the Reference Information Window (see left) by picking the icon.

The parent-child relationship is one of the most powerful aspects of parametric design. When a parent feature is modified, its children are automatically recreated to reflect the changes in the parent feature's geometry. It is essential to reference feature dimensions so that design modifications are correctly propagated through the model/part. As an example, the modification to the length of a part is automatically propagated through the part (Fig. 3.12) and will affect all children of the modified feature. The part's features have been modified to create two different versions as shown in Figure 3.13.

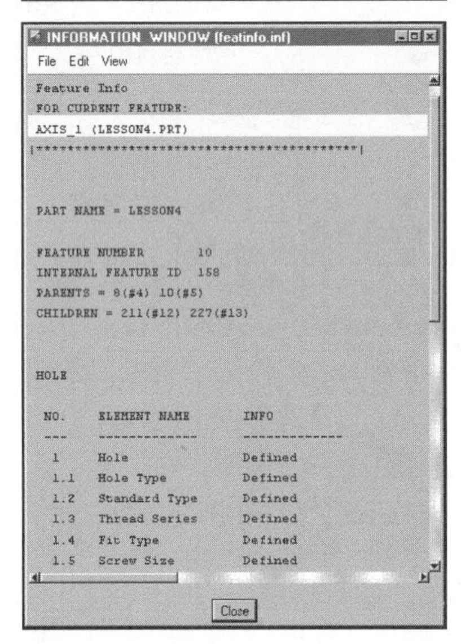

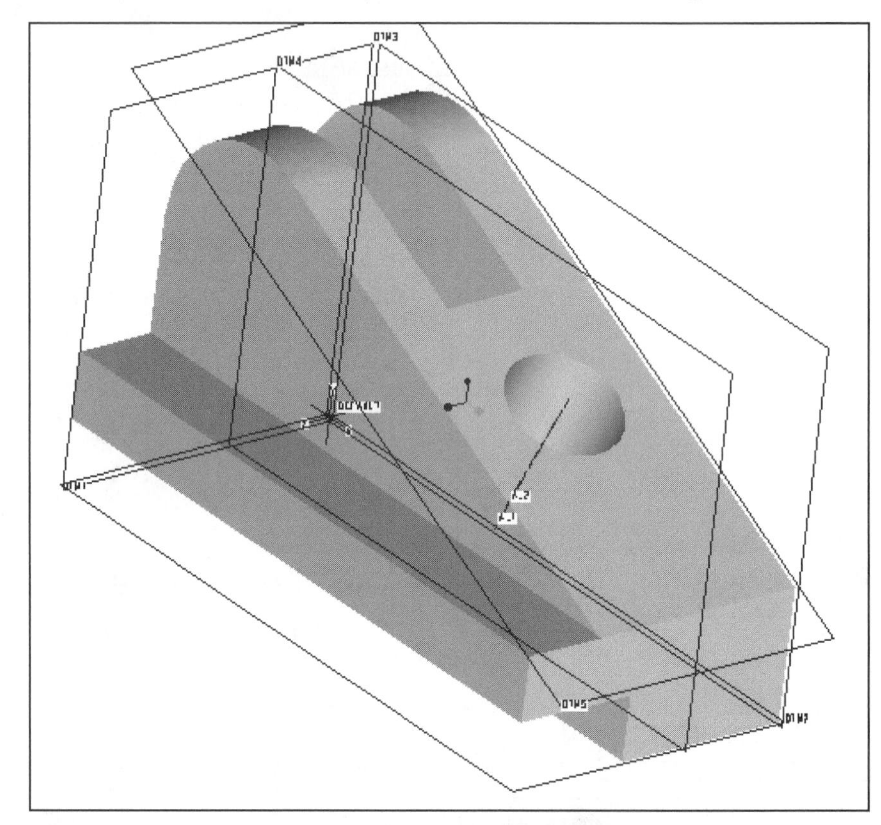

Figure 3.12
Original Design

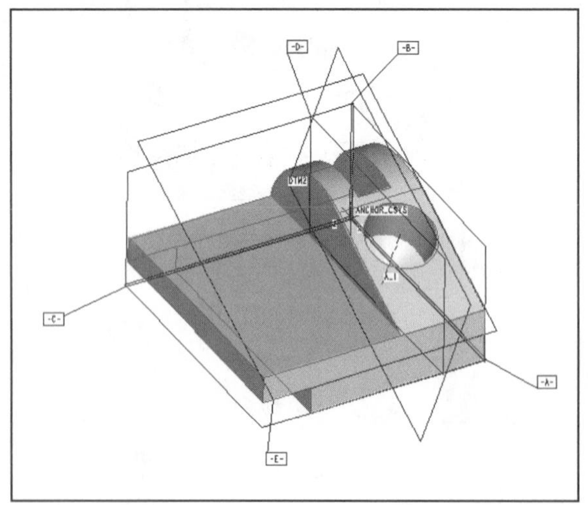

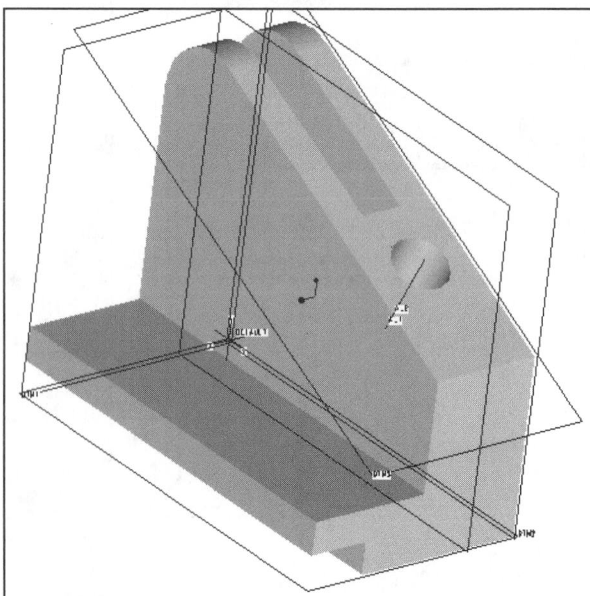

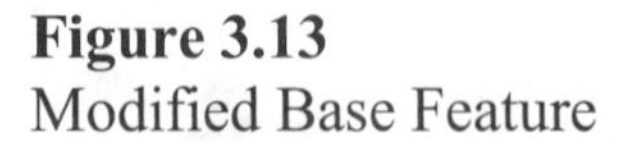

Figure 3.13
Modified Base Feature

Capturing Design Intent

Family of Parts

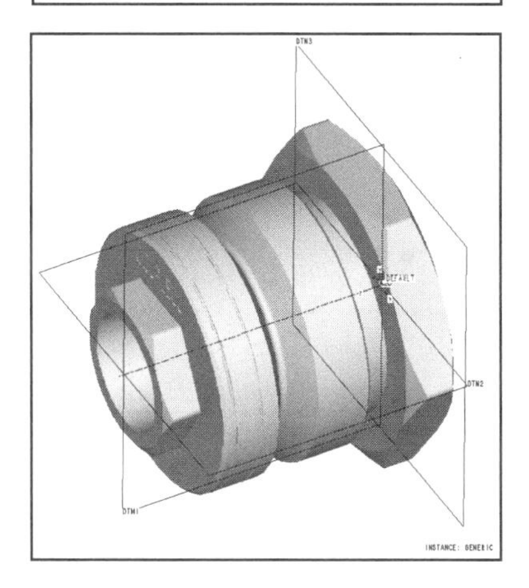

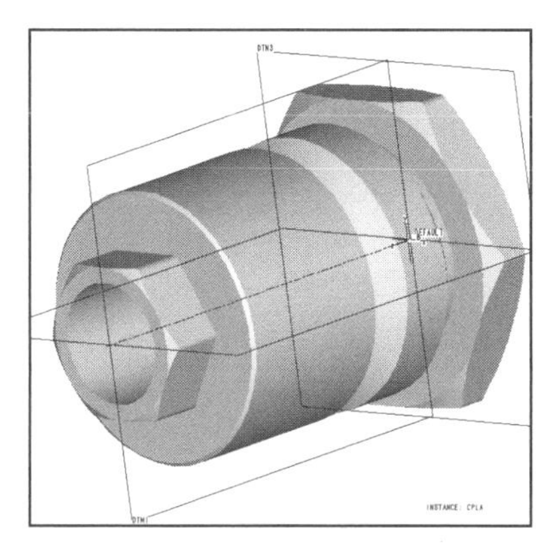

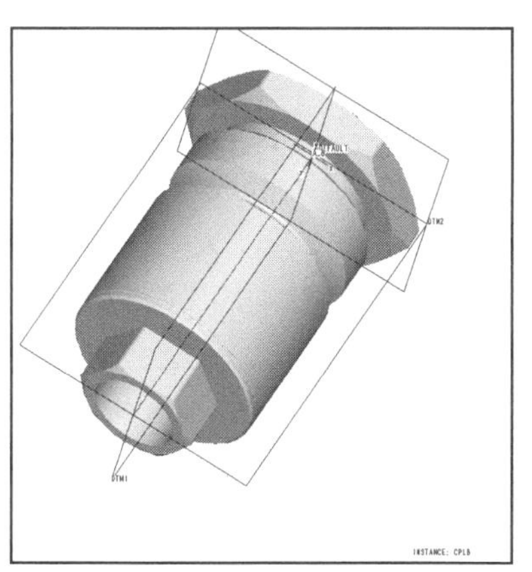

A valuable characteristic of any design tool is its ability to **render** the design and at the same time capture its **intent**. Parametric methods depend on the sequence of operations used to construct the design. The software maintains a *history of changes* the designer makes to specific parameters. The point of capturing this history is to keep track of operations that depend on each other. Whenever Pro/E is told to change a specific dimension, it can update all operations that are referenced to that dimension.

For example, a circle representing a bolt hole circle may be constructed so that it is always concentric to a circular slot. If the slot moves, so does the bolt circle. Parameters are usually displayed in terms of dimensions or labels and serve as the mechanism by which geometry is changed. The designer can change parameters manually by changing a dimension or can reference them to a variable in an equation (**relation**) that is solved either by the modeling program itself or by external programs such as spreadsheets.

Parametric modeling is particularly useful in modeling whole **families** of similar parts and in rapidly modifying complex 3D designs. It is most effective in working with designs where changes are likely to consist of dimensional changes rather than radically different geometry.

Feature-based modeling is the construction of geometry as a combination of **form features**. The designer specifies features in engineering terms, such as holes, slots, or bosses, rather than in geometric terms, such as circles or boxes.

Features can also store nongraphical information. This information can be used in activities such as drafting, numerical control (**NC**), finite-element analysis (**FEA**), and kinematics analysis.

Capturing design intent is based on incorporating engineering knowledge into a model by establishing and preserving certain geometric relationships. The wall thickness of a pressure vessel, for example, should be proportional to its surface area and should remain so, even as its size changes. Parametric design captures these relationships in several ways:

Implicit Relationships Implicit relationships occur when new model geometry is sketched and dimensioned relative to existing features and parts. An implicit relationship is established, for instance, when the section sketch of a tire (Fig. 3.14) uses rim edges as a reference.

Tire is sketched using rim edges to define sidewall.

Tire automatically adjusts as rim width is modified.

Figure 3.14
Tire and Rim

Patterns Design features often follow a geometrically predictable pattern. Features and parts are patterned in parametric design by referencing either construction dimensions or existing patterns. One example of patterning is a wheel hub with spokes (Fig. 3.15). First, the spoke holes are radially patterned. The spokes can then be strung by referencing this pattern.

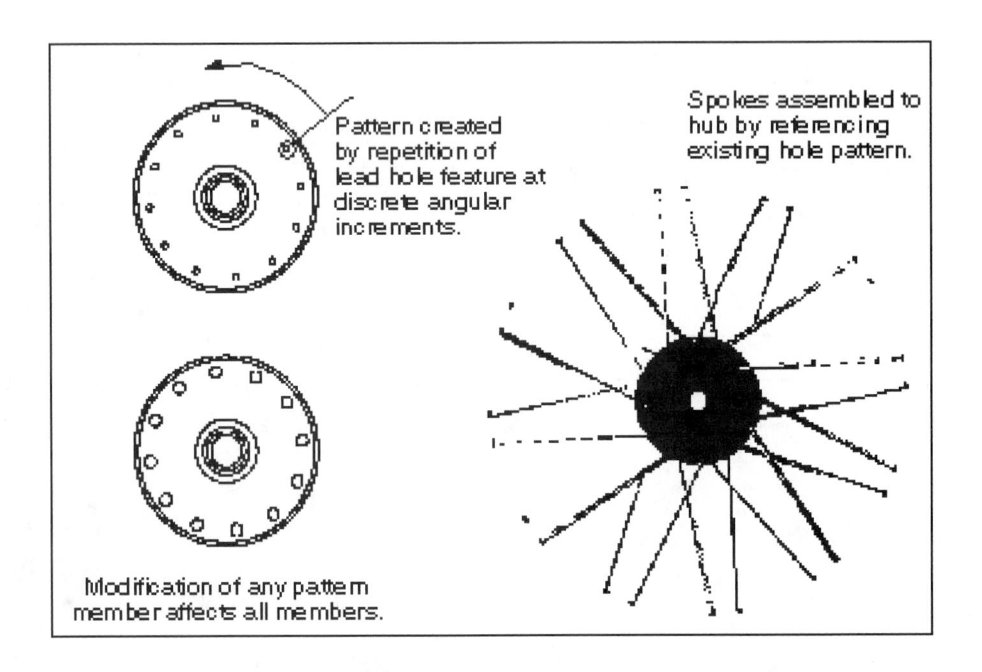

Figure 3.15
Patterns

Modification to a pattern member affects all members of that pattern. This helps capture design intent by preserving the duplicate geometry of pattern members.

Explicit Relations Whereas implicit relationships are implied by the feature creation method, an explicit relation is mathematically entered by the user. This equation is used to relate feature and part dimensions in the desired manner. An explicit relation might be used, for example, to ensure that any number of spoke holes will be evenly spaced around a wheel hub (Fig. 3.16).

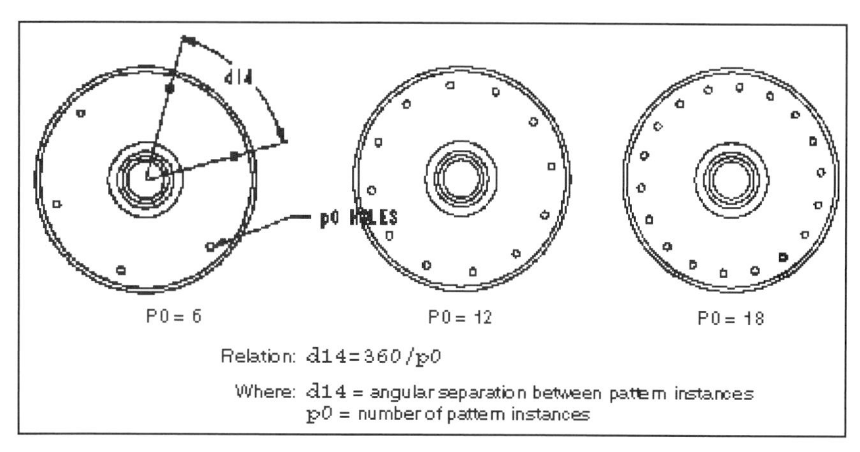

Figure 3.16
Adding Relations

Family Tables Family tables are used to create part families from generic models by tabulating dimensions or the presence of certain features or parts. A family table might be used, for example, to catalog a series of wheel rims with varying width and diameter as shown in Figure 3.17.

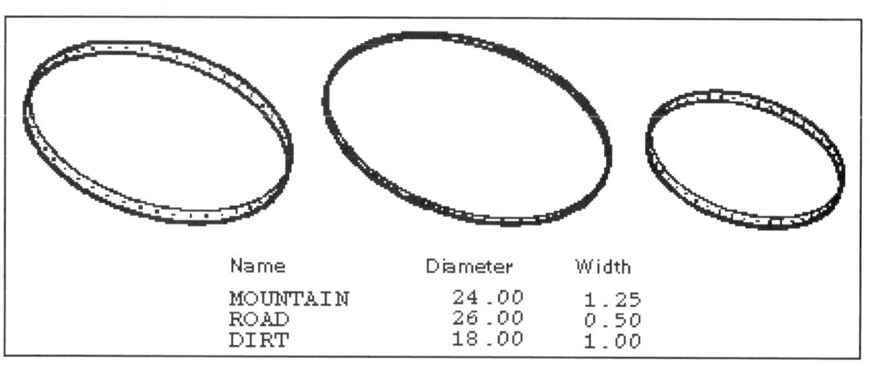

Name	Diameter	Width
MOUNTAIN	24.00	1.25
ROAD	26.00	0.50
DIRT	18.00	1.00

Figure 3.17
Family Table

The modeling task is to incorporate the features and parts of a complex design while properly capturing design intent to provide flexibility in modification. Parametric design modeling is a synthesis of physical and intellectual design (Fig. 3.18).

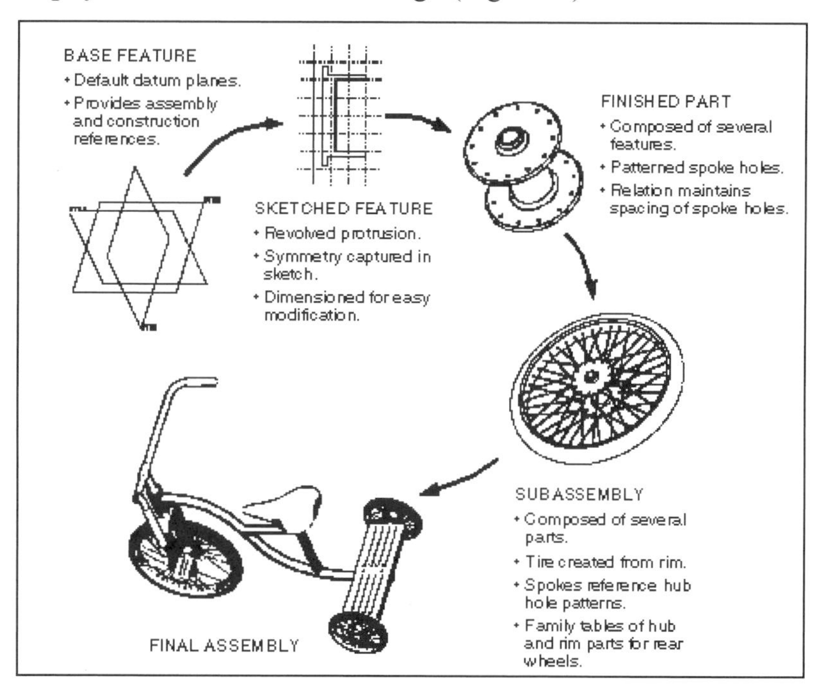

Figure 3.18
Relations

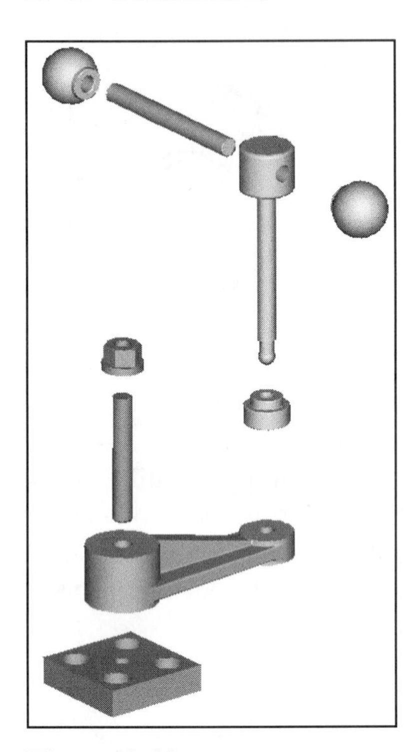

Figure 3.19
Clamp Assembly and Model Tree

Assemblies

Just as parts are created from related features, **assemblies** are created from related parts. The progressive combination of subassemblies, parts, and features into an assembly creates parent-child relationships based on the references used to assemble each component (Fig. 3.19).

The *Assembly* functionality is used to assemble existing parts and subassemblies. During assembly creation, you can:

* Simplify a view of a large assembly by creating a simplified representation
* Perform automatic or manual placement of component parts
* Create an exploded view of the component parts
* Perform analysis, such as mass properties and clearance checks
* Modify the dimensional values of component parts
* Define assembly relations between component parts
* Create documentation drawings of the assembly
* Create a shaded view of the assembly
* Use the assembly as a subassembly
* Perform automatic interchange of component parts
* Create parts in Assembly mode
* Create sssembly features

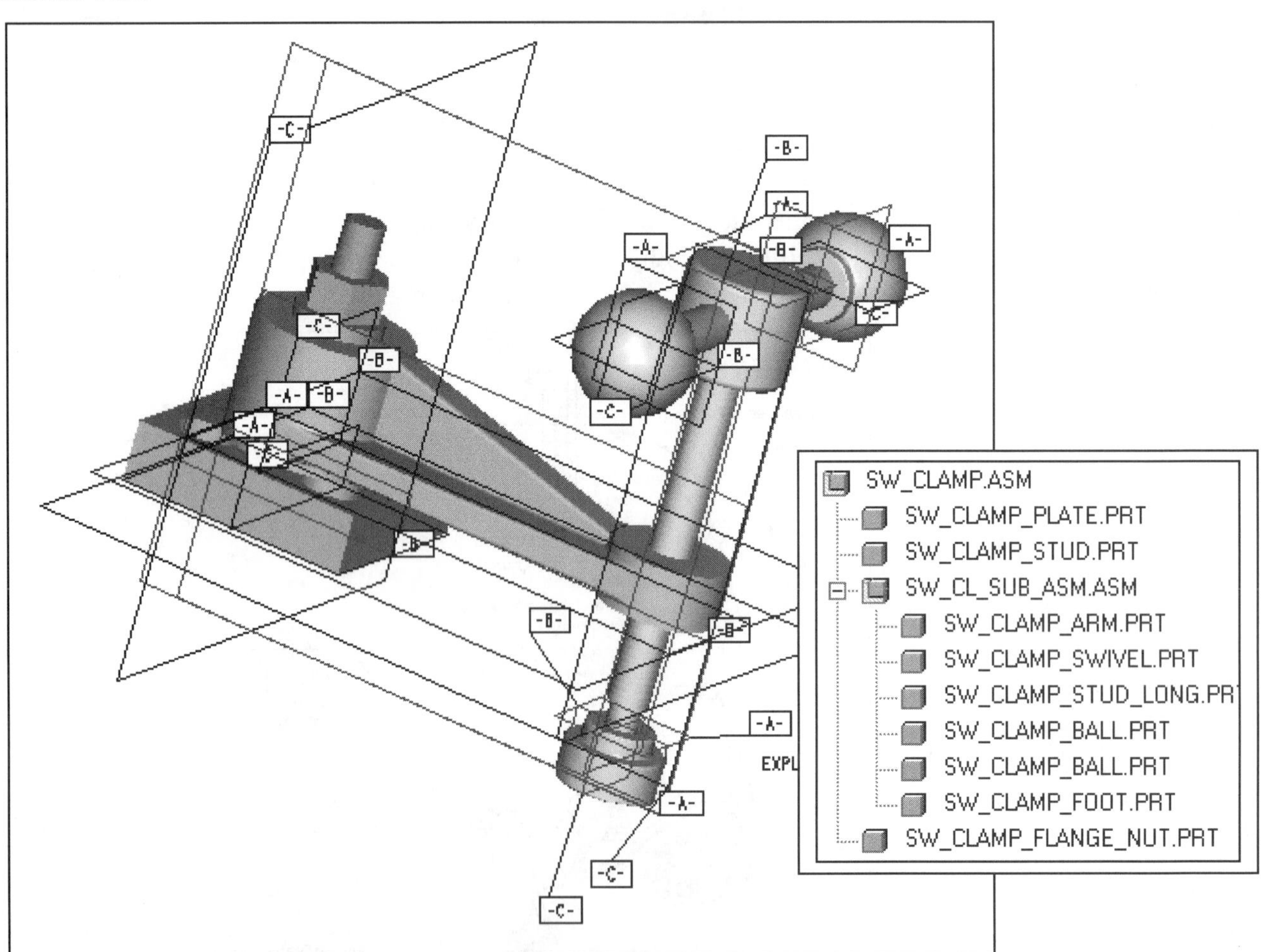

Just as features can reference part geometry, parametric design also permits the creation of parts referencing assembly geometry. **Assembly mode** allows the designer both to fit parts together and to design parts on the basis of how they should fit together.

In Figure 3.19, the assembly of the clamp was shown. **Info ⇒ BOM** is chosen from the pull-down Info menu (Fig. 3.20). A *Bill of Materials* report is generated for the assembly in Figure 3.21.

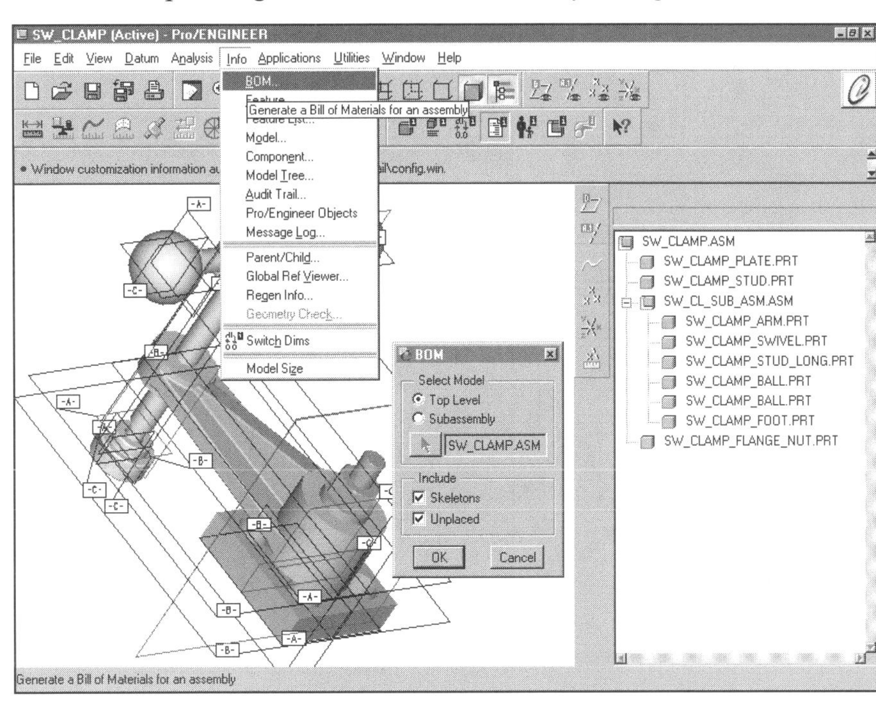

Figure 3.20
Clamp Assembly and BOM Command

Figure 3.21 Clamp Assembly and BOM

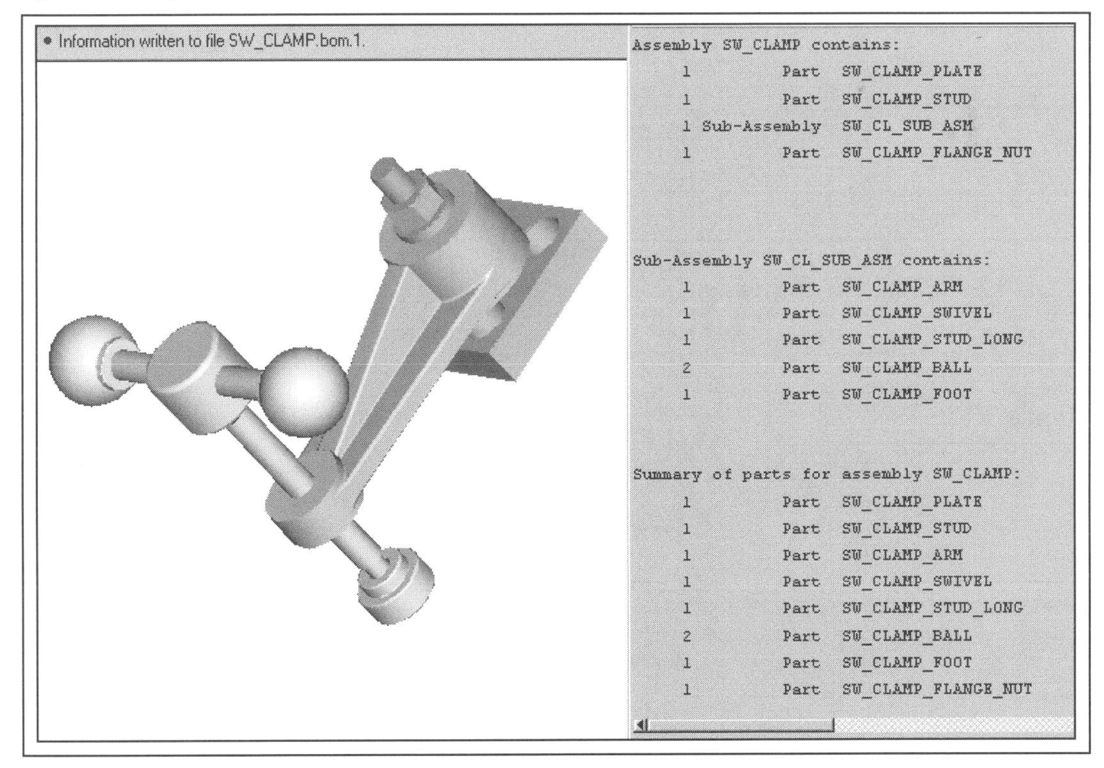

• Information written to file SW_CLAMP.bom.1.

```
Assembly SW_CLAMP contains:
     1          Part  SW_CLAMP_PLATE
     1          Part  SW_CLAMP_STUD
     1 Sub-Assembly  SW_CL_SUB_ASM
     1          Part  SW_CLAMP_FLANGE_NUT

Sub-Assembly SW_CL_SUB_ASM contains:
     1          Part  SW_CLAMP_ARM
     1          Part  SW_CLAMP_SWIVEL
     1          Part  SW_CLAMP_STUD_LONG
     2          Part  SW_CLAMP_BALL
     1          Part  SW_CLAMP_FOOT

Summary of parts for assembly SW_CLAMP:
     1          Part  SW_CLAMP_PLATE
     1          Part  SW_CLAMP_STUD
     1          Part  SW_CLAMP_ARM
     1          Part  SW_CLAMP_SWIVEL
     1          Part  SW_CLAMP_STUD_LONG
     2          Part  SW_CLAMP_BALL
     1          Part  SW_CLAMP_FOOT
     1          Part  SW_CLAMP_FLANGE_NUT
```

Drawings

You can create **drawings** of all parametric design models or create or retrieve some of them by importing files from other systems. All model views in the drawing are **associative:** if you change a dimensional value in one view, other drawing views update accordingly. Moreover, drawings are associated with their parent models. Any dimensional changes made to a drawing are automatically reflected in the model. Any changes made to the model (e.g., addition of features, deletion of features, dimensional changes, and so on) in Part, Sheet Metal, Assembly, or Manufacturing modes are also automatically reflected in their corresponding drawings.

The *Drawing* functionality is used to create annotated drawings of parts and assemblies. During drawing creation, you can:

* Add views of the part or assembly.
* Show existing dimensions.
* Incorporate additional driven or reference dimensions.
* Create notes to the drawing.
* Display views of additional parts or assemblies.
* Add sheets to the drawing.
* Create draft entities on the drawing.

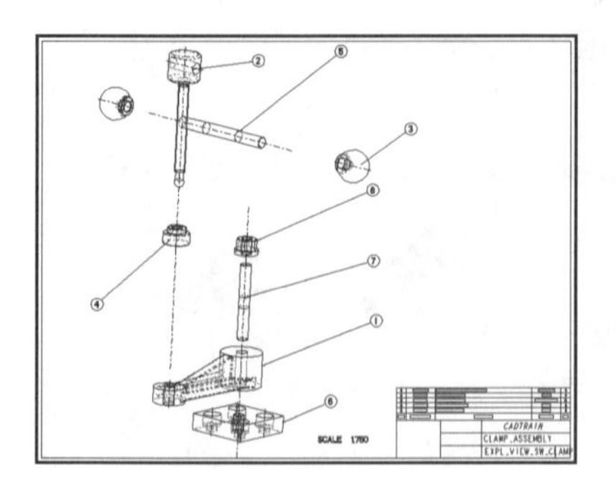

8	SW100-22FLN	.500-13 X HEX FLANGE NUT	PURCHASED	1
7	SW100-21ST	.500-13 X 3.50 DOUBLE END STUD	PURCHASED	1
6	SW100-20PL	SWING CLAMP PLATE	1020 CRS	1
5	SW101-9STL	.500-13 X 5.00 DOUBLE END STUD	PURCHASED	1
4	SW101-8FT	SWING CLAMP FOOT	NYLON	1
3	SW101-7BA	SWING CLAMP BALL	BLACK PLASTIC	2
2	SW101-6SW	SWING CLAMP SWIVEL	STEEL	1
1	SW101-5AR	SWING CLAMP ARM	STEEL	1
ITEM	PT NUM	DESCRIPTION	MATERIAL	QTY

TOOL ENGINEERING CO.

DRAWN

ISSUED

2.000 SW_CL_ASSEMBLY

SHEET 1 OF 1

A

2 1

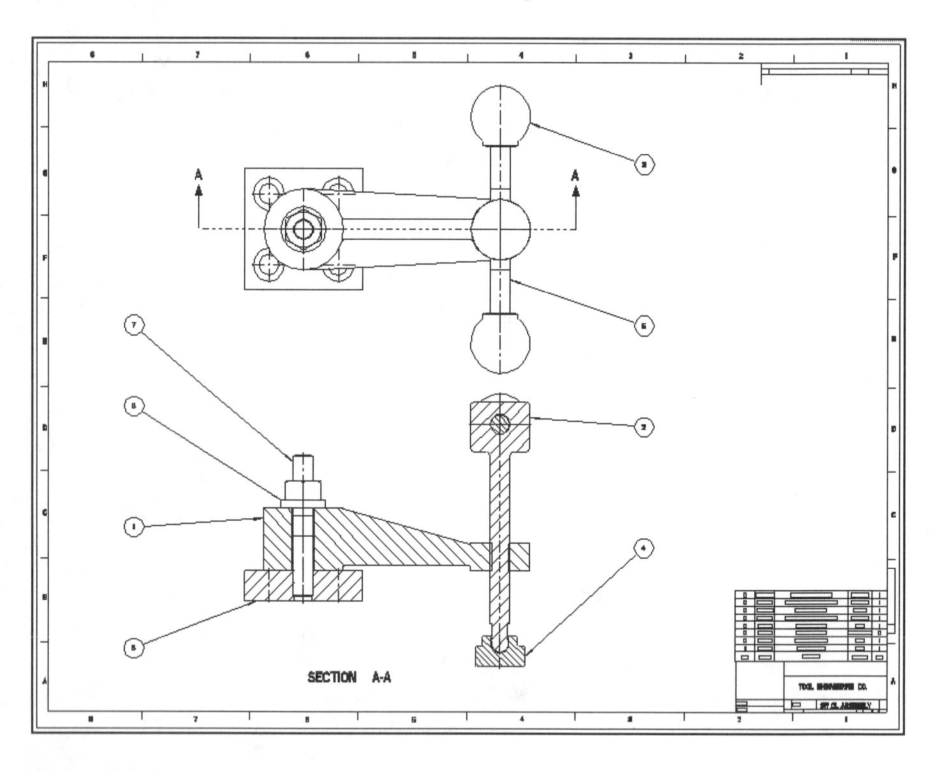

Figure 3.22
Assembly Drawing

Drawing Mode and Basic Parametric Design

Drawing mode in parametric design provides you with the basic ability to document solid models in drawings that share a two-way associativity (Fig. 3.22).

Changes that are made to the model in Part mode or Assembly mode will cause the drawing to update automatically and reflect the changes. Any changes made to the model in Drawing mode will be immediately visible on the model in Part and Assembly modes. The part shown in Figure 3.23 has been detailed in Figure 3.24.

Basic Pro/E allows you to create drawing views of one or more models in a number of standard views with dimensions.

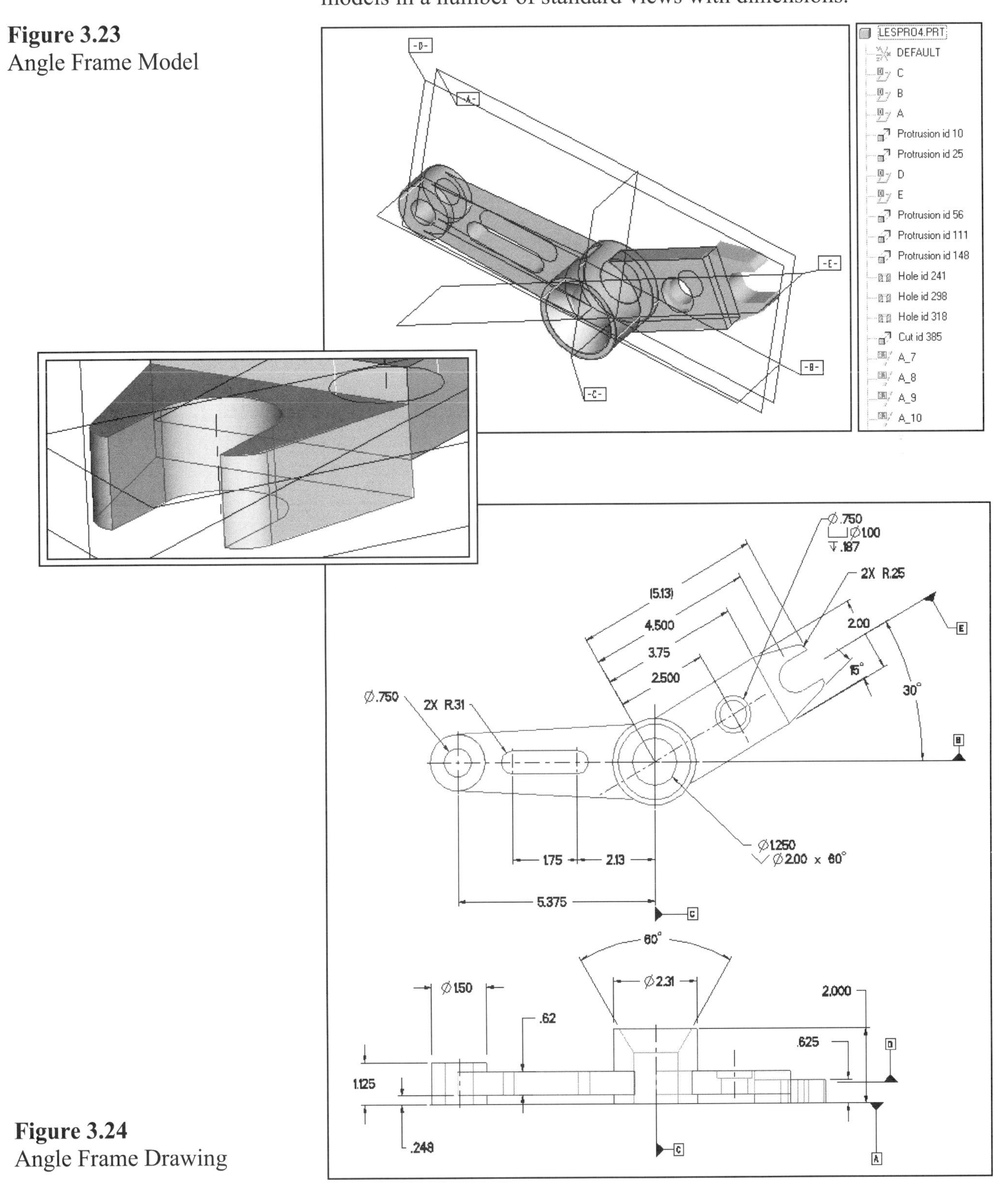

Figure 3.23
Angle Frame Model

Figure 3.24
Angle Frame Drawing

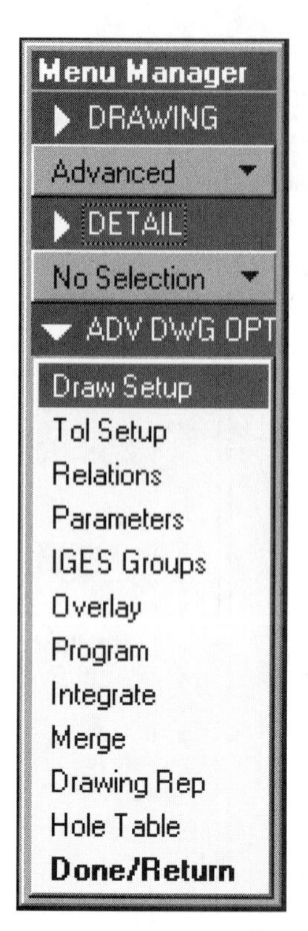

Figure 3.25
Setting **.dtl** Drawing Preferences

You can annotate the drawing with notes, manipulate the dimensions, and use layers to manage the display of different items on the drawing. The module **Pro/DETAIL** can be used to extend the drawing capability or as a stand-alone module allowing you to create, view, and annotate models and drawings.

Pro/DETAIL supports additional view types and multi-sheets and offers commands for manipulating items in the drawing and for adding and modifying different kinds of textural and symbolic information. In addition, the abilities to customize engineering drawings with sketched geometry, create custom drawing formats, and make numerous cosmetic changes to the drawing are available.

Drawing parameters are saved with each individual drawing and drawing format. Drawing parameters determine the height of dimension and note text, text orientation, geometric tolerance standards, font properties, drafting standards, and arrow lengths. Drawing parameter values are stored by Pro/E in files with **.dtl** extensions. The drawing parameters can be altered by picking **Advanced ⇒ Draw Setup** (Fig. 3.25).

When you regenerate a drawing, the drawing and the model that it represents are recreated, not simply redrawn. This means that if any of the model's dimension values are changed while you are in drawing mode, regenerating the drawing causes the model to update these changes. The regenerated drawing displays the updated model and any changes that were made to it.

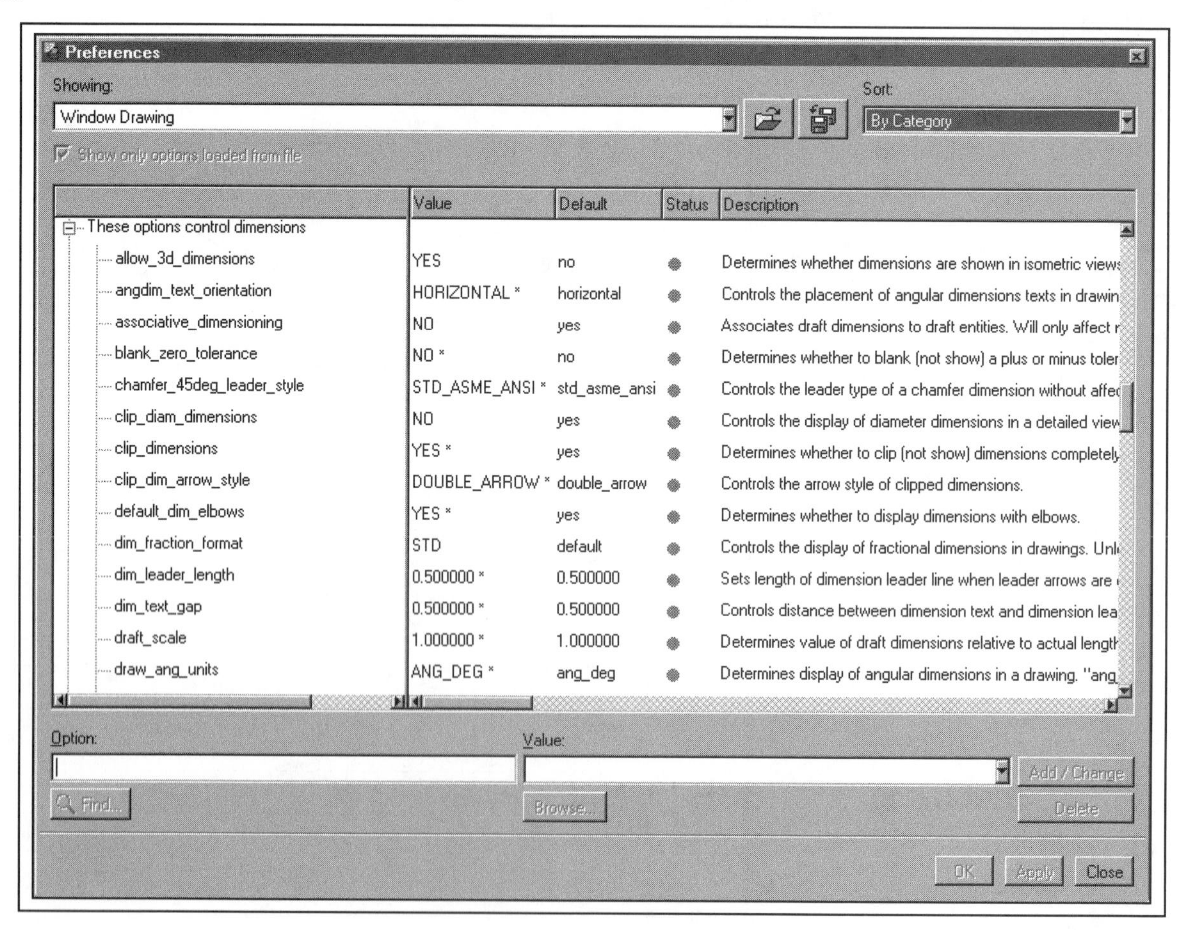

Manufacturing and Parametric Design

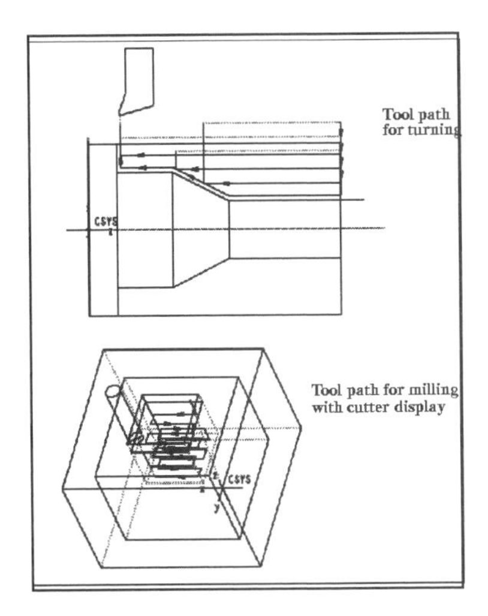

Figure 3.26
Numerical Control Machining Simulation

Parametric design systems also provide the tools to program and simulate numerical control manufacturing processes (Fig. 3.26). The information created can be quickly updated if the engineering design model changes. Numerical control (**NC**) programs in the form of ASCII cutter location (**CL**) data files, tool lists, operation reports, and in-process geometry can be generated.

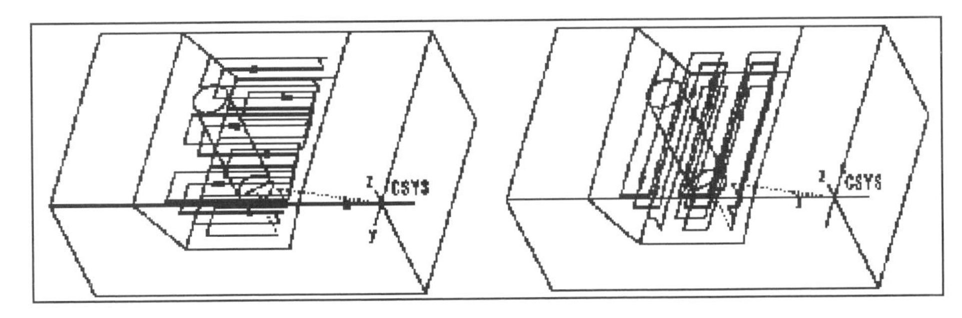

Pro/NC

Pro/NC software creates the data necessary to drive an NC machine tool to machine a part. It does this by providing the tools to let the manufacturing engineer follow a logical sequence of steps to progress from a design model to an ASCII CL data file that can be post-processed into NC machine data (Fig. 3.27). Figure 3.28 shows the work flow sequence for NC machining using Pro/NC.

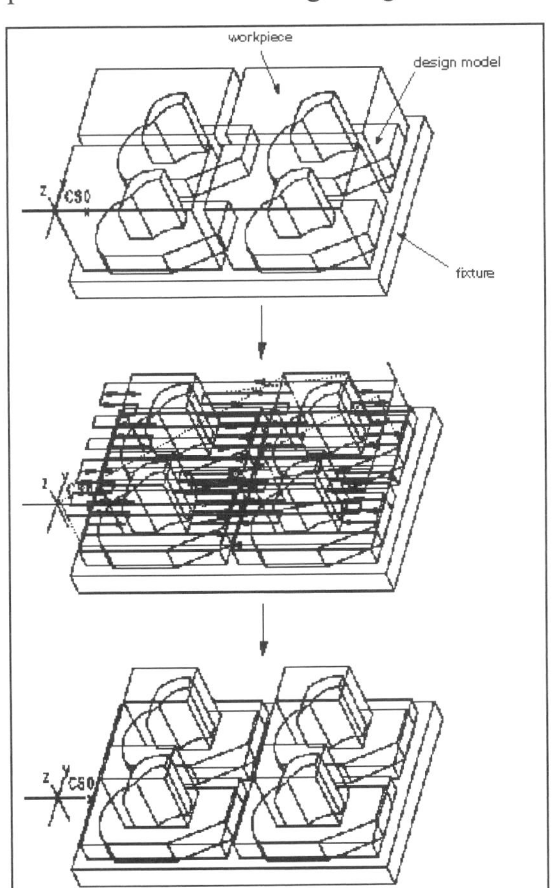

Figure 3.27
Multiple-Part Machining Using Pro/NC

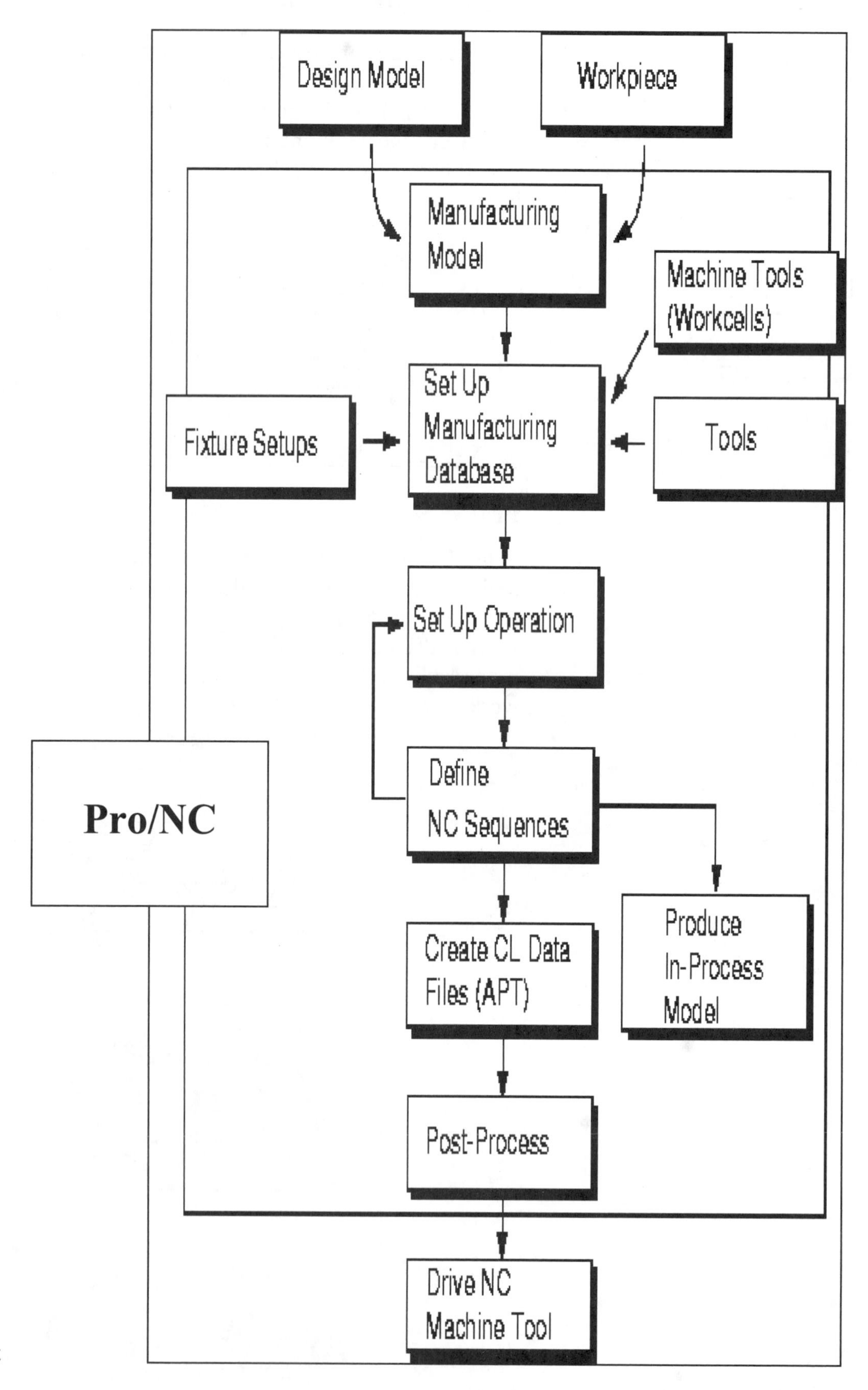

Figure 3.28
Pro/NC Flowchart

Section 4

Modes and File Management

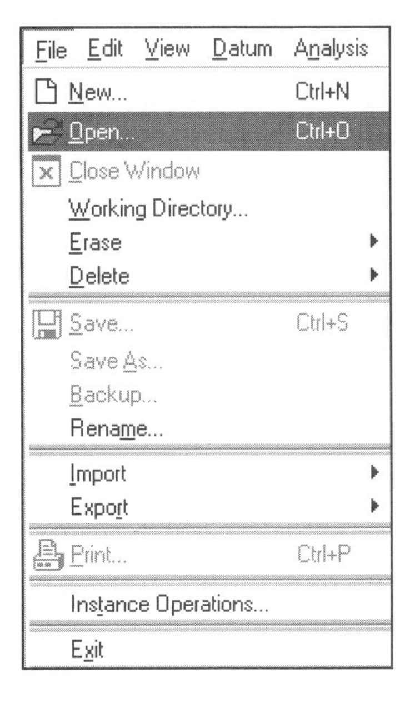

The available *mode options* on your system are determined by the licensed modules that you have. This text covers **part**, **assembly**, **drawing**, **format**, **markup, manufacturing,** and **sheetmetal** modes (Fig. 4.1). You can **Preview** a file before opening it (Fig. 4.1). *Previews are not available for Pro/E objects created prior to Ver. 18.*

Several *file management utilities* are available during the design of parts, assemblies, and drawings. The ability to access and change directories, save objects, delete objects, erase in session objects, rename objects, and save objects under different names is essential to the efficient and accurate creation of project databases.

Operating Modes

More than one module can be open in a Pro/E session at one time. You can have a part, the assembly where it is used, and its drawing on the screen simultaneously. In Figure 4.2, part, assembly, and drawing modes have been opened. You can move among the windows and work on the part, the assembly, or the drawing. *Only one widow can be active at any one time.* The part database essentially drives the assembly and the drawing. Because Pro/E is *parametric*, any editing or modifications of the part will be reflected in the Assembly and Drawing modes after regeneration. The same is true of Assembly mode and Drawing mode changes.

Figure 4.1
Modes

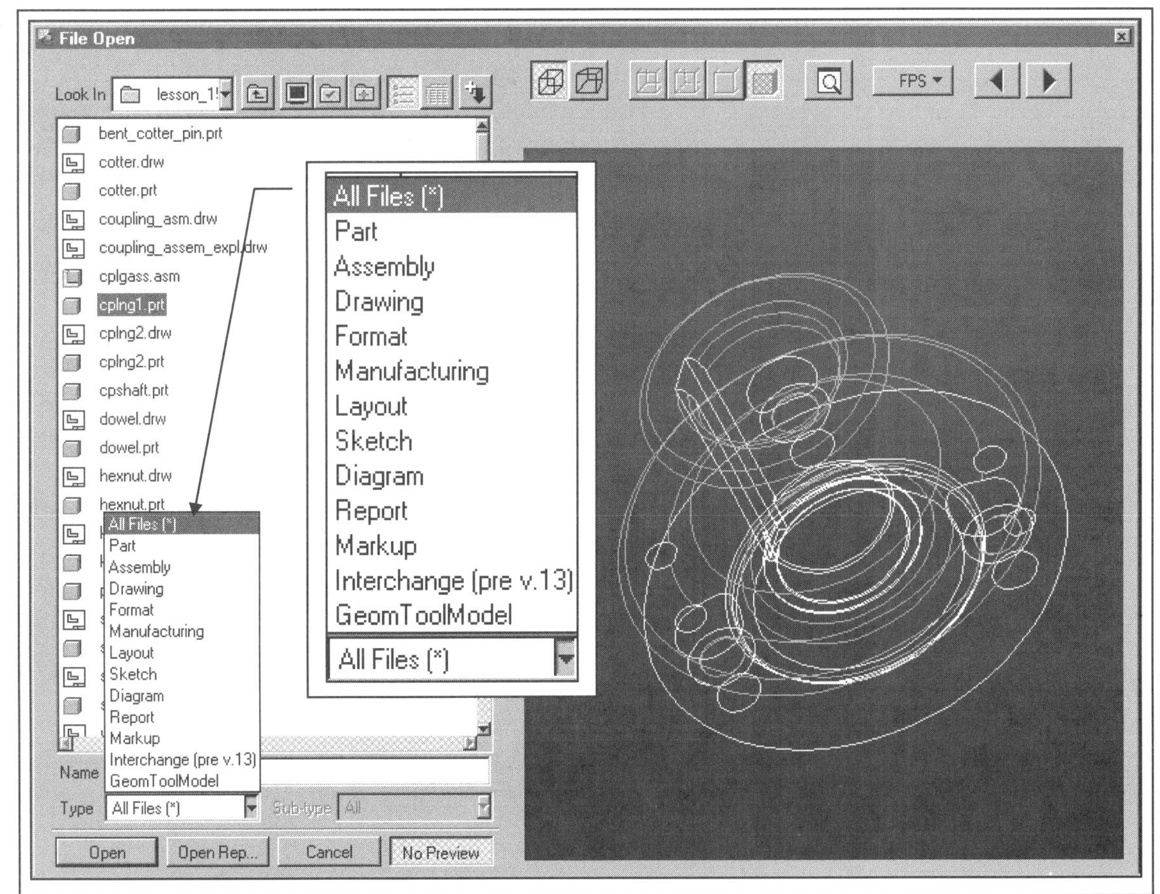

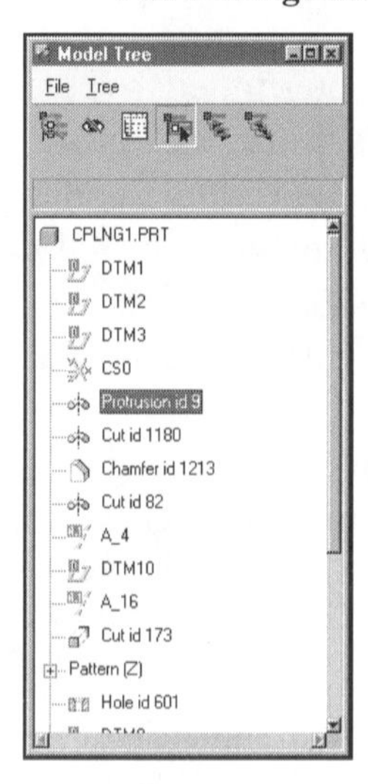

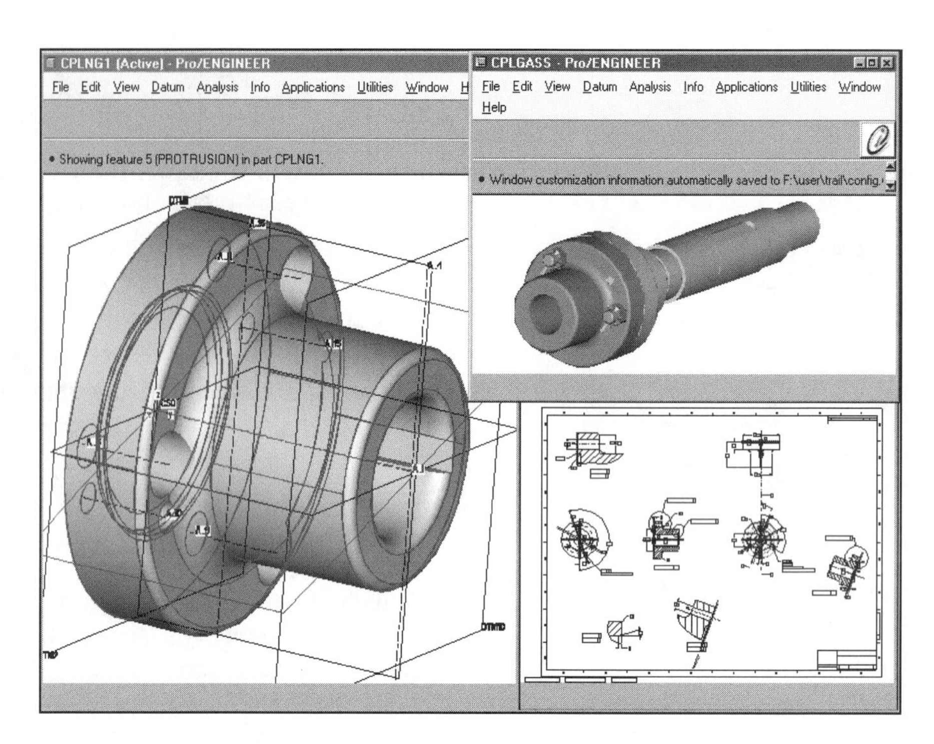

Figure 4.2
Part, Assembly, and Drawing Modes Open in the Same Pro/E Session

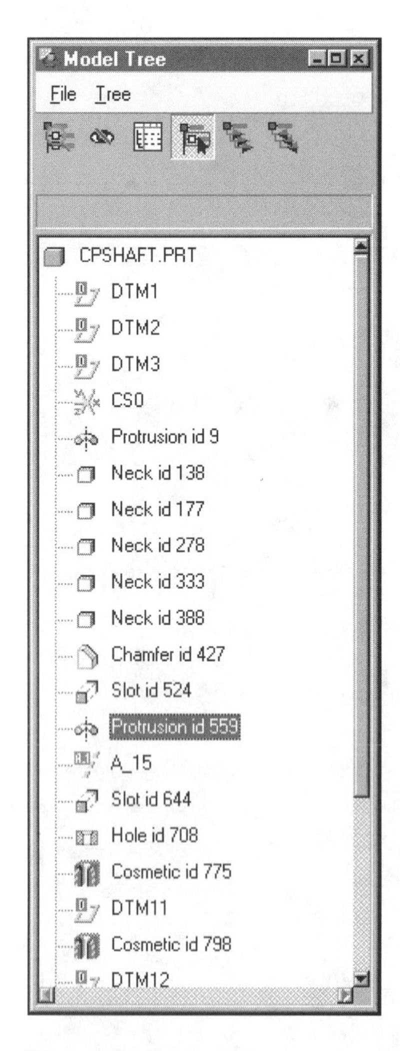

Figure 4.3
Part Mode

The *primary* modules covered in this text are:

Part This mode is used to create parts from *section sketches*, modify parts, and adding features to parts (Fig. 4.3). It is the module that you will be working in most of the time. This mode is covered in Lessons 1 through 13 and Lesson 21.

Assembly This mode is used to assemble components and is covered in Lessons 14 and 15.

Drawing This mode is used to create fully dimensioned drawings of parts and/or assemblies. The Drawing mode is covered in Lessons 16 through 20.

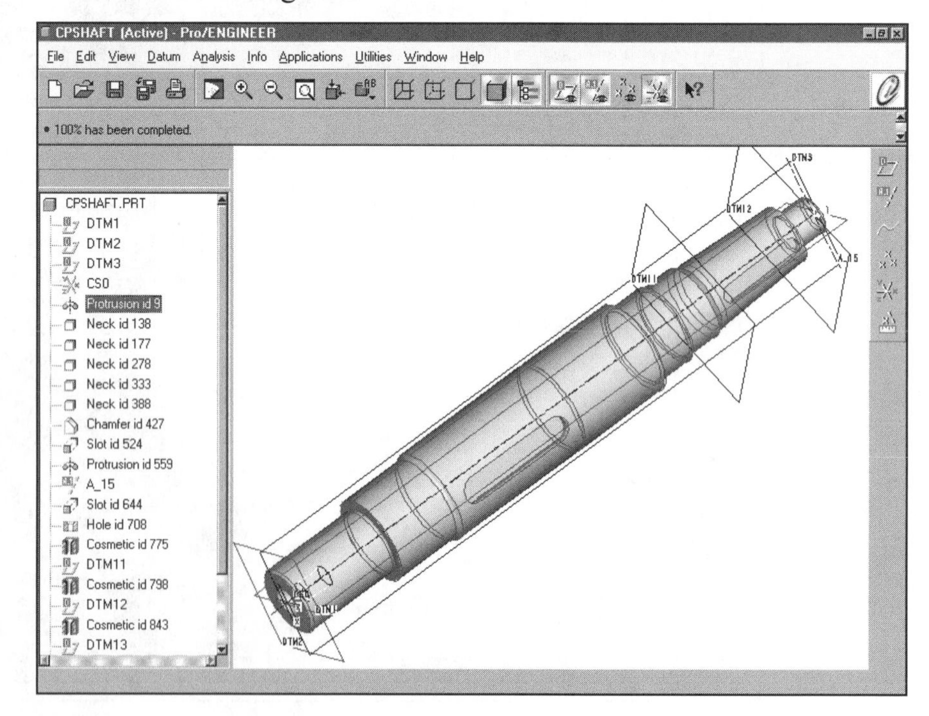

File Handling

File Naming

The rules for file naming, such as file name length and the effects of uppercase/lowercase names, are dependent on the operating system. Be aware of these rules before you file "objects" in Pro/E.

The following requirements apply when creating and performing operations on Pro/E objects/files:

* Names for all Pro/E files are restricted to a maximum of 31 characters. An object with more than 31 characters in its name cannot be created or retrieved.
* Brackets ([,], (, or)) and periods (.) cannot be used in file and directory names.
* Nonalphanumeric characters (such as @, #, and %) cannot be used in file or directory names.
* For Pro/E file and directory names, use only lowercase characters. Pro/E always saves objects and files on disk with lowercase file names.
* Because Pro/E adds a period (.) between the file name entered and the appropriate extension, do not use a period (.) in any file names you create.

The following is an example of default-naming conventions that are used by Pro/E to aid file management.

xxx.sec	Cross section or sketch
s2d####.sec	Default sketch
xxx.prt	Part
prt####.prt	Default part
xxx.drw	Drawing
drw####.drw	Default drawing
xxx.frm	Drawing format
xxx.asm	Assembly
asm####.asm	Default assembly
color.map	Shaded view color settings
names.inf	File used for Names INFO listing
rels.inf	Temporary file for relations listing
partname.inf	Temporary file used for Part INFO listing
partname.m_p	Temporary file used for Part Mass Properties listing
assemblyname.inf	Temporary file used for Assembly INFO listing
feature.inf	Temporary file used for Feature INFO listing
config.pro	Default configuration options file
trail.txt	Default name for Trail file
xxxx.cfg	Model Tree Settings file
config.win	Default customize dialog box options file

File Management

Within the Pro/E environment, files (**objects**; where an "object" may be a part, assembly, drawing, layout, sketch, or manufacturing model) are managed through the **File** menu (Fig. 4.4). The submenu under **File** has a variety of options, including:

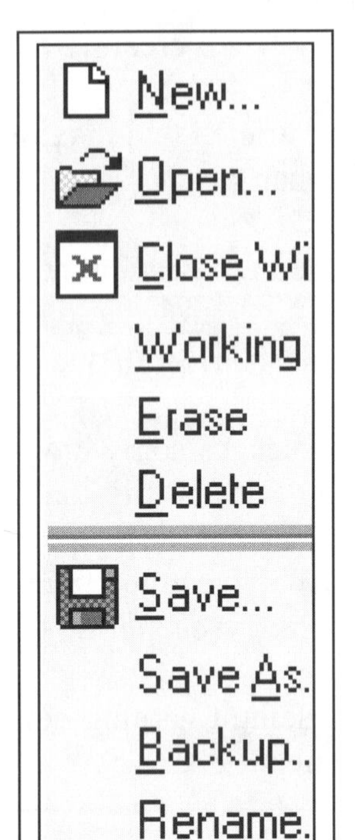

New Creates a new sketch, part, assembly, manufacturing, drawing, format, report, diagram, layout, or markup file.
Open Retrieves files of all types and sub-types from the current session (in memory) as well as from disk.
Close Window Close window and keep object in session.
Working Directory Selects a different working directory.
Erase

⇒ **Current** Removes an object from working memory.
⇒ **Not Displayed** Removes undisplayed objects from workstation working memory.

Delete

⇒ **Old Version** Every time you save an object using the **Save** command, you create a new version of the object in memory and write the previous version to disk. Pro/E numbers each version of an object file consecutively (for example; box.sec.1, box.sec.2, box.sec.3).

This command can be used to free up disk space and remove old, unnecessary versions of objects.

⇒ **All Versions** Deletes all versions of an object from the working memory *and from the disk.*

Save Files the object to the disk.
Save As Saves a copy to the same directory with a new name or to a different directory with the same name or with a new name.
Backup You can create a backup copy of an object.
Rename Changes the file name of the current object to a new name. *Have the part, assembly, and its drawing active (in memory) before changing names!*

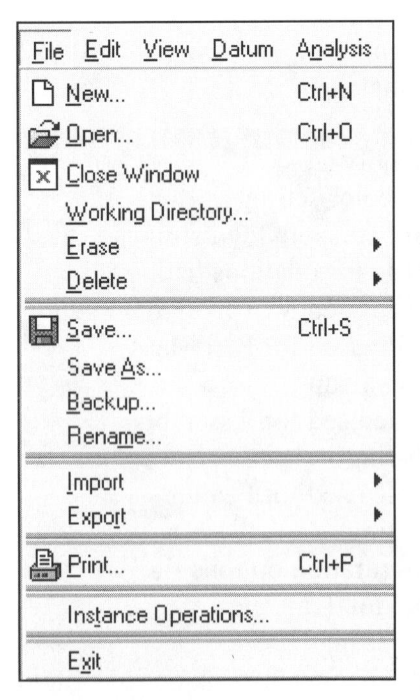

Figure 4.4
File Menu

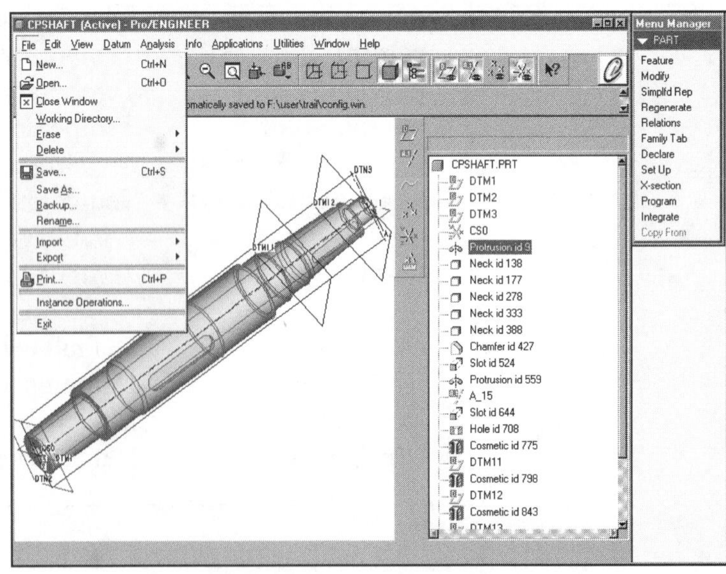

Creating New Files (Objects)

A new sketch, part, assembly, manufacturing, drawing, format, report, diagram, layout, or markup file in Pro/ENGINEER can be created using the **New** dialog box.

To open the New dialog box, from the **File** pull-down menu, choose **New**. You use this dialog box to (Fig. 4.5):

* Select the type of file to create, such as sketch, part, or report.
* If required, select a sub-type, such as **Solid** for **Part.**
* Enter a name for your file or use the default name.
* Choose a default *template file* to use in creating a new model.

File ⇒ New or pick on the Create new object icon:

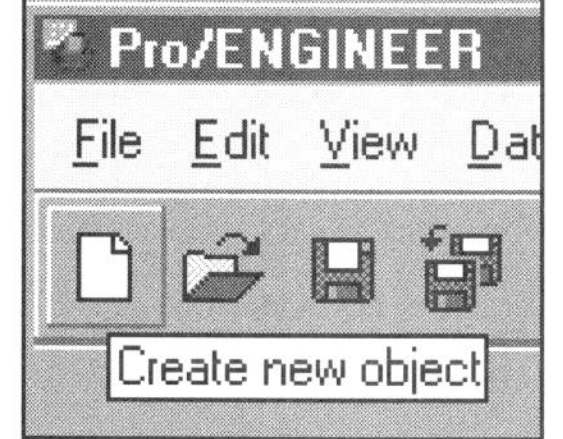

File sub-type

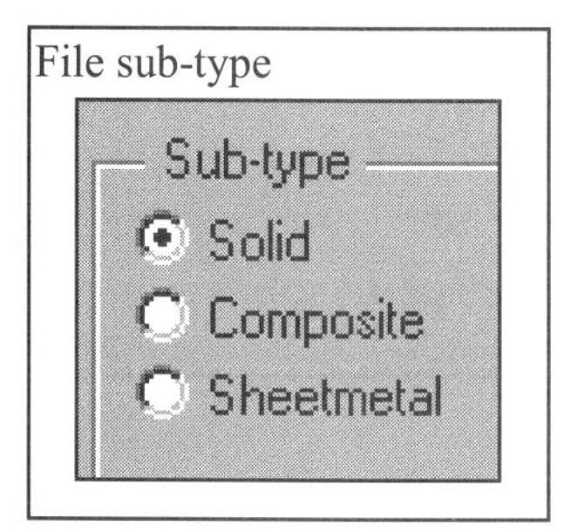

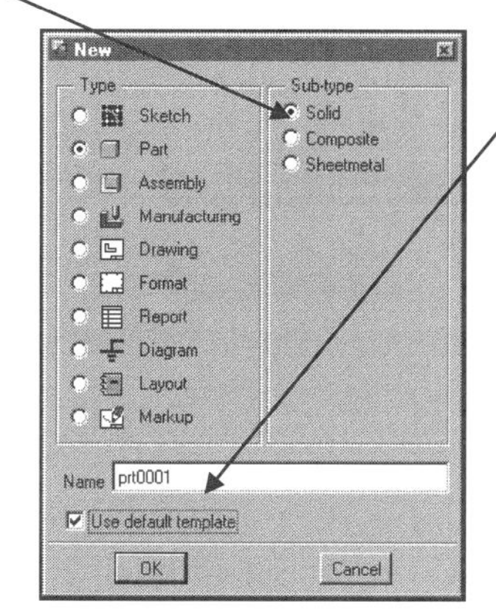

Default Part Name: **prt0001**

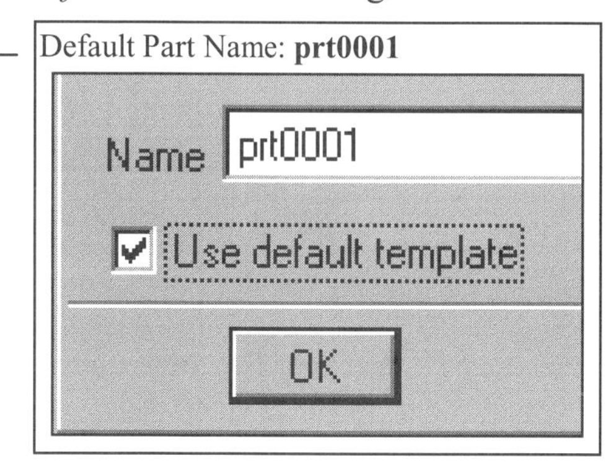

Figure 4.5
Dialog Box for Opening a New Object

Opening Files

You can retrieve files of all types and sub-types from the current session (in memory) as well as from disk using the **File Open** dialog box. To open the File Open dialog box, from the **File** menu, choose **Open**. You use this dialog box (Fig. 4.6) to:

* Navigate the disk and directory structure.
* Select and open a file.
* Create or select a simplified representation of a file.

Figure 4.6
Opening an Existing Object

List of objects in current directory **(lesson_15)**

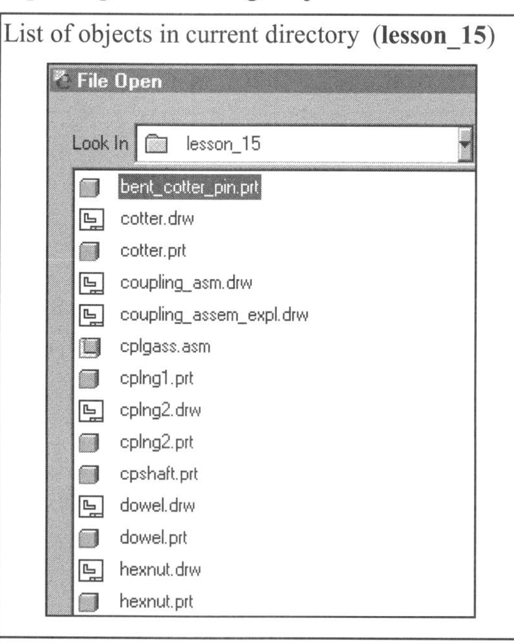

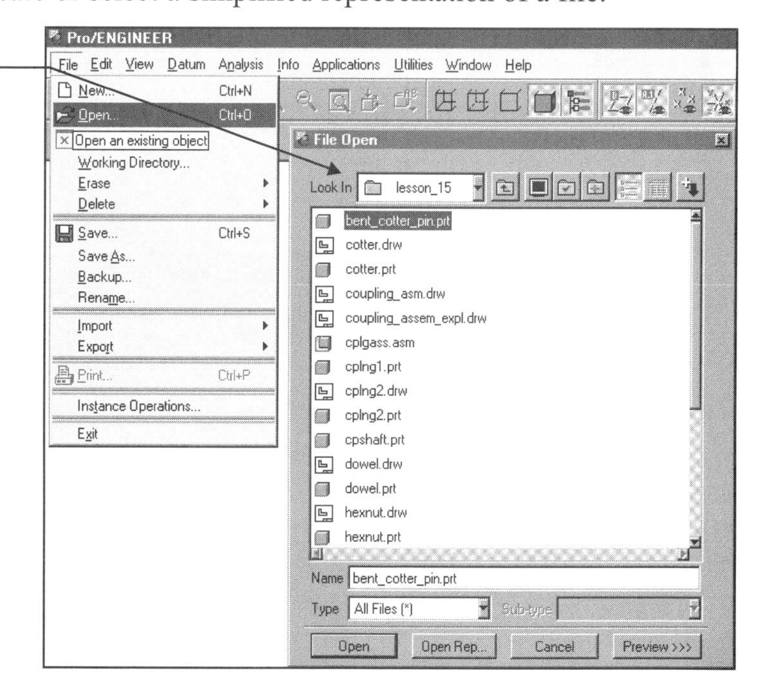

Working Directory

You can select a different working directory while in a Pro/E session using the **Working Directory** dialog box. When you retrieve files on your system, Pro/E generally retrieves them from a directory structure. The working directory is a directory that you set up for your Pro/E files. You need to have read/write access to this directory. You usually start Pro/E from your working directory (default directory). To open the Working Directory dialog box (Fig. 4.7) from the **File** pull-down menu, choose **Working Directory.** You use this dialog box to select the desired working directory. All new models that you create and save will be stored in this working directory (default directory).

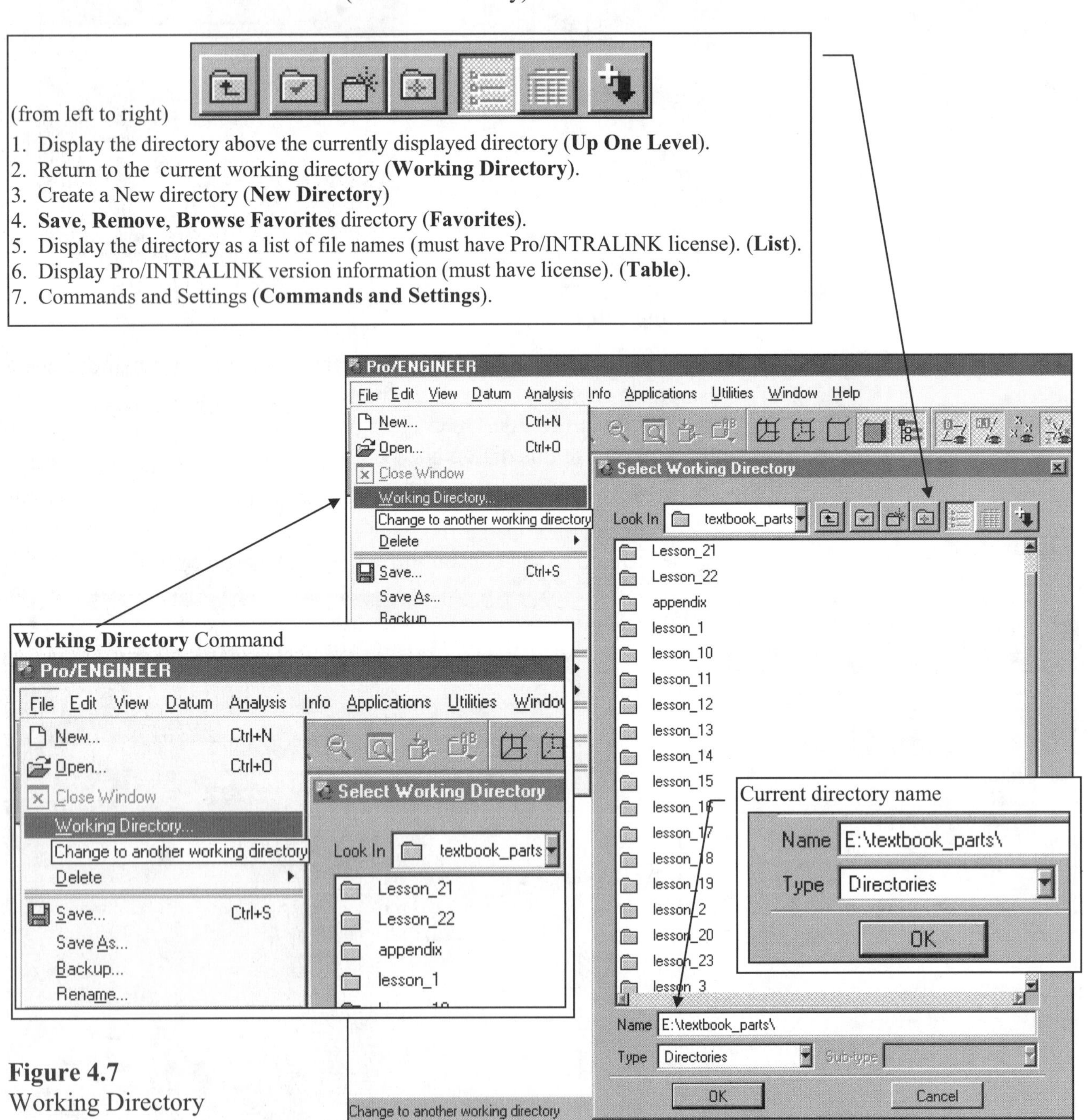

Figure 4.7
Working Directory

Erasing

You can erase the *current object* using the **Erase** dialog box. To open the Erase dialog box (Fig. 4.8) from the **File** pull-down menu, choose **Erase ⇒ Current.**

For assemblies (Fig. 4.8), manufacturing models, and drawings, the Erase dialog box opens. You use the Erase dialog box to remove the object or objects and its associated objects from memory.

For parts, formats, layouts, markups, and other types of objects, a confirmation dialog box opens.

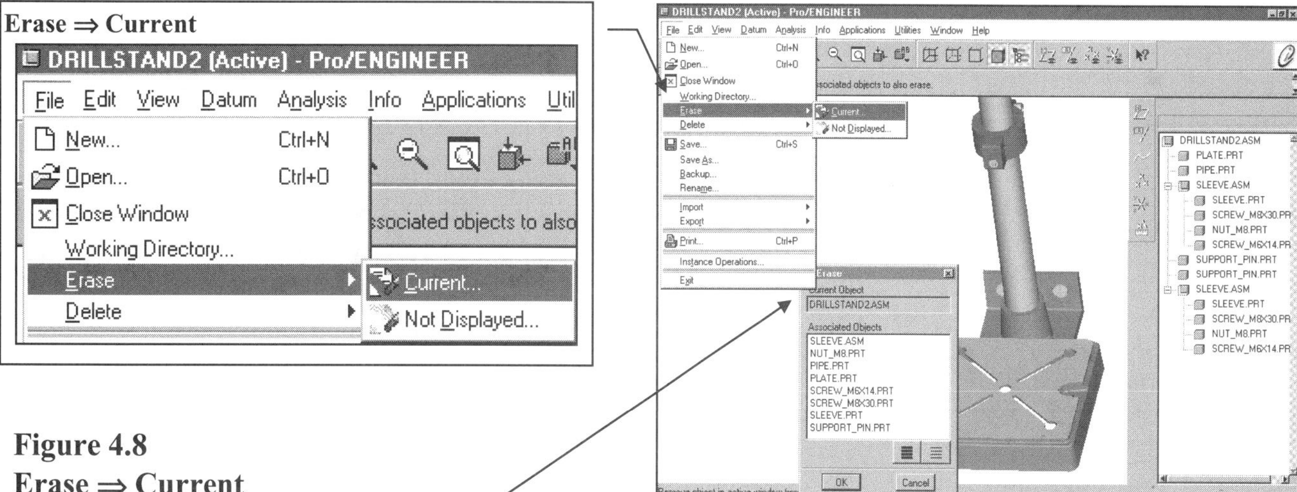

Figure 4.8
Erase ⇒ Current

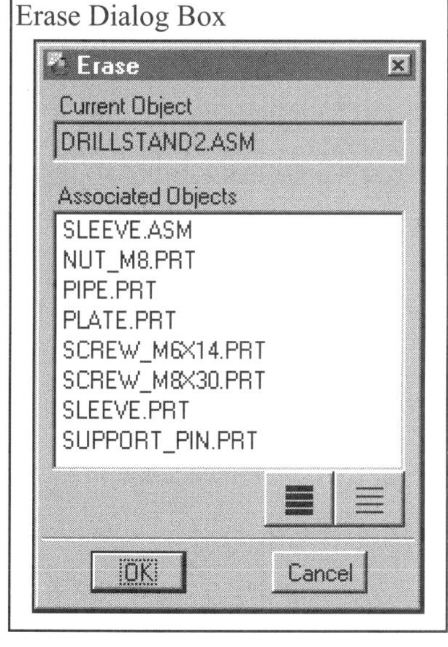

You can also erase objects using the **Erase Not Displayed** dialog box. This dialog box allows you to erase from the current session all objects except for those that are currently displayed and any objects referenced by the displayed objects. For example, if an assembly instance is displayed, the instance's generic and its components are not erased. To open the Erase Not Displayed dialog box (Fig. 4.9) from the **File** menu, choose **Erase ⇒ Not Displayed.** You use this dialog box to erase undisplayed objects.

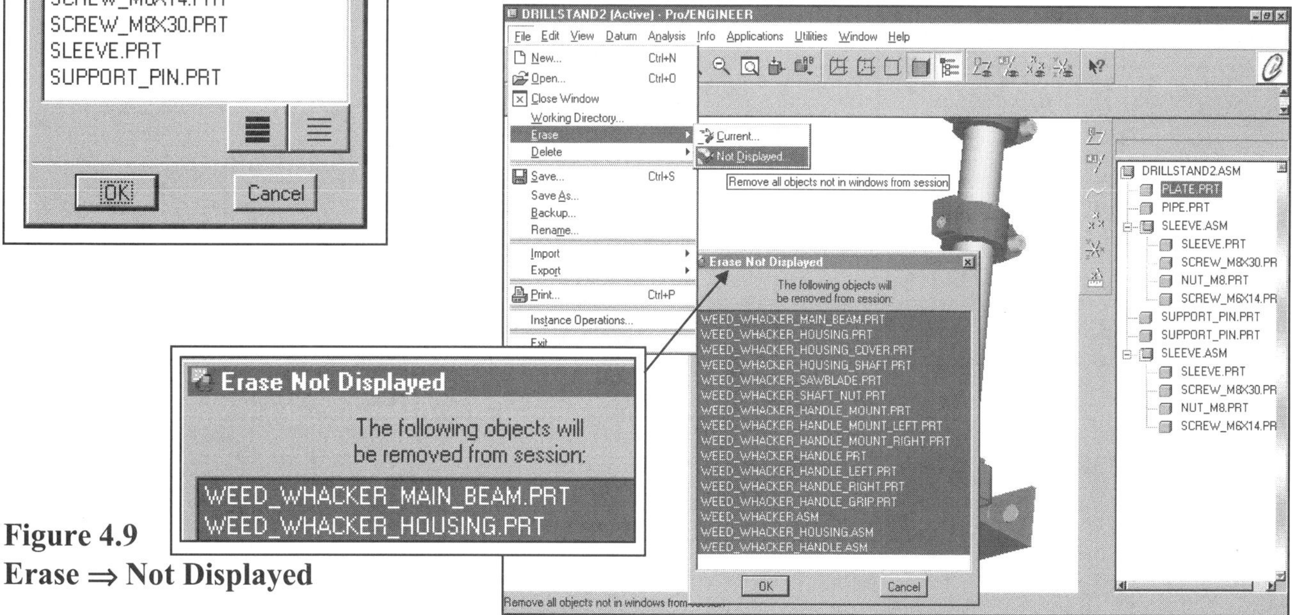

Figure 4.9
Erase ⇒ Not Displayed

Deleting Files (Objects)

Every time you save an object using the **Save** command from the **File** menu, you create a new ***version*** of the object in memory and write the previous version to disk. Pro/E numbers each version of an object file consecutively (for example, cylinder.prt.1, cylinder.prt.2, cylinder.prt.3).

The **Delete** command is used to free up disk space and remove old, unnecessary versions of object files.

You cannot delete a part or subassembly that has been used in an assembly or drawing during the current working session until the assembly or drawing has been deleted.

Deleting Old Versions of a File (Object) from Disk

You can *purge* objects from the disk using the **Delete** command from the **File** menu. To use this function from the **File** pull-down menu, choose **Delete ⇒ Old Versions.** A prompt appears in the message area with the name of the current file (Fig. 4.10). If you press the keyboard **Enter** or **Return** key or click the green check mark **✔,** you purge from the disk all versions of the specified object file except the most recent version.

Figure 4.10
Delete ⇒ Old Versions

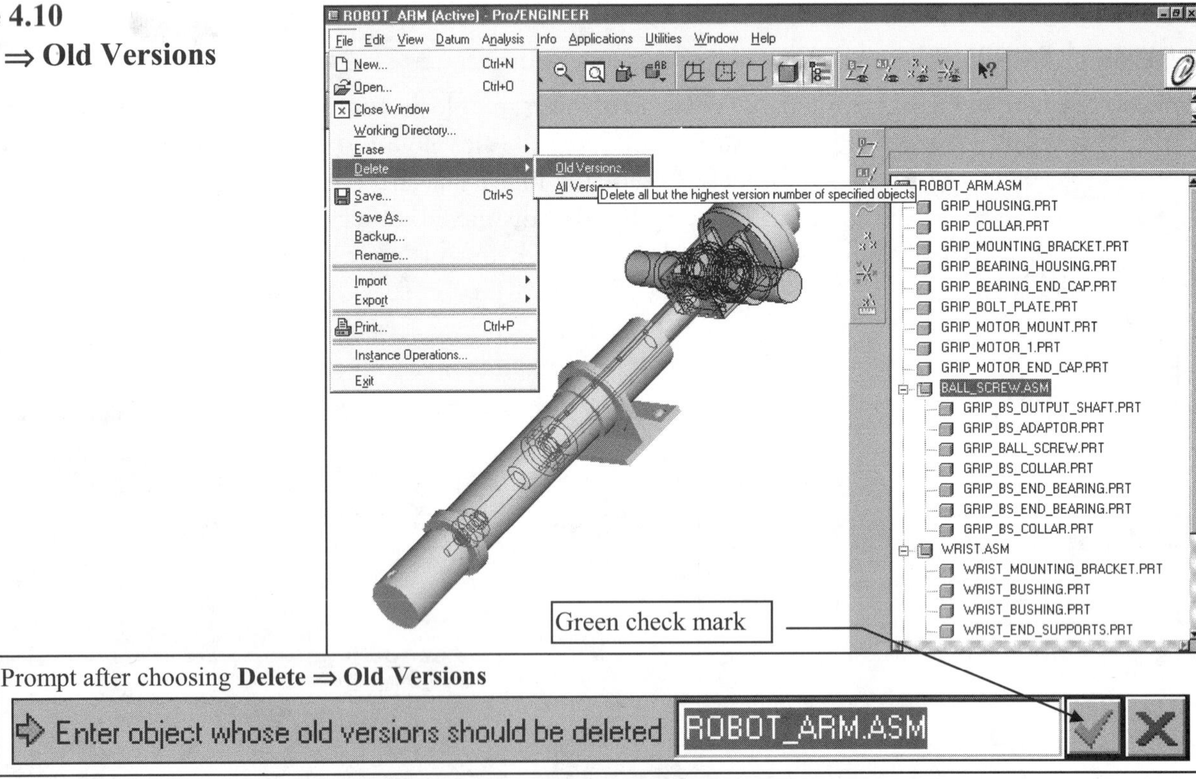

Deleting All Versions of a File (Object) from Disk

You can delete all versions of an object on the disk using the **Delete** command from the **File** menu. To use this function from the **File** pull-down menu, choose **Delete ⇒ All Versions**. You will be prompted with a warning after choosing this command (Fig. 4.11).

For assemblies, manufacturing models, drawings, parts, formats, layouts, markups, and other types of objects, a confirmation dialog box opens. You use the **Delete All Confirm** dialog box to remove the object and its associated objects from disk.

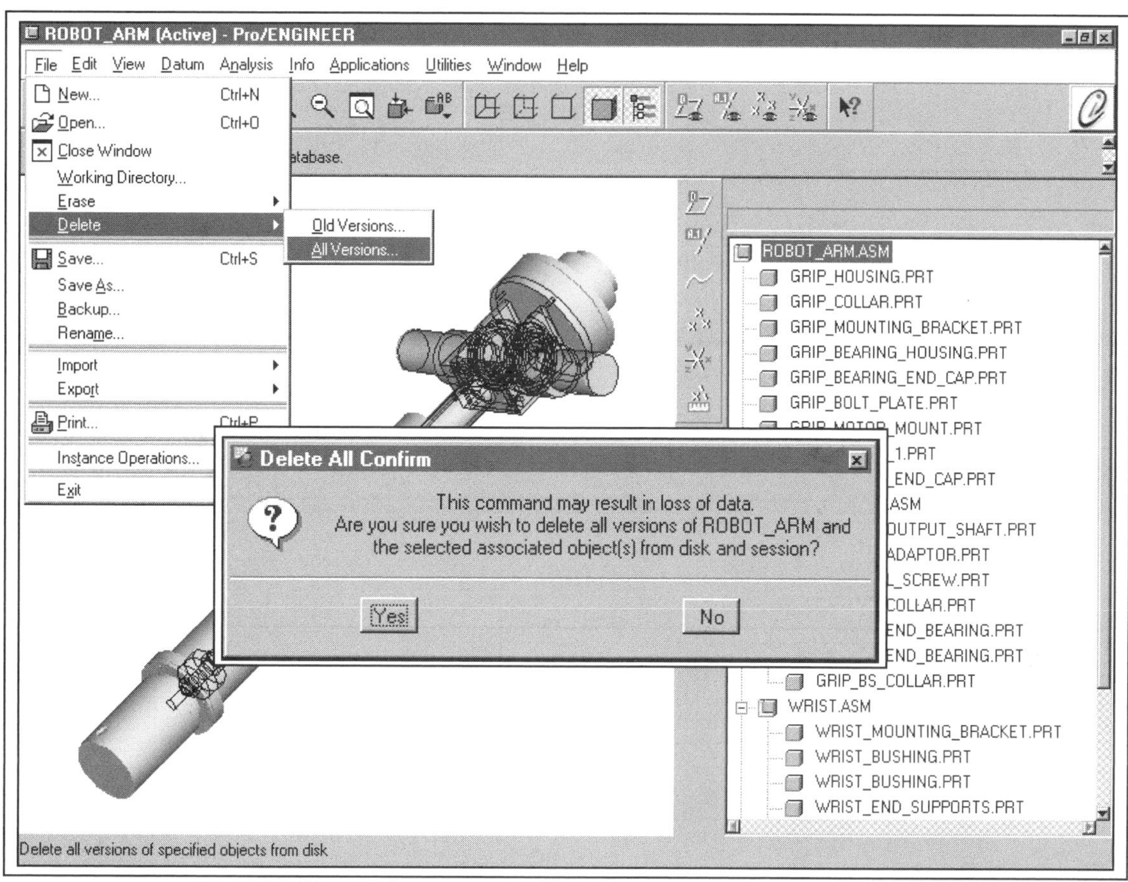

Figure 4.11
Delete ⇒ All Versions

CONFIG.PRO
Set your config.pro file to:

prompt_on_exit yes

and you will be prompted to save when you exit a Pro/E session.

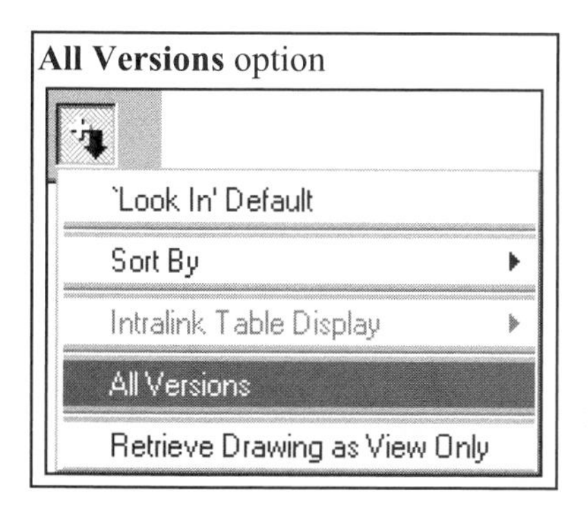

Saving Changes

Saving your files (objects) is the most important file management procedure. You can save changes in an object at any time by selecting **Save** from the **File** pull-down menu (Fig. 4.12). *It is important always to save and to save often.* Pro/E does not automatically save files for you as some software programs do. ***It is your responsibility to save the work you have completed.*** One good practice is to save after every important change to the object on which you are working. When saving an object, Pro/E creates a new version of it by increasing the version number. The following is an example:

Original part	**clamp.prt.1**
Part after 1st save	**clamp.prt.2**
Part after 2nd save	**clamp.prt.3**

Selecting the green check mark ✔ saves the object with the current default name (Fig. 4.12).

To retrieve an old version, specify the version number in the retrieval name. To display the version numbers in the File Open dialog box, use the **All Versions** option (see left column).

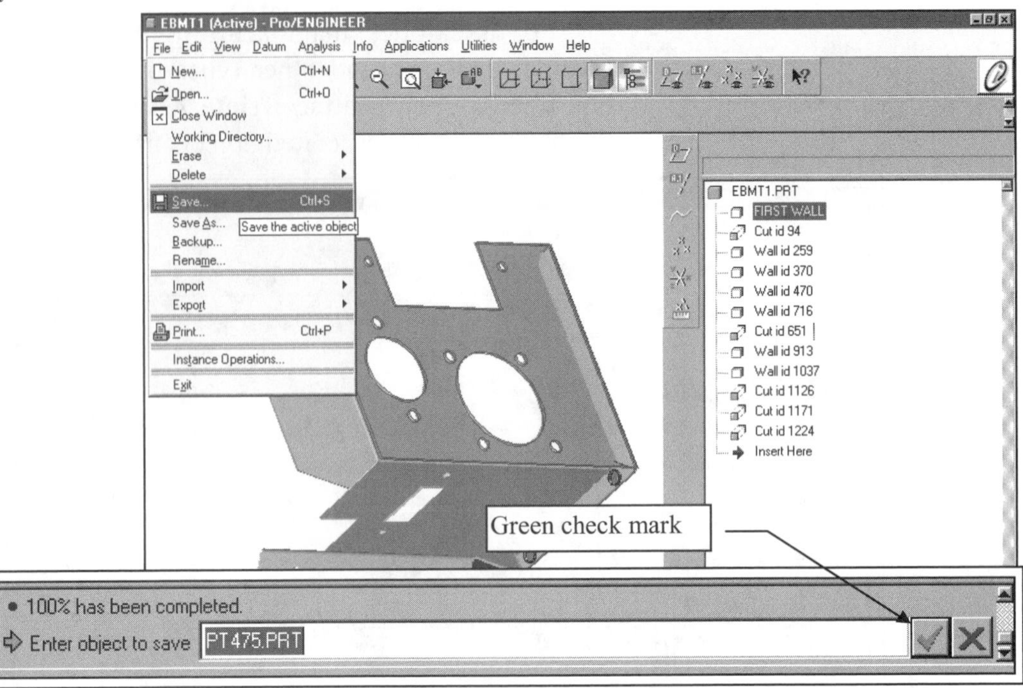

Figure 4.12
File ⇒ Save

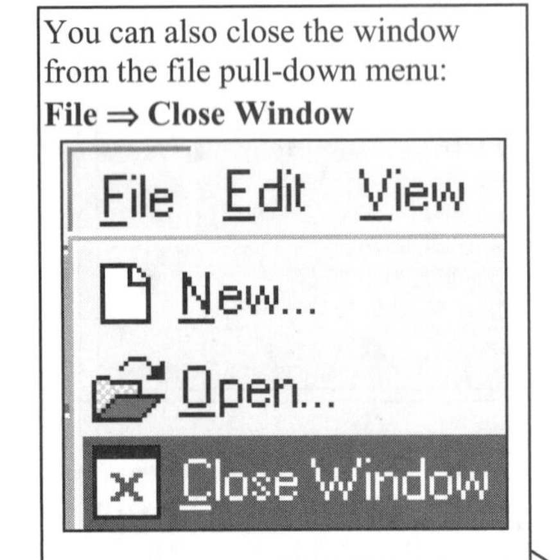

Closing Windows

To close a subwindow you can use the **Close Window** option in the **Window** pull-down menu (Fig. 4.13), or use **File ⇒ Close Window.** *Closing the window in this way does not remove the model from the current session of Pro/ENGINEER.* The model still occupies space in the current working memory of the computer. If multiple windows are in session, this may reduce the speed of using the software. When the model is no longer required in memory (RAM), erase it from memory by choosing **File ⇒ Erase ⇒ Current.** To erase models that are in session but are not displayed in the main working window, choose **File ⇒ Erase ⇒ Not Displayed.**

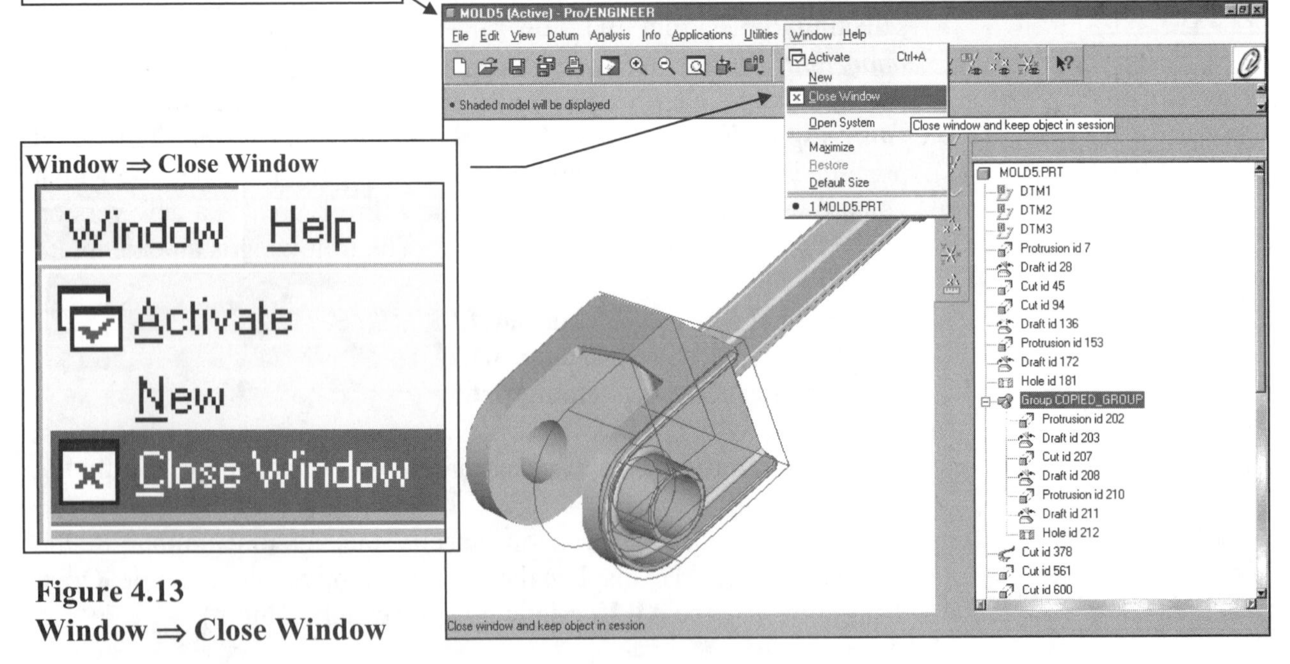

Figure 4.13
Window ⇒ Close Window

Using Multiple Pro/ENGINEER Windows

Each Pro/E object that you open or create appears in a window containing an integrated menu bar, toolbars, and message area in addition to the traditional graphics area. You can perform many operations from the top-level menus in multiple windows simultaneously without aborting pending operations.

If you have multiple windows open, only one of them is active at a time; however, you can still perform some functions in the inactive windows. To activate a window, from the **Window** menu, choose **Activate.** To switch between active objects; pick **Window** ⇒ (and then select the object you wish to work on), as shown in Figure 4.14.

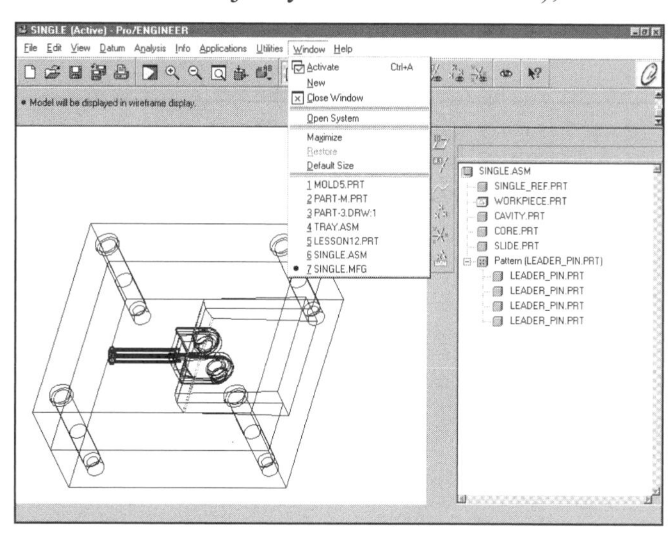

Figure 4.14
Active Objects

Application Manager

NOTE
This text is using the Windows NT version, so the Application Manager will not be displayed.

Pro/E automatically displays the Application Manager (Fig. 4.15) when it runs the Startup command. ***For Microsoft Windows systems, the Application Manager is not present; use the Windows taskbar tray area.*** The advantage of using the Application Manager is in the ease of window management. Windows do not become lost behind each other as you work with four or five different applications.

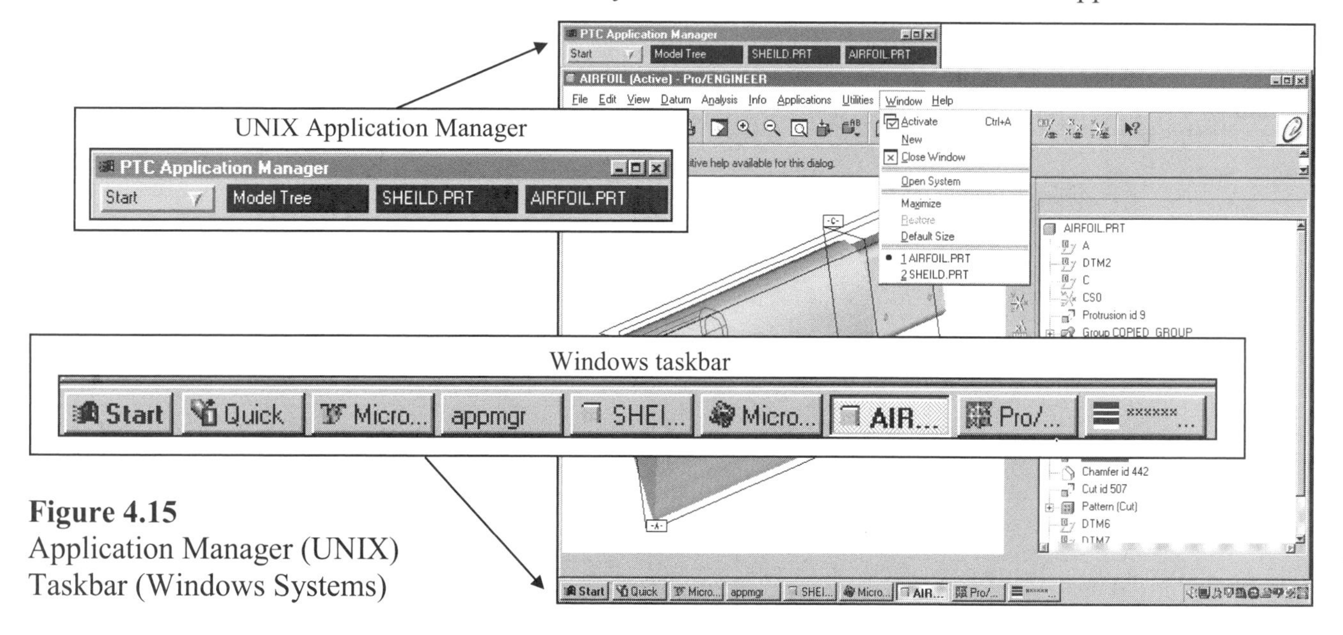

Figure 4.15
Application Manager (UNIX)
Taskbar (Windows Systems)

Printing

You can print an output file containing a picture of the object using the Print dialog box. From the **File** pull-down menu, choose **File ⇒ Print** (Fig. 4.16).

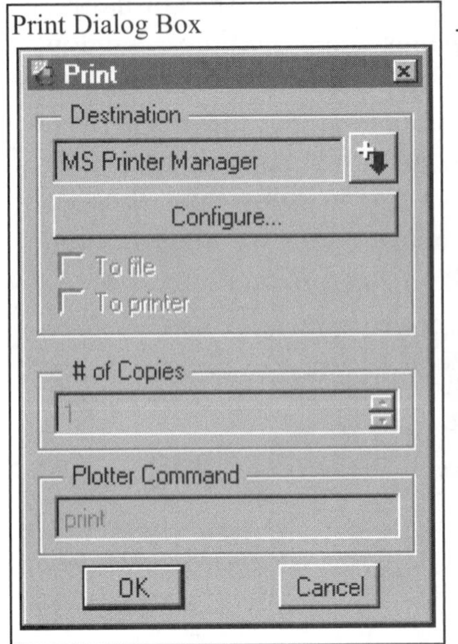

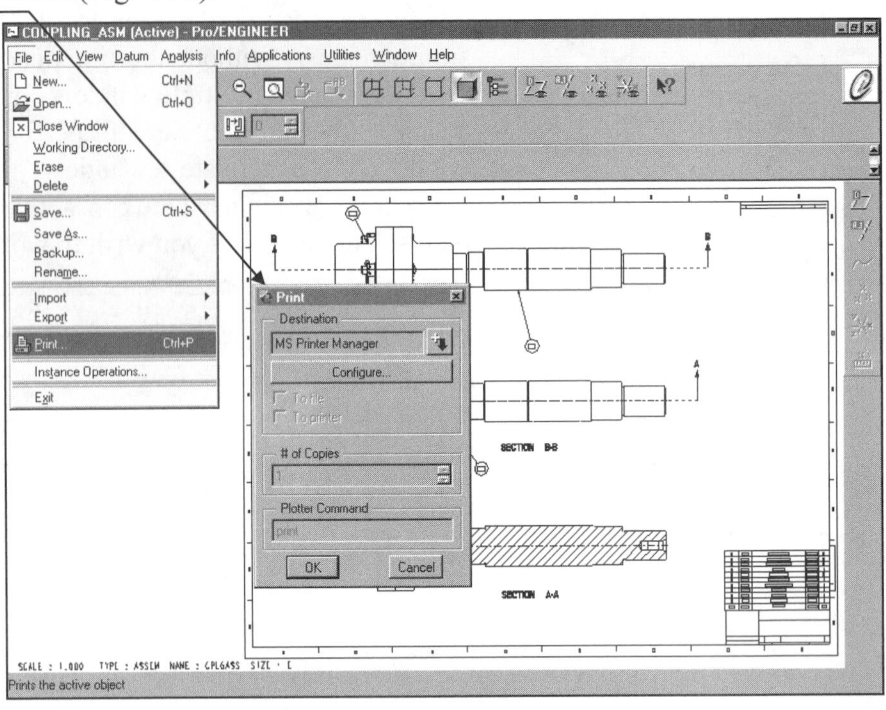

Figure 4.16
File ⇒ Print

Configuring Printers

You can configure your printer using the **Printer Configuration** dialog box. To open the Printer Configuration dialog box, select Configure in the **Print** dialog box. You use this dialog box to configure the printer for printing an object (Fig. 4.17). If you are printing a shaded image, the Shaded Image Configuration dialog box appears instead of the Printer Configuration dialog box.

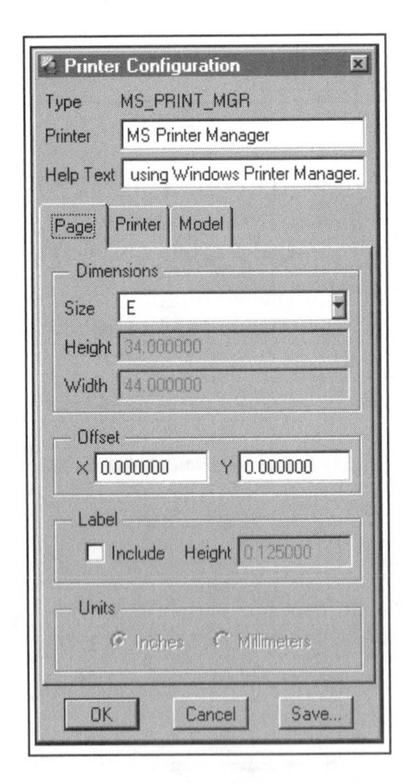

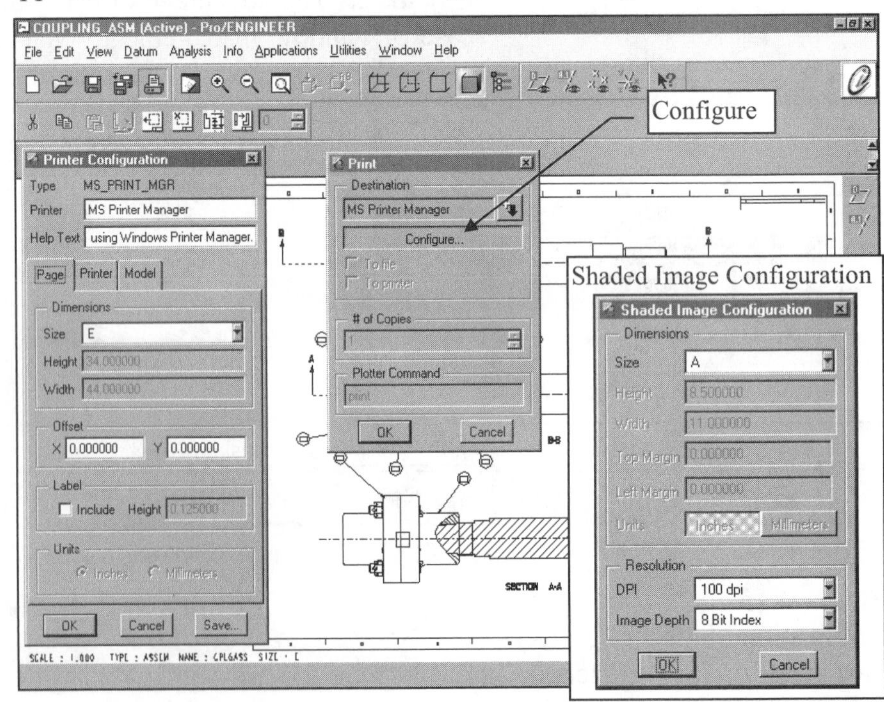

Figure 4.17
File ⇒ Print ⇒ Configure

Section 5

Interface

This section illustrates the basic skills required to begin a Pro/E session. You will learn how to interact with the Pro/E User Interface by performing some basic view manipulation operations.

When you start Pro/E, the *Main Window* opens on your desktop. This is the window in which you do most of your graphics work. It is composed of four areas:

* The ***pull-down menus*** on the Menu Bar at the top of the window contain options that are used for all modes.
* The ***Toolbar*** *(toolchest)* below the pull-down menus contains buttons that you can use as shortcuts to common Pro/E commands.
* The ***display area*** displays the model graphics.
* The ***message area*** displays system messages that prompt you for required information.

Pro/E uses a windows interface on the workstation to display a series of interactive work areas (Fig. 5.1). The Pro/E environment is divided into several main areas: the **Main (Working) Window** (the graphics window), the **Menu Manager** (Panel), the **Message Area,** the **Command Dialog Box,** and the **Model Tree.**

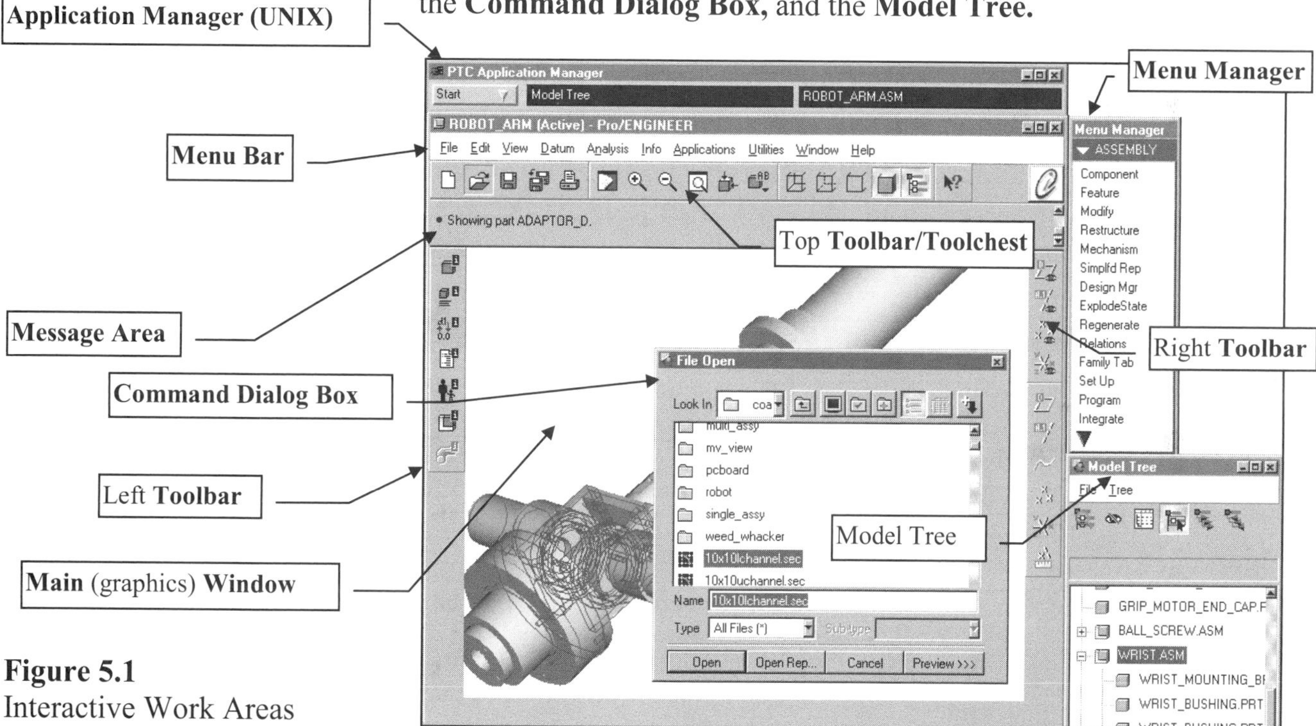

Figure 5.1
Interactive Work Areas

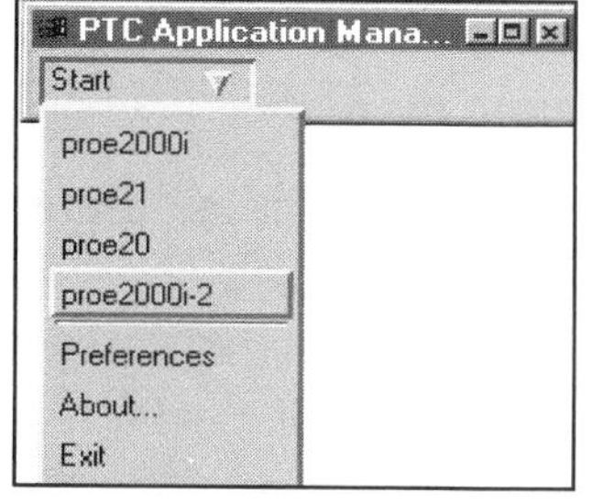

Figure 5.2
Application Manager (UNIX)

If you have a UNIX system, the **Application Manager** will be displayed along the top of the screen (default) (Fig. 5.2). As you can see, a lot is happening on your screen at the same time. You must be *visually ambidextrous* to master Pro/E!

The Main (Working) Window

The largest area is the Main Window. This is where the object is displayed and manipulated. This area is much like the screen on a television set. You have control over the scale and the angle at which you view your object. This Main Window is also the "parent" window in terms of the window manager. If you *iconify* the Main Window, all other Pro/E windows automatically disappear (until you restore the Main Window).

Pull-Down Menus

You can use the Pro/E *pull-down menus* at the top of the Main Window for all modes of the software (Fig. 5.3). Menu choices change according to the mode in which you are working, and others may be unavailable. In most modes, you have access to the following:

File commands for manipulating files.
Edit commands for editing your object.
View commands for controlling the model display and display performance.
Datum commands for creating datum features.
Analysis commands to analyze model properties.
Info commands for performing queries or reports.
Applications providing access to Pro/E modules.
Utilities commands for customizing your working environment.
Window commands for managing Pro/E windows.
Help commands for accessing online documentation.

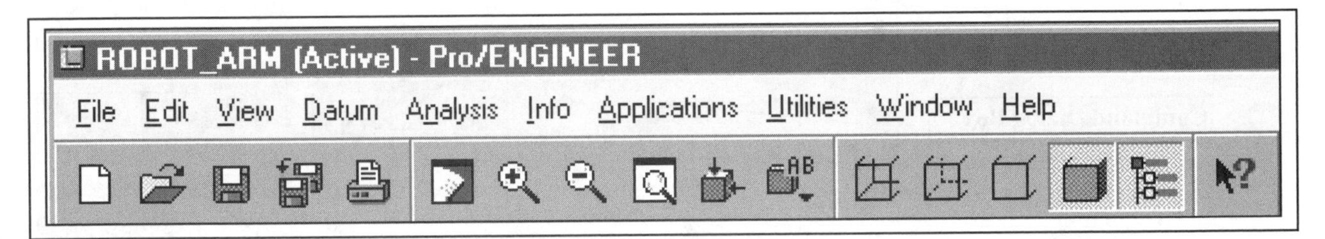

Figure 5.3
Menu Bar for Pull-down Menus

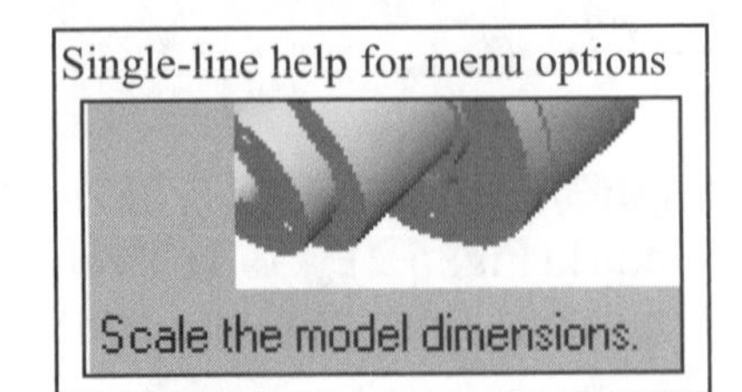

Message Area

The *message area* (Fig. 5.4), performs multiple functions, including:

* single-line help for menu options (bottom of main window)
* status information for an operation
* icons to represent warnings or status prompts
* ability to query for information to complete a command
* if needed, prompts for additional information by sounding a bell

Information is input in the *message area* by using the mouse or entered by the keyboard (if a white text box appears). Pro/E provides a default selection highlighted in *blue.* You can accept it by pressing **enter,** use the mouse to change the input parameters, or cancel the data entry operation by pressing the **<Esc>** key. To view old messages, you can use the scroll bar located on the right (Fig. 5.4).

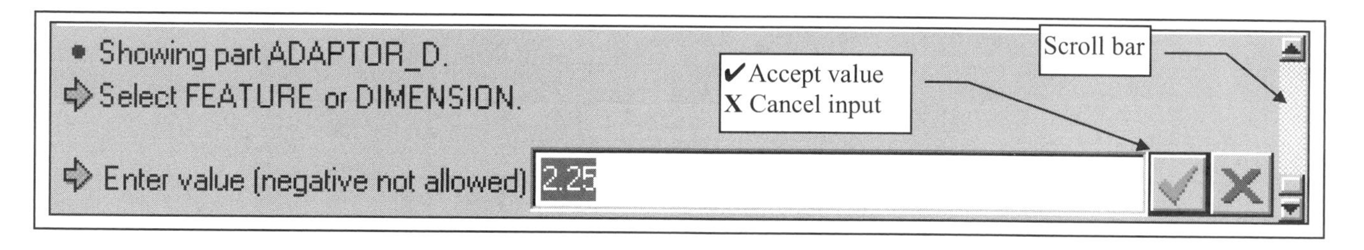

Figure 5.4
Message Area

The Menu Manager (panel)

You interact with Pro/E via a series of hierarchical menus. Menus are displayed in the **Menu Manager**. When you make a choice from this menu, Pro/E cascades a submenu, which offers you more choices, as shown on the right side of Figure 5.5.

Understanding Menu Selections

It is also important to understand how Pro/E displays options in the **Menu Manager**. When a menu first appears with an option already highlighted (in **REVERSE**), it is the ***default*** option and therefore will be automatically selected, as shown in Figure 5.5.

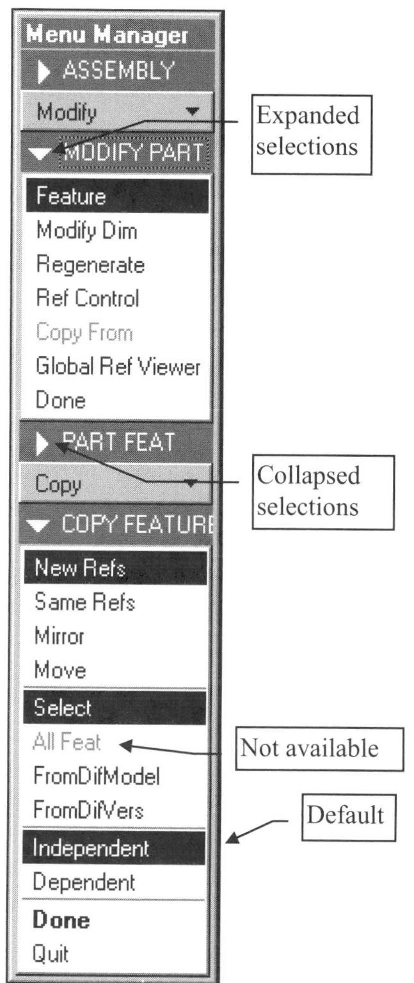

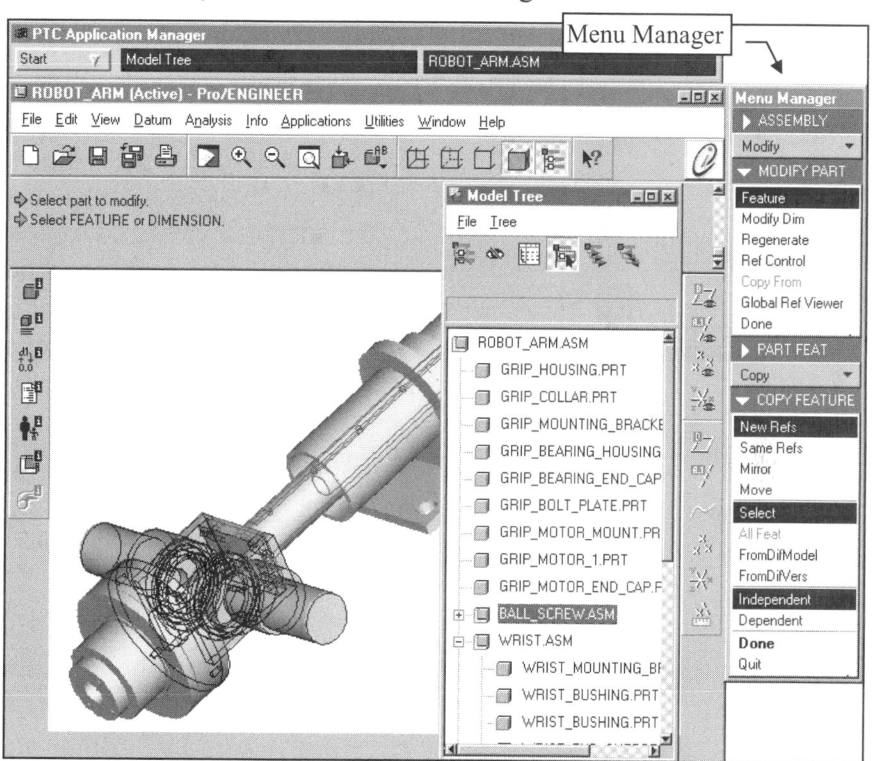

Figure 5.5
Menu Manager

When an option is *gray,* it is not available (Fig. 5.5). In this example, the **Copy From** option in the MODIFY PART menu is not available.

When a series of menus cascade, Pro/E expects you to pick a menu option from the lowest menu. However, you can select options from any displayed menu.

Some menus have multiple lists of options, separated by horizontal division lines. In general, you work your way down through the lists, choosing an option from each division.

In Figure 5.6, you must select one option from each of the first two lists. Here, **Extrude** and **Solid** have been chosen. You must make yet a *third* selection **(Done)** in order to proceed to the next set of menus (which will be displayed in place of the current set of menus).

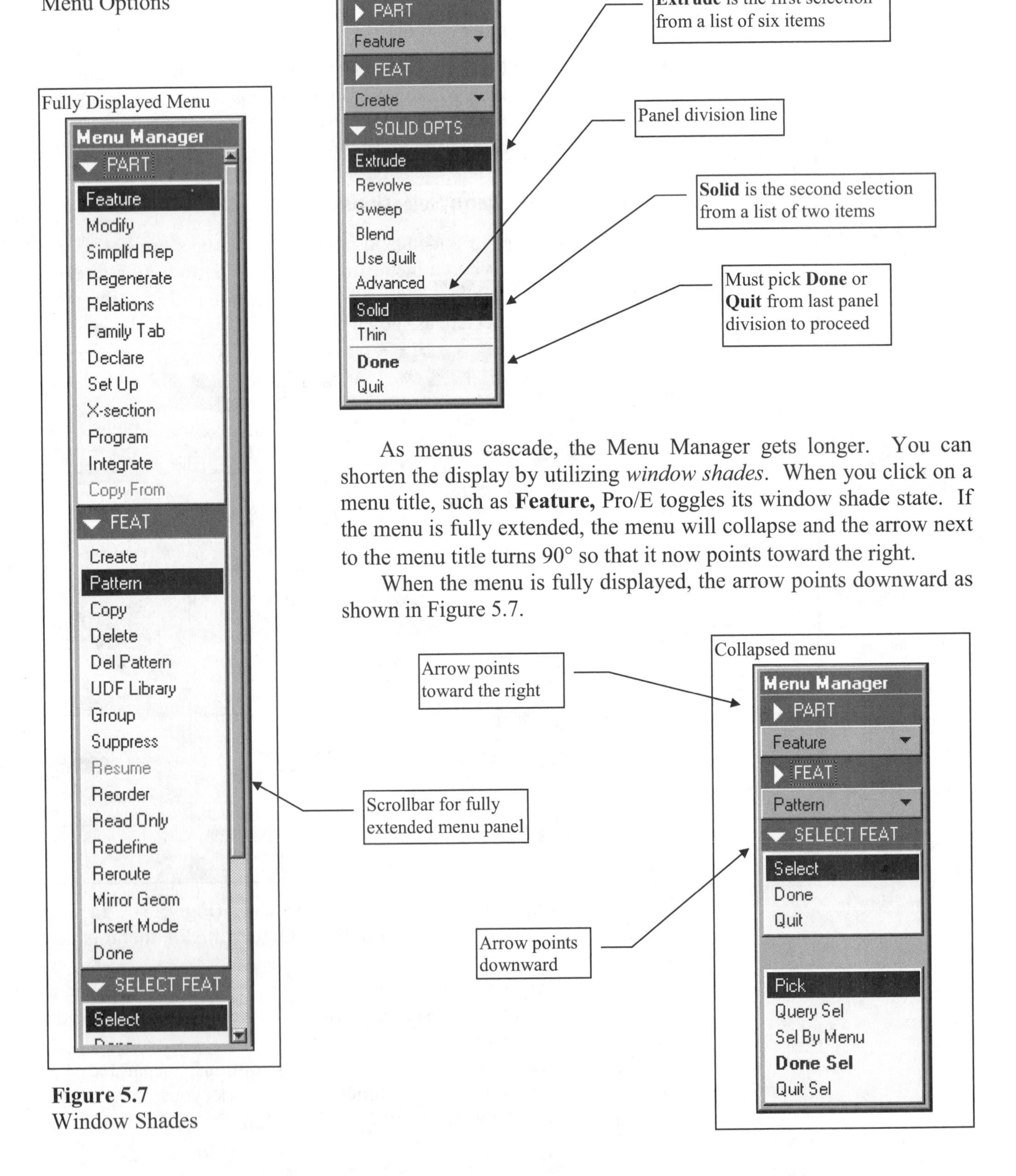

Figure 5.6
Menu Options

As menus cascade, the Menu Manager gets longer. You can shorten the display by utilizing *window shades*. When you click on a menu title, such as **Feature,** Pro/E toggles its window shade state. If the menu is fully extended, the menu will collapse and the arrow next to the menu title turns 90° so that it now points toward the right.

When the menu is fully displayed, the arrow points downward as shown in Figure 5.7.

Figure 5.7
Window Shades

To move on to the next set of menus in Pro/E, you usually must select **Done** or **Done Sel.** To terminate an operation, you usually select **Quit** or **Quit Sel**. To go back up a level in the menu structure, you usually select **Done, Done-Return,** or **Done/Return.** Keep in mind that these are only general rules, for you will discover that there are times when these rules are broken.

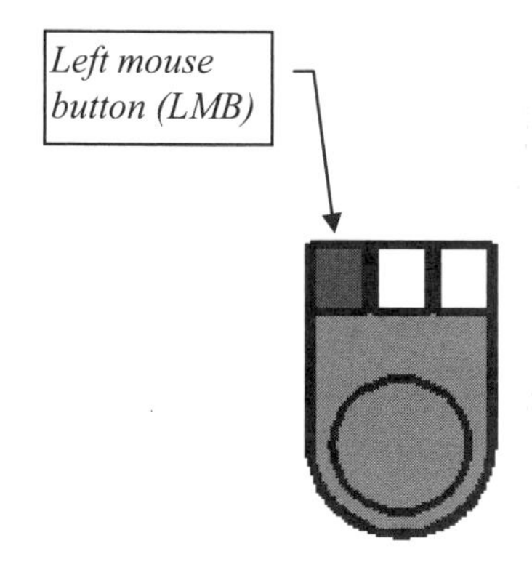

Selection of Items

Pro/E makes use of all three buttons on your mouse. Most of the time, however, you will use the left mouse button (**LMB**). The left mouse button is used to choose menu options, as well as to select objects in the Main (working) Window. You will learn more about the functionality of the other mouse buttons in another section.

Using the Message Area

As you move the mouse cursor over the menu options, they become momentarily highlighted (Fig. 5.8). If you press and release ("click") the left mouse button on an option, it will be selected and highlighted in *black.*

You can get additional information about Pro/E commands via menu options. Placing the mouse cursor on top of a menu option displays a one-line help message or hint about the menu option. The *hint* is displayed at the bottom, below the Message Area in Figure 5.8, explaining what that command will accomplish.

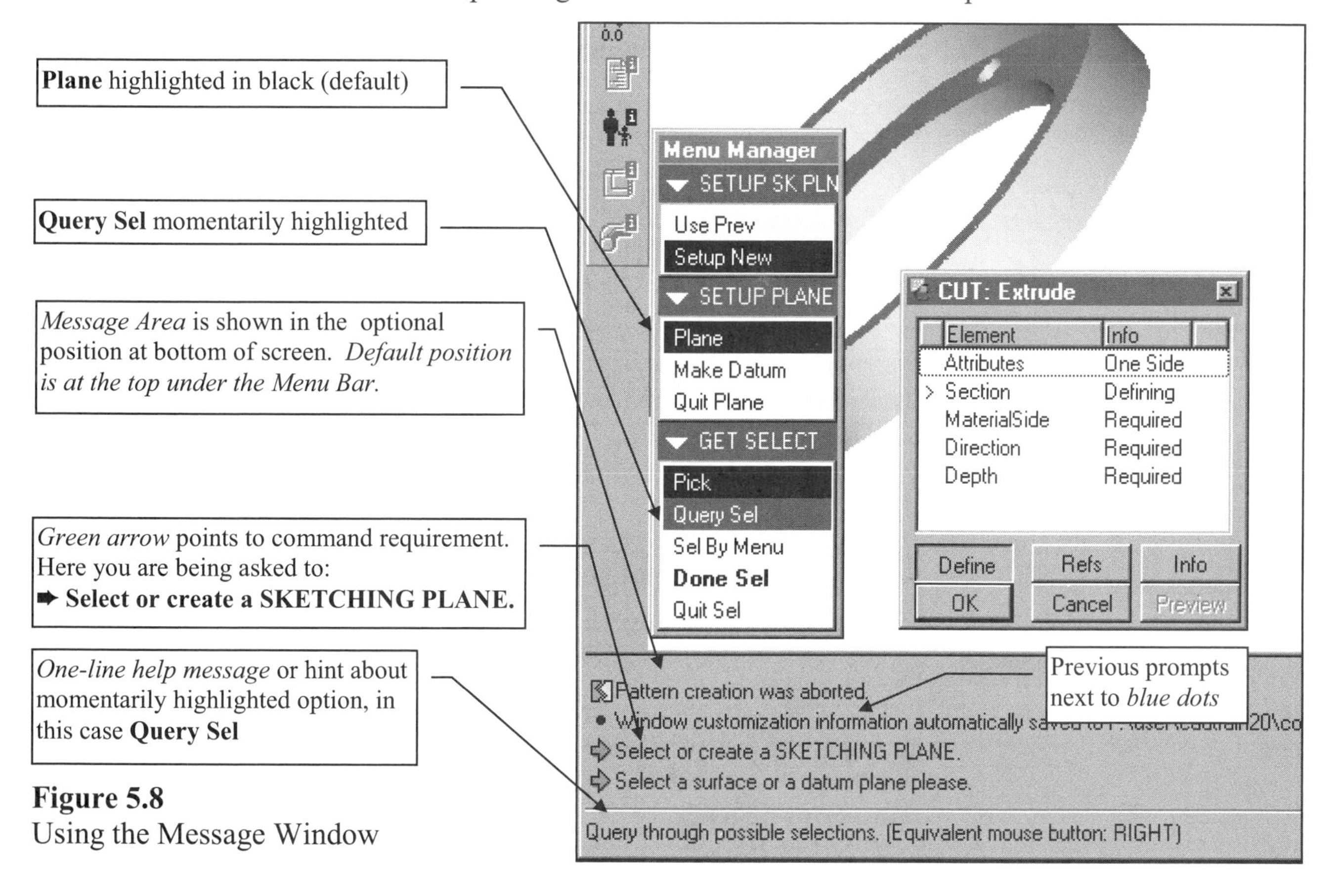

Figure 5.8
Using the Message Window

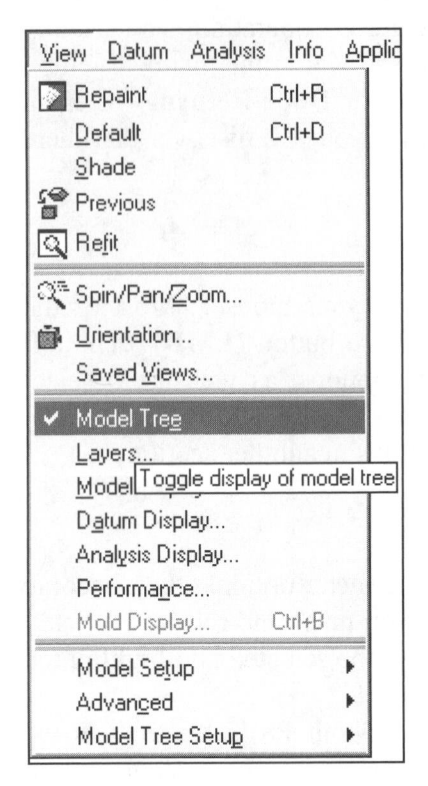

Figure 5.9
Model Tree

Reading the Message Area is very important. Not only does Pro/E display the menu hints as you move the mouse cursor over the menu options; it also displays a prompt that is related to the current operation (Fig. 5.8). The current prompt message is displayed next to the *green* arrow. The previous prompts are scrolled upwards in the Message Area and can be examined if necessary. This allows you to review the operations you have completed.

The Model Tree

When you actually retrieve an object, Pro/E displays a window called the **Model Tree.** The **Model Tree** is used to navigate through, and select objects in, a Pro/E model. You will learn more about how to use this tree in Section 9. You can turn it on and off by choosing the **View** pull-down menu from the Menu Bar and selecting ✔ **Model Tree.**

The expanded Pro/E Model Tree (Fig. 5.9) window tool presents a breakdown of the model structure, feature by feature, in the order in which the features were created. Using this tool, you can select features for various tasks such as **Modify** and **Delete** and change its format to display various forms of information. Icons appear beside features in the Model Tree to denote the item type and its current status. The Model Tree can expand to show the **Feat**(ure) **Name, Feat**(ure) **Type, Status,** and a variety of other information.

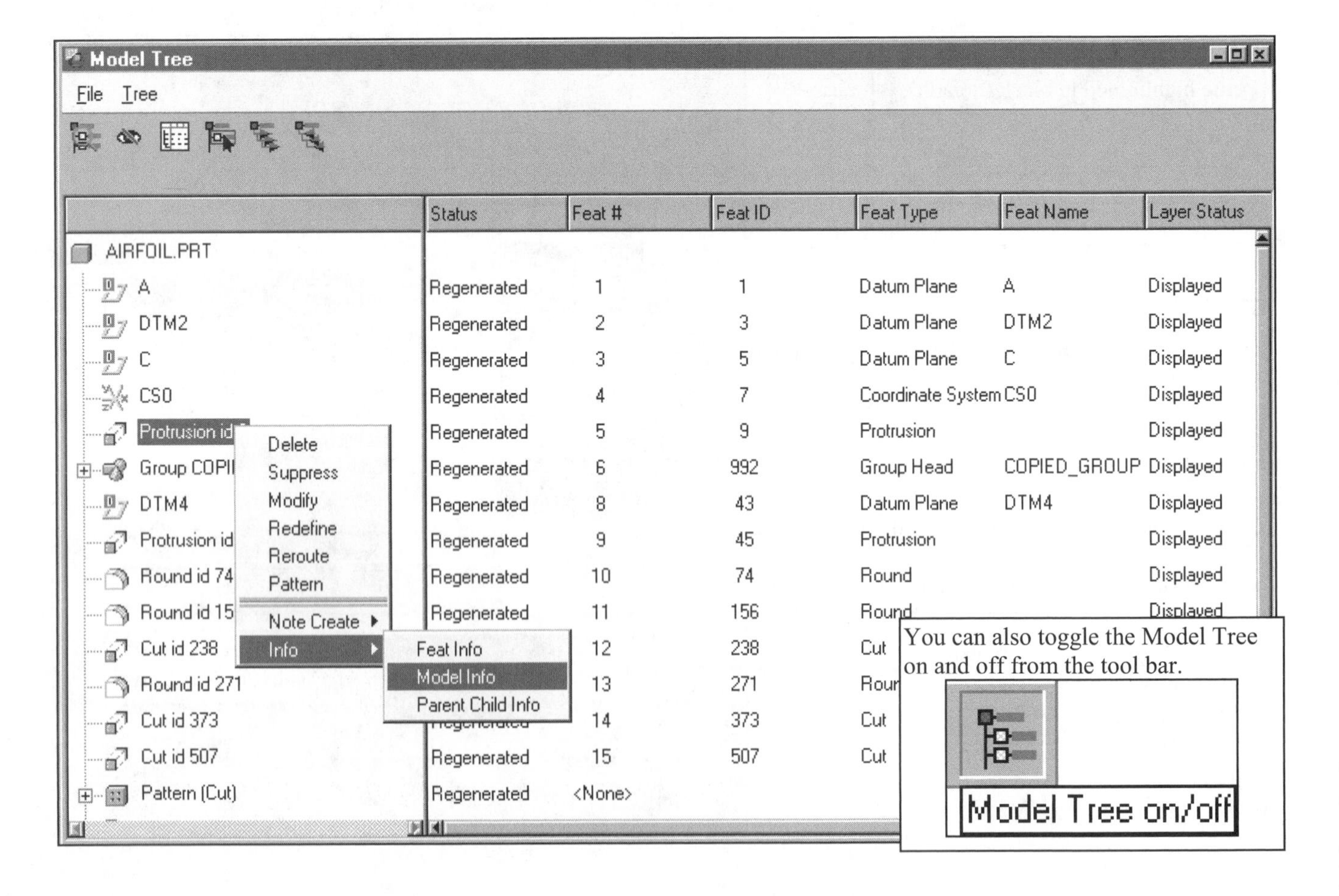

Orientation

You can think of the Main Window as being just like a camera. It displays a picture of the current object on your screen, based on the current orientation of the object and on how close the "camera" is to the object (Fig. 5.10).

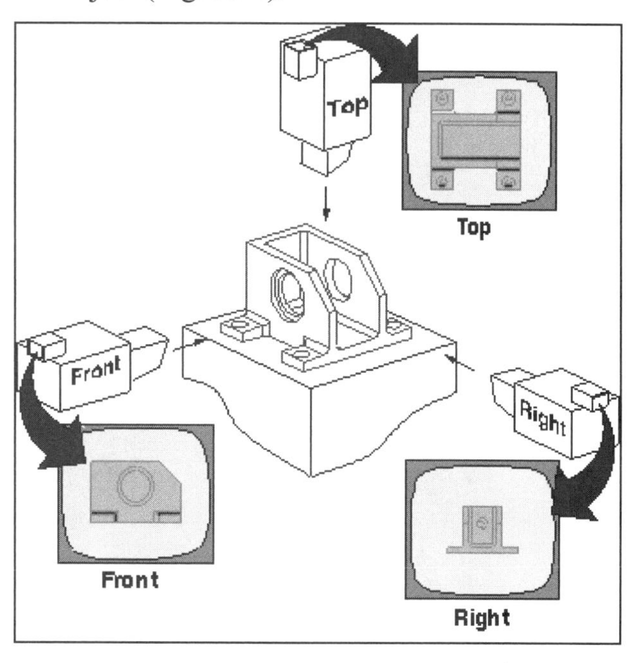

Figure 5.10
View Orientation

The function you use to view the object from different angles is called **Orientation.** This term is analogous to the technique of *rotating* the camera in photography to look at the objects in the view finder from a different angle.

The function you use to move closer to, or farther away from, the object is called **Zoom.** This term is analogous to the photographic technique of *zooming* in or out to make the objects in the picture appear closer or farther away.

You can also move the camera in such a way that it does not change the orientation or its distance from the object. You can **Pan** the view from side to side or up and down. Keep in mind that you are moving the camera, not the object itself (Fig. 5.11).

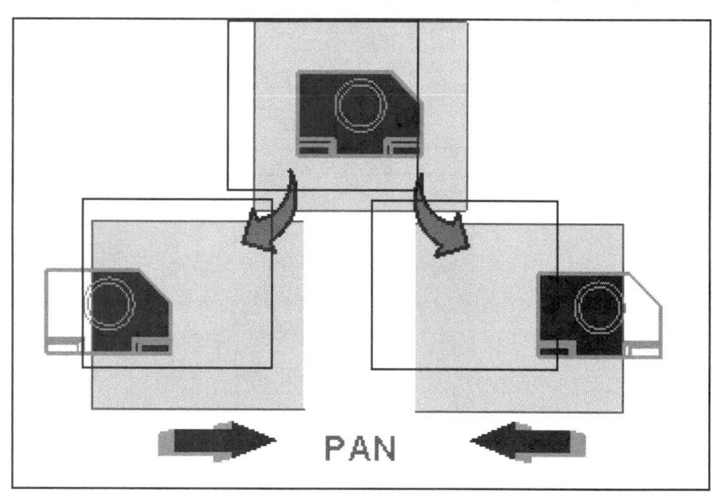

Figure 5.11
Pan

When you look at an object in a 3D view, while it is displayed in **Wireframe** mode, it often "flips" inside out. By default, Pro/E helps you visualize objects by displaying them in **Hidden Line** mode (Fig. 5.12). In **Hidden Line** mode, edges that are *behind* the object appear dimmer. This setting can be changed in the Toolbar.

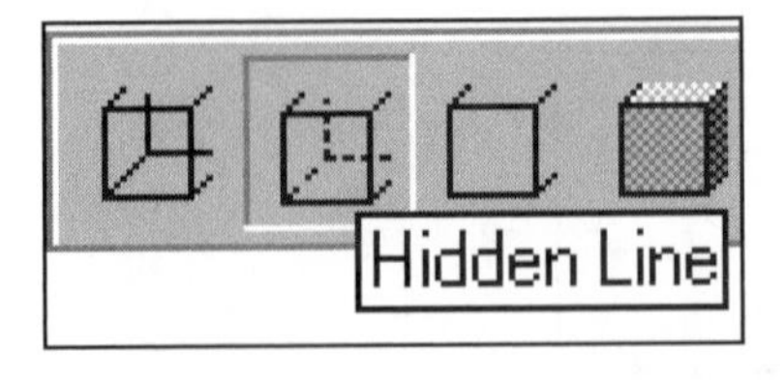

Figure 5.12
Hidden Line Mode

Orienting the View of an Object

You can create customized views of any object or use the built-in **Default** view [either the **Isometric** or the **Trimetric** (Fig. 5.13) view of the object]. You can also define 3D views of an object by rotating it with specified angles or by rotating it dynamically (using the **Spin** option).

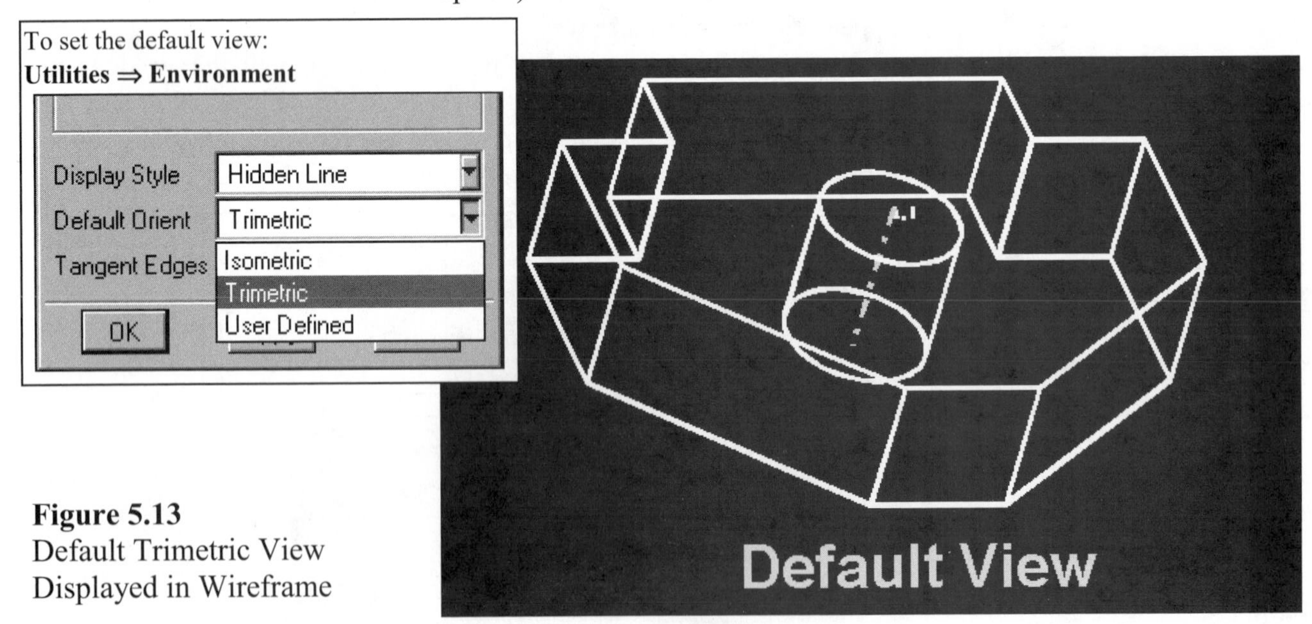

Figure 5.13
Default Trimetric View
Displayed in Wireframe

To define an orthographic view, use the **Orientation** function to specify model faces (or planes) that represent the front, back, top, bottom, right, or left face of the object. *Two* items are required to define any view: **Right** and **Front** (Fig. 5.14) or **Top** and **Right.**

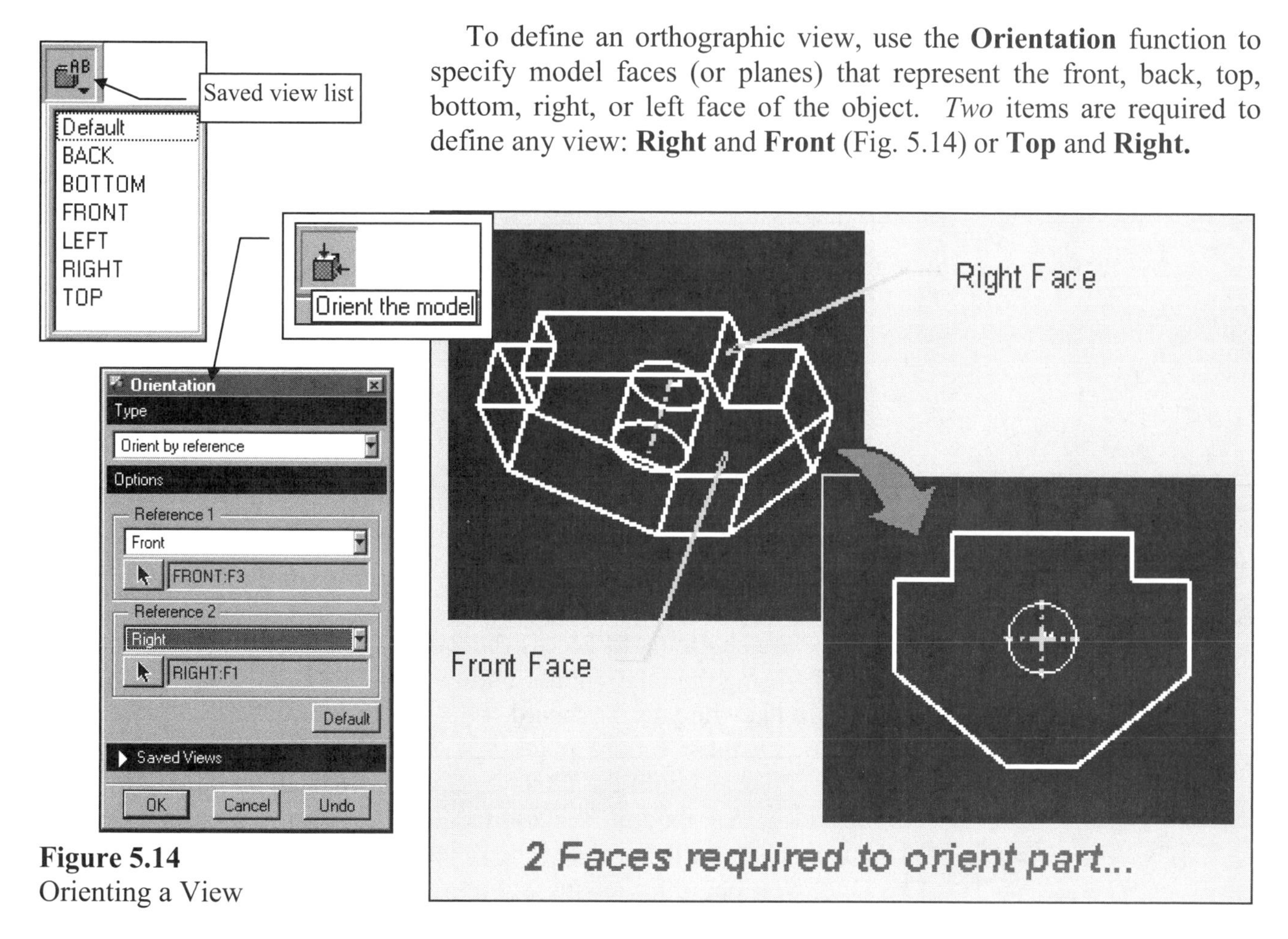

Figure 5.14
Orienting a View

The Basic Viewing Functions

The functions in Pro/E that change the view of an object are **Zoom, Spin,** and **Pan.** These functions can be accessed from the default **Toolbar** (Fig. 5.15). You can use the viewing functions even while you are in the middle of another operation (e.g., creating a feature):

> **Zoom In/Zoom Out** Increases or decreases the size of the object on the screen. *Zoom does not change the size of the object, only its display size on the screen.*
> **Spin** Moving the cursor spins the object around on the screen, relative to the current defined spin center of the model.
> **Pan** Moving the cursor drags the object. Basically, you can "pick up" the object with the cursor and drag it to anywhere you want within the Main Window.

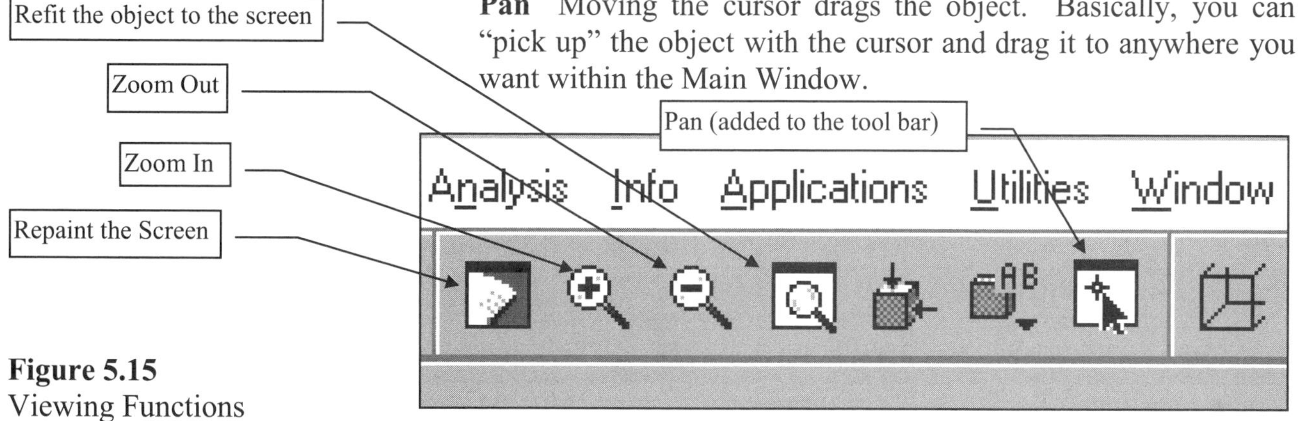

Figure 5.15
Viewing Functions

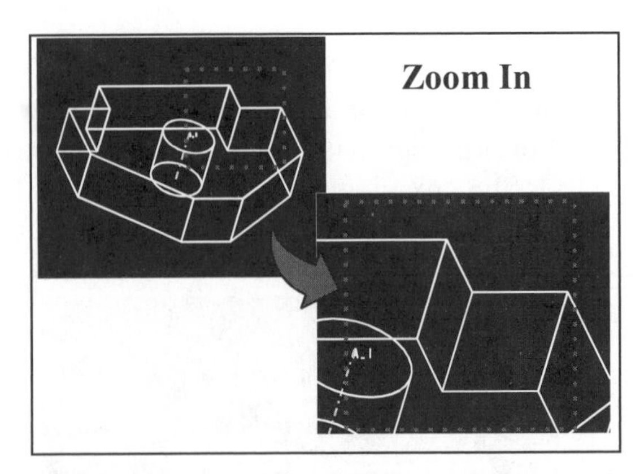

Figure 5.16
Zoom In

To perform a **Zoom In**, you place the cursor at one of the diagonal corners of the zoom box you want to define (Fig. 5.16). **Click** (do not hold) the **left** mouse button (LMB). Move the cursor to the other diagonal point (drawing a box), and click the **left** mouse button (LMB) again--the **Zoom** will immediately take effect.

Zoom/Spin/Pan Shortcuts

There are shortcuts to access the **Zoom**, **Spin**, and **Pan** functions in Pro/E--using the three mouse buttons (Fig. 5.17), in combination with the **Ctrl** key on your keyboard.

To use these mouse mappings, simply move the cursor into the Main Window, hold the **Ctrl** key down, and click and hold down the desired mouse button (for **Zoom, Spin,** or **Pan**). Then moving the cursor changes the view of the object, according to the view option. Once you begin a viewing action, it continues until you release the mouse button.

The mouse **Zoom** functionality is slightly different from the menu options. **Zoom In/Zoom Out** works like this: moving the cursor down or to the left zooms in on the object, and moving the cursor up or to the right zooms out. To perform a mouse **Rectangle Zoom,** hold down the **Ctrl** key and place the cursor at one of the diagonal corners of the zoom box you want to define. **Click** (do not hold) the **left** mouse button quickly. Move the cursor to the other diagonal point, and click the **left** mouse button again to finish zooming in.

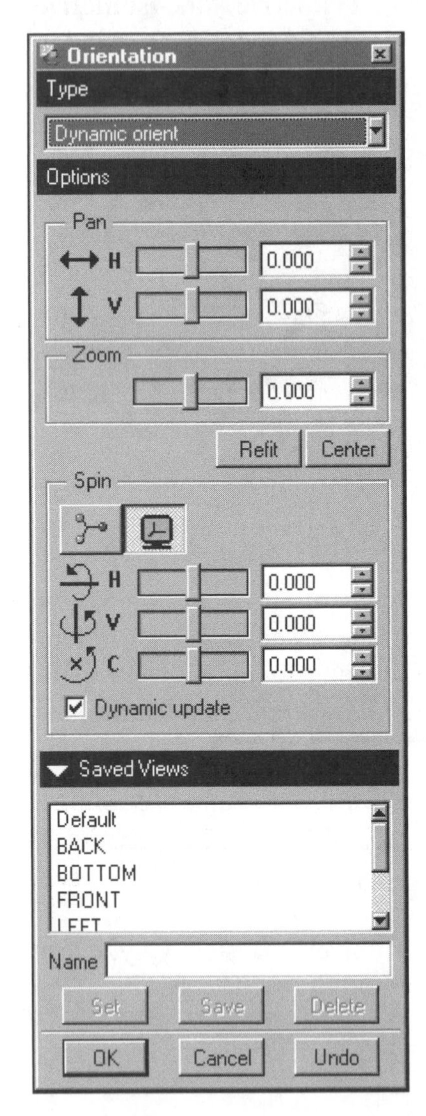

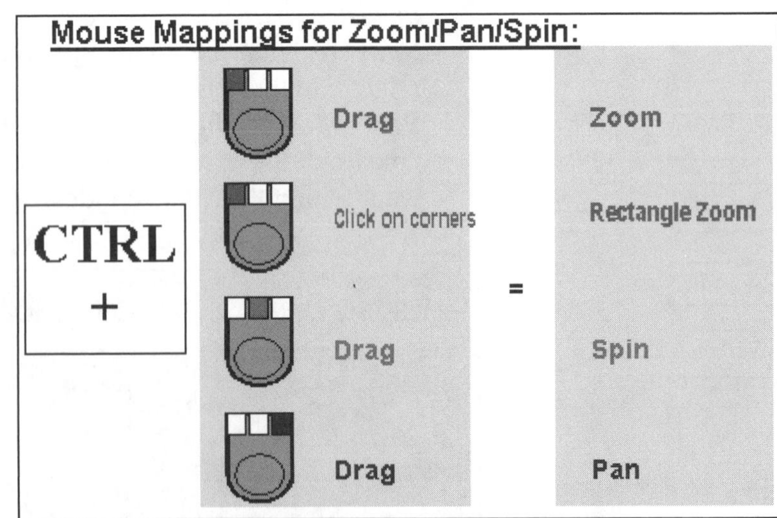

Figure 5.17
Mouse Functions

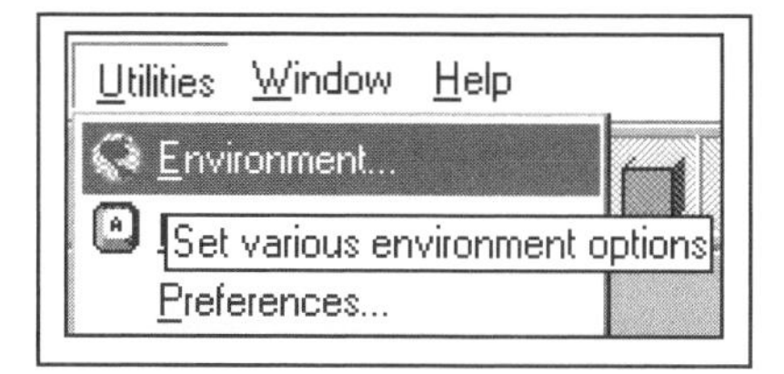

Pro/E allows you to manipulate the view of the active object. You can control the view display of the object in the active window by altering its scale and/or viewing angle using the **Zoom, Spin,** and **Pan** functions.

After opening an object you can manipulate the view. For a part called "**RADIO_MASTER**" displayed in the Main Window, the Title area of the Main Window will read (Fig. 5.18):

RADIO_MASTER (Active) Pro/ENGINEER

The Application Manager (UNIX) will also display the object name (as will the Model Tree) as seen in Figure 5.18. The Default view is displayed in the Main Window as shown in Figure 5.18. The **Default Orient** view will be either **Trimetric** or **Isometric,** depending on the **Environment** setting. The Model Tree window appears after the part model is retrieved or activated (Fig. 5.19). Notice that the part is displayed in the **Shading** Display Style.

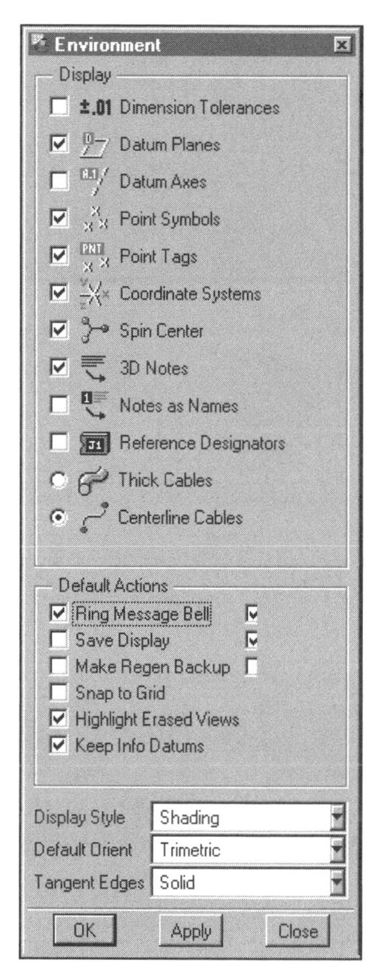

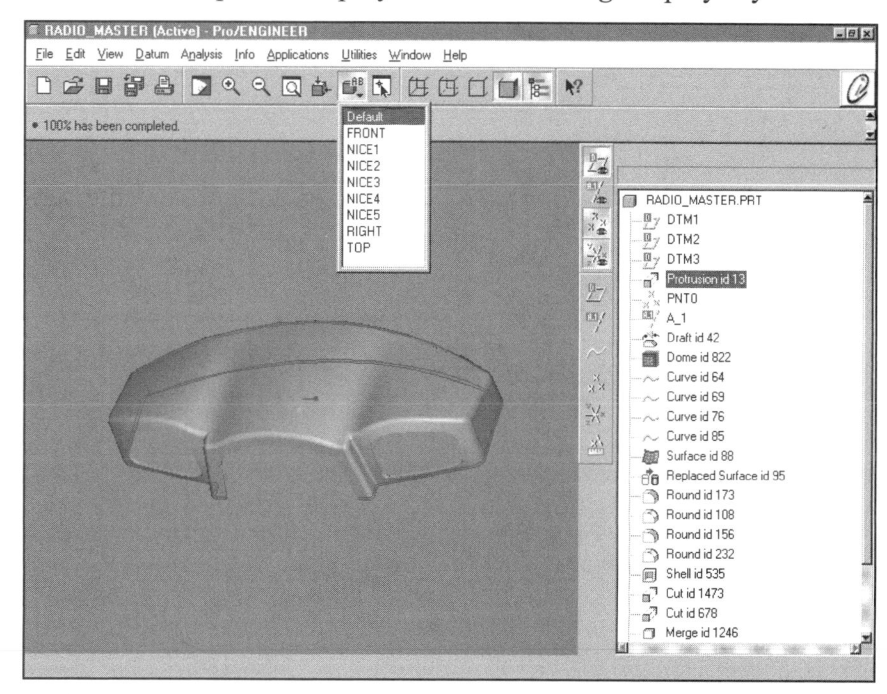

Figure 5.18
Default View with Shading

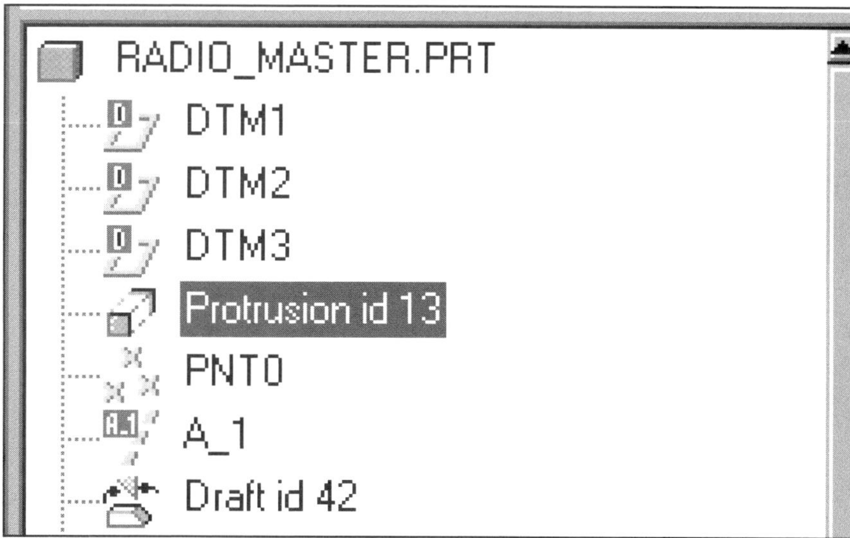

Figure 5.19
Model Tree

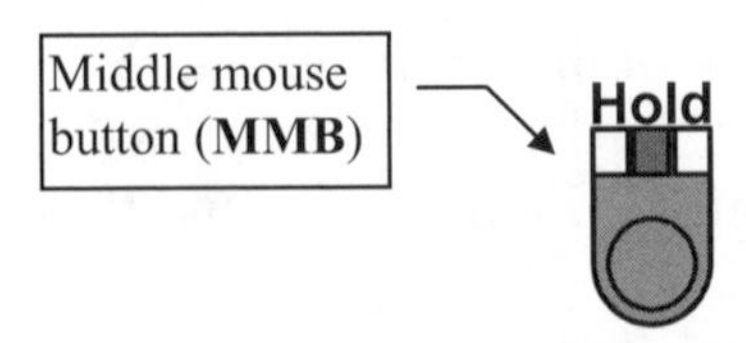

To change the model orientation using the mouse, move the cursor over the graphics area until the pointer is somewhere near the object (Fig. 5.20). Hold down the **Ctrl** key and hold down the **middle** mouse button (**MMB**).

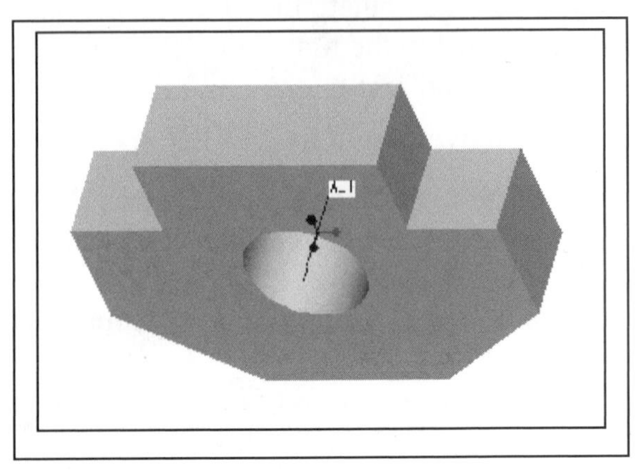

Figure 5.20
Rotating a View Using the Mouse

Move the cursor around on the screen until the object spins around as in Figure 5.21. Release the **middle** mouse button (MMB) to stop spinning the object.

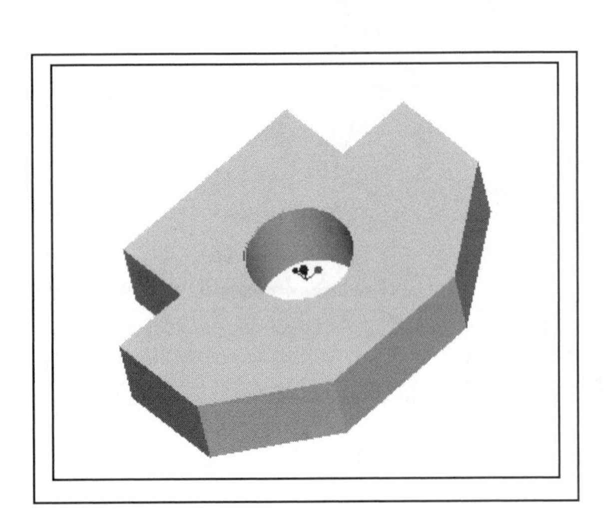

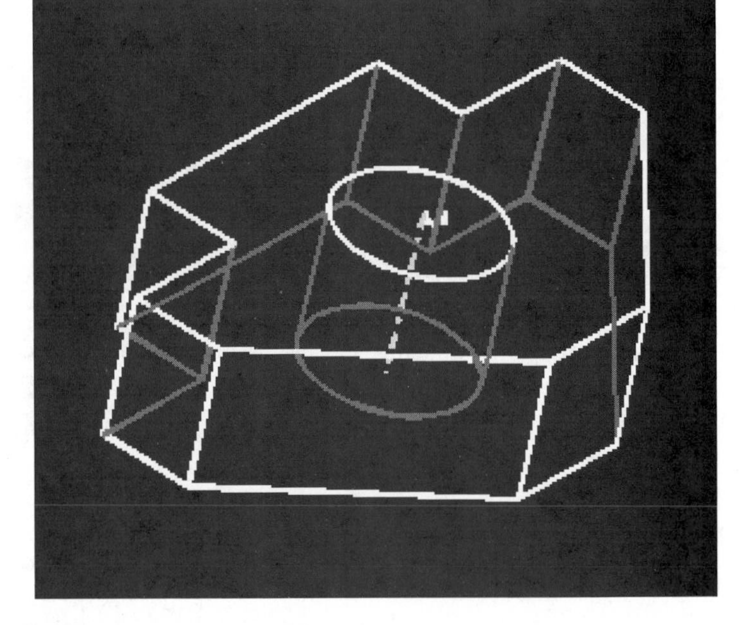

Figure 5.21
Rotated View

Using Dialog Boxes

In Pro/E, you use dialog boxes for such procedures as model manipulation, feature creation, and information. *General* dialog boxes enable you to perform general functions such as saving, viewing, and interrogating, whereas *modeling* dialog boxes enable you to create and manipulate model geometry.

General Dialog Box

General dialog boxes may contain (Fig. 5.22):

Title The name shown at the top of the dialog box.
List Box A box that contains a list of options.
Slider A bar that allows changes to values or scrolls through lists of information.
Drop-Down Arrow An arrow for expandable lists.
Text Box An area in which you can enter information.
Tab A label for multiple pages of a layered dialog box.
Check Box A box that you can select to choose an option.

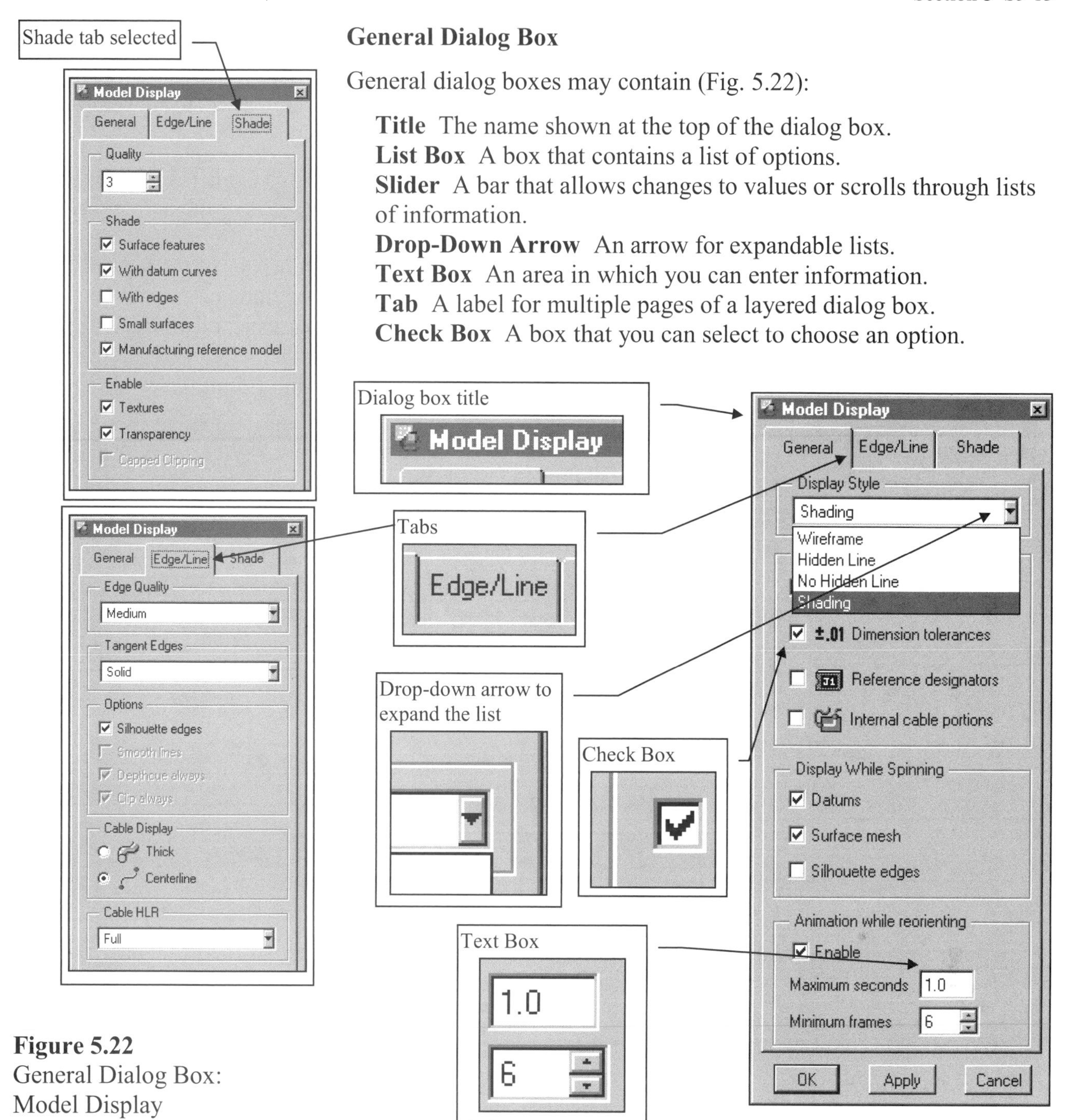

Figure 5.22
General Dialog Box:
Model Display

Model Dialog Box

Model dialog boxes are used to create and modify model geometry and have many of the properties contained in general dialog boxes, but they also include other important options such as required and optional elements. *Required* elements are modifiable properties of a Pro/E feature that you can change at any time during the design cycle. You must specify these elements to complete the definition of a feature. *Optional* elements are additional operations that you can perform on a feature, but you do not need to define them in order to complete the feature.

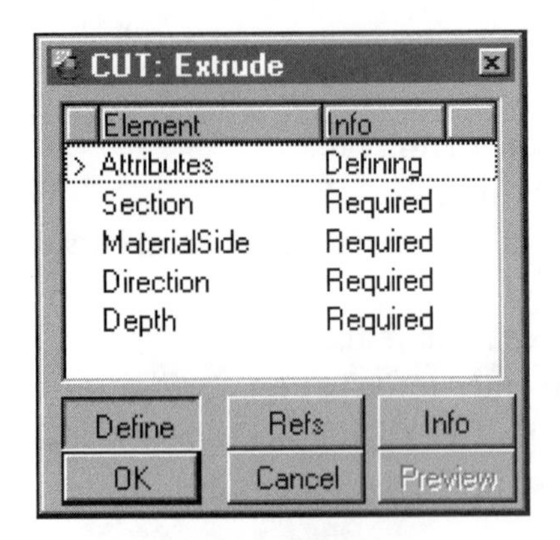

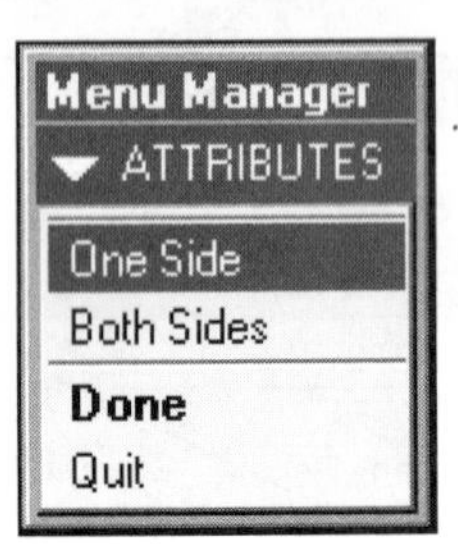

Model dialog boxes (Fig. 5.23) have command buttons that allow you to perform certain functions while creating and changing features. You must understand these functions in order to develop good modeling practices:

Define Allows for redefinition of selected elements in dialog box.
Show Refs Displays the external references of the current selected element.
Feat Info Generates a listing of the properties of the feature.
Feat Refs Displays the external parent references.
OK Completes the definition of the elements, creating the feature or model entity.
Cancel Cancels the current feature or model entity.
Preview Allows you to check the geometry before completing the feature definition. It is not available until you have defined all required elements.
Resolve Rectifies the failure of defined elements by changing elements.

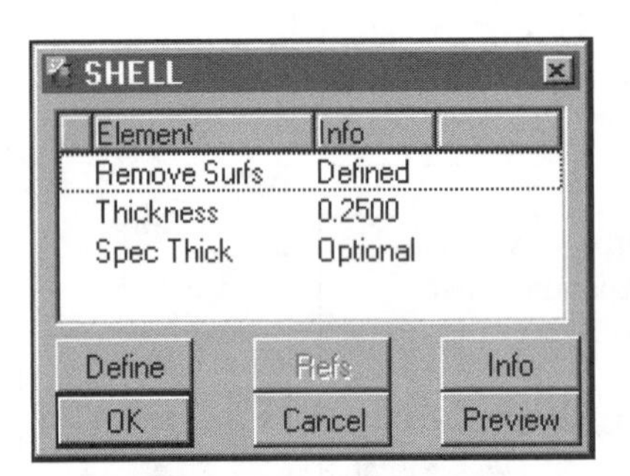

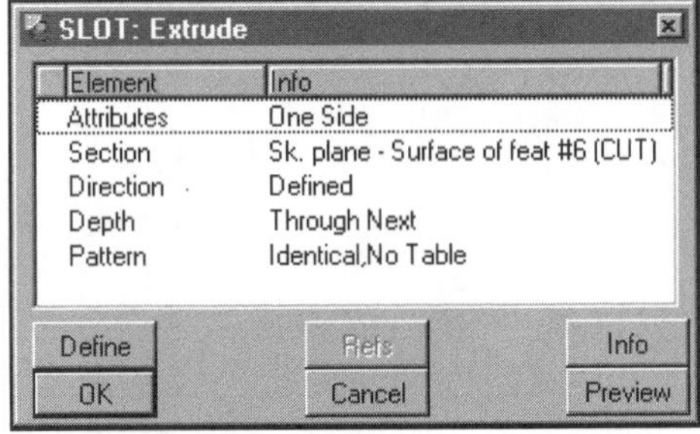

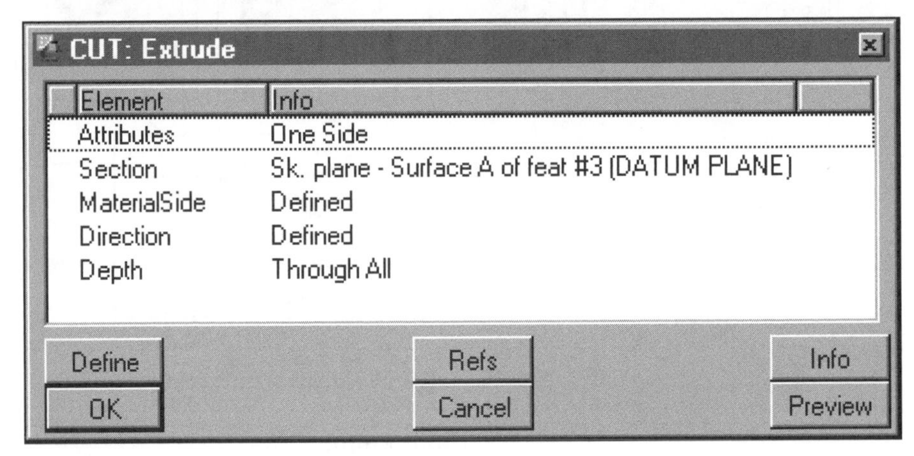

Figure 5.23
Model Dialog Boxes

Menu Help

As explained previously, when you hold your mouse cursor over a menu option, Pro/E provides a one-line help hint on the bottom of the Main Window that explains its function. For additional help, highlight the menu option with the mouse, press down the right mouse button (**RMB**), and select **Get Help** from the pop-up menu. This allows you to view online documentation from the section of the user's guide that discusses the uses of that particular menu option.

Section 6

Utilities

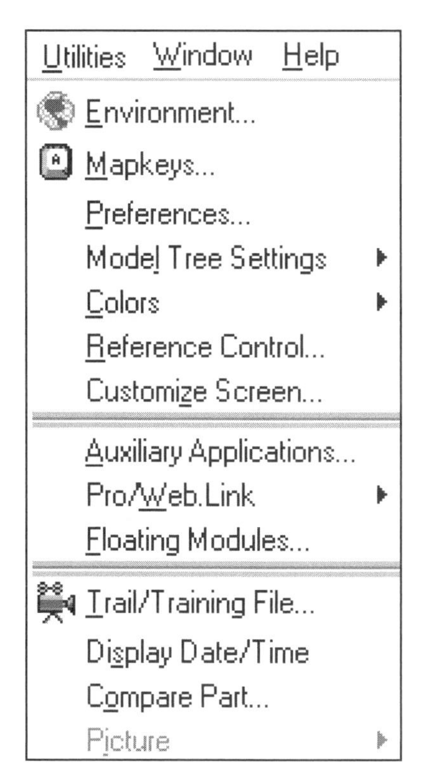

The **Utilities** pull-down menu (from the Menu Bar) is used to control your working environment, user interface, and color scheme and to perform other miscellaneous functions, including the following:

Controlling the Pro/E Environment
Defining Mapkeys
Editing and Loading Configuration Files (Preferences)
Establishing Model Tree Settings
Customizing the User Interface, Entity Colors, and System Colors
Setting Reference Controls
Customizing Menus, Mapkeys, and Toolbars
Running Auxiliary Applications
Setting Pro/Web.Link Security
Running Trail and Training Files
Displaying Date and Time
Comparing Part Files
Saving and Retrieving a Picture of the Model

Configuration Files (*config.pro*)

One of the most important tools in setting up and efficiently using Pro/E is the **configuration file.** The configuration file allows you to configure (set up) a Pro/E working session to your project or company requirements. A configuration file can set school or company standards for storage, formatting, and establishing default units for new parts (such as millimeters instead of inches). Configuration files are important in establishing the path to the location of directories that contain your library items.

Each user can have individual configuration files, but establishing a system-wide configuration file is a way to ensure compliance with project standards. Figure 6.1 shows a typical personalized configuration file.

You can control the environment in which Pro/E runs by specifying the way your objects are oriented, plotting configurations, environment settings, table editor, system colors, and so on. There are two ways to set the working environment: *editing the configuration file* and *using the Environment dialog box*.

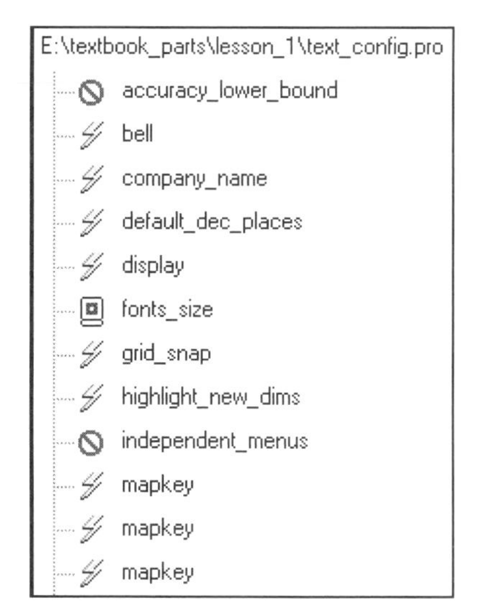

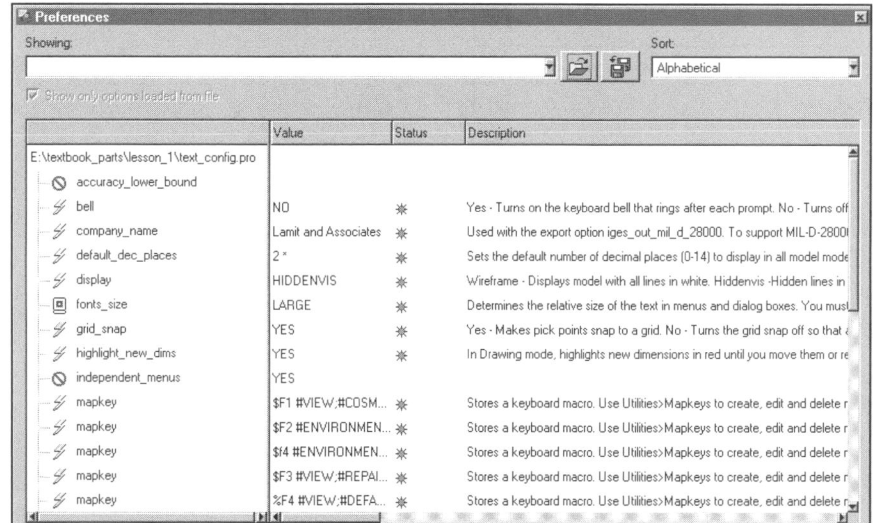

Figure 6.1 Configuration File
Utilities ⇒ Preferences

The default configuration file name is *config.pro*. You can enter the working environment settings into this configuration file before starting a new Pro/E session **(Utilities ⇒ Preferences).** These settings will then be activated every time you start a new session. If you do not enter a particular setting in the configuration file and there is no master configuration file, Pro/E will use its own default settings.

Using the Environment Menu

Frequently used working environment settings are included in the Environment dialog box **(Utilities ⇒ Environment).** This enables you to change the settings easily during the current session (Fig. 6.2). When you enter Pro/E, the values of the environment options are set to those in your *config.pro* file; otherwise, they are set to the default. To change a setting, just pick the item; a check (✔) will appear to activate this capability. To remove a setting, pick the item and the check will disappear.

Figure 6.2
Environment Options Used to Alter Configuration Settings

Setting Configuration Files

All the environment options and other global settings can be preset, before starting Pro/E, by entering the relevant options and their settings into the configuration file. Configuration files can also include settings for tolerance display, formats, calculation accuracy, and the number of digits used in Sketcher dimensions. Pro/E assumes default settings for variables that can be specified in the configuration file but are not.

As an example, *Bell* turns the keyboard bell, that rings after each prompt, on (yes) or off (no). You can override this setting at running time by using the Environment dialog box on the Utilities menu as in Figure 6.3.

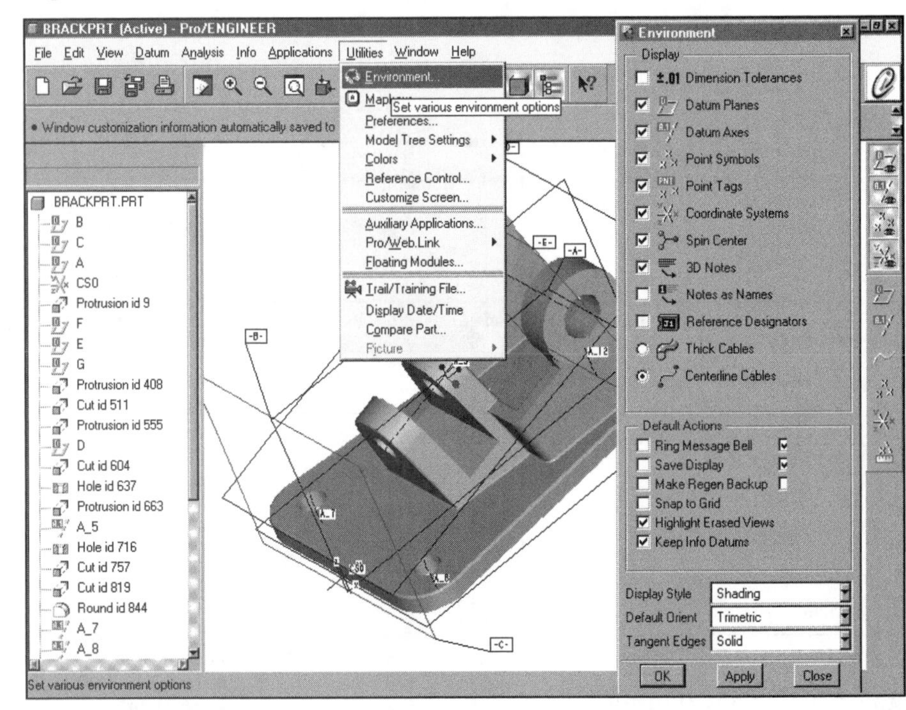

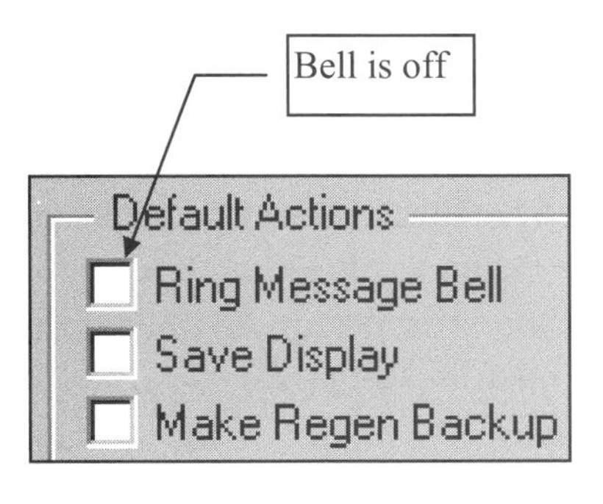

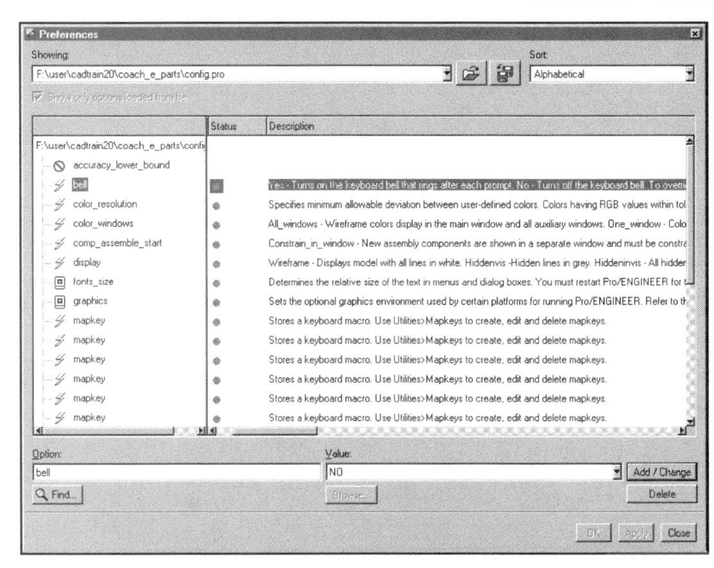

Figure 6.3
Configuration File - Bell

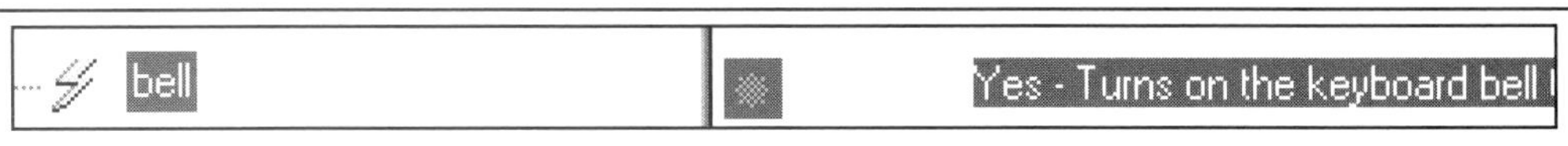

Pro/E reads configuration files automatically from several areas. If a particular option is present in more than one configuration file, the latest value is the one used by the Pro/E.

At startup, Pro/E first reads from a protected system configuration file called config.sup. It then searches for and reads in configuration files from the following directories in the following order:

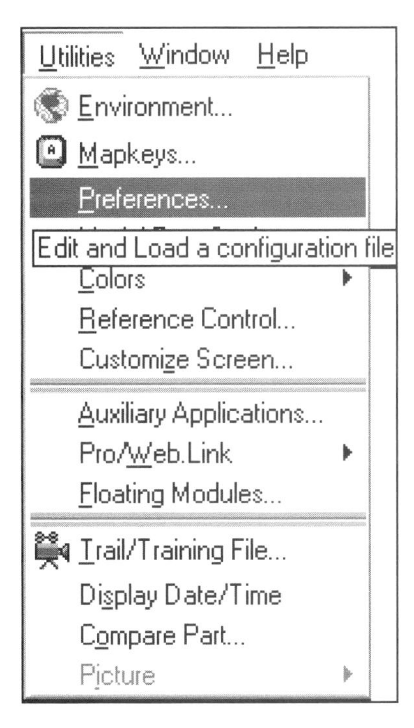

1. *loadpoint/text* (loadpoint is the Pro/E installation directory) - Your system administrator may have put a configuration file in this location to support company standards for formats and libraries. Any user starting Pro/E from this loadpoint uses the values in this file.
2. *Login directory* - This is the home directory for your login ID. Placing your configuration file here lets you start Pro/E from any directory without having a copy of the file in each directory.
3. *Startup directory* - This is your current or working directory when you start Pro/E.

Note: The local config.pro file (in your startup directory) is the last to be read; therefore, it overrides any conflicting configuration file option entries. It does not, however, override any config.sup entries.

Editing a Configuration File

To edit a configuration file during a Pro/E session from the Menu Bar, choose **Utilities ⇒ Preferences.** *Config.pro* is the default. To select a different config.pro, pick the Open button as shown in the left-hand column and navigate through your files and directories. Select the desired *config.pro* and pick **Open.**

Pro/E will display the selected config.pro in the Preferences dialog box (Fig. 6.4).

Using the Preferences Dialog Box

Use the Preferences dialog box to:

* Edit values for configuration options in Pro/E configuration files.
* Toggle display of all configuration options and values.
* Determine the source config file for each config option displayed.
* Save copies of configuration files with custom settings.

To access the Preferences dialog box, click **Utilities ⇒ Preferences**. The **Showing:** pull-down list box shows the last config.pro file read, and the left pane shows the cumulative off-default settings that have been read from any config.pro files found in the search path while Pro/E is loading. The value, status, and source are shown for each option. The status is either *applied* (solid dot) or *conflicting* values (crosshatched dot). Conflicting values means that two config files have differing values for the same option. In this case, the Pro/E default setting is used until you reset the value for the session. The source is the config file and path that the option and value come from. You can use the (**Showing:**) pull-down list to show only the configuration options from each of the sources.

Figure 6.4
Selecting a New *Config.pro*

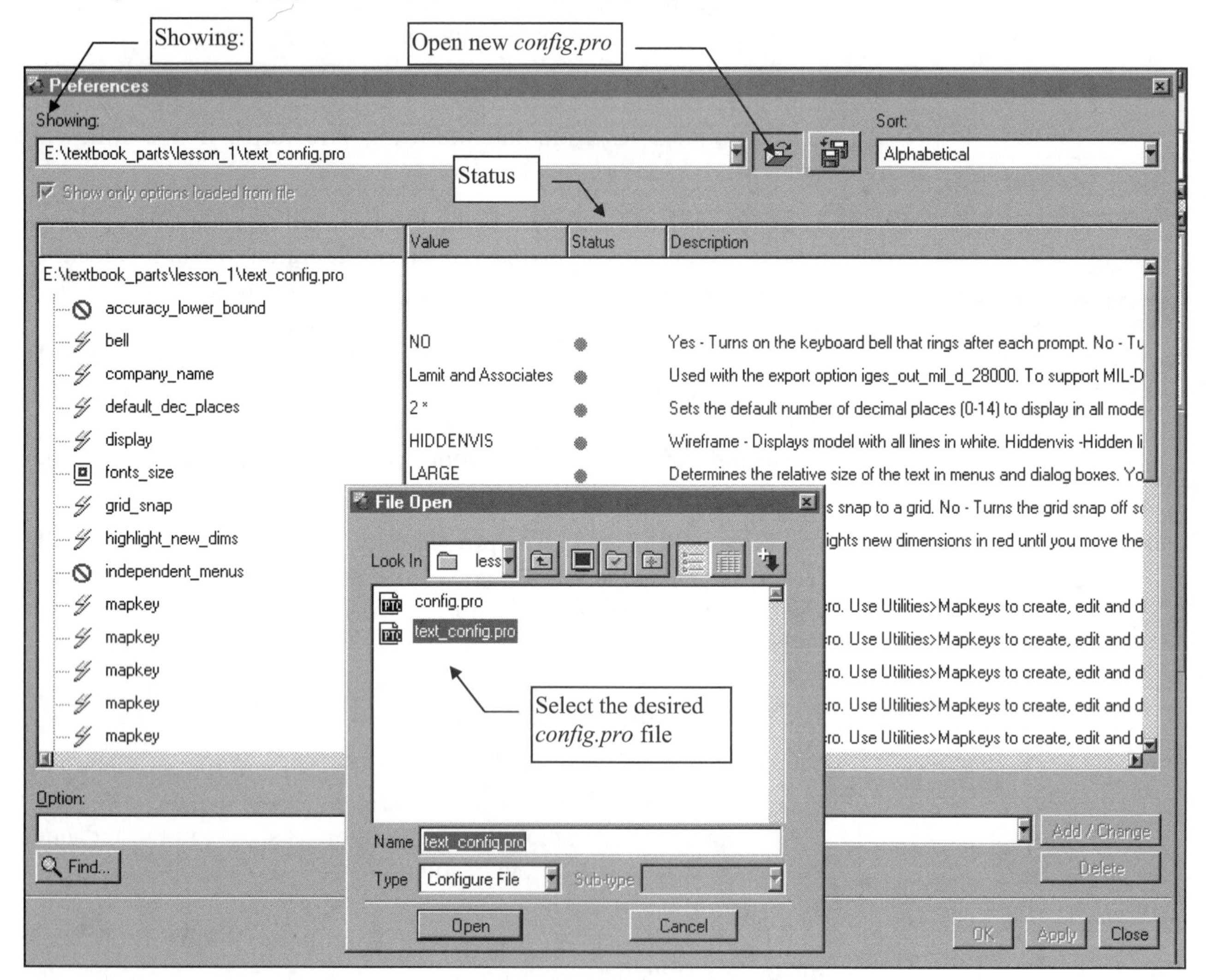

Use online Help (Fig. 6.5) to investigate the config.pro options that are available.

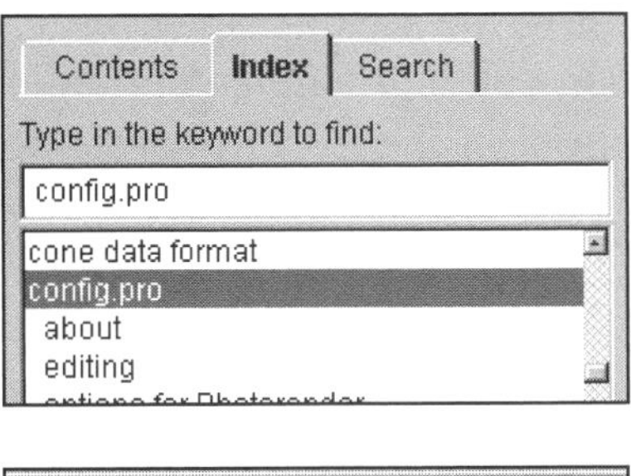

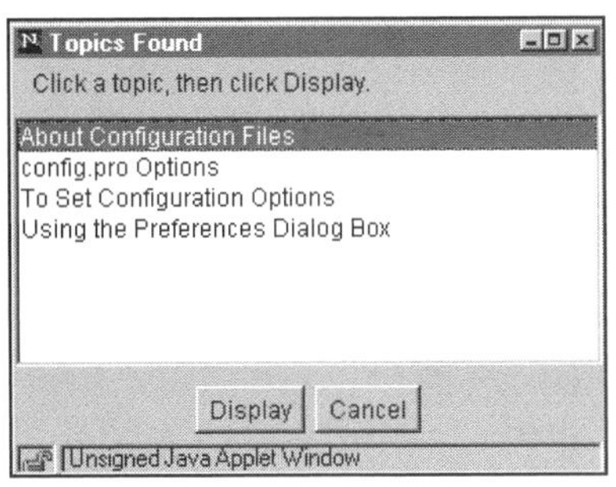

Figure 6.5
Config.pro Help

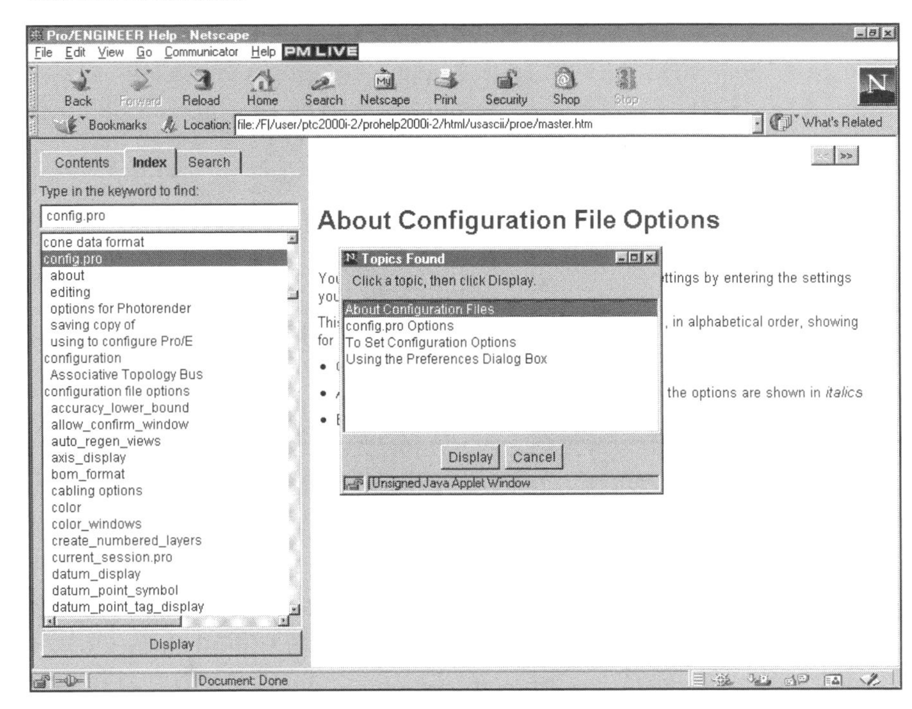

Application Effective Icons

The icon to the left of each option indicates whether a change is applied immediately or on the next startup (Fig. 6.6). The lightning icon means immediate effect. The wand icon means the change is applied to the next object created. The screen icon means the change is applied in the next session. The circle with the slash though it means an incomplete option or an obsolete option.

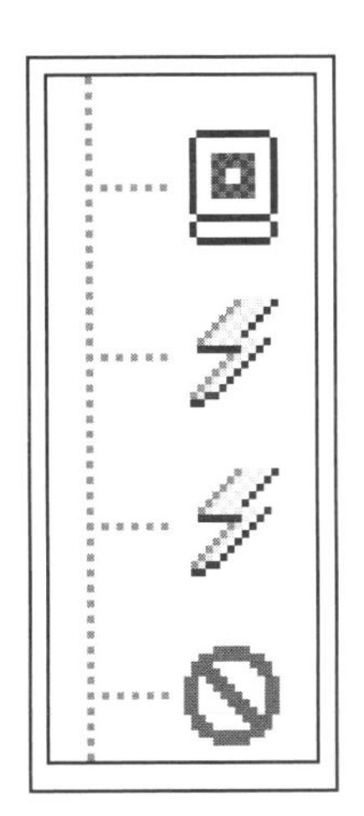

Figure 6.6
Application Effective Icons

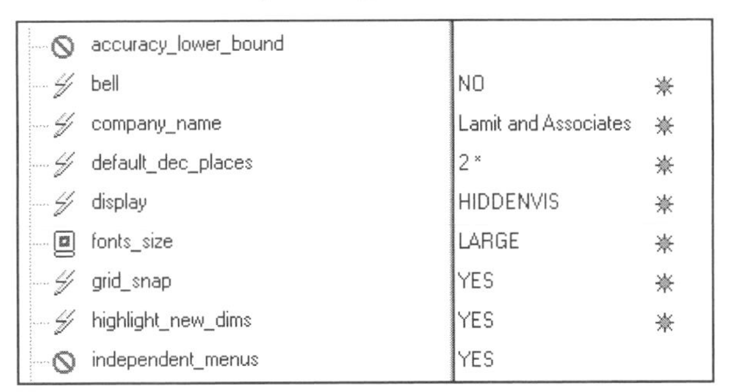

Editing Options and Values

When you make edits to options or values, the edits are saved to the file currently displayed in the **Showing:** pull-down menu when you click **Apply.** If **Current Session** is displayed, the changes are saved to a file named *current_session.pro*, which is automatically created in the current working directory. To edit a value, select it in the list window. The option appears in the **Option:** text field.

The value appears in the **Value:** field. If the value is fixed, for example **Yes** or **No**, use the pull-down list box to select from the choices for the option. If the value requires an integer, type it in. To reset a value to Pro/E default, select the value and click **Delete.**

To Set Configuration Options

1. Click **Utilities ⇒ Preferences.** The Preferences dialog box opens. Settings from the most recently loaded config file are displayed.
2. Locate the option you want to set. When you highlight an option in the list, it appears in the line labeled Option: in the lower part of the dialog box.
3. In the Value field opposite the Option field, enter the new value. If the value entries are fixed, for example **Yes** or **No,** you can use the pull-down menu to see what values are available. If the value requires an integer, type it in.
4. When you have edited the value, click **Add/Change.** The new value is applied in the options list window. The status icon for the affected line changes to show an edit to the default has been made.
5. When you have finished making edits to the config option values, click either **Apply** or **OK.** Changes that can be applied immediately appear in the user interface.

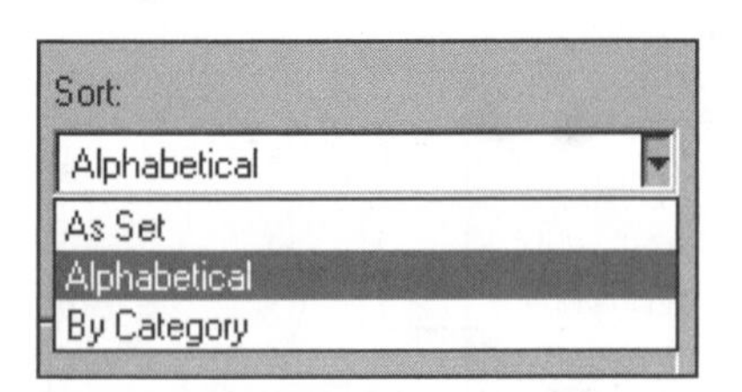

Additionally, a configuration file called *current_session.pro* is automatically created in the working directory. This file contains only the configuration options you changed, with their new settings. If the file already existed, it is edited or appended with the new settings.

Navigating the Options

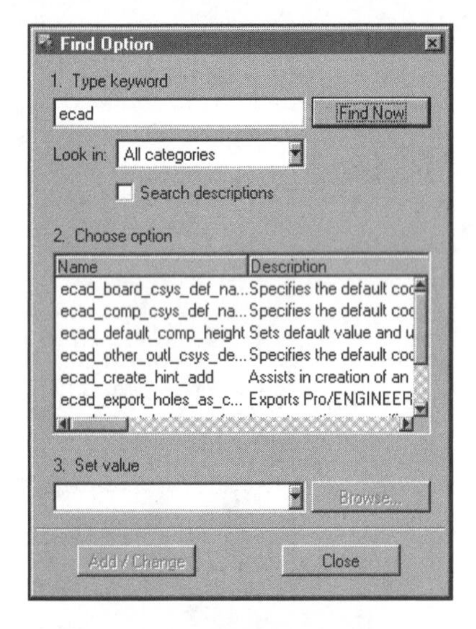

Use the **Sort:** pull-down list box to sort the options alphabetically or by category. You can also use the **Find** button to search for options using a text string. For example, if you type *ecad* in the **Type keyword** field, the dialog box lists all options using *ecad* in their strings. You can also search the descriptions for keywords. (You must still click **Apply** in the Preferences dialog box to save changes to the session.)

Mapkeys

You can create keyboard macros, which map frequently used command sequences to certain keyboard keys. The macros are saved in the configuration file. Each macro begins on a new line. Keyboard macros require the following format:

mapkey keyname #command;#command; . . .

As an example, the following shows a keyboard macro command statement to begin creating cosmetic threads mapped to the keyboard character sequence **ct:**

```
ct #FEATURE;#CREATE;#COSMETIC;#THREAD;
```

Entering **ct** from the keyboard during a current Pro/E session initiates the commands **Feature ⇒ Create ⇒ Cosmetic ⇒ Thread.**

In Figure 6.7, a macro to begin the creation of a round for the current model is associated with function key **cr:**

mapkey cr #FEATURE;#CREATE;#ROUND;#DONE;#DONE

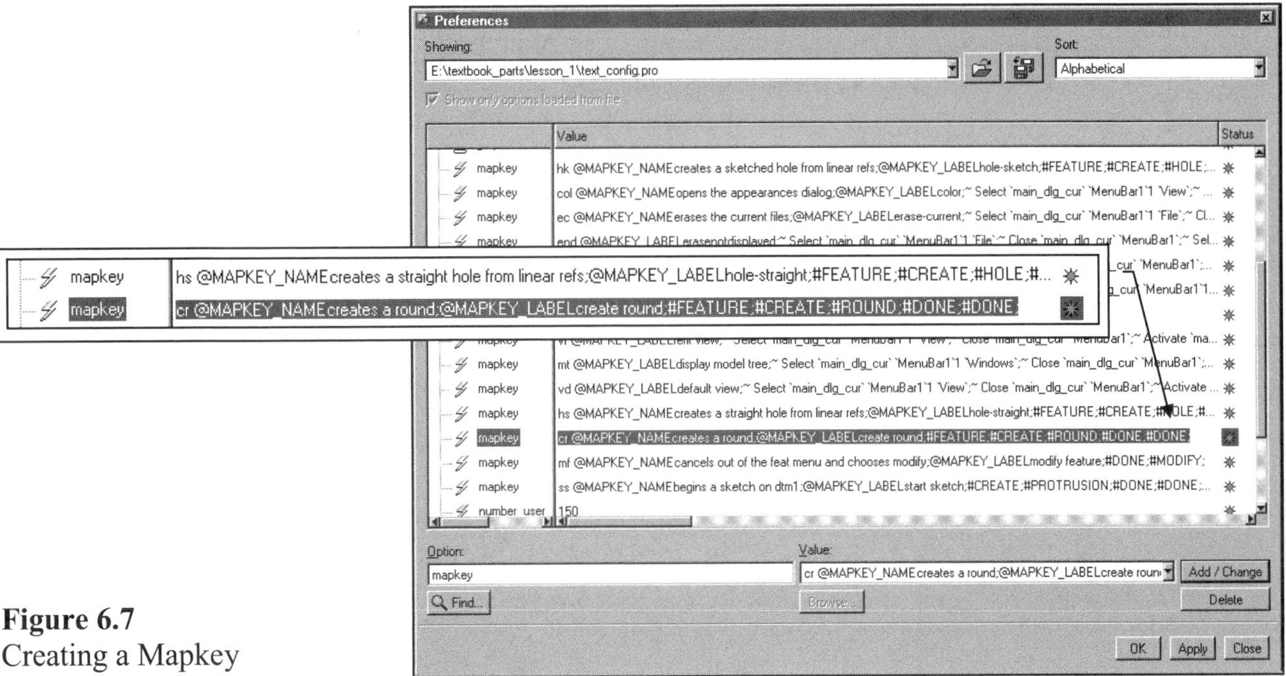

Figure 6.7
Creating a Mapkey

Defining Mapkeys

With the **Mapkeys** dialog box, you can create keyboard macros, which map frequently used command sequences to certain keyboard keys or sets of keys. By adding custom mapkeys to your toolbar, you can automate your work flow in a visible way. You can use Mapkeys with a single menu command. You use the Mapkeys dialog box to:

- Define new mapkeys.
- View, modify, and delete existing mapkeys.
- Execute a mapkey chosen from the list.
- Save the mapkeys to a configuration file.

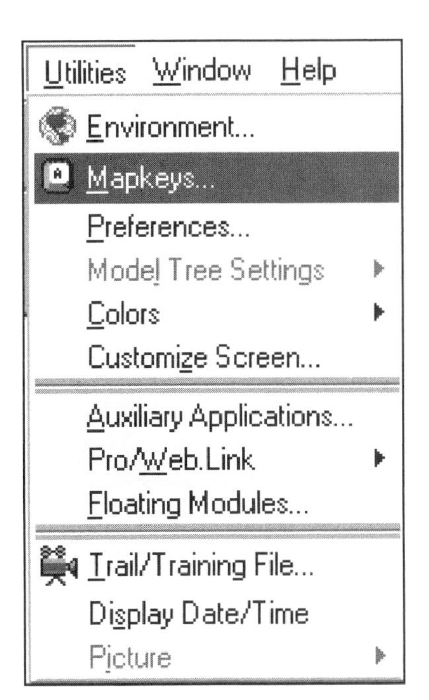

To open the Mapkeys dialog box, choose **Mapkeys** from the Utilities menu. The following are available:

New	Define a new mapkey.
Modify	Modify the mapkey highlighted in the list.
Run	Run the mapkey highlighted in the list.
Delete	Delete the mapkey highlighted in the list.
Save	Save the mapkeys to a configuration file.
Changed	Save only the mapkeys changed in the current session.
All	Save all the mapkeys.

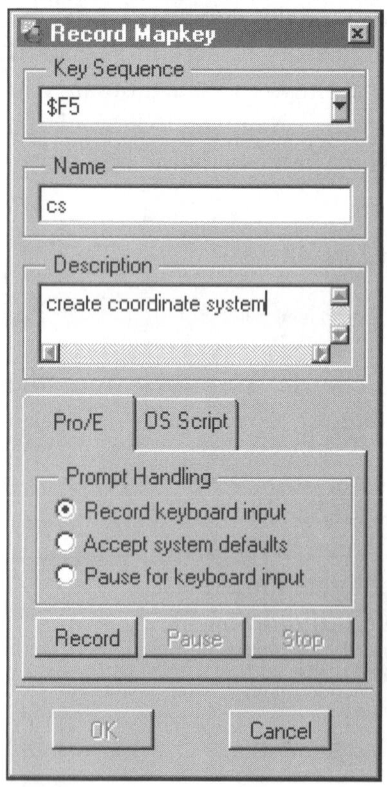

Adding User-Defined Mapkeys

To define a mapkey, you use the **Record Mapkey** dialog box to type a key sequence, a name, and a description for the mapkey. If the Mapkey is added to the Toolbar, the label appears on the button added, while the description appears as one-line help when you let the mouse cursor remain over that toolbar button.

After you define the mapkey, a corresponding button appears in the **Customize Toolbars** dialog box under the **Mapkeys** category. You can then drag the mapkey onto the toolbar just like the Pro/E-supplied buttons. Mapkeys include the ability to pause for user interaction, handle message window input flexibly, and run operating system commands.

When you define a mapkey, Pro/E automatically records a pause when you make screen selections, so that you can make new selections while the mapkey is running. Also, you can record a pause at any place in the mapkey along with a user-specified dialog prompt, which will appear at the corresponding point while the mapkey is running.

You can create mapkeys that run operating system scripts and commands. The Record Mapkey dialog box contains the OS Script tab, which runs OS commands rather than Pro/E commands. To define a new mapkey:

1. Choose **Utilities ⇒ Mapkeys.** The Mapkeys dialog box appears.
2. Pick **New.** The Record Mapkey dialog box opens.
3. Type the key sequence that is to be used to execute the mapkey in the Key Sequence text box. To use a function key, precede its name with a dollar sign (**$**). For example, to map to **F5**, type **$F5**.
4. Type the name and description of the mapkey in the appropriate text boxes (optional).
5. On the Pro/E tabbed page, specify how Pro/E will handle the prompts when running the mapkey:

    Record Record the keyboard input when defining the mapkey, and use it when running the macro.
    Accept Accept Pro/E defaults when running the macro.
    Pause Pause for keyboard input when running the macro.

6. Pick **Record** and start recording the macro by selecting menu commands in the appropriate order.
7. If desired, pick **Pause** to indicate where to pause while running the mapkey. Type the prompt in the Resume Prompt dialog box. Pick **Ok**. Then pick **Resume** and proceed recording the mapkey. When you run the macro, Pro/E will pause, display the prompt you typed, and give you the options to **Resume** running the macro or **Cancel.**
8. Pick **Stop** when finished recording the macro.

Figure 6.8 shows an example of creating a user-defined Mapkey. The command **Utilities ⇒ Mapkeys ⇒ New** is selected. The Mapkeys dialog box appears after you select **Utilities ⇒ Mapkeys** (Fig. 6.8). After you select **New**, the Record Mapkey dialog box appears as shown in Figure 6.9.

Figure 6.8
Creating a User-defined Mapkey

Figure 6.9
Record Mapkey Dialog Box

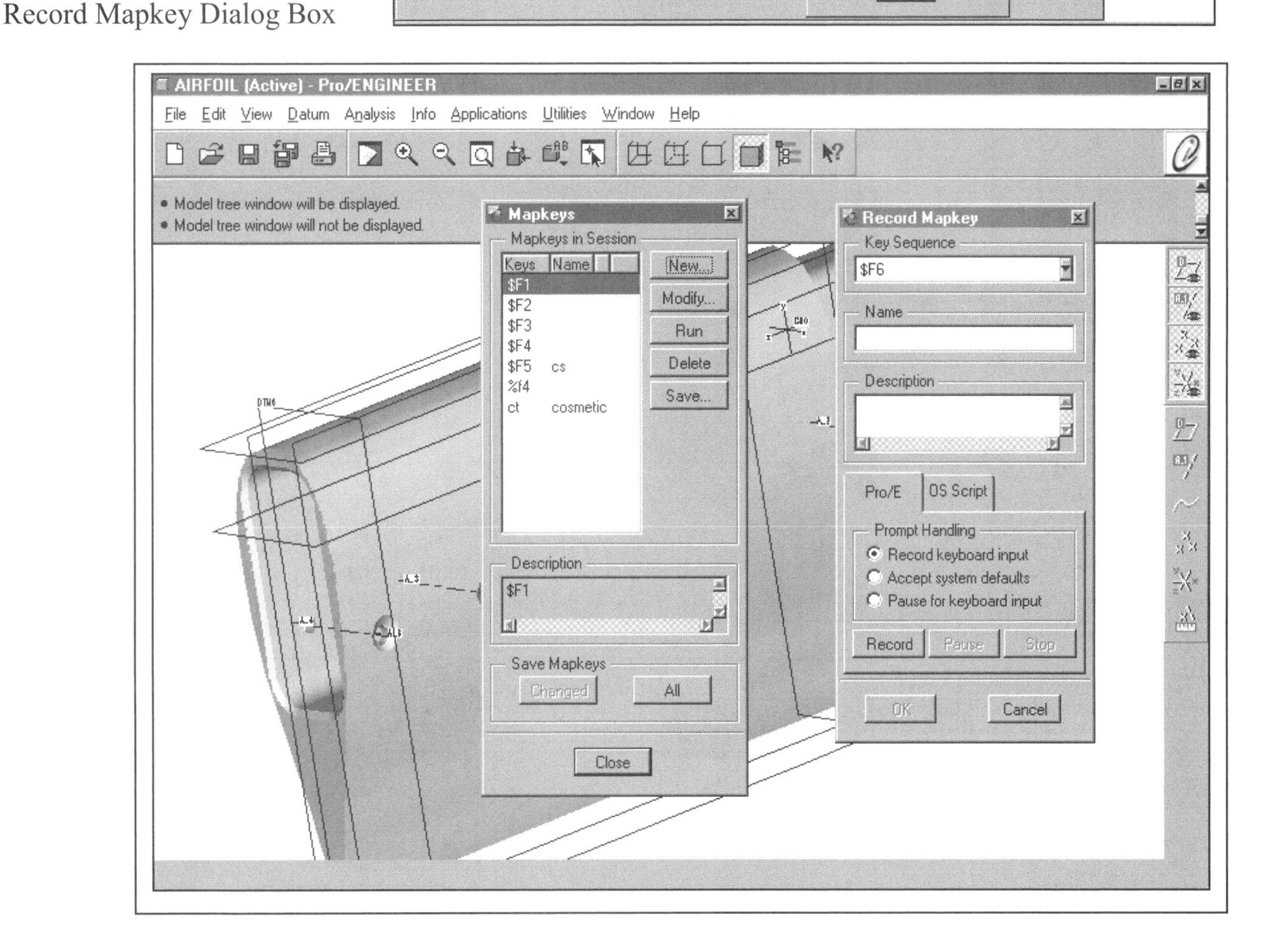

Controlling the Pro/E Environment

You control the environment in which Pro/E runs with the Environment dialog box (Fig. 6.10). You use the Environment dialog box to:

* Select how your models are displayed: default orientation, hidden line and tangent edge display styles.
* Establish whether or not Pro/E displays datums, spin center, model tree, and other optional entities.
* Specify whether or not Pro/E performs certain actions, such as ringing the bell or making backups.

To open the Environment dialog box, choose **Environment** from the Utilities menu.

Figure 6.10
Environment Dialog Box

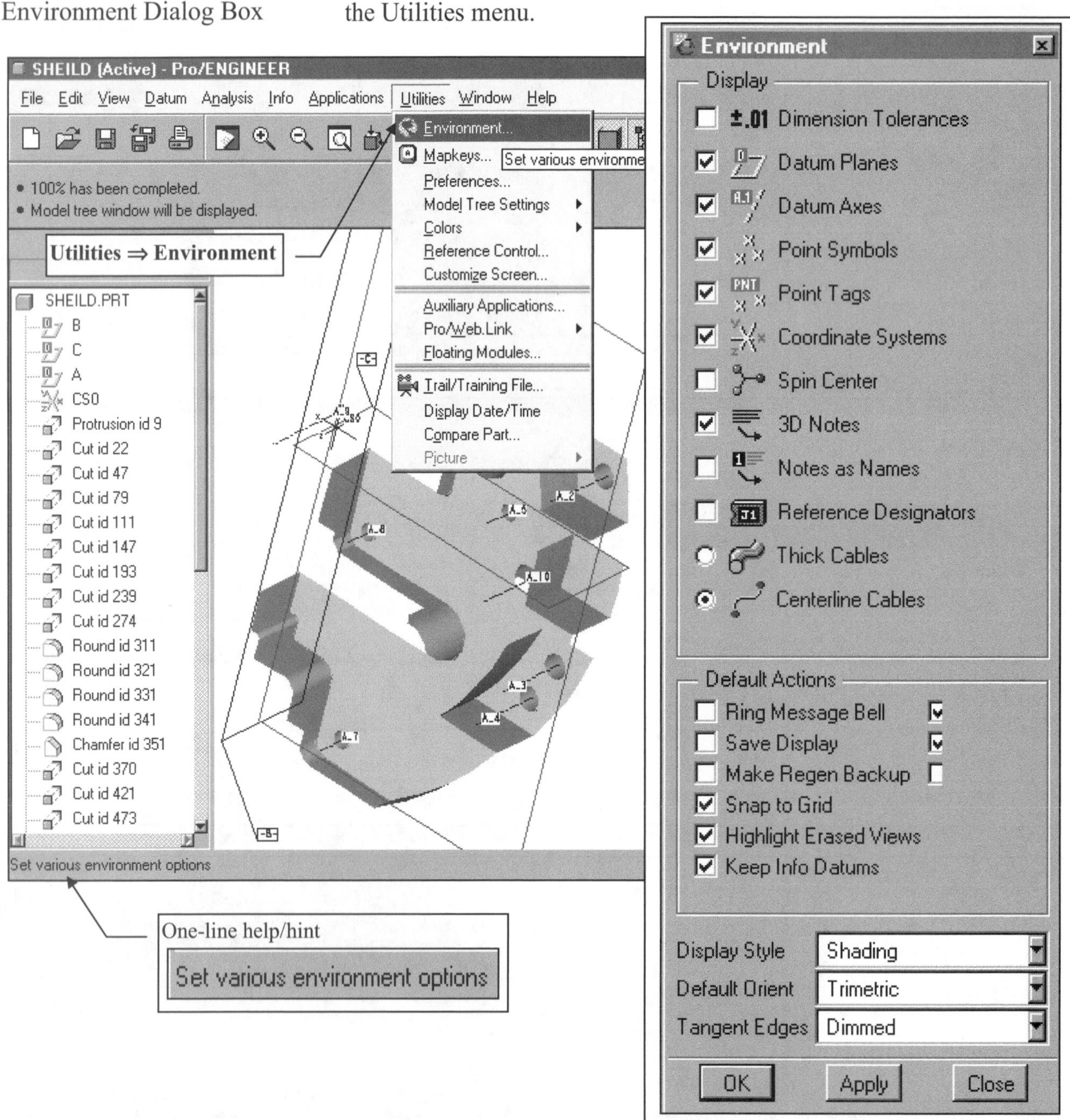

ENVIRONMENT MENU

Use	To Display
Dimension Tolerances	Display model dimensions with tolerances.
Datum Planes	Display the datum planes and their names. Note that when you turn off the datum plane display by clearing the checkbox, this will not affect a datum plane that is set as a reference datum for geometric tolerances.
Datum Axes	Display the datum axes and their names. In Drawing mode, when you turn the datum axis display off by clearing the checkbox, only the axis names are blanked.
Point Symbols	Display the datum points and their names.
Point Tags	Control the display of the datum point names separately from the datum point symbols. When you blank the datum point names by clearing the checkbox, the datum point locations are still marked by the **X** symbols.
Coordinate Systems	Display the coordinate systems and their names.
Spin Center	Display the spin center for the model.
3D Notes	Display notes created in Part or Assembly mode in the model window.
Notes as Names	If the preceding 3D Notes option is selected, lets you toggle between displaying the note text and displaying the note names.
Reference Designators	Make each assembly component designated as a connector in Pro/CABLING or imported as an ECAD component appear with its reference designator in 3D assembly views.
Thick Cables	Display a cable with 3D thickness. It can be shaded. This option and **Centerline Cables** are mutually exclusive.
Centerline Cables	Display a cable's centerline with location points shown in green. This option and **Thick Cables** are mutually exclusive.
Internal Cable Portions	Display cable portions that are hidden from view by other geometry.
Model Tree	Display the Model Tree window in Draw Mode.
Textures	Display textures on shaded models.
Snap Lines	Snaps to snap lines (Drawing and Report modes only).
Levels of Detail	Makes the use of **Levels of Detail** available in a shaded model during dynamic orientation (panning, zooming, and spinning).
2D Arc Centers	Display arc centers (appears in Drawing and Report modes only).
Highlight Erased Views	Show the box and the view name in place of erased drawing view.
Ring Message Bell	Ring bell (*beep*) after each prompt or system message.
Save Display	Save objects with their most recent screen display. This makes object retrieval in new sessions faster.

Default Actions	
Make Regen Backup	Make the system back up the current model(s) to disk before every regeneration: when you explicitly choose **Regenerate**, as well as whenever you start a function that ends in an implicit regeneration, such as **Feature ⇒ Redefine**. At the end of the session, Pro/ENGINEER automatically deletes the backup files that it creates.
Snap to Grid	Make points you select on the screen snap to a grid. This is particularly useful in Sketcher.
Highlight Erased Views	Show highlight box around erased views.
Snap to Snap Lines	Snap entities (such as dimensions, notes, symbols) onto snap lines (appears in Drawing and Report modes only).
Keep Info Datums	Control how the system treats datum planes, datum points, datum axes, and coordinate systems that you create "on the fly" under the **Info** functionality. If selected, the system includes them as features in the model. If cleared, the system erases them when you exit the Info functionality.
Orient Model	When entering sketcher, orient model so sketching is parallel to the screen.
Intent Manager	Start with Intent Manager on when sketching.
Use 2D Sketcher	Control the initial model orientation in Sketcher mode. If selected, the model is reoriented upon entering Sketcher to a 2 D orientation looking directly at the sketching plane. If cleared, the model orientation upon entering Sketcher is unchanged.
Display HLR	Make possible the hardware acceleration of dynamic spinning with hidden lines, datums, and axes.
Display Style	Choose one of the following display styles for your models: **Wireframe** Model is displayed in wireframe (no distinction between visible and hidden lines). **Hidden Line** Hidden lines are shown in *gray.* **No Hidden** Hidden lines are not shown. **Shading** All surfaces and solids are displayed as shaded.
Default Orient	Choose one of the following default orientations for your models: **Isometric** Standard isometric orientation. **Trimetric** Standard trimetric orientation. **User Defined** User-defined orientation (determined by the configuration file options *x_angle* and *y_angle*).
Tangent Edges	Choose one of the following display styles for tangent edges (edges formed by two tangent surfaces) in your models: **Solid** Display tangent edges as solid lines. **No Display** Blank tangent edges. **Phantom** Display tangent edges in phantom font. **Centerline** Display tangent edges in centerline font. **Dimmed** Display tangent edges in the Dimmed Menu system color. (Styles can be set with *tangent_edge_display* configuration option.)

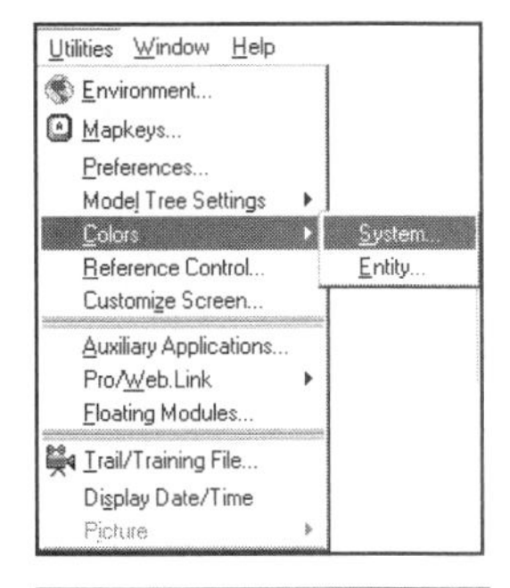

System and Entity Colors

From the Utilities menu you can access and change the system and entity colors used as defaults in a Pro/E session by choosing the commands **Utilities ⇒ Colors ⇒ System** or **Entity.**

Customizing System Colors

With the **System Colors** dialog box (Fig. 6.11), you can change the system colors used by Pro/E for displaying geometry, datums, text, background, and other entities. To open this dialog box, from the **Utilities** menu choose **Colors ⇒ System.**

For all of the color buttons, Pro/E displays the Color Editor dialog box. You change the color definition by moving the sliders corresponding to the ***Red, Green,*** and ***Blue*** contents. To change the default color scheme used by Pro/E, you can switch among several available schemes.

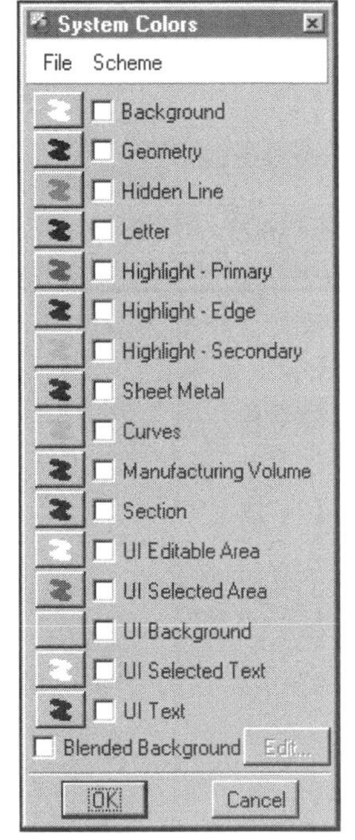

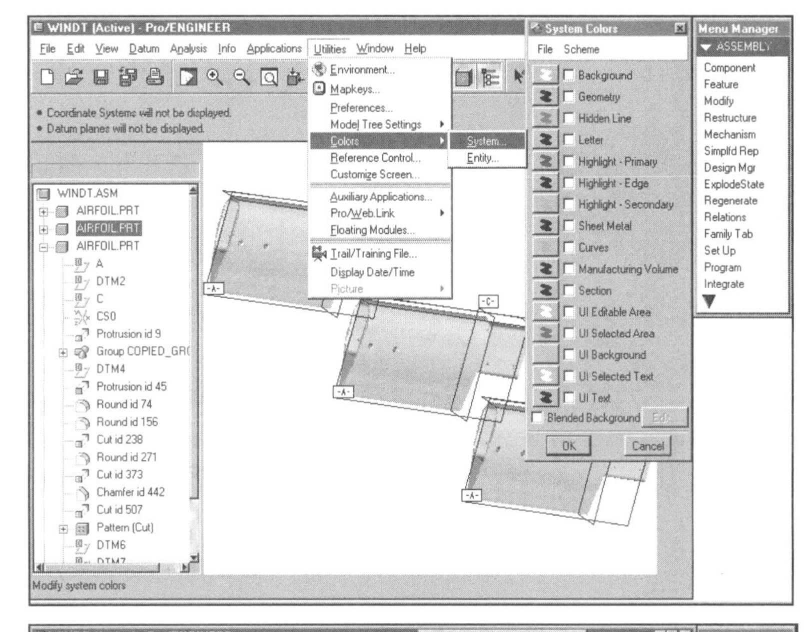

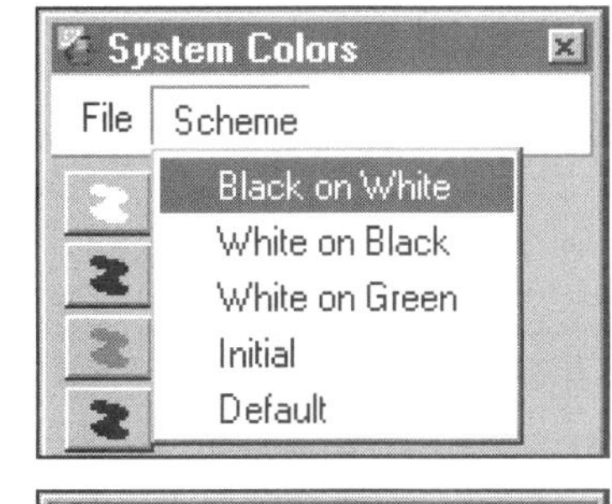

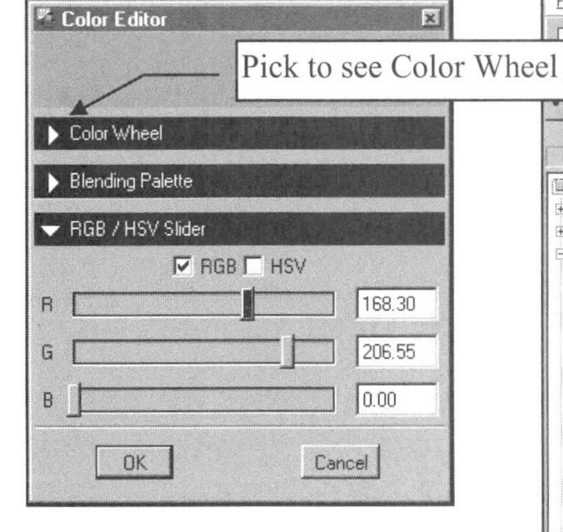

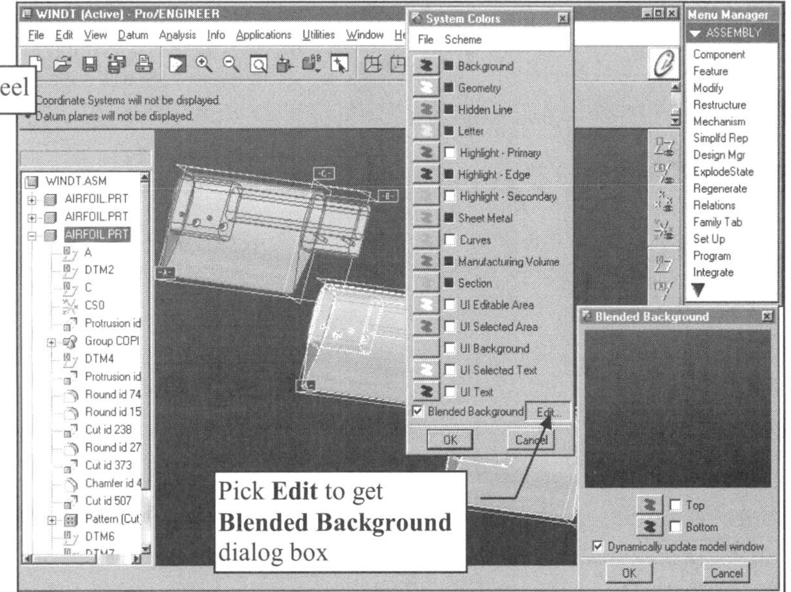

Figure 6.11
System Colors
Black on White (top)
Default (bottom)

Customizing the Entity Colors

With the **Entity Colors** dialog box (Fig. 6.12), you can change the colors that Pro/E uses for displaying certain entities, such as datum planes or quilt edges. To open the Entity Colors dialog box, from the Utilities menu choose **Colors ⇒ Entity.** This dialog box has two tabs, **Datum** and **Geometry.**

In the Datum tabbed page, you can change the color of planes, axes, points, and coordinate systems. In the Geometry tabbed page, you can change the color of edges, references, and surfaces.

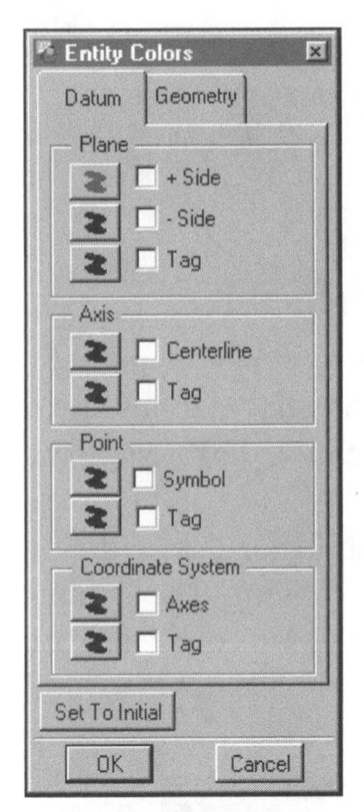

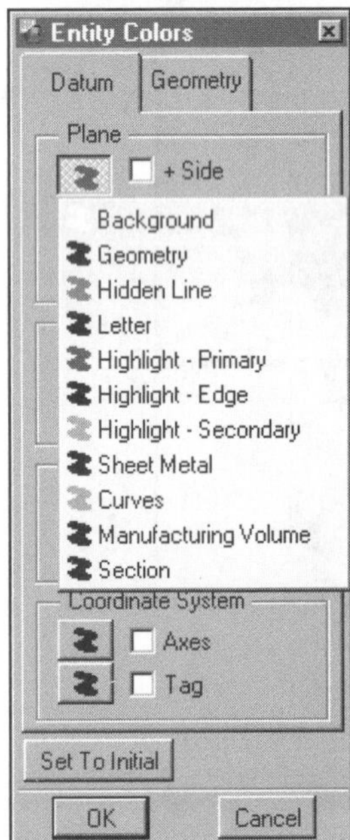

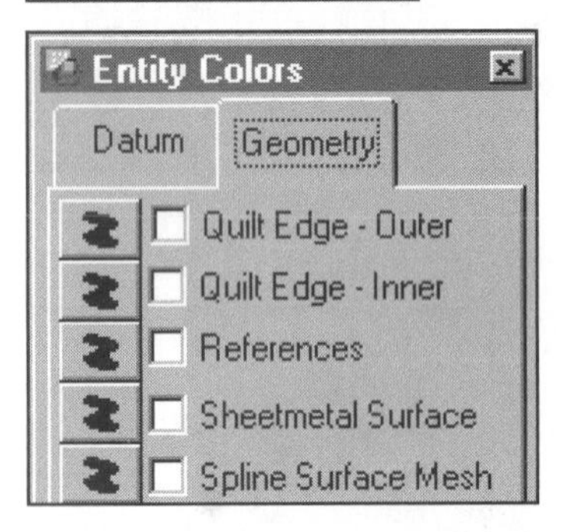

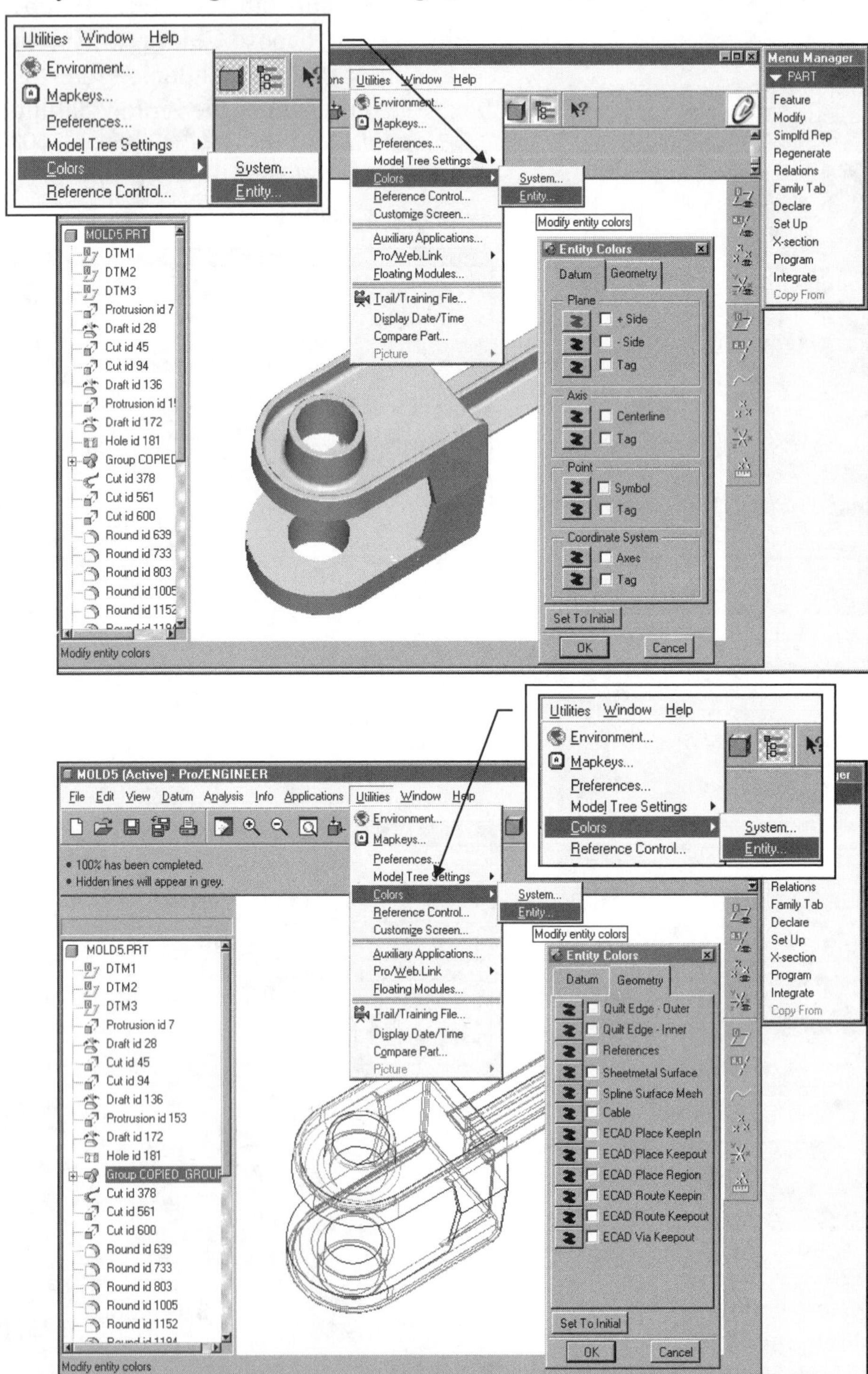

Figure 6.12
Entity Colors
Datum (top)
Geometry (bottom)

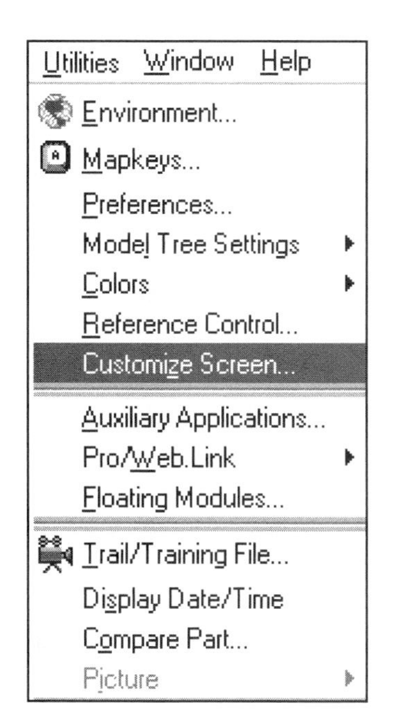

Customizing Your User Interface (IU)

You can customize your user interface in several ways. You can use the **Customize** command on the **Utilities** menu to:

* Add, delete, and move toolbar icons. Many toolbar icons are available in addition to those that appear initially.
* Set a menu command as a default choice for a menu.
* Add menu commands to menu bar menus.

You can use the **Mapkeys** command on the **Utilities** menu to:

* Create keyboard macros (copies of frequently used command sequences that are associated with defined key strokes).
* Add icons for the mapkeys to the toolbar and place labels on the icons for ease of use.
* Enter descriptions of each macro that help you keep track of the defined macros.

The **Customize** dialog box contains the following tabs (Fig. 6.13):

Toolbars Commands Options

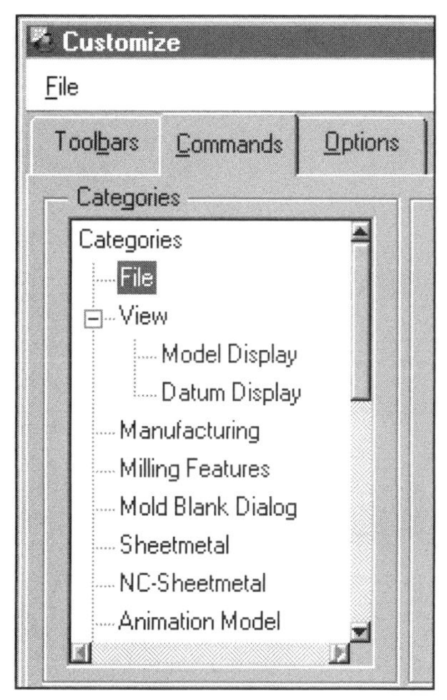

Figure 6.13
Customize Dialog Box

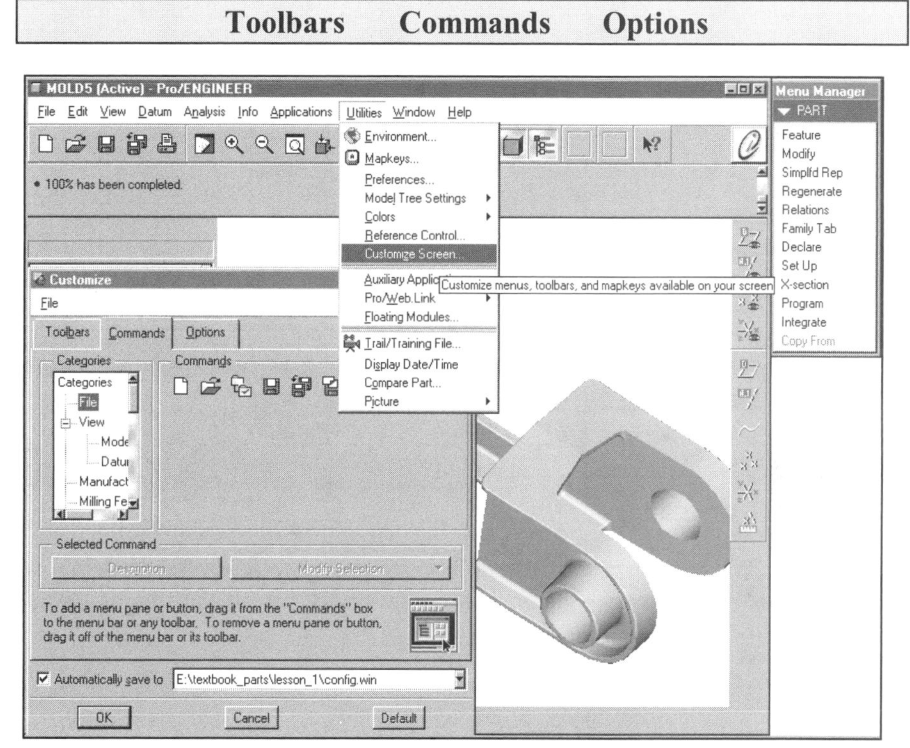

To customize your Toolbar, choose the following:

1. Choose **Utilities ⇒ Customize Screen.** The Customize dialog box appears (Figs. 6.13 and 6.14).
2. Click one of the tabs in the Customize dialog box.
3. Select a category, menu, or button and follow the instructions in the window.
4. To apply changes you made to the user interface, click **OK.** The Default button restores the default Pro/E user interface.

Commands

With the **Commands** page in the Customize dialog box (Fig. 6.14), you can create a toolbar that is specific for your needs. You can click, drag, and drop the buttons where you want them in the Pro/E toolbar.

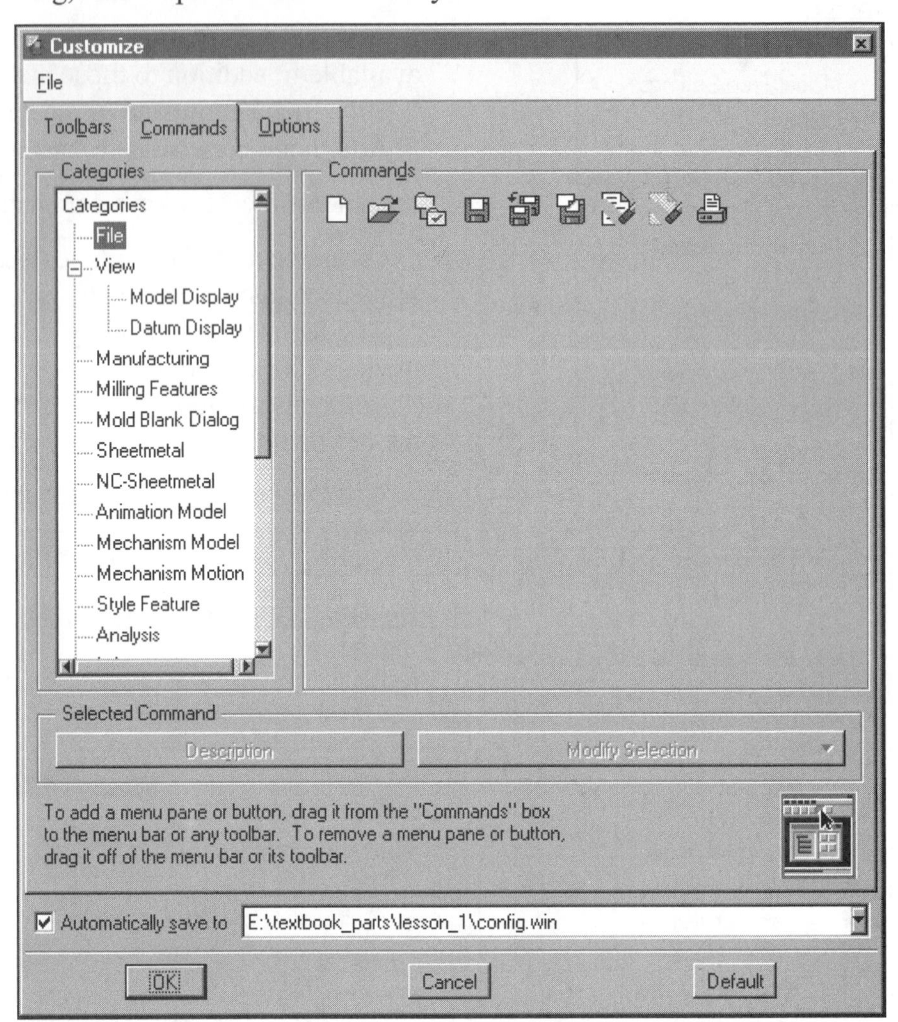

Figure 6.14
Customize Commands

The Commands Categories consist of commands that are most helpful in the core modules-- for example, Drawing, Sketcher, and Manufacturing. You can also move buttons that you have defined for mapkeys to the toolbar. To Customize screen commands, do the following:

1. Select a Pro/E category in the Categories list.
2. Pro/E displays the buttons available for this category under Commands.
3. Click a button to see its description in the Description box.
4. Drag the button to any toolbar in the current window.
5. To remove a button from the toolbar, drag it off the toolbar while the Customize dialog box is open.

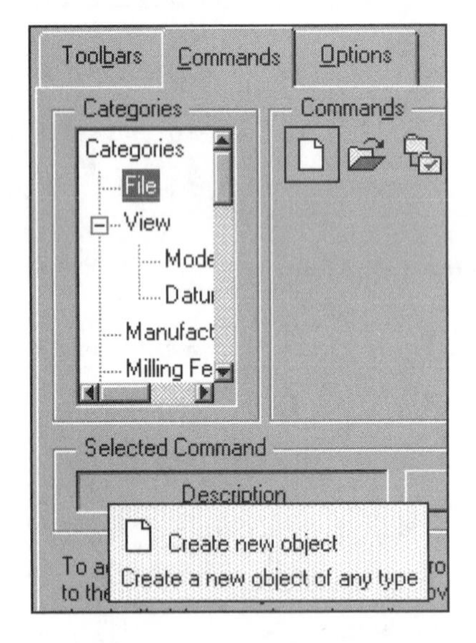

In Figure 6.15, the **Utilities** category was selected. The icons associated with **Utilities** appear to the right of the **Categories** area in the **Commands** area of the Customize dialog box.

Pick one of the icons (one at a time for this category, though in some cases the whole set of icons can be selected, dragged, and dropped), hold the left mouse button down, and drag the icon to a Toolbar. Release the left mouse button to drop the icon (Fig. 6.16).

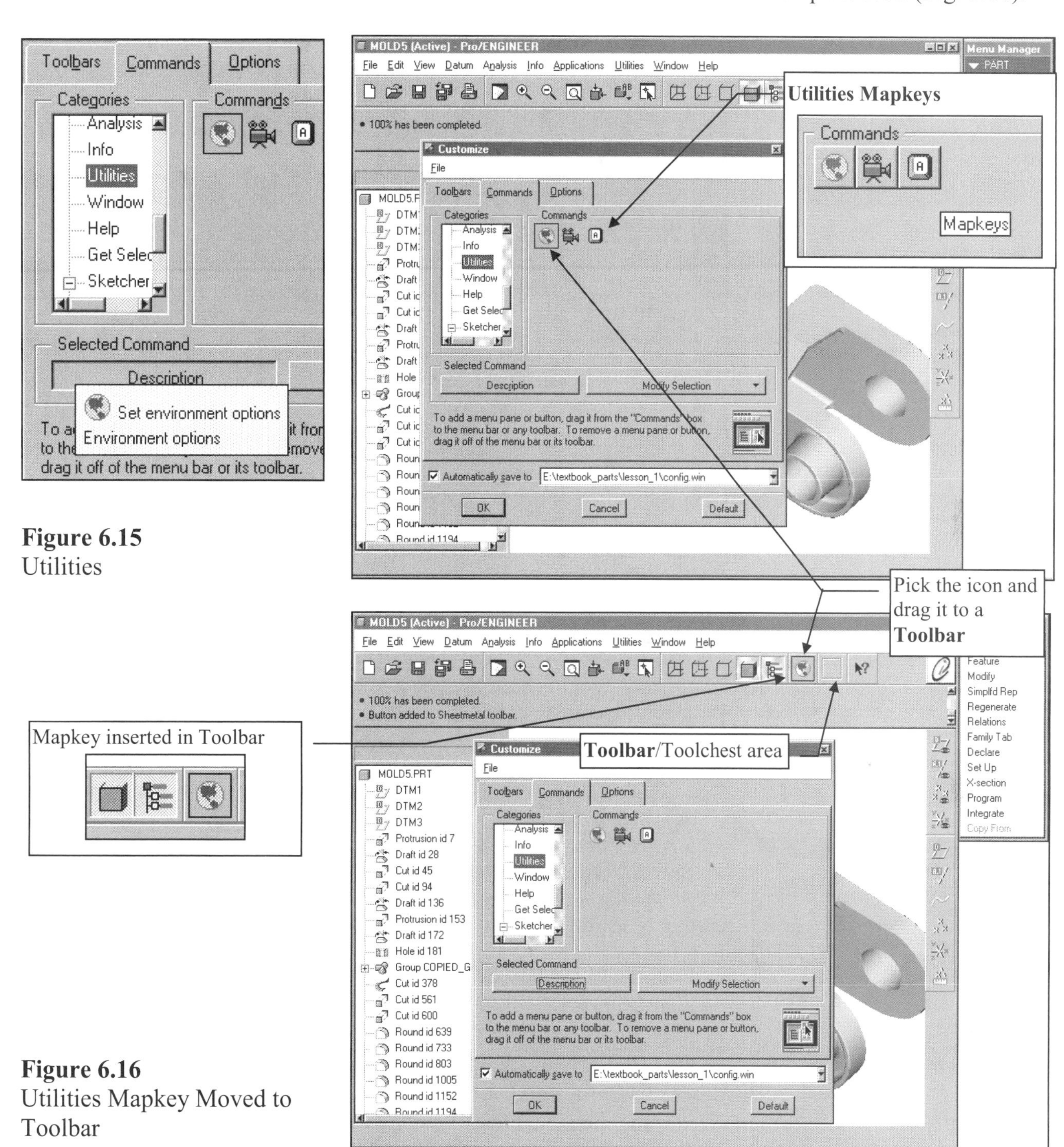

Figure 6.15
Utilities

Figure 6.16
Utilities Mapkey Moved to Toolbar

The **Commands** page in the Customize dialog box (Fig. 6.17) also lets you can customize menus by adding mapkeys. If there is no Pro/E default for a particular menu, you can set your own default using a *menu_def.pro* file. Like the configuration file, this file is loaded during Pro/E initialization. The Pro/E load point directory is searched first, then the current working directory. If a menu default entry has already been defined for a particular menu in the file *menu_def.pro*, any conflicting definitions in the local file are ignored.

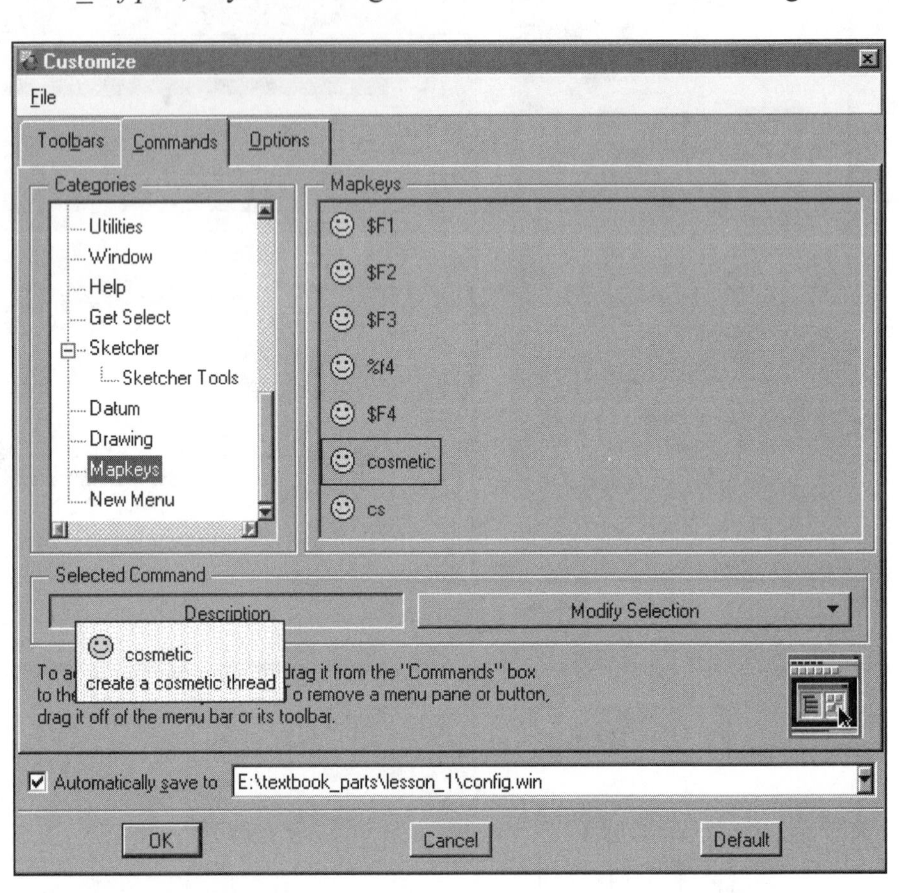

Figure 6.17
Customize Menus

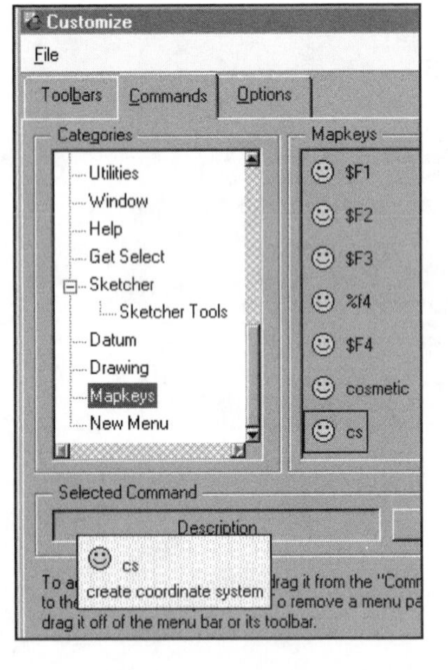

To customize menus, choose the following:

1. Select **Mapkeys** in the **Categories** list.
2. Select a mapkey to see its description in the Description box as shown in Figure 6.17 and in the left-hand column.

Previously, a Mapkey called **cs (Datum ⇒ Coord Sys ⇒ etc.)** was created (Fig. 6.18). When the Mapkey category was selected, the **cs** mapkey displayed in the Mapkeys area of the dialog box. To add the **cs** icon to Toolbar 1, click on the **cs** icon, hold the left mouse button down, drag the icon to the added Toolbar 1 (Fig. 6.19), and then release the mouse button.

To change the Button Image, position the cursor over the **cs** icon and press and hold the right mouse button (RMB) (Fig. 6.20). Select **Choose Button Image.** The Pick Mapkey Icon dialog box will display (Fig. 6.21). Select the image you want for your **cs** icon (mapkey). If you want to edit the image, select **Edit Button Image** (Fig. 6.22). Figure 6.23 shows another image being edited.

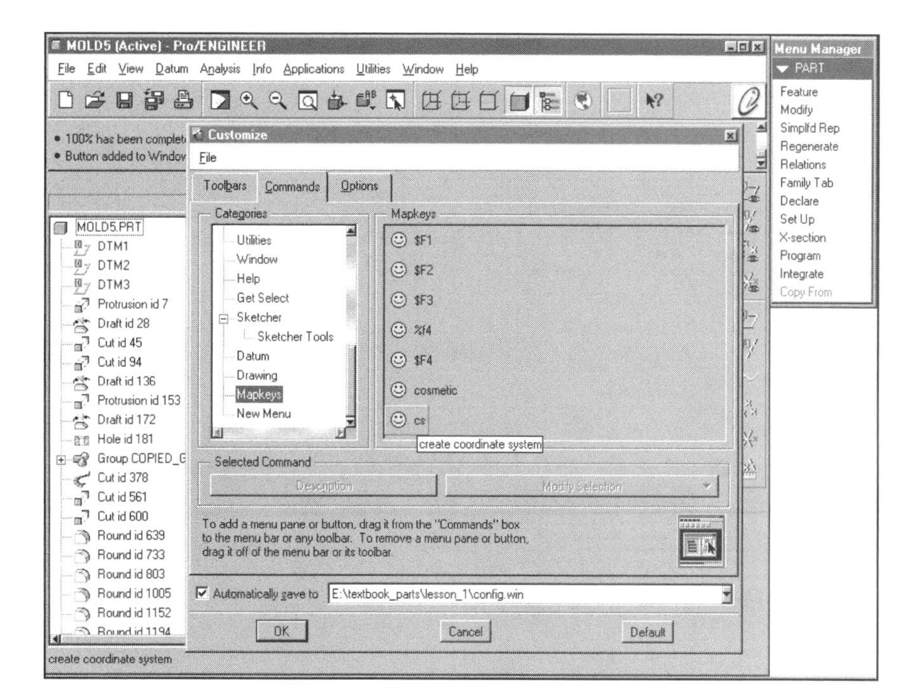

Figure 6.18
Customizing Tool Bars

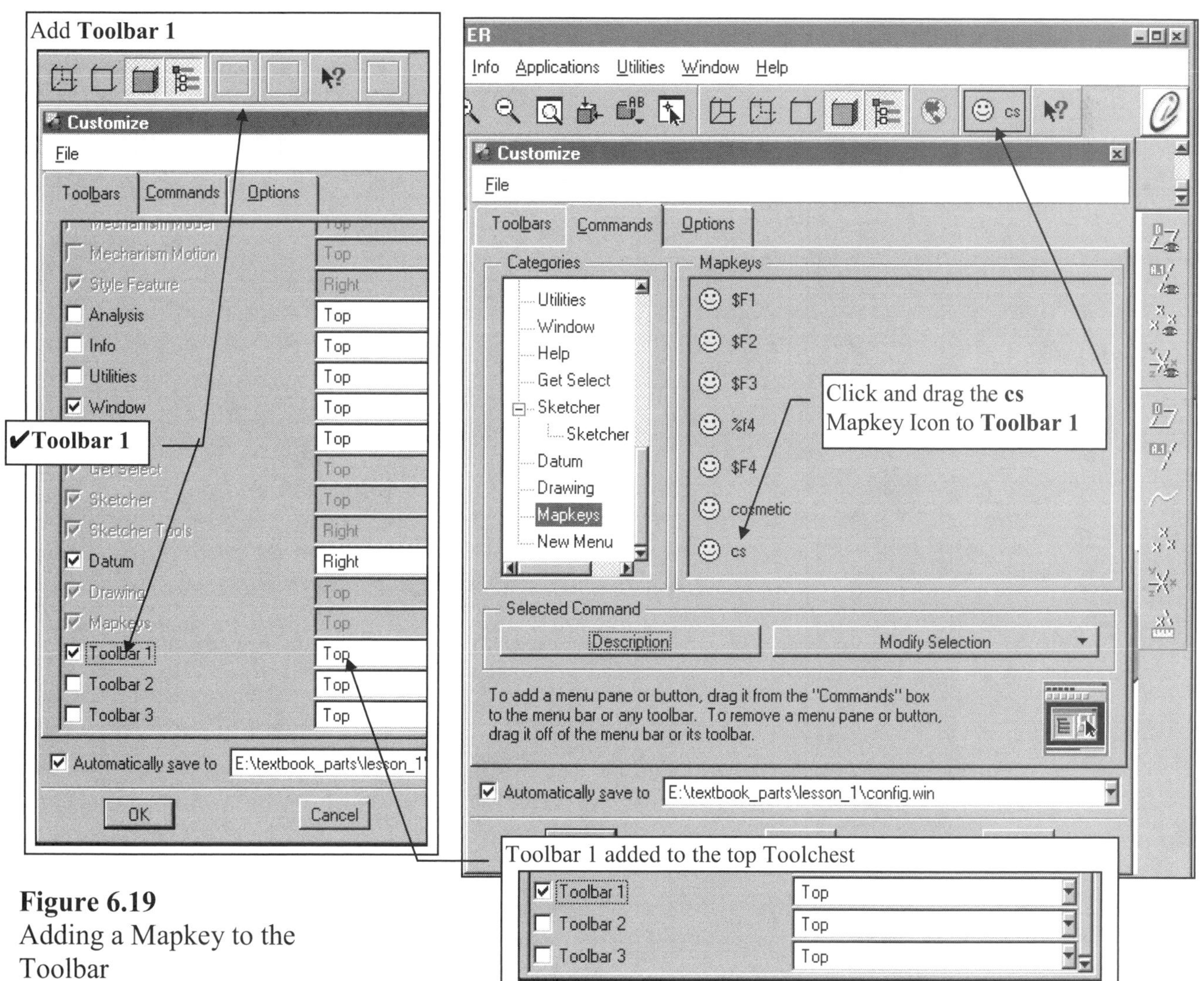

Figure 6.19
Adding a Mapkey to the Toolbar

Figure 6.20
Choosing a Button Image

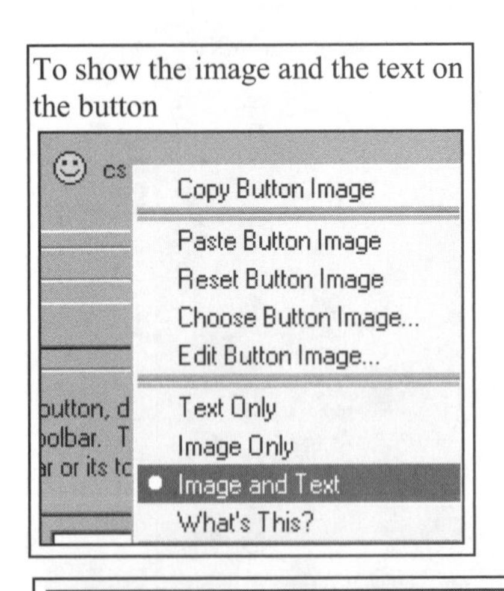

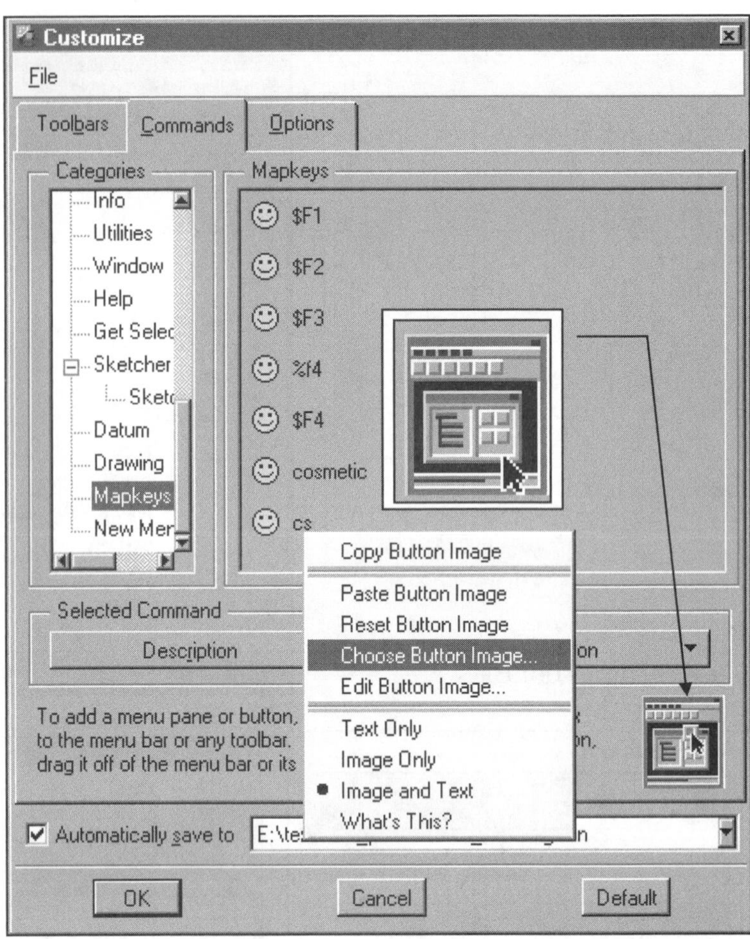

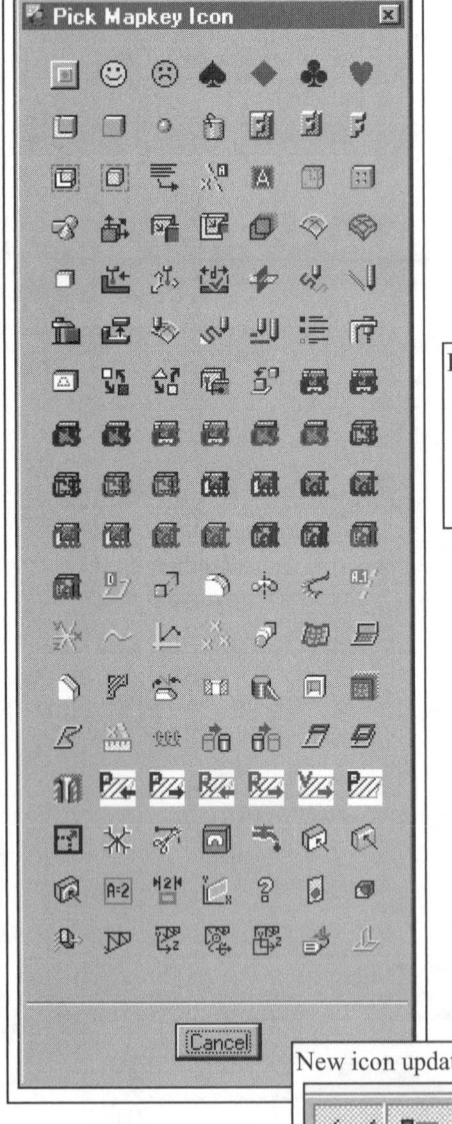

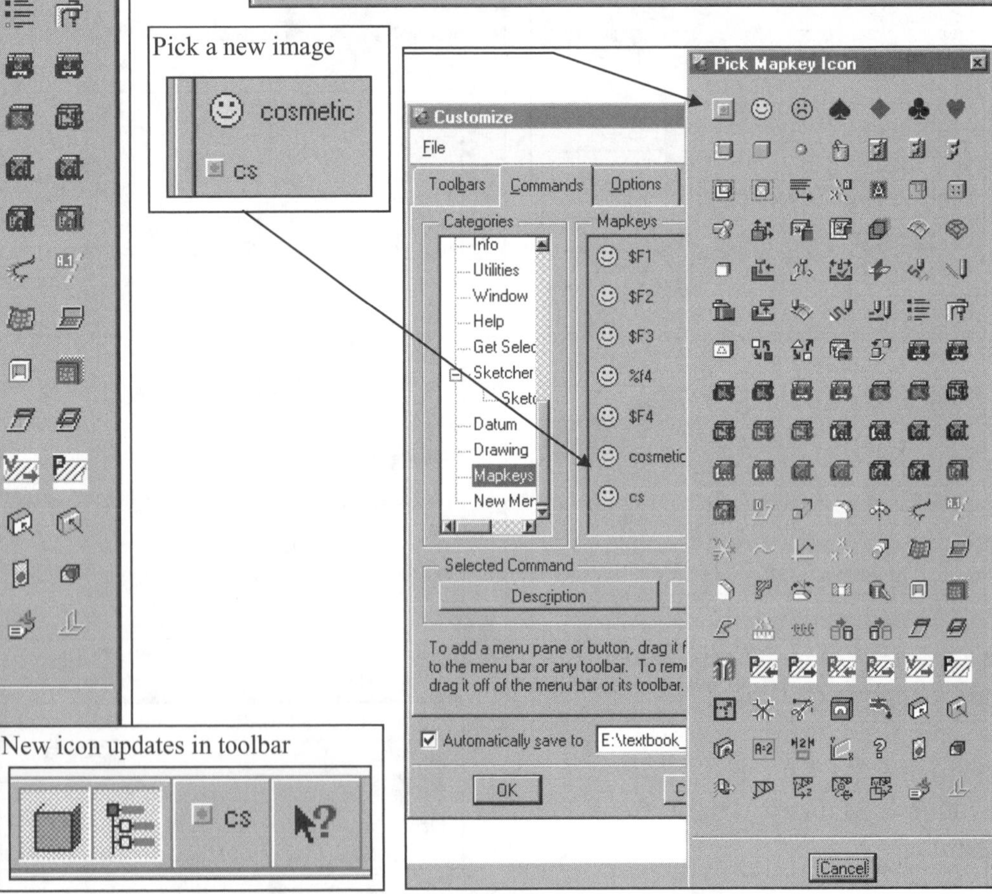

Figure 6.21
Pick Mapkey Icon

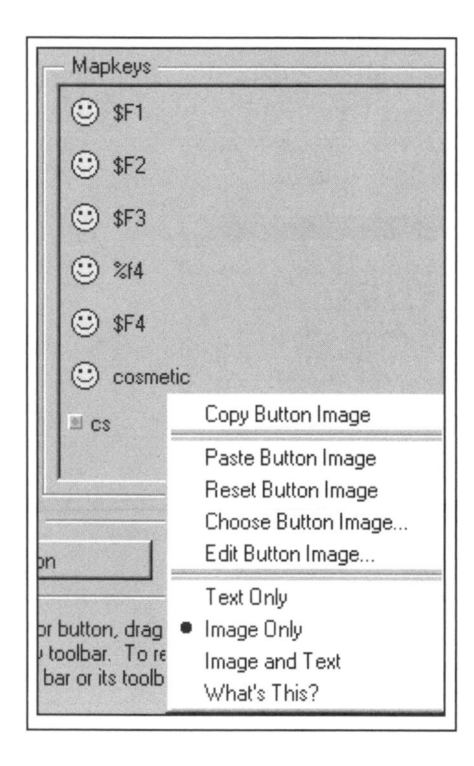

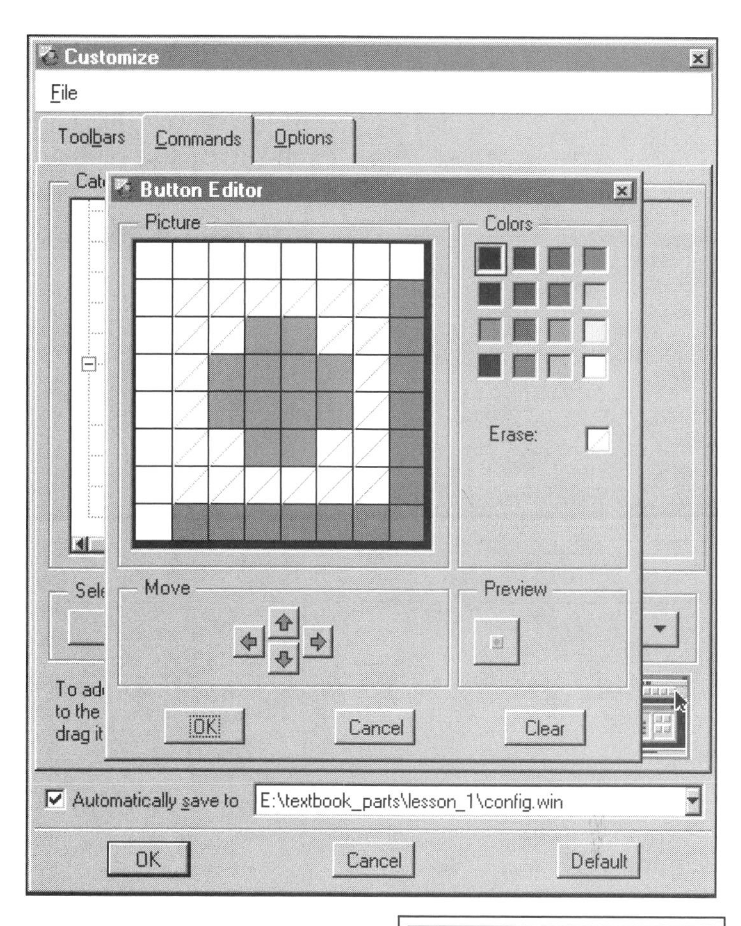

Figure 6.22
Edit Button Image

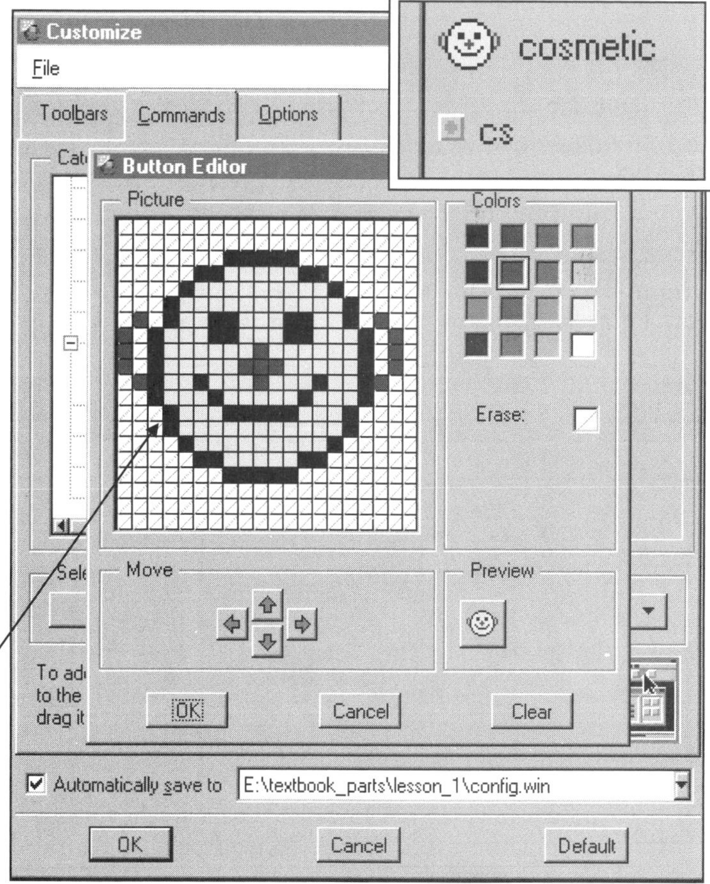

"Nose" and "ears" were added

Figure 6.23
Button Editor Dialog Box

Toolbars

Using the **Toolbars** page in the **Customize** dialog box (Fig. 6.24), you can customize menus by adding and removing function buttons to and from the toolbar on the *top*, *left side,* or *right side* of the Pro/E window. You can also add user-defined mapkeys to the toolbar.

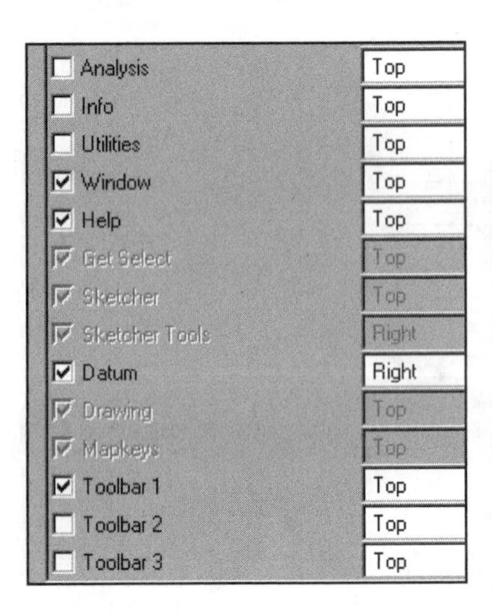

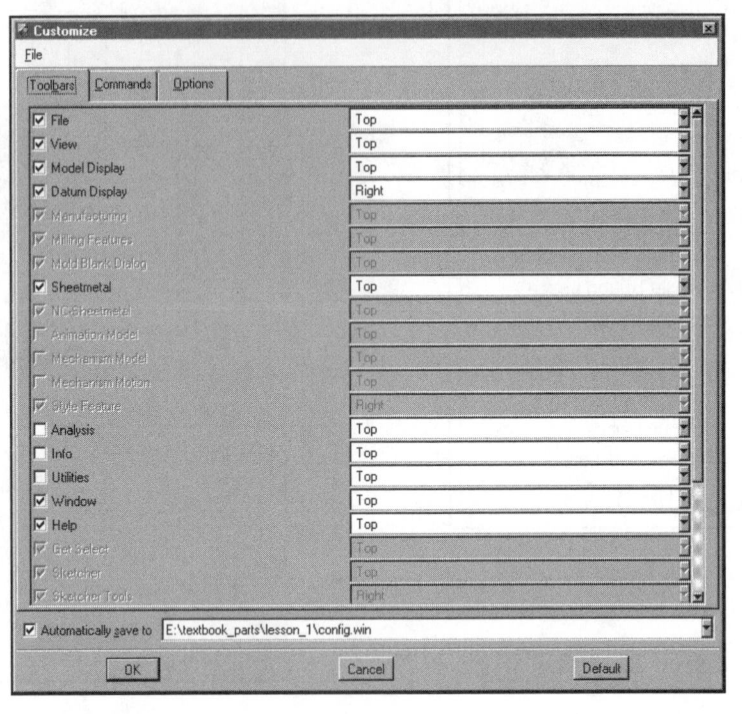

Figure 6.24
Customize Toolbars

Customizing the Toolbars

1. Select toolbar boxes that you want to display on the screen-- for example, **Info** (Fig. 6.25).
2. For each selected toolbar, specify its position in the Pro/E window (Top, Left, or Right).
3. Pro/E moves all the buttons in the appropriate toolbar category to the specified position. For example, if you specify Left for the **Info** toolbar, Pro/E moves the default set of buttons in the **Info** category to the left side of the window.

Figure 6.25
Info Icons Added to Left Side Toolbar
Left - Tool Bar (with Model Tree Displayed)
Right - Tool Bar (without Model Tree displayed)

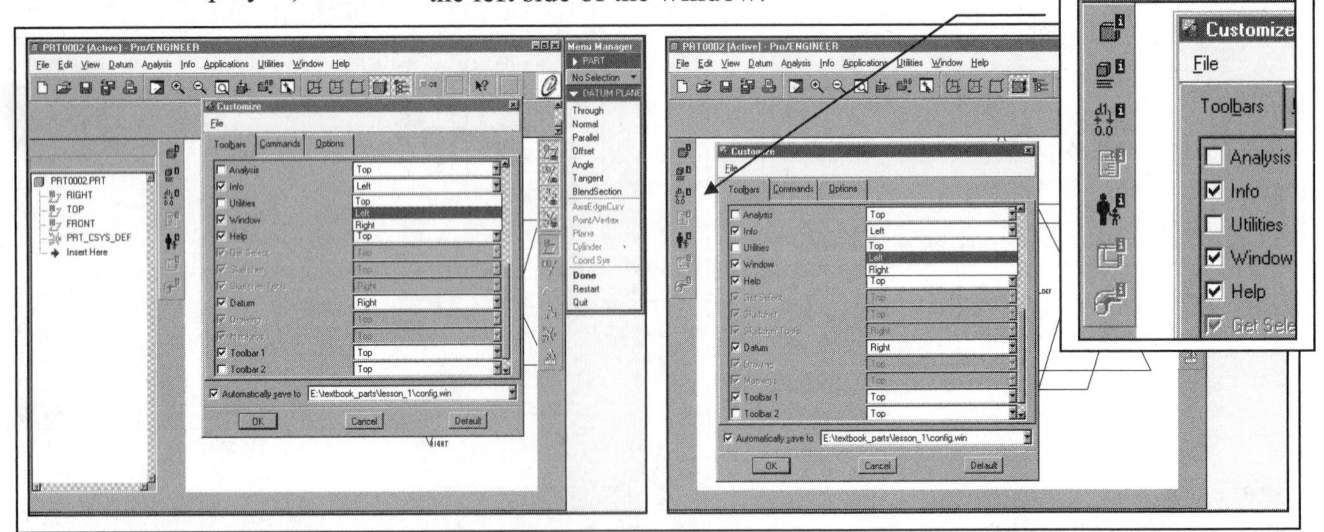

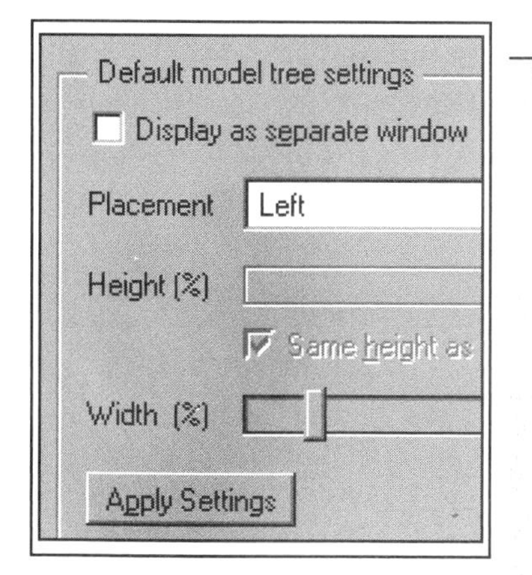

NOTE

Default position for the Model Tree is integrated into the left side of the Main Window. Most of the text illustrations display the Model Tree in a separate window, and moved to the lower right corner of the screen.

Options

With the **Options** tabbed page in the Customize dialog box, you can customize the position of the message area, the Model Tree position, the Secondary Windows size, and display icons in menu bar menus, as shown in Figure 6.26.

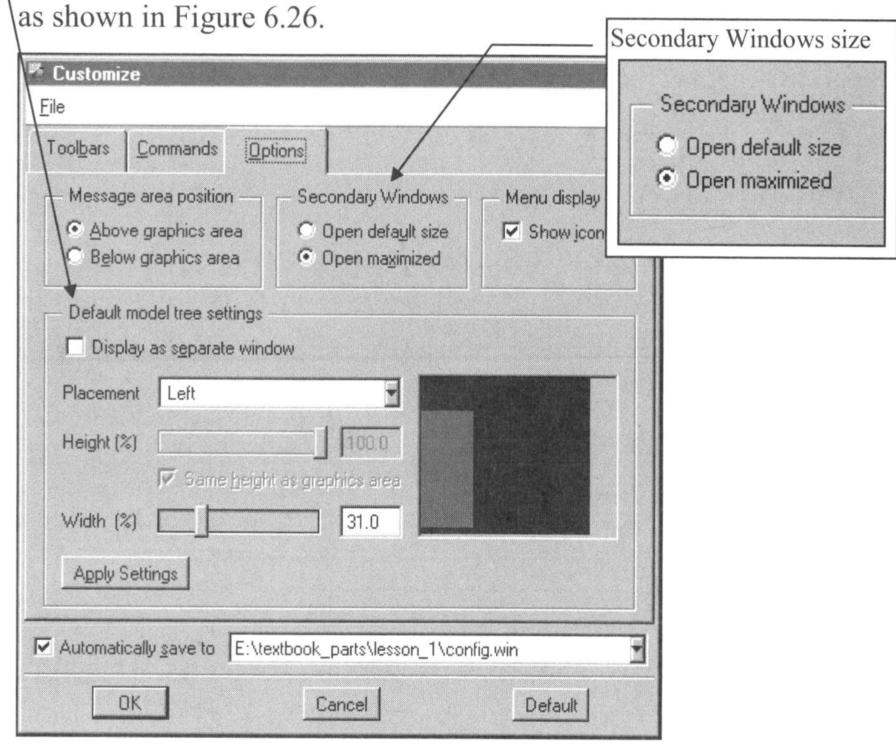

Figure 6.26
Options

Above graphics window (default)
Position the message area above the graphics window (Fig. 6.27).

Below graphics window
Position the message area below the graphics window.

The configuration file option ***windows_scale*** lets you scale the Pro/E windows with a given coefficient-- for example; **.95.**

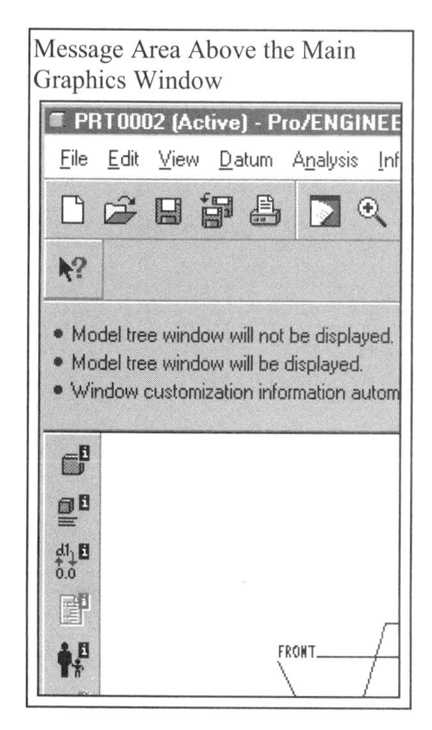

Figure 6.27
Message Area Positioned Above the Graphics Window (default)

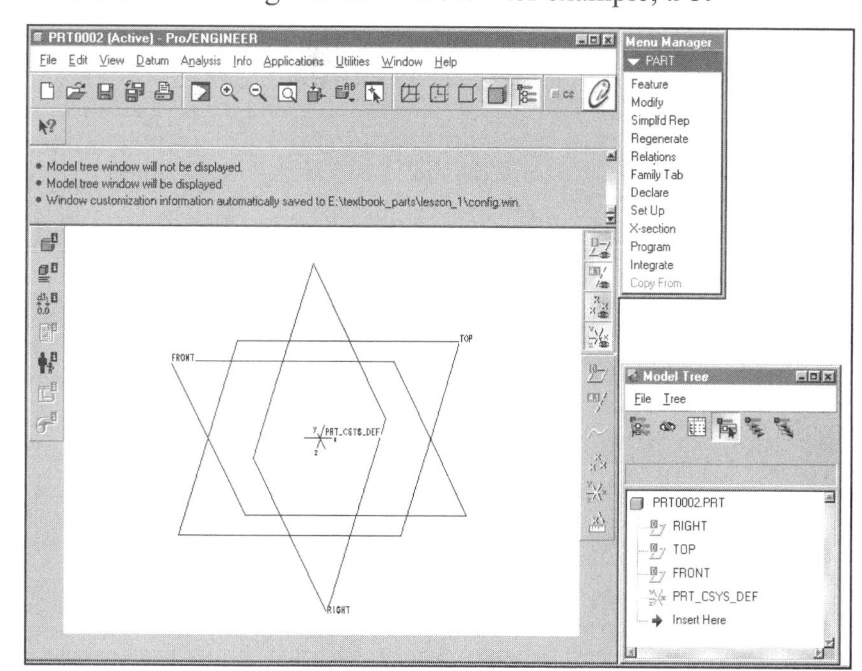

In Figure 6.28, the screen has been customized. Menu icons are located in **Top**, **Left**, and **Right Toolbars**, the Message Area is displayed above the Main Graphics Area, and the **Mapkey cs** is in the **Right Toolbar.**

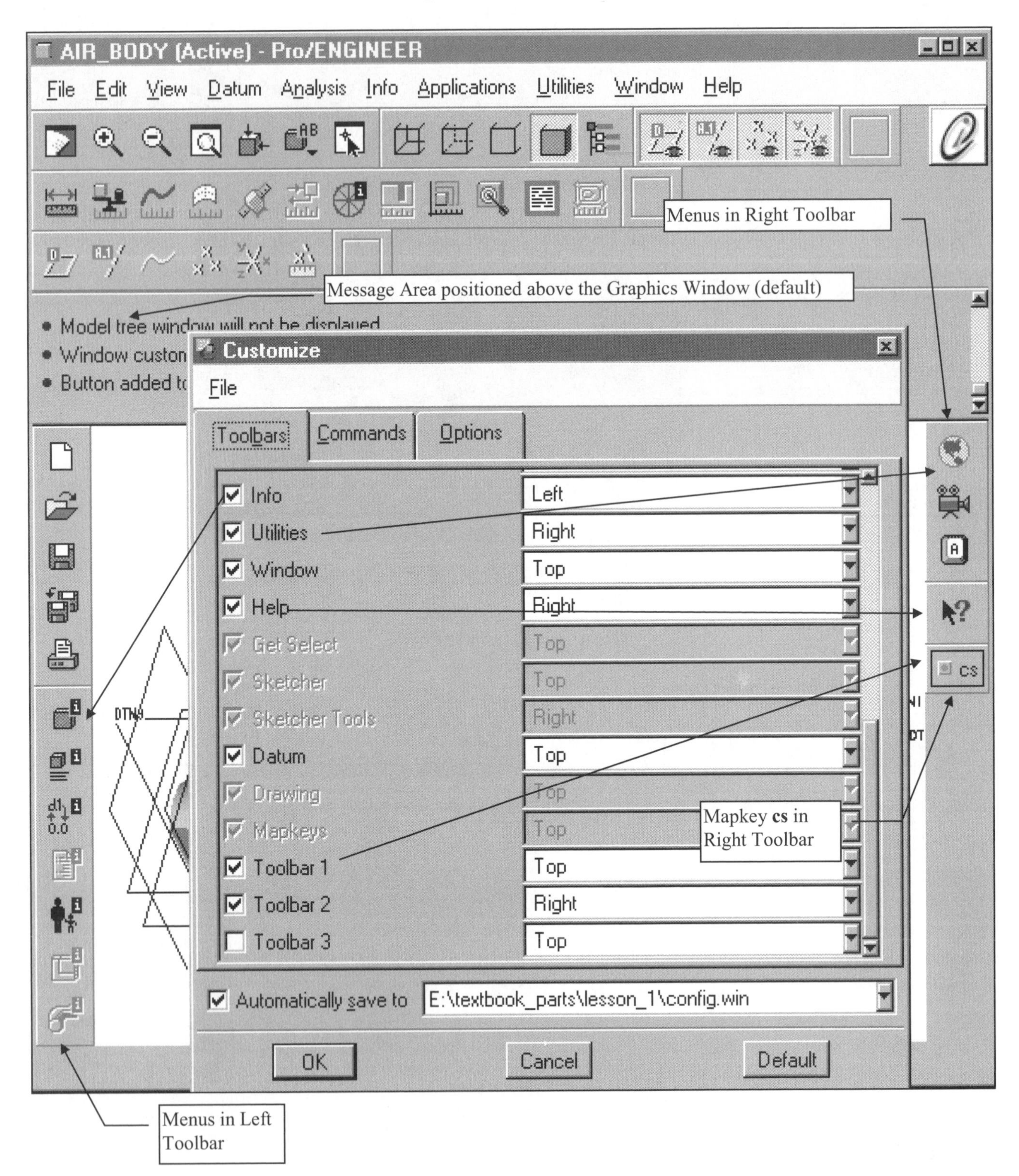

Figure 6.28
Customized Pro/E Session

Section 7

Setup, Information, and Analysis

Several capabilities are used during the design of parts, assemblies, and drawings. In this section, **Set Up, Info,** and **Analysis** functionality are covered. Before a project is started, whether it is a part or an assembly, you must set up the working standards for that project. Units must be selected. Dimension tolerance standards are set to ANSI (1982), ASME (1994), ISO, JIS, or DIN. Materials should be selected for the part. Parameters are input. Accuracy is set. Geometric tolerances are established, and other specific design settings are organized and input at this point.

The units, accuracy, and other settings established with the **Set Up** command will be reflected when information is requested about the model. At any time in the design process, **Info** (information) about the part or assembly can be requested from the pull-down menu. Mass properties (Figures 7.1 and 7.2), surface analysis, geometric tolerances, measurements, and a variety of information listings are available with the **Set Up, Info,** and **Analysis** commands.

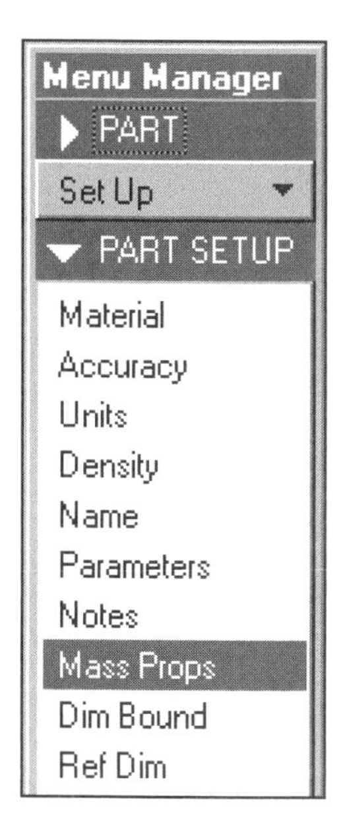

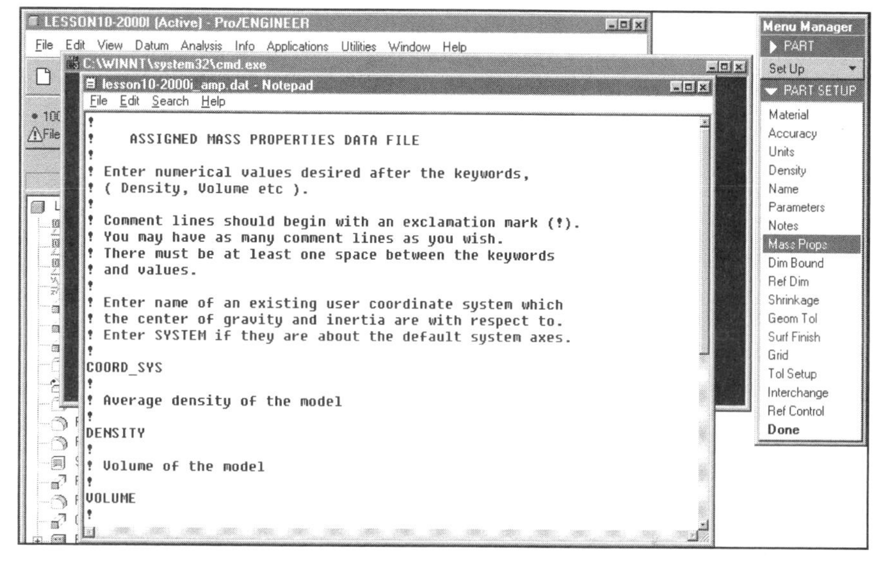

Figure 7.1
Setting Up Mass Properties: **Set Up ⇒ Mass Props**

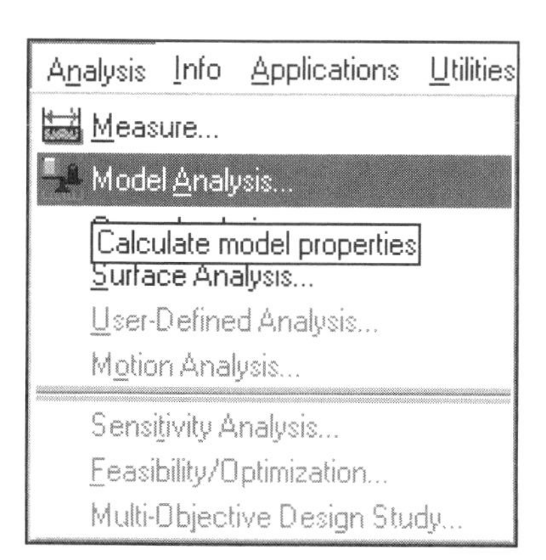

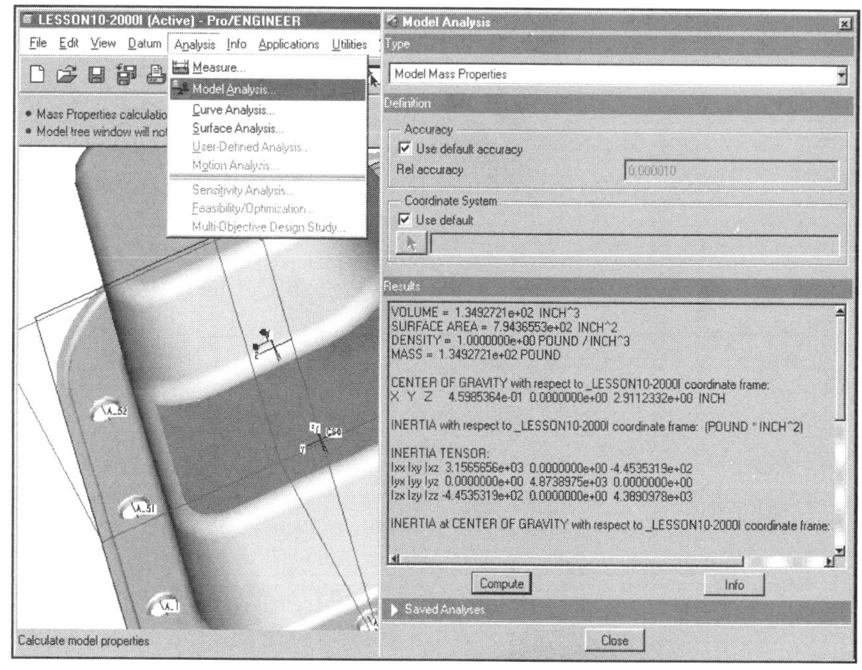

Figure 7.2
Analysis (from Menu Bar) ⇒ **Model Analysis ⇒ Model Mass Properties ⇒ Compute**

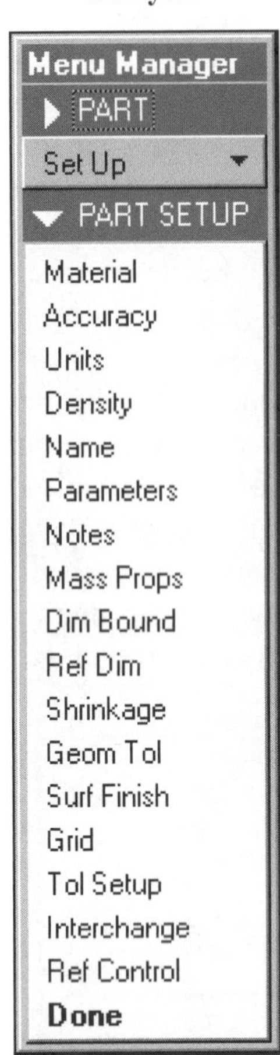

Set Up (Part and Assembly)

In *Part mode* and *Assembly mode*, the SETUP menu allows you to define various attributes of the model. The most important of these capabilities are discussed in detail in this section. The following is a listing of options:

Material Create and modify material data files.
Accuracy The **Accuracy** option in the PART SETUP menu lets you change the part computational accuracy. This lets you adjust the accuracy according to the ratio of the smallest edge of a feature to the overall size of the part. If you attempt to place a small feature on a very large part, you may get the error "Could not construct feature geometry." If you get this error, change the Accuracy to a smaller number. The range for Accuracy is from **.01** to **.0001**. The default Accuracy is **.0012**. To check the ratio, determine the largest overall size of the part using the **Info ⇒ Model Size** option, then multiply that number by **.0001**. This would give you the smallest edge for a feature you could create on the part. When you choose **Info ⇒ Model Size**, Pro/E displays a rectangular box around the part and a diagonal. It also displays the length of the diagonal.

Relative Modify relative part accuracy (Part mode). This is the computational accuracy of Pro/E geometry calculations and has a default value of **0.0012**. Part accuracy is relative to the size of the part. Increasing the part's accuracy increases regeneration time.
Absolute This functionality improves the matching of parts of different size or different accuracy (as an example, imported parts created on another system). In most cases, use relative accuracy.

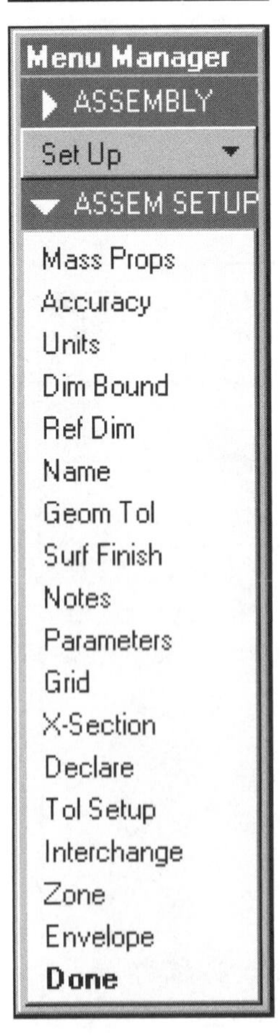

Units Specify the dimension units for the part or assembly.
Density Set a material density value to be assigned to the part.
Name Assign names to features, assembly members, and so on. Also used to name or rename datum planes and coordinate systems instead of using default names.
Parameters Allows editable user-defined parameters to be assigned to features, surfaces, models, etc.
Notes Create, modify, or remove notes associated with the model.
Mass Props Create a file of mass properties.
Dim Bound Change a dimension regeneration value from nominal to its upper, lower, or midpoint tolerance boundary.
Ref Dim Create reference dimensions for the model.
Shrinkage Modify shrinkage of part dimensions.
Geom Tol Specify geometric tolerances for surfaces and features.
Surf Finish Define surface finish symbols for the part model.
Grid Define a 3D grid for the model.
Tol Setup Specify tolerance standards (ANSI, ISO/DIN).
Interchange Shows information about, or removes references to, interchange groups.
Ref Control Controls creation of external references.
X-Section *(assembly)* Create, modify, and display cross sections.
Declare *(assembly)* Establish declarations to a layout.
Zone *(assembly)* Define a zone in an assembly.
Envelope *(assembly)* Create or modify an envelope component.

Material

The **Material** option is used to create and modify material data. To assign a specific material to the current part, choose **Material** from the PART SETUP menu. This calls up the material management menu, MATRL MGT, which includes the following options:

> **Define** Define the properties of a new material. The system editor displays a default specification file (Fig. 7.3) that must be edited in order to add the desired values for the material parameters. Different materials can be created by building new material files with the system editor (Fig. 7.4).

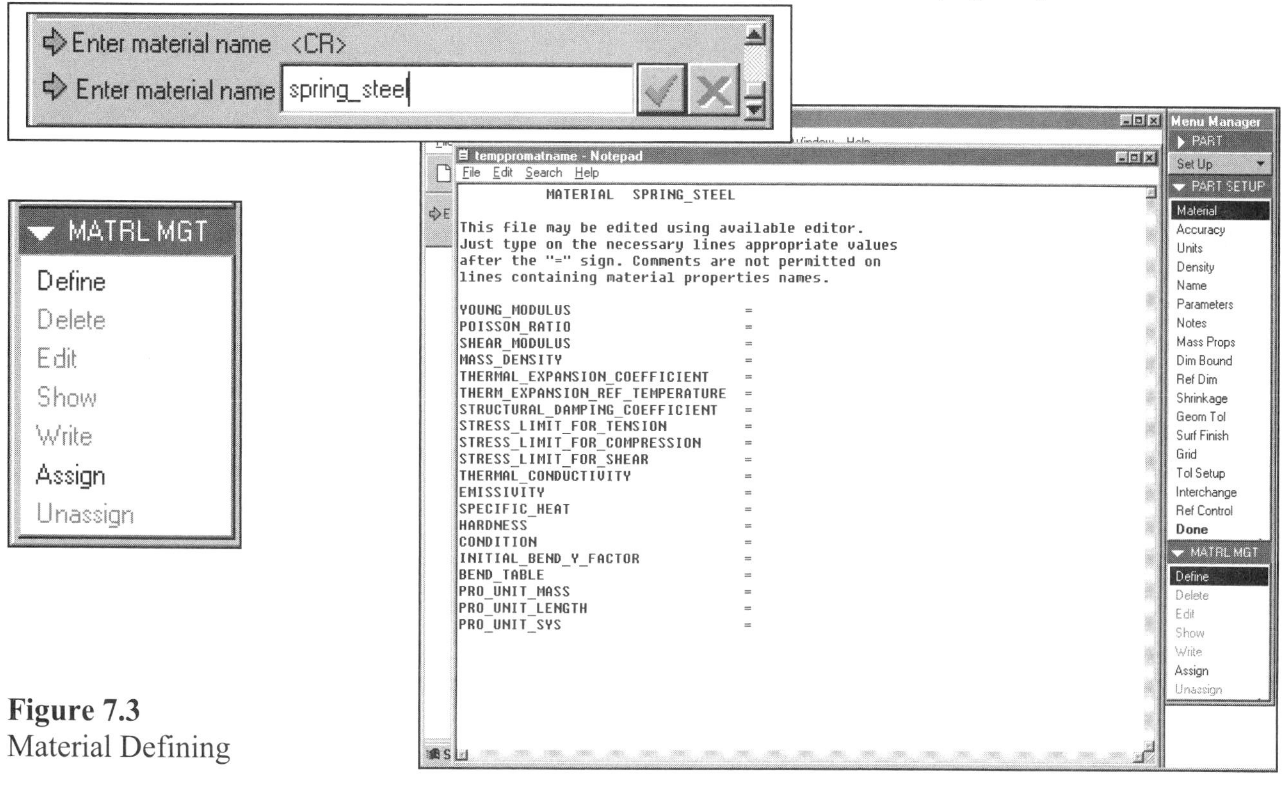

Figure 7.3
Material Defining

Material Steel_1040		
Young_Modulus	=	**29000000**
Poisson_Ratio	=	**0.27**
Shear_Modulus	=	**11000000**
Mass_Density	=	**0.00879**
Thermal_Expansion_Coefficient	=	**6.78**
Thermal_Expansion_Ref_Temperature	=	**32.0**
Structural_Damping_Coefficient	=	**0.01**
Stress_Limit_For_Tension	=	**36000**
Stress_Limit_For_ Compression	=	**36000**
Stress_Limit_For_ Shear	=	**36000**
Thermal_Conductivity	=	
Emissivity	=	
Specific_Heat	=	
Hardness	=	
Condition	=	
Initial_Bend_Y_Factor	=	
Bend_Table	=	**steel_1040**

Figure 7.4
Steel Material File

Delete Remove a material from the part's internal database. Pick the material from the MAT_LIST namelist menu.

Edit Edit the material specification file using the system editor. The material must be in the part's internal database. Select the material from the MAT_LIST namelist menu.

Show Displays the material specification file in an information window. The material must be in the part's internal database. Select the material from the MAT_LIST namelist menu.

Write Write material properties from the part to a disk file named *materialname.mat* (*materialname* is the name of the material). Select the material from the MAT_LIST namelist menu, and then specify the name under which the material will be stored. After defining the materials required in your design, create a materials library in the appropriate directory.

Assign Assign an existing material to the part. This material is used in all analysis calculations of the part. From the USE MATER menu, select the material list to choose from:

- **From Part** From the part's internal database, use the material as listed in the MAT_LIST namelist menu as shown in Figure 7.5.
- **From File** Use material that is stored in a disk file. Provide the name of the material to be retrieved.

Unassign Unassign the currently assigned part material.

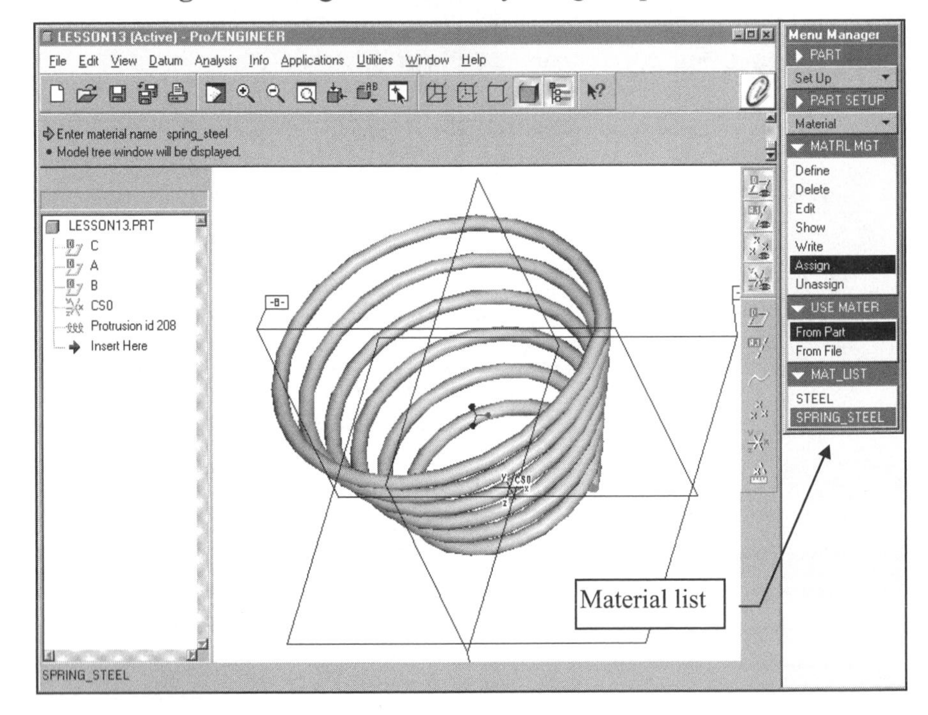

Figure 7.5
Assigning Material to a Part

Changing Material Parameters in a Part

You can specify or change material parameters by choosing **Material** ⇒ **Edit** to edit the material properties stored in the part's internal database or by choosing **Material** ⇒ **Assign** to change the material file assigned to the part. After regeneration, the material values are updated.

Units

The **Units** option in the PART SETUP menu enables you to specify the dimension units to be associated with the part (Fig. 7.6). The **Units** option in the ASSEM SETUP menu governs the units of assembly features, placement offsets, and explode distances.

Each model has a basic system of metric and nonmetric units to ensure that all material properties of that model are consistently measured and defined. All Pro/E models have length, mass/force, time, and temperature units defined. Pro/E comes with some predefined systems of units, one of which is a default system of units. You can change the assigned system of units, and you can define your own units and systems of units (called custom units and custom systems of units). You cannot change the predefined systems of units.

Use the **Units** command to set, create, change, review, or delete a system of units or a custom unit for your model. When you use the Units command, the **Units Manager** dialog box opens. This dialog box lists the predefined systems of units and any already-defined custom systems of units. Using this dialog box, you can create new custom units and systems of units. A *red* arrow indicates the current system of units for the model. Use custom units only if your model does not contain standard SI or British units or if your Material file (which you set up by using the **Material** command under **Set Up**) contains units that cannot be derived from the system of units or both. Pro/E uses the definition of a custom unit in interpreting material properties. You can also use custom units to create a new system of units.

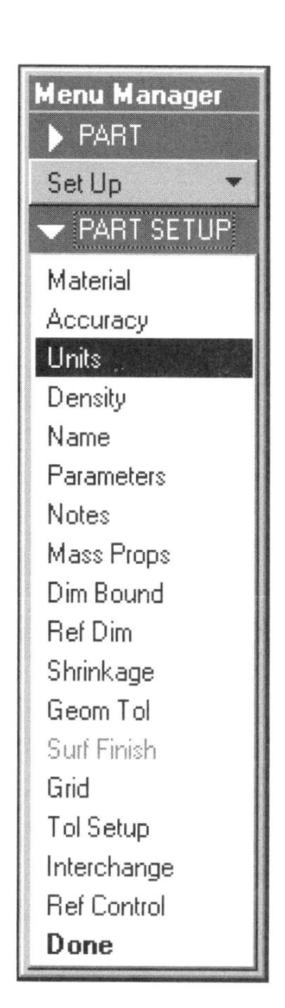

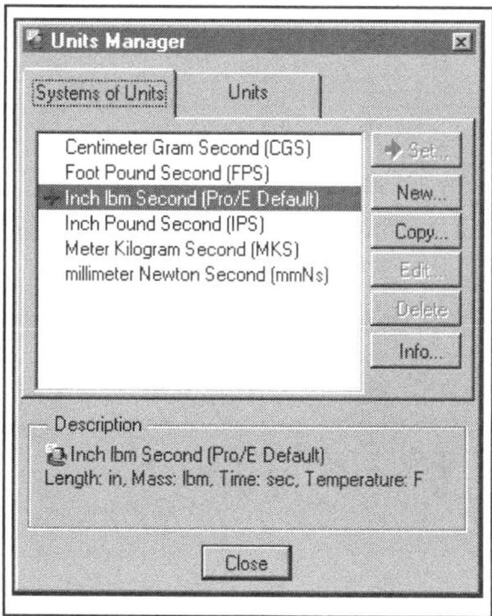

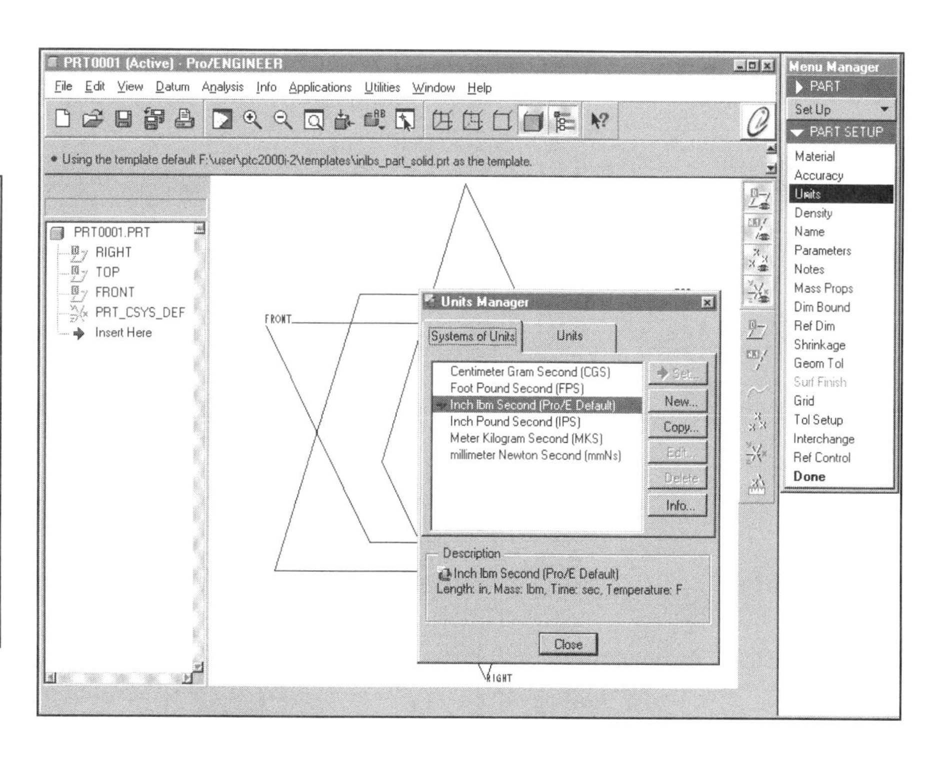

Figure 7.6
Units

Selecting the System of Units

The **Units Manager** dialog box (Fig. 7.7) is used to define a principal system of units for your model. To specify **Units:**

1. Choose **Set Up ⇒ Units.** The Units Manager dialog box appears.
2. From the System of Units list, select a system of units.

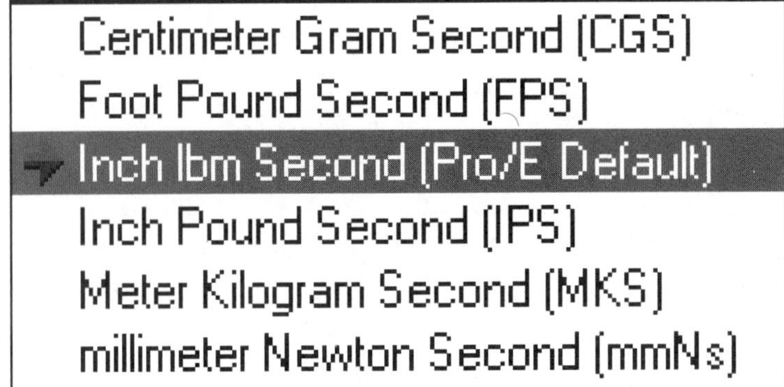

3. To change the system of units, select a different unit type from the list and then ➨**Set** (see left).

4. Select one of the following options:

   * **Convert Existing Numbers (Same Size)** Existing data is converted to the new system of units, but the physical size of the model remains the same.
   * **Interpret Existing Numbers (Same Dims)** Existing data is not converted to the new system of units. The physical size of the model changes, but the dimensions of the model remain the same.

5. Pick **OK** to return to the Units Manager dialog box.
6. You can create a new system of units by picking NEW and then editing the System of Units Definition dialog box (see lower left-hand column).

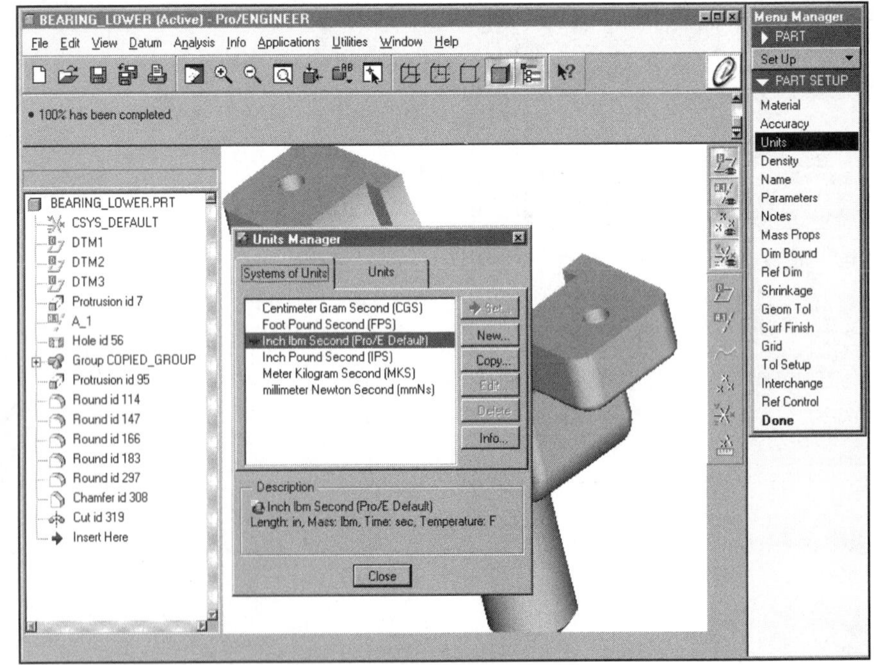

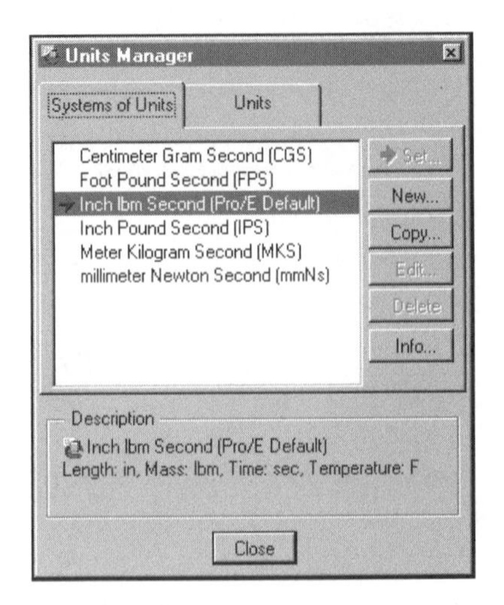

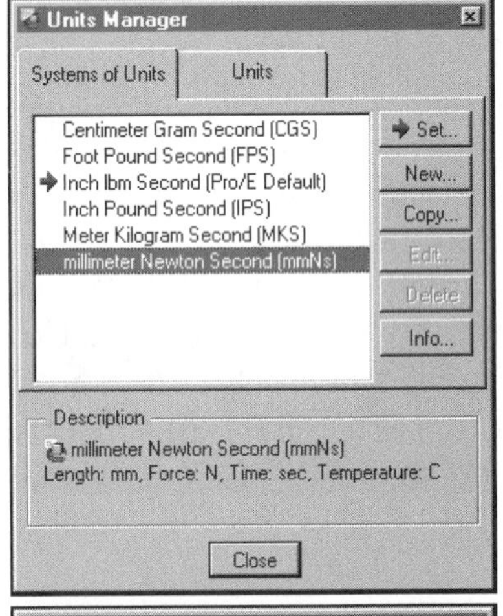

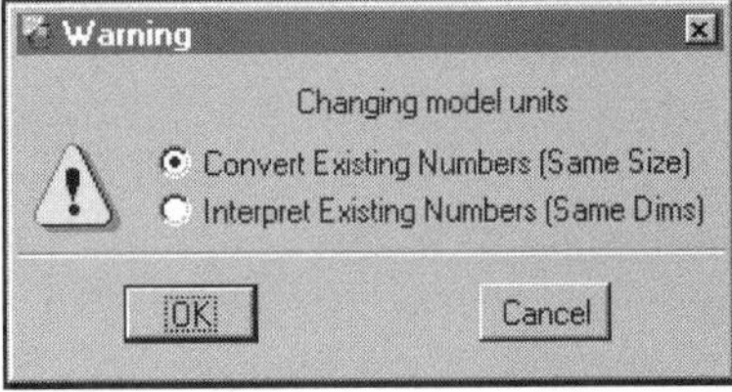

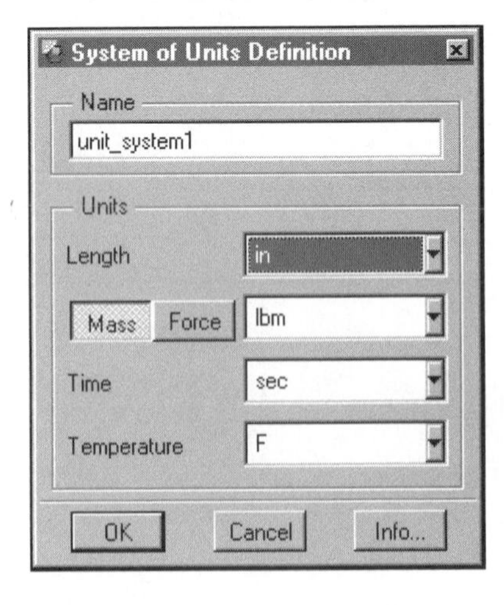

Figure 7.7
Units Manager

Dimension Tolerance Setup for Parts or Assemblies

The application of dimension tolerances is governed by either ANSI or ISO/DIN standards. By default, the configuration file option ***tolerance_standard*** is set to ***ANSI.*** ANSI dimension tolerances are assigned according to the number of digits specified. ISO/DIN dimension tolerances are driven by a set of ISO/DIN tolerance tables. To change to ISO/DIN tolerances, change the ***tolerance_standard*** option to ***ISO/DIN.***

Setting Up Dimension Tolerances

To use the **Tol Setup** command in Part or Assembly mode (Fig. 7.8), do the following:

1. Choose **Set Up** in the PART menu or ASSEMBLY menu.
2. Choose **Tol Setup** in the PART SETUP menu or ASSEM SETUP menu.
3. The TOL SETUP menu is displayed with the following options:

Standard Changes the tolerance standard of the current model from the TOL STANDARD menu.

ANSI Switches the standard from ISO/DIN to ANSI. New tolerances are determined for all dimensions on the basis of number of digits in the dimensions. Tolerance tables, if any, will be deleted.

ISO/DIN Switches the standard from ANSI to ISO/DIN. The system tolerance tables and any available user-defined tolerance tables are loaded.

Model Class Changes the tolerance class of the current model using TOL CLASSES menu options.

Tol Tables Specifies an action using the TOL TBL ACT menu options.

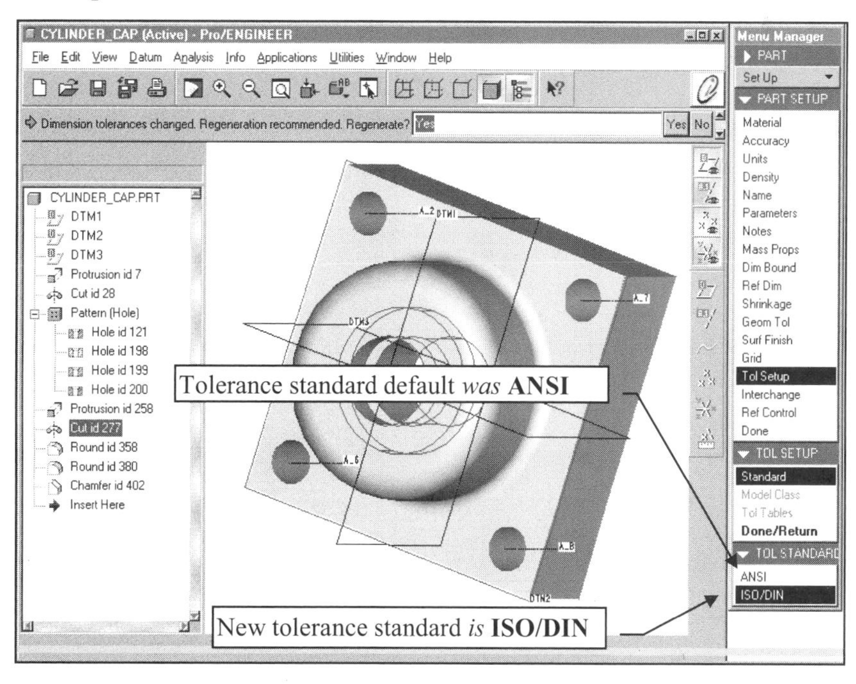

Figure 7.8
Tolerance Selection

Information

Choosing the **Info** menu from the Menu bar pull-down menu displays options to obtain information about the geometric properties of models, names of models and sections, and so on (Fig. 7.9). A variety of different types of information are available, including:

BOM (*Assembly mode*)
Feature
Feature list
Model *(part, assembly)*
Component (*Assembly mode*)
Audit trail
Pro/Engineer Objects
Message Log
Parent/Child
Global Ref_Viewer
Regen Info
Geometry Check
Switch Dims
Model Size

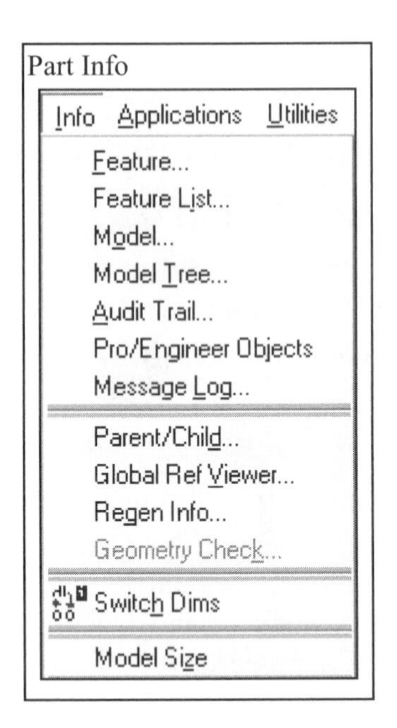

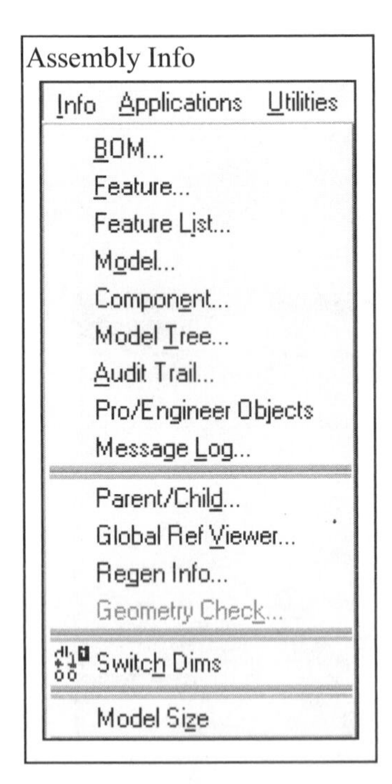

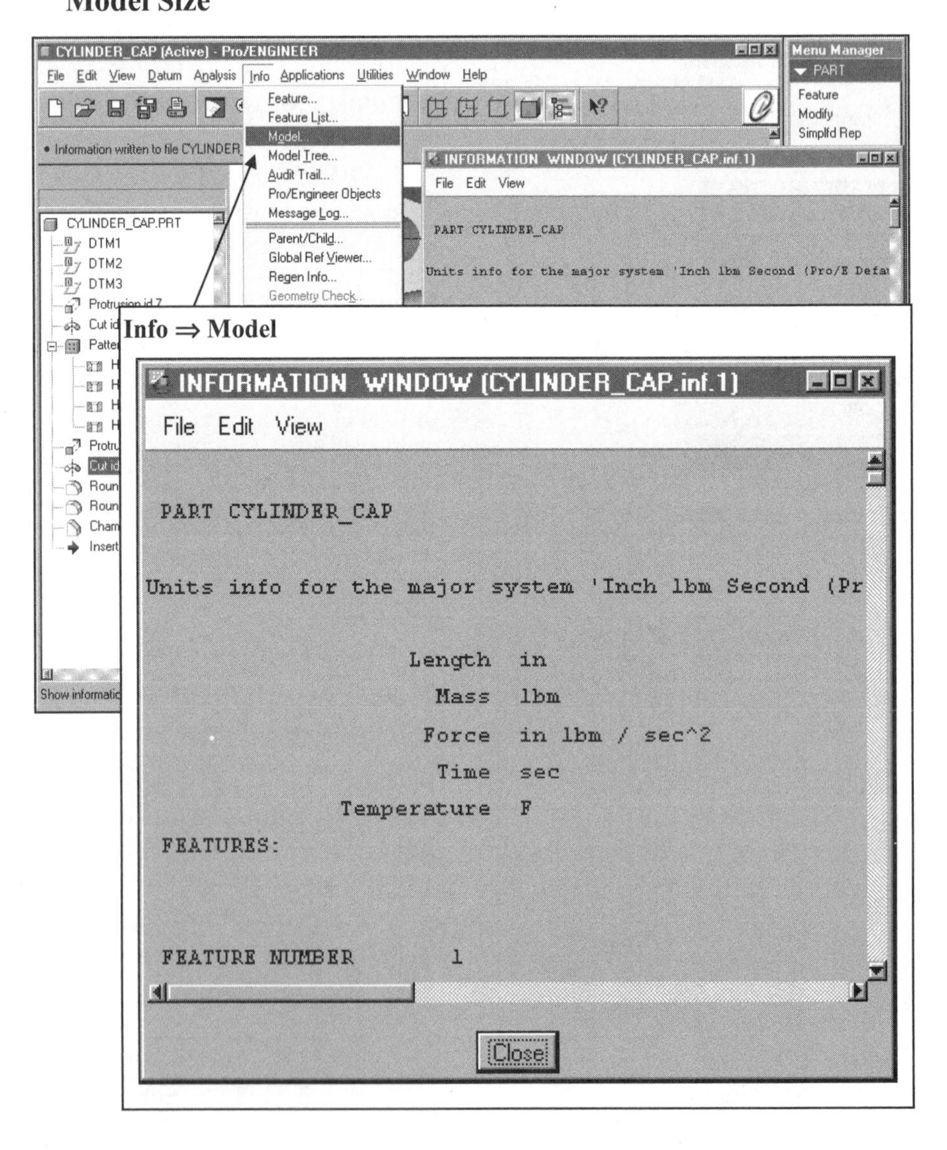

Figure 7.9
Model Information

The Analysis Pull-down Menu Options

Analysis on the Menu Bar can be used to compute mass properties; it can also be used to measure distances, clearances, and interference. Model information can be obtained by choosing **Info ⇒ Model.**

Mass properties for parts, assemblies, and cross sections can be computed by choosing **Model Analysis ⇒ Model Mass Properties ⇒ Compute** in the Part or the Assembly mode.

Measure

The **Measure** command is used to analyze model and draft geometry. It is available in all modes. Figure 7.10 shows the **Analysis ⇒ Measure ⇒ Area ⇒ Query Sel ⇒ Accept ⇒ Compute** commands given with the circular surface of the boss selected and accepted.

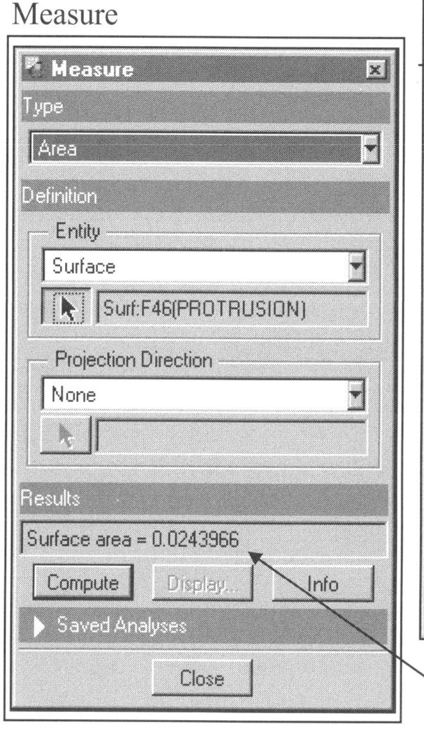

Figure 7.10
Measure

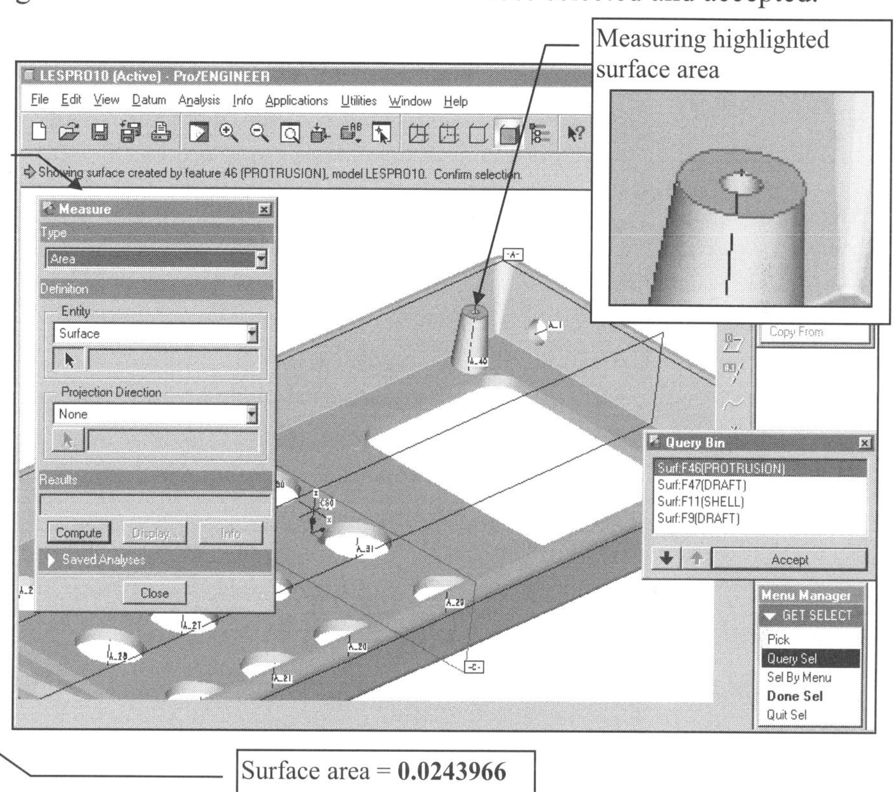

Curve/Edge

Part edges and datum curves are measured using the **Curve/Edge** option from the Measure dialog box. To measure model geometry, such as lengths, angles, and areas, from the Analysis menu, choose **Measure.** The Measure dialog box opens.

To measure an edge or datum curve:

1. Pick **Curve Length** from the Type menu.
2. Select the measurement type **(Curve/Edge** or **Chain).**
3. Select the edge or curve (Fig. 7.11a). Specify a chain by selecting two edges of a surface or two entities of a curve (Fig. 7.11b).

In Figure 7.11(a) the front curved edge of the part is measured as a length of **2.82649.** The Chain option was selected in Figure 7.11(b); the same edge was picked along with the corner curve. Pro/E then measures both curves and totals their combined lengths: **3.00511.**

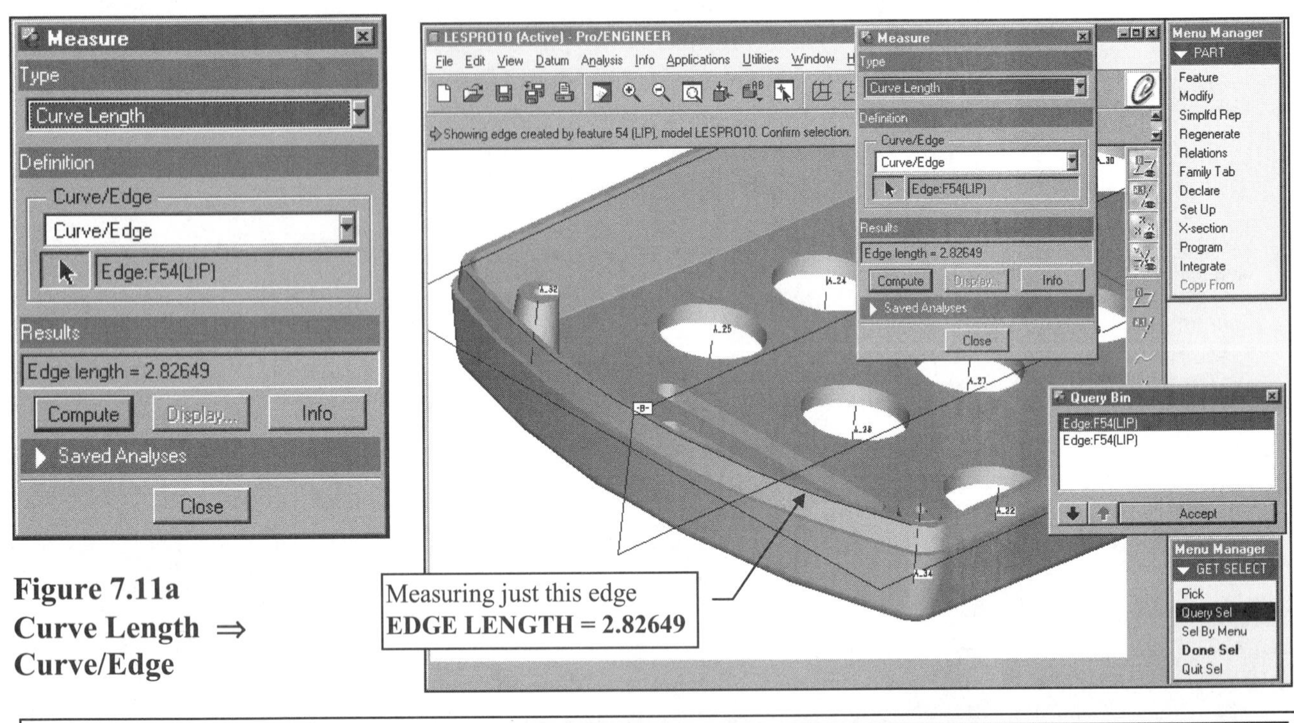

Figure 7.11a
Curve Length ⇒ Curve/Edge

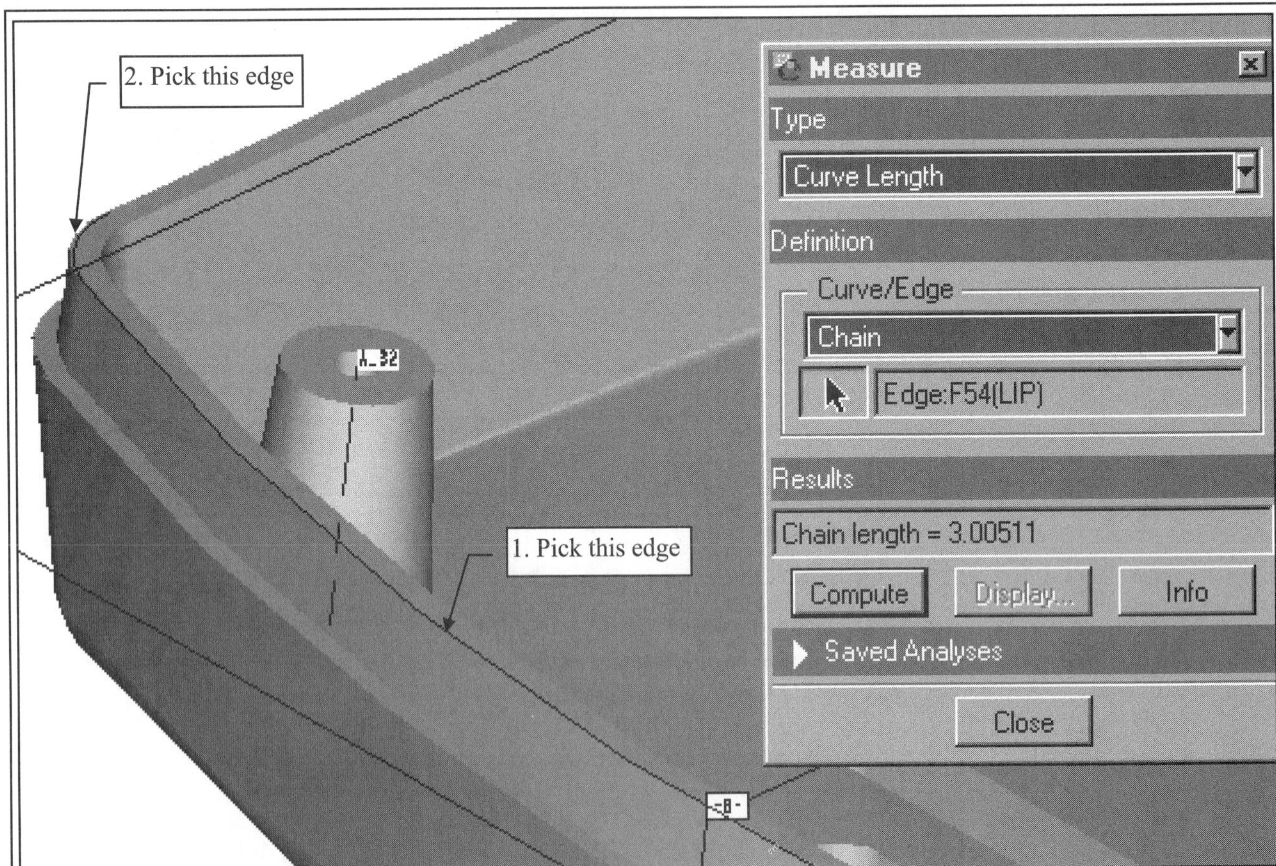

Figure 7.11b
Curve Length ⇒ Chain

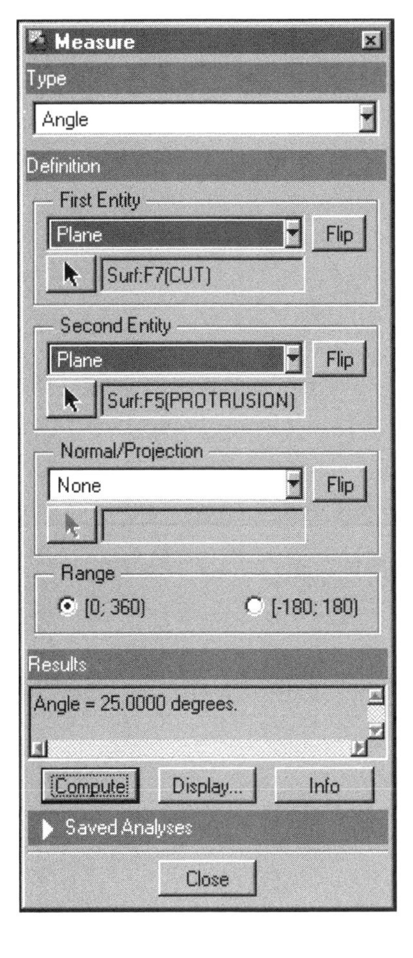

Angle

The **Angle** command is used to measure the angle between axes, planar curves, and planar nonlinear edges. The angle measured depends on where you select the edges. To measure angles between planes, do the following:

1. Pick **Angle** from the Type menu in the Measure dialog box.
2. Select **Curve/Edge, Axis, Plane,** or **Coordinate System**.

When computing the value of an angle between two planes, select the First Entity [**Surf:5(PROTRUSION)**] and then the Second Entity [**Surf:7(CUT)**], as shown in Figure 7.12. The angle measurement between the angled surface and the front ledge surface is **25.0000 degrees**. Two arrows are also displayed.

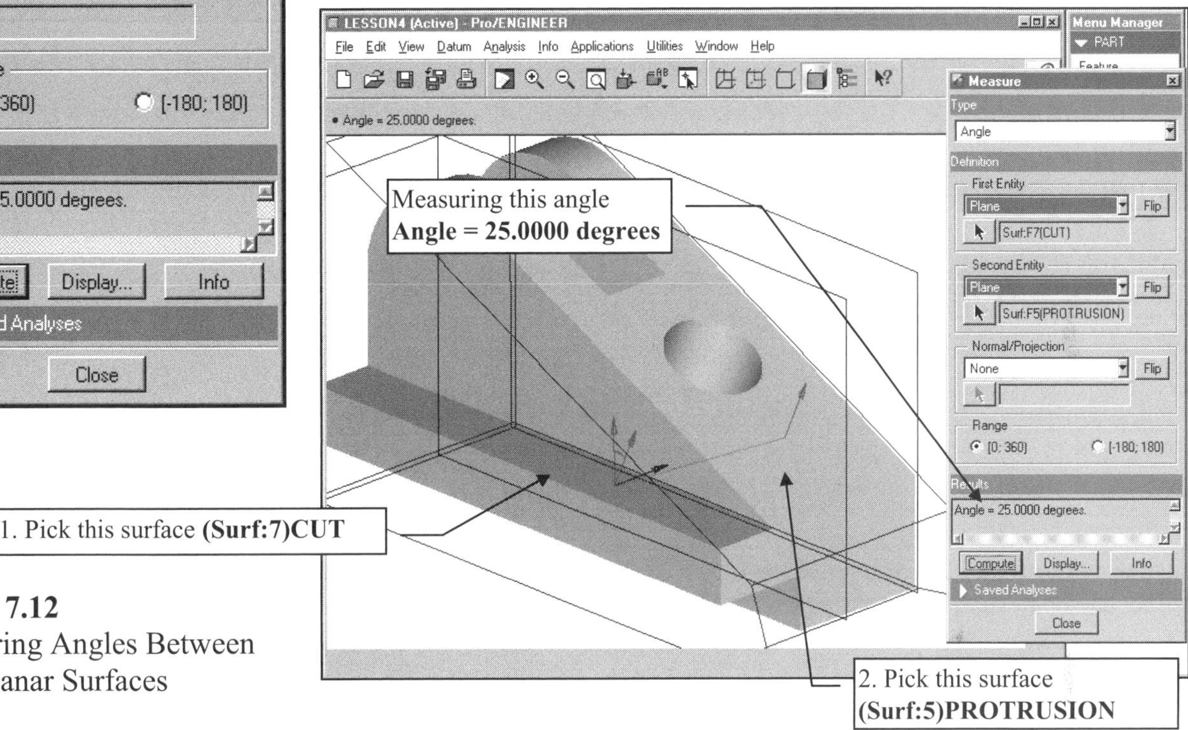

Figure 7.12
Measuring Angles Between Two Planar Surfaces

Distance

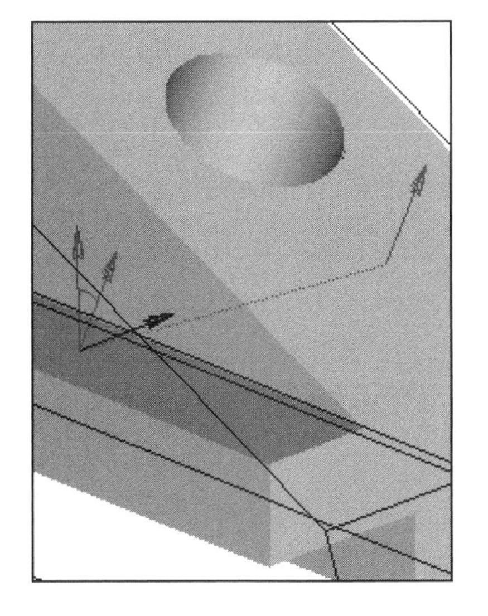

Distance is measured with respect to a basis entity. The *basis entity* is the one from which you measure, in other words, the first entity selected when you start measuring **Distance.** After the *basis entity* is selected, you can make as many measurements as you like by selecting various entities to measure to. Each distance will be calculated with respect to the first entity, until you restart the measuring process by selecting a new *basis entity*. After customizing your screen, you can initiate **Measure** commands from a Toolbar.

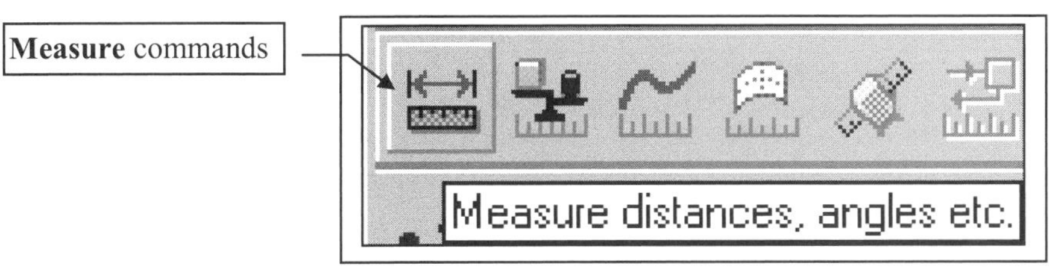

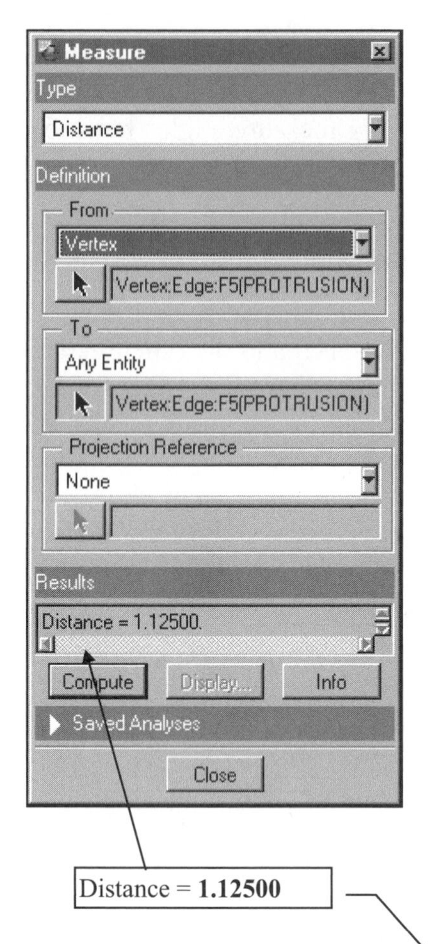

When measuring distances, you have to specify the entity type before selecting an entity from which, or to which, to measure. Some entity types are:

Any Entity Any selected entity.
Vertex Intersection of lines, edges, or surfaces (Fig. 7.13).
Point A point on the part surface or a datum point.
Coordinate System A coordinate system.
Line/Axis A line or axis.
Plane A planar part surface or datum plane.
Curve/edge An existing curve or edge.
Surface A planar surface.
Cable A cable.

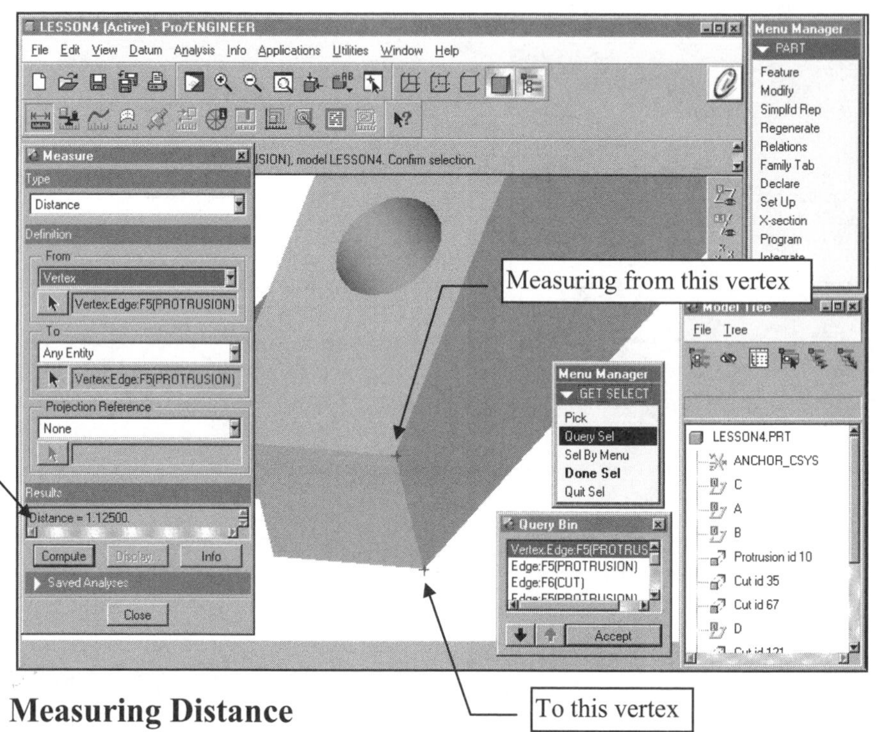

Figure 7.13
Measuring from a **Vertex** to a **Vertex**

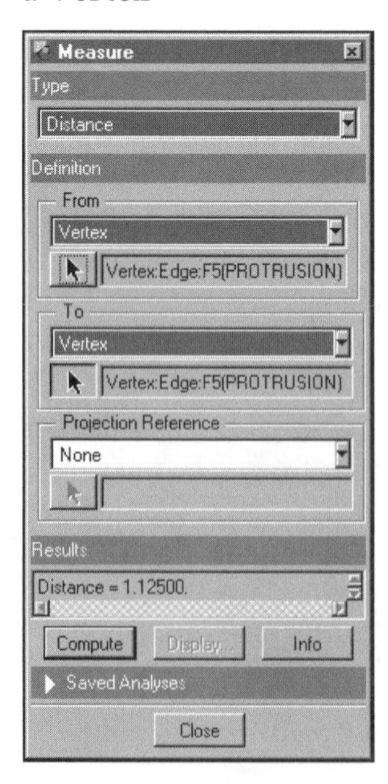

Measuring Distance

To measure the distance between any two entities:

1. Choose **Analysis** ⇒ **Measure** ⇒ Choose **Distance** from the Type drop-down menu.
2. Choose the appropriate From option to establish the *basis entity*.

Any Entity
Vertex
Point
Scan Point
Coordinate System
Line/Axis
Plane
Curve/Edge
Curve feature
Surface
Cable

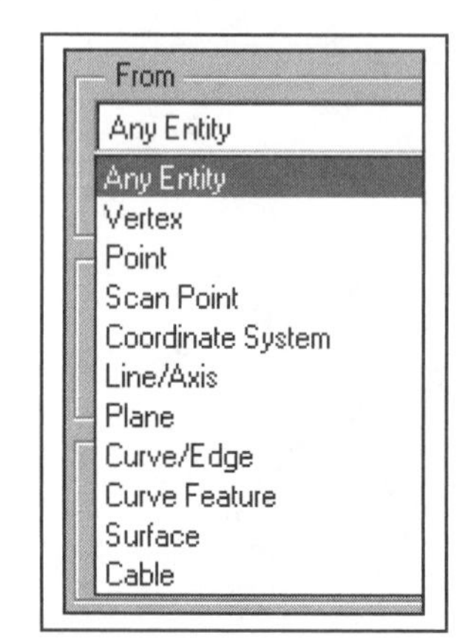

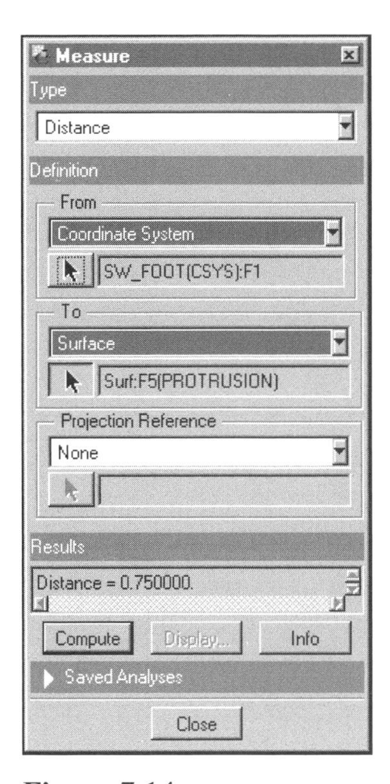

Figure 7.14
Measuring from the Coordinate System

3. Choose the appropriate To option.

Any Entity
Vertex
Point
Scan Point
Coordinate System
Line/Axis
Plane
Curve/Edge
Curve feature
Surface
Cable

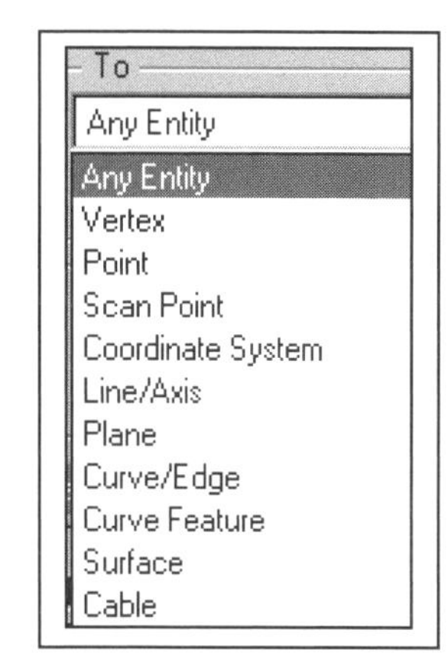

4. Make as many measurements as you want from the *basis entity*. You can change the type of entity you are measuring to at any time.
5. To restart the measuring process for a new *basis entity*, repeat the process from step 2, choosing a new *basis entity* option (e.g., From).
6. To end the measuring process, choose **Close** from the dialog box.

In Figure 7.14, **Measure** (from the Toolbar) ⇒ **Coordinate System** ⇒ (pick the coordinate system) ⇒ **Surface** ⇒ (use **Query Sel** to select the bottom circular plane of the part) was used to measure the **0.750000** distance.

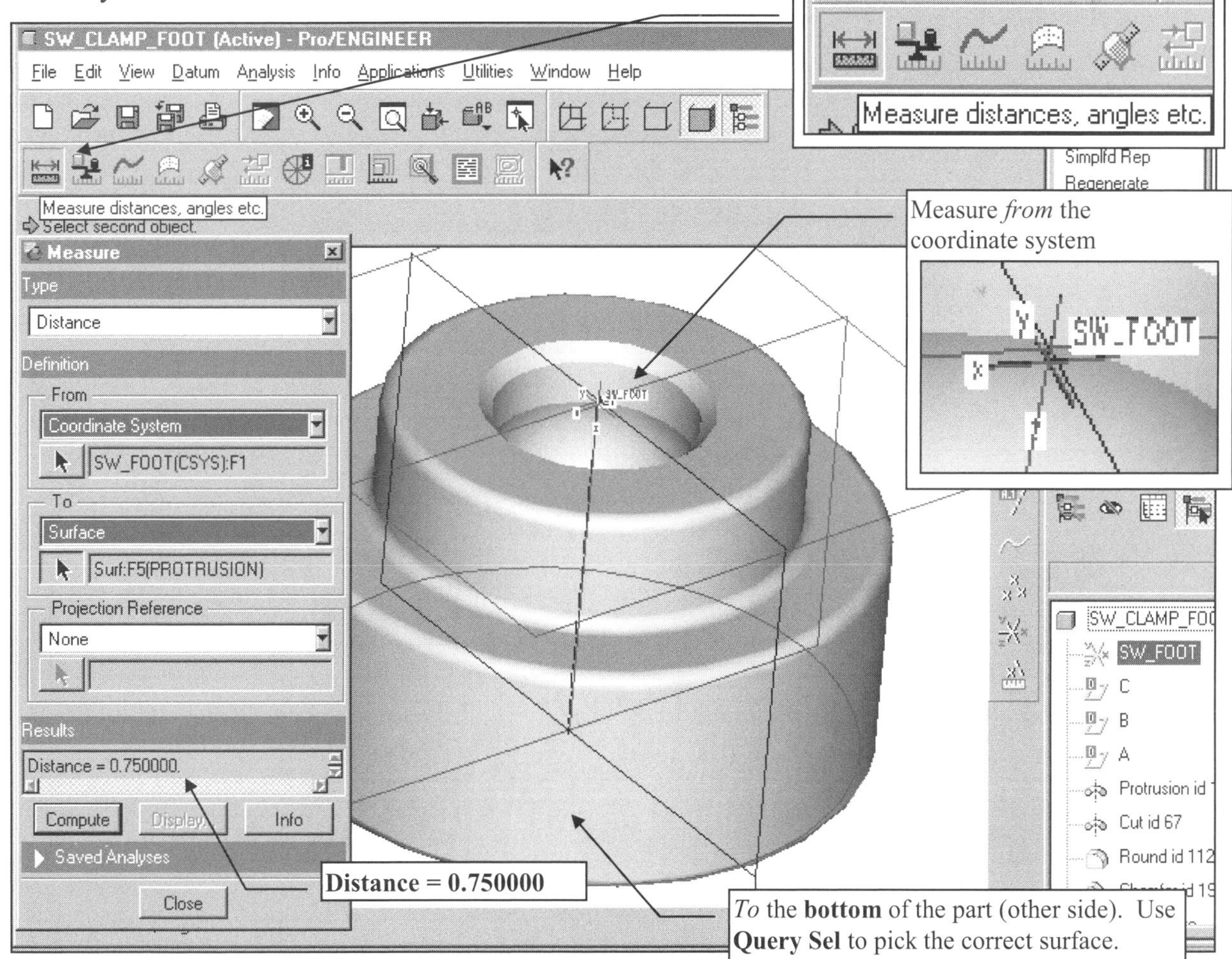

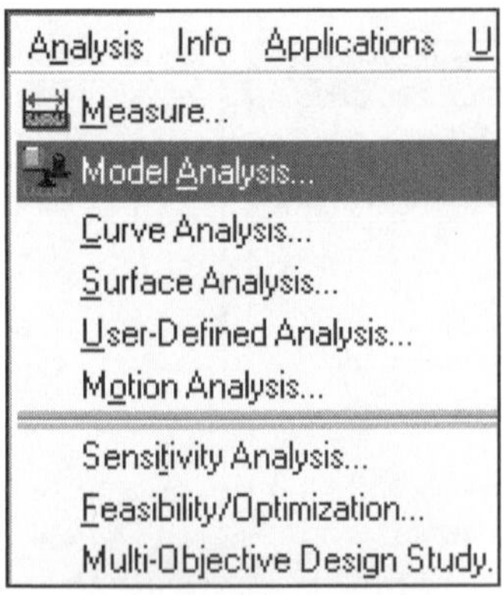

Model Analysis

The **Model Analysis** option calculates and displays a variety of information about your model. You can analyze model properties by using the Model Analysis dialog box. This dialog box includes a drop-down list offering the following types of analysis: **Assembly** (or Model) **Mass Properties**, **X-Section Mass Properties**, **One-Sided Volume**, **Pairs Clearance**, **Global Clearance**, **Volume Interference**, **Global Interference, Short Edge, Edge Type,** and **Thickness.**

Interferences and Clearances

If the objects selected do not interfere, the minimum clearance is displayed graphically as a red line. A small red circle with crosshairs will display at each end of the red line to identify the location at which the clearance is being measured. If there is interference between the two objects, Pro/E highlights the volume of interference and provides the value or highlights the curve or point of intersection, as appropriate for the items selected. The clearance or interference value is displayed in the dialog box. The following is the procedure to determine an interference:

1. Pick **Model Analysis** from the Analysis pull-down menu.
2. Select **Global Interference** from the Type menu.
3. Check **Parts Only** ⇒ **Exclude** or **Include** ⇒ **Exact Result** or **Quick Check** ⇒ **Compute**

The **Global Interference** ⇒ **Parts Only** ⇒ **Exclude** ⇒ **Exact Result** ⇒ **Compute** was chosen in Figure 7.15. Note that four pairs have interference. Pair 1 of these four is highlighted.

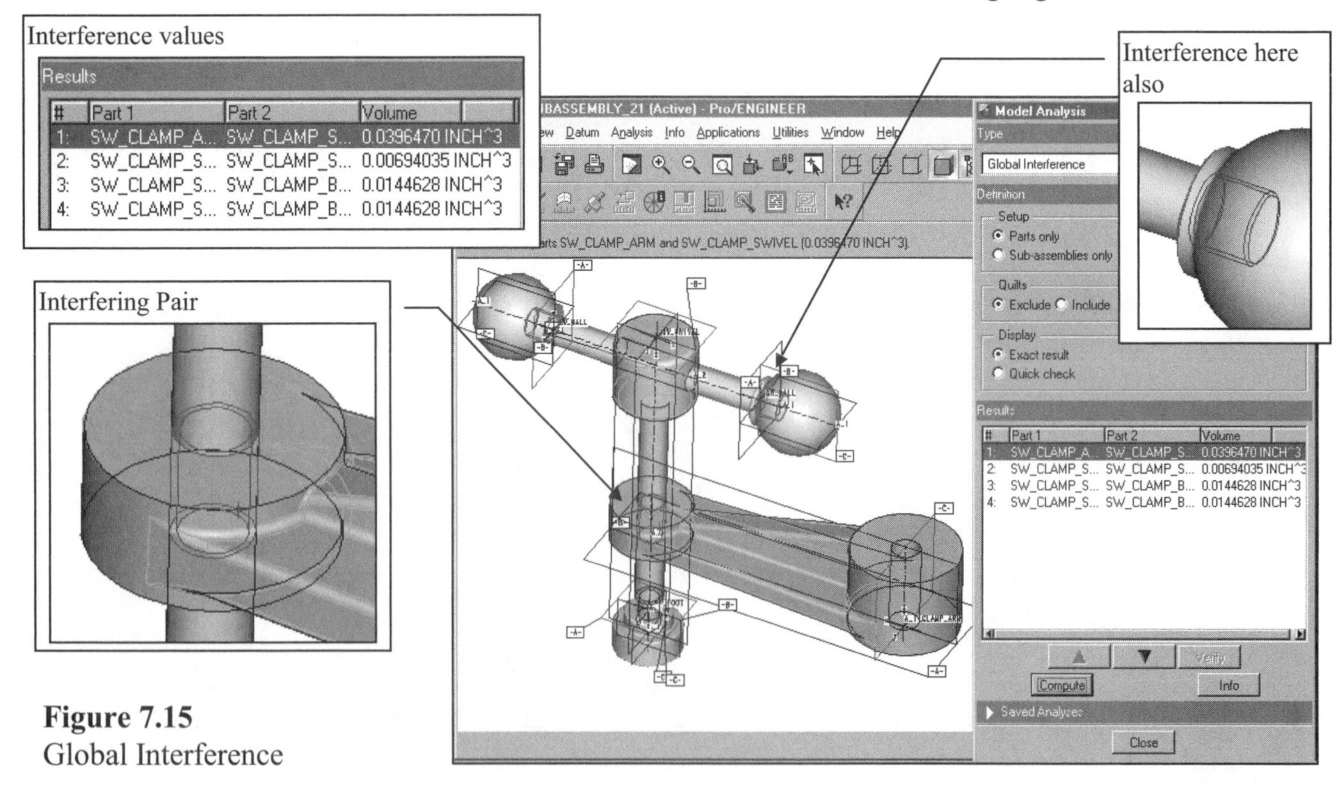

#	Part 1	Part 2	Volume
1:	SW_CLAMP_A...	SW_CLAMP_S...	0.0396470 INCH^3
2:	SW_CLAMP_S...	SW_CLAMP_S...	0.00694035 INCH^3
3:	SW_CLAMP_S...	SW_CLAMP_B...	0.0144628 INCH^3
4:	SW_CLAMP_S...	SW_CLAMP_B...	0.0144628 INCH^3

Figure 7.15
Global Interference

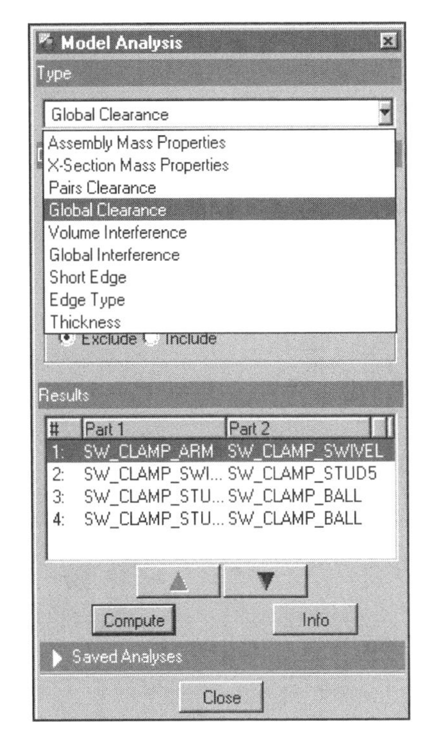

Global Clearance Within an Assembly

This measurement command is used to identify, in an entire assembly, all component parts or subassemblies for which clearances are less than or equal to a clearance distance that is specified in the design, as shown in Figure 7.16:

1. Pick **Model Analysis** from the Analysis pull-down menu.
2. Select **Global Clearance** from the Type menu.
3. Enter the **Clearance** value when applicable.
4. Check **Parts Only ⇒ Include ⇒ Compute.** Pro/E will determine whether any components of the assembly are within the specified clearance distance. All interferences are included.
5. Use the down arrow on the dialog box to step through the display of identified pairs (Fig. 7.16). To exit the process, choose **Close.**

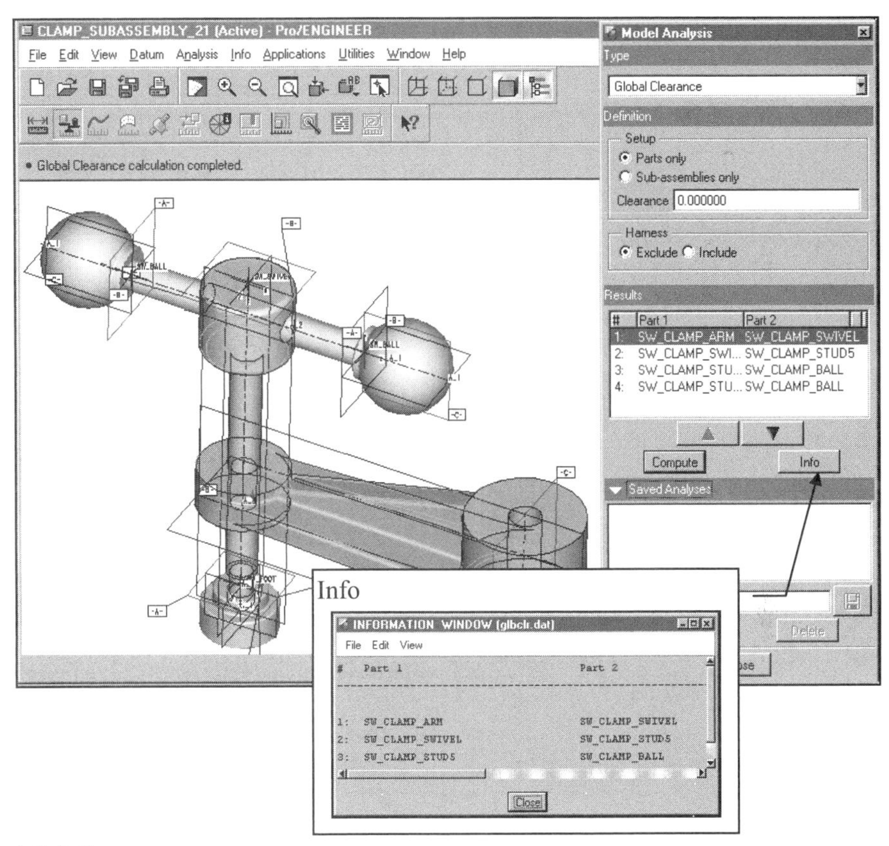

Thickness

Thickness measures the thickness of a part to determine if a specified region has a thickness that is greater than or less than a user-specified maximum or minimum value. The region is displayed as shown in Figure 7.17.

When the thickness check is complete, the cross section is highlighted: ***Yellow*** means that the thickness is between the specified maximum and minimum values, ***Red*** border means that the thickness exceeds the specified maximum value, and ***Blue*** border means that the thickness is below the specified minimum value.

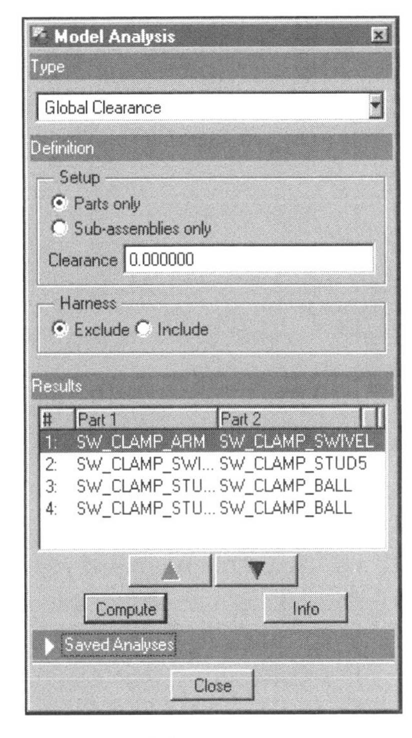

Figure 7.16
Global Clearance

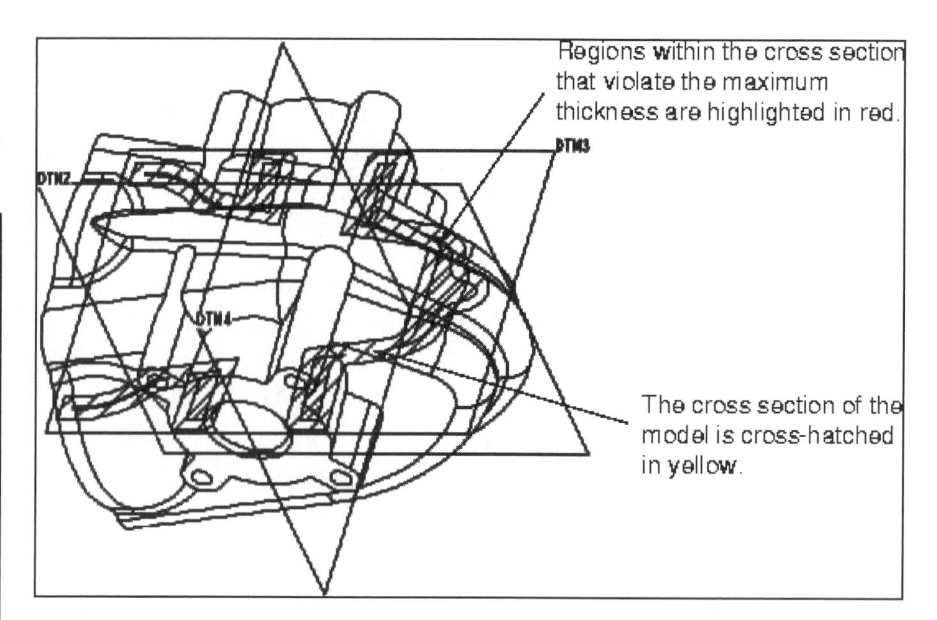

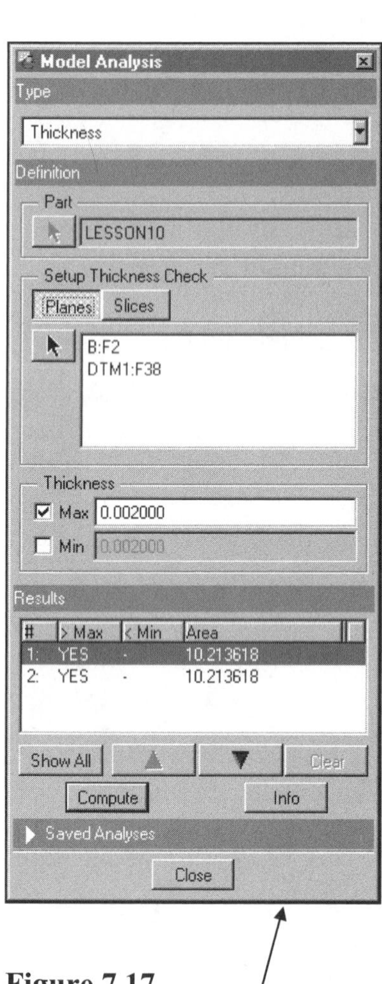

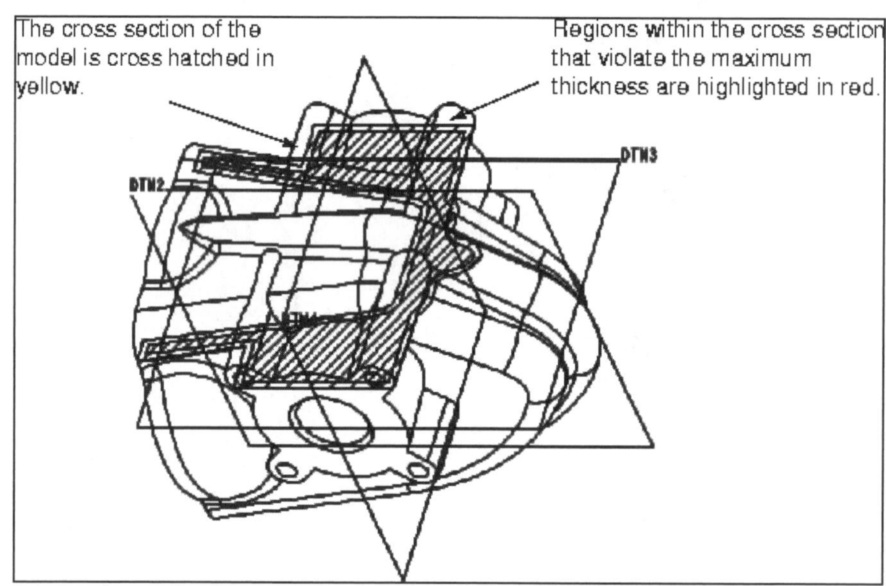

Figure 7.17
Thickness

Viewing Regeneration Information

In Part and Assembly modes, you can step through a regeneration of a model. You can skip some steps, get feature information on each component of the model, and even fix the model in the middle of the regeneration. To use this function, from the **Info** pull-down menu, choose **Regen Info**.

The **Regen Info** option lets you observe how a part was built (Fig. 7.18) and aids in the diagnosis of bad features in a part. You can use this option at any time or instead of the **Regenerate** option after you have modified the part.

To regenerate a part using the **Regen Info** option:

1. Choose the option **Regen Info** from the Info pull-down menu.
2. Select where to start regenerating the part by choosing one of the following options from the START OPTS menu:

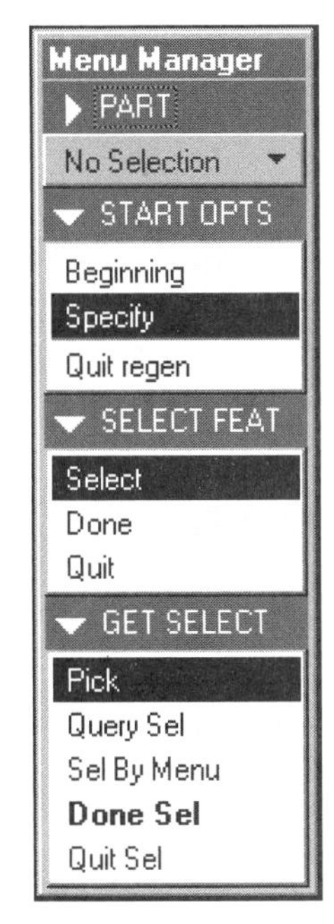

Beginning Start part regeneration with feature number 1.
Specify Specify the feature from which to start regenerating by entering its number at the prompt or selecting the desired feature.
Quit regen Return to the previous menu.

3. Pro/E regenerates and displays the specified feature of the part and then displays the INFO REGEN menu. The possible options are as follows:

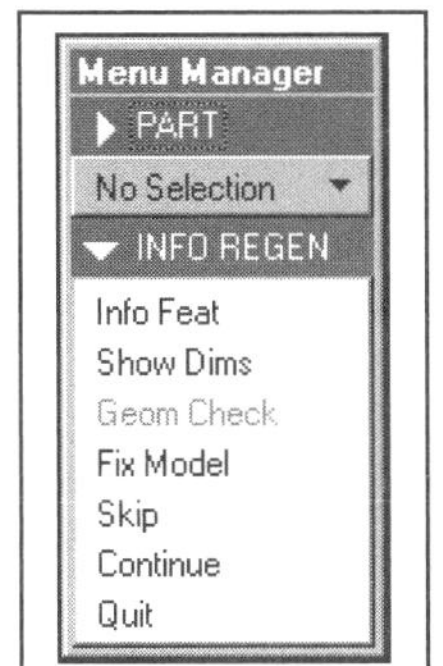

Info Feat Provide regular feature information about the last feature regenerated. If you choose this option, Pro/E displays a window with the feature information.
Show Dims Display the dimensions of the last regenerated feature.
Geom Check Allow you to investigate the geometry error for the feature just regenerated. This option is accessible only when a geometry error has been encountered.
Fix Model Activate the special Resolve functionality by forcing the feature to abort regeneration.
Skip Skip a designated number of features. The features regenerate without waiting and without allowing you to select any options from this menu. Enter the number of features to skip.
Continue Continue to the next feature for regeneration (Fig. 7.19).
Quit Complete the regeneration of the part without giving the option for further information.

4. If a feature intersects any other feature on the part, Pro/E displays the geometry of the feature in *red* wireframe before intersection.

Figure 7.18
Regen Info
Specify (pick the rib)

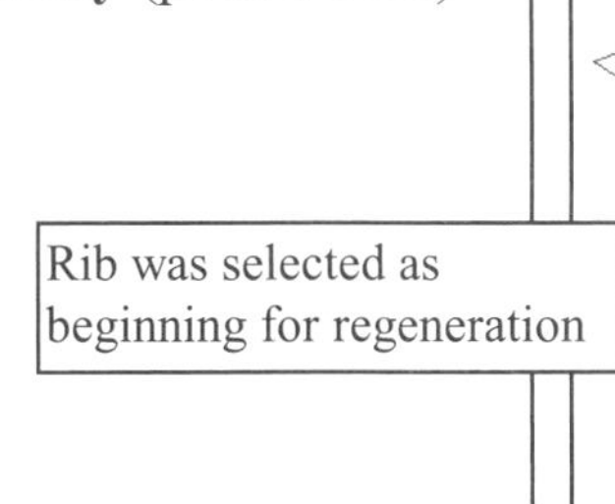

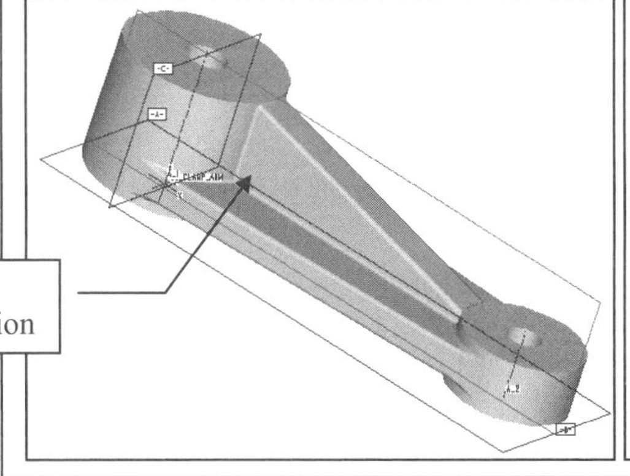

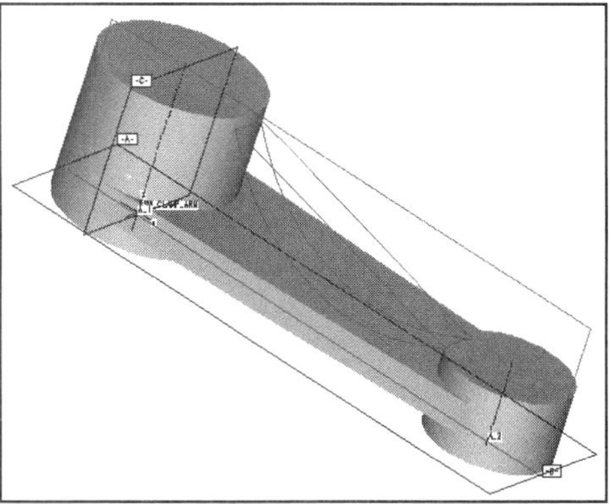

Figure 7.19
Continue (rib) (hole)

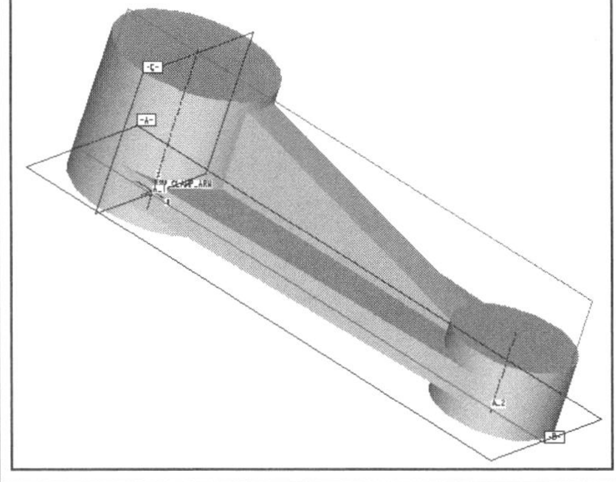

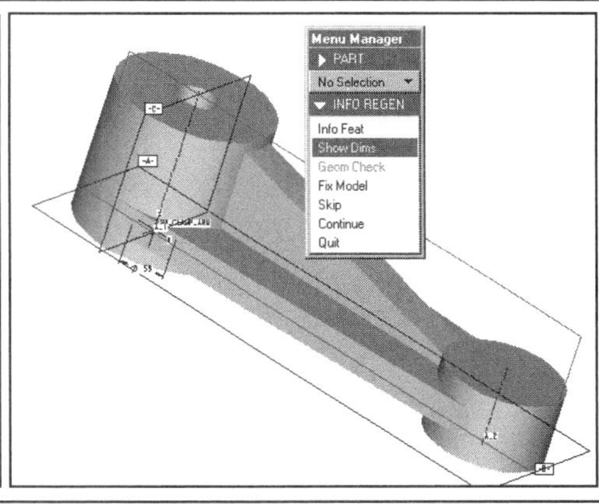

Model Information

Information about features can be accessed by choosing **Info** (from the pull-down Menu bar) ⇒ **Feature.** After choosing **Feature,** specify a feature by either picking it with the mouse from the Main Window or selecting it from Model Tree. In Figure 7.20, the rib was selected from the part. **Info** ⇒ **Feature** can be chosen in both Part and Assembly modes. Feature information is shown in the INFORMATION WINDOW. The Model Tree can also be used to get **Info** (see left-hand column).

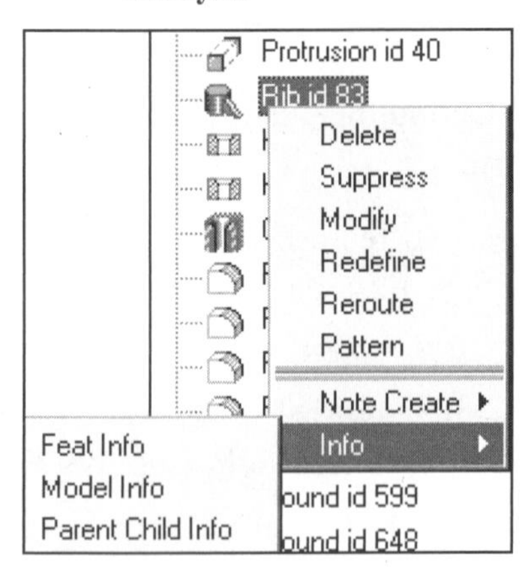

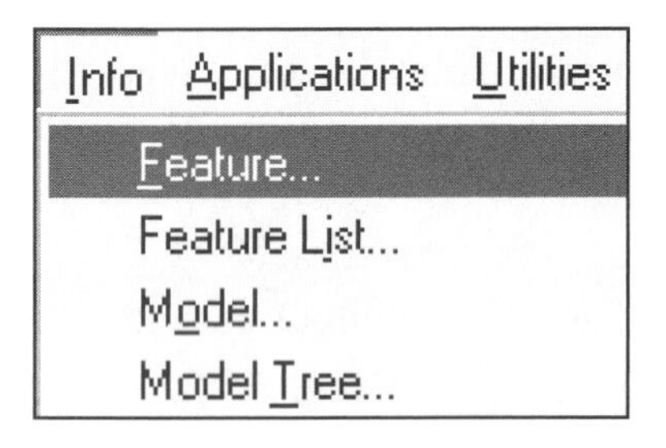

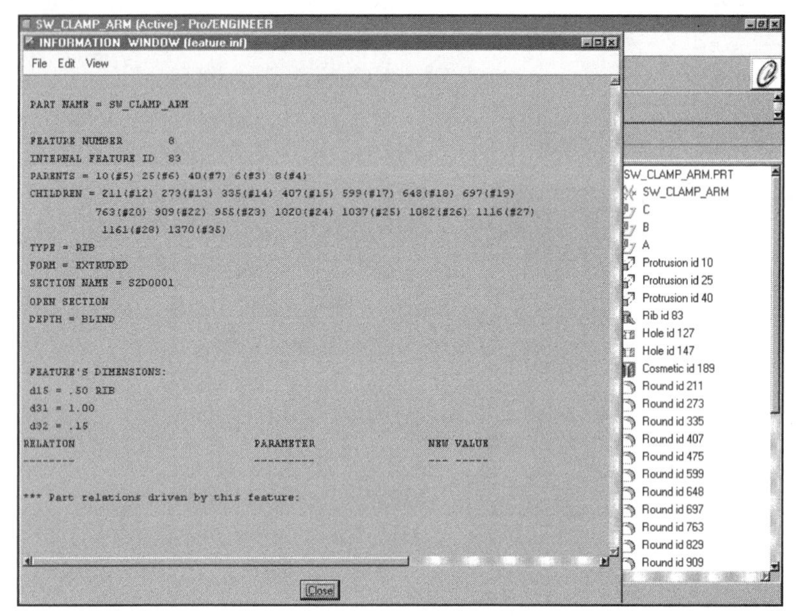

Figure 7.20
Feature Information

Feature List

When you choose **Feature List**, the INFORMATION WINDOW appears with a table listing the features in creation order and giving information, such as **Num**(ber), **ID, Name, Type, Sup**(pression) **Order,** and **Regen**(eration) **Status** (Regenerated, Unregenerated, Failed, and so on). An example is shown in Figure 7.21.

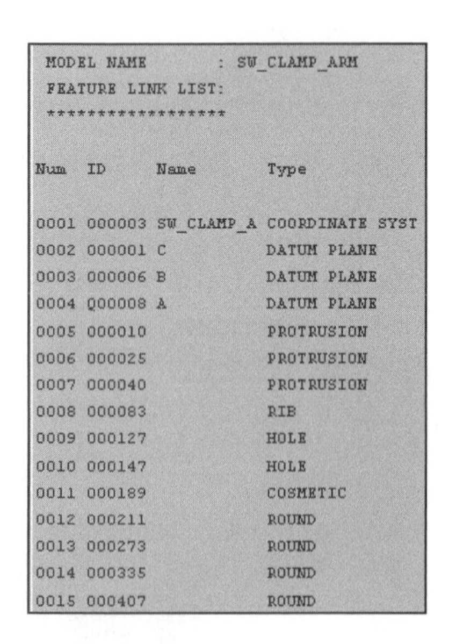

MODEL NAME : SW_CLAMP_ARM
FEATURE LINK LIST:
****

Num	ID	Name	Type
0001	000003	SW_CLAMP_A	COORDINATE SYST
0002	000001	C	DATUM PLANE
0003	000006	B	DATUM PLANE
0004	000008	A	DATUM PLANE
0005	000010		PROTRUSION
0006	000025		PROTRUSION
0007	000040		PROTRUSION
0008	000083		RIB
0009	000127		HOLE
0010	000147		HOLE
0011	000189		COSMETIC
0012	000211		ROUND
0013	000273		ROUND
0014	000335		ROUND
0015	000407		ROUND

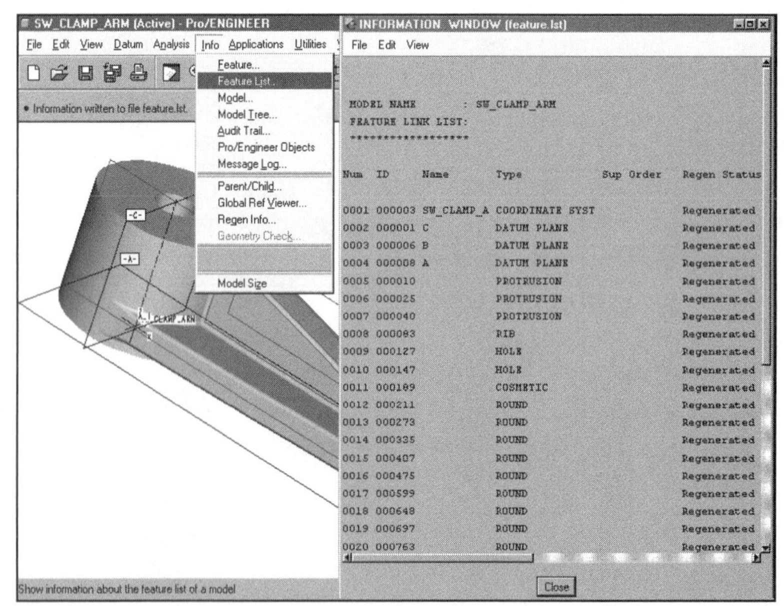

Figure 7.21
Info ⇒ Feature List

Part Information

Information about every feature of a part can be accessed by choosing **Info ⇒ Model.** In Part mode, the INFORMATION WINDOW will appear immediately after you choose **Info ⇒ Model** from the pull-down Menu Bar. You can also access the model information from the Model Tree by selecting on the parts name with the left or right mouse button, holding the right mouse button down, and then sliding the cursor until it activates the **Info ⇒ Model Info** choice (Fig. 7.22).

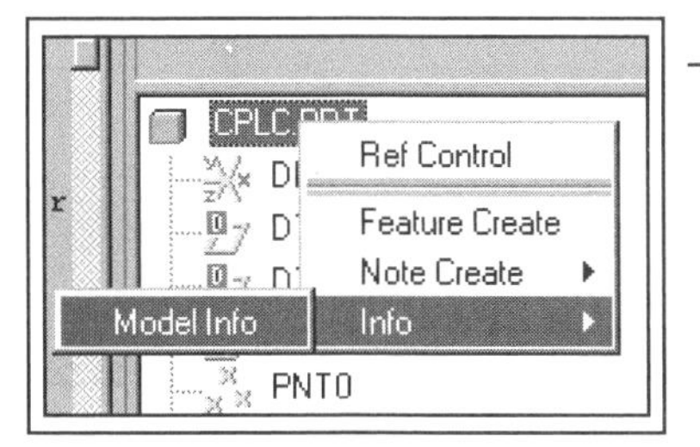

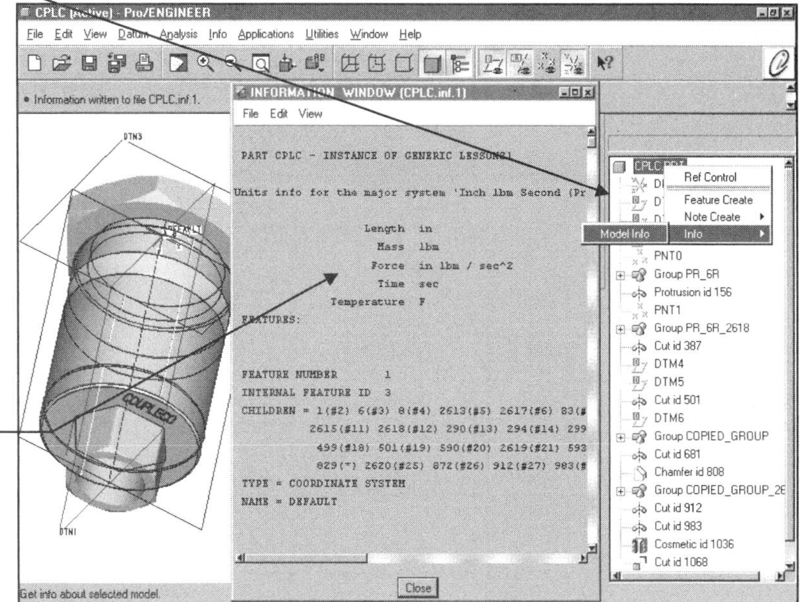

Information requested on the part

Figure 7.22
Part Model Information

Assembly Information

Assembly information can be accessed by choosing **Info ⇒ Model** and **Top Level** from the Model Info dialog box, then **Apply.** An INFORMATION WINDOW displaying the assembly information will then appear. The names of the components in the assembly are displayed in a hierarchical structure to show how it was assembled, as in Figure 7.23.

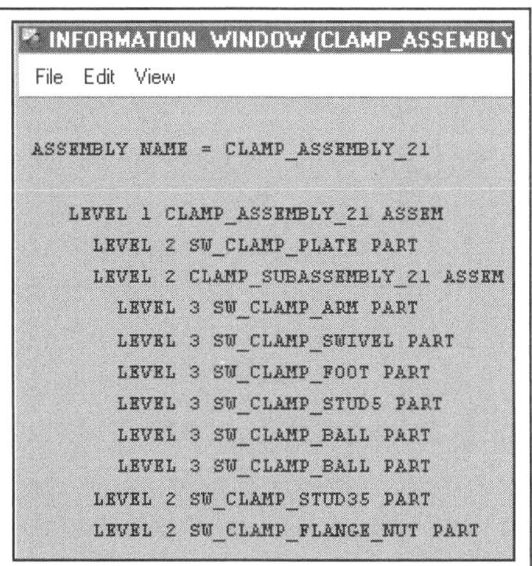

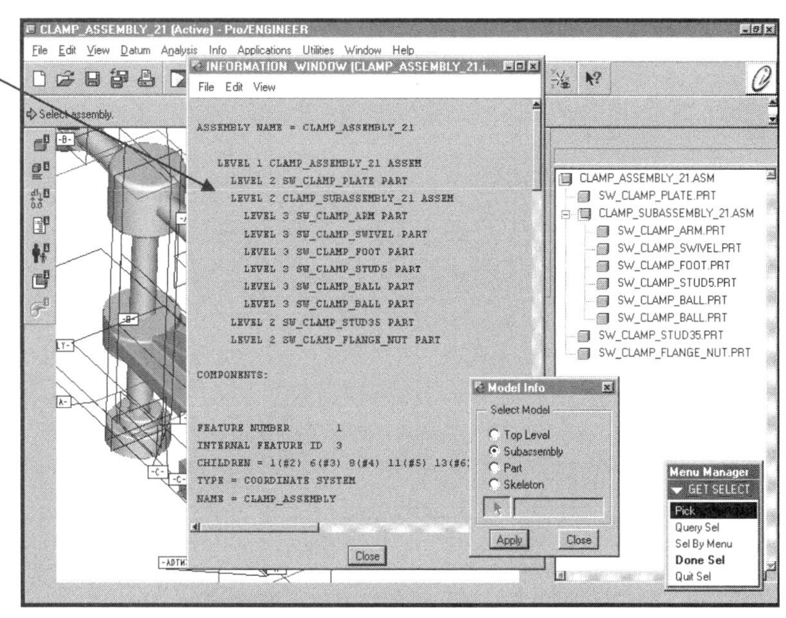

Figure 7.23
Assembly Model Information

Parent/Child Information

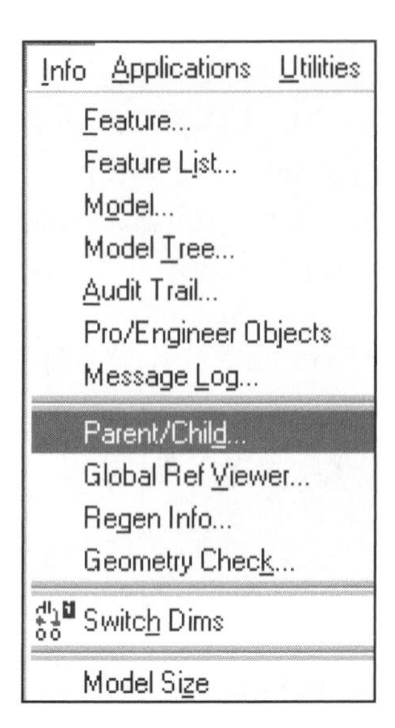

The **Parent/Child** option from the **Info** drop-down menu or Toolbar is used to highlight the relationships between features. After choosing **Parent/Child,** select the desired feature/object from the Main Window or the Model Tree. The Reference Information Window appears. Choose the button to obtain information on the selected feature/object.

In Figure 7.24, the cylindrical protrusion was chosen and its parents and children were listed. The **ActionMenu** provides a variety of types of information.

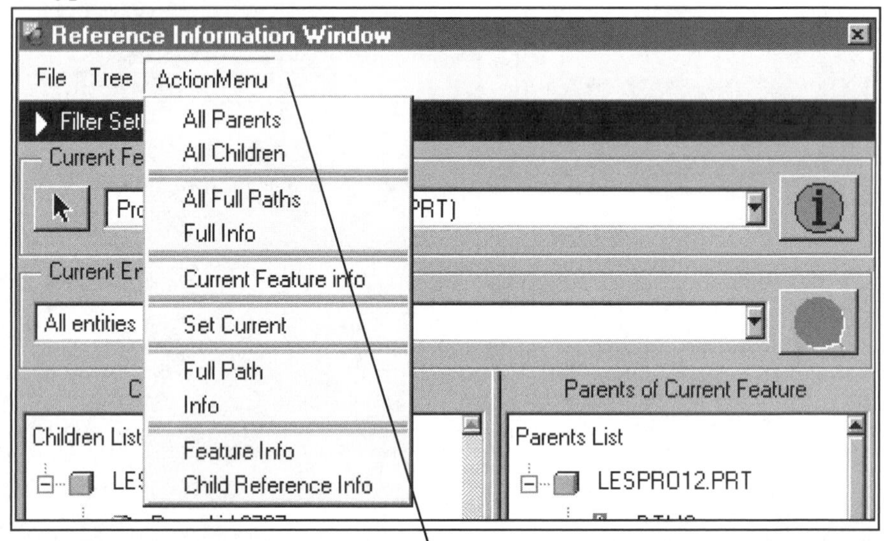

Figure 7.24
Parent/Child Information for a Part Feature

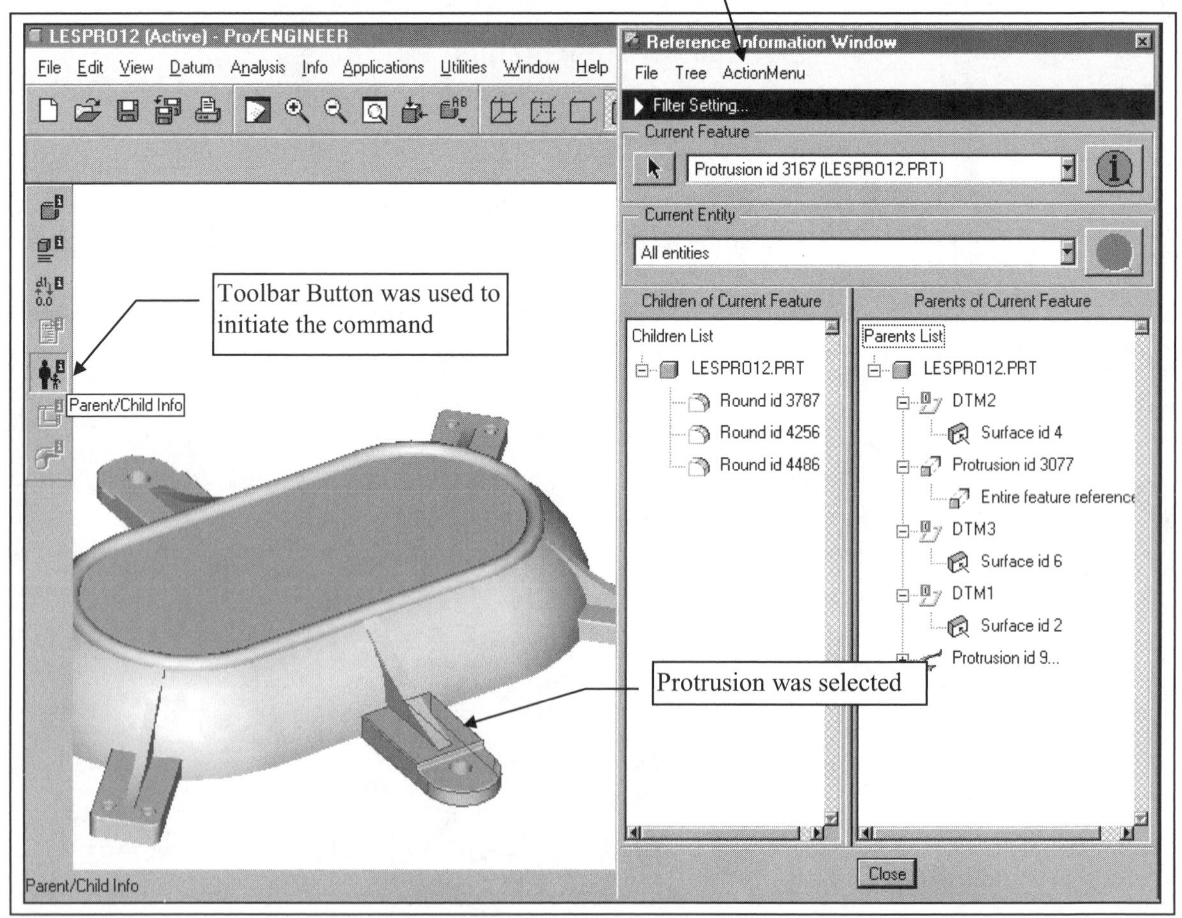

Displaying Parent/Child Information for Assemblies

This command is used to show parent/child relationships of a particular component (or feature) of an assembly:

1. Select **Parent/Child** from the Info drop-down menu.
2. Select the component or feature. In Figure 7.25, the swivel component of the assembly was selected.

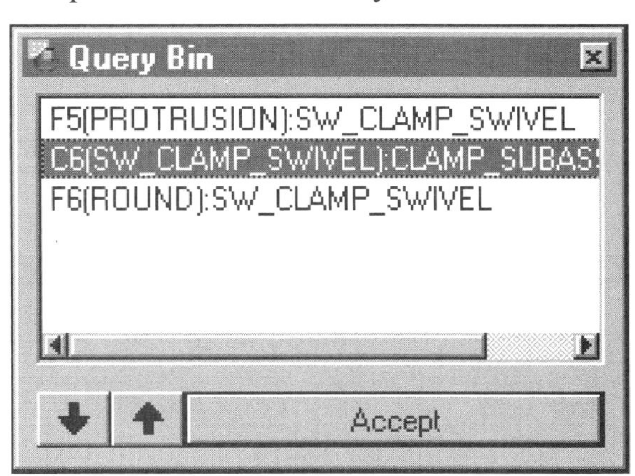

Figure 7.25
Parent Information

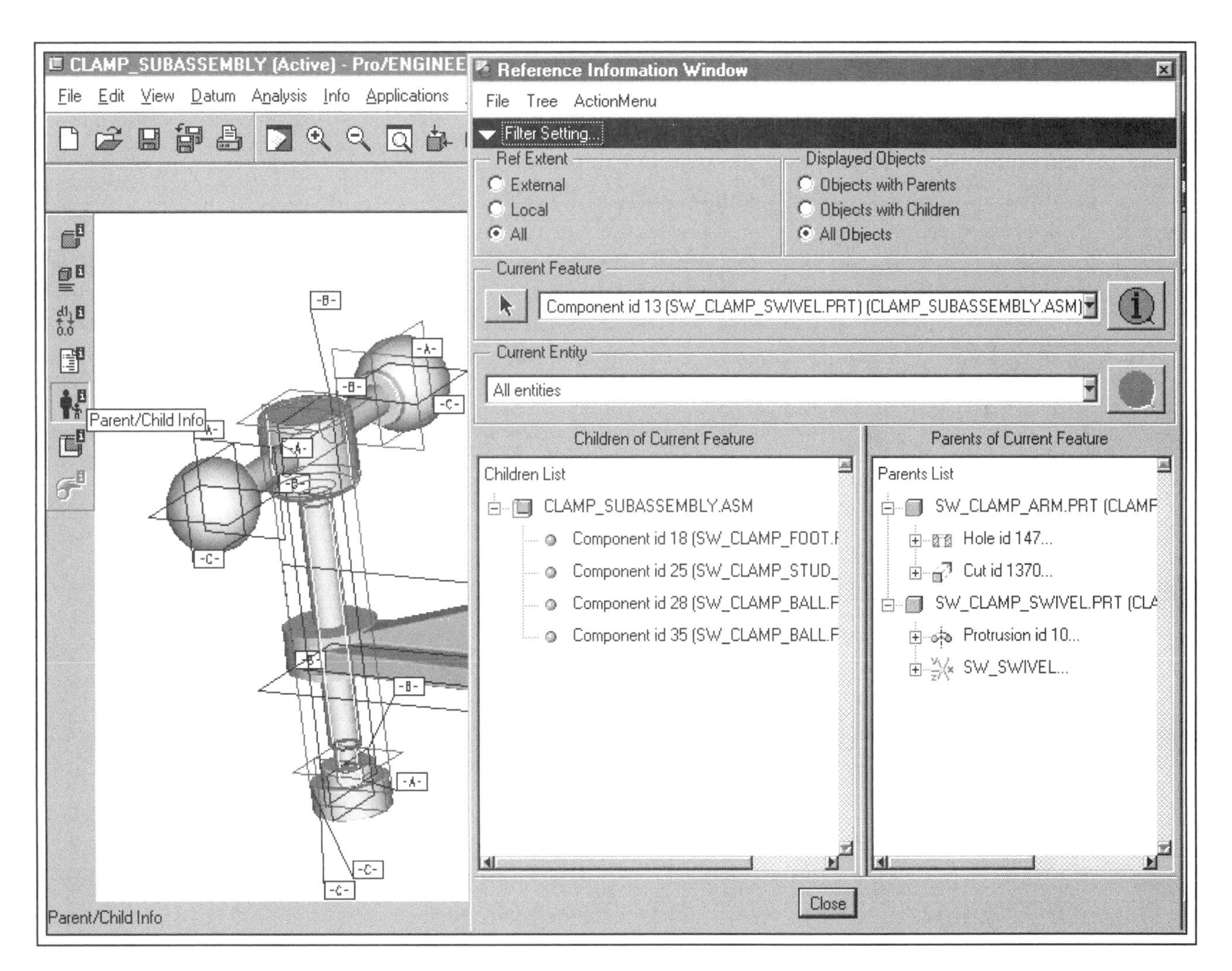

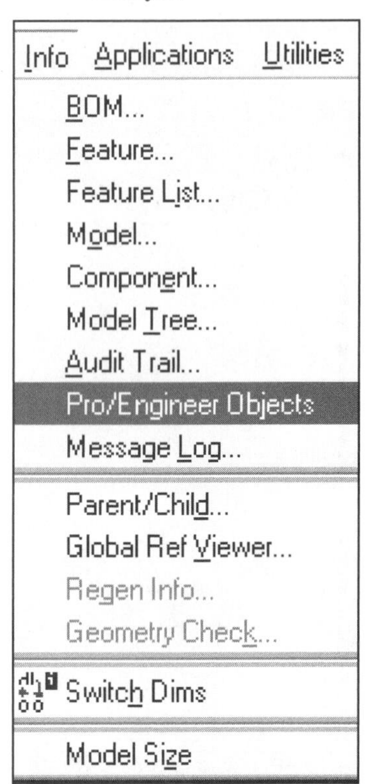

Figure 7.26
Pro/Engineer Objects

Pro/ENGINEER Objects

The Info menu option **Pro/Engineer Objects** displays a list of all stored files in an INFORMATION WINDOW (Fig. 7.26). The first portion of the listing shows parts, assemblies, drawings, layouts, and sections in memory. The second portion gives a complete listing of all Pro/ENGINEER objects in the current working directory.

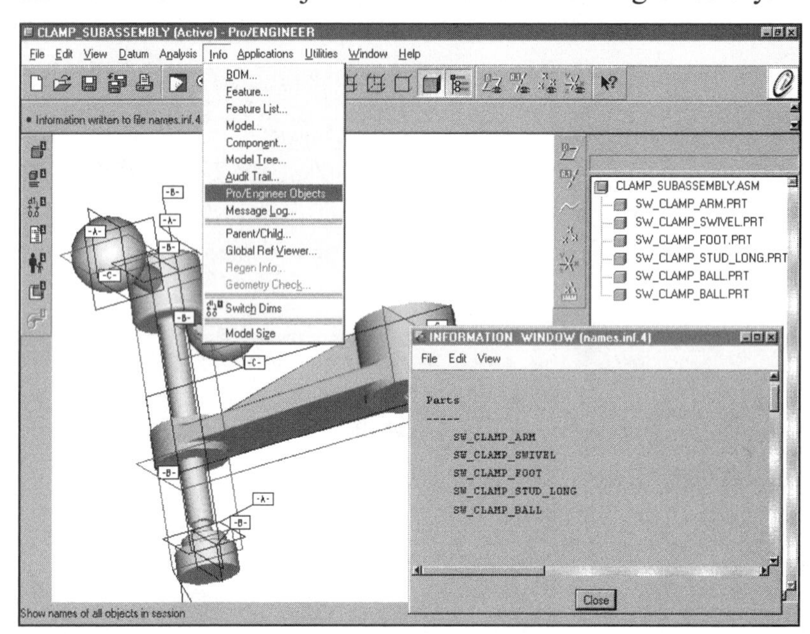

Bill of Materials (BOM)

A *bill of materials* (BOM) is a listing of all parts and part parameters in the current assembly (Fig. 7.27). You can customize the output formats to produce a particular form of presentation and content. BOM(s) can be created for assemblies in Assembly mode or from assembly drawings in Drawing mode.

Bill of materials

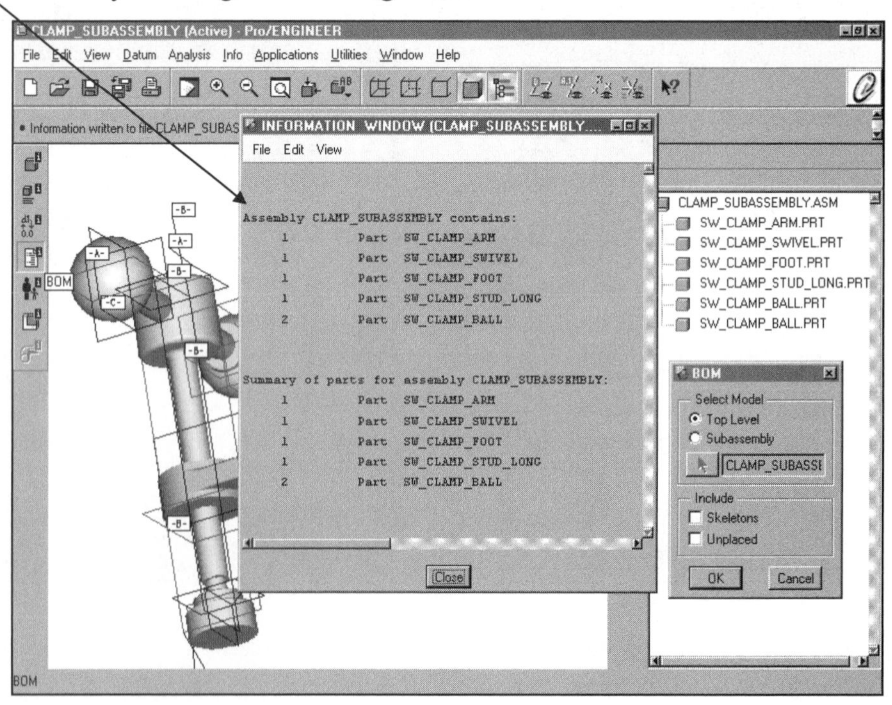

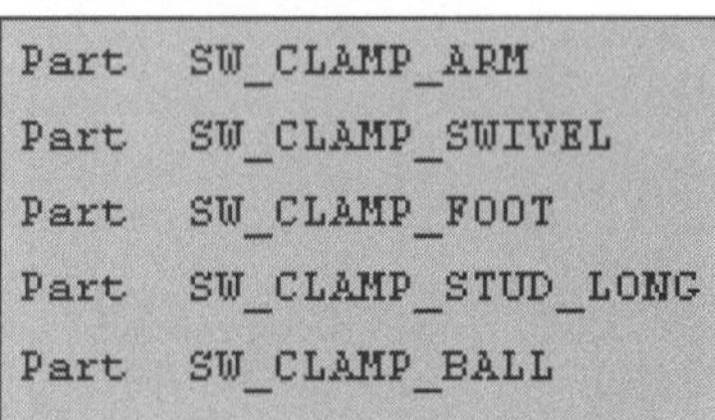
Part SW_CLAMP_ARM
Part SW_CLAMP_SWIVEL
Part SW_CLAMP_FOOT
Part SW_CLAMP_STUD_LONG
Part SW_CLAMP_BALL

Figure 7.27
BOM

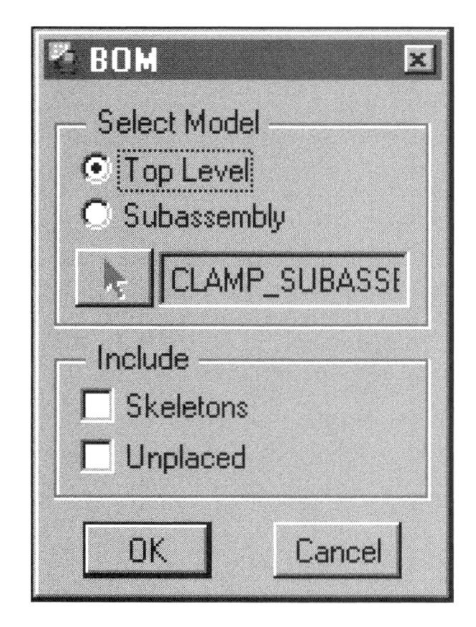

The information that follows shows how to create and format simple text BOM(s), which are stored as text files. The optional module Pro/REPORT provides functionality for creating BOM reports: graphical BOMs with complex formatting and indexing.

The source of the bill of materials output format can be configured by the configuration file. An example of the configuration file option for a user-defined format is:

bom_format ***bomcompany.fmt***

The default output format for the BOM is divided into two sections:

Breakdown Lists the name, type, and number of instances of each member and submember.
Summary Lists the total quantity of each part included in the assembly. It amounts to a "shopping list" of all the parts needed to build the assembly from the part level.
(Titles, Row These commands may be used within each section of the BOM to specify the column titles and the information to be included on the repeating rows of the BOM.)

The Bearing Assembly BOM was displayed by choosing **Info ⇒ BOM ⇒ Top Level ⇒ OK** (Fig. 7.28). Note that the **Model Tree** also displays information about an assembly.

A user-defined BOM output format specifies the formats of the breakdown section and the summary section. You can include one or both sections, but you must specify the column titles, row content, and display format for each included section.

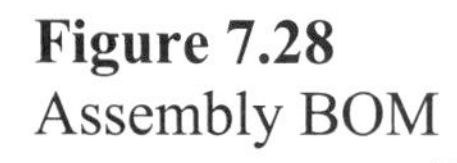
Figure 7.28
Assembly BOM

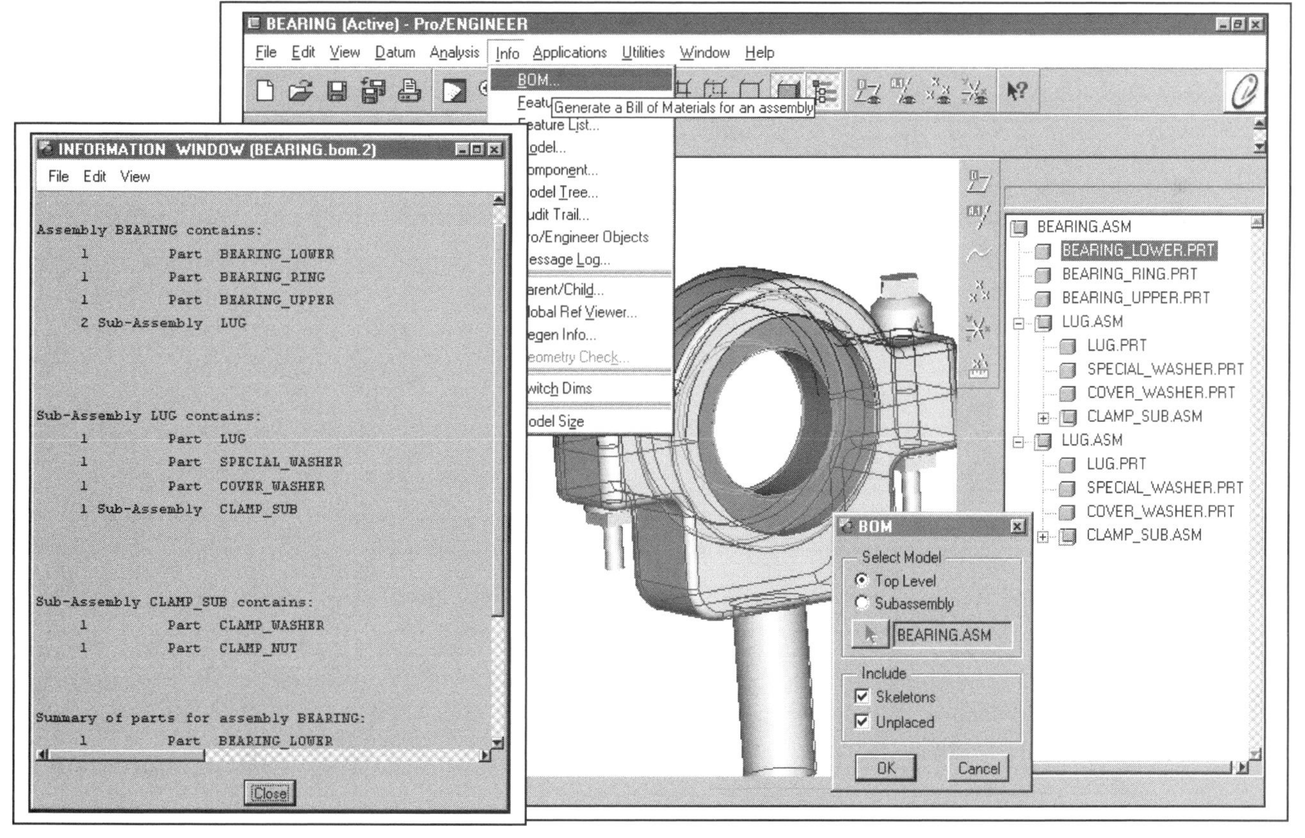

Importing and Exporting Files

To import or export information through Pro/E, the File pull-down menu is available (Fig. 7.29). The options that appear in this menu will change depending on the mode you are working in and the Pro/E modules that are available. Pro/E can import and export data in a variety of formats, including IGES, STEP, DXF, SET, VDA, CGM, STL, Plotter files, Shaded Images, Neutral files, PHOTORENDER, Render, Inventor, SUPRTB Geom, CatiaFacets, PDGS, ECAD, TIFF, CATIA, JPEG, EPS, and VRML.

To Create a Part or Assembly IGES File

1. Choose **File ⇒ Export ⇒ Model.** The EXPORT menu appears.
2. Choose **IGES** from the EXPORT menu.
3. Press **enter,** or type in the desired export file name and press **enter.** The Export IGES dialog box opens.
4. Complete the dialog box to specify the structure and contents of the output file.
5. If you want to export layers, click Customize Layers in the Export IGES dialog box. The Choose Layers dialog box opens.
6. Click Auto ID to assign layer IDs to layers that do not have IDs.
7. Make changes to the Choose Layers dialog box, then click **OK** to accept the changes made.
8. Click **OK** to create the part or assembly IGES file.

Figure 7.29
Export ⇒ Model

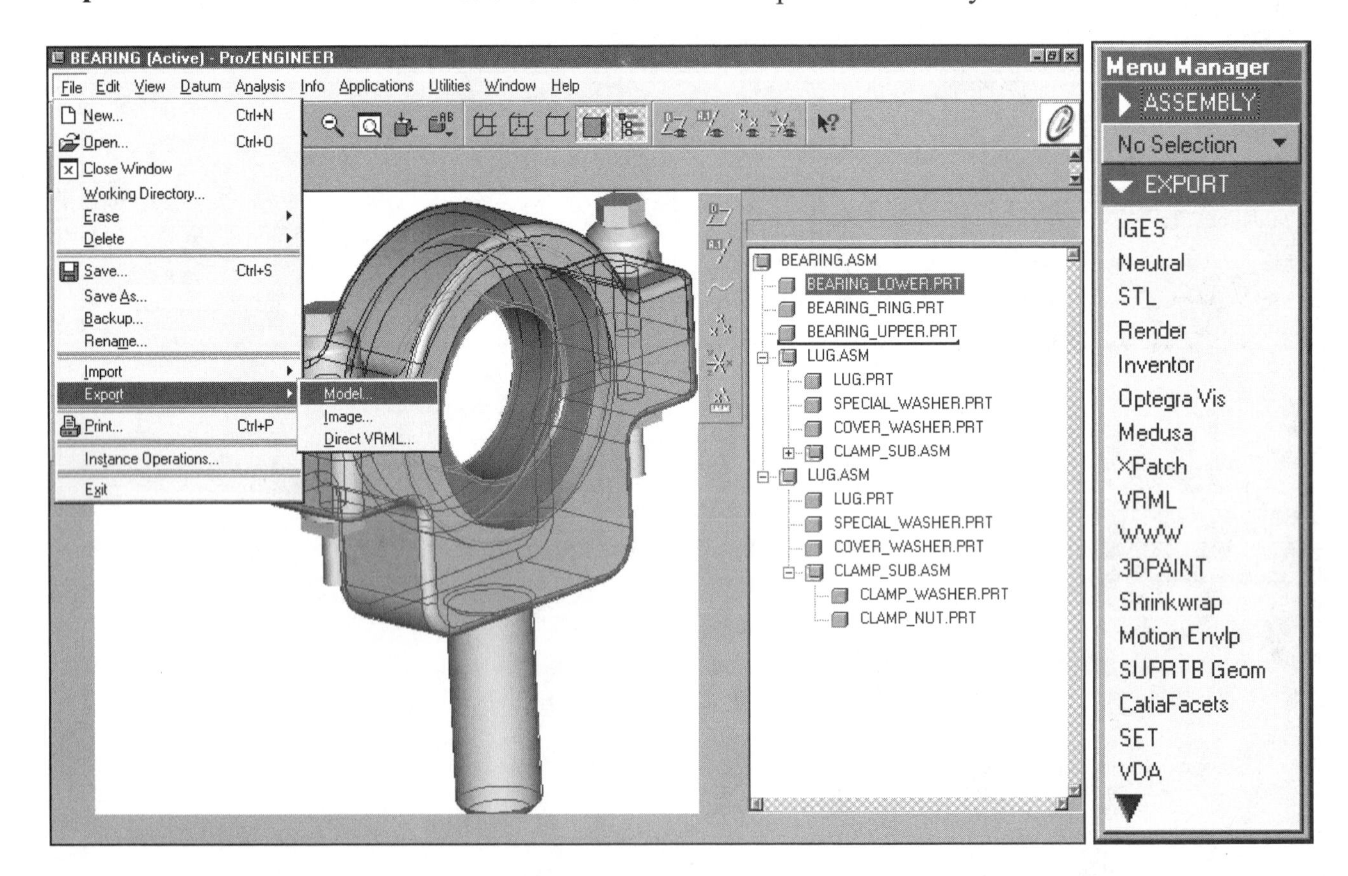

Section 8

Layers

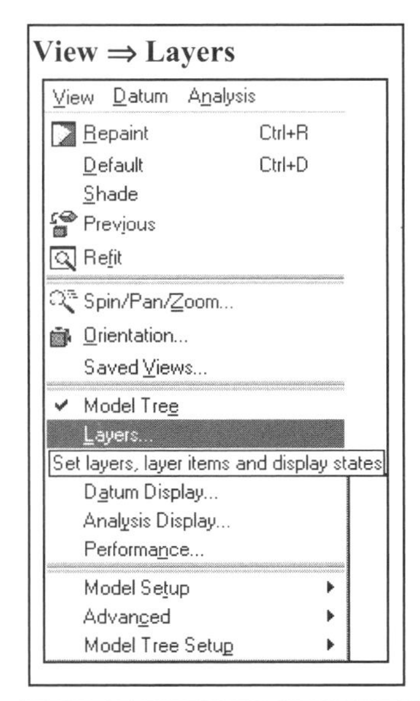

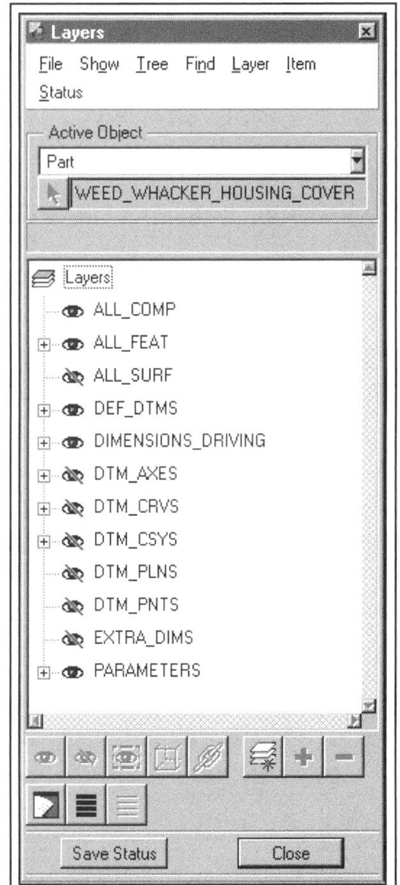

Figure 8.1
Layers

Layers are an essential tool for grouping items and performing operations on them, such as selecting, displaying or blanking, plotting, and suppressing. Any number of layers can be created. User-defined names are available, so layer names can be easily recognized. Most companies have a layering scheme that serves as a *default standard* so that all projects follow the same naming conventions and objects/items are easily located by anyone with access. Layer information, such as display status, is stored with each individual part, assembly, or drawing.

Controlling Layer Display

To open the Layer Display dialog box (Fig. 8.3), choose **Layers** from the View menu. Using the Layers dialog box, you can control layer display and manipulate layers, their items, and their display status. You use this dialog box to:

* View the layers in a part, assembly, or drawing.
* View the items (such as features, datums, and surfaces) assigned to each layer.
* Control how each layer is displayed in the model/object.

You can organize models into layers and then show or blank them selectively using the Layers dialog box. Showing or blanking layers does not affect model geometry. The **Layers** command is available while you are working in any mode (Part, Assembly, or Drawing and so on) so that you can manipulate layer display status or layer membership as required.

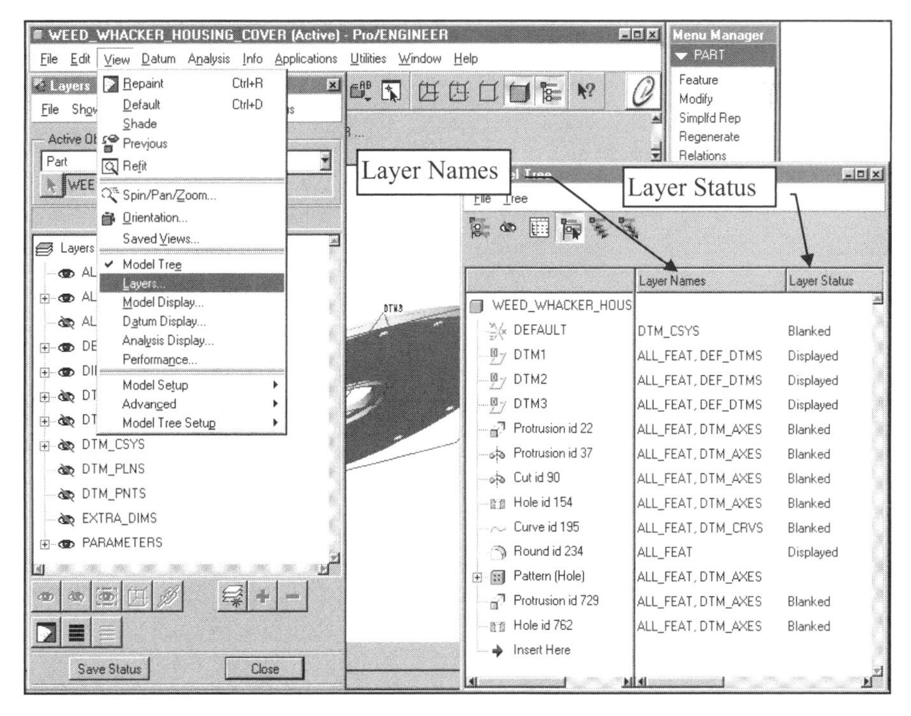

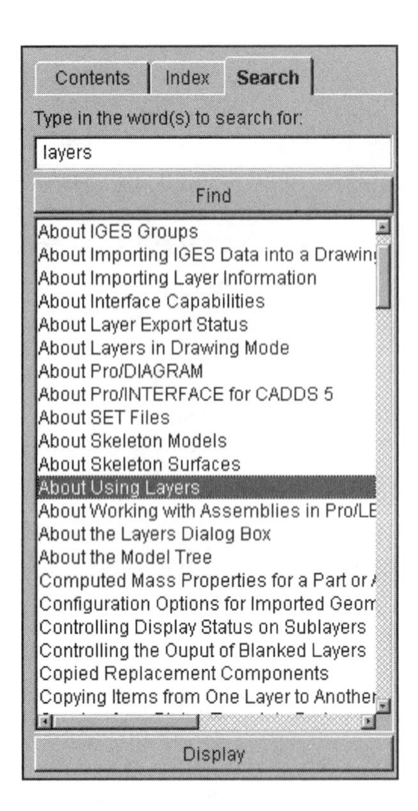

To use layers:

1. Set up the layers to which items will be added.
2. Add items to the specified layers:
   * Many items can be automatically added to layers as they are created, using configuration file options.
   * An item can be added by selecting its type and then picking the items themselves.
   * Features are also added by selecting an option that adds all features of a particular type to the layer.
   * A range of features can be added to a layer.
   * One layer can be added onto another layer.
   * Items can be copied from one layer to an existing layer or to a new layer.
3. Set the display status of the layers:
   * A layer's name can be picked from an Object name list box.
   * A layer status file can be retrieved that contains the desired layer status, which automatically sets each layer to the status specified in the file.
   * The current layer file can be edited. All changed layers will reflect the new status once the object is saved.

Layer Names

Layers are identified by name. Layer names can be expressed in numeric or alphanumeric form, with a maximum of 31 characters per name. When layers are displayed, numeric layer names are sorted first, then alphanumeric layer names. Layer names in alphabetic form are sorted alphabetically.

To get online documentation for **Layers,** use the context-sensitive help button in your *Toolbar* and pointing to the Layers dialog box [Fig. 8.2 (left column lower)] or using *Help* from the Menu bar. Figure 8.3 shows Help for **To Create a New Layer.**

Figure 8.2
Getting Information on the Layers Command

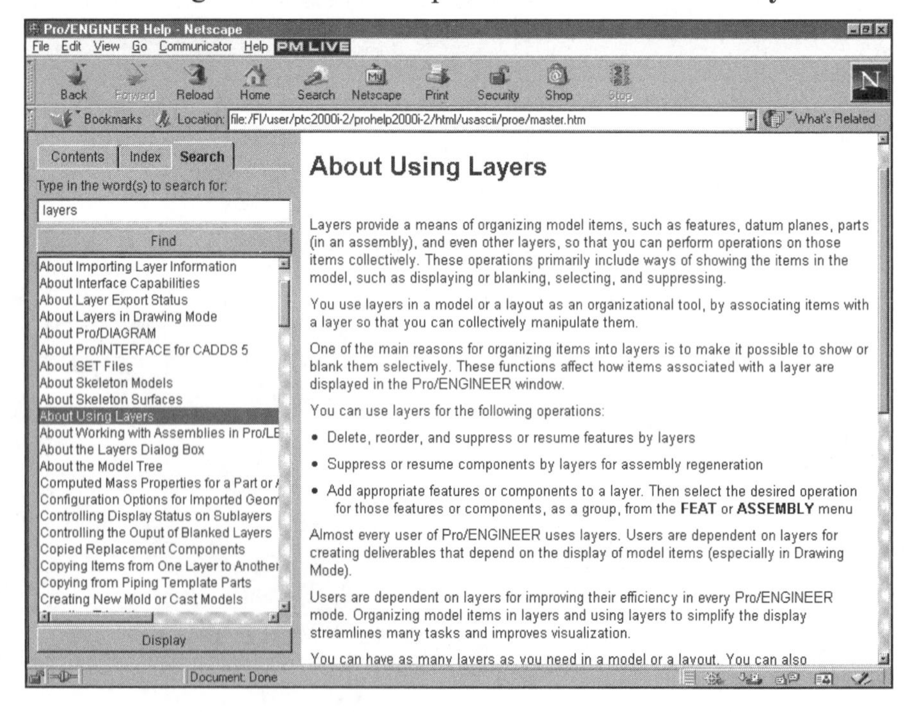

Layers Basics

Layers provide a means of organizing model items, such as features, datum planes, parts (in an assembly), and even other layers, so that you can perform operations on those items collectively. These operations primarily include ways of showing the items in the model, such as displaying or blanking, selecting, and suppressing.

You use layers in a model or a layout as an organizational tool by associating items with a layer so that you can collectively manipulate them.

You can have as many layers as you need in a model or a layout. You can also have items associated with more than one layer. For example, you could have an axis associated with several layers.

Layer display status is stored locally with its object. This means that the changes made to the layer display status of one object do not affect similarly named layers present in any other object active in the current session of Pro/E. However, changes to the layers in assemblies may affect layers in lower-level objects (subassemblies or parts).

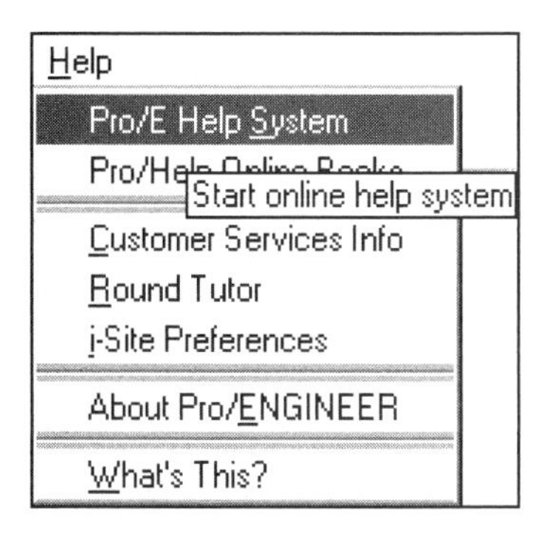

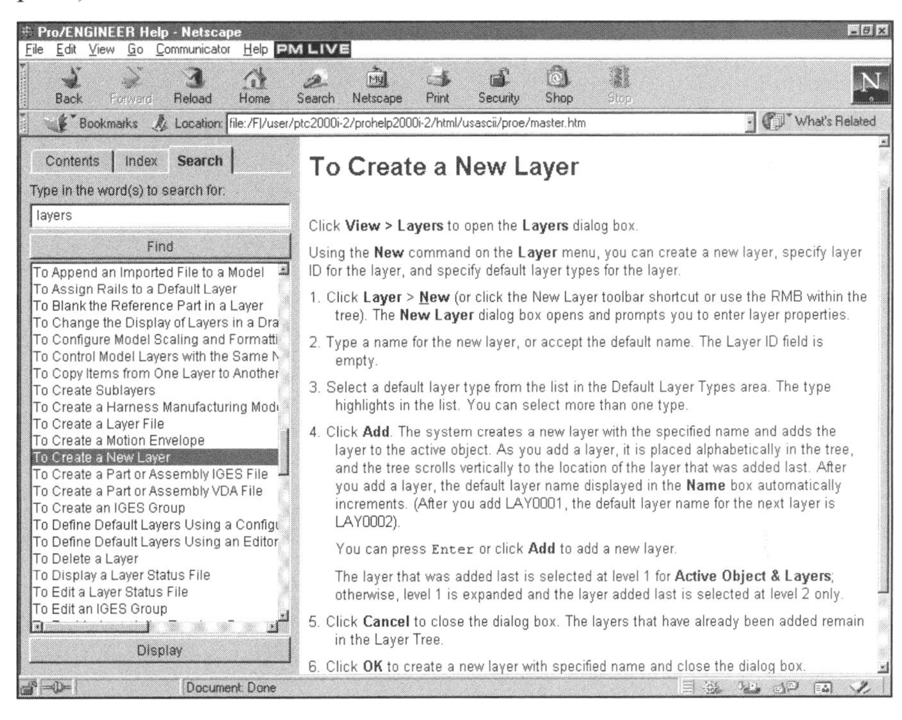

Figure 8.3
Layer Help
To Create a New layer

Creating a Layer

Before you can place items on a layer, you must first create the layer using the following method:

1. Pick **View** from the Menu bar.
2. Choose **Layers** (Fig. 8.4).
3. Select **Create new layer** icon as shown in Figure 8.5.
4. New Layer dialog box appears (Fig. 8.6).
5. **Add** to add the highlighted default layer name **LAY0001** or enter a new name of your choice. Continue to pick **Add** to add more layers (Fig. 8.7). After entering the last layer name, pick **OK.**

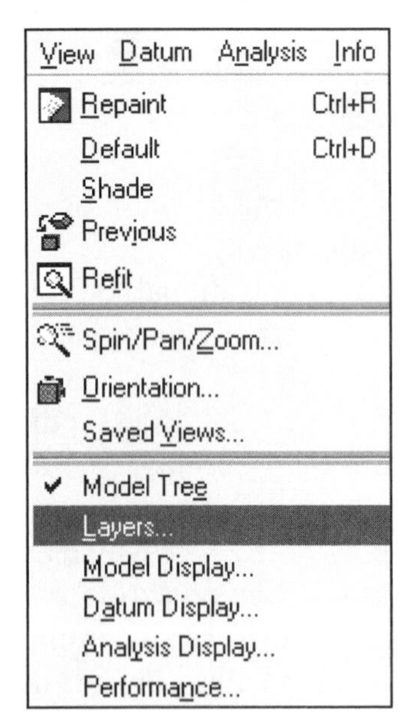

Figure 8.4
View ⇒ Layers

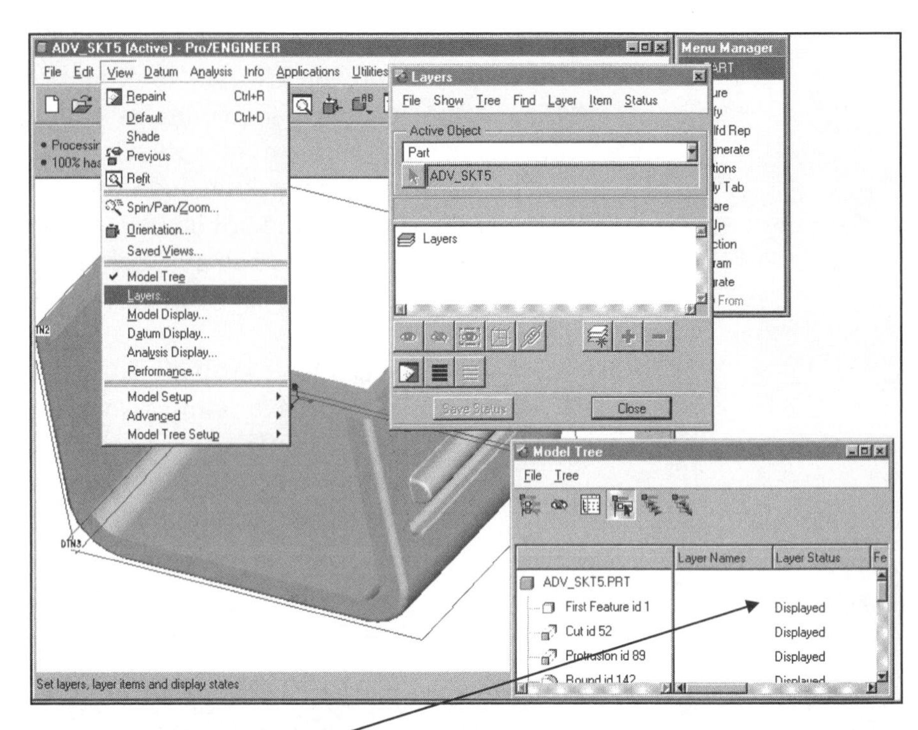

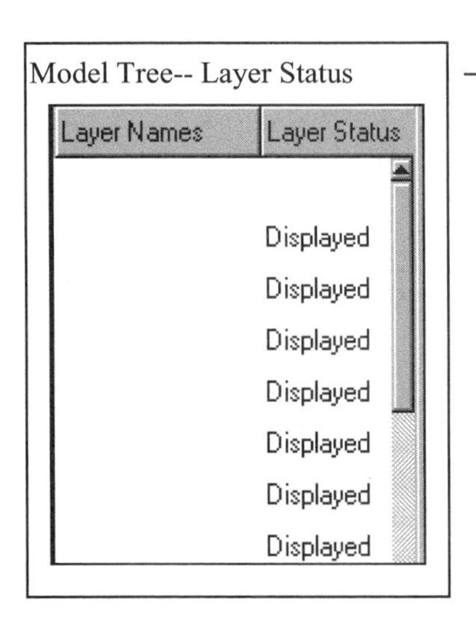

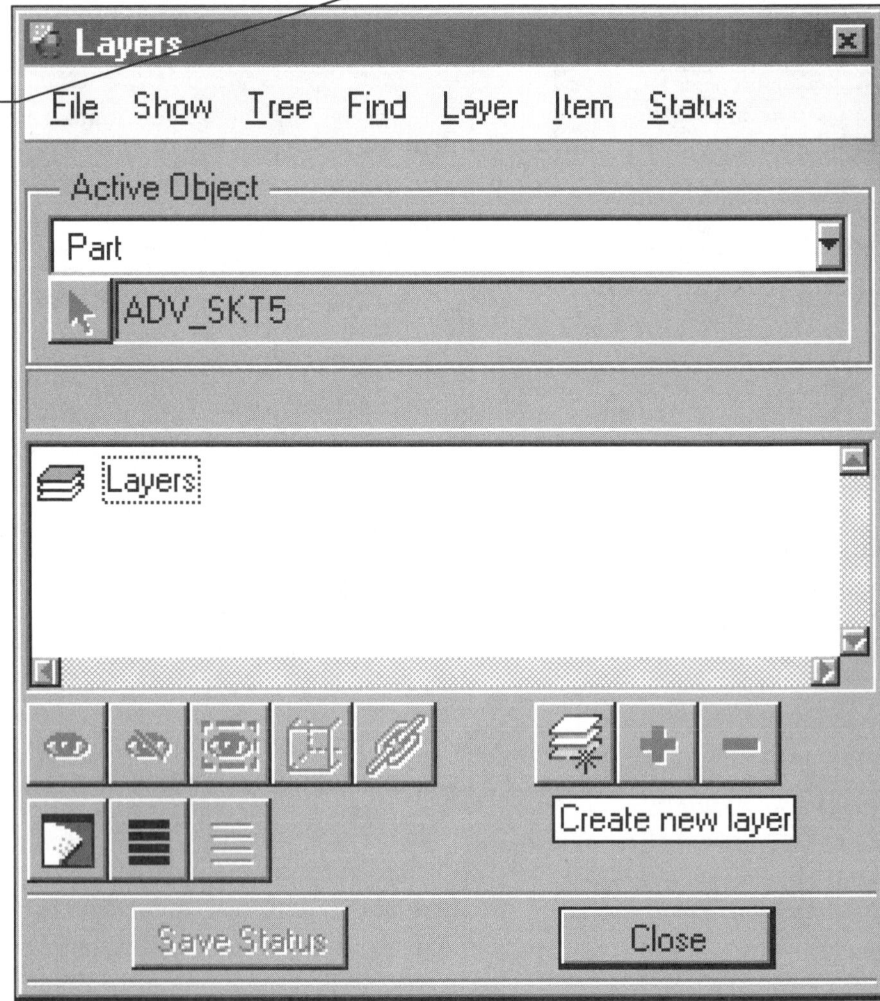

Figure 8.5
Create new layer Icon

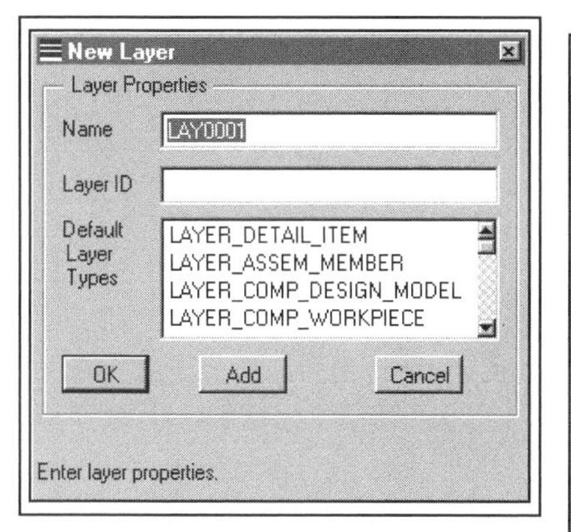

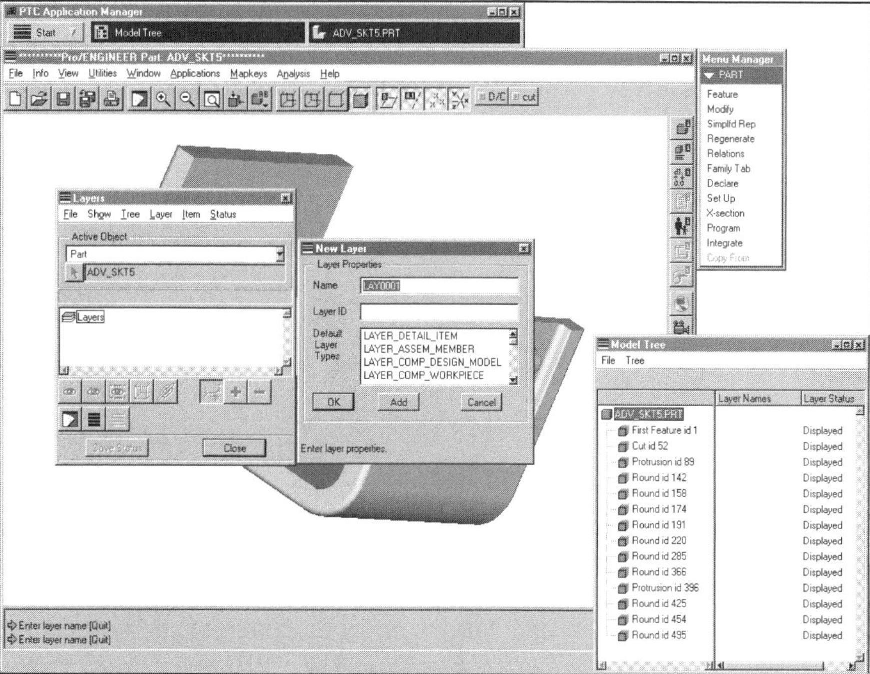

Figure 8.6
New Layer Dialog Box

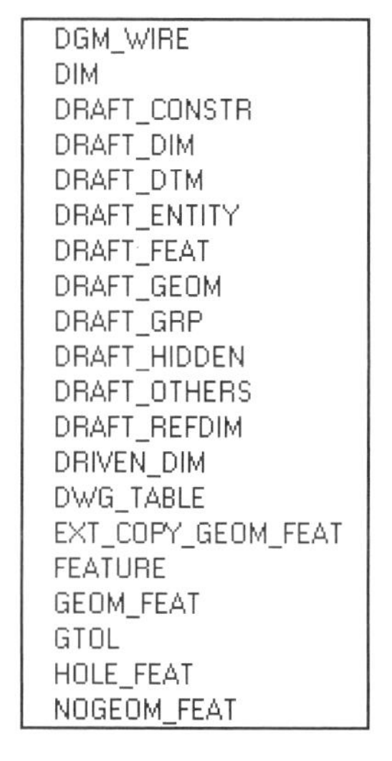

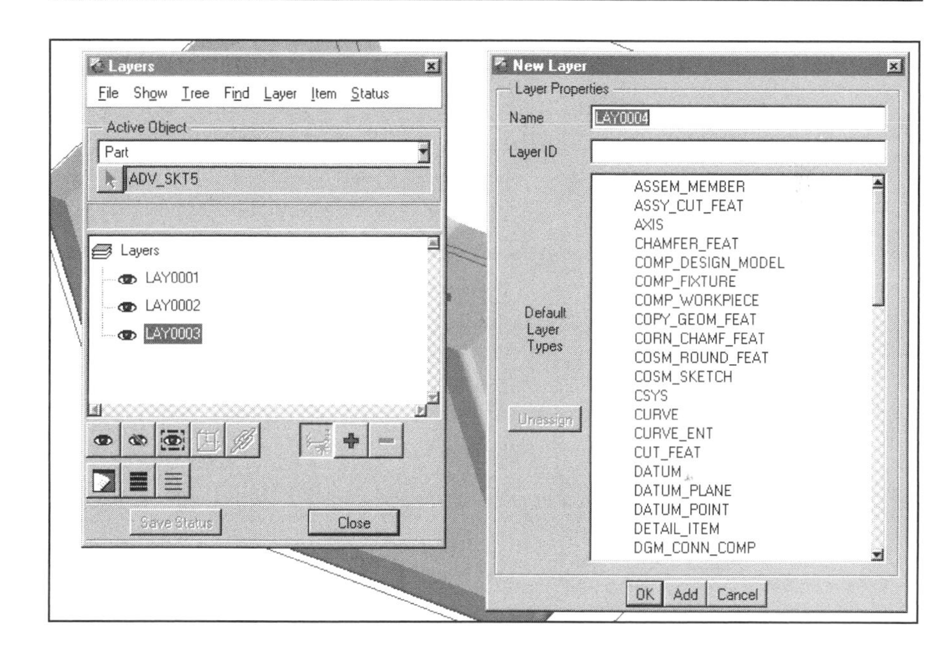

Figure 8.7
Creating New Layers

Adding Items to Layers

After you have created a layer, you can associate items to it by taking the following steps:

1. Highlight the layer name you wish to set items (Fig. 8.8).
2. Select **Add item to selected layers** icon from the Layers dialog box (Fig. 8.8).
3. The LAYER OBJ menu appears, with a list of possible item types. Select the desired item type.
4. The GET SELECT menu appears. Select the desired objects or features by picking them from the screen or from the model tree by highlighting each item or by selecting them by navigating through the menu structure (Fig. 8.9). Then choose **Done Sel.**

1. Pick a layer name to select (pick again to unselect)

2.

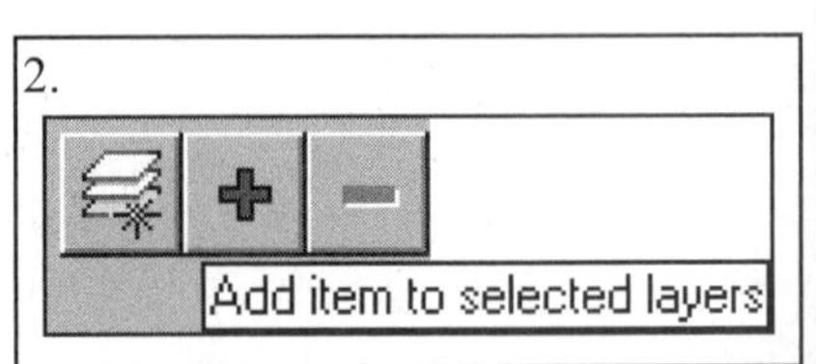

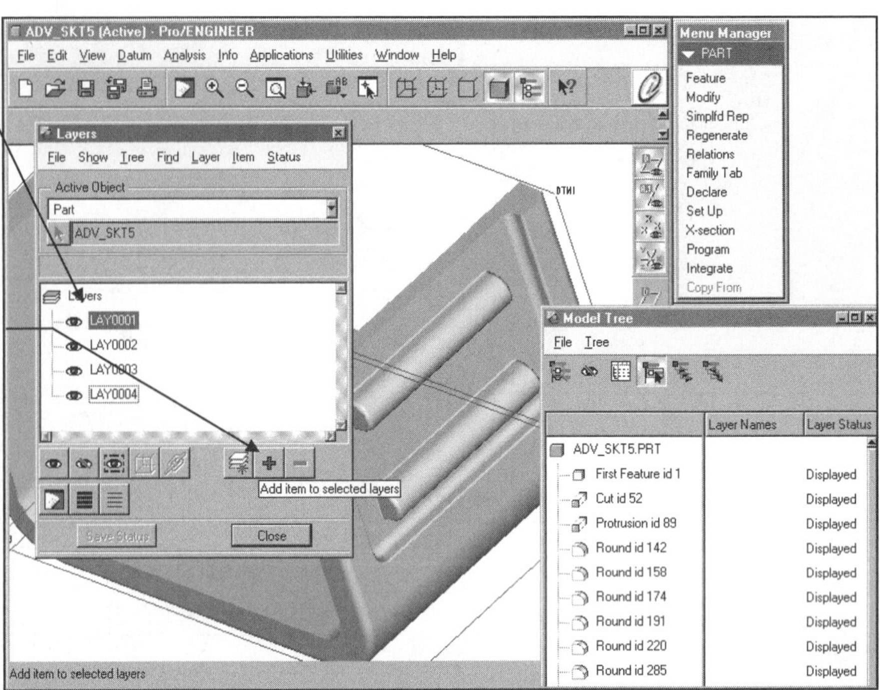

Figure 8.8
Adding Item to Selected Layers

To select all items from the Names dialog box, pick the icon

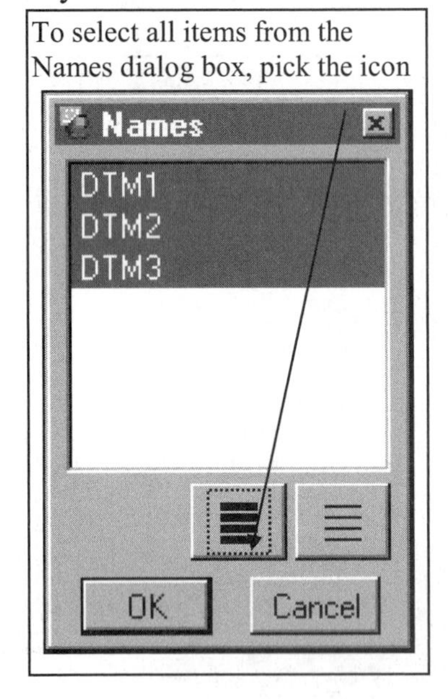

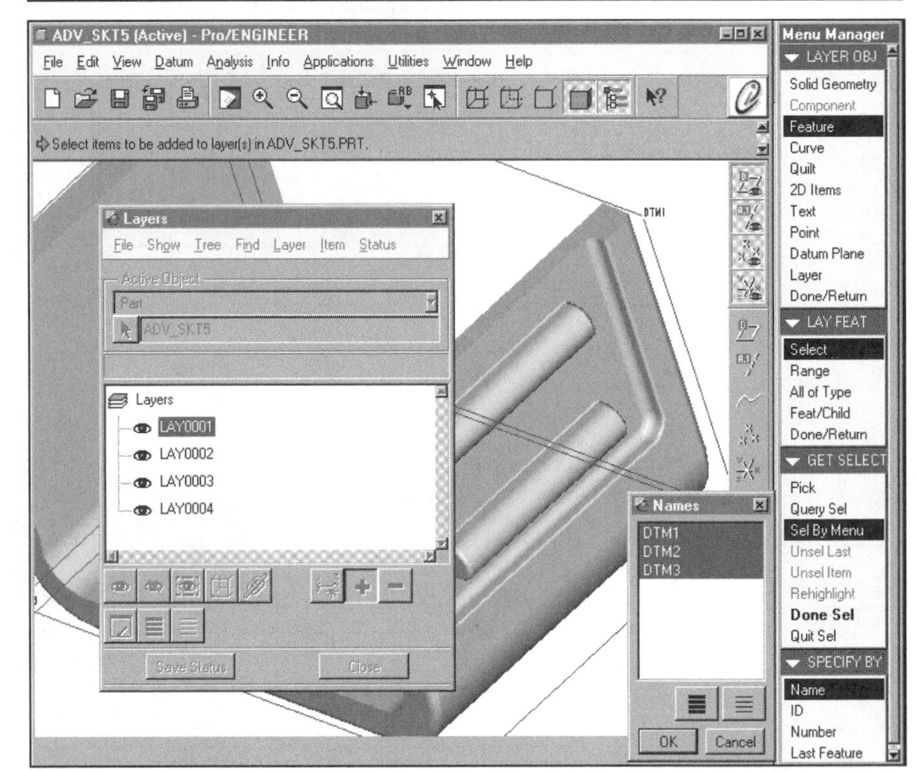

Figure 8.9
LAYER OBJ Menu and GET SELECT Menu

After the items have been set, you can verify that they have been placed by picking **View** ⇒ **Layers** ⇒ (select the layer name from the list) ⇒ **Layer** ⇒ **Info**, as shown in Figure 8.10.

Info displays an INFORMATION WINDOW showing display layer status, layer items, and active object information; contains a layer information block for each selected layer and an item information block for each selected item. You can save the Layer INFORMATION WINDOW with the default file name *layer_filename.inf* or enter a unique file name.

Figure 8.10
Layer ⇒ Info

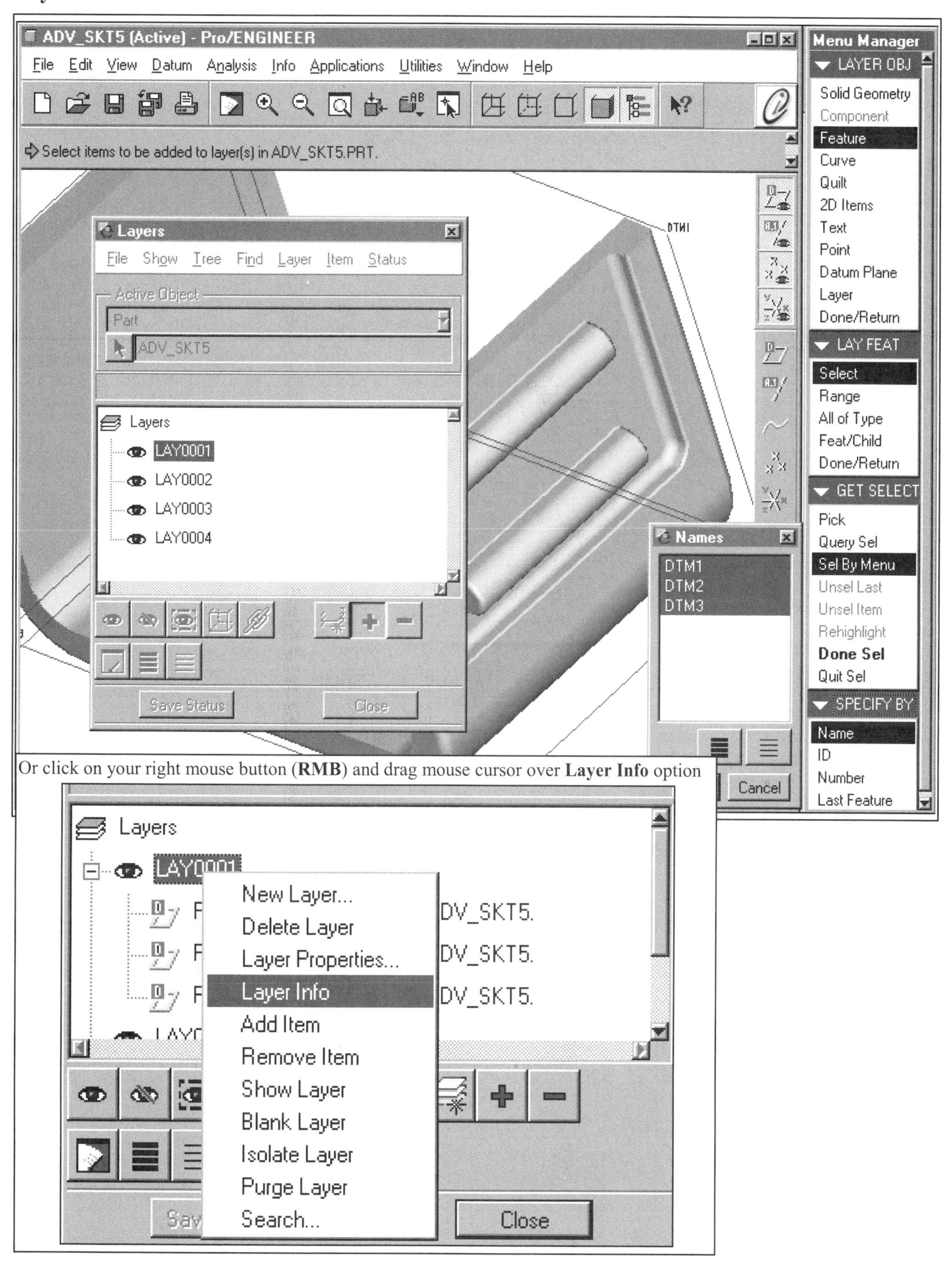

Removing Items from Layers

When you remove items from a layer in the active model, you disassociate them from the layer. The items in the Layers dialog box Tree (Fig. 8.11) were already shown using **View ⇒ Layers ⇒ Show ⇒ ✔ Layer Items.** You can select the items to remove directly from this tree using your left mouse button (Fig. 8.12); then pick the Remove button.

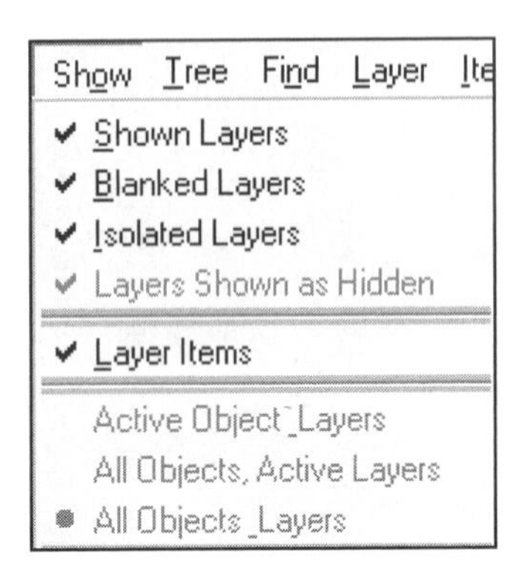

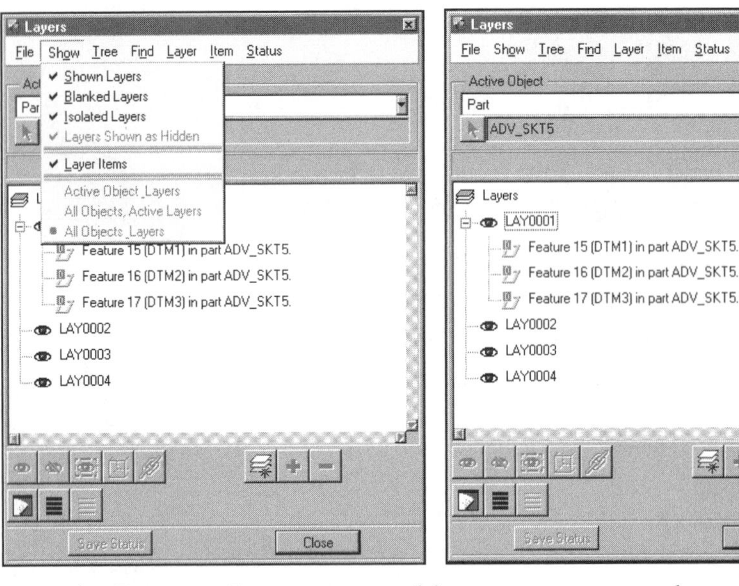

Figure 8.11
Layer Items

The **Remove Item** command has some commands available only in the pop-up menu. You can use these commands to remove selected item or items from all layers, from all 3D layers, or from all drawing layers. When you use these commands to remove items from a layer, the Confirmation dialog box prompts you to confirm before the Pro/E removes the selected item or items from the layer(s).

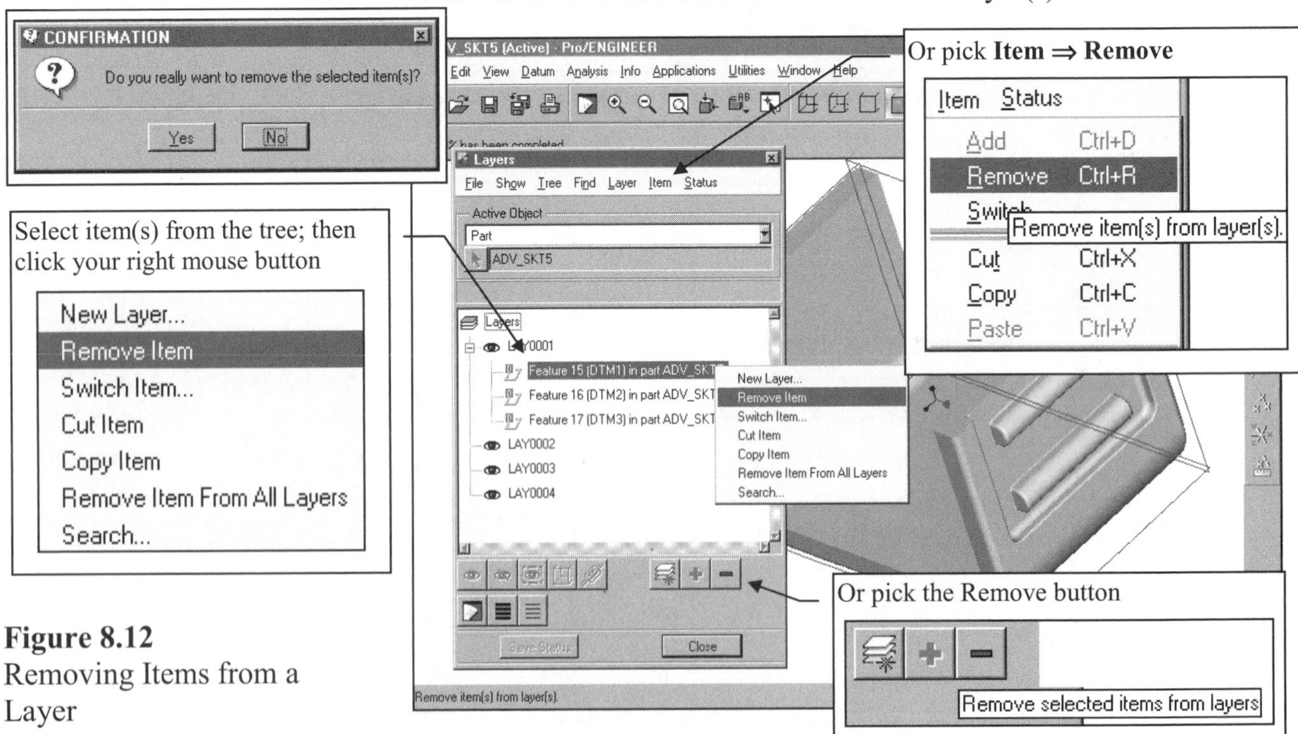

Figure 8.12
Removing Items from a Layer

Editing Layers

To delete an existing layer, select **View** ⇒ **Layers** ⇒ (select the layer from the list) ⇒ **Layer** ⇒ **Delete** (Fig. 8.13 left).

To rename an existing layer, select **View** ⇒ **Layers** ⇒ (select the layer from the list) ⇒ **Layer** ⇒ **Properties** (Fig. 8.13 right) ⇒ (type a new name) ⇒ **OK** (Fig. 8.14).

Figure 8.13
Deleting a Layer (left)
Layer Properties (right)

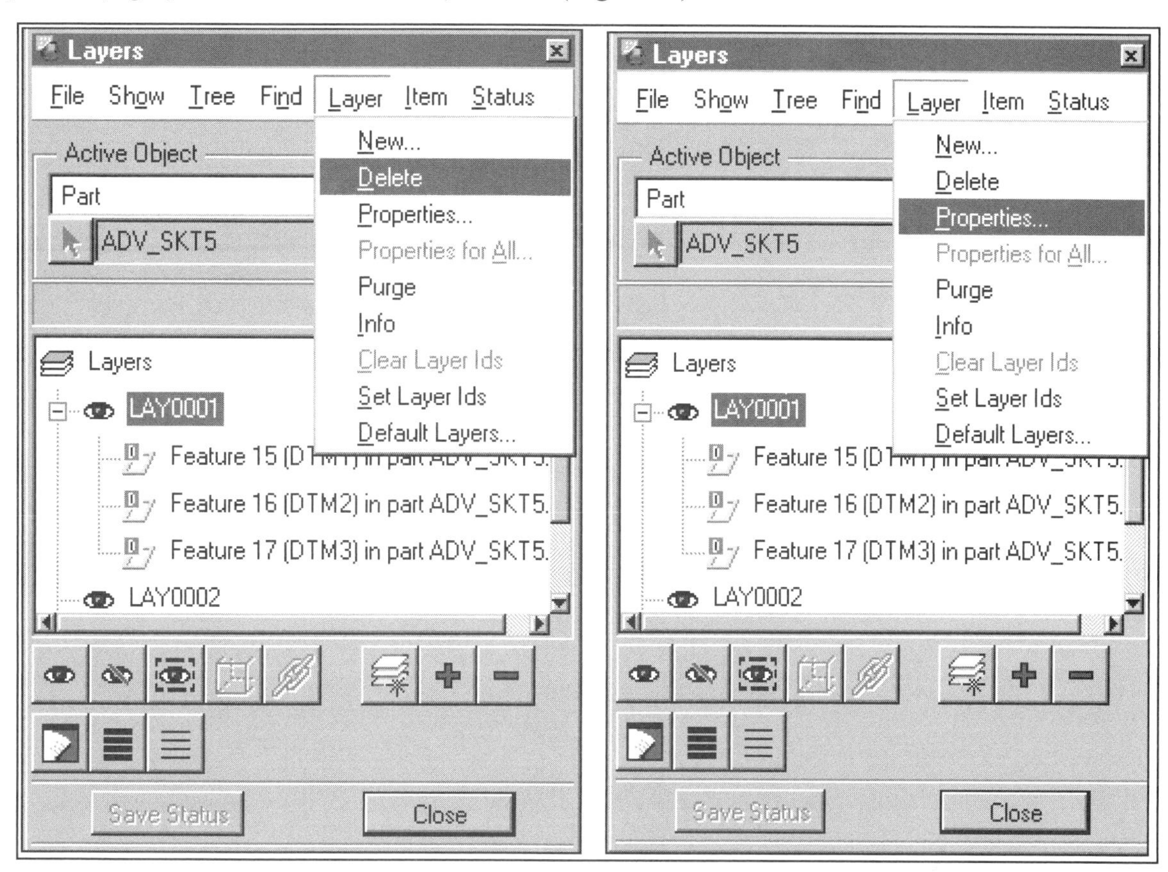

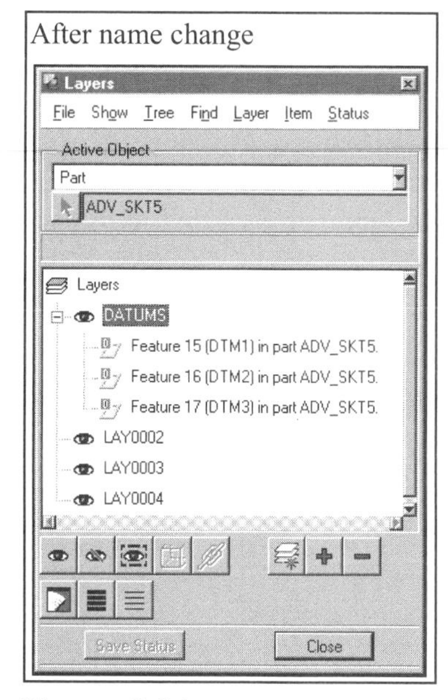

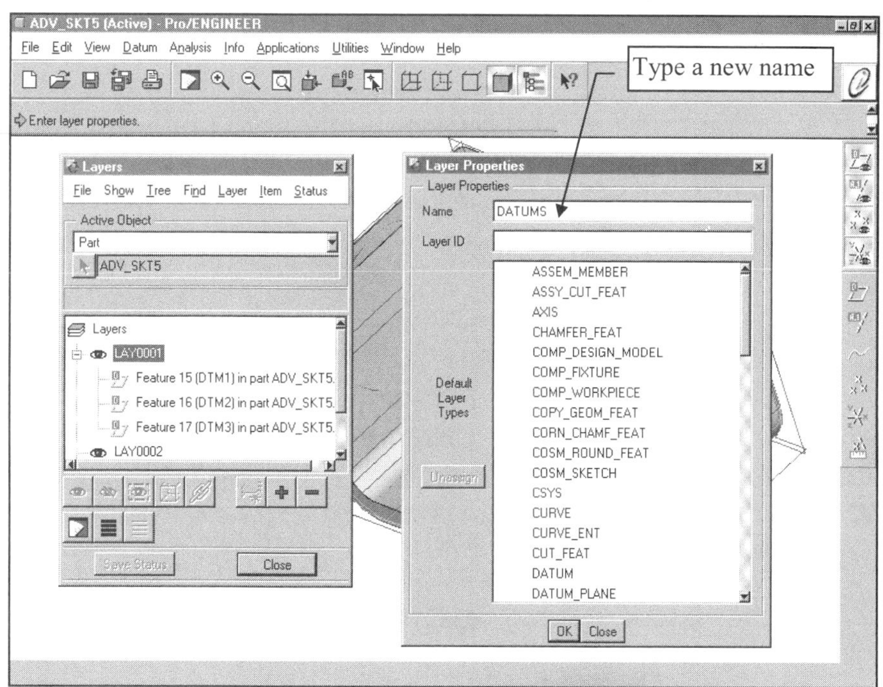

Figure 8.14
Renaming a Layer

System Default Layering

It is possible to have default layers to which certain items are automatically assigned. You can establish up to 32 default layers for a model. The following table lists valid type options for default layers. Pro/E automatically places certain types of items on specified layers when they are created, using the configuration file option ***def_layer.***

Type option	**Description**
Layer_assem_member	Assembly members
Layer_feature	All features
Layer_axis	Features with axes
Layer_geom_feat	Features with geometry
Layer_nogeom_feat	Features without geometry
Layer_cosm_sketch	Cosmetic sketches
Layer_surface	Surface features
Layer_datum	Datum planes
Layer_datum_point	Datum point features
Layer_slot_feat	Slot
Layer_dgm_highway	Diagram highways
Layer_dgm_rail	Diagram rails
Layer_shell_feat	Shell
Layer_assy_cut_feat	Assembly cut
Layer_chamfer_feat	Chamfer
Layer_corn_chamf_feat	Corner chamfer
Layer_cut_feat	Cut
Layer_draft_feat	Draft
Layer_hole_feat	Hole
Layer_protrusion_feat	Protrusion
Layer_rib_feat	Rib
Layer_round_feat	Round

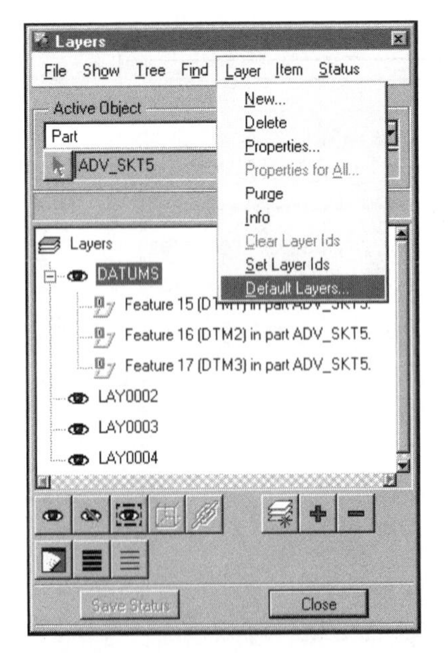

When you create an entity of one of these types, Pro/E will automatically add it to the specified default layer. If a feature (e.g., a hole) has an axis (other than a datum axis), that axis can be automatically placed on two default layers, one for features with axes (option **layer_axis**) and one for the particular type of feature (e.g., hole option **layer_hole_feat**).

As an alternative to editing the configuration file, you can use the **View ⇒ Layers ⇒ Layer ⇒ Default Layers** to edit the default layer table (Fig. 8.15). Layering options defined or changed here will *not* reflect back to the configuration file and are good only for *new* features created in the current Pro/E session in *all* models. It is a good practice to set up the default layering *before* you start work on a part, assembly, or drawing.

The Default Layer Table is initially empty when you start a new session and do not have any def_layer configuration file options.

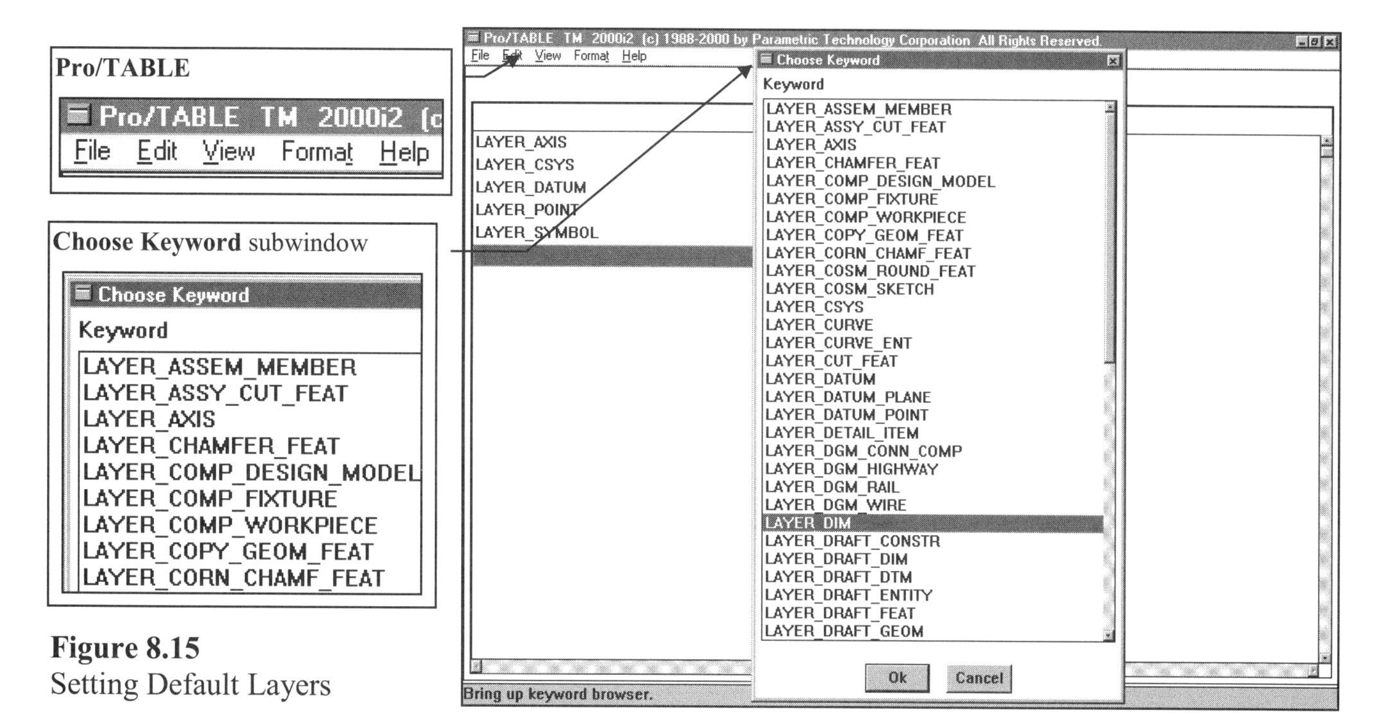

Figure 8.15
Setting Default Layers

The left column lists layer items. The right column lists the layers to which the layer items are assigned.

When you are adding layer items using Pro/TABLE, select the following:

1. Select an empty cell in the left column.
2. Press **F4** and the Choose Keyword NAMELIST menu opens.
3. Pick on an item in the box (here the **LAYER_DIM** keyword is highlighted) and the editor will copy it into the active cell.

Pro/E supplies a number of default layer names:

Layer_dim	**Def_dims**
Layer_corn_chamf_feat	**Def_chamfers**
Layer_protrusion_feat	**Def_protrusions**
Layer_axis	**Def_axis**
Layer_assem_member	**Def_components**
Layer_assy_cut_feat	**Def_features**
Layer_chamfer_feat	**Def_chamfers**

You can enter layer names into the right column by:

1. Selecting an empty cell in the right column adjacent to your previous entry.
2. Pick the **Choose Keywords** option in the EDIT menu. This will bring up the Choose Keyword NAMELIST menu of default layer names.
3. Pick a layer name. Choose **Ok** and the editor will copy it into the active cell. Alternatively, you can type in a layer name from the keyboard. When finished, choose **Exit** from the FILE menu.

The simplest way to use the default layering system is to give the following commands: **View** ⇒ **Layers** ⇒ **New** ⇒ [enter the layer name and choose a type from the list (Fig. 8.16)] ⇒ [Default layer types (select from the list **AXIS**)] ⇒ **Add** (this will make the layer name in the Name box the default for this type of entity--**AXIS**) ⇒ [*if you want the default layering to have the same name as the one you selected*, you must type it in the Name box (Fig. 8.17)] ⇒ **Cancel.**

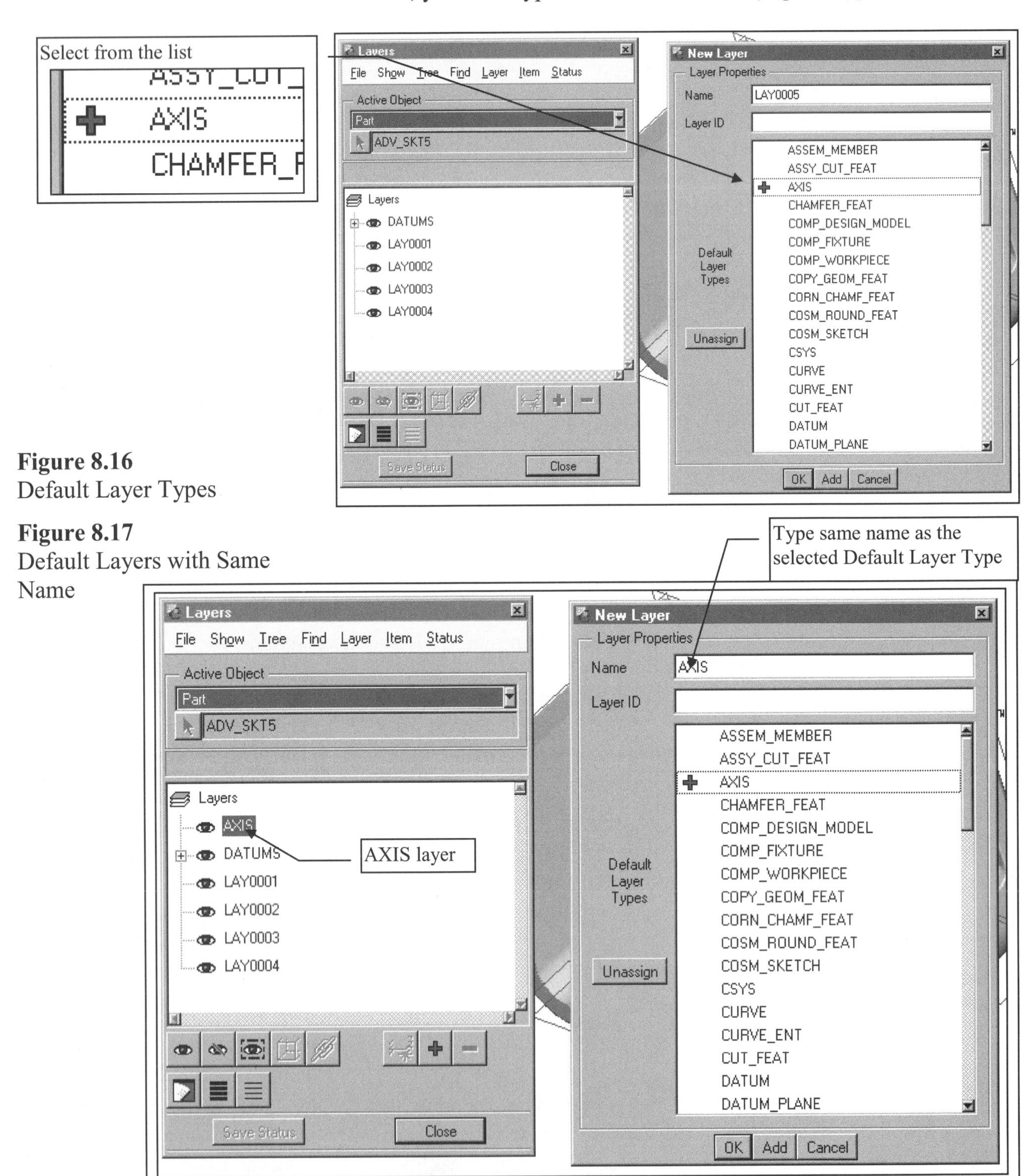

Figure 8.16
Default Layer Types

Figure 8.17
Default Layers with Same Name

Displaying Layers

Using the Environment dialog box **(Utilities ⇒ Environment)** or the Toolbar (left on, right off):

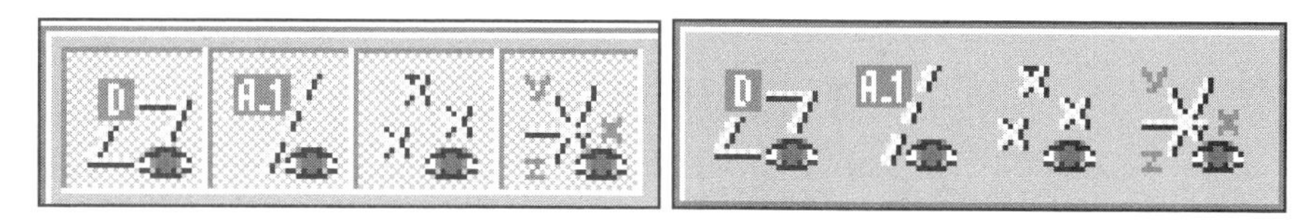

you can choose to turn on or off a variety of items, including datum planes, datum axes, datum points, and coordinate systems (Fig. 8.18). Another way to control the display of the items on the screen is to put groups of items on layers and show or blank them as needed. Later you will learn that the datums can be set as default geometric tolerance features (basic datums). Datum planes set in this way *cannot be turned off using environment settings*; therefore, when the object is displayed in Part or Assembly mode, the only way to control the basic datum display is to blank its layer. This will become obvious when you see assemblies with 10 to 50,000 components, all with basic datums displayed at once! Blanking layers becomes essential when this happens.

Figure 8.18
An Assembly with Multiple Layers

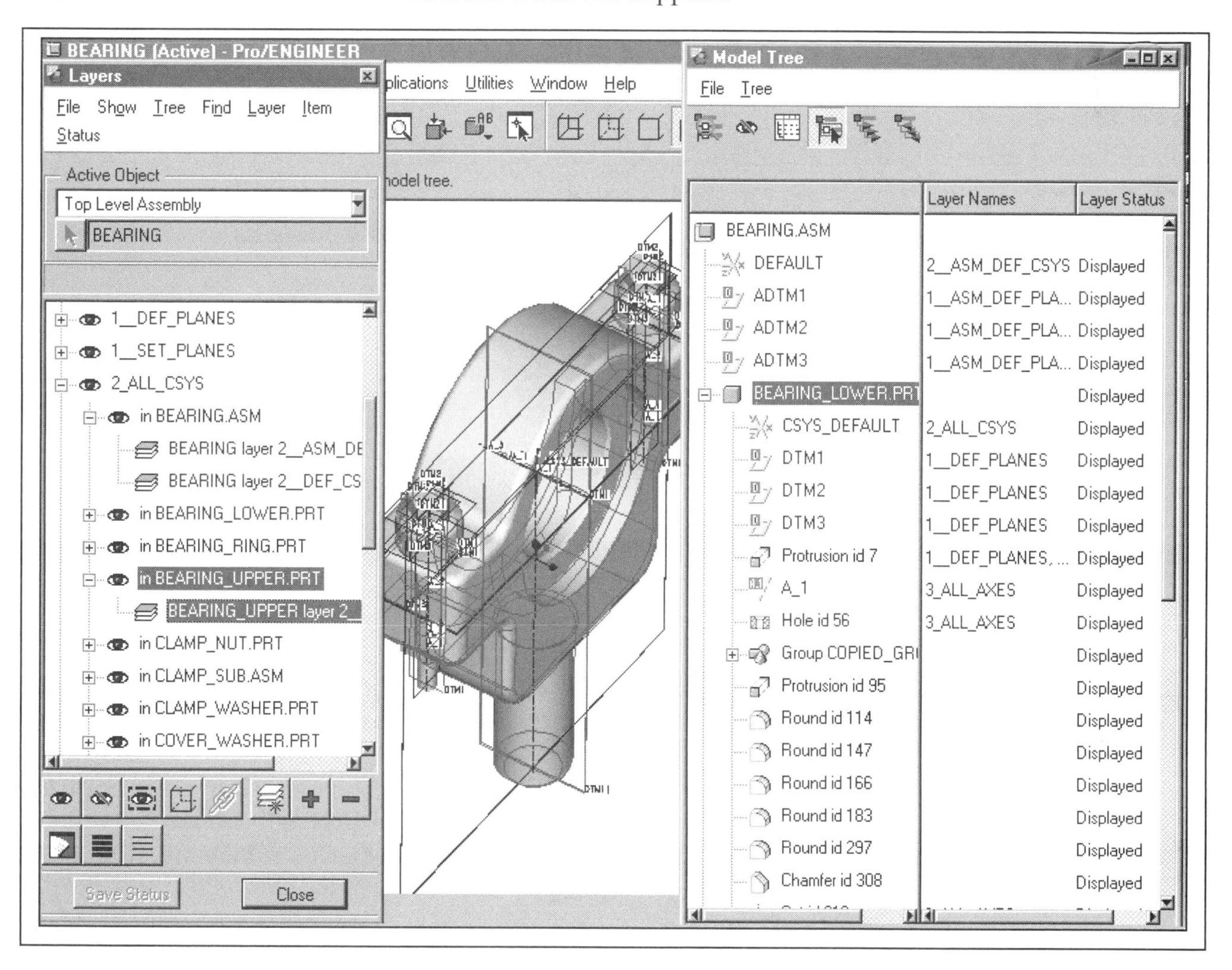

You use the Layers dialog box (Fig. 8.19) to view layers in a part, assembly, or drawing, to view items assigned to a layer, and to control how layers are displayed in the object. To open this dialog box, from the View menu choose **Layers.** One of the main reasons for organizing items into layers is to make it possible to show or blank the items selectively. These functions affect how items associated with a layer are displayed in the Pro/E main window.

Blanking or showing layers does not affect model geometry, because these functions affect only features that do not affect mass properties, such as datum planes, axes, and coordinate systems.

You can set the display status of a layer with one of the following *icons*:

Show (Default) Make selected layer shown (can be seen).
Blank Make selected layer blanked (cannot be seen).
Isolate Make selected layers shown and treat all nonisolated layers as blanked.
Hidden (Assembly mode only) Make components in hidden layers blanked in accordance with the Environment settings for hidden line display.

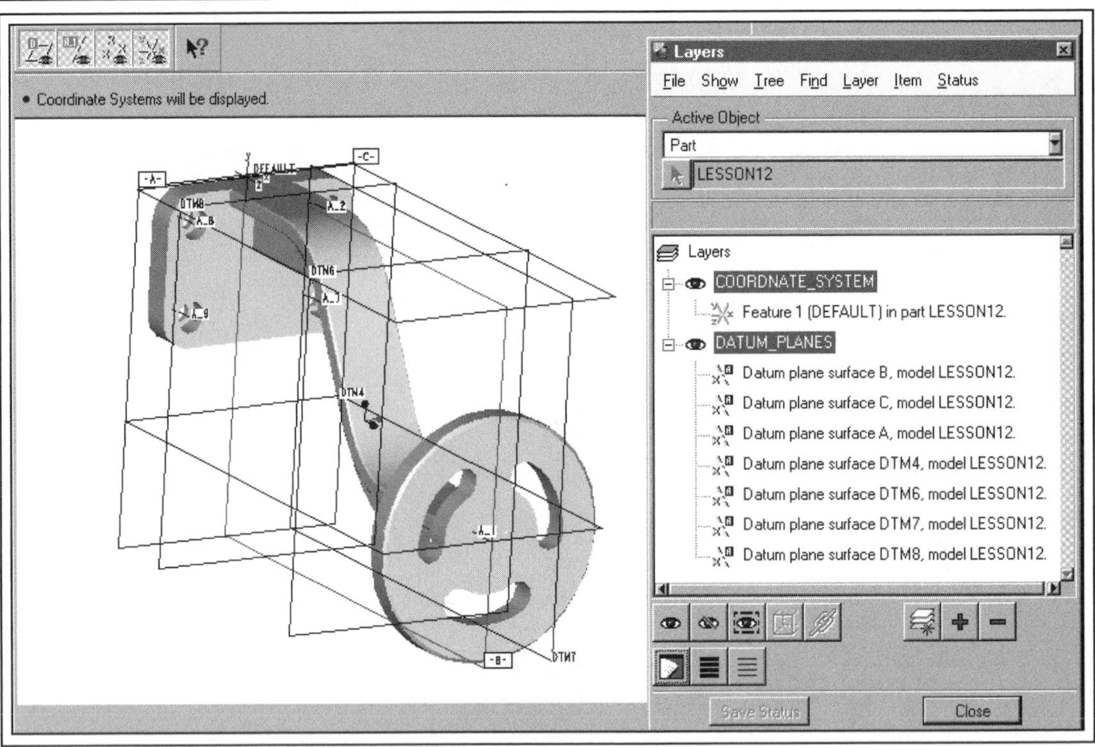

Figure 8.19
Layer Display Dialog Box

In Assembly mode, if you set a specific layer or layers to **Isolate,** Pro/E blanks all other layers. In addition, Pro/E blanks all other items that are not assigned to any layer.

In Assembly mode (Fig. 8.20), if you have components on layers that you then set to **Blank,** Pro/E blanks all nongeometry items (datum planes, datum axes, feature axes) even if they are also on the displayed layers.

Using **Tree ⇒ Expand** in the Layer dialog box allows you to expand the Model Tree list as shown in Figure 8.20.

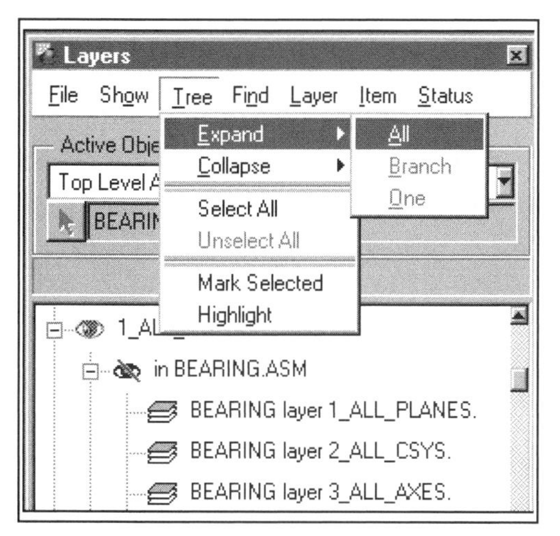

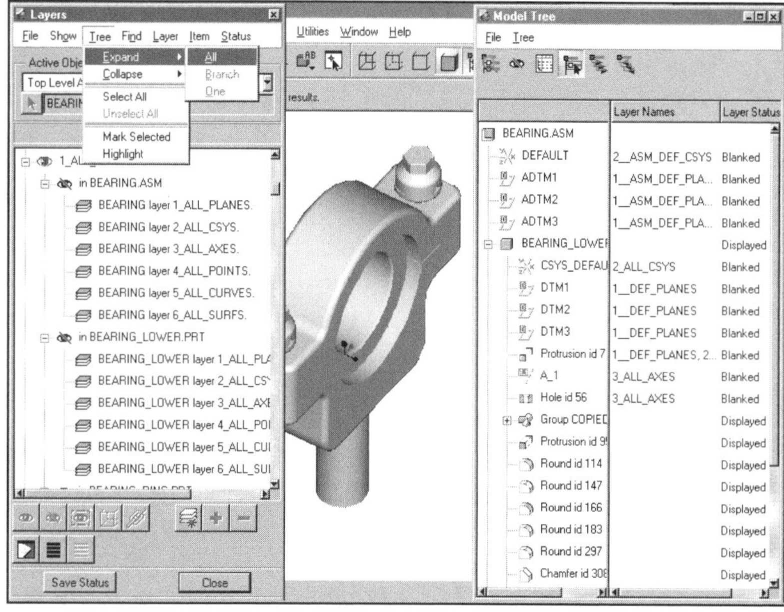

Figure 8.20
Expanded Layer Tree

In all other modes, if you set a specific layer or layers to **Isolate,** Pro/E also blanks all other layers. However, Pro/E continues to display items that are not assigned to any layer. You can blank the following types of items in any modeling mode (such as Part, Assembly, or Manufacturing): *Datum features*, such as planes, axes, curves, and points; *Feature axes*, such as axes for holes on the layer; *Cosmetic features; Quilts;* and *Notes.*

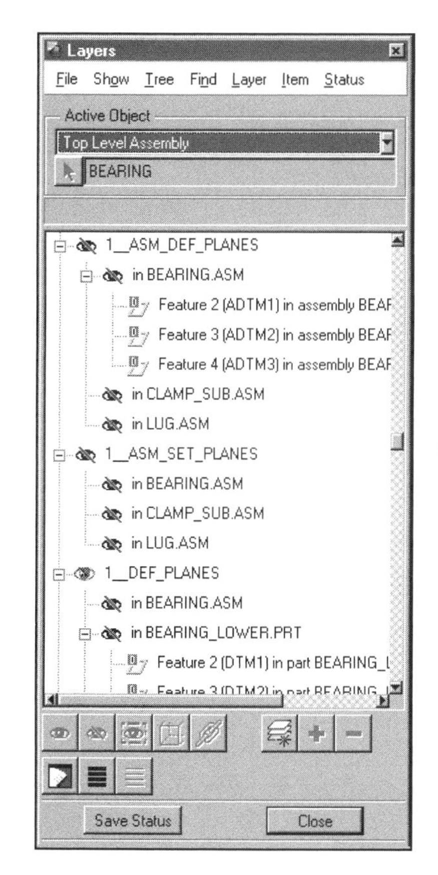

Isolate has priority over **Blank.** Therefore, if a member is on two layers, one set to **Isolate** and the other set to **Blank,** the member is shown. However, if a feature consists of several entities (for example, datum curves), individual entities are not shown if the entire feature is in a layer that is set to **Blank,** even if the entities themselves are in layers that are set to **Isolate.**

The only features on a layer that layer display operations affect are datum and surface features. Solid geometry is not affected. For example, in Part mode, if you put a hole on a layer and then blank the layer, only the hole's datum axis is blanked. *You can prevent the hole from being displayed by suppressing it directly or by suppressing its layer.* The only exception to this rule is that in an assembly solid, you can blank components.

If a new member is added to an assembly, and a default layer for it does not exist, Pro/E does not automatically add it to an existing layer. If a layer in the assembly is already set to **Isolate** when the member is added, the new member is not shown until you either add it to a layer that is explicitly isolated and repaint the display or clear all the isolated layers. In the latter case, which is the recommended action, all the members of the assembly are shown. In Assembly mode, **Show** affects the level of the member and levels above it; **Blank** affects the level of the member and levels below it.

Remember that the Environment and Toolbar settings affect only datum features that are not set as basic. In Figure 8.21, the datums have been turned off with the Toolbar. The Layer Dialog Box shows the two layers as displayed. Remember that all of the datums were placed on a layer. The datums that were not set as basic can be toggled on and off using the Environment settings or with the Toolbar selections. These two methods do not affect the basic datum planes. You control the display of the basic datum planes with the Layers dialog box. The **Eye** icon (Fig. 8.22) without the slash through it **Makes the selected layers Shown**. "Picking" the Eye icon with a slash through it **Makes the selected layers Blanked.**

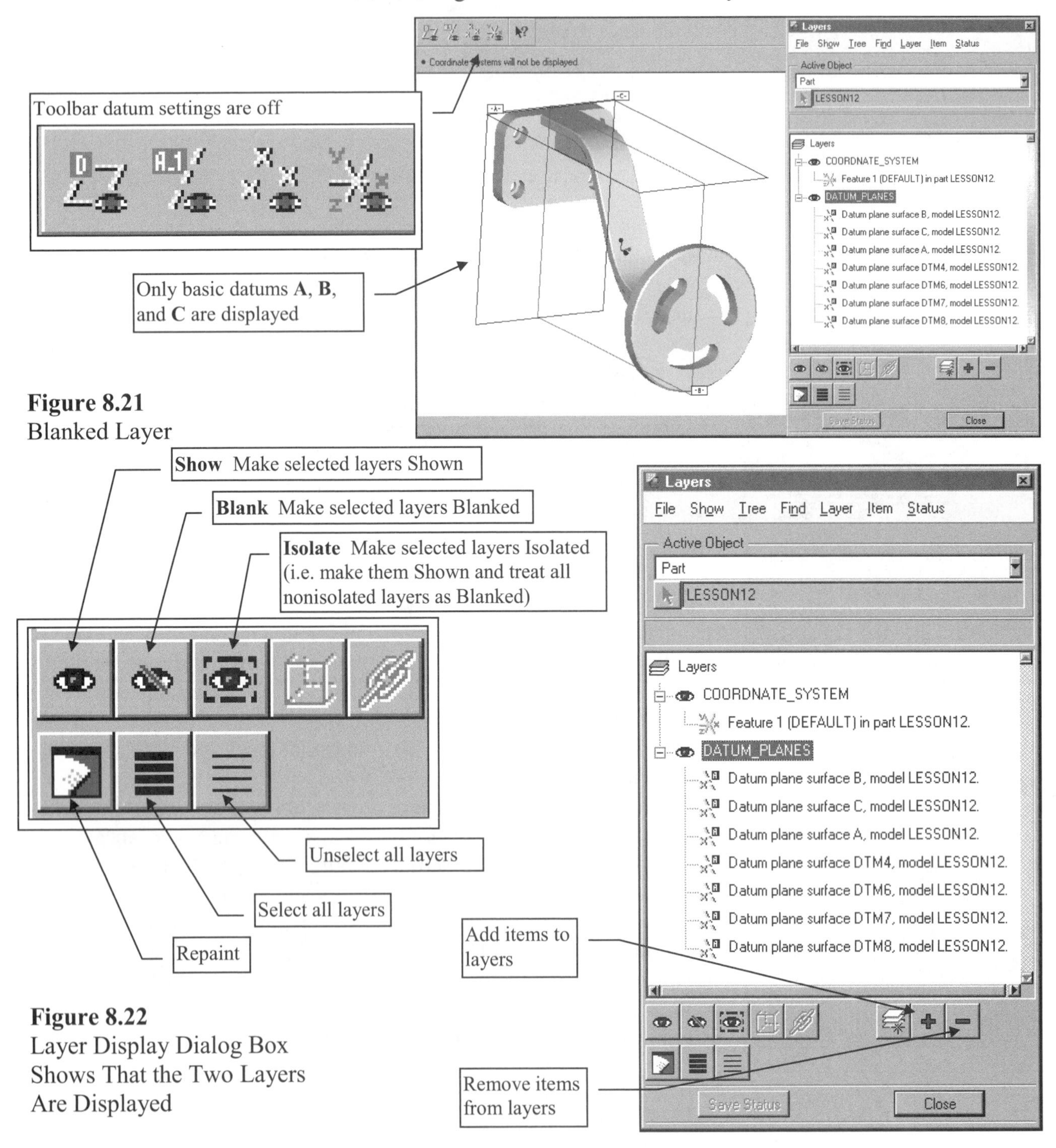

Figure 8.21
Blanked Layer

Figure 8.22
Layer Display Dialog Box Shows That the Two Layers Are Displayed

The results show that the part is now displayed without datum planes. Note that if you pick **Utilities** (from the Menu Bar pull-down menu) ⇒ **Environment**, the **Datum Planes** selection is still checked (on), as shown in Figures 8.23 and 8.24. The Blank Layer command overrides the *temporary* Environment and Toolbar settings for those datum planes not set as basic. After making changes in the Layer dialog box, you may need to repaint the screen before the changes take effect.

Figure 8.23
Blanking Layers

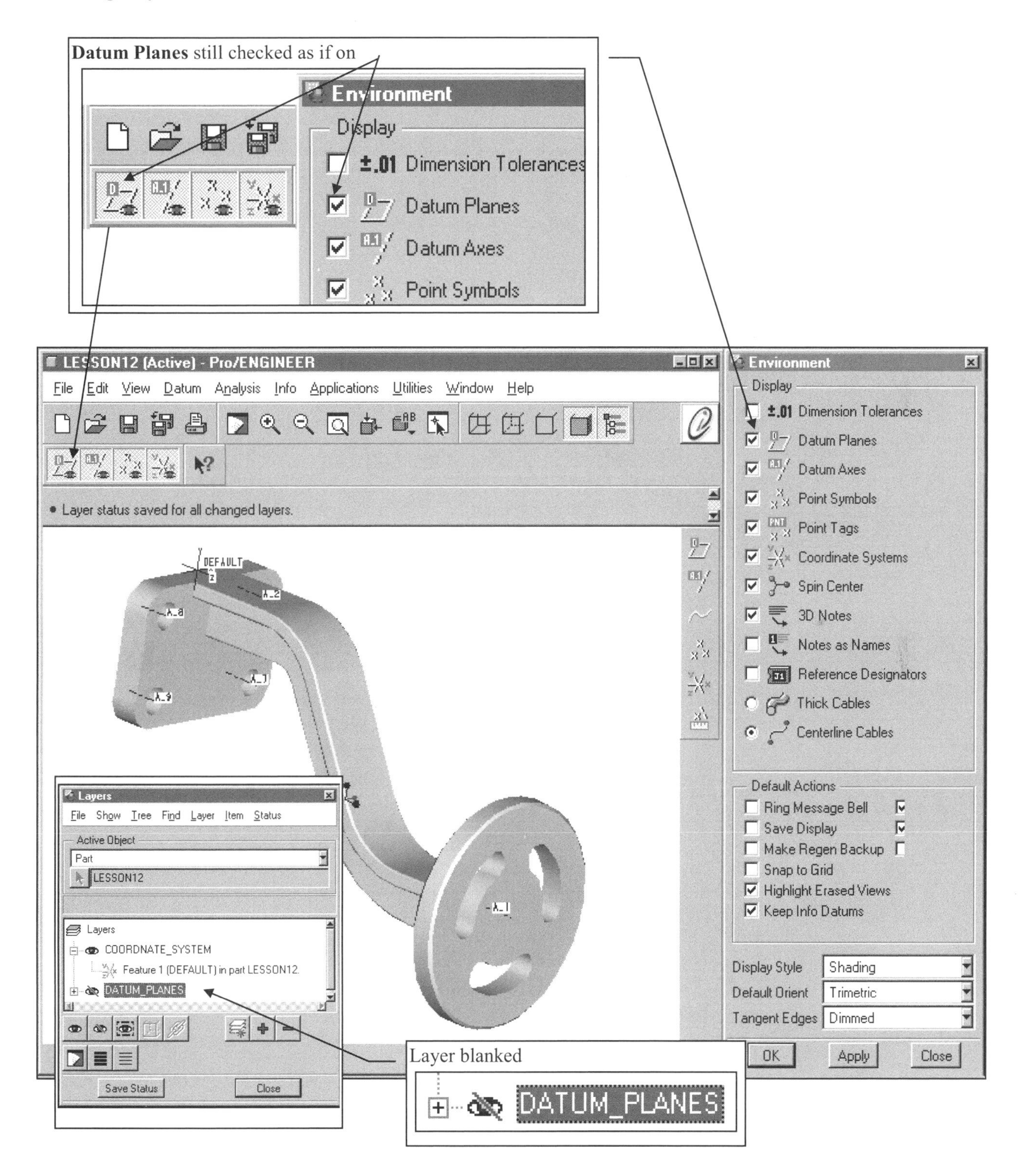

Figure 8.24
Blanking Layers Assembly Mode

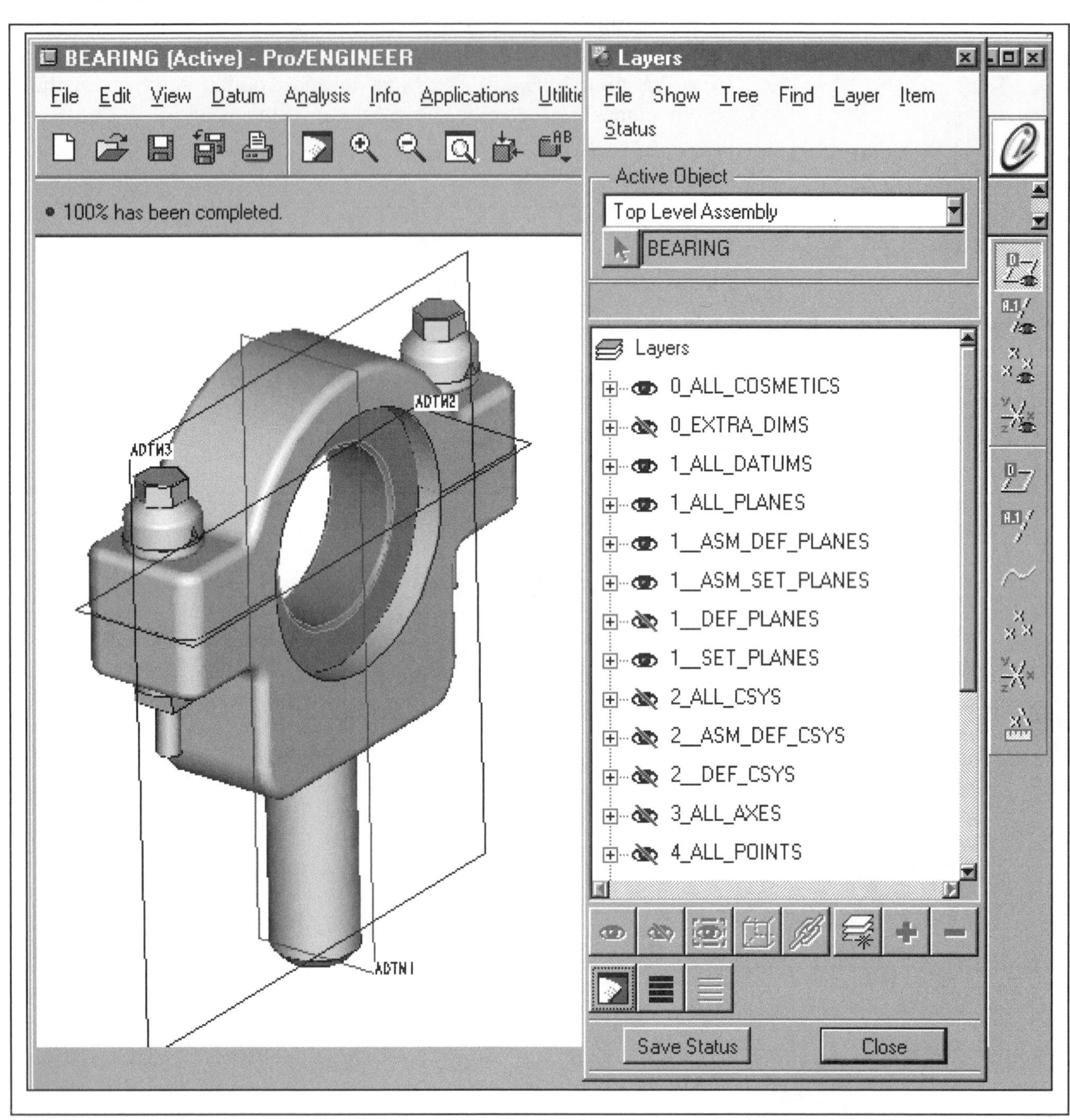

Section 9

Model Tree

Learning how to use the **Model Tree** (Fig. 9.1) is critical for using Pro/E effectively. As you or your workgroup develops more sophisticated models, you will find that the Model Tree provides both basic and sophisticated features for manipulating and managing the objects that you create. These features include the following:

* Displays a list of objects currently available in your Pro/E session. For example, if you open an assembly, the Model Tree displays all the constituent parts associated with that assembly.
* Allows selection of specific features or parts for various operations (e.g., you can select a feature in a part by selecting its name in the Model Tree without having to alter the current view of the part to make the feature easily selectable).
* Enables you to run commonly used commands for the features that you select (e.g., you can right-click a feature listed in the Model Tree and create a model note for it. In Assembly mode, you can select an assembly in the Model Tree and create a new part for that assembly).
* Displays detailed information about each object, including feature types, feature ID, and names.
* Permits creation of new information parameters for selected features.
* Provides sophisticated search capabilities so you can query all the objects in your Pro/E session for specific built-in or user-defined parameters.
* Using ➡ **Insert Here** you can move a feature or create a feature at new positions in the Model Tree sequence.

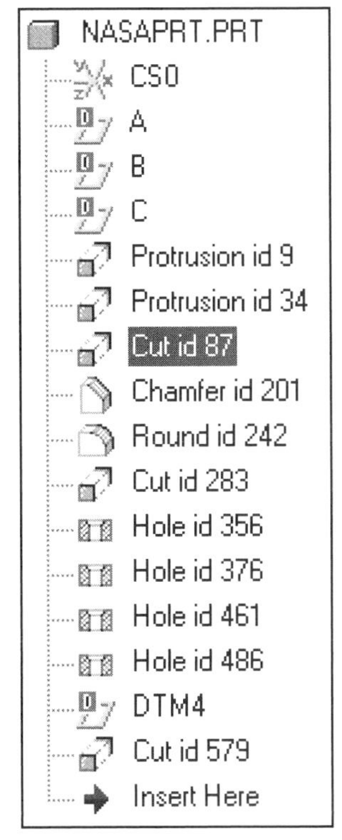

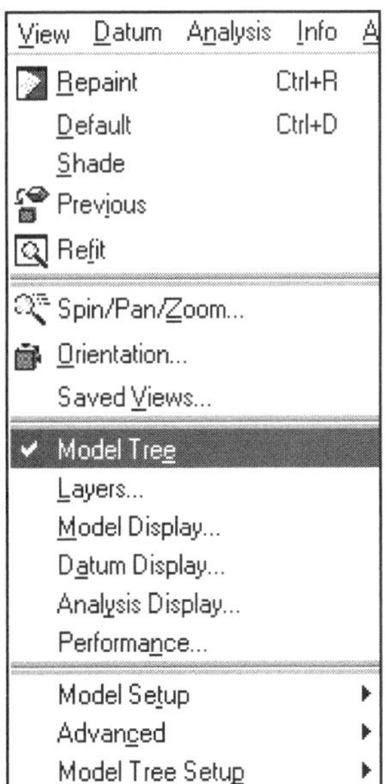

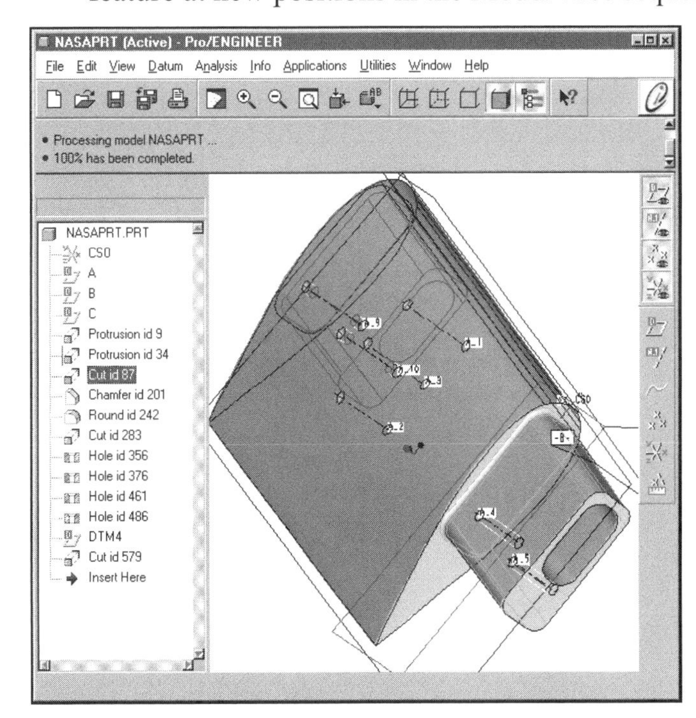

Figure 9.1
Part with Model Tree Displayed

To display the Model Tree and to enable the Model Tree display whenever the active window is displayed, select the Model Tree check box on the **View** pull-down menu (see left-hand column).

About the Model Tree

The Model Tree window lists every object in the design in a hierarchically ordered tree format. *The window can be set to "float" or to be embedded in the Main Working Window.* To change the Model Tree from the default position of embedded (Fig. 9.1), select **Utilities ⇒ Customize Screen ⇒ Options ⇒ ✔** Display as a separate window ⇒ **OK** (Fig. 9.2). Because the *embedded* option takes up space in the main graphics area, most of this text *floats* the Model Tree (Fig. 9.3) and positions it in the lower right-hand corner of the screen. The floating Model Tree has pull-down menus and a set of icons along the top of its window (see upper left-hand column).

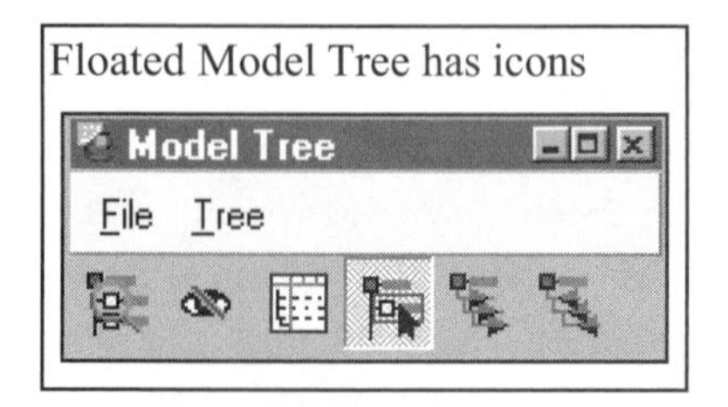

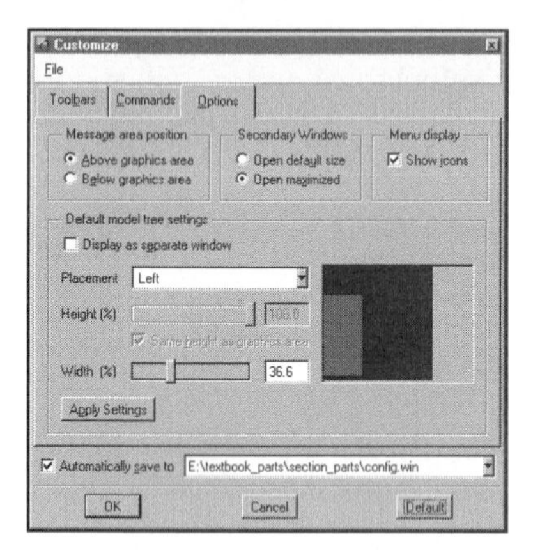

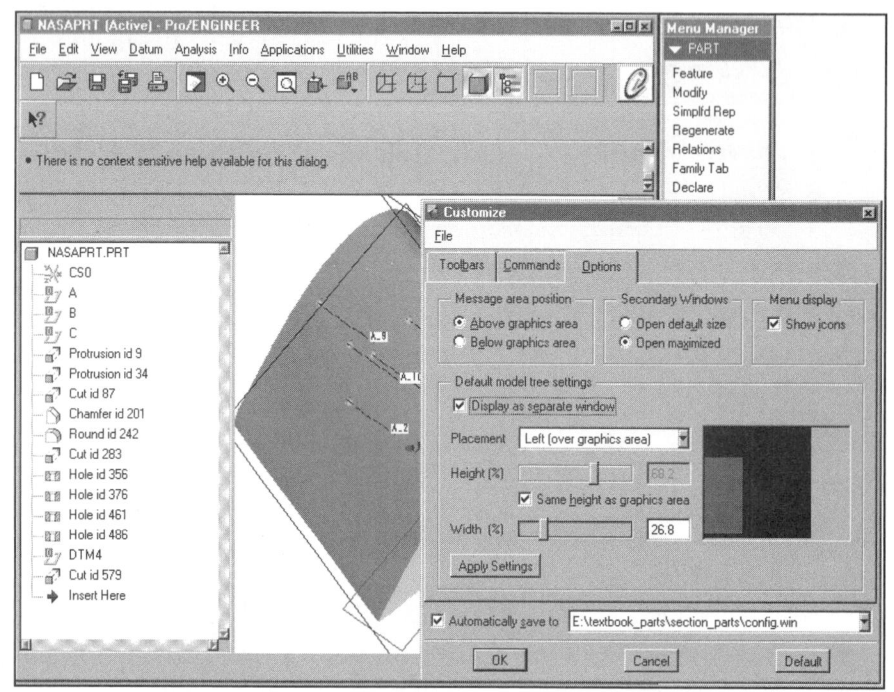

Figure 9.2
Customize Location of Model Tree

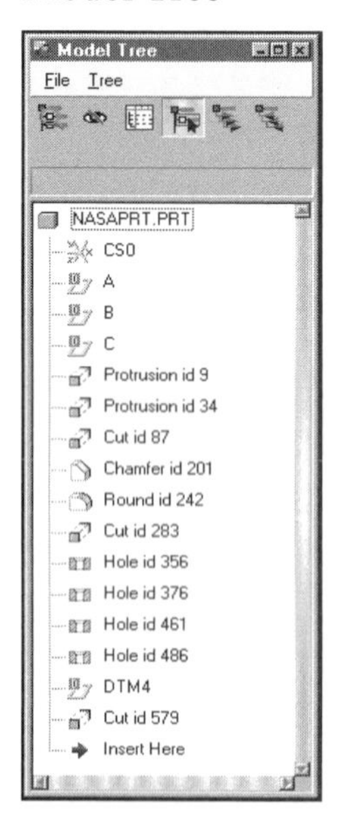

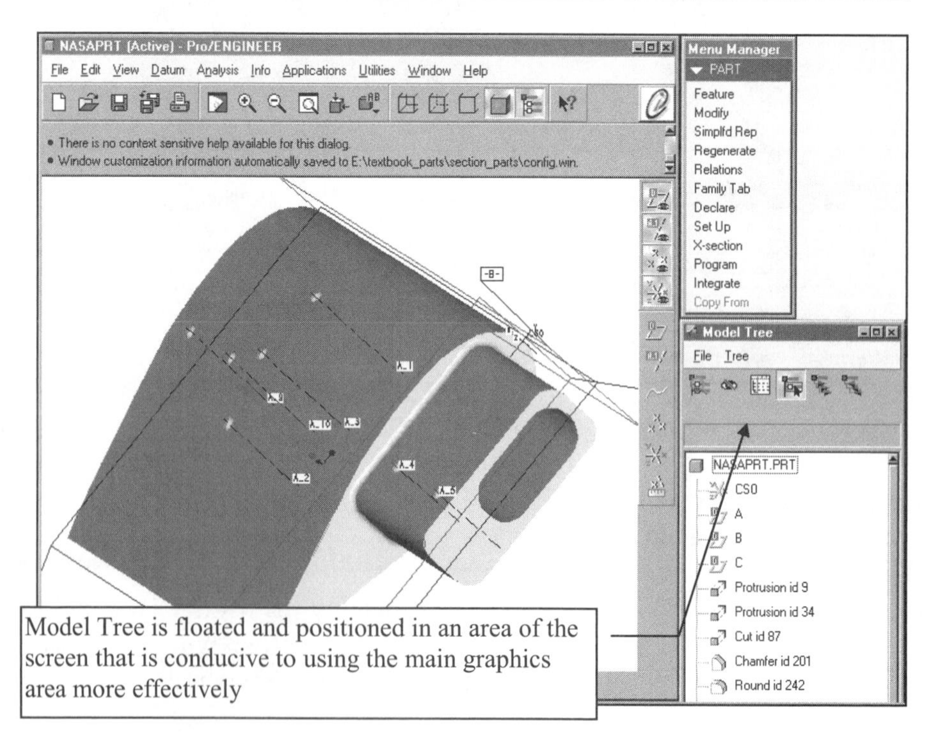

Figure 9.3
Floating Location of Model Tree

If you have multiple windows open, the Model Tree is active for the active window. You can filter the Model Tree display by item type or status, such as, showing or hiding datum objects, or suppressed objects.

You can use the Model Tree icon on the toolbar to show or hide the Model Tree

You can use an icon on the toolbar to show or hide the Model Tree (see left-hand column). You can also save and reuse the configuration settings for the model tree in a *.cfg* file.

The icon next to each item (see left-hand column) in the tree reflects its object type, such as, assembly, part, feature, or datum. The icon can also show the display or completion status, such as, suppressed or unregenerated.

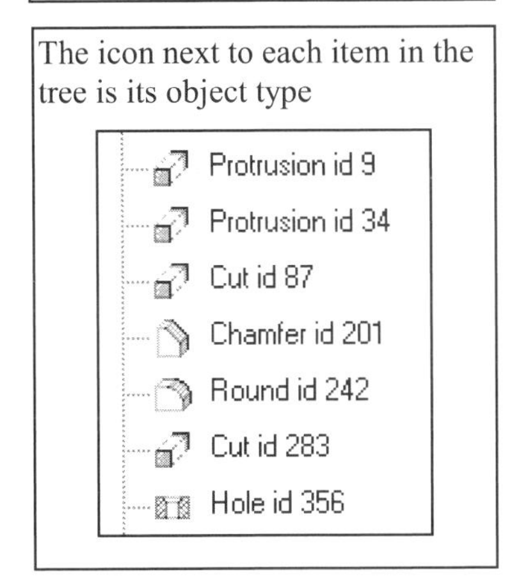

You can use the model tree to select features or parts for editing when they are not visible in the graphics area. When an item is selected, you can click the right mouse button (RMB) to select object-specific commands from a pop-up menu.

You can add informational columns to the model tree window, containing, for example, parameters and values, assigned layers, or feature name for each item. You can use the cells in the columns to perform context-sensitive edits and deletions.

You can reorder objects in the design by dragging them up or down in the model tree. If you move a child object to a level above its parent, the parent is moved also to retain the parent-child relationship.

You can show or hide items by type in the model tree **View ⇒ Model Tree Setup** (Fig. 9.4). You can toggle on and off a wide variety of items with this menu.

Figure 9.4
Model Tree Setup

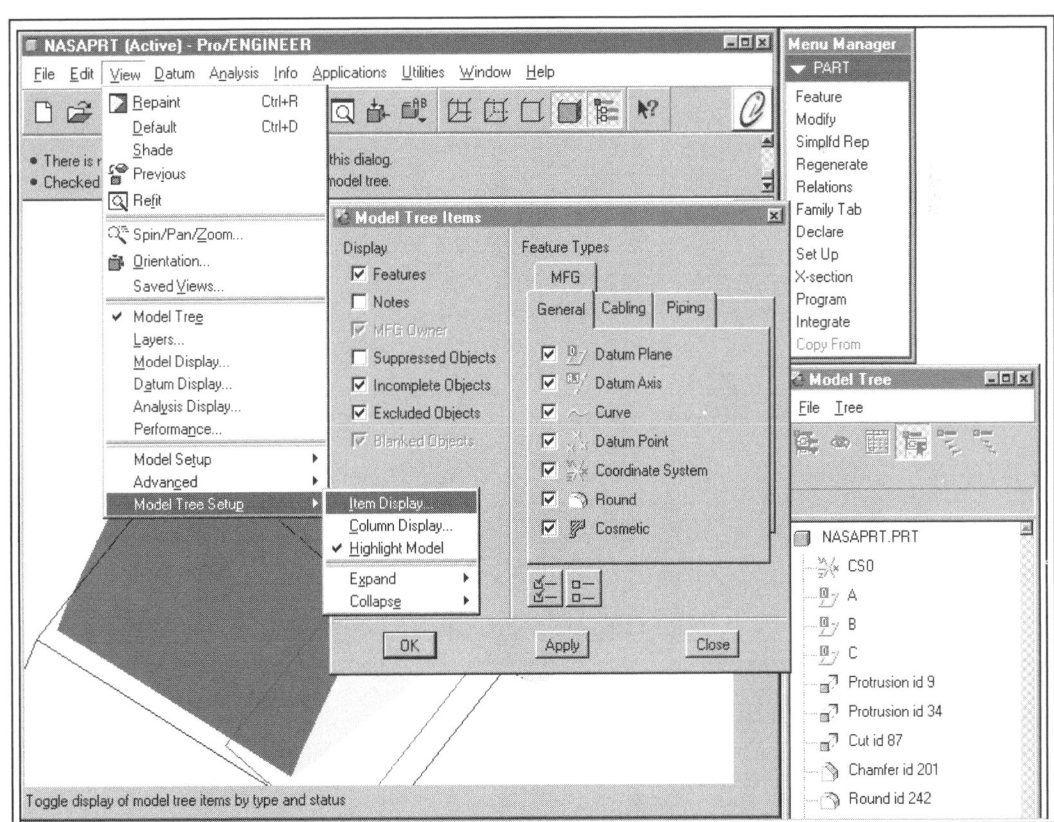

Model Tree Window in Assembly Mode

A graphical, hierarchical representation of the assembly is shown in the Model Tree window (Fig. 9.5). The nodes of the model tree represent the subassemblies, parts, and features that make up an assembly. The model tree can be used as a selection tool, allowing objects to be quickly identified and selected for various component and feature operations. Additionally, different types of information regarding components and features may be displayed in the model tree by toggling a variety of information columns.

Pro/E displays only one Model Tree window at a time--the tree for the active model. You can expand or compress the tree display by double-clicking with the left mouse button (LMB) on the name of the component.

Click and hold down the right mouse button (RMB) on the name or symbol of an entity in the Model Tree window to access a menu containing the following assembly operations:

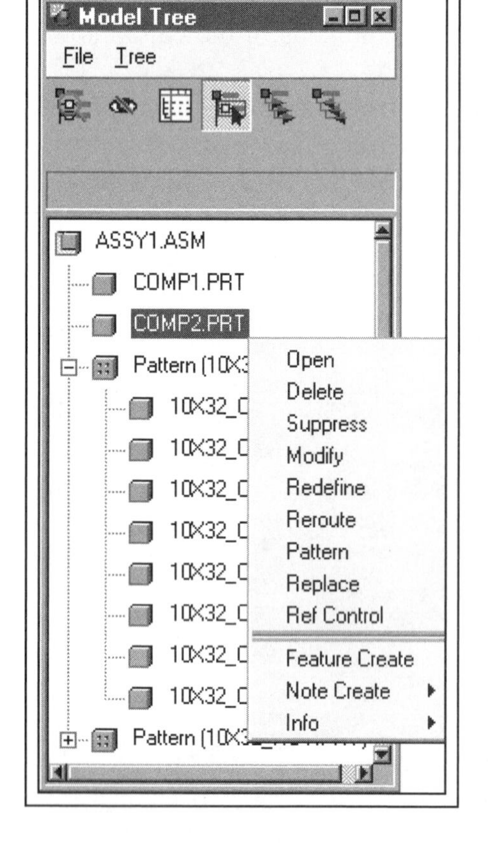

* Modify an assembly or any component in an assembly.
* Redefine component constraints.
* Reroute, delete, suppress, resume, replace, and pattern components.
* Create, assemble, or include a new component.
* Create assembly features.
* Create Notes (see left column).
* Control External References.
* Access model and component information.
* Modify the status of individual components in the current simplified representation.
* Redefine the display status of all components.
* Redefine the display status of individual components.

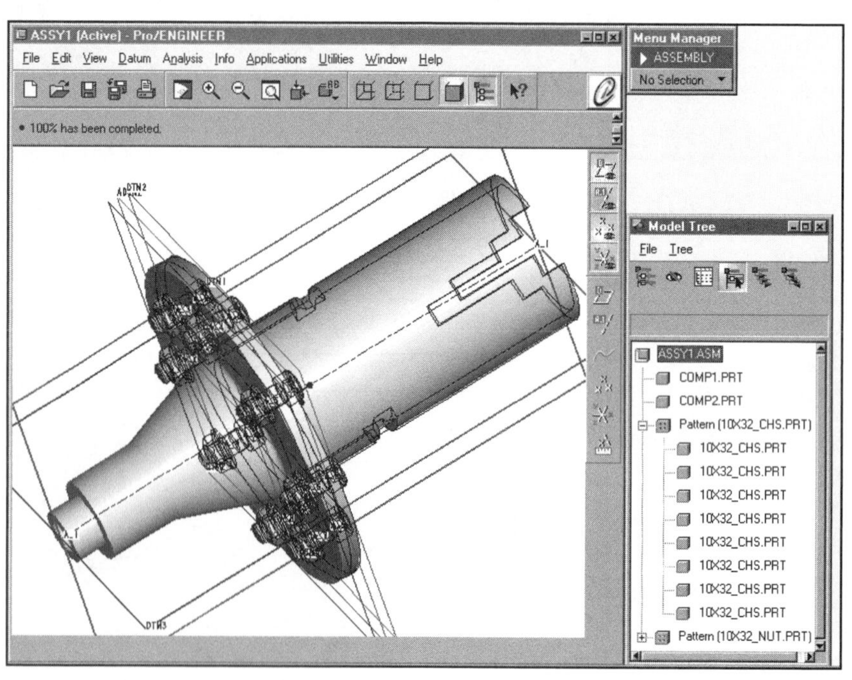

Figure 9.5
Assembly with Model Tree Displayed

Besides the feature or object listing in the model tree, a variety of information columns can also be displayed. Information columns can be added or removed and their format altered. The format can also be saved for later use. A Model Tree, in Assembly Mode, is shown in Figure 9.6. In Figure 9.7, columns for information about the Assembly model have been enabled by choosing **View ⇒ Model Tree Setup ⇒ Column Display ⇒** (select the desired option and then >>) **⇒ OK.**

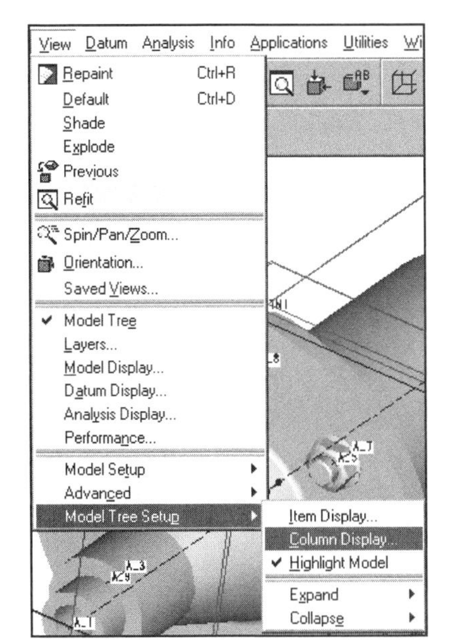

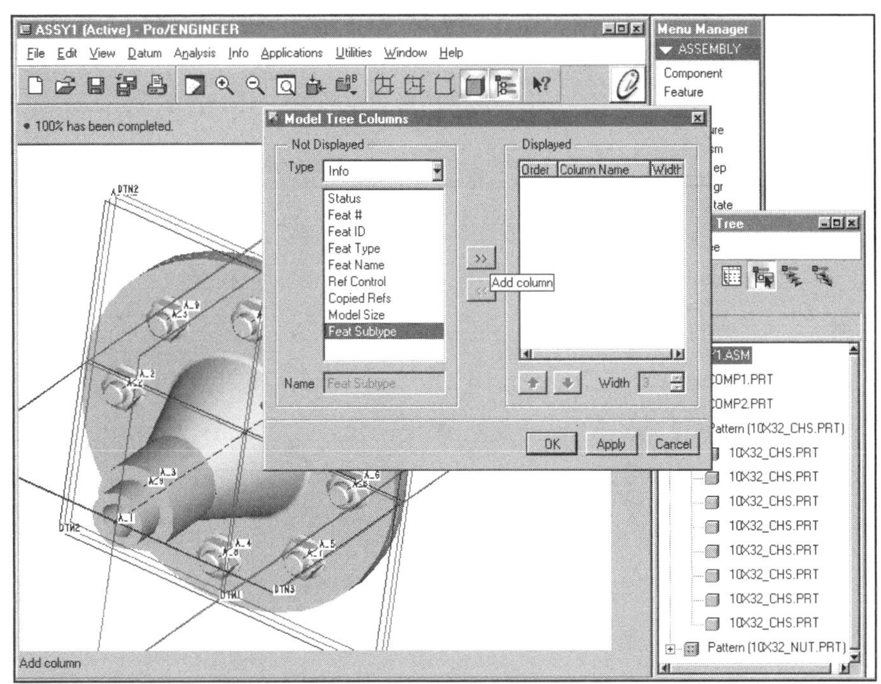

Figure 9.6
Adding Columns to the Model Tree in Assembly Mode

You can also alter the selections using **Tree ⇒ Column Display** from the Model Tree menu (see left-hand column).

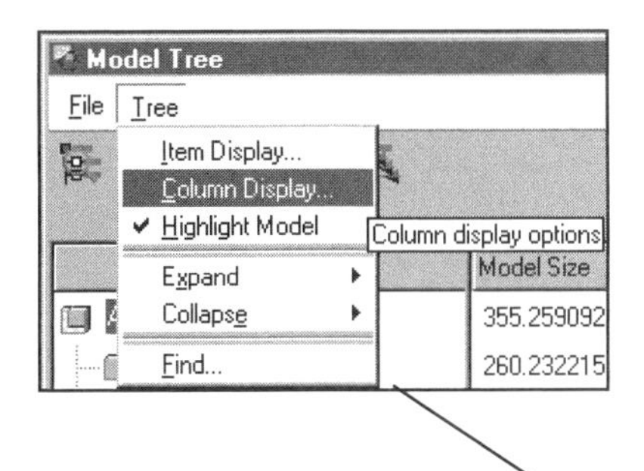

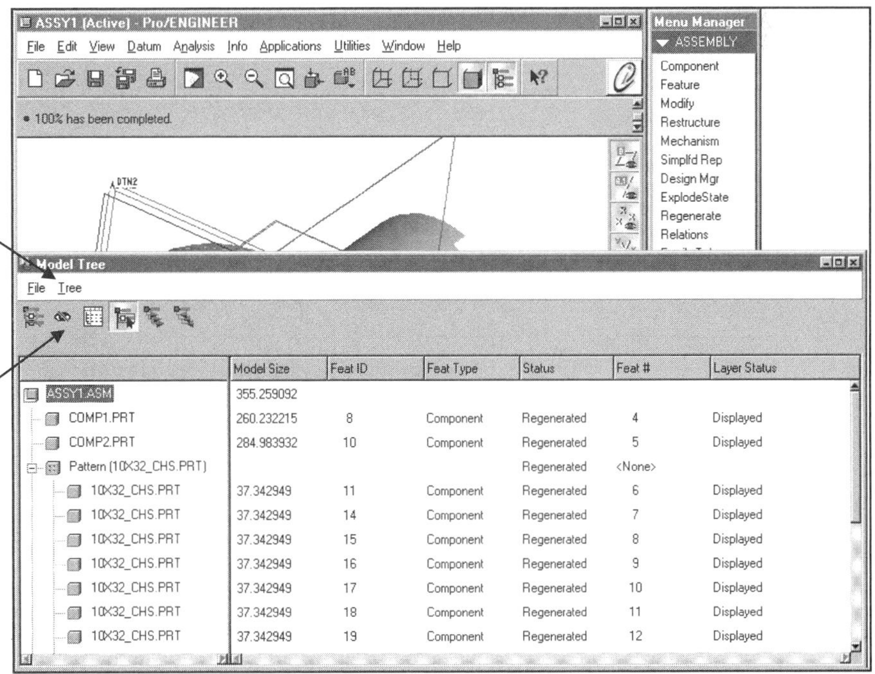

	Model Size	Feat ID	Feat Type	Status	Feat #	Layer Status
ASSY1.ASM	355.259092					
COMP1.PRT	260.232215	8	Component	Regenerated	4	Displayed
COMP2.PRT	284.983932	10	Component	Regenerated	5	Displayed
Pattern (10X32_CHS.PRT)				Regenerated	<None>	
10X32_CHS.PRT	37.342949	11	Component	Regenerated	6	Displayed
10X32_CHS.PRT	37.342949	14	Component	Regenerated	7	Displayed
10X32_CHS.PRT	37.342949	15	Component	Regenerated	8	Displayed
10X32_CHS.PRT	37.342949	16	Component	Regenerated	9	Displayed
10X32_CHS.PRT	37.342949	17	Component	Regenerated	10	Displayed
10X32_CHS.PRT	37.342949	18	Component	Regenerated	11	Displayed
10X32_CHS.PRT	37.342949	19	Component	Regenerated	12	Displayed

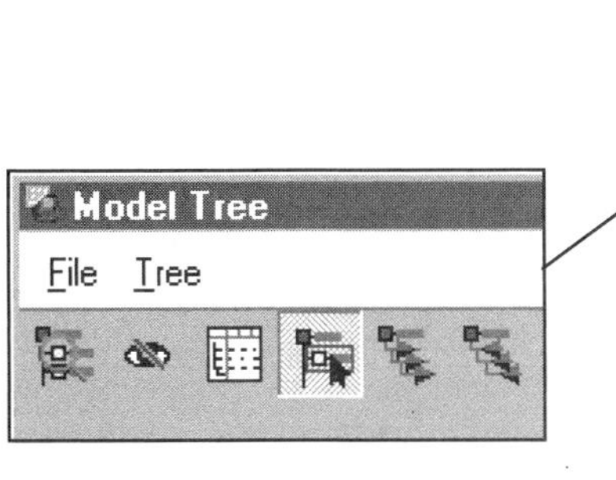

Figure 9.7
Columns Added to the Model Tree in Assembly Mode

A variety of information columns can also be displayed for the Part Mode. A Model Tree, in Part Mode, is shown in Figure 9.8. In Figure 9.9, columns for information about the part model have been enabled by choosing **View ⇒ Model Tree Setup ⇒ Column Display ⇒** (select the desired option and then >>) **⇒ OK.**

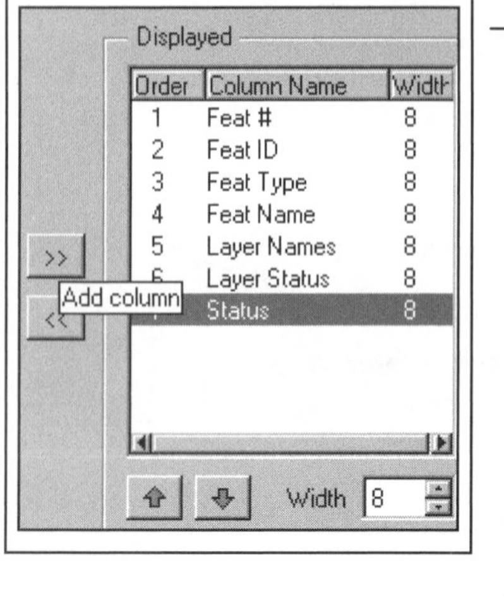

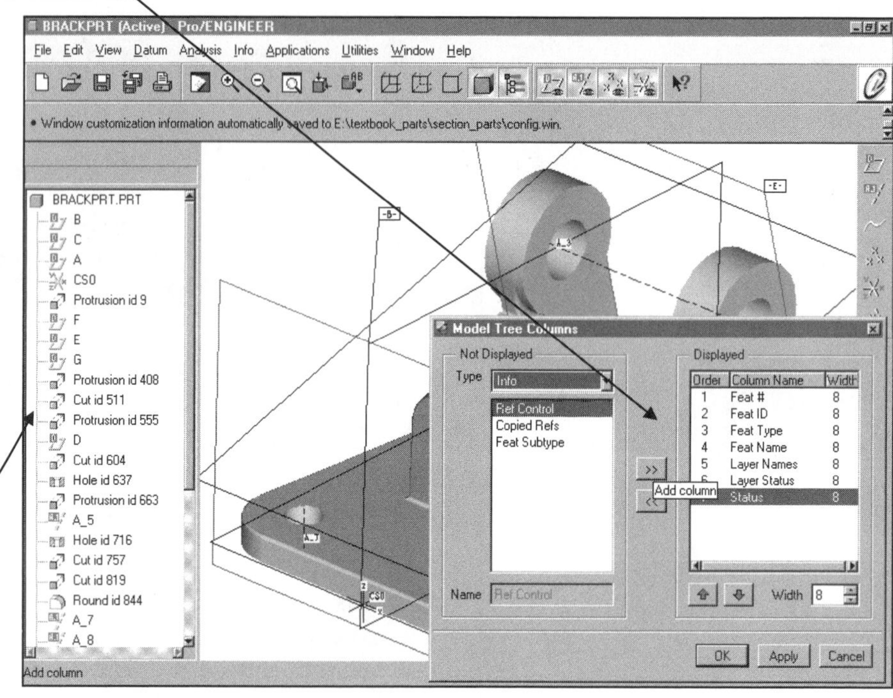

Model Tree is in the embedded, default position (option)

Figure 9.8
Adding Columns to Model Tree in Part Mode

Slide the divider bar to the right to see the Model Tree columns. You can also move the column dividers to see the Model Tree as shown below.

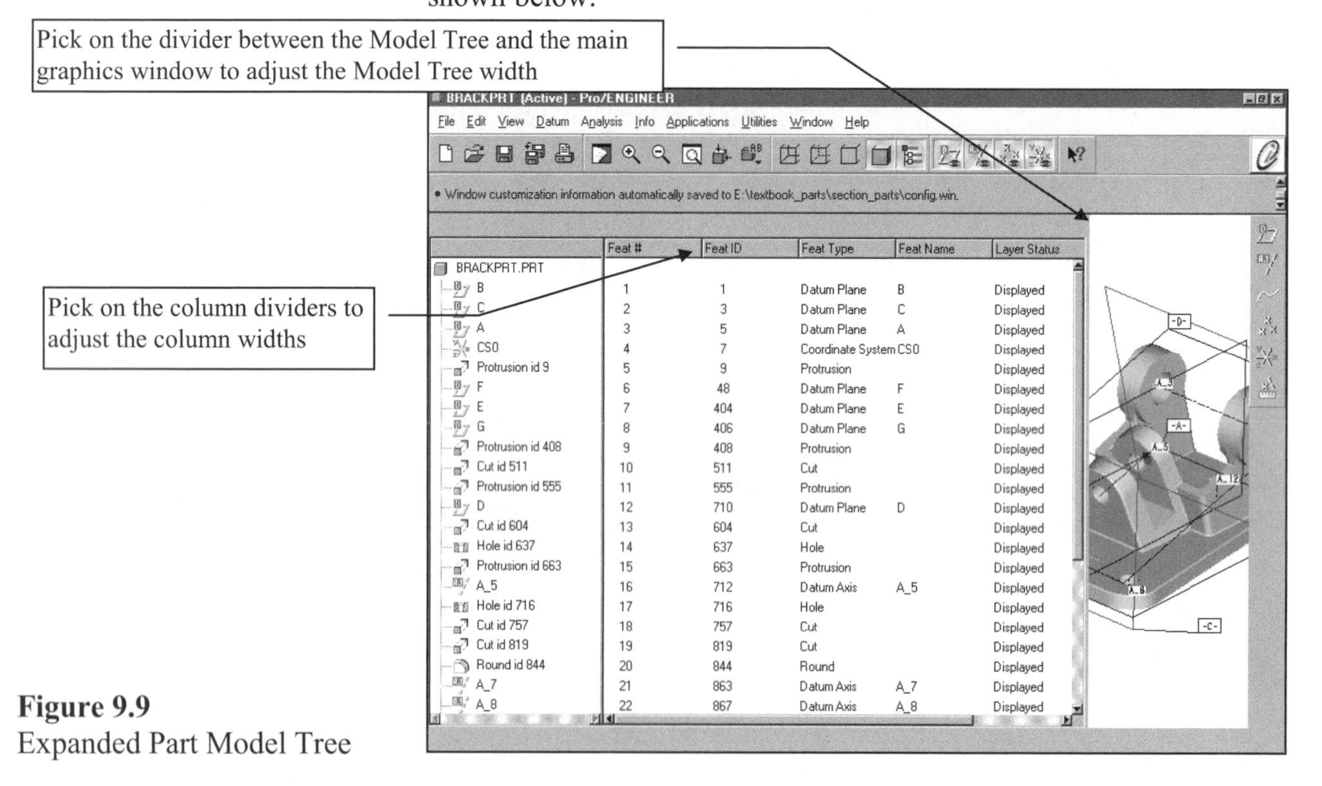

Figure 9.9
Expanded Part Model Tree

If the Model Tree is floating as in Figure 9.10, the columns can also be altered by picking **Column Display Options** button (as shown in the left-hand column) and changing the width of each category (Fig. 9.10). To save a Model Tree set up to your present working directory, choose **File ⇒ Save Settings ⇒** (enter a name) **⇒ Save.** To use a saved Model Tree, choose **File ⇒ Load Settings ⇒** (select the name) **⇒ Open.**

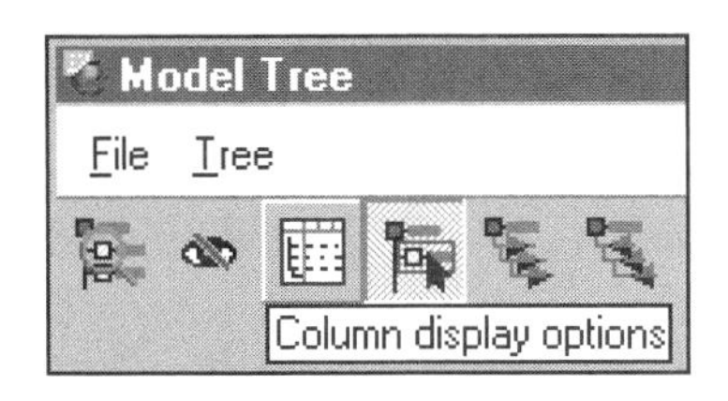

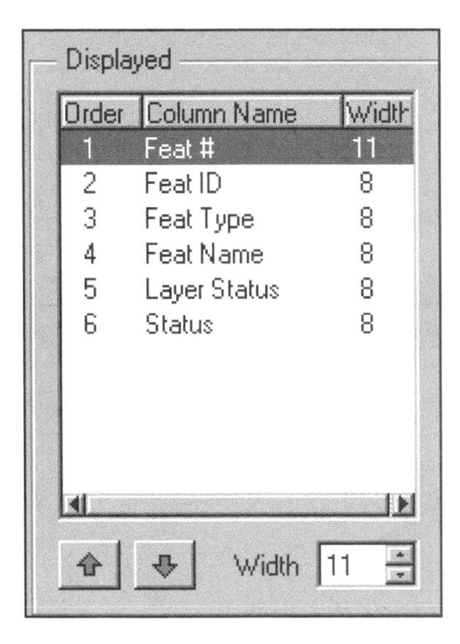

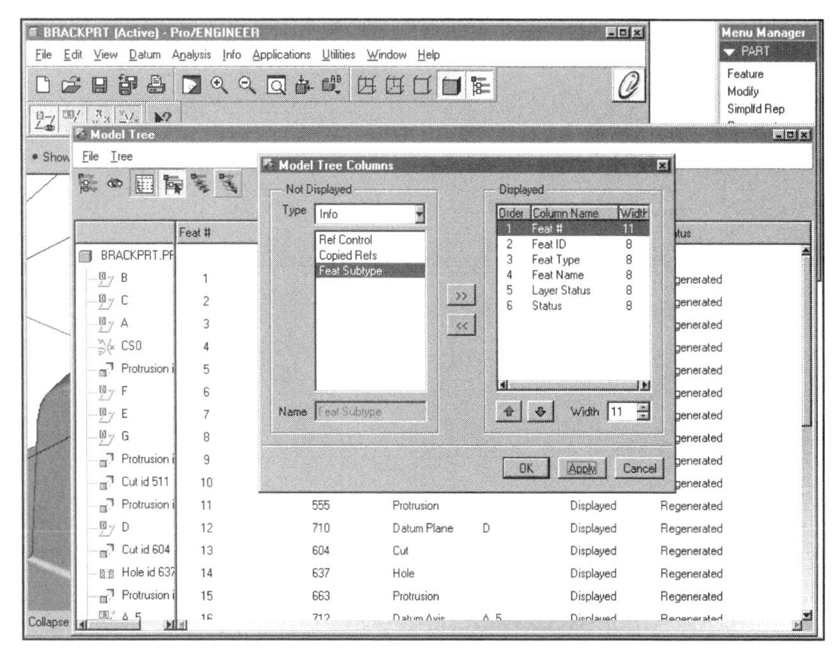

Figure 9.10
Changing the Column Width

The Model Tree window is interactive, so you can either select objects directly from it or select **Sel By Menu** from the GET SELECT menu and then pick a component or feature from a menu list.

Figure 9.11 is an example of selecting a *subassembly* from the assembly Model Tree. The **WRIST.ASM** has been selected and is shown highlighted.

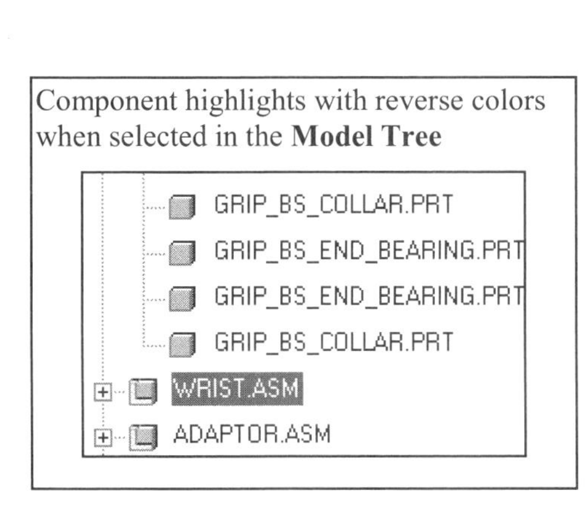

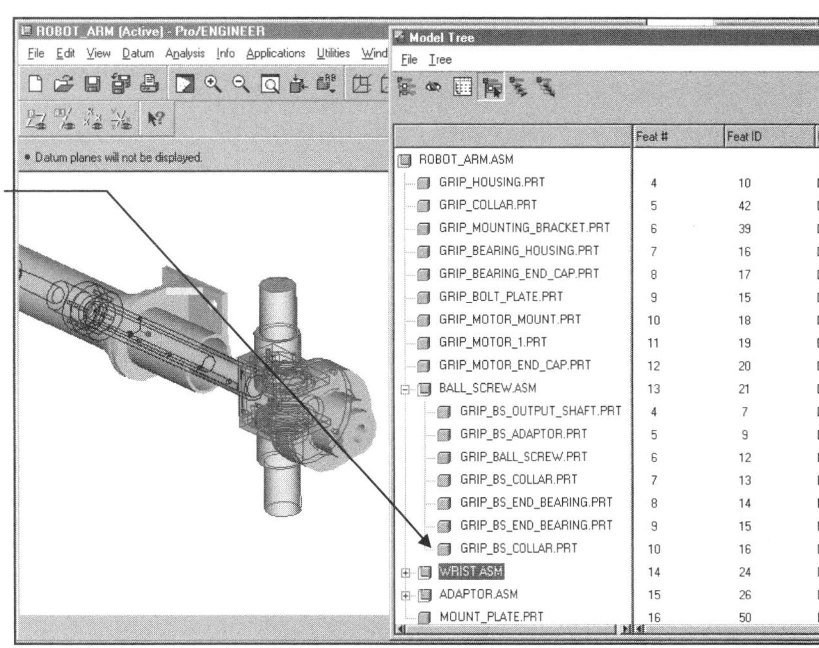

Figure 9.11
Selecting Model Tree Items

NOTE

Choose the feature from the Model Tree.

Highlight and select from the cascading choices.

The Model Tree is used to quickly access a variety of commands. Commands such as **Modify, Redefine, Reroute, Delete, Suppress, Replace, Feature Create, Pattern, Note Create, Ref Control, Open,** and **Info** can be chosen directly from the Model Tree. The feature is selected and the command chosen without choosing from the menu structure (Menu Manager) and without selecting the feature from the working window. The **Shell** feature (Fig. 9.12), is selected from the Model Tree with the *left mouse button*. The **Info** command and the **Feat Info** option were then highlighted and selected with the *right mouse button*. In Assembly mode (Fig. 9.13), the Model Tree can be expanded to include the selected component's features. In the assembly Model Tree you can also **Open** a part (Fig. 9.13).

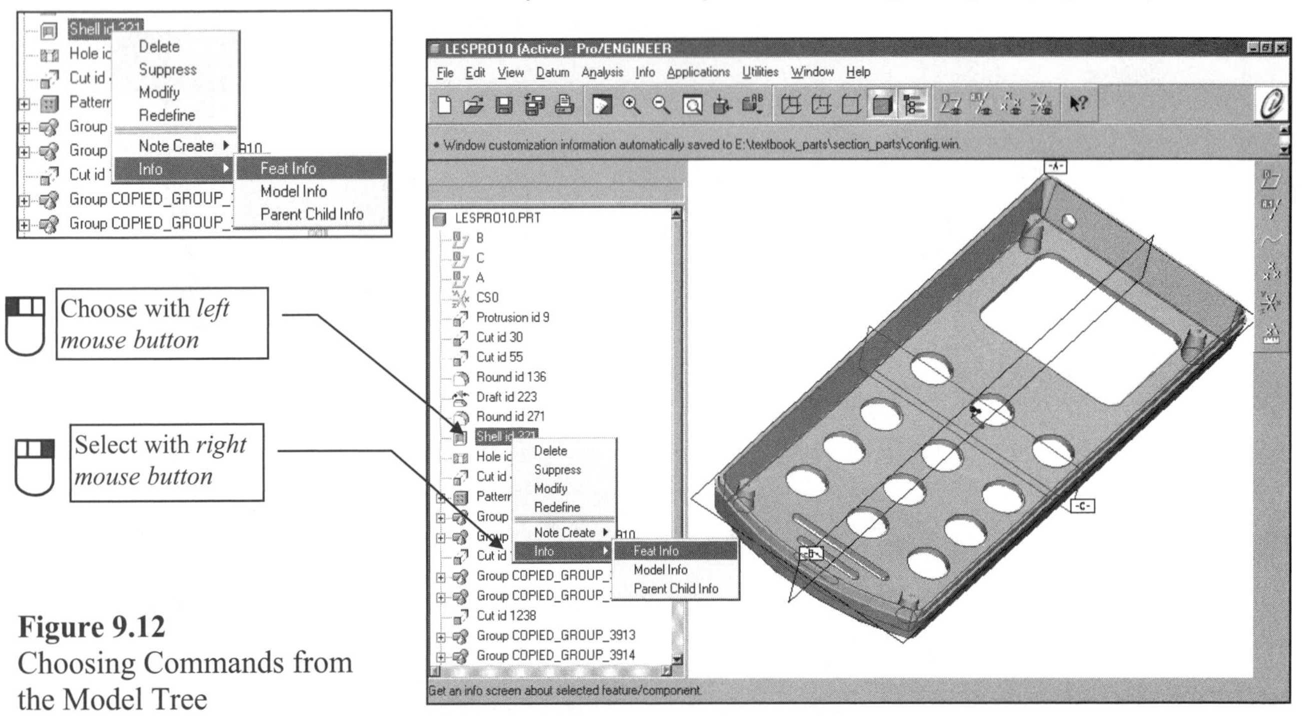

Figure 9.12
Choosing Commands from the Model Tree

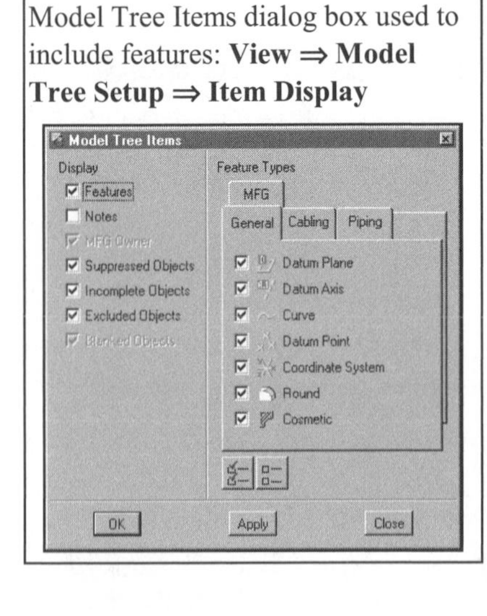

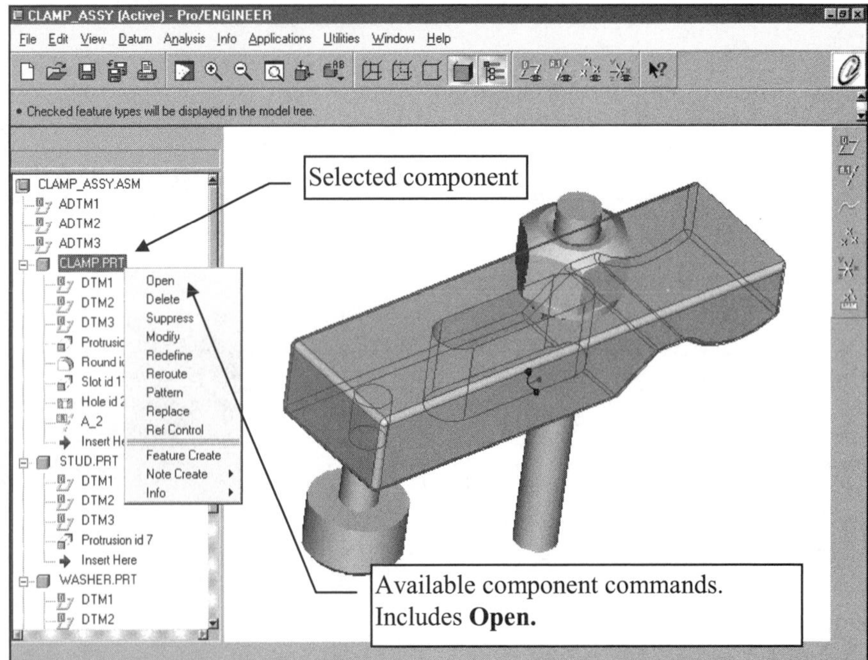

Figure 9.13
Selecting Items

Information from the Model Tree

As has been stated, information can be accessed directly from the Model Tree. Select a feature or part (in Assembly mode) in the Model Tree as shown in Figure 9.14, where the first feature is highlighted. Press the right mouse button (RMB) and slide the cursor down to **Info** and then highlight one of the two choices displayed, **Feat Info** or **Model Info**. The part's feature information is then displayed in a separate window (Fig. 9.15).

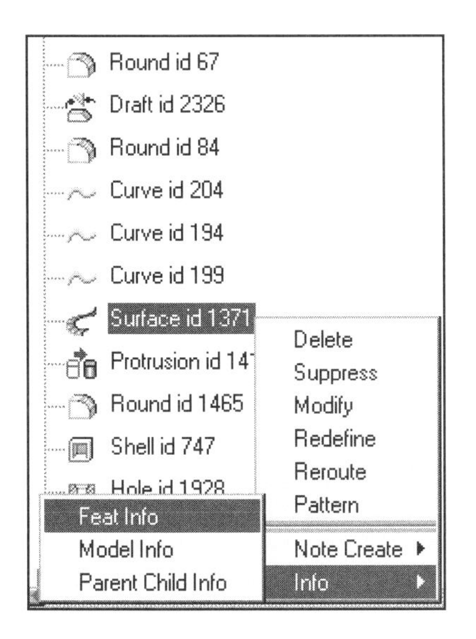

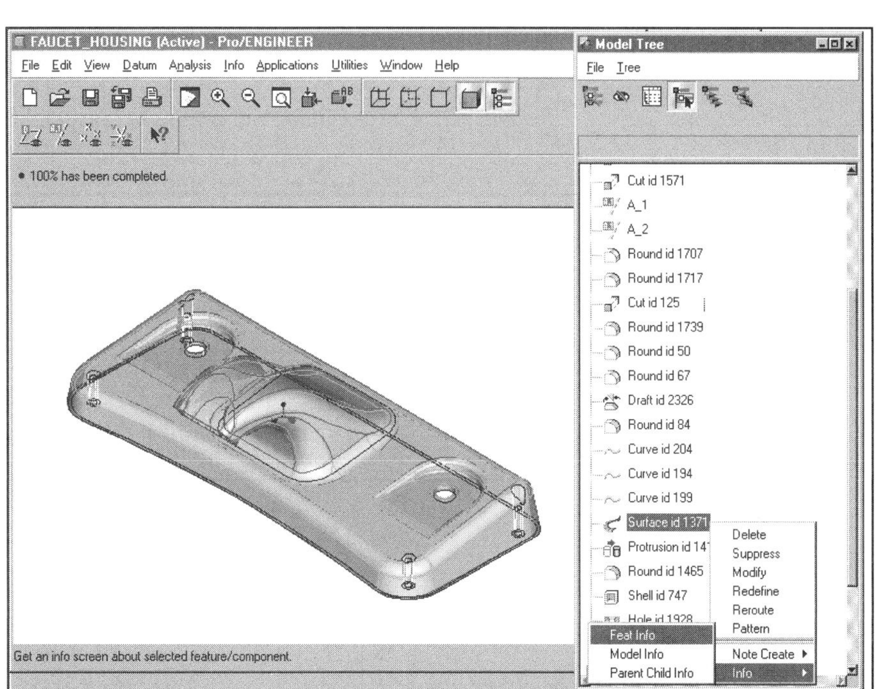

Figure 9.14
Information Accessed from the Model Tree

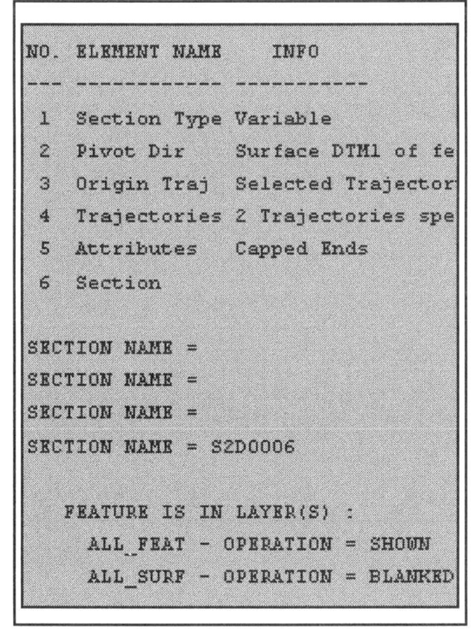

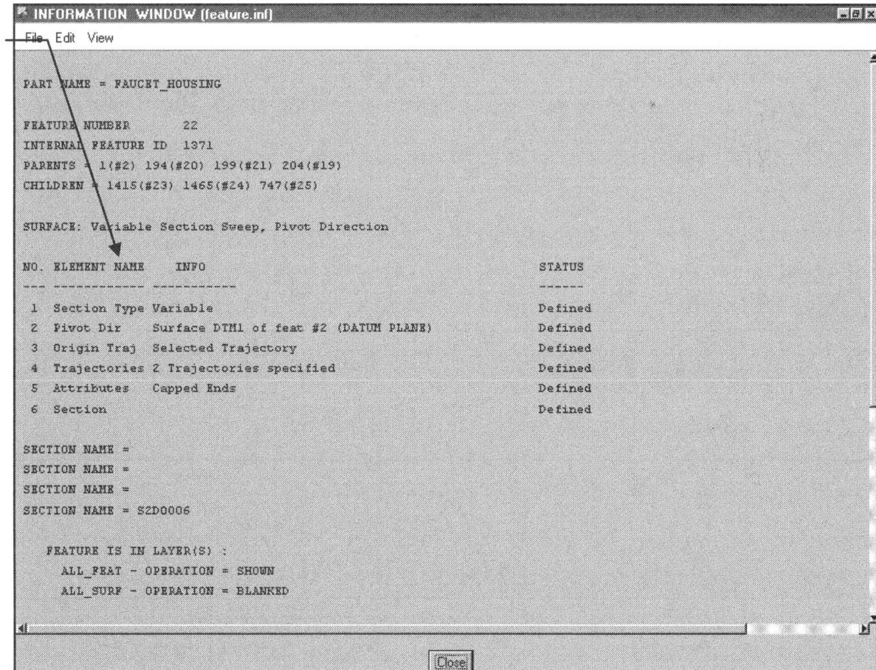

Figure 9.15
Feature Information

In Figure 9.16 the Model Tree was used to access information about an assembly.

Figure 9.16 Assembly Information from the Model Tree

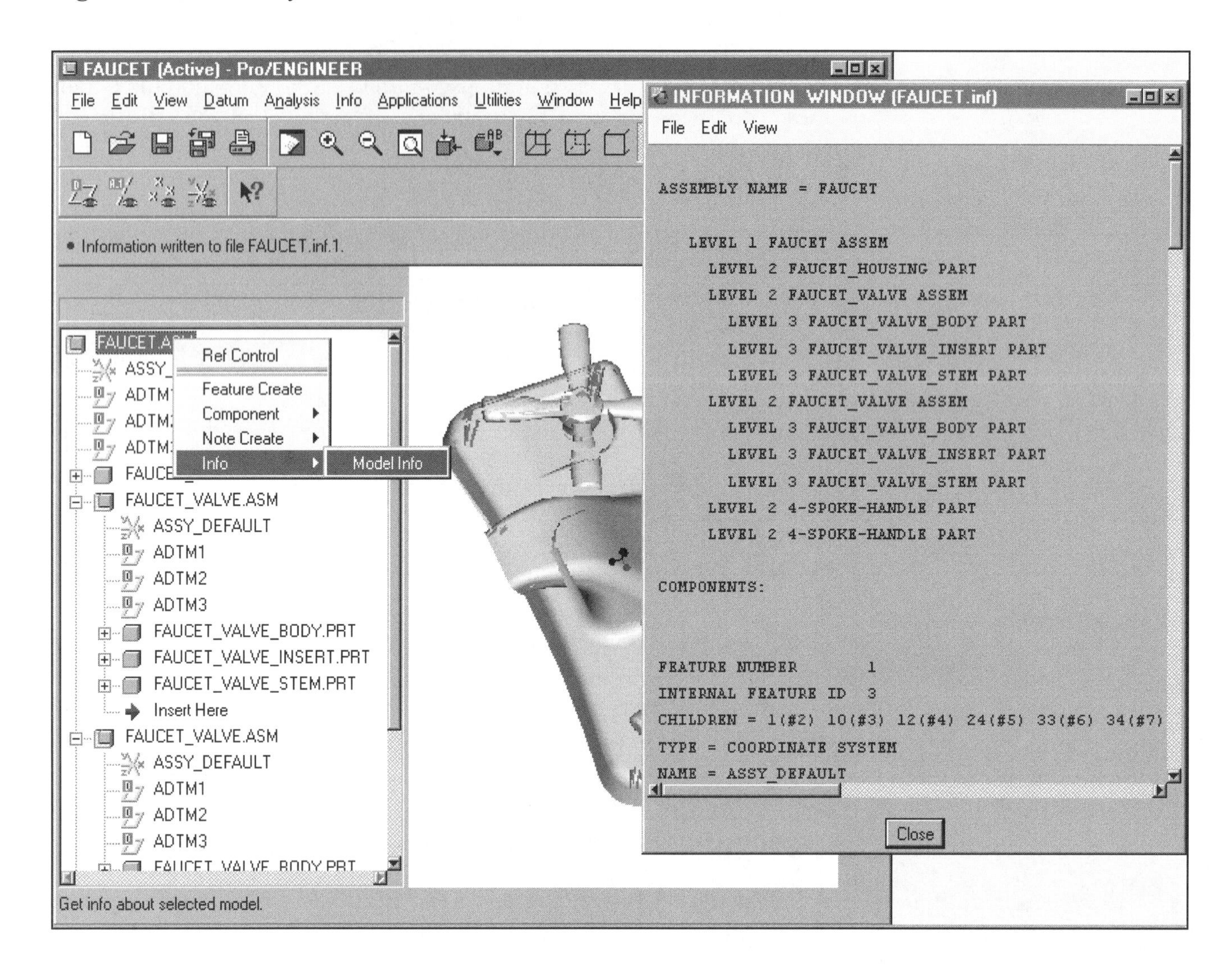

Section 10

Help

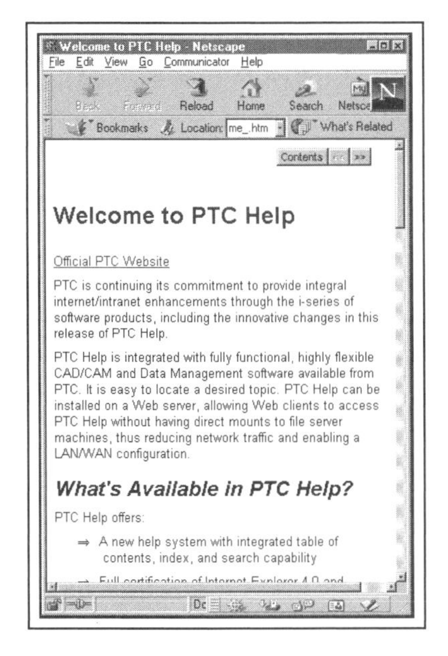

The **Help** menu is used to gain access to online information and customer service information about Pro/E. The following commands are available on the Pro/E Help menu:

Pro/Help System Displays the contents of Welcome to the PTC Help System in a browser window.
Pro/Help Online Books Displays contents of Welcome to release 2000i Pro/Help.
Customer Services Info Displays product information, including the release level, license information, installation date, and customer support contact information.
Round Tutor Makes available or unavailable the Pro/E Round Feature Tutor.
i-Site Preferences Pro/E connects you to related information, resources, and knowledge from PTC via the Internet.
About Pro/ENGINEER Displays Pro/E copyright information.
What's This? Makes context-sensitive Help mode available.

Starting Pro/HELP

Use Pro/HELP to gain access to the Pro/E online documentation in a browser window. You must have Pro/HELP installed to use this feature.

Start Pro/HELP by choosing **Pro/E Help System** from the Help pull-down menu (Fig. 10.1). When Pro/HELP starts, a browser appears, displaying the Welcome to the PTC Help System home page. For information on how to use Pro/HELP, click the About button in the browser window.

Figure 10.1
Toolbar Help Menu

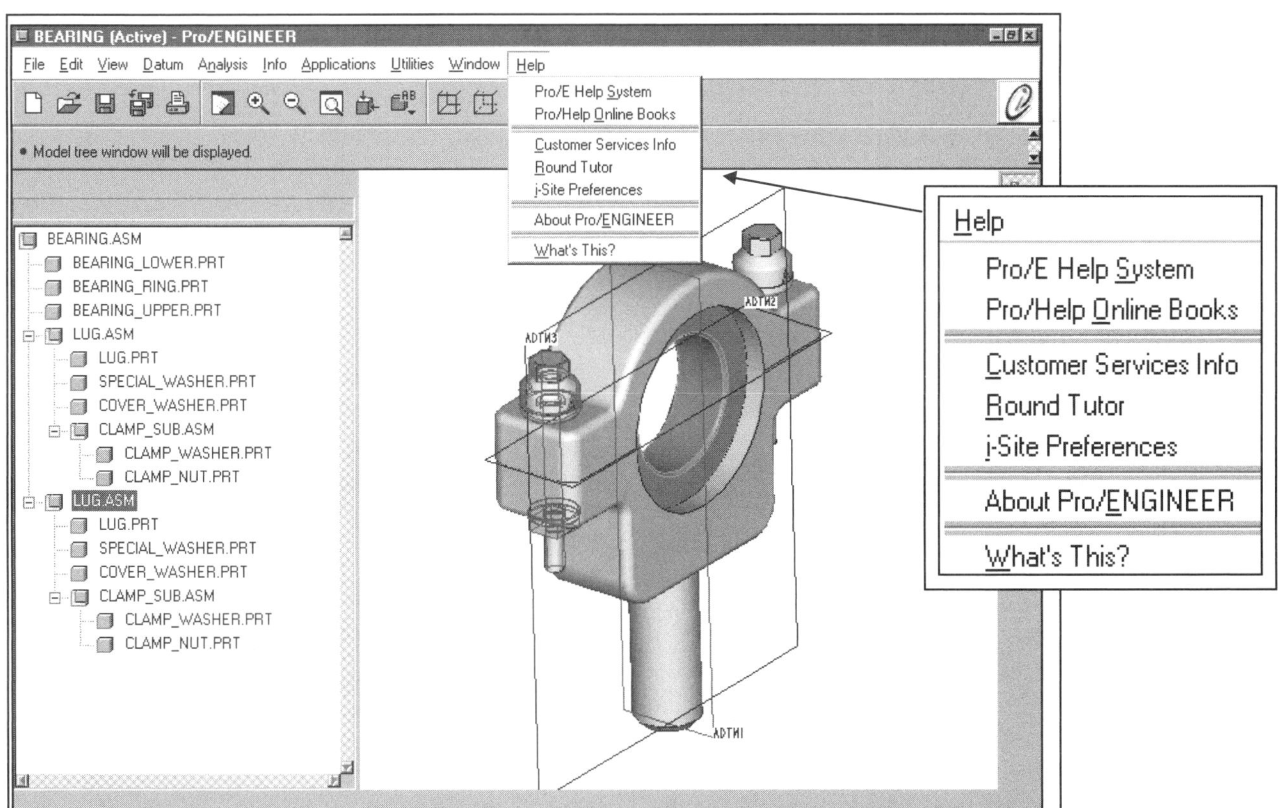

Using Online Help (Online Documentation)

When you scroll the cursor up and down over the menu items, a one-line description of the menu item that is lightly highlighted (not picked) appears at the bottom of the main working window. If more detailed help is needed you can access **online documentation.**

In Figure 10.2 (Assembly mode), the option **Modify** was highlighted by passing the cursor over the menu item and then by holding down and clicking the *right* mouse button; **? Get Help** then appears (see left column). Release the right mouse button to see the online documentation for that menu item. Scrolled pages of the **Assembly Modeling User's Guide** are brought to the screen in a separate browser window. You can scroll to other pages, or go to a table of contents or index, or do a search by scrolling to the buttons at the very top or bottom of the online documentation.

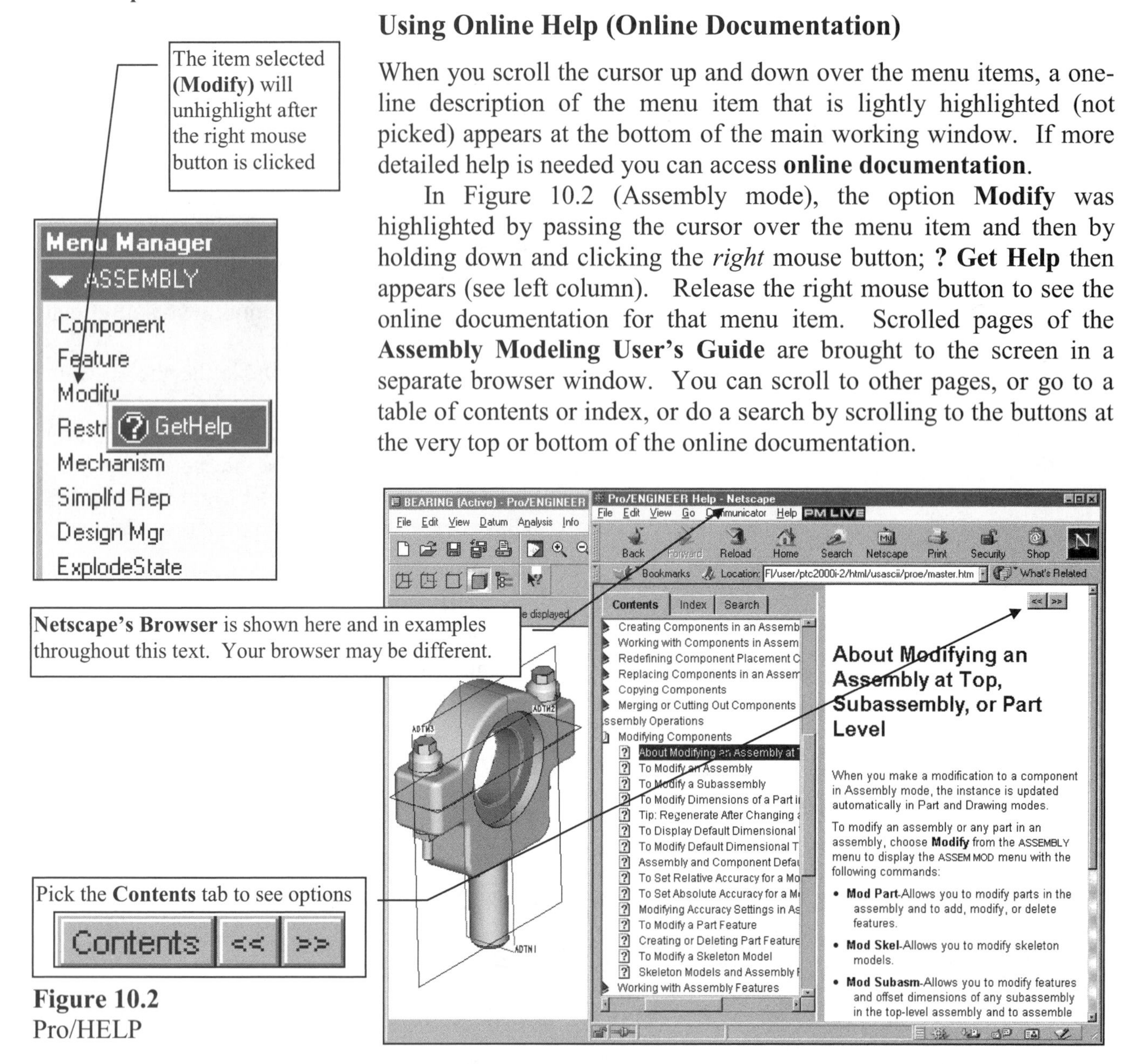

Figure 10.2
Pro/HELP

The Pro/E Help System is a tool that allows you to view the Pro/E user guides and support documentation on your computer screen. When installed and configured on your system, Pro/HELP opens to the appropriate document and page to provide information and help whenever the *right* mouse button is clicked on a menu item.

With this access to the various user guides and libraries installed on your system, you have the ability to:

Browse User's Guide documentation
Establish bookmarks
Search for keywords
Print pages (Fig. 10.3)

The Print dialog box is activated when you choose **Print,** found along the top of your browser (Netscape's Menu in this text's examples), as shown in Figure 10.3.

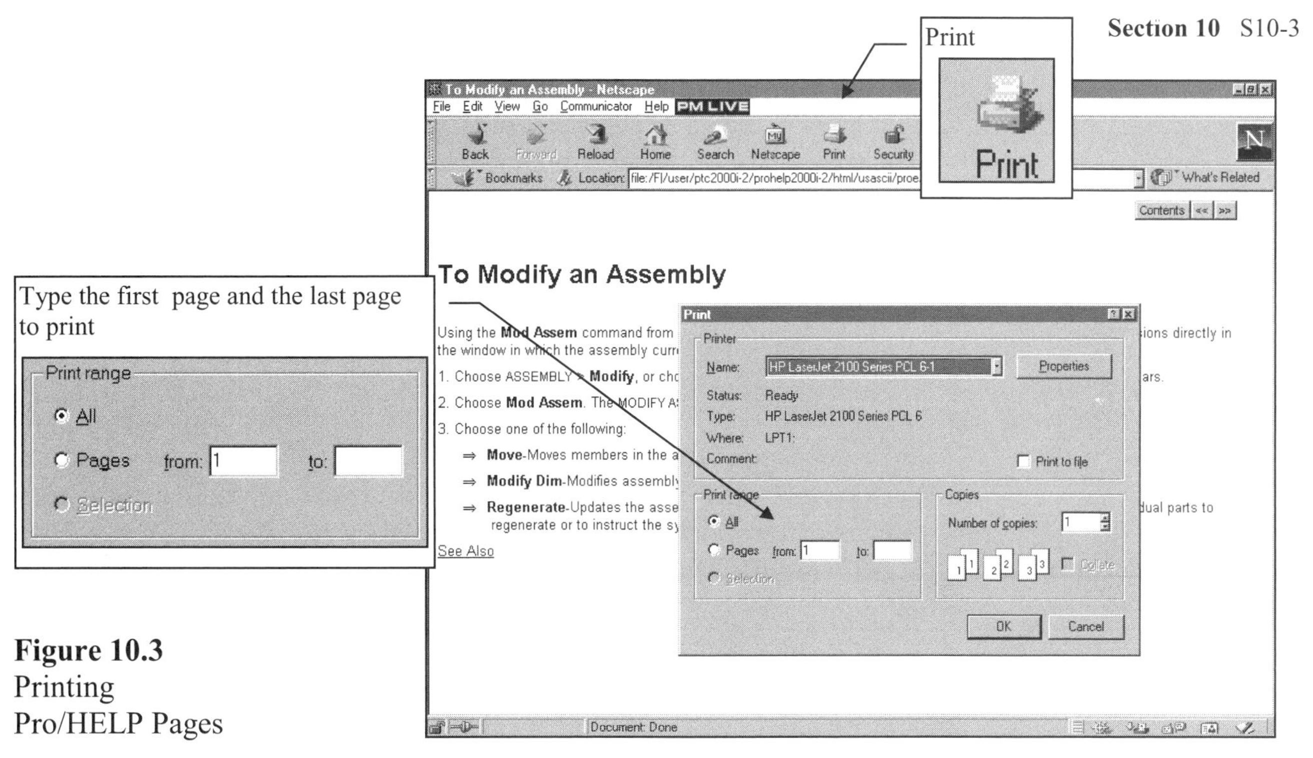

Figure 10.3
Printing
Pro/HELP Pages

Besides providing help, Pro/HELP can access any library that you have installed on your system. In Figure 10.4, the Tooling library has been accessed.

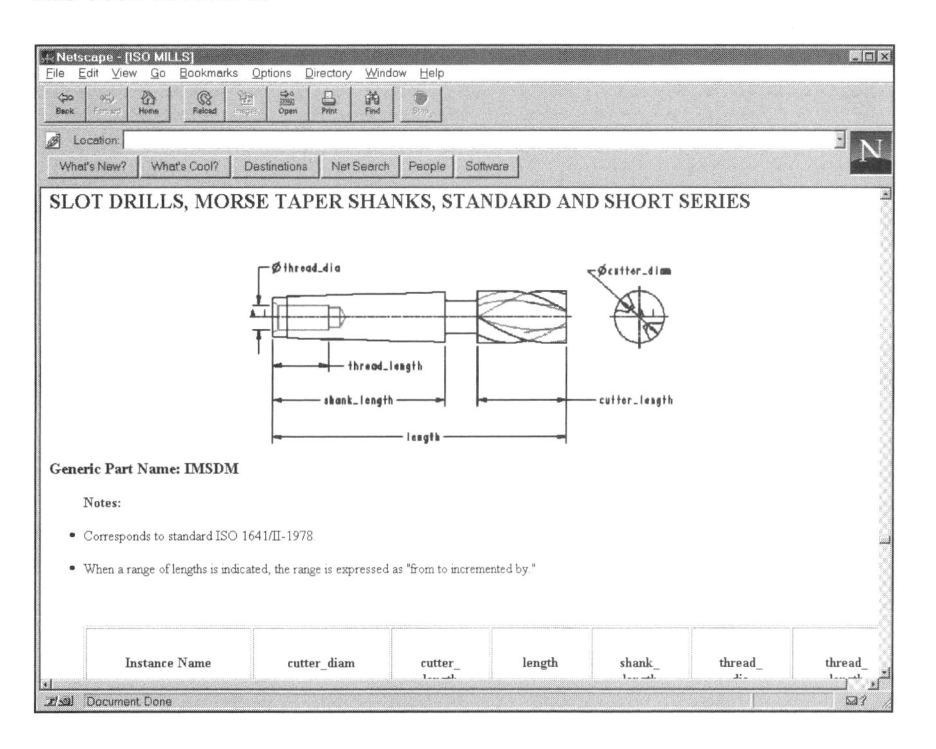

Figure 10.4
Pro/HELP
Tooling Library

Throughout the text, you will be prompted to access help for the menu items that you will be learning in each lesson. There is no way that this text can cover all the commands and capabilities of Pro/E. ***You must use part of every session looking up information on commands and discovering the wide range of options available in Pro/E.*** It is the only way to expand your understanding of Pro/E and become a more competent user.

What's This ? (Help)

You can display context-sensitive help for each top-level menu item and a variety of other areas by choosing **Help** from the menu bar and then selecting **What's This?**

What's This? Icon found on the Toolbar

1. Choose **Help ⇒ What's This?** (Fig. 10.5). The cursor changes to a question mark. You can also click the icon for What's This?
2. Move the pointer over the item or area about which you want help.
3. Click the left mouse button (LMB). Help appears in the (Netscape) browser window (Fig. 10.6).

In Figure 10.5, the pick was made in the Main Window. Online Documentation was displayed in a browser window describing *To Pre-Select Objects in the Graphics Window* (Fig. 10.6).

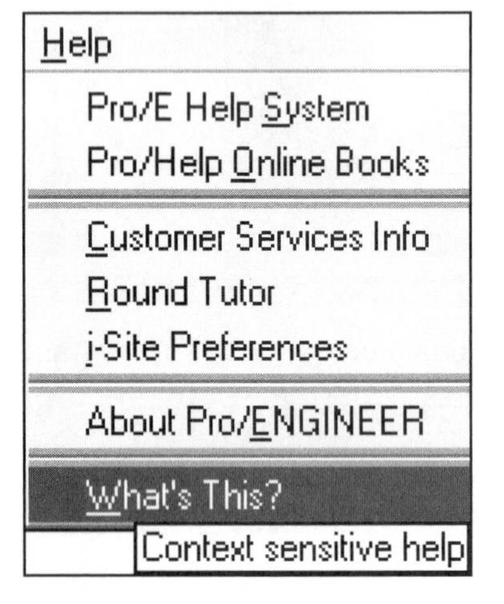

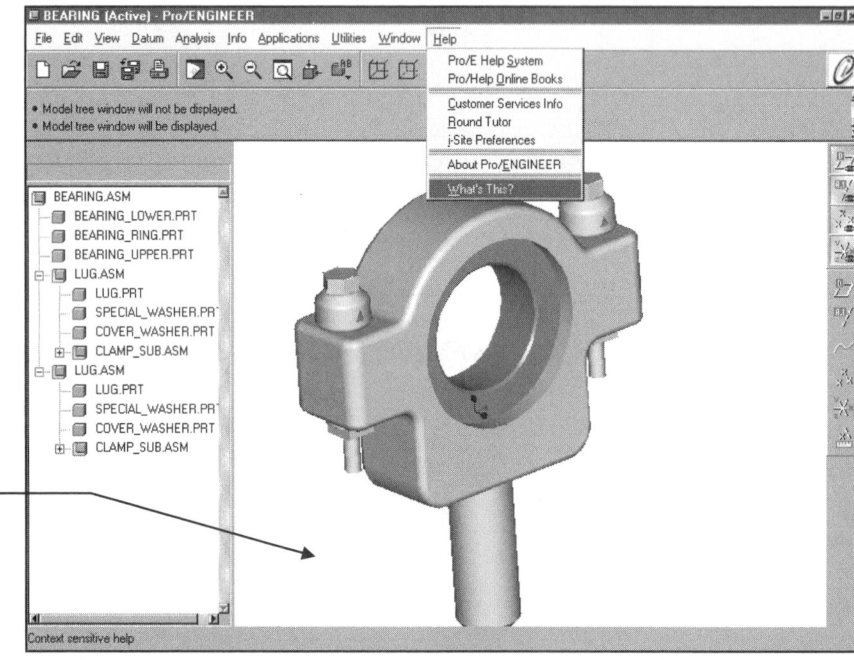

? Pick was made somewhere inside the Main Window

Figure 10.5
What's This? Help

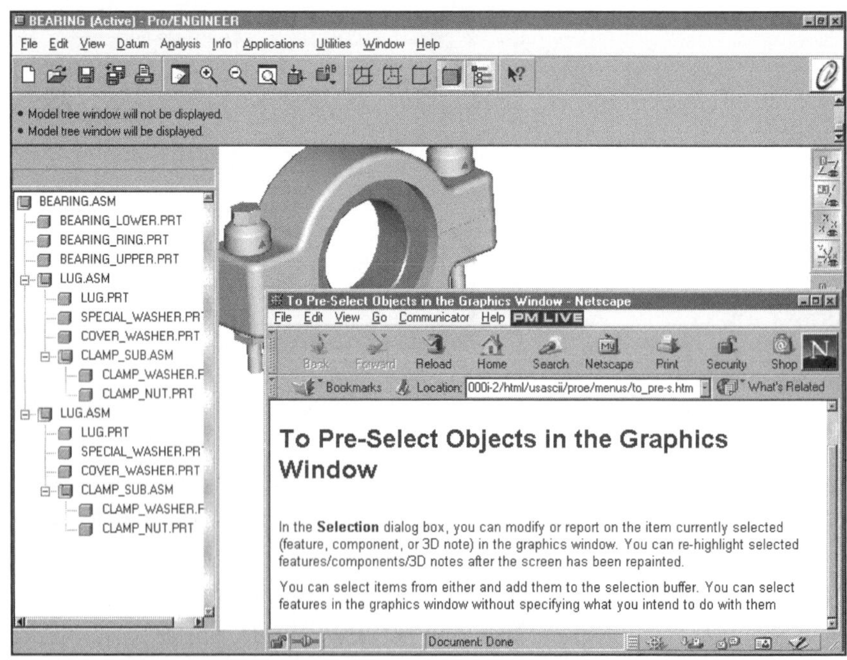

Figure 10.6
Online Documentation

Section 11

Intent Manager

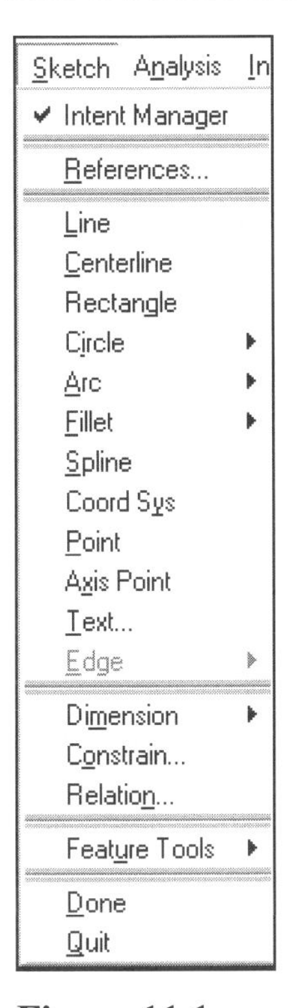

Figure 11.1 SKETCHER ENHANCEMENT - INTENT MANAGER

You can sketch geometry in two distinct ways: with the **Intent Manager** or the traditional Pro/E **Sketcher.**

The most important difference between the two methods of sketching is the fact that the sketch is always resolved (regenerated) when the ✔**Intent Manager** is on (default).

Every time you start Pro/E and enter Sketcher for the first time, a window appears announcing **Intent Manager** (Fig. 11.1). (If you would like Pro/E to stop displaying this window, add the following line to your *config.pro:sketcher_overview_alert no).* Selecting the **Overview** button will provide a description of the **Intent Manager** and its capabilities.

When you are in the Sketcher, the Utilities menu will give an option called Sketcher Preferences (see left-hand column).

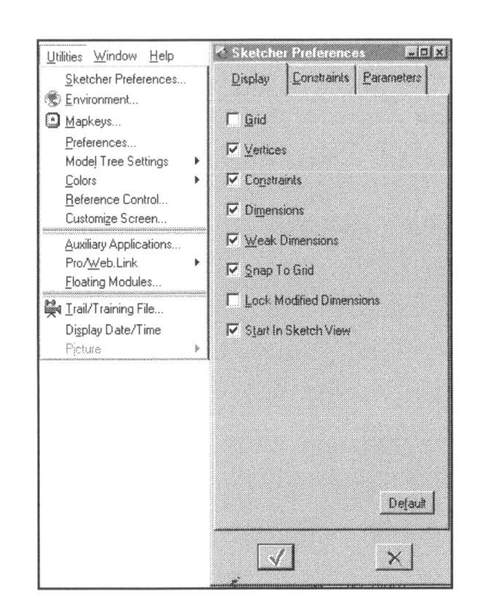

When you sketch curves (geometry) with the ✔ **Intent Manager** on, they are automatically dimensioned and constrained. In other words, after you draw a horizontal line, a dimension for its length is automatically created and it is assigned the horizontal constraint (**H**). You can also assign and remove geometric constraints to curves in the sketch. For example, you can specify that any number of radii are equal or that lines are either *collinear* or *of equal length.*

There is no **Mouse Sketch** option available when you are using the **Intent Manager.** The default option for sketching is **Line.** The **Line** option allows you to draw a line between two points or tangent to two curves. As each line is sketched, the constraints (e.g. **V**, **H**) are automatically displayed. When the sketch is completed, the **Intent Manager** resolves the sketch, automatically dimensions the geometry, and successfully regenerates the sketch.

It is important to investigate the Intent manager using Pro/E's online documentation:

Help ⇒ Pro/E Help System ⇒ Contents tab **⇒ -Pro/ENGINEER Foundation ⇒ -Using Sketcher ⇒ -Sketcher with Intent Manager ⇒ ?About Sketcher Mode with Improved Intent Manager** as in Figure 11.2

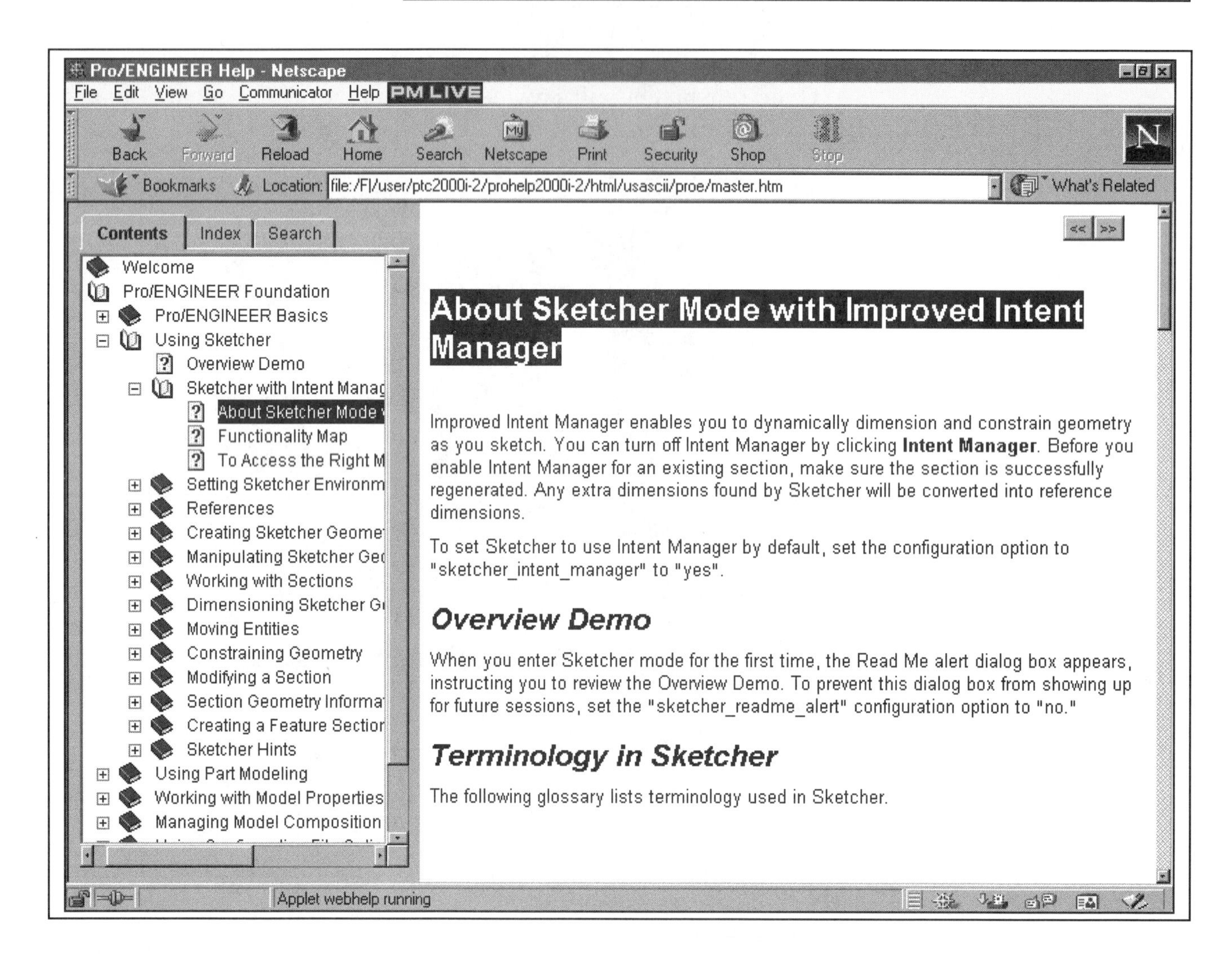

Figure 11.2
Online Documentation for Intent Manager

Sketcher Mode Using Intent Manager

The Intent Manager enables you to dimension and constrain geometry dynamically as you sketch. You can turn off Intent Manager by clicking Intent Manager from the Sketch pull-down menu. Before you enable Intent Manager for an existing section, make sure the section is successfully regenerated. Any extra dimensions found by Sketcher will be converted into reference dimensions.

To set Sketcher to use Intent Manager by default, set the configuration option *"sketcher_intent_manager"* to *"yes."*

Terminology in Sketcher

The following glossary lists terminology used in Sketcher.

Entity Any element of the section geometry (such as line, arc, circle, spline, conic, point, or coordinate system). You create entities when you sketch, divide, or intersect the section geometry or when you reference geometry outside the section.

Reference entity An entity of the section that is created in 3-D Sketcher when you reference geometry outside the section. The referenced geometry (for example, part edge) is "known" to Sketcher. For example, creating a dimension to a part edge creates a reference entity in the section which is the projection of that part edge onto the sketching plane.

Dimension A measurement of an entity or a relationship among entities.

Constraint A condition defining the geometry of the entity or a relationship among entities. A constraint symbol appears next to the entity to which the constraint is applied. For example, you can constrain two lines to be parallel. A parallel constraint symbol appears to indicate this.

Parameter An auxiliary numeric value in Sketcher.

Relation An equation relating dimensions and/or parameters. For example, a relation can be used to set the length of one line to be half the length of some other line.

Weak dimension or **constraint** A dimension or constraint is called "weak" if Sketcher can remove it when appropriate without any confirmation from the user. Dimensions created by Sketcher are weak. When you add a dimension, Sketcher can remove an extra weak dimension or constraint without any confirmation. Weak dimensions and constraints appear in *gray.*

Strong dimension or **constraint** A dimension or constraint is called "strong" if Sketcher cannot delete it automatically. Dimensions and constraints created by the user are always strong. If several strong dimensions or constraints are in conflict, Sketcher asks you to remove one. Strong dimensions and constraints appear in *yellow.*

Conflict Contradicting or redundant conditions of two or more strong dimensions or constraints. When this occurs, the conflict must be resolved immediately by removing an undesired constraint or dimension.

Using the Toolbar Icons

After you enter Sketcher, its toolbar displays the option icons:

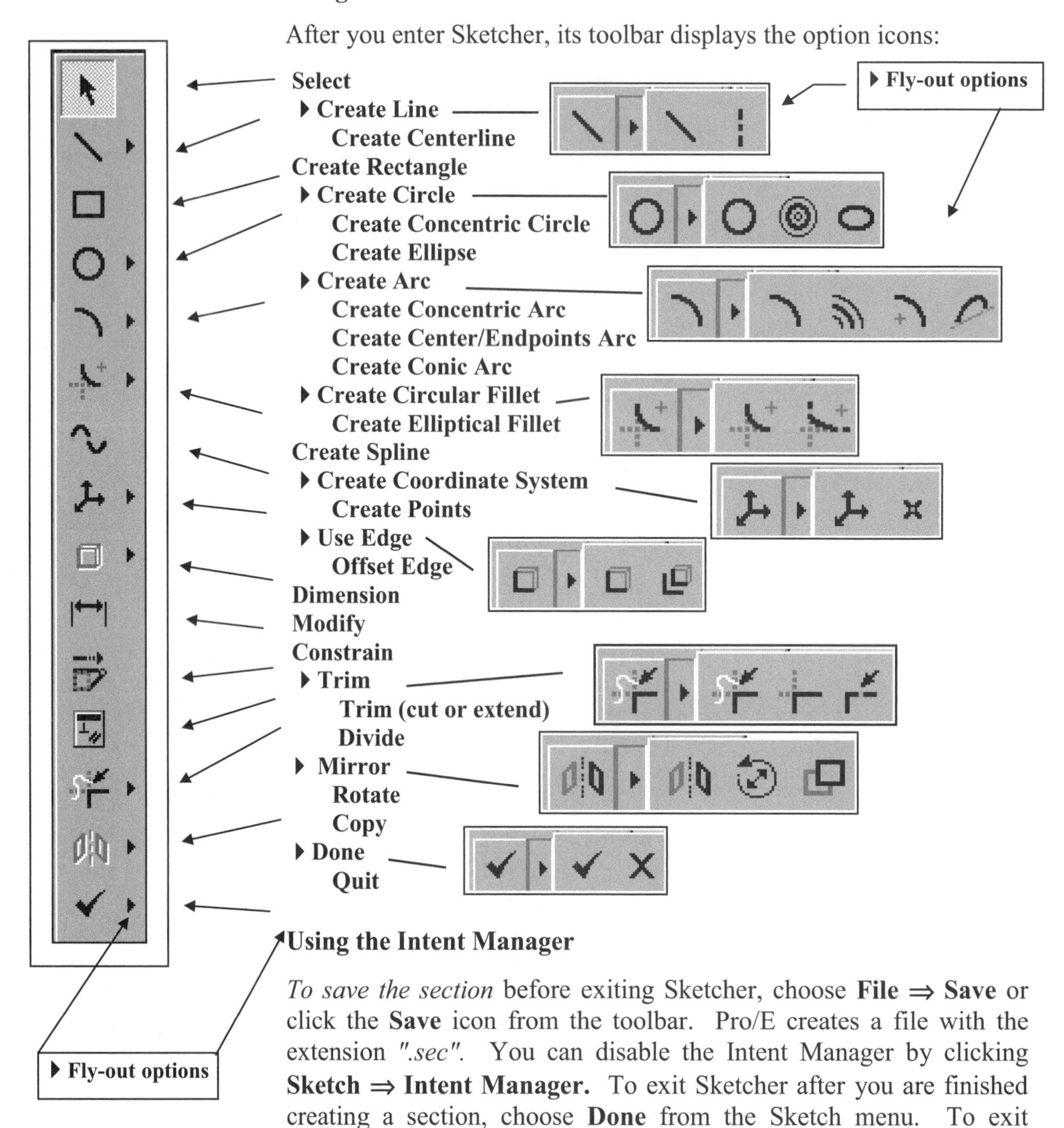

Using the Intent Manager

To save the section before exiting Sketcher, choose **File ⇒ Save** or click the **Save** icon from the toolbar. Pro/E creates a file with the extension *".sec"*. You can disable the Intent Manager by clicking **Sketch ⇒ Intent Manager.** To exit Sketcher after you are finished creating a section, choose **Done** from the Sketch menu. To exit Sketcher and discard any sketched geometry, choose **Quit** from the Sketch menu and **Yes** from the CONFIRMATION dialog box.

Using Shortcuts with the Right Mouse Button

You can access the most frequently used drafting operations by pressing the right mouse button. Additionally the right mouse button shortcut menu is context-sensitive. The right mouse button shortcut menu is divided into three areas. The top of the menu contains editing, manipulation, and selection commands. The middle portion of the menu contains creation commands, and the bottom portion of the menu always contains the **Undo** command.

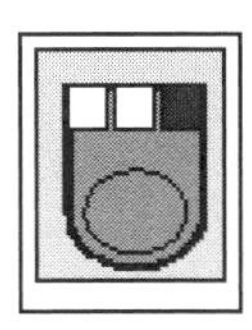

Disable Constraint
Delete
Modify...
Move
Toggle Construction
Line
Rectangle
Circle
3 Point / Tangent End
Centerline
Fillet
Dimension
Undo

Delete
Modify...
Move
Strong
Reference
Toggle Lock
Dimension
Undo

To Access the Mouse Button Shortcut Menus

The right mouse button (RMB) shortcut menu is context-sensitive. Sketcher dynamically assembles the shortcut menu, taking the following factors into consideration:

* What command is currently invoked
* What kind of entity is selected
* What is currently pre-highlighted

The top portion of the menu contains editing, manipulation, and selection commands. Some of the commands that the top portion may contain include:

- **Accept**
- **Delete**
- **Enable/Disable**
- **Modify**
- **Move**
- **Next**
- **Pick**
- **Previous**
- **Query Sel**
- **Strong**
- **Unlock**
- **Unselect Last**

The middle portion of the menu contains creation commands. Generally the following commands are available:

- **Line**
- **Rectangle**
- **Circle**
- **3 Point/Tangent End**
- **Centerline**
- **Fillet**

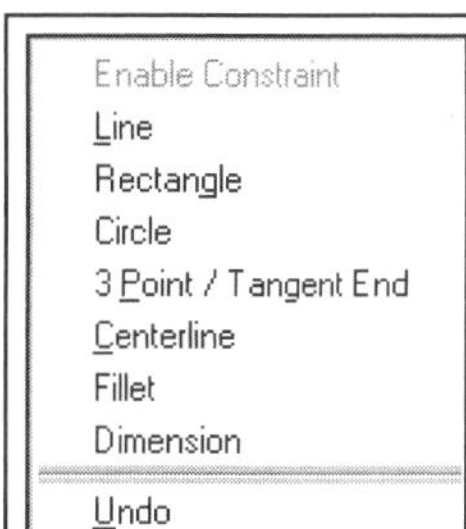

The menu, in its lower portion, lists only the following command:

- **Undo**—Undo the most recent operation.

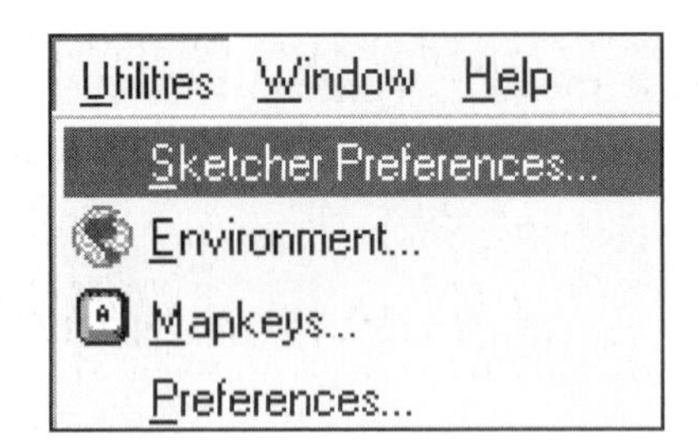

To Set Sketcher Parameters

1. In Sketcher mode, click **Utilities ⇒ Sketcher Preferences.** The Sketcher Preferences dialog box opens (Fig. 11.3).
2. Click the **Parameters** tab.
3. The **Parameters** tabbed page lists the following options:

Grid-- You can modify the grid **Origin**, **Angle** and **Type.**
Grid Spacing-- You can change the spacing of both the Cartesian and Polar grids.
Accuracy-- You can modify the number of decimal places that Pro/E displays for dimensions. In addition, you can change the relative accuracy for Sketcher solving.

4. Click the ✔ button to apply the changes and close the dialog box.

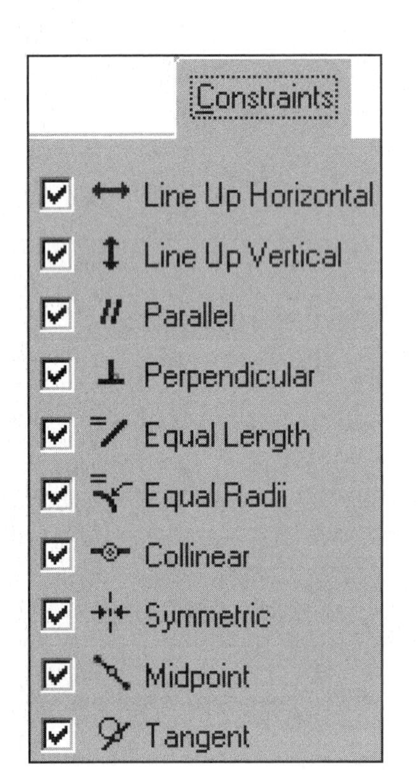

To Set Constraint Preferences (see left-hand column)

1. In Sketcher mode, click **Utilities ⇒ Sketcher Preferences.** Pro/E displays the Sketcher Preferences dialog box (Fig. 11.3).
2. Click the **Constraints** tab.
3. The **Constraints** tabbed page lists the following constraints. You can control the constraints that Sketcher assumes by placing or removing a check mark **✔.**

To Set Display Preferences

1. In Sketcher mode, click **Utilities ⇒ Sketcher Preferences.** Pro/E displays the Sketcher Preferences dialog box (Fig. 11.3).
2. The **Display** tabbed page lists the following options that you can switch on and off by placing or removing a check mark **✔:**

Grid-- display of the screen grid.
Vertices-- display of vertices. You can control the display of vertices by setting the configuration option *sketcher_disp_vertices.*
Constraints-- display of constraints. You can control the display of constraints by setting the configuration option *sketcher_disp_constraints.*
Dimensions-- display of all section dimensions.
Weak Dimensions-- display of weak dimensions.
Snap To Grid-- engage or disengage the snap to grid option.
Lock Modified Dimensions-- lock or unlock modified dimensions.
Start In Sketch View-- orient model so that the sketching plane is parallel to the screen (2D orientation).

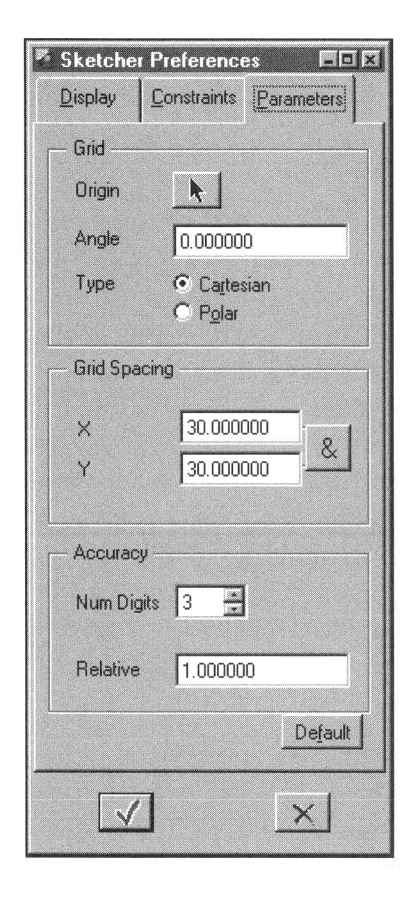

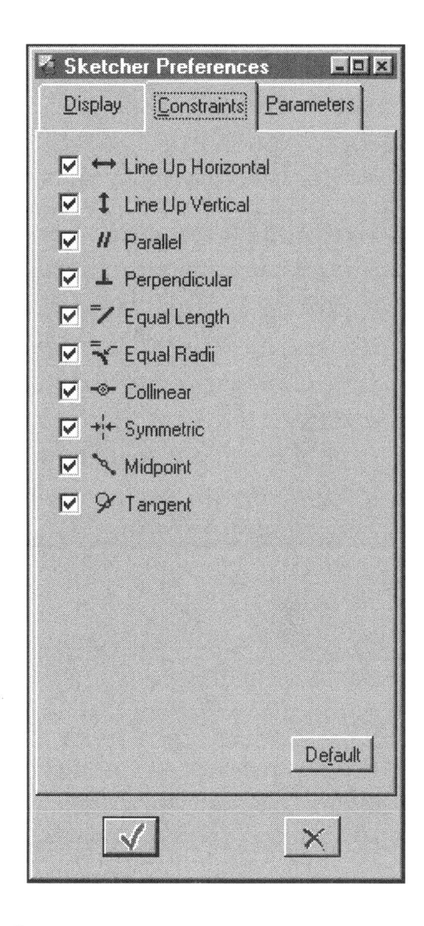

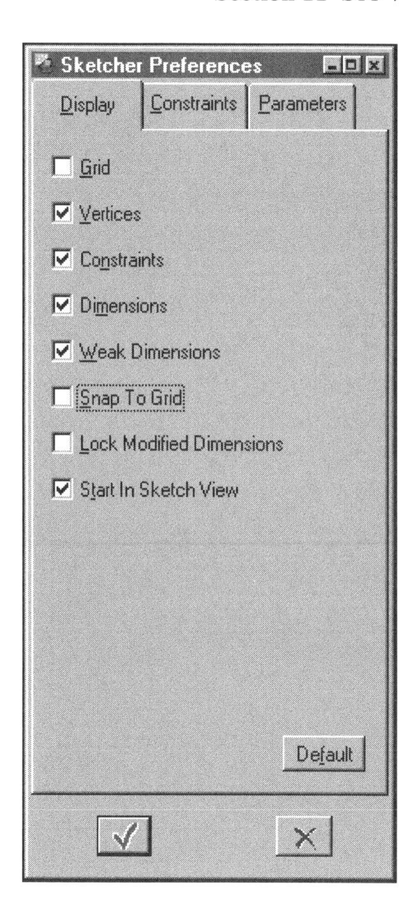

Figure 11.3
Sketcher Preferences

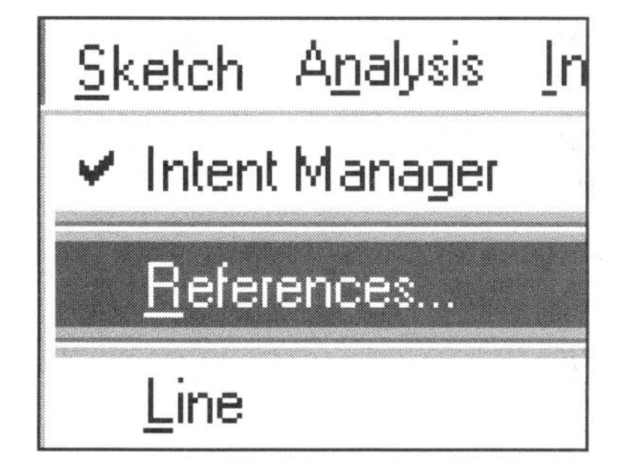

References

To dimension and constrain geometry, Pro/E requires you to create or accept references. References can be created through the References dialog box. To open the References dialog box, click **Sketch ⇒ References** (see left-hand column). When you create a new feature, Pro/E automatically selects default Sketcher references. You can change these references or create new ones in the References dialog box. Pro/E prompts you to create references in the following situations:

* When you create a new feature, the References dialog box opens (see left-hand column). Pro/E prompts you to select a perpendicular surface, edge, or vertex relative to which the section will be dimensioned and constrained.
* When you redefine a feature that is missing references.
* When you do not have enough references to place a section.

About Creating Geometry in Sketcher

To start sketching, select an option from the Sketcher toolbar or the Sketch menu. Create entities by clicking points inside the Sketcher window. As you move the mouse pointer, Sketcher determines applicable constraints and displays them; Pro/E displays the active constraint in *red*.

As you create geometry, it snaps to satisfy these constraints (for example, horizontal or vertical line constraint). After the entities are sketched, you can apply additional constraints by choosing the **Constrain** option in the Sketch menu.

You use the mouse in Sketcher in many ways:

* Use the left mouse button to pick points on the screen and the middle mouse button to abort the current action.
* SHIFT key and click the left mouse button to switch between circle and ellipse creation. You can use the same mouse operation to switch between circular fillet and elliptical fillet creation.
* While you are sketching, you can disable the current constraint (shown in *red*) by pressing the right mouse button and lock the constraint by pressing SHIFT key and the right mouse button.
* SHIFT key and click the left mouse button to gather selected items.
* You can click the right mouse button menu for a shortcut menu with frequently used sketching commands (while you are not in the rubberband mode).

Pro/E automatically dimensions geometry, as you sketch entities, by adding only those dimensions that are necessary to solve the section (Fig. 11.4). Pro/E dimensions are called "weak" dimensions (they appear in *gray*), because Pro/E can remove or change them without your input. Use the Dimension option in the Sketch menu to add "strong" dimensions (they appear in *yellow*).

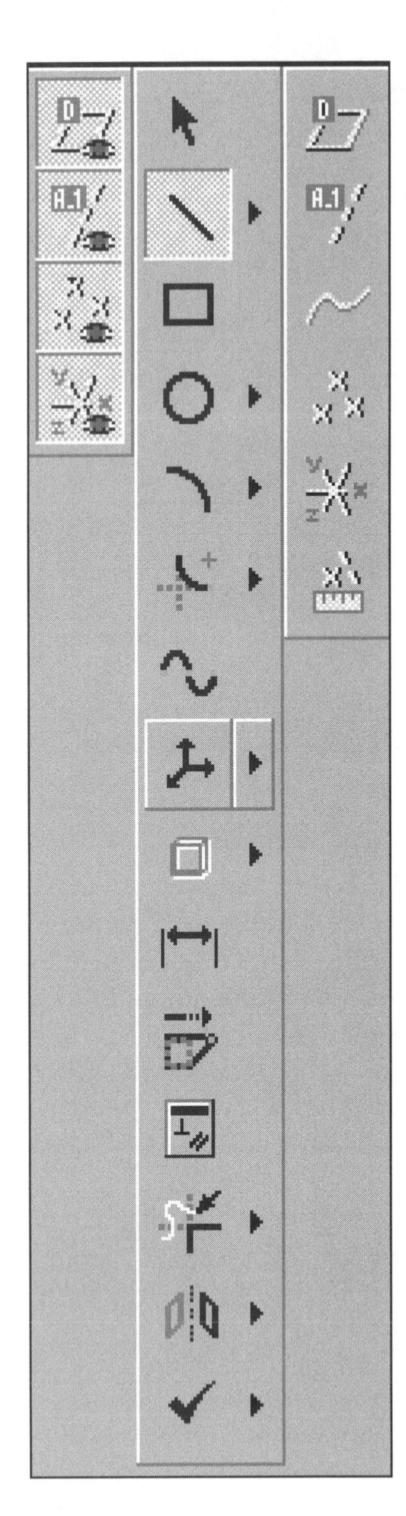

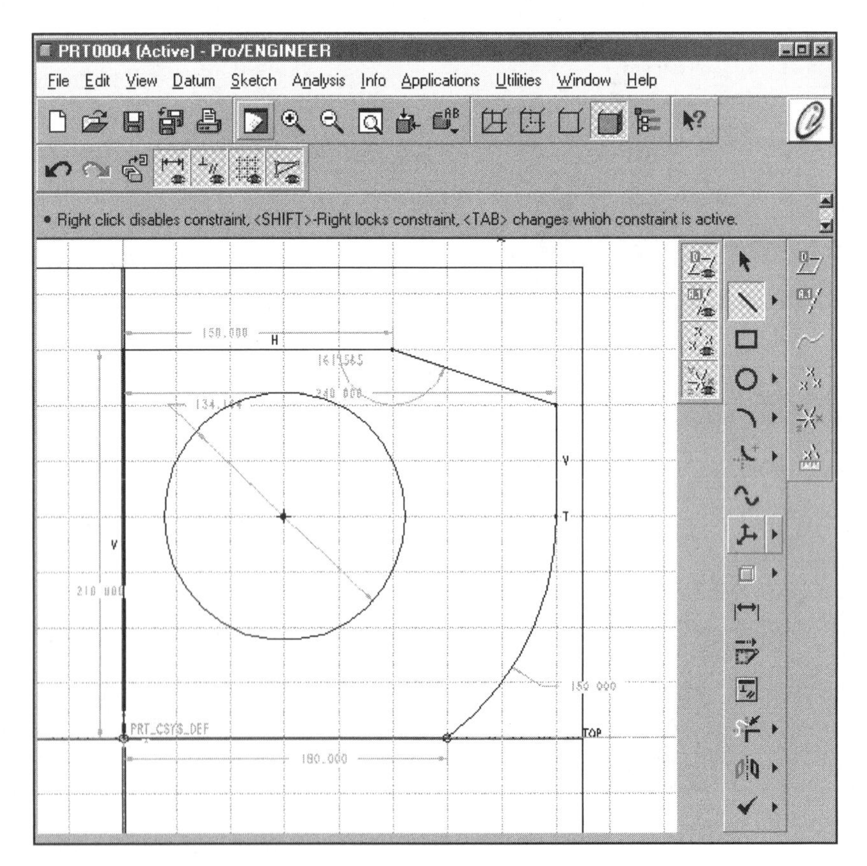

Figure 11.4
Sketching with Intent Manager On

Dimensioning

Sketcher makes sure that the section is adequately constrained and dimensioned at any stage of the section creation. As you sketch a section, Pro/E automatically dimensions the geometry. These dimensions are called **"weak"** dimensions, because Pro/E creates and removes them without warning. Weak dimensions appear in gray. You can also add your own dimensions to create the desired dimensioning scheme. User dimensions are considered **"strong"** dimensions by Pro/E. As you add strong dimensions, Pro/E automatically removes unnecessary weak dimensions and constraints. It is good practice to strengthen weak dimensions that you intend to keep in a section before you exit Sketcher. This ensures that Pro/E does not delete these dimensions without your input. If adding a dimension leads to a conflict or redundancy in the dimensioning scheme and constraints, Sketcher issues a warning and lets you resolve the conflict.

Weak dimensions cannot be deleted. They are automatically removed when you create strong dimensions that make the weak dimensions unnecessary.

You can selectively turn weak dimensions into strong dimensions by using the **Strong** command in the Convert to menu.

1. Click a dimension to strengthen.
2. Click **Edit ⇒ Convert to ⇒ Strong**. The dimension changes from *gray* to *yellow*. Or simply pick the dimension, press the right mouse button, and select **Strong** (Fig. 11.5).

When you modify a value of a weak dimension or use it in a relation, that dimension becomes strong.

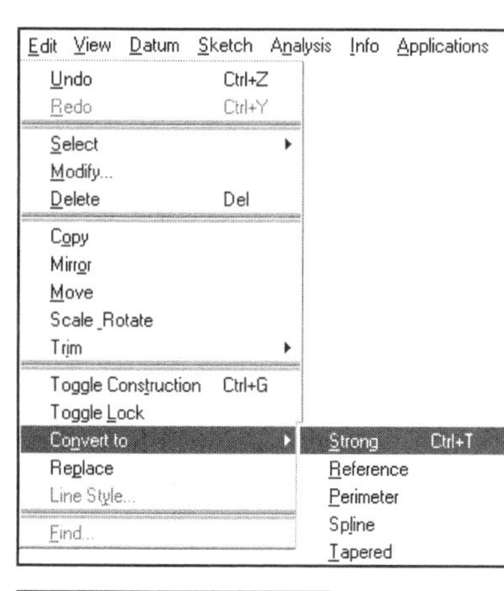

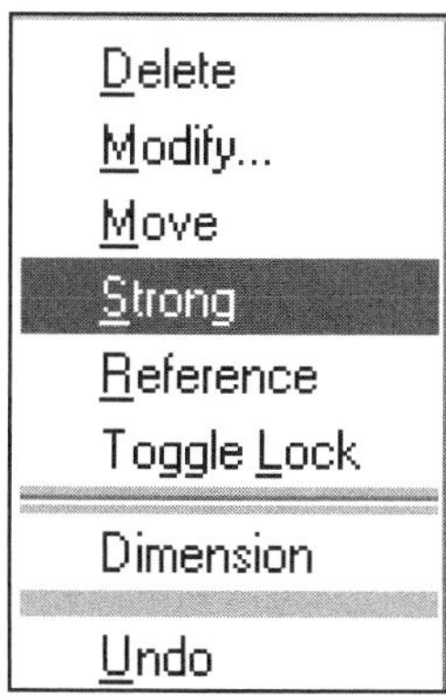

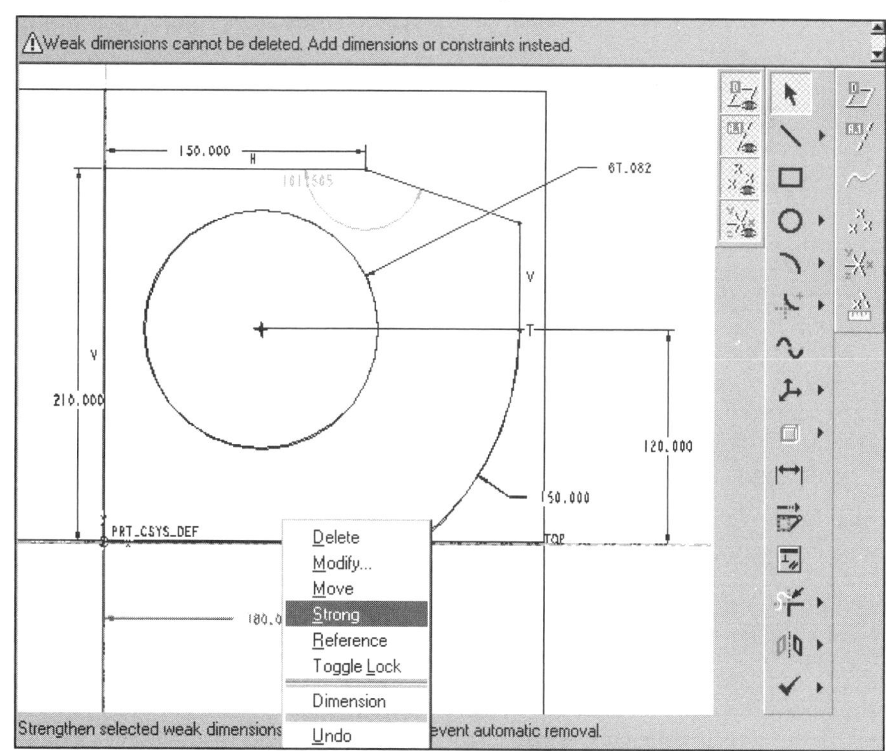

Figure 11.5 Converting Weak Dimensions to Strong Dimensions

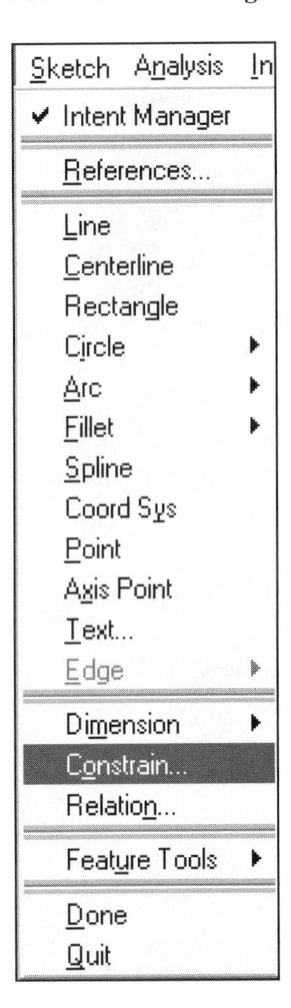

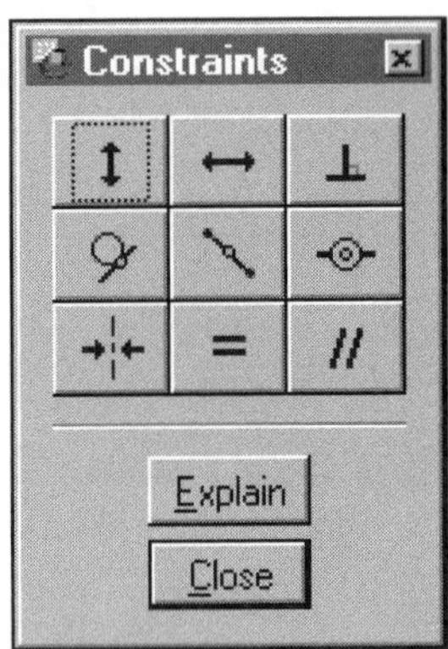

There are hundreds more capabilities and techniques available with the Intent Manger on. Throughout the text you will be introduced to a number of them. It is up to you to explore and master this tool using the text and Pro/E Help.

Sketcher Constraints

When you sketch geometry, Pro/E uses certain assumptions to help you locate geometry. When the cursor comes within the tolerance of some constraints, Pro/E snaps to that constraint and shows its graphical symbol next to that entity. Before you pick the location with the left mouse button (LMB), you can:

* Disable a constraint by pressing the right mouse button. To enable it again, press the right mouse button again.
* Lock in a constraint by holding the Shift key and pressing the right mouse button. To unlock the constraint, repeat this action.
* When more than one constraint is active, you can change the active constraint by using the Tab key.
* Constraints that appear in *gray* are called "weak" constraints. They can be removed by Pro/E without warning. You can add your own constraints with the Constrain option in the Sketch menu.

Pro/E shows constraints as follows:

* Current constraint--*red*
* Weak constraint--*gray*
* Strong constraint--*yellow*
* Locked constraint--enclosed in a circle
* Disabled constraint--with a line crossing the constraint symbol

The following table lists constraints with the corresponding graphical symbols.

Constraint	Symbol
Midpoint	M
Same points	o
Horizontal entities	H
Vertical entities	V
Point on entity	-O- - -
Tangent entities	T
Perpendicular entities	⊥
Parallel lines	$//_1$
Equal radii	R with an index in subscript
Line segments with equal lengths	L with an index in subscript (for example, L1)
Symmetry	—▸◂—
Entities are lined up horizontally or vertically	-- ¦
Collinear	=̶
Alignment	Symbol for the appropriate alignment type.
Use Edge/Offset Edge	—— o

Section 12 will cover the Sketcher without the Intent Manager on. You can switch back and forth between using and not using the Intent Manager.

Section 12

The Sketcher

As we have stated in Section 11, in Pro/ENGINEER, *sketching* is done in the **Sketcher**. *In this section we will discuss aspects of the Sketcher that apply when you sketch with the Intent Manager off.* Almost all traditional lines, circles, arcs, and their variations can be accomplished on the screen without creating exact and perfectly constructed geometry. Pro/E will assume a variety of conditions, such as tangency, similar sizes for same-type geometry, parallelism, perpendicularity, verticals, horizontals, coincident endpoints, tangent points, and symmetry.

The Sketch Mode can be entered directly by picking **File ⇒ New** (from the Menu Bar or from the Toolbar) ⇒ ● **Sketch ⇒ OK ⇒ Sketch ⇒ Intent Manager** (off) (Fig. 12.1). If you go directly into the Sketcher, you will be creating sections that can be recalled later during a part feature creation. The sections are like sheets of graph paper with sketched geometry on them. They can be saved for later use in any feature creation where a section is called for. Instead of a part name, Pro/E will display **S2D0001 (Active)** in the window bar. The sketches are numbered sequentially. If you choose to save a sketch (in Sketch mode or any other mode), you need to give it a unique name or use the default.

You are automatically put into the Sketcher when creating most geometry in Part mode. The section in Figure 12.2 is an example of a section created in Part mode. A section created in Part mode or any other mode can also be saved for later use as well as being used to create the geometry needed for the present feature. Sketcher techniques are used in many areas of parametric design. The purpose of the Sketcher is to enable the quick and simple creation of geometry for your model. The Sketcher requires you to create and dimension this geometry, but during the sketching process you need not be concerned with exact dimension sizes or the creation of perfect and accurate geometry. The Sketcher has four command icons that are automatically displayed by default when you enter this mode.

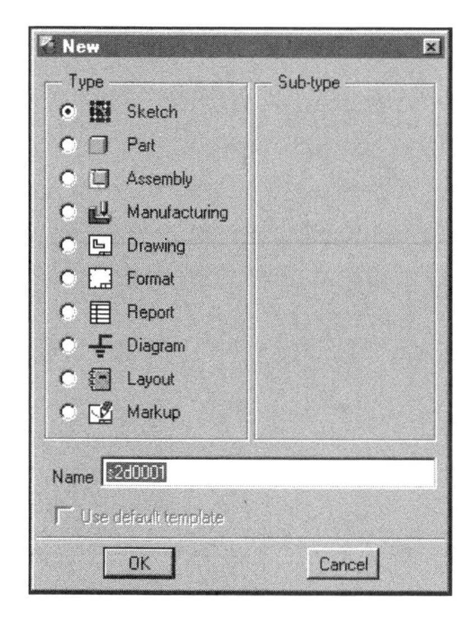

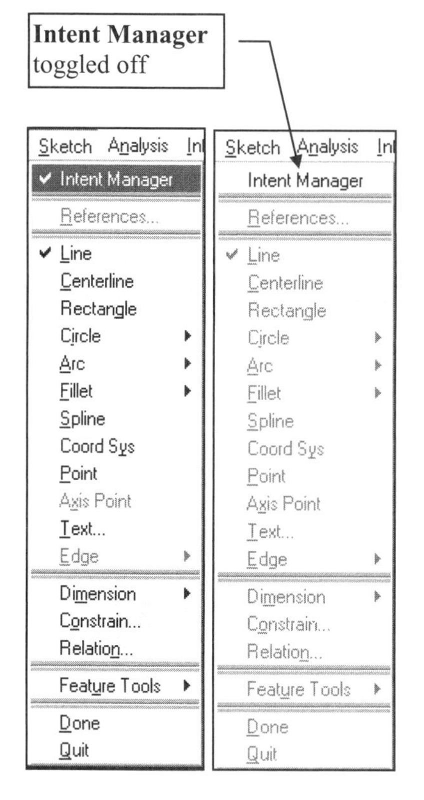

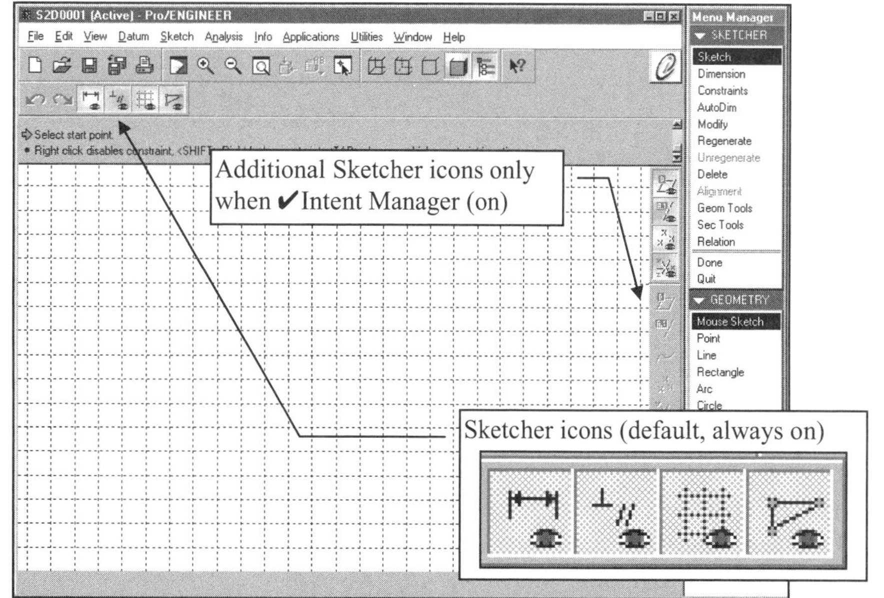

Figure 12.1
Sketcher Mode

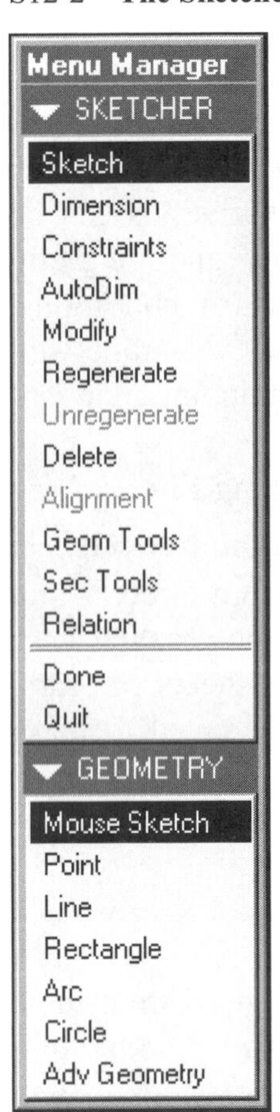

Creating sections within the SKETCHER menu is not difficult. There are only a few steps to remember:

1. **Sketch** Sketch the section geometry. Use SKETCHER tools to create the section geometry (Fig. 12.2).
2. **Dimension** Dimension the section. Use a dimensioning scheme that you want to see in a drawing. Dimension to control the characteristics of the section geometry (Fig. 12.3).
3. **Alignment** Align the section geometry to a datum feature or to a part feature (Fig. 12.3).
4. **Regenerate** Regenerate the section. Regeneration solves the section sketch on the basis of your dimensioning scheme (Fig. 12.4).
5. **Relations** Add section relations. Add relations to control the parametric behavior of your section.

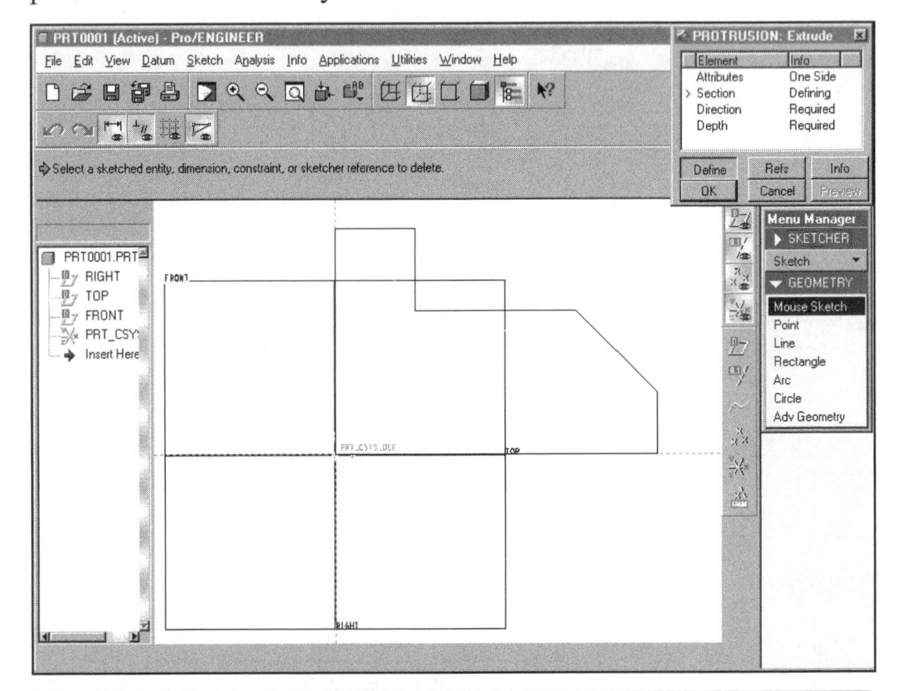

Figure 12.2
Sketch the Geometry

NOTE
Approximately 50% of the texts commands are demonstrated with the Intent Manager off. Later in Part Lessons 8-13, many commands are left for you to complete without direction. In these cases, you may choose to use or not to use the Intent Manager.

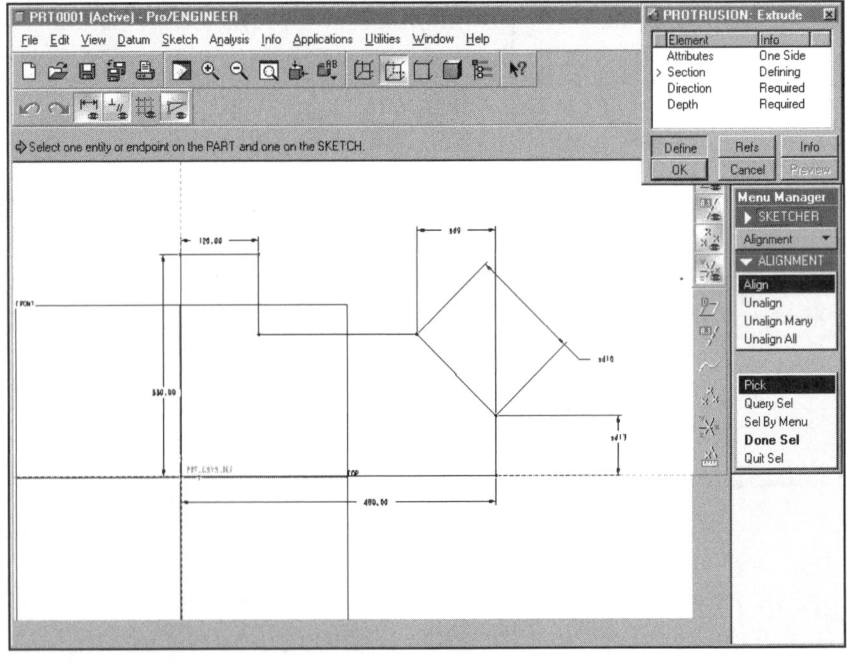

Figure 12.3
Dimension and
Alignment of the Sketch

6. **Modify** Change the sketch dimension values to the required design sizes.
7. **Regenerate** Regenerate the section again.

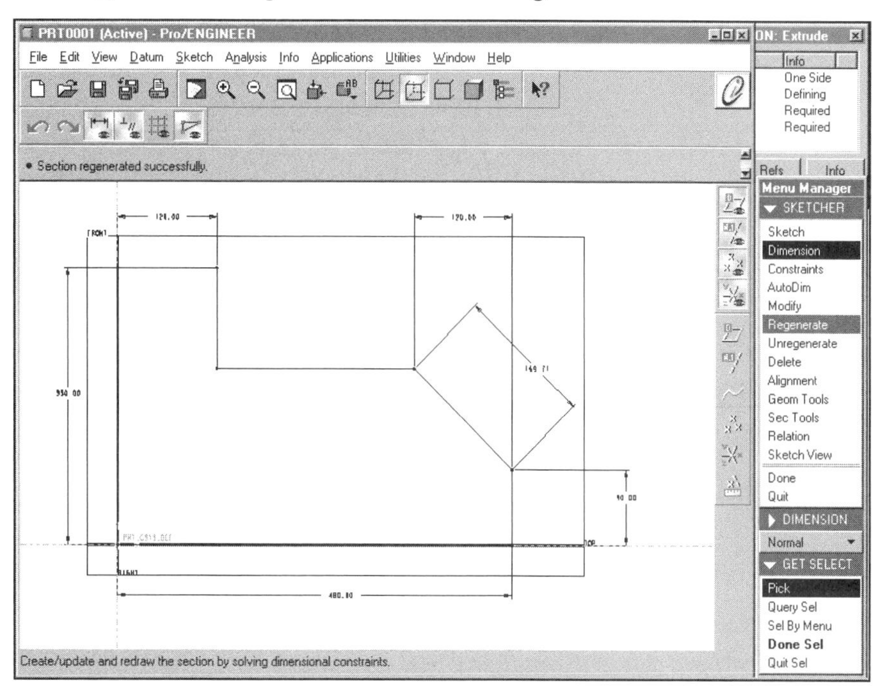

Figure 12.4
Regenerate the Section

After Pro/E regenerates the sketch, the feature can be modified and completed. In Figure 12.5 the sketch dimensions were changed to the design size (after regenerating) and then regenerated again. More features can then be added, using *sketched* features or *pick-and-place* features such as holes and chamfers.

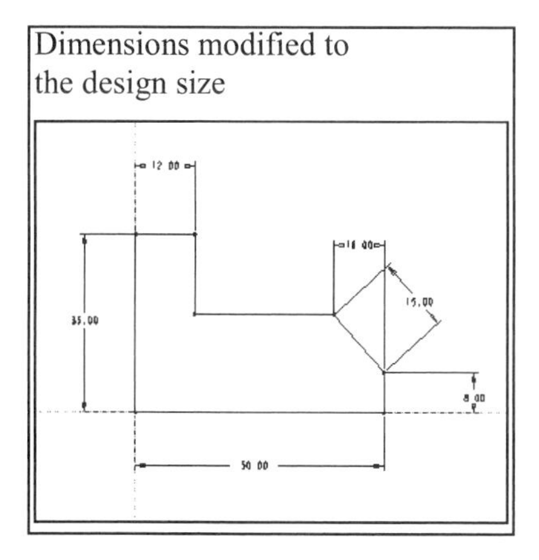

Figure 12.5
Extruded Protrusion Feature Created from the Sketch

Sketcher and the Mouse

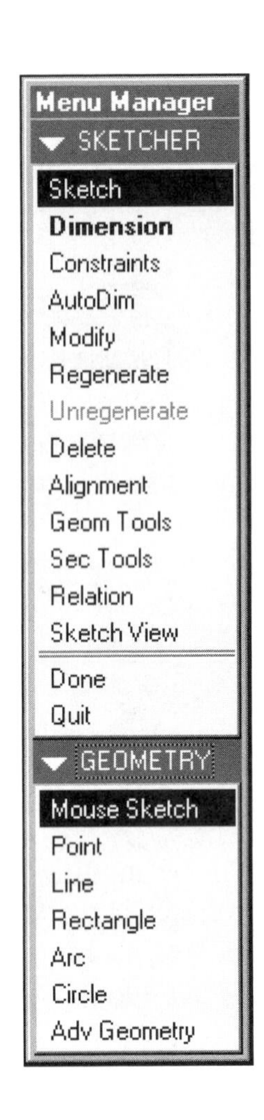

The **SKETCHER** is used to establish 2D sections that are the basis for the 3D feature being created. In order to understand just how powerful the Sketcher is in a parametric design system, you need only look at what sketching has been throughout the ages: *Sketching is a process of simply and efficiently establishing the basic design and intent of a designer-engineer* on paper, and that is now possible with the mouse.

In the Sketcher (with **Mouse Sketch** option), much of the section geometry can be created using the three buttons on the mouse.

LEFT button Used to create lines. The line command in the Sketcher chains lines together. This button also aborts the creation of circles.

MIDDLE button Used to create circles. The first pick is the center of the circle; the second pick is a location on the diameter of the circle. Also used to end line creation.

RIGHT button Used to create tangent arcs. *There must be an existing line, arc, or spline to reference for tangency.* Place the cursor near the tangent entity, and press the right button to start the arc. Press the right button again to set the endpoint of the arc.

These functions are available when the **Sketch** option is selected from the SKETCHER menu and **Mouse Sketch** (Fig. 12.6) is left as the default in the GEOMETRY menu. Also, there are other geometry functions in the GEOMETRY menu.

A **Sketched Hole** command will automatically bring up the Sketcher so you can create the profile of *one half* of the hole geometry

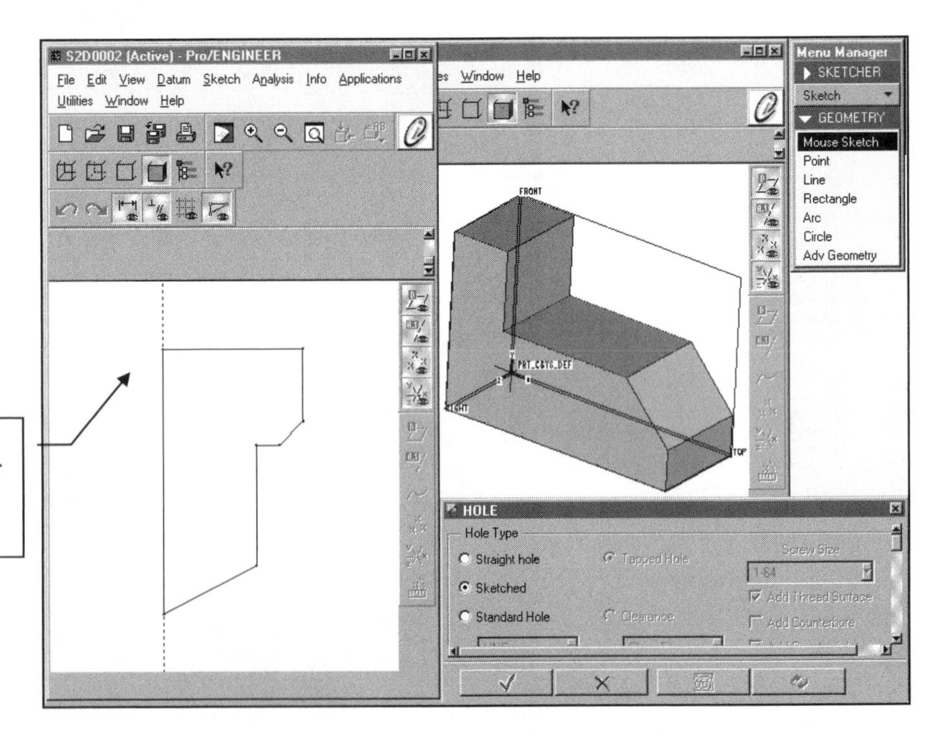

Figure 12.6
Mouse Sketch

Regenerating a Section Sketch

In the past, designers have sketched on paper, showing lines, arcs, circles, and other geometric forms in rough, simplified outline and internal forms. The sketched shapes are assumed to be what they *sort of* look like. Round shapes approximating a circle are assumed by the reader of the sketch to be circles, curved shapes are assumed to be arcs, and lines drawn straight up or down are assumed to be vertical. Lines drawn left to right are assumed to be horizontal. Lines sketched at an angle are straight lines that are angled. Dimensions roughly sketched on a less-than-perfect drawing of a part are assumed to represent the exact, perfect shape desired by the person sketching. All this seems obvious to most people involved in engineering design. With the introduction of parametric design, we can now sketch on the screen and allow Pro/E to make all the assumptions that were traditionally made by a person creating a sketch or reading a sketch. These assumptions include, but are not limited to, the following: *symmetry*, *tangency*, *parallelism*, *perpendicularity*, *equal angles*, *same-size arcs* and *circles*, and *coincident centers*.

The sketch started in Figure 12.7 was completed without the **Snap to Grid** activated. Note that the lines and arcs are not sketched perfectly. After regeneration, Pro/E will straighten the lines and align the features according to a set of assumptions or rules. Almost-vertical lines will be vertical, close-to-horizontal lines will become horizontal, and so on. We suggest that you keep the grid snap on (✔**Snap to Grid**) most of the time while sketching the first features of a part; experienced users normally keep it off and trust the Pro/E assumptions to clean up any sketching inconsistencies.

Because our screen is set to display the commands and icons at a larger size than yours, three of the options are not displayed with their names:

- ❑ Use 2D Sketcher
- ❑ Sketcher Intent Manager
- ❑ Use Fast HLR

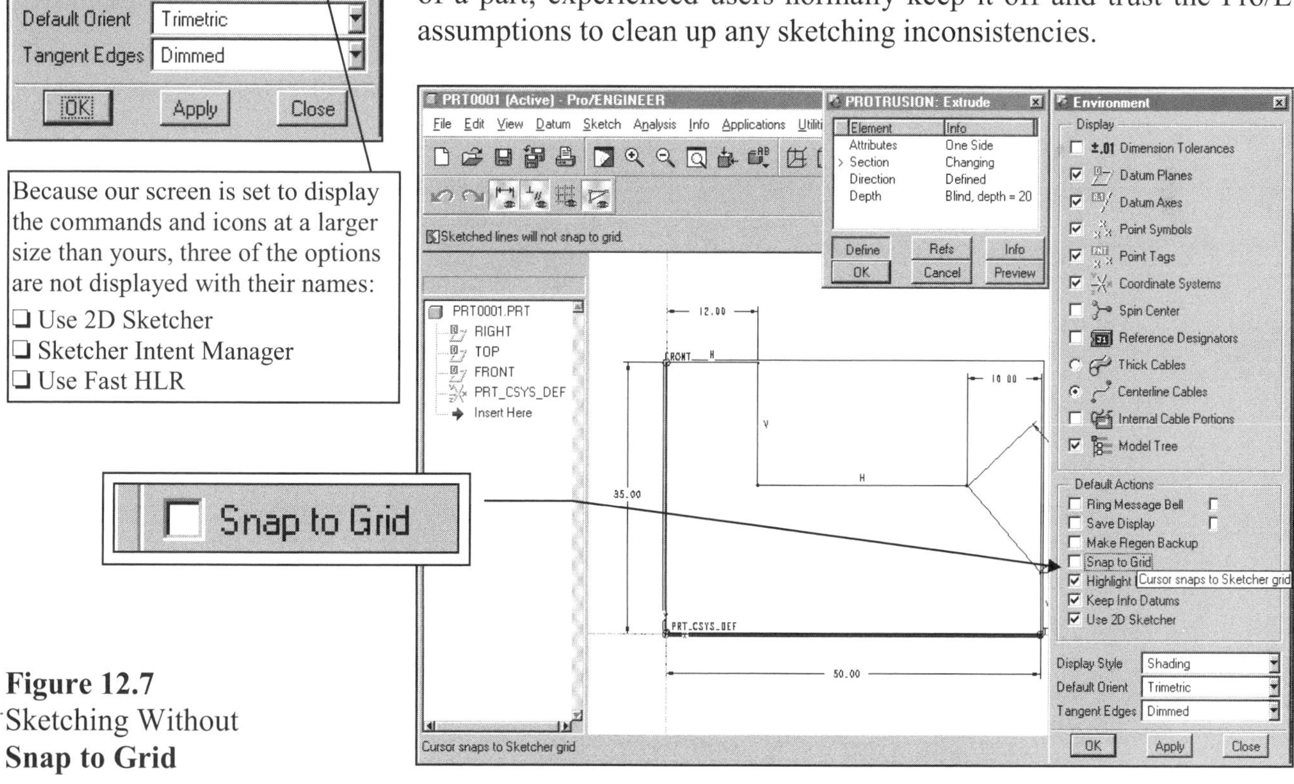

Figure 12.7
Sketching Without
Snap to Grid

During regeneration, Pro/E checks to make sure that it understands your dimensioning scheme and that you have created a complete and independent set of parameters. Pro/E analyzes your section based on the geometry you have sketched and the dimensions you have created. In the absence of explicit dimensions, implicit information based on the sketch is used. You can quickly understand and control the assumptions made in solving the sketch. This streamlines the process of sketch creation and behavior diagnosis.

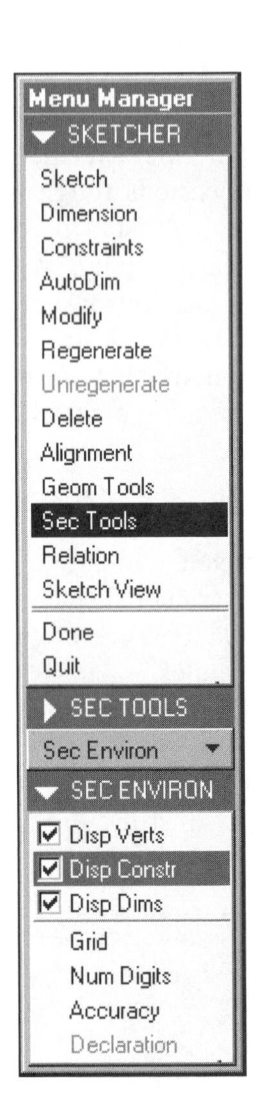

Modifications made in sketches are *animated* over a brief time during regeneration. If a sketch fails, it changes shape up to the point of failure, allowing you to view the section at the point of failure with the option of restoring the dimensions to the old values. This provides an understanding of how the sketch fails by showing the point of failure and by displaying, through animation, how that point was achieved. Corrective actions can then be taken.

Here is a list of implicit information that Pro/ENGINEER uses to regenerate a section (these rules are applied to *all* Pro/ENGINEER sketches):

RULE: Equal radius/diameter
DESCRIPTION: If two or more arcs or circles are sketched with approximately the same radius, they are assigned the same radius value.
RULE: Symmetry
DESCRIPTION: Entities sketched symmetrically about a centerline are assigned equal values with respect to the centerline.
RULE: Horizontal and vertical lines
DESCRIPTION: Lines that are approximately horizontal or vertical are considered to be exactly horizontal or vertical.
RULE: Parallel and perpendicular lines
DESCRIPTION: Lines that are sketched approximately parallel or perpendicular are considered to be exactly parallel or perpendicular.
RULE: Tangency
DESCRIPTION: Entities sketched approximately tangent to arcs or circles are assumed to be tangent.
RULE: 90°, 180°, 270° arcs
DESCRIPTION: Arcs are considered to be multiples of 90° if they are sketched with approximately horizontal or vertical tangents at the endpoints.
RULE: Collinearity
DESCRIPTION: Segments that are approximately collinear are considered to be exactly collinear.
RULE: Equal segment lengths
DESCRIPTION: Segments of unknown length are assigned a length equal to that of a known segment of approximately the same length.
RULE: Point entities lying on other entities
DESCRIPTION: Point entities that lie approximately on lines, arcs, or circles are considered to be exactly on them.
RULE: Centers lying on the same horizontal
DESCRIPTION: Two centers of arcs or circles that lie approximately along the same horizontal direction are set to be exactly horizontally aligned.
RULE: Centers lying on the same vertical
DESCRIPTION: Two centers of arcs or circles that lie approximately along the same vertical direction are set to be exactly vertically aligned.
RULE: Equal coordinates
DESCRIPTION: Endpoints and centers of the arcs may be assumed to have the same X- or the same Y-coordinates.

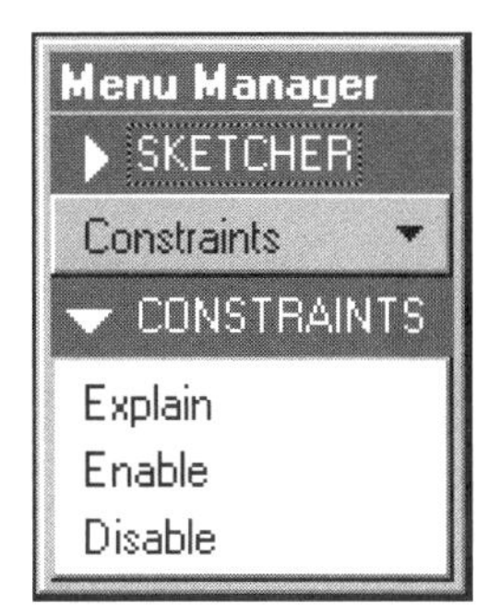

Sketcher Mode and Constraints

Constraints used in solving a sketch are displayed graphically on the sketch with the aid of *small symbols* that appear next to the entities to which they apply (Fig. 12.8). You can turn off the display of these symbols. You can also click on the symbols to disable or enable the constraints and to obtain a brief explanation. Also, endpoints of section entities are highlighted with the aid of small dot symbols. This graphical display of symbols for constraints replaces the old user interface of a Constraints dialog box.

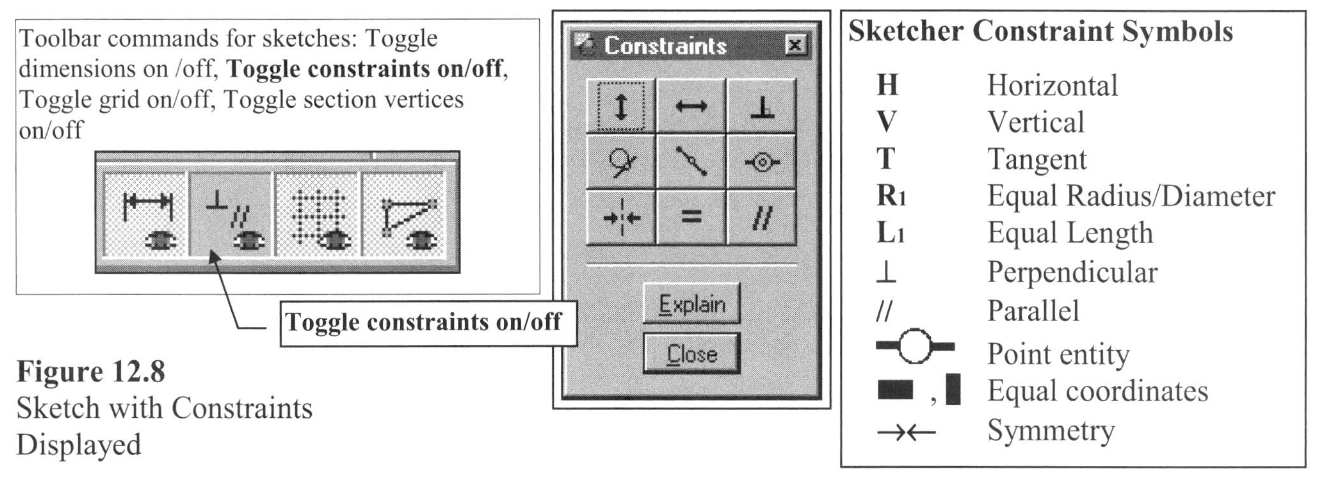

Sketcher Constraint Symbols

Symbol	Meaning
H	Horizontal
V	Vertical
T	Tangent
R1	Equal Radius/Diameter
L1	Equal Length
⊥	Perpendicular
//	Parallel
—○—	Point entity
▬ , ▮	Equal coordinates
→←	Symmetry

Figure 12.8
Sketch with Constraints Displayed

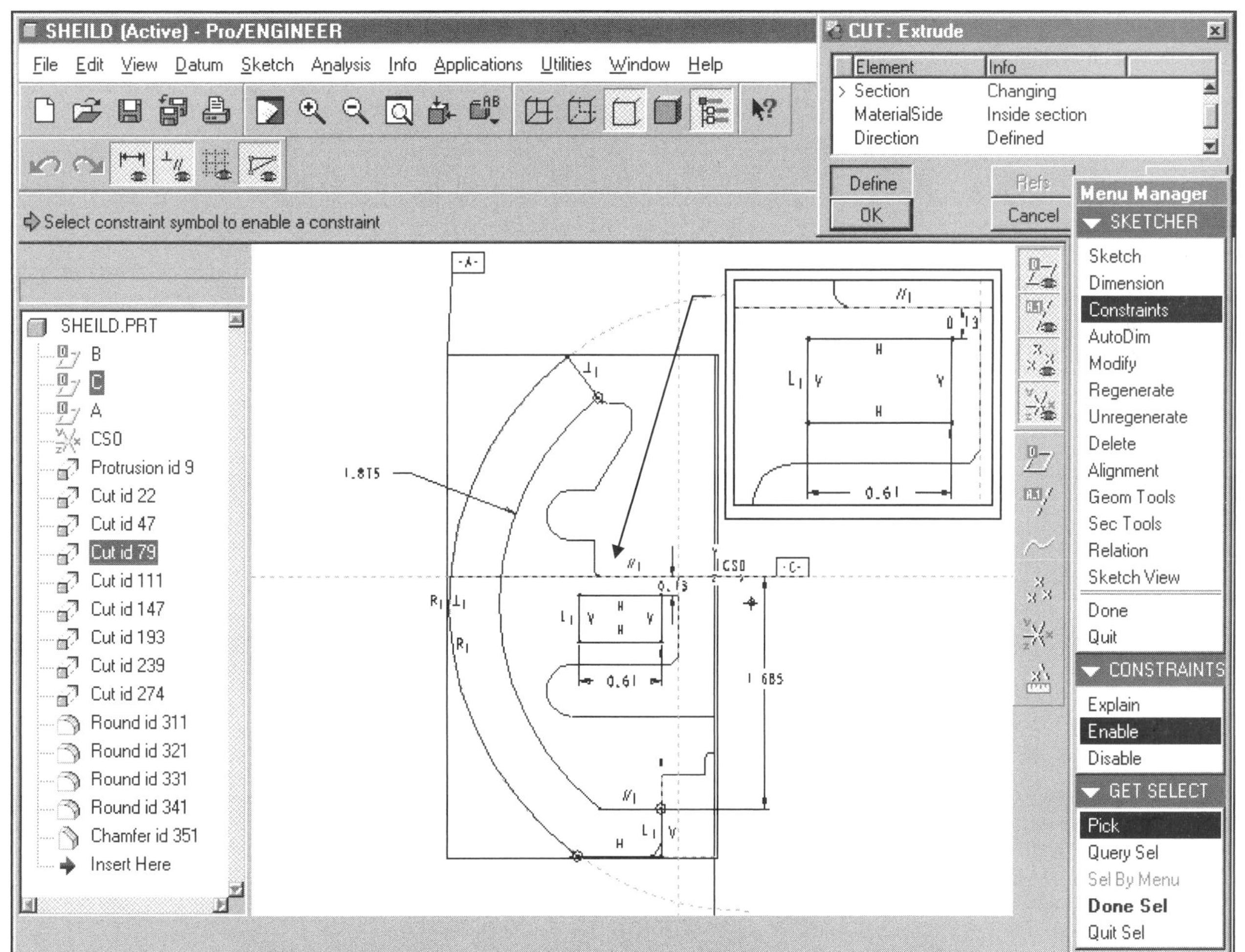

Sketching Lines

You can create two types of **lines** with Pro/E: geometry lines and centerlines. **Geometry lines** are used to create *feature geometry*. **Centerlines** are used to define the *axis of revolution* of a revolved feature, to define a line of symmetry within a section, or to create construction lines. In Figure 12.9, the section contains lines, centerlines, arcs, and a circle shown in 2D **Sketch View.** You can also reorient the sketch into a 3D **Default** view orientation, as shown in Figure 12.10.

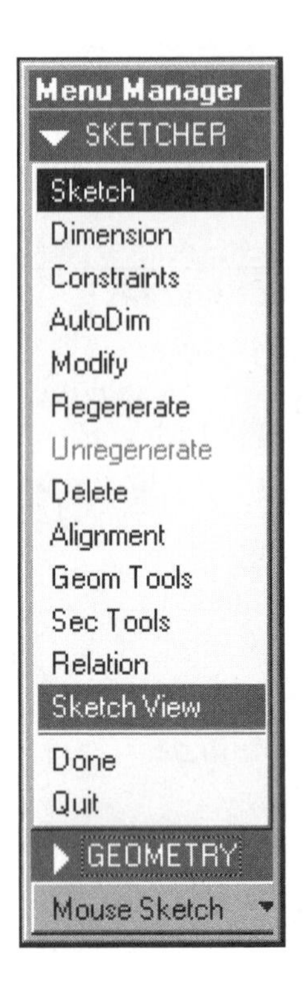

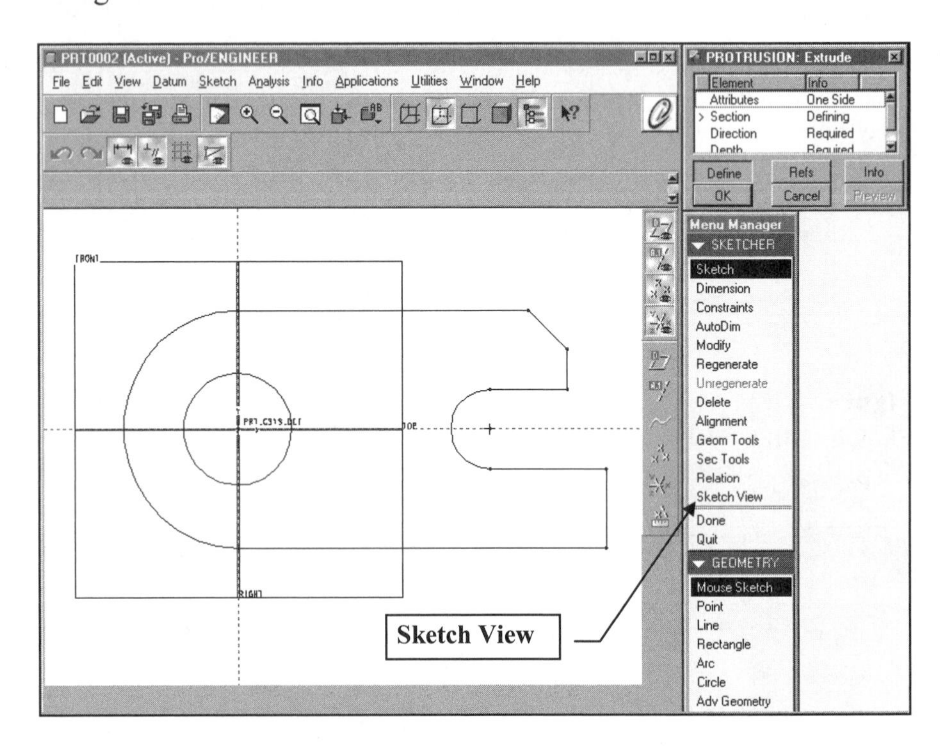

Figure 12.9
2D Sketch View

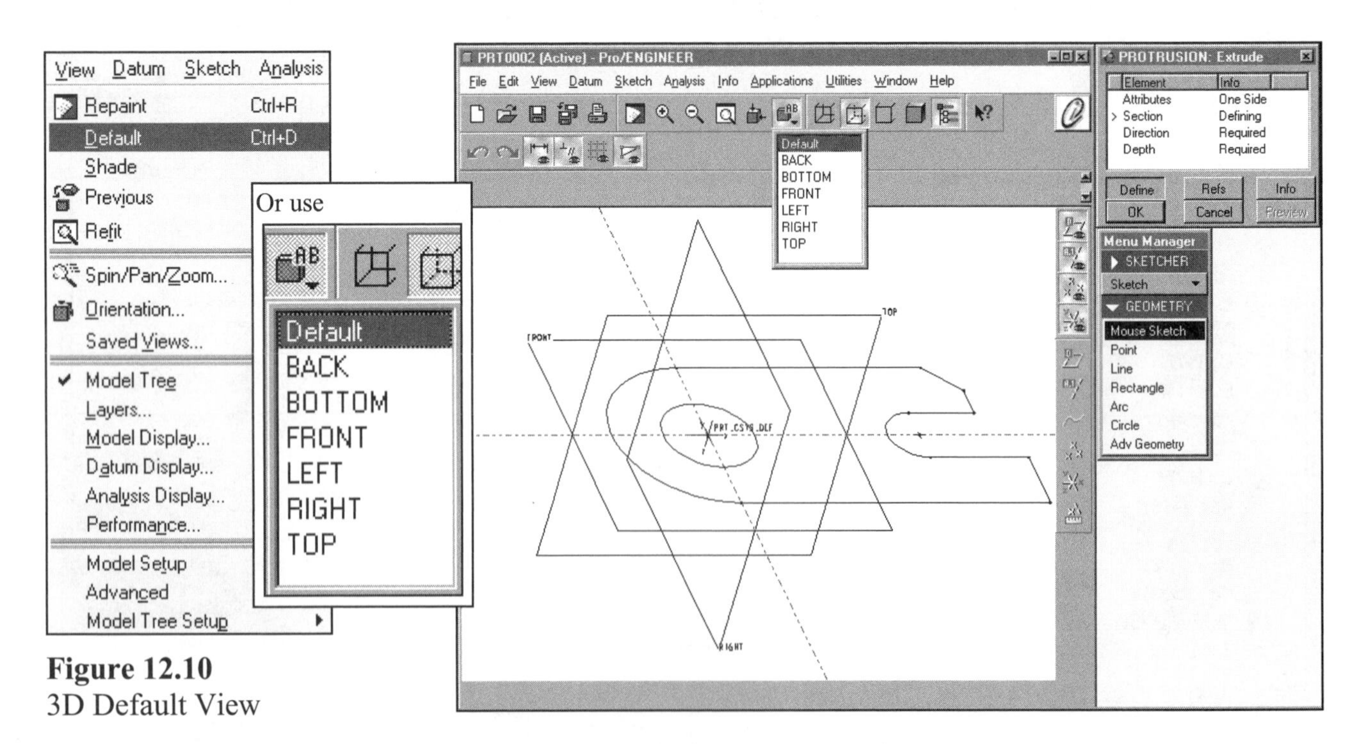

Figure 12.10
3D Default View

To sketch **geometry lines** or **centerlines,** do the following:

1. Choose **Line** from the GEOMETRY menu. The LINE TYPE menu appears.
2. Choose **Geometry** or **Centerline** from the top portion of the menu to indicate the type of line that you want.
3. Choose an option from the bottom portion of the menu to indicate how to create the *line* or *centerline*:

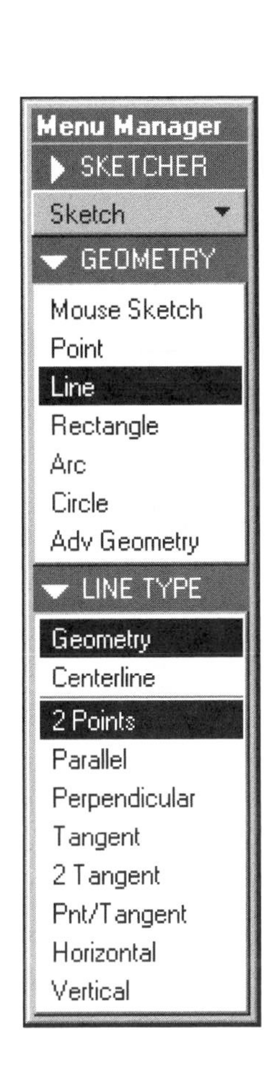

2 Points Create a line by picking the start point and endpoint. Geometry lines created using this command will automatically be chained together.

Parallel Pick an existing line to determine the new line's direction; then pick the start point and endpoint. For a *centerline*, only a single pick is needed to determine the parallel placement of the *centerline*, and the ends of the *centerline* will be chosen to fit model or section outlines.

Perpendicular Pick an existing line to determine the new line's direction; then pick the start point and endpoint. For a *centerline*, only a single pick is needed to determine the perpendicular placement of the *centerline*, and the ends of the *centerline* will be chosen to fit model or section outlines.

Tangent Pick an endpoint of an arc or spline to start the new line and determine its direction; then pick the endpoint of the line. For a *centerline*, only a single pick is needed to determine the tangent placement of the *centerline*, and the ends of the *centerline* will be chosen to fit model or section outlines.

2 Tangent Pick two arcs, splines, or circles to determine the direction of the new line. The line is automatically created between the selected entities, and *it splits the entities at the tangency points*. A 2 Tangent line created to construction entities will not split the entity. A 2 Tangent *centerline*, created as a 2 Tangent line defined using two circles, will not split the circles.

Pnt/Tangent Pick a point anywhere in the current section; then pick an arc, spline, or circle to which the line must be tangent.

Horizontal Create a line that is horizontal relative to the orientation of the section. For a geometry line, the endpoint is automatically the starting point of a chained vertical line. For a *centerline*, only a single pick is needed to determine the horizontal location of the *centerline*.

Vertical Create a line that is vertical relative to the orientation of the section. For a geometry line, the endpoint is automatically the starting point of a chained horizontal line. For a *centerline*, only a single pick is needed to determine the vertical location of the *centerline*.

Circles

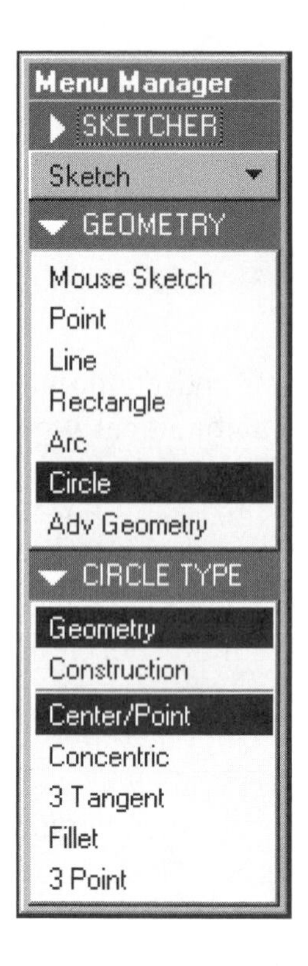

A variety of **geometry circles** and **construction circles** can be created in the Sketcher (Fig. 12.11). Geometry circles are used to create feature geometry, whereas construction circles serve as guides and references but do not create feature geometry. Construction circles (Fig. 12.11) are displayed in the same color as circles, but with a *phantom line font* rather than the *solid font* that is used for circles and other geometry.

To sketch geometry and construction circles, choose the following commands:

1. Select **Circle** from the GEOMETRY menu.
2. Pick **Geometry** or **Construction** from the top part of the menu.
3. Choose one of the following options from the bottom part of the menu:

Center/Point Same as creating a circle with the **Mouse Sketch** option and the middle button of the mouse, except that here the left button is used for the picks.
Concentric Pick an existing circle or arc; then pick the radius of the new circle.
3 Tangent Create a circle connecting three existing reference entities. These can be centerlines, construction features, or geometry features.
Fillet Create a circle tangent to two existing entities.
3 Point Pick three points you wish to define the circle's circumference.

Figure 12.11
Geometry Circles and Construction Circles

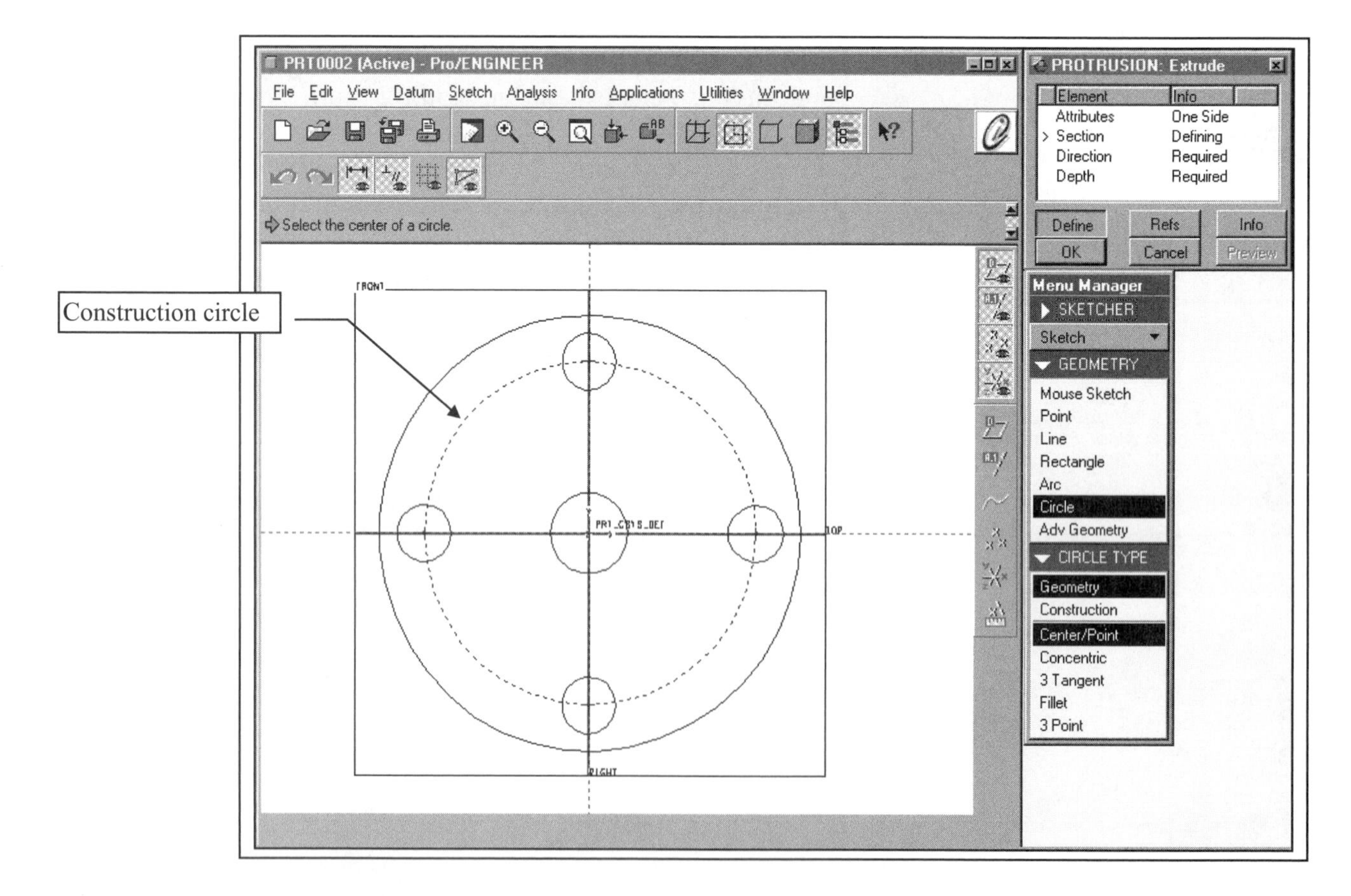

Arcs

Arcs are sketched using the menu or the mouse. To sketch arcs:

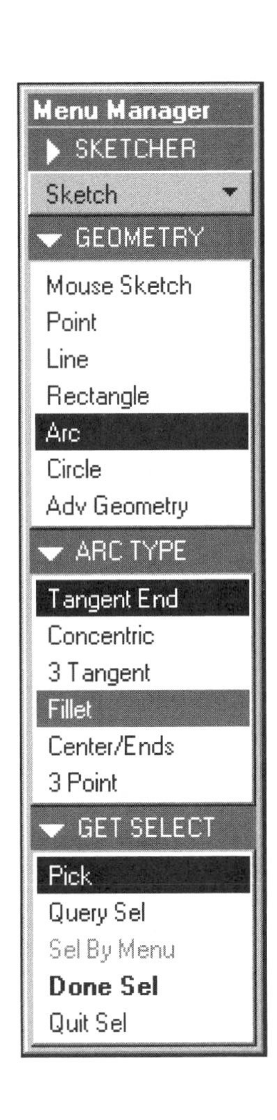

1. Select **Arc** from the GEOMETRY menu. The ARC TYPE menu appears.
2. Pick one of the following options from the ARC TYPE menu:

    Tangent End Same as creating an arc using **Mouse Sketch**, except that you must use the left mouse button. Pick an end of an existing entity to determine tangency; then pick the desired endpoint of the arc.
    Concentric Pick an existing circle or arc as a reference, then pick the endpoints of the new arc. As you create the arc, a radial line will appear through its center to assist you in aligning the endpoint.
    3 Tangent Select three entities for the new arc to be tangent to. The arc will be created in the same direction as the reference picks.
    Fillet Pick two entities to connect by creating a tangent arc (Figure 12.12).
    Center/Ends Pick the center point of the arc; then pick the arc's endpoints.
    3 Points Pick the endpoints of the arc; then pick a point on the arc.

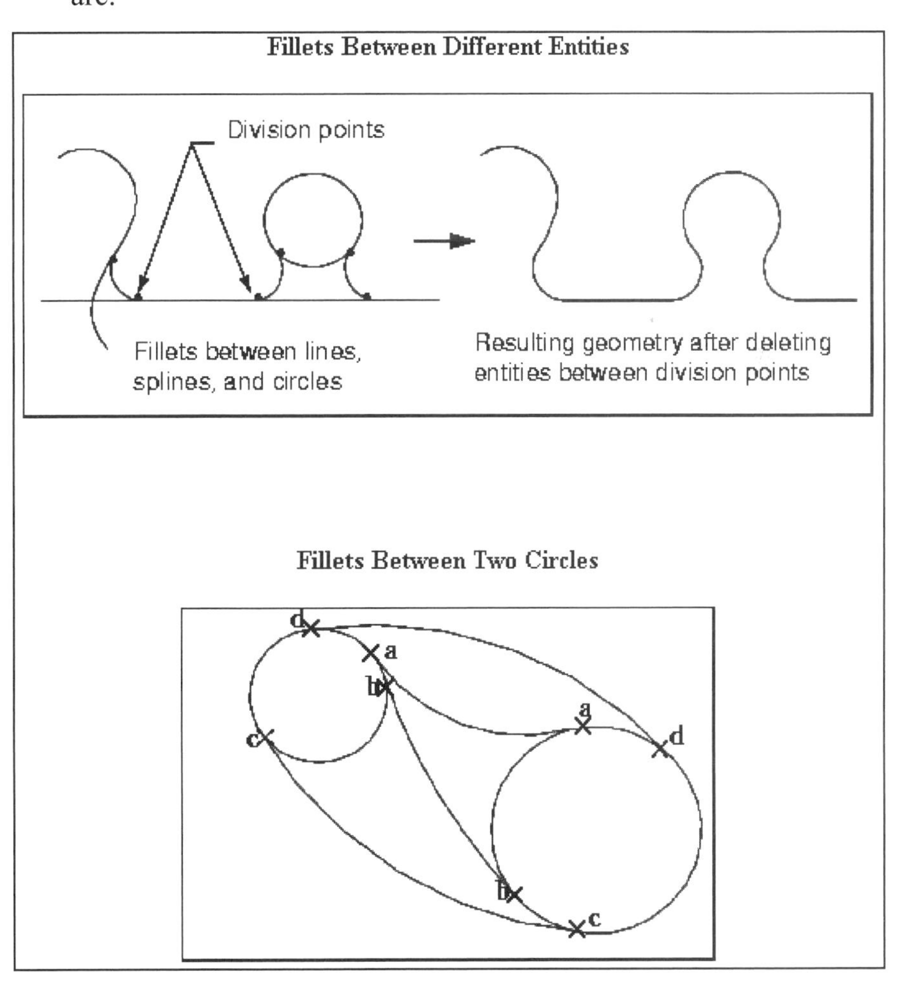

Figure 12.12
Arc Fillets and Fillets Between Circles

Advanced types of geometry, such as conics (Fig. 12.13) and splines (Fig. 12.14), can also be created in the Sketcher.

The **Conic** option allows you to sketch a conic. To create a conic, choose the following commands:

1. Choose **Conic** from the ADV GEOMETRY menu.
2. Pick the first endpoint for the conic using the left mouse button.
3. Pick the second endpoint for the conic using the left mouse button.
4. Pick the location for the shoulder using the left mouse button. The conic will rubberband as you do this.

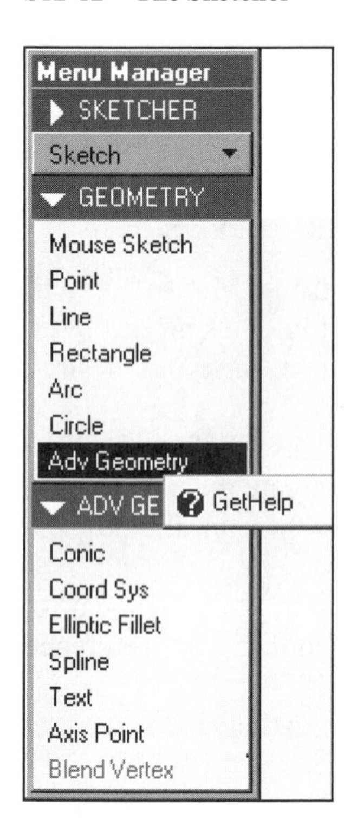

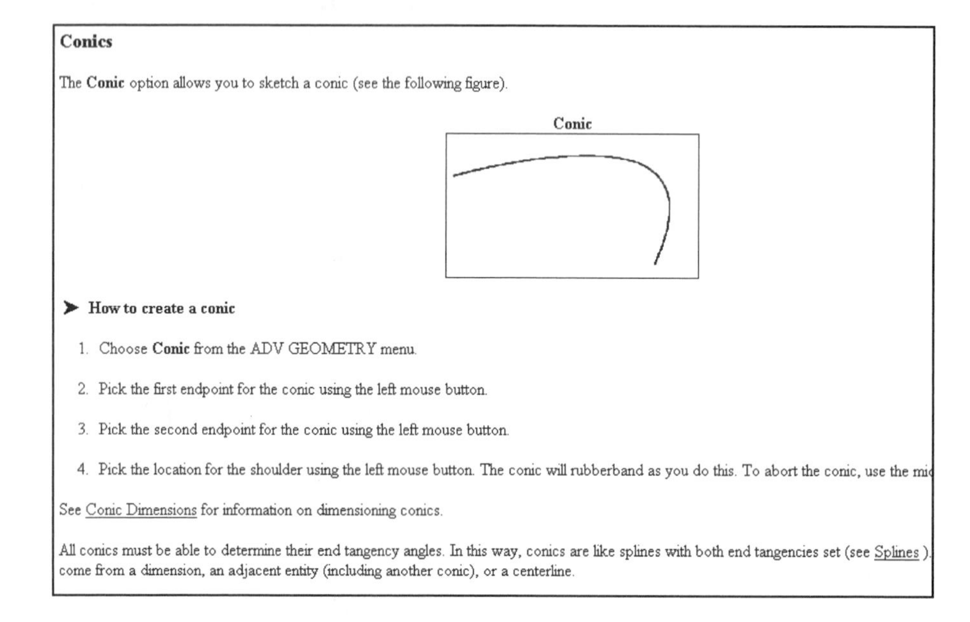

Conics

The **Conic** option allows you to sketch a conic (see the following figure).

➤ **How to create a conic**

1. Choose **Conic** from the ADV GEOMETRY menu.
2. Pick the first endpoint for the conic using the left mouse button.
3. Pick the second endpoint for the conic using the left mouse button.
4. Pick the location for the shoulder using the left mouse button. The conic will rubberband as you do this. To abort the conic, use the mi

See Conic Dimensions for information on dimensioning conics.

All conics must be able to determine their end tangency angles. In this way, conics are like splines with both end tangencies set (see Splines). come from a dimension, an adjacent entity (including another conic), or a centerline.

Figure 12.13
Sketching **Conics**

Splines are curves that smoothly pass through any number of intermediate points. The tangency angle and radius of curvature can be set at the ends of a spline to control its shape further.

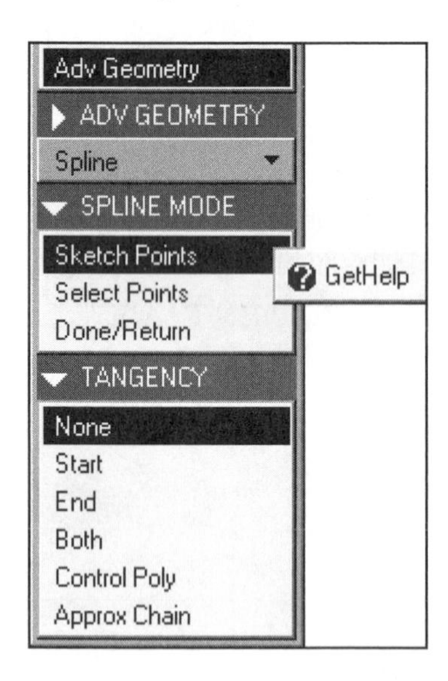

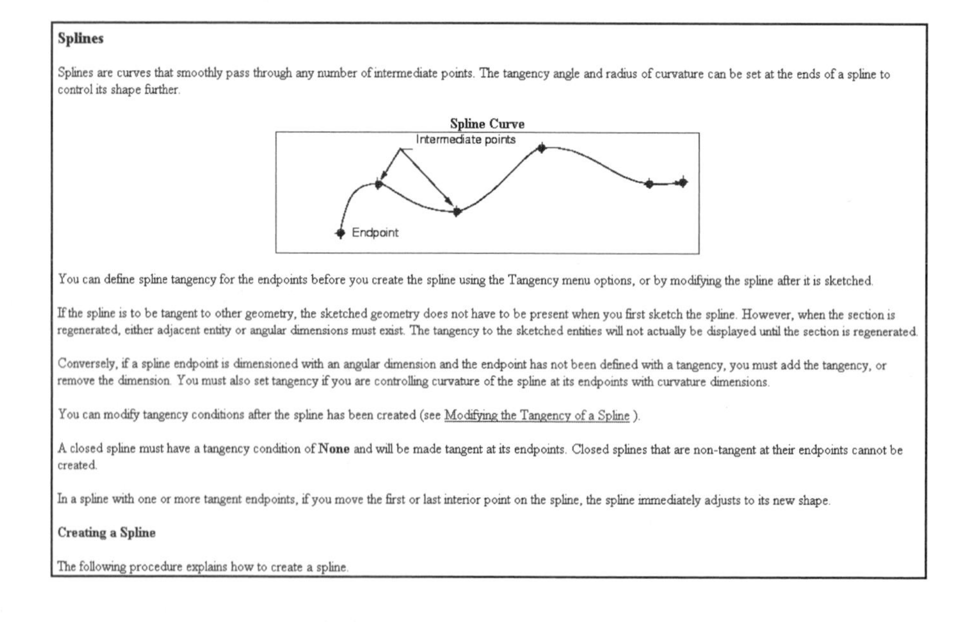

Splines

Splines are curves that smoothly pass through any number of intermediate points. The tangency angle and radius of curvature can be set at the ends of a spline to control its shape further.

You can define spline tangency for the endpoints before you create the spline using the Tangency menu options, or by modifying the spline after it is sketched.

If the spline is to be tangent to other geometry, the sketched geometry does not have to be present when you first sketch the spline. However, when the section is regenerated, either adjacent entity or angular dimensions must exist. The tangency to the sketched entities will not actually be displayed until the section is regenerated.

Conversely, if a spline endpoint is dimensioned with an angular dimension and the endpoint has not been defined with a tangency, you must add the tangency, or remove the dimension. You must also set tangency if you are controlling curvature of the spline at its endpoints with curvature dimensions.

You can modify tangency conditions after the spline has been created (see Modifying the Tangency of a Spline).

A closed spline must have a tangency condition of **None** and will be made tangent at its endpoints. Closed splines that are non-tangent at their endpoints cannot be created.

In a spline with one or more tangent endpoints, if you move the first or last interior point on the spline, the spline immediately adjusts to its new shape.

Creating a Spline

The following procedure explains how to create a spline.

Figure 12.14
Splines-- Sketch Points

Dimensioning Sections

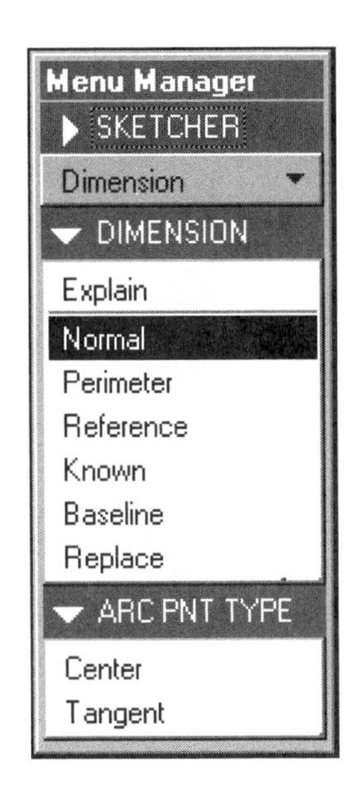

To regenerate a sketch successfully, it must be properly dimensioned. The Sketcher provides the ability to dimension a sketch with just a push of a button. If any references are required to locate a section, you are prompted to select the desired references and then complete the dimensioning scheme. There are two steps in dimensioning an entity: pick the entity or entities with the ***left mouse button (LMB);*** then place the dimension at the desired location using the ***middle mouse button (MMB).***

Linear Dimensions

Linear dimensions indicate the length of a line segment or the distance between two entities. The dimension value (Fig. 12.15) is displayed as a symbol until the sketch is successfully regenerated. Only horizontal and vertical dimensions are allowed when you create a dimension between two arc or circle extents (tangency points). The dimension is created to the tangency point closest to the pick point.

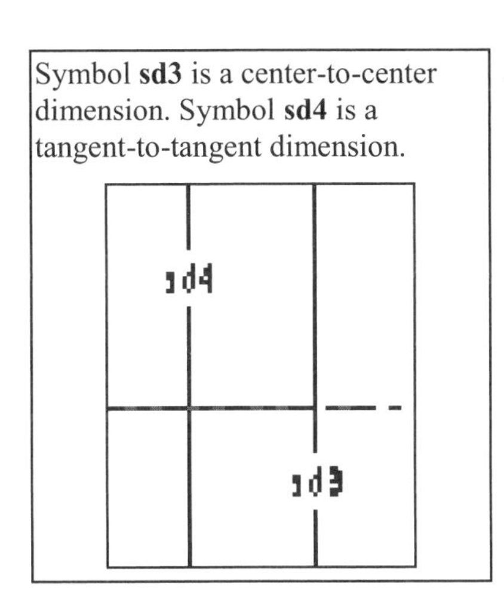

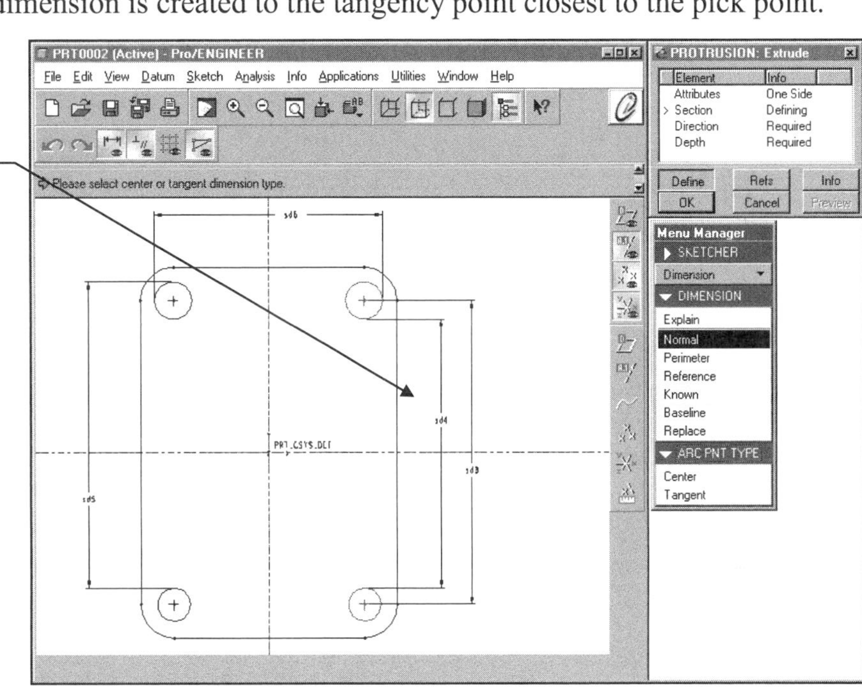

Figure 12.15
Linear Dimensioning

Linear dimensions (Fig. 12.15) let you do the following:

* Dimension the explicit length of a line: pick the line and then place the dimension.
* Dimension the distance between two parallel lines: pick the two lines and then place the dimension.
* Dimension the distance between a point and a line: pick the line, pick the point, and place the dimension.
* Create a dimension between two points (centerpoints and coordinate systems are included, but vertices are excluded): pick the points and location for the dimension. The distance between the points or, as in Figure 12.16, the shortest distance between a point and a circle is established based on where you pick the dimension location.

HINT
You can specify critical dimensions to identify key design intent, and the rest of the dimensioning can be completed automatically using **AutoDim.**

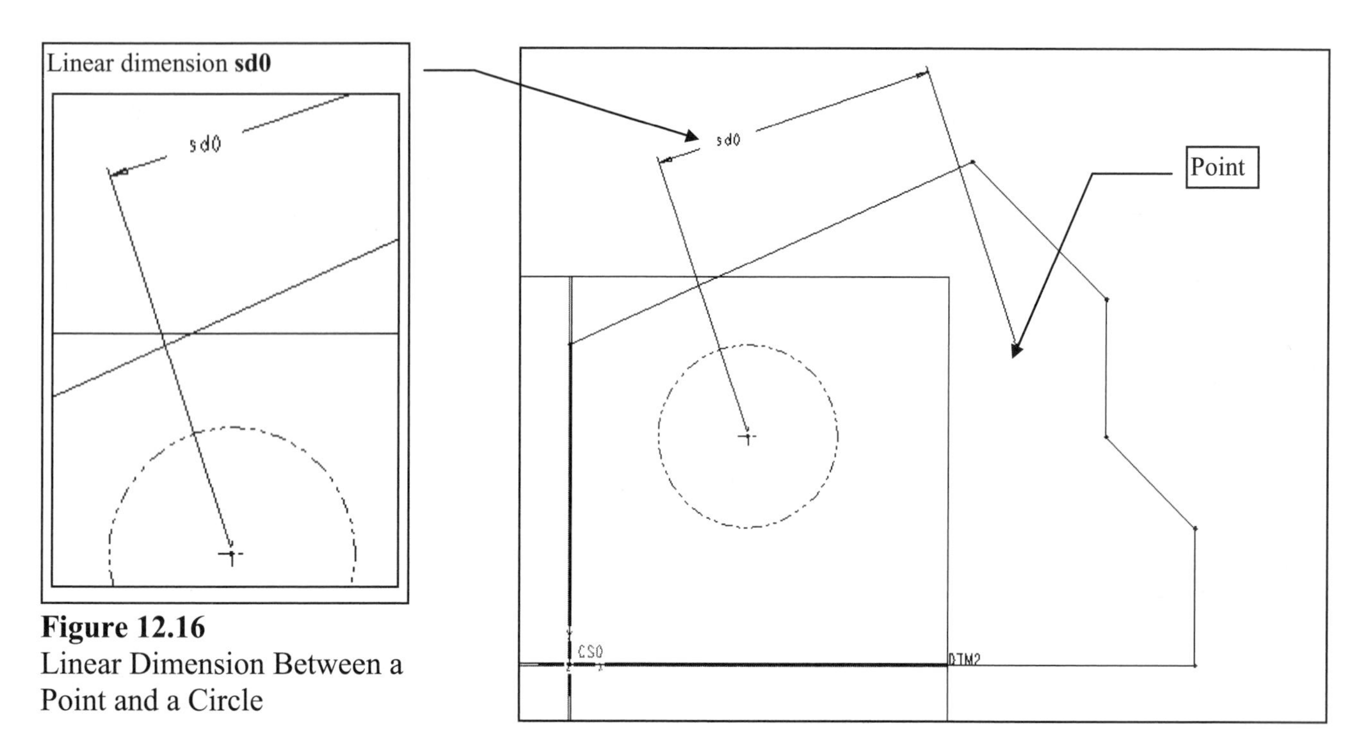

Figure 12.16
Linear Dimension Between a Point and a Circle

To dimension the distance between a line and a circle or arc:

1. Pick the line.
2. Pick the arc or circle.
3. Place the dimension [with the middle mouse button (MMB)].
4. The ARC PNT TYPE menu will appear, with the following options:

   Center Use to dimension between the arc or circle center and the line (Fig. 12.17).

   Tangent Use to dimension between a line and the point of nearest tangency on the arc or circle.

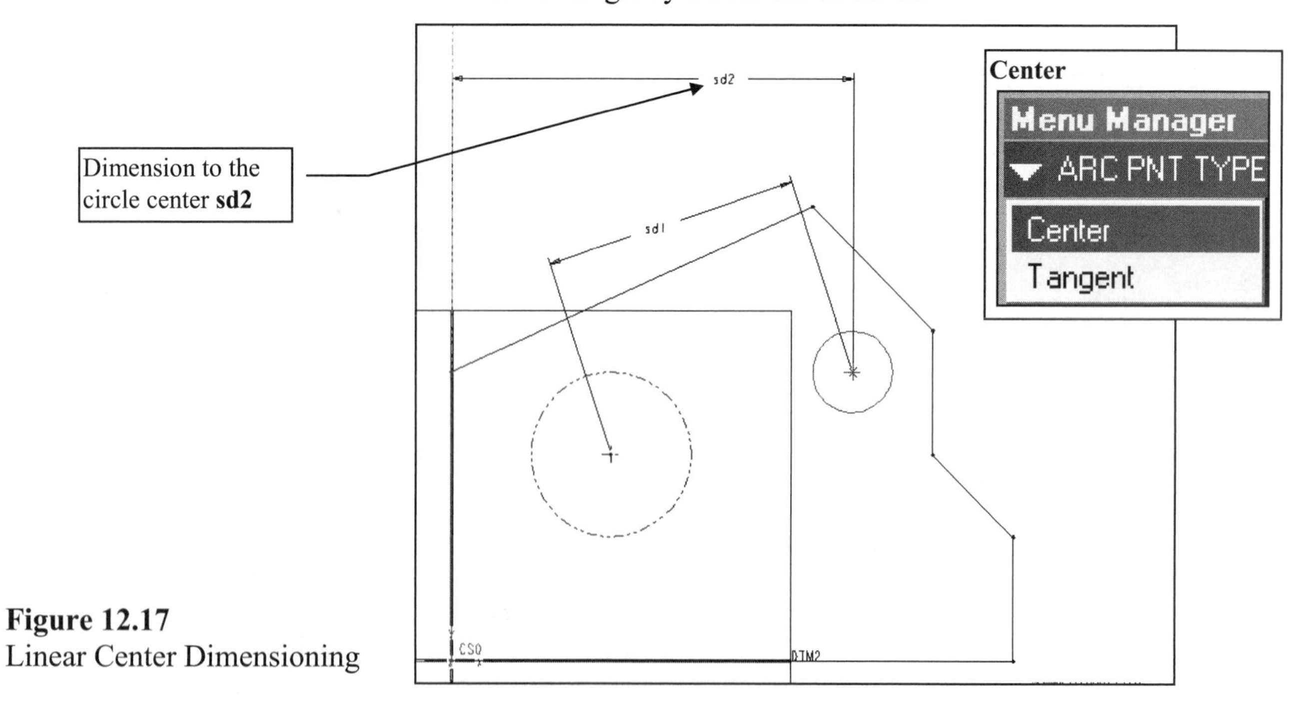

Figure 12.17
Linear Center Dimensioning

Figure 12.18 shows the dimension between a line and a circle at its tangency. To dimension between tangencies (Fig. 12.19):

1. Pick the first arc or circle.
2. Pick the second arc or circle.
3. Place the dimension.
4. Select **Tangent** from the ARC PNT TYPE menu.
5. Select either **Vert** or **Horiz** for the proper orientation.

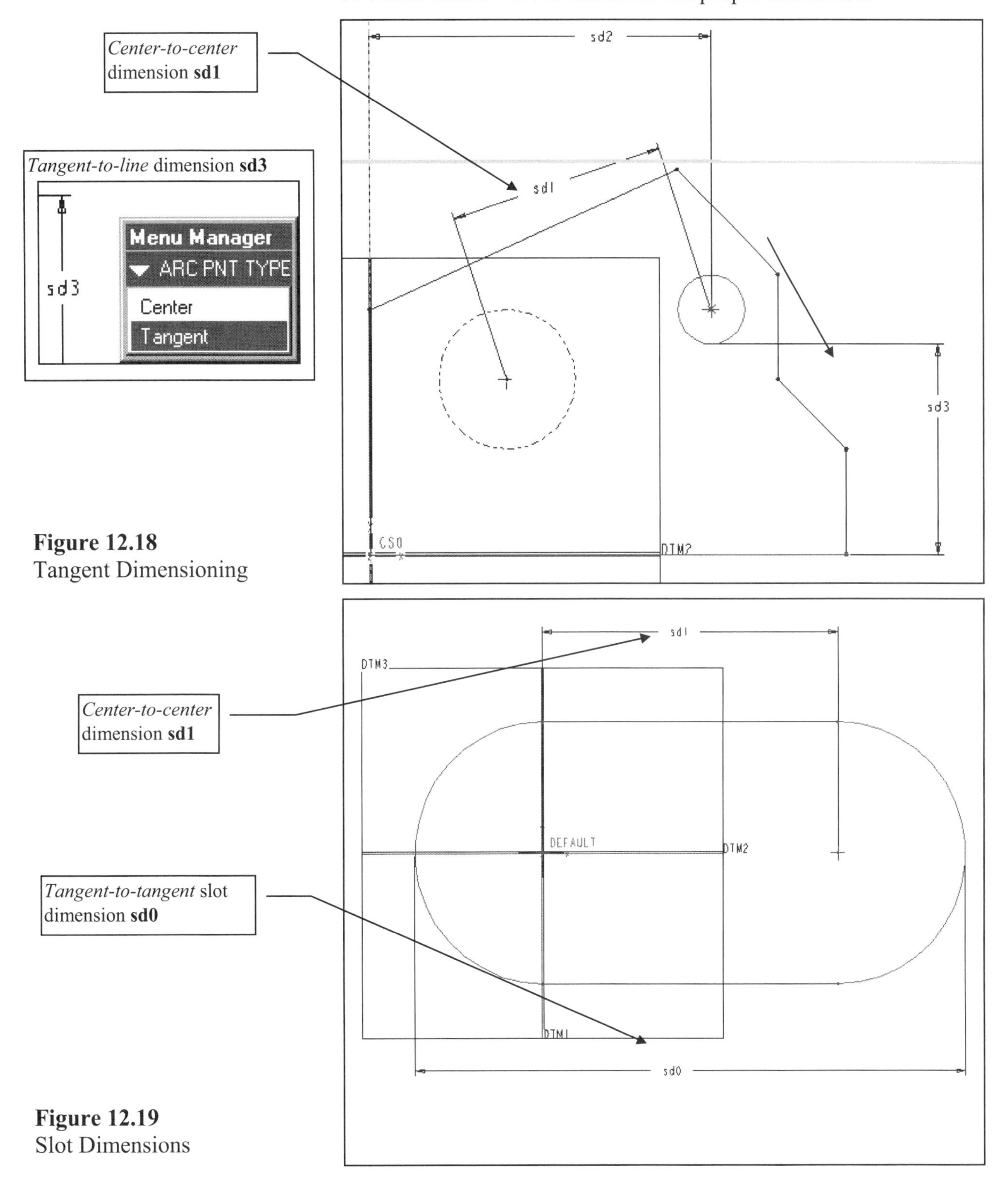

Figure 12.18
Tangent Dimensioning

Figure 12.19
Slot Dimensions

Diameter Dimensions

Diameter dimensions measure the diameters of sketched circles and arcs or the diameters for sketching sections about a centerline axis.

To create a diameter dimension for an arc or a circle, pick on the arc or circle twice, and then place the dimension (Fig. 12.20).

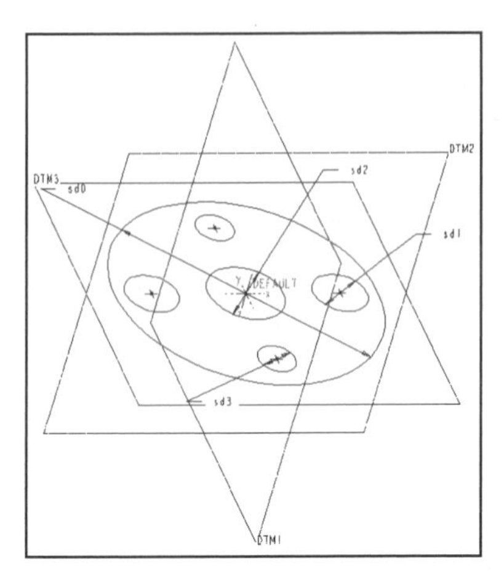

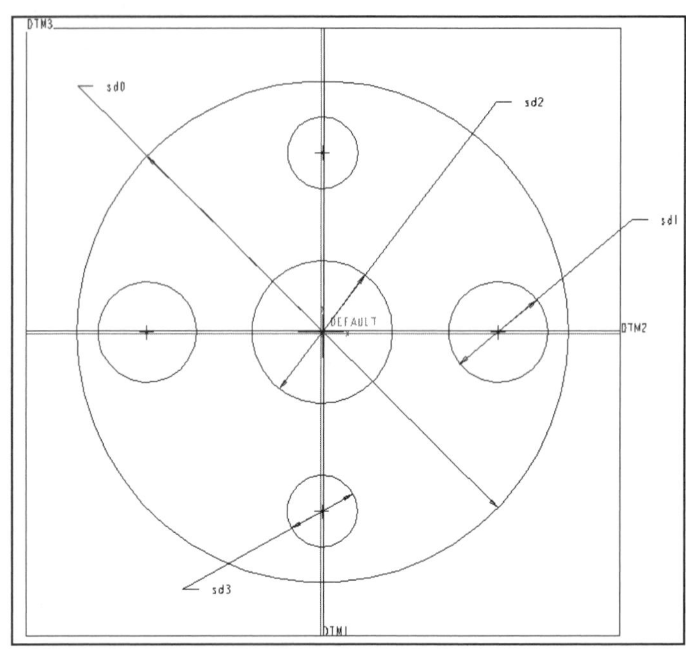

Figure 12.20
Diameter Dimensions

The diameter dimension for a revolved feature will extend beyond the centerline, indicating that it is a diameter dimension and not a radius dimension (Fig. 12.21). To create a diameter dimension for a section that will be revolved:

1. Select the entity to be dimensioned.
2. Pick the centerline that will be the axis of revolution.
3. Pick the entity again.
4. Place the dimension.

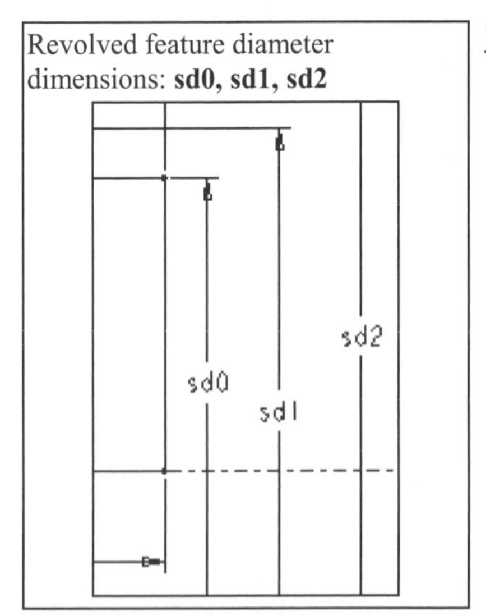

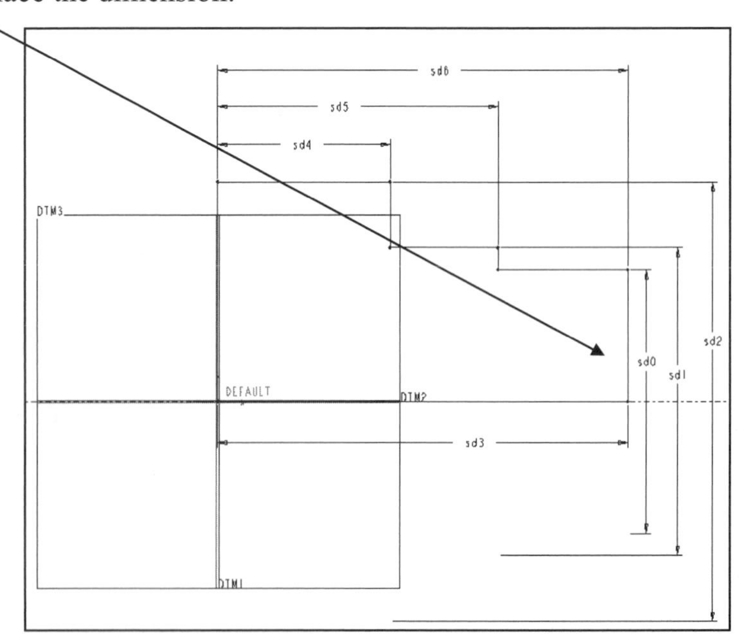

Figure 12.21
Revolved Feature Diameter Dimensioning

Radial Dimensions

Radial dimensions measure the radii of circles and arcs and the radii of circles and arcs created by revolving a sketched section about an axis. To create a radial dimension for an arc or circle, pick on the arc or circle and then place the dimension. In general, circles are dimensioned as diameters and arcs as radii (Fig. 12.22).

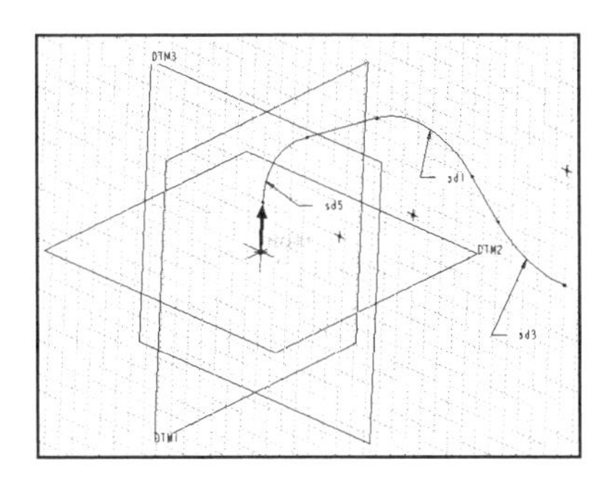

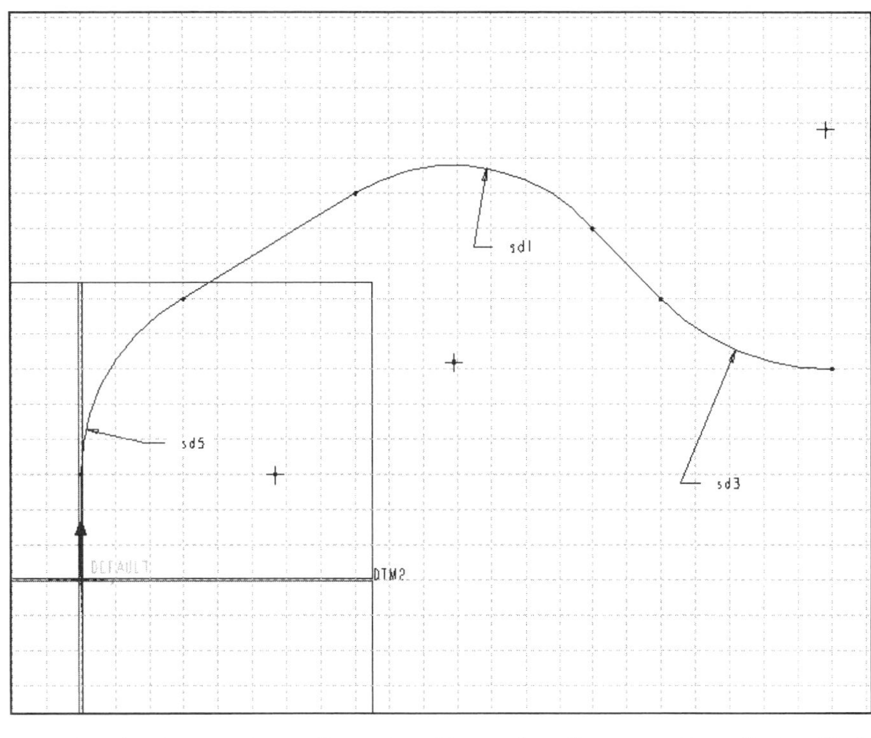

Figure 12.22
Arc Dimensioning

Again, to create a diameter dimension for a section that will be revolved, pick on the entity, pick on the centerline axis, pick on the entity, and then place the dimension. The example in Figure 12.23 was dimensioned in the 3D **Default** view instead of 2D **Sketch View.**

Diameter dimension **sd3** shown in 3D **Default** view

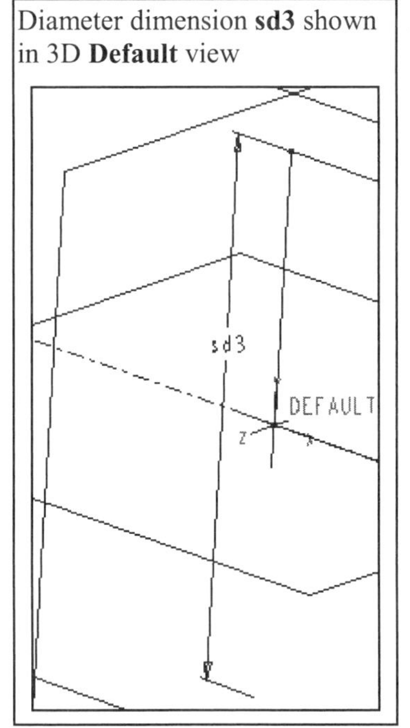

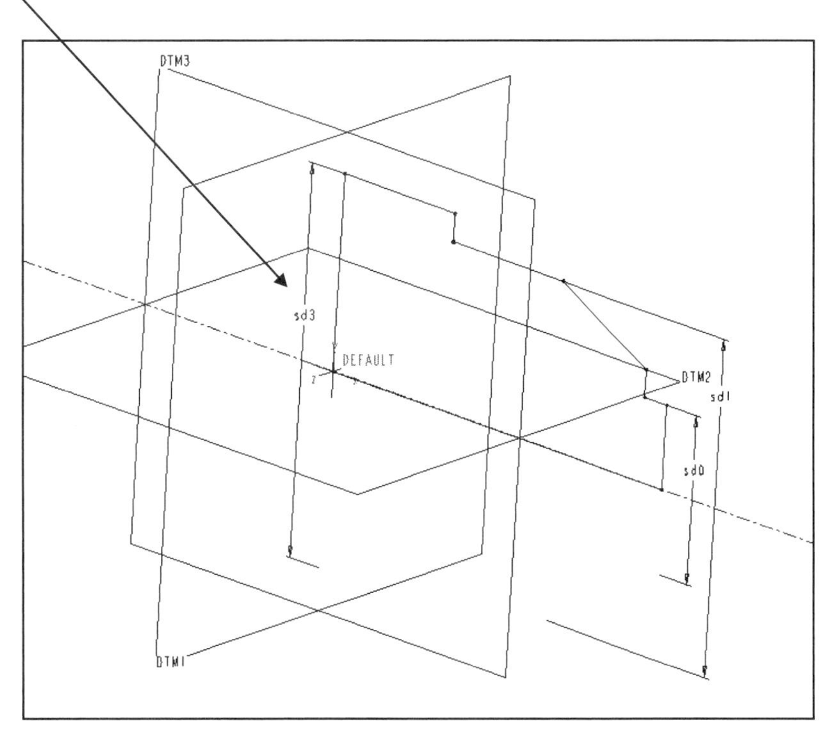

Figure 12.23
3D View
Diameter Dimensioning

Angular Dimensions

Angular dimensions measure the angle between two lines (Fig. 12.24) or the angle of an arc between its endpoints.

To create an angular dimension between lines, pick the first line, pick the second line, and then place the dimension. Where you place the dimension determines how the angle is measured (either acute or obtuse).

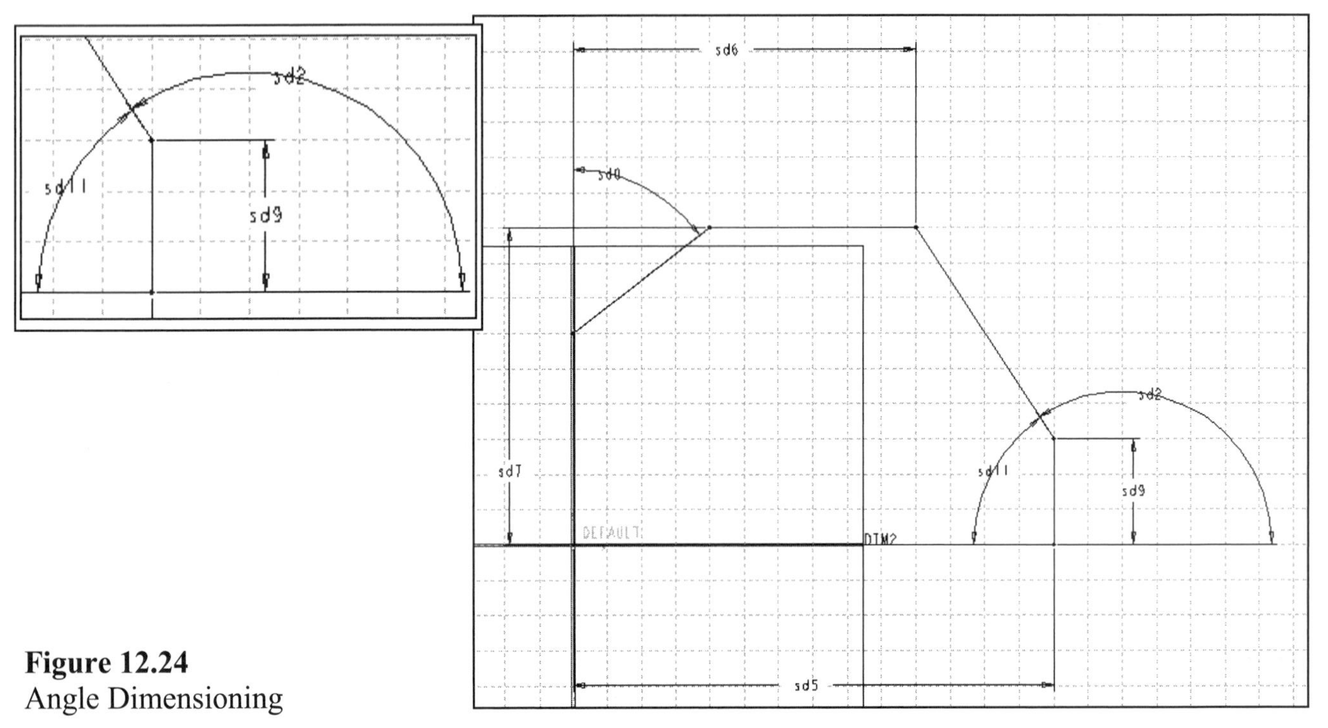

Figure 12.24
Angle Dimensioning

To create an arc angle dimension, pick one endpoint of the arc, pick the other endpoint of the arc, pick on the arc, and then place the dimension (Fig. 12.25).

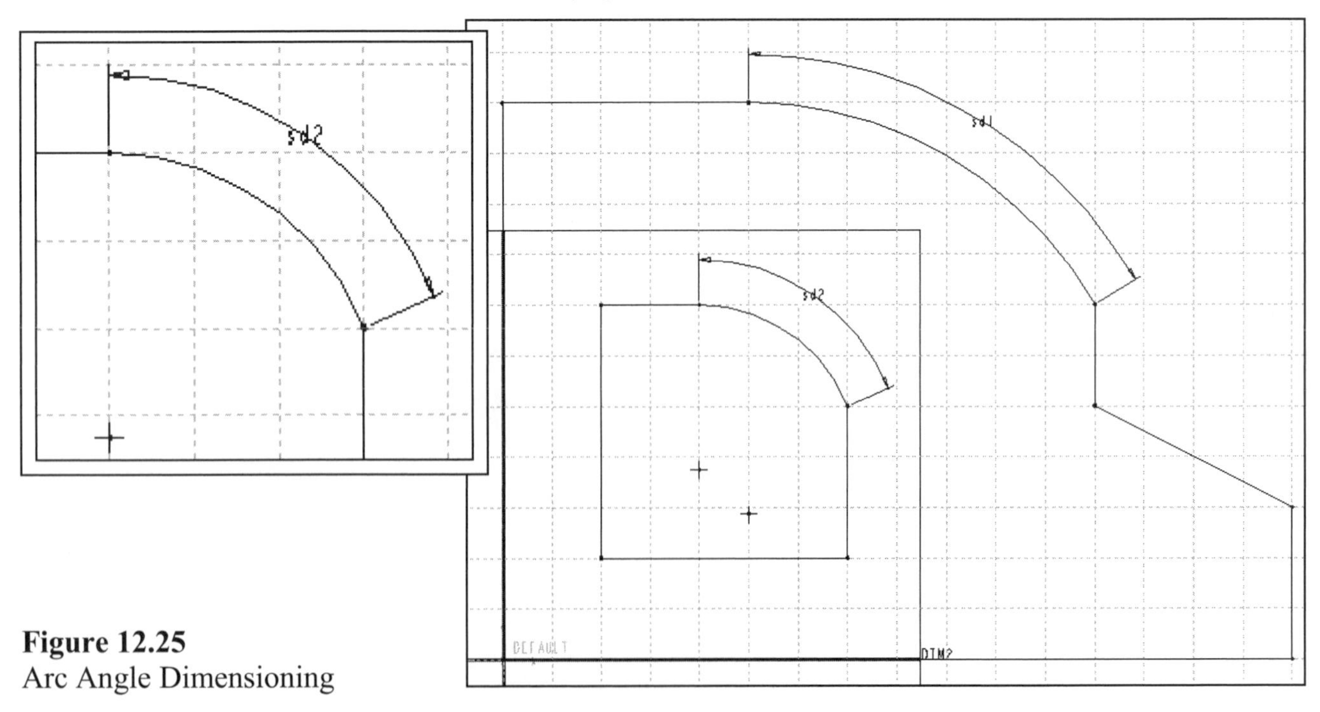

Figure 12.25
Arc Angle Dimensioning

Summing up, the aim of the Sketcher in parametric feature-based design is to create quick and simple geometry for your model. The sketching process enables you to create and dimension the geometry for a feature or set of features based on your design. Remember that during the sketching process, you need not concern yourself with creating perfect geometry or exact dimensions. You can modify your dimensions later in the design process (Figs. 12.26 and 12.27).

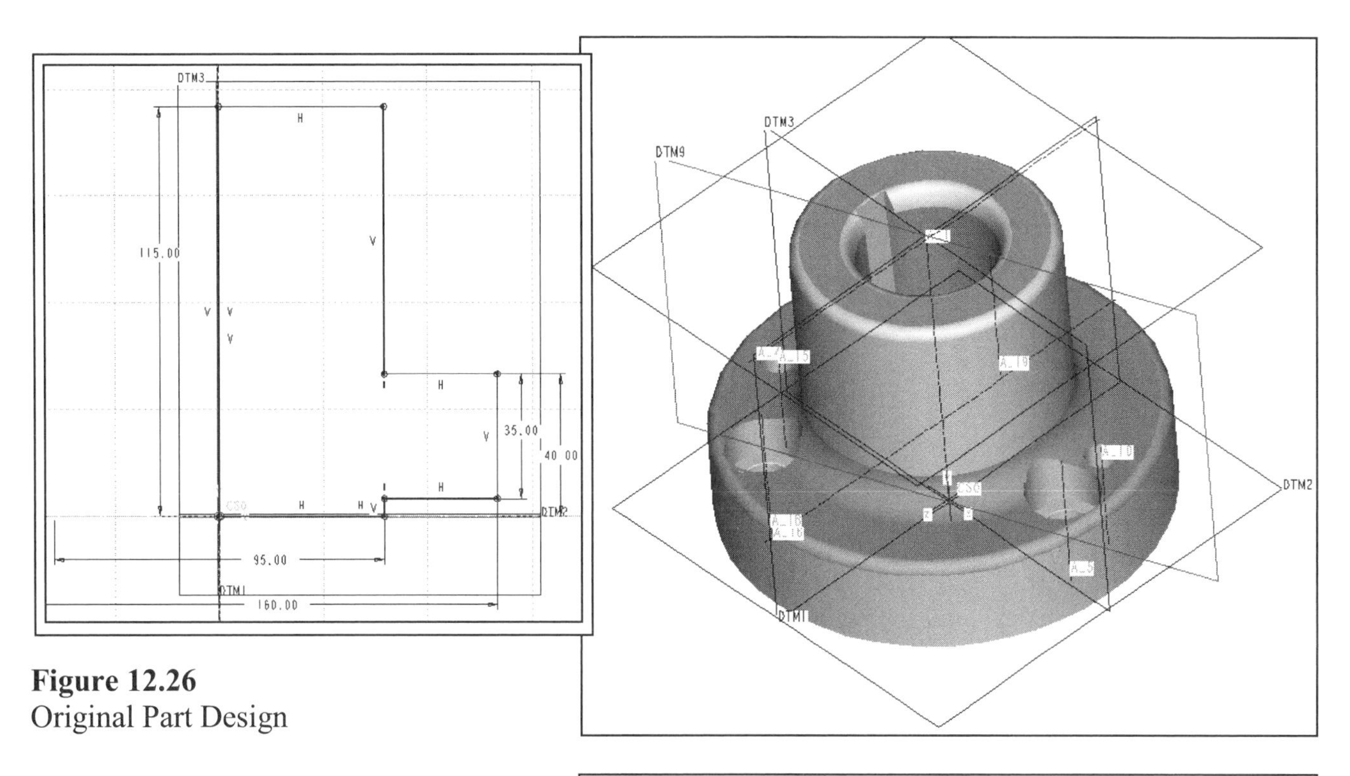

Figure 12.26
Original Part Design

Figure 12.27
Modified Final Design

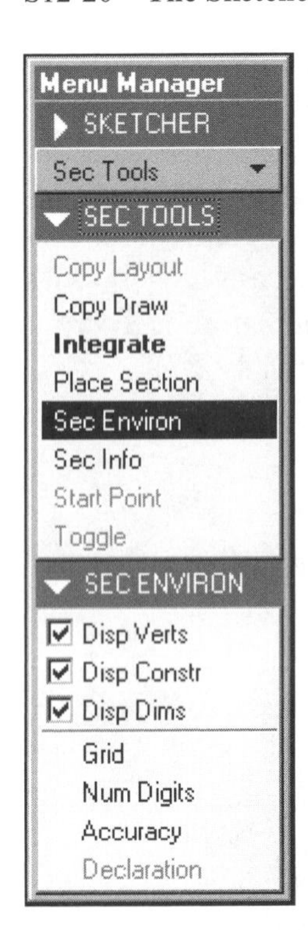

Figure 12.28
Sec Tools ⇒ Sec Environ

Sketcher Environment Options

The following procedure explains how to set the Sketcher environment. To set the Sketcher environment options, choose the following options (Fig. 12.28):

1. Choose **Sec Tools** from the SKETCHER menu.
2. Choose **Sec Environ** from the SEC TOOLS menu.
3. The SEC ENVIRON menu lists the following options:

Disp Verts Toggle the display of vertices by placing or removing a check mark. You can preset the display of vertices by setting the configuration option "*sketcher_disp_vertices*".
Disp Constr Toggle the display of constraints by placing or removing a check mark. You can preset the display of constraints by setting the configuration option "*sketcher_disp_constraints*".
Disp Dims Toggle the display of sketcher dimensions by placing or removing a check mark.
Grid Access the grid options.
Num Digits Change the number of decimal places for Sketcher dimensions you will enter.
Accuracy Change Sketcher accuracy.
Declaration Declare a sketch to the section.

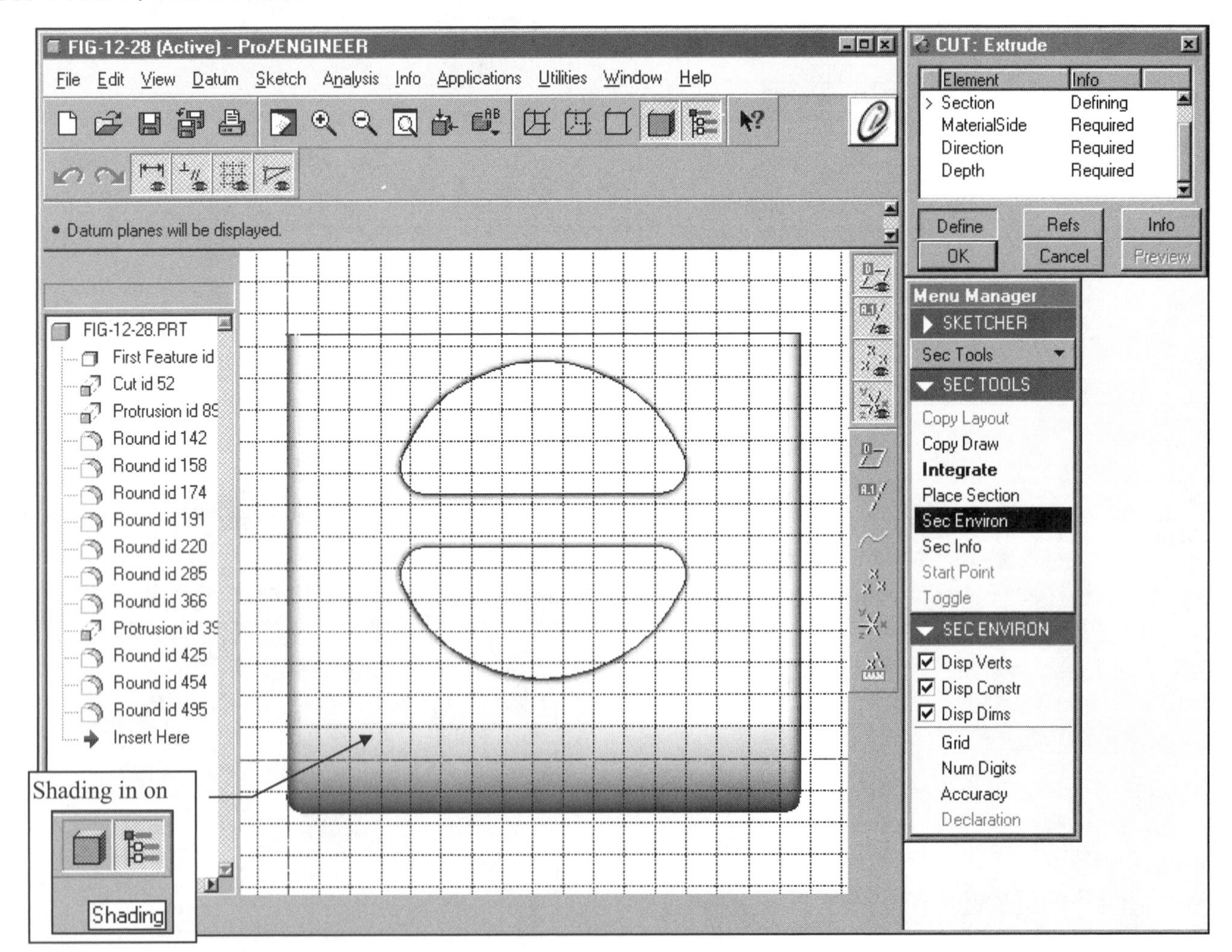

Grid

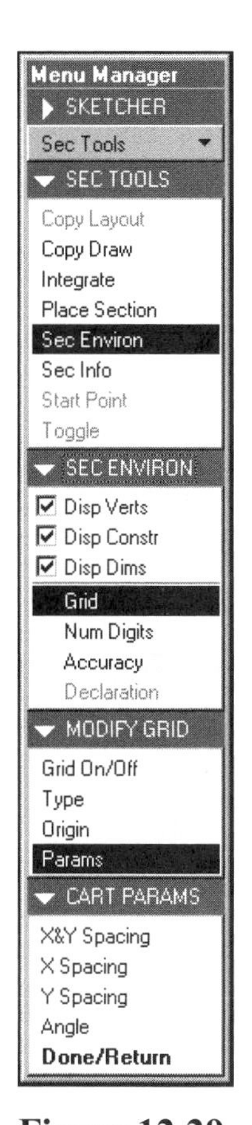

Sketcher mode supports both Cartesian and Polar grids. When you first enter Sketcher mode, Pro/E displays a Cartesian grid (Fig 12.29).

For the first feature section of a part and for auxiliary sketches such as blind holes, the grid has a spacing equal to one model unit. For example, sketching a box **4x6** grid spaces creates a box measuring **4x6** units.

Additional section sketches for a model use a grid for reference only. You can modify this grid spacing, but the first grid displayed is scaled for the current part size and does not have a value of one unit between grid lines.

Modifying the Grid

You can modify the grid spacing to suit the intended size of your sketch, making it easier to adjust to actual dimension values when you are done. To modify the grid, use the command sequence **Sec Tools ⇒ Sec Environ ⇒ Grid**, and choose an option from the MODIFY GRID menu.

The MODIFY GRID menu lists the following options:

Grid On/Off Toggle the display of the grid on or off.
Type Modify the type of the grid.
Origin Modify the origin of the grid.
Params Modify the parameters of the grid.

Figure 12.29
Cartesian Grid

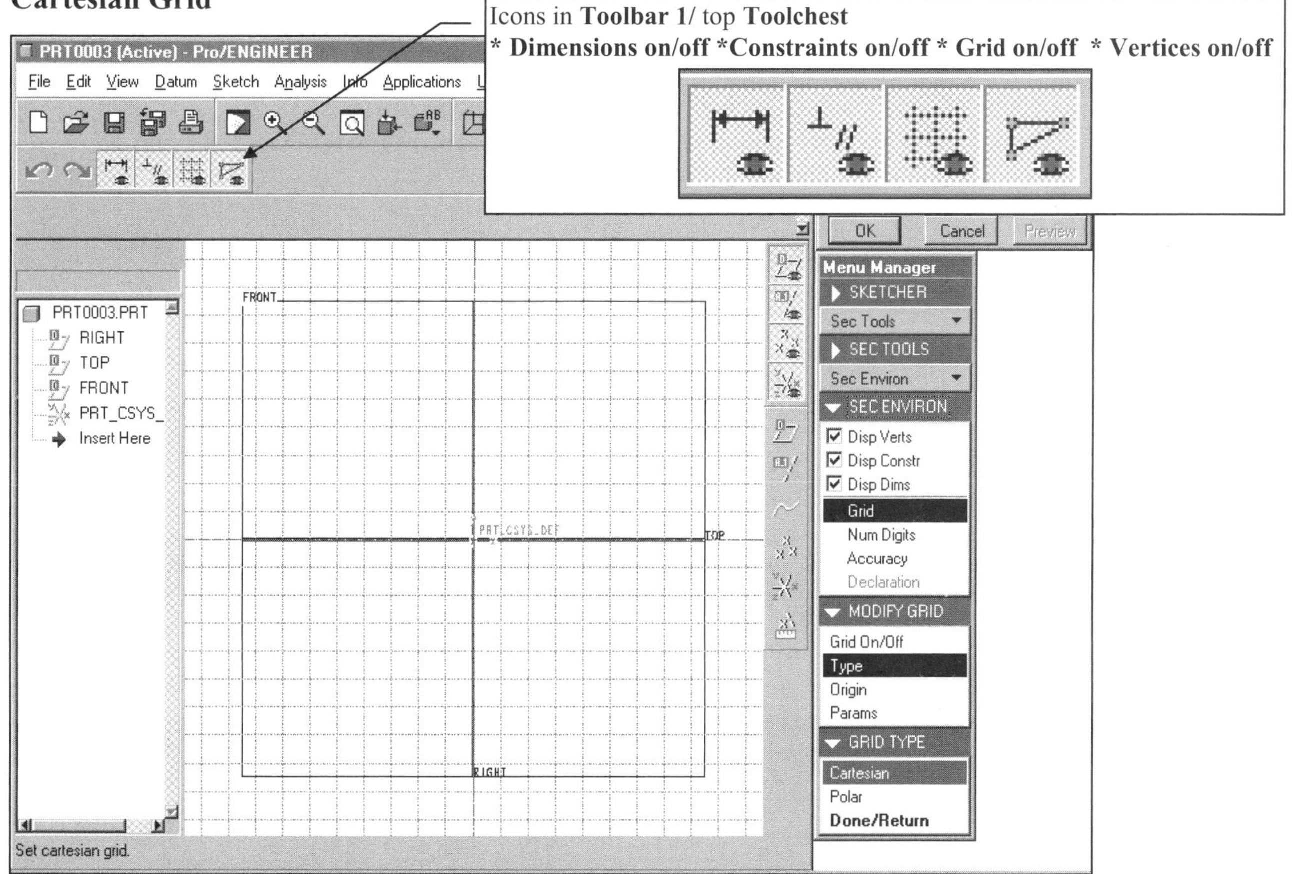

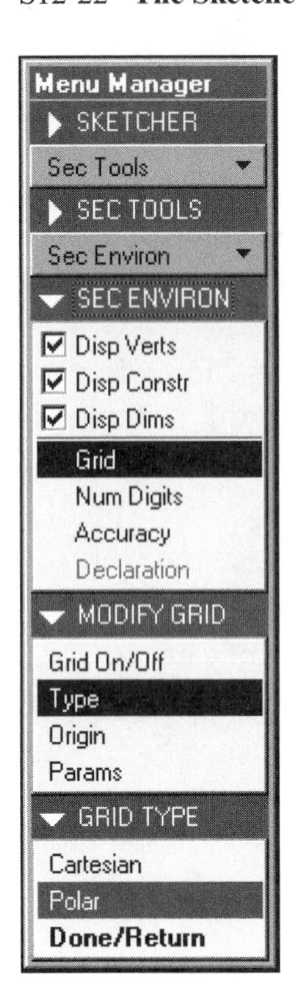

Displaying the Grid

To toggle the display of the grid, choose **Grid On/Off** from the MODIFY GRID menu or from Toolbar 1 (Fig. 12.29). *Note that this does not affect the snapping of sketched entities to the grid intersections.*

Moving the Grid Origin

You can set the grid intersection at the following locations:

* Sketched entity endpoint and center.
* Sketched point and coordinate system.
* Datum point and coordinate system.
* Edge or curve vertex.

To do so, choose **Origin** from the MODIFY GRID menu. Select the appropriate geometry to locate the origin.

Modifying the Type of Grid

To change the type of grid being used, choose **Type** from the MODIFY GRID menu, and then choose one of the GRID TYPE menu options:

Cartesian Use a Cartesian grid.
Polar Use a Polar grid (Fig. 12.30).

Figure 12.30
Polar Grid

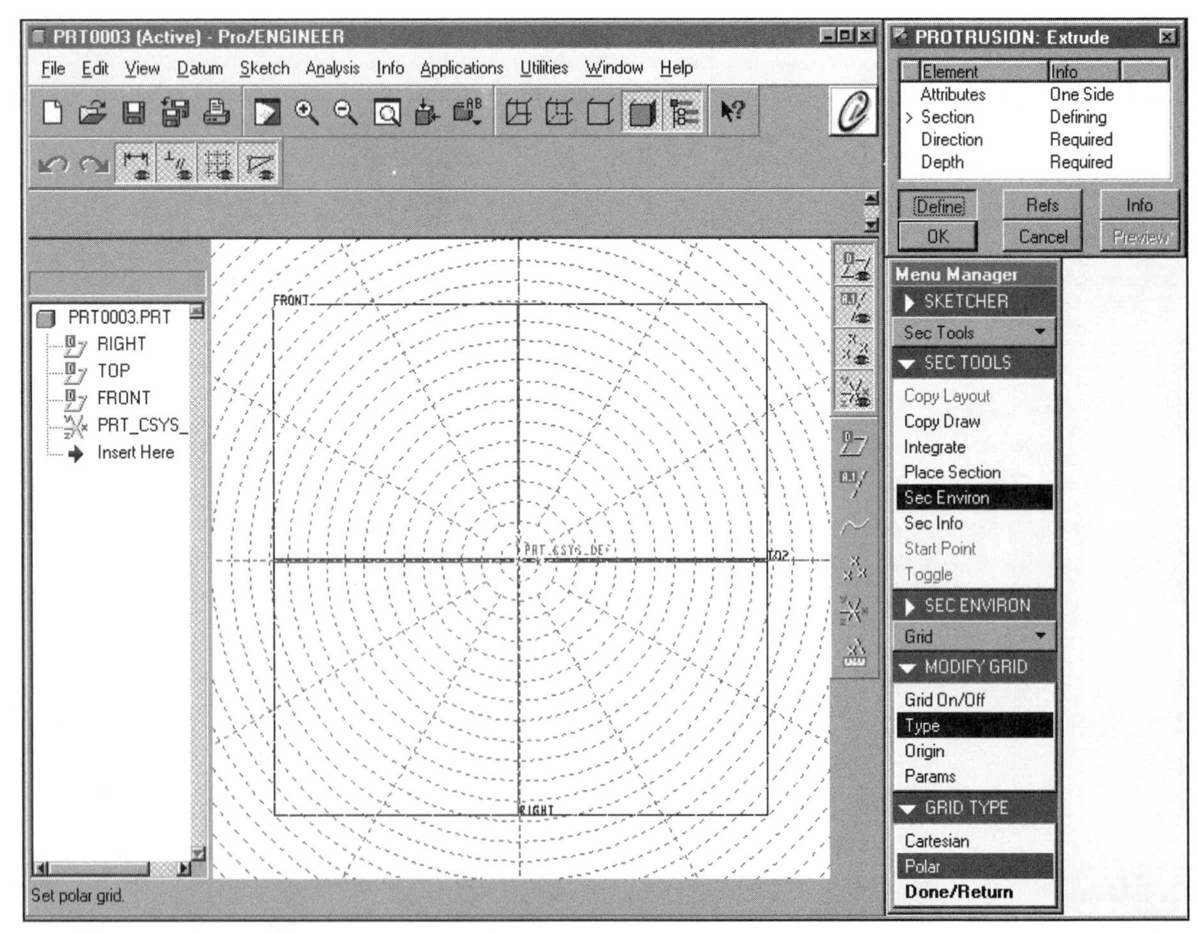

Modifying the Grid Spacing

The **Params** option allows you to modify the grid spacing and angle. You can use this option when you start a sketch (before any geometry has been created) to control the approximate size of the section. For example, if you have a blank sketch and a **20x17** Cartesian grid in your window, and you change the **X** and **Y** spacing from **1.0** to **0.5,** then instead of seeing a **40x34** grid, you will have decreased the size of the sketching area to **10x8.5** units. If you have sketched at least one entity, and then you modify the grid spacing, the grid spacing changes while the sketched entities remain unchanged.

To modify the grid spacing or angle, choose **Params** from the MODIFY GRID menu. The options available depend on the type of grid.

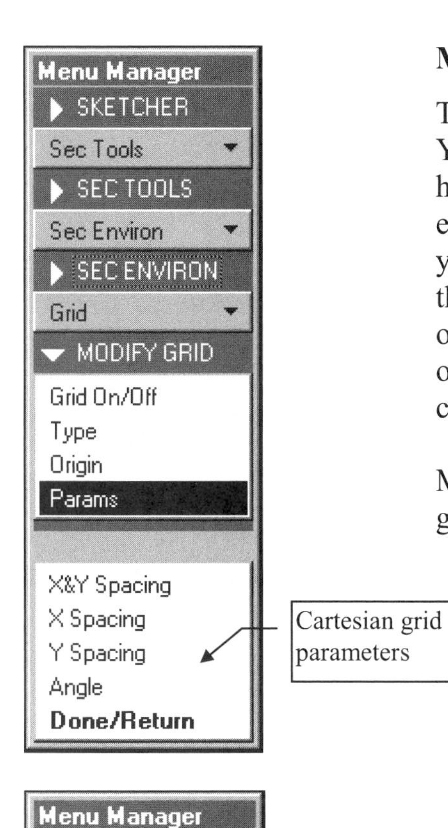

For a *Cartesian grid*, the available options are as follows:

X&Y Spacing Set the spacing in both the **X** and **Y** directions to the same value (in Fig. 12.31 the grid spacing was set to **15**).
X Spacing Set the **X**-direction spacing only.
Y Spacing Set the **Y**-direction spacing only.
Angle Modify the angle between the horizontal and the **X**-direction grid.

For a *Polar grid*, the available options are as follows:

Ang Spacing Set the angular spacing between radial lines. The specified value must divide evenly into **360.**
Num Lines Set the number of radial lines. The angular spacing is **360** divided by the number of lines.
Rad Spacing Modify the spacing of the circular grid.
Angle Modify the angle between the horizontal and the **0** degree radial line.

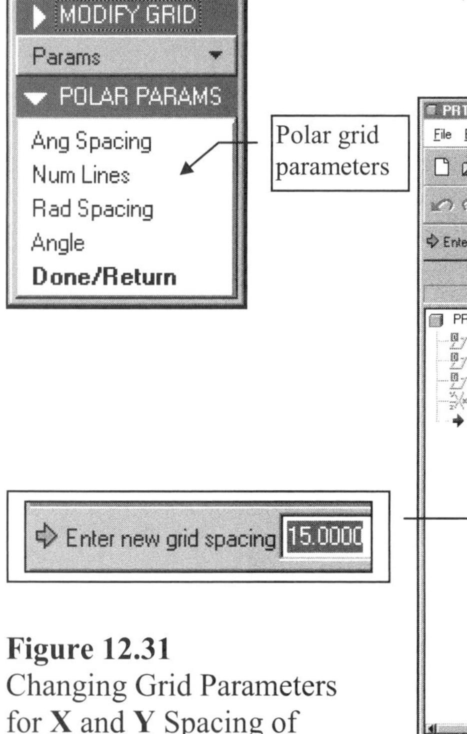

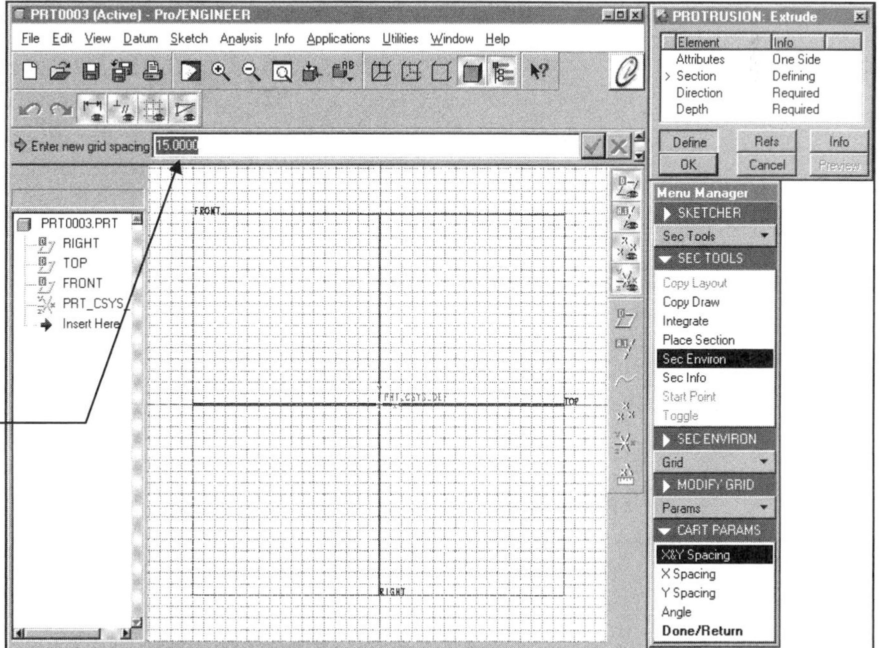

Figure 12.31
Changing Grid Parameters for **X** and **Y** Spacing of Cartesian Grid

Setting the Grid to Snap

With ✔**Snap To Grid** turned on, each pick of the mouse causes Pro/E to snap a point to the intersection of grid lines. To enable grid snapping, choose **Utilities** ⇒ **Environment** ⇒ ✔ **Snap To Grid** or to disable-- ❑ **Snap To Grid.** You can also turn on grid snapping by default by setting the *"grid_snap"* configuration option to *"yes."* Snapping to the grid helps when sketching in 3D [using the **Default** or rotated view (Fig. 12.32)].

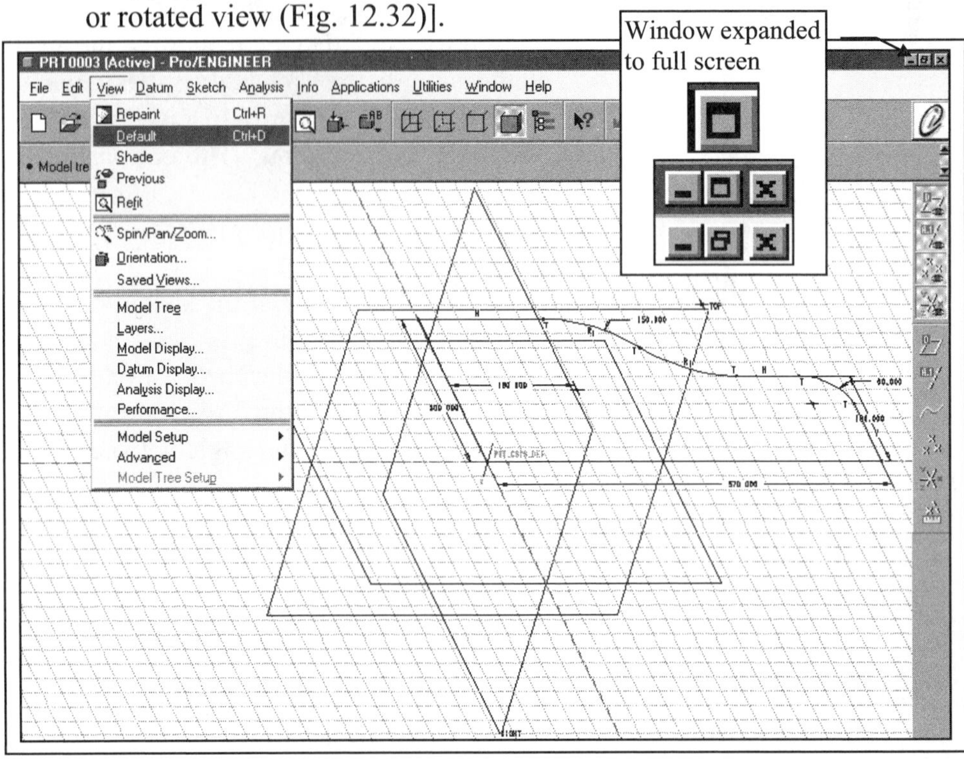

Figure 12.32
Sketching in (3D) Default View with ✔**Snap To Grid**

Sketcher Accuracy

Modify the Sketcher accuracy to help solve certain section regeneration problems. For example, if Pro/E issues the message that says an entity has zero length (because its length is less than the accuracy used by the Sketcher), increasing the accuracy by entering a smaller number makes the length appear to be nonzero. Pro/E can then solve the section (assuming that no other problems arise as a result of increasing the accuracy).

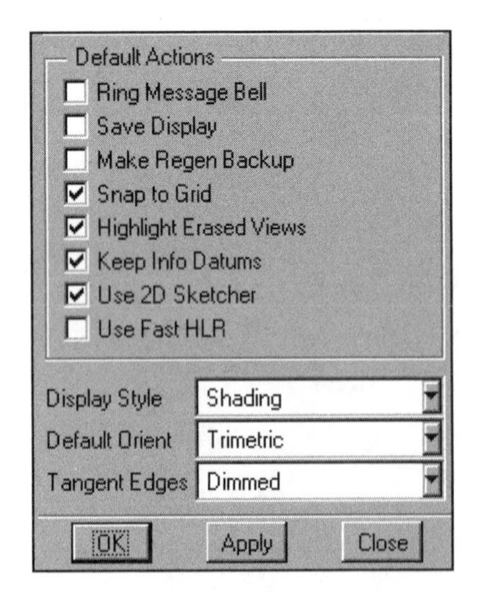

How to change the Sketcher accuracy:

1. Choose **Sec Environ** from the SEC TOOLS menu.
2. Choose **Accuracy** from the SEC ENVIRON menu.
3. At the prompt for the new accuracy, enter a value between **1.0E-12 (0.000000000001)** and **1.0**.
4. **Regenerate** the section.

If the section still fails to regenerate successfully, try increasing the accuracy again (entering a smaller number), or evaluate the section for other problems.

Pro/E remembers the relative accuracy of each section when you redefine a feature that contains the section.

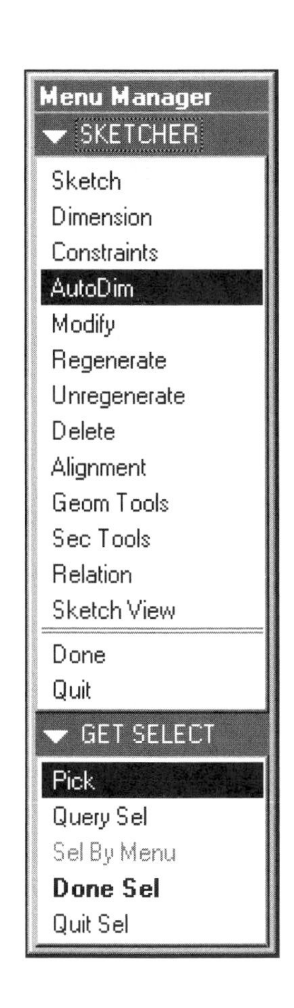

Automatic Dimensioning

After you sketch a section, you can automatically dimension it by choosing **AutoDim** from the SKETCHER menu. Pro/E adds all dimensions that are necessary to constrain the section and then regenerates it. You can also use **AutoDim** on a partially dimensioned section.

Once Pro/E places all dimensions, you can move them to the desired location by using **Move Entity** from the GEOM TOOLS menu and then **Dimension** from the MOVE ENTITY menu.

For information on how to use **AutoDim** with sections created in a part with existing geometry, see Using Automatic Dimensioning.

Using Automatic Dimensioning

With automatic dimensioning, Pro/E adds dimensions to your sketch so that it is fully constrained and then regenerates it (Fig. 12.33). When you create a section in a part with existing geometry, Pro/E must locate the section with respect to the part. For this, Sketcher uses "known" geometry. There are several ways to make geometry "known":

When you specify horizontal and vertical references for orienting the sketching plane, Pro/E uses these as known geometry.

You can make geometry known by referencing it before you use **AutoDim** (for example, by aligning it, using a model edge for sketching, or dimensioning to it.)

When you align the section to the part geometry at Pro/E alignment query, that geometry becomes "known." To make any known geometry unknown, use the **Unalign** command.

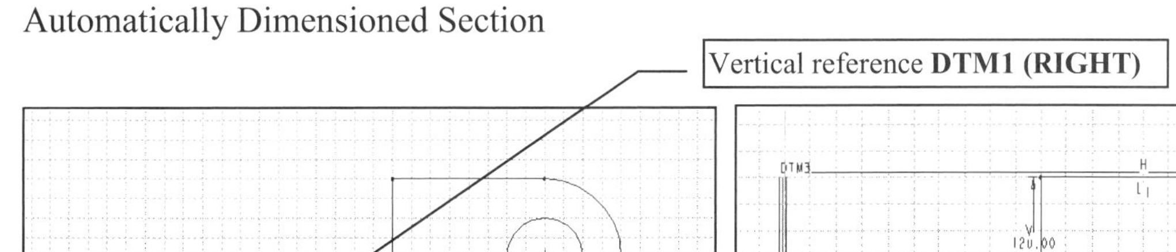

Figure 12.33
Automatically Dimensioned Section

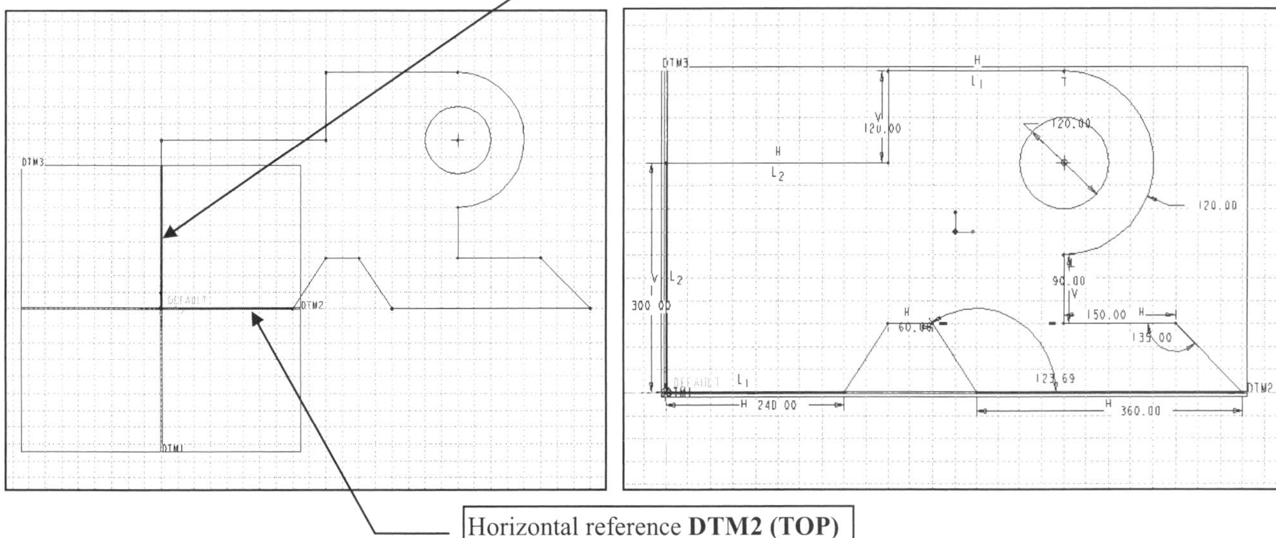

Figure 12.34
Final Regenerated Sketch

To dimension a section automatically:

1. Sketch geometry and choose **AutoDim**.
2. If Pro/E needs additional references for locating the section with respect to the part geometry, it prompts you to select these references by picking edges and vertices. When finished, choose **Done Sel**(ect) from the GET SELECT menu or press the middle mouse button.
3. Pro/E adds all necessary dimensions to constrain the section. Notice that projections of known geometry onto the sketching plane appear in orange with a dotted font.
4. Pro/E checks whether the section can be aligned to the existing part geometry. If such alignment is possible, Pro/E brings up the query menu so you can select the desired action. Choose one of the following options:

   Align Align geometry as prompted in the alignment query.
   Don't Align Do not align geometry for that alignment query.
   DontAlignAny Do not align any geometry in the current automatic dimensioning operation.

5. After Pro/E regenerates successfully, you can move dimensions to the desired location by using **Move Entity** from the GEOM TOOLS menu and then **Dimension** from the MOVE ENTITY menu.

The sketch shown in Figure 12.33 was cleaned up in Figure 12.34 by moving dimensions. The required design dimensions were also added, and the unneeded dimensions displayed using **AutoDim** were removed. Dimensions shown using **AutoDim** are sometimes ambiguous and useless, but other times it can help you solve a sketch.

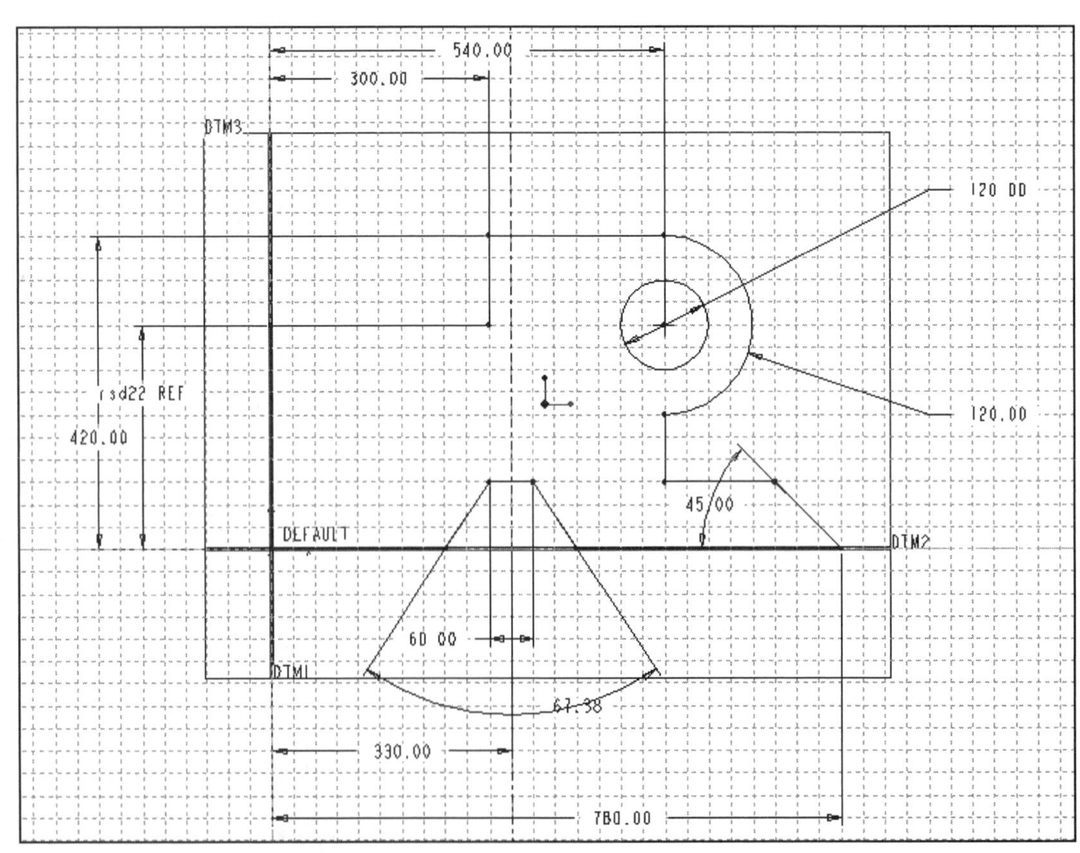

Place Section

Saved sketches are referred to as **Sections**. They always have a file extension of **.sec.** You can retrieve predefined section files in Pro/E, allowing you to reuse common sketches easily. This is accomplished while you are in the Sketcher by using the **Sec Tools ⇒ Place Section** function.

After working with Pro/E for a while, you will most likely find that you are redoing the same sketch repeatedly. If you have sections that you want to reuse, you can save them (using the **File ⇒ Save As** function to give them meaningful names) into a "common sections" directory.

You can reuse a sketch in many different models, using the **Sec Tools ⇒ Place Section** function in the Sketcher (Fig. 12.35). Pro/E will prompt you to enter a section name.

If you have already created several features in the part, Pro/E also allows you to list all of the sections that are **In Session**. Recall that as a part of the process of creating a new feature (for example, with the **Protrusion** or **Cut** option), Pro/E puts you into Sketch mode.

In Session allows you to select any of the sketches that you have created during the current session, no matter which part they were originally created in.

Only those sketches that you have created (or worked on) in the current session will be available with the **In Session** option. Sketches created in previous sessions will not be available. To reuse a sketch (not saved to disk) that you created in a previous session, you must first **Open** the part, then **Redefine** the **Sketch,** and then store it to disk (using the **File ⇒ Save As** option).

When you reuse sections in this manner, there is no associativity between the features created from the same section file.

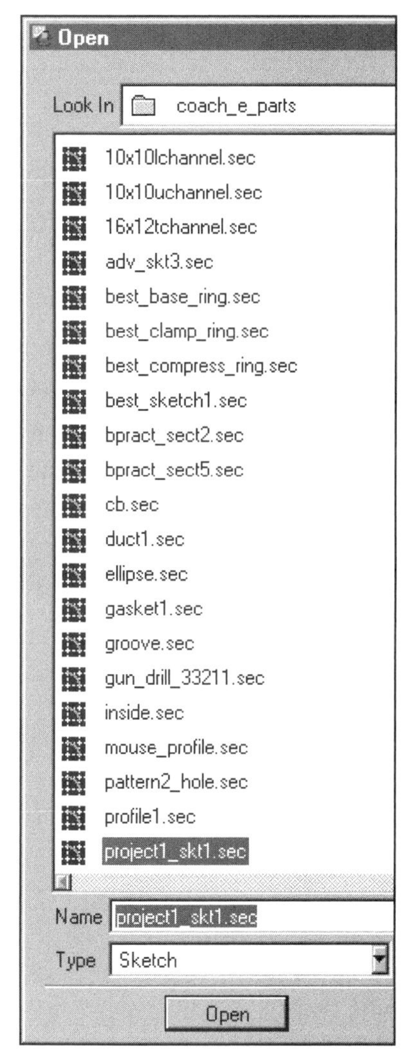

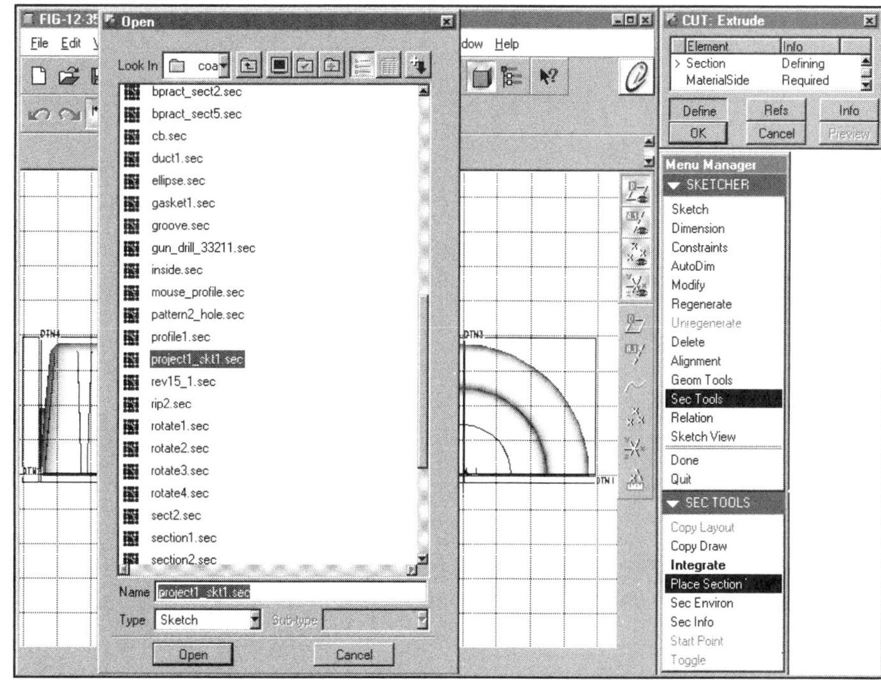

Figure 12.35
Using **Place Section**

Retrieving an Existing Section

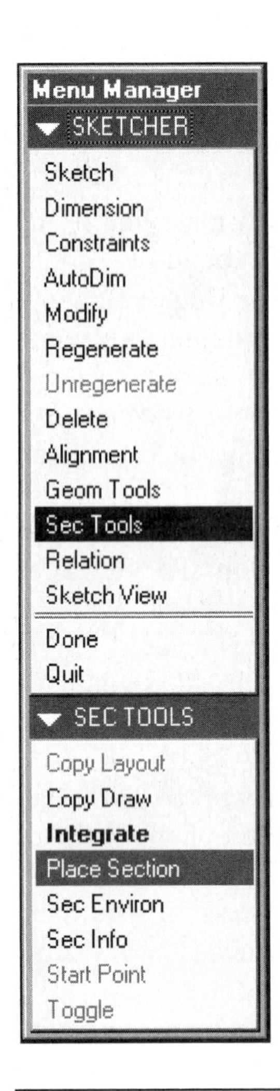

The **Place Section** option in the SEC TOOLS menu allows you to retrieve a section from disk or from memory and place it on the current sketch as an independent copy of the original section. The target section can be empty or can contain existing entities (and dimensions). Placing a section does not alter other sketched geometry.

The **Place Section** option copies the entities and relations (if any) of the original section without reference to the original context in which they were created. Thus the accuracy, grid parameters, and units of measure are those of the current model.

The placed section behaves as a regular sketched section. After you place the section, it is no longer associated with the source section. Figures 12.35 through 12.39 show the creation of a cut using an existing section. To retrieve an existing section:

1. Choose **Place Section** from the SEC TOOLS menu (Fig. 12.35).
2. Select a section file in the Open dialog box.
3. Pro/E retrieves the section and displays it in a subwindow as shown in Figure 12.36.
4. If the section sketch is being placed on a part sketching plane (not an auxiliary section, such as for a sketched blind hole or a shaft), you can modify the location, orientation, and scaling of the section.
5. Enter a rotation angle for the sketch. Be aware that some dimensioning schemes may change because of the change of sketch orientation (Fig 12.36).
6. Select an origin point on the sketch for scaling. During scaling of the section using the mouse, the origin point remains stationary.
7. Select a drag point on the sketch. This is the point that will follow the mouse during positioning. The drag point cannot be coincident with the scaling point.
8. Enter a preliminary scale factor for the sketch (**.25** was used here).
9. Move the mouse from the subwindow to your part window. The section appears in red and follows your mouse pointer as it moves around the screen (Fig. 12.37).
10. Using the mouse, you can do any of the following:
    * Click the left mouse button to place the section. The section changes from *red* to the normal section color, and Pro/E displays any dimensions.
    * Click the middle mouse button to abort the section placement. This returns you to step 5.
    * Click the right mouse button to switch between scaling and drag modes. When you are scaling, your scale origin remains stationary, and moving the mouse increases or decreases the size of the section. Returning to drag mode causes the drag origin to follow the mouse again (maintaining the new scale).
11. Locate the section with respect to the part by dimensioning and/or aligning (Fig. 12.38). In Figure 12.39, the cut is shown **Preview**(ed).

To place an existing section while using the Intent Manager, choose the following commands:

File ⇒ Import ⇒ Append to Model ⇒ with the LMB move the section to the proper location using rotate and translate (⊗)

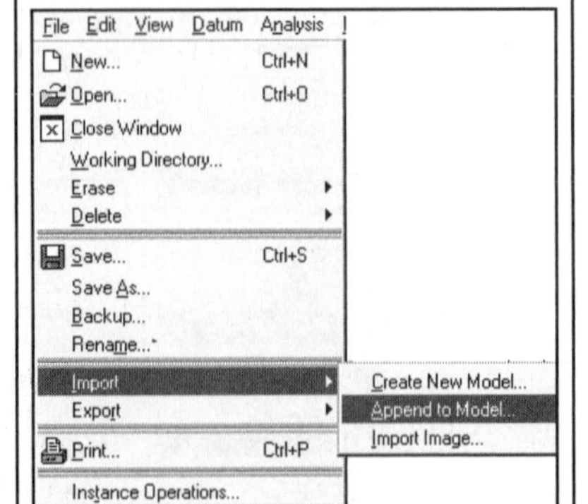

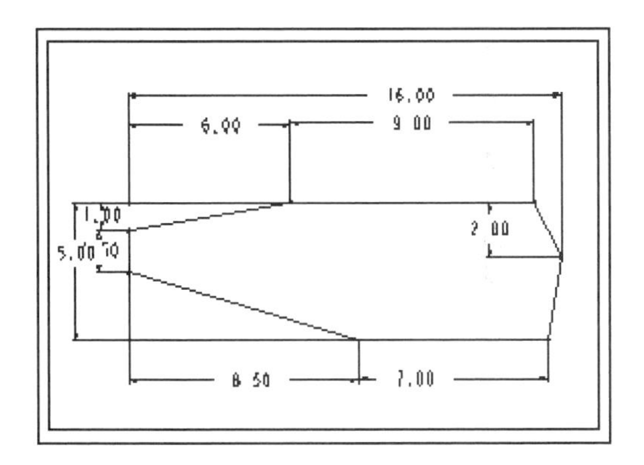

Figure 12.36
Enter Rotating Angle

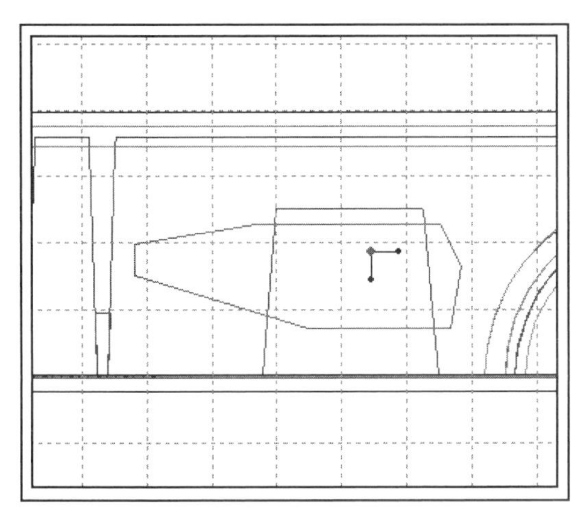

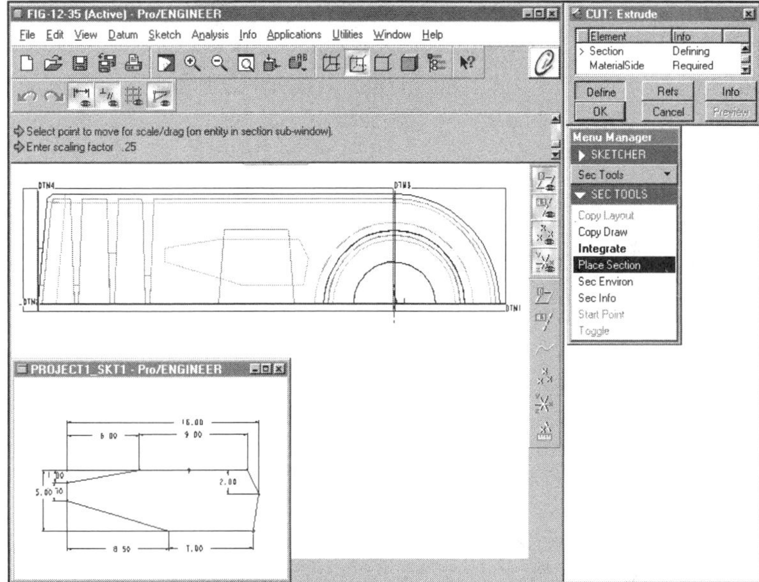

Figure 12.37
Place Section on Part
(**.25** Scale)

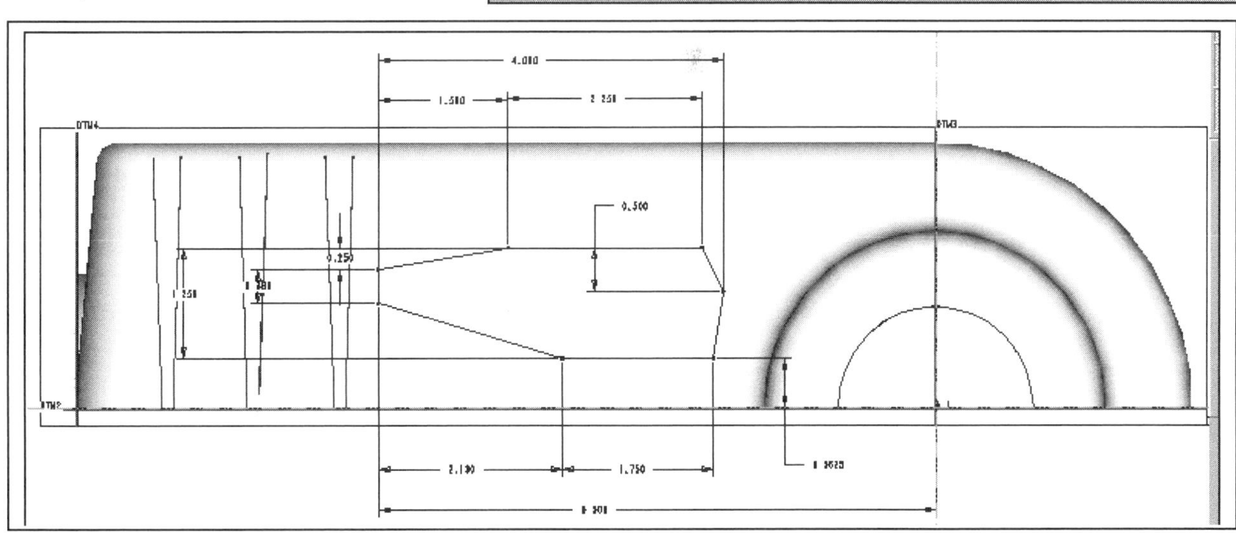

Figure 12.38
Section Placed and Located

Figure 12.39
Completed Cut Previewed

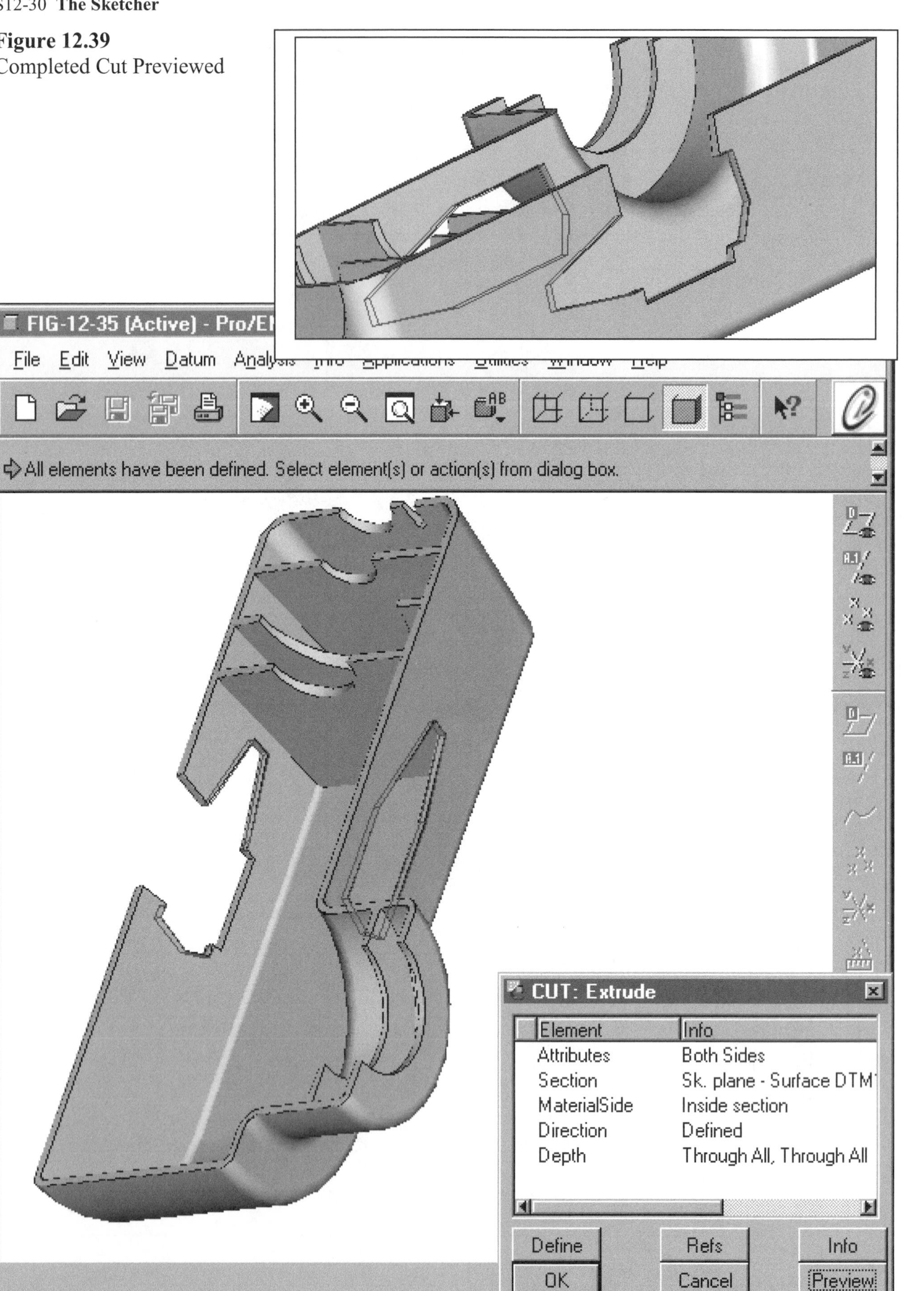

Part One

Creating Parts

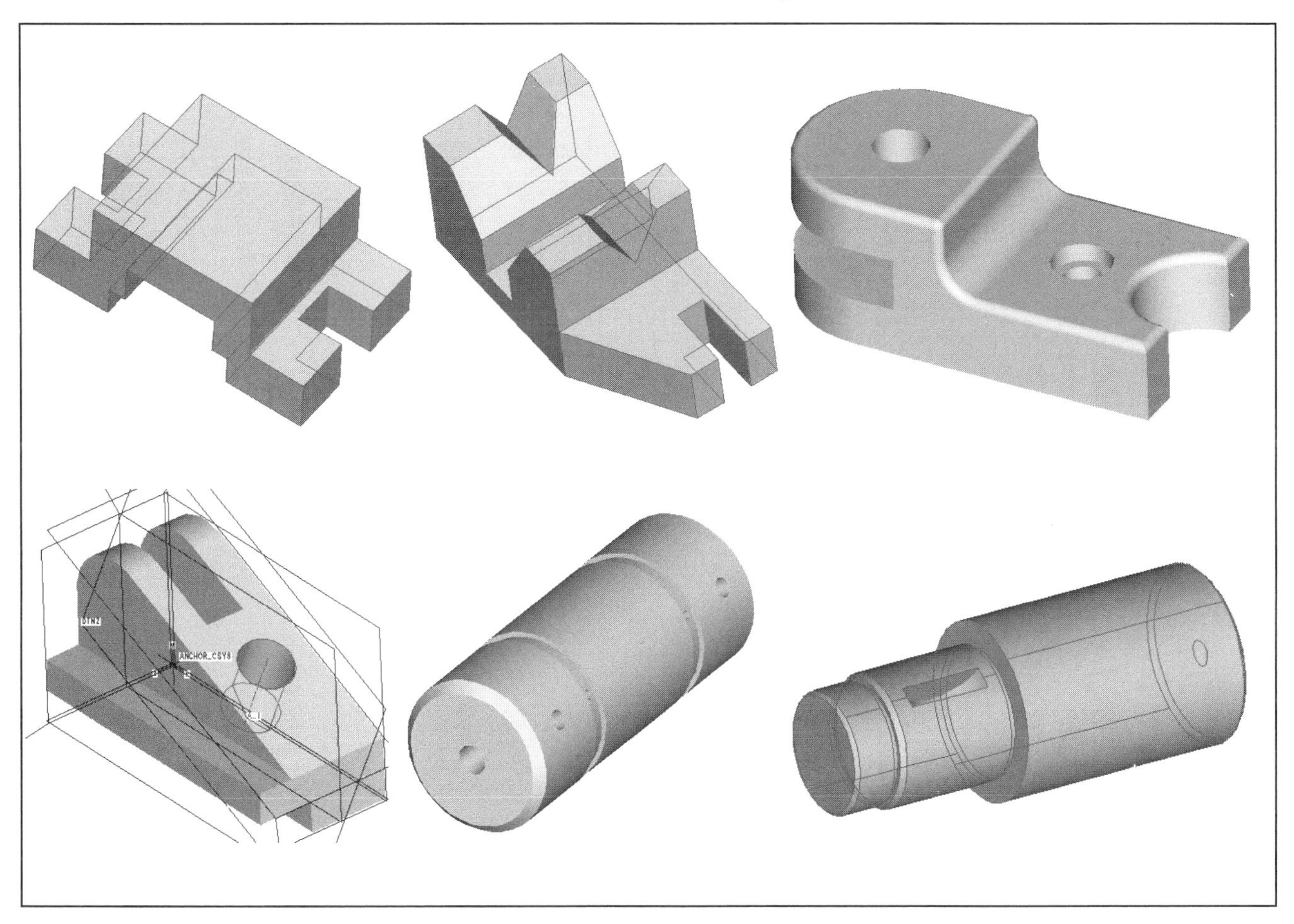

Lesson Parts 1 through 6

Creating Parts

The *design intent* of a feature, a part, or an assembly (and even a drawing) should be established before any work is done with Pro/ENGINEER. Skipping this step in the design process is a recipe for disaster. In industry, there are thousands of stories of how a designer created a *graphically correct* part or assembly that *"looked"* visually precise.

Upon closer examination, the model or assembly had too many or too few datum planes, parent-child relationships that were glaring examples of the designer's incorrect use of Pro/E, and massive feature failures that resulted when minor ECOs were introduced after the original design was complete. Pro/E is only as good as the drafter, designer or engineer using it.

Without proper process planning, organization, and a well-defined design intent, the part model is useless. In most cases, such poor design habits result in the parts being remodeled, because it would take more time to reorder, modify, redefine, and reroute. In fact, most poor designs can't be fixed.

Part Design Philosophy: Design Intent

The **design intent** of a project must be understood before modeling geometry is started. Use the Design Intent Planning Sheets provided in Appendix D to sketch and analyze your part before modeling.

The *dimensioning scheme* will establish the dimensions that are critical for the design: What dimensions on the part might be modified during an *ECO*? What dimensions are required for *manufacturing* the part economically and to the correct *tolerances*? Are there any dimensional *relationships* that must be established and maintained? Will the part be a member of a *family of similar parts?* How does the part relate to other *parts in the assembly?*

Lesson Parts 7 through 13

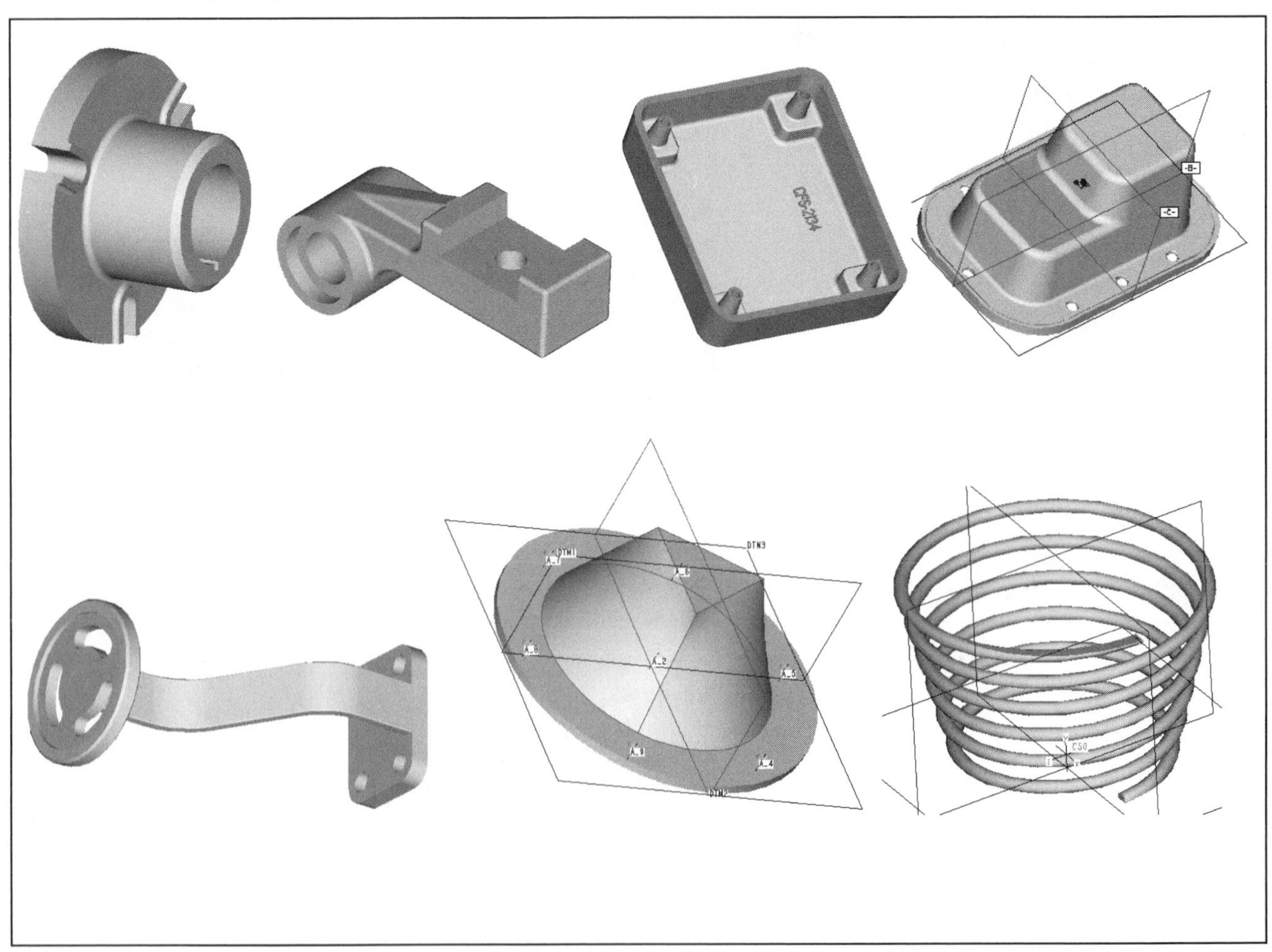

Use the following basic guidelines to create a typical part:

1. Establish the Pro/E environment that you will work in for the project, including environment setting and *config.pro* settings.
2. Use setup to establish the material and units.
3. Establish the datum planes and coordinate system.
4. Create a layering scheme, and set the datum planes and the coordinate system on a layer.
5. Rename the datum planes per the part and geometric tolerance requirements.
6. Determine the base feature and protrusion type. Sketch the base feature on the appropriate datum plane.
7. Establish the dimensioning scheme for the feature.
8. Determine what construction features should be used on the part.
9. Build a construction feature using a dimensioning scheme, keeping relationship requirements in mind.
10. Add relations to control the feature where desired, per the design intent.
11. Adjust dimension cosmetics as desired.
12. Create new layers, and establish a layering scheme for dimensions and features.
13. Add reference dimensions required for documentation.
14. Repeat steps 7-14 for each feature of the part until the part is completed.

Lesson Project Parts

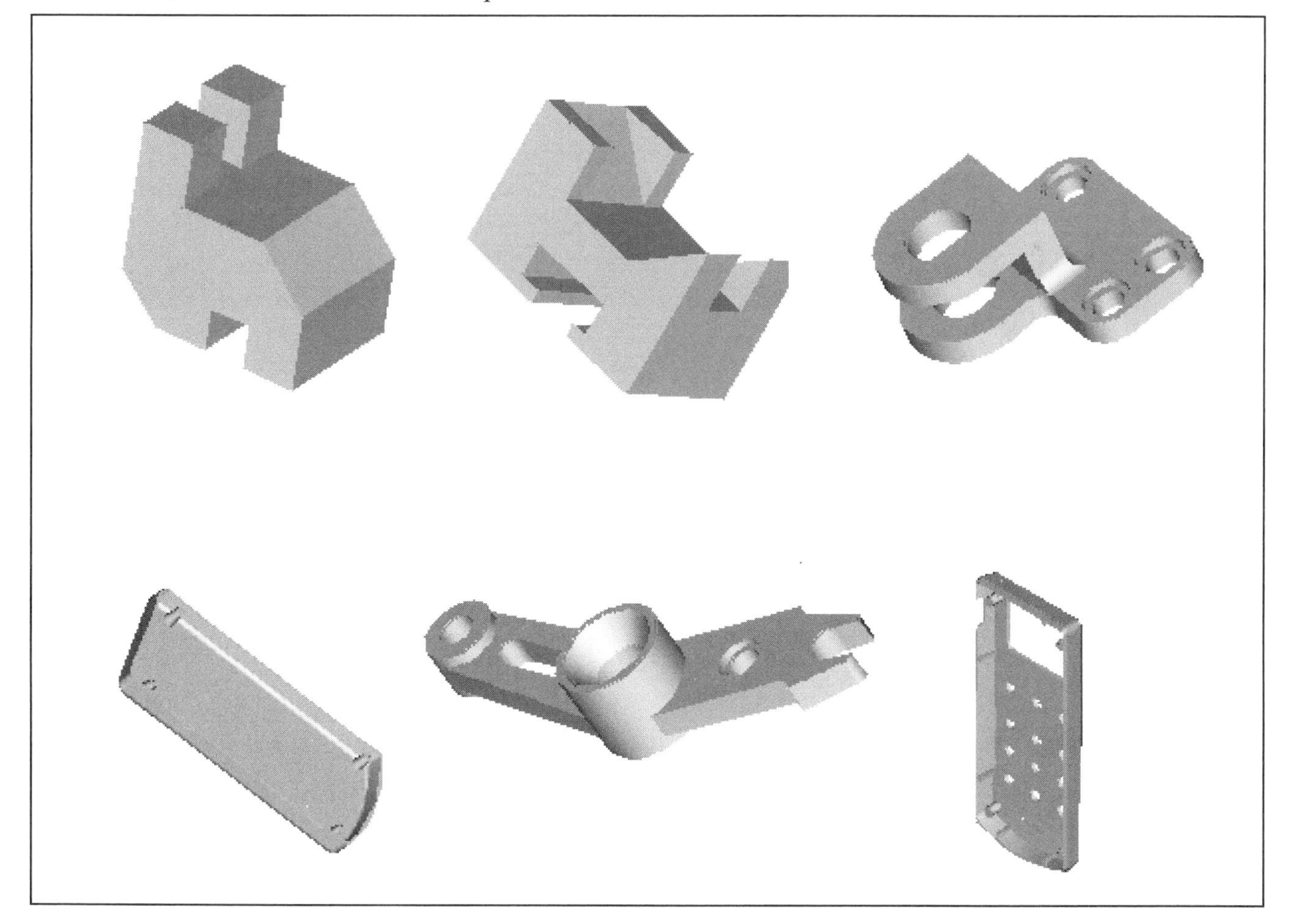

A wide variety of features can be created with Pro/ENGINEER. The following part-creation lessons incorporate most of the following capabilities:

Protrusions (Lesson 1) Part features that add material
Cuts (Lesson 1) Features that remove material from a part
Holes (Lesson 3) Creates different types of holes: through, counterbored (sketched), blind, etc.
Rounds (Lesson 3) Creates many types of rounds
Chamfers (Lesson 6) Creates edge and corner chamfers, which remove flat sections of material to create a beveled surface
Cosmetic features (Lesson 6) Creates cosmetic features, sketched, thread, groove, and user-defined
Ribs (Lesson 8) Creates a rib, which creates a thin fin or web that is attached to a part
Tweak features (Lesson 9) Creates drafts and free-form features
Shells (Lesson 10) Creates a shell feature, which removes a surface or surfaces from the solid, and then hollows out the inside of the solid, leaving a shell of a specified wall thickness
Pipes Creates a pipe, which is a three-dimensional centerline that represents the centerline of a pipe

COAch for Pro/ENGINEER
Protrusions

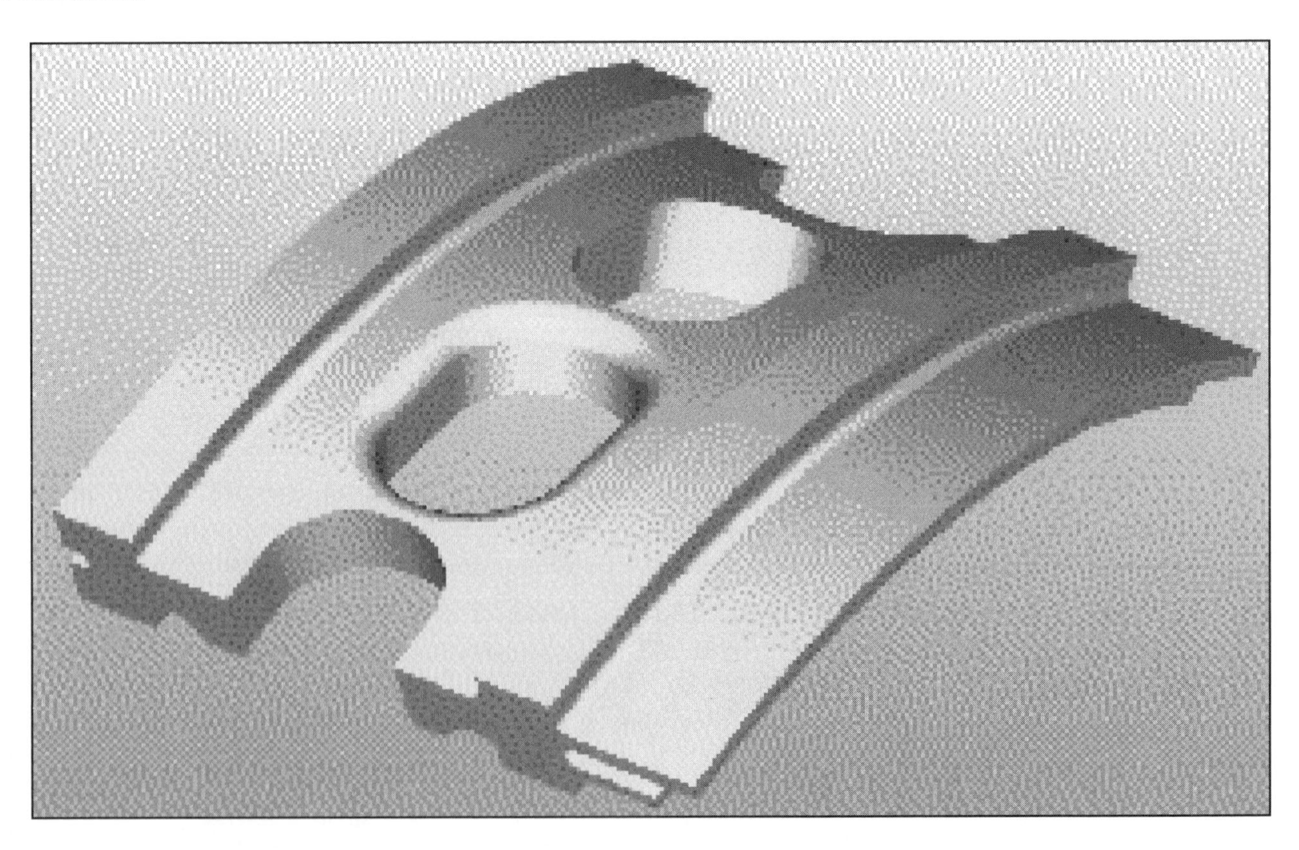

Lesson 1

Protrusions and Cuts

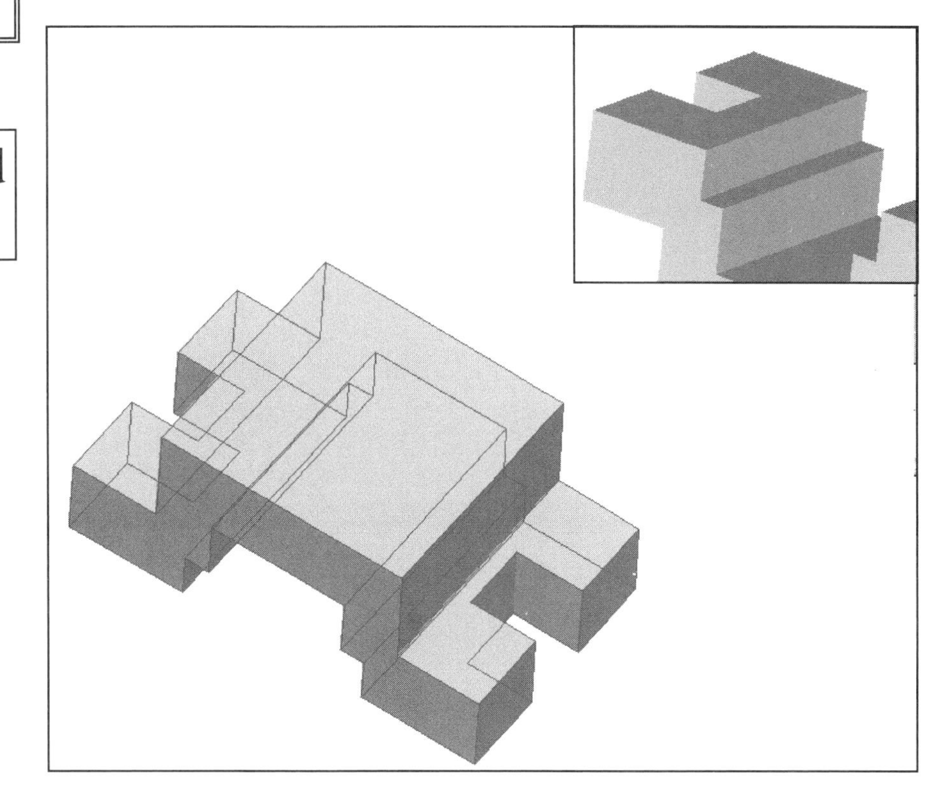

Figure 1.1
Clamp

OBJECTIVES

1. **Create a base feature using an extruded protrusion**
2. **Understand setup and environment settings**
3. **Define and set a material type**
4. **Create and use datums**
5. **Sketch a protrusion and a cut feature in the Sketcher**
6. **Understand the feature dialog box**
7. **Align sketch geometry**
8. **Shade a part**
9. **Copy a cut feature**
10. **Save and remove old versions of a part file**

EGD REFERENCE
Fundamentals of Engineering Graphics and Design
by L. Lamit and K. Kitto
Read Chapters 5 and 10.
See pages 111, 192-193, 302, and 409.

COAch™ for Pro/ENGINEER
If you have **COAch for Pro/ENGINEER** on your system, go to SEARCH and do the Segments shown in Figure 1.4 and Figure 1.6.

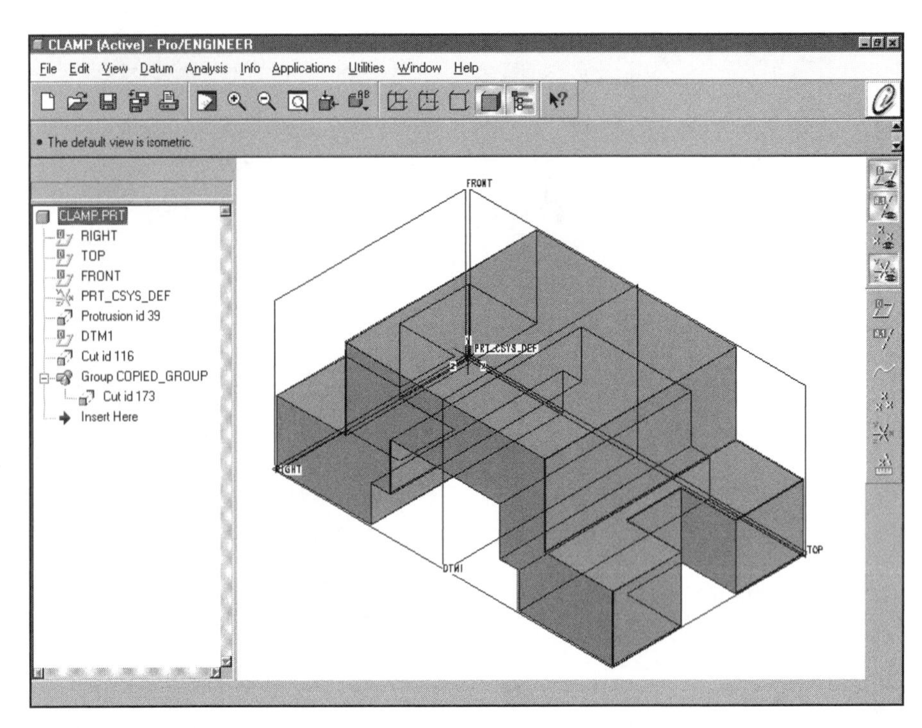

Figure 1.2
Clamp Part Showing the Datum Planes and the Model Tree

> ✔ ***READ THIS!!***
> The first few pages of each lesson contain general information (online documentation, references to COAch, etc.) on the commands of features introduced in that lesson. *The actual step-by-step sequence comes after this material.* You will see the **START HERE ➡ ➡ ➡** (shown below) where the sequence of steps and procedures actually starts:
>
> **Lesson procedures & commands**
> **START HERE ➡ ➡ ➡**
>
> ***This lesson's steps and procedures start on page L1-7, not here!***

PROTRUSIONS AND CUTS

A **protrusion** is a part feature that adds material. You can sketch different geometry by combining a variety of form options and attributes during creation of the protrusion feature. **Cuts** are used to remove material from existing solid features. Figures 1.1 and 1.2 show a simple protruded part with identical cuts on both sides.

Protrusions

A protrusion (Fig.1.3) is *always the first solid feature created.* This can be the **base feature** or the first feature created after a base feature of datum planes.

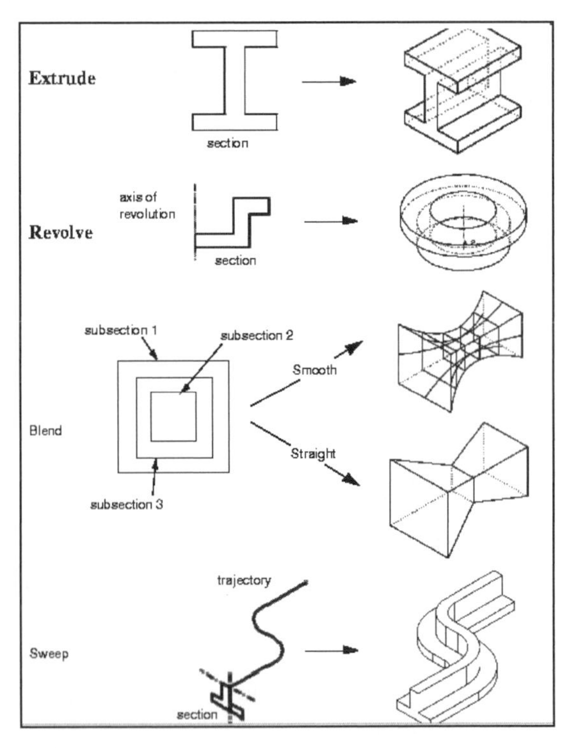

Figure 1.3
Protrusions

Menu Manager
PART
Feature
FEAT
Create
FEAT CLASS
Solid
Surface
Sheet Metal
Composite
Cosmetic
User Defined
CDRS In
ICEM In
DESKTOP In
Data Sharing
Done/Return
SOLID
Hole
Round
Chamfer
Cut
Protrusion
Rib
Shell
Pipe
Tweak
Intersect

The first solid feature is always a **Protrusion**

To create an extruded protrusion:

1. Choose **Feature** from the PART menu, and then choose **Create** from the FEAT menu.
2. Choose **Protrusion** from the SOLID menu.
3. Pick **Extrude ⇒ Solid ⇒ Done** from the SOLID OPTS menu.
4. Pro/E displays the PROTRUSION: Extrude dialog box, which lists the elements needed for creating this type of protrusion.
5. Pro/E displays the ATTRIBUTES menu, listing these options:
   One Side Creates the feature on one side of the sketching plane.
   Both Sides Feature is created on both sides of the sketch plane.
6. Pick **One Side** or **Both Sides ⇒ Done** from ATTRIBUTES menu.
7. Select the sketching plane and the sketch orientation reference.
8. **Sketch** the protrusion (Fig. 1.4).
9. Align (**Alignment**) and **Dimension** the section geometry.
10. **Regenerate** the section.
11. **Modify** and **Regenerate** the section sketch. Choose **Done.**
12. Specify the depth of the protrusion and choose **OK**.

Cuts

The Cut feature can be used to create a **Thru All** or **Thru Next** cut or to create the user-defined features (UDF) used for notches and punches. The techniques used to create a cut in Sheet Metal mode are similar to those used in Part mode. To create a solid-class cut feature, use the SOLID OPTS menu to choose one of the following:

* **Solid** Displays a dialog box to sketch a solid cut, without thickness.
* **Thin** Displays a dialog box to create a simple line sketch for a cut that has uniform thickness (Fig. 1.5).

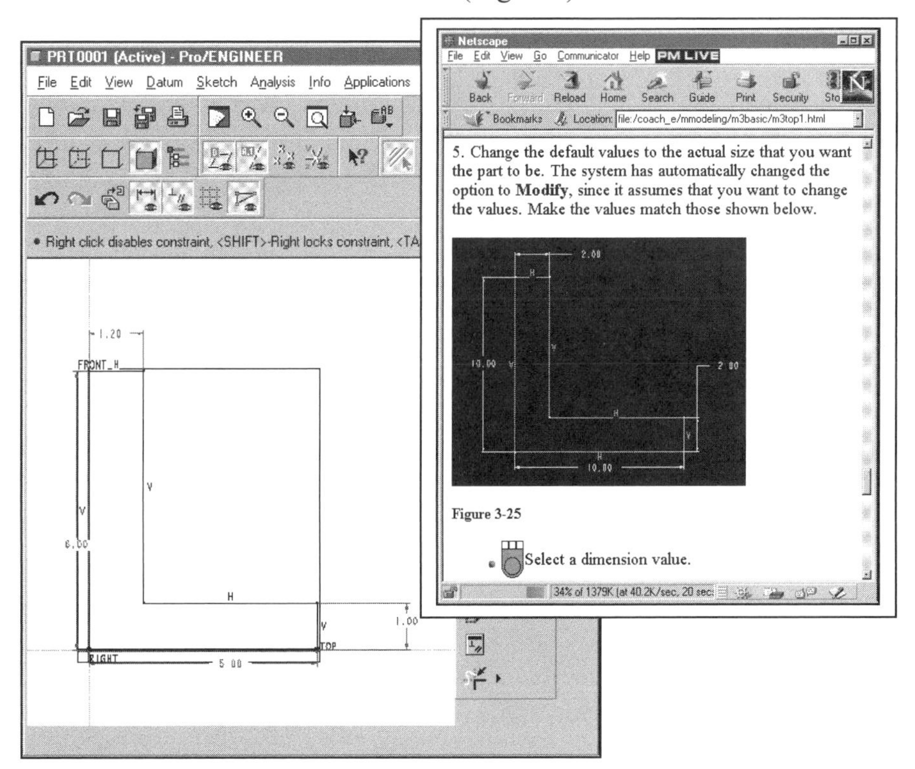

Figure 1.4
COAch for Pro/ENGINEER, Basic Modeling (Creating a Sketch)

The **Slot** option on the SOLID menu is available when the configuration file option is set ***allow_anatomic_features yes.***

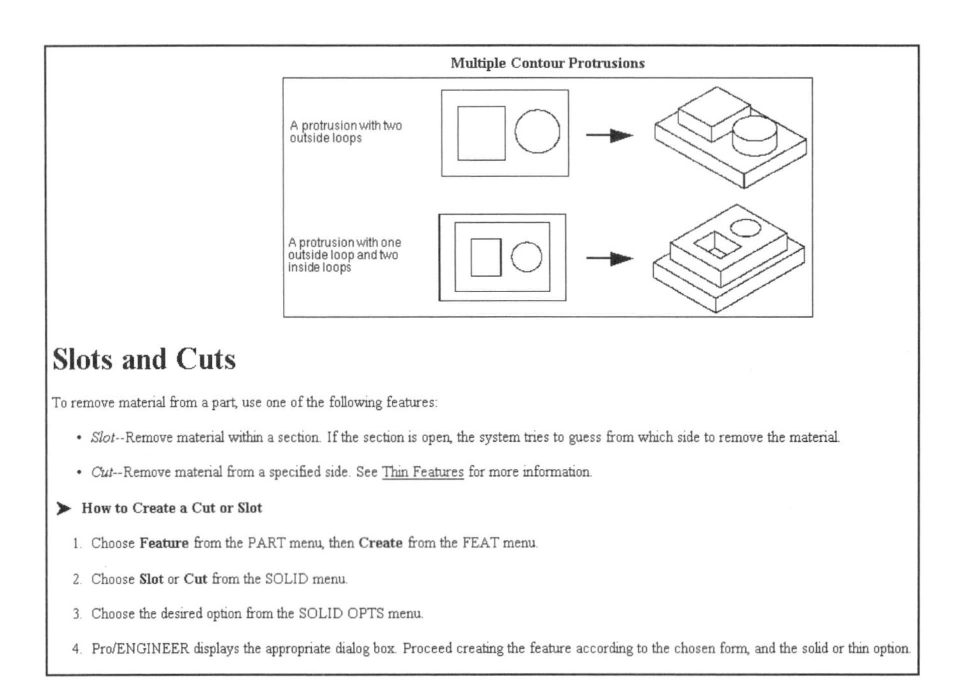

Slots and Cuts

To remove material from a part, use one of the following features:

- *Slot*--Remove material within a section. If the section is open, the system tries to guess from which side to remove the material.
- *Cut*--Remove material from a specified side. See Thin Features for more information.

➤ **How to Create a Cut or Slot**

1. Choose **Feature** from the PART menu, then **Create** from the FEAT menu.
2. Choose **Slot** or **Cut** from the SOLID menu.
3. Choose the desired option from the SOLID OPTS menu.
4. Pro/ENGINEER displays the appropriate dialog box. Proceed creating the feature according to the chosen form, and the solid or thin option.

Figure 1.5
Cuts and Slots

To create a cut (Fig. 1.6), do the following:

1. Choose **Feature** from the PART menu, and then choose **Create** from the FEAT menu.
2. Choose **Cut** or **Slot** from the SOLID menu.
3. Choose **Extrude** ⇒ **Solid** ⇒ **Done** (SOLID OPTS menu).
4. The appropriate dialog box is displayed.
5. Choose **One Side** or **Both Sides** ⇒ **Done** (ATTRIBUTES menu).
6. Select the sketching plane on the part and the part's orientation.
7. **Sketch, Alignment, Dimension,** and **Regenerate** the section.
8. **Accept** the cut direction or flip the arrow.
9. Determine the depth of the cut or slot.

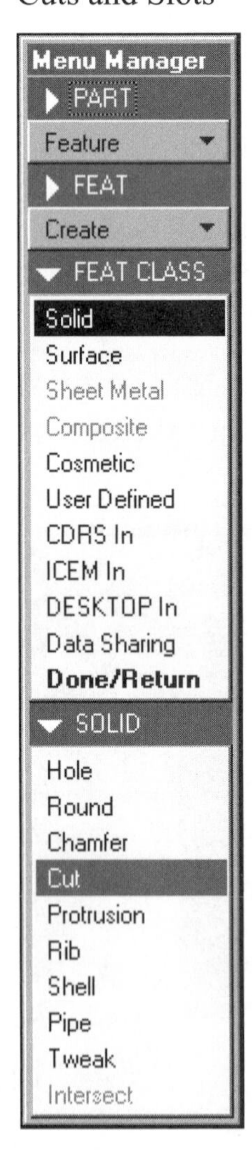

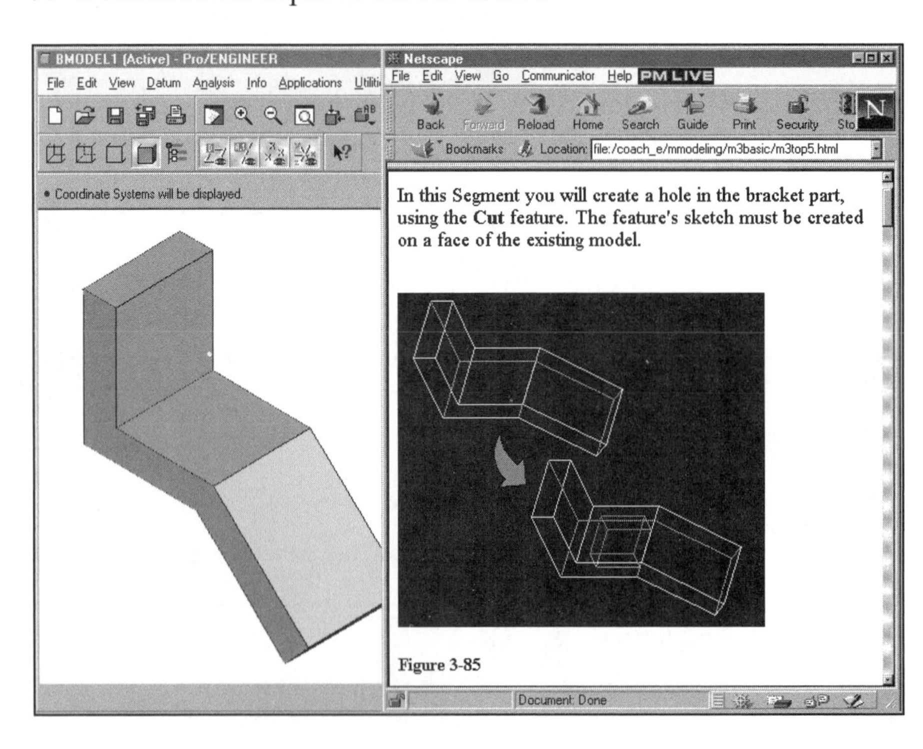

Figure 1.6
COAch for Pro/ENGINEER, Basic Modeling (Sketching on a Face)

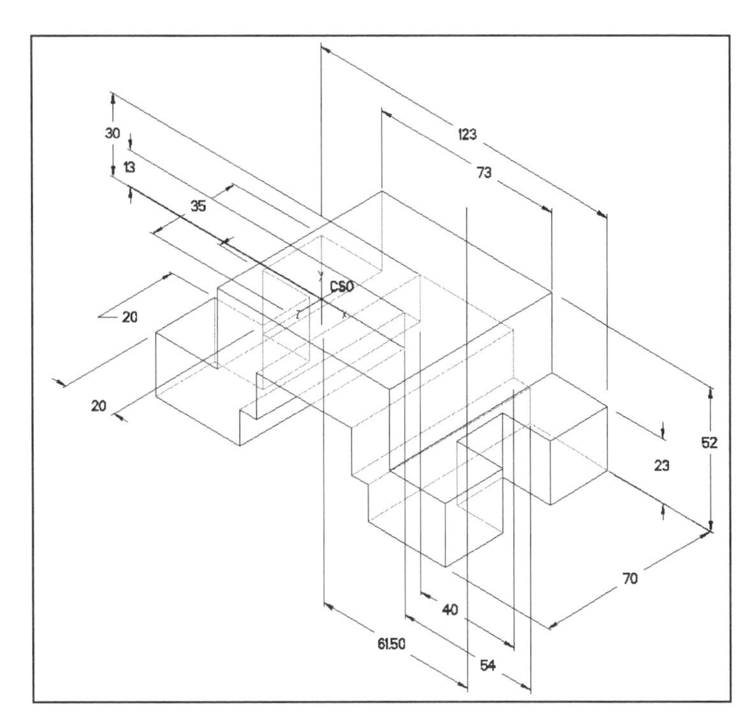

Figure 1.7
Clamp Dimensions

Clamp

The clamp (Fig. 1.7) will be our first ***lesson part.*** It is composed of a simple protrusion and a cut. A number of things need to be established before you actually start modeling. These include setting up the *environment*, selecting the *units*, and establishing the *material* for the part.

Before you begin any part using Pro/E, you must plan the design. The **design intent** will depend on a number of things that are out of your control and on a number that you can establish. Asking yourself a few questions will clear up the design intent you will follow: Is the part a component of an assembly? If so, what surfaces or features are used to connect one part to another? Will geometric tolerancing be used on the part and assembly? What units are being used in the design, SI or decimal inch? What is the part's material? What is the primary part feature? How should I model the part? And what features are best used for the primary protrusion (the first solid mass)? On what datum plane should I sketch to model the first protrusion? These and many other questions will be answered as you follow the step-by-step lesson part. But you must answer many of the questions on your own when completing the ***lesson project***, which does not come with step-by-step instructions.

Using the appropriate ***Design Intent Planning Sheet*** for each **Lesson Project** will increase your chance of having a part, assembly, or drawing with the appropriate design intent and project sequence. For the lesson part step-by-step designs and the lesson projects found in Lessons 1-13, you should use **DIPS 1**, **2**, **7**, and **8**. Block out some trial feature sequences to establish the parent-child relationships required by your design intent using **DIPS 1** and **2**. Use **DIPS 7** and **DIPS 8** for your feature geometry sketches and dimensioning scheme.

? Pro/HELP

Remember to use the help available on Pro/E by highlighting a command and pressing the right mouse button.

HINT

Before you start modeling with Pro/E, copy the Design Intent Planning Sheets **(DIPS)** from Appendix D to have them available for planning parts, feature sketches, assemblies, and drawings.

If you wish, you can purchase pads of engineering grid paper with ¼" spacing (isometric and Cartesian) for sketching your projects and laying out your design intent.

After you have started your workstation and loaded Pro/ENGINEER, you can set up the environment. The screen on your system will look similar to Figure 1.8. We will be concerned with the **Environment** command first, to set up certain Pro/E defaults.

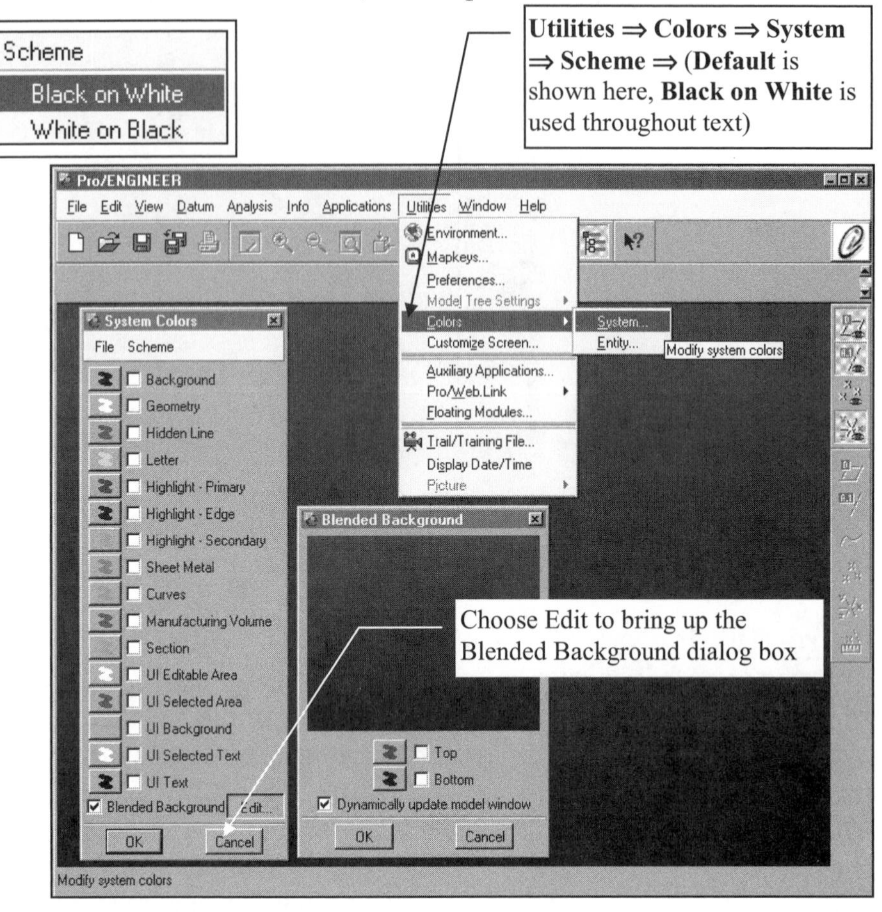

Figure 1.8
System Colors

Your screen will have a *blended dark blue background* (by default). For most of this text, an alternative system color will be used for the background color. The text illustrations of screen captures show a *white* MAIN WINDOW color, not a solid *black* one as in Figure 1.8.

When you are starting a new session, the first thing you should always do is check your environment settings. The following command block will appear at the beginning of each of the early lessons of the text (you must do this without being prompted when starting the lesson projects). The command block will also prompt you to *choose the* **Set Up** *command after the Part mode, Assembly mode, or Drawing mode has been activated.* You can change the environment settings at any time during an active Pro/E session.

NOTE- Do not choose these commands!

This is a *SAMPLE* **SETUP AND ENVIRONMENT** command block. Your part will start on the next page ***NOT HERE!***

SETUP AND ENVIRONMENT

Set Up ⇒ **Units** ⇒ Units Manager ⇒ **Inch lbm Second (Pro/E Default)** ⇒ **Close** ⇒ **Material** ⇒ **Define** ⇒ (type **STEEL**) ⇒ ✔ ⇒ (table of material properties, change or add information) ⇒ **File** ⇒ **Save** ⇒ **File** ⇒ **Exit** ⇒ **Assign** ⇒ (pick **STEEL**) ⇒ **Accept** ⇒ **Done**
Utilities ⇒ **Environment** ⇒ ✔ **Snap to Grid**
Display Style **Hidden Line** Tangent Edges **Dimmed** ⇒ **OK**

Lesson procedures & commands
START HERE ➡ ➡ ➡

The first command you will choose in this lesson will be to set the *environment* for the lesson part to be created. Using your mouse, pick **Utilities ⇒ Environment** command. Choose it by clicking the left mouse button on **Utilities** in the menu bar. This will bring up the Environment dialog box (Fig. 1.9). To activate an environment setting, pick on the small box ❑ to the left of the selections. A ✔ means that the choice is activated. To deactivate a choice, pick on a box ❑ to remove the ✔. To set the model visibility, choose **Hidden Line** and **Dimmed** (Fig. 1.9). Also, activate the grid snap (**✔Snap to Grid**) if it is not already on by default. When you are finished with your selections, pick **OK** from the Environment dialog box.

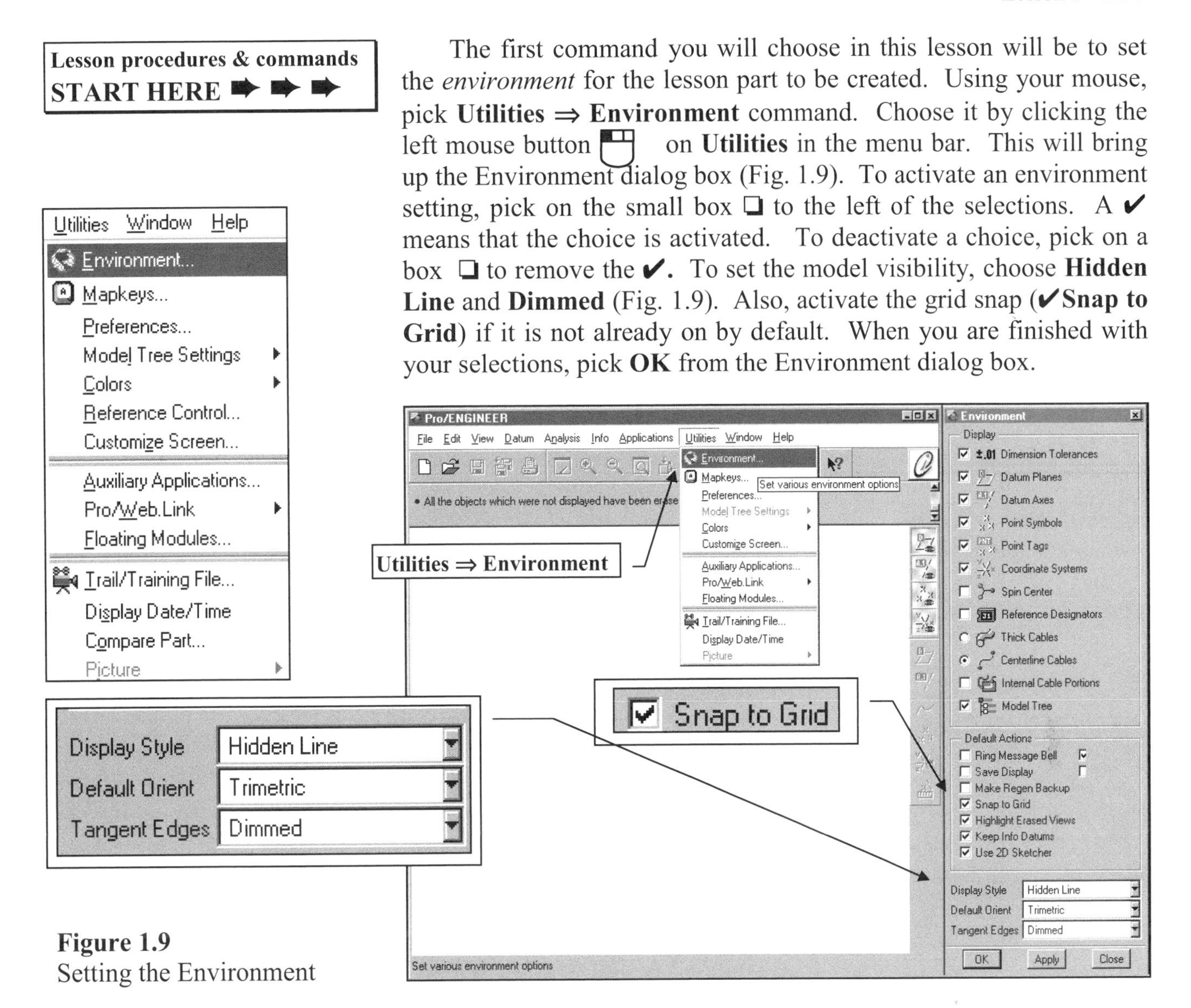

Figure 1.9
Setting the Environment

To start the actual part, assembly, drawing etc., you must choose **File ⇒ New** from the menu bar and then select the mode (Type). The **Part** mode is used in Lessons 1-13 and Lesson 21. In Lessons 14 and 15 the **Assembly** mode will be used, and in Lessons 16-20 the **Drawing** mode is accessed. Lesson 22 utilizes the **Manufacturing** mode. Lesson 23 uses the **Part** mode with the sub-type **Sheetmetal.**

Choose the following commands:

File (from the menu bar) ⇒ **New** ⇒ **●Part** ⇒ (type the part name **CLAMP) ⇒ ✔ Use default template ⇒ OK**

The screen will show the **Model Tree** on the left side of the main window in its default position (Fig. 1.10). You can remove (or move and resize) the Model Tree out of the working area (the main window) of the screen.

Notice that the part name appears along the top of the main window. The part name also appears in the Model Tree [and in the Application Manager (UNIX)].

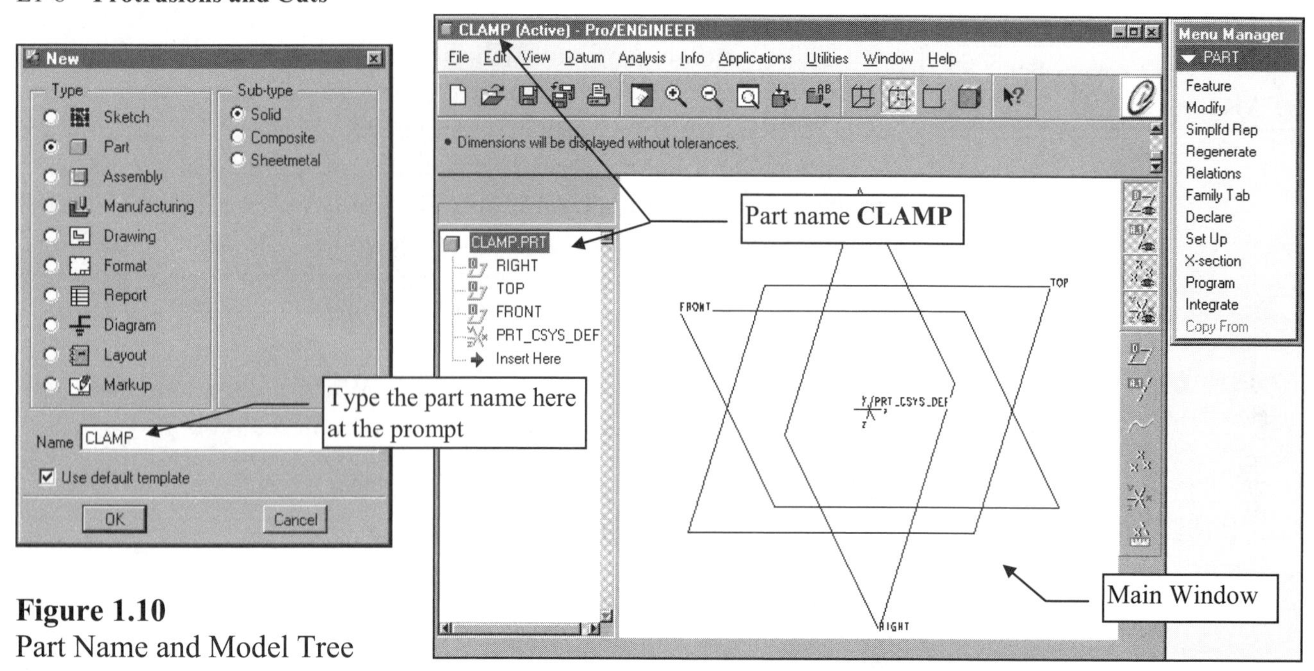

Figure 1.10
Part Name and Model Tree

The next step is to set up the part. Normally, the text will prompt you to do this using the **Set Up** and **Environment** command box shown earlier. Before you do the setup, move the Model Tree from its default location in the main window then choose:

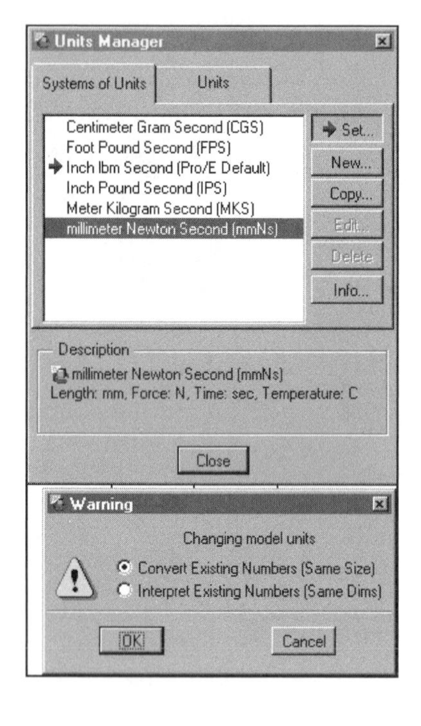

Utilities ⇒ **Customize Screen** ⇒ **Options** ⇒ **✔Display as a separate window** ⇒ **▼Lower right** ⇒ **OK** ⇒ **Set Up** ⇒ **Units** ⇒ Units Manager **millimeter Newton Second** ⇒ **Set** ⇒ **●Convert existing numbers (Same Size)** ⇒ **OK** ⇒ **Close** ⇒ **Material** ⇒ **Define** ⇒ [type **STEEL** at the prompt and pick **✔**. A material *Table* appears on the screen--enter the values (Fig. 1.11)] ⇒ **File** (from material table) ⇒ **Save** ⇒ **File** ⇒ **Exit** ⇒ **Assign** ⇒ (pick **STEEL**) ⇒ **Accept** ⇒ **Done** (Fig. 1.12)

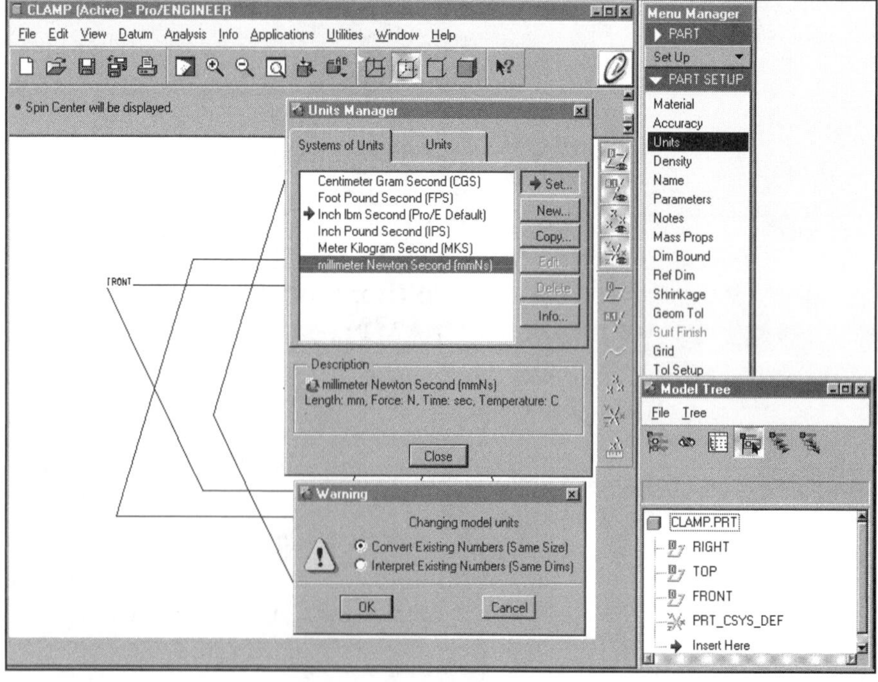

Figure 1.11
Setting Up the Part Units

Your text editor may differ.

The **Material** command brings up the Material Table using the default text editor (Fig. 1.12). You can fill in the information if it is available or just save and exit the table. **Assign** the material to the part after exiting the table.

File ⇒ Save ⇒ File ⇒ Exit

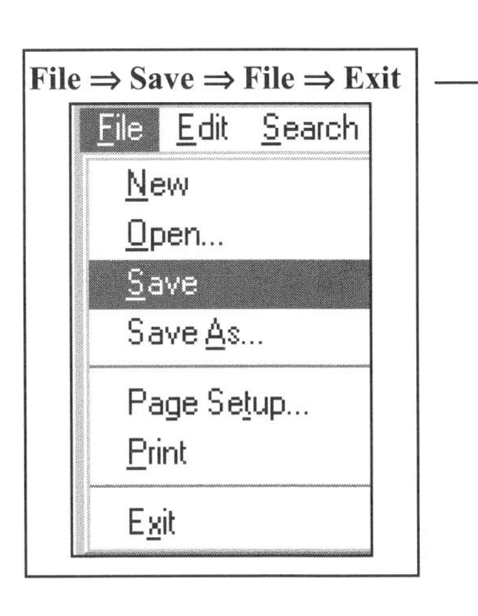

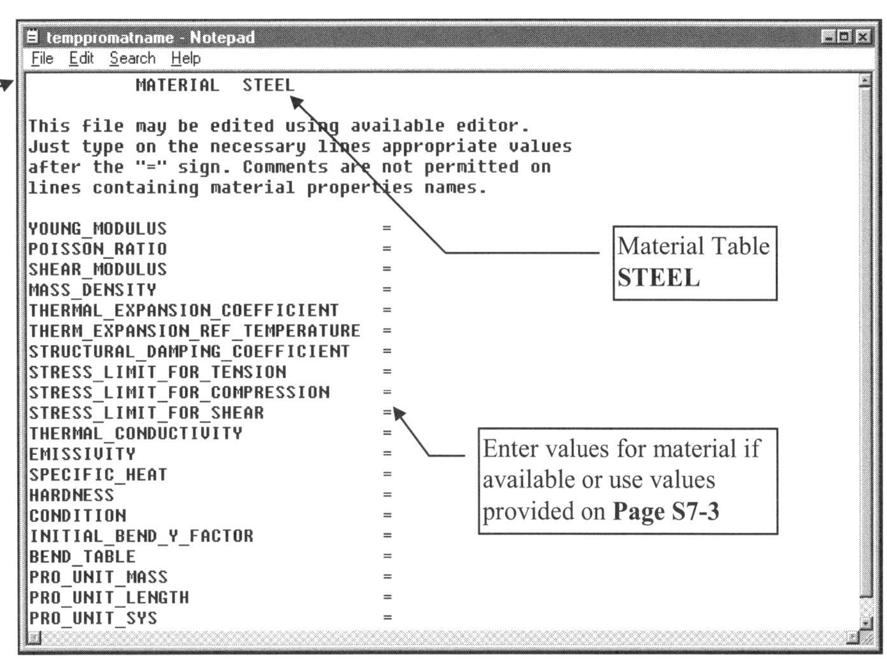

Figure 1.12
Setting Up the Material

Menu Manager
PART
Set Up
PART SETUP
Material
MATRL MGT
Assign
USE MATER
From Part
From File
MAT_LIST
STEEL
Accept

At this point, it may seem that there is a lot to do before any modeling takes place. You are building not just a solid model but a *complete integrated database* that controls all aspects of the design, including the part's material properties, units, geometry, and other important parameters that capture the design intent of the project.

The three default datum planes and default coordinate system are automatically displayed in the main window and the Model Tree (see Figure 1.13) because ✔ **Use default template** was used.

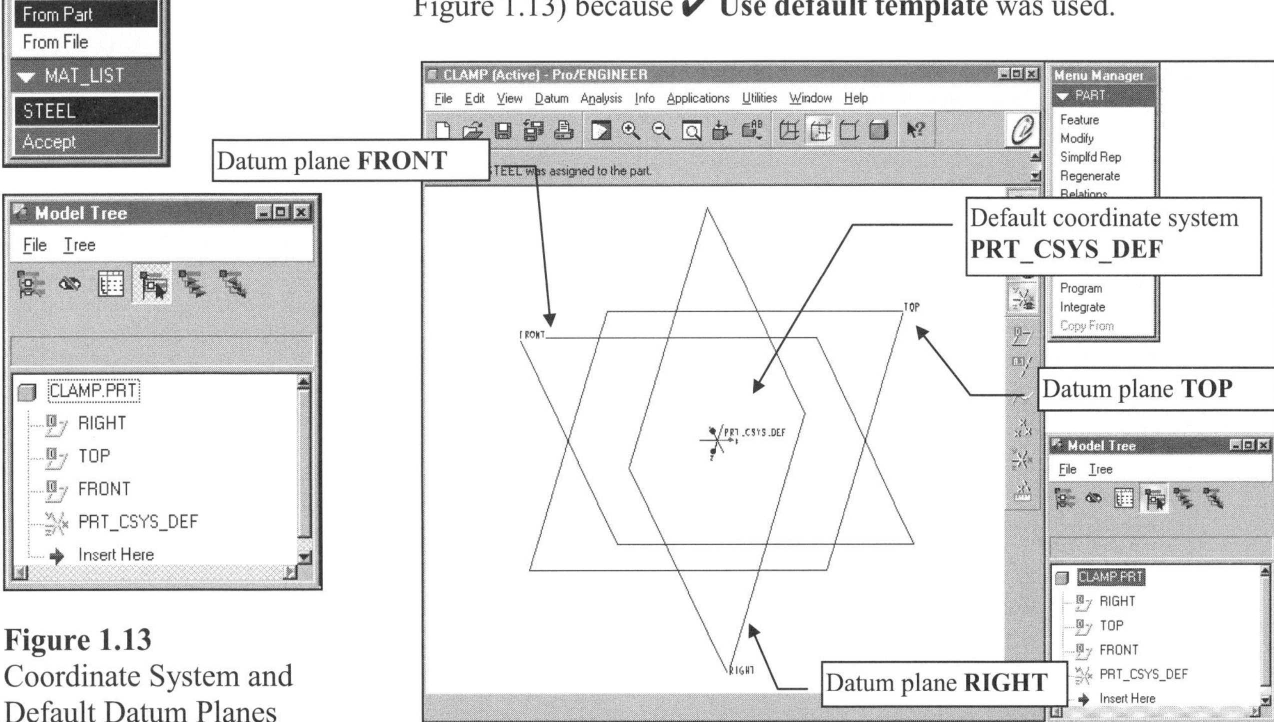

Figure 1.13
Coordinate System and Default Datum Planes

The **default datum planes** and the **default coordinate system** will be the first features on all parts *and* assemblies. The datum planes will be used to sketch on and orient the part's section geometry. Because an isometric view of three datum planes shows ambiguous geometry, the **Trimetric** setting is the default in the Environment dialog box. *After you create the first protrusion for a part, go back to the Environment dialog box and choose* ***Isometric*** *for a more lifelike view of the part (or assembly).* The datum planes and the coordinate system show in the Model Tree (Fig. 1.13). They are the first features of the part. These features become the *parents* of any feature tied to them through sketching or alignment. Having datum planes as the first features of a part, instead of the first protrusion, gives the modeler more flexibility later in the design.

The following commands create the first protrusion (Fig. 1.14):

Feature ⇒ Create ⇒ Solid ⇒ Protrusion ⇒ Extrude ⇒ Solid ⇒ Done ⇒ One Side (default from ATTRIBUTES menu) **⇒ Done ⇒** (pick datum **FRONT** as the sketching plane) **⇒ Okay** (for selecting the direction of protrusion projection--*red* arrow)

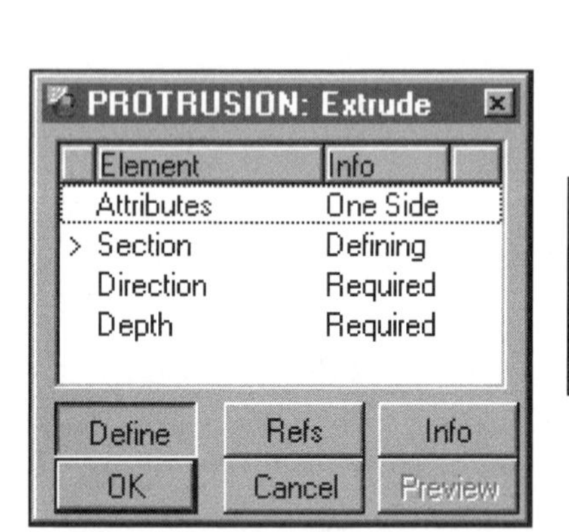

Figure 1.14
Selecting the Sketching Plane

By picking on datum **FRONT,** as shown in Figure 1.14, this becomes the plane on which you will be sketching the first protrusion's section geometry. The protrusion's feature dialog box appears on the screen at this time. For many of the actions you will be performing on this part, you will use the dialog box buttons.

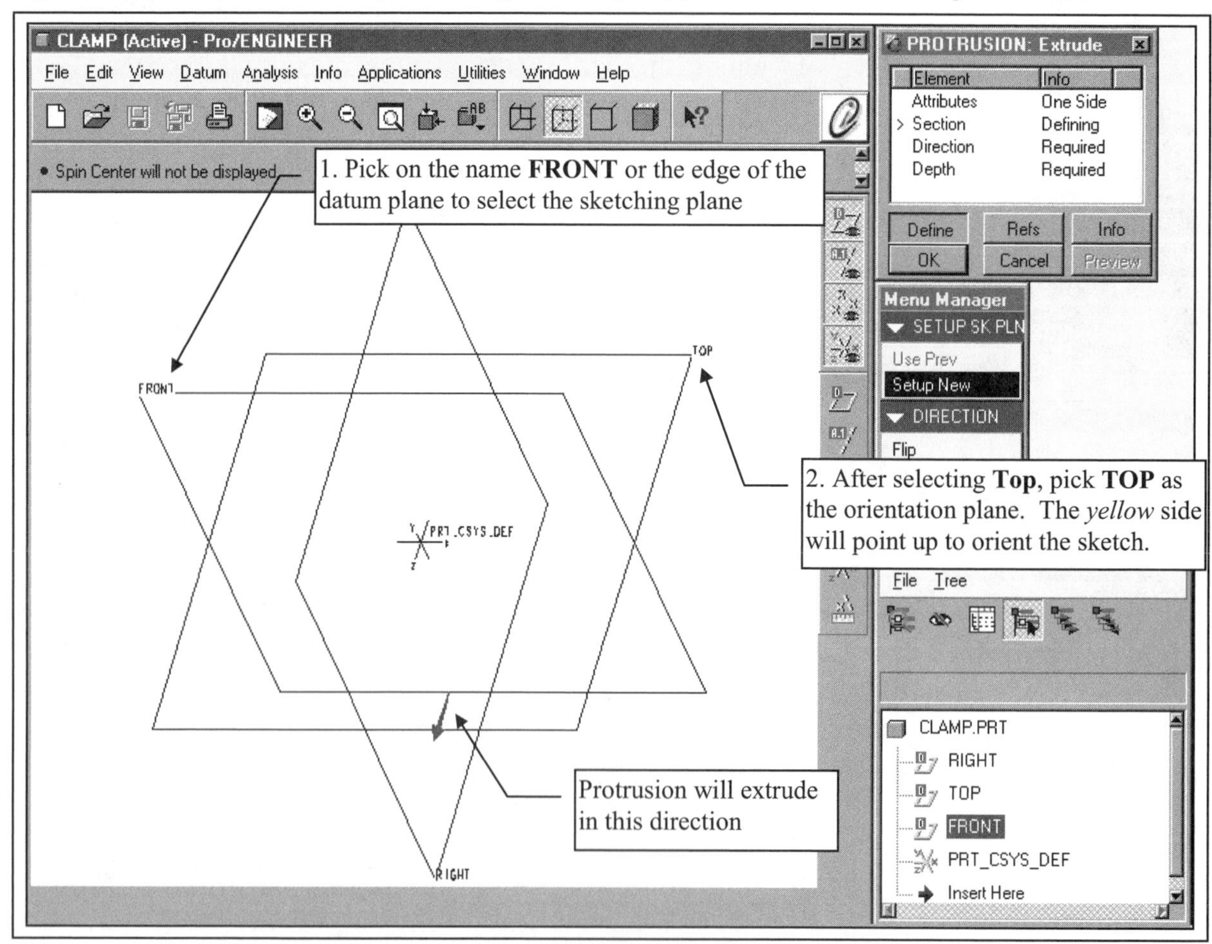

NOTE

Most, but not all, of your sketches will be created on **FRONT** and aligned to **RIGHT** and **TOP.**

HINT

Datum **FRONT** (yellow side) is facing you and datum **TOP** (*yellow* side) is facing the top of the screen (you are seeing it as an edge).

You must still orient the sketch. Complete the process by choosing:

> **Top** ⇒ (pick datum **TOP)** ⇒ **Close** (SKETCHER ENHANCEMENT) ⇒ **Sketch** (menu bar) ⇒ **Intent Manager** (off)

The sketch is now oriented as shown in Figure 1.15. The sketch is displayed in 2D, with the coordinate system at the middle of the sketch, where datum **RIGHT** and datum **TOP** intersect. The **X** coordinate arrow points to the right and the **Y** coordinate arrow points up. The **Z** arrow is coming toward you (out from the screen). The square box you see is the limited display of **FRONT**, which is like a piece of graph paper you will be sketching on when you create the protrusion's geometry. The coordinate system is used for some commands and for Pro/NC. Pro/E is not a coordinate-based software, so you need not enter geometry with **X**, **Y**, and **Z** coordinates as with many other CAD systems. In reality, you don't need to see the coordinate system, but many modelers prefer to see the axes displayed on the screen.

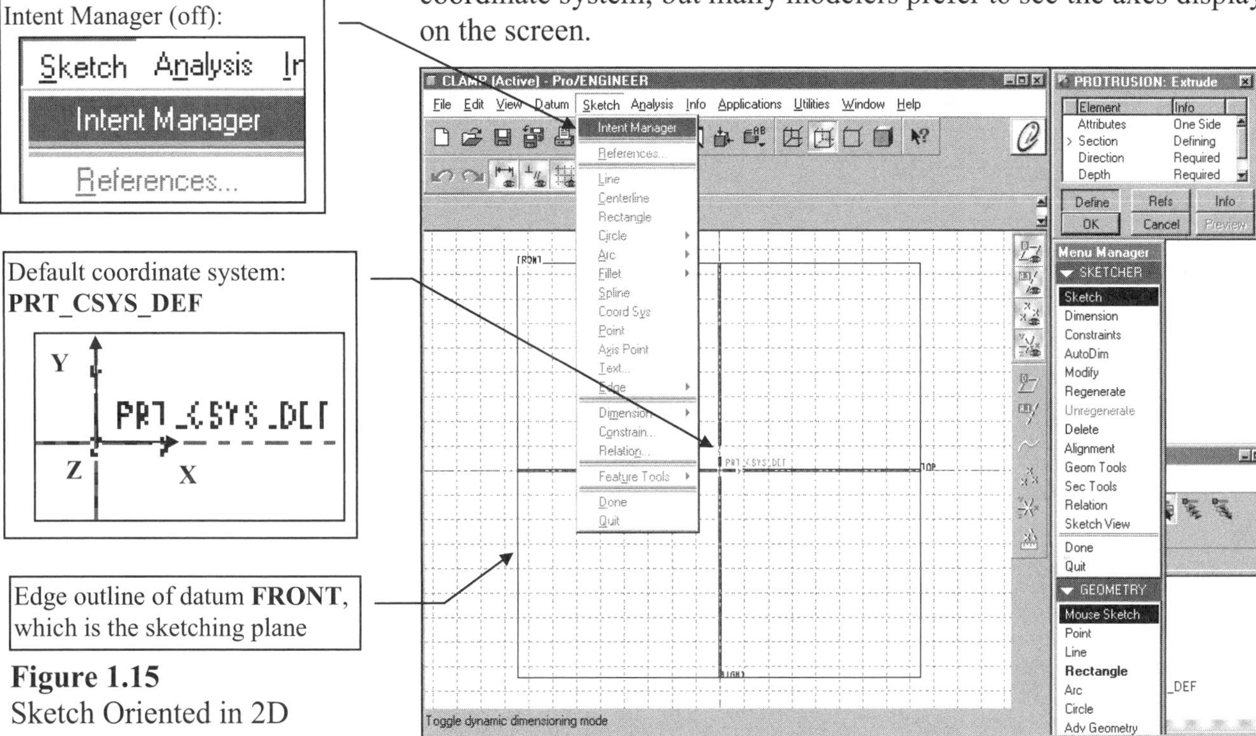

Figure 1.15
Sketch Oriented in 2D

Reread Section 11, Intent Manager, and Section 12, The Sketcher, at this time. It will help you understand some of the Sketcher's capabilities and what is required before a successful feature can be completed. In general, many of the part base protrusion features you will be modeling start with sketching in the first quadrant (Fig. 1.16). To make this more convenient and to have a greater sketching area, we need to change the position and size of the sketch, as shown in Figure 1.16. To resize the sketch on the screen, place the mouse cursor in the center of the screen and hold down the control key (**Ctrl**) while simultaneously holding down the left mouse button (LMB); moving the mouse will enlarge or shrink (*zoom*) the display of the sketch.

Zooming the graphics in the Main Window with the mouse (while holding down the **Ctrl** key)

Move the mouse up or down until the sizes of the sketch and grid are correct. While you press **Ctrl**, the right mouse button *pans* the sketch across the screen and the middle button *rotates* the sketch in 3D. You can sketch in 2D or 3D with Pro/E; for now, let's stick with 2D sketching. If you rotated your sketch and wish to return to the 2D orientation, pick **Sketch View** from the SKETCHER menu.

HINT

With the mouse in the Main Window, and with the **Ctrl** key depressed, if you quickly click the left mouse button (LMB) without moving the mouse, you start a zoom box. Now move the cursor to the desired end location of the zoom box and click again with the left mouse button (LMB). To cancel, pick the middle mouse button (MMB).

You will be sketching in the first quadrant

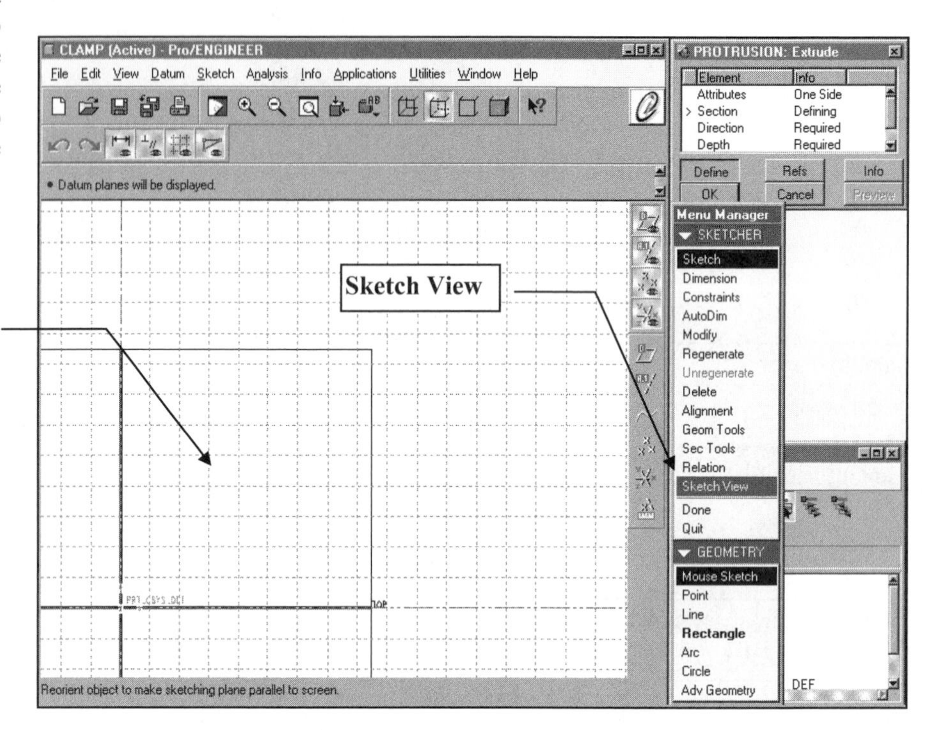

Figure 1.16
2D Sketch

Because you checked ✔**Snap to Grid** in the Environment dialog box, you can now sketch by simply picking grid points representing the part's geometry (outline). Because this is a sketch in the true sense of the word, you need only create geometry that *approximates* the shape of the feature; the sketch does not have to be accurate as far as size or dimensions are concerned. Pro/E constrains the geometry according to rules. In Section 9, the rules for the **constraints** are listed. They include the following:

? Pro/HELP

Highlight the **Constraints** command and press the right mouse button

RULE: Symmetry
DESCRIPTION: Entities sketched symmetrically about a centerline are assigned equal values with respect to the centerline.
RULE: Horizontal and vertical lines
DESCRIPTION: Lines that are approximately horizontal or vertical are considered to be exactly horizontal or vertical.
RULE: Parallel and perpendicular lines
DESCRIPTION: Lines that are sketched approximately parallel or perpendicular are considered to be exactly parallel or perpendicular.
RULE: Tangency
DESCRIPTION: Entities sketched approximately tangent to arcs or circles are assumed to be exactly tangent.

READ THIS!
Do not draw lines or other entities on top of each other. Pro/E will not regenerate the sketch, because it will think you have two sections, not one. It is better to **Delete ⇒ Delete All** and start the sketch again. Later you will feel more comfortable troubleshooting a failed sketch.

The **Mouse Sketch** default is active in the GEOMETRY menu, so we can begin the sketching process using the mouse buttons without selecting another command. The left button on the mouse creates lines; the middle button, circles; and the right mouse button, arcs that are tangent to the end of an existing line/arc.

The outline of the part's primary feature is sketched using a set of connected lines. The part's dimensions and general shape are provided in Figure 1.7. Because much of the part can be defined by sketching an outline similar to its front view, you can complete most of the part's geometry with one protrusion. The cuts on the sides will be the second feature created. The part will have its base (bottom horizontal edge) aligned with the edge of **TOP**, its left edge aligned with **RIGHT**, and, because you are sketching on **FRONT**, its back face aligned with that datum. If you are at all confused about this, reread Sections 3, 11, and 12. It is important not to create any unintended constraints while sketching. Therefore, remember to exaggerate the sketch geometry and not to align edges that have no relationship. Pro/E is very smart: If you draw two lines at the same horizontal level, Pro/E thinks they are horizontally aligned even if you later dimension them differently! Figure 1.17 shows how to sketch vertical lines that *are not in line with each other.*

Place the mouse near the center of the coordinate system at the intersection of the **RIGHT** and **TOP** datum planes and click the left button. Continue picking until you have sketched an outline (see Figure 1.17) approximating the primary feature of the clamp part as shown in Figures 1.2 and 1.7.

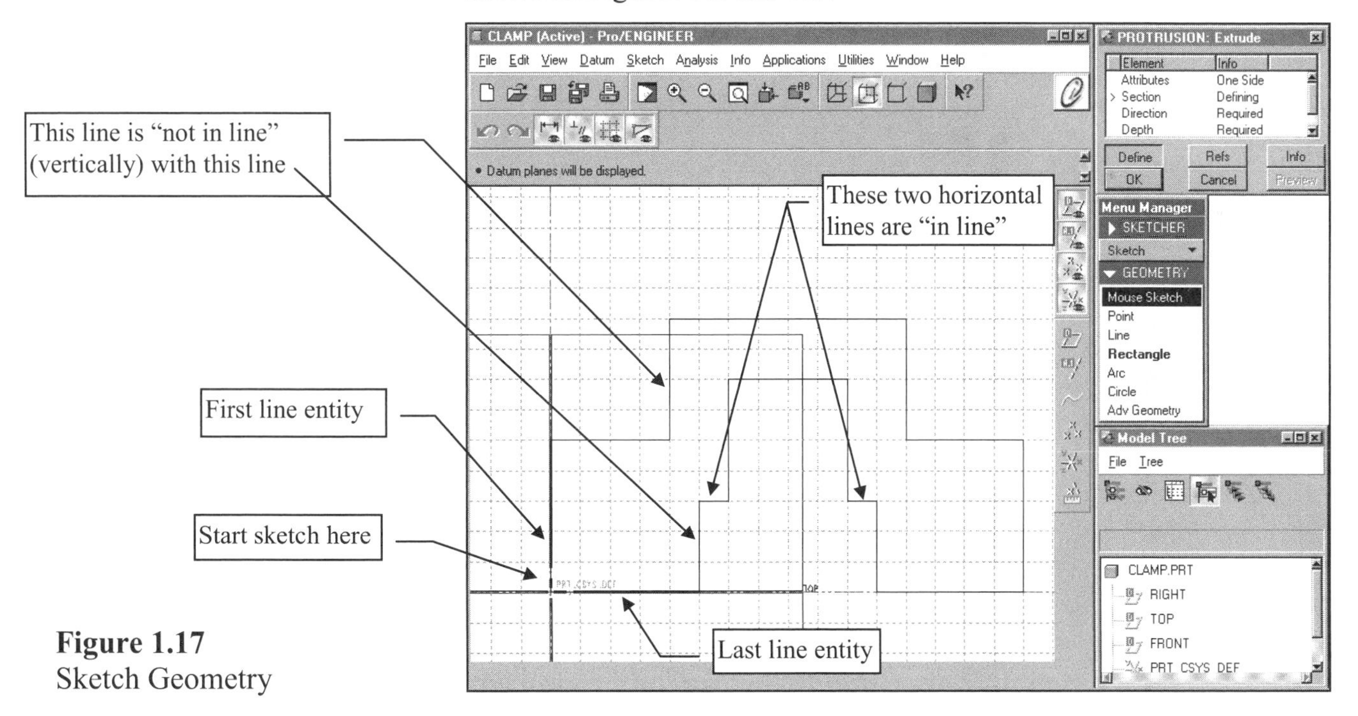

Figure 1.17
Sketch Geometry

After completing the sketch outline (section), select **Regenerate** from the SKETCHER menu. The endpoints of section entities are highlighted with the aid of small colored dot symbols (•).

HINT

Choose **View** ⇒ **Default** to orient the sketch in 3D.

Sketcher constraints can be turned on or off (enabled or disabled) while you are sketching. A Sketcher *constraint symbol* appears next to the entity that is controlled by that constraint. An **H** next to a line means horizontal; a **T** means tangent. To disable a constraint, pick **Disable** from the CONSTRAINTS menu and then pick a symbol on the screen. To obtain a brief explanation, pick **Explain** from the CONSTRAINTS menu and then pick a symbol on the screen. If the constraints are not displayed on the sketch, select **Constraints** from the SKETCHER menu (Fig. 1.18). Check **Enable** in the CONSTRAINTS menu. The constraints will appear on the section. They will also be displayed when the sketch is oriented in 3D.

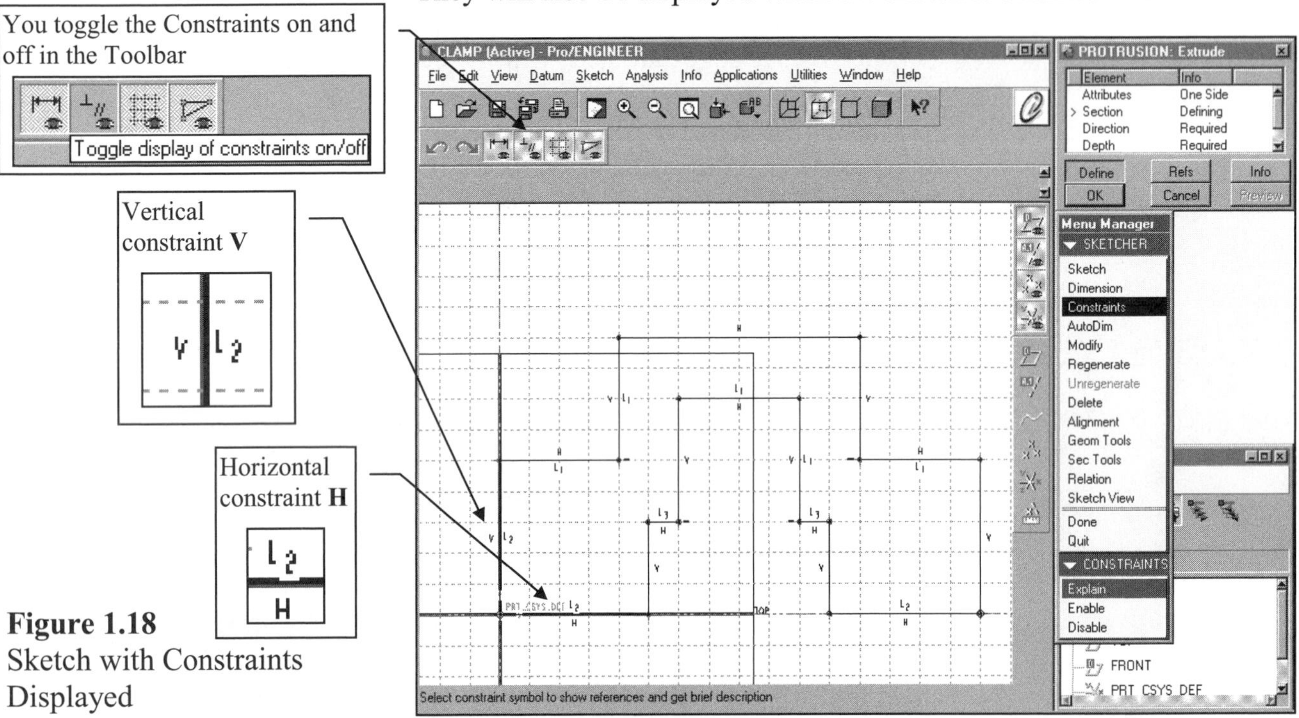

Figure 1.18
Sketch with Constraints Displayed

In general, the following six steps are used when sketching a section:

HINT

Memorize this:

Sketch
Regenerate
Alignment
Regenerate
Dimension
Regenerate
(Relations
Regenerate)
Modify
Regenerate

1. **Sketch** Sketch the section geometry. Use Sketcher tools to create the section geometry.
2. **Alignment** Align the section geometry to a datum feature or to an existing part feature (if applicable).
3. **Dimension** Dimension the section. Use a dimensioning scheme that you want to see in a drawing. Dimension to control the characteristics of the section.
4. **Regenerate** Regenerate the section. Regeneration solves the section sketch on the basis of your dimensioning scheme.
5. ***Relations*** Add section relations to control the behavior of your section. *This step depends on the feature.*
6. **Modify** Modify the dimension values. Change the dimension values to reflect the design intent. **Regenerate** again after modifying.

Next, add a centerline to the sketched section (Fig. 1.19):

Sketch ⇒ Line ⇒ Centerline ⇒ Vertical (Pick on a vertical grid line running through the center of the sketch) ⇒ **Regenerate**

This is where your sketching skills come in handy. Did you sketch the section so that it is symmetrical? Did you leave an even number of grid squares on each side of the section's center? It's not mandatory, but when you are learning Pro/E, it helps to simplify the sketch and pay attention to details. You do not have the necessary skills, at this time, to get the sketch to regenerate easily.

Read carefully!

If you did not sketch the section symmetrically, turn the grid snap off or change the grid spacing, and then redraw the centerline:

Utilities ⇒ Environment ⇒ ❑ Snap To Grid ⇒ OK
(***or*** change the grid spacing using **Sec Tools ⇒ Sec Environ ⇒ Grid ⇒ Params ⇒ X&Y Spacing ⇒** (type **15**) **⇒ ✔ ⇒ Done/Return)**

Grid is now twice as dense because we changed the spacing from the default of **30** units to **15** units

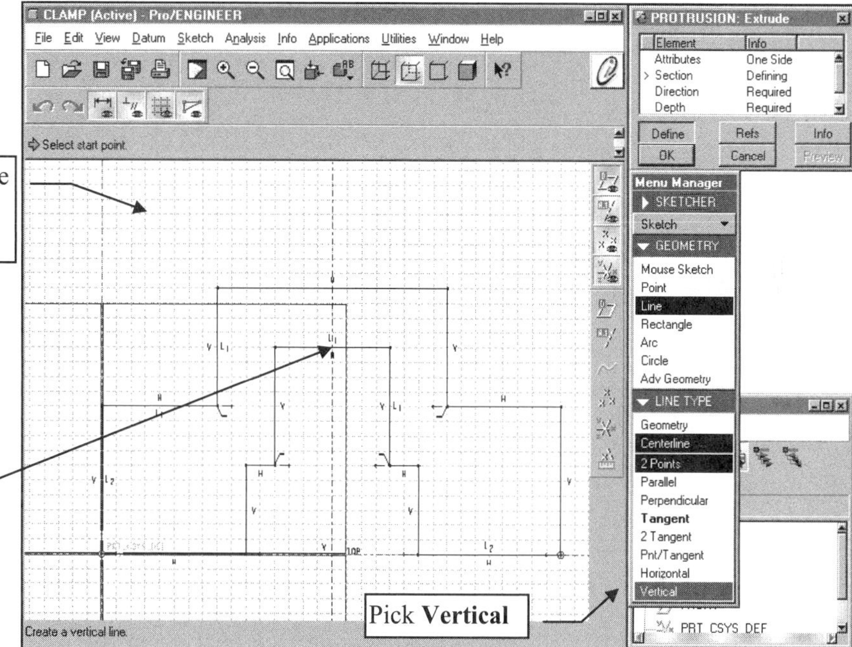

Sketch a centerline through the middle of the section

Figure 1.19
Sketching a Centerline

It is now time to *align* the sketch (Fig. 1.20). This simple procedure sometimes confuses the beginner. Even though you have sketched the base and left edge against **TOP** and **RIGHT,** you still need to tell Pro/E what portion of the sketch will be aligned with **RIGHT** and **TOP.** Choose the following:

Alignment [(from the SKETCHER menu), then pick twice with the left mouse button on datum **TOP** and on the lower horizontal line of the sketch (you need not move the cursor between the picks)] ⇒ (repeat with datum **RIGHT** and the left-edge vertical line of the sketch) ⇒ (the message line says **ALIGNED) ⇒ Regenerate**

Because the horizontal line and edge of datum **RIGHT** are coincident, you need only click twice with the left mouse button on exactly the same spot. The alignment of the sketch tells Pro/E that the bottom edge of the part will be on datum **TOP** and that the left side of the part will be against datum **RIGHT.**

In Lesson 4, you will find out how to set the datum planes, according to **ASME Y14.5M 1994** geometric tolerance standards, as *basic* datums and how to rename the datum planes to **A, B,** and **C.**

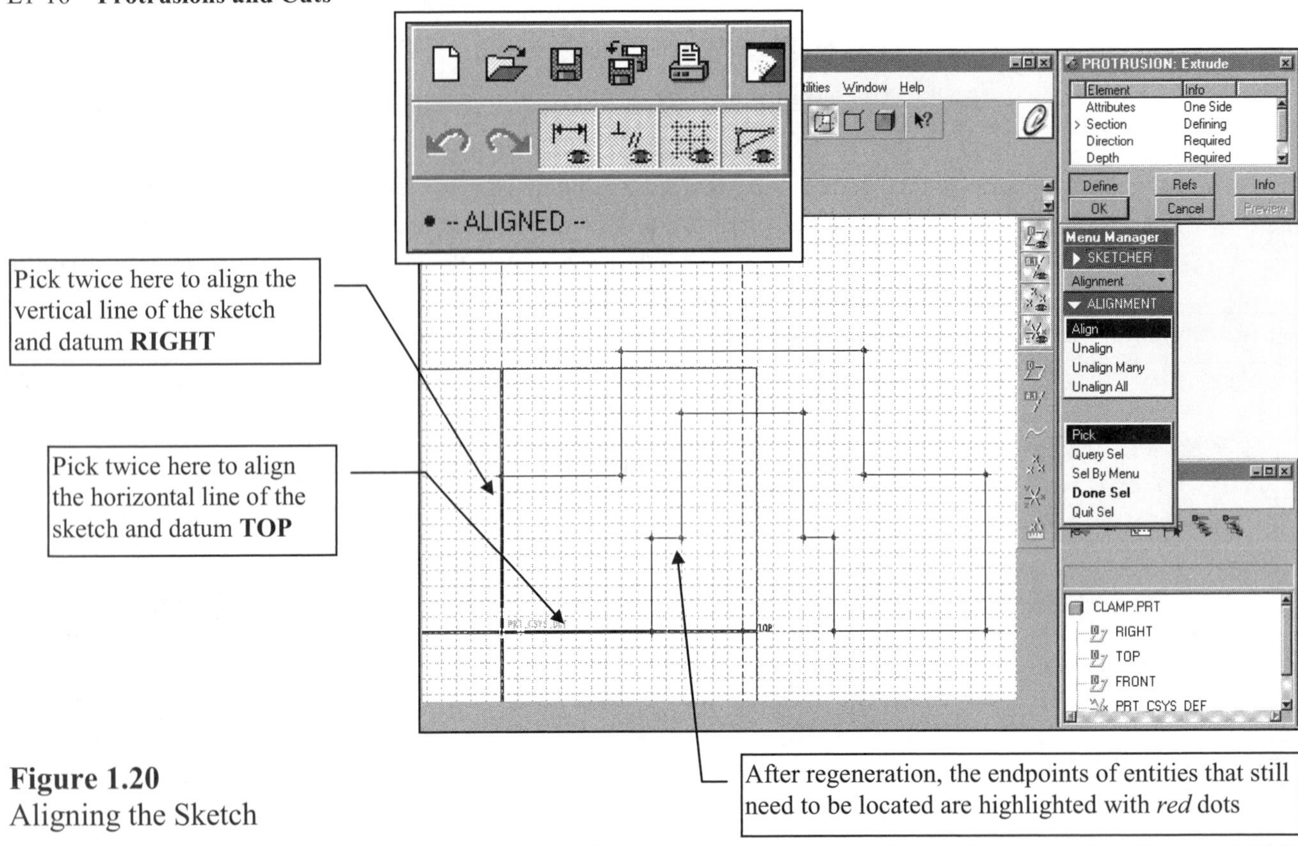

Figure 1.20
Aligning the Sketch

Look at the part again (Figures 1.1, 1.2, and 1.7). Datum **TOP** could be datum **A**, **FRONT** datum **B**, and **RIGHT** datum **C**, if geometric tolerancing were used on the part.

The sketch now needs to be dimensioned according to the design intent of the part. Look at Figures 1.7 and 1.21 to see the dimensioning scheme used to create the part's geometry. *Pro/E can automatically dimension a sketch, but the design intent may not be what you want.*

NOTE

On the job, ***never*** change the design intent of the model you are working on without getting the permission of the project engineer (or instructor for a class project)!

Your sketch dimensions may be placed above or below or to the right or the left of the object's view. The important factor is that the dimensions are correct as per the features being dimensioned, not the placement location on your sketch.

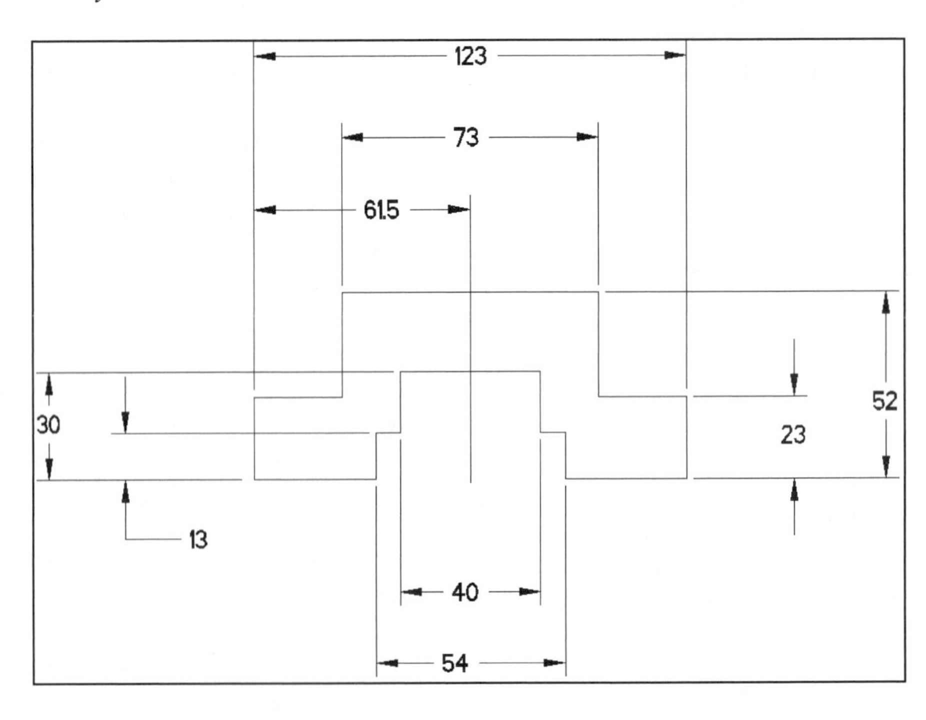

Figure 1.21
Part Dimensions

Place the dimensions as shown in Figure 1.22. Do not be concerned with the perfect positioning of the dimensions, but try to follow the spacing and positioning standards found in the **ASME Geometric Tolerancing and Dimensioning** standards. (This will save you time when you create a drawing of the part).

Dimension (from the SKETCHER menu) (see Fig. 1.22 for picks)

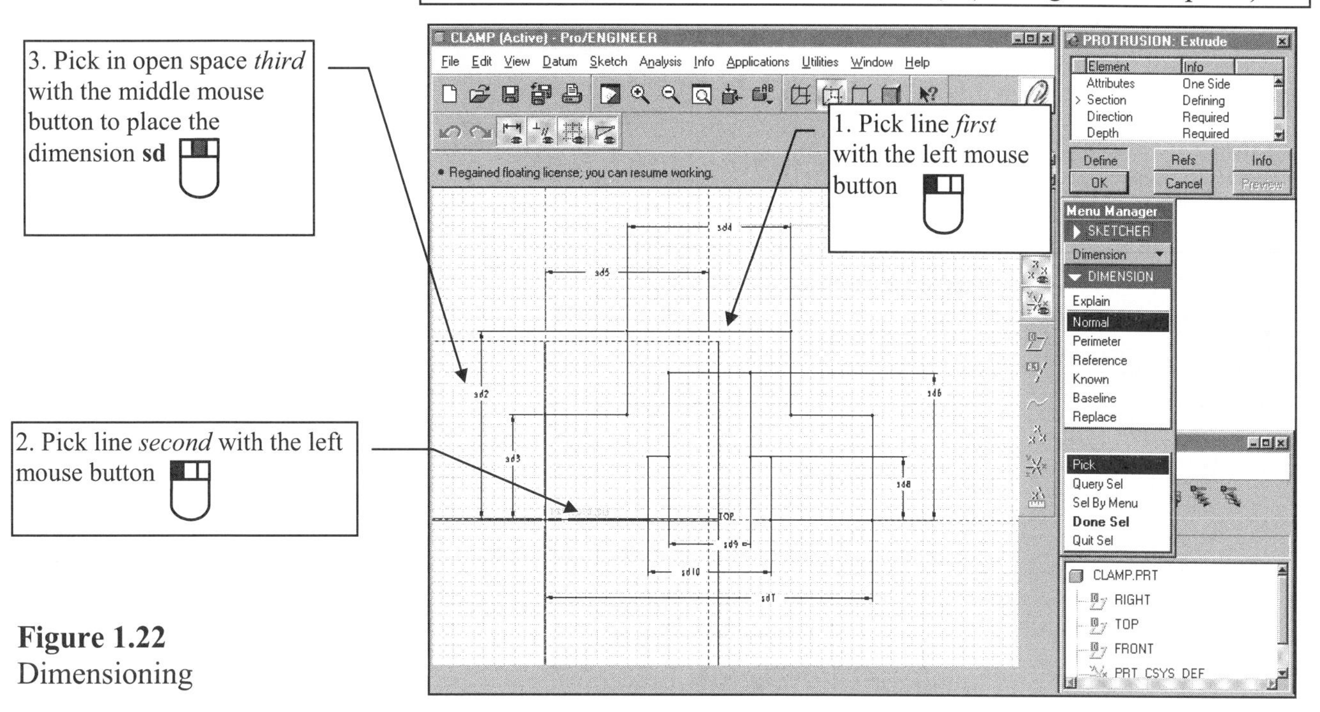

Figure 1.22
Dimensioning

HINT Again, your sketch dimensions may be placed above or below or to the right or the left of the object's view. The important factor is that the dimensions are correct as per the features being dimensioned, not the placement location on your sketch.

HINT
If you pick too many lines or choose an incorrect sequence of entities, simply pick the **Dimension** command to restart the dimensioning process.

Dimensions placed at this stage of the design process are displayed on the drawing document by simply asking to show all the dimensions. The most important thing is to get enough dimensions on the sketch to regenerate. The dimensions should be ones required to manufacture the part. To dimension between two lines, simply pick the lines with the left mouse button and place the dimension value with the middle button. At this stage of the dimensioning process, the dimensions are displayed as **sd** symbols: **sd1**, **sd2**, **sd3**, and so on.

It is now time to see if the sketch and dimensions will regenerate. Choose **Regenerate** from the SKETCHER menu (Fig. 1.23). The command line says **Section regenerated successfully**. The **Modify** command is automatically initiated at the successful regeneration of the section. You can now change/modify the dimensions to the *design sizes*. Note that the regenerated dimensions are based on a grid size of **15** units, here **15** millimeters (originally **30 X 30**). All the dimensions are too big. Pick on each dimension individually and type the correct value at the prompt, using the correct dimensions from Figures 1.7 and 1.21. Make sure you change every dimension correctly. If you change all but the center cut dimension and leave it as a value greater than the part's width, the section will fail at the next regeneration. If this happens, immediately pick **Unregenerate** from the SKETCHER menu and correct your neglected dimension(s).

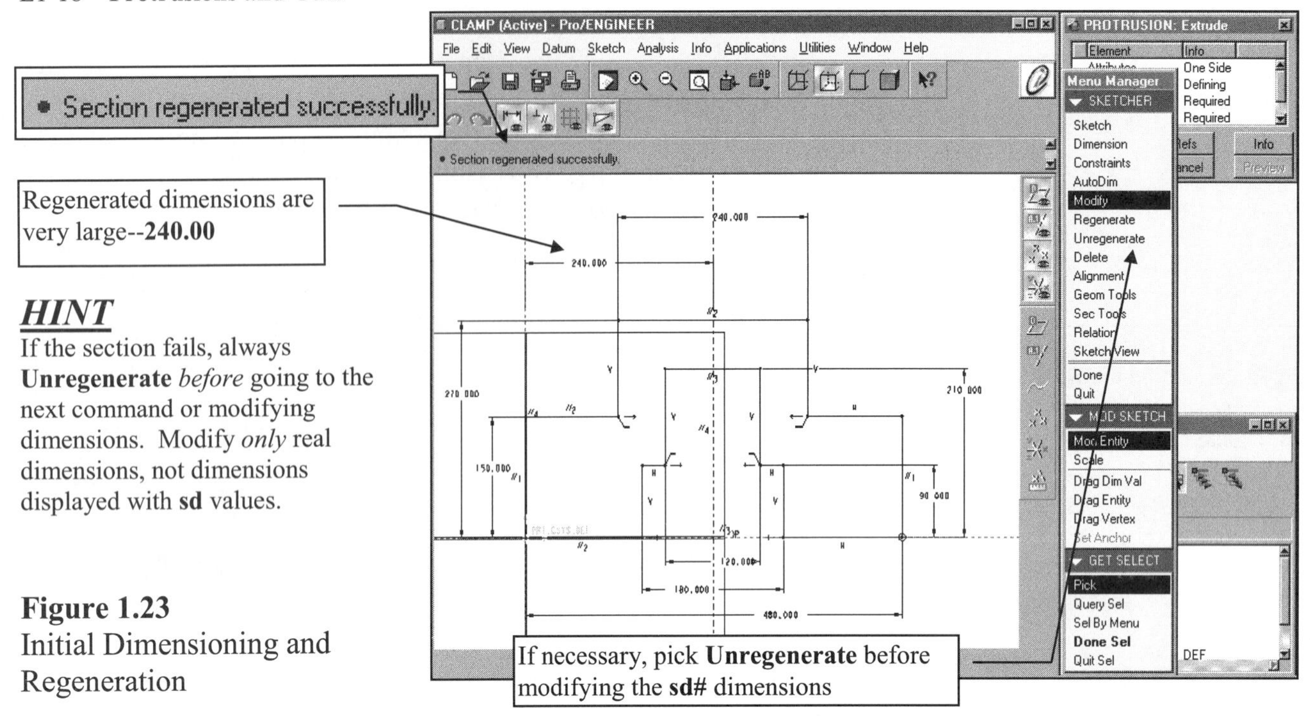

HINT

If the section fails, always **Unregenerate** *before* going to the next command or modifying dimensions. Modify *only* real dimensions, not dimensions displayed with **sd** values.

Figure 1.23
Initial Dimensioning and Regeneration

For a failed section, do not try to change the dimensions before unregenerating, because initiating certain Pro/E commands will block the use of **Unregenerate**. The **Unregenerate** choice becomes *dimmed* on the menu when it is unavailable (Fig. 1.24).

Make corrections and **Regenerate** the sketch. The correct sizes for the features are now displayed according to the design intent dimensions, and the section is complete. Turn off the datum planes, constraints, and grid to see the sketch more clearly.

Figure 1.24
Correct Design Intent Dimensions (see Fig. 1.21 for dimensions)

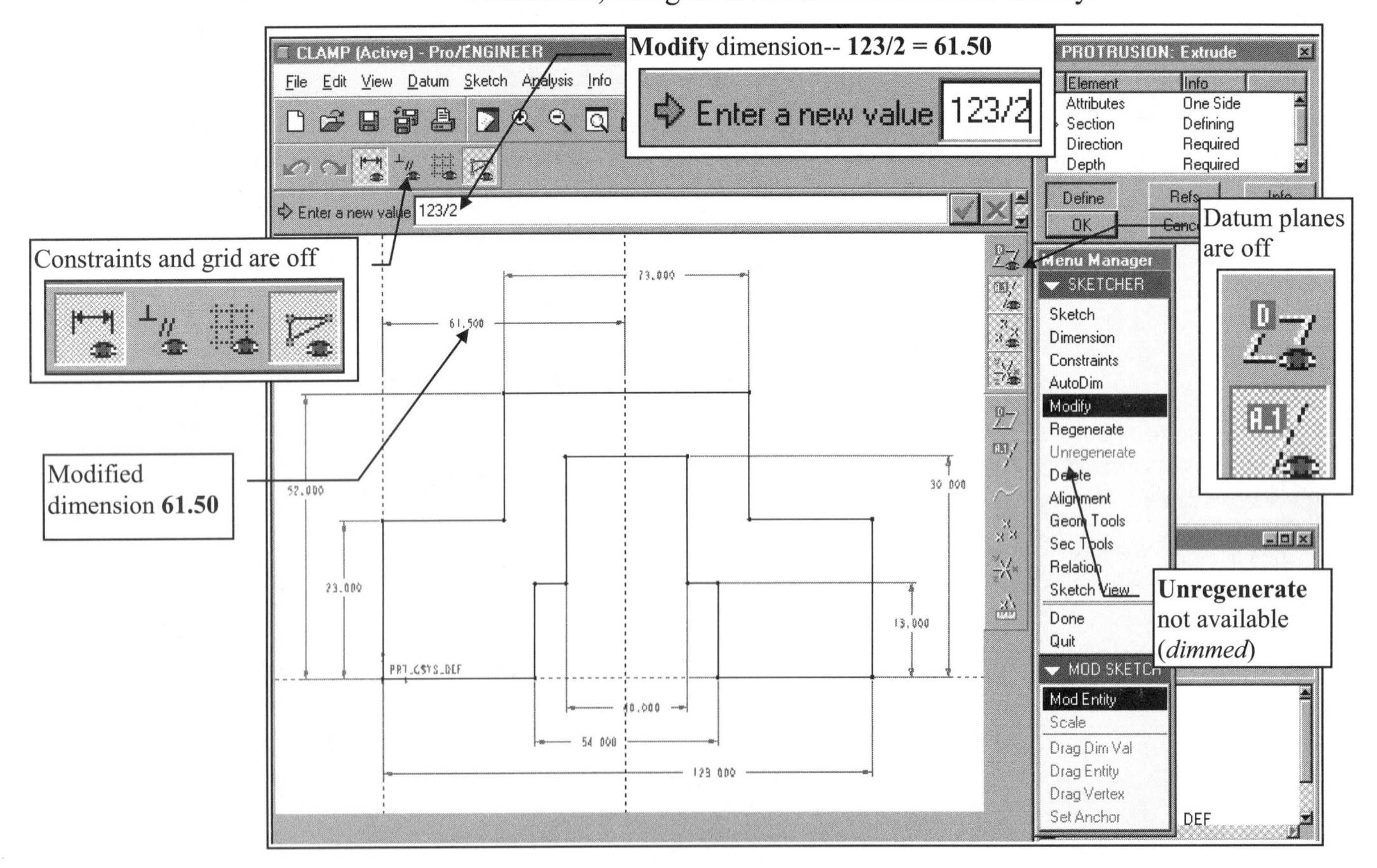

Complete the protrusion by choosing:

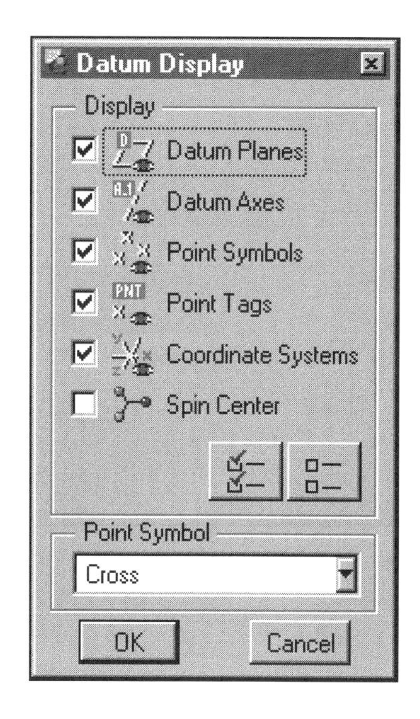

Regenerate ⇒ **Done** (in the Sketcher) ⇒ **View** (from the menu bar) ⇒ **Datum Display** ⇒ ✔ **Datum Planes** ⇒ **OK** ⇒ **View** ⇒ **Default** (to see the sketch in 3D trimetric)

Pro/E now prompts for the depth in the *direction* of protrusion creation and displays an arrow ➞ pointing in the direction previously set when **FRONT** was selected as the sketch plane (Fig. 1.25).

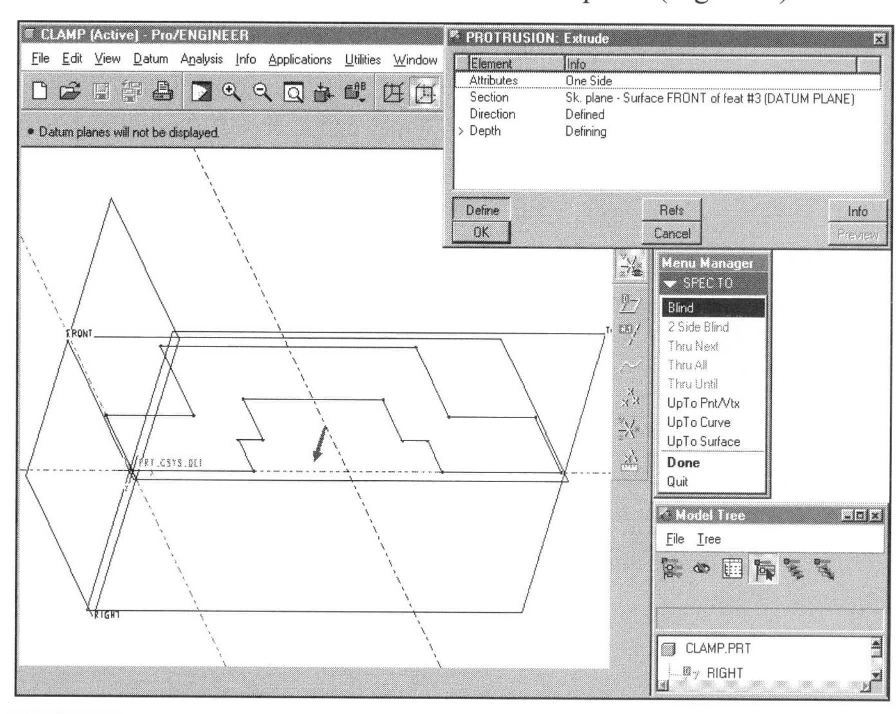

Figure 1.25
3D View of Sketch Showing the Arrow Direction

Blind (from the SPEC TO menu) ⇒ **Done** (type the depth dimension of the part at the prompt **(70)** ⇒ ✔ ⇒ **Preview** [from the dialog box to see the feature (Fig. 1.26)]

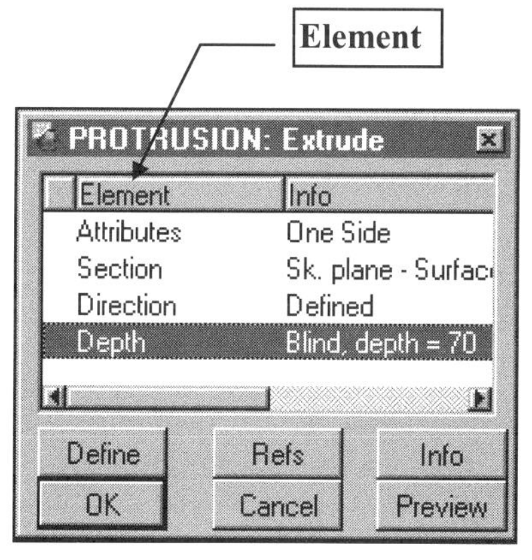

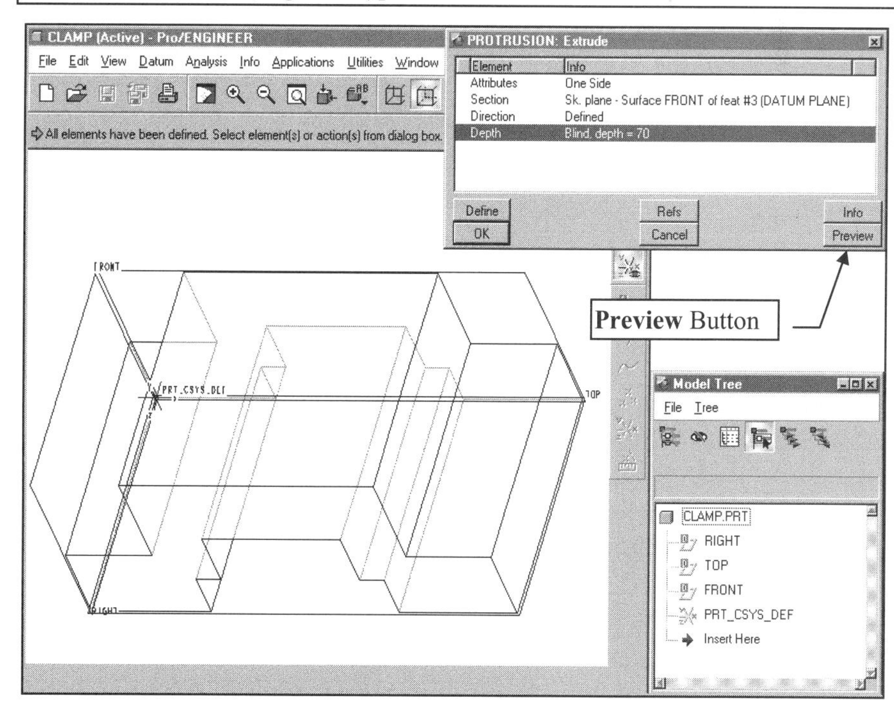

Figure 1.26
Preview of Part Protrusion

At this point, you could pick one of the elements from the dialog box and **Define** to change anything completed up to this point. Instead, pick **OK** from the dialog box. To see the model in a shaded state (Fig. 1.27), choose:

OK ⇒ **Utilities** ⇒ **Environment** ⇒ Display (uncheck all boxes) ⇒ (make **Isometric** and **Shading** defaults) ⇒ **Apply** ⇒ **OK**

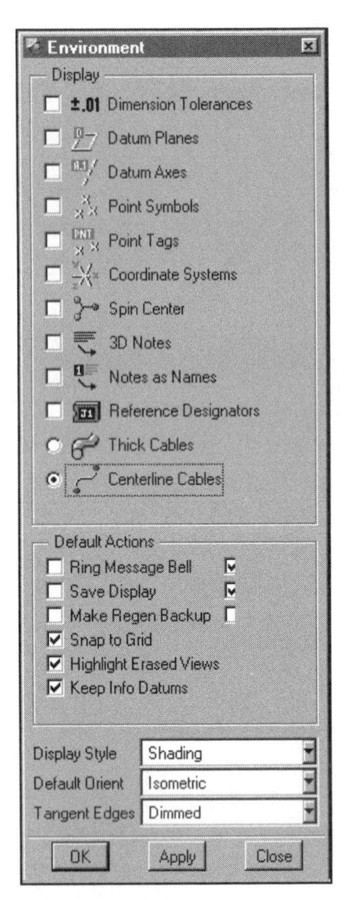

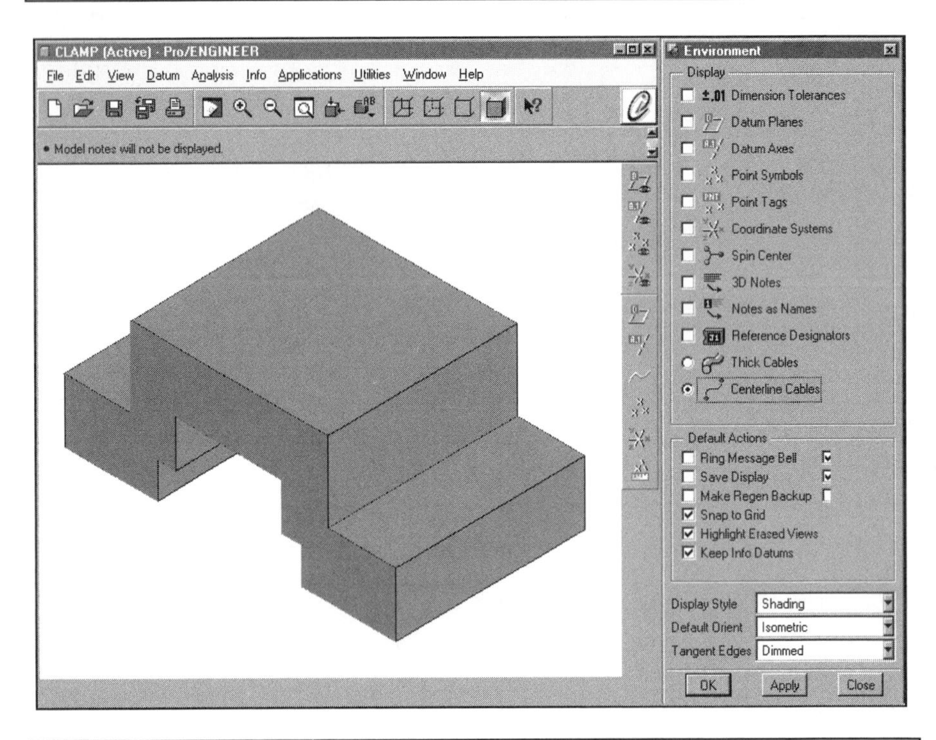

Figure 1.27
Shaded Isometric Part

File (from the menu bar) ⇒ **Save** (this will save the part at its present stage) ⇒ ✔ ⇒ **File** ⇒ **Delete** ⇒ **Old Versions** ⇒ ✔

Use **File** ⇒ **Delete** ⇒ **Old Versions** ⇒ ✔ after the **Save** command to eliminate previous versions of the part that may have been saved. This is the first time you have saved this part, so **Delete** ⇒ **Old Versions** ⇒ ✔ is not necessary. But it is a good habit to get into so that you do not fill up your hard drive with versions of the same part saved at different stages in the design process.

The next feature will be a cut (Fig. 1.28). The **20 X 20** centered cut is on both sides of the part. Before we create this feature, change the **Environment** back to **Hidden Line** and turn on the settings for datum planes, datum axes, etc. Leave the view set as isometric. Because the cut feature is identical on both sides of the part, we can mirror the cut after it has been created. The cut is made on one side and mirrored about a datum plane to the other side of the protrusion. Start by creating a new datum plane that is offset from datum **RIGHT**. Choose the following commands (Fig. 1.29):

Datum (from the menu bar) ⇒ **Plane** ⇒ **Offset** (pick datum **RIGHT** as the plane to offset from) ⇒ **Enter Value** (type the distance at the prompt) **123/2** **(123** divided by **2**; Pro/E will do the higher math: **123/2 = 61.5)** ⇒ ✔ ⇒ **Done**

The new protrusion shows in the Model Tree-- **Protrusion id**

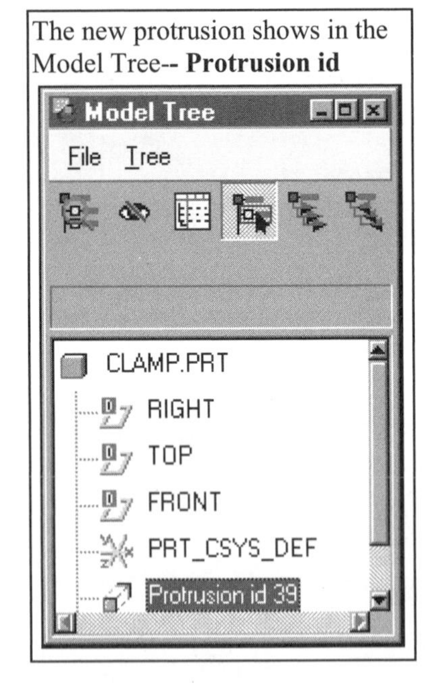

HINT
Use **Trimetric** before the first protrusion is created; then use **Isometric** as the default orient after the first protrusion is completed.

HINT

Never complete the math yourself if Pro/E can complete it for you. Sometimes you may think that you have completed the math correctly without noticing that you have rounded the values. For instance, **5.255** divided by **2** is **2.6275**. If you round to **2.628,** it will not be correct because it is not one-half of **5.255.**

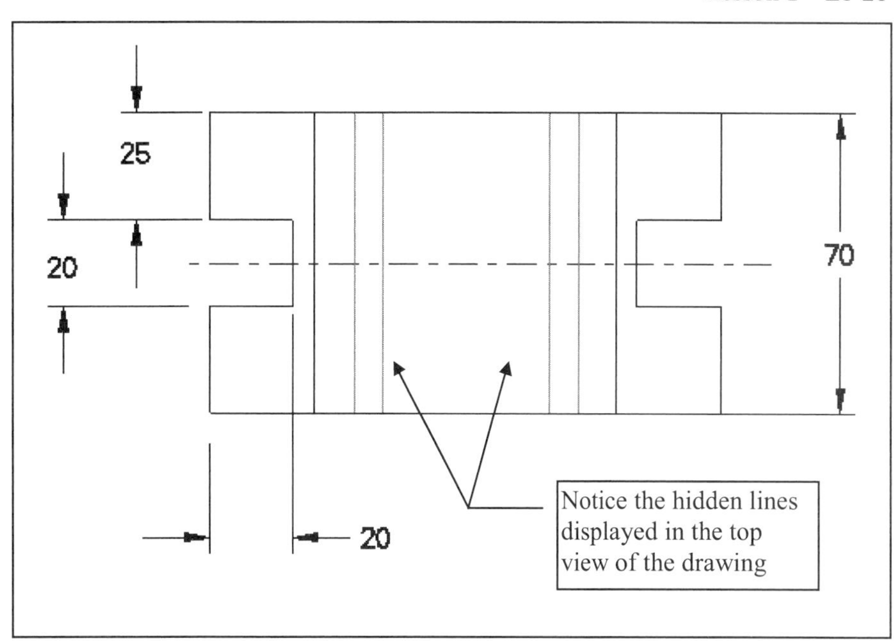

Figure 1.28
Cut Dimensions

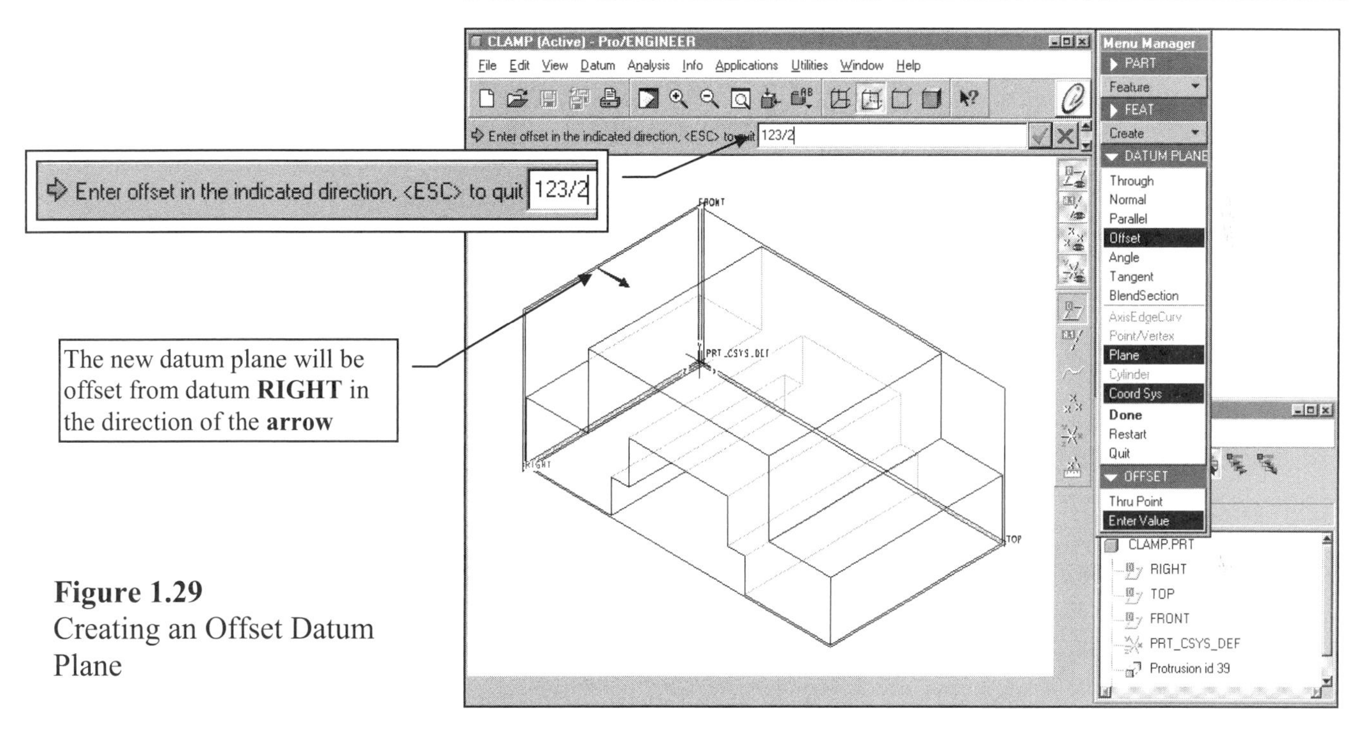

Figure 1.29
Creating an Offset Datum Plane

The datum plane is offset from **RIGHT,** which is its *parent feature*. The new datum plane is shown on the screen passing through the center of the part and also appears in the Model Tree (Fig. 1.30). **DTM1** can now be used in the construction of other features as required by the design.

The cut will be a sketched feature similar to the first protrusion, except that it will remove material. Choose the following commands:

Feature ⇒ **Create** ⇒ **Cut** ⇒ **Extrude** (default) ⇒ **Solid** (default) ⇒ **Done** ⇒ **One Side** (default) ⇒ **Done** (pick **TOP** as the sketching plane) ⇒ **Flip** ⇒ **Okay** ⇒ **Top** (pick **FRONT** to orient the sketch)

You can initiate a variety of commands by picking on the feature in the Model Tree and then pressing the right mouse button (RMB)

Model Tree
File Tree
CLAMP.PRT
RIGHT
TOP
FRONT
PRT_CSYS_DEF
Protrusion id 39
DTM1
Insert
Delete
Suppress
Modify
Redefine
Reroute
Pattern
Note Create
Info

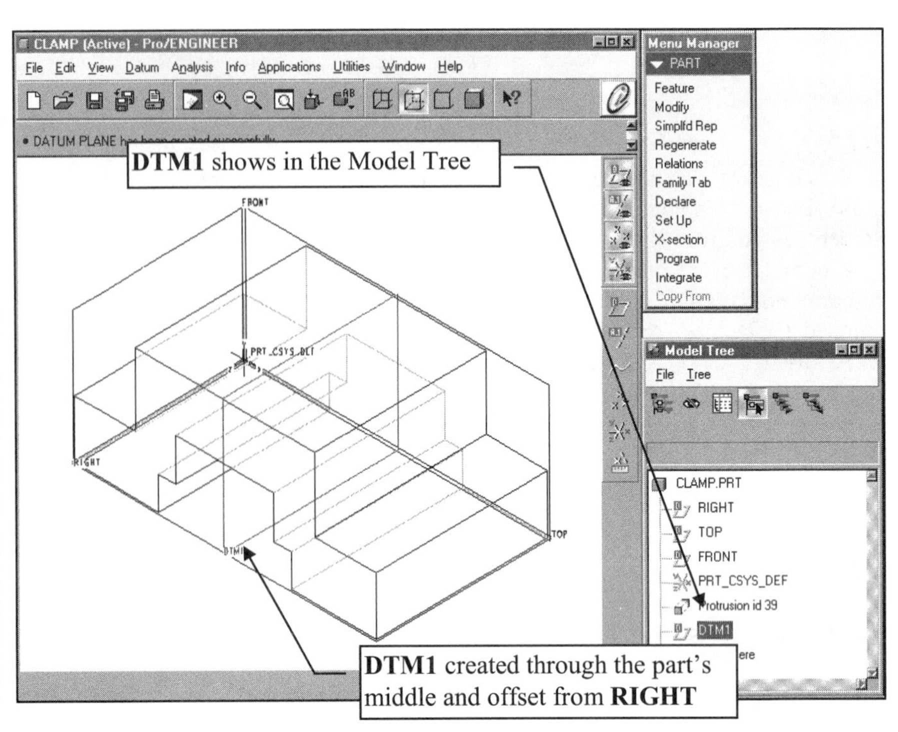

Figure 1.30
Offset Datum Plane **DTM1**

In Figure 1.31, you are being prompted for the **Direction of Feature Creation**. Flipping the arrow changes the direction of the feature creation so that it will cut the part protrusion and not *air!*

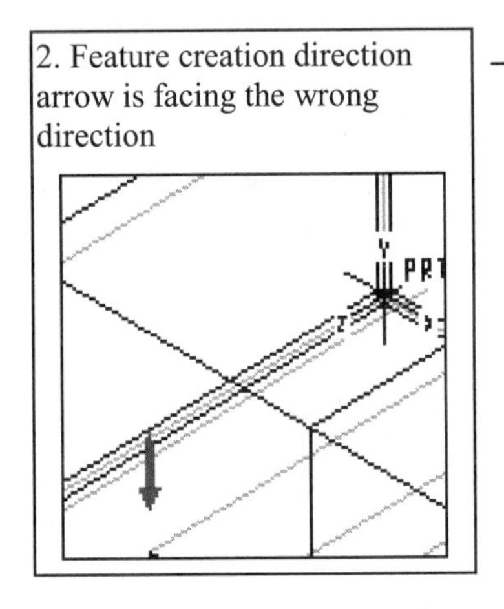

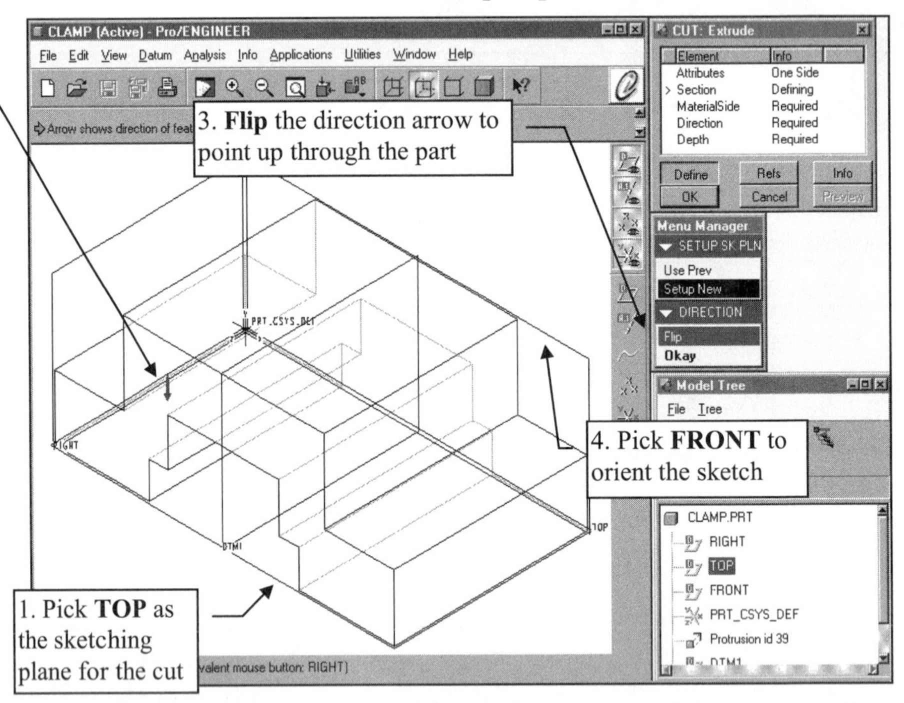

Figure 1.31
Pick **TOP** as the Sketching Plane and **Flip** the Arrow for the Feature Creation Direction

In Figure 1.32, you are looking at the *bottom of the part*. At first, it is hard to see the orientation of the part. To get visually oriented, note the location of the coordinate system. **RIGHT** and **FRONT** show as edges and will be used to dimension the sketch. The sketch is composed of three lines forming an *open section*. The endpoints of the two lines that touch **RIGHT** will be aligned with that datum plane. The dimensioning scheme and values will be the same as those shown in Figure 1.28 (which shows the top view of the part on the drawing). The edge of the cut is dimensioned from **FRONT.**

Turn the grid snap off in the Environment dialog box, and continue the commands (Fig. 1.32):

Utilities ⇒ **Environment** ⇒ ❑ **Snap to Grid** ⇒ **OK** ⇒ **Close** (to keep the **References)** ⇒ (click the *create line icon* and pick on **RIGHT** to start the line)

Complete the three continuous lines by picking four endpoints, as shown in Figure 1.32. After the second pick, the **Sketcher** toggles from horizontal to vertical line option automatically.

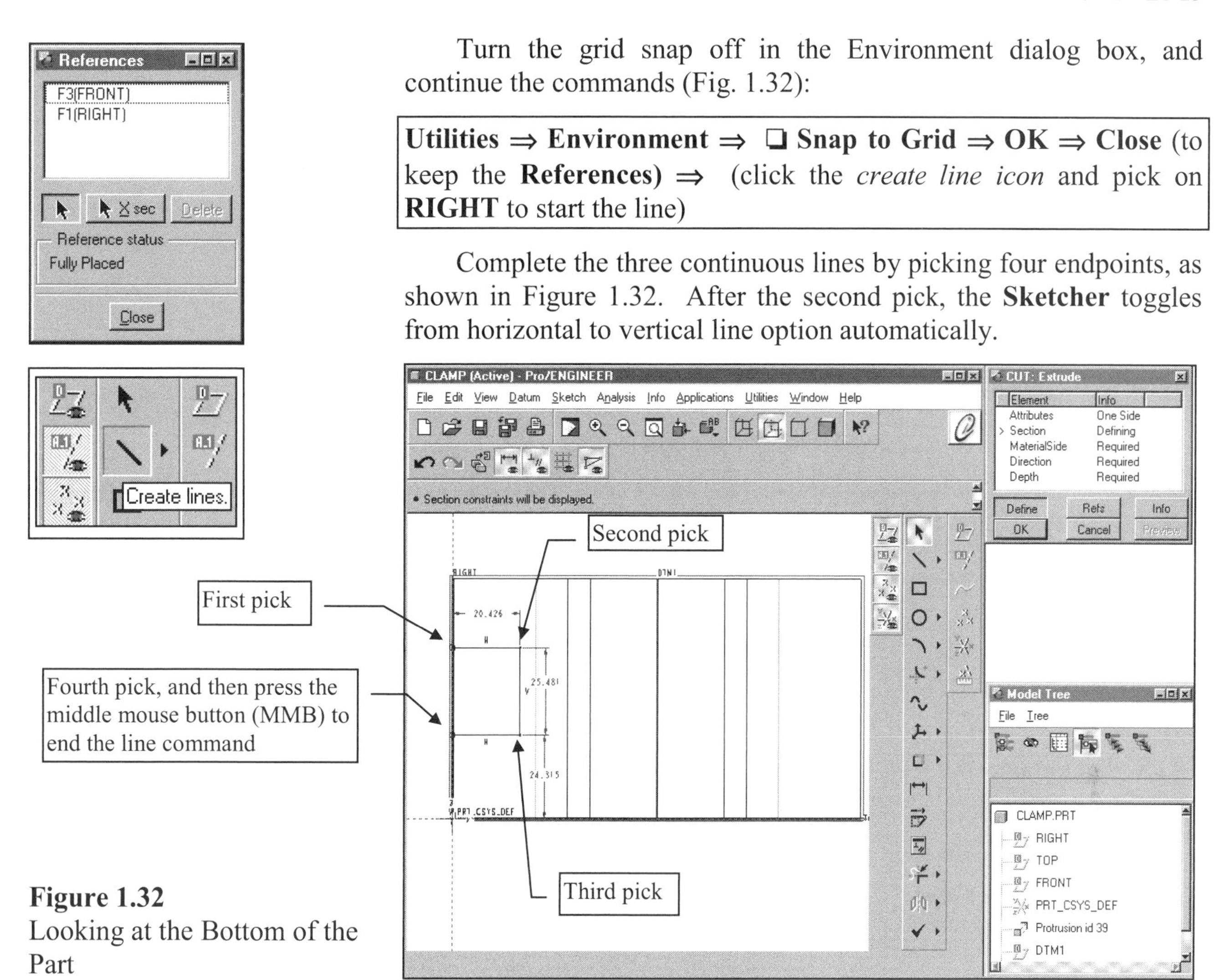

Figure 1.32
Looking at the Bottom of the Part

The view shown in Figure 1.33 was panned and zoomed.

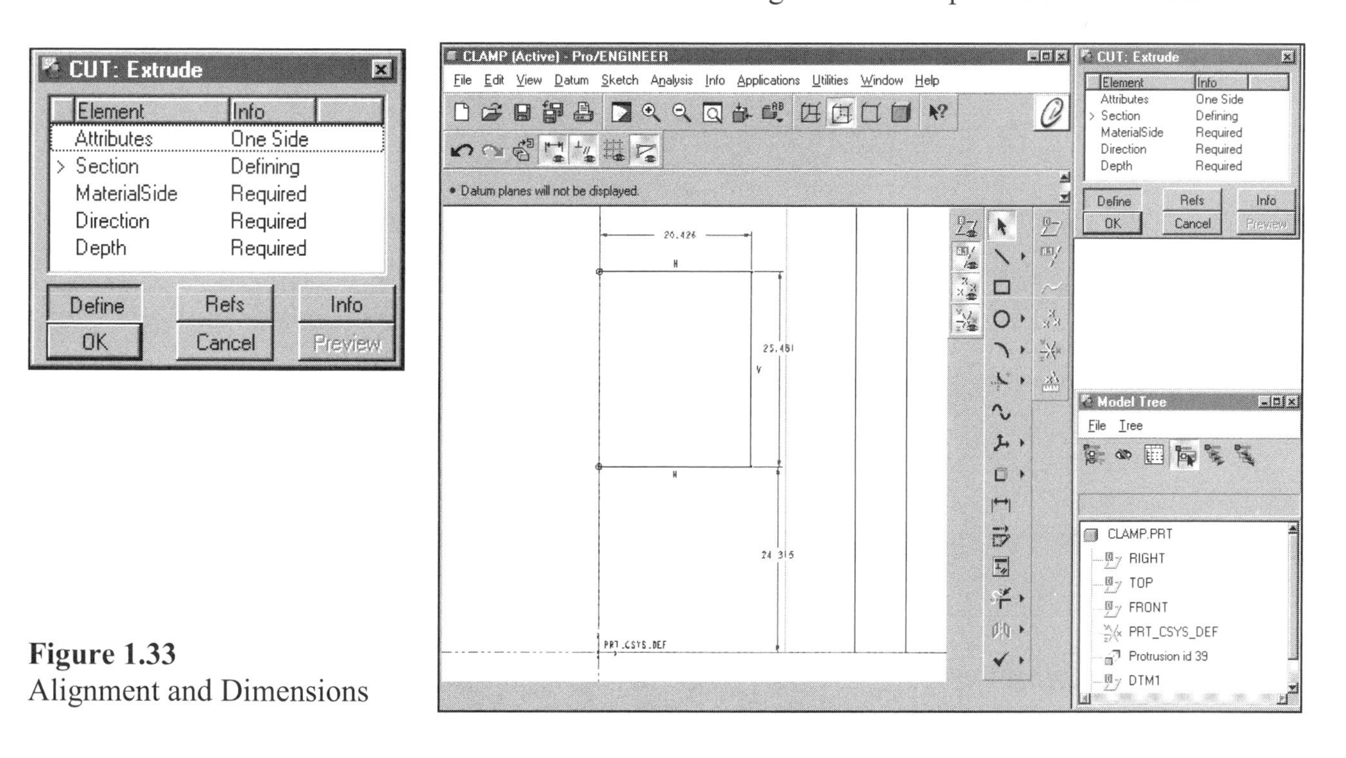

Figure 1.33
Alignment and Dimensions

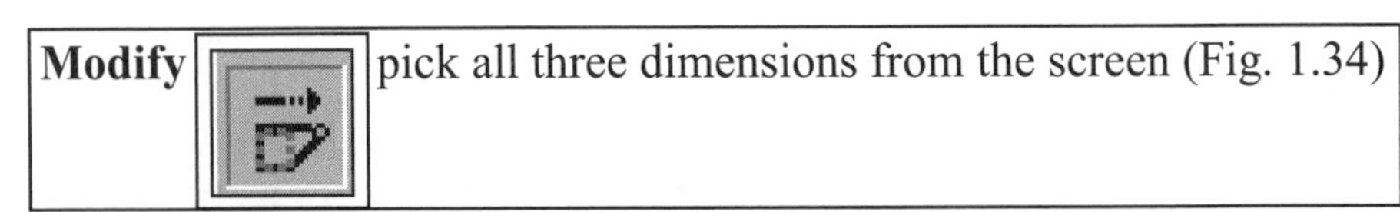

Modify pick all three dimensions from the screen (Fig. 1.34)

Before

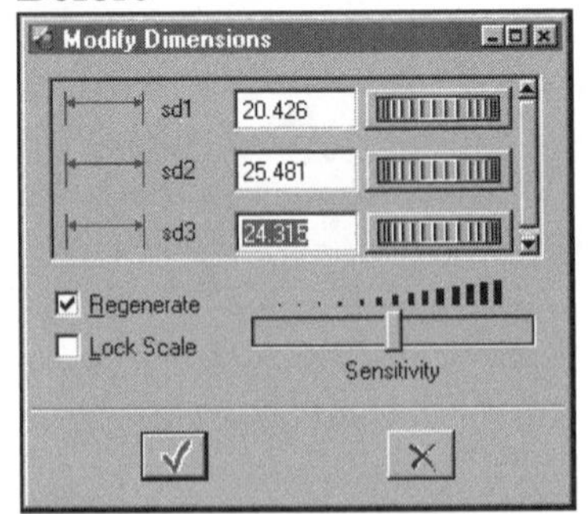

After

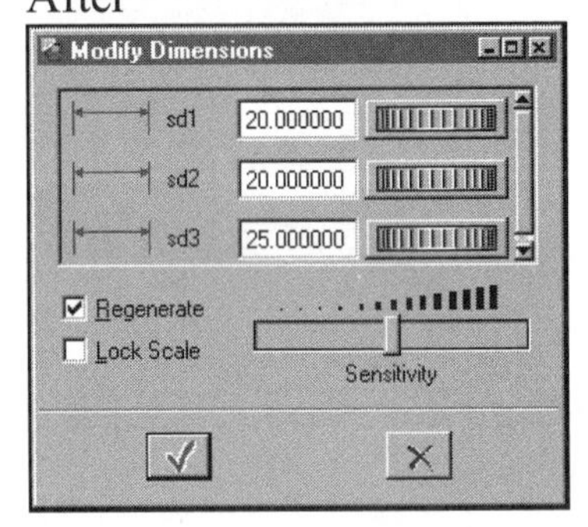

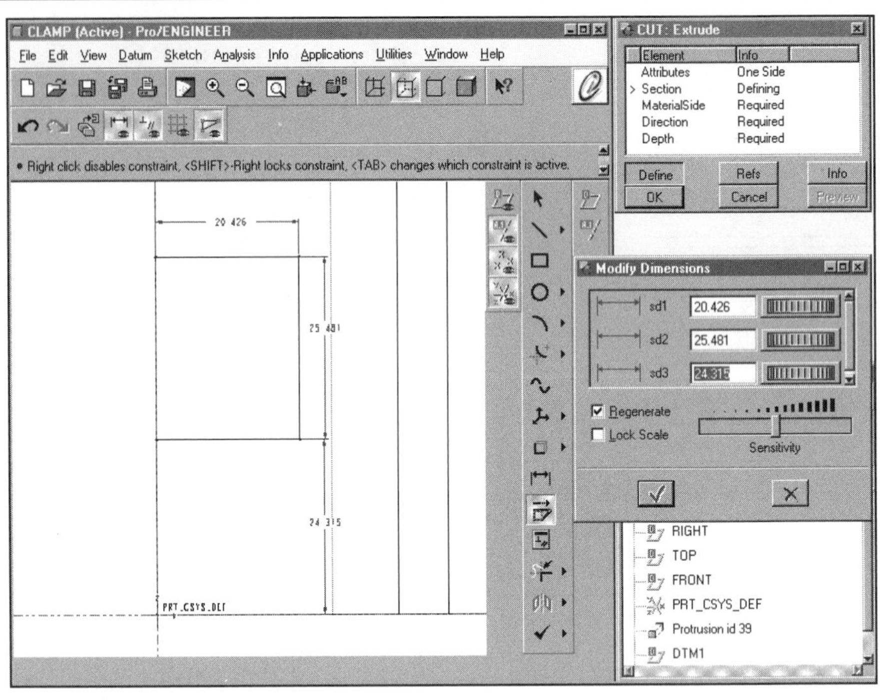

Figure 1.34
Modifying dimensions

(Input the correct design value for each dimension) ⇒ ✔ ⇒ L [left-click and hold down on a dimension and move it to a better position off of the object [the cut's section is now complete (Fig. 1.35)] ⇒ ✔

Figure 1.35
Modified Sketch

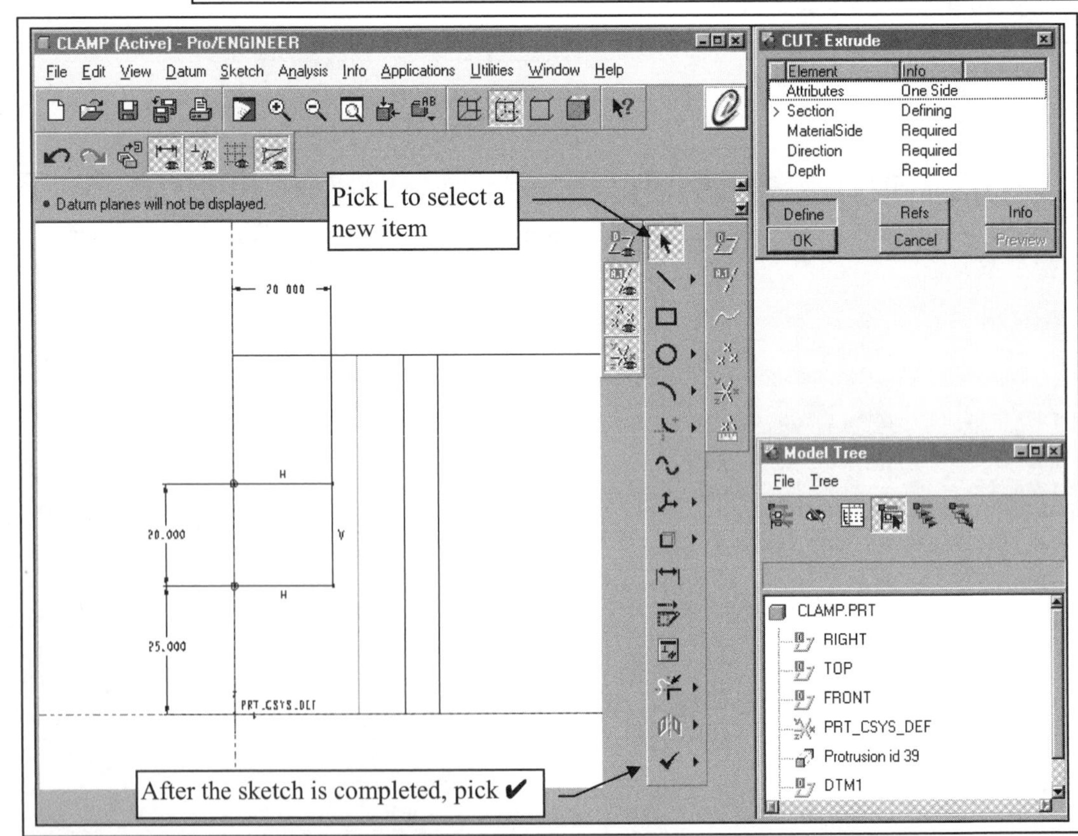

The direction of material removal is required at this stage of the cut creation. An arrow is displayed showing the direction in which the material will be removed (Fig. 1.36). Here, the direction arrow points toward the area to be removed for the cut; therefore, pick **Okay** from the DIRECTION menu. If you select the incorrect direction for the material removal direction arrow, the whole part will be cut away and the small **20 X 20** square piece will be all that's left of your part!

HINT

If the removal direction was pointing away from the cut and all that was left after the command was completed was a small square block of material, you should pick **MaterialSide** and **Define** from the dialog box and redo the command.

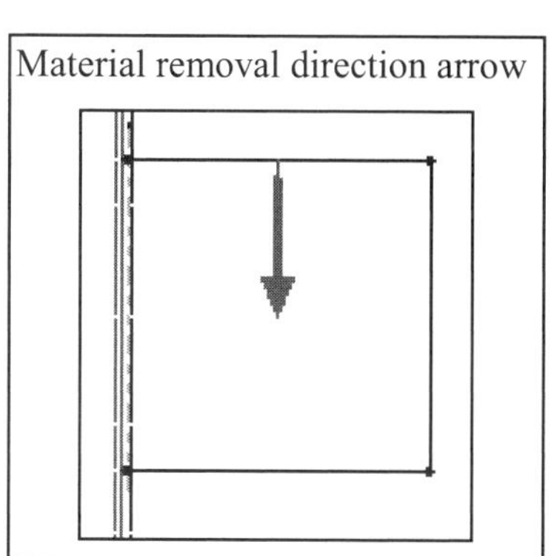

Pick **MaterialSide** ⇒ **Define** to redo (if needed)

Pick **Okay**

Figure 1.36
Direction of Material Removal

We recommend changing to a 3D view orientation in order to verify the correct cut-depth direction (Fig. 1.37).

Choose **View** ⇒ **Default** (make sure the cut-depth direction arrow points *into* the cut area to be removed) ⇒ **Thru All**

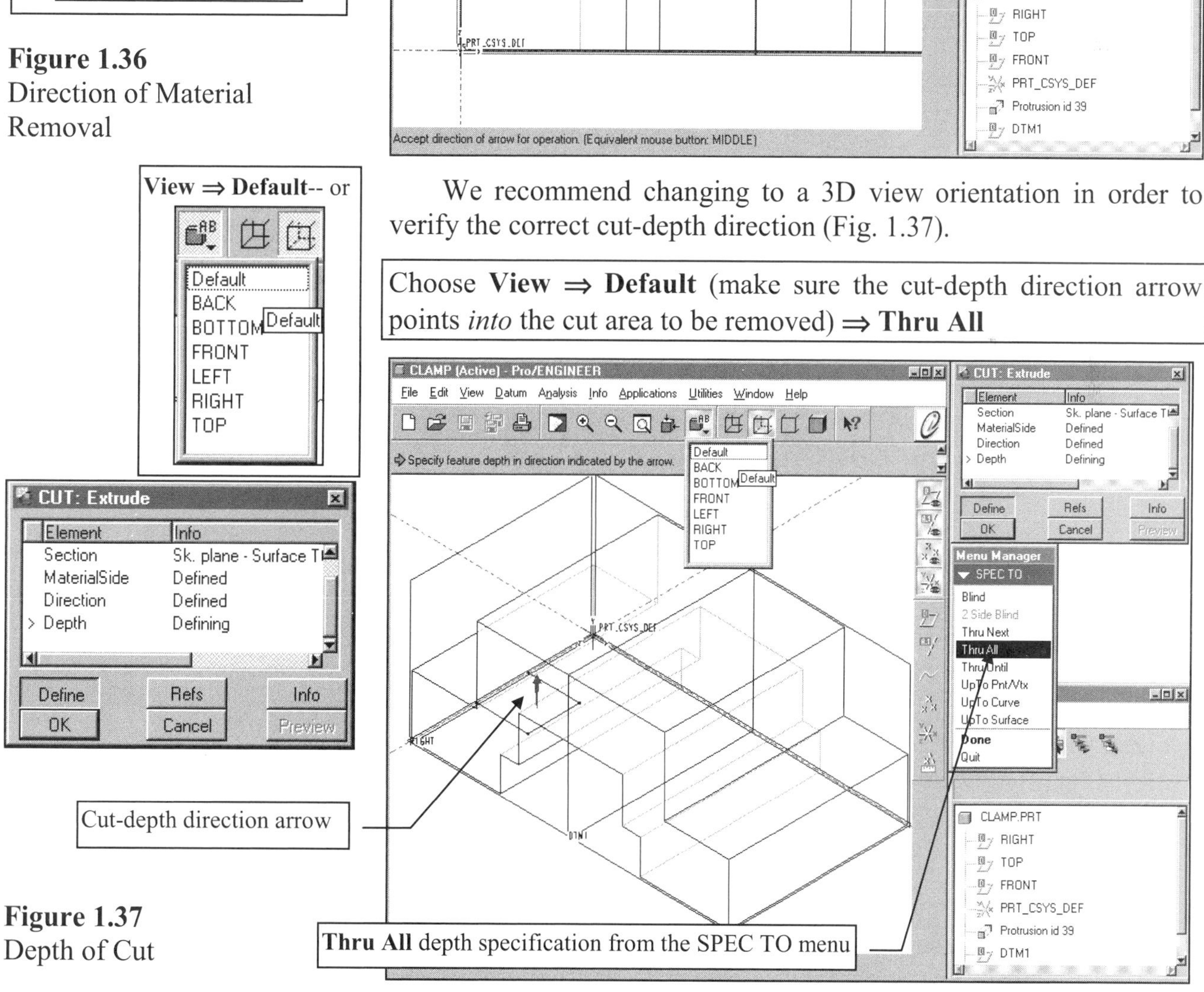

Figure 1.37
Depth of Cut

Done from the SPEC TO menu ⇒ **Preview** from the dialog box ⇒ **OK** from the dialog box (Fig. 1.38) ⇒ **File** ⇒ **Save** ⇒ ✔

Newest feature is displayed in the model tree: **Cut id**

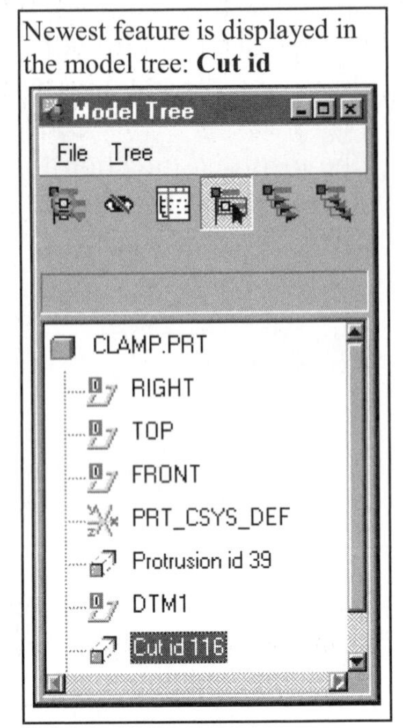

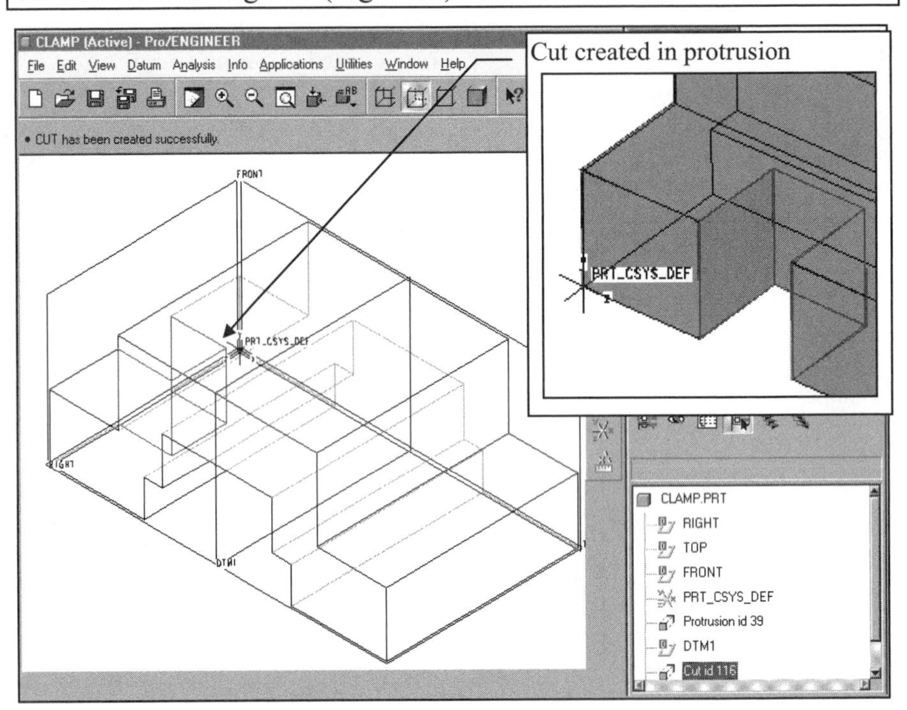

Figure 1.38
Cut Created Successfully

The last step is to copy the cut feature to the right side of the part. The feature is hidden behind the protrusion, so you must use **Query Sel** to filter through to the cut feature. It will be highlighted when selected. From the FEAT menu, choose the following (Fig. 1.39):

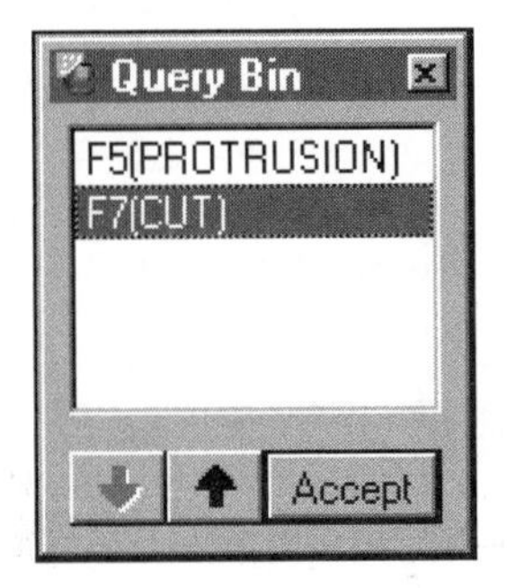

Copy ⇒ **Mirror** ⇒ **Independent** ⇒ **Done** ⇒ **Query Sel** (pick on the cut feature) ⇒ (pick the **CUT** from the **Query Bin)** ⇒ **Accept** ⇒ **Done Sel** ⇒ **Done**

HINT
Instead of selecting from the Query Bin, use your *right mouse button* to query to the next selection and the *middle mouse button* to accept the selection.

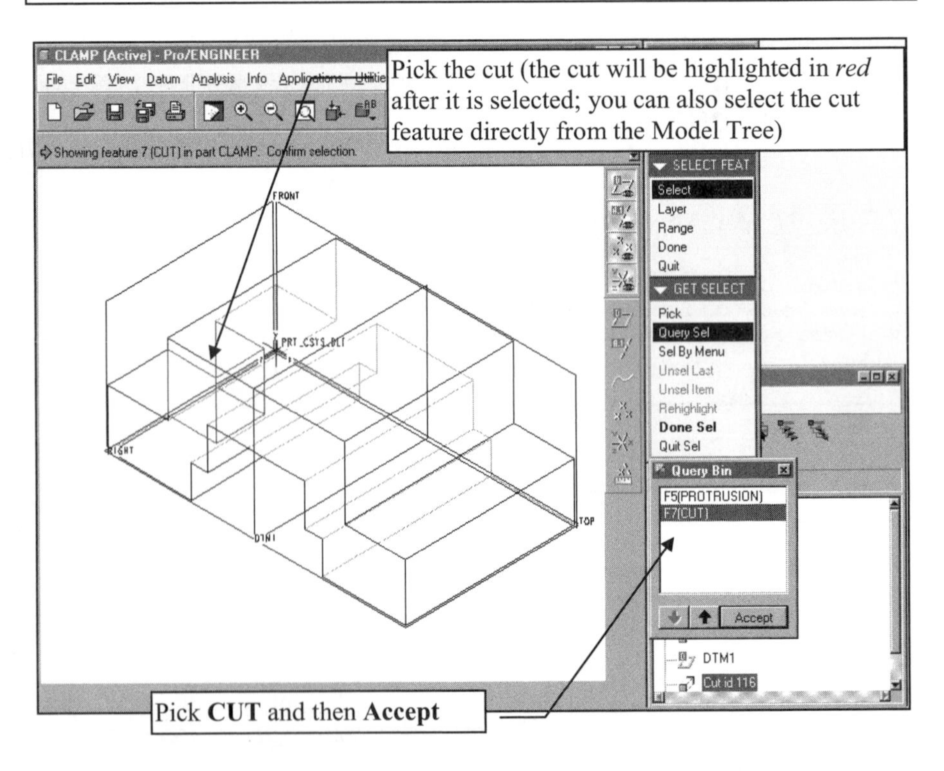

Figure 1.39
Copying the Cut

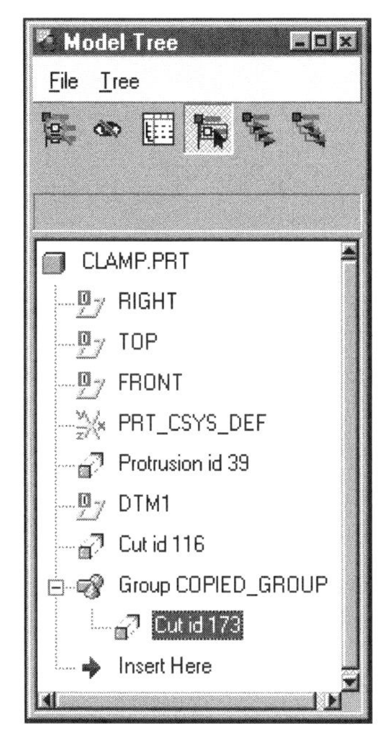

Figure 1.40
Copying the Cut by Mirroring it About **DTM1**

Pro/E now prompts you to select the plane or datum to use as the mirroring plane (Fig. 1.40). The cut will be mirrored about **DTM1.**

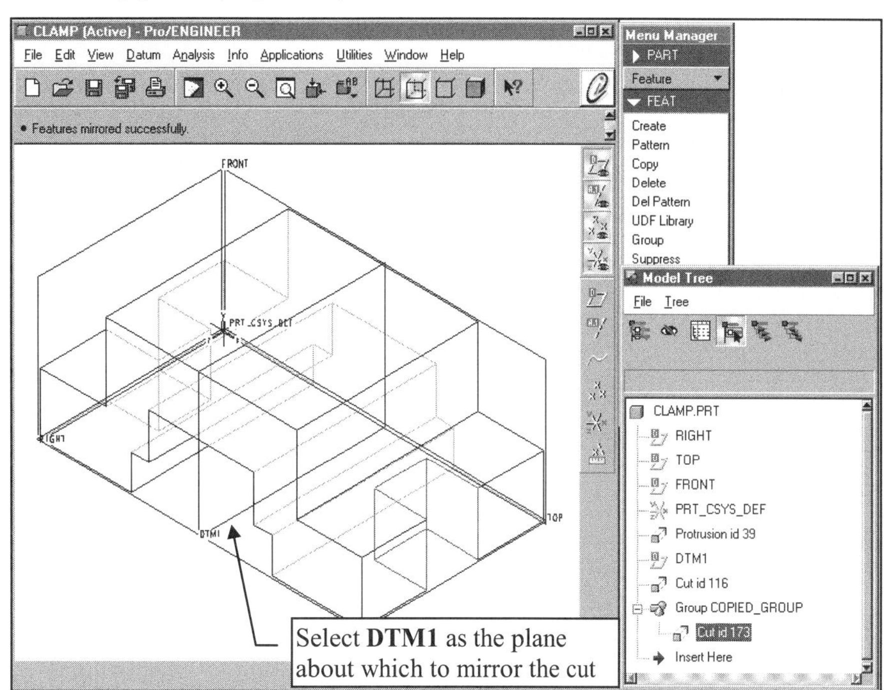

Pick on the edge or name of datum plane **DTM1.** Once you **Pick** (or **Accept** after **Query Sel**), the cut is mirrored automatically. Pro/E completes the feature and displays the part. You have completed your first Pro/E part! To see the protrusion and the cuts clearly, rotate and shade the part, make the Main Window fill the screen, and use the default settings for the placement of the Model Tree (Fig. 1.41):

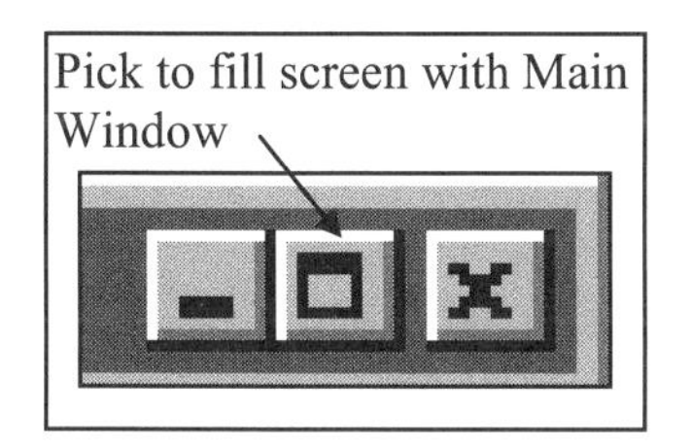

Utilities ⇒ Customize Screen ⇒ Options ⇒ Default ⇒ OK ⇒ Utilities ⇒ Environment ⇒ Shading ⇒ OK ⇒ View ⇒ Default ⇒ File ⇒ Save ⇒ ✔ ⇒ File ⇒ Delete ⇒ Old Versions ⇒ ✔

HINT

Hold down the **Ctrl** key and press the right (**Pan**), left (**Zoom**), or middle (**Rotate**) mouse button to orient the part on the screen.

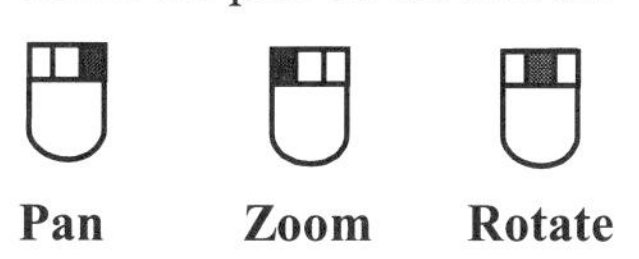

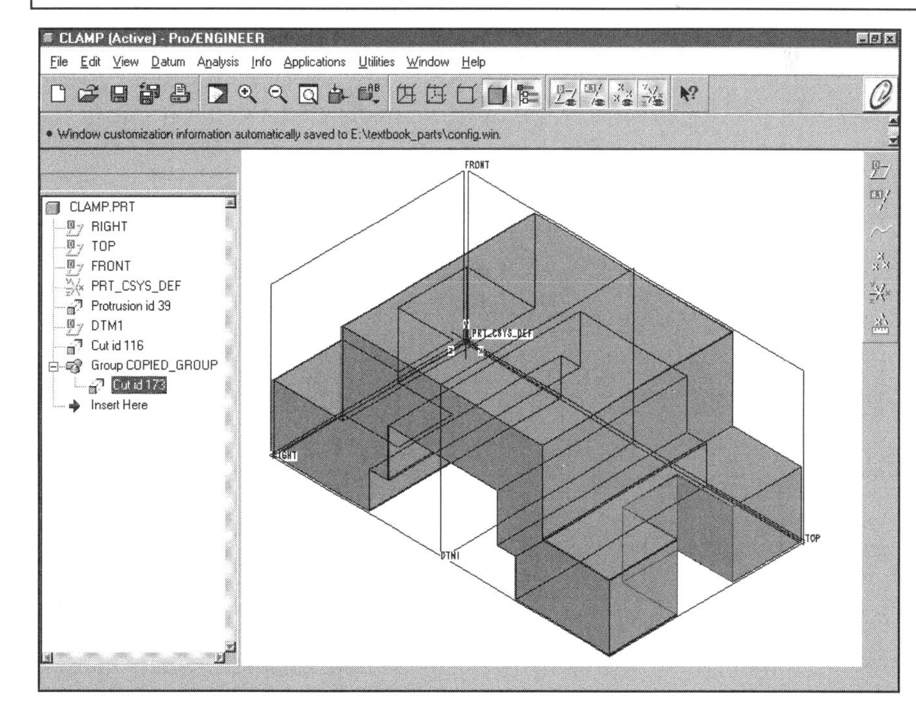

Figure 1.41
Rotated and Shaded Clamp

Lesson 1 Project

Angle Block

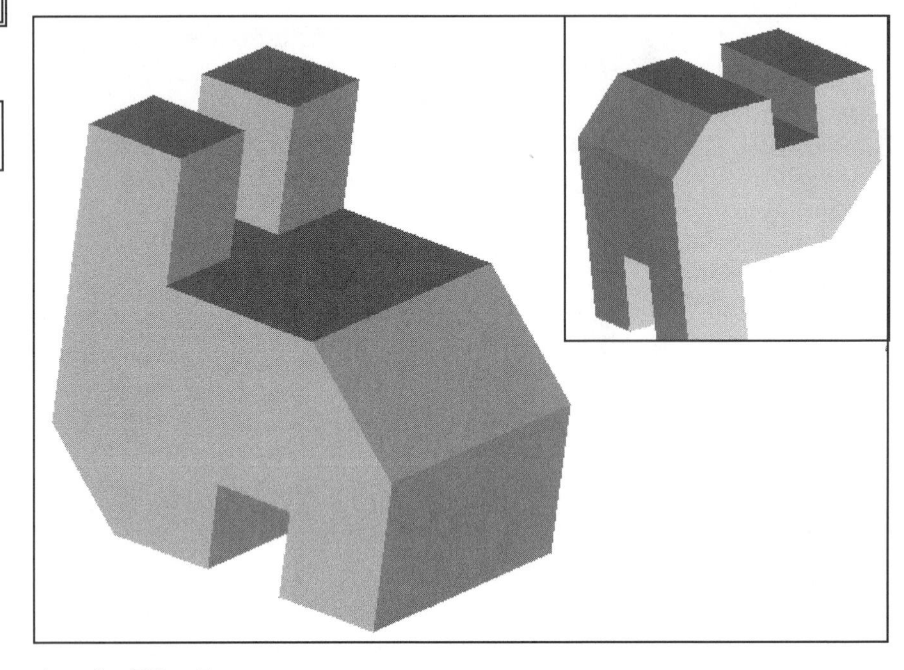

Figure 1.42
Angle Block

Angle Block

The first **lesson project** is a simple block that requires many of the same commands as the **Clamp.** Using the default datums, create the part shown in Figures 1.42 through 1.46. The datum planes used here are named **DTM1, DTM2,** and **DTM3** instead of **RIGHT, TOP,** and **FRONT.** Sketch the protrusion on **DTM3** (your datum **FRONT**). Sketch the cut on **DTM1** (your datum **RIGHT**) and align it to the upper surface/plane of the protrusion, as shown in Figure 1.43.

? Pro/HELP
Remember to use the help available on Pro/E by highlighting a command and pressing the right mouse button (RMB)

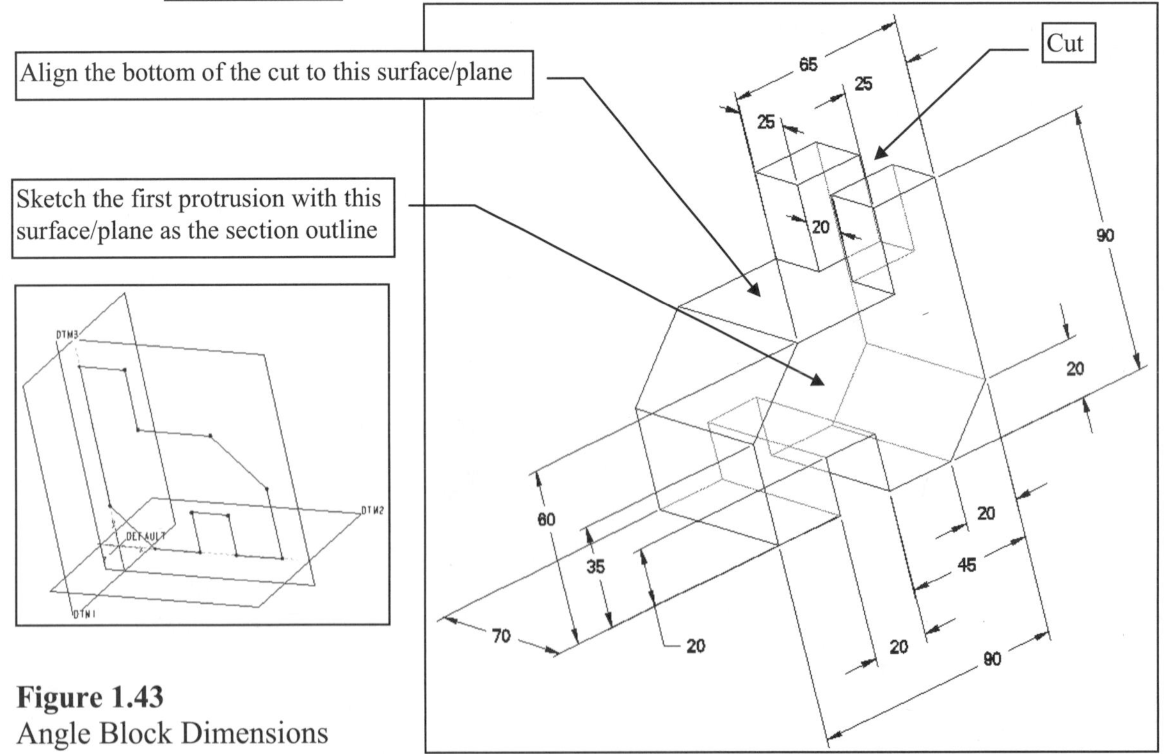

Figure 1.43
Angle Block Dimensions

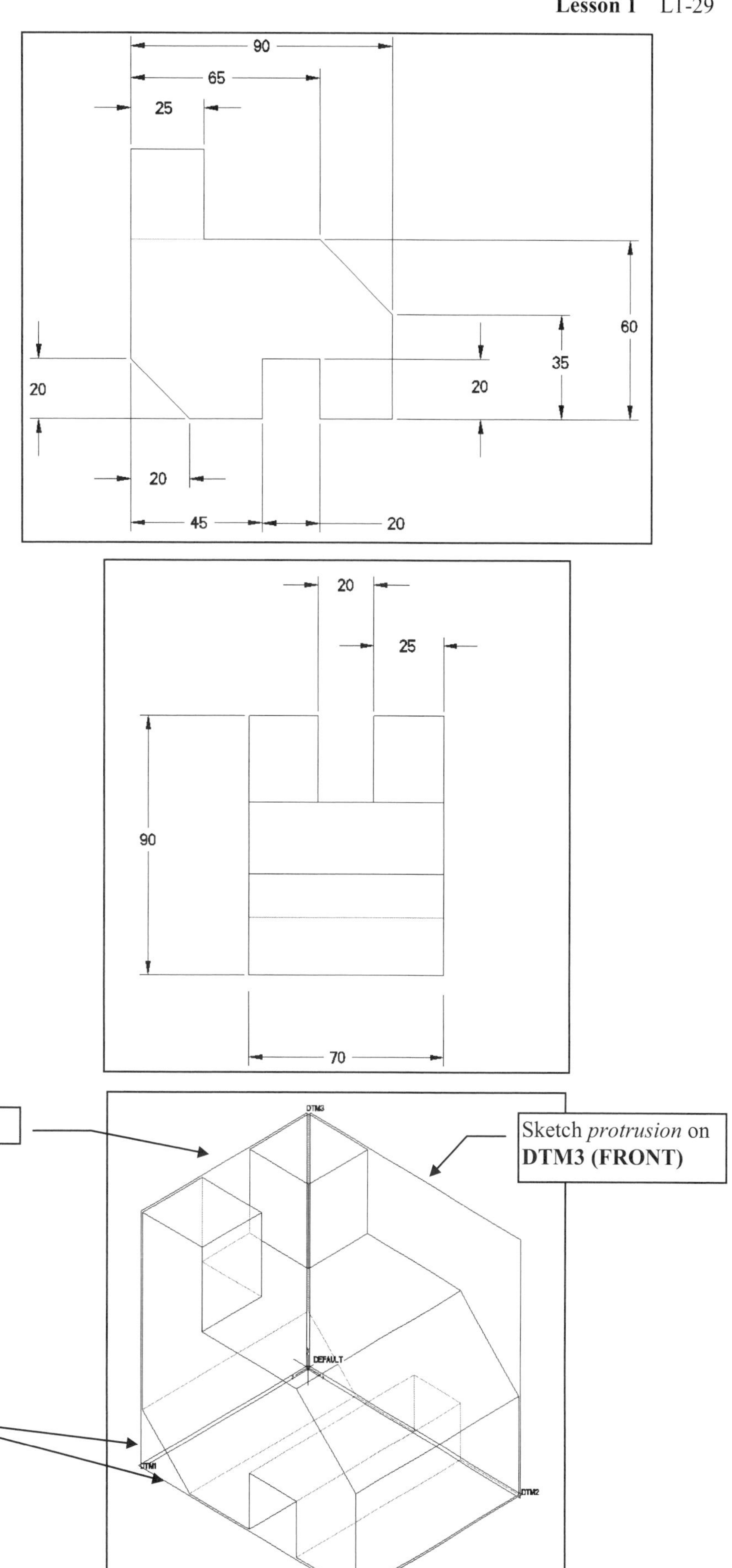

Figure 1.44
Angle Block Protrusion
Sketch Dimensions

File ⇒ Save ⇒ ✔
File ⇒ Delete ⇒
Old Versions ⇒ ✔

Figure 1.45
Angle Block Cut Sketch
Dimensions

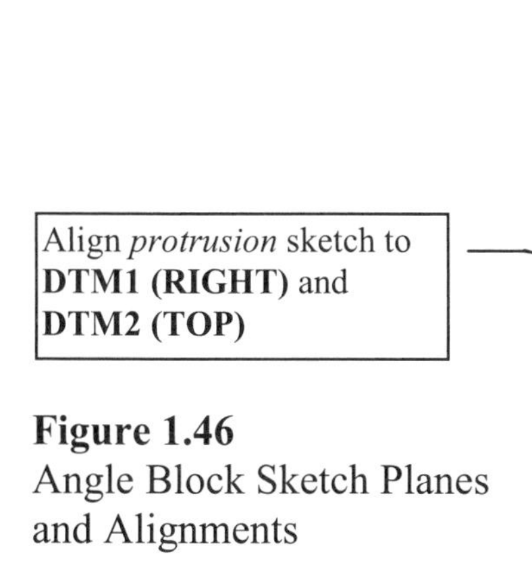

Figure 1.46
Angle Block Sketch Planes
and Alignments

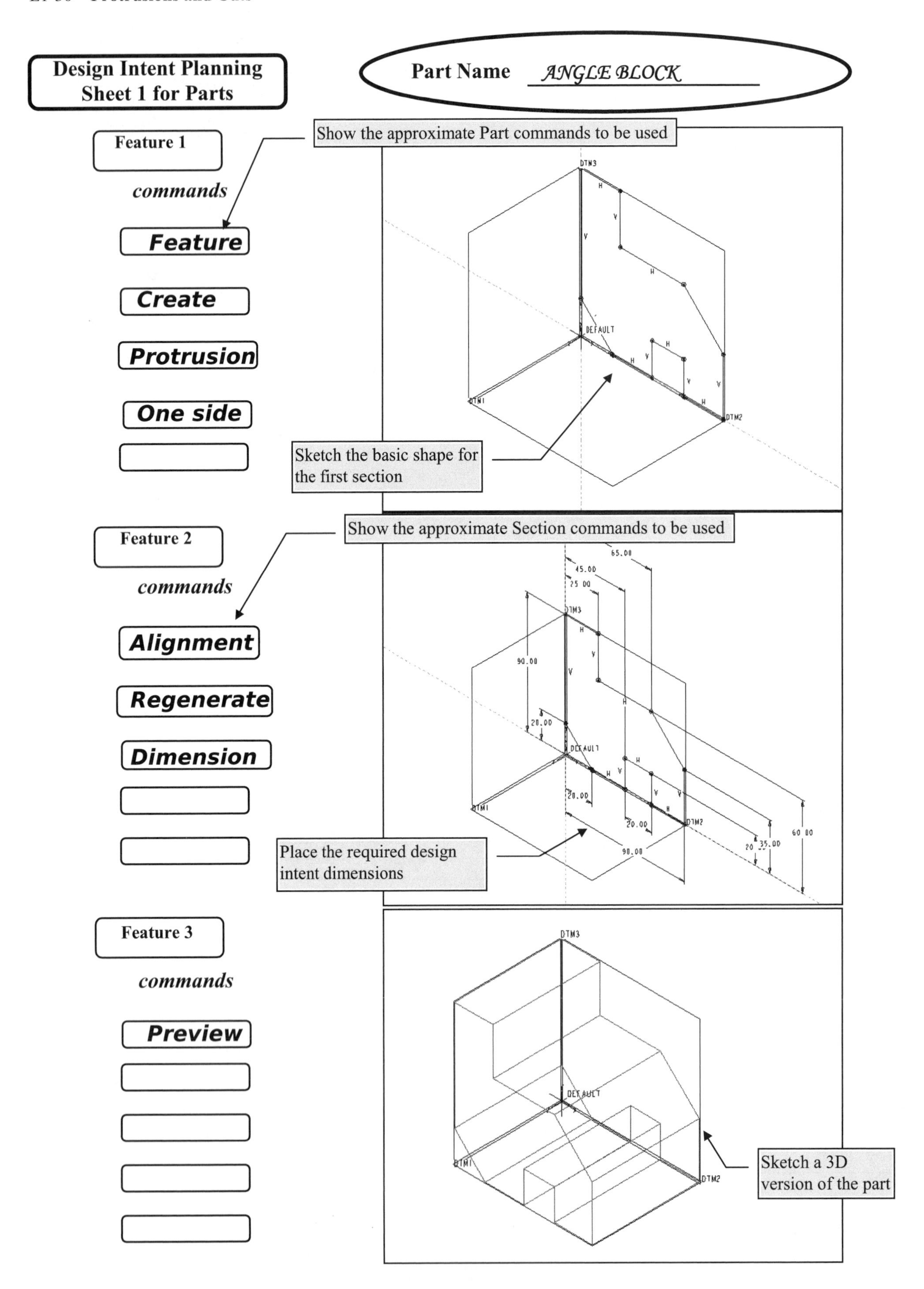

Design Intent Planning Sheet 1 for Parts
Part Name ANGLE BLOCK
Feature 1
commands
Show the approximate Part commands to be used
Feature
Create
Protrusion
One side
Sketch the basic shape for the first section
Feature 2
commands
Show the approximate Section commands to be used
Alignment
Regenerate
Dimension
Place the required design intent dimensions
Feature 3
commands
Preview
Sketch a 3D version of the part

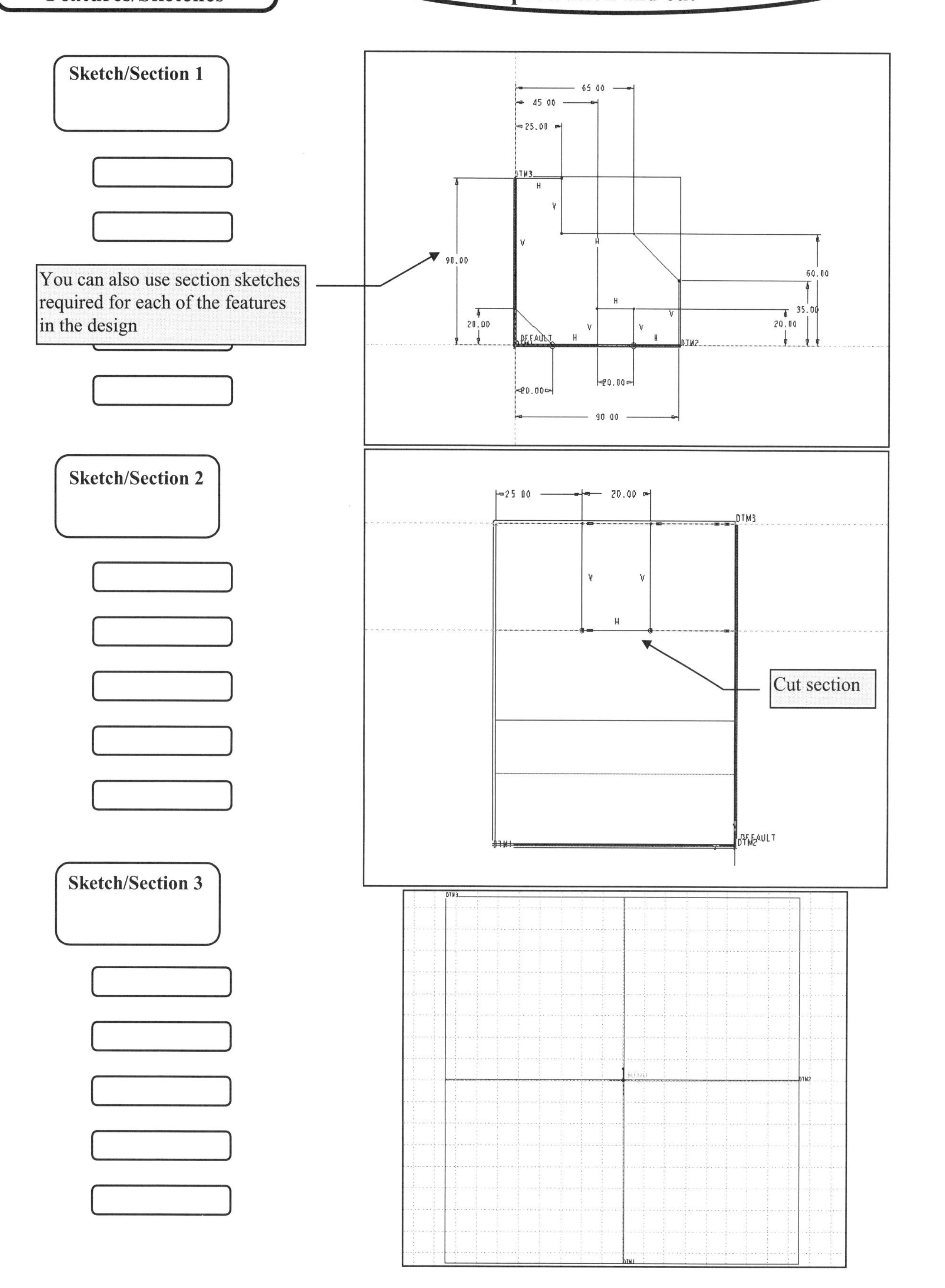
Design Intent Planning Sheet 7 for Features/Sketches
Part/Section Name ANGLE BLOCK
protrusion and cut
Sketch/Section 1
You can also use section sketches required for each of the features in the design
Sketch/Section 2
Cut section
Sketch/Section 3

Appendix D

Design Intent Planning Sheets (DIPS)

In **Appendix D** you will find more **DIPS.** You may use, with your instructor's permission, engineering grid paper instead of the DIPS. Metric, isometric, and ¼-inch grid sheets should all be purchased so that you have a variety of choices for your planning and sketching.

The appendix provides a variety of sketching formats for planning your design. The ***design intent*** of a feature, a part, or an assembly (or even a drawing) should be established before any work is started with Pro/ENGINEER. A number of different formats are provided in which to sketch and plan your feature, part, assembly, or drawing. Copy the sheets so that you have a number available for each lesson in the text. ***Inch, metric, and isometric engineering grid paper can be substituted for the DIPS.***

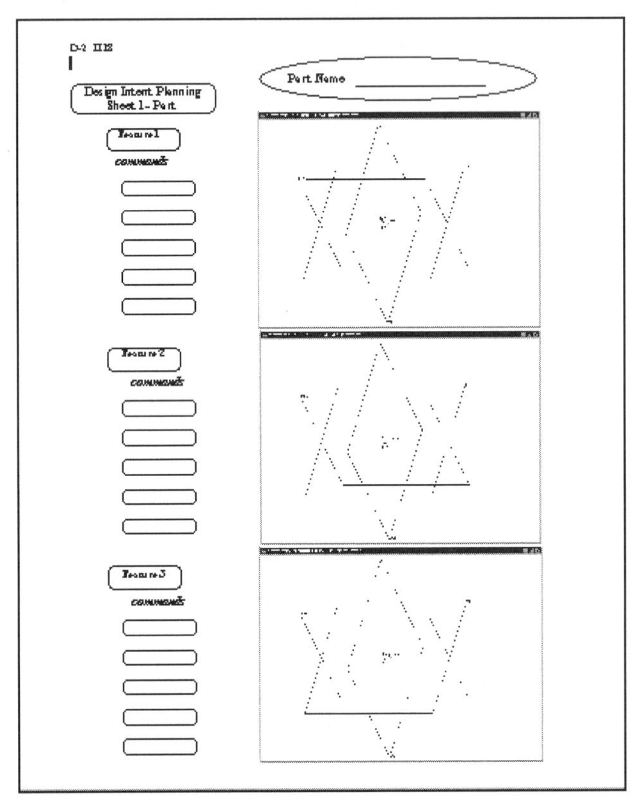

This appendix contains the following sheets:

DIPS1 Part Planning Sheet (trimetric)

DIPS2 Part Planning Sheet (pictorial with axes)

DIPS3 Assembly Planning Sheet (trimetric)

DIPS4 Assembly Planning Sheet (pictorial with axes)

DIPS5 Drawing Planning Sheet (no format)

DIPS6 Drawing Planning Sheet (format)

DIPS7 Feature/Sketch Planning Sheet (three sketches)

DIPS8 Feature/Sketch Planning Sheet (two sketches)

DIPS9 Part Model Trees & Assembly Model Trees

READ THIS!!

We cannot emphasize enough the importance of manually sketching your project before starting the actual modeling. Planning your parent-child relationships, the features required for the part, etc. will in the end cut the actual time required to complete a project by 30%. We guarantee this! Most of the time you normally waste redoing, rerouting, modifying, and redefining your part will be reduced or eliminated!

Make sure you use these sheets or engineering grid paper for every Lesson Project (and if you're smart, for every Lesson). Do the step-by-step Lessons exactly as stated and they will never fail. But if you use the DIPS, you will see better ways to model the project (hey, we're not perfect!). If you have the time, recreate the model with your planned-out modeling sequence and new design intent.

The front inside cover has a check-off list that includes the category DIPS. In my class I always say: "NO DIPS NO GRADE!"

Lesson 2

Modify and Redefine

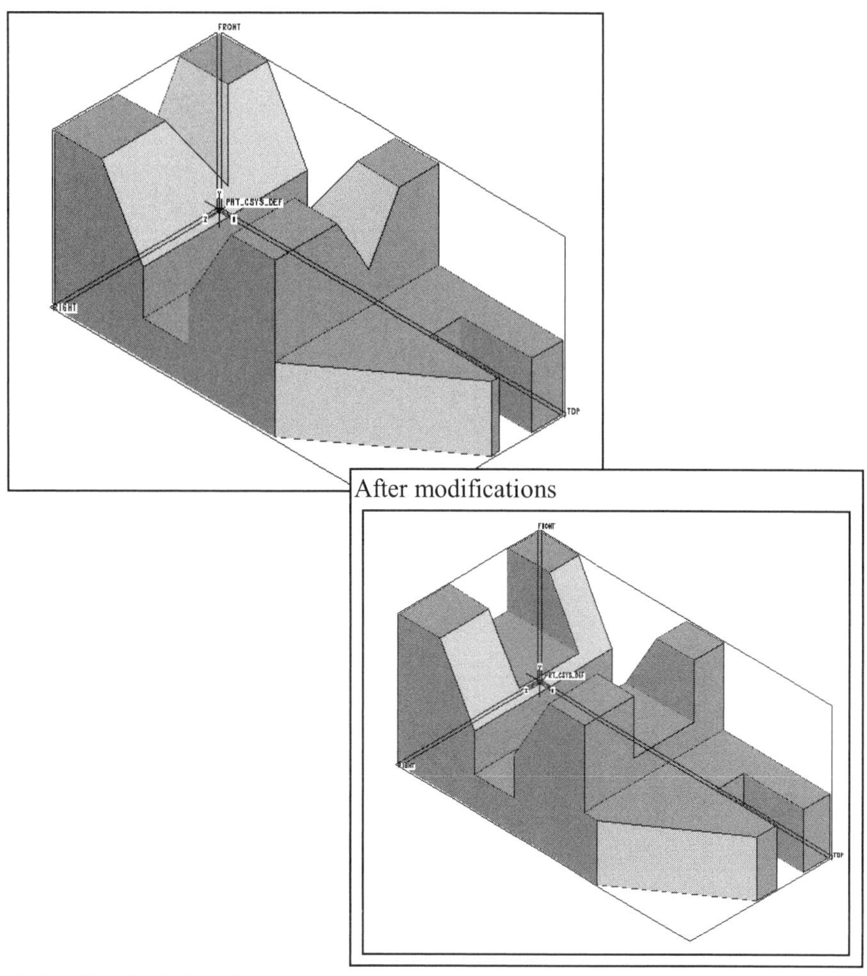

Figure 2.1
Base Angle Versions

OBJECTIVES

1. **Change existing features by modifying dimensions**
2. **Alter the view consideration in dimension display**
3. **Redefine a part's features**
4. **Modify the number of decimal places of dimensions**
5. **Regenerate modified and redefined parts**
6. **Input a *config.pro* change**
7. **Use the Info command to get feature and model information**
8. **Use the Model Tree to modify, redefine, and get information**

EGD REFERENCE
Fundamentals of Engineering Graphics and Design
by L. Lamit and K. Kitto
Read Chapter 10.
See pages 298 and 399.

COAch™ for Pro/ENGINEER

If you have **COAch for Pro/ENGINEER** on your system, go to SEARCH and do the Segments shown in Figures 2.3 and 2.5.

Figure 2.2
Base Angle Showing Datums, and Model Tree

MODIFY AND REDEFINE

Both **Modify** and **Redefine** are important capabilities in the design of parts (Fig. 2.1). With **Modify**, you can change any dimension used in the creation of a feature. With **Redefine**, the feature's attribute, the placement plane, the placement/orientation references, and the size and configuration of the feature can be redone. **Modify** is for simple dimensional changes, and **Redefine** is for more comprehensive changes to the model's features. You will be creating the **BASE ANGLE** shown in Figure 2.2 and then modifying and redefining some of its features. If you have **COAch for Pro/ENGINEER**, complete the appropriate segment (Fig. 2.3).

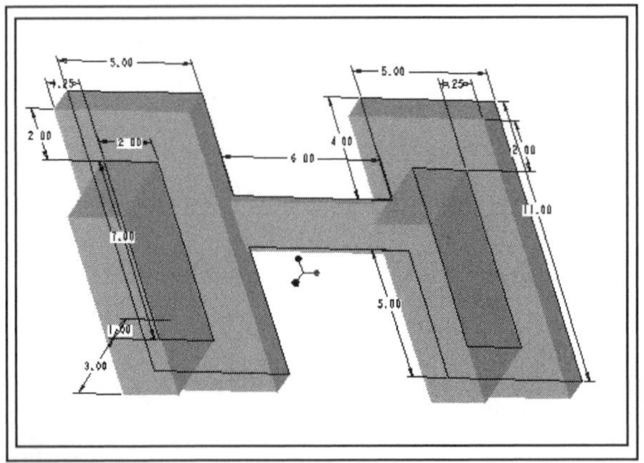

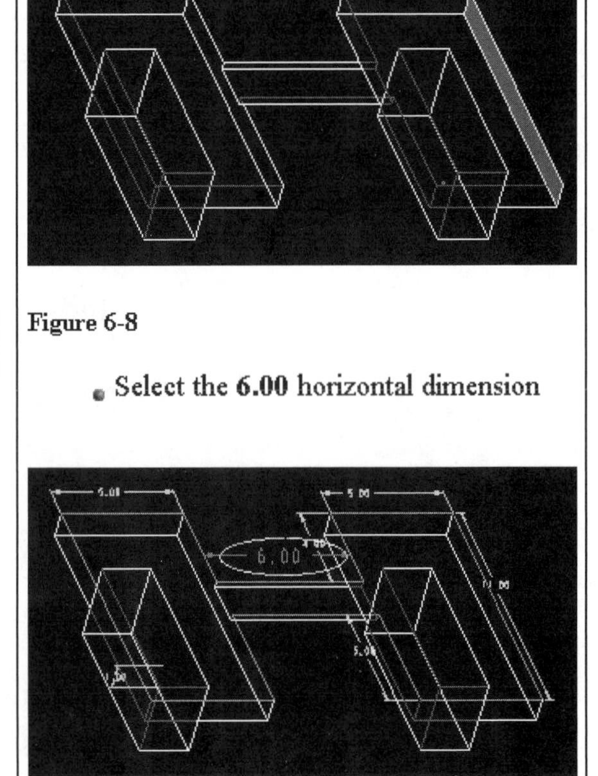

Figure 2.3
COAch for Pro/E, Basic Modifications (Changing Parameters)

Modify

To modify dimensions, choose **Modify**, then **Value** (default) from the MODIFY menu, and select a feature. Pro/E displays all the dimensions associated with the selected feature. If you pick an edge that is shared by two features while you are using the **Query Sel** option, Pro/E highlights the associated features in turn. Pro/E displays the **Query Bin**, which, after you pick ↓ (Next) or ↑ (Previous), lets you step back and forth through the highlighted features in order to choose the one you want.

View Orientation When Displaying Dimensions

The view orientation of a part can be adjusted to improve clarity in viewing dimensions. The need to change the view orientation becomes apparent when the dimensions overlap, the dimensions are in planes that are perpendicular to the current view, or Pro/E displays the section dimensions on the original sketching plane. Hold down the **Ctrl** key and press the middle button of the mouse to rotate the model to the desired position. Figure 2.4 shows a portion of the available help for the modify-view consideration.

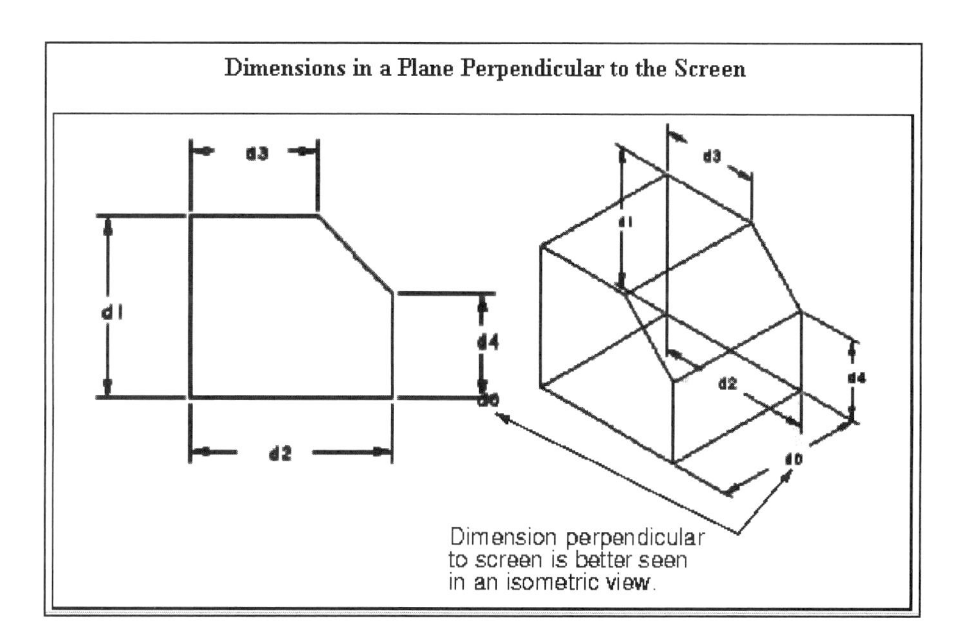

Figure 2.4
Modify- View Consideration

Modifying Dimension Values

When you modify the value of a dimension, you can enter a new number or a **Relation.** Pro/E supports the use of negative dimensions. The value entered depends on the displayed sign of the dimension. By default, Pro/E displays all dimensions as positive values, and entering a negative value tells Pro/E to create the section geometry on the opposite side, but the *direction* of a feature creation cannot be changed by entering a negative number. Use the **Redefine** option to redefine the direction of the feature.

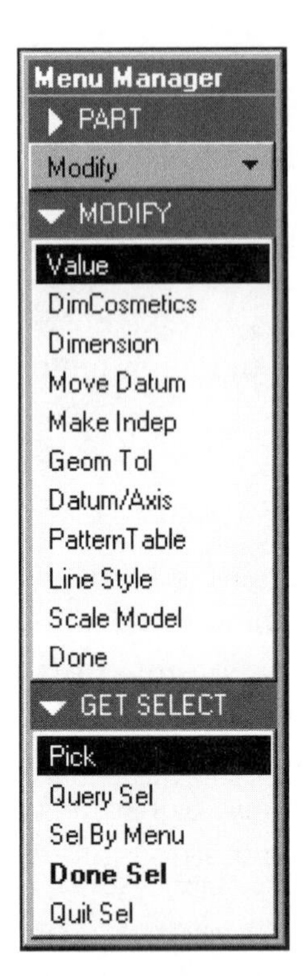

To modify a dimension value with the **Intent Manager** off:

1. Choose **Modify** from the PART menu.
2. Display the dimensions of a feature by picking on any surface of the desired feature.
3. Pick the *yellow* dimension to change. The value highlights in *red*, and Pro/E displays a prompt in the message area.
4. Enter a new value, or accept the current value by pressing **enter.** In many cases, this value can be negative. This new value displays in *white,* replacing the old value.
5. Modify other dimensions as required.
6. When you have completed all the changes, choose **Regenerate** to recalculate the part using the new dimension values.

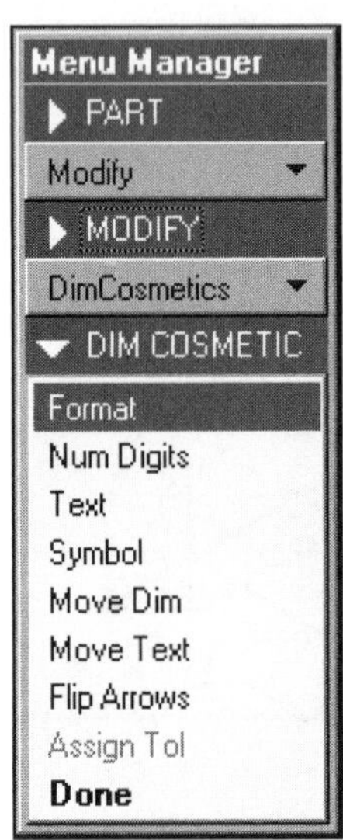

Modifying the Number of Decimal Places for Dimensions

The default number of decimal places for dimensions is two. To increase the precision of a particular dimension, enter a new value with the desired precision. Modifying the number of decimal places, for a dimension, rounds the value of the dimension.

To decrease the precision of a particular dimension:

1. Turn the tolerances on by choosing **✔± .01 Dimension Tolerances** from the Environment dialog box.
2. Choose **Modify ⇒ DimCosmetics ⇒ Format.**
3. Choose **Nominal** from the DIM FORMAT menu; then pick a dimension and **Done Sel.** Its tolerance display changes to nominal.
4. To modify the number of decimal places to display for one or more dimensions (including reference dimensions), from the DIM COSMETIC menu choose **Num Digits.** Enter the number of desired significant digits.
5. Select the dimensions whose display is to be changed.

Redefining Features with Elements

You redefine a variety of features using the Feature Definition dialog box to change the elements with which they were created:

Protrusions	Cuts	Slots
Shells	Rounds	Holes
Some surface features	Drafts	Draft offsets
Some datum curves	Shafts	

To redefine a feature that has elements:

1. Choose the **Redefine** option from the FEAT menu, and pick the feature to be redefined.
2. Pro/E displays the Feature Definition dialog box. Each element and its current value are listed. Select the element to redefine; then select the **DEFINE** button. Pro/E prompts for the information needed to redefine the element.

Redefining Features

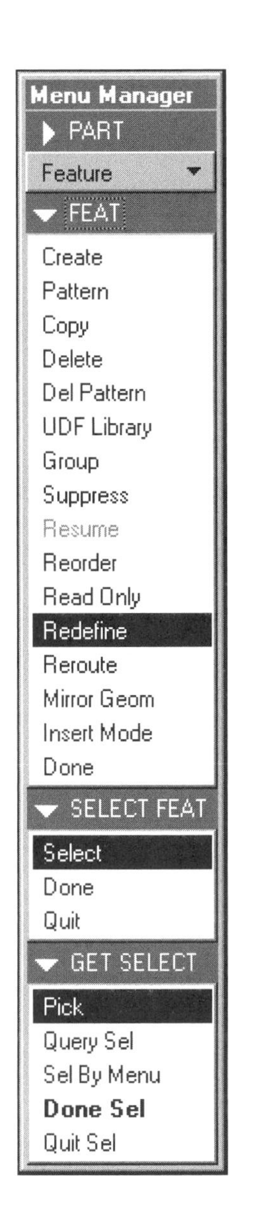

The **Redefine** option in the FEAT menu allows you to change the way a feature is created, including section geometry. The types of changes you can make depend on the selected feature.

A cut has a *section;* therefore, it can be redefined. If you have **COAch for Pro/ENGINEER** on your system, go to SEARCH and complete the appropriate Segment shown in Figure 2.5.

Pro/E recreates the feature using the new feature definitions. When feature sections are redefined, you may need to redimension any child feature whose reference edge or surface was replaced. Redimension the child feature using the options **Redefine** and **Scheme,** or **Reroute.** If you make any changes to the feature that cause the feature creation to abort, you enter the **Resolve** environment.

When you are using the **Redefine** option for a feature created with the options **Copy, Mirror,** and **Dependent,** Pro/E issues a warning message stating that the selected feature is a dependent copy of the highlighted feature. If **Confirm** is chosen from the CONFIRMATION menu, Pro/E will remove the feature dependency and display the Feature Definition dialog box with the elements **Attributes, Section, Direction,** and **Depth.** For example, if you choose **Section** after you select the option to redefine, the section of the selected feature can be changed, yet upon completion it is independent of the referenced feature.

When you apply the redefinition, Pro/E removes the feature geometry and creates temporary geometry for your changes. When you exit from the user interface, Pro/E regenerates the part.

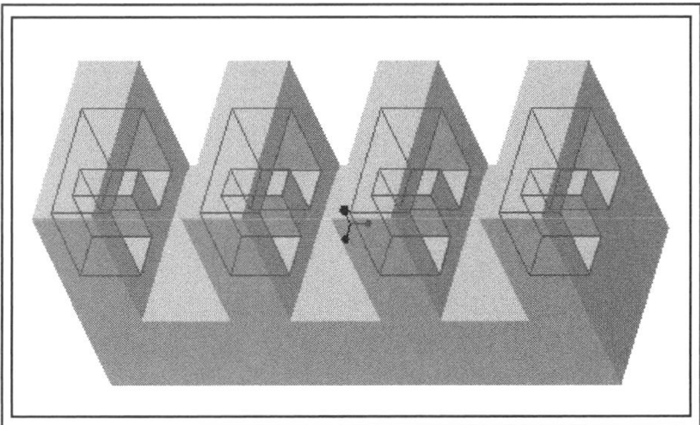

Figure 2.5
COAch for Pro/E, Basic Modifications (Redefining the Depth)

Edit the **Depth** attributes of the provided part.

1. Retrieve **edit1.prt**.

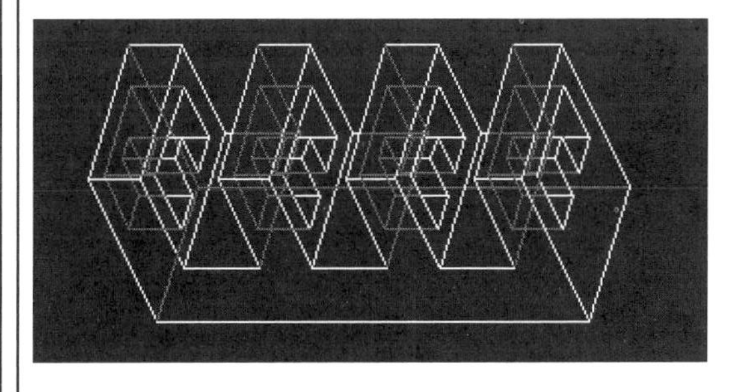

Figure 6-29

2. Change the depth of the "U-shaped" cut out to a different face.

- Choose **Feature**

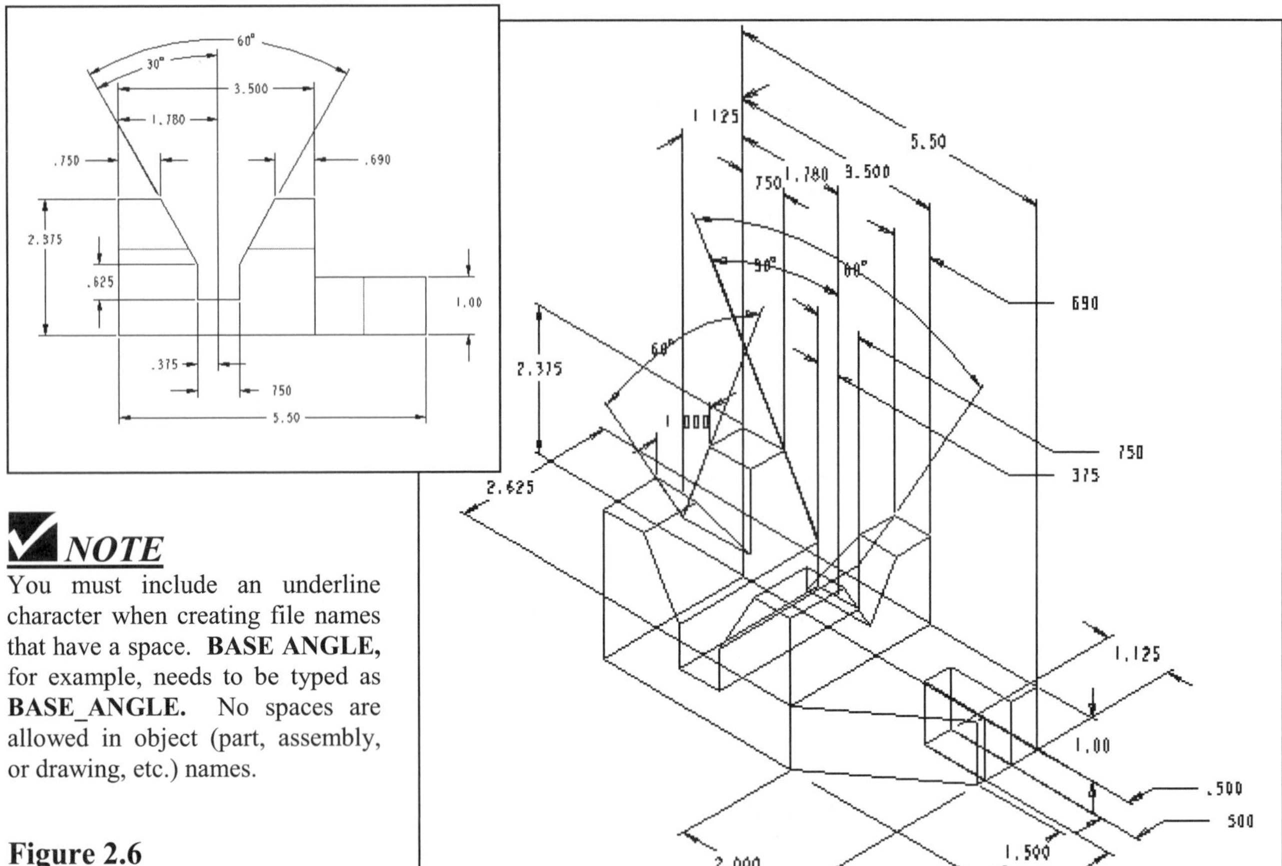

✔ *NOTE*

You must include an underline character when creating file names that have a space. **BASE ANGLE,** for example, needs to be typed as **BASE_ANGLE.** No spaces are allowed in object (part, assembly, or drawing, etc.) names.

Figure 2.6
Base Angle in Isometric

Base Angle

The base angle (Fig. 2.6) will be our second ***lesson part***. It is composed of one protrusion and three cuts. Along with creating a new part, you will *modify* and *redefine* the part after it is completed.

Create a new part by choosing the following commands:

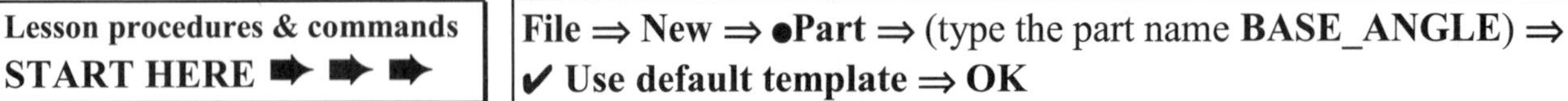

Lesson procedures & commands START HERE ➡ ➡ ➡

File ⇒ **New** ⇒ ●**Part** ⇒ (type the part name **BASE_ANGLE**) ⇒ ✔ **Use default template** ⇒ **OK**

As in Lesson 1 (and for all parts), the *environment*, *units*, and *material* for the part need to be established:

SETUP AND ENVIRONMENT

Utilities ⇒ **Environment** ⇒ ✔ **Snap to Grid**
Display Style **Hidden Line** Tangent Edges **Dimmed** ⇒ **OK** ⇒
Set Up ⇒ **Units** ⇒ Units Manager ⇒ **Inch lbm Second** ⇒ **Close** ⇒ **Material** ⇒ **Define** ⇒ (type **ALUMINUM**) ⇒ ✔ ⇒ (table of material properties, change or add information) ⇒ **File** ⇒ **Save** ⇒ **File** ⇒ **Exit** ⇒ **Assign** ⇒ (pick **ALUMINUM**) ⇒ **Accept** ⇒ **Done** ⇒ **File** (menu bar Pro/E window) ⇒ **Save** ⇒ ✔

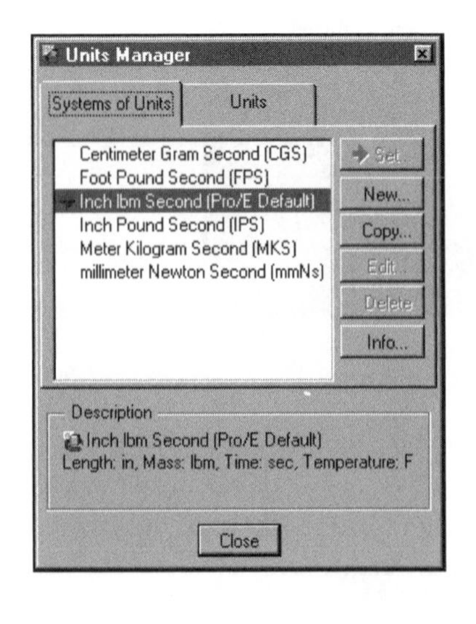

The three default datum planes and the default coordinate system are shown in Figure 2.7.

> ***CONFIG.PRO***
>
> ***default_dec_places*** sets the default number of decimal places to be displayed in all model modes for nonangular dimensions; it does not affect the *previously displayed* number of digits as modified using NUM DIGITS.
>
> ***sketcher_dec_places*** controls the number of digits displayed when you are in the Sketcher.

The default number of digits for the sketcher is **2.** You can set the number of digits using the configuration file option ***sketcher_dec_places*** (a value in the range **0** to **14**).

> **Utilities** ⇒ **Preferences** ⇒ Showing: ▼ ⇒ **Current Session** ⇒ ❑ **Show only options loaded from file** ⇒ (slide the bar down to the option or type the Option ***default_dec_places)*** ⇒ (type **3** in the Value line) ⇒ **Add/Change** ⇒ (slide the bar down to the option or type the option ***sketcher_dec_places)*** ⇒ (type **3** in the Value line) ⇒ **Add/Change** ⇒ **Apply** ⇒ **Close**

The number of decimal places was changed to be three; therefore Pro/E will round four places to three (**1.0625** becomes **1.063**).

The next commands create the first protrusion (Fig. 2.7):

> **Feature** ⇒ **Create** ⇒ **Solid** ⇒ **Protrusion** ⇒ **Extrude** ⇒ **Solid** ⇒ **Done** ⇒ **One Side** ⇒ **Done** ⇒ (pick **FRONT** as the sketching plane) ⇒ **Okay** (for selecting the *direction* of protrusion projection) ⇒ **Top** ⇒ (pick **TOP** as the *orientation* plane) ⇒ **Close**

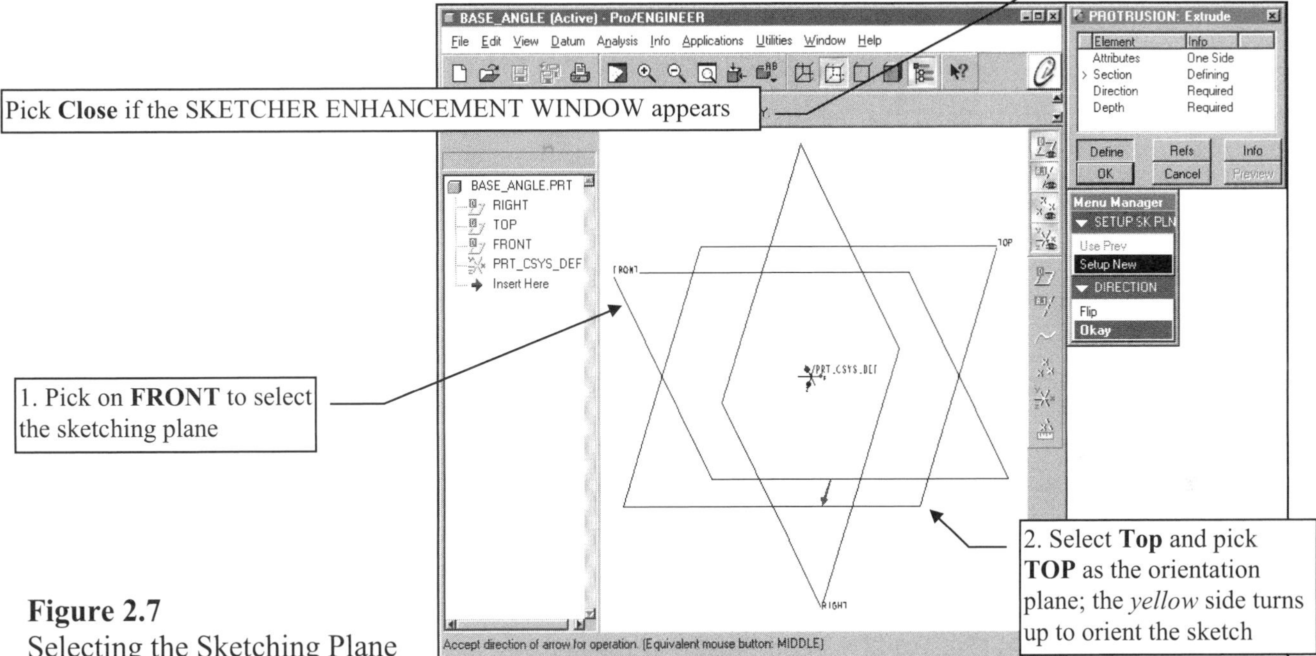

Figure 2.7
Selecting the Sketching Plane

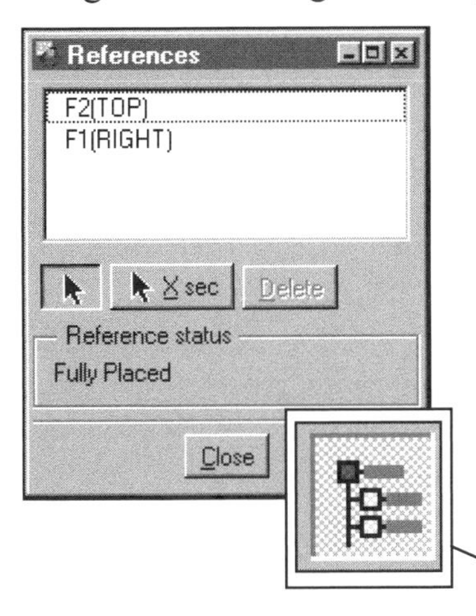

The ✔**Intent Manager** is on by default. We will use the Intent Manager for the protrusion. After you select TOP as the reference plane, Pro/E puts you into the Sketcher. The References dialog box is up and **F2(TOP)** and **F1(RIGHT)** are listed as the references. The words Fully Placed appear in the Reference status area. You can close the dialog and start sketching (unless you wish to add or delete references, which we don't want to do for this project). The main window shows the **2D Sketch** orientation and includes the datum planes and orange phantom lines that indicate your existing references. Sketch entities that touch these references will be automatically aligned. Select **Close** from the References dialog box.

Toggle the Model Tree off with the Toolbar icon (see left).

> **Read this carefully!**
>
> In some places, early in the text, we have not enclosed the required commands in a command box. In the first few lessons, there are many instances where we must explain information in detail as you model; command boxes would be too unwieldy and cumbersome.
>
> In most cases, commands are placed inside of command boxes. Regardless, ***you must read all of every lesson to complete the work!***

Sketch the outline (Fig. 2.8) of the part's primary feature using a set of connected lines, as was done in Lesson 1. The part's general shape and dimensions are provided in Figures 2.6 and 2.11. Most of the part can be defined by sketching an outline similar to its front view. Therefore, you can complete most of the part's geometry with one protrusion. The cuts on the top and side will be created later.

Create a vertical centerline down the middle of the slot. Do not create any constraints that you do not want as part of the design. Make sure that lines not intended to be at the same level (horizontal) are not sketched in line and that features of differing sizes are sketched with different lengths. Make sure that lines at the same angle are sketched that way, that symmetrical features are drawn equally about the centerline, and so on. Add design dimensions to the sketch with the scheme in Figure 2.9. As you add a dimension the Sketcher will remove unneeded dimensions. Note that the dimensioning scheme has more dimensions than were originally created by the Sketcher. Make existing dimensions that you wish to keep **Strong.**

Lines of differing dimensional lengths are sketched with different grid spacing. Though it would be considered poor practice in industry, it is OK for you to count the grid spacing for this feature's sketch.

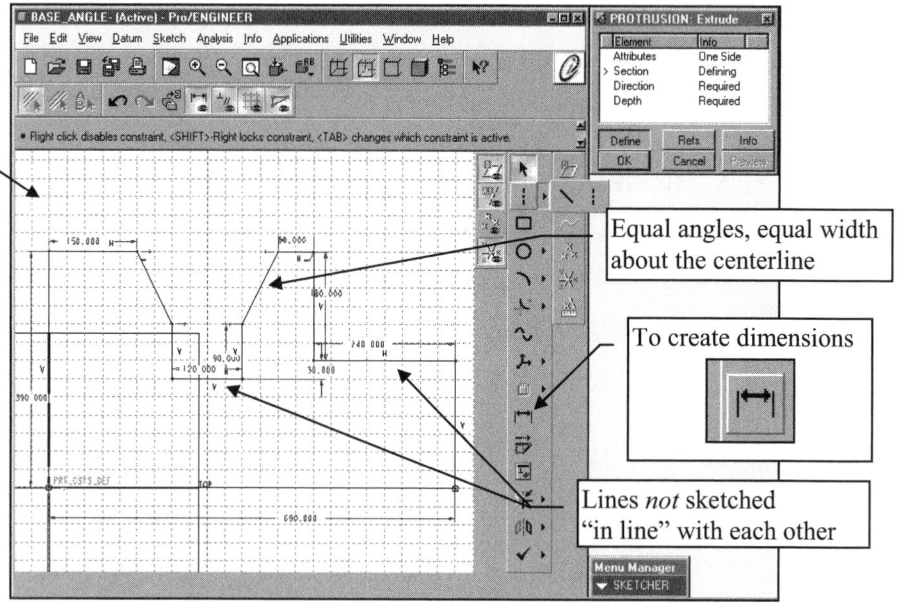

Figure 2.8
Sketching the First Protrusion's Section Geometry

HINT
The Sketcher with or without the Intent Manager is a complicated tool. Go back and reread Sections 11 and 12 at this time.

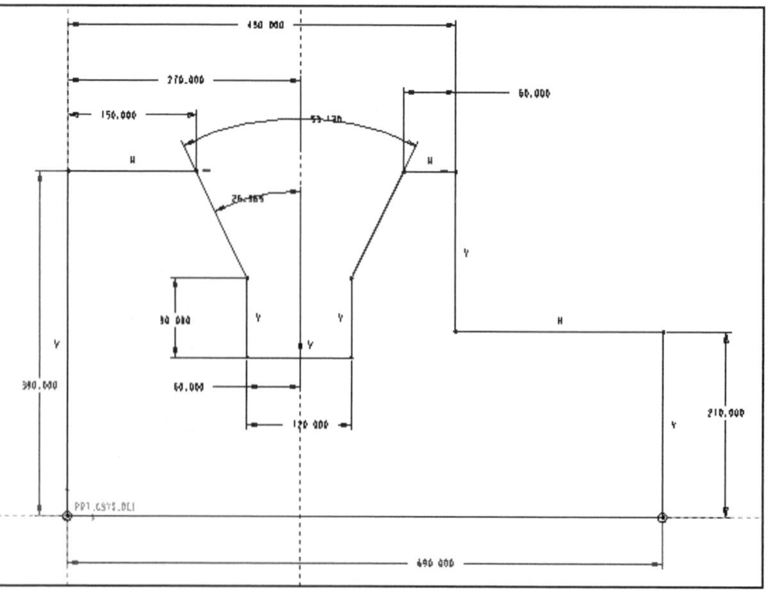

Figure 2.9
Add Dimensions to the Sketch Section

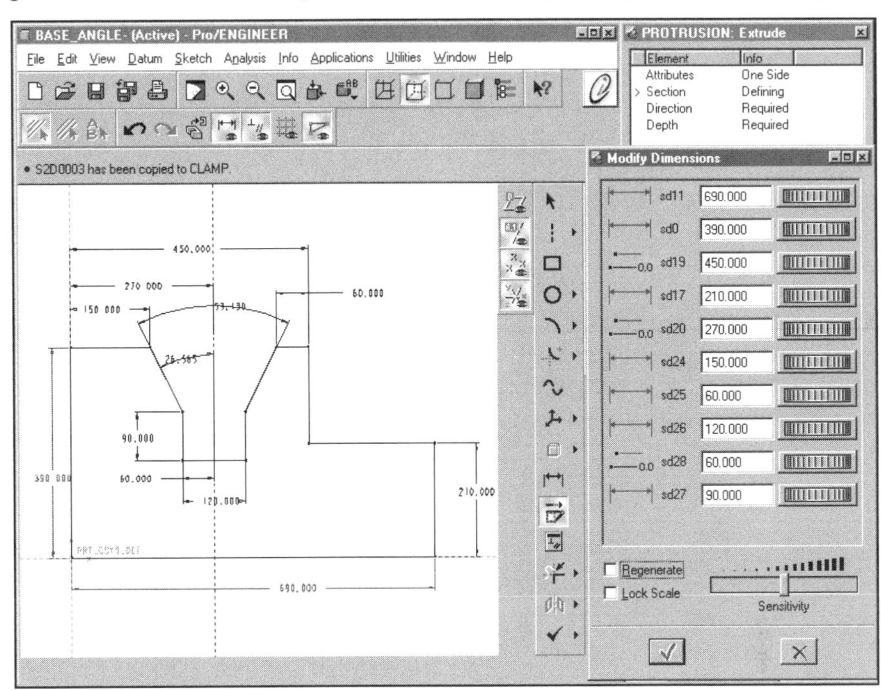

Pick the **Modify** icon and ***add all linear dimensions*** to the Modify Dimensions dialog box with ❑ **Regenerate** (off) (Fig. 2.10). Change the dimensions using the design values (Fig. 2.11) and then pick ✔**Regenerate (View ⇒ Refit).** If the dimensions are correct, pick ✔ from the Modify Dimensions dialog box (see left column).

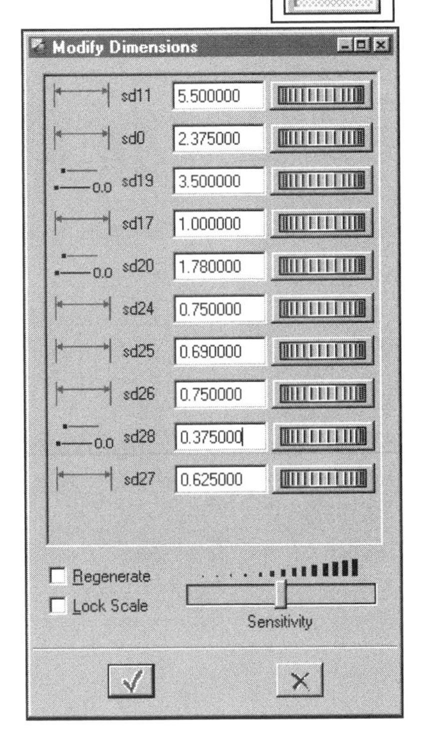

Figure 2.10
Modify Linear Dimensions

Figure 2.11
Part Dimensions

Modify the angle dimensions to the design values shown below

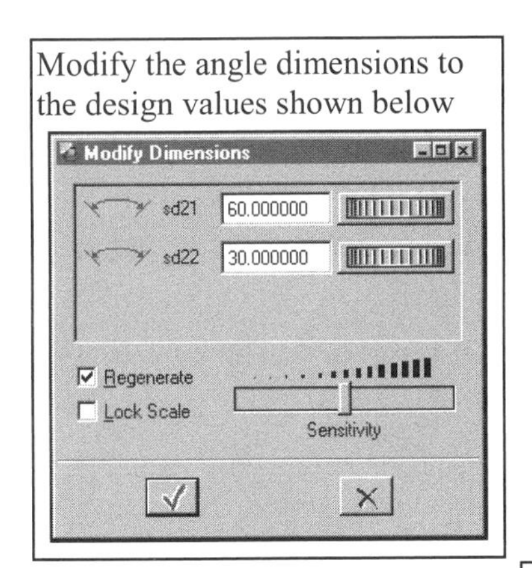

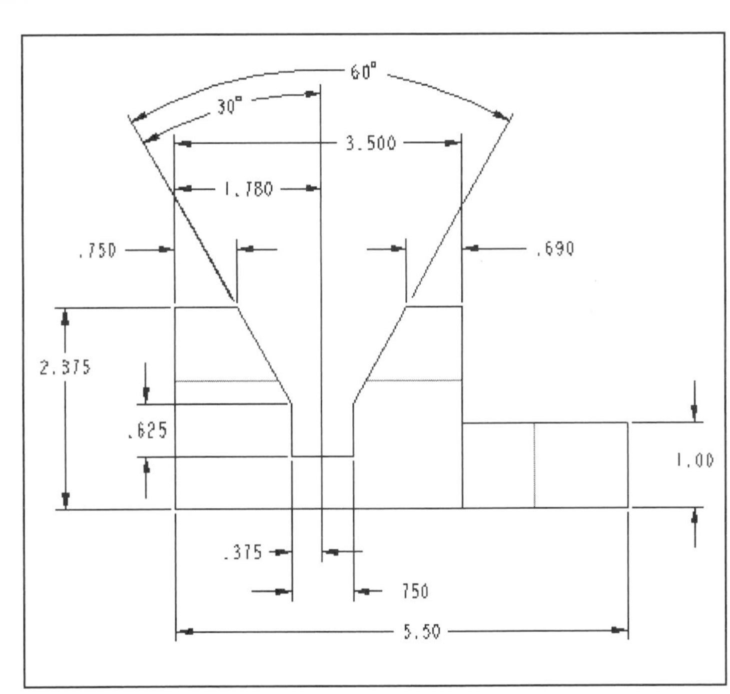

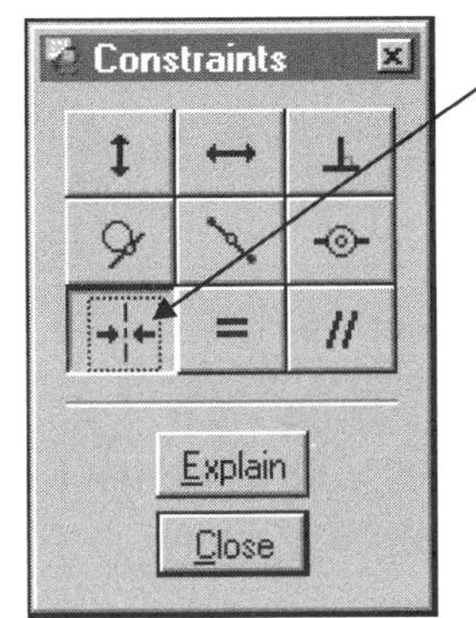

Symmetric icon

Now repeat the process and modify the angle dimensions.

Sketch ⇒ **Constrain** ⇒ (pick symmetric symbol from Constraints dialog box) ⇒ (pick centerline and then pick two vertices to be symmetric) ⇒ [the Resolve Sketch dialog box displays (Fig. 2.12)] ⇒ (pick **1. Dimension sd# = 0.375)** ⇒ **Delete** ⇒ **Close** ⇒ ✔

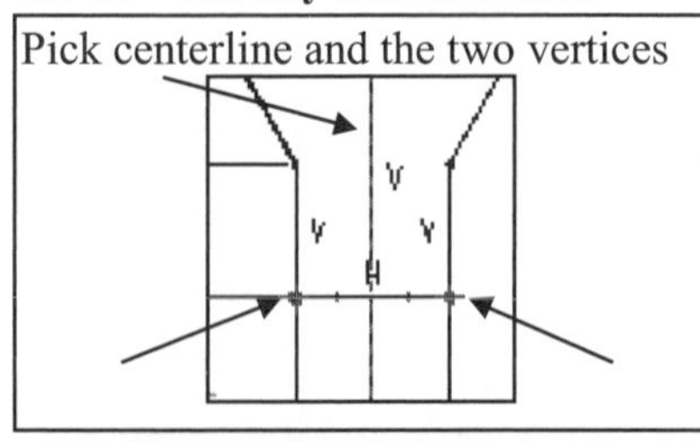

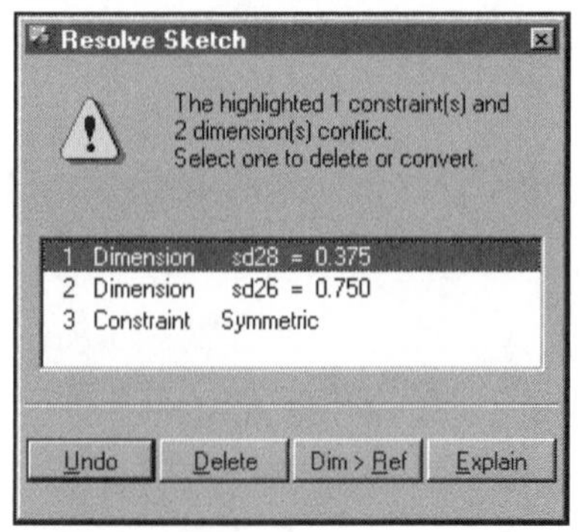

Figure 2.12
Regenerated Section

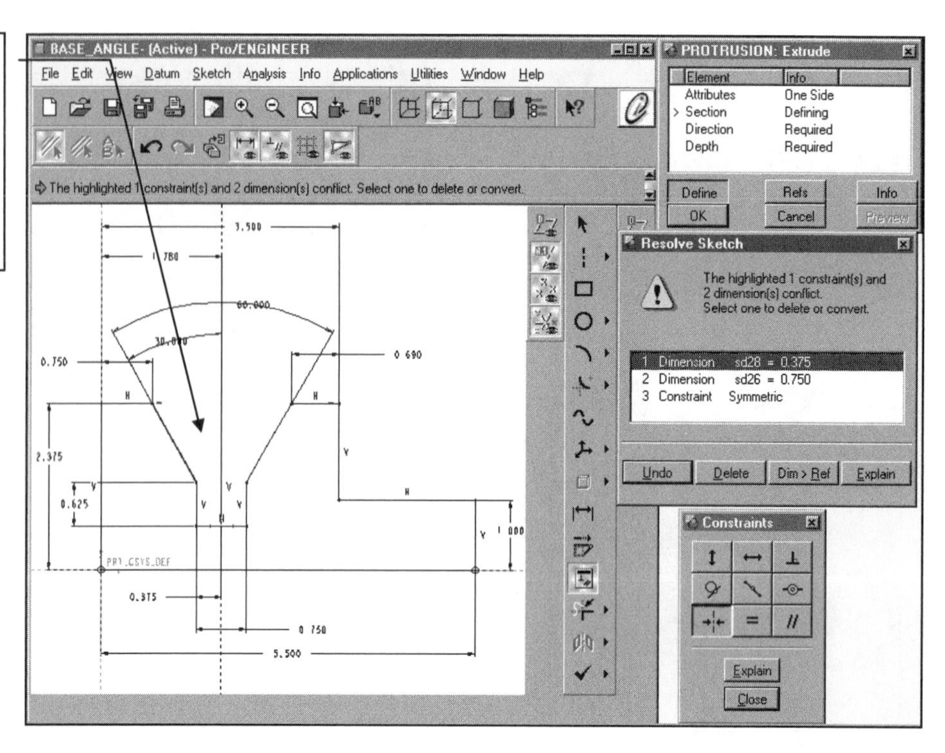

Continue with the commands:

View ⇒ **Default** ⇒ **Blind** ⇒ **Done** ⇒ (type **2.625** for part's depth) ⇒ ✔ ⇒ (turn on the datums and coordinate system using the icons) ⇒ **Preview** (Fig. 2.13) ⇒ **OK** ⇒ **Done** ⇒ **File** ⇒ **Save** ⇒ ✔

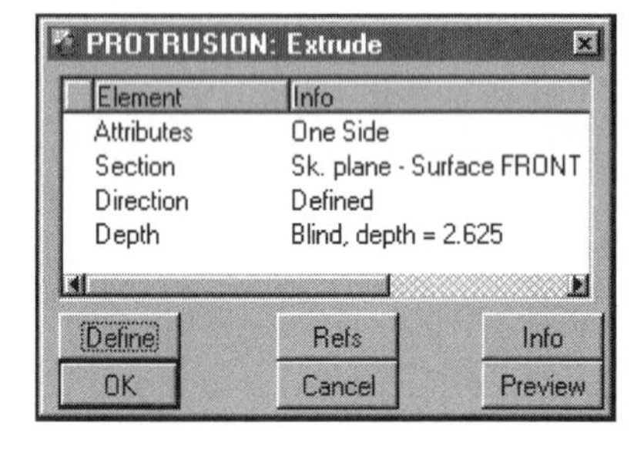

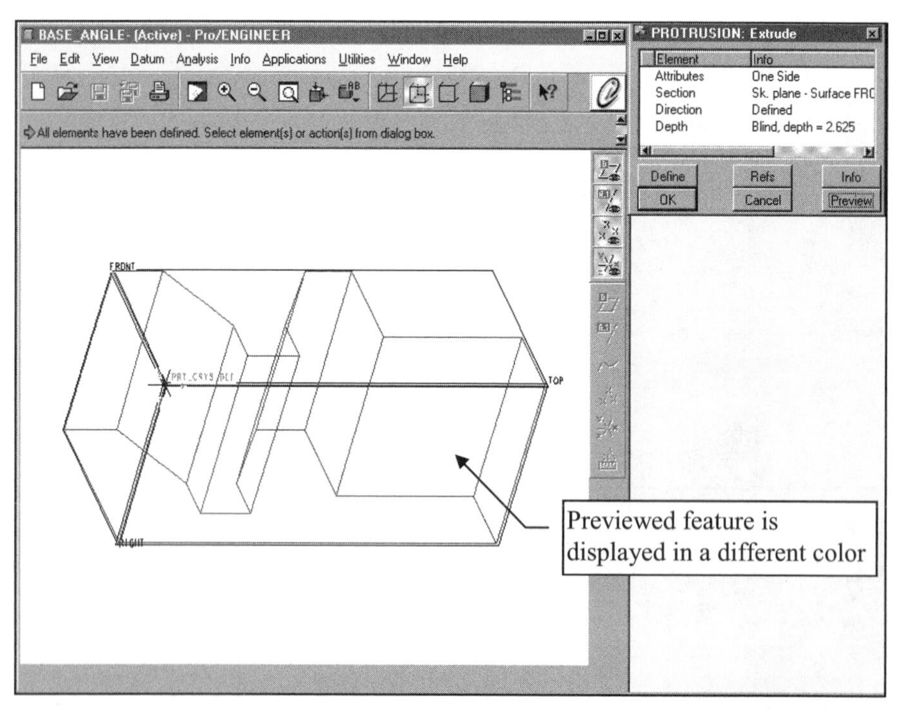

Figure 2.13
Preview of Part Feature

Three separate cuts will be the next features to be created.

Figure 2.14 shows the right side view of the part with dimensions for the **V**-shaped cut, and Figure 2.15 shows the top view with dimensions for two more cuts.

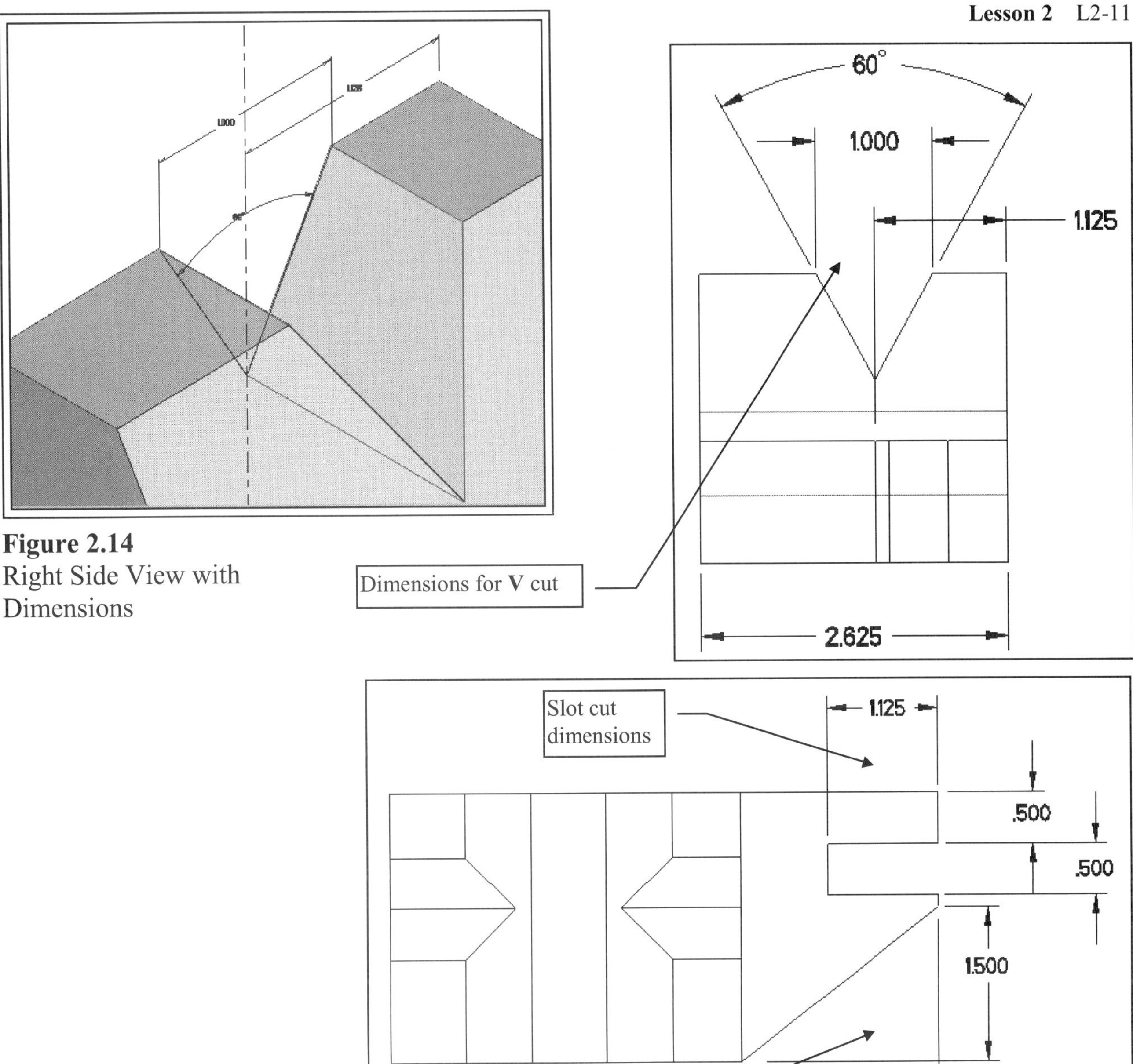

Figure 2.14
Right Side View with Dimensions

Figure 2.15
Dimensioned Top View

The next feature that will be created is the V-shaped cut. The dimensions for the cut are shown in Figure 2.14. The **V** cut is sketched similar to the first protrusion except that it removes material.

Choose the following commands:

Turn the Model Tree back on using the icon; also turn on the grid snap:

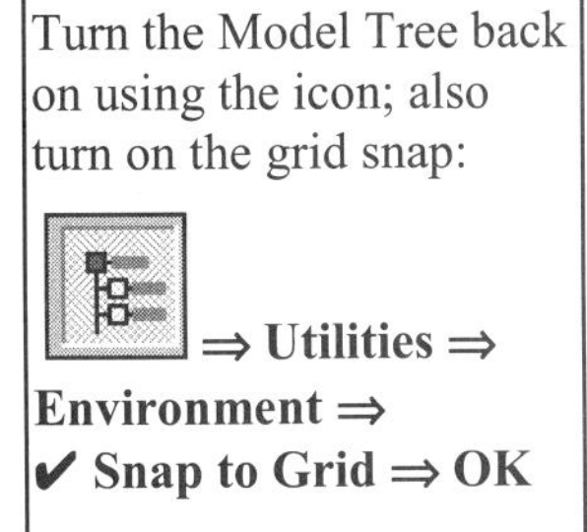

⇒ **Utilities** ⇒ **Environment** ⇒ ✔ **Snap to Grid** ⇒ **OK**

Feature ⇒ **Create** ⇒ **Solid** ⇒ **Cut** ⇒ **Extrude** (default) ⇒ **Solid** (default) ⇒ **Done** ⇒ **One side** (default) ⇒ **Done** (if necessary, use **Query Sel** to pick the datum planes) ⇒ (pick **RIGHT** from the Model Tree as the sketching plane) ⇒ **Flip** *(if needed)* ⇒ **Okay** ⇒ **Top** (pick **TOP** from the Model Tree to orient the sketch) ⇒ **Close** (the SKETCHER ENHANCEMENT if it appears)

Pro/E now enters the **Sketcher** automatically and displays the part and datum planes as shown in Figure 2.16.

You are looking at the left side of the part in this orientation of the part. The coordinate system is in the lower left, **TOP** is on the bottom, and **FRONT** is along the left. Delete the **TOP** as a reference and add ↖ the top surface/edge of the part as shown in Fig. 2.16.

References
F3(FRONT)
Surf:F5(PROTRUSION)
X sec
Delete
Reference status
Fully Placed
Close

Add the top surface/edge of the part as a reference

Edge of **FRONT**

PRT_CSYS_DEF coordinate system

Delete **TOP** as a reference

Figure 2.16
Sketcher View for **V** Cut

Pick **Close** from the References dialog box. Sketch the **V** cut symmetrically about a vertical centerline (Fig. 2.17). If desired, adjust the grid size or turn the grid snap off (❑ **Snap to Grid).** Turn off the model tree to have more usable space in your main window.

If the sketch is not drawn symmetrically, you may need to add the **30°** half-angle dimension.

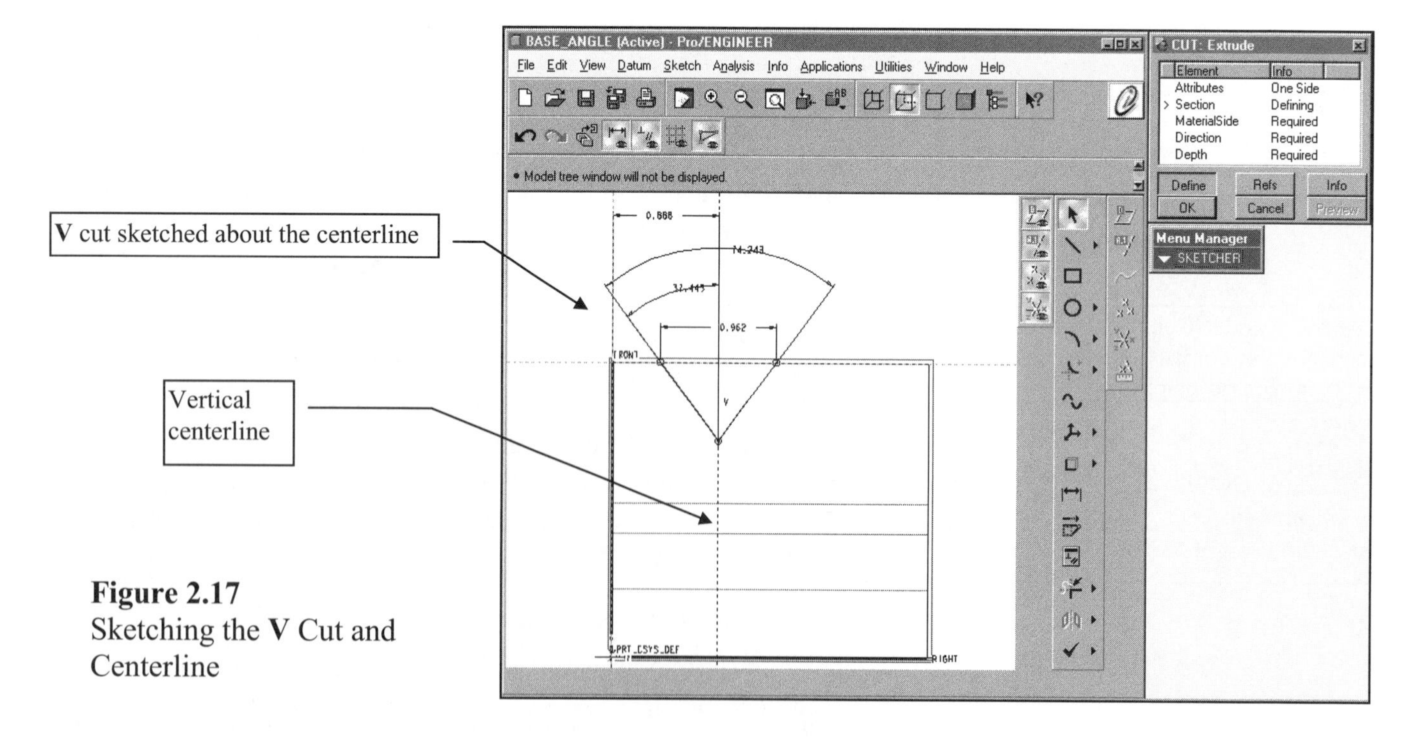

Figure 2.17
Sketching the **V** Cut and Centerline

Modify the sketch by clicking on the Modify icon in the Sketcher icon bar. Pick all the dimensions on the sketch to add them to the Modify Dimensions dialog box (Fig. 2.18).

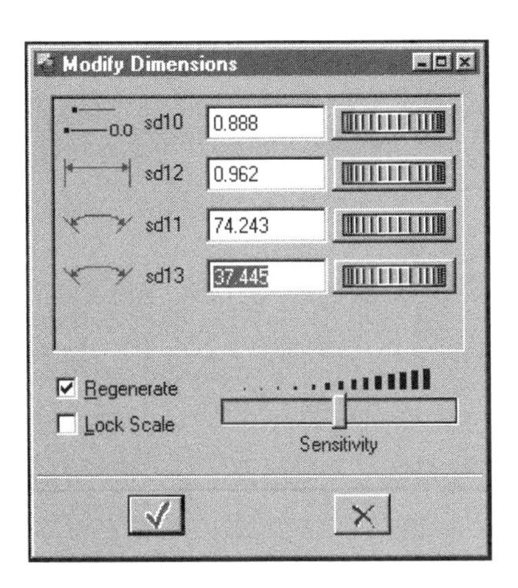

Figure 2.18
Modify

Modify ⇒ ✔ **Regenerate** (Fig. 2.19) the dimensions using the design values in Figure 2.14 ⇒ ✔ (in the Modify Dimensions dialog box)

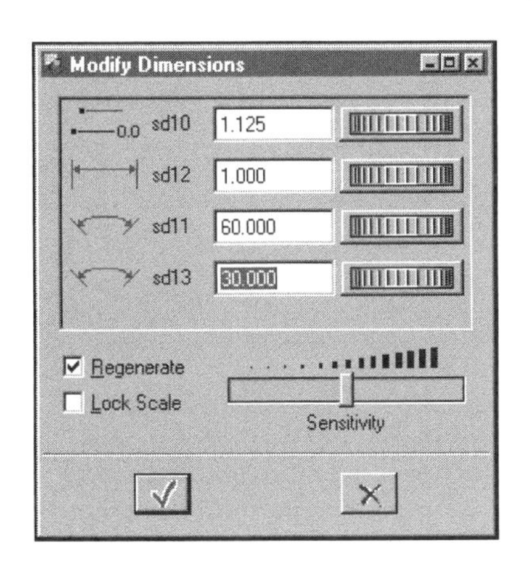

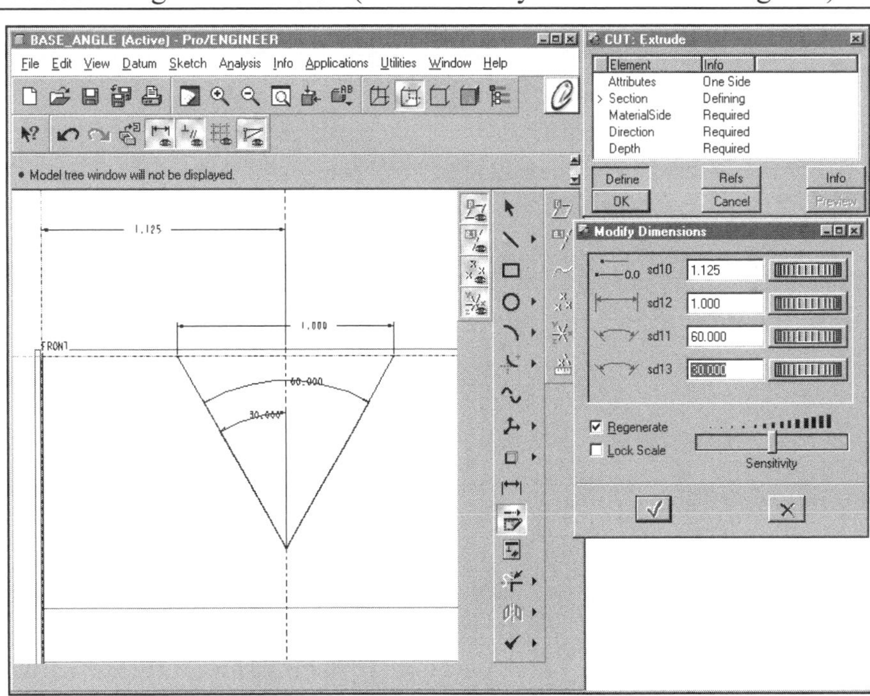

Figure 2.19
Sketch with Design Dimensions

Or pick the Constraints icon

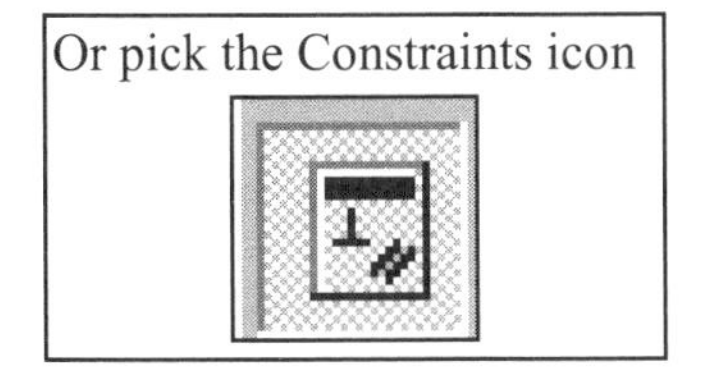

Sketch ⇒ **Constrain** ⇒ (pick the symmetric button) ⇒ [select the centerline on the section sketch and the two upper endpoints of the V cut (Fig. 2.20)] ⇒ **Delete** (the **30** dimension from the Resolve Sketch dialog box) ⇒ **Close** (Constraints dialog box) ⇒ ✔ (Sketcher icon) ⇒ **Okay** (to accept the *material removal direction)*

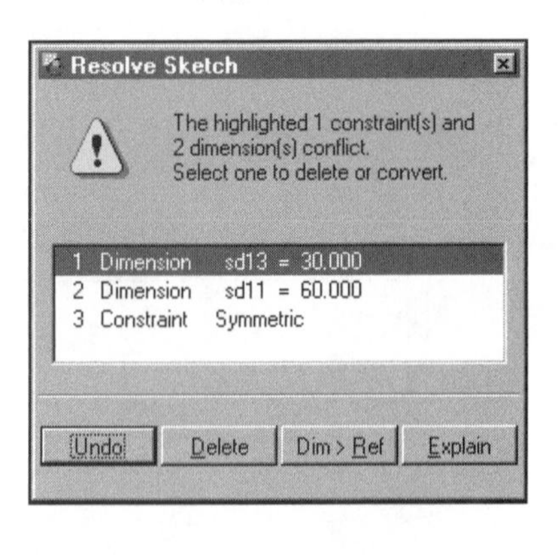

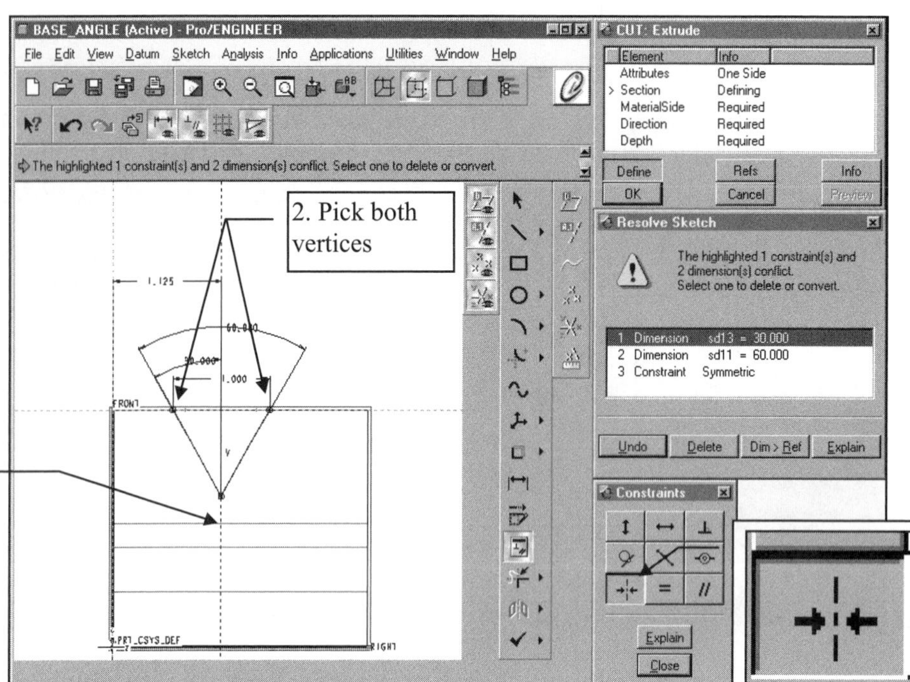

Figure 2.20
Symmetry Constraint

You are prompted with **Arrow points TOWARD area to be REMOVED.** Pick **Flip** or **Okay.** Arrow points in the direction of material removal.

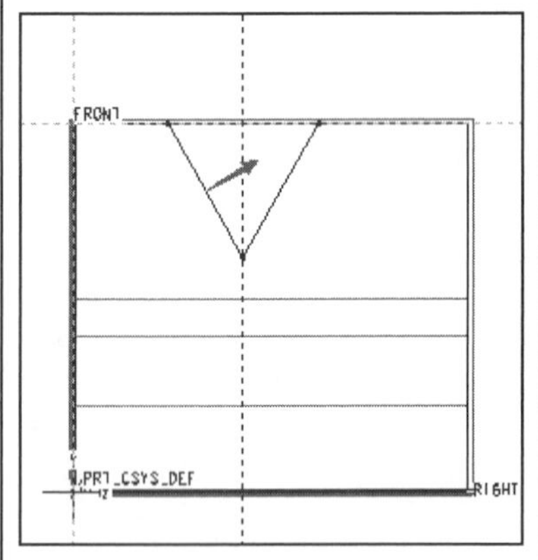

View ⇒ Default ⇒ Thru Until ⇒ Done ⇒ (pick the vertical surface/plane shown in Figure 2.21)

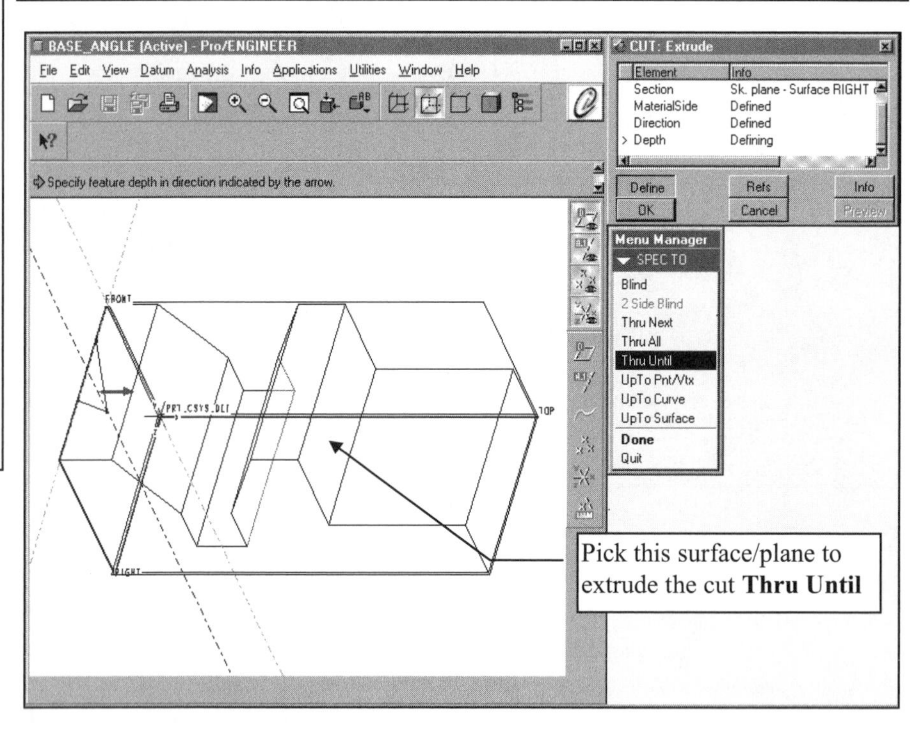

Figure 2.21
Defining the Depth of the Cut

Pick **Preview** from the dialog box to display the cut, as shown in Figure 2.22. Instead of accepting the design, pick the **Depth** element in the dialog box and then pick **Define**. Choose **Thru Next** from the SPEC TO menu (Fig. 2.23), then **Done**. Shade the part **(View ⇒ Shade)** and then choose **Preview** from the dialog box.

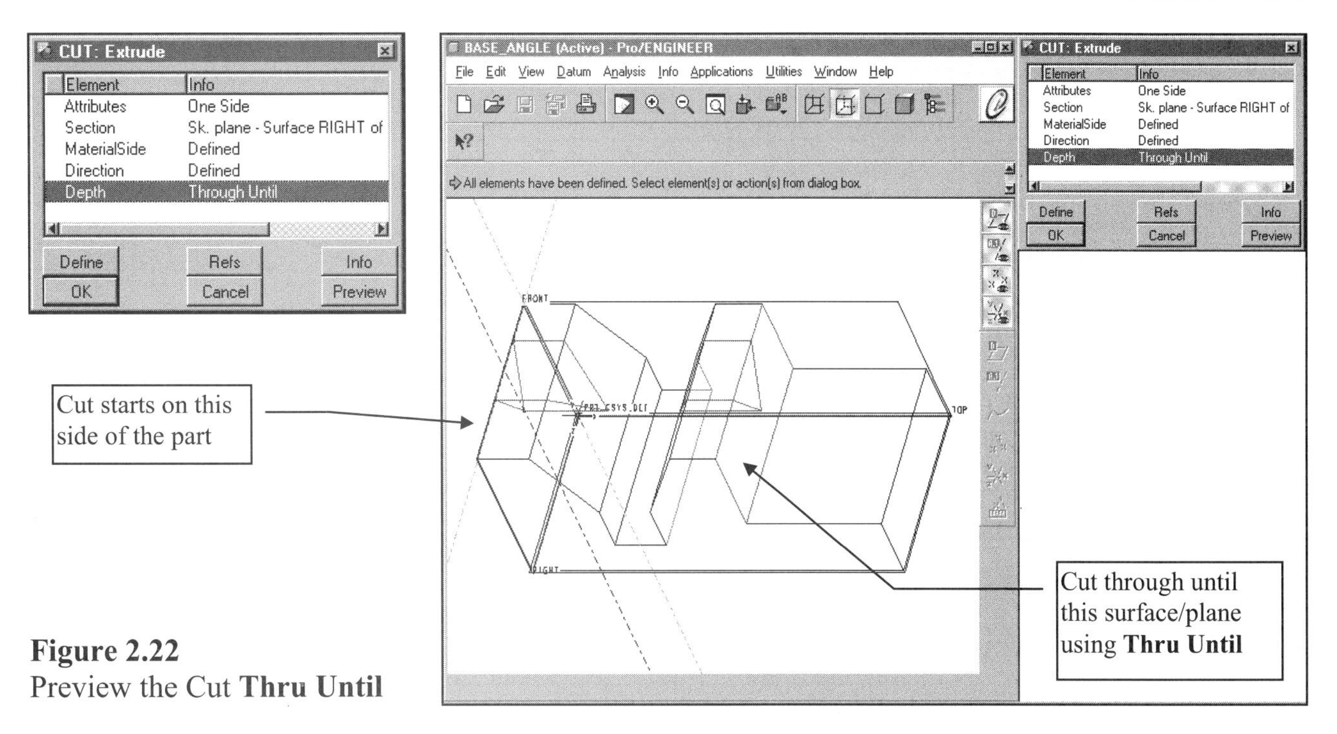

Figure 2.22
Preview the Cut **Thru Until**

The cut now goes up to the next feature it encounters and stops there (Fig. 2.23). You may need to repaint the screen to see the cut correctly **(View ⇒ Repaint).**

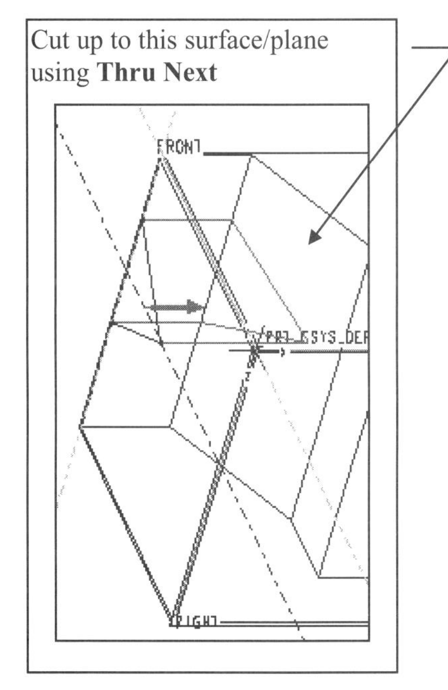

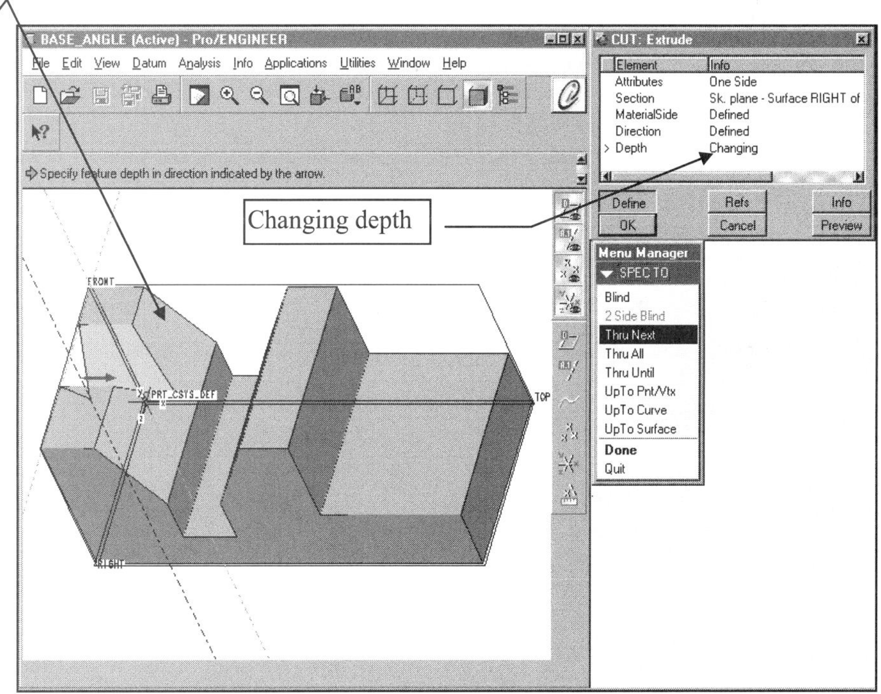

Figure 2.23
Preview the Cut **Thru Next**

Pick **Depth** and **Define** from the dialog box again, select **UpTo Surface**, then **Done,** and pick the same surface as in the first example when you selected **Thru Until** ⇒ **Preview** ⇒ **OK** (from the dialog box) ⇒ **Done** ⇒ **File** ⇒ **Save** ⇒ ✔

The part should look like Figure 2.24.

Figure 2.24
Completed Cut

The cut is now complete. Change the part's environment, shading, orientation, and position on the screen to those shown in Figure 2.25. Maximize the Main Window and turn on the Model Tree. Change the environment by choosing the following commands:

Default Actions
- ☐ Ring Message Bell
- ☐ Save Display
- ☐ Make Regen Backup
- ☐ Snap to Grid
- ☑ Highlight Erased Views
- ☑ Keep Info Datums
- ☑ Use 2D Sketcher
- ☐ Use Fast HLR

Display Style: Shading
Default Orient: Isometric
Tangent Edges: Isometric / Trimetric / User Defined
OK

Utilities ⇒ Environment ⇒ ❑ Coordinate Systems ⇒ Shading ⇒ Isometric (or rotate the part-- press and hold down the Ctrl key and click and hold down the mouse button) ⇒ **OK**

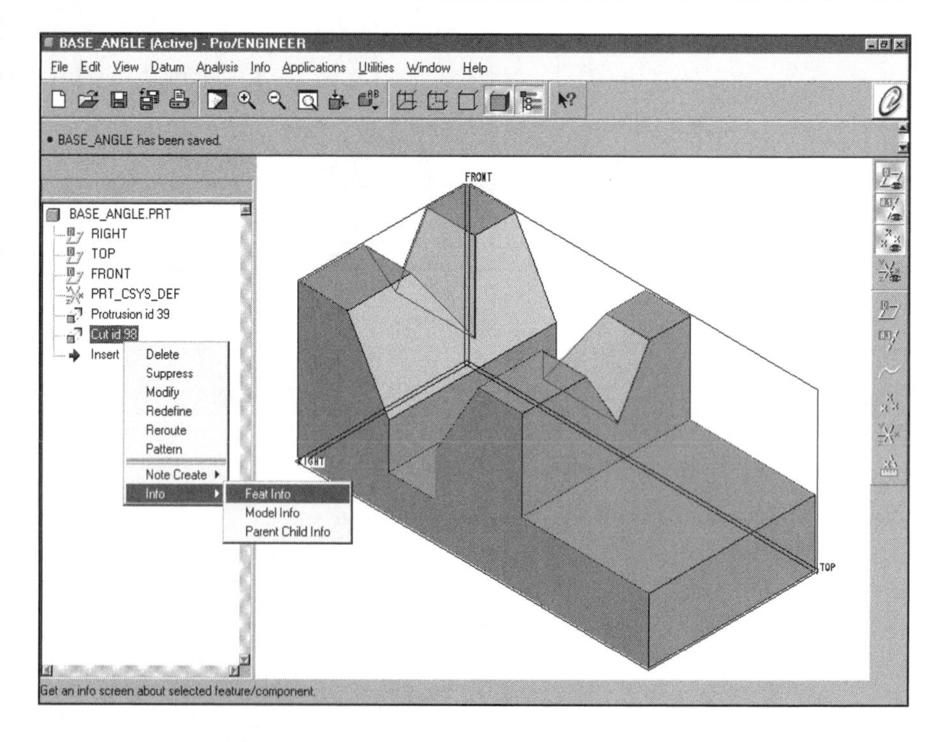

Figure 2.25
Isometric View of Shaded Part

Access the feature's information by selecting the **CUT** in the Model Tree and holding down the RMB. Select **Info ⇒ Feat Info** (and release the mouse button) ⇒ **Close** (Information Window).

Before creating the next cut, resize your Main Window. Then, set the environment and view with the following commands:

HINT

You can make only *one open* section cut at a time. You can make *multiple closed* sections as one feature, but this is considered poor design practice.

Utilities ⇒ Environment ⇒ ✔Coordinate Systems ⇒ Hidden Line ⇒ Isometric ⇒ OK ⇒ View ⇒ Repaint ⇒ View ⇒ Default

The next feature will be another cut. Two cuts are still required for the completion of the part. They can be created together using closed sections or separately with one open section at a time. It is better design intent to make the cuts separately, because they do not have any particular relationship except that they cut the same direction and start on the same surface/plane. A *closed section* is a sketched set of entities that start and end at the same position, like a square □ **shape**. An *open section* does not form a closed figure; an example would be a ∩ or ∪ **shape.**

Start the angled cut by choosing the following commands:

Feature ⇒ Create ⇒ Solid ⇒ Cut ⇒ Extrude ⇒ Solid ⇒ Done ⇒ One Side ⇒ Done ⇒ (pick **TOP** as the sketching/placement plane) **⇒ Flip** [change the direction of the cut to pass through the part, not out into space (Fig. 2.26)] **⇒ Okay ⇒ Top ⇒** (pick **FRONT** as the orientation plane)

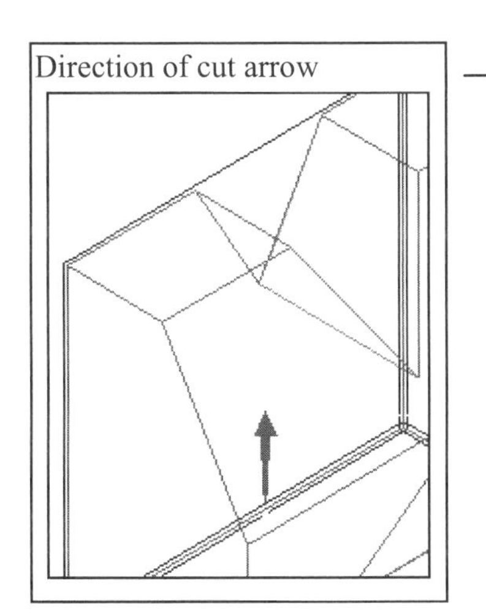

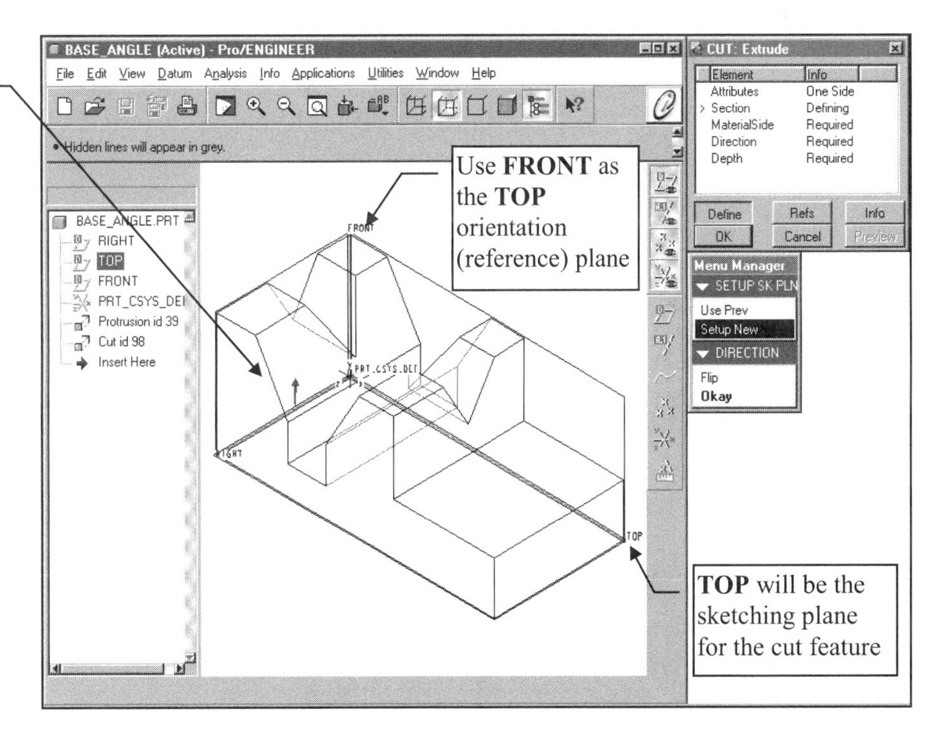

Figure 2.26
Establishing the Cut Placement Plane and the Direction of the Cut

If it's still on, turn off the grid snap in the Environment dialog box: **Utilities ⇒ Environment ⇒ ❑ Snap to Grid ⇒ OK.** Your screen should show the part in the Sketcher looking from the bottom, through the part, as in Figure 2.27. Because you are looking at the bottom, the cut will be sketched on the upper right of the part.

Add the top surface/edge of the part to the References dialog box; then pick **Close.**

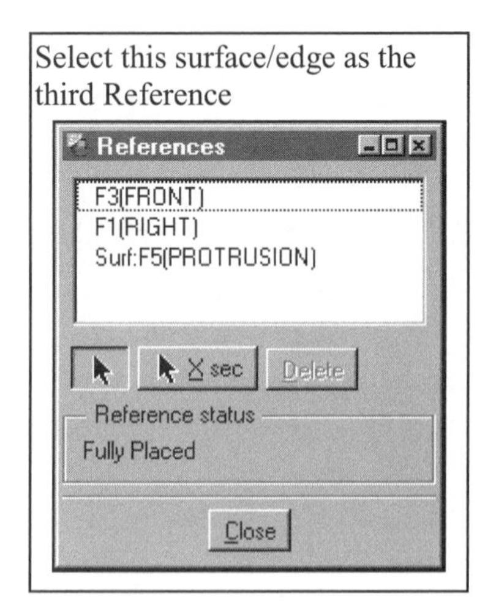

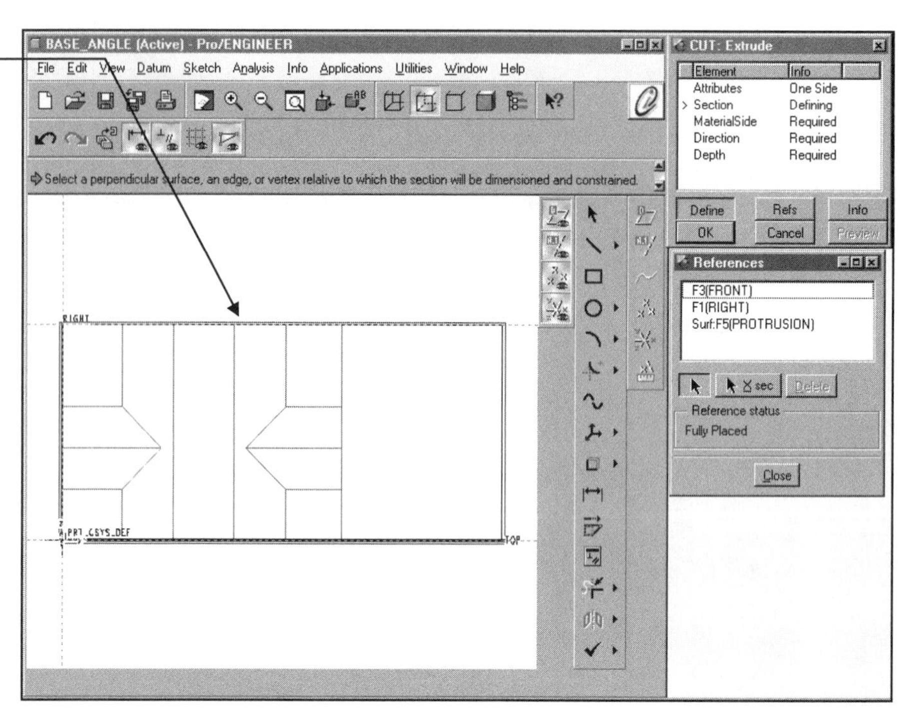

Figure 2.27
Sketching the Cut

Do not sketch the endpoint of the line near the left vertical edge (Fig. 2.27) of the part feature, in order to avoid an unwanted assumption that the point and the part edge are exactly at the same position. This will allow the cut to be modified later by changing the value of the cut dimension. Turn off the Model Tree.

After sketching the one line, you are prompted **Should the highlighted entities be aligned?**, pick **Yes.** Add dimensions as needed.

Modify (Fig. 2.28) the dimensions using the design values from Figure 2.15. Complete the sketch by picking ✔ ⇒ ✔. After you pick ✔, the side of material removal must be selected (Fig. 2.29).

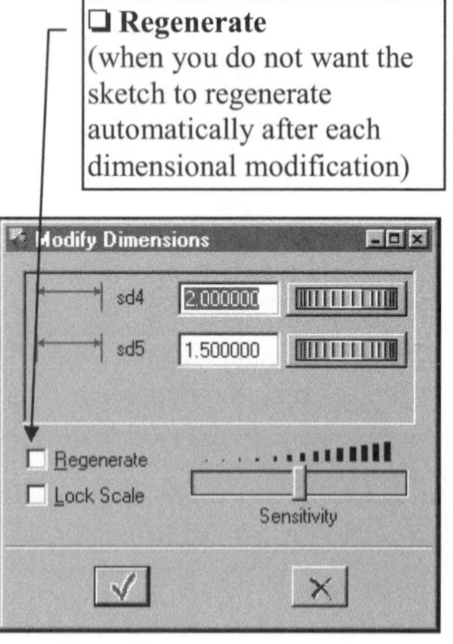

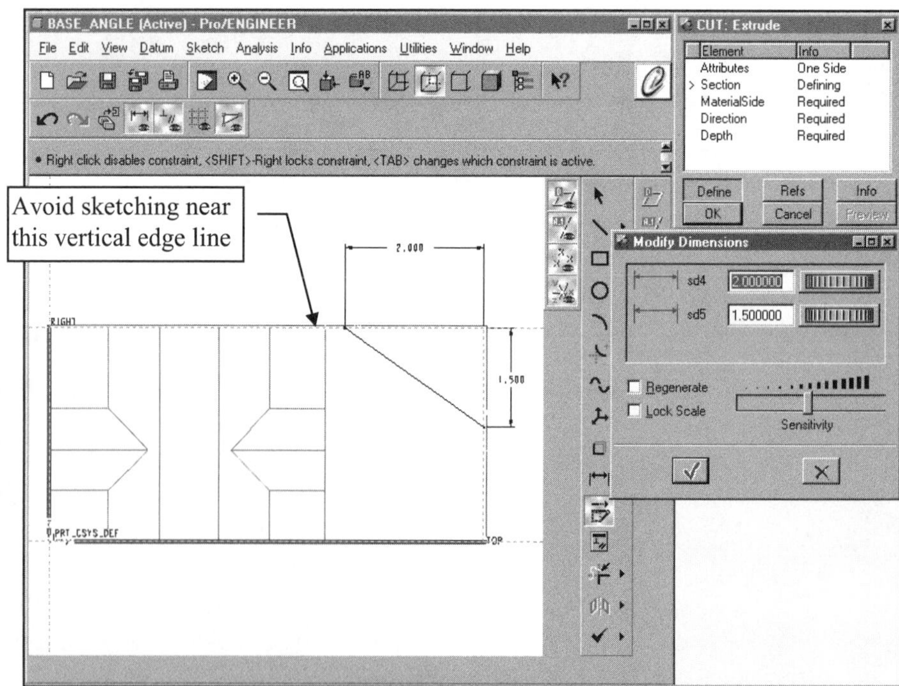

Figure 2.28
Cut Sketched, Dimensioned, and Modified
(❑ Regenerated)

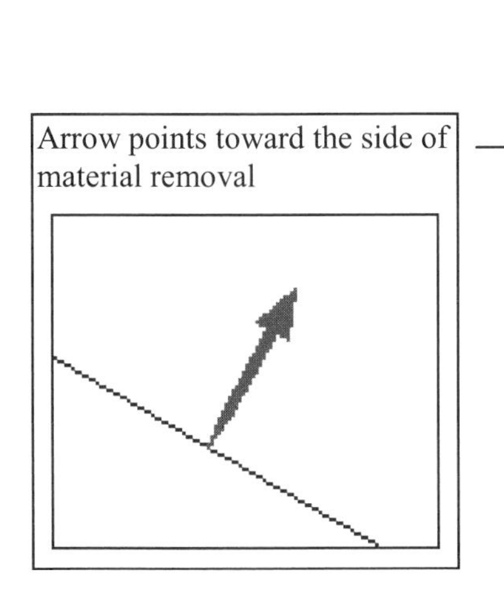

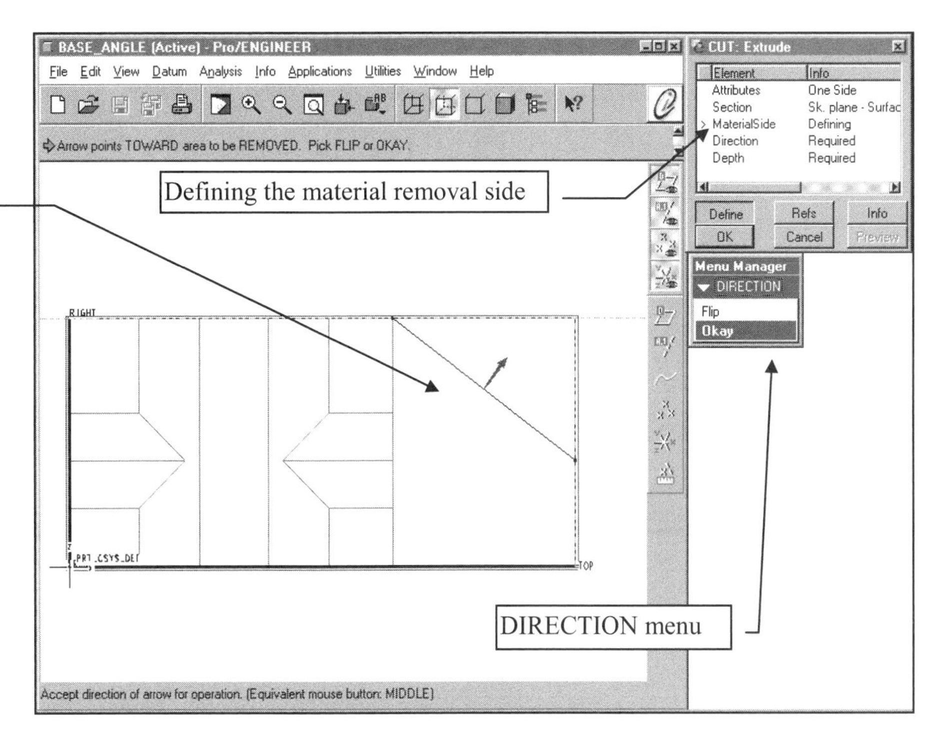

Figure 2.29
Material Removal Side of the Cut

To complete the cut, choose the following commands:

Okay (to accept the material removal side) ⇒ **View** ⇒ **Default** ⇒ (rotates the model to see the cut direction) ⇒ **Thru All** (Fig. 2.30) ⇒ **Done** ⇒ **Preview** ⇒ **OK** (from the dialog box) ⇒ **Done**

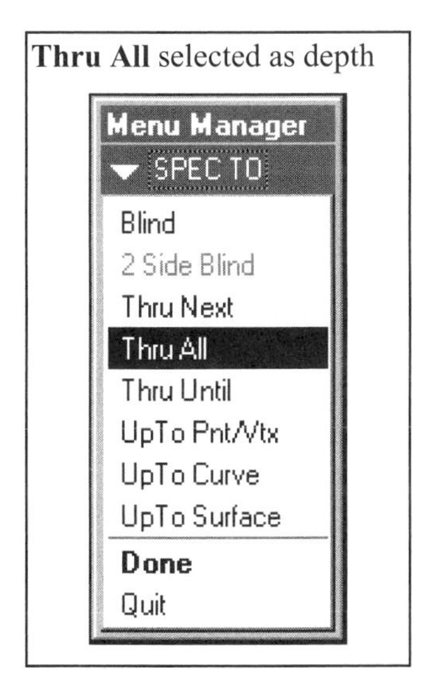

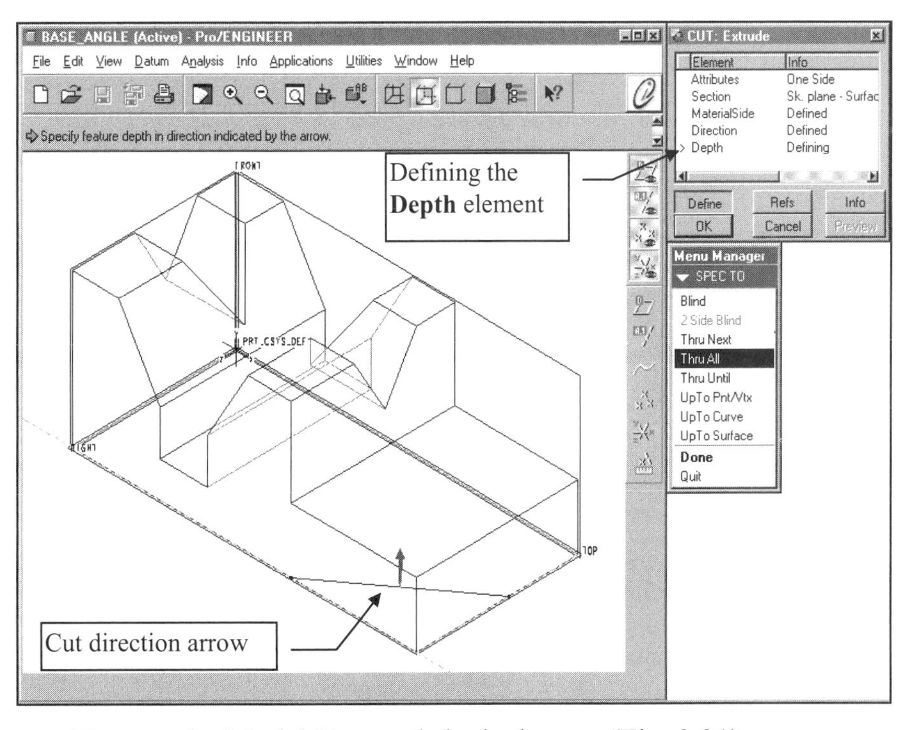

Figure 2.30
Defining Depth

Turn on the Model Tree and shade the part (Fig. 2.31):

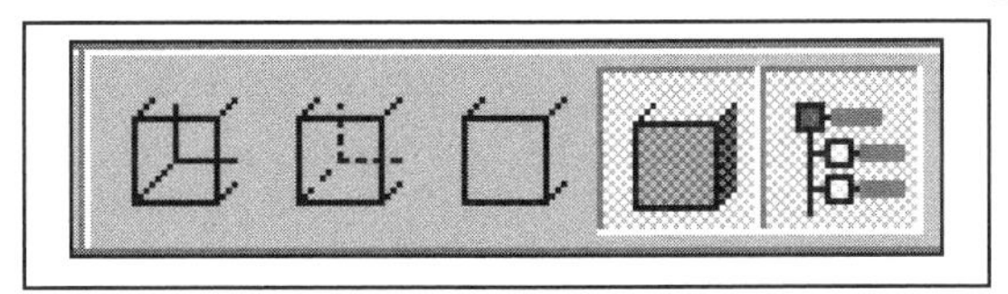

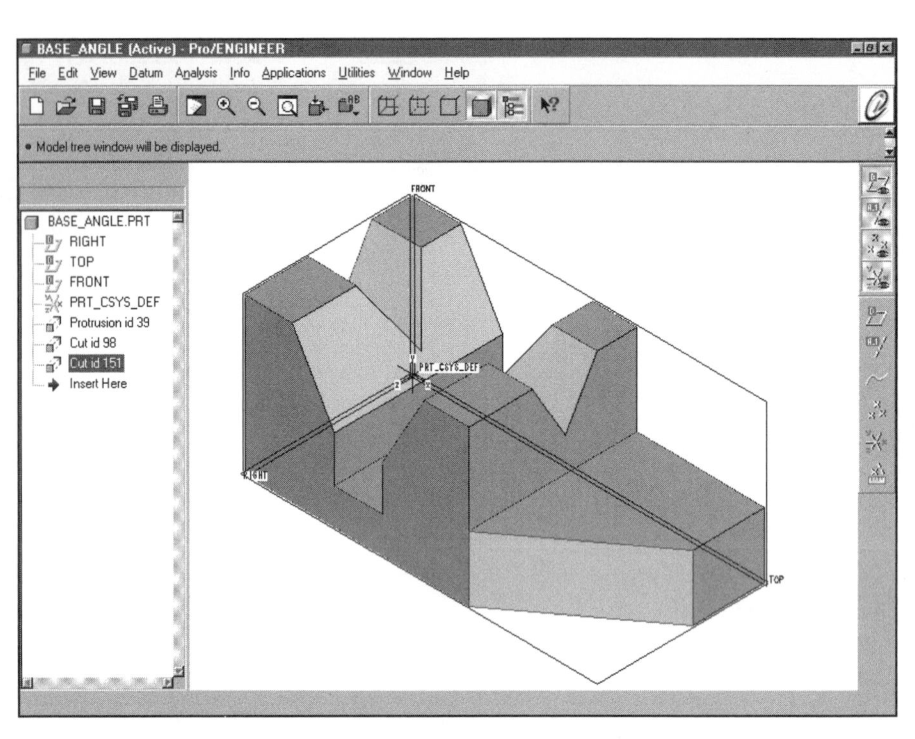

File ⇒ Save ⇒ ✔
File ⇒ Delete ⇒
Old Versions ⇒ ✔

Figure 2.31
Completed Cut

The last feature of the part is the **U**-shaped cut. Use the previous cut's placement plane and orientation. Choose these commands:

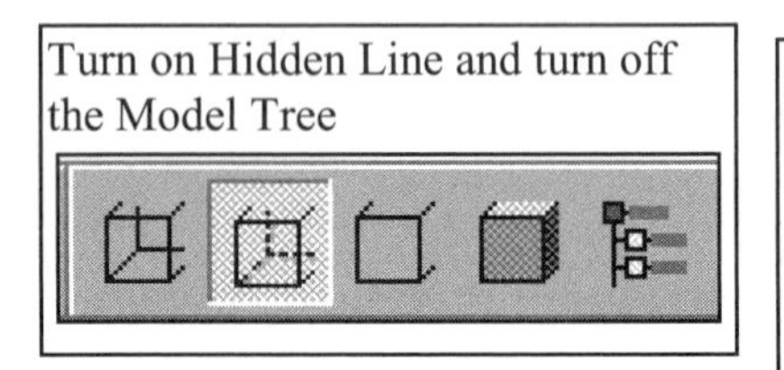

Turn on Hidden Line and turn off the Model Tree

Feature ⇒ Create ⇒ Solid ⇒ Cut ⇒ Extrude ⇒ Solid ⇒ Done ⇒ One Side ⇒ Done ⇒ Use Prev (sketching plane and reference plane) ⇒ **Flip** (change the direction of the cut to pass through the part, not out into space) ⇒ **Okay** ⇒ (from the References dialog box highlight) **RIGHT** ⇒ **Delete** ⇒ ↖ [add the right vertical surface/edge of the model to the References (Fig. 2.32)] ⇒ **Close**

Sketch the three lines (Fig. 2.33) and **Modify** the dimensions using the design values (from Fig. 2.15).

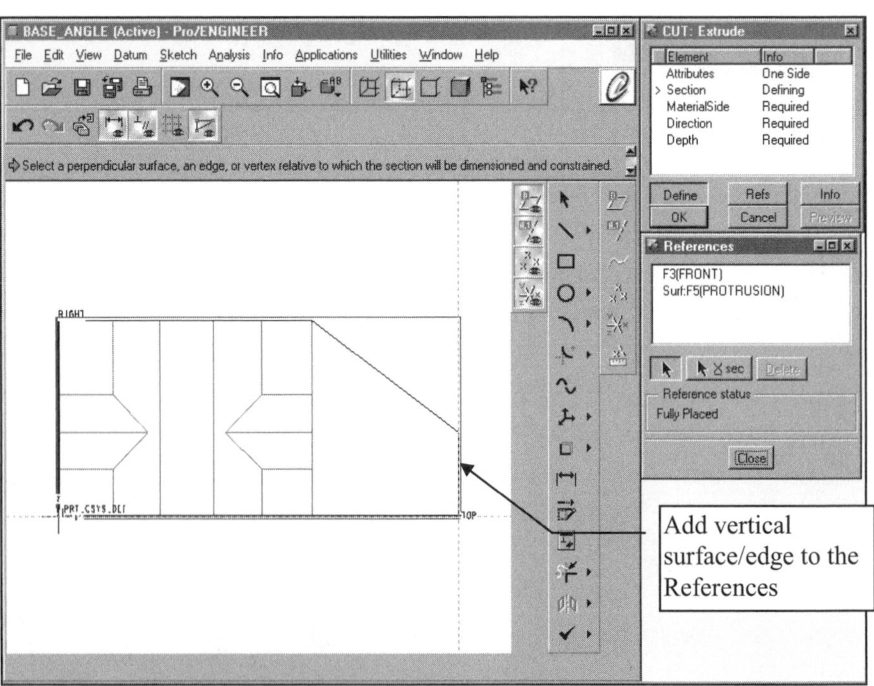

Figure 2.32
References for the **U**-Shaped Cut

The cut is dimensioned, modified, and regenerated as shown in Figure 2.34.

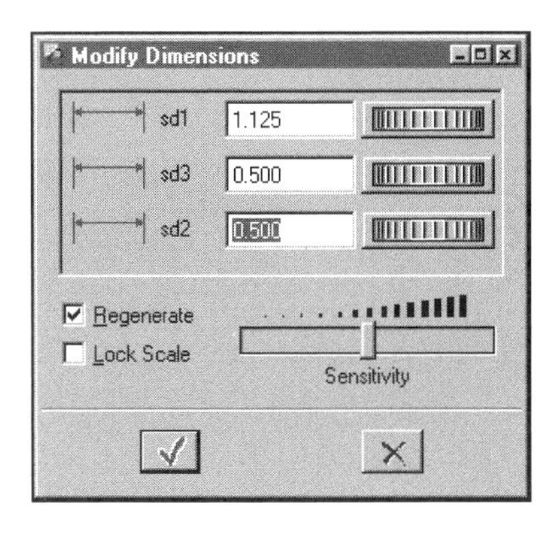

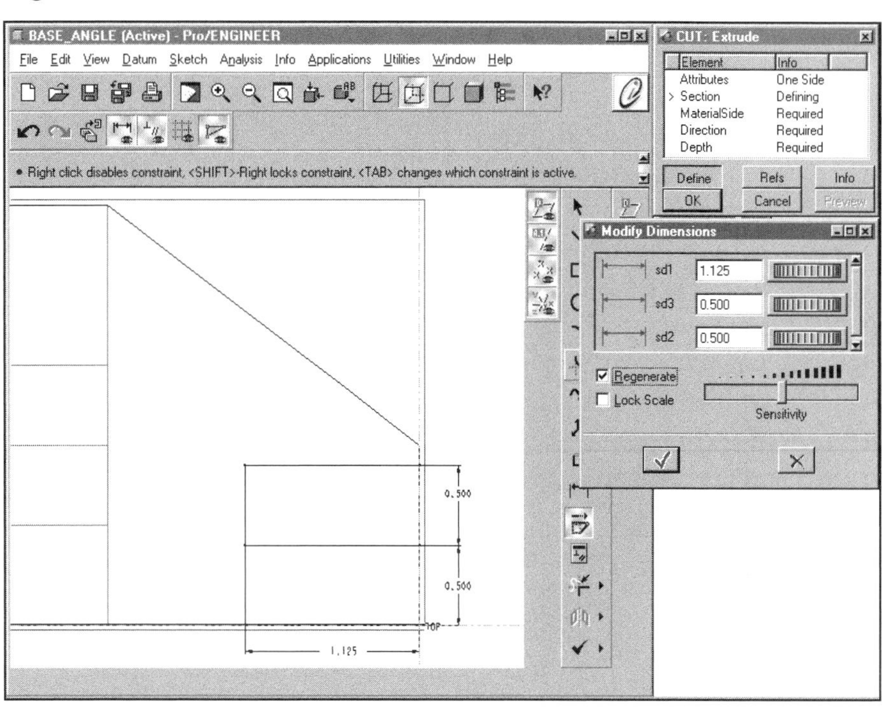

Figure 2.33
U-Shaped Cut

Complete the part with the following commands (Fig. 2.34):

✔ ⇒ **Okay** (accepts material removal side) ⇒ **Thru All** ⇒ **Done** ⇒ **OK** (from dialog box) ⇒ **View** ⇒ **Default** ⇒ **File** ⇒ **Save** ⇒ ✔

BASE_ANGLE.PRT
RIGHT
TOP
FRONT
PRT_CSYS_DEF
Protrusion id 39
Cut id 98
Cut id 151
Cut id 187
Insert Here

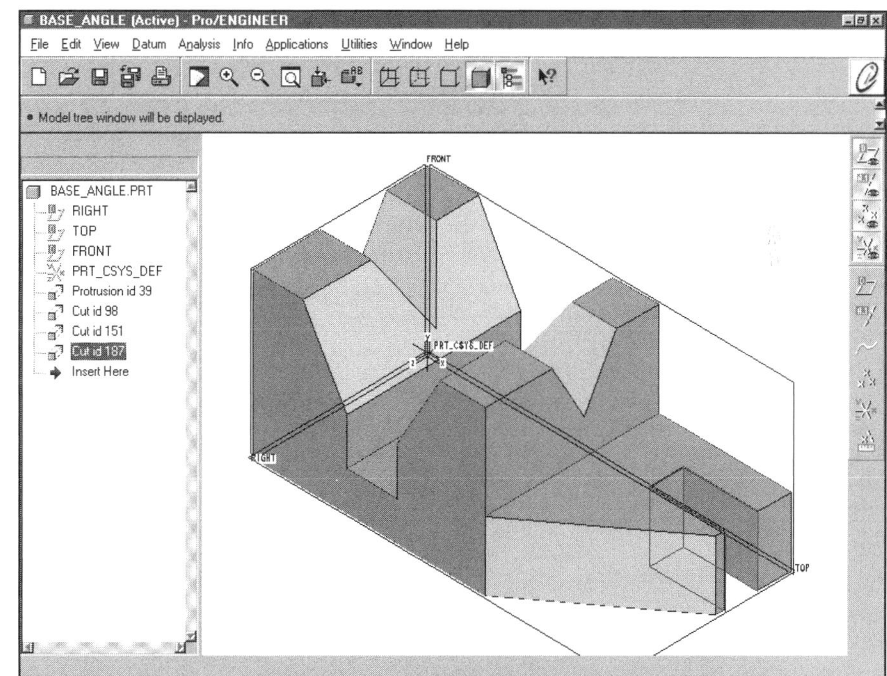

Figure 2.34
Completed **BASE_ANGLE**

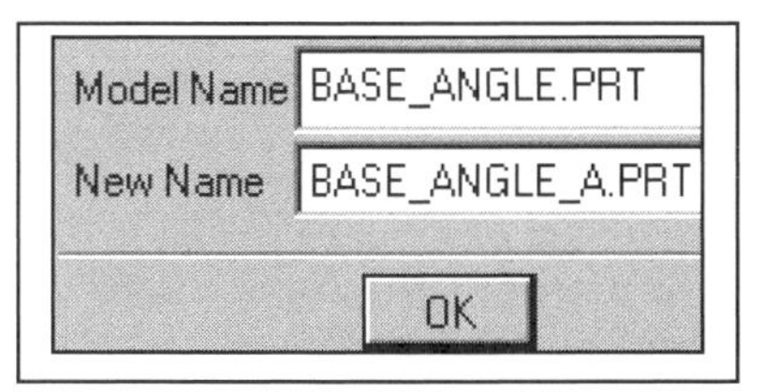

Because you will be modifying and redefining the part, save a version under another name:

File ⇒ **Save As** ⇒ New Name **BASE_ANGLE_A** ⇒ **OK** ⇒ **File** ⇒ **Close Window** ⇒ **File** ⇒ **Erase** ⇒ **Not Displayed** ⇒ **OK**

Very few projects make it through the design, engineering, and manufacturing phases without changes. Changes can be simple dimensional changes in the part's size or more extreme changes in the part's configuration. For simple dimensional changes, the **Modify** command is used; when configuration changes are needed, the **Redefine** command is called upon. Engineering or Manufacturing departments normally release an Engineering Change Order **(ECO)** (Fig. 2.35). **Open** the part you **Saved As** (Fig. 2.36):

File ⇒ Open ⇒ (pick **BASE_ANGLE_A**) **⇒ Preview >>> ⇒ Open**

ECO

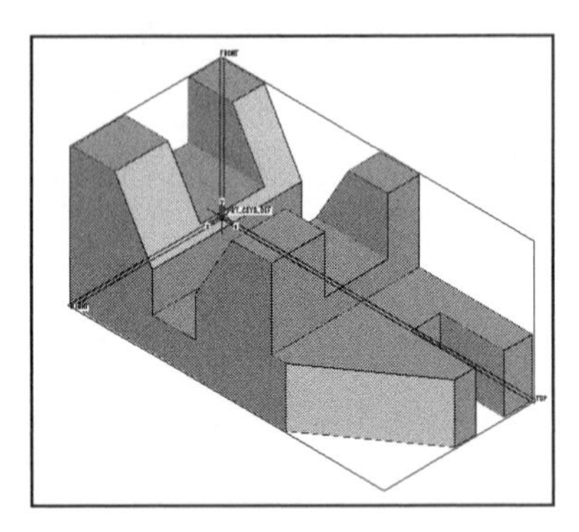

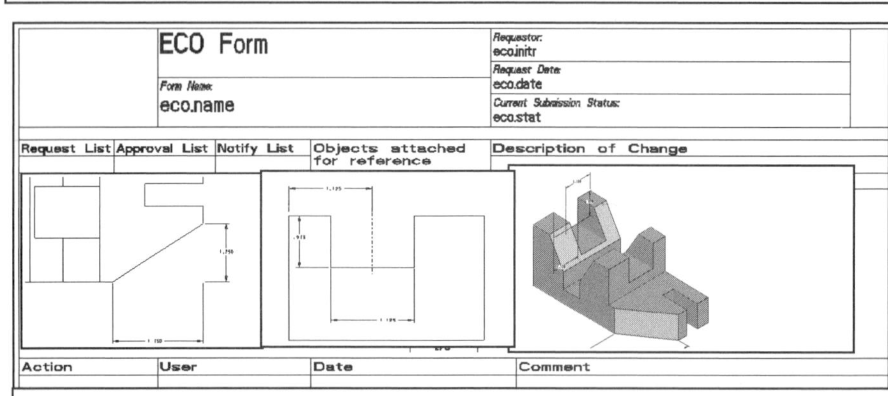

ECO Form
Requestor: eco.initr
Request Date: eco.date
Form Name: eco.name
Current Submission Status: eco.stat

Request List	Approval List	Notify List	Objects attached for reference	Description of Change

Action	User	Date	Comment

New design requirements for the cuts
Change the size of the angled cut from **2.00** to **1.75** and the **1.50** dimension to **1.25.** Alter the **V** cut to be a **U-shaped** slot, **1.125** wide by **.975** deep.

Figure 2.35
ECO Changing the Angled Cut Size and the **V** Cut

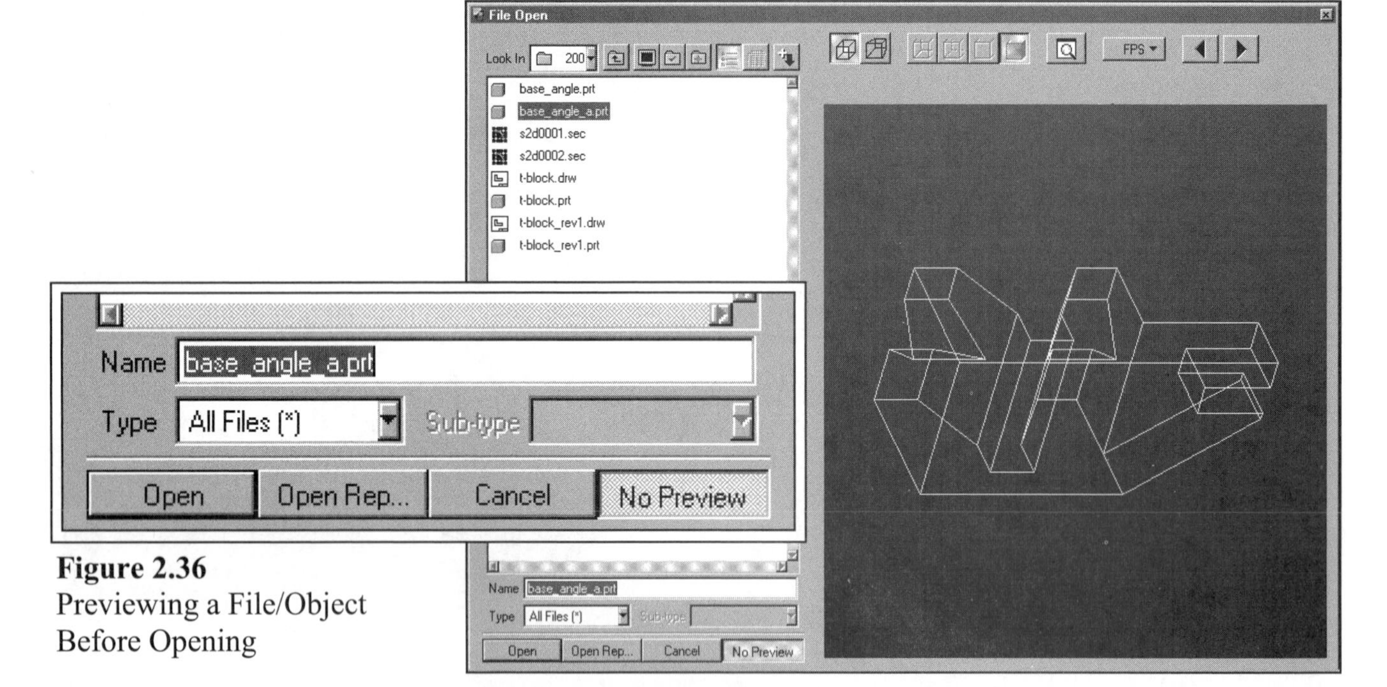

Figure 2.36
Previewing a File/Object Before Opening

You can **Modify** and **Redefine** from the menu structure or from the Model Tree. Use the **Model Tree** for both ECO changes as shown in Figure 2.37.

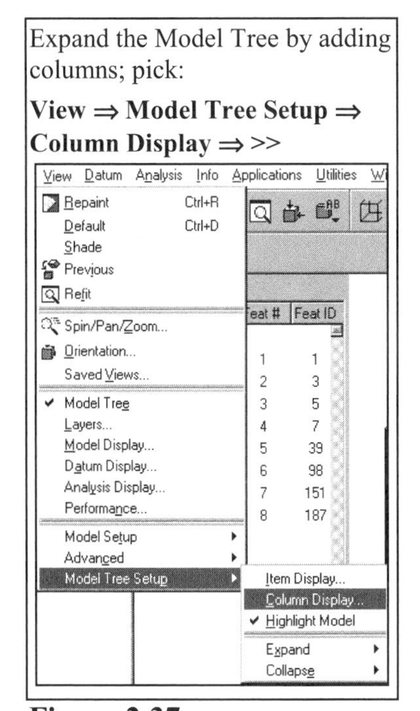

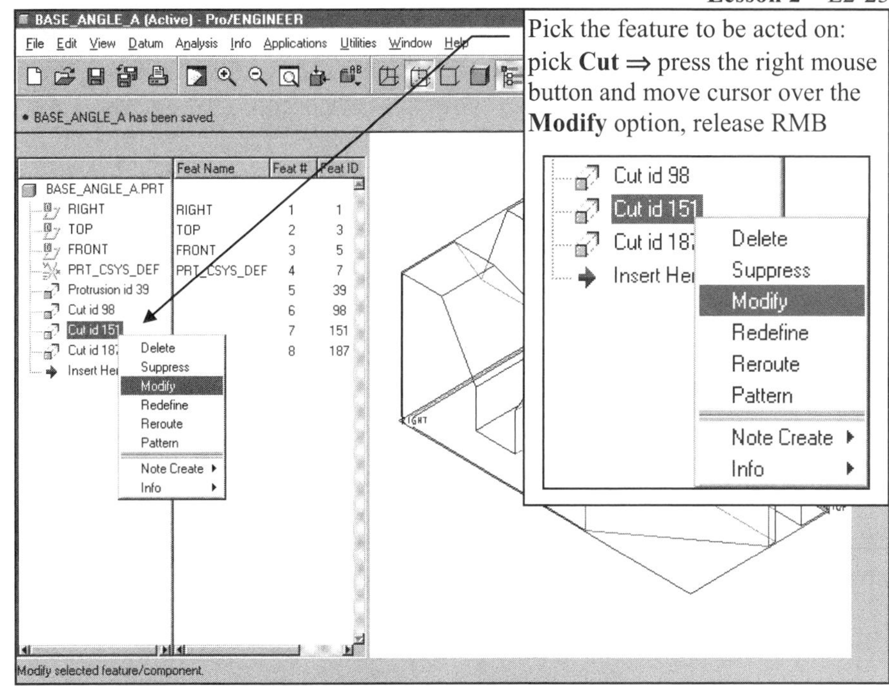

Figure 2.37
Press RMB to See the Pull-down Menu

After selecting the feature to be modified from the Model Tree, press and hold down the right mouse button to display the pull-down menu (the mouse cursor must be somewhere in the Model Tree for this to work). Move the mouse to highlight the **Modify** command and then release RMB. Pick the **2.00** dimension, type the new value in the **Enter value** command line (**1.75**), and ✔; do likewise to the **1.50** dimension to change it to **1.25** (Fig. 2.38). **Regenerate** the part to see the changes. **Shade** and rotate the part.

Figure 2.38
Modify the Angled Cut Dimensions from **2.00 X 1.75** to **1.50 X 1.25**

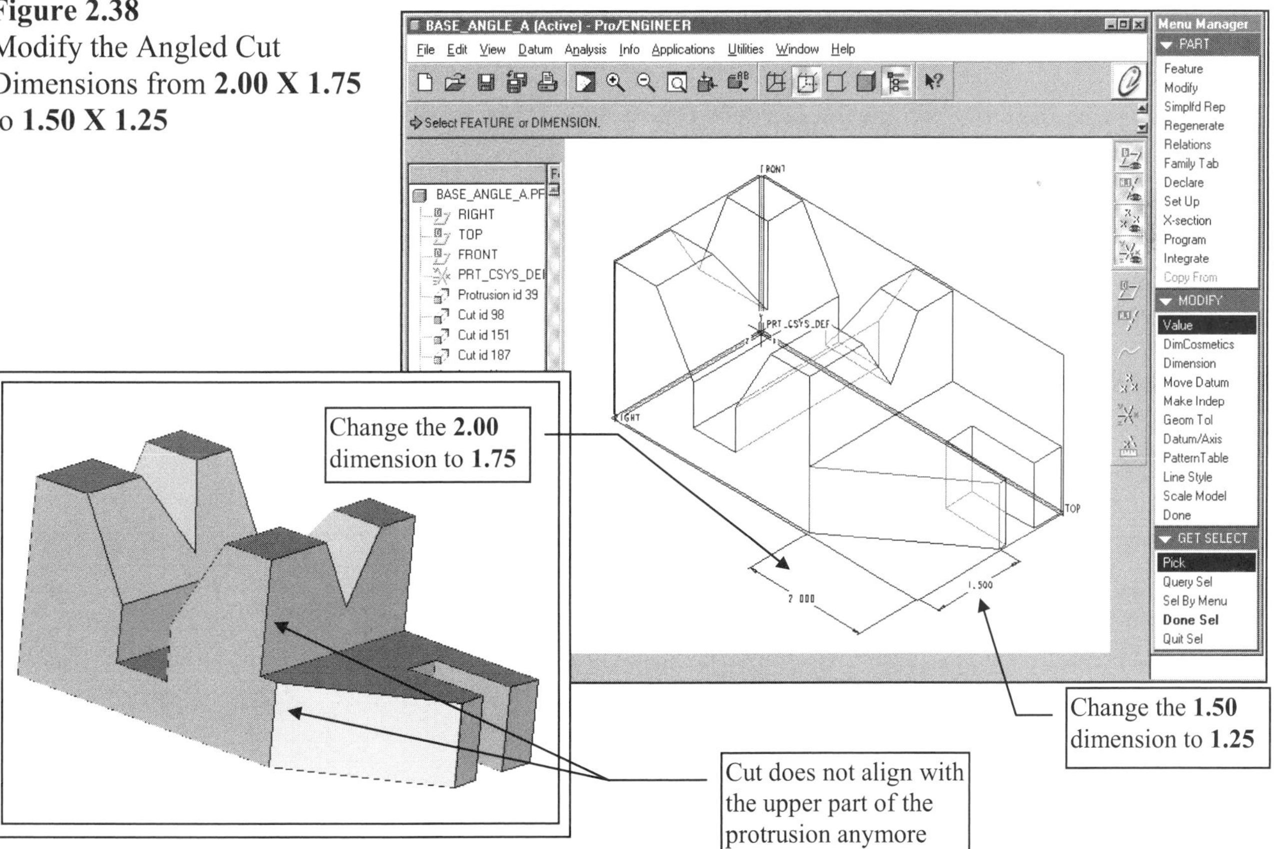

The other requirement from the ECO was to change the shape and the size of the **V** cut. Alter the **V** cut to be a **U**-shaped slot **1.125** wide by **.975** deep. Again, let's use the Model Tree to redefine the **V** cut feature (Fig. 2.39).

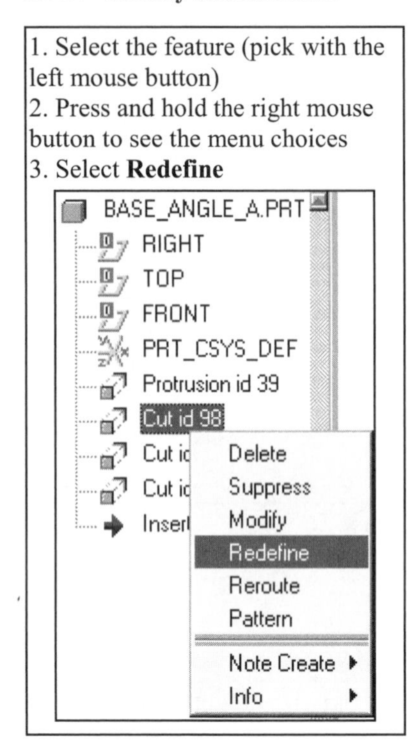

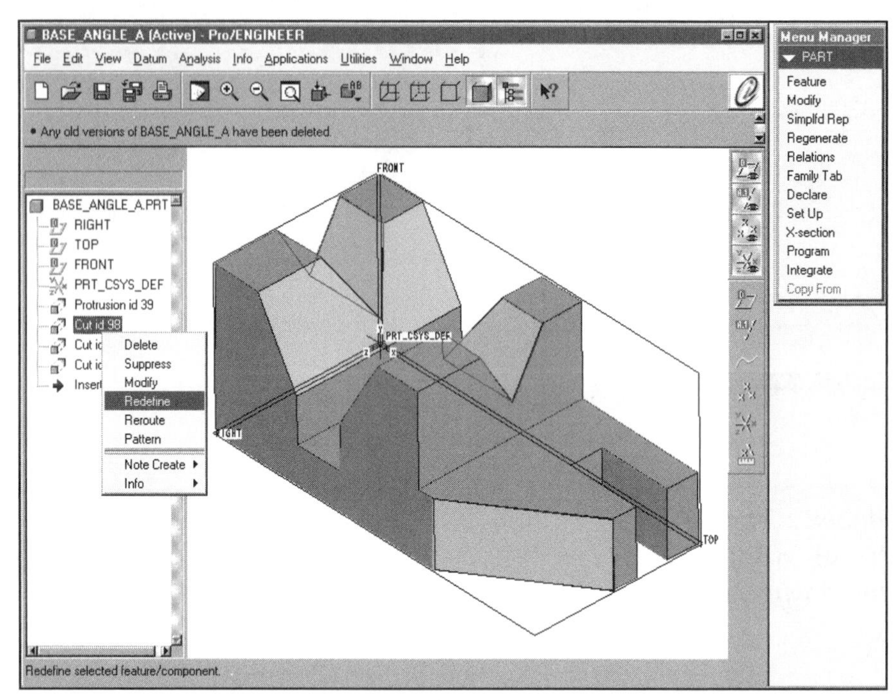

Figure 2.39
Redefine the **V** Cut

From the dialog box, pick **Section** and then choose **Define**, as shown in Figure 2.40. You are now able to redefine the section.

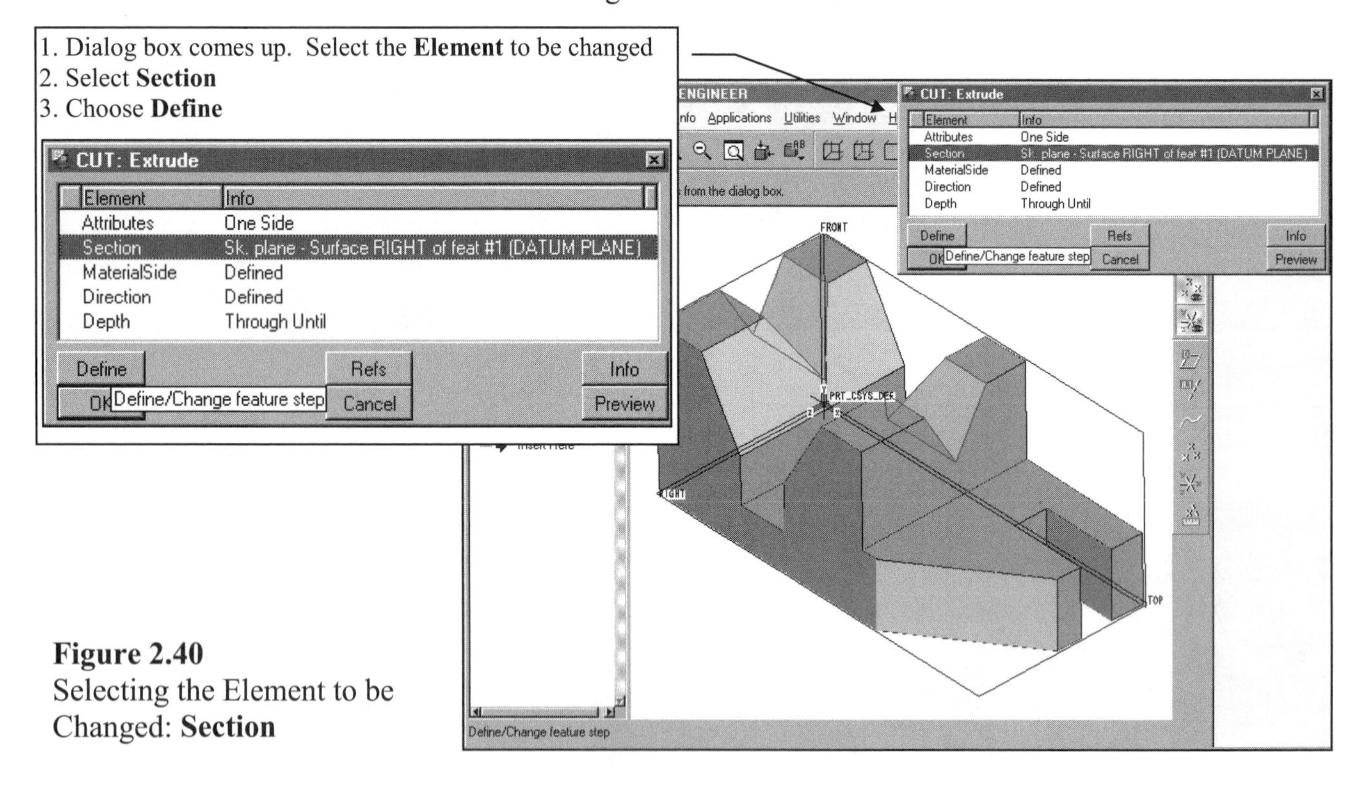

Figure 2.40
Selecting the Element to be Changed: **Section**

Choose **Sketch** from the SECTION menu. The section sketch will be displayed as in Figure 2.41. **Close** the SKETCHER ENHANCEMENT if it appears.

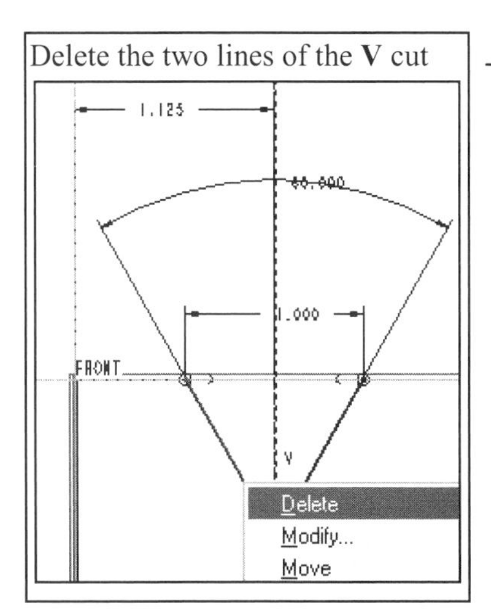

Figure 2.41
Section Sketch Showing Original Design **V** Cut

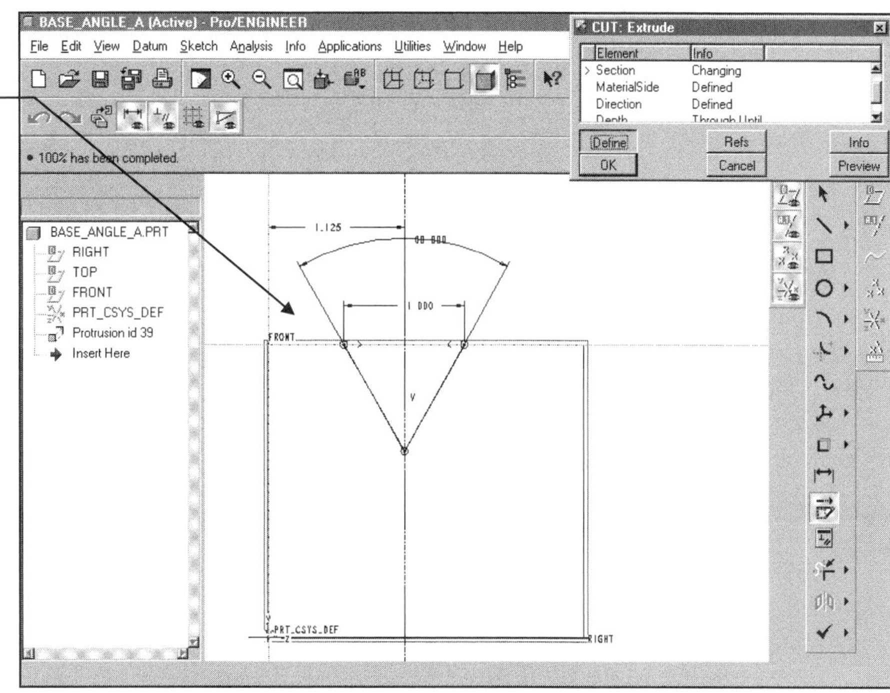

Use the same centerline for the redefined cut section. Delete the two lines forming the **V.** Select the line (it turns *red*); click your RMB to display the pop-up menu options; pick **Delete.** Repeat this process and delete both lines. Sketch three new lines of the slot as shown in Figure 2.42. Modify the dimensions using the **ECO** sizes of **1.125** wide by **.975** deep (Fig. 2.42). Pick ✔ ⇒ ✔ ⇒ **OK** (from the dialog box). **Shade** and spin the part (Fig. 2.43). Save the modified and redefined part.

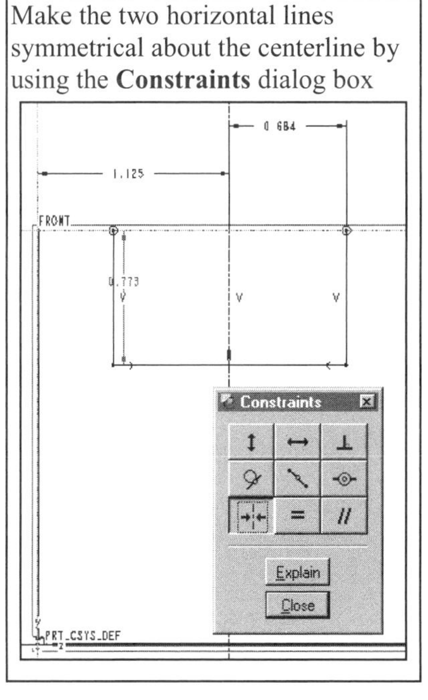

Figure 2.42
Redefined Sketch

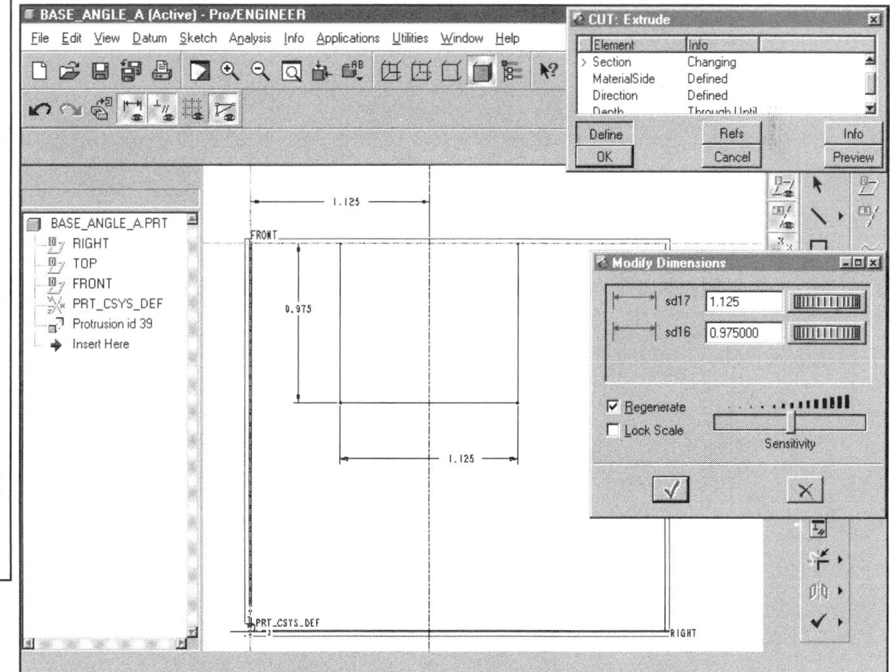

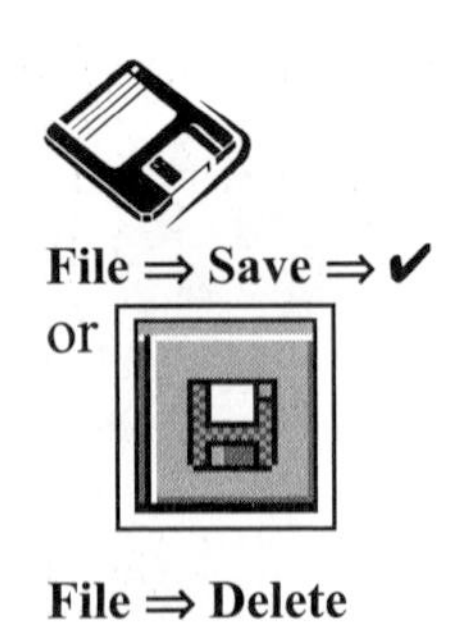

File ⇒ Save ⇒ ✔
or

File ⇒ Delete
Old Versions ⇒ ✔

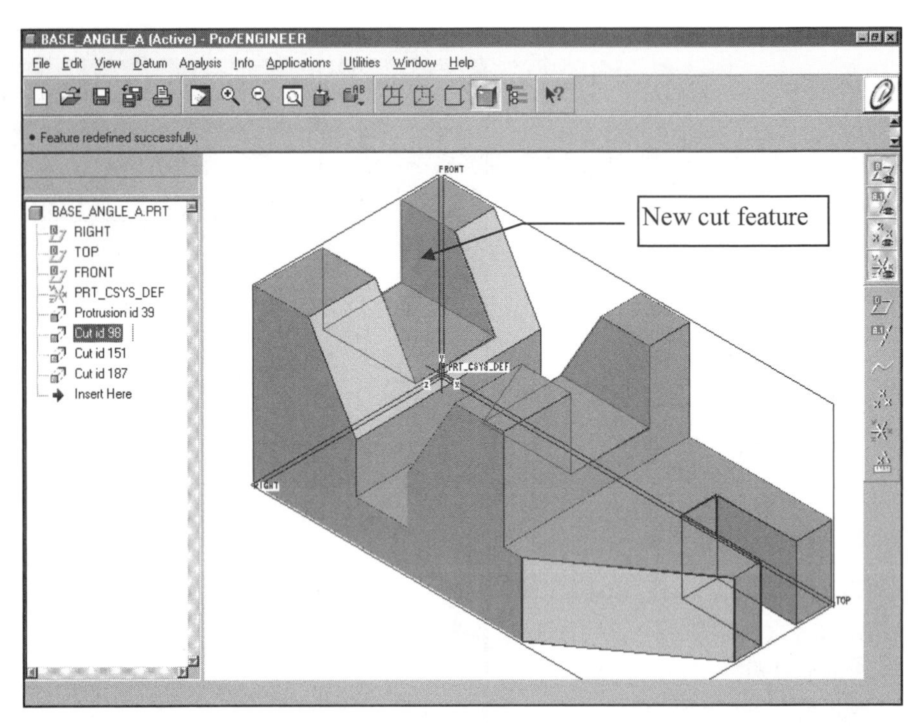

Figure 2.43
ECO Changes Completed

You can get information regarding your model at any time in the design or redesign process. Using the Model Tree, request information on the part feature just redefined (Fig. 2.44). A variety of information is displayed on the screen, including the **Feature's Dimensions**, **Feature Number**, **Internal Feature ID Number**, and **Parents** (Fig. 2.45).

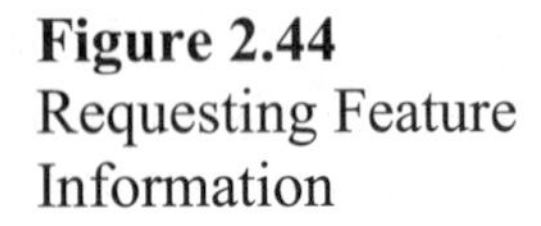

Figure 2.44
Requesting Feature Information

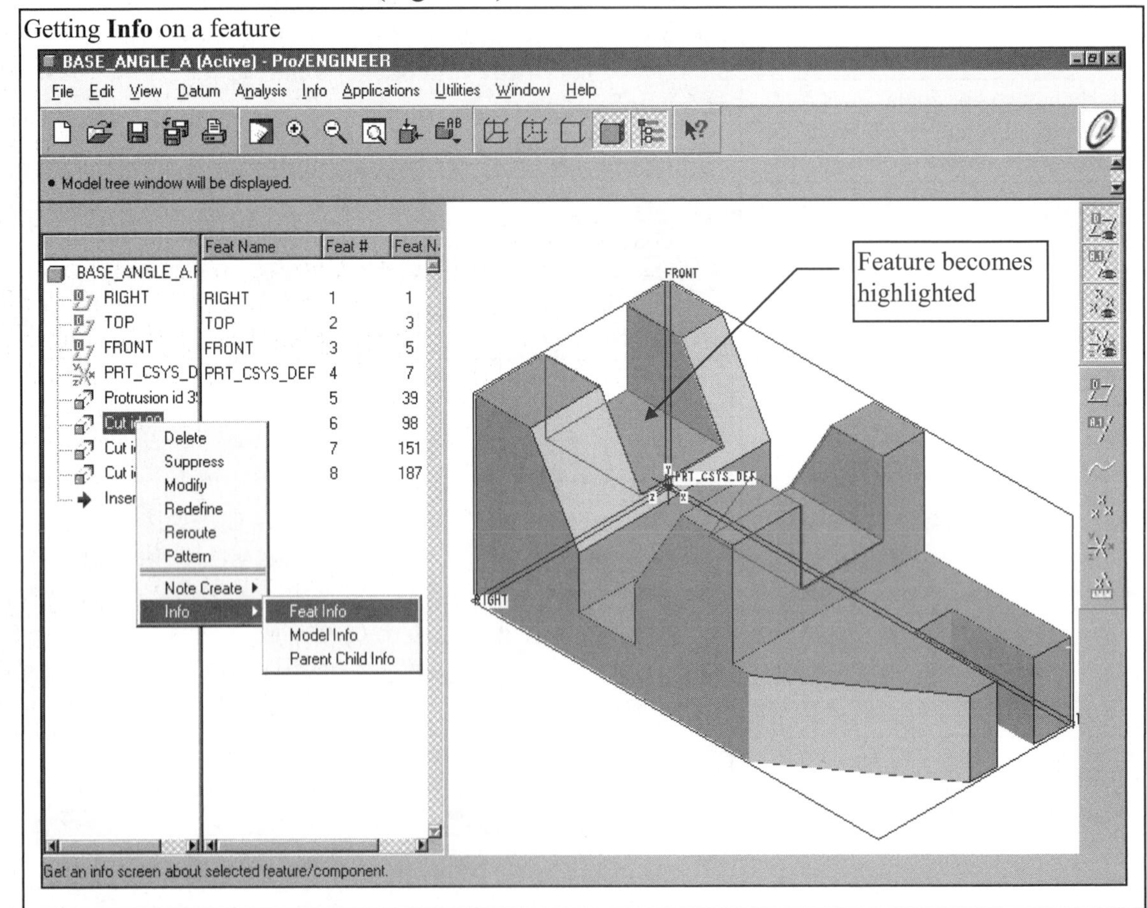

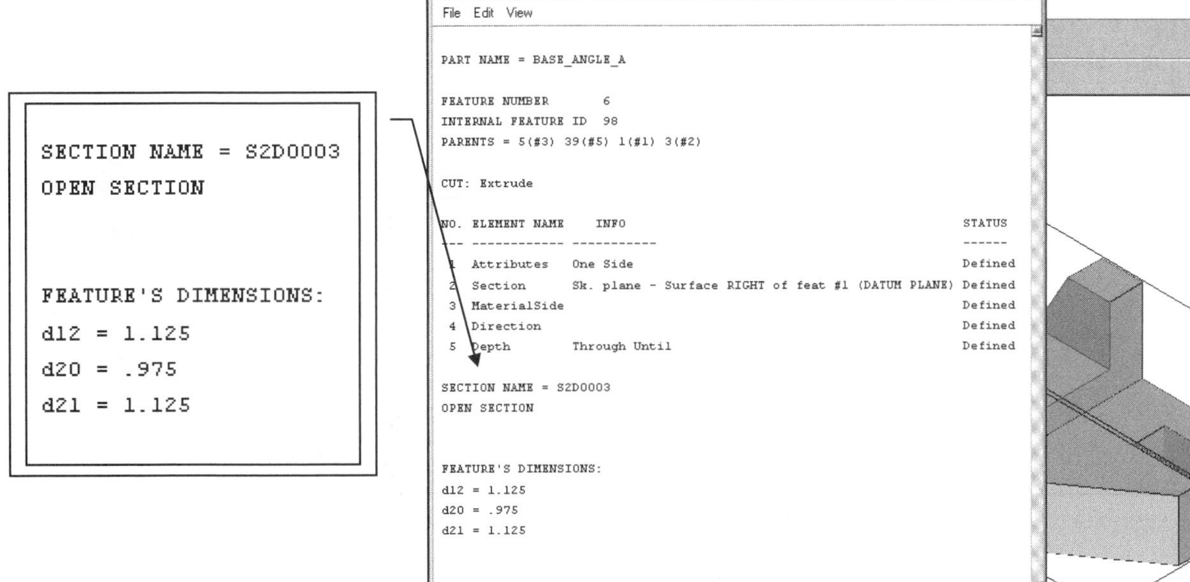

Figure 2.45
Feature Information

Besides feature information, you can extract information on the whole model by choosing:

Info (from the menu bar) ⇒ **Model** (Fig. 2.46)

Figure 2.46
Model Information

BASE_ANGLE_A (Active) - Pro/ENGINEER
File Edit View Datum Analysis Info Applications Utilities Window
Feature...
Feature List...
Model...
Model Show information about a specified model
Audit Trail...
Pro/Engineer Objects
Message Log...
Parent/Child...
Global Ref Viewer...
Regen Info...
Geometry Check...
Switch Dims
Model Size
Information written to file BASE_ANG
Show information about a specified model

INFORMATION WINDOW (BASE_ANGLE_A.inf.2)
File Edit View
PART BASE_ANGLE_A
MATERIAL FILENAME: ALUMINUM
Units info for the major system 'Inch lbm Second (Pro/E
Length in
Mass lbm
Force in lbm / sec^2
Time sec
Temperature F
FEATURES:
FEATURE NUMBER 1
INTERNAL FEATURE ID 1
CHILDREN = 39(#5) 98(#6) 151(#7) 187(#8)
TYPE = DATUM PLANE
NAME = RIGHT
FEATURE IS IN LAYER(S) :
01___PRT_ALL_DTM_PLN - OPERATION = SHOWN
01___PRT_DEF_DTM_PLN - OPERATION = SHOWN
Close

Lesson 2 Project

T-Block

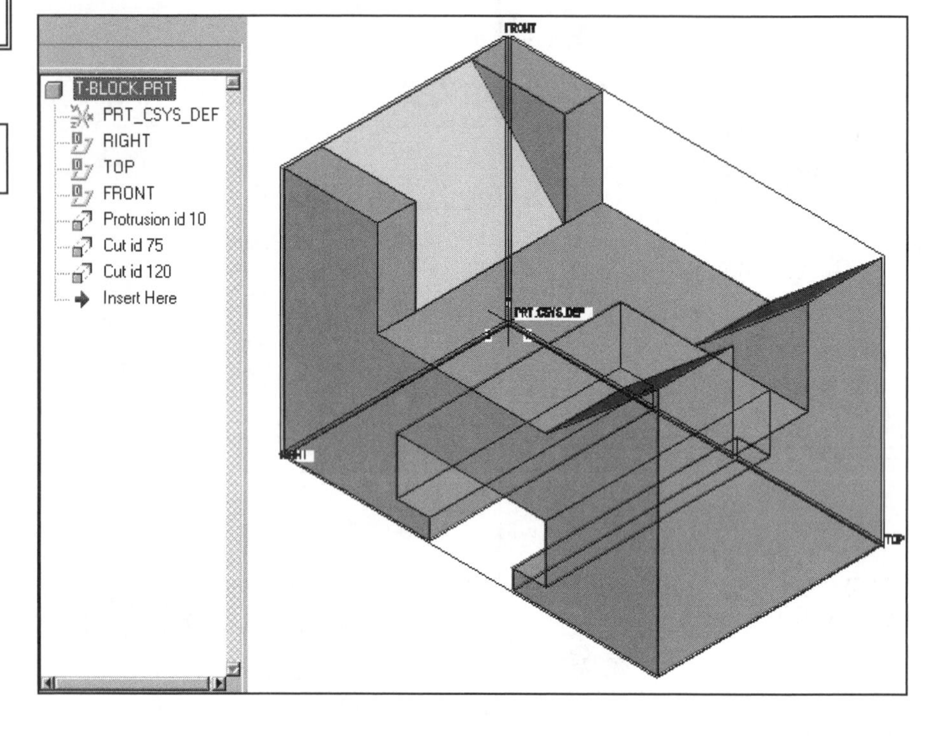

Figure 2.47
T-Block

T-Block

The second **lesson project** is a block (Figs. 2.47 through 2.51) that is created with types of commands similar to those for the **BASE_ANGLE.**

Sketch the protrusion on **FRONT.** After the T-Block is modeled, you will be prompted to modify and redefine a number of its features from an **ECO**.

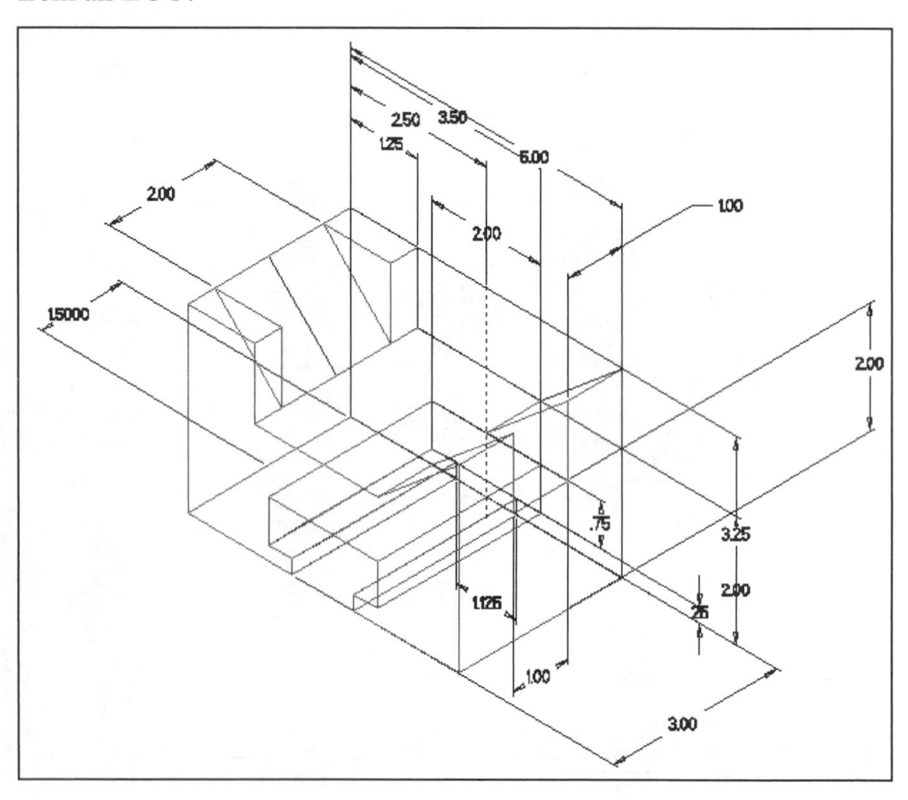

Figure 2.48
T-Block Dimensions

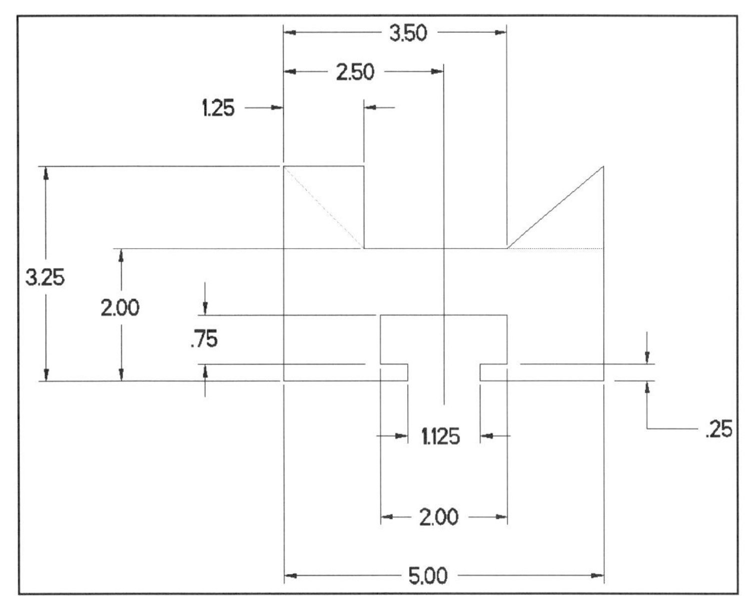

Figure 2.49
Front View

NOTE
The **T**-shaped slot is symmetrical and is located at the center of the part. Later, when completing the ECO, be careful not to assume that this condition still applies!

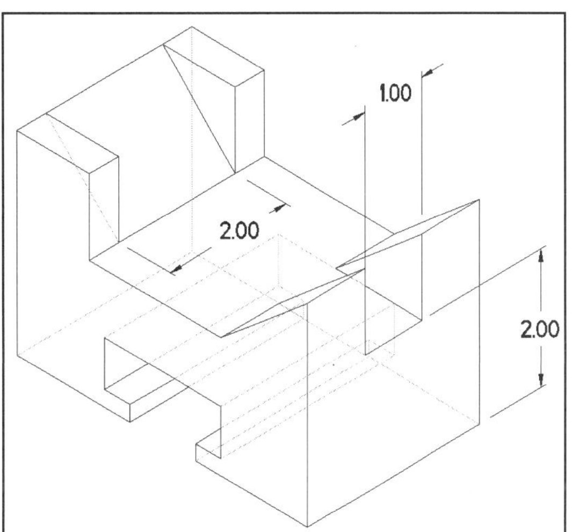

Figure 2.50
Pictorial View

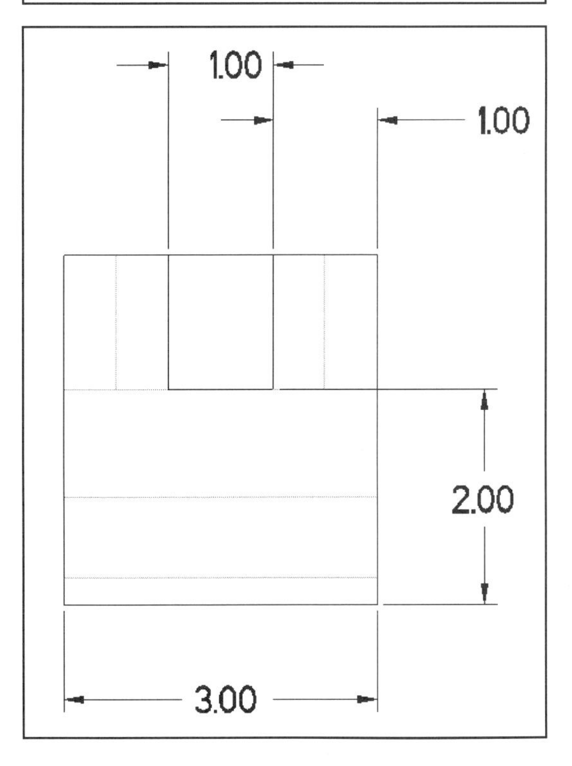

Figure 2.51
Right Side View

Redefine and **Modify** the part after you save it to another name. Figures 2.52 through 2.57 provide the ECO and feature redefinition requirements.

File ⇒ Save ⇒ ✔
File ⇒ Delete ⇒
Old Versions ⇒ ✔

ECO Form

Form Name:
eco.name

Requestor:
eco.initr

Request Date:
eco.date

Current Submission Status:
eco.stat

Request List	Approval List	Notify List	Objects attached for reference	Description of Change

Action	User	Date	Comment

Change the cut to the configuration shown in Figures 2.53 and 2.54.

Change the height and the depth of the part to 4.00.

Change the T slot to that shown in Figures 2.54 through 2.56.

Figure 2.52
ECO

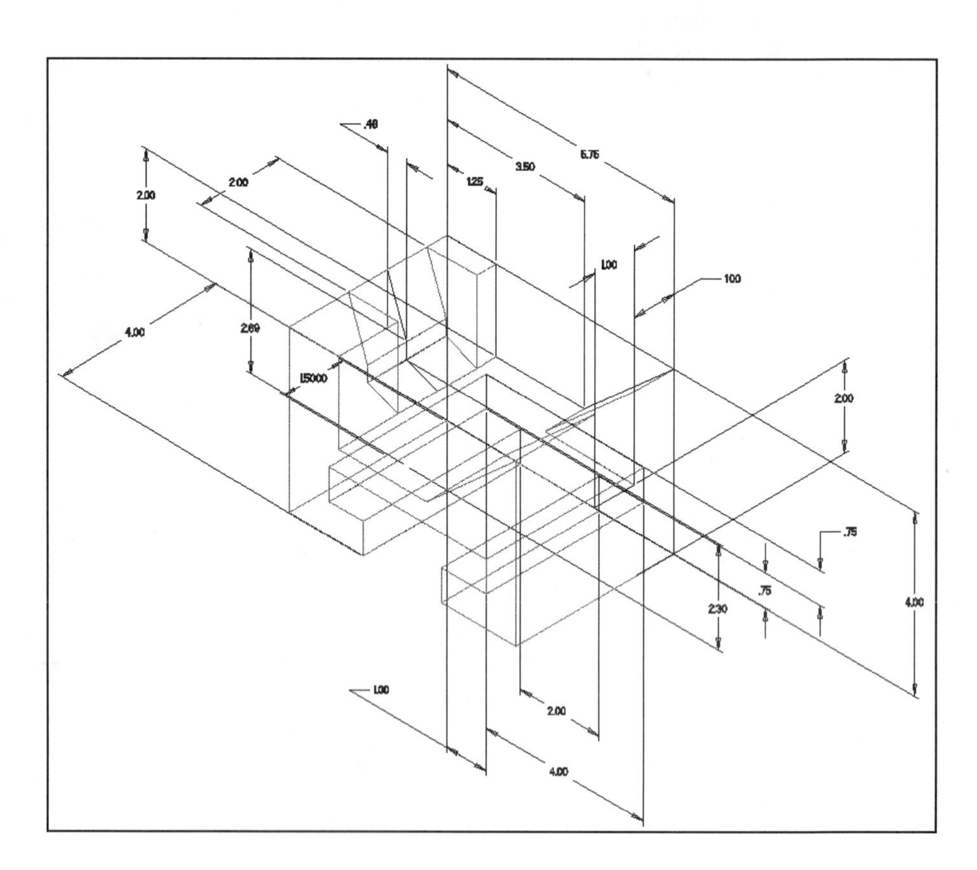

Figure 2.53
Modified Dimensions for T-Block

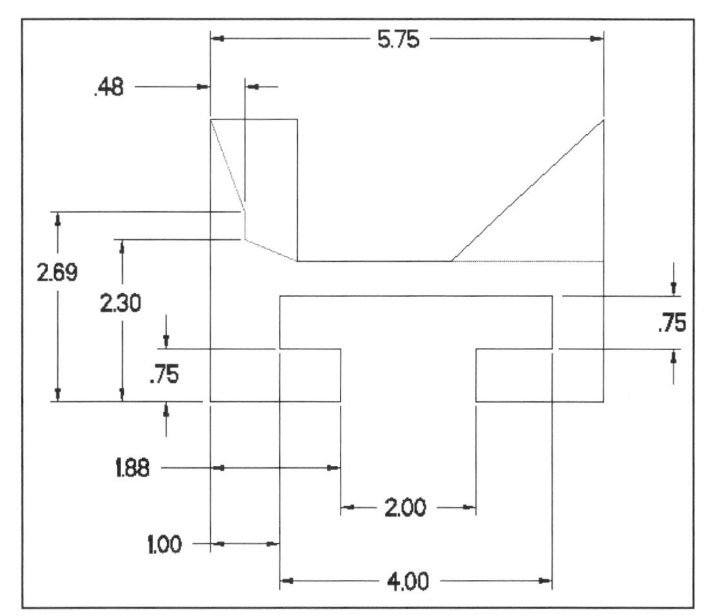

Figure 2.54
Front View

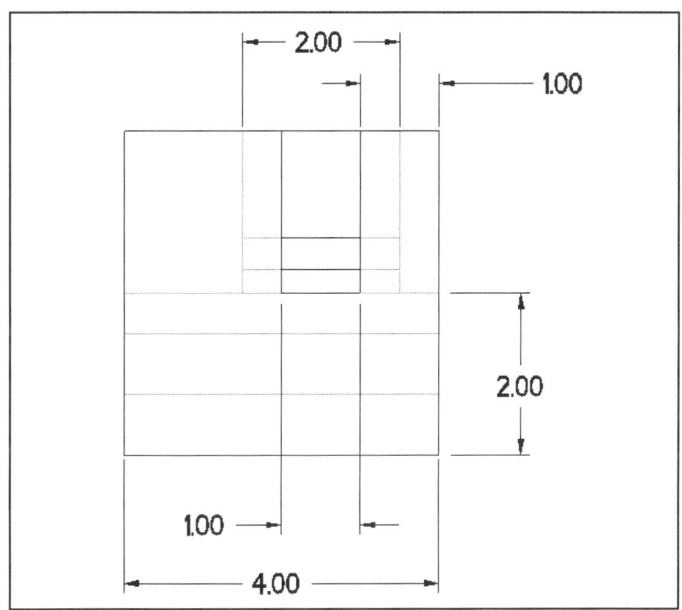

Figure 2.55
Right Side View

File ⇒ Save ⇒ ✔
File ⇒ Delete ⇒
Old Versions ⇒ ✔

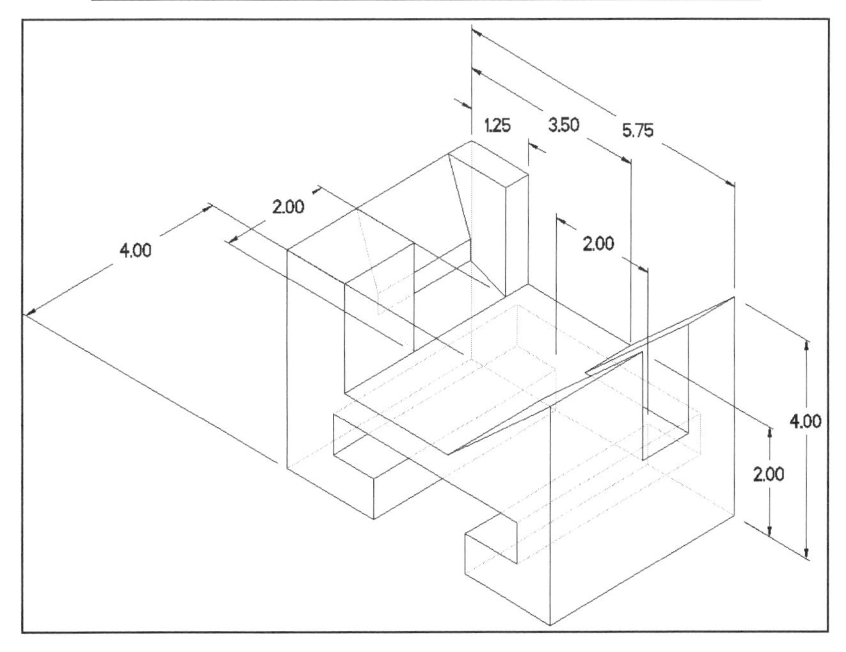

Figure 2.56
Pictorial View

Figure 2.57
Completed ECO

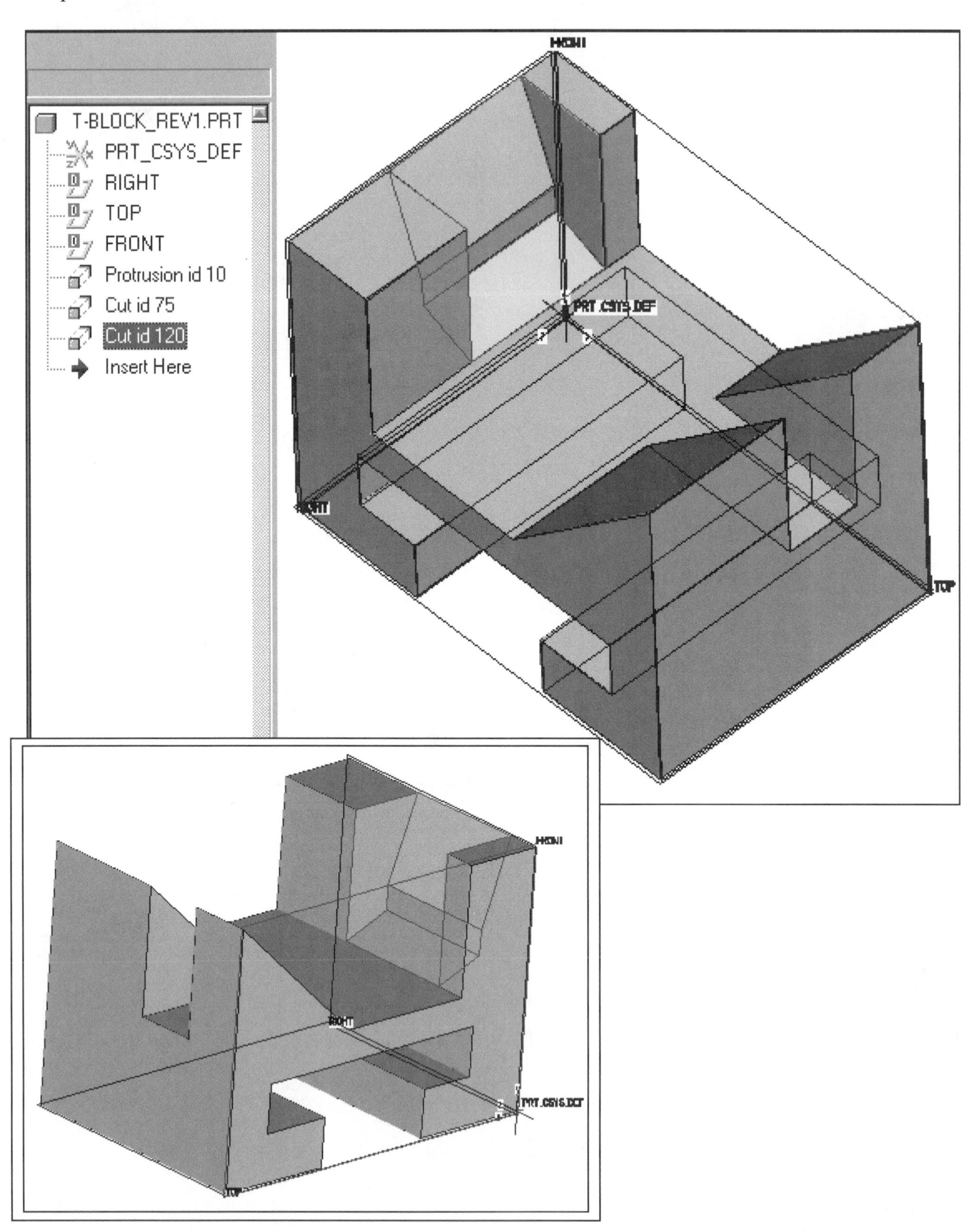

Lesson 3

Holes and Rounds

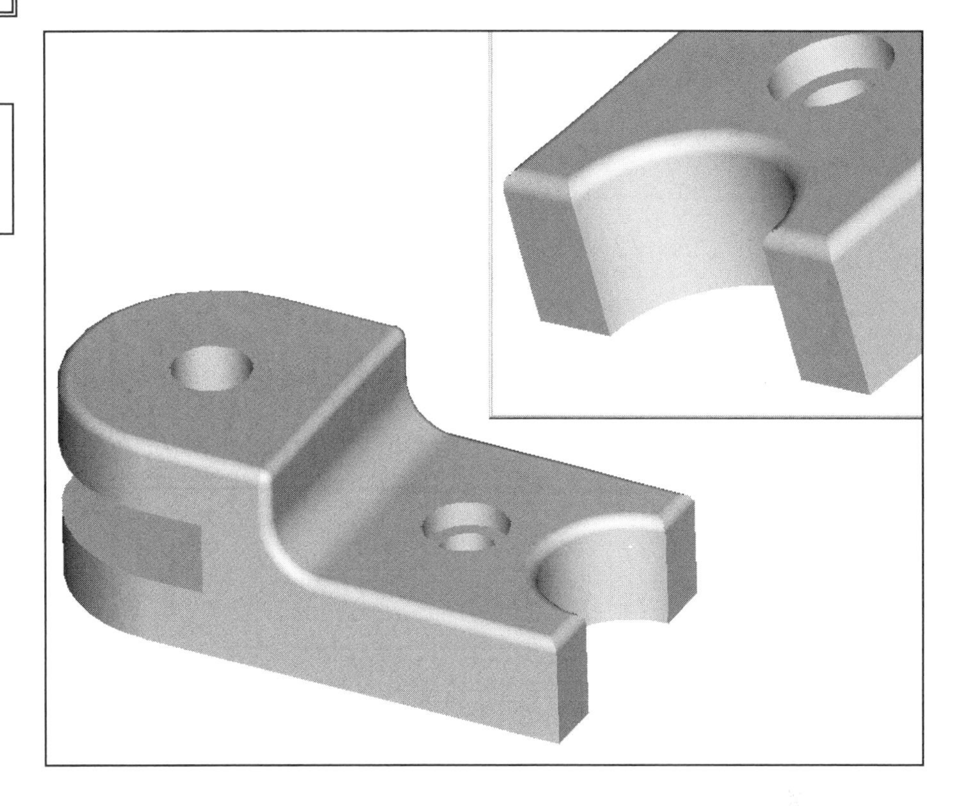

Figure 3.1
Breaker

EGD REFERENCE
Fundamentals of Engineering Graphics and Design
by L. Lamit and K. Kitto
Read Chapter 10.
See pages 286-341, 364, and 570.

COAch™ for Pro/ENGINEER

If you have **COAch for Pro/ENGINEER** on your system, go to SEARCH and do the Segments shown in Figure 3.4 and Figure 3.6.

OBJECTIVES

1. **Create simple rounds along model edges**

2. **Sketch arcs on sections**

3. **Create a straight hole through a part**

4. **Complete a sketched hole**

5. **Understand the difference between sketched and pick-and-place features**

6. **Understand the options for specifying hole depth, including Blind, Thru Next, Thru All, and Thru Until**

7. **Use the Info command to extract feature information**

8. **Understand the types of round creation options**

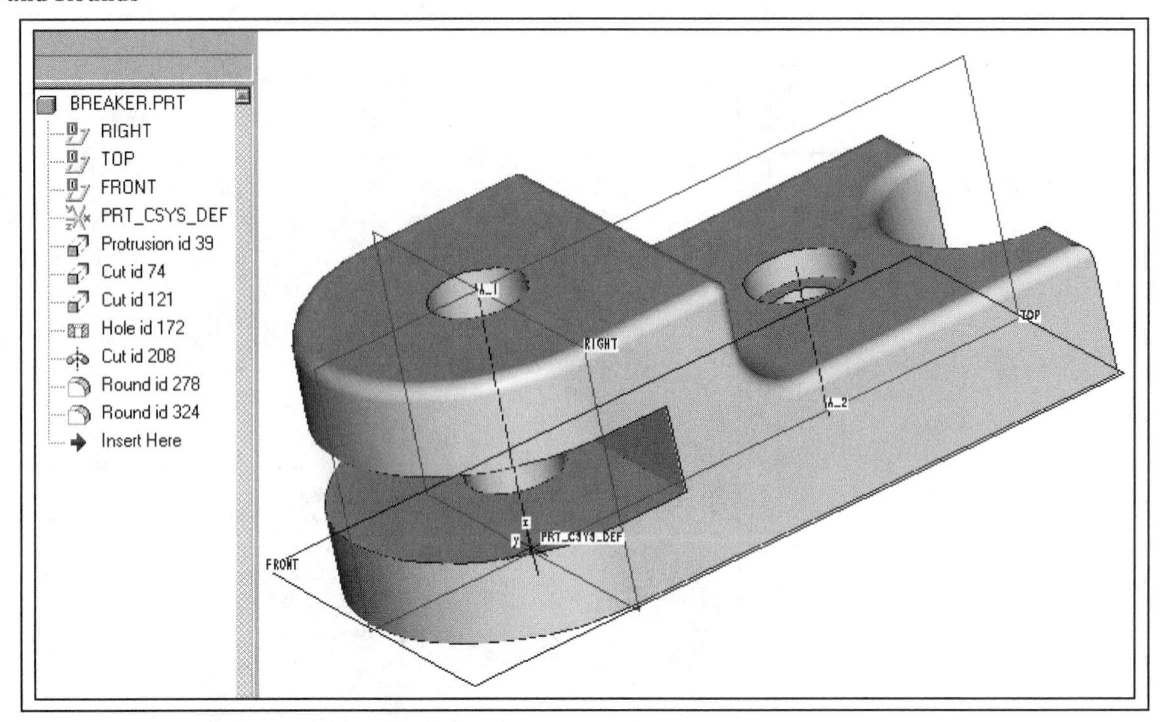

Figure 3.2
Breaker with Axes, Datums, Coordinate System, and Model Tree

HOLES AND ROUNDS

A variety of geometric shapes and constructions are accomplished automatically with Pro/E, including *holes* and *rounds*. These features are called *pick-and-place* features, because they are created automatically from your input and then placed according to prompts by Pro/E. A hole can also be created using **Cut,** but it must be sketched. In general, pick-and-place features are not sketched (except for the **Sketched** option when you are creating a complex hole shape such as a countersink hole or counterbore, as in Figs. 3.1 and 3.2). **Round** creates a fillet, or a round on an edge, that is a smooth transition with a circular profile between two adjacent surfaces.

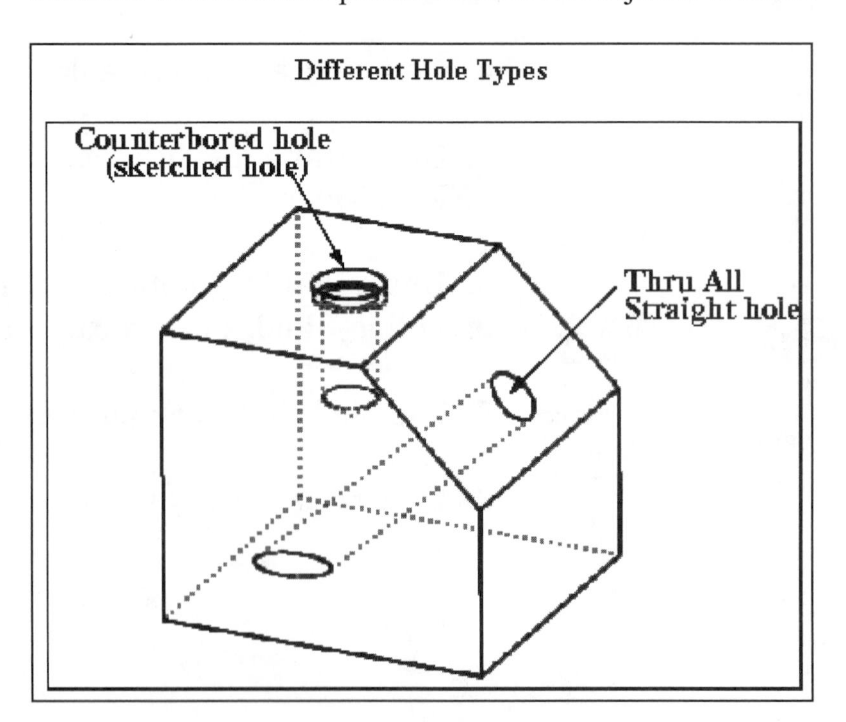

Figure 3.3
Online Documentation, Holes

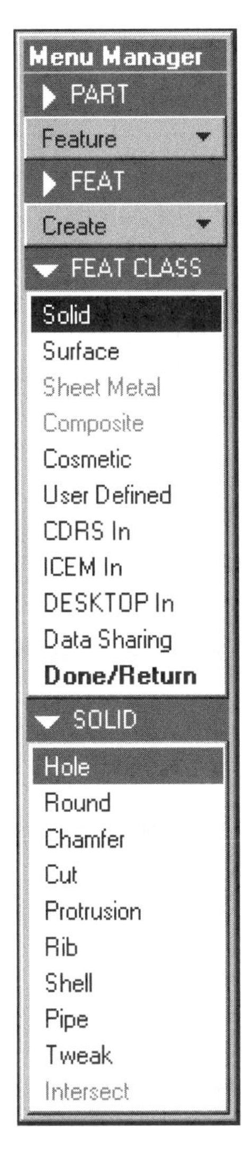

Holes

The **Hole** option creates a variety of holes. Figure 3.3 shows a small sample of online help available for the **Hole** command. Types of hole geometry include:

> **Straight hole** An extruded slot with a circular section. Passes from the placement surface to the specified end surface.
> **Sketched hole** A revolved feature defined by a sketched section. Counterbore and countersink holes, for example, are created as sketched holes.
> **Standard hole** A revolved feature created with UNC, UNF, or ISO standards.

Straight Holes

All straight holes are created with a constant diameter. To create a straight hole:

1. Choose **Hole** from the SOLID menu.
2. HOLE dialog box is displayed. Default is ●**Straight hole.**
3. Select the placement plane.
4. Choose one of the options: **Linear, Radial, Diameter,** or **Coaxial.**
5. Select the first reference ↖ (pick an edge, axis, planar surface, or datum).
6. Enter the distance from the first reference ⇒ ✔.
7. Select the second reference ↖.
8. Enter the distance from the second reference ⇒ ✔.
9. Select the extent to which the hole will be created. The Depth list box includes:

   Variable Creates a hole with a flat bottom at a specified depth.
   Thru Next Creates a hole that continues until it reaches the next part surface.
   Thru All Creates a hole that intersects all the surfaces.
   Thru Until Creates a hole that goes through all the surfaces until it intersects with the specified surface.
   To Reference Creates a hole with a flat bottom that continues until it reaches the specified point or vertex.
   None (Depth Two only) No extrusion in Side Two direction.
   Symmetric (Depth Two only) Creates a two-sided hole of equal depth on both sides of the sketch plane.

10. Enter the depth or select the reference, if necessary.
11. Repeat steps 9 and 10 for the other direction.
12. Enter the diameter of the hole.
13. Select ✔ button in the dialog box to create the hole.

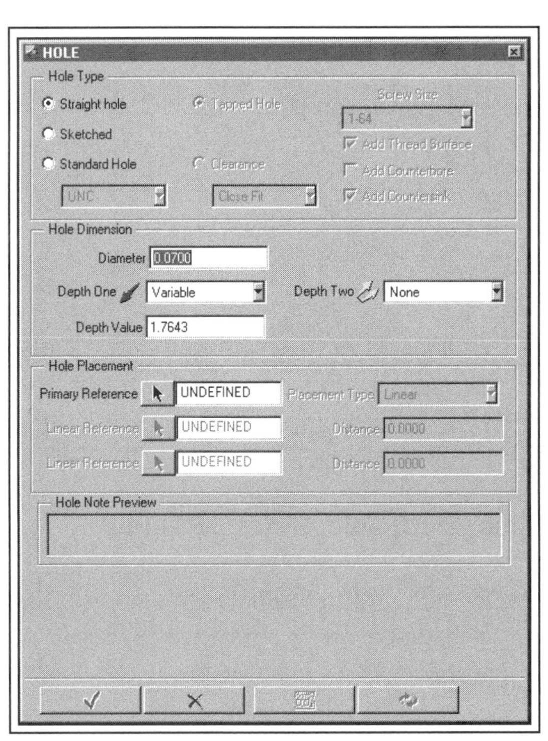

Sketched Holes

A *sketched hole* is created by sketching a section for revolution and then placing the hole on the part. Sketched holes are always blind and one-sided. Sketched holes must have a vertical centerline (*axis of revolution*), with at least one of the entities sketched normal to the axis centerline. Pro/E aligns the normal entity with the placement plane. The remainder of the sketched feature is cut from the part, as with a revolved cut. You can also use the revolved cut command to create holes. If you have **COAch for Pro/ENGINEER** on your system, go to SEARCH and do the Sketched Holes segment as shown in Figure 3.4.

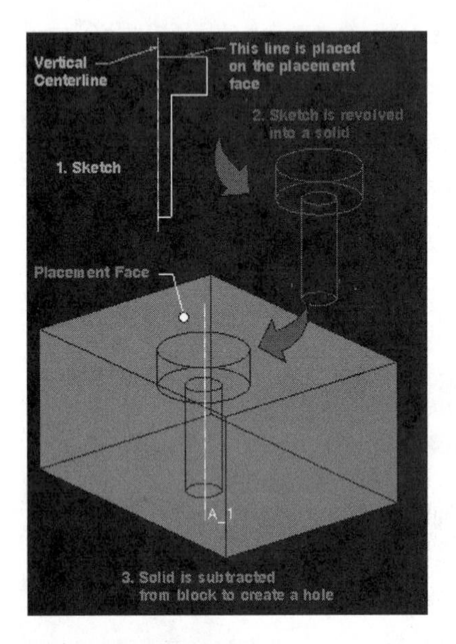

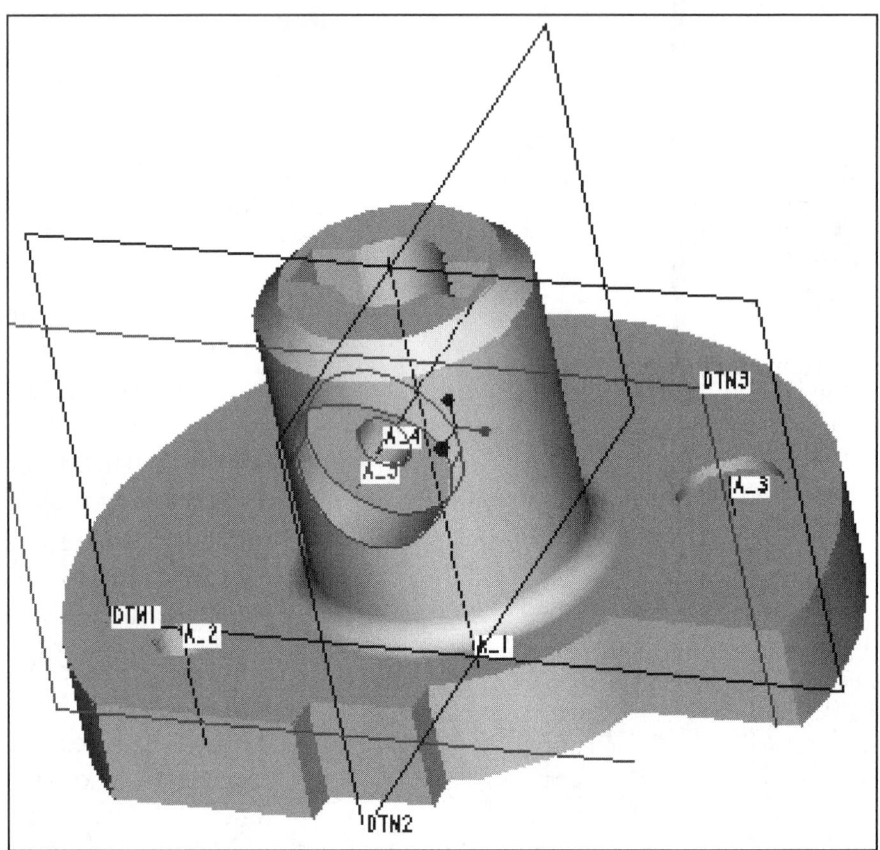

Figure 3.4
COAch for Pro/E, Holes (Sketched Holes)

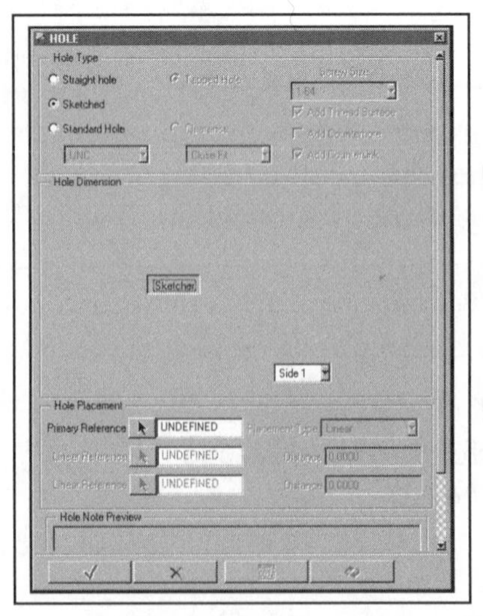

To create a sketched hole:

1. Choose **Hole** from the SOLID menu. Choose ● **Sketched** from the HOLE dialog box.
2. Sketcher subwindow displays.
3. Sketch and dimension the hole geometry (sketched holes must have a vertical centerline); then ✔.
4. Choose the placement scheme for the hole: **Linear, Radial, Diameter,** or **Coaxial.**
5. Select the Primary Reference↖ and fill in the Distance as required.
6. Linear Reference↖**.** Enter the distance for the linear references.
7. Choose ✔ to complete the feature.

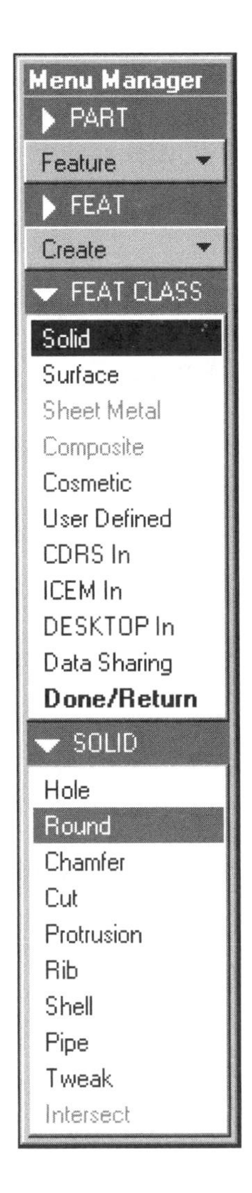

Rounds

Rounds (Fig. 3.5) are created at selected edges of the part. Tangent arcs are introduced as rounds between two adjacent surfaces of the solid model. There are cases in which rounds should be added early, but in general, wait until later in the design process to add the rounds. Introducing rounds into a complex design early in the project can cause a series of failures later in the design. You may choose to place all rounds on a layer and suppress that layer to speed up your working session. There are a variety of rounds to consider, including **Edge Chain, Surf**(ace)**-Surf**(ace), **Edge-Surf**(ace), and **Edge Pair.**

Reference Type	Original Geometry	Rounded Geometry
(a) Edge Chain/ One By One	Select these edges.	Resulting round
(b) Edge Chain/ Tangnt Chain	An edge from a tangent chain	Resulting round
(c) Surf-Surf	Select these two surfaces	Resulting round

Figure 3.5
Rounds

Two categories of rounds are available: simple and advanced. Much of the time, you will create *simple* rounds. These rounds smooth the hard edges between two adjacent surfaces. If you have **COAch for Pro/ENGINEER** on your system, go to SEARCH and do the appropriate segment for Rounds, shown in Figure 3.6.

Creating a Simple Round

Figures 3.7 (edge pair) and 3.8 (advanced variable) show some of the online documentation available on rounds. You should read your manual, or highlight the **Round** option in the menu, press the right mouse button, and choose **? GetHelp**.

The basic steps to create a simple round include:

1. Choose **Round** from the SOLID menu. (Note that the **Rounds Tutor** may open. See the left column on the next page.)
2. A dialog box appears, listing elements of the round feature.
3. Choose **Simple** and **Done** from the ROUND TYPE menu.
4. The Attributes element is selected by default. Use the RND SET ATTR menu options to specify the round's attributes.

Choose the command **Help ⇒ ✔Round Tutor** from the menu bar. Round Tutor will automatically open when you choose the **Round** command: **Feature ⇒ Create ⇒ Round.**

Each time you use the **round** option, spend some time investigating the Round Tutor and Online Documentation to deepen your knowledge of rounds.

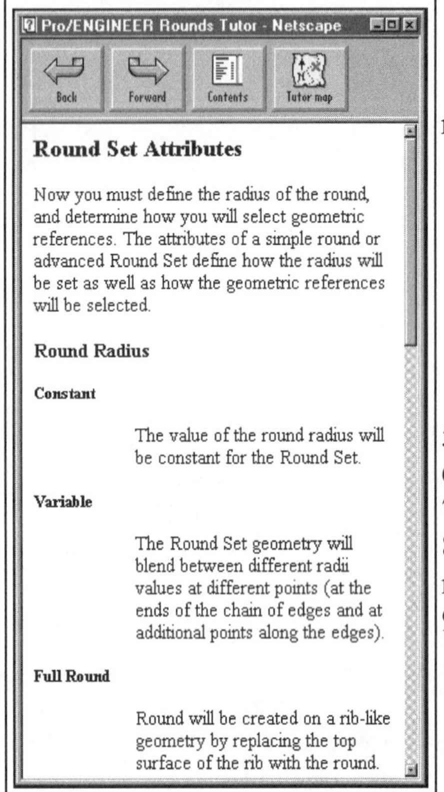

Specify the type of round by selecting one of these options:

Constant Creates a round between two sets of surfaces with a constant radius.
Variable Creates a round between two sets of surfaces with variable radii. Specify radii at the ends of the chain of edges or at the ends of the spine (when the spine is required) and, optionally, at additional points along the edges or along the spine.
Full Round Creates a round by removing a surface; the consumed surface becomes a round.
Thru Curve Creates a round through a selected curve.

Select one of the following options to specify the type of references for placing the round:

Edge Chain Places a round by selecting a chain of edges. To select the chain, use options in the CHAIN menu.
Surf-Surf Places a round by selecting two adjacent surfaces.
Edge-Surf Places a round by specifying an edge and a tangent surface.
Edge Pair Places a full round by specifying a pair of edges.

5. Choose **Done** from the RND SET ATTR menu.
6. Pro/E prompts you to select the placement references.
7. For other than a full round, enter the radius for the round.
8. For other than a full round, define the extension boundaries of the round by specifying the **Round Extent** element (optional).
9. If required, define the **Attach Type** element.
10. Choose **OK** from the dialog box.

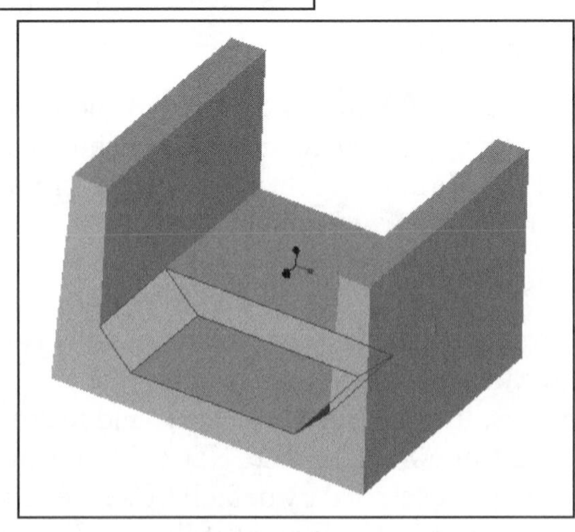

Figure 3.6
COAch for Pro/E, Rounds (Overview)

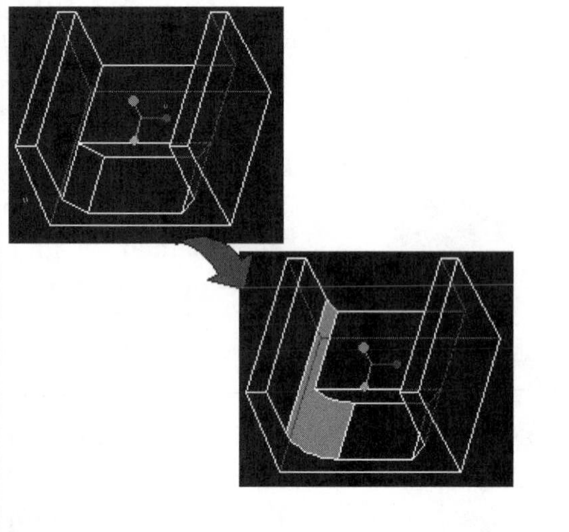

Figure 5-5

Surf-Edge Rounds are created between a surface and an edge - only one face will exist after the Round is created. This is an effective way of creating "cliff-edge" Rounds...

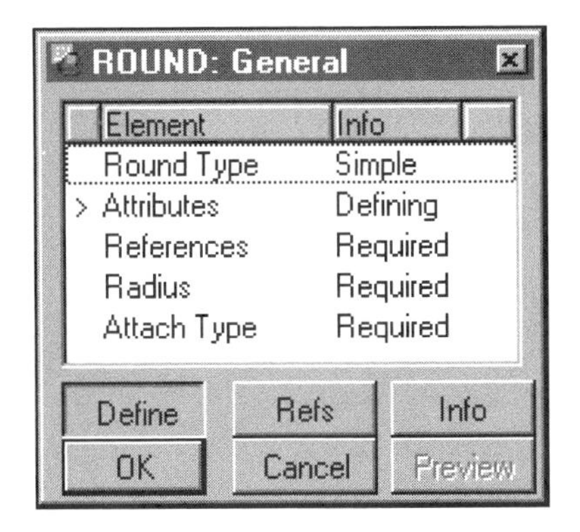

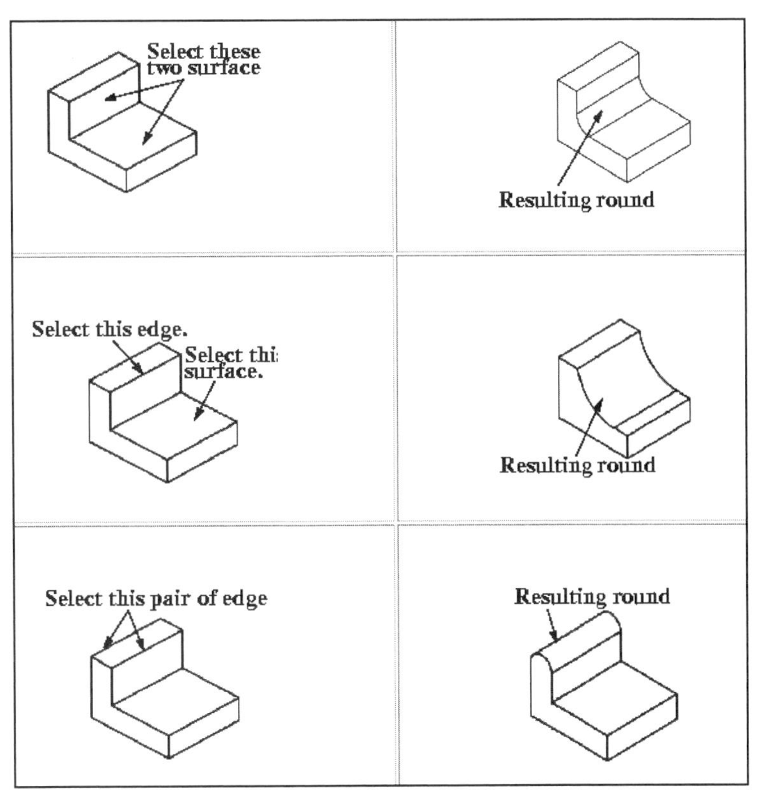

Figure 3.7
Online Documentation, Rounds

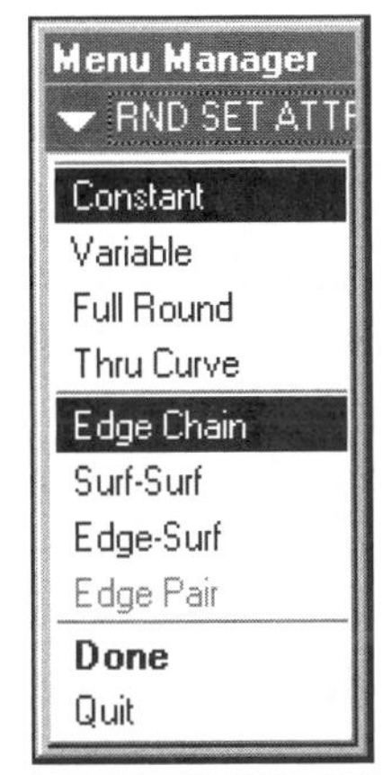

Figure 3.8
Variable-Radius Rounds

Using the Chain Menu Options

When you select reference edges with the **Edge Chain** option, Pro/E displays the CHAIN menu. Note that you can choose more than one option: choose an option, select the references as prompted by Pro/E, and then choose the next option. The CHAIN menu lists the following options:

One By One Define a chain one at a time by selecting individual edges and curves.
Tangnt Chain Define a chain by selecting an edge. All tangent edges are included in the selection.
Surf Chain Define a chain of edges by selecting a surface.
Intent Chain Define a chain by selecting multiple associated edges or surfaces.
Unselect Unselect references from one of the preceding options.

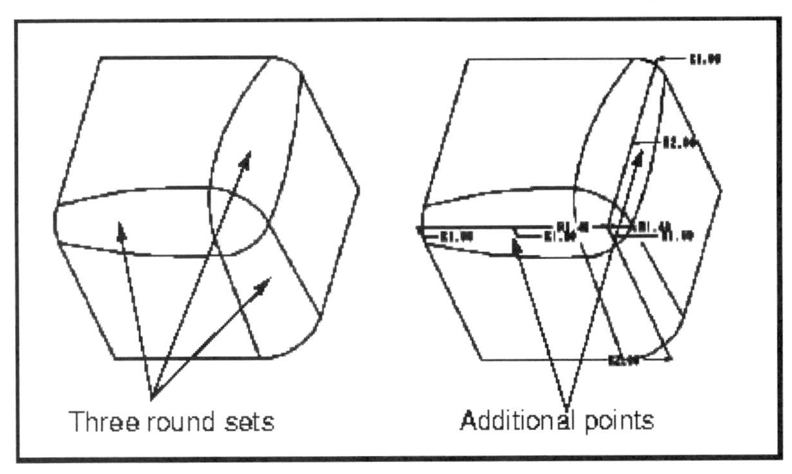

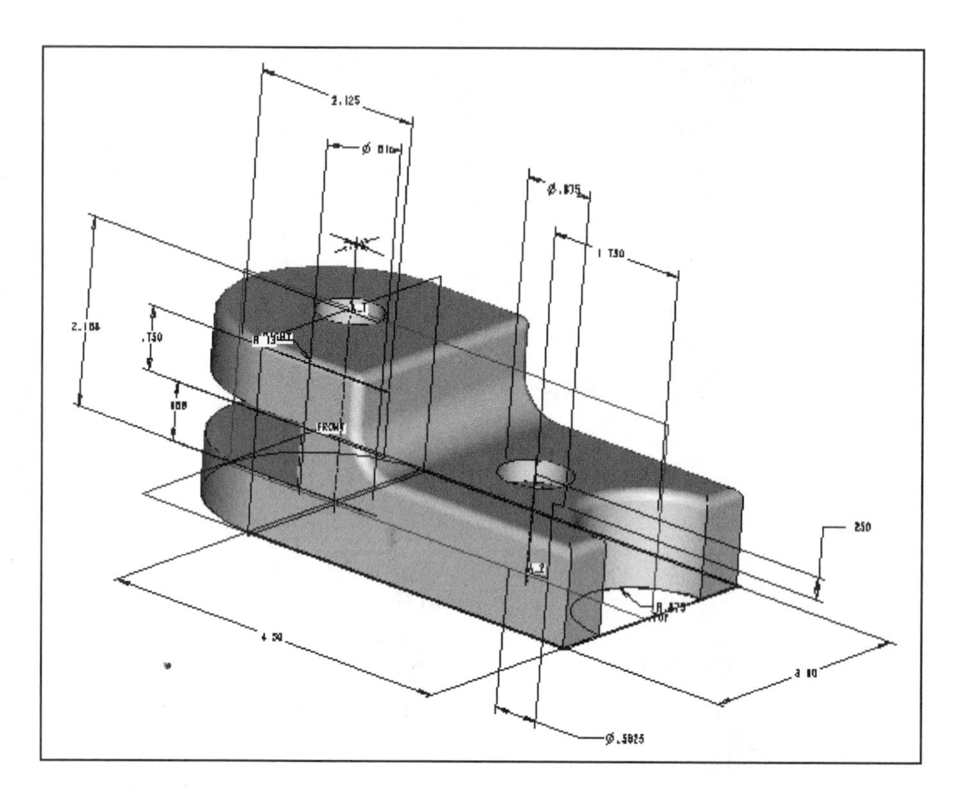

Figure 3.9
Breaker

Breaker

Lesson procedures & commands
START HERE ➡ ➡ ➡

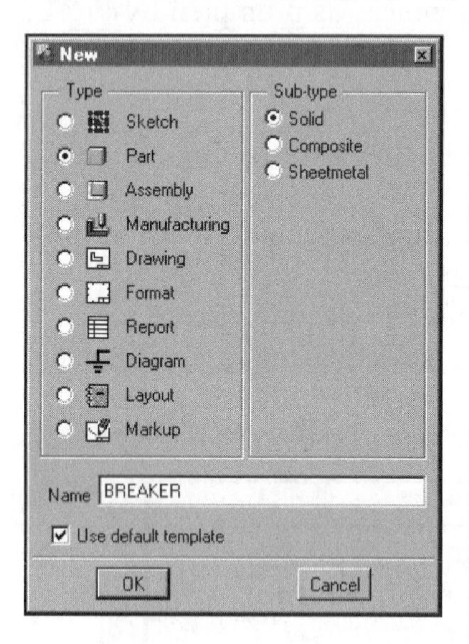

The **Breaker** (Fig. 3.9) is the third ***lesson part.*** This part introduces two new features, *holes* and *rounds*. Also, in the Sketcher, the **Arc** command will be used to create the rounded end and the half-circle cut of the Breaker. Choose **Part** from the Type options. Choose the following commands:

File ⇒ **New** ⇒ Type **●Part** ⇒ Sub-type **●Solid** ⇒ Name **BREAKER** ⇒ **✔Use default template** ⇒ **OK**

As always, the *environment*, *units*, and *material* for the part need to be established:

SETUP AND ENVIRONMENT

Set Up ⇒ **Units** ⇒ Units Manager **Inch lbm Second** (Pro/E default) ⇒ **Close** ⇒

Material ⇒ **Define** ⇒ (Type **ALUMINUM**) ⇒ **✔**⇒ (table of material properties, change or add information) ⇒ **File** ⇒ **Save** ⇒ **File** ⇒ **Exit** ⇒ **Assign** ⇒ (pick **ALUMINUM**) ⇒ **Accept** ⇒ **Done**

Utilities ⇒ **Environment** ⇒ **✔ Snap to Grid**
Display Style **Hidden Line**
Tangent Edges **Solid** (try this for a different look) ⇒ **OK** ⇒

Utilites ⇒ **Preferences** ⇒ Option: (type in) ***sketcher_dec_places*** ⇒ Value: (type in) **3** ⇒ **Add/Change** ⇒ **Apply** ⇒ **Close**

The first protrusion for the Breaker is created using an extruded protrusion and sketching (on **FRONT)** the outline of the part as seen from its top. Choose the following commands to set up the section for the sketch:

Feature ⇒ Create ⇒ Protrusion ⇒ Extrude ⇒ Solid ⇒ Done ⇒ One Side ⇒ Done ⇒ (pick **FRONT** as the sketching plane) **⇒ Okay** (for selecting the *direction* of feature creation) **⇒ Top ⇒** (pick **TOP** as the *orientation* plane) **⇒ Close** (to accept default References) **⇒ Sketch** (menu bar) **⇒ Intent Manager** (off)

Though it is not really necessary for the sketching of this section, sometimes the grid spacing needs to be altered to a different size. Change the size of the grid spacing by choosing the following commands:

Sec Tools ⇒ Sec Environ ⇒ Grid ⇒ Params ⇒ X&Y Spacing ⇒ (type **15** at the prompt) **⇒ ✔ ⇒ Done/Return**

Your screen should look similar to Figure 3.10. The grid will now be twice as dense.

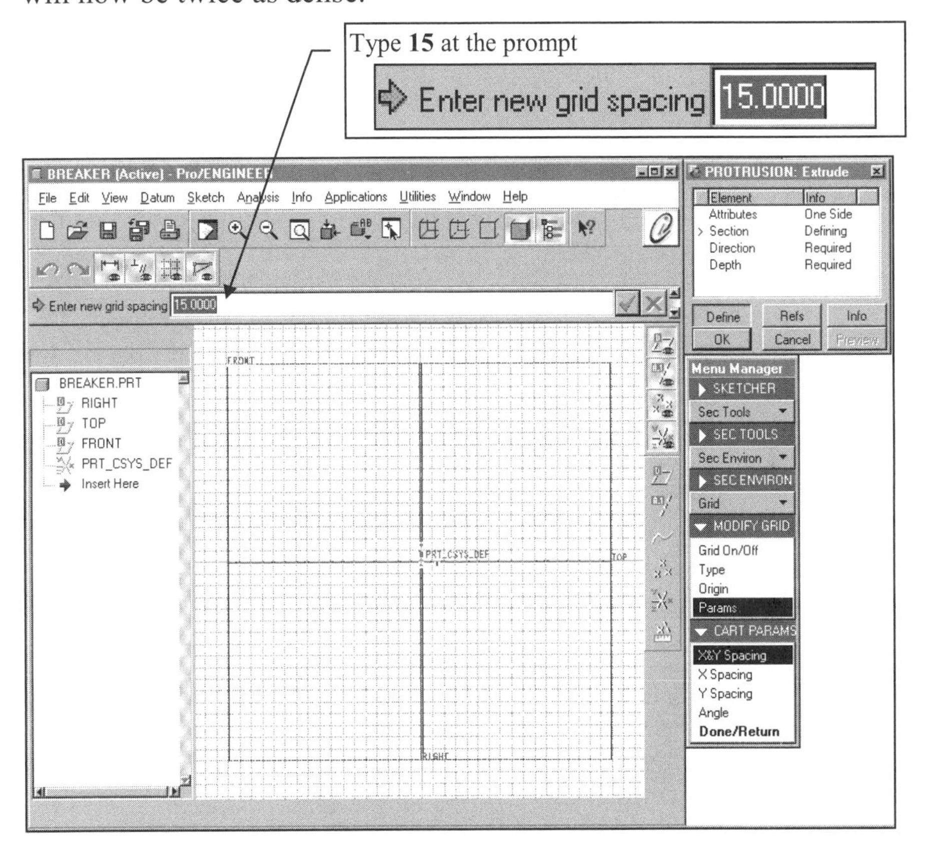

Figure 3.10
Grid Size Changed to **15** Units

You will sketch the section by creating an arc representing the part's curved end. The origin of the arc will be at the intersection of **RIGHT** and **TOP**, as shown in Figure 3.11. The section will be composed of two arcs, four lines, and a horizontal centerline. Align the large arc with **RIGHT** and **TOP**, and align the small arc and centerline to **TOP.**

Create the arcs first (Fig. 3.11), and then add the lines and the centerline (Fig. 3.12). Choose the following commands:

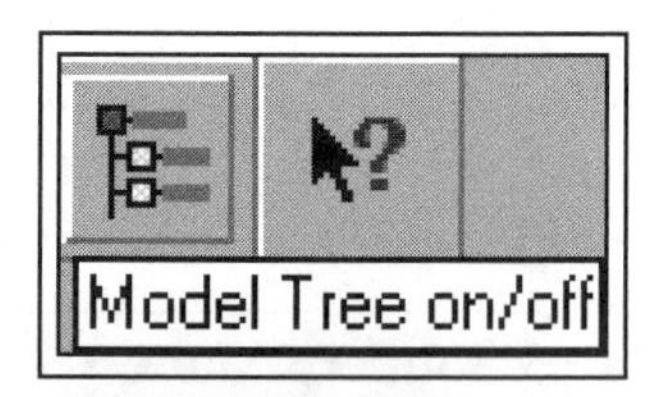

Model Tree (toggle icon off) ⇒ **Sketch** ⇒ **Arc** ⇒ **Center/Ends** (pick the intersection of **RIGHT** and **TOP** as the arc's center; then pick the starting and ending points; repeat the process for the smaller arc) ⇒ **Sketch** ⇒ **Mouse Sketch** ⇒ (create the lines and centerline of the section geometry) ⇒ **Regenerate**

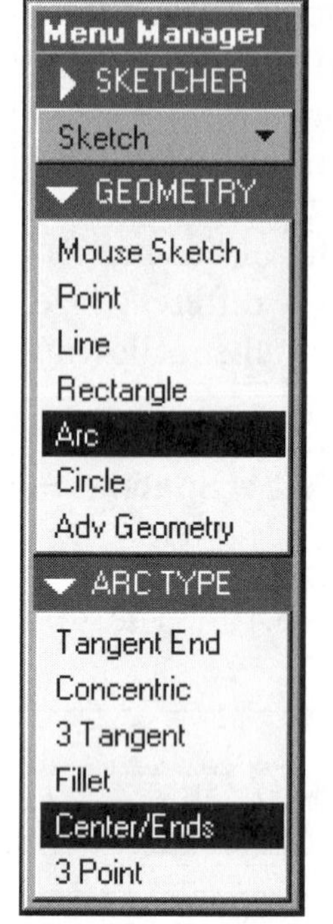

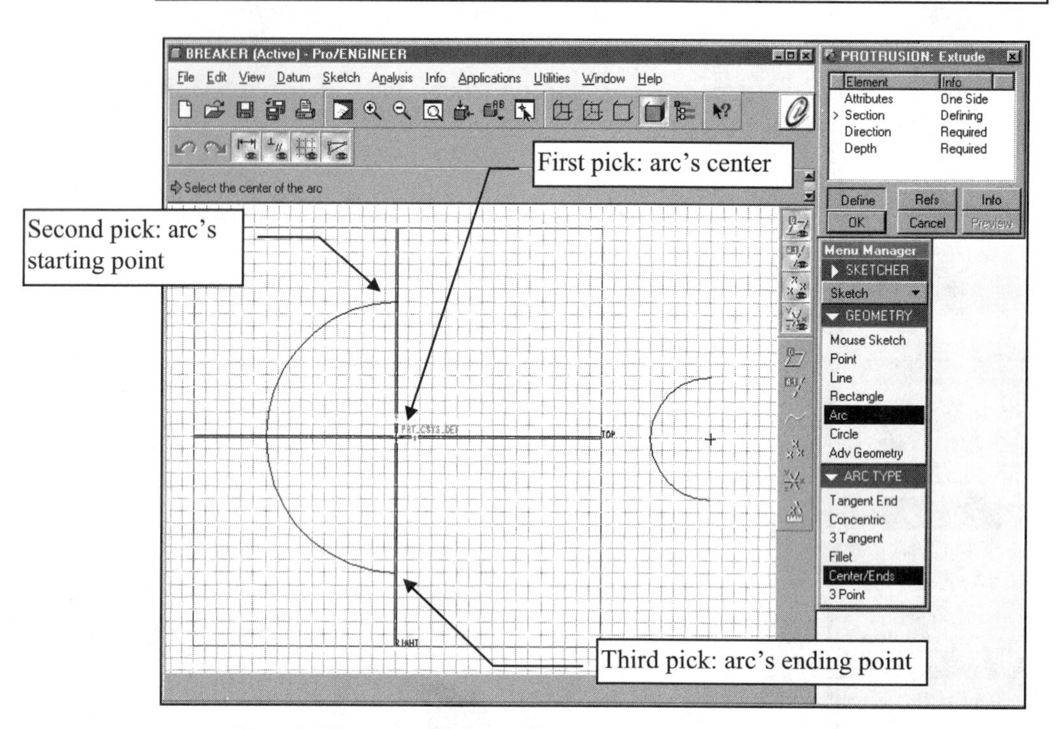

Figure 3.11
Sketched Arcs

Align (**Alignment**) the centerline by double-picking on the entity and datum plane **TOP.** To align the arcs, pick on the arc (or its center point) and then on the datum with which it will be aligned. **Regenerate** again.

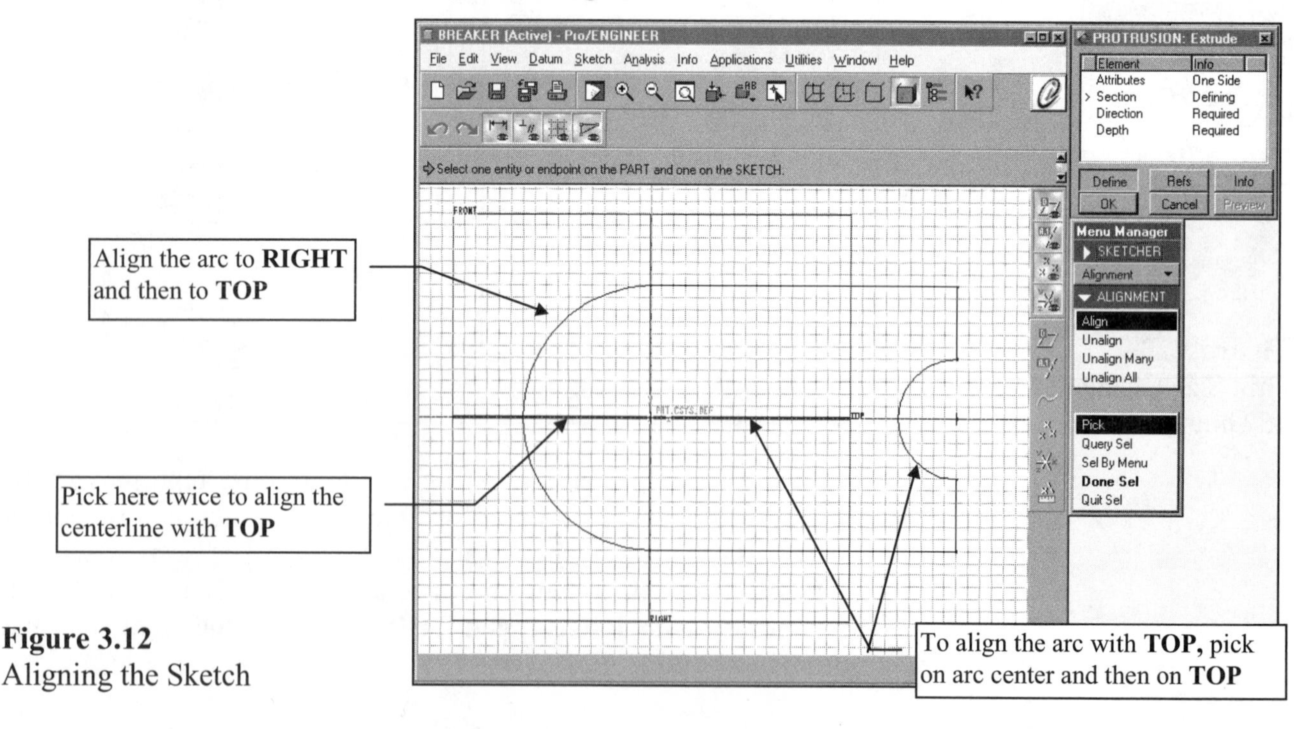

Figure 3.12
Aligning the Sketch

Use the dimensions and dimensioning scheme provided in Figure 3.13. **Dimension**, **Regenerate**, **Modify**, and **Regenerate** the sketch as shown in Figure 3.14. Only three dimensions are required for the successful regeneration of the sketch.

? Pro/HELP

Remember to use the help available on Pro/E by highlighting a command and pressing the right mouse button.

Figure 3.13
Design Intent Dimensions

HINT

You can change the grid size at any point in the sketching process. After the sketch is regenerated, the **15.00** inch grid zooms out of view beyond the **5.00** inch long part model. Change the grid **X** and **Y** spacing to **.25.**

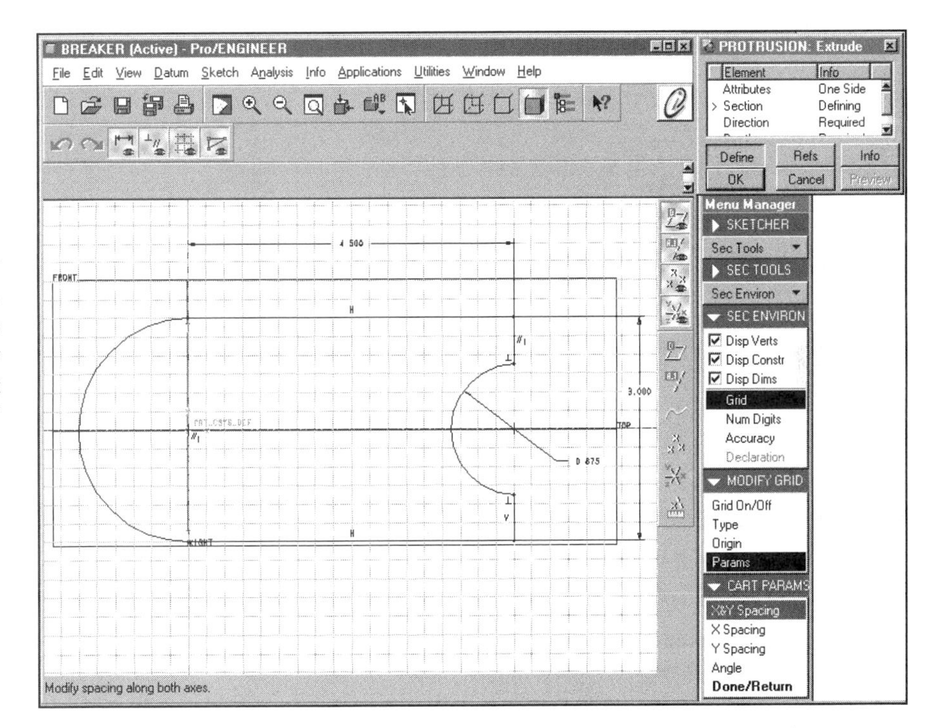

Figure 3.14
Modified and Regenerated Sketch

To complete the feature, choose the following commands:

Regenerate ⇒ **Done** ⇒ **Blind** ⇒ **Done** ⇒ **2.188** (as the depth of protrusion) ⇒ ✔ ⇒ **OK** (from the dialog box) ⇒ **View** ⇒ **Default** (to view the protrusion as in Fig. 3.15) ⇒ **Done** ⇒ **View** ⇒ **Shade** (or use the Shade icon to display the part *and the datum planes* as in Figure 3.16)

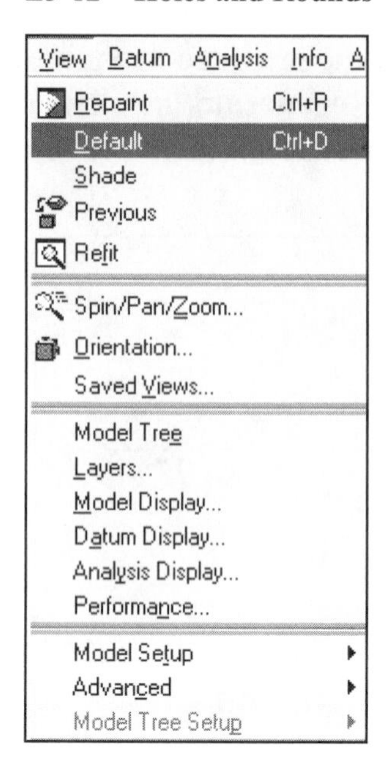

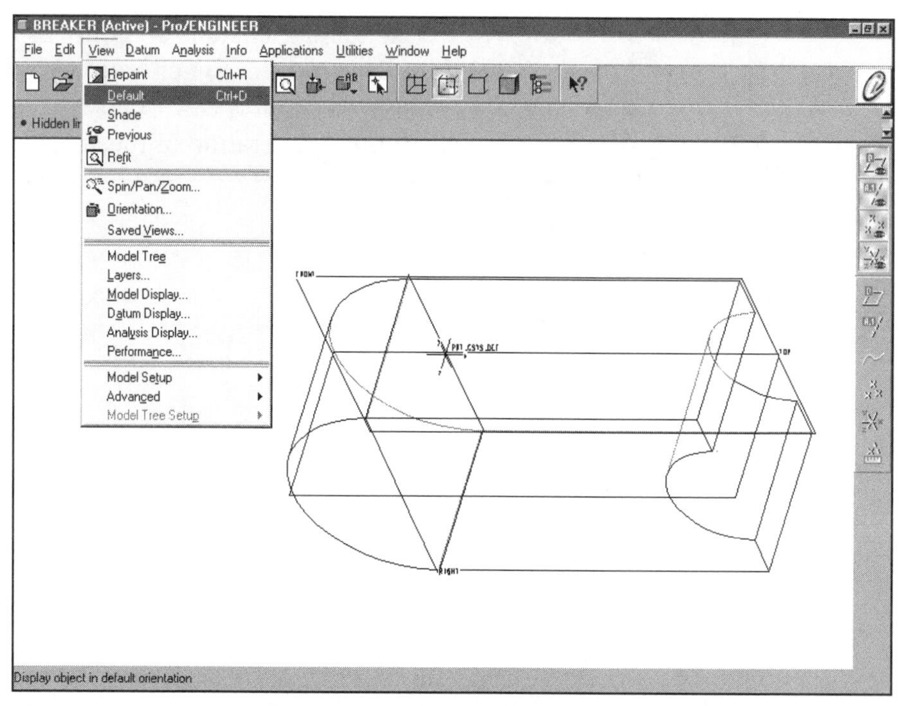

Figure 3.15
Completed Protrusion

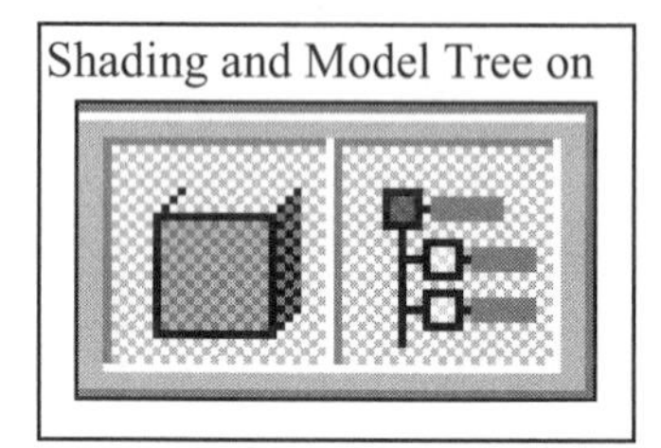

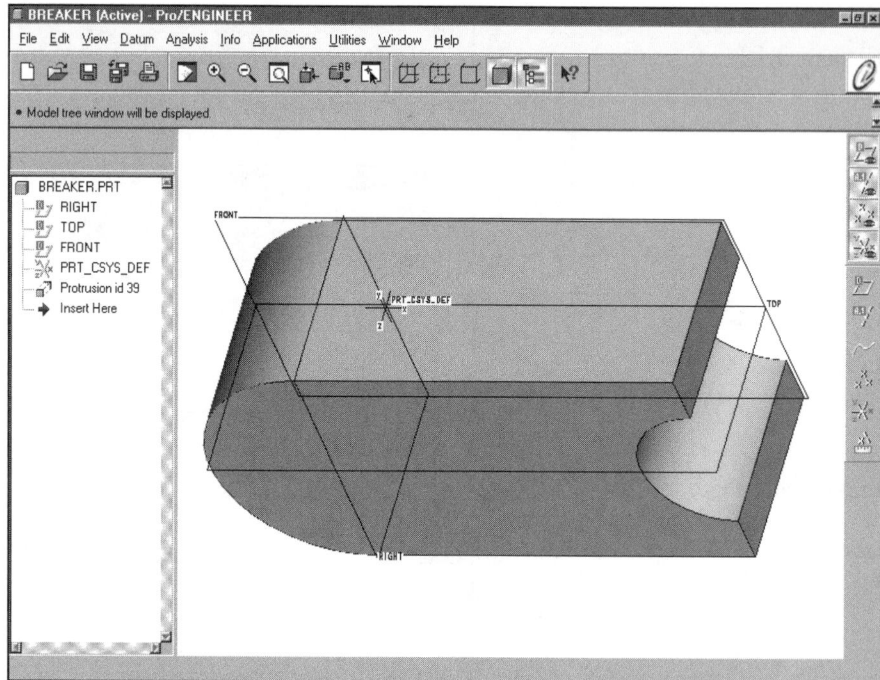

Figure 3.16
Shaded Protrusion

Unless you get in the habit of saving every time you create a feature, you will someday lose your file due to a system crash, etc. Develop good habits now, instead of learning from a bad experience.

File ⇒ Save ⇒ ✔
File ⇒ Delete ⇒ Old Versions ⇒ ✔

The next features will be the cuts created to remove portions of the protrusion. The cuts will complete the primary features of the part. **TOP** is used as the sketching plane for both cuts, and the extruded cuts are made using the option **Both Sides.**

In general, leave *holes* and *rounds* as the final features of the part. Most holes are *pick-and-place* features that are added to the model at a similar step, such as when they are drilled, reamed, or bored during actual manufacturing. In most cases, this means after most of the machining has been completed. Rounds are the very last features created. A good many model failures happen when a set of rounds is being created. Leaving them as the final features reduces the effort needed to resolve modeling problems.

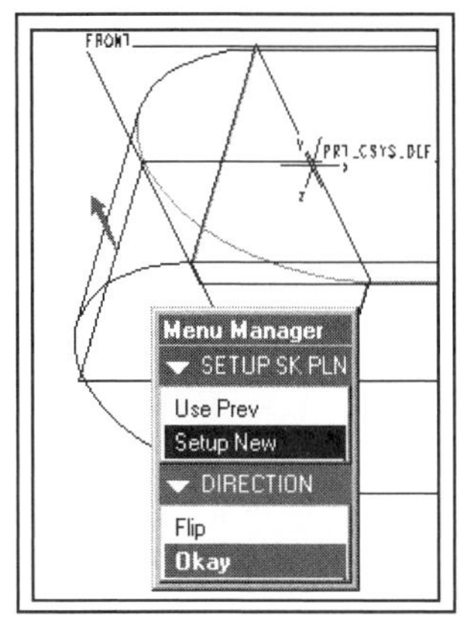

The next feature that will be created is the cut on the top of the part. The dimensions for the cut are shown in Figure 3.17. Choose the following commands:

Hidden Line icon on ⇒ **Feature** ⇒ **Create** ⇒ **Cut** ⇒ **Extrude** ⇒ **Done** ⇒ **Both Sides** ⇒ **Done** (pick **TOP** as the sketching plane) ⇒ **Flip** ⇒ **Okay** ⇒ **Top** (pick **FRONT** to orient the sketch) ⇒ **Close**

If necessary, use **Query Sel** to pick the datum planes (Fig. 3.18). Pro/E now enters the **Sketcher** and displays the part and datum planes, as shown in Figure 3.19.

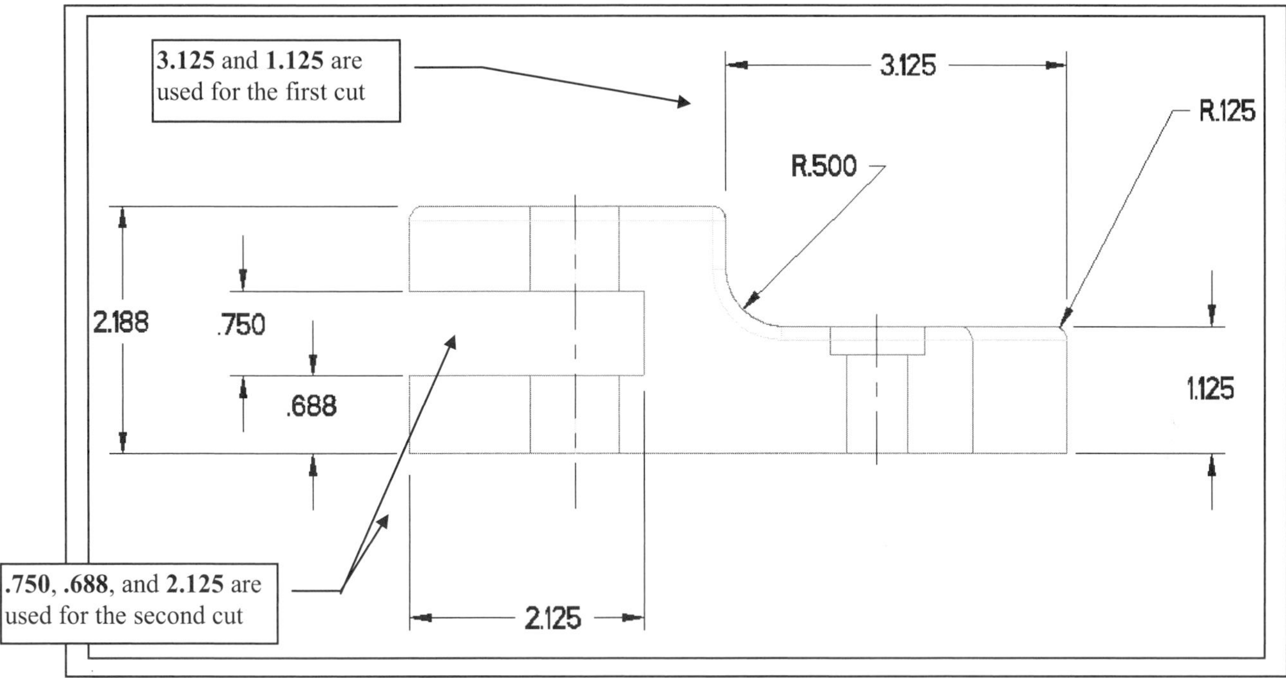

Figure 3.17
Front View of Part

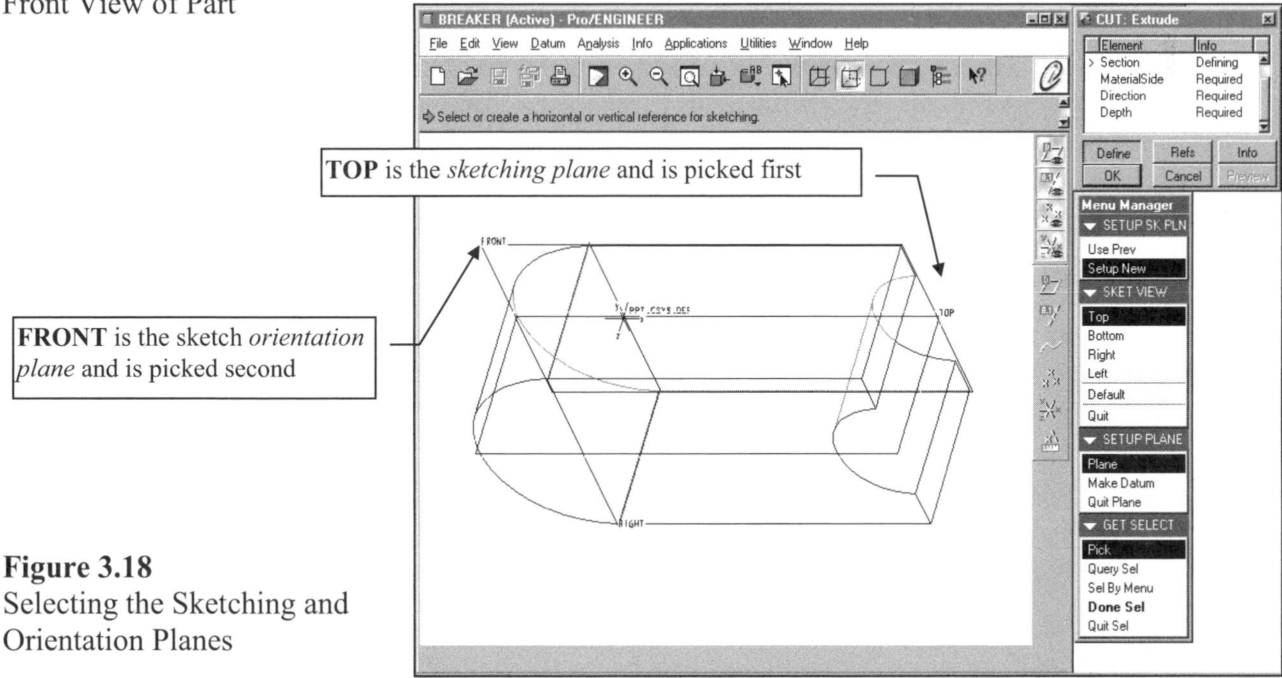

Figure 3.18
Selecting the Sketching and Orientation Planes

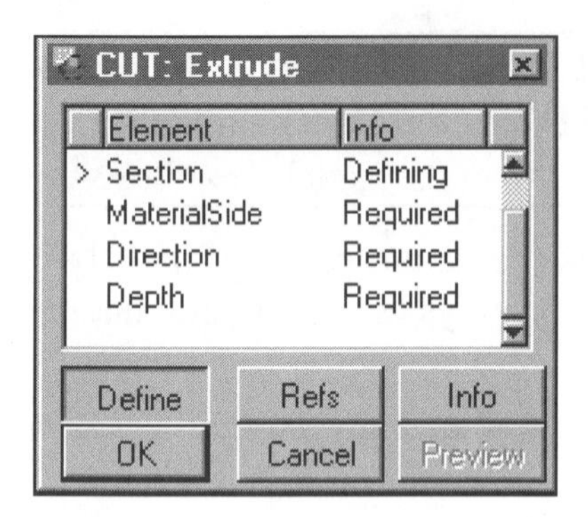

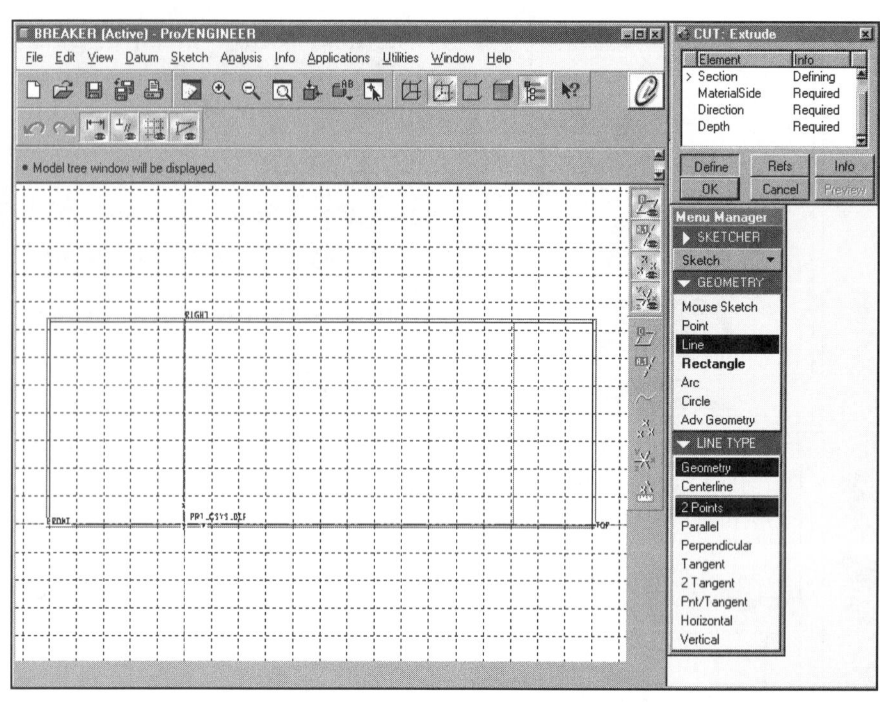

Figure 3.19
Sketcher

If it is still on, turn off the **Spin Center**:
Utilities ⇒ Environment ⇒

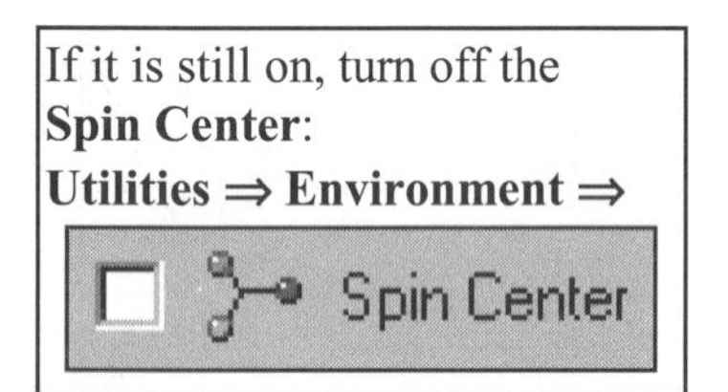

Before you start sketching, turn off the grid display (use the icon) and the Spin Center. Sketch the three endpoints of the two lines as shown in Figure 3.20. Align the two endpoints that touch the part's edges, and **Regenerate** the sketch. **Dimension** (use the dimensions shown in Figure 3.17), **Regenerate**, **Modify**, and **Regenerate** to complete the sketch (Fig. 3.21). Choose the following commands:

Utilities ⇒ Environment ⇒ ❑ Snap to Grid (turn off the Grid Snap) **⇒ OK ⇒ Sketch** (menu bar) **⇒ Intent Manager** (off) **⇒ Sketch ⇒ Line ⇒ Vertical** (pick the three endpoints of the lines) **⇒ Regenerate ⇒ Alignment ⇒ Regenerate ⇒ Dimension ⇒ Regenerate ⇒ Modify ⇒ Regenerate**

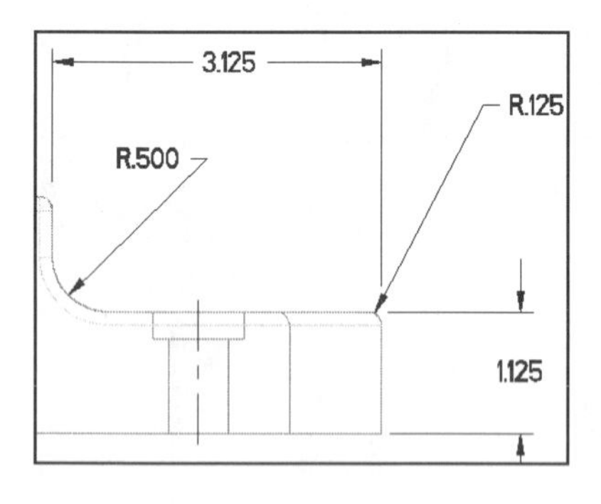

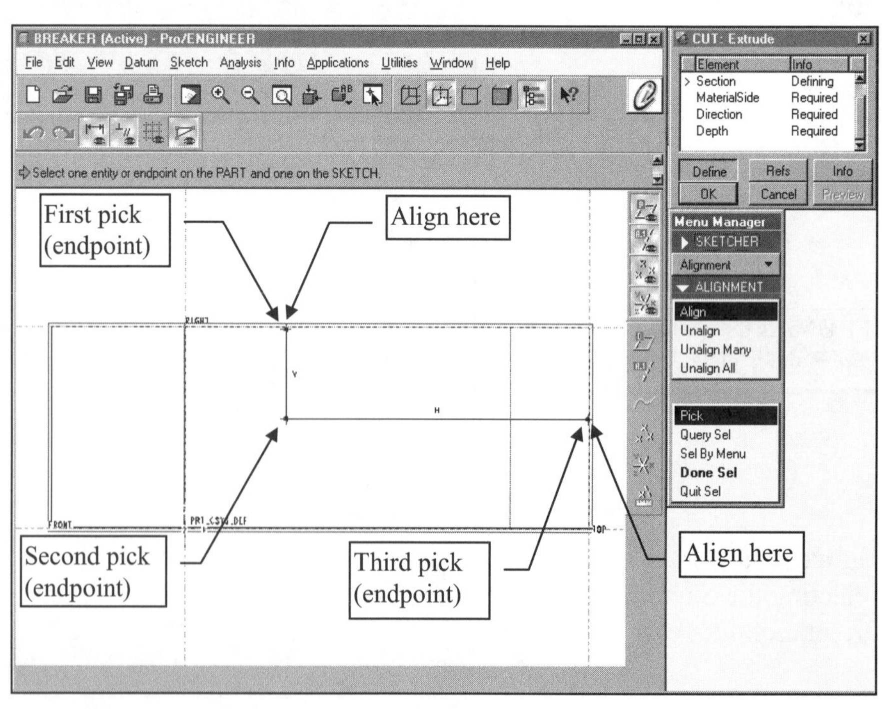

Figure 3.20
Section Sketch of Cut

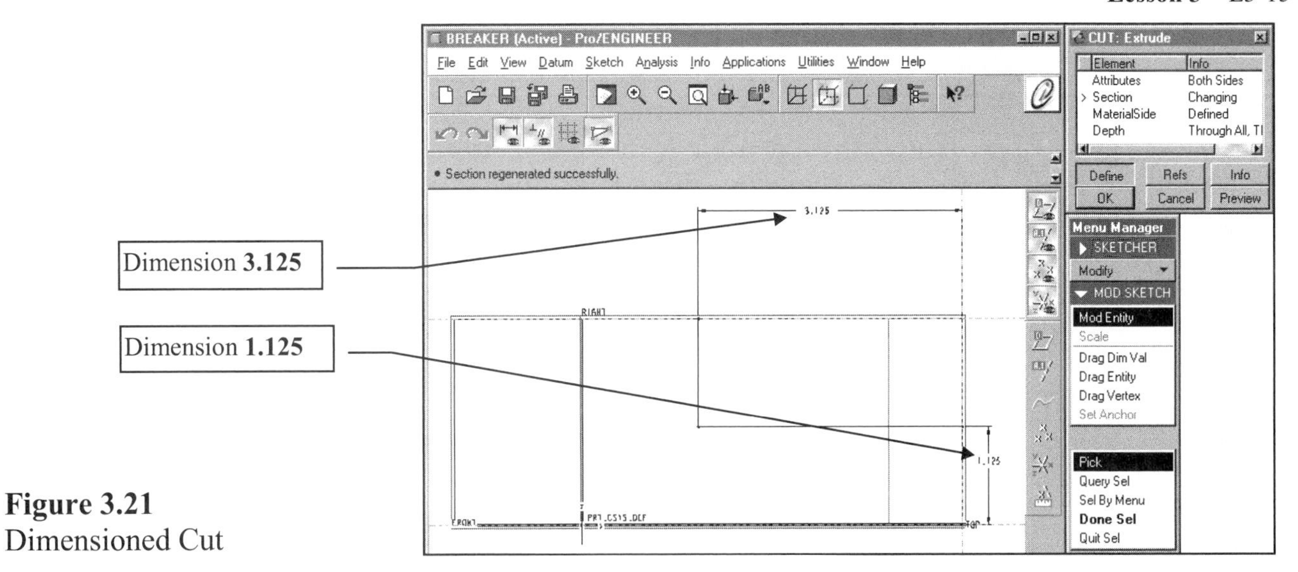

Figure 3.21
Dimensioned Cut

Complete the cut: **Done** (from SKETCHER menu) ⇒ **Okay** (for the material removal direction, as shown in Fig. 3.22) ⇒ **View** ⇒ **Default** ⇒ **Thru All** (first side) ⇒ **Done** ⇒ **Thru All** (second side) ⇒ **Done** ⇒ **Preview** ⇒ **OK** ⇒ **Utilities** ⇒ **Environment** ⇒ **Shading** ⇒ **Isometric** ⇒ **OK** (Fig. 3.23)

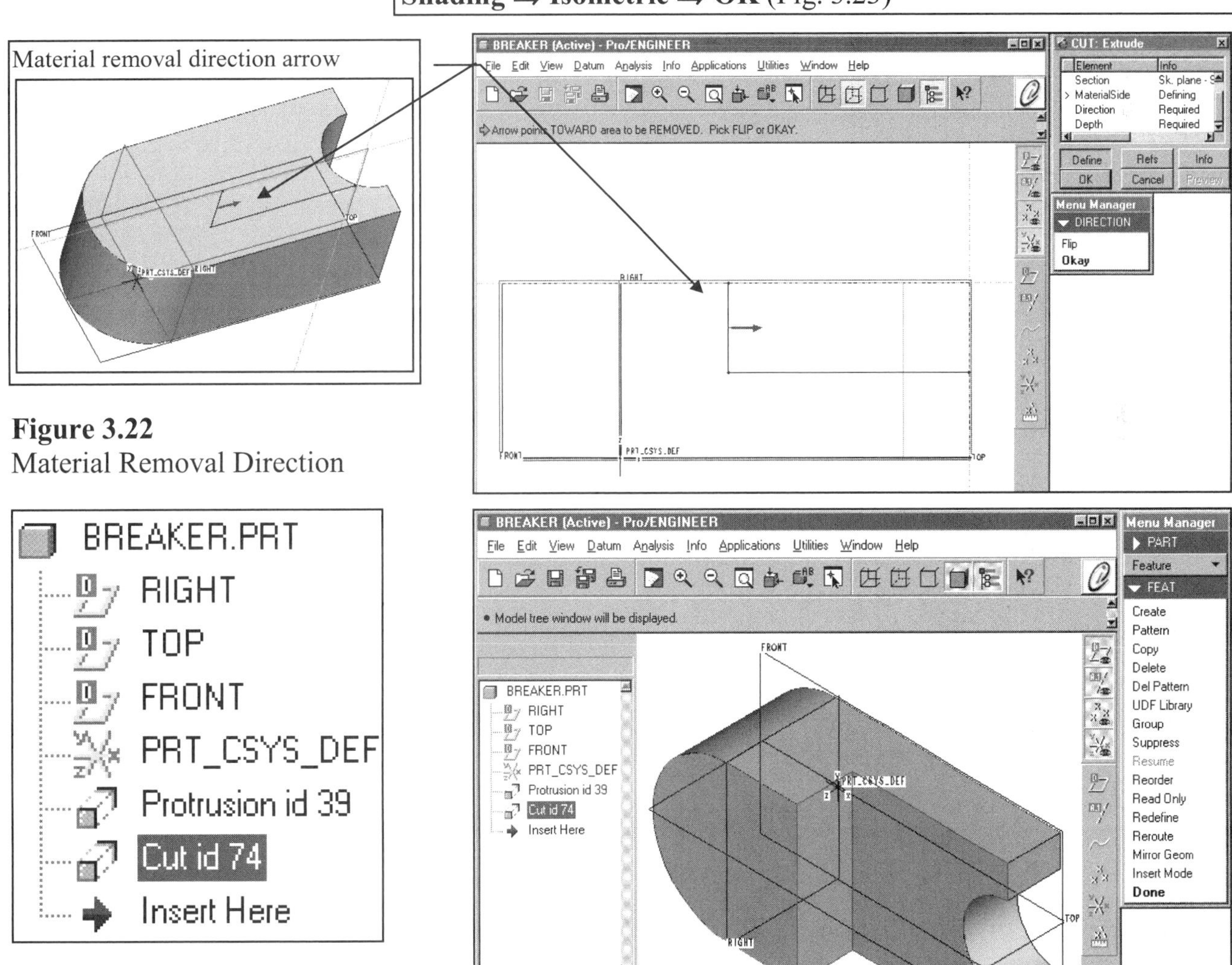

Figure 3.22
Material Removal Direction

Figure 3.23
Completed Cut

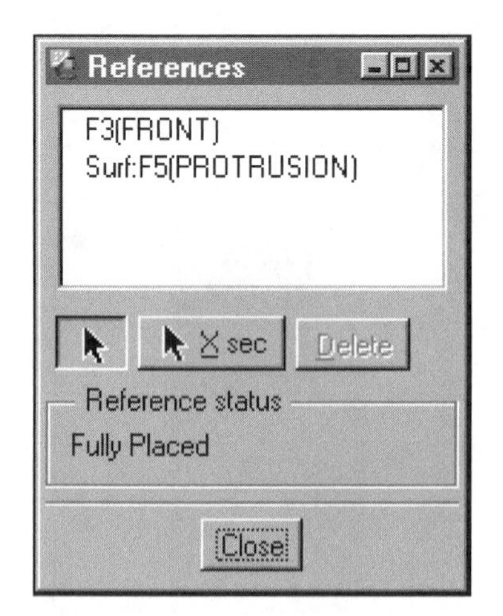

The second cut will use the same sketching plane and orientation. Choose the following command sequence (Figs. 3.24 and 3.25):

Utilities ⇒ **Environment** ⇒ **Hidden Line** ⇒ **OK** ⇒ **Create** ⇒ **Cut** ⇒ **Extrude** ⇒ **Solid** ⇒ **Done** ⇒ **Both Sides** ⇒ **Done** ⇒ **Use Prev** ⇒ **Flip** ⇒ **Okay** ⇒ (Delete **RIGHT** as a Reference and add the circular edge **Surf:F5)** ⇒ **Close** ⇒ (sketch the three lines and modify each by double-clicking on each dimension and changing it to the design value shown in Figure 3.17) ⇒ ✔ ⇒ **Okay** (for the material removal direction) ⇒ **View** ⇒ **Default** ⇒ **Thru All** (first side) ⇒ **Done** ⇒ **Thru All** (second side) ⇒ **Done** ⇒ **OK** ⇒ (Shade icon on) ⇒ **Ctrl** (rotate the part) (Fig. 3.25) ⇒ **Done**

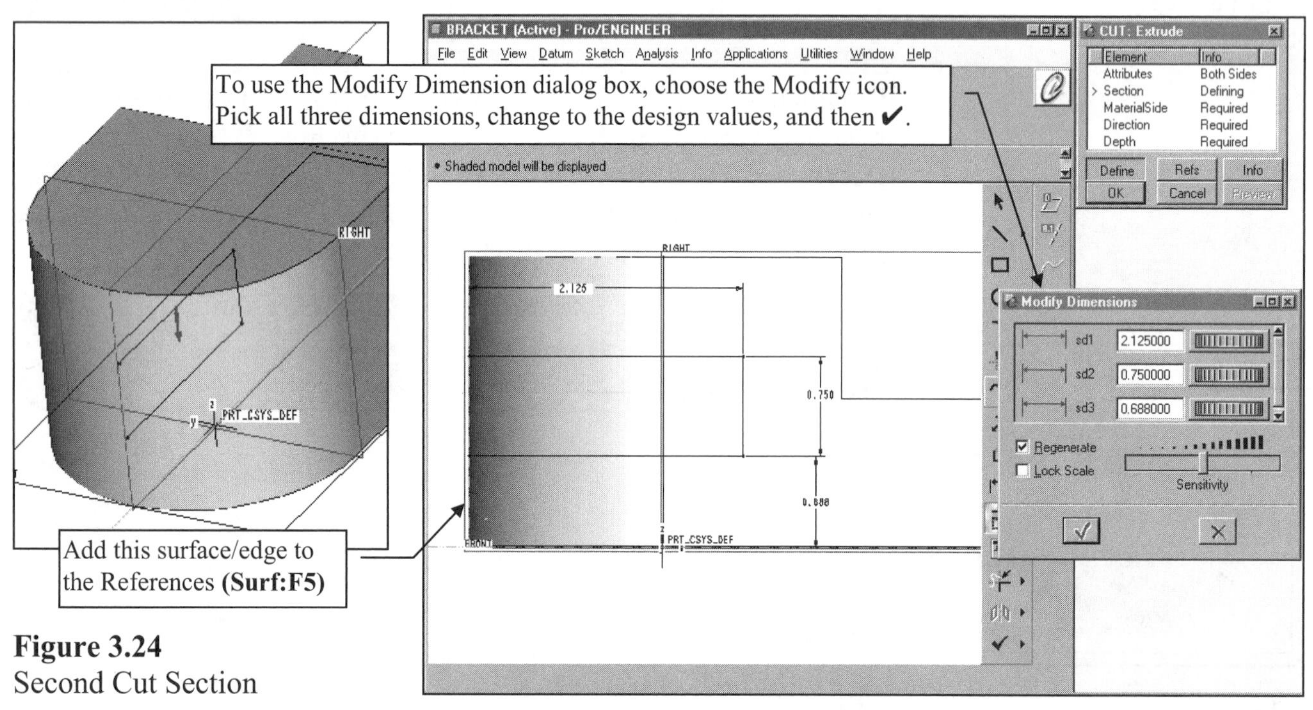

Figure 3.24
Second Cut Section

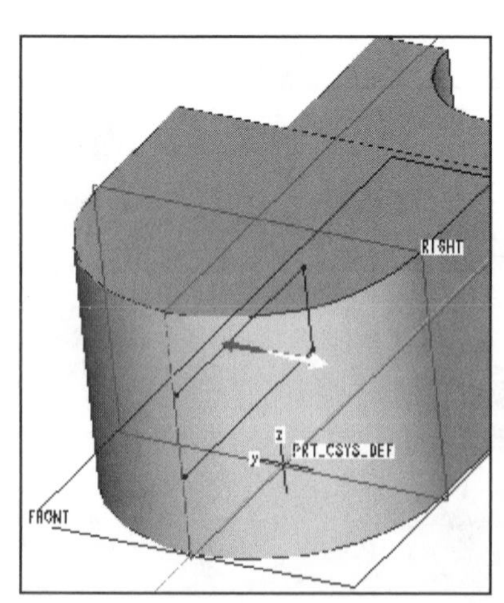

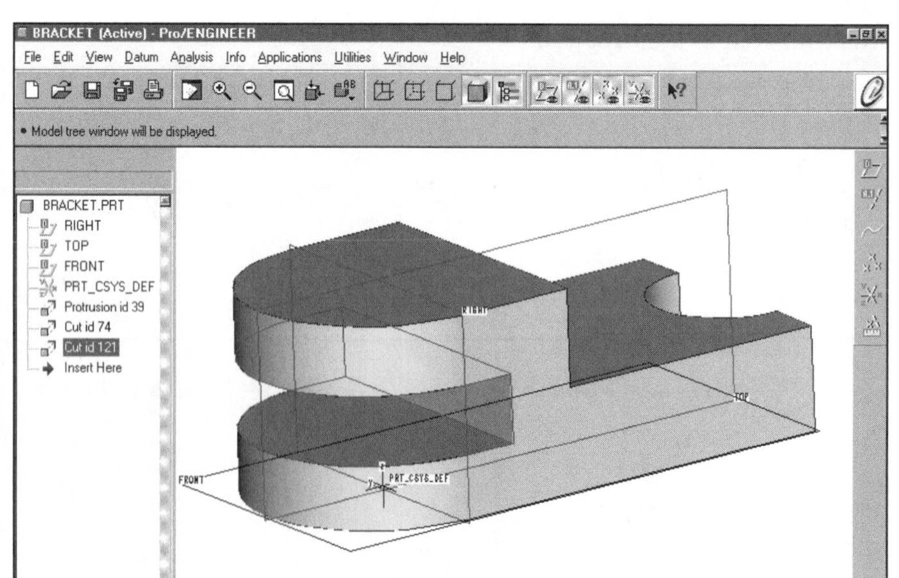

Figure 3.25
Second Cut Completed

Create a Mapkey command for Straight Linear holes:

MKEY ***TOOLCHEST***

Utilities ⇒ Mapkeys ⇒ New ⇒ Key Sequence- **$F**(number 1-12) ⇒ Name ⇒ **LHOLE ⇒** Description **creates a hole ⇒ Record** ⇒ **Feature ⇒ Create ⇒ Hole ⇒ ●Straight hole ⇒ Done ⇒ Stop ⇒ OK ⇒** Save Mapkeys **All ⇒ Ok ⇒ Close ⇒ Utilities ⇒ Customize Screen ⇒ Commands ⇒** (Categories **Mapkey**) ⇒ (pick the new mapkey and press the RMB) ⇒ **●Image and Text ⇒** (pick **Choose Button Image** and or **Edit Button Image** and select and/or edit the image) ⇒ (pick the new Mapkey, drag to the Toolbar, and drop) ⇒ **OK**

LHOLE
Copy Button Image
Paste Button Image
Reset Button Image
Choose Button Image.
Edit Button Image...
Text Only
● Image and Text
Image Only

The next feature to be created is a hole. This is a *pick-and-place* feature that does not require a sketch. The placement plane is the part's top surface; the dimensioning edges/planes are **RIGHT** and **TOP** (Fig. 3.26). The hole will be at the intersection of the two datum planes; therefore, the distance from both will be **0** inches.

Choose the following commands:

Feature ⇒ Create ⇒ Solid ⇒ Hole ⇒ ● Straight hole ⇒
(from the HOLE dialog box--Diameter **.8125,** Depth one-- **Thru All)**
Primary Reference ↖-- [pick the top of the part (Fig. 3.26)]
Placement Type- **Linear**
Linear Reference ↖-- (pick **TOP**) Distance **0**
Linear Reference ↖-- (pick **RIGHT**) Distance **0**
✔ (Fig. 3.27) ⇒ **Done ⇒ File ⇒ Save ⇒ ✔**

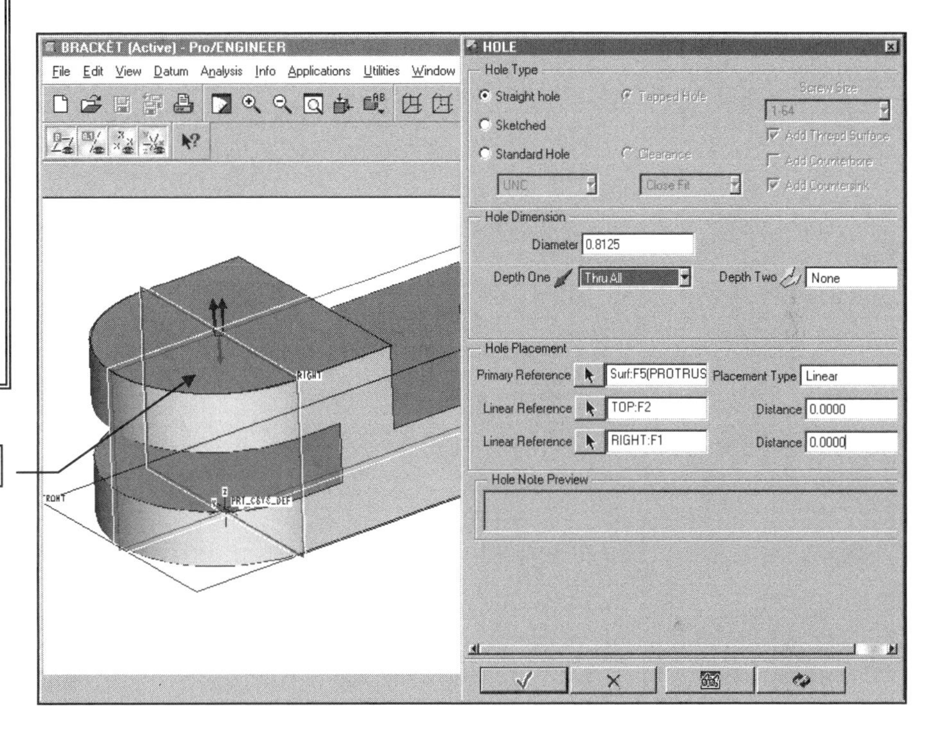

Placement plane

Figure 3.26
Hole Placement Plane and Measuring Edges

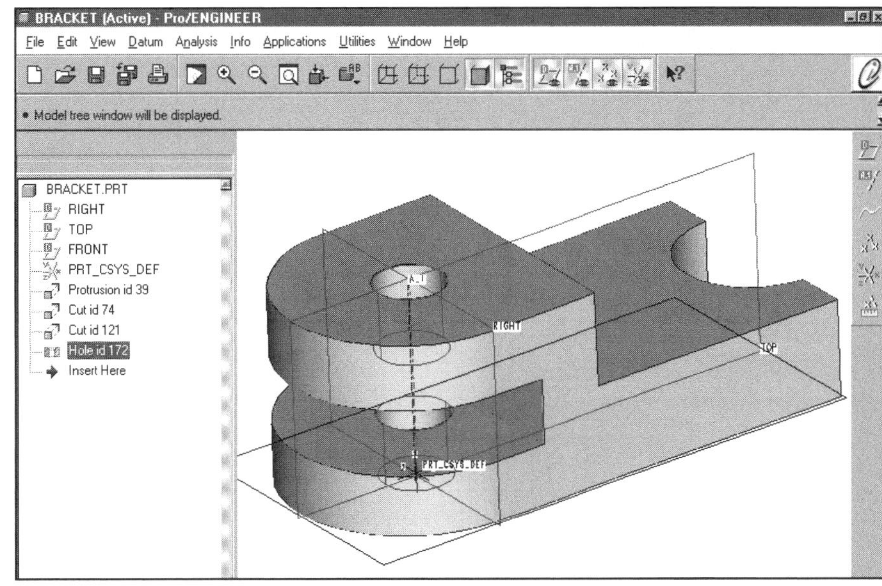

? Pro/HELP

To get more information about holes, highlight the **Hole** command and press the right mouse button and **? GetHelp**

Figure 3.27
Completed Hole

Instead of using the **Hole** command, we will create the next hole with a *revolved cut*. Sketched holes are really nothing more than revolved cuts. When we use this part to complete Lesson 17, the revolved cut will be used to demonstrate the modification of feature dimension id's to create parametric notes. This feature could be created with a **Standard Hole.**

NOTE

All entities of a revolved sketch must be on one side of a *vertical* centerline, and the section must be closed. You cannot use a horizontal centerline.

You should reread Sections 11 and 12 at this time.

Sketched holes and revolved cuts are created with a section sketch. ***The section must have a vertical centerline, and all entities must be on one side of that centerline.*** Always start by sketching the vertical centerline first. Next, sketch the entities required to describe the hole's shape (half of the shape). No alignment is necessary, but dimensions are required. ***The section must be closed.*** Choose the following commands:

Feature ⇒ **Create** ⇒ **Cut** ⇒ **Revolve** ⇒ **Solid** ⇒ **Done** ⇒ **One Side** ⇒ **Done** ⇒ (select **TOP** as the sketching plane) ⇒ **Flip** ⇒ **Okay** ⇒ **Top** ⇒ (pick **FRONT** as reference-orientation plane) ⇒ **Delete RIGHT as a Reference** ⇒ ↖ [select the right surface/edge as a Reference (Fig. 3.28)] ⇒ **Close** ⇒ (sketch and modify)

Figure 3.28
References

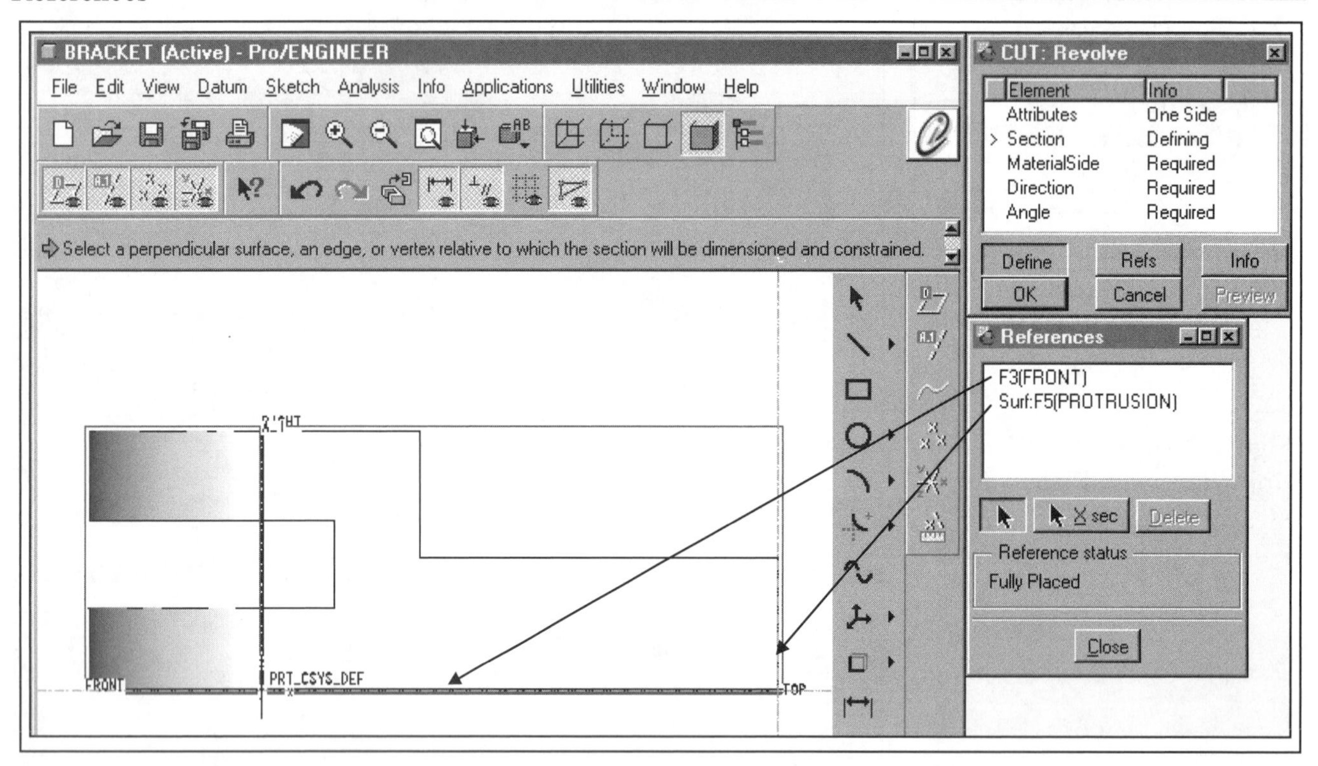

The diameter of a hole is dimensioned by choosing the Dimension icon and picking the centerline with the LMB (left mouse button), then the edge to be dimensioned with the LMB, and then the centerline a second time with the LMB. Place the dimension by picking a position with the MMB (middle mouse button). Start with a vertical centerline, create the required *closed* section, dimension, and modify the sketch as in Figure 3.29. (If prompted to align the top horizontal sketched line to the part; pick "**Yes.**")

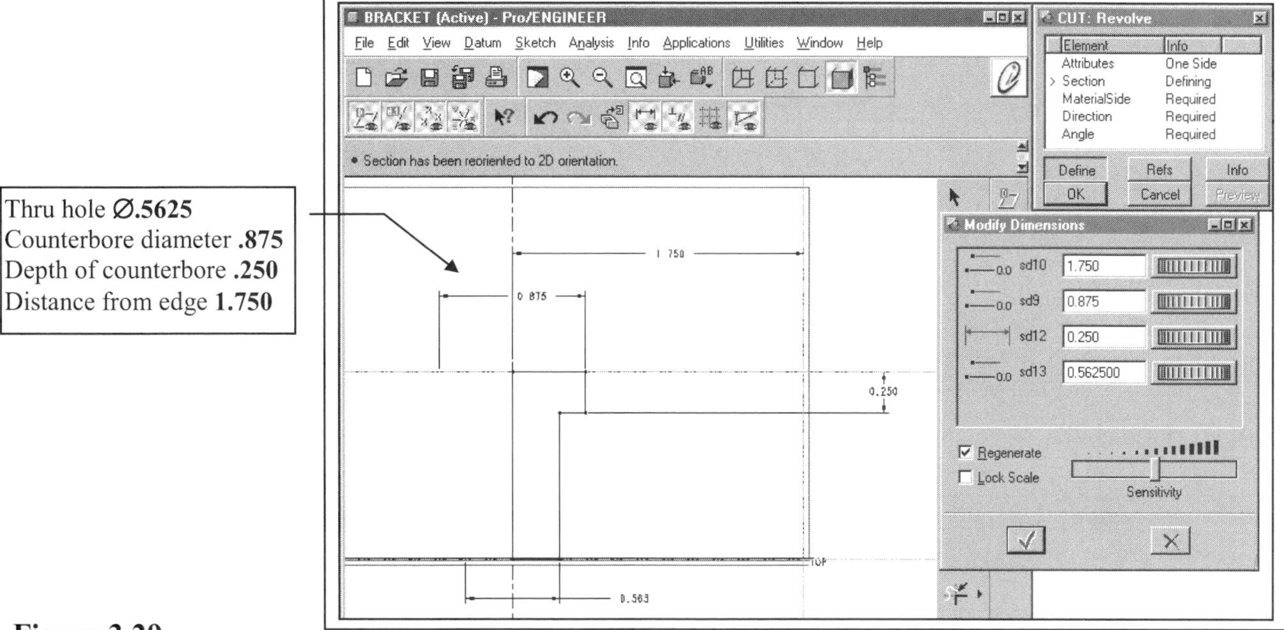

Figure 3.29
Modify Section Dimensions

Add the remaining dimensions; then **Modify** (Fig. 3.29). The counterbore diameter is **.875** and the thru hole diameter is **.5625.** The depth of the counterbore is **.250.**

Figure 3.30
Counterbore

✔ (continue with the current section) ⇒ **Okay** (Material Side) ⇒ **360** ⇒ **Done** ⇒ **Preview** ⇒ **OK** (Fig. 3.30) ⇒ **File** ⇒ **Save** ⇒ ✔

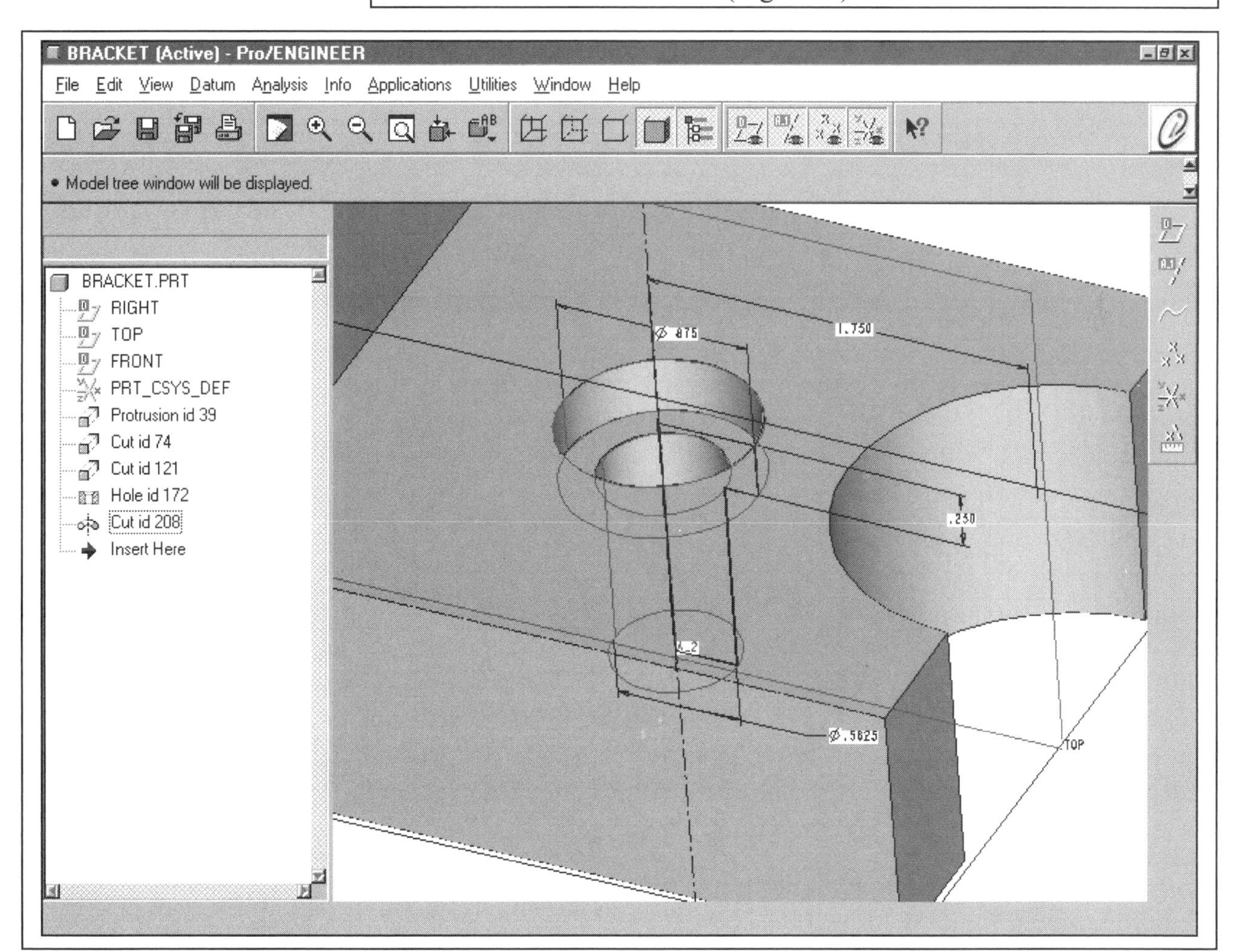

Both holes are now completed (Fig. 3.31). The Straight hole shows as a Hole in the Model Tree and the counterbored hole shows as a Cut. In Lesson 17, we will detail this part in a drawing.

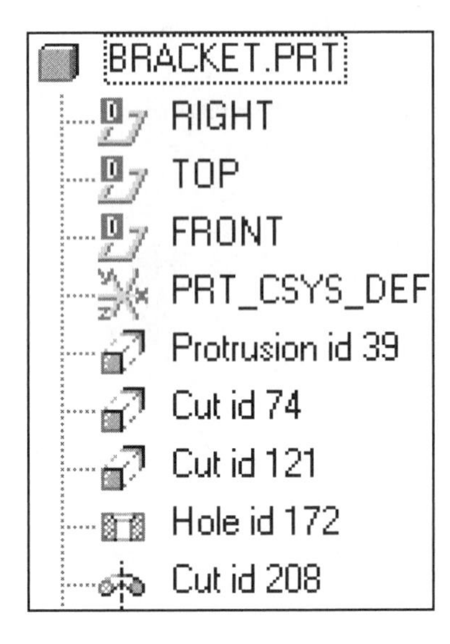

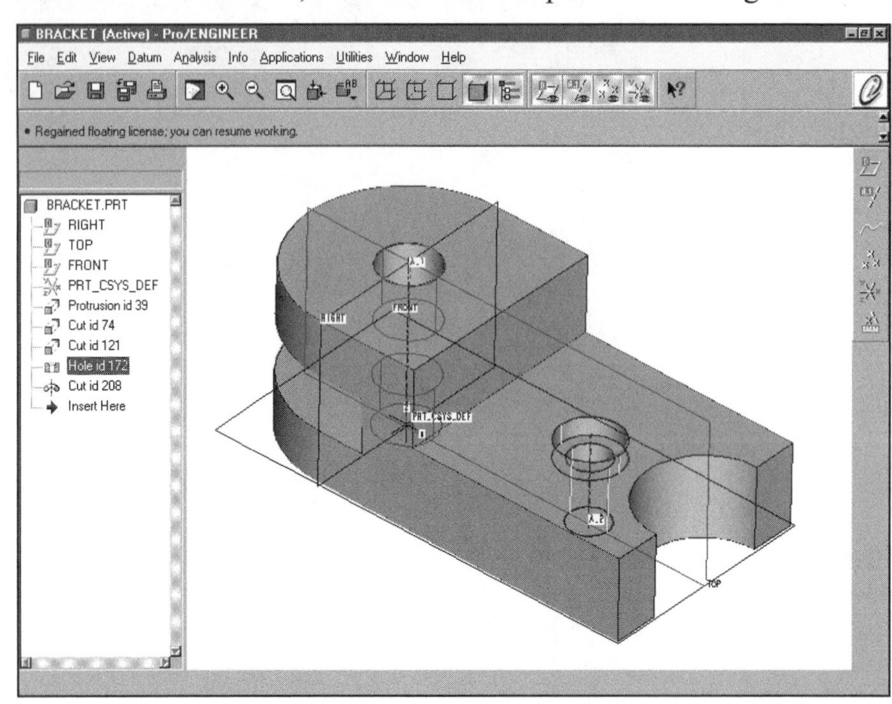

Figure 3.31
Completed Holes

If the counterbore were for a standard fastener, you could have created it with the Hole command. The Standard Hole option (Fig. 3.32) allows the varying of the counterbore diameter and depth but does not permit the thru hole diameter to be altered for the screw shaft size Clearance (Close Fit or Free Fit).

Figure 3.32
Standard Hole Option

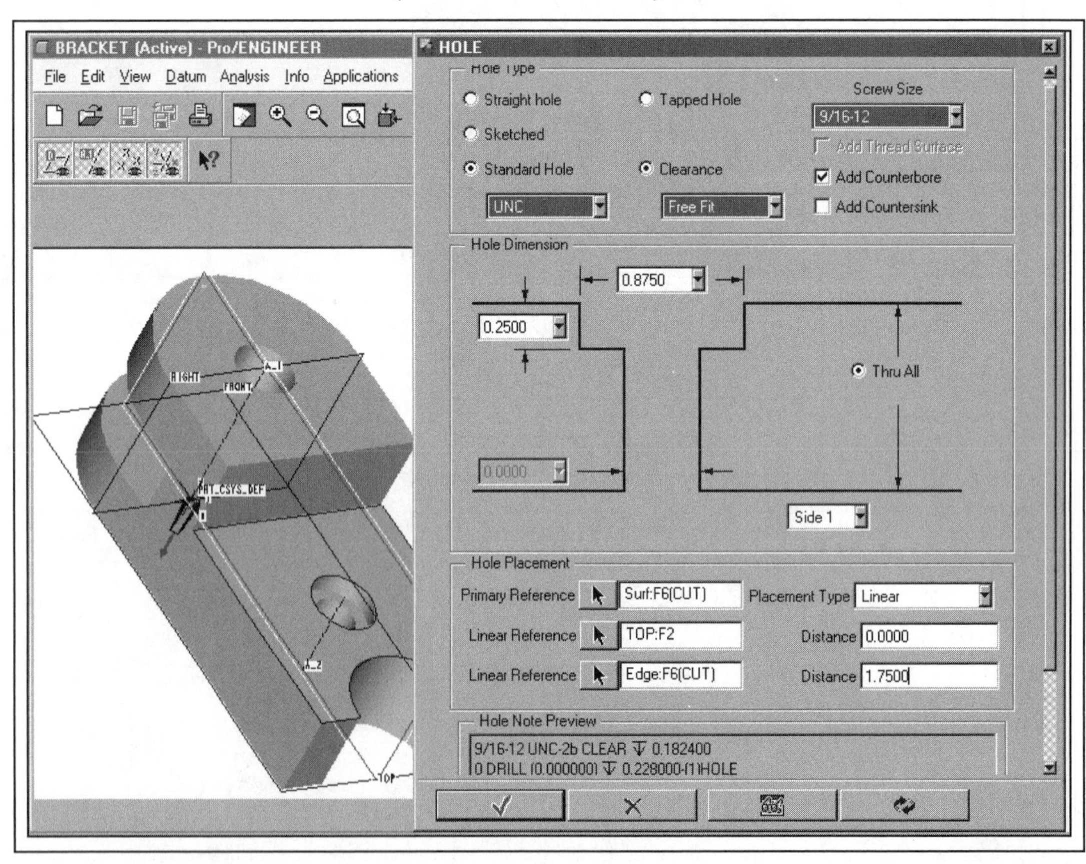

Open a new window and zoom in on the counterbore (Fig. 3.33). Highlight the feature in the Model Tree and press the RMB. Select **Info ⇒ Feat Info.** The INFORMATION WINDOW will appear on your screen, as in Figure 3.34.

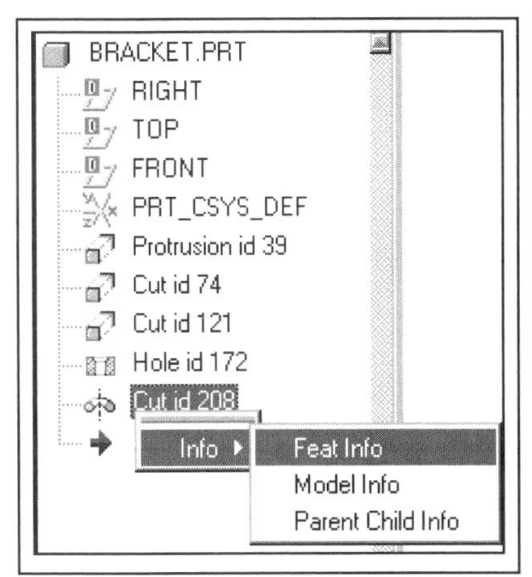

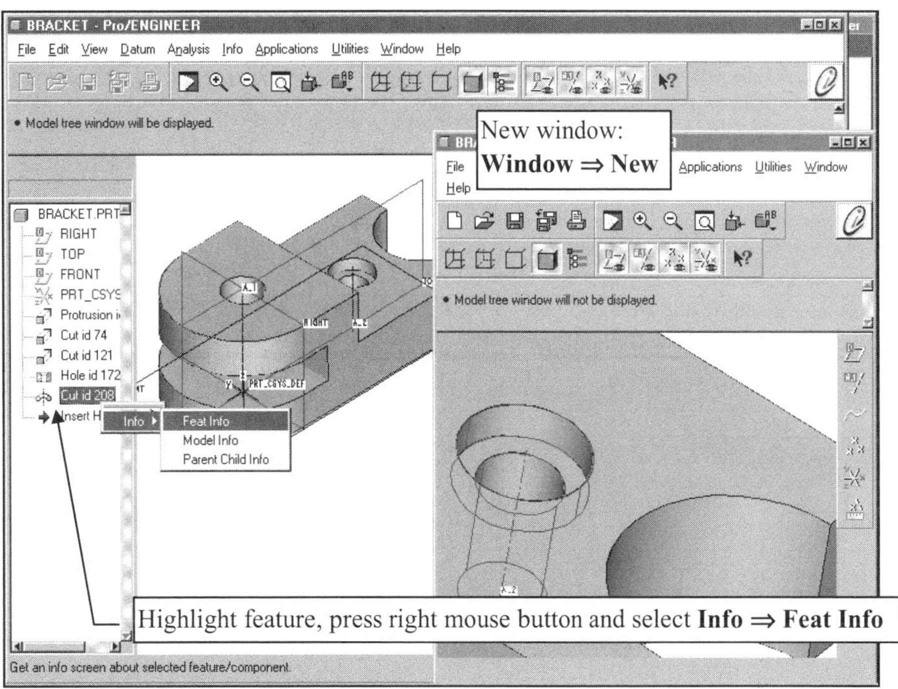

Figure 3.33
New Window and Feature Information Requested from the Model Tree

Information about a feature, part, assembly, etc. can be requested as needed during a Pro/E Session.

To close an INFORMATION WINDOW, and close the new window, and activate the first window, choose:

Close ⇒ Window ⇒ Close Window ⇒ Window ⇒ Activate

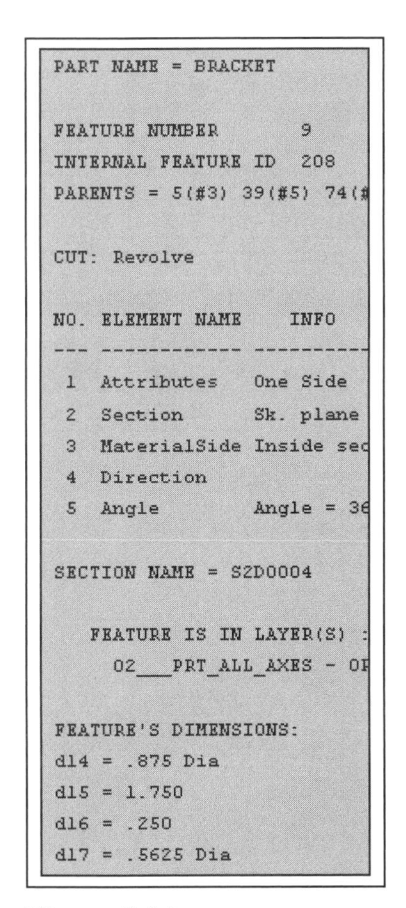
```
PART NAME = BRACKET

FEATURE NUMBER       9
INTERNAL FEATURE ID  208
PARENTS = 5(#3) 39(#5) 74(#

CUT: Revolve

NO. ELEMENT NAME    INFO
--- ------------ ---------
 1  Attributes    One Side
 2  Section       Sk. plane
 3  MaterialSide  Inside sec
 4  Direction
 5  Angle         Angle = 36

SECTION NAME = S2D0004

   FEATURE IS IN LAYER(S) :
     02___PRT_ALL_AXES - OP

FEATURE'S DIMENSIONS:
d14 = .875 Dia
d15 = 1.750
d16 = .250
d17 = .5625 Dia
```

Figure 3.34
Feature Information

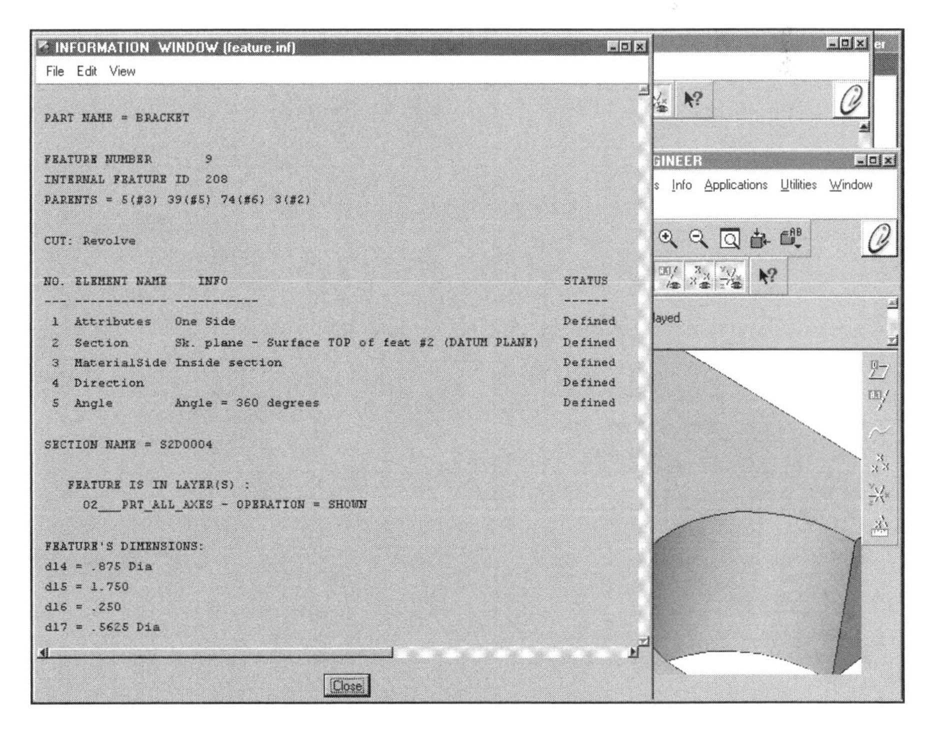

Create a Mapkey command for a **Round**:

MKEY *TOOLCHEST*

Utilities ⇒ **Mapkeys** ⇒ **New** ⇒ Key Sequence- **$F(#)** ⇒ Name ⇒ **RND** ⇒ Description **creates a round** ⇒ **Record** ⇒ **Feature** ⇒ **Create** ⇒ **Round** ⇒ **Simple** ⇒ **Done** ⇒ **Constant** ⇒ **Edge Chain** ⇒ **Done** ⇒ **One By One** ⇒ **Stop** ⇒ **OK** ⇒ Save Mapkeys **All** ⇒ **OK** ⇒ **Close** ⇒ **Utilities** ⇒ **Customize Screen** ⇒ **Commands** ⇒ (Categories **Mapkeys**) ⇒ (pick the new mapkey and press right mouse button) ⇒ ●**Image and Text** ⇒ (pick **Choose Button Image** and or **Edit Button Image** and select and or edit the image) ⇒ (pick the new Mapkey, drag to the Toolbar and drop) ⇒ **OK**

RND
Copy Button Image
Paste Button Image
Reset Button Image
Choose Button Image
Edit Button Image...
Text Only
● Image and Text
Image Only

To complete the part, a number of rounds need to be created. The first round is an edge round between the vertical and horizontal faces of the first cut. Choose the following commands:

Feature ⇒ **Create** ⇒ **Round** ⇒ **Simple** ⇒ **Done** ⇒ **Constant** ⇒ **Edge Chain** ⇒ **Done** ⇒ **One By One** ⇒ (pick the edge as shown in Fig. 3.35) ⇒ **Done Sel** ⇒ **Done** ⇒ **New Value** ⇒ (type the round radius value at the prompt, **.50**) ⇒ ✔ ⇒ **Preview** ⇒ **OK** (round will appear as in Fig. 3.36) ⇒ **Done** ⇒ **File** ⇒ **Save** ⇒ ✔

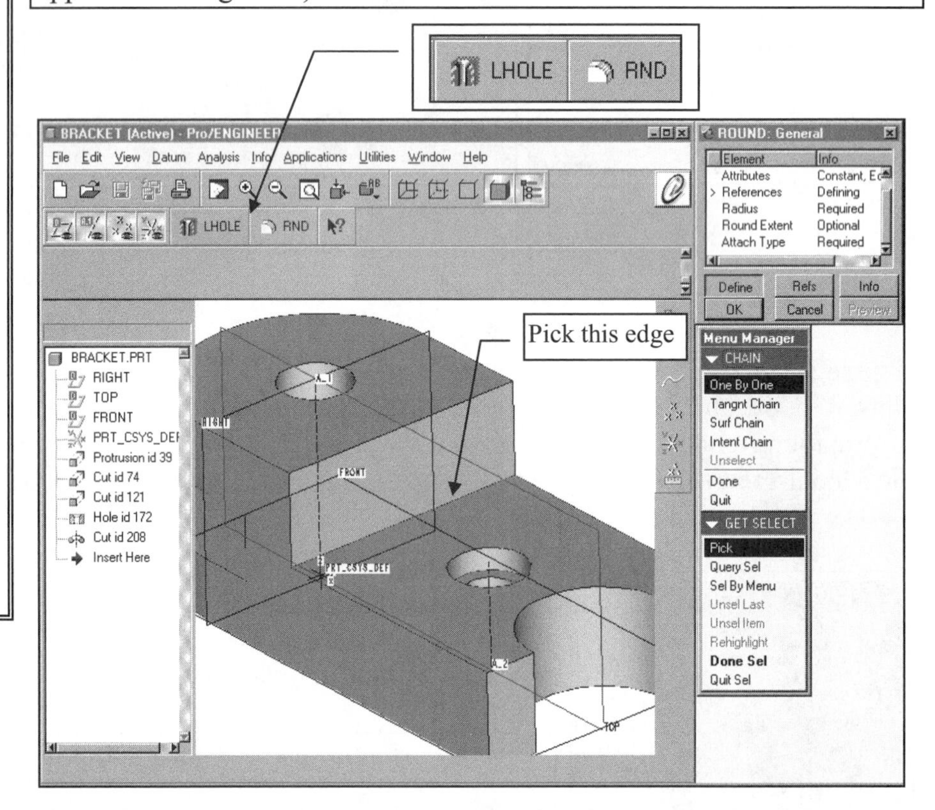

Figure 3.35
Edge to Be Rounded

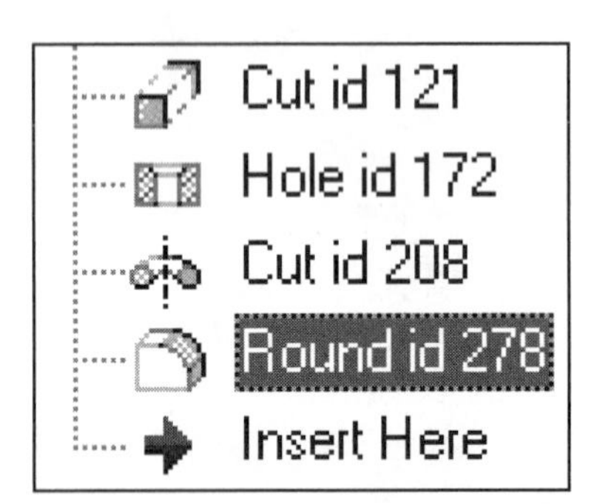

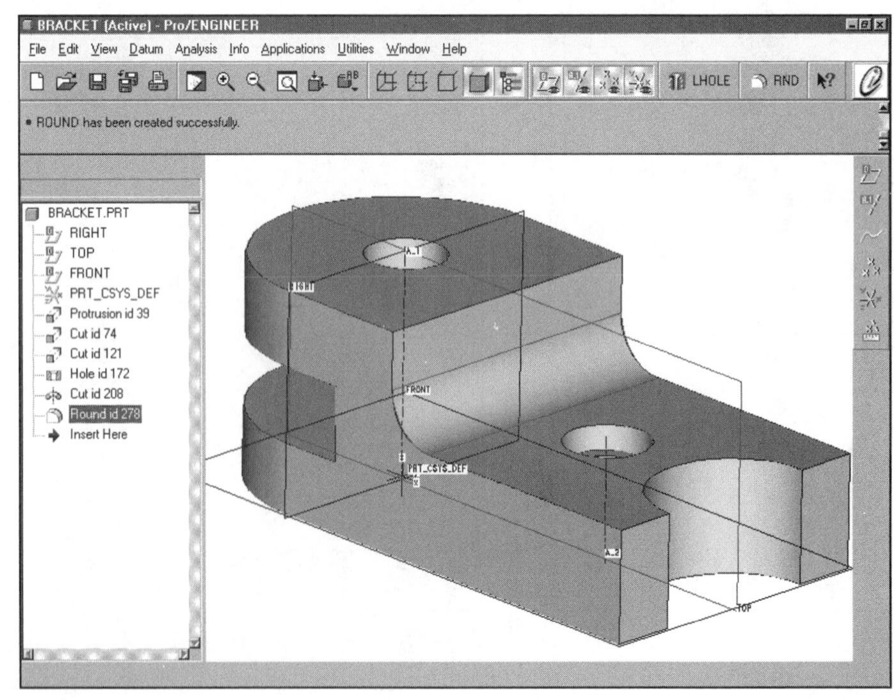

Figure 3.36
Round

If you created a mapkey for making a round pick it and continue with the command:

RND
creates a round
⇒ **Tangnt Chain** (pick the edges as shown in Figure 3.37) ⇒ **One By One** ⇒ (pick the remaining edges as shown in Figure 3.37) ⇒ **Done Sel** ⇒ **Done** ⇒ **New Value** ⇒ (type the round radius value at the prompt, **.125**) ⇒ ✔ ⇒ **Preview** (Fig. 3.38) ⇒ **OK**

Finally, the last round is a set that can be created all at the same time. Choose the following commands:

Feature ⇒ **Create** ⇒ **Round** ⇒ **Simple** ⇒ **Done** ⇒ **Constant** ⇒ **Edge Chain** ⇒ **Done** ⇒ **Tangnt Chain** (pick the edges as shown in Figure 3.37) ⇒ **One By One** ⇒ (pick the remaining edges as shown in Figure 3.37) ⇒ **Done Sel** ⇒ **Done** ⇒ **New Value** ⇒ (type the round radius value at the prompt, **.125**) ⇒ ✔ ⇒ **Preview** (Fig. 3.38) ⇒ **OK** ⇒ **File** ⇒ **Save** ⇒ ✔ **File** ⇒ **Delete** ⇒ **Old Versions** ⇒ ✔

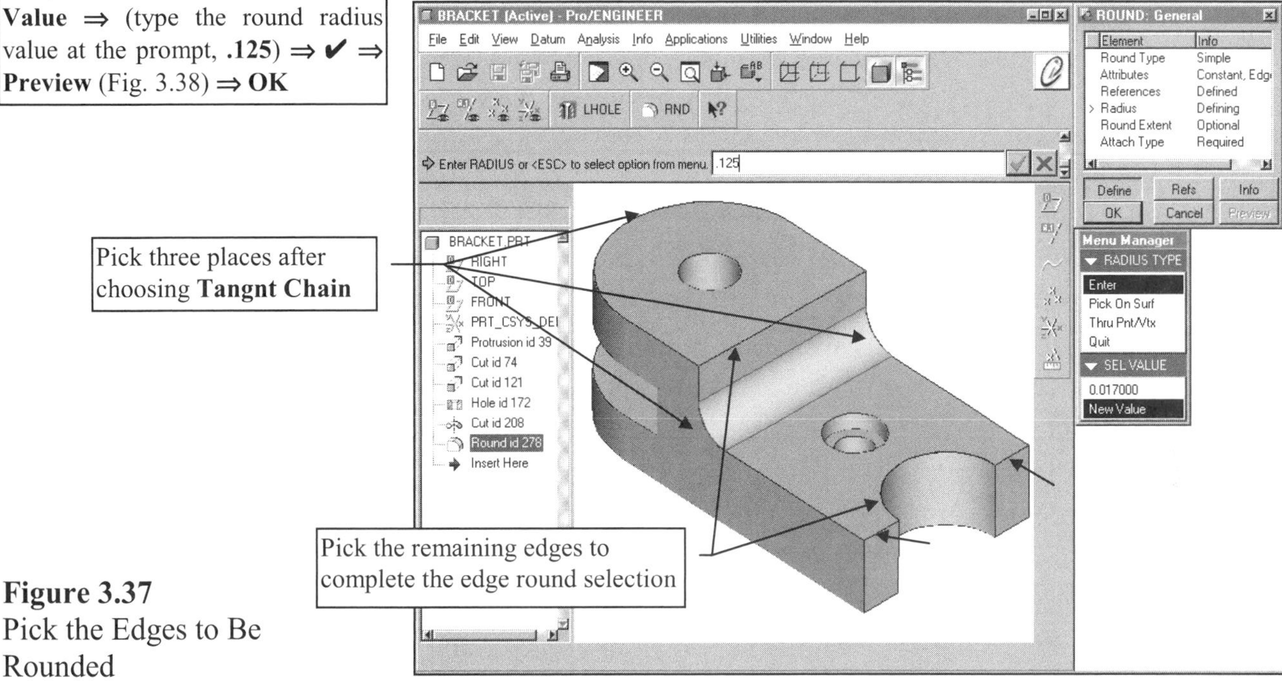

Figure 3.37
Pick the Edges to Be Rounded

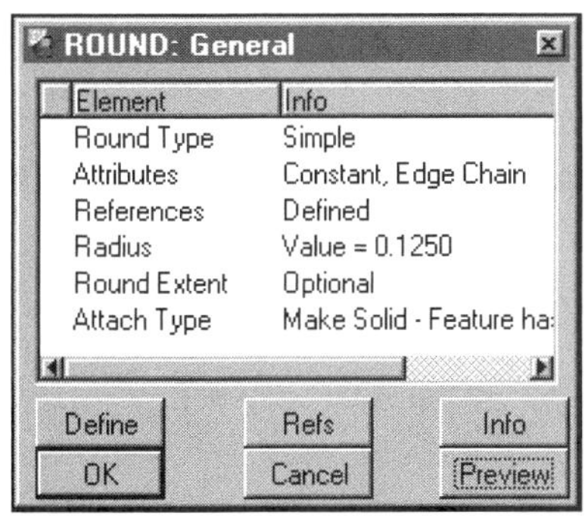

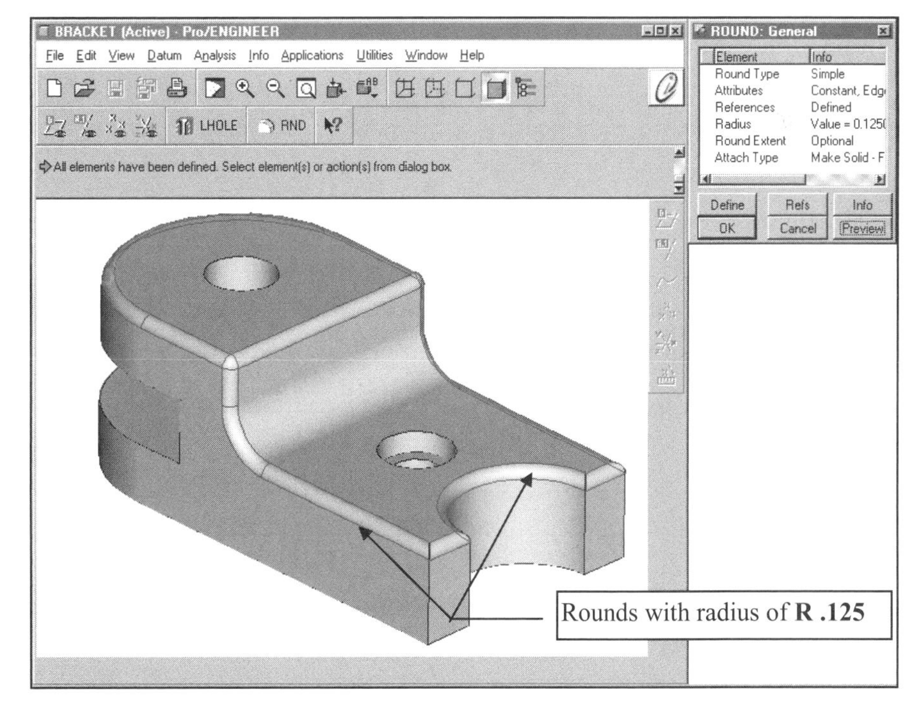

Figure 3.38
Completed Part

Figure 3.39 and Figure 3.40 provide more information on rounds.

Figure 3.39
How to Navigate the User Interface

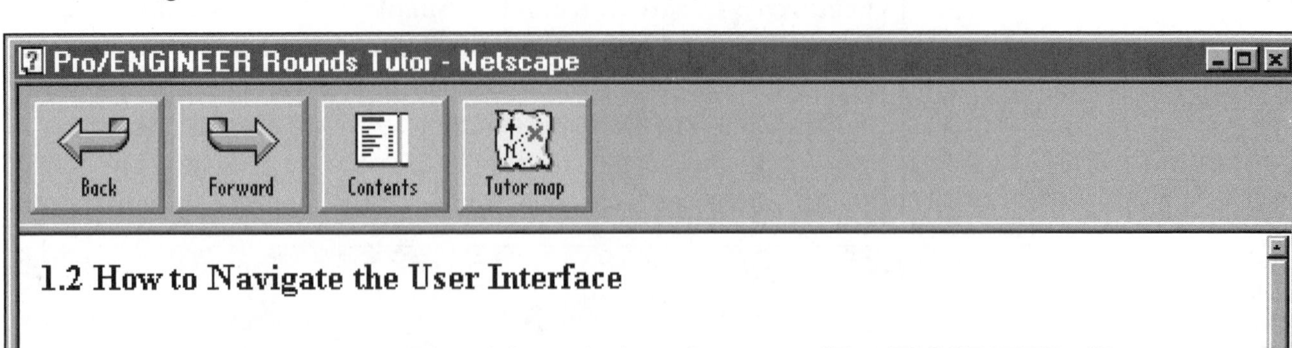

1.2 How to Navigate the User Interface

The Round feature is one of the pick-and-place features of Pro/ENGINEER. You can create rounds as a solid or surface feature.
The Round feature has option driven user interface, which means that to define a round you have to define a set of elements. All the elements are listed in the dialog box, and each element opens another dialog box, or a set of menus. The common elements for rounds are Type, Shape, References, Radius, and Attachment. Some selections (like, Simple/Advanced, Constant/Variable) change the set of elements defining the feature. Some elements are required (References), some are optional (Transitions), and some are predefined (Attachment).
The most common way to define round is to define required elements in the order they are listed, and press **OK** in the dialog box. However, you have an option to come back and redefine any element at any time, change predefined elements, or select and define optional elements. Status of each element (Required, Defining, Defined, Optional) is also listed in the dialog box. Additionally, you have an option of pressing **OK** even when not all required elements are defined, creating an incomplete feature (no geometry is created). Incomplete feature remembers all of the previously defined elements and serves as a placeholder in the feature list, so that you can later come back and complete the definition of the feature.
During creation of a Round feature, Pro/ENGINEER displays in yellow the temporary geometry as it is being defined. This visual feedback is often important, and we strongly recommend you to pay close attention to it. By investigating the temporary geometry, you will see when additional references are needed, the Round Extent element should be used, Transitions should be defined, or where Radius values should be modified.
When all the required elements are defined, the Preview button becomes available in the dialog box. You can use it to see the final geometry of the Round feature.
You can use Tutor Mode for online help with the user interface.

Figure 3.40
Round Tutor Map

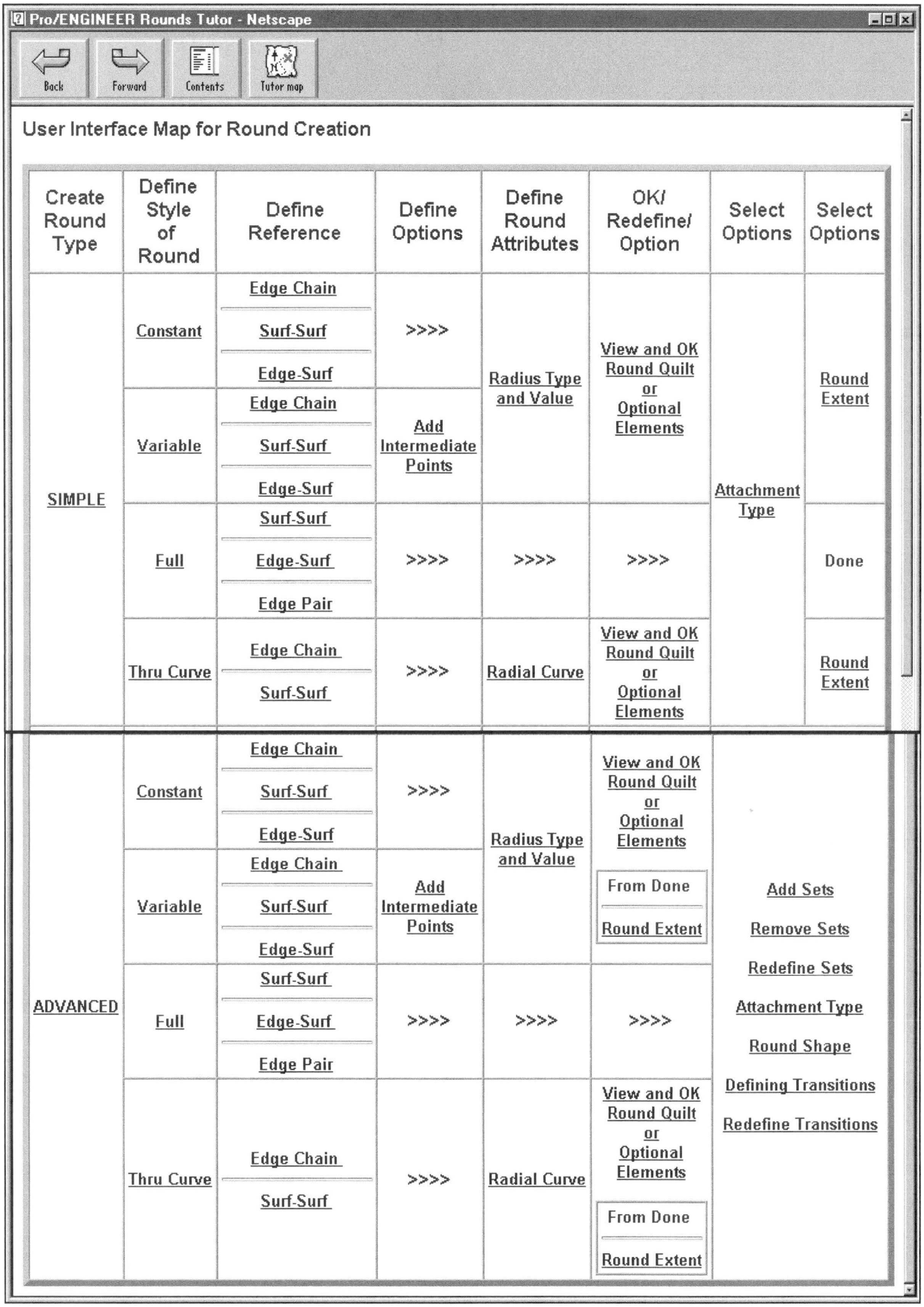

Pro/ENGINEER Rounds Tutor - Netscape

Back Forward Contents Tutor map

User Interface Map for Round Creation

Create Round Type	Define Style of Round	Define Reference	Define Options	Define Round Attributes	OK/ Redefine/ Option	Select Options	Select Options
SIMPLE	Constant	Edge Chain Surf-Surf Edge-Surf	>>>>	Radius Type and Value	View and OK Round Quilt or Optional Elements	Attachment Type	Round Extent
	Variable	Edge Chain Surf-Surf Edge-Surf	Add Intermediate Points				
	Full	Surf-Surf Edge-Surf Edge Pair	>>>>	>>>>	>>>>		Done
	Thru Curve	Edge Chain Surf-Surf	>>>>	Radial Curve	View and OK Round Quilt or Optional Elements		Round Extent
ADVANCED	Constant	Edge Chain Surf-Surf Edge-Surf	>>>>	Radius Type and Value	View and OK Round Quilt or Optional Elements	Add Sets Remove Sets Redefine Sets Attachment Type Round Shape Defining Transitions Redefine Transitions	
	Variable	Edge Chain Surf-Surf Edge-Surf	Add Intermediate Points		From Done Round Extent		
	Full	Surf-Surf Edge-Surf Edge Pair	>>>>	>>>>	>>>>		
	Thru Curve	Edge Chain Surf-Surf	>>>>	Radial Curve	View and OK Round Quilt or Optional Elements From Done Round Extent		

Lesson 3 Project

Guide Bracket

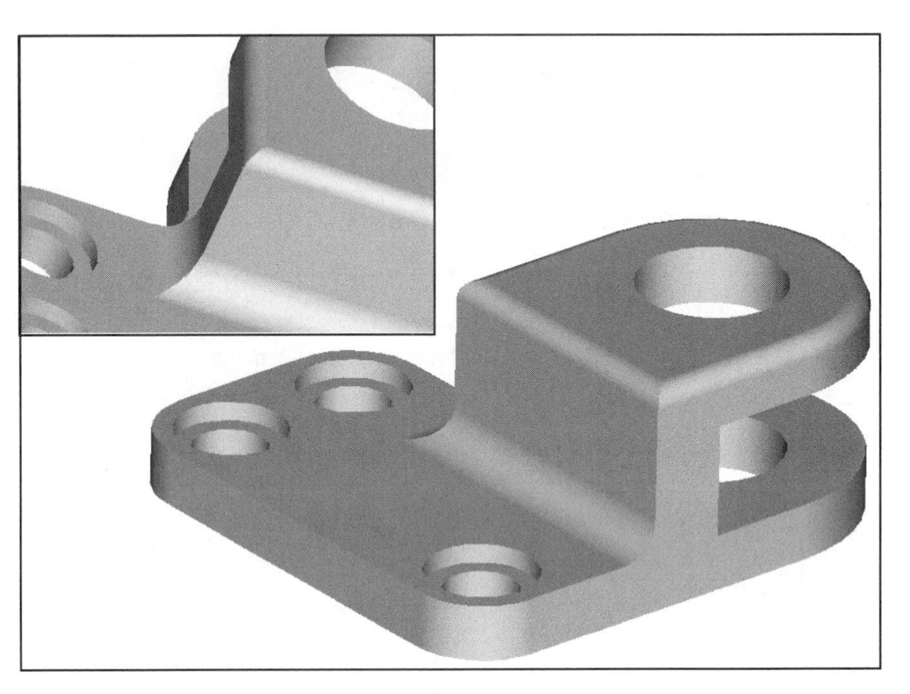

NOTE
Use the **DIPS** to plan out your feature creation sequence and the selection of datum sketching planes.

Figure 3.41
Guide Bracket

Guide Bracket

The third **lesson project** is a machined part that requires commands similar to the **Breaker**. Simple rounds and straight and sketched holes are part of the exercise. Create the part shown in Figures 3.41 through 3.47. At this stage in your understanding of Pro/E, you should be able to analyze the part and plan out the steps and features required to model it. You do not have to use the exact design intent shown here in the lesson project. You must use the same dimensions and dimension scheme, but the choice and quantity of datum planes and the sequence of modeling features can be different.

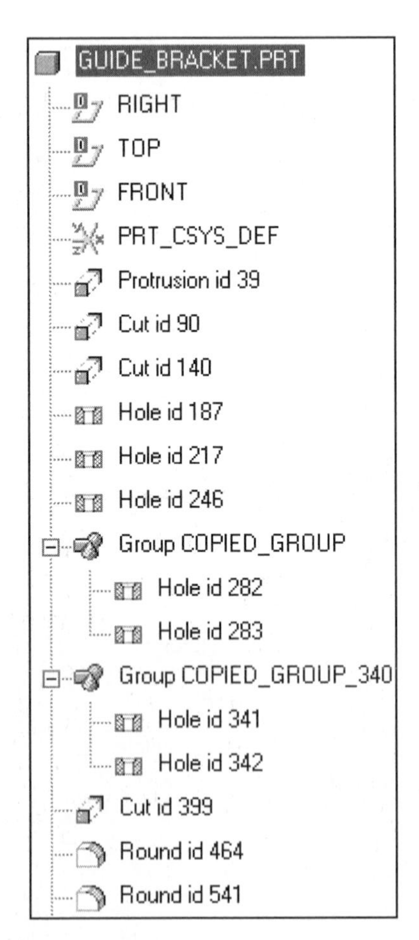

Figure 3.42
Guide Bracket Model Tree

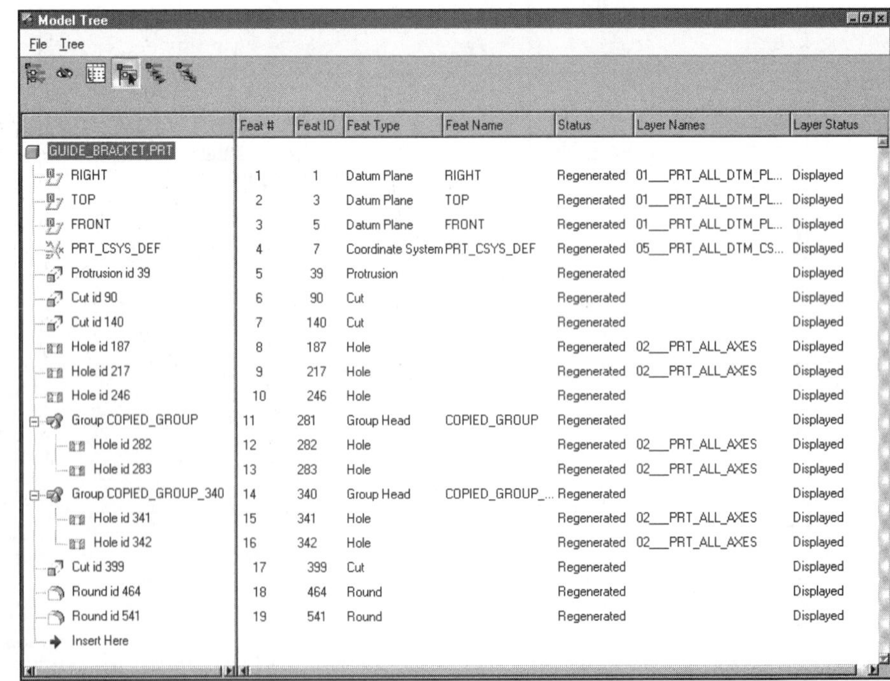

	Feat #	Feat ID	Feat Type	Feat Name	Status	Layer Names	Layer Status
GUIDE_BRACKET.PRT							
RIGHT	1	1	Datum Plane	RIGHT	Regenerated	01__PRT_ALL_DTM_PL...	Displayed
TOP	2	3	Datum Plane	TOP	Regenerated	01__PRT_ALL_DTM_PL...	Displayed
FRONT	3	5	Datum Plane	FRONT	Regenerated	01__PRT_ALL_DTM_PL...	Displayed
PRT_CSYS_DEF	4	7	Coordinate System	PRT_CSYS_DEF	Regenerated	05__PRT_ALL_DTM_CS...	Displayed
Protrusion id 39	5	39	Protrusion		Regenerated		Displayed
Cut id 90	6	90	Cut		Regenerated		Displayed
Cut id 140	7	140	Cut		Regenerated		Displayed
Hole id 187	8	187	Hole		Regenerated	02__PRT_ALL_AXES	Displayed
Hole id 217	9	217	Hole		Regenerated	02__PRT_ALL_AXES	Displayed
Hole id 246	10	246	Hole		Regenerated		Displayed
Group COPIED_GROUP	11	281	Group Head	COPIED_GROUP	Regenerated		Displayed
Hole id 282	12	282	Hole		Regenerated	02__PRT_ALL_AXES	Displayed
Hole id 283	13	283	Hole		Regenerated	02__PRT_ALL_AXES	Displayed
Group COPIED_GROUP_340	14	340	Group Head	COPIED_GROUP_...	Regenerated		Displayed
Hole id 341	15	341	Hole		Regenerated	02__PRT_ALL_AXES	Displayed
Hole id 342	16	342	Hole		Regenerated	02__PRT_ALL_AXES	Displayed
Cut id 399	17	399	Cut		Regenerated		Displayed
Round id 464	18	464	Round		Regenerated		Displayed
Round id 541	19	541	Round		Regenerated		Displayed
Insert Here							

Figure 3.43
Guide Bracket

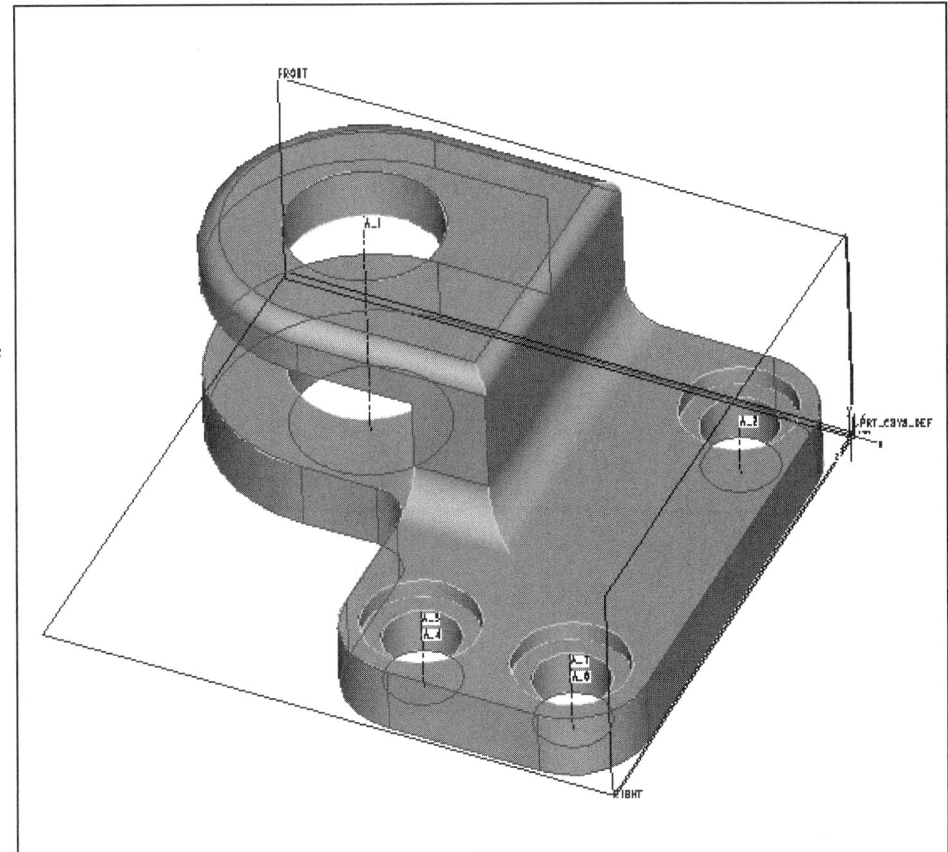

 ***NOTE***

Use the **DIPS** to plan out your feature creation sequence and the selection of datum sketching planes.

Figure 3.44
Guide Bracket Drawing

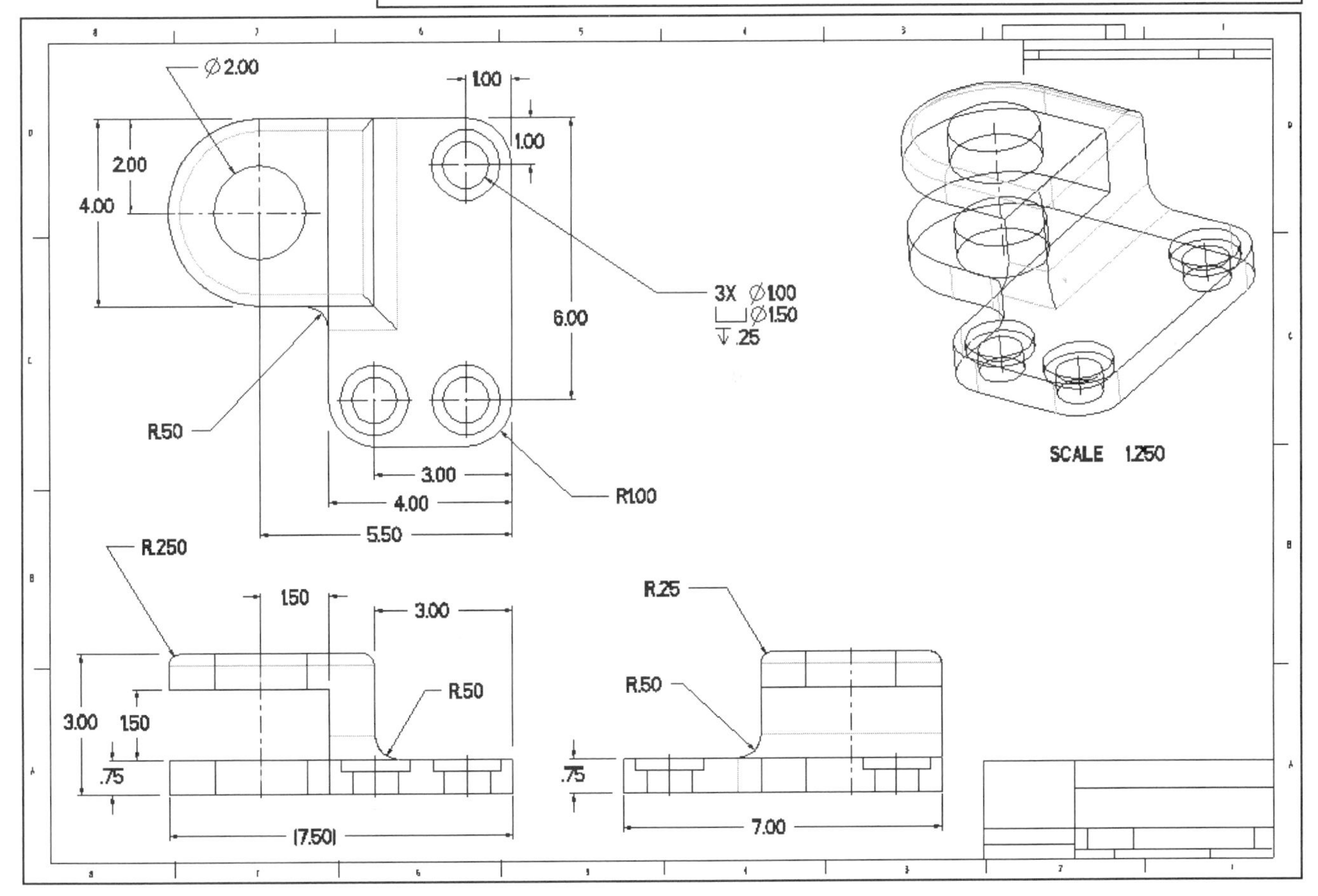

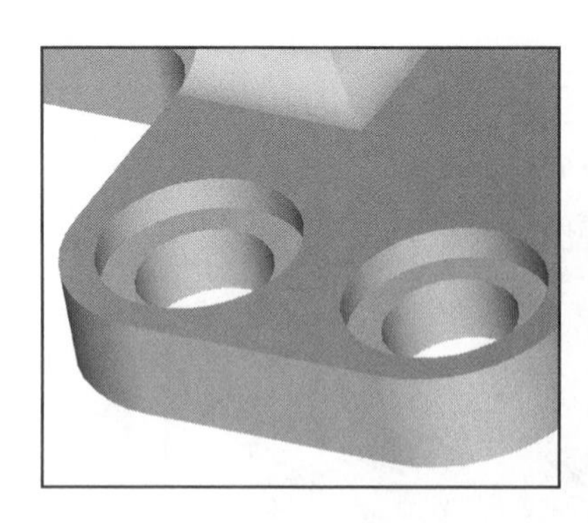

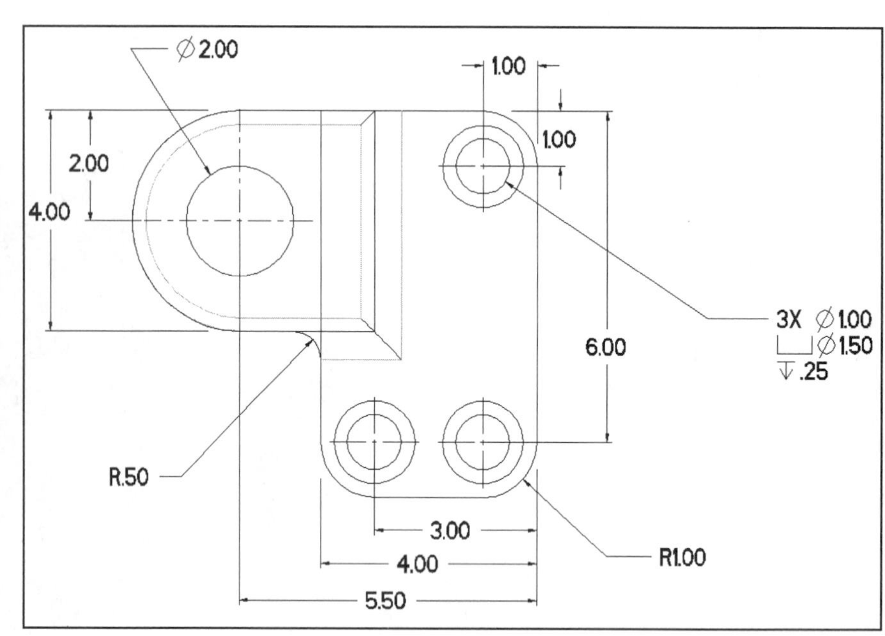

Figure 3.45
Guide Bracket Drawing,
Top View

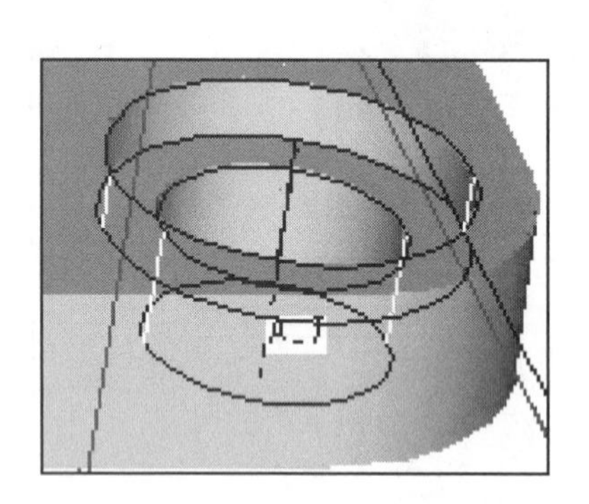

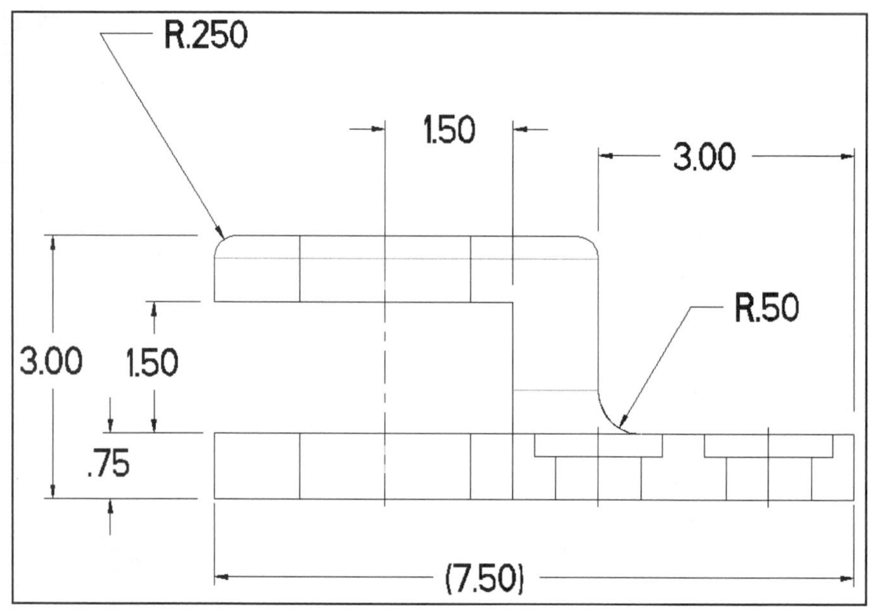

Figure 3.46
Guide Bracket Drawing,
Front View

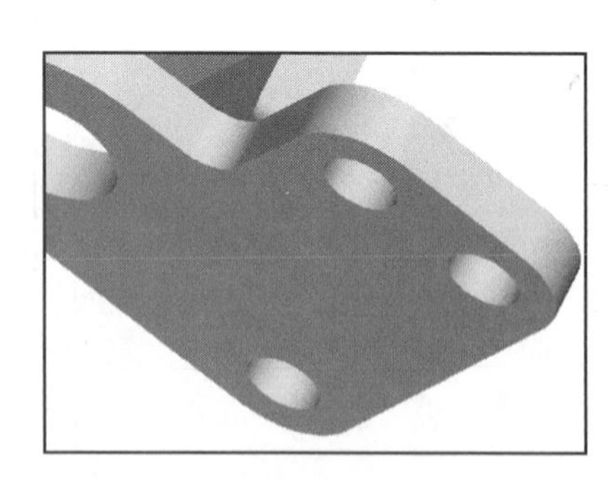

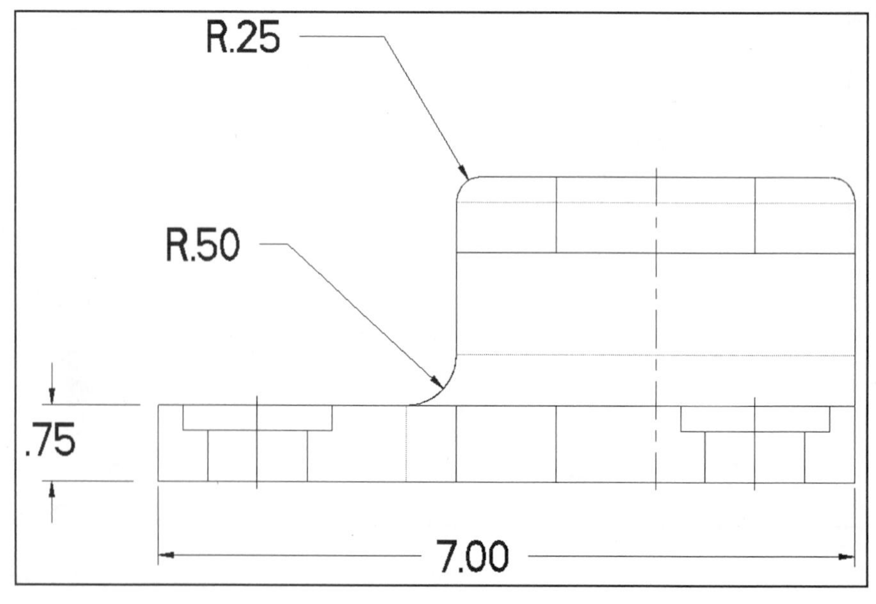

Figure 3.47
Guide Bracket Drawing,
Right Side View

Lesson 4

Datums and Layers

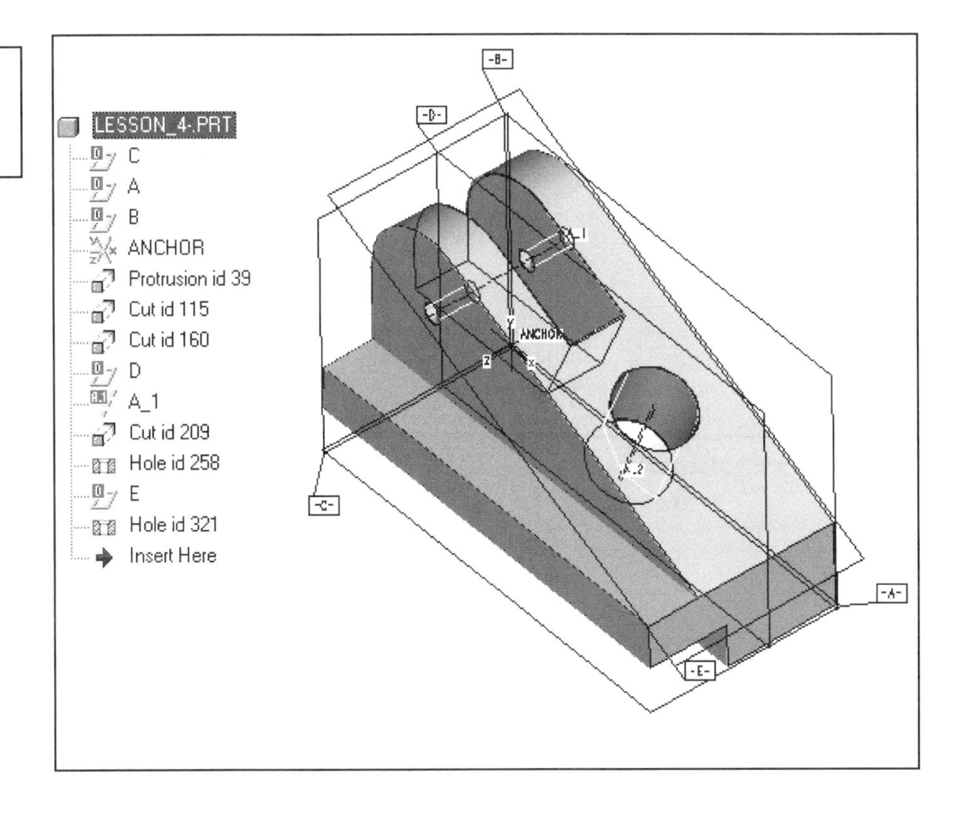

Figure 4.1
Anchor

OBJECTIVES

1. **Create datums to locate features**
2. **Use layers to organize part features**
3. **Set datum planes for geometric tolerancing**
4. **Rename datums and other features**
5. **Add a simple relation to control a feature**
6. **Use datum planes to establish sections**
7. **Reroute a features references**
8. **Use Info command to get layer information**
9. **Learn how to change the color and shading of models**

EGD REFERENCE
Fundamentals of Engineering Graphics and Design
by L. Lamit and K. Kitto
Read Chapters 10 and 25.
See pages 388 and 549.

COAch™ for Pro/ENGINEER

If you have **COAch for Pro/ENGINEER** on your system, go to SEARCH and do the Segment shown in Figure 4.3.

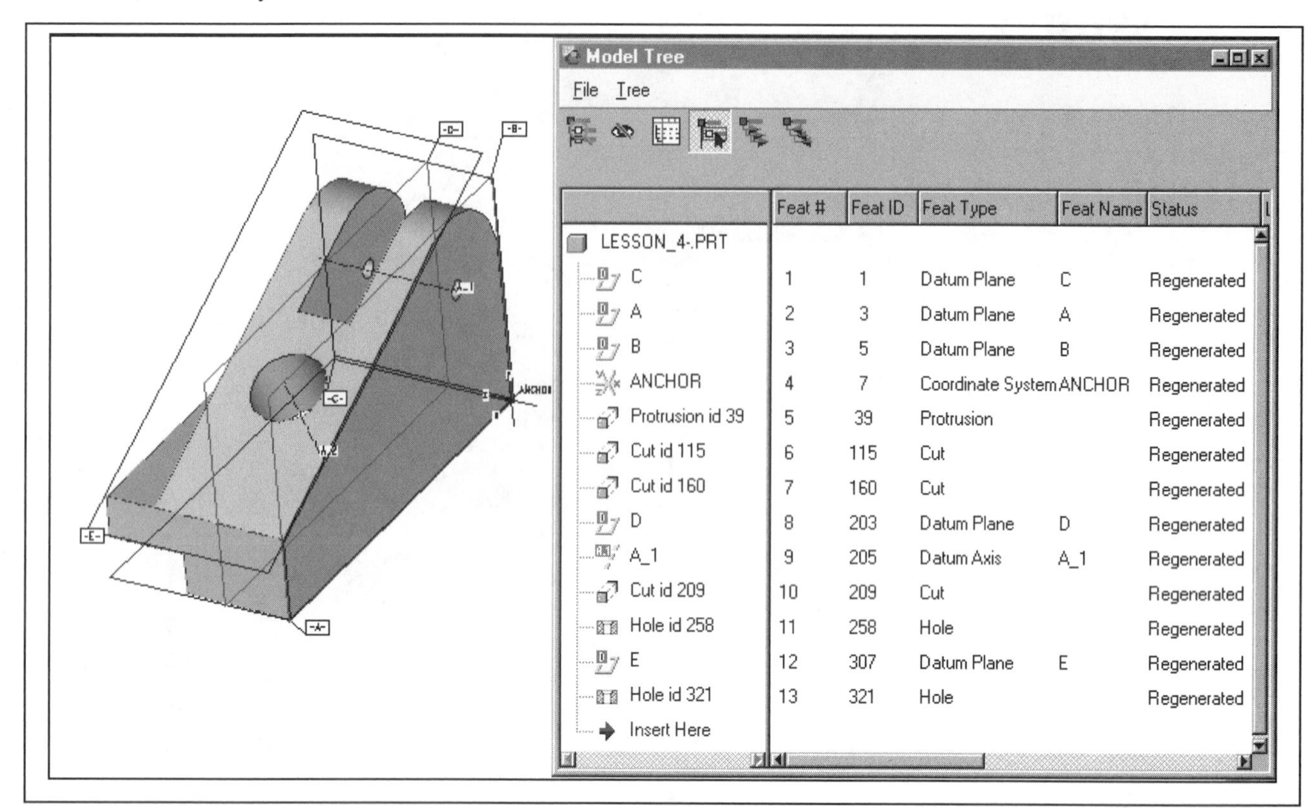

Feat #	Feat ID	Feat Type	Feat Name	Status
1	1	Datum Plane	C	Regenerated
2	3	Datum Plane	A	Regenerated
3	5	Datum Plane	B	Regenerated
4	7	Coordinate System	ANCHOR	Regenerated
5	39	Protrusion		Regenerated
6	115	Cut		Regenerated
7	160	Cut		Regenerated
8	203	Datum Plane	D	Regenerated
9	205	Datum Axis	A_1	Regenerated
10	209	Cut		Regenerated
11	258	Hole		Regenerated
12	307	Datum Plane	E	Regenerated
13	321	Hole		Regenerated

Figure 4.2
Anchor with Datums

DATUMS AND LAYERS

Datum planes and layers are two of the most useful mechanisms for creating and organizing your design (Figs. 4.1 and 4.2). ***Layers** were covered in detail in Section 8, and that section should be reread at this time.* Datum features such as *datum planes* and *datum axes* are essential for the creation of all parts, assemblies, and drawings using Pro/E. If you have **COAch for Pro/ENGINEER** on your system, do the appropriate segment (Fig. 4.3, Datums).

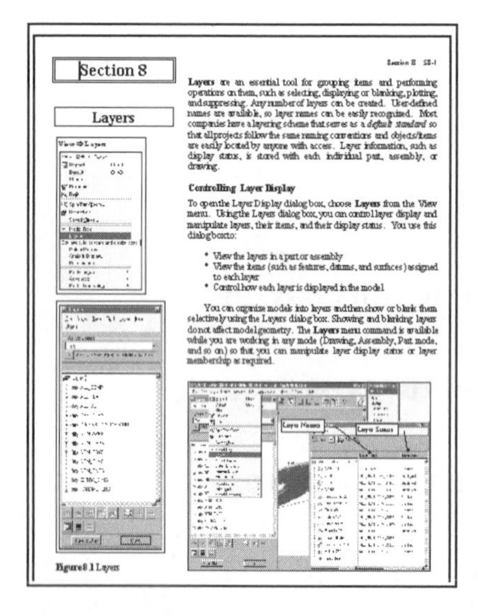

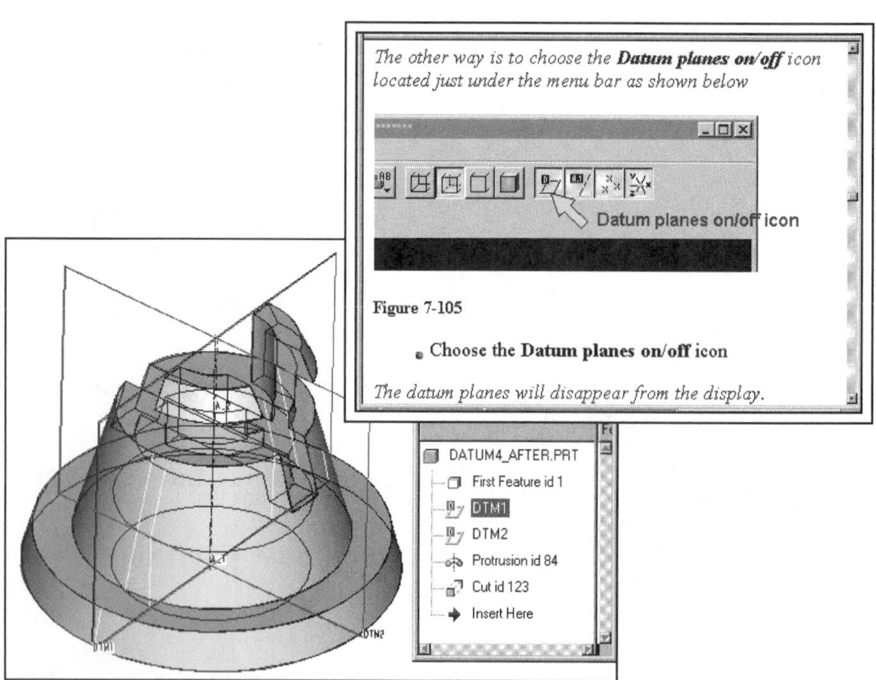

Figure 4.3
COAch for Pro/E, Datums (Datum Axis)

Datum Planes

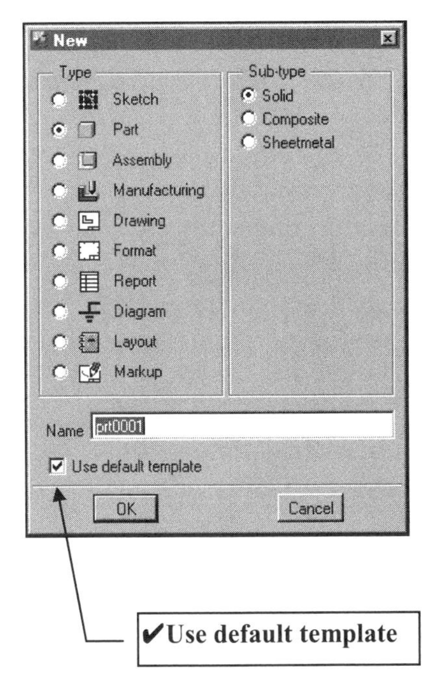

As previously mentioned, **Datum planes** are used to create a reference for a part that does not already exist. For example, you can sketch or place features on a datum plane when there is no appropriate planar surface; you can also dimension to a datum plane as though it were an edge. When you are constructing an assembly, you can use datums with assembly commands. All datums have a *red* side and a *yellow* side so that you know on which side you are working.

In all models, the first features created are the three default datums (Fig. 4.4). When you use ✔ **Use default template,** the default datums are named **FRONT, TOP,** and **RIGHT.** These three datums and the coordinate system are the parents of all other features. The base construction feature (a protrusion) is created using these datums. Datum features can be created as needed to complete the design. Datum planes can be used as references, as sketching planes, and as parent features for a variety of nonsketched part features.

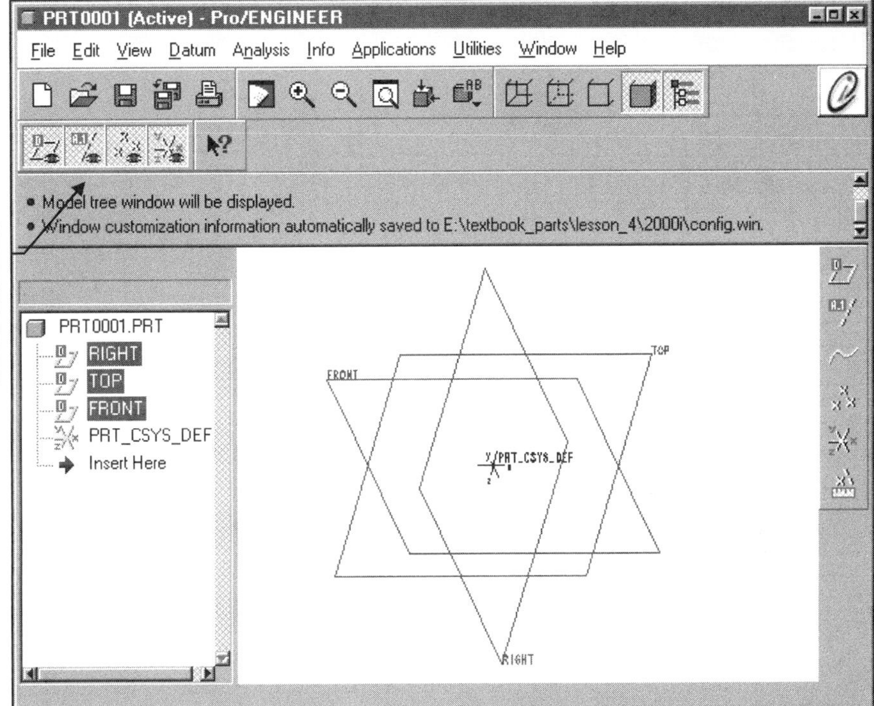

Datum planes on/off

Figure 4.4
Default Datums

A nondefault datum is created by specifying one or more **constraints** that locate it with respect to existing geometry. As an example, a datum plane might be created to pass **Tangent** to a cylinder and **Parallel** to a planar surface. Figure 4.5 shows an angled datum used to model the part. The following list provides the available datum plane constraints:

Through Creates a datum plane coincident with a planar surface.
Normal Creates a datum plane that is normal to an object that has been selected.

Parallel Creates a datum plane that is parallel to a selected object.
Offset Creates a datum plane that is parallel to a selected planar surface and is offset from that surface by a specified distance.
Angle Creates a datum plane that is at an angle to an object that has been selected.
Tangent Creates a datum plane that is tangent to a selected object.
BlendSection Creates a datum plane through the section that was used to make a feature. When sections exist, as for a blend, you will be prompted for the section number.

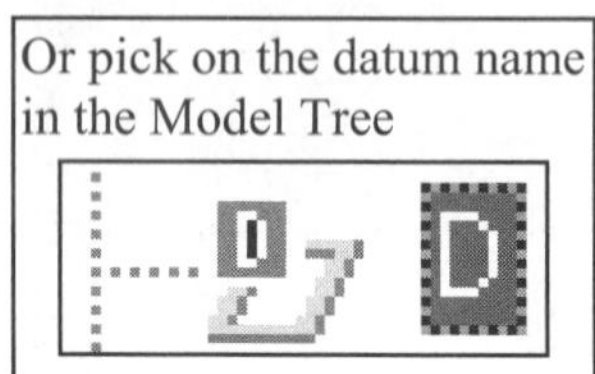

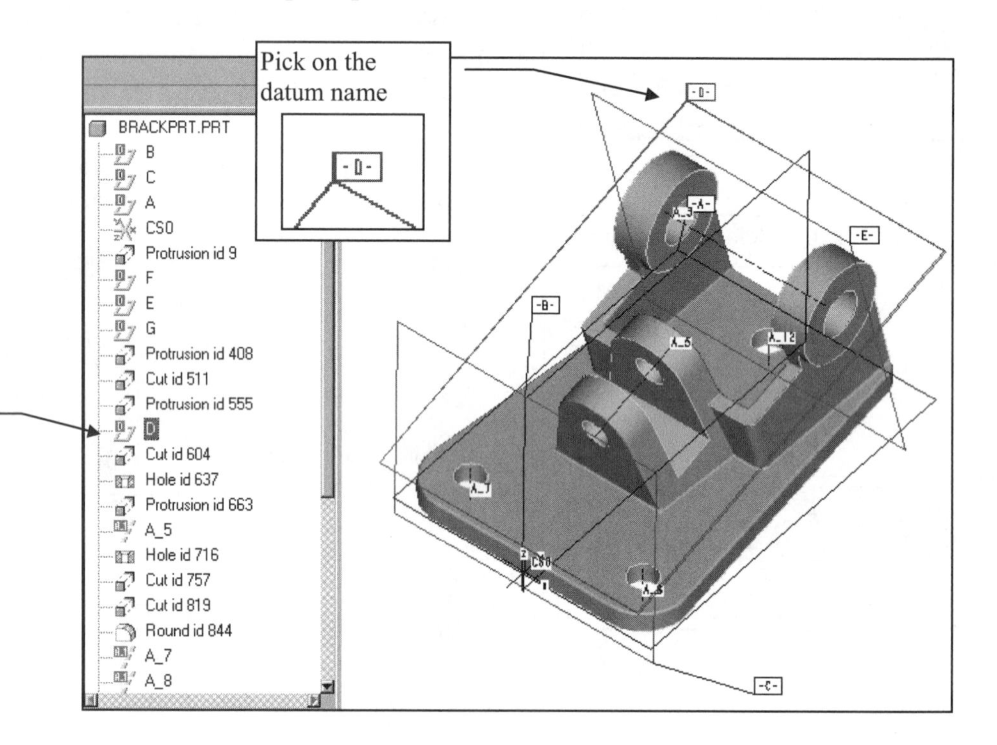

Figure 4.5
Non-default Datums

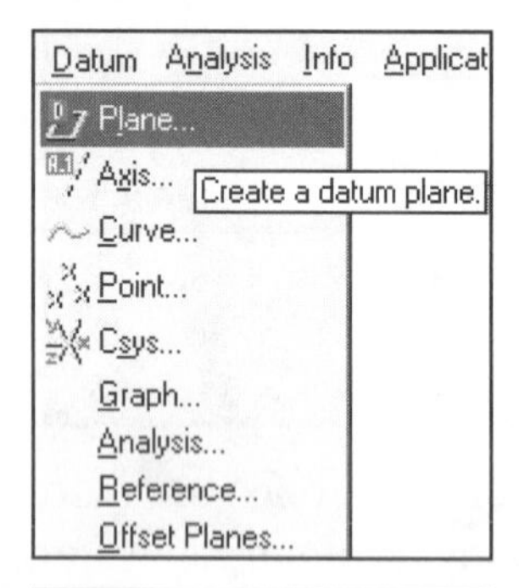

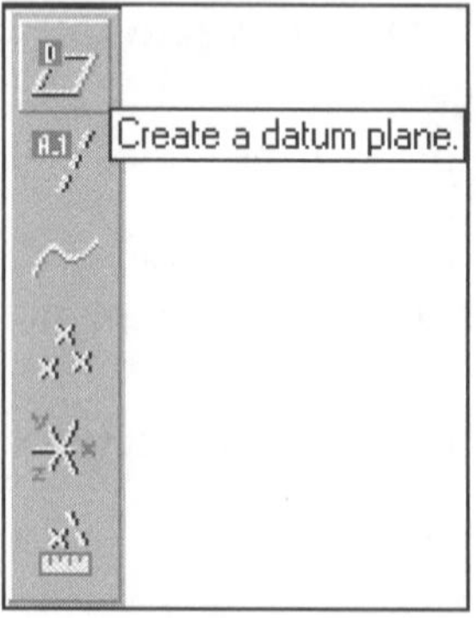

To create a datum plane:

1. Choose **Datum** from the menu bar [or **Make Datum** from SETUP PLANE menu (sometimes called a ***datum on the fly***)]; then choose **Plane** (see left-hand column). You can also pick the *Create a datum plane icon* to initiate the command.
2. Choose the desired constraint option from the DATUM PLANE menu. All appropriate geometry options in the lower section of the menu will be selected automatically. *To limit the items to select, click on highlighted menu options to unhighlight them.*
3. Pick the necessary references.
4. Repeat steps 2 and 3 until the required constraints have been established. When the maximum number of constraints have been specified, Pro/E notifies you by *dimming out* all options except **Done**, **Restart**, and **Quit**. Although they are actually *infinite planes*, datum planes are displayed scaled to the model size. To select a datum plane, you can pick on its name or select one of its boundaries, or you can select it by picking on its name in the **Model Tree** (Fig. 4.5).

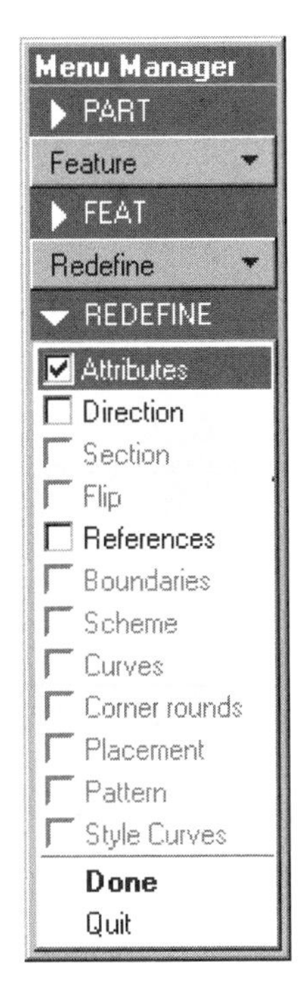

Figure 4.6
Datums

The size of a displayed datum plane changes with the dimensions of a part (Fig. 4.6). All datum planes, except those made *on the fly* (within other commands using **Make Datum**), can be sized to specific geometry using **Redefine ⇒ ✔Attributes ⇒ Done**. You can make a datum plane as big as the model or as small as an edge or surface on the model.

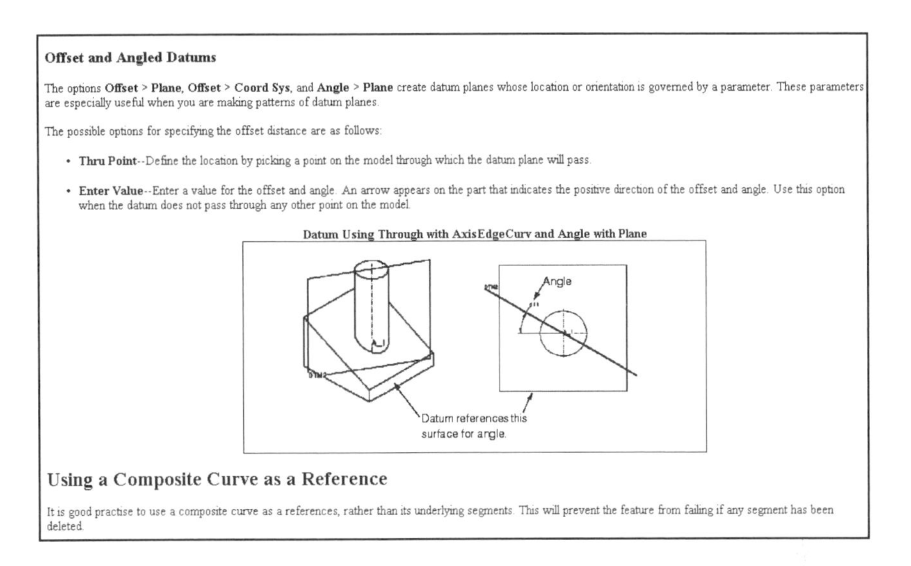

Offset and Angled Datums

The options **Offset** > **Plane**, **Offset** > **Coord Sys**, and **Angle** > **Plane** create datum planes whose location or orientation is governed by a parameter. These parameters are especially useful when you are making patterns of datum planes.

The possible options for specifying the offset distance are as follows:

- **Thru Point**--Define the location by picking a point on the model through which the datum plane will pass.
- **Enter Value**--Enter a value for the offset and angle. An arrow appears on the part that indicates the positive direction of the offset and angle. Use this option when the datum does not pass through any other point on the model.

Datum Using Through with AxisEdgeCurv and Angle with Plane

Using a Composite Curve as a Reference

It is good practise to use a composite curve as a references, rather than its underlying segments. This will prevent the feature from failing if any segment has been deleted.

The options available for sizing the datum plane (see left-hand column outline) are:

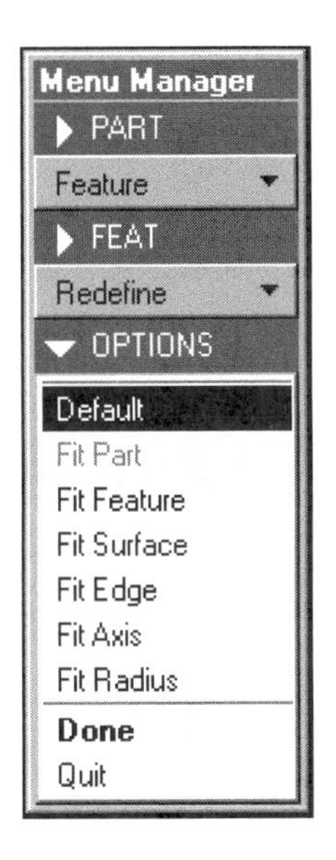

Default The datum plane is sized to the model (part or assembly).

Fit Part Sizes the datum plane to a part in Assembly mode.

Fit Feature Sizes the datum plane to a part or assembly feature.

Fit Surface Sizes the datum plane to any surface.

Fit Edge Sizes the datum plane to fit an edge.

Fit Axis Sizes the datum plane to fit an axis.

Fit Radius Sizes the datum plane to fit a specified radius, centering itself within the constraints of the model.

Datum Axes

Datum Axes (Fig. 4.7) can be used as references for feature creation, such as the coaxial placement of a hole. They are particularly useful for making datum planes, for placing items concentrically, and for creating radial patterns. Axes can be used to measure from, place coordinate systems, and place specific features. The angle between a feature and an axis, the distance between an axis and a feature, and so on can be determined using the **Info** command. Axes (appearing as centerlines) are automatically created for:

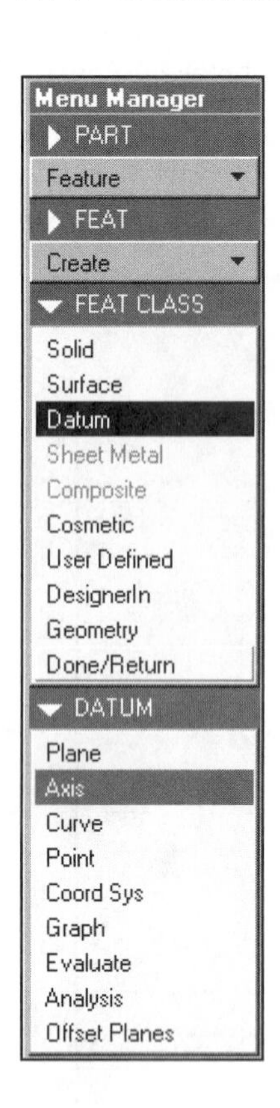

Revolved features All features whose geometry is revolved, including revolved base features, holes (Fig 4.8), shafts, revolved slots, cuts, and circular protrusions (Fig. 4.9).

Extruded circles An axis is created for every extruded circle in any extruded feature.

Extruded arcs An axis can be created automatically for extruded arcs only when you set the configuration option *(show_axes_for_extr_arcs yes).*

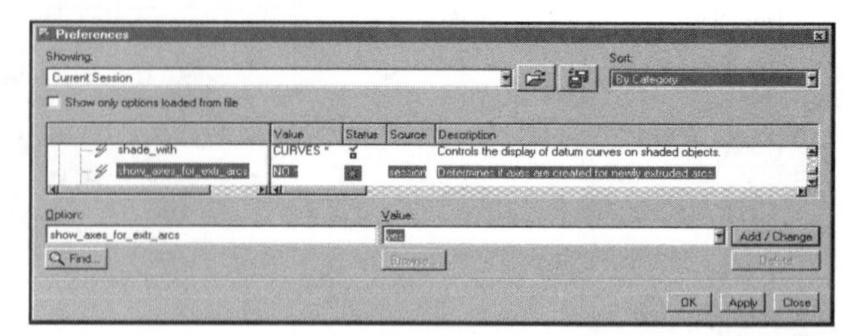

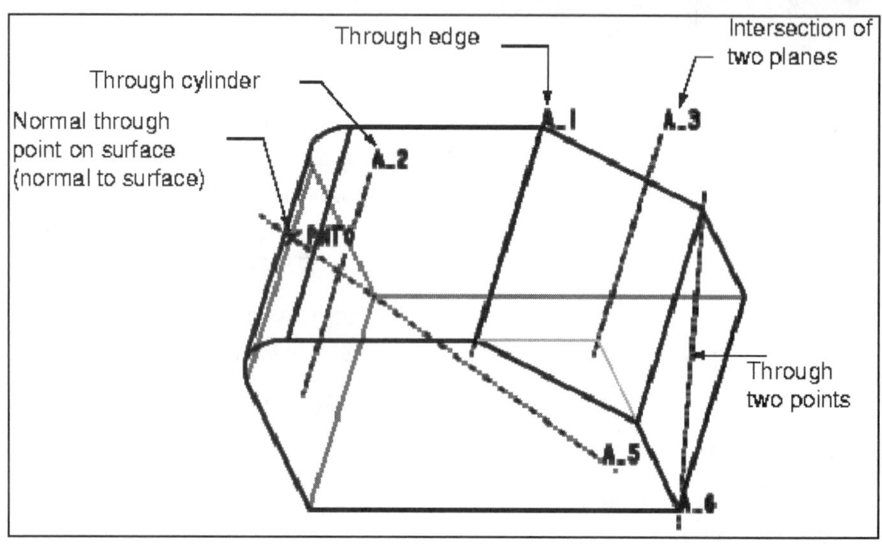

Figure 4.7
Datum Axes

To create a datum axis:

1. Choose the datum axis icon (see left-hand column) or pick **Datum** from the menu bar and then **Axis.**
2. Choose the desired constraint option from the DATUM AXIS menu:

Thru Edge Creates a datum axis through a straight edge. Select the edge.

Norm Pln Creates an axis that is normal to a surface, with linear dimensions locating it on that surface.

Pnt Norm Pln Creates an axis through a datum point and normal to a specified plane.

Thru Cyl Creates an axis through the "imaginary" axis of any surface of revolution (where an axis does not already exist). Select a cylindrical surface or a revolved surface. Note that some features that only appear to be cylindrical, such as a revolved feature round, cannot be selected.

Two Planes Creates a datum axis at the intersection of two planes (datum planes or surfaces). Select two nonparallel planes (they need not be shown to intersect on the screen).

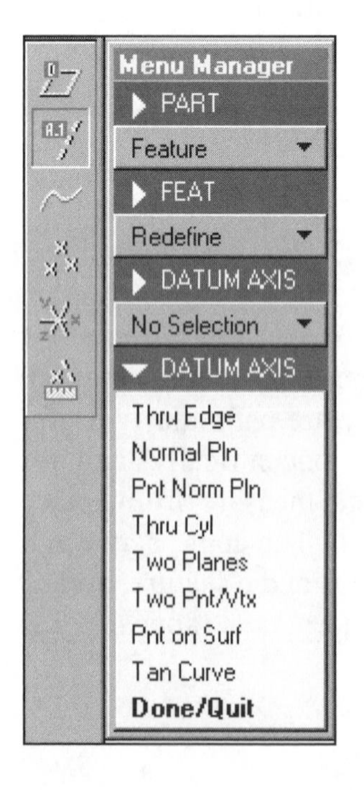

Two Pnt/Vtx Creates an axis between two datum points or edge vertices. Select datum points or edge vertices.

Pnt on Surf Create an axis through any datum point located on a surface; the point does not need to have been created using **On Surface**. The axis will be normal to the surface at that point.

Tan Curve Creates an axis that is tangent to a curve or edge at its endpoint. Select the curve/edge for it to be tangent to, and then select an endpoint of the curve/edge.

3. Pick the necessary references for the selected option.

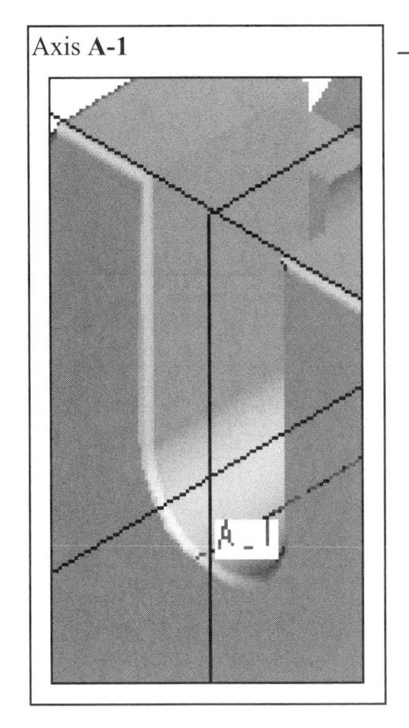

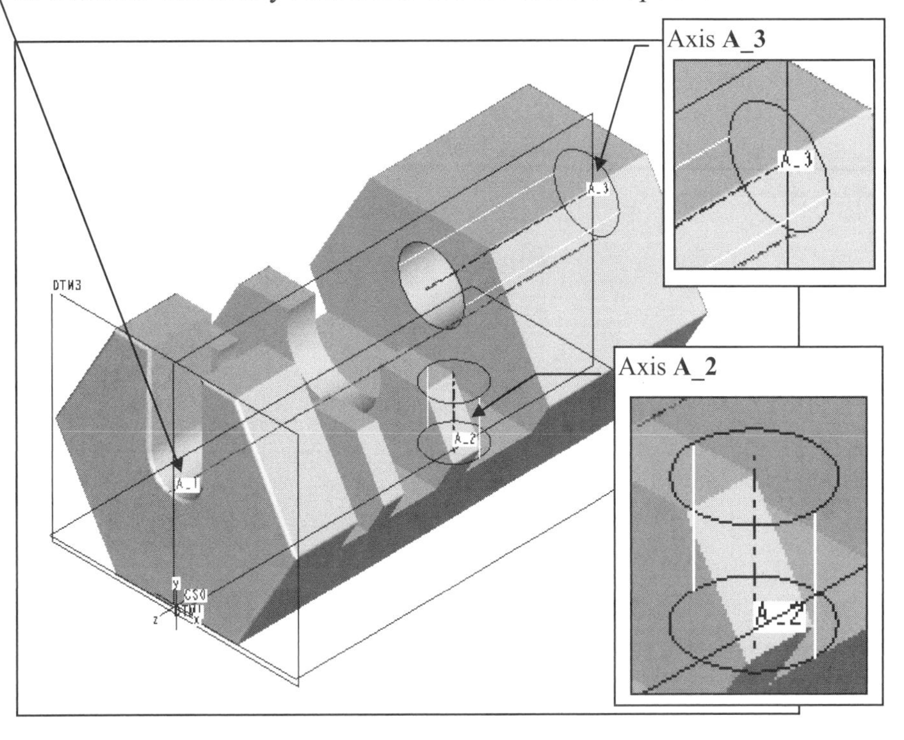

Figure 4.8
Datum Planes and Datum Axes

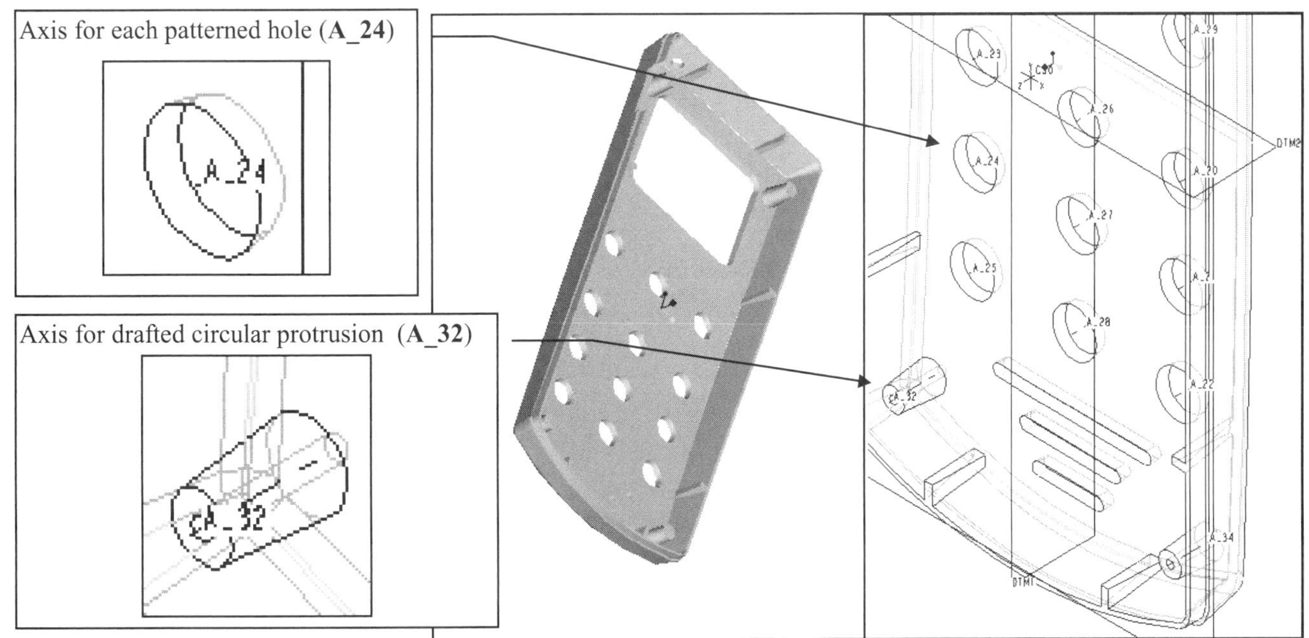

Figure 4.9
Datum Axes for Holes and Circular Protrusions

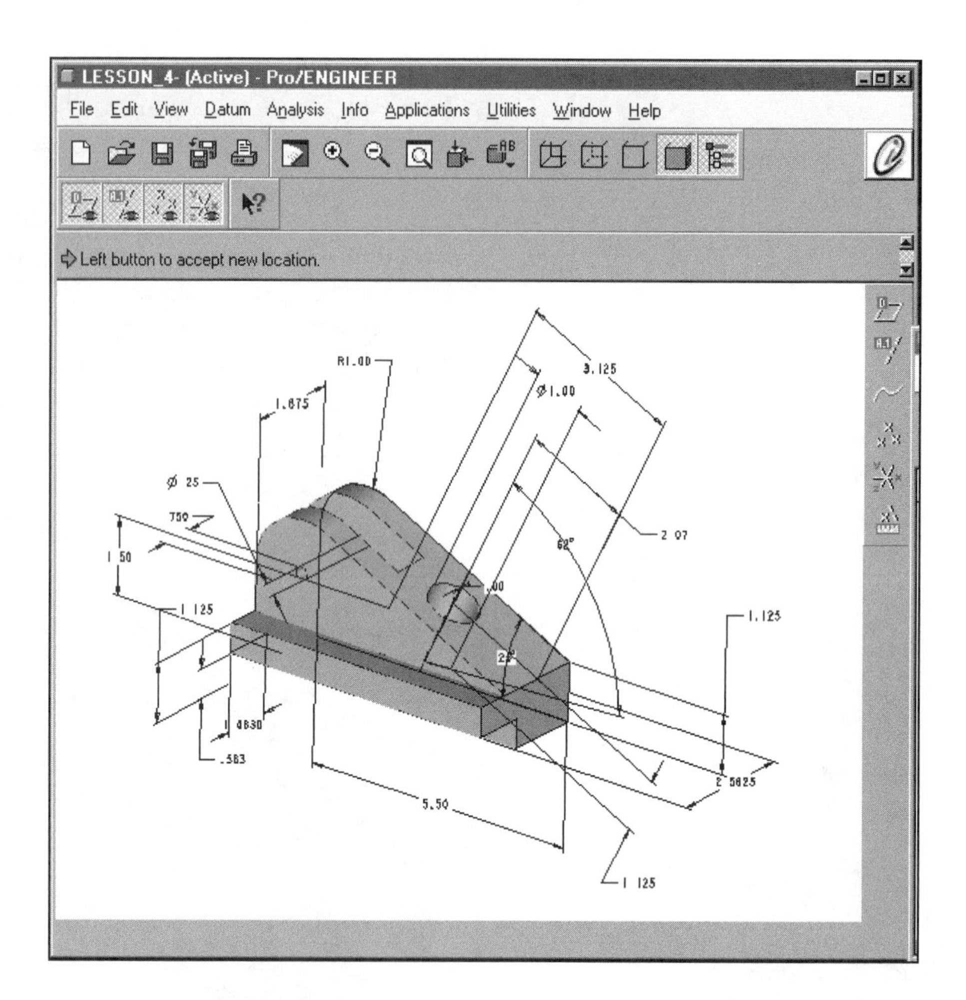

Figure 4.10
Anchor

Anchor

Lesson procedures & commands
START HERE ➡ ➡ ➡

The **Anchor** is the fourth ***lesson part*** (Fig. 4.10). Though default datums have been sufficient in previous lessons, Lesson 4 requires the creation of nondefault datums and the assignment of datums to layers. The datum planes will be set as geometric tolerance features and put on a separate layer. Reread Section 8, Layers, at this time. Choose the following commands:

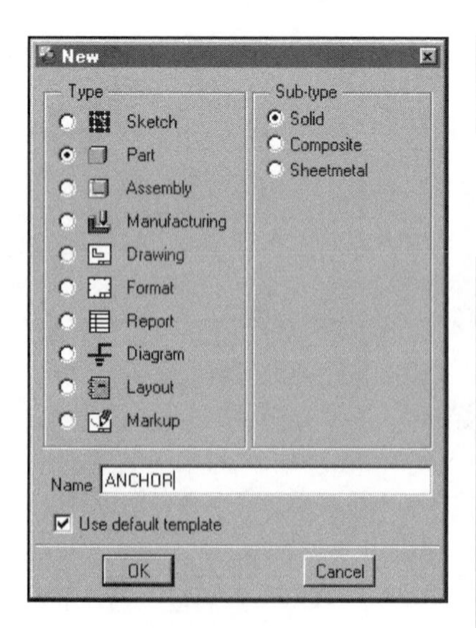

File ⇒ New ⇒ Part ⇒ (type the part name **ANCHOR**) **⇒ ✔ Use default template ⇒ OK**

Set up the units and the environment:

SETUP AND ENVIRONMENT

Set Up ⇒ Units ⇒ Units Manager **⇒ Inch lbm Second ⇒ Close ⇒ Material ⇒ Define ⇒** (type **STEEL**) **⇒ ✔ ⇒** (table of material properties, change or add information) **⇒ File ⇒ Save ⇒ File ⇒ Exit ⇒ Assign ⇒** (pick **STEEL**) **⇒ Accept ⇒ Done ⇒ Set Up ⇒ Name ⇒** (pick on the default coordinate system name in the **Model Tree**) **⇒** (type **CSYS_ANCHOR**) **⇒ ✔ ⇒ Done ⇒ ✔**

Utilities ⇒ Environment ⇒
❑ Spin Center ✔ Snap to Grid
Display Style **Hidden Line ⇒ OK**

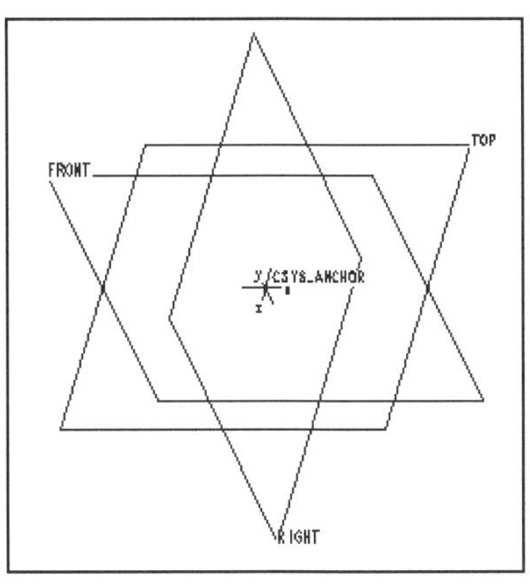

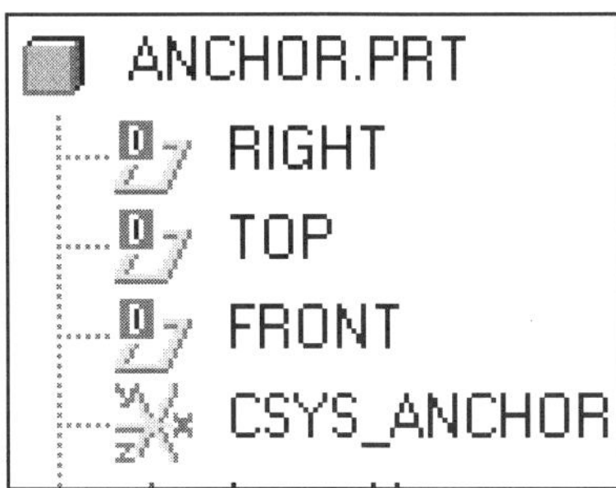

As in previous lessons, the first protrusion will be sketched on **FRONT.** Use Figure 4.11 for the protrusion dimensions, and sketch the outline in Figure 4.12. Use only **5.50**, **R1.00**, **1.125**, and **25°** dimensions. Choose the following commands:

Feature ⇒ **Create** ⇒ **Protrusion** ⇒ **Extrude** ⇒ **Solid** ⇒ **Done** ⇒ **One Side** ⇒ **Done** ⇒ (pick **FRONT** as the *sketching* plane) ⇒ **Okay** (for selecting the *direction* of feature creation) ⇒ **Top** ⇒ (pick **TOP** as *orientation* plane) ⇒ **Sketch** ⇒ **Intent Manager** (off) ⇒ **Sketch** ⇒ **Mouse Sketch** (complete the sketch as shown in Figure 4.12) ⇒ **Constraints** (to toggle on the constraints)

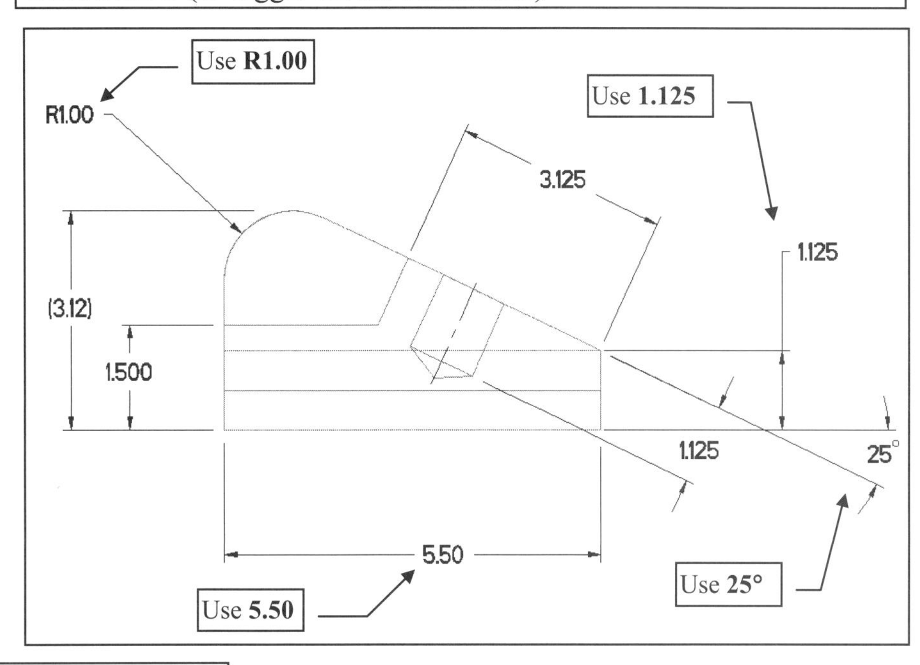

Figure 4.11
Protrusion Dimensions (Use only the four dimensions)

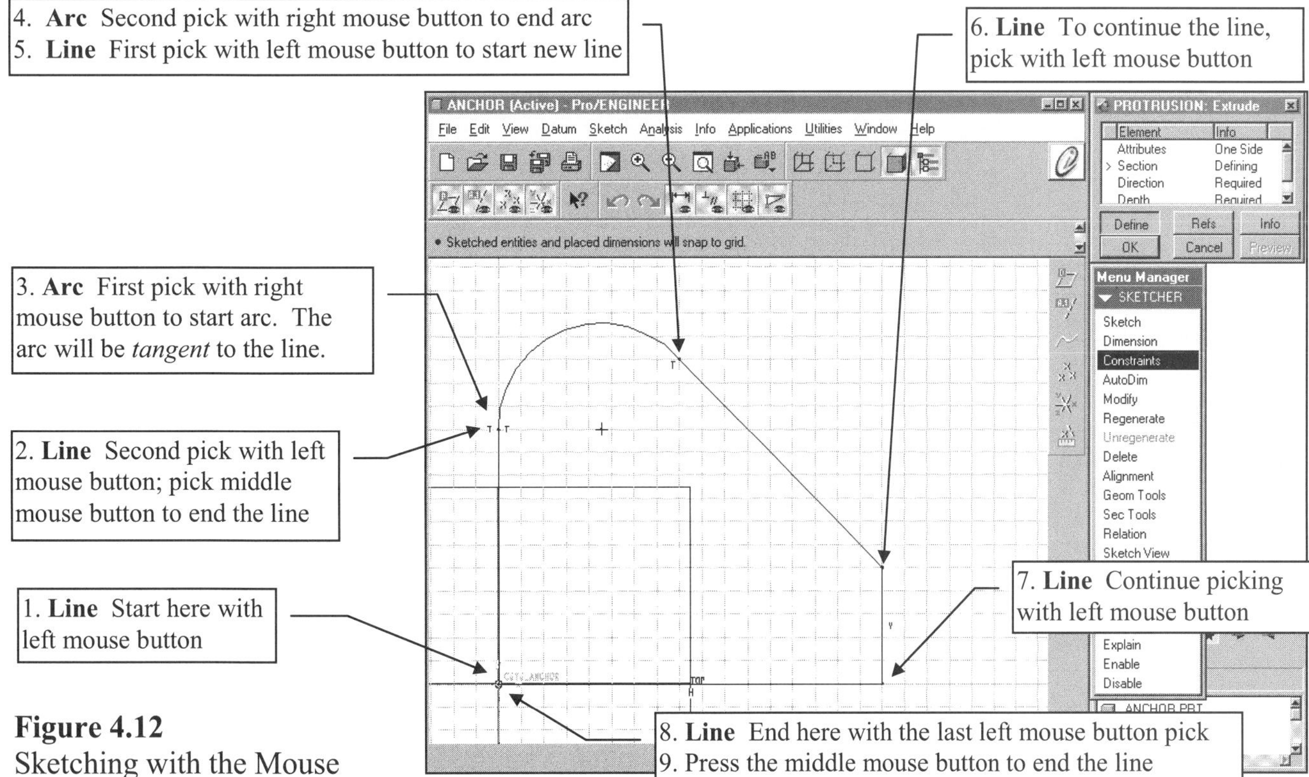

Figure 4.12
Sketching with the Mouse

? Pro/HELP

Use online documentation to understand the use of the mouse buttons better.

Sketching with the mouse is a fast and efficient method of creating geometry. The **left mouse button** is used to create a continuous set of *lines*. The **middle mouse button** is used for creating *circles*, and the **right mouse button** creates *tangent arcs* when the first pick is near the end of an existing line or arc.

Start the sketch (Fig. 4.12) by picking the first endpoint of the vertical line with the left mouse button (1). Pick the second endpoint at a position needed to create a vertical line (2). Use the middle mouse button to end the line. Next, using the right mouse button, pick near the last endpoint created for the line to start an arc (3). The arc will rubberband tangent from the line. Pick with the right mouse button to finish the arc (4). Now use the left mouse button to finish the section sketch lines (5-8). The sketch must be closed and composed of single entities. Do not draw lines or arcs on top of one another or the sketch will fail to regenerate. The sketch does not have to have exactly the same dimensional scale as the physical part (Fig. 4.13). As long as the outline is similar, Pro/E corrects the sketch when the dimensions are modified *(an alternative technique would have you creating the four lines and then creating the arc using **Arc ⇒ Fillet.** This method ensures that the arc and the lines are tangent).*

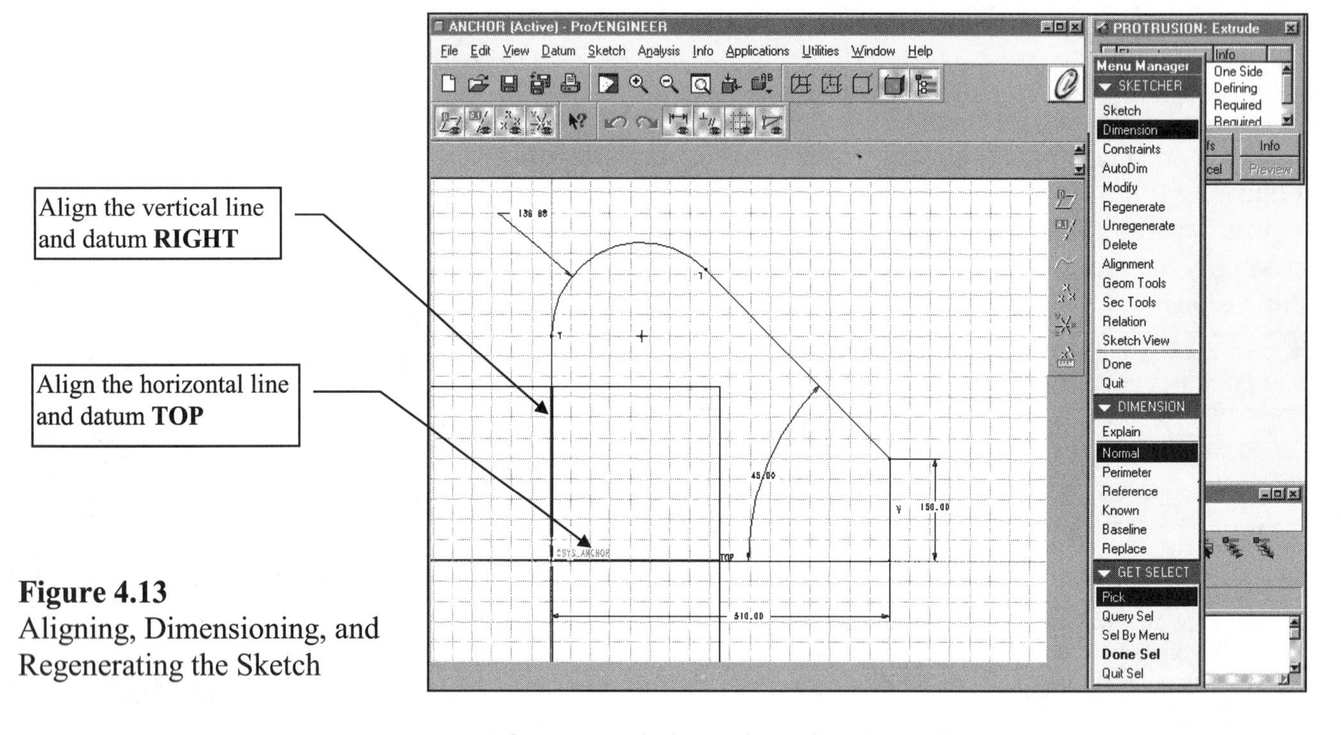

Figure 4.13
Aligning, Dimensioning, and Regenerating the Sketch

After completing the sketch and getting it to regenerate successfully, modify the values to the design dimensions shown in Figure 4.11. Use the following commands to complete the feature:

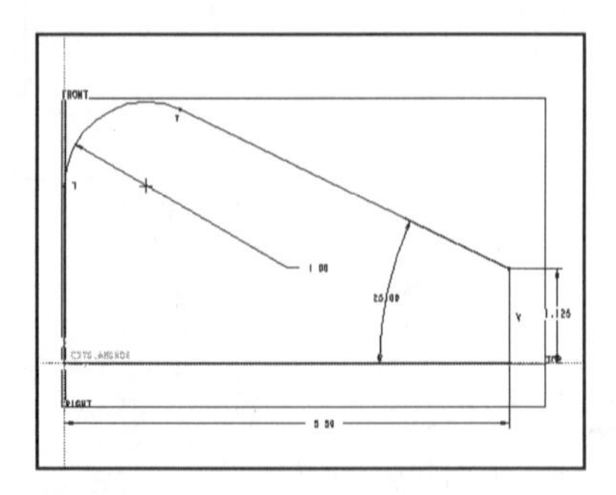

Alignment (align the datums with the vertical and horizontal lines) ⇒ **Regenerate** ⇒ **Dimension** [add the four dimensions (Fig. 4.13)] ⇒ **Regenerate** ⇒ **Modify** (see Fig. 4.11 for dimensions) ⇒ **Regenerate** (see left-hand column) ⇒ **Done** ⇒ **View** ⇒ **Default** ⇒ **Blind** ⇒ **Done** (type the depth dimension at the prompt: **2.5625**) ⇒ ✔ ⇒ **OK** (from the dialog box) ⇒ **Done** ⇒ **View** ⇒ **Shade** (Fig. 4.14)

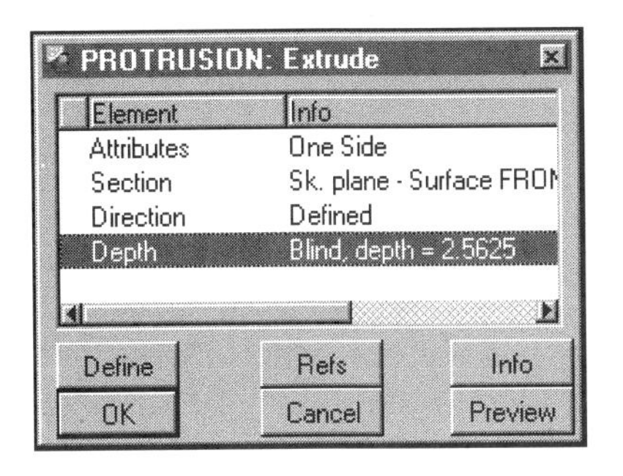

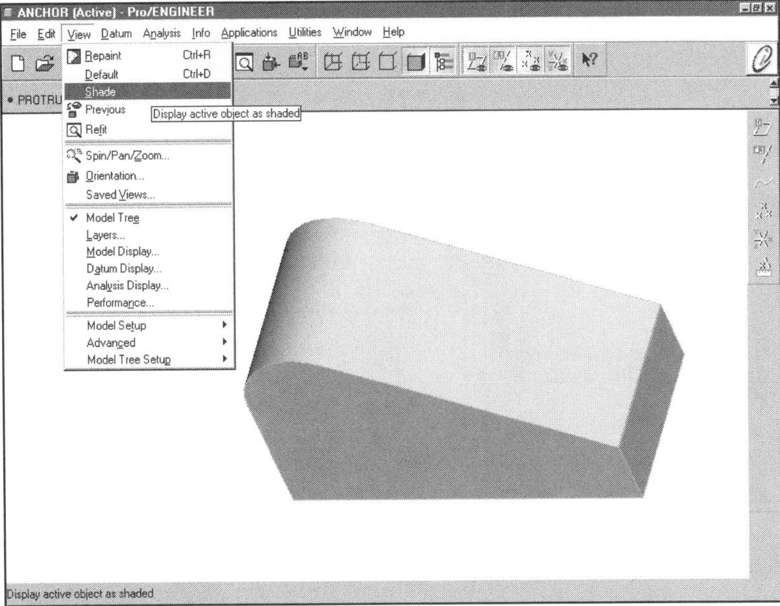

Figure 4.14
Shaded First Protrusion

At this point, let's change the color of the model. In general, try to avoid the colors that are used as defaults. Because feature and entity highlighting is defaulted to *red*, colors similar to *red* should be avoided. Datum planes have *red* and *yellow* sides; therefore, *yellow* should be avoided. Shades of *blue* and *green* work well. It is really up to you to choose colors that are pleasant to look at when you're gazing at the monitor for hours on end!

File ⇒ **Save** ⇒ ✔ ⇒ **View** ⇒ **Model Setup** ⇒ **Color & Appearances** [activates the Appearances dialog box (Fig. 4.15)] ⇒ **Add** ⇒ Pick on the color box to initiate the **Color Editor** ⇒ Click and hold on the slide bars to attain the desired color (Fig. 4.16)]

1. Appearances dialog box

2. Pick **Add**

3. Appearance Editor dialog box

4. Pick on the color box to activate the Color Editor

5. Color Editor dialog box

▼ **Color Wheel** or
▼ **blending Palette**
to use instead of the slide bars

6. Slide bars control the color

Figure 4.15
Appearance Editor

1. Move the cursor over the vertical bar, press and hold the *left* mouse button, and slide the bars to change the color. Also try creating colors with the **Color Wheel** or the **Blending Palette.**

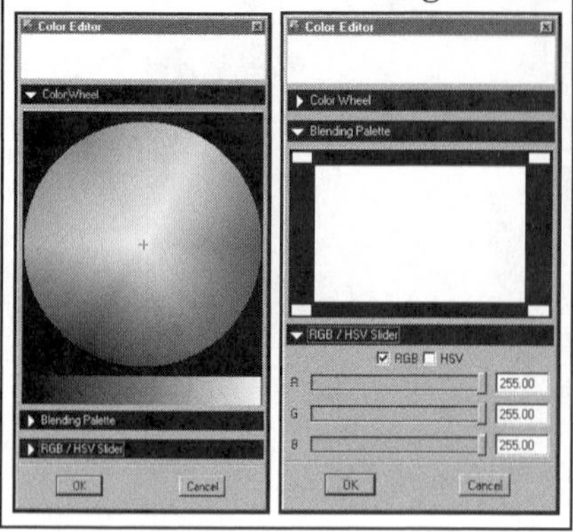

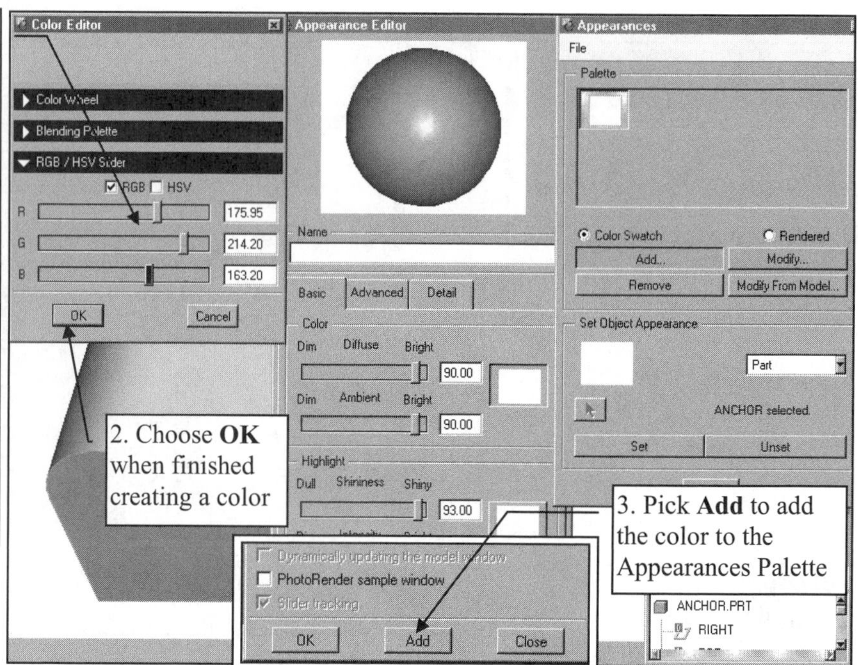

Figure 4.16
Color Editor

Slide the RGB bars to create useful colors. Add a color to the Appearances Palette by choosing **OK** from the Color Editor and then **Add** from the Appearance Editor (Fig. 4.17). Continue making a few more colors, adding each to the palette. Finally, select the **Close** button from the Appearance Editor.

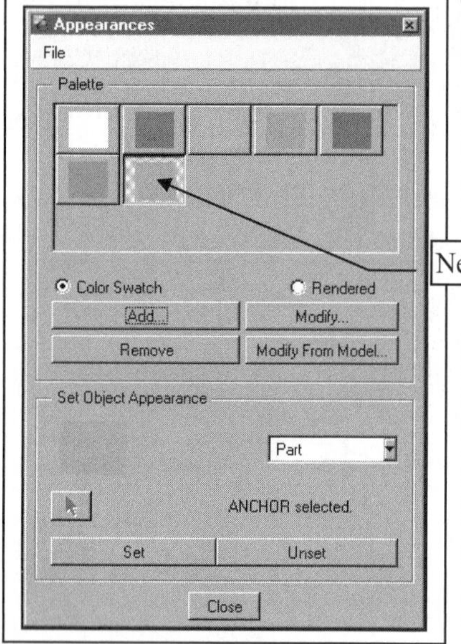

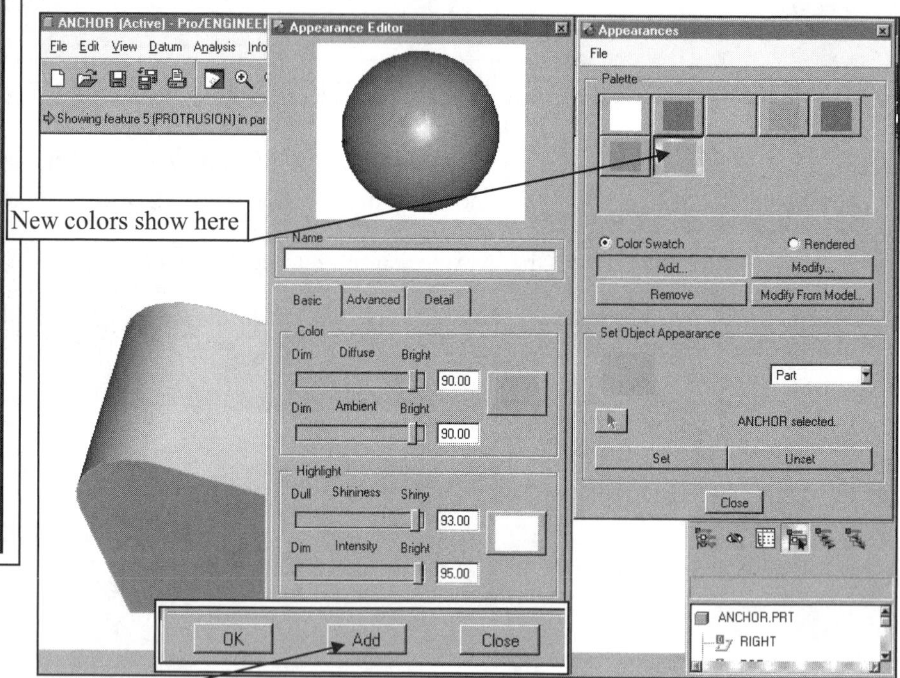

Figure 4.17
Adding Colors

Pick **Add** to include a color on the Appearances Palette

To change the color of the *part*, select a color from the Appearances Palette, then choose **Part** from Set Object Appearance drop-down menu (Fig. 4.18), and then pick **Set.** The part will now be the new selected color. To change the color of an object's *surface*, select a color from the Appearances Palette and then **Surfaces** from Set Object Appearance menu. Pick the surface of the model to be changed (Fig. 4.19), and then pick **Set.** Select **Close** when finished.

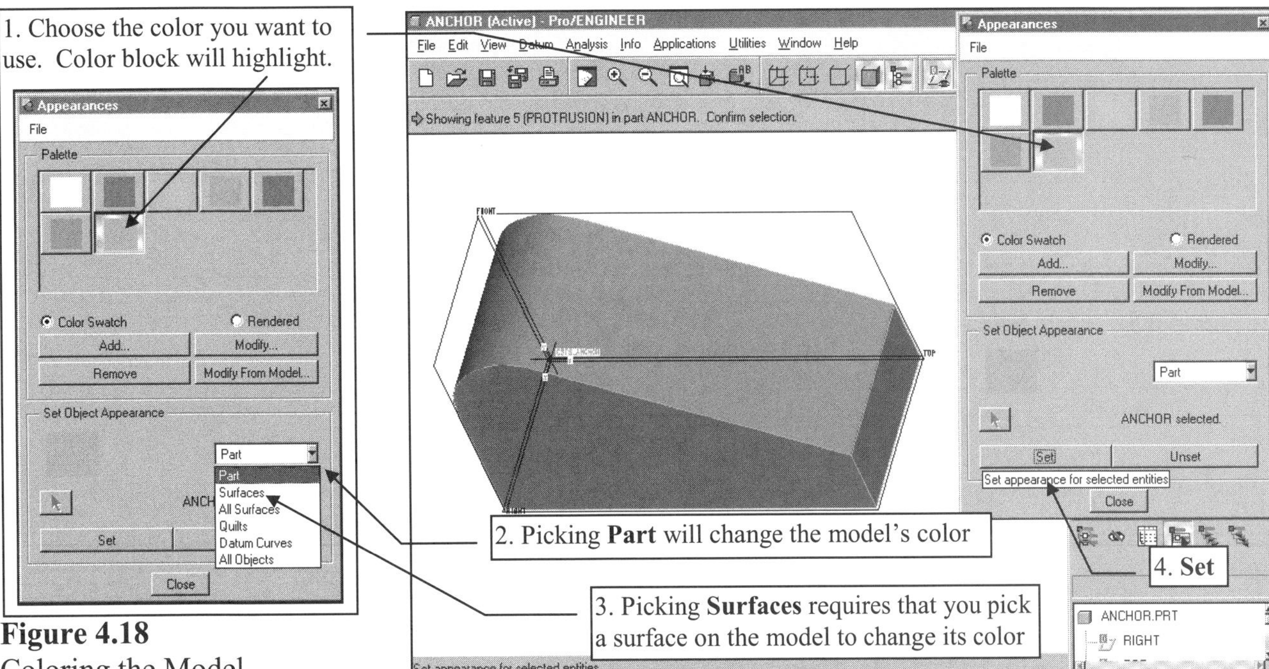

Figure 4.18
Coloring the Model

To change the color of an object's *surface:*

Select a color from the Appearances Palette ⇒ **Surfaces** from Set Object Appearance drop-down menu ⇒

⇒ Pick the surface of the model ⇒ **Done Sel** ⇒ **Done** ⇒ **Set** (from the Appearances dialog box) ⇒ **Close**

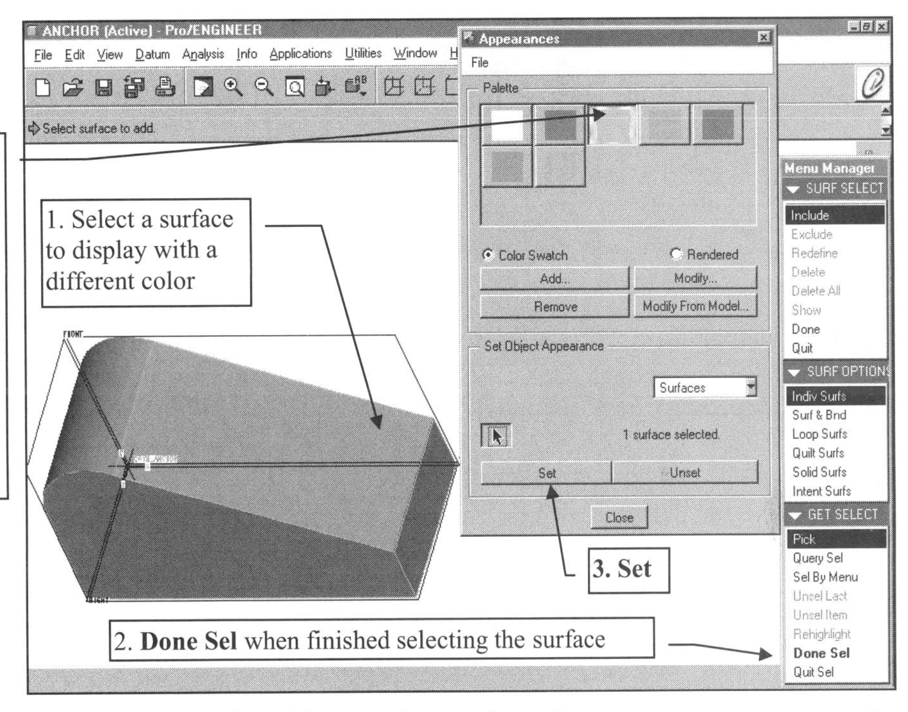

Figure 4.19
Model with New Colors

HINT

Create a set of colors for your project before you model the first protrusion. Set the part color and surface colors as you model.

You can save your colors by picking **File** ⇒ **Save** from the Appearances dialog box.

File ⇒ **Open** will recall a saved palette ***(color.map)***.

Unfortunately, this text is not in color, so you cannot see the changes to the model (Fig. 4.19). Experimenting with the colors is most students' idea of fun. Enjoy yourself, but don't create too many colors; they really aren't necessary at this stage of the project.

The next two features will be cuts. For both of the cuts, use **RIGHT** as the sketching plane, use **TOP** as the reference plane, and give **Thru All** as the depth. Each cut requires just two lines, two alignments, and two dimensions.

Use the first cut's sketching/placement plane and reference/orientation (**Use Prev**) for the second cut. Create the cuts separately, each as an open section. Figure 4.20 shows the dimensions for each cut. Choose the following commands:

1.875
.750
.938
TO CENTER OF SLOT
First cut
1.125
.563
1.063
Second cut

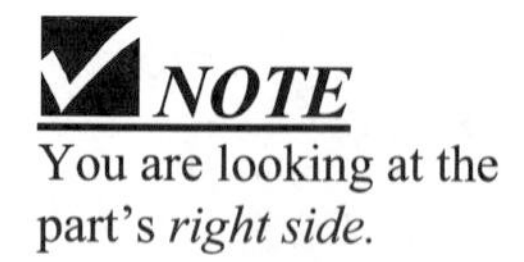

NOTE
You are looking at the part's *right side.*

Figure 4.20
Dimensions for the Two Cuts

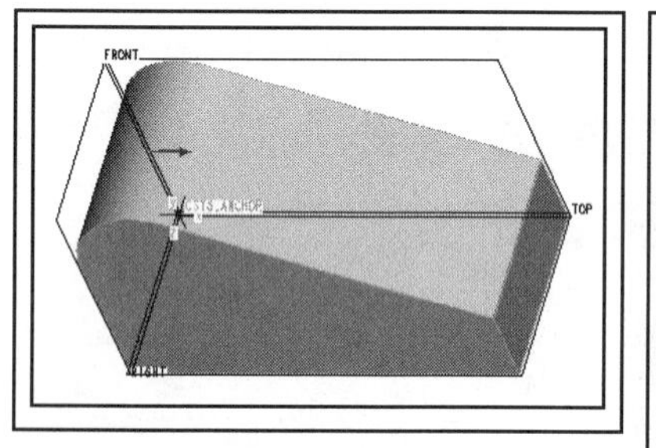

NOTE
You are looking at the part's *left side.*

Utilities ⇒ Environment ⇒ ❑ Snap to Grid ⇒ OK ⇒ View ⇒ Repaint ⇒ Feature ⇒ Create ⇒ Cut ⇒ Extrude ⇒ Solid ⇒ Done ⇒ One Side ⇒ Done ⇒ (pick **RIGHT** as the sketching/placement plane) ⇒ **Flip** (change the direction of the cut to pass through the part (see at left) ⇒ **Okay ⇒ Top ⇒** (pick **TOP** as the orientation plane; you are looking at the left side) ⇒ **Sketch ⇒ Intent Manager** (off) ⇒ **Line ⇒ Vertical** (sketch the two lines) ⇒ **Alignment** (align the endpoint of the vertical line with the top of the part and the end point of the horizontal line with the right-side edge) ⇒ **Regenerate ⇒ Dimension** (add the dimensions) ⇒ **Regenerate ⇒ Modify** (change the dimensions to **1.125** and **1.875**) ⇒ **Regenerate** (Fig. 4.21) ⇒ **Done**

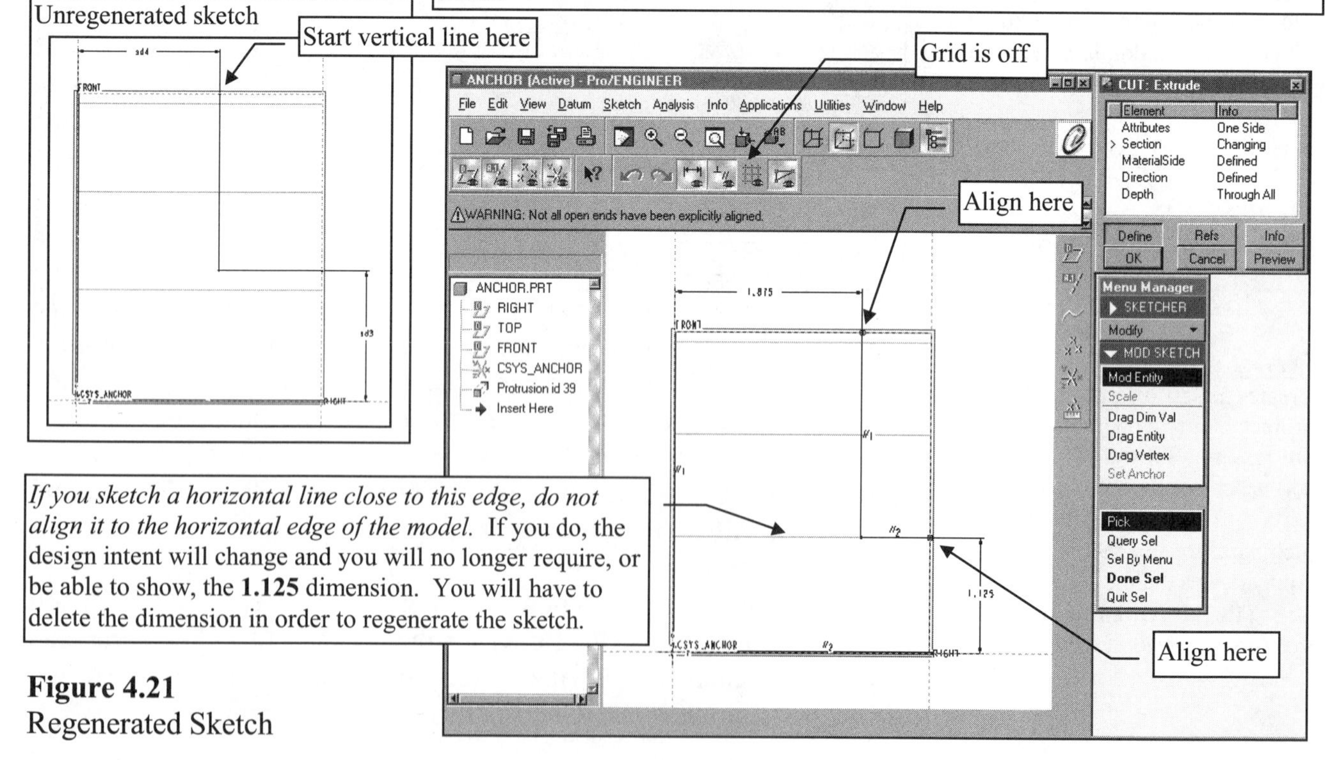

Figure 4.21
Regenerated Sketch

Complete the cut using the following commands:

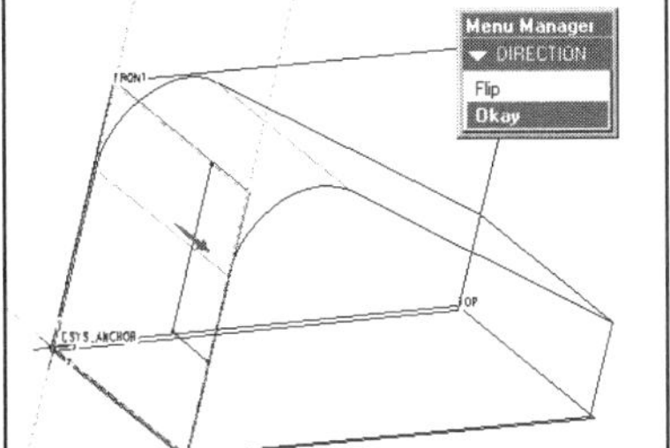
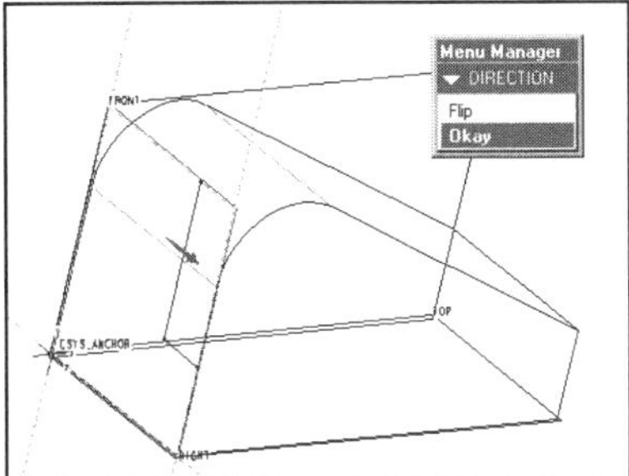

Okay ⇒ Thru All ⇒ Done ⇒ View ⇒ Default ⇒ OK ⇒ View ⇒ Shade ⇒ Done (Fig. 4.22) **⇒ File ⇒ Save ⇒ ✔**

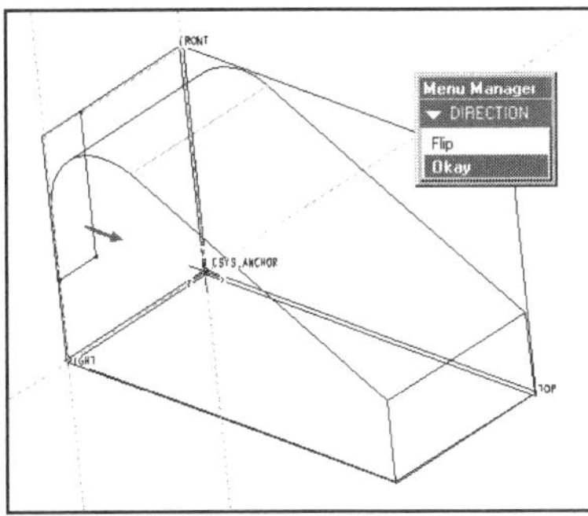

Figure 4.22
Completed Cut

Complete the second cut using the last two command blocks (Figure 4.23). Everything will be the same except for the dimensions **.563** and **1.063**. You can expedite things by selecting **Use Prev** for the sketching and orientation planes. *You may want to try and complete the feature with the* ***✔ Intent Manager.***

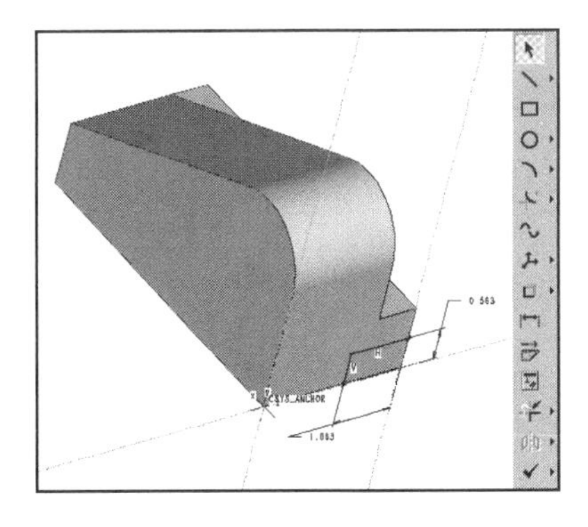

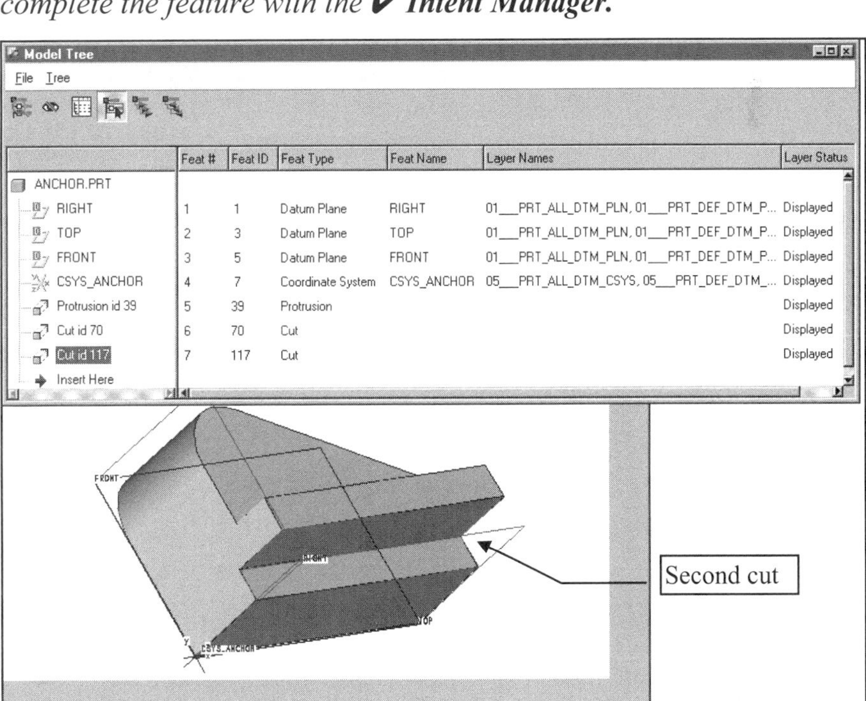

Figure 4.23
Second Cut

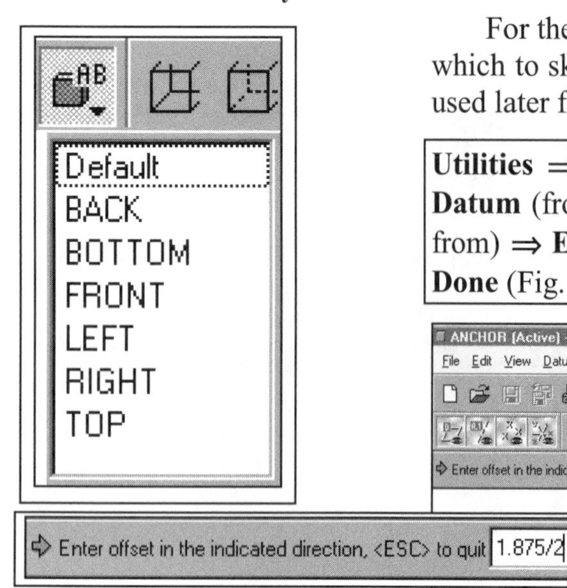

For the next feature, we will need to create a new datum plane on which to sketch. A datum axis will also be created at this time, to be used later for the creation of a hole. Choose the following commands:

Utilities ⇒ **Environment** ⇒ Default Orient **Isometric** ⇒ **OK** ⇒ **Datum** (from menu bar) ⇒ **Plane** ⇒ **Offset** (pick **FRONT** to offset from) ⇒ **Enter Value** (type the offset distance of **1.875/2**) ⇒ ✔ ⇒ **Done** (Fig. 4.24) ⇒ **File** ⇒ **Save** ⇒ ✔

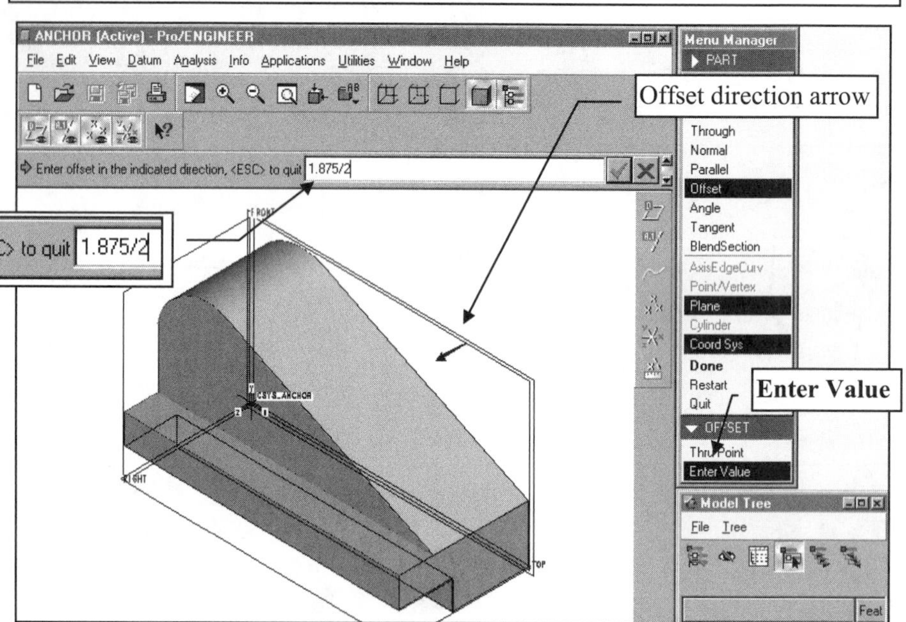

Figure 4.24
Creating an Offset Datum Plane

NOTE

An axis can be inserted through any curved feature. Holes and circular features automatically have axes when they are created. Features created with arcs, fillets, and so on need to have axes added afterward, unless set in the *configuration file* as the default.

The new datum plane is **DTM1.** New datum planes are by default numbered sequentially. A datum axis will now be created through the curved top of the part, using the following commands:

Datum (from the menu bar) ⇒ **Axis** ⇒ **Thru Cyl** (pick as shown in Fig. 4.25) ⇒ **Done**

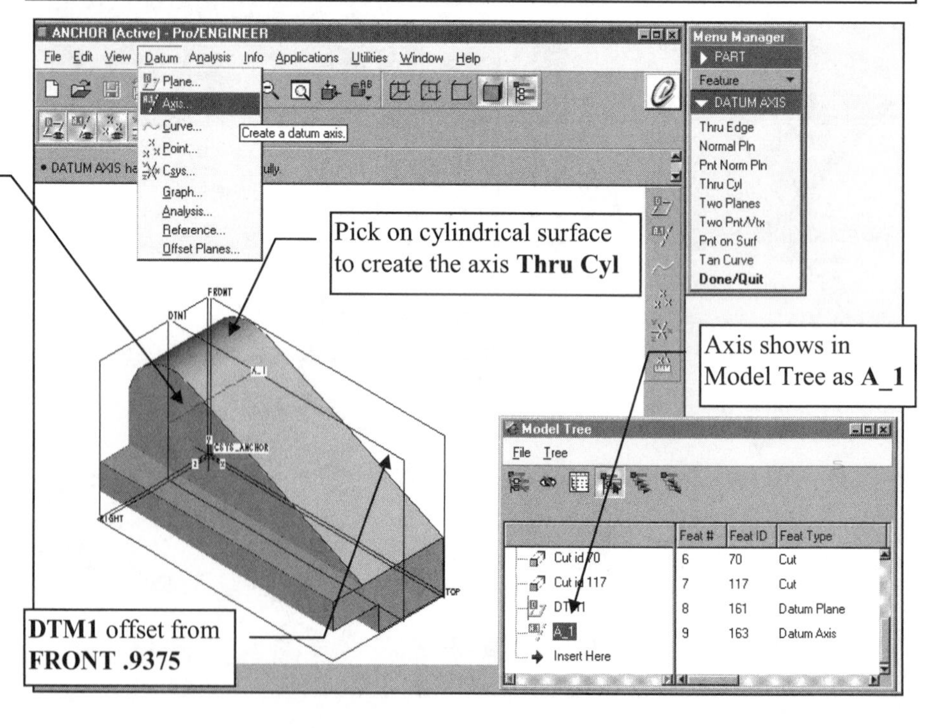

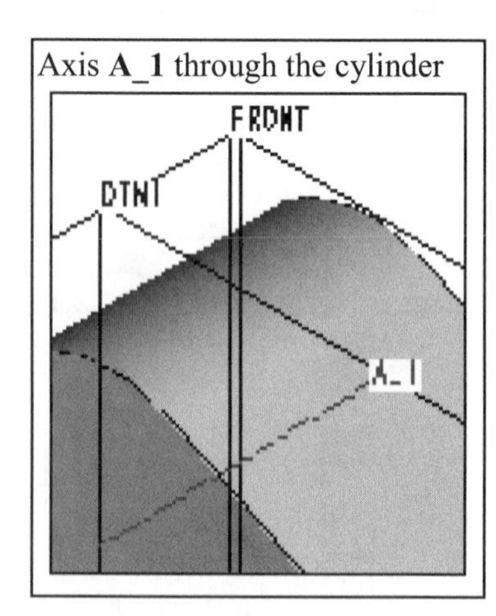

Figure 4.25
Creating a Datum Axis

The cut on the top of the part is created by sketching on **DTM1** and projecting it toward both sides. Use the Model Tree to select the appropriate datum planes. Choose the following commands:

> **Feature ⇒ Create ⇒ Solid ⇒ Cut ⇒ Extrude ⇒ Solid ⇒ Done ⇒ Both Sides ⇒ Done ⇒** (pick **DTM1** as the sketching/placement plane) **⇒ Okay ⇒ Top ⇒** (pick **TOP** as the orientation plane) **⇒ Sketch** (from menu bar) **⇒ Intent Manager** (off) **⇒ Sketch ⇒ Line ⇒ Perpendicular** (pick the angled edge and then the starting and ending points of the line) **⇒ Line ⇒ Horizontal ⇒** (pick the endpoints of the horizontal line as shown in Fig. 4.26) **⇒ Sketch ⇒ Point** (place a sketch point at the intersection of the angled edge and the vertical edge) **⇒ Regenerate ⇒ Alignment** (align the angled line's endpoint with the angled edge of the part and the horizontal line with the left-side edge. Align the point vertically and with the angled edge) **⇒ Regenerate ⇒ Dimension** (add the dimensions) **⇒ Regenerate ⇒ Modify** (change the dimensions to **3.125** and **1.50**, as in Figure 4.11) **⇒ Regenerate** (Figure 4.26)

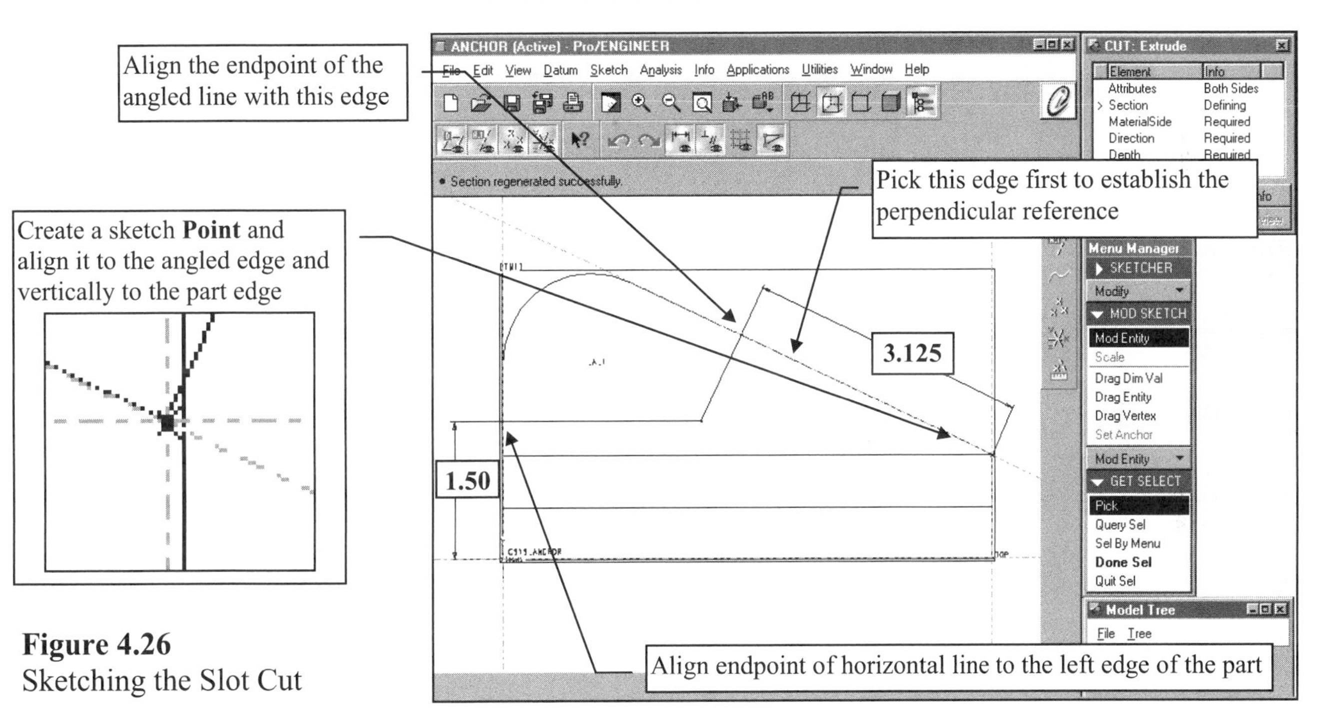

Figure 4.26
Sketching the Slot Cut

Complete the cut with the following commands (Fig. 4.27):

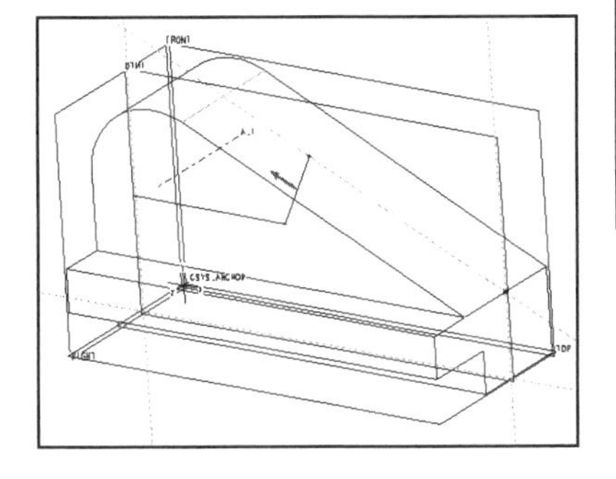

> **Done ⇒ Okay** (for the material removal side) **⇒ Blind ⇒ Done ⇒** (type the full width of **.750** for the slot) **⇒ ✔ ⇒ OK ⇒ Done ⇒ Window ⇒ New ⇒ View ⇒ Default ⇒ View** (in new window) **⇒ Shade ⇒ Window ⇒ Close Window ⇒ Window ⇒ Activate**

The hole drilled in the angled surface appears to be aligned with **DTM1.** Upon closer inspection, it can be seen that the hole is at a different distance from the edge (**.875** from **FRONT**) and is not in line with the slot and datum plane ***(DTM1*** *was created* ***.9375*** *from* ***FRONT).*** Create the feature using a sketched hole (Fig. 4.28).

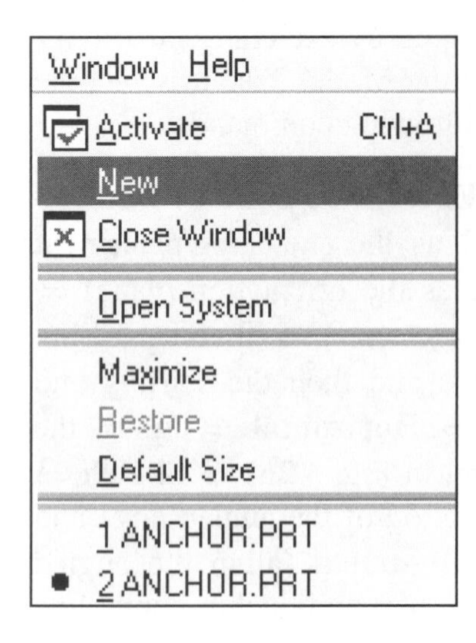

Figure 4.27
Cut Created on Both Sides of **DTM1**

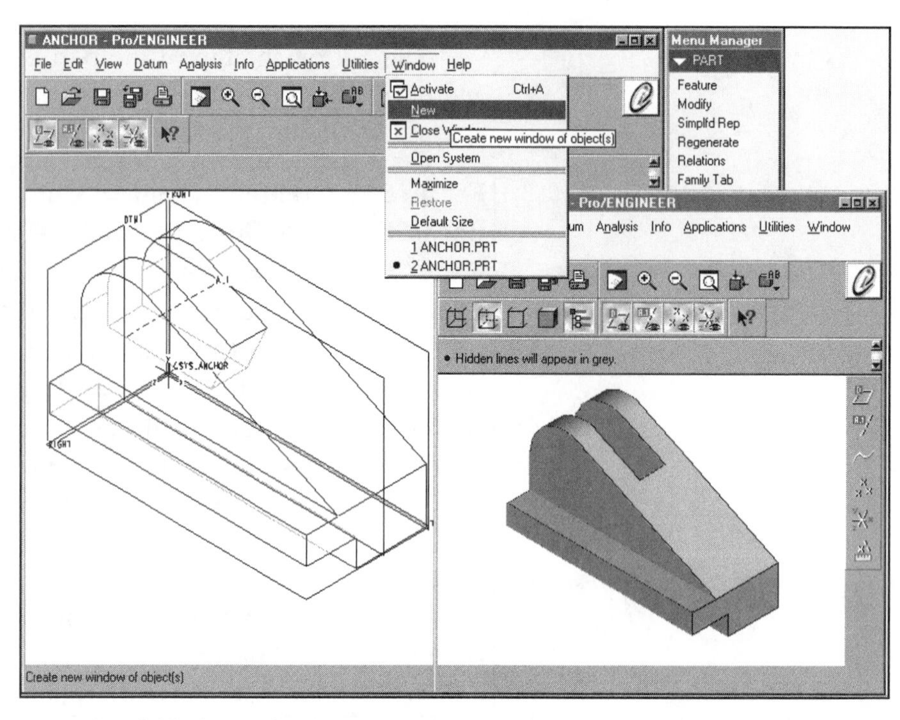

The drill tip at the bottom of the hole needs to be modeled so the hole is created as a sketched hole (Fig. 4.28). Use the following commands:

HINT

Remember that the diameter dimension is created by picking the rightmost vertical line. Then pick the centerline, and finally pick the rightmost line again before placing the dimension on top.

Utilities ⇒ **Environment** ⇒ ✔ **Snap to Grid** ⇒ **OK** ⇒ **Feature** ⇒ **Create** ⇒ **Hole** ⇒ (from the HOLE dialog box-- ● Sketched) ⇒ (sketch a *vertical* centerline) ⇒ (sketch the four lines to create the *closed section* on one side of the centerline) ⇒ **View** ⇒ **Repaint** ⇒ Dimension and modify as required [**1.125** depth, **62°** (one-half of the drill tip angle), and **Ø1.00** (Fig. 4.28)] ⇒ ✔ (Sketcher command icon) ⇒ Primary Reference ↖[pick the angled surface (Fig. 4.29)] ⇒ Placement Type **Linear** (default)

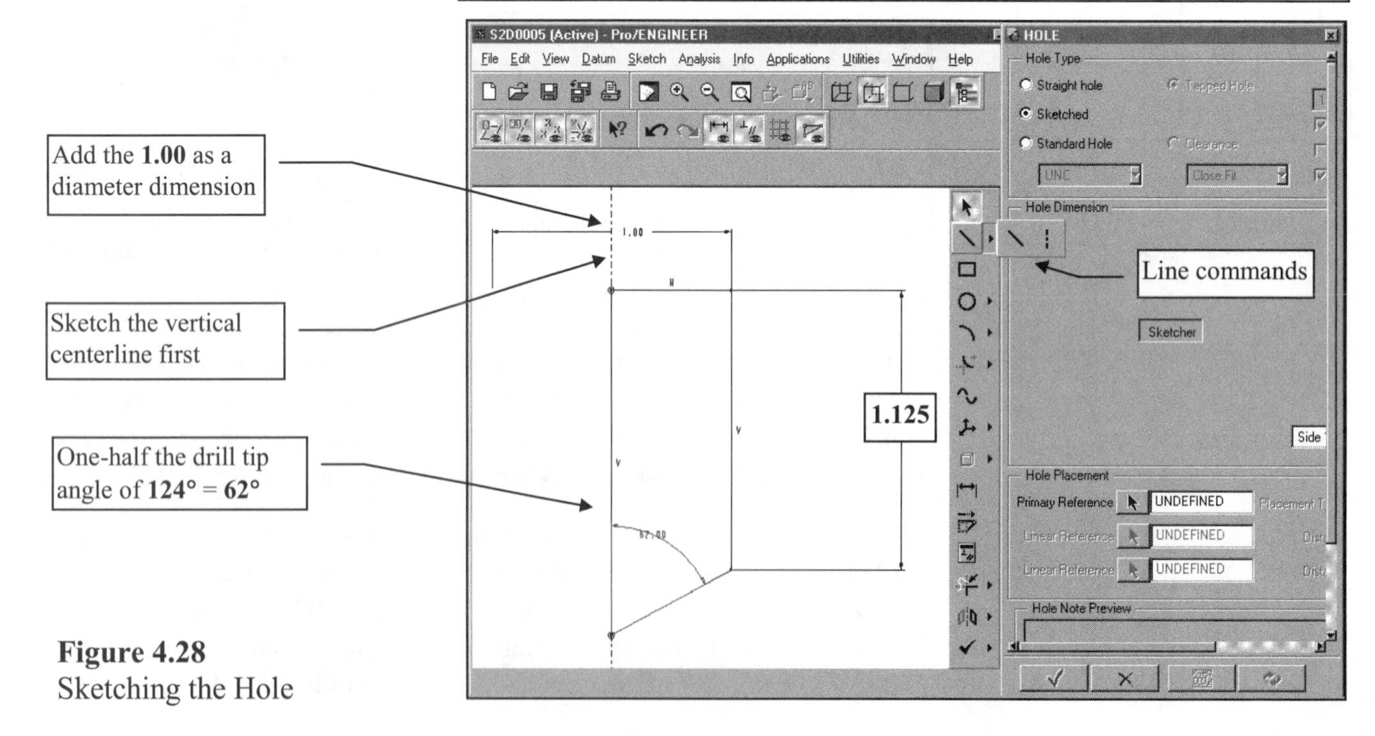

Figure 4.28
Sketching the Hole

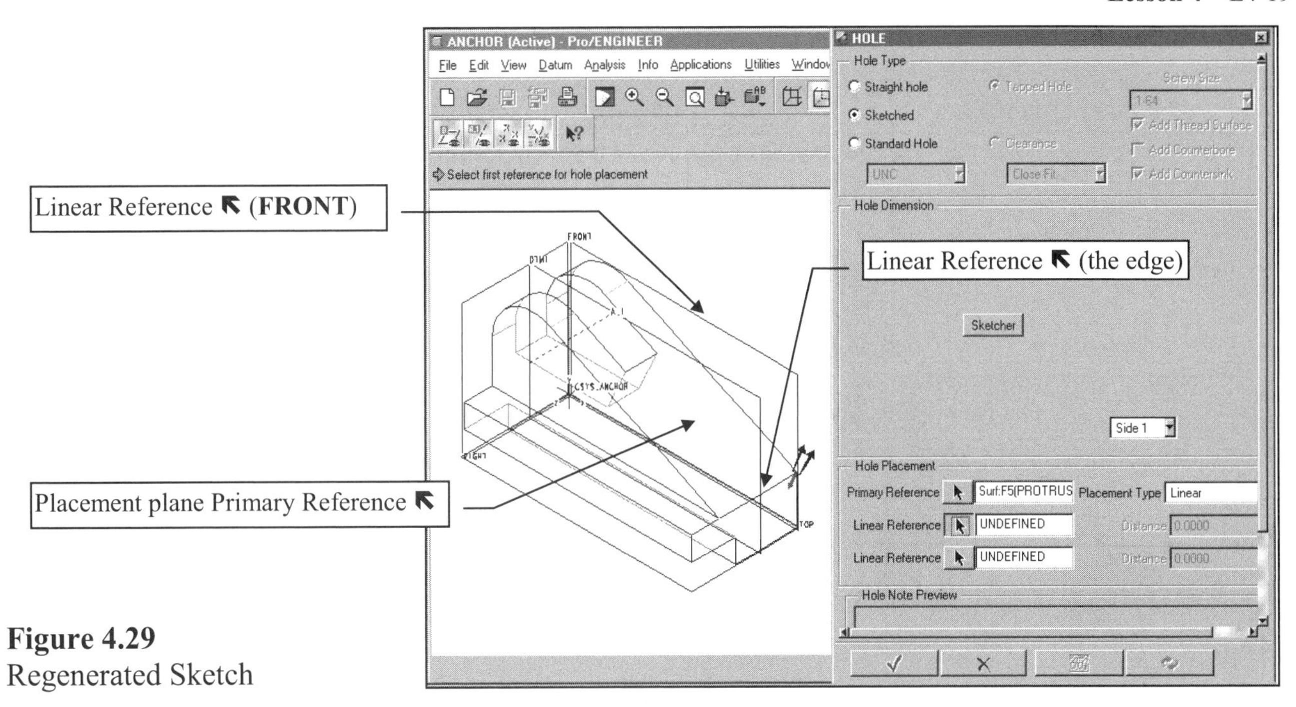

Figure 4.29
Regenerated Sketch

To complete the hole, choose the following commands:

Linear Reference ↖ (pick **FRONT**) ⇒ Distance-- type **.875** ⇒ **enter** ⇒ Linear Reference ↖ (pick the edge at the bottom of the angled surface) ⇒ Distance-- type **2.0625** ⇒ **enter** ⇒ Shade icon ⇒ **Preview** [from the Hole dialog box (Fig. 4.30)] ⇒ ✔ [from the Hole dialog box (Build feature)]

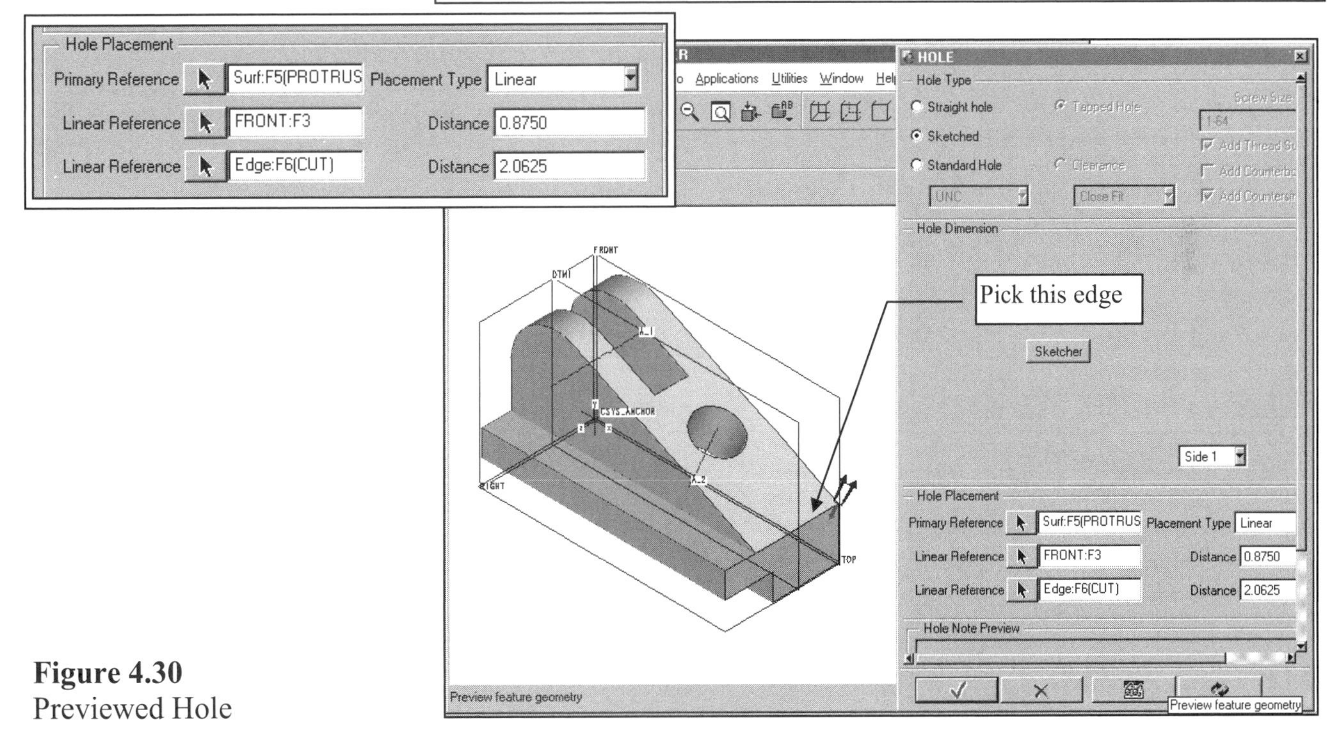

Figure 4.30
Previewed Hole

Next, a new datum plane will now be created that passes through the angled surface. All datum planes are to be *set* as geometric tolerancing features (basic).

Create the datum plane though the angled surface (Fig. 4.31) using the following commands:

File ⇒ Save ⇒ ✔ ⇒ File ⇒ Delete ⇒ Old Versions ⇒ ✔ ⇒ View ⇒ Default ⇒ Datum ⇒ Plane ⇒ Through ⇒ (pick the angled surface) **⇒ Done ⇒ Done**

Pick the angled surface to create the datum plane *through*

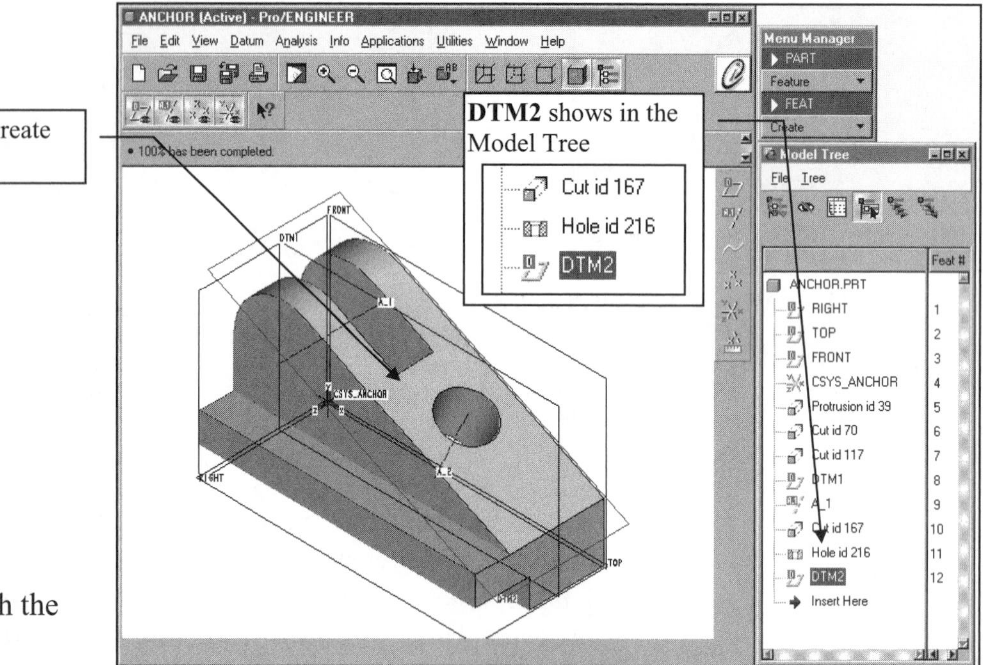

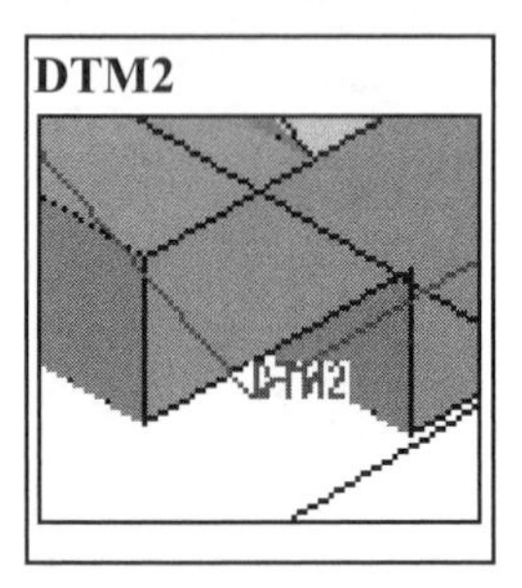

Figure 4.31
DTM2 Created Through the Angled Surface

HINT

Datums used in geometric tolerancing (basic):

A (primary--three-point contact),
B (secondary--two-point contact),
C (tertiary--one-point contact),

should be established on your ***DIPS*** before you start the Lesson Project models.

Set (Fig. 4.32) and rename the datum planes as geometric tolerancing features using the following commands:

Utilities ⇒ Customize Screen ⇒ Options ⇒ Default ⇒ OK ⇒ Set Up (from PART menu) **⇒ Geom Tol ⇒ Set Datum ⇒** (pick **TOP** from Model Tree) ⇒ Name (change name from **TOP** to **A**) ⇒ **OK**

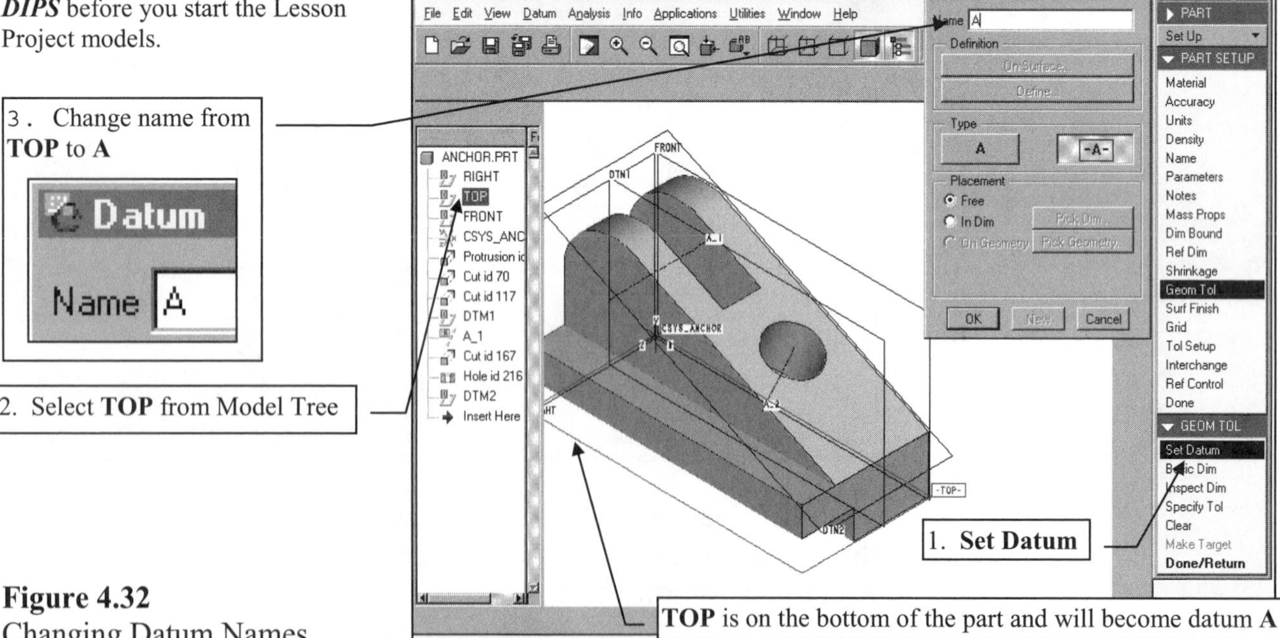

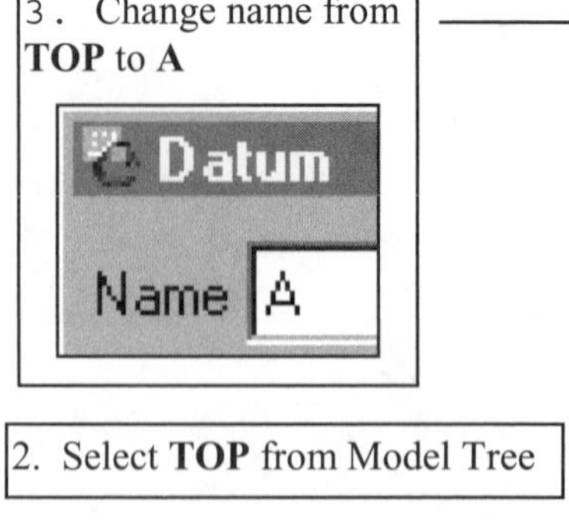

Figure 4.32
Changing Datum Names

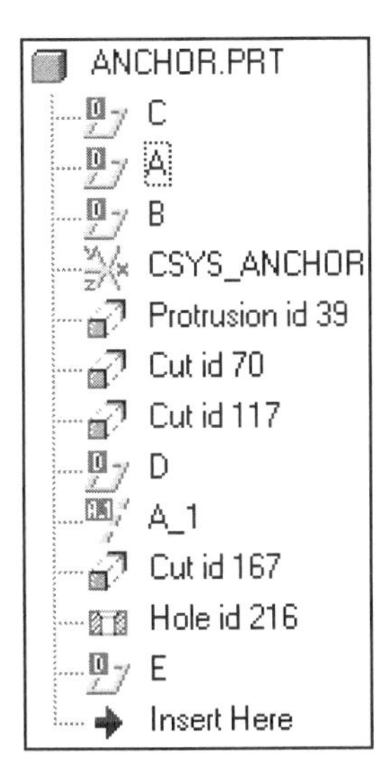

Repeat to complete the renaming and setting of all five datum planes (**A** through **E**) (Fig. 4.33). Move the positions of the names of the datums by choosing the following commands (Fig. 4.33):

Done/Return ⇒ **Done** ⇒ **Modify** (from the PART menu) ⇒ **Move Datum** ⇒ (pick the datum name, edge, or name from the Model Tree) ⇒ (pick a new position on the screen for the name) ⇒ **Done**

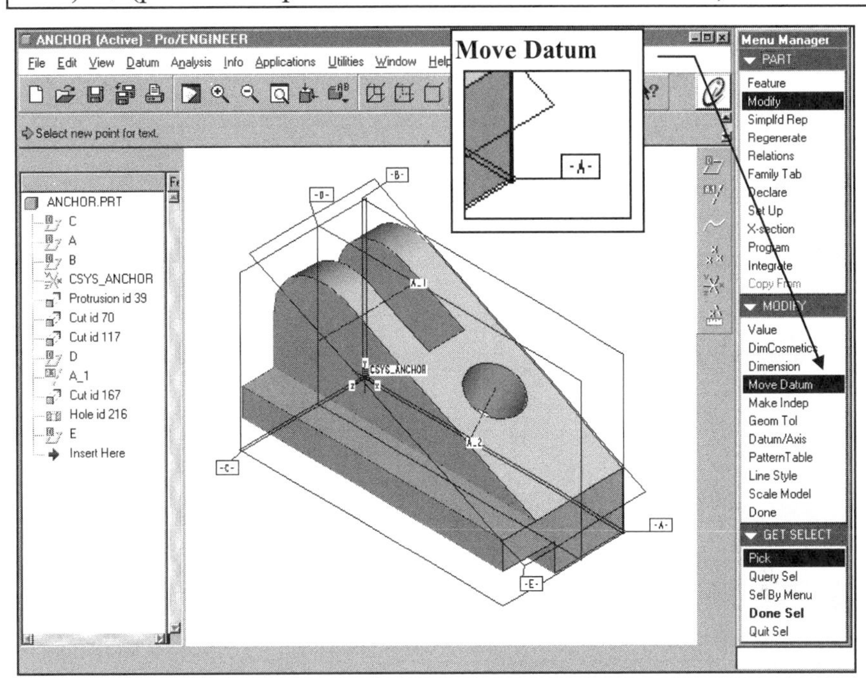

Figure 4.33
The Five Datum Plane Names in New Positions, Away from the Part Model View

Next you will create a section to be used when you are detailing the model in Lesson 18. The section will pass through the part lengthwise using datum plane **D**. The section will be named **A** and will show as **Section A-A** when you are detailing the view in **Drawing Mode**. Figure 4.34 shows the section.

HINT

Create sections in **Part Mode** to be used later in **Drawing Mode** when you are detailing the part model.

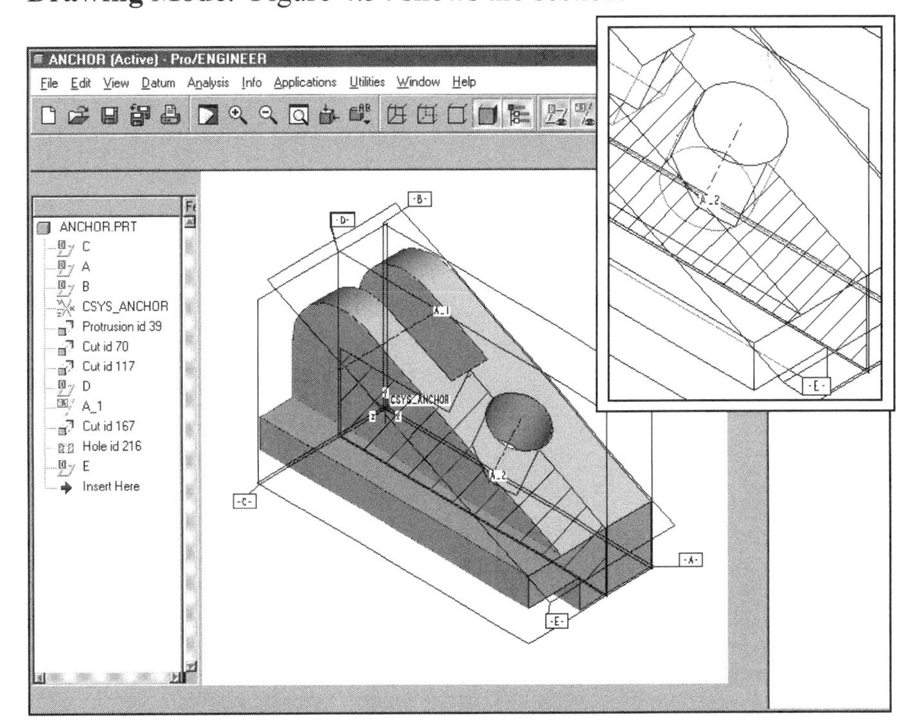

Figure 4.34
Creating a Section

HINT

If you type **AA** as the section name you will get a section called **AA-AA!** Therefore type only one letter.

Choose the following commands to create the section (Fig. 4.34):

X-section (from the PART menu) ⇒ **Create** ⇒ **Planar** ⇒ **Single** ⇒ **Done** ⇒ (enter the **NAME** for the cross section at the prompt: **A)** ⇒ ✔ ⇒ (select the planar surface of datum plane: **D)** ⇒ **Done/Return**

The section passes through the slot and the hole, but it doesn't pass through the center of the hole. The "boss" has a "suggestion" and provides you with the following ECO (Fig. 4.35):

E C O

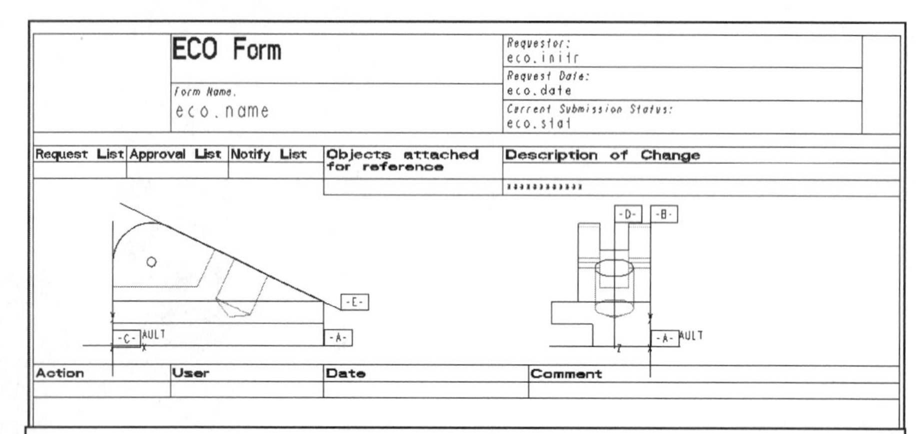

ECO Form

Form Name:
eco.name

Requestor:
eco.initr
Request Date:
eco.date
Current Submission Status:
eco.stat

Request List	Approval List	Notify List	Objects attached for reference	Description of Change

Action	User	Date	Comment

1. Change the dimensioning reference for the **Ø1.00** hole to be **.00** from datum plane **D** instead of **.875** from datum plane **B**.

2. Write a relation to keep datum plane **D** centered about the upper portion of the part.

3. Add a **Ø.250** hole coaxially through the cylindrical surface.

Figure 4.35
ECO

Choose the following commands to change the dimensioning reference for the hole:

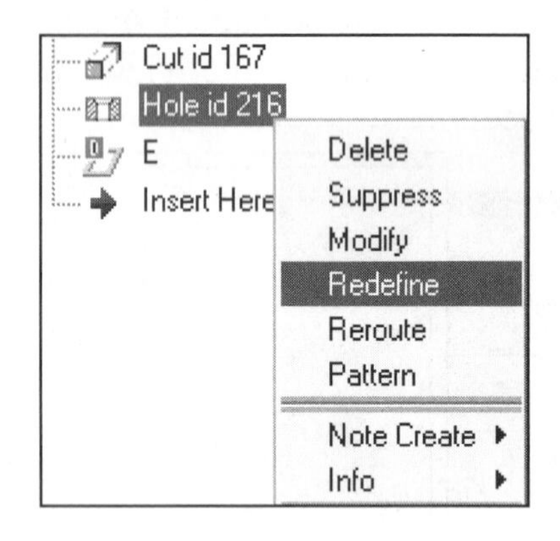

[highlight the hole in the Model Tree by picking it with the left mouse button] ⇒ [click the right mouse button and then slide the cursor down to **Redefine** and release the button (see left-hand column)] ⇒ Linear Reference ↖ (pick datum **D)** ⇒ (double click in the Distance box and type **0)** ⇒ **enter** (to accept the value) ⇒ **Preview** (Fig. 4.36) ⇒ ✔ (Build feature)

The hole and the slot are now ***children*** of datum **D**. If datum **D** moves, so will the slot and the hole. In order to ensure that the plane stays through the middle of the upper portion (the hole stays centered), create a relation to control the location of datum **D:**

NOTE

With **Info ⇒ Switch Dims** the dimensions will change from their numeric value to their parameter value. The parameter value *may be different* on your model.

Modify (PART menu, Fig. 4.37) ⇒ [pick datum **D** (from Model Tree) and the *upper front cut* to display the dimensions] ⇒ **DimCosmetics** ⇒ **Move Dim** (pick on a dimension and move it to see it more clearly) ⇒ **Info** ⇒ **Switch Dims** ⇒ **Relations** (from the PART menu) ⇒ **Add** ⇒ (type **d13=d10/2** at the prompt--your **d** symbols may be different) ⇒ ✔ ⇒ ✔ ⇒ **Done Sel** ⇒ **Done**

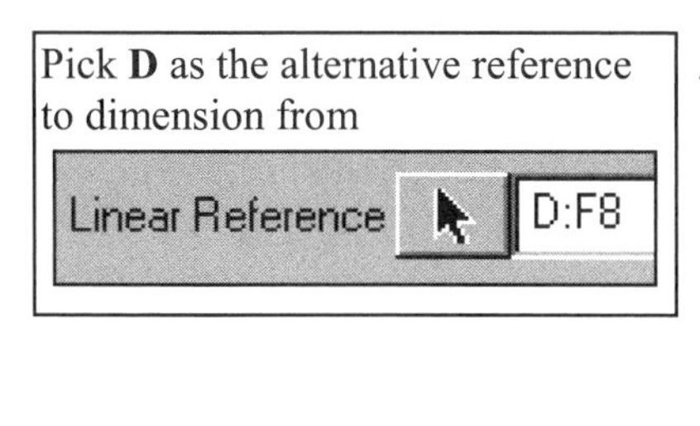

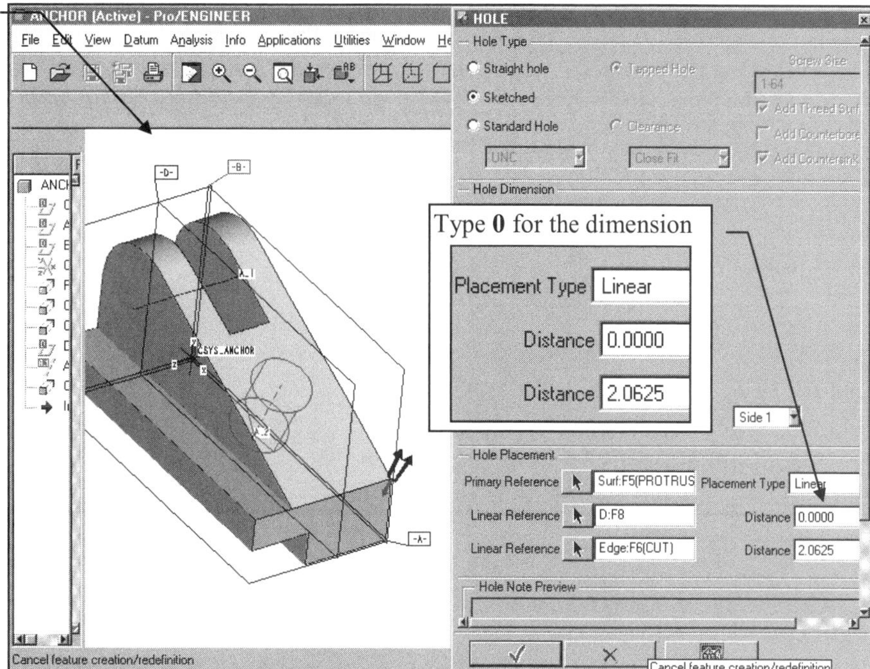

Figure 4.36
Changing the Reference of a Feature

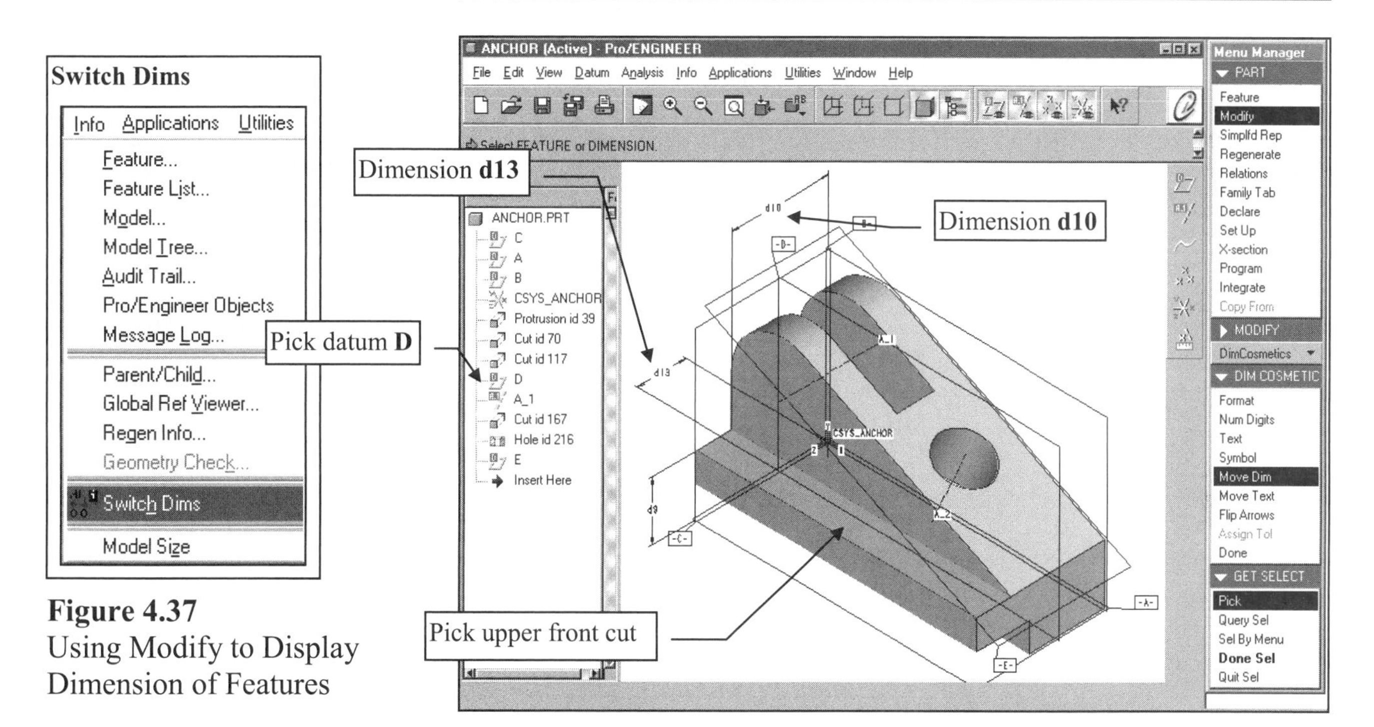

Figure 4.37
Using Modify to Display Dimension of Features

The relation (Fig. 4.38) states that the distance (**d13**) from datum **B** to datum **D** will be one-half the value of the distance from datum **B** to the cut surface (**d10**). If the thickness of the upper portion of the part (**1.875**) changes, datum **D** will remain centered, as will the slot cut and the **∅1.00** hole. If your *dimension symbols* do not show on the screen after you select them during the **Modify** command (Fig. 4.37), pick **Info ⇒ Switch Dims** from the menu bar *(or **Switch Dims** from the RELATIONS menu)*. To see the new relation, select **Show Rel** before completing the command (Fig. 4.39) or choose:

Relations ⇒ Show Rel ⇒ Close ⇒ Done Sel ⇒ Done

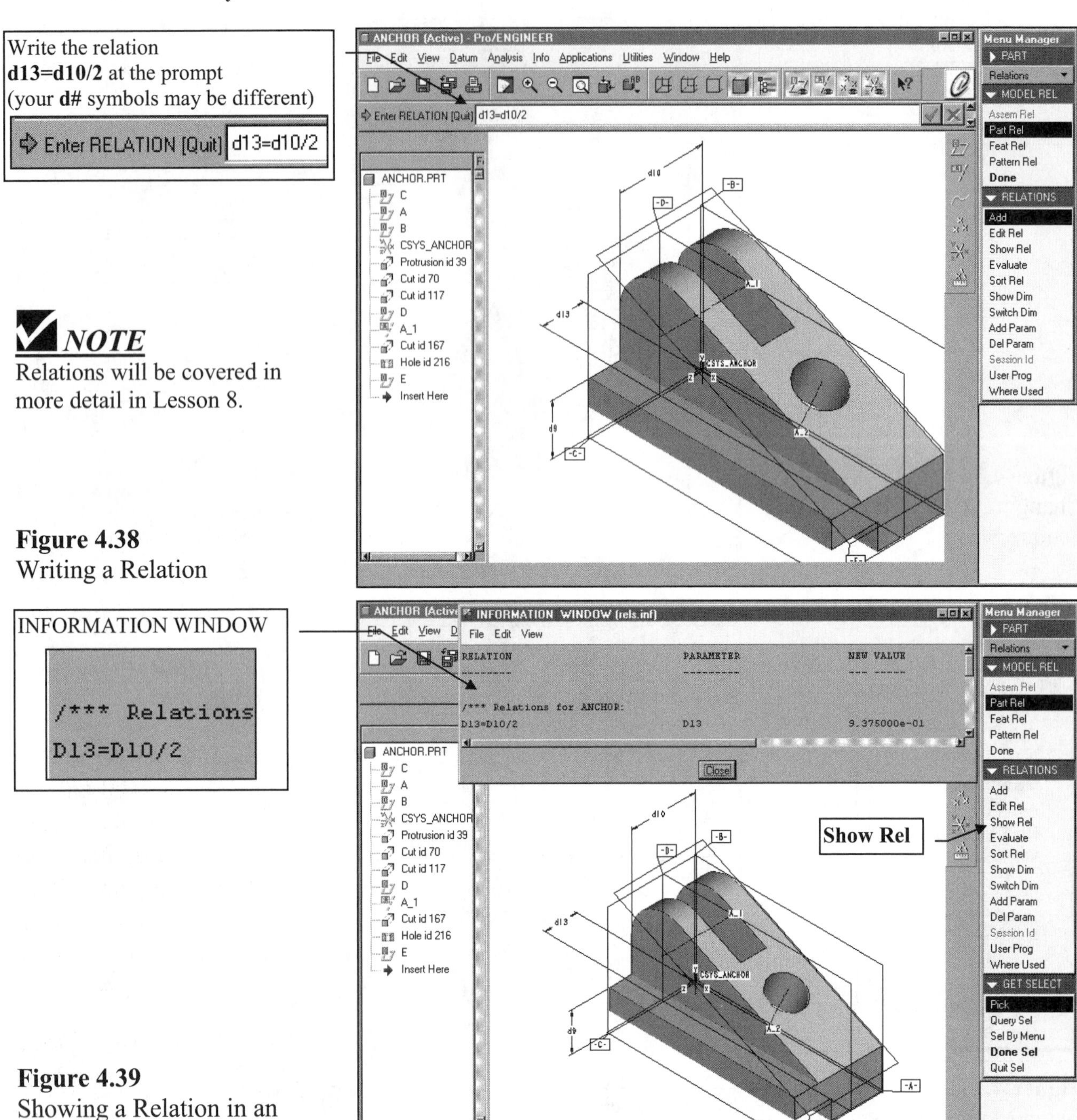

NOTE
Relations will be covered in more detail in Lesson 8.

Figure 4.38
Writing a Relation

Figure 4.39
Showing a Relation in an INFORMATION WINDOW

The last feature to create is a **Ø.250** hole to be placed coaxially with **A_1**. Choose the following commands:

File ⇒ Save ⇒ ✔ ⇒ Feature ⇒ Create ⇒ Solid ⇒ Hole ⇒ [fill in the required information-- Diameter **.250**, Depth One **Thru All**, Primary Reference ↖ [pick the front surface (Fig. 4.40) as the placement plane, Placement Type **Coaxial**, Axial Reference ↖ (pick on axis **A_1)] ⇒ ✔ ⇒ Done**

Many layers are created by default with Pro/E. A new layer is required that contains the part's holes. This layer can be suppressed to blank the holes from the display.

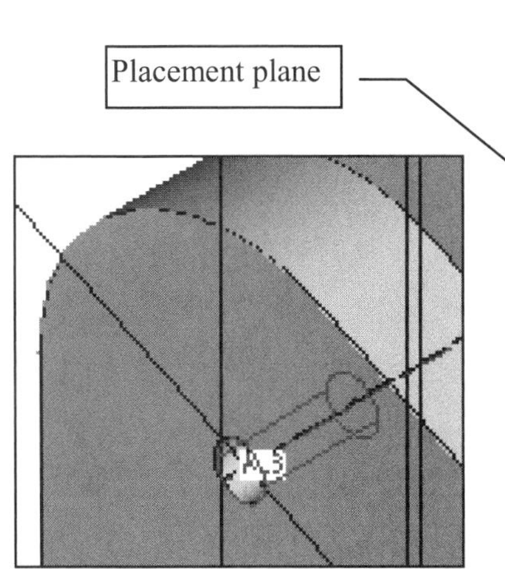

Figure 4.40
Creating a Coaxial Hole

Create a layer and add the holes to it (reread Section 8, Layers, at this time), using these commands:

View (menu bar) ⇒ **Layers** ⇒ **Show** ⇒ **Layer Items** (Fig. 4.41) ⇒ **Layer** ⇒ **New** ⇒ (type name of new layer at prompt: **HOLE**) ⇒ **OK** (from New Layer dialog box) ⇒ **Item** (from Layers dialog box) ⇒ **Add** ⇒ **Feature** ⇒ [pick the two holes from the Model Tree (Fig 4.42)] ⇒ **Done Sel** ⇒ **Done/Return** ⇒ **Done/Return** ⇒ pick **HOLE** (from Layers dialog box) ⇒ **Layer** (Fig. 4.43) ⇒ **Info** ⇒ **Close** ⇒ **Close** ⇒ **Feature** ⇒ **Suppress** ⇒ **Layer** ⇒ ✔ **HOLE** (Fig. 4.44) ⇒ **Done Sel** ⇒ **Done Sel** ⇒ **Done** ⇒ **Done** ⇒ **File** ⇒ **Save** ⇒ ✔

Pro/E created a number of default layers as you were modeling and added features to it automatically

- Layers
  - 01__PRT_ALL_DTM_PLN
    - Feature 1 (C) in part ANCHOR.
    - Feature 2 (A) in part ANCHOR.
    - Feature 3 (B) in part ANCHOR.
    - Feature 8 (D) in part ANCHOR.
    - Feature 12 (E) in part ANCHOR.
  - 01__PRT_DEF_DTM_PLN
    - Feature 1 (C) in part ANCHOR.
    - Feature 2 (A) in part ANCHOR.
    - Feature 3 (B) in part ANCHOR.
  - 02__PRT_ALL_AXES
    - Feature 9 (A_1) in part ANCHOR.
    - Feature 11 (HOLE) in part ANCHOR
  - 03__PRT_ALL_CURVES
  - 04__PRT_ALL_DTM_PNT
  - 05__PRT_ALL_DTM_CSYS
    - Feature 4 (CSYS_ANCHOR) in part A
  - 05__PRT_DEF_DTM_CSYS
    - Feature 4 (CSYS_ANCHOR) in part A
  - 06__PRT_ALL_SURFS

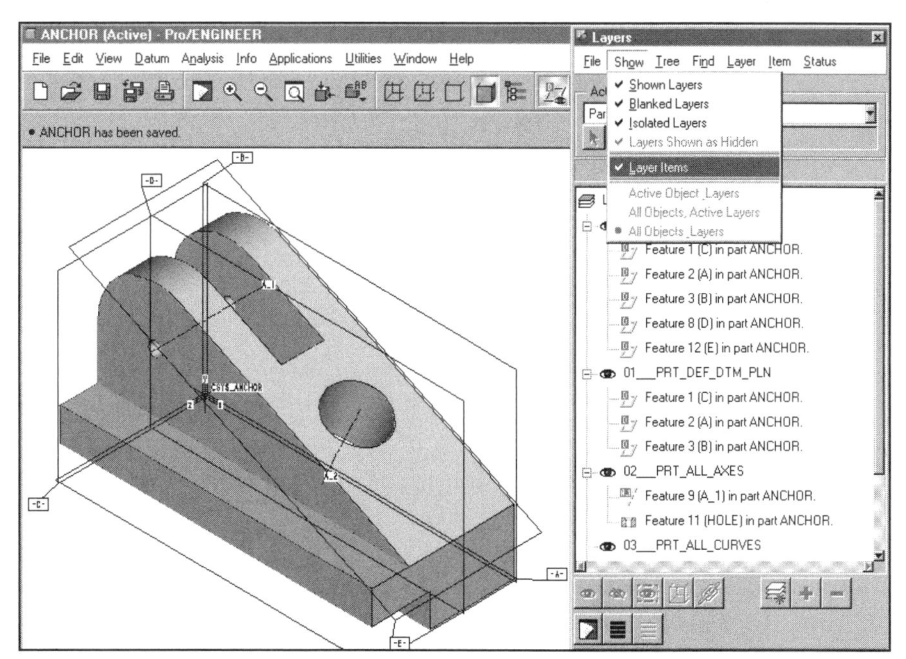

Figure 4.41
Default Layering

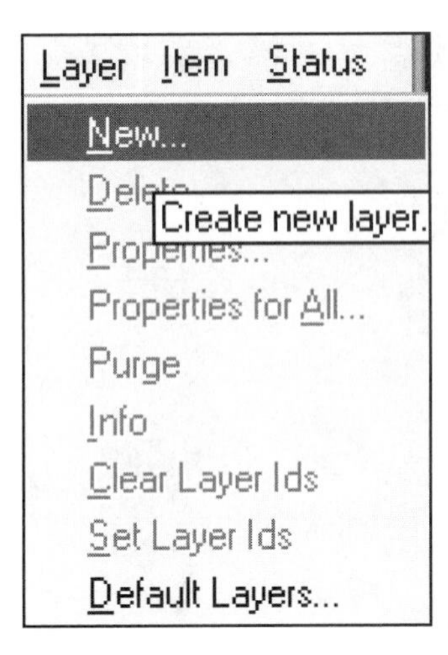

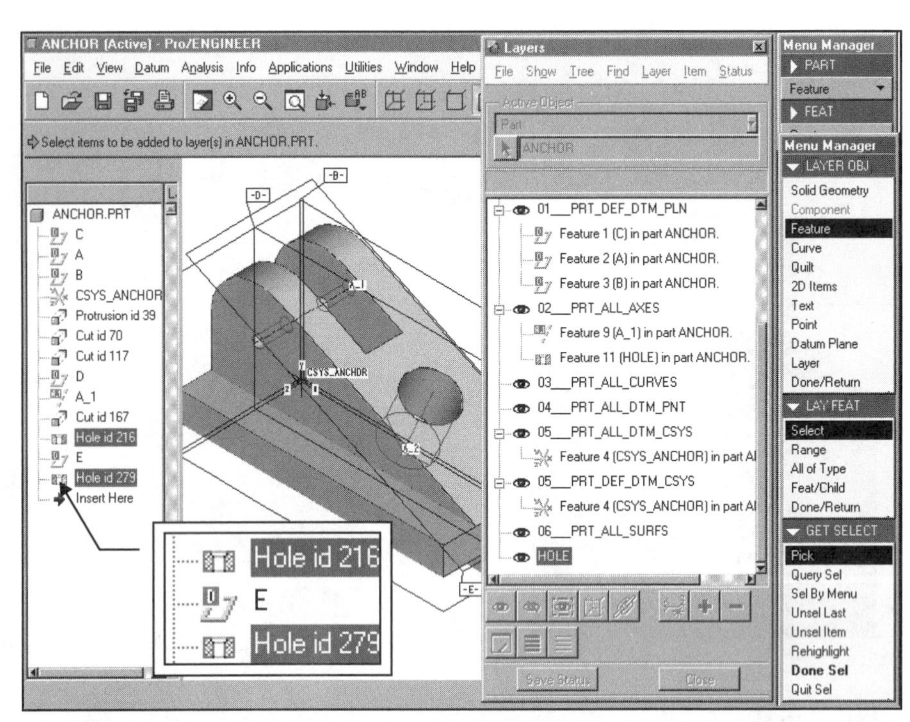

Figure 4.42
Adding the Holes to the New Layer **HOLE**

Figure 4.43
Layer Info

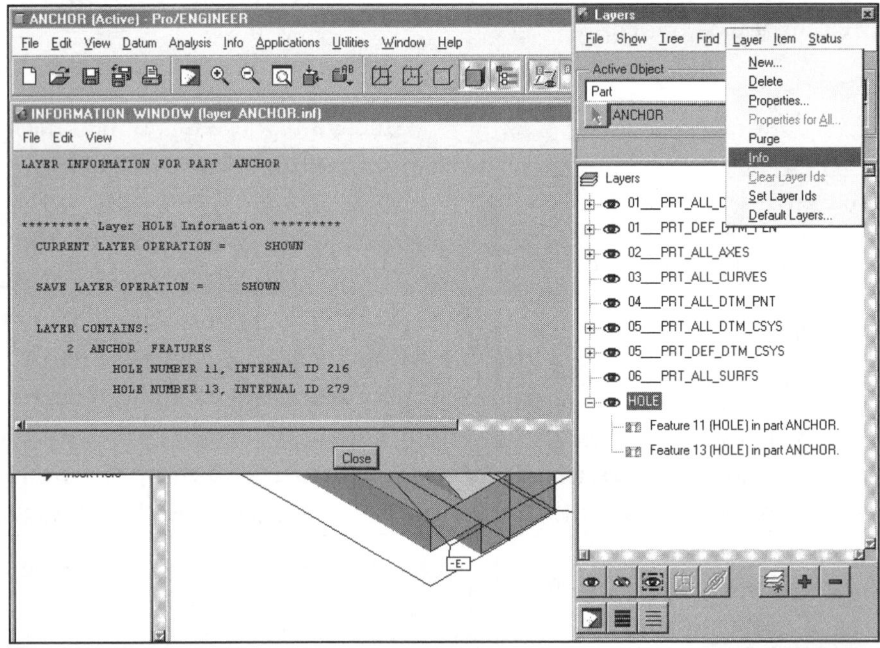

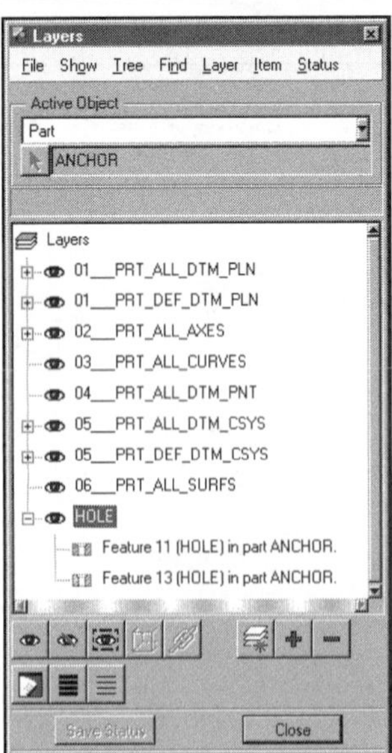

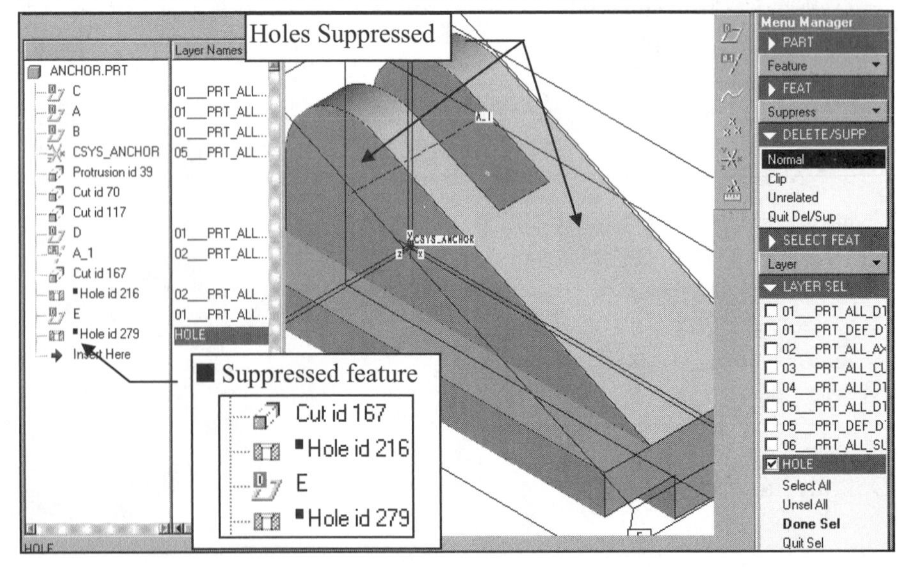

Figure 4.44
Suppressing Layers

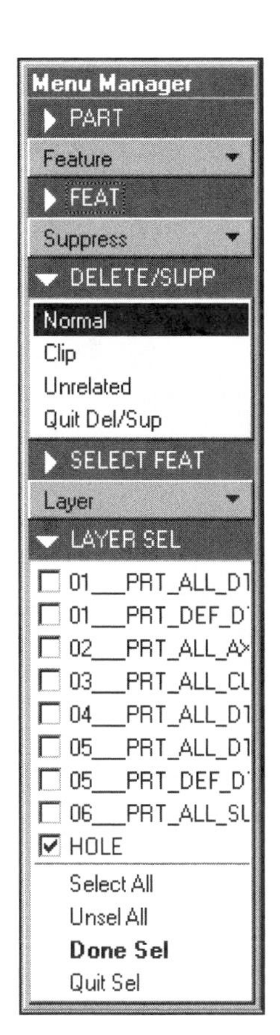

The features on the layer **HOLE** are now suppressed. To complete the part, select:

Feature ⇒ Resume ⇒ All ⇒ Done ⇒ Done ⇒ X-section ⇒ Show ⇒ A ⇒ Done/Return

Notice that the first hole is on two layers, a default layer and the layer you created (HOLE) (Fig. 4.45).

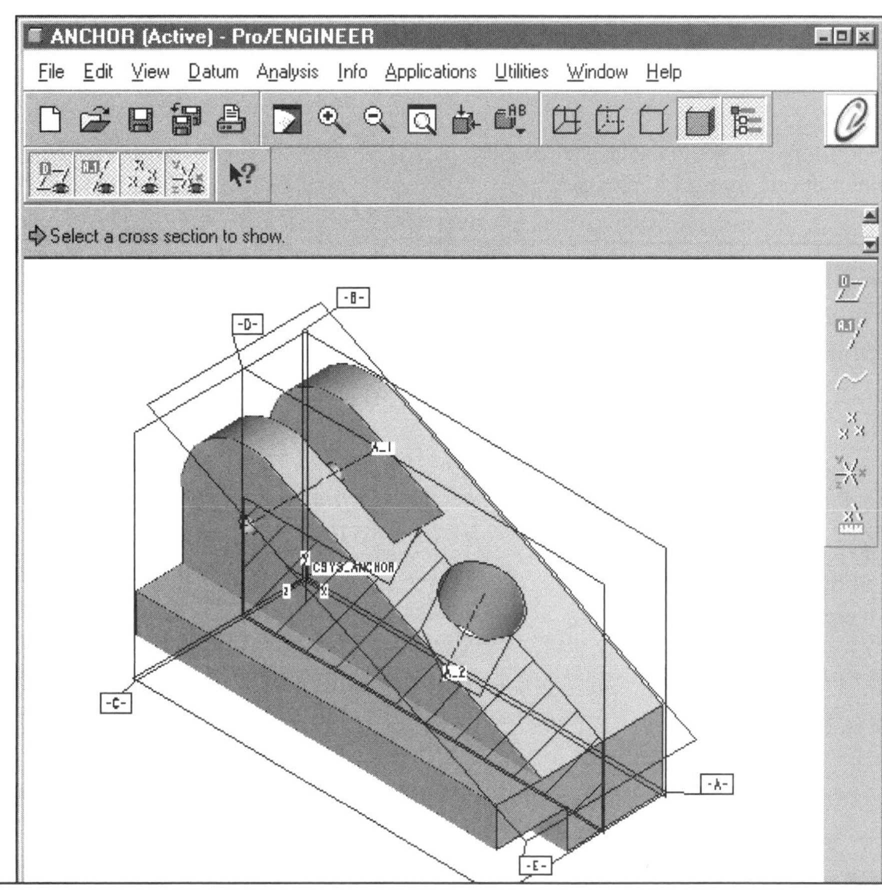

NOTE
Reread Section 8, Layers, and Section 9, Model Tree, at this time.

Figure 4.45
Resumed Features

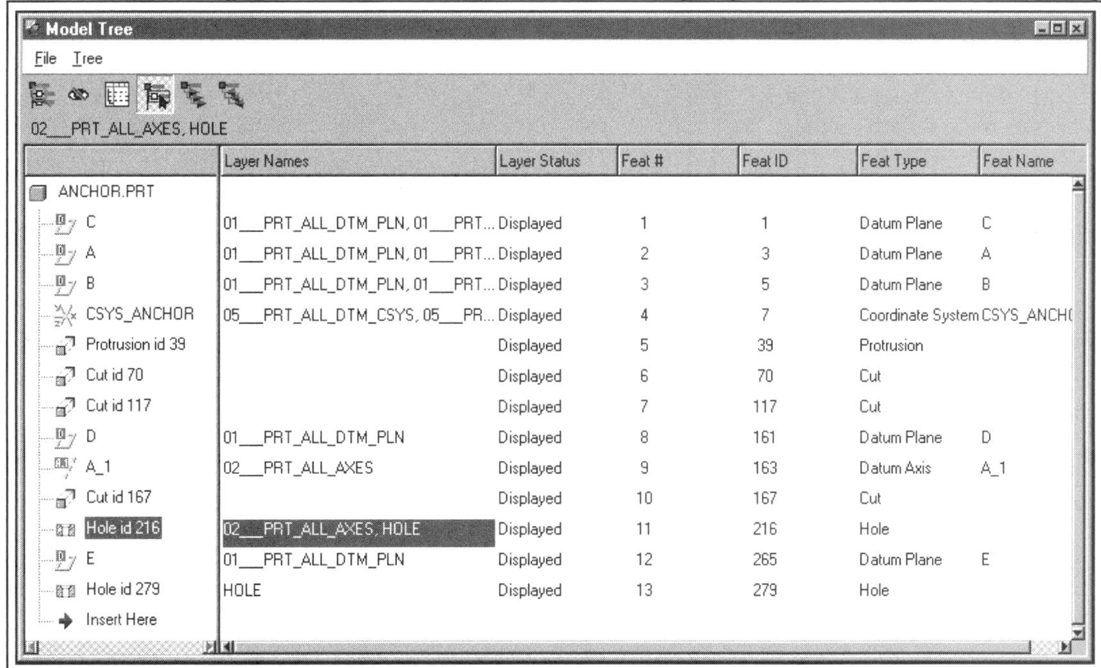

	Layer Names	Layer Status	Feat #	Feat ID	Feat Type	Feat Name
ANCHOR.PRT						
C	01__PRT_ALL_DTM_PLN, 01__PRT...	Displayed	1	1	Datum Plane	C
A	01__PRT_ALL_DTM_PLN, 01__PRT...	Displayed	2	3	Datum Plane	A
B	01__PRT_ALL_DTM_PLN, 01__PRT...	Displayed	3	5	Datum Plane	B
CSYS_ANCHOR	05__PRT_ALL_DTM_CSYS, 05__PR...	Displayed	4	7	Coordinate System	CSYS_ANCH(
Protrusion id 39		Displayed	5	39	Protrusion	
Cut id 70		Displayed	6	70	Cut	
Cut id 117		Displayed	7	117	Cut	
D	01__PRT_ALL_DTM_PLN	Displayed	8	161	Datum Plane	D
A_1	02__PRT_ALL_AXES	Displayed	9	163	Datum Axis	A_1
Cut id 167		Displayed	10	167	Cut	
Hole id 216	02__PRT_ALL_AXES, HOLE	Displayed	11	216	Hole	
E	01__PRT_ALL_DTM_PLN	Displayed	12	265	Datum Plane	E
Hole id 279	HOLE	Displayed	13	279	Hole	
Insert Here						

Lesson 4 Project

Angle Frame

NOTE
Don't forget to set the units and the material (aluminum) for the part.

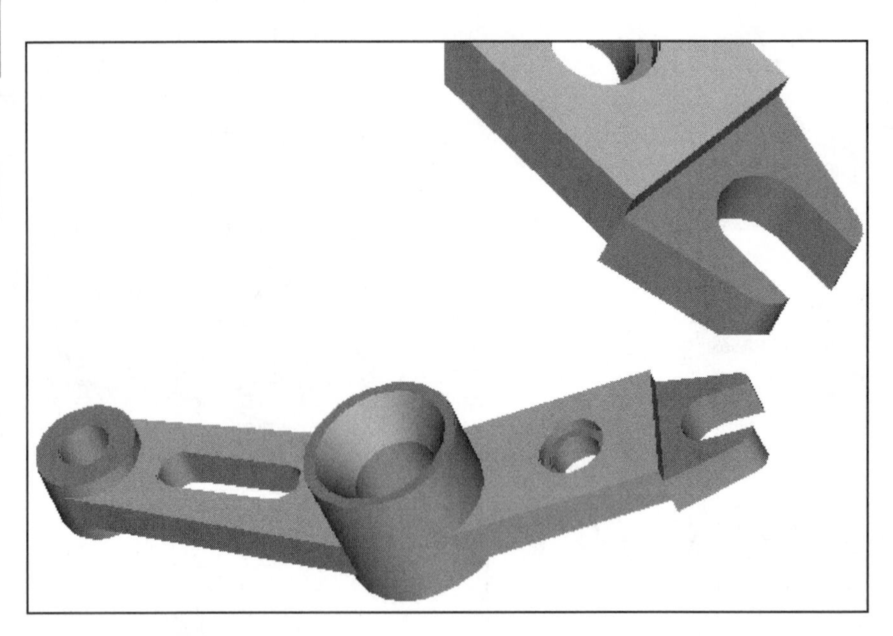

Figure 4.46
Angle Frame

Angle Frame

The fourth **lesson project** is a machined part that requires the use of a variety of datum planes and a layering scheme. You will also add a relation to control the depth of the large countersink hole at the part's center. Analyze the part and plan out the steps and features required to model it. Use the **DIPS** in Appendix D to establish a feature creation sequence before you start modeling. Create the part shown in Figures 4.46 through 4.55.

EGD REFERENCE
Fundamentals of Engineering Graphics and Design
by L. Lamit and K. Kitto
See pages 311, 461, and 548.

NOTE
Set the datums using **Geom Tol**, rename all three default datum planes to **A**, **B**, **C**, and so on.

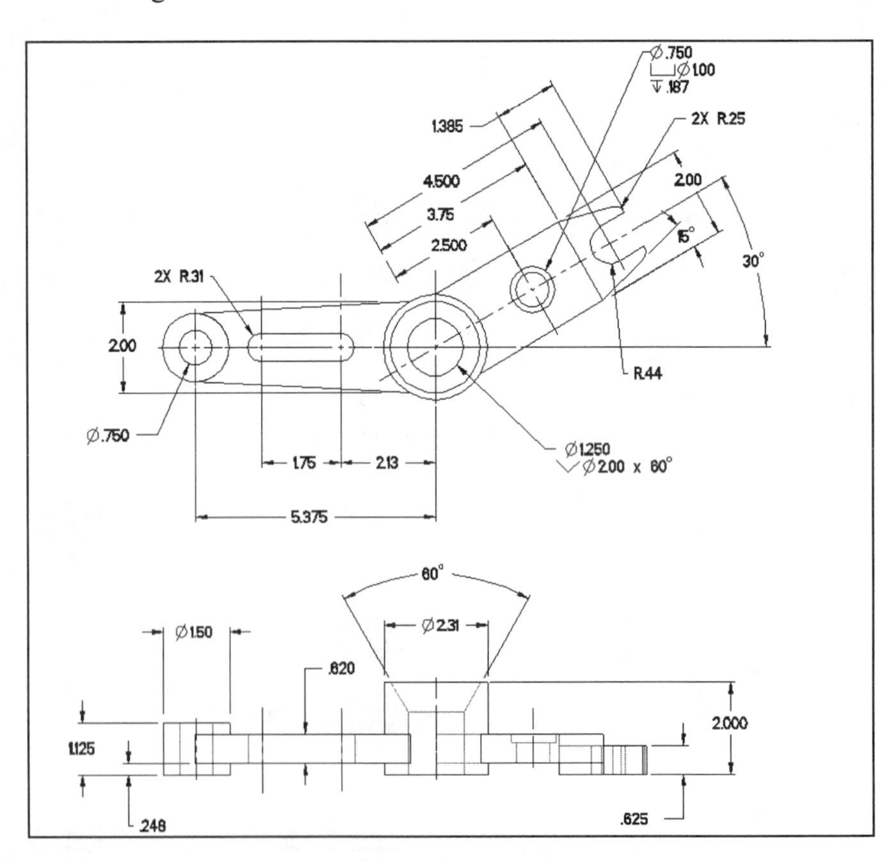

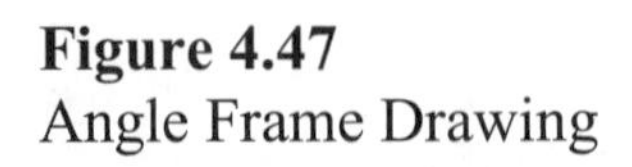

Figure 4.47
Angle Frame Drawing

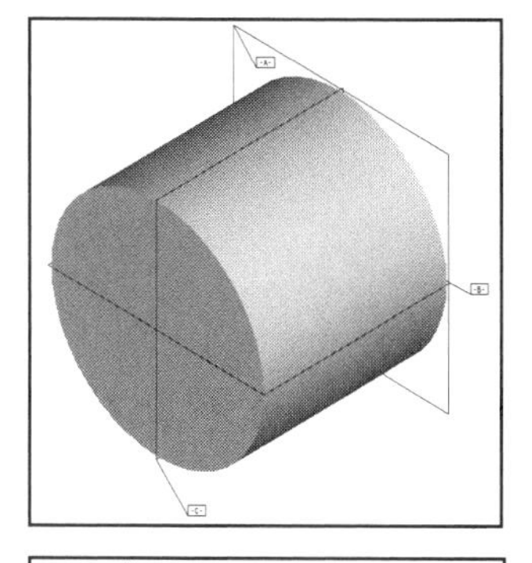

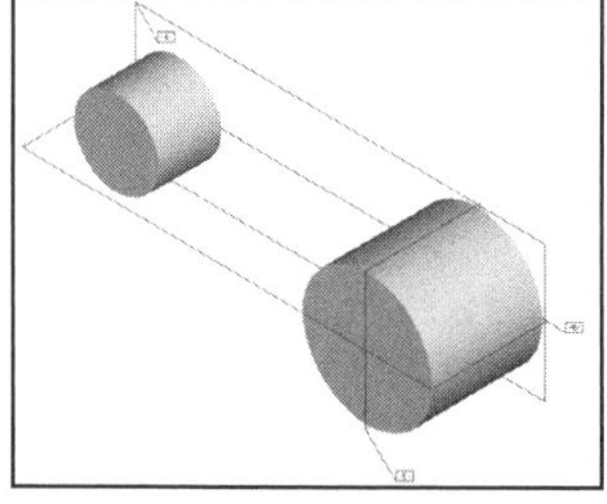

Figure 4.48
Angle Frame, Top View

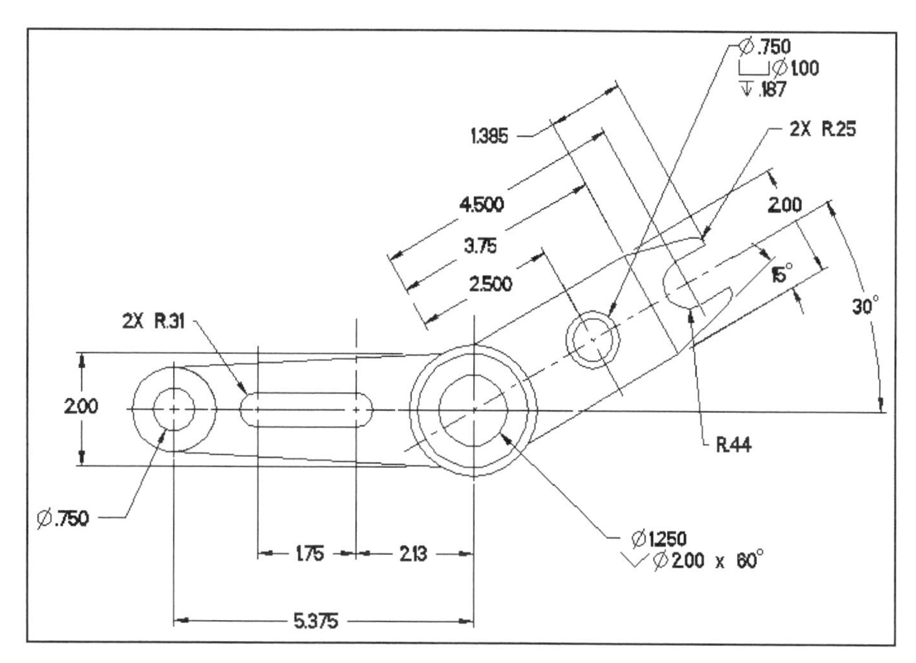

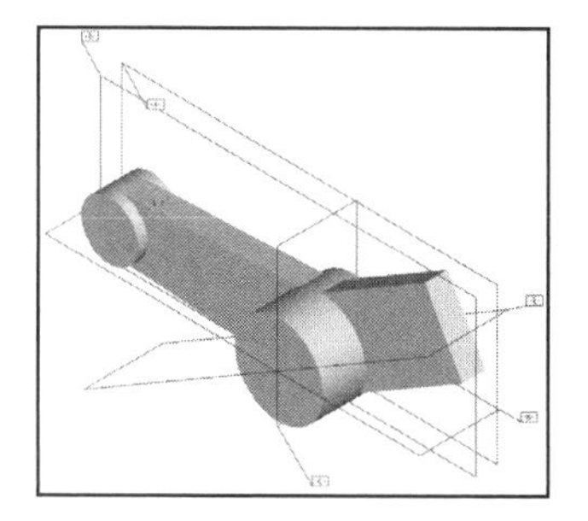

Figure 4.49
Angle Frame, Front View

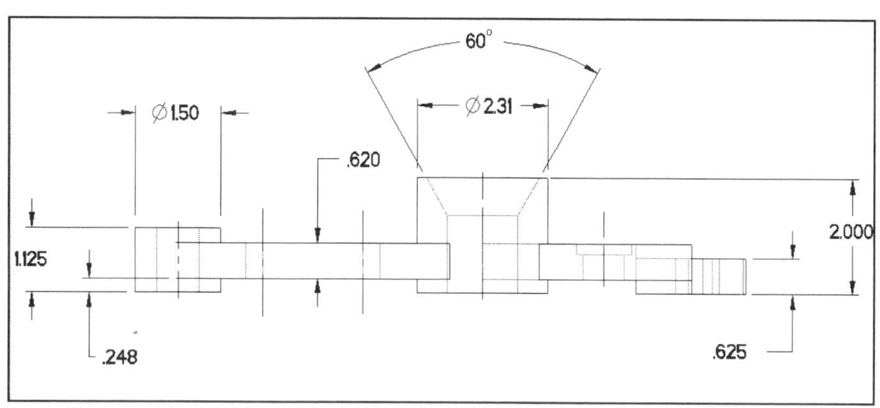

Set the datums with the appropriate geometric tolerance names: **A, B, C,** and so on.

Create two *sections* through the Angle Frame to be used later in a Drawing Lesson. For the sections, use datum planes **B** and **E**, which pass vertically through the center of the part. Name the cross sections **A** (**SECTION A-A**) and **B** (**SECTION B-B**).

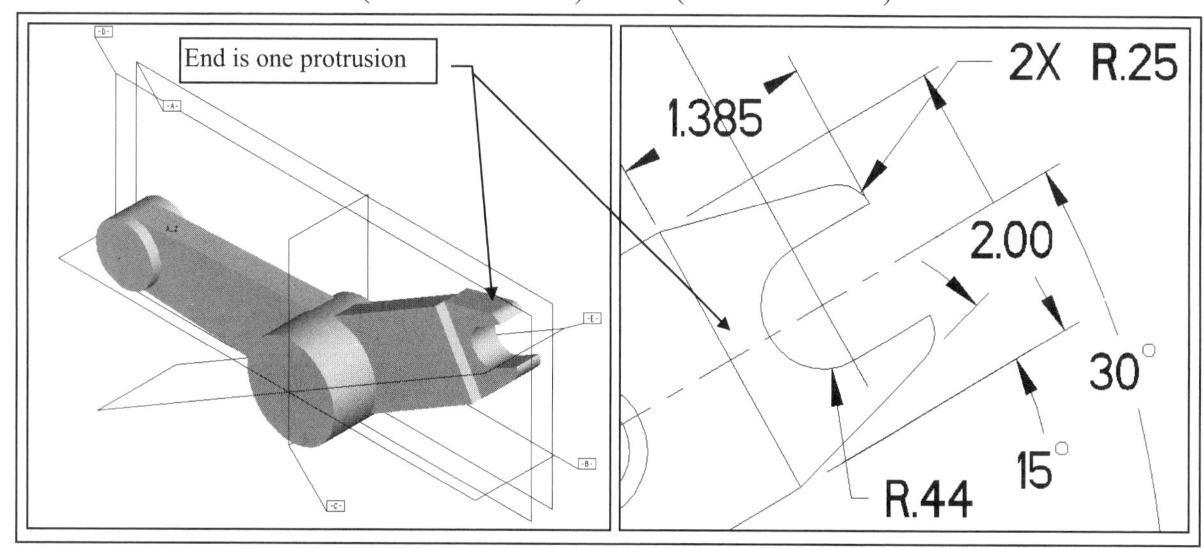

Datum and Section Naming

RIGHT - C
TOP - B (SECTION A-A)
FRONT - A
datum - **D**
datum - **E (SECTION B-B)**

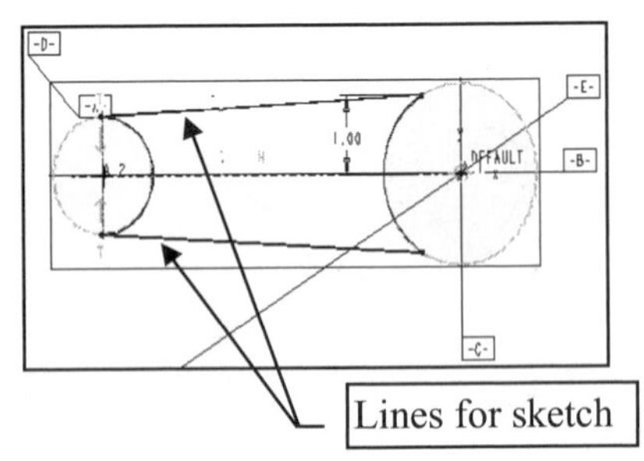

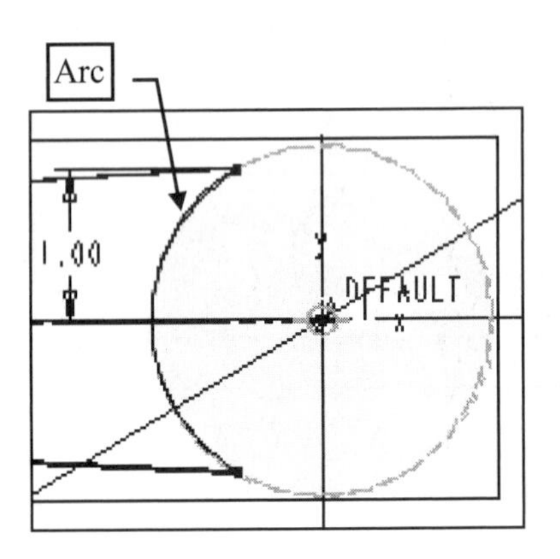

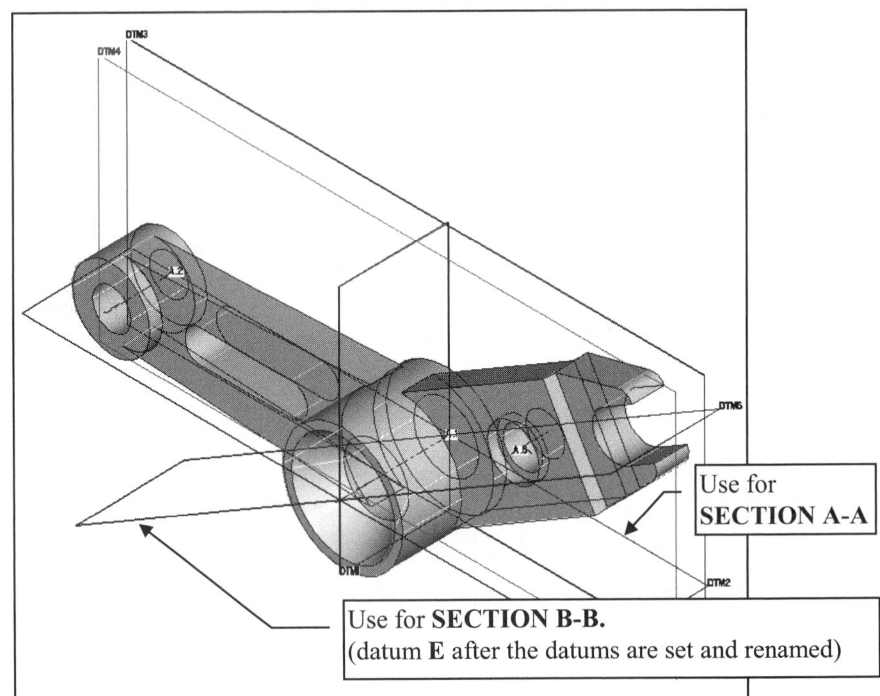

Figure 4.50
Datums

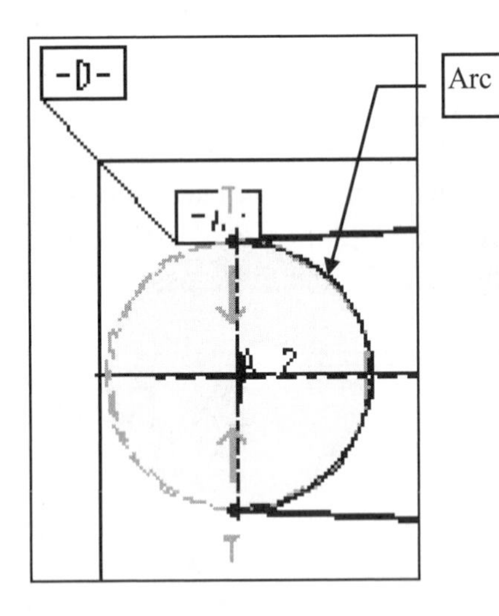

HINT
You can format, add, or remove columns of feature information about the model to the Model Tree

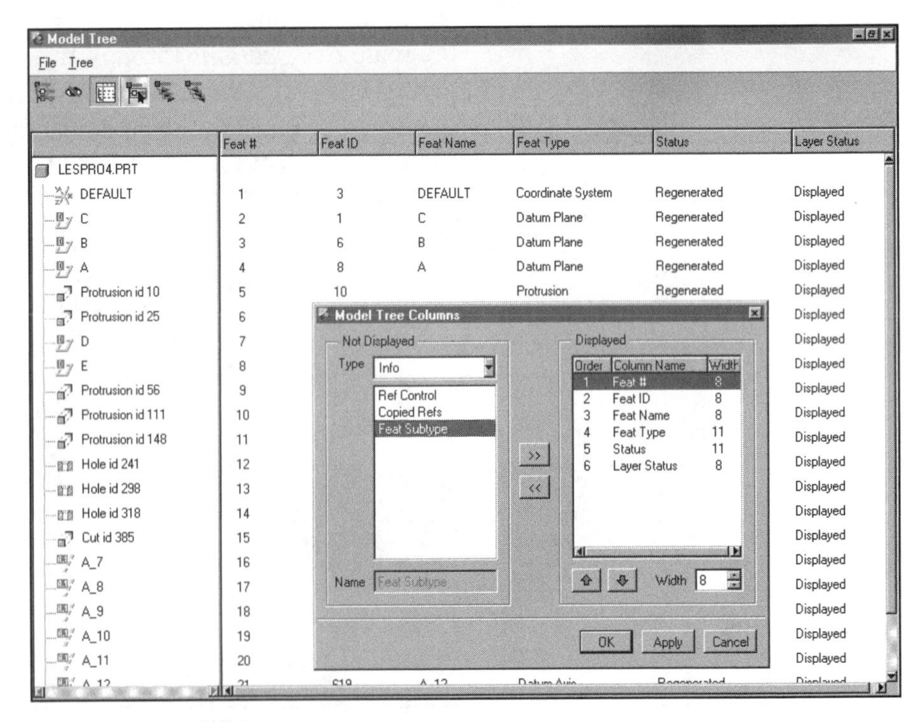

Figure 4.51
Model Tree

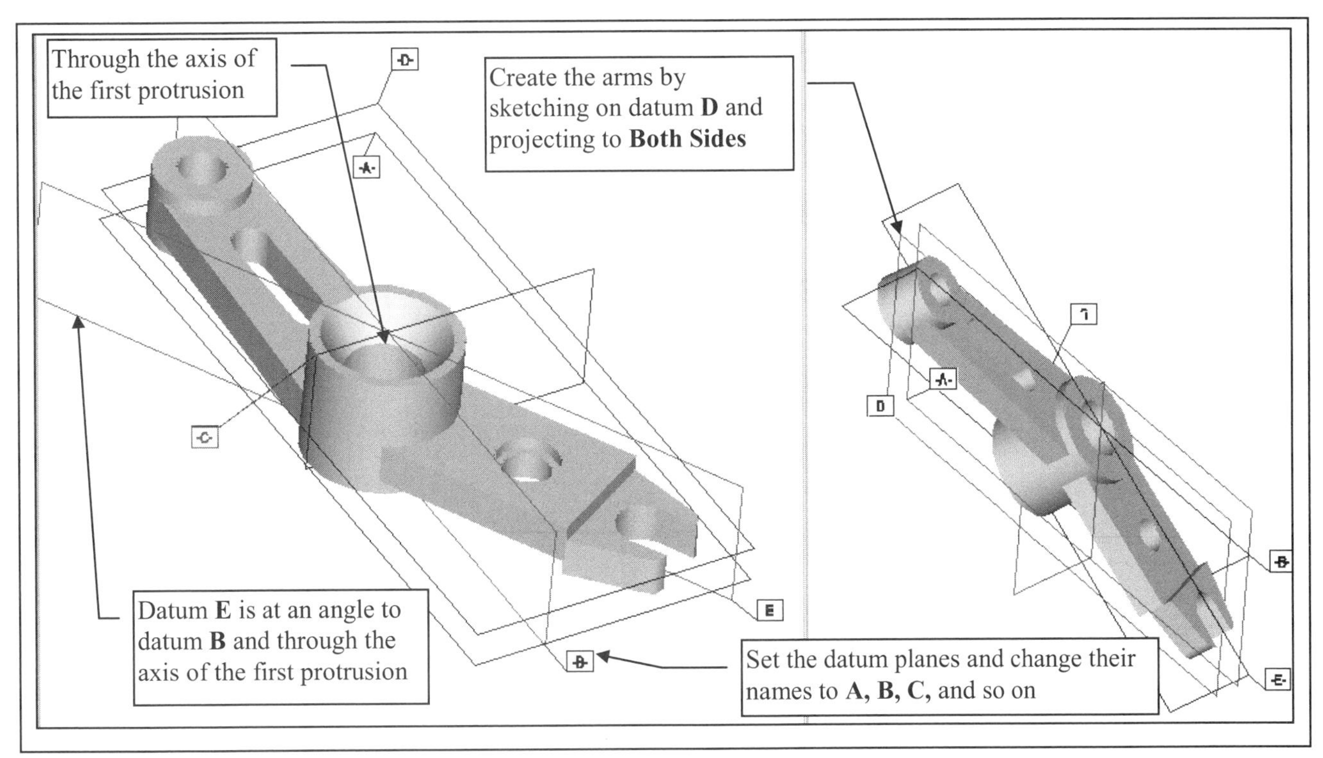

Figure 4.52
Using the Datums to Create Features

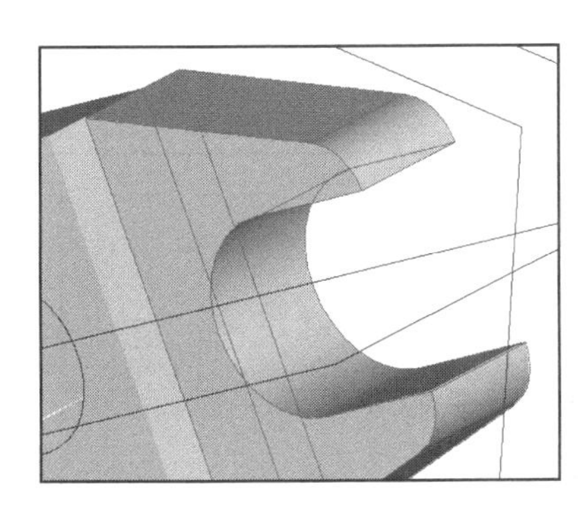

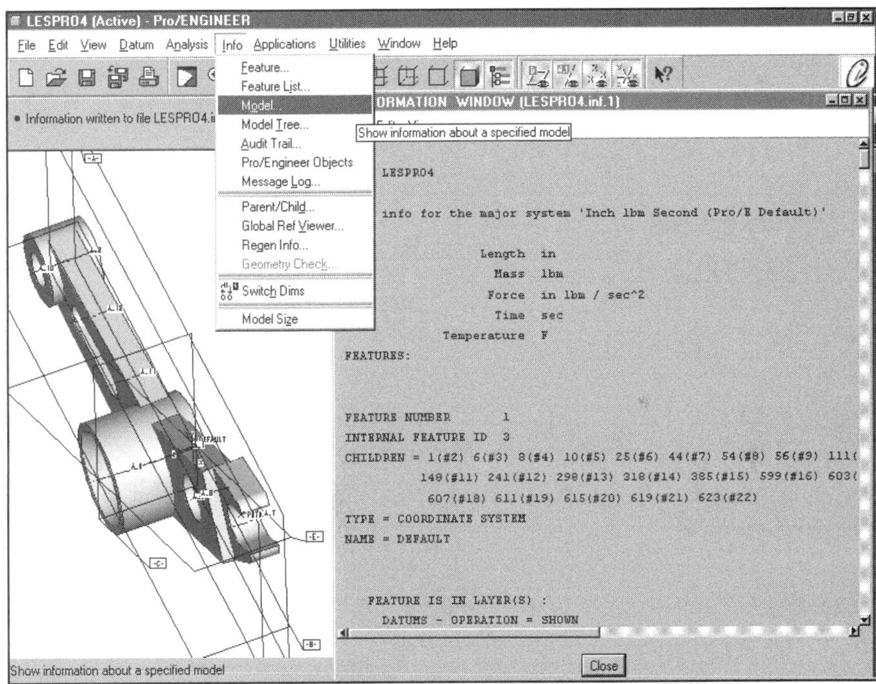

Figure 4.53
Info ⇒ Model

Modify the thickness of the boss from **2.00** to **2.50.** Note that the hole does not go through the part. Modify the boss back to the original design dimension of **2.00.** Add a relation to the hole that says the depth of the hole should be equal to the thickness of the boss (**d43=d6**), as shown in Figure 4.54.

HINT
Your **d#** symbols will probably be different from the ones shown here.

Now change the thickness of the boss (original protrusion) to see that the hole still goes through the part. No matter what the boss thickness dimension changes to, the hole will always go completely through it. This relation controls the *design intent* of the hole.

NOTE

Relations are used to control features and preserve the design intent of the part. Lesson 8 will cover relations in more detail.

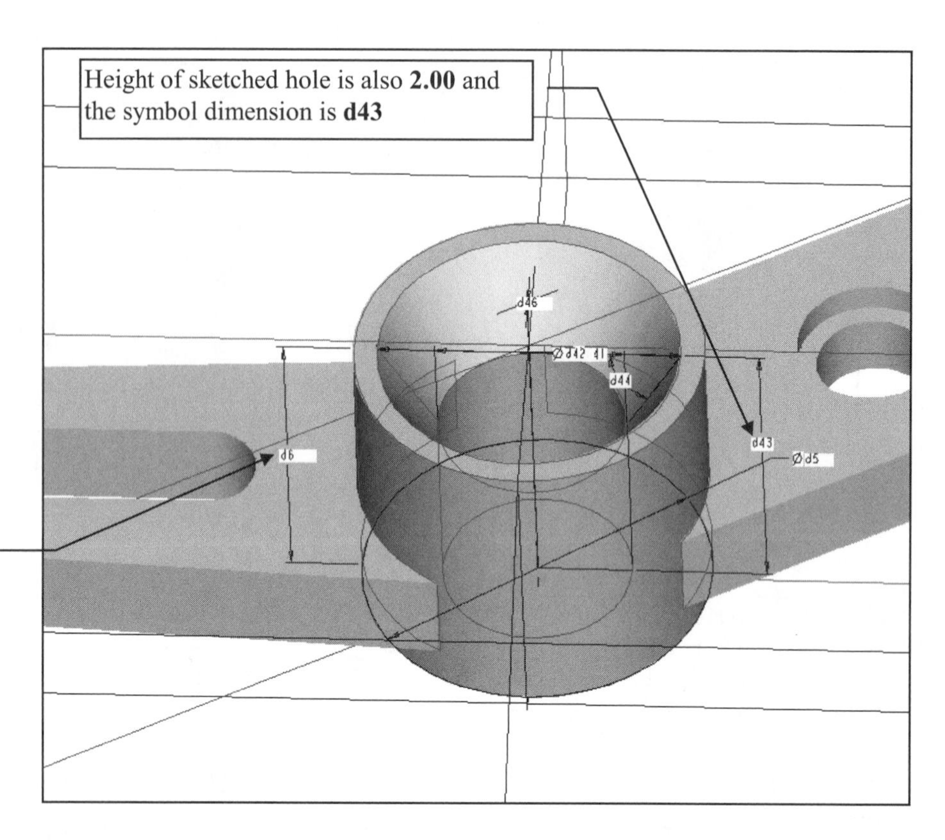

Add a relation that says **d43=d6** (your **d#** symbols may be different)

Boss protrusion has a height of **2.00** and has a symbol dimension of **d6**

Figure 4.54
Adding a Relation to Control the Hole Depth

Create the sections required to describe the part while you are in **Part Mode** so they will be available for use when you are detailing the part in **Drawing Mode.**

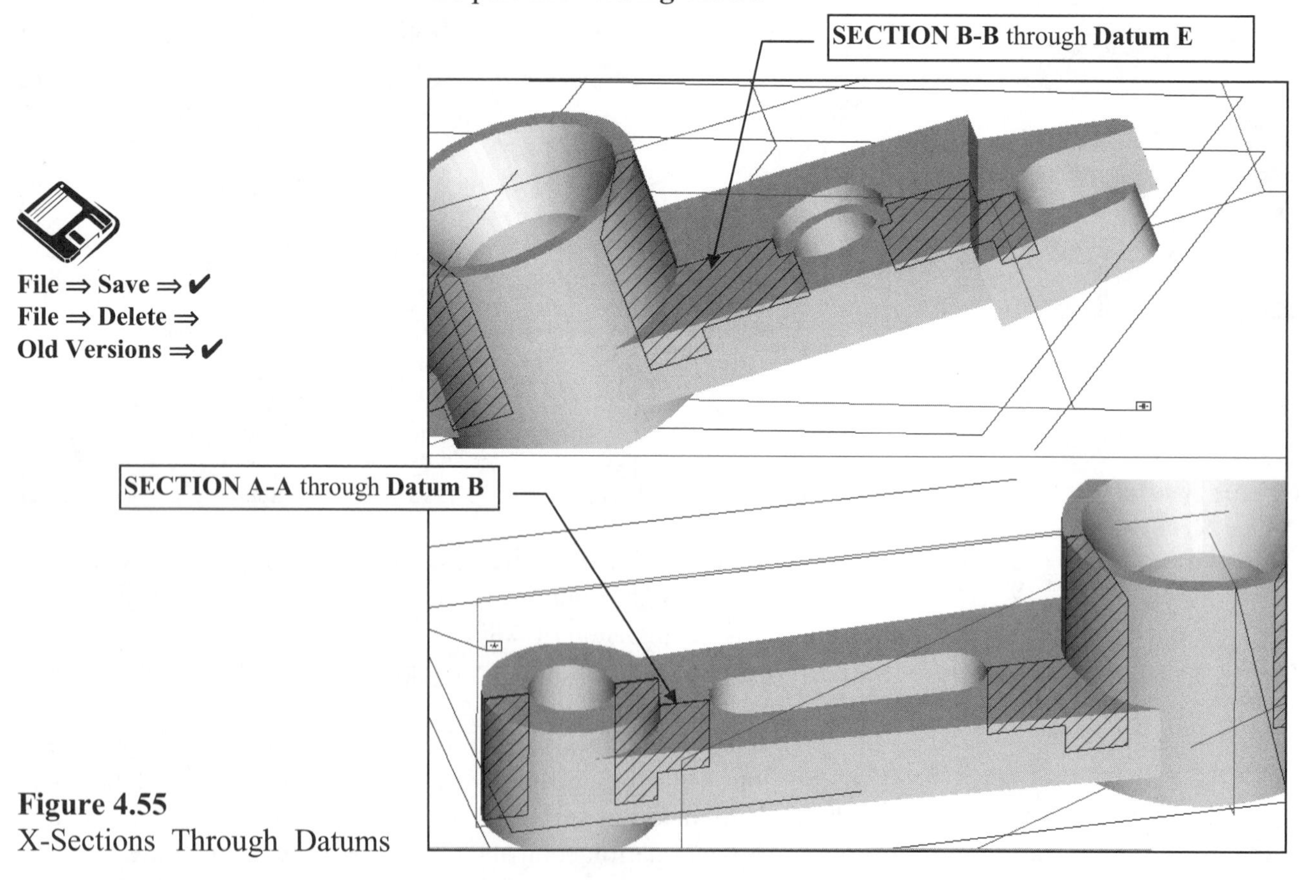

File ⇒ Save ⇒ ✔
File ⇒ Delete ⇒
Old Versions ⇒ ✔

Figure 4.55
X-Sections Through Datums

Lesson 5

Revolved Protrusions and Revolved Cuts

CONFIG.PRO
You can reinstate a variety of pre Pro/ENGINEER 2000i commands (**Lip**, **Shaft**, **Neck**, **Slot**, **Flange** etc.) by using:
allow_anatomic_features *Yes*

Utilities ⇒ Preferences ⇒ *allow_anatomic_features* *Yes* ⇒ Add/Change ⇒ Apply ⇒ Close

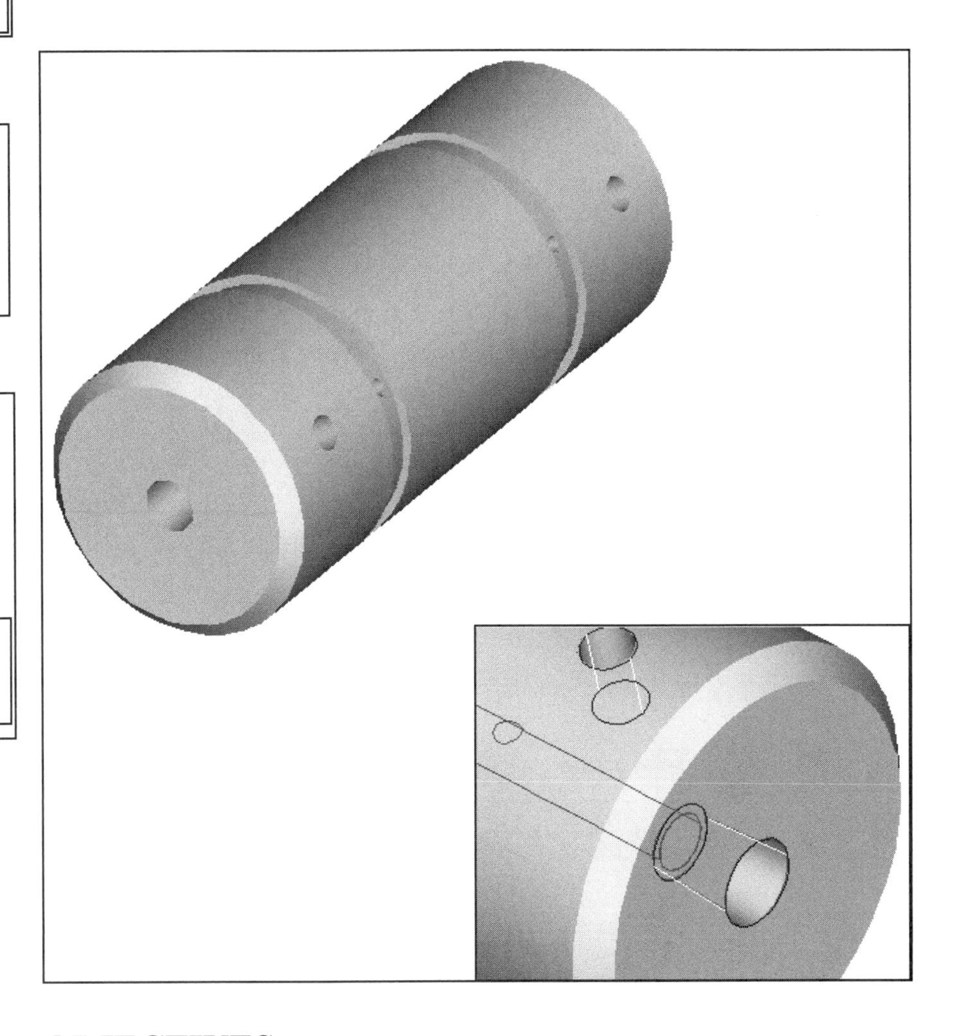

Figure 5.1
Pin

EGD REFERENCE
Fundamentals of Engineering Graphics and Design
by L. Lamit and K. Kitto
Read Chapters 8, 14, and 26.
See pages 239, and 486.

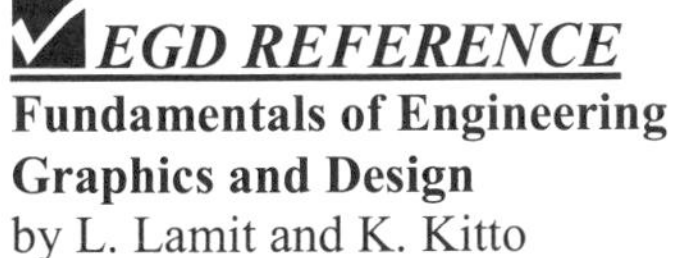
COAch™ for Pro/ENGINEER
If you have **COAch for Pro/ENGINEER** on your system, go to SEARCH and do the Segment shown in Figure 5.4.

OBJECTIVES

1. **Create a revolved protrusion**

2. **Understand the angle options used to create revolved features**

3. **Use datums to locate holes**

4. **Cut necks/grooves in revolved protrusions**

5. **Create a conical revolved cut**

6. **Use the Info command to measure a revolved feature**

7. **Get a hard copy using the Interface command**

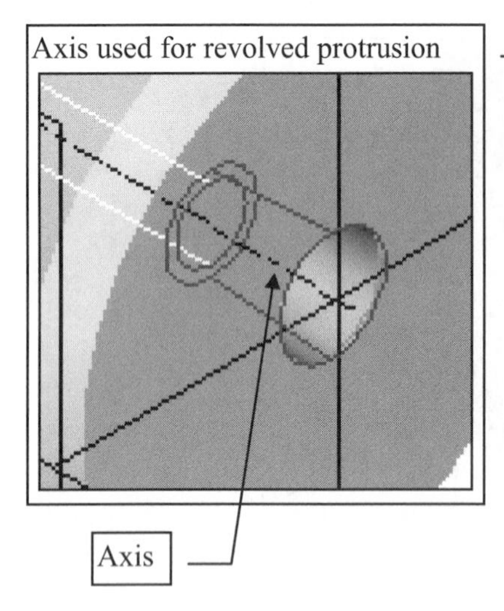

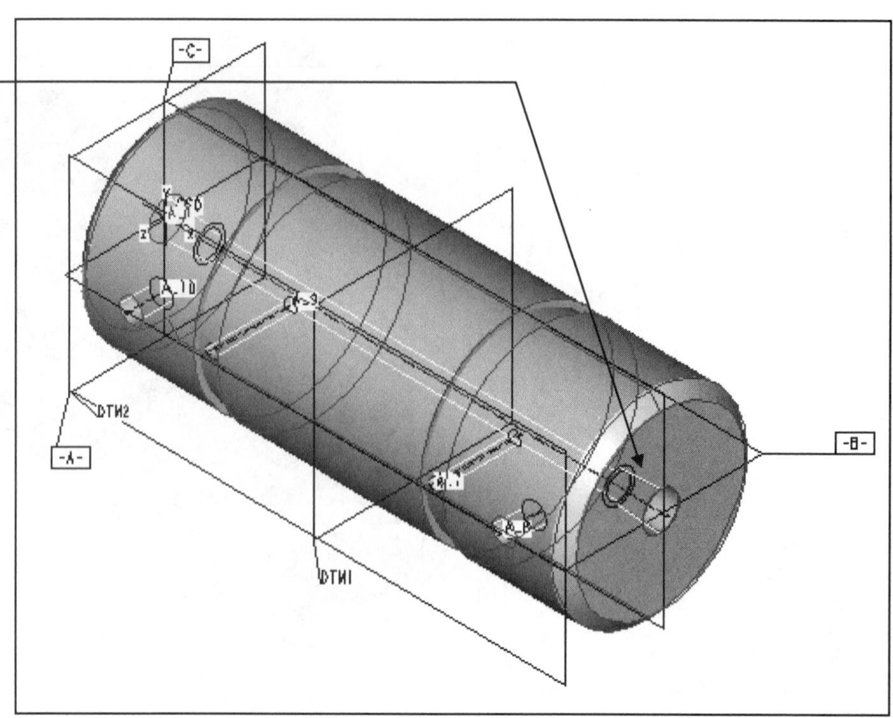

Figure 5.2
Pin

REVOLVED PROTRUSIONS AND REVOLVED CUTS

The **revolve** option creates a feature by revolving the sketched section around a centerline from the sketching plane into the part (Fig. 5.1). You can have any number of centerlines in your sketch, but the first centerline will be the one used to rotate your section geometry.

When you are sketching the feature to be revolved, the first centerline sketched is the *axis of revolution* (Fig. 5.2). The section geometry must be closed and must lie on one side of this centerline as shown in Figure 5.3. **Revolved Protrusions** and **Revolved Cuts** can be created with this method.

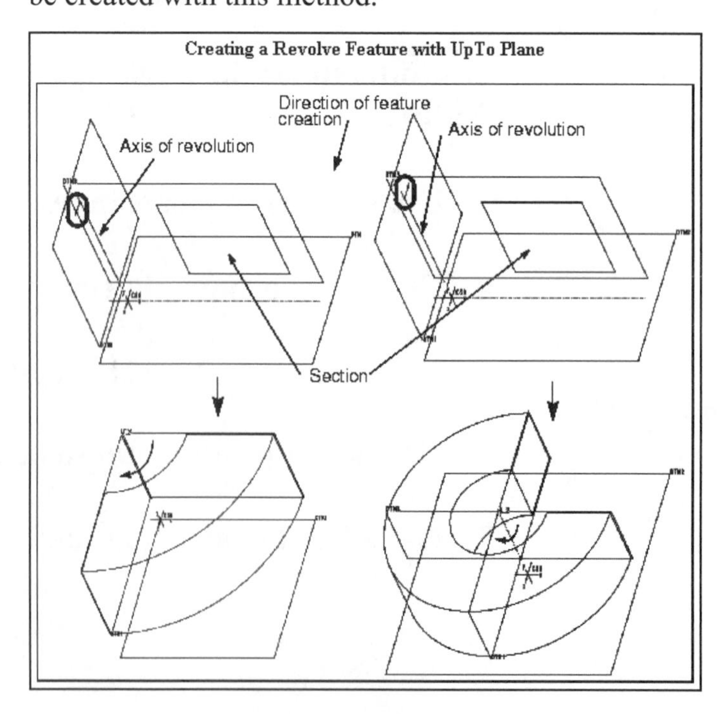

Figure 5.3
Revolved Protrusions

A revolved feature can be created either entirely on one side of the sketching plane or symmetrically on both sides of the sketching plane. The **One Side** and **Both Sides** options are available. If you choose **Both Sides**, the feature will be revolved symmetrically in each direction for one-half of the angle specified in the REV TO menu.

After successfully regenerating the revolved section, select **Done** and the REV TO menu appears. This menu allows you to specify the value of the feature's angle of revolution. You can choose the **Variable** option for a user-defined angle of revolution, or you can choose one of four preset angles: **90**, **180**, **270**, and **360**.

If you choose **Variable**, the angle can be specified and modified after the section is created. This angle must be greater than **0°** and less than **360°**. The angle is controlled by a dimension that appears when you are modifying the part and in drawings. A corresponding dimension will not appear if a preset angle is chosen. The base feature of the pin was created with a **360°** revolved section.

If you have **COAch for Pro/ENGINEER** on your system, go to SEARCH and do the appropriate Segment, shown in Figure 5.4.

Figure 5.4
COAch for Pro/E, Basic Modeling (Using Centerlines)

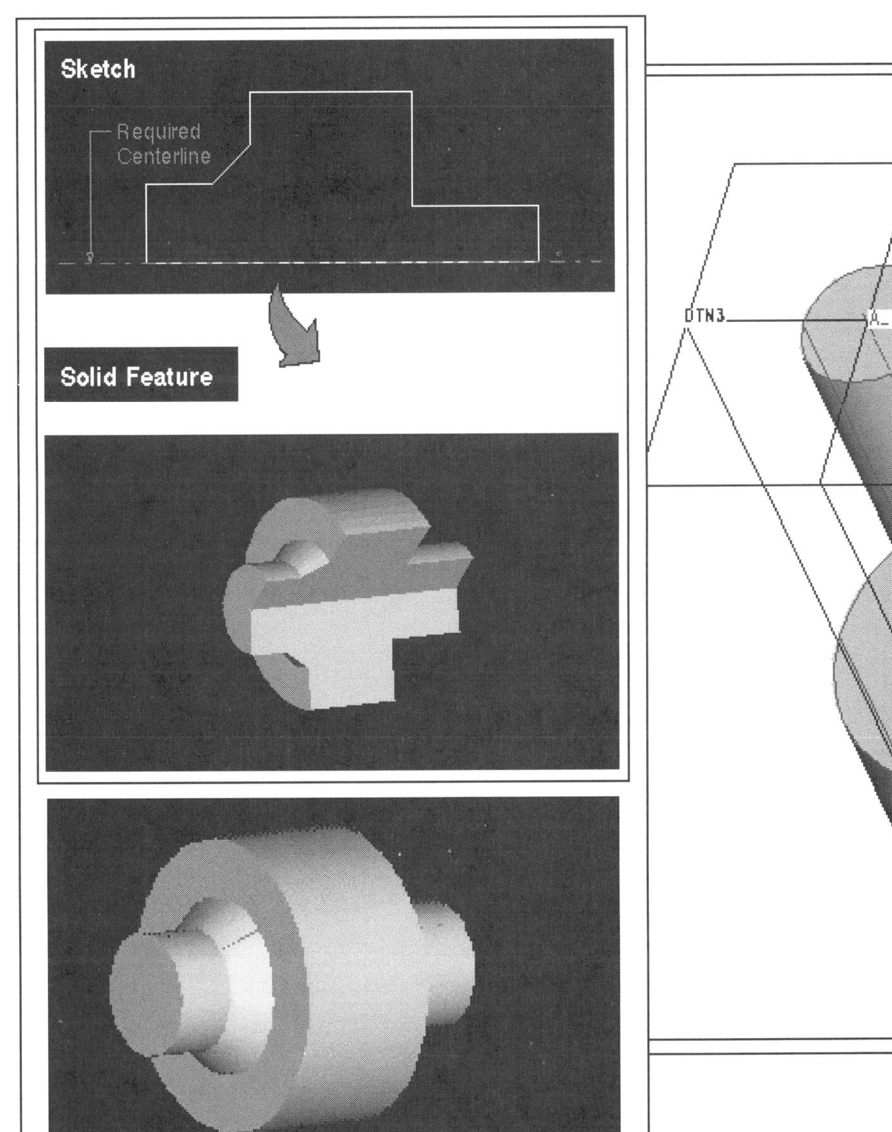

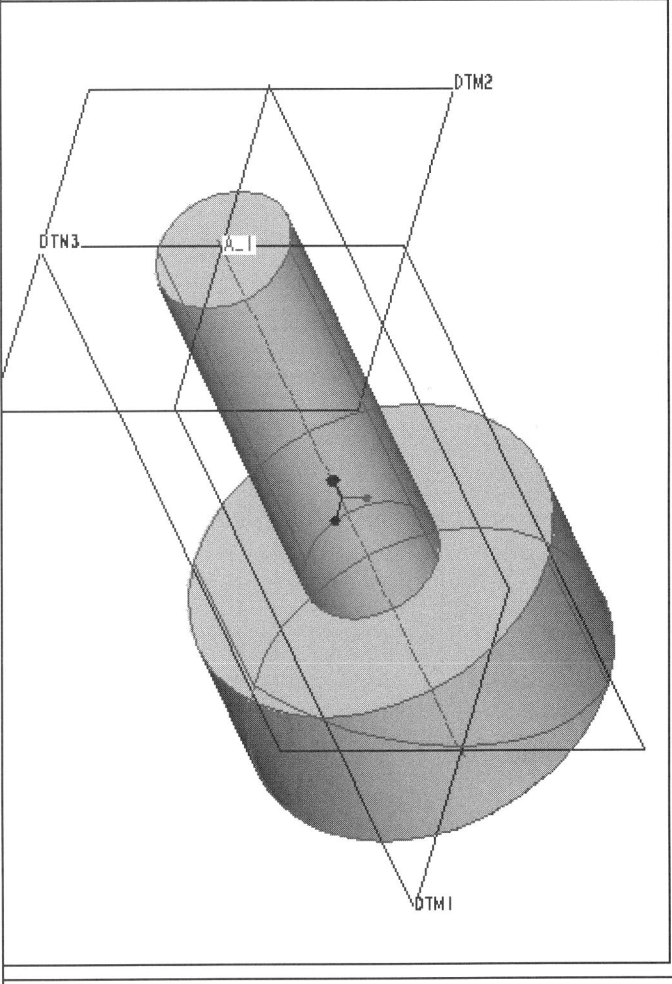

Sketching the Revolved Feature Section

To create a revolved section, create a centerline and the geometry that will be revolved about that centerline. The rules for sketching a revolved feature are:

* The revolved section must have a centerline. If you use more than one centerline in the sketch, Pro/ENGINEER uses the first centerline sketched as the axis of rotation.
* Geometry must be sketched on one side of the axis of revolution.
* The section must be closed for a protrusion but can be open for a cut.

Specifying the Angle of Revolution

Use the options in the REV TO menu to specify the angle of revolution of the feature and whether that angle is to be measured entirely on one side of the sketching plane or symmetrically on both sides of the sketching plane.

In creating a **Revolved Cut**, Pro/E revolves the section around the part to the specified angle measurement, thereby removing the material inside the section (Fig. 5.5).

Figure 5.5 Revolved Cut

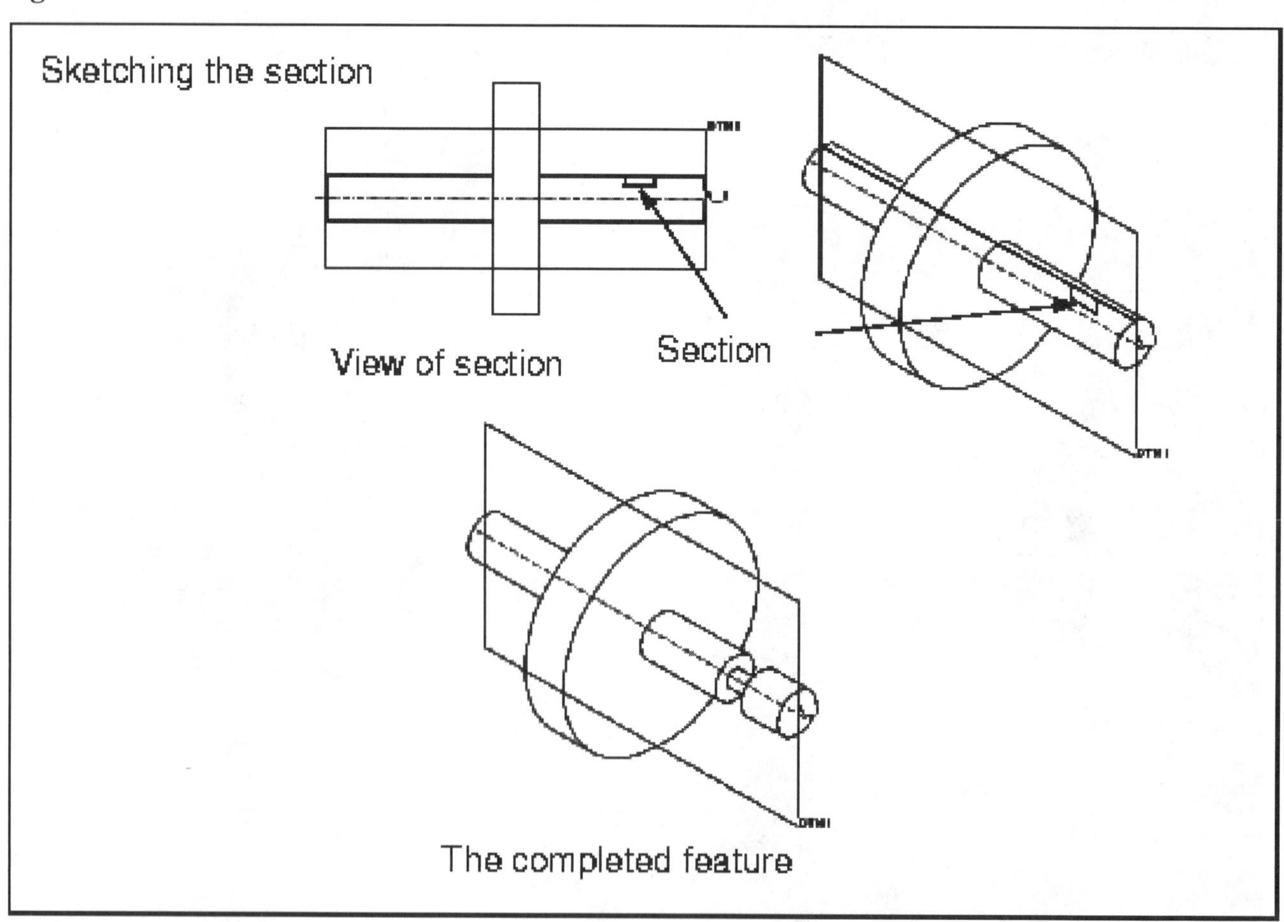

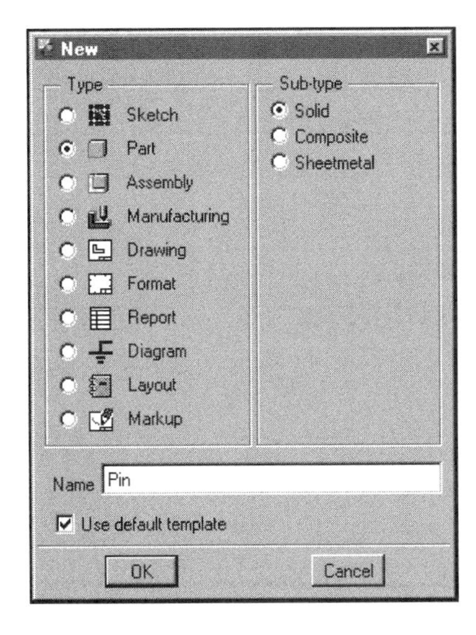

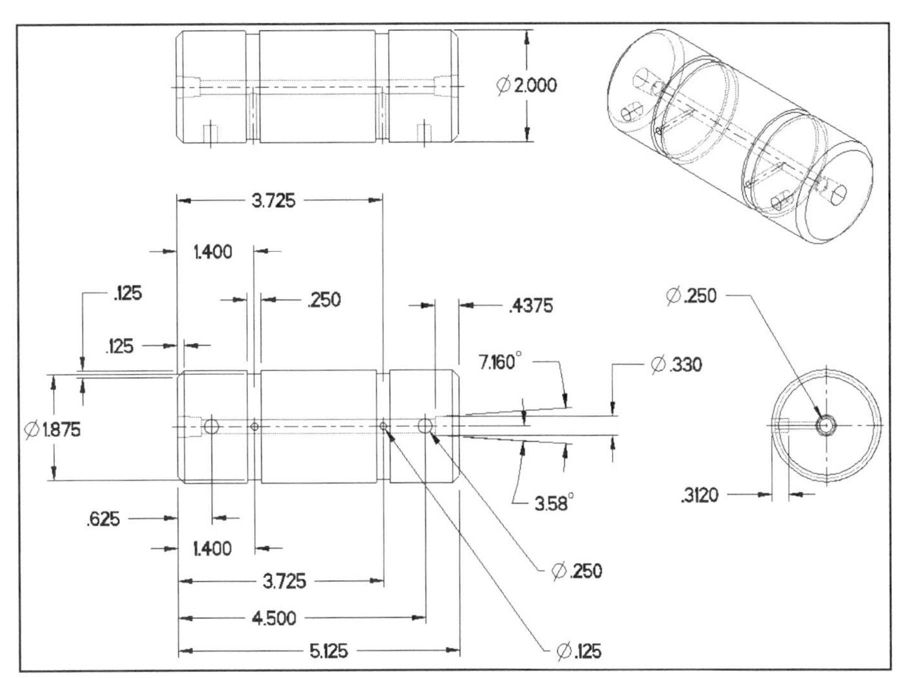

Figure 5.6
Pin Dimensions

Pin

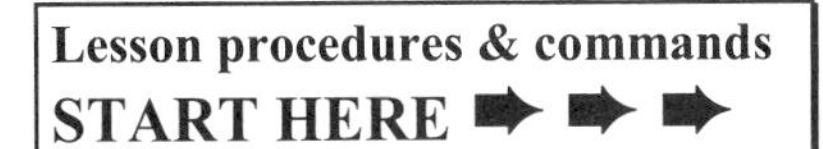

The pin is an example of a part created by revolving one section about a centerline (Fig. 5.6). The pin was created as a **revolved protrusion**. The chamfers are created on the first revolved protrusion. The grooves were created with revolved cuts. The holes were added using datum axes and a new datum plane.

The pin's complete geometry (with the exception of the holes) could have been created with one revolved protrusion. In general, this is poor design practice, because it limits the flexibility of modifying the geometry later in the design process. For most parts, the basic revolved shape should be the first protrusion, followed by the most important secondary features (such as secondary protrusions and cuts). The holes required for the part are then created. Lastly, the rounds and chamfers are created where required. Start the part:

File ⇒ New ⇒ •Part ⇒ (type the part name, **Pin**) **⇒ ✔Use default template ⇒ OK**

SETUP AND ENVIRONMENT

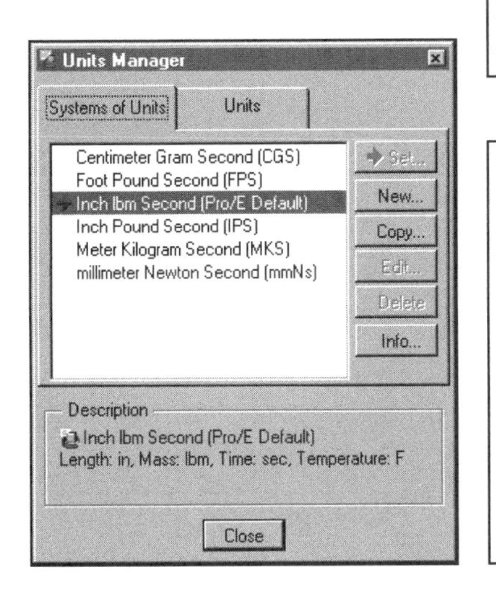

Set Up ⇒ Units ⇒ Units Manager **Inch lbm Second** (Pro/E default) **⇒ Close ⇒ Material ⇒ Define ⇒** (Type **STEEL**) **⇒ ✔⇒** (table of material properties, change or add information) **⇒ File ⇒ Save ⇒ File ⇒ Exit ⇒ Assign ⇒** (pick **STEEL**) **⇒ Accept ⇒ Done**

Utilities ⇒ Environment ⇒
Spin Center
✔ Snap to Grid
Display Style **Hidden Line**
Tangent Edges **Dimmed ⇒ OK**

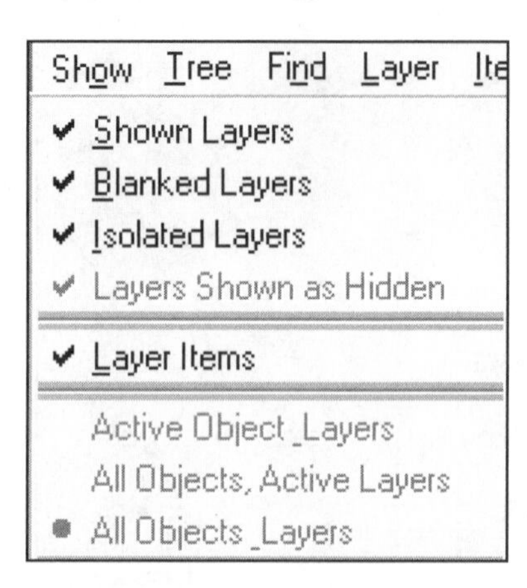

Start the pin by viewing the default layering system Note that the default datums, default coordinate system, and the default layers were created automatically by using the ✔ **Use default template** option. Choose the following commands to view the default layering (see the left-hand column):

View ⇒ Layers ⇒ Show ⇒ ✔ Layer Items ⇒ (expand the list for **-01_PRT_ALL_DAT_PLN** to see the **TOP, FRONT,** and **RIGHT** datum planes) **⇒ Close**

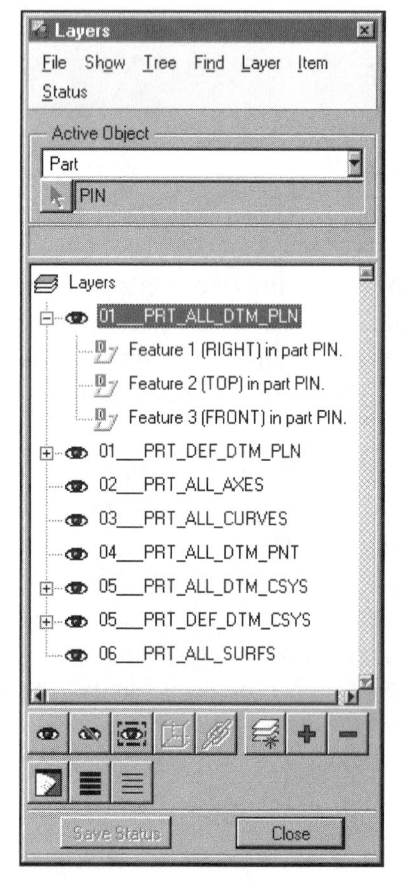

The first protrusion is a revolved protrusion. Use the front view of the pin in Figure 5.6 to sketch the revolved protrusion's section in Figure 5.7. Create the chamfers and the main body of the pin with the first protrusion. *In Lesson 6, you will see that chamfers, like rounds, are normally added near the end of the modeling sequence.* The commands, sketch, references, and so on are very similar to those in the previous lessons. ✔ **Intent Manager** is the default:

Feature ⇒ Create ⇒ Solid ⇒ Protrusion ⇒ Revolve ⇒ Solid ⇒ Done ⇒ One Side ⇒ Done ⇒ (pick **FRONT** as sketching plane) **⇒ Okay ⇒ Top** (pick **TOP** as the horizontal reference) **⇒ Close** (References) **⇒ Model Tree Off** (icon) **⇒** (now hold down **Ctrl** key and press RMB to pan the screen as shown in Figure 5.7) ⇒ [create a **Centerline** (sketch the horizontal centerline on datum **TOP** as the first centerline) ⇒ (sketch the remaining section geometry using **Lines**) ⇒ [add the diameter dimension and the chamfer dimensions (Fig. 5.8), the weak dimensions that are not required are automatically deleted by the Sketcher] ⇒ ↖ (icon)

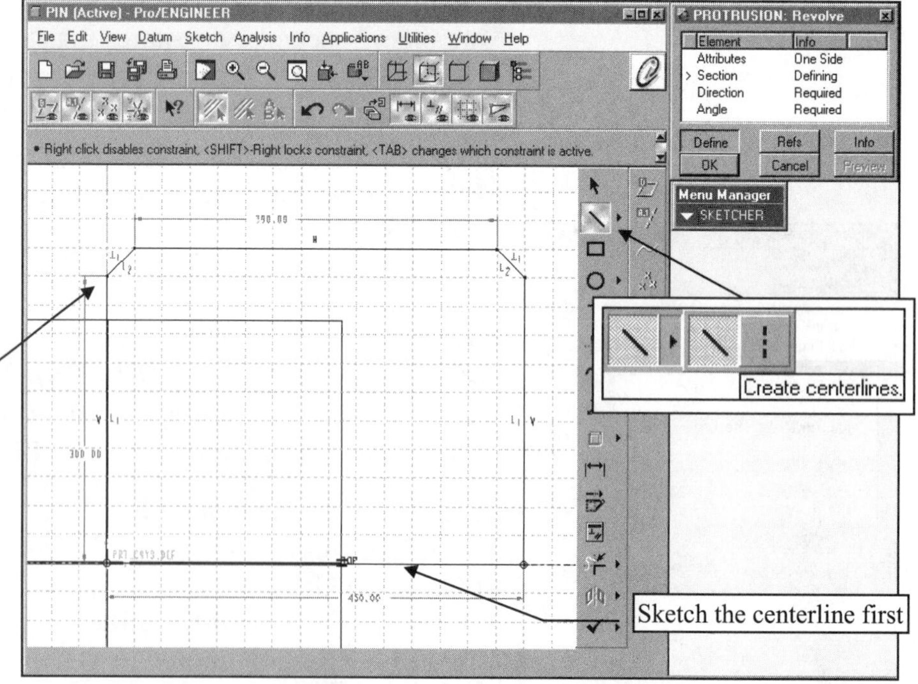

Figure 5.7
Section Sketch for Revolved Protrusion

Toggle off the grid from the Toolbar

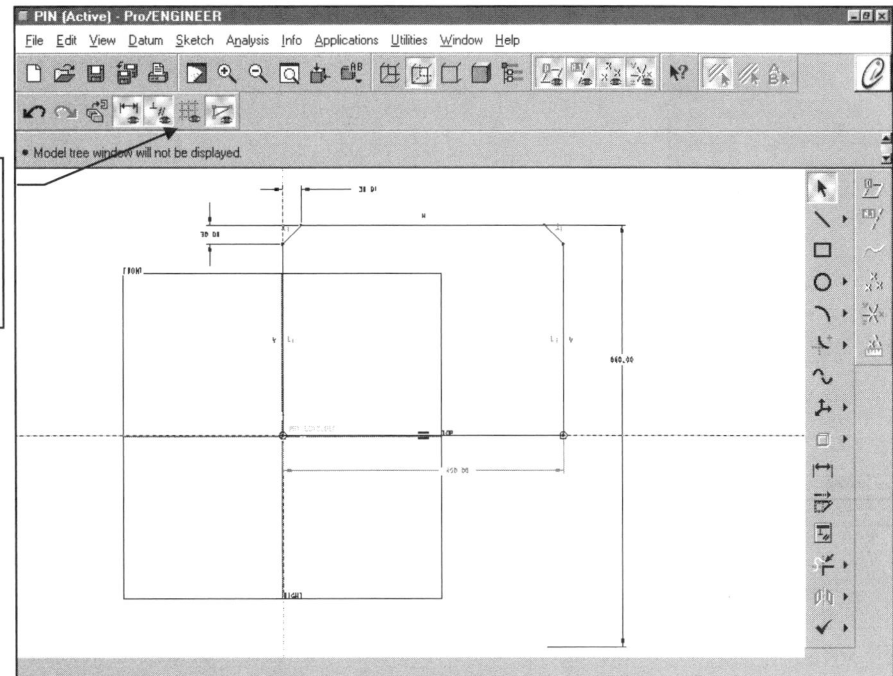

Figure 5.8
Dimensioned Section Sketch

[Pick on each dimension individually, after the dimension highlights, press the RMB for the pop-up options and select **Modify.** Change the dimensions to the design values (see Figure 5.6 for dimensions)] ⇒ (Figure 5.9 shows the updated sketch) ⇒ **Repaint**

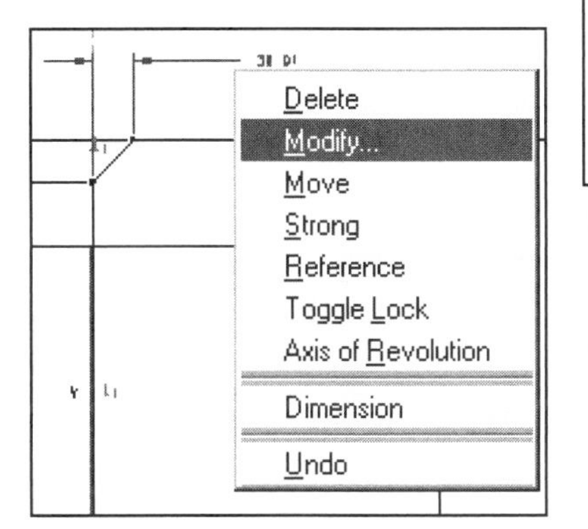

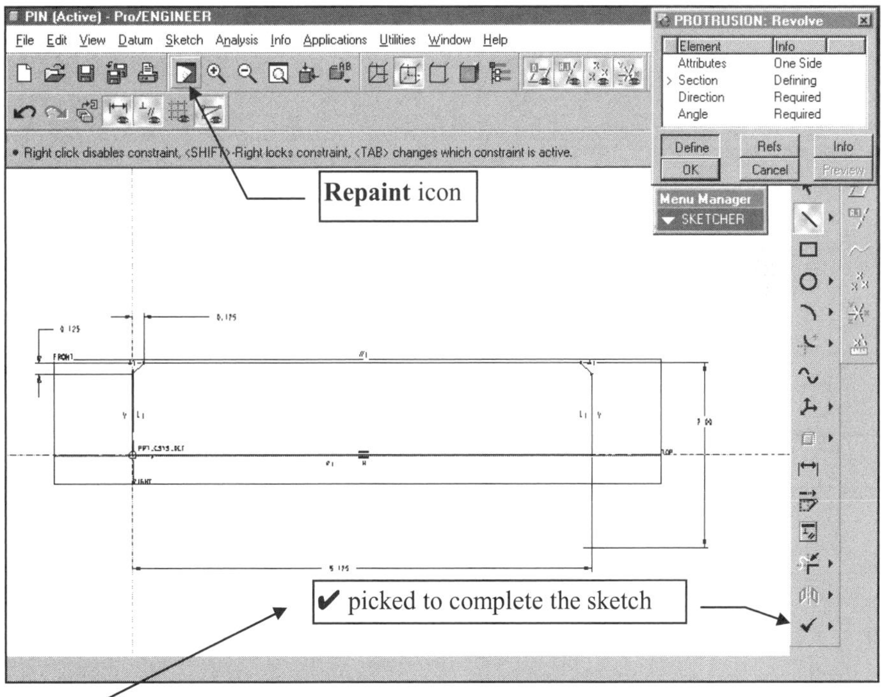

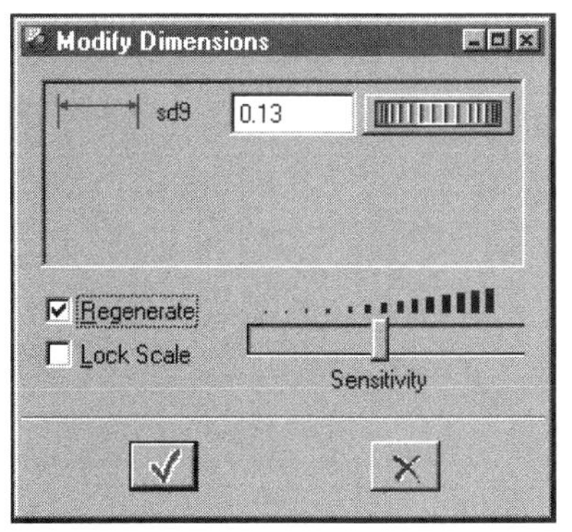

Figure 5.9
Design Dimensions

✔ ⇒ **360** (from the REV TO menu) ⇒ **Done** ⇒ **View** (from the menu bar) ⇒ **Default** ⇒ shade the model from the Toolbar (Shading)

⇒ **Preview** (Fig. 5.10) ⇒ **OK** ⇒ **Done** ⇒ **File** ⇒ **Save** ⇒ ✔

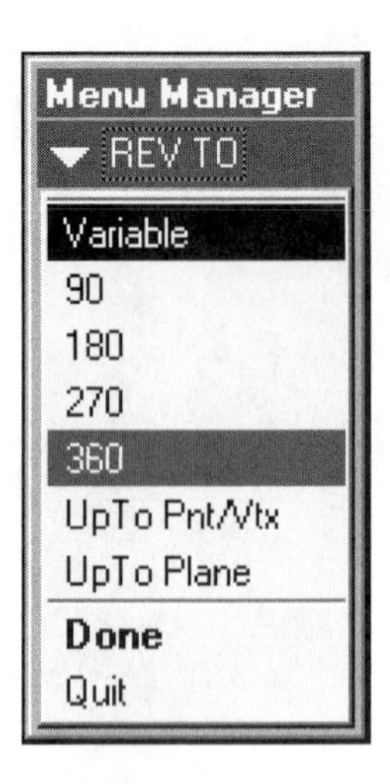

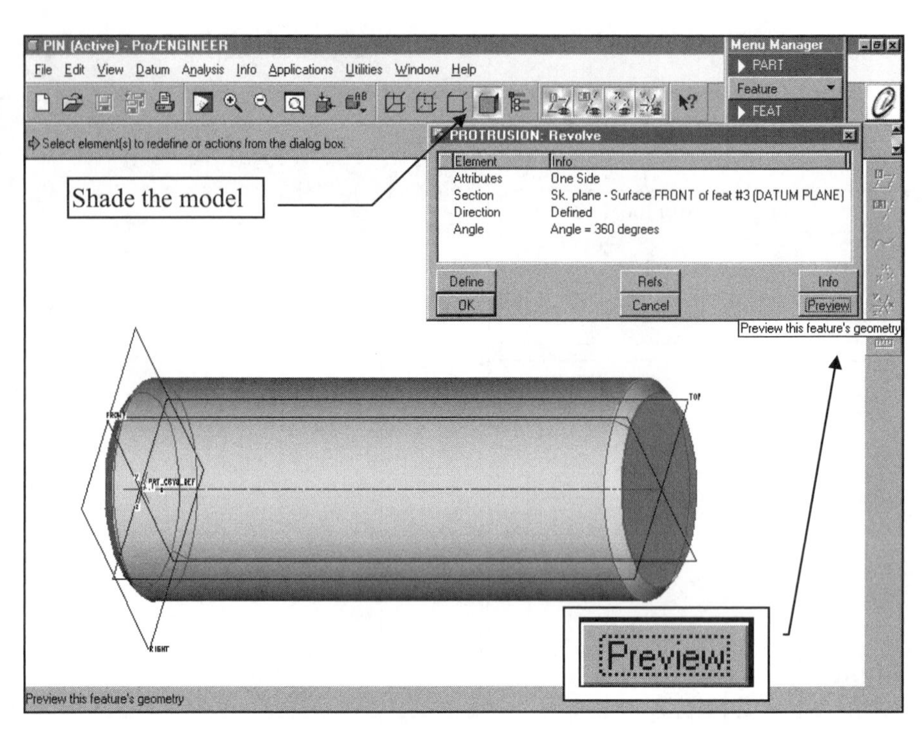

Figure 5.10
Preview of Revolved Protrusion

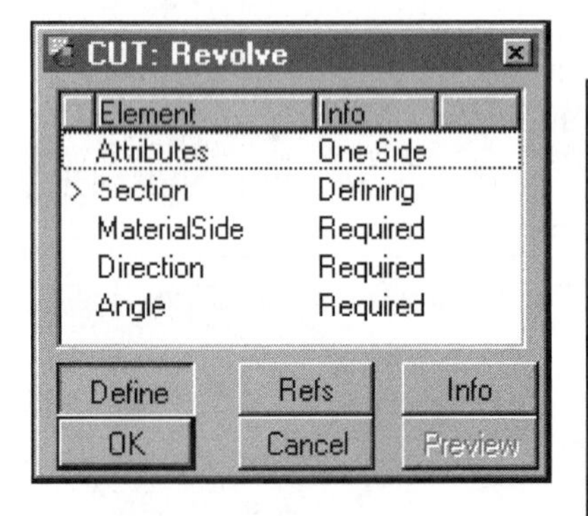

The revolved cuts are created next. Exaggerate the cut's size:

Utilities ⇒ Environment ⇒ ❑ Snap to Grid ⇒ Hidden Line ⇒ OK ⇒ Feature ⇒ Create ⇒ Cut ⇒ Revolve ⇒ Solid ⇒ Done ⇒ One Side ⇒ Done ⇒ Use Prev ⇒ Okay ⇒ Sketch ⇒ (off) **Intent Manager ⇒ Sketch ⇒ Line ⇒ Centerline ⇒ Horizontal** (sketch the centerline on datum **TOP** to be used for revolving the cut) **⇒ Vertical** (sketch the vertical centerline to establish the middle of the cut) **⇒ Sketch ⇒ Line ⇒ Vertical** (sketch the three lines of the cut) **⇒ Regenerate ⇒ Alignment** (align the two endpoints of the vertical lines with the top of the part, and align the *horizontal* centerline with datum **TOP**) **⇒ Regenerate ⇒ Dimension** (add dimensions) **⇒ Regenerate** (Fig. 5.11) **⇒ Modify** [change the sketch dimensions to the design dimensions (Fig. 5.6)] **⇒ Regenerate ⇒ Done ⇒ Okay ⇒ 360 ⇒ Done ⇒ OK ⇒ Done ⇒ File ⇒ Save ⇒ ✔**

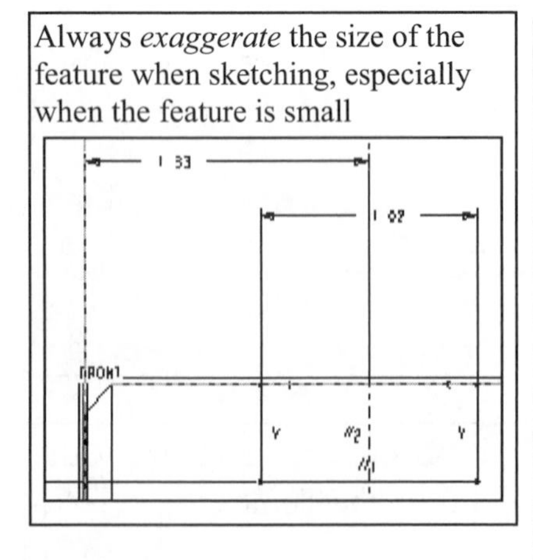

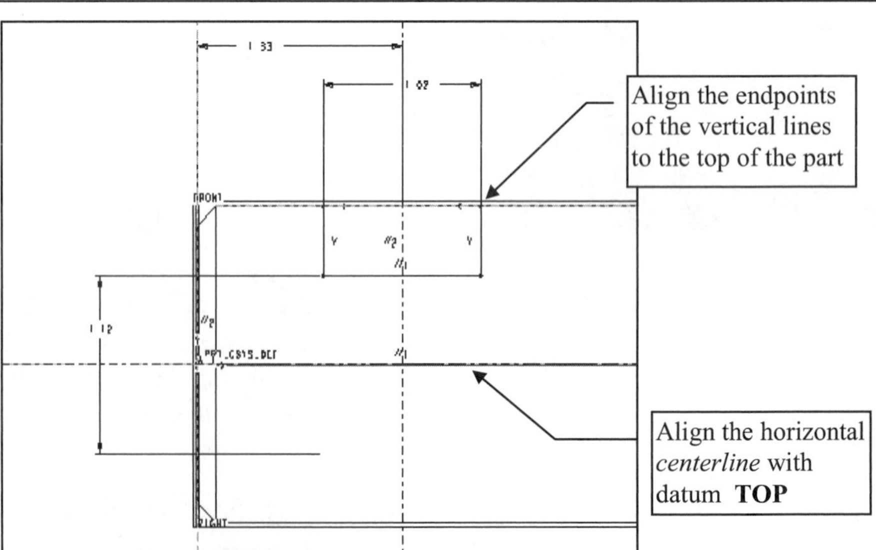

Figure 5.11
Revolved Cut Sketch

The revolved cut is now complete. Change the view, turn the Model Tree on, and shade and color the part (Fig. 5.12).

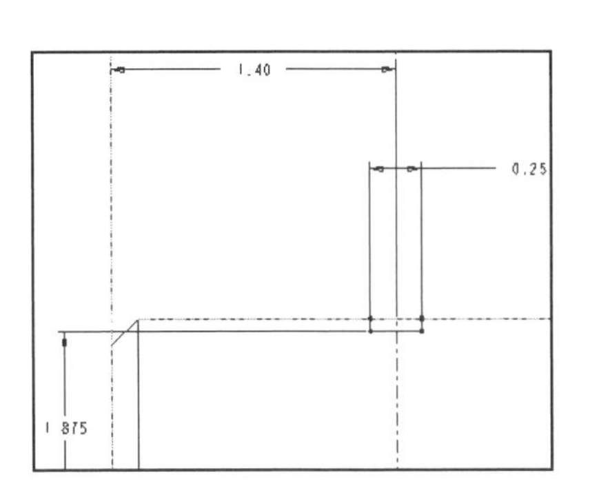

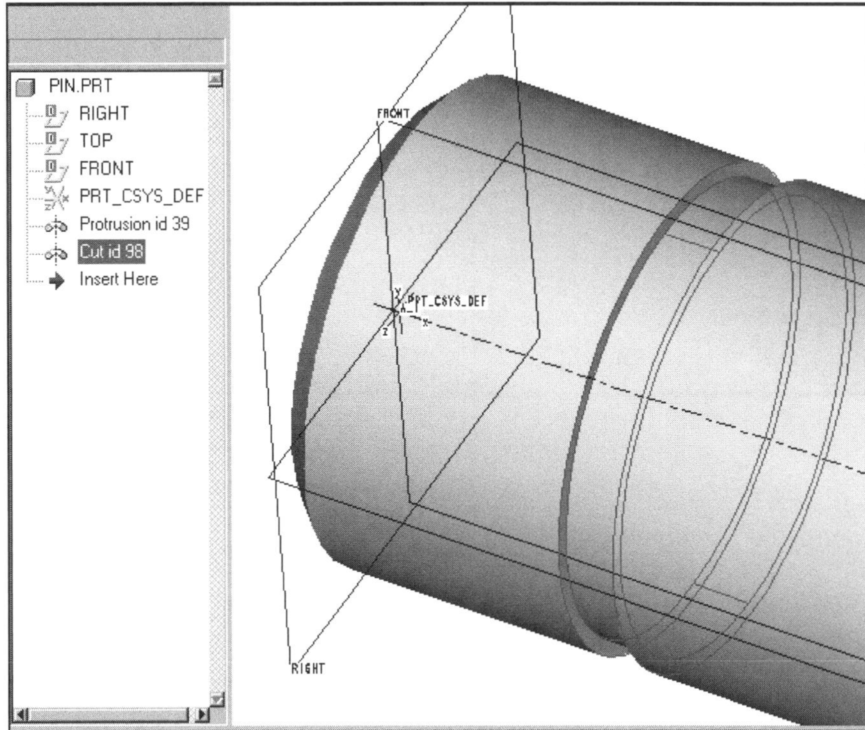

Figure 5.12
Completed Revolved Cut

Create the other revolved cut using the same steps (Fig. 5.13). Try sketching with the shading on, as shown in Figure 5.14. Rotate the part to see the cuts, as in Figure 5.14. Color the revolved cuts differently from the base part. In industry, the cut would have been copied instead of created again, but it is good to practice repeating commands at this stage of your understanding of Pro/E.

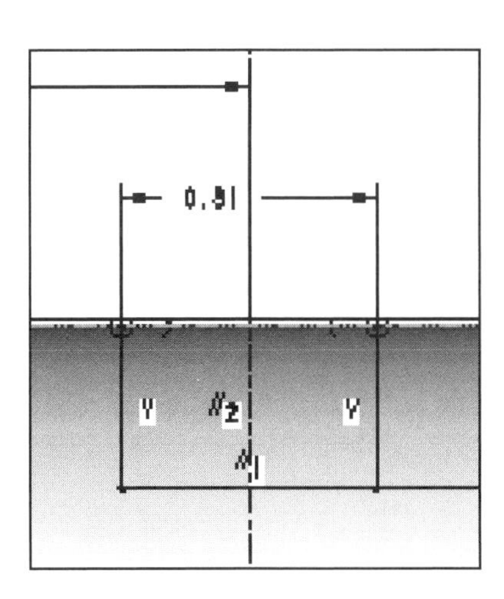

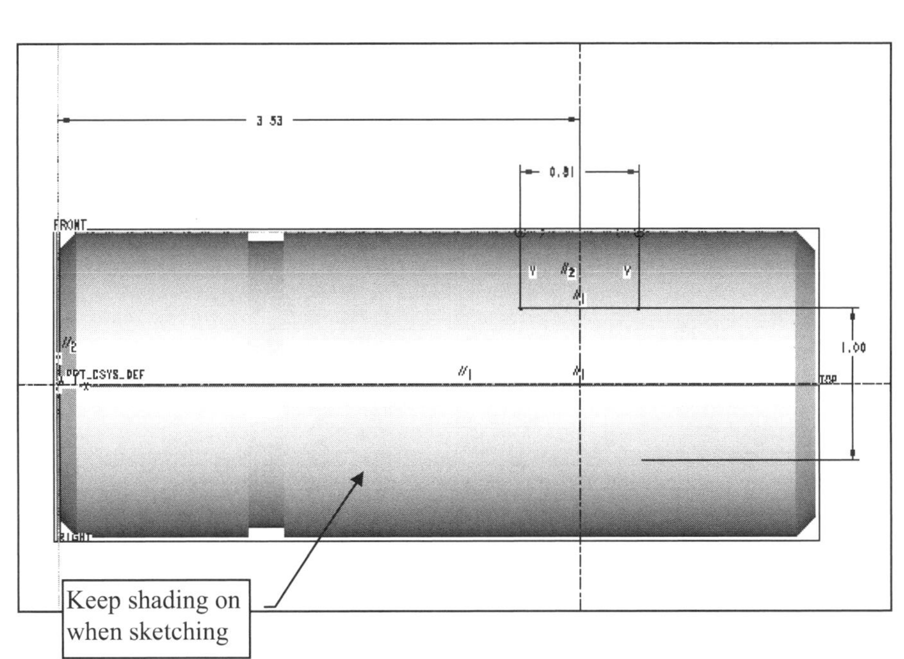

Figure 5.13
Sketching Second Cut with Shading On

NOTE

Set as basic and rename the datum planes and coordinate system:

Set Up ⇒ Geom Tol ⇒ Set Datum

RIGHT = datum **A**
TOP = datum **B**
FRONT = datum **C**

Set Up ⇒ Name

PRT_CSYS_DEF = CS0

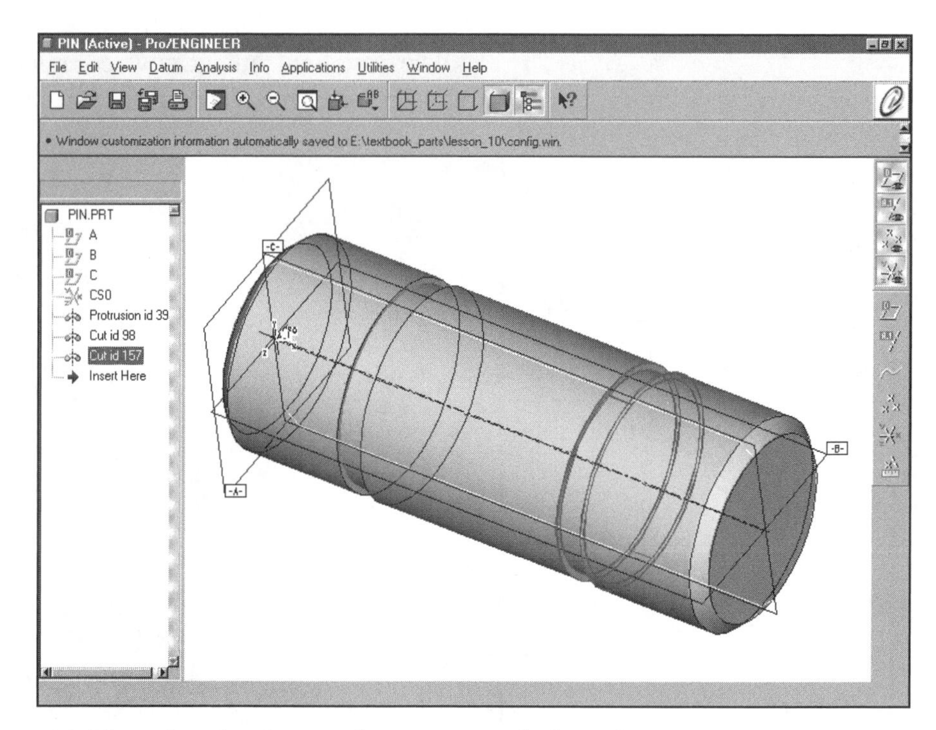

Figure 5.14
Completed Revolved Cuts

Note that the datums in our example have been *set*.

The remaining features are all holes. The first hole to create is the **∅.250** hole through the center (coaxial) of the pin (Fig. 5.15):

File ⇒ Save ⇒ ✔
File ⇒ Delete ⇒ Old Versions ⇒ ✔

Feature ⇒ Create ⇒ Solid ⇒ Hole ⇒ [fill in the required information--(Diameter **.250**, Depth One **Thru All,** Primary Reference ↖--(pick the left side of the revolved protrusion as the placement plane), Placement Type **Coaxial**- pick on axis **A_1)] ⇒ ✔ ⇒ Done ⇒ File ⇒ Save ⇒ ✔**

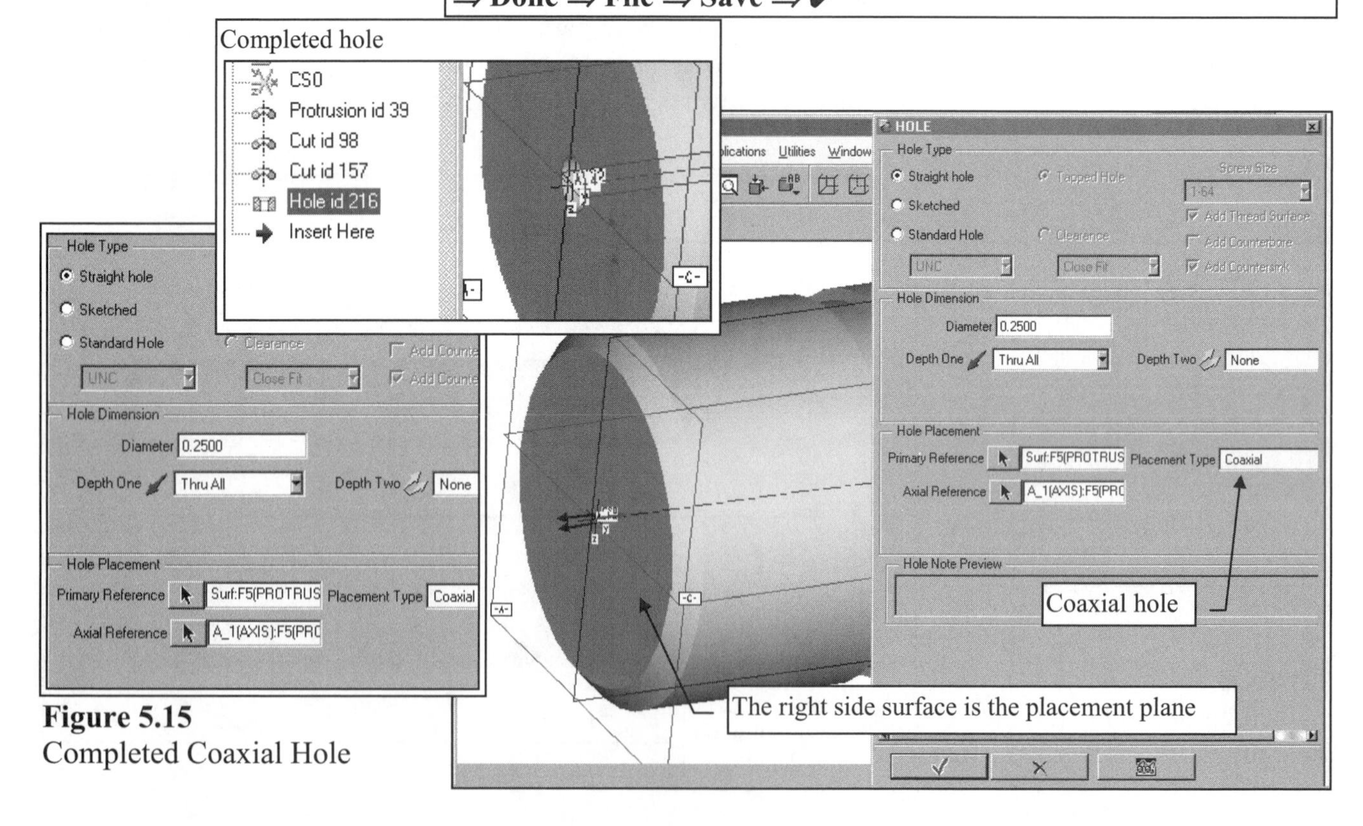

Figure 5.15
Completed Coaxial Hole

The pin has a conical cut (hole) at both ends that is coaxial with the **Ø.250** hole and axis **A_1**. Make the conical feature with a *sketched hole* placed on the datum plane **A** side of the pin (Fig. 5.16):

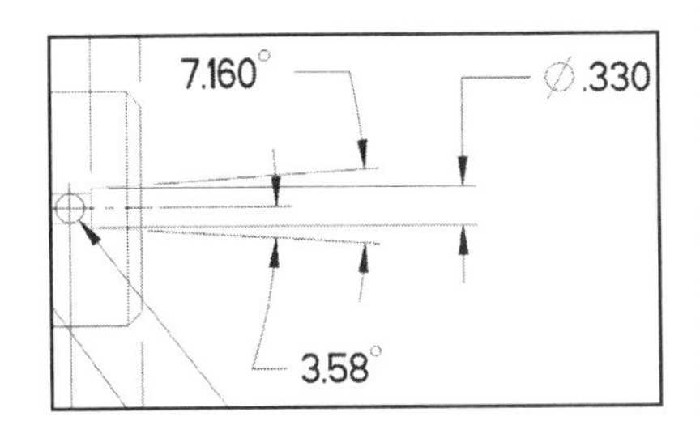

Utilities ⇒ Environment ⇒ ✔ Snap to Grid ⇒ OK ⇒ Feature ⇒ Create ⇒ Hole ⇒ (from the HOLE dialog box-- ● Sketched), ⇒ (sketch a *vertical* centerline) ⇒ (sketch the four lines of the *closed* section on one side of the centerline) ⇒ [add the required diameter and angle dimensions and remove any unwanted dimensions (Fig. 5.16)] ⇒ [modify the values of the dimensions as shown in the left-hand column (Depth = **.4375**, Diameter at small end = **.330**, and Angle = **3.58°**)] ⇒ **Sketch** (menu bar) ⇒ **Done ⇒** Primary Reference ↖pick the cylinders end plane that is coincident with datum **A**, Placement Type **Coaxial**) ⇒ ✔[(Build feature) (Fig. 5.17)]

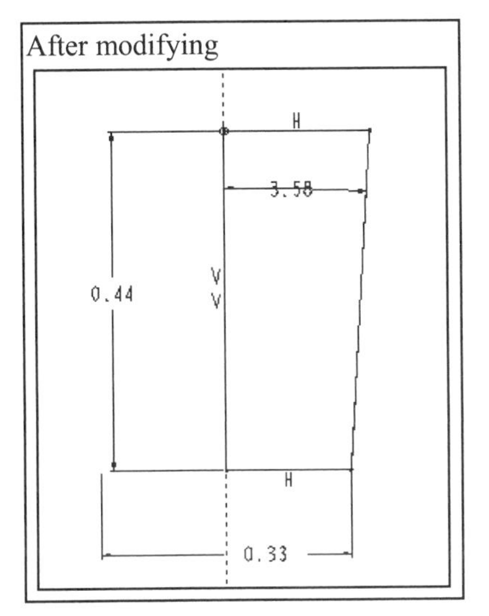

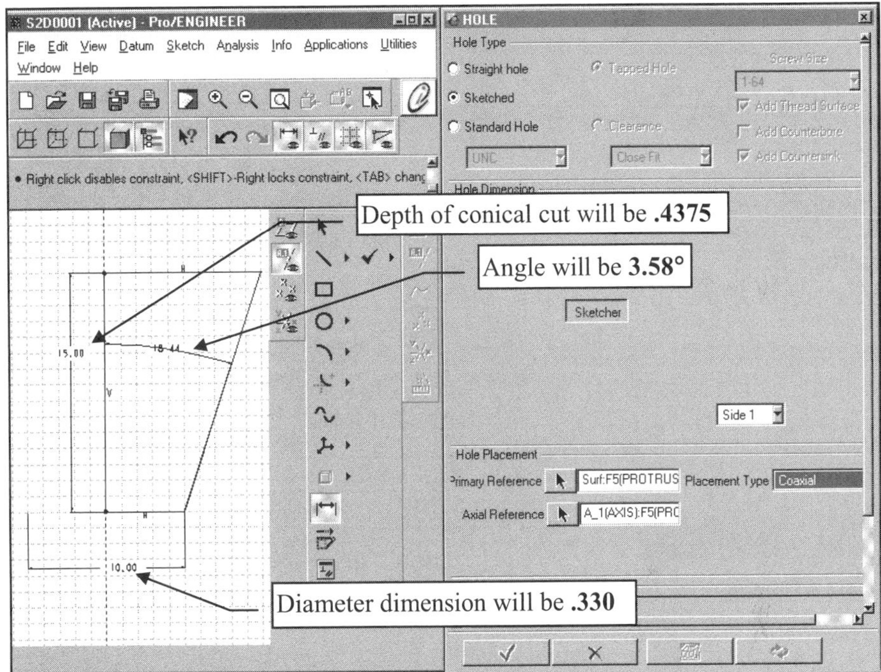

Figure 5.16
Sketch of Hole Before Modifying Dimensions

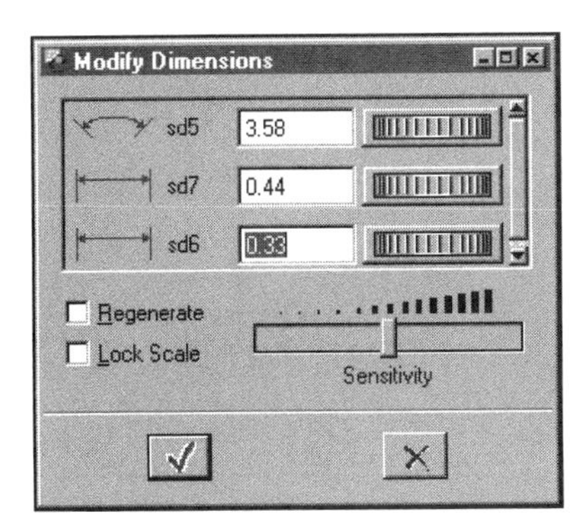

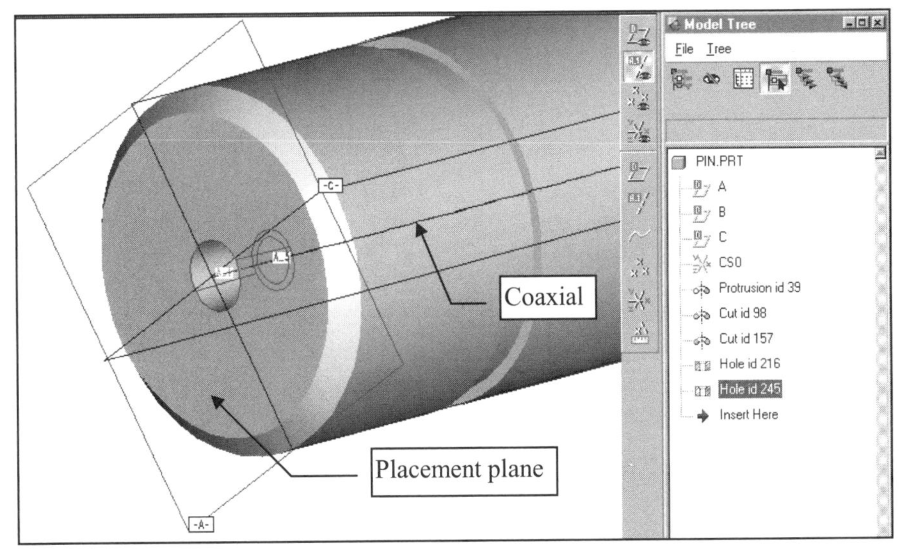

Figure 5.17
Coaxial Hole

File ⇒ Save ⇒ ✔
File ⇒ Delete ⇒
Old Versions ⇒ ✔

The conical hole is on both sides of the pin. Copy and mirror the hole using the following commands:

Feature ⇒ Copy ⇒ Mirror ⇒ Select ⇒ Dependent ⇒ Done ⇒ [select the conical hole to be mirrored; use **Query Sel** (or rotate the model to see the hole)] **⇒ Accept ⇒ Done Sel ⇒ Done ⇒ Make Datum ⇒ Offset ⇒** (pick datum plane **A** to offset from) **⇒ Enter Value** (type **5.125/2** at prompt) **⇒ ✔ ⇒ Done** (Fig. 5.18) **⇒ Done**

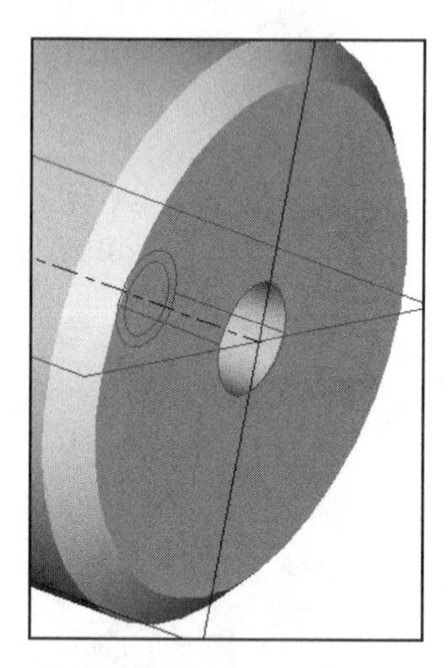

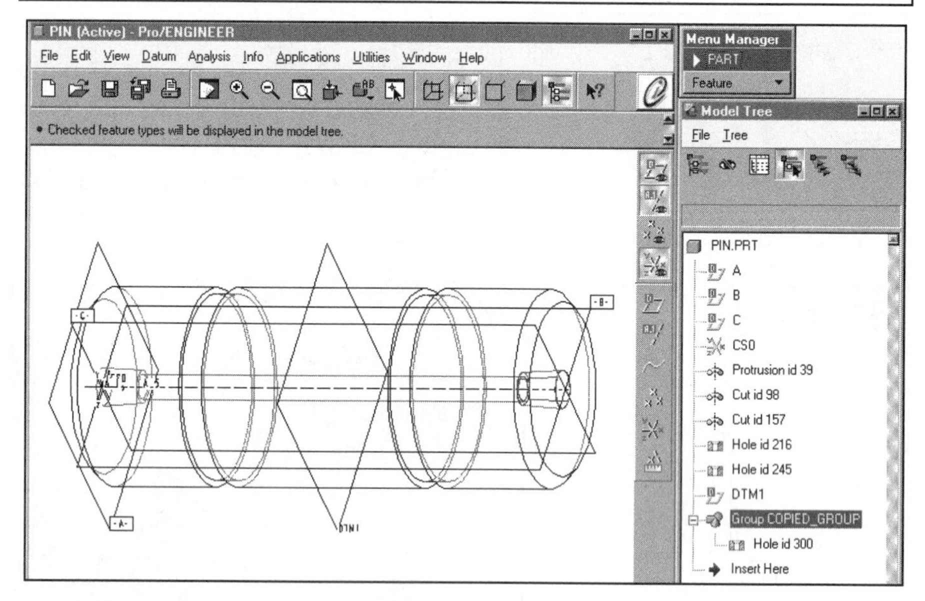

Figure 5.18
Mirrored Conical Hole

You just created your first ***datum-on-the-fly*** (**Make Datum**).

Create a new datum plane tangent to the pin's outside diameter and parallel to datum plane **C**:

[from the menu bar (or Toolbar icon)] **Datum ⇒ Plane ⇒ Tangent** (pick the left front of the pin's cylinder) **⇒ Parallel** (pick datum plane **C**, as shown in Fig. 5.19) **⇒ Done**

Datum Analysis Inf
Plane...
Axis...
Curve...
Point...
Csys...
Graph...
Analysis...
Reference...
Offset Planes...

Create the new datum plane *tangent* to the pin's cylindrical surface

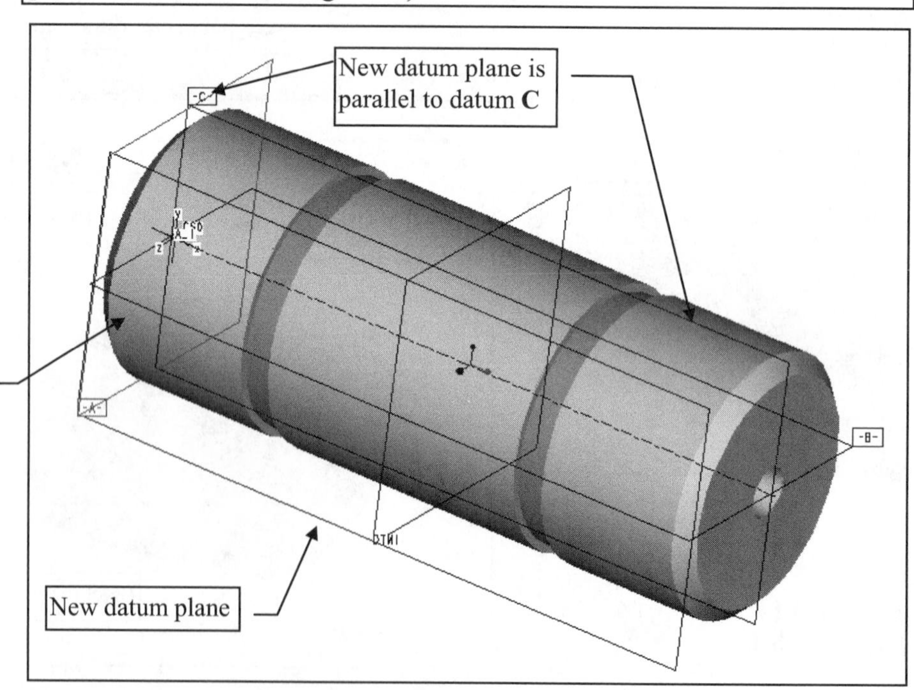

Figure 5.19
New Datum Plane Created Tangent to the Pin's Revolved Protrusion

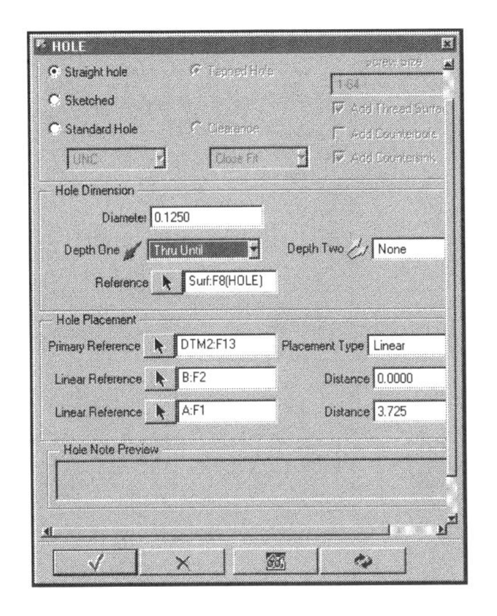

Using this new datum plane, create the two **Ø.125** holes using the new datum plane as the placement plane and the **Ø.250** coaxial hole running through the pin as the ending surface.

Feature ⇒ Create ⇒ Solid ⇒ Hole ⇒ ● Straight hole ⇒ [from the HOLE dialog box-- Diameter **.125,** Depth one **Thru Until** (pick the **Ø.250** hole by using **Query Sel** to filter through the part to get to the hole)],

Primary Reference ↖ [pick the tangent datum plane (**DTM2**)]

Placement Type **Linear**

Linear Reference ↖(pick datum **B**) Distance **0.00**

Linear Reference ↖(pick datum **A**) Distance **3.725** ⇒ ✔ (Fig. 5.20)

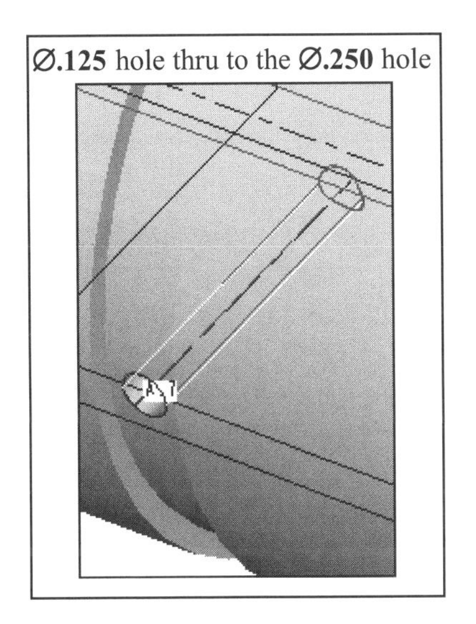

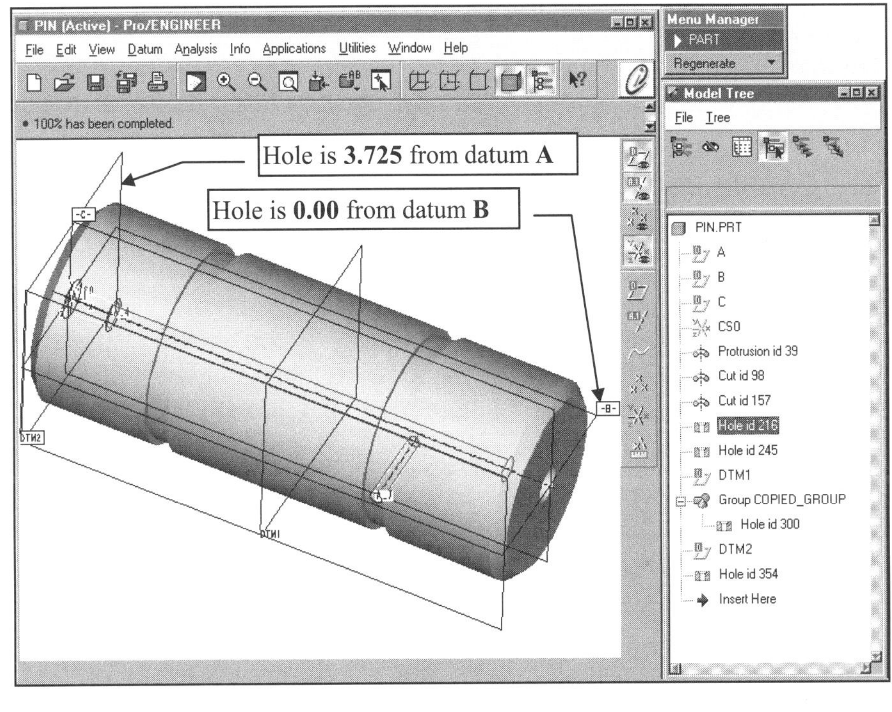

Figure 5.20
Ø.125 Hole

Now create the **Ø.250** hole on the part's side (you will mirror and copy both holes later). The command is exactly the same as the previous command, for the **Ø.125** hole, except that the diameter is **.250**, its distance from datum plane **A** is **4.50**, and it is a *"blind"* (Variable) hole with a Depth Value of **.3120.**

The holes can now be copied and mirrored (Fig. 5.21) about the same datum plane that was used for mirroring the conical hole (Fig. 5.6). Use the following commands:

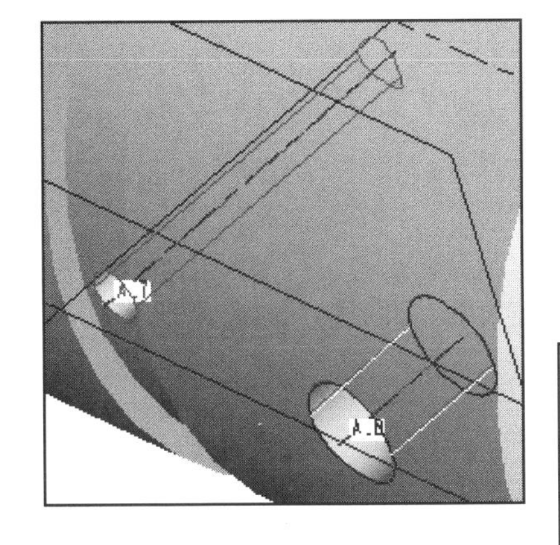

Feature ⇒ Copy ⇒ Mirror ⇒ Dependent ⇒ Done ⇒ (select both the **Ø.125** and the **Ø.250** holes just created) **⇒ Done Sel ⇒ Done ⇒** (pick **DTM1** as the plane to mirror about) **⇒ Done** (Fig. 5.21) **⇒ File ⇒ Save ⇒ ✔ ⇒ File ⇒ Delete ⇒ Old Versions ⇒ ✔**

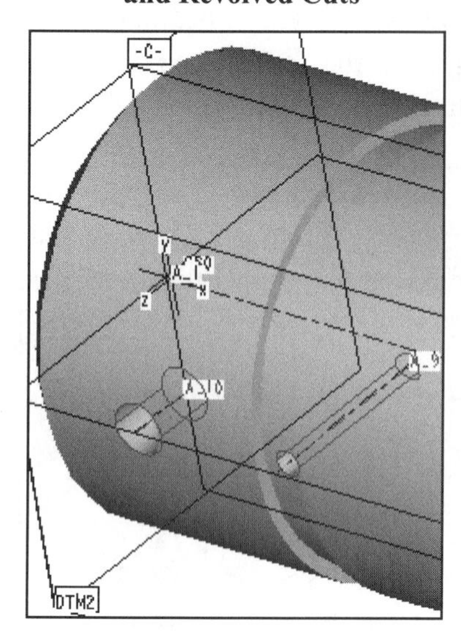

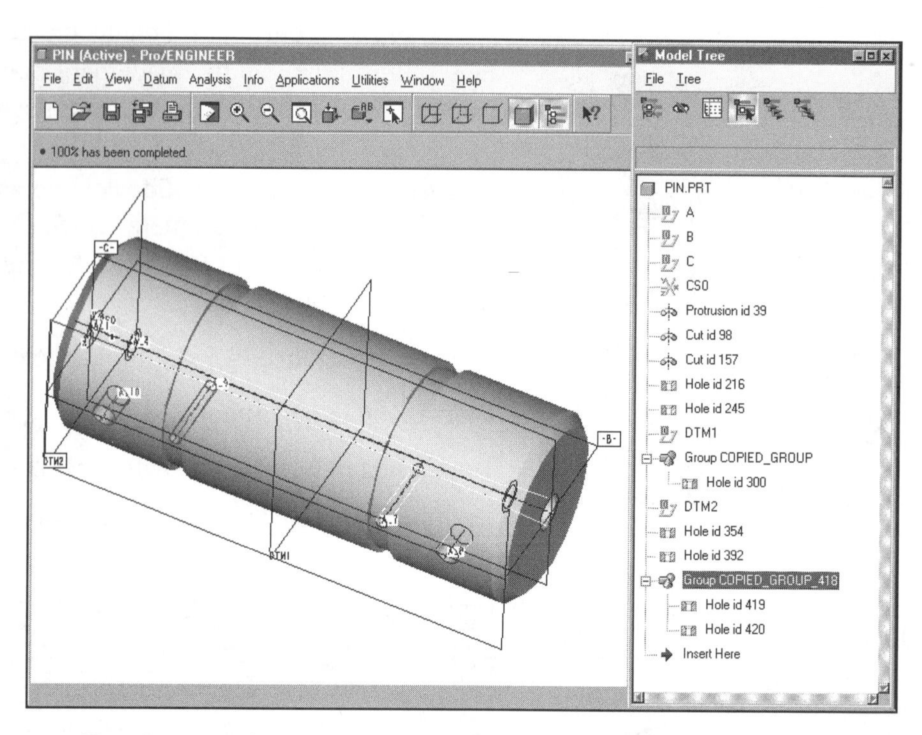

Figure 5.21
Mirrored Holes

To plot your part or drawing, choose **File** (from menu bar) ⇒ **Print** ⇒ **OK** ⇒ **OK.** For more information on plotting, read Section 6, Utilities.

To get information about your part, select the part name from the Model Tree and then press and hold down the RMB. Slide the cursor to the **Info** selection and, then slide the cursor to the **Model Info** selection. Release the RMB to see information displayed in the INFORMATION WINDOW (Fig. 5.22).

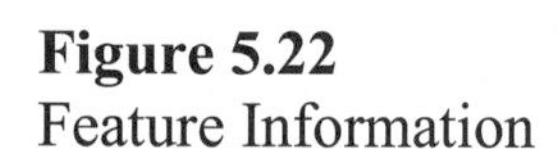

Figure 5.22
Feature Information

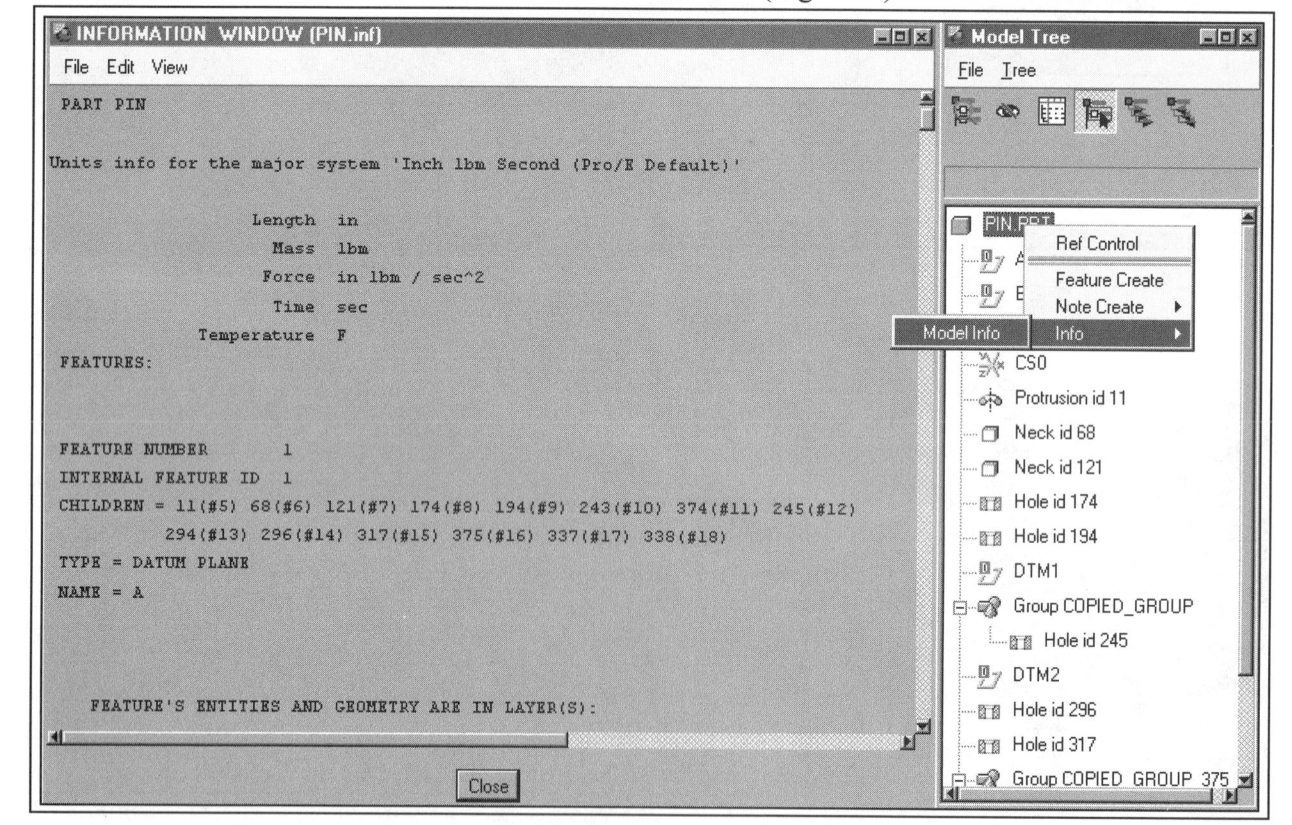

Lesson 5 Project

Clamp Foot and Clamp Swivel

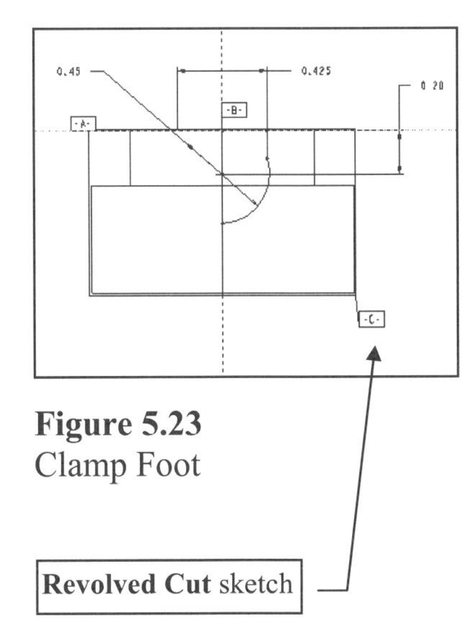

Figure 5.23
Clamp Foot

Revolved Cut sketch

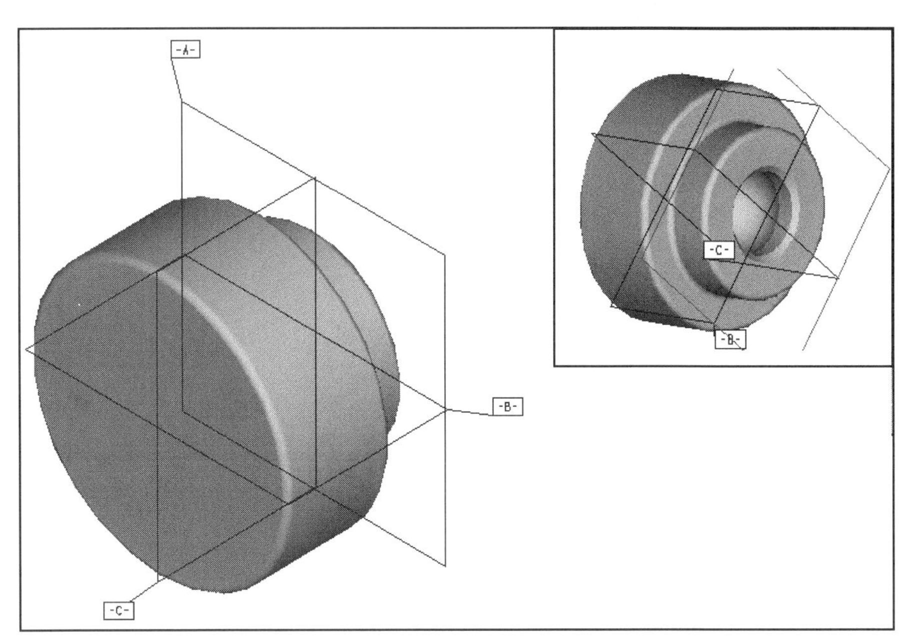

Clamp Foot and Clamp Swivel

Two **lesson projects** are provided in Lesson 5. You will use both of these parts in Lessons 14 and 15 when creating an assembly. Both the Foot and the Swivel are simple revolved protrusions. The Foot is nylon and the Swivel is steel. The Swivel fits inside the Foot.

Analyze each part and plan out the steps and features required to model it. Use the **DIPS** in Appendix D to establish a feature creation sequence before the start of modeling. Remember to set up the environment, establish datum planes, and set them on layers.

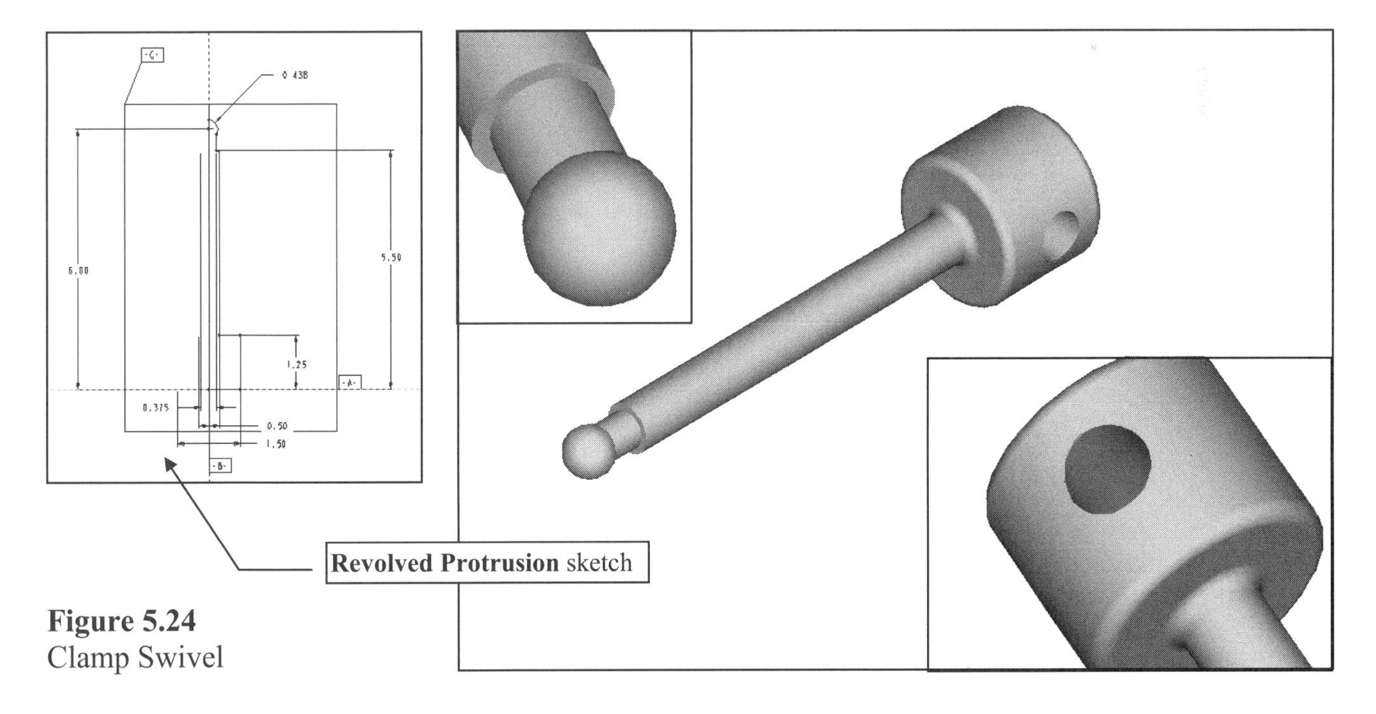

Revolved Protrusion sketch

Figure 5.24
Clamp Swivel

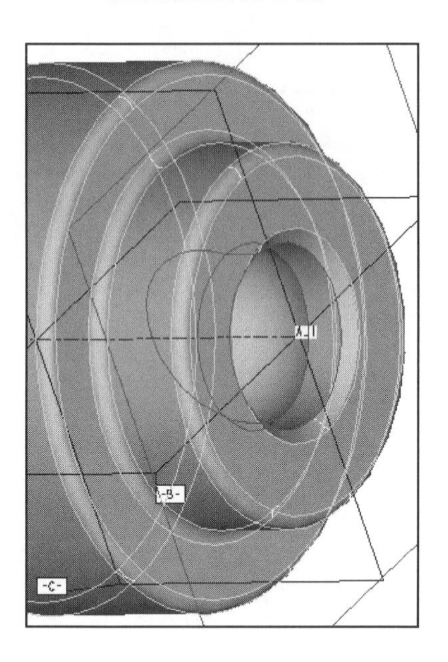

Figure 5.25
Clamp Foot Dimensions

Create the two parts (Figs. 5.23 through 5.30) with revolved protrusions using datum **C** (**RIGHT**) as the sketching plane. Create the internal cut on the **Foot** with a *revolved cut*. Add the rounds on both parts at the end of the modeling process; do not include them on the first revolved protrusions.

Figure 5.26
Clamp Foot Datums

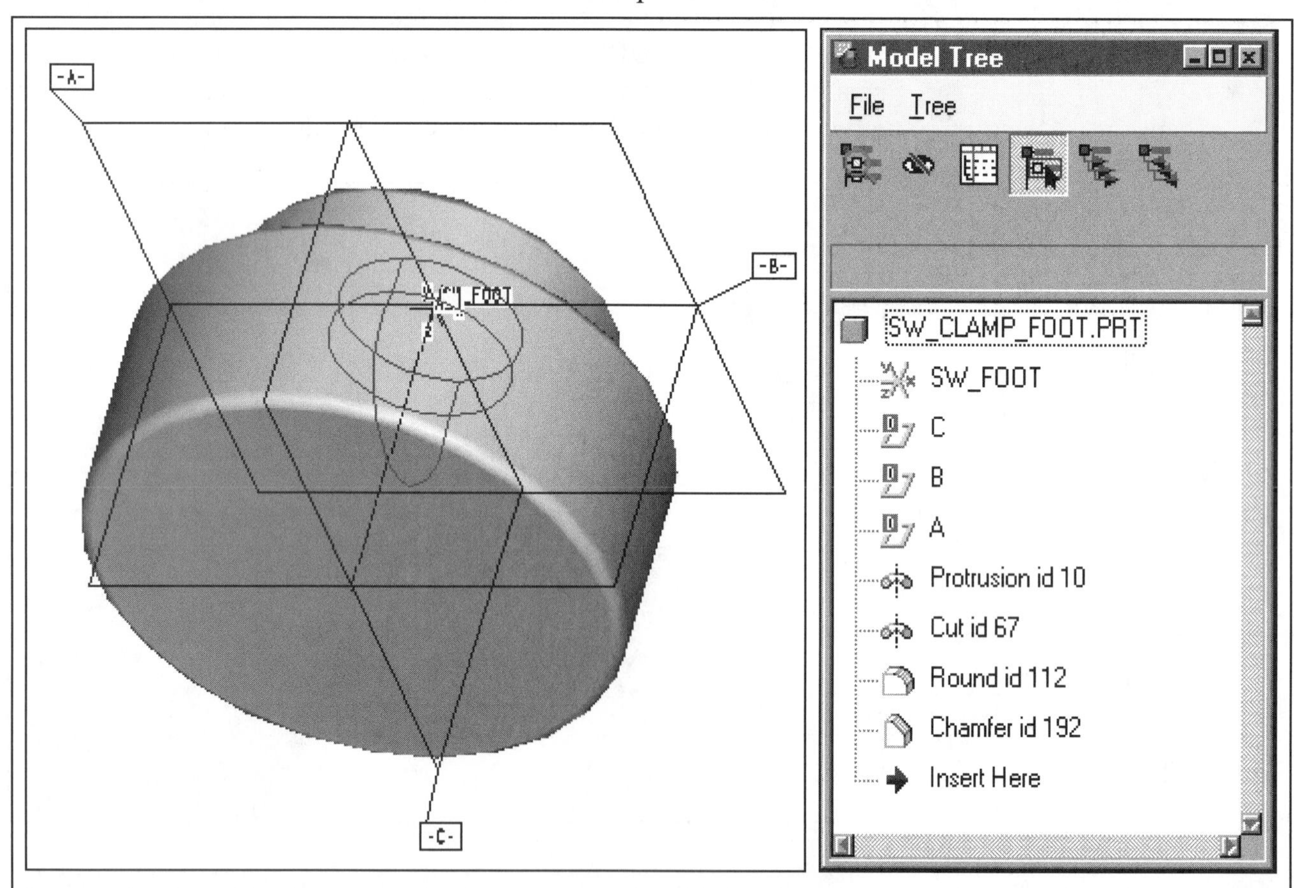

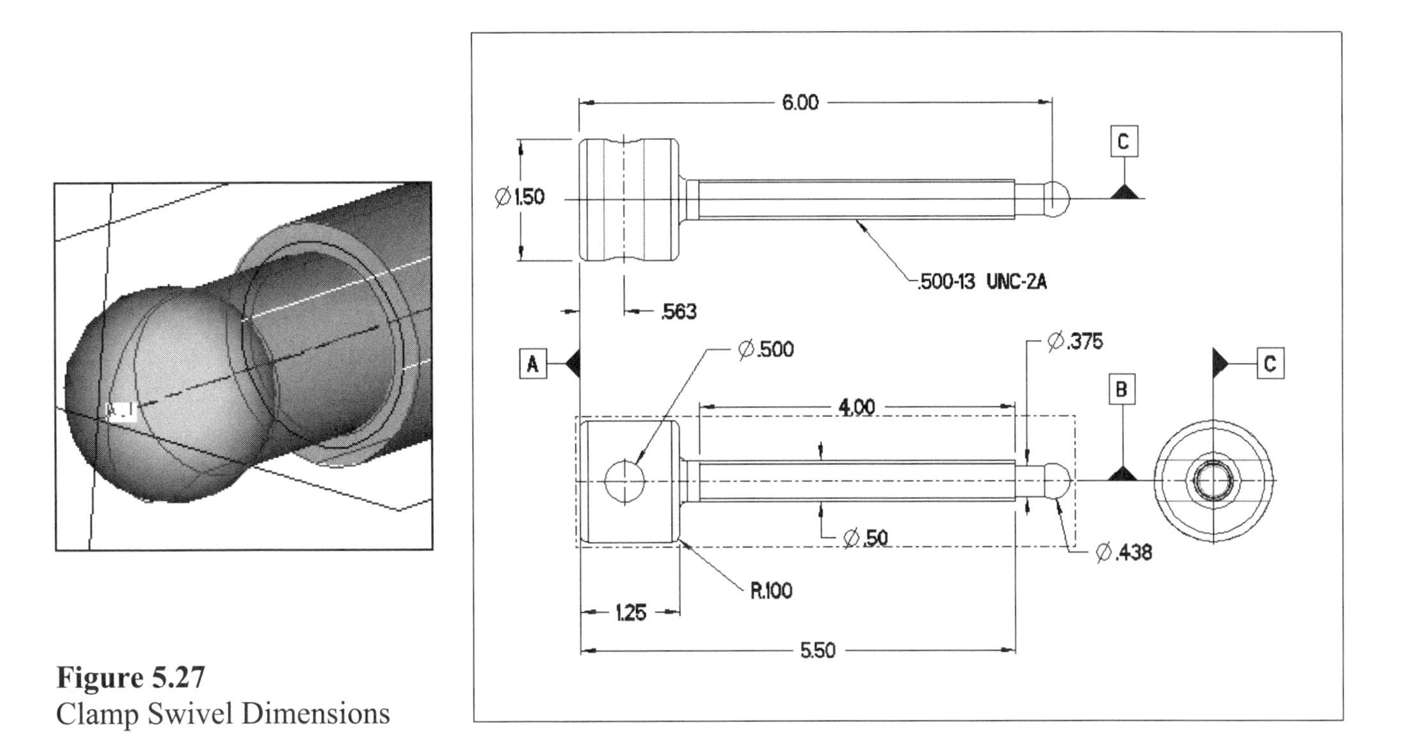

Figure 5.27
Clamp Swivel Dimensions

Figure 5.28 Clamp Swivel with Model Tree and Datums Displayed

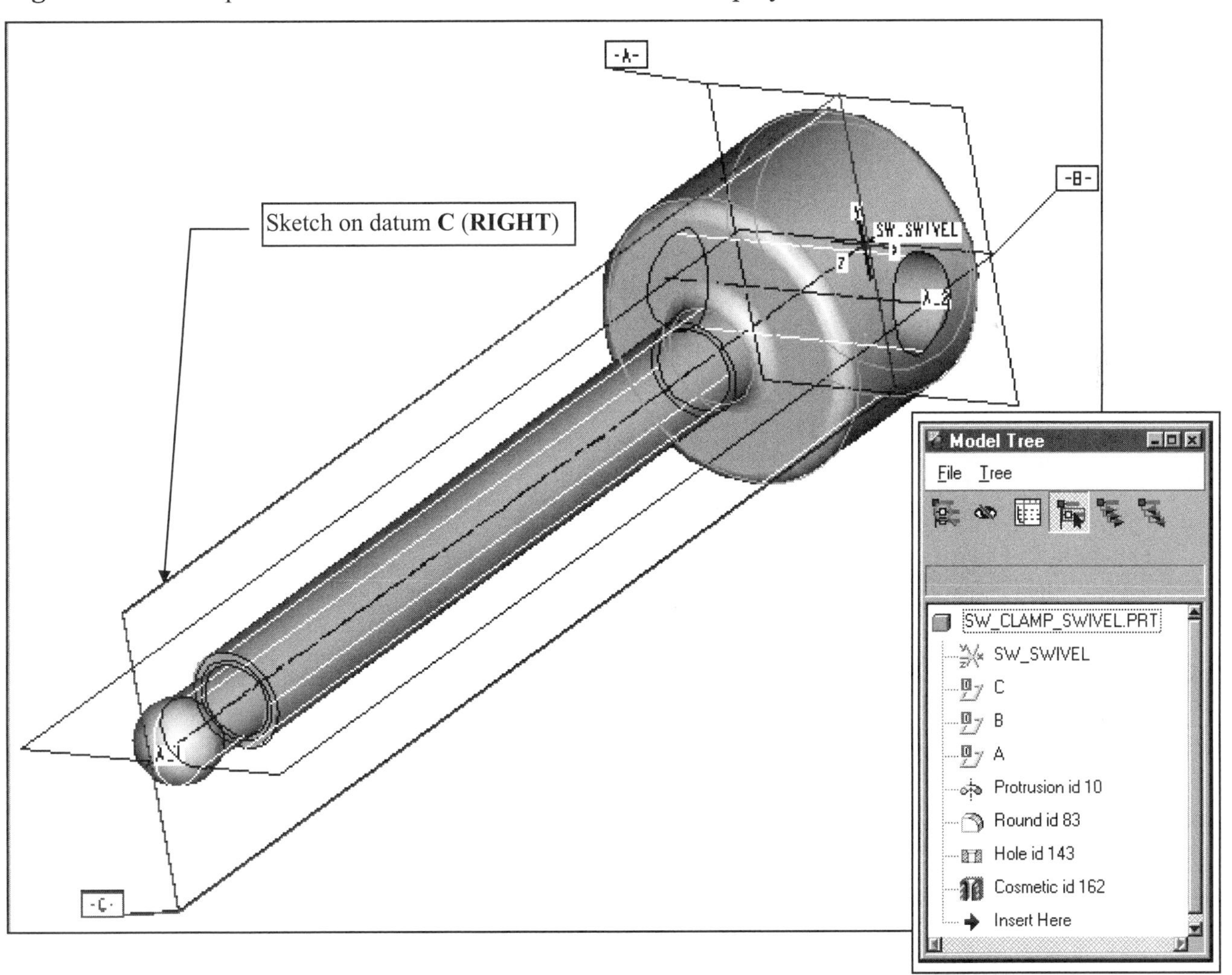

Figure 5.29 Clamp Foot

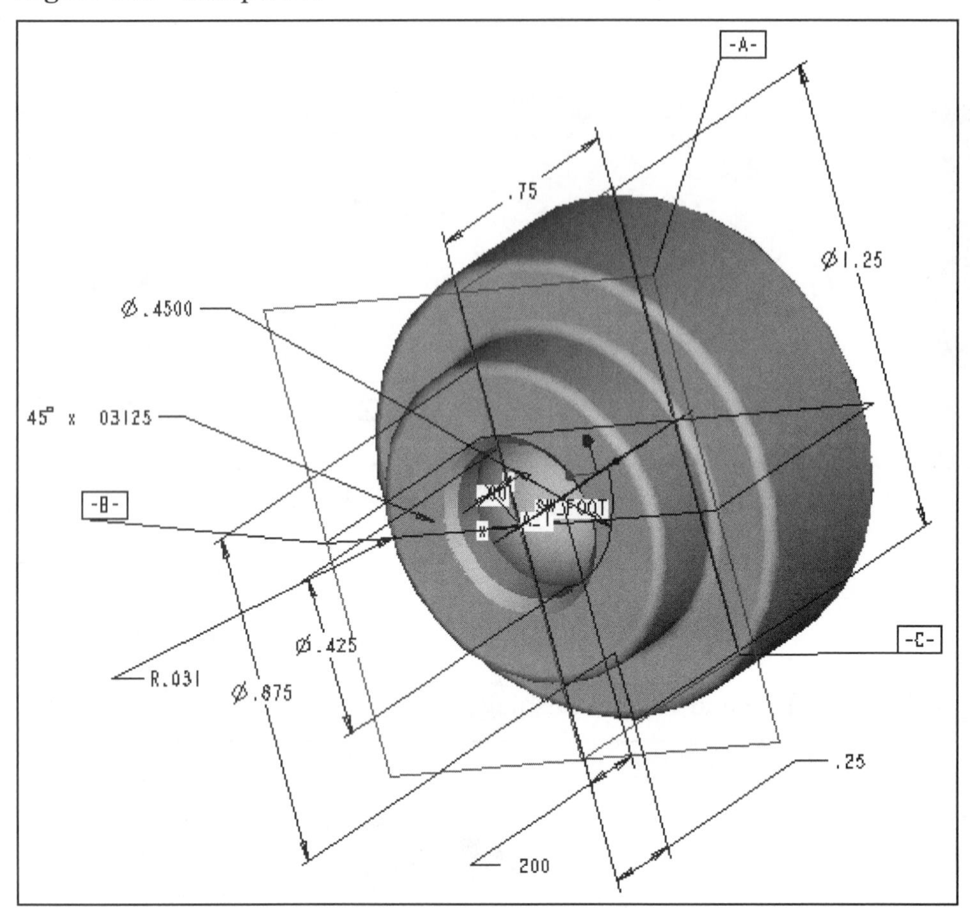

Figure 5.30 Clamp Swivel

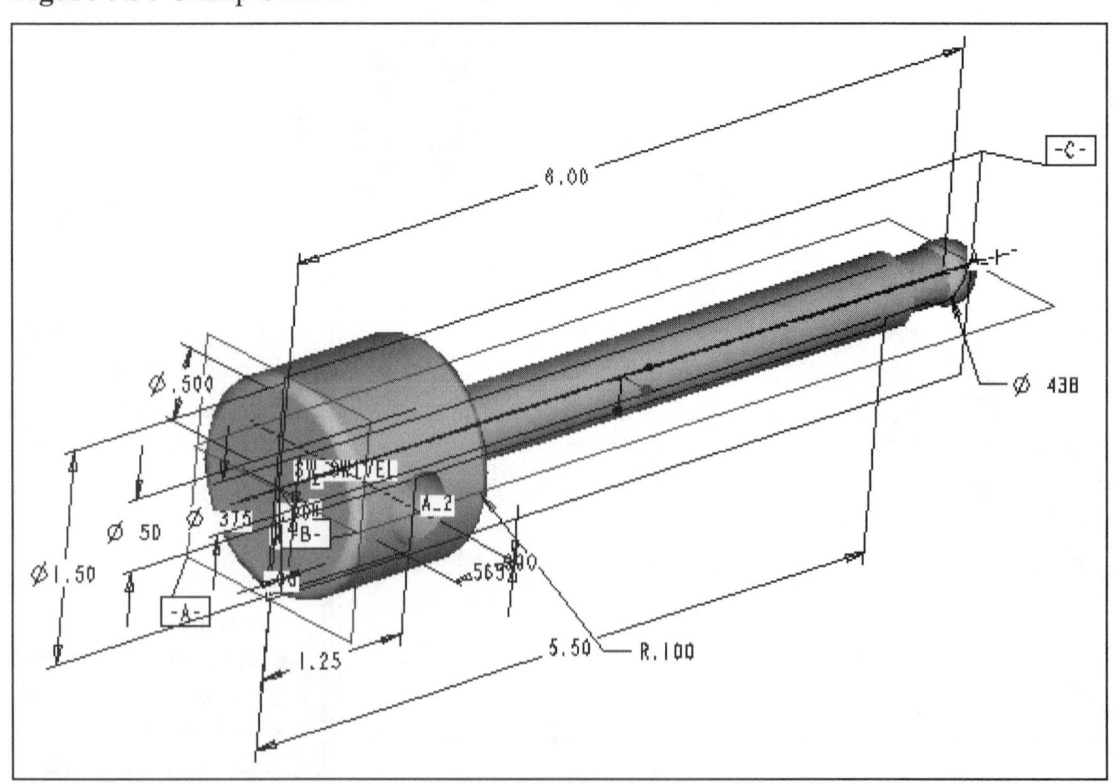

Lesson 6

Chamfers and Cosmetic Threads

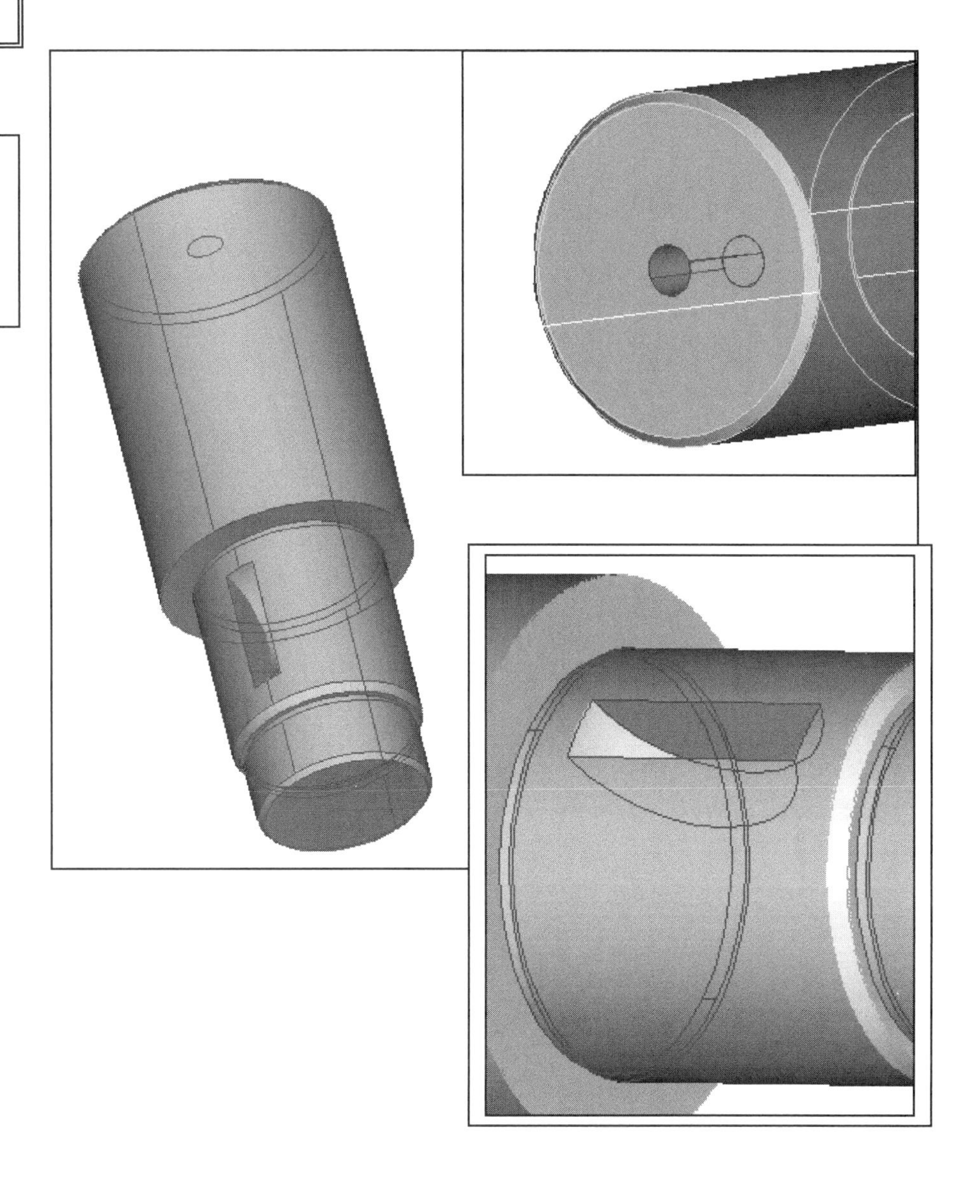

Figure 6.1
Cylinder Rod

OBJECTIVES

1. **Create simple chamfers along part edges**

2. **Learn how to sketch in 3D**

3. **Dynamically modify sketch dimension values**

4. **Create standard tapped holes**

5. **Create cosmetic threads**

6. **Complete tabular information for threads**

EGD REFERENCE
Fundamentals of Engineering Graphics and Design
by L. Lamit and K. Kitto
Read Chapters 14 and 17.
See pages 497, 540-541, and 674-678.

COAch™ for Pro/ENGINEER

If you have **COAch for Pro/ENGINEER** on your system, go to SEARCH and do the Segment shown in Figure 6.3.

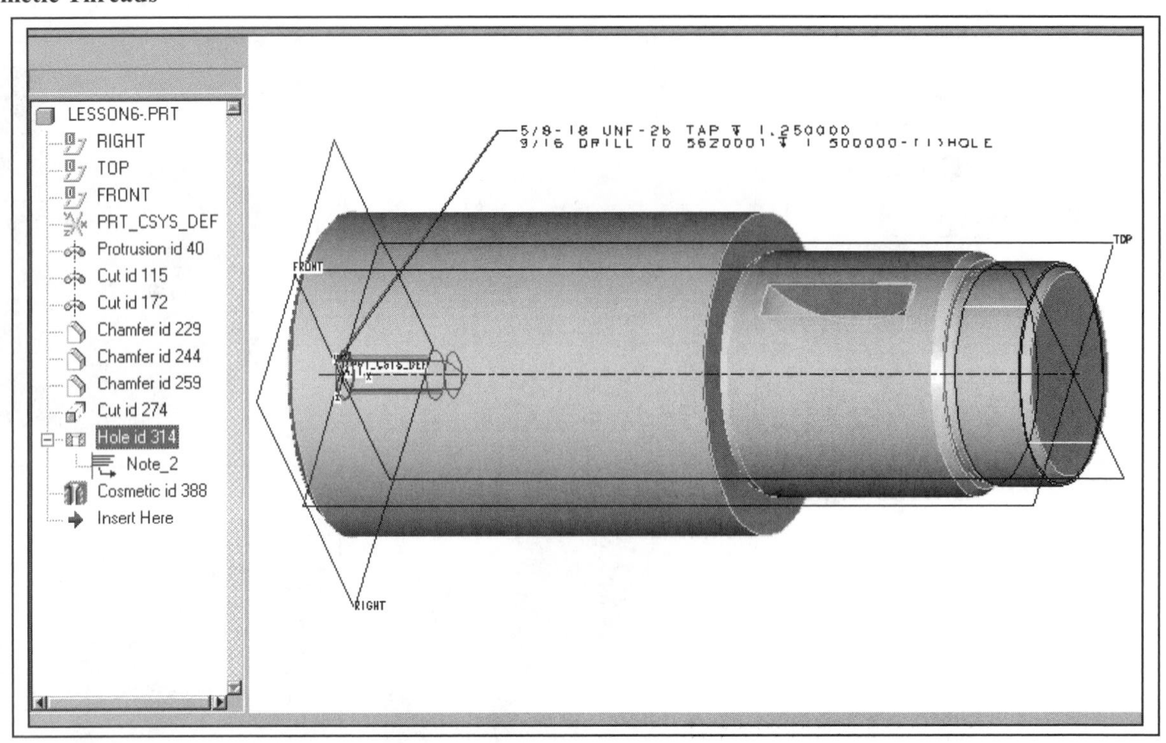

Figure 6.2
Cylinder Rod with Datums and Model Tree

CHAMFERS AND COSMETIC THREADS

A variety of geometric shapes and constructions are accomplished with a CAD system using parametric modeling. For instance, **chamfers** are created at selected edges of the part (Figs. 6.1 and 6.2). Chamfers are *pick-and-place* features (Fig. 6.3).

Threads are usually a *cosmetic feature* representing the *nominal diameter* or the *root diameter* of the thread. Information can be embedded in the feature. Threads show as a unique color *(magenta)*. By putting cosmetic threads on a separate layer, you can display, blank, or suppress them as required.

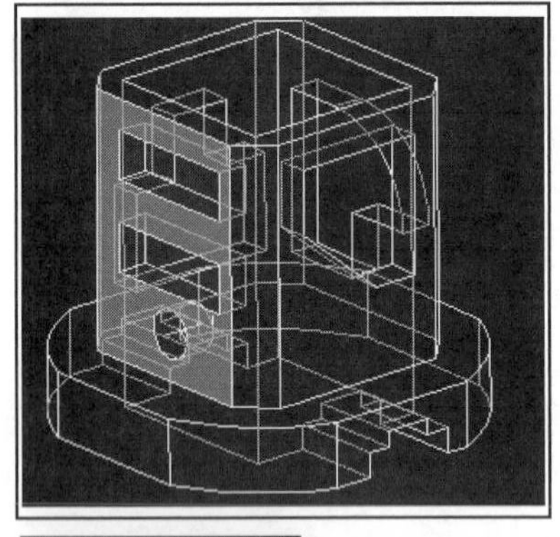

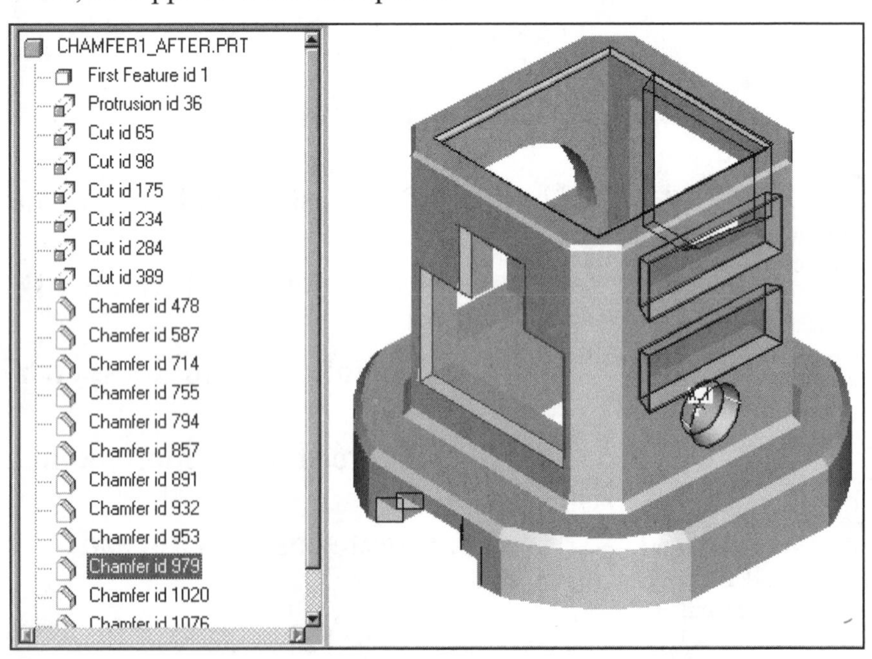

Figure 6.3
COAch for Pro/E, More Features (Chamfer Feature)

Chamfers

Chamfers are created between abutting edges of two surfaces on the solid model. An **edge chamfer** removes a flat section of material from a selected edge to create a beveled surface between the two original surfaces common to that edge (Fig. 6.4). Multiple edges can be selected.

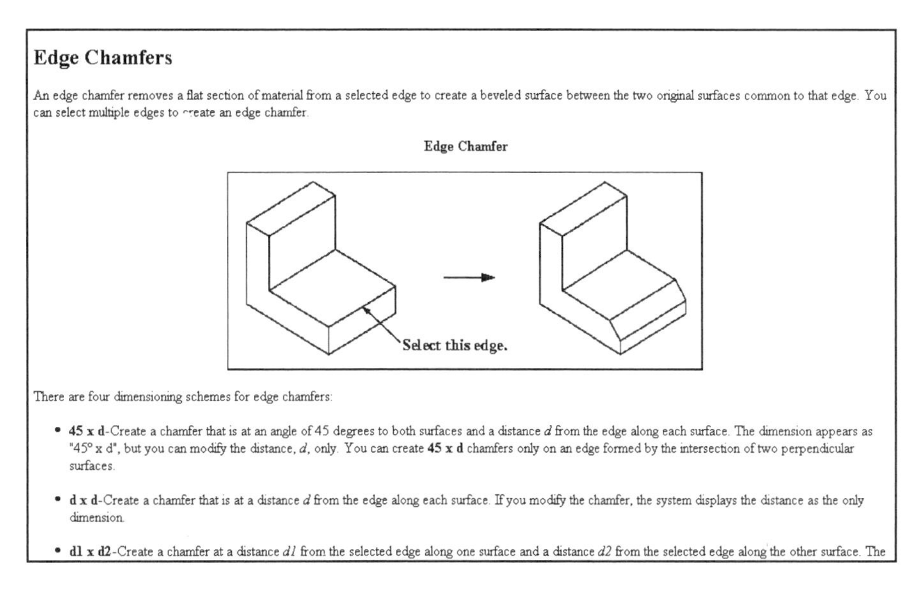

Edge Chamfers

An edge chamfer removes a flat section of material from a selected edge to create a beveled surface between the two original surfaces common to that edge. You can select multiple edges to create an edge chamfer.

Edge Chamfer

There are four dimensioning schemes for edge chamfers:

- **45 x d**-Create a chamfer that is at an angle of 45 degrees to both surfaces and a distance *d* from the edge along each surface. The dimension appears as "45° x d", but you can modify the distance, *d*, only. You can create **45 x d** chamfers only on an edge formed by the intersection of two perpendicular surfaces.
- **d x d**-Create a chamfer that is at a distance *d* from the edge along each surface. If you modify the chamfer, the system displays the distance as the only dimension.
- **d1 x d2**-Create a chamfer at a distance *d1* from the selected edge along one surface and a distance *d2* from the selected edge along the other surface. The

Figure 6.4
Edge Chamfers

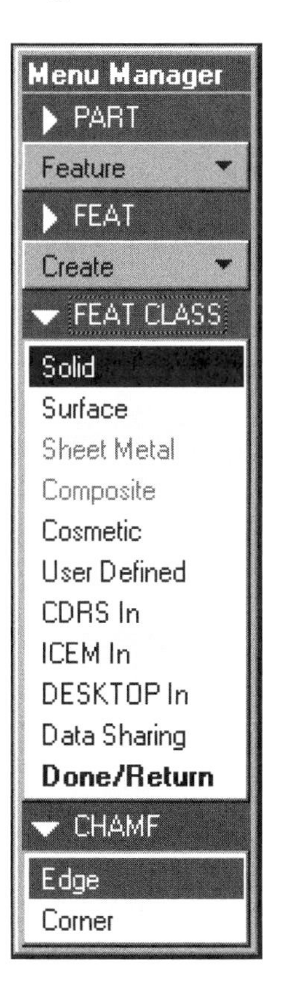

There are four dimensioning schemes for edge chamfers, as shown in Figure 6.5:

45 x d Creates a chamfer that is at an angle of **45°** to both surfaces and a distance **d** from the edge along each surface. The distance is the only dimension to appear when modified. **45 x d** chamfers can be created only on an edge formed by the intersection of two *perpendicular* surfaces.

d x d Creates a chamfer that is a distance **d** from the edge along each surface. The distance is the only dimension to appear when modified.

d1 x d2 Creates a chamfer at a distance **d1** from the selected edge along one surface and a distance **d2** from the selected edge along the other surface. Both distances appear along their respective surfaces when modified.

Ang x d Creates a chamfer at a distance **d** from the selected edge along one adjacent surface at a specified angle to that surface.

The dimensioning schemes appear as options in the SCHEME menu. The SCHEME menu appears after **Edge** has been chosen from the CHAMF menu.

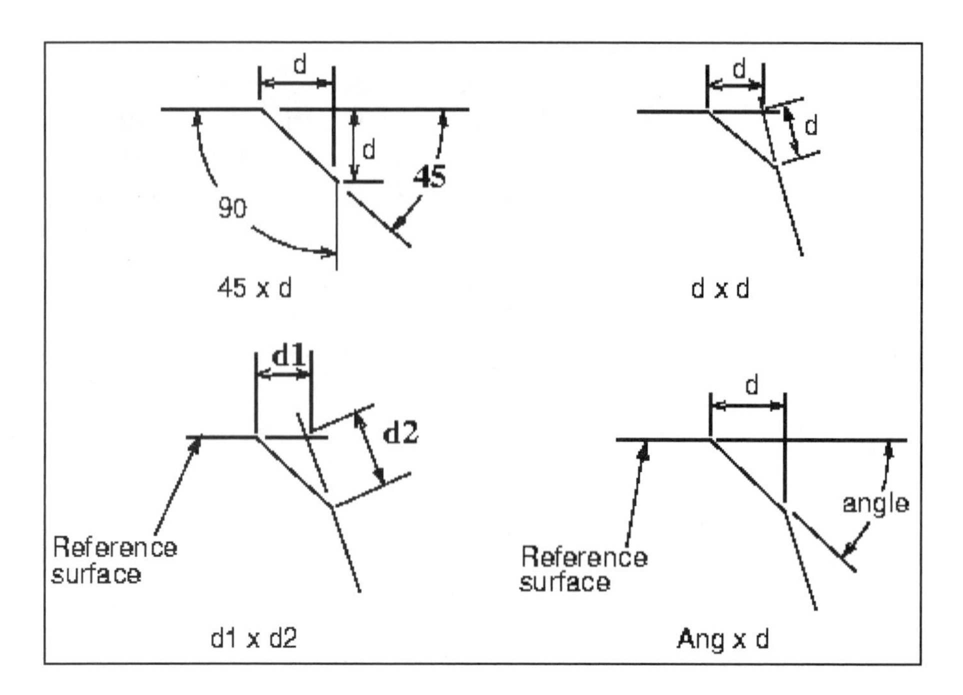

Figure 6.5
Online Documentation, Chamfer Dimensioning Schemes

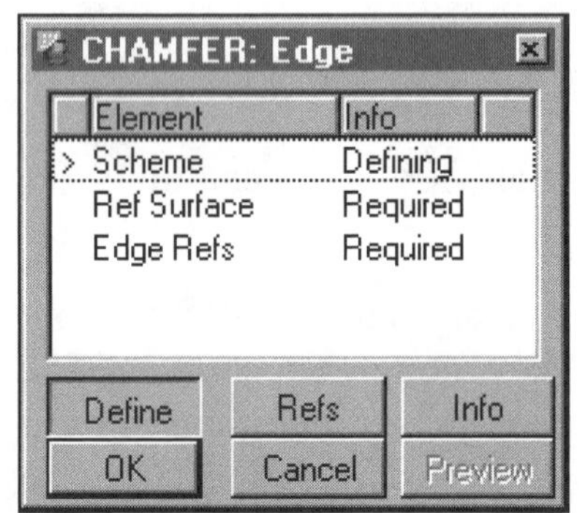

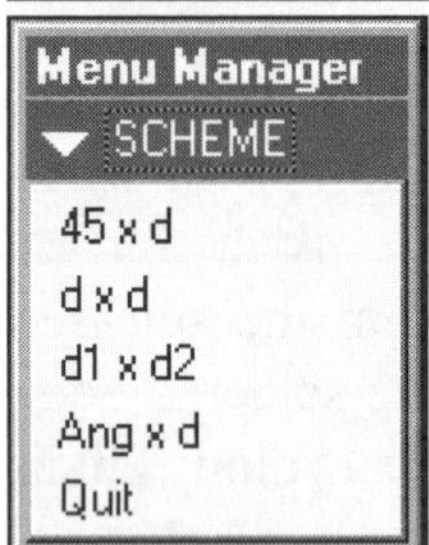

To create a **45 x d** and **d x d** edge chamfer:

1. Choose **Chamfer** from the SOLID menu.
2. Choose **Edge** from the CHAMF menu.
3. Choose the **45 x d** or **d x d** option.
4. Enter the chamfer dimension.
5. Select the edges to chamfer. Remember that for a **45 x d** edge chamfer, the surfaces bounding an edge must be at **90°** to each other.

To create a **d1 x d2** chamfer:

1. Choose **Chamfer** from the SOLID menu.
2. Choose **Edge** from the CHAMF menu.
3. Choose the **d1 x d2** option.
4. Input a distance along a surface to be selected.
5. Input a second distance.
6. Pick the surface along which the first distance will be measured; pick the edge(s) to chamfer.

To create an **Ang x d** chamfer:

1. Choose **Chamfer** from the SOLID menu.
2. Choose **Edge** from the CHAMF menu.
3. Choose the **Ang x d** option.
4. Input distance.
5. Input an angle from a surface to be selected.
6. Select the reference surface from which the values will be measured.
7. Pick the edge(s) to chamfer.

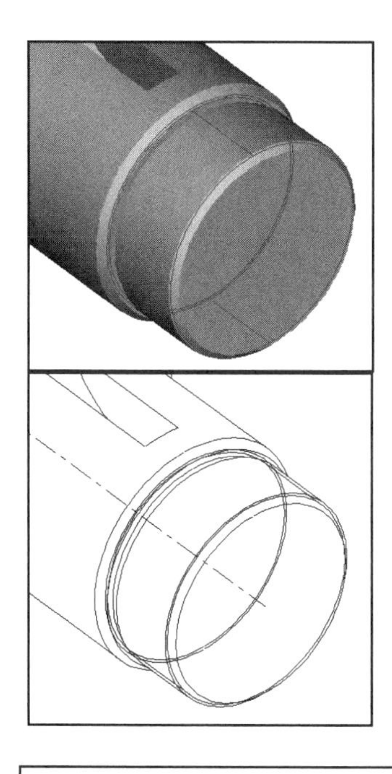

To create a corner chamfer (Fig. 6.6):

1. Choose **Chamfer** from the SOLID menu, and then choose **Corner** from the CHAMF menu.
2. Select the corner you want to chamfer.
3. Pro/E displays the PICK/ENTER menu, which allows you to specify the location of the chamfer vertex on the highlighted edge. The PICK/ENTER menu options are as follows:
   Pick Point Pick a point on the highlighted edge to define the chamfer distance along that edge.
   Enter-input Type in a value for the chamfer distance along the highlighted edge.
4. Pick or enter values to describe the chamfer lengths along the edge. After you have selected the first vertex, Pro/E highlights the other edges, one at a time, so you can place the other two vertices.
5. Select the **OK** button in the dialog box.

➤ **How to Create a Corner Chamfer**

1. Choose **Chamfer** from the SOLID menu, then choose **Corner** from the CHAMF menu.
2. Select the corner you want to chamfer.
3. The system displays the PICK/ENTER menu, which allows you to specify the location of the chamfer vertex on the highlighted edge. The PICK/ENTER menu options are as follows:
   - **Pick Point**--Pick a point on the highlighted edge to define the chamfer distance along that edge.
   - **Enter-input**--Type in a value for the chamfer distance along the highlighted edge.
4. Pick or enter values to describe the chamfer lengths along the edge. After you have selected the first vertex, Pro/ENGINEER highlights the other edges, one at a time, so you can place the other two vertices
5. To create the chamfer, click **OK** in the dialog box.

The following figure shows a corner chamfer.

Corner Chamfer

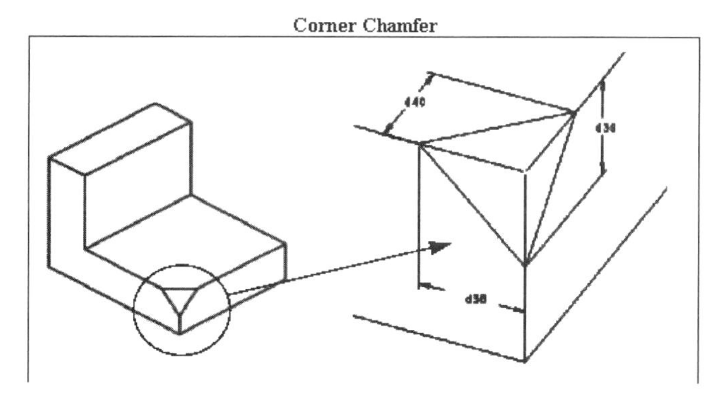

Figure 6.6
Corner Chamfer

Threads

Cosmetic threads (Fig. 6.7) are displayed with *magenta* lines and circles. Cosmetic threads can be external or internal, blind or through. In the Rod part, one end has external blind threads and the opposite end has internal blind threads.

A thread has a set of supported parameters that can be defined at its creation or later, when the thread is added.

The following parameters can be defined for a thread:

PARAMETER DESCRIPTION	PARAMETER NAME	PARAMETER VALUE
Thread major diameter	MAJOR_DIAMETER	Number
Threads per inch (1/pitch)	THREADS_PER_INCH	Number
Thread form	THREAD_FORM	String
Thread class	CLASS	Number
Thread placement (A-external, B-internal)	PLACEMENT	A or B
Thread is Metric	METRIC	TRUE/FALSE

Menu Manager
PART
Feature
FEAT
Create
FEAT CLASS
Solid
Surface
Sheet Metal
Composite
Cosmetic
User Defined
CDRS In
ICEM In
DESKTOP In
Data Sharing
Done/Return
COSMETIC
Sketch
Thread
Groove
ECAD Areas

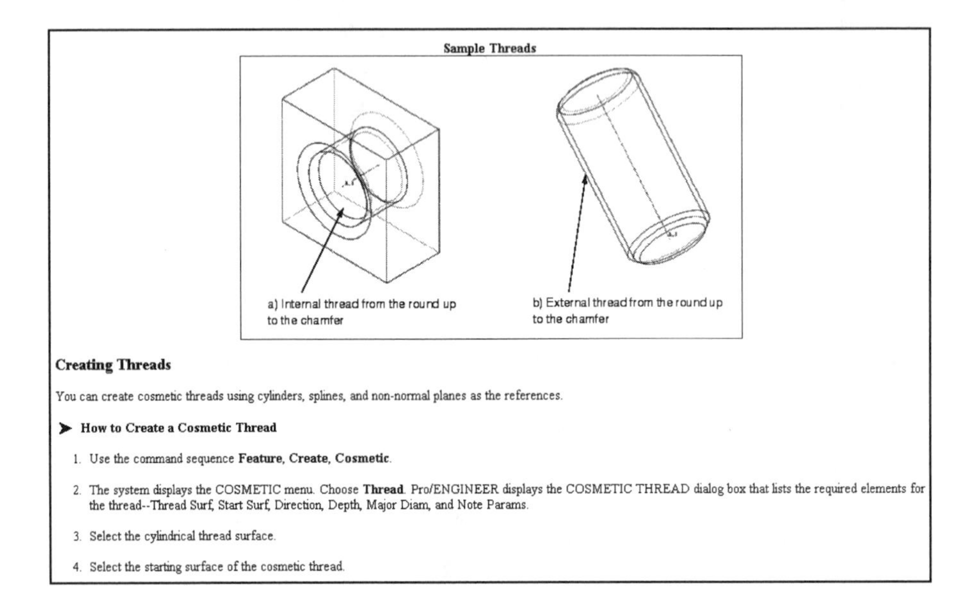
Sample Threads

Creating Threads

You can create cosmetic threads using cylinders, splines, and non-normal planes as the references.

How to Create a Cosmetic Thread

1. Use the command sequence **Feature, Create, Cosmetic**.
2. The system displays the COSMETIC menu. Choose **Thread**. Pro/ENGINEER displays the COSMETIC THREAD dialog box that lists the required elements for the thread--Thread Surf, Start Surf, Direction, Depth, Major Diam, and Note Params.
3. Select the cylindrical thread surface.
4. Select the starting surface of the cosmetic thread.

Figure 6.7
Online Documentation, Cosmetic Threads

To create a cosmetic thread feature:

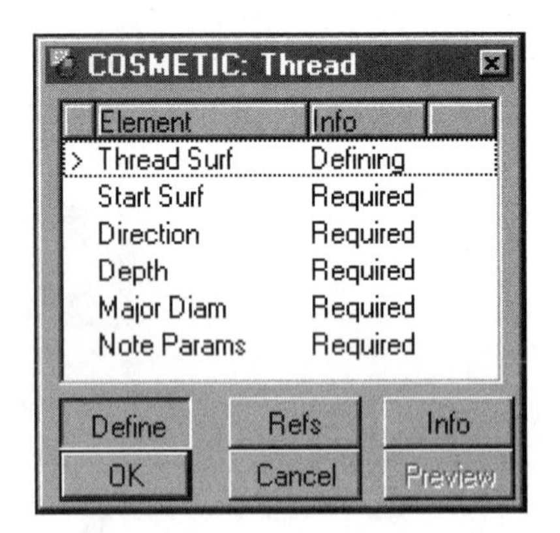

1. Choose **Feature ⇒ Create ⇒ Cosmetic ⇒ Thread.**
2. Pick the circular internal or external thread surface at the prompt.
3. Pro/E automatically knows whether the threads are internal or external, based on the feature selected.
Select the thread start surface and then **Flip** or **Okay** for the direction as needed.
4. From the SPEC TO menu, pick **Blind**, **UpTo Pnt/Vtx**, **UpTo Curve**, or **UpTo Surface** and then **Done.**
5. Follow the prompts, which differ depending on step 4.
6. Enter the diameter at the prompt. (If *external threads*, give the *minor diameter*; if *internal threads*, give the *major diameter*).
7. From the FEAT PARAM menu, select one of the options. In general, you will be picking **Mod Params** to enter the thread parameters into a file using Pro/TABLE. Select from the FEAT PARAM menu:

Retrieve Retrieves a previously created and saved thread file.

Save After completing the table, to save your thread file.
Mod Params Modifies thread parameters in the Pro/TABLE environment [Fig. 6.8(a)].
Show Displays a set of thread parameters in the INFORMATION WINDOW.
Done/Return To complete the process, exit from this menu.

8. When finished, choose **Done/Return** to continue, then **OK**.

Thread parameters can be manipulated like other user-defined parameters: they can be added, modified, deleted, or displayed using menu options.

Information is extracted using the commands **Info ⇒ Feature ⇒** (select the thread). Information on the thread appears in an INFORMATION WINDOW [Fig. 6.8(b)].

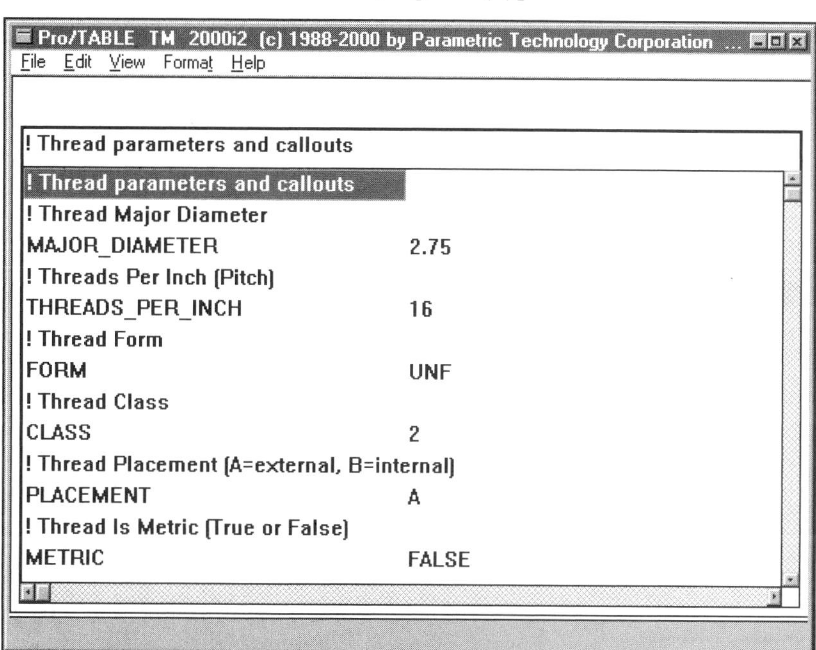

Figure 6.8(a)
Using Pro/TABLE to Input Thread Parameters

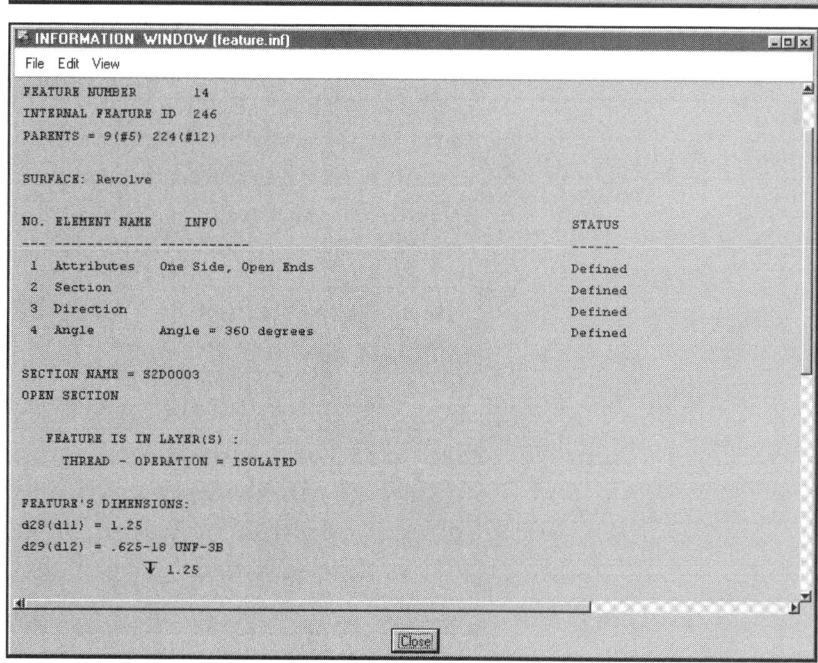

Figure 6.8(b)
Information on a Thread

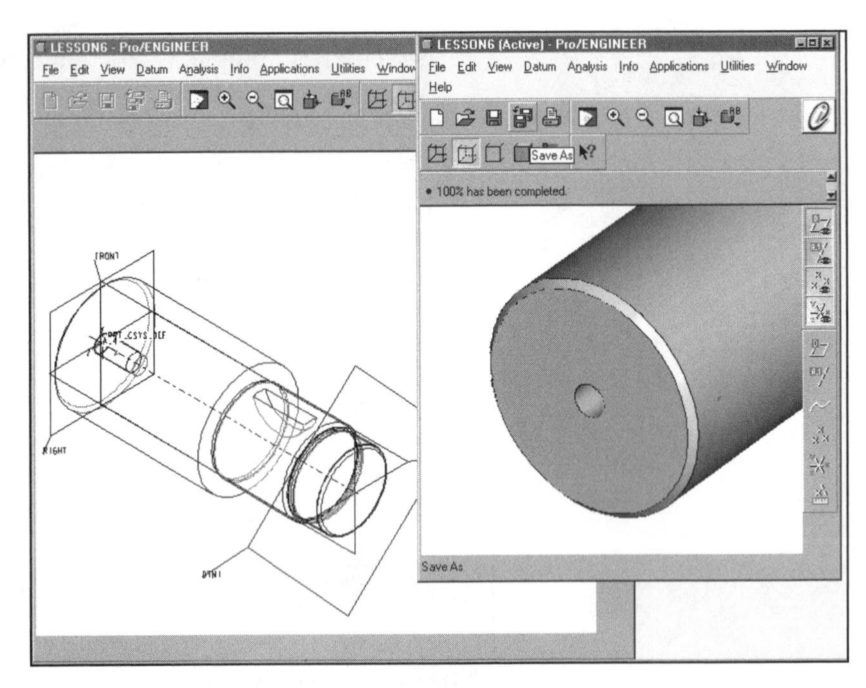

Figure 6.9
Cylinder Rod

Cylinder Rod

Lesson procedures & commands
START HERE ➡ ➡ ➡

The **Cylinder Rod** is modeled by creating a revolved protrusion, similar to the pin in Lesson 5. The geometry of the revolved feature is shown in Figure 6.9 (also see Fig. 6.38). After the revolved protrusion (base feature) is created, the necks (reliefs), the chamfers, the key seat, and the tap drill hole are modeled. In this lesson, we will create our first revolved protrusion by sketching in 3D.

Three chamfers are required for this part. The **45 x d** option was used to chamfer the left side (**Ø4.00**) of the part (**45° x .125**). A **45° x .09** chamfer is added to the right side (**Ø2.75**) of the rod, and a **30° x .14** chamfer is used on the **Ø3.00** step of the rod at the relief. Two necks are required; both are **.120 x .045 DEEP**.

The cosmetic threads for the external threaded shaft end and the internal hole threads are added last. They are created by specifying the minor or major diameter (for external or internal threads, respectively), a starting plane (for the external threads; a **DTM4** was used for the external threads starting plane), and a thread length or ending edge. The internal threaded hole is **.625-18 UNF-3B** by **1.25 DEEP.** The external threaded shaft is **2.75-16 UN-2A**.

Start the part with the usual commands (Fig. 6.10):

CONFIG.PRO

sketcher_dec_places	3
sketcher_disp_constraints	YES *
sketcher_disp_dimensions	YES *
sketcher_disp_grid	YES *
default_dec_places	3

File ⇒ New ⇒ ●Part ⇒ CYLINDER_ROD ⇒ ✔Use default template ⇒ OK

SETUP AND ENVIRONMENT

Set Up ⇒ Units ⇒ Units Manager **⇒ Inch lbm Second ⇒ Close ⇒ Material ⇒ Define ⇒** (type **ALUMINUM**) **⇒ ✔ ⇒** (table of material properties, change or add information) **⇒ File ⇒ Save ⇒ File ⇒ Exit ⇒ Assign ⇒** (pick **ALUMINUM**) **⇒ Accept ⇒ Done**
Utilities ⇒ Environment ⇒ ✔ Snap to Grid

❑ **Use 2D Sketcher** (we will be sketching in 3D)
Display Style **Hidden Line** Tangent Edges **Dimmed** ⇒ **OK**

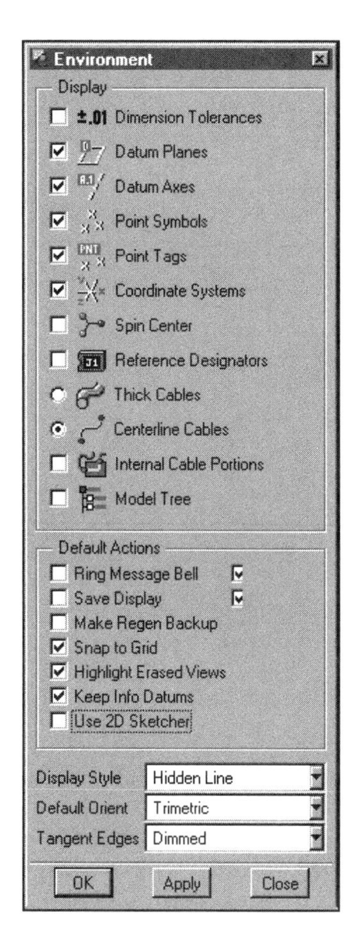

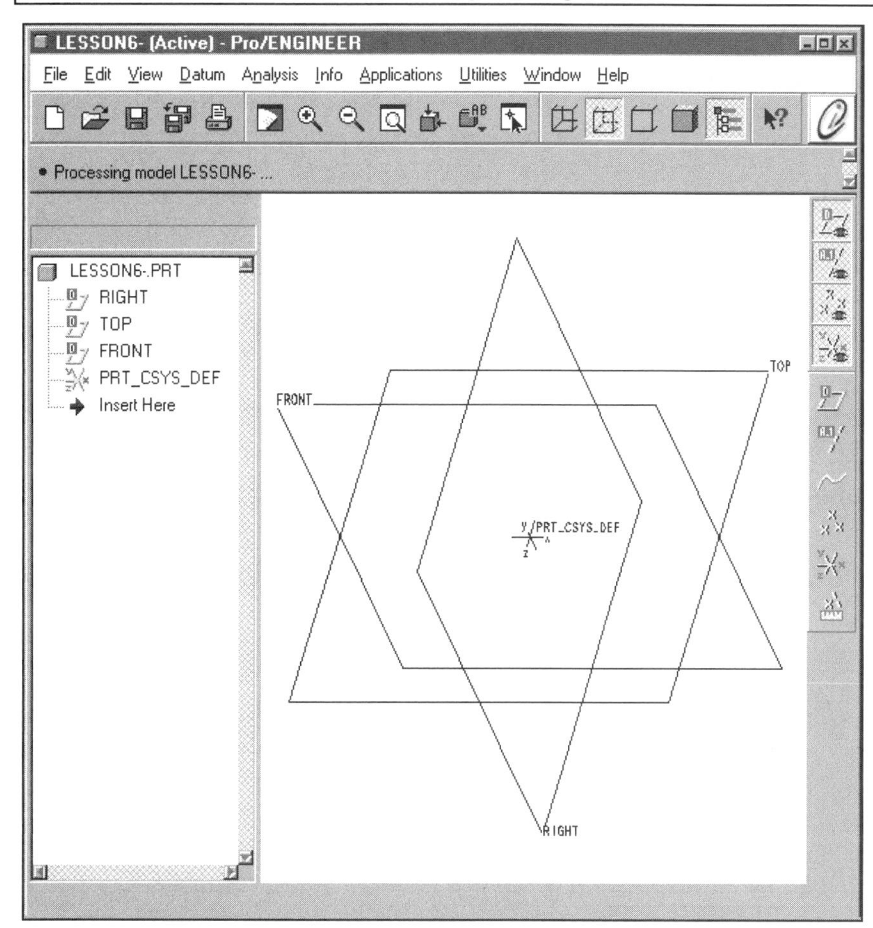

Figure 6.10
Default Datum Planes and Default Coordinate System

Default Layering

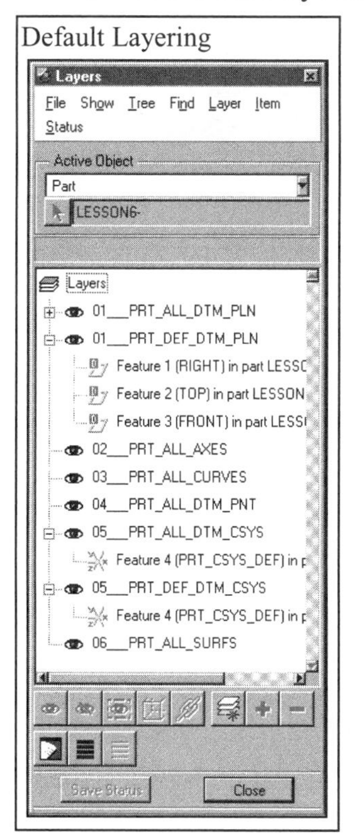

The first protrusion is a revolved feature, you will be sketching the section in 3D, and the **Intent Manager** is turned off:

Feature ⇒ **Create** ⇒ **Solid** ⇒ **Protrusion** ⇒ **Revolve** ⇒ **Solid** ⇒ **Done** ⇒ **One Side** ⇒ **Done** ⇒ (pick **FRONT** as sketching plane) ⇒ **Setup New** ⇒ **Okay** (DIRECTION options) ⇒ **Top** (pick **TOP** as horizontal reference for the reference plane) ⇒ **View** ⇒ **Model Tree** (Model Tree off) ⇒ **View** ⇒ **Default** (to place the model in 3D, *if* it is displayed in 2D orientation) ⇒ [using the Ctrl key and the right mouse button, pan the screen (Fig. 6.11)] ⇒ **References** (Fully Placed) ⇒ **Close** ⇒ **Sketch** (menu bar) ⇒ **Intent Manager** (toggled off) ⇒ **Sketch** ⇒ **Line** ⇒ **Centerline** ⇒ **Horizontal** (*first*--pick once to draw the horizontal centerline thru the coordinate system) ⇒ **Sketch** ⇒ **Mouse Sketch** (*second*-- sketch the section geometry) ⇒ **Constraints** (Fig. 6.11) ⇒ **Regenerate** ⇒ **Alignment** (align the horizontal centerline with datum **TOP**, the horizontal line with **TOP**, and the left side vertical line with datum **RIGHT**, as in Fig. 6.12) ⇒ **Done Sel** ⇒ **Regenerate** ⇒ **Dimension** [use dimension scheme--remember to create diameter dimensions (Fig. 6.13)] ⇒ **Regenerate** ⇒ **Modify** (modify to design dimensions shown in Figure 6.13 and Figure 6.38) ⇒ **Regenerate**

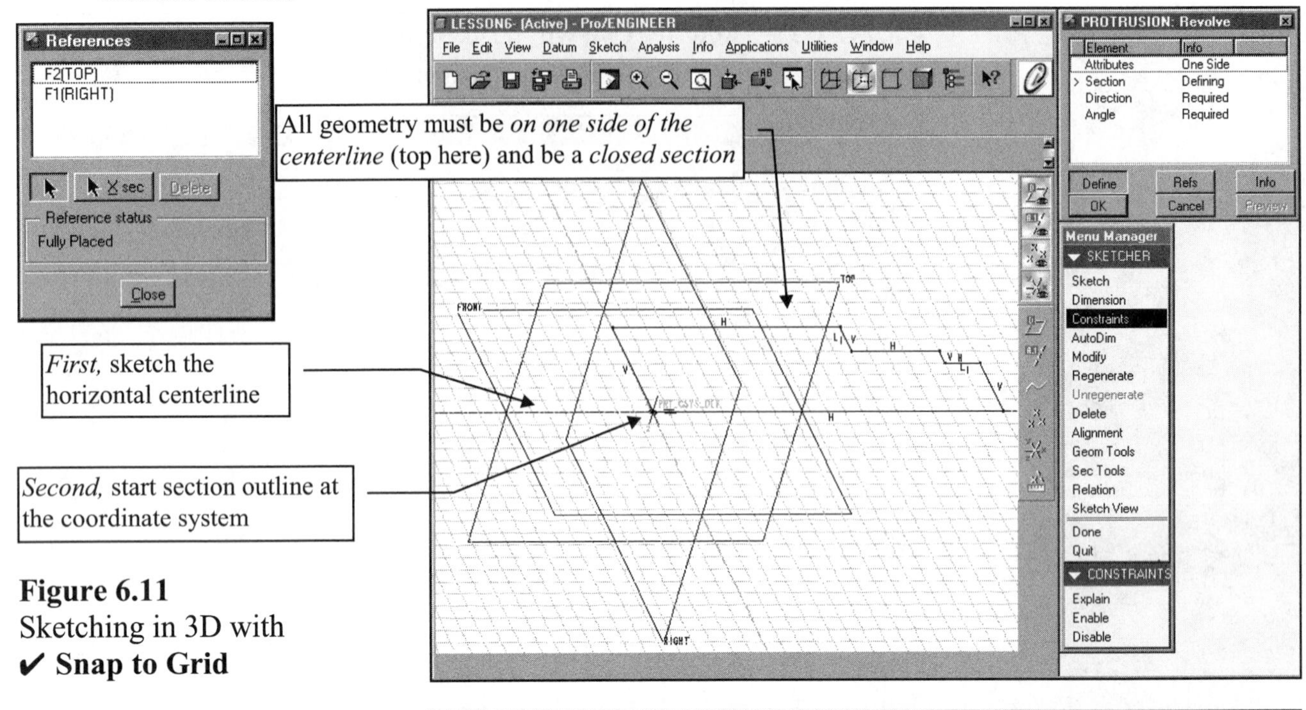

Figure 6.11
Sketching in 3D with
✔ **Snap to Grid**

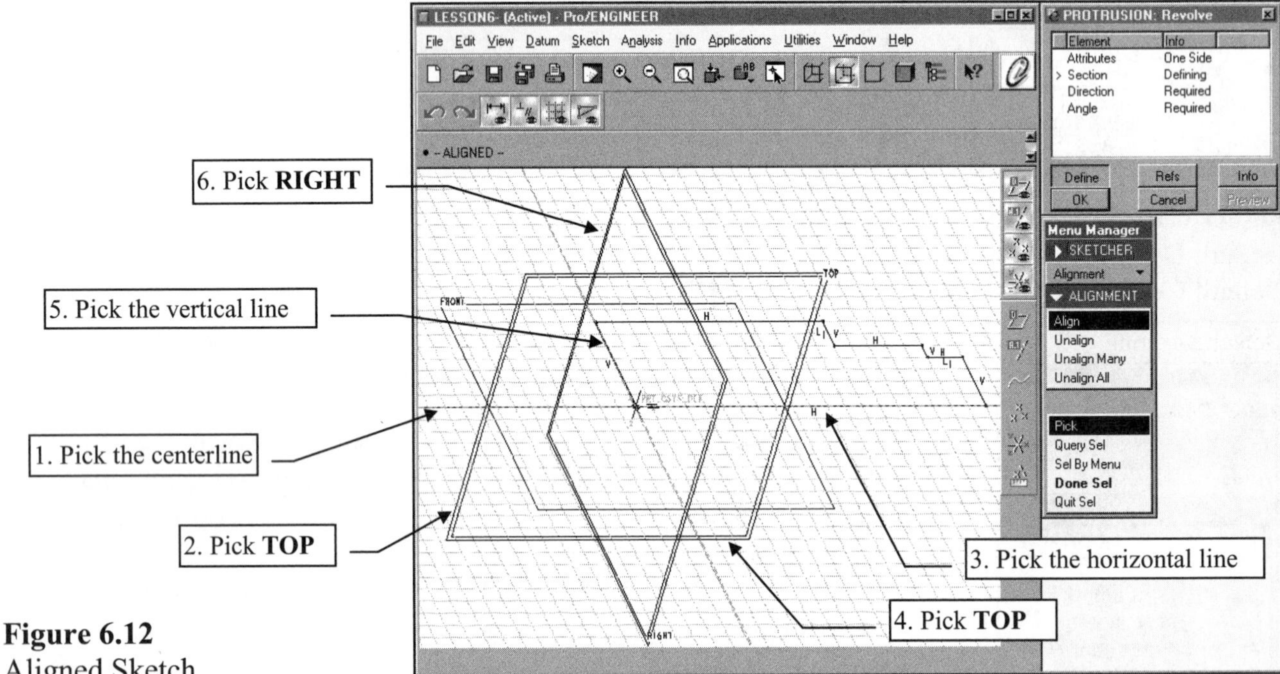

Figure 6.12
Aligned Sketch

HINT

You can toggle between sketching in 3D and in 2D. If you want to go to 2D, choose **Sketch View.**

You can dynamically change any dimension on the sketch. This capability is available for 3D and 2D sketching. After the sketch has been regenerated with the design dimensions (Figures 6.13 and 6.38), choose the following commands:

Modify ⇒ **Drag Dim Val** ⇒ (pick the **9.94** dimension) ⇒ **Done Sel** ⇒ (the **Modify Dims** *thermotool* will appear on the screen) ⇒ [move the mouse pointer into the region under the title "**linear**," pick once with the LMB, slide the linear indicator to the right until the dimension is about **12.25**, as shown in Fig. 6.14, and then pick once more with the LMB (in the region) to finish]

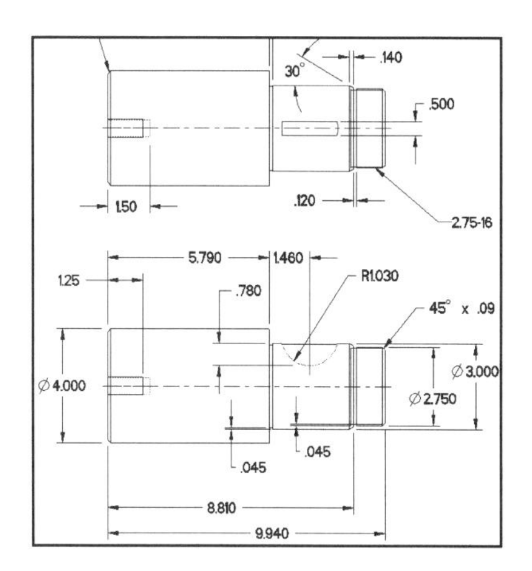

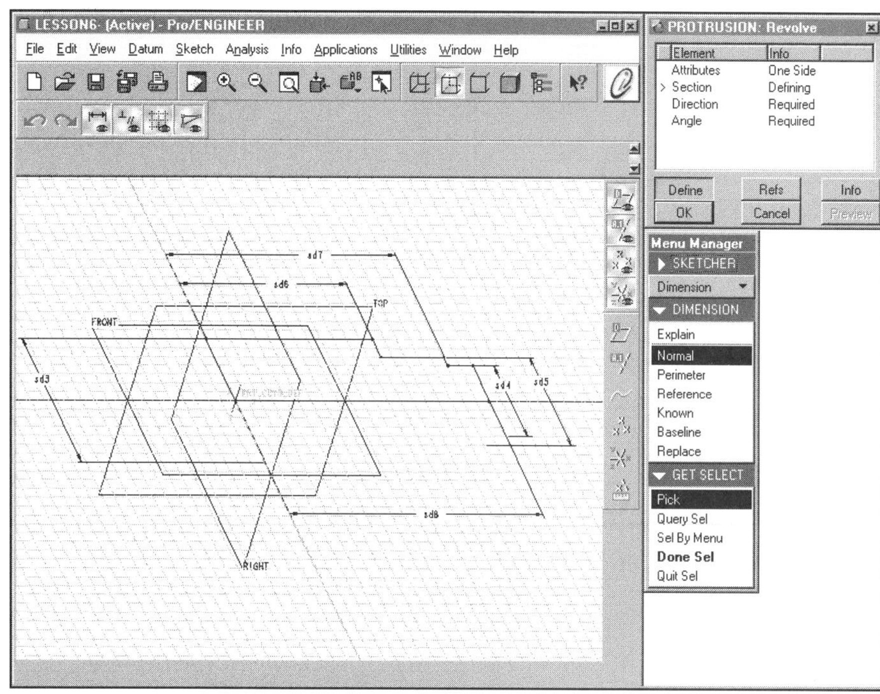

Figure 6.13
Section with Sketch Dimensions

Thermotool with the **Modify Dims** sliders and the **Scale** slider. Slide the **linear** indicator bar to the right to increase the dimension size.

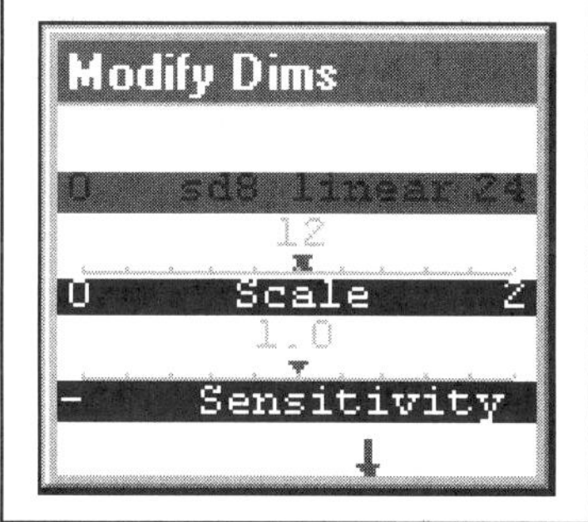

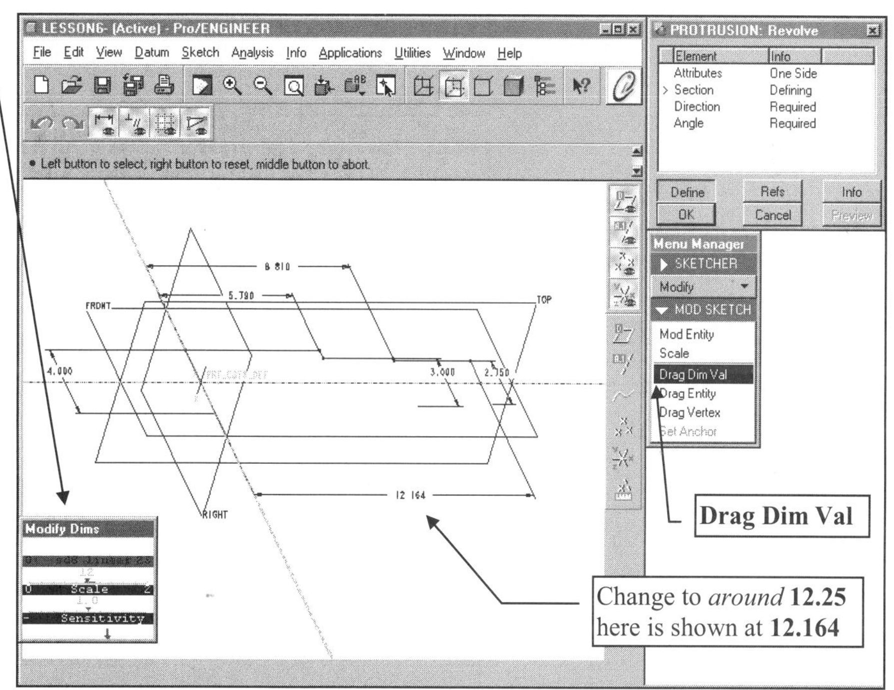

Figure 6.14
Dynamically Changing a Dimension Value

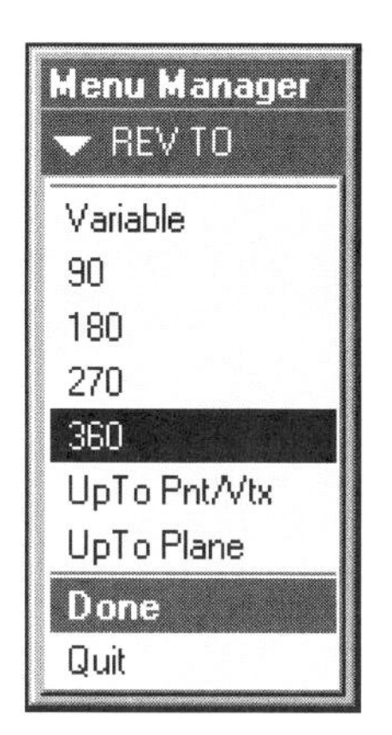

The sketch will dynamically change with the length value. After experimenting with this capability, complete the protrusion by choosing the following commands:

Regenerate (the sketch again) ⇒ **Modify** (the length back to the design dimension of **9.94)** ⇒ **Regenerate** ⇒ **Done** ⇒ **360** (from the REV TO menu) ⇒ **Done** ⇒ **OK** ⇒ **View** ⇒ **Default** ⇒ **View** ⇒ **Shade** ⇒ **View** ⇒ **✔Model Tree** (on) ⇒ **Done** (Fig. 6.15)

or **Default** view, **Shade**, and **Model Tree** on

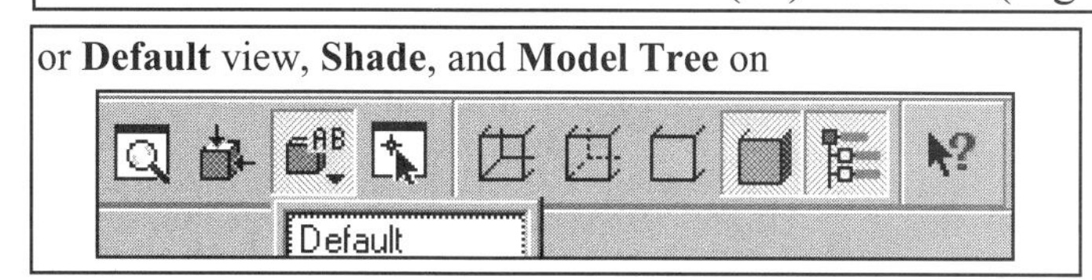

Pick on Protrusion in the Model Tree to highlight the geometry

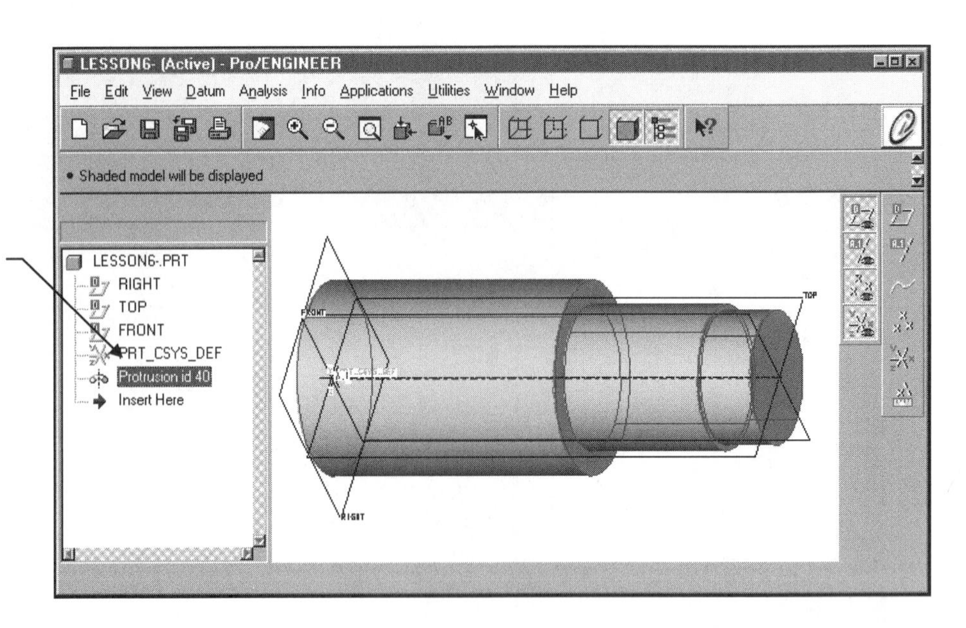

Figure 6.15
Revolved Protrusion

The next two features are revolved cuts similar to those for the Pin in Lesson 5:

File ⇒ Save ⇒ ✔

View ⇒ Repaint ⇒ Shade (icon) **⇒ Utilities ⇒ Environment ⇒ ✔ Use 2D Sketcher ⇒ ❑ Snap to Grid ⇒ OK ⇒ Feature ⇒ Create - ⇒ Cut ⇒ Revolve ⇒ Solid ⇒ Done ⇒ One Side ⇒ Done ⇒ Use Prev ⇒ View ⇒ Model Tree** (off) **⇒ Okay ⇒ References** dialog box (pick the vertical shoulder edge/surface of the cylinder rod as shown in Fig. 6.16) **⇒ Close ⇒** (sketch the centerline along the rod's axis) **⇒** [sketch the three lines of the **U-***shaped* cut (Fig. 6.17)] **⇒ Yes** (if asked should the highlighted entities to be aligned) **⇒ Modify** the *sketch* dimensions to *design* dimensions **⇒ ✔ Regenerate ⇒ ✔** (Sketcher icon) **⇒ Okay ⇒ 360 ⇒ Done ⇒ OK ⇒ Done ⇒ View ⇒ Default ⇒ View ⇒ ✔ Model Tree** (Fig. 6.18)

Pick the shoulder edge as a third Reference

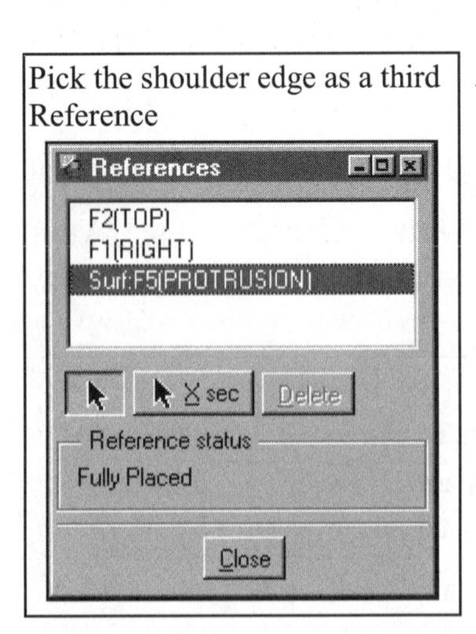

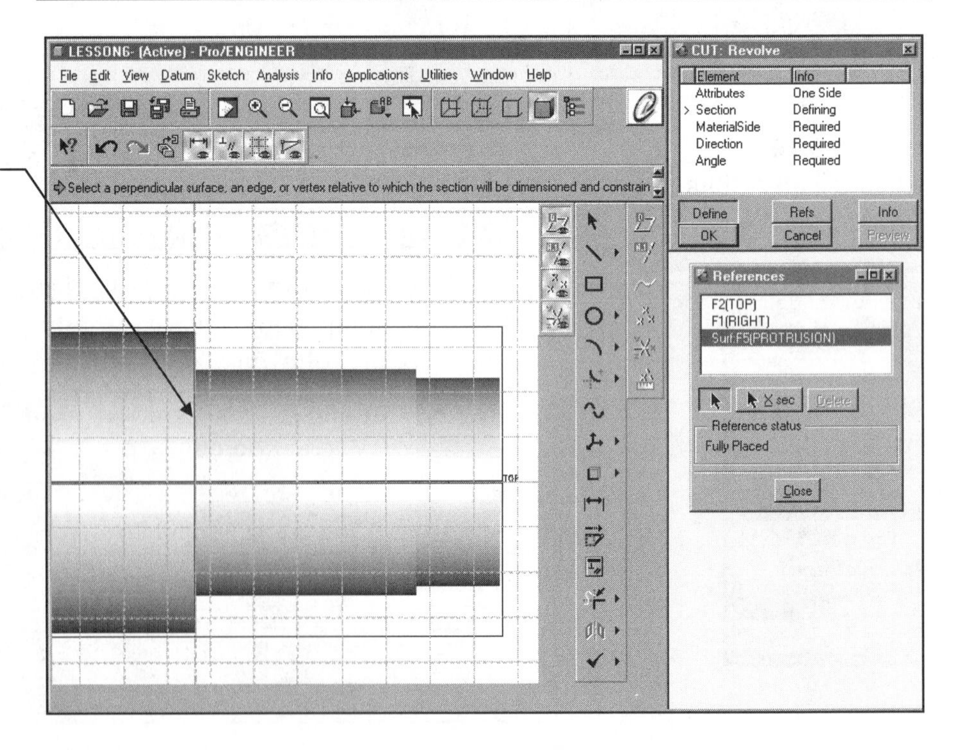

Figure 6.16
Adding a Reference

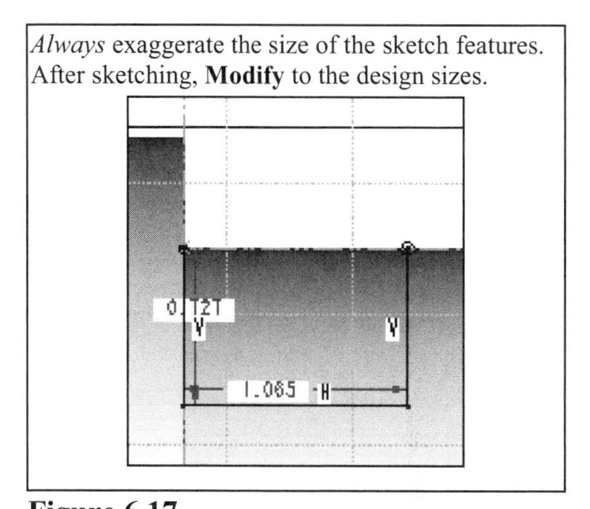

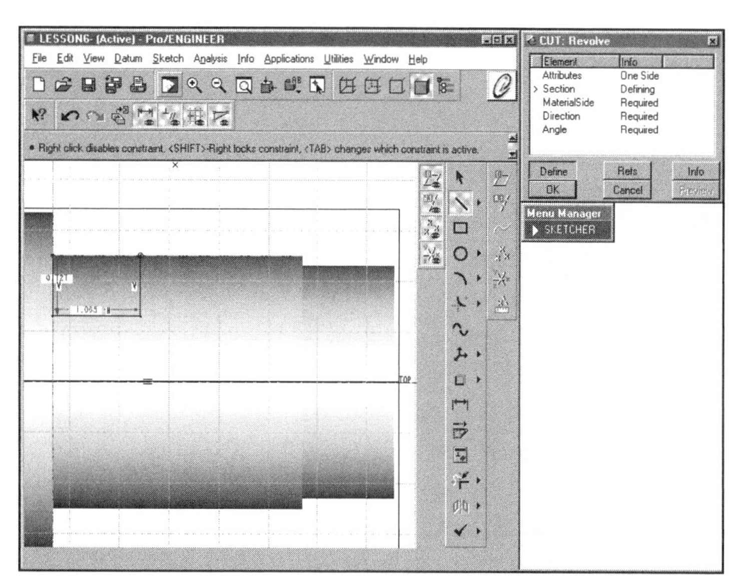

Figure 6.17
Sketching the Revolved cut

After the first revolved cut is successfully created, model the second revolved cut using the same procedure (Fig. 6.19).

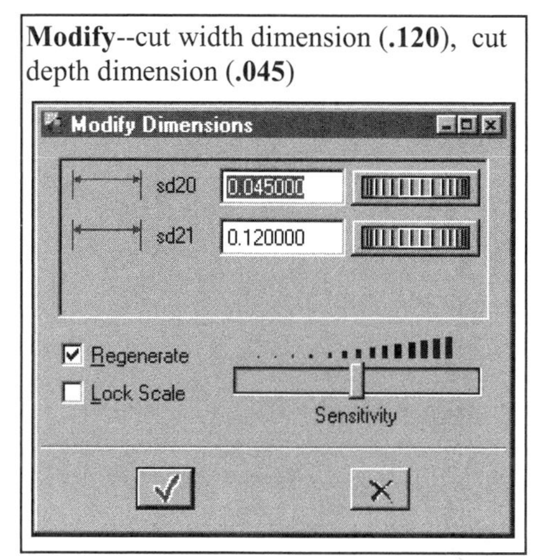

Figure 6.18
.120 X .045 (Revolved Cut)

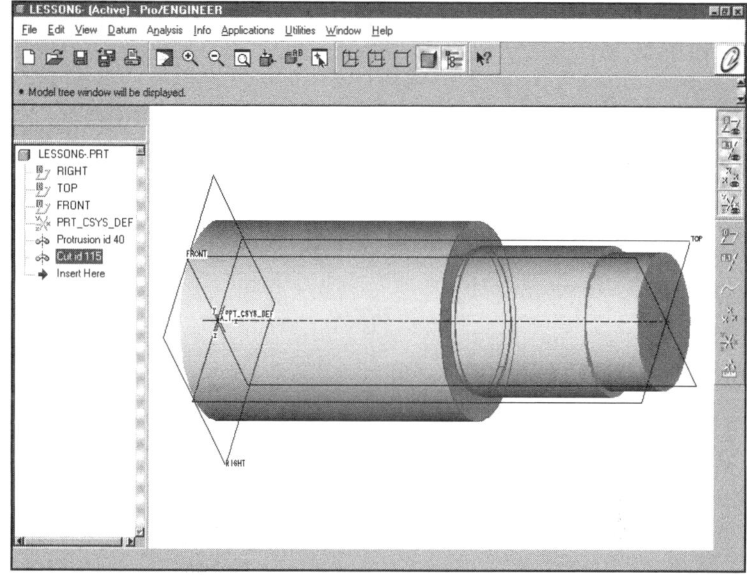

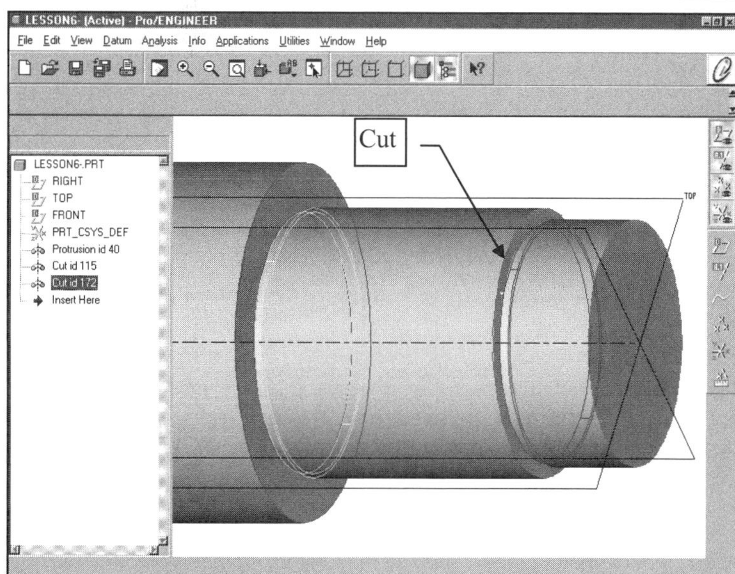

Figure 6.19
Second **.120 X .045** (Cut)

The chamfers will be created next. Choose the following commands:

File ⇒ Save ⇒ ✔ ⇒ Feature ⇒ Create ⇒ Solid ⇒ Chamfer ⇒ Edge ⇒ 45 x d ⇒ (type **.120** at the prompt) **⇒ ✔ ⇒** (select the edge to be chamfered as shown in Figs. 6.20 and 6.21) **⇒ Done Sel ⇒ Done Refs ⇒ Preview ⇒ OK ⇒ Done**

CHAMFER dialog box

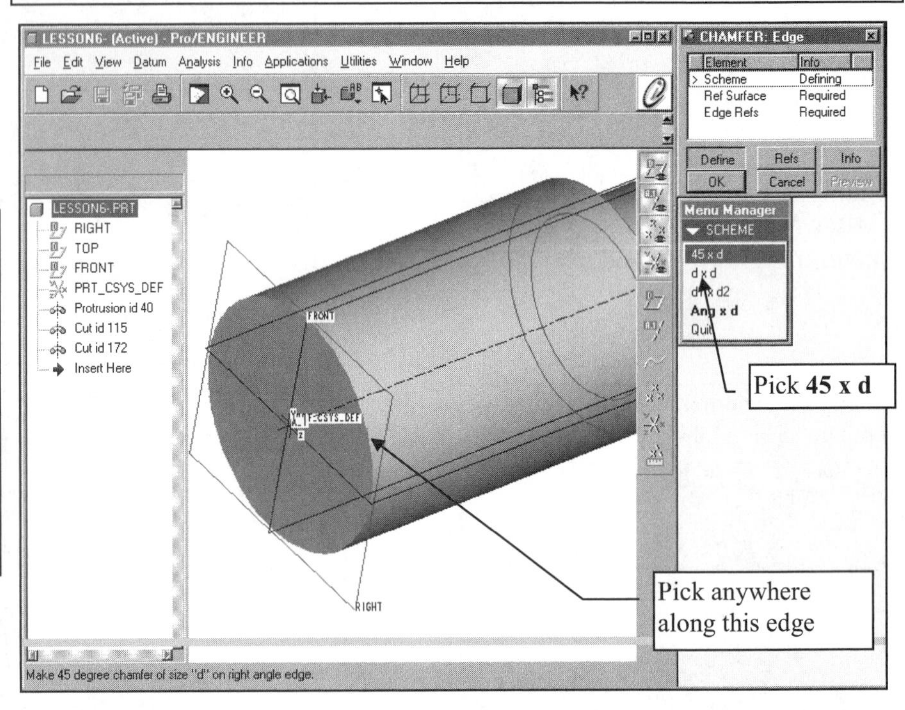

Figure 6.20
Chamfer Command

Open new window (**Window ⇒ New**) to view your part from different angles and zoom states. You can work in either window by choosing **Window ⇒ Activate** from the menu bar in the window you will work in. To close the new window, choose **Window ⇒ Close Window.** Make sure that the smaller window is the active window before you close it. Choose **Window ⇒ Activate** to work in the original window.

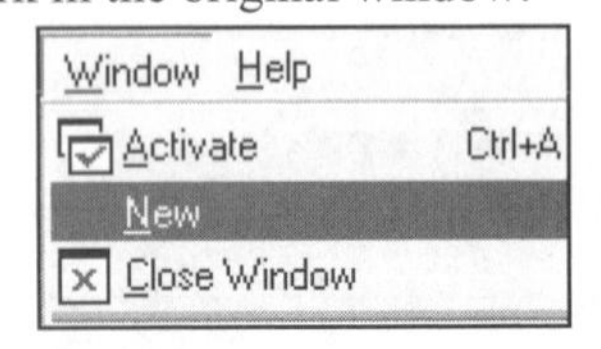

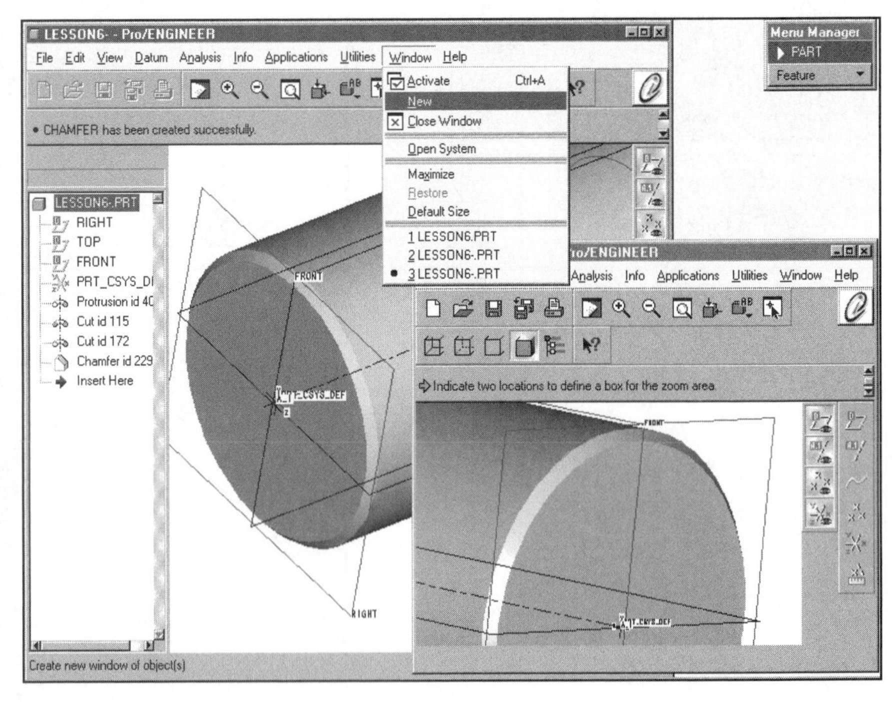

Figure 6.21
Chamfer

Create the **45° X .09** chamfer on the **∅2.75** side of the part using the same commands. The **30° X .14** chamfer requires a slightly different choice of commands (Fig. 6.22):

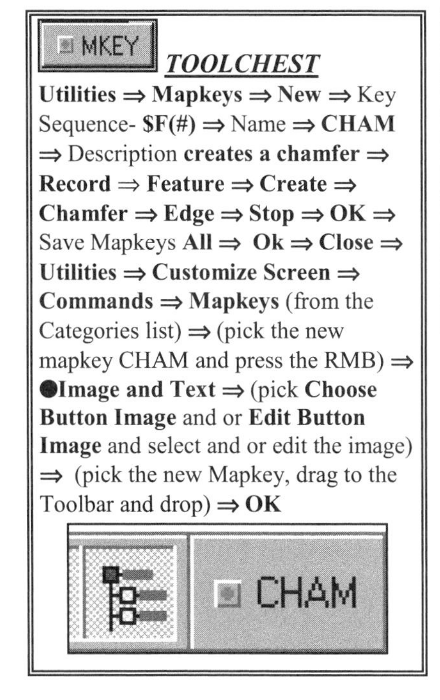

TOOLCHEST

Utilities ⇒ **Mapkeys** ⇒ **New** ⇒ Key Sequence- **$F(#)** ⇒ Name ⇒ **CHAM** ⇒ Description **creates a chamfer** ⇒ **Record** ⇒ **Feature** ⇒ **Create** ⇒ **Chamfer** ⇒ **Edge** ⇒ **Stop** ⇒ **OK** ⇒ Save Mapkeys **All** ⇒ **Ok** ⇒ **Close** ⇒ **Utilities** ⇒ **Customize Screen** ⇒ **Commands** ⇒ **Mapkeys** (from the Categories list) ⇒ (pick the new mapkey CHAM and press the RMB) ⇒ ●**Image and Text** ⇒ (pick **Choose Button Image** and or **Edit Button Image** and select and or edit the image) ⇒ (pick the new Mapkey, drag to the Toolbar and drop) ⇒ **OK**

Utilities ⇒ **Environment** ⇒ Default Orient **Isometric** ⇒ **Apply** ⇒ **OK** ⇒ **Feature** ⇒ **Create** ⇒ **Solid** ⇒ **Chamfer** ⇒ **Edge** ⇒ **Ang x d** ⇒ (type **.14** when prompted for the distance dimension) ⇒ ✔ ⇒ (type **30** when prompted for angle) ⇒ ✔ ⇒ [select the cylinder's revolved surface as the reference surface (Fig. 6.22)] ⇒ [select the edge to be chamfered (Fig. 6.22)] ⇒ **Done Sel** ⇒ **Done Refs** ⇒ **OK** (Fig. 6.23) ⇒ **Done** ⇒ (from the Model Tree, highlight the Chamfer, press your RMB and activate **Info)** ⇒ **Feat Info** ⇒ **Close**

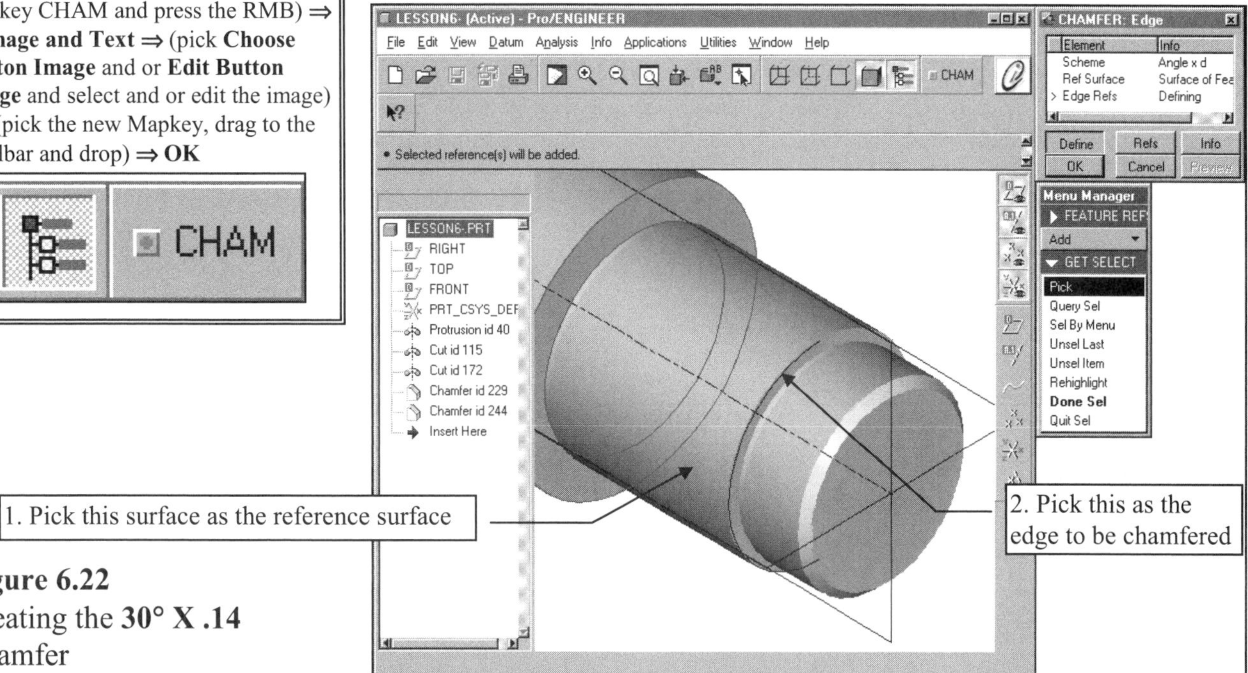

Figure 6.22
Creating the **30° X .14** Chamfer

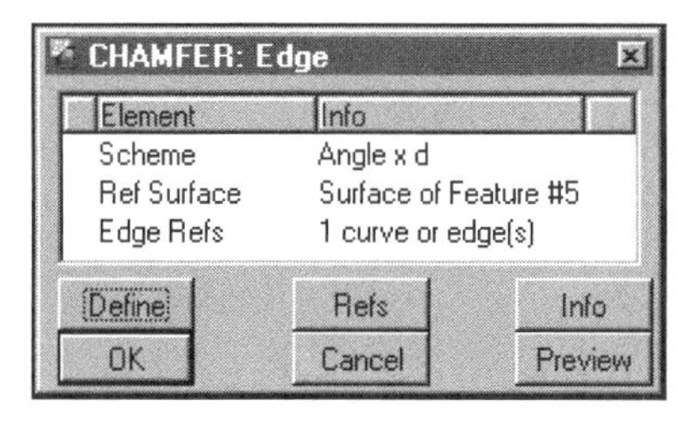

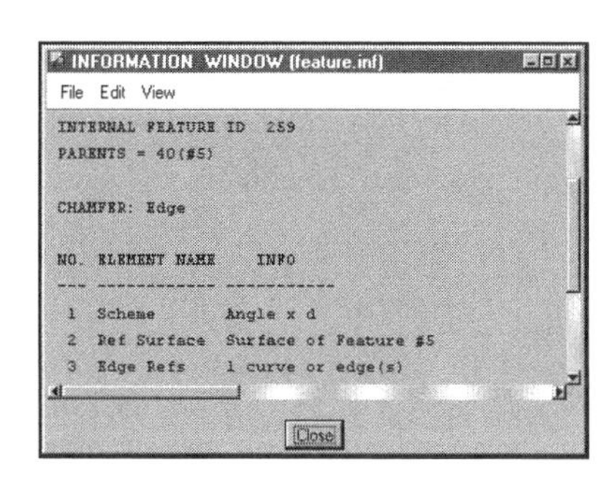

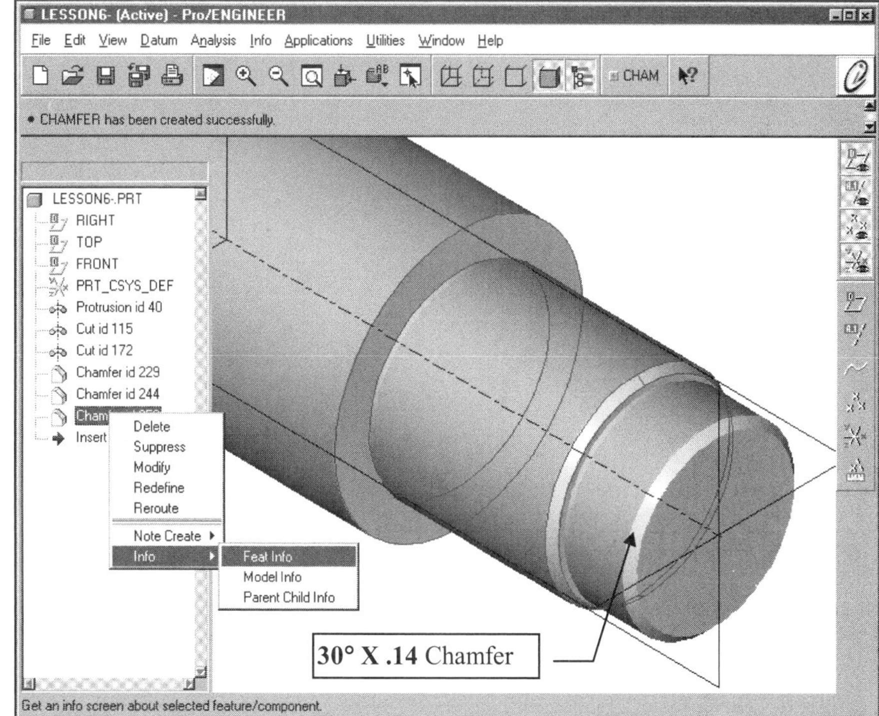

Figure 6.23
Chamfer **30° X .14**

Create the key seat using an extruded cut:

View ⇒ Default ⇒ Feature ⇒ Create ⇒ Cut ⇒ Extrude ⇒ Solid ⇒ Done ⇒ Both Sides ⇒ Done ⇒ Use Prev ⇒ Okay ⇒ Sketch (menu bar) ⇒ **Intent Manager** (off) ⇒ **Sketch View ⇒ Arc ⇒ Center/Ends** ⇒ [sketch the arc (Fig. 6.24)] ⇒ **Sketch ⇒ Line ⇒ Centerline ⇒ Vertical** [sketch the vertical centerline through the arc's center point, (Fig. 6.25)] ⇒ **Regenerate ⇒ Alignment** [align the endpoints of the arc to the upper edge of the part (Fig. 6.25)] ⇒ **Regenerate ⇒ Dimension** (add the three dimensions) ⇒ **Regenerate ⇒ Modify** (to the design values of **.780** deep, **R1.030**, and **1.460** from the edge) ⇒ **Regenerate** (Fig. 6.26) ⇒ (if you want to see it rotated, choose **View ⇒ Default)** ⇒ **Done ⇒ Okay ⇒ Blind ⇒ Done ⇒** (type **.500** at the prompt for the width of the keyseat) ⇒ ✔ ⇒ **Preview ⇒ OK** (Fig. 6.27) ⇒ **Done ⇒ View ⇒ Default**

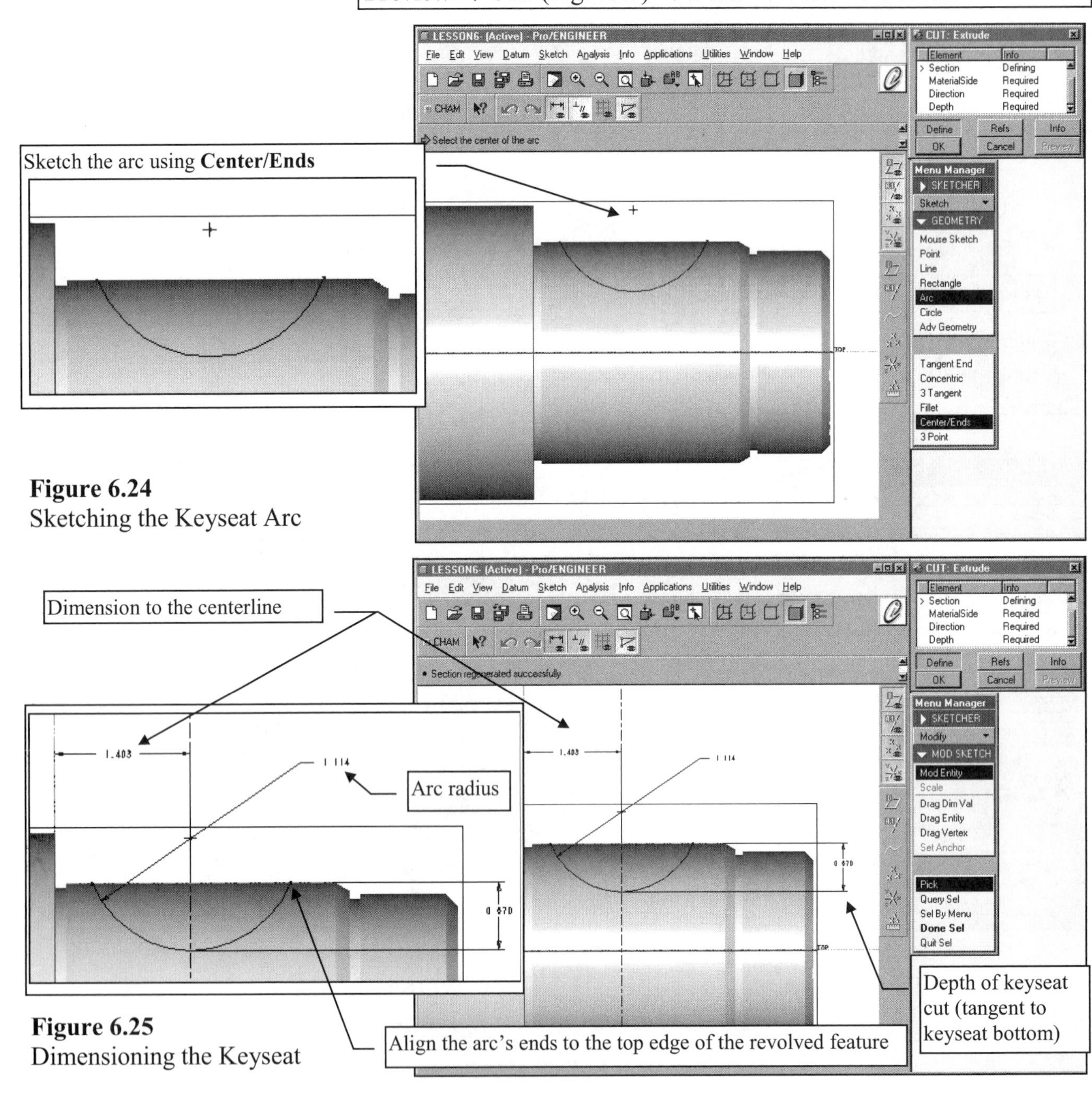

Figure 6.24
Sketching the Keyseat Arc

Figure 6.25
Dimensioning the Keyseat

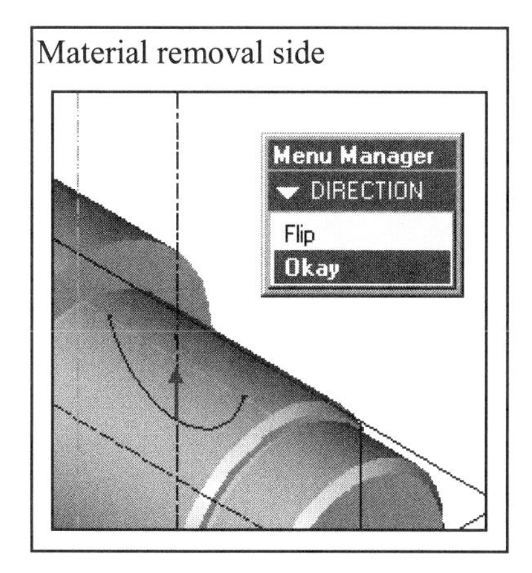

Keyseat will be cut on **Both Sides** of datum **FRONT**

Figure 6.26
Keyseat Dimensions

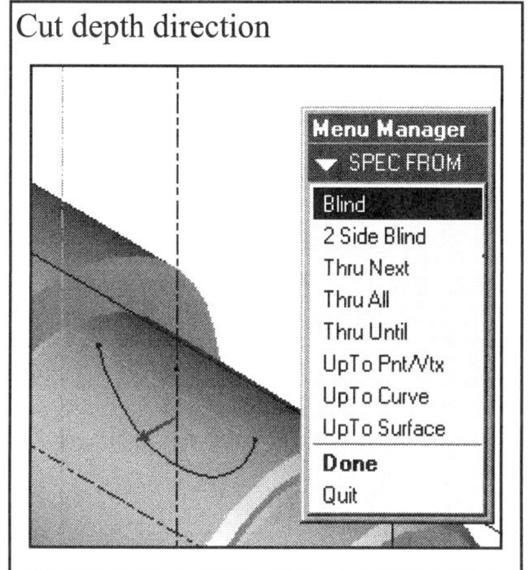

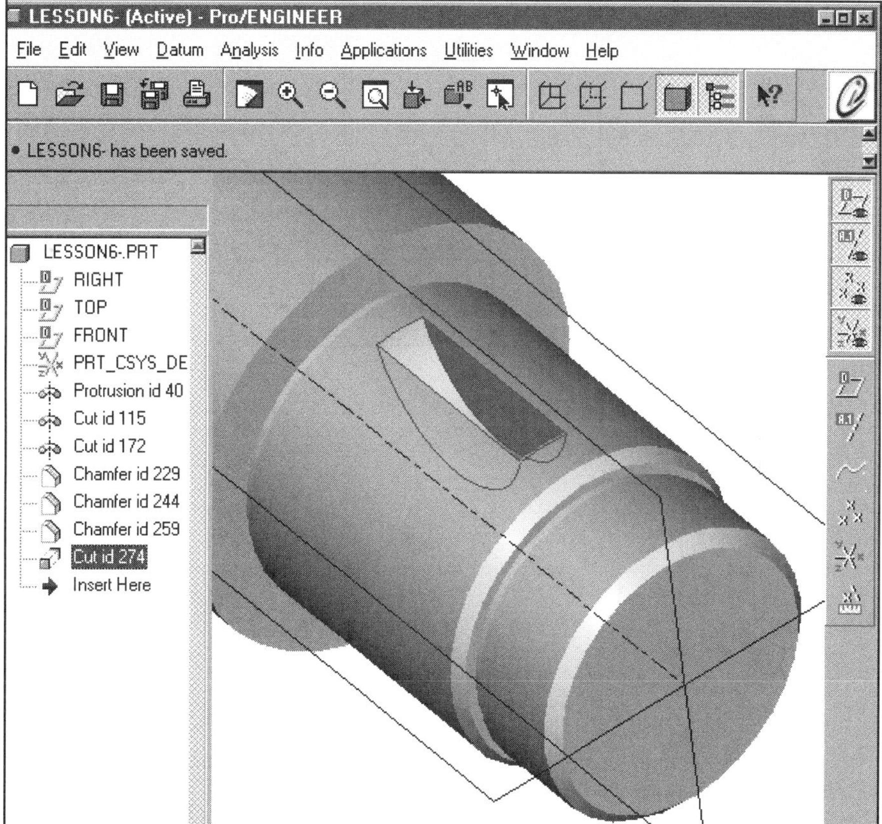

File ⇒ Save ⇒ ✔
File ⇒ Delete ⇒
Old Versions ⇒ ✔

Figure 6.27
Keyseat

Next, add the hole at the large-diameter end of the part using a standard hole. A *standard hole* is a combination of sketched and extruded geometry. It is based on industry-standard fastener tables. You can calculate either the tapped or clearance diameter appropriate to the selected fastener. You can use system-supplied standard lookup tables for these diameters or create your own.

Standard holes are either tapped holes or clearance holes that have a basic shape. To create a tapped hole, choose the following:

File ⇒ **Save** ⇒ ✔ ⇒ **Feature** ⇒ **Create** ⇒ **Hole** [HOLE dialog box displays, see left-hand column (Fig. 6.28)] ⇒ [in the Hole Type area, •**Standard Hole** Fig. 6.29)] ⇒ (next, select **UNF)** ⇒ •**Tapped Hole** ⇒ (in the Screw Size area select **5/8-18)** ⇒ ✔**Add Thread Surface** ⇒ [in the Hole Dimension area select •**Variable** and type **1.50** (for the tap drill depth)] ⇒ •**Variable** and then type **1.25** (for the thread depth) ⇒ [in Hole Placement area, Primary Reference ↖--(pick the end of the part) ⇒ Placement Type (select **Coaxial** from the drop-down list) ⇒ Axial Reference ↖ (pick the part's axis)] ⇒ [verify the note in the Hole Note Preview (Fig 6.30)] ⇒ ✔ (Build Feature) ⇒ **View** ⇒ ✔ **Model Tree** ⇒ **View** ⇒ **Model Tree Setup** ⇒ **Item Display** ⇒ ✔ **Notes** [Model Tree Items dialog box (Fig. 6.31 and 6.32)] ⇒ **OK** ⇒ **Done**

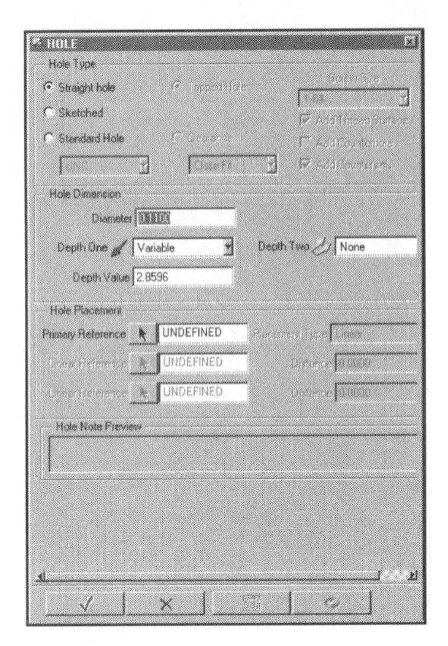

Figure 6.28 (above) HOLE dialog box

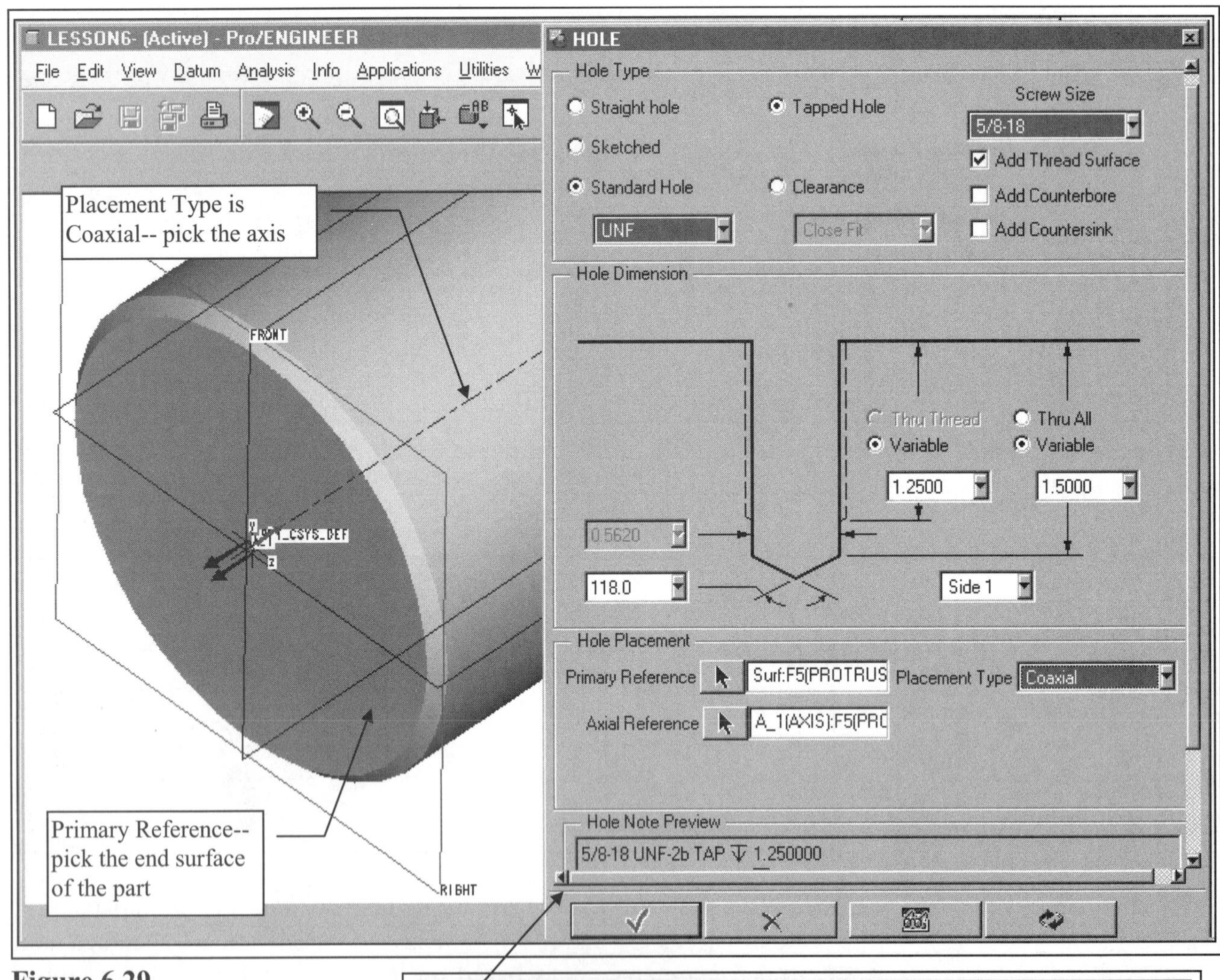

Figure 6.29
5/8-18 UNF-2B
Threaded Depth
1.50 Tap Drill Depth

Hole Note Preview
5/8-18 UNF-2b TAP ⟟ 1.250000
9/16 DRILL (0.562000) ⟟ 1.500000-(1)HOLE

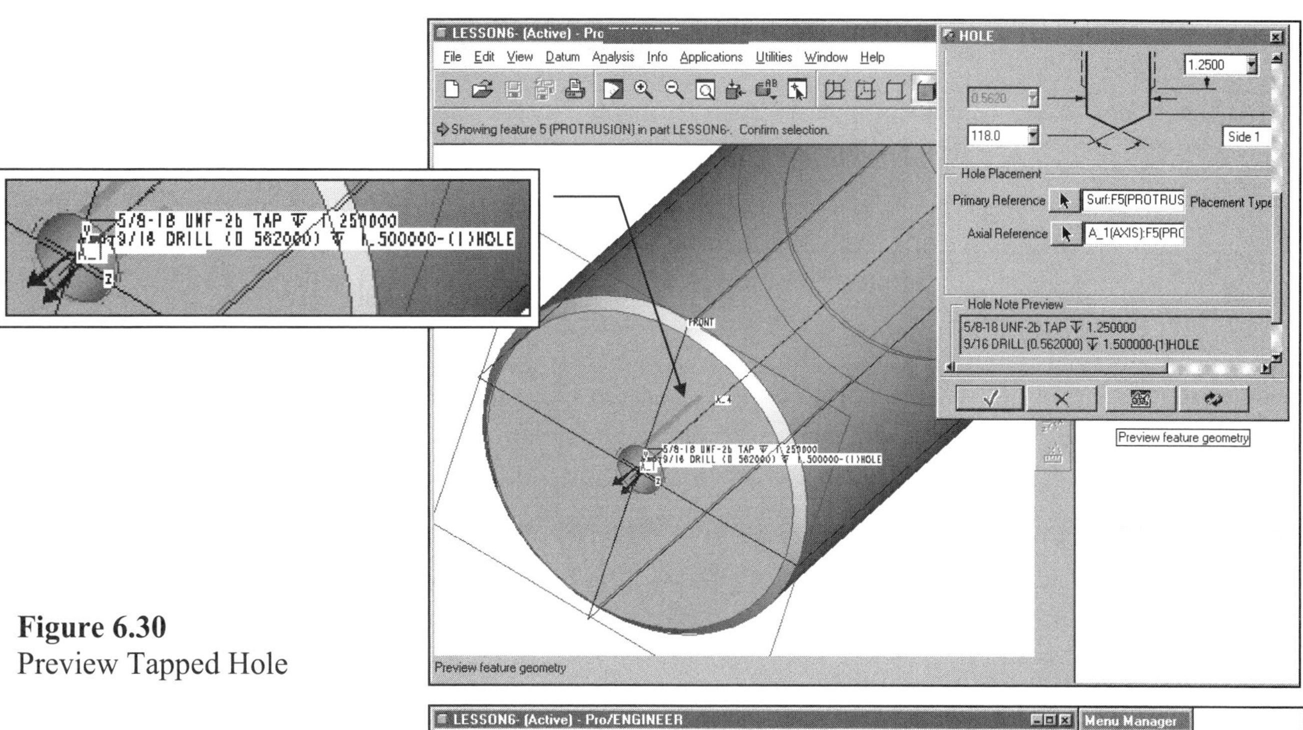

Figure 6.30
Preview Tapped Hole

Figure 6.31
View ⇒ ✔ Model Tree ⇒ View ⇒ Model Tree Setup ⇒ Item Display ⇒ ✔ Notes

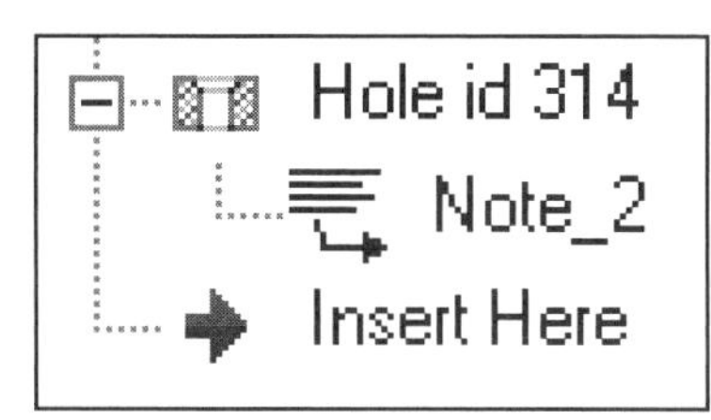

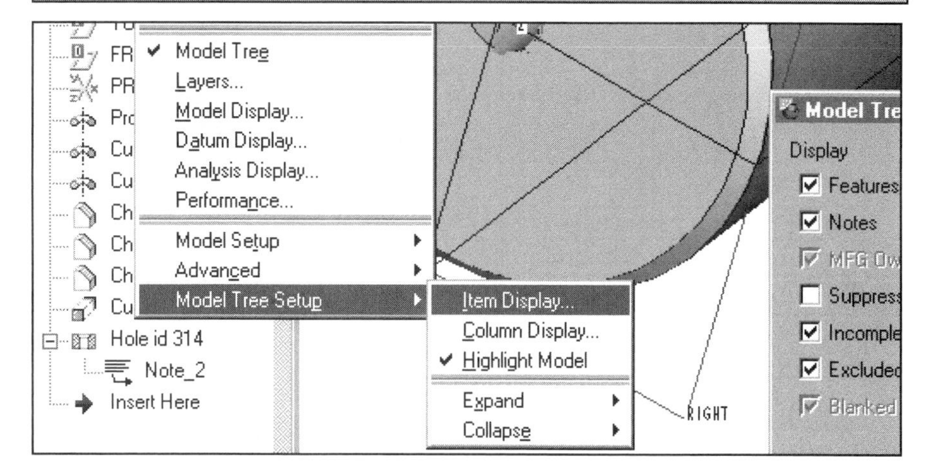

Figure 6.32
Note in Model Tree

HINT

Internal cosmetic threads were created automatically when you created the Standard (Tapped) Hole. For threaded shafts you must create the cosmetic threads.

External cosmetic threads represent the *root diameter*. After creating the cosmetic thread, edit the thread table. The thread size of an external thread must be changed to the nominal size from the root diameter defaulted on the Table.

Create an external cosmetic thread using the **∅2.75** surface. The thread starts at the "neck cut" and goes to the edge of the chamfer:

Feature ⇒ Create ⇒ Cosmetic ⇒ Thread ⇒ [pick the cylindrical surface (Fig. 6.33)] ⇒ (pick the Start Surf--the edge lip surface created by the "neck" cut) ⇒ **Query Sel ⇒ Accept ⇒ Okay** (arrow must point toward the end of the part) ⇒ **UpTo Surface ⇒ Done ⇒ Make Datum ⇒ Through ⇒** (unselect by picking on the default items in the DATUM PLANE Menu Manager except **AxisEdgeCurv) ⇒ Query Sel ⇒ Accept** [pick the edge of the chamfer (Figure 6.34)] ⇒ **Done ⇒** (type the diameter of the cosmetic thread (root diameter, **2.6875) ⇒ ✔ ⇒ Mod Params** (Fig. 6.35) ⇒ (edit the table)

Figure 6.33 Starting Surface For Thread

If you use **Blind,** you must first measure the distance from the lip of the neck cut to the edge of the chamfer before creating the cosmetic thread: **Analysis ⇒ Measure ⇒ Distance ⇒** From **Curve/Edge ⇒** (select the edge of the chamfer) ⇒ To **Curve/Edge** ⇒ (select the edge of the neck cut)

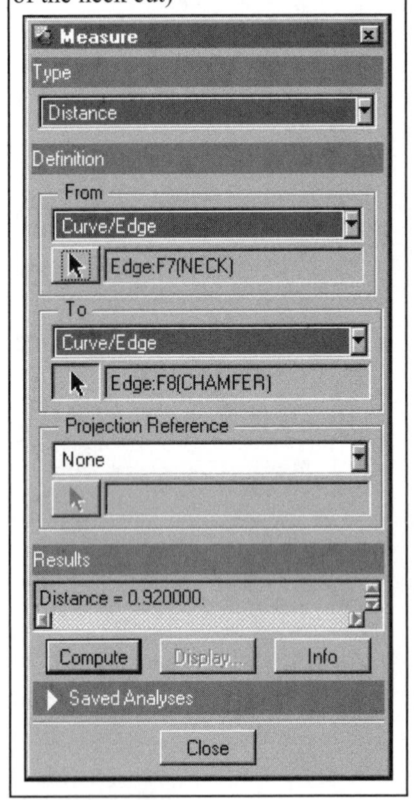

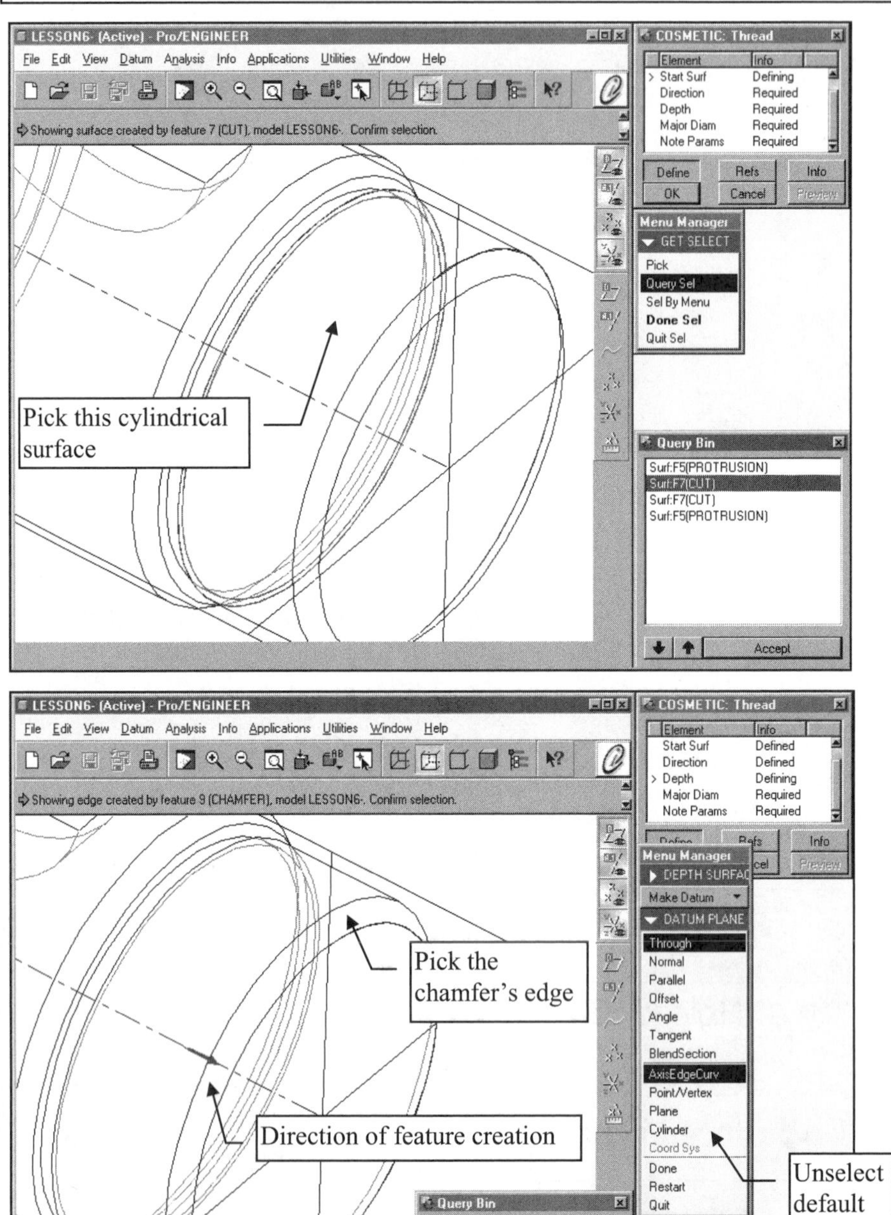

Figure 6.34
Direction Ending Surface for the Cosmetic Thread

MAJOR_DIAMETER	**2.75**
THREADS_PER_INCH	**16**
FORM	**UN**
CLASS	**2**
PLACEMENT	**A**
METRIC	**False**

The Thread Table information (Fig. 6.35) shows the diameter as **2.6875,** because the *magenta*-colored cylinder representing the thread on your screen will be **Ø2.688.** Because you are cosmetically representing the *root diameter* of the thread on the model, the *thread diameter* is *smaller* than the *nominal* thread size. Change the **2.6875** dimension to **2.75** on the Thread Table. After editing, choose:

File ⇒ Save ⇒ File ⇒ Exit ⇒ Show (Fig. 6.36) **⇒ Close** (Fig. 6.37) **⇒ Done/Return ⇒ OK** (Fig. 6.38) **⇒ Done**

Figure 6.35
Thread Table **2.75-16 UN-2A**

You must edit this number from the root diameter of **2.6875** to the nominal size of **2.75**

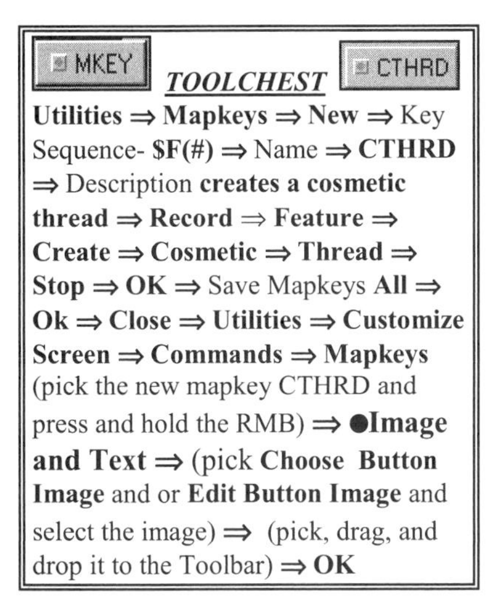
MKEY *TOOLCHEST* CTHRD

Utilities ⇒ Mapkeys ⇒ New ⇒ Key Sequence- **$F(#) ⇒** Name **⇒ CTHRD ⇒** Description **creates a cosmetic thread ⇒ Record ⇒ Feature ⇒ Create ⇒ Cosmetic ⇒ Thread ⇒ Stop ⇒ OK ⇒** Save Mapkeys **All ⇒ Ok ⇒ Close ⇒ Utilities ⇒ Customize Screen ⇒ Commands ⇒ Mapkeys** (pick the new mapkey CTHRD and press and hold the RMB) **⇒ ●Image and Text ⇒** (pick **Choose Button Image** and or **Edit Button Image** and select the image) ⇒ (pick, drag, and drop it to the Toolbar) **⇒ OK**

Figure 6.36
Showing Feature Parameters

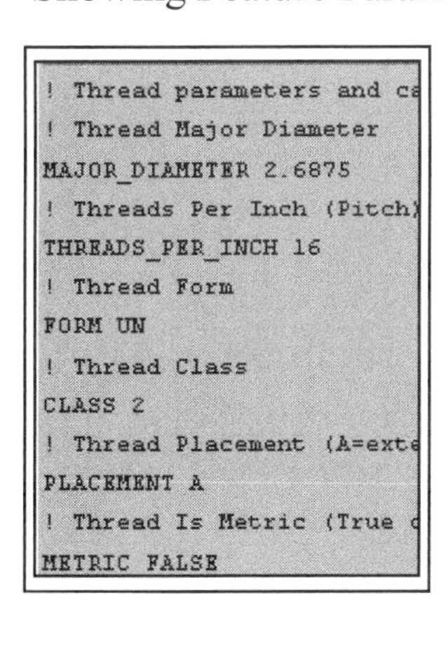
! Thread parameters and ca
! Thread Major Diameter
MAJOR_DIAMETER 2.6875
! Threads Per Inch (Pitch)
THREADS_PER_INCH 16
! Thread Form
FORM UN
! Thread Class
CLASS 2
! Thread Placement (A=exte
PLACEMENT A
! Thread Is Metric (True c
METRIC FALSE

Because it is a datum on the fly **(Make Datum), DTM1** will not display on the model or in the Model Tree after you choose **OK**

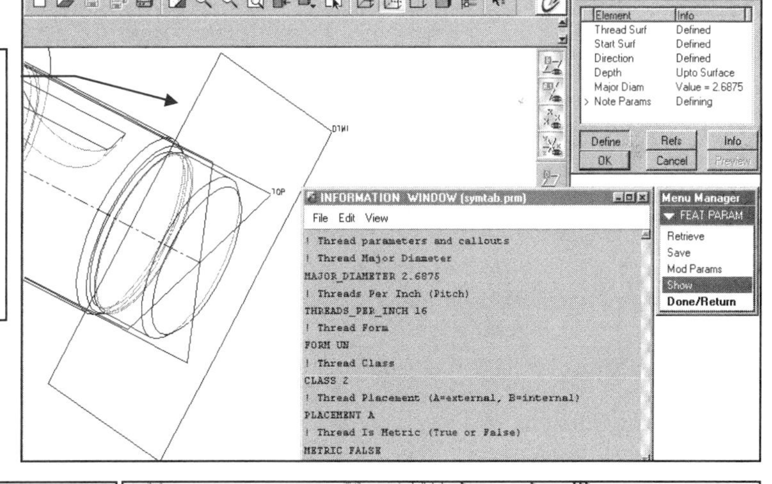

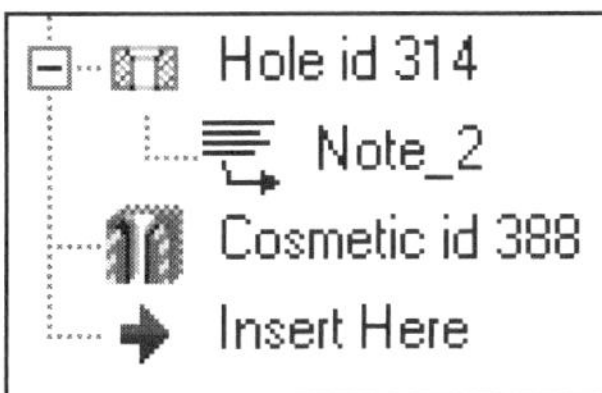

Figure 6.37
Cosmetic Thread Displays as a *Magenta* Cylinder

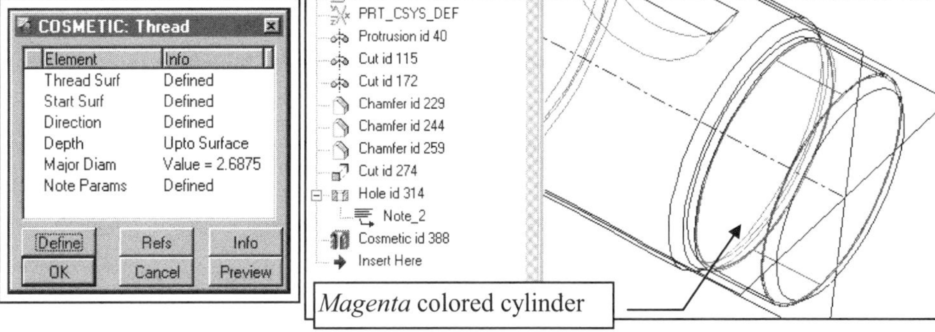

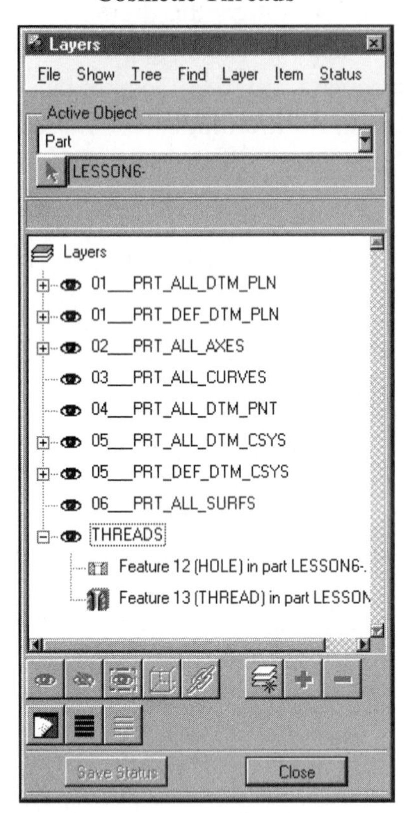

Utilities ⇒ Environment ⇒
(to change the display of the note on the screen to **NOTE_2) ⇒ OK ⇒ File ⇒ Save ⇒ ✔⇒ File ⇒ Delete ⇒ Old Versions ⇒ ✔**

Create a layer (called **THREADS**). Add the cosmetic thread and the hole to the **THREADS** layer (see left-hand column); then **File ⇒ Save.** Figure 6.38 shows a drawing of the completed cylinder rod.

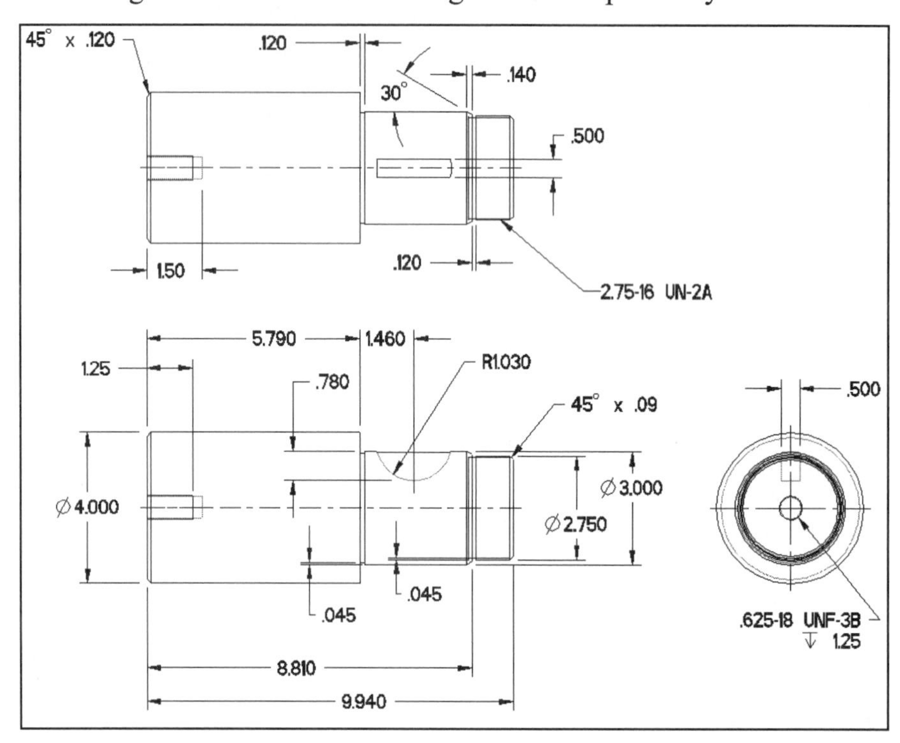

Figure 6.38
Cylinder Rod Drawing

Before going on to the Lesson 6 Project, complete the **ECO** shown in Figure 6.39 for the pin part from Lesson 5.

ECO

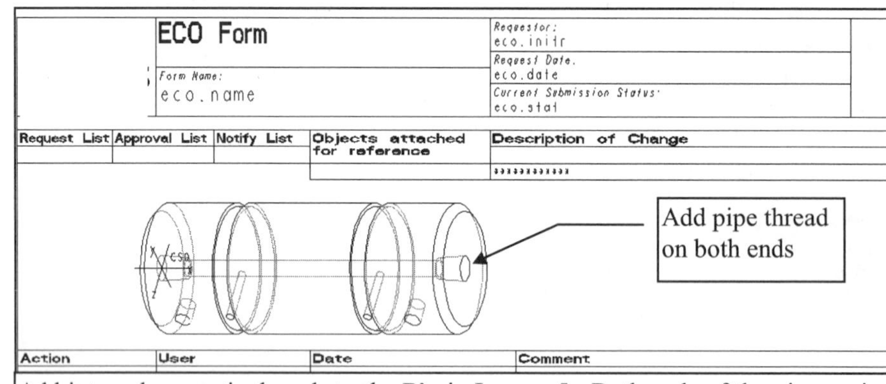

NOTE

Look up the geometry sizes for the pipe thread in your **Machinery's Handbook**, a drafting text, or **Fundamentals of Engineering Graphics and Design** by L. Lamit and K. Kitto, Chapter 17, Threads and Fasteners. See pages 661-662.

Figure 6.39
ECO for the Pin in Lesson 5

Add internal cosmetic threads to the **Pin** in Lesson 5. Both ends of the pin require *pipe threads*. Pro/E does not create a *conical* pipe cosmetic thread. Therefore, the cosmetic thread created will not be correct as far as the visual geometry is concerned--it will be *cylindrical.*
The pipe thread for both ends is: **.125-27 NPTF**
.44 DEEP
The diameter at the small end of the tapered thread is **.3339** and the angle of taper is **1.48°** (taper of thread on diameter is *3/4 inch per foot*) from the centerline. ***Modify the existing conical cut before you create the cosmetic threads.***

Lesson 6 Project

Clamp Ball and Coupling Shaft

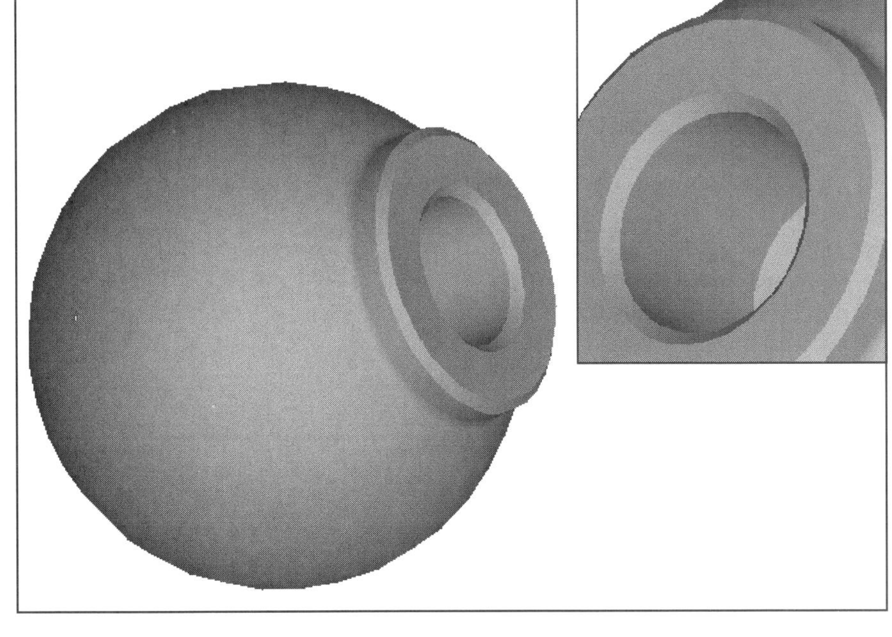

Figure 6.40
Clamp Ball

Clamp Ball and Coupling Shaft

As with Lesson 5, two **lesson projects** are provided here in Lesson 6 (Figs. 6.40 through 6.53). You will use both parts in Lessons 14 and 15 when creating different assemblies. Both the **Clamp Ball** (decimal inch) and the **Coupling Shaft** (SI units) are revolved protrusions. The Clamp Ball is simpler and easier to complete. The Ball is *black plastic* and the Shaft is *steel*. Create all cosmetic threads required on each part. The two parts are used on different assemblies.

Analyze the parts and plan out the steps and features required to model them. Use the **DIPS** in Appendix D to establish a feature creation sequence before the start of modeling. Remember to set up the environment, establish datum planes, and set them on layers.

EGD REFERENCE
Fundamentals of Engineering Graphics and Design
by L. Lamit and K. Kitto
Read Chapter 23.
See pages 865-866.

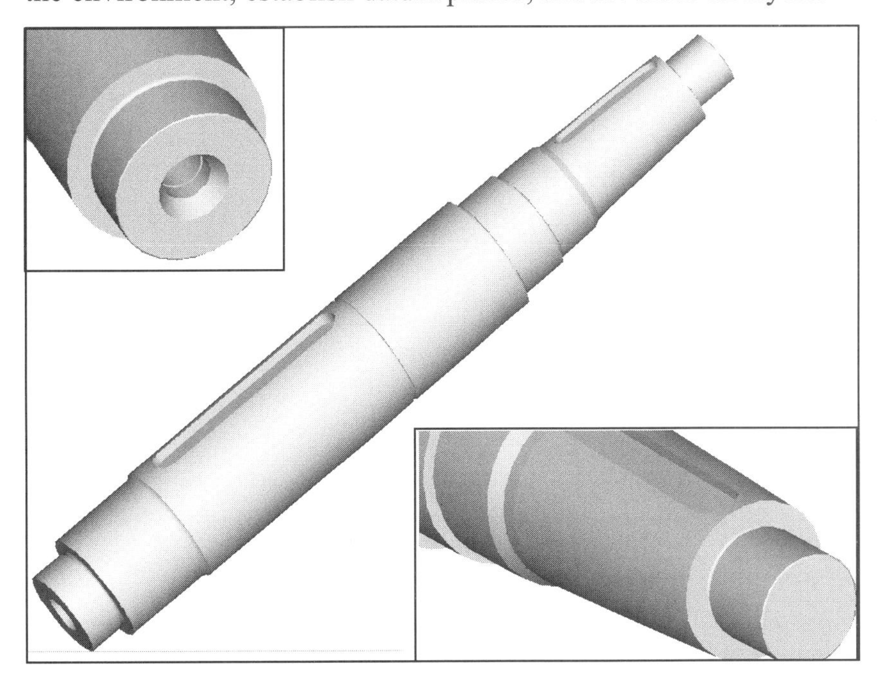

Figure 6.41
Coupling Shaft

L6-24 **Chamfers and Cosmetic Threads**

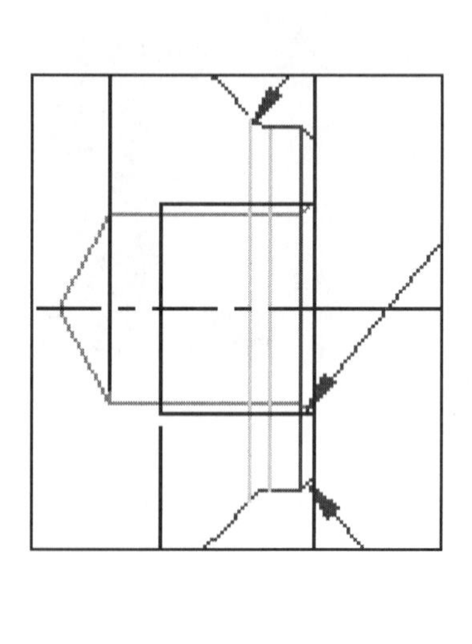

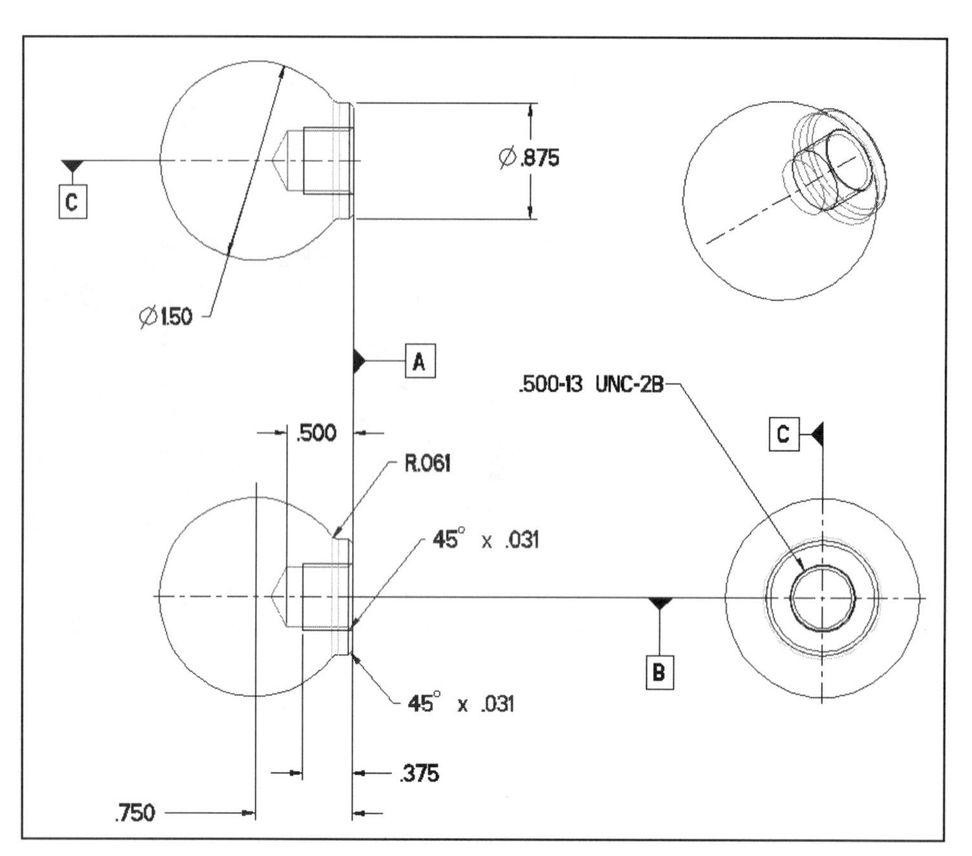

Figure 6.42
Clamp Ball Dimensions

Figure 6.43 Coupling Shaft Drawing, Sheet One

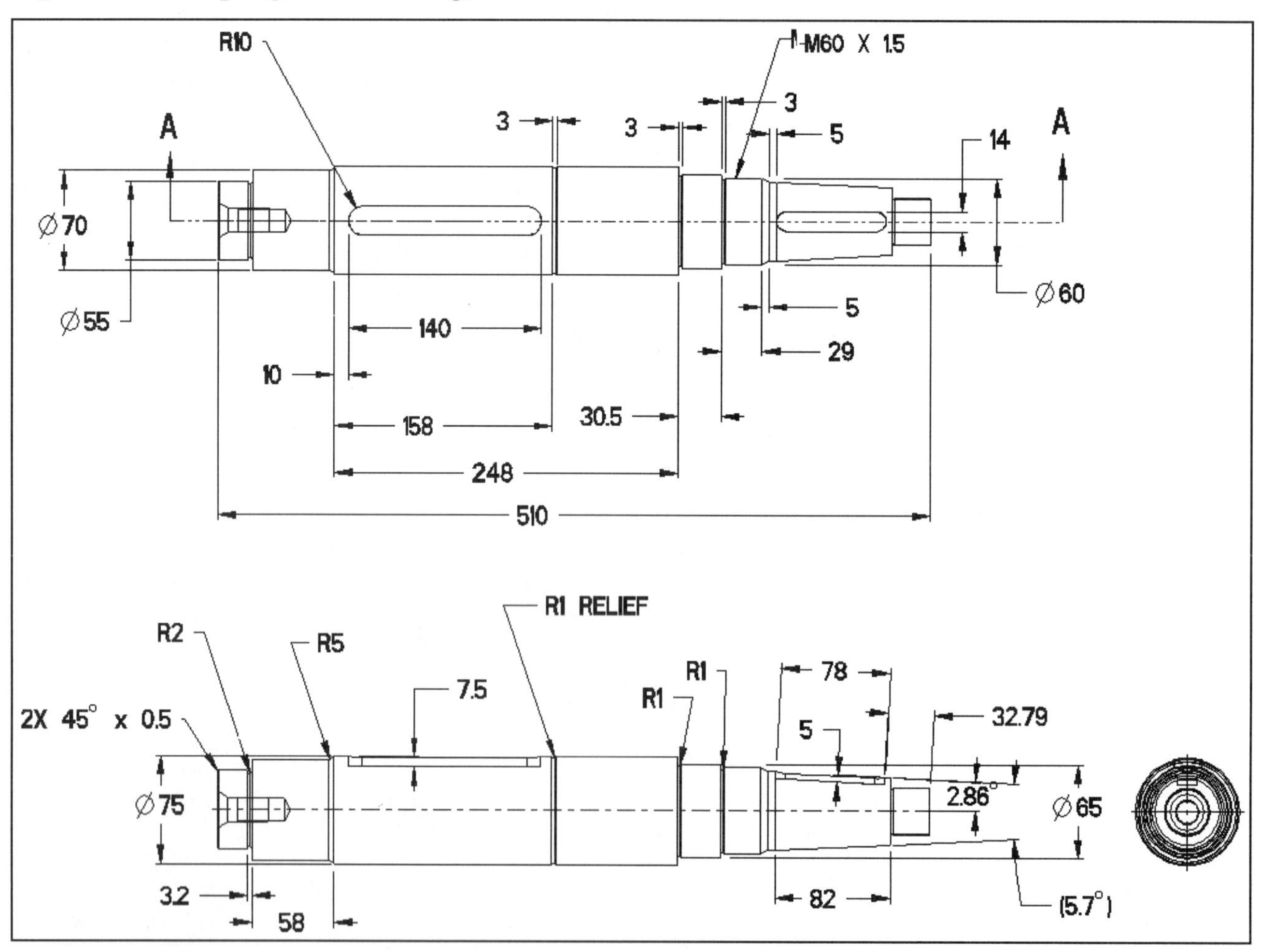

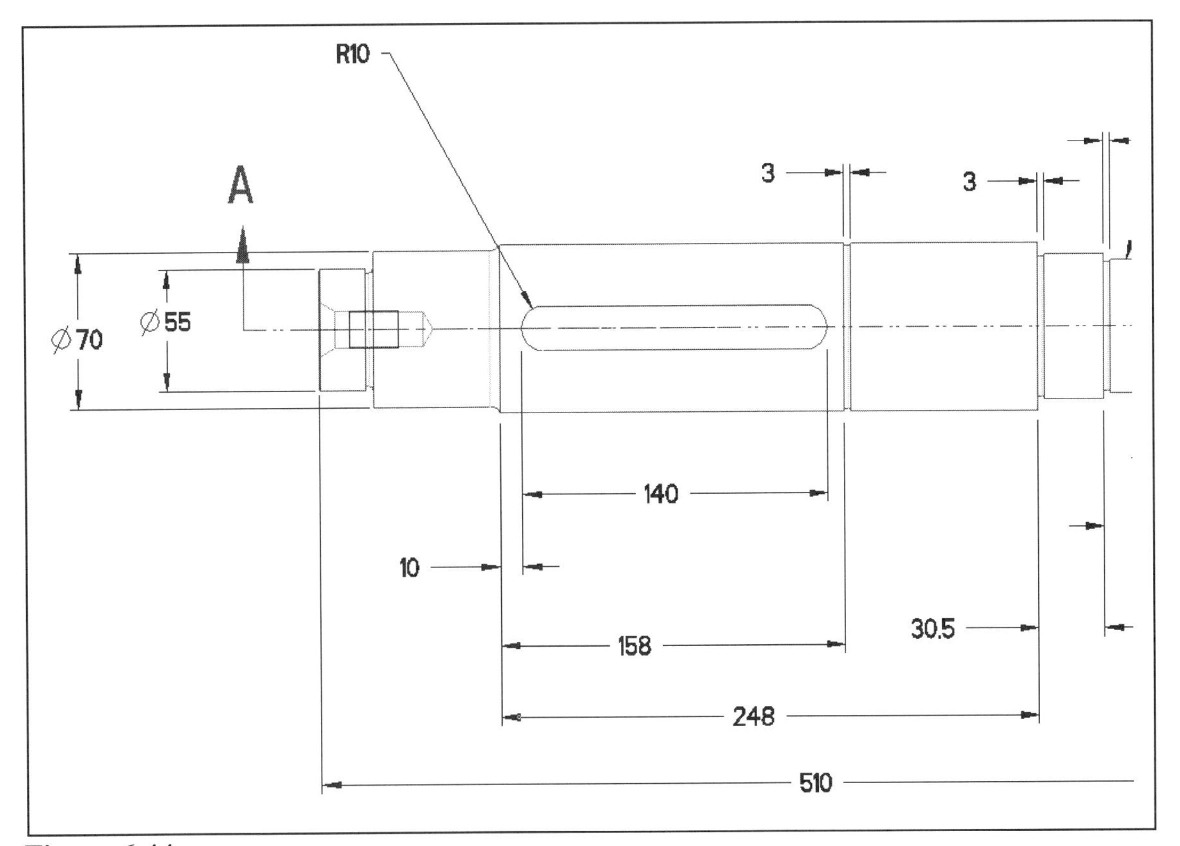

Figure 6.44
Coupling Shaft Drawing,
Top View, Left Side

Figure 6.45
Coupling Shaft Drawing,
Sheet Two

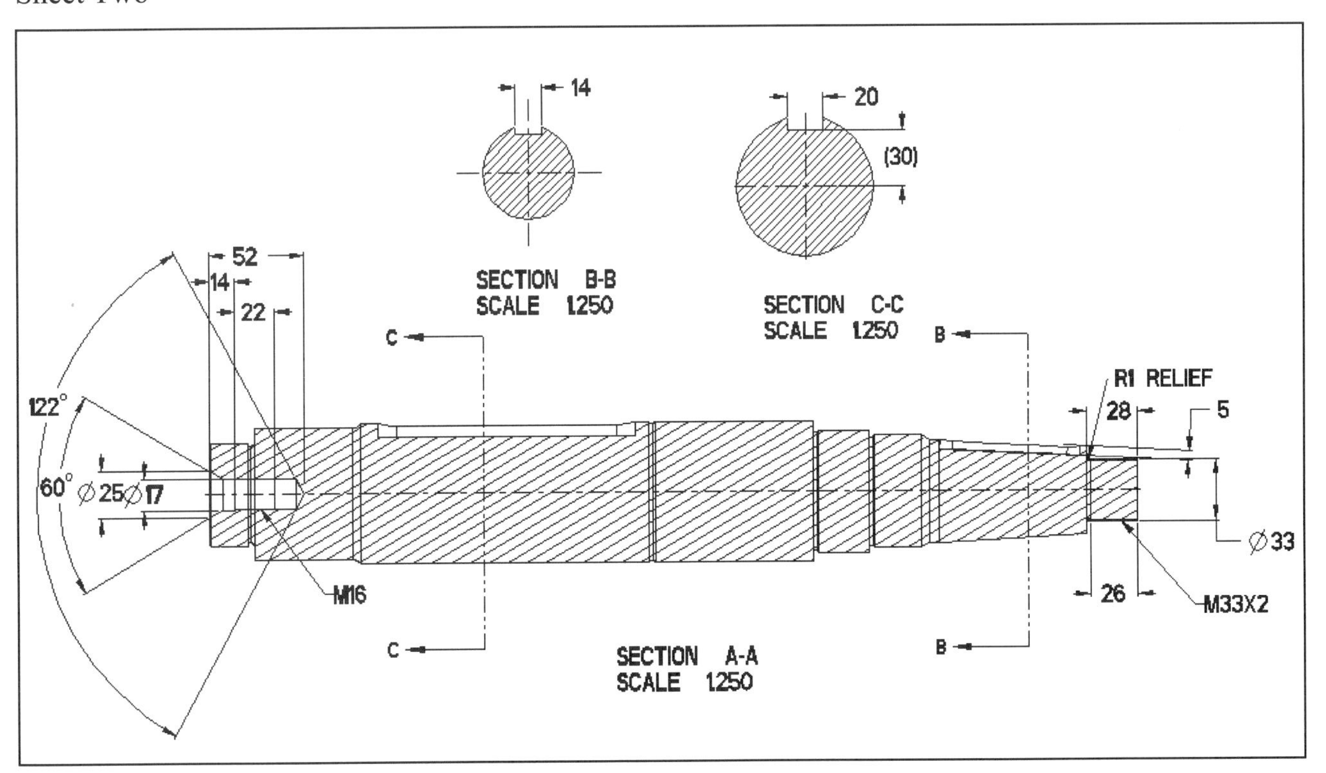

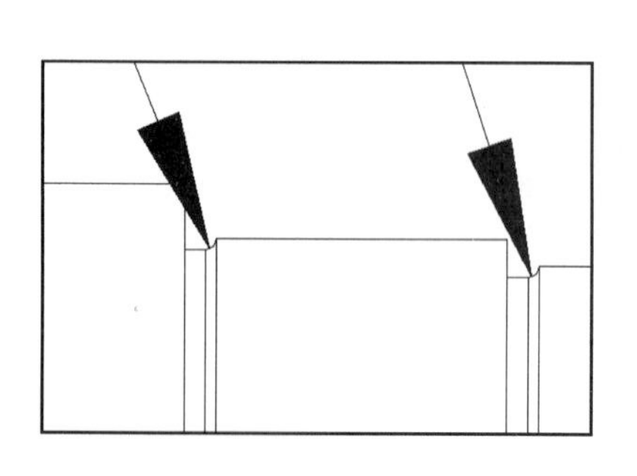

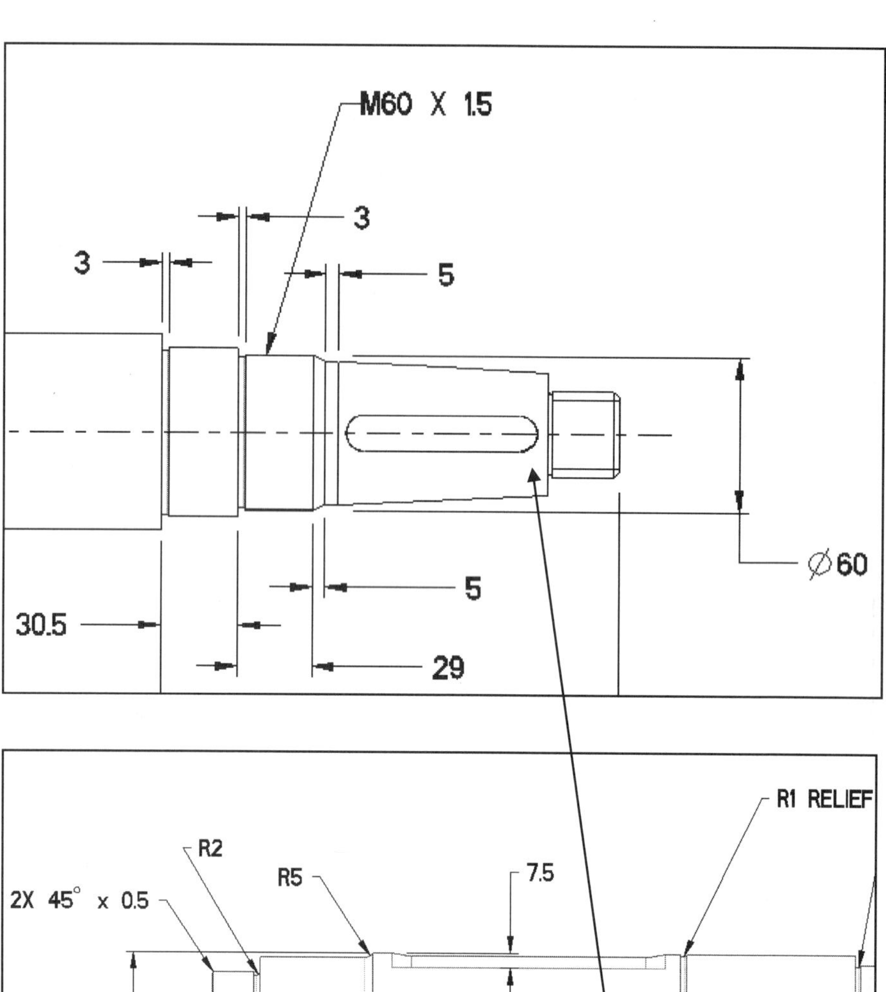

Figure 6.46
Coupling Shaft Drawing, Top View, Right Side

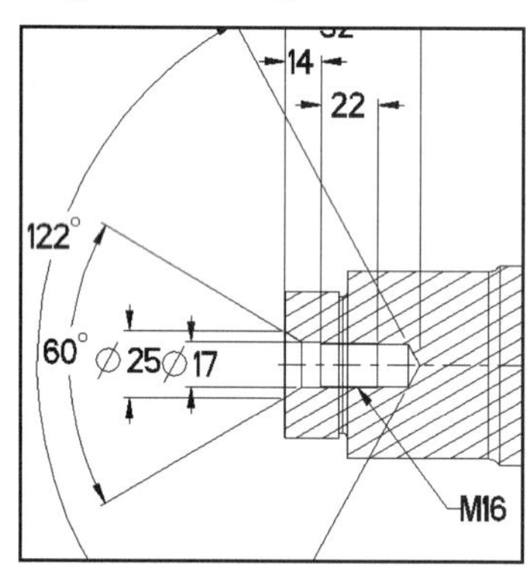

Figure 6.47
Coupling Shaft Drawing, Front View, Left Side

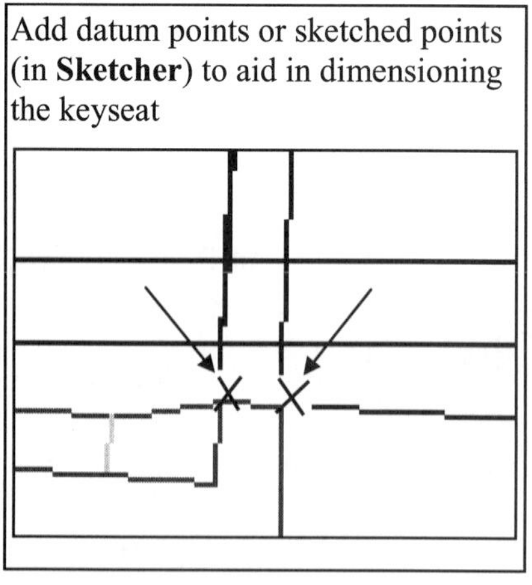

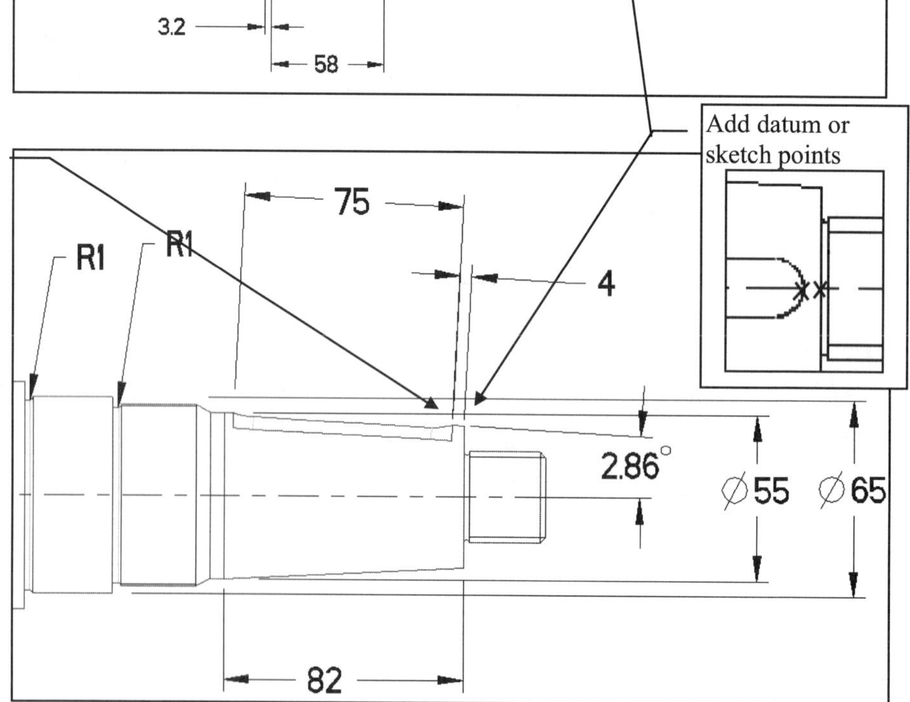

Figure 6.48
Coupling Shaft Drawing, Front View, Right Side

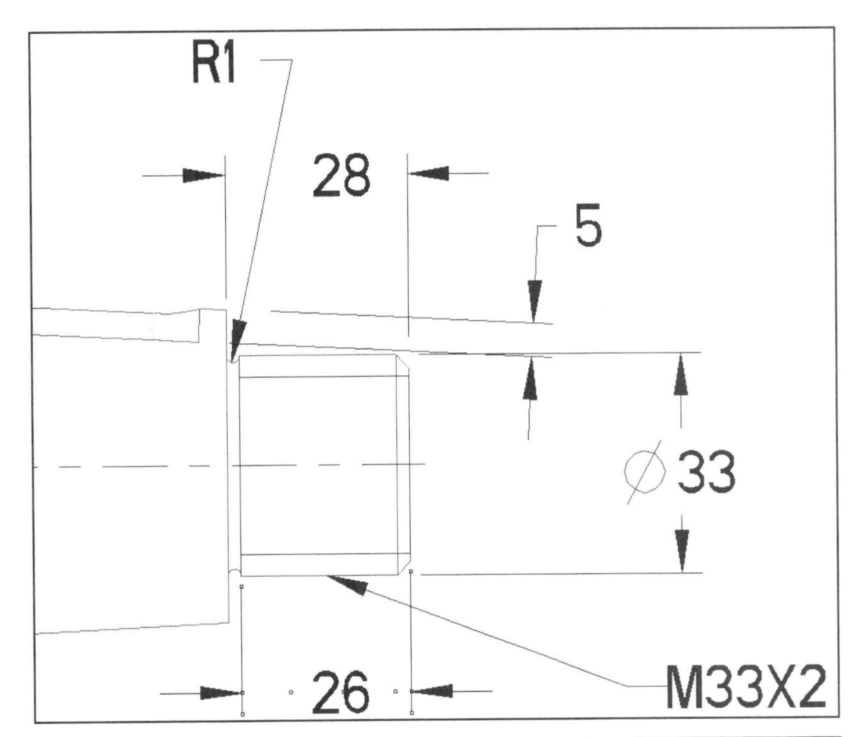

Figure 6.49
Coupling Shaft Drawing,
M33 X 2 Threads

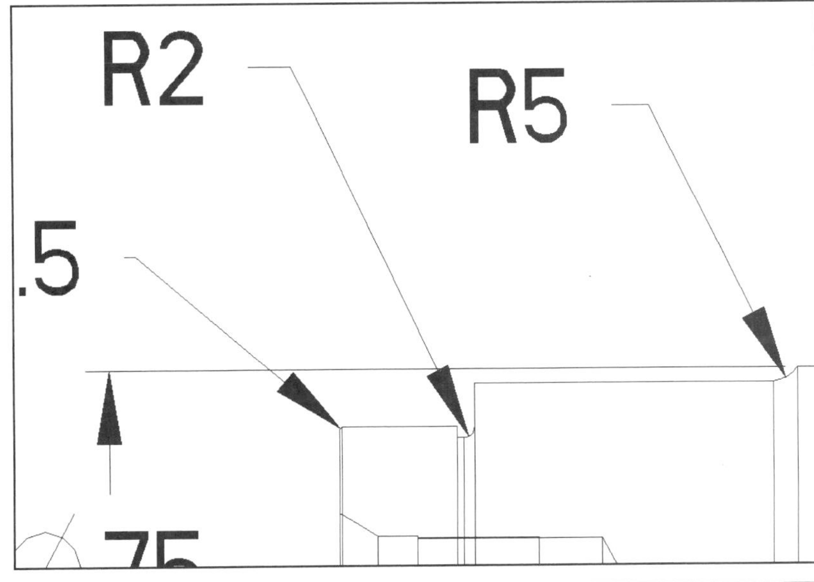

Figure 6.50
Coupling Shaft Drawing,
Reliefs

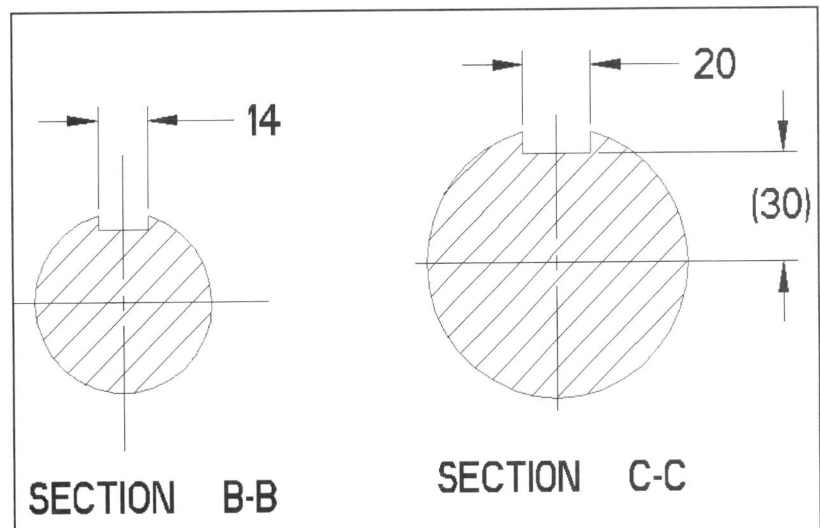

Figure 6.51
Coupling Shaft Drawing,
Sheet Two,
SECTION B-B and
SECTION C-C

File ⇒ Save ⇒ ✔
File ⇒ Delete ⇒
Old Versions ⇒ ✔

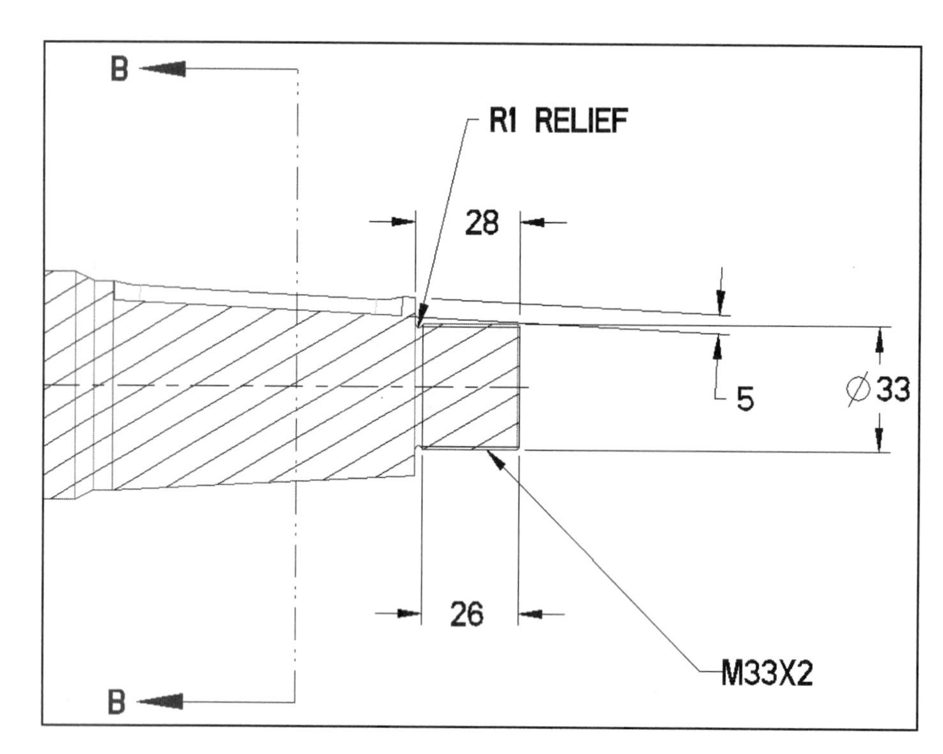

Figure 6.52
Coupling Shaft Drawing, Sheet Two, **SECTION A-A** Right Side

Figure 6.53
Coupling Shaft Drawing, Sheet Two, **SECTION A-A** Left Side

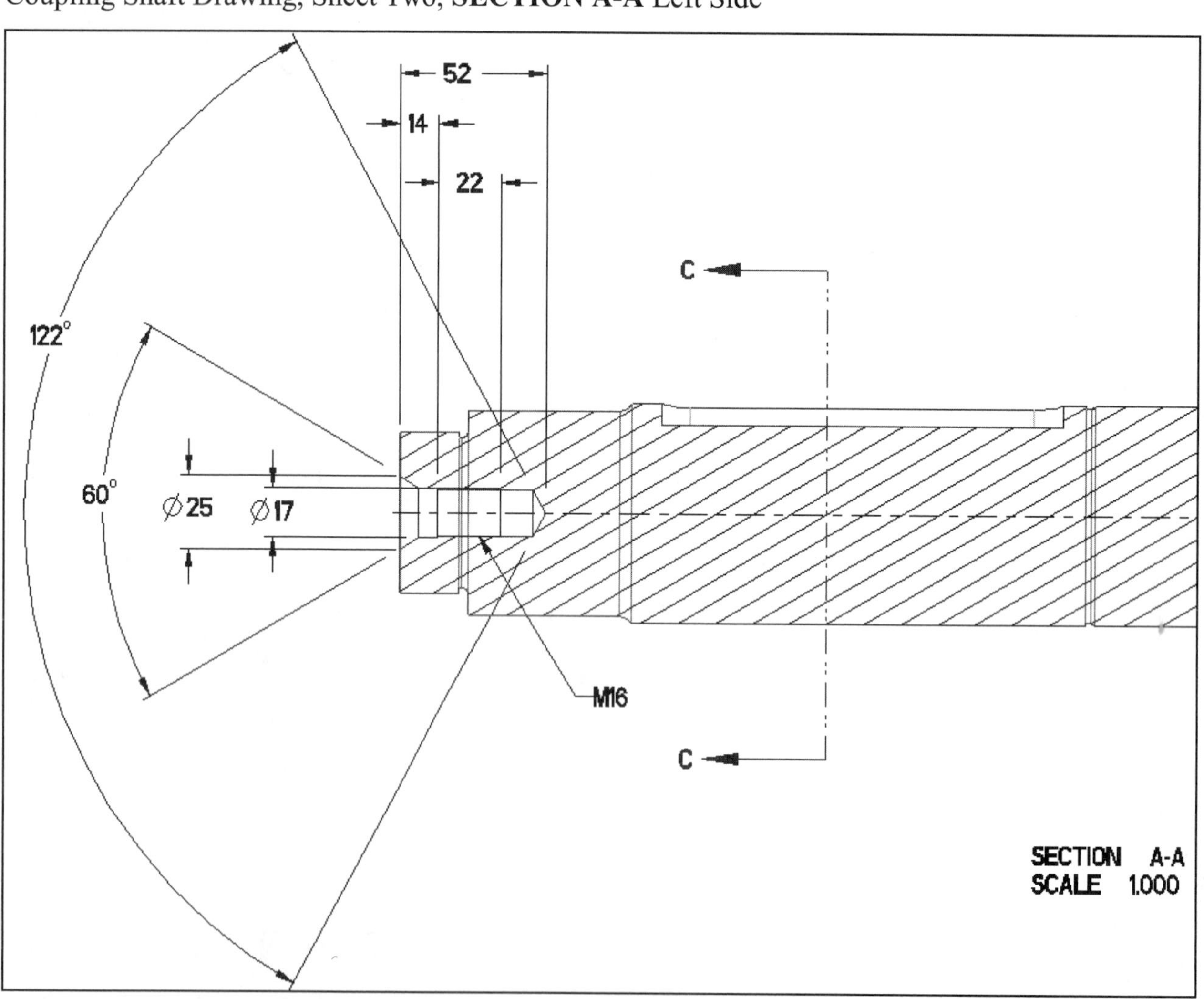

Lesson 7

Pattern and Group

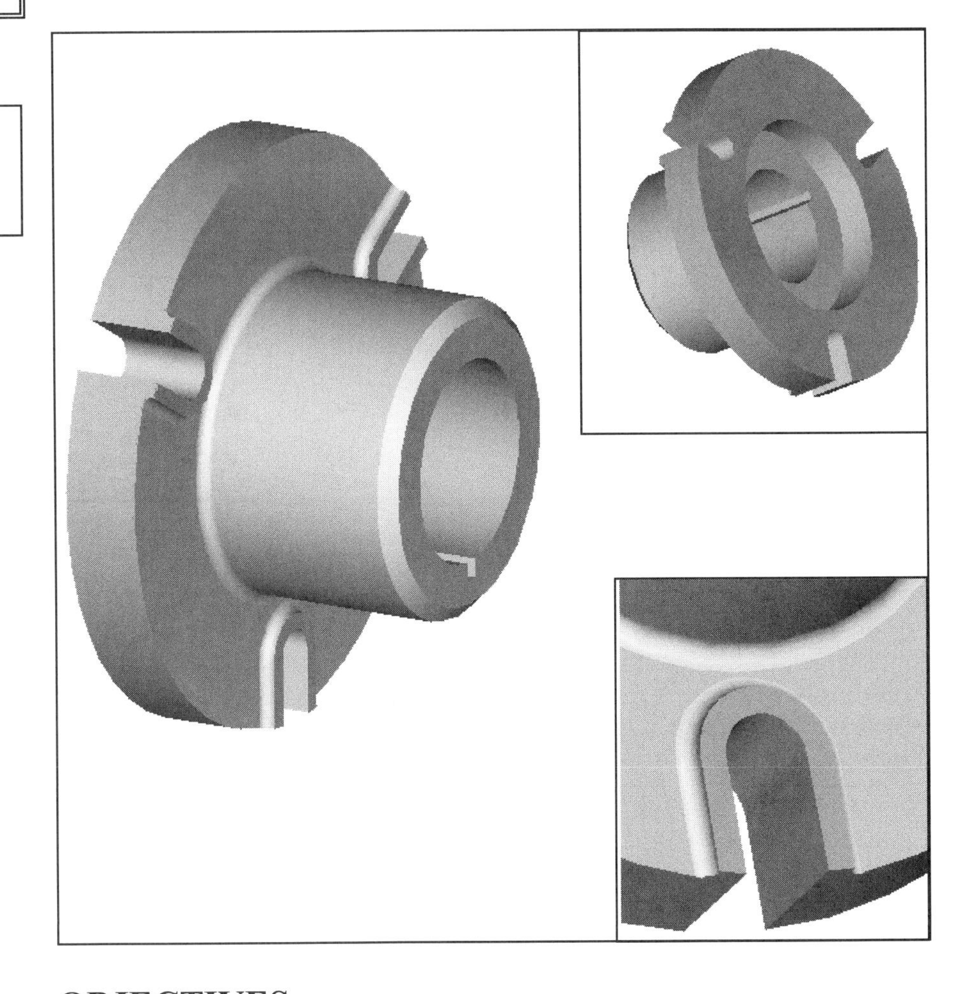

Figure 7.1
Post Reel

OBJECTIVES

1. Change dimension cosmetics and move part dimensions

2. Use offset edge to create sketch entities from existing geometry

3. Pattern features

4. Create and manipulate grouped features

5. Understand how to pattern groups

EGD REFERENCE
Fundamentals of Engineering Graphics and Design
by L. Lamit and K. Kitto
Read Chapter 10.
See page 296.

COAch™ for Pro/ENGINEER

If you have **COAch for Pro/ENGINEER** on your system, go to SEARCH and do the Segment shown in Figure 7.3.

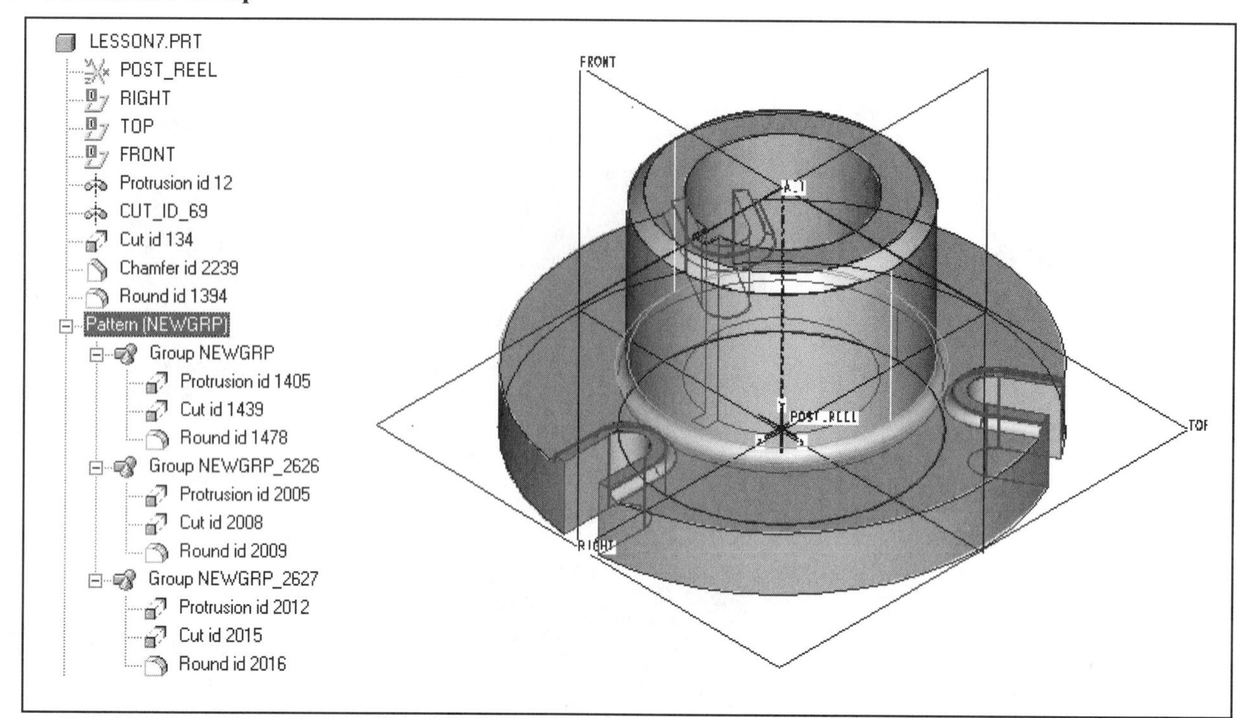

Figure 7.2
Post Reel Showing Model Tree

PATTERN AND GROUP

To create multiple features (Fig. 7.1) from a single feature (or group of features), the **Pattern** option can be used (Figs. 7.2 and 7.3). After it is created, a pattern behaves as if it were a single feature. When you create a pattern, you create instances (copies) of the selected feature (or group of features). The **Group** option unites a series of features. Creating a pattern (Fig. 7.4) is a quick way to reproduce a feature or a set of features that are related and grouped for easy manipulation. Manipulating a pattern may be more advantageous for you than operating on individual features.

CADTRAIN
Duplicating Features
Modeling I

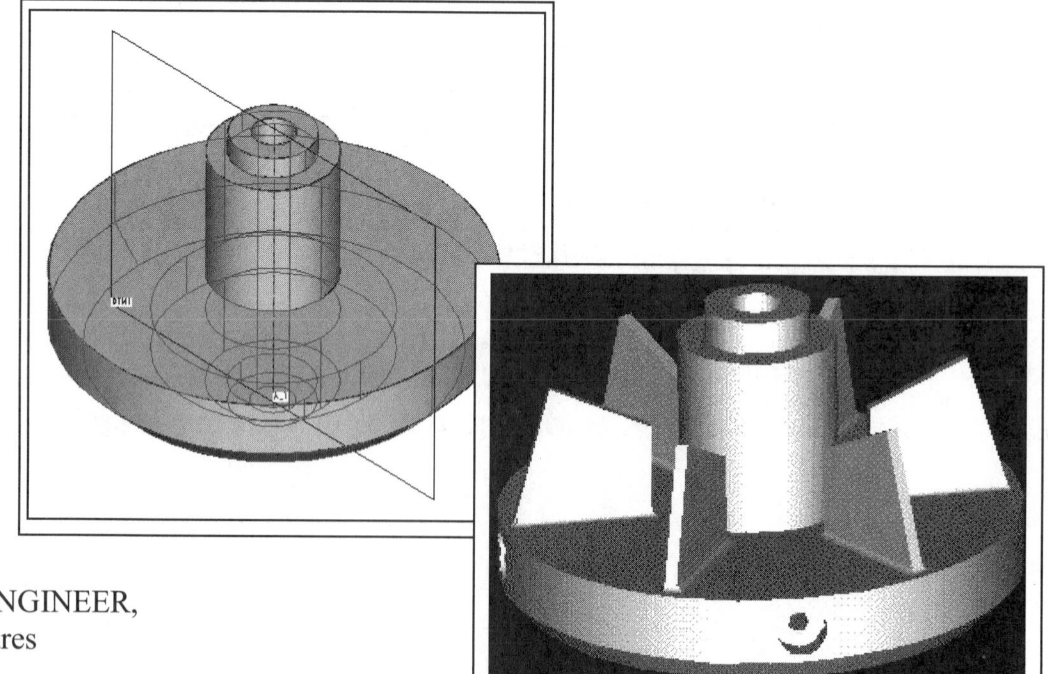

Figure 7.3
COAch for Pro/ENGINEER, Duplicating Features (Circular Pattern)

Group

When you create a *local group*, you must select the features in the consecutive (sequential) order of the regeneration list. A quick way to do this is to select the intended group by range. If there are features between the specified features in the regeneration list, Pro/E asks whether you want to group all the features in between.

Features that are already in other groups cannot be grouped again.

To create a local group, do the following:

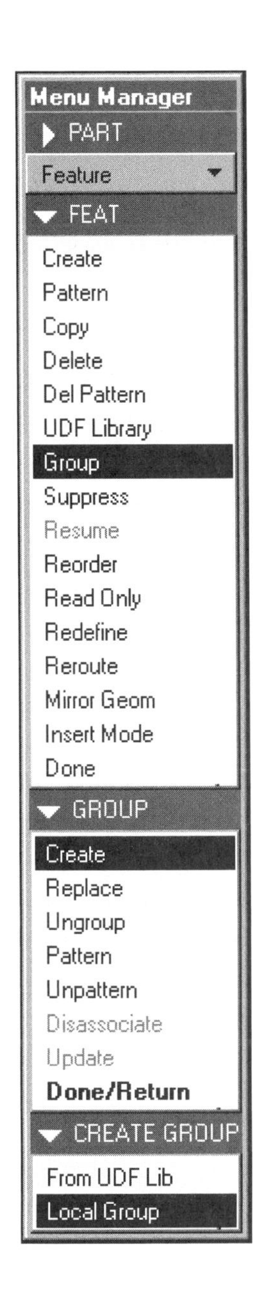

1. Choose **Create** from the GROUP menu and **Local Group** from the CREATE GROUP menu.
2. Select each feature to include in the local group.

   or

3. Answer **yes** to the prompt asking if you want to group all the features in the between. If you answer **no** to this prompt, Pro/E does not create the local group.

Pattern

Modifying patterns is more efficient than modifying individual features. In a pattern, when you change the dimensions of the original feature, the whole pattern will be updated automatically. A pattern is parametrically controlled. Therefore, a pattern can be modified by changing pattern parameters, such as the number of instances, the spacing between instances, and the original feature dimensions.

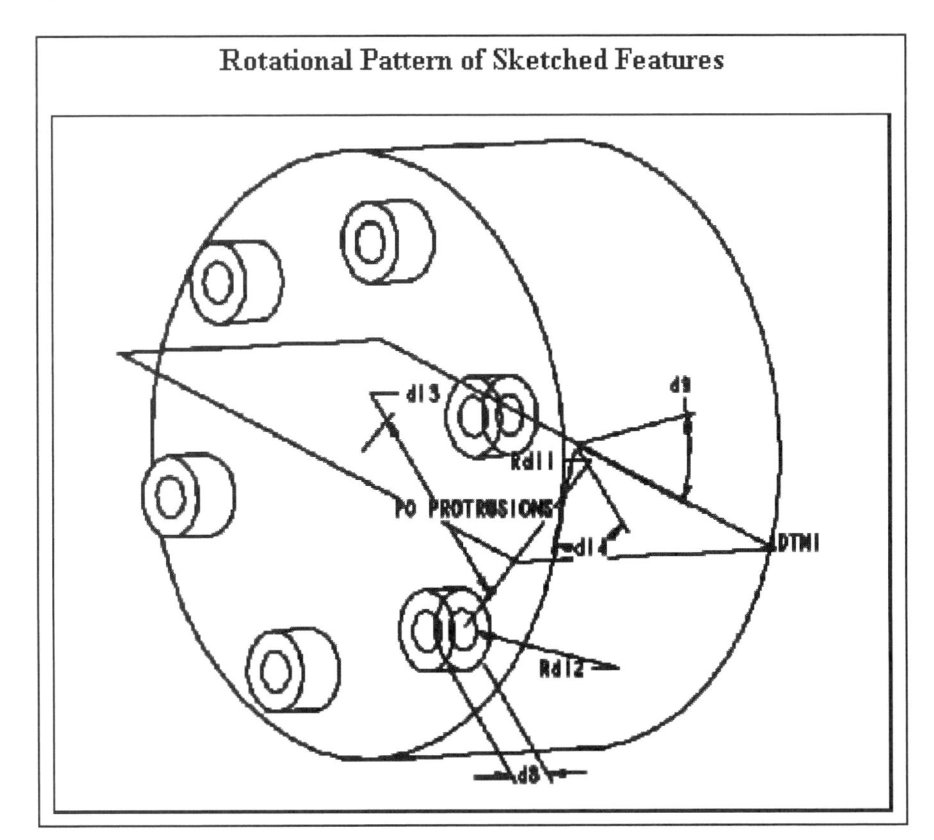

Figure 7.4
Patterns

MKEY *TOOLCHEST*

Utilities ⇒ Mapkeys ⇒ New ⇒ Key Sequence- **$F(#) ⇒ Name ⇒ GRP ⇒** Description **creates a group ⇒ Record ⇒ Feature ⇒ Group ⇒ Local Group ⇒ Stop ⇒ OK ⇒** Save Mapkeys **All ⇒ Ok ⇒ Close ⇒ Utilities ⇒ Customize Screen ⇒ Commands ⇒** (Category **Mapkeys**) ⇒ (pick the new Mapkey **GRP**) ⇒ **Modify Selection ⇒ Choose Button Image ⇒** (select a new image) ⇒ (pick the new Mapkey, drag to the Toolbar and drop) ⇒ **OK**

MKEY *TOOLCHEST*

Utilities ⇒ Mapkeys ⇒ New ⇒ Key Sequence- **$F(#) ⇒ Name ⇒ GRP_PAT ⇒** Description **patterns a group ⇒ Record ⇒ Feature ⇒ Pattern ⇒ Stop ⇒ OK ⇒** Save Mapkeys **All ⇒ Ok ⇒ Close ⇒ Utilities ⇒ Customize Screen ⇒ Commands ⇒** (Category **Mapkey**) ⇒ (pick the new Mapkey **PAT_GRP**) ⇒ **Modify Selection ⇒ Choose Button Image ⇒** (select a new image- see illustration below) ⇒ (pick the new Mapkey, drag to the Toolbar and drop) ⇒ **OK**

When you create a pattern, Pro/E assumes it is a "single" feature. Creating a pattern is a quick way to reproduce a feature. Patterning is an easier and more effective way to perform a single operation on the multiple features contained in a pattern, rather than on the individual features. For example, you can easily suppress a pattern or add it to a layer.

You can pattern most features using the FEAT menu's **Pattern** option. A thin feature "remembers" the surface to which it is attached and patterns to that surface.

Pro/E allows you to pattern a single feature only. However, you can pattern several features as if they were a single feature by arranging them in a **Local Group** and then patterning the group (Fig. 7.5). After the pattern is created, you can unpattern and ungroup the instances and then make them independently modifiable using the option **Make Indep.** There are two ways to pattern a feature using the PRO PAT TYPE menu:

Dim Pattern Controls the pattern using driving dimensions to determine the incremental changes to the pattern. This is the case for all of the pattern.

Ref Pattern Controls the pattern by referencing another pattern. The dimension pattern must exist before you can create the next pattern type. For example, counterbore one hole in the pattern and then have Pro/E pattern the counterbore. It automatically makes a reference pattern.

When you are working with features or components for which it does not make sense to have both **Dim Pattern** and **Ref Pattern**, Pro/E does not display the PRO PAT TYPE menu.

The features that can be patterned are:

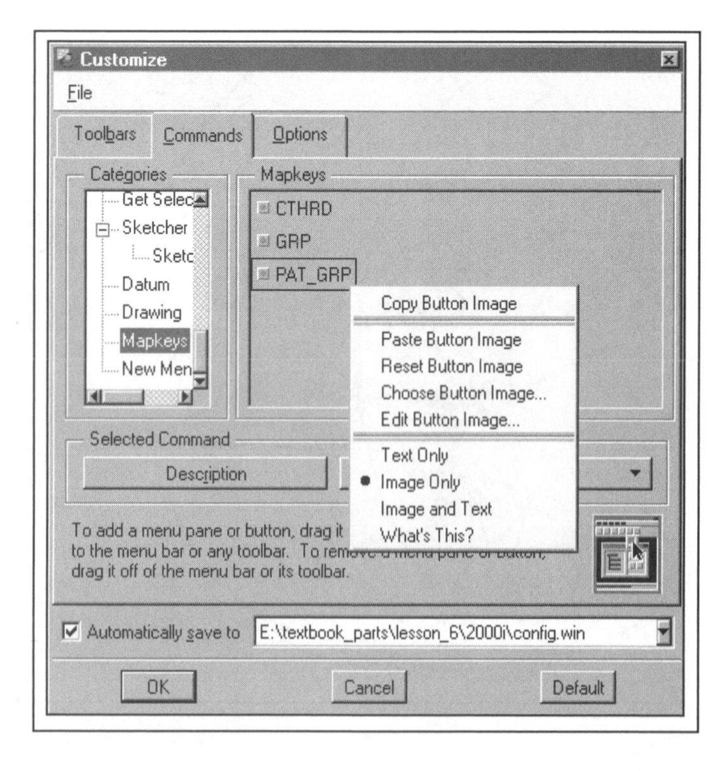

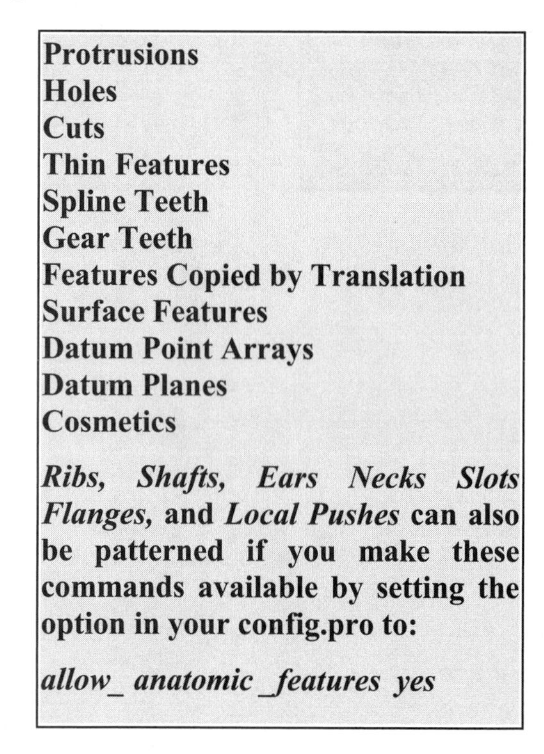

Protrusions
Holes
Cuts
Thin Features
Spline Teeth
Gear Teeth
Features Copied by Translation
Surface Features
Datum Point Arrays
Datum Planes
Cosmetics

***Ribs, Shafts, Ears Necks Slots Flanges,* and *Local Pushes* can also be patterned if you make these commands available by setting the option in your config.pro to:**

allow_ anatomic _features yes

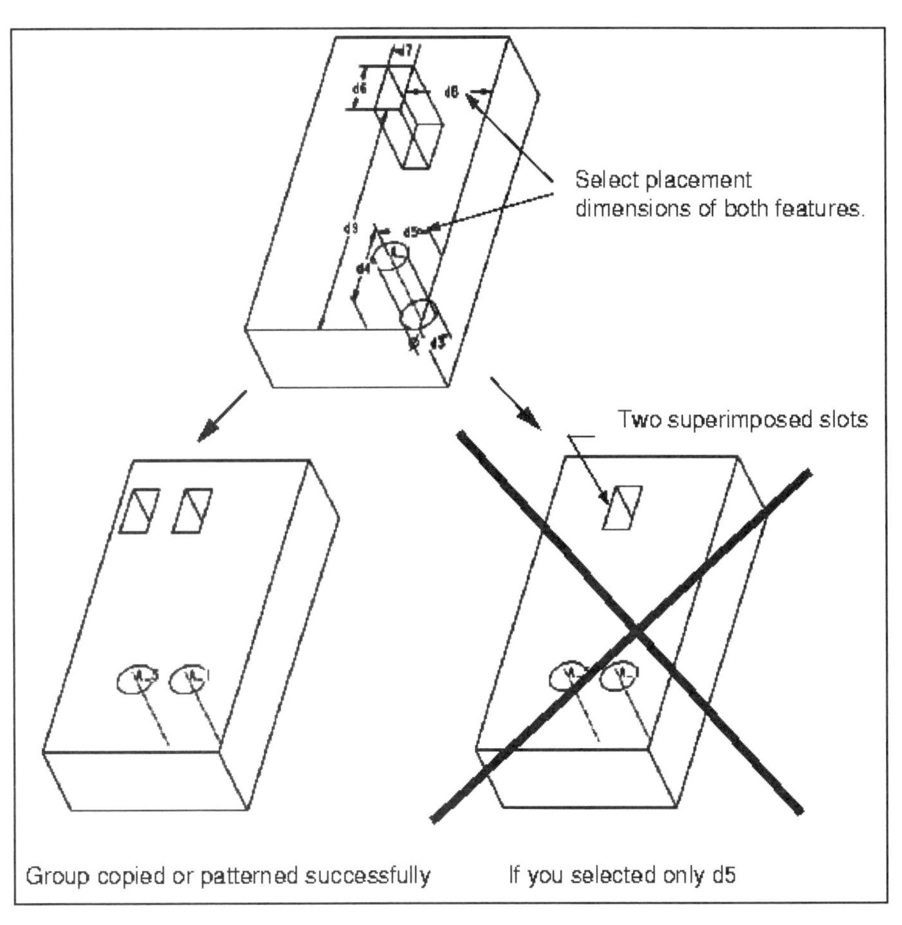

Figure 7.5
Patterns

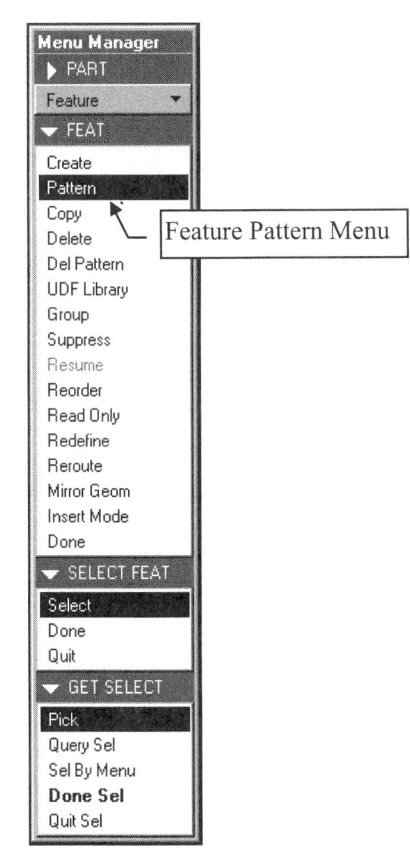

Figure 7.6
Patterning Groups

Patterning a Group

You can pattern groups created from UDFs (user-defined features) and local groups using the GROUP menu option **Pattern**. This option differs from the FEAT menu option **Pattern** in that the GROUP menu option treats an entire group as a single entity. The FEAT menu **Pattern** option is used to pattern one feature at a time.

You can select all the dimensions in the selected group (Fig. 7.6), except those used to create a feature pattern within the group, as incremental dimensions. When you create a patterned group, one member represents the whole group. When regenerating, however, Pro/E regenerates all the features individually.

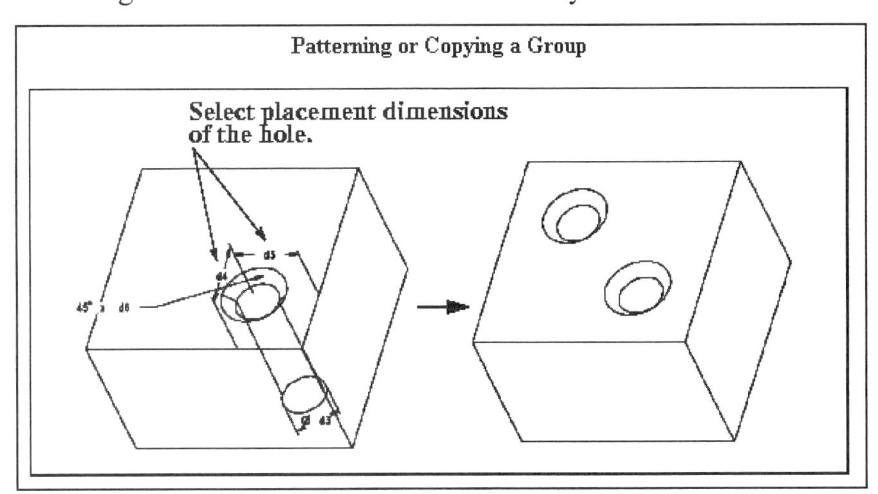

HINT

To pattern a group, you must first name and group two or more features into a local group.

Figure 7.7
Modifying Patterns

To pattern a group, do the following:

1. Choose **Pattern** from the GROUP menu.
2. Using the SELECT FEAT menu, select the group to be patterned.
3. Specify the variable dimensions, increments, and number of instances.

When you pattern or copy a group, be careful which placement dimensions you select to increment or vary. If a feature in a group references another for placement (for example, a chamfer references the edge of a hole), you need to change only the placement dimensions of the referenced feature.

If you place features in a group separately, you must change the placement dimensions of ***each member.*** Otherwise, features with unchanged dimensions will have several copies superimposed onto one another.

Patterns can be modified as shown in Figure 7.7. Table-driven patterns can also be created (Fig. 7.8).

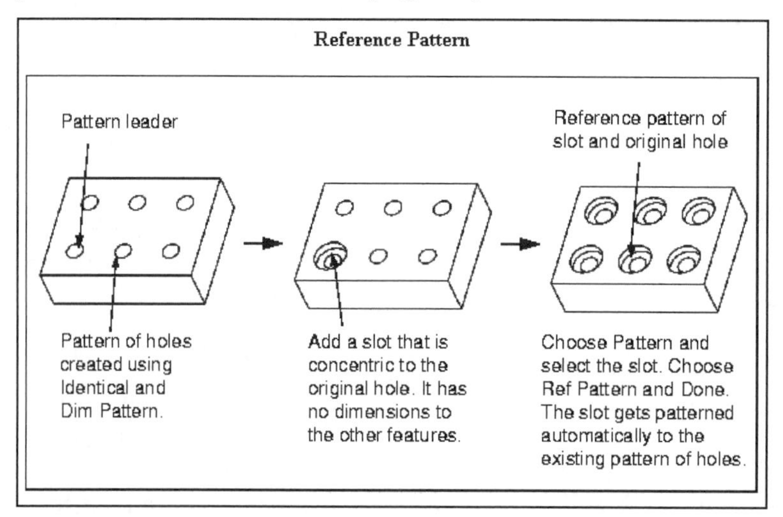

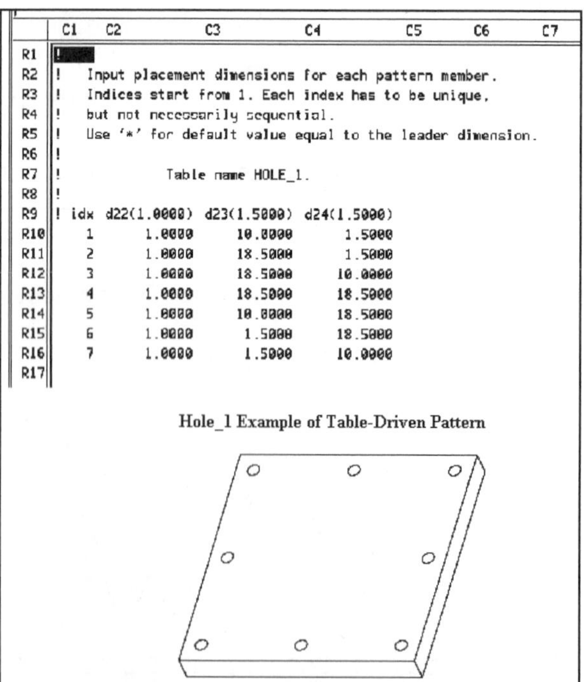

```
     C1   C2          C3          C4          C5     C6     C7
R1  !
R2  !   Input placement dimensions for each pattern member.
R3  !   Indices start from 1. Each index has to be unique,
R4  !   but not necessarily sequential.
R5  !   Use '*' for default value equal to the leader dimension.
R6  !
R7  !           Table name HOLE_1.
R8  !
R9  ! idx  d22(1.0000)  d23(1.5000)  d24(1.5000)
R10    1       1.0000      10.0000       1.5000
R11    2       1.0000      18.5000       1.5000
R12    3       1.0000      18.5000      10.0000
R13    4       1.0000      18.5000      18.5000
R14    5       1.0000      10.0000      18.5000
R15    6       1.0000       1.5000      18.5000
R16    7       1.0000       1.5000      10.0000
R17
```

Hole_1 Example of Table-Driven Pattern

Figure 7.8
Table-Driven Patterns

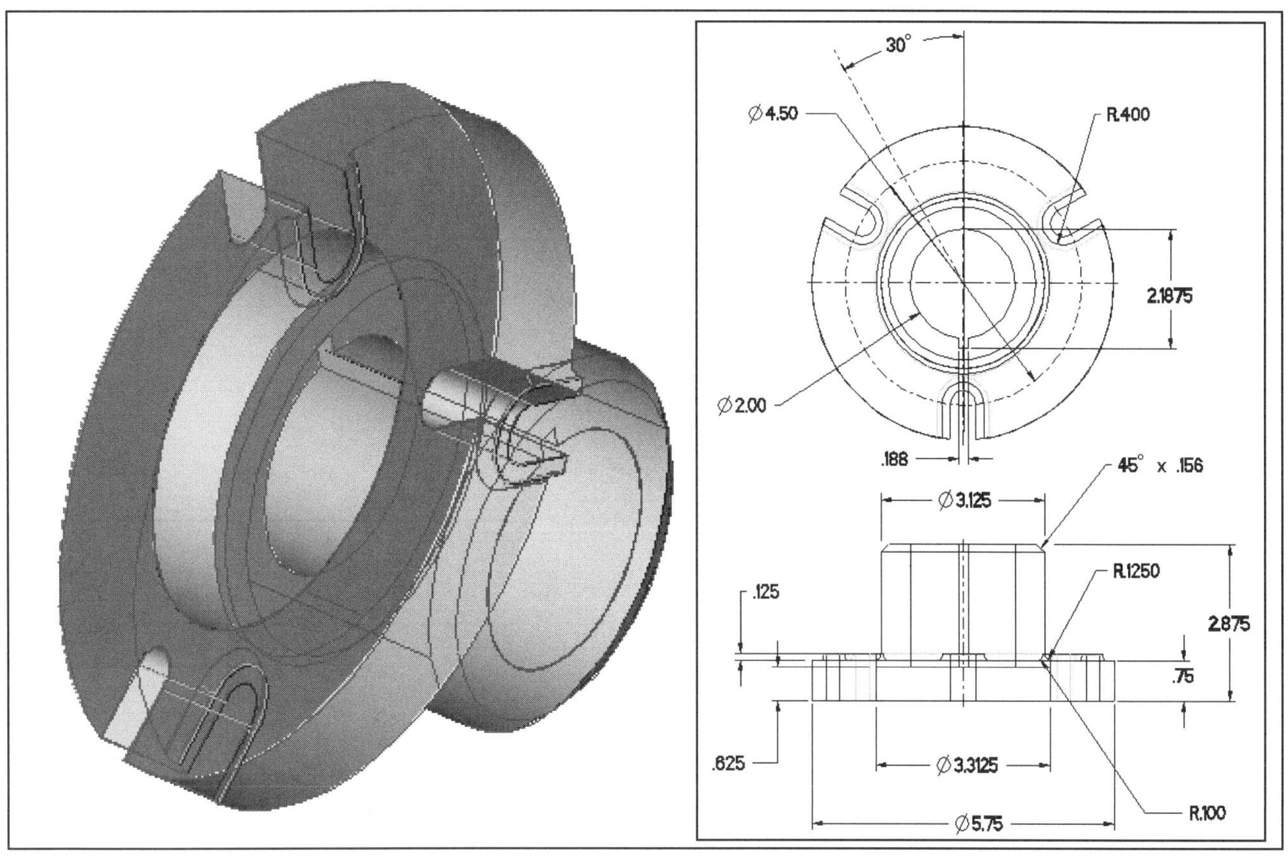

Figure 7.9
Post Reel Part Model and Detail Drawing

Lesson procedures & commands
START HERE ➡ ➡ ➡

Post Reel

The **Post Reel** (Fig. 7.9) is created with a *revolved protrusion*, as in Lesson 5 and Lesson 6. The internal geometry of the Post Reel can be created with a *revolved cut* instead of two holes of differing diameters or a sketched hole. The chamfers and the rounds are simple pick-and-place features. One boss and one slot (Fig. 7.10) are created using a *datum on the fly* (**Make Datum**), then grouped and patterned to complete the part. A detailed set of instructions will be supplied only for the boss and slot, because the other geometry is similar to that in previous lessons. The dimensions for the part are provided in Figures 7.9 through 7.12. Set up the part using the following commands:

File ⇒ **New** ⇒ ●**Part** ⇒ Name **POST_REEL** ⇒ ✔**Use default template** ⇒ **OK**

SETUP AND ENVIRONMENT

Set Up ⇒ **Units** ⇒ Units Manager **Inch lbm Second** ⇒ **Close** ⇒ **Material** ⇒ **Define** ⇒ (type **STEEL_1020**) ⇒ ✔ ⇒ (table of material properties, change or add information) ⇒ **File** ⇒ **Save** ⇒ **File** ⇒ **Exit** ⇒ **Assign** ⇒ (pick **STEEL _1020**) ⇒ **Accept** ⇒ **Done**
Utilities ⇒ **Environment** ⇒ ✔ **Snap to Grid**
❑ **Use 2D Sketcher** (sketch in 3D)
Display Style **Hidden Line** Tangent Edges **Dimmed** ⇒ **OK**
(Rename the default coordinate system = **POST_REEL)**

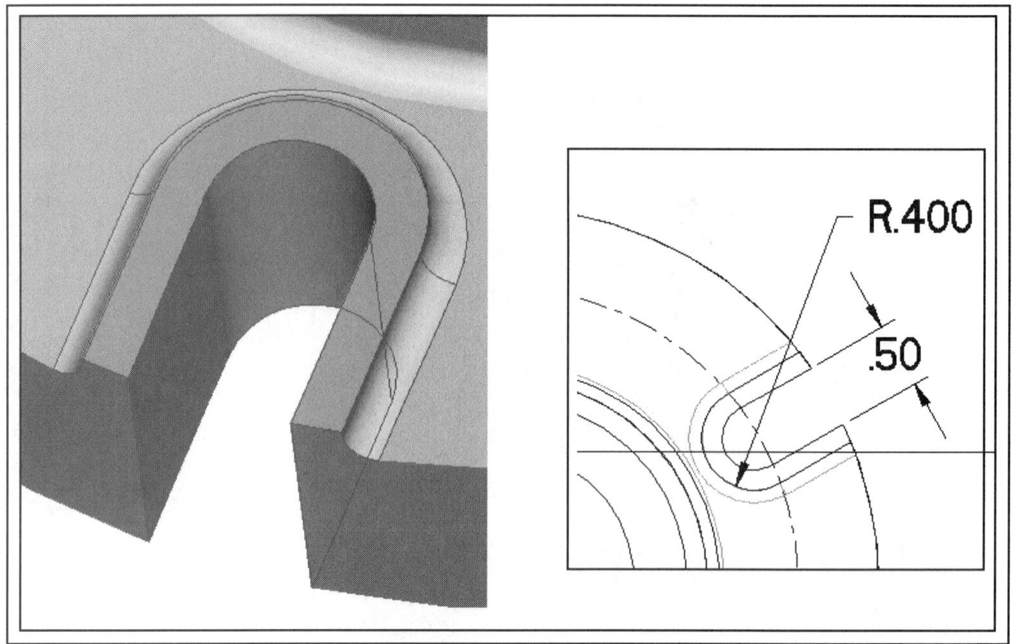

Figure 7.10
Slot and Boss

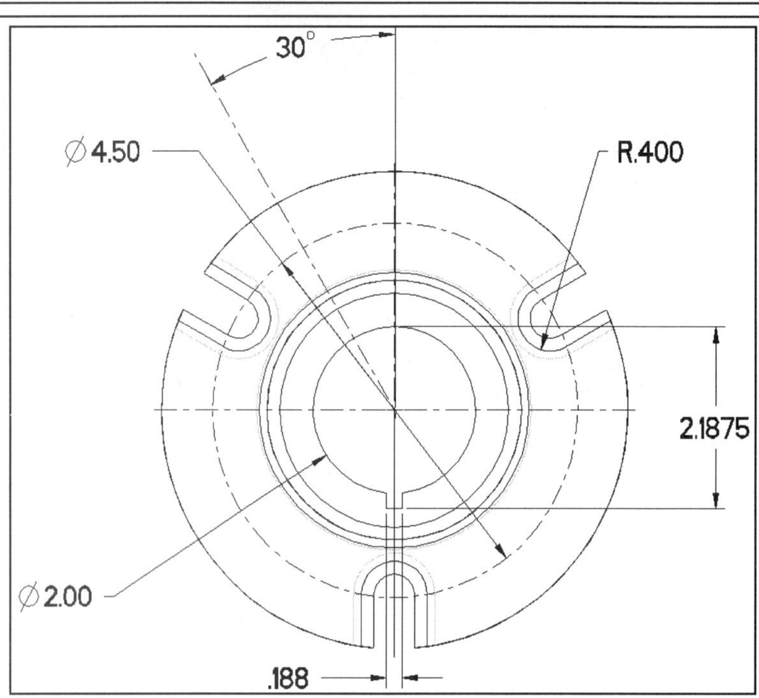

Figure 7.11
Post Reel Drawing,
Top View

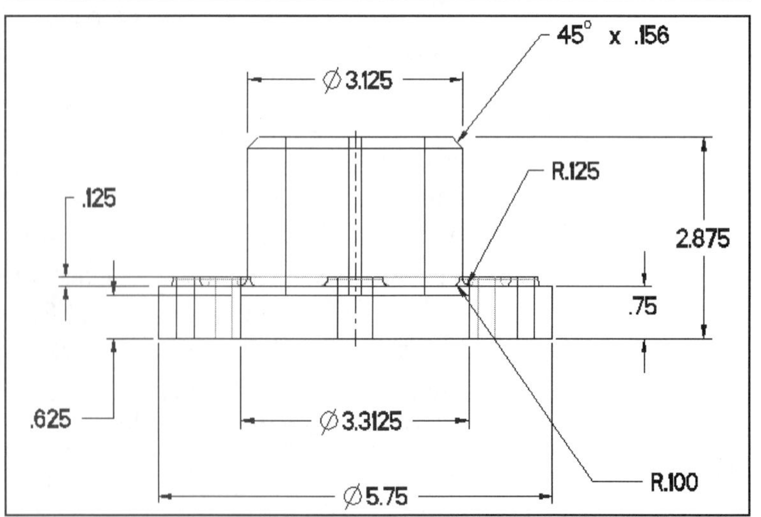

Figure 7.12
Post Reel Drawing,
Front View

HINT
Remember to create a palette of colors and use them on the model as needed.

The first protrusion is revolved and consists of the flange-like geometry shown in Figure 7.13. Sketch on **FRONT**, and revolve the section **360°** about a *vertical axis*.

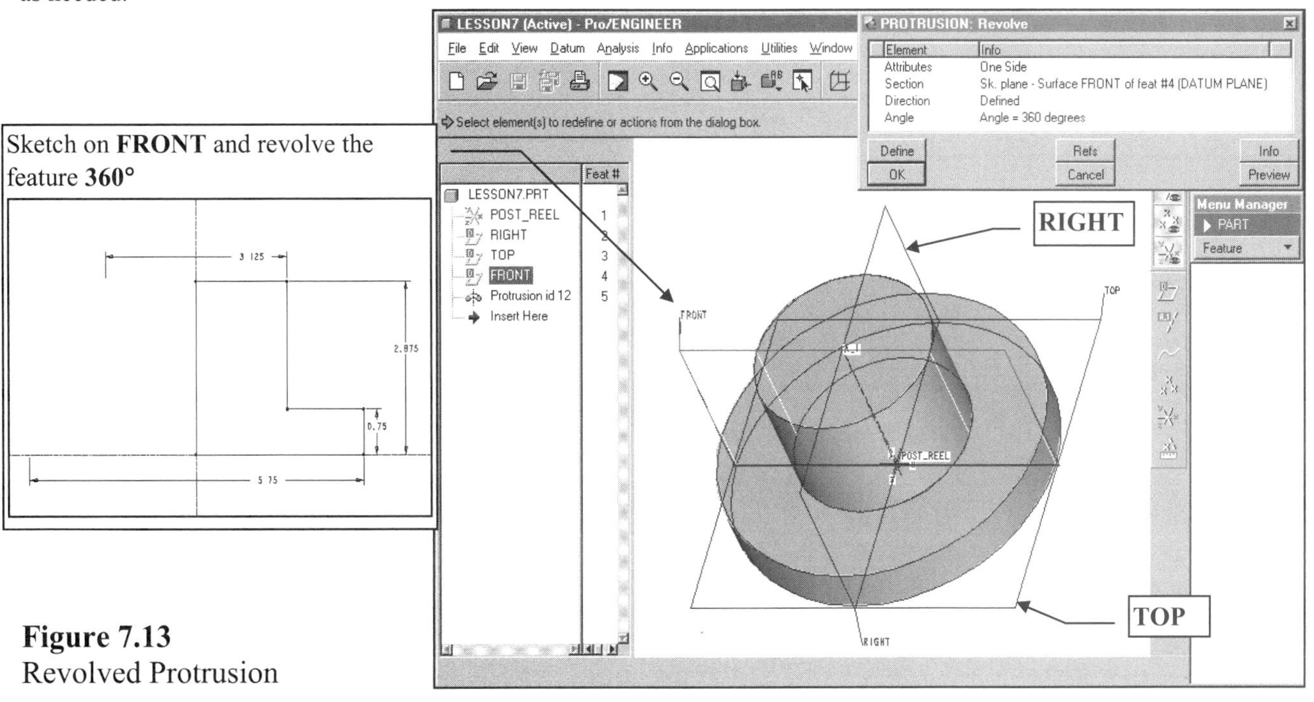

Figure 7.13
Revolved Protrusion

The second feature will be the internal cut shown in Figure 7.14. You can use the previous sketching plane and references.

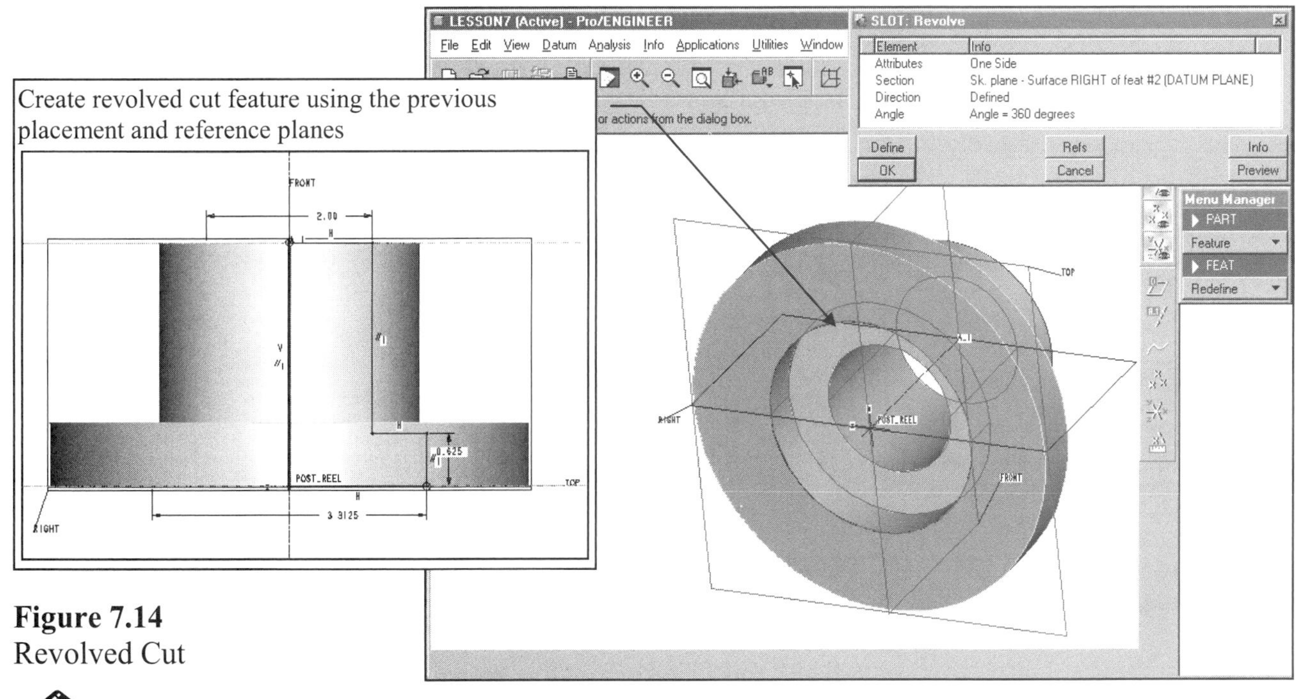

Figure 7.14
Revolved Cut

File ⇒ Save ⇒ ✔
File ⇒ Delete ⇒
Old Versions ⇒ ✔

The keyseat can be created with a cut from the end of the Post Reel using three lines sketched on **TOP** (or the top face), or it can be created with one line from the side, sketching on **RIGHT** and projecting to both sides (Fig. 7.15). Be careful to use the proper dimensioning scheme.

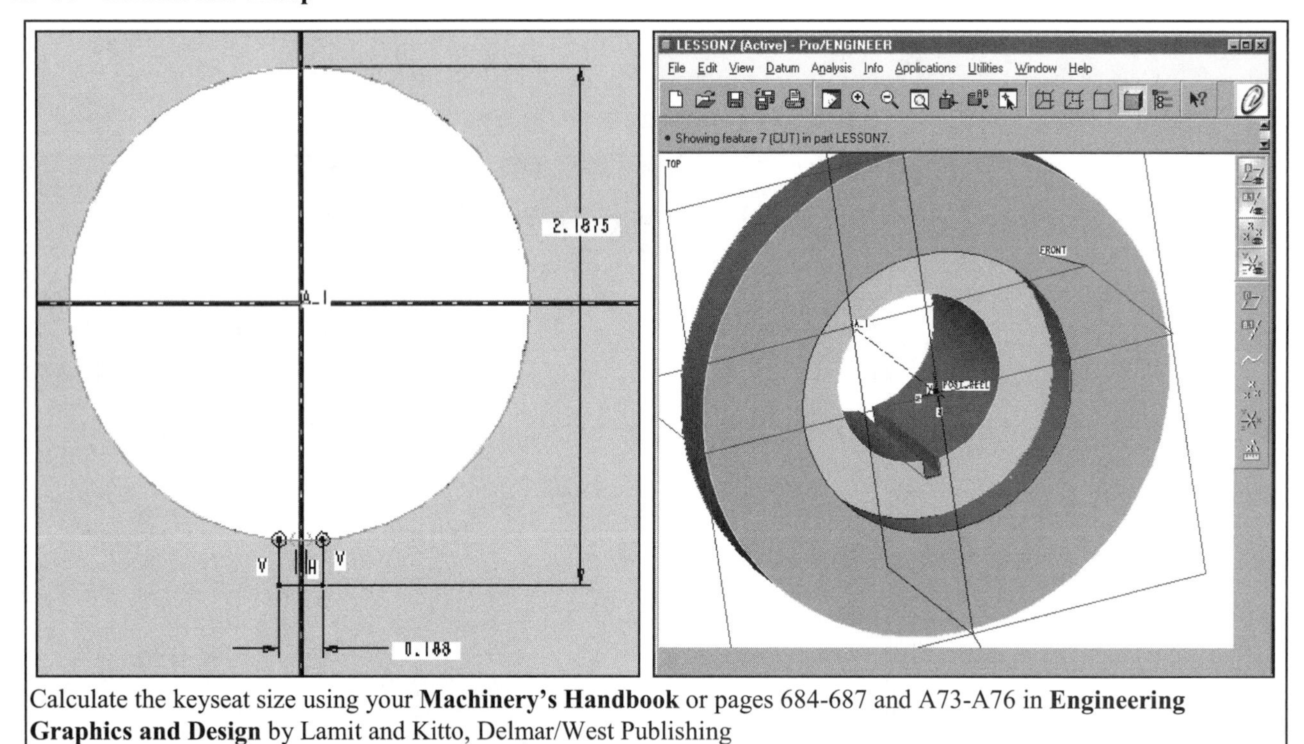

Calculate the keyseat size using your **Machinery's Handbook** or pages 684-687 and A73-A76 in **Engineering Graphics and Design** by Lamit and Kitto, Delmar/West Publishing

Figure 7.15
Keyseat

Keyseats are normally dimensioned by giving the width of the key slot and the distance between the top of the keyseat cut and the tangent edge of the shaft hole. The size of the keyseat is driven by the shaft size.

Time permitting, after you complete Lesson 8, return to the Post Reel part and create a relation that controls the relationship of the keyseat size as per any modified shaft size.

Add the **45° X .156** chamfer to the top of the Post Reel and the **R.125** round, as shown in Figure 7.16.

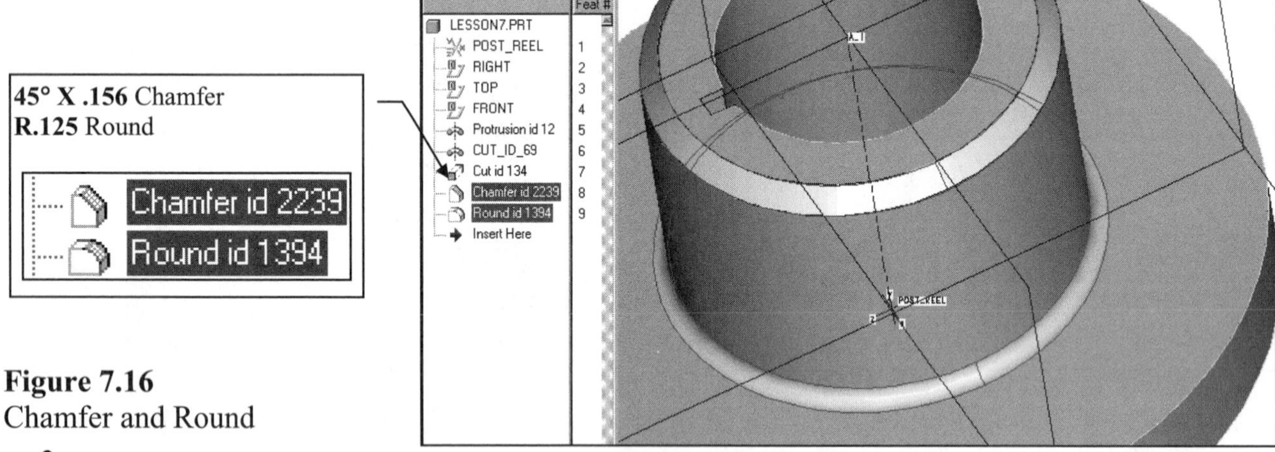

Figure 7.16
Chamfer and Round

File ⇒ Save ⇒ ✔
File ⇒ Delete ⇒
Old Versions ⇒ ✔

The final feature creation sequence consists of a boss protrusion, a slot cut, and a series of rounds. The three types of features are grouped, and the group is then patterned. The protrusion and the slot must be created with a *datum on the fly* (**Make Datum** within the SETUP SK PLN menu during feature creation). **Make Datum** is used here to provide the direction of rotation and is to orient the sketch plane for the feature sketch.

Create the boss protrusion using the following commands:

Feature ⇒ **Create** ⇒ **Protrusion** ⇒ **Extrude** ⇒ **Solid** ⇒ **Done** ⇒ **One Side** ⇒ **Done** ⇒ [pick top of the flange as the placement surface (Fig. 7.17)] ⇒ **Okay** (for the direction of feature creation) ⇒ **Bottom** ⇒ **Make Datum** ⇒ **Through** (pick axis **A_1**) ⇒ **Angle** (pick **FRONT**) ⇒ **Done** ⇒ **Enter Value** (type **30**) ⇒ ✔ (Fig. 7.18)

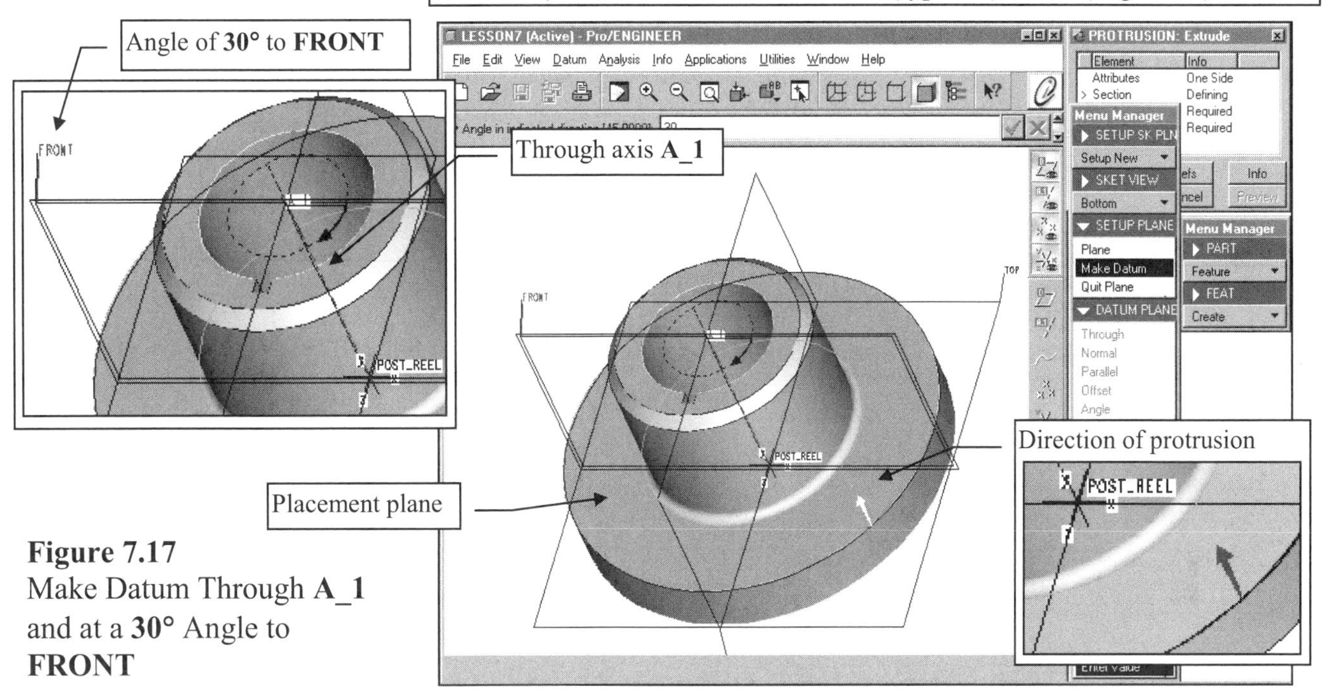

Figure 7.17
Make Datum Through **A_1** and at a **30°** Angle to **FRONT**

Turn on the grid snap and change the grid spacing:

Close (References) ⇒ **Yes** ⇒ **Sketch** ⇒ **Intent Manager** (off) ⇒ **Utilities** ⇒ **Environment** ⇒ ✔ **Snap to Grid** ⇒ **OK** ⇒ **Sec Tools** ⇒ **Sec Environ** ⇒ **Grid** ⇒ **Params** ⇒ **X&Y Spacing** ⇒ (type **.20** at the prompt) ⇒ ✔ ⇒ [zoom and pan the sketch (Fig. 7.19)]

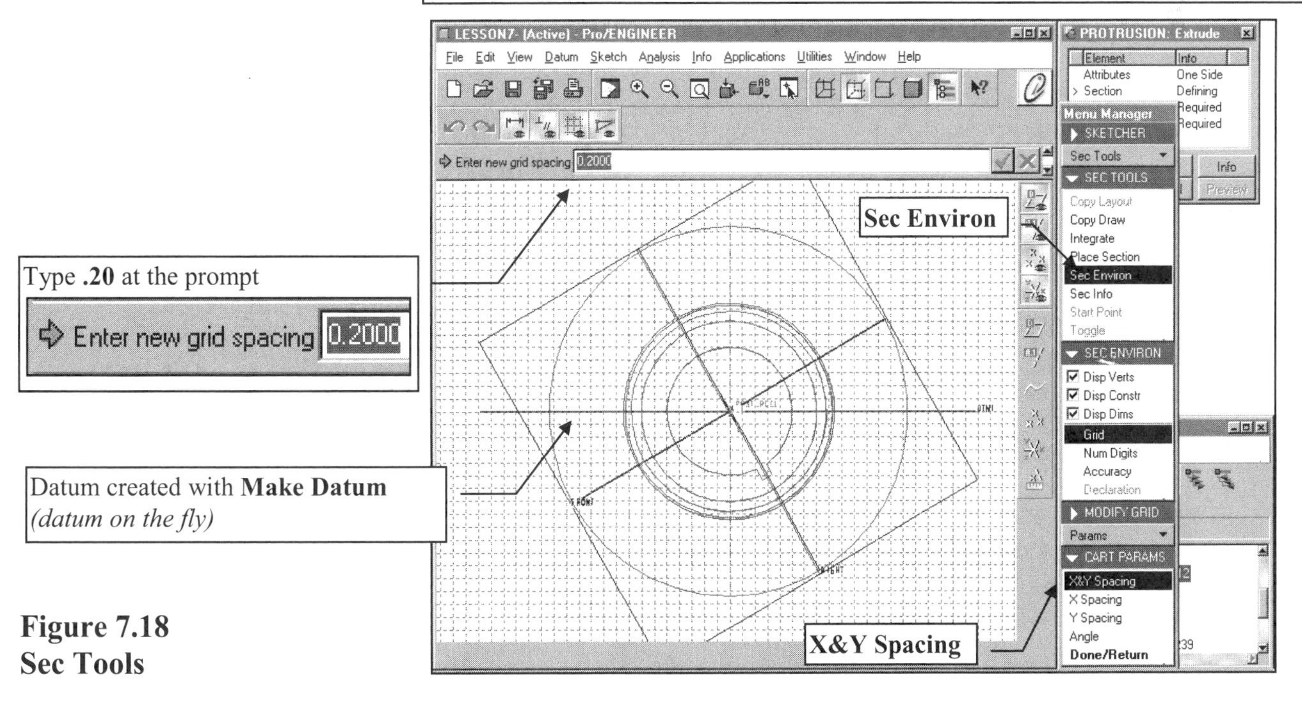

Figure 7.18
Sec Tools

Utilities ⇒ **Environment** ⇒ **✔Snap to Grid** ⇒ **Sketch** ⇒ [sketch the two horizontal lines and the two arcs *(the section must be closed)]* ⇒ [Create a construction circle to locate the center of the boss arc (Fig. 7.19)] **Sketch** ⇒ **Circle** ⇒ **Construction** ⇒ **Concentric** ⇒ (pick any of the circles composing the part and then pick the center of the boss arc twice) ⇒ **Regenerate** ⇒ **Alignment** [align the endpoints of the lines to the flange curve and the centerpoint of the boss arc to **DTM1** *(the datum on the fly).* Align the arc to the outside circle of the part and align the construction circle with the center of the part (**POST_REEL**)] ⇒ **Regenerate** ⇒ **Dimension** (the boss protrusion with an arc radius. Dimension the diameter of the construction circle.) ⇒ **Regenerate**

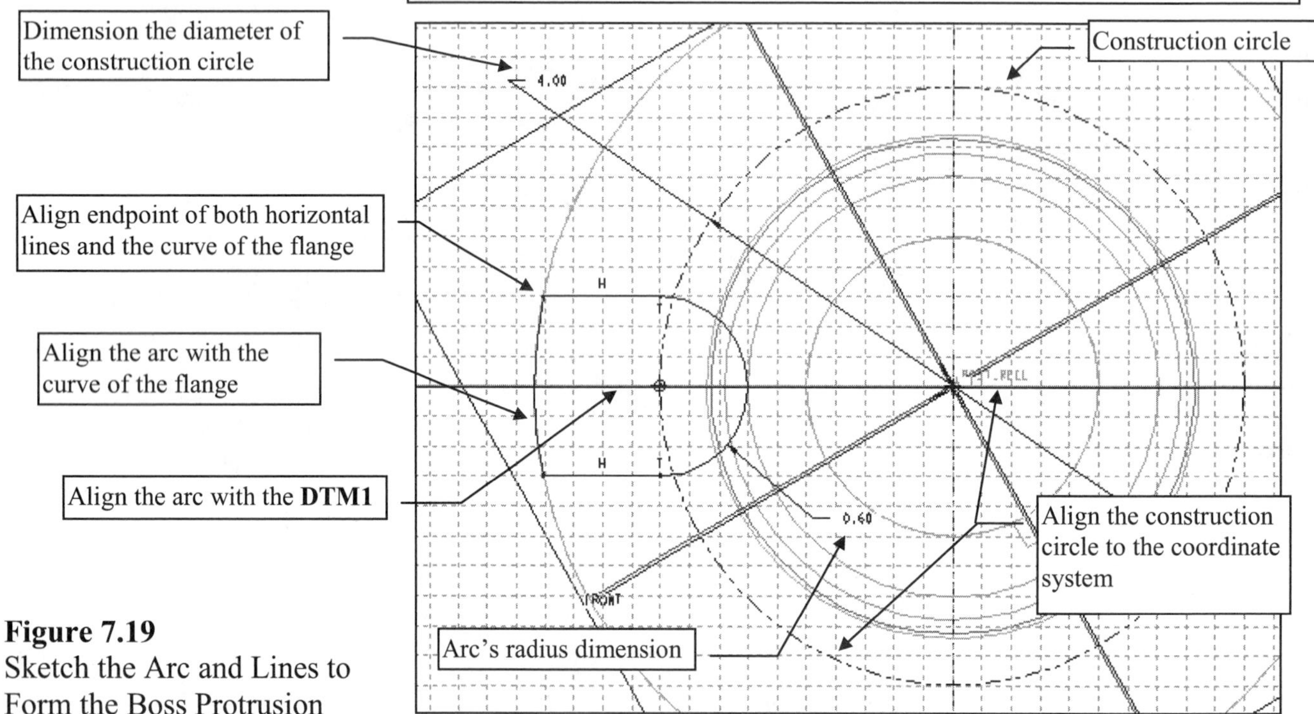

Figure 7.19
Sketch the Arc and Lines to Form the Boss Protrusion

If *the sketch does not regenerate, add a horizontal dimension from the center of the boss's arc to the center of the part, or add a sketch point at the intersection of the construction circle and the boss arc's center (Fig. 7.20).*

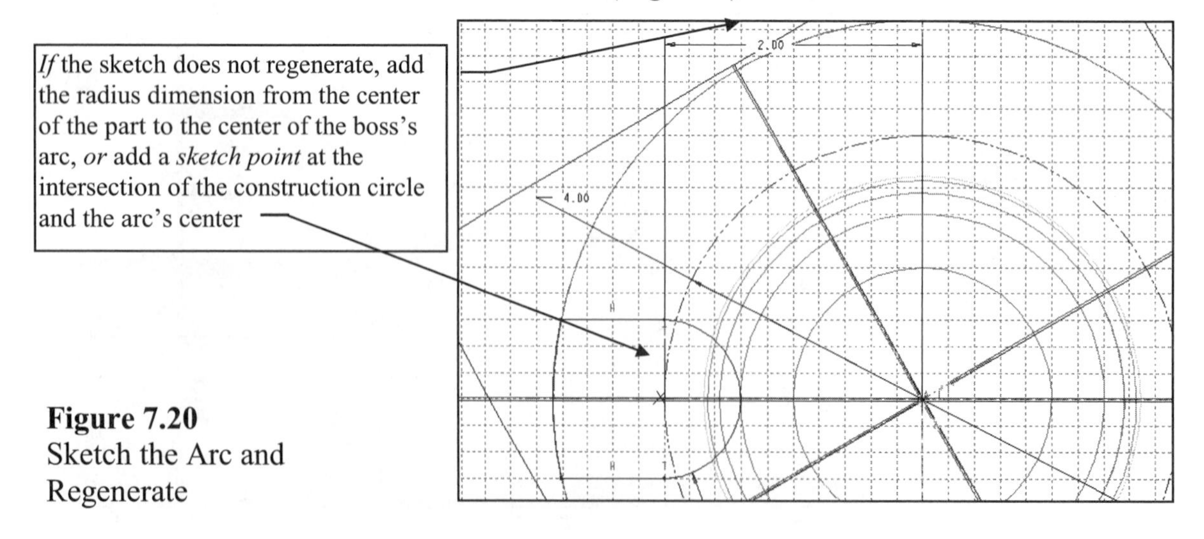

Figure 7.20
Sketch the Arc and Regenerate

After the sketch regenerates successfully, **Modify** the dimensions and **Regenerate** the sketch as shown in Figure 7.21.

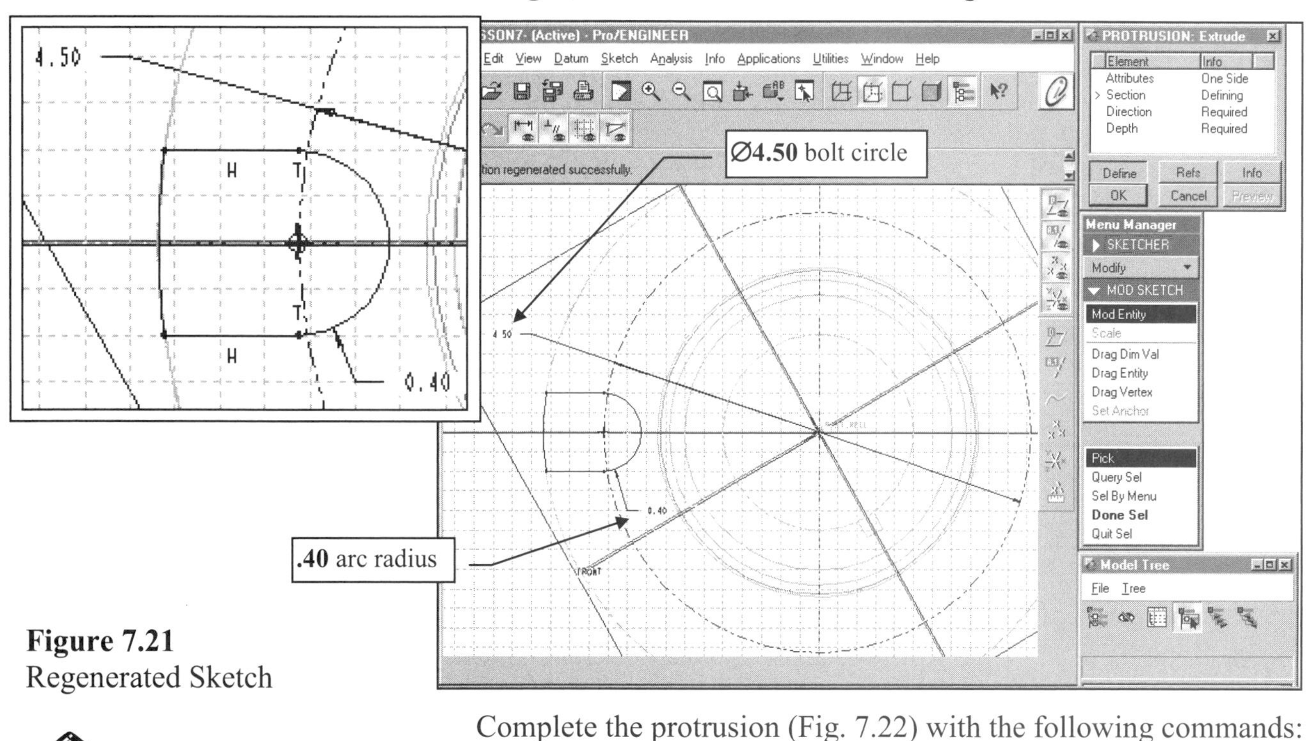

Figure 7.21
Regenerated Sketch

Complete the protrusion (Fig. 7.22) with the following commands:

Done ⇒ **Blind** ⇒ **Done** ⇒ (enter the depth of **.125**) ⇒ ✔ ⇒ (spin the model as in Figure 7.22, **Ctrl** key and the middle mouse button) ⇒ **Preview** ⇒ **OK** ⇒ **Done**

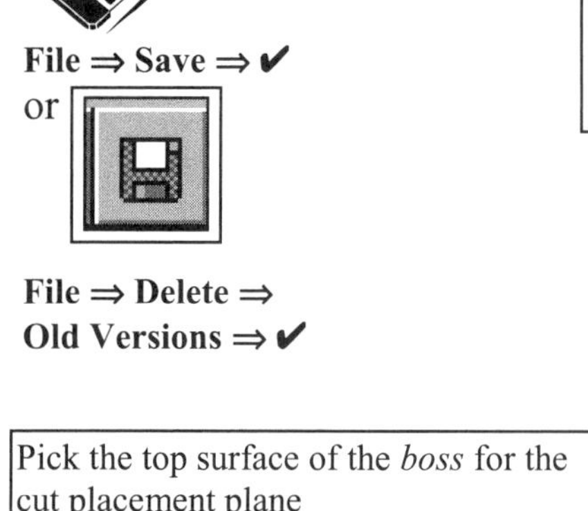

File ⇒ **Save** ⇒ ✔
or

File ⇒ **Delete** ⇒ **Old Versions** ⇒ ✔

Pick the top surface of the *boss* for the cut placement plane

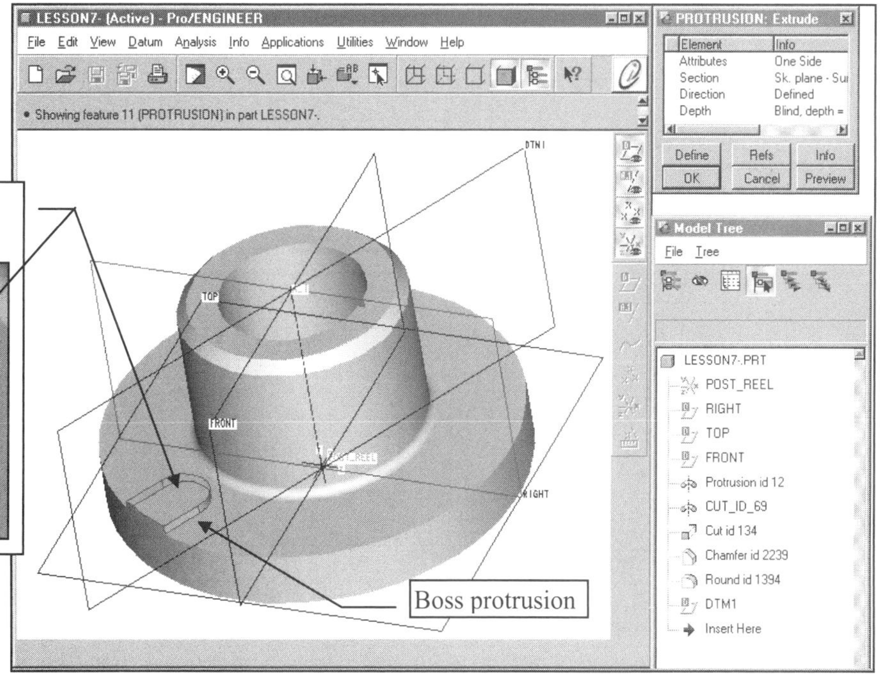

Figure 7.22
Regenerated Sketch

The next feature is a cut through the boss. The boss's surface will be the placement plane (Fig. 7.22). *The reference/orientation plane is again created using* **Make Datum.** Choose the following commands:

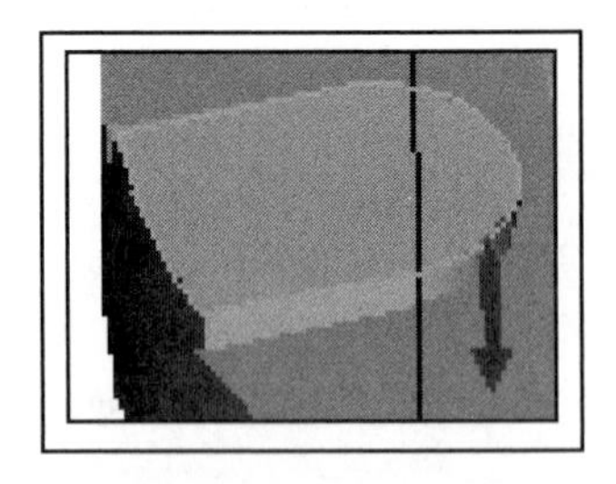

Feature ⇒ Create ⇒ Cut ⇒ Done ⇒ One Side ⇒ Done ⇒ (pick the top of the Boss's surface; see Fig. 7.22) **⇒ Okay ⇒ Bottom ⇒ Make Datum ⇒ Through** (pick axis **A_1**) **⇒ Angle** (pick **FRONT**) **⇒ Done ⇒ Enter Value** (type **30** at the prompt) ⇒ ✔ (Fig. 7.23)

Sketch the lines and arc using the GEOM TOOLS option **Offset Edge** (Fig. 7.23 and Fig. 7.24):

Close (References) **⇒ Yes ⇒ Sketch ⇒ Intent Manager** (off) **Utilities ⇒ Environment ⇒ ❑ Snap to Grid ⇒ OK ⇒ Geom Tools ⇒ Offset Edge ⇒** (pick the lower line) **⇒ Done/Return ⇒** (type **-.15** as the offset distance) ⇒ ✔ ⇒ (pick the arc) **⇒ Done/Return ⇒** (type **-.15** as the offset distance) ⇒ ✔ ⇒ (pick the upper line) **⇒ Done/Return ⇒** (type **-.15** as the offset distance) ⇒ ✔

HINT

Type a *plus* **.15** or a *minus* **.15** based on the direction of the arrow showing the side on which the geometry will be created. Draw the lines and the arc to the inside of the boss.

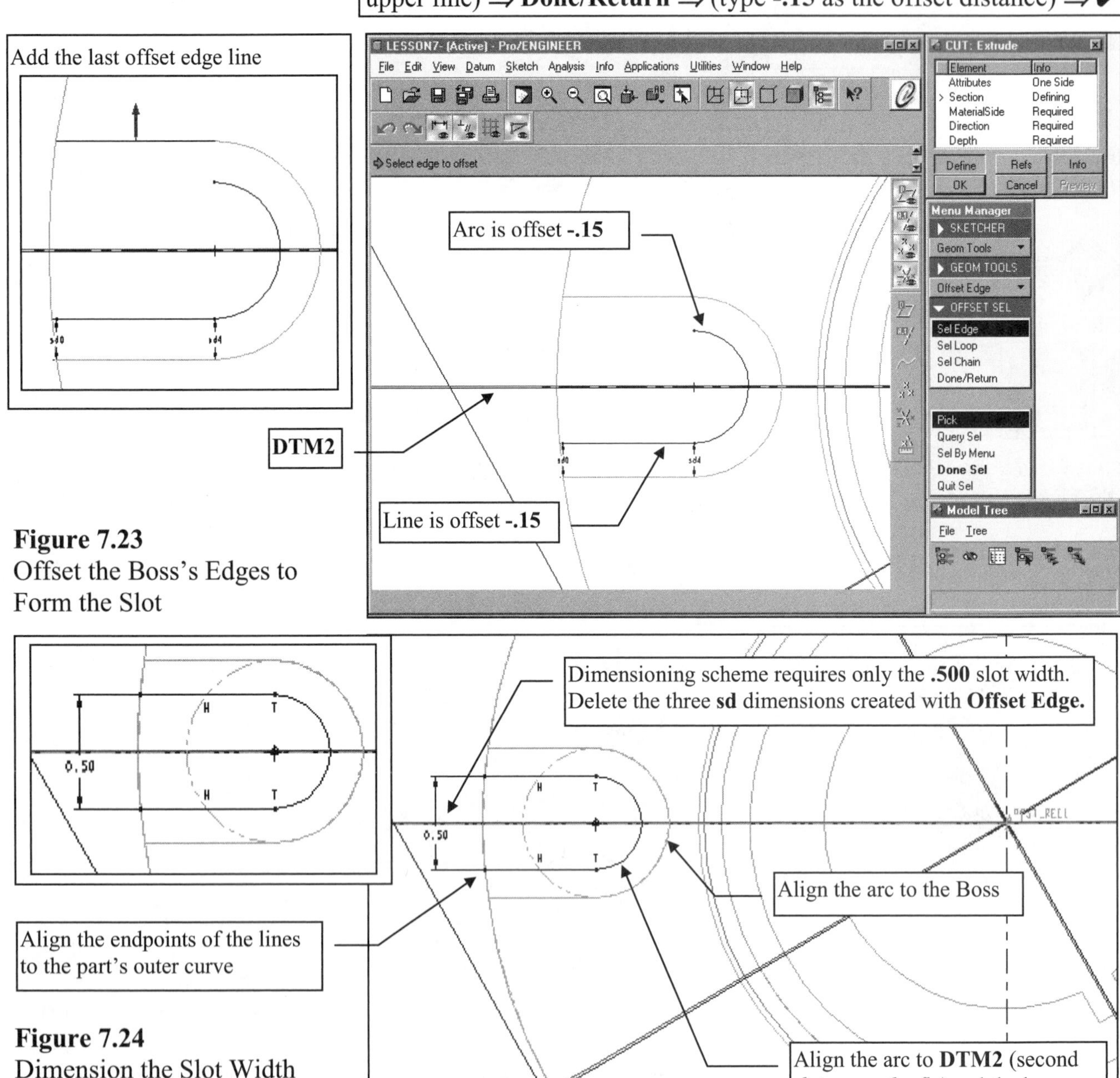

Figure 7.23
Offset the Boss's Edges to Form the Slot

Figure 7.24
Dimension the Slot Width **(.500)**

Delete all three offset dimensions (**sd**) and repaint the screen to see the slot sketch geometry (Figs. 7.24 and 7.25). Align the endpoints of the horizontal lines to the outer curve of the part. Align the arc to the boss arc and to **DTM2.** **Regenerate** and add the dimension for the slot width. **Modify** the width (**.500**). *There is no need to modify the value if it was created at the correct design size.*

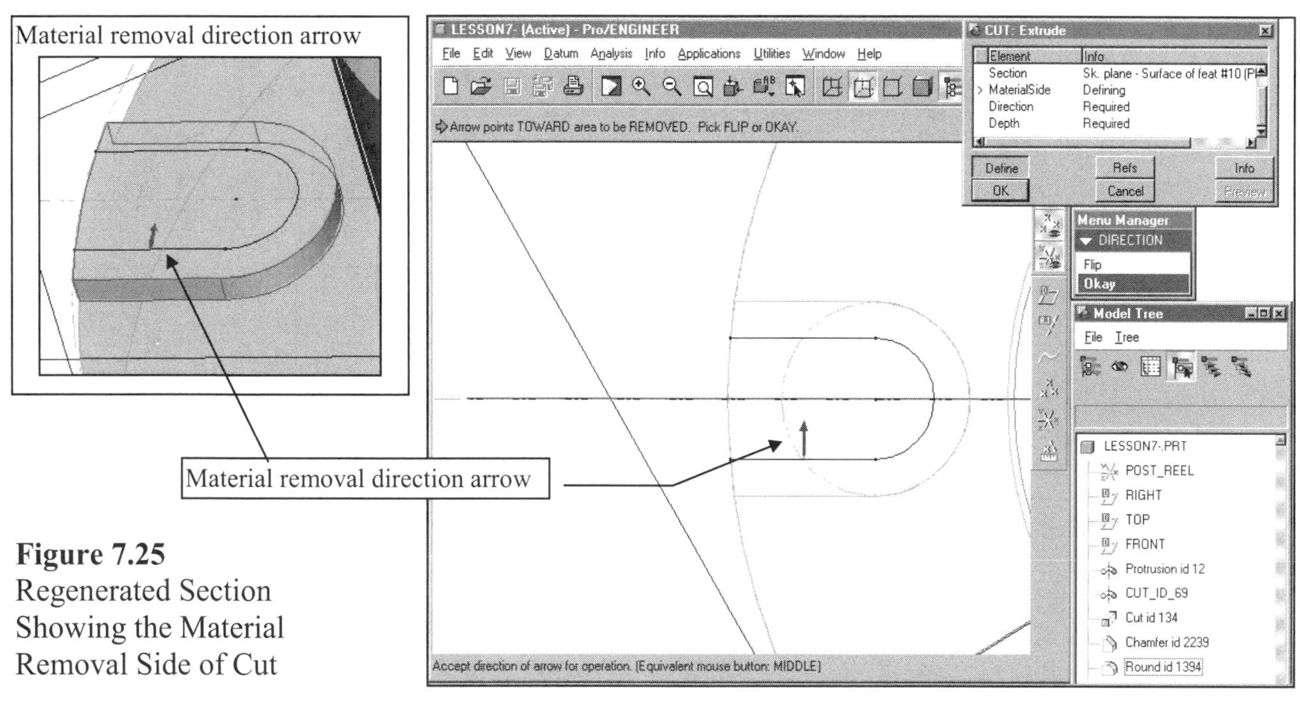

Figure 7.25
Regenerated Section Showing the Material Removal Side of Cut

Regenerate the sketch, choose **Done**, and choose **Okay** for the material removal direction. Make sure the direction of cut is correct. The depth is **Thru All**. Change the sketch orientation by rotating or using the default view. *If* the cut does not intersect the part, pick **Direction** and **Define** from the dialog box and flip the arrow; then choose **OK** and **Done** (Fig. 7.26).

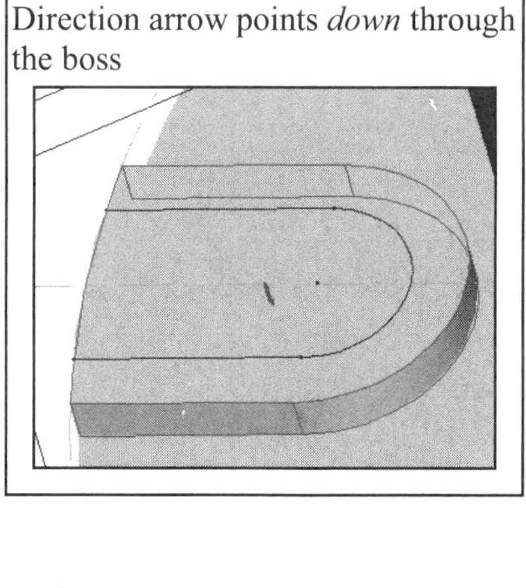

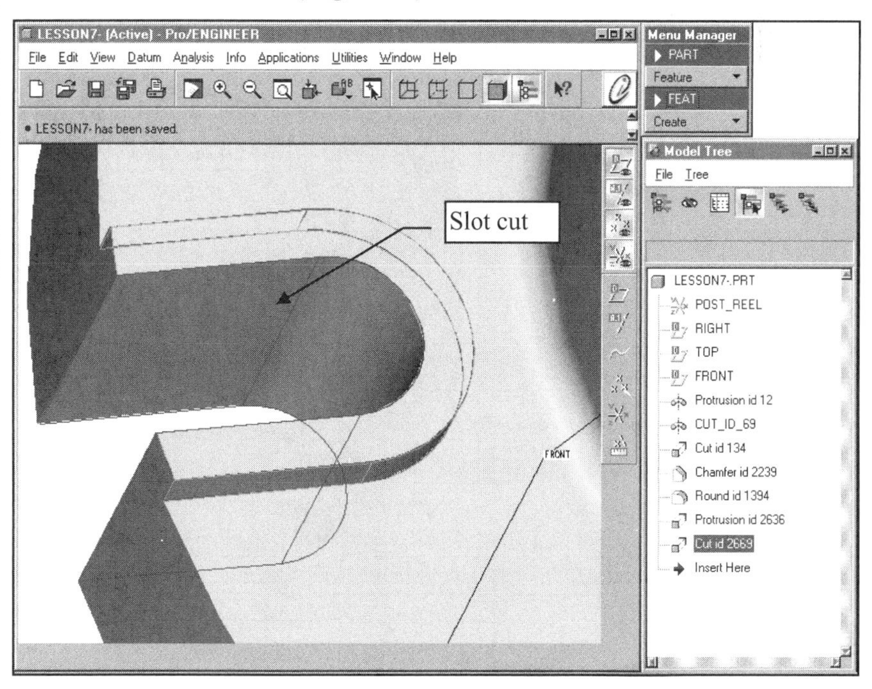

File ⇒ Save ⇒ ✔
File ⇒ Delete ⇒
Old Versions ⇒ ✔

Figure 7.26
Slot Cut

Create a set of rounds around the boss, as shown in Figure 7.27:

Feature ⇒ **Create** ⇒ **Round** ⇒ **Simple** ⇒ **Done** ⇒ **Done** ⇒ **Tangnt Chain** ⇒ (pick the line at the edge of the boss; both lines and the arc will highlight) ⇒ **Done Sel** ⇒ **Done** ⇒ (type the radius value of **.10** at the prompt) ⇒ ✔ ⇒ **Preview** ⇒ **OK** ⇒ **Done**

Pick the *line* at the edge of the boss; both lines and the arc will highlight

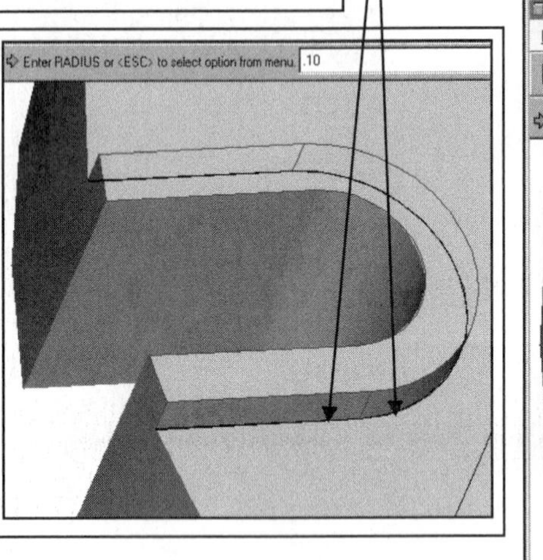

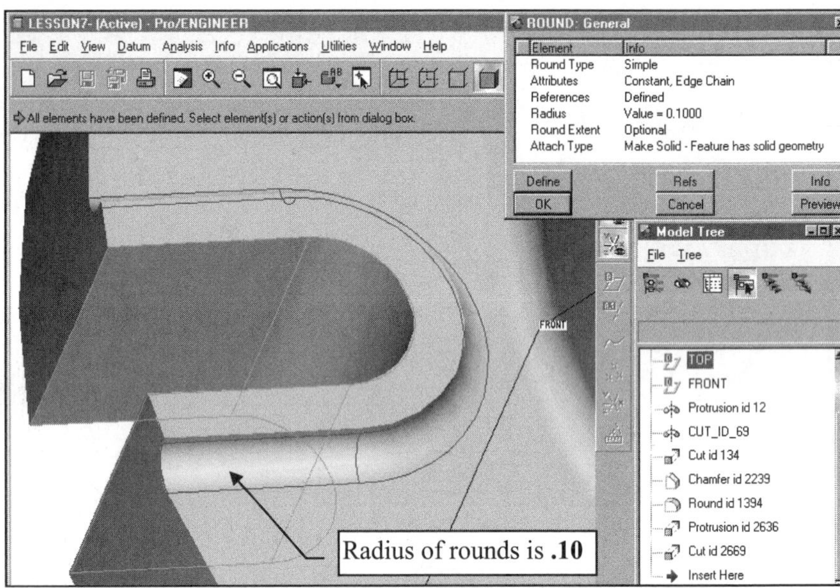

Figure 7.27
Rounds

Next, group (Fig. 7.28) the three features together using the following commands:

If you created a Mapkey for the group command, use it!

GRP
PAT
creates a group

Feature ⇒ **Group** ⇒ **Create** ⇒ **Local Group** ⇒ (enter the group name at the prompt: **BOSS**) ⇒ ✔ ⇒ (select the features for the group: the *boss protrusion*, the *slot*, and the *rounds*) ⇒ **Done Sel** ⇒ **Done** ⇒ (Pro/E will respond with "**Group BOSS has been created**") ⇒ **Done/Return** ⇒ **Done**

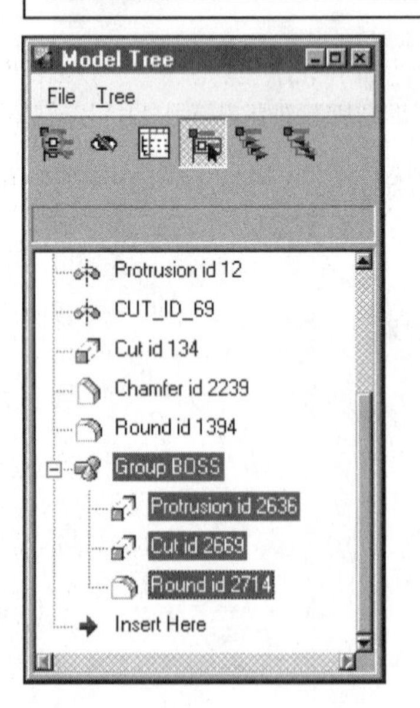

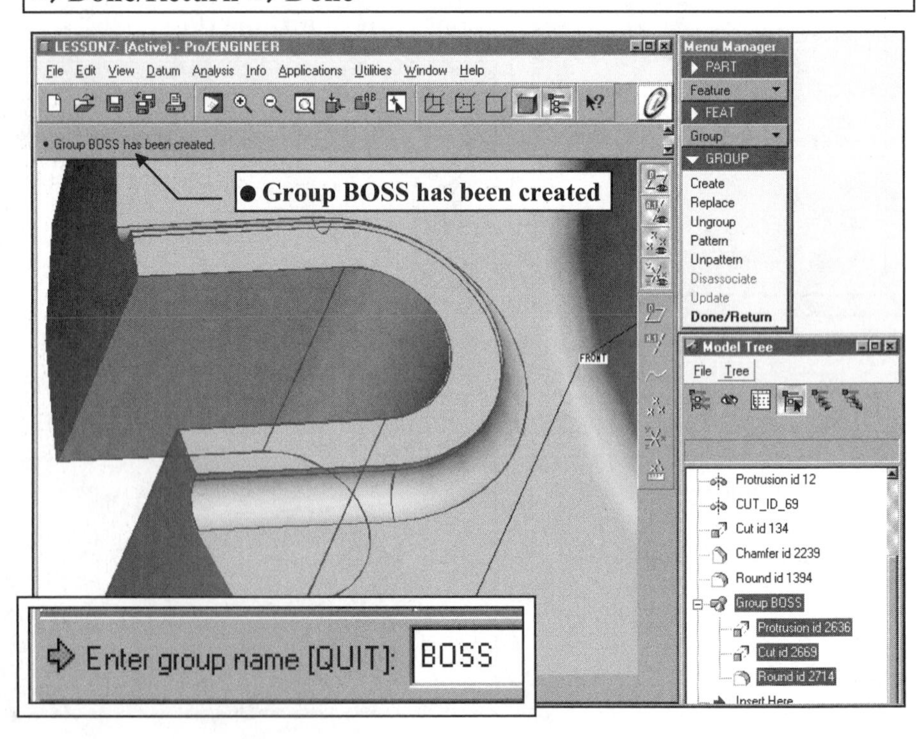

Figure 7.28
Group

File ⇒ Save ⇒ ✔
File ⇒ Delete ⇒
Old Versions ⇒ ✔

Because the dimensions for the features that you will be using to pattern the group are the same for the angle of the boss protrusion and for the angle of the cut, you need to move one of the **30°** dimensions. This must be done before the group is patterned (Fig. 7.29). Use the following commands:

Modify ⇒ (pick the boss and the cut) ⇒ **DimCosmetics** ⇒ **Move Dim** ⇒ (pick the original position of the **30°** angle) ⇒ (pick the new position for the angle dimension) ⇒ **Done Sel** ⇒ **Done** ⇒ **Done** ⇒ **View** ⇒ **Repaint**

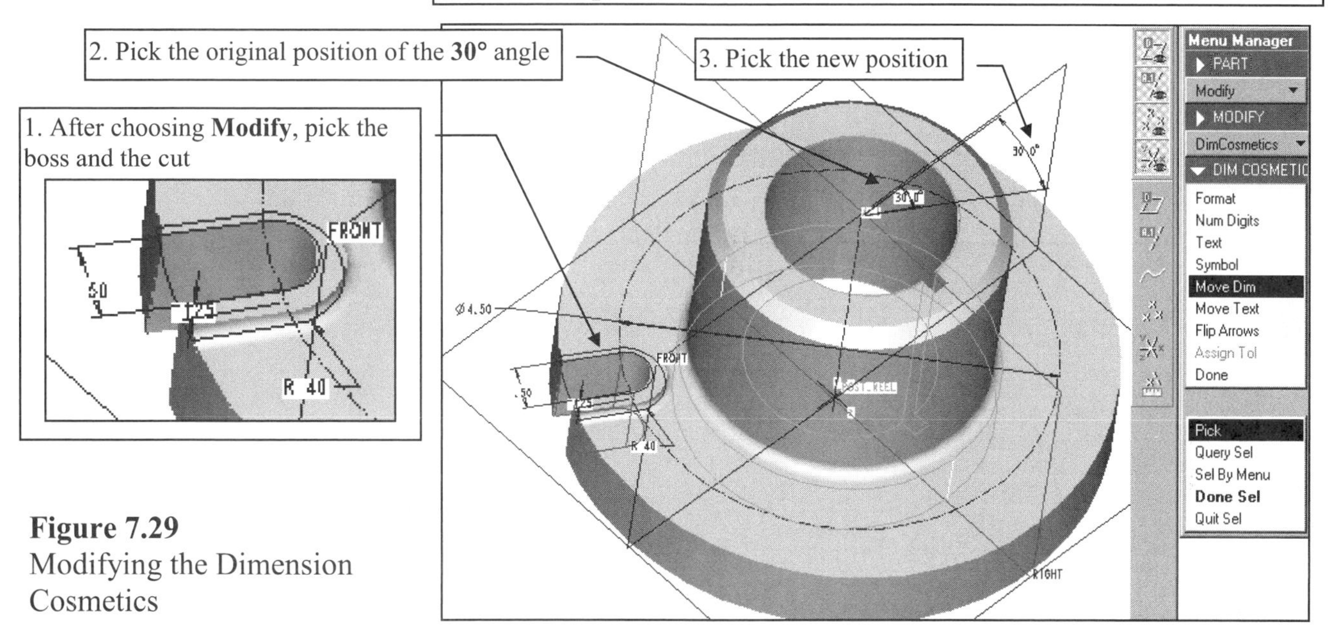

Figure 7.29
Modifying the Dimension Cosmetics

The BOSS group now needs to be patterned. To pattern a group, you need to choose **Pattern** from the GROUP menu choices, not from the FEAT menu. Choose the following commands (Fig. 7.30):

Feature ⇒ **Group** ⇒ **Pattern** ⇒ (select the Group **BOSS** from the Model Tree; notice that *two* **30°** angle dimensions show on the screen, one for the slot and another for the protrusion)

Menu Manager
PART
Feature
FEAT
Group
GROUP
Create
Replace
Ungroup
Pattern
Unpattern
Disassociate
Update
Done/Return
SELECT FEAT
Select
GET SELECT
Pick

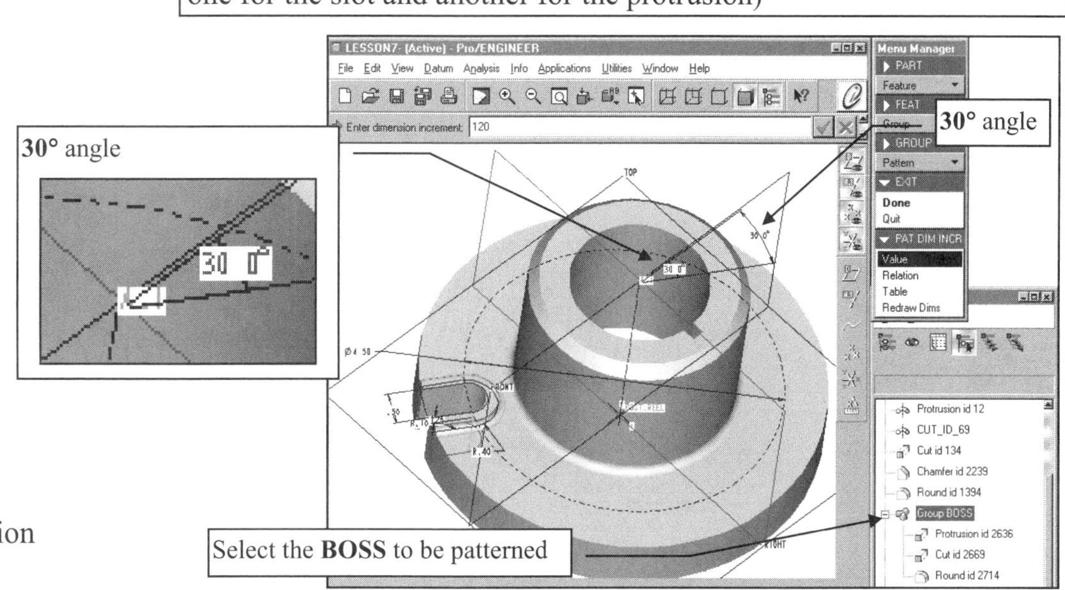

Figure 7.30
Modified Dimension Cosmetics

Pro/E will respond with **"Select pattern dimensions for FIRST direction, or increment type."** Continue with the following commands (Fig. 7.30):

(pick one of the **30°** angles) ⇒ (type **120**) ⇒ ✔ ⇒ (pick the other **30°** angle) ⇒ (type **120**) ⇒ ✔⇒ **Done** ⇒ (Pro/E responds with **"Enter TOTAL number of instances in this direction (including original):"**) ⇒ (type **3**) ⇒ ✔ ⇒ **Done** (Figs. 7.31 and 7.32) ⇒ **Done/Return** ⇒ **Done** ⇒ **File** ⇒ **Save** ⇒ ✔ ⇒ **File** ⇒ **Delete** ⇒ **Old Versions** ⇒ ✔

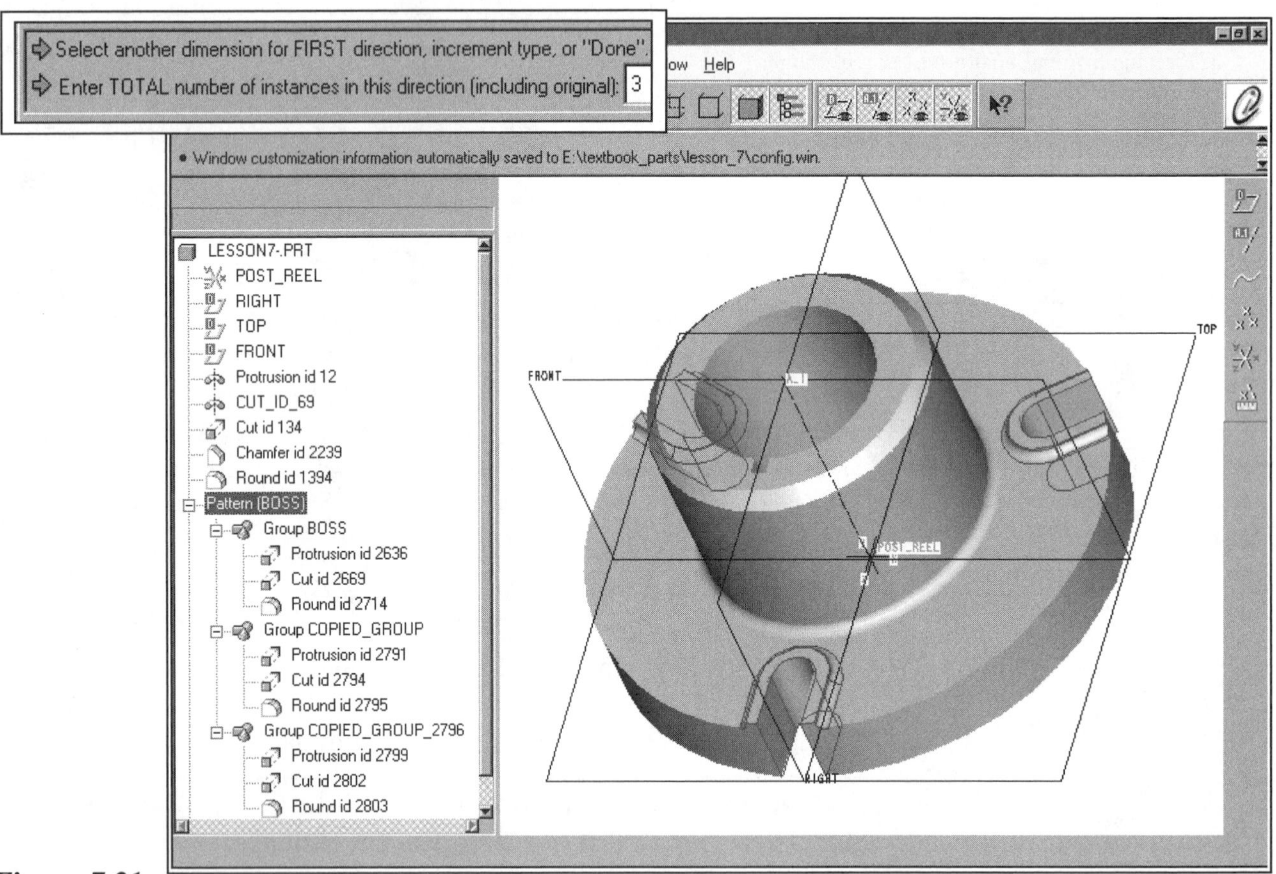

Figure 7.31
Patterning the Group

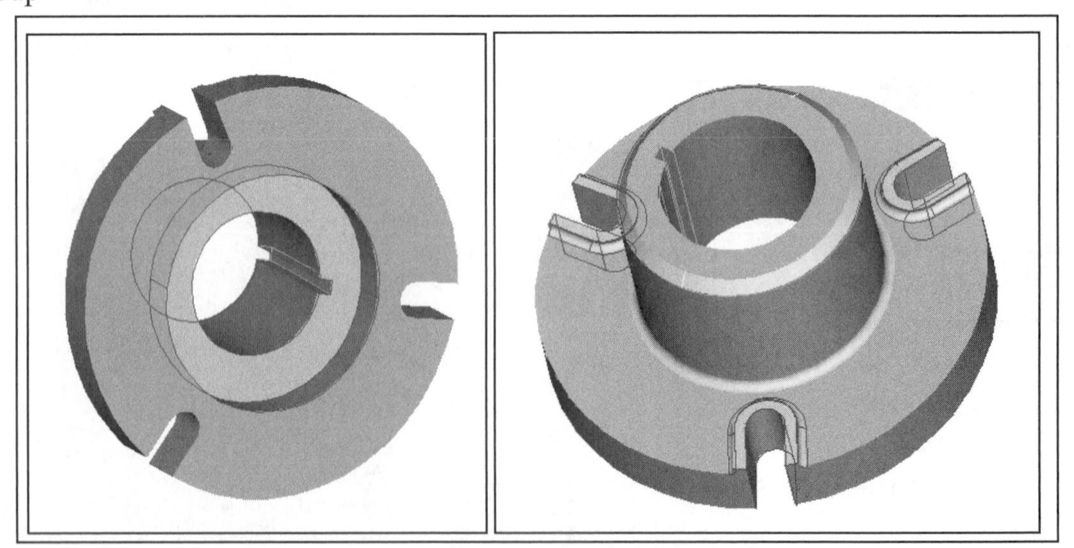

Figure 7.32
Post Reel

Lesson 7 Project

Taper Coupling

EGD REFERENCE
Fundamentals of Engineering Graphics and Design
by L. Lamit and K. Kitto
Read Chapter 23.
See pages 865-866.

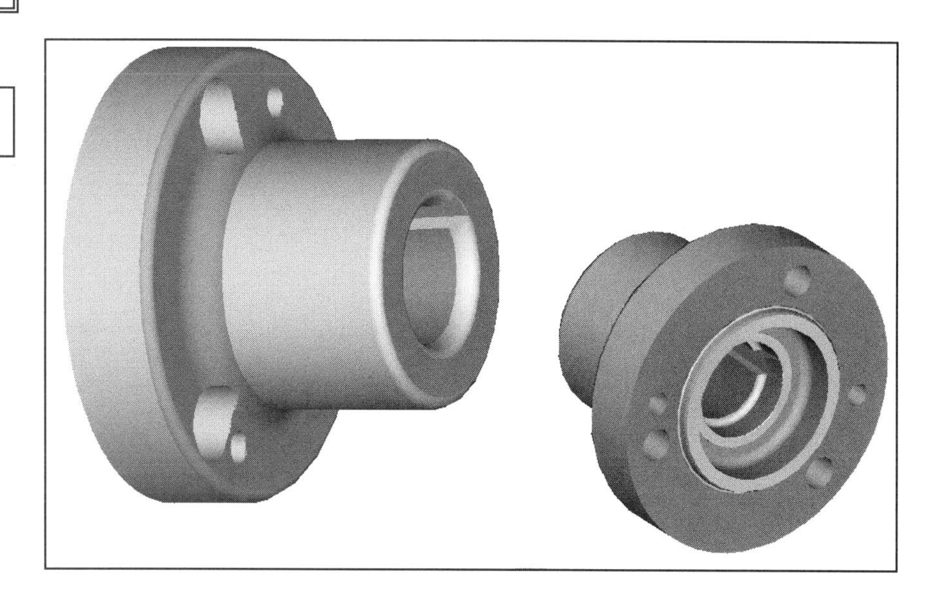

Figure 7.33
Taper Coupling

Taper Coupling

The seventh **lesson project** is a machined part that requires commands similar to the **Post Reel**. Create the part shown in Figures 7.33 through 7.47. This part will be used in Lessons 14 and 15.

At this stage in your understanding of Pro/E, you should be able to analyze the part and plan out the steps and features required to model it. Use the **DIPS** in Appendix D to plan out the feature creation and the parent-child relationships for the part. The coupling will be used in an assembly in Lessons 14 and 15. The machined face of the coupling mates with and is fastened to a similar surface when assembled. Plan your geometric tolerancing requirements accordingly, and set the datums to anticipate the mating surfaces.

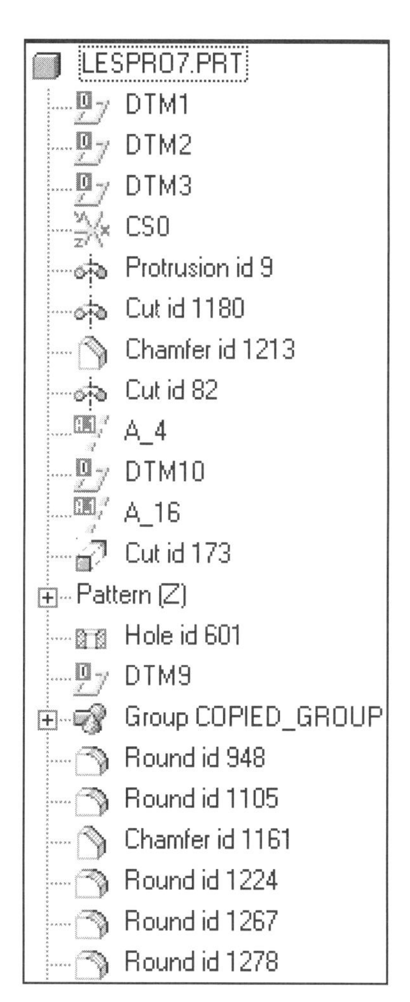

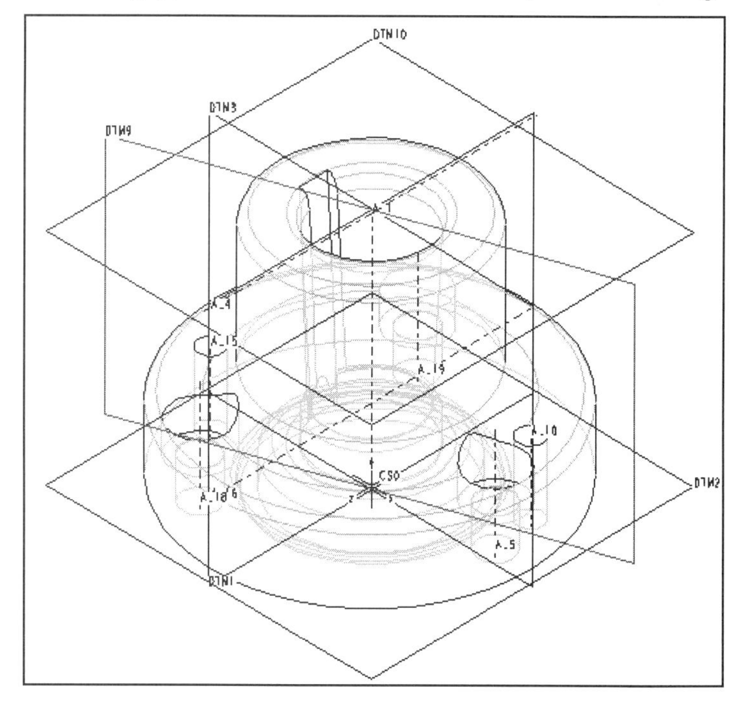

Figure 7.34
Taper Coupling Model with Datum Planes

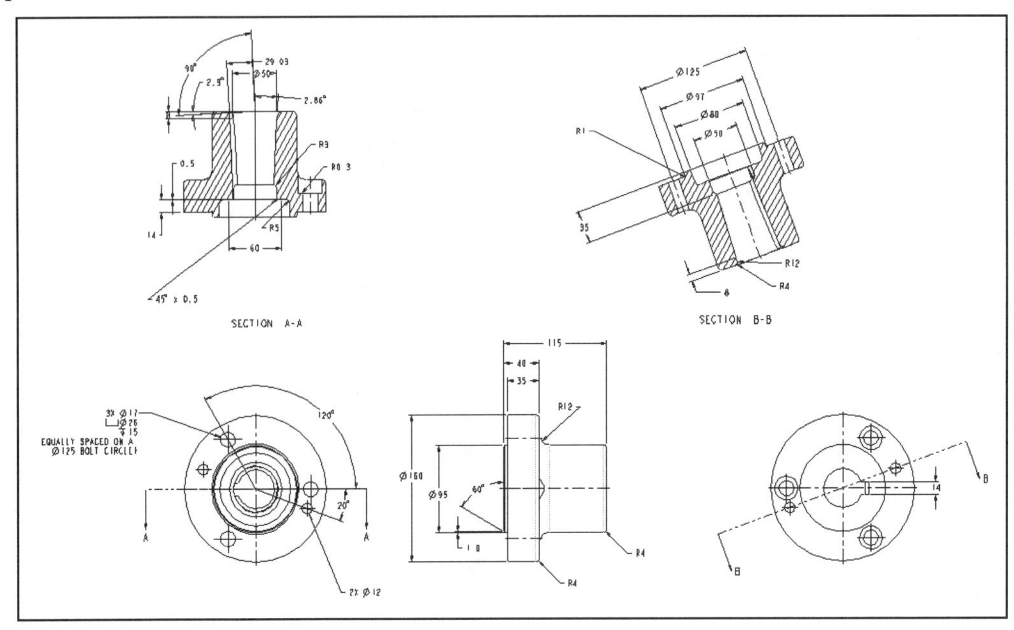

Figure 7.35
Taper Coupling Drawing

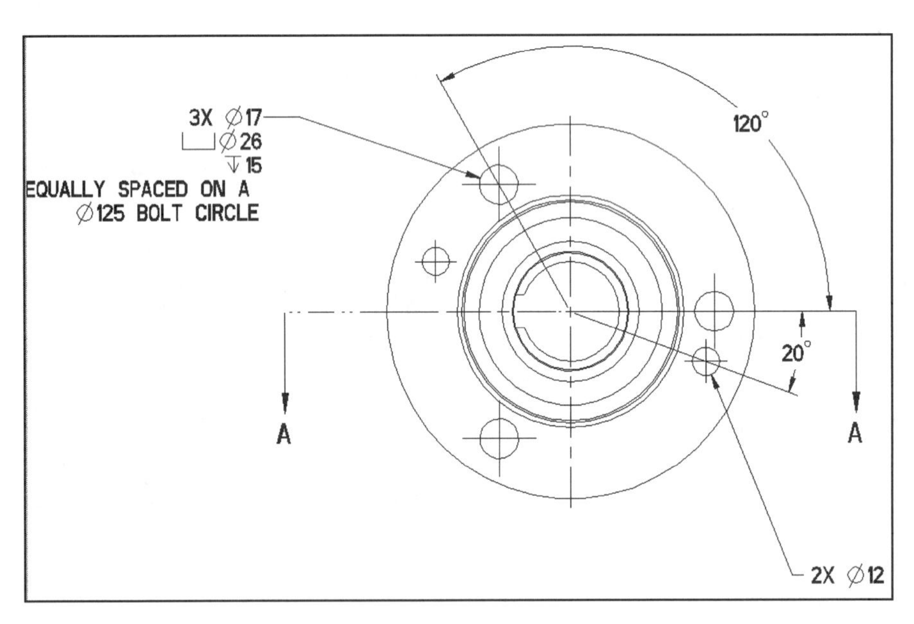

Figure 7.36
Taper Coupling Drawing, Bottom

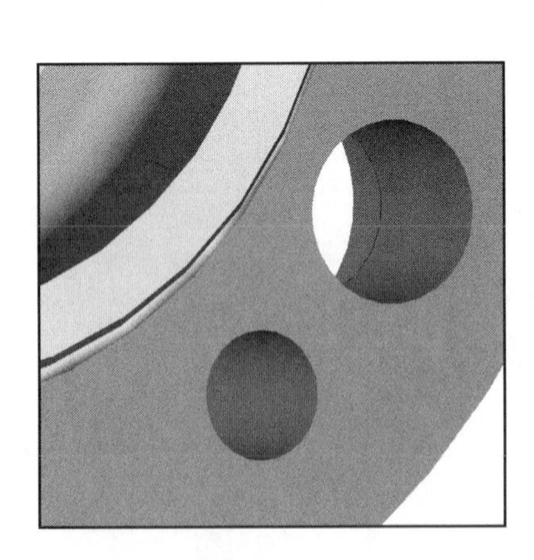

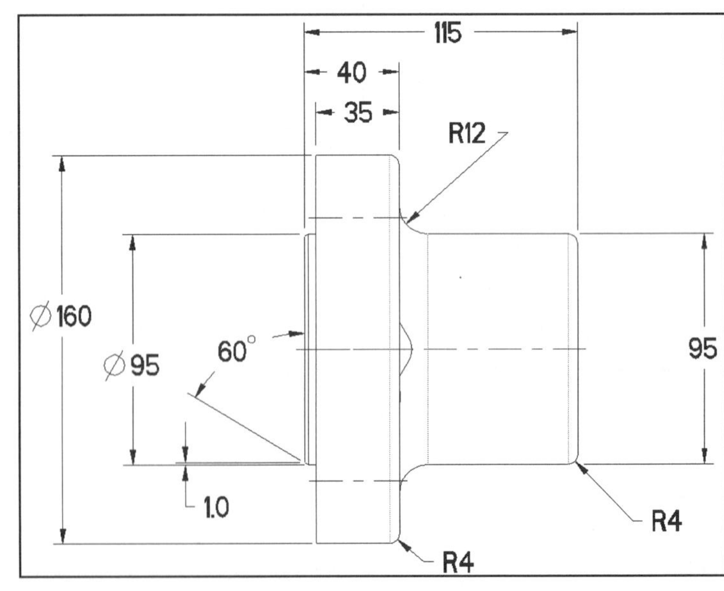

Figure 7.37
Taper Coupling Drawing, Side View

After completing Lesson 8, return to this project. Write a relation that will keep this dimension equal to the depth of the counterbore plus the radius of the large round (**R12**).

Dim=15+Radius
d18=d9+d6

Your dim symbols (**d#'s**) will differ.

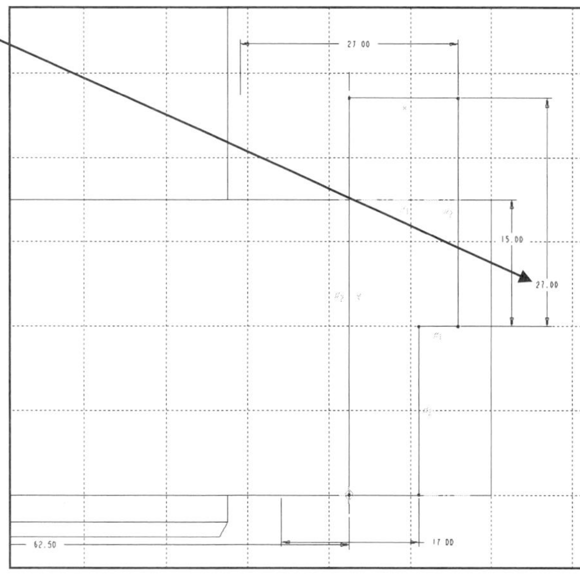

Figure 7.38
Taper Coupling Section, Counterbore

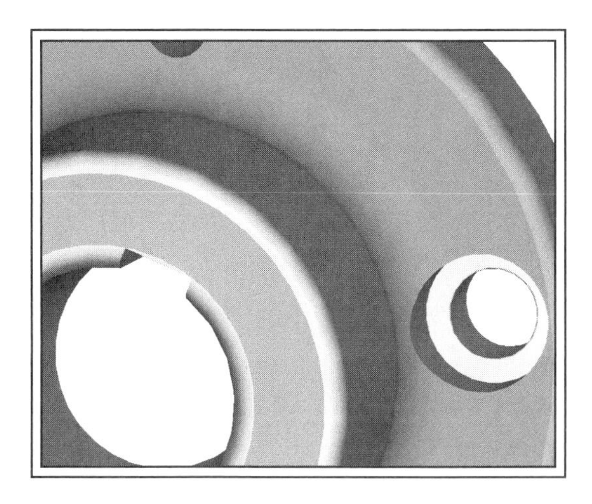

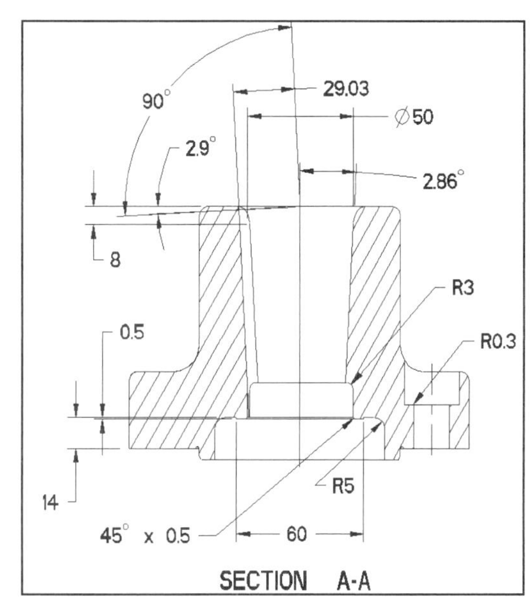

Figure 7.39
Taper Coupling Drawing,
SECTION A-A

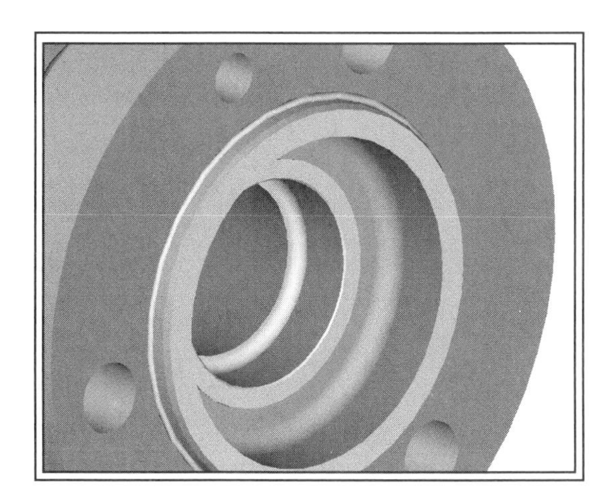

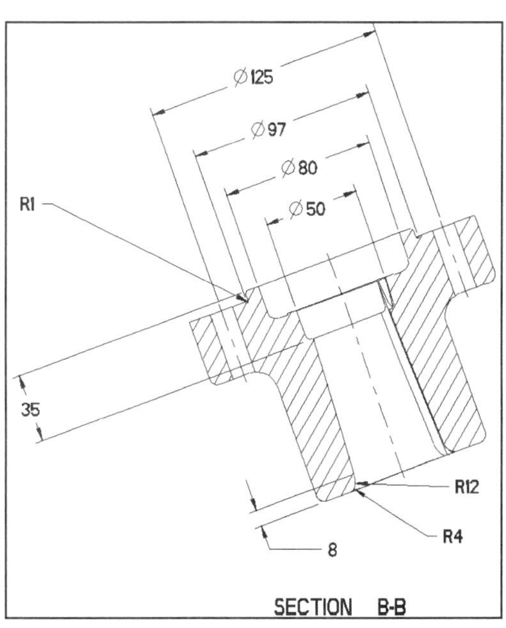

Figure 7.40
Taper Coupling Drawing,
SECTION B-B

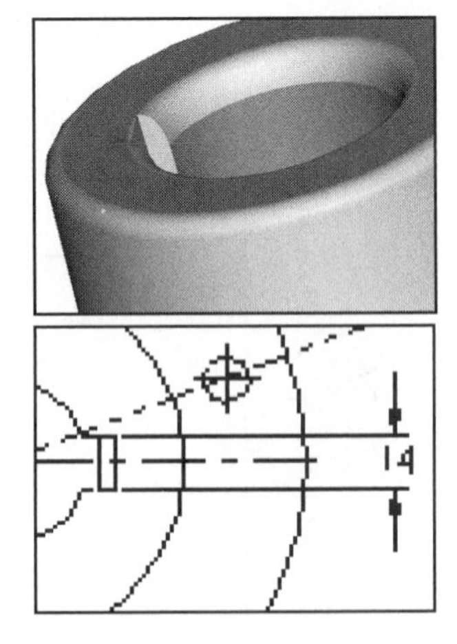

Figure 7.41
Taper Coupling Drawing,
SECTION A-A,
Taper Angle

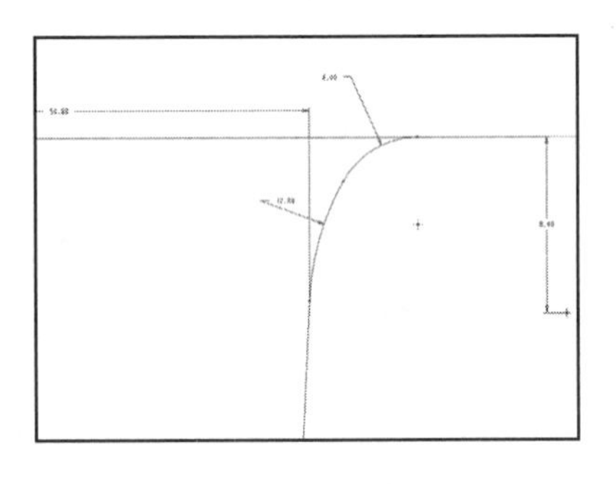

Figure 7.42
Taper Coupling Drawing,
SECTION A-A,
Close-up

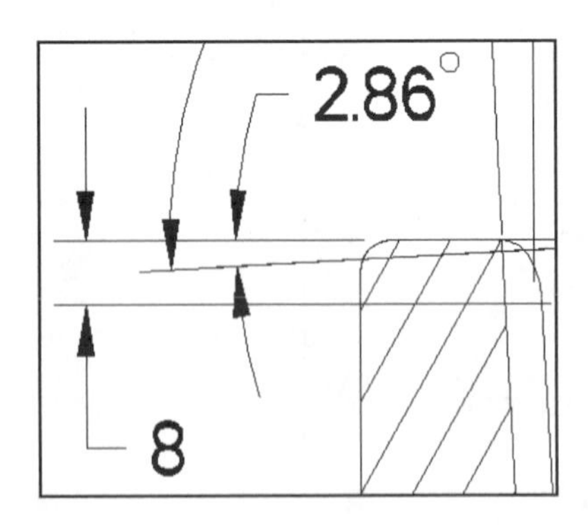

Figure 7.43
Taper Coupling Drawing,
SECTION B-B,
Close-up

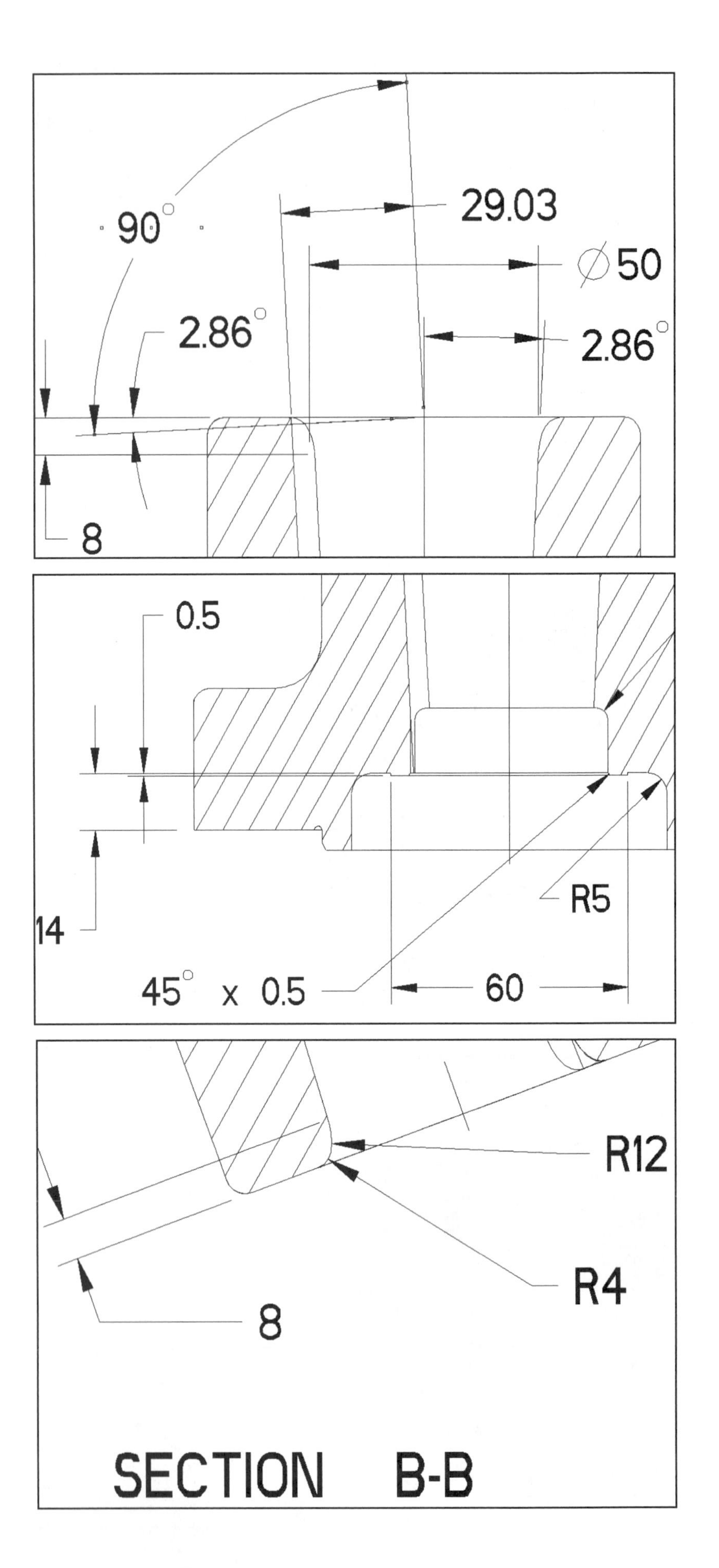

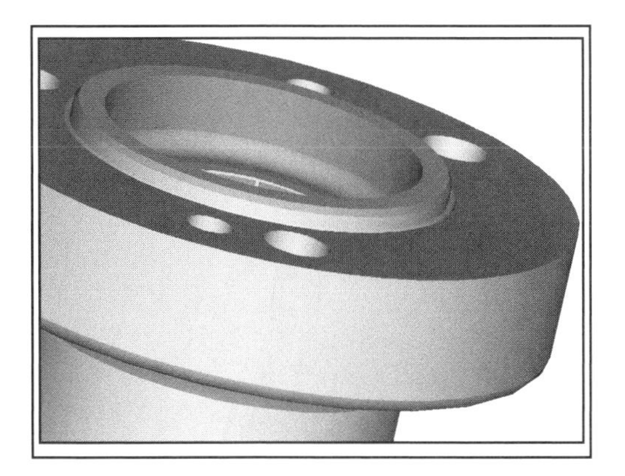

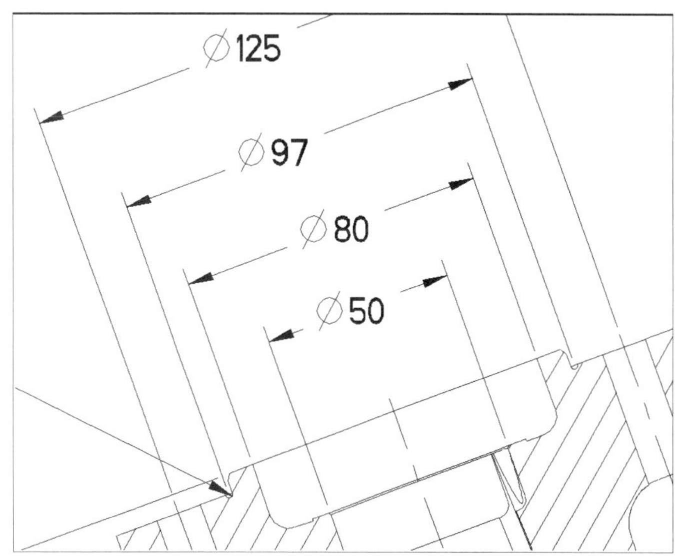

Figure 7.44
Taper Coupling Drawing,
SECTION B-B,
Mating Diameters

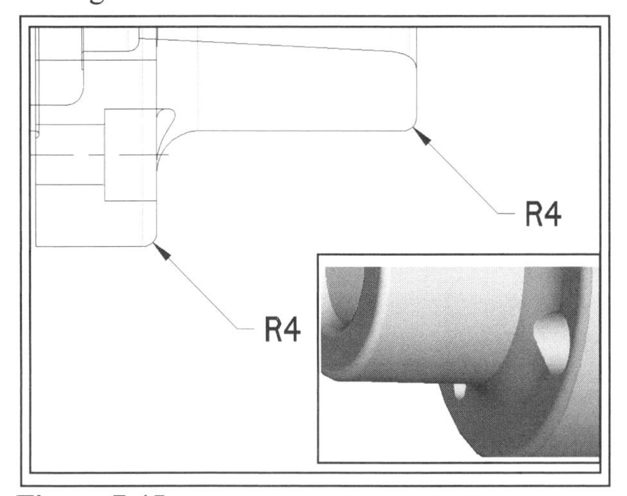

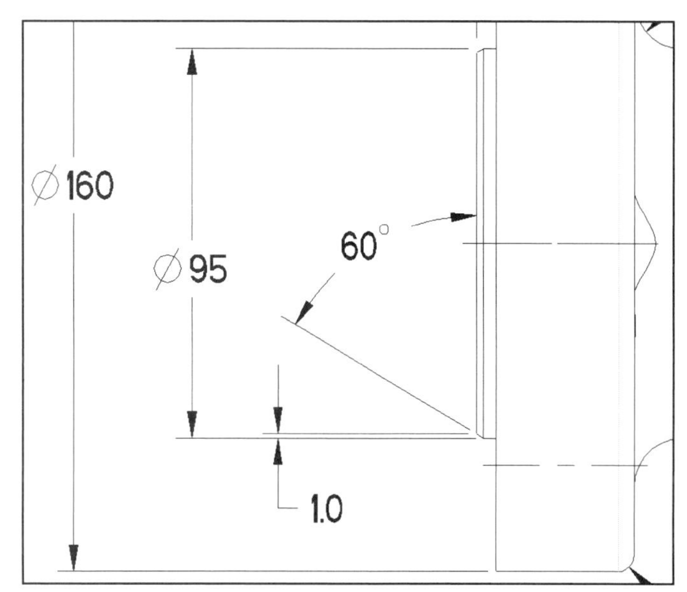

Figure 7.45
Taper Coupling Drawing,
Side View, Close-up

File ⇒ Save ⇒ ✔
File ⇒ Delete ⇒
Old Versions ⇒ ✔

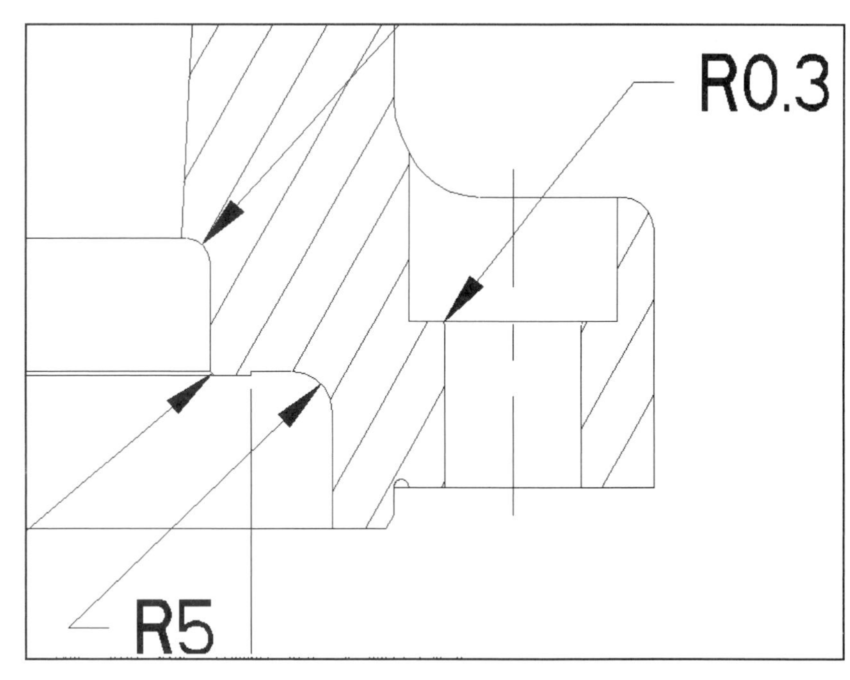

Figure 7.46
Taper Coupling Drawing,
SECTION A-A,
Close-up of Radii

Figure 7.47 Taper Coupling

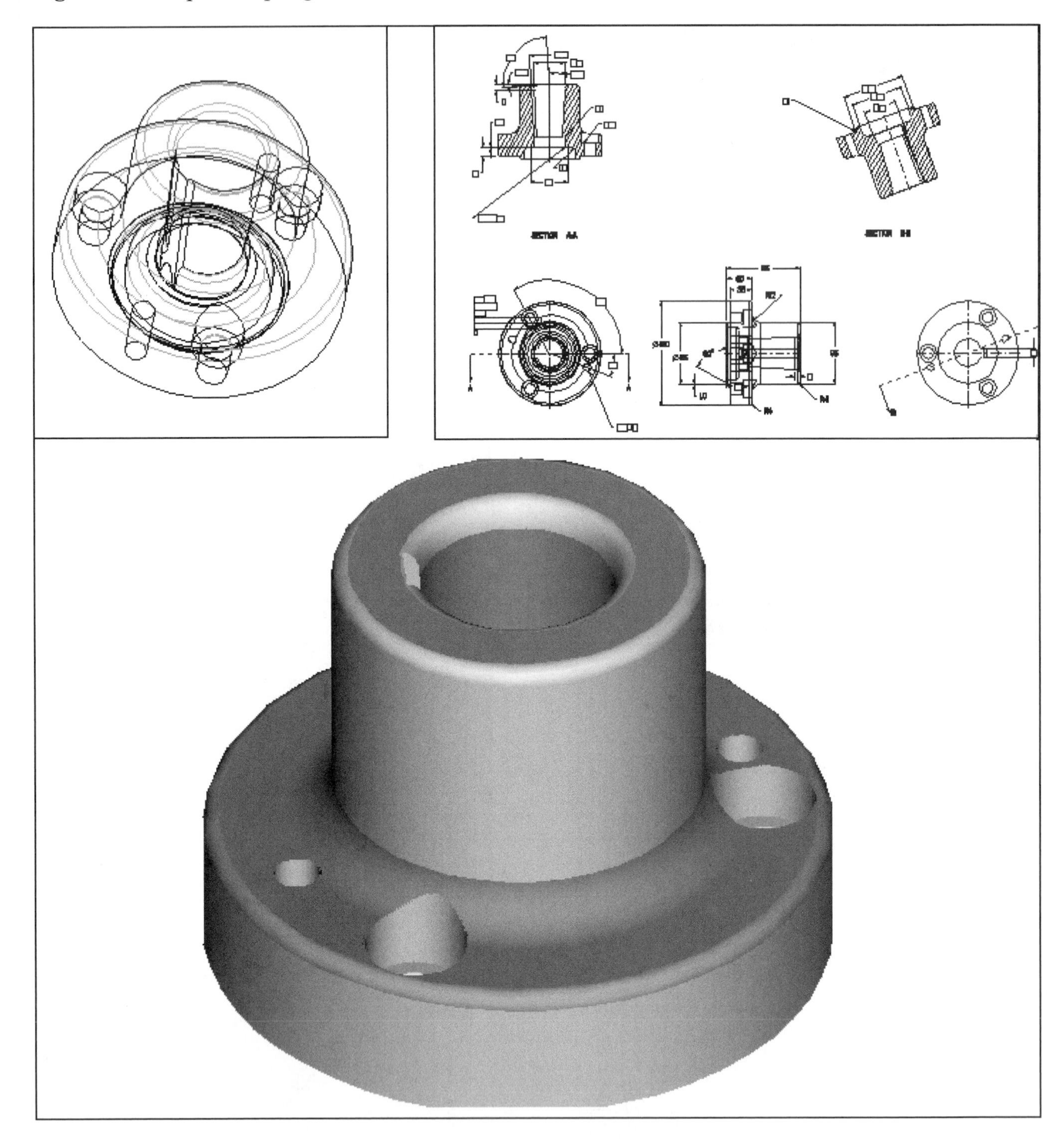

Lesson 8

Ribs, Relations, and Failures

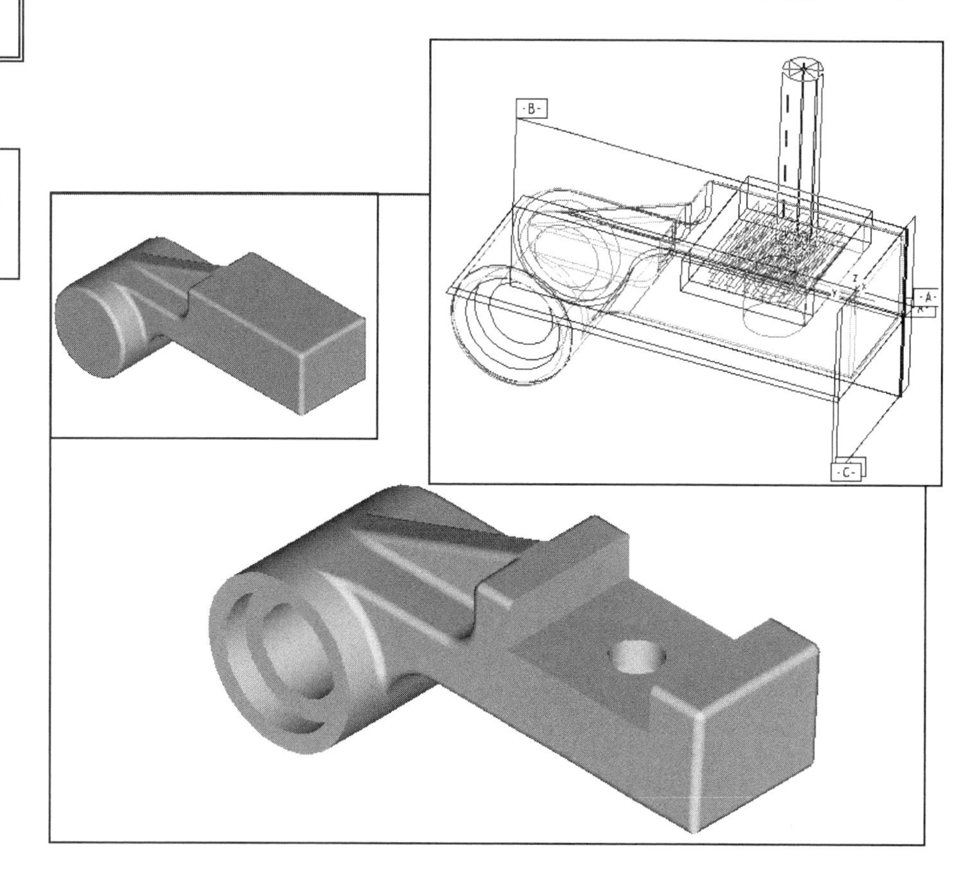

Figure 8.1
Adjustable Guide, Casting, and Machine Part

EGD REFERENCE
Fundamentals of Engineering Graphics and Design
by L. Lamit and K. Kitto
Read Chapters 11 and 14.
See pages 498 and 852.

COAch™ for Pro/ENGINEER

If you have **COAch for Pro/ENGINEER** on your system, go to SEARCH and do the Segments shown in Figures 8.3, 8.6, and 8.9.

OBJECTIVES

1. **Understand parameters and relations**
2. **Create straight ribs**
3. **Troubleshoot and resolve failures**
4. **Write relations to control features**
5. **Create a manufacturing model**
6. **Model a workpiece and a design part to be used in Pro/NC**
7. **Establish parameters for a part**
8. **Create equality and comparison relations**
9. **Understand the types of parameter symbols**

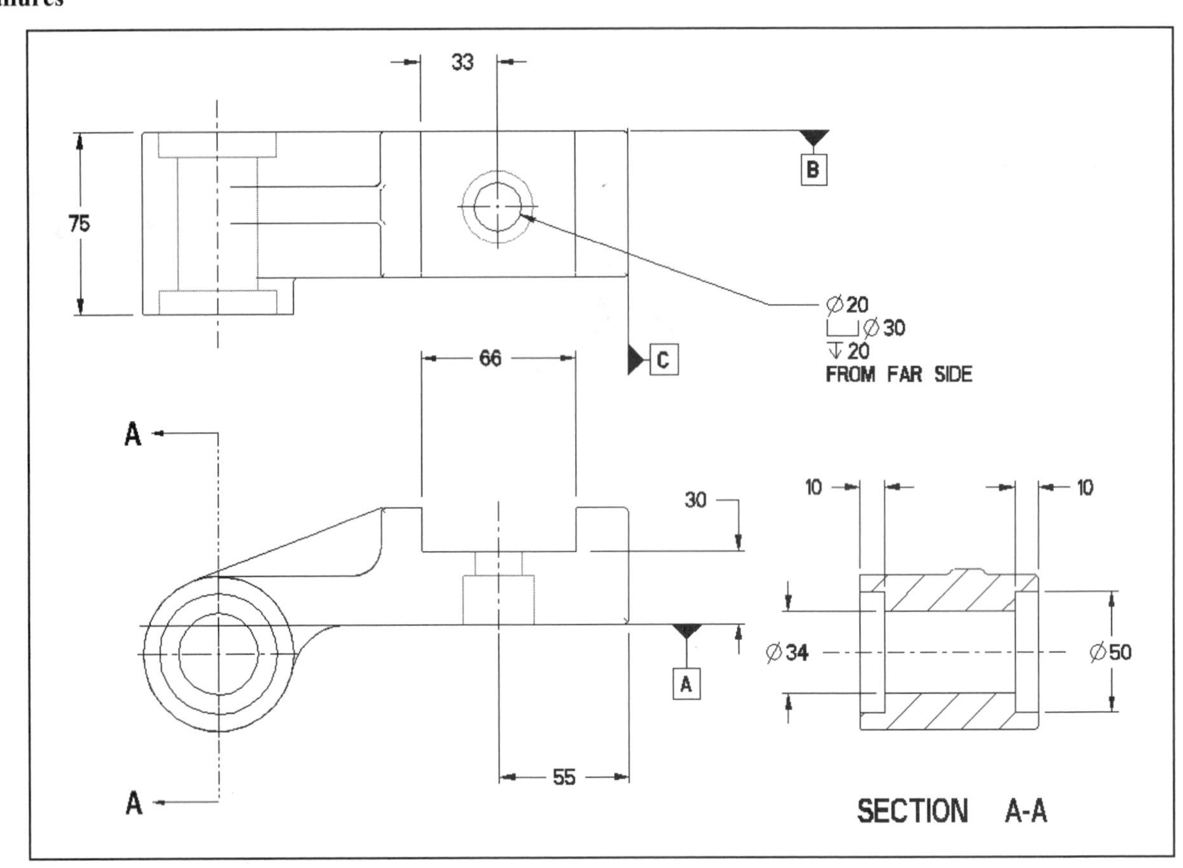

Figure 8.2
Adjustable Guide Machining Drawing

RIBS, RELATIONS, AND FAILURES

A **Rib** is a special type of protrusion designed to create a thin fin or web that is attached to a part (Figs. 8.1 through 8.4).

Relations are equations written between symbolic dimensions and parameters. By writing relations between dimensions in a part or an assembly, we can control the effects of modifications.

Failures happen when the model cannot be regenerated. You need to know how to avoid and how to resolve part and assembly failures.

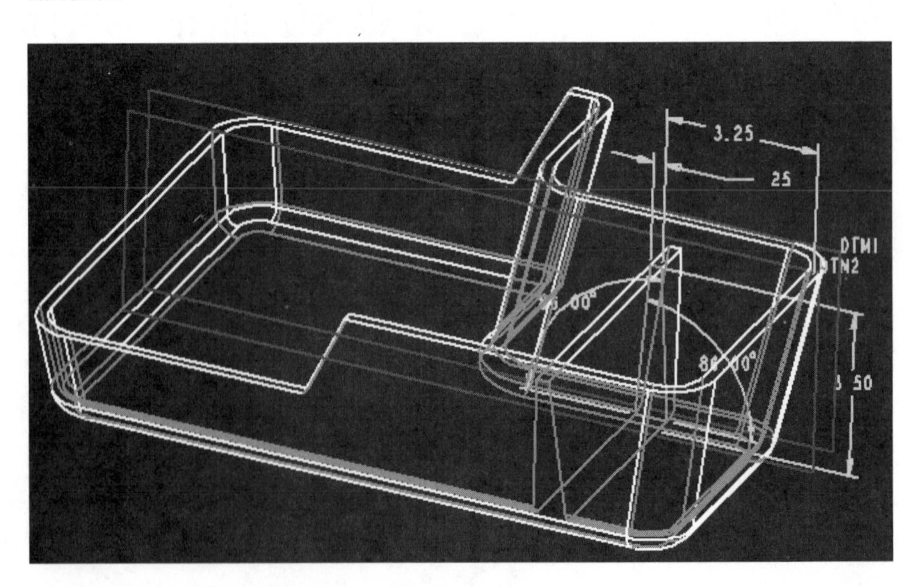

Figure 8.3
COAch for Pro/E, More on Modeling

A rib is always sketched from a side view, and it grows symmetrically about the sketching plane. Because of the way ribs are attached to the parent geometry, they are always sketched as open sections. A rib must "see" material everywhere it attaches to the part; otherwise, it becomes an unattached feature. There are two types of ribs, *straight* and *rotational*. We will confine our discussion to straight ribs. For more information on ribs and, in particular, on rotational ribs, use online documentation (see Fig. 8.4).

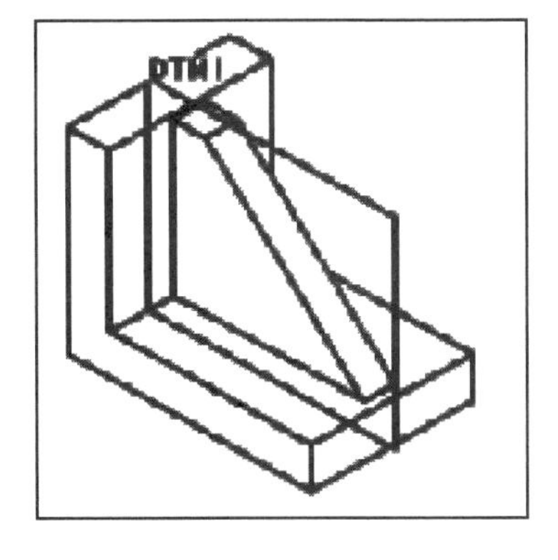

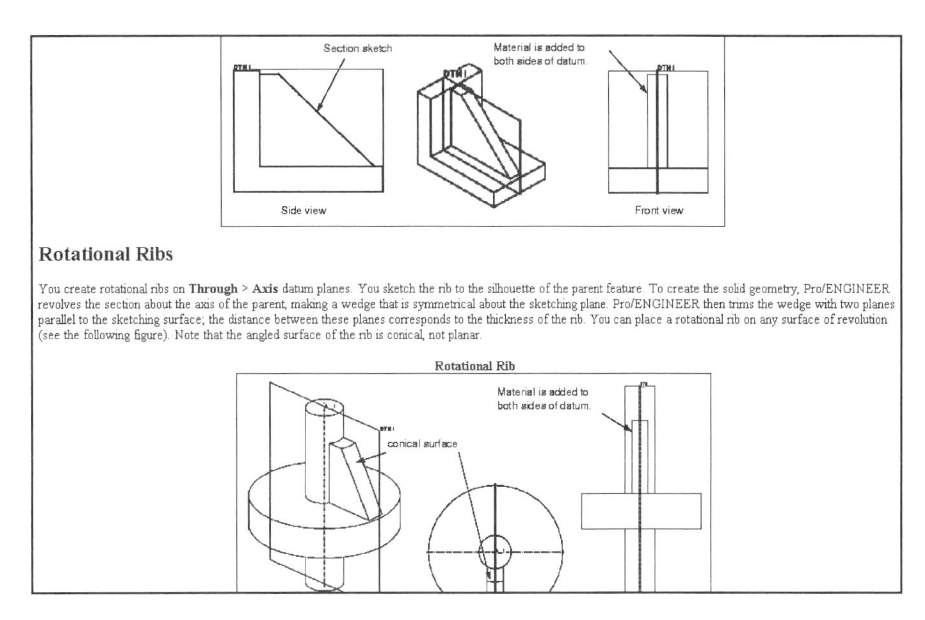

Rotational Ribs

You create rotational ribs on **Through** > **Axis** datum planes. You sketch the rib to the silhouette of the parent feature. To create the solid geometry, Pro/ENGINEER revolves the section about the axis of the parent, making a wedge that is symmetrical about the sketching plane. Pro/ENGINEER then trims the wedge with two planes parallel to the sketching surface; the distance between these planes corresponds to the thickness of the rib. You can place a rotational rib on any surface of revolution (see the following figure). Note that the angled surface of the rib is conical, not planar.

Rotational Rib

Figure 8.4

Straight Ribs

Ribs that are not created on **Through/Axis** datum planes are extruded symmetrically about the sketching plane. You must still sketch the ribs as open sections. Because you are sketching an open section, Pro/E may be uncertain about which side to add the rib on. Pro/E displays the DIRECTION menu after the rib section has been regenerated. Pro/E adds all material in the direction of the arrow. If the incorrect choice is made, correct the direction, pick **Okay**, and enter the value for the thickness.

Relations

Relations (Figs. 8.5 and 8.6) can be used to provide a value for a dimension. But they can also be used to notify you when a condition has been violated, such as when a dimension exceeds a certain value.

There are two basic types of relations, *equality* and *comparison*. An *equality relation* equates a parameter on the left side of the equation to an expression on the right side. This type of relation is used for assigning values to dimensions and parameters.

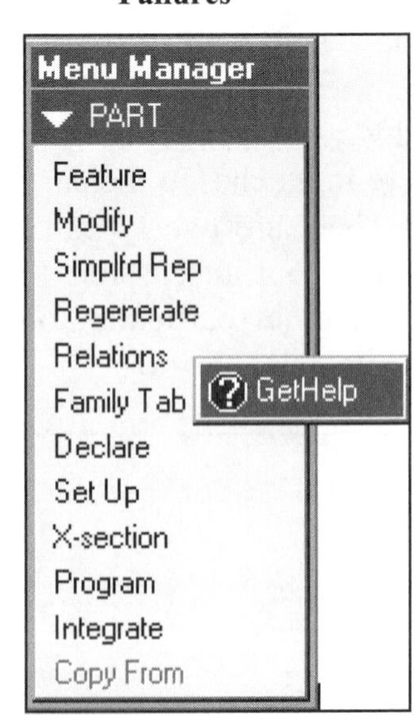

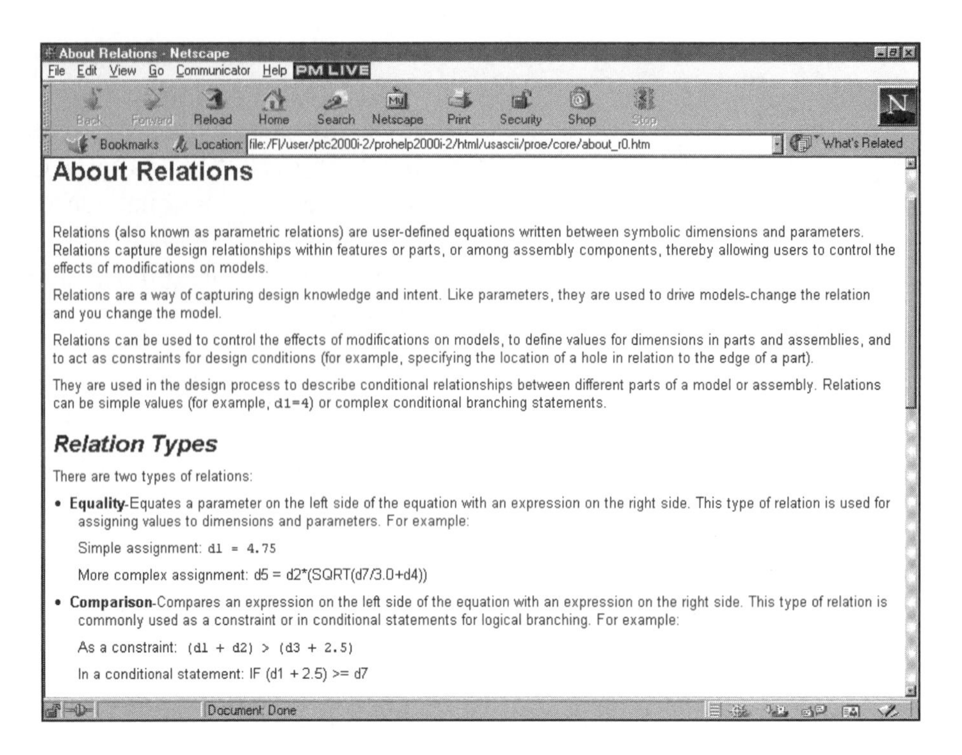

About Relations

Relations (also known as parametric relations) are user-defined equations written between symbolic dimensions and parameters. Relations capture design relationships within features or parts, or among assembly components, thereby allowing users to control the effects of modifications on models.

Relations are a way of capturing design knowledge and intent. Like parameters, they are used to drive models-change the relation and you change the model.

Relations can be used to control the effects of modifications on models, to define values for dimensions in parts and assemblies, and to act as constraints for design conditions (for example, specifying the location of a hole in relation to the edge of a part).

They are used in the design process to describe conditional relationships between different parts of a model or assembly. Relations can be simple values (for example, d1=4) or complex conditional branching statements.

Relation Types

There are two types of relations:

- **Equality**-Equates a parameter on the left side of the equation with an expression on the right side. This type of relation is used for assigning values to dimensions and parameters. For example:

  Simple assignment: d1 = 4.75

  More complex assignment: d5 = d2*(SQRT(d7/3.0+d4))

- **Comparison**-Compares an expression on the left side of the equation with an expression on the right side. This type of relation is commonly used as a constraint or in conditional statements for logical branching. For example:

  As a constraint: (d1 + d2) > (d3 + 2.5)

  In a conditional statement: IF (d1 + 2.5) >= d7

Figure 8.5
Online Documentation, Relations

The following are a few examples of equality relations:

d2 = 25.500
d8 = d4/2
d7 = d1+d6/2
d6 = d2*(sqrt(d7/3.0+d4))

A *comparison relation* compares an expression on the left side of the equation to an expression on the right side. This type of relation is commonly used as a constraint or as a conditional statement for logical branching.

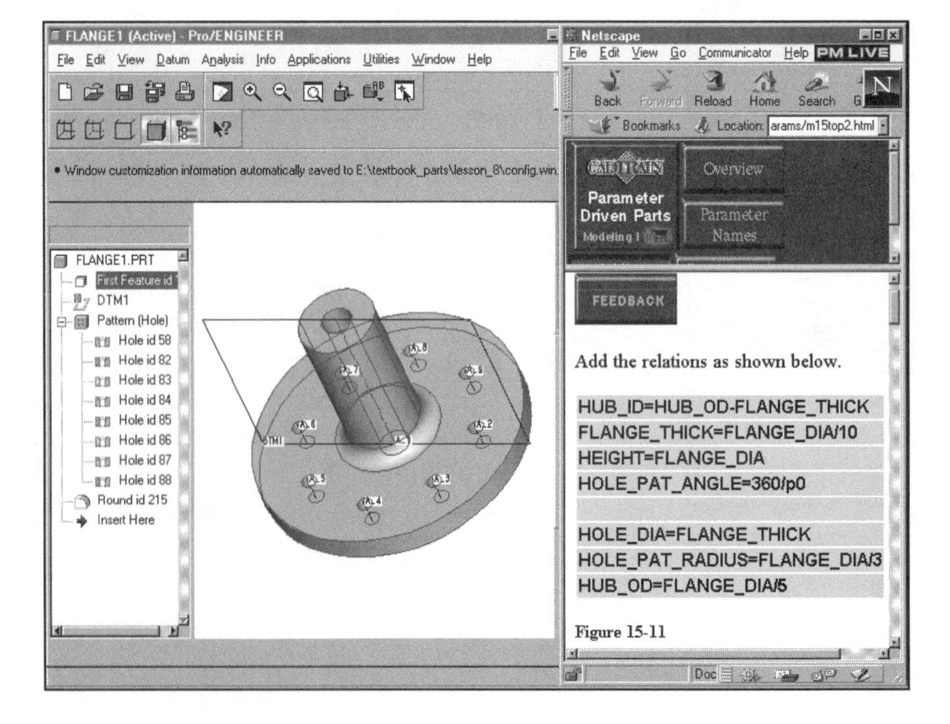

Figure 8.6
COAch for Pro/E, Parameter Driven Parts (Adding Relations)

The following are examples of comparison relations:

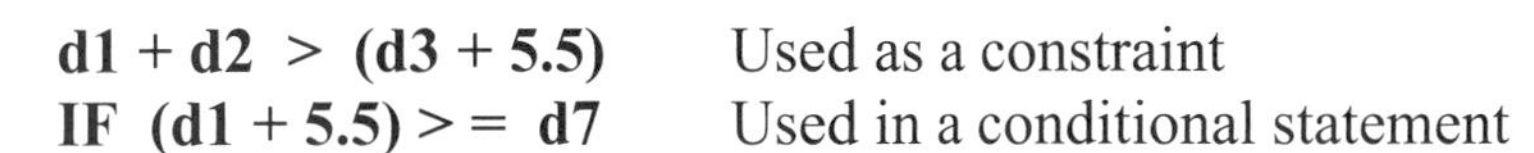

d1 + d2 > (d3 + 5.5) Used as a constraint
IF (d1 + 5.5) > = d7 Used in a conditional statement

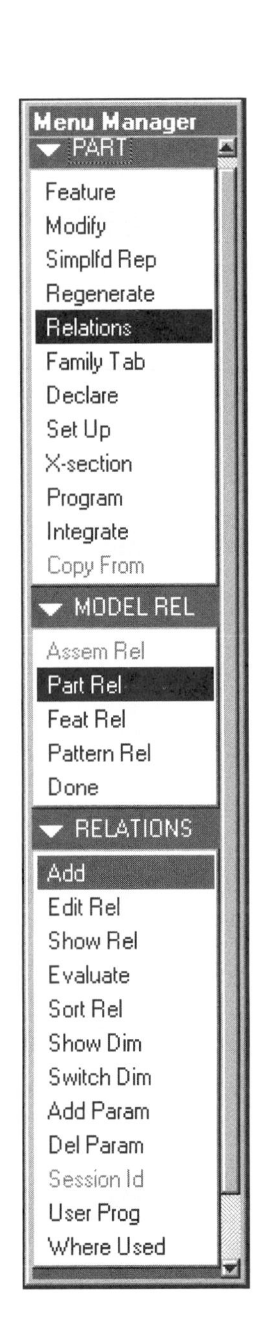

Parameter Symbols

Four types of parameter symbols are used in relations:

Dimensions These are dimension symbols, such as **d8**, **d12**.
Tolerances These are parameters associated with ± symmetrical and plus-minus tolerance formats. These symbols appear when dimensions are switched from numeric to symbolic.
Number of Instances These are integer parameters for the number of instances in a direction of a pattern.
User Parameter These can be parameters defined by adding a parameter or a relation (e.g., **Volume = d3 * d4 * d5**).

Operators and Functions

The following operators and functions can be used in equations and conditional statements:

Arithmetic Operators

+ Plus, for addition
- Minus, for subtraction
/ Divided by, for division
* Times, for multiplication
^ Exponentiation
( ) Parentheses, for grouping; for example, **d10 = (d5 - d6)*d7**

Assignment Operator

The equals sign is an assignment operator that equates the two sides of an equation. Obviously, when it is used, the equation can have only a single parameter on the left side.

= Equal to

Comparison Operators

Comparison operators are used wherever a TRUE/FALSE value can be returned. For example, the equation **d1 >= .625** will return TRUE whenever **d1** is greater than or equal to **.625**. It will return FALSE whenever **d1** is less than **.625**.

The following comparison operators can be used:

=   Equal to
\>   Greater than
\>=   Greater than or equal to
!=, < >, ~ =   Not equal to
<   Less than
<=   Less than or equal to
| Or
&   And
~, !   Not

The |, **&**, !, and ~ operators extend the use of comparison relations by allowing several conditions to be set in a single statement.

Functions

Mathematical Functions Relations may also include the following mathematical functions:

cos ( ) cosine
tan ( ) tangent
sin ( ) sine
sqrt ( ) square root
asin ( ) arcsine
acos ( ) arccosine
atan ( ) arctangent
sinh ( ) hyperbolic sine
cosh ( ) hyperbolic cosine
tanh ( ) hyperbolic tangent

Failures

Sometimes model geometry cannot be constructed because features that have been modified or created conflict with or invalidate other features. For example, this can happen when the following occurs:

* A protrusion is created that is unattached and has a one-sided edge.
* New features are created that are unattached and have one-sided edges.
* A feature is resumed that now conflicts with another feature (such as having two chamfers on the same edge).
* The intersection of features is no longer valid because dimensional changes have moved the intersecting surfaces.
* A relation constraint has been violated.

Resolving Feature Failures During Creation/Redefinition

Depending on the type of environment used to create a feature (e.g., whether the feature uses the dialog box interface--see Fig. 8.7), Pro/E handles feature failures that may occur during feature creation or redefinition in two different ways:

* *For features that do not use the dialog box interface* If the feature fails, Pro/E brings up the FEAT FAILED menu.

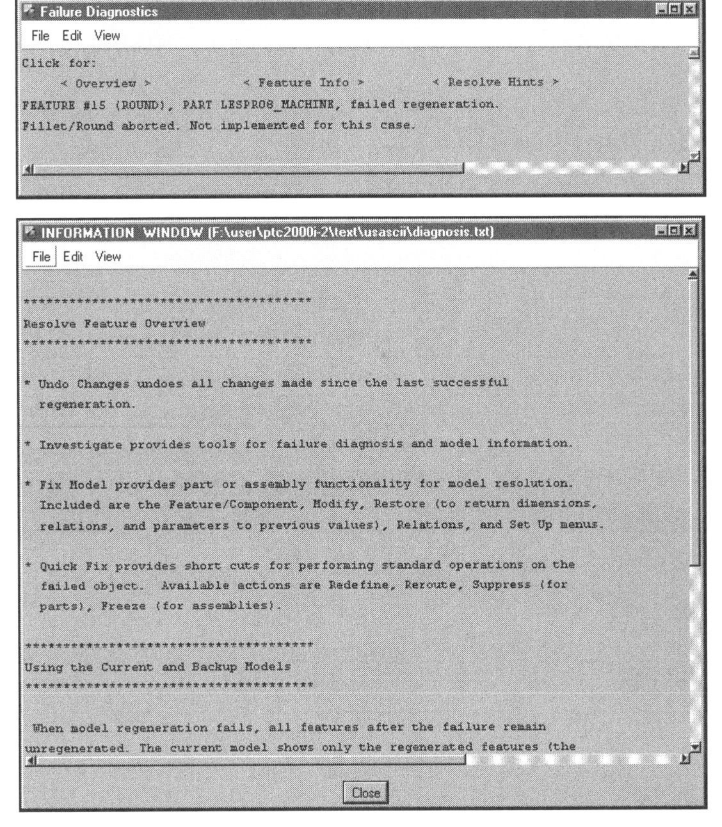

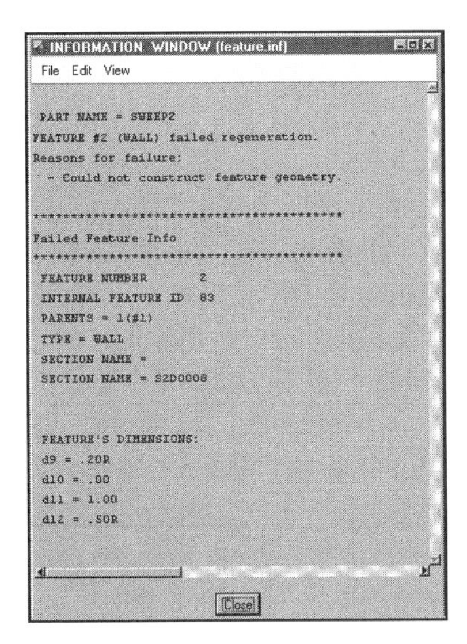

Figure 8.7
Feature Failure Window:
Failure Diagnostics
(top right)
Resolve Overview
(bottom right)
Failed Feature Information
(top left)

* *For features that use the dialog box interface* If the feature fails after you choose **OK** or **Preview**, the **Resolve** button appears in the feature creation dialog box. You can either stay in the dialog box environment and redefine feature elements with the **Define** button, or click on the **Resolve** button to access the Resolve environment so you can obtain diagnostics or make changes to other parts of the model.

Using the FEAT FAILED Menu

If a feature fails during creation and it does not use the dialog box interface, Pro/E displays the FEAT FAILED menu, with the following options:

Redefine the feature.
Show Ref Display the SHOW REF menu so you can see the references of the failed feature. Pro/E displays the reference number in the MESSAGE WINDOW.
Geom Check for problems with overlapping geometry, misalignment, and so on. This command may be *dimmed.* If a shell, offset surface, or thickened surface fails, Pro/E stores information about the surfaces that could not be offset. The GEOM CHECK menu displays a list of features with failed geometry and a **Restore** command.
Feat Info Get information about the feature.

Using the Resolve Button in the Dialog Box

If a feature fails after you choose **Preview** or **OK** from a dialog box, the **Resolve** button appears in the dialog box, enabling you to enter the "fix model" environment. To resolve a feature failure by using the **Resolve** environment, follow these steps:

1. After a feature creation fails, choose **Resolve** from the dialog box to access the "fix model" environment.
2. The INFORMATION WINDOW appears, listing features that failed regeneration. Select an option in the RESOLVE FEAT menu to resolve the problem.
3. After you have fixed the problem, choose **Preview** or **OK** in the dialog box.

Working in the Resolve Environment

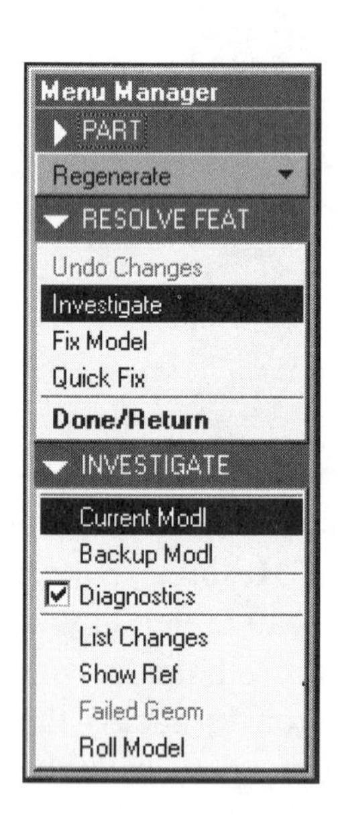

When a model regeneration fails, you must resolve the problem before continuing with normal model processing. Pro/E provides a special error resolution environment (the **Resolve** environment) for recovering from changes that have caused the model to fail regeneration.

As soon as a regeneration fails, Pro/E enters the **Resolve** environment, where the following occurs:

* The **File** command is unavailable and the model cannot be saved.
* The failed feature and all subsequent features remain unregenerated. The current model displays only the regenerated features as they were at the last successful regeneration. Pro/E displays a message that indicates the problem in the MESSAGE WINDOW.
* Pro/E displays the RESOLVE FEAT menu and the (failed-feature) Failed Diagnostic window.

The **Resolve** environment (Fig. 8.8) allows you to do the following:

* Undo all the changes made since the last successful regeneration.
* Diagnose the cause of the model failure.
* Fix the problems within this special environment while using standard part or assembly functionality.
* Attempt a quick fix of the problems using shortcuts for performing standard operations on the failed feature, including **Redefine, Reroute, Suppress** (for parts), **Freeze** (for assemblies), **Clip Suppress,** and **Delete.**

For both diagnosing and fixing the problem, you can choose to work on the current (failed) model or on the backup model. The backup model shows all features in their preregenerated state and can be used to modify or restore dimensions of the features that are not displayed in the current (failed) model.

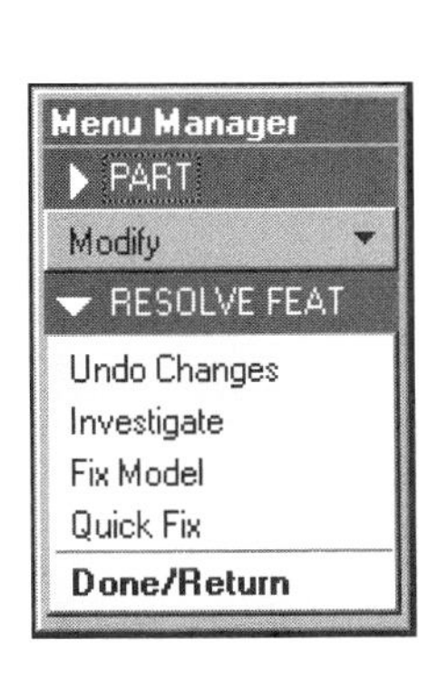

Using the RESOLVE FEAT Menu

The RESOLVE FEAT menu options are as follows:

Undo Changes Undo the changes that caused the failed regeneration attempt, and return to the last successfully regenerated model. Pro/E displays the CONFIRMATION menu.
Investigate Investigate the cause of the regeneration failure using the INVESTIGATE submenu.
Fix Model Enter the environment to fix the cause of the regeneration failure.
Quick Fix Use the QUICK FIX menu to perform the specified option on the current model immediately. Use these options:
- **Redefine** Redefine the failed feature.
- **Reroute** Reroute the failed feature.
- **Suppress** Suppress the failed feature and its children.
- **Clip Supp** Suppress the failed feature and all the features after it.
- **Delete** Delete the failed feature and its manage its children.

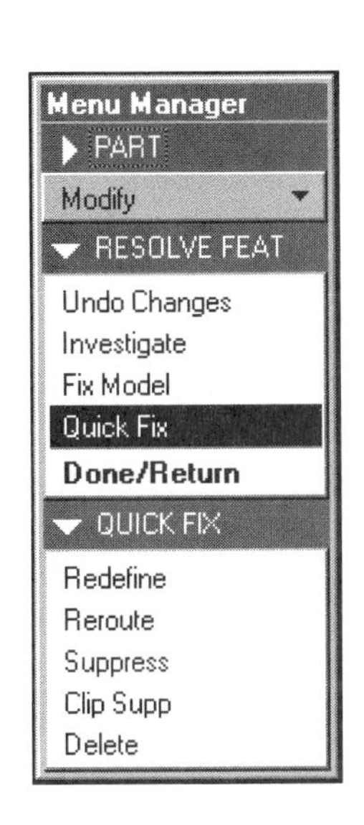

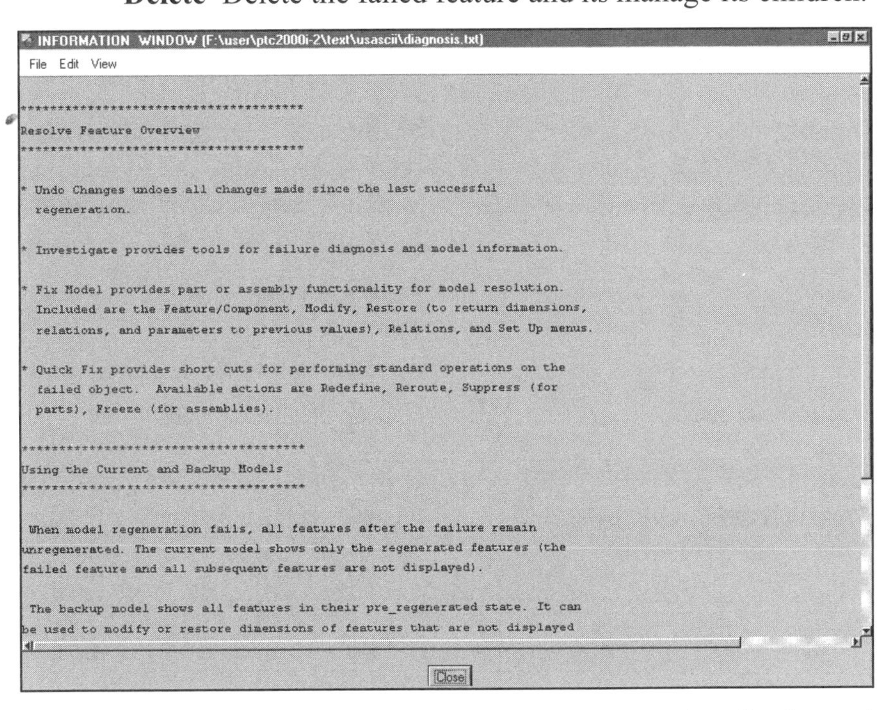
INFORMATION WINDOW (F:\user\ptc2000i-2\text\usascii\diagnosis.txt)
File Edit View

```
*************************************
Resolve Feature Overview
*************************************

* Undo Changes undoes all changes made since the last successful
  regeneration.

* Investigate provides tools for failure diagnosis and model information.

* Fix Model provides part or assembly functionality for model resolution.
  Included are the Feature/Component, Modify, Restore (to return dimensions,
  relations, and parameters to previous values), Relations, and Set Up menus.

* Quick Fix provides short cuts for performing standard operations on the
  failed object.  Available actions are Redefine, Reroute, Suppress (for
  parts), Freeze (for assemblies).

*************************************
Using the Current and Backup Models
*************************************

 When model regeneration fails, all features after the failure remain
unregenerated. The current model shows only the regenerated features (the
failed feature and all subsequent features are not displayed).

 The backup model shows all features in their pre_regenerated state. It can
be used to modify or restore dimensions of features that are not displayed
```

Close

Figure 8.8
Resolve Overview

If you choose the **Investigate** option, Pro/E displays the INVESTIGATE menu, which includes the following options:

Current Modl Perform operations on the current (failed) model.
Backup Modl Perform operations on the backup model, displayed in a separate window (current model displayed in active window).

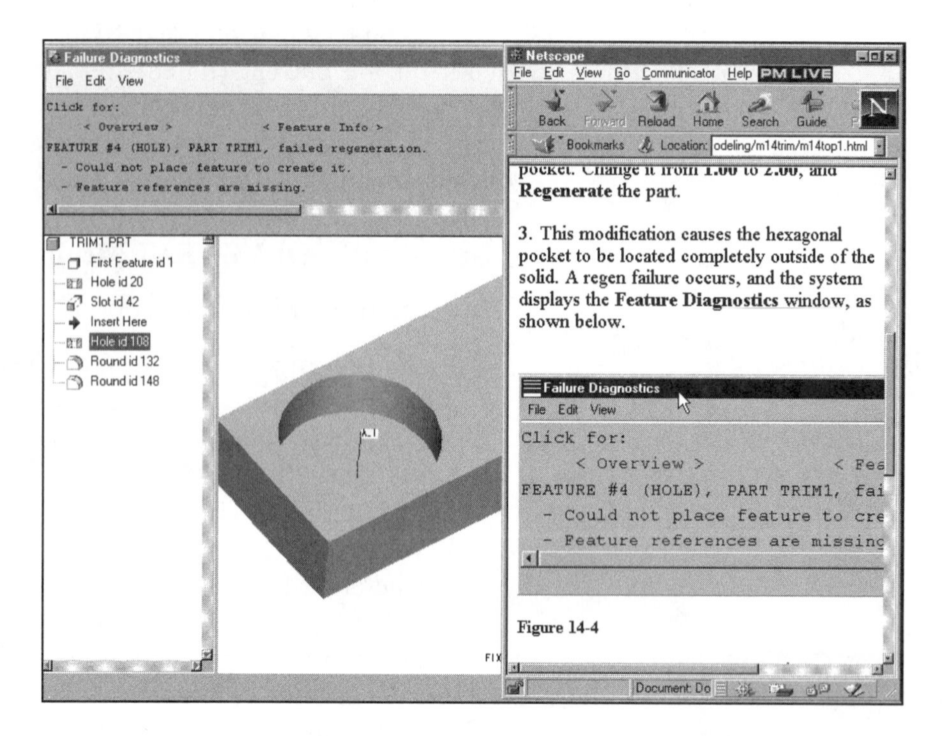

Figure 8.9
COAch for Pro/E, Modeling I, Resolve Features (Undoing Changes)

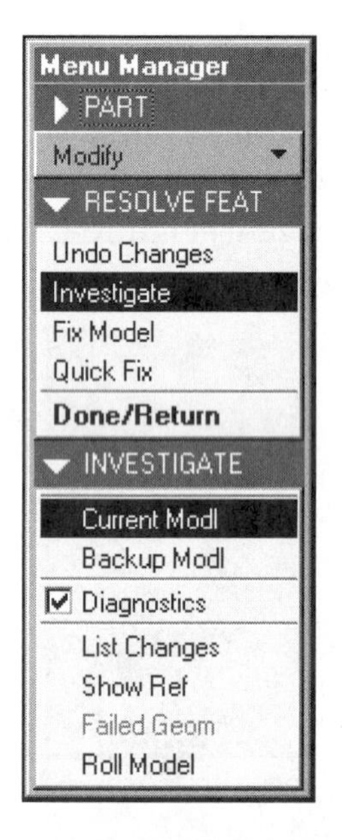

Diagnostics Toggle on or off the display of the (failed-feature) Failure Diagnostic window (Fig. 8.9).

List Changes Show the modified dimensions in the Main Window and in a preregenerated model window (Review Window), if available. Also, a table is displayed that lists all the modifications and changes.

Show Ref Display the SHOW REF menu to show all the references for the failed feature in the models, in both the Review Window and the Main Window. Pro/E displays the Reference Information Window (see left-hand column).

Failed Geom Display the invalid geometry of the failed feature. This command may be unavailable. The FAILED GEOM menu displays a list of features with failed geometry and a **Restore** command.

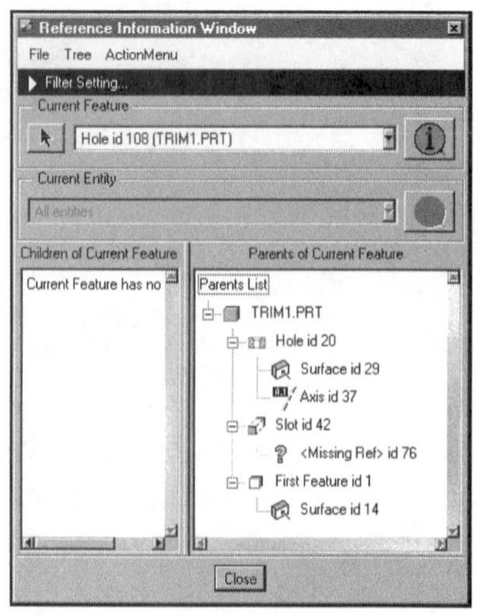

Roll Model Roll the model back to the option selected in the ROLL MDL TO submenu. The options are as follows:

> **Failed Feat** Roll the model back to the failed feature (for the backup model only).
> **Before Fail** Roll the model back to the feature just before the failed feature.
> **Last Success** Roll the model back to the state it was in at the end of the last successful feature regeneration.
> **Specify** Roll the model back to the specified feature.

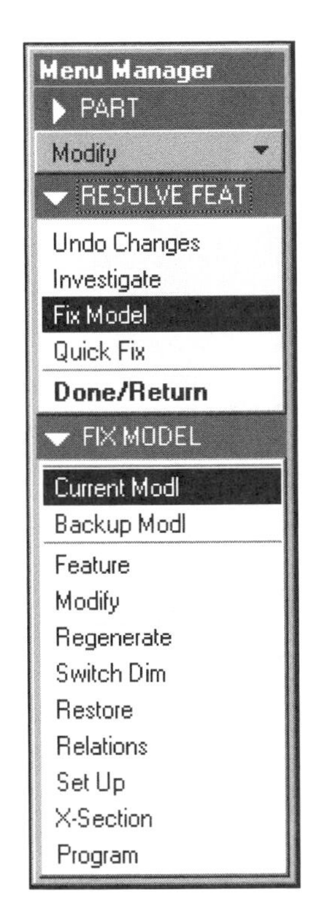

If you choose the **Fix Model** option, Pro/E displays the FIX MODEL menu (see left-hand column), which includes the following options (for Part mode):

Current Modl Perform operations on the current active (failed) model.

Backup Modl Perform operations on the backup model, displayed in a separate window from the current model in the active window.

Feature Perform feature operations on the model using the standard FEAT menu. Pro/E displays the CONFIRMATION menu, so you can confirm or cancel the request only if the **Undo Changes** option is not possible. However, the **Undo Changes** option is not always possible if you used the **Regen Backup** option in the Environment window dialog box.

Modify Modify dimensions using the standard MODIFY menu.

Regenerate Regenerate the model.

Switch Dim Switch the dimension display from symbols to values, or vice versa.

Restore Display the RESTORE menu so that you can restore dimensions, parameters, relations, or all of these to their values prior to the failure (see left-hand column).

Relations Add, delete, or modify relations, as necessary, to be able to regenerate the model, using the MODEL REL and RELATIONS menus.

Set Up Display the standard PART SETUP menu to perform additional part setup procedures.

X-Section Create, modify, or delete a cross-sectional view using the CROSS SEC menu.

Program Access Pro/PROGRAM capabilities by using the PROGRAM menu.

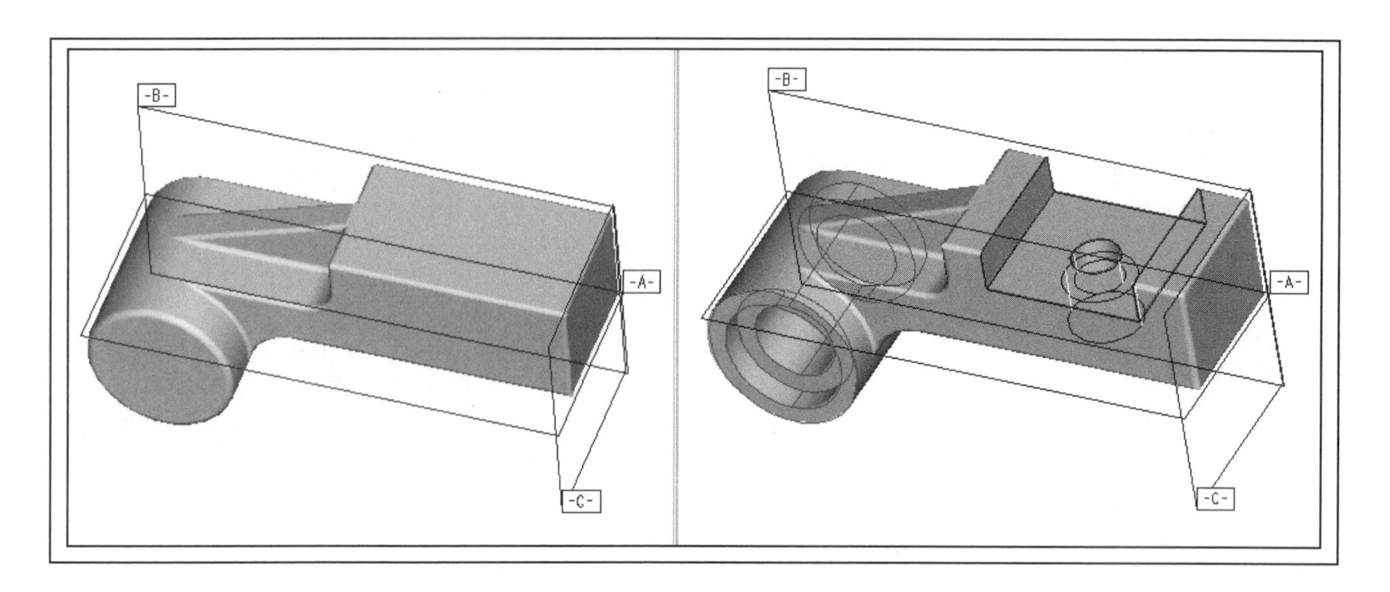

Figure 8.10
Adjustable Guide Casting, and Machined Part

Lesson procedures & commands
START HERE 

Adjustable Guide

Model the casting first (Fig. 8.10, left) and save it when it is completed. After saving the casting under a different name, use the part model to create the machined **Adjustable Guide** (Fig. 8.10, right). By having a *casting* (which is called a ***workpiece*** in **Pro/NC**) and a separate but almost identical *machined part* (which is called a ***design part*** in **Pro/NC**), you can create an operation for machining and an NC sequence. During the manufacturing process, *you merge the workpiece into the design part* and create a ***manufacturing model*** (Fig. 8.11). The difference between the two files is the difference between the volume of the casting and the volume of the machined part. The removed volume can be seen as *material removal* when you are performing an **NC Check** operation on the manufacturing model. The manufacturing model is *green;* the cuts completed in NC Check show as *yellow* until the machine tool reaches the design size, and then they show *magenta.* If the machining process gouges the part, the gouge will be displayed as *cyan.* The cutter location (CL) can also be displayed as an animated machining process, as shown in Figure 8.12.

The rib created in the casting model will have a **relation** added to it to control its location. The relation will keep the rib centered on the rectangular side of the part.

The Adjustable Guide is a simple part, so the process of describing step-by-step procedures will start with the creation of the rib. The rounds are added late in the modeling process, for they can cause the model to fail in many cases. The process of fixing the part so that the rounds do not make the regeneration fail is also described. The machined version of the part can be finished with two cuts: a sketched counterbore hole (use a revolved cut) and a thru hole with two counterbores (also created with a revolved cut).

NOTE
From this lesson on, you will be required to set up your parts without step-by-step commands.

Set up the Adjustable Guide:

- Material = STEEL
- Units = millimeters
- Hidden Line
- Dimmed
- ✔ Snap to Grid
- Set datums with Geom Tol
- Rename datum planes and coordinate system :

Datum **TOP** = **A**
Datum **FRONT** = **B**
Datum **RIGHT** = **C**
Coordinate System = **DEFAULT**

HINT

* Create the part with **ADJ_GUIDE_MACHINE** as the part name.
* Model only the casting dimensions.
* Save the part, using **File ⇒ Save As,** with the name of **ADJ_GUIDE_CAST**.
* Continue modeling the machined features.
* Save the part under its original name.

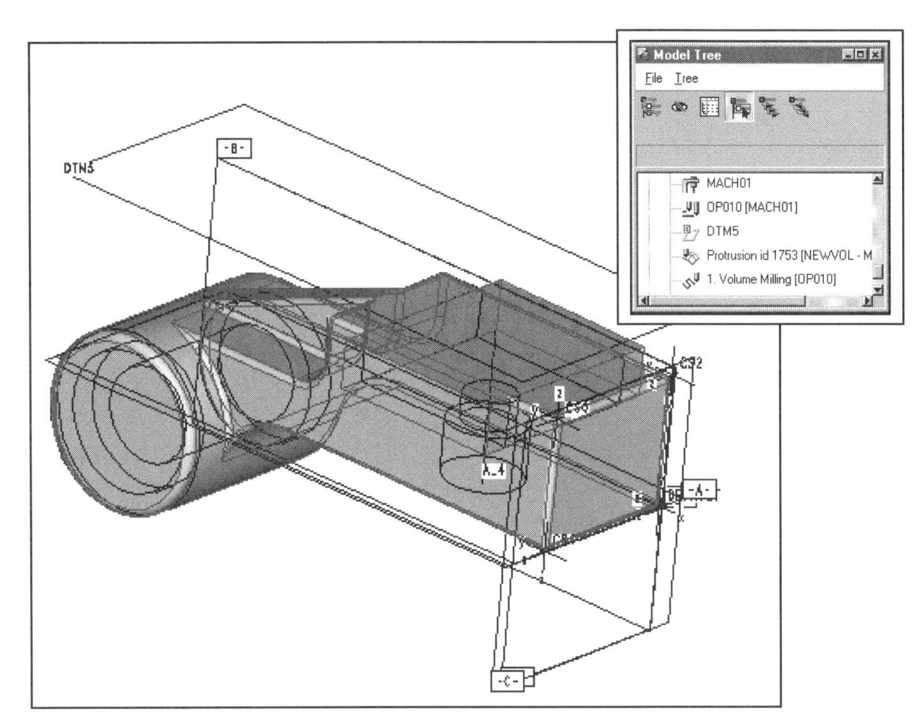

Figure 8.11
Manufacturing Model

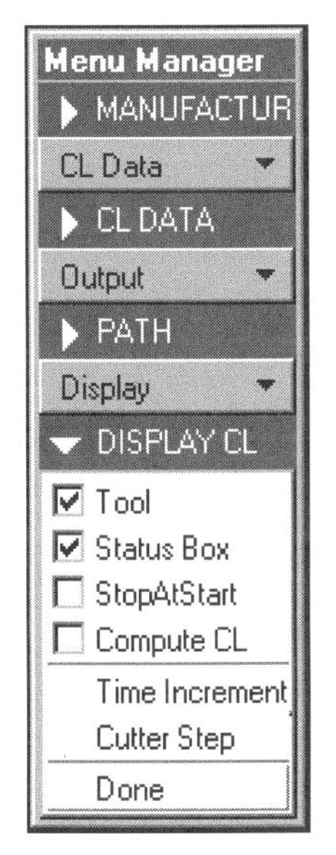

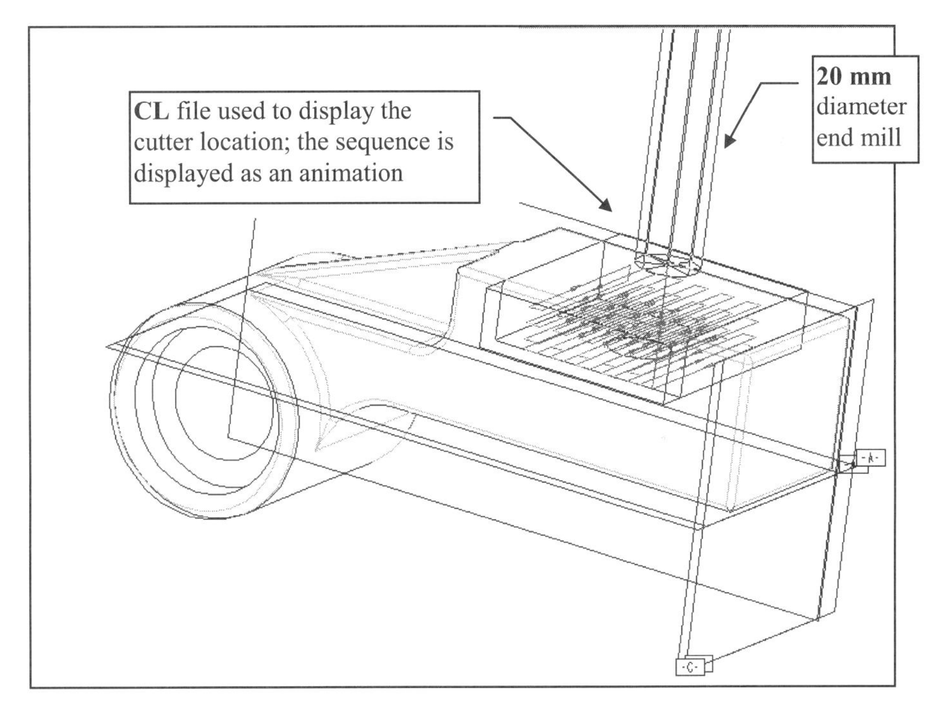

Figure 8.12
Pro/NC
NC Sequence and a **CL** File

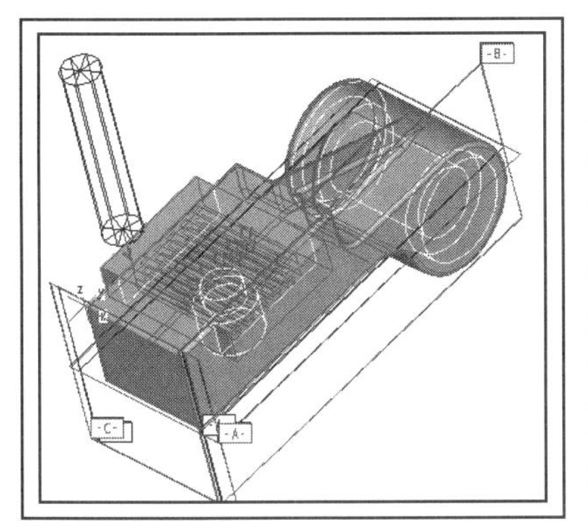

You will begin this process by creating a part called **ADJ_GUIDE_MACHINE**. Use Figure 8.13 for the modeling of the Adjustable Guide casting and Figure 8.14 for the Adjustable Guide machined part. Remember to create the casting first and save it using **File ⇒ Save As ⇒** (type a new name, such as **ADJ_GUIDE_CAST)** ⇒ ✔ (to save). Continue working on the part by adding the machined features as required by the design. In this edition of this text, Pro/NC is presented in Lesson 22. If your company or school has this module, you may wish to create a manufacturing model and machine it.

NOTE:
ALL RADII TO BE R2 UNLESS OTHERWISE NOTED

30
15
∅64
175
105
R12
(60)
48
12
R24
80
60
A
B
C

Figure 8.13
Adjustable Guide, Casting Detail

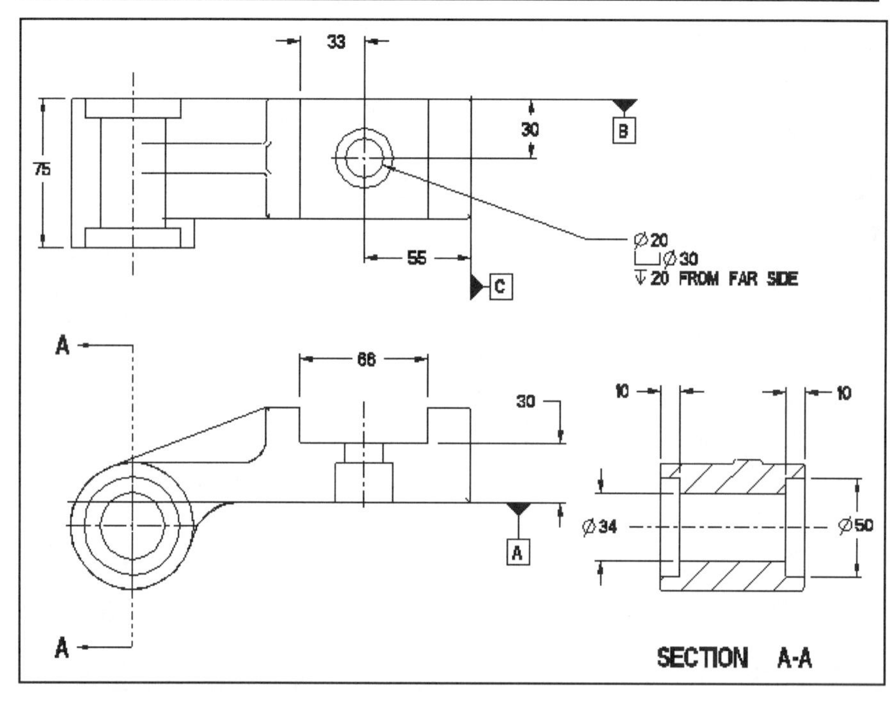

Figure 8.14
Adjustable Guide, Machining Detail

Use the same datum planes shown in the following figures. The sketching plane for the part is **FRONT** and is on the back side of the part (datum **B** in Fig. 8.15). Datum **A** was originally **TOP** and runs along the bottom of the part. A *jig and fixture* assembly will be used to hold, locate, and clamp the part for machining. Datum **C** was originally **RIGHT** and runs along the right side of the part.

The first protrusion is shown in Figure 8.15. The circular protrusion is extended in Figure 8.16. The first round set is created in Figure 8.17. This round is needed to create a tangent rib. Normally, all rounds would be added last. Create the round by choosing the following commands:

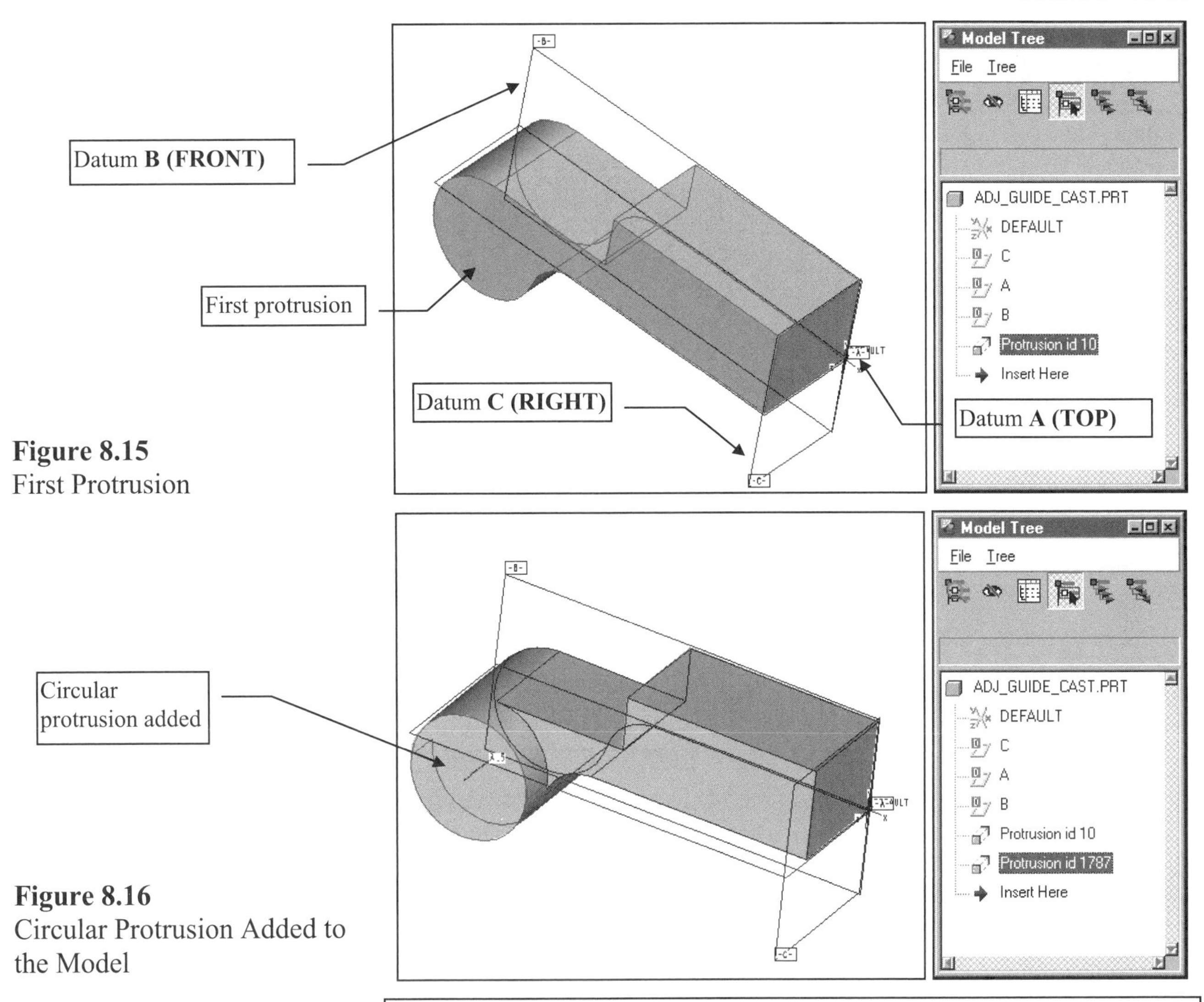

Figure 8.15
First Protrusion

Figure 8.16
Circular Protrusion Added to the Model

Feature ⇒ Create ⇒ Round ⇒ Simple ⇒ Done ⇒ Constant ⇒ Edge Chain ⇒ Done ⇒ Surf Chain ⇒ (pick the top surface of the part as shown in Fig. 8.17) **⇒ Select All ⇒ Done Sel ⇒ Done ⇒** (type **2**) **⇒ ✔ ⇒ Preview ⇒ OK ⇒ Done ⇒ File ⇒ Save ⇒ ✔**

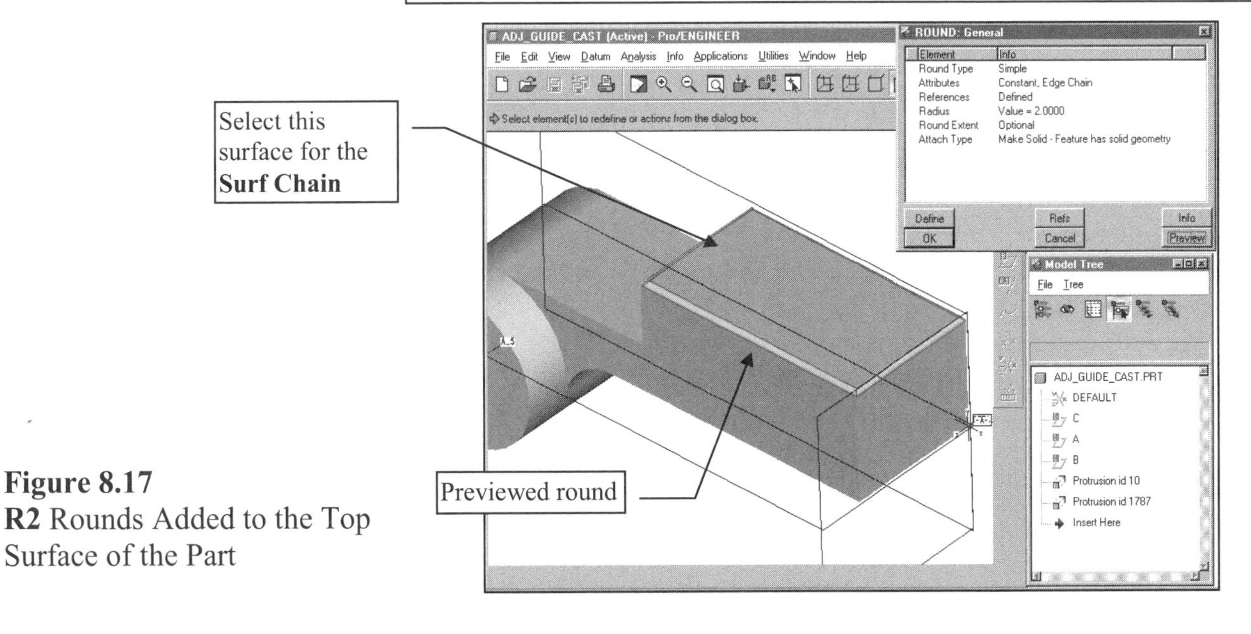

Figure 8.17
R2 Rounds Added to the Top Surface of the Part

The rib will be constructed using a *datum on the fly* (**Make Datum**). The sketch of the rib will require just one tangent line between the large circular protrusion and the round just created. Choose the following commands:

MKEY ***TOOLCHEST***
Utilities ⇒ Mapkeys ⇒ New ⇒ Key Sequence- **$F(#) ⇒** Name **⇒ RIB ⇒** Description **creates a rib ⇒ Record ⇒ Feature ⇒ Create ⇒ Rib ⇒ Stop ⇒ OK ⇒** Save Mapkeys **All ⇒ Ok ⇒ Close ⇒ Utilities ⇒ Customize Screen ⇒ Commands ⇒ Mapkeys** (from the Categories list) **⇒** (pick the new mapkey and press the RMB) **⇒ ●Image and Text ⇒** (with the RMB pick **Choose Button Image** and/or **Edit Button Image** and select and/or edit the image) **⇒** (pick the new Mapkey, drag to the Toolbar and drop) **⇒ OK**

RIB [or **Feature ⇒ Create ⇒ Rib] ⇒ Make Datum ⇒ Offset ⇒** (pick datum **B**--originally datum **FRONT**) **⇒ Enter Value ⇒** (type **30** at the prompt) **⇒ ✔ ⇒ Done ⇒ Top ⇒** (pick datum **A**--originally **TOP) ⇒ Sketch ⇒ Intent Manager ⇒ Sketch ⇒ Geom Tools ⇒ Use Edge** [pick the circular edge and the round arc (Fig. 8.18)] **⇒ Done Sel ⇒ Sketch ⇒ Line ⇒ 2 Tangent ⇒** [pick the circle and the arc at the approximate points of tangency (Fig. 8.19)] **⇒ Delete ⇒ Delete Item** (delete the *three* arcs--Pro/E broke the arc and the circle at the points of tangency, so you must zoom in and make sure only the line is left after deleting the arcs) **⇒ Regenerate ⇒ Alignment ⇒** [pick the end of the line and the circular edge for one alignment and the other end of the line and the circular edge for the second alignment (Fig. 8.19)] **⇒ Regenerate ⇒ Done ⇒ Flip** [if needed, flip the arrow to point toward the part so as to add material in the correct direction (Fig. 8.20)] **⇒ Okay ⇒** (input the rib thickness by typing **15** at the prompt) **⇒ ✔ ⇒ Done** (Fig. 8.21)

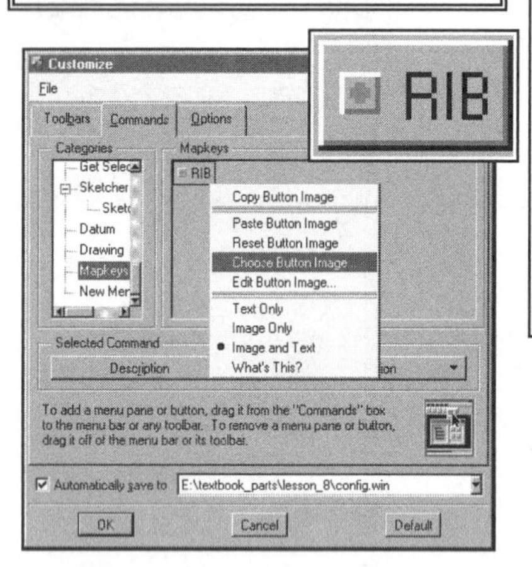

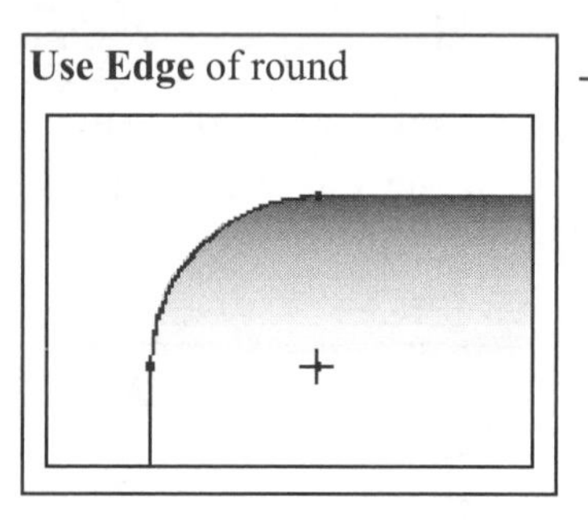

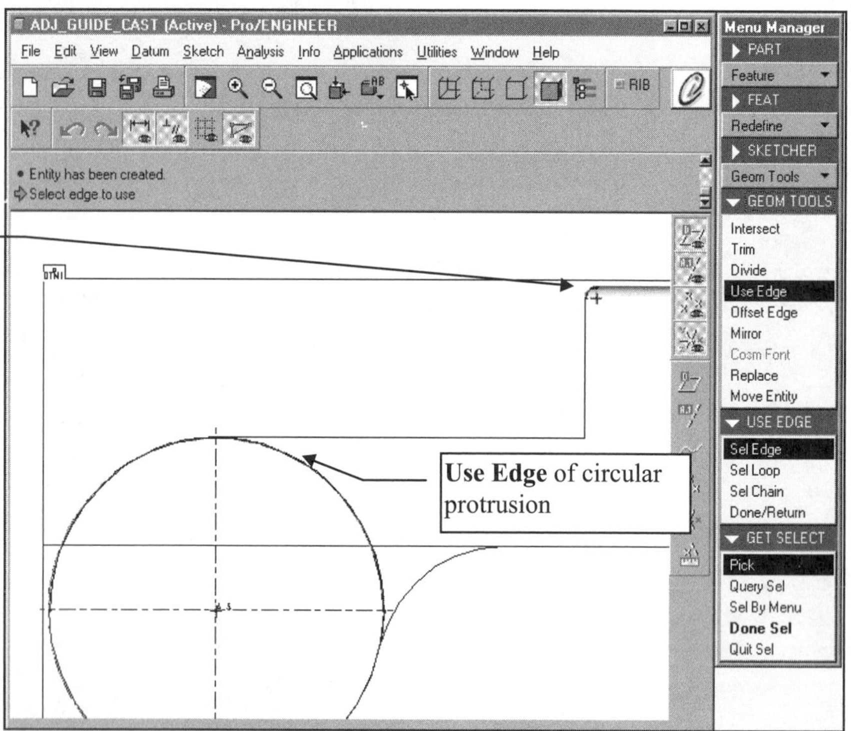

Figure 8.18
Use Edge

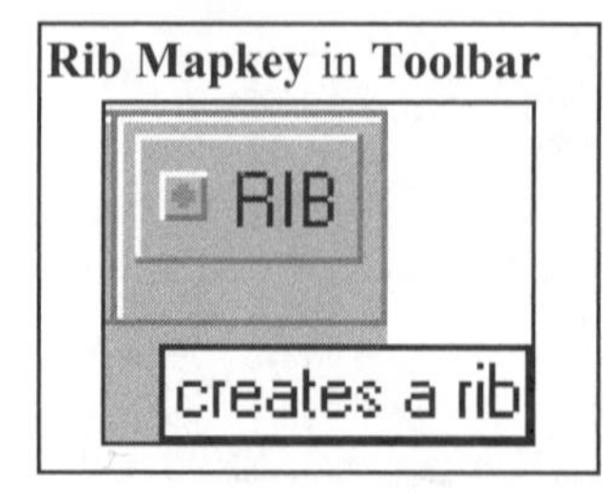

Before continuing with the modeling process, write a relation that will control the position of the rib. You used a *datum on the fly* to locate the center of the rib, so you need to write a relation that will say that the rib's center will remain at the center of the rectangular protrusion no matter how wide the protrusion becomes during a possible **ECO** change. The datum plane used to sketch on was offset **30 mm** from datum **B (FRONT).**

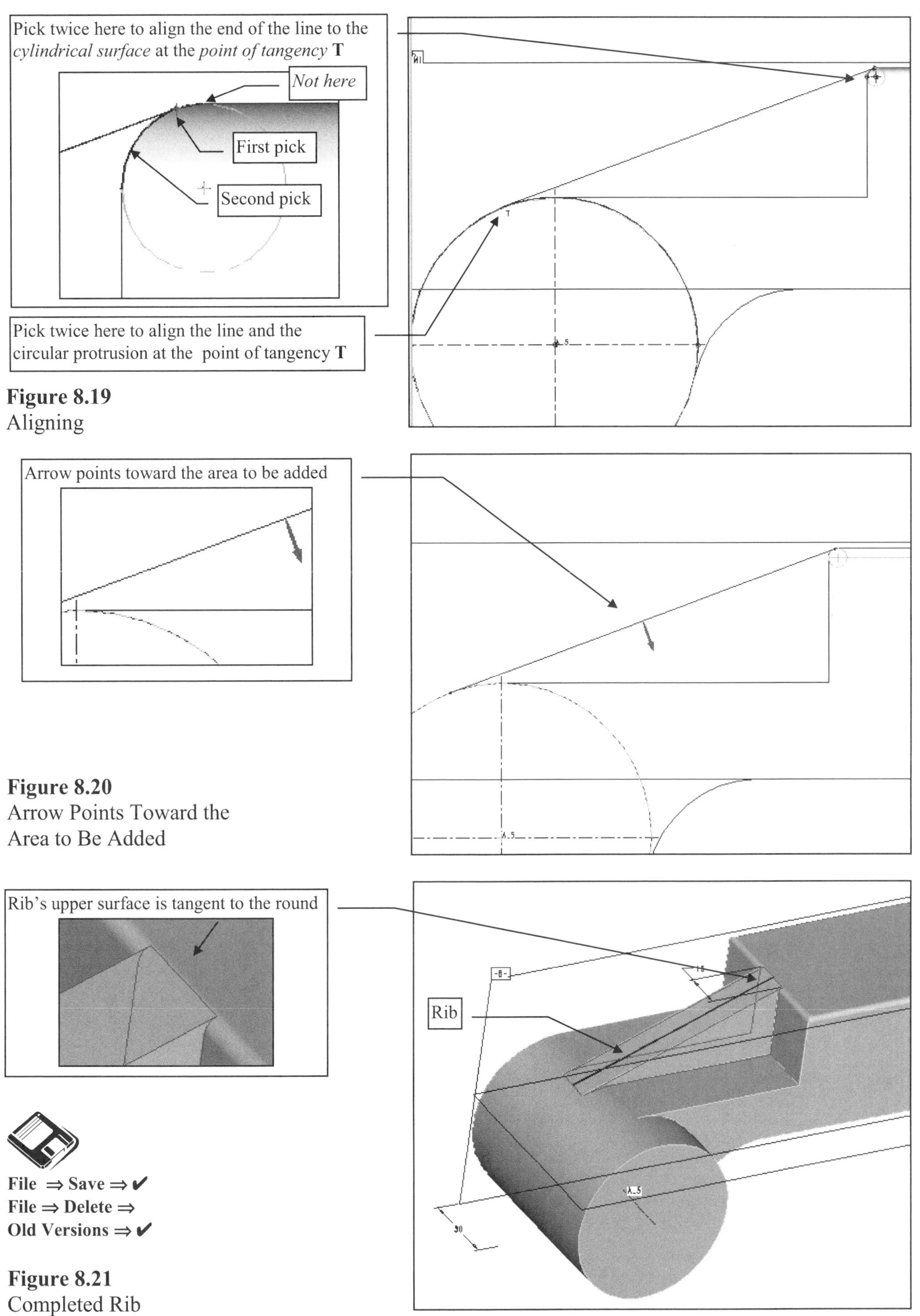

Figure 8.19
Aligning

Figure 8.20
Arrow Points Toward the Area to Be Added

File ⇒ Save ⇒ ✔
File ⇒ Delete ⇒
Old Versions ⇒ ✔

Figure 8.21
Completed Rib

The relation should control the distance from datum **B** to the datum created on the fly by saying that the distance equals one-half the width of the protrusion. You may need to use **Modify** ⇒ **DimCosmetics** ⇒ **Move Dim** ⇒ (select the desired feature and move the dimensions to new positions) to see the dimensions clearly. ***Your dimension symbol numbers may be different.*** Choose the following commands:

Relations ⇒ **Show Dim** ⇒ **enter** ⇒ (pick the rib and the protrusion as shown in Figure 8.22) ⇒ **Add** ⇒ (at the prompt, **Enter RELATION [Quit]: d101=d11/2**) ⇒ ✔ ⇒ ✔⇒ **Show Rel** (Fig. 8.23) ⇒ **(Close** to quit INFORMATION WINDOW) ⇒ **Done**

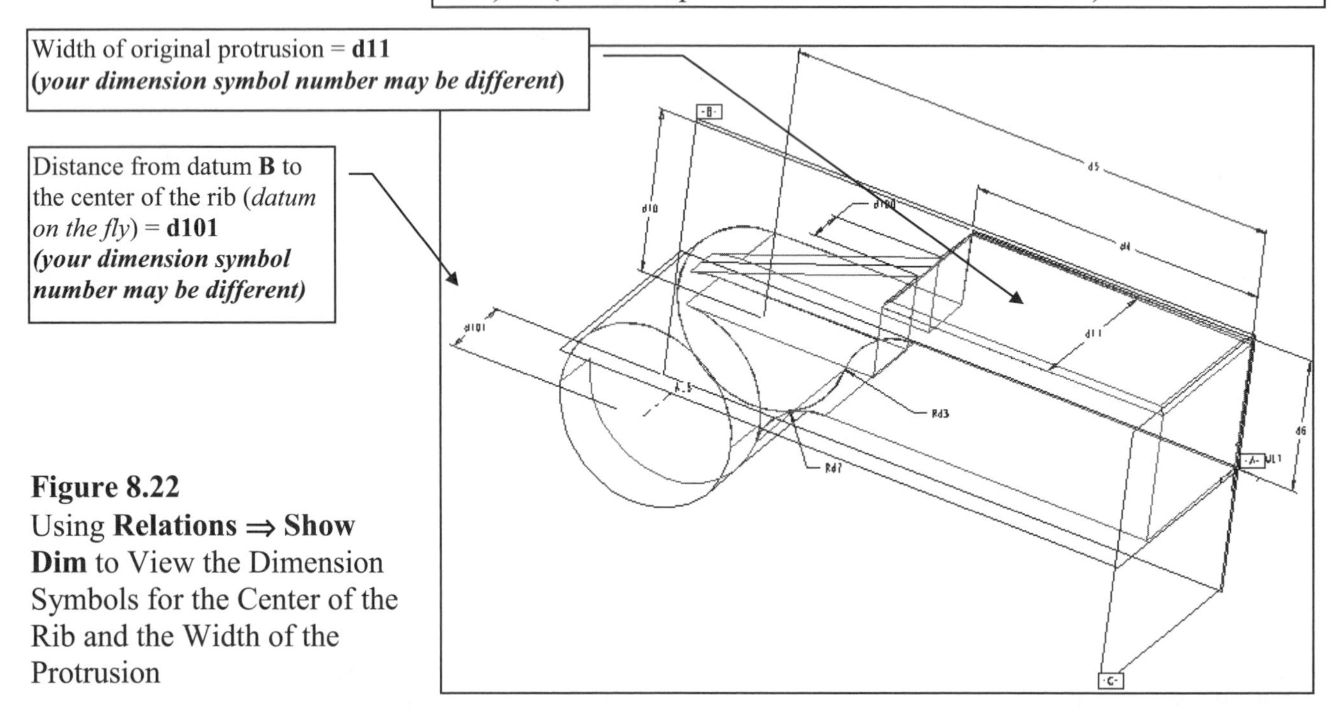

Figure 8.22
Using **Relations** ⇒ **Show Dim** to View the Dimension Symbols for the Center of the Rib and the Width of the Protrusion

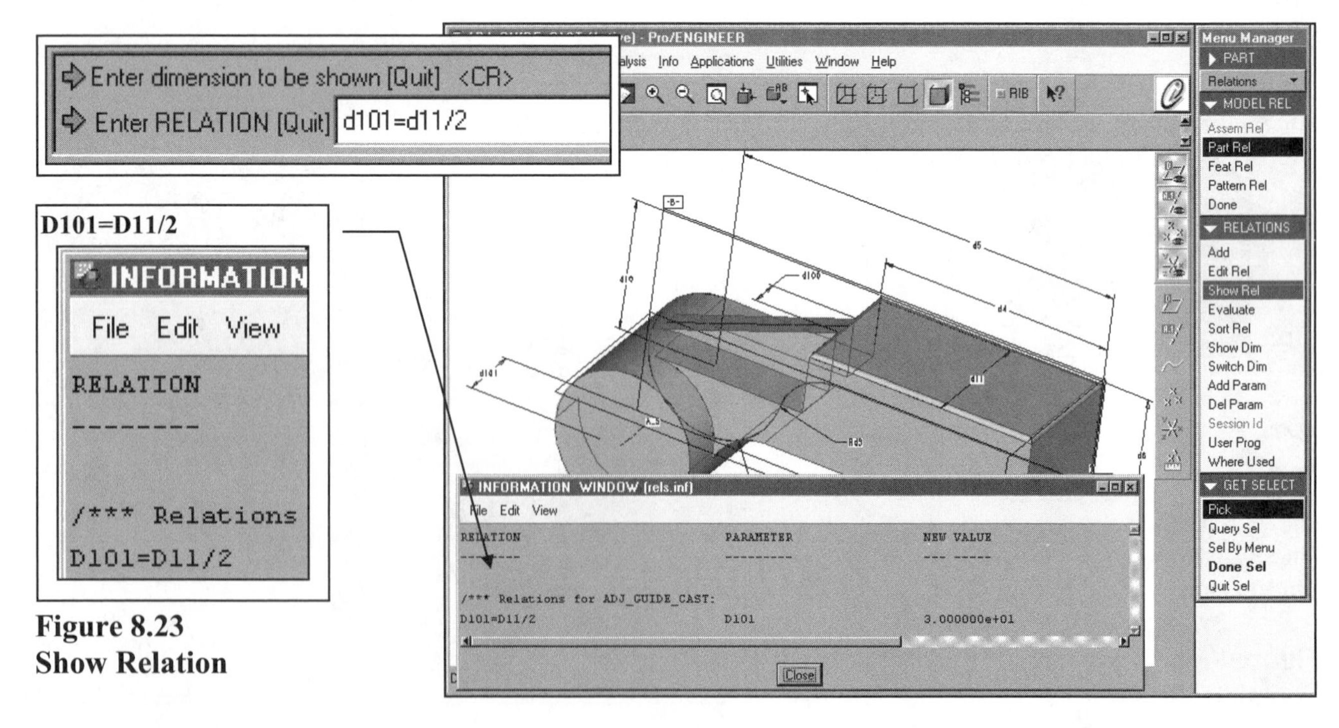

Figure 8.23
Show Relation

Create the rounds for the rest of the part. If you pick too many edges using **One By One**, you may get a failure, as shown in Figure 8.24, where the **Feature Info** choice was selected after the failure to explain the error. Redo the round, working one small set of connected edges at a time using **One By One** and **Tangnt Chain**, as shown in Figure 8.25.

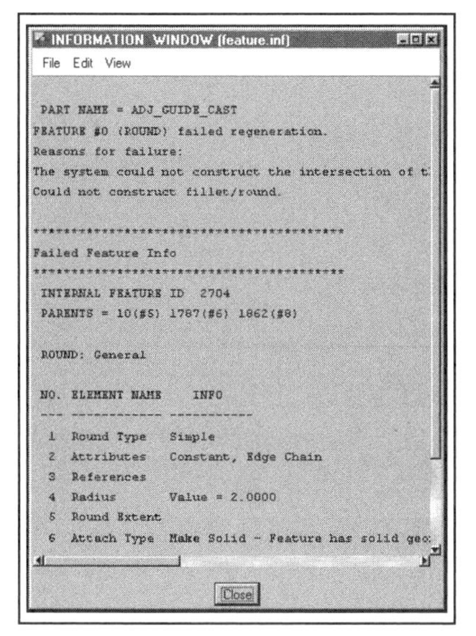

Figure 8.24
Feature Info During Failed Round

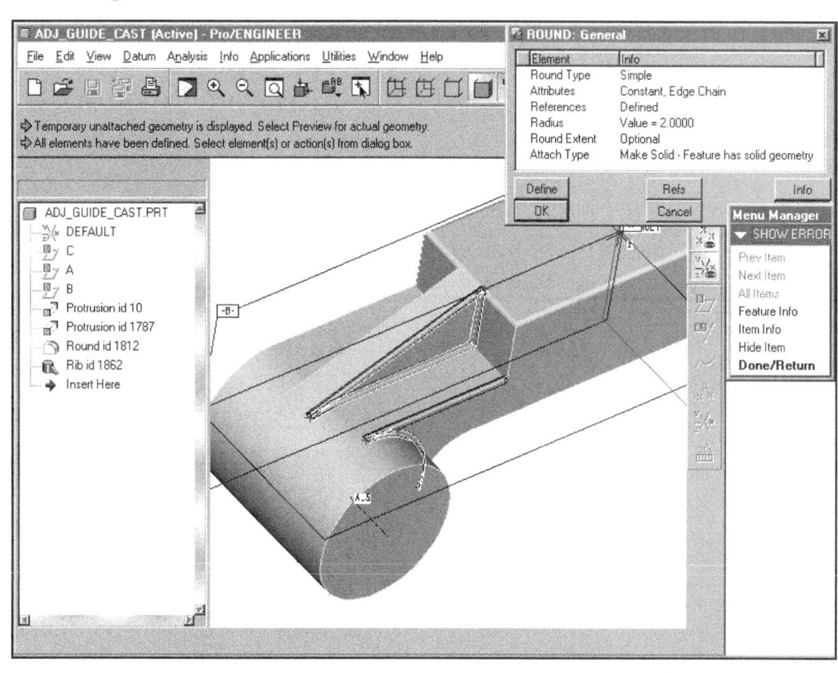

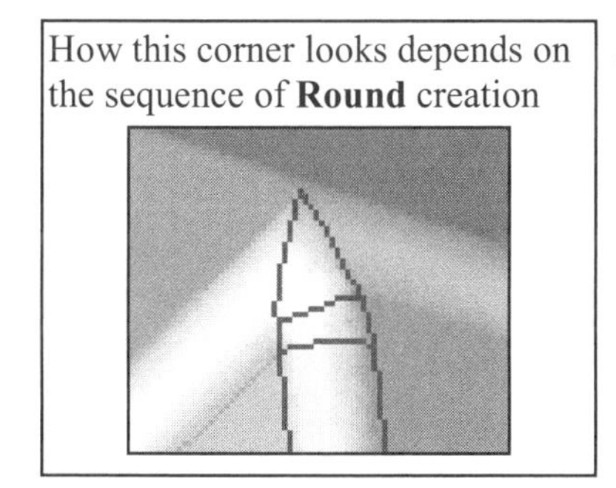

Use **Tangnt Chain** for remaining rounds that use this surface

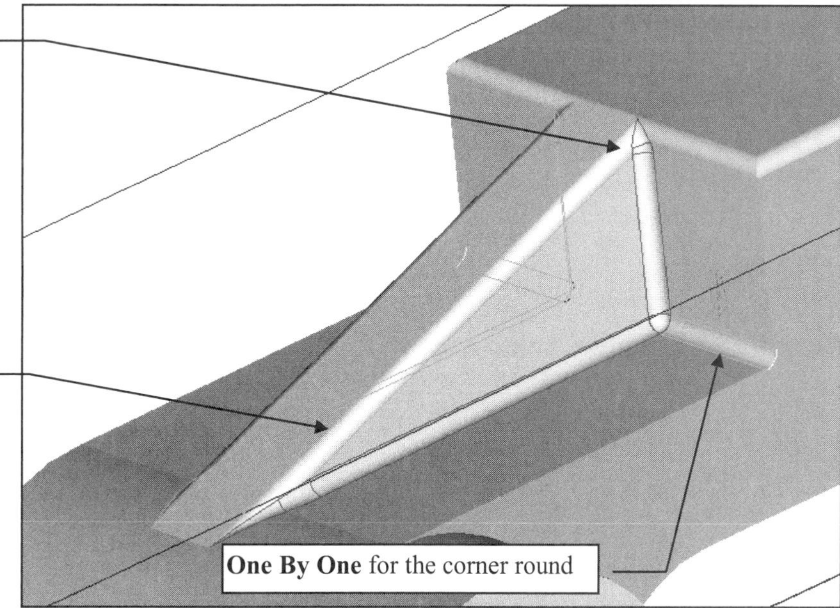

Figure 8.25
Rounds Created Successfully Using **One By One** and **Tangnt Chain**

If you get a failure as shown in Figure 8.26, choose:

Item Info (see what Pro/E suggests to solve the problem) ⇒ **Close** ⇒ **Done/Return** ⇒ **Resolve** ⇒ **Undo Changes** (if you use **Quick Fix** ⇒ **Delete** ⇒ **Confirm** instead, you may remove other rounds that you want to keep) ⇒ **Confirm**

Try other pick sequences for the round selection (Figure 8.27 and Figure 8.28).

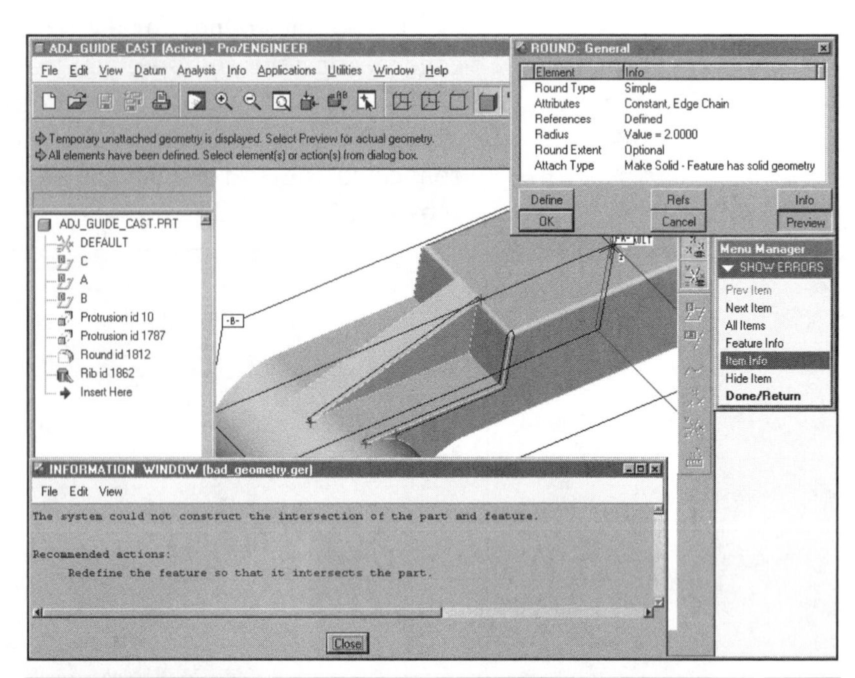

Figure 8.26
Round Failure and **Item Info**

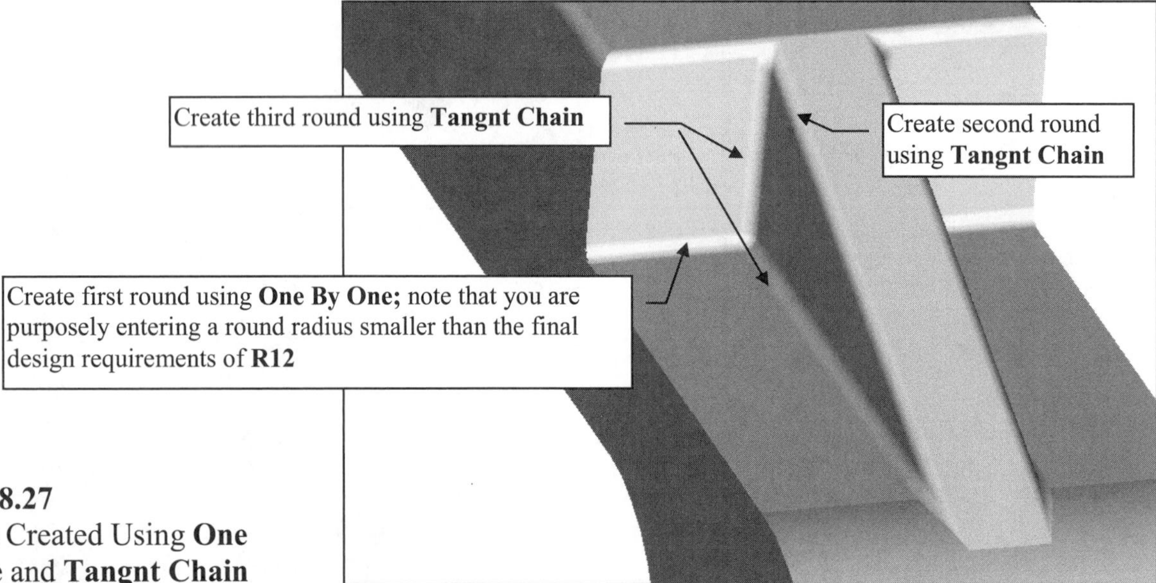

Figure 8.27
Rounds Created Using **One By One** and **Tangnt Chain**

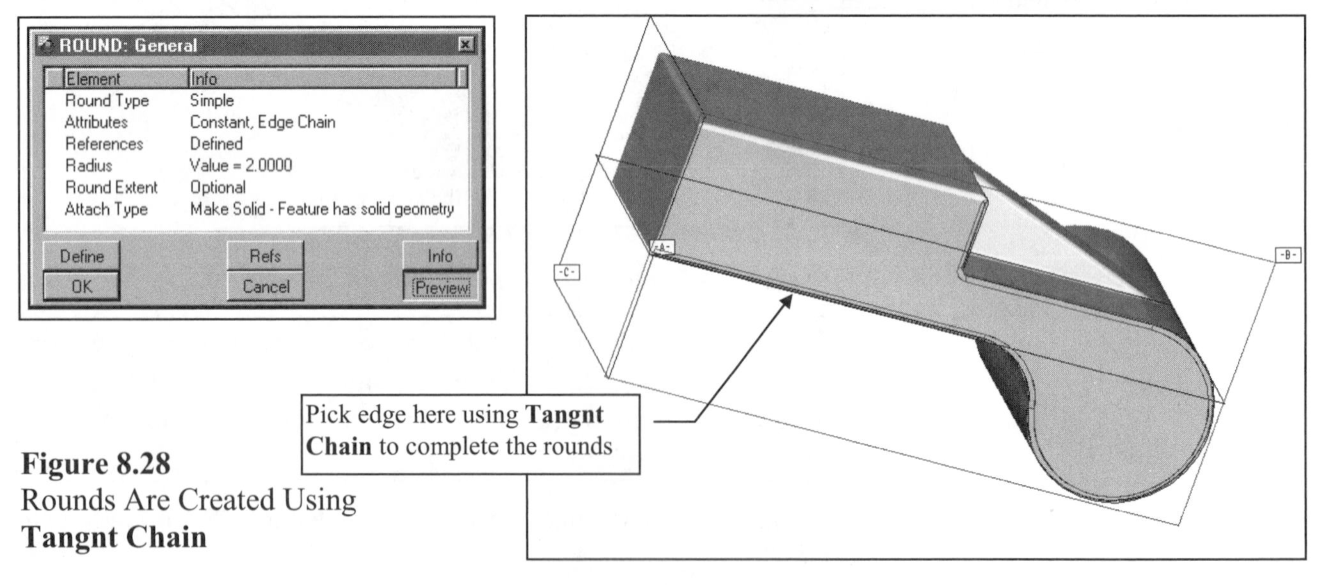

Figure 8.28
Rounds Are Created Using **Tangnt Chain**

It will take some trial and error (the whole purpose for this lesson!), but it is possible to get all the rounds done correctly, as shown in Figure 8.29.

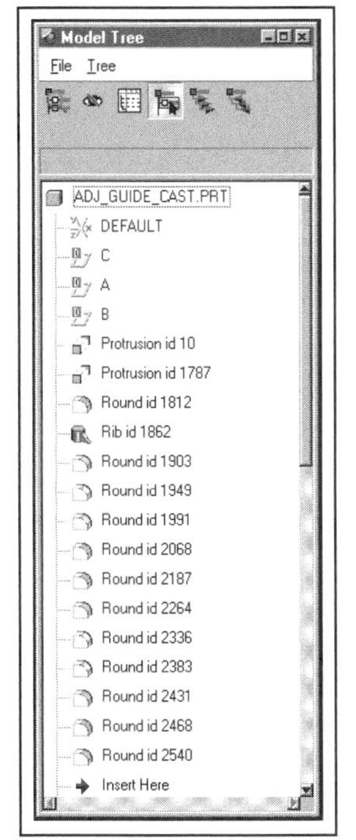

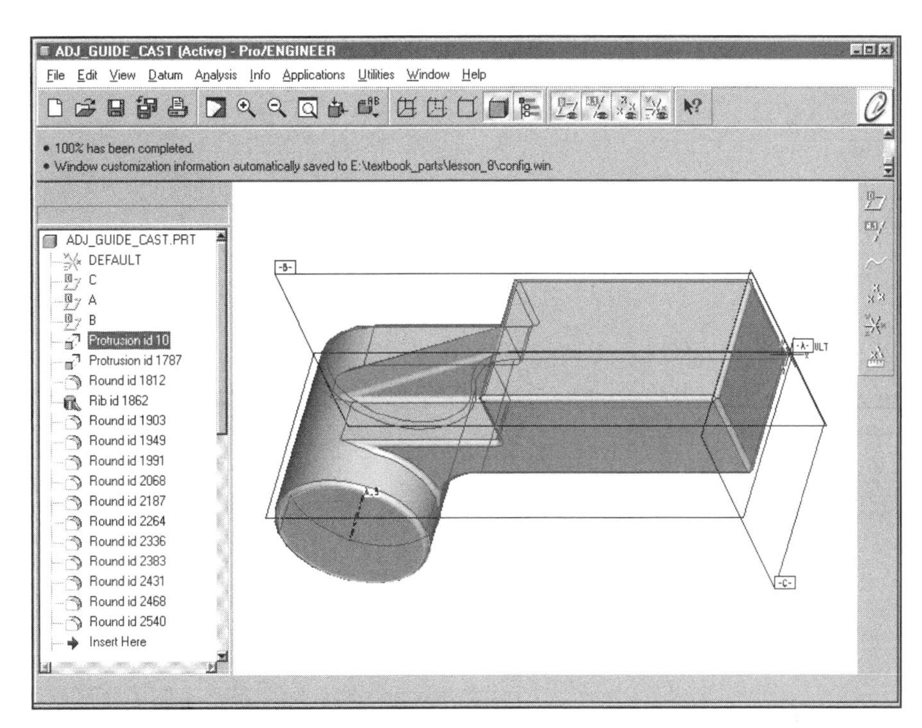

Figure 8.29
Completed Casting Part
(Workpiece)

File ⇒ Save ⇒ ✔
File ⇒ Delete ⇒
Old Versions ⇒ ✔
File ⇒ Save As ⇒
ADJ_GUIDE_CAST ⇒ OK

Modify the **R2** round to the design size of **R12** for the feature shown in Figure 8.30. The radius for the round must be updated on the casting and machine part to reflect the design requirements.

Use **File ⇒ Save As ⇒** (type new name for the casting **ADJ_GUIDE_CAST**) **⇒ ✔** to save the casting separate from the part file on which you will continue to model. Complete the machined part using the dimensions shown in Figure 8.14, and save.

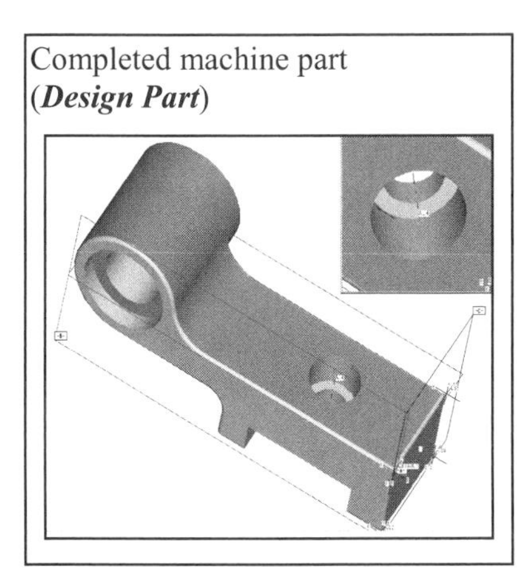

Figure 8.30
Modify the Round to **R12**

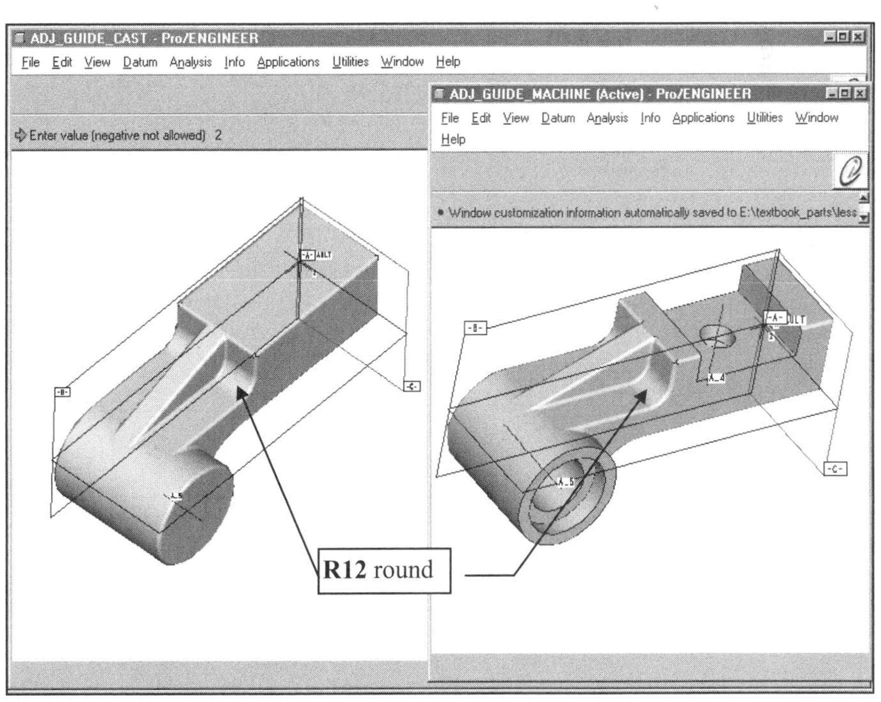

Lesson 8 Project

Clamp Arm

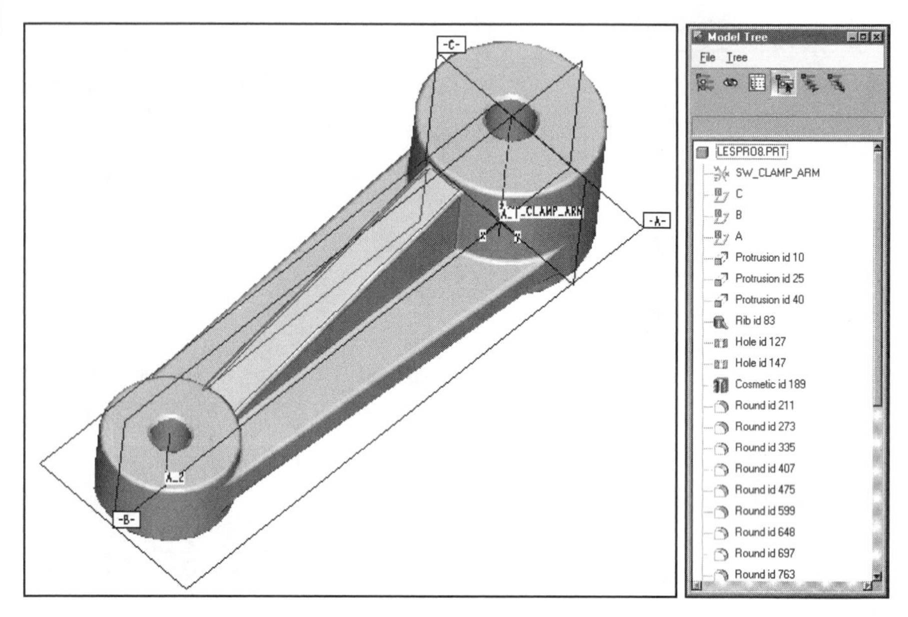

Figure 8.31
Clamp Arm (casting-workpiece)

Clamp Arm

The eighth **lesson project** is a cast part that requires commands similar to those for the **Adjustable Guide**. Create the part shown in Figures 8.31 through 8.37. Analyze the part and plan out the steps and features required to model it. Use the **DIPS** in Appendix D to plan out the feature creation sequence and the parent-child relationships for the part. The Clamp Arm is used in an assembly (Fig. 8.32) in Lessons 14 and 15. Create two versions of the Arm, one with all cast surfaces and one with machined ends.

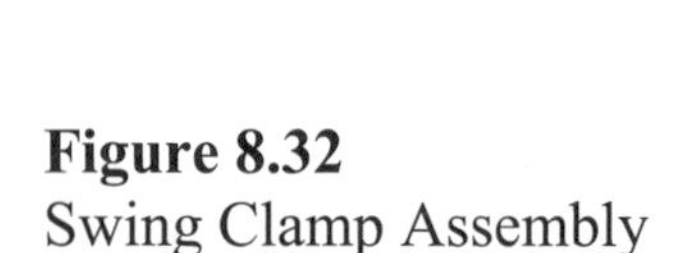

Figure 8.32
Swing Clamp Assembly

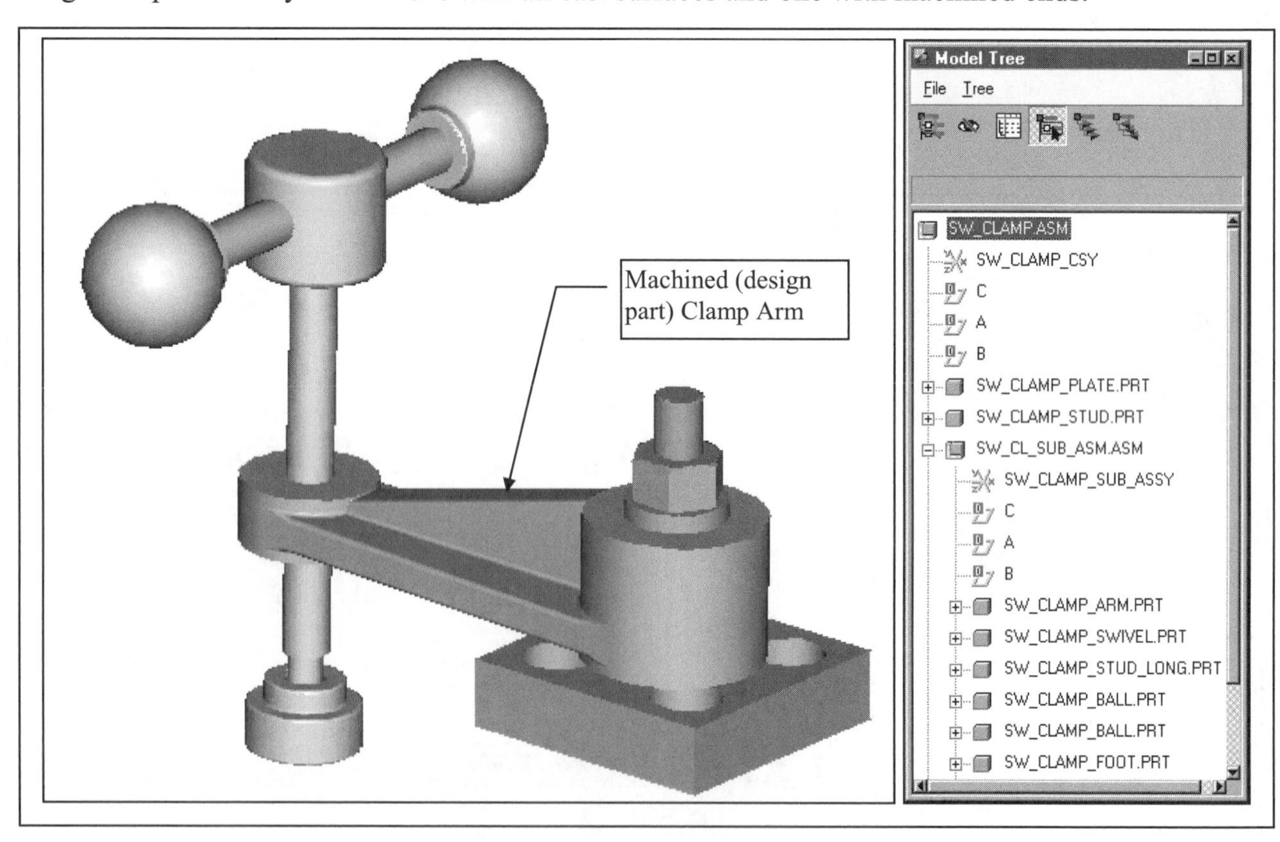

Write a relation to control the position of the horizontal ribs (Fig. 8.33). They must remain at the center of the smaller circular end.

Save a casting version of the part (no holes or machined surfaces) and a machined version. Include cosmetic threads where required. Machine the top and bottom of both round protrusions (**.10** off either end).

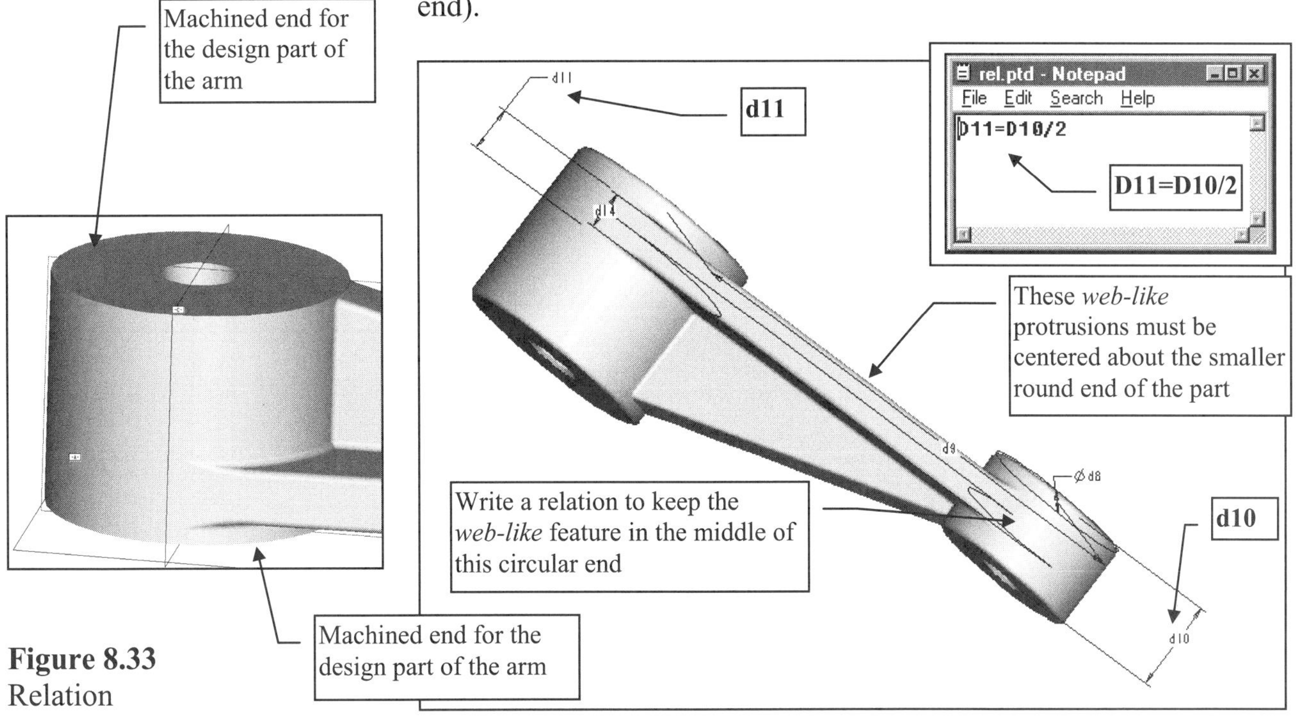

Figure 8.33
Relation

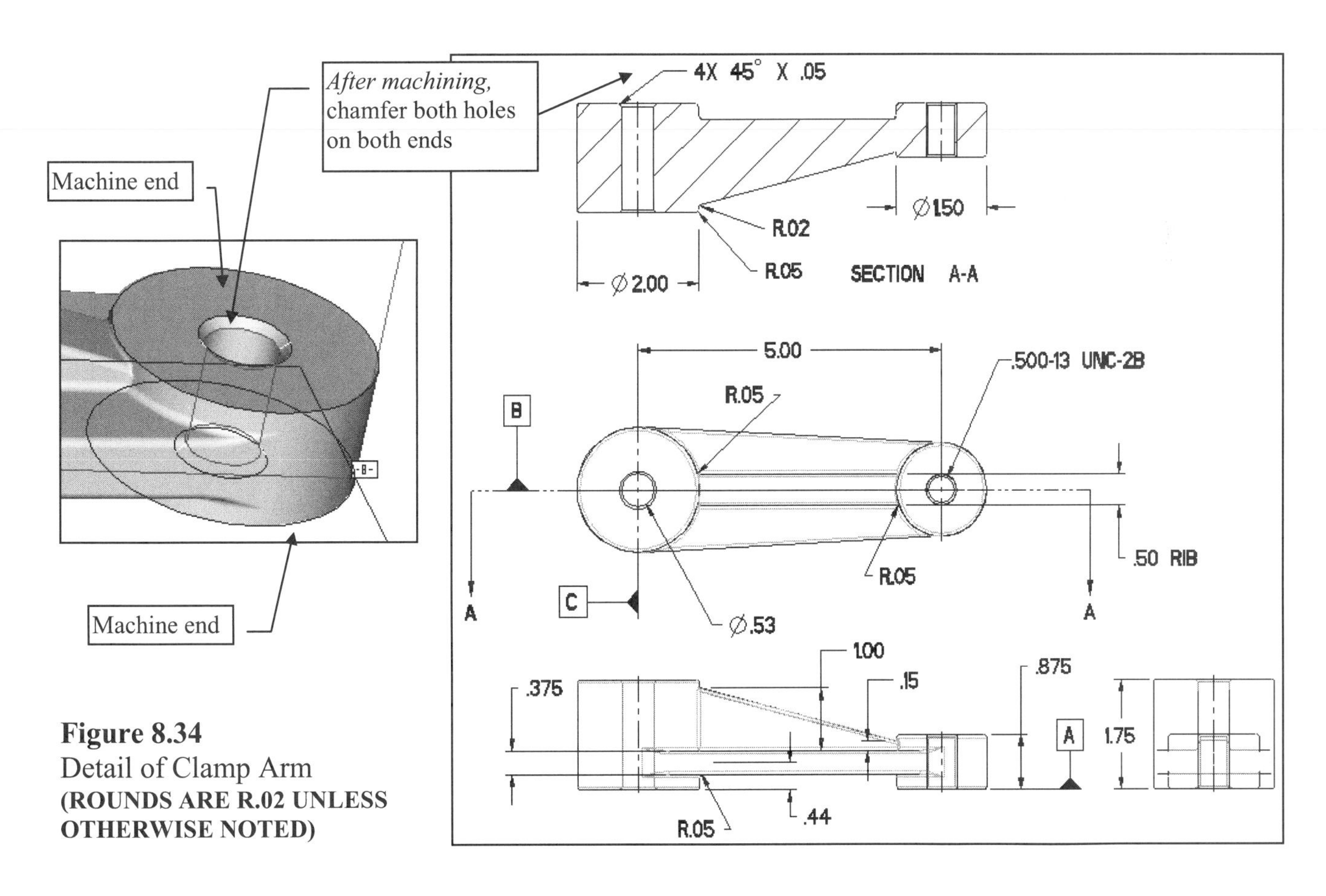

Figure 8.34
Detail of Clamp Arm
(ROUNDS ARE R.02 UNLESS OTHERWISE NOTED)

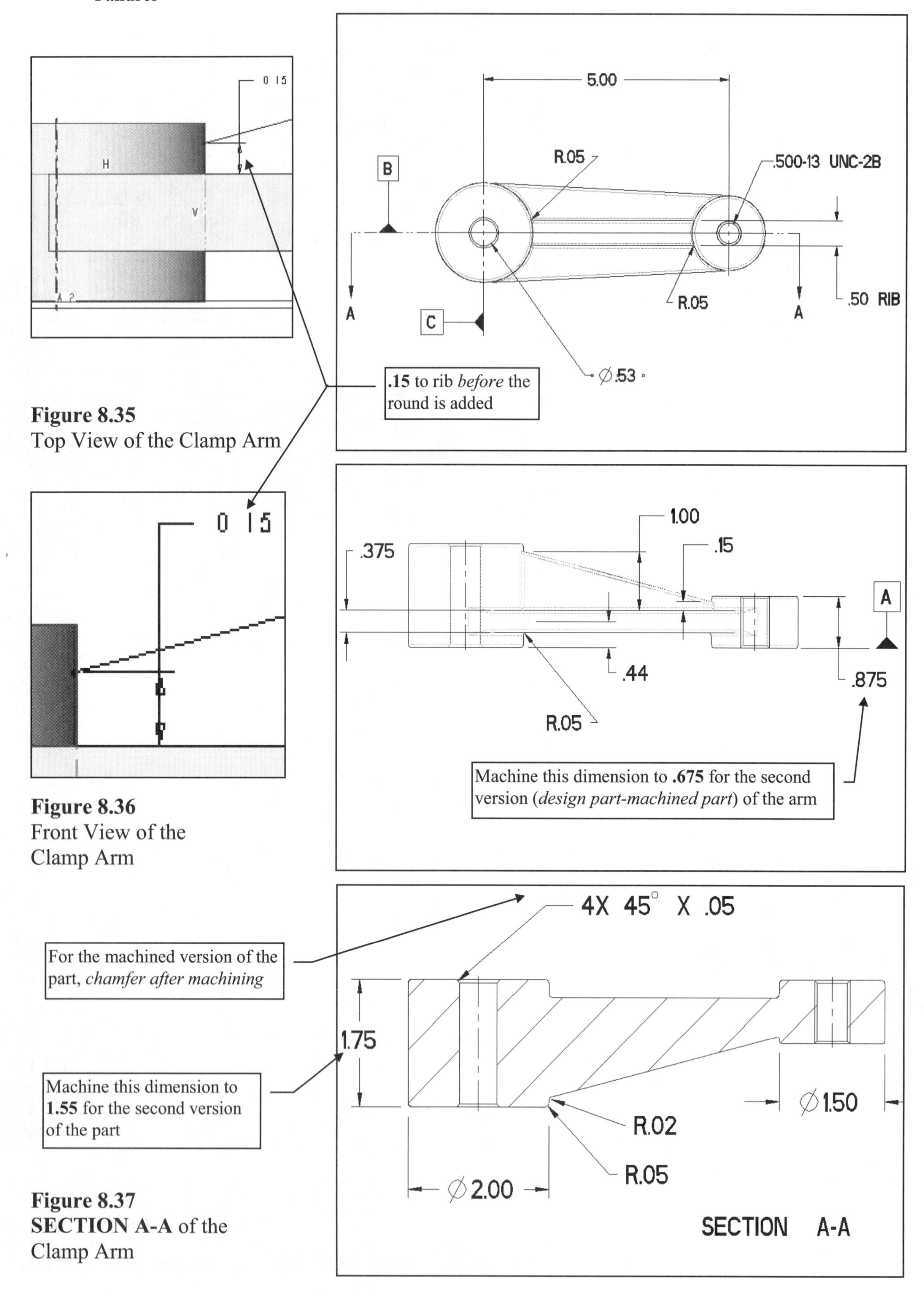

Figure 8.35
Top View of the Clamp Arm

Figure 8.36
Front View of the Clamp Arm

Figure 8.37
SECTION A-A of the Clamp Arm

Lesson 9

Drafts, Suppress, and Text Protrusions

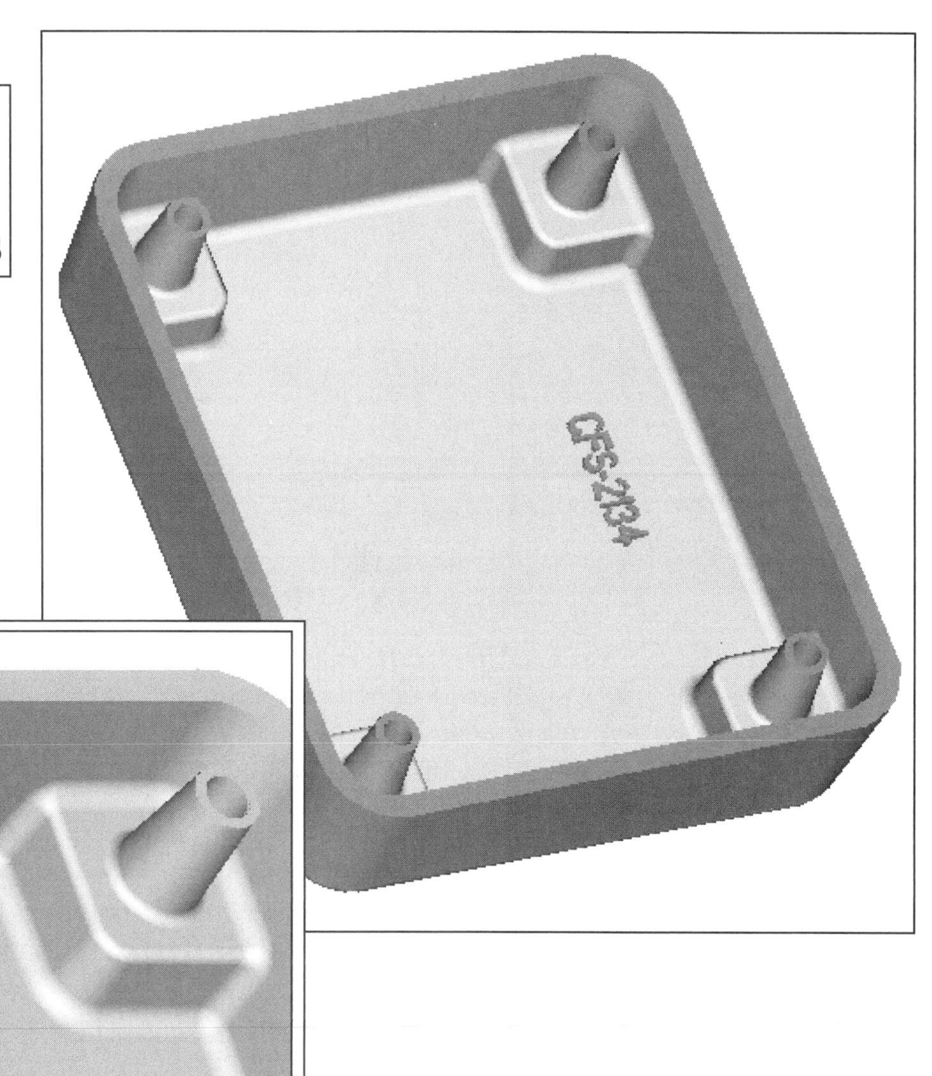

Figure 9.1
Enclosure

EGD REFERENCE
Fundamentals of Engineering Graphics and Design
by L. Lamit and K. Kitto
Read Chapter 14.
See pages 493-498.

COAch™ for Pro/ENGINEER

If you have **COAch for Pro/ENGINEER** on your system, go to SEARCH and do the Segment shown in Figure 9.3.

OBJECTIVES

1. **Understand the terminology, theory, and capability of the draft command**

2. **Create draft features**

3. **Shell a part**

4. **Suppress features to decrease regeneration time**

5. **Resume a set of suppressed features**

6. **Create text protrusions on parts**

Figure 9.2
Enclosure Dimensions

DRAFTS, SUPPRESS, AND TEXT PROTRUSIONS

The **Draft** feature adds a draft angle between surfaces. A wide variety of parts incorporate drafts into their design. Casting, injection mold, and die parts normally have drafted surfaces. The ENCLOSURE in Figure 9.1 and Figure 9.2 is a plastic injection-molded part.

Suppressing features by using the **Suppress** command temporarily removes them from regeneration. Suppressed features can be "unsuppressed" (**Resume**) at any time. It is sometimes convenient to suppress text protrusions and rounds to speed up regeneration of the model.

Text can be included in a sketch for extruded protrusions and cuts, trimming surfaces, and cosmetic features. To decrease regeneration time of the model, text can be suppressed after it has been created. Text can also be drafted.

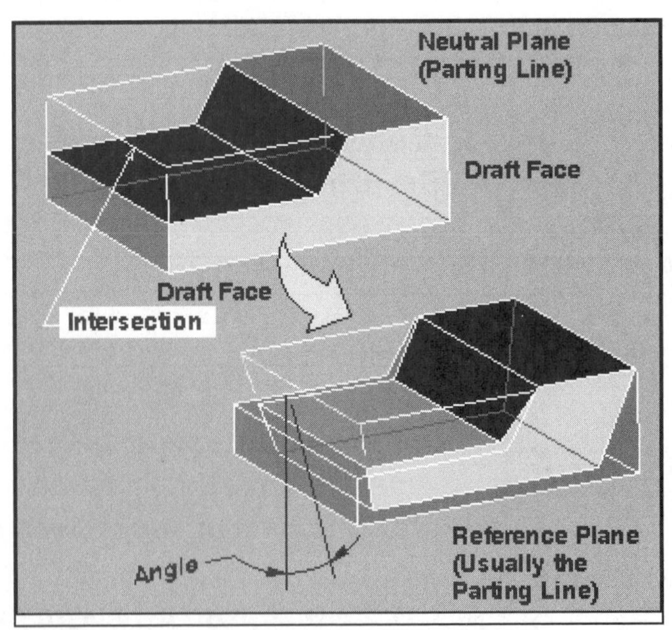

Figure 9.3
COAch for Pro/E, More Features (Draft Features)

Drafts

The **Draft** feature adds a draft angle between two individual surfaces or to a series of selected planar surfaces (Fig. 9.3). The following terminology is used in draft creation:

Draft surfaces are the selected surfaces of the model designated for drafting.

Neutral plane defines the pivot plane (Fig. 9.4). The intersection of the neutral plane and drafted surfaces defines the *axis of rotation.*

Neutral curve defines the curve on the surfaces to be drafted. Drafted surfaces are rotated about the neutral curve.

Draft direction indicates the direction in which material is added. The draft direction is defined as a normal to the reference plane that you specify; it is shown on the screen as a *green* arrow. The normal also indicates the direction from which the draft angle is measured.

Draft angle is measured between the draft direction and the resulting drafted surfaces. If the draft surfaces are split, you can define two independent angles for each portion of the draft.

Direction of rotation defines how surfaces are rotated with respect to the neutral plane or neutral curve.

Split areas of the draft allow you to divide draft surfaces so different draft angles can be applied.

When creating drafts, you can draft only surfaces that are formed by tabulated cylinders or planes. The reference plane used to establish the draft direction must be parallel to the neutral plane.

Surfaces with fillets around the edge boundary cannot be drafted. Therefore, draft the surfaces first, and then fillet the edges as required.

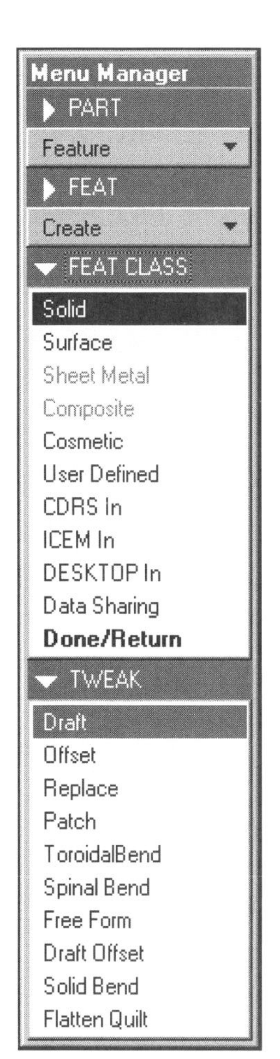

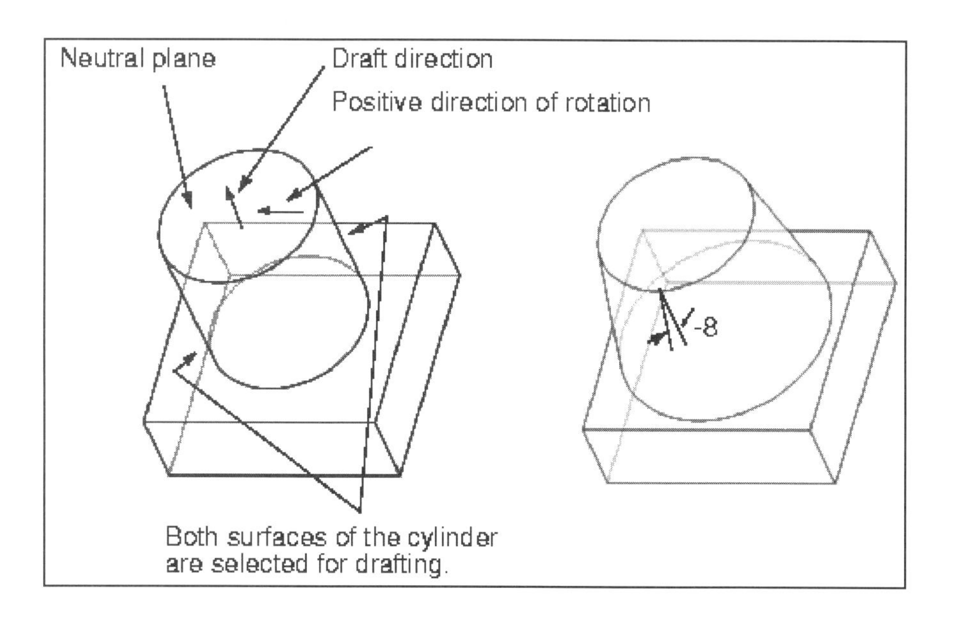

Figure 9.4
Drafts

Suppressing and Resuming Features

Suppressing a feature is similar to removing the feature from regeneration temporarily (Fig. 9.5). You can "unsuppress" **(Resume)** suppressed features at any time. Features on a part can be suppressed to simplify the part model and decrease regeneration time. For example, while you work on one end of a shaft, it may be desirable to suppress features on the other end of the shaft. Similarly, while working on a complex assembly, you can suppress some of the features and components for which the detail is not essential to the current assembly process.

Unlike other features, the base feature cannot be suppressed. If you are not satisfied with your base feature, you can redefine the section of the feature, or you can delete it and start over again.

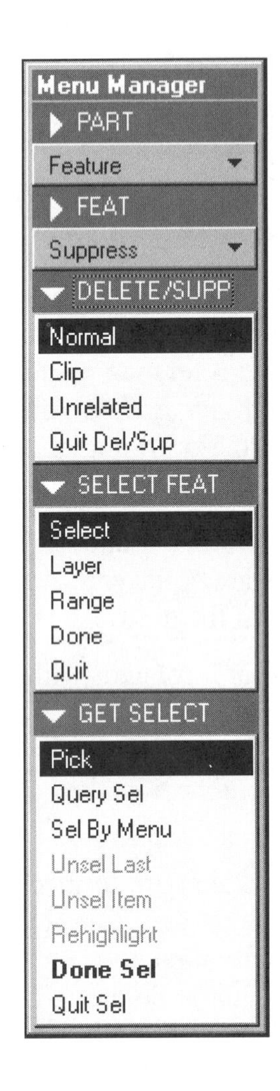

Suppressing Features

To suppress features, do the following:

1. Choose **Suppress** from the FEAT menu. Pro/E displays the SELECT FEAT and GET SELECT menus.
2. Choose one of the following options from the DELETE/SUPP menu:

   Normal Suppress the selected feature and all its children.
   Clip Suppress the selected feature and all the features that follow.
   Unrelated Suppress any feature that is not a child or parent of the selected feature.
3. Select a feature to suppress by picking on it, selecting from the Model Tree, specifying a *range*, entering its *feature number* or *identifier*, or using *layers*.

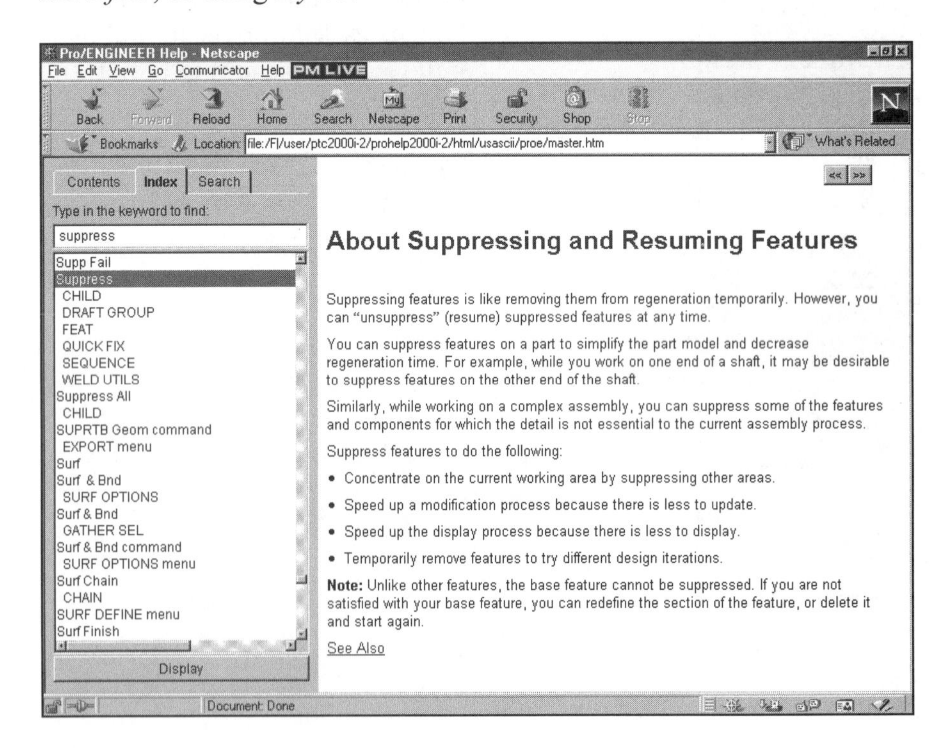

Figure 9.5
Suppressing Features Online Documentation

You can use **Suppress** and **Resume** to simplify the part before inserting features such as text protrusions. Also, you may wish to suppress the text protrusion if there is other work to be done on the part. Text protrusions take time to regenerate, because they increase the file size considerably.

Text Protrusions

When you are modeling, **Text** can be included in a sketch for extruded protrusions and cuts, trimming surfaces, and cosmetic features (Fig. 9.6). The characters that are in an extruded feature use the font **font3d** as the default. Other fonts are available. For cosmetic features, any font may be used, and this is done by modifying the text after creating the sketch (Intent Manager off).

Figure 9.6
Text Protrusions

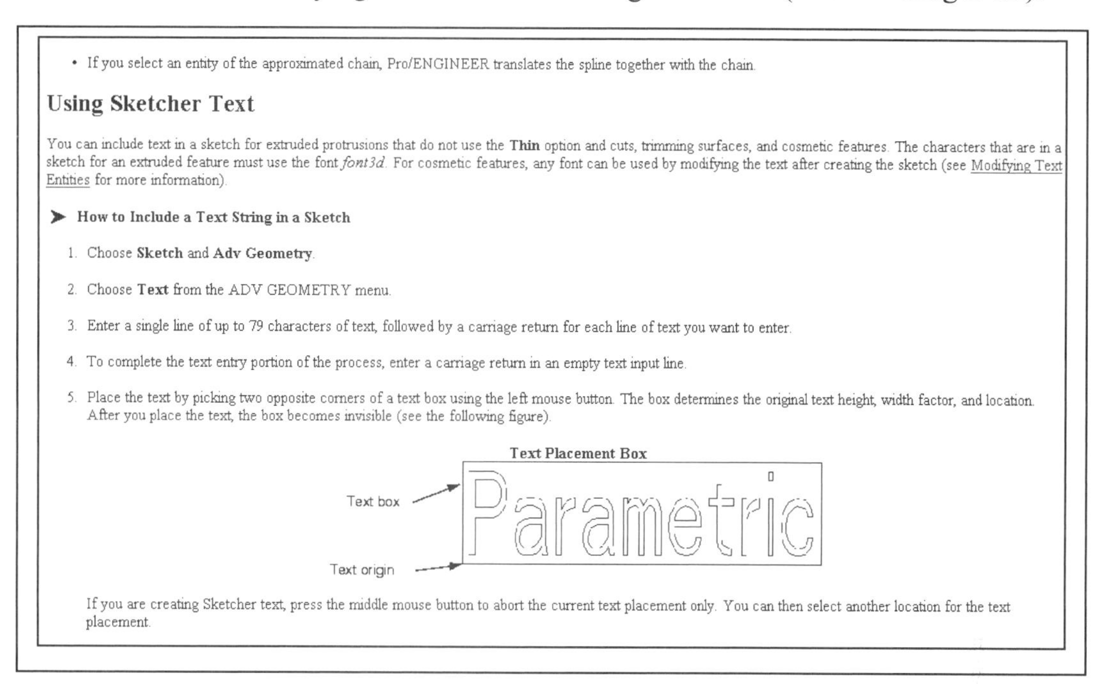

- If you select an entity of the approximated chain, Pro/ENGINEER translates the spline together with the chain.

Using Sketcher Text

You can include text in a sketch for extruded protrusions that do not use the **Thin** option and cuts, trimming surfaces, and cosmetic features. The characters that are in a sketch for an extruded feature must use the font *font3d*. For cosmetic features, any font can be used by modifying the text after creating the sketch (see Modifying Text Entities for more information).

➤ How to Include a Text String in a Sketch

1. Choose **Sketch** and **Adv Geometry**.
2. Choose **Text** from the ADV GEOMETRY menu.
3. Enter a single line of up to 79 characters of text, followed by a carriage return for each line of text you want to enter.
4. To complete the text entry portion of the process, enter a carriage return in an empty text input line.
5. Place the text by picking two opposite corners of a text box using the left mouse button. The box determines the original text height, width factor, and location. After you place the text, the box becomes invisible (see the following figure).

If you are creating Sketcher text, press the middle mouse button to abort the current text placement only. You can then select another location for the text placement.

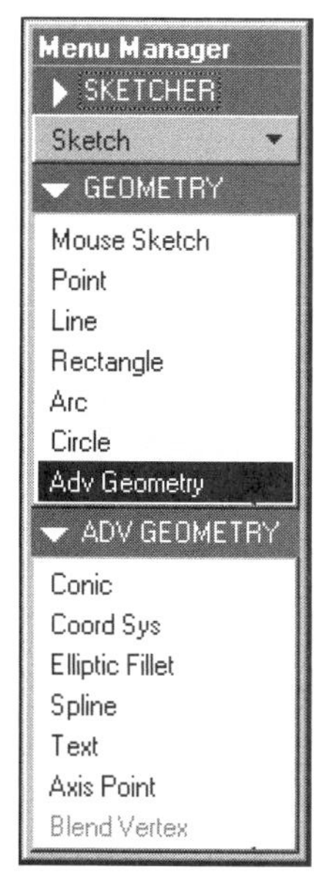

To include a text entry in a sketch (Intent Manager off):

1. Choose **Sketch ⇒ Adv Geometry.**
2. Choose **Text** from the ADV GEOMETRY menu.
3. Enter a single line of text. For the Enclosure, the part lot number, **CFS-2134**, was added to the model, as shown in Figure 9.1 and Figure 9.2.
4. Place the text by picking two opposite corners of a text box. The box determines the original text height, width factor, and location.
5. Dimension the text to the part or Sketcher geometry. To dimension the text, choose **Dimension** from the SKETCHER menu, pick anywhere on the text, pick a geometry entity, and place the dimension. The dimension will be created from the text origin (the lower left corner of the text box).

Figure 9.7
Enclosure

Lesson procedures & commands
START HERE ➡ ➡ ➡

NOTE

Set up the Enclosure:

- Material = PLASTIC
- Units = Inch lbm Second
- ✔ Snap to Grid
- Set datums with Geom Tol
- Rename the coordinate system: **ENCLOSURE**

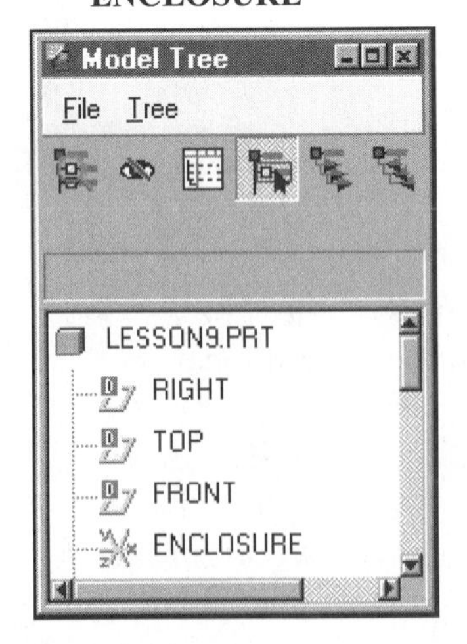

Figure 9.8
Enclosure Detail Drawing

Enclosure

The **Enclosure** is a plastic injection-molded part. A variety of drafts will be used in the design of this part. A *raised text protrusion* will be modeled on the inside of the Enclosure, as shown in Figure 9.7. The dimensions for the part are provided in Figures 9.8 through 9.13.

The first protrusion has a draft angle of **5°**, as well as the pad and cylindrical protrusions. The holes have a **.3°** draft angle.

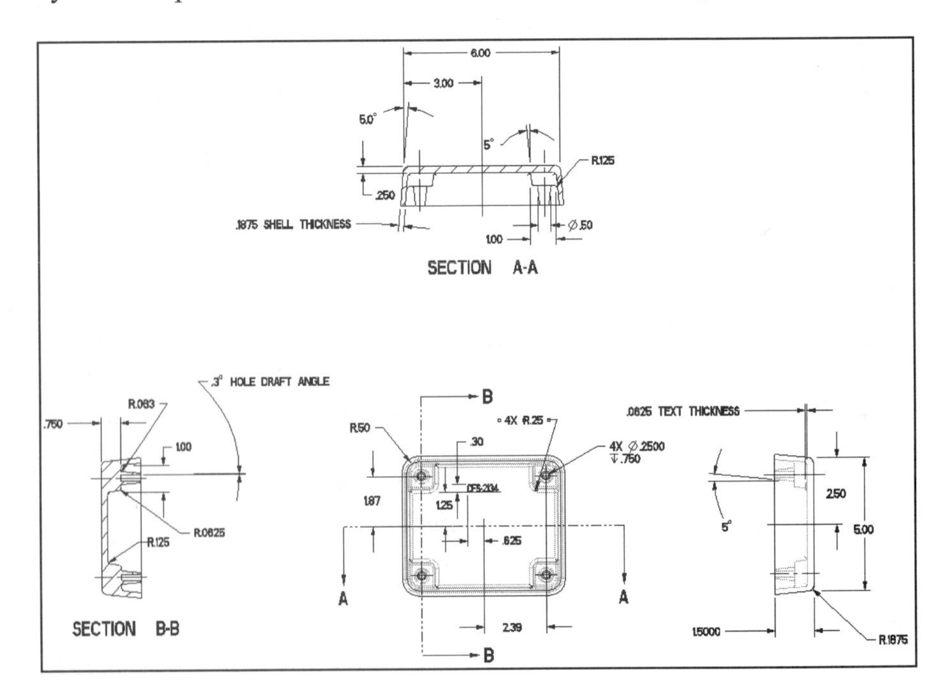

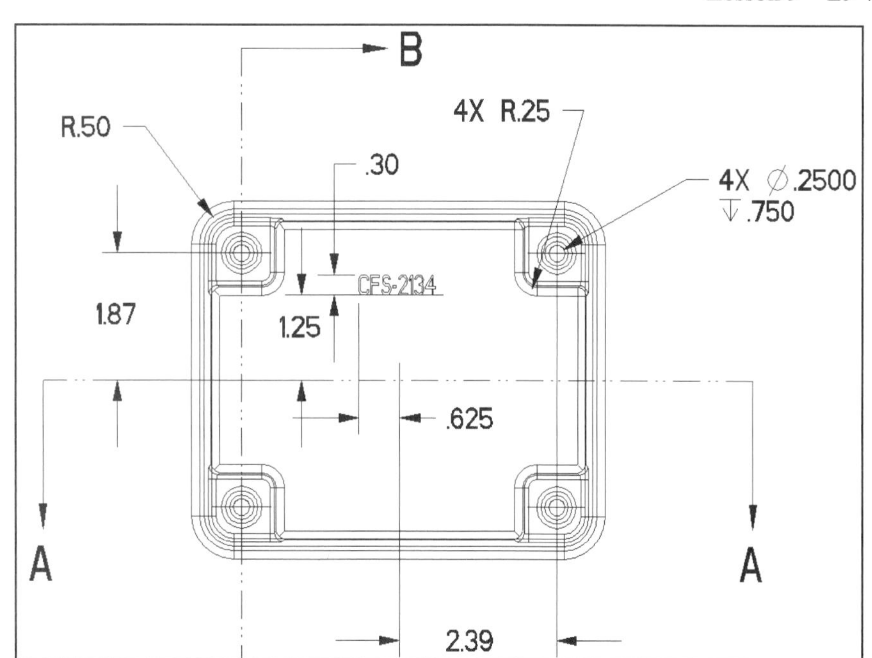

Figure 9.9
Enclosure Drawing,
Front View

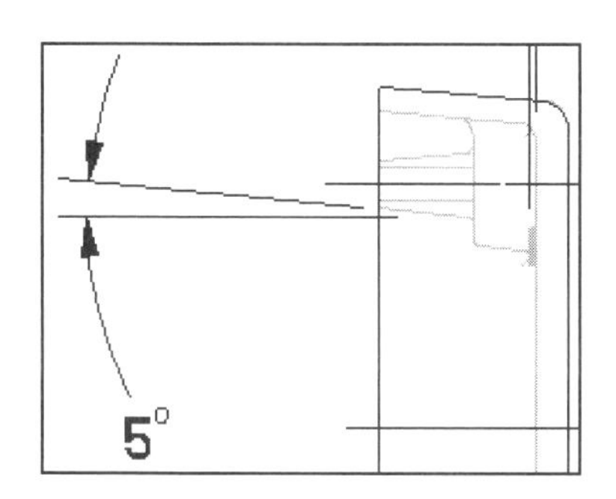

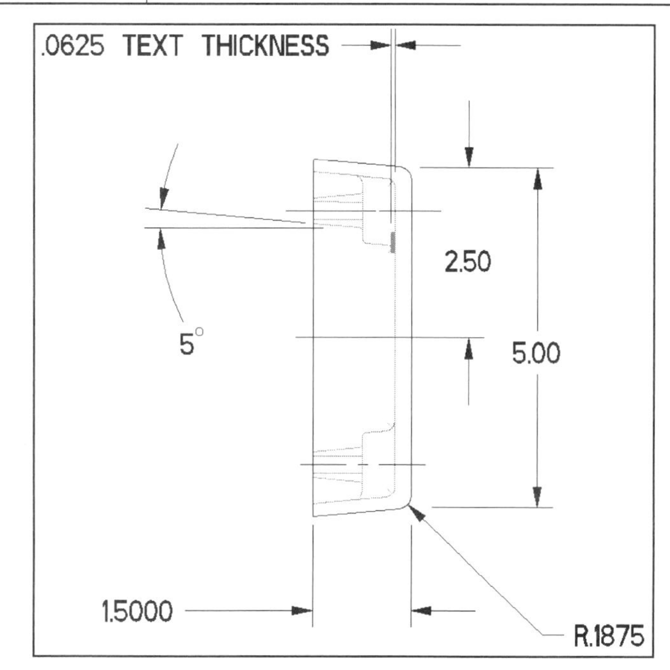

Figure 9.10
Enclosure Drawing,
Right Side View

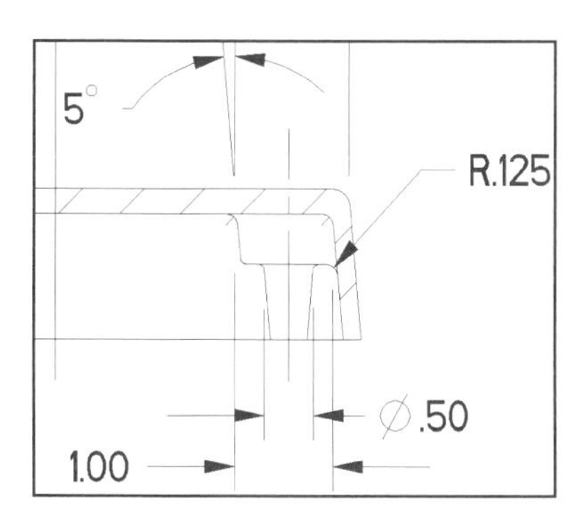

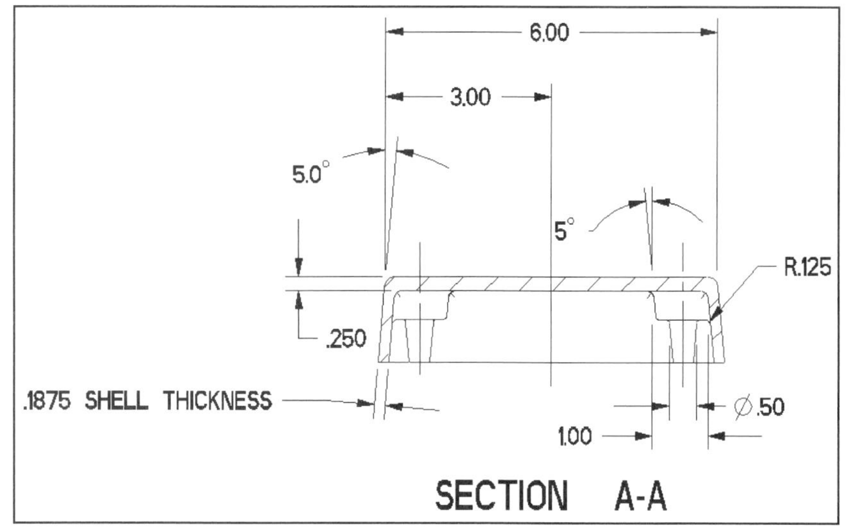

Figure 9.11
Enclosure Drawing,
SECTION A-A

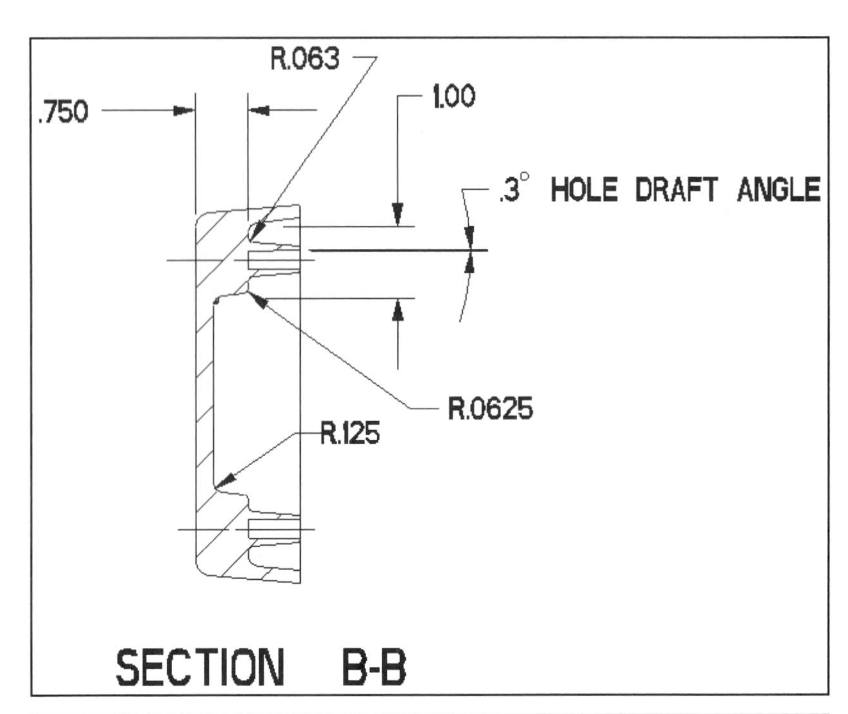

Figure 9.12
Enclosure Drawing,
SECTION B-B

CFS-2134

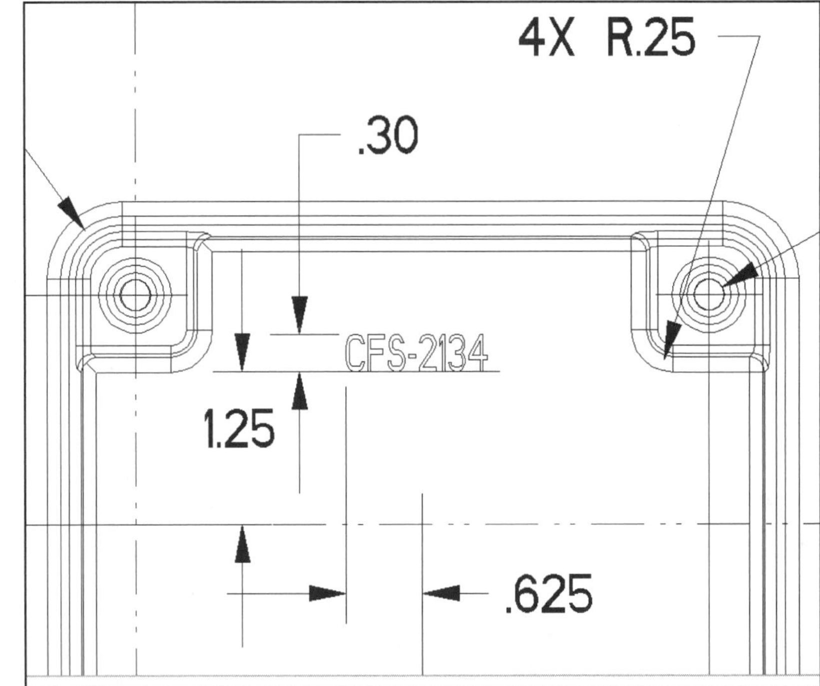

Figure 9.13
Enclosure Drawing,
Text Protrusion Location

Start the part with the ✔ **Use default template** and the ✔ **Intent Manager** (on). Make first protrusion **6.00** (height) **X 5.00** (width) **X 1.50** (depth), with **R.50** rounds [include the rounds **(Arc ⇒ Fillet)** in the sketch].

Sketch on datum plane **FRONT**, and center the first protrusion horizontally on datum **TOP** and vertically on datum **RIGHT**, as shown in Figure 9.14. Remember to add the fillets to the sketch rather than after the first protrusion is complete (Fig. 9.14). Add the drafts to the protrusion before it is shelled (Fig. 9.15). The protrusion is shelled to a thickness of **.1875** and **Spec Thick .25** on the bottom. Only the **Shell** command is introduced in this lesson. A complete description of shell capabilities appears in Lesson 10.

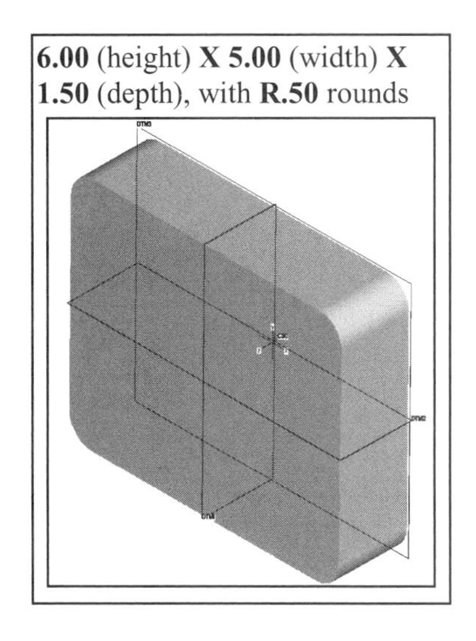

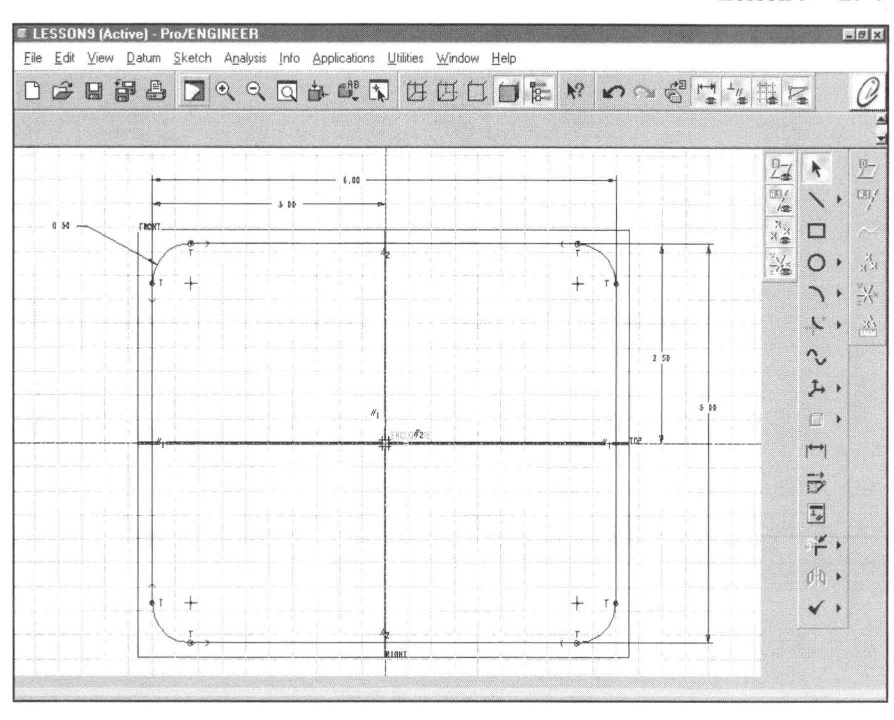

Figure 9.14
First Protrusion with Design Dimensions Added

Create the draft for the lateral surfaces of the protrusion using the following commands:

Feature ⇒ Create ⇒ Tweak ⇒ Draft ⇒ Neutral Pln ⇒ Done ⇒ No Split ⇒ Constant ⇒ Done ⇒ Include ⇒ Indiv Surfs ⇒ (pick all *eight lateral surfaces* of the protrusion, as shown in Fig. 9.15) ⇒ **Done Sel ⇒ Done ⇒** (select **FRONT** as the neutral plane) ⇒ (select **FRONT** as the plane the direction will be perpendicular to, as shown in Figure 9.15) ⇒ (specify the draft angle by typing **5** at the prompt) ⇒ ✔ ⇒ **Preview** (Fig. 9.16) ⇒ **OK ⇒ Done ⇒ File ⇒ Save ⇒ ✔**

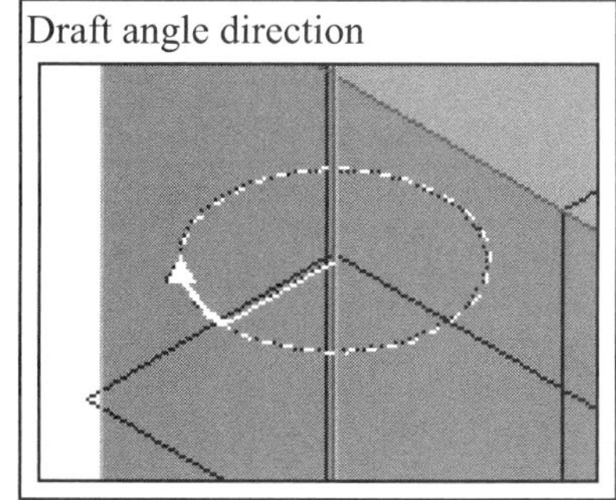

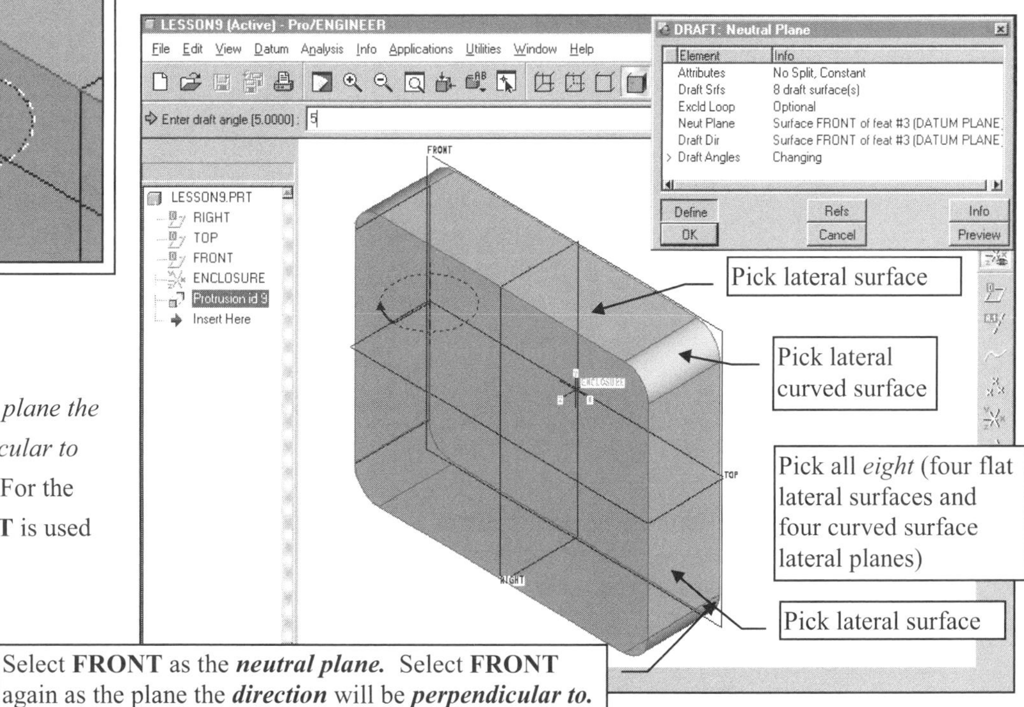

NOTE

The *neutral plane* and the *plane the direction will be perpendicular to* are often the same plane. For the draft created here, **FRONT** is used for both.

Figure 9.15
Draft Creation

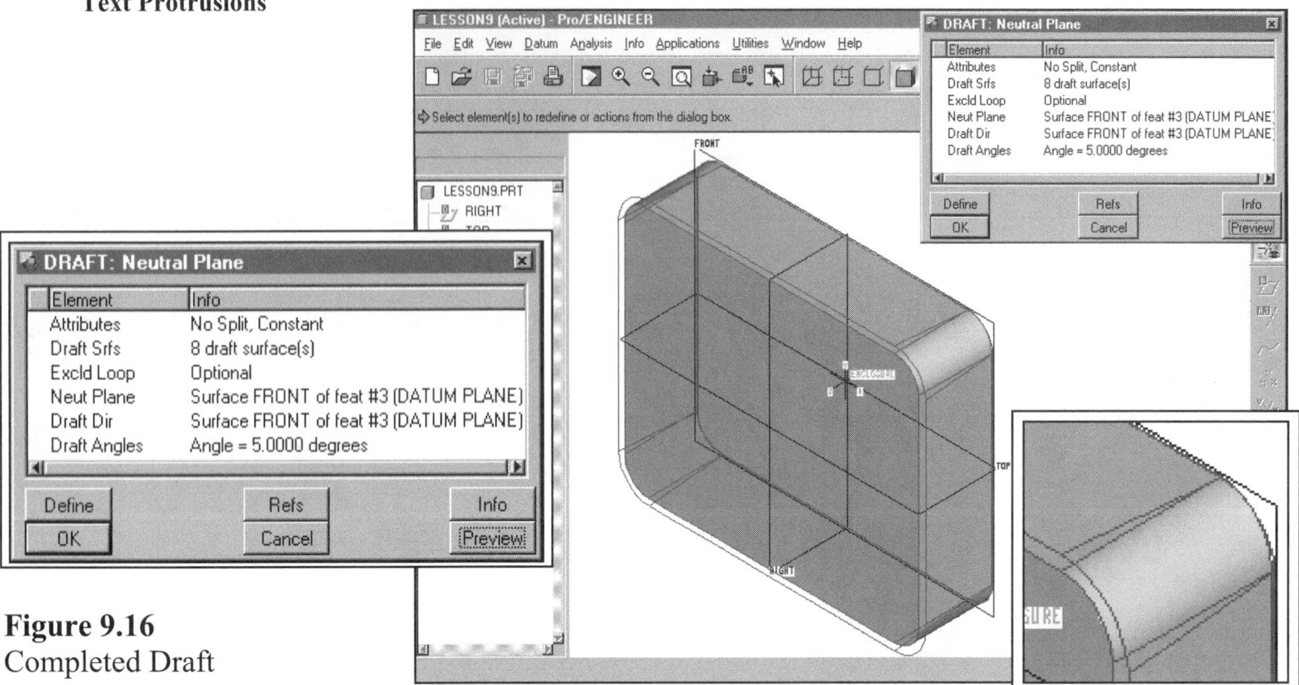

Figure 9.16
Completed Draft

Next you will shell the part using the following commands:

Feature ⇒ **Create** ⇒ **Shell** ⇒ [select the surface to remove by picking the vertical face (Fig. 9.17)] ⇒ **Done Sel** ⇒ **Done Refs** ⇒ (enter the thickness by typing **.1875** at the prompt, as shown in Figure 9.17) ⇒ ✔ ⇒ **Preview** (Fig. 9.18)

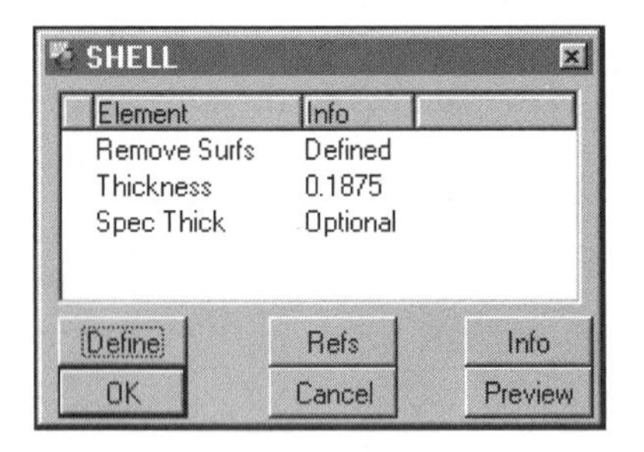

Shell thickness **.1875**

Pick this surface to shell

Figure 9.17
Selecting the Surface to Shell

Select **Spec Thick** from the dialog box ⇒ **Define** ⇒ **Set Thickness** ⇒ **Query Sel** (select the back surface as shown in Figure 9.18) ⇒ **Accept** ⇒ (Type **.250**) ⇒ ✔ ⇒ **Done Sel** ⇒ **Done** ⇒ **OK** ⇒ **Done**

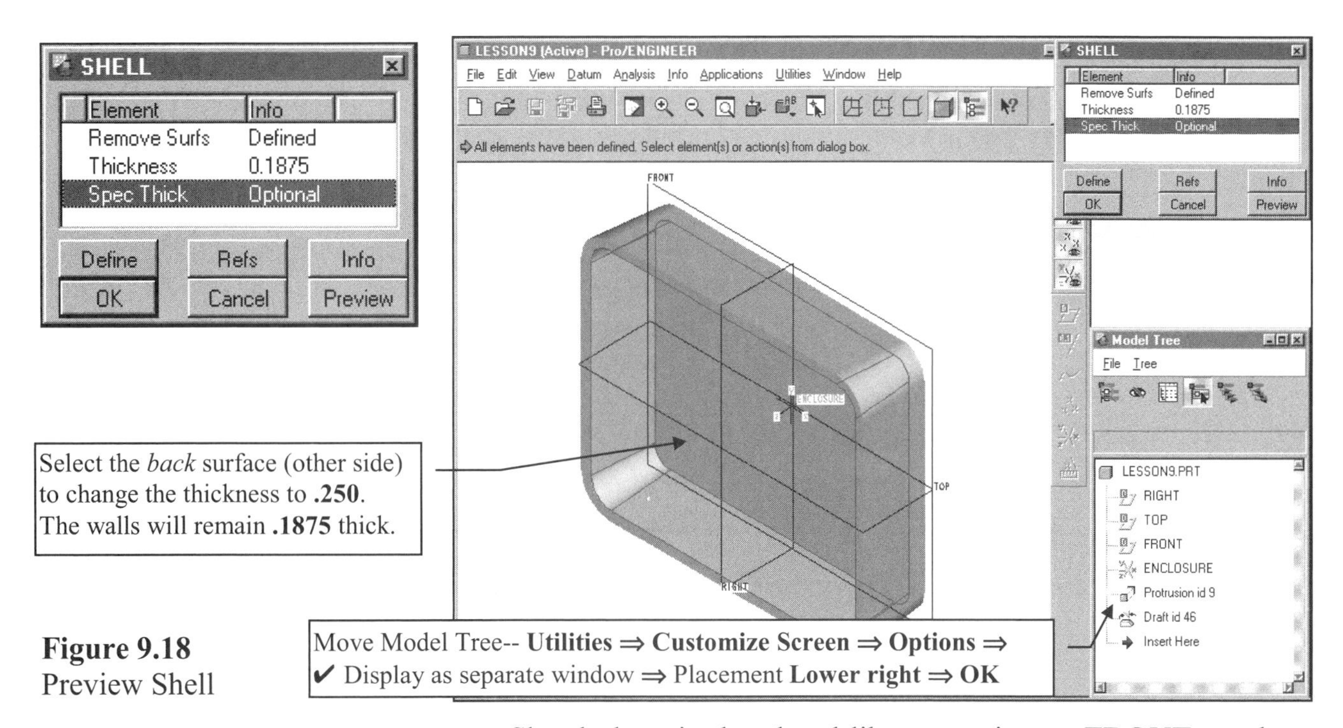

Figure 9.18
Preview Shell

Sketch the raised pedestal-like protrusion on **FRONT**, as shown in Figure 9.19. In most cases, the protrusion would be sketched on the inside of the shelled surface, but for this design we will project from the datum plane. Use the existing ***internal shelled edge*** and the dimensioning scheme shown in the figure and in the detail drawings provided earlier. Figure 9.20 shows the completed protrusion.

Figure 9.19
Sketch of the Second Protrusion

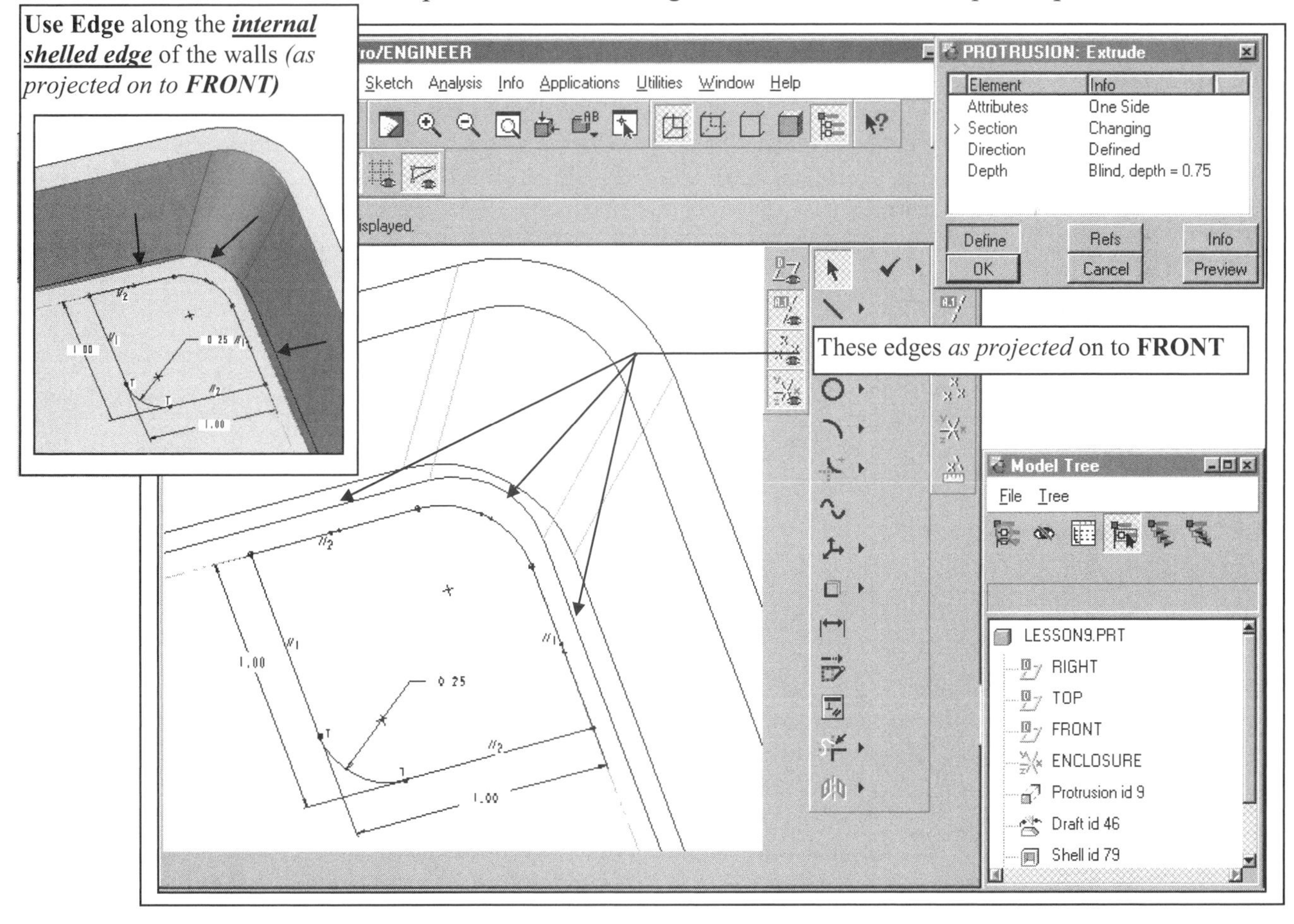

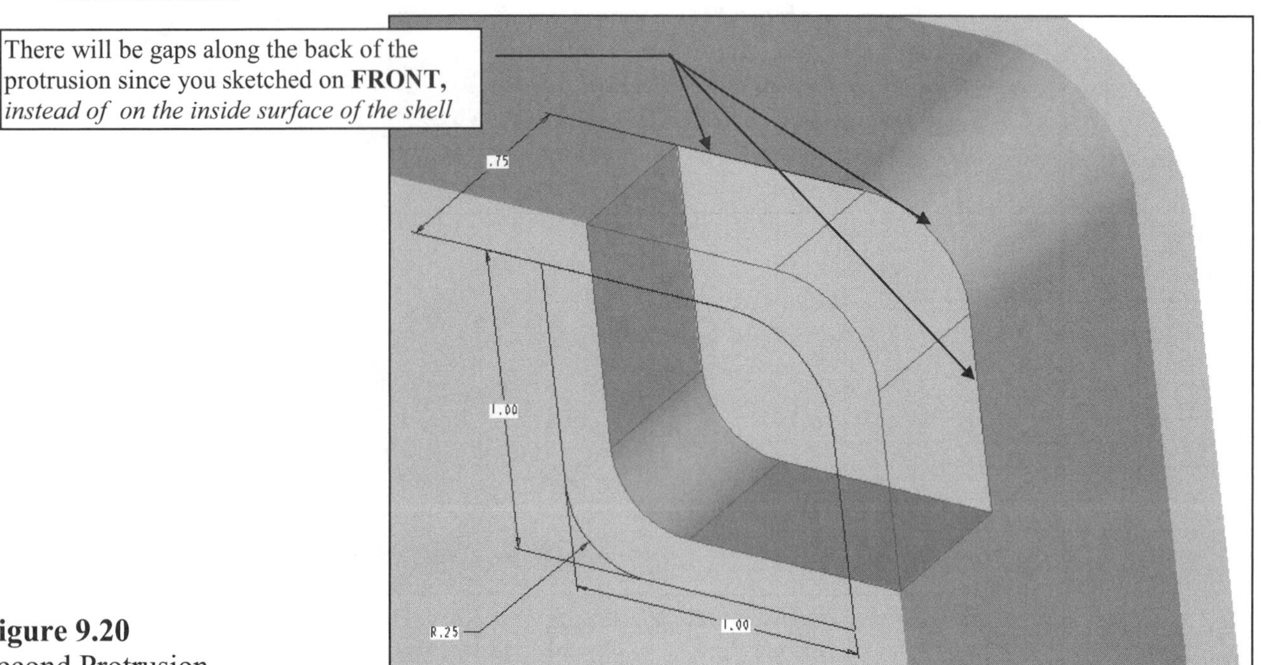

Figure 9.20
Second Protrusion

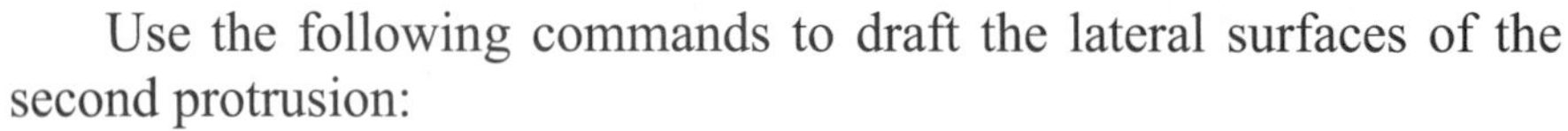

Use the following commands to draft the lateral surfaces of the second protrusion:

File ⇒ **Save** ⇒ ✔ ⇒ **File** ⇒ **Delete** ⇒ **Old Versions** ⇒ ✔ ⇒ **Feature** ⇒ **Create** ⇒ **Tweak** ⇒ **Draft** ⇒ **Neutral Pln** ⇒ **Done** ⇒ **No Split** ⇒ **Constant** ⇒ **Done** ⇒ **Include** ⇒ **Indiv Surfs** ⇒ (pick all ***three*** *lateral surfaces* of the protrusion, as shown in Fig. 9.21) ⇒ **Done Sel** ⇒ **Done** ⇒ (select **FRONT** as the neutral plane) ⇒ [select **FRONT** as the plane the direction will be perpendicular to (or pick **Use Neut Pln**)] ⇒ (specify the draft angle by typing **-5** at the prompt) ⇒ ✔ ⇒ **Preview** (Fig. 9.21) ⇒ **OK** ⇒ **Done**

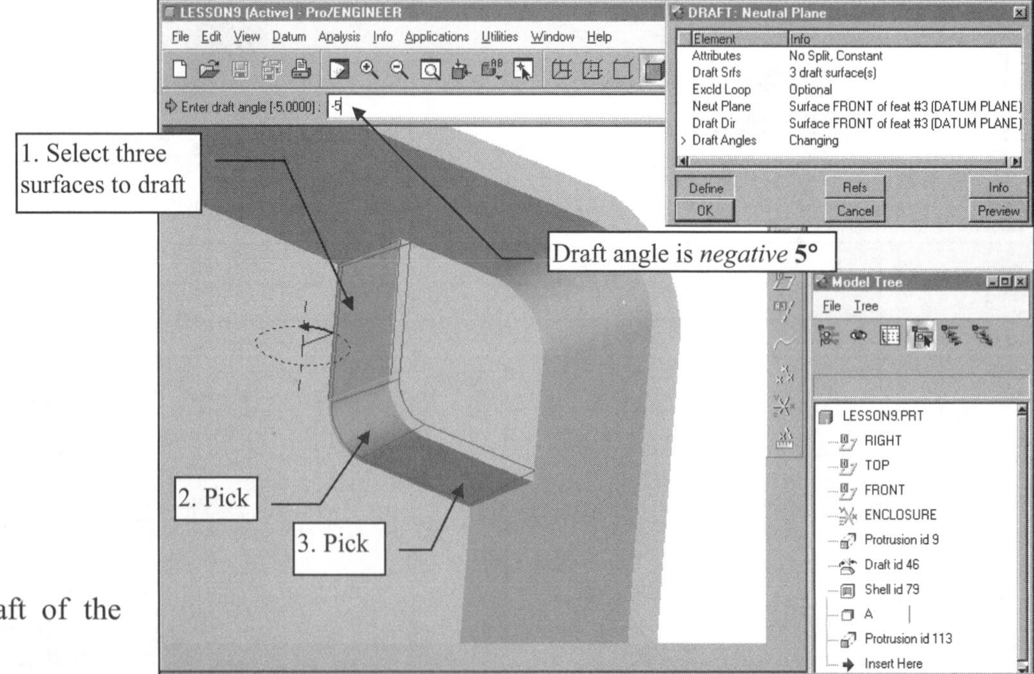

Figure 9.21
Preview of the Draft of the Second Protrusion

File ⇒ Save ⇒ ✔

Model the circular protrusion (Fig. 9.22). Draft the new protrusion at **-5°** (Fig. 9.23). Create a hole (Fig. 9.24), and add an internal draft of **-.3°.**

Protrusion goes **UpTo Surface**

Sketching plane

Figure 9.22
Circular Protrusion

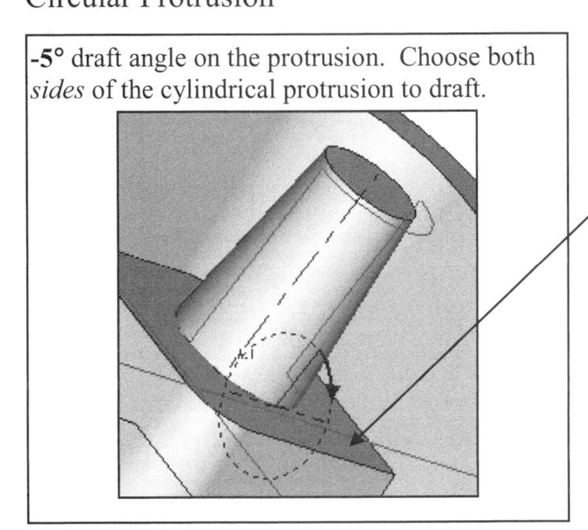

-**5°** draft angle on the protrusion. Choose both *sides* of the cylindrical protrusion to draft.

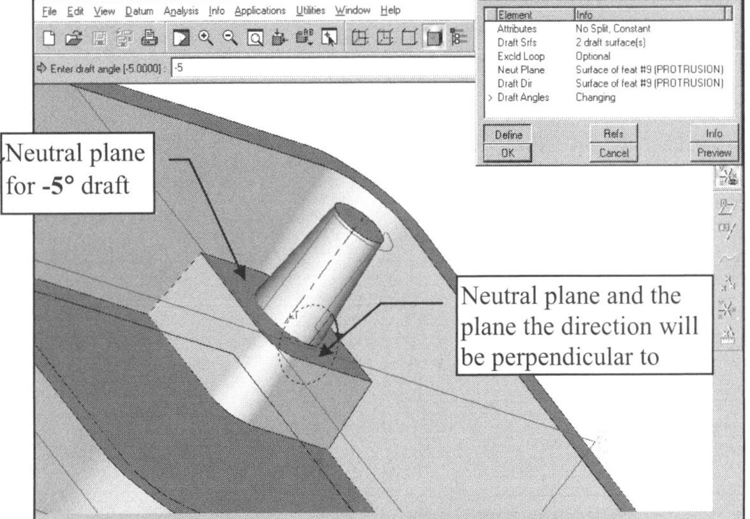

Neutral plane for **-5°** draft

Neutral plane and the plane the direction will be perpendicular to

Figure 9.23
Draft

Draft of **-.3°** for the hole.
Remember to choose both halves of the hole for the draft.

Neutral plane and the plane the direction will be perpendicular to

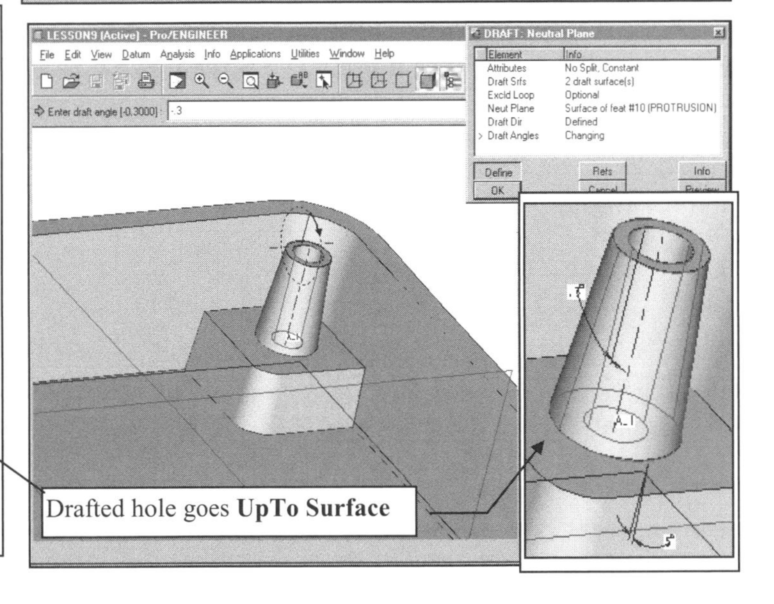

Drafted hole goes **UpTo Surface**

Figure 9.24
Hole

Create the rounds as shown in Figure 9.25. **Group** the protrusions, hole, and rounds shown in Figure 9.25 and **Copy** ⇒ **Mirror** in both directions, as in Figures 9.26 and 9.27.

File ⇒ Save ⇒ ✔
File ⇒ Delete ⇒
Old Versions ⇒ ✔

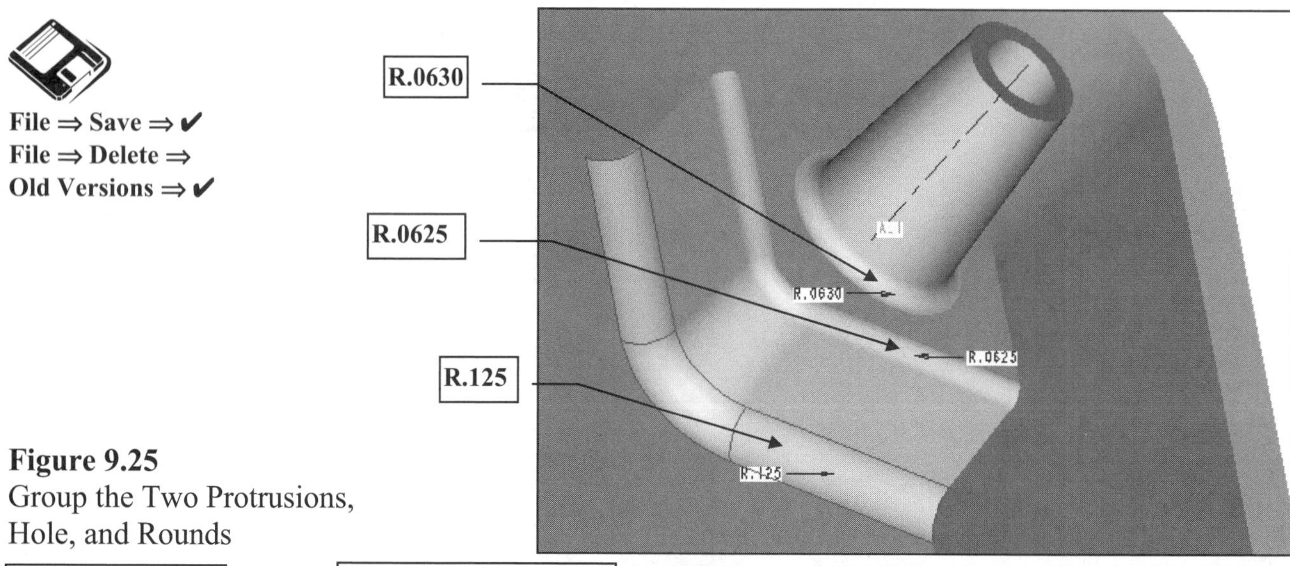

Figure 9.25
Group the Two Protrusions, Hole, and Rounds

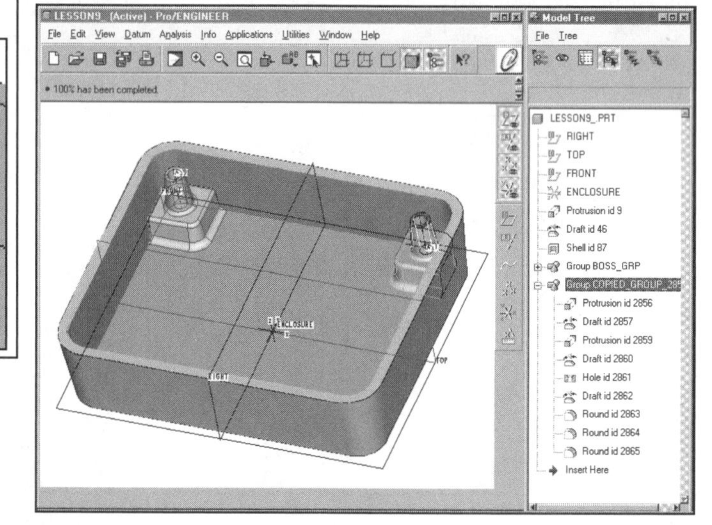

Figure 9.26
Copy ⇒ Mirror the **Group** About **RIGHT** (Change the colors of the surfaces to show them more clearly)

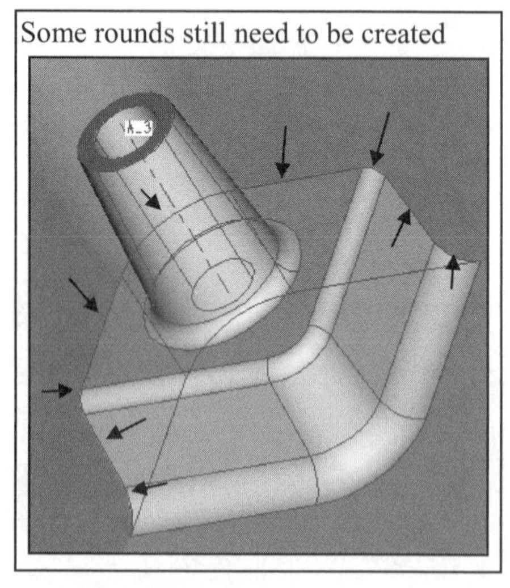

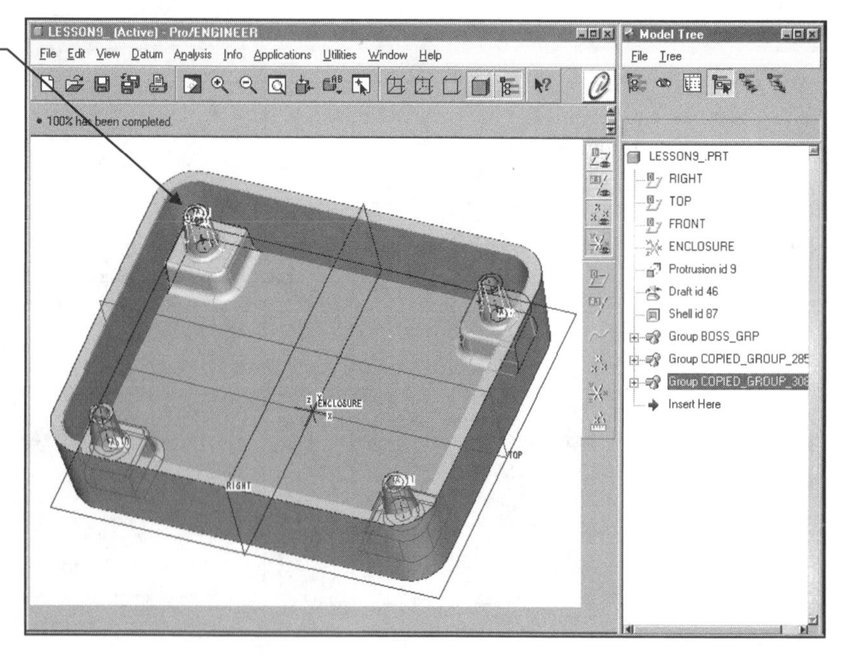

Figure 9.27
Mirror and Copy the Groups About **TOP**

Create the missing rounds (Fig. 9.27) by using the following commands:

Feature ⇒ Create ⇒ Round ⇒ Simple ⇒ Done ⇒ Constant ⇒ Surf-Surf ⇒ Done ⇒ (pick the first surface) ⇒ (pick the second surface) ⇒ (enter the radius **.125** at the prompt) ⇒ ✔ ⇒ **Preview** (Fig. 9.28) ⇒ **OK ⇒ Done ⇒ File ⇒ Save ⇒ ✔**

Select the first surface

Select the second surface

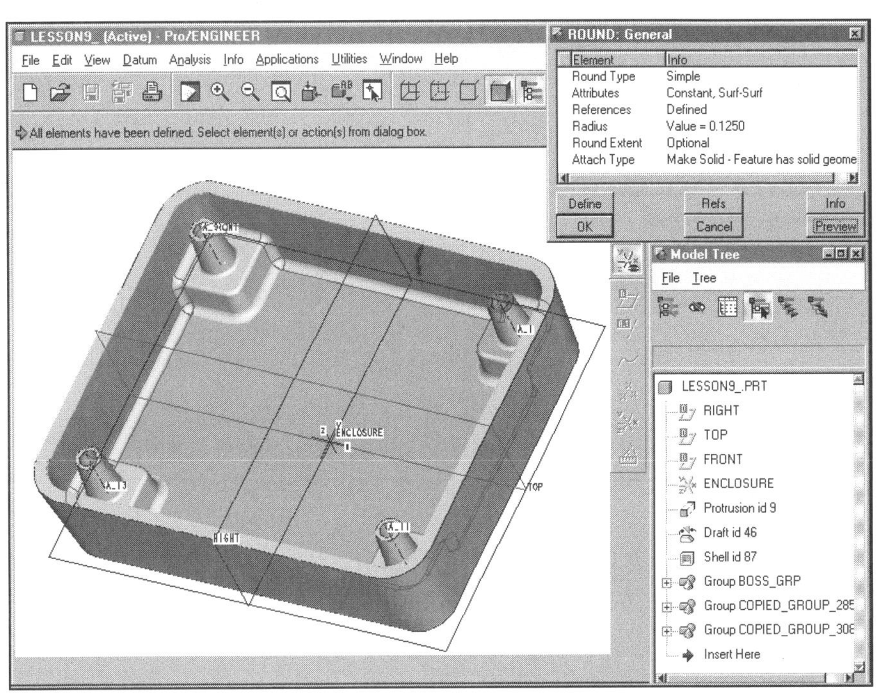

Figure 9.28
Create the Last Internal Rounds

Create the external round on the part, as shown in Figure 9.29. Try to **Reorder** the round to come before the shell. If the model fails, **Undo Changes ⇒ Confirm.**

File ⇒ Save ⇒ ✔
File ⇒ Delete ⇒
Old Versions ⇒ ✔

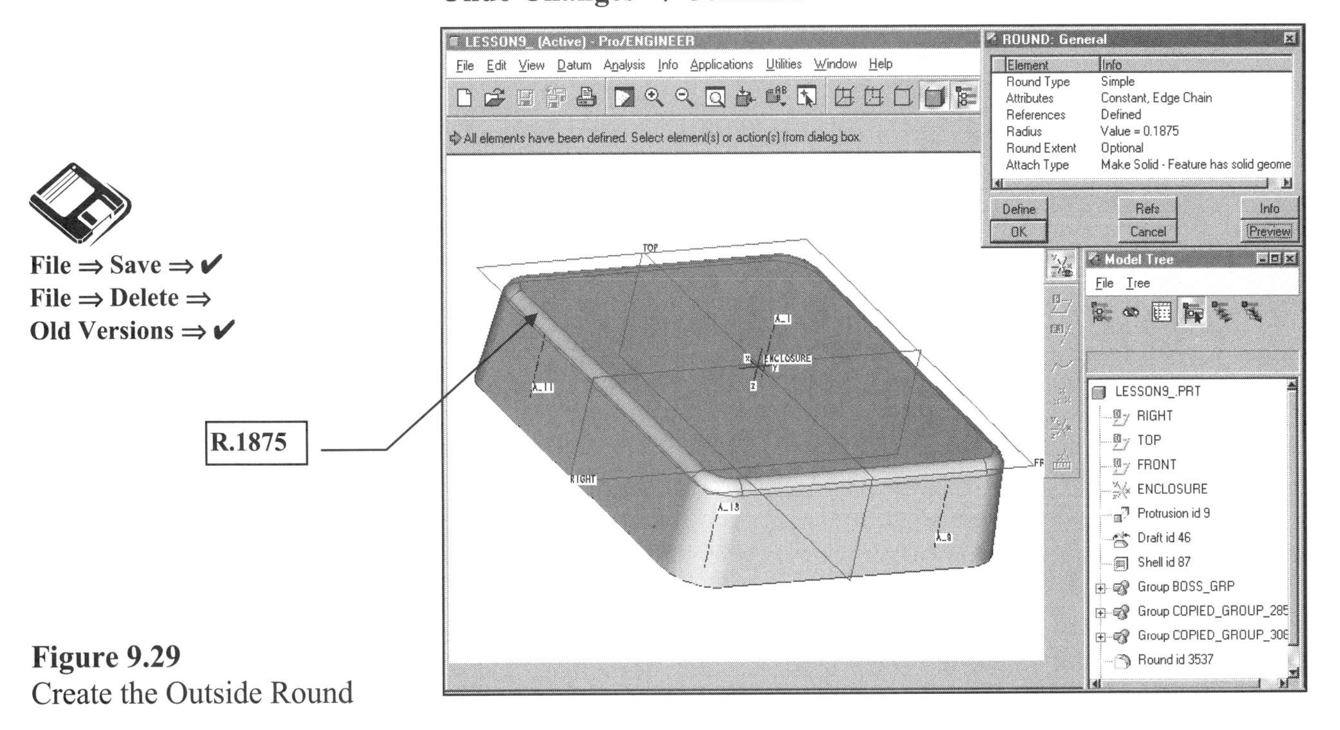

Figure 9.29
Create the Outside Round

Before creating the text protrusion, **Suppress** all the features after the shell command. Expand the Model Tree to include the feature number and status. Use the following commands:

Feature ⇒ **Suppress** ⇒ **Range** ⇒ (enter the number of the feature following the shell; type **8** at the prompt--*your number for this feature may be different*) ⇒ ✔ ⇒ (enter the regeneration number of the ending feature; type **45** at the prompt--*your number for this feature may be different*) ⇒ ✔ (Fig. 9.30) ⇒ **Done Sel** ⇒ **Done** ⇒ **Done**

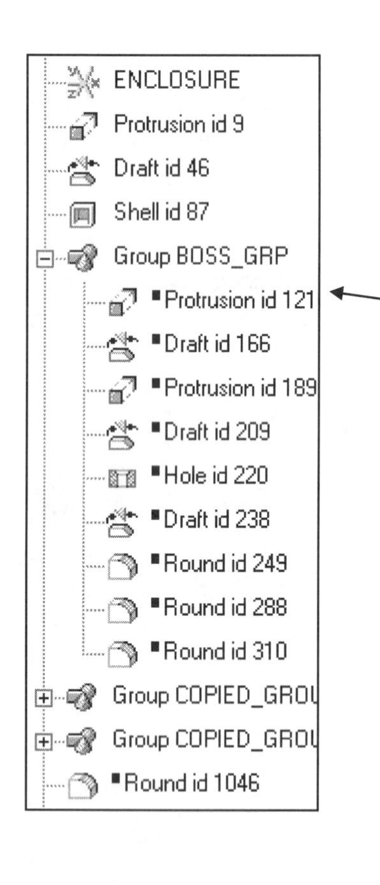

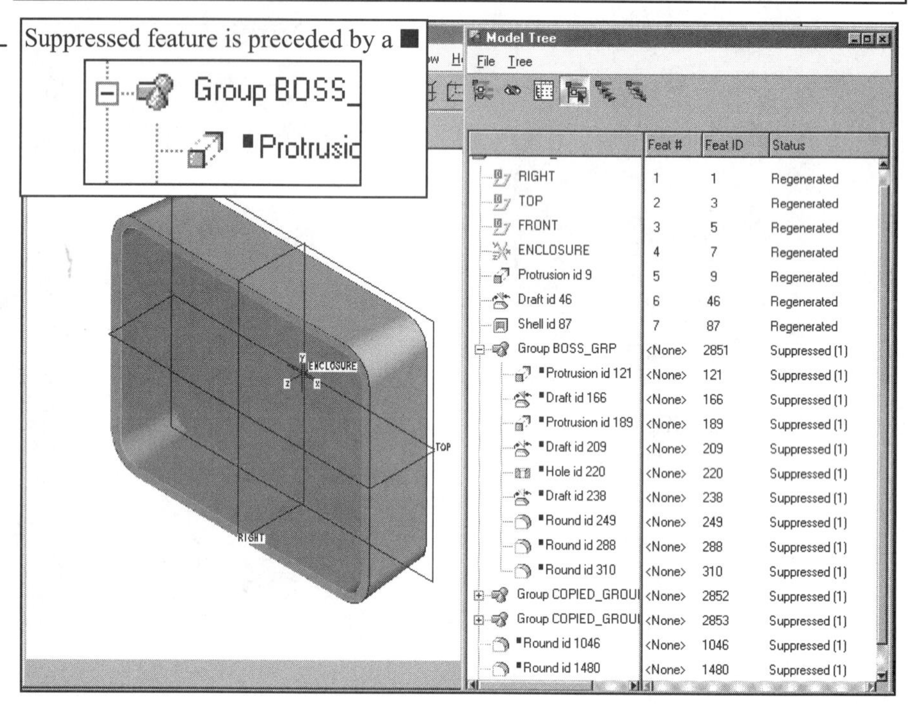

Figure 9.30
Suppressing Model Features

The regeneration time for your model will now be much shorter. Add the text protrusion shown in Figures 9.9 and 9.13 using the following commands:

Feature ⇒ **Create** ⇒ **Protrusion** ⇒ **Done** ⇒ **Done** ⇒ (pick the inside of the enclosure for the sketching plane as in Fig. 9.31) ⇒ **Okay** (for direction) ⇒ **Top** ⇒ (pick **TOP**) ⇒ ✔ **Snap to Grid** ⇒ ✔ **Intent Manager** ⇒ **Close** (References) ⇒ **Sketch** ⇒ **Text** ⇒ Select *start point* of line to determine text height and orientation ⇒ Select *second point* of line to determine text height and orientation (see left-hand column) ⇒ Text Line **CFS-2134** (Fig. 9.32 and Fig. 9.33) ⇒ ✔ ⇒ **Modify** the dimensions (Fig. 9.34) ⇒ **Sketch** ⇒ **Done** ⇒ **Blind** ⇒ **Done** (type **.0625** at the prompt) ⇒ ✔ ⇒ **OK** (Figure 9.35)

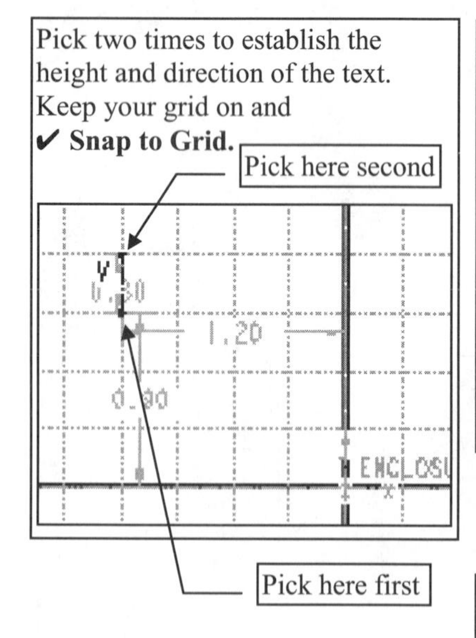

After the text protrusion is completed, choose:

Feature ⇒ **Resume** ⇒ **All** ⇒ **Done** [from the RESUME menu (Fig. 9.36)] ⇒ **Regenerate** (it now takes longer to regenerate) ⇒ **File** ⇒ **Save** ⇒ ✔ ⇒ **File** ⇒ **Delete** ⇒ **Old Versions** ⇒ ✔

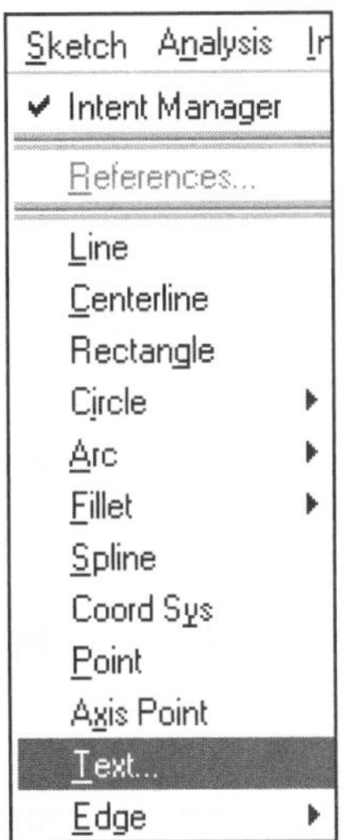

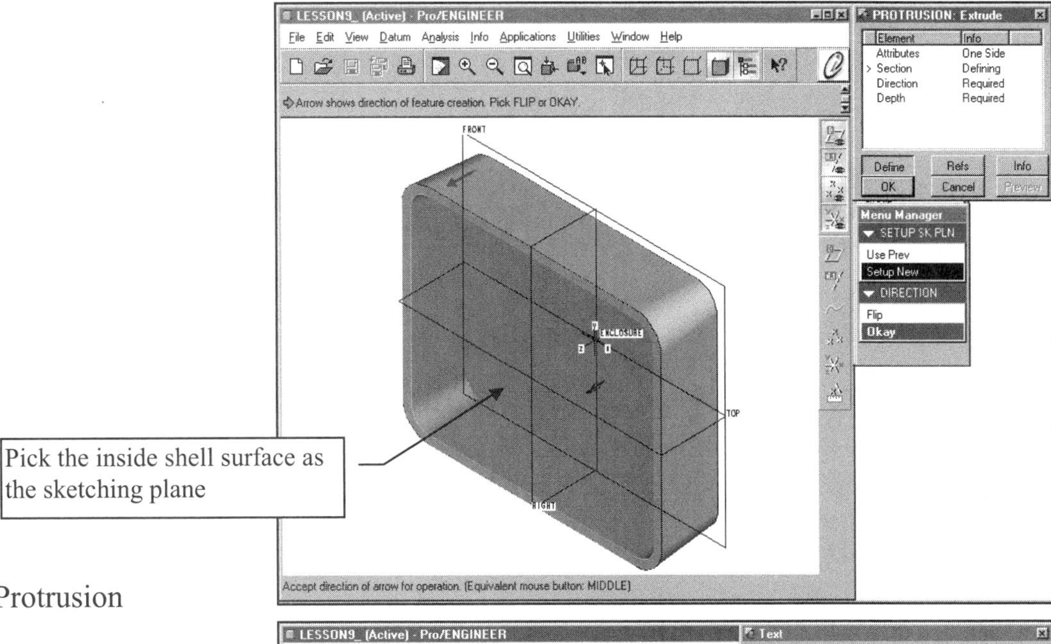

Figure 9.31
Direction of Text Protrusion

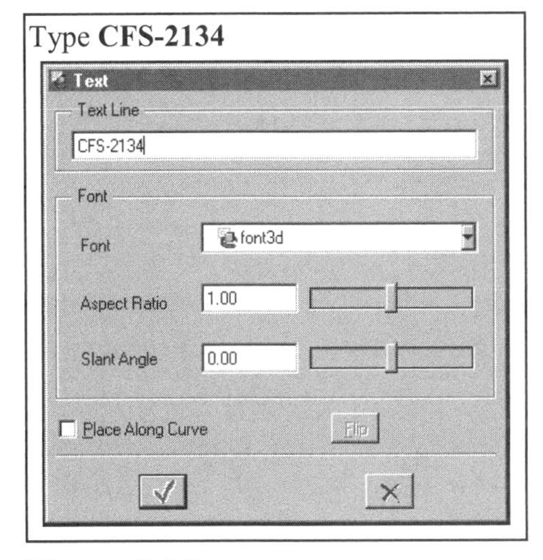

Figure 9.32
Sketch Text Line

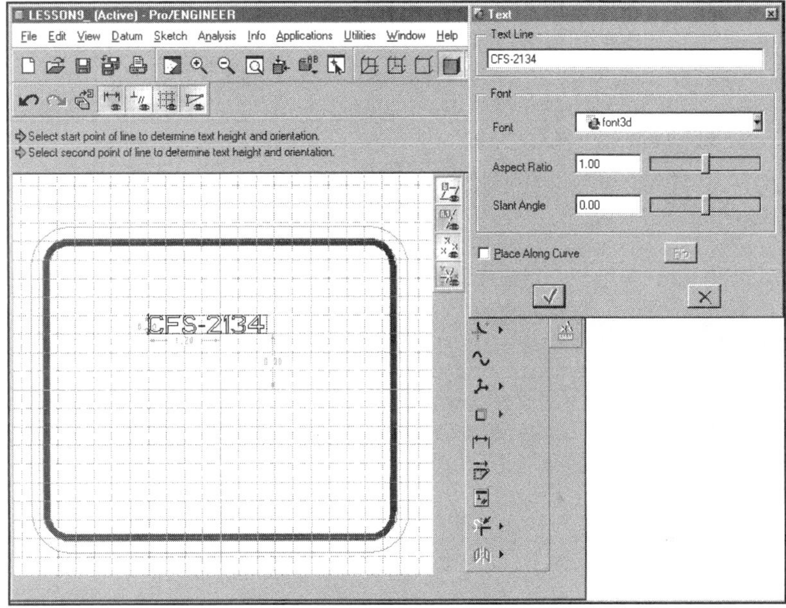

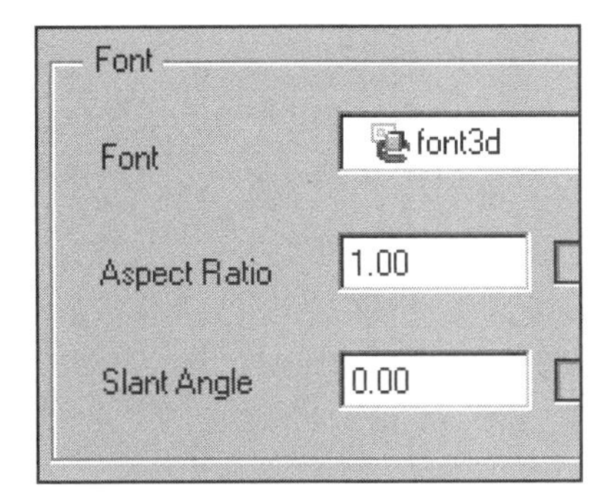

Figure 9.33
Creating Text

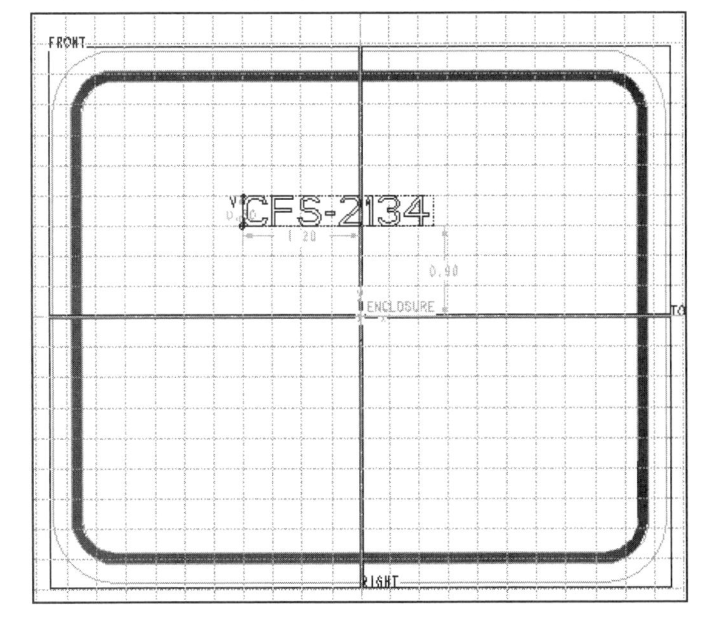

Modify the dimensions as shown

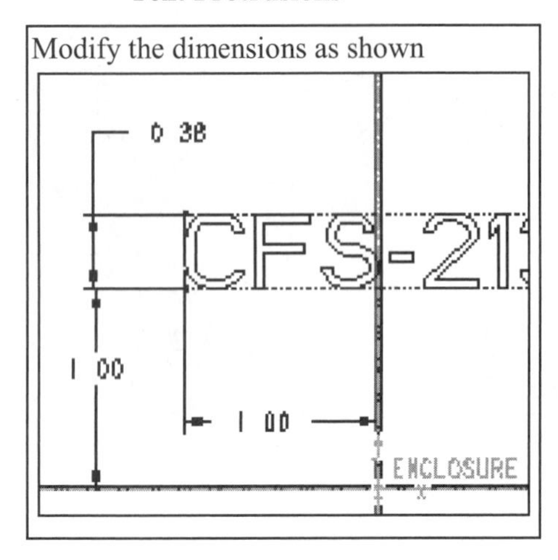

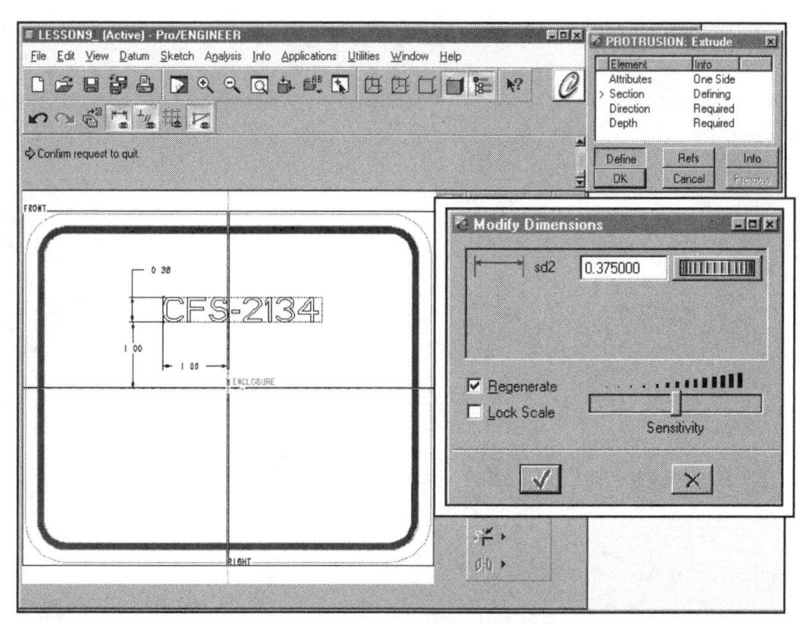

Figure 9.34
Dimension the Sketch Text

Modify the depth and location of the text protrusion: **Modify** ⇒ (pick the text) ⇒ (change the depth to **.25**) ⇒ **Regenerate.** If you have not resumed the model, do so now: **Resume** ⇒ **All** ⇒ **Done.**

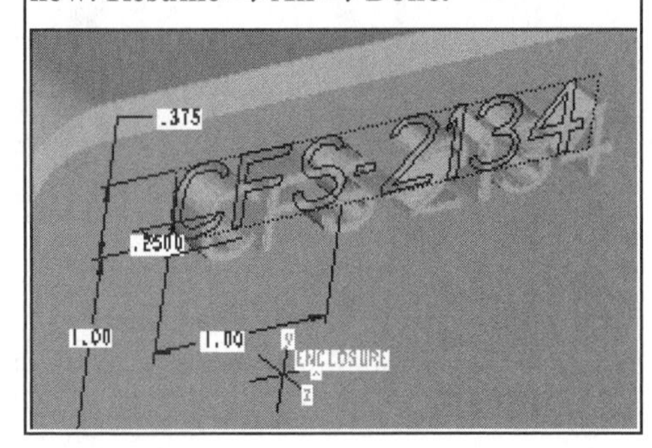

Figure 9.35
Text Protrusion

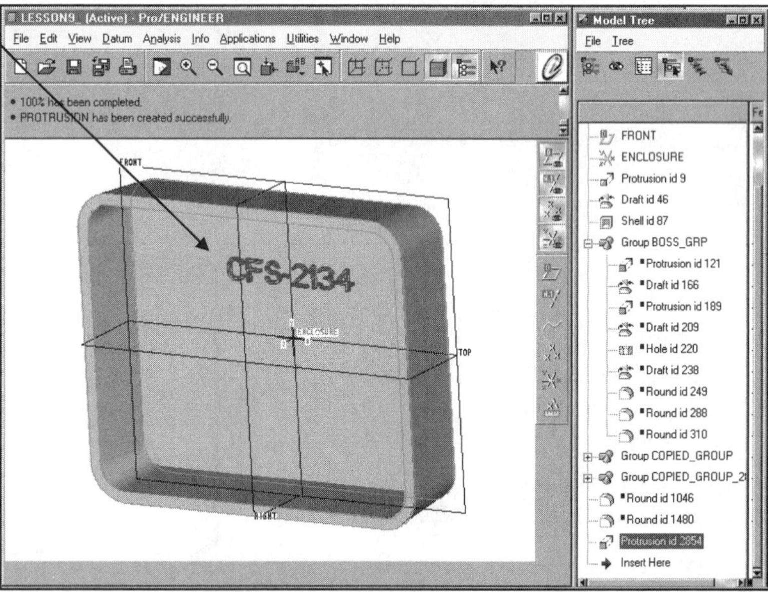

Regenerated (**Resume**) features

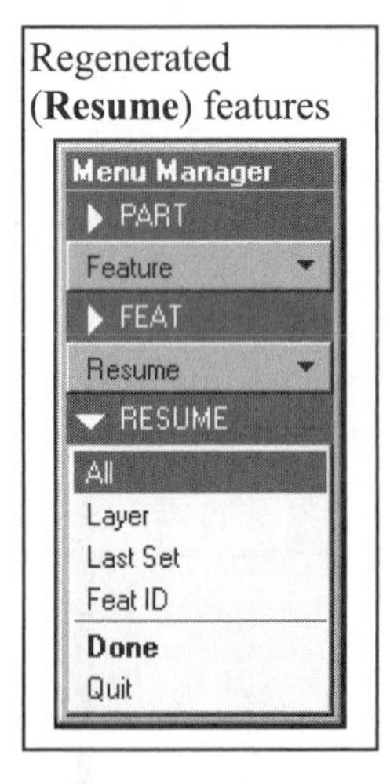

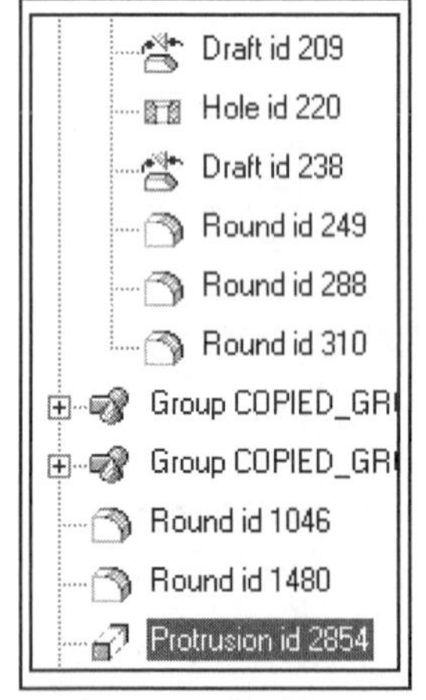

Figure 9.36
Resume

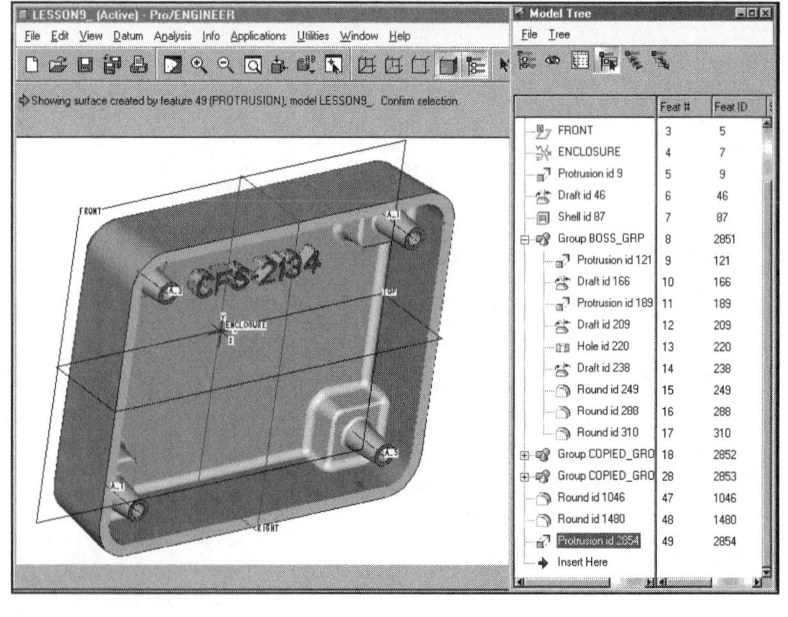

Lesson 9 Project

Cellular Phone Bottom

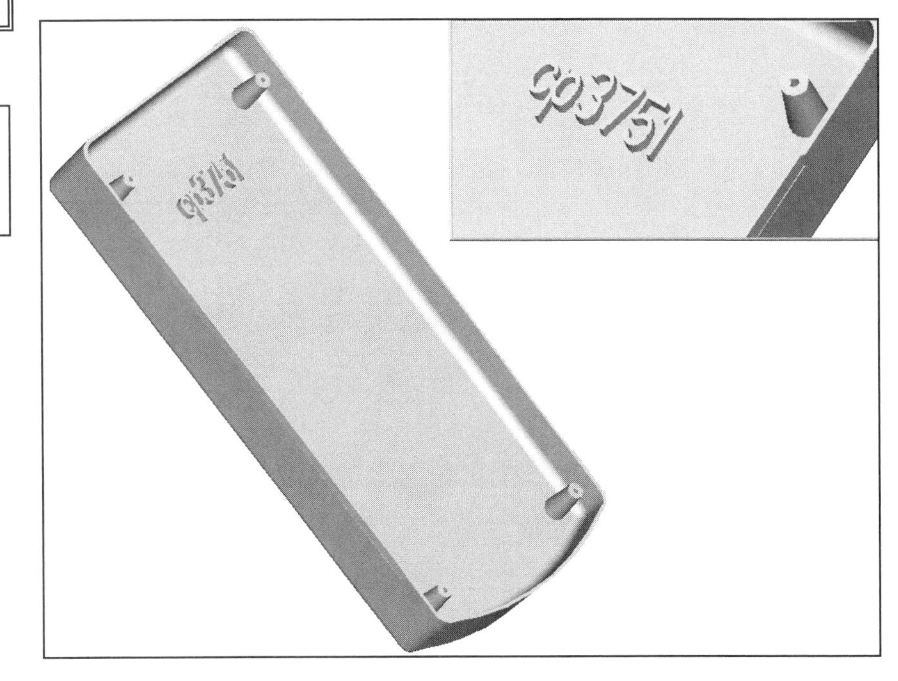

Figure 9.37
Cellular Phone Bottom

Cellular Phone Bottom

The ninth **lesson project** is a die-cast plastic part that requires commands similar to those for the **Enclosure**. Create the part shown in Figures 9.37 through 9.47. Analyze the part and plan out the steps and features required to model it. Use the **DIPS** in Appendix D to plan out the feature creation sequence and the parent-child relationships for the part. The top of the cellular phone is created in the Lesson 10 Project. Shell the Cellular Phone Bottom **.04** on its sides and **Spec Thick .500** on its inside bottom.

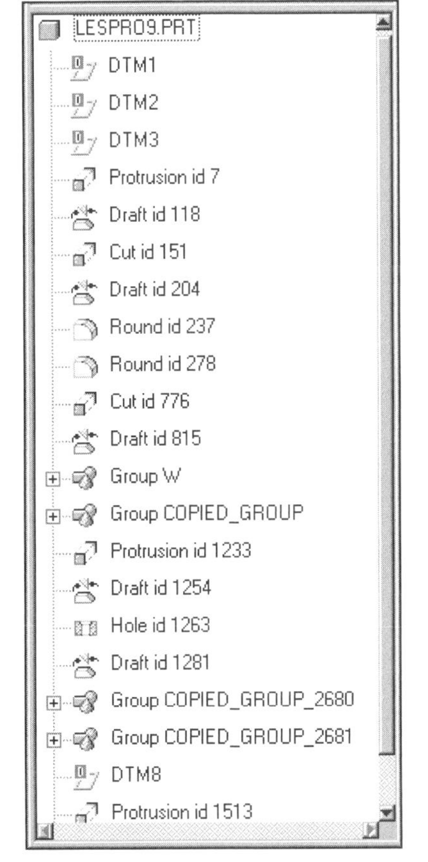

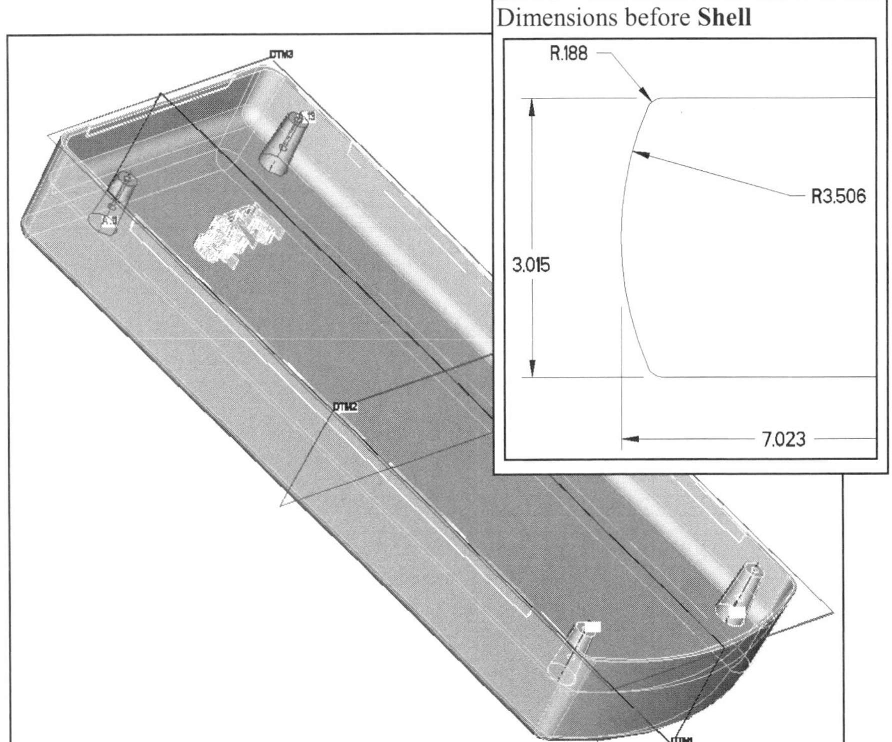

Figure 9.38
Cellular Phone Bottom
Showing Datum Planes

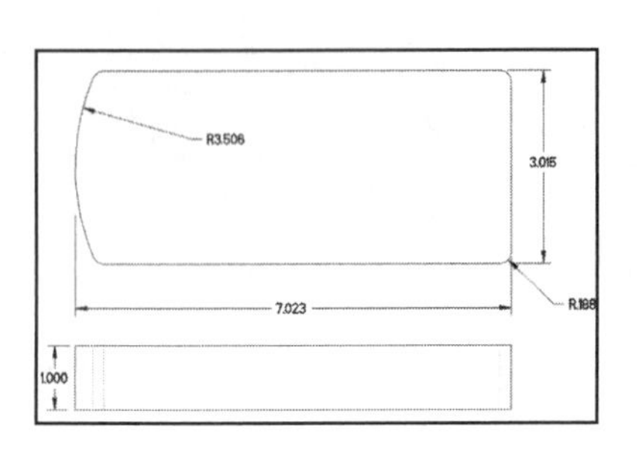

Figure 9.39
Cellular Phone Drawing

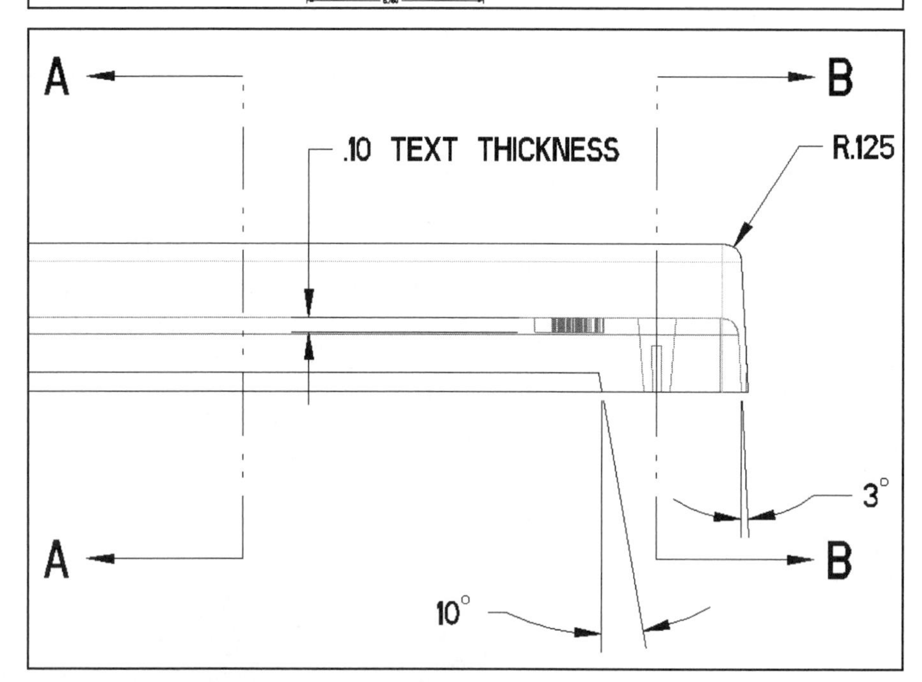

Figure 9.40
Cellular Phone Drawing, Front View

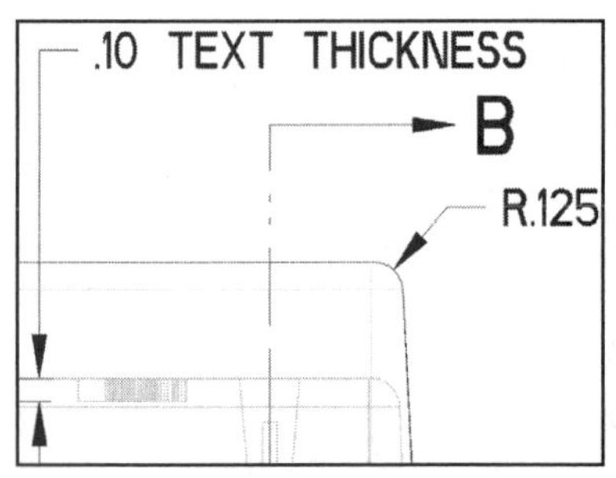

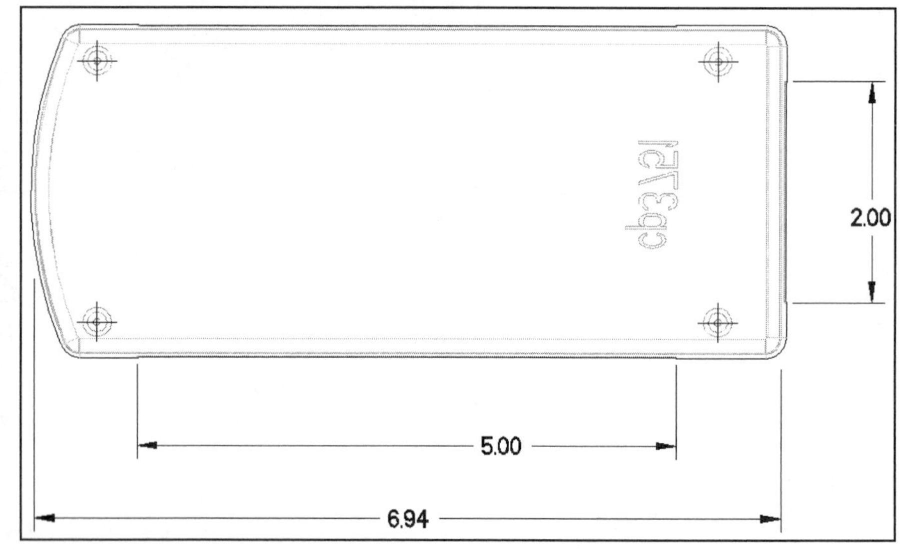

Figure 9.41
Cellular Phone Drawing, Top View

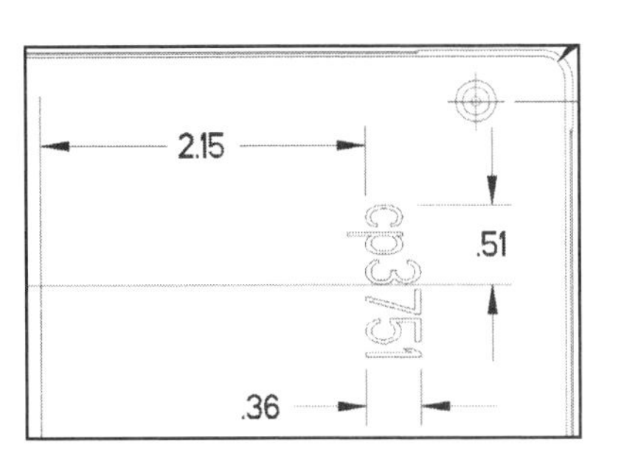

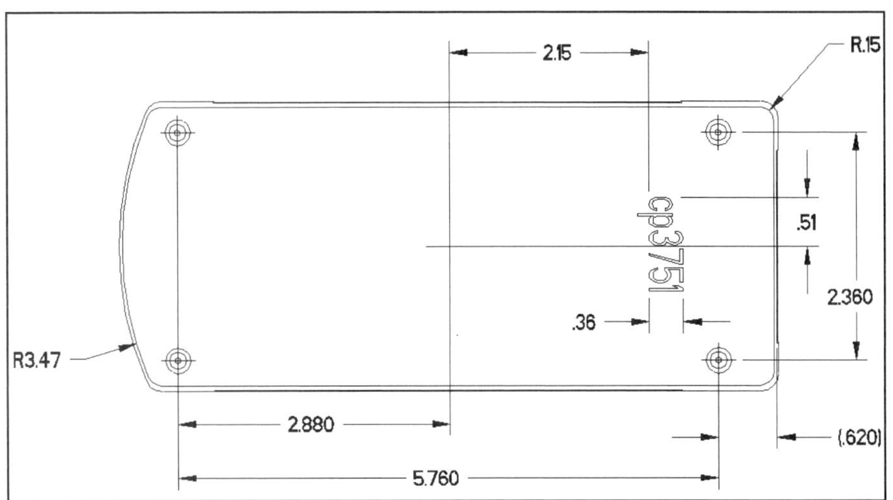

Figure 9.42
Cellular Phone Drawing,
Bottom View

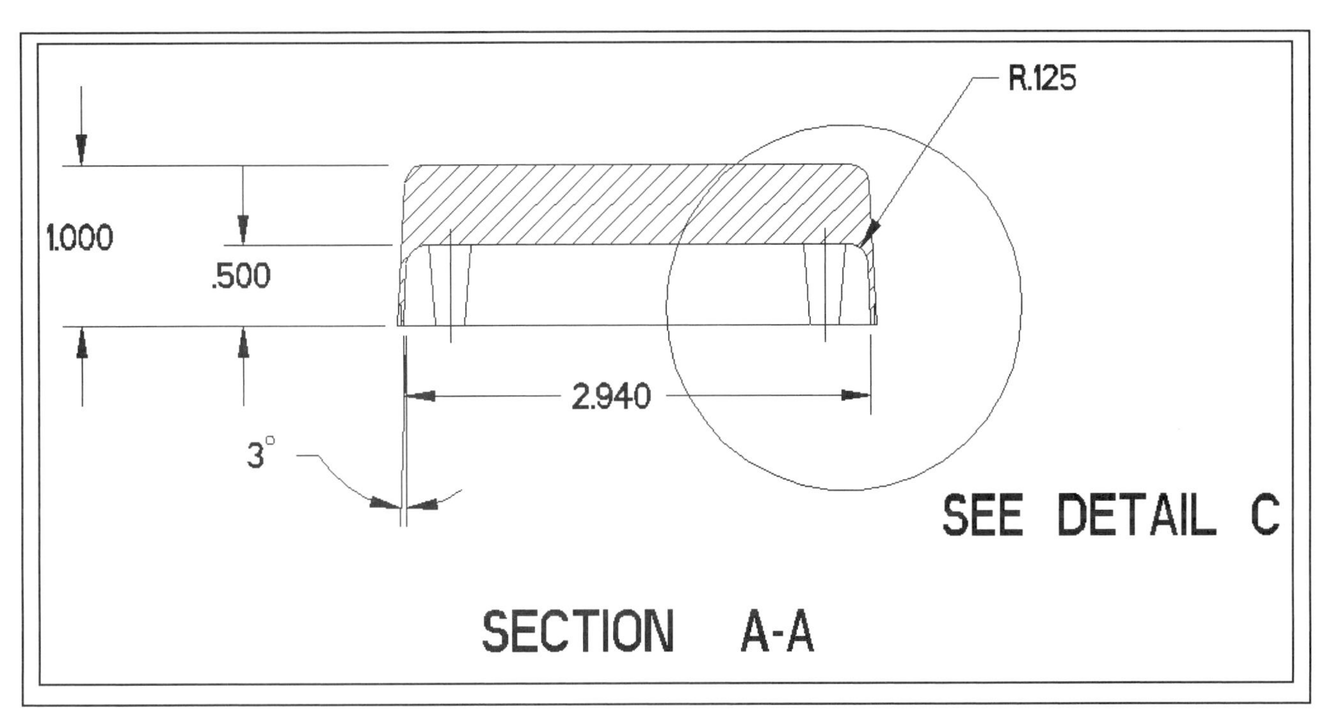

Figure 9.43
Cellular Phone Drawing,
SECTION A-A

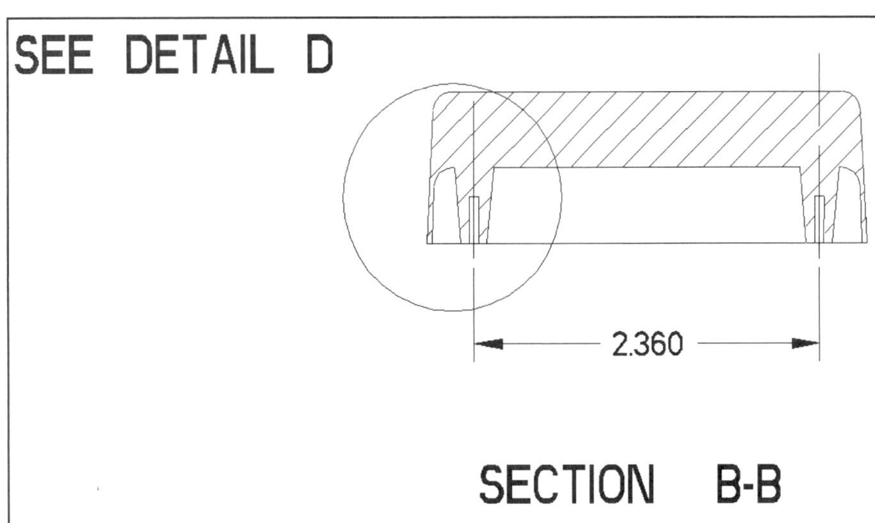

Figure 9.44
Cellular Phone Drawing,
SECTION B-B

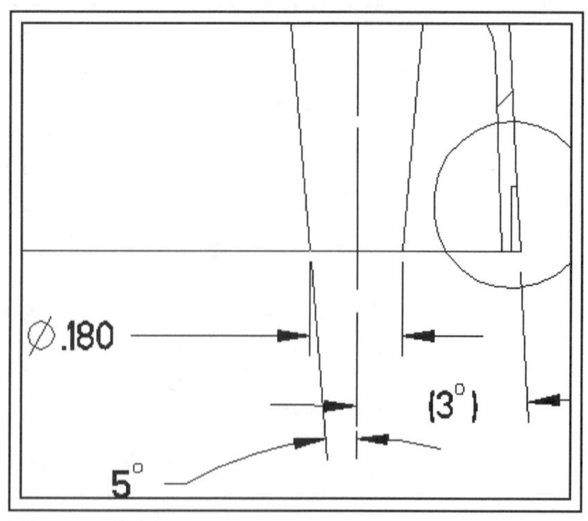

Figure 9.45
Cellular Phone Drawing,
DETAIL C

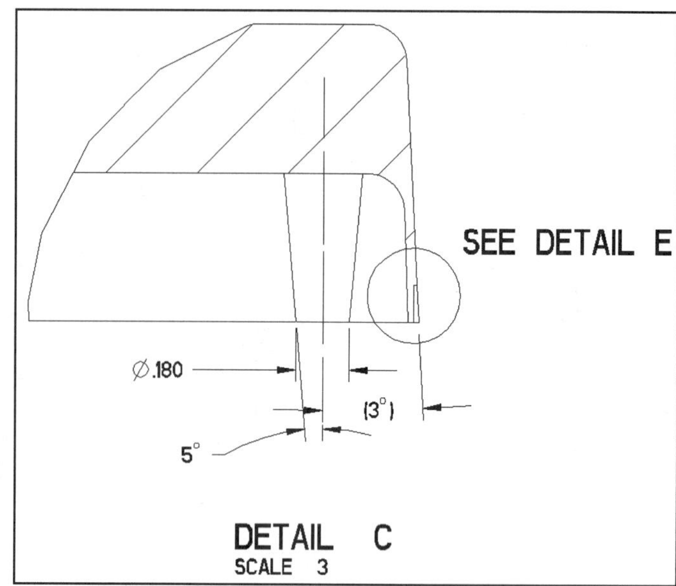

.312
(3°)
∅.180
∅.0625
.5° HOLE DRAFT ANGLE
DETAIL D
SCALE 3

Figure 9.46
Cellular Phone Drawing,
DETAIL D

File ⇒ Save ⇒ ✔
File ⇒ Delete ⇒
Old Versions ⇒ ✔

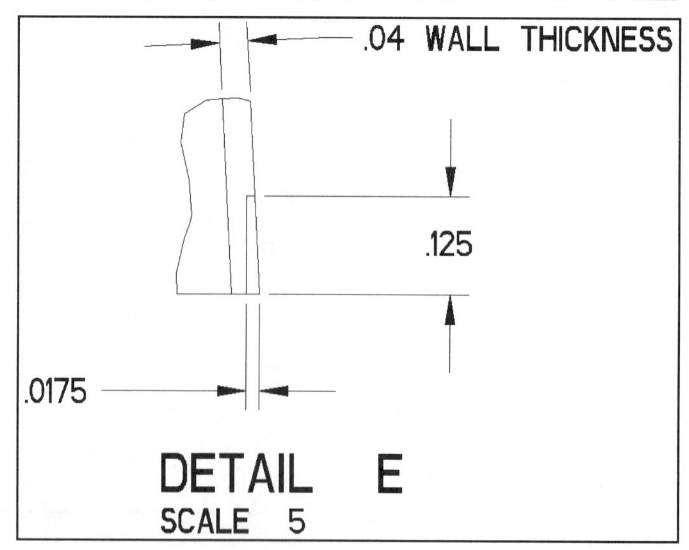

Figure 9.47
Cellular Phone Drawing,
DETAIL E

Lesson 10

Shell, Reorder, and Insert Mode

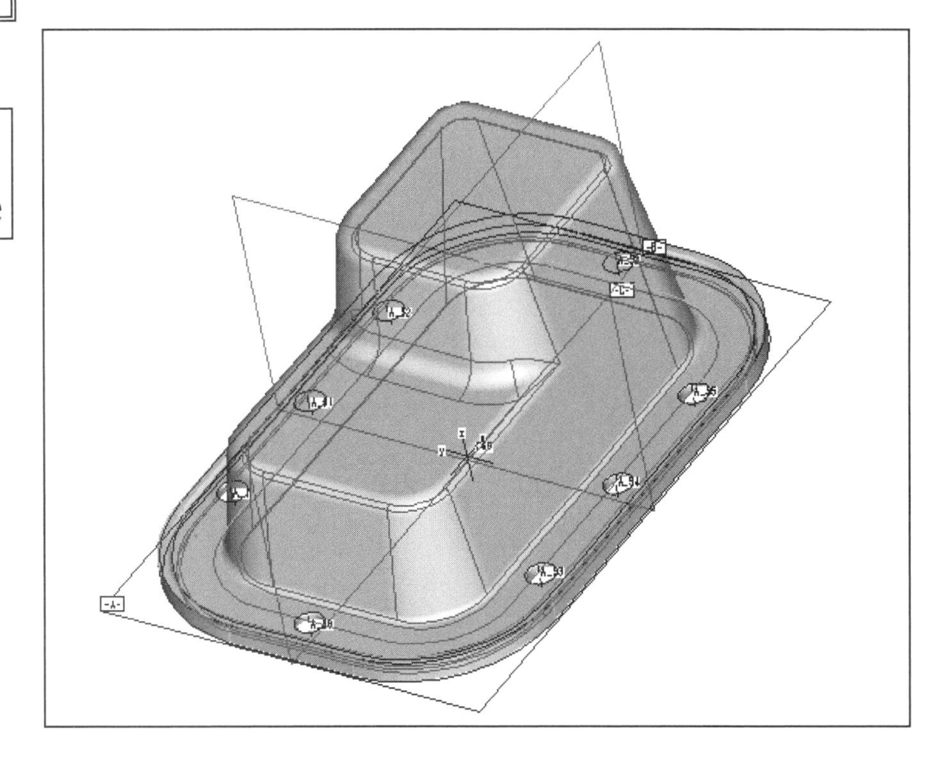

Figure 10.1
Oil Sink

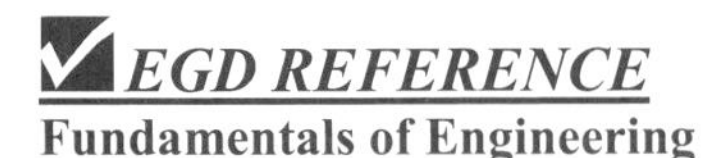

EGD REFERENCE
Fundamentals of Engineering Graphics and Design
by L. Lamit and K. Kitto
Read Chapter 14.

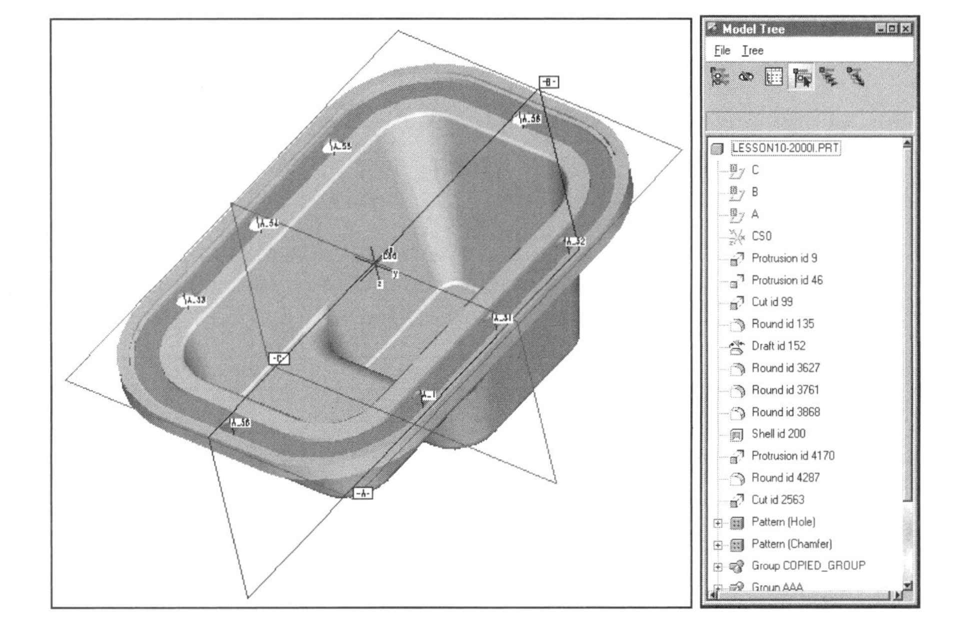

Figure 10.2
Oil Sink with Model Tree

COAch™ for Pro/ENGINEER

If you have **COAch for Pro/ENGINEER** on your system, go to SEARCH and do the Segments shown in Figures 10.4, 10.6, and 10.7.

OBJECTIVES

1. **Shell out a part**
2. **Alter the creation sequence of a feature with Reorder**
3. **Insert a feature at a specific point in the design order**

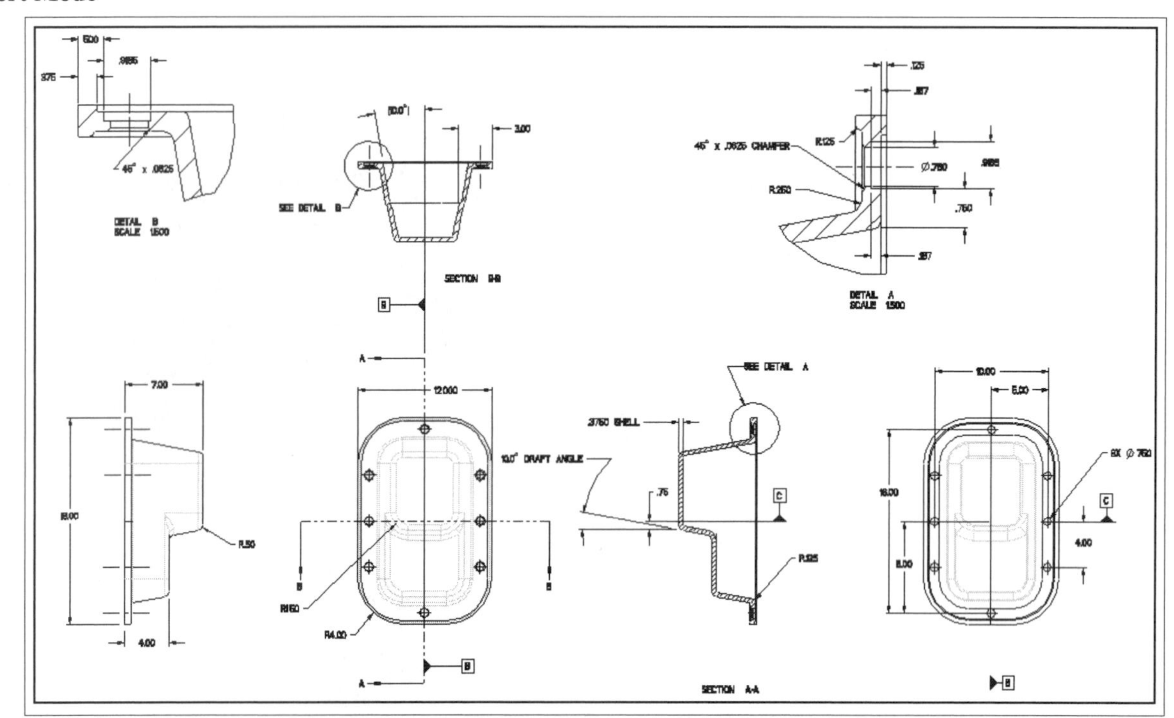

Figure 10.3
Oil Sink Dimensions

SHELL, REORDER, AND INSERT MODE

The **Shell** option removes a surface or surfaces from a solid and then hollows out the inside of the solid, leaving a shell of a specified wall thickness, as in the **Oil Sink** shown in Figures 10.1 through 10.3. When Pro/E makes the shell, all the features that were added to the solid before you chose **Shell** are hollowed out (Fig. 10.4). Therefore, the *order of feature creation* is very important when you use **Shell**. You can alter the feature creation order by using the **Reorder** option. Another method of placing a feature at a specific place in the feature/design creation order is to use the **Insert Mode** option.

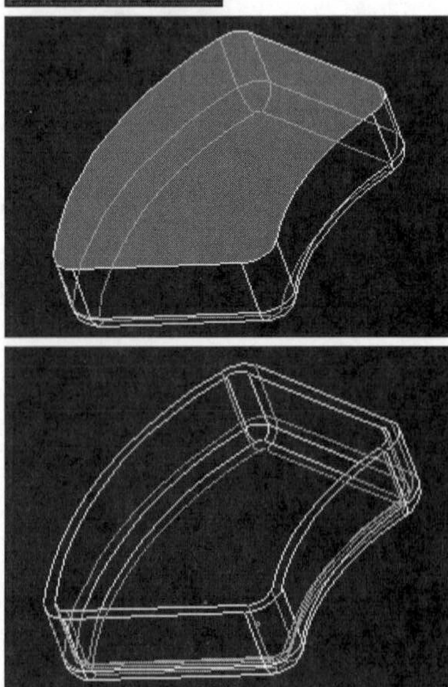

Figure 10.4
COAch for Pro/E, More Features (Shell Feature)

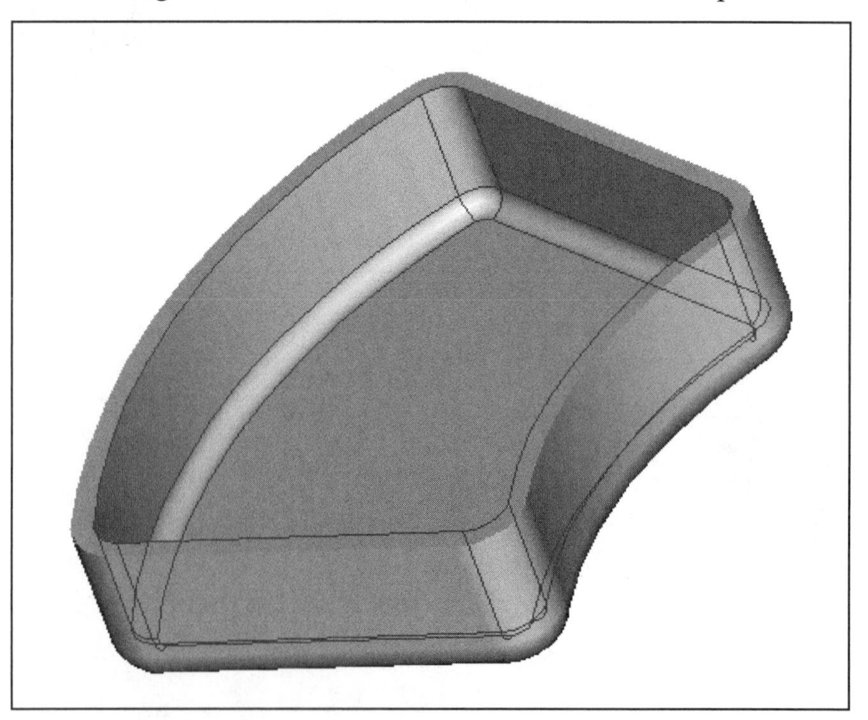

Creating Shells

To create a shell (Fig. 10.5), do the following:

1. Choose **Shell** from the SOLID menu.
2. Pro/E displays the SHELL feature creation dialog box. If you wish to, select the optional element **Spec Thick** to specify thickness individually. Select the **Define** button from the dialog box.
3. Select a surface or surfaces to be removed. When you have finished, choose **Done Refs** from the FEATURE REFS menu.
4. Enter the thickness of the wall. This thickness applies to all surfaces except those to which you assign a different thickness.
5. If you choose the **Spec Thick** element, Pro/E displays the SPEC THICK menu, which lists the following options:

   Set Thickness Set thickness for the individual surface.
   Reset to Def Reset the surfaces to the default thickness.

Choose **Set Thickness**. Select a surface and enter the thickness. Continue this process until you have specified all the surfaces you want. When you have finished, choose **Done** from the SPEC THICK menu.

6. To create the shell, select **OK** from the dialog box. If you entered a positive value for the thickness, material will be removed, leaving the shell thickness "inside" the part. However, if you entered a negative value, the shell thickness will be added to the "outside" of the part.

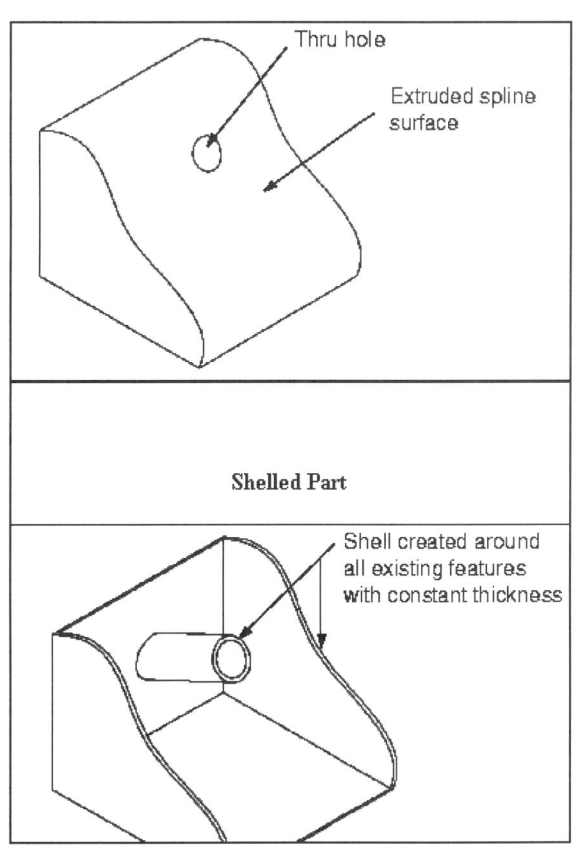

Figure 10.5
Online Documentation, Shell

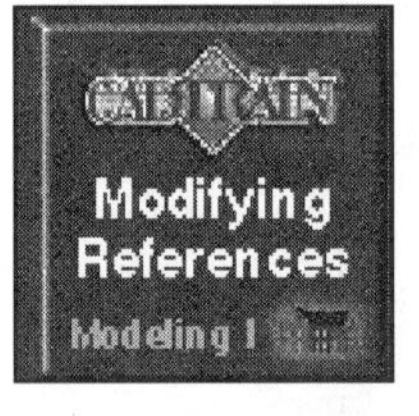

Reordering Features

You can move features forward or backward in the feature creation (regeneration) order list, thus changing the order in which they are regenerated (Fig. 10.6). You can reorder multiple features in one operation, as long as these features appear in *consecutive* order.

Feature reorder *cannot* occur under the following conditions:

Parents cannot be moved so that their regeneration occurs after the regeneration of their children.
Children cannot be moved so that their regeneration occurs before the regeneration of their parents.

To reorder a feature, do the following:

1. Use the command sequence **Feature ⇒ Reorder.**
2. Specify the selection method by choosing an option from the SELECT FEAT menu:

Select Select features to reorder by picking on the screen and/or from the Model Tree. You can also choose **Sel By Menu** to enter the feature number. Choose **Done Sel** when finished.
Layer Select all features from a layer by selecting the layer. Choose **Done Sel** from the LAYER SEL menu when finished.
Range Specify the range of features by entering the regeneration numbers of the starting and ending features.

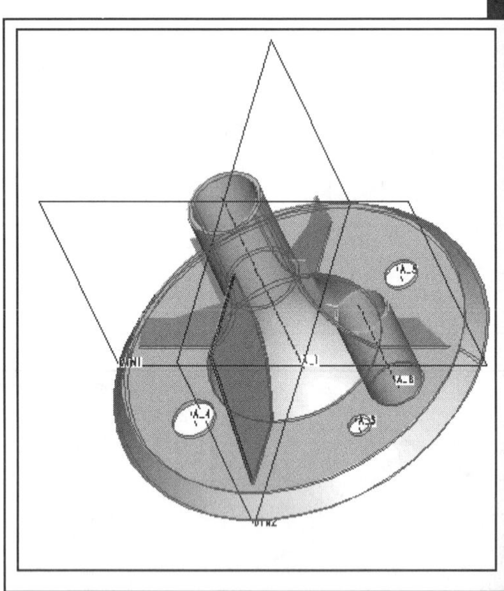

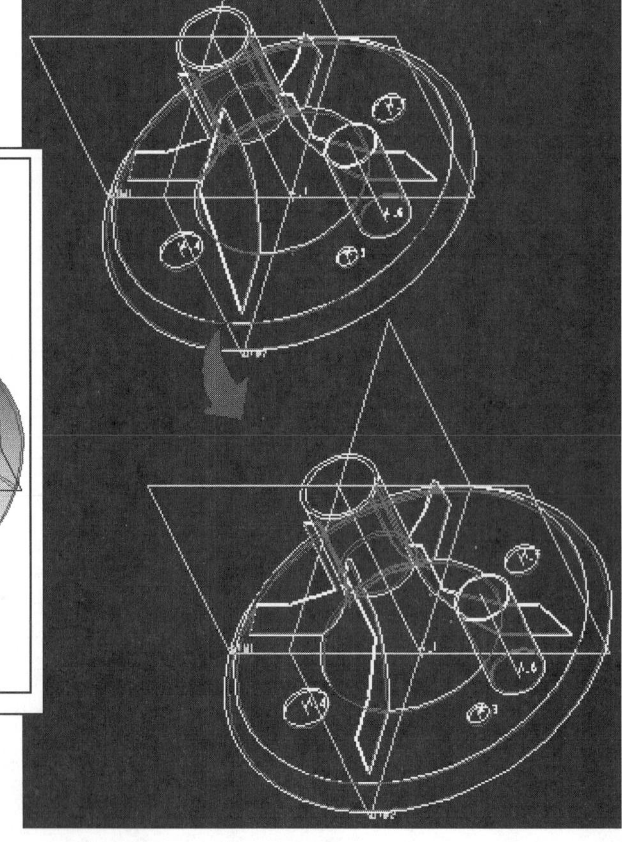

Figure 10.6
COAch for Pro/E, Modifying References (Reorder Features)

You can **Reorder** features in the **Model Tree** by dragging one or more features to a new location in the feature list. If you try to move a child feature to a higher position than its parent feature, the parent feature moves with the child feature in context, so the parent/child relationship is maintained.

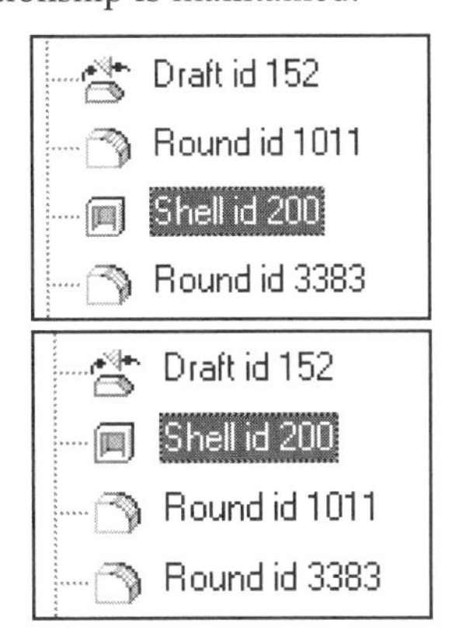

You can also **Insert** features using the Model Tree. There is an arrow-shaped icon on the Model Tree that indicates where features will be inserted upon creation. By default, it is always at the end of the Model Tree. You may drag the location of the *arrow* higher or lower in the tree to insert features at a different point. When the *arrow* is dropped at a new location, the model is rolled backward or forward in response to the insertion *arrow* being moved higher or lower.

Figure 10.7
COAch for Pro/E, Modifying References (Insert Features)

3. A Pro/E message lists the selected features for reorder and states the valid ranges for the new insertion point.
4. Choose **Done** from the SELECT FEAT menu.
5. Choose one of the options in the REORDER menu:

   Before Insert the feature before the insertion point feature.
   After Insert the feature after the insertion point feature.

   You can pick a feature indicating the insertion point, or choose **Sel By Menu** to enter the feature number.

Inserting Features

Normally, Pro/E adds a new feature after the last existing feature in the part, including suppressed features. **Insert Mode** (Fig. 10.7) allows you to add new features at any point in the feature sequence, except before the base feature or after the last feature.

To insert features, do the following:

1. Choose **Insert Mode** from the FEAT menu; then choose **Activate**.
2. Select a feature after which the new features will be inserted. All features after the selected one will be automatically suppressed.
3. Choose **Create** and create the new features as usual.
4. Cancel **Insert Mode** by choosing **Resume** from the FEAT menu and select to resume the features that were suppressed when you activated **Insert Mode**, or choose **Cancel** from the INSERT MODE menu. Pro/E asks you whether to resume the features that were suppressed when you activated **Insert Mode.** Then it automatically regenerates the part.

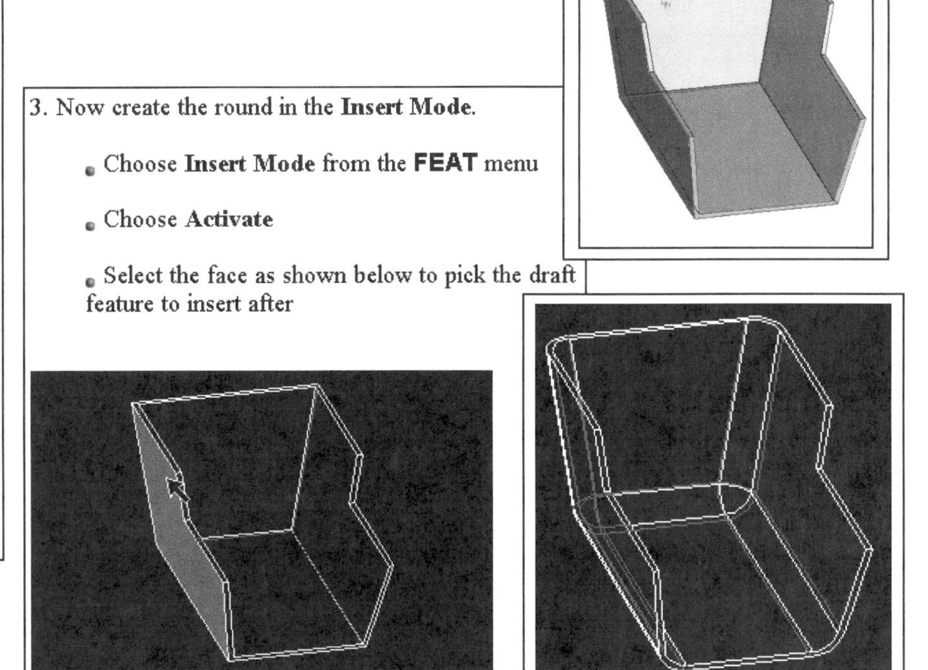

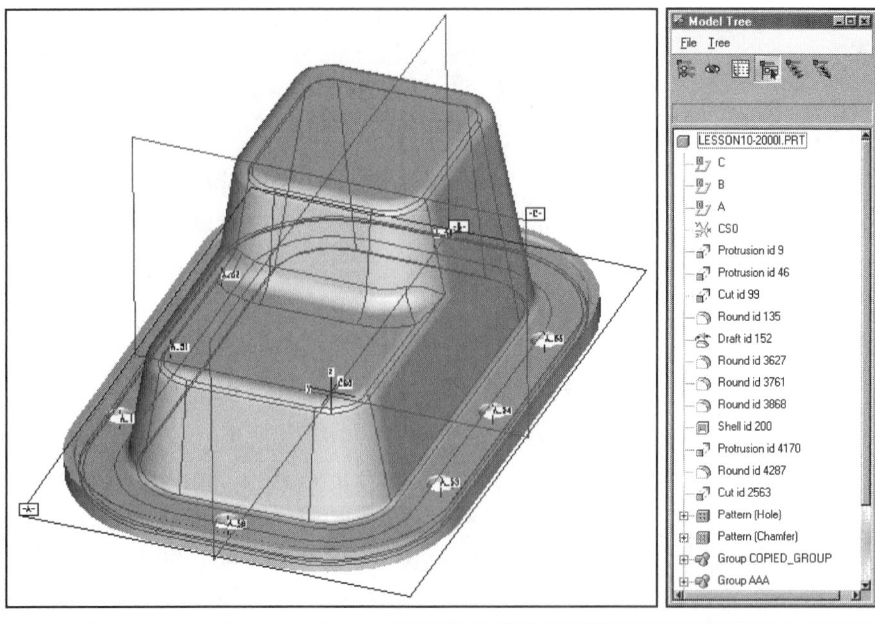

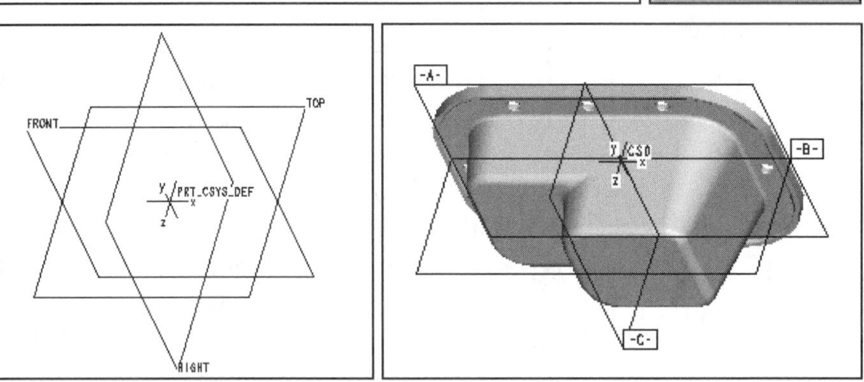

Figure 10.8
Oil Sink
Datum **TOP** = **B**
Datum **FRONT** = **A**
Datum **RIGHT** = **C**
Coordinate System = **CS0**

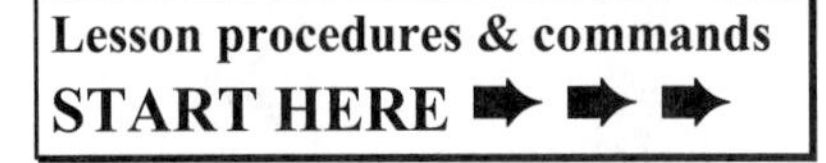

Oil Sink

The **Oil Sink** (Fig. 10.8) requires the use of the **Shell** option. The shelling of a part should be done after the desired protrusions and most rounds have been modeled. This lesson part will have you create a protrusion, a cut, and a set of rounds. Some of the required rounds will be left off the part model on purpose. Pro/E's **Insert Mode** option enables you to insert a set of features at a earlier stage in the design of the part. In other words, you can create a feature after or before a selected existing feature even if the whole model has been completed. You can also *move the order in which a feature was created* and therefore have subsequent features affect the reordered feature. A round created after a shell operation can be reordered to appear before the shell, to have the shell be affected by the round.

In this lesson, you will also insert a round or two before the existing shell feature using **Insert Mode**. The rounds will be shelled after the **Resume** option is picked, because the rounds now appear before the shell feature.

The first protrusion and cut for the Oil Sink are to be modeled by using the dimensions provided in Figures 10.9 through 10.16.

Set up the Oil Sink:

- Material = STEEL
- Units = Inch lbm Second
- Layers = **DATUM_LAYER**
- Hidden Line
- ✔ Snap to Grid

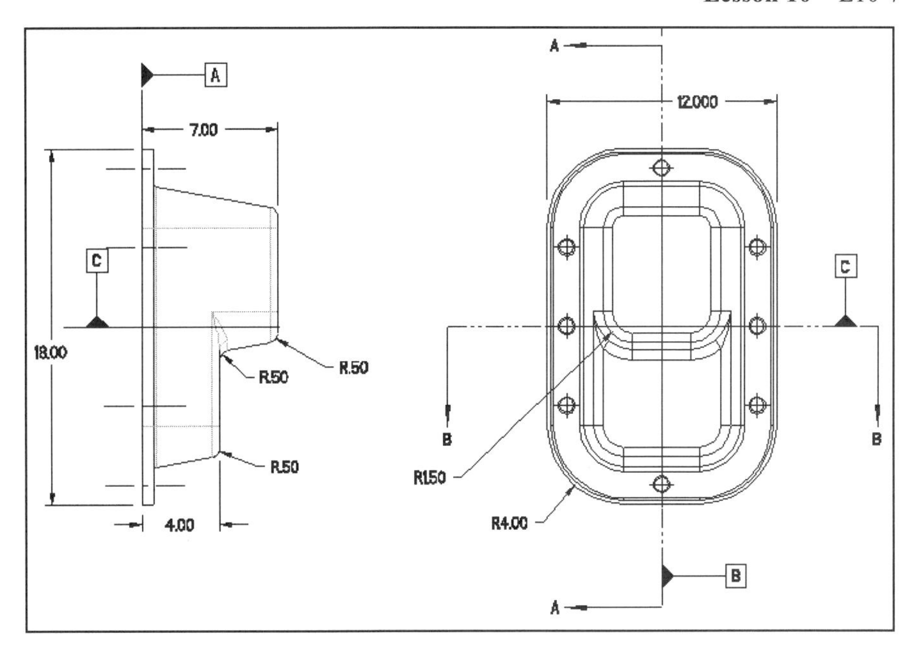

Figure 10.9
Oil Sink Detail Drawing

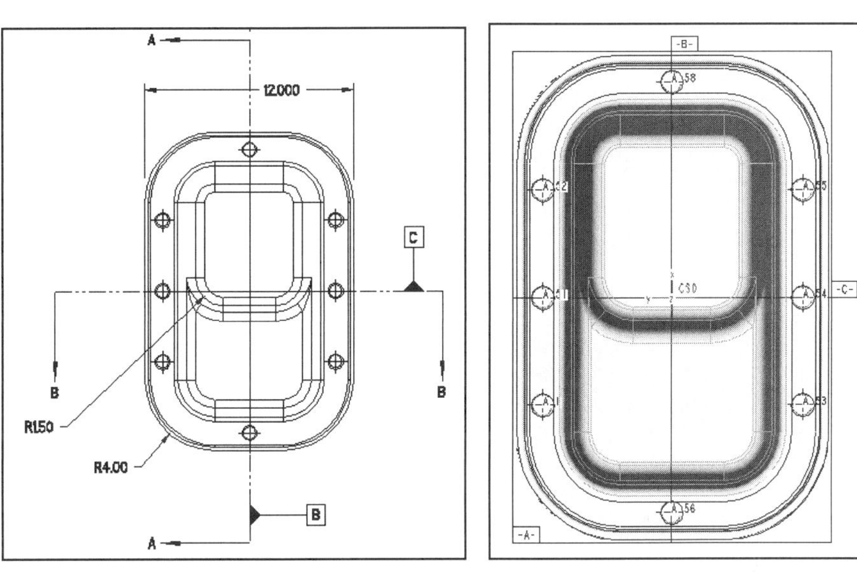

Figure 10.10
Oil Sink Drawing,
Front View

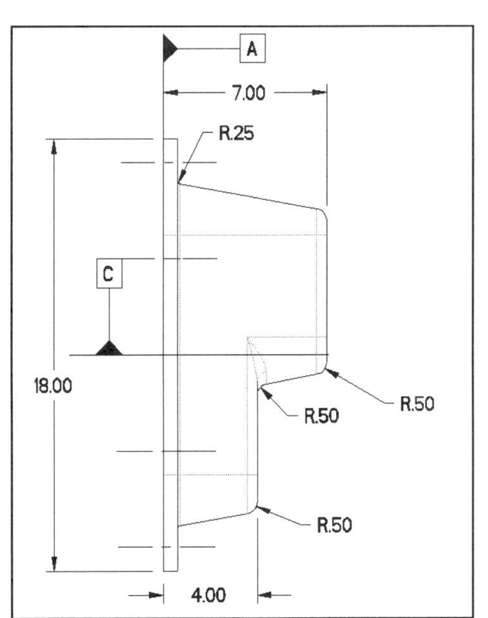

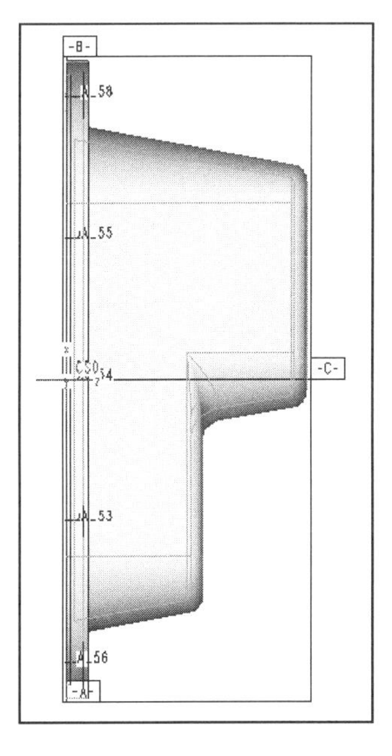

Figure 10.11
Oil Sink Drawing,
Left Side View

SEE DETAIL A
.3750 SHELL
10.0° DRAFT ANGLE
.75
C
R.125
SECTION A-A

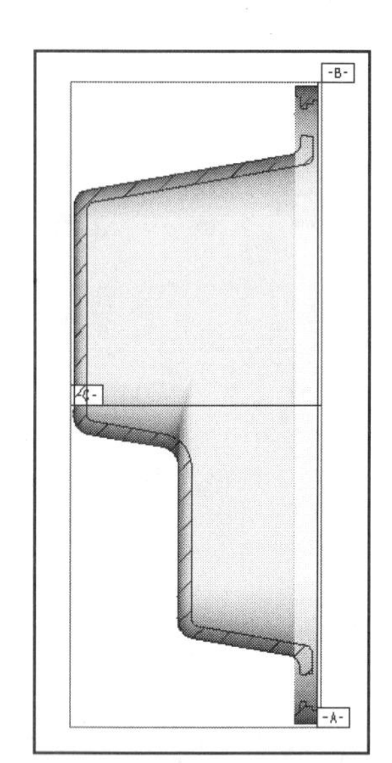

Figure 10.12
Oil Sink Drawing,
SECTION A-A

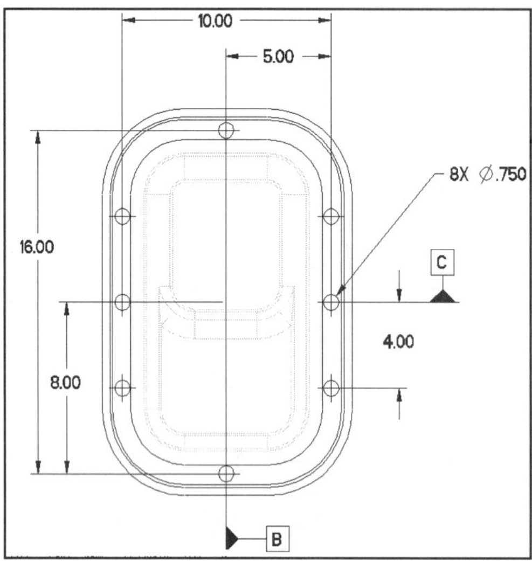

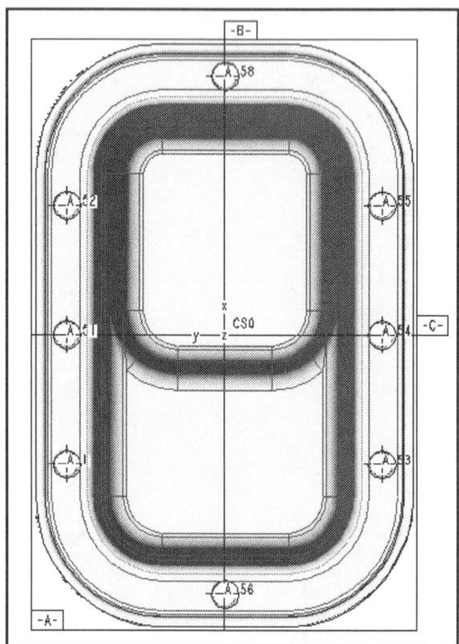

Figure 10.13
Oil Sink Drawing,
Bottom View

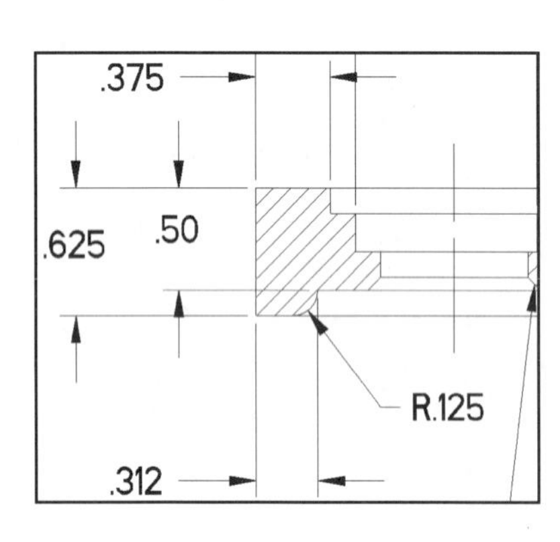

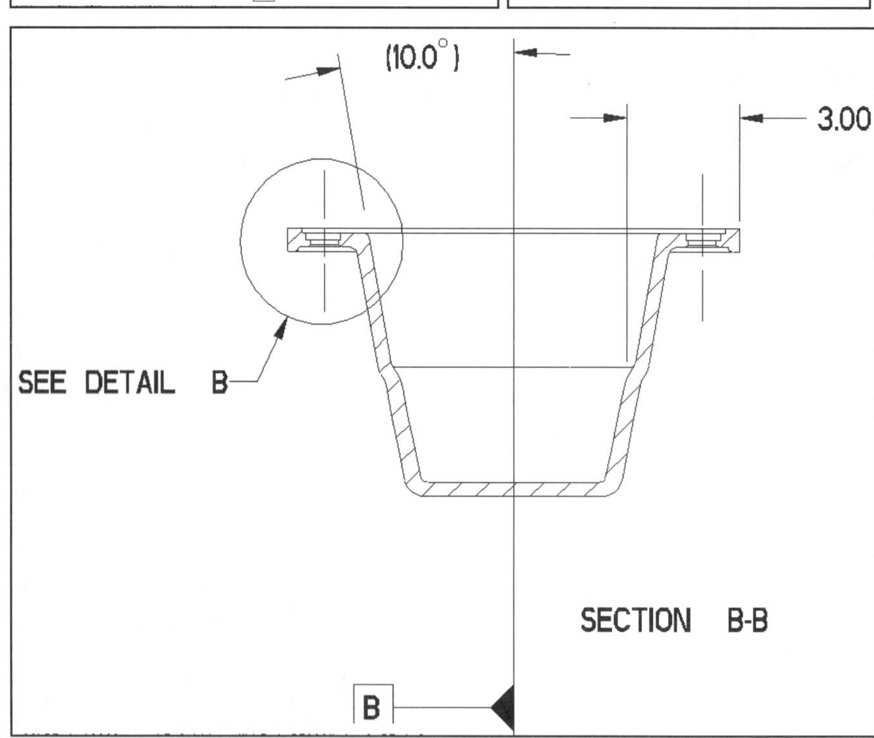

Figure 10.14
Oil Sink Drawing,
SECTION B-B

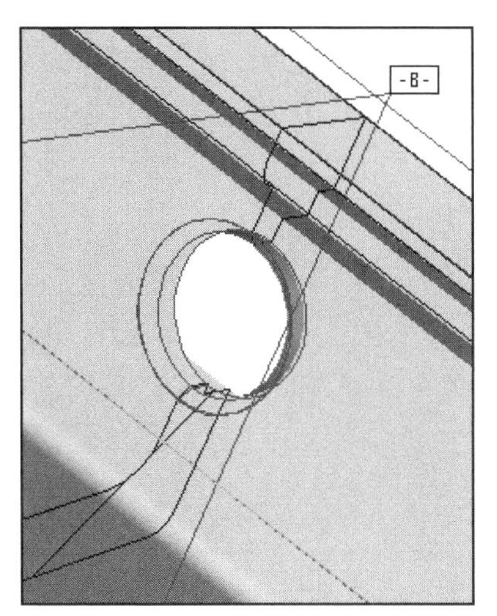

Figure 10.15
Oil Sink Drawing,
DETAIL A

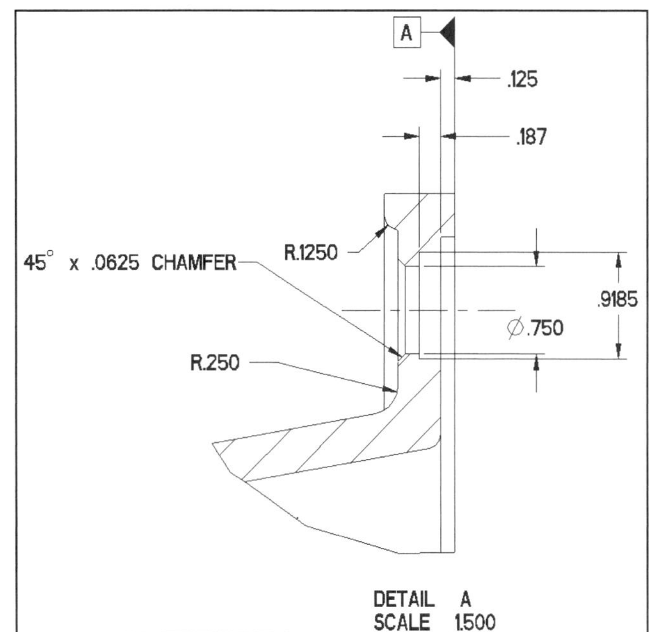

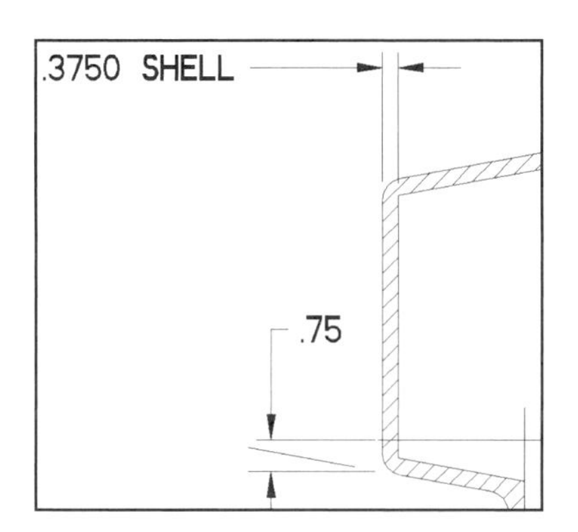

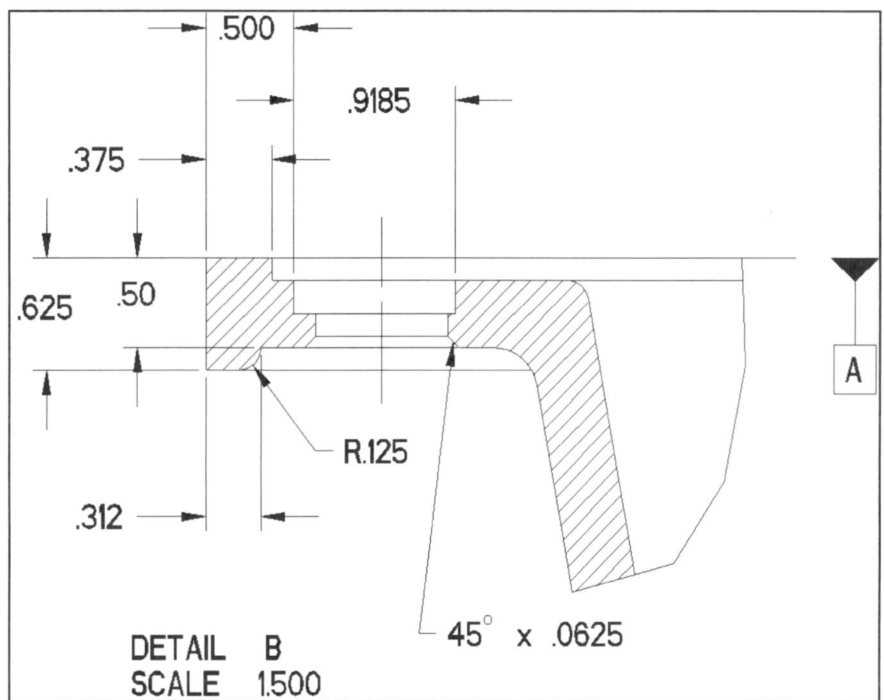

Figure 10.16
Oil Sink Drawing,
DETAIL B

After renaming the datum planes and coordinate system, model the first protrusion shown in Figure 10.17. Next, the large protrusion forming the Oil Sink's tub-like shape is modeled (Figure 10.18). In Figure 10.19, a cut is added to the model. The **R1.50** rounds are added next (Fig. 10.20). The lateral sides of the protrusion are drafted at an angle of **-10°**, as shown in Figure 10.21. Shell the part with the following commands (Fig. 10.22):

Feature ⇒ **Create** ⇒ **Shell** ⇒ (select one or more surfaces--pick the bottom surface) ⇒ **Done Sel** ⇒ **Done Refs** ⇒ (enter thickness of **.375** at the prompt) ⇒ ✔ ⇒ **OK** ⇒ **Done**

L10-10 **Shell, Reorder, and Insert Mode**

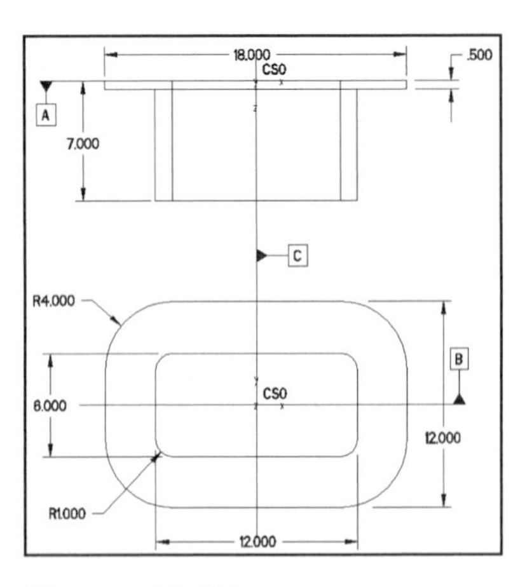

Figure 10.17
Oil Sink's First Protrusion

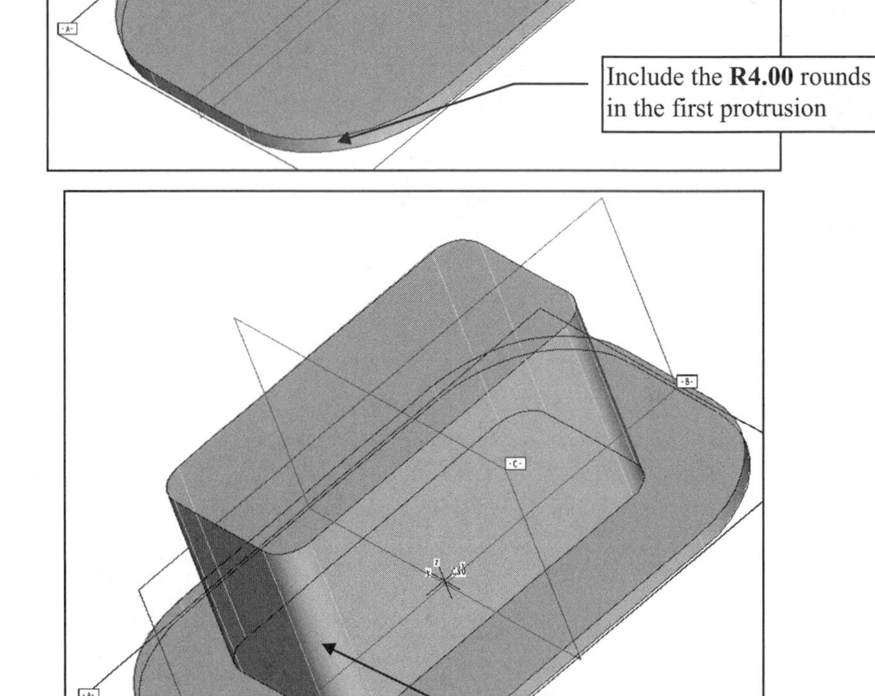

Figure 10.18
Oil Sink's Second Protrusion

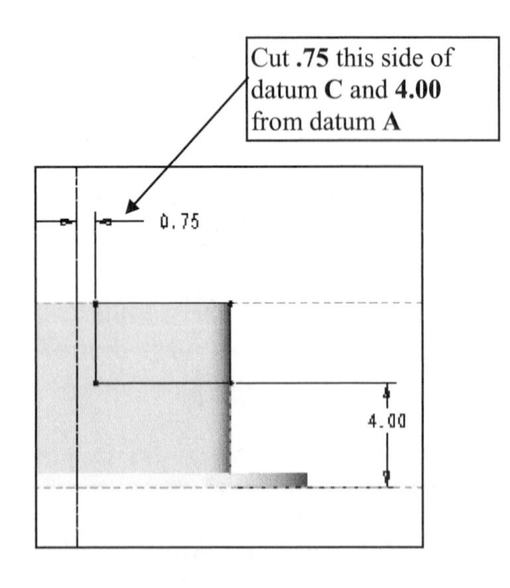

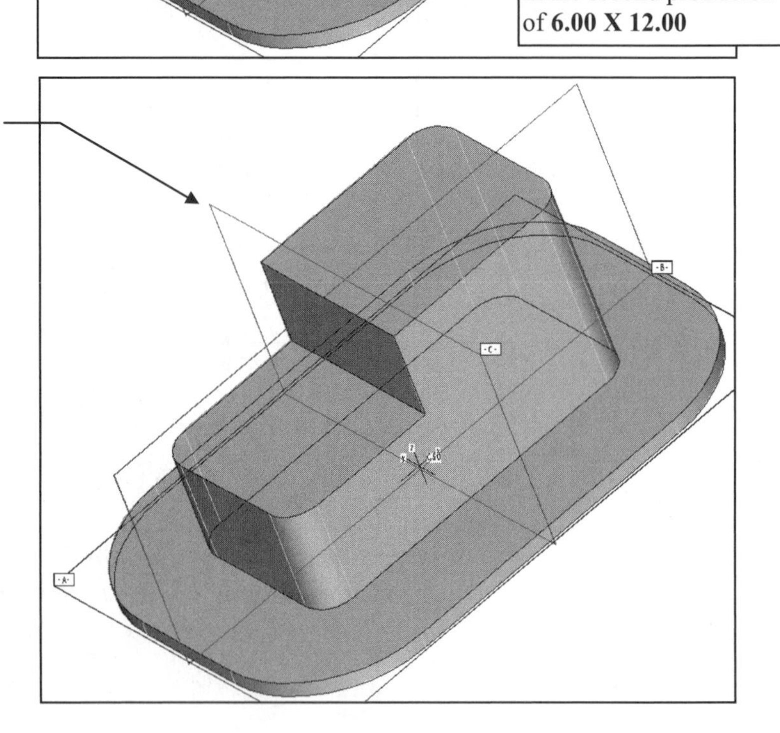

Figure 10.19
Oil Sink's First Cut

File ⇒ Save ⇒ ✔

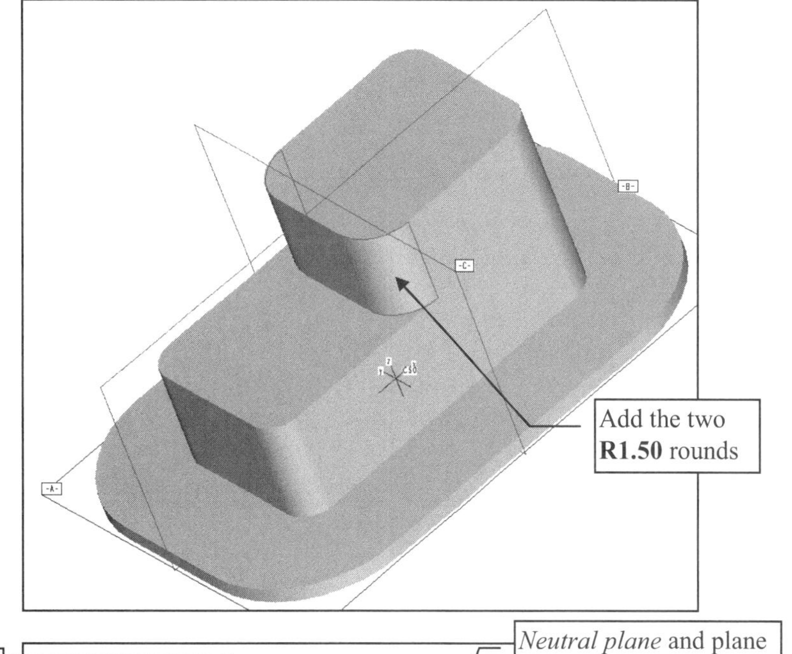

Figure 10.20
Add Rounds

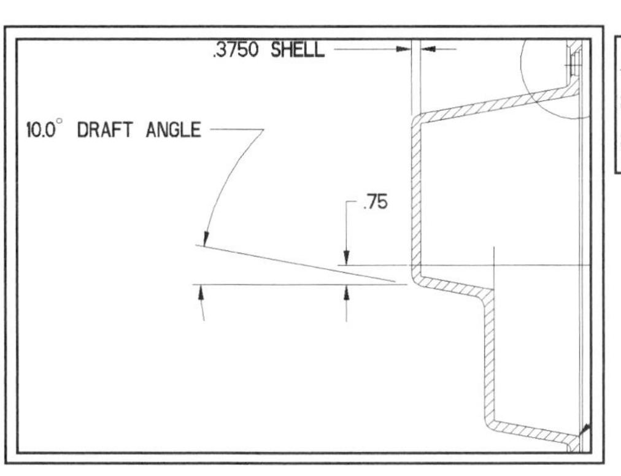

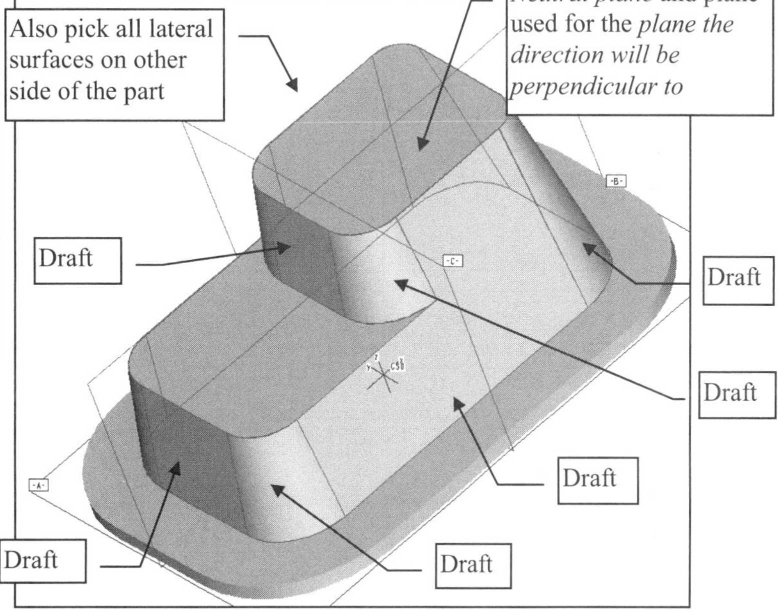

Figure 10.21
Draft *All* Upper Lateral Surfaces of the Oil Sink (**-10°**)

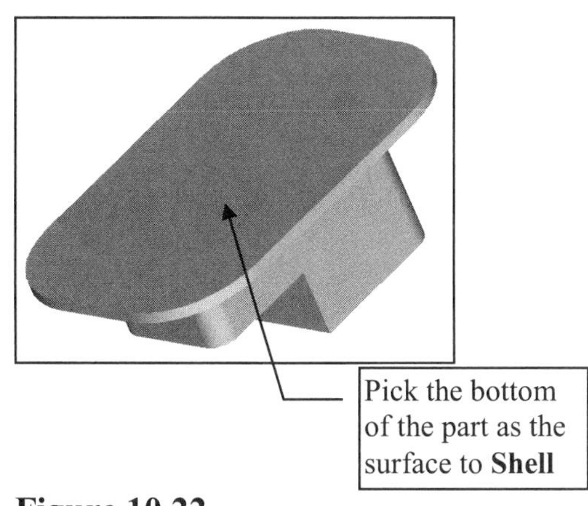

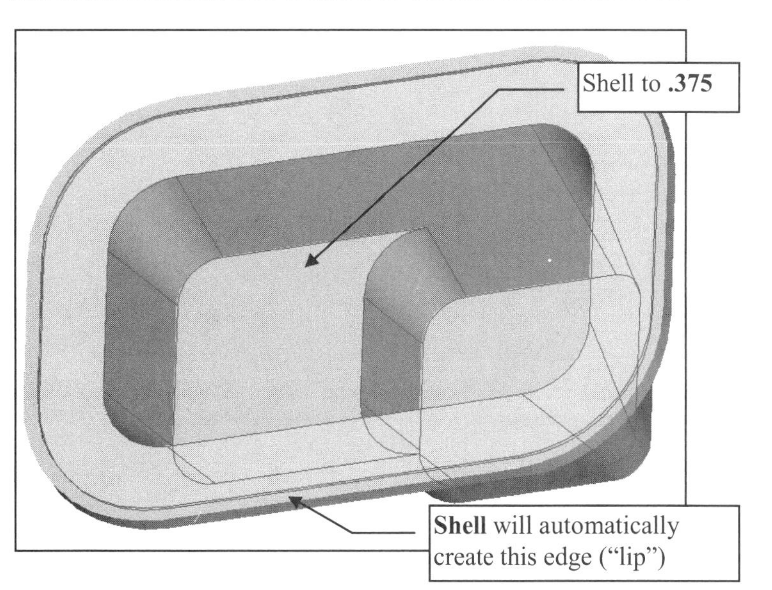

Figure 10.22
Shell the Part

CONFIG.PRO
You can reinstate a variety of pre Pro/ENGINEER 2000i commands (**Lip**, **Shaft**, **Neck**, **Slot**, **Flange** etc.) by using:
allow_anatomic_features yes

Utilities ⇒ Preferences ⇒ ***allow_anatomic_features yes*** **⇒ Add/Change ⇒ Apply ⇒ Close**

The next feature you need to create is a *"lip"* (Fig. 10.23) around the part using a protrusion and a round. Also, you may wish to try creating the *"lip"* with a **Sweep Protrusion** after reviewing Lesson 12 or by reinstating "anatomic" commands (**Lip**, **Shaft**, **Neck**, **Slot**, **Flange,** etc.) from pre Pro/ENGINEER 2000i (see left-hand column).

Here we will use a protrusion and a round (Fig. 10.24).

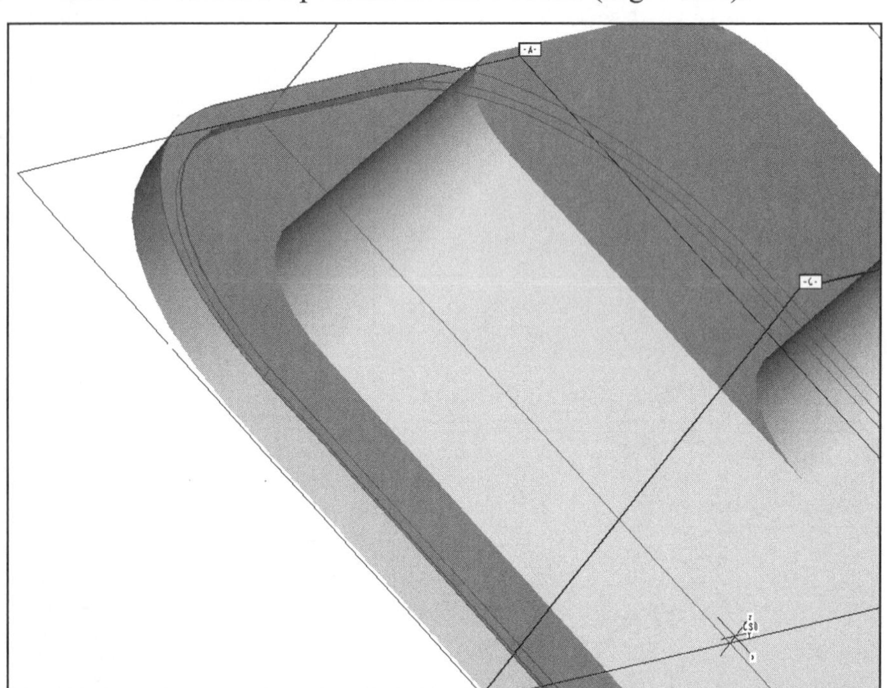

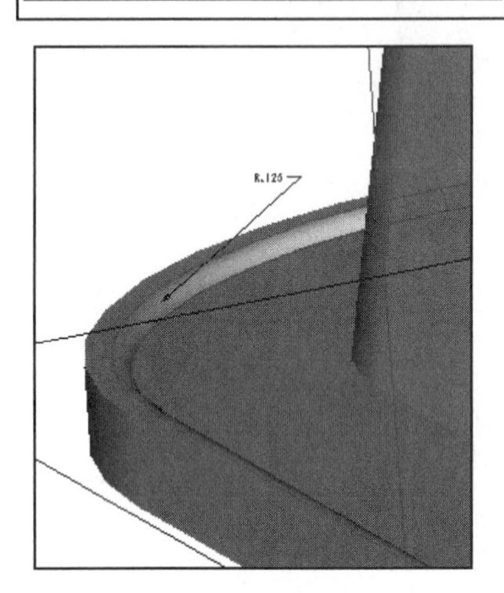

Figure 10.23
"Lip" (Protrusion)

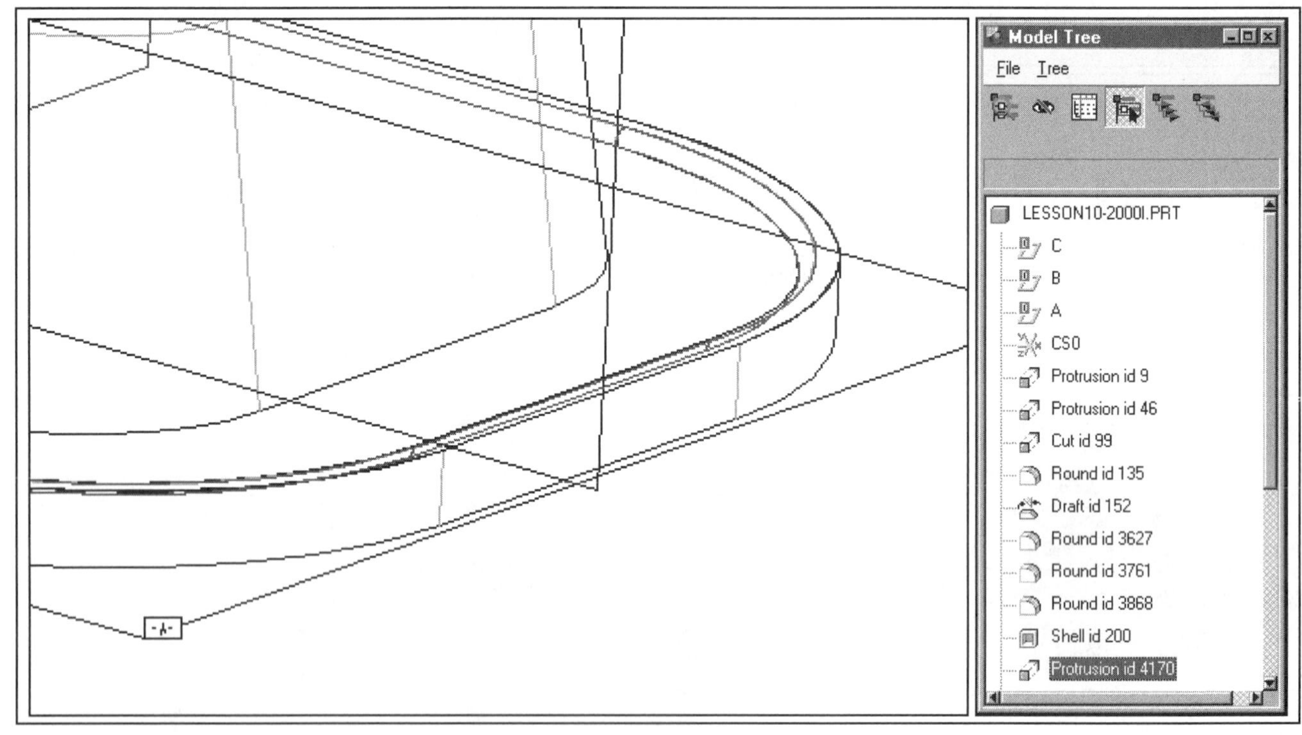

Figure 10.24
Protrusion and Round Required for the *"Lip"*

Create the *"lip"* with the following [Intent Manager (off)]:

Feature ⇒ **Create** ⇒ **Solid** ⇒ **Protrusion** ⇒ **Extrude** ⇒ **Solid** ⇒ **Done** ⇒ **One Side** ⇒ **Done** ⇒ [select the top plane as the sketching plane (Fig. 10.25)] ⇒ **Okay** [to accept the protrusion direction (see left column)] ⇒ **Bottom** ⇒ (select datum **B**) ⇒ **Sketch** (on menu bar) ⇒ **Edge** ⇒ **Use** (from fly-out menu) ⇒ **Sel Loop** ⇒ (select the outside edge of the part to define the entity one loop) ⇒ **Accept** ⇒ **Edge** ⇒ **Offset Edge** ⇒ **Sel Loop** (select the outside edge) ⇒ **Accept** ⇒ [Enter offset in the direction of the arrow **-.3125** (see inset)] ⇒ ✔ ⇒ **Regenerate** ⇒ **Done** ⇒ **Blind** ⇒ **Done** ⇒ Enter depth **.125** ⇒ ✔ ⇒ **OK** ⇒ **Done** ⇒ **File** ⇒ **Save** ⇒ ✔

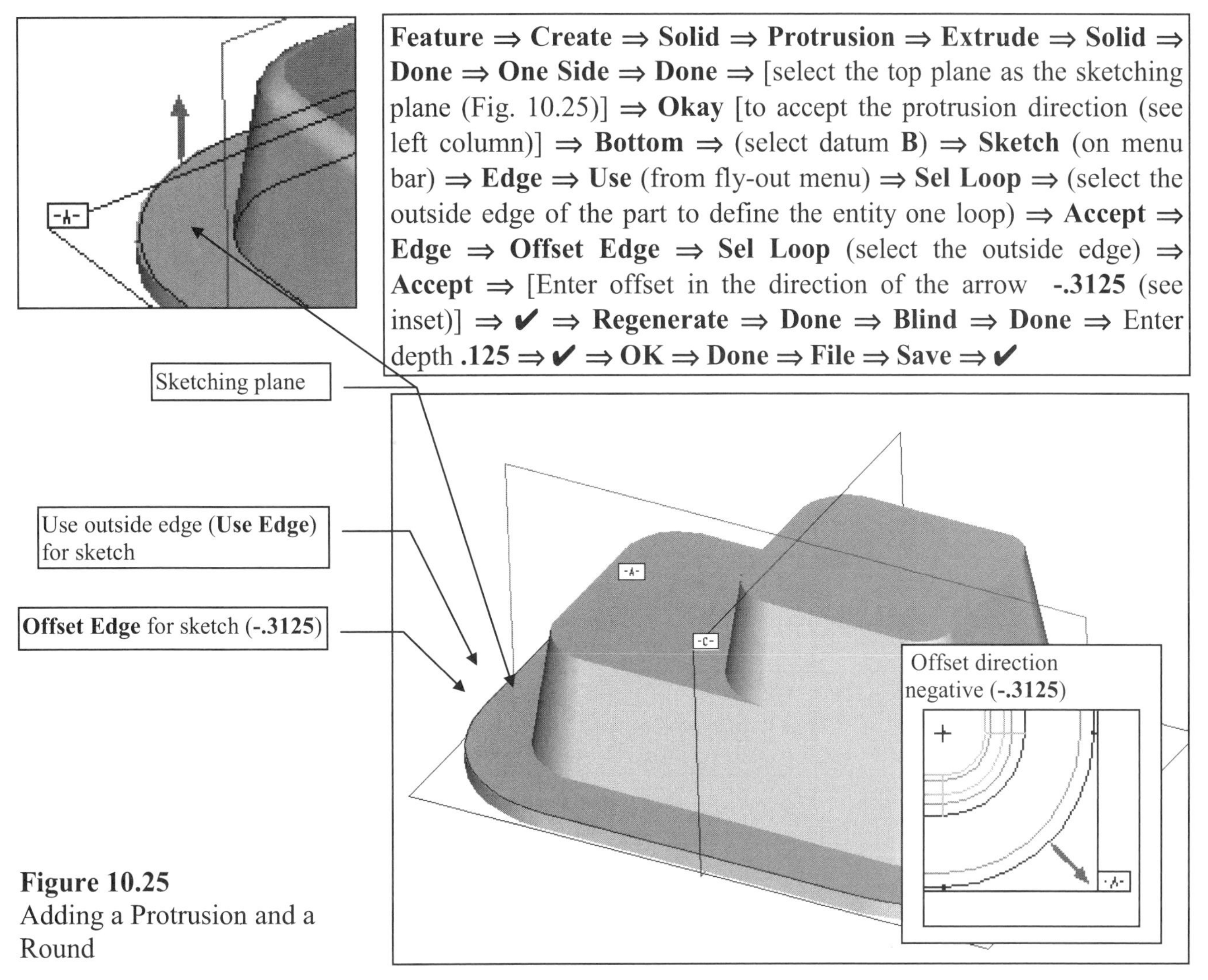

Figure 10.25
Adding a Protrusion and a Round

Add the **R.125** round to the inside of the *"lip"* (Fig. 10.26).

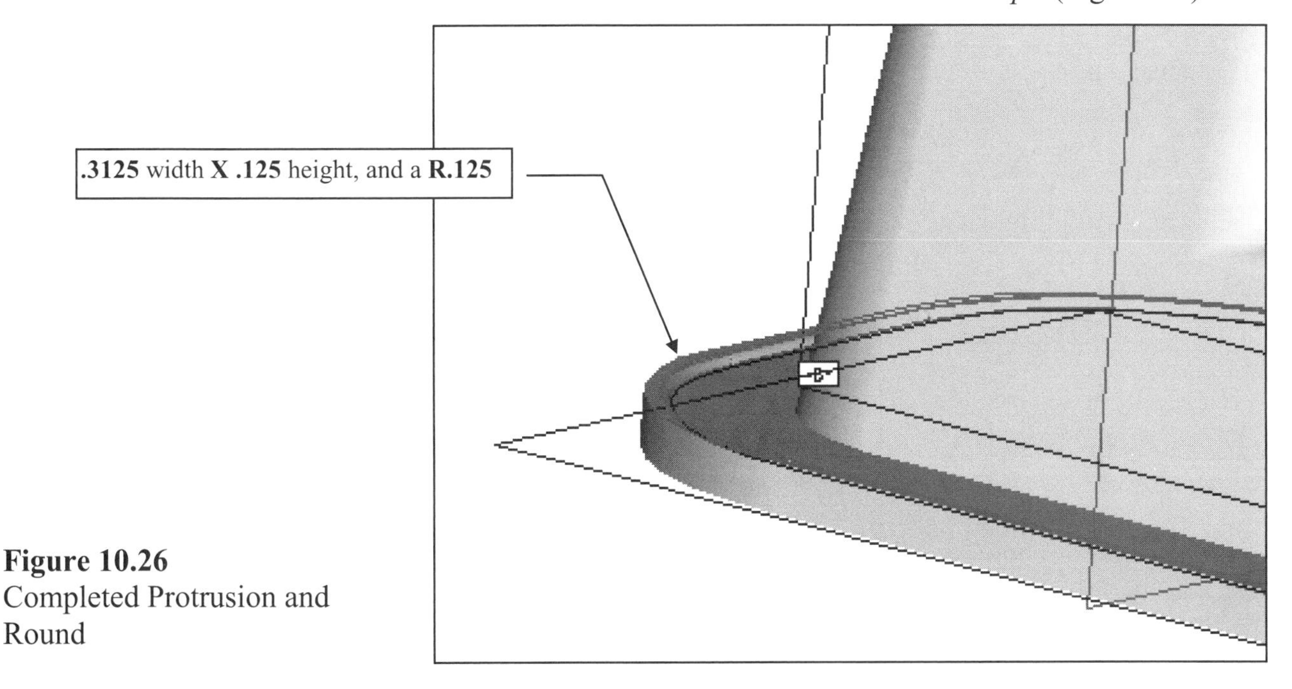

Figure 10.26
Completed Protrusion and Round

The next feature is a cut measuring **.9185** wide by **.187** deep (see Figs. 10.15 and 10.16). Figures 10.27 and 10.28 show the cut. Add the countersunk holes, the chamfers, and the rounds (Fig. 10.29).

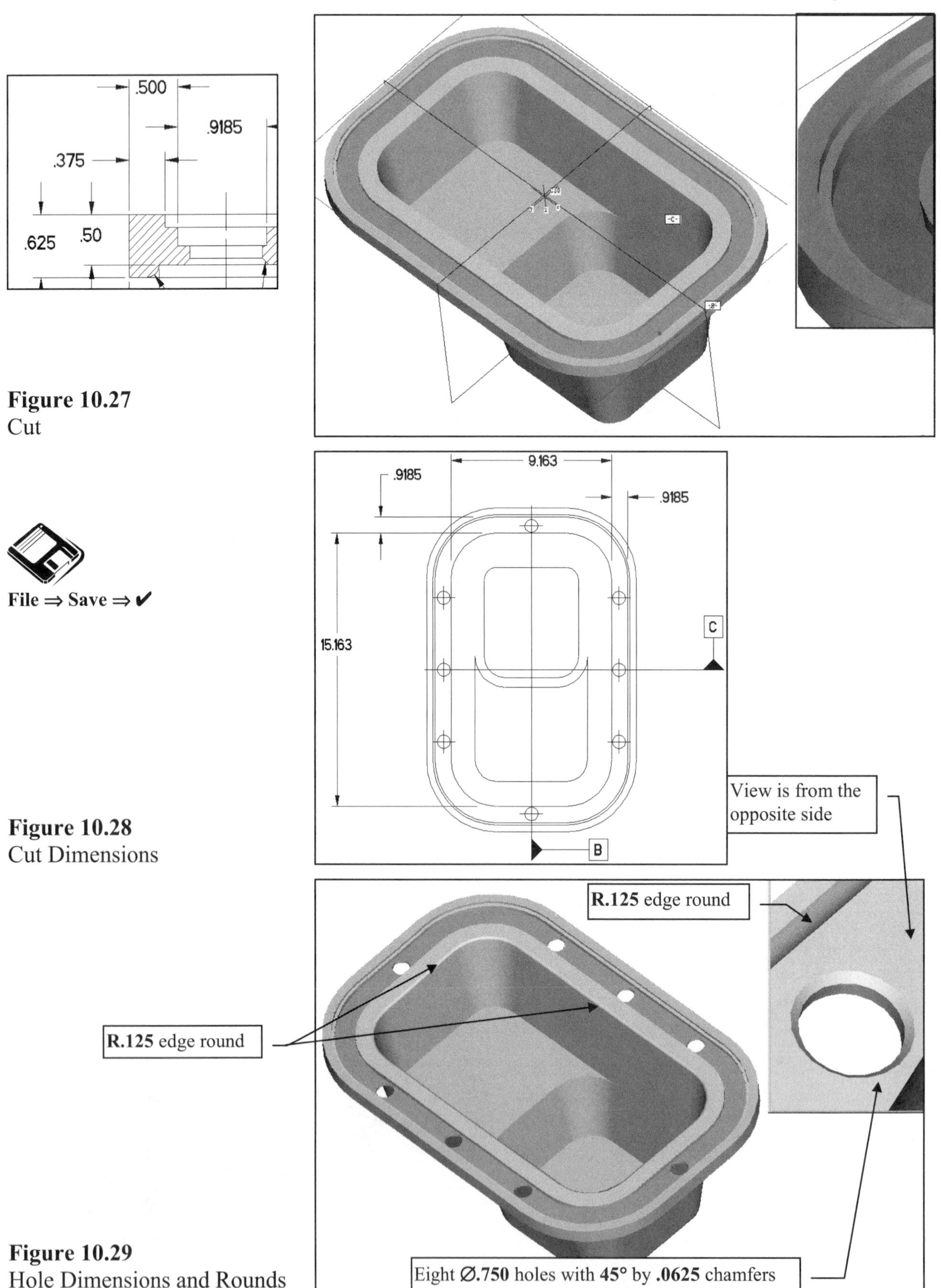

Figure 10.27
Cut

Figure 10.28
Cut Dimensions

Figure 10.29
Hole Dimensions and Rounds

The next series of features will purposely be created at the wrong stage of the project. Create an **R.50** round as shown in Figure 10.30. *Because the design intent should have been to have a constant thickness for the part, the round should have been created before the shell.* Use the **Reorder** option to change the position of this round in the design sequence. *Your prompts will have different feature numbers.* Choose the following commands:

File ⇒ Save ⇒ ✔

Feature ⇒ Reorder ⇒ (pick the **R.50** round) **⇒ Done Sel ⇒ Done ⇒ Before ⇒** (Pro/E will prompt with: ***Feature # can be inserted before features [10-26] or [30-34]. Please, select feature.***) **⇒** (from the Model Tree, select the **Shell id**) **⇒ Done** (Fig. 10.31)

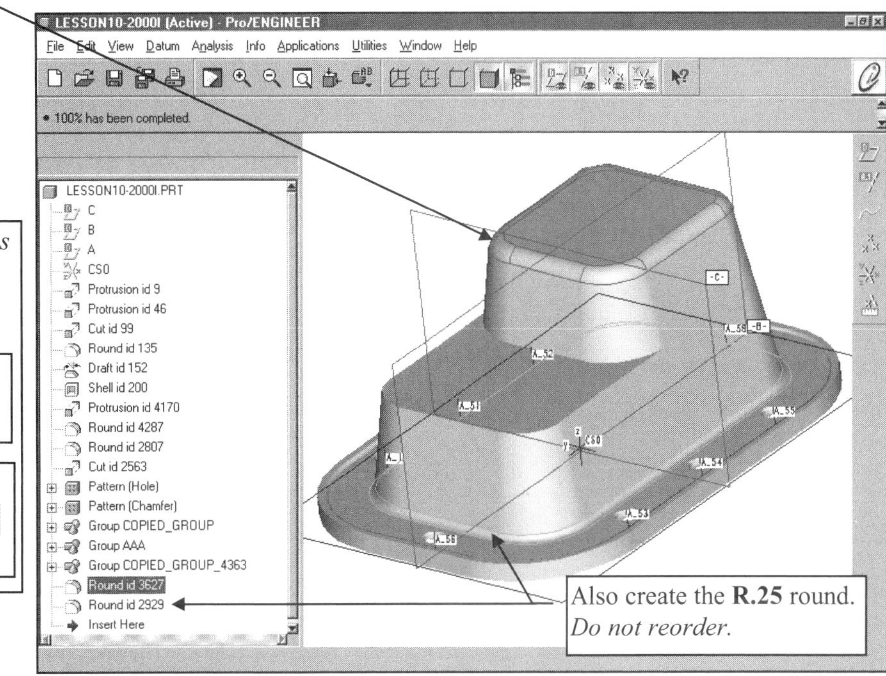

*The new **Reorder** capability completes the operation within the Model Tree.* Pick and drag the Round to a new location in the feature list:

Drag and drop

Draft id 152
Shell id 200

new position

Draft id 152
Round id 3627
Shell id 200

Figure 10.30
R.50 Round

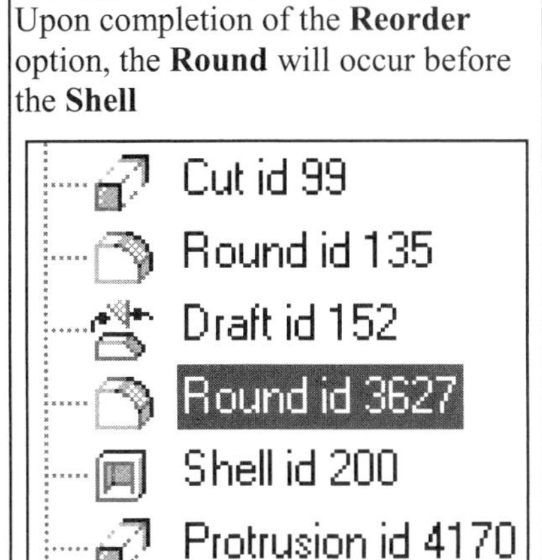

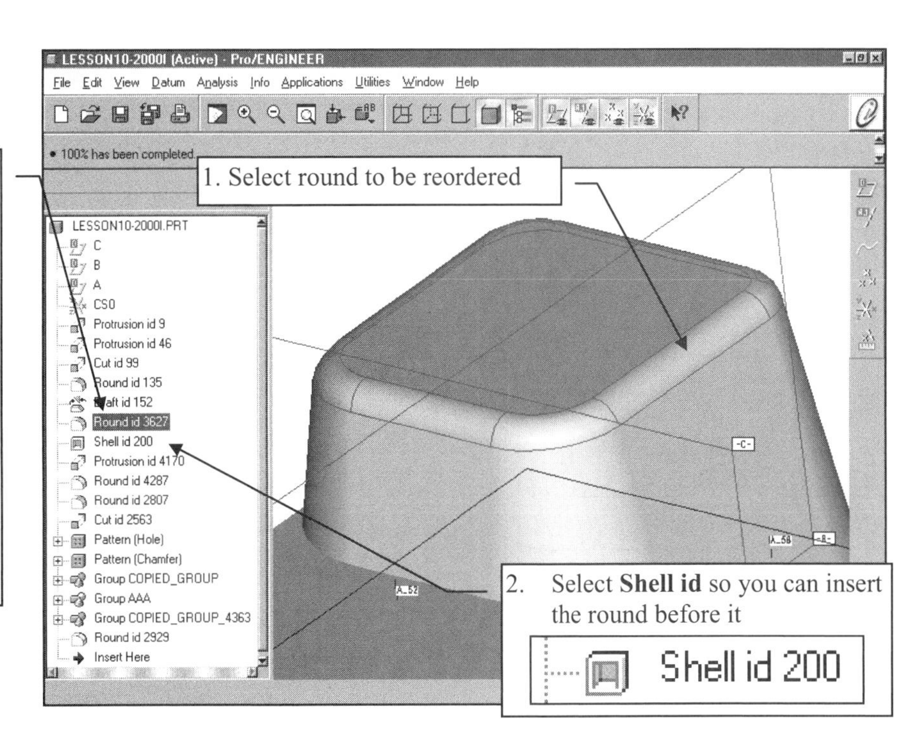

Figure 10.31
Reorder

*You can also **Insert** features using the Model Tree.* The arrow-shaped icon on the Model Tree that indicates where features will be inserted upon creation is by default at the end of the Model Tree.

Drag the location of the node (**➡Insert Here)** higher so that its new position is before the **Shell id 200.** When the *node* is dropped at a new location, the model is rolled backward or forward in response to the insertion node being moved higher or lower.

From **Shell id 200** on down, the Model Tree displays a small square ■ next to the features that are not active (suppressed) during **Insert Mode.**

The previous round was created at the wrong stage in the design sequence and then reordered. To eliminate the reordering of a feature, the **R.50** rounds will be created using **Insert Mode. Insert Mode** allows you to insert a feature at a previous stage of the design sequence. This is like going back into the past and doing something you wish you had done before--not possible with life, but with Pro/E less of a problem. Add the additional **R.50** rounds. Choose the following commands to enter **Insert Mode,** create the rounds, and then use **Resume** to return to the previous stage in the design as shown in Figure 10.32):

Feature ⇒ Insert Mode ⇒ Activate ⇒ (select a feature to insert after; from the Model Tree, pick the round you just reordered, as shown in Fig. 10.33) **⇒ Done**

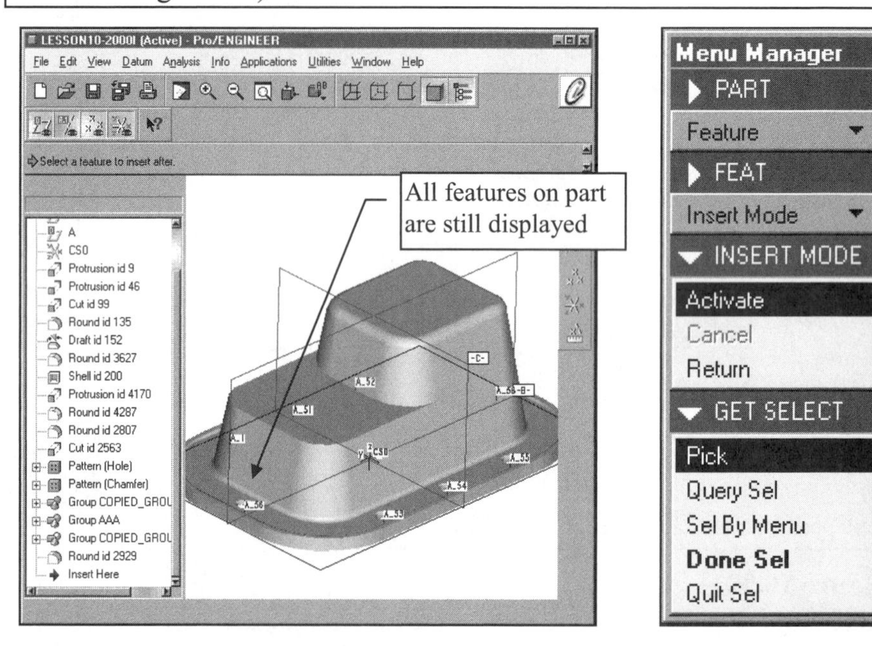

Figure 10.32
Before **Insert Mode** Is Activated by Selecting a Feature

Show suppressed features
View ⇒ Model Tree Setup ⇒ Item Display ⇒ ✔Suppressed Objects ⇒ OK

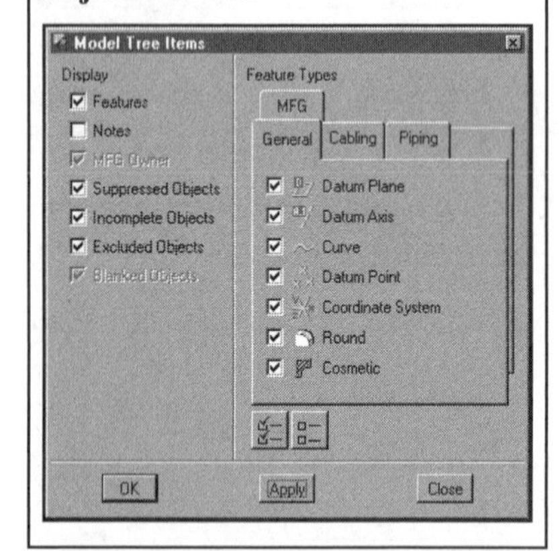

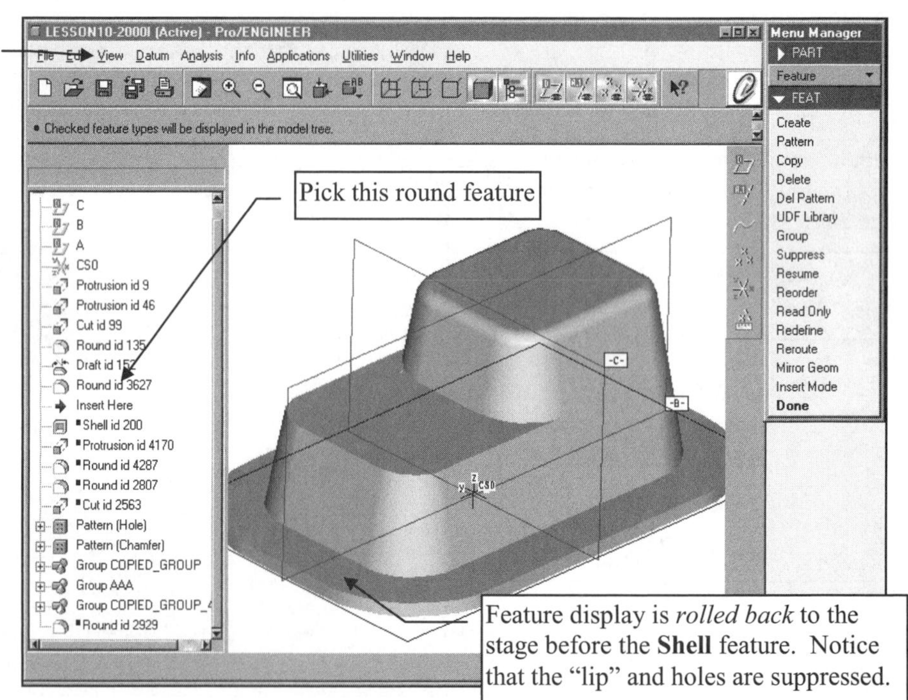

Figure 10.33
Insert Mode Is Now Active and the Model Is Rolled Back

The rounds can now be created using the following commands:

Feature ⇒ Create ⇒ Round ⇒ Simple ⇒ Done ⇒ Constant ⇒ Edge Chain ⇒ Done ⇒ Tangnt Chain ⇒ (pick the edge shown in Fig. 10.34) **⇒ Done Sel ⇒ Done ⇒** (enter the radius value of **.50** at the prompt) **⇒ ✔ ⇒ OK ⇒ Done ⇒** (*repeat the same commands,* with **R.50,** and pick the edge shown in Fig. 10.35) **⇒ Feature ⇒ Resume ⇒ All** (Fig. 10.36) **⇒ Done** (Fig. 10.37) **⇒ Done**

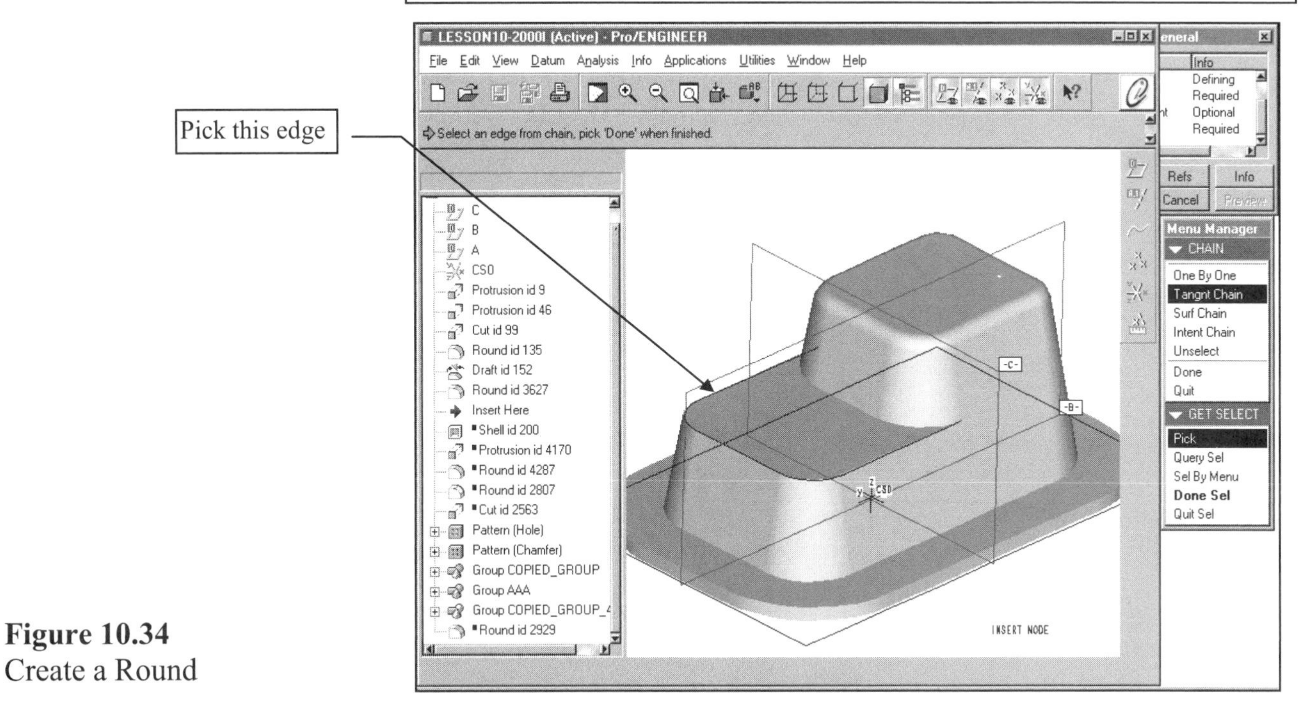

Figure 10.34
Create a Round

Pick this edge for the next round

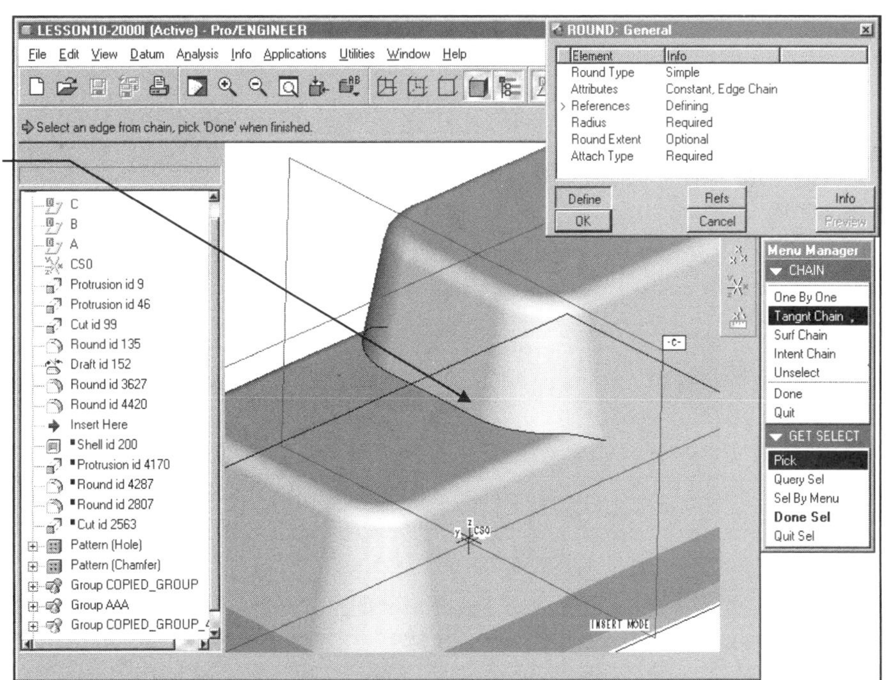

Figure 10.35
Create Another Round

File ⇒ Save As ⇒ etc.

The rounds are now in the proper design sequence for the shell feature to create a constant thickness. After the part is finished, save it under a new name and complete the **ECO** (Fig. 10.38).

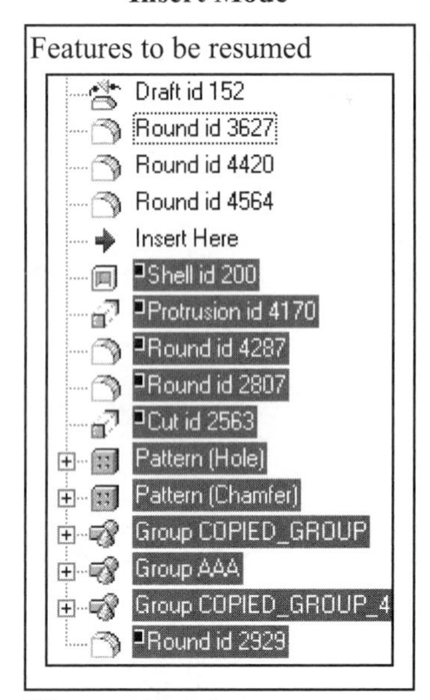

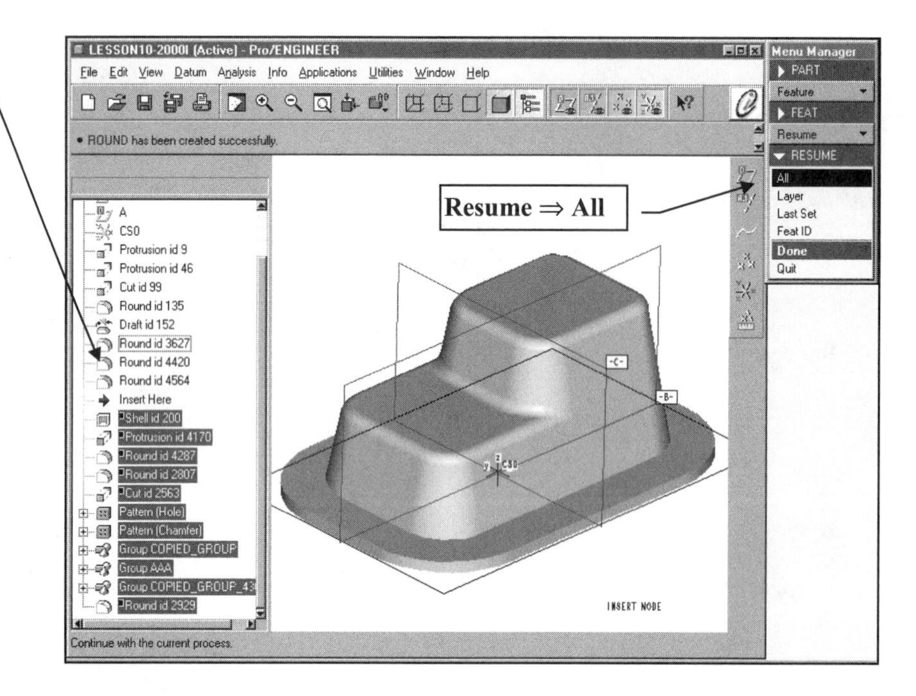

Figure 10.36
Resume ⇒ All

File ⇒ Save ⇒ ✔

Figure 10.37
Resumed Part

E C O

File ⇒ Save ⇒ ✔
File ⇒ Delete ⇒
Old Versions ⇒ ✔

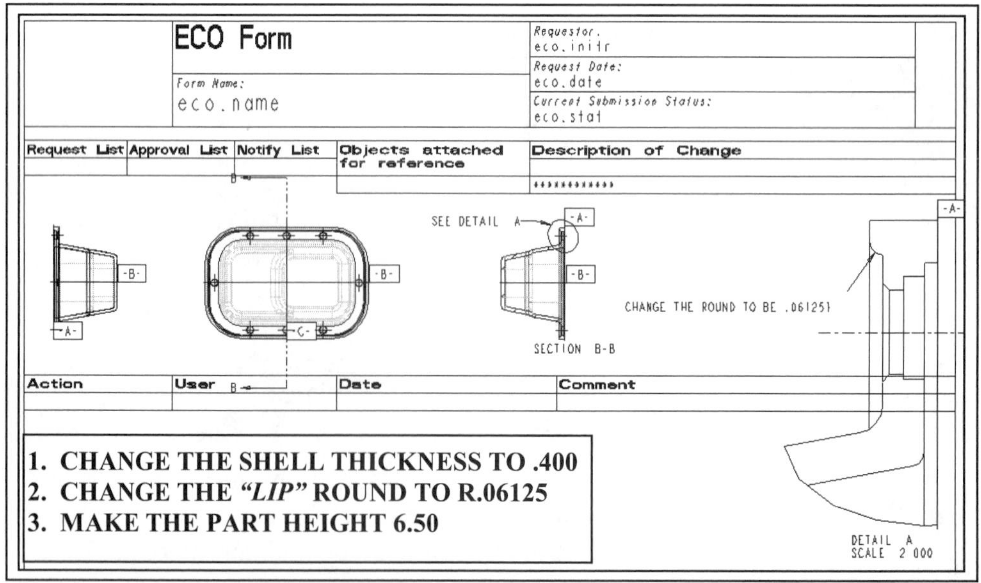

Figure 10.38
ECO

Lesson 10 Project

Cellular Phone Top

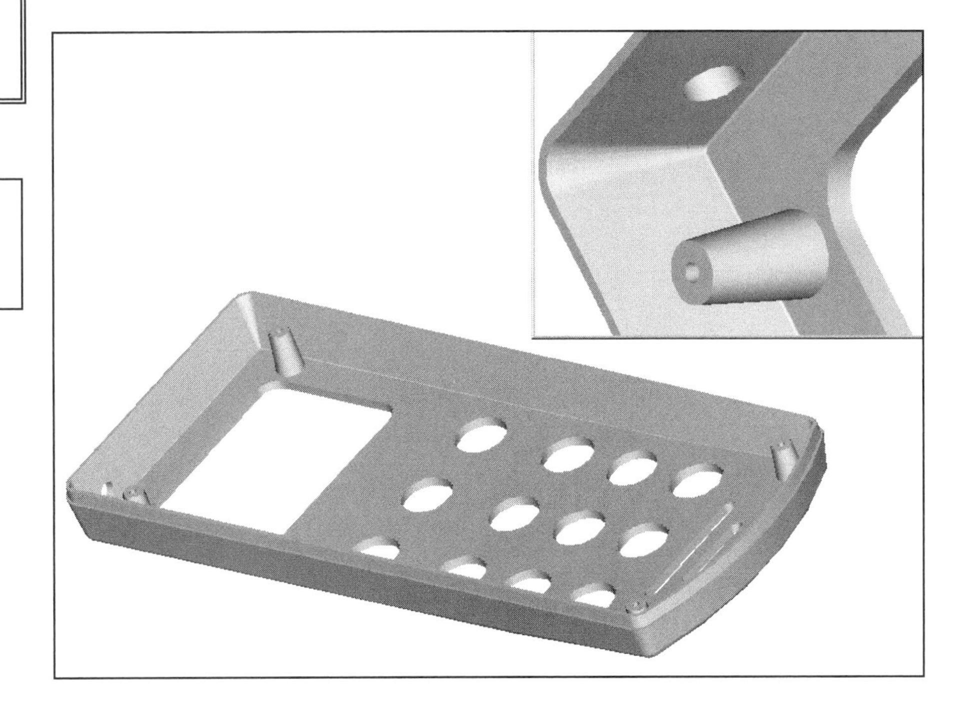

Figure 10.39
Cellular Phone Top

Cellular Phone Top

The Cellular Phone Top (Figures 10.39 through 10.49) is one of two major components for a cellular phone. You created the other as a project in Lesson 9. The part is made of the same plastic as the Cellular Phone Bottom. If time permits, try to assemble the two pieces after completing Lesson 14, Assembly Constraints (you may need to modify the parts).

Analyze each part and plan out the steps and features required to model it. Use the **DIPS** in Appendix D to establish a feature creation sequence before the start of modeling.

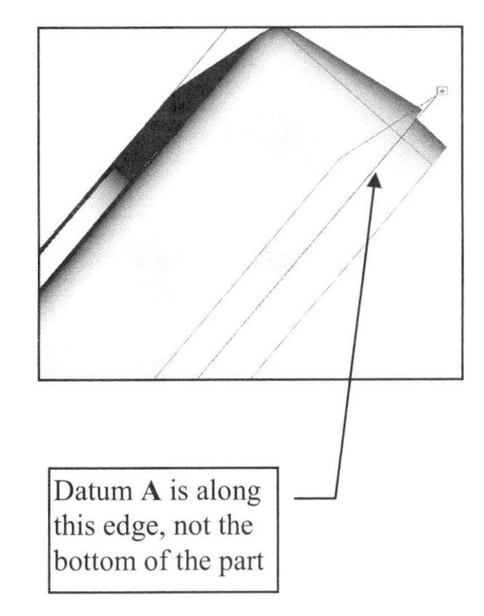

Datum **A** is along this edge, not the bottom of the part

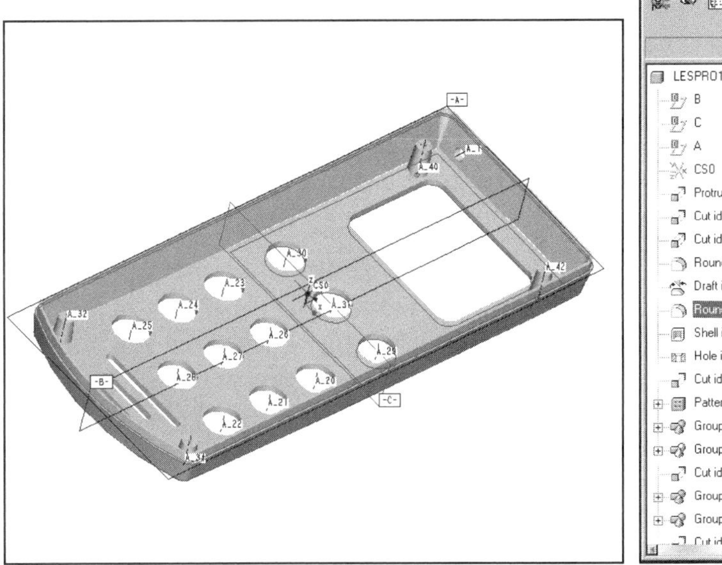

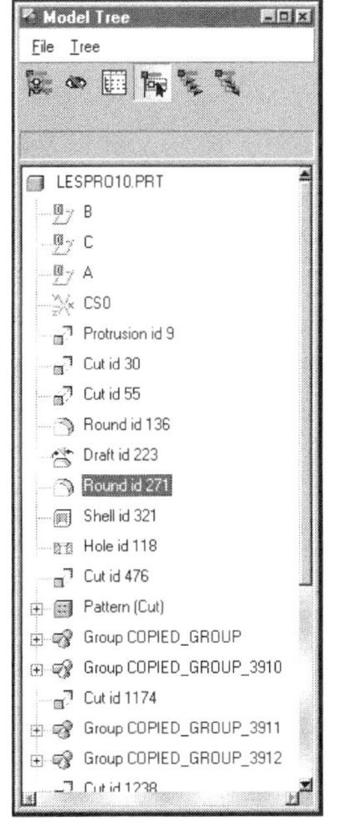

Figure 10.40
Cellular Phone Top Showing Datum Planes and Model Tree

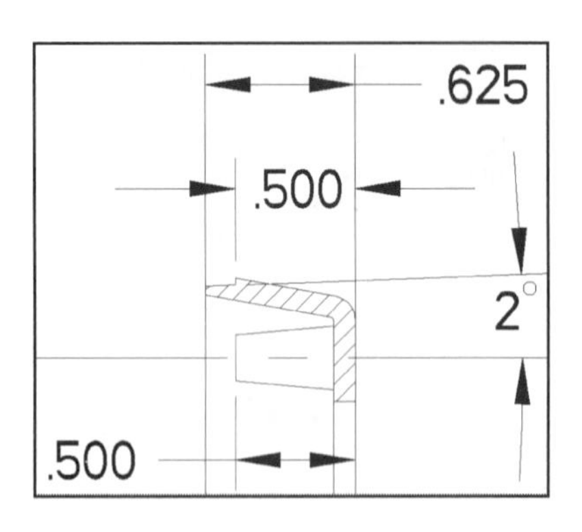

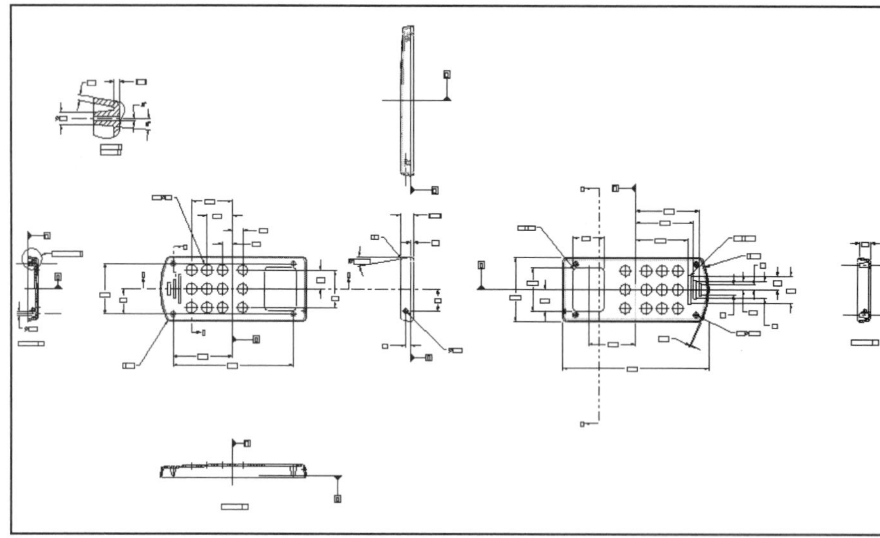

Figure 10.41
Cellular Phone Top, Detail Drawing

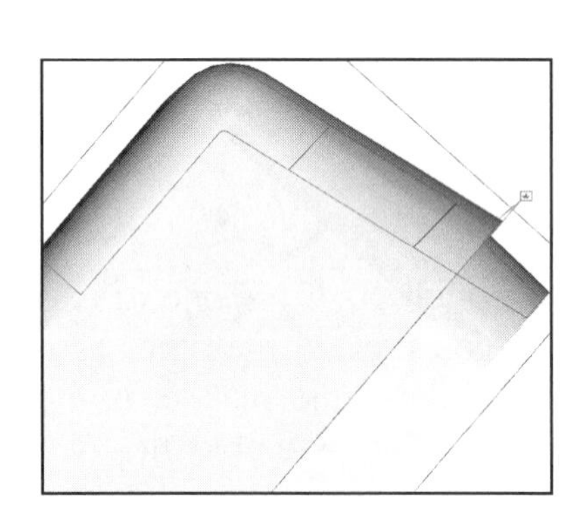

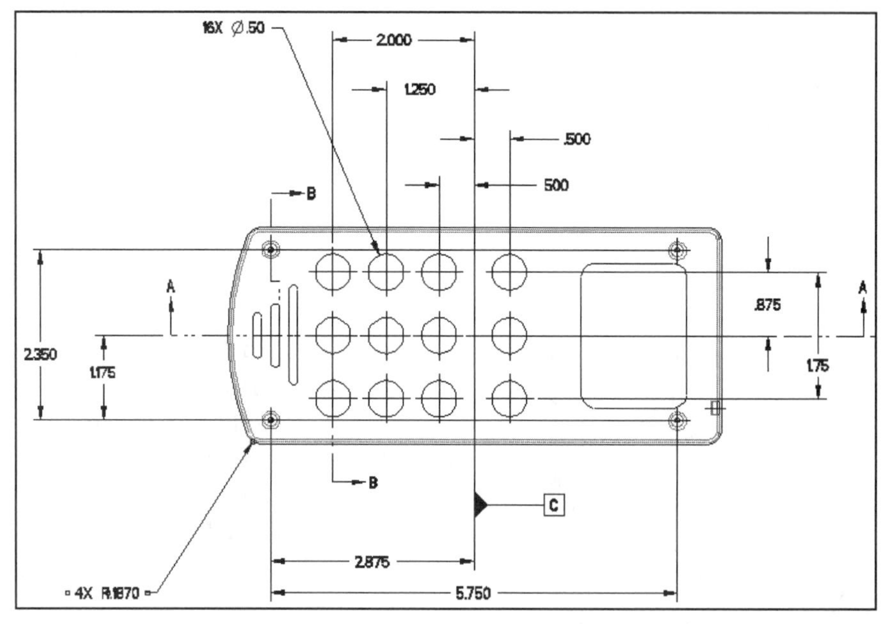

Figure 10.42
Cellular Phone Top, Front View

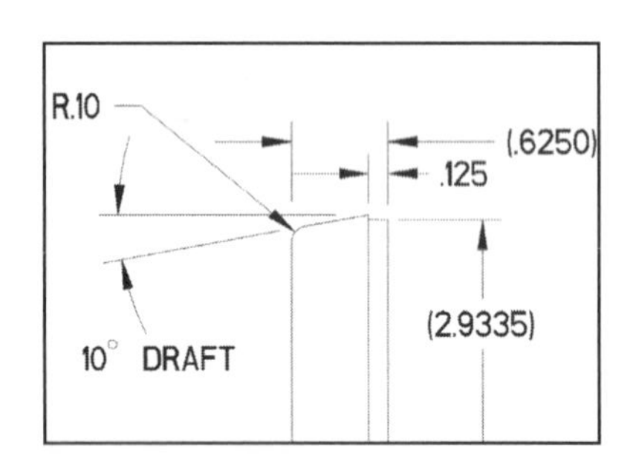

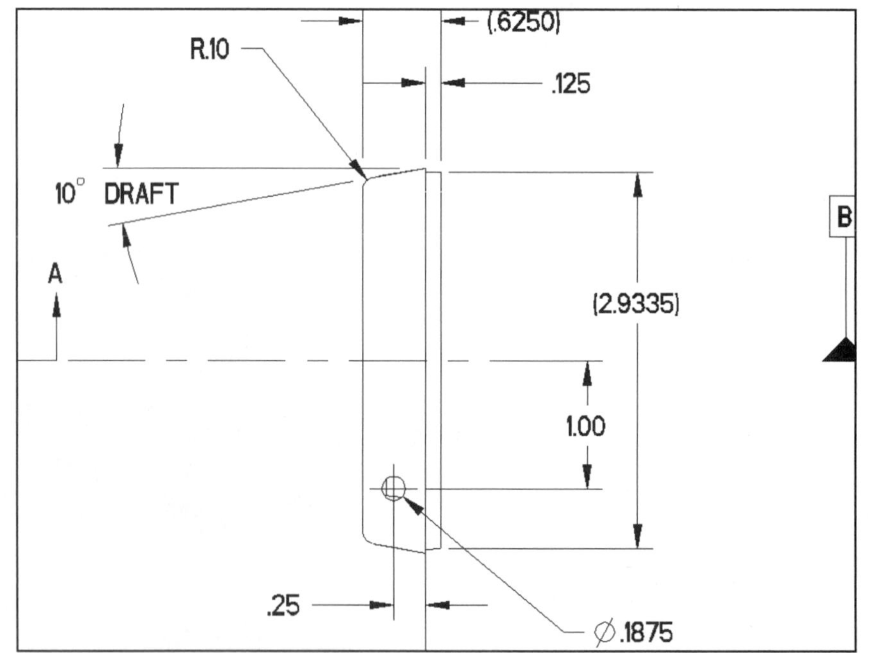

Figure 10.43
Cellular Phone Top, Right Side View

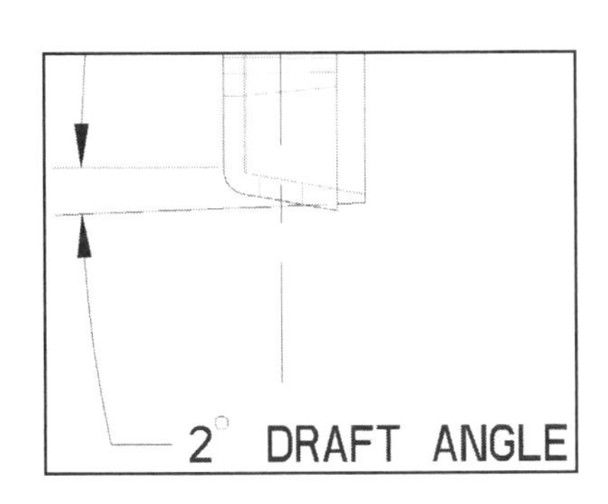

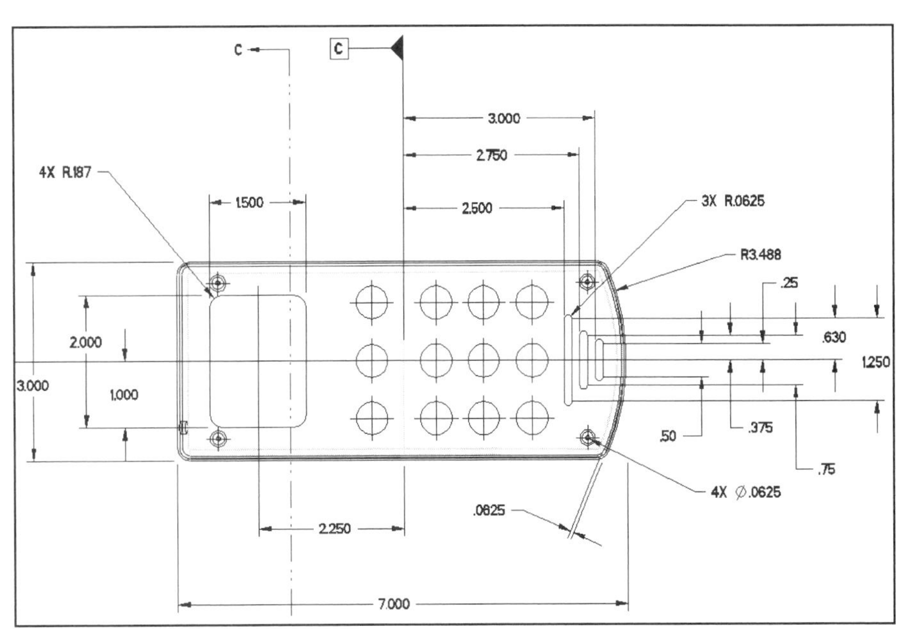

Figure 10.44
Cellular Phone Top,
Bottom View

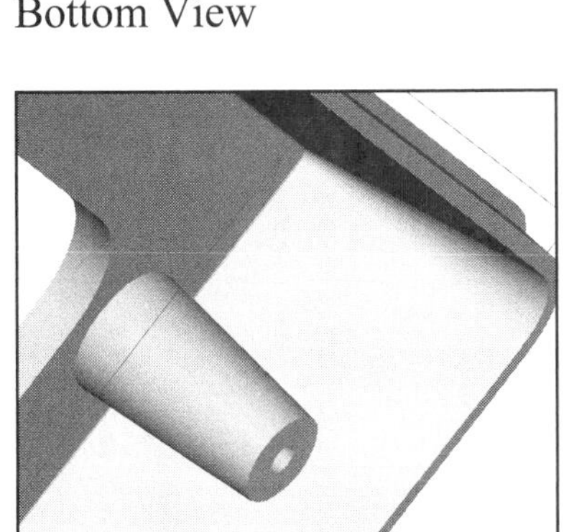

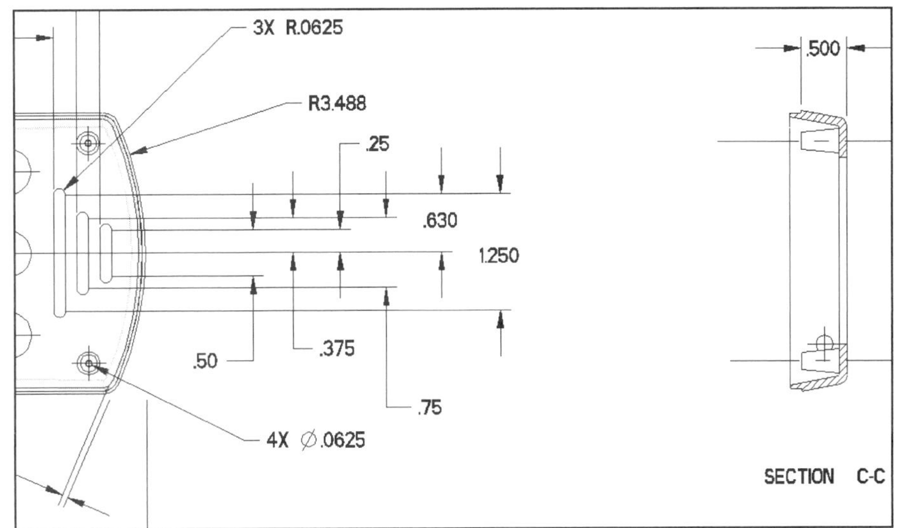

Figure 10.45
Cellular Phone Top,
SECTION C-C

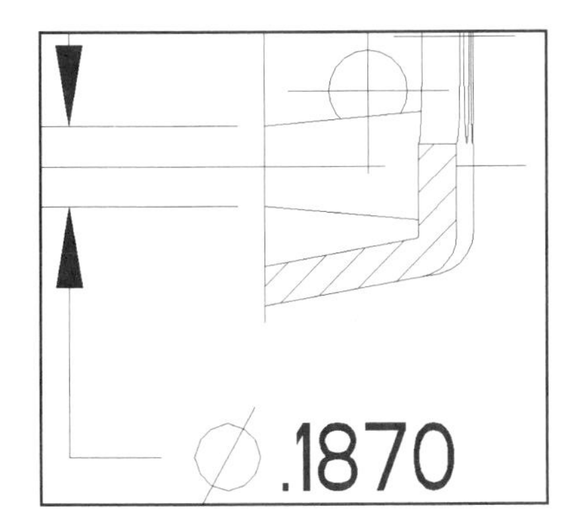

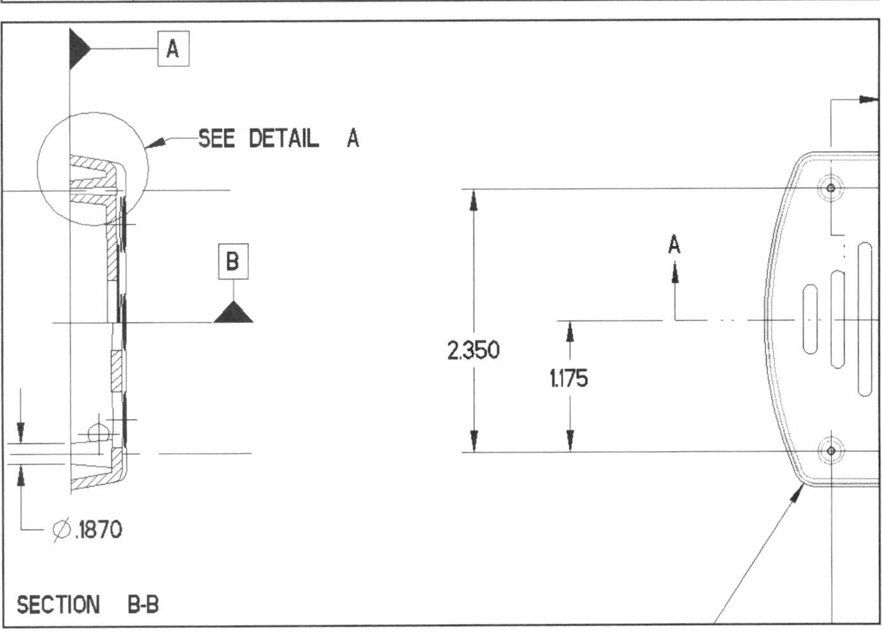

Figure 10.46
Cellular Phone Top,
SECTION B-B

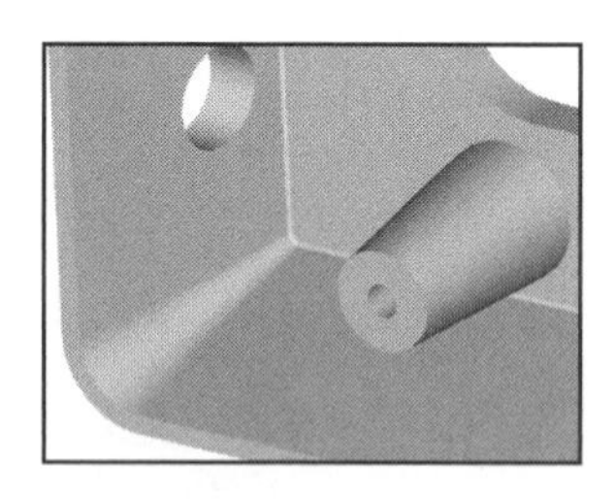

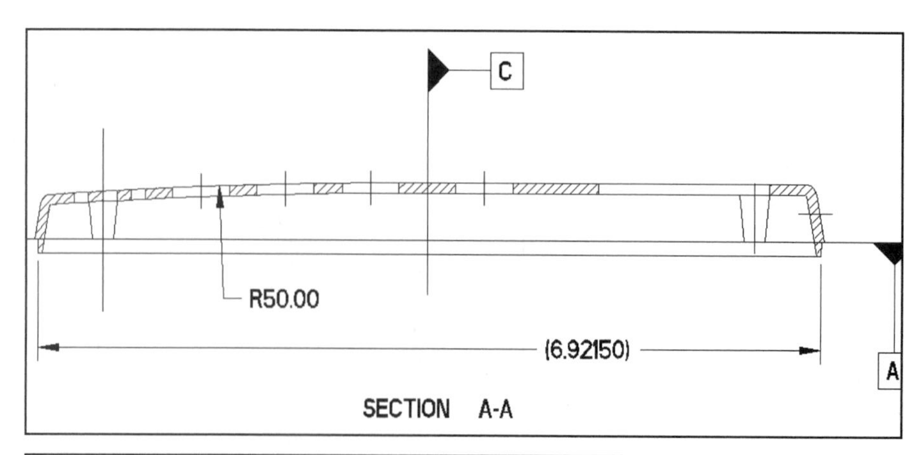

Figure 10.47
Cellular Phone Top,
SECTION A-A

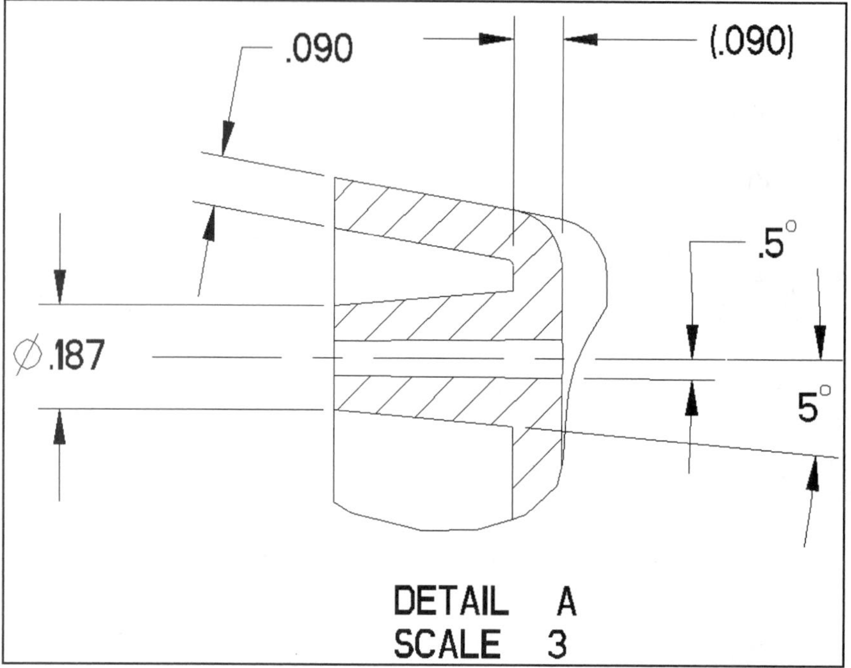

Figure 10.48
Cellular Phone Top,
DETAIL A

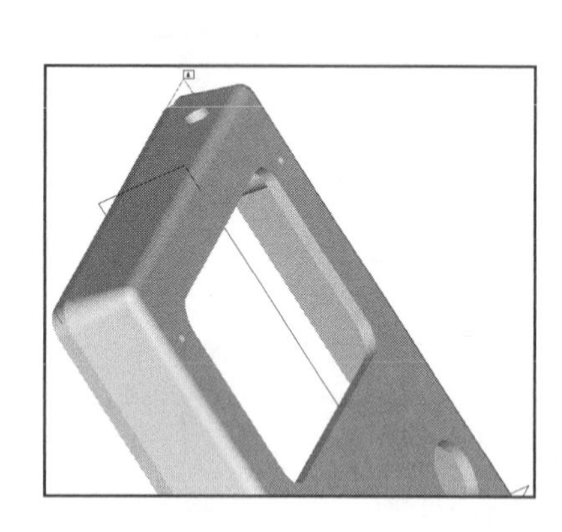

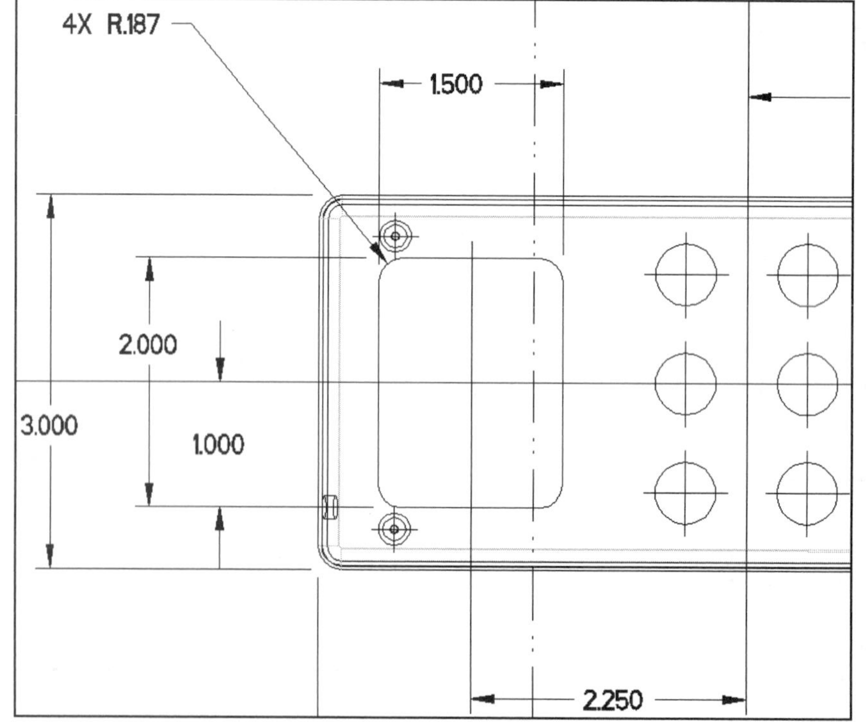

Figure 10.49
Cellular Phone Top,
Opening

Lesson 11

Sweeps

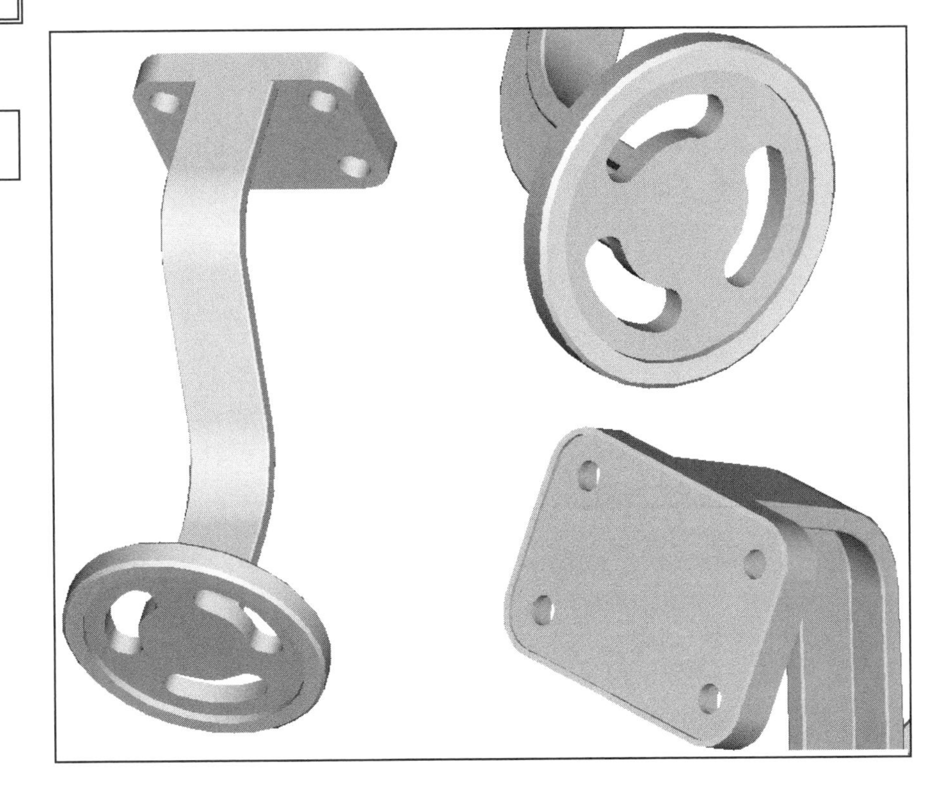

Figure 11.1
Bracket

OBJECTIVES

1. **Create a constant-section swept feature**

2. **Sketch a trajectory for a sweep**

3. **Sketch and locate a sweep section**

4. **Understand the difference between adding and not adding inner faces**

5. **Be able to redefine a sweep**

6. **Understand the difference between a sketched and a selected trajectory**

7. **Create variable sweeps**

EGD REFERENCE
Fundamentals of Engineering Graphics and Design
by L. Lamit and K. Kitto
Read Chapter 20.
See pages 378 and 382.

COAch™ for Pro/ENGINEER
If you have **COAch for Pro/ENGINEER** on your system, go to SEARCH and do the Segments shown in Figures 11.3 through 11.5.

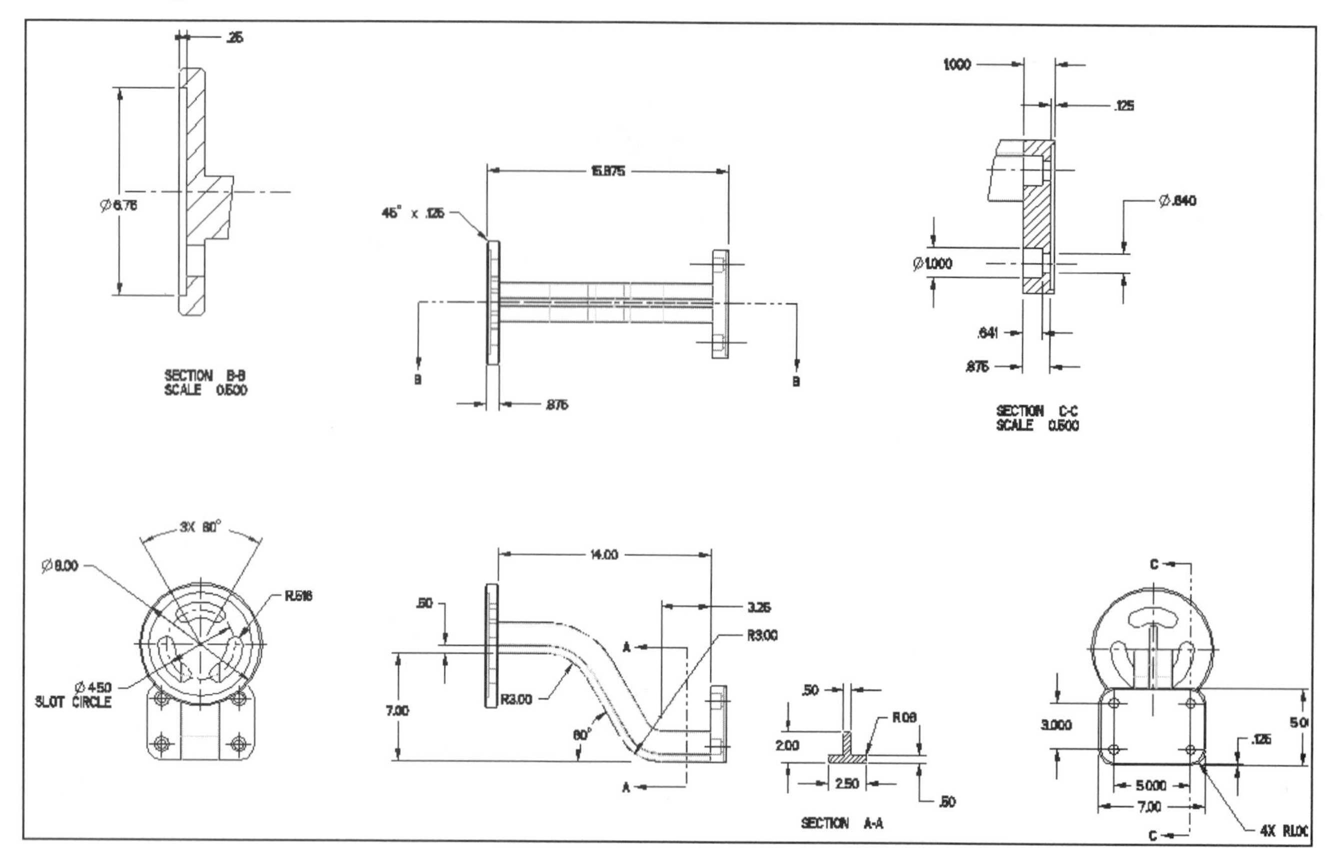

Figure 11.2
Bracket Detail

SWEEPS

A **Sweep** is created by sketching or selecting a *trajectory* and then sketching a *section* to follow along it. The **Bracket** shown in Figures 11.1 and 11.2 uses a simple sweep in its design.

Defining a Trajectory

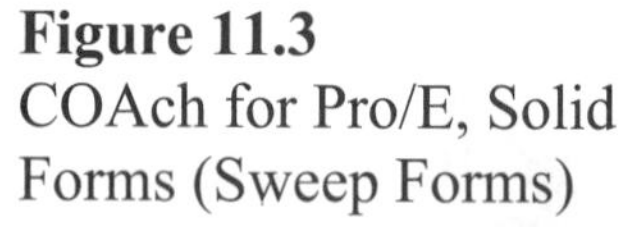

Figure 11.3
COAch for Pro/E, Solid Forms (Sweep Forms)

A *constant-section sweep* can use either trajectory geometry sketched at the time of feature creation or a trajectory (Fig. 11.3) made up of selected datum curves or edges.

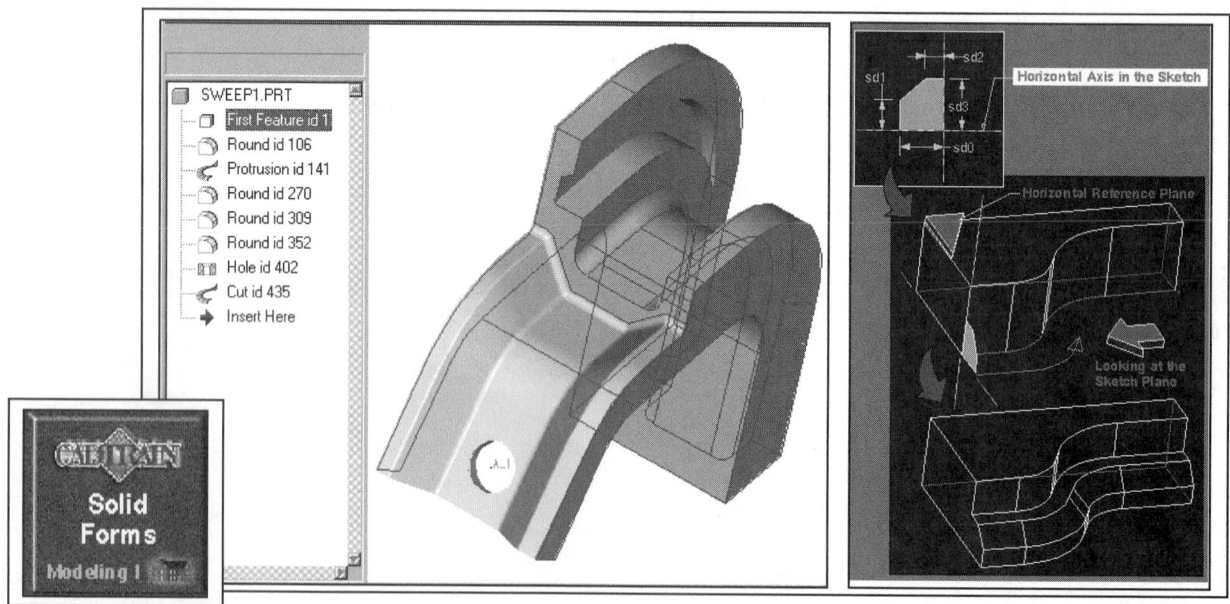

The trajectory must have adjacent reference surfaces or be planar (Fig. 11.4). When defining a sweep (Fig. 11.5), Pro/E checks the specified trajectory for validity and establishes normal surfaces. When ambiguity exists, Pro/E prompts you to select a normal surface.

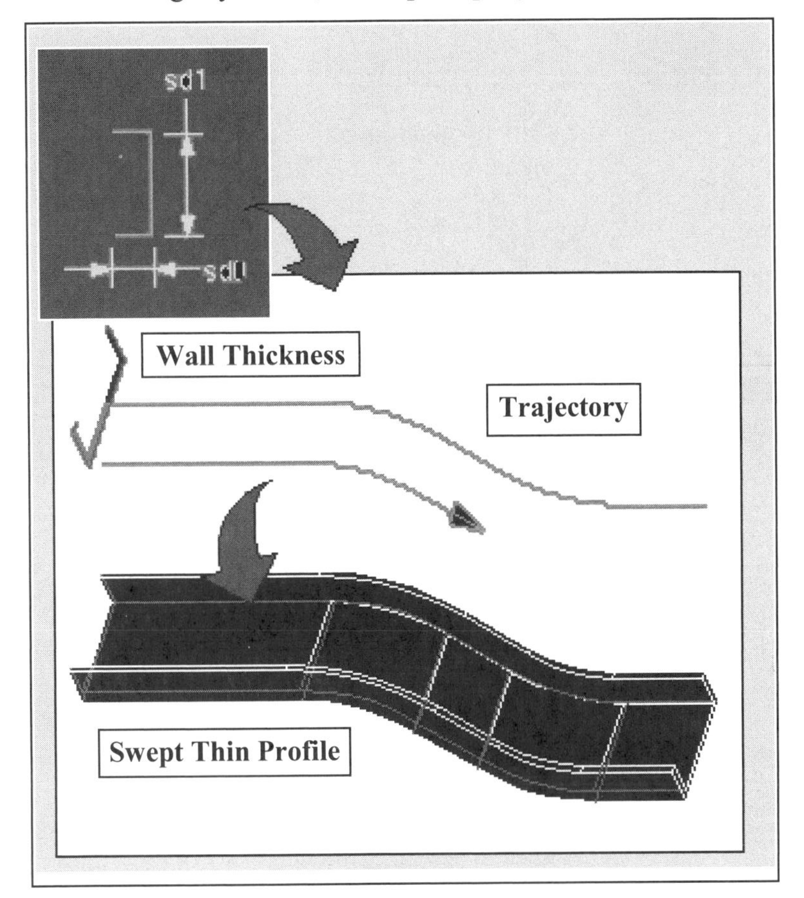

Figure 11.4
COAch for Pro/E, Solid Forms (Sweep Forms)

Creating a Swept Feature

To create a sweep, do the following:

1. **Feature ⇒ Create ⇒ Solid ⇒ Protrusion**.
2. Choose **Sweep** and **Done** from the SOLID OPTS menu.
3. The feature creation dialog box for sweeps is displayed.

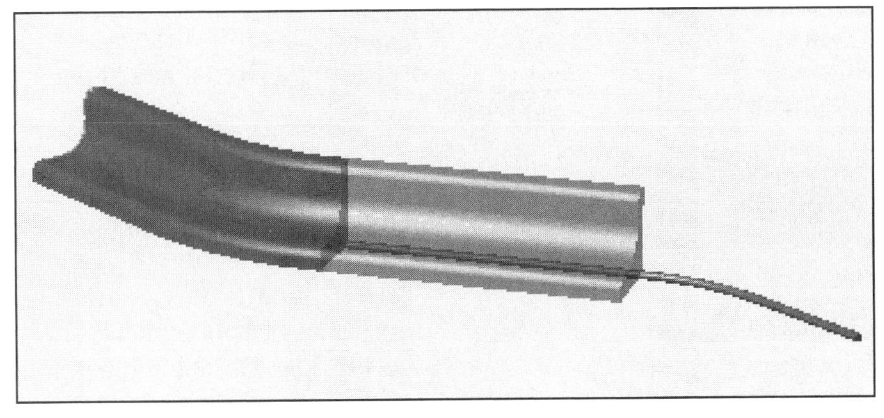

Figure 11.5
COAch for Pro/E, Solid Forms (Sweep Forms)

4. Sketch or select an open or closed trajectory (Fig. 11.6), using a SWEEP TRAJ menu option. The following options are available:

Sketch Traj Sketch the sweep trajectory using Sketcher mode.
Select Traj Select a chain of existing curves or edges as the sweep trajectory. The CHAIN menu allows you to select the desired trajectory.

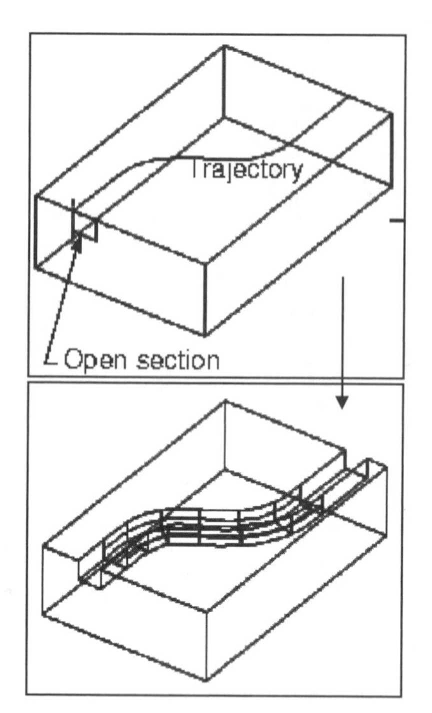

Figure 11.6
Sweeps

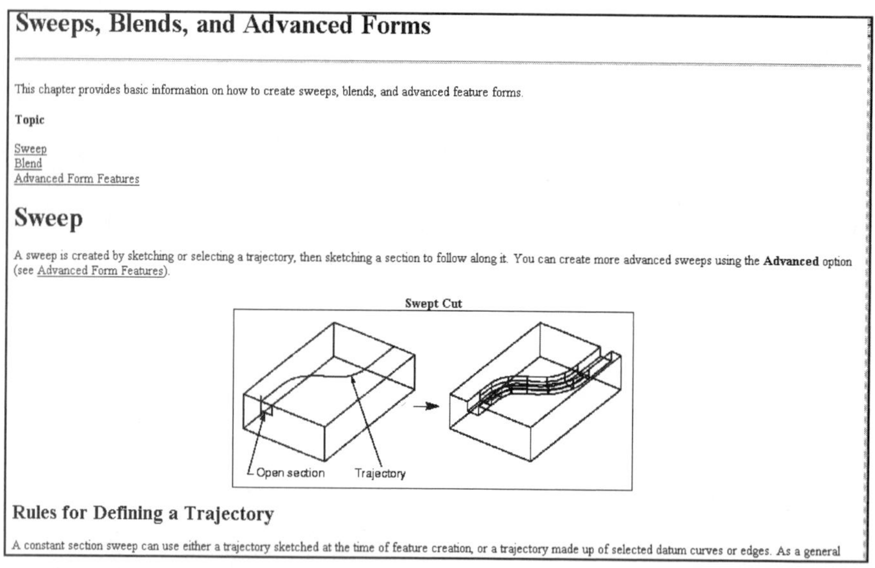

Sweeps, Blends, and Advanced Forms

This chapter provides basic information on how to create sweeps, blends, and advanced feature forms.

Topic

Sweep
Blend
Advanced Form Features

Sweep

A sweep is created by sketching or selecting a trajectory, then sketching a section to follow along it. You can create more advanced sweeps using the **Advanced** option (see Advanced Form Features).

Swept Cut

Open section Trajectory

Rules for Defining a Trajectory

A constant section sweep can use either a trajectory sketched at the time of feature creation, or a trajectory made up of selected datum curves or edges. As a general

5. If the trajectory lies in more than one surface, such as a trajectory defined by a datum curve created using **Intr Surfs**, Pro/E prompts you to select a normal surface for the sweep cross section. Pro/E orients the **Y** axis of the cross section to be normal to this surface along the trajectory.
6. Create or retrieve the section to be swept along the trajectory, and dimension it relative to the *crosshairs* displayed on the trajectory. Choose **Done**.
7. If the trajectory is open (the start point and endpoint of the trajectory do not touch) and you are creating a solid sweep, choose an ATTRIBUTES menu option; then choose **Done**. The possible options are as follows:

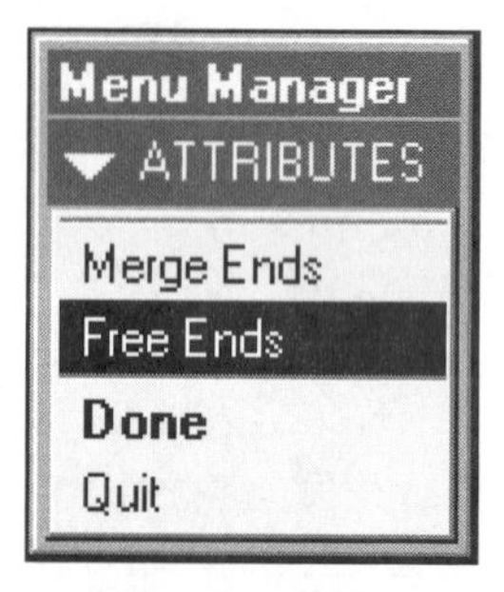

Merge Ends Merge the ends of the sweep, if possible, into the adjacent solid. For you to do this, the sweep endpoint must be attached to part geometry.
Free Ends Do not attach the sweep end to adjacent geometry.

8. If the sweep trajectory is closed (Fig. 11.7), choose one of the following SWEEP OPT menu options and then **Done**:

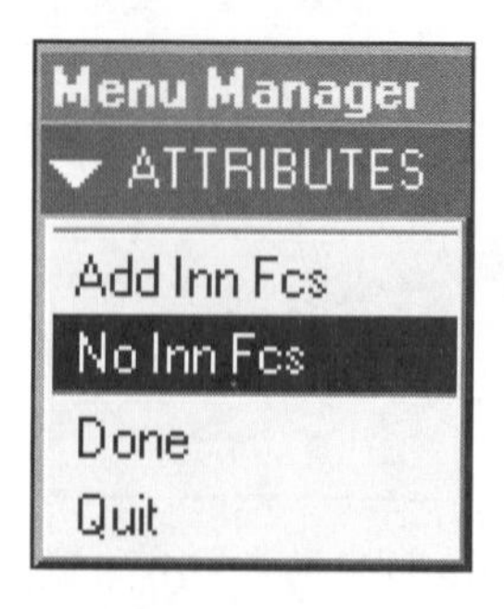

Add Inn Fcs For open sections, add top and bottom faces to close the swept solid (planar, closed trajectory, and open section). The resulting feature consists of surfaces created by sweeping the section and has two planar surfaces that cap the open ends.
No Inn Fcs Do not add top and bottom faces.

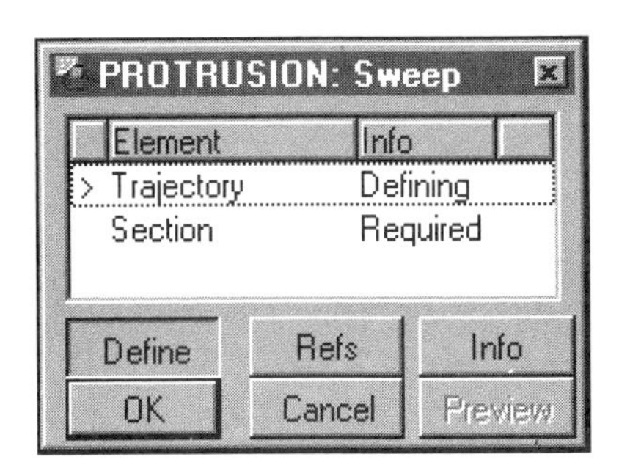

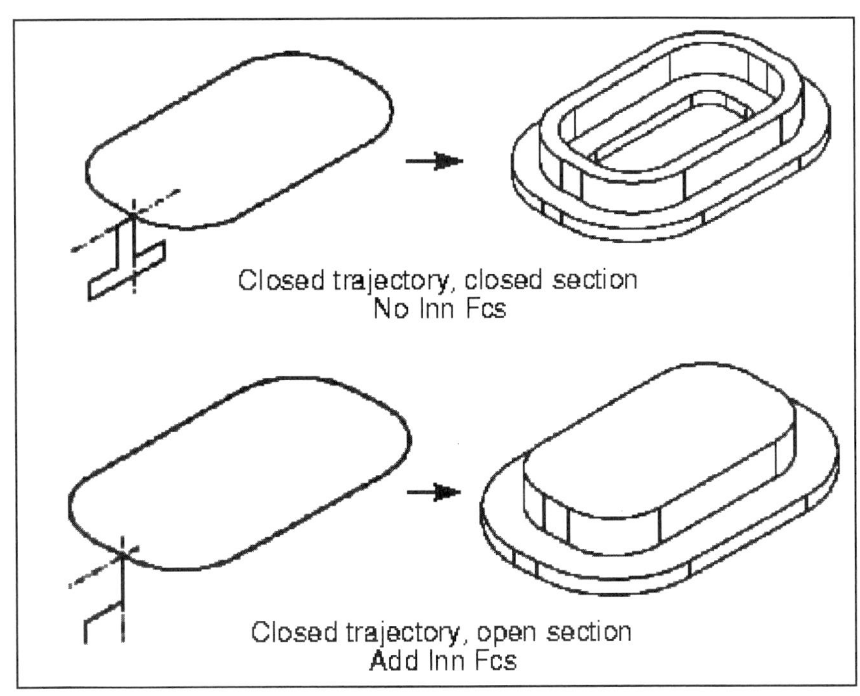

Figure 11.7
Online Documentation, Sweep Trajectories and Sections--**No Inn Fcs** and **Add Inn Fcs**

9. Choose **Flip**, if desired, and then **Okay** from the DIRECTION menu to select the side on which to remove material for swept cuts.
10. Pro/E issues a message stating that all the elements have been defined.
11. Select the **OK** button in the dialog box to create the sweep.

Feature ⇒ Redefine

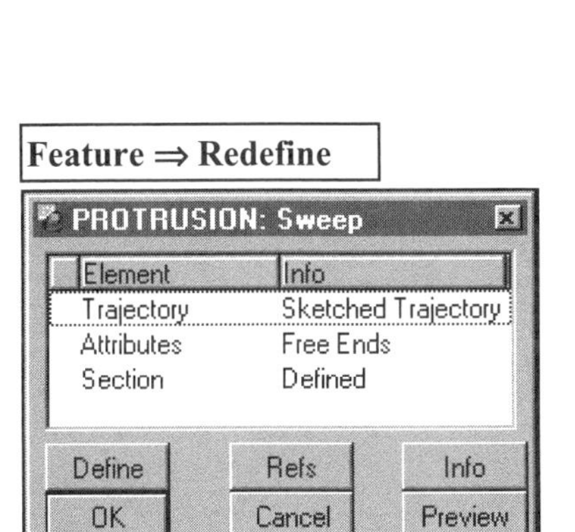

To redefine sweep sections or trajectories after the feature is created, choose **Redefine** and then select the sweep. Pro/E will display the feature creation dialog box. You can redefine the trajectory or section elements.

Variable-Section Sweeps

A solid sweep feature using one or more longitudinal trajectories and a single variable section can also be created (Fig. 11.8). The parameters of the section can vary as the section moves along the sweep trajectories. These sweeps are called **variable-section sweeps**.

Every variable-section sweep requires one *longitudinal "spine" trajectory*. You can define a sweep for which the **X** axis of the section follows the **X**-vector trajectory while remaining normal to the spine at all times as it sweeps along the spine (**Nrm to Spine**). To do so, you must also specify an **X**-vector trajectory to orient the section as it sweeps along the spine. The section plane is always normal to the spine trajectory at the point of their intersection. The **X** axis of each section's coordinate system is defined by the direction from the point of intersection of the plane and the spine to the point of intersection of the plane and the **X**-vector trajectory for that section. You can also define a variable-section sweep for which the **Y** axis of the section remains constant. The section will follow the spine such that it is normal to the selected pivot plane. The **X** axis and **Z** axis will still follow the spine and **X**-vector trajectories.

Variable-section sweeps need the following trajectories:

Spine trajectory The trajectory along which the section is swept (Fig. 11.9). If you choose **Nrm To Spine**, the origin of the section (crosshairs) is always located on the spine trajectory, with the **X** axis pointing toward the **X**-vector trajectory.
X-vector trajectory Sweeps created using **Nrm To Spine** need this additional trajectory. It defines the orientation of the **X** axis of the section coordinate system. *The X-vector and spine trajectories cannot intersect.*

After the direction is specified, Pro/E displays the VAR SEC SWP menu so you can define a trajectory. You can use a composite curve as a trajectory.

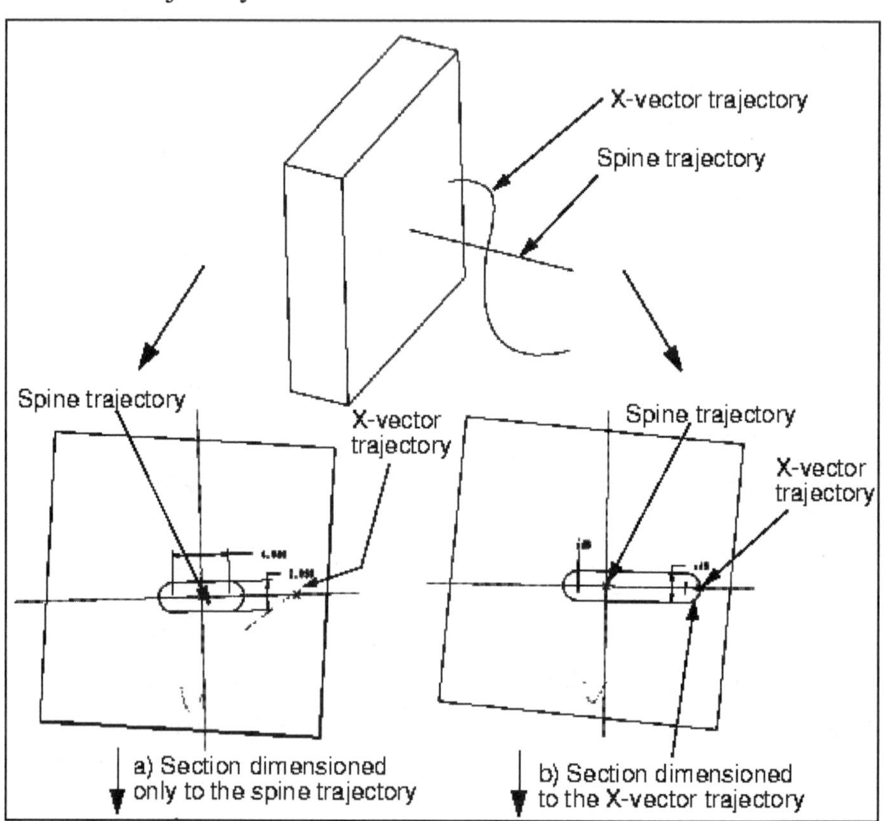

Figure 11.8
Variable-Section Sweeps

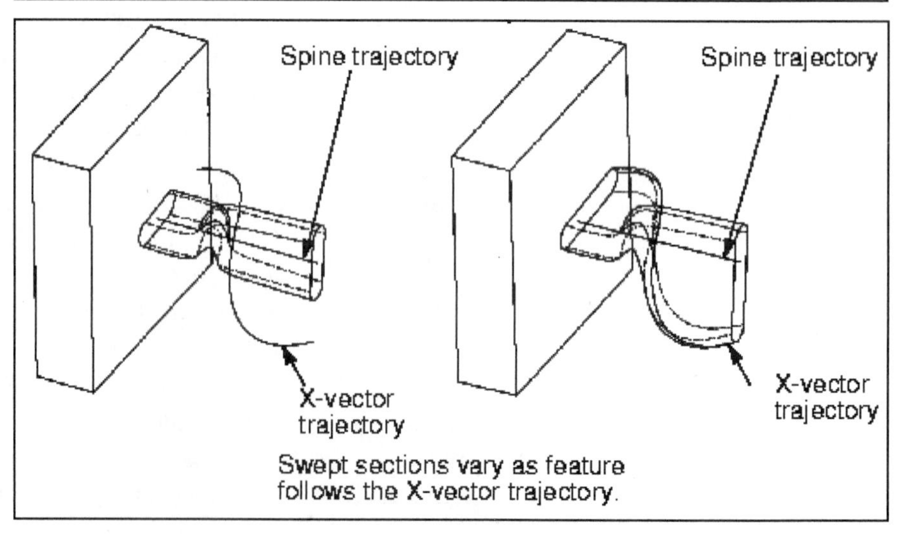

Figure 11.9
Spine Trajectories

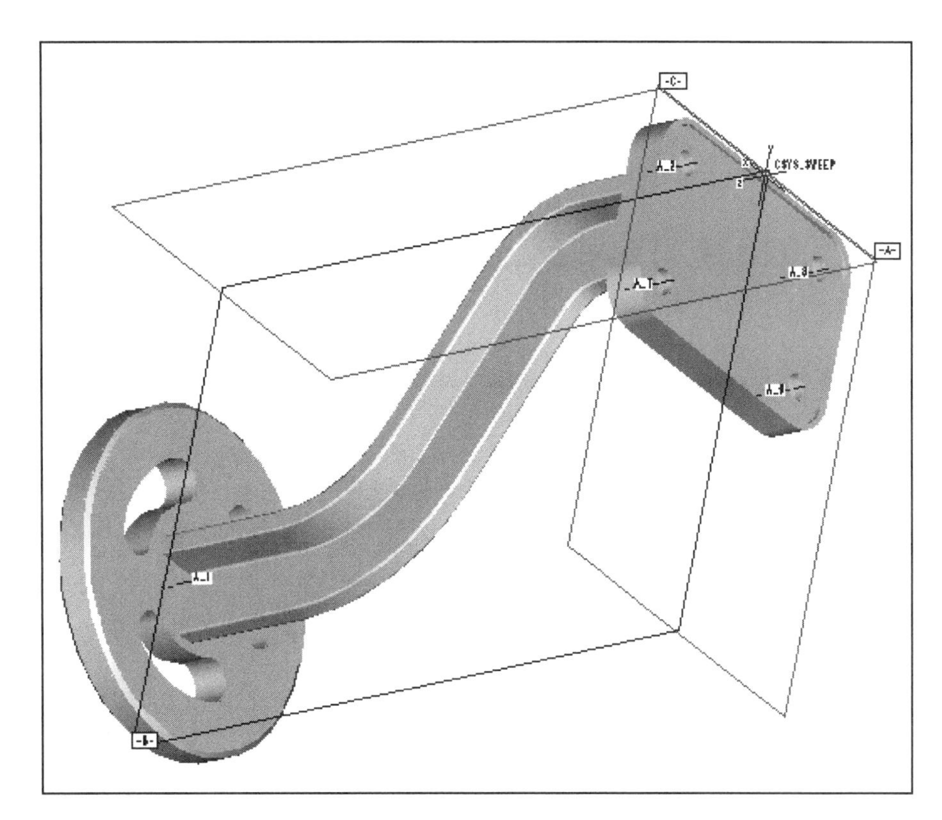

Figure 11.10
Bracket Showing Set Datum Planes and Coordinate System

Bracket

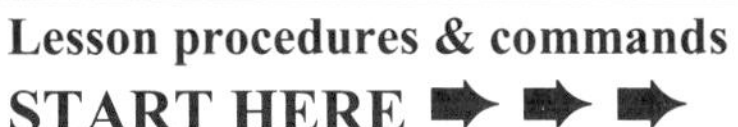

The **Bracket** (Figs. 11.10 through 11.39) requires the use of the **Sweep** command. The T-shaped section is swept along the selected *trajectory*. The protrusions on both sides of the swept feature are to be created with the dimensions given in Figures 11.11 through 11.17. Step-by-step commands are provided only for the sweep trajectory and its cross section.

NOTE

Set up the Bracket:

- Material = CAST IRON
- Units = Inch lbm Second
- Hidden Line
- ✔ Snap to Grid
- Datum **TOP** = **C**
- Datum **FRONT** = **A**
- Datum **RIGHT** = **B**
- Coordinate System = **CSYS_SWEEP**

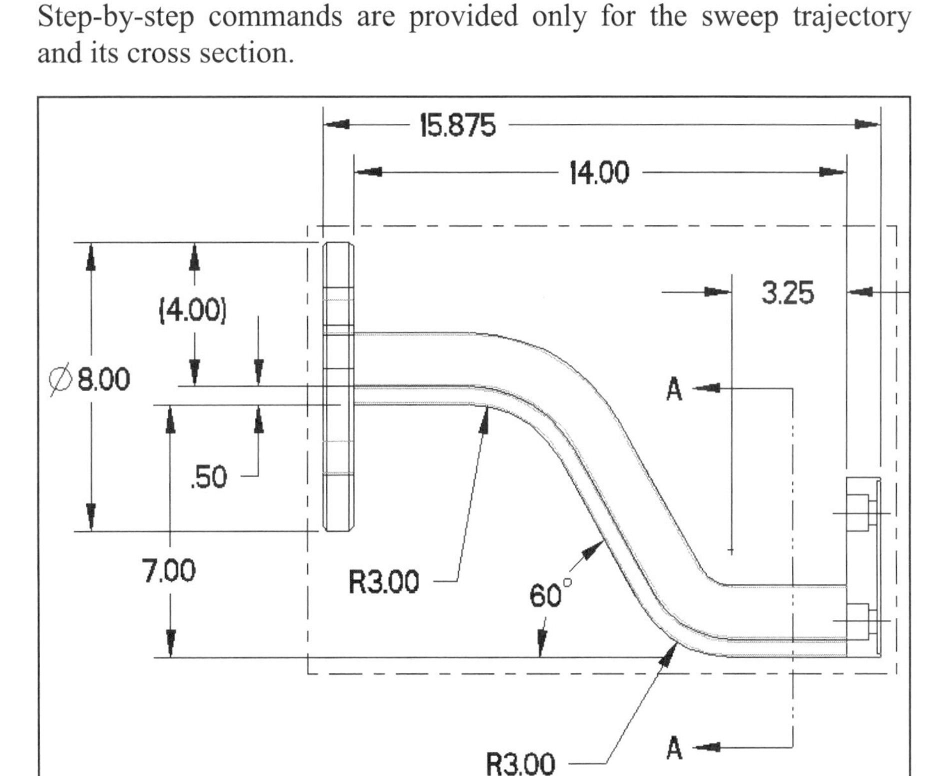

Figure 11.11
Bracket Drawing, Front View

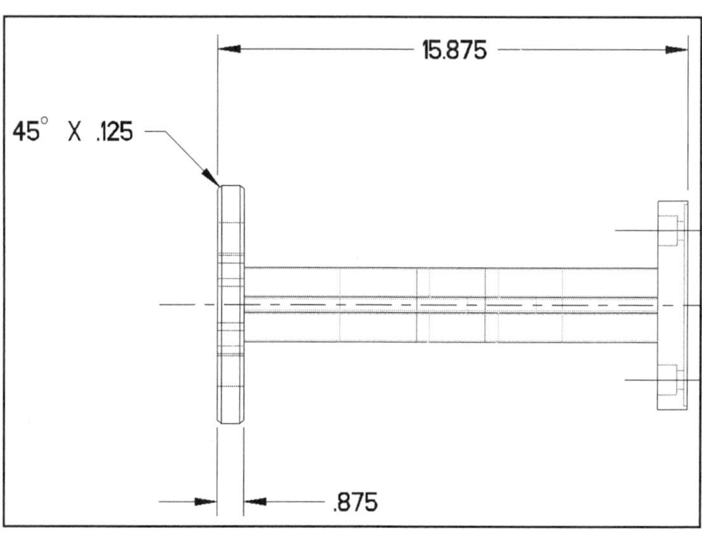

Figure 11.12
Bracket Drawing,
Top View

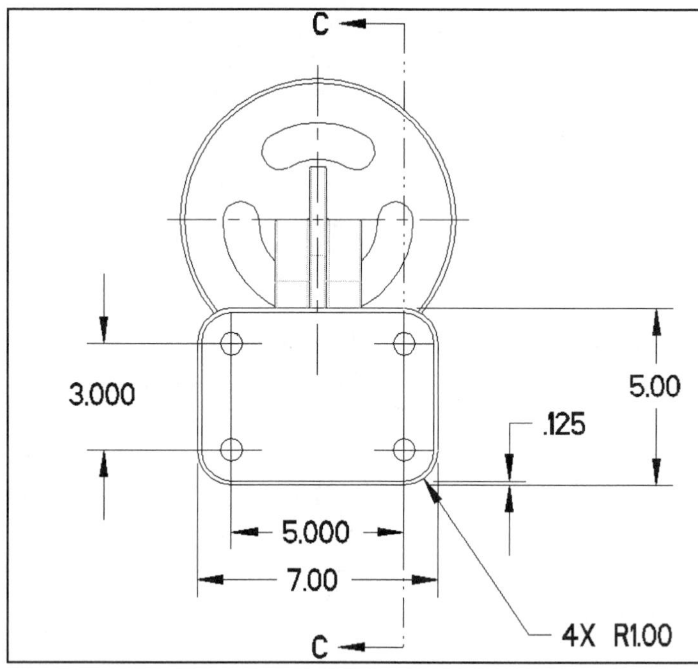

Figure 11.13
Bracket Drawing,
Right Side View

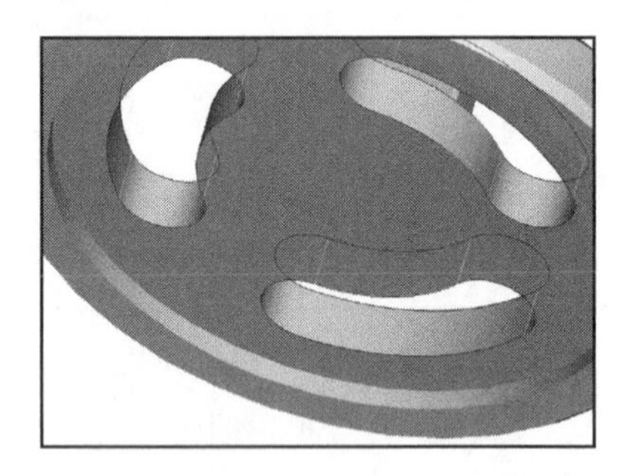

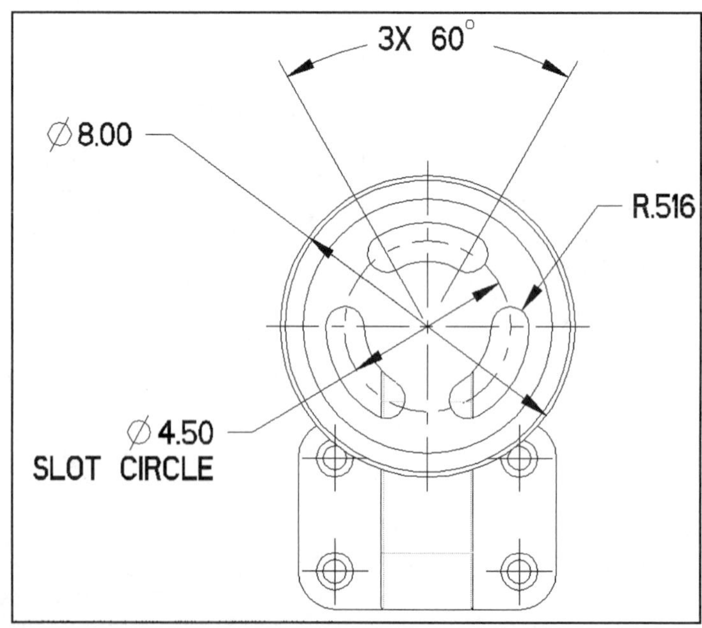

Figure 11.14
Bracket Drawing,
Left Side View

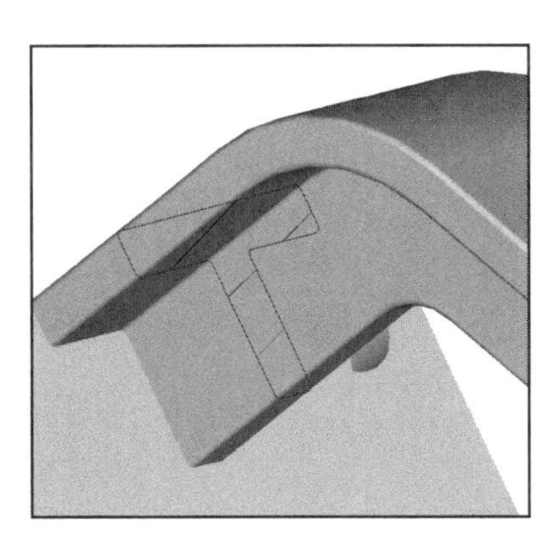

Figure 11.15
Bracket Drawing,
SECTION A-A

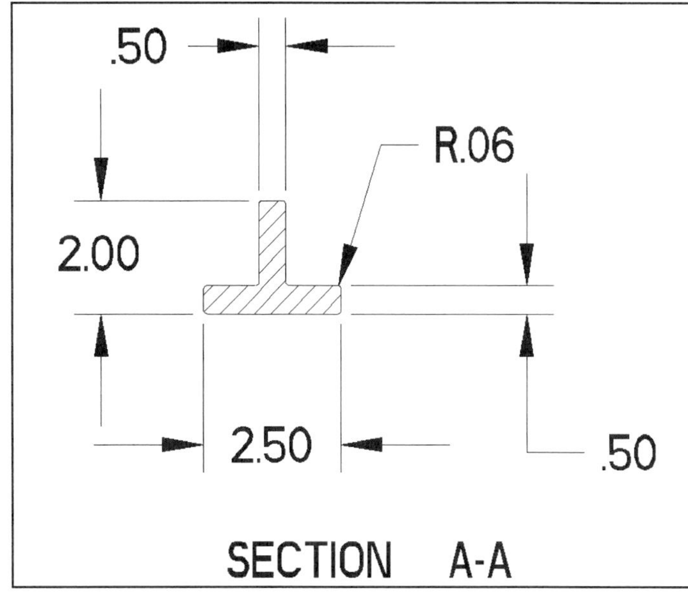

Figure 11.16
Bracket Drawing,
SECTION B-B

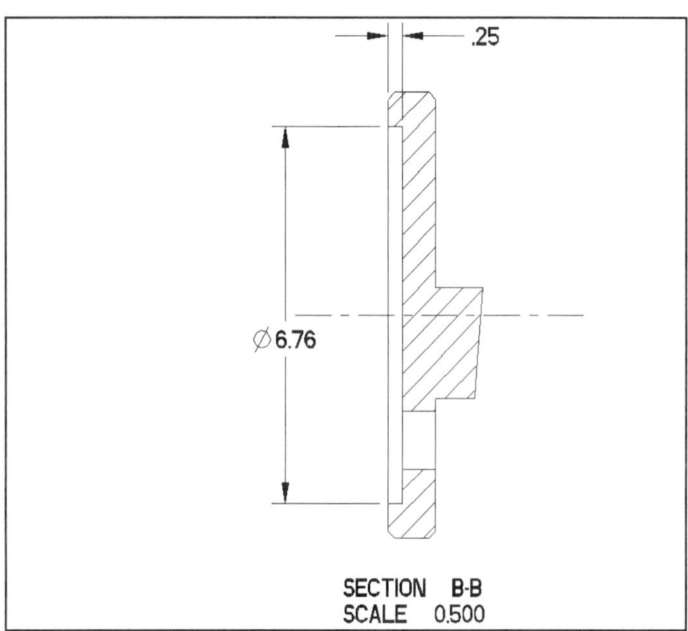

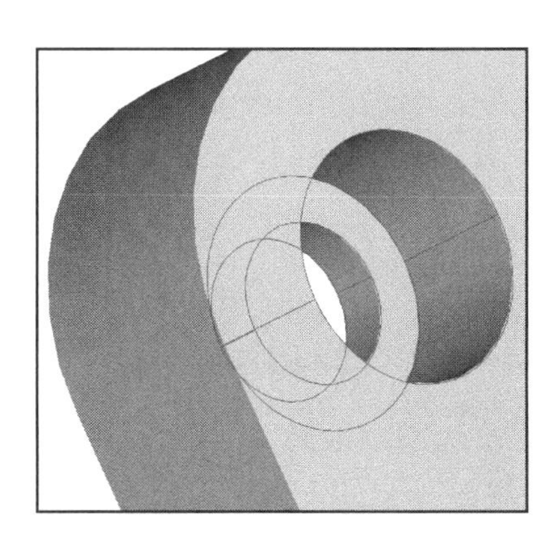

Figure 11.17
Bracket Drawing,
SECTION C-C

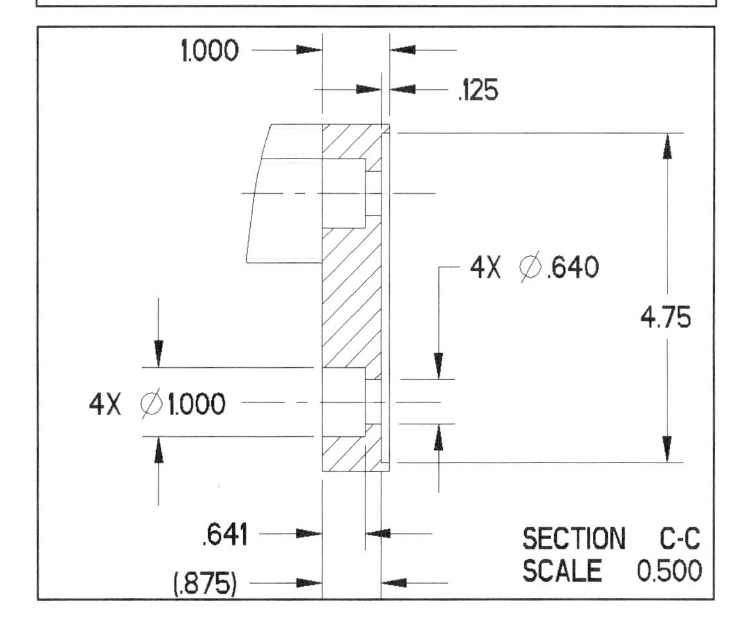

The bracket is started by modeling the protrusion shown in Figure 11.18. This protrusion will be used to establish the sweep's position in space. Sketch the protrusion on datum **A**.

The second protrusion is a sweep and is shown in Figure 11.19.

NOTE

- Datum TOP = **C**
- Datum FRONT = **A**
- Datum RIGHT = **B**
- Coordinate System = **CSYS_SWEEP**

File ⇒ Save ⇒ ✔

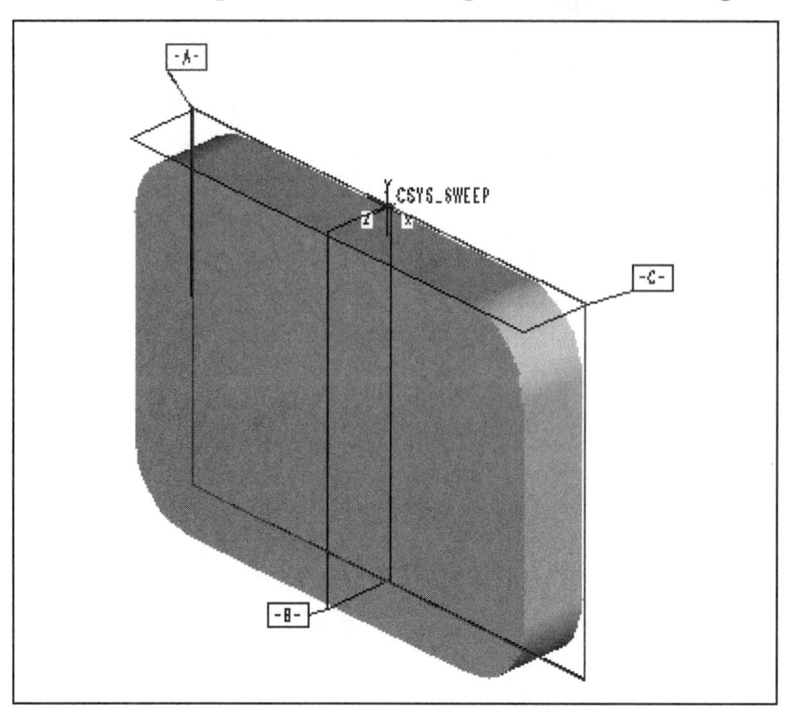

Figure 11.18
Bracket's First Protrusion

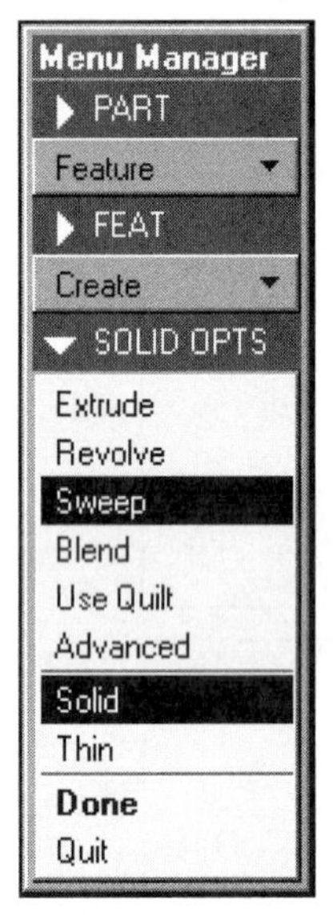

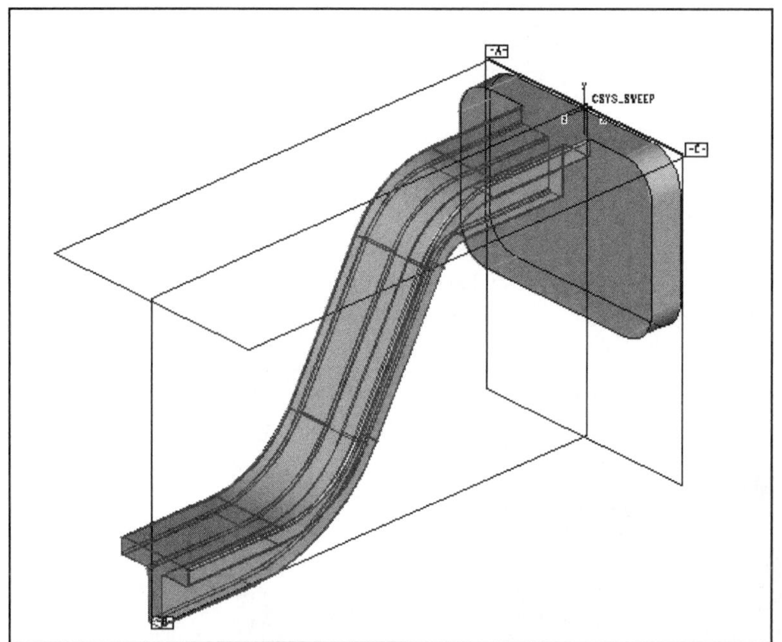

Figure 11.19
Swept Protrusion

Start the sweep by choosing the following commands:

Utilities ⇒ Environment ⇒ ❑ Snap to Grid ⇒ OK ⇒ Feature ⇒ Create ⇒ Protrusion ⇒ Sweep ⇒ Done ⇒ Sketch Traj ⇒ (pick datum **B** as the sketching plane for the trajectory) **⇒ Okay ⇒ Top ⇒** (pick datum **C** as the orientation plane) **⇒ Intent Manager** (your choice--use it or not use it) **⇒ (Sketch**, **Dimension**, and **Modify** the trajectory as shown in Figs. 11.20 through 11.25)

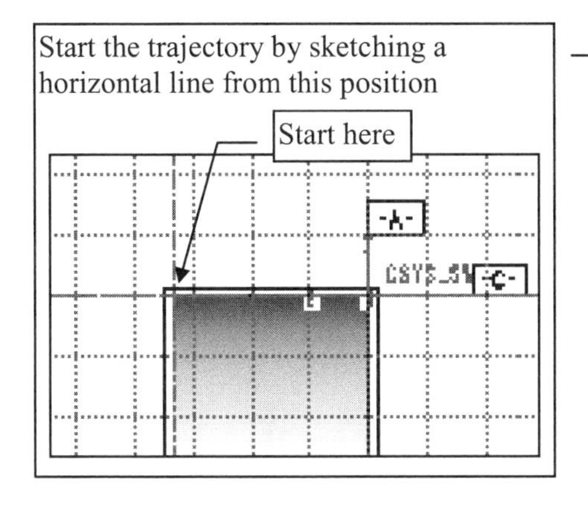

Figure 11.20
Starting the Sweep Trajectory Sketch

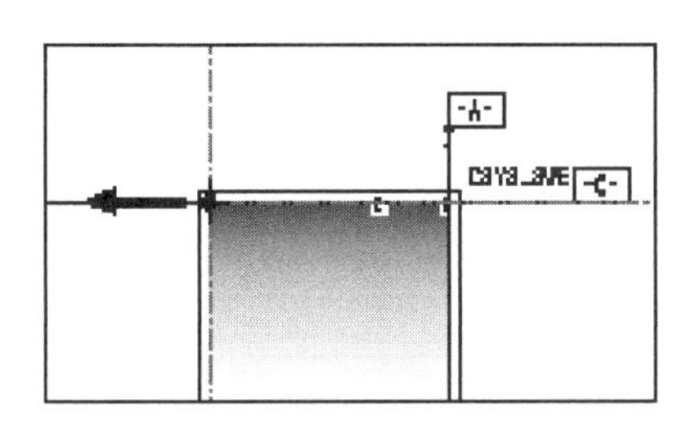

Start point

Figure 11.21
Sketch the Three Lines

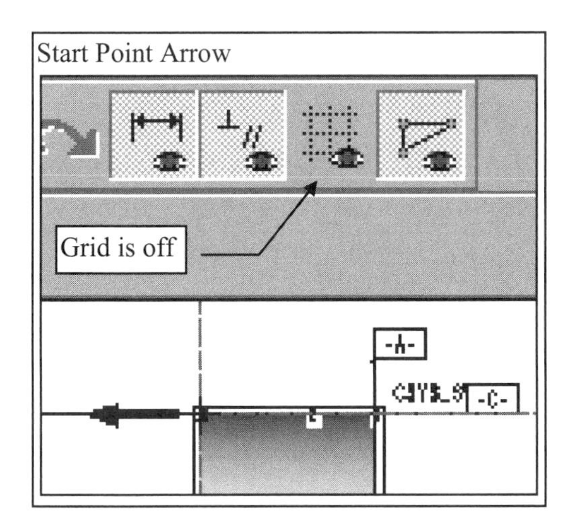

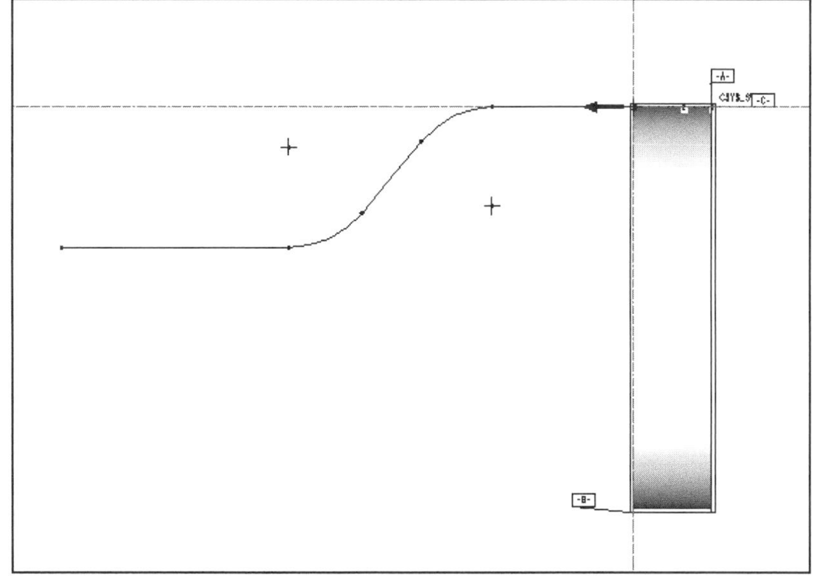

Figure 11.22
Add the Arc Fillets

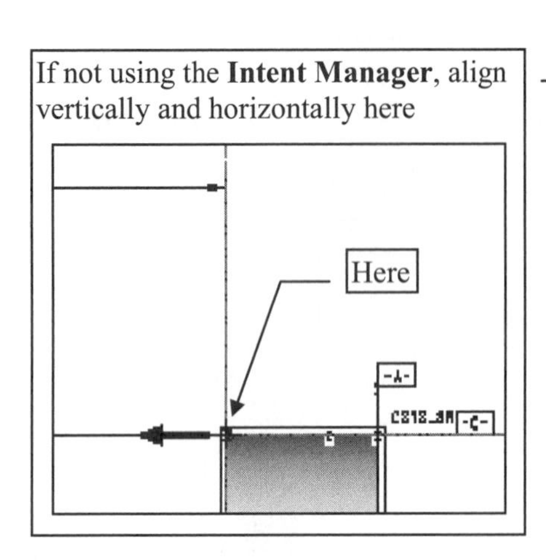

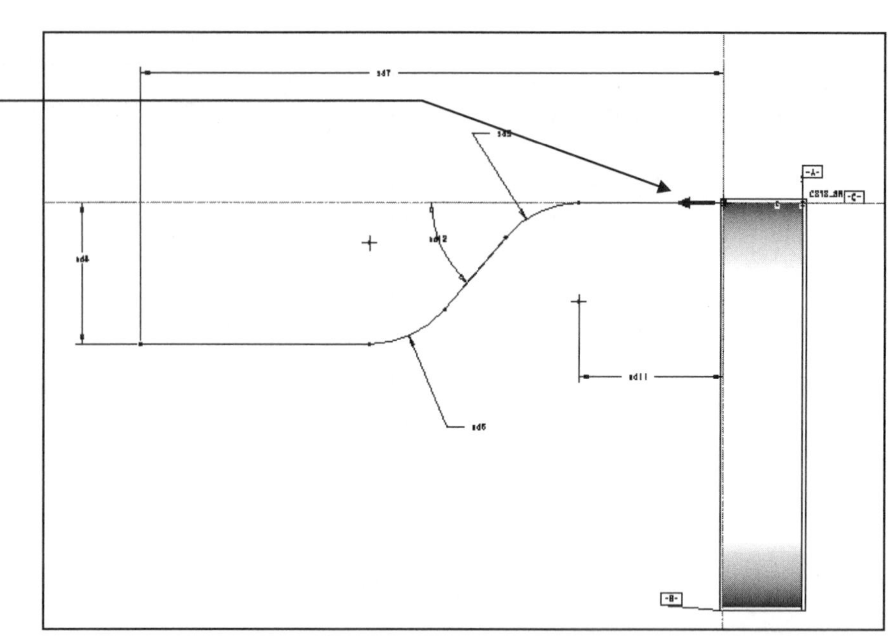

Figure 11.23
Alignment and **Dimension**

Figure 11.24
Modify and **Regenerate** the Sketch

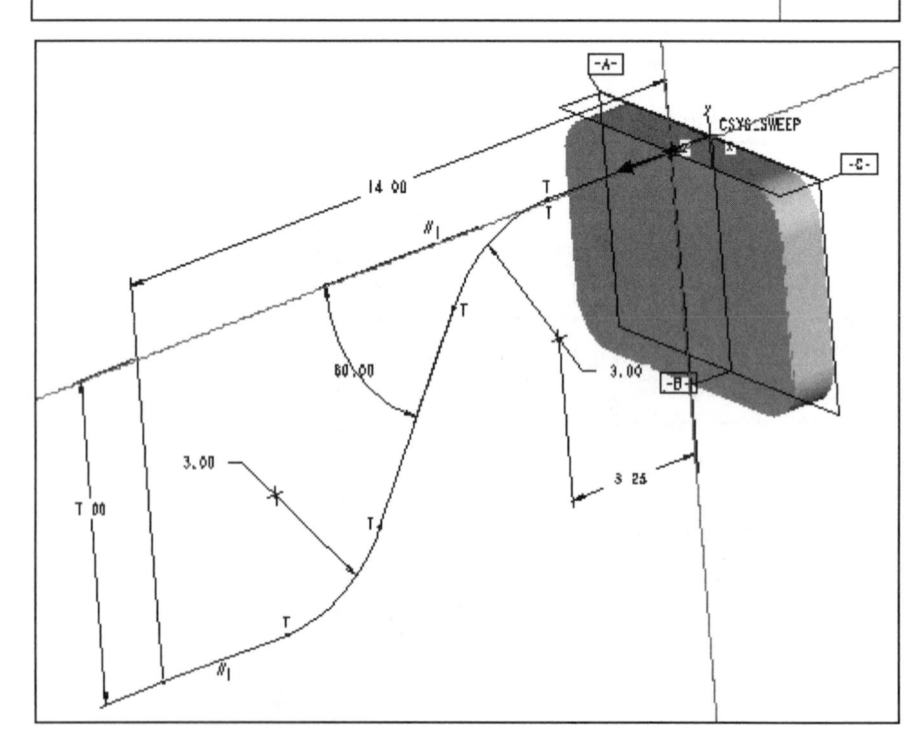

Figure 11.25
Completed Sketch

After the trajectory has been completed, finish the sweep using:

Free Ends ⇒ **Done** ⇒ **Utilities** ⇒ **Environment** ⇒ ✔ **Snap to Grid** ⇒ **OK** ⇒ **Sketch** ⇒ **Intent Manager** (off) ⇒ **Regenerate** ⇒ **Done** ⇒ **Sec Tools** ⇒ **Sec Environ** ⇒ **Grid** ⇒ **Params** ⇒ **X&Y Spacing** ⇒ (type **.25** at prompt) ⇒ ✔ ⇒ [now sketch a section (Figs. 11.26 and 11.27)] ⇒ [add *eight* fillets (Fig. 11.28)] ⇒ **Regenerate** ⇒ **Alignment** [align the left vertical line of the sketch with datum **C,** and align the horizontal centerline to datum **B** (Fig. 11.29)]

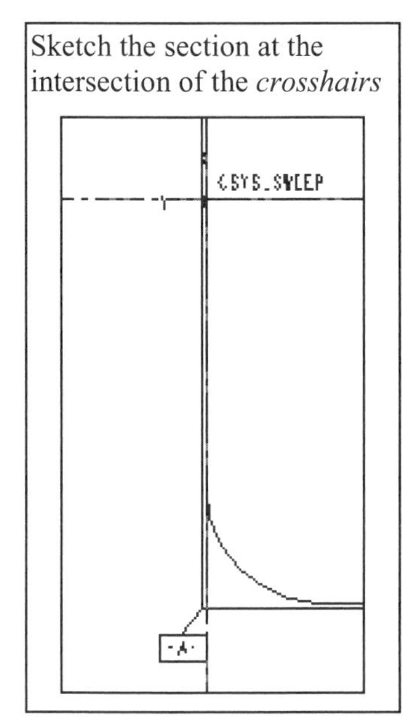

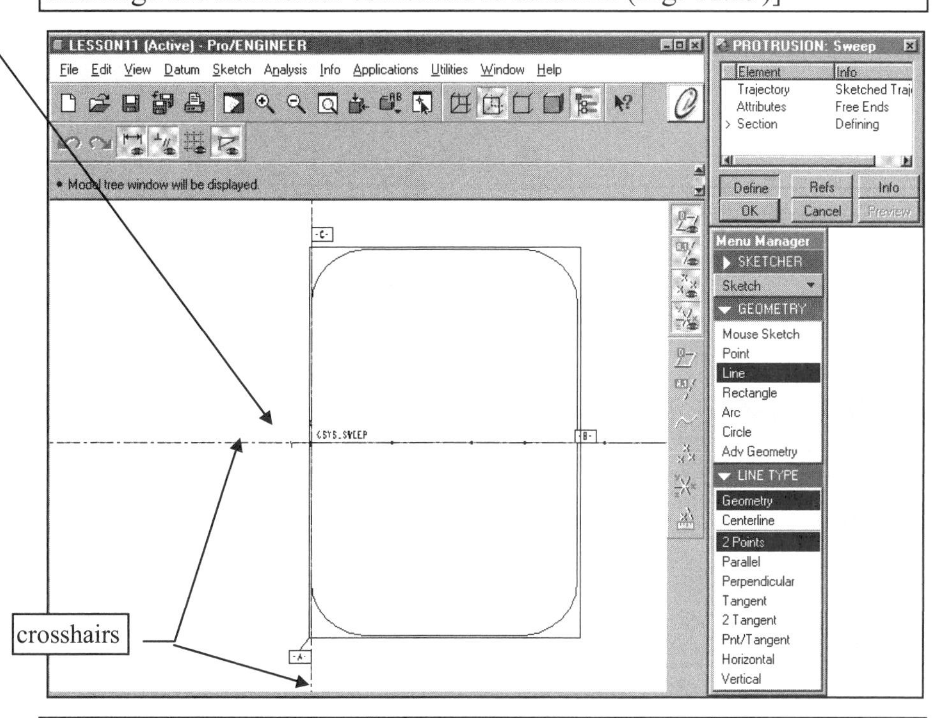

Figure 11.26
Sketch the Section

Regenerate ⇒ **Dimension** (Fig. 11.29) ⇒ **Regenerate** (Fig. 11.30) ⇒ **Modify** (Fig. 11.31) ⇒ **Regenerate** ⇒ **Done** ⇒ **Preview** (Fig. 11.32) ⇒ **OK** ⇒ **Done** (Fig. 11.33) ⇒ **File** ⇒ **Save** ⇒ ✔

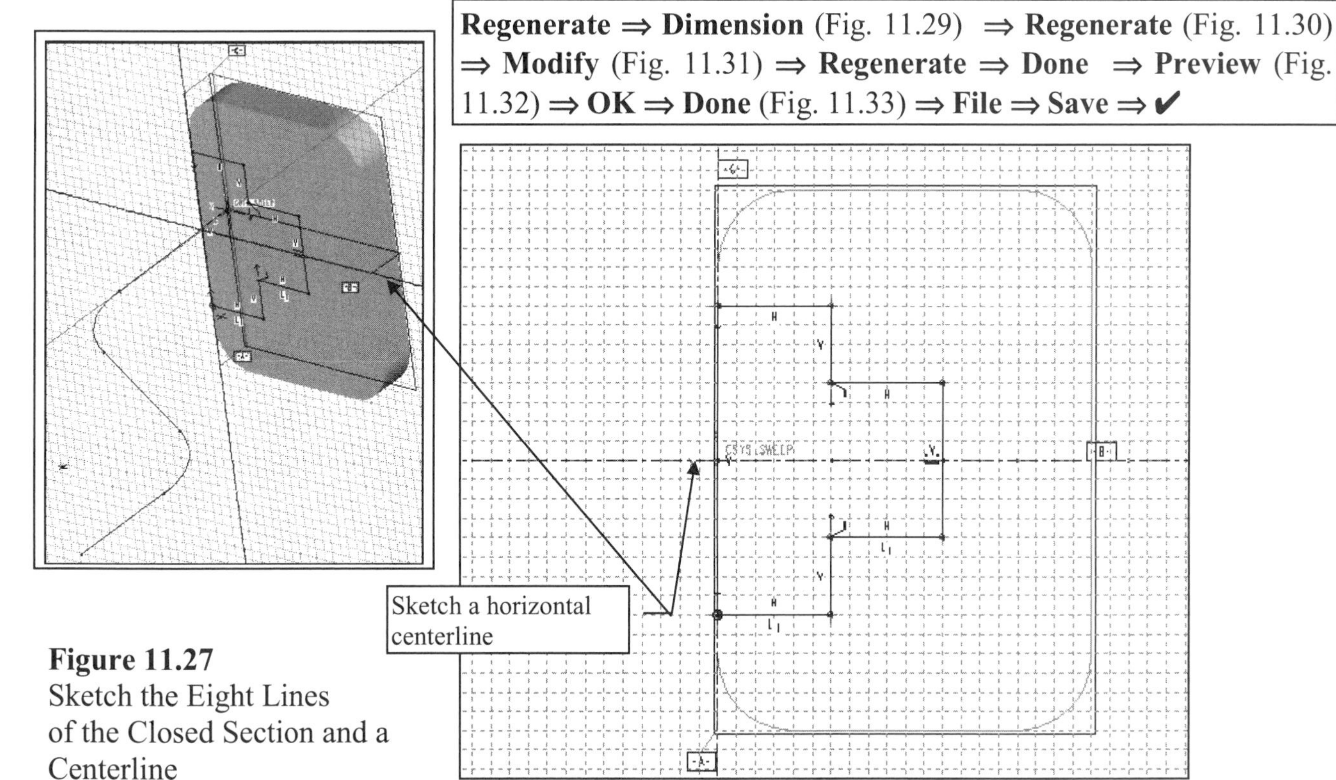

Figure 11.27
Sketch the Eight Lines of the Closed Section and a Centerline

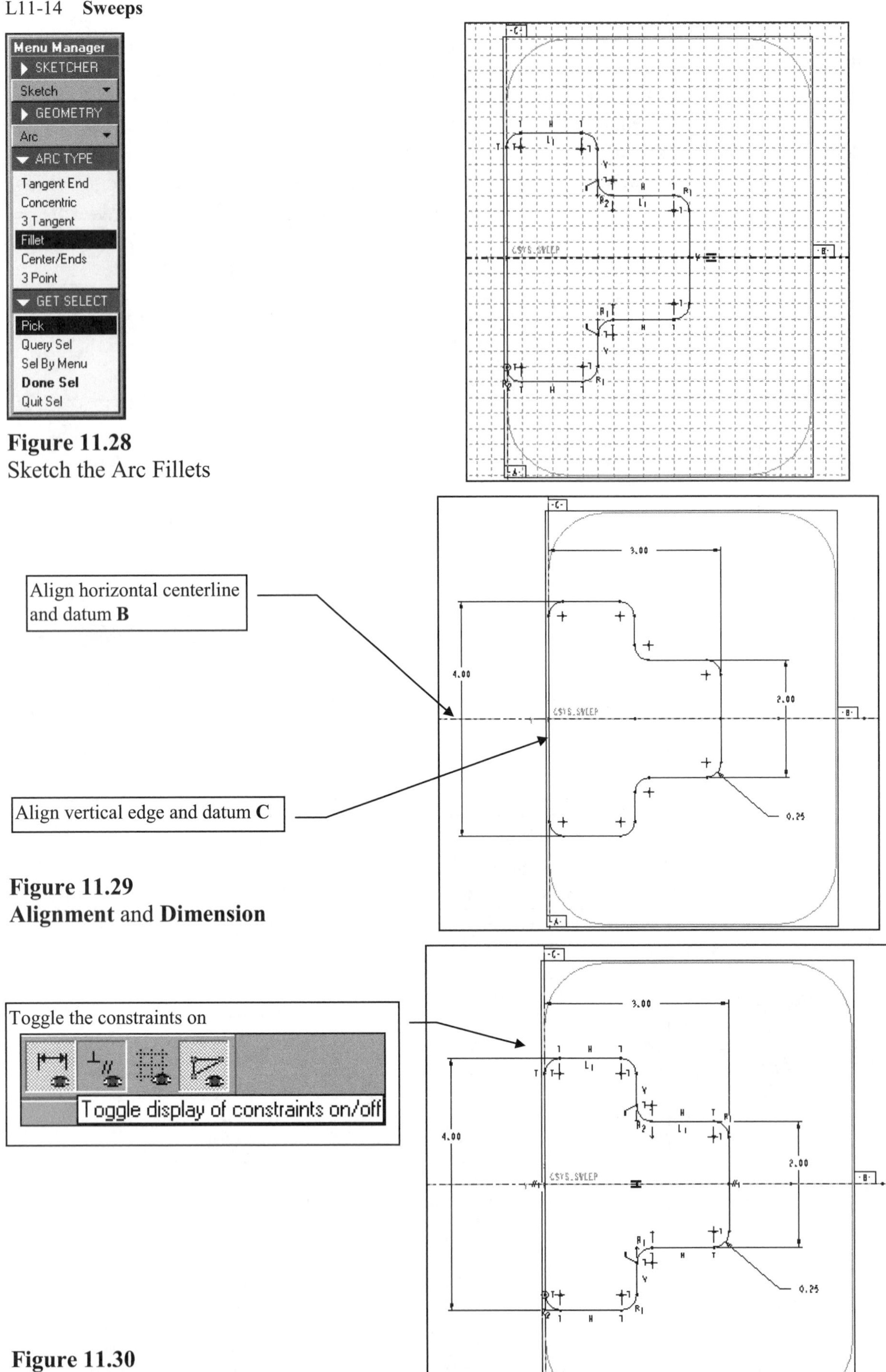

Figure 11.28
Sketch the Arc Fillets

Figure 11.29
Alignment and **Dimension**

Figure 11.30
Regenerate the Sketch

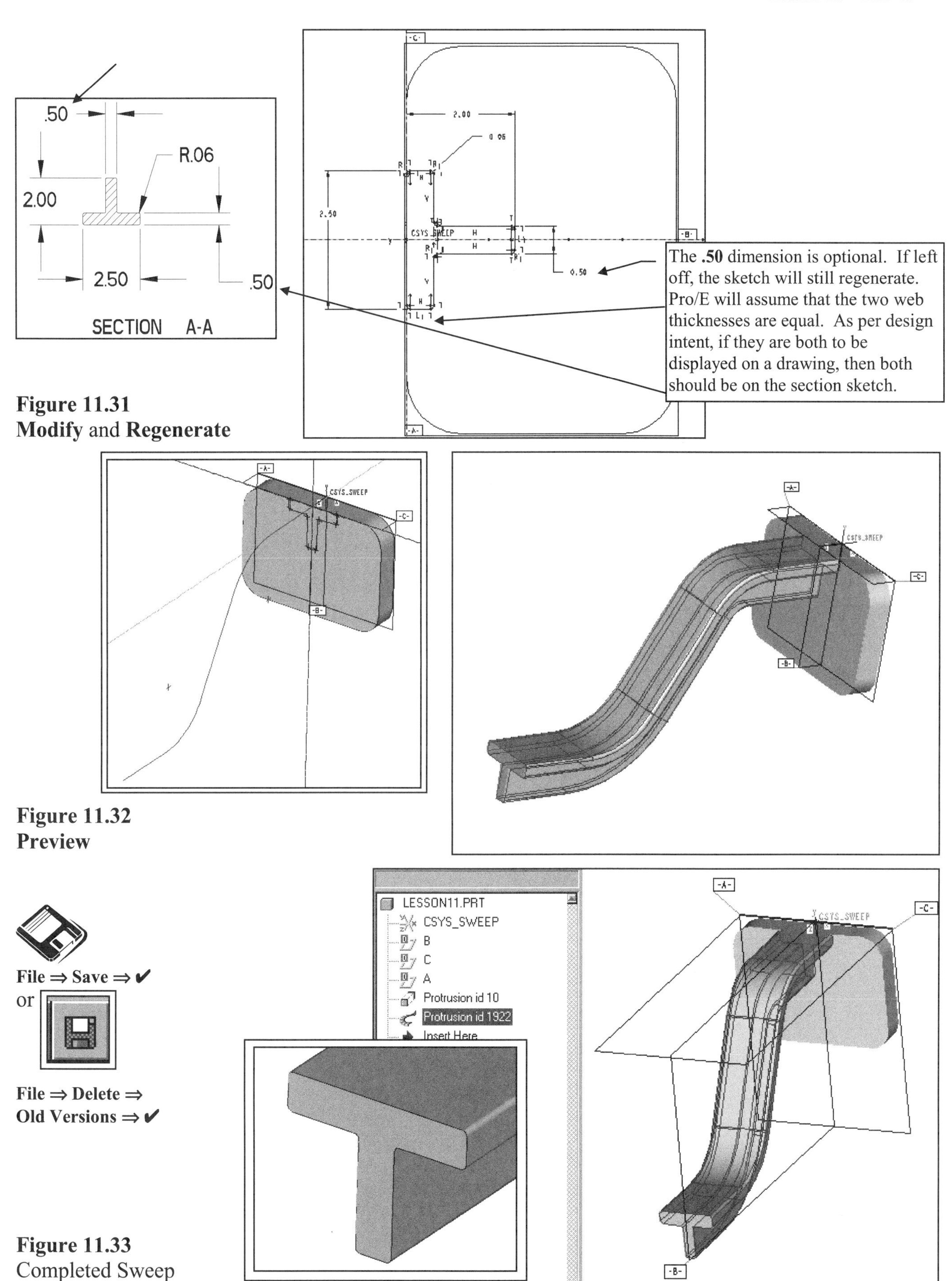

Figure 11.31
Modify and **Regenerate**

Figure 11.32
Preview

Figure 11.33
Completed Sweep

Model the third protrusion (Fig. 11.34), make the cuts (Figs. 11.35 and 11.36), and create and pattern the counterbore holes (Fig. 11.36). Complete the part by modeling the chamfers and the slots (Figs. 11.35 through 11.39).

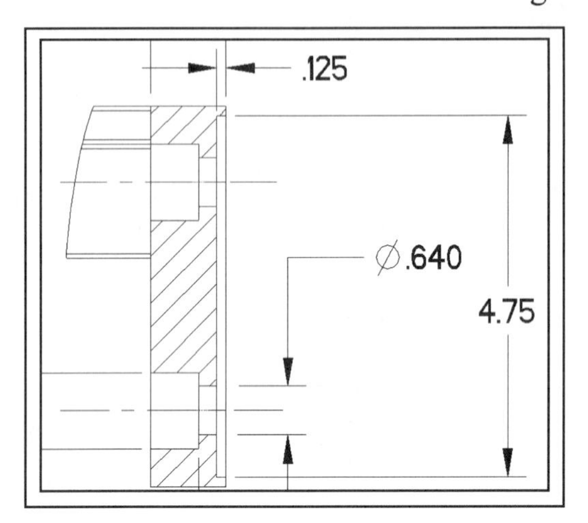

Figure 11.34
Third Protrusion

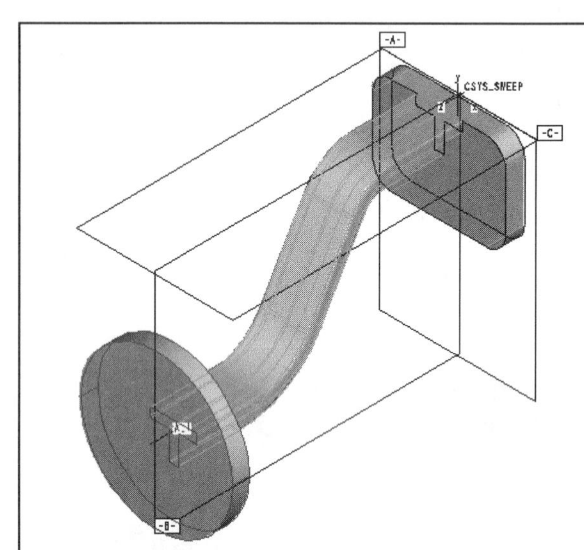

File ⇒ Save ⇒ ✔

45° X .125 chamfer both sides

Figure 11.35
Add the Cuts and the Chamfers

Figure 11.36
Create the Cut; Model, and Pattern the Counterbore Holes

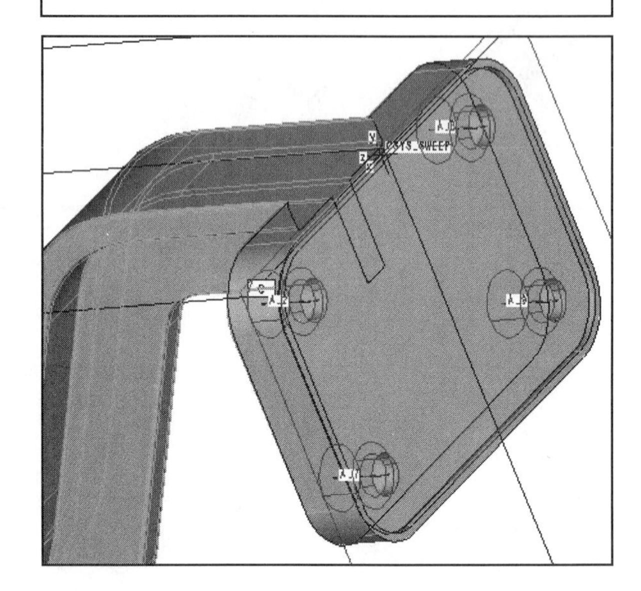

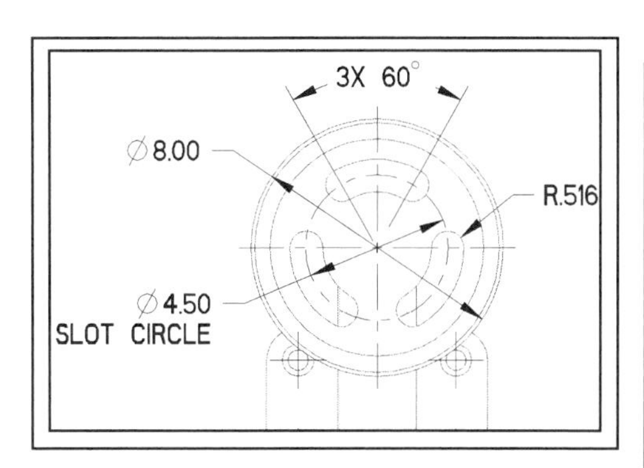

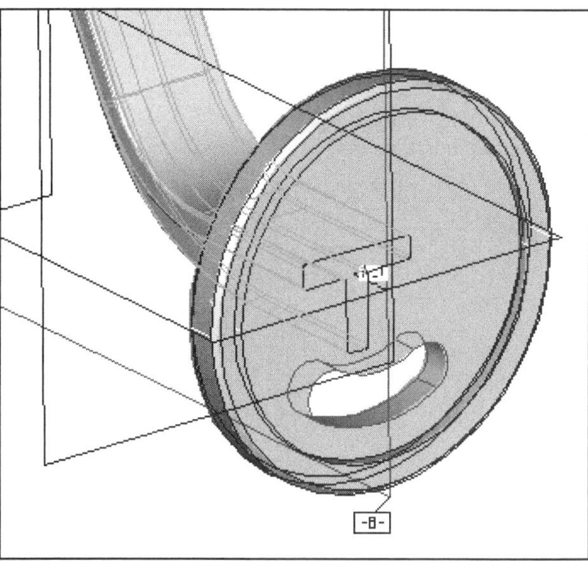

Figure 11.37
Slot Cut. [In order to pattern the slot feature, create a *datum on the fly* (**Make Datum**) as the reference plane (see Lesson 7).]

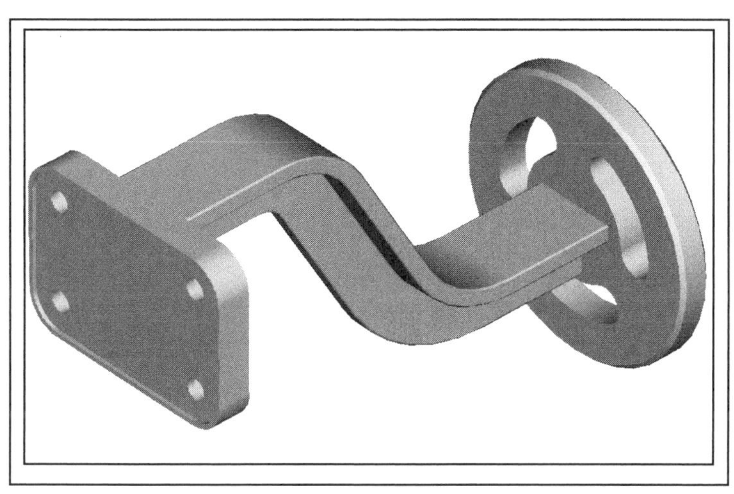

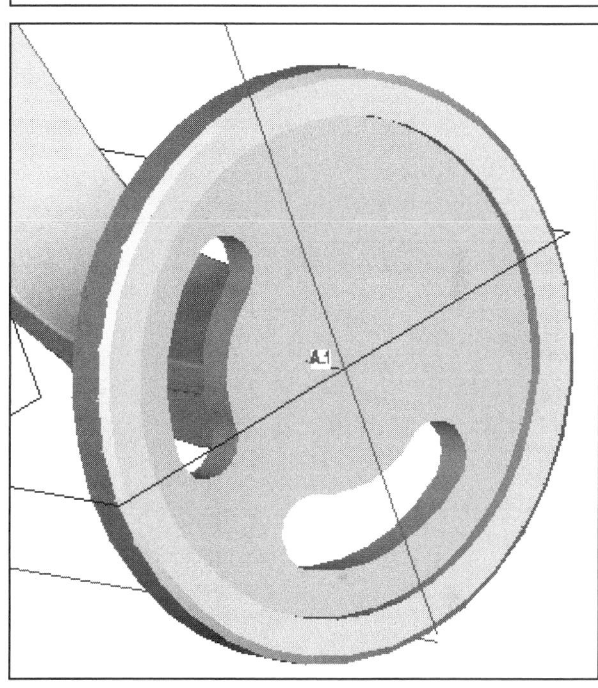

Figure 11.38
Pattern the Slots

File ⇒ Save ⇒ ✔
File ⇒ Delete ⇒
Old Versions ⇒ ✔

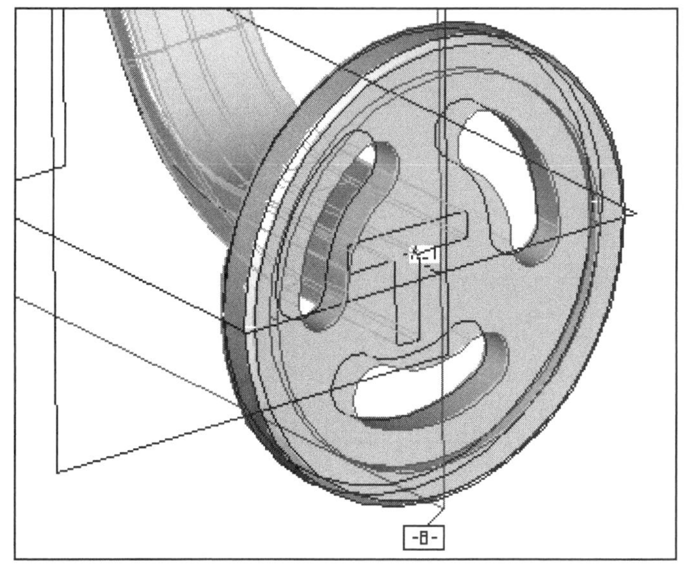

Figure 11.39
Completed Pattern and Part

Lesson 11 Project

Cover Plate

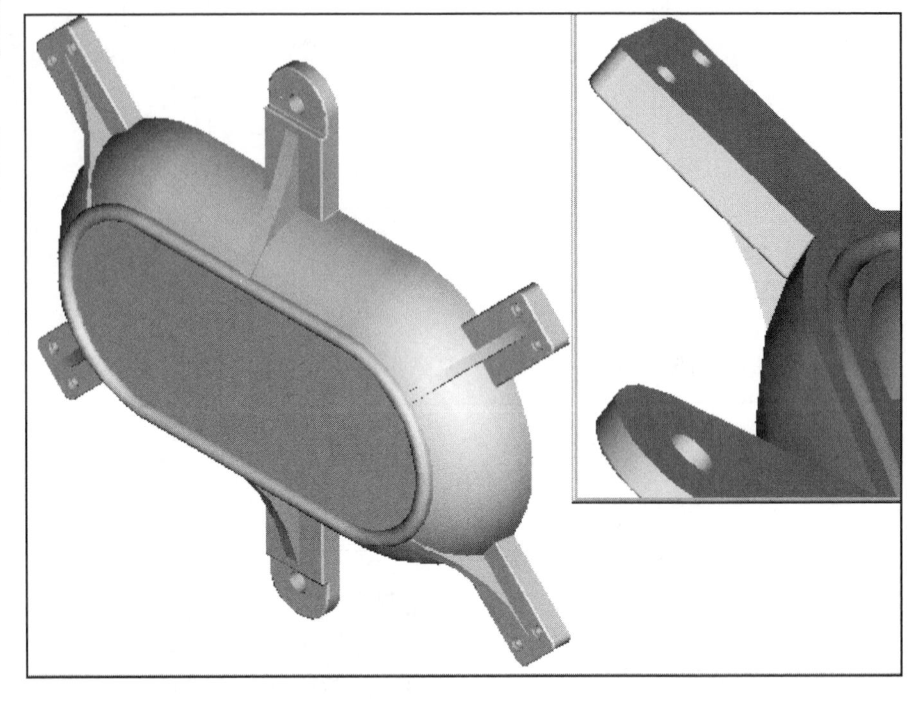

Figure 11.40
Cover Plate

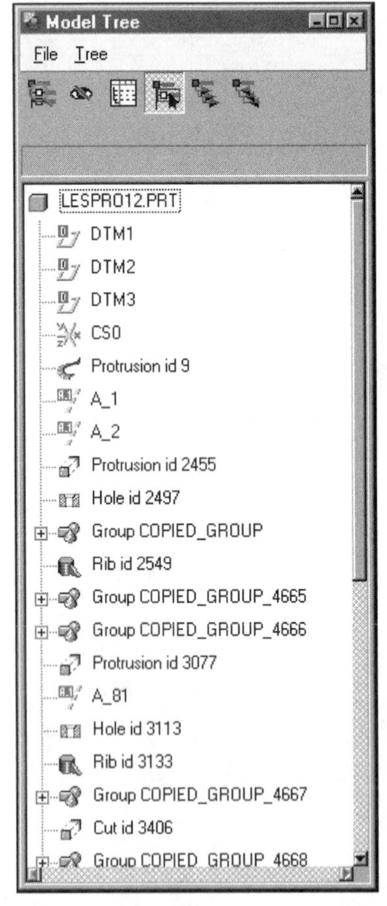

Figure 11.41
Cover Plate with Model Tree, and Datum Planes

Cover Plate

The eleventh **lesson project** is a cast-iron part. Create the part shown in Figures 11.40 through 11.53. The sweep will have a *closed trajectory* with *inner faces included.* Analyze the part and plan out the steps and features required to model it. Use the **DIPS** in Appendix D to plan out the feature creation sequence and the parent-child relationships for the part. Add rounds on all nonmachined edges.

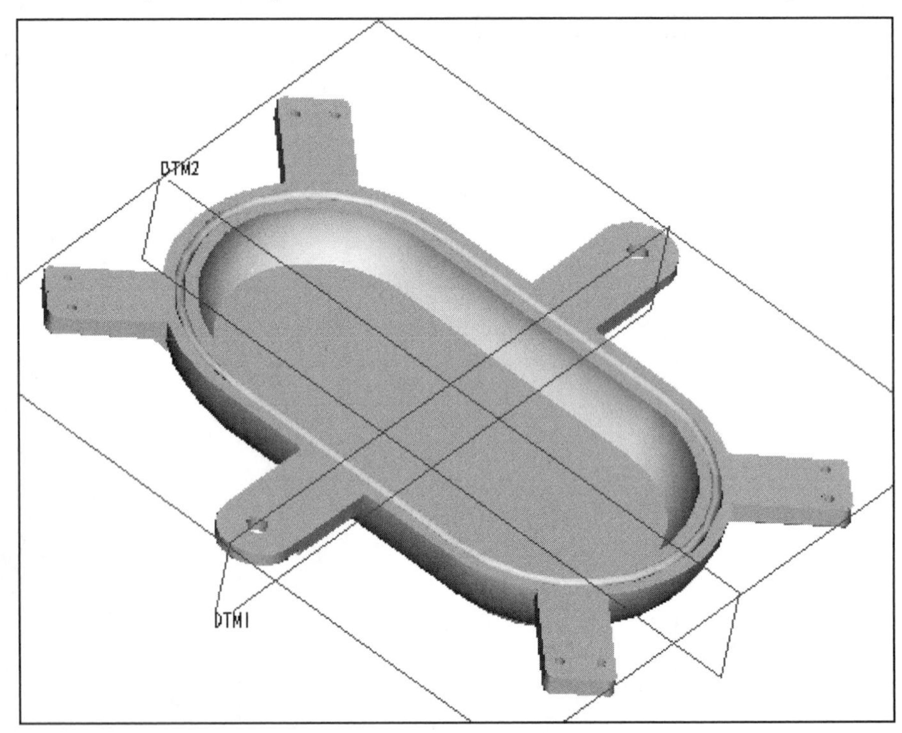

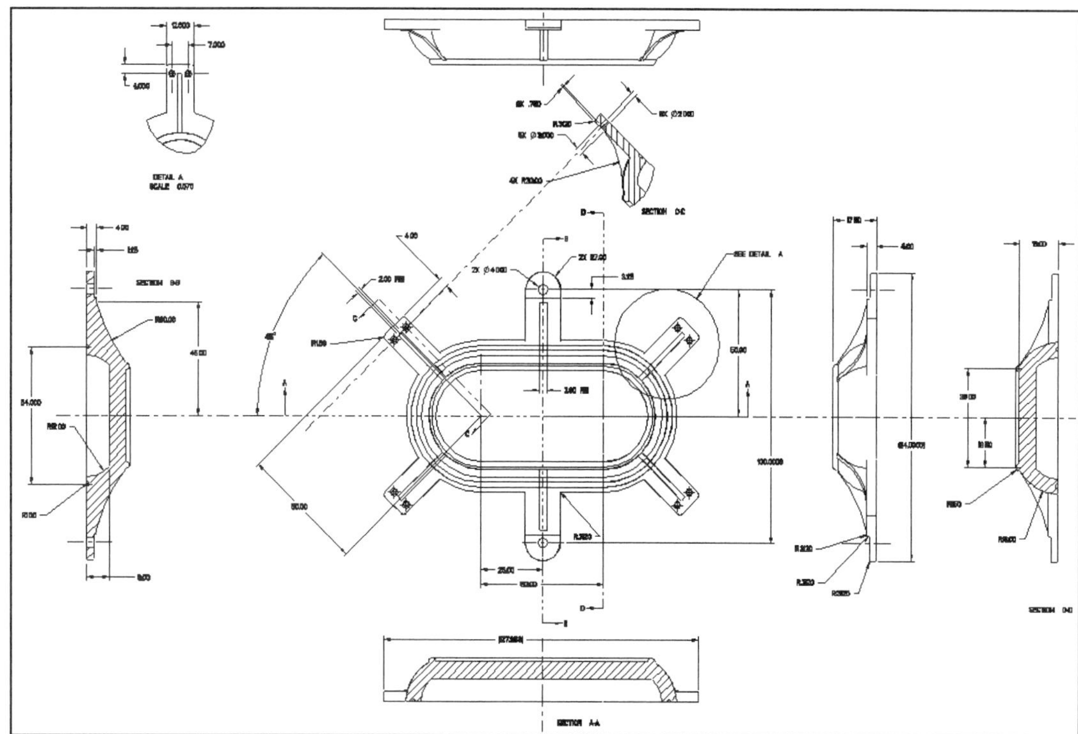

Figure 11.42
Cover Plate Detail Drawing, Sheet One

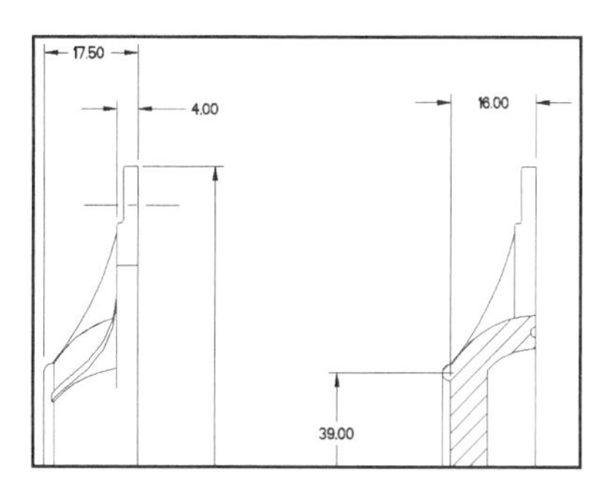

Figure 11.43
Cover Plate Drawing, Sheet Two, Bottom View

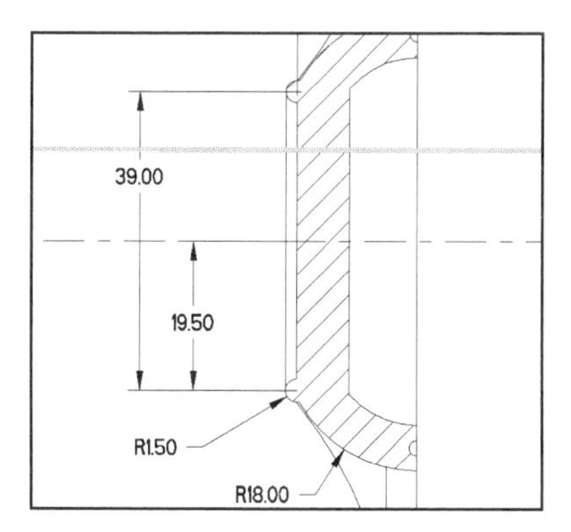

Figure 11.44
Cover Plate Drawing, Top View

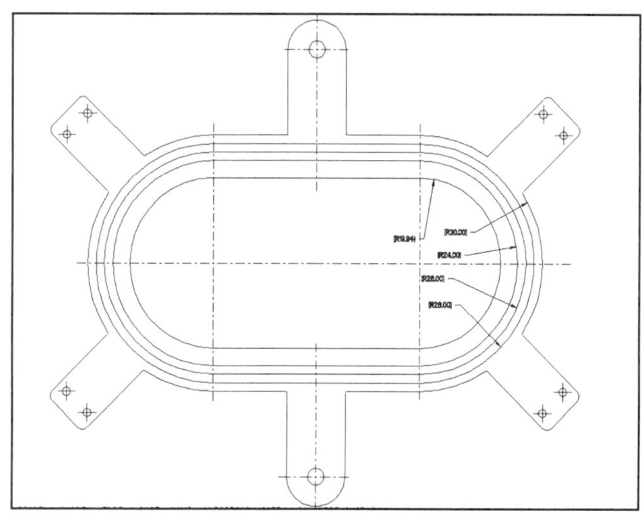

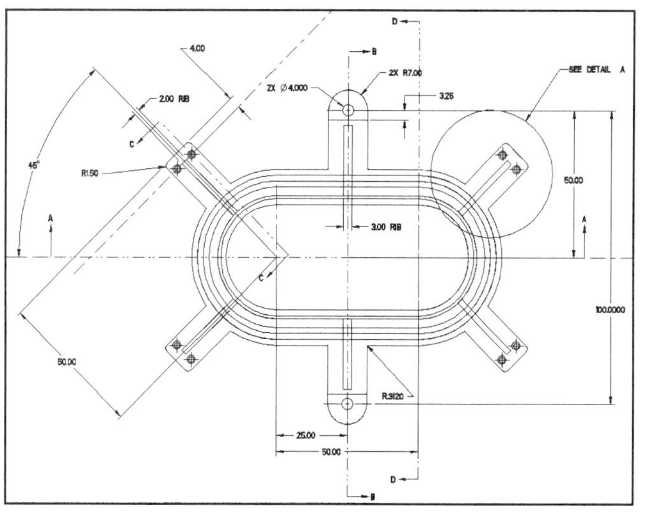

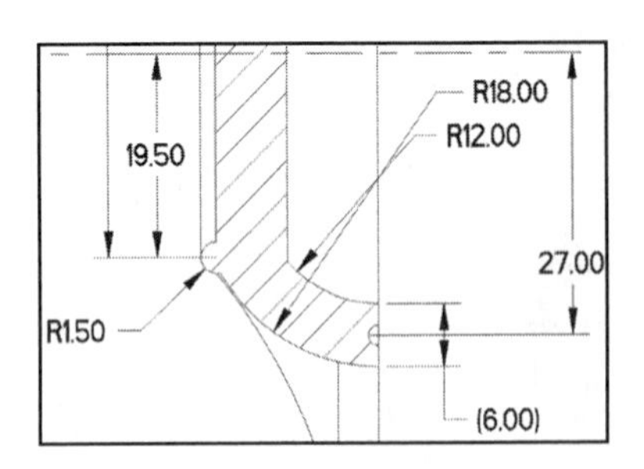

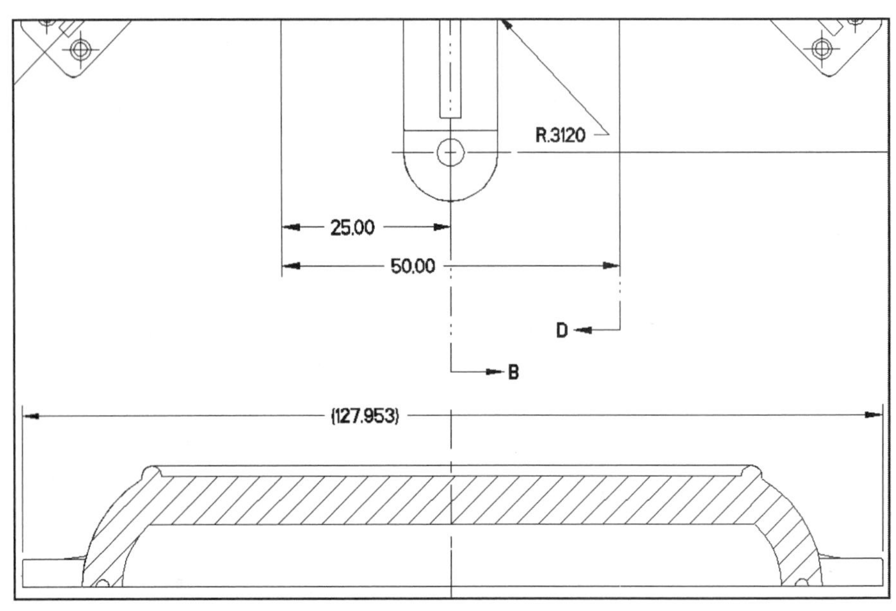

Figure 11.45
Cover Plate Drawing,
Front View,
SECTION A-A

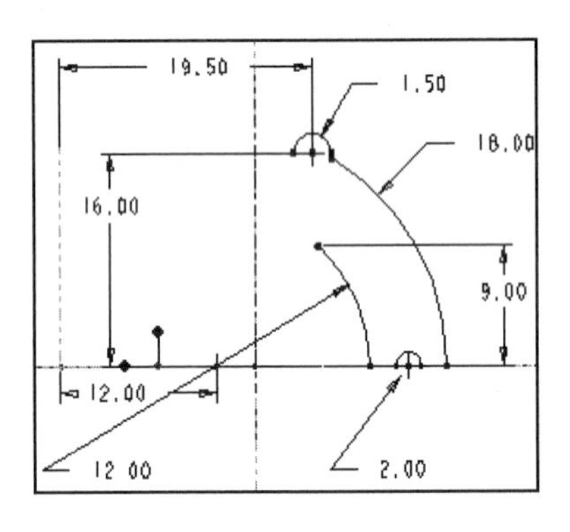

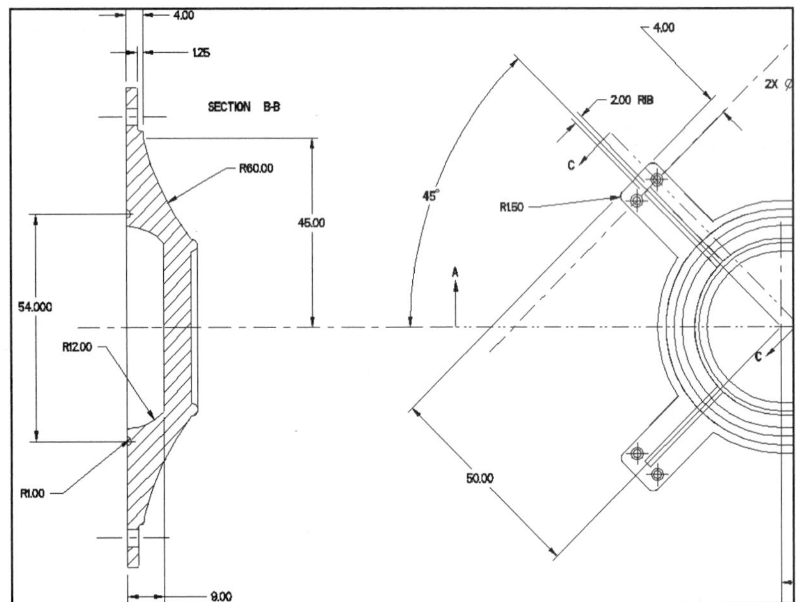

Figure 11.46
Cover Plate Drawing,
Left Side View,
SECTION B-B

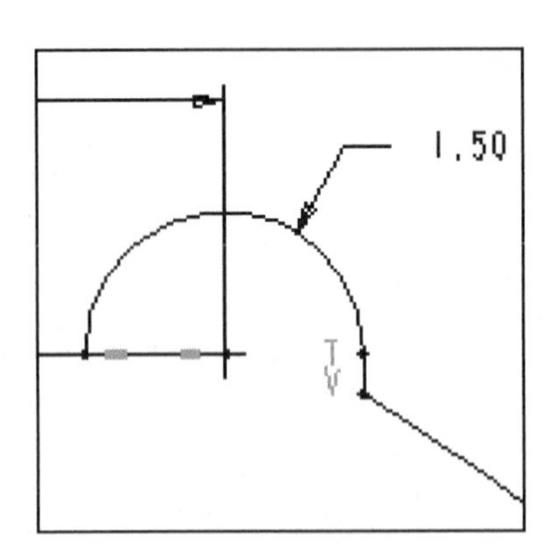

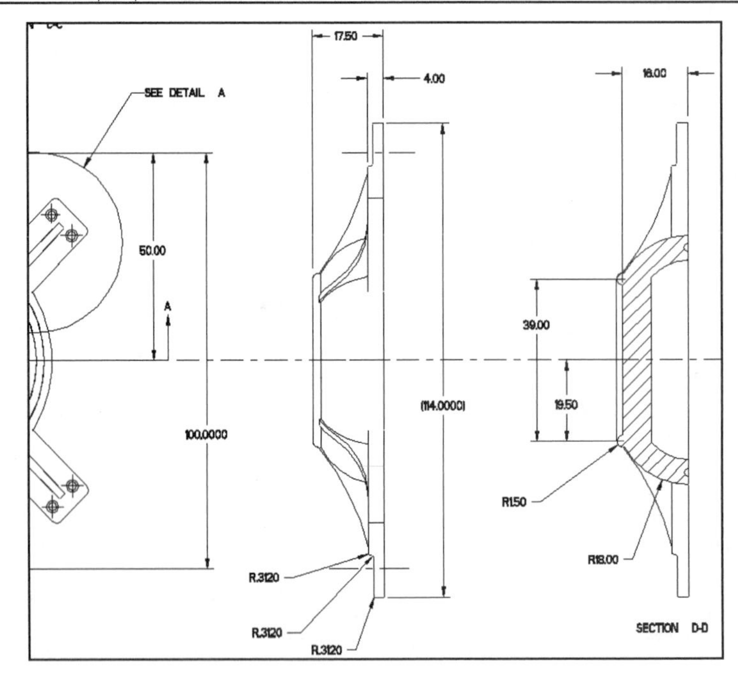

Figure 11.47
Cover Plate Drawing,
Right Side View,
SECTION D-D

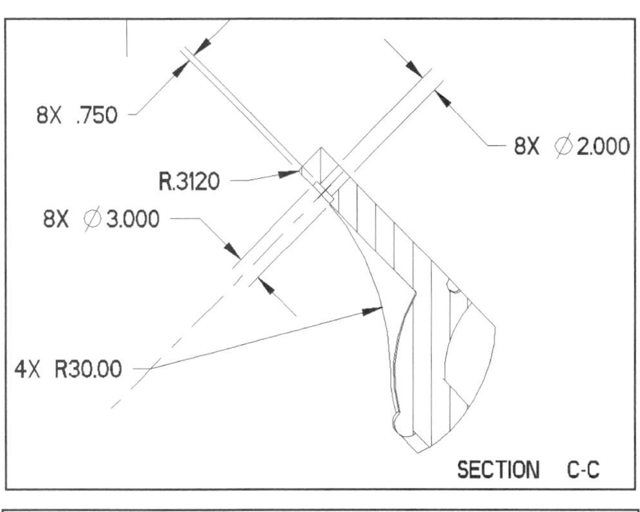

Figure 11.48
Cover Plate Drawing,
SECTION C-C

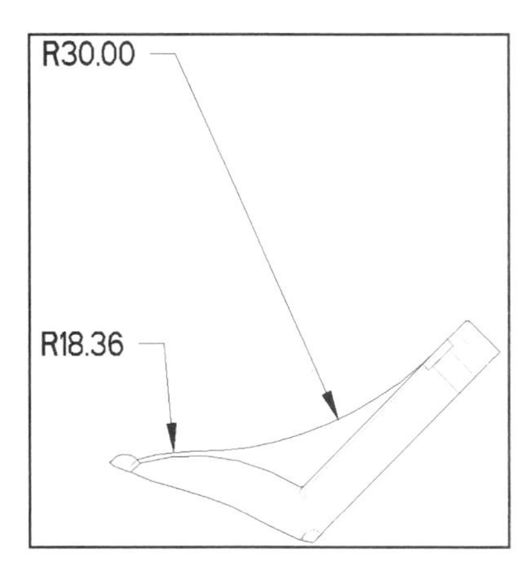

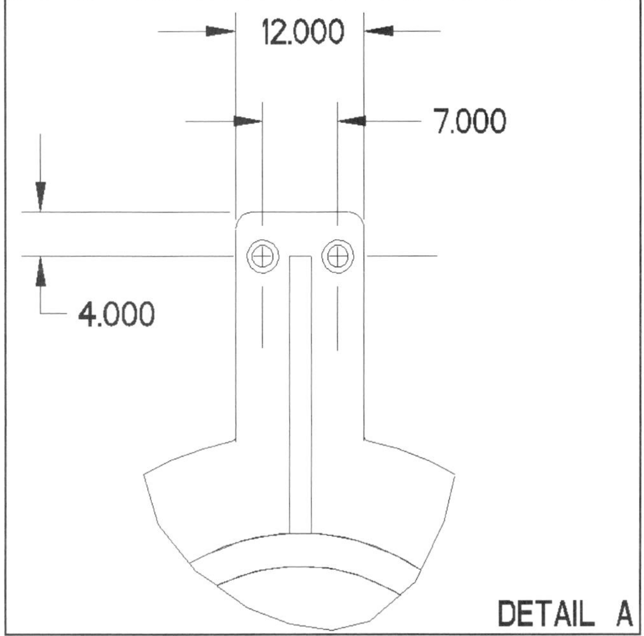

Figure 11.49
Cover Plate Drawing,
DETAIL A

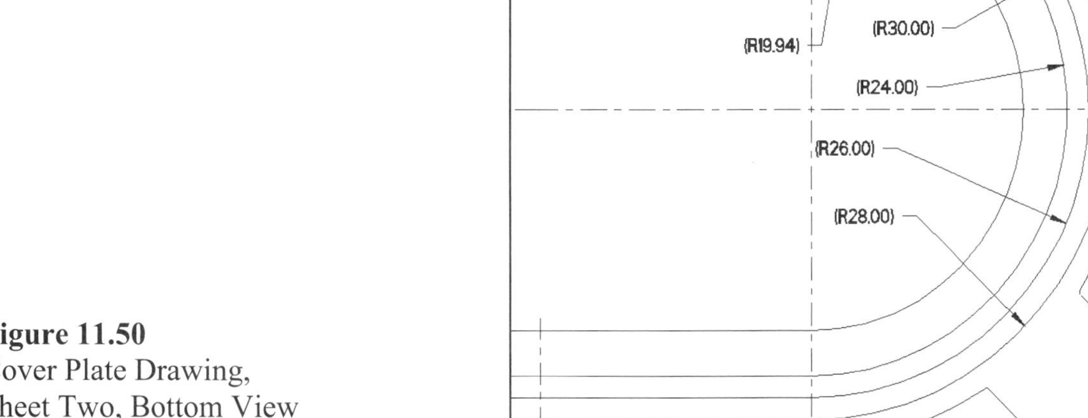

Figure 11.50
Cover Plate Drawing,
Sheet Two, Bottom View

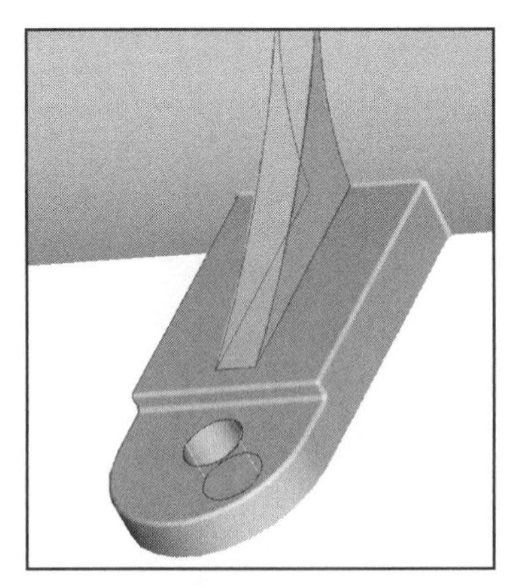

Figure 11.51
Cover Plate Drawing,
Close-up of Top Right

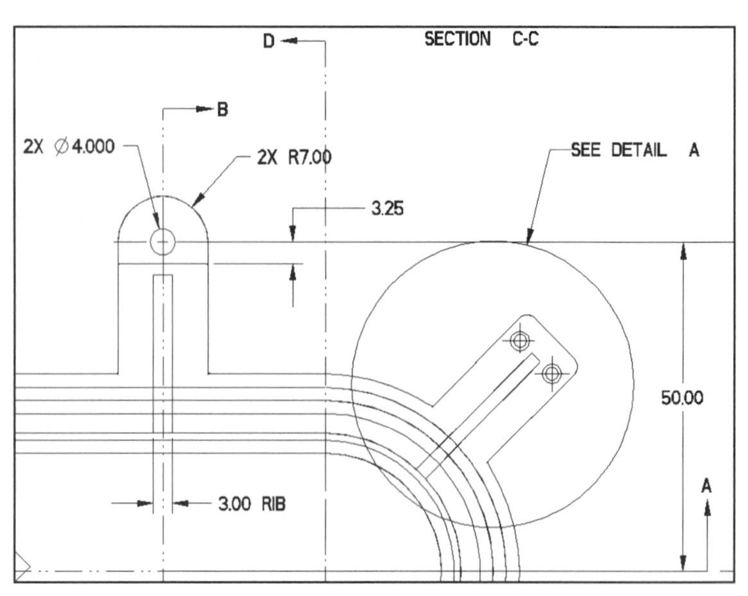

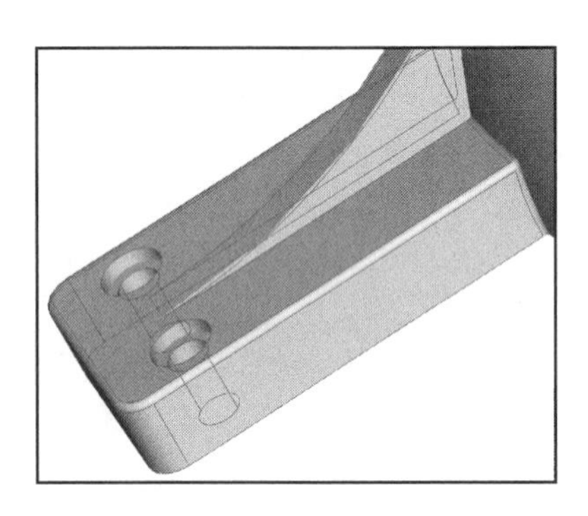

Figure 11.52
Cover Plate Drawing,
Close-up of Top Left

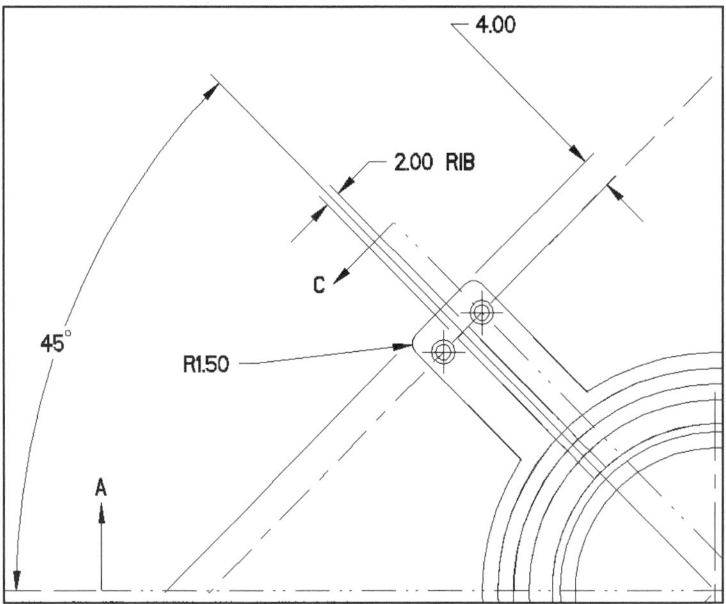

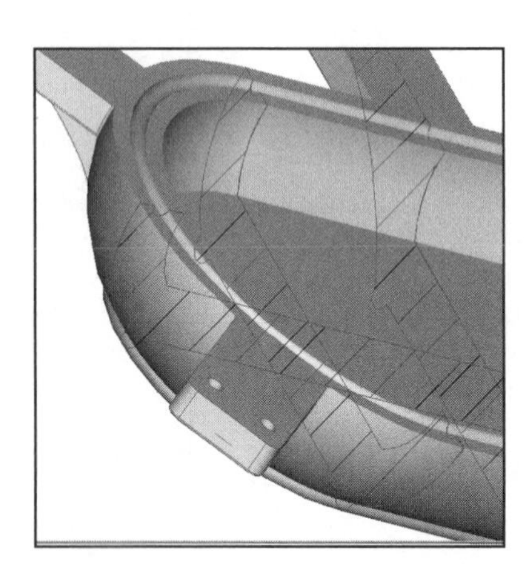

Figure 11.53
Cover Plate Drawing,
Close-up of **SECTION B-B**

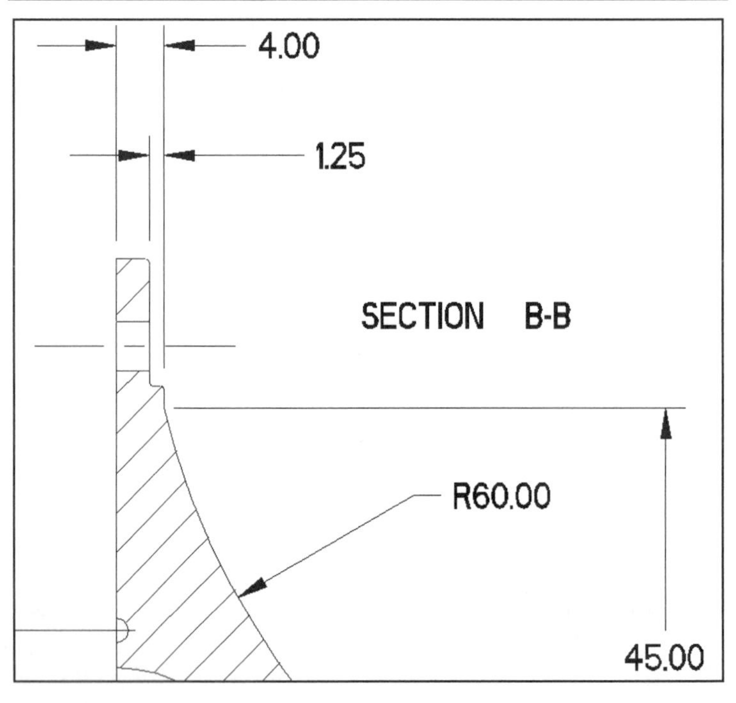

Lesson 12

Blends and Splines

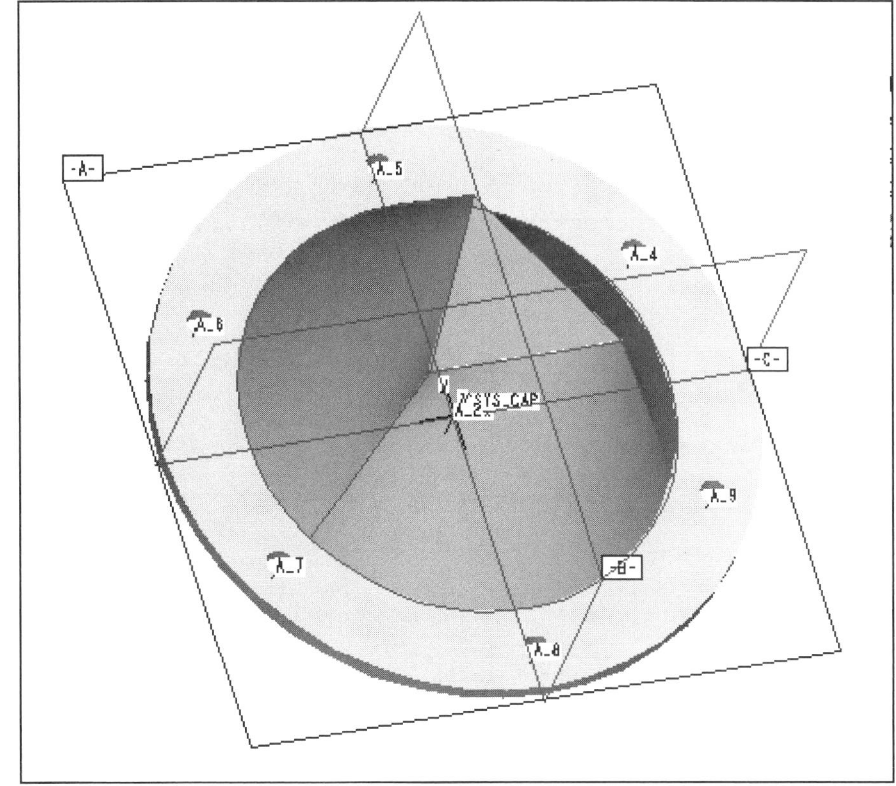

Figure 12.1
Cap

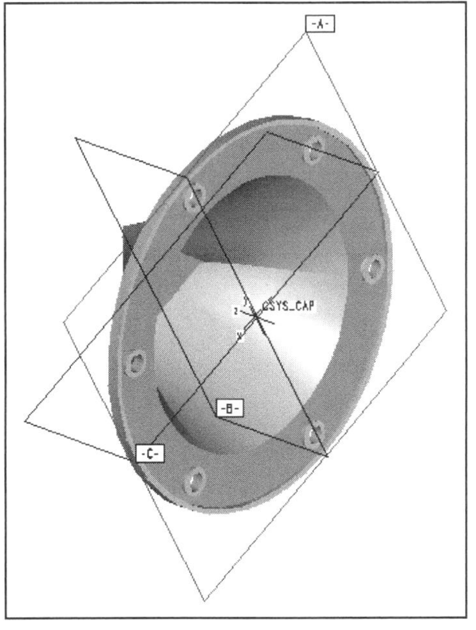

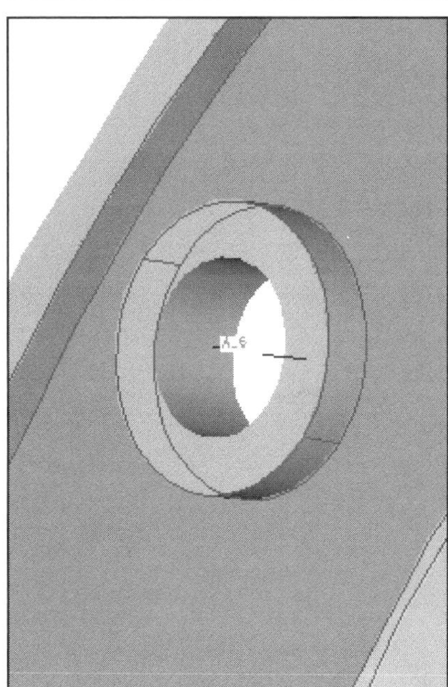

EGD REFERENCE
Fundamentals of Engineering Graphics and Design
by L. Lamit and K. Kitto
Read Chapter 12.
See pages 364-366.

Figure 12.2
Cap with Datum Planes Displayed

COAch™ for Pro/ENGINEER

If you have **COAch for Pro/ENGINEER** on your system, go to SEARCH and do the Segment shown in Figure 12.5.

OBJECTIVES

1. **Create a parallel blend feature**
2. **Shell a blend feature**
3. **Create a spline and use it in a swept blend**
4. **Create a swept blend feature**
5. **Create sections in Part mode**

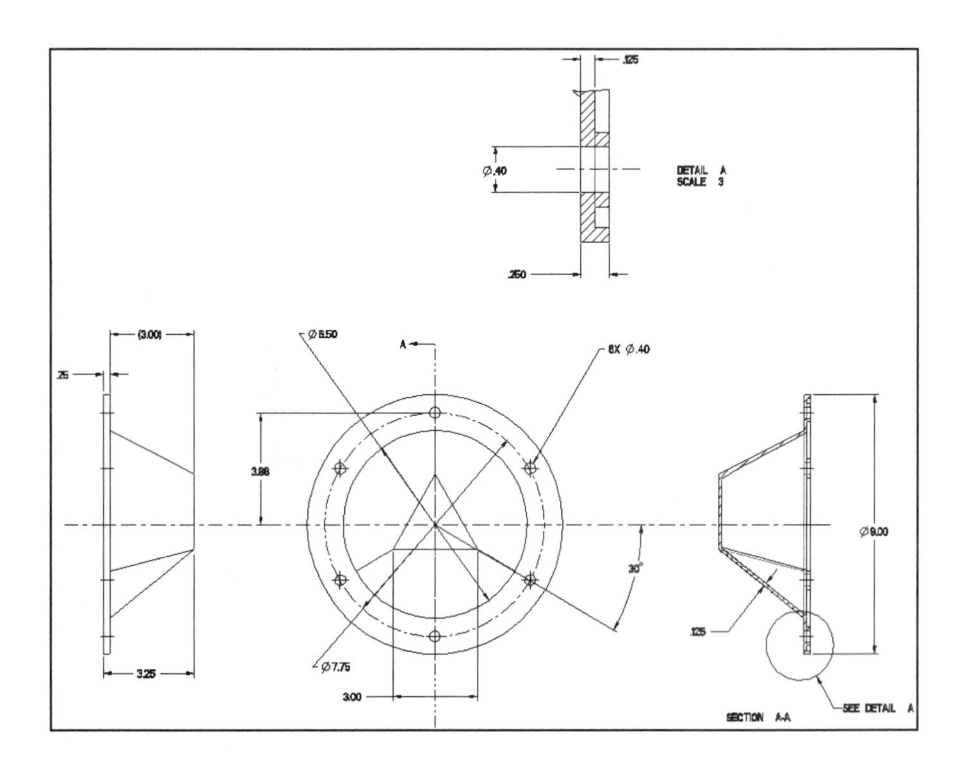

Figure 12.3
Cap Detail

BLENDS AND SPLINES

A blended feature consists of a series of at least two planar sections that are joined together at their edges with transitional surfaces to form a continuous feature. The Cap in Figures 12.1 through 12.3 uses a blend feature in its design. A **Blend** can be created as a **Parallel Blend** (Figs. 12.4 and 12.5), or you can construct a **Swept Blend**.

A **spline** is similar to an **irregular** curve and is used in a variety of industrial designs.

Figure 12.4
Blends

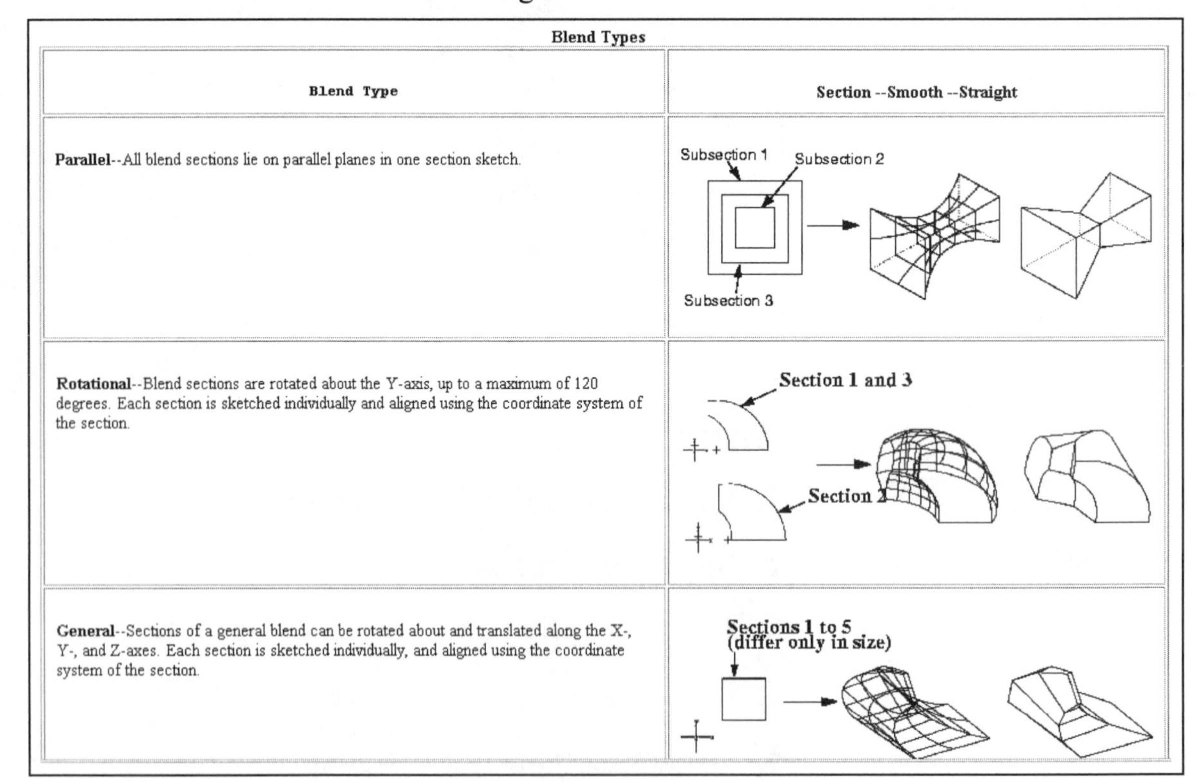
Blend Types

Blend Type	Section --Smooth --Straight
Parallel--All blend sections lie on parallel planes in one section sketch.	Subsection 1, Subsection 2, Subsection 3
Rotational--Blend sections are rotated about the Y-axis, up to a maximum of 120 degrees. Each section is sketched individually and aligned using the coordinate system of the section.	Section 1 and 3, Section 2
General--Sections of a general blend can be rotated about and translated along the X-, Y-, and Z-axes. Each section is sketched individually, and aligned using the coordinate system of the section.	Sections 1 to 5 (differ only in size)

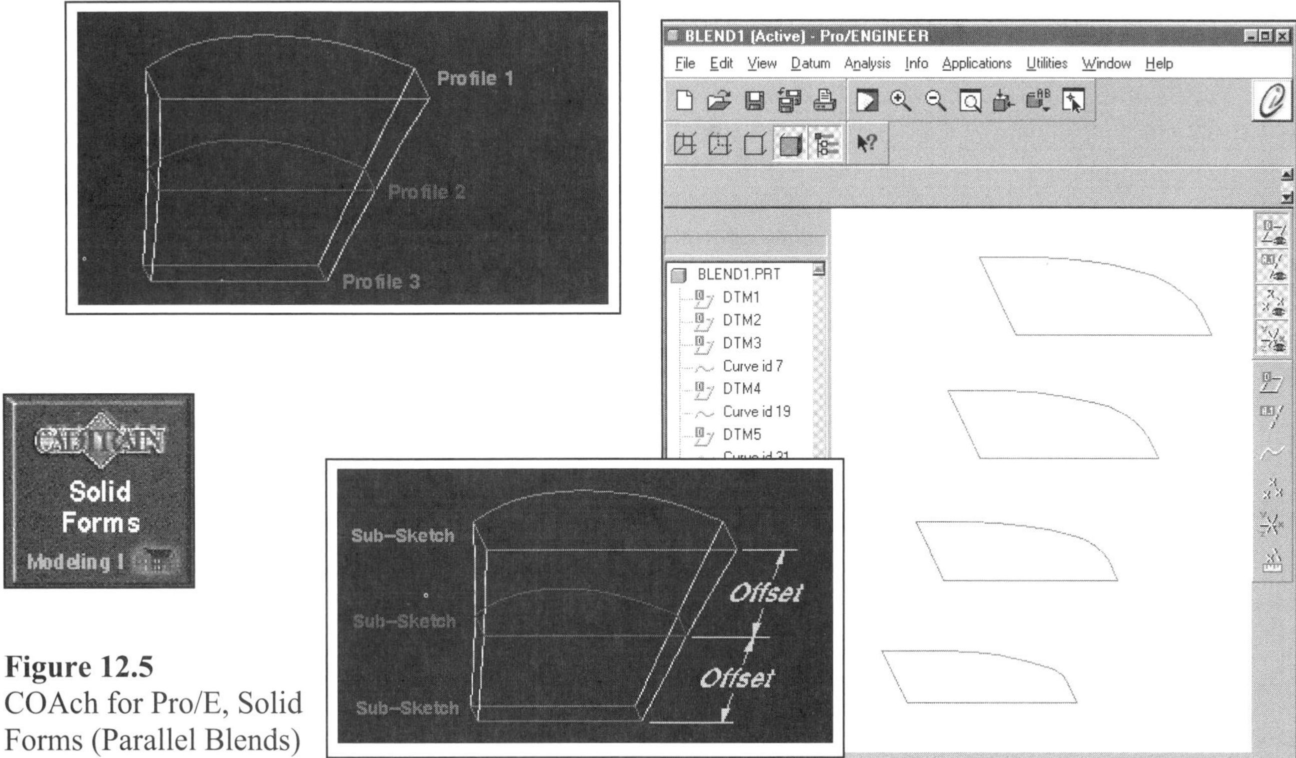

Figure 12.5
COAch for Pro/E, Solid Forms (Parallel Blends)

Blend Sections

Figure 12.5 shows a parallel blend for which the *section* consists of four *subsections*. Each segment in the subsection is matched with a segment in the following subsection; the blended surfaces are created between the corresponding segments.

HINT
For the most parts, blends must have the same number of entities in each subsection. The only exception is a capped blend.

Starting Point of a Section

To create the transitional surfaces, Pro/E connects the *starting points* of the sections and continues to connect the vertices of the sections in a clockwise manner. By changing the starting point of a blend section, you can create blended surfaces that twist between the sections.

The default starting point is the first point sketched in the subsection. You can position the starting point to the endpoint of another segment by choosing the option **Start Point** from the SEC TOOLS menu and selecting the new position.

Straight and Smooth Attributes

Blends use one of the following transitional surface ATTRIBUTES menu options:

Straight Create a straight blend by connecting vertices of different subsections with straight lines. Edges of the sections are connected with ruled surfaces.

Smooth Create a smooth blend by connecting vertices of different subsections with smooth curves. Edges of the sections are connected with ruled (spline) surfaces.

Creating Blends

To create a blend, choose the following:

1. **Feature ⇒ Create ⇒ Solid ⇒ Protrusion.**
2. Choose **Blend** and **Solid** or **Thin** from the SOLID OPTS menu, then choose **Done**.
3. Choose options from the BLEND OPTS menu; then choose **Done**.

The BLEND OPTS menu options are as follows:

Parallel All blend sections lie on parallel planes in one section sketch.
Rotational The blend sections are rotated about the **Y** axis, up to a maximum of **120°**. Each section is sketched individually and aligned using the coordinate system of the section.
General The sections of a general blend can be rotated about and translated along the **X**, **Y**, and **Z** axes. Sections are sketched individually and aligned using the coordinate system of the section.
Regular Sec The feature will use the regular sketching plane.
Project Sec The feature will use the projection of the section on the selected surface. This is used for parallel blends only.
Select Sec Select section entities (not available for parallel blends).
Sketch Sec Sketch section entities.

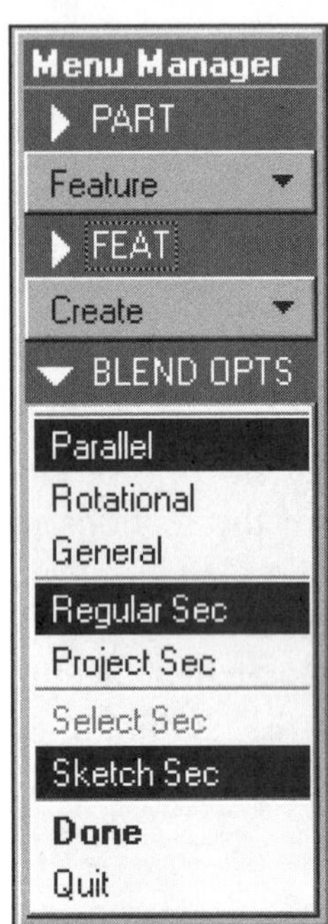

Figure 12.6
Parallel Blends

Parallel Blends

You create parallel blends (Fig. 12.6) using the **Parallel** option in the BLENDS OPTS menu. A parallel blend is created from a single section (Fig. 12.7) that contains multiple sketches called *subsections*. A first or last subsection can be defined as a point or a blend vertex.

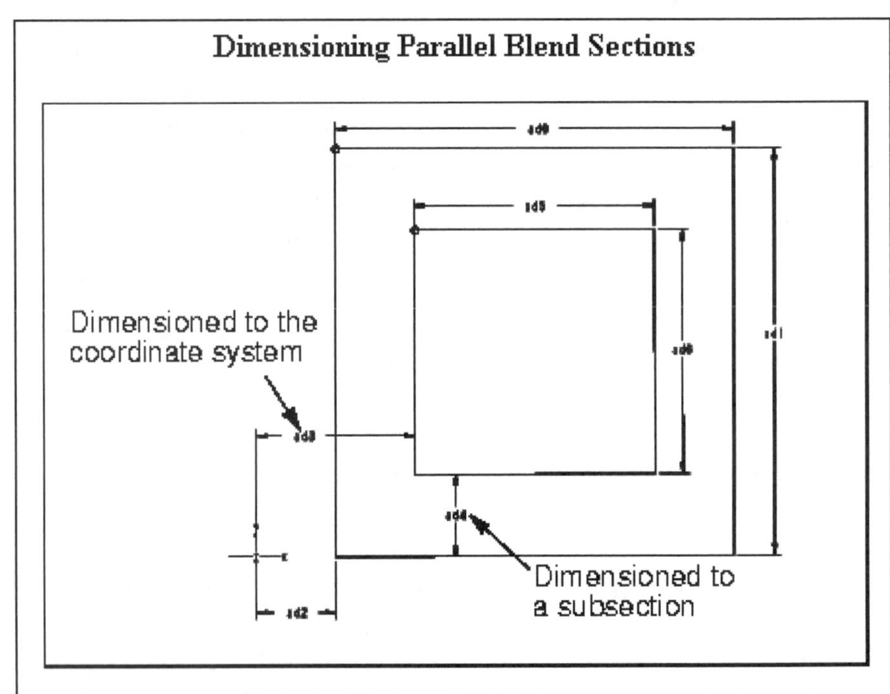

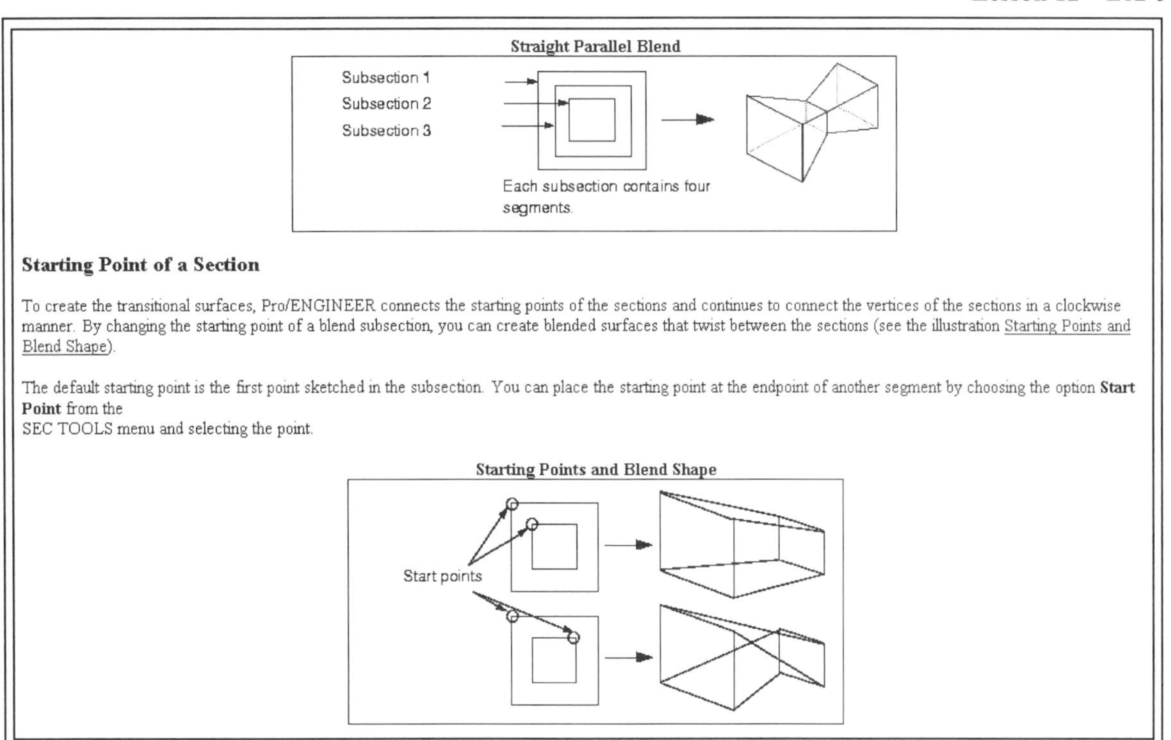

Straight Parallel Blend

Subsection 1
Subsection 2
Subsection 3

Each subsection contains four segments.

Starting Point of a Section

To create the transitional surfaces, Pro/ENGINEER connects the starting points of the sections and continues to connect the vertices of the sections in a clockwise manner. By changing the starting point of a blend subsection, you can create blended surfaces that twist between the sections (see the illustration Starting Points and Blend Shape).

The default starting point is the first point sketched in the subsection. You can place the starting point at the endpoint of another segment by choosing the option **Start Point** from the
SEC TOOLS menu and selecting the point.

Starting Points and Blend Shape

Start points

Figure 12.7
Blend Sections

Creating a Parallel Blend

To create a parallel blend, choose the following:

1. When you choose **Done** from the BLEND OPTS menu, Pro/E displays the feature creation dialog box and the ATTRIBUTES menu. Choose either **Straight** or **Smooth**.

2. Create the first subsection using the Sketcher. You determine the direction of feature creation as you set up the sketching plane. Dimension and regenerate each subsection sketch to ensure the validity of the dimensioning scheme. A parallel blend requires more than one subsection, so after successfully regenerating this section, choose **Sec Tools** from the SKETCHER menu.
3. Choose **Toggle** from the SEC TOOLS menu. The first subsection turns gray and becomes inactive.
4. Choose **Sketch** and sketch the second subsection. Make sure its starting point corresponds to the starting point of the first subsection in the manner that you intend. Dimension and regenerate it.
5. If you are sketching more than two subsections, choose **Toggle** repeatedly until all the current geometry is gray; then sketch the subsection. Repeat this step until all subsections have been sketched.

6. To modify an existing subsection, toggle through until the subsection you want is active. While you can place or move the starting point of a subsection only when it is active, you can modify the dimensions of any subsection at any time.
7. When you have sketched all the subsections and regenerated, choose **Done** from the SKETCHER menu. When prompted, enter the distances between each of the subsections.
8. Select the **OK** button to create the feature.

Swept Blends

A swept blend (Figs. 12.8 and 12.9) is created using a single trajectory (a spine) and multiple sections. You create the spine of the swept blend by sketching or selecting a datum curve or an edge. Spines can be created with splines. You sketch the sections at specified segment vertices or datum points on the spine. Each section can be rotated about the **Z** axis with respect to the section immediately preceding it.

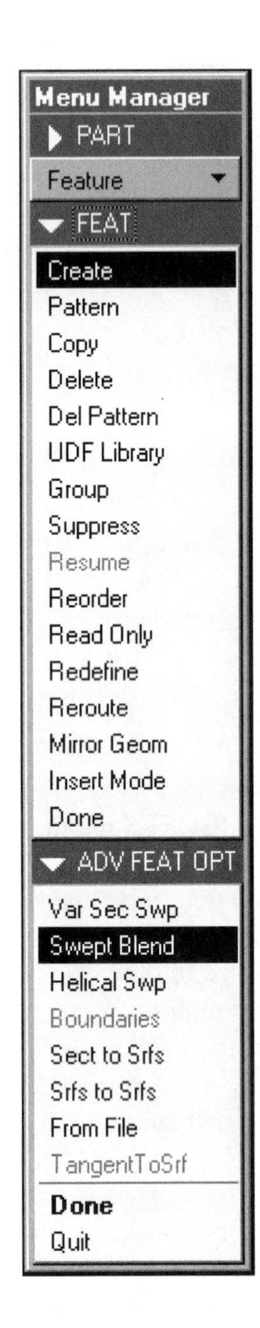

Note the following restrictions:

* A section cannot be located at a sharp corner in the spine.
* For a closed trajectory profile, sections must be sketched at the start point and at least one other location. Pro/E uses the first section at the endpoint.
* For an open trajectory profile, you must create sections at the start point and endpoint. You cannot skip placement of a section at those points.
* Sections cannot be dimensioned to the model, because modifying the trajectory would invalidate those dimensions.
* A composite datum curve cannot be selected for defining sections of a swept blend (**Select Sec**). Instead, you must select one of the underlying datum curves or edges for which a composite curve is determined.
* If you choose **Pivot Dir** and **Select Sec**, all selected sections must lie in planes that are parallel to the pivot direction.
* You cannot use a nonplanar datum curve from an equation as a swept blend trajectory.

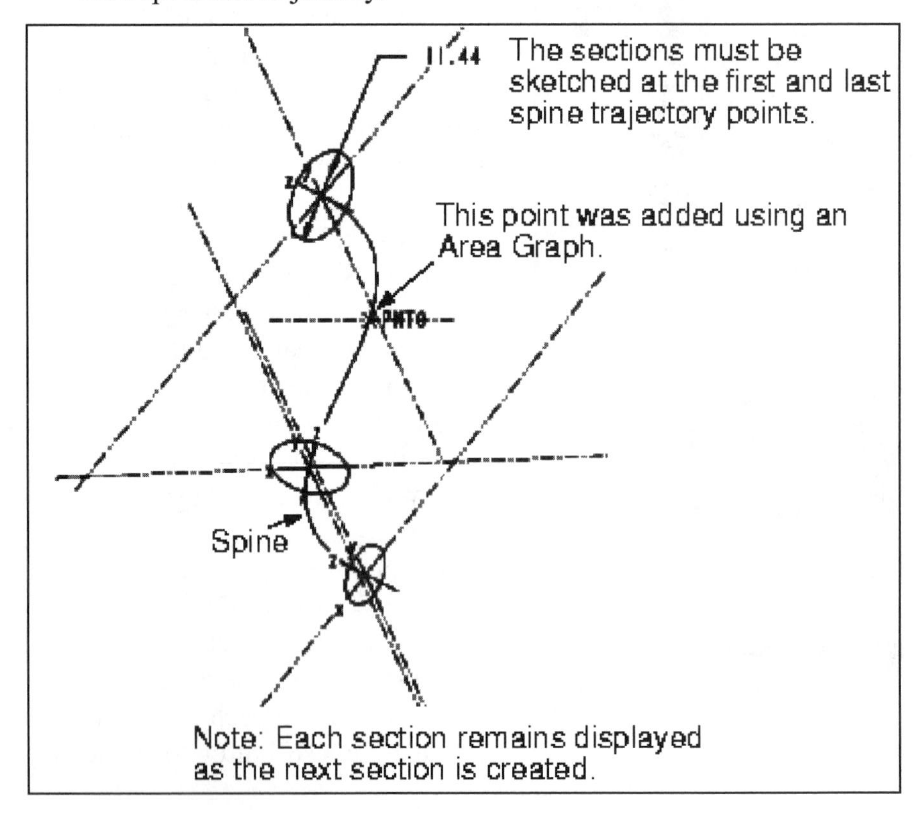

Figure 12.8
Creating a Swept Blend

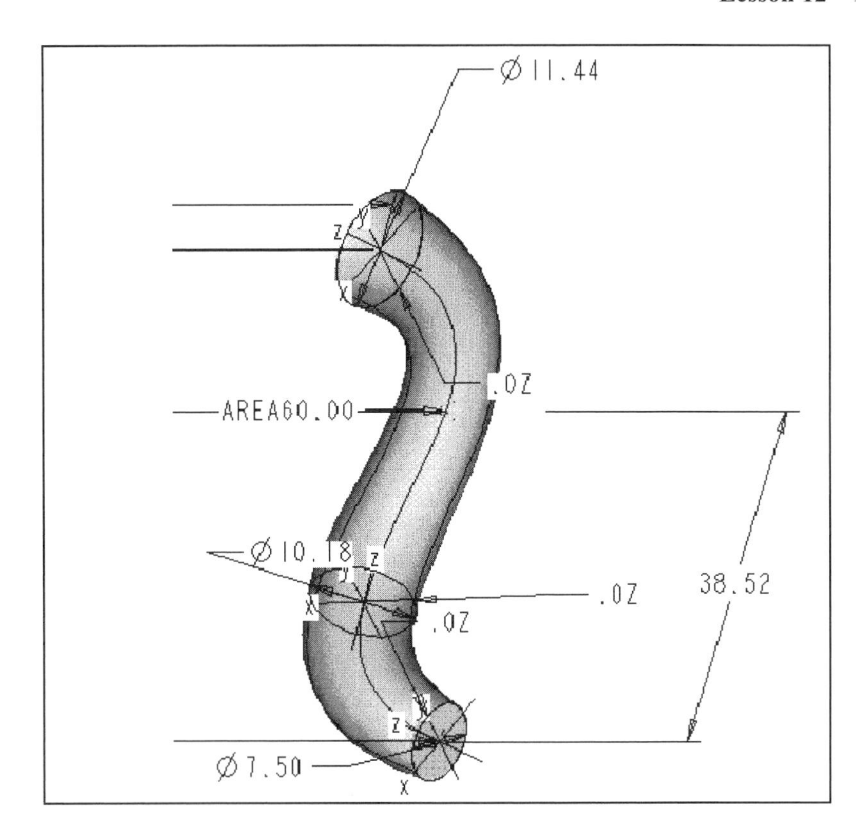

Figure 12.9
Swept Blends

Creating a Swept Blend

To create a **Swept Blend** (Figs. 12.9 and 12.10), you can define the trajectory by sketching a trajectory or by selecting existing curves and edges and extending or trimming the first and last entities in the trajectory. Use the following procedure:

1. Choose **Advanced** and **Done** from the SOLID OPTS menu and **Swept Blend** and **Done** from the ADV FEAT OPT menu.
2. Choose the desired options from the BLEND OPTS menu, presented in mutually exclusive pairs, and then choose **Done** from the BLEND OPTS menu. The possible options are as follows:

Select Sec Select existing curves or edges to define each section, using the CRV SKETCHER menu.
Sketch Sec Sketch new section entities to define each section.
NrmToOriginTraj The section plane remains normal to the *Origin Trajectory* throughout its length. The generic **Sweep** behaves this way.
Pivot Dir The section remains normal to the *Origin Trajectory* as it is viewed along the *Pivot direction.* The upward direction of the section remains parallel to the *Pivot Direction.*
Norm To Traj Two trajectories must be selected to determine the location and the orientation of the section. The *Origin Trajectory* determines the origin of the section along the length of the feature. The section plane remains normal to the *Normal Trajectory* along the length of the feature.

3. Choose a SWEEP TRAJ menu option. The options include:

   Sketch Traj Sketch a spine. The spine can have sharp corners (a discontinuous tangent to the curve), except at the endpoint of a closed curve. At nontangent vertices, Pro/E miters the geometry as in constant section sweeps.
   Select Traj Define the spine trajectory using existing curves and edges. Pro/E displays the CHAIN menu. Choose **Select**, define the chain, and then choose **Done**.

NOTE
A section cannot be located at a sharp corner in the spine.

4. Sketch or select the trajectory of the spine.
5. Use the CONFIRM menu options to choose the points at which to define any additional sections. As appropriate to the defined trajectory, those points may include endpoints of spine entities, spine entity control points, and any existing datum points on spine entities (if you select a spine). The CONFIRM SEL menu options are:

   Accept Sketch or select a section at this highlighted location.
   Next Bypass this highlighted location and go to the next point.
   Previous Bypass this highlighted location and return to the previous point.

6. For each section, specify the rotation angle about the **Z** axis (with a value between **-120°** and **+120°**).
7. Select or sketch the entities for each section, depending on whether you choose **Select Sec** or **Sketch Sec**. Choose **Done** from the SKETCHER menu.
8. When all sections have been sketched or selected, unless you chose the **Area Graph** or **Blend Control** element, select the **OK** button in the dialog box to generate the swept blend feature.

Figure 12.10
Spines, Set Perimeter Option

- **Set Perimeter**-Control the shape of the feature by controlling its perimeter between the sections. If two consecutive sections have equal perimeters, the system attempts to maintain the same cross-section perimeter between these sections (see the following illustration, Using the Set Perimeter Option). For sections that have different perimeters, the system uses smooth interpolation along each curve of the trajectory to define the perimeter of the feature between its sections.

  Note:

  You cannot specify both perimeter control and tangency conditions for the swept blend-only one of these conditions is allowed.

- **Area Graph**-Control the shape of the feature through control points and area values (see Modifying Swept Blend Geometry Using an Area Graph).
- **None**-Do not set any blend control for the feature.
- **Center Crv**-Show a curve connecting the centroids of the feature's cross-sections. This option is available only with the Set Perimeter option.

Using the Set Perimeter Option

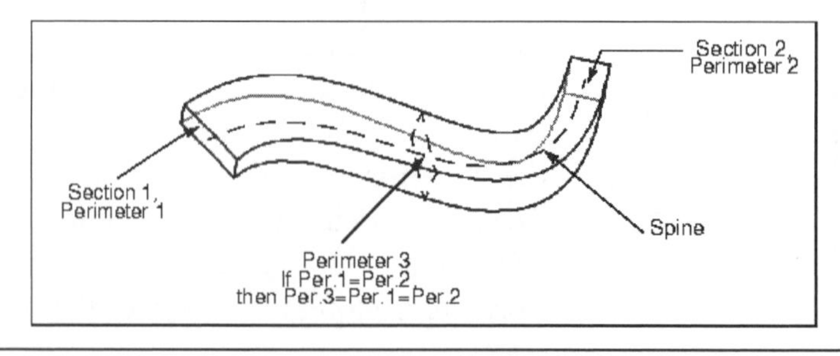

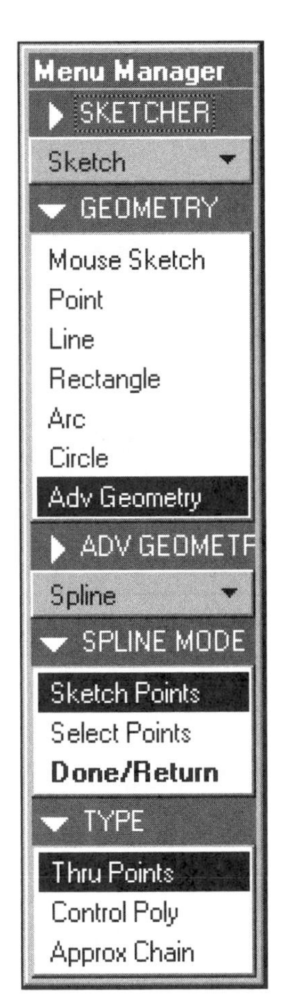

Splines

Sketching a **Spline** is similar to drawing an **irregular** curve (Fig. 12.11). Splines (Fig. 12.12) are created by picking or sketching a series of specific points.

To create a spline:

1. Choose **Spline** from the ADV GEOMETRY menu.
2. The SPLINE MODE menu appears, with the following options:

Sketch Points Create a spline by picking screen points for the spline to pass through.
Select Points Create a spline by selecting existing Sketcher points. Once the point has been selected, there is no further link between the point and the spline.

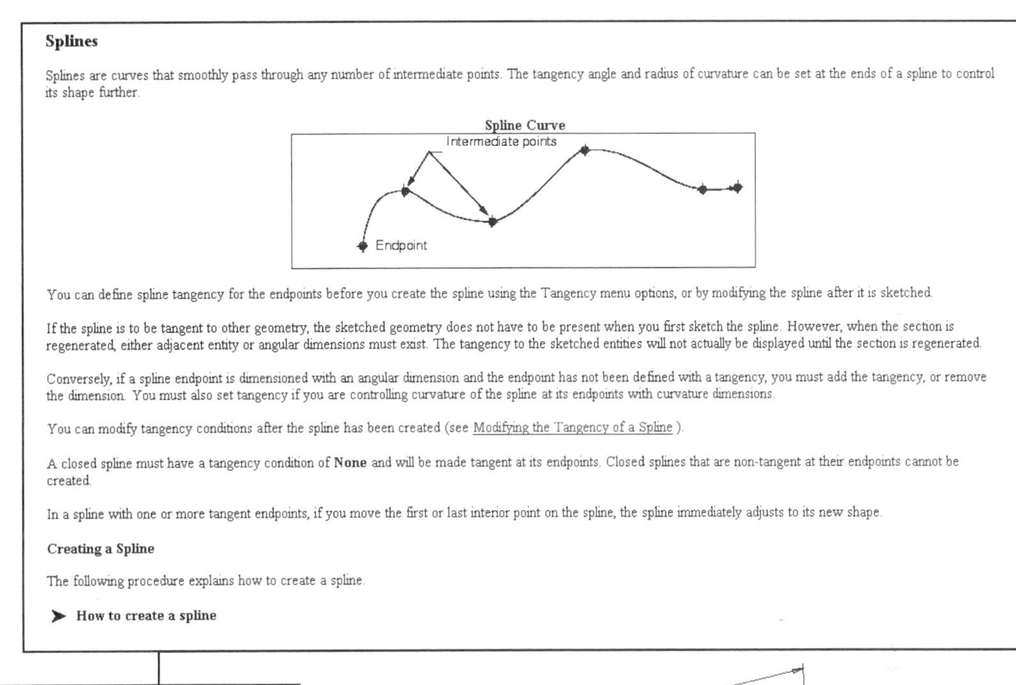

Splines

Splines are curves that smoothly pass through any number of intermediate points. The tangency angle and radius of curvature can be set at the ends of a spline to control its shape further.

You can define spline tangency for the endpoints before you create the spline using the Tangency menu options, or by modifying the spline after it is sketched.

If the spline is to be tangent to other geometry, the sketched geometry does not have to be present when you first sketch the spline. However, when the section is regenerated, either adjacent entity or angular dimensions must exist. The tangency to the sketched entities will not actually be displayed until the section is regenerated.

Conversely, if a spline endpoint is dimensioned with an angular dimension and the endpoint has not been defined with a tangency, you must add the tangency, or remove the dimension. You must also set tangency if you are controlling curvature of the spline at its endpoints with curvature dimensions.

You can modify tangency conditions after the spline has been created (see Modifying the Tangency of a Spline).

A closed spline must have a tangency condition of **None** and will be made tangent at its endpoints. Closed splines that are non-tangent at their endpoints cannot be created.

In a spline with one or more tangent endpoints, if you move the first or last interior point on the spline, the spline immediately adjusts to its new shape.

Creating a Spline

The following procedure explains how to create a spline.

➤ **How to create a spline**

Figure 12.11
Splines

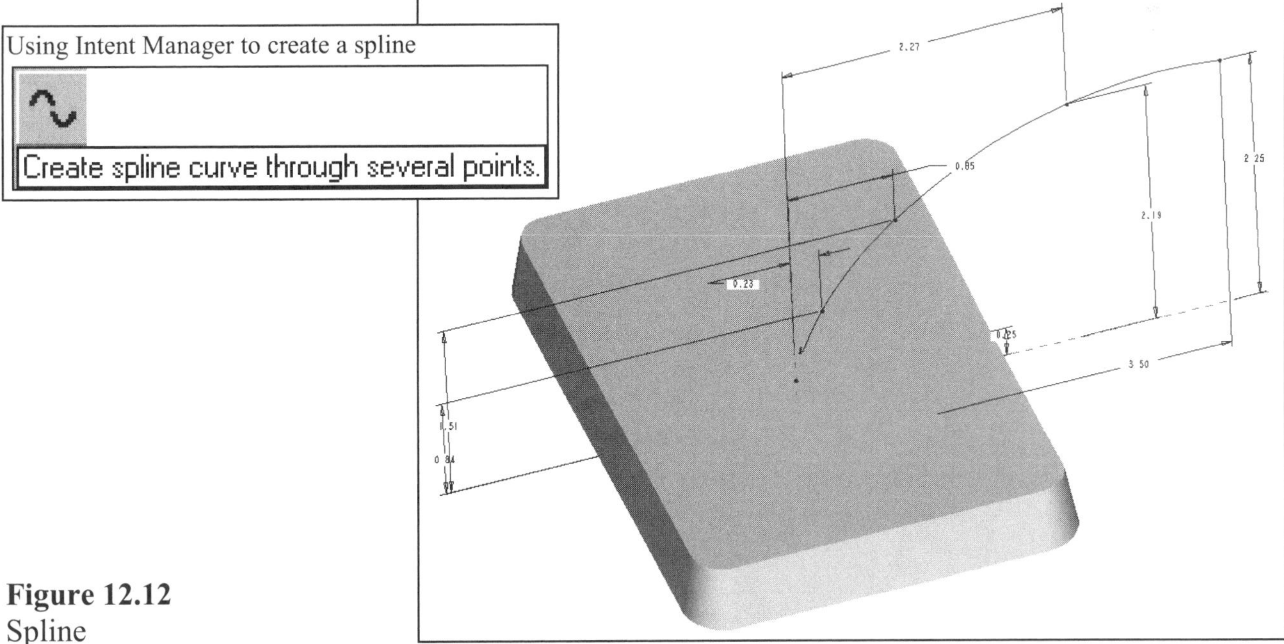

Figure 12.12
Spline

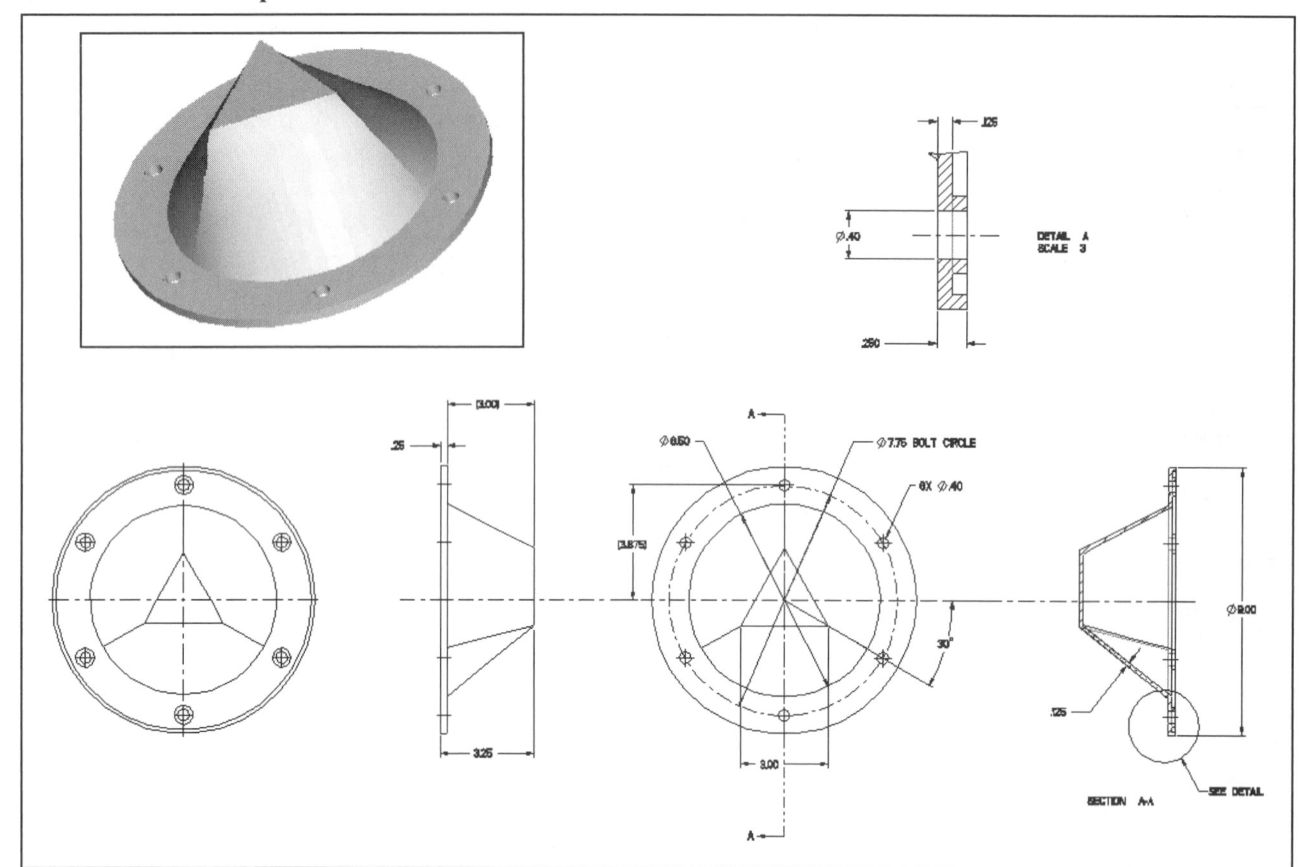

Figure 12.13
Cap Part and Detail Drawing

Lesson procedures & commands START HERE ➡ ➡ ➡

NOTE

Set up the Cap:

- Material = PLASTIC
- Units = Inch lbm Second
- Hidden Line
- Tan Dimmed
- Datum **TOP** = **C**
- Datum **FRONT** = **A**
- Datum **RIGHT** = **B**
- Coordinate System = **CSYS_CAP**

✔ Snap to Grid
✔ ± Dimension Tolerances

CONFIG.PRO

angular_tol	***0***
tol_mode	***plusminus***
sketcher_dec_places	***3***
default_dec_places	***3***

Cap

The **Cap** is a part created with a **Parallel Blend** (Figs. 12.13 through 12.19). The blend sections are a circle and a triangle. The circle is actually three equal arcs, because *the sections of a blend must have equal segments*.

The part is shelled as the last feature in its creation. The **Shell** command will create *bosses* around each hole as it hollows out the part.

For this part, a cross section will be created in the Part mode to be used when you are detailing the Cap in the Drawing mode.

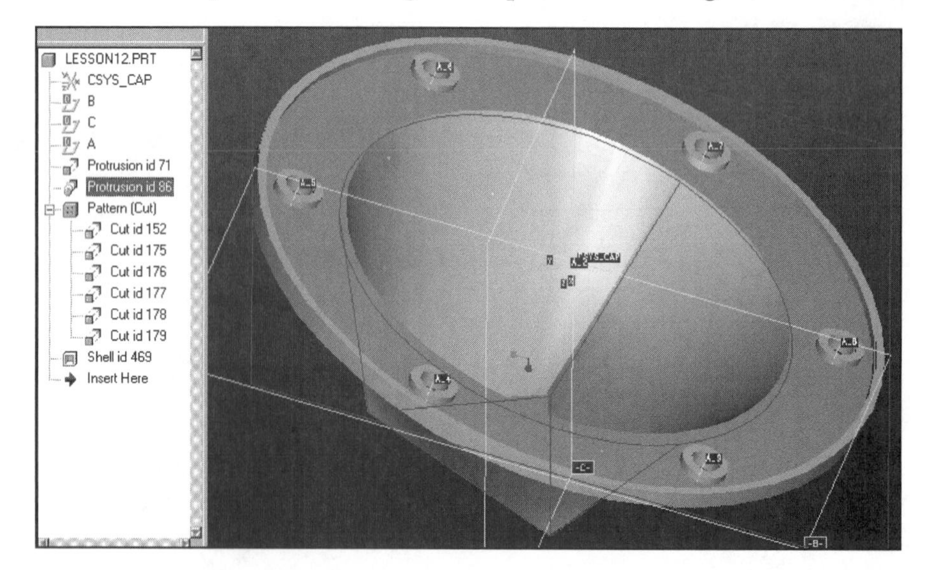

Figure 12.14
Cap and Model Tree

	Value
Current Session	
angular_tol	0
default_dec_places	3
sketcher_dec_places	3
tol_mode	plusminus

```
X.X    +-0.1
X.XX   +-0.01
X.XXX  +-0.001
ANG    +-0.5
```

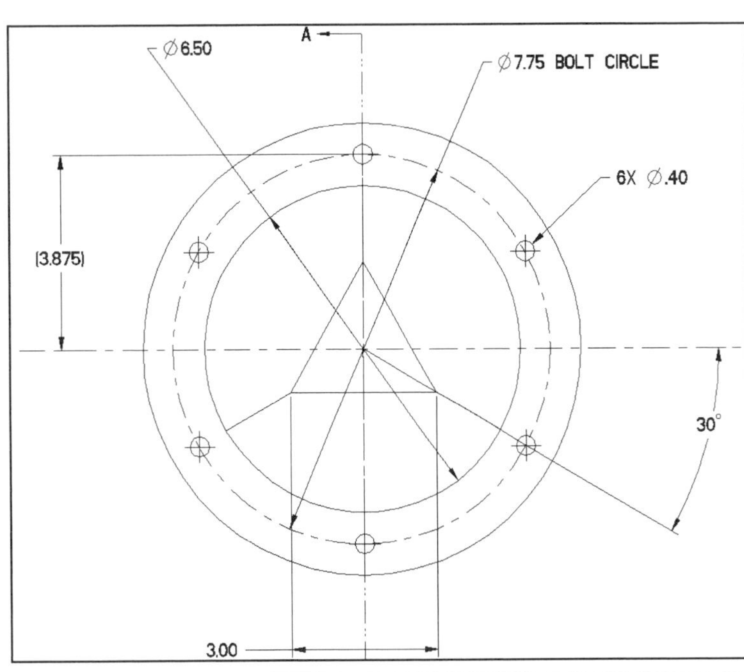

Figure 12.15
Cap, Top View

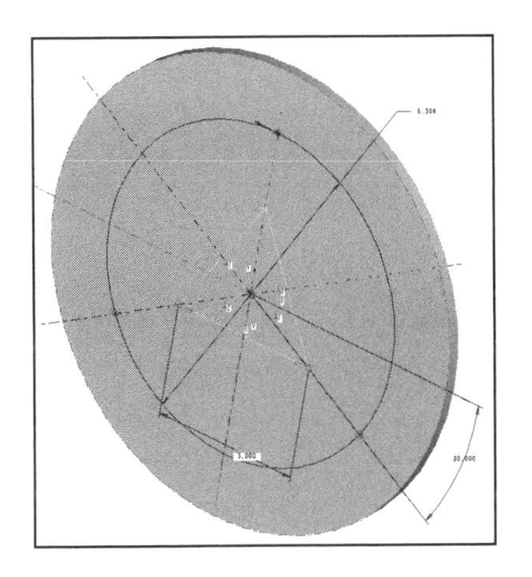

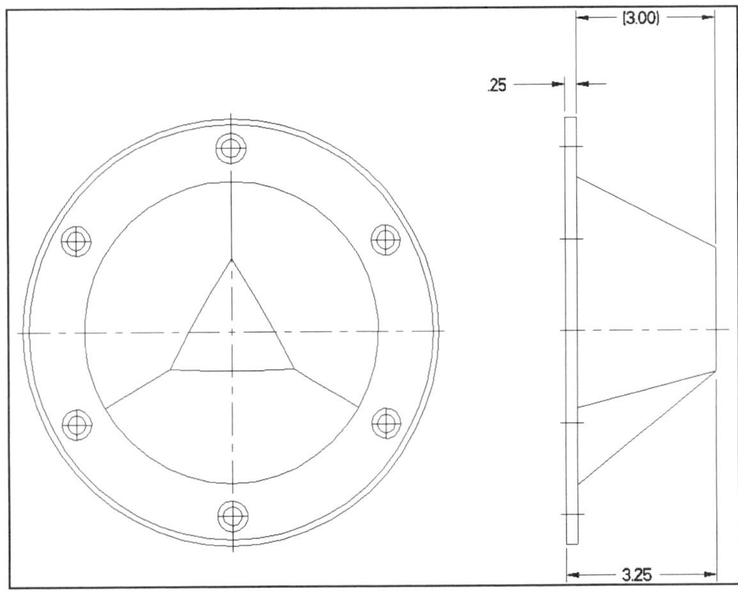

Figure 12.16
Cap, Left Side View, and Back View

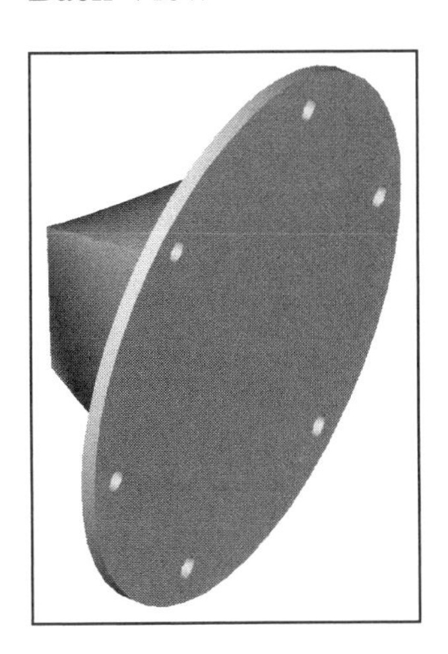

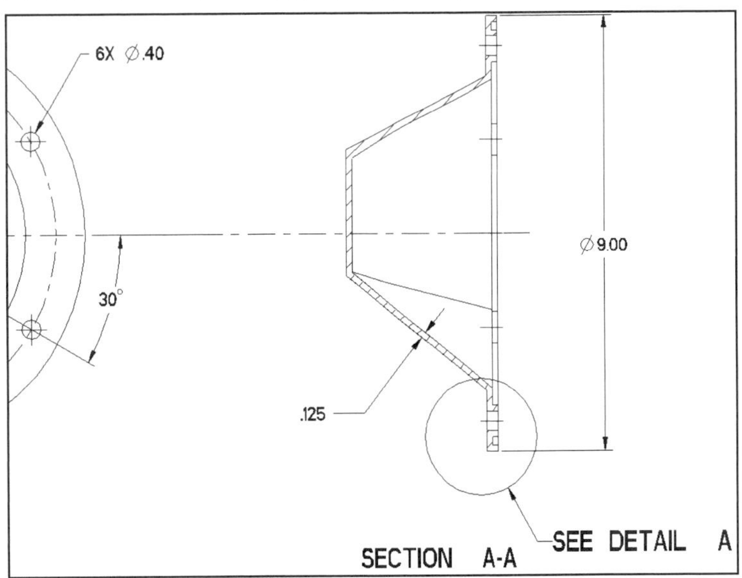

Figure 12.17
Cap, **SECTION A-A**

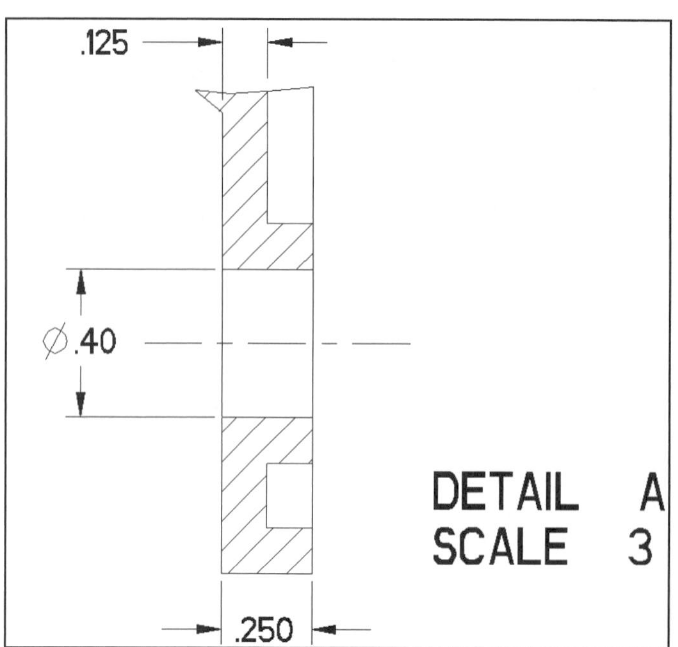

Figure 12.18
Cap, **DETAIL A**

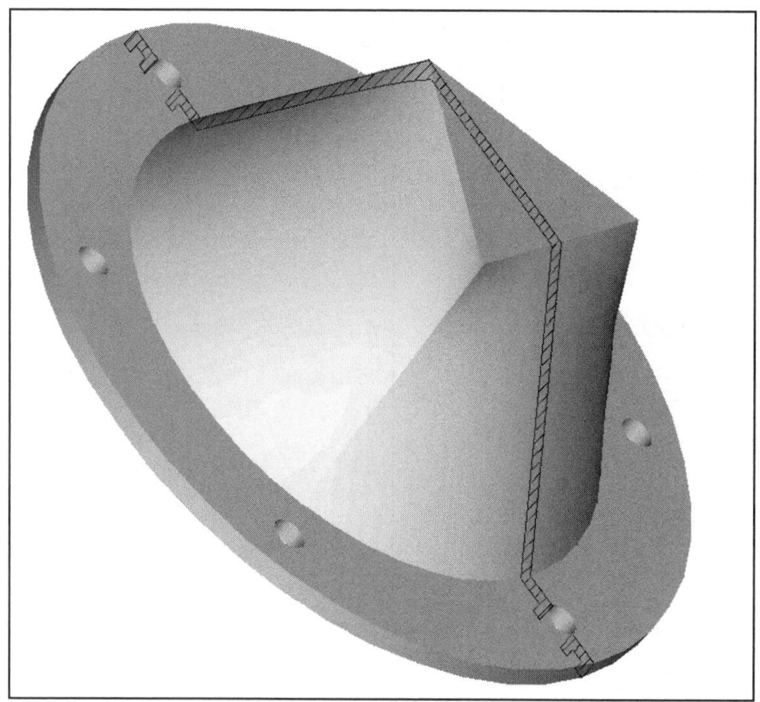

Figure 12.19
Cap, Part Section

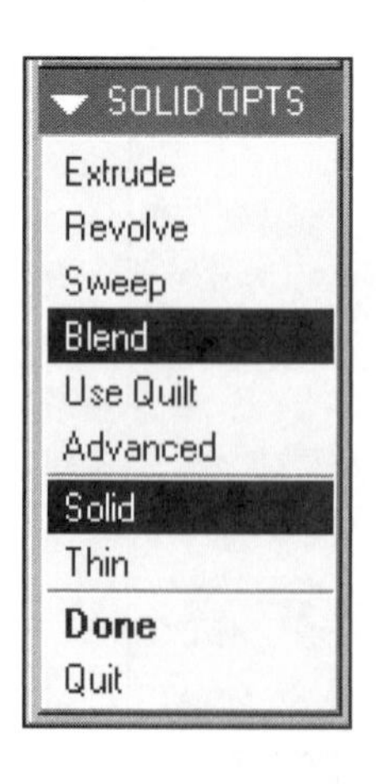

Start the cap by modeling the circular protrusion that is **Ø9.00** by **.25** thick shown in Figure 12.20. Sketch the first protrusion on datum **A (FRONT)** and centered on **B (RIGHT)** and **C (TOP).** The **Blend** feature is modeled next (Fig. 12.21). Choose the following commands:

Feature ⇒ **Create** ⇒ **Solid** ⇒ **Protrusion** ⇒ **Blend** ⇒ **Solid** ⇒ **Done** ⇒ **Parallel** ⇒ **Regular Sec** ⇒ **Sketch Sec** ⇒ **Done** ⇒ **Straight** ⇒ **Done** ⇒ [pick the top surface of the first protrusion (Fig. 12.22)] ⇒ **Okay** (to confirm direction of feature creation) ⇒ **Top** ⇒ (choose datum **C** as the orientation plane) ⇒ **Close** (References dialog box) ⇒ **Sketch** ⇒ **Intent Manager** (off)

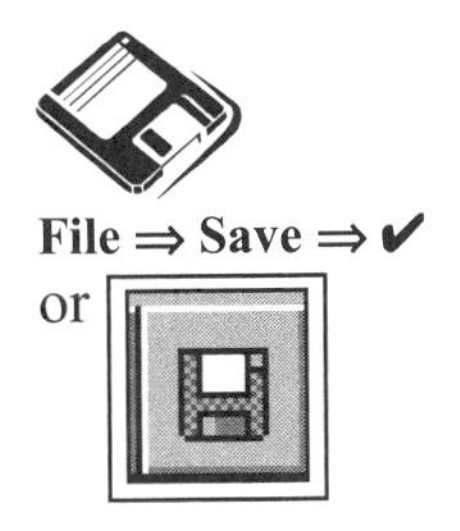

File ⇒ Save ⇒ ✔
or

Figure 12.20
First Protrusion

Menu Manager
ATTRIBUTES
Straight
Smooth
Done
Quit

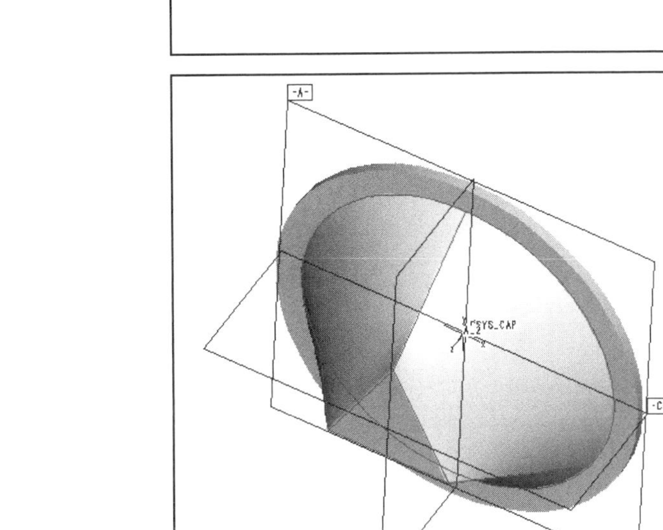

Figure 12.21
Blend Feature

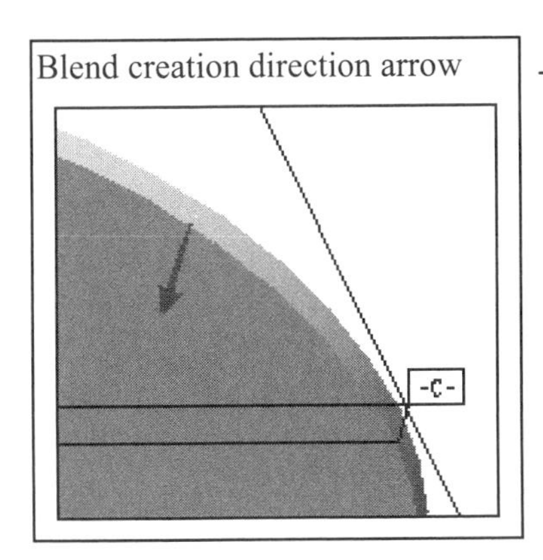

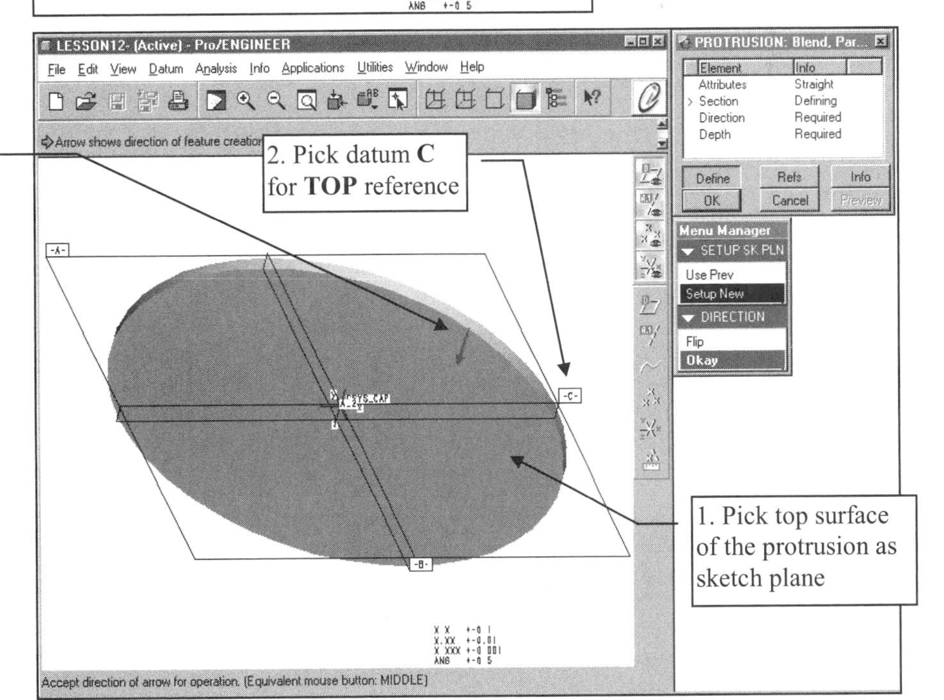

Figure 12.22
Blend Feature Starting Surface and Direction of Creation

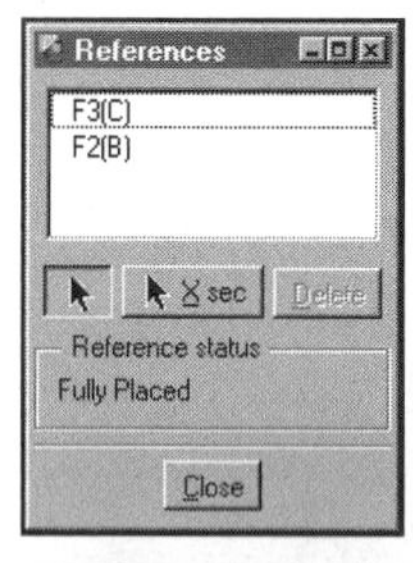

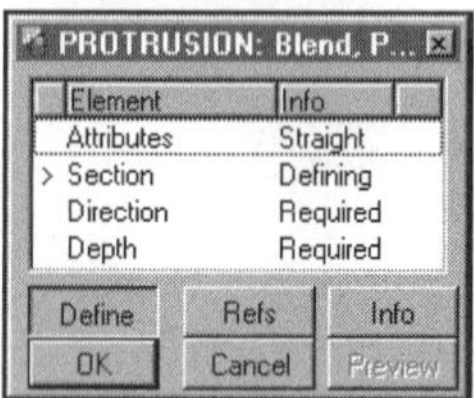

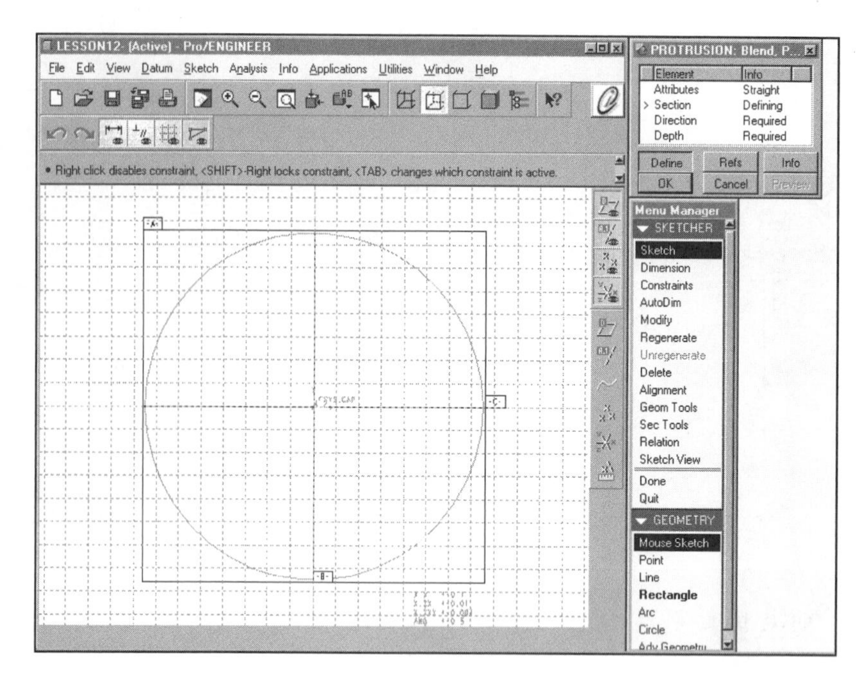

Figure 12.23
Sketcher Showing Cartesian Grid

NOTE

See page L12-16 and Sections 11 and 12 for information on setting preferences while the Intent Manager is activated.

The section grid would be better utilized if it were a **Polar** grid rather than a **Cartesian** grid (Fig. 12.23). Change the grid type and size, and continue with the blend commands:

Sec Tools ⇒ **Sec Environ** ⇒ **Grid** ⇒ **Type** ⇒ **Polar** ⇒ **Done/Return** ⇒ **Params** ⇒ **Ang Spacing** ⇒ ✔ (to accept **30°** default) ⇒ **Rad Spacing** ⇒ [type **.50** at the prompt (Fig. 12.24)] ⇒ ✔ ⇒ **Done/Return**

Sketch ⇒ **Arc** ⇒ **Center/Ends** ⇒ [sketch the first section of the blend by creating three equal **120°** arcs, sketch each arc in a counter-clockwise direction (Fig. 12.25)] ⇒ **Regenerate** ⇒ **Alignment** (select each arc and the default coordinate system) ⇒ **Regenerate** ⇒ **Dimension** (add the diameter dimension) ⇒ **Regenerate**

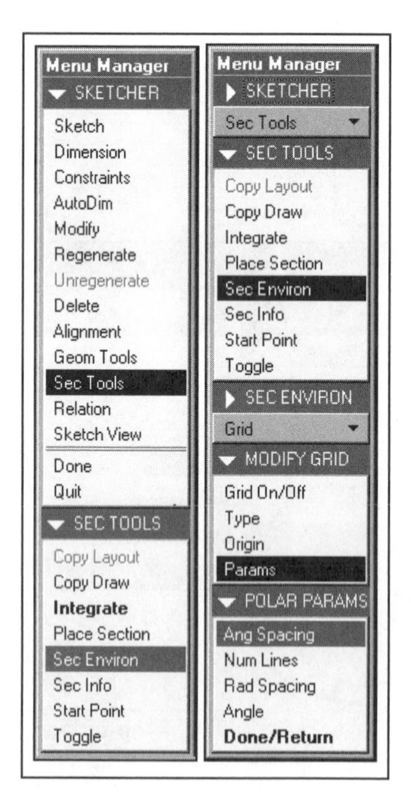

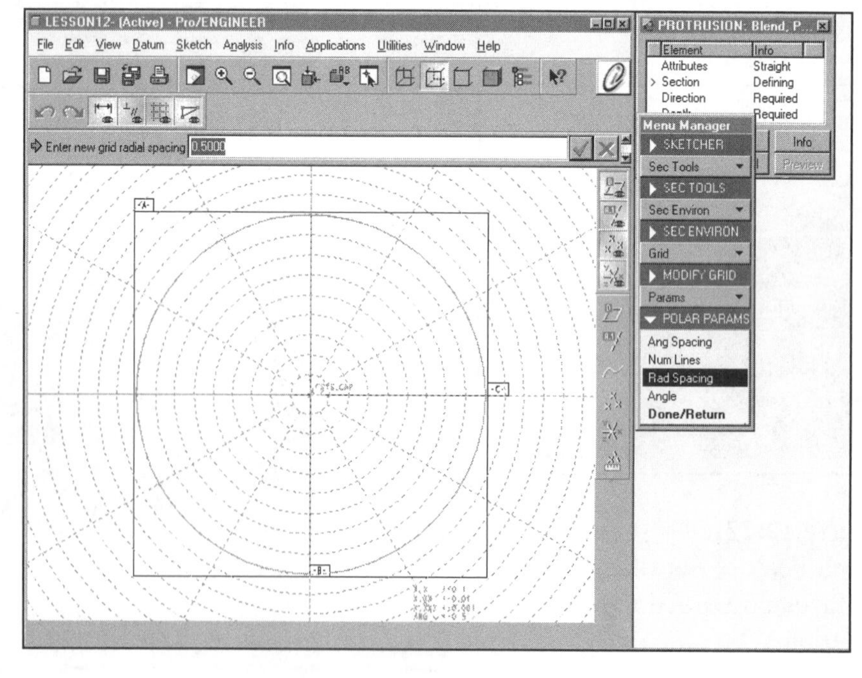

Figure 12.24
Sketcher Showing Polar Grid

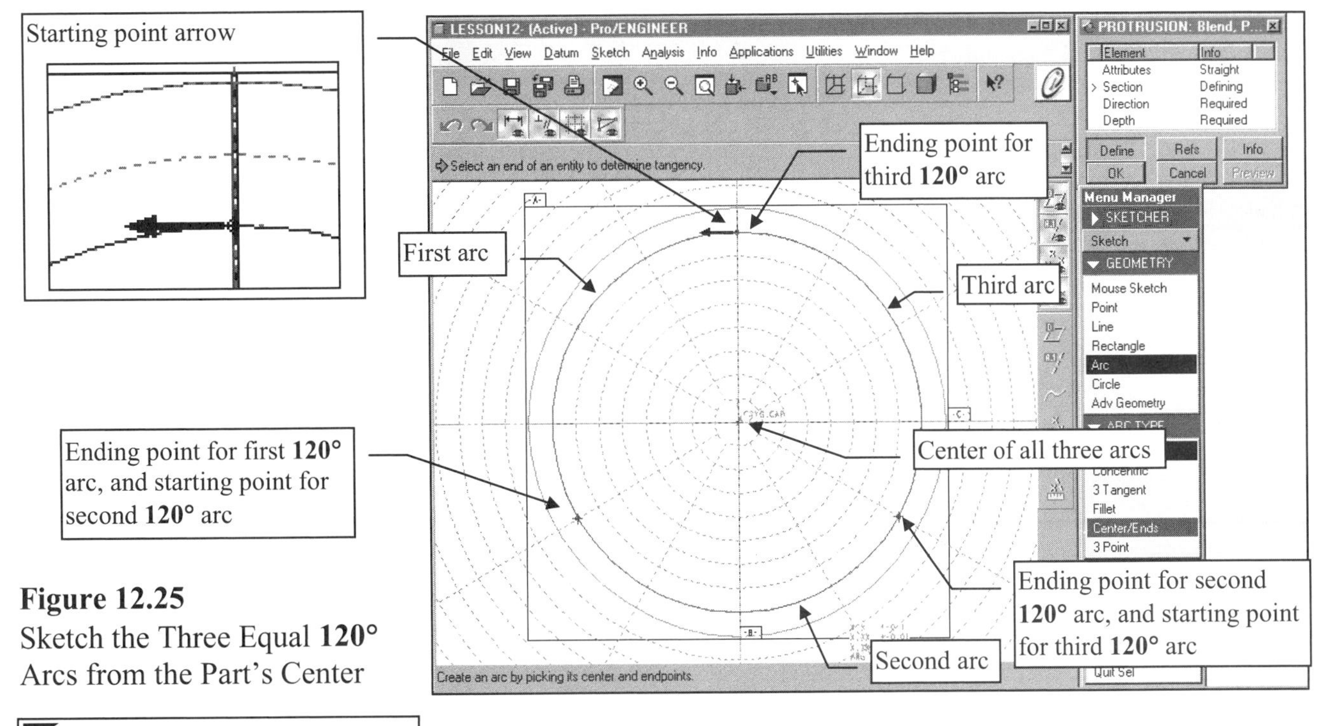

Figure 12.25
Sketch the Three Equal **120°** Arcs from the Part's Center

NOTE

As you toggle between sections, the active section shows in *cyan* (light blue), and the inactive section is *grayed.* You can toggle both directions. Remember that each section is separate and needs its own unique dimensioning scheme, as required by the design intent. *Each section must have the same total number of entity segments.* Any number of subsections can be created, depending on the design requirements. Here, only two subsections are incorporated into the design, and each subsection has three segments: three arcs in one subsection and three lines in the other.

Sketch ⇒ **Line** ⇒ **Centerline** (Add the two centerlines to locate the arcs' ends) ⇒ **Dimension** (dimension the centerlines from the vertical datum plane with **120°** angles) ⇒ **Regenerate** (Fig. 12.26) ⇒ **Sec Tools** ⇒ **Toggle** [to sketch the second parallel section (the first section is *grayed out*] ⇒ **Sketch** ⇒ **Line** ⇒ [sketch the three lines of the triangle starting at the top of datum **B** and picking points *in the same direction* in which the arcs were created (Fig. 12.27)] ⇒ **Alignment** (align the two angled lines to datum **B**)

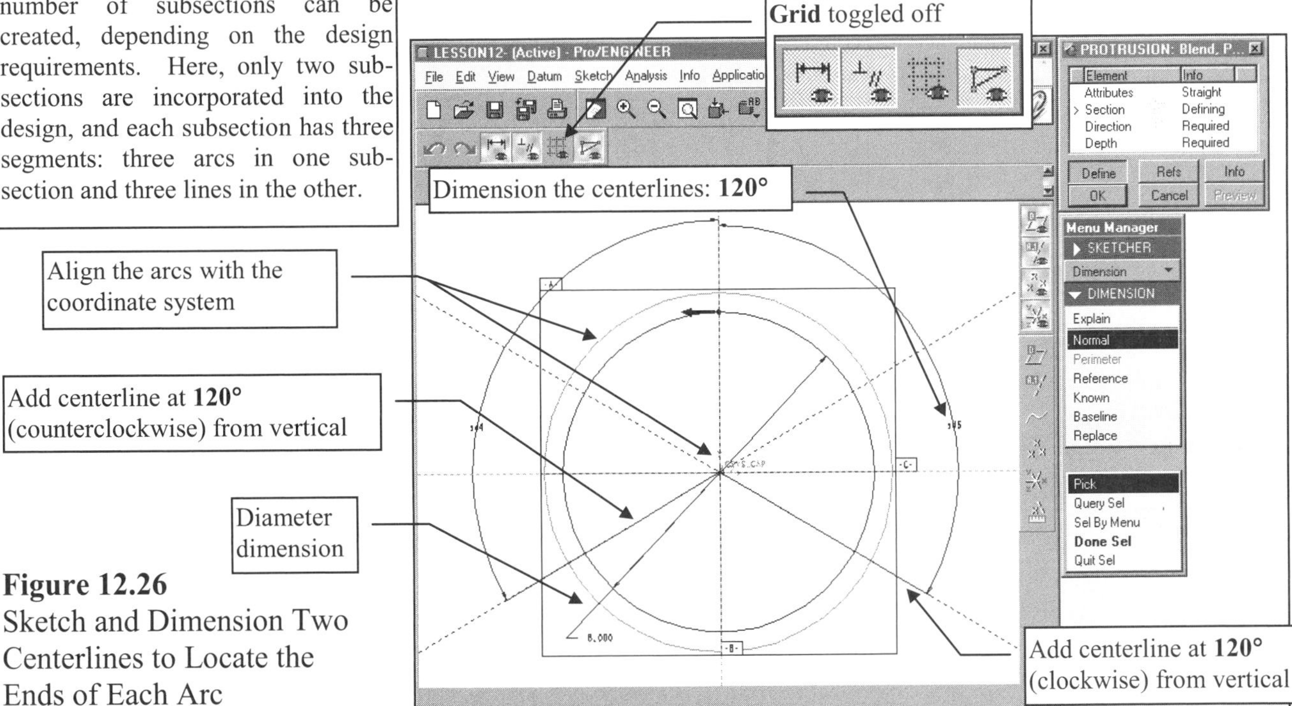

Figure 12.26
Sketch and Dimension Two Centerlines to Locate the Ends of Each Arc

Start at the *vertical position* and create the section in the *same direction* as the arcs. Make the triangle subsection *smaller* than the arc subsection. The two start arrows must point in the same direction.

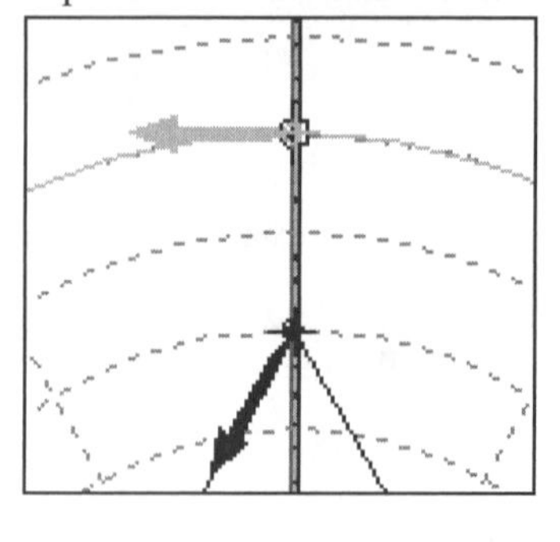

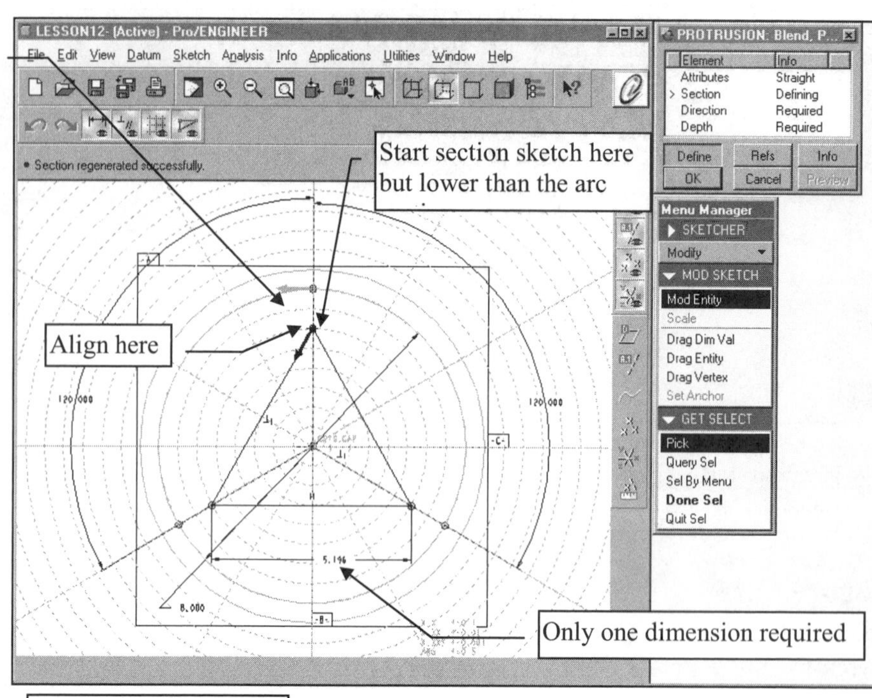

Figure 12.27
Sketch and Dimension the Three Lines

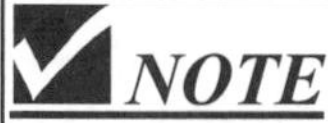

NOTE

You have been working without the **Intent Manager** on. If the Intent Manager was activated, you can change the **Grid** type by picking **Utilities ⇒ Sketcher Preferences ⇒ Parameters** tab ⇒ Grid ● **Polar**

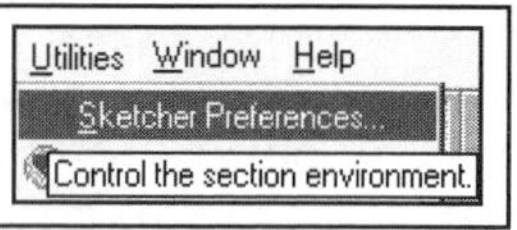

Regenerate ⇒ Dimension ⇒ Regenerate (Fig. 12.27) ⇒ **Modify** (change the diameter dimension to **7.75**, the leg of the triangle dimension to **3.00,** and the two angles to **120°**) ⇒ ✔ ⇒ **Regenerate** (Fig. 12.28) ⇒ **Done ⇒ View ⇒ Default ⇒ Blind ⇒ Done ⇒** (type **3.00** at the prompt) ⇒ ✔ ⇒ **Preview** (Fig. 12.29) ⇒ **OK ⇒ Done**

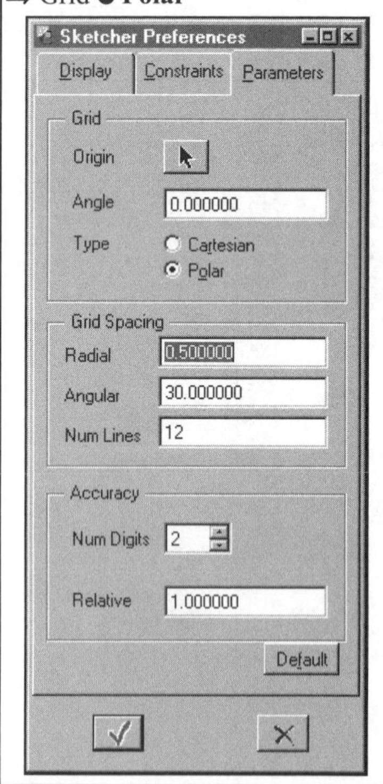

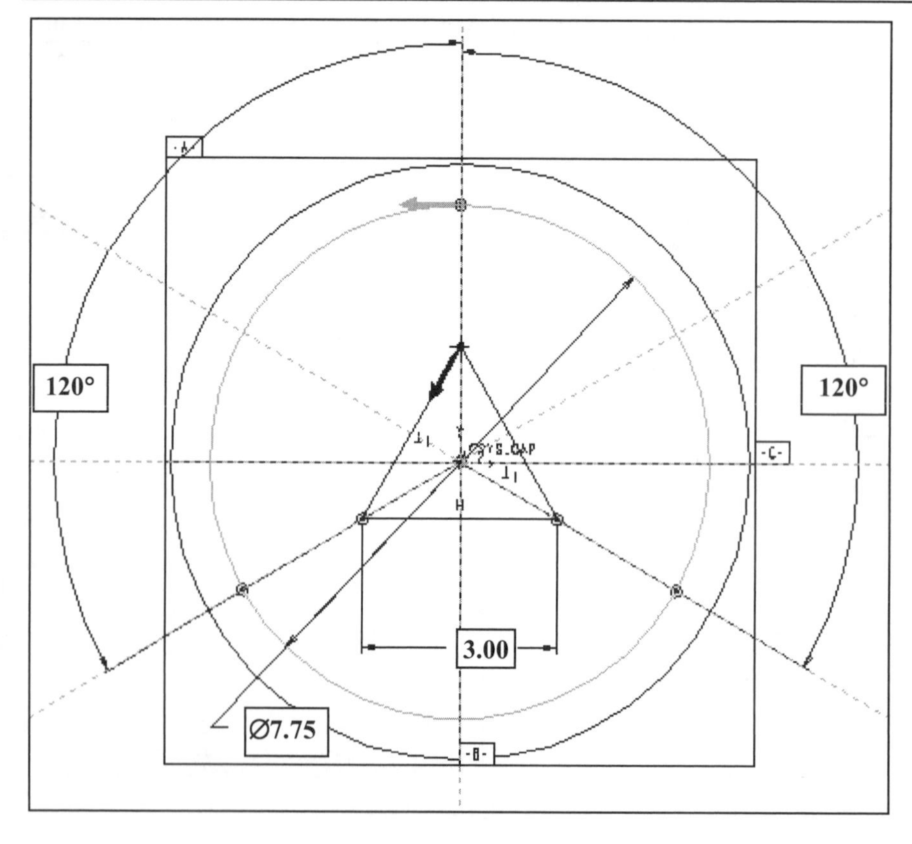

Figure 12.28
Modified and Regenerated Sketch

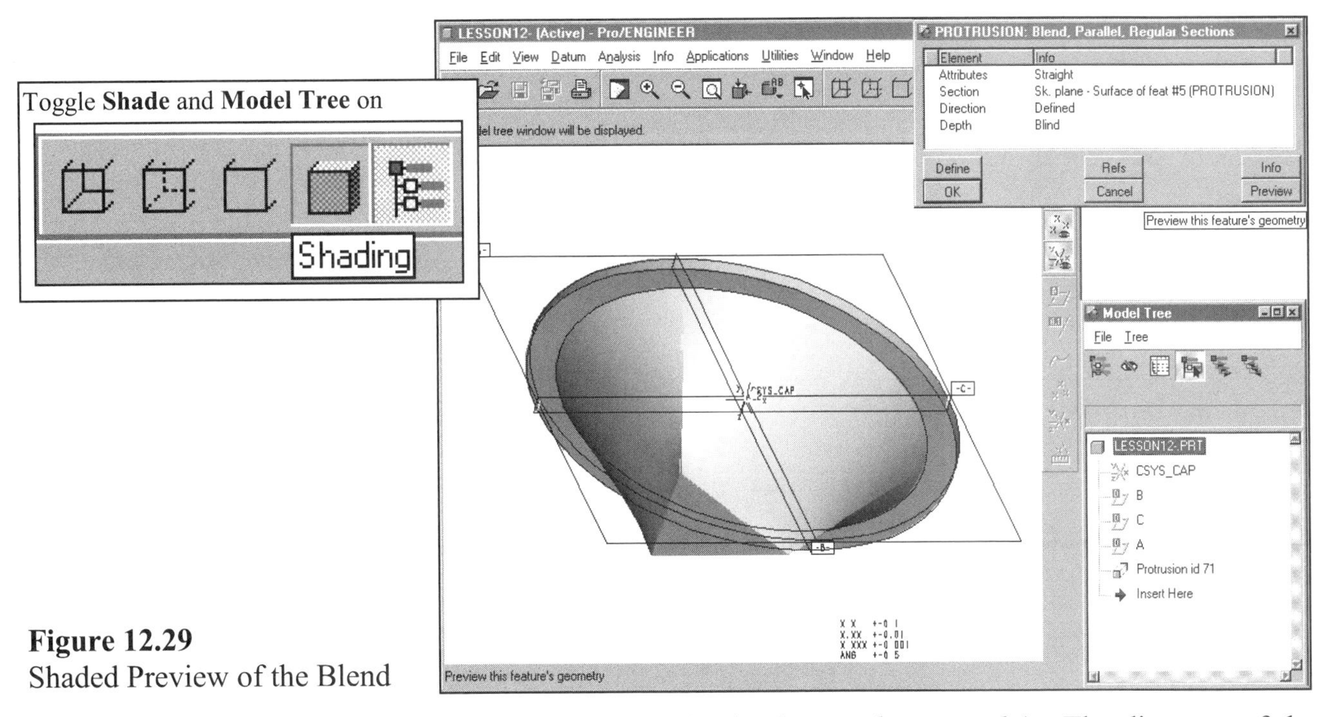

Figure 12.29
Shaded Preview of the Blend

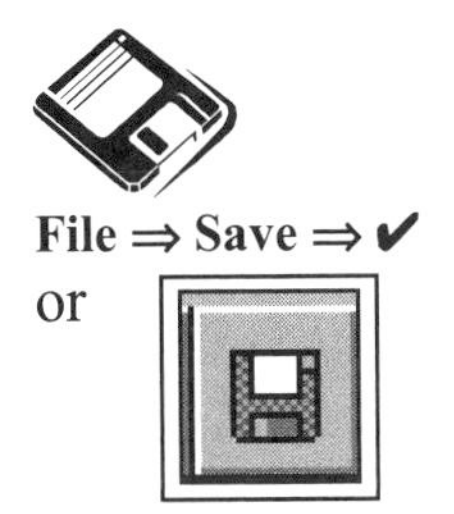

File ⇒ Save ⇒ ✔
or

You made a mistake (*we made a mistake*). The diameter of the blend was supposed to be **6.50,** not **7.75**, which is the diameter of the bolt circle.

Modify ⇒ (pick the blend protrusion from the Model Tree, as shown in Fig. 12.30) ⇒ (pick the **7.75** dimension) ⇒ (change the value to **6.50**; see Fig. 12.15) ⇒ ✔ ⇒ **Regenerate** (Fig. 12.31)

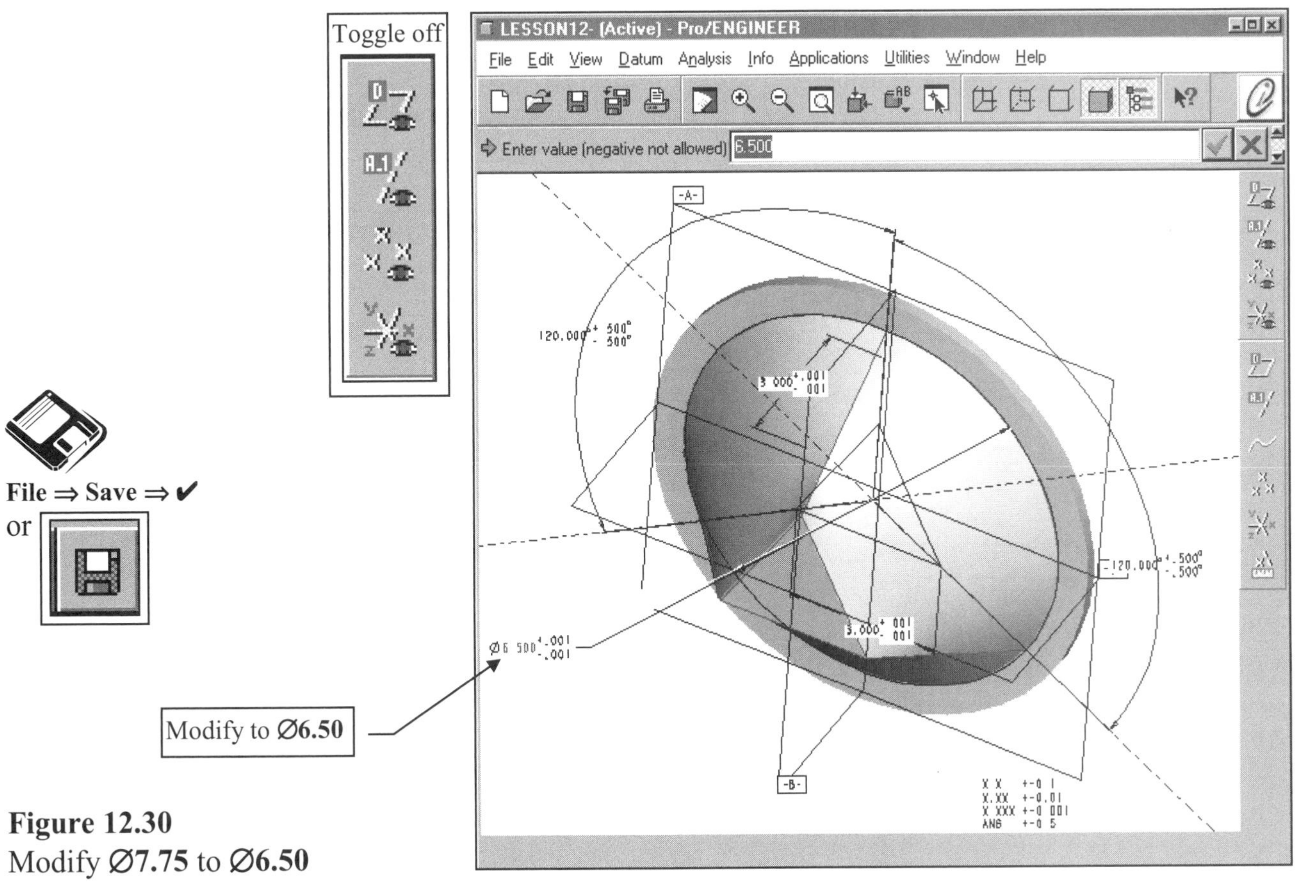

File ⇒ Save ⇒ ✔
or

Figure 12.30
Modify **Ø7.75** to **Ø6.50**

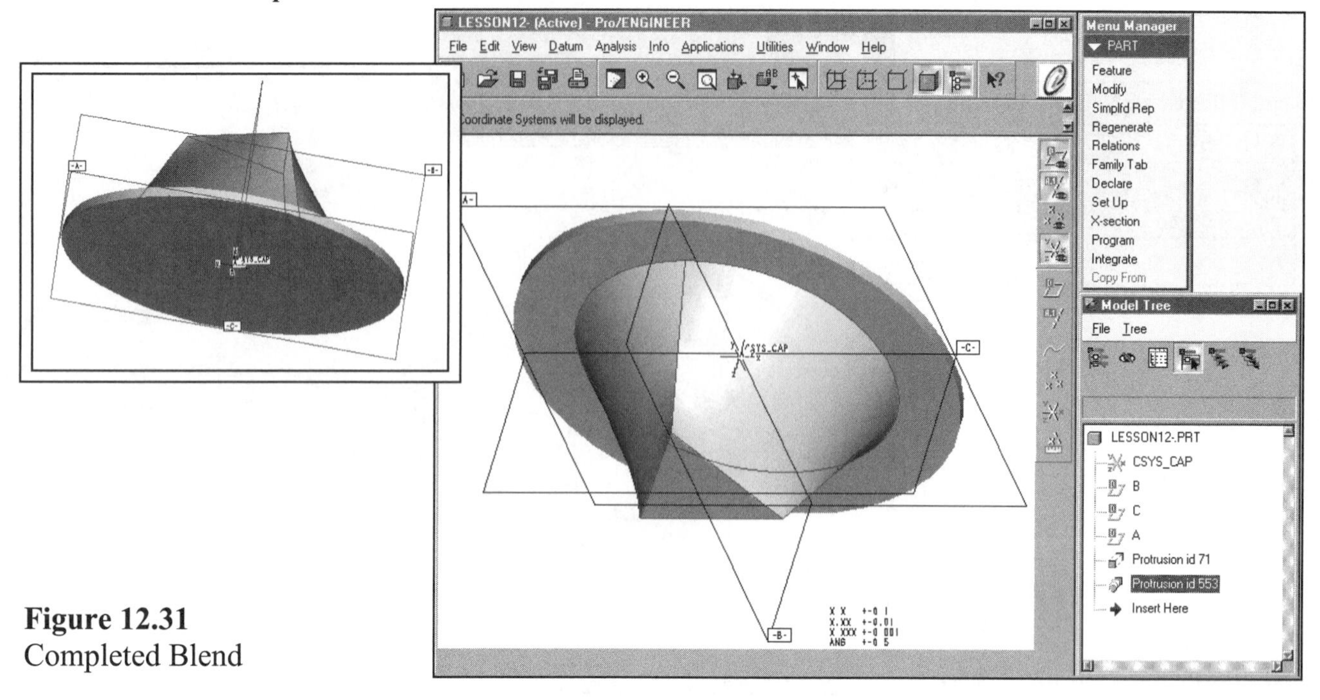

Figure 12.31
Completed Blend

Create and pattern the holes as shown in Figure 12.32. Next, **Shell** the part using the following commands:

Feature ⇒ **Create** ⇒ **Shell** ⇒ (pick the bottom surface of the part as the surface to remove, as shown in Fig. 12.33) ⇒ **Done Sel** ⇒ **Done Refs** ⇒ (type a thickness of **.125** at the prompt) ⇒ ✔ ⇒ **OK** (Fig. 12.34) ⇒ **Done**

Before shelling, create and pattern the six holes **(∅.400)** with a **∅7.75** bolt circle. Use the **Diameter** option so that the hole can be patterned. The **Diameter** option will create a diameter dimension that can be displayed on the drawing of the part.

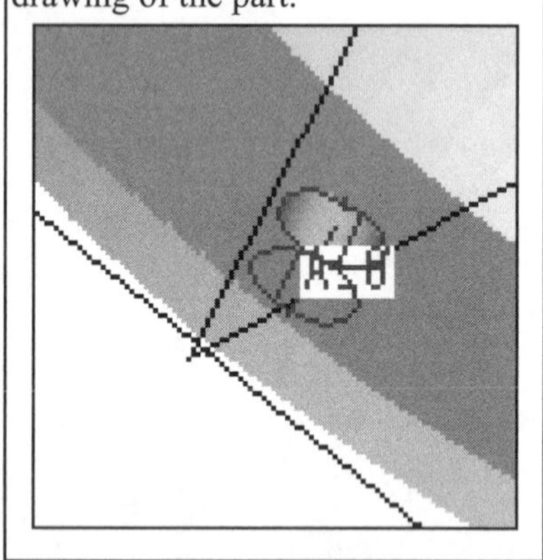

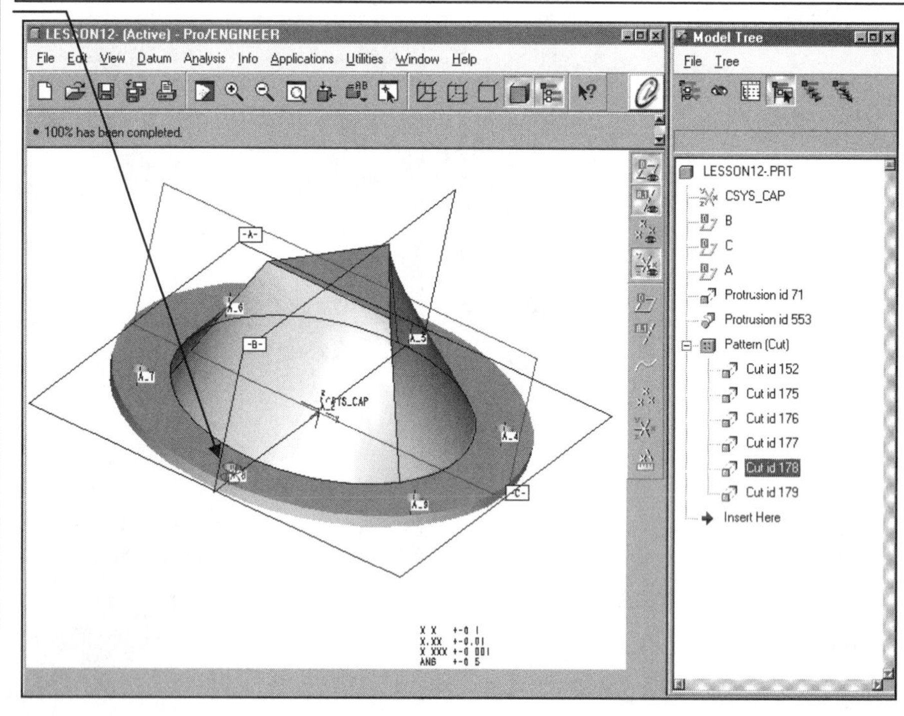

Figure 12.32
Create the Hole Pattern

The **Shell** will automatically create the bosses (Figs. 12.34 and 12.35), because the **.125** thickness is left around all previously created features. If the bosses were not desired, you would simply **Reorder** the holes to come after the shell feature.

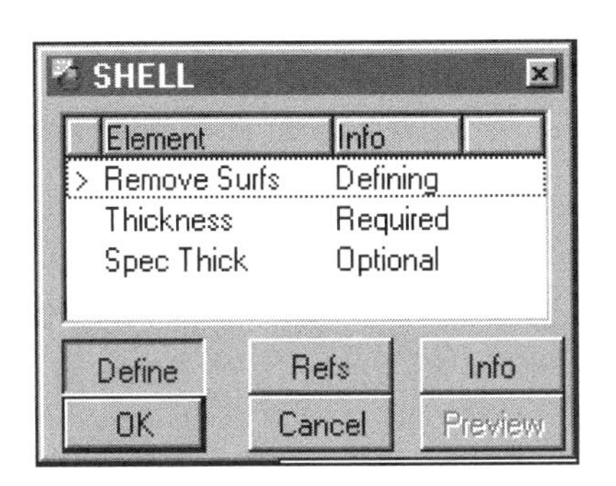

Pick the bottom surface to remove

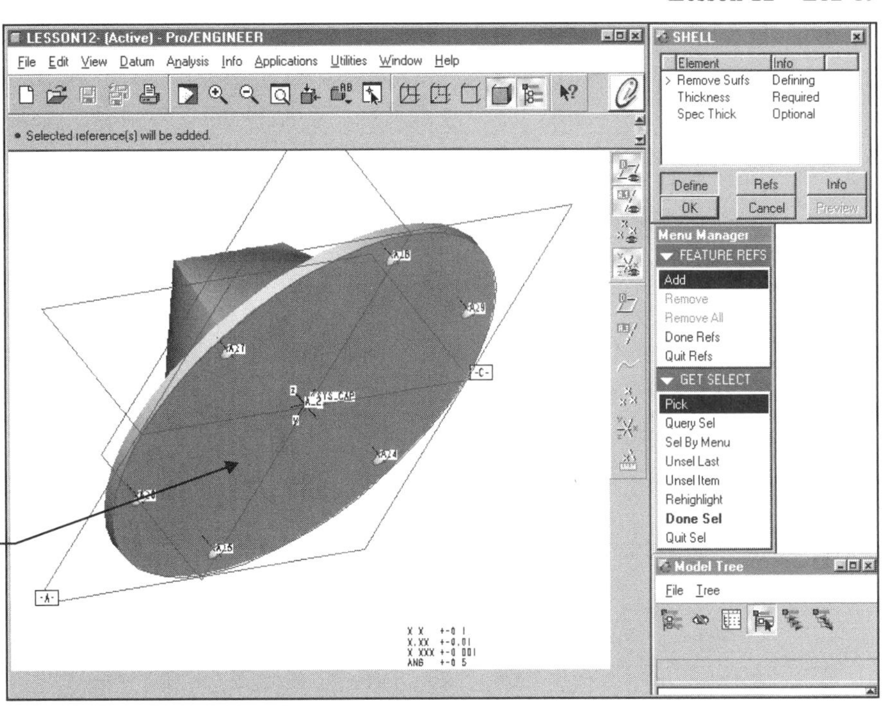

Figure 12.33
Shell the Part

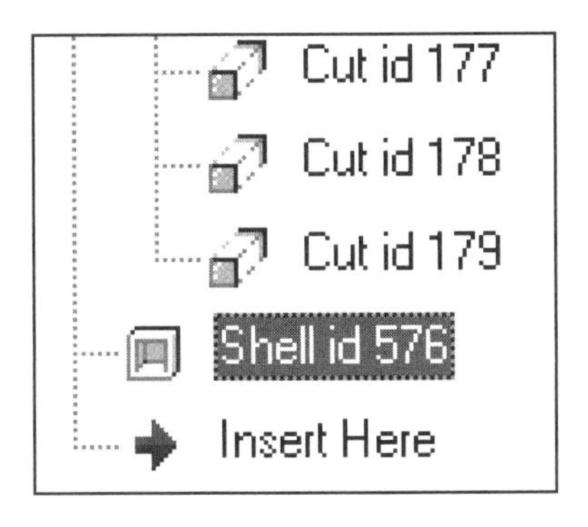

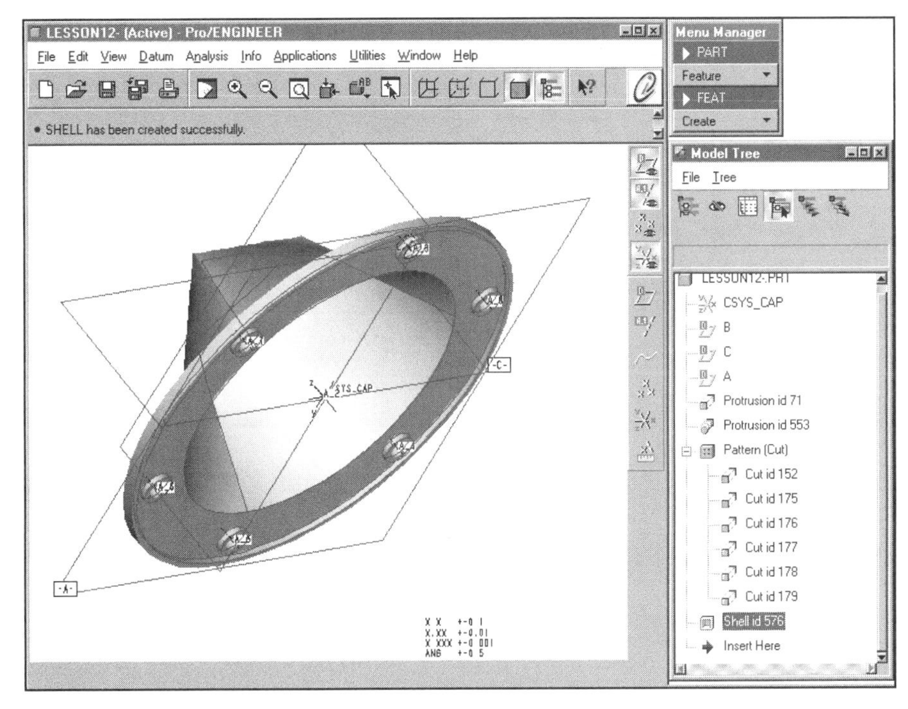

Figure 12.34
Shelled Part

NOTE
The *bosses* around the holes are created automatically at **.250** larger than the holes:
(**.125 + .125 + .400 = Ø.650**)

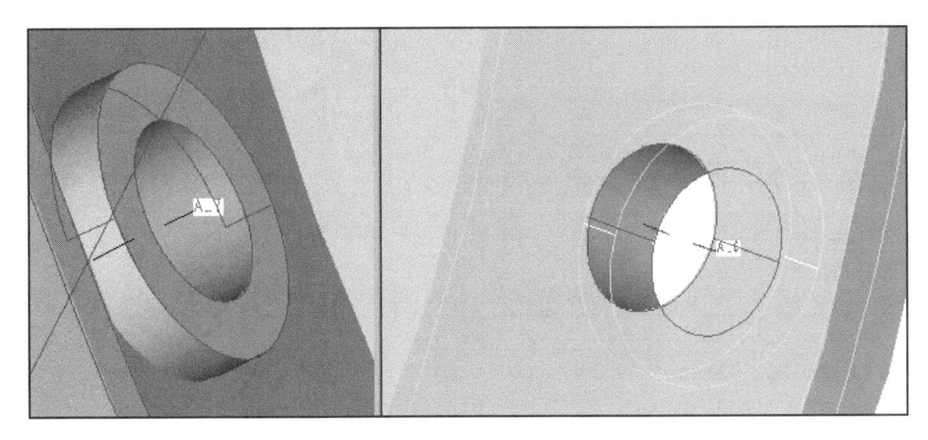

Figure 12.35
Hole and Boss Shown from Both Sides

Measure the size of the boss:

Analysis ⇒ **Measure** ⇒ (pick the arrow to open the **Type** drop down selections) ⇒ **Diameter** ⇒ (pick the boss as shown in Fig. 12.36, and observe the Message area) ⇒ **Close**

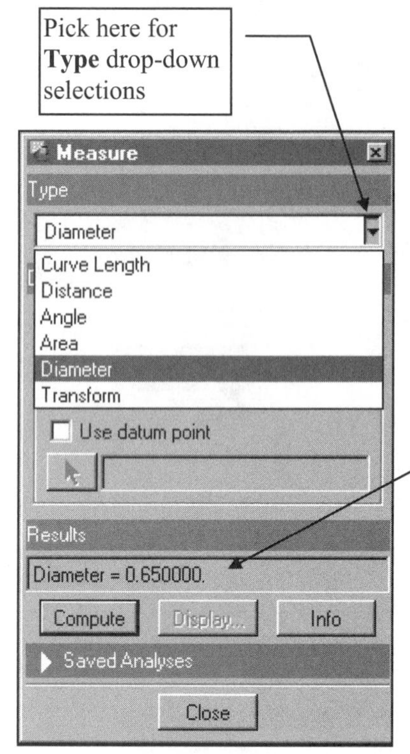

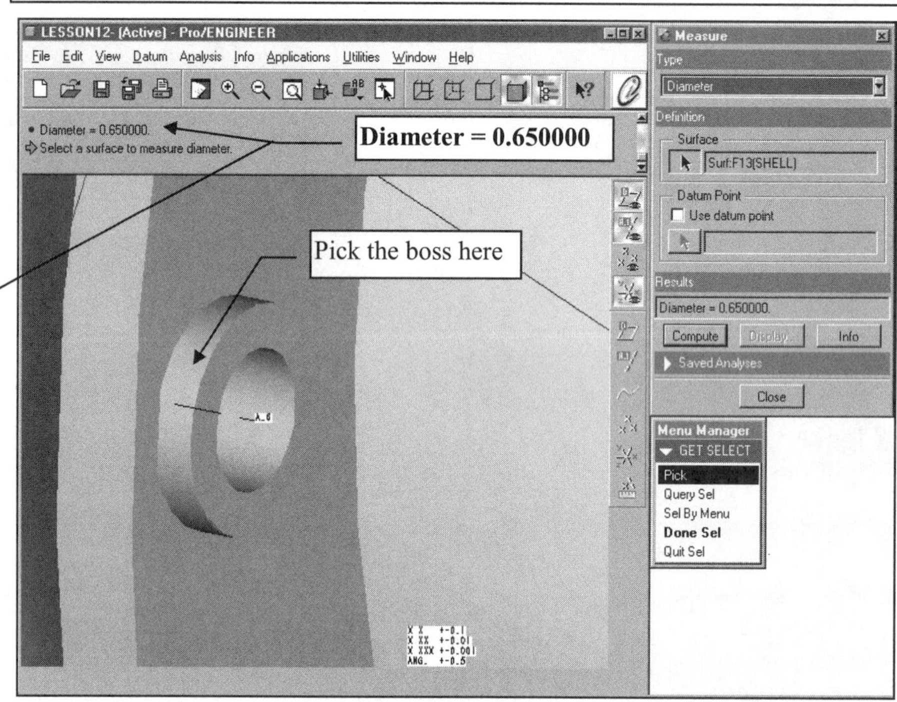

Figure 12.36
Measuring the Boss

Now create a cross section (Fig. 12.37) to be used in the Drawing mode when detailing the cap. Choose the following commands:

X-section (from the PART menu) ⇒ **Create** ⇒ **Planar** ⇒ **Single** ⇒ **Done** ⇒ (type **A** at the prompt) ⇒ ✔ ⇒ (select planar surface or datum plane; pick datum **B** from the Model Tree) ⇒ **Done/Return**

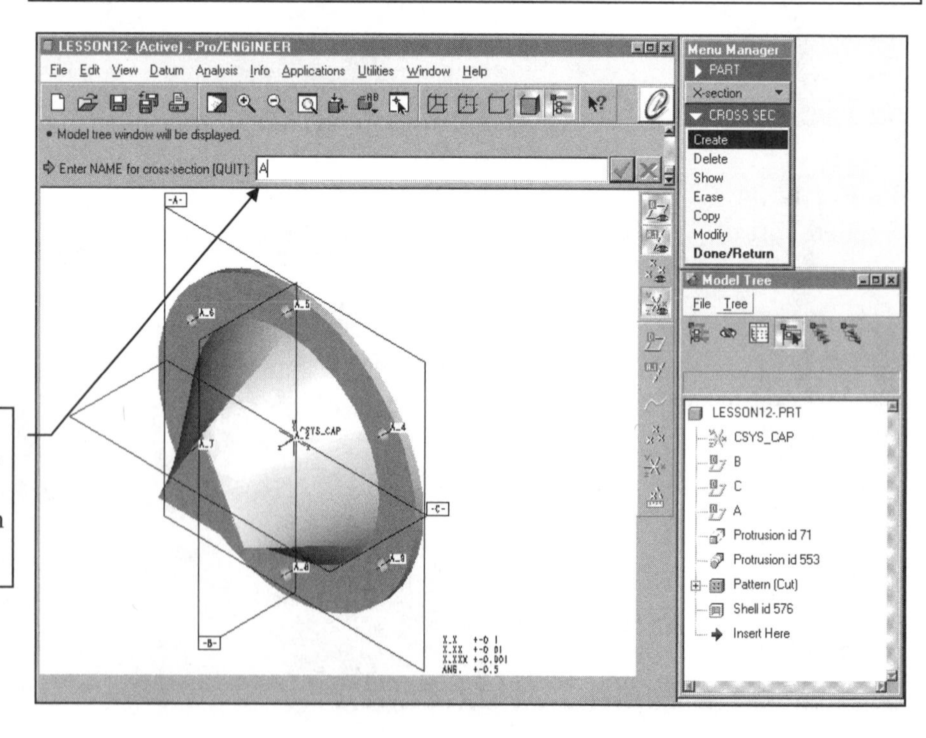

Type only *one letter* to establish a section name. Typing **A** will create a section called **SECTION A-A**. Typing *two* **A**'s will create a section called **SECTION AA-AA!**

Figure 12.37
Creating a **X-section**

Save the cap under a new name, and complete the **ECO** (Figures 12.38 and 12.39) to establish a second part with different dimensional sizes and features.

ECO

File ⇒ Save ⇒ ✔
File ⇒ Delete ⇒
Old Versions ⇒ ✔
File ⇒ Save As ⇒ (type **CAP_PART_ONE) ⇒ OK**

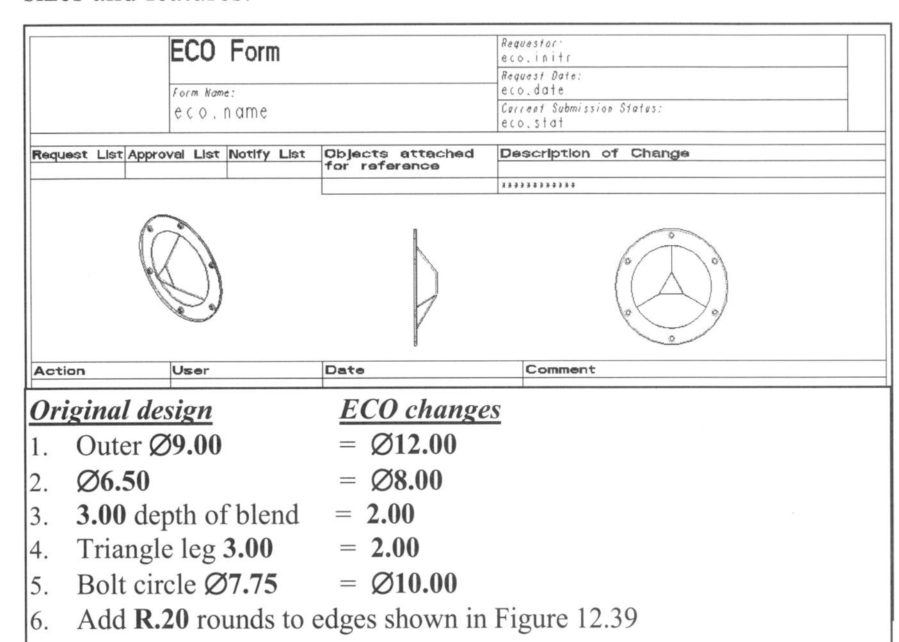

ECO Form

Form Name: eco.name

Requestor: eco.initr

Request Date: eco.date

Current Submission Status: eco.stat

Request List	Approval List	Notify List	Objects attached for reference	Description of Change

Action	User	Date	Comment

Original design	*ECO changes*
1. Outer **Ø9.00**	= **Ø12.00**
2. **Ø6.50**	= **Ø8.00**
3. **3.00** depth of blend	= **2.00**
4. Triangle leg **3.00**	= **2.00**
5. Bolt circle **Ø7.75**	= **Ø10.00**
6. Add **R.20** rounds to edges shown in Figure 12.39	

Figure 12.38
ECO

Figure 12.39
ECO Changes to Model

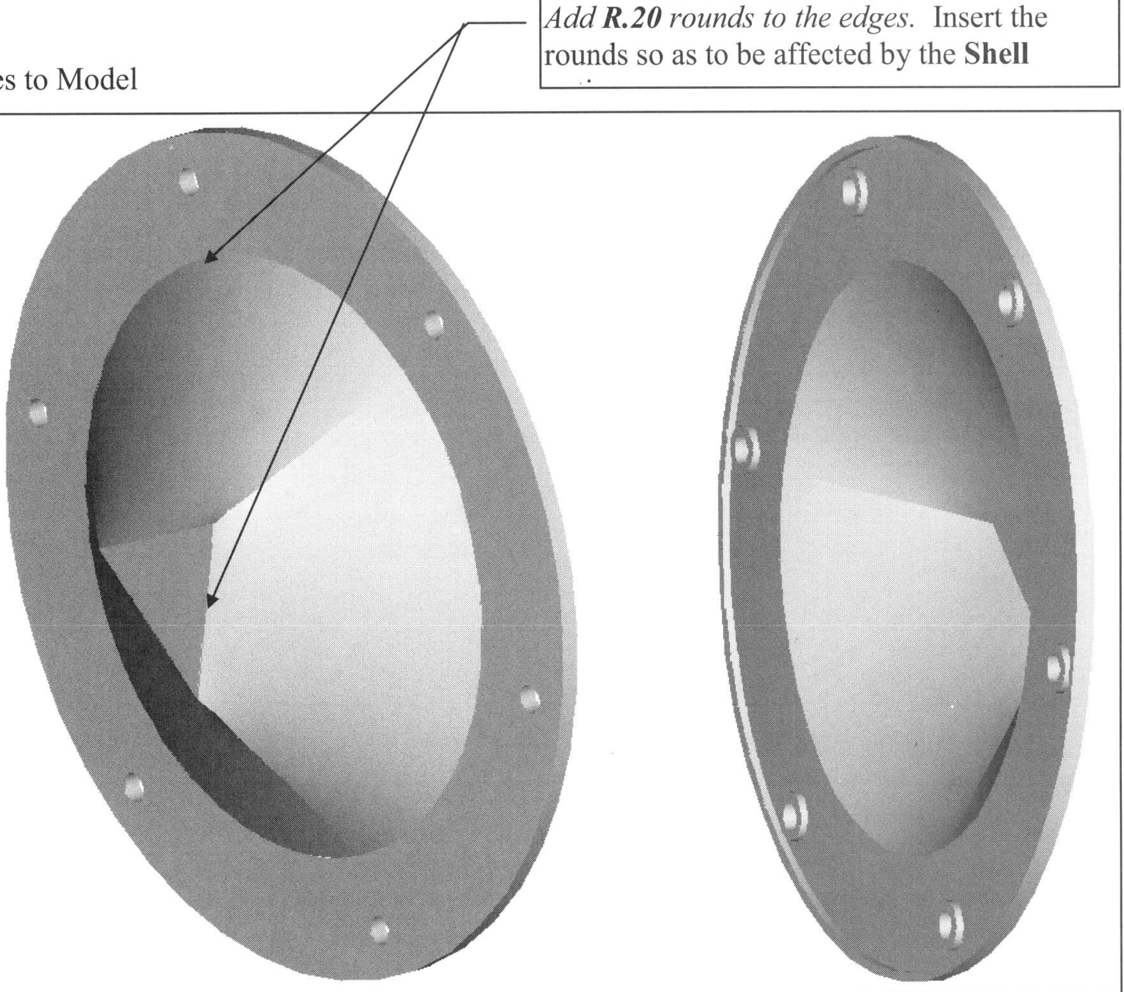

Lesson 12 Project

Bathroom Faucet

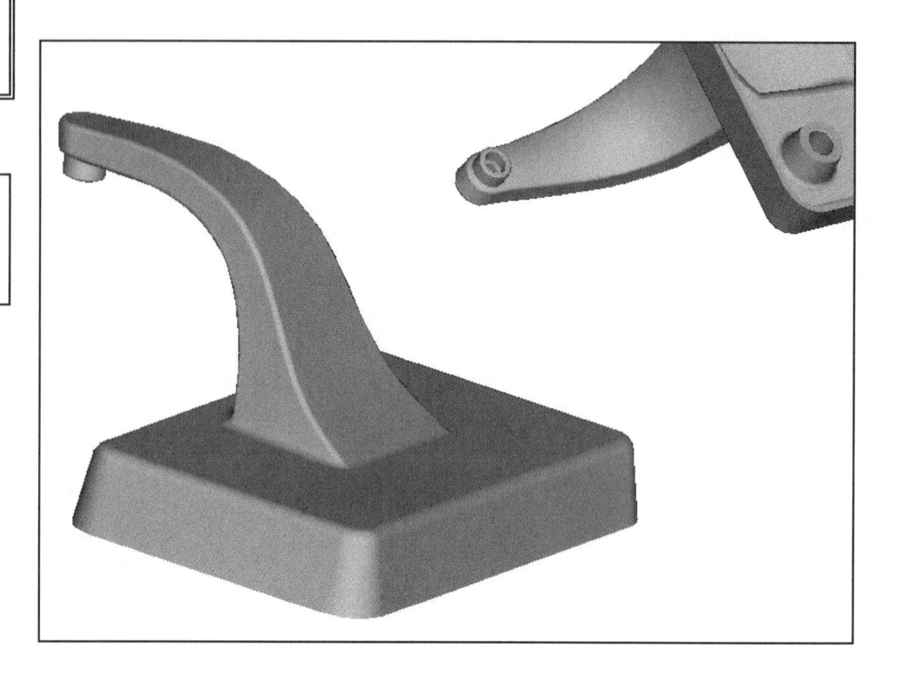

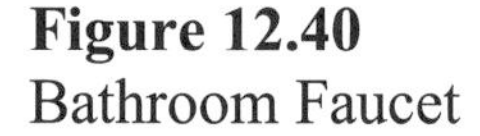

Figure 12.40
Bathroom Faucet

Bathroom Faucet

This is an advanced **lesson project.** Because you have created over 20 parts, you should be able to use that knowledge to model the **Swept Blend** required to create the **Bathroom Faucet** (Figs. 12.40 through 12.82). Some instructions accompany this lesson project, but you will be required to research online documentation **(Pro/HELP)** and learn about **Splines** and **Swept Blends.**

After the model is complete, create a number of sections that can be used in Drawing mode when you are detailing the Faucet.

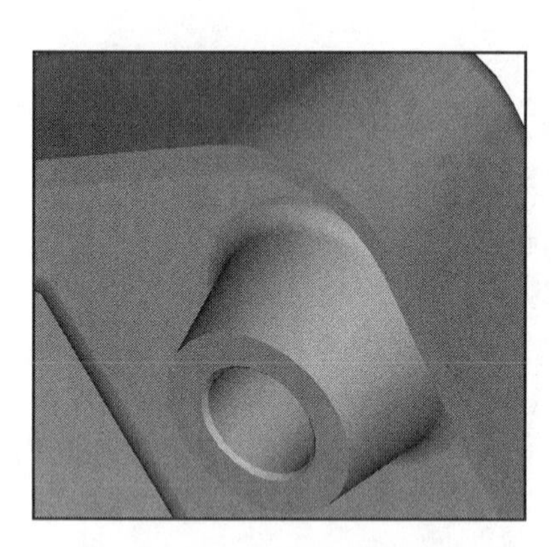

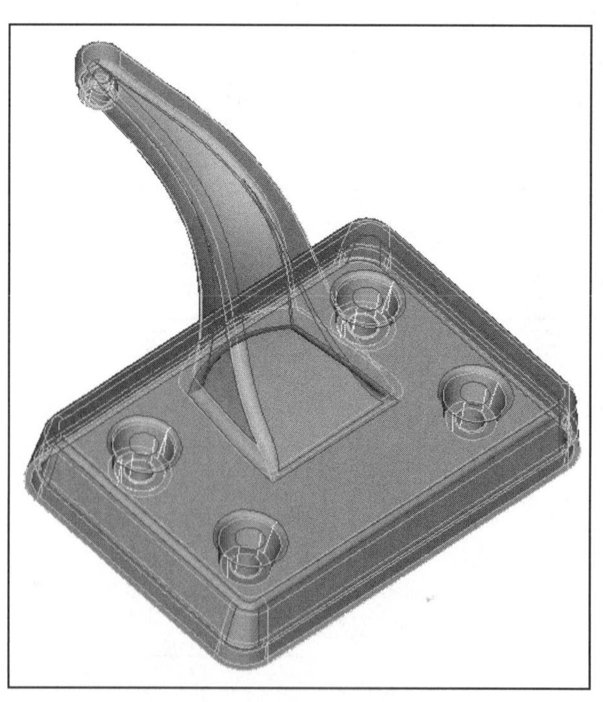

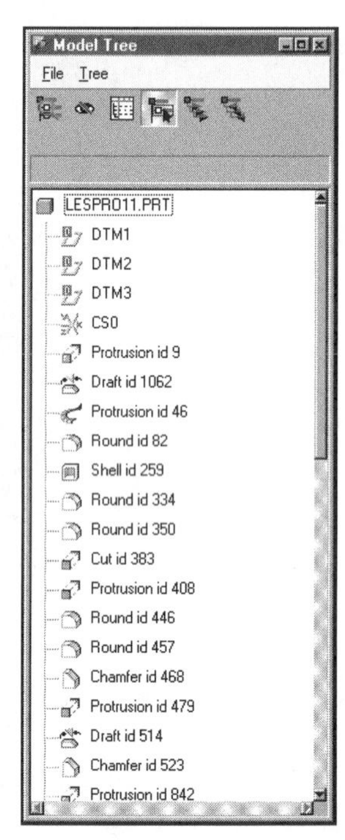

Figure 12.41
Bathroom Faucet with Model Tree

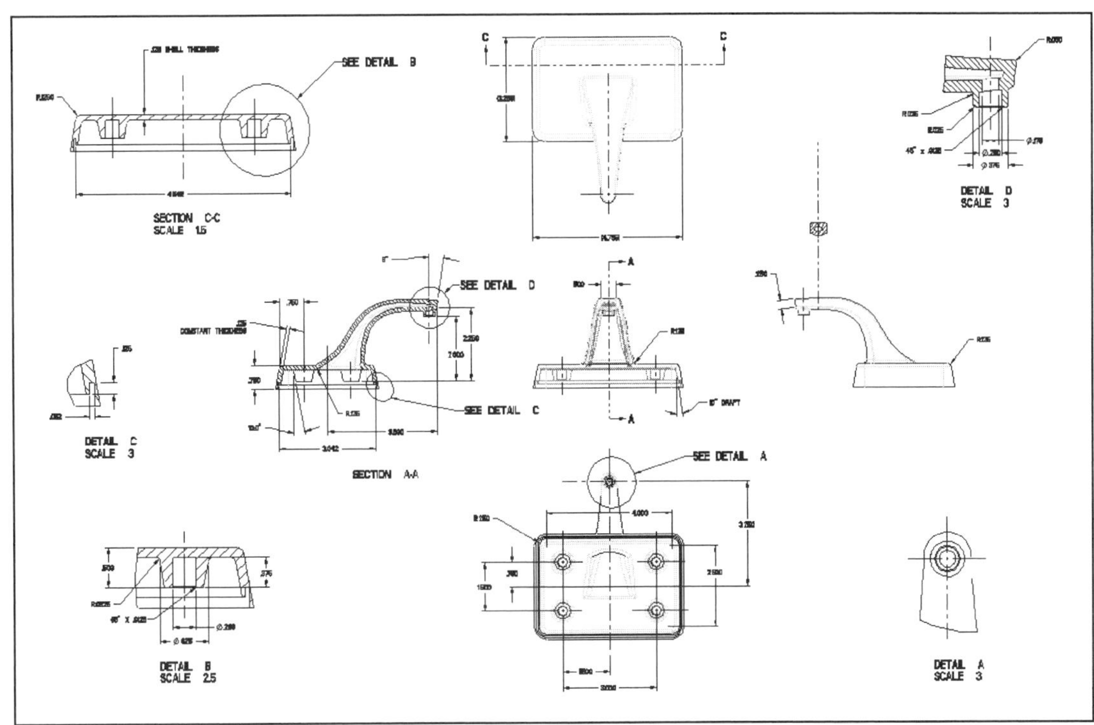

Figure 12.42
Bathroom Faucet,
Detail Drawing

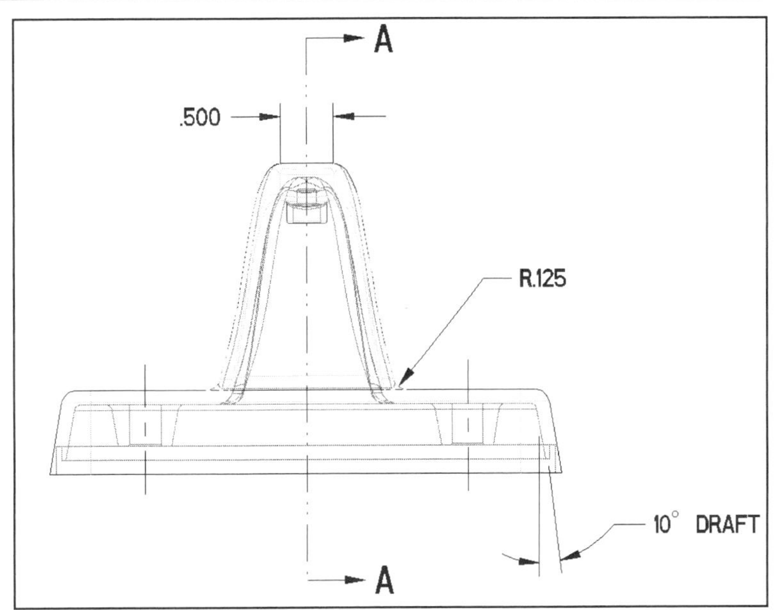

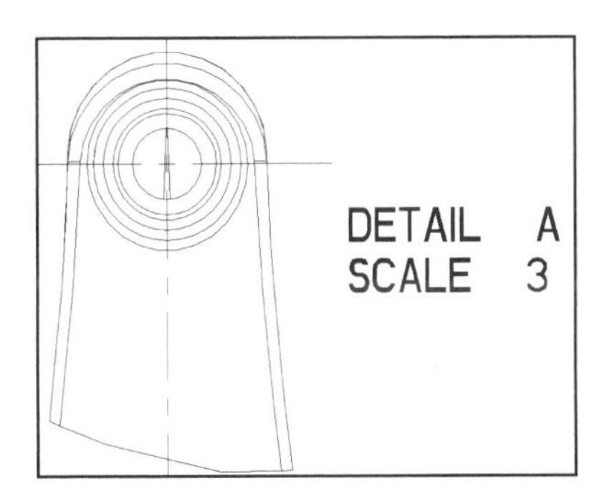

Figure 12.43
Bathroom Faucet Drawing,
Front View

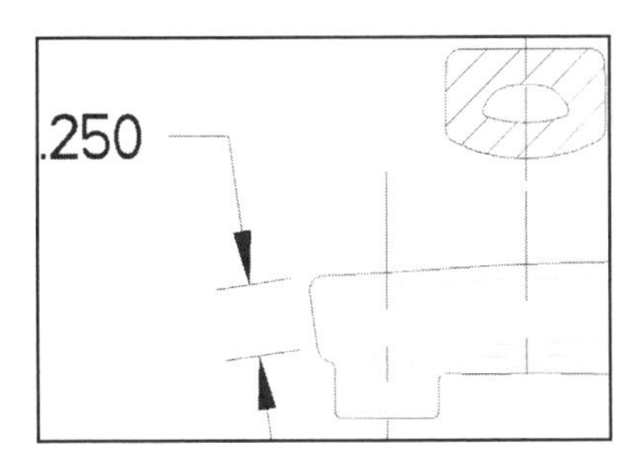

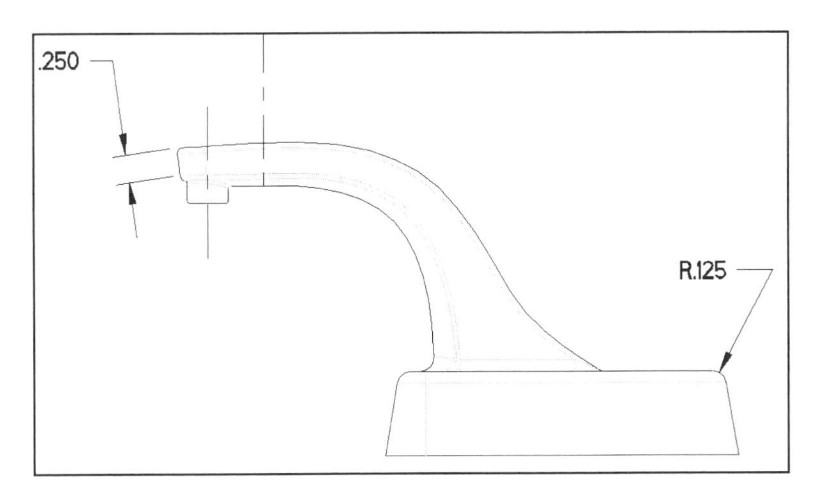

Figure 12.44
Bathroom Faucet Drawing,
Right Side View

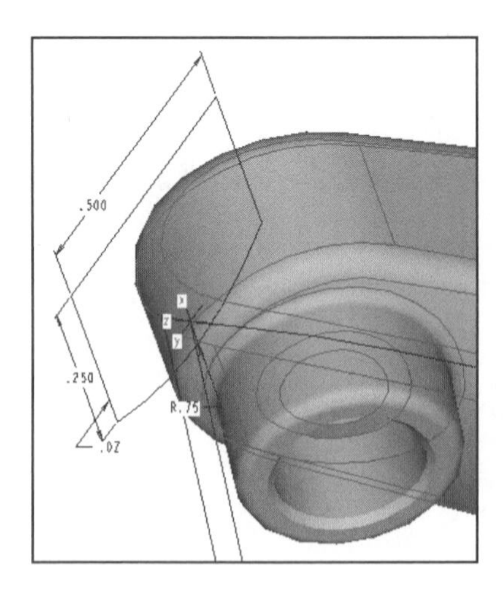

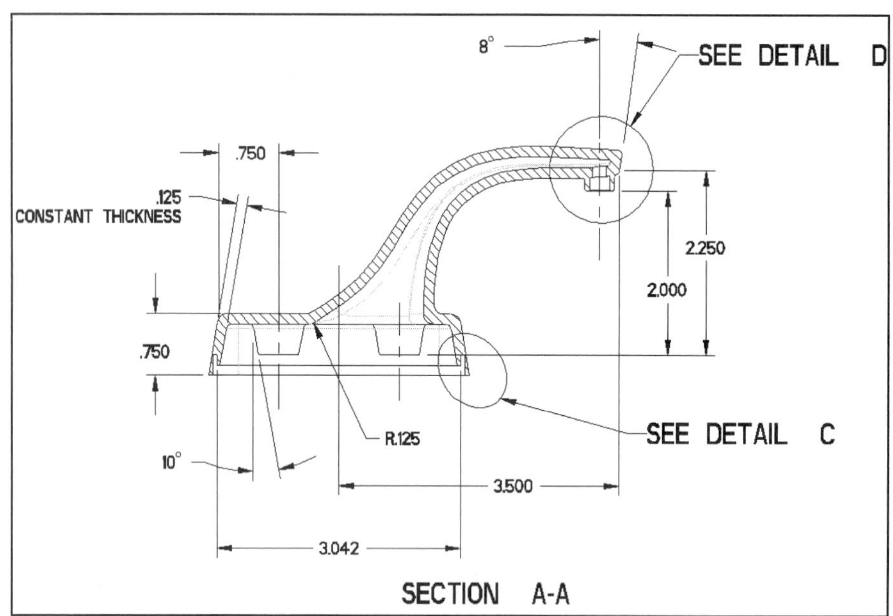

Figure 12.45
Bathroom Faucet Drawing, **SECTION A-A**

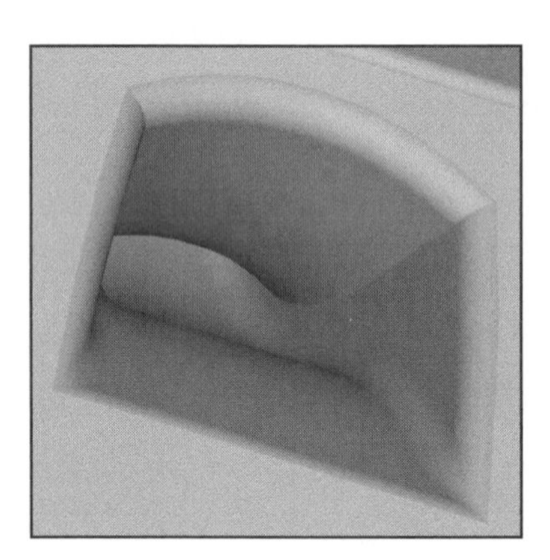

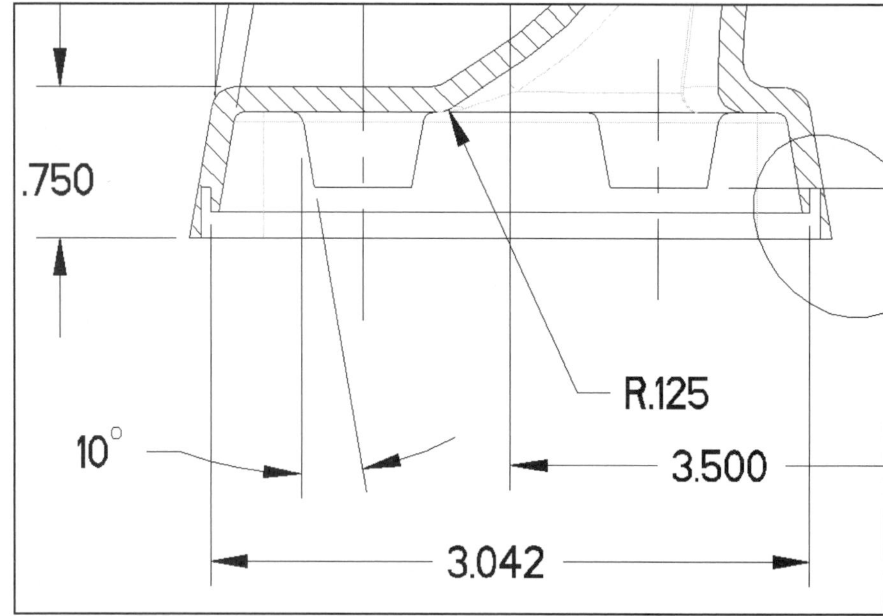

Figure 12.46
Bathroom Faucet Drawing, Lower **SECTION A-A**

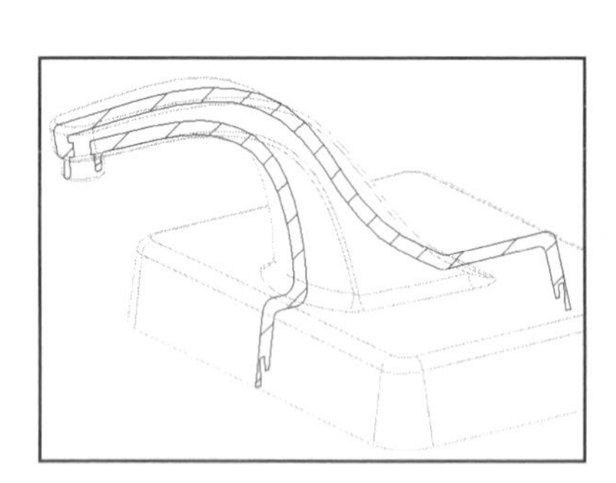

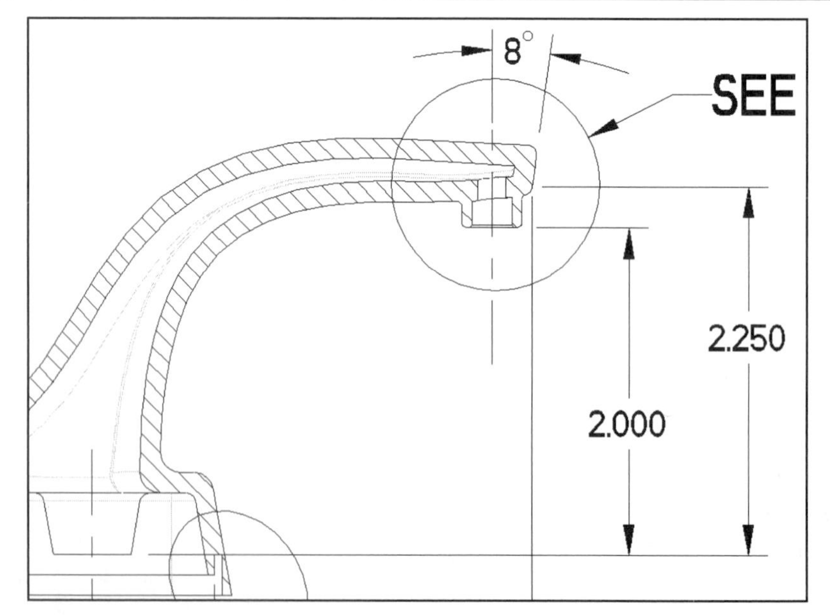

Figure 12.47
Bathroom Faucet Drawing, Upper **SECTION A-A**

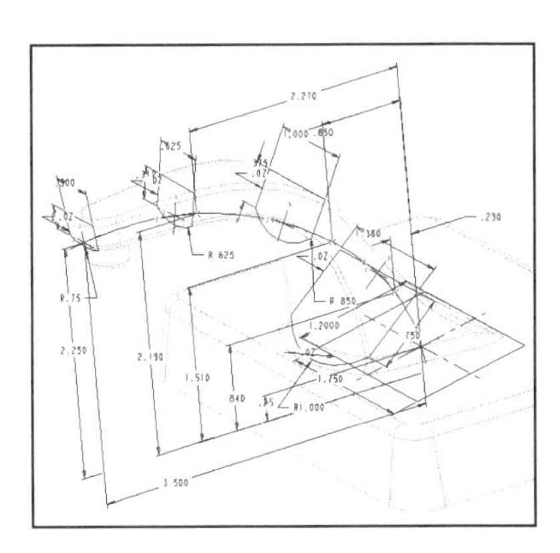

Figure 12.48
Bathroom Faucet Drawing, Top View

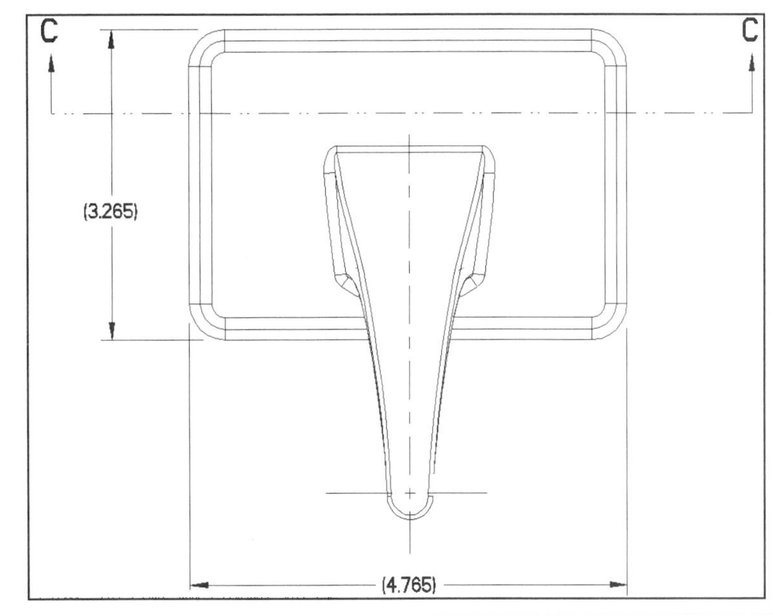

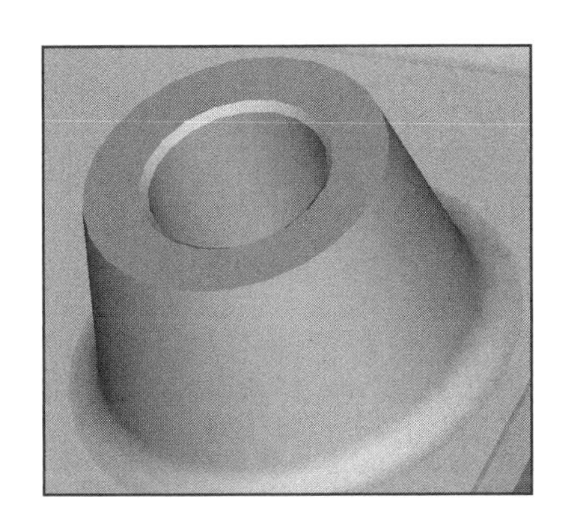

Figure 12.49
Bathroom Faucet Drawing, **DETAIL B**

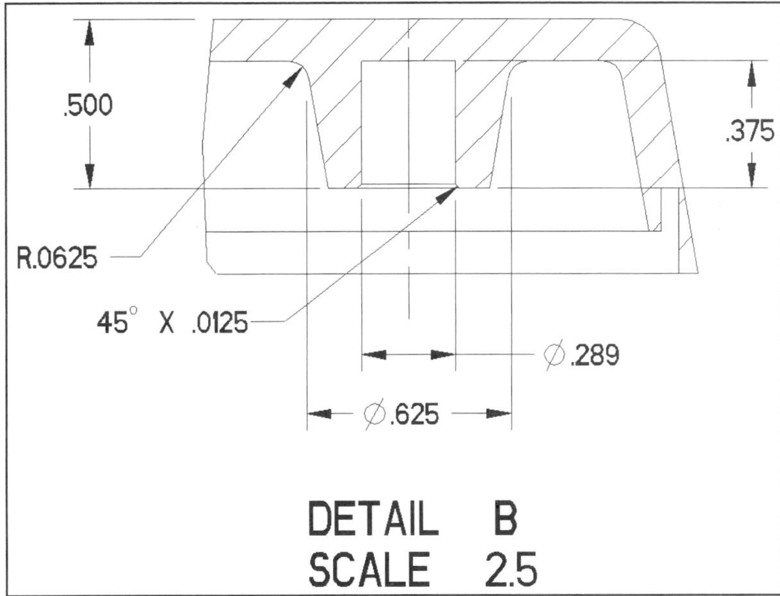

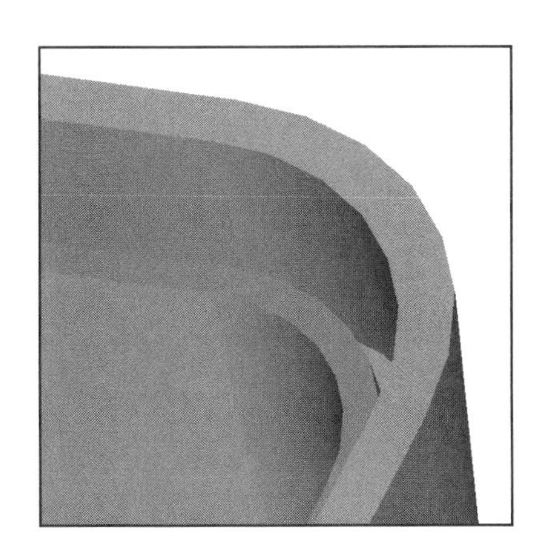

Figure 12.50
Bathroom Faucet Drawing, **DETAIL C**

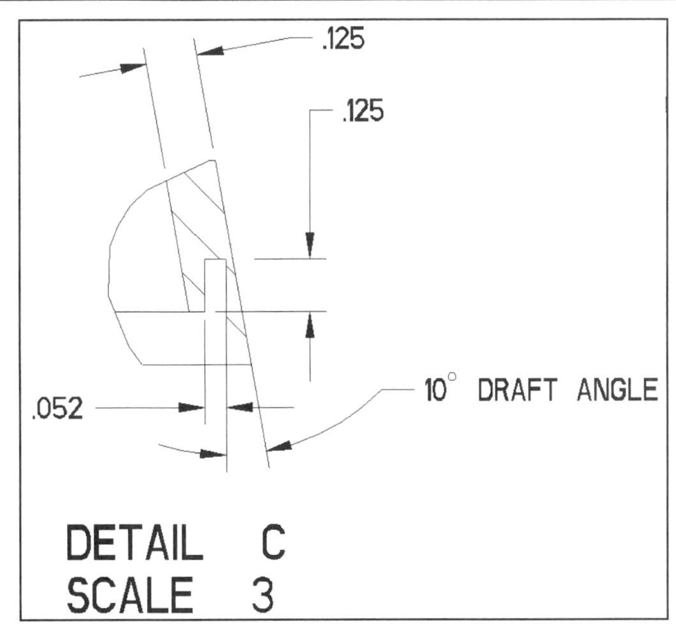

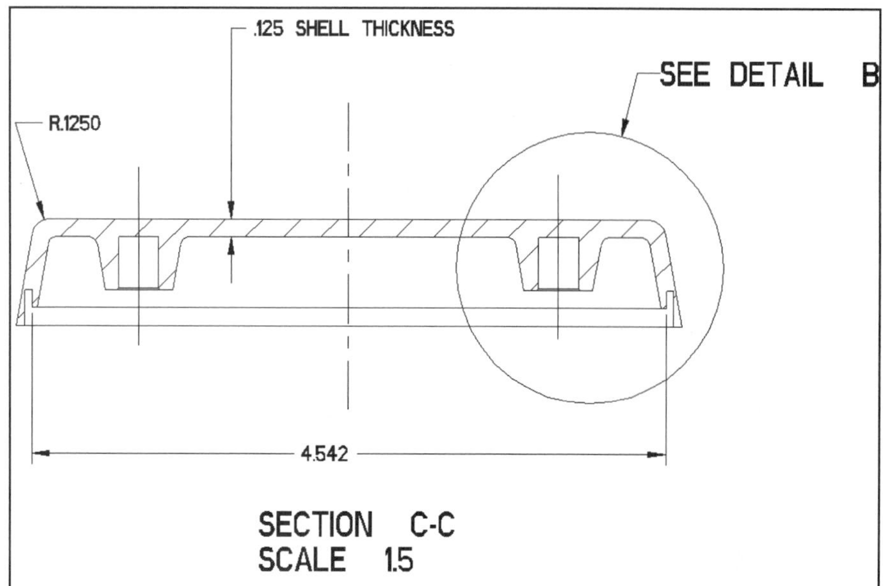

Figure 12.51
Bathroom Faucet Drawing,
SECTION C-C

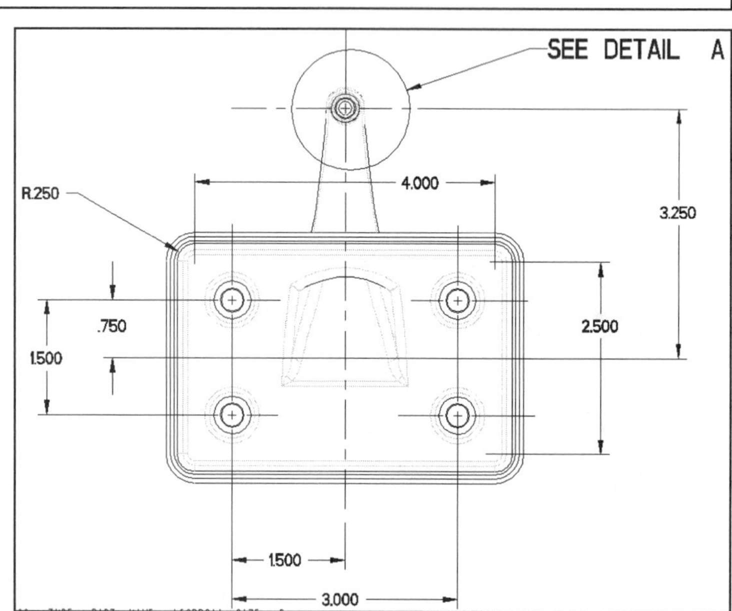

Figure 12.52
Bathroom Faucet Drawing,
Bottom View

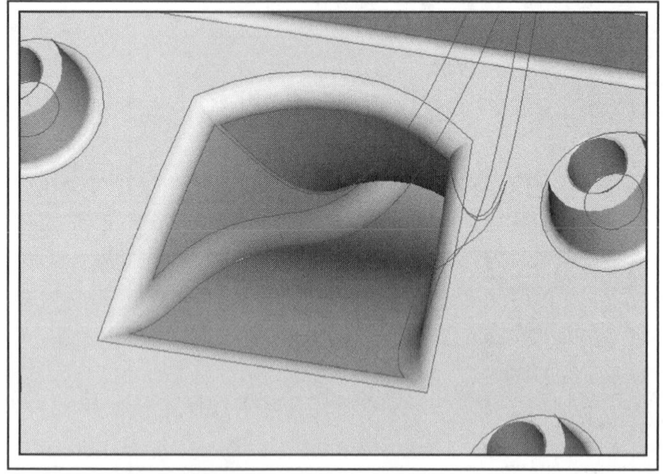

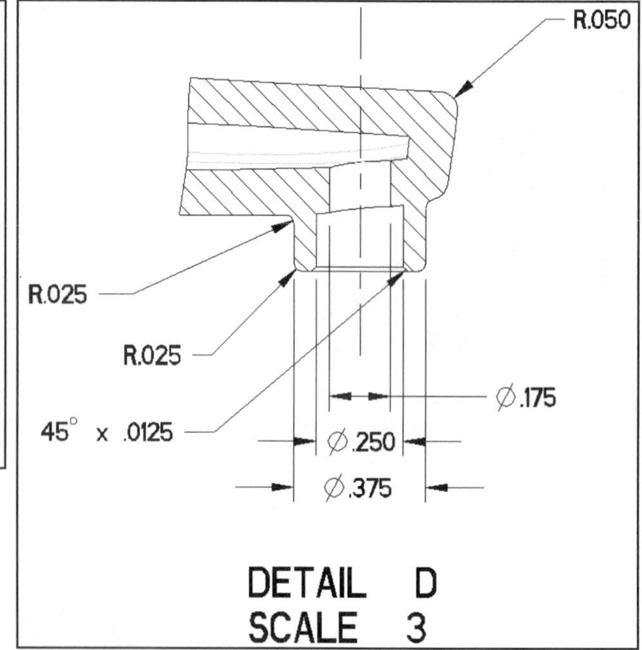

Figure 12.53
Bathroom Faucet Drawing,
DETAIL D

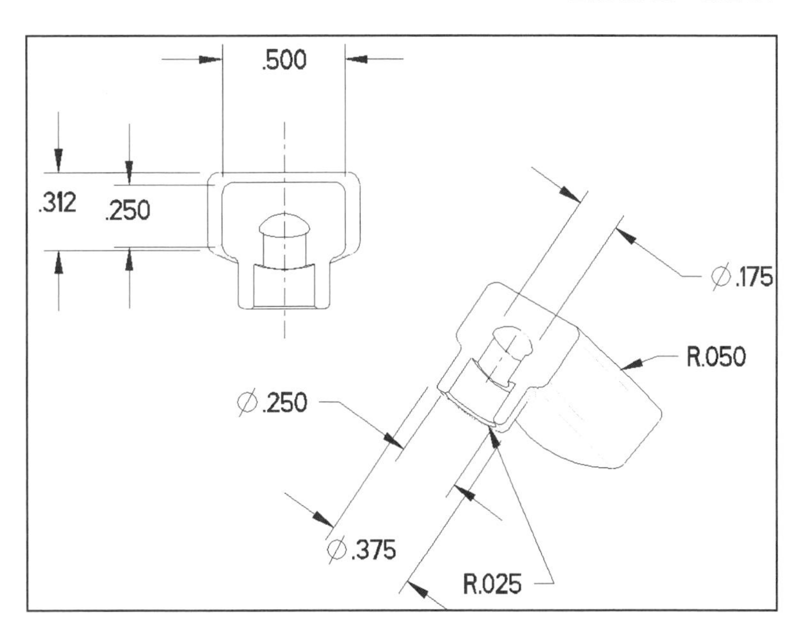

Figure 12.54
Bathroom Faucet Drawing,
Spout Section

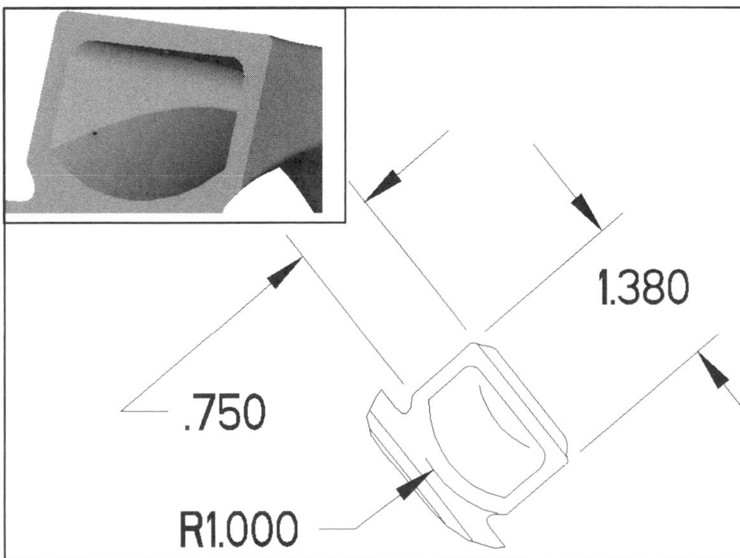

Figure 12.55
Bathroom Faucet Drawing,
Second Blend Section

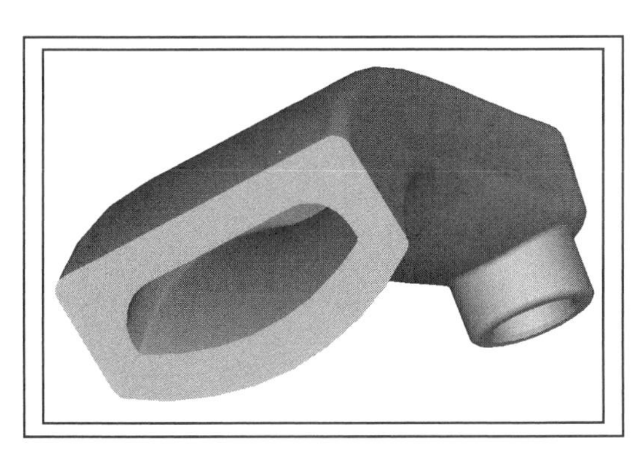

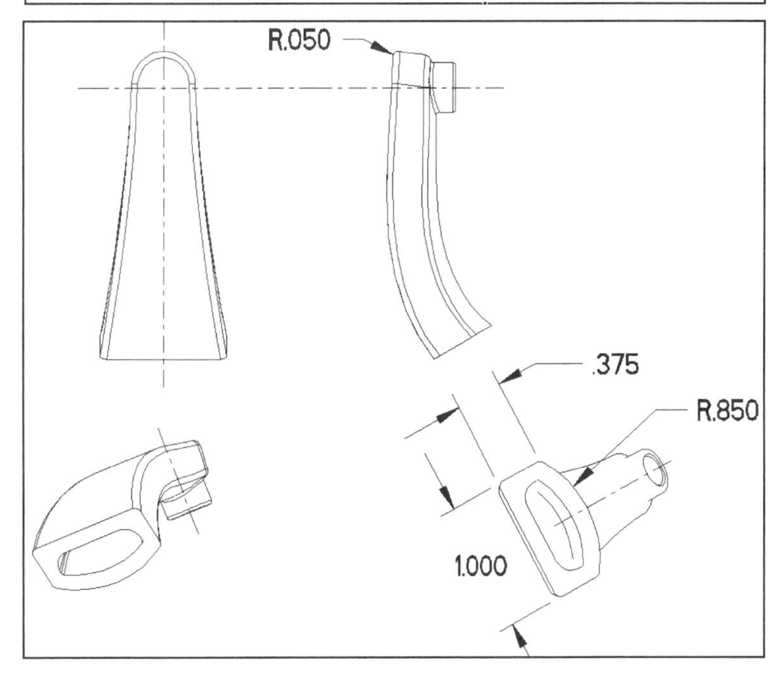

Figure 12.56
Bathroom Faucet Drawing,
Third Blend Section

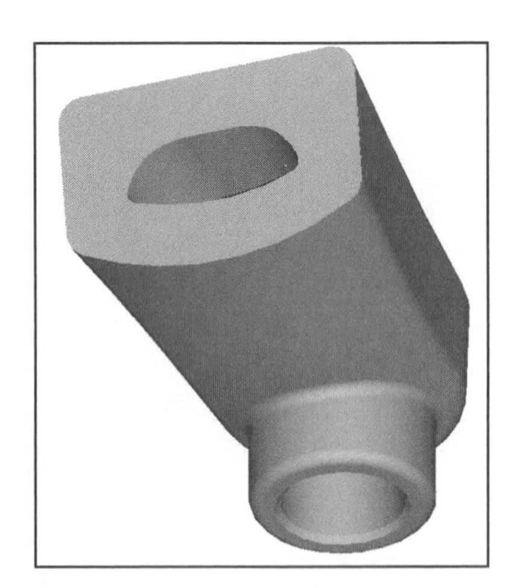

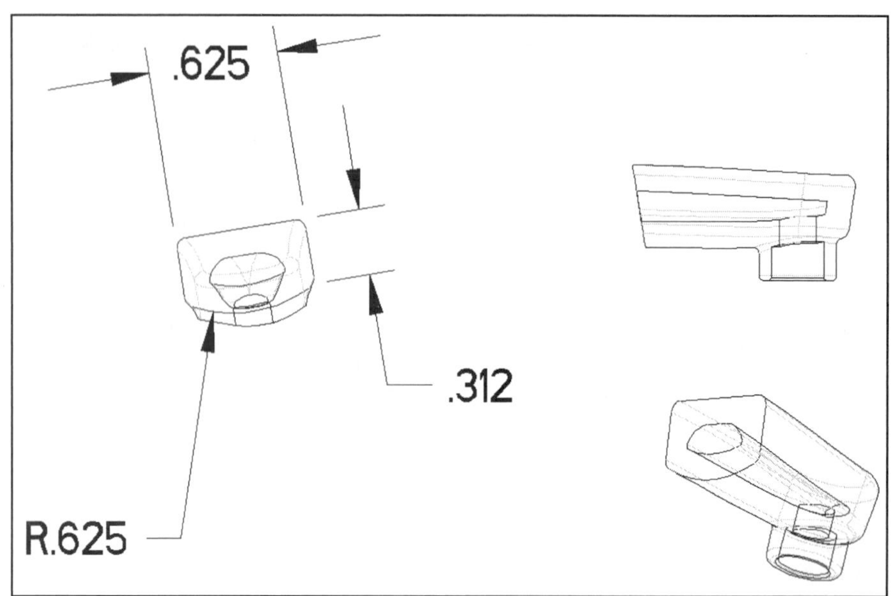

Figure 12.57
Bathroom Faucet Drawing, Fourth Blend Section

Figure 12.58
Bathroom Faucet Drawing, Fifth Blend Section

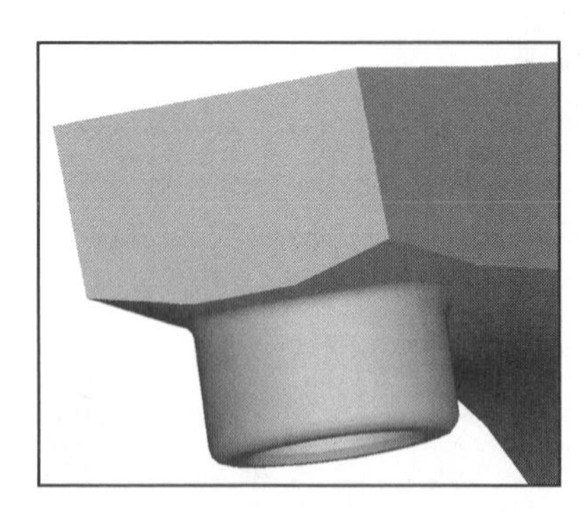

Figure 12.59
Bathroom Faucet Drawing, Sections

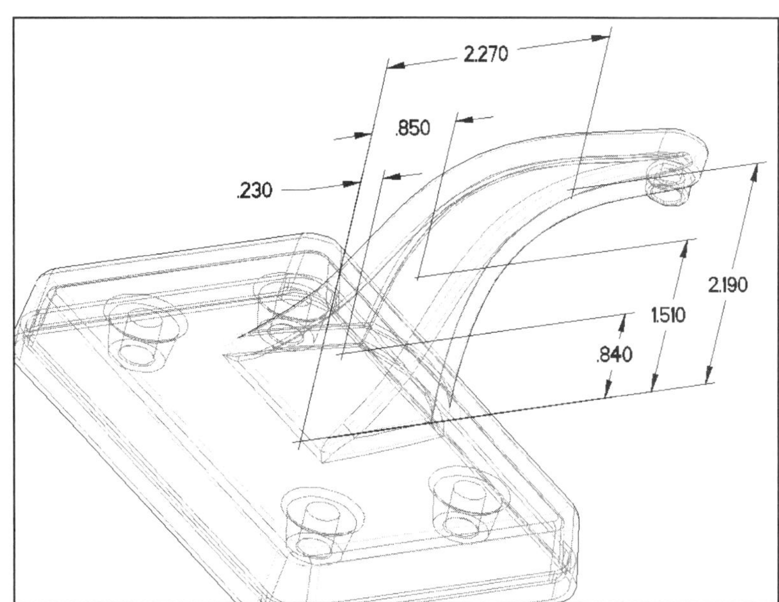

Figure 12.60
Bathroom Faucet Drawing, First Three Blend Section Locations

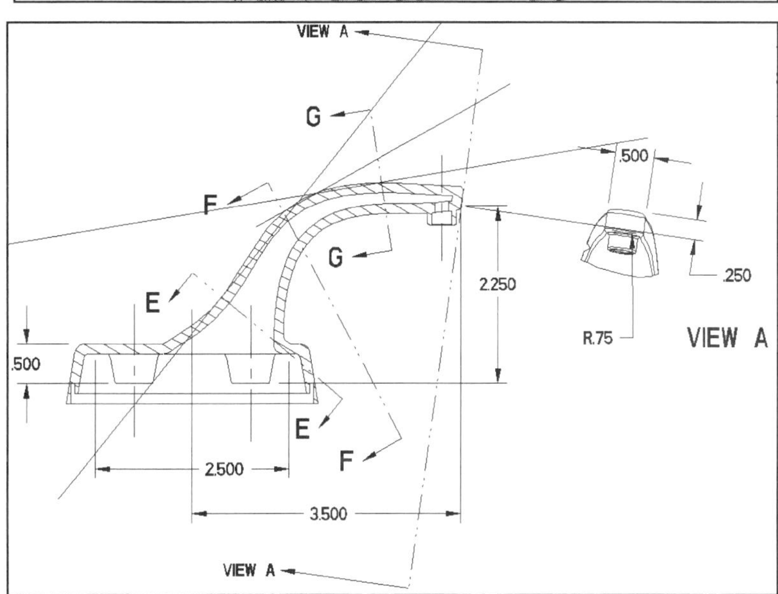

Figure 12.61
Bathroom Faucet Drawing, **SECTION A-A**

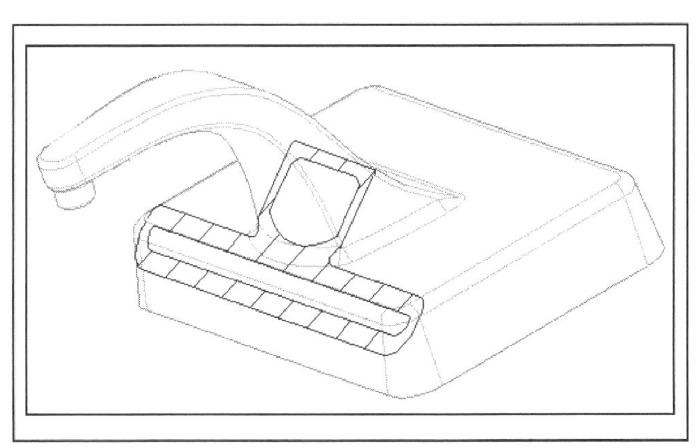

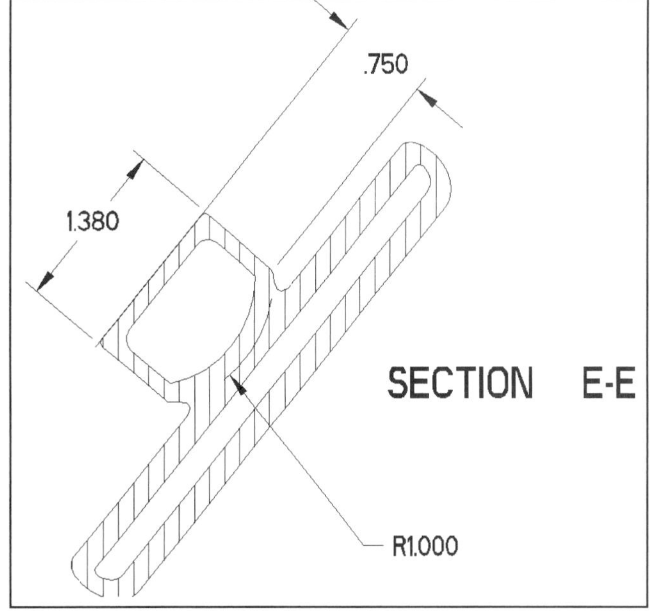

Figure 12.62
Bathroom Faucet Drawing, **SECTION E-E**

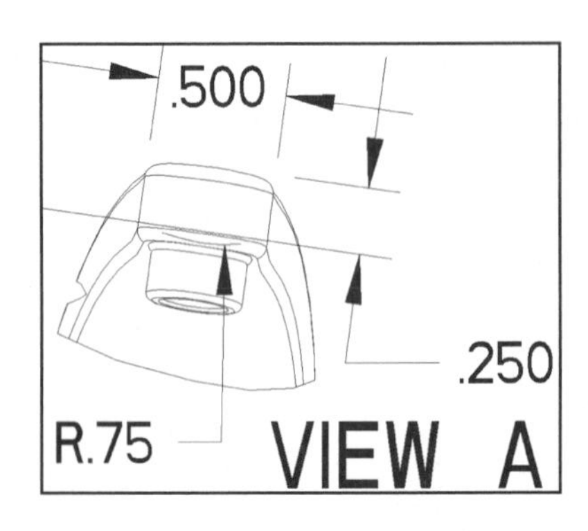

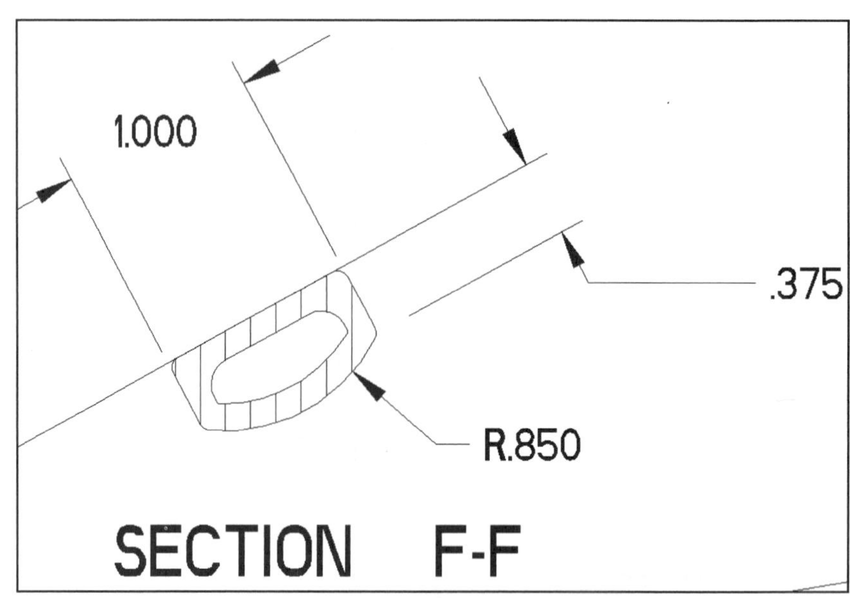

Figure 12.63
Bathroom Faucet Drawing,
SECTION F-F

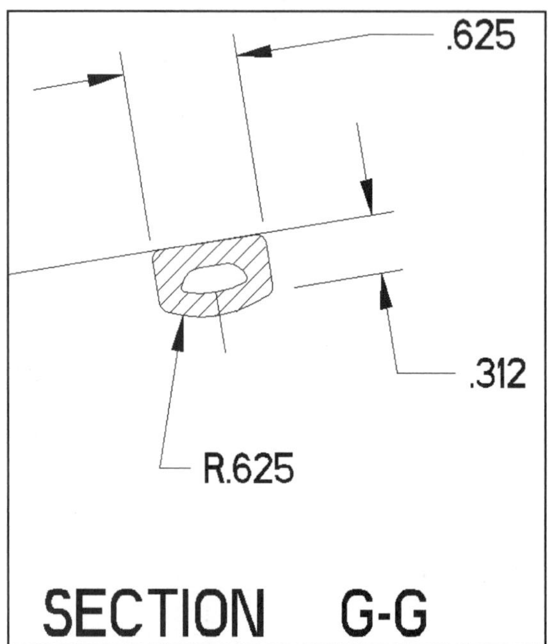

Figure 12.64
Bathroom Faucet Drawing,
SECTION G-G

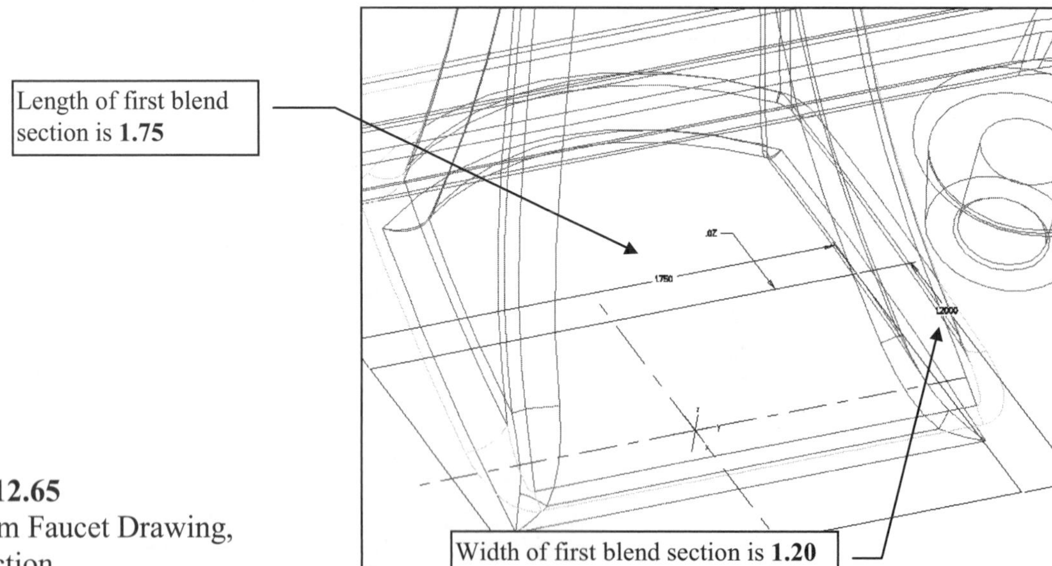

Figure 12.65
Bathroom Faucet Drawing,
First Section

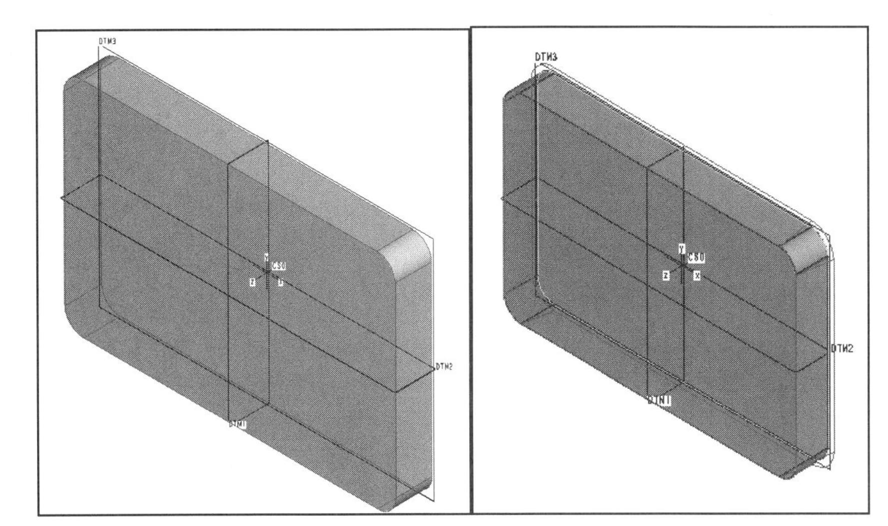

Figure 12.66
Bathroom Faucet, First Protrusion and Draft

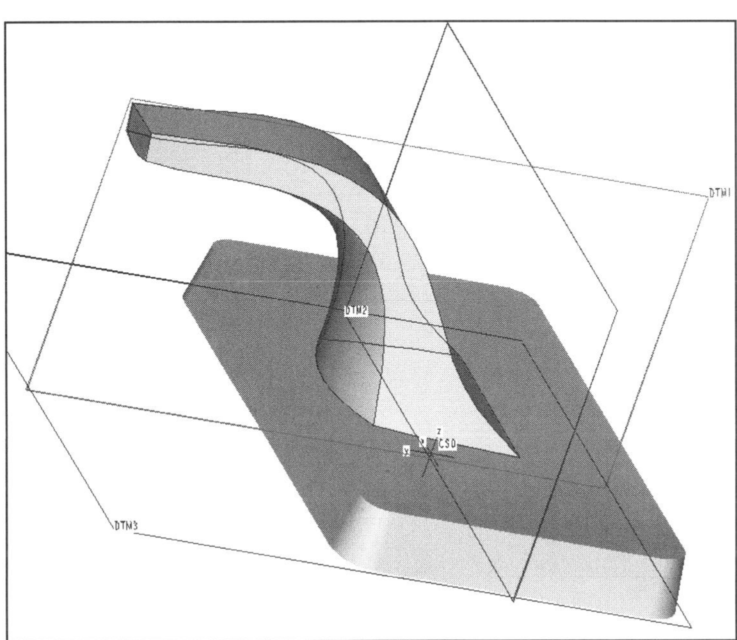

Figure 12.67
Bathroom Faucet, Swept Protrusion

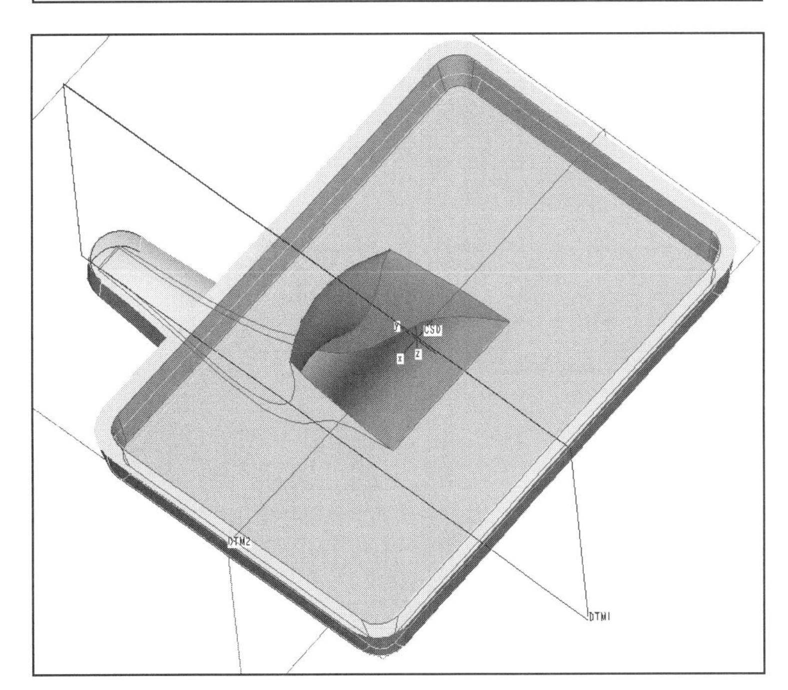

Figure 12.68
Bathroom Faucet, **Shell**

HINT
Each section in the **Swept Blend** has four entities. Keep the start point direction arrow of each section in the same region and facing the same direction.

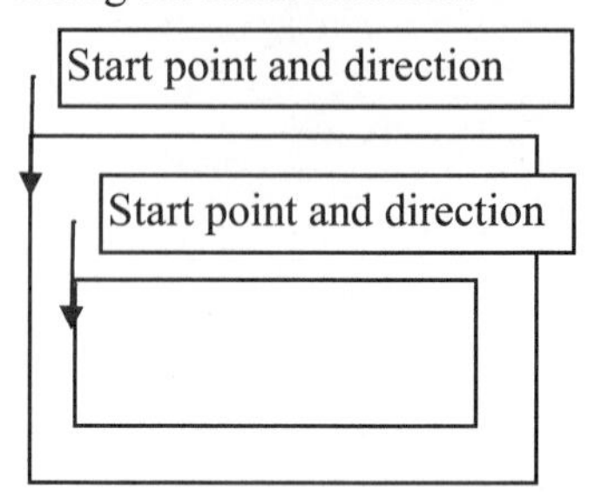

HINT
DTM1 = RIGHT
DTM2 = TOP
DTM3 = FRONT

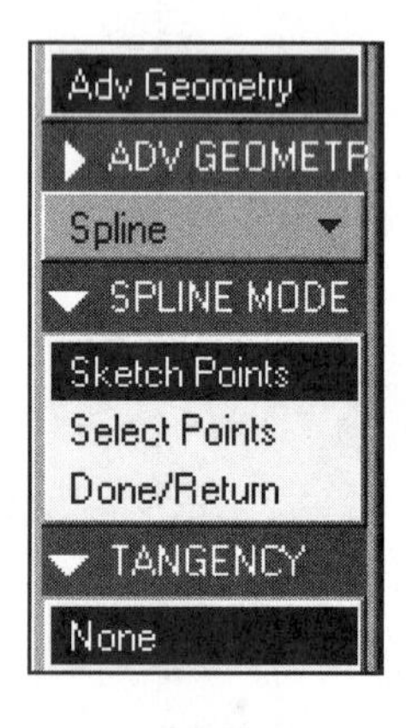

Create the first protrusion and draft as shown in Figure 12.66. Next, a **Swept Blend** will be used to create the geometry for the **Bathroom Faucet** as shown in Figures 12.69 through 12.82. The default options of **Sketch Sec** and **NrmToOriginTraj** will be used. When prompted for the trajectory, choose **Sketch Traj**. Sketch a trajectory (spine) with the **Spline** option in the **Adv Geometry** menu. Create a total of five points along the trajectory. These will locate the five sections of the blend. Before you begin to sketch any sections, you are prompted for where the sections are to be located. A total of five sections will be sketched for this protrusion. Two will be at the endpoints (mandatory) and the other three at the datum points. The first location to be highlighted will be the second point of the sketched trajectory. Select **Accept** from the CONFIRM menu. The next point will then be highlighted, so **Accept** this location. Choose **Accept** until all points have been accepted.

To sketch the first section, accept the default **Z**-axis rotation of zero. Sketch the section as shown in Figure 12.72. Be aware of the *start point* location. When finished with the first section, Pro/E will prompt you for the next **Z**-axis rotation. Accept the default value and sketch the second section (Fig. 12.73). Again, keep track of the start point; it must match up with the first section. Sketch the third section (Fig. 12.75), again using a **Z**-axis rotation value of zero. Sketch the fourth section (Fig. 12.77), again using a **Z**-axis rotation value of zero. When finished with the fourth section, sketch the fifth section (Fig. 12.79). Again, accept the default of zero for the **Z**-axis rotation.

The completed feature should look like Figure 12.81. **Preview** your feature elements before choosing **OK**, to check for twist in the swept blend because of start points that do not line up. If there is a problem, use the Sections element from the dialog box and choose **Define** to change the start point(s) so that they are aligned.

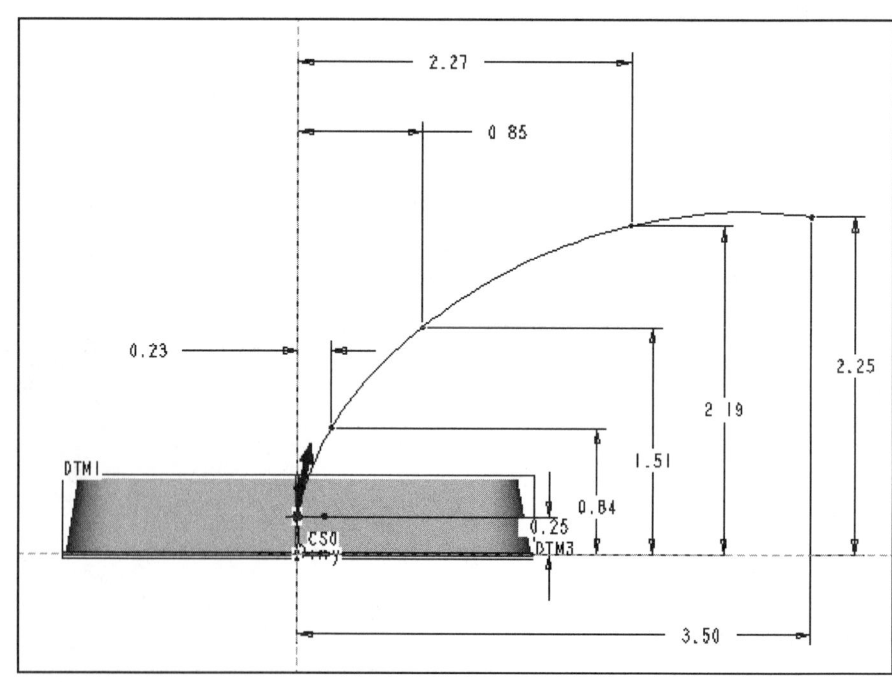

Figure 12.69
Bathroom Faucet, Trajectory Dimensions

Create the **Swept Blend** with the following commands:

Feature ⇒ Create ⇒ Protrusion ⇒ Advanced ⇒ Solid ⇒ Done ⇒ Swept Blend ⇒ Done ⇒ Sketch Sec ⇒ NrmToOriginTraj ⇒ Done ⇒ Sketch Traj ⇒ (pick **DTM1**) **⇒ Okay ⇒ Right ⇒** (pick **DTM2**) **⇒ Sketch ⇒ Intent Manager** (off) **⇒ Adv Geometry ⇒ Spline ⇒ Sketch Points ⇒ Thru Points ⇒** [sketch the trajectory (Fig. 12.69)] **⇒ Regenerate ⇒ Alignment** (align endpoint to **DTM2**) **⇒ Regenerate ⇒ Dimension** (pick the spline to display the points, dimension as shown in Fig. 12.69) **⇒ Regenerate ⇒ Modify ⇒ Regenerate ⇒ Done ⇒ Accept** (all three middle points will be highlighted one at a time as you accept them. The first and last points are automatically accepted) **⇒ ✔** (accept the default of **0°**) ⇒ (sketch the first section centered about the light *blue* crosshairs--starting point--each section will have one) **⇒ Regenerate ⇒ Alignment ⇒ Regenerate ⇒ Dimension ⇒ Regenerate ⇒ Modify ⇒ Regenerate ⇒ Done ⇒** (continue creating the last four sections) **⇒ Done ⇒ Preview ⇒ OK**

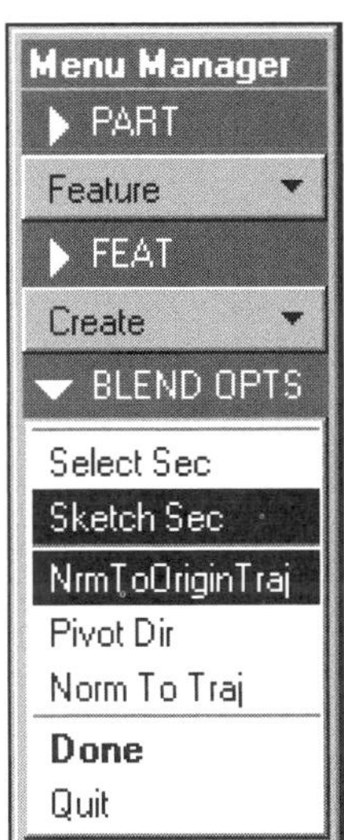

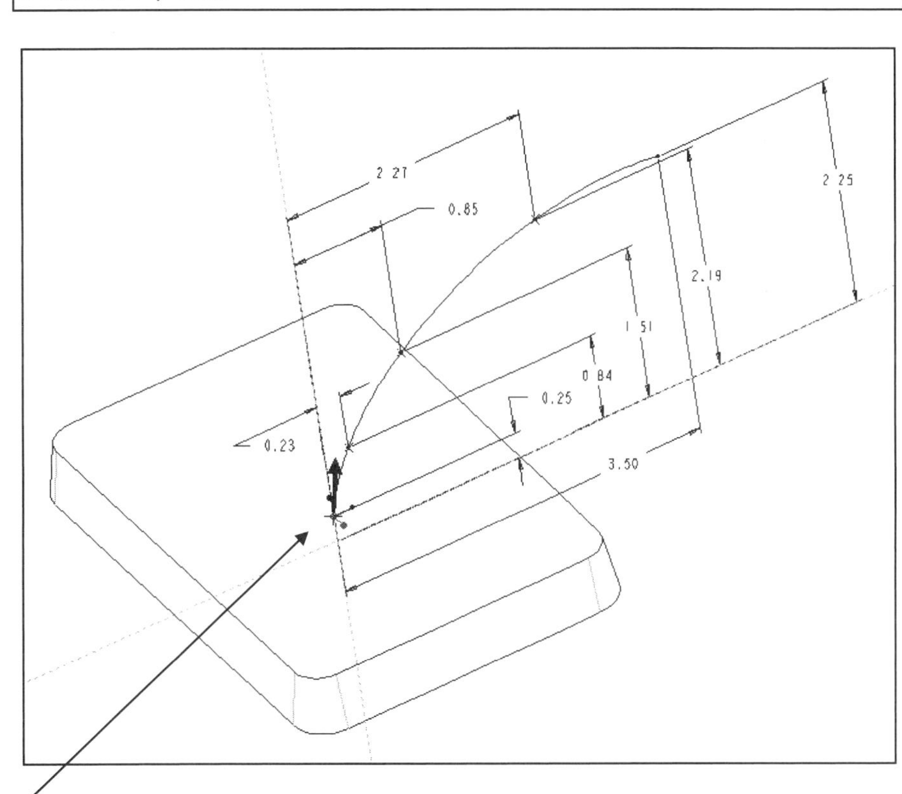

Figure 12.70
Bathroom Faucet,
First Section: Pictorial

The *first section* is **1.75 X 1.20,** shown in Figures 12.71 and 12.72. The *second section* is **1.380 X .75 X R1.00,** shown in Figures 12.55, 12.73, and 12.74. The *third section* is **1.00 X .375 X R.850,** shown in Figures 12.56, 12.75, and 12.76. The *fourth section* is **.625 X .312 X R.625,** shown in Figures 12.57, 12.77, and 12.78. The *fifth section* is **.500 X .25 X R.75,** shown in Figures 12.58, 12.79, and 12.80. Figure 12.81 shows the final part design.

HINT

Create the first section approximately centered about its respective colored crosshairs (starting point). Use centerlines when possible.

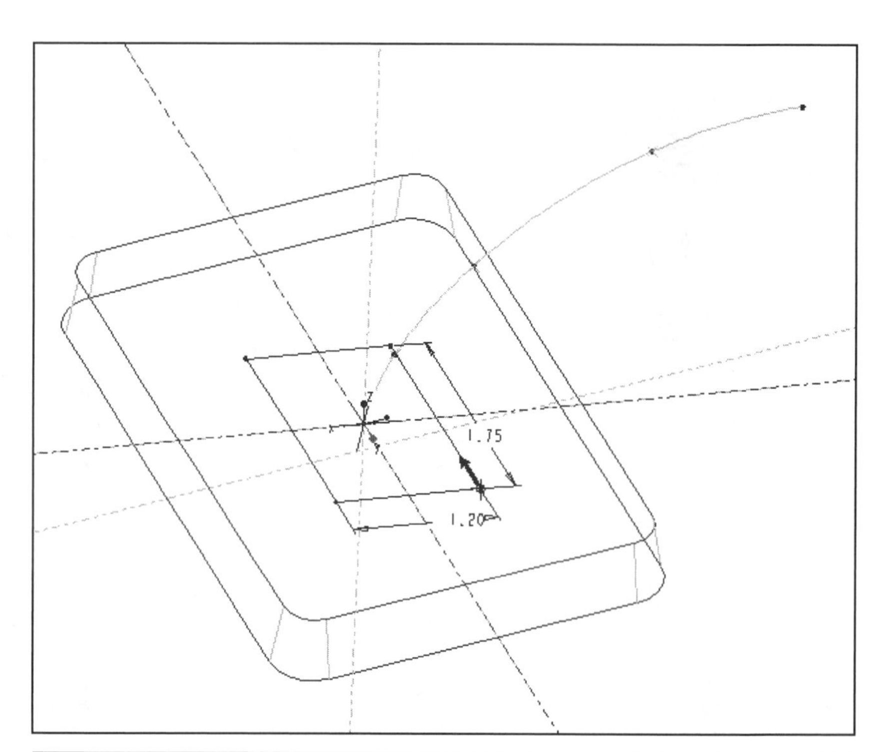

Figure 12.71
Bathroom Faucet,
First Section Dimensions

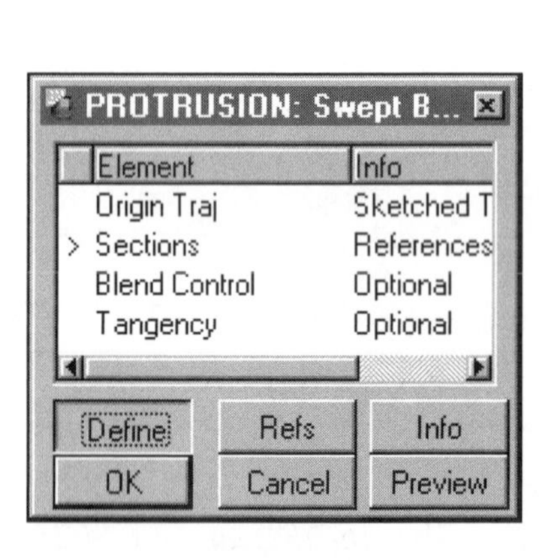

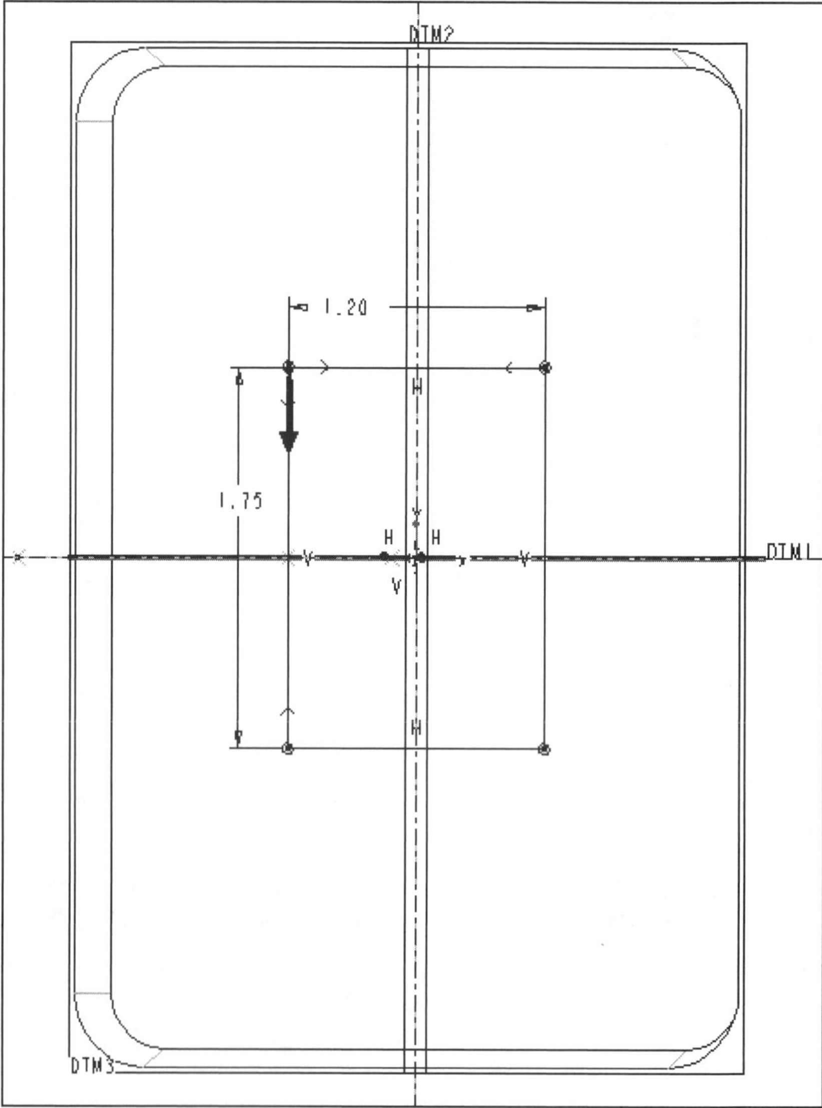

Figure 12.72
Bathroom Faucet,
First Section: **1.75 X 1.20**

HINT

For the second section, align the arc center to the colored crosshairs (starting point).

Figure 12.73
Bathroom Faucet,
Second Section:
1.380 X .75 X R1.00

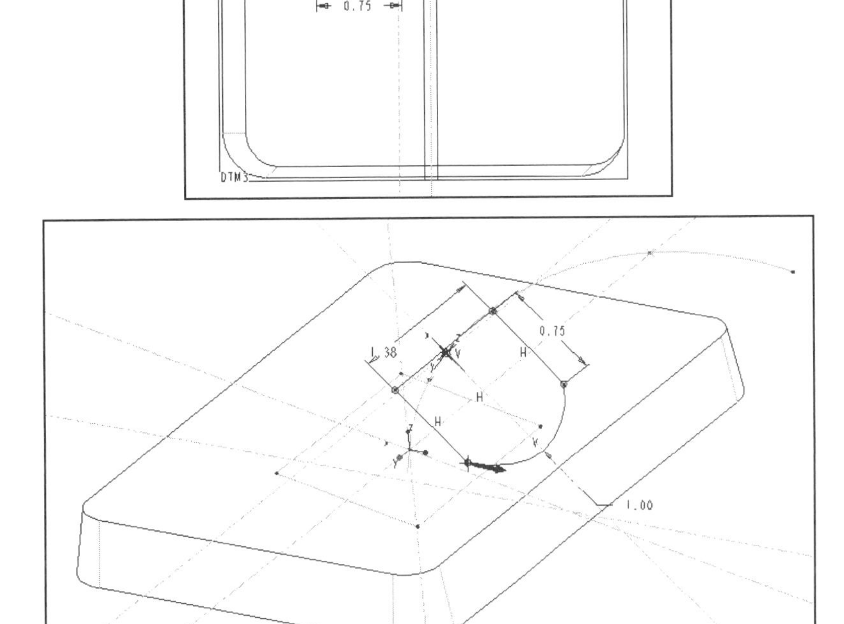

Figure 12.74
Bathroom Faucet,
Second Section: Pictorial

HINT

Note where all endpoints are in relation to the colored crosshairs (starting point).

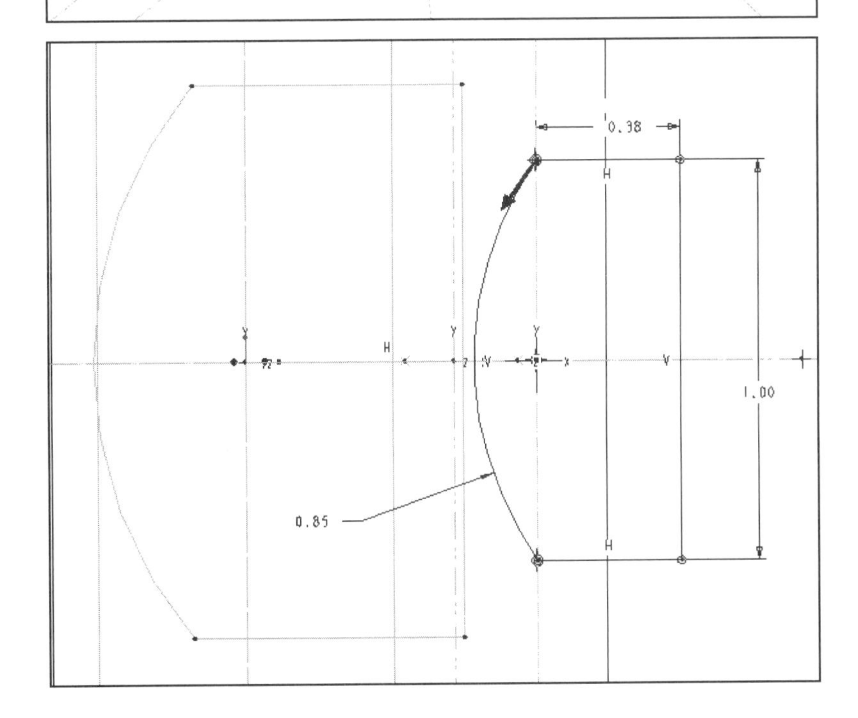

Figure 12.75
Bathroom Faucet,
Third Section:
1.00 X .375 X R.850

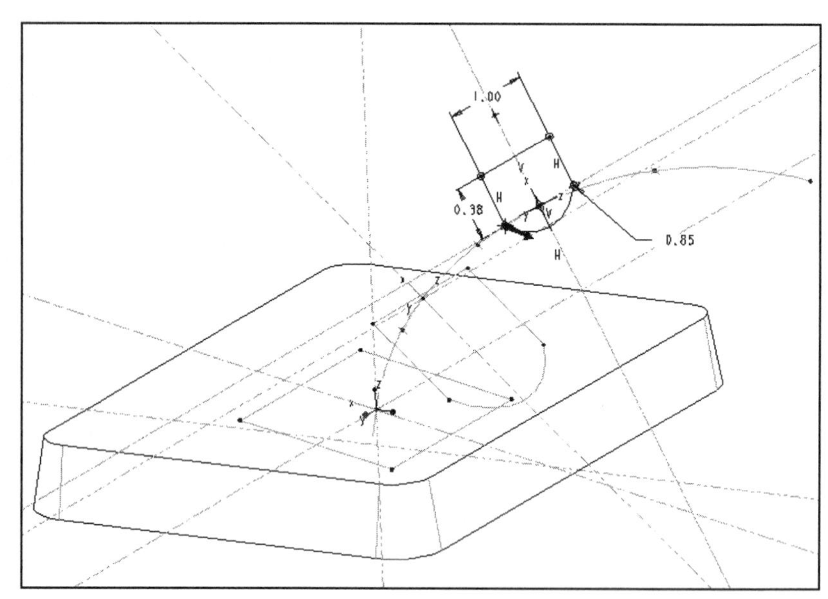

Figure 12.76
Bathroom Faucet,
Third Section: Pictorial

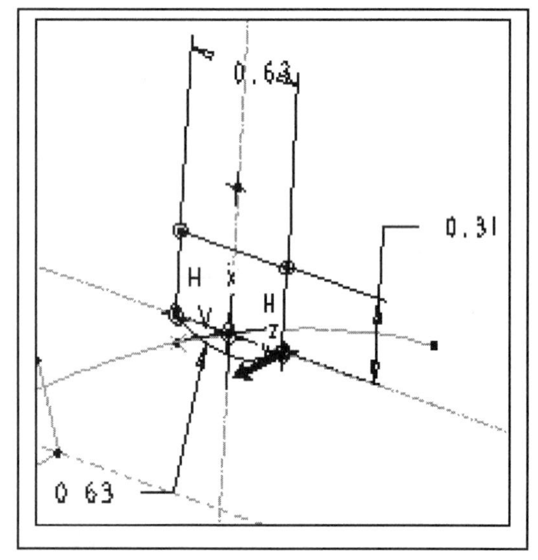

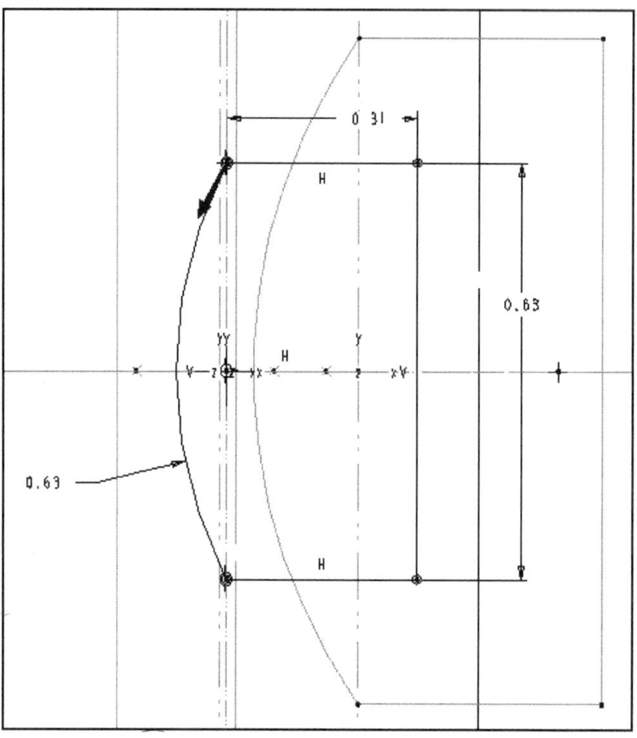

Figure 12.77
Bathroom Faucet
Fourth Section:
.625 X .312 X R.625

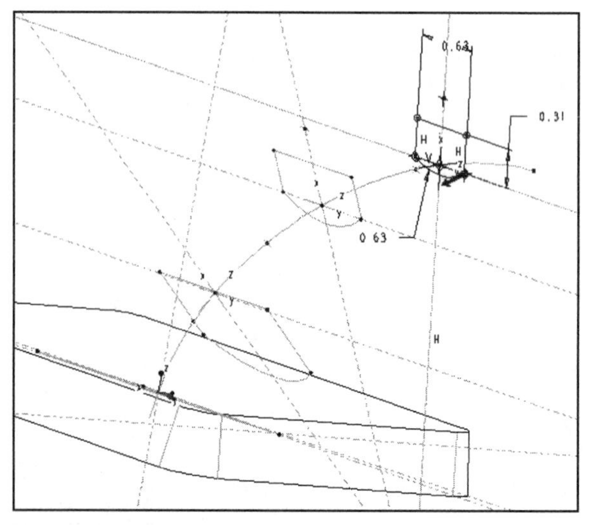

Figure 12.78
Bathroom Faucet,
Fourth Section: Pictorial

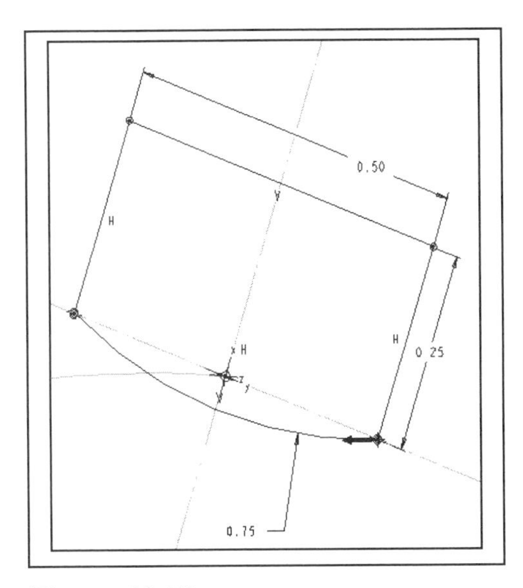

Figure 12.79
Bathroom Faucet,
Fifth Section:
.500 X .25 X R.75

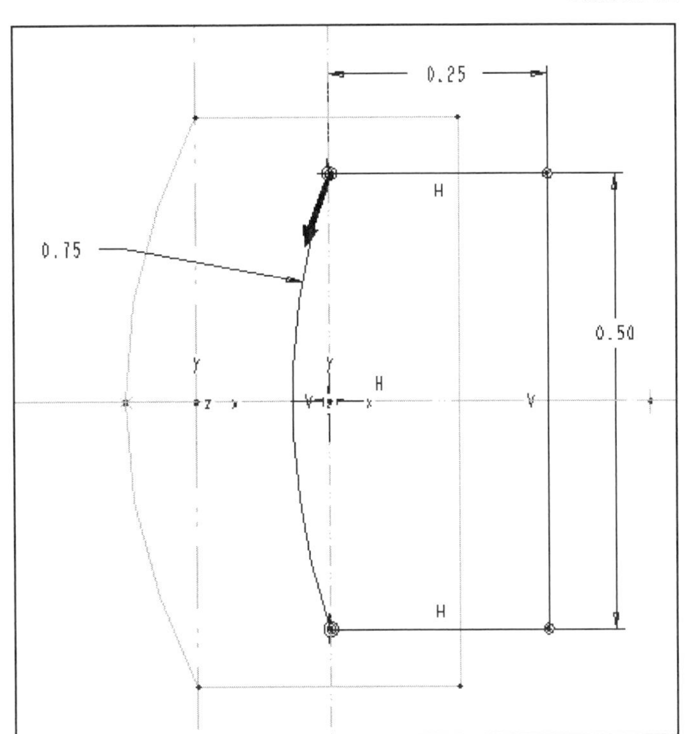

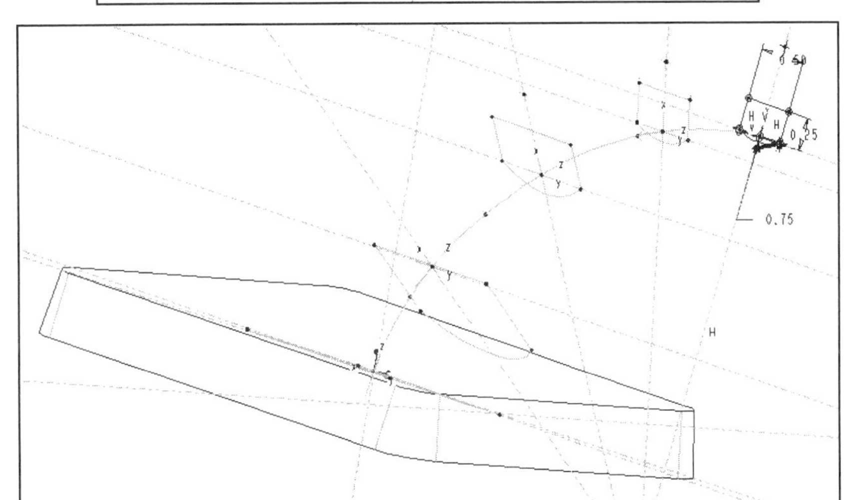

Figure 12.80
Bathroom Faucet,
Fifth Section: Pictorial

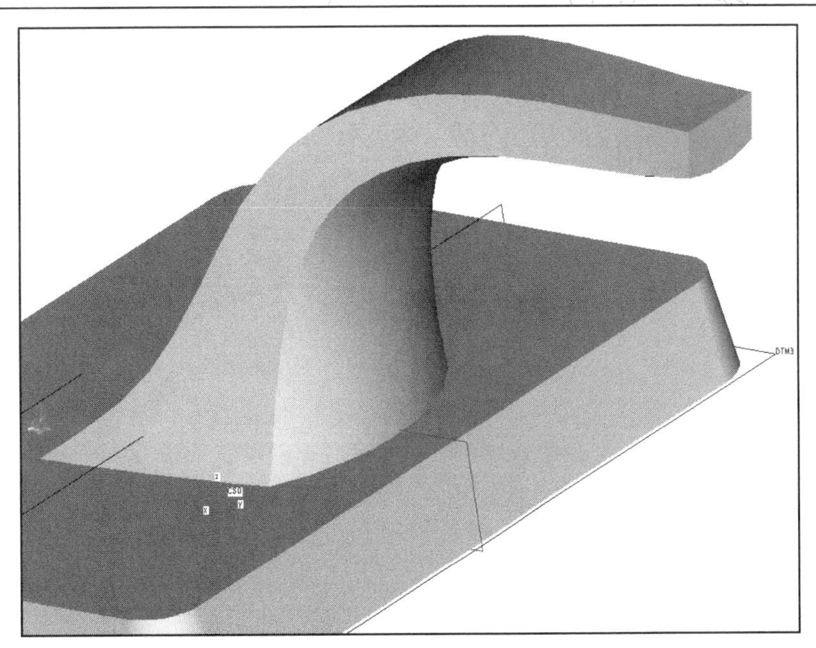

Figure 12.81
Bathroom Faucet,
Completed Swept Blend

Figure 12.82 Bathroom Faucet

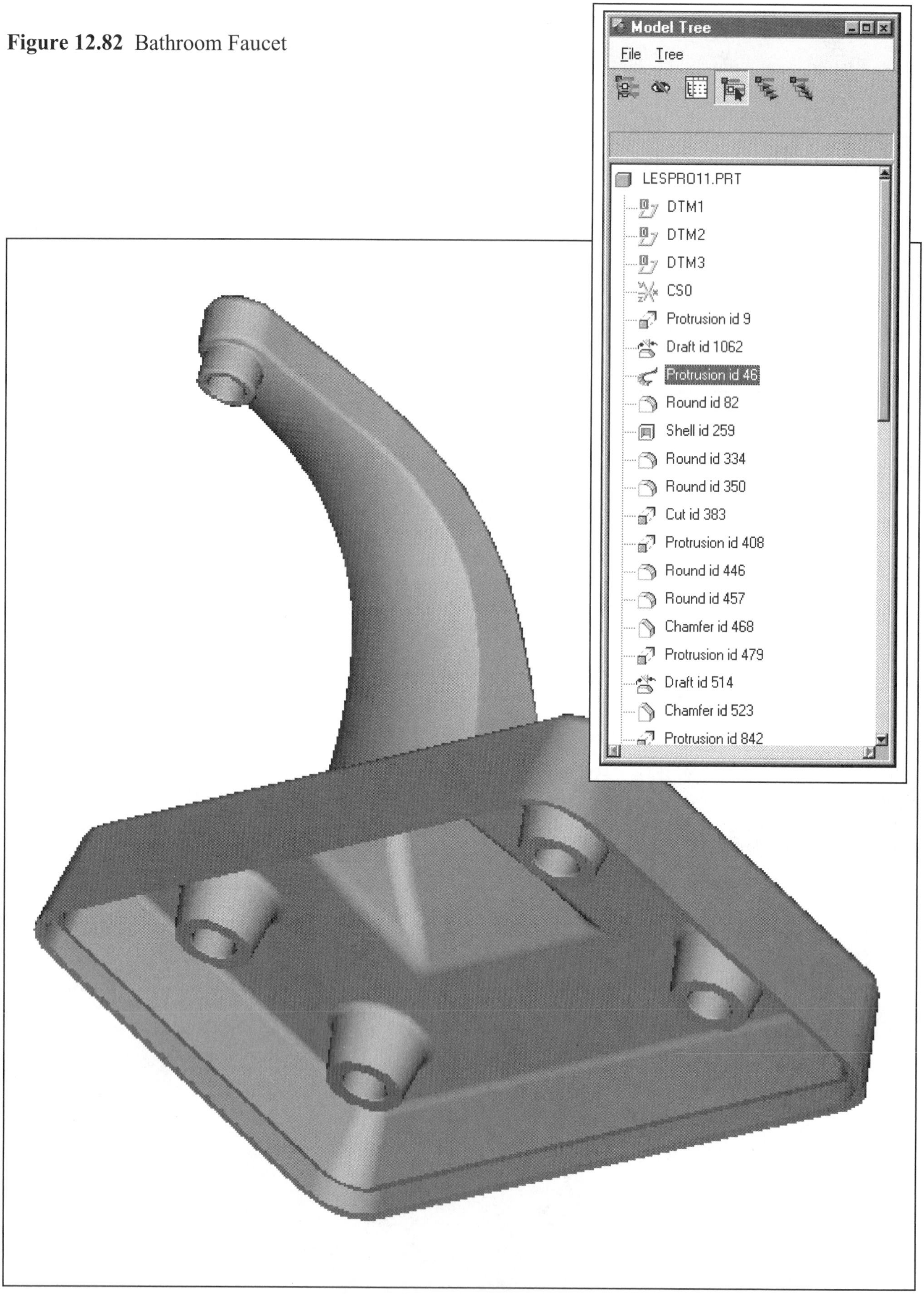

Lesson 13

Helical Sweeps and 3D Notes

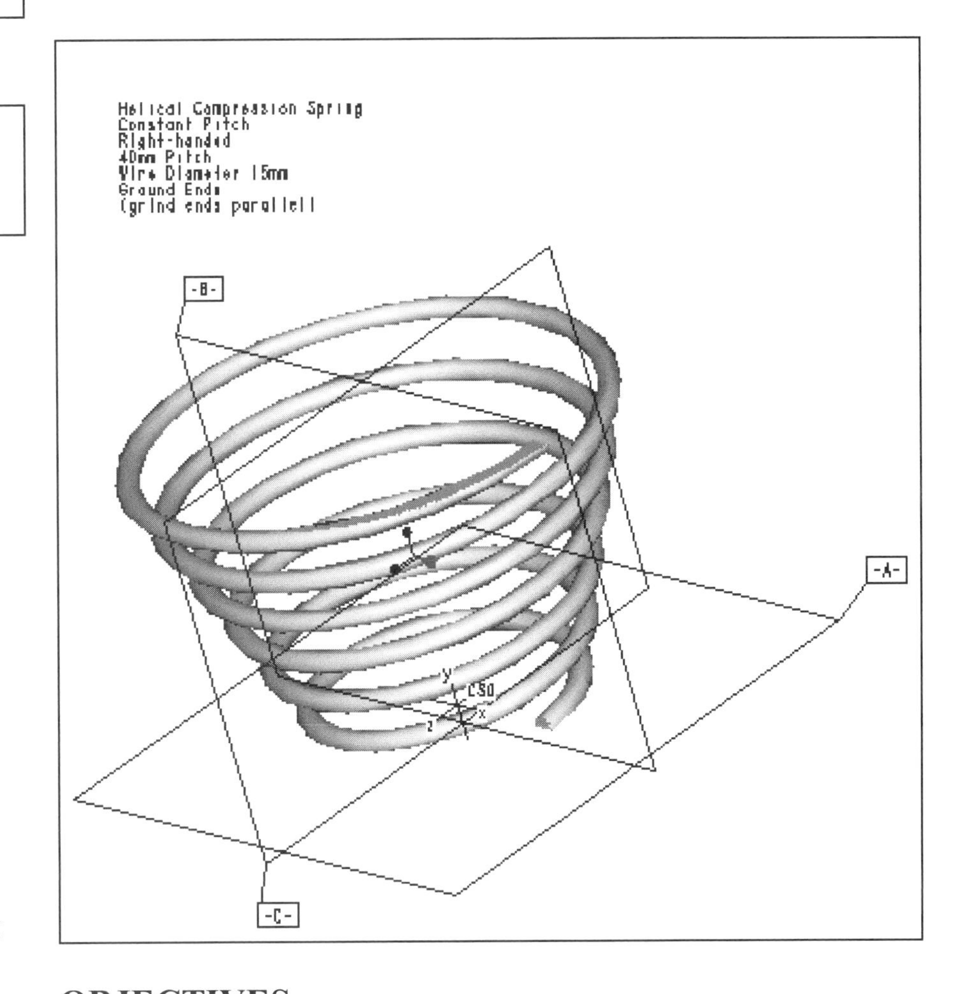

Figure 13.1
Helical Compression Spring

EGD REFERENCE
Fundamentals of Engineering Graphics and Design
by L. Lamit and K. Kitto
Read Chapter 18.
See pages 702-704.

OBJECTIVES

1. **Create springs with a helical sweep**

2. **Model a helical compression spring**

3. **Use sweeps to create hooks on extension springs**

4. **Design an extension spring with a machine hook**

5. **Create plain ground ends on a spring**

6. **Model a convex spring**

7. **Create 3D Notes**

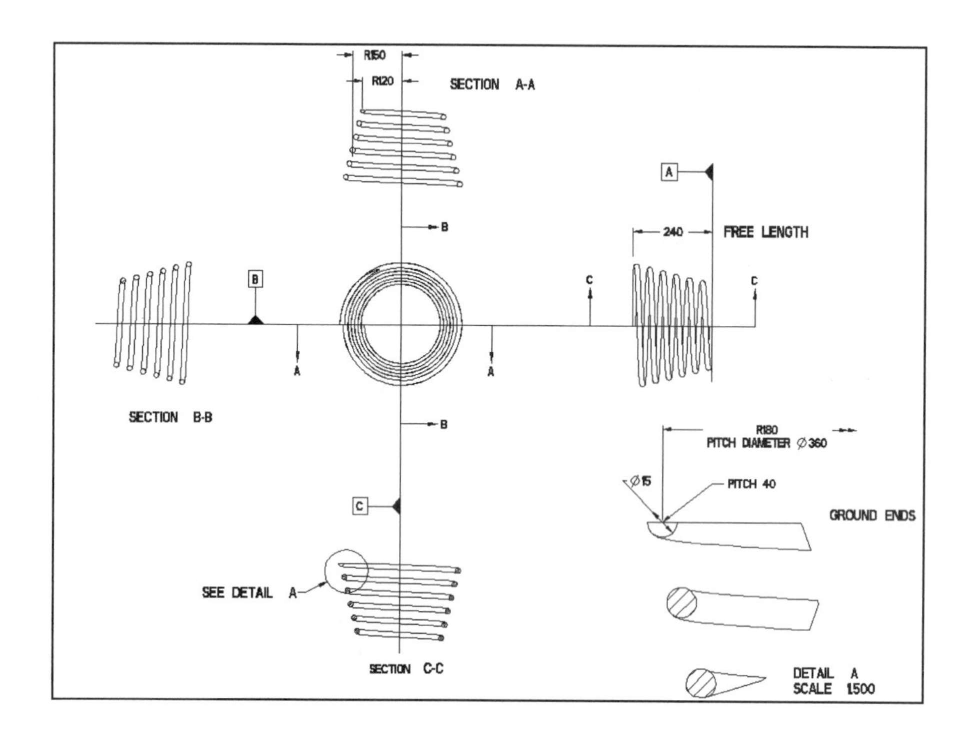

Figure 13.2
Helical Compression Spring Dimensions

HELICAL SWEEPS AND 3D NOTES

A **helical sweep** (Figs. 13.1 through 13.4) is created by sweeping a section along a helical *trajectory*. The trajectory is defined by both the *profile* of the *surface of revolution* (which defines the distance from the section origin of the helical feature to its *axis of revolution*) and the *pitch* (the distance between coils). The trajectory and the surface of revolution are construction tools and do not appear in the resulting geometry.

Model Notes are pieces of text or links to World Wide Web pages that you can attach to objects in Pro/E. **Model Notes** increase the usefulness of this feature by letting you attach any number of notes to any object in your model.

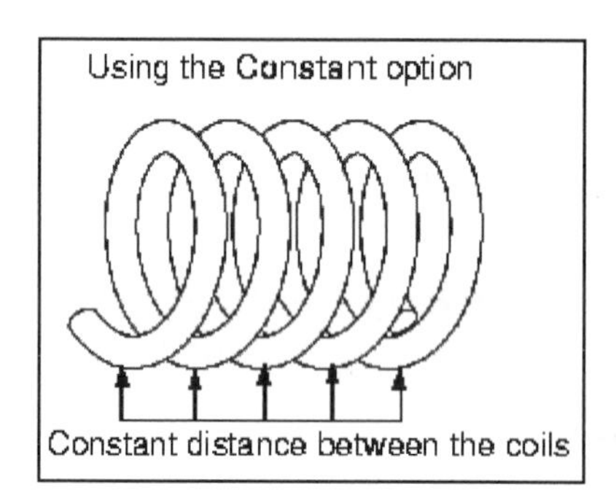

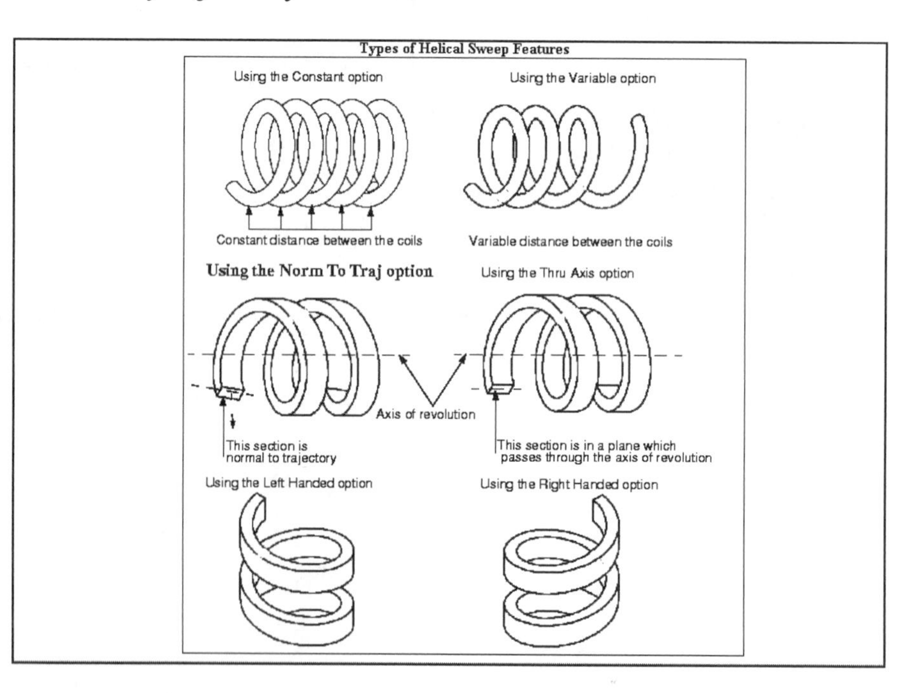

Figure 13.3
Helical Sweeps

Helical Sweeps

The **Helical Swp** option in the ADV FEAT OPT menu is available for both solid and surface features. Use the following ATTRIBUTES menu options, presented in mutually exclusive pairs, to define the helical sweep feature:

> **Constant** The pitch is constant.
> **Variable** The pitch is variable and defined by a graph.
> **Thru Axis** The section lies in a plane that passes through the axis of revolution.
> **Norm To Traj** The section is oriented normal to the trajectory (or surface of revolution).
> **Right Handed** The trajectory is defined by the right-hand rule.
> **Left Handed** The trajectory is defined by the left-hand rule.

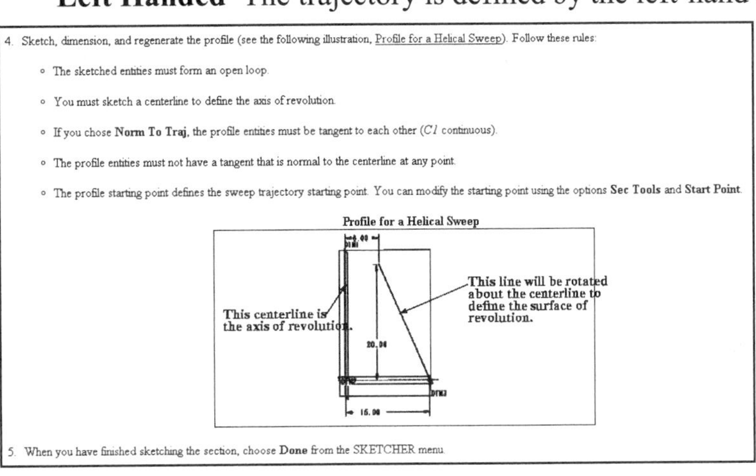
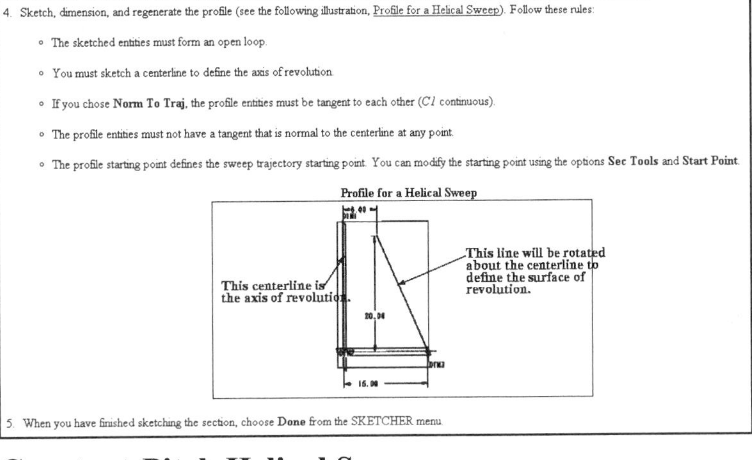
4. Sketch, dimension, and regenerate the profile (see the following illustration, Profile for a Helical Sweep). Follow these rules:

- The sketched entities must form an open loop.
- You must sketch a centerline to define the axis of revolution.
- If you chose **Norm To Traj**, the profile entities must be tangent to each other (*C1* continuous).
- The profile entities must not have a tangent that is normal to the centerline at any point.
- The profile starting point defines the sweep trajectory starting point. You can modify the starting point using the options **Sec Tools** and **Start Point**.

Profile for a Helical Sweep

5. When you have finished sketching the section, choose **Done** from the SKETCHER menu.

Figure 13.4
Profile for a Helical Sweep

Constant-Pitch Helical Sweeps

To create a helical sweep with a constant pitch value, choose the following command sequence:

1. Choose **Advanced ⇒ Done** from the SOLID OPTS menu, then **Helical Swp ⇒ Done.** Pro/E displays the feature creation dialog box.
2. Define the feature by selecting from the ATTRIBUTES menu; then choose **Done.**
3. Specify the sketching plane and its orientation. Then Pro/E places you in Sketcher mode. Sketch, dimension, and regenerate the surface of revolution profile. The sketched entities must form an *open loop*. As with a revolved feature, you must *sketch a centerline* to define the **axis of revolution**.
4. Enter the pitch value.
5. Pro/E places you in another Sketcher orientation. Sketch, dimension, and regenerate the section profile to be swept along the trajectory.

If you choose **Norm To Traj**, the profile entities must be tangent to each other (continuous). The profile entities should not have a tangent that is normal to the centerline at any point. The profile starting point defines the sweep trajectory starting point. Modify the starting point using the options **Sec Tools** and **Start Point.**

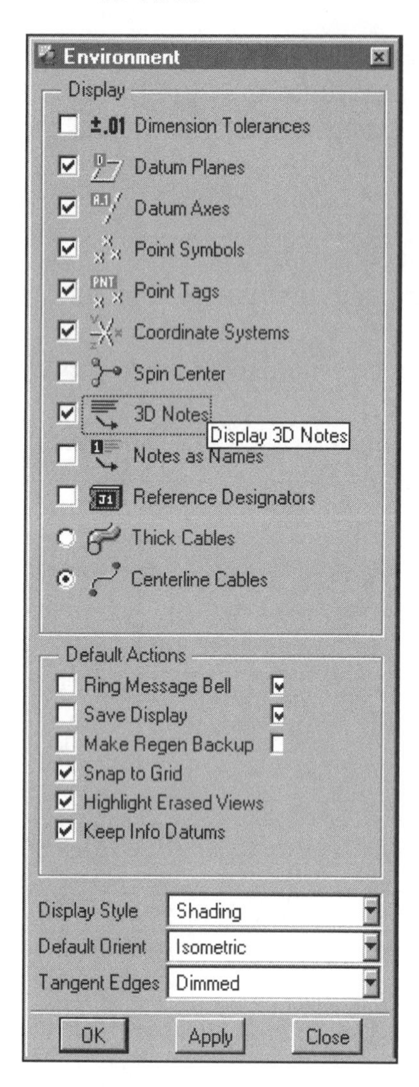

3D Model Notes (Notes)

Model Notes let you attach any number of notes to any object in your model. You can use **Model Notes** to do the following:

* Inform other members of your design workgroup how to review or use a model that you have created.
* Explain how you approached or solved a design problem when defining features of a model.
* Explain changes made to features of a model.
* Embed information about the model.

Creating a Model Note

You can attach a **Model Note** to a feature (or object) by using either the **Model Tree** or the **Set Up** command (Figs. 13.5 and 13.6).

1. To create a model note by using the **Set Up** command, from the PART menu choose **Set Up ⇒ Notes ⇒ New**.
2. [To create a model note by using the **Model Tree,** make the **Model Tree** window active, select the object in the **Model Tree** list to which you want to attach a note, and right-click the item in the **Model Tree** list to display a menu of commands associated with that object (Figure 13.5).] Choose **Note Create ⇒ Part** from the menu.
3. Specify the name for the note in the **Name** field or use the default.
4. Type the text for the note in the **Text** field or click **Insert** to insert text from a file or from another note.
5. (Optional) Select **Symbols** to select from a gallery of symbols to add to the note.
6. (Optional) Select a **Placement** button to specify an option for attaching the note to the displayed object.
7. Click **OK** to create the note and attach the note to the object.
8. You can toggle on and off the note with an **Environment** setting.

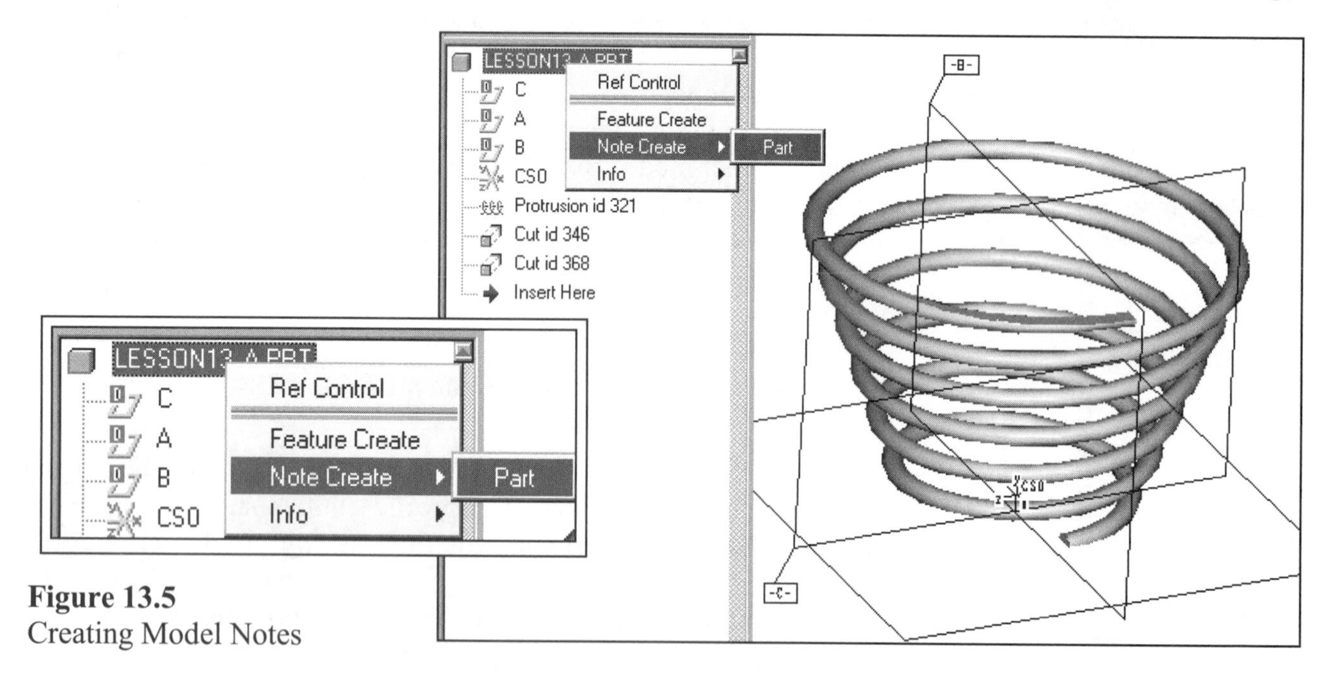

Figure 13.5
Creating Model Notes

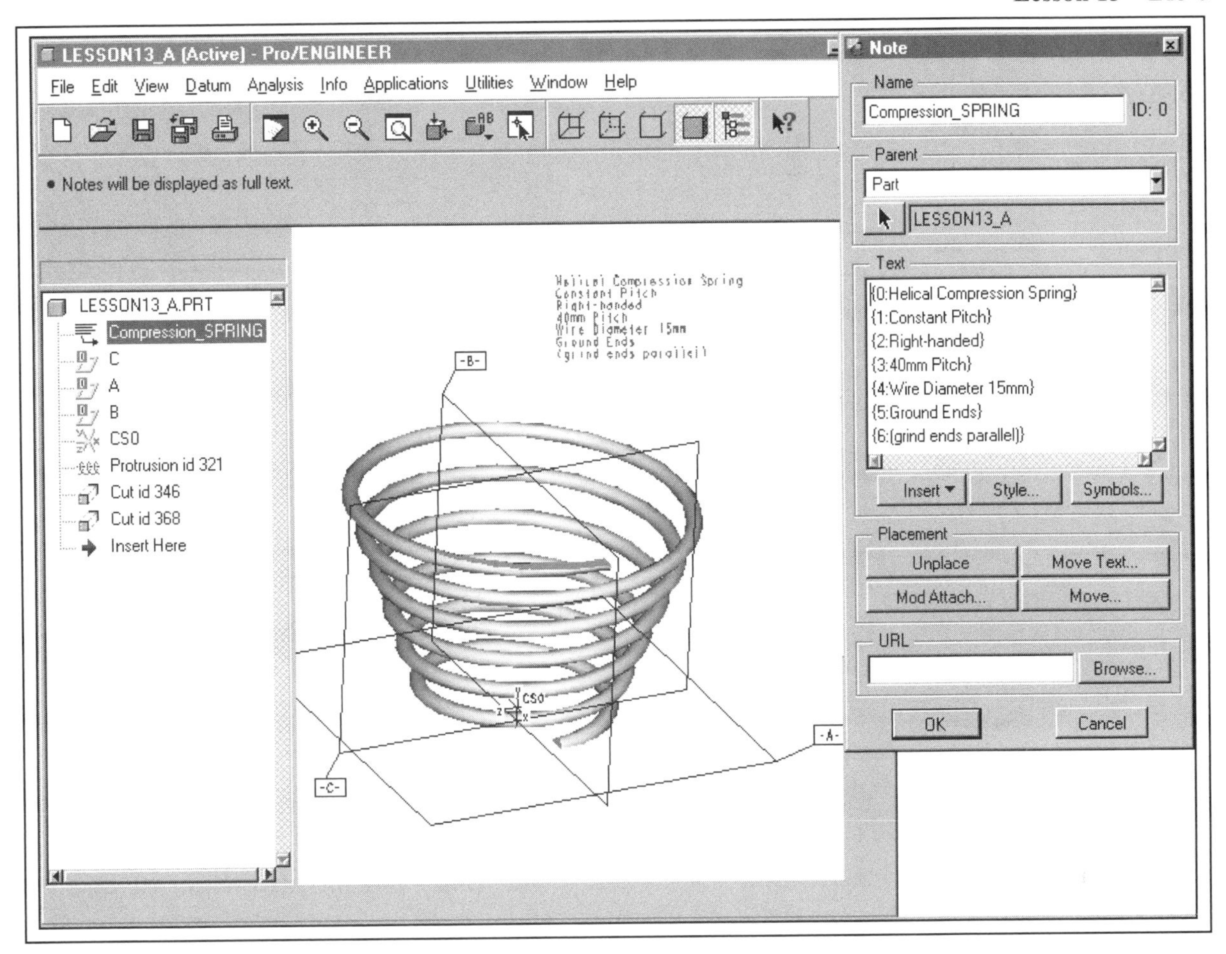

Figure 13.6
Model Notes

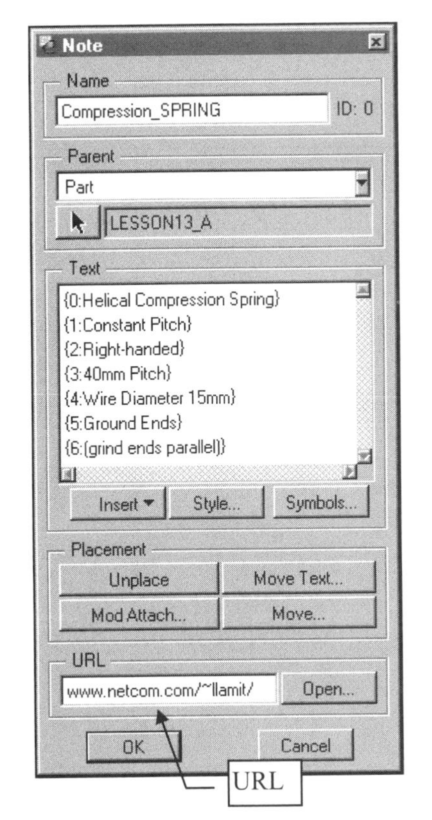

Attaching a URL to an Object

To attach a URL to an object:

1. Follow the procedure to attach a model note to an object.
2. Enter a Web address or Website name in the URL field.
3. (Optional) Click **Open** to start your default Web browser and to load the specified URL into the browser.
4. Click **OK** to create the note.
5. To see the address, choose **Set Up** ⇒ **Notes** ⇒ **Open URL**, and then select a note in the graphics window.

Modifying a Model Note

You can modify, delete, or display **Model Notes** as you would any other object displayed in the **Model Tree**.

1. Make the **Model Tree** window active.
2. Select a **Model Note** in the list.
3. Right-click the note to display a menu of commands.
4. Choose **Delete** to delete the selected **Model Note**.
5. Choose **Modify** to display the dialog box used to create the **Model Note**.
6. Modify the **Model Note** and click **OK** to save the modifications.
7. Choose **Info** ⇒ **Model Info** to display feature information about the model.

Figure 13.7
Helical Compression Spring with Datum Planes and Model Tree

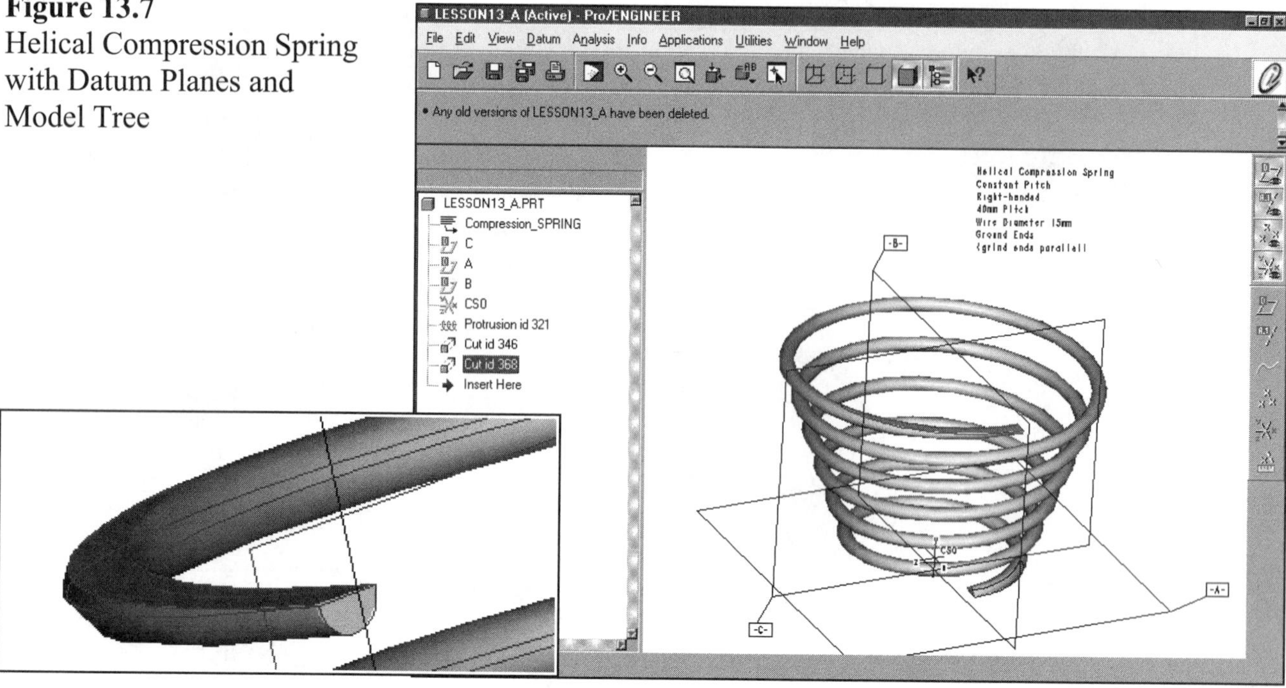

Helical Compression Spring

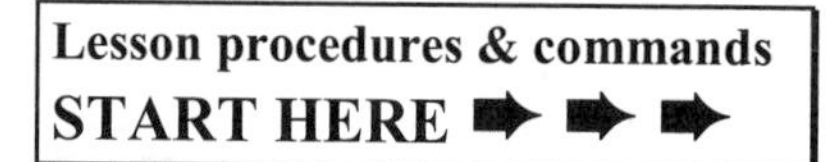

Springs (Fig. 13.7) and other helical features are created with the **Helical Swp** command. A helical sweep is created by sweeping a *section* along a *trajectory* that lies in the *surface of revolution:* the trajectory is defined by both the *profile* of the surface of revolution and the distance between coils. The model for this lesson is a *constant-pitch right-handed helical compression spring with ground ends, a pitch of **40 mm**, and a wire diameter of **15 mm*** (Figs. 13.8 through 13.11).

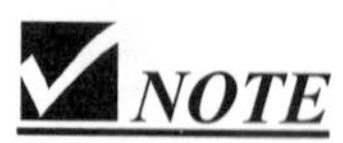

Set up the Helical Compression Spring:

- Material = SPRING_STEEL
- Units = millimeters
- Hidden Line
- No Display
- ✔ Snap to Grid
- Datum TOP = **A**
- Datum FRONT = **B**
- Datum RIGHT = **C**
- Coordinate System = **CS0**

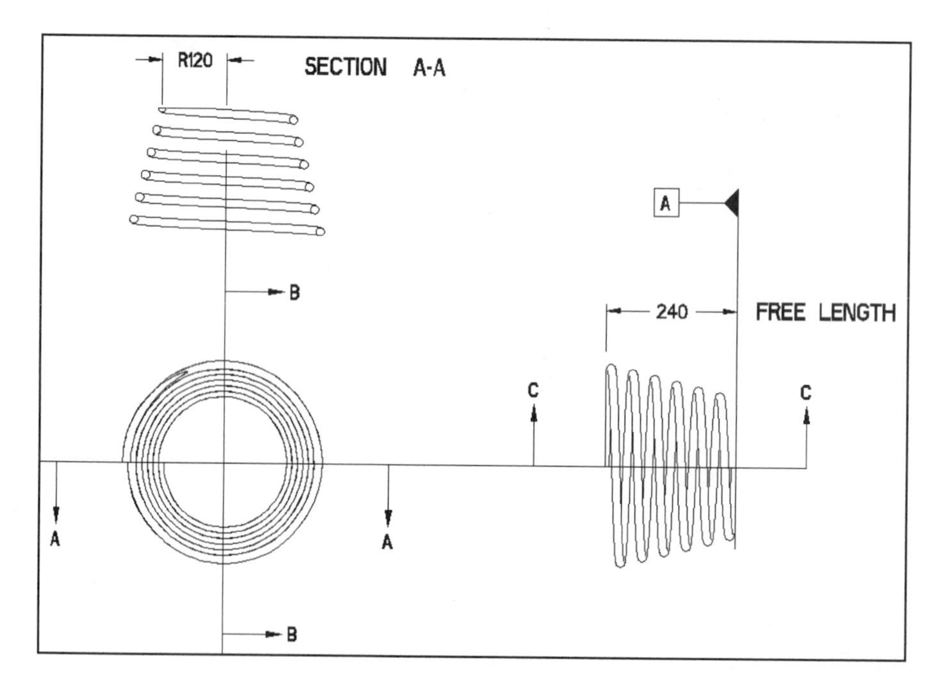

Figure 13.8
Helical Compression Spring

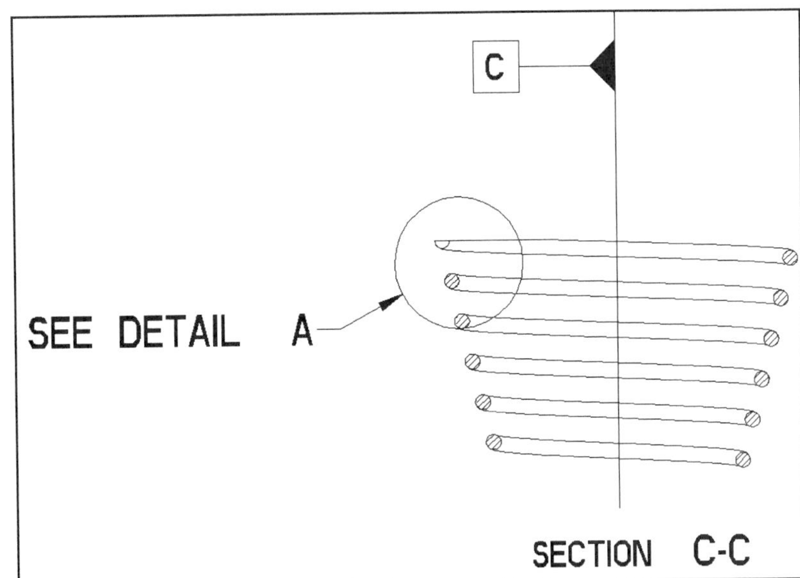

Figure 13.9
Helical Compression Spring Drawing, **SECTION C-C**

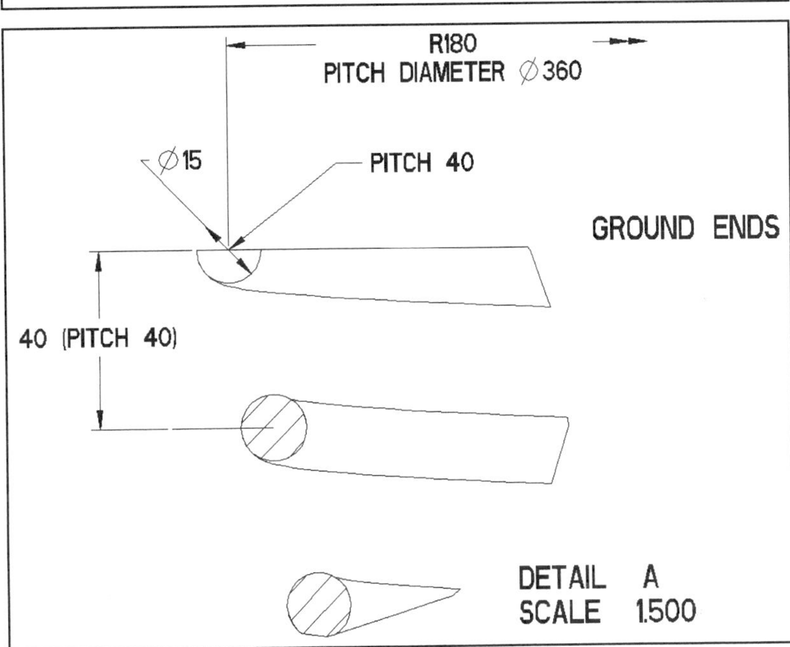

Figure 13.10
Helical Compression Spring Drawing, **DETAIL A**

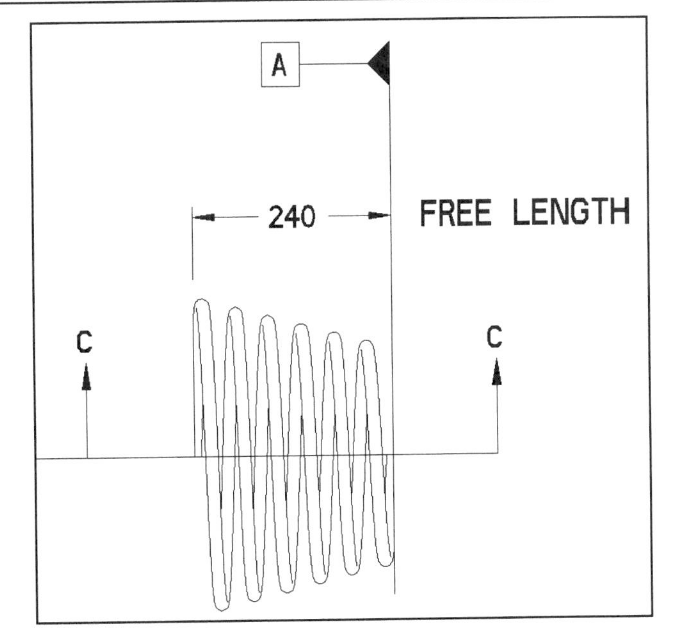

Figure 13.11
Helical Compression Spring Drawing, Right Side View

Helical Swp

?GetHelp

The only protrusion needed for the spring is created with an advanced protrusion command, **Helical Swp**. Start the part with the usual default datum planes and coordinate system (Fig. 13.12). Choose the following commands to create the first protrusion:

Feature ⇒ Create ⇒ Protrusion ⇒ Advanced ⇒ Solid ⇒ Done ⇒ Helical Swp ⇒ Done ⇒ Constant ⇒ Thru Axis ⇒ Right Handed ⇒ Done ⇒ (pick datum **B** (**FRONT**) **⇒ Okay ⇒ Top** (pick datum **A**, originally **TOP**) **⇒ Sketch** (menu bar) **⇒ Intent Manager** (off) **⇒ Sketch ⇒** [sketch a line (Fig. 13.13)] **⇒ Regenerate ⇒ Alignment** [align the lower end of the line to datum **A** (**TOP**)]

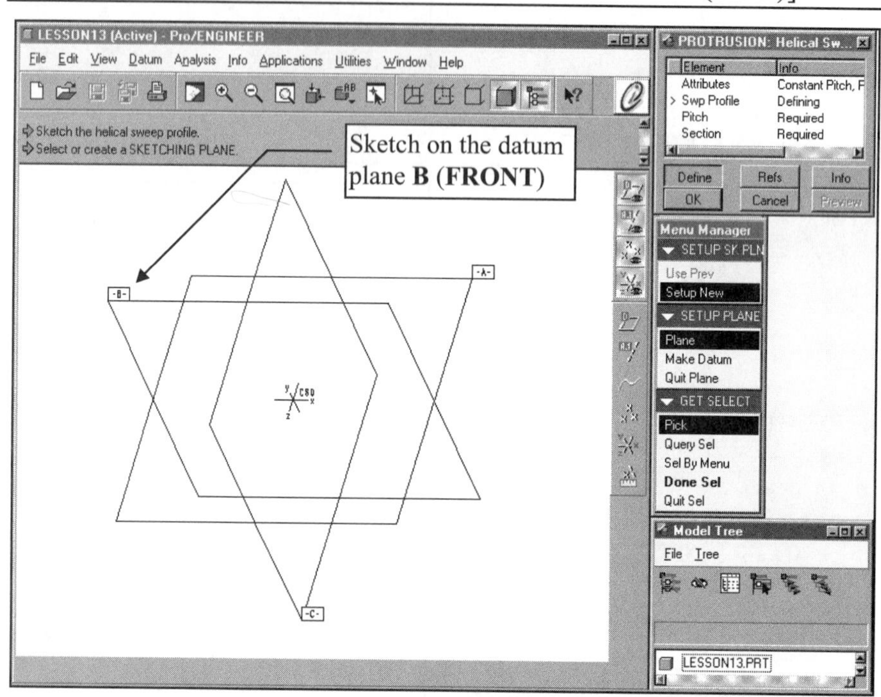

Figure 13.12
Default Datum Planes and Coordinate System

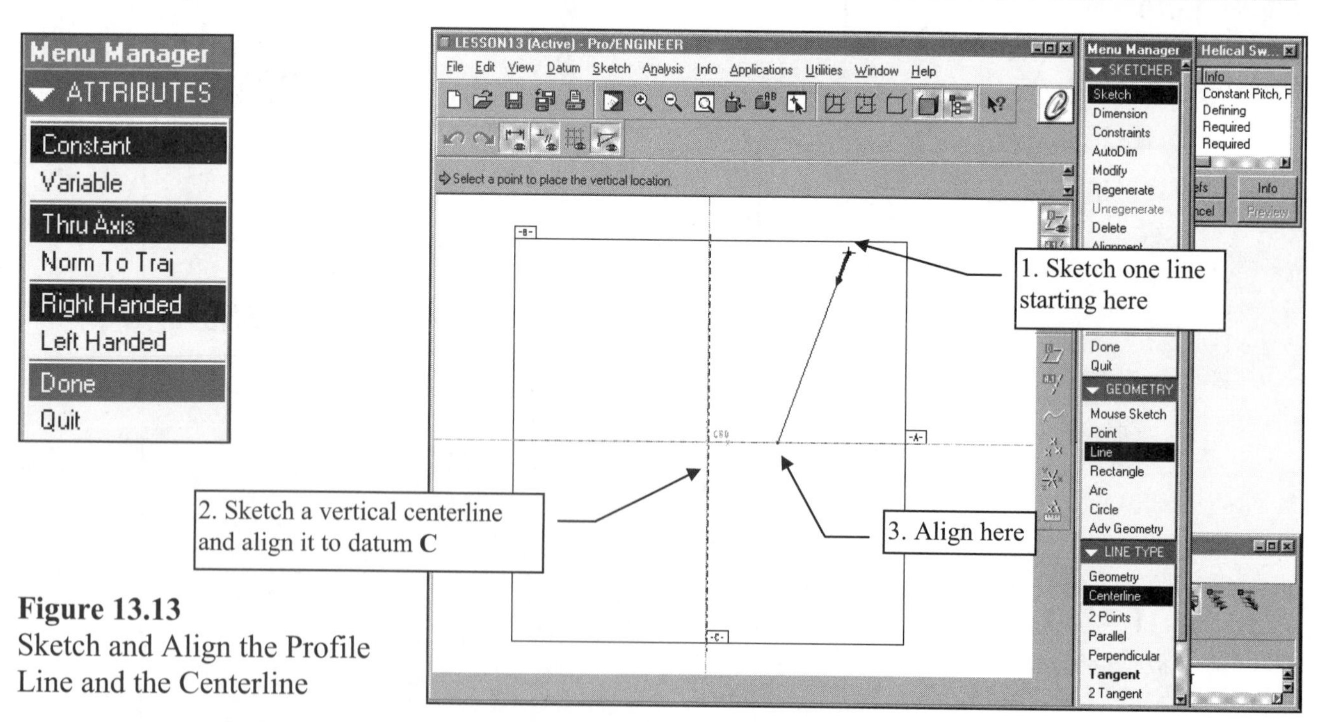

Figure 13.13
Sketch and Align the Profile Line and the Centerline

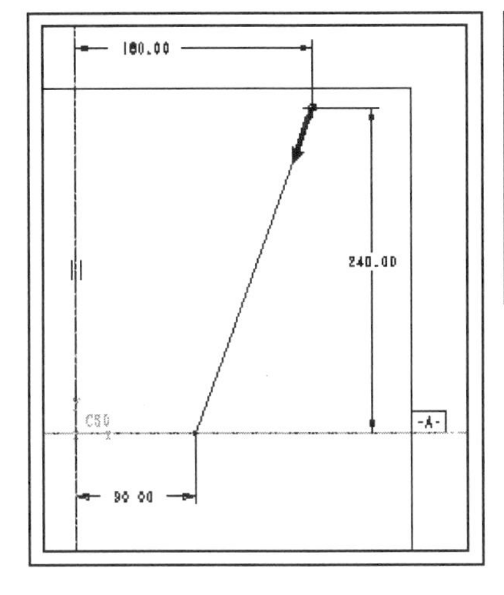

Regenerate ⇒ Sketch ⇒ Line ⇒ Centerline ⇒ Vertical ⇒ (add a vertical axis line along datum **C** (**RIGHT**)] ⇒ **Alignment** (align the centerline with datum **C**) ⇒ **Regenerate ⇒ Dimension** (add dimensions as shown in Figure 13.14) ⇒ **Regenerate ⇒ Modify** (change the values to the design sizes) ⇒ **Regenerate ⇒ Done ⇒** (enter the pitch value **40** at the prompt, as shown in Fig. 13.15) ⇒ ✔

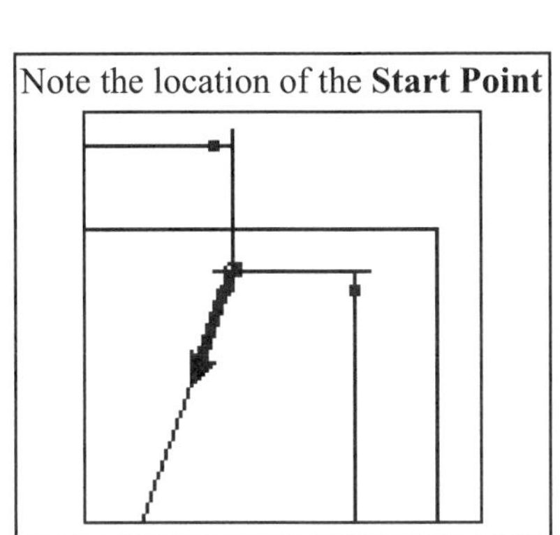

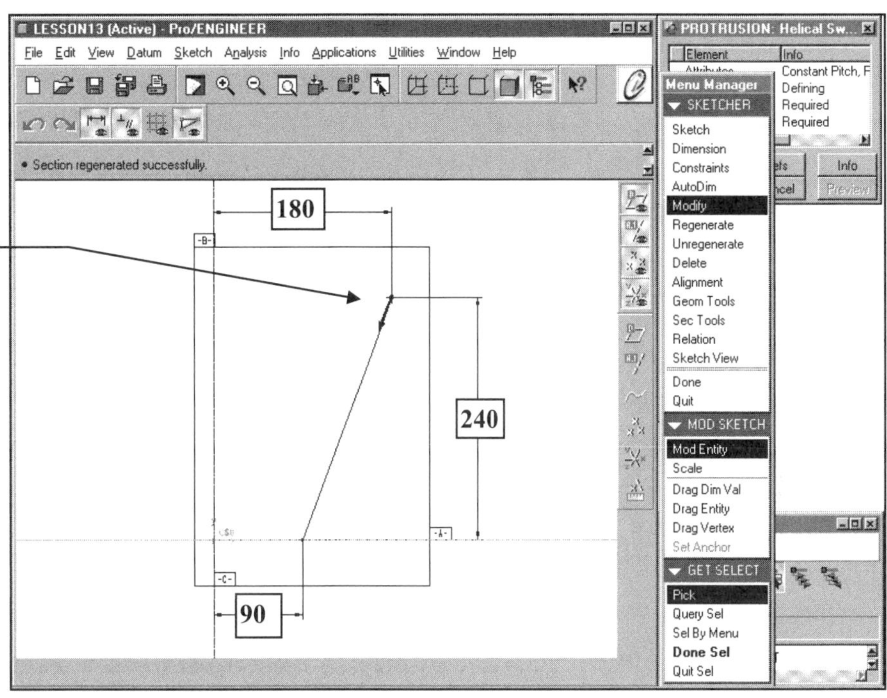

Figure 13.14
Modify the Sketch

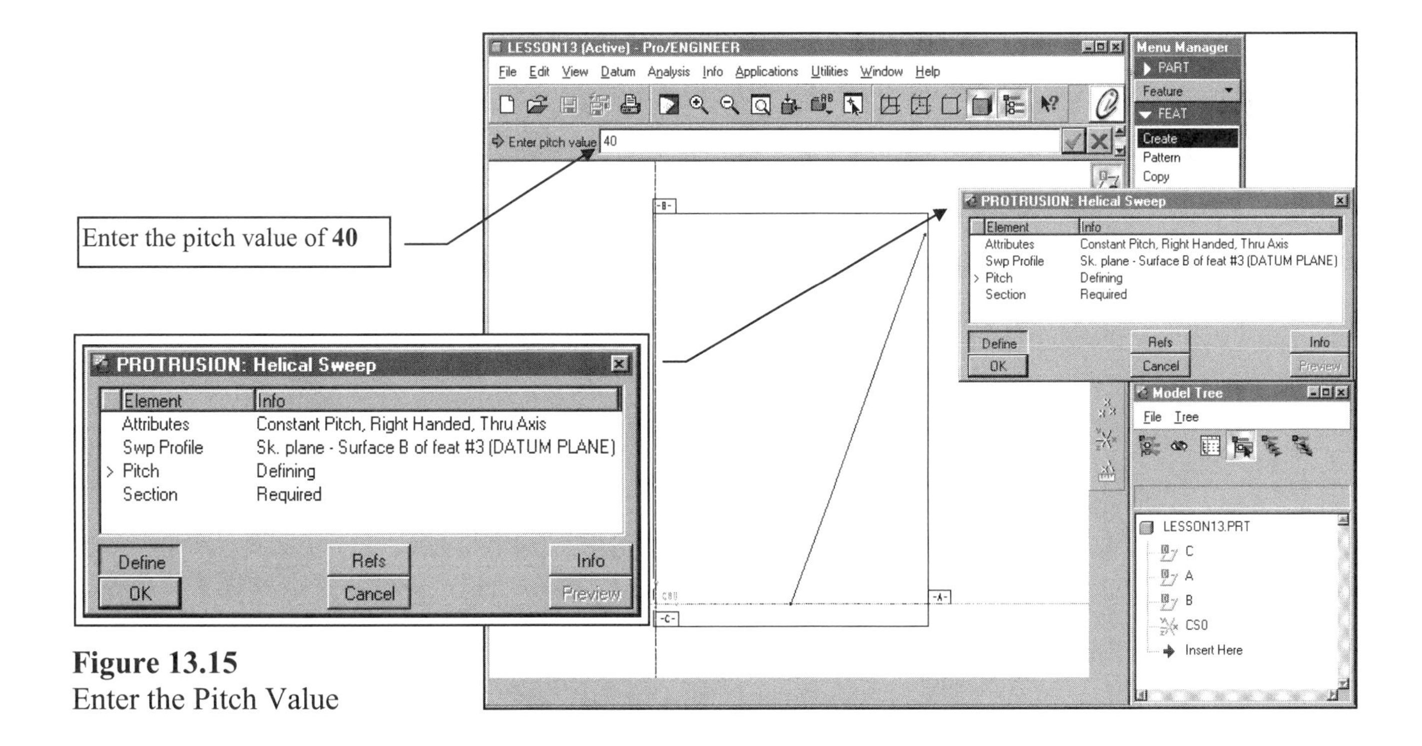

Figure 13.15
Enter the Pitch Value

Sketch the section geometry of the spring (here it is a circle):

Sketch ⇒ **Intent Manager** ⇒ **Sketch** ⇒ **Circle** ⇒ **Center/Point** (Fig. 13.16) ⇒ **Regenerate** ⇒ **Alignment** (align the center of circle with the *crosshairs*) ⇒ **Regenerate** ⇒ **Dimension** (add wire diameter) ⇒ **Regenerate** ⇒ **Modify** (type **15**) ⇒ ✔ ⇒ **Regenerate** ⇒ **Done** ⇒ **OK** ⇒ **View** ⇒ **Default** ⇒ **Shade** icon (Fig. 13.17)

Sketch a circle as the section geometry (wire diameter)

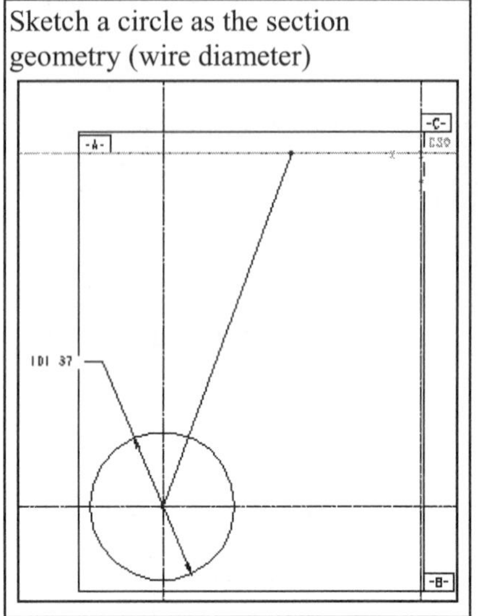

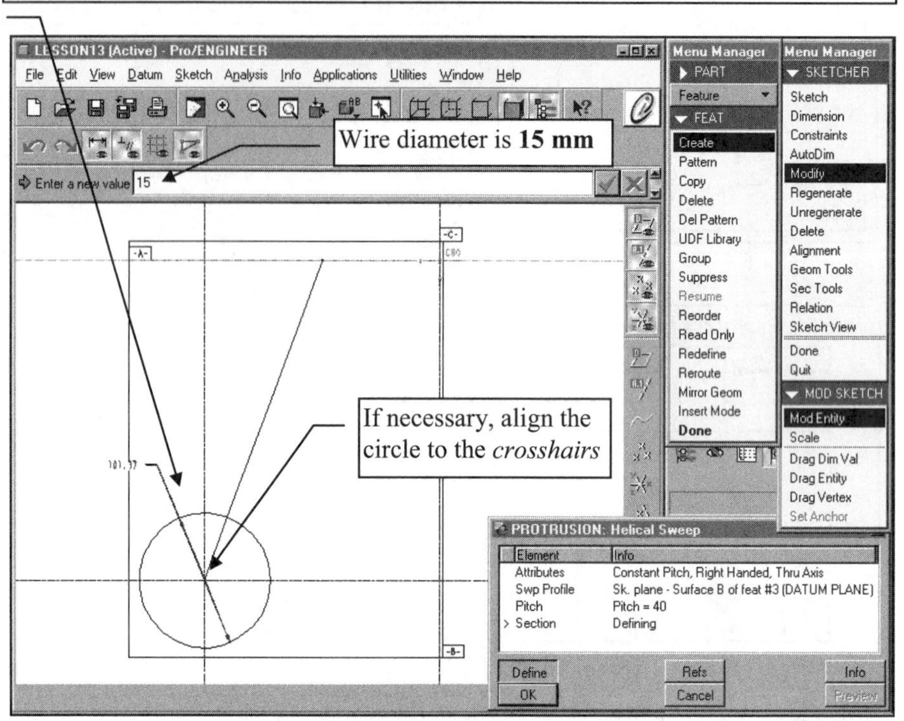

Figure 13.16
Sketching the Circle

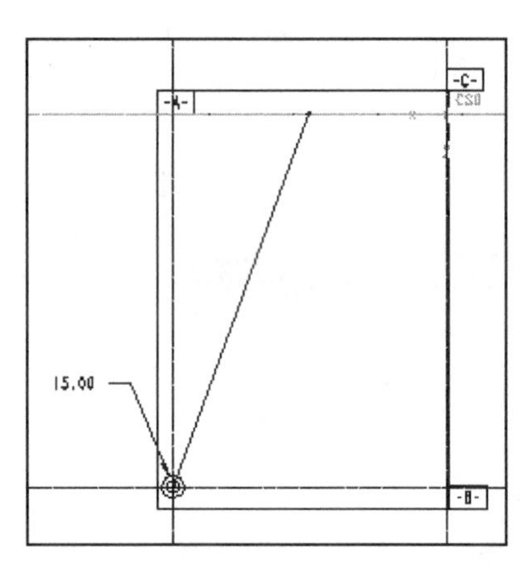

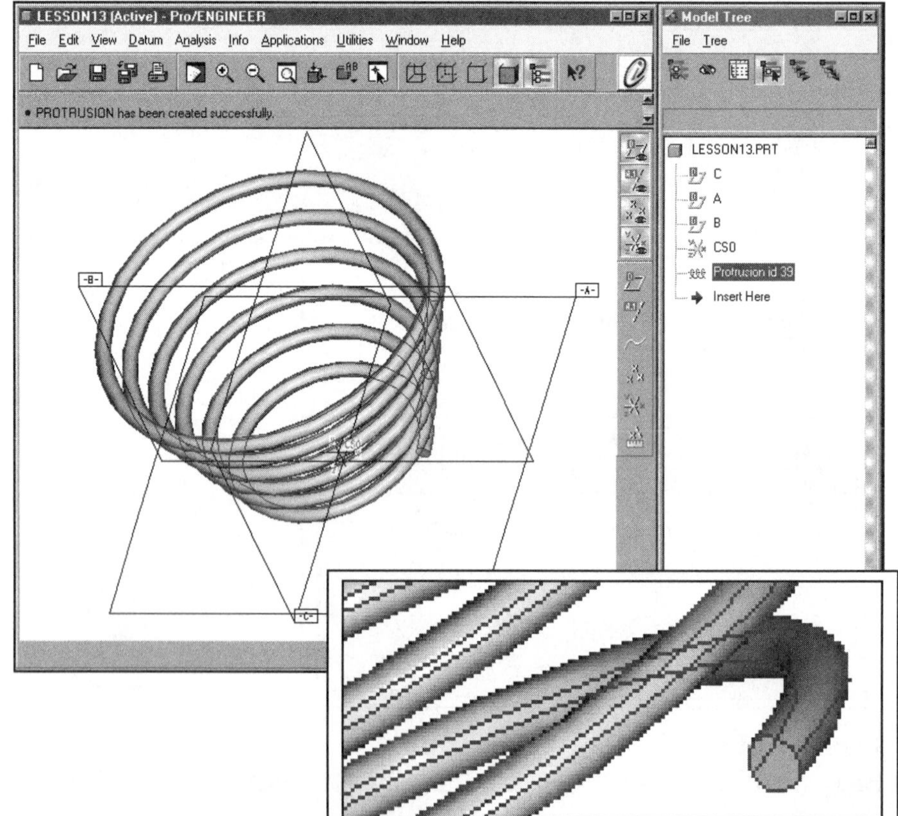

File ⇒ **Save** ⇒ ✔

Figure 13.17
Completed Helical Sweep

The spring is almost complete (Fig. 13.18). Add the cut line to create the *ground end* as shown in Figure 13.19:

File ⇒ Save ⇒ ✔ ⇒ Create ⇒ Cut ⇒ Extrude ⇒ Solid ⇒ Done ⇒ Both Sides ⇒ Done ⇒ Plane (pick datum **C**) **⇒ Okay ⇒ Bottom** [pick datum **A** (Fig. 13.18)] **⇒ Sketch ⇒ Intent Manager** (off) **⇒ Sketch ⇒ Line ⇒ Horizontal ⇒ Regenerate ⇒ Alignment** (align the line to datum **A**, and the end of the line to datum **B**) **⇒ Regenerate ⇒ Dimension ⇒ Regenerate** (there is no need to modify the dimension) **⇒ Done ⇒ Flip** (if necessary) **⇒ Okay ⇒ Thru All ⇒ Done ⇒ Thru All ⇒ Done ⇒ OK** (Fig. 13.20)

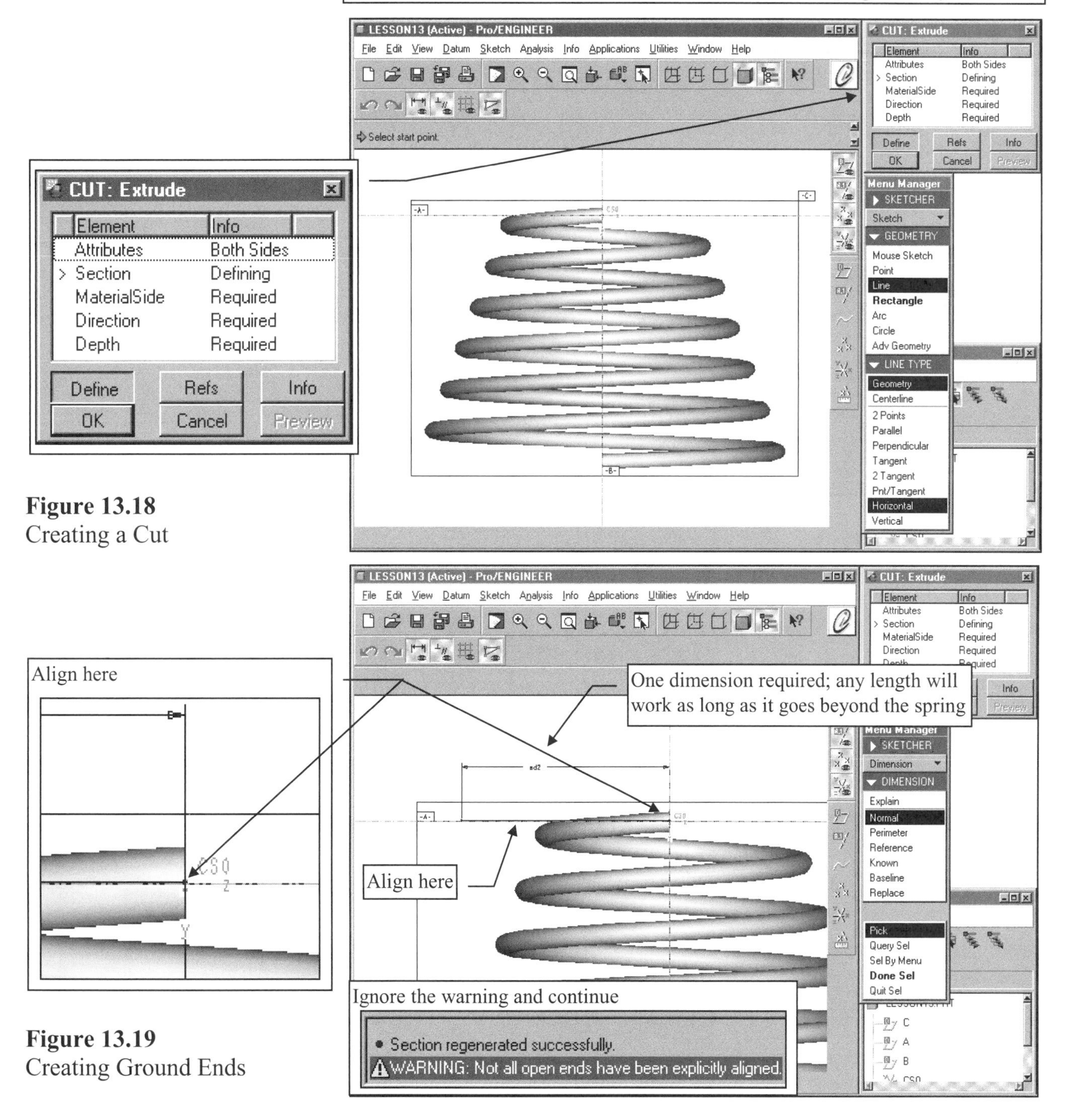

Figure 13.18
Creating a Cut

Figure 13.19
Creating Ground Ends

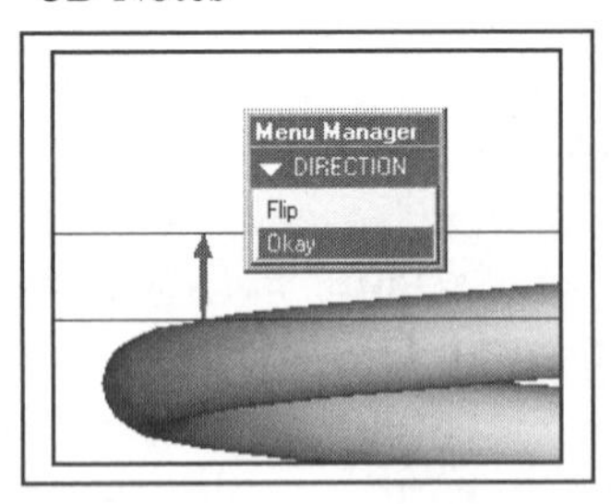

You will get a message that says **"Not all open ends have been explicitly aligned."** Ignore this warning and continue (Fig. 13.19).

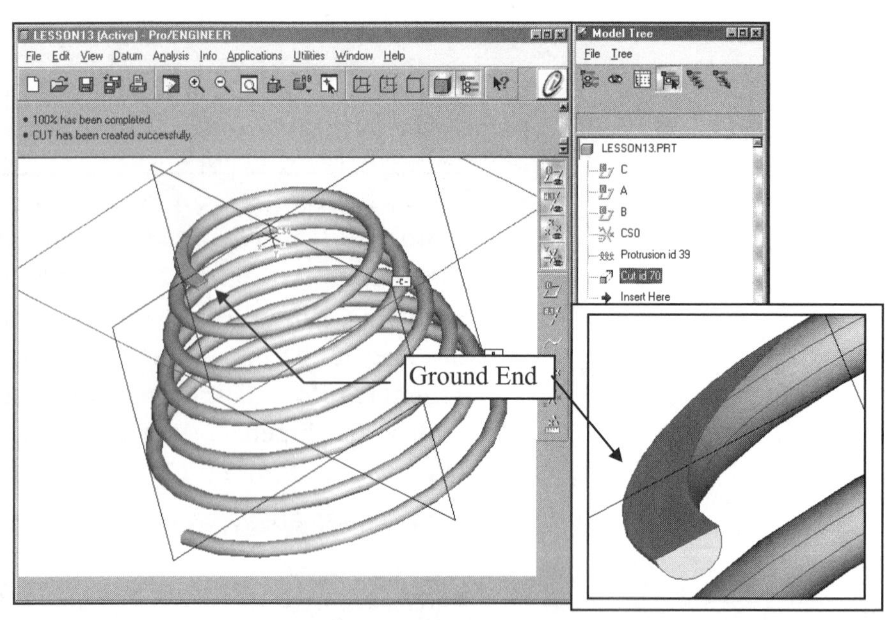

Figure 13.20
Completed Ground End

The second ground end is cut in Figure 13.21 using similar commands. The completed spring is shown in Figure 13.22.

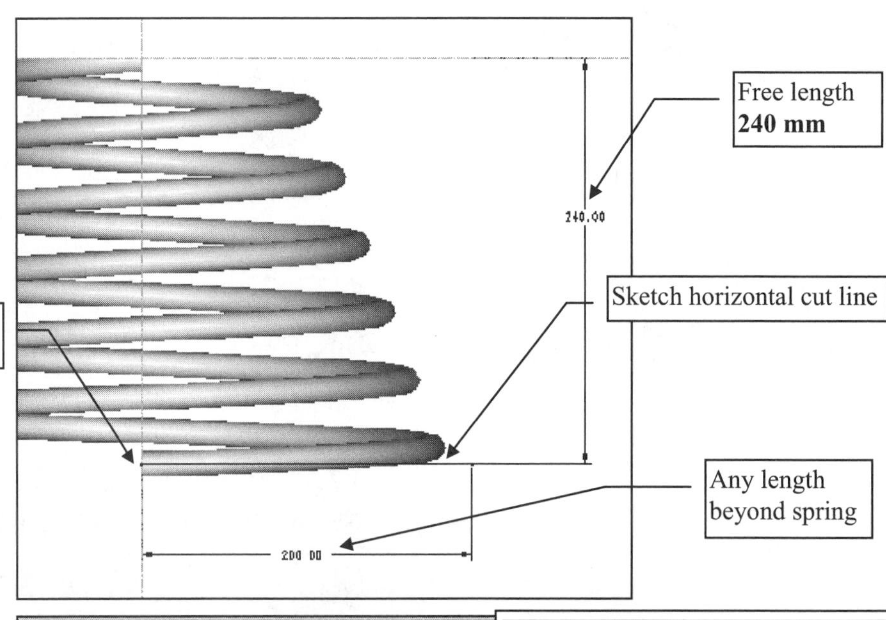

Figure 13.21
Creating the Second Ground End

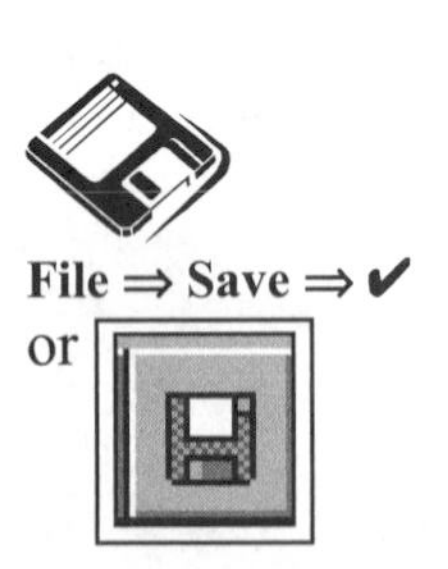

File ⇒ Save ⇒ ✔
or

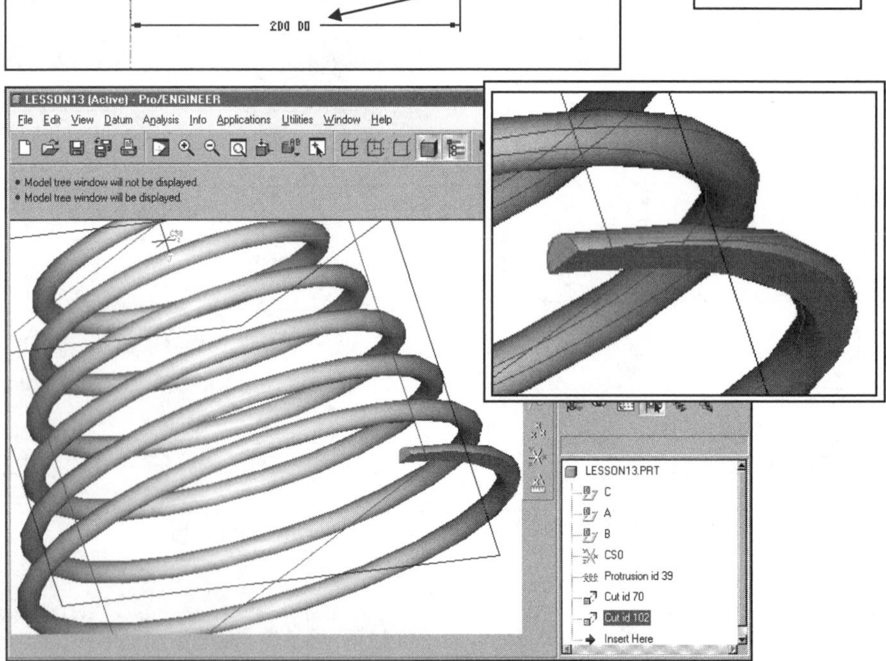

Figure 13.22
Completed Spring

Save the Helical Compression Spring by choosing the commands **File ⇒ Save As** and giving it a different name--for example, **HEL_COMP_SPR_GRND_ENDS**. Rename the file you are working on by using **File ⇒ Rename**, and give it a name such as **HEL_EXT_SPR_MACH_ENDS**. Figure 13.23 provides an **ECO** for the new spring. Delete the existing ground ends and modify the pitch to **10**. Figure 13.24 shows the wire diameter changed to **7.5 mm** and the pitch changed to **10 mm.** The ground ends are deleted.

Complete the extension spring as shown in Figures 13.25 through 13.33. The free length is to be **120 mm.** The large radius will now be **90 mm**, and the small radius is **60 mm**.

E C O

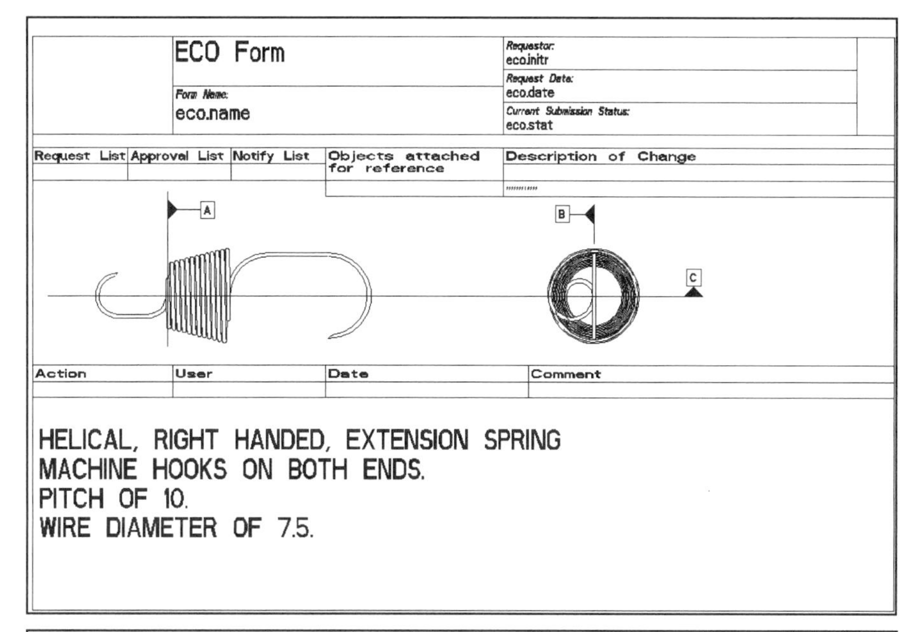

ECO Form

Form Name: eco.name

Requestor: eco.initr

Request Date: eco.date

Current Submission Status: eco.stat

Request List	Approval List	Notify List	Objects attached for reference	Description of Change

Action	User	Date	Comment

HELICAL, RIGHT HANDED, EXTENSION SPRING
MACHINE HOOKS ON BOTH ENDS.
PITCH OF 10.
WIRE DIAMETER OF 7.5.

Figure 13.23
ECO to Create a Helical Extension Spring

File ⇒ Save ⇒ ✔

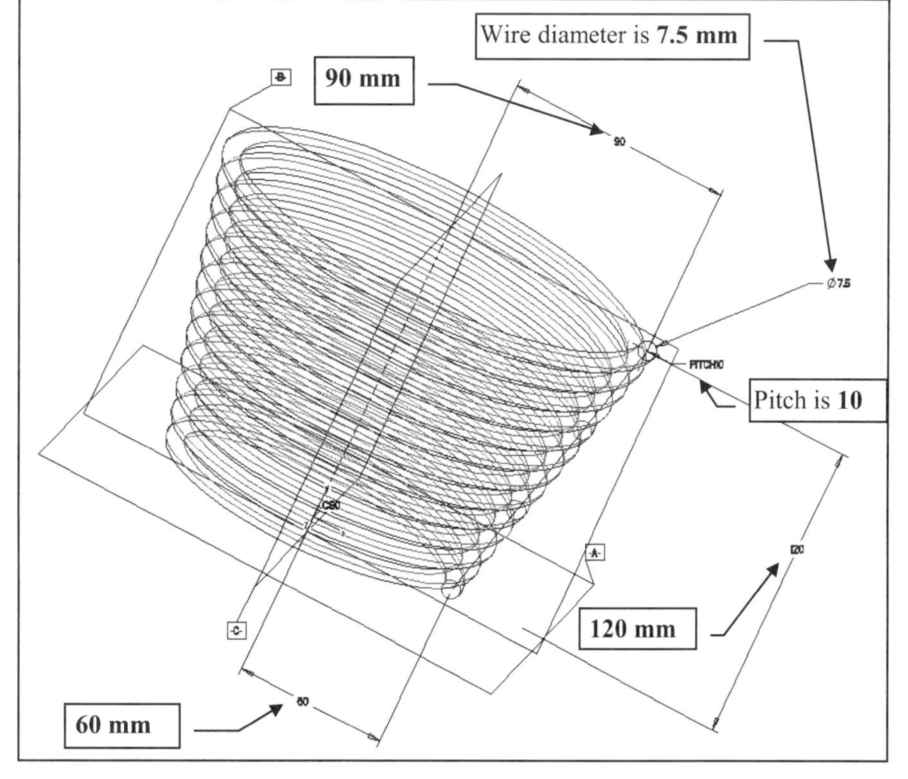

Figure 13.24
ECO Changes

Figure 13.25
Helical Extension Spring with Machine Hook Ends

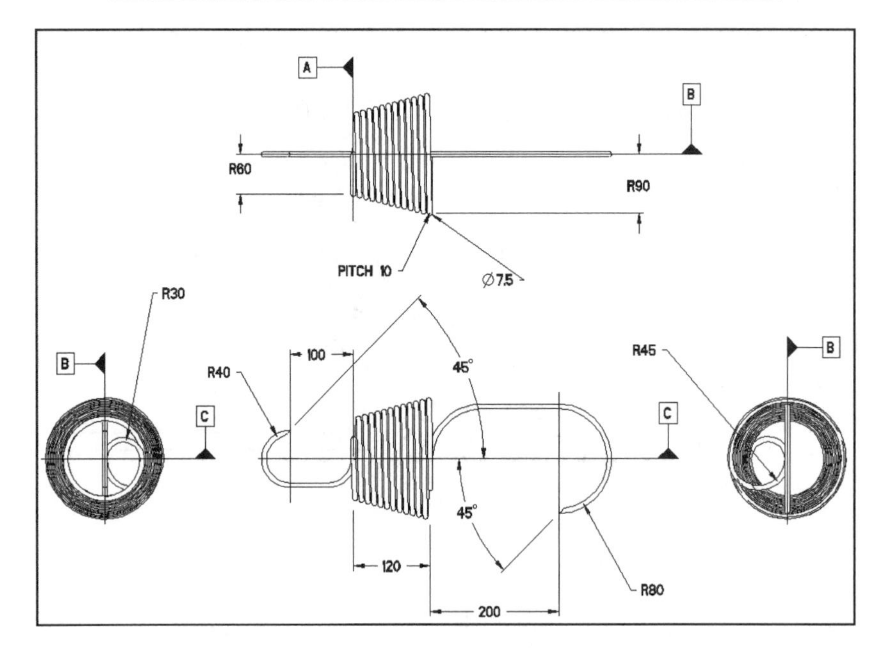

Figure 13.26
Detail Drawing of Helical Extension Spring with Machine Hook Ends

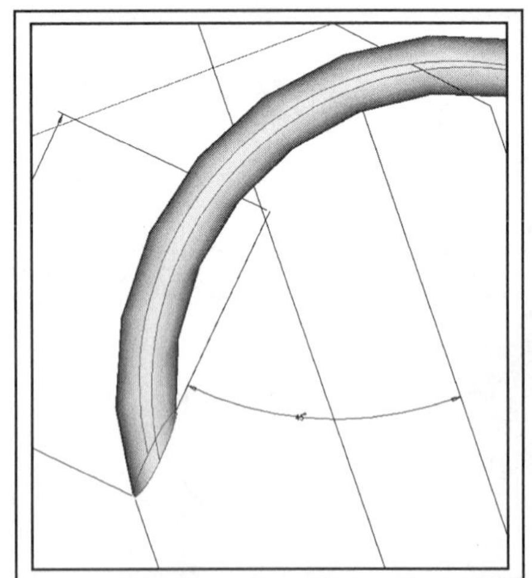

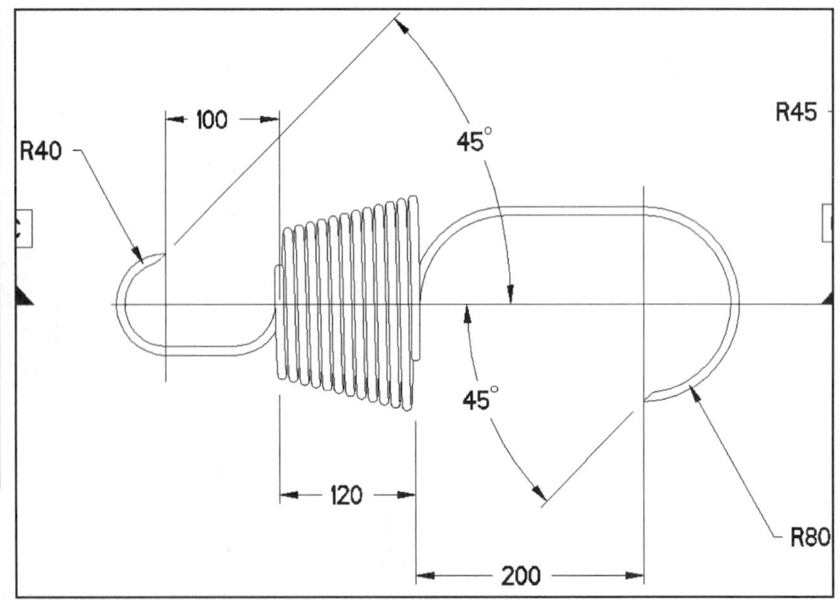

Figure 13.27
Front View

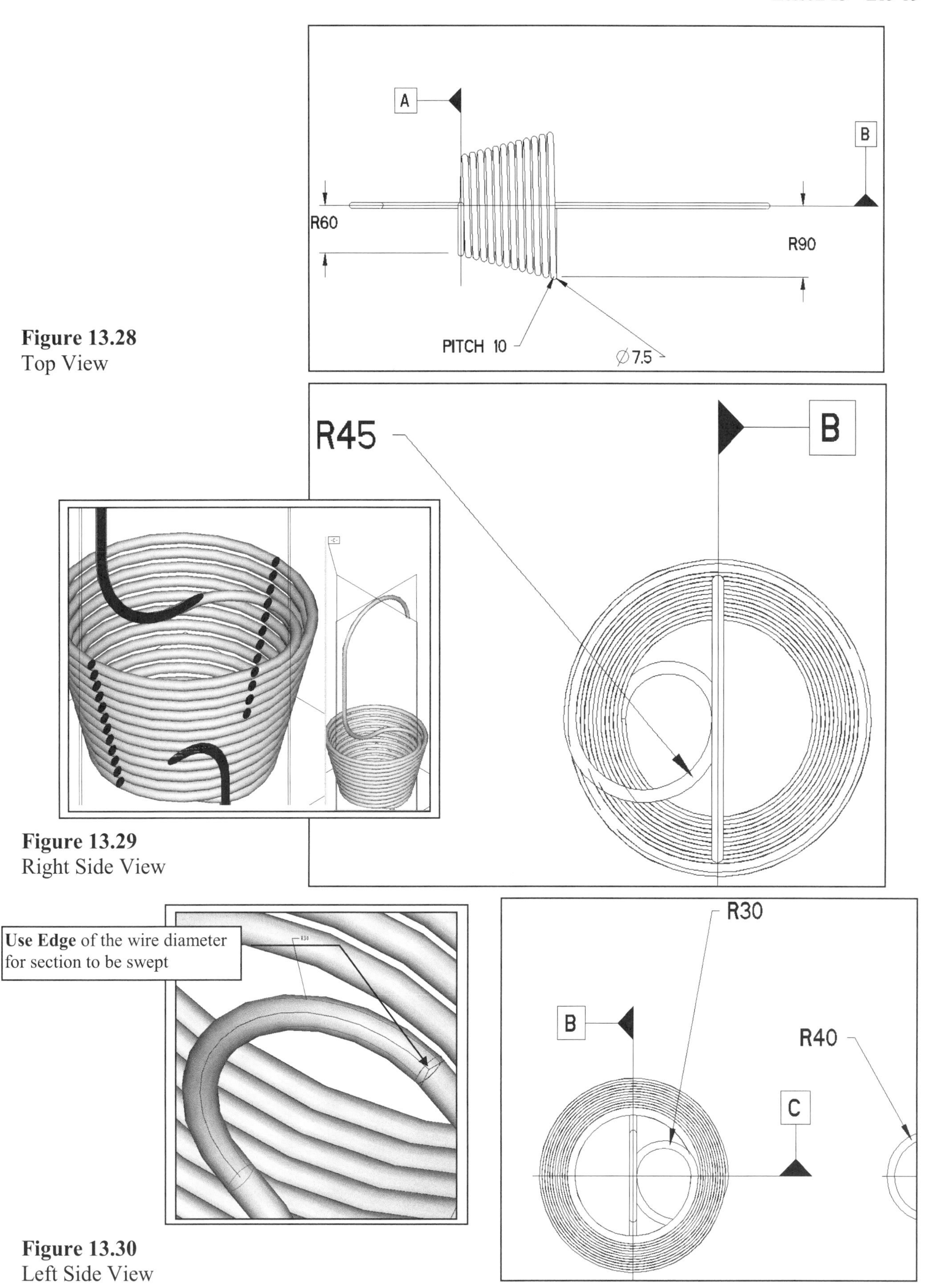

Figure 13.28
Top View

Figure 13.29
Right Side View

Figure 13.30
Left Side View

Create the machine hooks using simple sweeps and cuts, as shown in Figures 13.31 through 13.33.

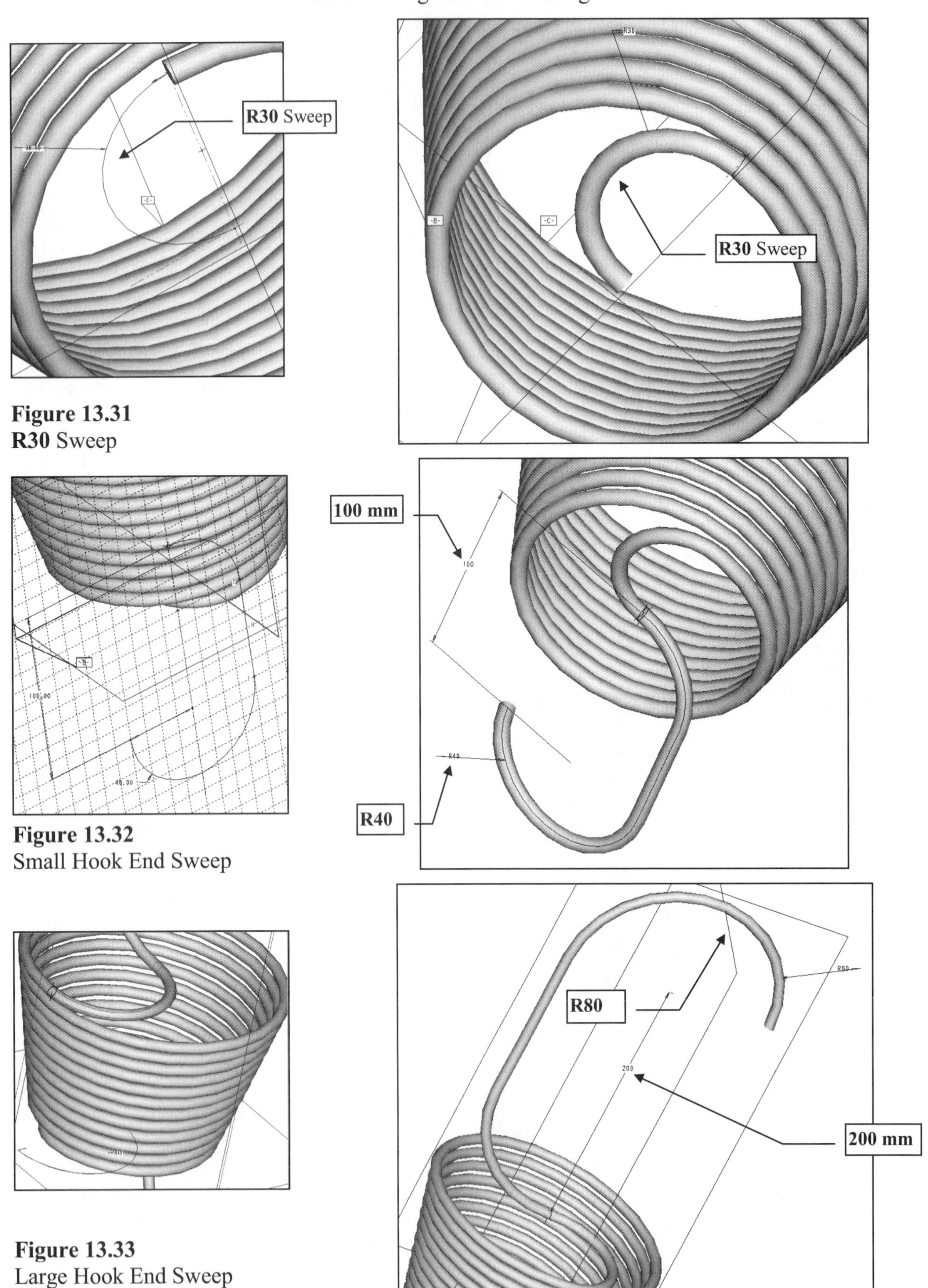

Figure 13.31
R30 Sweep

Figure 13.32
Small Hook End Sweep

Figure 13.33
Large Hook End Sweep

Model Notes

When you attach a note to an object, the object is considered the "parent" of the note. Deleting the parent deletes all of the notes of the parent. You can attach model notes anywhere in the model; they do not have to be attached to a parent.

Here we will add a note to the part and describe the spring. Open the saved spring file that has the ground ends (Fig. 13.34). Choose the following commands:

Set Up ⇒ **Notes** ⇒ **New** ⇒ (type **Compression_SPRING** as the name of the note; no spaces are allowed in the name) ⇒ pick in the **Text** area and type the note (Fig. 13.35):

Helical Compression Spring
Constant Pitch
Right-handed
40 mm Pitch
Wire Diameter 15 mm
Ground Ends
(grind ends parallel)

⇒ **Place** ⇒ **No Leader** ⇒ **Standard** ⇒ **Done** (pick a place on the screen to place the note (Fig. 13.36) ⇒ **OK** ⇒ **Done/Return** ⇒ **Done**

You can toggle the note on and off with the Environment setting ✔**3D Notes** (Fig. 13.37). You can make the note appear in the Model Tree by choosing **Tree** (from the Model Tree) ⇒ **Item Display** ⇒ ✔**Notes** ⇒ **OK** ⇒ **File** ⇒ **Save** ⇒ ✔.

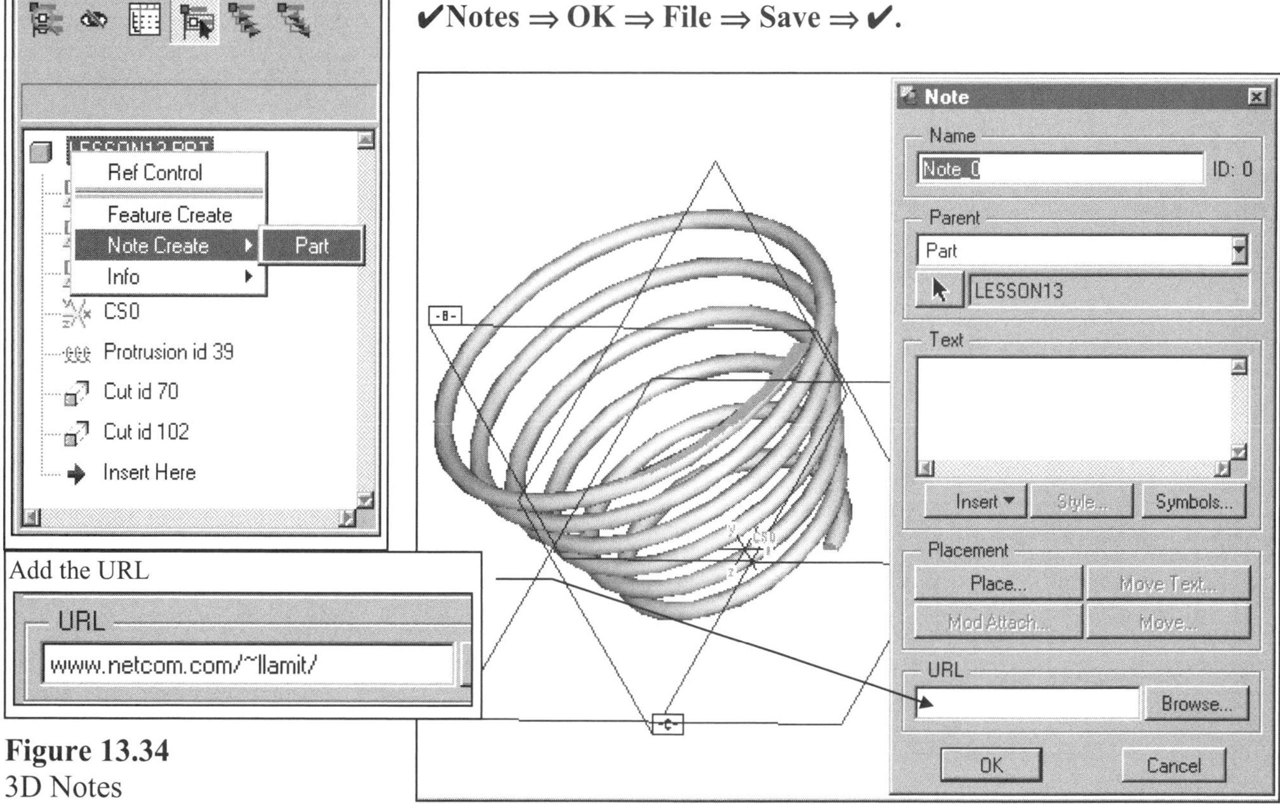

Figure 13.34
3D Notes

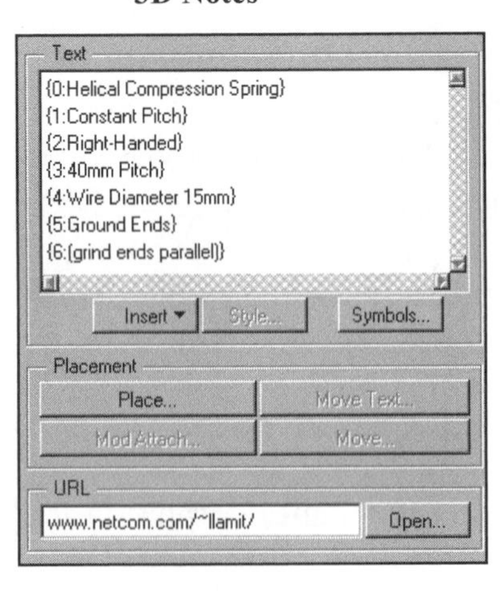

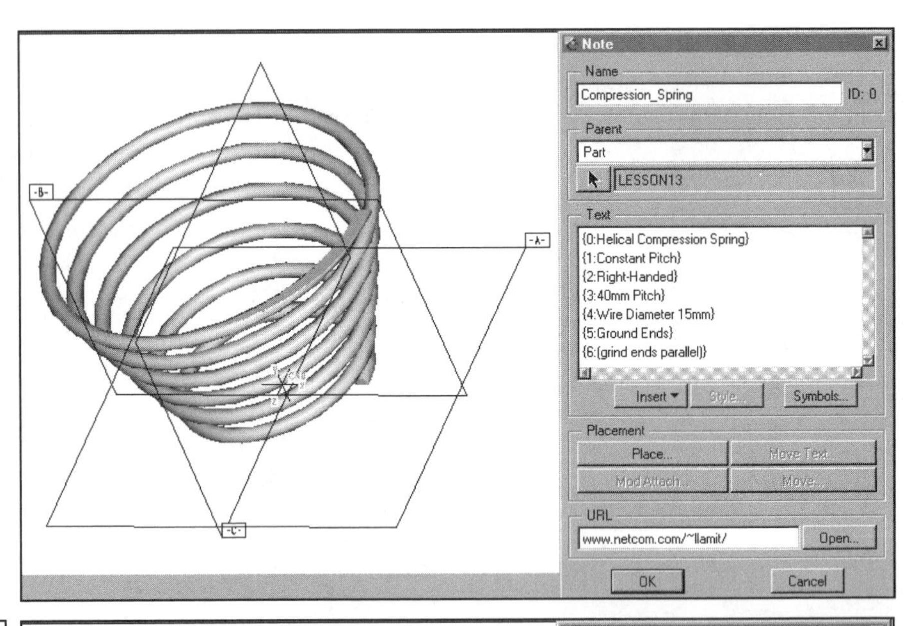

Figure 13.35
Creating the Note

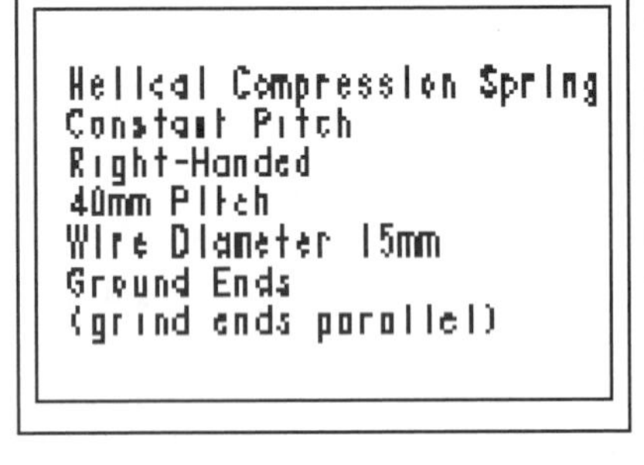

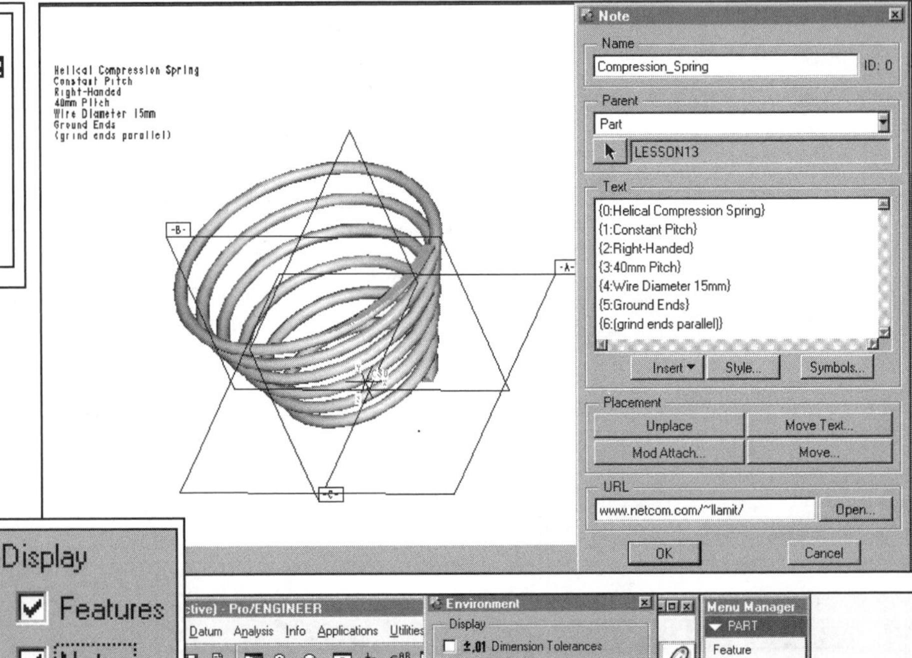

Figure 13.36
Placing the Note

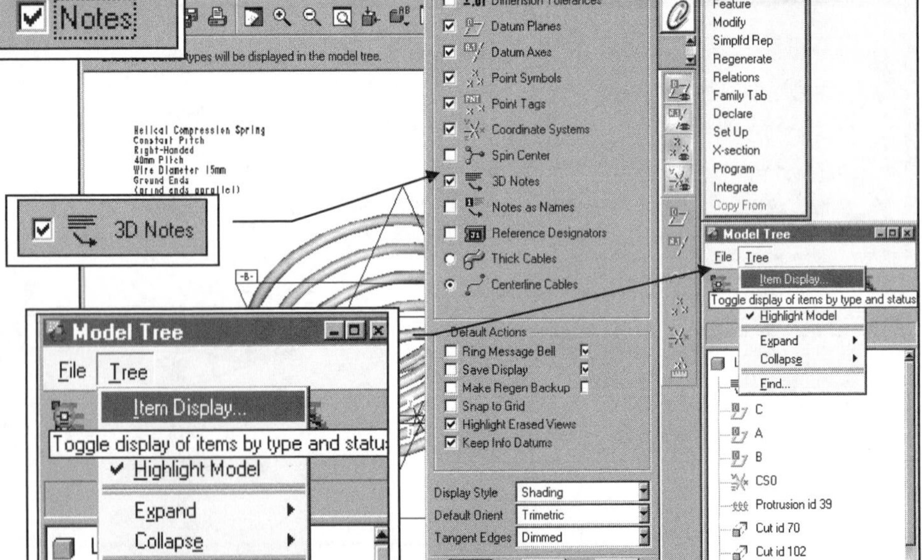

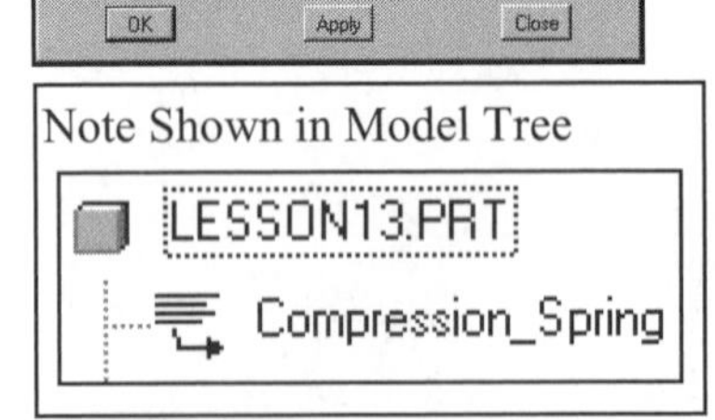

Figure 13.37
Environment Window with **3D Notes** Checked and Model Tree Showing Notes

Lesson 13 Project

Convex Compression Spring

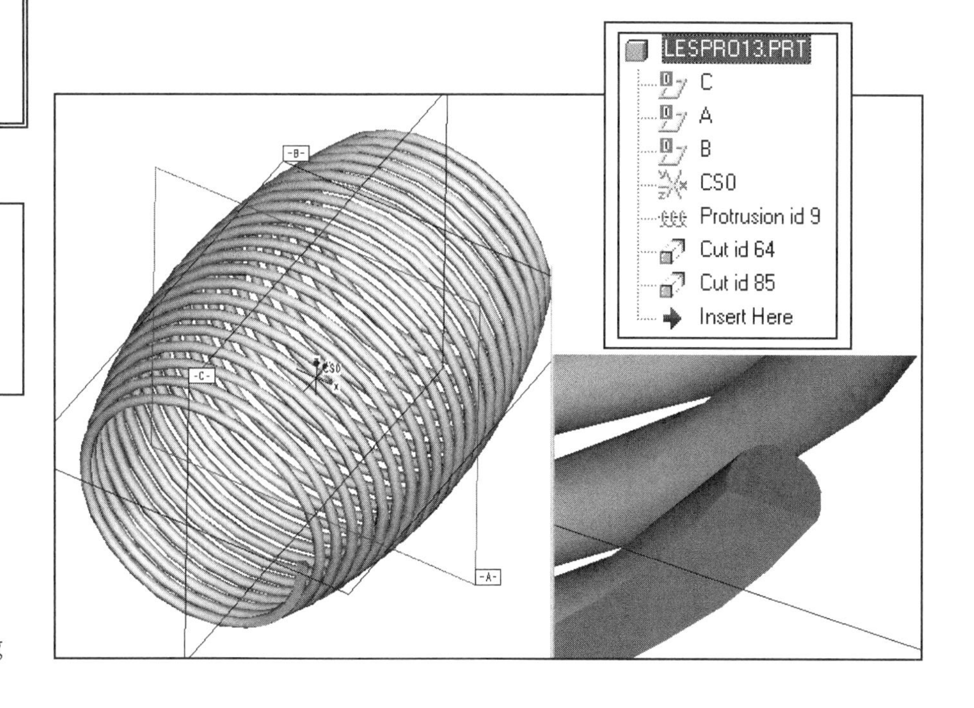

Figure 13.38
Convex Compression Spring

Convex Compression Spring

This **lesson project** is a **Convex Compression Spring**. This project uses commands similar to those for the **Helical Compression Spring**. Create the part shown in Figures 13.38 through 13.46. Analyze the part and plan out the steps and features required to model it. Use the **DIPS** in Appendix D to plan out the feature creation sequence and the parent-child relationships for the part. The spring is made of *spring steel*. Add **3D Notes** describing the spring.

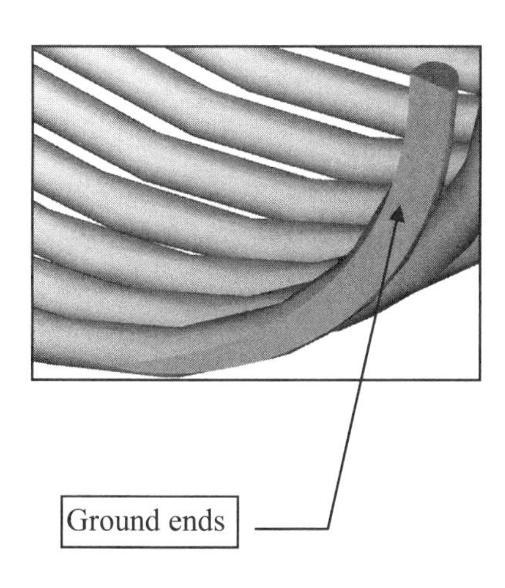

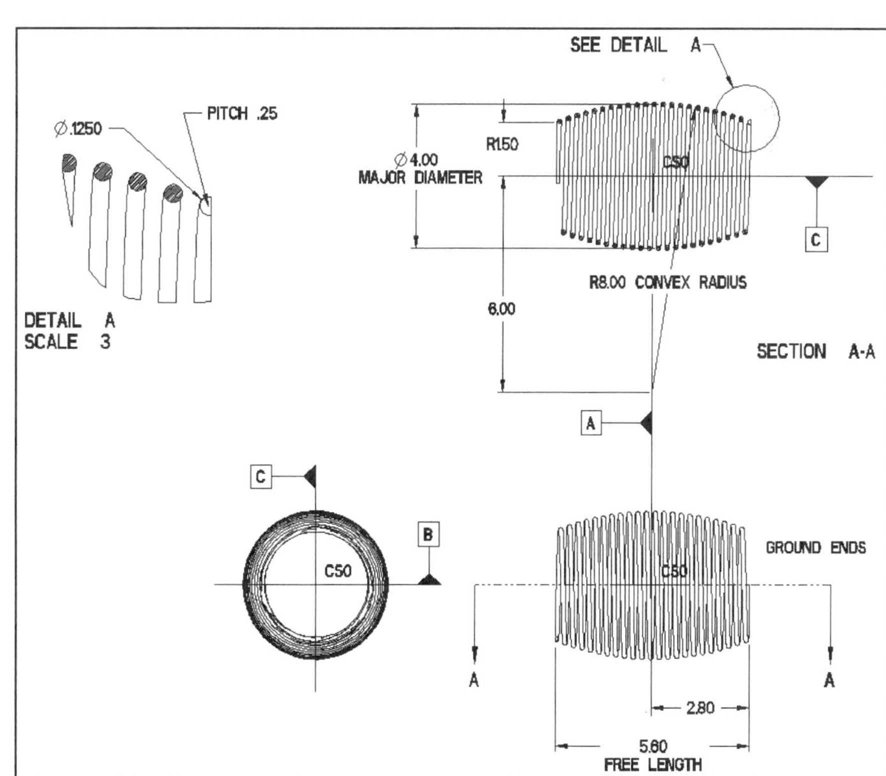

Figure 13.39
Convex Compression Spring, Detail Drawing

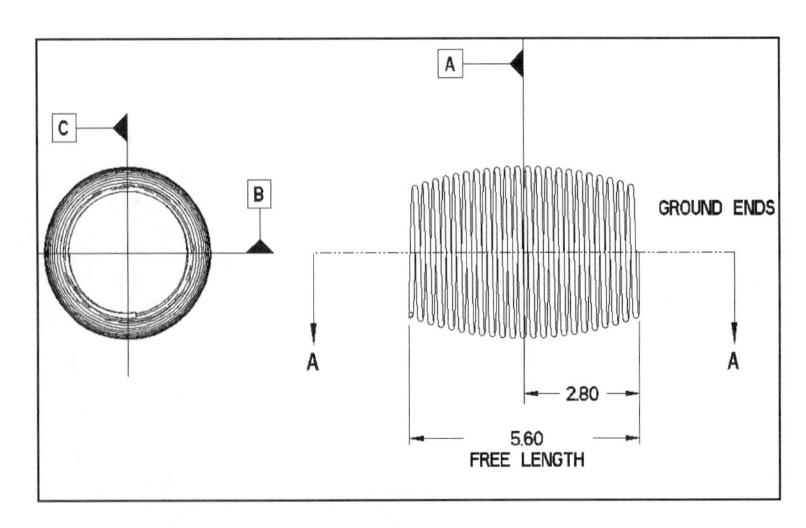

Figure 13.40
Convex Compression Spring Drawing, Front and Left Side Views

SEE DETAIL A
R1.50
∅4.00
MAJOR DIAMETER
C
R8.00 CONVEX RADIUS
6.00
SECTION A-A
A

Figure 13.41
Convex Compression Spring Drawing, Top View

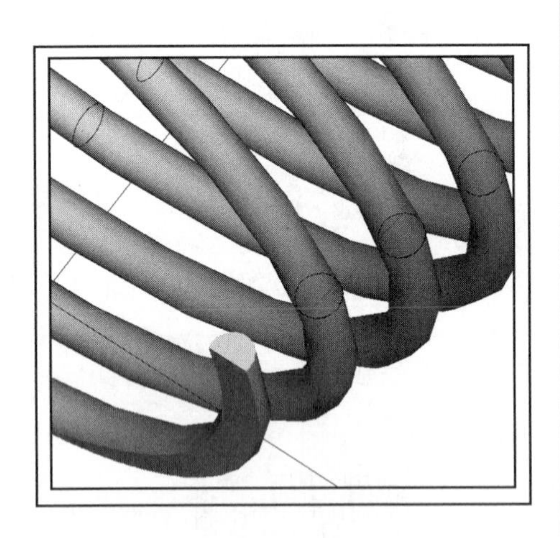

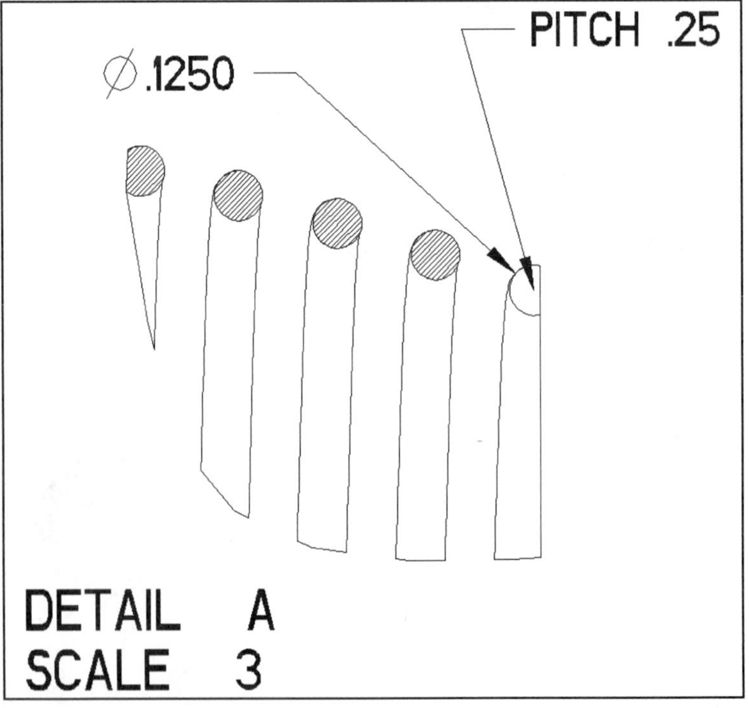

Figure 13.42
Convex Compression Spring Drawing, **DETAIL A**

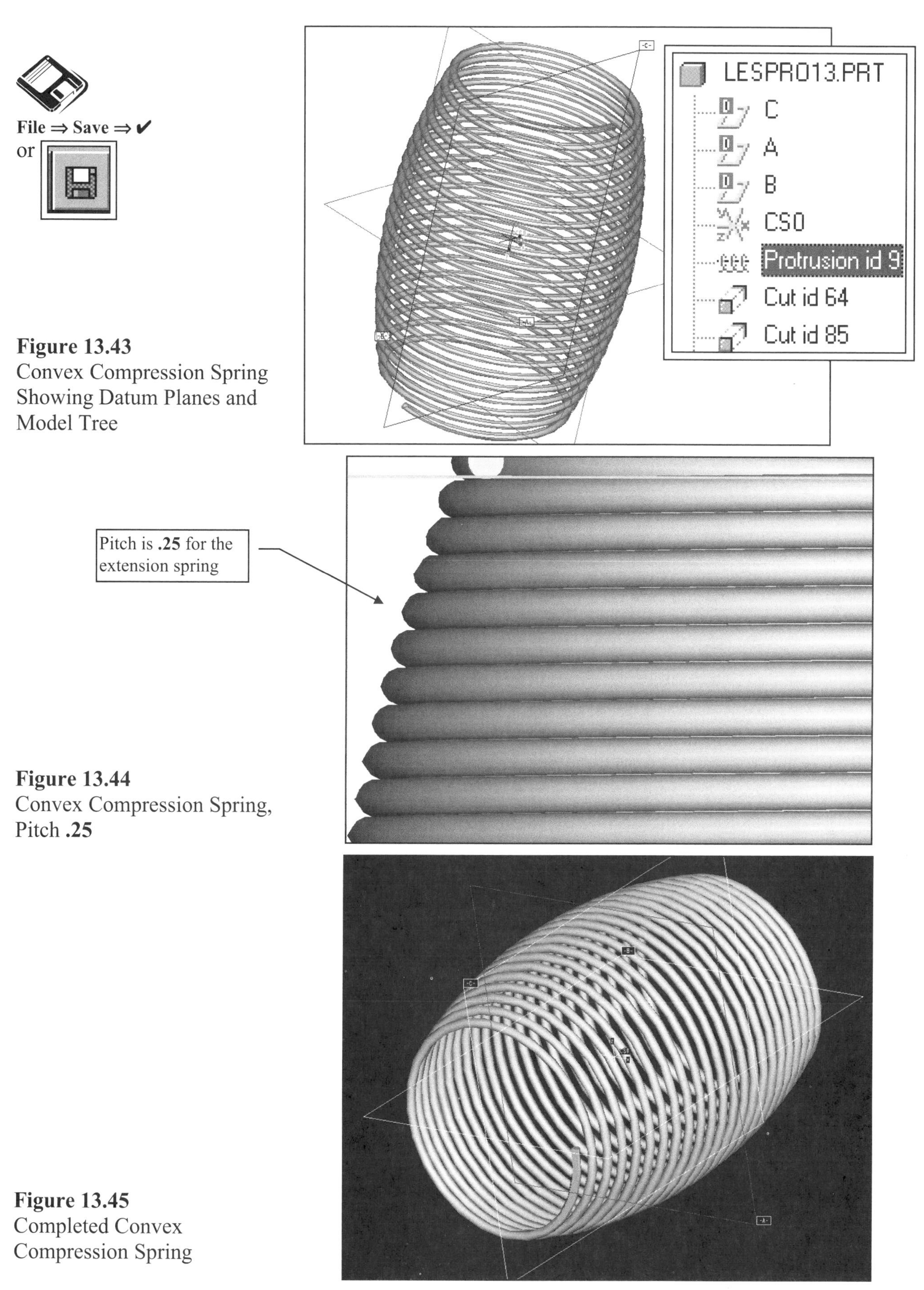

Figure 13.43
Convex Compression Spring Showing Datum Planes and Model Tree

Figure 13.44
Convex Compression Spring, Pitch **.25**

Figure 13.45
Completed Convex Compression Spring

The ECO in Figure 13.46 does *not* alter the Convex Compression Spring you just created. *The ECO requests that a different extension spring be designed with hook ends instead of ground ends.* The same size and dimensions are to be used for the new spring that were required in the Convex Compression Spring.

Save the Convex Compression Spring under a new name--for example, **CONVEX_COM_SPR_GRND_ENDS**. Rename the active part to something like **CON_EXT_SPR_HOOK_ENDS**. Delete the ground ends, and design machine hooks for both ends of the new spring. Refer to a Machinery's Handbook or your engineering graphics text for acceptable design options and dimensions for the hook ends.

Create **3D Notes** to describe the spring.

EGD REFERENCE
Fundamentals of Engineering Graphics and Design
by L. Lamit and K. Kitto
Read Chapter 18.
See pages 702-704.

Figure 13.46 ECO for New Convex *Extension* Spring

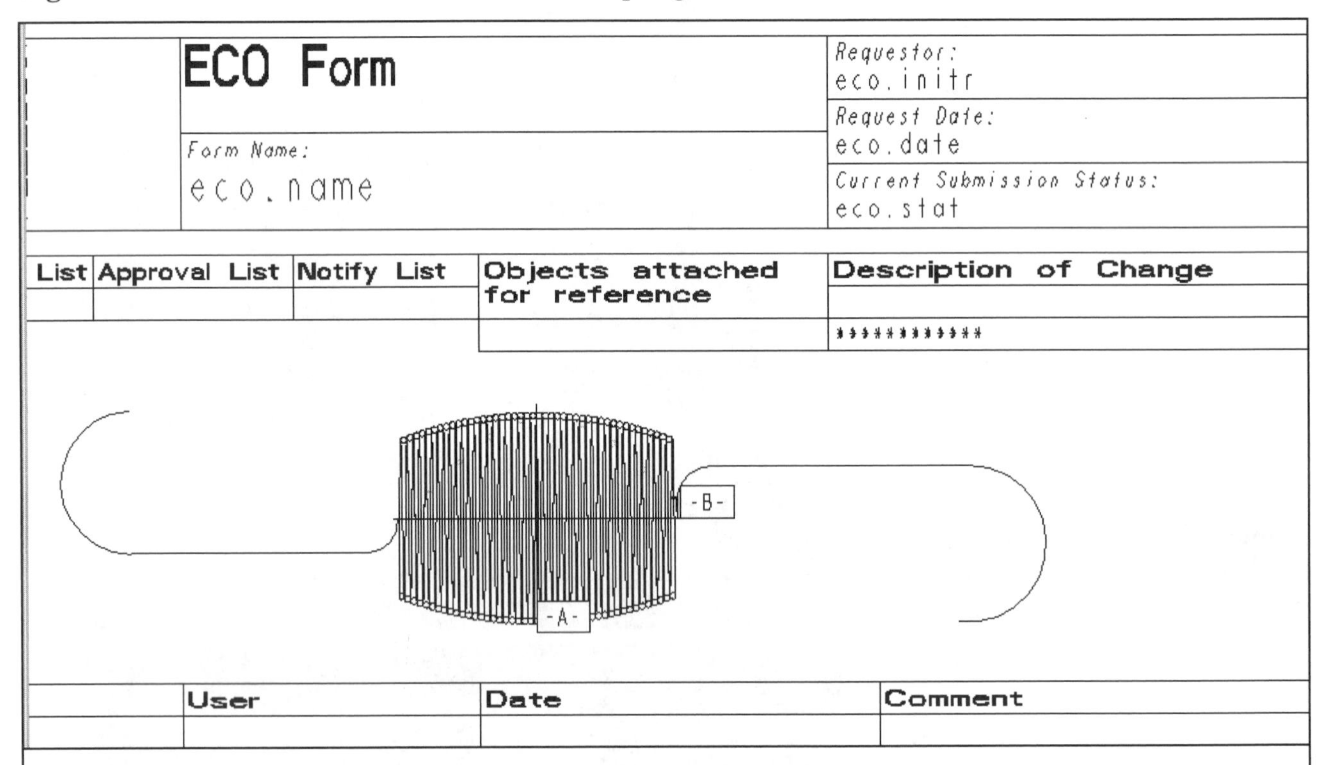
ECO Form
Requestor:
eco.initr
Request Date:
eco.date
Form Name:
eco.name
Current Submission Status:
eco.stat

List	Approval List	Notify List	Objects attached for reference	Description of Change
				************

User	Date	Comment

Design a **convex extension spring** with machine hook ends.

Use the same dimensions as those for the Convex *Compression* Spring (Figs. 13.39 through 13.45).

Design the hook ends using your own dimensions.

Pitch to be **.125.**

Part Two

Assemblies

Lesson 14 Assembly Constraints
Lesson 15 Exploded Assemblies

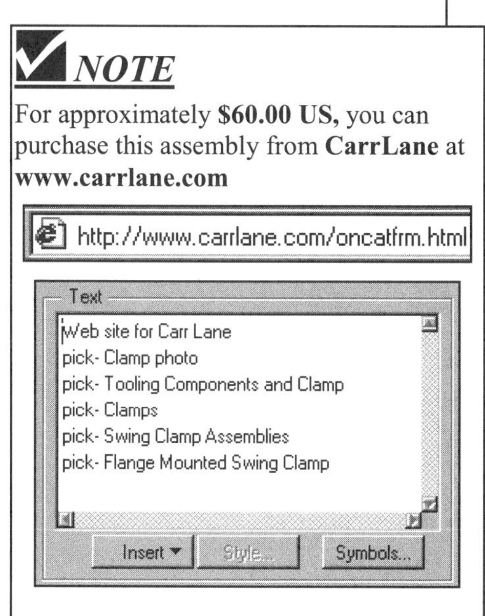

NOTE

For approximately **$60.00 US,** you can purchase this assembly from **CarrLane** at **www.carrlane.com**

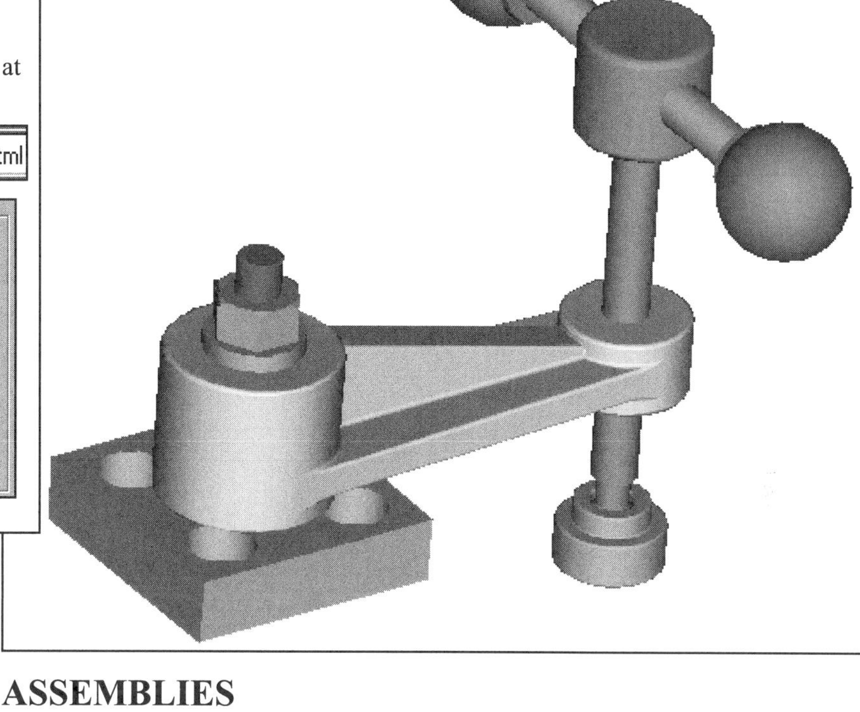

Swing Clamp Assembly

ASSEMBLIES

The **Assembly mode** allows you to place together component parts and subassemblies to form assemblies. Assemblies can then be modified, reoriented, analyzed, and documented. Assembly mode is used for the following functions:

* Placing components into assemblies (*bottom-up* assembly design)
* Exploding views of assemblies
* Altering the display settings for individual components
* Designing in Assembly mode (*top-down* assembly design)
* Part modification, including feature construction
* Analysis of assemblies

With **Pro/ENGINEER,** you can:

* Assemble component parts and subassemblies to form assemblies
* Delete or replace assembly components
* Modify assembly placement offsets, and create and modify assembly datum planes, coordinate systems, and sectional views
* Modify parts directly in Assembly mode
* Get assembly engineering information, perform viewing and layer operations, create reference dimensions, and work with interfaces

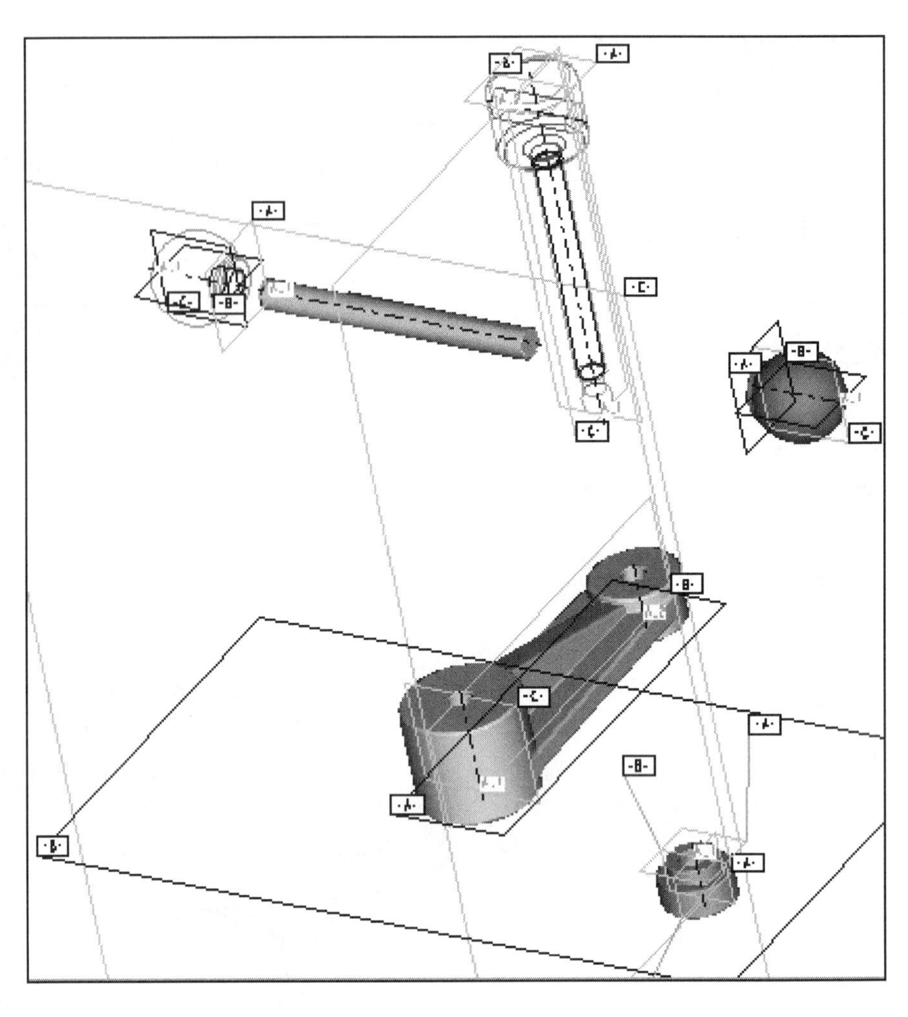

Exploded Swing Clamp Subassembly

With **Pro/ASSEMBLY,** you can:

* Create new parts in Assembly mode
* Create sheet metal parts in Assembly mode
* Mirror parts in Assembly mode (create a new part)
* Replace components automatically by creating interchangeable groups. Create assembly features, existing only in Assembly mode and intersecting several components
* Create families of assemblies, using the family table
* Simplify the assembly representation
* Use the **Move** and **Copy** commands for assembly components
* Use Pro/PROGRAM to create design programs that allow users to respond to program prompts to alter the design model

The process of creating an assembly is accomplished by adding models (parts/subassemblies) to a base component (parent part/subassembly) using a variety of constraints. A **placement constraint** specifies the relative position of a pair of surfaces on two components. The **Mate**, **Align**, **Insert**, and **Orient** commands and their variations are used to accomplish this task.

Lesson 14

Assembly Constraints

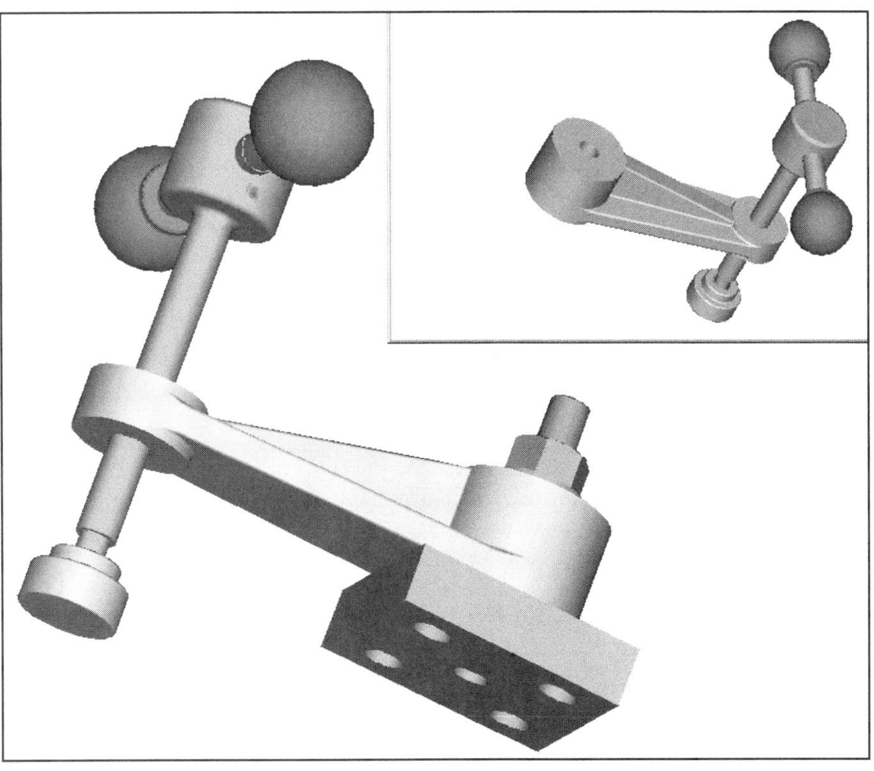

Figure 14.1
Swing Clamp

EGD REFERENCE

Fundamentals of Engineering Graphics and Design
by L. Lamit and K. Kitto
Read Chapter 23.
See pages 832-838 and 845-846.

COAch™ for Pro/ENGINEER

If you have **COAch for Pro/ENGINEER** on your system, go to SEARCH and do the Segments shown in Figures 14.3 through 14.5.

OBJECTIVES

1. **Assemble components to form an assembly**
2. **Create a subassembly**
3. **Understand and use a variety of assembly constraints**
4. **Redefine a component constraint**
5. **Modify a constraint**
6. **Check for clearance and interference**

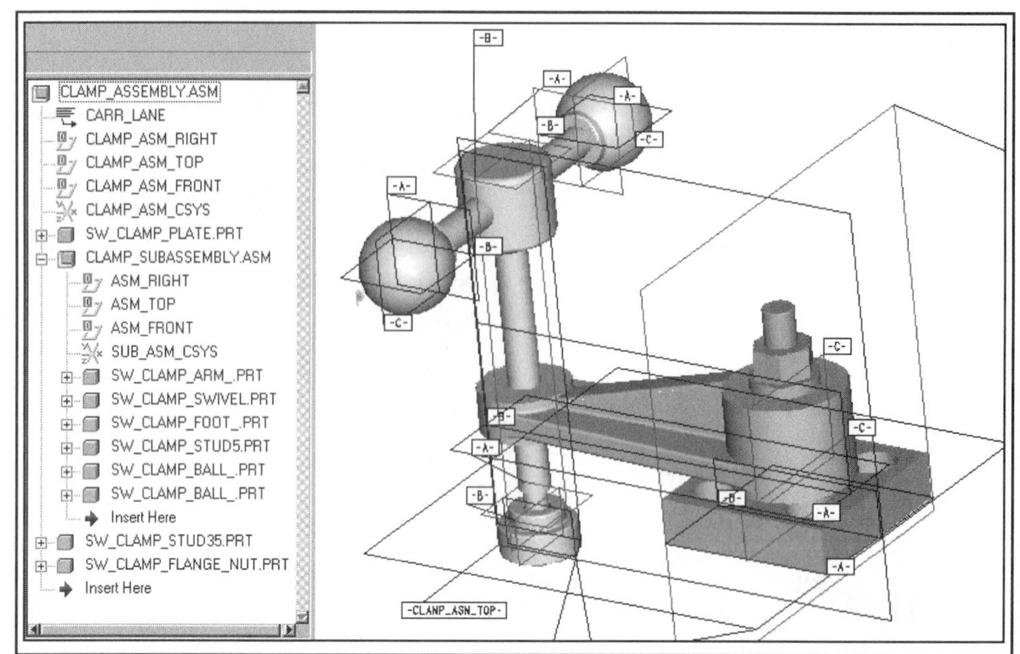

Figure 14.2
Swing Clamp Assembly with Model Tree

ASSEMBLY CONSTRAINTS

Assembly mode allows you to place together component parts and subassemblies to form an **assembly** (Figs. 14.1 and 14.2). Assemblies (Fig. 14.3) can be modified, reoriented, documented, or analyzed. An assembly can be assembled into another assembly, thereby becoming a **subassembly**.

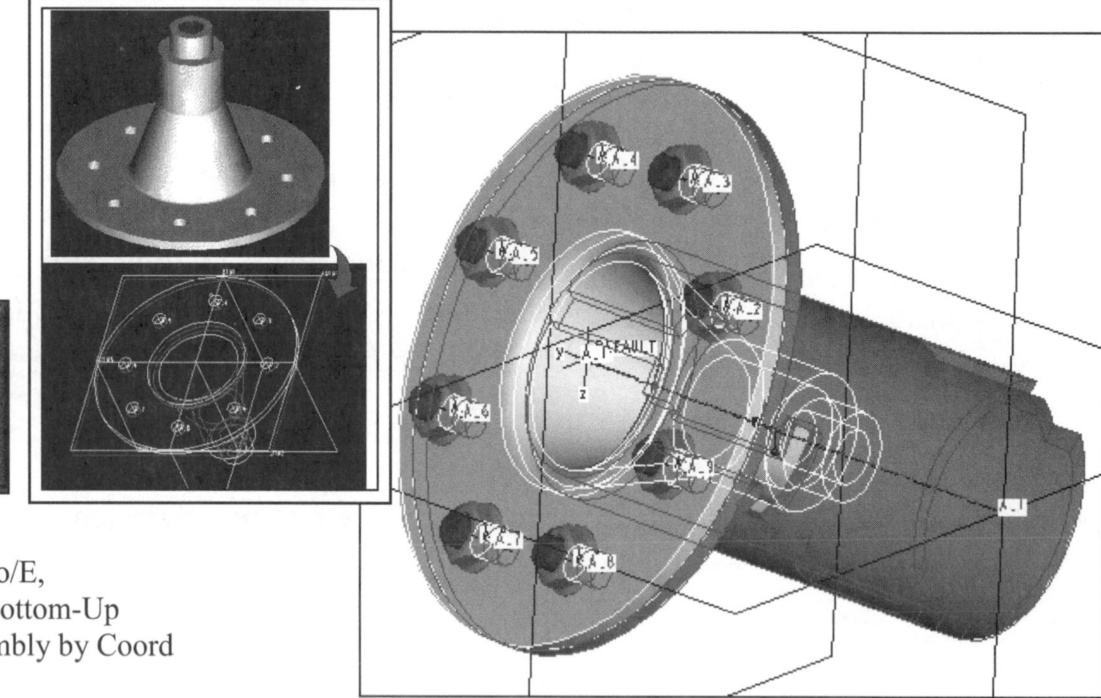

Figure 14.3
COAch for Pro/E, Assemblies, Bottom-Up Design (Assembly by Coord System)

Assembly mode is used for the following functions:

* Placing components in assemblies (Fig. 14.4)
* Exploding views of assemblies
* Part modification, including feature construction
* Analysis

With **Pro/E,** you can:

* Place together component parts and subassemblies to form assemblies
* Remove or replace assembly components
* Modify assembly placement offsets, and create and modify assembly datum planes, coordinate systems, and cross sections
* Modify parts directly in Assembly mode
* Get assembly engineering information, perform viewing and layer operations, create reference dimensions, and work with interfaces

With **Pro/ASSEMBLY,** you can:

* Create new parts in Assembly mode
* Create sheet metal parts in Assembly mode (using Pro/SHEETMETAL)
* Mirror parts in Assembly mode (create a new part)
* Replace components automatically by creating interchangeable groups. Create assembly features, existing only in Assembly mode, that intersect several components
* Create families of assemblies, using the family table
* Simplify the assembly representation
* Use **Move** and **Copy** for assembly components
* Use Pro/PROGRAM to create design programs that allow users to respond to program prompts to alter the design model

Figure 14.4
COAch for Pro/E, Assemblies, Top-Down Design (Datum Planes and Layers)

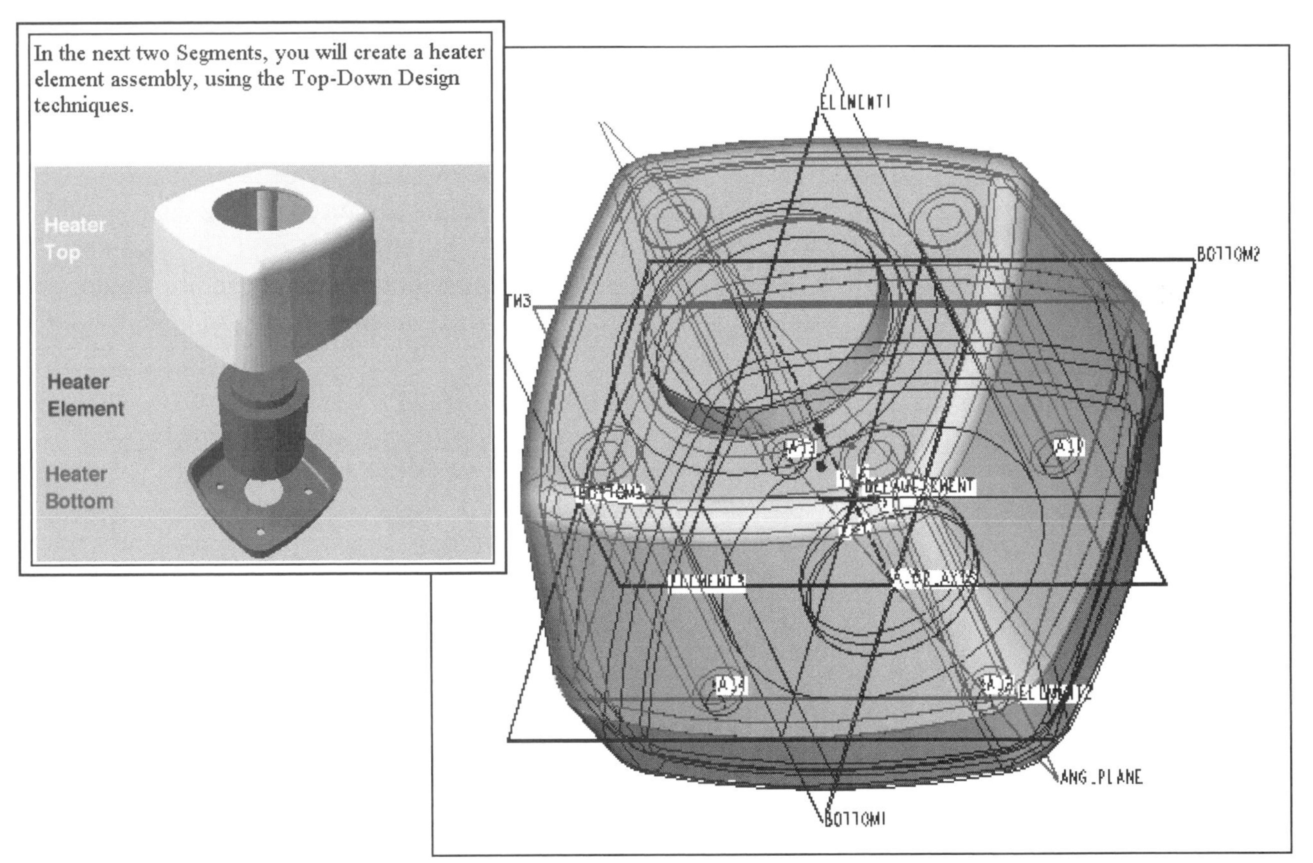

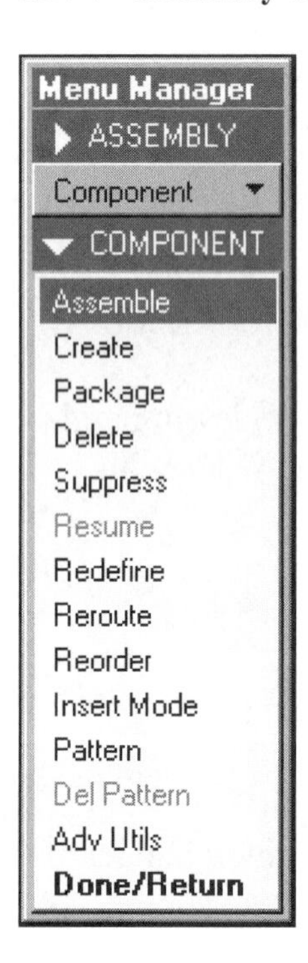

Assembling Components

The process of creating an assembly involves adding components (parts/subassemblies) to a base component (parent part/subassembly) using a variety of constraints (Fig. 14.5). Components can also be created in Assembly mode using existing components as references.

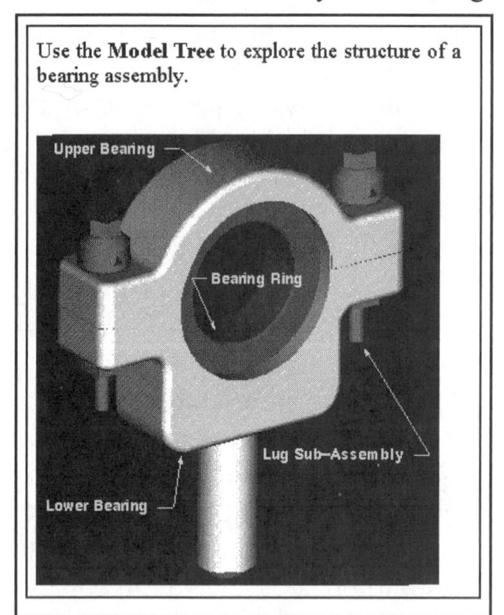

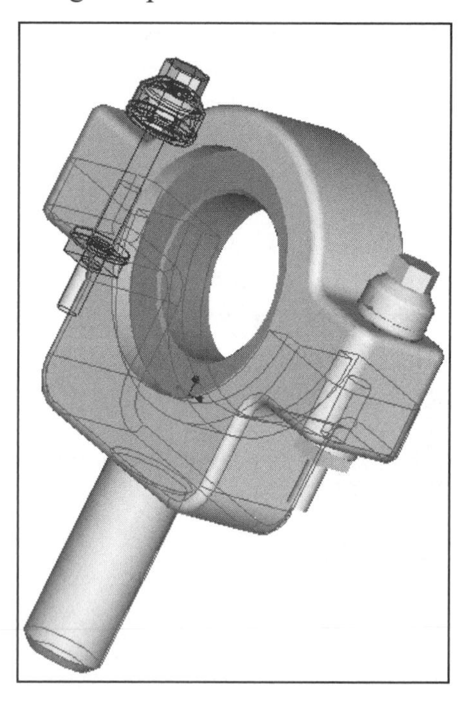

Figure 14.5
COAch for Pro/E, Assemblies, Bottom-Up Design

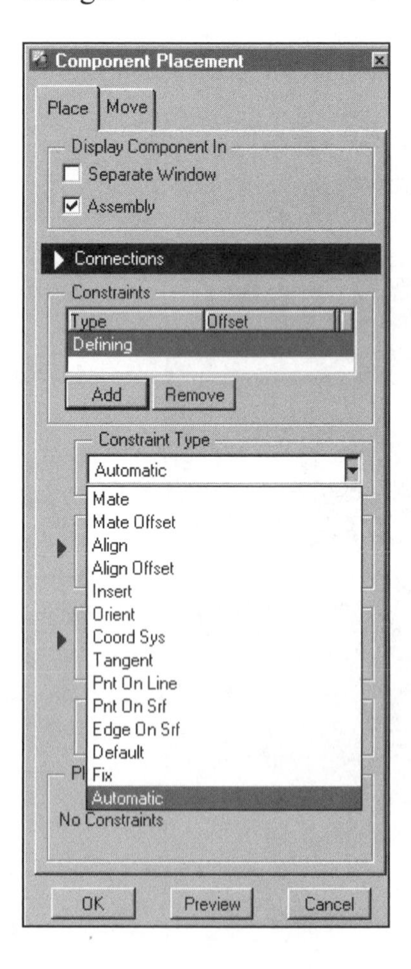

A **placement constraint** specifies the relative positions of a pair of references on two components. The **Mate**, **Align**, **Insert**, and **Orient** commands and their variations are used to accomplish this task. The general principles to apply during constraint placement are as follows:

* The two surfaces must be of the same type (for example, plane-plane, revolved-revolved). The term *revolved surface* means a surface created by revolving a section or by extruding an arc or a circle. Only the following surfaces are allowed: plane, cylinder, cone, torus, sphere.
* If you put a placement constraint on a datum plane, you should specify which side of it, *yellow* or *red,* you are going to use.
* When using **Mate Offset** or **Align Offset,** you will be shown the positive offset direction. If you need an offset in the opposite direction, enter a negative value.
* When a model surface is selected in one window, another window may become hidden, in which case it will need to be brought forward.
* Constraints are added one at a time.
* Placement constraints are used in combinations, in order to specify placement and orientation completely. For example, one pair of surfaces may be constrained to mate, another pair to insert, and a third pair to orient.

Mate [Fig. 14.6 (top)] is used to make two surfaces touch one another: coincident and facing each other. When you are using datums, this means that two *yellow* sides, or two *red* sides, will face each other. **Mate Offset** [Fig. 14.6 (bottom)] makes two planar surfaces parallel and facing each other. The offset value determines the distance between the two surfaces.

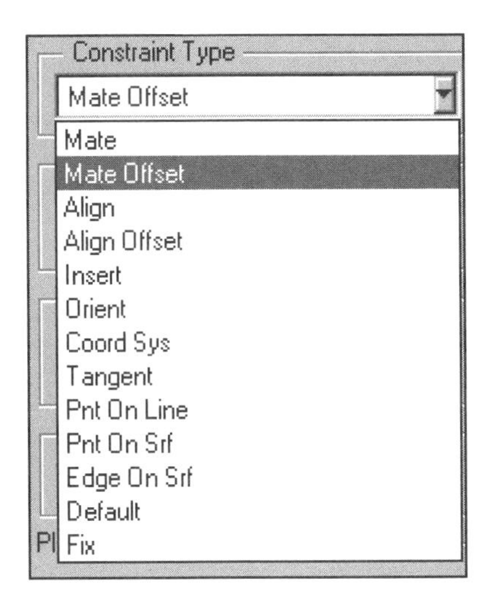

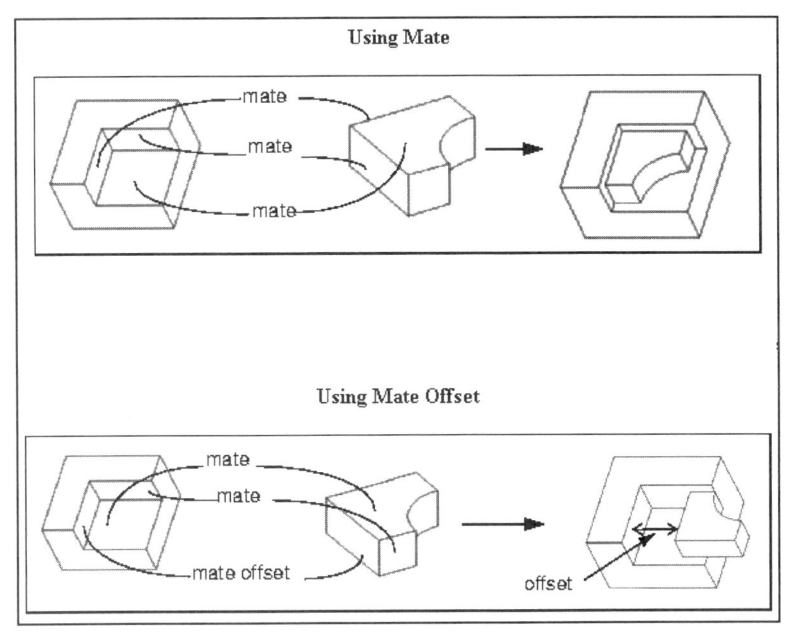

Figure 14.6
Mate and **Mate Offset**

The **Align** command (Fig. 14.7) makes two planes coplanar: coincident and facing in the same direction. The **Align** constraint also aligns revolved surfaces or axes to make them coaxial. You can also align two datum points, vertices, or curve ends; selections on both parts must be of the same type (that is, if a datum point is selected on one part, only a datum point can be selected on another part). The **Align Offset** constraint [Fig. 14.8 (top)] aligns two planar surfaces at an offset: parallel and facing in the same direction.

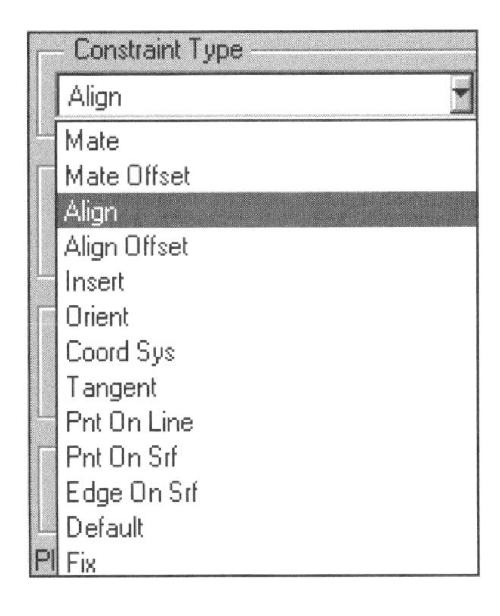

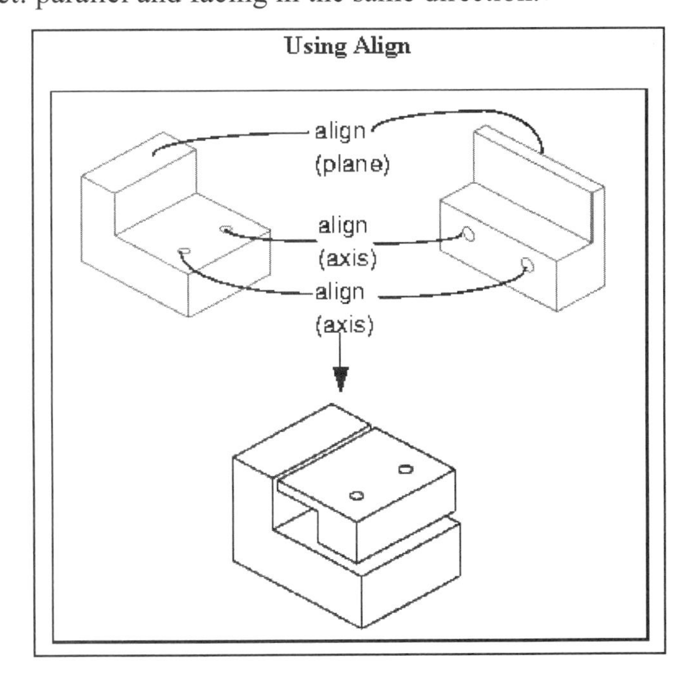

Figure 14.7
Pro/HELP
Online Documentation,
Align

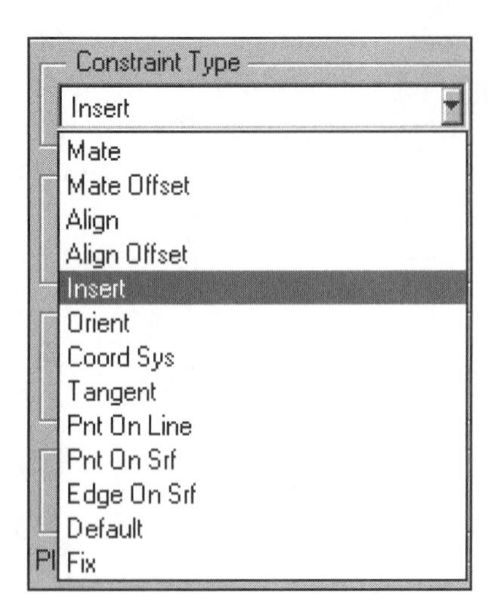

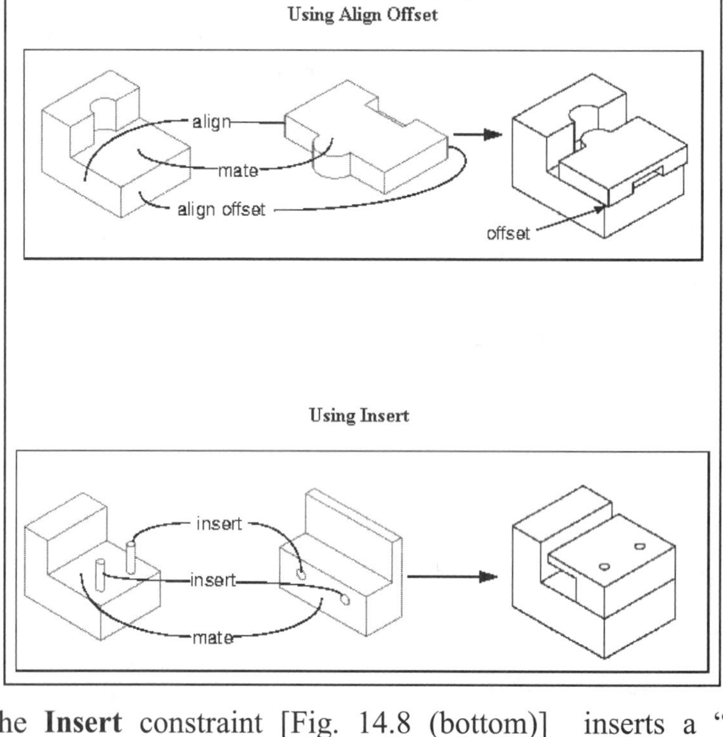

Figure 14.8
Pro/HELP
Online Documentation,
Align Offset and **Insert**

The **Insert** constraint [Fig. 14.8 (bottom)] inserts a "male" revolved surface into a "female" revolved surface, aligning axes.

The **Orient** constraint [Fig. 14.9 (top)] orients two planar surfaces so that they are parallel and facing in the same direction; offset is not specified.

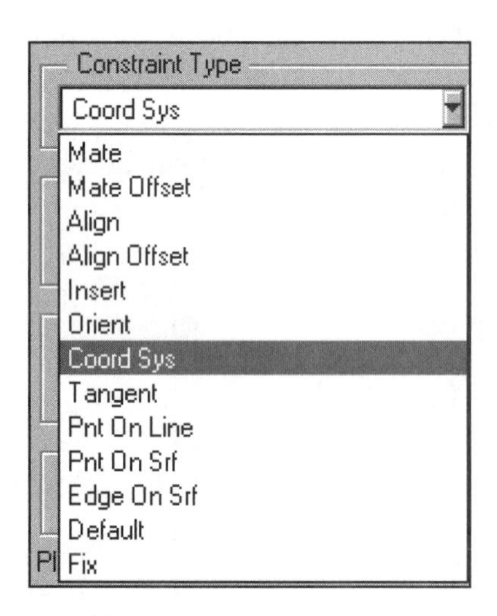

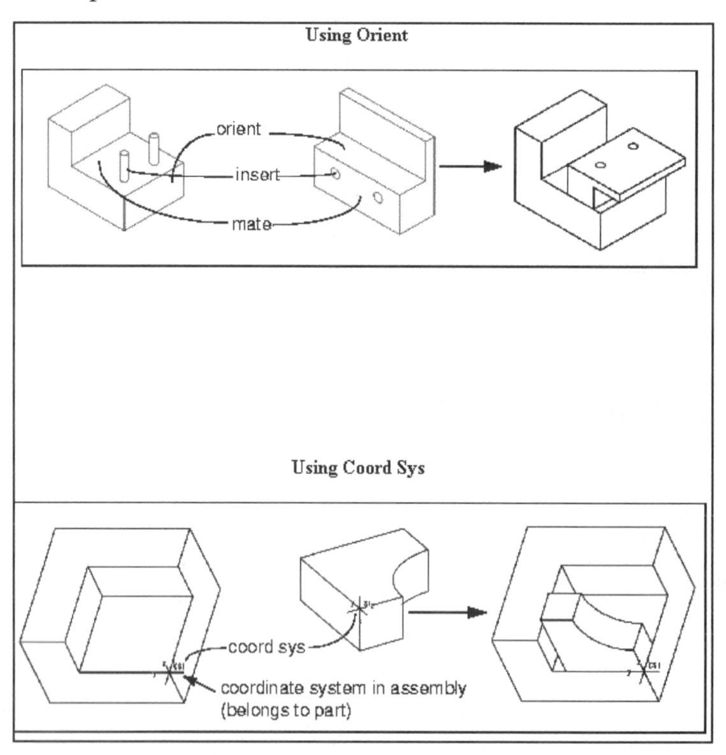

Figure 14.9
Pro/HELP
Online Documentation,
Orient and **Coord Sys**

The **Coord Sys** constraint [Fig. 14.9 (bottom)] places a component in an assembly by aligning its coordinate system with a coordinate system in the assembly (both assembly and part coordinate systems can be used).

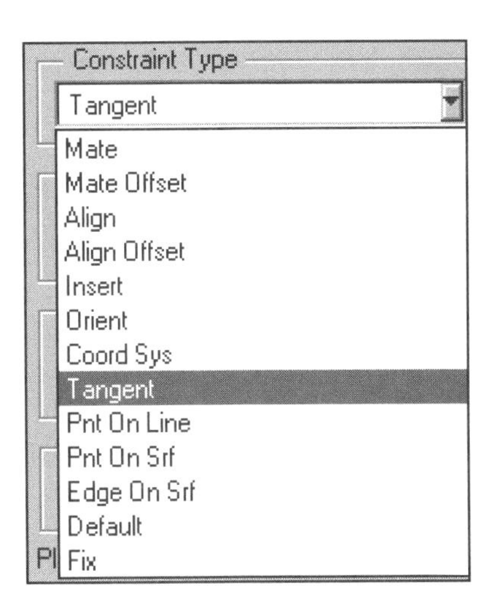

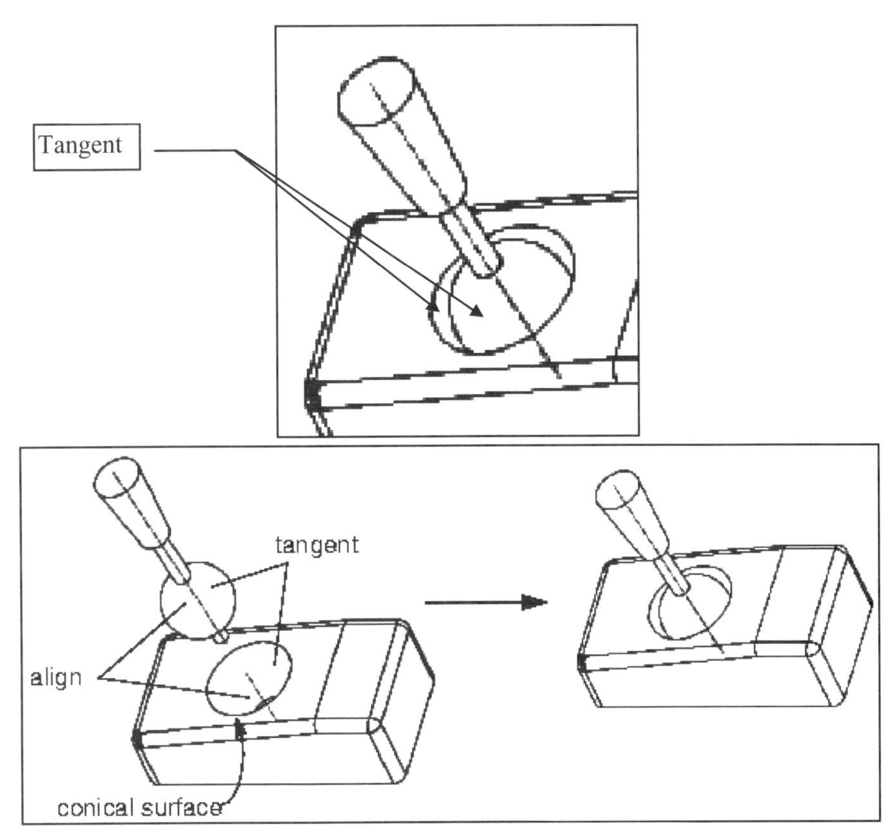

Figure 14.10
Pro/HELP
Online Documentation,
Using **Tangent**

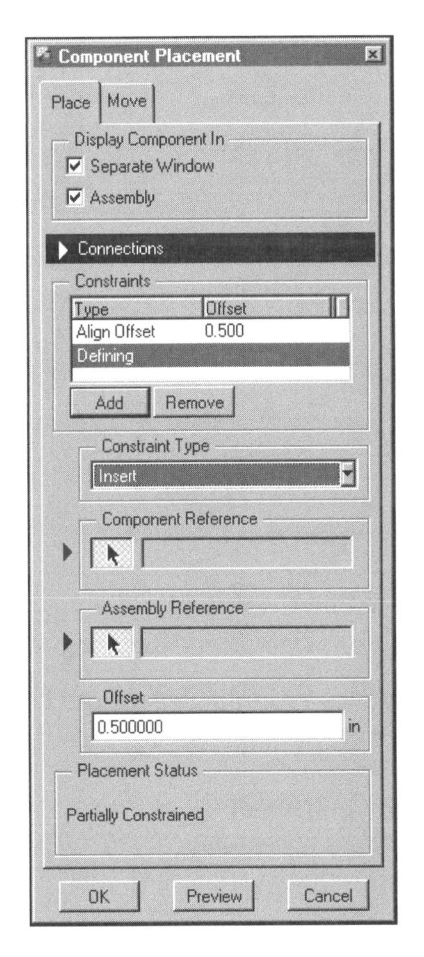

Coordinate systems can be picked or selected by name from namelist menus. The components will be assembled by aligning the **X**, **Y**, and **Z** axes of the selected coordinate systems.

The **Tangent, Pnt On Line, Pnt On Srf,** and **Edge On Srf** constraints are used to control the contact of a surface at the tangency of another surface, at a point, or at an edge (Fig. 14.10). An example of the use of these placement options is the contact surface or point between a cam and its actuator.

In most cases, a combination of constraints will be required. **Mate** and **Insert** are necessary to constrain the two parts shown in Figure 14.11. **Mate**, **Insert**, and **Orient** are another possibility, depending on the parts.

Assemblies can also be displayed *exploded.* The exploded components can be placed anywhere in 3D space. An exploded cosmetic view of the assembly can be displayed with or without the reference planes, coordinate system, and hidden lines. The completed assembly can be modified by redefining the constraints. The last stage of most projects involves putting the assembly into Drawing mode and displaying the appropriate views and information.

The Automatic Placement Constraint

The **Automatic** placement constraint is selected by default when a new component is introduced into an assembly for placement in the assembly window. Do one of the following:

1. Select a reference on the component and a reference on the assembly, in either order, to define a placement constraint.

After you select a pair of valid references from the assembly and the component, Pro/E automatically selects a constraint type appropriate to the specified references.

2. Before selecting references, you can change the type of constraint by selecting a type from the Constraint Type list. After you specify a constraint type, you cannot change it back to **Automatic.** However, when you create another new constraint, the constraint type again defaults to **Automatic.**

Component Placement

When you choose **Assemble** or **Redefine** from the COMPONENT menu, a *Component Placement dialog box* appears (Fig. 14.11 and left; next page). This interactive window includes the *Constraints*, the corresponding *Component References* and *Assembly References*, and the *Offset* (if any) and a variety of other information and prompts.

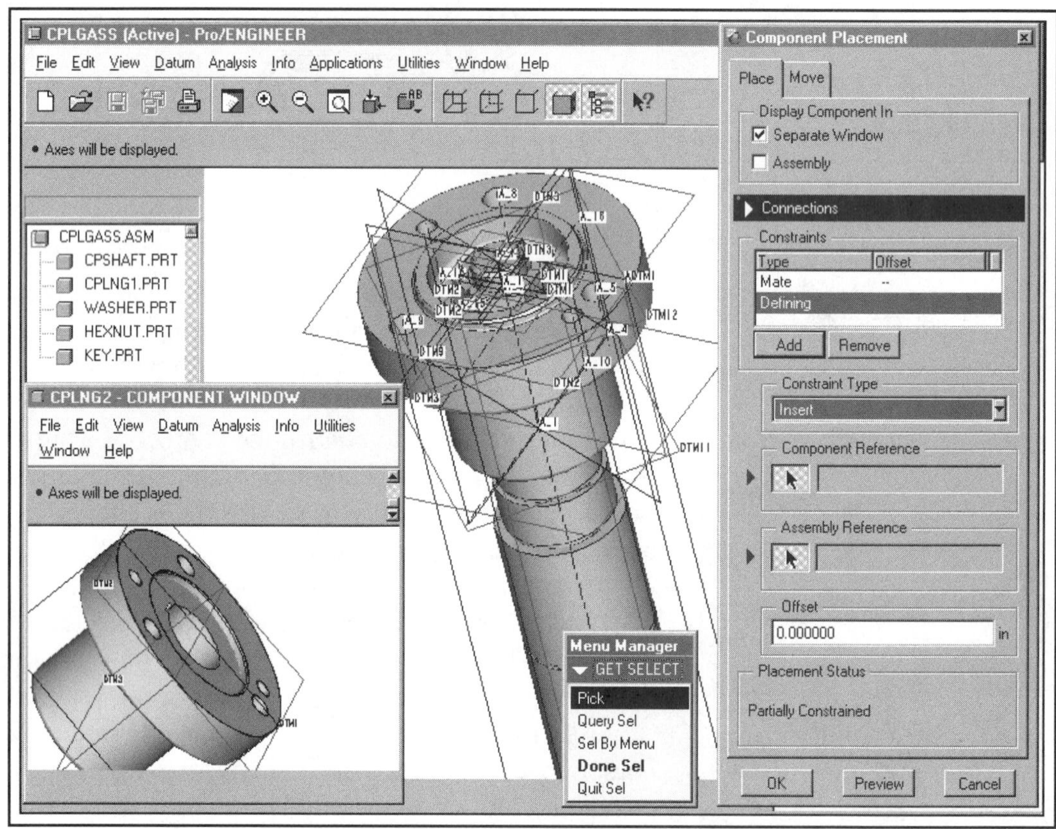

Figure 14.11
Component Windows

To assemble a component parametrically by defining placement constraints:

1. From ASSEMBLY choose **Component ⇒ Assemble**, or choose the assembly name in the Model Tree, then RMB and choose **Component ⇒ Assemble** from the pop-up menu.
2. Select the component. The **Component Placement** dialog box opens, and the component appears in the assembly window.
3. Under the **Display Component In** area of the dialog box, specify the screen window in which the component is displayed while you position it. You can change windows at any time.

Separate Window Displays the component in its own window (Fig. 14.11) while you specify its constraints.
Assembly Displays the component in the main assembly graphics window while you specify its constraints.

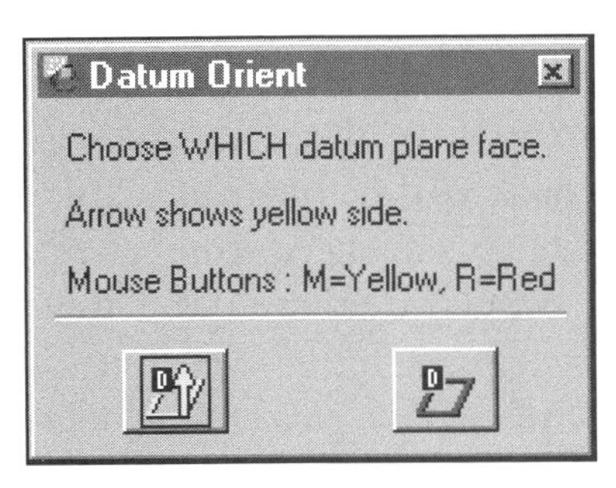

4. Select a reference on the component and a reference on the assembly, in either order, to define a placement constraint for the default constraint type, e.g. **Automatic.** Before selecting references, you can change the type of constraint by selecting a type from the **Constraint Type** list.

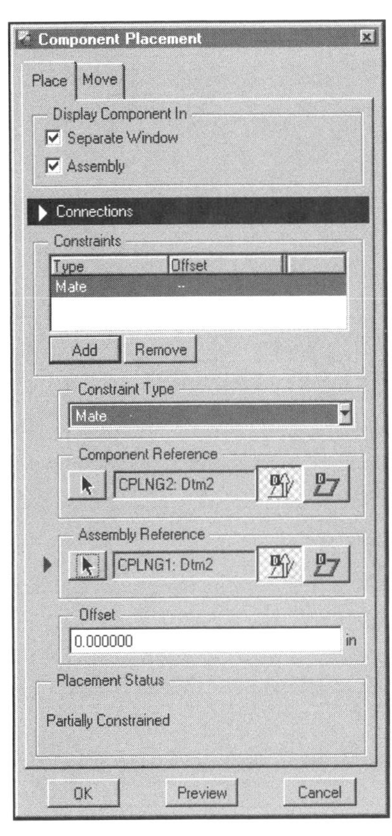

As you define constraints, each constraint is listed under the **Constraints** area. Until you define the first constraint, the **Type** is shown as **Defining.** If you have selected the **Assembly** option, the component is displayed in the main assembly graphics window, and *component placement is updated as you specify constraints.* As you select references, the current status of the component is reported in the **Placement Status** area. Initially, **No Constraints** is displayed.

When you select a mating surface or datum plane on the model, Pro/E prompts you to select ***Yellow*** or ***Red*** to specify (in the **Datum Orient** dialog box shown at left top) which side of the datum plane should face the direction indicated by the arrow. The name of the selected reference is then shown in the **Component Reference** or the **Assembly Reference** text box, along with a *yellow* and a *red* button (shown at left bottom), each with an arrow indicating direction. You can choose these buttons to flip sides from here, or you can return to the assembly window to flip sides.

When you create a **Mate Offset** or **Align Offset** constraint, Pro/E prompts you to enter a value to define the offset from the reference. The value that you enter is shown in the **Offset** text box. You can then change the value in the text box without having to redefine the value in the assembly window.

5. After you define a constraint, **Add** is automatically selected, and you can repeatedly define another constraint. You can define as many constraints as you want (up to the Pro/E limit of 50 constraints).
6. You can select one of the constraints listed in the **Constraints** area at any time and change the constraint type, flip sides, or modify the offset value.

You can select **Remove**, **Retr Refs**, or **Preview** at any time. To delete a placement constraint for the component, select one of the constraints listed in the **Constraints** area, and then choose **Remove**.

The **Retr Refs** option appears when you redefine a component in a simplified representation, and that component depends on components that are not in the current simplified representation. Choose **Retr Refs** to retrieve any other components that define the location of the component.

Choose **Preview** to show the location of the component as it would be with the current placement constraints.

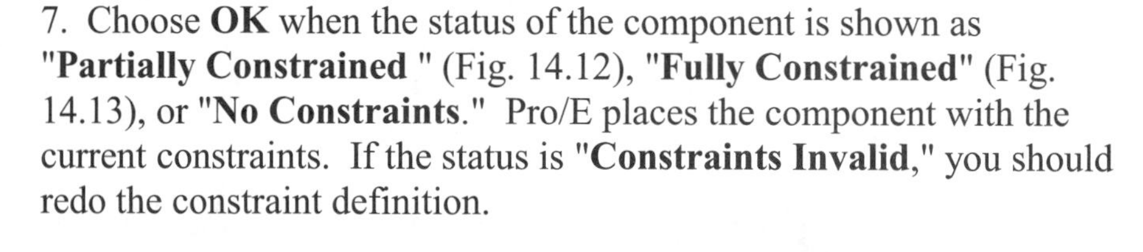

7. Choose **OK** when the status of the component is shown as "**Partially Constrained** " (Fig. 14.12), "**Fully Constrained**" (Fig. 14.13), or "**No Constraints**." Pro/E places the component with the current constraints. If the status is "**Constraints Invalid**," you should redo the constraint definition.

NOTE

If components are *packaged* but not placed (*fully constrained*), you cannot create children that reference them.

If constraints are incomplete, you can leave the component as packaged. Because the components are packaged but not placed, you cannot create children that reference them. Packaged components follow the behavior dictated by the configuration file option ***package_constraints.***

If constraints are conflicting, you can restart or continue placing the component. Restarting erases all previously defined constraints for the component.

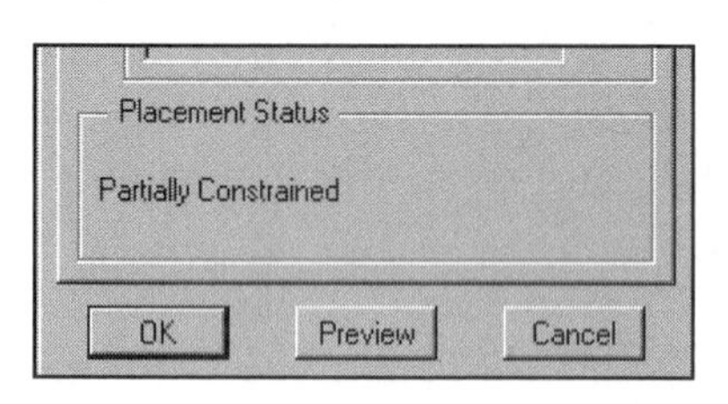

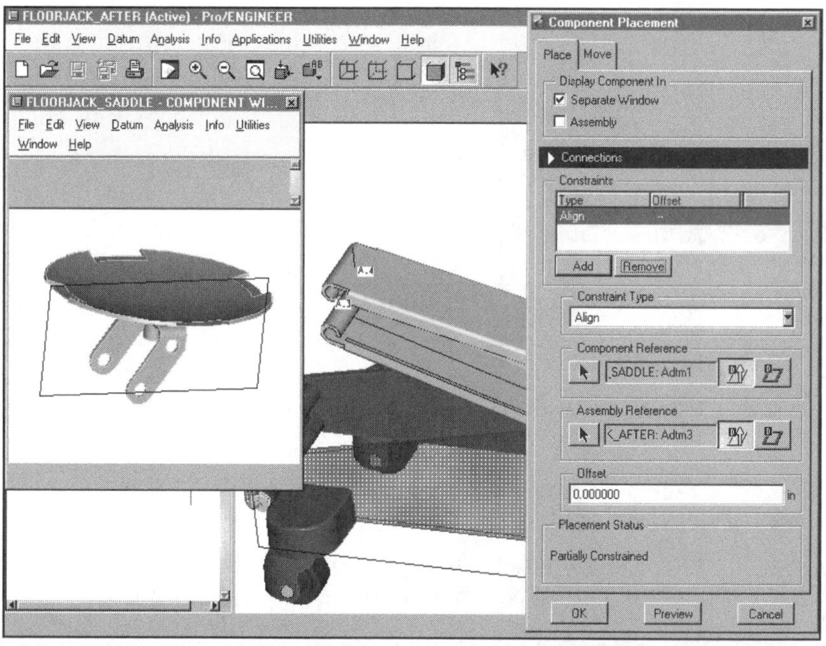

Figure 14.12
Partially Constrained

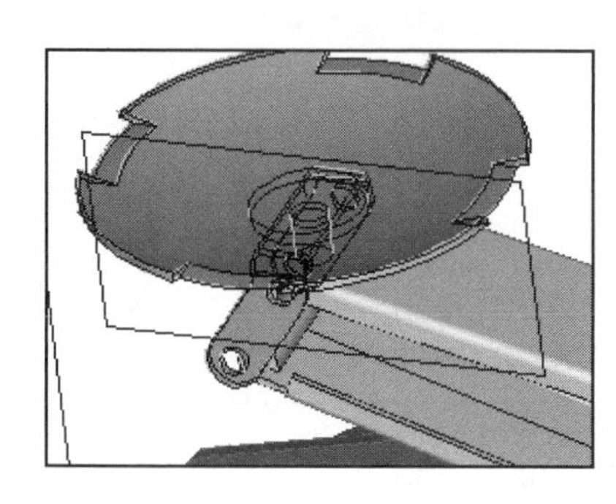

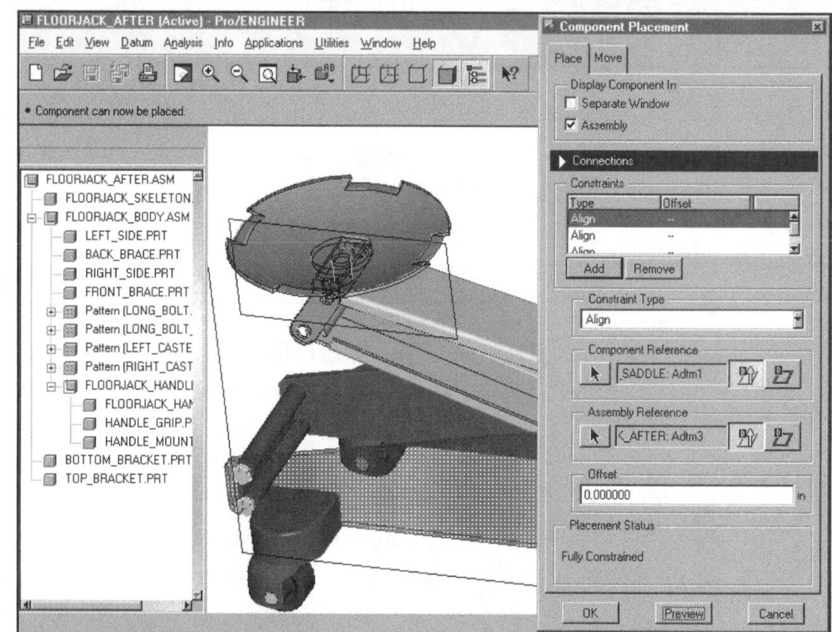

Figure 14.13
Fully Constrained

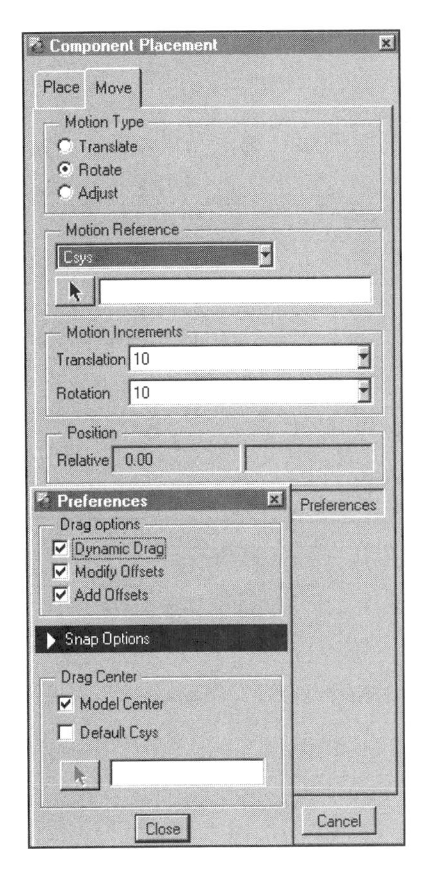

Figure 14.14
Using the Freeform Mouse-Driven Manipulation

Freeform Mouse-Driven Component Manipulation

Whenever the Component Placement dialog box is available for placing a component or redefining placement constraints, a spin center for the active component is always visible, and you can manipulate the active component using a combination of mouse and keyboard commands ***(Ctrl+Alt**, pressed at the same time).*

Manipulating a component is easier than moving to a separate tab in the Component Placement dialog box to package-move a component around the screen. You can switch between full view navigation and component manipulation easily with the **Alt** key. You can perform translation and rotation (Fig. 14.14) adjustments while you establish constraints. The component motion respects any constraints as they are established, as is the case with *Move tab* functionality. The spin icon appears during the entire component placement operation and defaults to the bounding box center. You can modify this location, using the Preferences dialog box, accessed from the Move tab on the Component Placement dialog box. Because of the orthographic projection used by Pro/E, motion in the Z-axis, or screen normal, is noticeable only if objects intersect. Thus, while a component is moving in the Z-axis, the camera angle is adjusted to provide a noticeable effect. In **Rotate mode**, when View Plane is the selected motion reference, the spin center does not use the screen normal spinning, but rather the two orthogonal axes (within the screen plane). Pro/E automatically sets the drag origin. Pro/E uses the screen selection point as the origin of rotation.

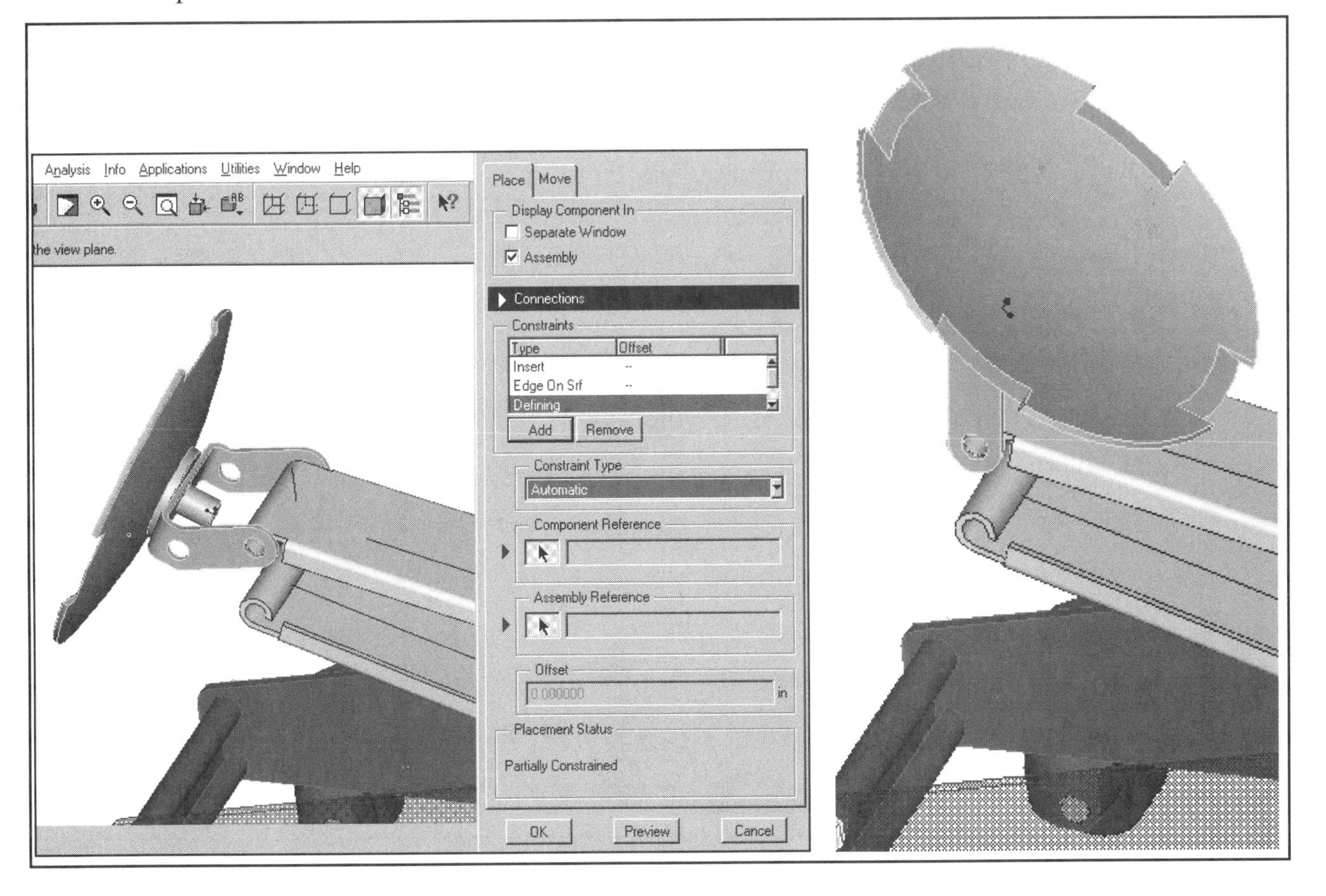

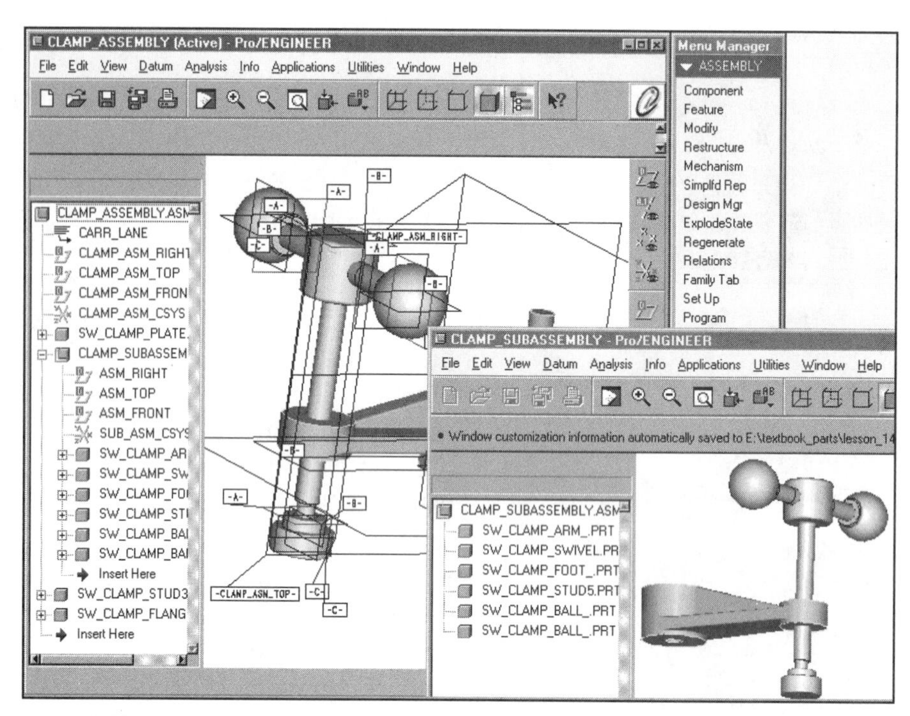

Figure 14.15
Swing Clamp Assembly with Model Tree and Subassembly

Swing Clamp Assembly

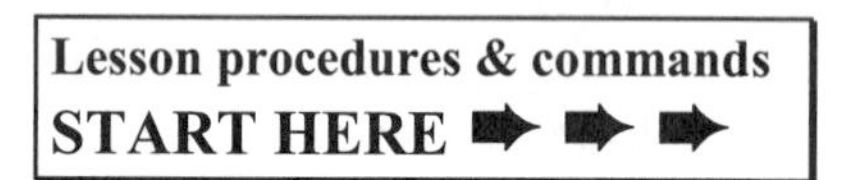

Most of the parts required in this lesson are lesson projects from this text. The **Clamp Foot** and **Clamp Swivel** are from Lesson 5. The **Clamp Ball** is from Lesson 6, and the **Clamp Arm** was created in Lesson 8. If you have not modeled these parts previously, please do so before you start the following step-by-step instructions. The other parts required for the assembly (Fig. 14.15) are standard off-the-shelf hardware items that you can get from Pro/E by accessing the library. If your system does not have a Pro/LIBRARY license for the Basic and Manufacturing libraries, model the parts using the detail drawings provided here. The **Flange Nut**, the **3.50 Double-ended Stud**, and the **5.00 Double-ended Stud** are standard parts (Fig. 14.16).

Set up the **Swing Clamp** assembly:

- Units = Inch lbm Second
- Shading
- No Display

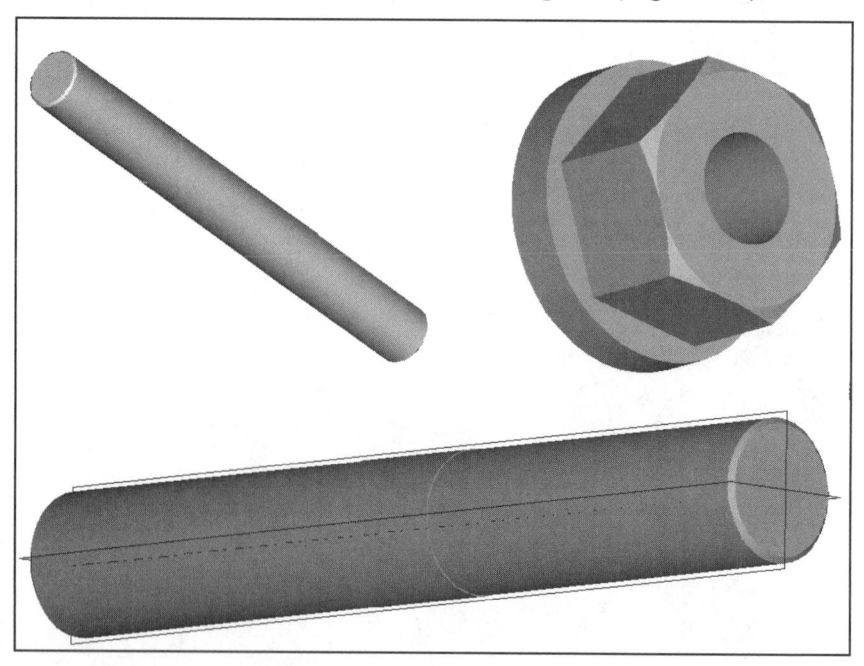

Figure 14.16
Double-Ended Studs and a Flange Nut

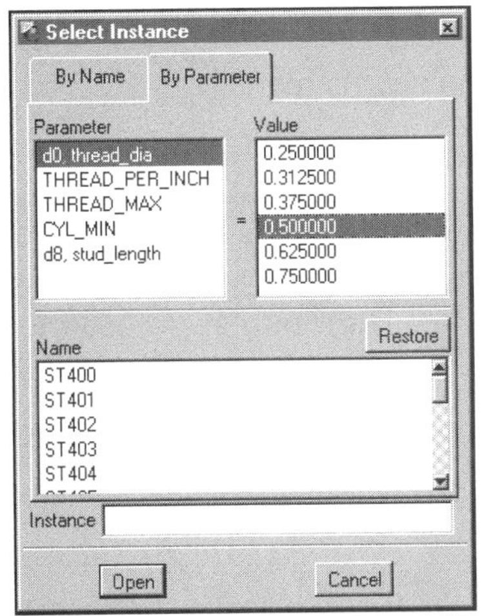

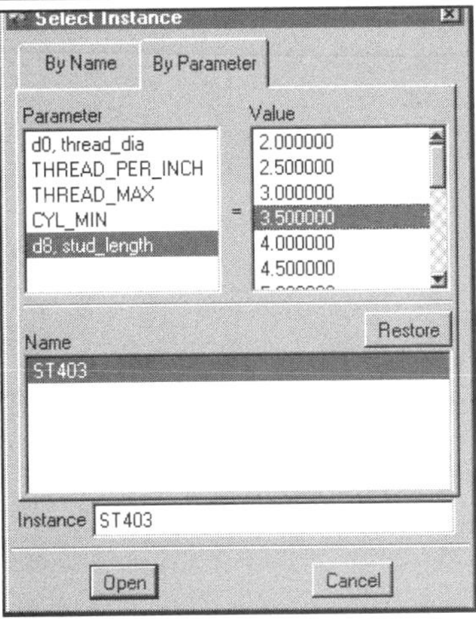

Figure 14.17
∅3.50
Double-Ended Stud

Before starting the assembly, you will be modeling each part or retrieving *standard parts* from the library and saving them under unique names in *your* directory. ***Unless instructed to do so by your teacher, do not use the library parts directly in the assembly.*** Start this process by choosing the following commands to access the existing standard double-ended studs in the library:

DOUBLE-ENDED STUD ∅.500 by 3.50 length

File ⇒ Open ⇒ ▼ ⇒ prolibrary ⇒ mfglib ⇒ Open ⇒ fixture_lib ⇒ Open ⇒ nuts_bolts_screws ⇒ Open ⇒ st.prt ⇒ Open ⇒ By Parameter (from Select Instance dialog box) **⇒ d0,thread_dia ⇒ .500 ⇒ d8,stud_length ⇒ 3.500 ⇒ Open ⇒ File ⇒ Save As ⇒ SW_CLAMP_STUD35 ⇒ OK ⇒ File ⇒ Erase ⇒ Current ⇒ Yes**

INSTANCE = ST403 (Fig. 14.17)

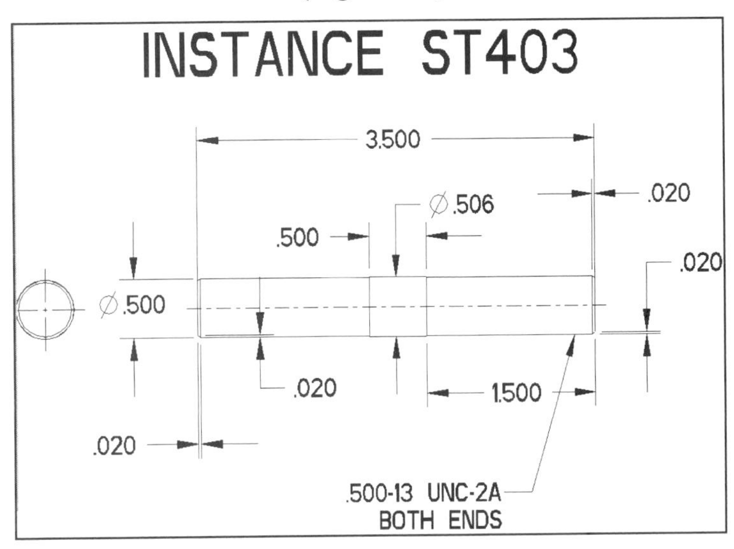

DOUBLE_ENDED STUD ∅.500 by 5.00 length

File ⇒ Open ⇒ ▼ ⇒ prolibrary ⇒ mfglib ⇒ Open ⇒ fixture_lib ⇒ Open ⇒ nuts_bolts_screws ⇒ Open ⇒ st.prt ⇒ Open ⇒ By Parameter (from Select Instance dialog box) **⇒ d0,thread_dia ⇒ .500 ⇒ d8,stud_length ⇒ 5.00 ⇒ Open ⇒ File ⇒ Save As ⇒ SW_CLAMP_STUD5 ⇒ OK ⇒ File ⇒ Erase ⇒ Current ⇒ Yes**

INSTANCE = ST406 (Fig. 14.18)

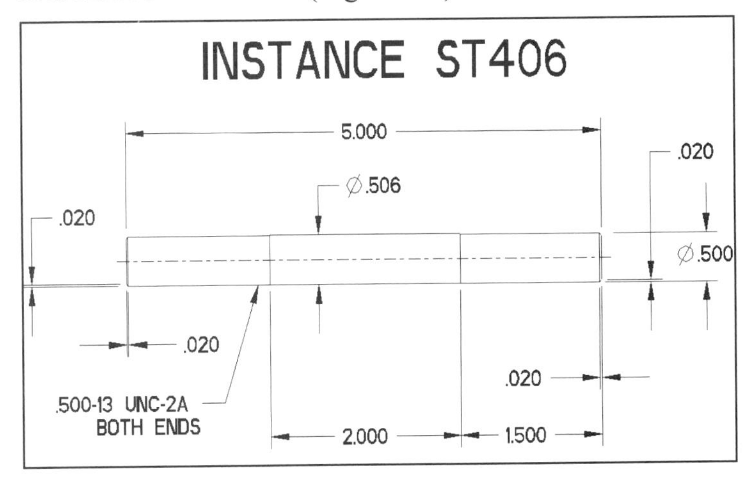

NOTE
Save the library part with a new name.

Figure 14.18
∅5.00
Double-Ended Stud

The last standard item used for this assembly is a flange nut (Fig. 14.19). Choose the following commands to access the part:

prolibrary ⇒

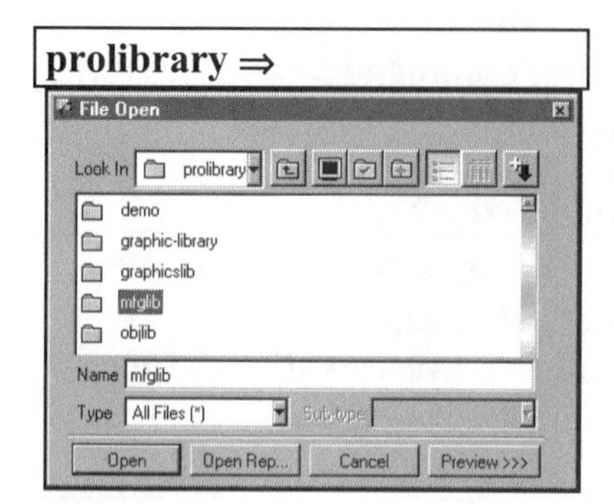

FLANGE NUT

File ⇒ Open ⇒ ▼ ⇒ prolibrary ⇒ mfglib ⇒ Open ⇒ fixture_lib ⇒ Open ⇒ nuts_bolts_screws ⇒ Open ⇒ fn.prt ⇒ Open ⇒ By Parameter (from Select Instance dialog box) **⇒ d4,thread_dia ⇒ .500 ⇒ Open ⇒ File ⇒ Save As ⇒ SW_CLAMP_FLANGE_NUT ⇒ OK ⇒ File ⇒ Erase ⇒ Current ⇒ Yes**

mfglib ⇒

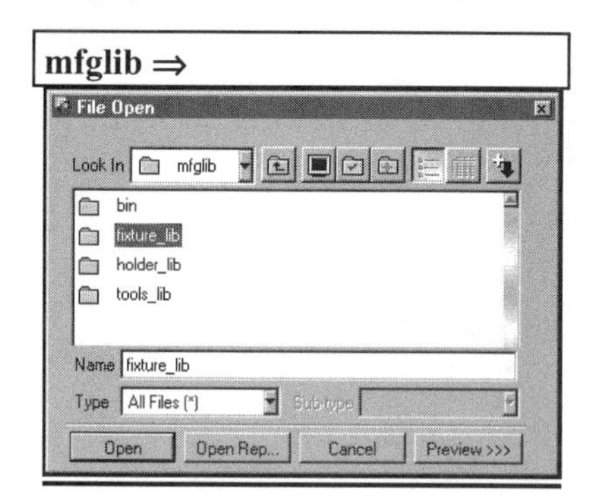

INSTANCE = FN7

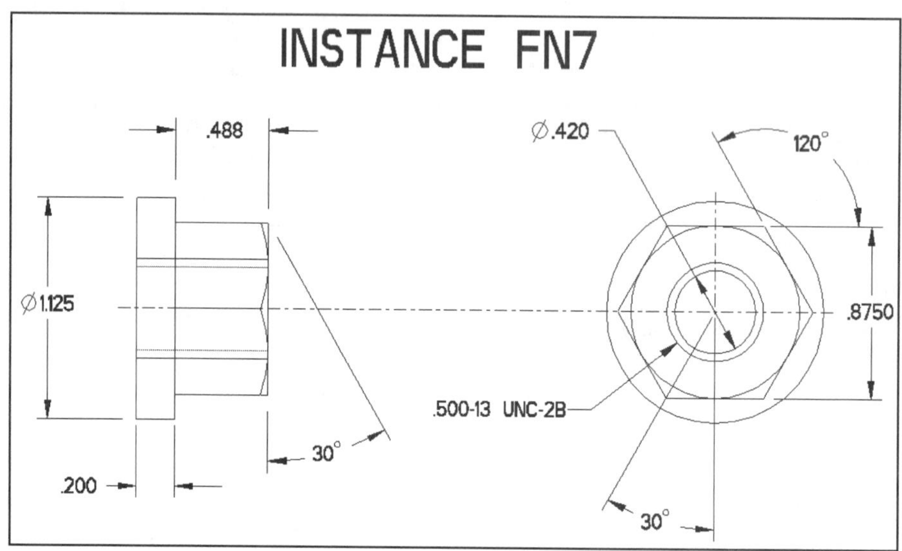

Figure 14.19
Flange Nut

fixture_lib ⇒

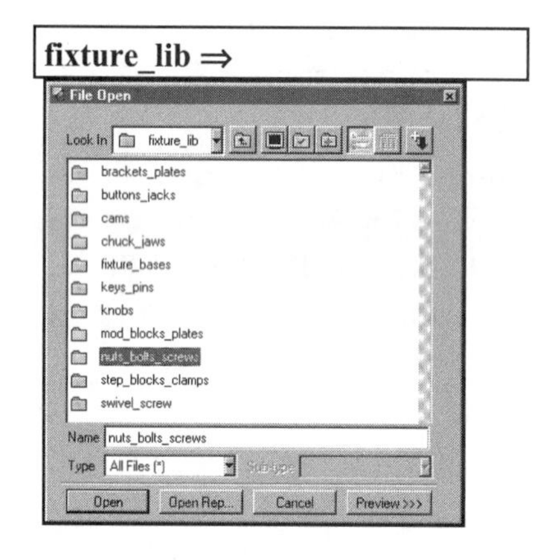

nuts_bolts_screws ⇒

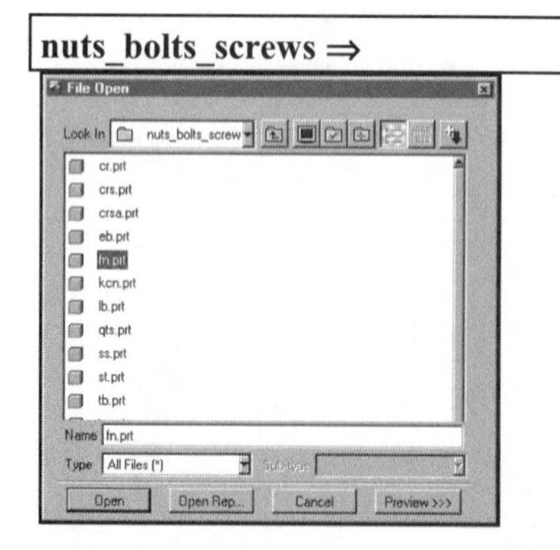

Because you will be creating an assembly using the bottom-up design approach, all the components must be available before the assembling starts. *Bottom-up design* means that existing parts are assembled one by one until the assembly is complete. The assembly starts with a set of datum planes and a coordinate system. The parts are added to the datum features of the assembly. The sequence of assembly will determine the parent-child relationships between components.

Top-down design is the design of an assembly whose component parts are created in Assembly mode as the design unfolds. Some existing parts are available, such as standard components and a few modeled parts. The remaining design evolves during the assembly process.

Regardless of the design method, the assembly datums and coordinate system should be on their own separate *assembly layer*. Each part should also be placed on separate assembly layers; the part's datum features should already be on *part layers*.

The **Clamp Plate** component is the first component of the main assembly. The Plate is a new part (Fig. 14.20). You must model the plate before creating the assembly. Figure 14.21 provides the dimensions necessary to model the **Clamp Plate.**

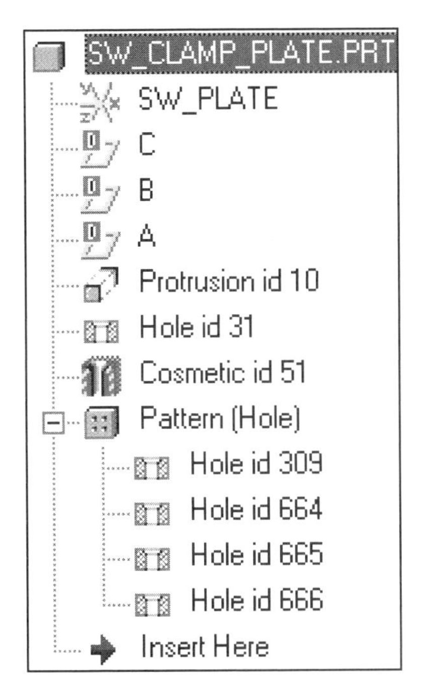

Figure 14.20
Clamp Plate

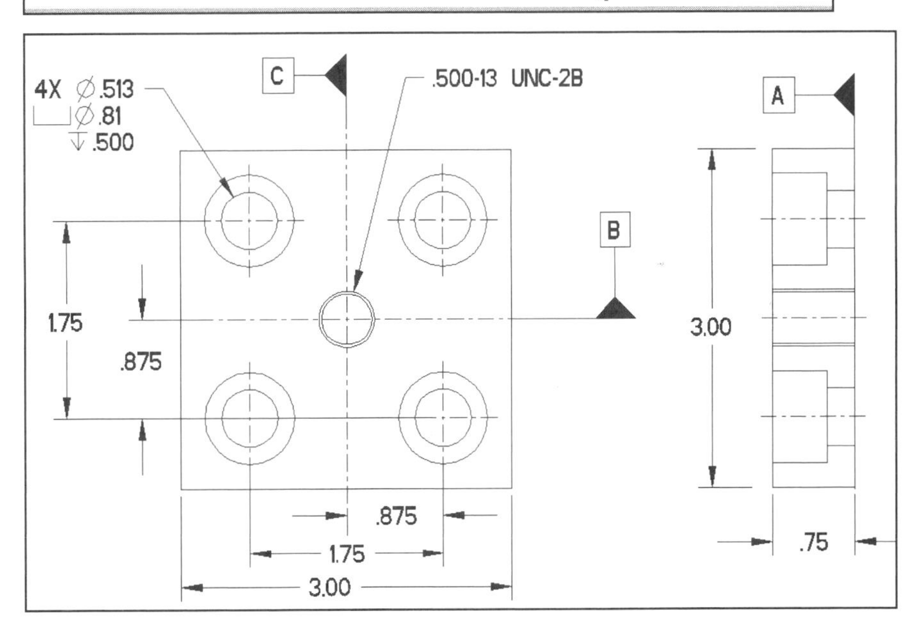

Figure 14.21
Clamp Plate Detail Drawing

You now have all nine parts (two identical Ball components are used) required for the assembly. *A subassembly will be created first.* The main assembly is created second. The subassembly is assembled to the main assembly. Start the subassembly (Fig. 14.22) using the following commands:

File ⇒ New ⇒ ●Assembly ⇒ CLAMP_SUBASSEMBLY ⇒ ✔ Use default template ⇒ OK ⇒ View ⇒ Model Tree Setup ⇒ Item Display ⇒ Display **✔Features ✔Notes ⇒ OK**

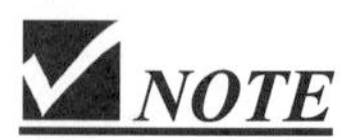

NOTE

It is very important to go back and reread Section 8, Layers, at this time!

Datum planes and the coordinate system are created automatically as a default. The datums will have default names, **ASM_RIGHT**, **ASM_TOP**, and **ASM_FRONT,** provided by Pro/E.

Change the coordinate system name:

Set Up ⇒ Name ⇒ Feature ⇒ (pick the coordinate system) ⇒ (type new name **SUB_ASM_CSYS**) ⇒ ✔ ⇒ **Done ⇒ File ⇒ Save ⇒ ✔**

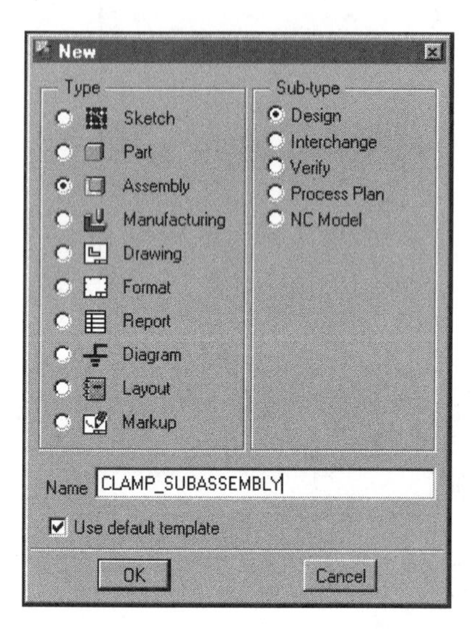

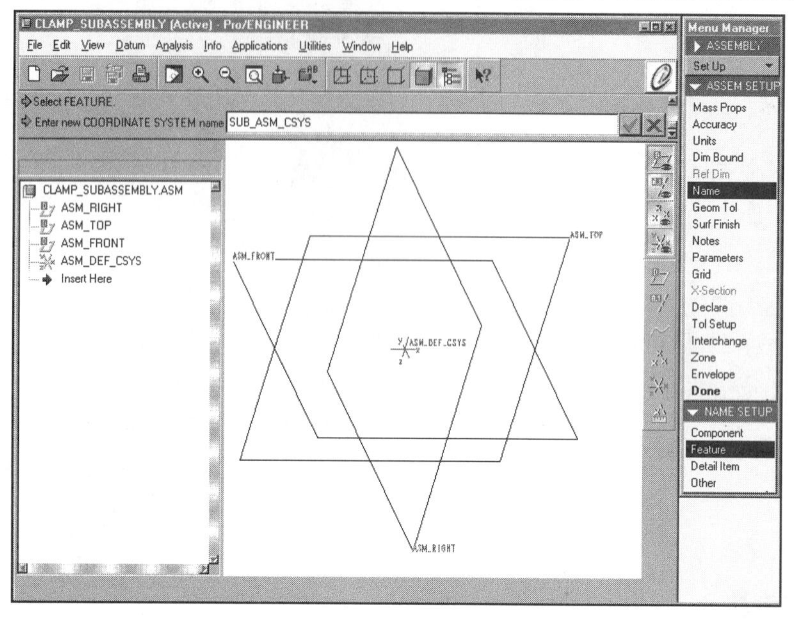

Figure 14.22
Subassembly Datums

Layers (Fig. 14.23) for the datum planes and coordinate system are also created by default. You can create your own layering system, using unique names, or use the ones provided here or by Pro/E.

View ⇒ Layers ⇒ Show ⇒ Layer Items ⇒ pick **- 05_ASM_DEF_DTM_CSYS ⇒ Layer ⇒ Info ⇒ Close** **File ⇒ Save Status File ⇒ ✔⇒ OK ⇒ Close**

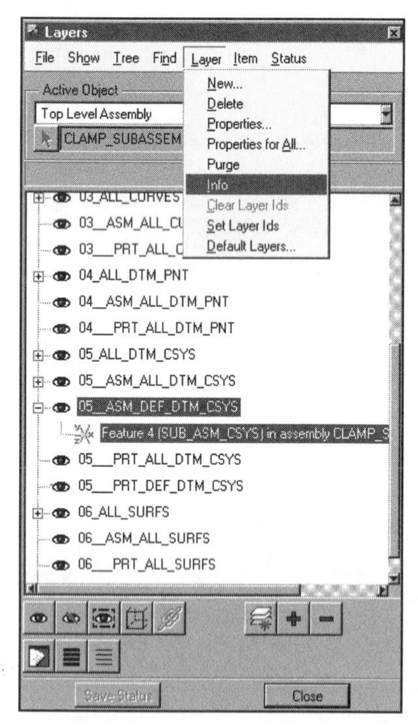

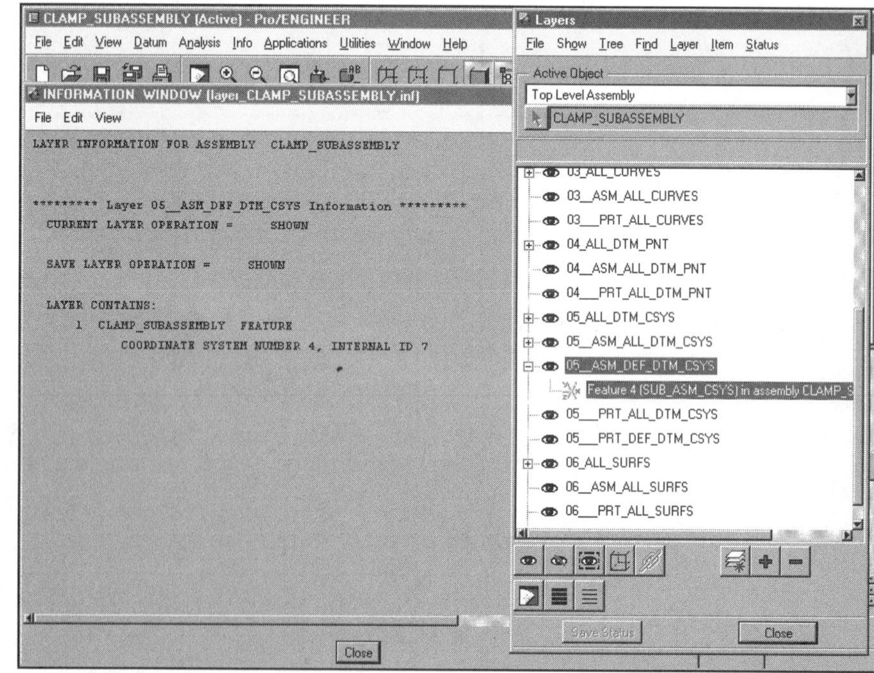

Figure 14.23
Layer ⇒ Info

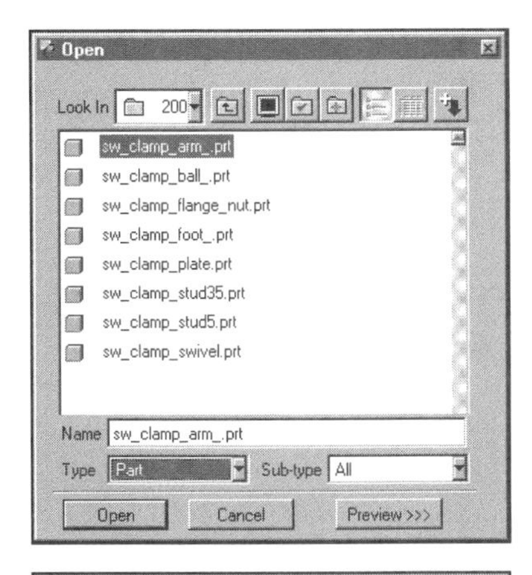

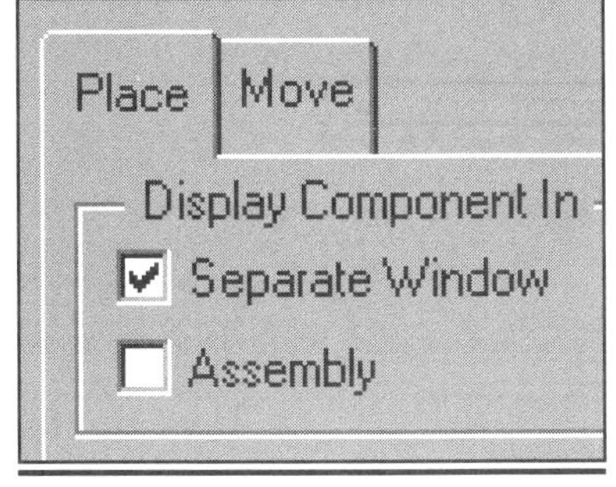

The first component to be assembled to the subassembly is the Swing Clamp Arm (Fig. 14.24). The simplest and quickest method of adding a component to an assembly is to match the coordinate systems. The first component assembled is usually where this *constraint* is used, because after the first component is established, few if any of the remaining components are assembled to the assembly coordinate system (with the exception of top-down design) or, for that matter, other parts' coordinate systems.

The **Coord Sys** option (Fig. 14.25) places a component in an assembly by aligning its coordinate system with a coordinate system in the assembly. Before you start the assembly process, both coordinate systems should exist on their respective models (Pro/E refers to both assemblies and parts as *models* or *objects*). Make sure all your models are in the same working directory before you start the assembly process. ***If you named your models something that is not listed here, pick the appropriate part model as requested.***

Choose the following commands (Figs. 14.25 and 14.26):

Component ⇒ **Assemble** ⇒ [pick the **sw_clamp_arm** (the machined part, not the casting) from the Open dialog box] ⇒ **Open** ⇒ ✔ **Separate Window** ❑ **Assembly** ⇒ (you may need to resize your windows) ⇒ ▼ **Coord Sys** (from the Constraint Type drop-down selections) ⇒ (↖pick the coordinate systems of the part model and then ↖pick the assembly model) ⇒ **OK** ⇒ **Done/Return**

After the component window appears, rotate the component so that you can see more clearly the features to be used for constraining. When a feature in one window is selected [the *active window* (**Active**)], another window may become hidden.

? Pro/HELP

Remember to use the help available on Pro/E by highlighting a command and pressing the right mouse button.

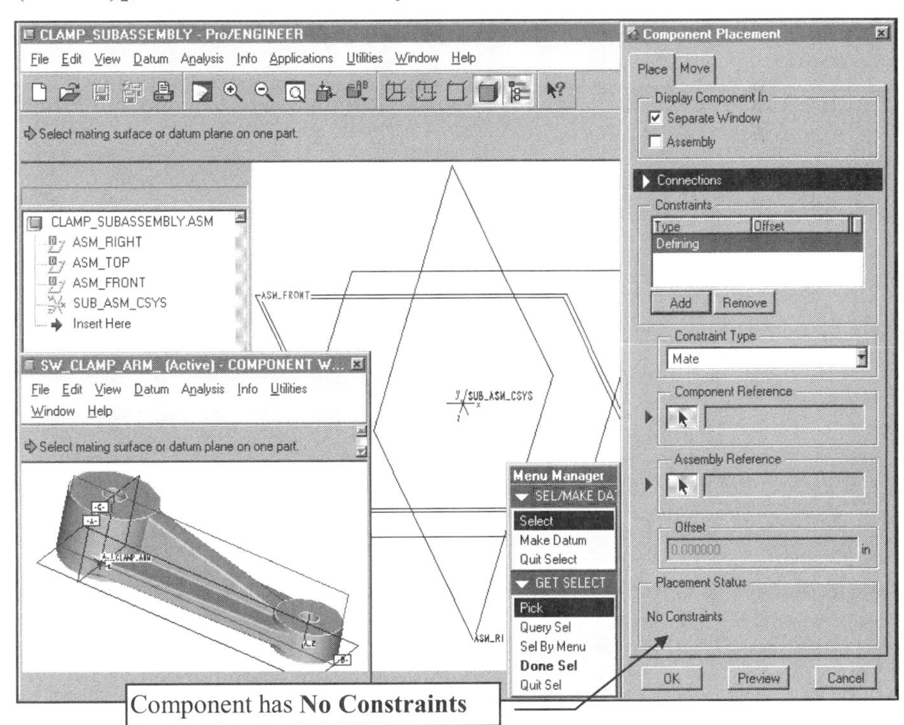

Figure 14.24
COMPONENT WINDOW and Component Placement Dialog Box (the Component Window is "**Active**")

Figure 14.25
Component is
Fully Constrained

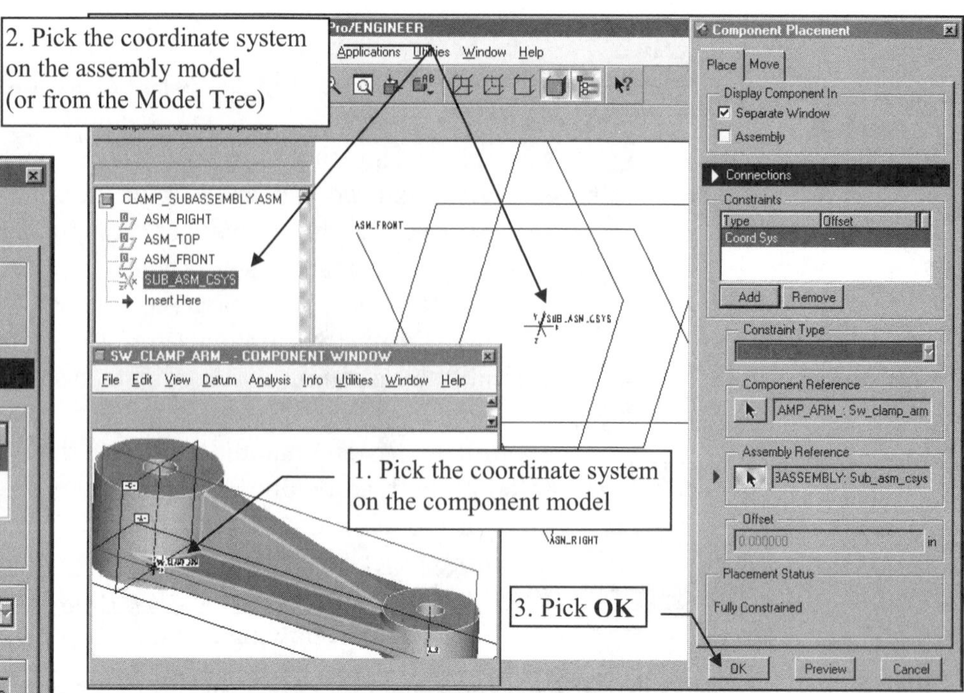

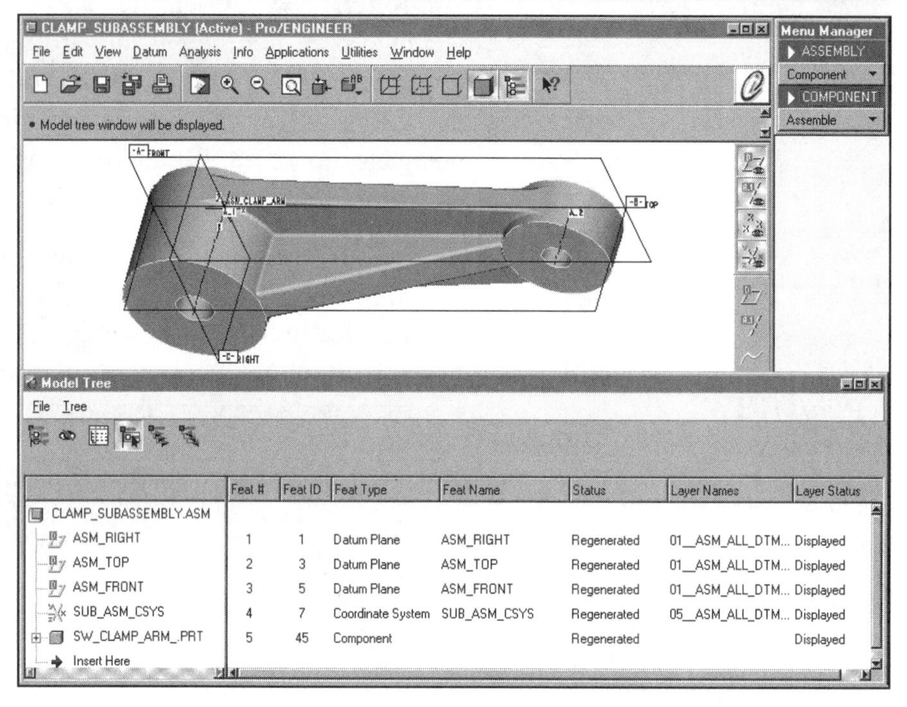

Figure 14.26
Subassembly Now Has One Component (the Model Tree has been displayed as a separate window and has a number of columns added)

The next component to be assembled is the Swing Clamp Swivel. Two constraints will be used with this component: **Insert** and **Mate Offset**. *Placement constraints* are used to specify the relative position of a *pair of surfaces/references* between two components. The **Mate**, **Align**, **Insert**, and **Orient** commands are placement constraints. The two surfaces/references must be of the same type. When using a datum plane as a placement constraint, specify which side to use, *yellow* or *red*. When using **Mate Offset** or **Align Offset**, enter the offset distance. The *offset direction* is displayed with a large arrow. If you need an offset in the opposite direction, enter a *negative value*. Choose the following commands to assemble the next component (Fig. 14.27):

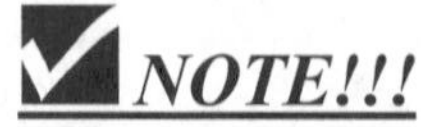
NOTE!!!
Follow these instructions exactly as provided, regardless of what you might think is the correct placement. There is a reason for the selections made. If you correct these instructions, you will be unable to experience some of the later commands involving editing.

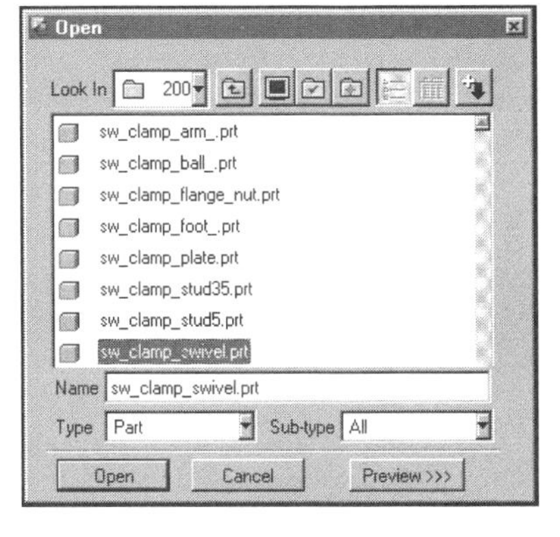

File ⇒ **Save** ⇒ ✔ ⇒ **Component** ⇒ **Assemble** ⇒ (pick the **sw_clamp_swivel** from the Open dialog box list) ⇒ **Open** ⇒ ▼ ⇒ **Insert** ⇒ [↖pick the shaft of the Swivel (Component Reference) and then ↖pick the hole of the Arm (Assembly Reference)]

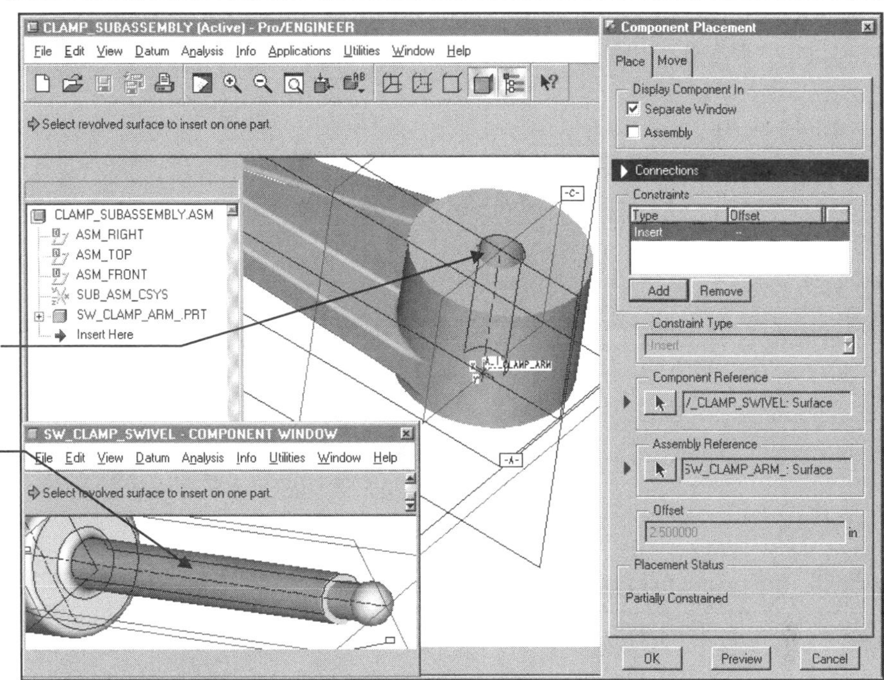

Figure 14.27
Assembling the Sw_Clamp_Swivel

The next constraint is **Mate Offset**:

Mate Offset ⇒ [↖pick the lower surface of the Swivel and then ↖pick the upper surface on the assembly model (Figure 14.28 and Figure 14.29)] ⇒ [**Offset (ins) in indicated direction 0.0000** (type) **2.00**] ⇒ ✔ ⇒ **OK** ⇒ **Done/Return** ⇒ **File** ⇒ **Save** ⇒ ✔

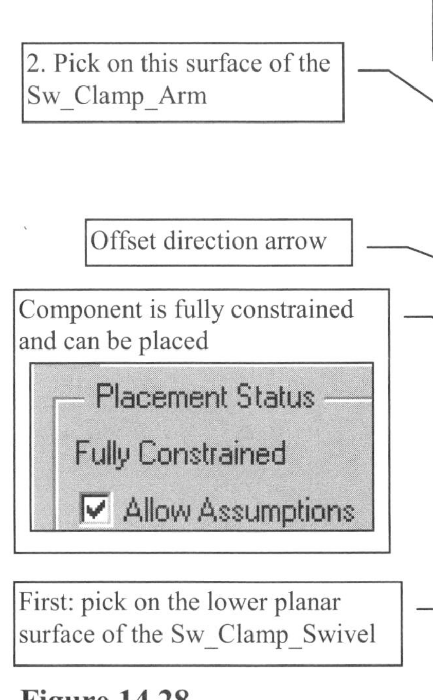

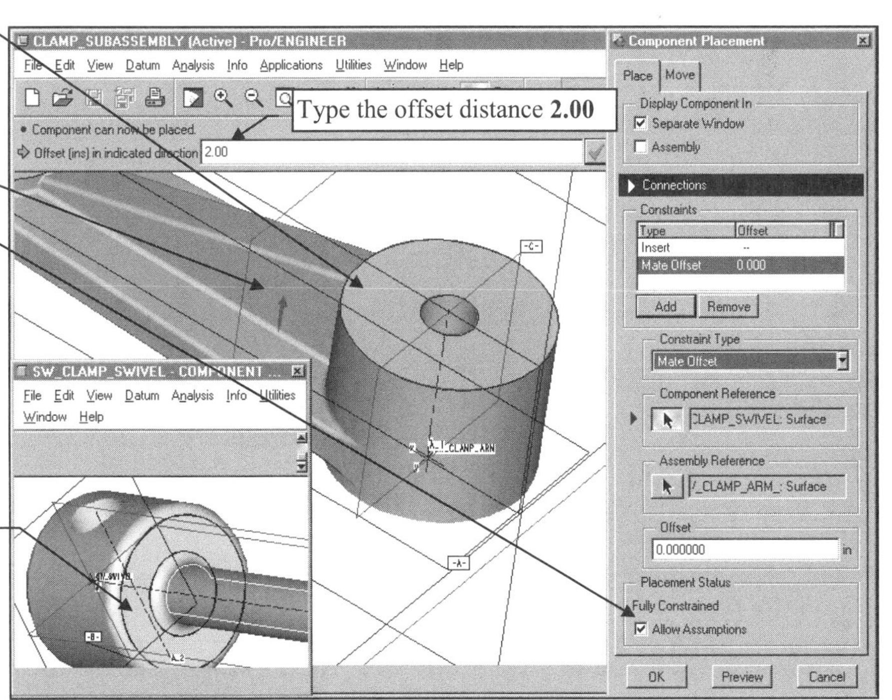

Figure 14.28
Constrained Sw_Clamp_Swivel

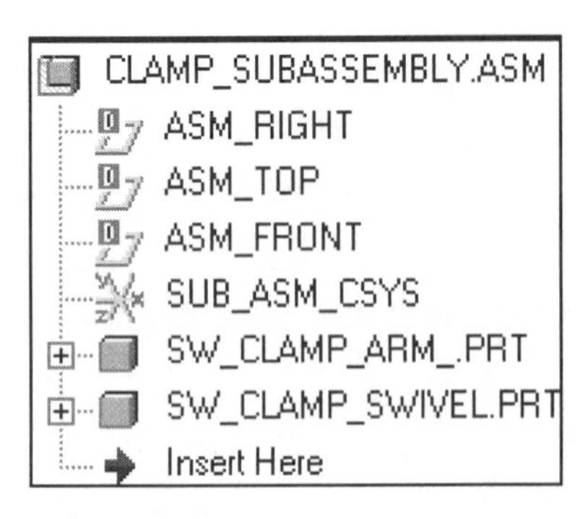

Figure 14.29
Subassembly with Sw_Clamp_Arm and Sw_Clamp_Swivel

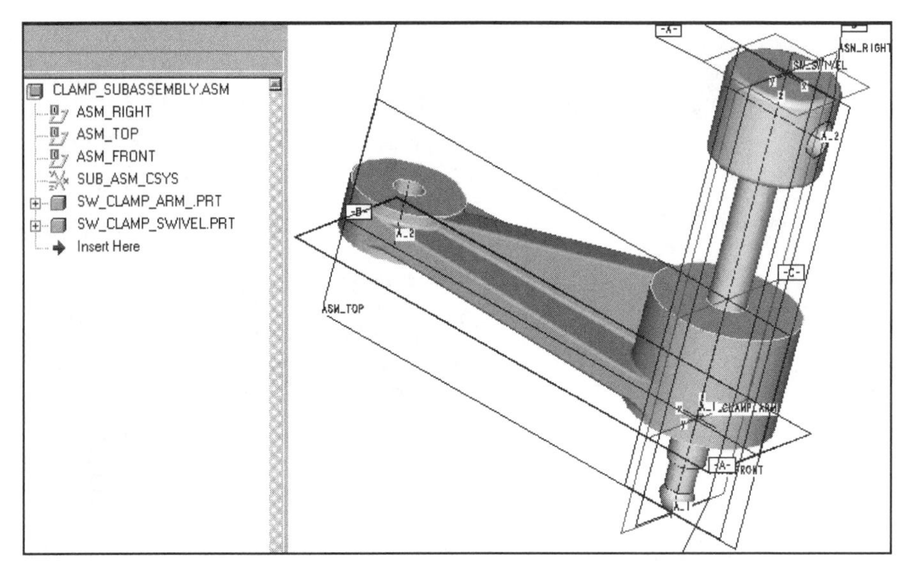

Change the offset distance to **1.50** using the following commands:

Modify ⇒ **Mod Assem** ⇒ **Modify Dim** ⇒ **Value** ⇒ (pick the Swivel) ⇒ [pick the **2.00** dimension from the model (Fig. 14.30)] ⇒ (type **1.50** at the prompt) ⇒ ✔ ⇒ **Done Sel** ⇒ **Done** ⇒ **Done** ⇒ **Done/Return** ⇒ **Regenerate** ⇒ **Automatic** ⇒ (Fig. 14.31)

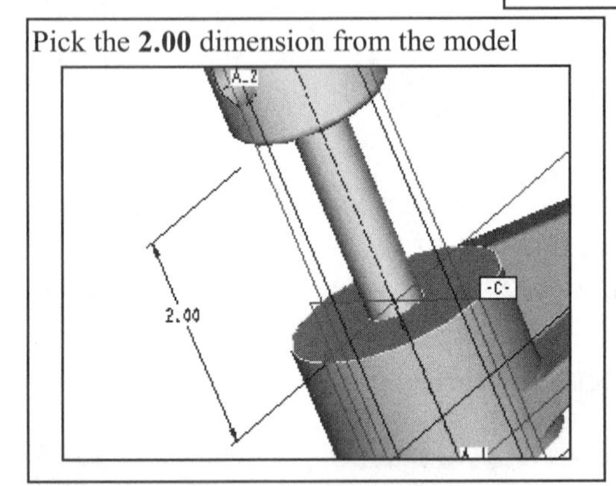

Figure 14.30
Modify the Offset Distance to **1.50**

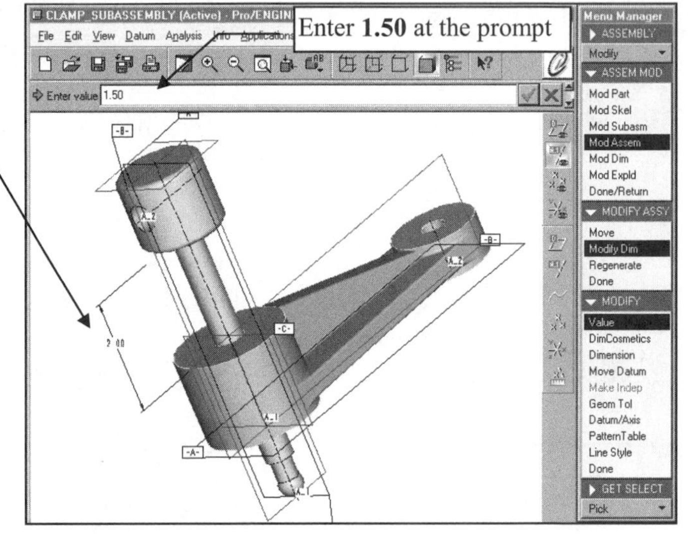

File ⇒ **Save** ⇒ ✔
File ⇒ **Delete** ⇒
Old Versions ⇒ ✔

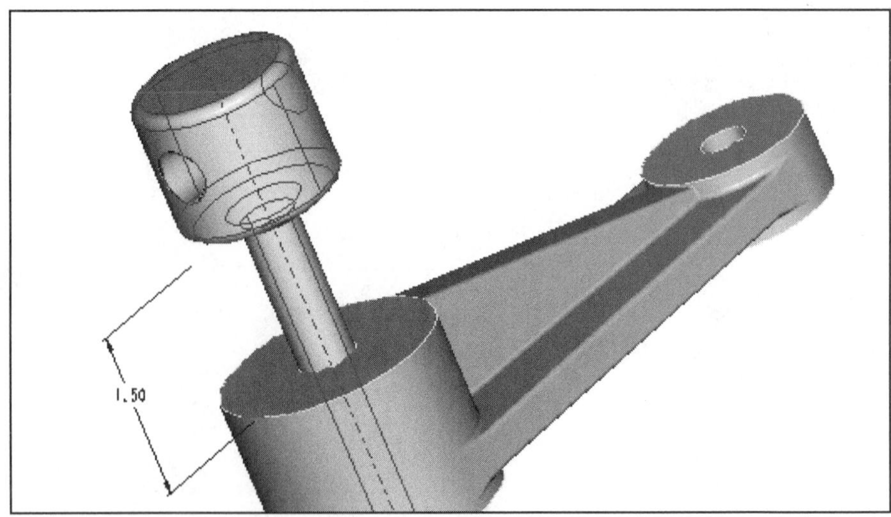

Figure 14.31
Regenerated Model with **1.50** as Offset

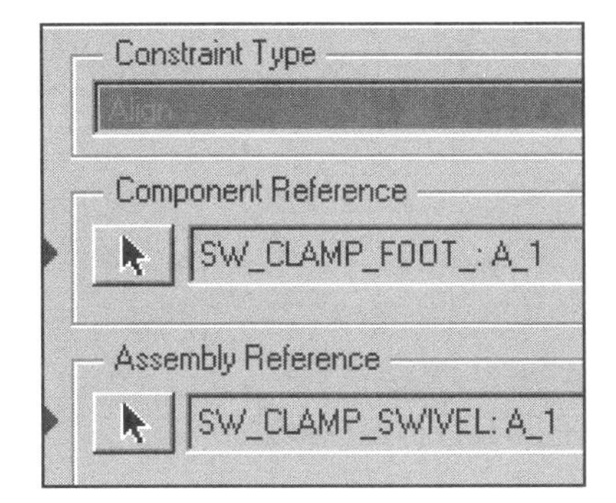

The next component to be assembled is the Sw_Clamp_Foot component. Choose the following commands:

> **Component** ⇒ **Assemble** ⇒ (pick **sw_clamp_foot** from the **Open** dialog box) ⇒ **Open** ⇒ ✔**Separate Window** ❑ **Assembly** ⇒ **Align** ⇒ [⇖pick the *axis* of the Sw_Clamp_Foot and then ⇖pick the *axis* on the Sw_Clamp_Swivel (Fig. 14.32)] ⇒ **Mate** ⇒ [⇖pick the spherical hole of the Sw_Clamp_Foot and then pick the ⇖spherical end on the Sw_Clamp_Swivel (Fig. 14.33)] ⇒ **Preview** ⇒ **OK**

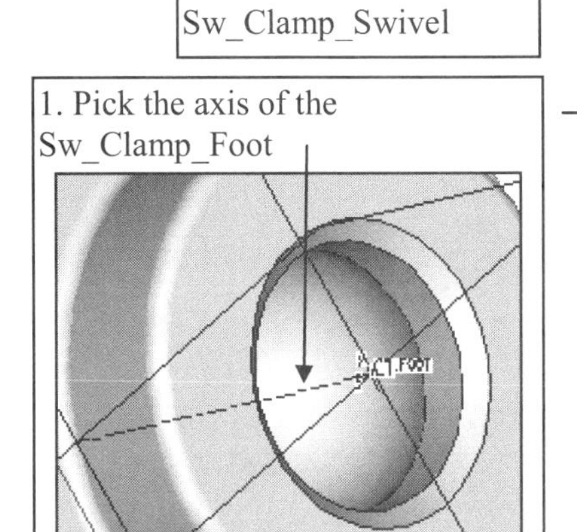

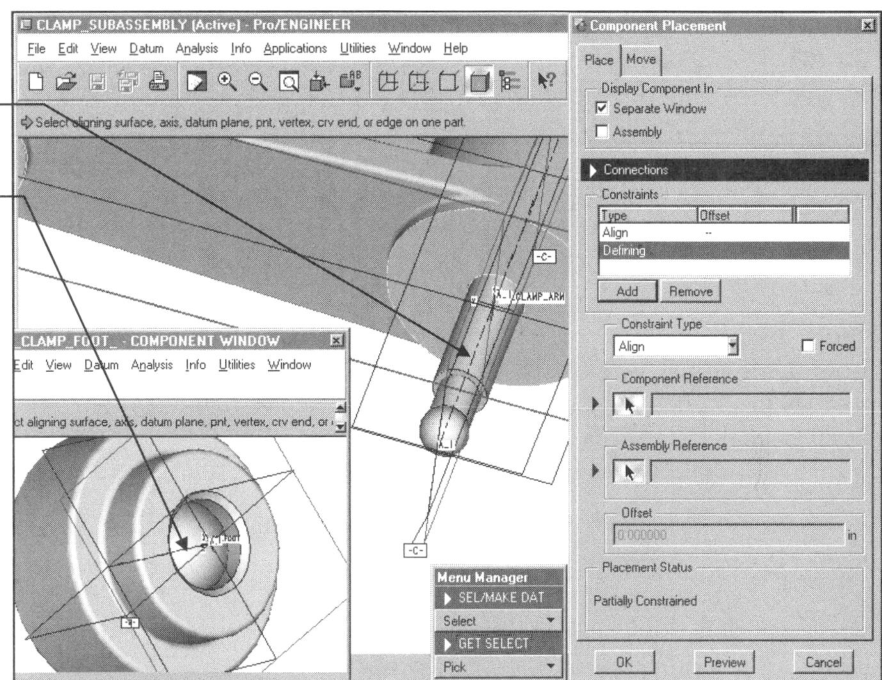

Figure 14.32
Using **Align** as a Constraint

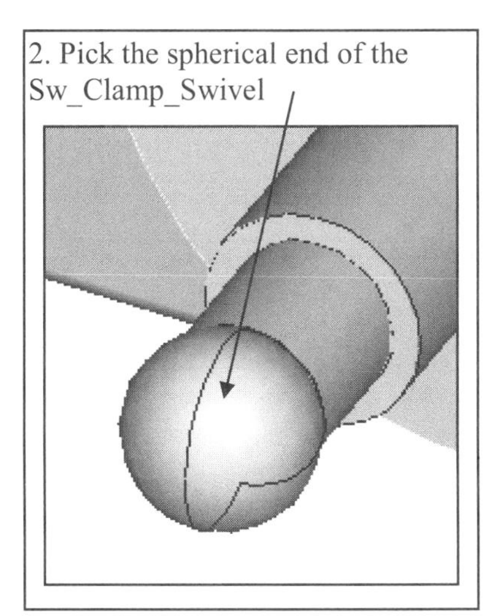

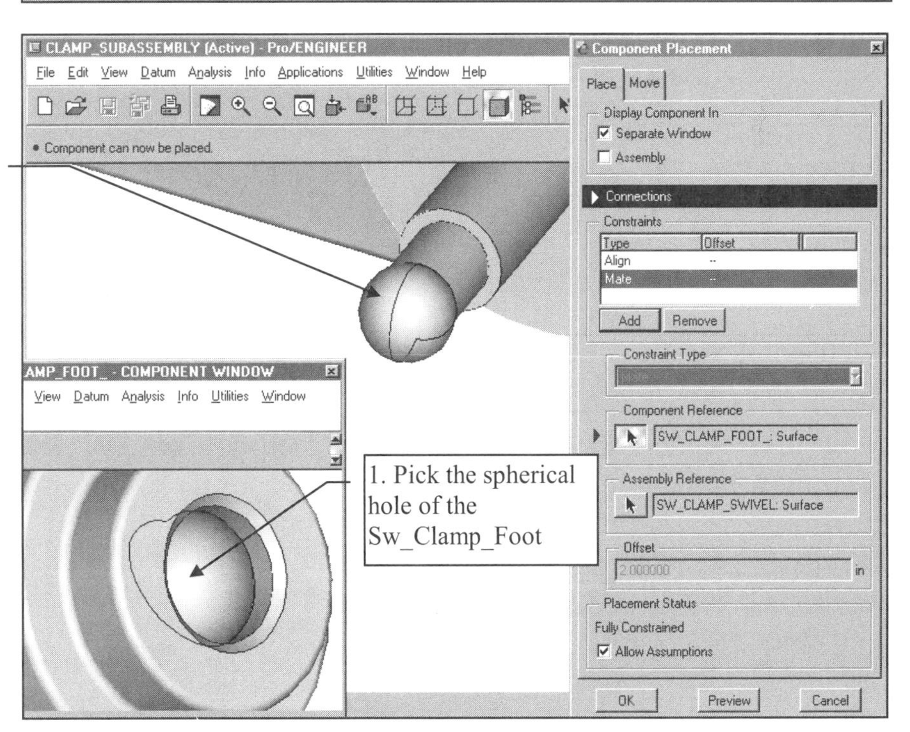

Figure 14.33
Using **Mate** as a Constraint

Well, it doesn't look exactly right yet (Fig. 14.34). A third constraint needs to be added to orient the Sw_Clamp_Foot correctly. Choose the following commands (Fig. 14.35):

Redefine ⇒ (pick Sw_Clamp_Foot) ⇒ **Add** ⇒ **Orient** ⇒ [↖pick lower surface of the Sw_Clamp_Arm (Fig. 14.36)] ⇒ **Query Sel** ⇒ (↖pick top surface of the Sw_Clamp_Foot) ⇒ **Next** ⇒ **Accept** ⇒ **Preview** (Fig. 14.37) ⇒ **OK** ⇒ **Done/Return** (Fig. 14.38)

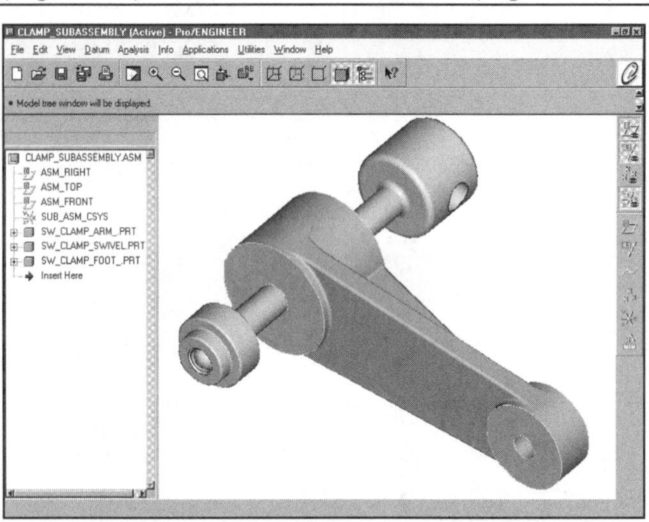

Figure 14.34
Foot Facing in the Wrong Direction

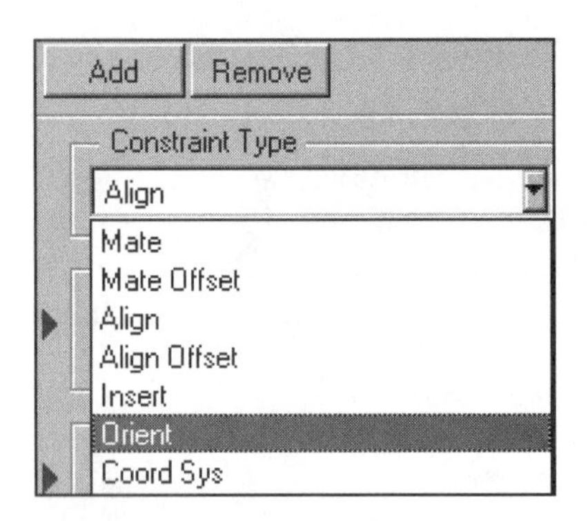

Figure 14.35
Adding **Orient** as a Constraint

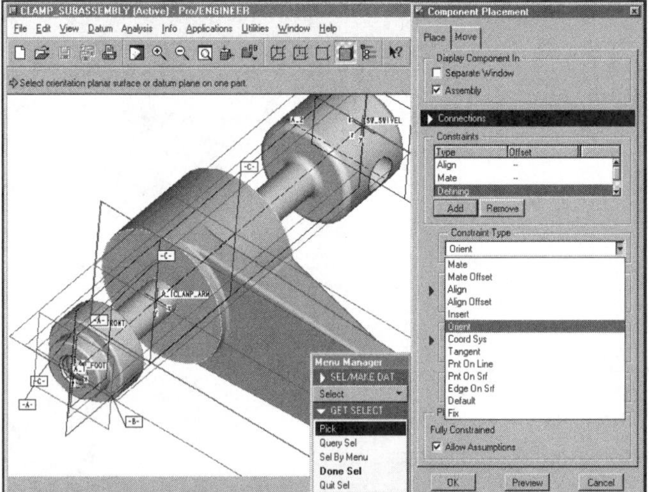

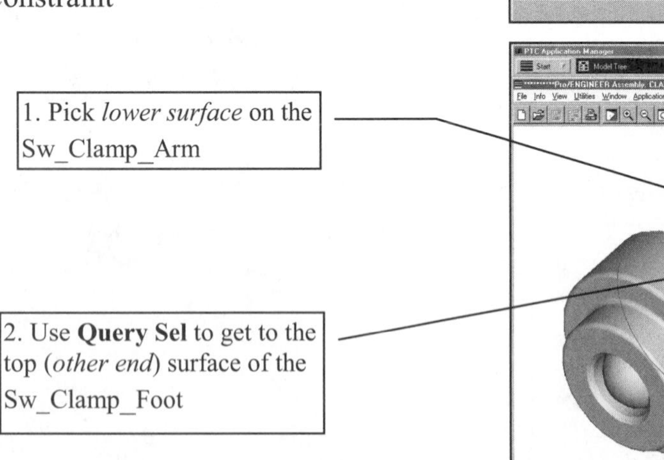

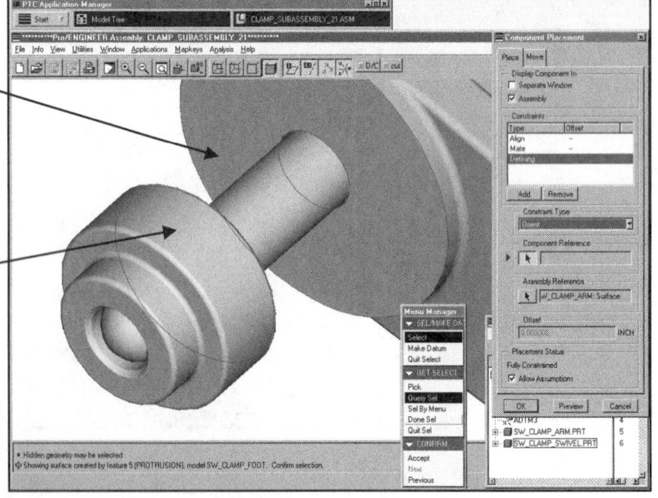

Figure 14.36
Picking Surfaces for **Orient**

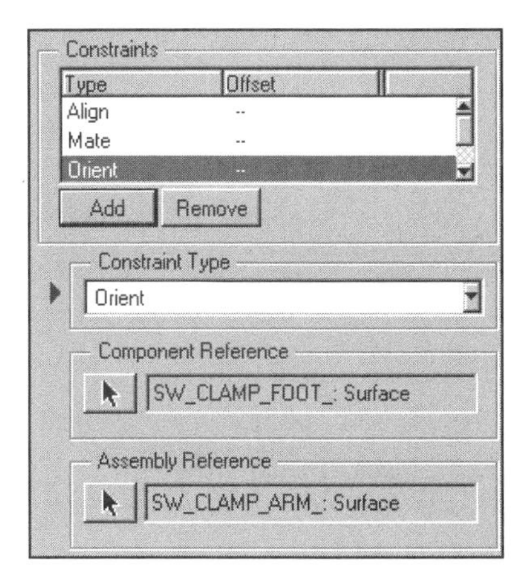

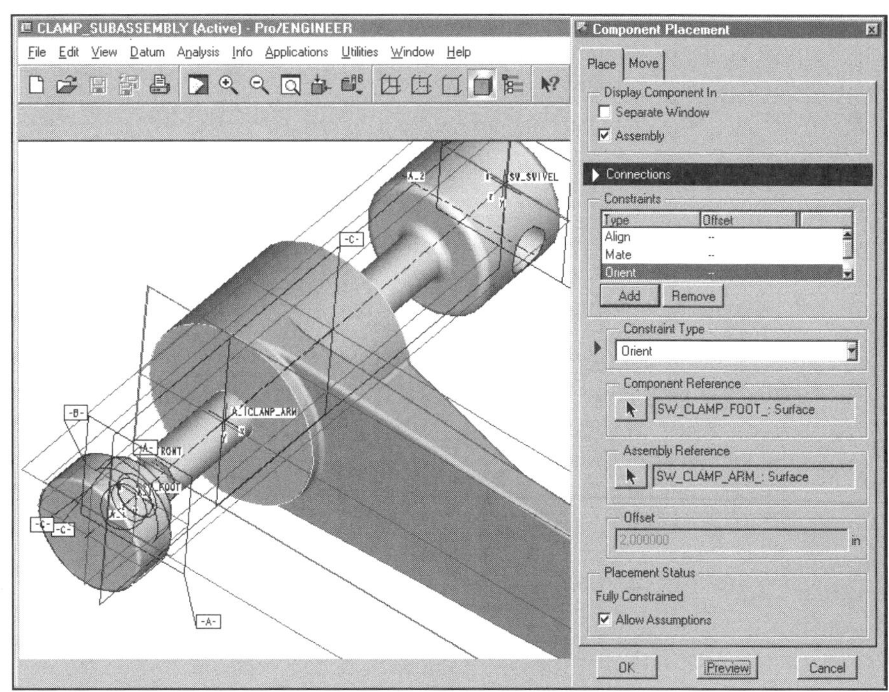

Figure 14.37
Preview of Redefined Foot

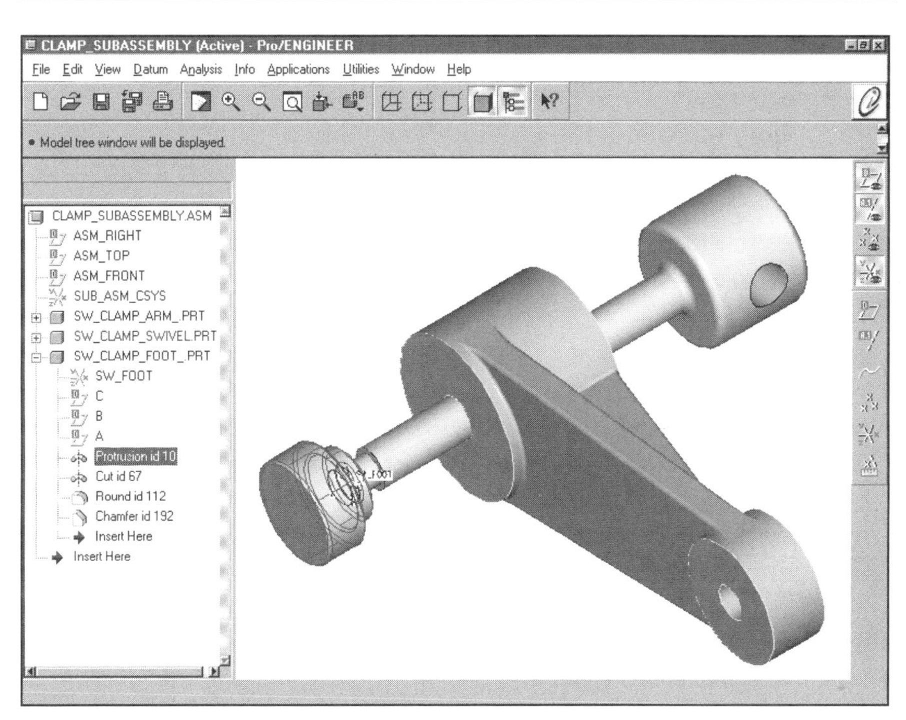

Figure 14.38
Redefined Component

Here comes your boss again. Unfortunately, *you* (Hey, don't blame us! We can't help it if you follow instructions well.) put the Swivel and the Foot in the wrong hole! Before the oversight gets any undue attention, redefine its placement by choosing:

Component ⇒ **Redefine** ⇒ (select the Sw_Clamp_Swivel from the Model Tree) ⇒ (select the **Insert** constraint you added before from the Component Placement dialog box) ⇒ (select Assembly Reference ⇒ (pick the other hole surface as shown in Figure 14.39) ⇒ **Preview** (Fig. 14.40)

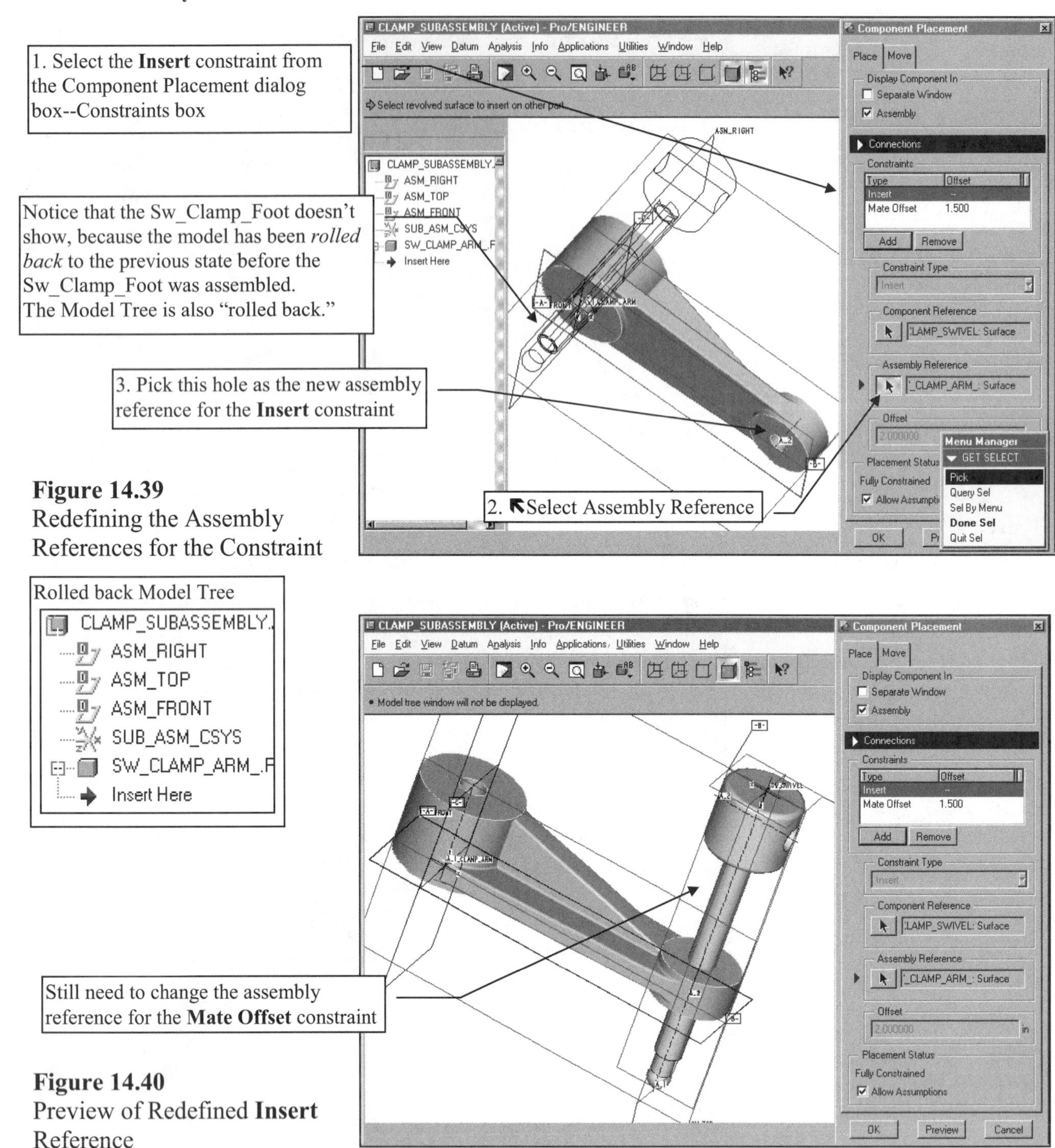

Figure 14.39
Redefining the Assembly References for the Constraint

Figure 14.40
Preview of Redefined **Insert** Reference

The Swivel is now in the correct hole, but the **Mate Offset** assembly reference is incorrect, because the bottom of the Swivel head is offset from the top surface of the large-diameter boss on the other side of the part. Continue the redefinition:

(select the **Mate Offset** constraint from the Component Placement dialog box) ⇒([icon] select Assembly Reference ⇒ [select offset mating surface (Fig. 14.41)] ⇒ (type **2.50** in the Offset box) ⇒ **enter** ⇒ **Preview** ⇒ **OK** ⇒ **Done/Return** (Fig. 14.42)

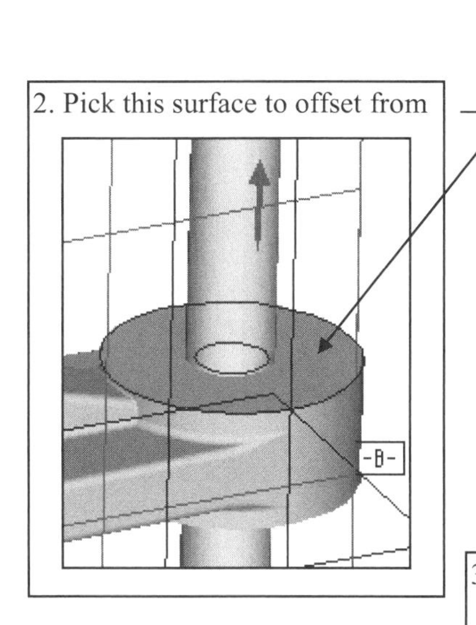

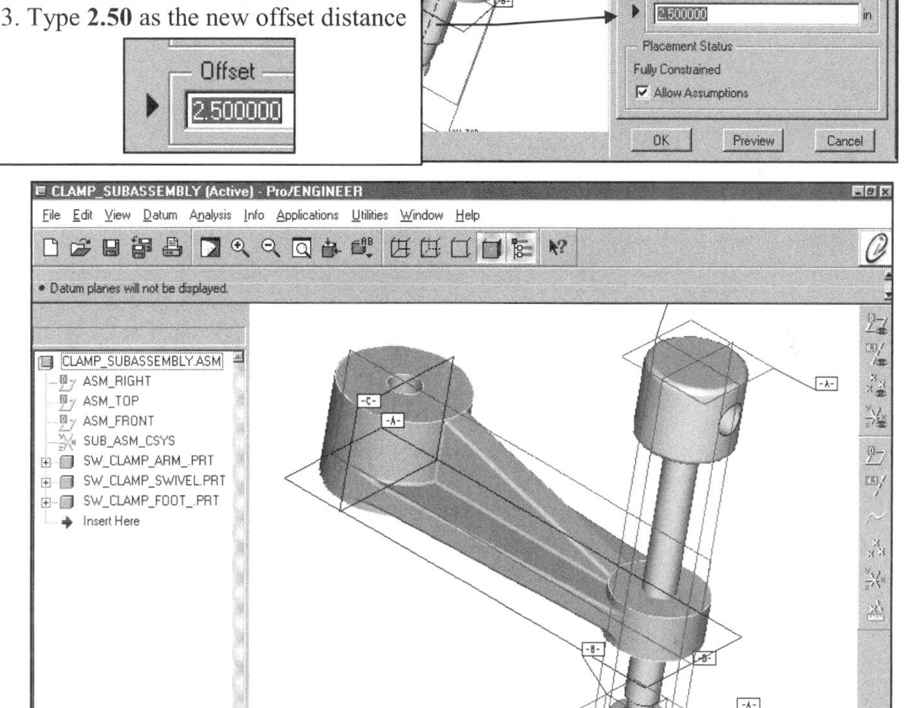

Figure 14.41
Redefined **Mate Offset** Reference

NOTE

Both **Feature ⇒ Redefine** and **Component ⇒ Redefine** are available. You want to redefine the placement of the component, not an assembly feature; therefore, use **Component ⇒ Redefine**.

Figure 14.42
Completed Redefinition

The next step is to assemble the **5.00** double-ended stud:

Component ⇒ Assemble ⇒ sw_clamp_stud5 ⇒ Open ⇒ Automatic ⇒ ✔ Assembly ❑ Separate Window ⇒ *(the component will "float" in the main wind; hold down* ***Ctrl Alt*** *and your mouse buttons to adjust the component location using* ***Driven Component Manipulation****)* ⇒ [↖pick the *revolved surface* of the Sw_Clamp_Stud5 as the Component Reference, and then ↖pick the *revolved surface hole* on the Sw_Clamp_Swivel as the Assembly Reference (Fig. 14.43)] ⇒ **Automatic** ⇒ (use *Freeform Mouse* to adjust the component's position) ⇒ (↖pick the end of the Sw_Clamp_Stud5 and then ↖pick datum **C** of the Sw_Clamp_Swivel from the Model Tree) ⇒ ***Yellow*** (Fig. 14.44) ⇒ (type an offset of **-5.00/2** at the prompt) ⇒ **✔⇒ OK** (Fig. 14.45) ⇒ **Done/Return**

NOTE

When the part was modeled, the datum planes were set as geometric tolerance features (Basic Datums) and renamed:

TOP = B
FRONT = A
RIGHT = C

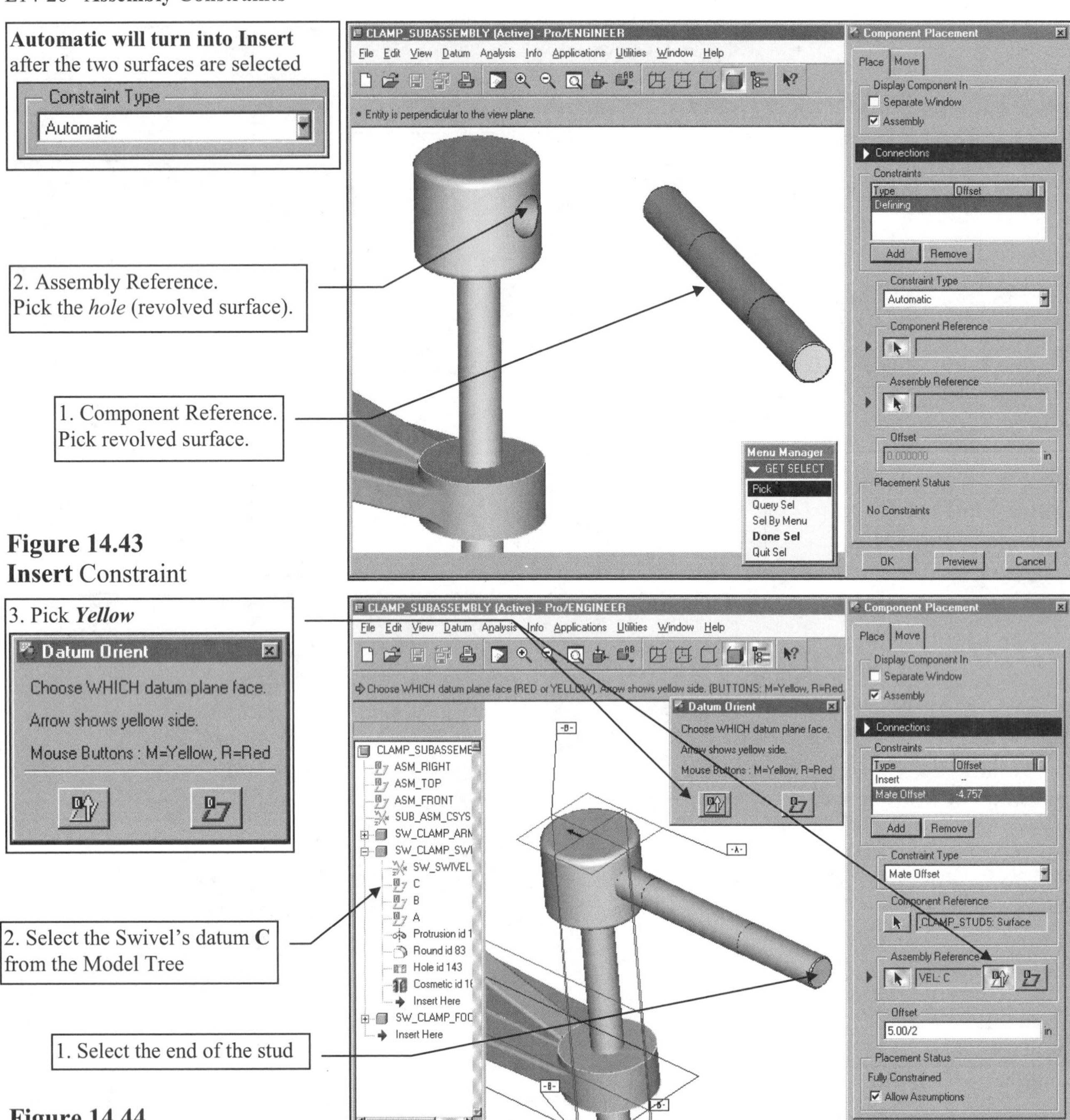

Figure 14.43
Insert Constraint

Figure 14.44
Align Offset Constraint

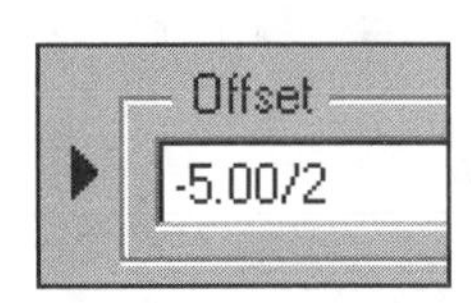

turns into this

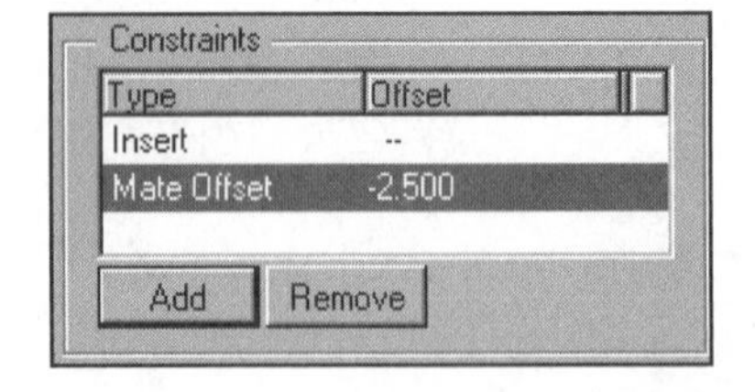

Automatic turns into **Insert** after the two surfaces are selected, and **Automatic** turns into **Mate Offset** after the end surface and datum **C** are selected. The two Ball handles are assembled next. Instructions are provided for assembling one Ball; you must assemble the other on your own. Choose the following commands:

File ⇒ **Save** ⇒ ✔ ⇒ **Component** ⇒ **Assemble** ⇒ **(sw_clamp_ball)** ⇒ **Open** ⇒ (use *Freeform Mouse* to adjust components position) ⇒ **Insert** ⇒ [pick the *revolved hole* of the Ball as the Component Reference and then the *revolved surface* on Stud5 as the Assembly Reference (Fig. 14.46)]

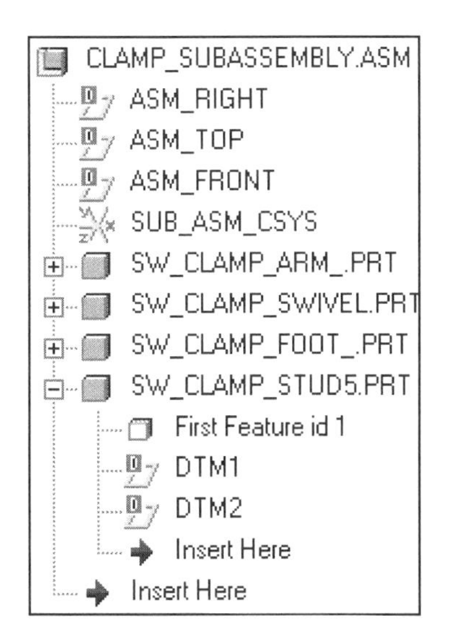

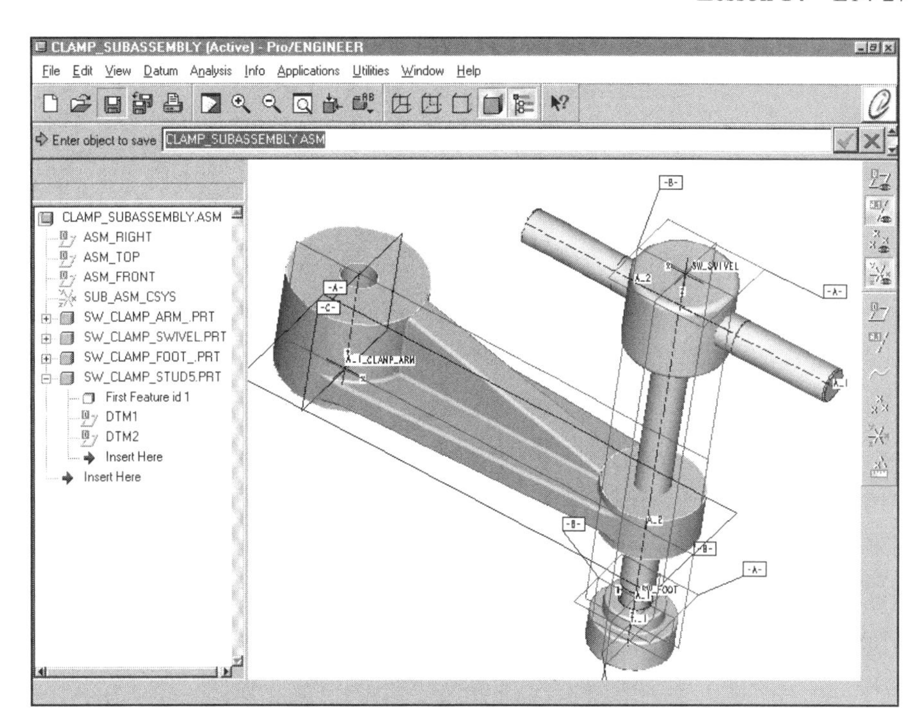

Figure 14.45
Assembled Double-Ended Stud

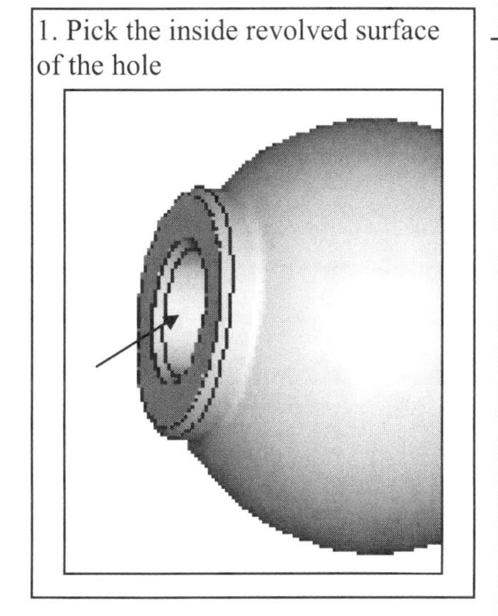

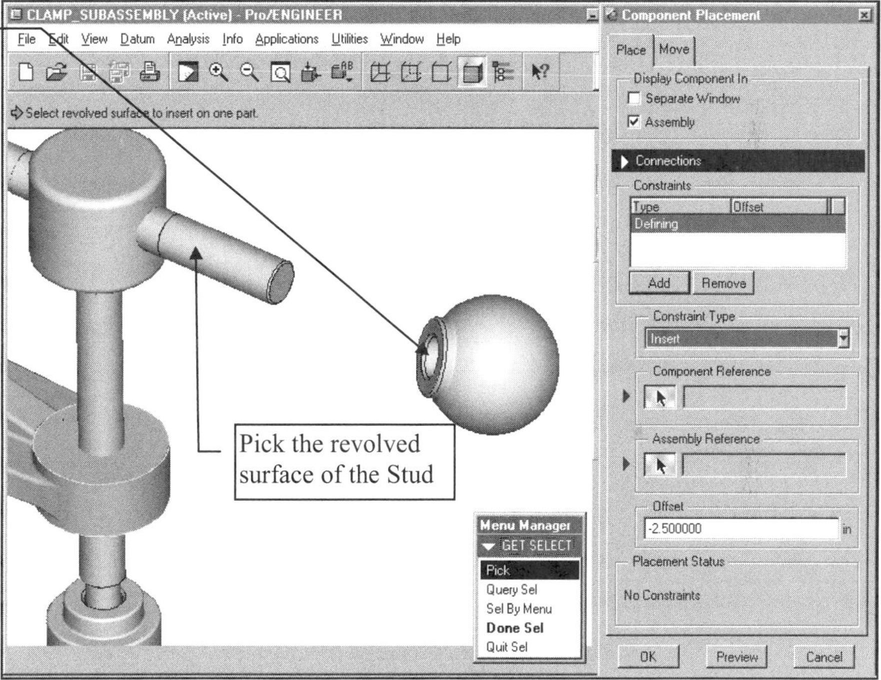

Figure 14.46
Assembling the Ball

Mate Offset ⇒ (use *Freeform Mouse* to adjust component position) ⇒ (pick the flat end of the Ball and then the flat end of Stud5, as in Figure 14.47) ⇒ **(Offset [ins] in indicated direction)** ⇒ (type **-.50** at the prompt) ⇒ ✔⇒ **OK** (Figure 14.48) ⇒ **Done/Return** ⇒ **File** ⇒ **Save** ⇒ **File** ⇒ ✔ ⇒ **Delete** ⇒ **Old Versions** ⇒ ✔

Assemble the second Ball as shown in Figure 14.49. The subassembly is now complete. **Save** the model. The subassembly will be added to the assembly in the next set of steps.

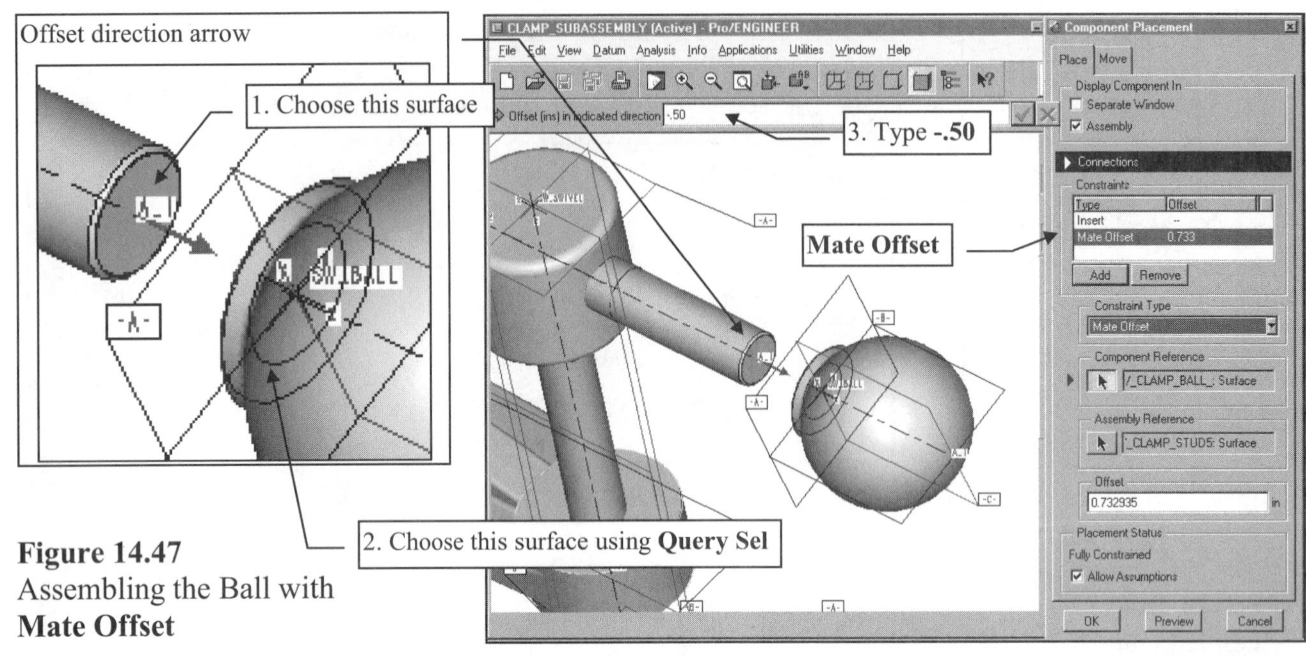

Figure 14.47
Assembling the Ball with **Mate Offset**

Figure 14.48
Ball Assembled on Stud5

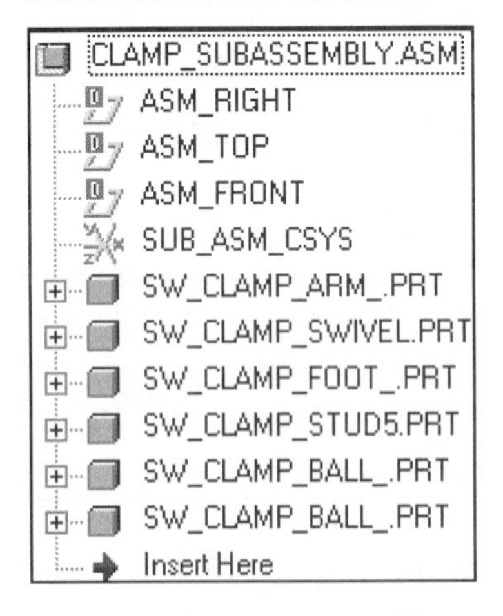

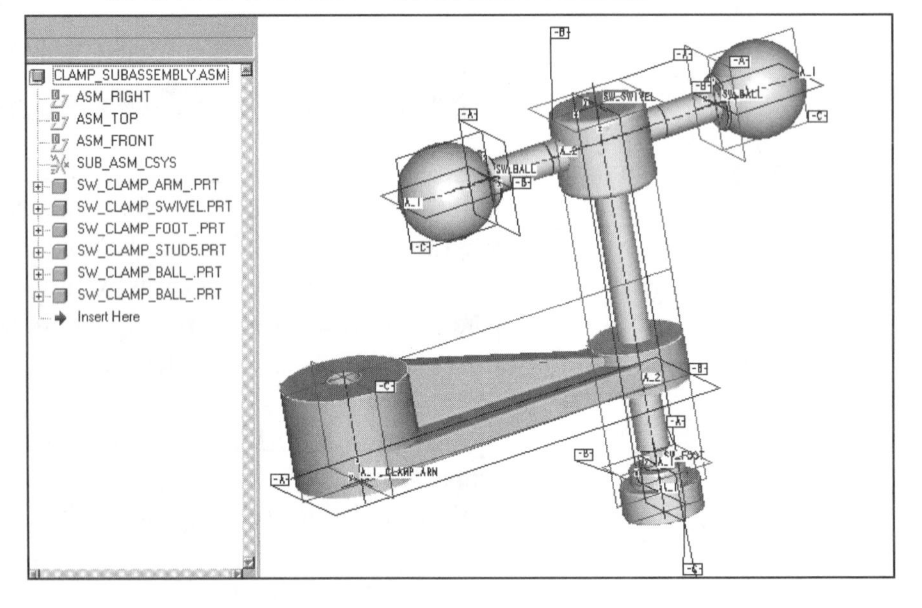

Figure 14.49
Completed Subassembly

ECO

As a minor **ECO,** redefine or modify the Ball component as offset **.44** from the end of the Shaft so that Stud5 does not bottom out in the Ball's hole.

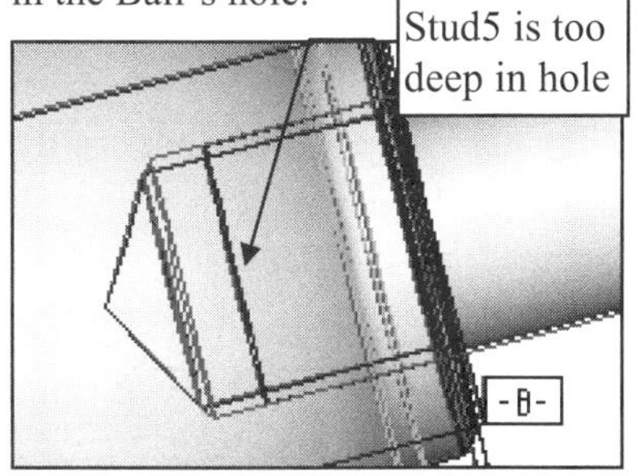

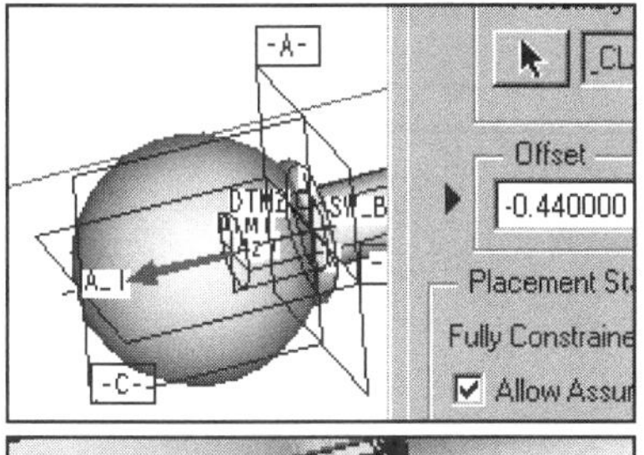

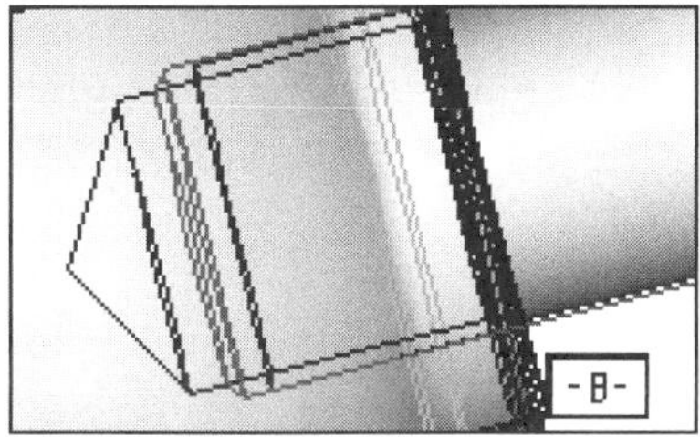

Figure 14.50
Sw_Clamp_Assembly

The first features for the next assembly will be the appropriate datum planes and coordinate system. The first part assembled on the main assembly will be the Sw_Clamp_Plate. When you start another assembly, a new window will open. Leave the subassembly in session and resize the windows as needed. Choose the following commands to start the assembly (Fig. 14.50):

Utilities ⇒ Environment ⇒ Trimetric ⇒ OK ⇒ File ⇒ New ⇒ ●Assembly ⇒ CLAMP_ASSEMBLY ⇒ ✔ Use default template ⇒ OK ⇒ Set Up ⇒ Geom Tol ⇒ Set Datum (rename, set, and move datum identifiers) **⇒ OK ⇒ Done/Return ⇒ Done**

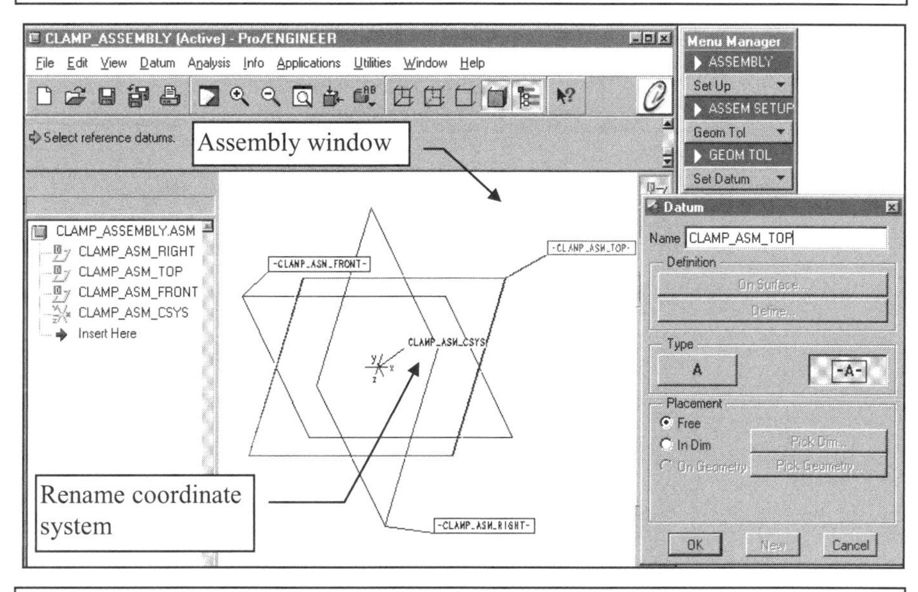

Component ⇒ Assemble ⇒ (sw_clamp_plate) ⇒ Open ⇒ ✔Separate Window ❑ Assembly ⇒ Coord Sys ⇒ [pick the coordinate system on the Sw_Clamp_Plate (Fig. 14.51) and then on **CLAMP_ASM_CSYS** on the assembly] **⇒ OK**

New component window

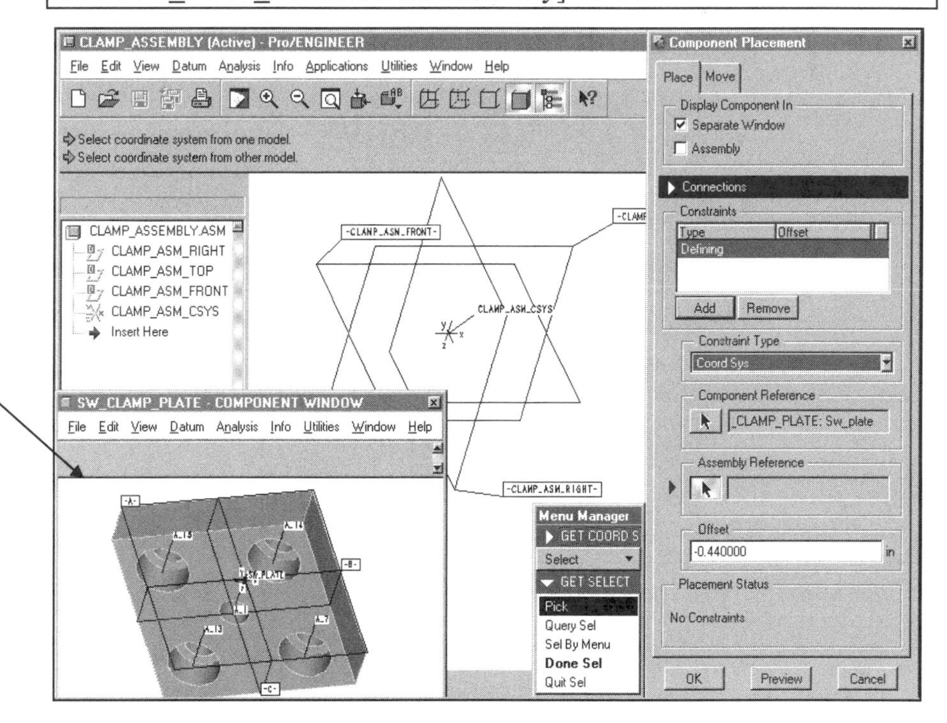

Figure 14.51
Use **Coord Sys** as the Only Constraint for the Plate

HINT
Your datum planes may have different names. Pick the *yellow* side of both datum planes. To **Orient**, pick a vertical datum on the assembly and subassembly.

Also, you may have to pick on the **CLAMP_ASSEMBLY** window to bring it forward.

Assemble ⇒ (pick the **clamp_subassembly** from the Open dialog box) ⇒ **Open** ⇒ **Mate** (Fig. 14.52) ⇒ [pick the top surface of the Sw_Clamp_Plate (Assembly Reference) and then the bottom circular planar surface on the Sw_Clamp_Arm of the subassembly (Component Reference)] ⇒ **Insert** ⇒ (pick the hole in the Sw_Clamp_Plate and then the hole in the Sw_Clamp_Arm) ⇒ **Add** ⇒ **Orient** ⇒ [pick a vertical datum plane of the assembly (**CLAMP_ASM_TOP**)] ⇒ ***Yellow*** ⇒ [pick the vertical datum plane (ASM_TOP) of the subassembly (Fig. 14.53)] ⇒ ***Yellow*** ⇒ **Preview** ⇒ **OK** ⇒ **Done/Return** (Fig. 14.54) ⇒ **File** ⇒ **Save** ⇒ ✔

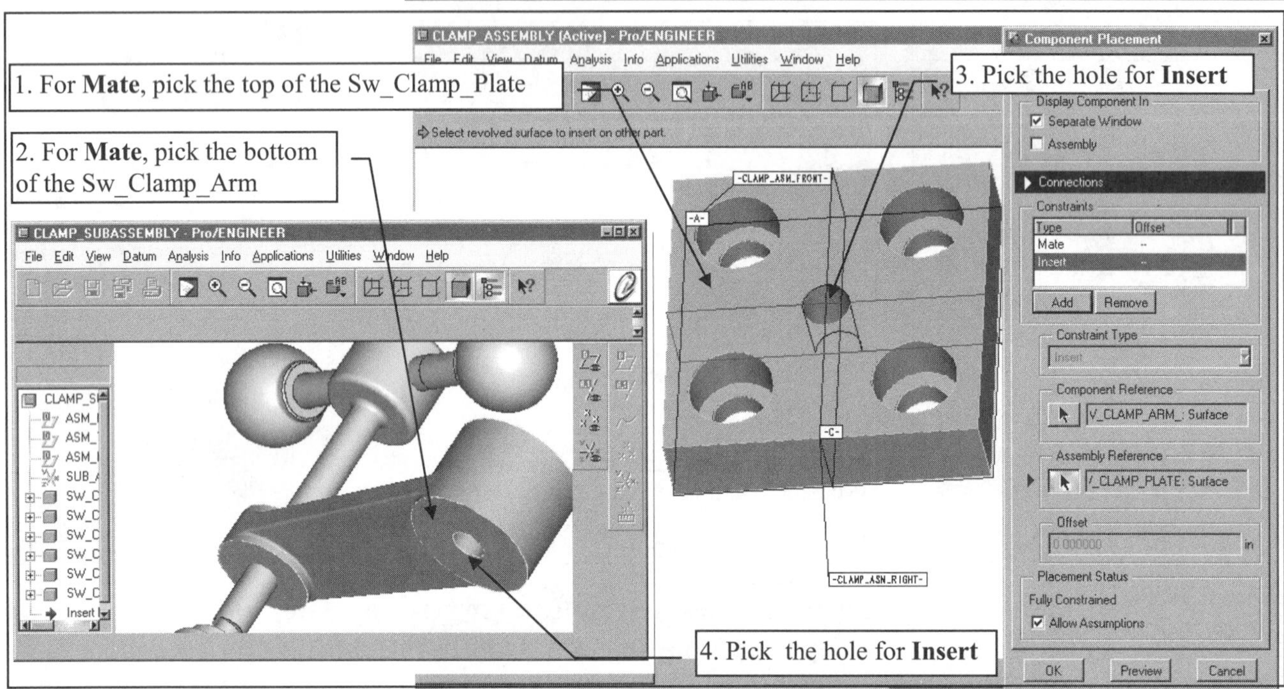

Figure 14.52
Using **Mate** and **Insert**

1. Pick the datum plane **CLAMP_ASM_TOP** to **Orient**

2. Pick the vertical datum plane **ASM_TOP** to **Orient**

Figure 14.53
Using **Orient**

Using constraints learned in this lesson, assemble the short stud (Sw_Clamp_Stud35) and the Sw_Clamp_Flange_Nut (Figure 14.55 and Figure 14.56) to complete the assembly.

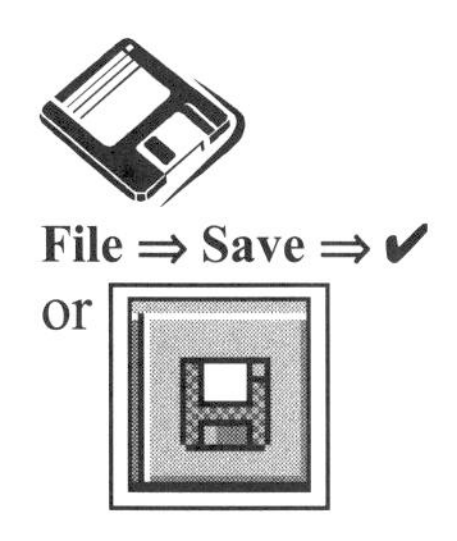

File ⇒ Save ⇒ ✔
or

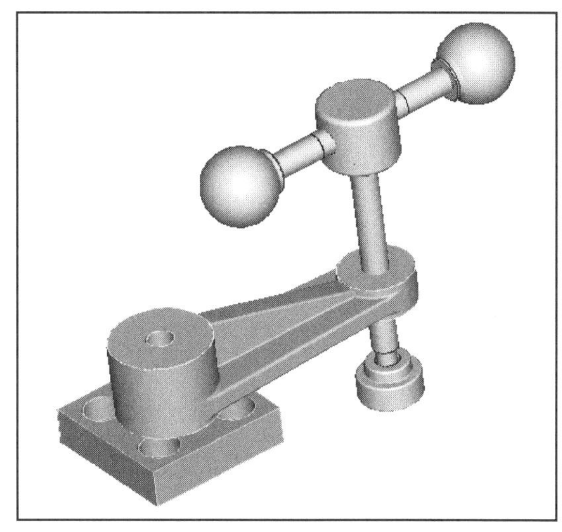

Figure 14.54
Assembled Plate and Subassembly

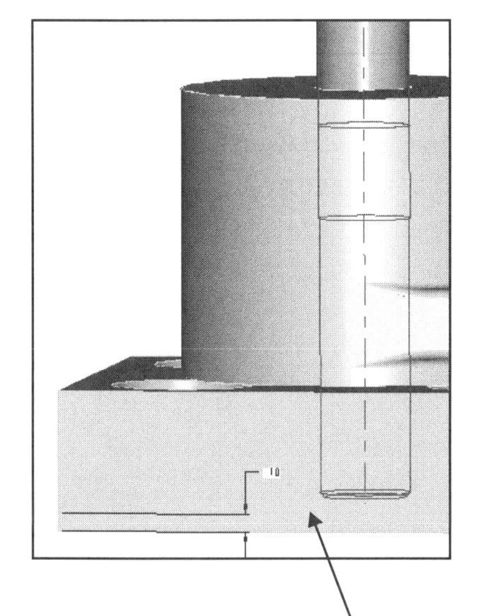

Figure 14.55
Assembled Short Stud

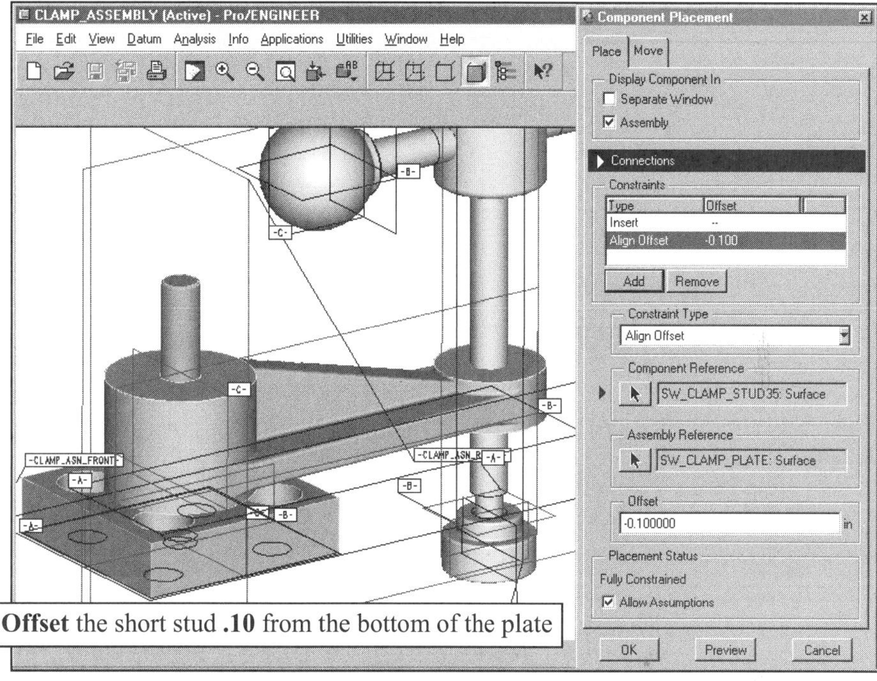

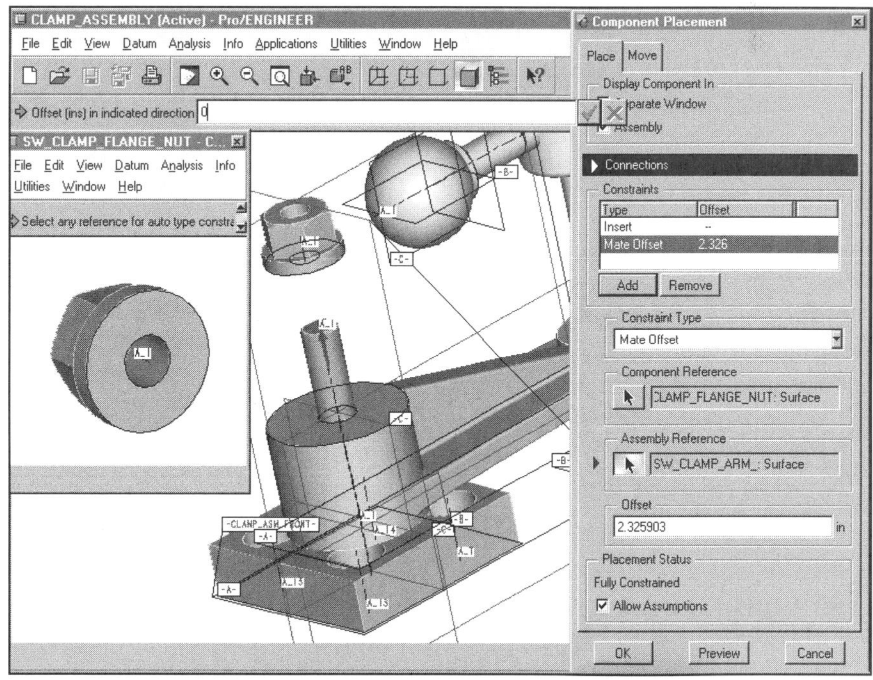

File ⇒ Save ⇒ ✔
File ⇒ Delete ⇒
Old Versions ⇒ ✔

Figure 14.56
Assembling the Swing Clamp Flange Nut

The final thing you will want to do before going on to the next lesson is to check the assembly using the **Analysis** command as shown in Figure 14.57.

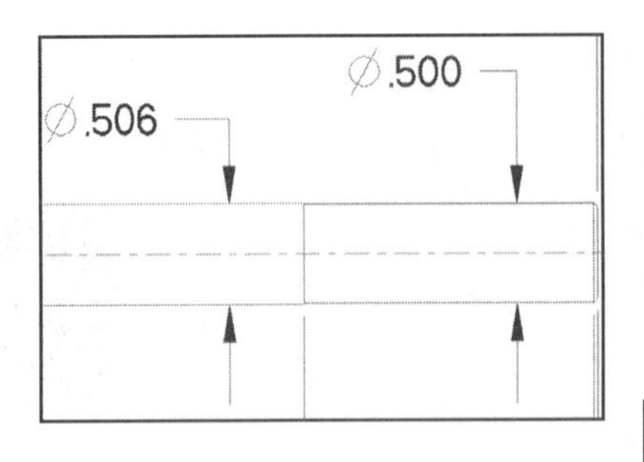

If you look at the long **5.00** stud's detail drawing (see Figure 14.18), you will see that the shaft diameter is greater than **.500** at the center of the stud and that each end has a **.500-13 UNC** thread. The hole in the swivel is **.500**. This means that there should be a slight interference between the two components. Check the clearance and interference between these components using the following commands (Fig. 14.57):

Analysis ⇒ Model Analysis ⇒ Pairs Clearance ⇒ Whole Part ⇒ Quilts ●**Exclude** ⇒ (pick the Swivel) ⇒ (pick the **5.00** Stud)

The **Analysis** command provides you with a statement about the status of the two parts:

Volume of Interference is 0.00694034.

The assembly is now complete. In the next lesson, you will learn how to move and rotate components in the assembly, establish views for use in the Drawing mode, create exploded views of the assembly, generate a bill of materials, and change the component visibility.

Figure 14.57
Analysis Command Used to Verify Clearance and Interference

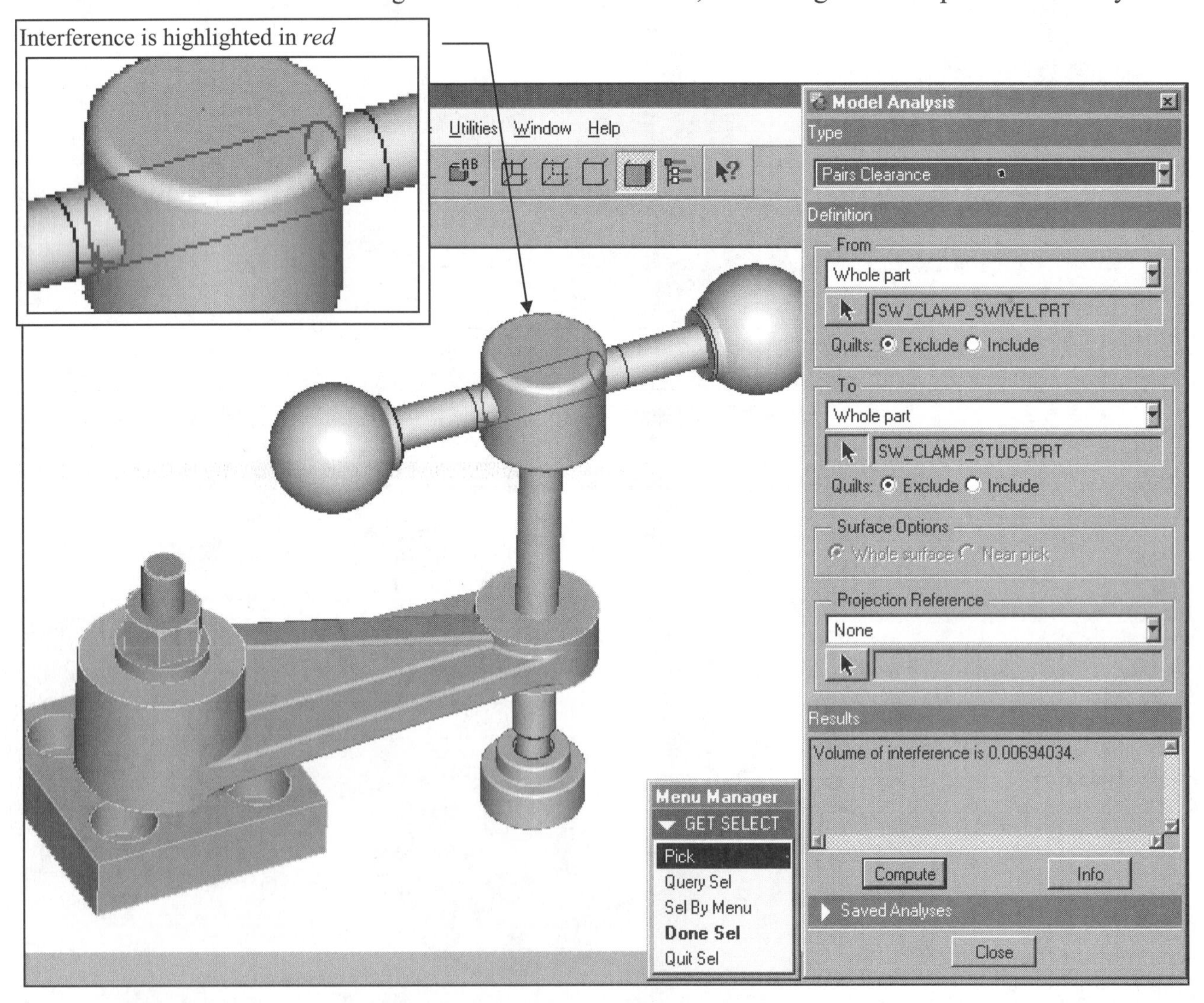

Lesson 14 Project

Coupling Assembly

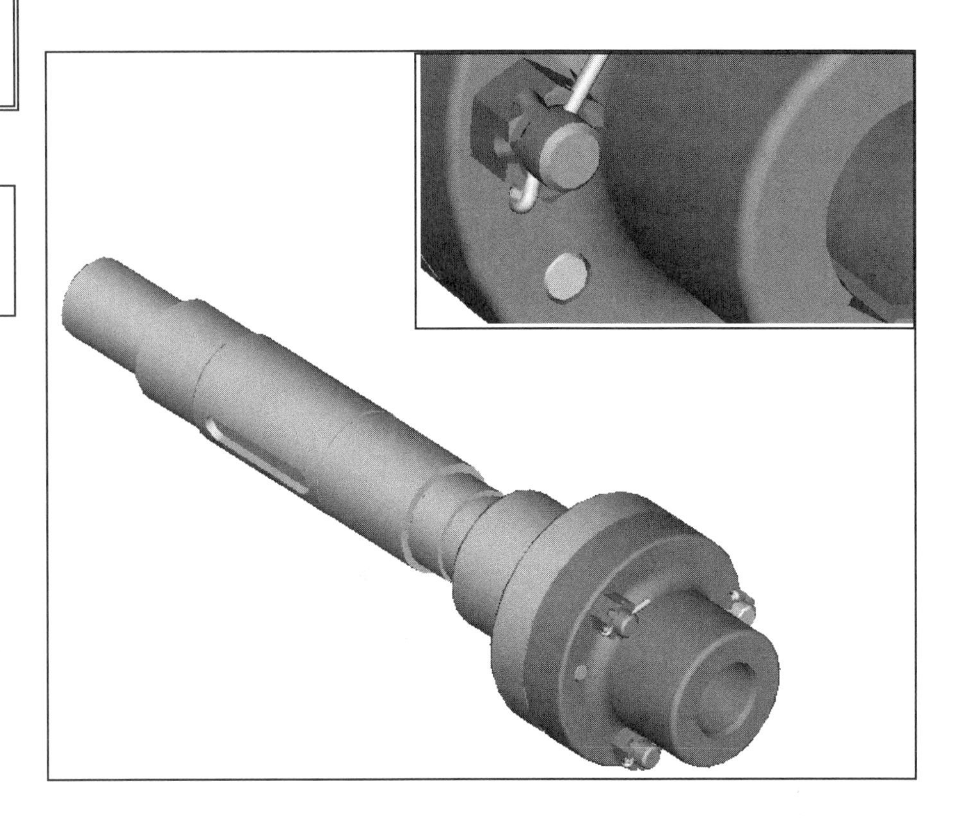

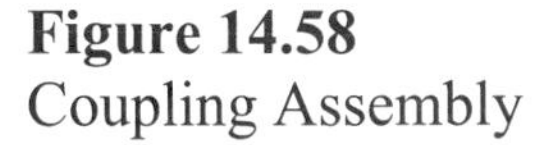

Figure 14.58
Coupling Assembly

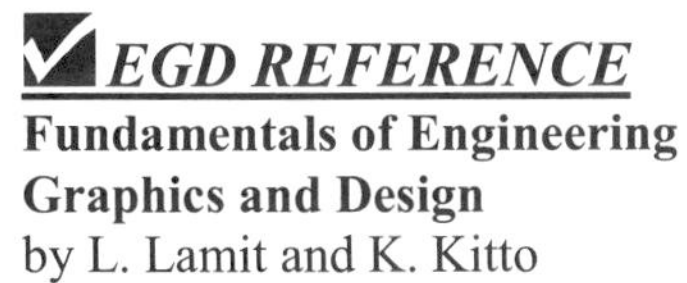

EGD REFERENCE
Fundamentals of Engineering Graphics and Design
by L. Lamit and K. Kitto
Read Chapter 23.
See pages 865-866.

NOTE
For this project, do not assemble the library parts directly from the library. Save each library part in your own directory with a new name, and then use the new part names in the assembly.

Coupling Assembly

The fourteenth **lesson project** is an assembly that requires commands similar to those for the **Swing Clamp**. Model the parts and create the assembly shown in Figures 14.58 through 14.81. Analyze the assembly and plan out the steps required to assemble it. Use the **DIPS** in Appendix D to plan out the assembly component sequence and the parent-child relationships for the assembly.

You will use the **Coupling Shaft** from the Lesson 6 Project. The Coupling Shaft should be the first component assembled. The **Taper Coupling** from the Lesson 7 Project is also used in the assembly. The detail drawings for the second coupling are provided here, in this lesson project. Model this second **Coupling** *before* you start the assembly. Depending on the library parts available on your system, you may need to model the **Key**, the **Dowel**, and the **Washer**.

Because not all organizations purchase the libraries, details are provided for all the components required for the assembly, including the standard off-the-shelf parts available in Pro/E's library. Pro/LIBRARY commands to access the standard components are provided for those of you who have them loaded on your systems. The instance number is given for every standard component used in the assembly. The **Slotted Hex Nut**, **Socket Head Cap Screw**, **Hex Jam Nut**, and **Cotter Pin** are all standard parts from the library. The Cotter Pin is in *inch* units, and the remaining items are *metric*.

Redesign the length of the threaded end to accommodate the washer and nut. *You* decide the new length based on the combined thickness of the two components.

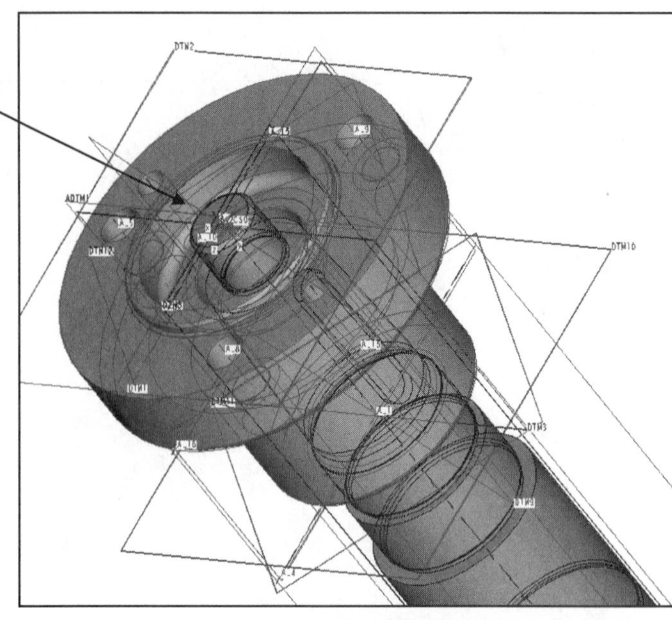

Figure 14.59
Assembling the Coupling Shaft and Taper Coupling

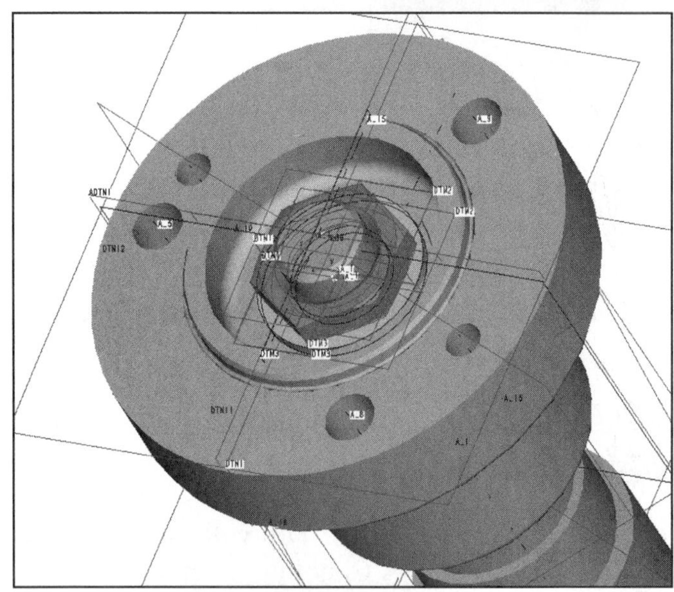

Figure 14.60
Hex Jam Nut and Washer

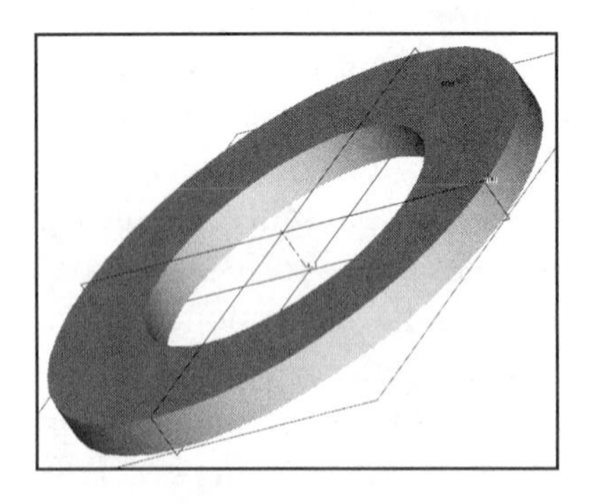

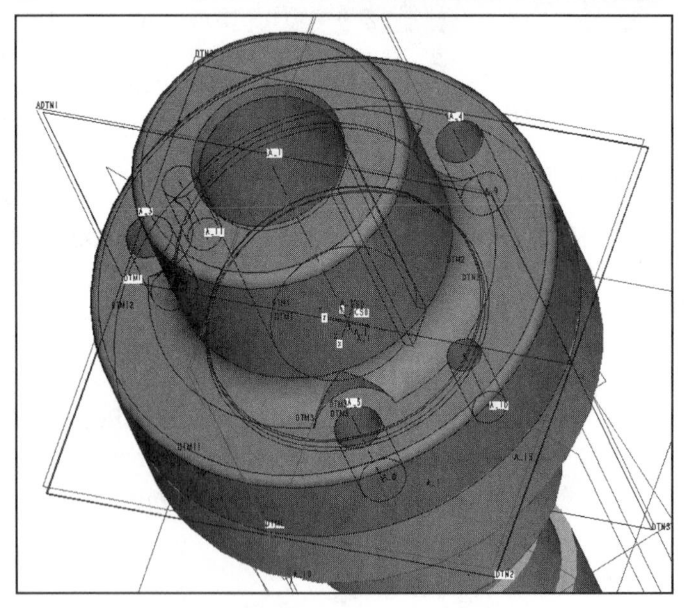

Figure 14.61
Second Coupling

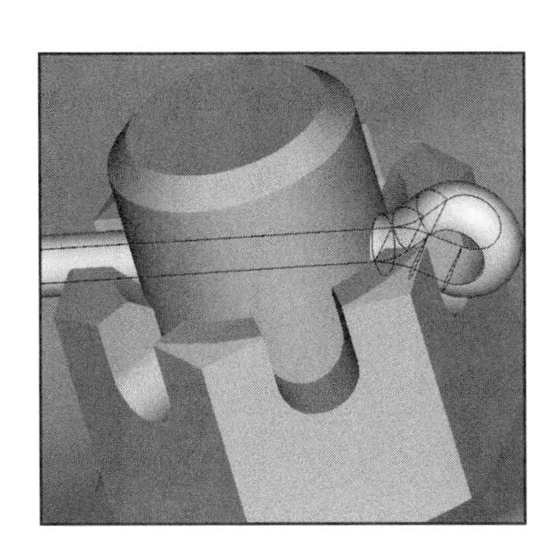

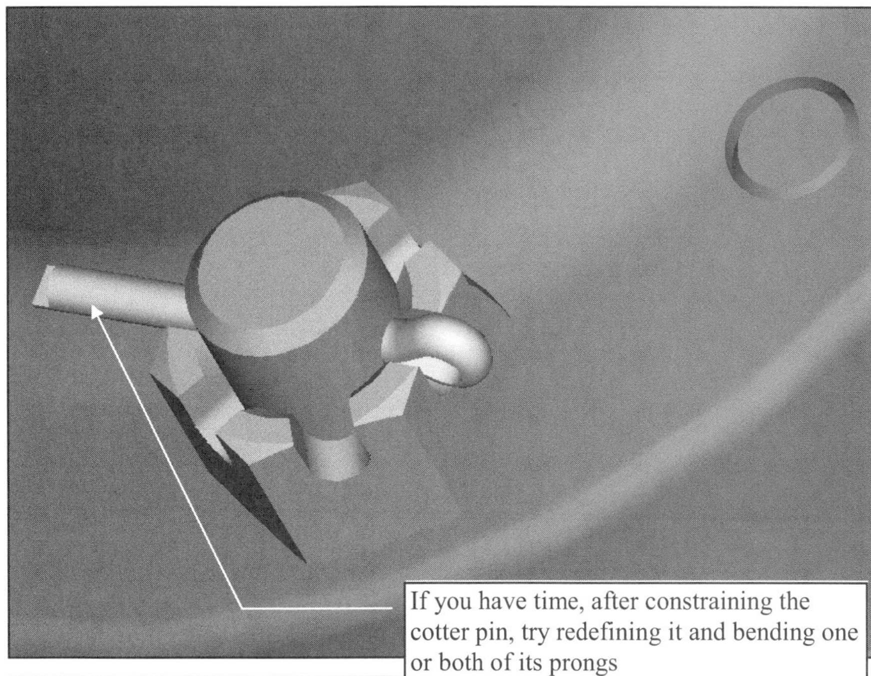

Figure 14.62
Dowel, Slotted Hex Nut, and Cotter Pin

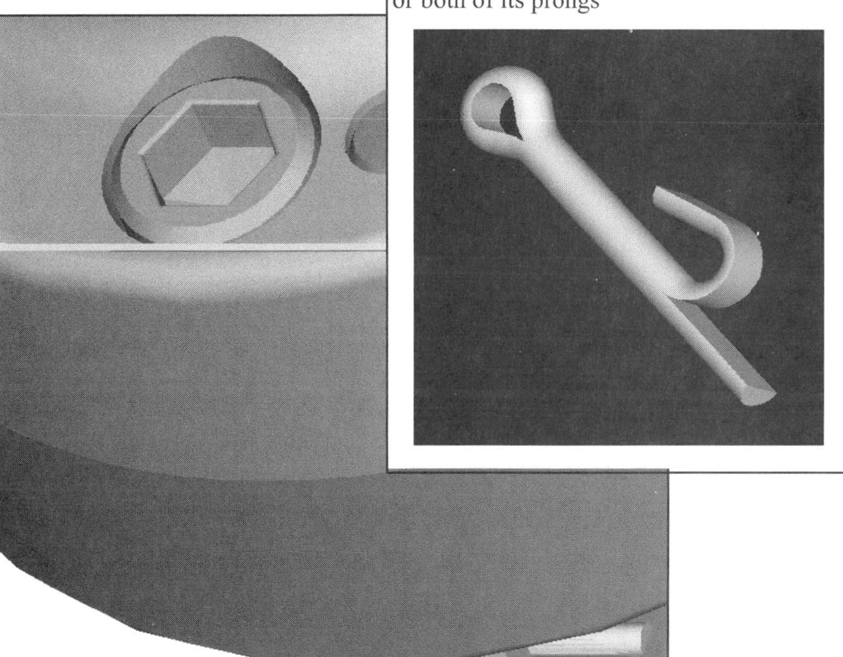

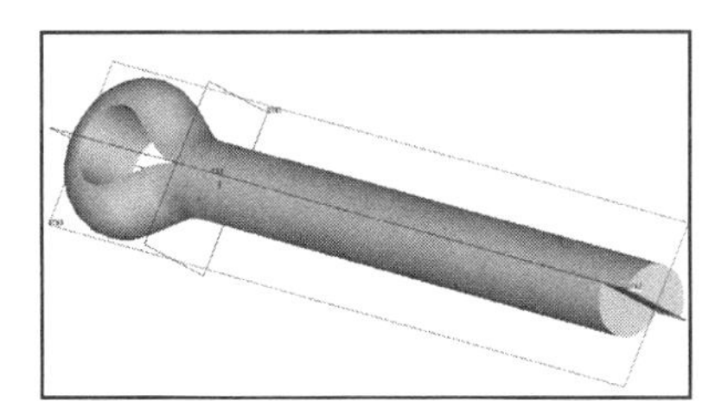

Figure 14.63
Socket Head Cap Screw

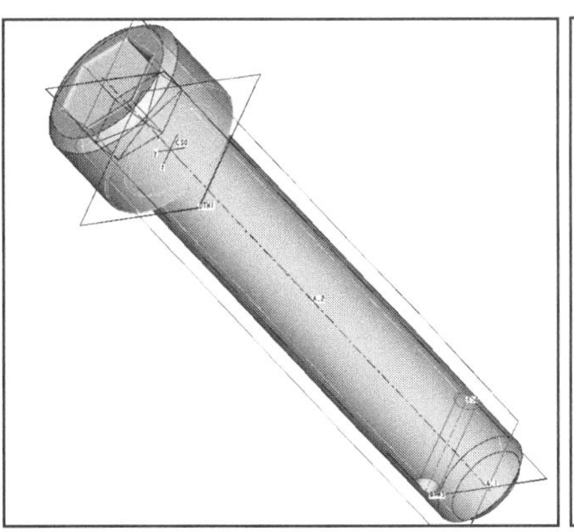

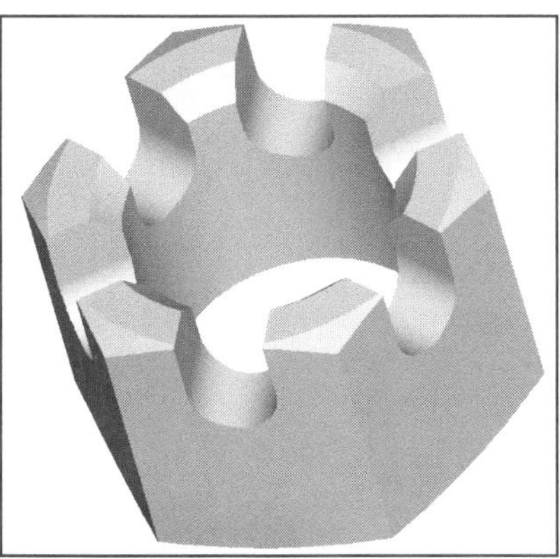

Figure 14.64
Socket Head Cap Screw and Slotted Hex Nut

Figure 14.65
Second Coupling, Part Model

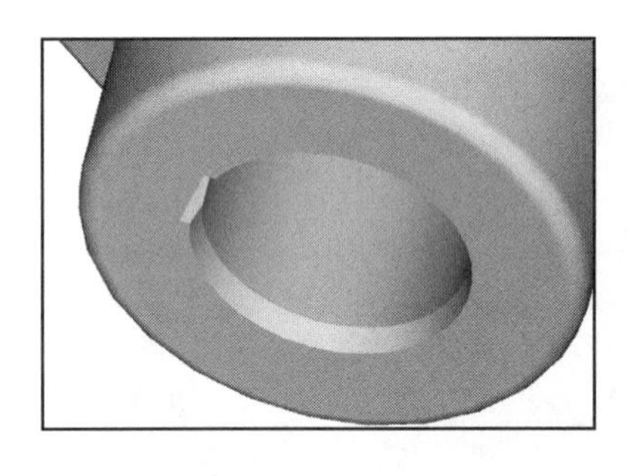

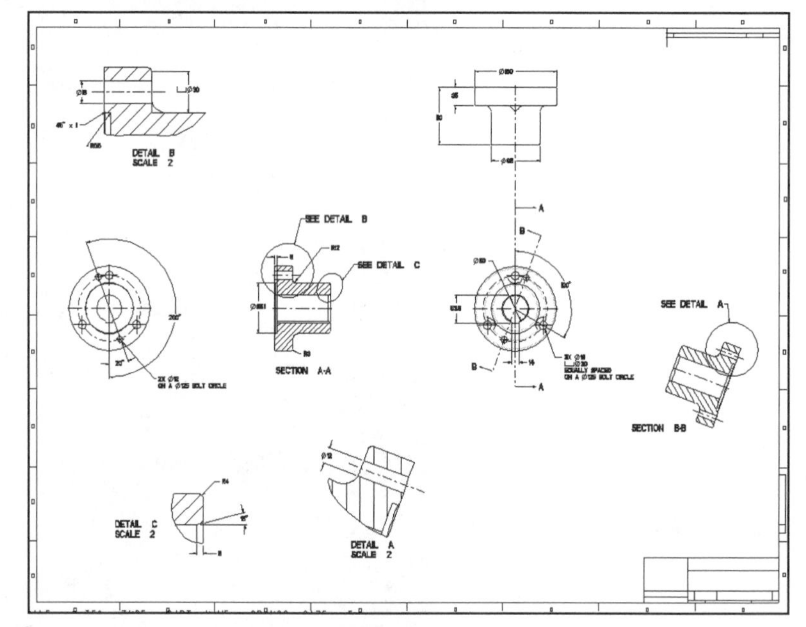

Figure 14.66
Second Coupling, Detail Drawing

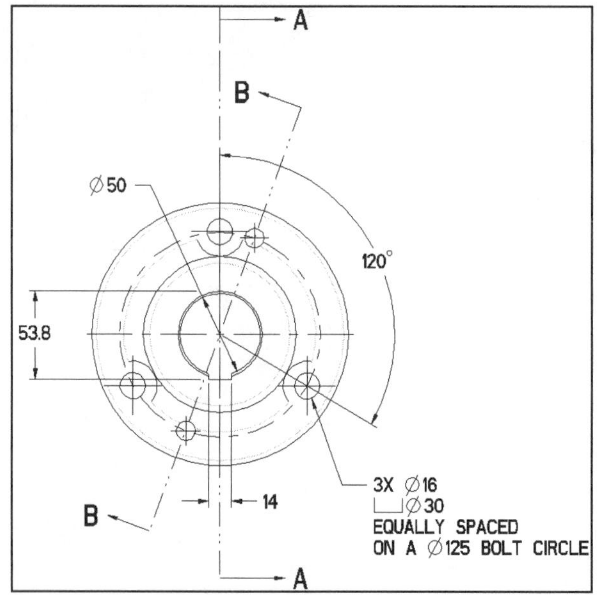

Figure 14.67
Second Coupling, Detail Drawing, Front View

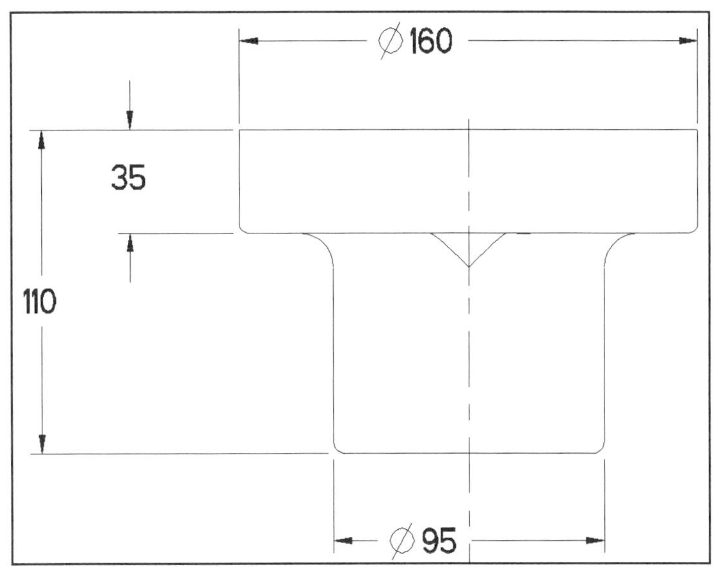

Figure 14.68
Second Coupling, Detail Drawing, Top View

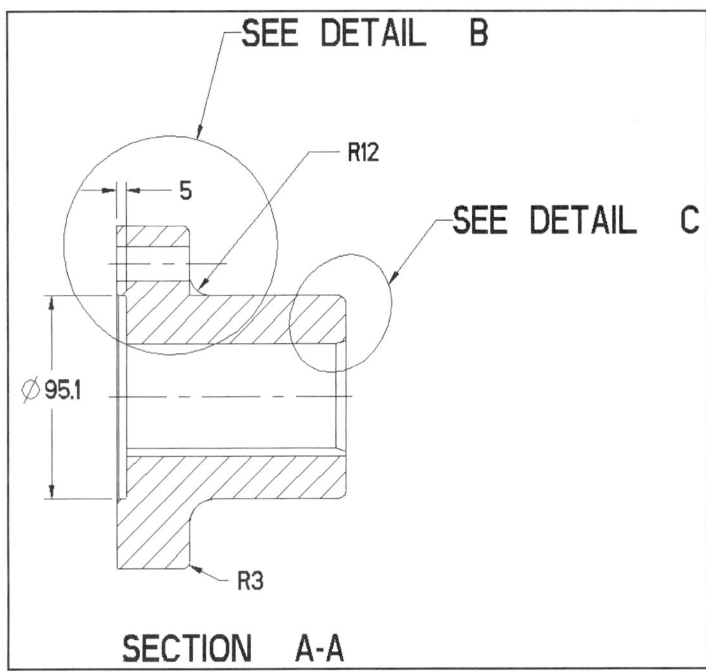

Figure 14.69
Second Coupling, Detail Drawing, **SECTION A-A**

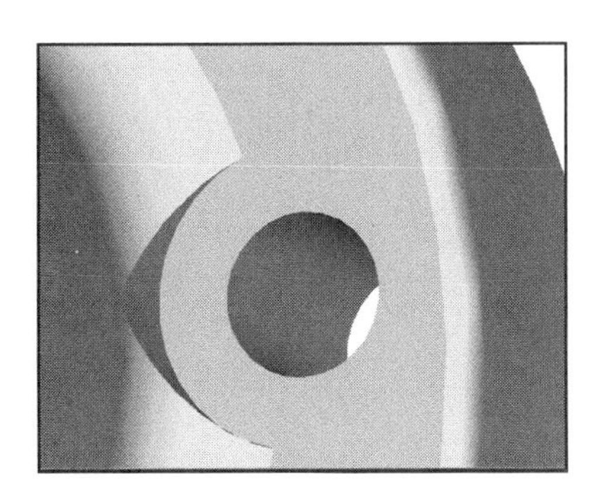

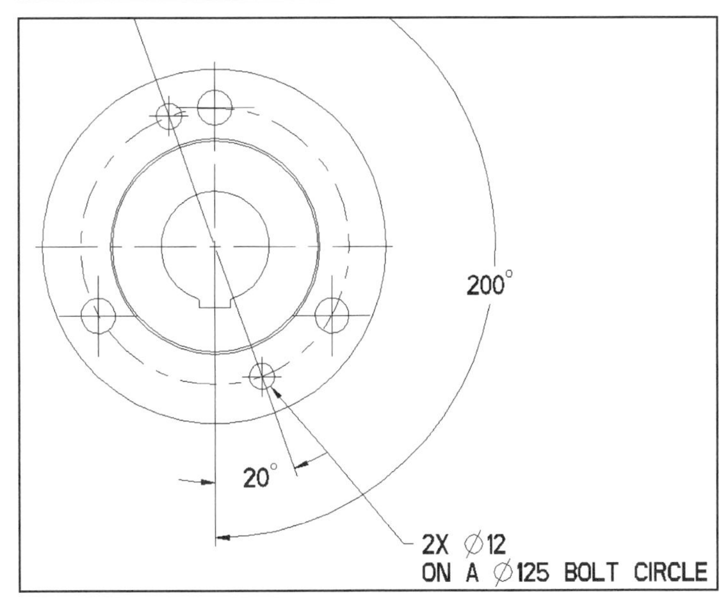

Figure 14.70
Second Coupling, Detail Drawing, Back View

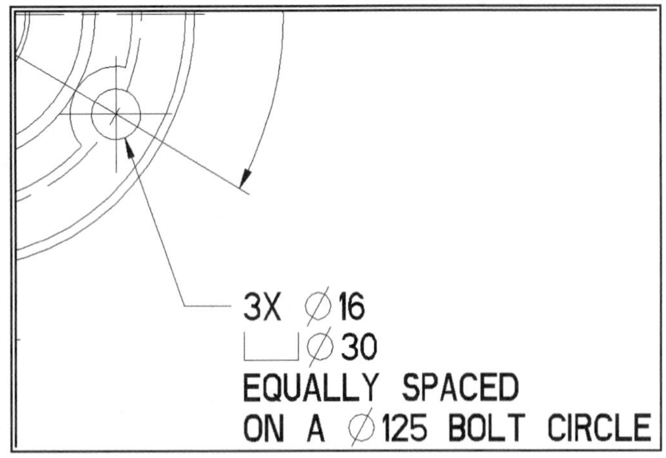

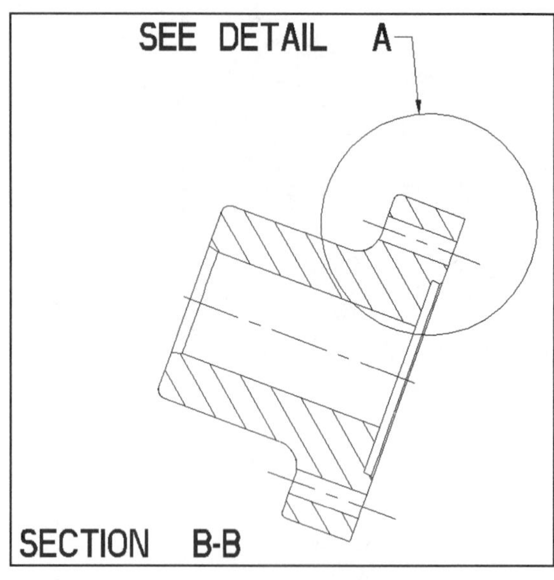

Figure 14.71
Second Coupling, Detail Drawing, **SECTION B-B**

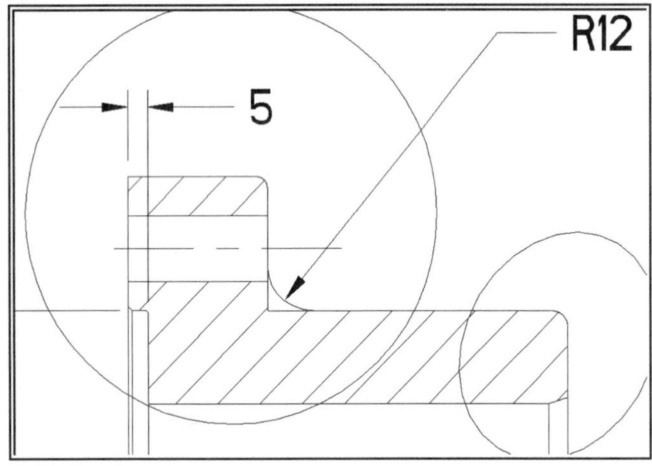
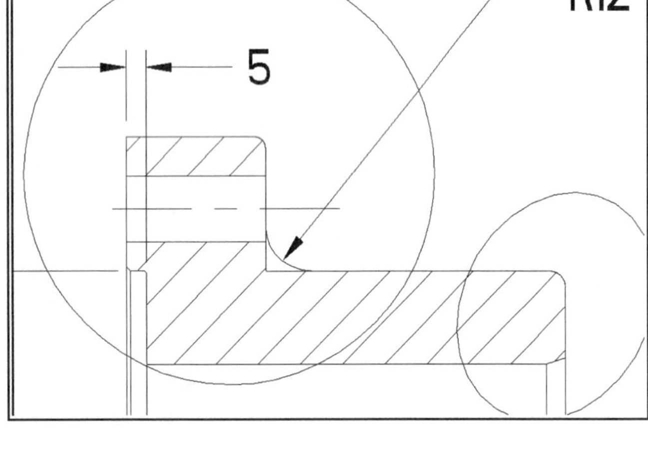

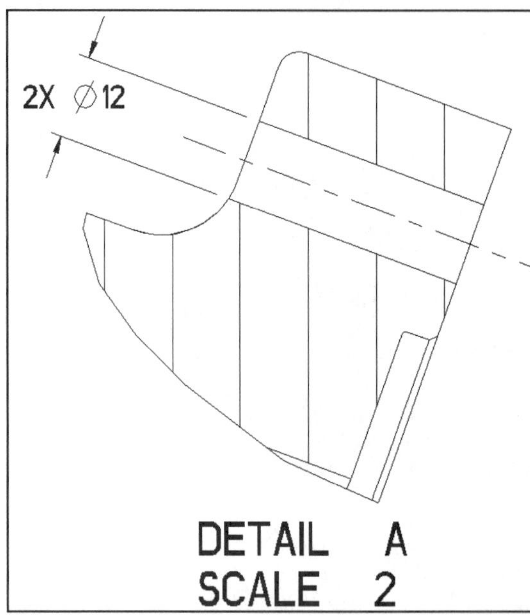

Figure 14.72
Second Coupling, Detail Drawing, **DETAIL A**

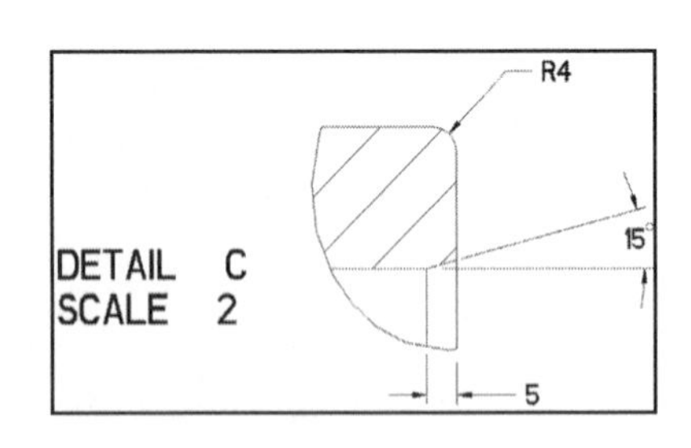

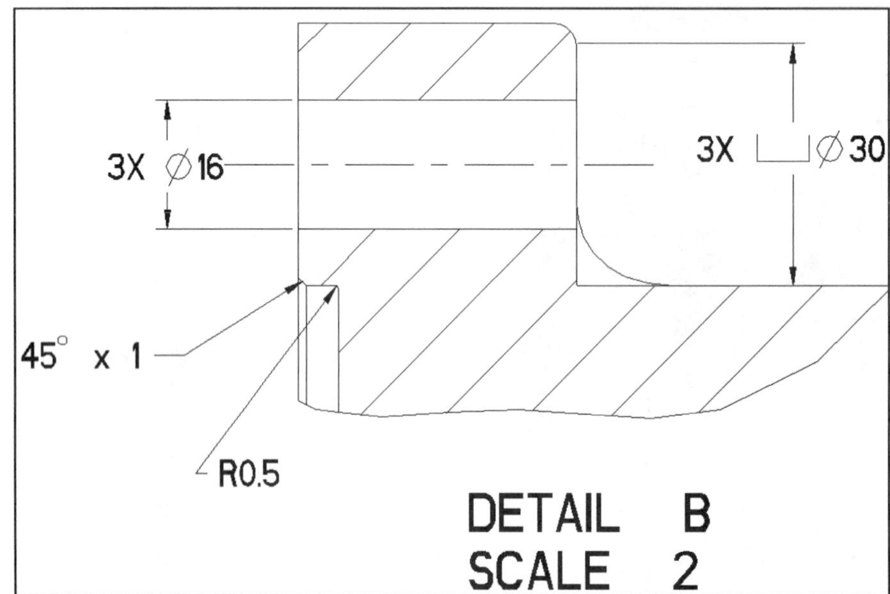

Figure 14.73
Second Coupling, Detail Drawing, **DETAIL B**

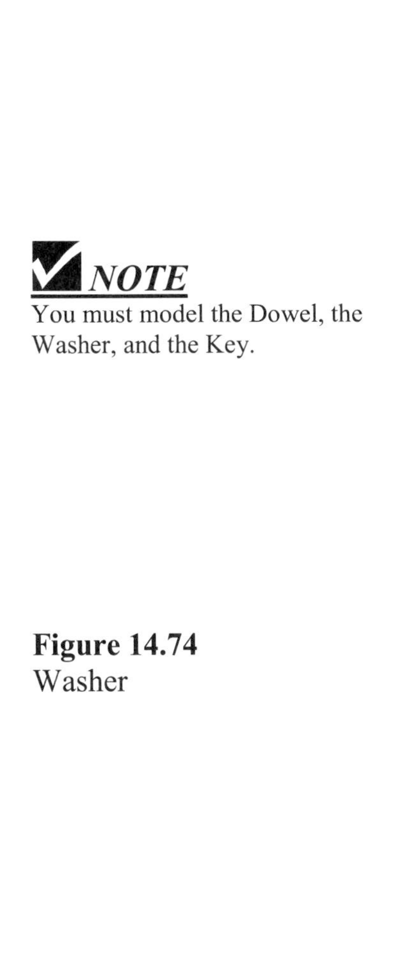

NOTE

You must model the Dowel, the Washer, and the Key.

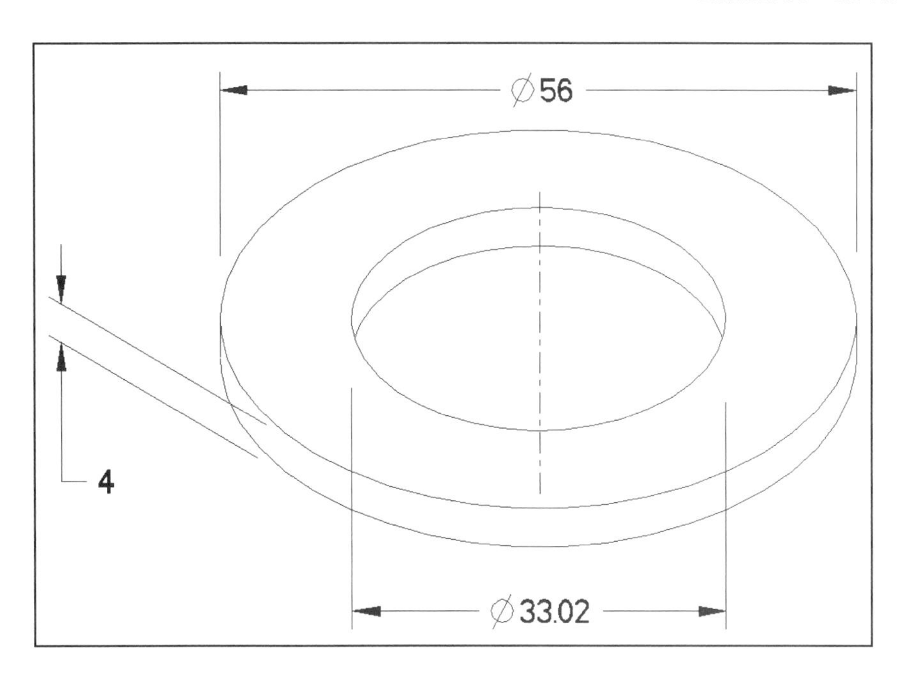

Figure 14.74
Washer

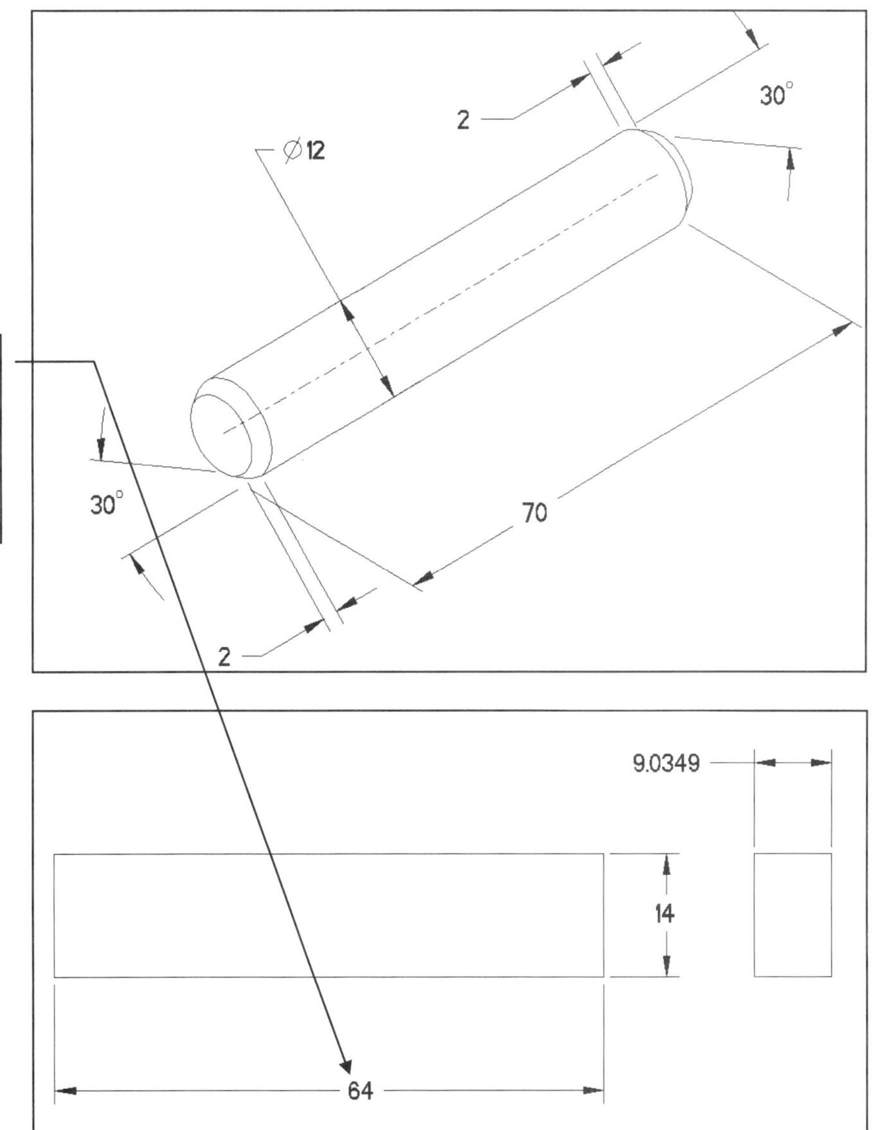

After completing the assembly, do an **Analysis** using **Global Interference.** If there is an interference between the shaft and the key, modify the key to the correct size.

Figure 14.75
Dowel

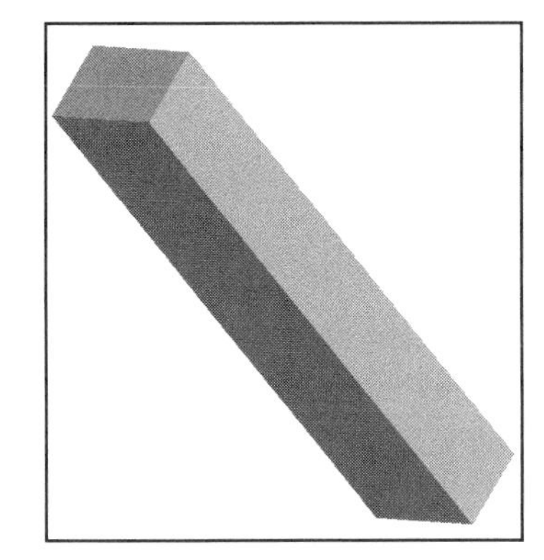

Figure 14.76
Key

SOCKET HEAD CAP SCREW

File ⇒
Open ⇒
pro/library ⇒
objlib ⇒ metriclib ⇒
sock_hd_scr ⇒ mscs.prt ⇒
By Parameter ⇒
NOMINAL_SIZE_THR_PITCH ⇒
M16X2 ⇒ Open
By Parameter ⇒
d5,length ⇒
80.000 ⇒ Open

INSTANCE = MSCS1210

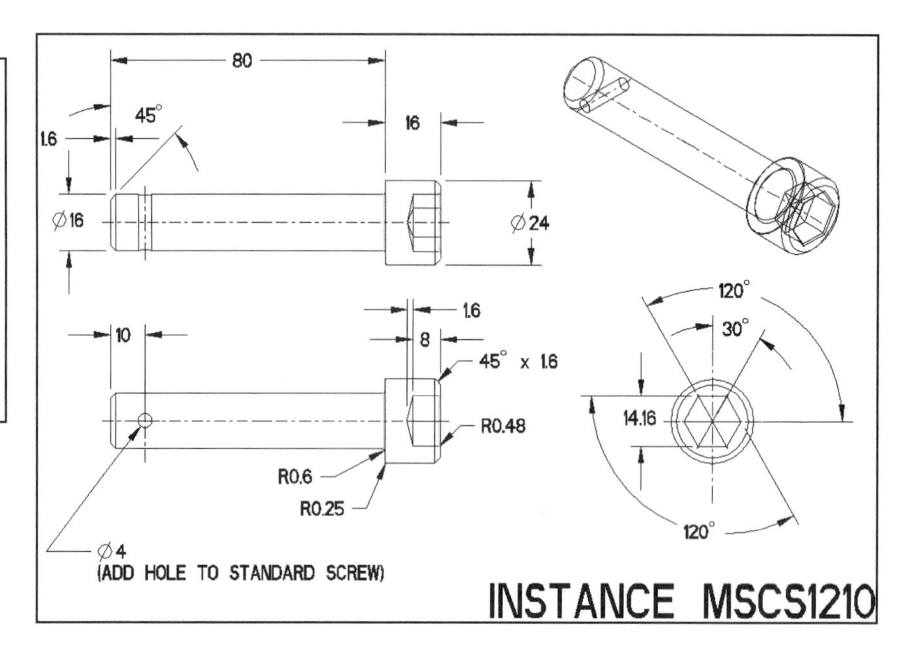

Figure 14.77
Socket Head Cap Screw

SLOTTED HEX NUT

File ⇒
Open ⇒
pro/library ⇒
objlib ⇒ metriclib ⇒
hex_nuts ⇒
mshn.prt ⇒
By Parameter ⇒
NOMINAL_SIZE_THR_PITCH ⇒
M16X2 ⇒ Open

INSTANCE = MSHN07

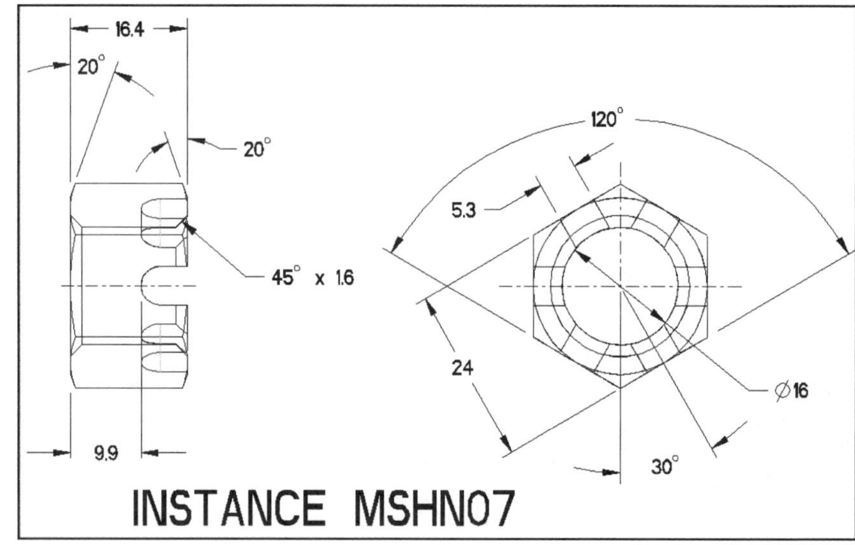

Figure 14.78
Slotted Hex Nut

COTTER PIN

File ⇒
Open ⇒
pro/library ⇒
objlib ⇒ eng_part_lib ⇒
cot_clvs_pin ⇒ Pina.prt ⇒
By Parameter ⇒
NOMINAL_SIZE ⇒
.1562 ⇒ Open
By Parameter ⇒
d13,1 ⇒
1.250 ⇒ Open

INSTANCE = PNA09L05

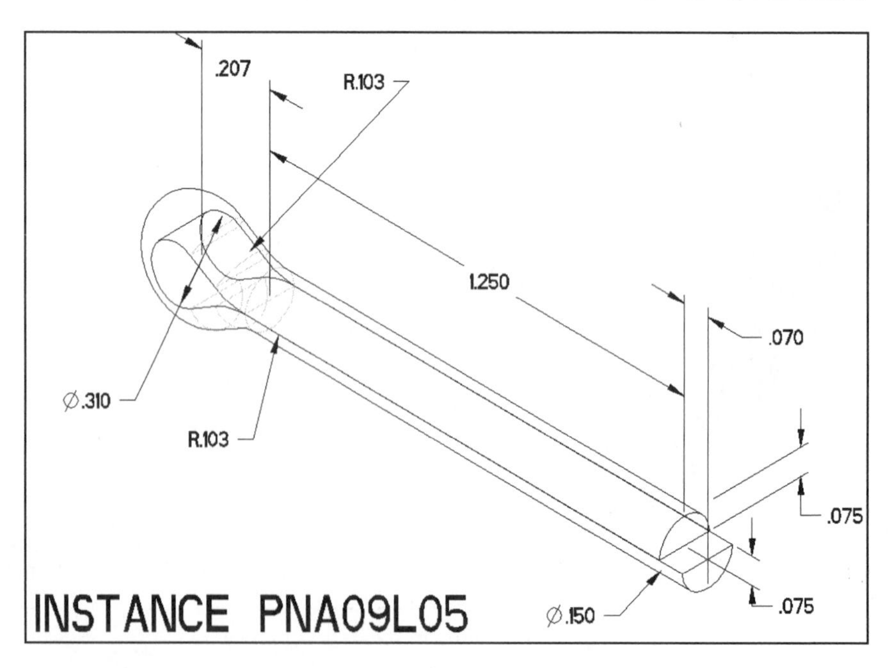

Figure 14.79
Cotter Pin

HEX JAM NUT

File ⇒
Open ⇒
pro/library ⇒
objlib ⇒ metriclib ⇒
hex_nuts ⇒
mhjn.prt ⇒
By Parameter ⇒
NOMINAL_SIZE_THR_PITCH ⇒
M30X3.5 ⇒ Open

INSTANCE = MHJN10

- Modify the thickness of the nut to **10 mm.**

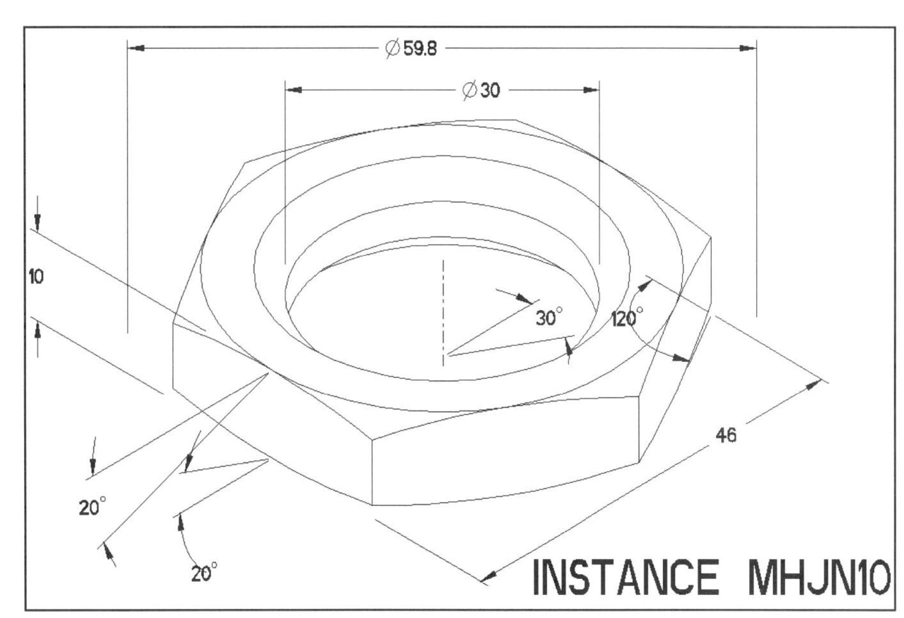

Figure 14.80
Hex Jam Nut

Figure 14.81
Coupling Assembly

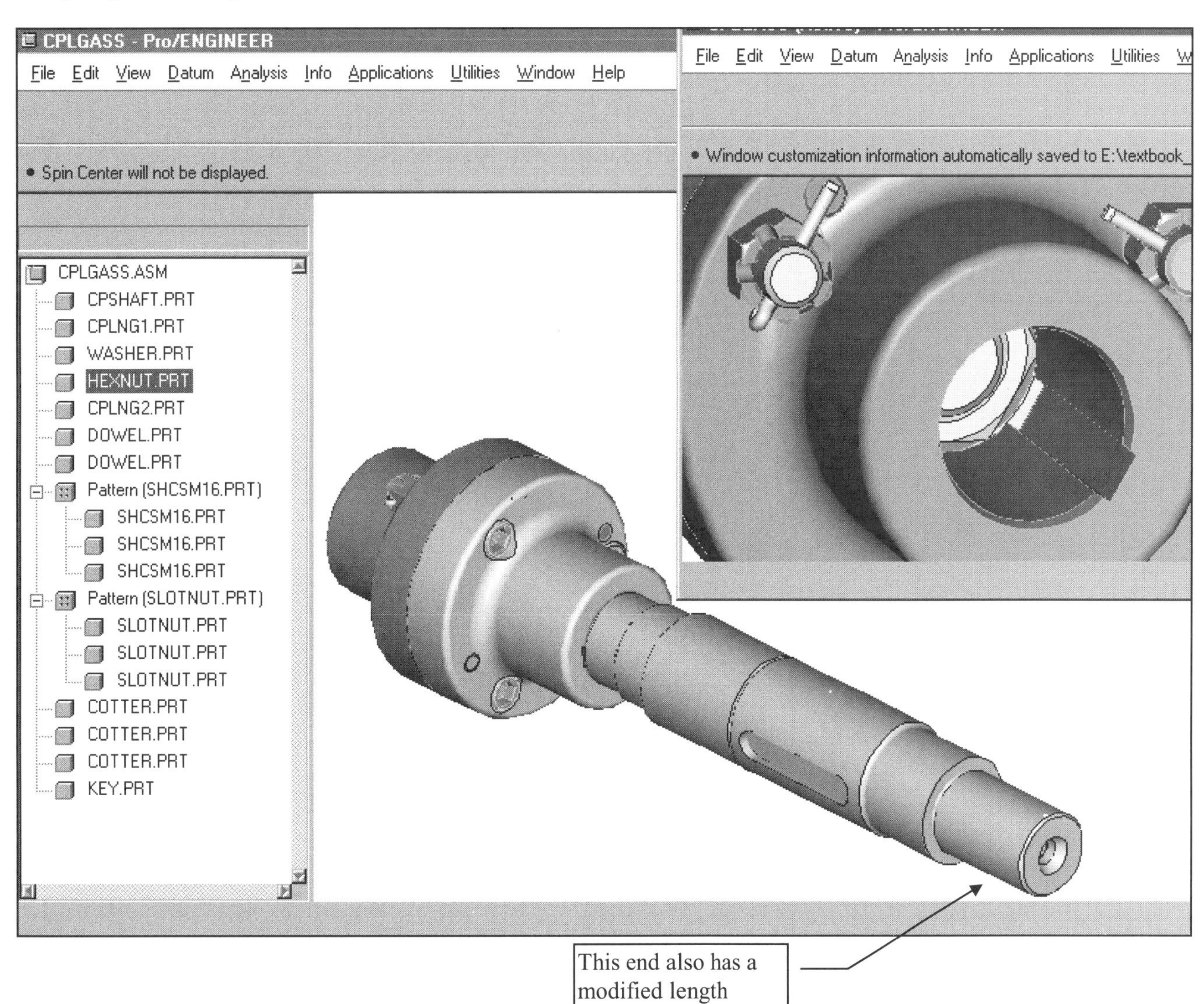

Lesson 15

Exploded Assemblies

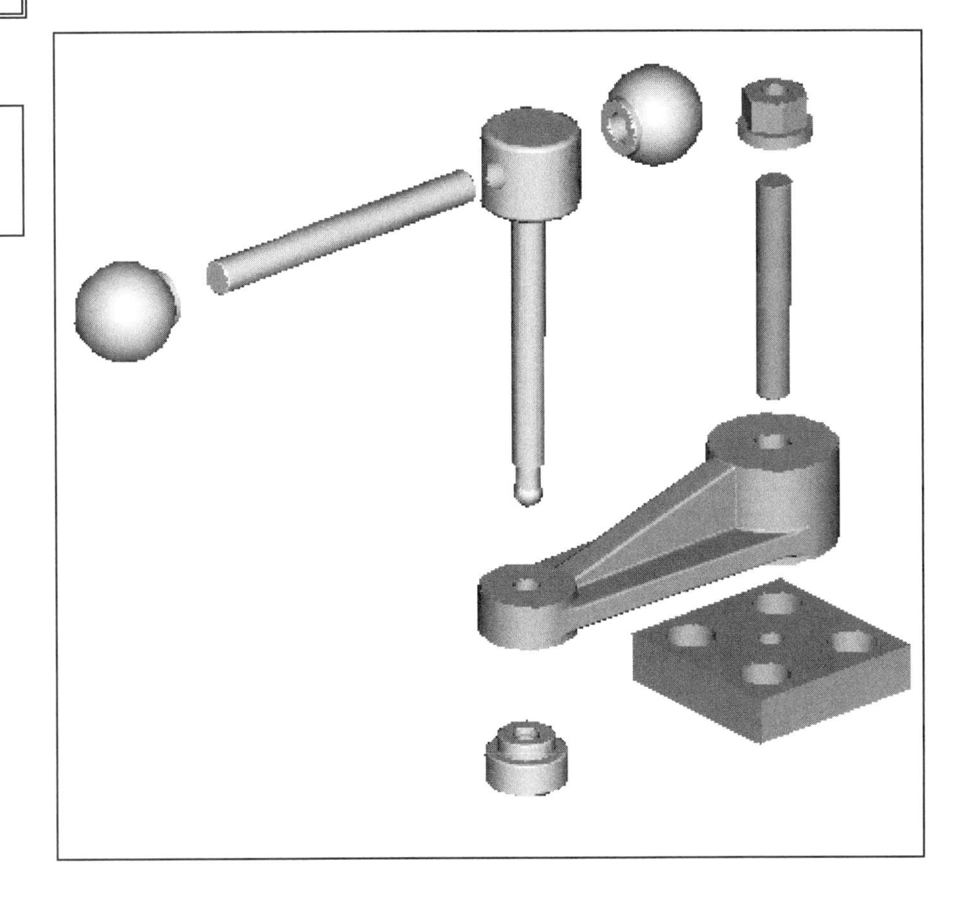

Figure 15.1
Exploded Swing Clamp Assembly

OBJECTIVES

1. **Create exploded views**
2. **Create Explode States**
3. **Edit exploded views**
4. **Create unique component visibility settings**
5. **Move and rotate components in an assembly**
6. **Create shaded and pictorial views of an assembly**
7. **Save named views to use later in Drawing mode**

EGD REFERENCE
Fundamentals of Engineering Graphics and Design
by L. Lamit and K. Kitto
Read Chapter 23.
See pages 865-866.

COAch™ for Pro/ENGINEER

If you have **COAch for Pro/ENGINEER** on your system, go to SEARCH and do the Segment shown in Figure 15.3.

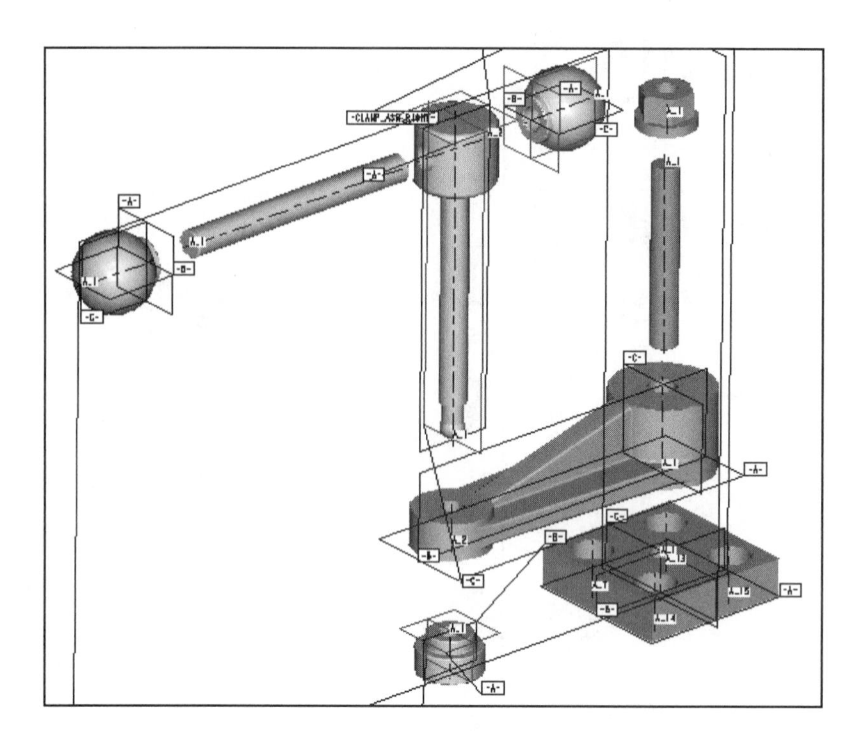

Figure 15.2
Exploded Swing Clamp

EXPLODED ASSEMBLIES

Pictorial illustrations, such as exploded views, are generated directly from the 3D model database (Figs. 15.1 through 15.7). The model can be displayed and oriented in any position. Each component in the assembly can have a different display type: wireframe, hidden line, no hidden, and shading. You can select and orient the part to provide the required view orientation to display the part from underneath or from any side or position. Perspective projections are made with selections from menus. The assembly can be spun around, reoriented, and even clipped to show the interior features. When assemblies are illustrated, you have the choice of displaying all components and subassemblies or any combination of parts in the design.

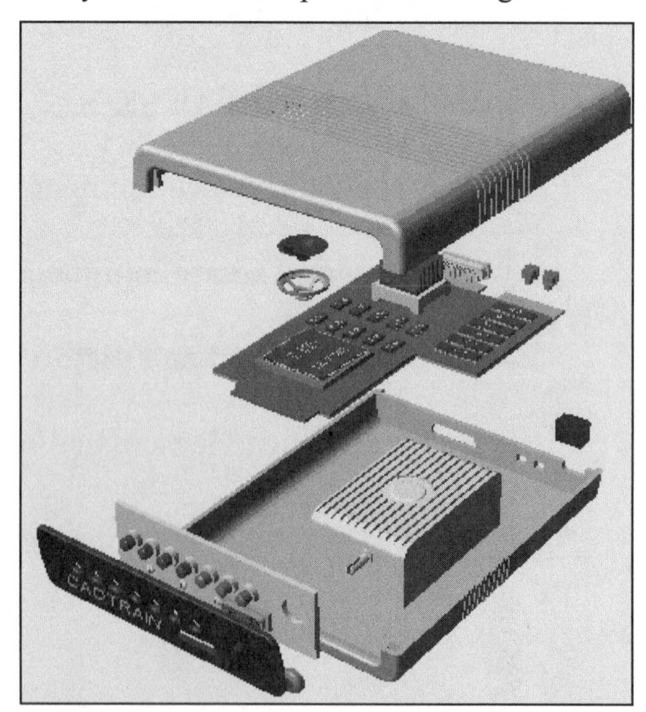

Figure 15.3
COAch for Pro/E, Exploded Assemblies (Overview)

Creating Exploded Views

Using the **ExplodeState** option in the ASSEMBLY menu, you can create an exploded view of an assembly. Exploding an assembly affects only the display of the assembly; it does not alter actual distances between components. Exploded states are created to define the exploded positions of all components. For each explode state, you can toggle the explode status of components, change the explode locations of components, and create explode offset lines. To access this functionality, choose the **ExplodeState** option in the ASSEMBLY menu.

You can define multiple explode states for each assembly and then explode the assembly using any of these explode states at any time. You can also set an explode state for each drawing view of an assembly. Pro/E gives each component a default explode position determined by the placement constraints. By default, the reference component of the explode is the parent assembly (top-level assembly or subassembly).

To explode components, you use a drag-and-drop user interface similar to the **package** functionality. You select one or more components and the motion reference and then drag the outlines to the desired positions. The component outlines drag along with the mouse cursor. You control the type of explode motion using a Preferences setting. Two types of explode instructions can be added to a set of components. The children components follow the parent component being exploded or they do not follow it. Each explode instruction consists of a set of components, explode direction references, and dimensions that define the exploded position from the final (installed) position with respect to the explode direction references.

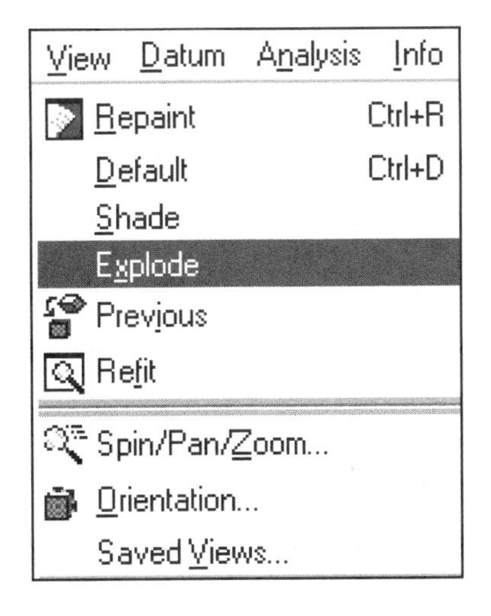

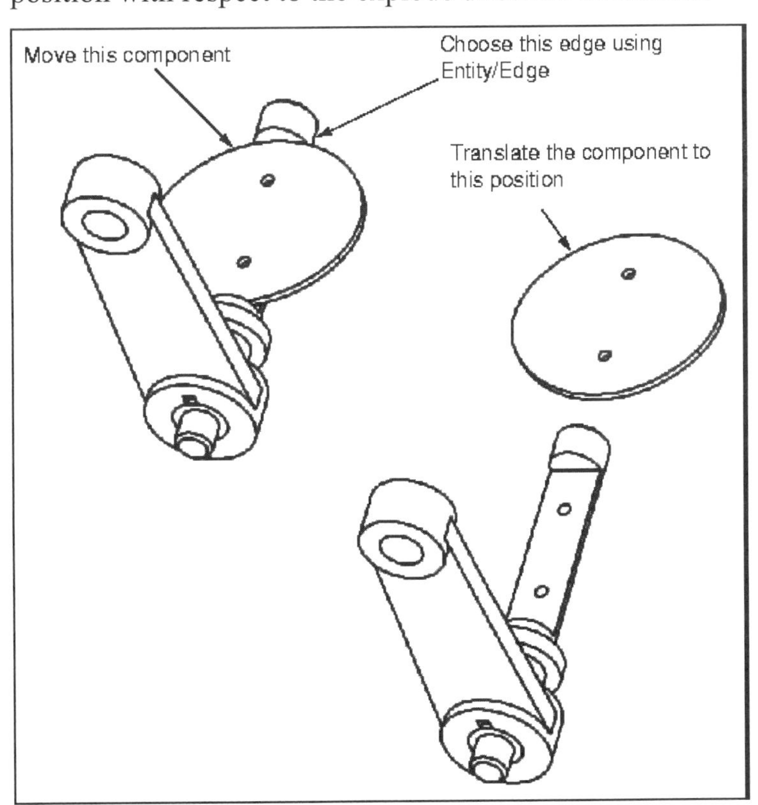

Figure 15.4
Exploded Views

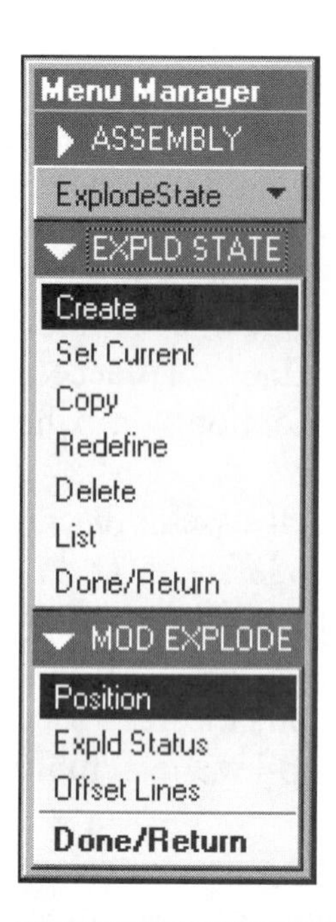

To Create Exploded Views

To create, set, and modify the explode states of the assembly; use the following commands:

1. From the ASSEMBLY menu, pick **ExplodeState.** The EXPLD STATE menu opens.
2. Click **Create**, and then enter a name for the state. The Explode Position dialog box opens. You can select a type of motion for the explode position. Motion types are:

⇒ **Translate** Select a component on which to base the motion
⇒ **Copy Pos** Select a component and copy its position
⇒ **Default Explode** Place selected components in the default position
⇒ **Reset** Reset the position of selected components

3. Select from the list of entities in the Motion Reference field to select an entity as a reference to use in the explode motion.
4. Select from the list of increments in the Motion Increments field. This determines how smooth the explode motion will be.
5. Click **Preferences** to open the Preferences dialog box with the following commands:

⇒ **Move One** Moves one component at a time
⇒ **Move Many** Moves multiple components simultaneously
⇒ **Move With Children** Highlights the children of the selected component and moves the entire group as one unit

6. Click a command from the Preferences dialog box.
7. Pick a component(s) to move. As you move it, the relative position is dynamically updated in the Position field of the dialog box.
8. You can **Undo** the explode motion.
9. Click **OK** to successfully explode the assembly components.

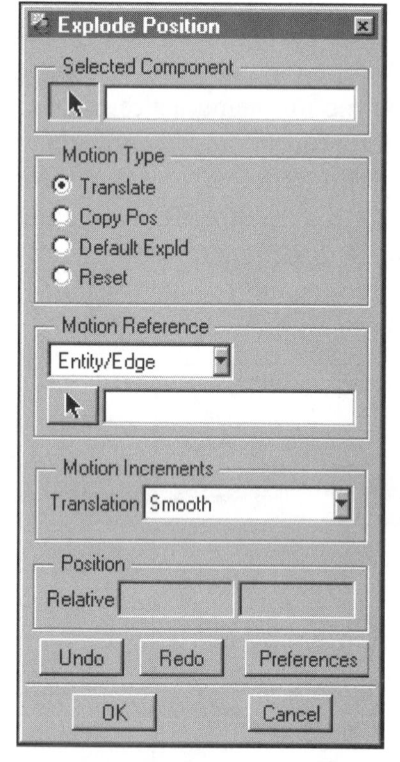

Explode Position Dialog Box Commands

The Explode Position dialog box has several commands with multiple options, and it contains several fields that you fill in by selecting an entity or by moving a selected entity.

* The Selected Component field displays the name of a component that you select to move in the explode assembly.
* The **Motion Type** command has the following options:

⇒ **Translate** Translates the selected component
⇒ **Copy Pos** Copy the position of the selected component
⇒ **Default Expld** Place the selected component in the default explode position
⇒ **Reset** Reset the position of the selected component

* The **Motion Reference** command allows you to select a reference type to define the direction of the explode motion:

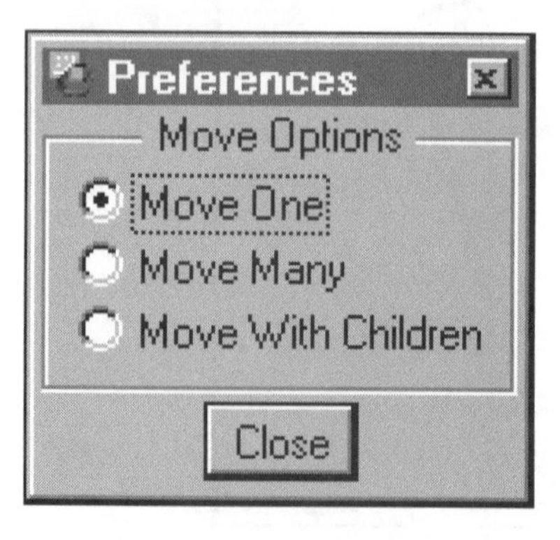

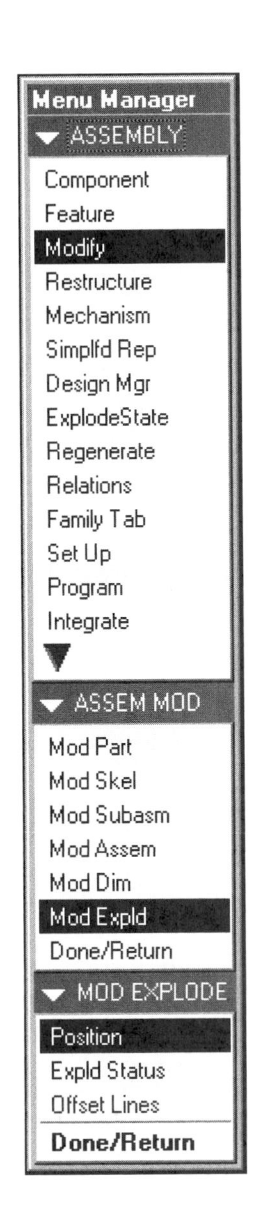

⇒ **Entity/Edge** Translation occurs about the reference line in the plane that is normal to it and contains the drag origin point.
⇒ **View Plane** Translation occurs about the component's drag origin in the view plane.
⇒ **Sel Plane** Translation occurs about the component's drag origin in the selected plane.
⇒ **Plane Normal** Translation occurs about the component's drag origin in the plane that contains the drag origin point and is parallel to the reference plane.
⇒ **2 Points** Translation occurs about the reference line in the plane that is normal to it and contains the drag origin point.
⇒ **Csys** Translation occurs about the axis in the plane that is normal to it and contains the drag origin point.

* The Motion Increments command allows you to define the smoothness of the explode motion by selecting from a pull-down list of options.
* Position Relative window updates dynamically as you move components.
* Preferences allows setup preferences for moving components.

Undo Undoes the last motion.
Redo Redoes the last motion that was undone.

Setting Display Modes for Components

Using **View** (menu bar) ⇒ **Model Setup** ⇒ **Component Display**, you can set different visualization (display) modes for components in an assembly (Fig. 15.5). Wireframe, hidden line, no hidden, shaded, or blanked display modes can be assigned to components. The components will be displayed according to their display status in the current display state (that is, blanked, shaded, drawn in hidden line color, and so on). The setting, in the Environment menu, controls the display of unassigned components. As a result, they appear according to the current mode of the environment.

Figure 15.5
Component Display Status

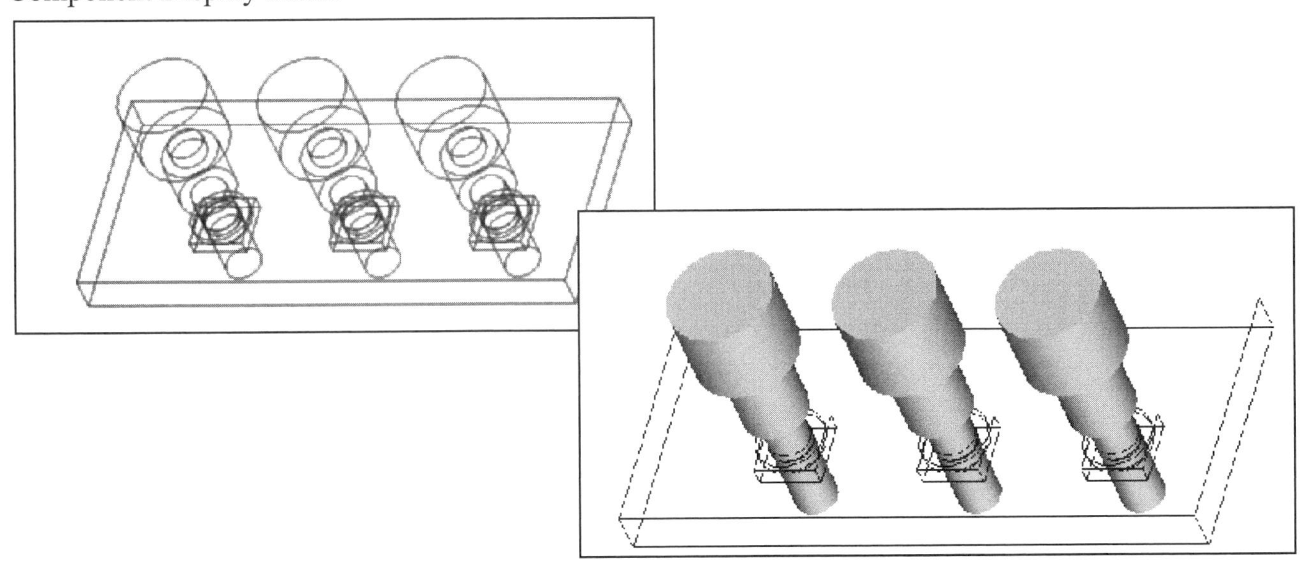

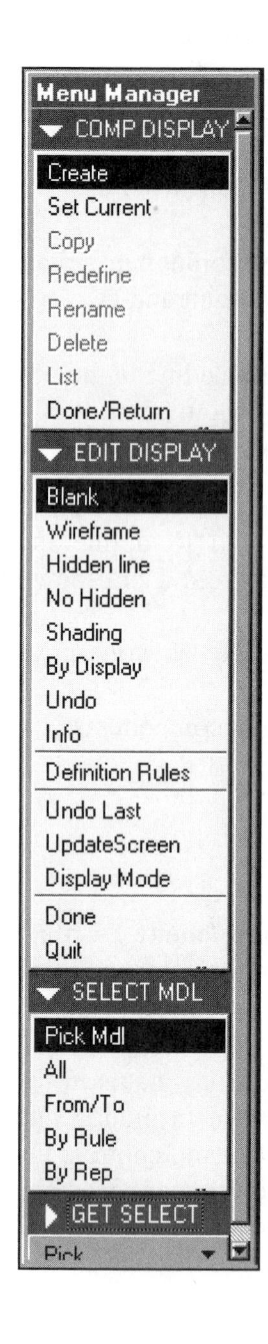

To Set Component Display State

1. Choose **View ⇒ Model Setup ⇒ Component Display.** The COMP DISPLAY menu appears, with the following commands:

⇒ **Create** Creates a new component display state.
⇒ **Set Current** Sets the current component display state.
⇒ **Copy** Copies an existing component display state.
⇒ **Redefine** Edits an existing component display state.
⇒ **Rename** Renames an existing component display state.
⇒ **Delete** Removes an existing component display state.
⇒ **List** Allows you to view existing component display states.

2. Choose **Create**, and enter a name for the component display state.
3. The system displays the Model Tree window and the EDIT DISPLAY menu. Choose one of these menu options from the EDIT DISPLAY menu to set the display status:

⇒ **Blank** Blanks components in the current component display state.
⇒ **Wireframe** Applies Wireframe display state to assigned components.
⇒ **Hidden line** Applies Hidden line display state to assigned components.
⇒ **No Hidden** Applies No Hidden line display state to assigned components.
⇒ **Shading** Applies Shading display state to assigned components.
⇒ **By Display** Uses the display state of a subassembly in this display state. After you select a subassembly, the By Rep dialog box displays a list of possible component display states for that subassembly.
⇒ **Undo** Removes the display operation on selected components. Resets the display of selected components to normal state.
⇒ **Info** Displays the information window showing all assigned display states for a selected component.
⇒ **Definition Rules** Sets or changes the rules that define the content and settings of the components in the display state.
⇒ **Undo Last** Undoes the last action applied.
⇒ **UpdateScreen** Displays the current state as it is defined.
⇒ **Display Mode** Sets the rule for component display in the Model Tree window. Displays the DISPLAY MODE menu.

4. Choose a command from the SELECT MDL menu to assign the specified display status state to assembly components.
5. Select one or more components to display in the specified display state. In the Model Tree, the current display state is indicated next to specified components in the display editing column.
6. Choose **Done** from the EDIT DISPLAY menu, and **Done/Return** from the COMP DISPLAY menu.

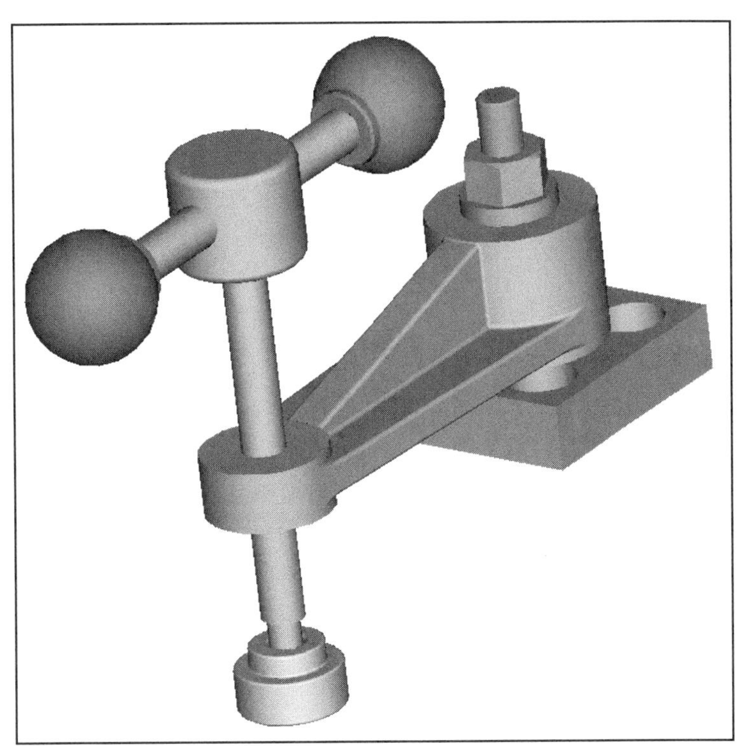

Figure 15.6
Swing Clamp Assembly, Original (Default) State

Set up the **Swing Clamp** Assembly Exploded View:

- ⊙ Units = Inch lbm Second
- ⊙ Shading
- ⊙ No Display
- ❑ Snap to Grid

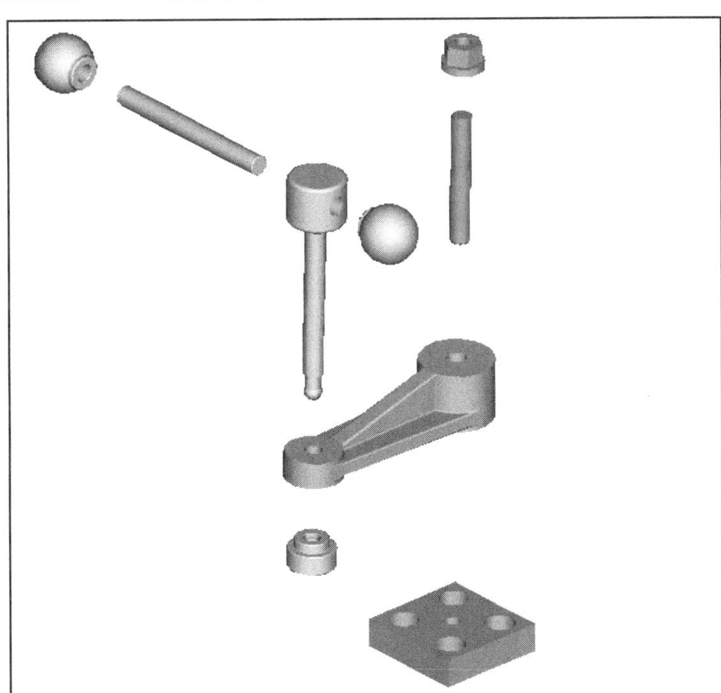

Figure 15.7
Exploded Swing Clamp Assembly with Rotated Swivel

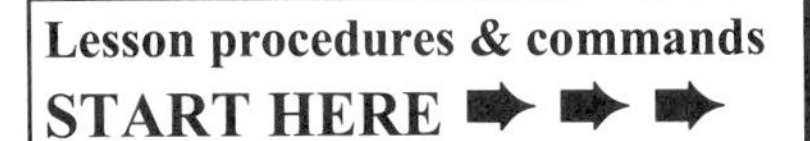

Exploded Swing Clamp

In this lesson, you will use the subassembly and assembly created in Lesson 14 to establish and save new views, exploded views, and views with component display states that differ from one another. You will also be required to move and rotate components of the assembly before cosmetically displaying the assembly in an exploded state. The creation and assembly of new components will not be required. A bill of materials will also be displayed using the **Info** command.

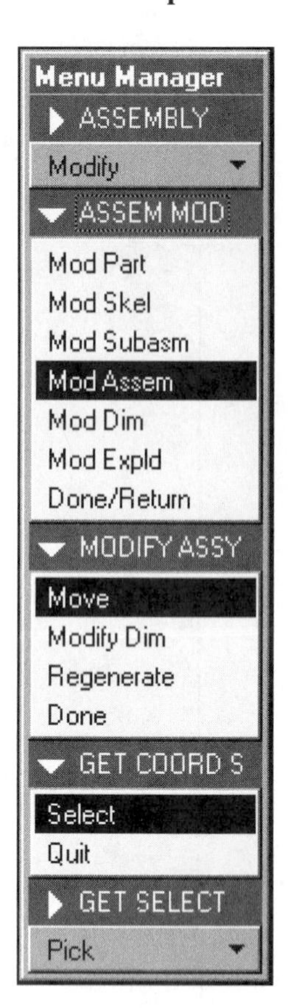

Rotating Components of an Assembly

To rotate an existing component or set of components of an assembly (or subassembly), you select a coordinate system to use as a reference, pick one or more components, and give the rotation angle about a chosen axis of a coordinate system. The **Swing Clamp** subassembly shown here has a **Swivel**, **Foot**, **Stud**, and two **Ball** components that can be rotated about the **Arm** during normal operation of the assembly. You will rotate these components so that the Stud and Balls are perpendicular to the Arm. This position looks better when you are displaying the assembly as exploded (and in its unexploded state). The components will be rotated in the subassembly, and the change will be propagated to the assembly.

The following sequence of commands are used to rotate the Swivel, Foot, Stud, and two Ball components about the Arm (Fig. 15.8) (all of these make up the *subassembly*):

File ⇒ **Open** ⇒ select **clamp_subassembly** or your subassembly name ⇒ **Open** ⇒ **Modify** ⇒ **Mod Assem** ⇒ **Move** ⇒ (select the Swivel's coordinate system) ⇒ (select components to move; pick the Swivel and the Foot *only*) ⇒ **Done Sel** ⇒ **Rotate** ⇒ **Z Axis** ⇒ (input the angle about the **Z** direction, **90°**) ⇒ ✔ ⇒ **Done Move** ⇒ **Regenerate** ⇒ **Automatic** ⇒ **Done** ⇒ **Done/Return**

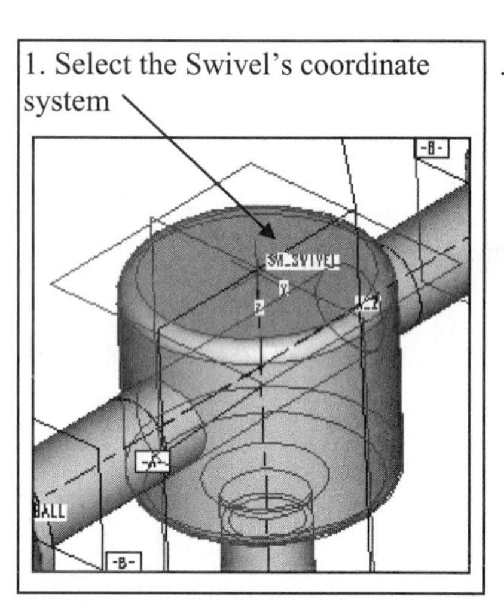

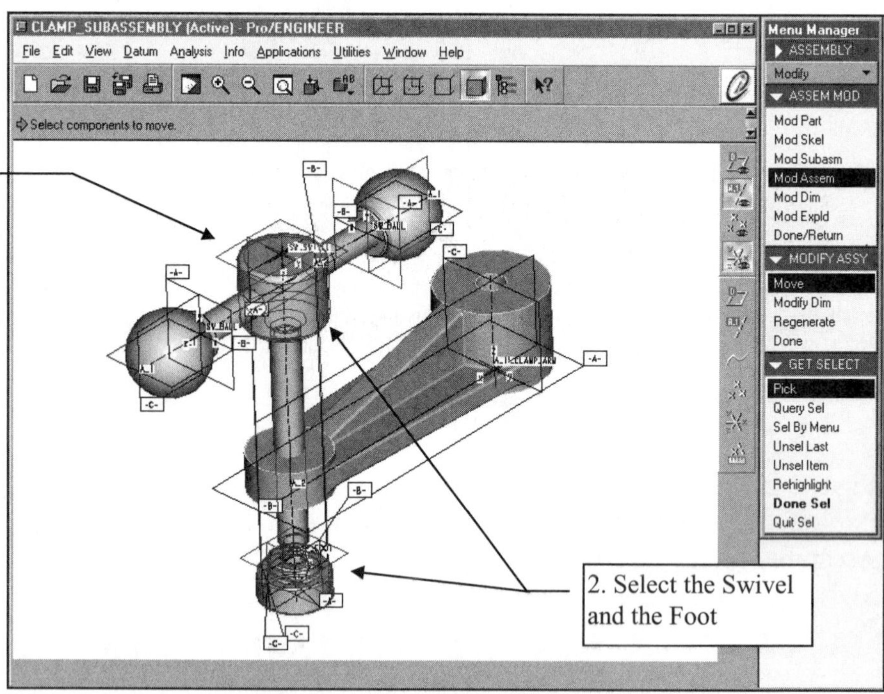

Figure 15.8
Modifying the Subassembly

The Swivel, Foot, Balls, and Stud are now rotated **90°** (Fig. 15.9). Capture this view with the following commands:

View (from the menu bar) ⇒ **Saved Views** ⇒ (type the name **Position 1**) ⇒ **Save** ⇒ **Close**

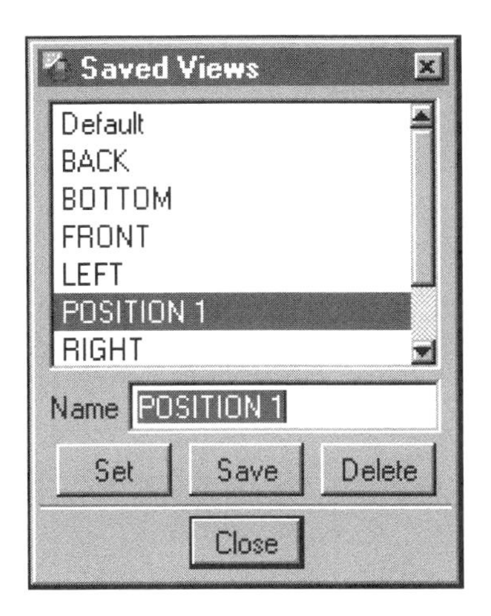

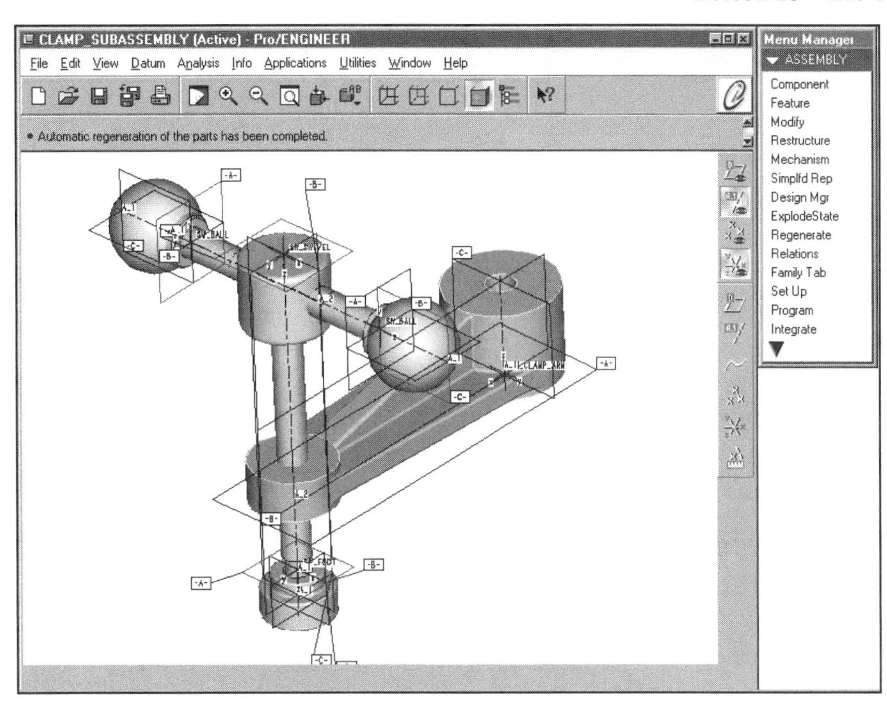

Figure 15.9
Subassembly with Rotated Components

Views: Perspective, Exploded, and Altered Component Displayed State Views

You will now create a variety of cosmetically altered views. Cosmetic changes to the assembly do not affect the model itself, only the way it is displayed on the screen. One type of view that can be created is the *perspective view*. Choose the following commands to create a perspective view (Fig. 15.10):

View ⇒ **Advanced** ⇒ **Perspective** ⇒ (create a view to your liking) **OK** ⇒ **View** ⇒ **Saved Views** ⇒ (type the name **Perspective 1**) ⇒ **Save** ⇒ **Close** ⇒ **File** ⇒ **Save** ⇒ ✔ ⇒ **Window** ⇒ **Close Window**

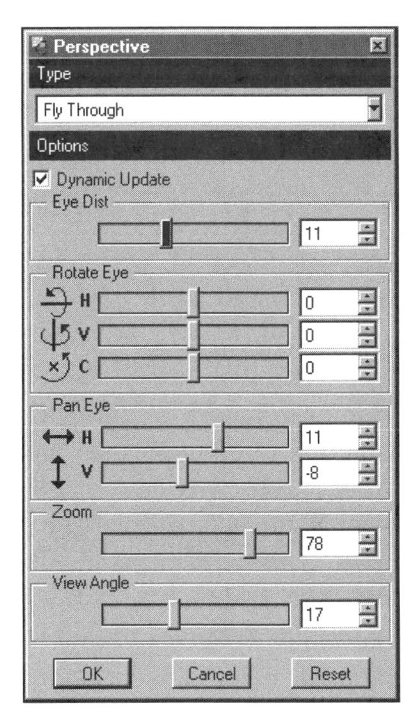

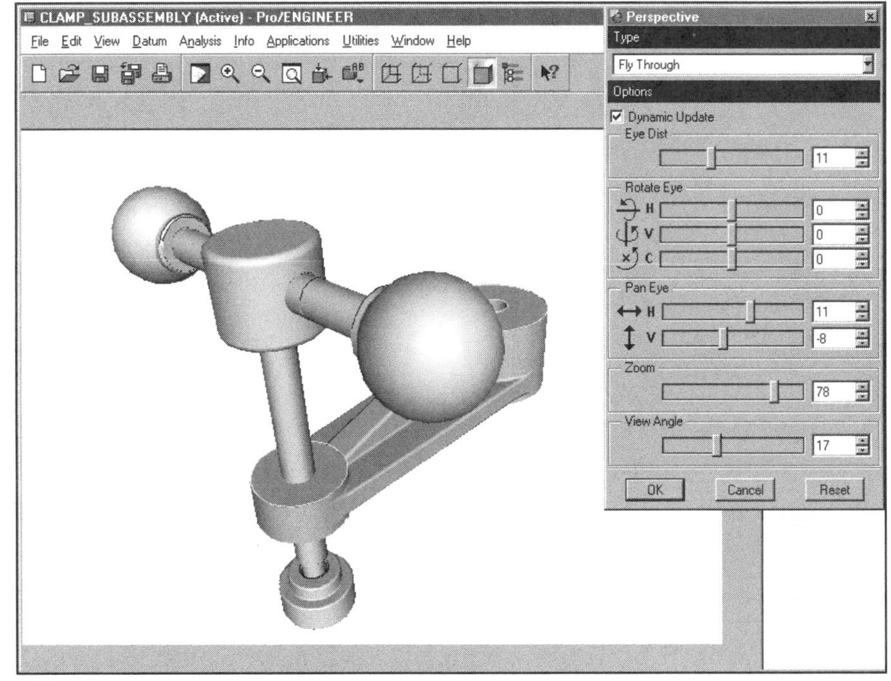

Figure 15.10
Assembly Perspective View

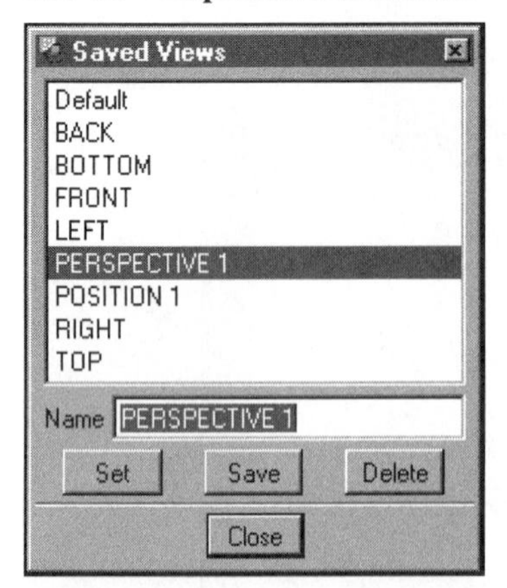

When you create an *exploded view*, Pro/E moves apart the components of an assembly to a set default distance. Your exploded view will look different from the example shown in Figure 15.11.

Create and save the default exploded view of the full assembly (Fig. 15.11) using the following commands:

File ⇒ **Open** ⇒ **clamp_assembly** ⇒ **Open** ⇒ **View** ⇒ **Explode** ⇒ **Shade** (to remove datums, etc.) ⇒ **View** ⇒ **Saved Views** ⇒ type NAME: **EXPLODE 1** (Fig. 15.11) ⇒ **Save** ⇒ **Close**

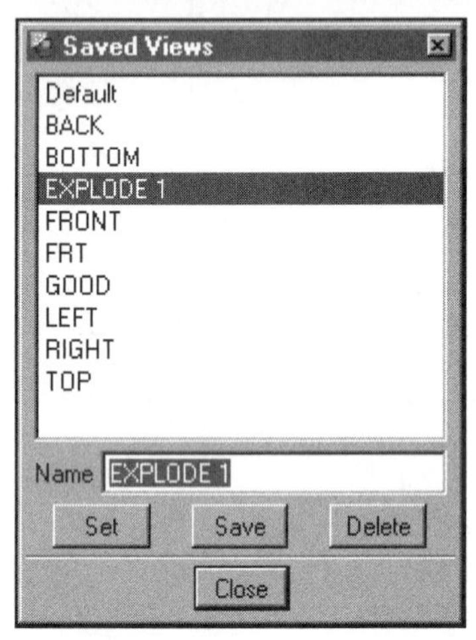

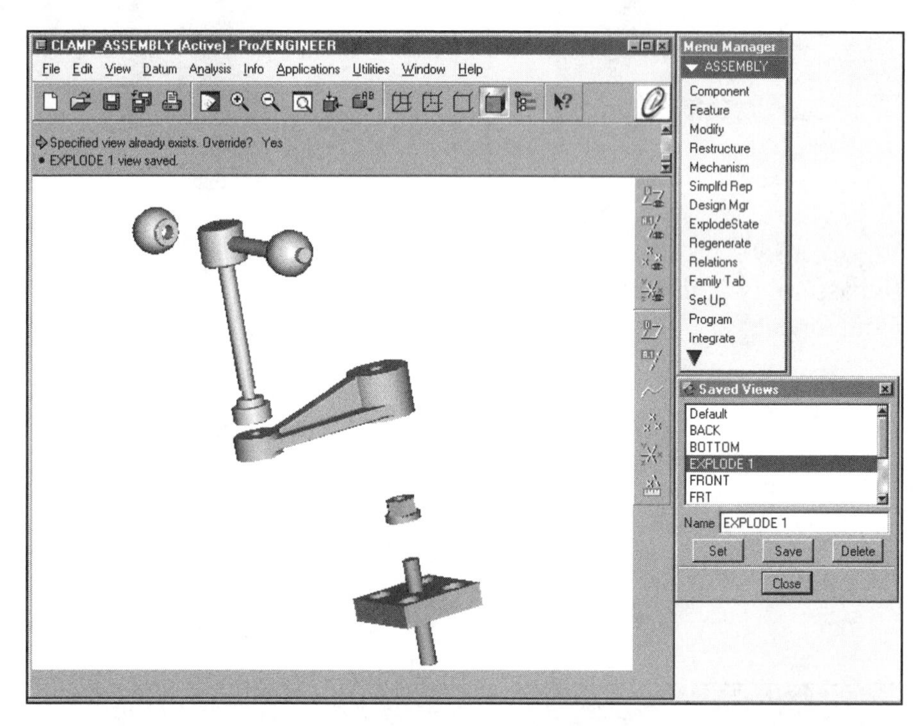

Figure 15.11
Default Exploded Assembly
(Yours May Look Different)

Because the default exploded view is seldom perfect, Pro/E has a wide variety of commands available to adjust, change, modify, and reorient the exploded view. Using **Modify** from the ASSEMBLY menu, choose the following commands:

Modify ⇒ **Mod Expld** ⇒ **Position** ⇒ Motion Reference **Entity/Edge** ⇒ **[Select an axis or straight edge as the motion reference.** ↖pick a vertical edge on the plate (Fig. 15.12)] ⇒ **[Select component(s) to move.** ↖pick the Foot, and slide it to a new position (Fig. 15.13)] ⇒ (pick once with the left mouse button to place the component) ⇒ (continue selecting and moving the components you wish to adjust vertically)

After you are through moving the components vertically to better positions; start the process again:

●**Translate** ⇒ **Entity/Edge** ⇒ **Preferences** ⇒ ●**Move Many** ⇒ **Close** ⇒ ↖pick the horizontal edge of the plate as the reference edge ⇒ ↖pick the Ball and the Stud and move to new positions (Fig. 15.14) ⇒ **OK** ⇒ **Done/Return** ⇒ **Done/Return** ⇒ **View** ⇒ **Saved Views** ⇒ type Name: **Explode 2** ⇒ **Save** (Fig. 15.15) ⇒ **Close** ⇒ **File** ⇒ **Save** ⇒ ✔ ⇒ **File** ⇒ **Delete** ⇒ **Old Versions** ⇒ ✔

NOTE You can move more than one component at a time by selecting: **Preferences** ⇒ **Move Many** (or **Move With Children**)

Preferences
Move Options
Move One
Move Many
Move With Children
Close

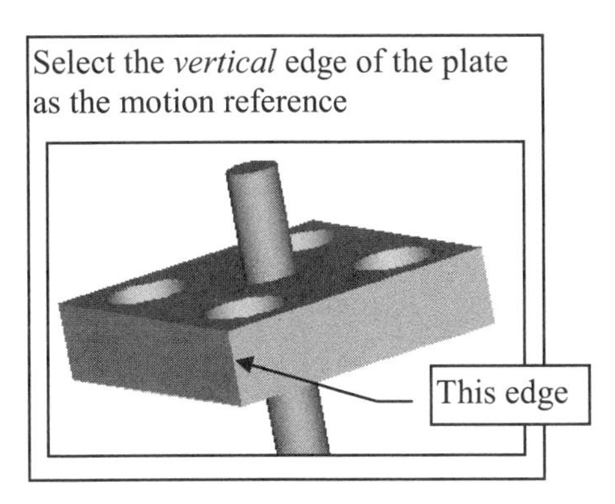

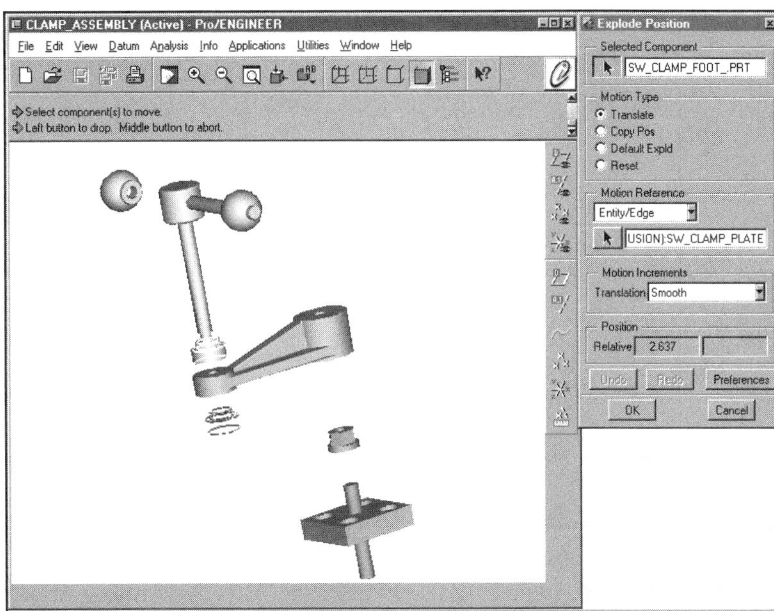

Figure 15.12
Modifying an Exploded Assembly

Foot is in dynamic motion

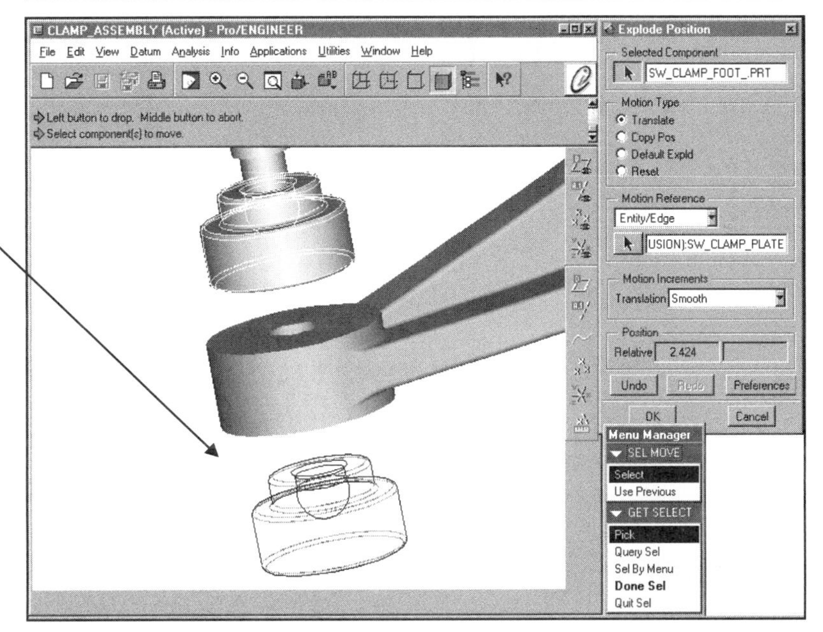

Figure 15.13
New Vertical Positions

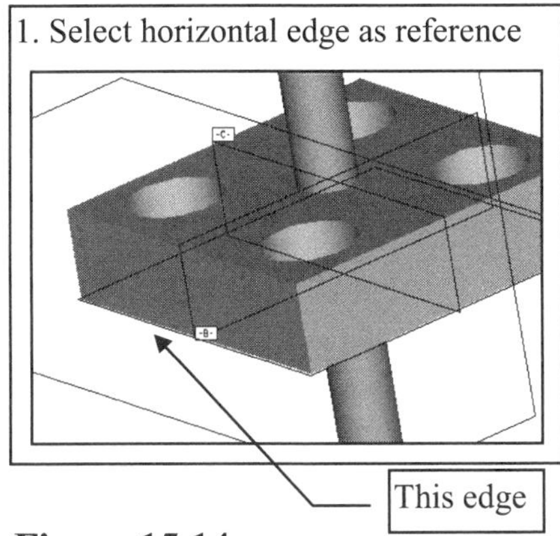

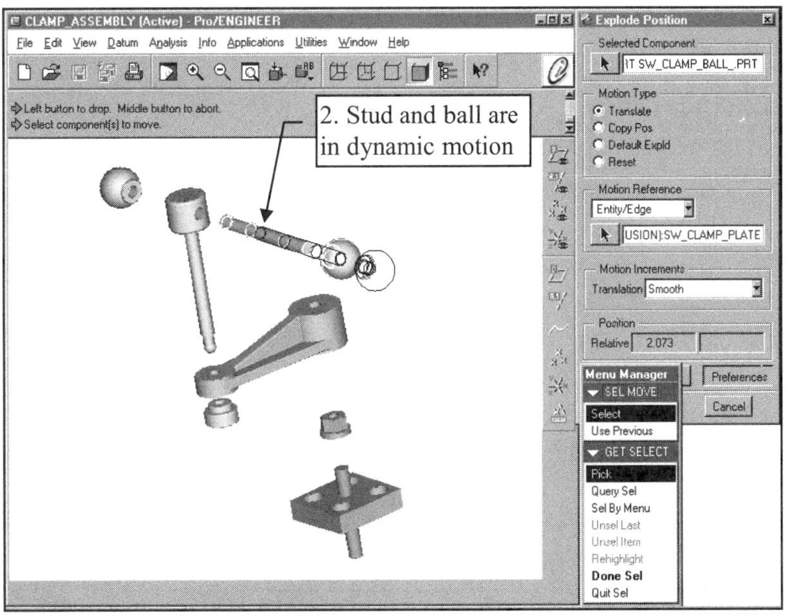

Figure 15.14
Moving Components Horizontally

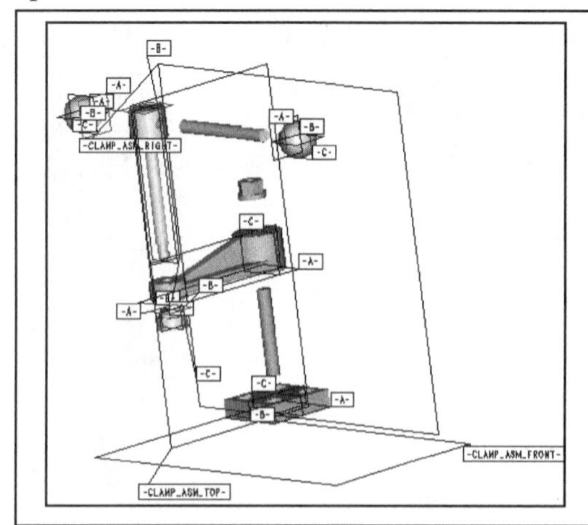

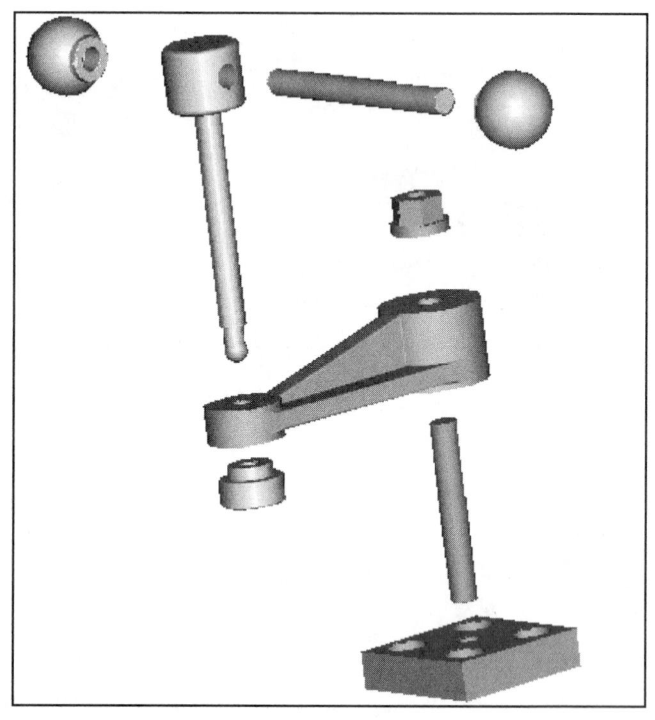

Figure 15.15
Save the Exploded View's New Component Positions (Hopefully, your final exploded view will look better than ours)

The components of an assembly (whether exploded or not) can be displayed individually with **Wireframe**, **Hidden line**, **No Hidden**, or **Shading** using **Component Display.** Change the component display of the exploded view (Fig. 15.16) with the following commands:

Utilities ⇒ **Customize Screen** ⇒ **Options** ⇒ ❑ **Display as a separate window** ⇒ **OK** ⇒ **View** ⇒ **Model Setup** ⇒ **Component Display** ⇒ **Create** ⇒ (**VIS0001**) ⇒ ✔ (to accept the default) ⇒ **Hidden line** ⇒ (pick the Swivel) ⇒ **No Hidden** ⇒ (pick the right Ball) ⇒ **Wireframe** ⇒ (pick the Foot) ⇒ **UpdateScreen** ⇒ **Done** ⇒ **Set Current** ⇒ **VIS0001** (Fig. 15.17) ⇒ **OK** ⇒ **Utilities** ⇒ **Environment** ⇒ Display Style **Shading** ⇒ **OK** ⇒ **File** ⇒ **Save** ⇒✔

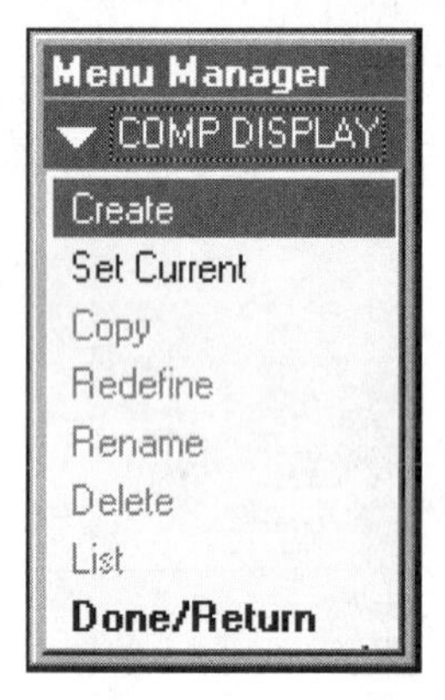

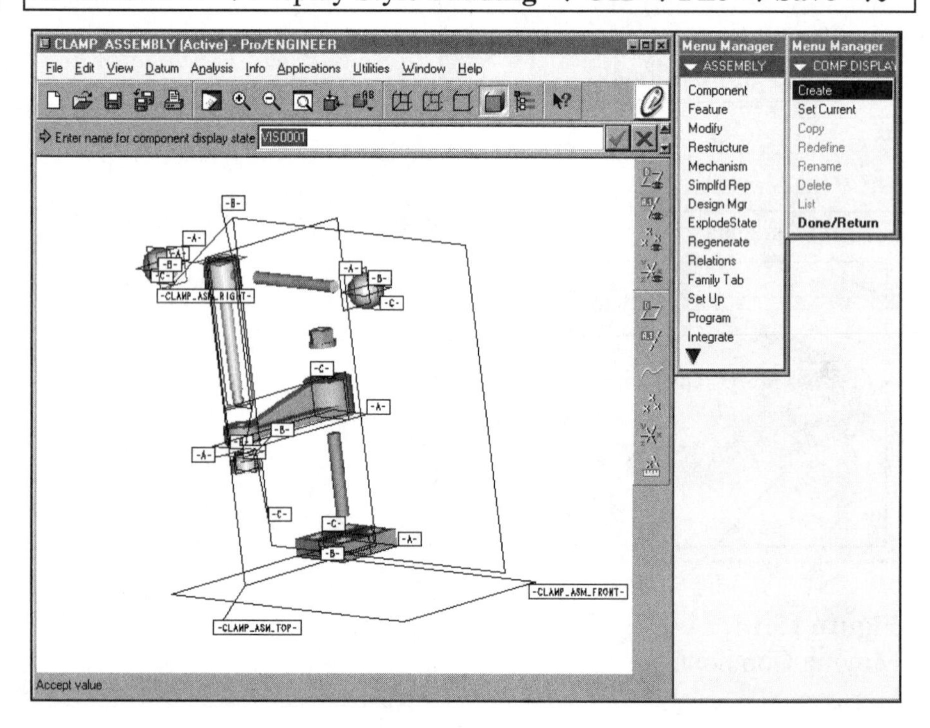

Figure 15.16
Changing the Component Display State

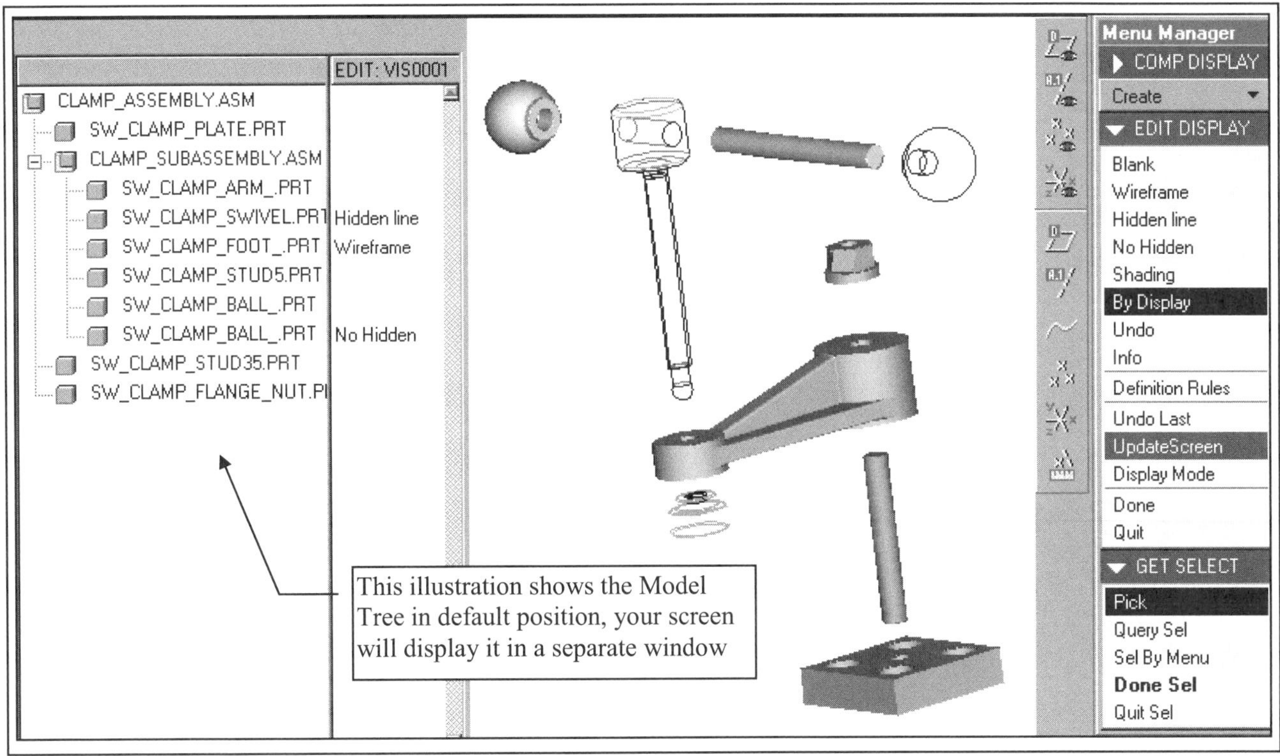

Figure 15.17
Component Display State

You can see the new component display states (Fig. 15.18) in the Model Tree by choosing the following:

Tree (in the Model Tree) ⇒ **Column Display** ⇒ ▼ ⇒ **Info** ⇒ >> (**add/column** 8-10 times) ⇒ ▼ ⇒ **Visual Modes** ⇒ >> (**Add column** twice) ⇒ **Apply** ⇒ **OK** ⇒ (you will need to adjust your column format to see all the information on the screen)

You can change the component display using the Model Tree by selecting the component display type and then ▼

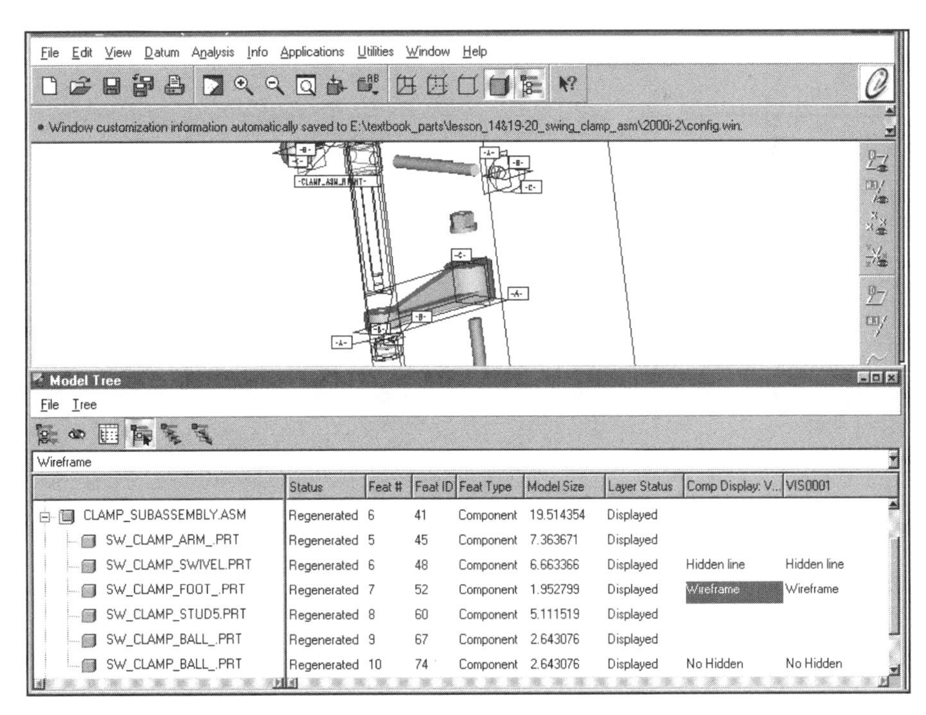

Figure 15.18
Components Display

Multiple **Explode States** can be created for your assembly. In Figure 15.19 a new explode state is being created using:

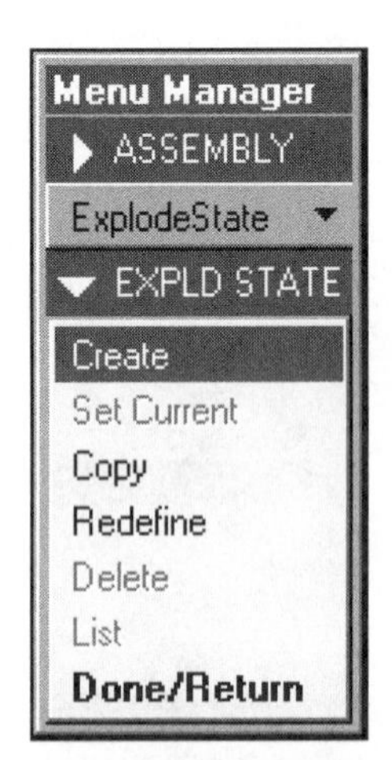

View ⇒ Unexplode ⇒ ExplodeState ⇒ Create ⇒ Enter name for new Exploded State **EXP0001 ⇒ ✔** (to accept default name) ⇒ (adjust positions of components for new state) **⇒ OK ⇒ View ⇒ Saved Views ⇒** (provide view name) **⇒ Save ⇒ Close ⇒ Done/Return ⇒ Done/Return ⇒ File ⇒ Save ⇒ ✔**

After you have created a number of explode states, you can activate them by choosing the following command:

ExplodeState ⇒ Set Current ⇒ (select **✔ EXP0002) ⇒ Done ⇒**

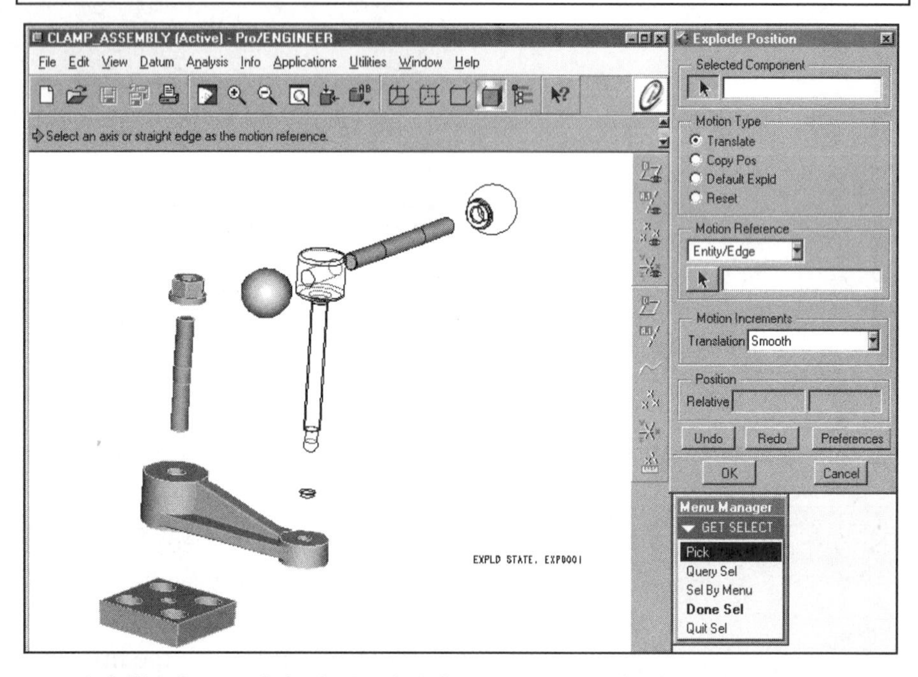

Figure 15.19
Explode States

A bill of materials (BOM) (Fig. 15.20) can be seen by choosing:

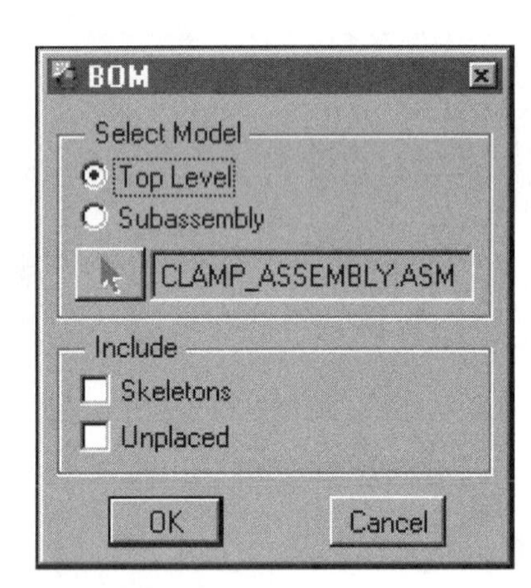

Info ⇒ BOM ⇒ Top Level ⇒ OK ⇒ Close ⇒ File ⇒ Save ⇒ ✔

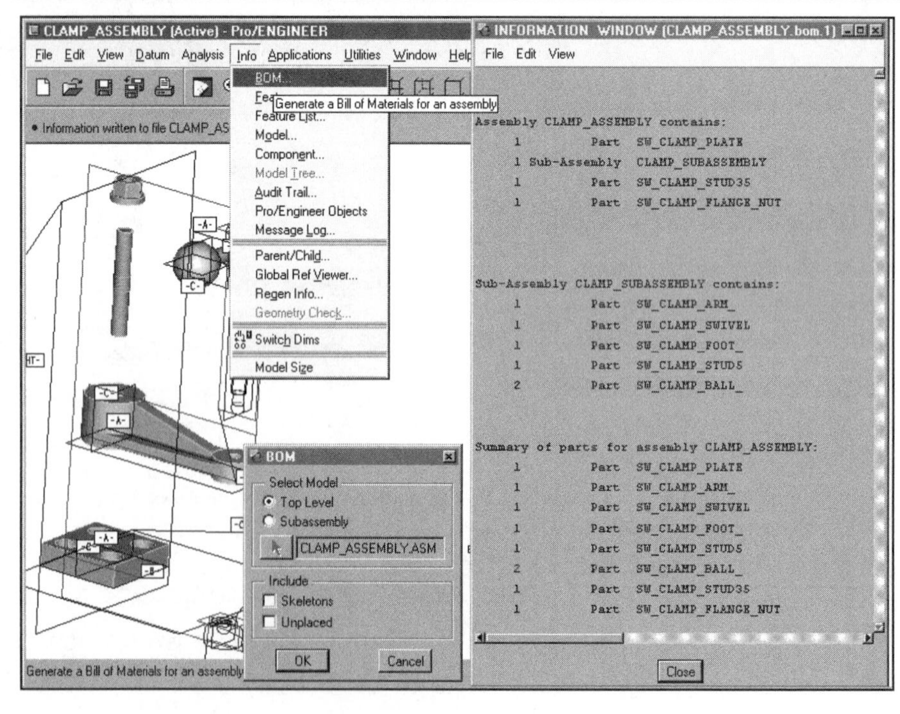

Figure 15.20
BOM

Since the assembly you created is a standard clamp from CARR LANE Manufacturing Co.; you can go to CARR LANE's Website and see the assembly, order it (about $60.00 U.S.) and even download a 3D IGES model!

Attach a 3D note to the assembly identifying the manufacturer and Website information using the following commands:

Set Up ⇒ **Note** ⇒ **New** ⇒ Name CARR_LANE ⇒ Text (see Fig. 15.21) ⇒ URL **www.carrlane.com** ⇒ **Place** ⇒ **No Leader** ⇒ **Done** ⇒ **(Select LOCATION for note:** pick near the assembly on the screen) ⇒ **OK** ⇒ **Done/Return** ⇒ **Done** ⇒ **File** ⇒ **Save** ⇒ ✔ ⇒ **File** ⇒ **Delete** ⇒ **Old Versions** ⇒ ✔

Figure 15.21
Adding a 3D Note

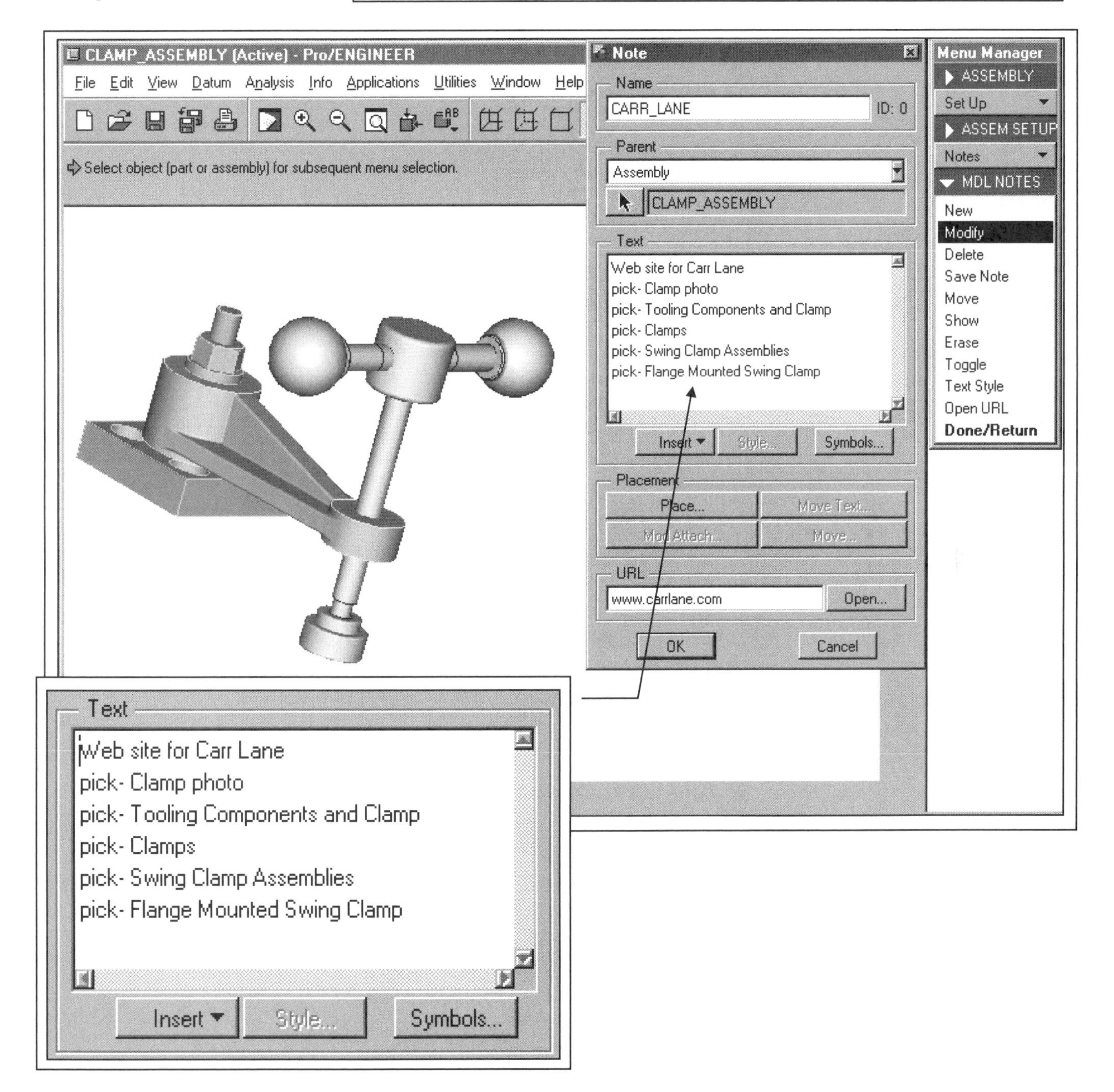

You can also open a URL note from the Model Tree:

Highlight the note in the Model Tree with the (RMB) and then select **Open URL.**

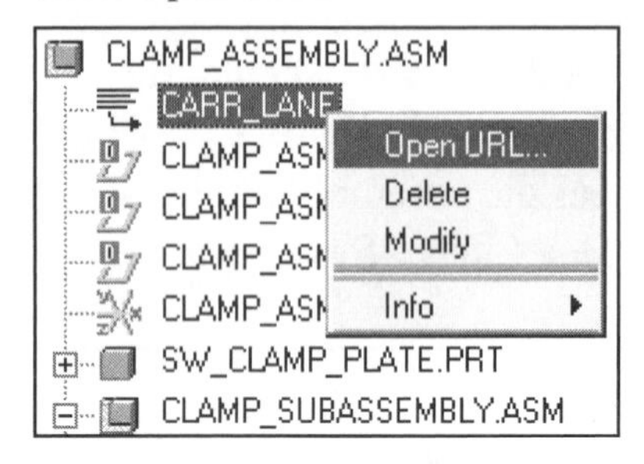

To Open a URL in a Model Note

To launch a Web browser and go to a World Wide Web URL (Universal Resource Locator) associated with a model note, use the **Set Up** command from the PART or ASSEMBLY menu.

1. Click **Set Up ⇒ Notes ⇒ Open URL**. The GET SELECT menu opens.
2. Select a note. Your Web browser opens, displaying the URL you selected.

Open the URL and follow the Text instructions to navigate the site to the Swing Clamp using the following commands:

Set Up ⇒ Notes ⇒ Modify (pick on the note in the Model Tree or from the screen to see the note before opening the URL) ⇒ **URL Open** (from the Note dialog box) ⇒ (navigate the site to the Swing Clamp)

Figure 15.22
Opened CARR LANE URL

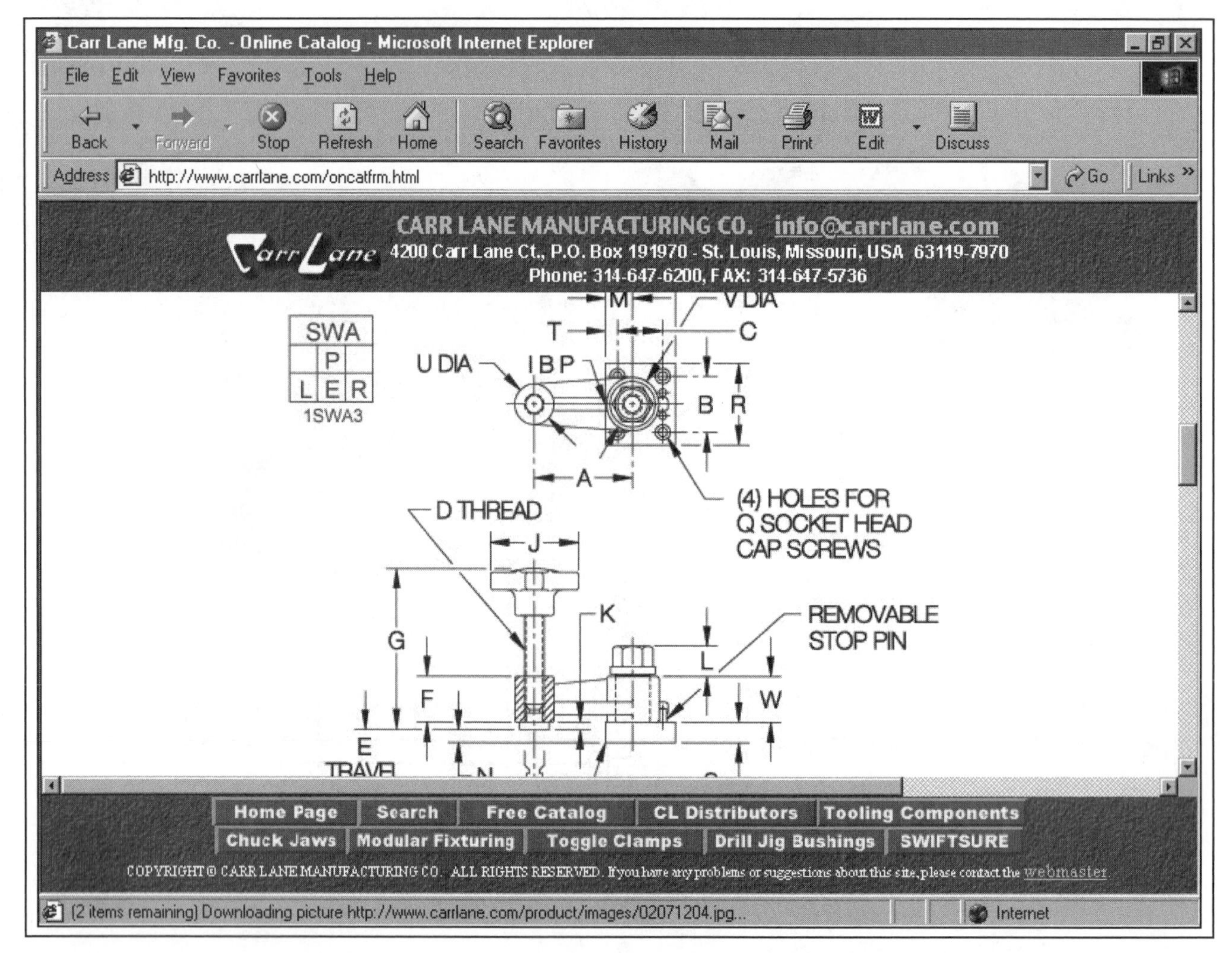

Lesson 15 Project

Exploded Coupling Assembly

EGD REFERENCE
Fundamentals of Engineering Graphics and Design
by L. Lamit and K. Kitto
Read Chapter 23.
See pages 865-866.

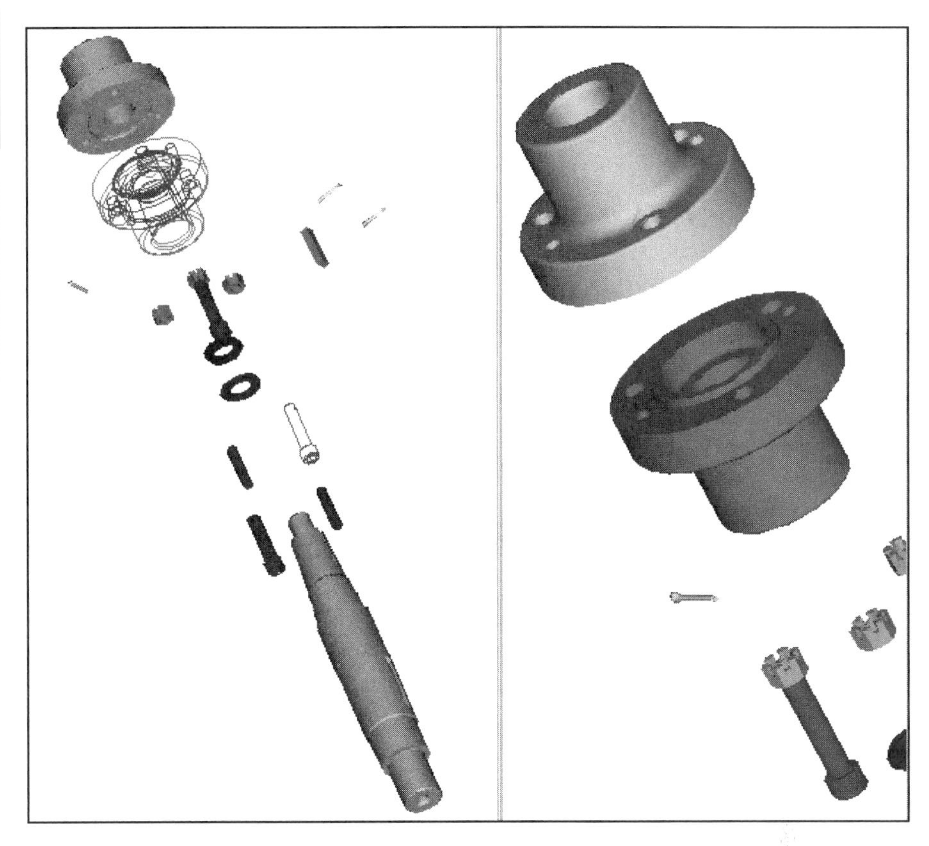

Figure 15.23
Exploded Coupling Assembly

Exploded Coupling Assembly

The fifteenth **lesson project** uses the assembly created in the Lesson 14 Project. An exploded view needs to be created and saved for use later in Drawing mode for the Lesson 20 Project. A variety of other views are suggested, including a section of the assembly, a perspective view, and an exploded view with a different component display variation. Each component should have its own color. If you did not color the components during the part creation, bring up each part in Part mode, define, and set the component with a color. Save three or four unique exploded states (view positions) and component display (visibility) states. You do not need to match the examples (Figs. 15.23 through 15.26) provided in this lesson project.

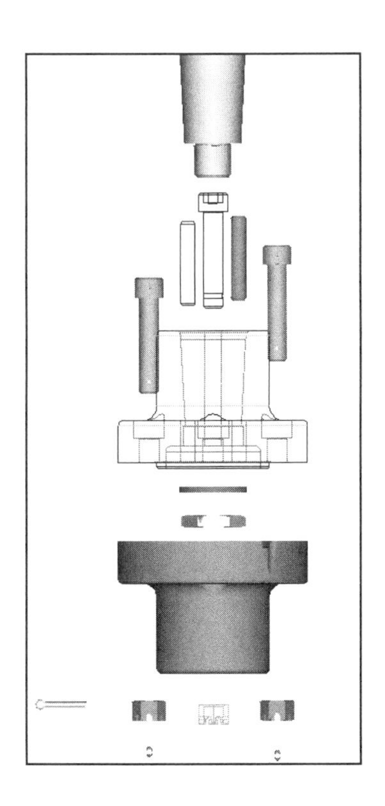

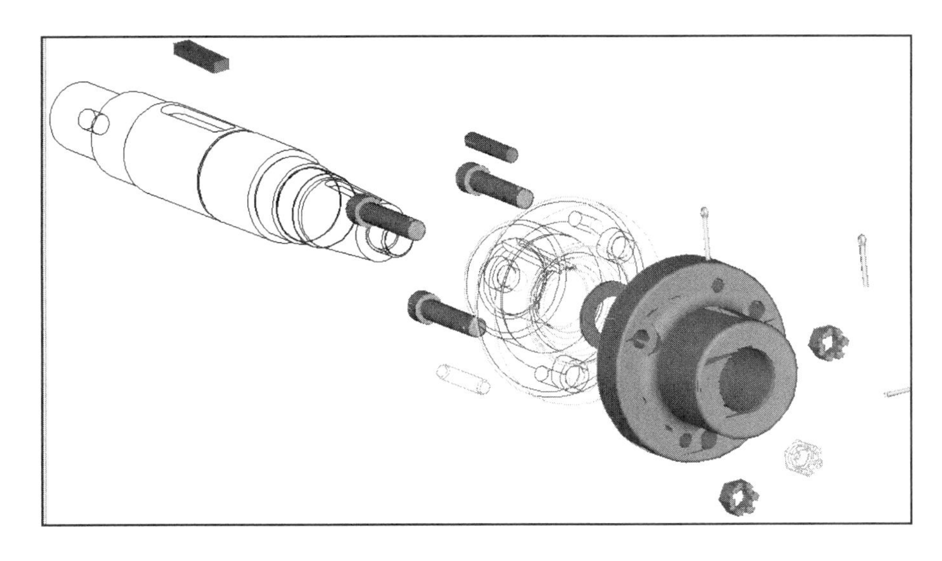

Figure 15.24
Perspective View of Exploded Coupling Assembly with a Variety of Component Display States

Figure 15.25
Shaded Exploded Coupling Assembly

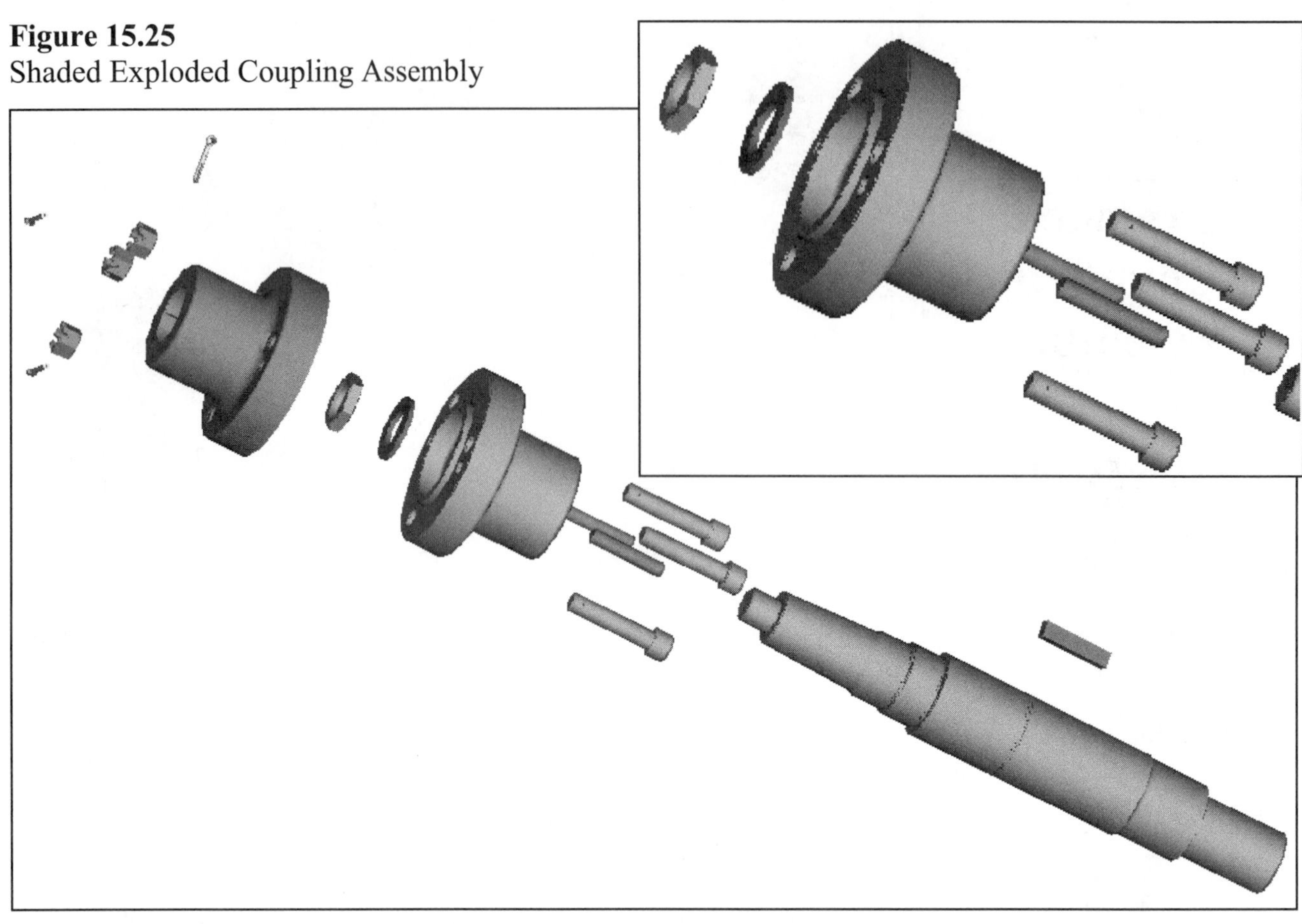

Figure 15.26
Exploded Coupling Assembly with Different Component Display States

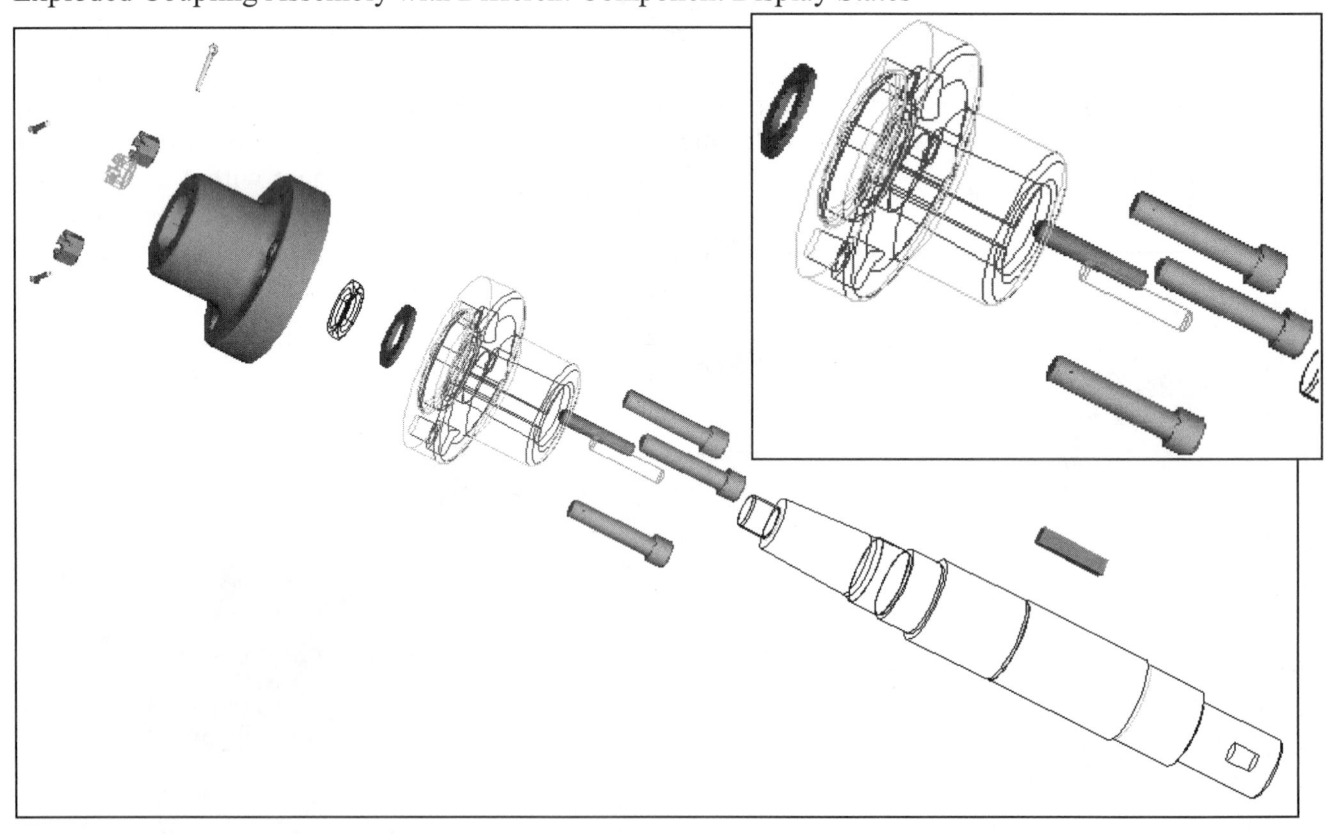

Part Three

Generating Drawings

Lesson 16 Formats, Title Blocks, and Views
Lesson 17 Detailing
Lesson 18 Sections and Auxiliary Views
Lesson 19 Assembly Drawings and BOM
Lesson 20 Exploded Assembly Drawings

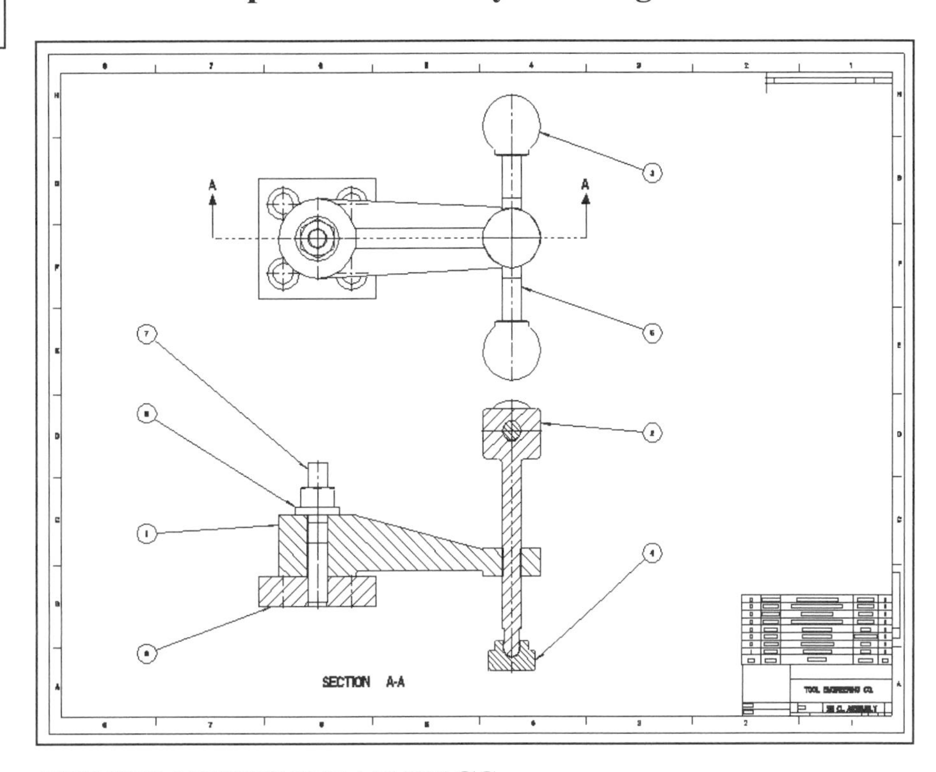

Assembly Drawing

GENERATING DRAWINGS

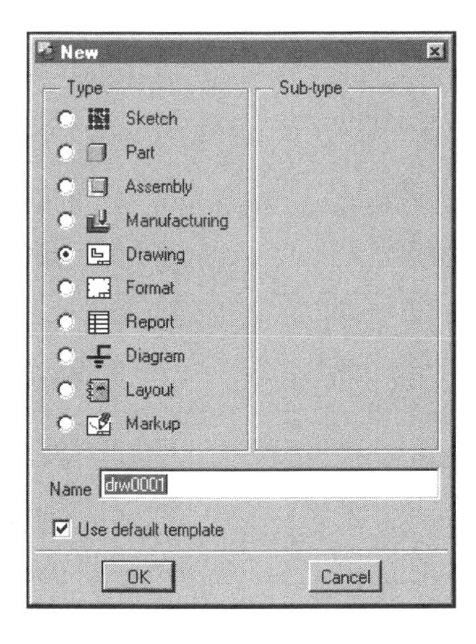

The drawing functionality in Pro/E is used to create annotated drawings of parts and assemblies and has a variety of options, including:

* Use default views created on template drawings
* Add views of the part or assembly to the drawing
* Display existing design dimensions
* Create additional driven or reference dimensions
* Create and insert notes on the drawing
* Add views of additional parts or assemblies
* Add multiple sheets to the drawing
* Create a BOM and balloon the assembly
* Add draft entities to the drawing

Construction of Drawings

Drawings can be created of parts and assemblies. Drawings can be multiview or pictorial and can include section, auxiliary, detailed, exploded, and broken views. With Pro/DETAIL, ANSI, ISO, DIN, or JIS, standard drawings can be created.

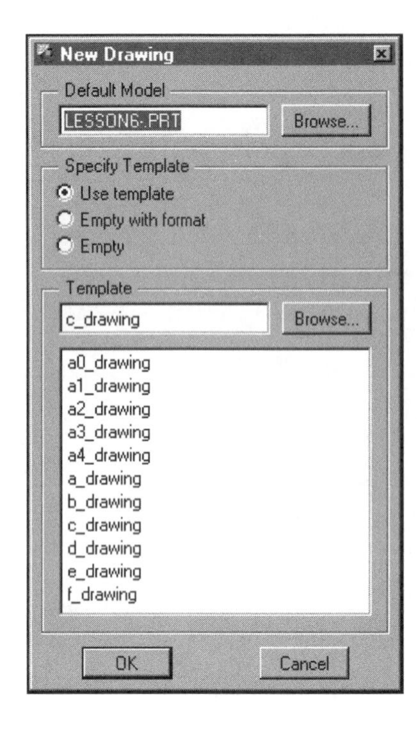

The **Drawing Module** is designed to allow you to create drawings, add views, dimension, and document the part or assembly. Pro/E offers several methods for creating views on the drawing. All methods are based on the rules of orthographic projection.

Orthographic projection allows you to project a view from an existing view based on either first-angle or third-angle projection.

The projected view, by default, is created at the same scale as the view it is projected from. You can alter the scale of the view, if necessary, after it is placed.

When you create a Drawing, Pro/E creates a new file to hold your drawing. Drawing files all have a **.drw** file extension. The new drawing is displayed in its own graphics window.

The Drawing Module is very parameter-intensive. There are over 60 separate **parameters,** which control the display of drafting annotation, views, formats, etc. The user interface would become too cumbersome if all of these parameters were controlled by menu options, and because virtually all of these parameters are actually company standards (that you do not change constantly), the parameters are defined and modified in the Drawing **Setup** file.

You can create and utilize several Setup files at one time. When you are using the Drawing Module, you can change the file at any time. When you change files, Pro/E regenerates the drawing to display based on the settings in the new file. You can also edit the current file while you are working on a drawing; however, if you do, you should be conscious of the fact that you may be altering your drawing in such a way that it deviates from company standards.

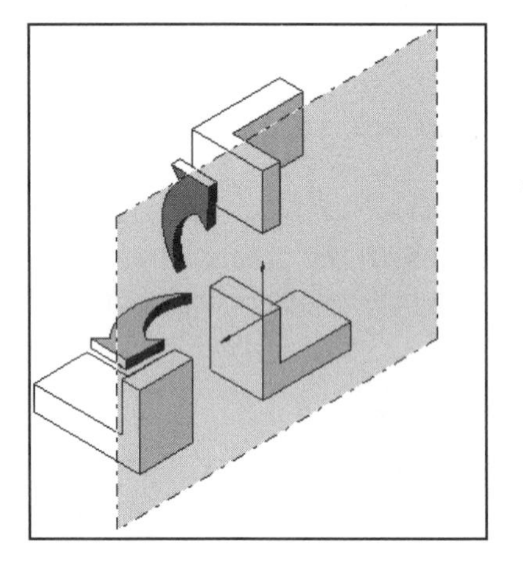

In the following Drawing Module Lessons, there will be references to pertinent Setup Parameters. When you see a parameter, or its value in text, it will be shown in a different font and color: *projection_type*. There is a complete listing of all Setup Parameters in the Pro/ENGINEER Drawing User's Guide. The listing is in its Appendix A.

Every Drawing contains certain attributes. All drawings have a **Name,** a **Size** (Height by Length), a **Scale,** a **Projection Angle,** and a **Drawing Unit.**

The **Name** is the name of the file containing the drawing. The size defines, in Drawing Units, how much space is available on the drawing to place views and annotation. The size is also used to plot the drawing.

The **Scale** establishes the default size at which views are placed on the drawing. This scale does not affect the size of annotation on the drawing.

The **Projection Angle** controls how views are projected. Views are projected using one of two conventions: **First-Angle** or **Third-Angle.** The Projection Angle is controlled by a drawing Setup Parameter called *projection_type*. The default value of this parameter is *third_angle.*

Lesson 16

Formats, Title Blocks, and Views

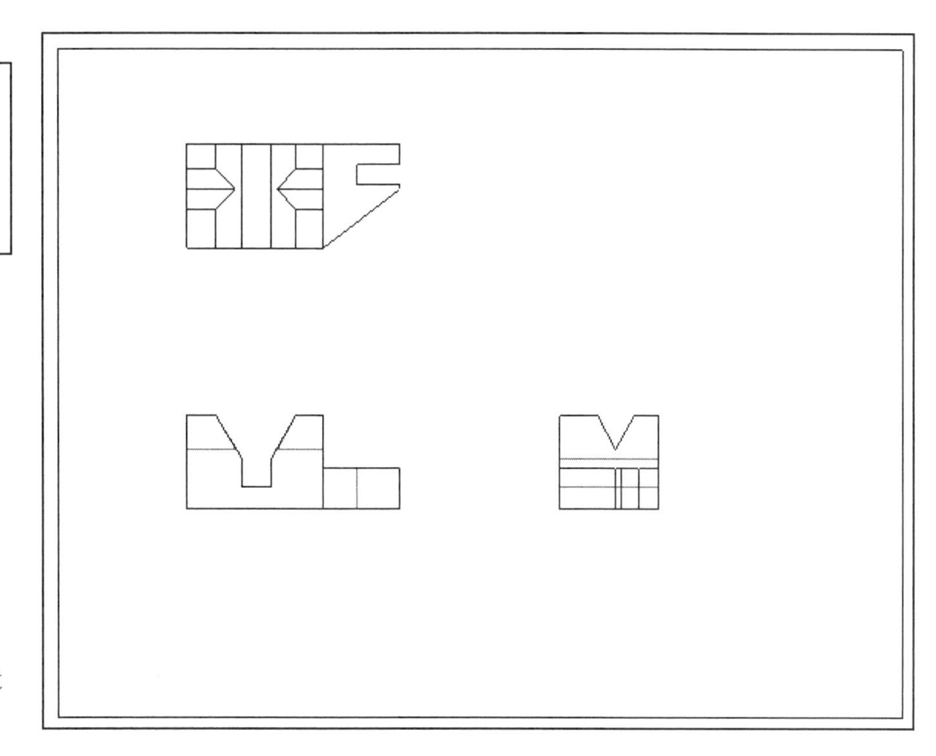

Figure 16.1
Base Angle Drawing Without Drawing Format

EGD REFERENCE
Fundamentals of Engineering Graphics and Design
by L. Lamit and K. Kitto
Read Chapters 10 and 23.
See pages 314-316.

COAch™ for Pro/ENGINEER

If you have **COAch for Pro/ENGINEER** on your system, go to SEARCH and do the Segments shown in the many figures taken from COAch in this lesson. Note that *COAch is not necessary to use this text.* Having access to CADTRAIN's COAch tutorials *will expedite* your mastering of Pro/ENGINEER.

OBJECTIVES

1. **Create drawings with views**
2. **Create and save title blocks and sheet formats**
3. **Change the scale of a view**
4. **Display views for detailing a project**
5. **Move, erase, and delete views**
6. **Specify a standard-format paper size and units**
7. **Retrieve formats from the format library**

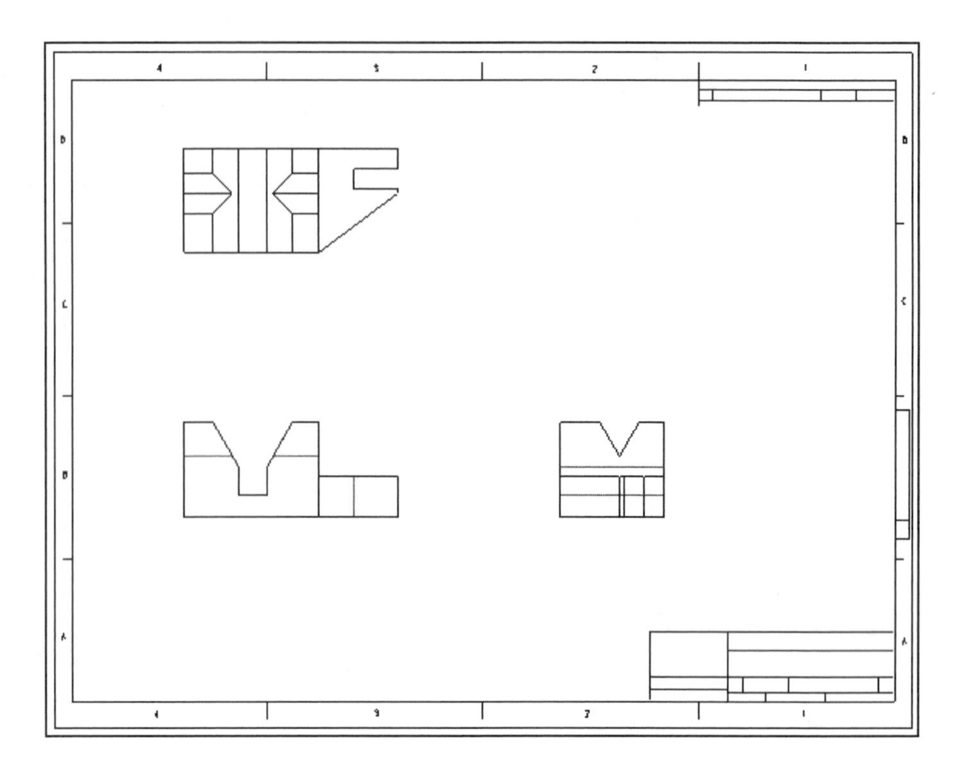

Figure 16.2
Formatted Base Angle Drawing

FORMATS, TITLE BLOCKS, AND VIEWS

Drawing formats are user-defined drawing sheet layouts (Figs. 16.1 and 16.2). A drawing format can be used in any number of drawings. It can also be modified or replaced in a Pro/E drawing at any time. **Title Blocks** are standard or sketched line entities that can contain parameters so that the part name, tolerances, scale, and so on will show when the drawing format is retrieved (Fig. 16.3). **Views** created by Pro/E are exactly like views constructed manually by a designer on paper. The same rules of projection are applied; the only difference is that you choose commands in Pro/E to create the views as needed.

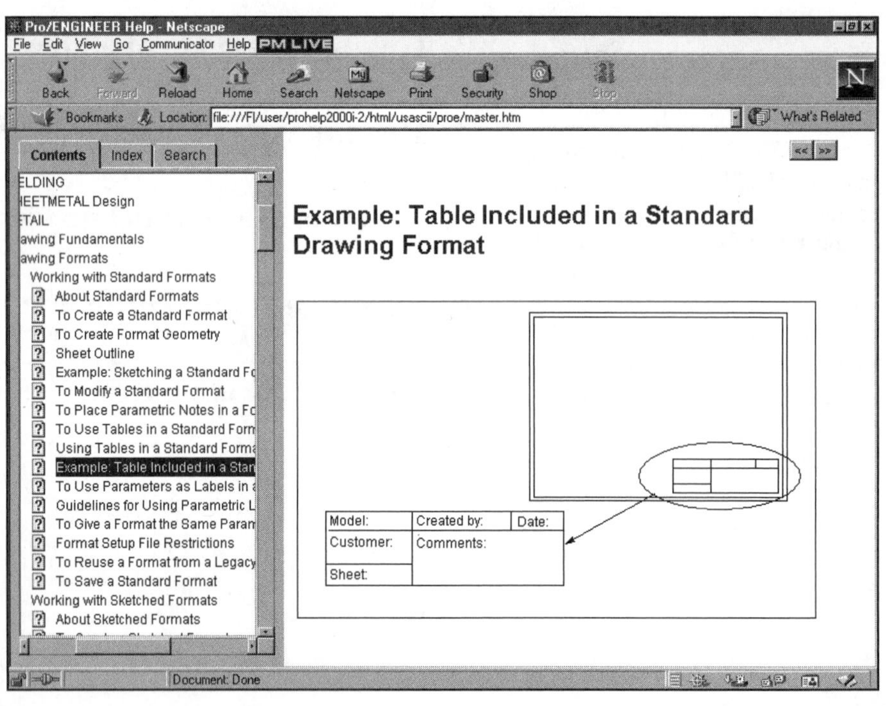

Figure 16.3
Online Documentation, Formats

Drawing Formats

There are two types of drawing formats: standard (Fig. 16.2) and sketched. **Drawing formats** consist of draft entities, not model entities. You can select the desired format size from a list of standard sizes (**A-F** and **A0-A4**) or create a new size by entering values for length and width.

Sketched formats created in Sketcher mode (Fig. 16.4) may be parametrically modified, enabling you to create nonstandard-size formats or families of formats.

Formats can be altered to include note text, symbols, tables, and drafting geometry, including drafting cross sections and filled areas.

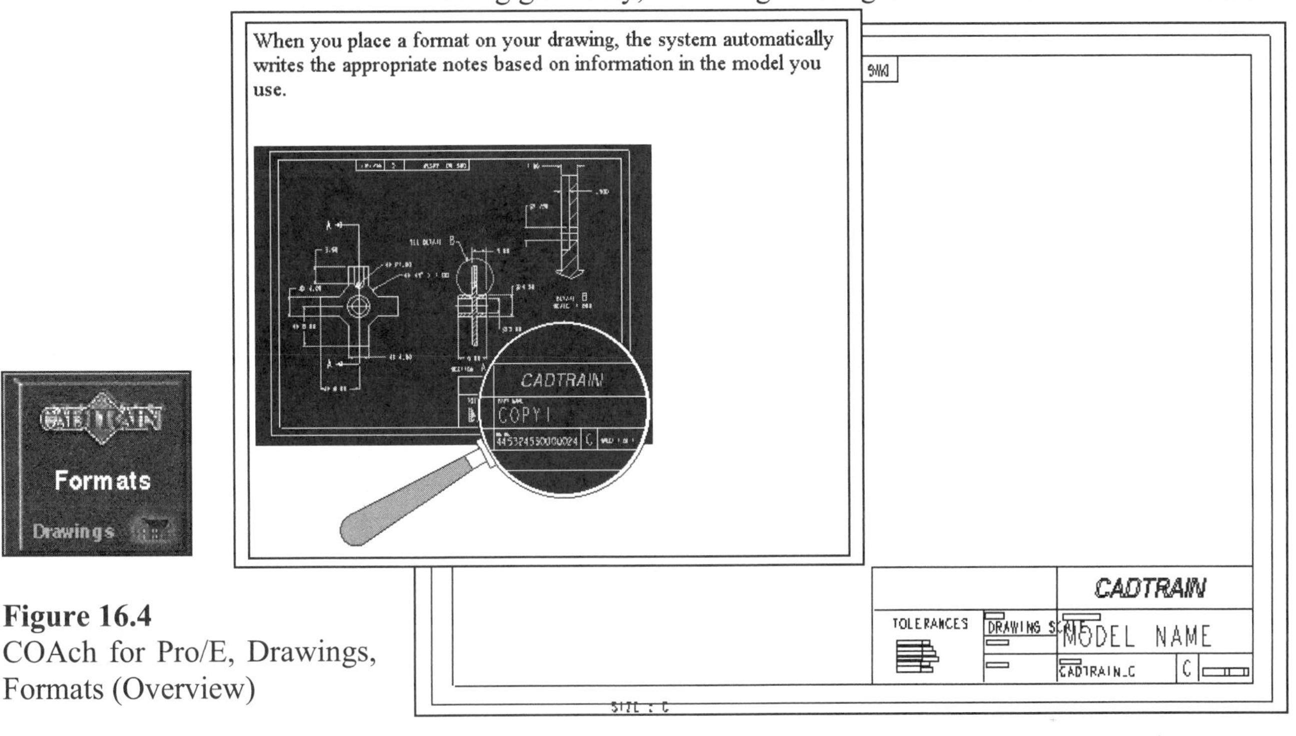

Figure 16.4
COAch for Pro/E, Drawings, Formats (Overview)

With Pro/E (you must have a license for Pro/DETAIL) you can do the following in **Format** mode:

* Create draft geometry, notes.
* Move, mirror, copy, group, translate, and intersect geometry.
* Use and modify the draft grid.
* Enter user attributes (Fig. 16.5).
* Create drawing tables.
* Use interface tools to create plot, DXF, SET, and IGES files.
* Import IGES, DXF, and SET files into the format.
* Create user-defined line styles.
* Create, use, and modify symbols.
* Include drafting cross sections in a format.

Whether you use a standard format or a sketched format, the format is added to a drawing that is created for a set of specified views of a parametric 3D model.

Figure 16.5
COAch for Pro/E, Drawings, Formats (Modifying Text)

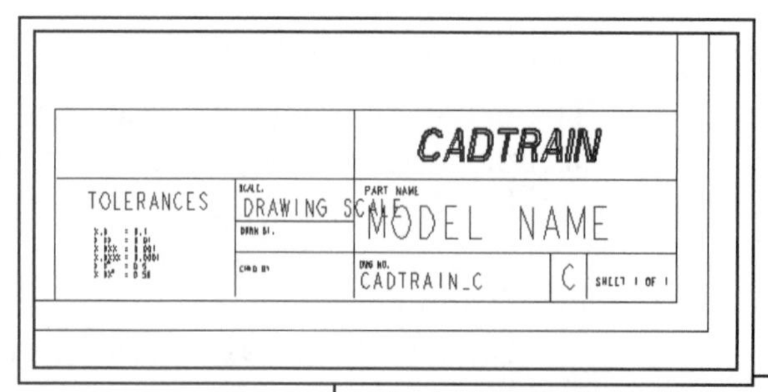

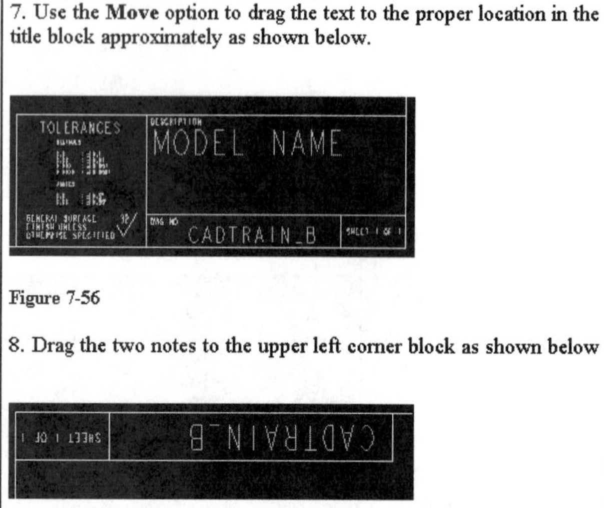

7. Use the **Move** option to drag the text to the proper location in the title block approximately as shown below.

Figure 7-56

8. Drag the two notes to the upper left corner block as shown below

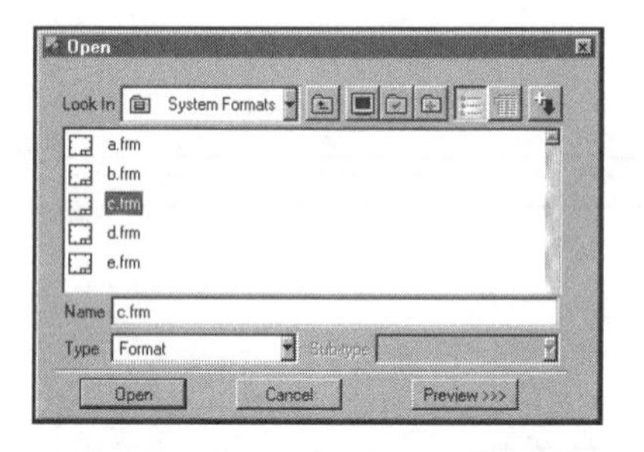

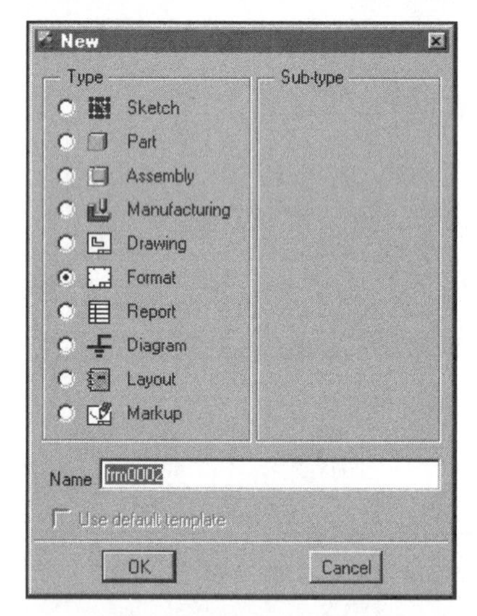

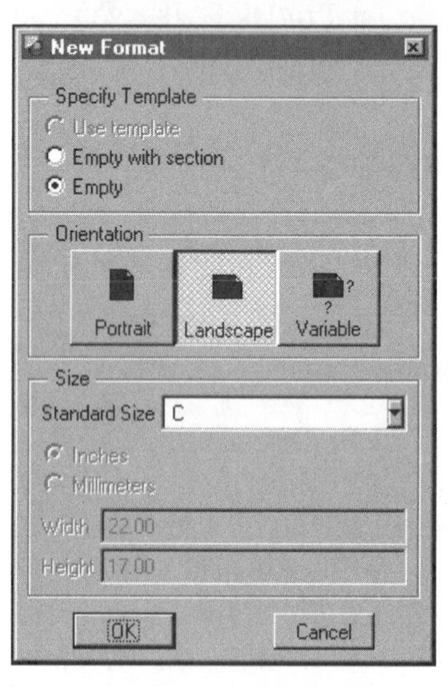

Retrieving Pro/E Formats from the Format Library

You can retrieve formats from a format library directory within Pro/E. To enter the format directory:

1. Choose **File** from the Menu Bar
2. Choose **Open** ⇒ pick **Look In ▼**
3. Choose **System Formats**
4. Select a name
5. Choose **Open**

After specifying a format size, you enter **Format** mode; a sheet outline appears on the screen. If necessary, turn on the drawing grid by choosing **Modify** from the DETAIL menu, **Grid** from the MODIFY DRAW menu, and **Grid On** from the GRID MODIFY menu. The grid size depends on the units for the format.

Specifying the Format Size and Units

You establish the format size by selecting one from a list of standard format sizes or by creating your own size by entering values for the width and length of the format. The main grid spacing and format text units depend on the units you select for the format.

To select the format units, choose **Inches** or **Millimeters** from the **New Format** dialog box. The text height units are in the units of the format: inches or millimeters. The major grid spacing is an inch if the units are inches; it is a centimeter if the units are millimeters.

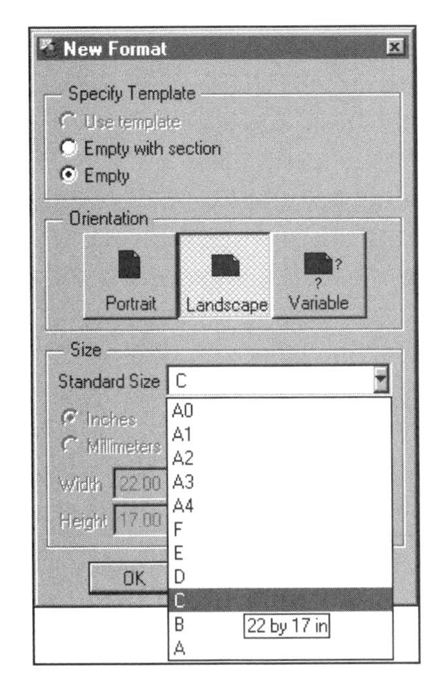

To specify a standard-format paper size and units:

1. Choose an option from the **New Format** dialog box (see left) to determine the orientation of the format sheet:

   Portrait Uses the larger of the dimensions of the sheet size for the format's height; uses the smaller for the format's width.
   Landscape Uses the larger of the dimensions of the sheet size for the format's width; uses the smaller for the format's height.
   Variable Select the unit type, **Inches** or **Millimeters,** and then enter specific values for the **Width** and **Height** of the format.

2. If you select **Landscape** or **Portrait,** then choose the desired standard size:

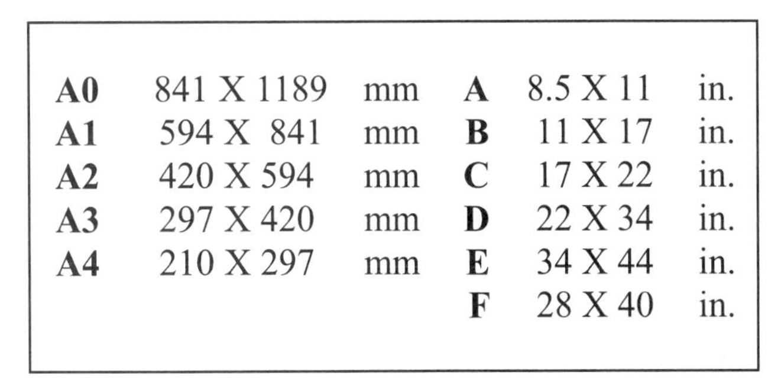

A0	841 X 1189	mm	**A**	8.5 X 11	in.
A1	594 X 841	mm	**B**	11 X 17	in.
A2	420 X 594	mm	**C**	17 X 22	in.
A3	297 X 420	mm	**D**	22 X 34	in.
A4	210 X 297	mm	**E**	34 X 44	in.
			F	28 X 40	in.

3. If **Variable** is selected as the Orientation, then choose the desired units (**Inches** or **Millimeters**) from the **New Format** dialog box.

Creating a Format

To create a format:

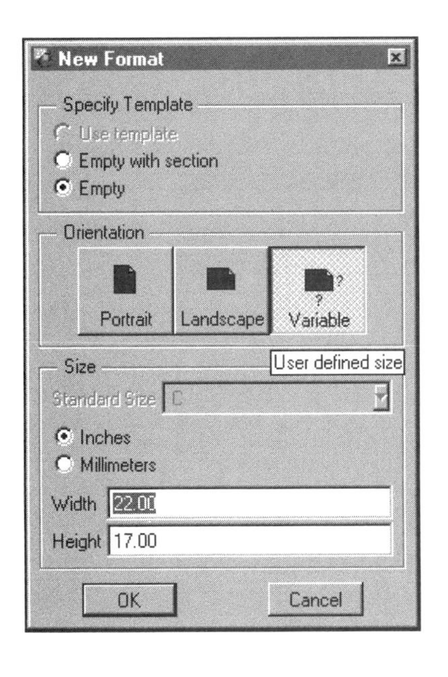

1. Choose **File ⇒ New** from the Menu Bar. The **New** dialog box is displayed.
2. From the **New** dialog box, choose **Format ⇒** (type a name) **⇒ OK ⇒ Empty.**
3. Choose **Variable** ⇒ select the unit type, **Inches** or **Millimeters,** and then enter specific values for the **Width** and **Height** of the format.
4. Draw the format.

Sketching the Format

To create format geometry, you use draft geometry. To sketch draft geometry, choose **Sketch** from the DETAIL menu, and any option from the DRAFT GEOM menu.

The sheet outline is the border of the standard drawing format you selected. Because it is the actual border, it may not show up on "pen plots" unless you use a paper size larger than the drawing size. Everything within the sheet outline border also plots, but you should make an allowance for the plotter's hold-down rollers.

Sketched Format

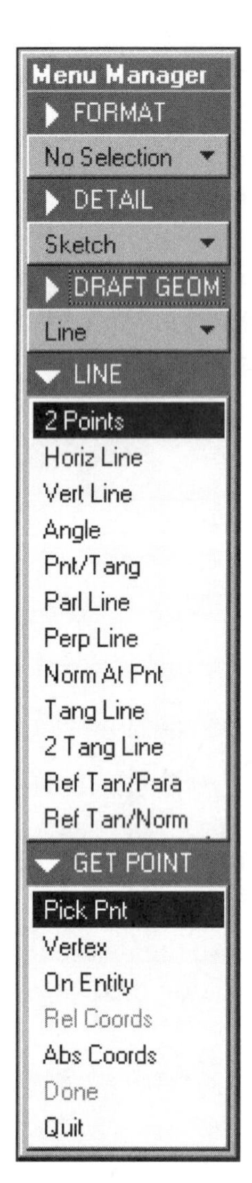

When you add a sketched format to a drawing, Pro/E aligns the lower left corner (the origin) of the format with the lower left corner (the origin) of the drawing, and centers all items in the drawing on the new sheet in the locations that correspond to their positions on the original sheet. If necessary, it adjusts the drawing scale, maintaining relative distances between items. You can draw lines, arcs, splines, etc. on a drawing format to represent the border and title block. Even though the option to draw geometry is called "sketch," you need to be aware that the **Format** option does not support the true **Sketcher**. No allowance is made for constraining the curves and regenerating them to a dimension-driven size. You must draw the curves exactly as they will appear on the format. To create a sketched format (Fig. 16.6):

1. Choose **File ⇒ New ⇒ Format ⇒** (name) **⇒ Variable ⇒ OK**
2. Sketch the boundary and the title block using **Sketch ⇒ Line** (etc.). File and save the format: **File ⇒ Save ⇒ ✔**

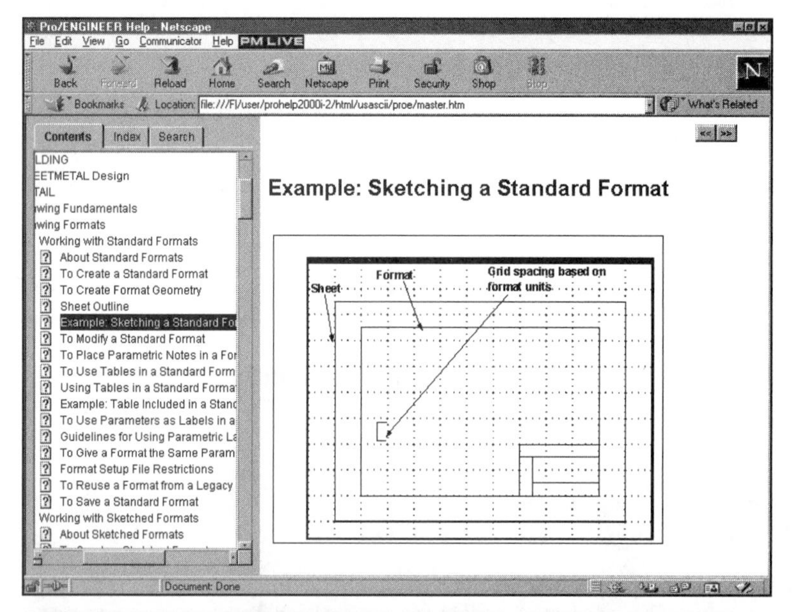

Figure 16.6
Online Documentation, Sketching Formats

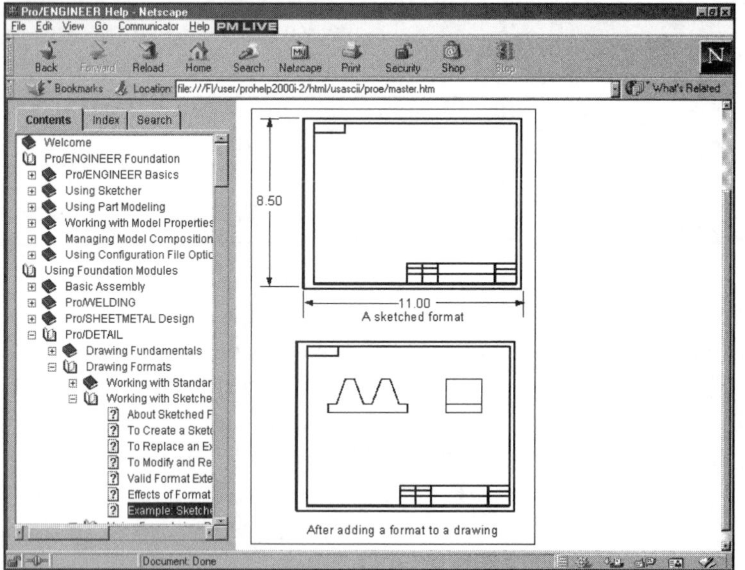

Figure 16.7
Online Documentation, Adding Views to a Sketched Format

Views

A wide variety of views (see Fig. 16.7) can be derived from the parametric model. Among the most common are projection views. Pro/E creates projection views by looking to the left of, to the right of, above, and below the picked view location to determine the orientation of a projection view (Fig. 16.8). When conflicting view orientations are found, you are prompted to select the view that will be the parent view. A view will then be constructed from the selected view.

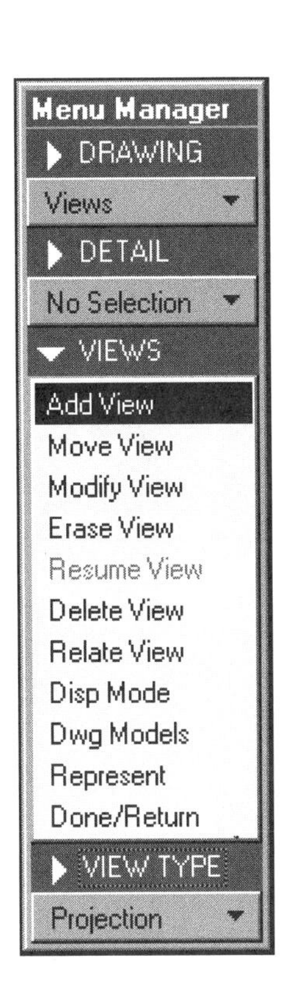

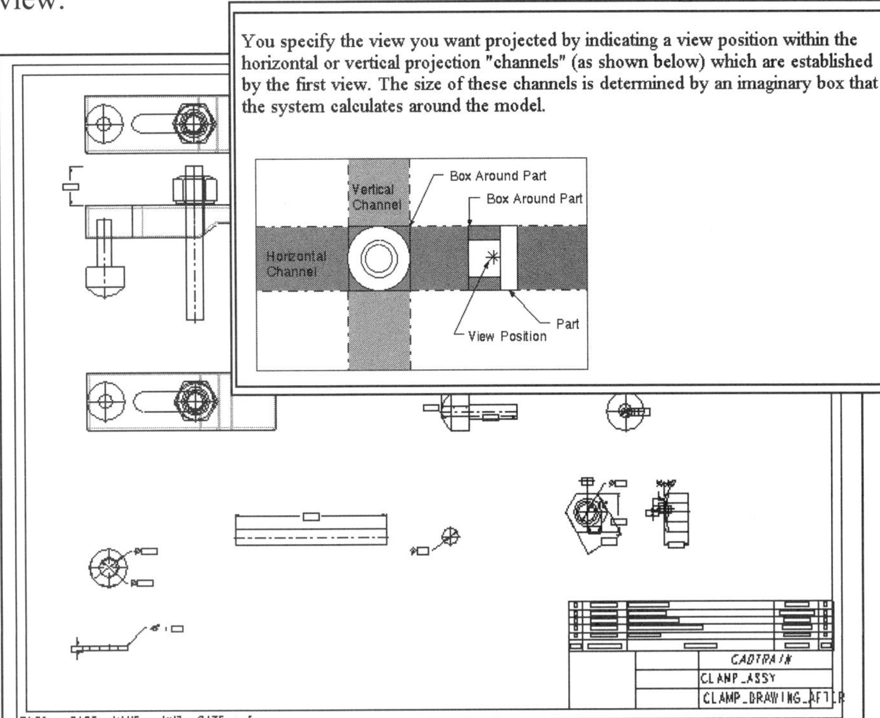

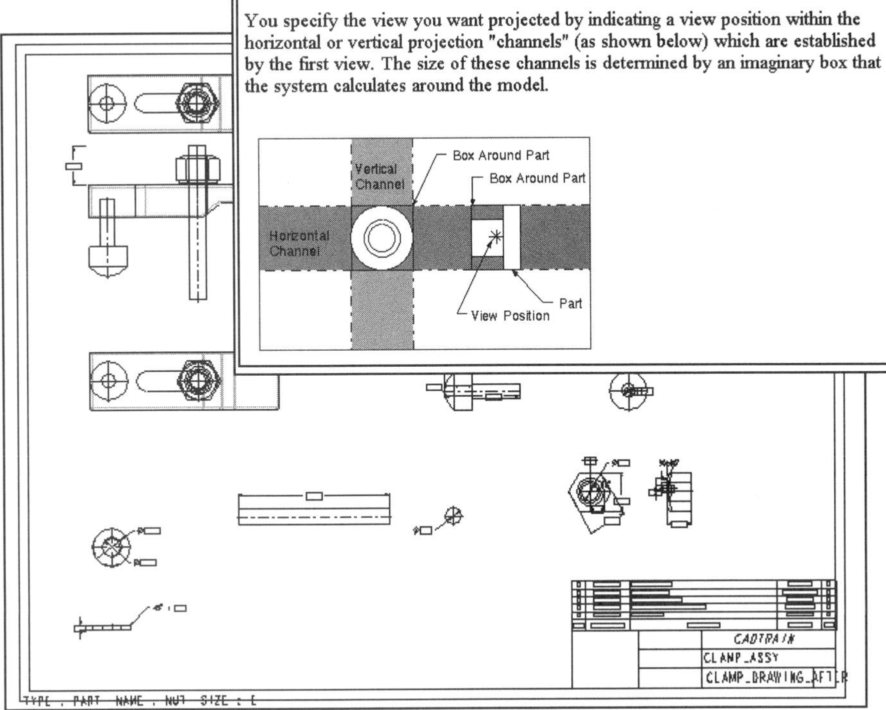

Figure 16.8
COAch for Pro/E, Creating a Drawing (Detail Views)

At the time when they are created, projection, auxiliary, detailed, and revolved views have the same representation and explosion offsets, if any, as their parent views. From that time onward, each view can be simplified, be restored, and have its explosion distance modified without affecting the parent view. The only exception to this is detailed views, which will always be displayed with the same explosion distances and geometry as their parent views. Here are some view types that are available:

Projection Creates a view that is developed from another view by projecting the geometry along a horizontal or vertical direction of viewing (orthographic projection). The projection type is specified by you in the drawing setup file and can be based on third-angle (default) or first-angle rules.

Auxiliary Creates a view that is developed from another view by projecting the geometry at right angles to a selected surface or along an axis. The surface selected from the parent view must be perpendicular to the plane of the screen (Fig. 16.9).

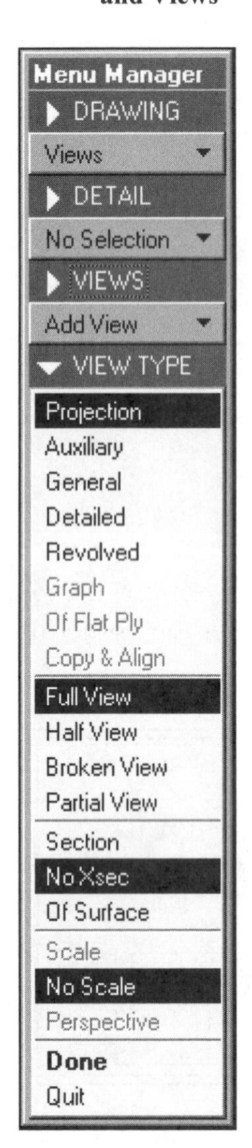

General Creates a view with no particular orientation or relationship to other views in the drawing. The model must first be oriented to the desired view orientation established by you.
Detailed Details a portion of the model appearing in another view. Its orientation is the same as that of the view it is created from, but its scale may be different so that the portion of the model being detailed can be better visualized.
Revolved Creates a planar area cross section from an existing view; the section is revolved **90°** around the cutting plane projection and offset along its length. A revolved view may be full or partial, exploded or unexploded.

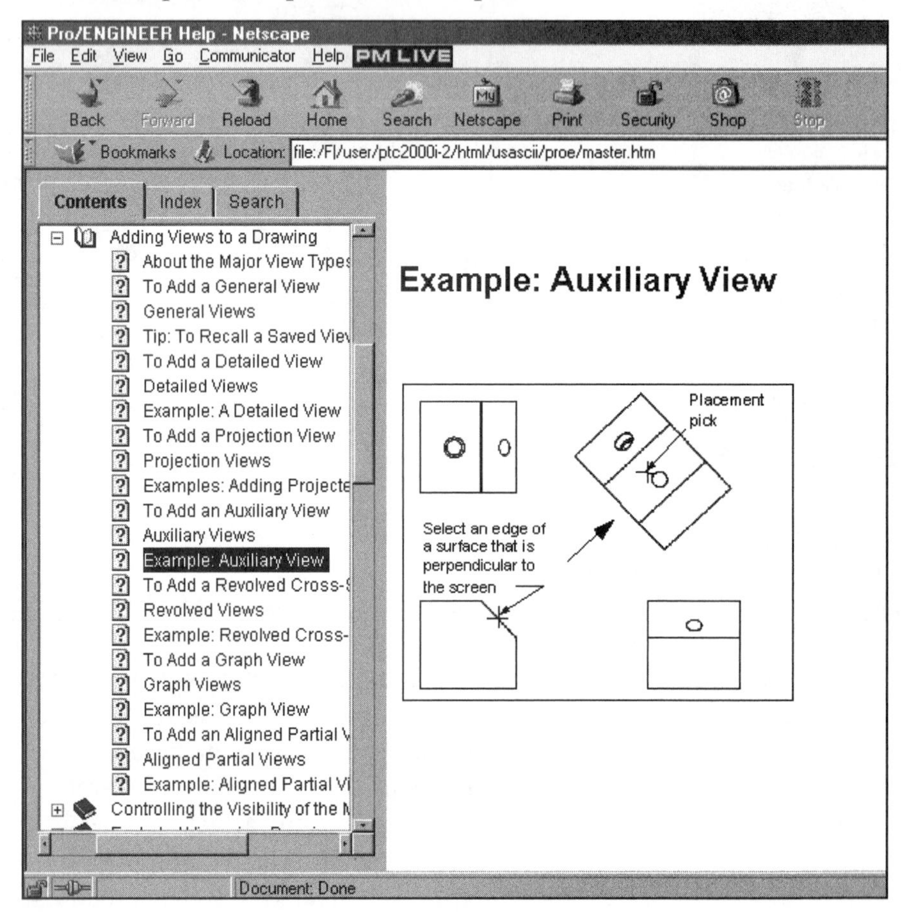

Figure 16.9
Online Documentation, Auxiliary Views

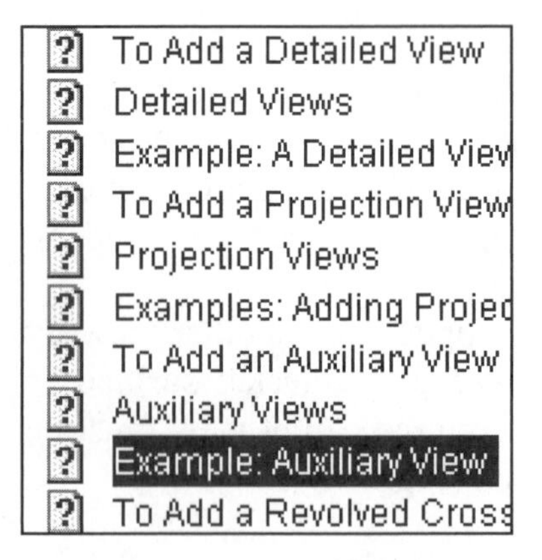

The view options that determine how much of the model is visible in the view are:

Full View Shows the model in its entirety.
Half View Removes a portion of the model from the view on one side of a cutting plane.
Broken View Removes a portion of the model from between two selected points and closes the remaining two portions together within a specified distance.
Partial View Displays a portion of the model in a view within a closed boundary. The geometry appearing within the boundary is displayed; the geometry outside of it is removed.

Figure 16.10
Section Views

The options that determine whether the view is of a single surface or has a cross section are:

Section Displays an existing cross section of the view if the view orientation is such that the cross-sectional plane is parallel to the screen (Figs. 16.10 and 16.11).
No Xsec Indicates that no cross section is to be displayed.
Of Surface Displays a selected surface of a model in the view. The single-surface view can be of any view type except detailed.

The options that determine whether the view is scaled are:

Scale Allows you to create a view with an individual scale shown under the view. When a view is being created, Pro/E will prompt you for the scale value. This value can be modified later. General and detailed views can be scaled.
No Scale A view will be scaled automatically using a pre-defined scale value that will be shown in the lower left corner of the screen as "SCALE."
Perspective Creates a perspective general view.

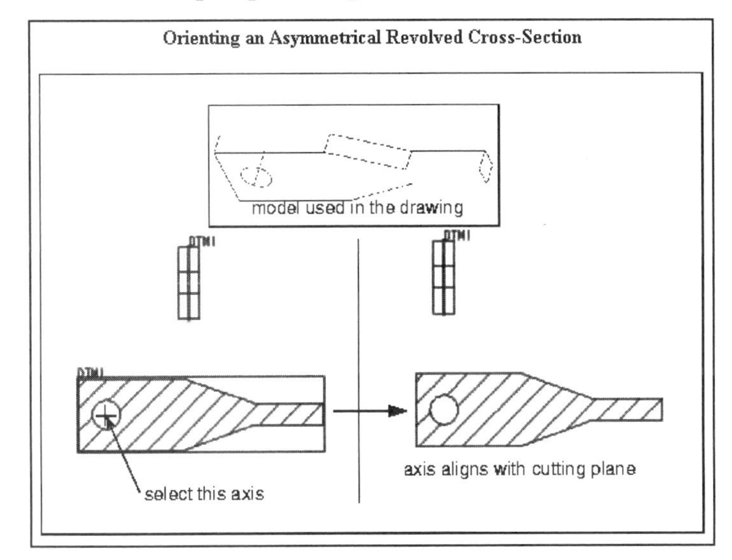

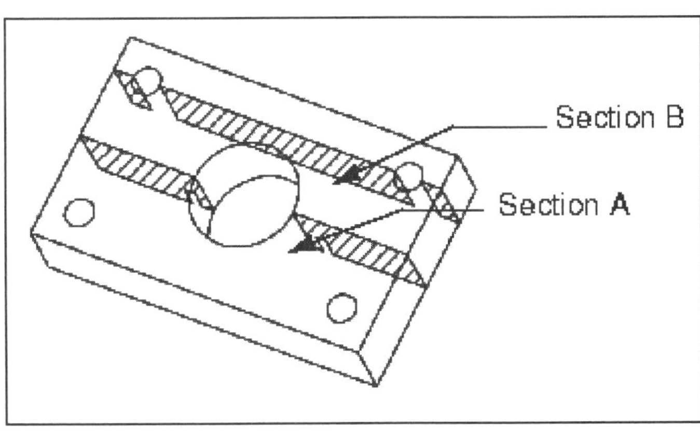

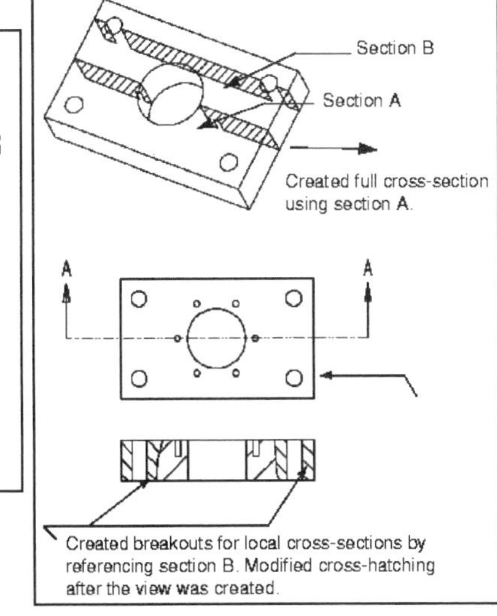

Figure 16.11
Sections

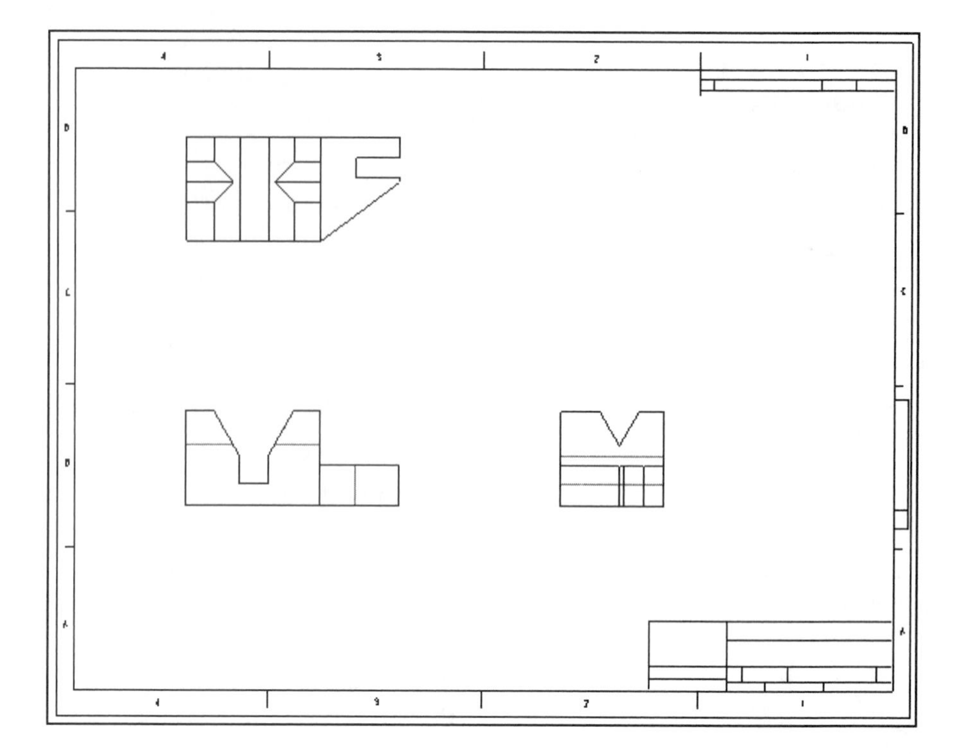

Figure 16.12
Base Angle Drawing on an ANSI/ASME Standard Drawing Format

Base Angle Drawing

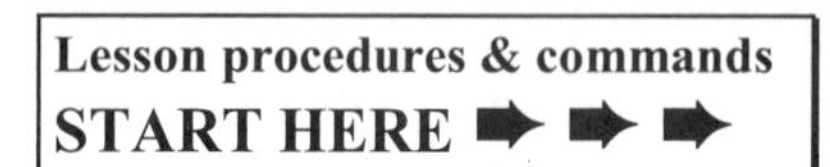

Pro/E allows you to create your own drawing formats to support your company or school standards. This includes the lines that make up the border and title block, as well as text. The text can also contain parameter information that refers to Pro/E-generated values, such as the drawing scale, the drawing name, and the model name.

In most cases, you will use the standard formats that come with Pro/E. A standard "**C**" size format has been added to the drawing of the Base Angle part in Figure 16.12.

When you add a format to your drawing, Pro/E can write the appropriate notes on the basis of information stored or defined in the model you use.

You can create two basic types of formats: a standard format, which is *locked* to a specific drawing size, or a sketched format, which is parametrically linked to the size of the drawing and thus changes as the drawing size changes.

This lesson deals with altering an existing standard format to match your format requirements. After you master the techniques required to make this type of format, you can easily extend these principles to using the Sketcher-like commands to draw the border outlines (to make a sketched parametric format).

Pro/E provides a subset of the 2D Drafting functionality to allow you to draw lines, arcs, splines, and so on to define a drawing border. The **Format** function also supports a complete range of text functions to create notes on the drawing that serve as title block information or standard notes.

Much of this lesson has been adapted from CADTRAIN's COAch for Pro/ENGINEER, with the company's permission.

Altering Pro/E Formats from the Format Library

One of the easiest ways to create a drawing format is to use an existing format from the Pro/E format library, save it in your directory under a different name, and alter it as required. This method will keep the sheet an ANSI/ASME standard size and format with a standard title block. You can add parameters to the sheet so as to display the appropriate information in the title block and at other locations on the sheet where appropriate. Formats have an **.frm** extension and are read-only files.

Retrieve and save (for later modification) a "**C**" size format using the following commands (Fig. 16.13):

File ⇒ Open ⇒ ▼ ⇒ System Formats ⇒ c.frm ⇒ Open ⇒ File ⇒ Save As (type a unique name for your format: **FORMAT_C**) **⇒ OK ⇒ File ⇒ Erase ⇒ Current ⇒ Yes ⇒ File ⇒ Open ⇒** Pick **FORMAT_C ⇒ Open**

You can make any number of personal formats using this method. Adding parameters to the sheet is the next step in creating your own library of formats that will automatically display the part's name, scale, tolerances, and so on.

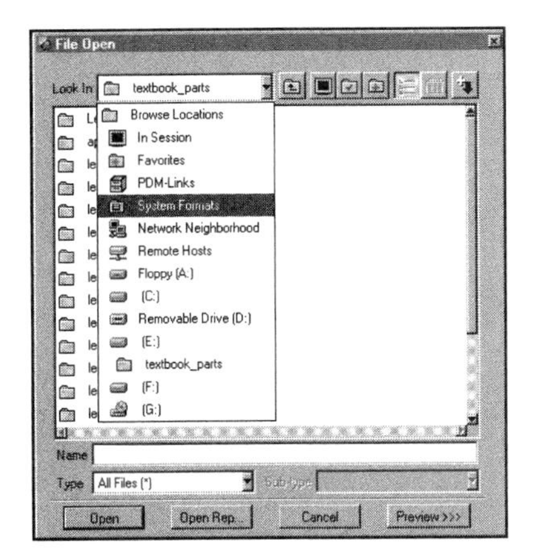

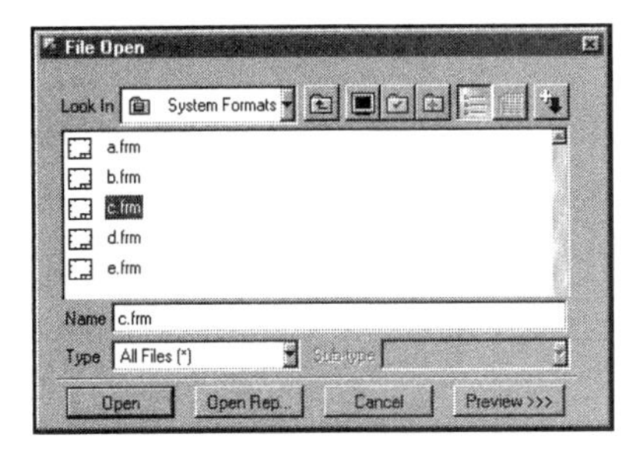

Figure 16.13 FORMAT_C

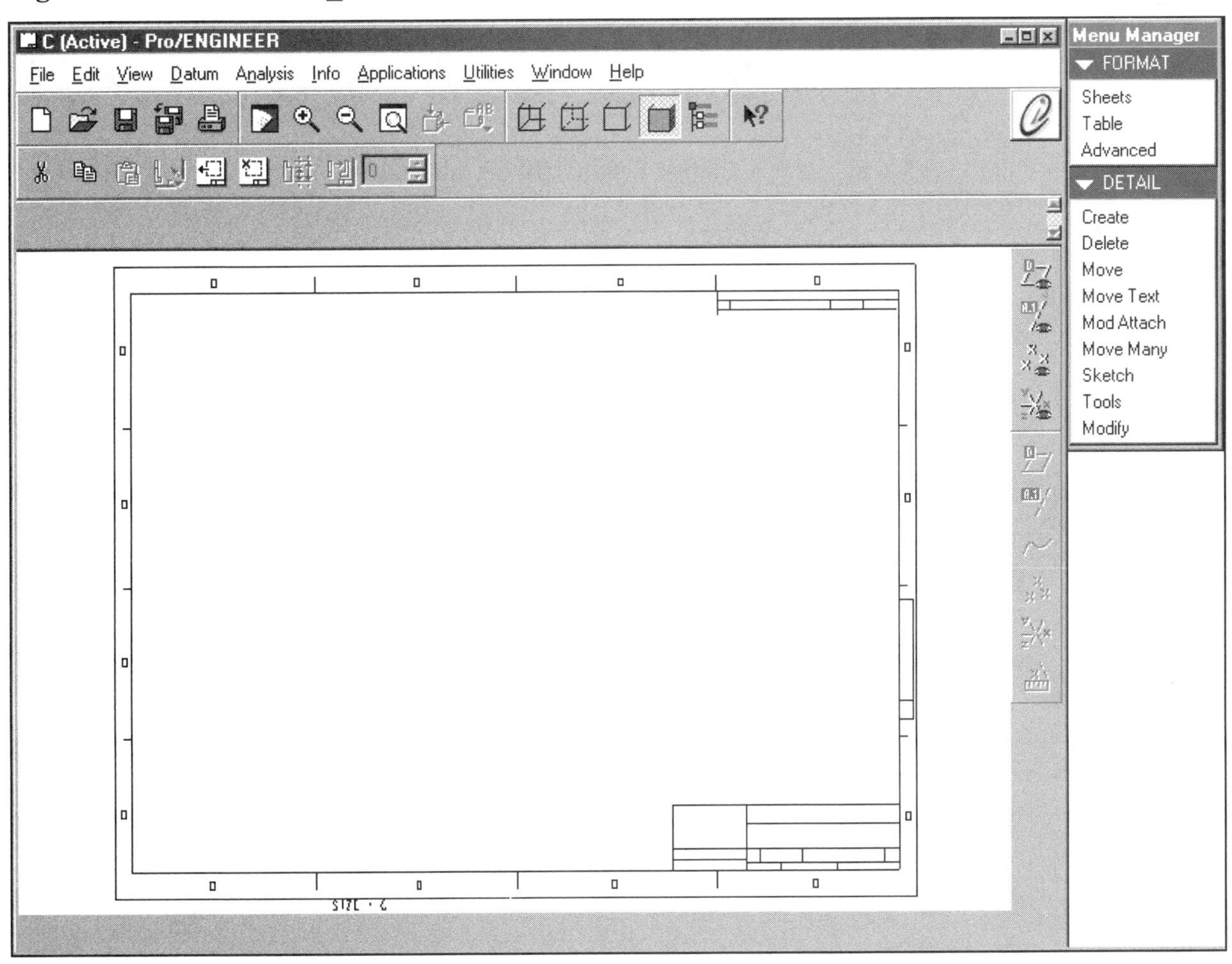

NOTE
If you do not wish to create your own format, (or if you have already done this part of the exercise using CADTRAIN), but instead want to use one of the standard formats and alter it, follow the steps on the previous page and then skip to page L16-18, **Adding Text to the Title Block.**

Sketching the Format (Using Draft Geometry)

This part of the lesson leads you through creating a "**C**" size, inches format (Fig. 16.14). It is the only coverage of using draft entities in this text.

You can substitute metric values to create the equivalent-size metric format if desired.

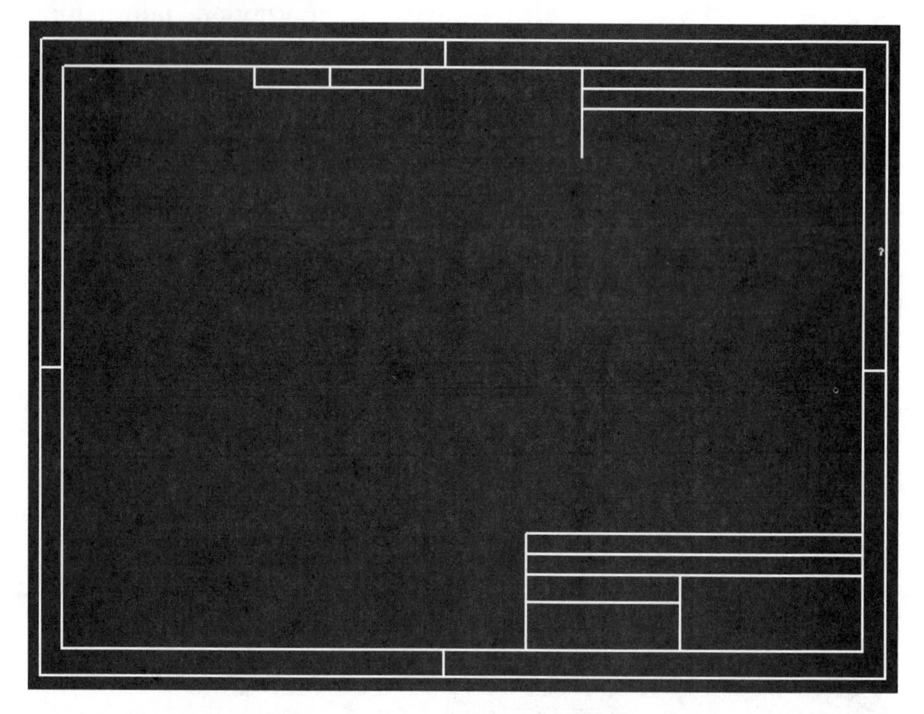

Figure 16.14
CADTRAIN, Production Drawing (Formats)

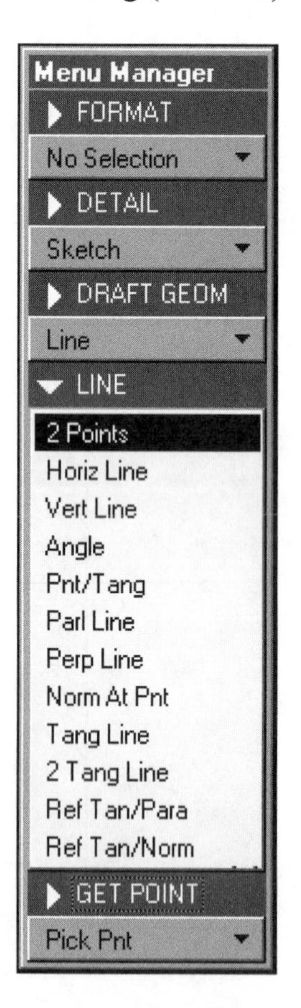

You can draw lines, arcs, splines, and so on using a drawing format to represent the border and title block. Even though the option to draw geometry is called "sketch," you need to be aware that the **Format** mode does not support the true Sketcher. No allowance is made for constraining the curves (entities) and regenerating them to a dimension-driven size. You **must** draw the curves exactly as they will appear on the format.

You use the basic **2D Drafting** functionality of Pro/E to create the geometry you need. Therefore, you do not have implied constraints such as horizontal or vertical lines. You also do not have implied endpoint connection (even if the curves are close to each other, they are not connected).

You can, however, draw curves using the **On Entity** option, which will force picks near an endpoint to snap to that endpoint. You can also use the **Start Chain** option to make sure that endpoints connect.

You use either the **Horizontal** or the **Vertical** option to *force* lines to be created correctly for the format border. You can also **Trim** curves after you have drawn them. When you **Trim** a curve, you pick the bounding curve first, then the curve you want to trim. You pick a new bounding curve by choosing the **Bound** option again.

When you pick the curve to trim, you should always pick on the portion of the curve you want to keep (Fig. 16.15).

? Pro/HELP

Remember to use the help available on Pro/E by highlighting a command and pressing the right mouse button.

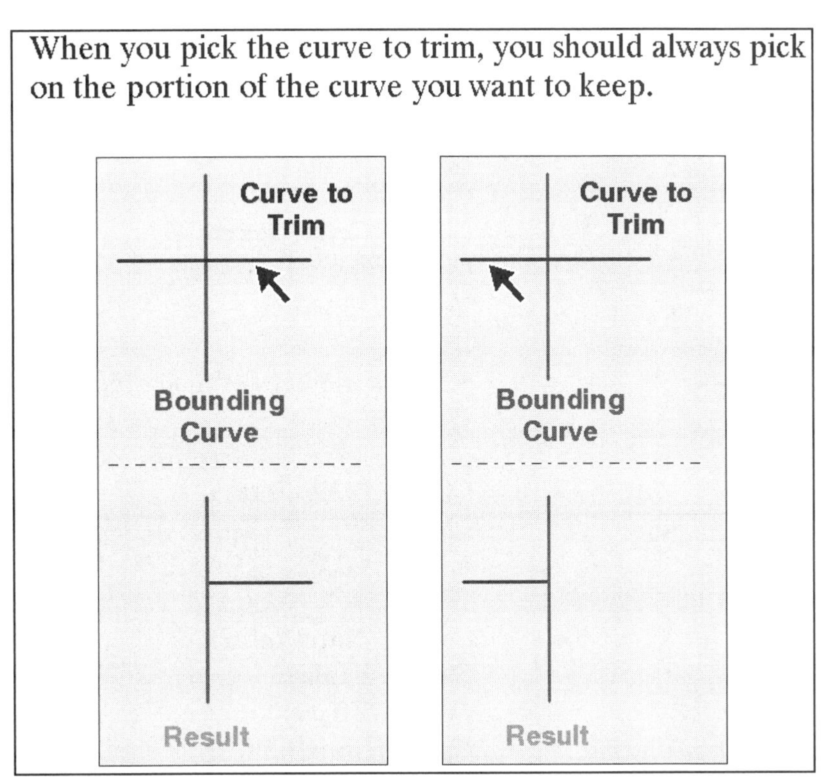

Figure 16.15
CADTRAIN, Production Drawing (Trimming)

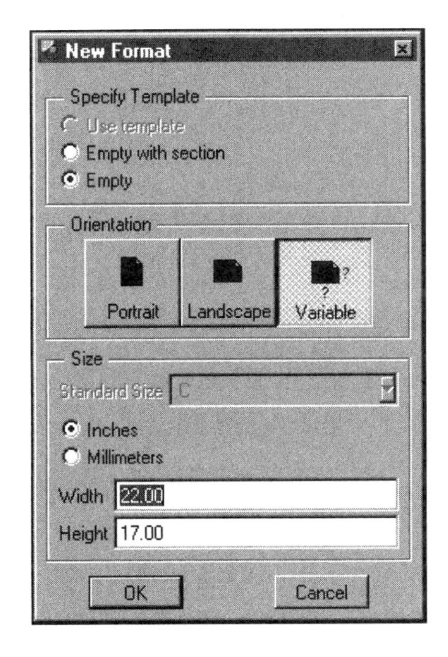

Draw the lines of a standard **"C"** size format using the following commands (Fig. 16.16):

1. Create a new format file.

Choose **File ⇒ New ⇒ Format**
Type **my_format_** and ***your initials*** **⇒ OK**
Choose **C** from the list of sizes or
Choose **Variable ⇒ Inches ⇒ 22.00 ⇒ 17.00 ⇒ OK**

HINT

The box that is displayed represents the edge of the paper. Many companies allow for about **.500** inch inside this box for the border of the format.

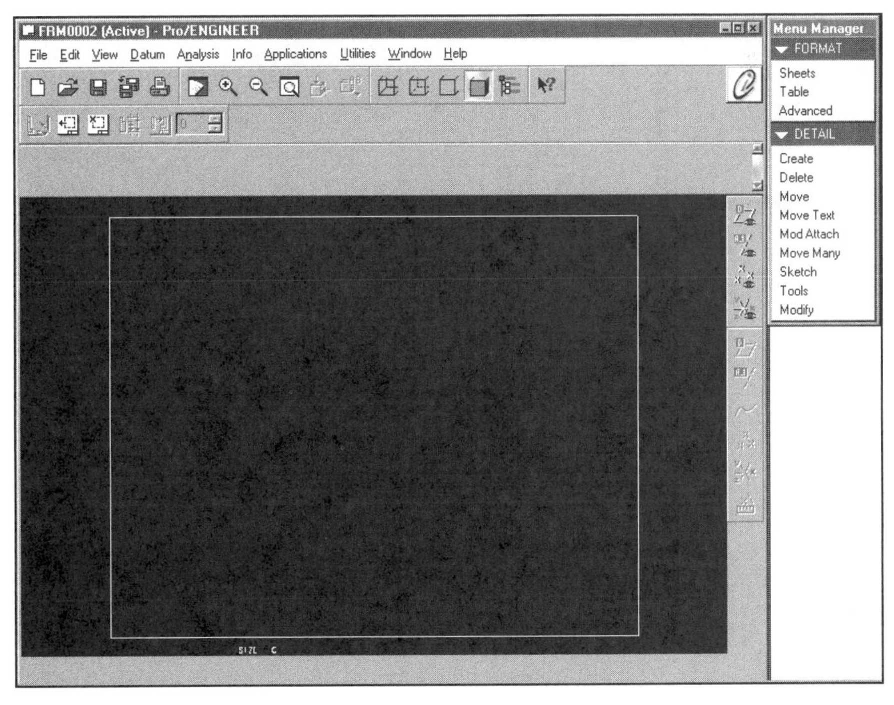

Figure 16.16
"C" Size Drawing Format (Edge of the Paper)

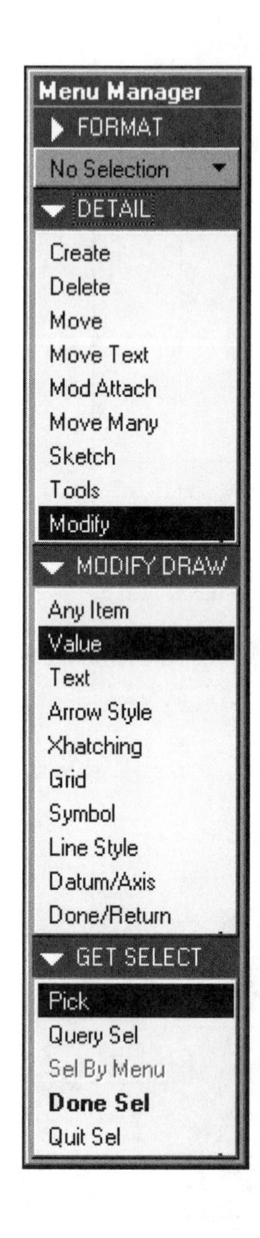

2. Turn the **GRID** on and change the spacing to **.50**.

Choose **Modify** from the DETAIL menu
Choose **Grid**
Choose **Grid On**
Choose **Grid Params**
Choose **X&Y Spacing**
Type **.5** ⇒ ✔

3. Draw the lines of the border (Fig. 16.17).

Choose **Done/Return**
Choose **Done/Return**
Choose Zoom Out from the Toolbar to Zoom Out
Choose **Done/Return**
Choose **Tools** (under the DETAIL menu)
Choose **Offset**
Choose **Ent Chain**
Select the four lines on the screen
Choose **Done Sel**
Type **-.5** ⇒ ✔ (the boundary is zero with negative defined as inward)
Choose

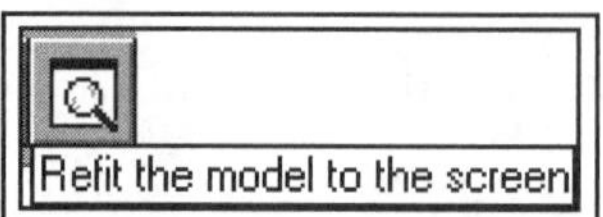

from the Toolbar to refit the format to the screen

Figure 16.17
CADTRAIN, Production Drawing (Four Lines of the Border)

4. Create the lines for the title block.

Select the second-to-bottom horizontal line using the **left** mouse button (Fig. 16.18)
Click the **middle** mouse button to finish selection
Type **-.5** ⇒ ✔
Select the line you just created with the **left** mouse button
Click the **middle** mouse button
Type **-1.5** ⇒ ✔
Select the second-to-right vertical line as shown in Figure 16.18
Click the **middle** mouse button
Type **-1** ⇒ ✔
Select the line you just created
Click the **middle** mouse button
Type **-3** ⇒ ✔
Select the line you just created
Click the **middle** mouse button
Type **-2** ⇒ ✔

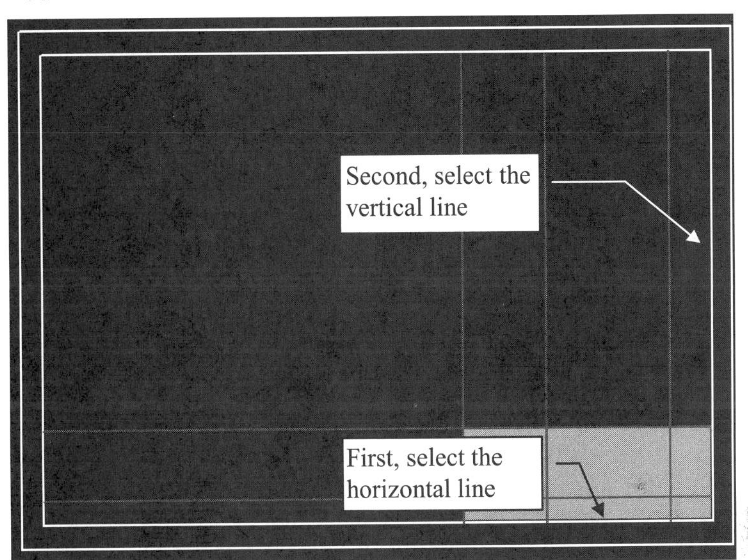

Figure 16.18
CADTRAIN, Production Drawing (Create Lines for the Title Block)

5. **Trim** the lines to form the finished title block (Fig. 16.19).

Figure 16.19
CADTRAIN, Production Drawing (Finished Title Block)

6. Draw the lines for the drawing number that is inverted at the top left corner of the format.

Choose **Offset** ⇒ **Ent Chain**
Select second the left vertical line (Fig. 16.20)
Click the **middle** mouse button
Type **-1** ⇒ ✔
Select the line you just created
Click the **middle** mouse button
Type **-3** ⇒ ✔
Select the horizontal line (Fig. 16.20)
Click the **middle** mouse button
Type **-.5** ⇒ ✔

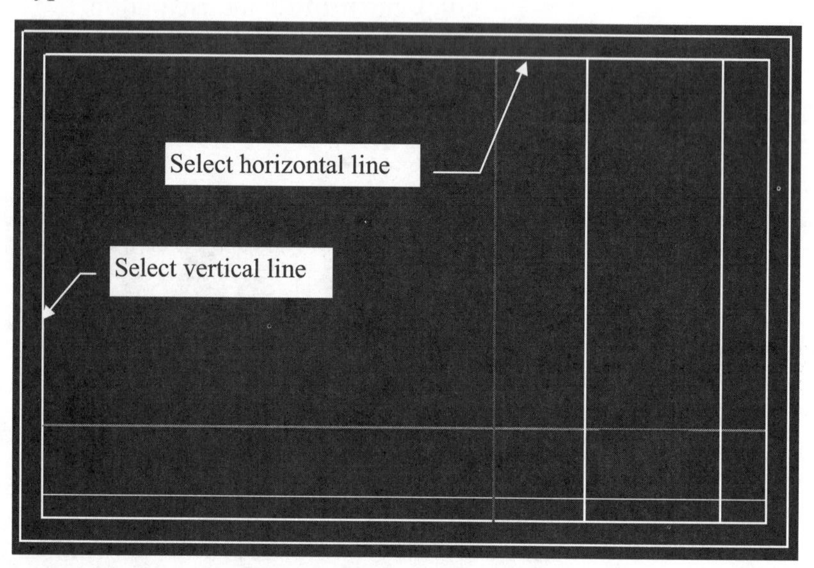

Figure 16.20
CADTRAIN, Production Drawing

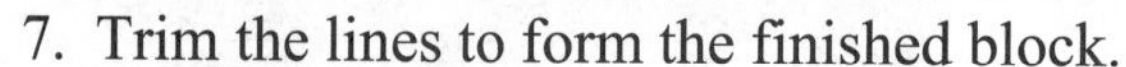

7. Trim the lines to form the finished block.

Choose **Trim**
Select the line as shown in Figure 16.21 as the bounding curve
Select the *two* curves to be trimmed as shown in Figure 16.22

Menu Manager
FORMAT
No Selection
DETAIL
Tools
TOOLS
Trim
DRAFT TRIM
Bound
Length
Increm
Corner
GET SELECT
Pick

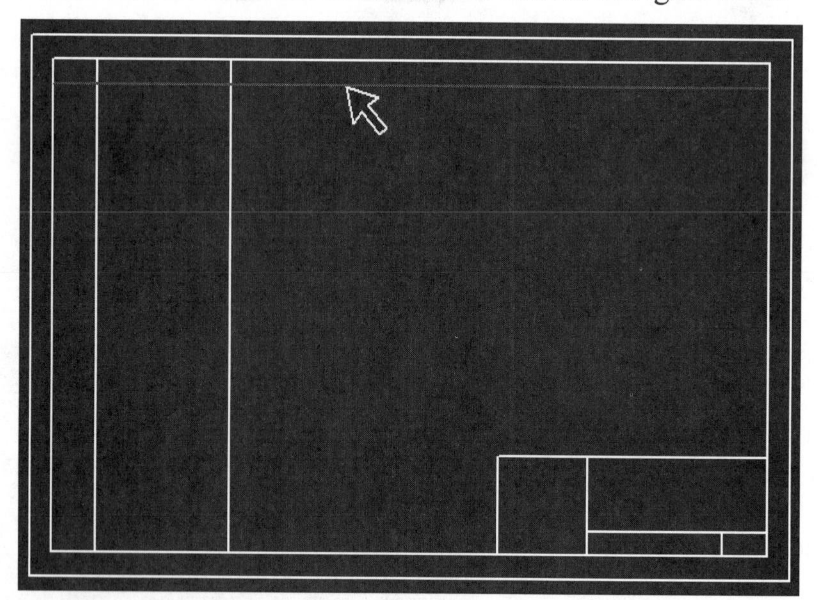

Figure 16.21
CADTRAIN, Production Drawing (Bounding Curve)

Choose **Bound**
Select the rightmost vertical line of the inverted drawing number block (Figs. 16.22 and 16.23) as the bounding curve
Select the curve to trim as shown in Figure 16.23
Choose **Done/Return**
Choose **View ⇒ Repaint ⇒ File ⇒ Save ⇒ ✔**

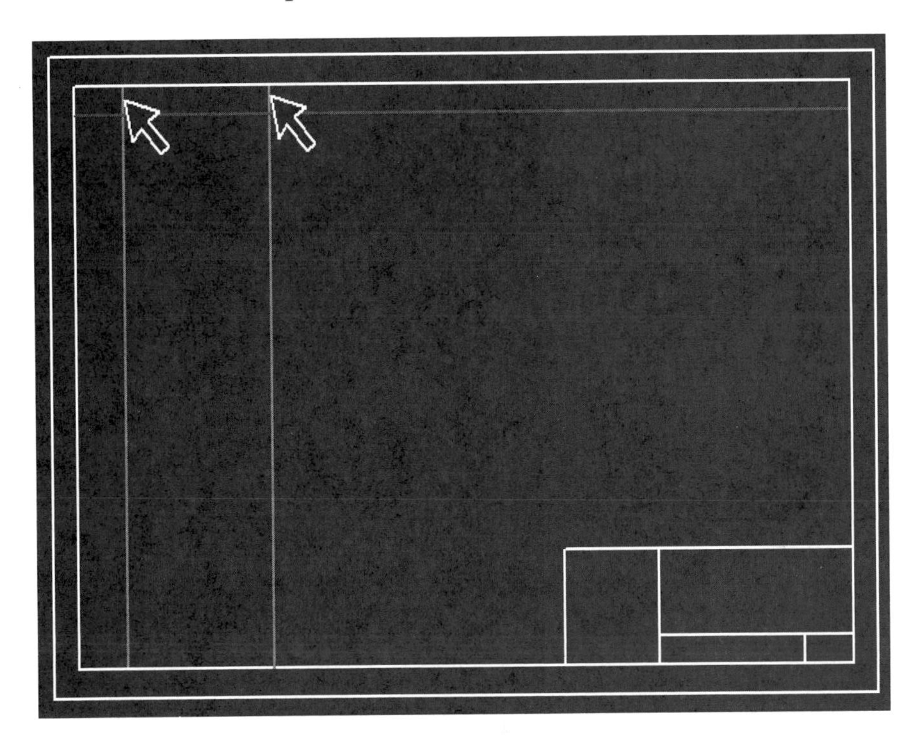

Figure 16.22
CADTRAIN, Production Drawing (Trimming the Block)

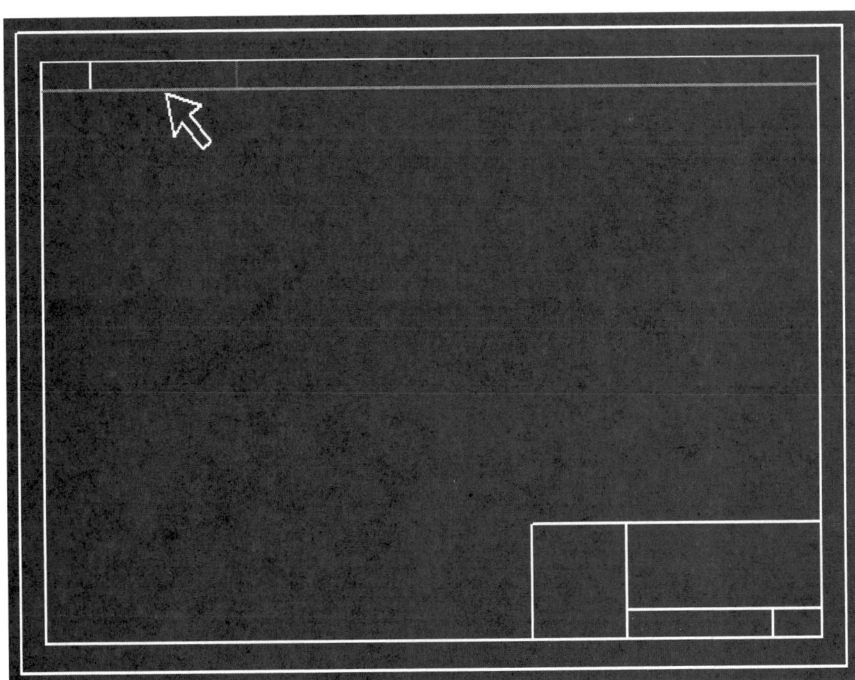

Figure 16.23
CADTRAIN, Production Drawing (Trimming the Block, Final Trim Curve)

Next, you will add *parameters* to the sheet.

Adding Text to the Title Block (Border and Format)

For every drawing, the title block is filled in with standard data or data unique to that design. **Parameters** for the title block information are created and saved in a format in **Format** mode. *Parameter text* will automatically reflect the design when the format is added to a drawing (in **Drawing** mode). Examples of *parameter text* include tolerance standards and sheet numbering (Fig. 16.24). Besides parameter text, *plain text,* which does not change when the format is added, can also be added in the format. The company name might be an example of *plain text* on a format (Fig. 16.24).

User-defined parameters can be included in the format, but these user-defined parameters must also be embedded in the part or assembly model. After the model is inserted on the drawing, views made and placed, and the format added to the drawing, then the user-defined parameters will be read by the format parameters and the information displayed on the drawing. An example of a user-defined parameter is the DESCRIPTION of a component used in a bill of material (BOM) on an assembly drawing.

NOTE
If user-defined parameters are not defined in the model (part or assembly), then the format will display the *parameter titles* created in the format mode, instead of converting the parameters to model values.

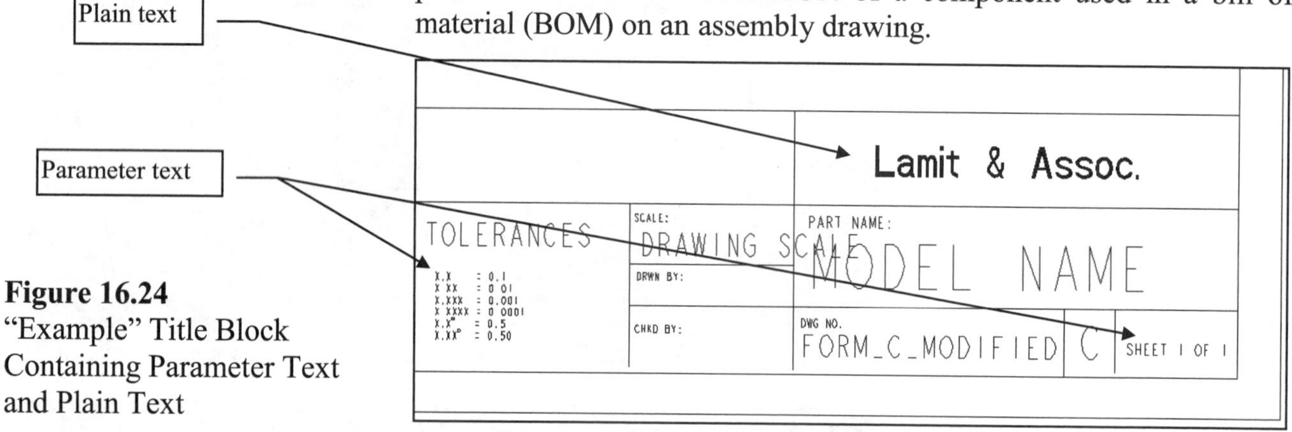

Figure 16.24
"Example" Title Block Containing Parameter Text and Plain Text

You can then create and place the notes almost anywhere on the format, because you will nearly always want to alter and move the parameter and plain text after they have been created.

A format has its own parameter **.dtl** file. The **.dtl** entries listed apply to a *format*. A *drawing* **.dtl** file will have these entries and more (see Lesson 19). You may want to establish a standard **.dtl** file for use only with formats. This is especially true if your standards call for text parameters (font size, arrow size, etc.) that are different from those used when you detail a drawing.

You may also want to create items such as your company logo and any other special symbols you plan to use (e.g., projection angle and inspection symbols) before you create your formats. You can then add these symbols without having to redraw them for each format size (**A**, **B**, **C**, etc.).

You should also be aware that you can make copies of text as you create the format. However, when you make a copy, any parameters in the copy become simple text. This means that the copy will *not* utilize the parametric value.

NOTE
The **format** will have a **.dtl** file associated with it, and the **drawing** will have a different **.dtl** file associated with it. *They are separate .dtl files and have nothing to do with each other.* When you activate a drawing and then add a format, the **.dtl** for the format controls the font etc., for the format only. The drawing **.dtl** file still needs to be established.

As an example, if you changed the **.dtl** file in the format to have a filled font, the format will show as filled font when used on a drawing. The drawing dimensions and text will *not* show as filled because it uses a separate **.dtl** file!

Create the notes you need on the drawing (Fig. 16.25). The notes initially will all have the same text parameters; you can edit them later. You also do not need to place the notes anywhere special. You will find it easier to move them into place later in the process.

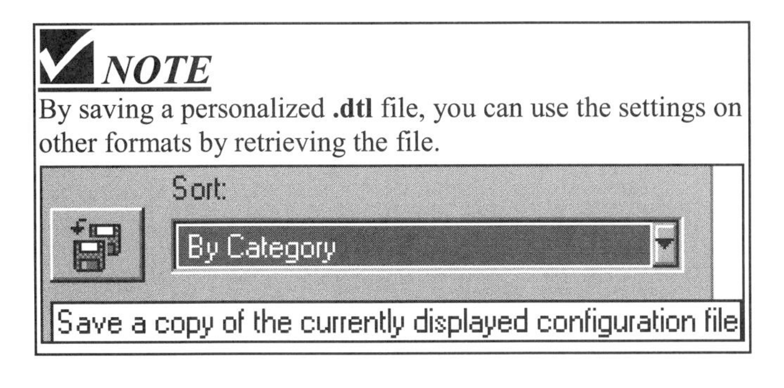

Figure 16.25
Creating Drawing Format Notes

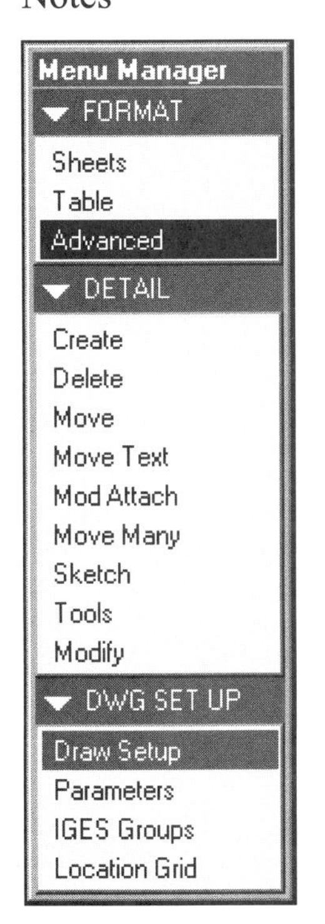

You will be working on one of the formats you previously created or one that you saved under a different name from the Pro/E format directory. When you save this file, it will become a format (**.frm**) and can be used to replace a blank format for a project or as the original format retrieved from your directory.

Retrieve the **cadtrain.dtl** file if you have CADTRAIN's COAch for Pro/ENGINEER installed on your system, or use a **.dtl** file provided by your instructor. You may also alter or create your own **.dtl** file (Fig. 16.26) by choosing:

Advanced ⇒ Draw Setup ⇒ Option ***draw_arrow_style filled, default_font filled*** **⇒ Add/Change ⇒ Apply ⇒ Save** icon **⇒ Close**

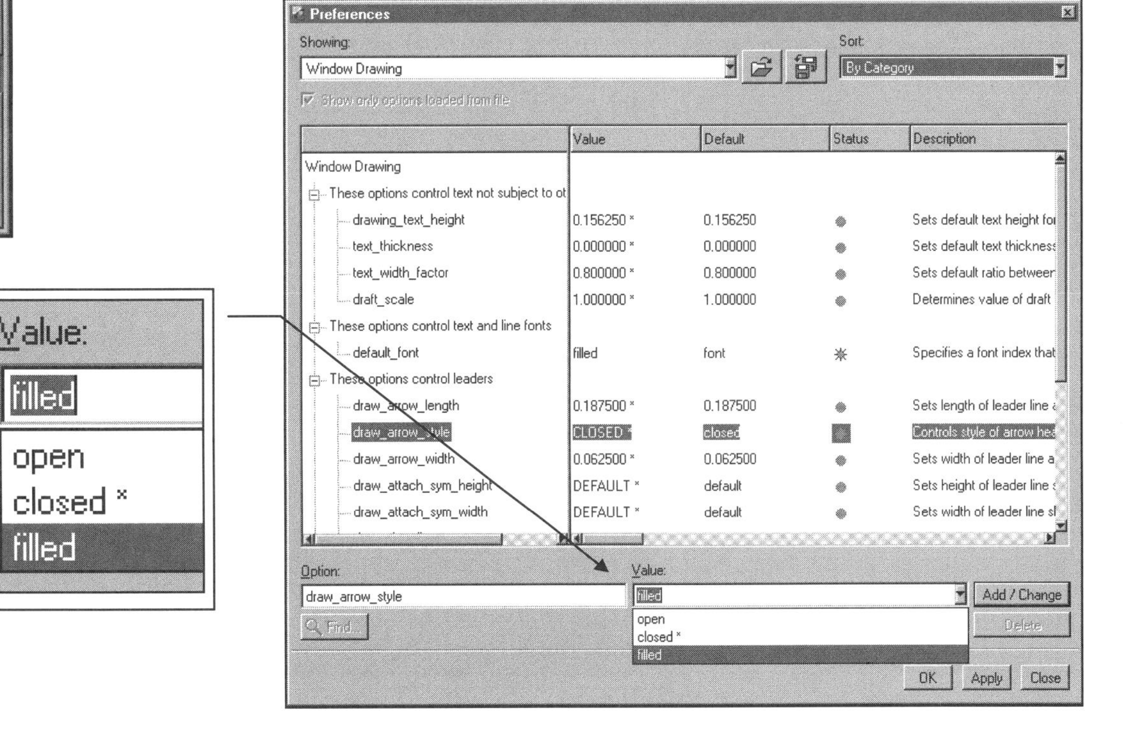

Figure 16.26
.dtl File

1. Create the plain text notes (Fig. 16.27).

NOTE

If you have CADTRAIN, retrieve the **cadtrain.dtl** file: **Advanced ⇒ Draw Setup**

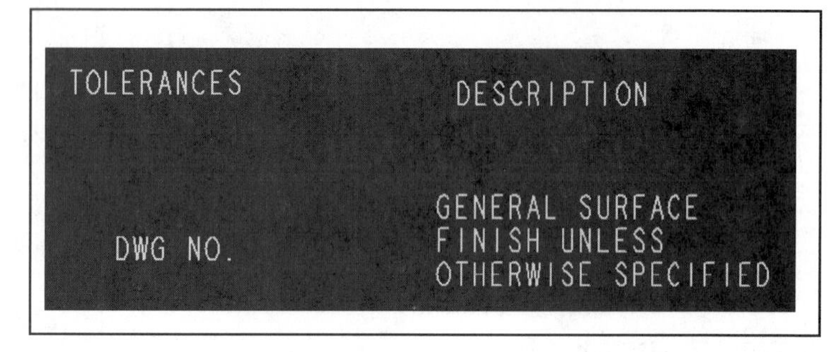

Figure 16.27
Plain Text Notes

Press your ***Caps Lock*** key to turn it on.

Choose **Create** from the DETAIL menu
Choose **Note** (accept all defaults)
Choose **Make Note**
Indicate the note origin anywhere on the screen (pick once with the *left* mouse button)
Type **TOLERANCES** and press **enter** twice
Choose **Make Note**
Indicate the note origin anywhere on the screen
Type **DESCRIPTION** and press **enter** twice
Choose **Make Note**
Indicate the note origin anywhere on the screen
Type **DWG NO.** and press **enter** twice
Choose **Make Note**
Indicate the note origin anywhere on the screen
Type **GENERAL SURFACE** and press **enter**
FINISH UNLESS and press **enter**
OTHERWISE SPECIFIED and press **enter** twice

NOTE

A variety of Pro/E standard drawing and format parametric notes are available, including:

&dwg_name
&todays_date
&scale
&type (drawing model type)
&format (format size)
&linear_tol_0_0 through ***&linear_tol_0_000000***
&angular_tol_0_0 through ***&angular_tol_0_000000***
¤t_sheet
&total_sheets
&det_scale (detail scale)

HINT

Set your font and arrow style to filled:

default_font	***filled***
draw_arrow_style	***filled***

2. Create the tolerances note (Fig. 16.28).

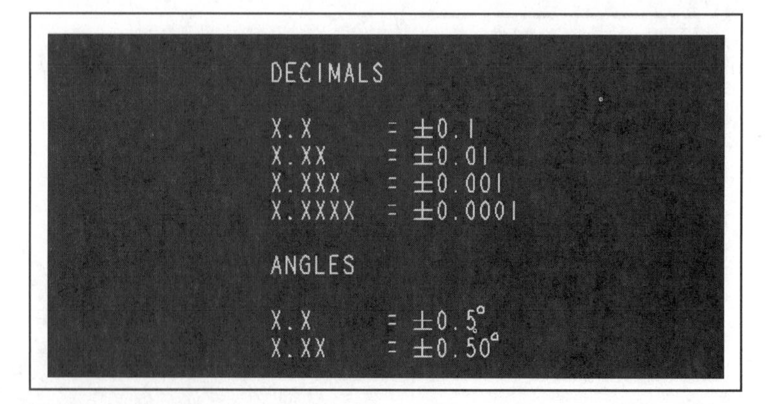

Figure 16.28
Tolerances Note

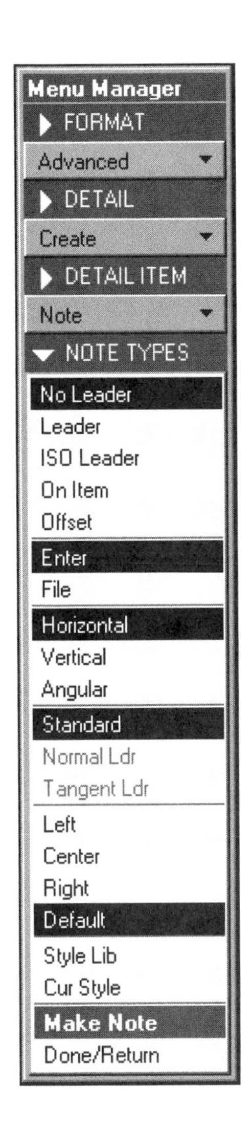

Choose **Make Note**
Indicate the note origin anywhere on the screen
Type **DECIMALS ⇒ enter**

You now need a *blank* line. *If* you just press **enter** again, Pro/E will assume that you are finished making the note, which you are not.

Press the ***Spacebar*** and then press **enter**

The ***Spacebar*** counts as a character on the current line and makes Pro/E think that you have entered actual text. Therefore, **enter** is not taken to mean that you are finished making the note.

Now you must input the parameterized variables.

Make sure ***Caps Lock*** is off

This is very important: *You cannot enter the name of a Pro/E parameter (such as* ***&linear_tol_0_0****) in capital letters.* Pro/E will not recognize it as a parameter.

Type **X.X**
Press the ***Spacebar*** five times
Type = ***Spacebar***
Choose the ***plus and minus*** symbol ± from the palette
Type **&linear_tol_0_0 ⇒ enter**
Type **X.XX**
Press the ***Spacebar*** four times
Type = ***Spacebar***
Choose the ***plus and minus*** symbol ± from the palette
Type **&linear_tol_0_00 ⇒ enter**
Type **X.XXX**
Press the ***Spacebar*** three times
Type = ***Spacebar***
Choose the ***plus and minus*** symbol ± from the palette
Type **&linear_tol_0_000 ⇒ enter**
Type **X.XXXX**
Press the ***Spacebar*** twice
Type = ***Spacebar***
Choose the **plus and minus** symbol ± from the palette
Type **&linear_tol_0_0000 ⇒ enter**

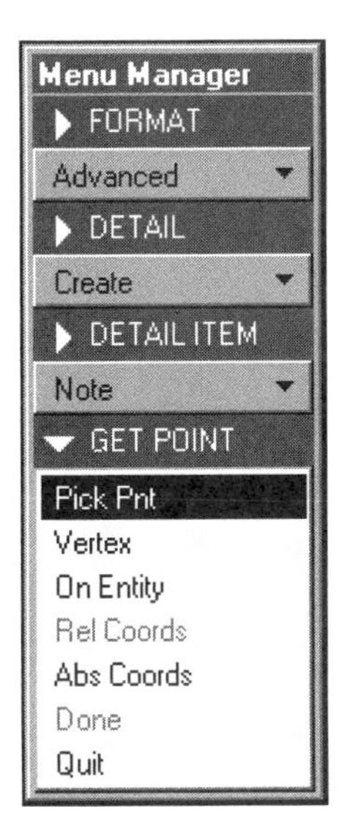

You now need another *blank* line. *If* you press **enter** again, Pro/E will assume that you are finished making the note, and you are not.

Press the ***Spacebar*** and then press **enter**
Type **ANGLES ⇒ enter**
Press the ***Spacebar*** and press **enter**
Type **X.X**
Press the ***Spacebar*** five times
Type = ***Spacebar***
Choose the ***plus and minus*** symbol ± from the palette
Type **&angular_tol_0_0** and press **enter**

Type **X.XX**
Press the ***Spacebar*** four times
Type = ***Spacebar***
Choose the **plus and minus** symbol ± from the palette
Type **&angular_tol_0_00** and press **enter** (and press **enter** again to finish creating the note)

3. Create the note that will display the model name (Fig. 16.29) in the description area in the title block. Because there is no model at the moment, the text of the name of the parameter is displayed. ***When this format is placed on a drawing and a view is added (thus adding a model), this text will reflect the name of the model.***

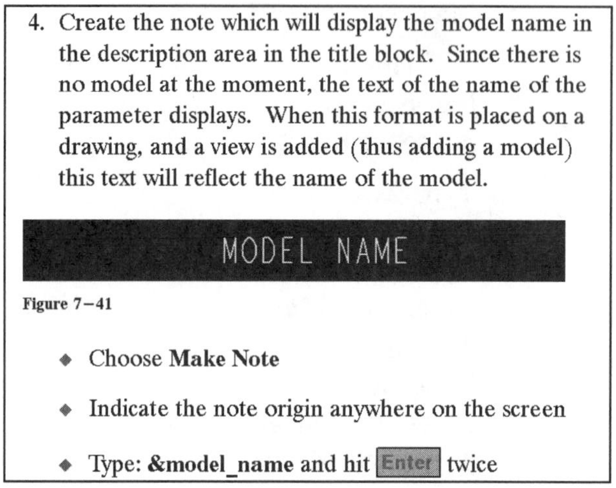

4. Create the note which will display the model name in the description area in the title block. Since there is no model at the moment, the text of the name of the parameter displays. When this format is placed on a drawing, and a view is added (thus adding a model) this text will reflect the name of the model.

MODEL NAME

Figure 7–41

- Choose **Make Note**
- Indicate the note origin anywhere on the screen
- Type: **&model_name** and hit Enter twice

Figure 16.29
Model Name

Choose **Make Note**
Indicate the note origin anywhere on the screen
Type **&model_name** and press **enter** twice

4. Create the note that will display the drawing name in the drawing number area in the title block (Fig. 16.30). Your version of this note will actually show the name of the format that is the current format name. When this format is placed on a drawing, the drawing name will replace this text.

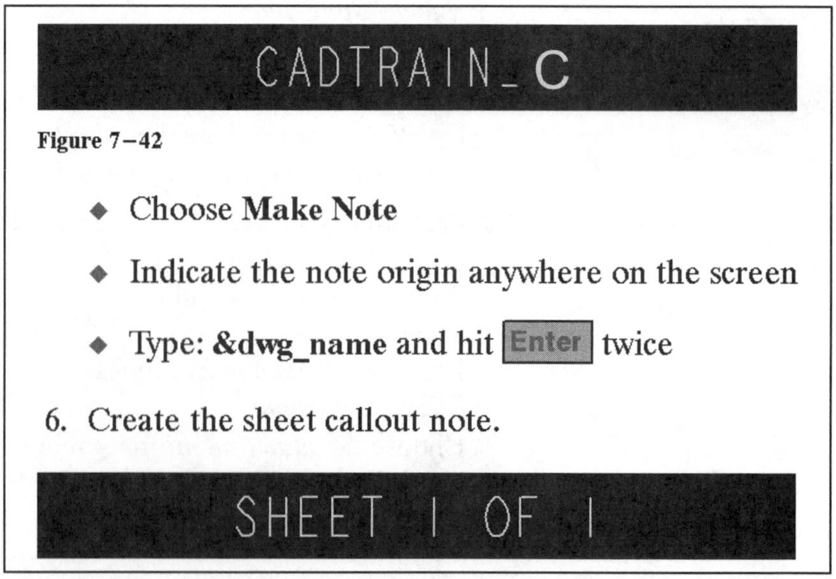

CADTRAIN_C

Figure 7–42

- Choose **Make Note**
- Indicate the note origin anywhere on the screen
- Type: **&dwg_name** and hit Enter twice

6. Create the sheet callout note.

SHEET 1 OF 1

Figure 16.30
Model Name and Sheet Callout

Choose **Make Note**
Indicate the note origin anywhere on the screen
Type **&dwg_name** and press **enter** twice

5. Create the sheet callout note (Fig. 16.30).

Choose **Make Note**
Indicate the note origin anywhere on the screen
Type **SHEET** (***Spacebar***) **¤t_sheet** (***Spacebar***) **OF** (***Spacebar***) **&total_sheets** and press **enter** twice

6. Create another copy of the note that will display the drawing name (Fig. 16.30). This one will be rotated and will be placed in the box in the upper left corner of the drawing (see Fig. 16.25).

Choose **Make Note**
Indicate the note origin anywhere on the screen
Type **&dwg_name** and press **enter** twice

7. Create another copy of the sheet callout note (see Fig. 16.30).

Choose **Make Note**
Indicate the note origin anywhere on the screen
Type **SHEET** (***Spacebar***) **¤t_sheet** (***Spacebar***) **OF** (***Spacebar***) **&total_sheets** and press **enter** twice

8. Create the note that will display the model scale

Choose **Make Note**
Indicate the note origin anywhere on the screen
Type **&scale** and press **enter** twice

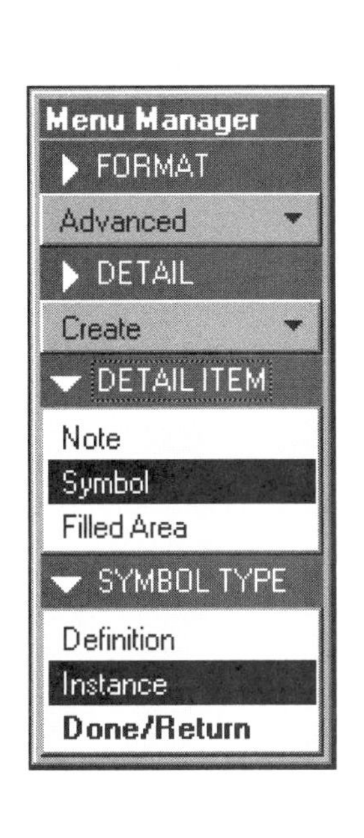

9. Add a finish symbol to go with the note you created earlier.

Choose **Done/Return**
Choose **Symbol**
Choose **Instance**

Pro/E displays the **Symbol Instance** dialog box (Fig. 16.31).

Figure 16.31
Symbol Instance Dialog Box

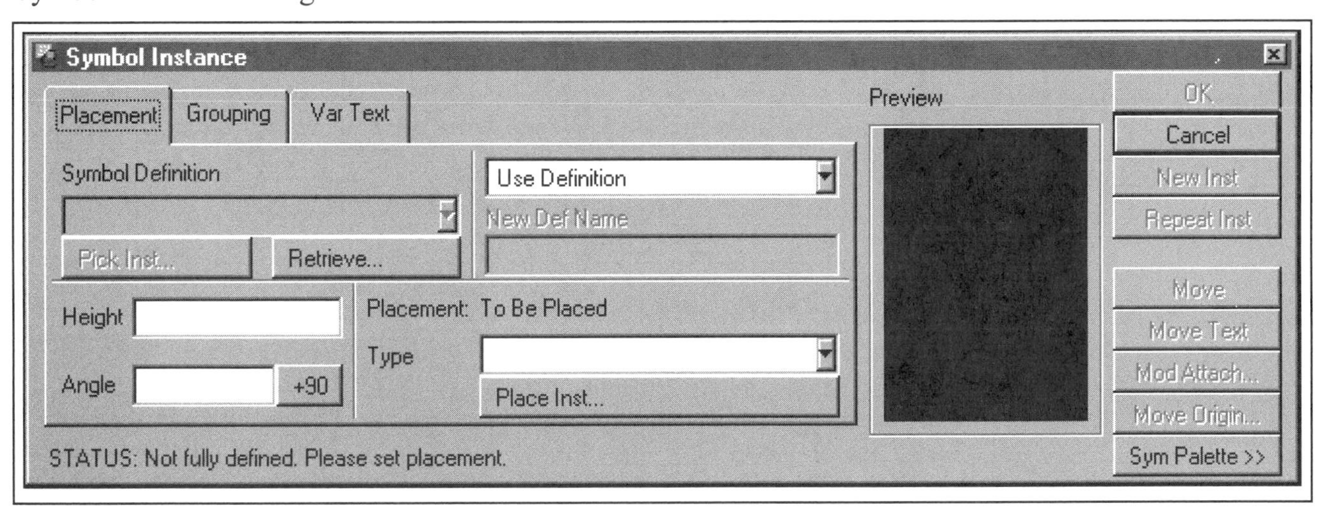

Choose the **Retrieve** button in the dialog box (Fig. 16.32)
Choose **System Syms** from the Look In menu, not the dialog box
Choose **surftextsymlib ⇒ Open** from the dialog box
Choose **surftexture.sym ⇒ Open** from the dialog box
Type **.5** for the symbol **Height** in the dialog box (Fig. 16.33)
Choose the **Grouping** tab in the dialog box

Pro/E now changes the left side of the dialog box so it contains the settings for the finish symbol (Fig. 16.34).

Figure 16.32
Left Side of Symbol Instance Dialog Box

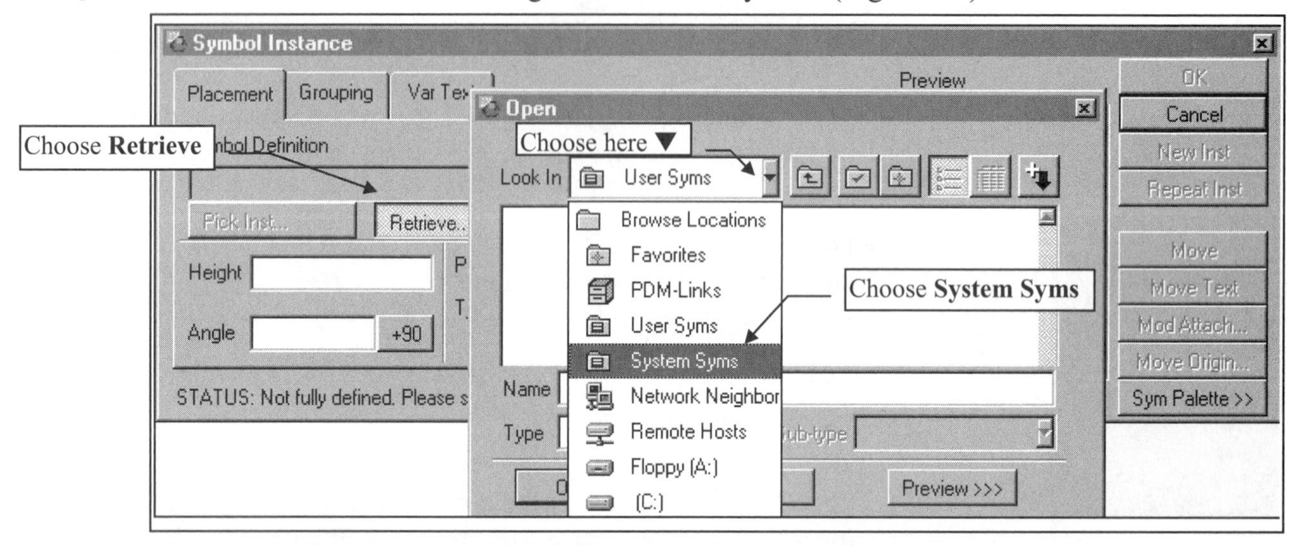

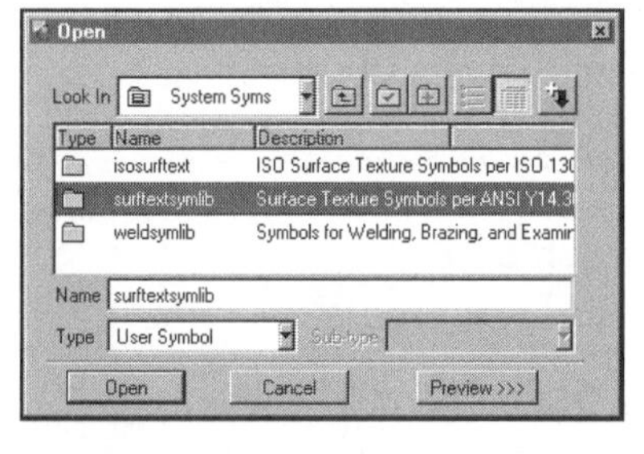

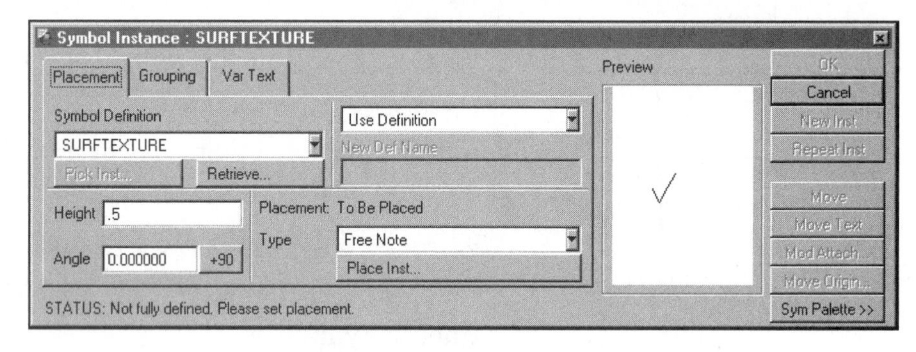

Figure 16.33
Setting the Height

Click on the + just to the left of the text **UNSPECIFIED**

Pro/E now expands the list under **UNSPECIFIED** to show the options available (Fig. 16.34).

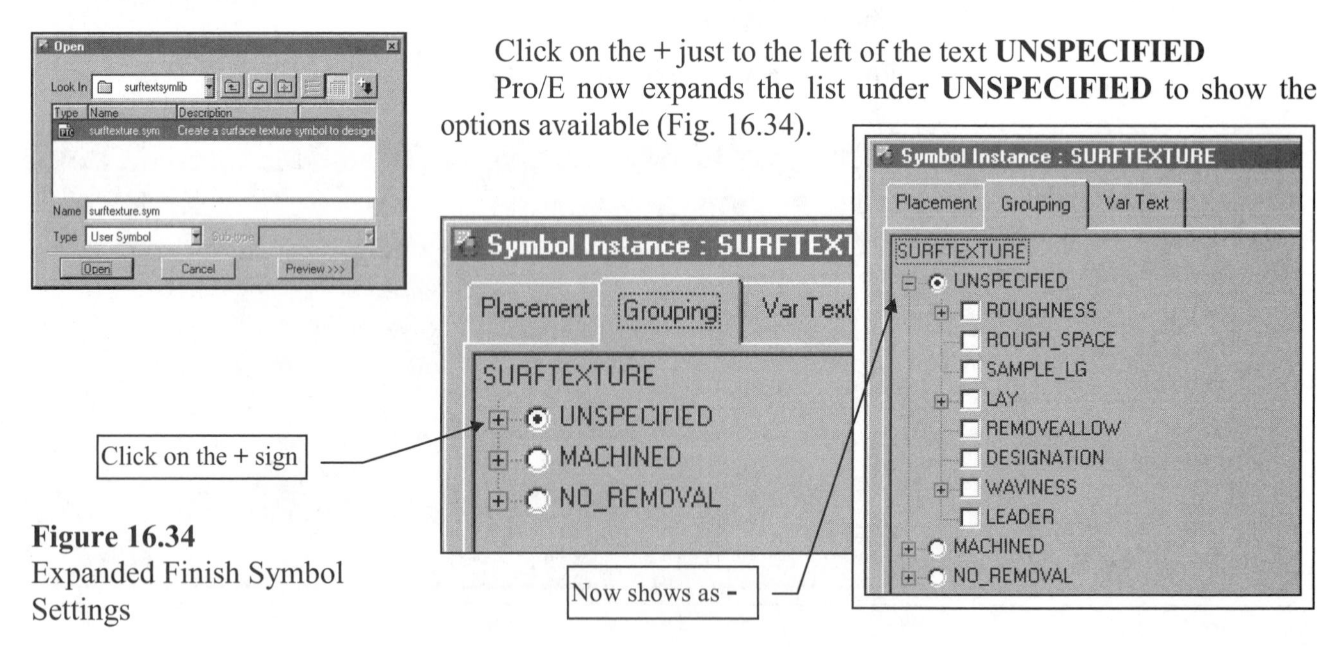

Figure 16.34
Expanded Finish Symbol Settings

Figure 16.35
Expanded Finish Symbol Roughness

Click on the toggle box ❑ just to the left of the **ROUGHNESS** text (Fig. 16.35).

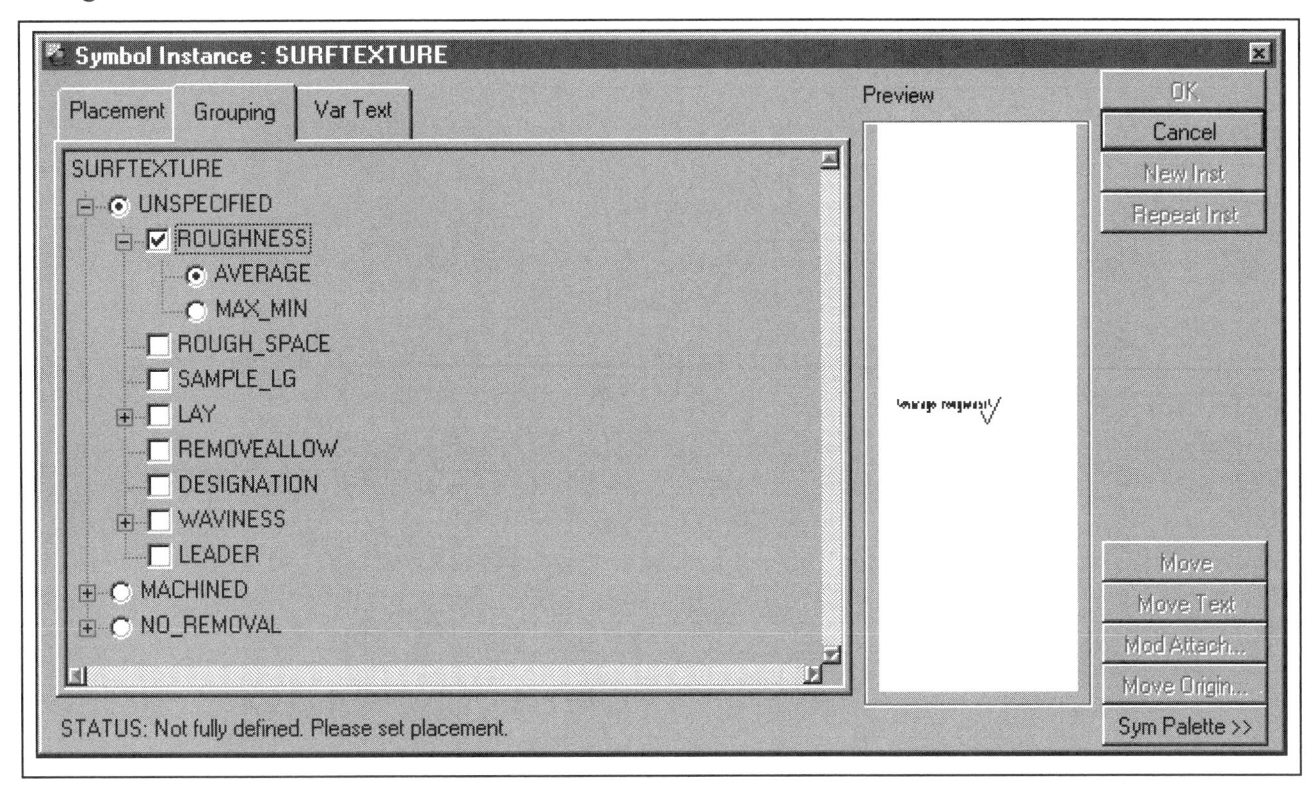

Choose the **Var Text** tab (Fig. 16.36)

Figure 16.36
Finish Symbol Variable Text

Pro/E now changes the left side of the dialog box to display the variable text entries for the finish symbol (Fig. 16.36).

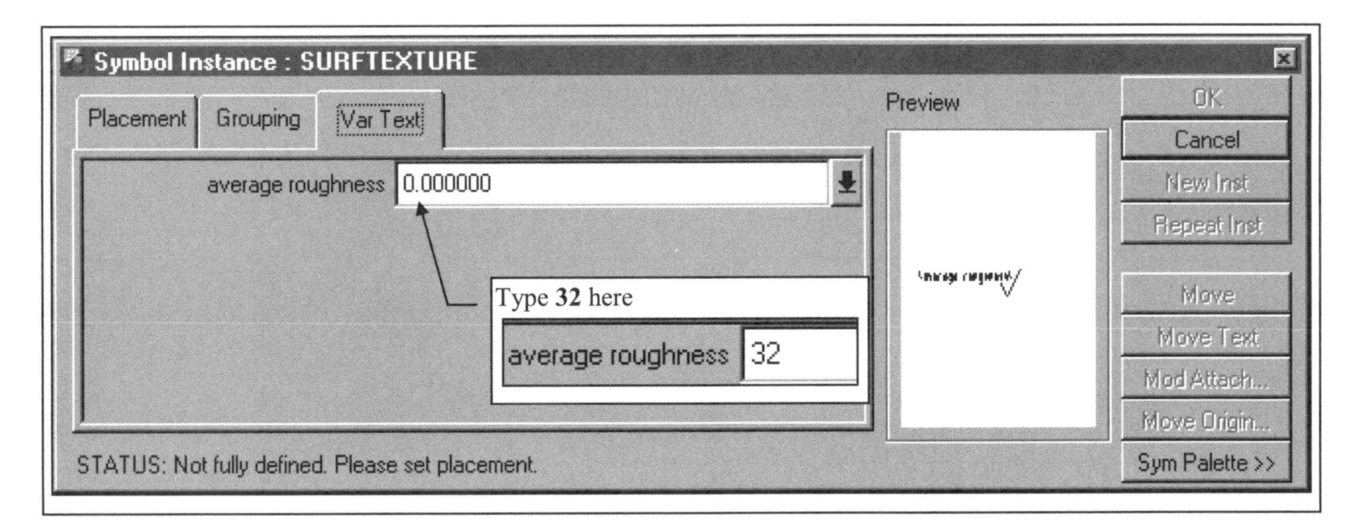

If this symbol had been defined with a series of default parameters, you would be able to select them from an option menu. Because this symbol has the roughness value only as a single number, you need to type in the desired value. Figure 16.37 shows the Symbol Instance dialog box with SURFTEXTURE option.

Click in the text area containing **0.000000**, and type **32** for the roughness value (Fig. 16.36)
Choose the **Placement** tab
Choose the ▼ (drop-down arrow) to show placement types
Choose **Free Note**
Indicate the origin of the symbol anywhere on the screen
Choose **OK** (from the **Symbol Instance** dialog box as shown in Figure 16.37)
Choose **Done/Return**

HINT
If you did not get the correct roughness number:
Choose **Modify** (from the DETAIL menu)
Choose **Value**
Pick the value "**0**" on the format
Type **32** ⇒ ✔
Choose **Done Sel**
Choose **Done/Return**

Figure 16.37
Symbol Instance: Surface Texture

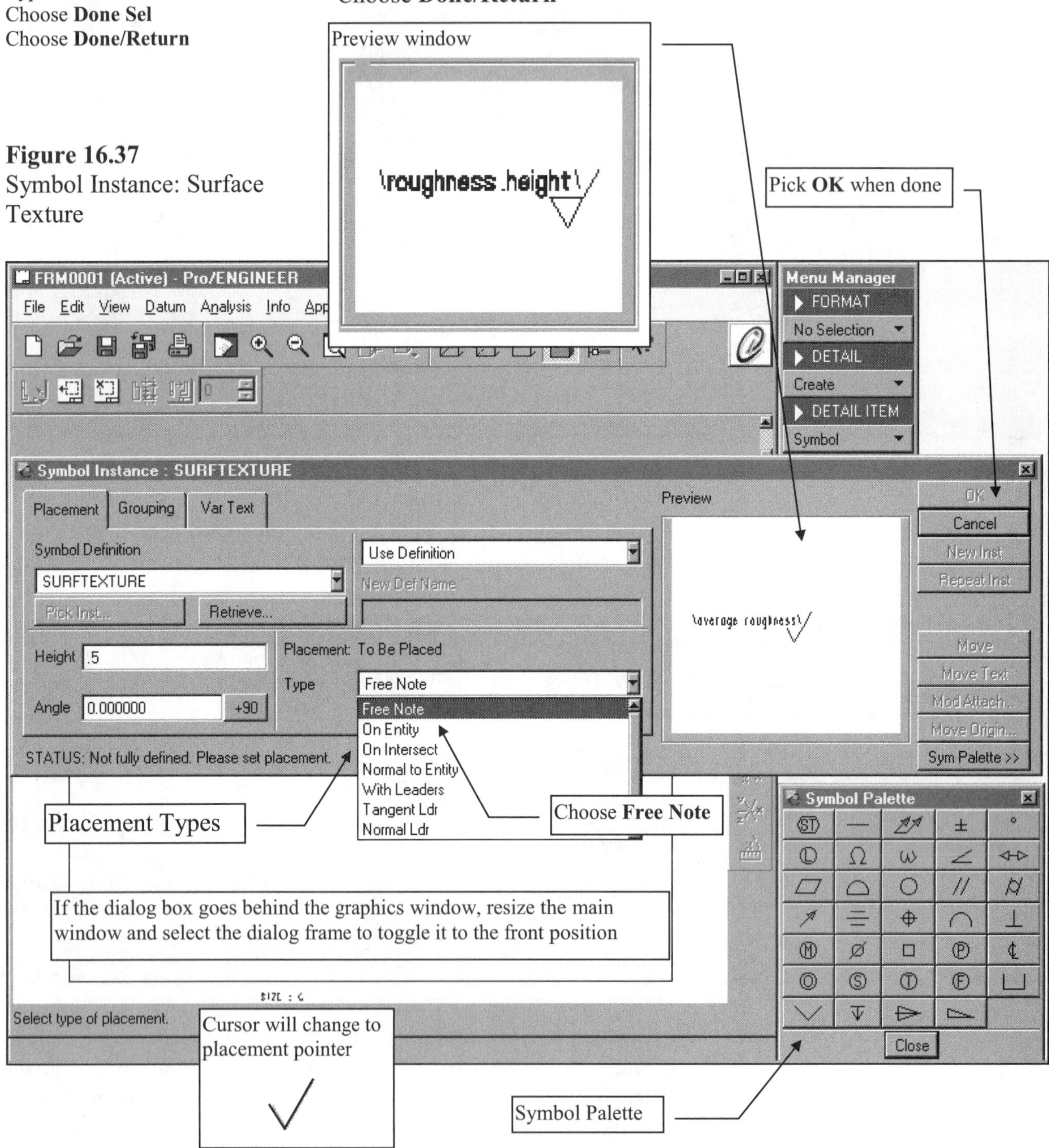

NOTE

You can **Modify Text** in either the **Format** mode or the **Drawing** Mode. The **Format** mode is demonstrated here.

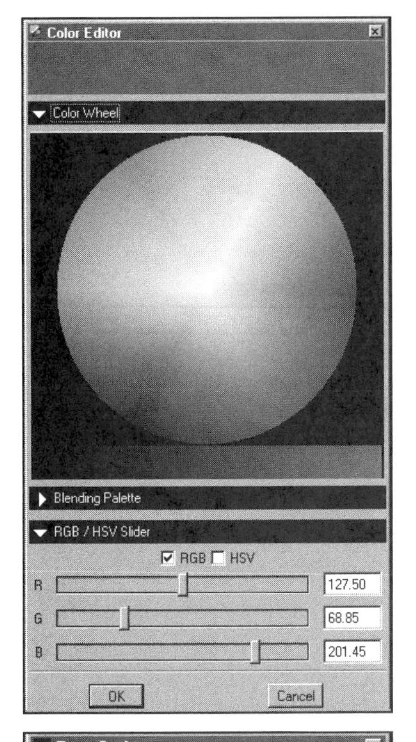

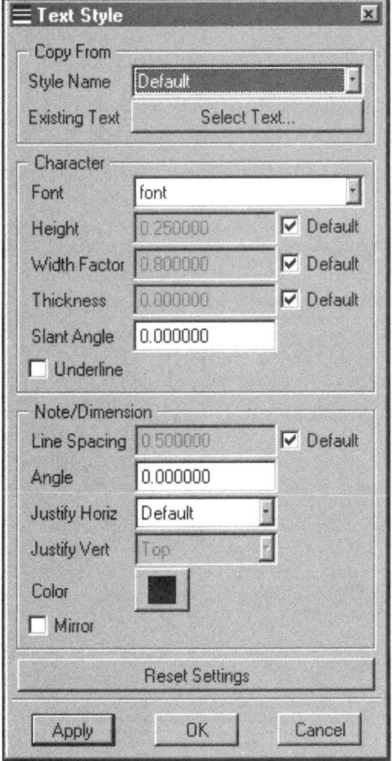

To modify Colors, pick here

Figure 16.38
Text Style Dialog Box

Modifying Text

After you create the text, you often need to modify the text parameters. The most common modification is to the text height. When you modify the parameters of text, Pro/E treats each text segment (**{1:text1}**) as a separate entity. This allows you to set the text height, for example, of different entity sets within a text string.

Because this would make selection quite time-consuming for complex notes, the **Pick Many** option allows you to pick the entire note using a rectangle.

You can also use the **Angle** option to rotate text. This allows you to write text upside down along the top of a format or at **90°** or **270°** to make it parallel to the right or left border. The **Mirror** option allows you to change text so that it can be read from the back side of the vellum (when you plot it).

Modify the text height of the notes and place them at the proper location in the format.

1. Change the height of all of the small title text to **.09.**

Choose **Modify**
Choose **Text**
Choose **Text Style**
Select the notes: **DESCRIPTION**, the entire **GENERAL SURFACE** note, and **DWG NO**.
Click the *middle* mouse button to stop selecting notes to change

(Pro/E now displays the **Text Style** dialog box, which contains all of the parameters that control how text displays, as shown in Figure 16.38.)

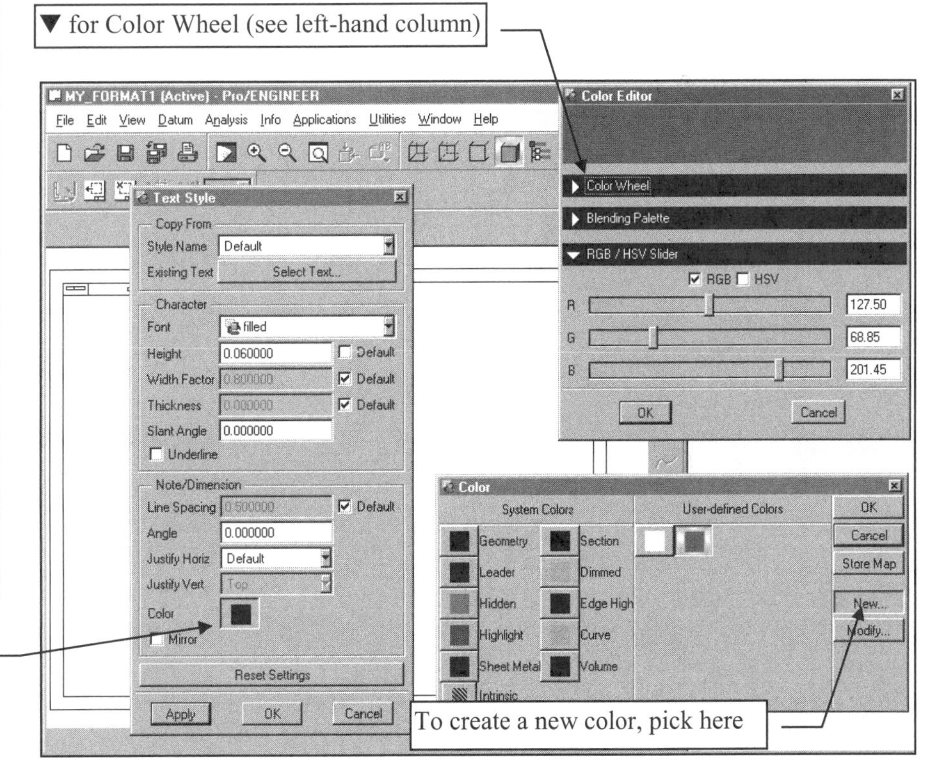

Click on the ❑ **Default** toggle box that is just to the right of the **Height** text area to select it (Fig. 16.39)

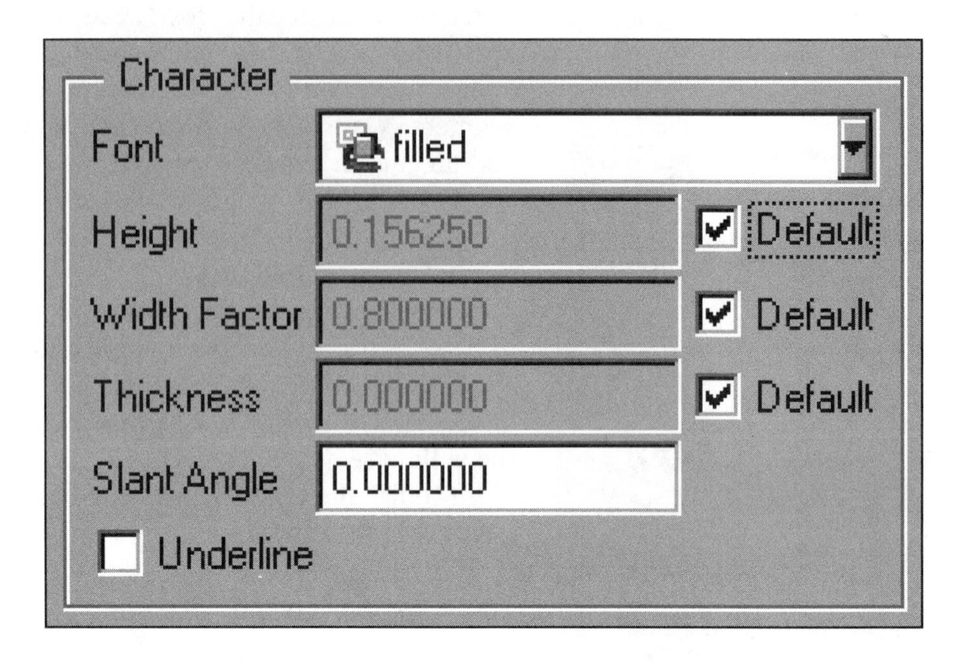

Figure 16.39
Text Height Use Default

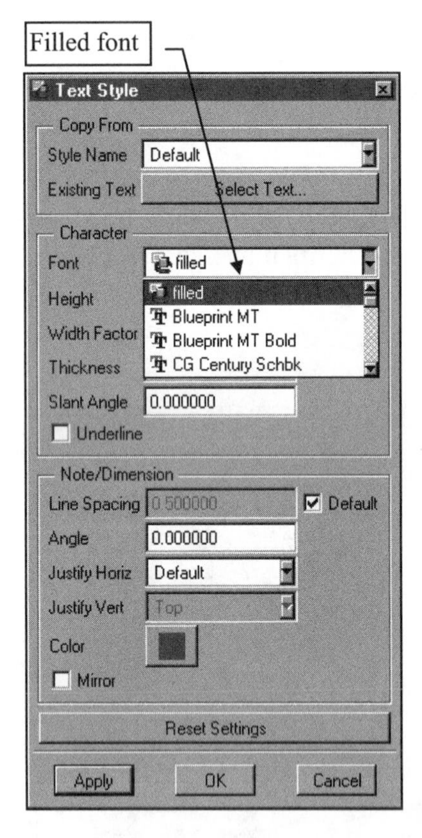

Double-click in the **Height** text area and type **.09**
Choose **Apply**
Choose **OK**

2. Change the size of the two **SHEET** callouts to **.09.**

Choose **Pick Many**
Drag a box around both the texts **"SHEET 1 OF 1"** to select them
Click the *middle* mouse button to stop selecting notes to change
Click on the ❑ **Default** toggle box that is just to the right of the **Height** text area to select it (Fig. 16.40).

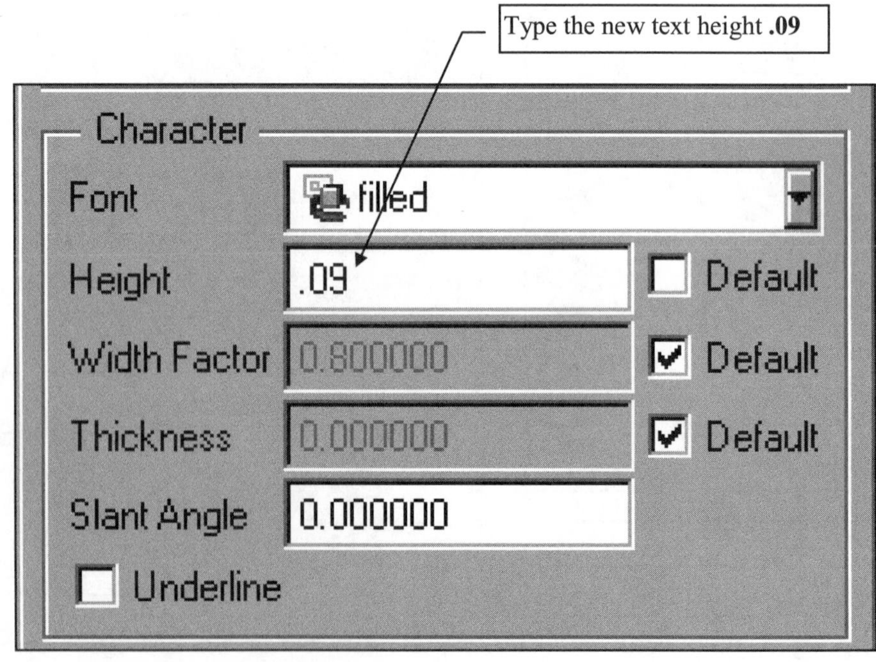

Figure 16.40
Text Height

Double-click in the **Height** text area and type **.09**
Choose **Apply**
Choose **OK**

3. Change the size of the two drawing number notes to **.25**.

Select the two notes that reflect the current name of the format (**MY_FORMAT**...)
Click the *middle* mouse button to stop selecting notes to change
Click on the ❑ **Default** toggle box that is just to the right of the **Height** text area to select it
Double-click in the **Height** text area and type **.25**
Choose **Apply**
Choose **OK**

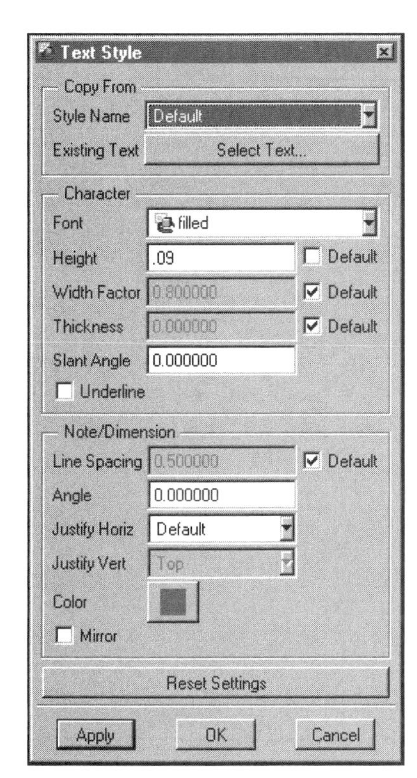

4. Change the size of the tolerances note.

Choose **Pick Many**
Drag a box around the complete tolerances note (not the text **TOLERANCES**)
Click the *middle* mouse button to stop selecting notes to change
Click on the ❑ **Default** toggle box that is just to the right of the **Height** text area to select it
Double-click in the **Height** text area and type **.06**
Choose **Apply**
Choose **OK**

5. Change the size of the **MODEL NAME** text to **.375**.

6. Rotate one of the drawing name notes and one of the sheet callouts.

Select one of the notes that reflects the current name of the format (**MY_FORMAT**) and the sheet callout note (**SHEET 1 OF 1**)
Click the *middle* mouse button to stop selecting notes to change
Double-click in the **Angle** text area (Fig. 16.41)

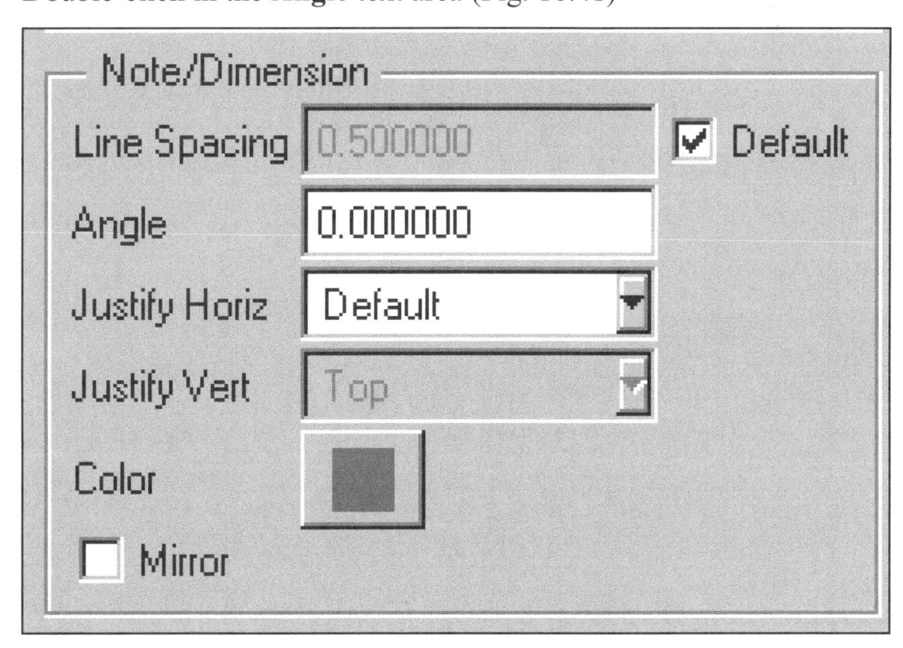

Figure 16.41
Text Angle

HINT

You can use the grid to line up and position text. Change the grid spacing to a convenient spacing.

In the **Format** mode and the **Drawing** mode, the **Grid** is 2D and is aligned with the lower left-hand corner of the sheet. You can also change the **Origin** of the grid. First, check to see whether your **Snap to Grid** is on or off in the **Environment** menu. Turn it on.

Second, to *see* the grid, you must turn it on using **Modify**. Choose: **Modify** ⇒ **Grid** ⇒ **Grid On** ⇒ **Grid Params** ⇒ **X&Y Spacing** ⇒ (type the new spacing value; the default is **.500**).

Type **180**
Choose **Apply**
Choose **OK**
Choose **Done/Return** ⇒ **Done/Return**

7. Use the **Move** option to drag the text to the proper location in the title block, approximately as shown in Figure 16.42. You can place the **&scale** note above the DWG NO. area or create a separate block.

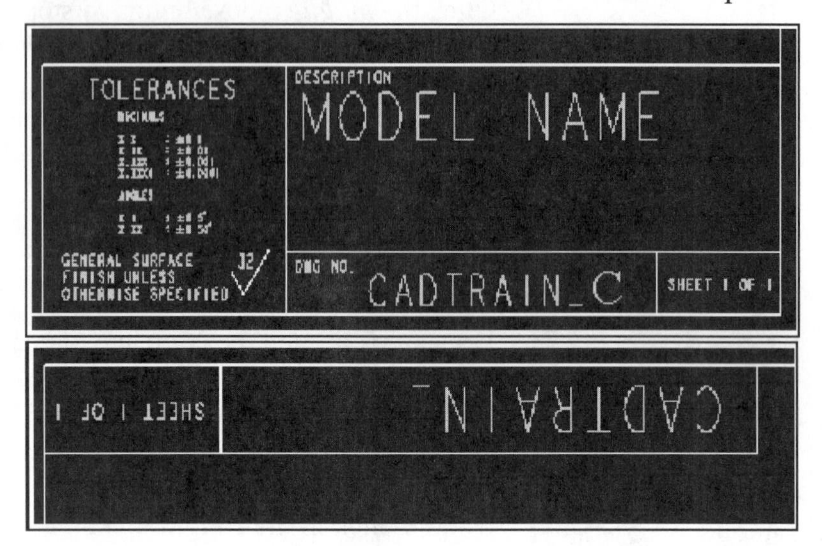

Figure 16.42
Moving the Text to the Title Block

8. Drag the two notes to the upper left corner block, as shown in Figure 16.42. Figure 16.43 shows another format that can be placed on a drawing.

9. Save the format.

Choose **File**
Choose **Save**
Choose ✔ (now you have a "**C**" size format with parameters)

As the model is added to the first drawing view (Drawing mode), the format will "read" embedded *parameter values* from models with the same parameters.

Choose **Window** ⇒ **Close Window** ⇒ **File** ⇒ **Save** ⇒ ✔

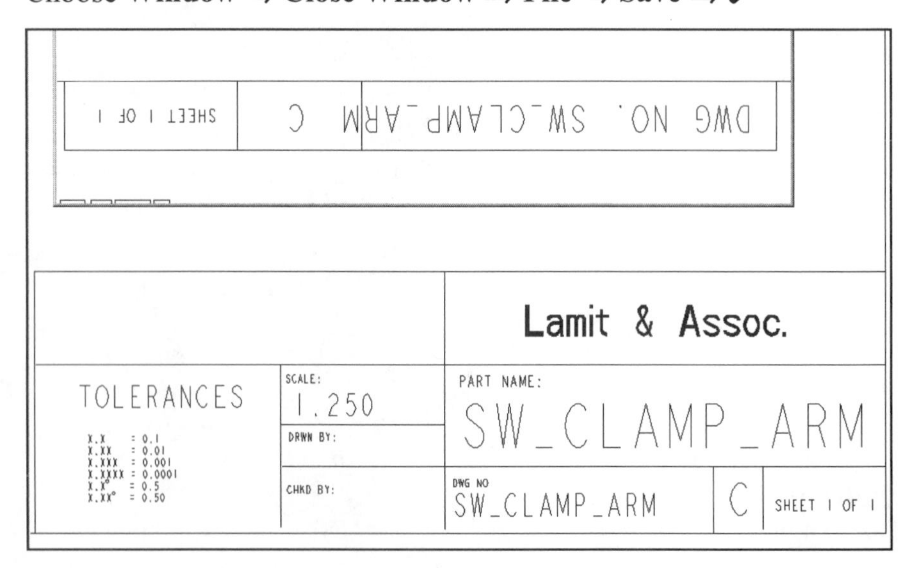

Figure 16.43
Company Formatted Sheet

Drawing Mode and Views

When you start a drawing, you can automatically have default views of the part or assembly displayed, or you can add the views of your liking to a blank drawing. In this lesson, you will use the ✔ **Use default template** setting and have the three primary views displayed for you. In Lesson 17, you will create the views by adding them to the drawing. ***You must be in the same directory as your part.*** It is also a good practice to bring up the part in a separate window.

Choose the following commands (Fig. 16.44):

File ⇒ **New** ⇒ ●**Drawing** ⇒ (type the name for your drawing: **BASE_ANGLE**) ⇒ ✔ **Use default template** ⇒ **OK** ⇒ (from the New Drawing dialog box, pick **Browse** ⇒ (pick **BASE_ANGLE**) ⇒ ●**Use template** ⇒ Template (pick **c_drawing** from the list) ⇒ **OK** ⇒ **File** (from the menu bar) ⇒ **Open** ⇒ (pick **BASE_ANGLE**) ⇒ **Open** ⇒ (move and resize the Part window) ⇒ **Window** (Drawing Window) ⇒ **Activate** ⇒ (Fig. 16.45) ⇒ **File** ⇒ **Save** ⇒ ✔

Figure 16.44
New Dialog Box
New Drawing Dialog Box

This will give you a drawing with the BASE_ANGLE part (before modifications) displayed in three standard views. Later, you may also wish to create a drawing for the ECO part (Fig. 16.46).

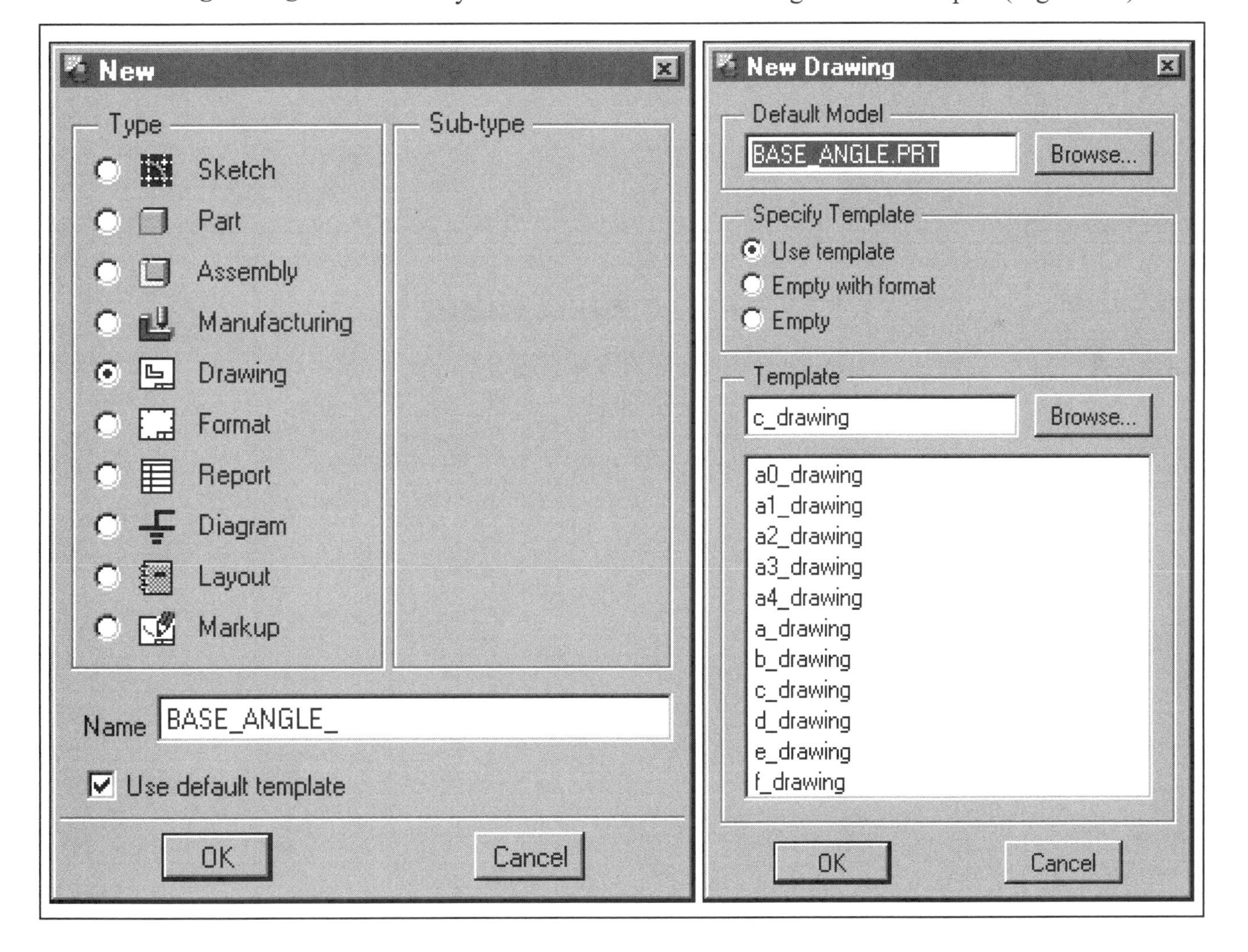

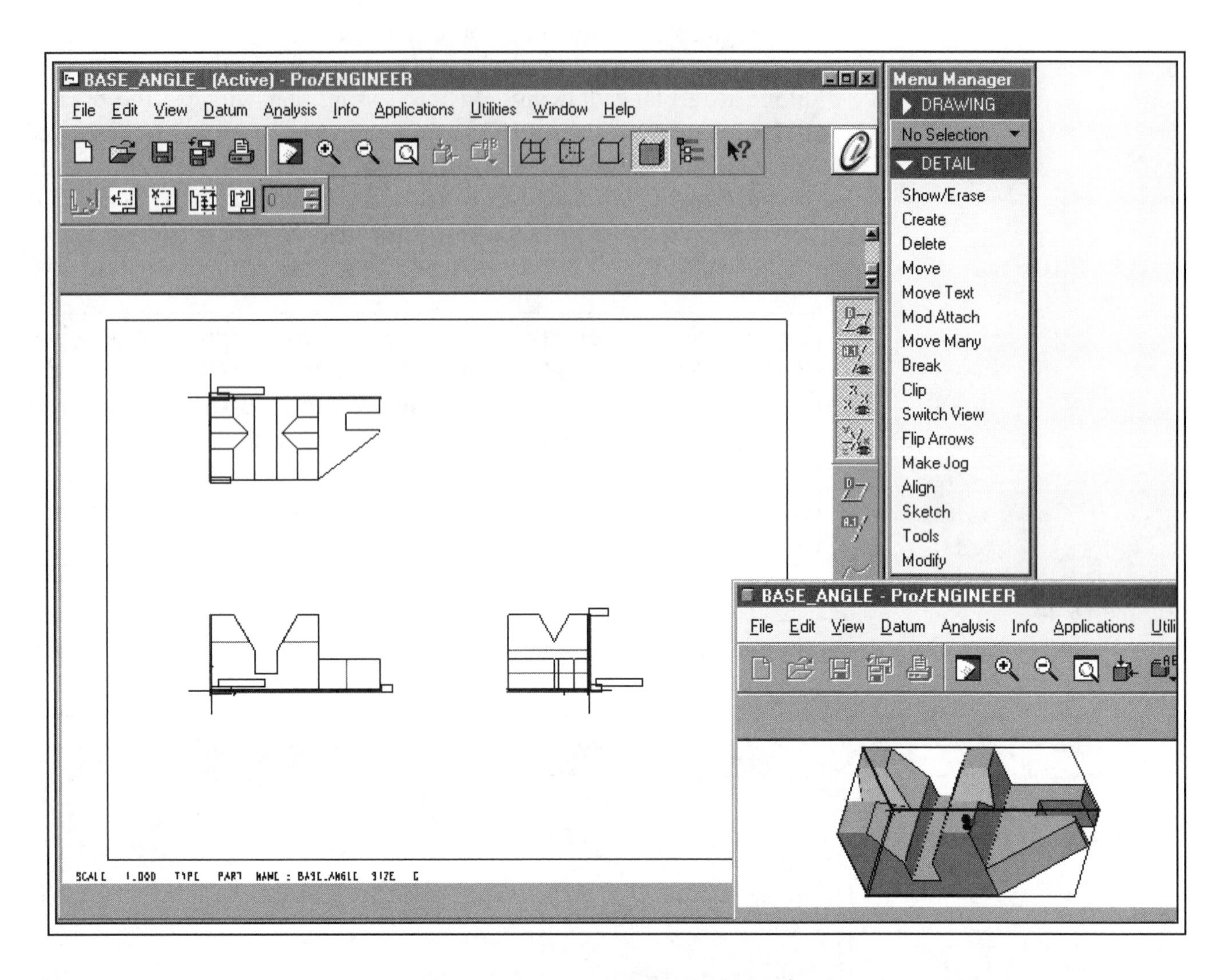

Figure 16.45
Base Angle Drawing of Base Angle Part

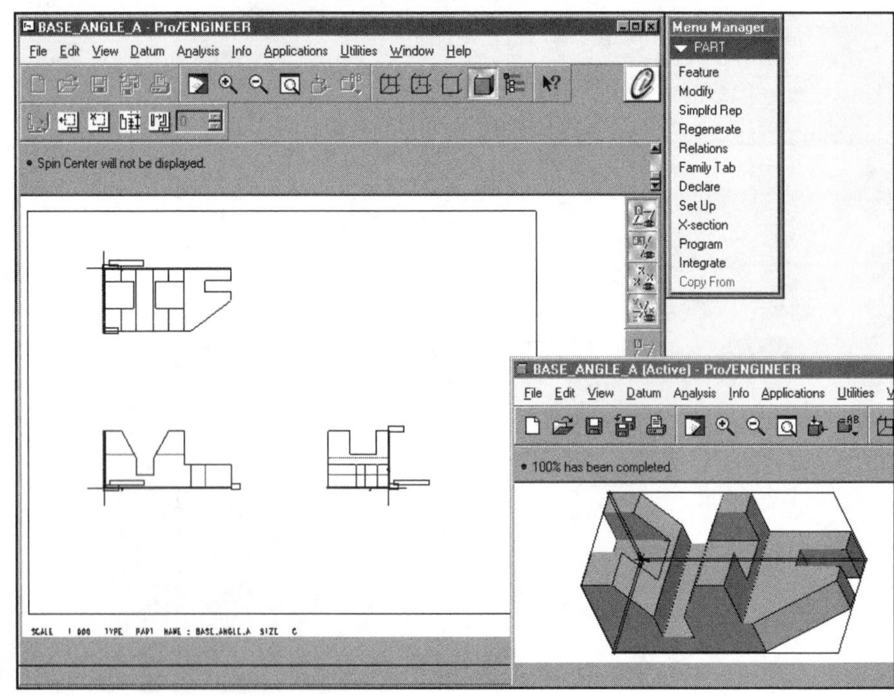

Figure 16.46
Base Angle A Drawing (modified and redefined part)

Add another projected view (Figs. 16.47 and 16.48):

Views ⇒ Add View ⇒ Projection ⇒ Full View ⇒ No Xsec ⇒ No Scale ⇒ Done ⇒ [pick the position for the new view (Fig. 16.47)] (Pro/E will prompt you with the following: **Conflict in parent view exists. Select parent view for making the projection.**) ⇒ [Select the right side view (Fig. 16.48)]

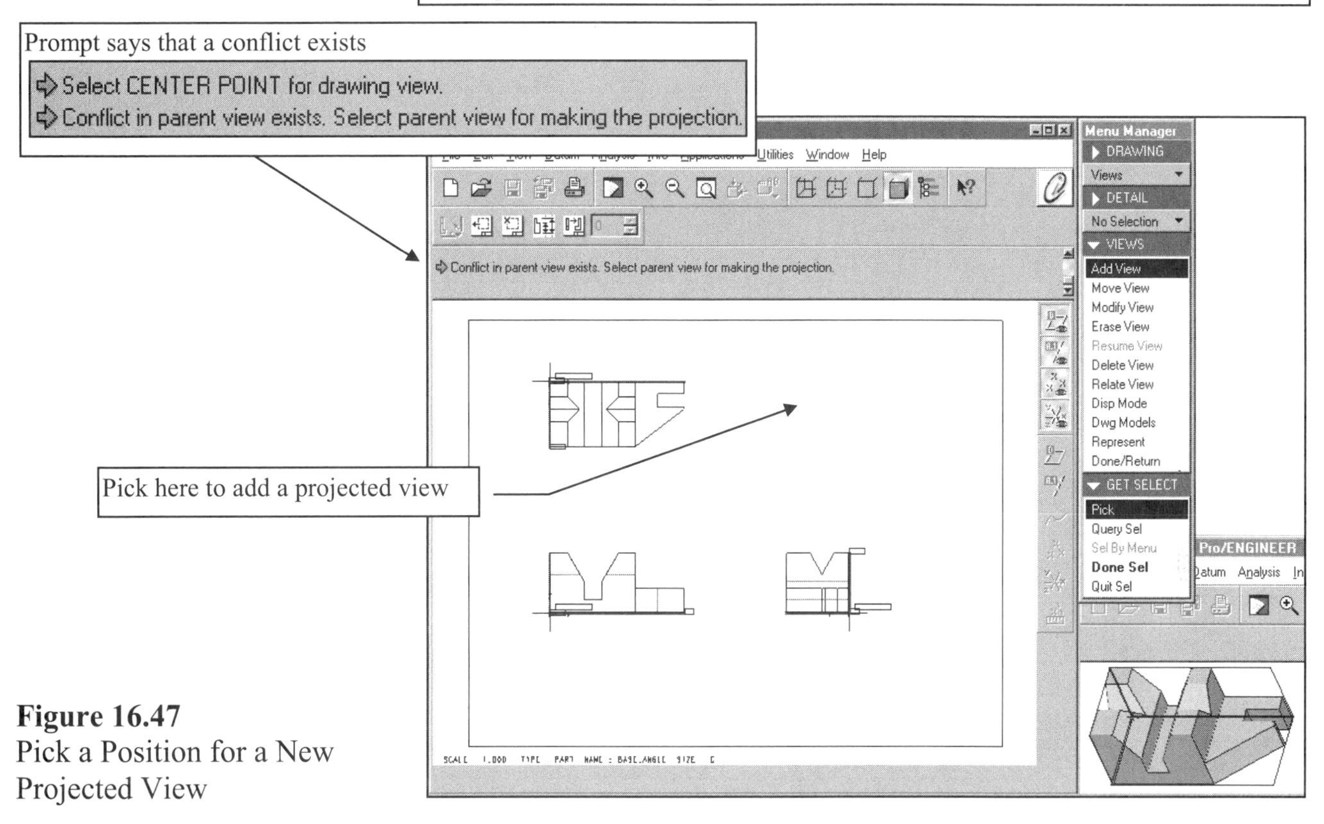

Figure 16.47
Pick a Position for a New Projected View

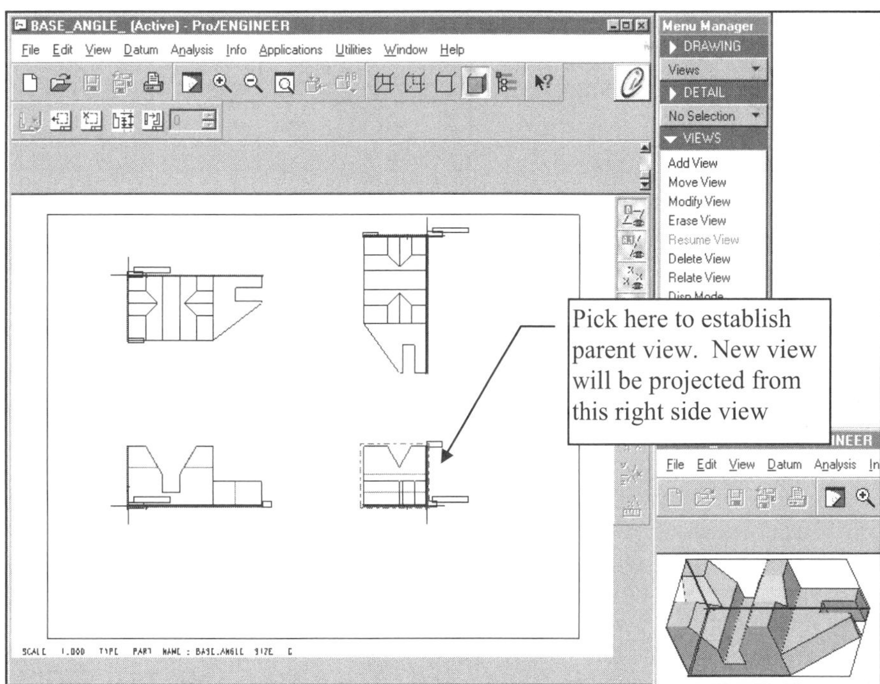

Figure 16.48
Pick the Parent View that the New View Will Be Projected From

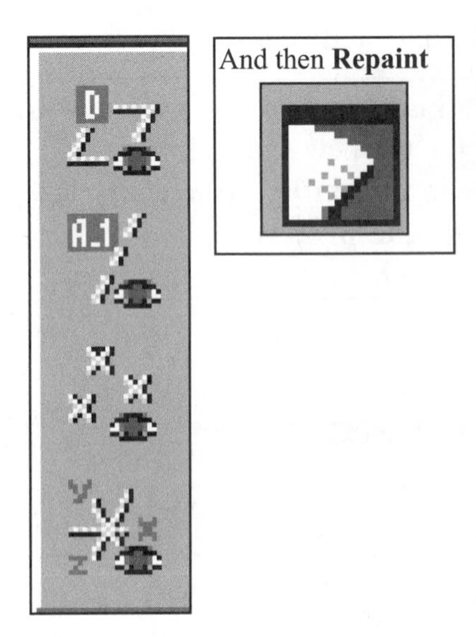

Figure 16.49
Datum Features Toggled Off

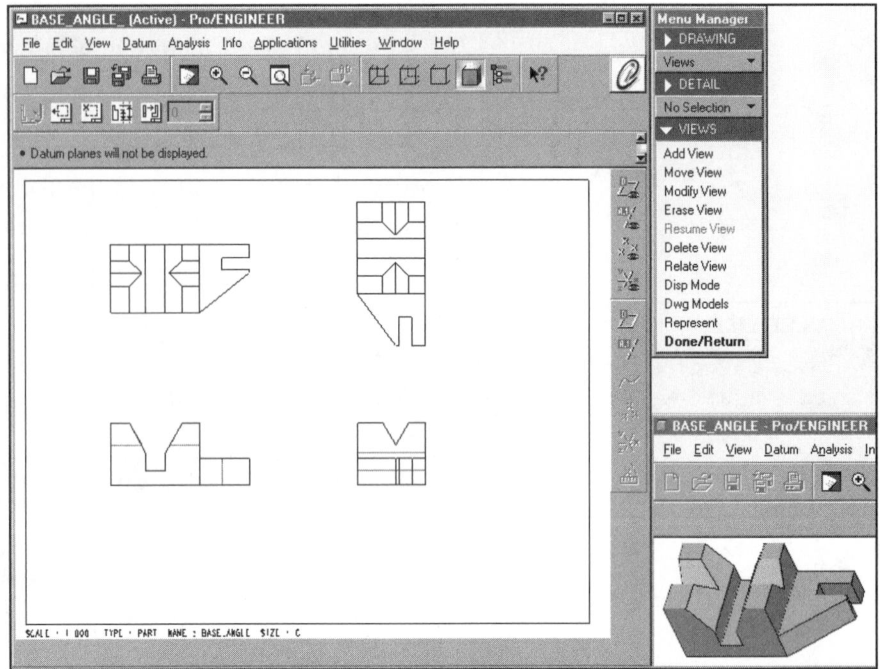

You do not need the view that was just created. Remove it using:

Delete View ⇒ (pick the view) ⇒ **Confirm** (Fig. 16.50) ⇒ **Done Sel** ⇒ **Done/Return**

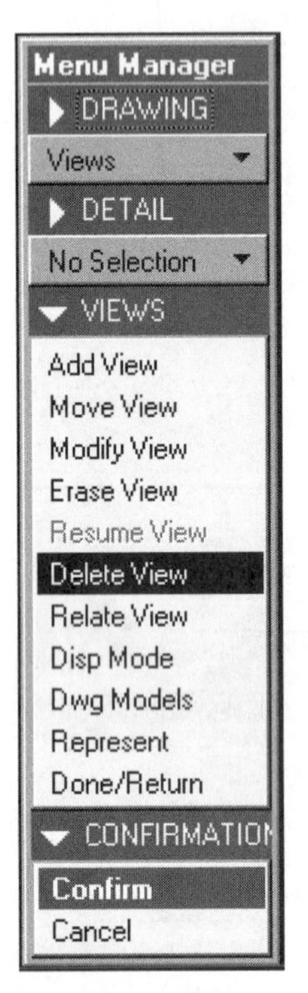

Figure 16.50
Delete the New View

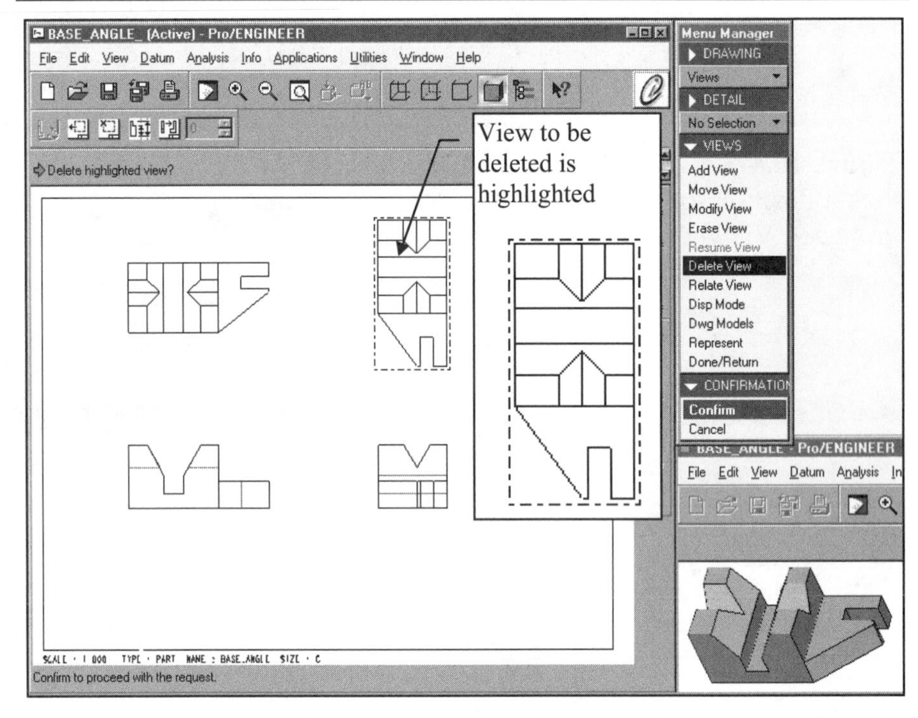

Next, reposition the views on the drawing. After you select the view to move, it will be highlighted with dashed lines (Fig. 16.51). If the view is a parent of another view, that view will also be highlighted and will move with the selected view. Choose :

Views ⇒ **Move View** ⇒ (pick the front view) ⇒ [move the view to a new position, try a variety of positions (Fig. 16.52)] ⇒ (pick to place view) ⇒ **Done Sel** ⇒ **Done/Return** ⇒ **File** ⇒ **Save** ⇒ ✔

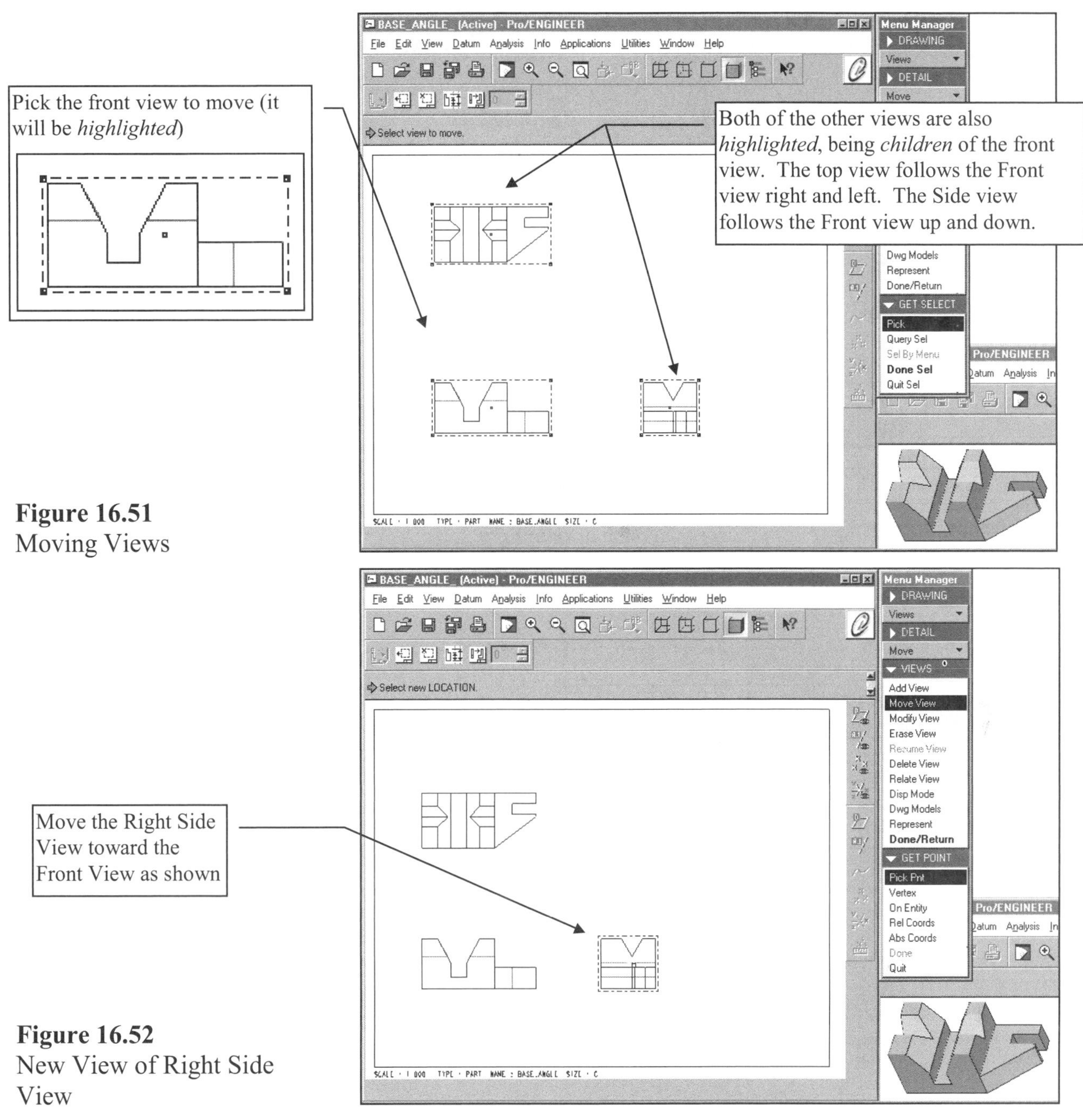

Figure 16.51
Moving Views

Figure 16.52
New View of Right Side View

Open
Look In: System Formats
a.frm
b.frm
c.frm
d.frm
e.frm
Name: c.frm
Type: Format
Open Cancel

Now add to the drawing sheet a standard ANSI "**C**" size format using the following commands (Fig. 16.53):

Sheets ⇒ Format ⇒ Add/Replace ⇒ System Formats ⇒ (pick: **c.frm**) (there are no parameters on this format) **⇒ Open ⇒ Done/Return**

Now change the drawing format to the one you previously created and saved as a **.frm** for use on your projects. This format should contain parameters you set up and saved with the format, as in Figure 16.54. Choose the following commands:

Sheets ⇒ Format ⇒ Add/Replace ⇒ (change to your file directory) **⇒** (pick ***your_format_name*.frm**) **⇒ Open ⇒ Done/Return**

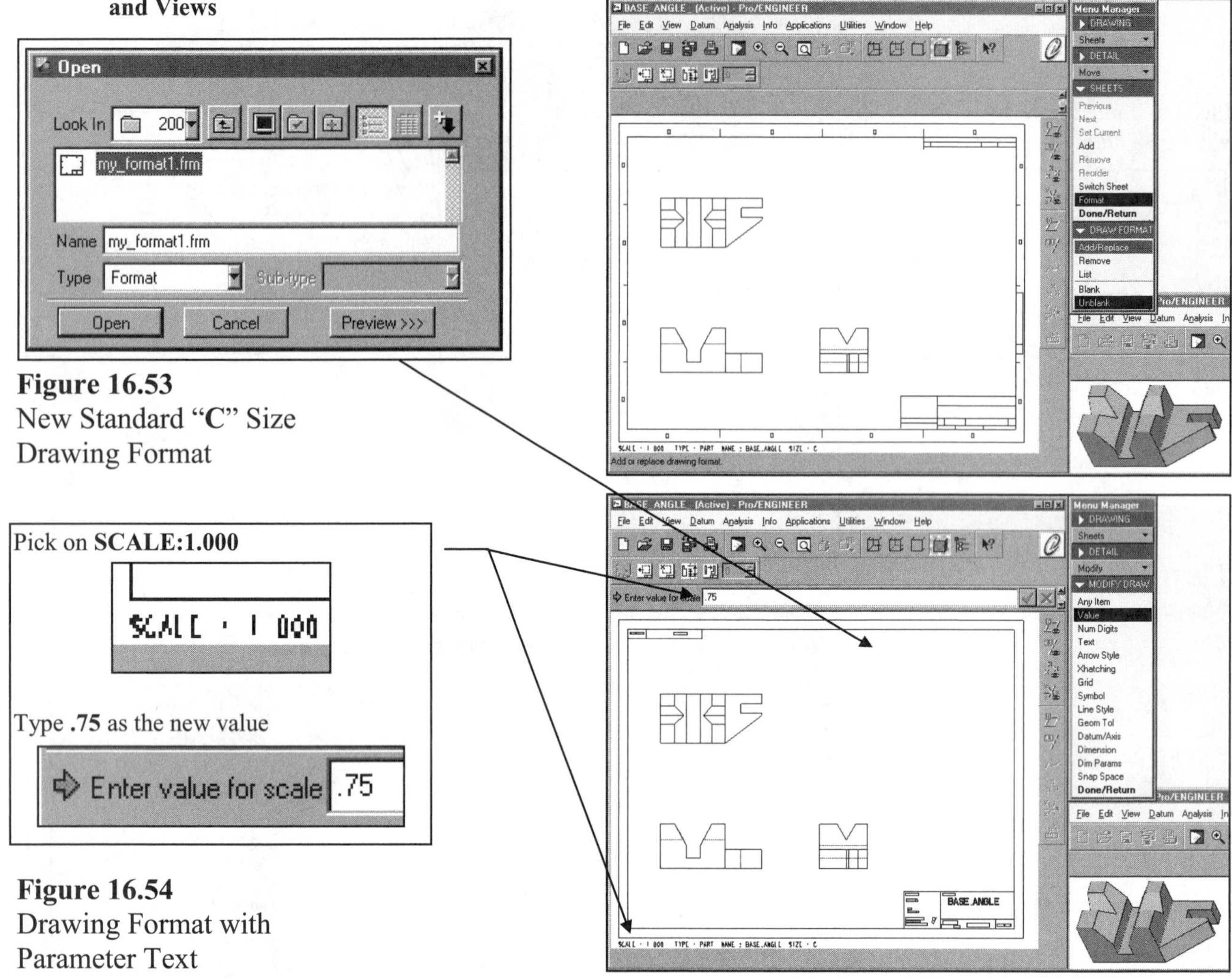

Figure 16.53
New Standard "**C**" Size Drawing Format

Figure 16.54
Drawing Format with Parameter Text

To change the scale (Fig. 16.54) of the drawing (*not the part*), pick **Modify** ⇒ [pick the **SCALE: 1.000** (bottom left edge of the Drawing Window)] ⇒ **.75** ⇒ ✔ (Fig. 16.55) ⇒ **File** ⇒ **Save** ⇒ ✔.

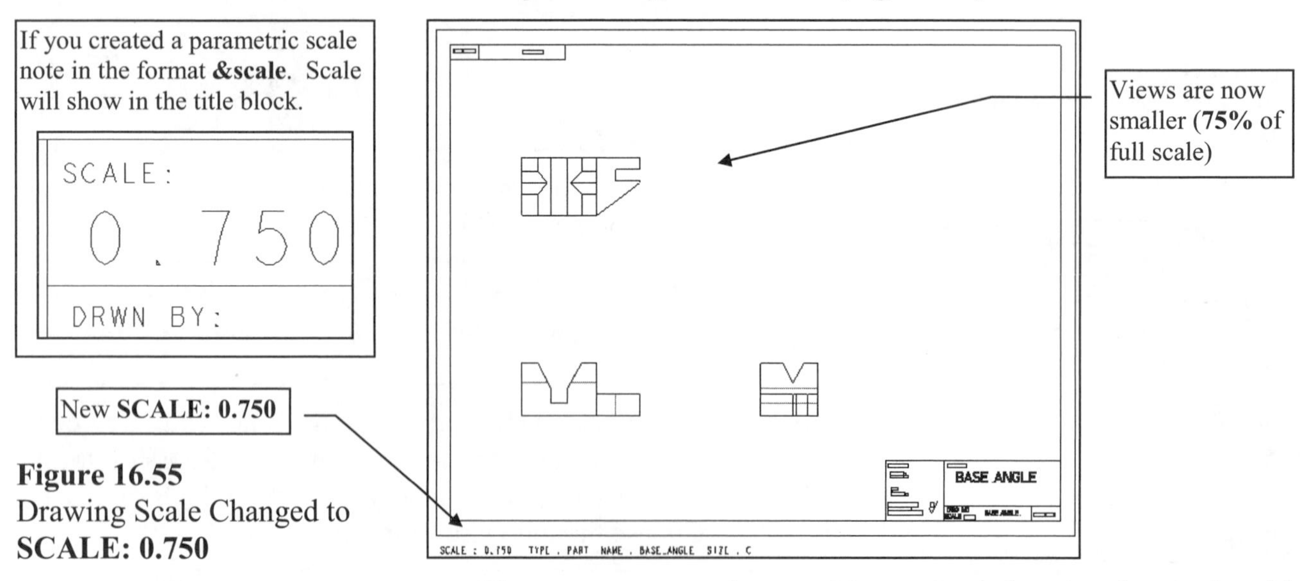

Figure 16.55
Drawing Scale Changed to **SCALE: 0.750**

You can now experiment with creating other drawings using this part or other parts; with adding, moving, and erasing views; and with creating, adding, and removing drawing formats.

Lesson 16 Project

Clamp Arm Drawing

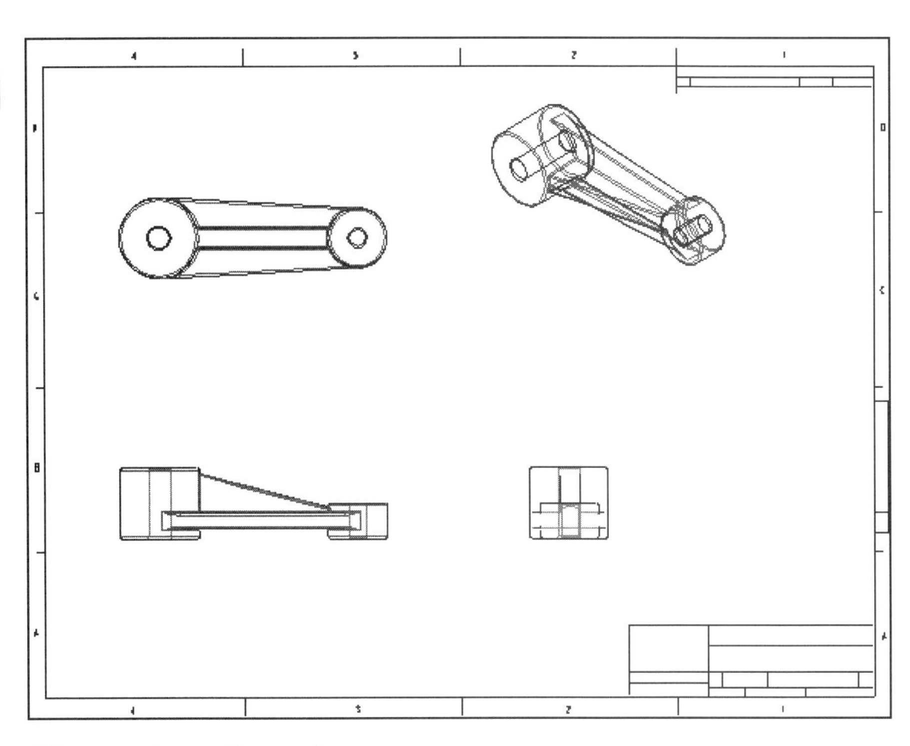

Figure 16.56
Clamp Arm Drawing Using ANSI (ASME) Standard Format

Clamp Arm Drawing

The first **lesson project** for Drawing mode will use the project modeled in Lesson 8. The drawing for the Clamp Arm (Fig. 16.56) is just one of many lesson parts and lesson projects that can be brought into Drawing mode and detailed.

Analyze the part, and plan out the sheet size and the drawing views required to display its features for detailing (Fig. 16.57). Use the format created in this lesson. Figure 16.58 shows a **C** format.

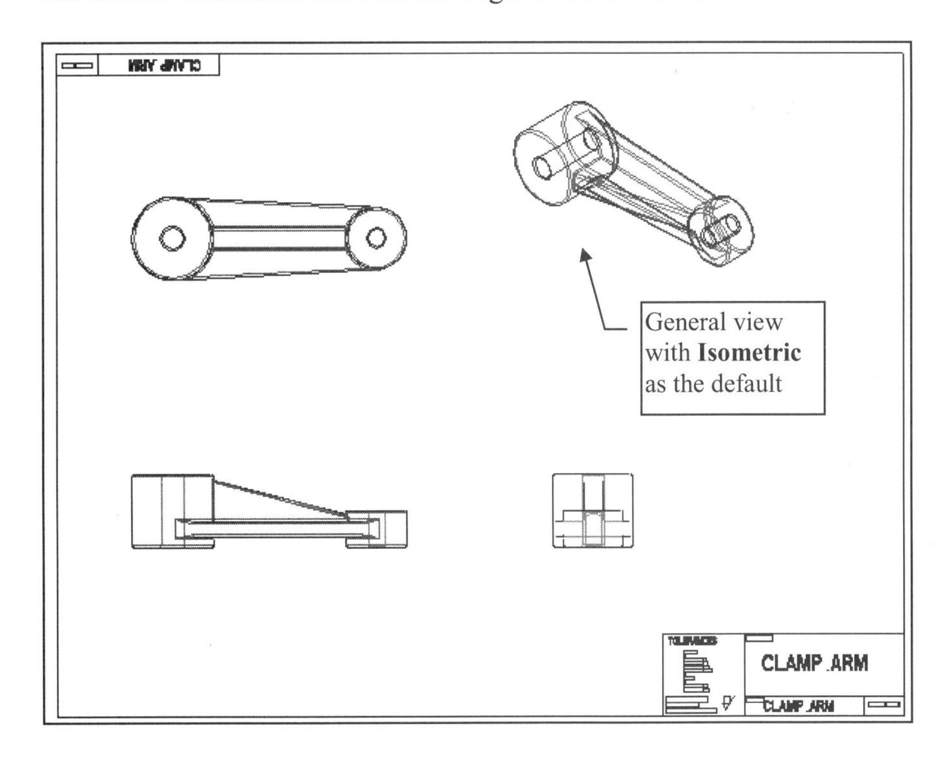

Figure 16.57
Base Angle Drawing Format with Parameters

Figure 16.58 C Format

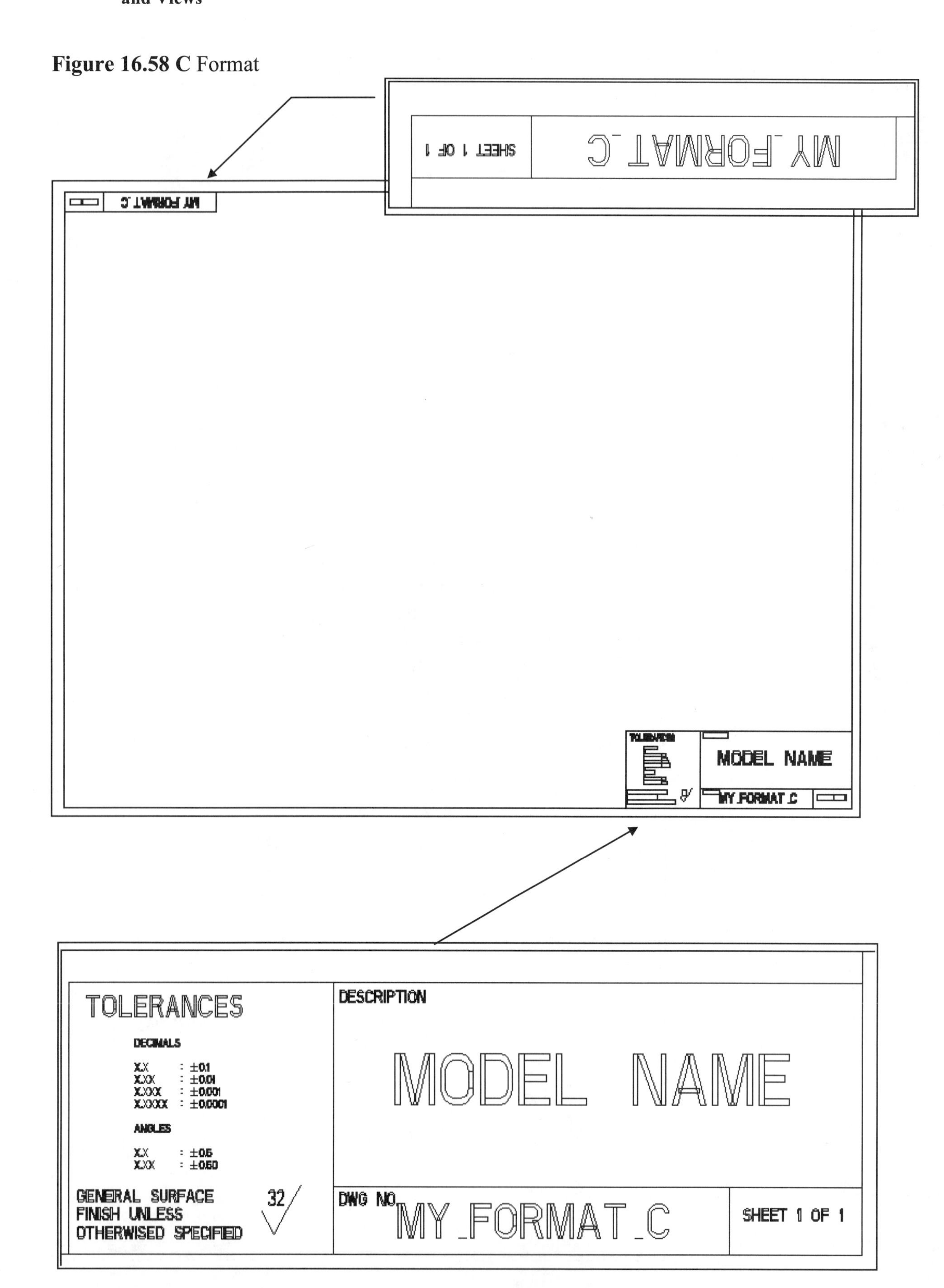

Lesson 17

Detailing

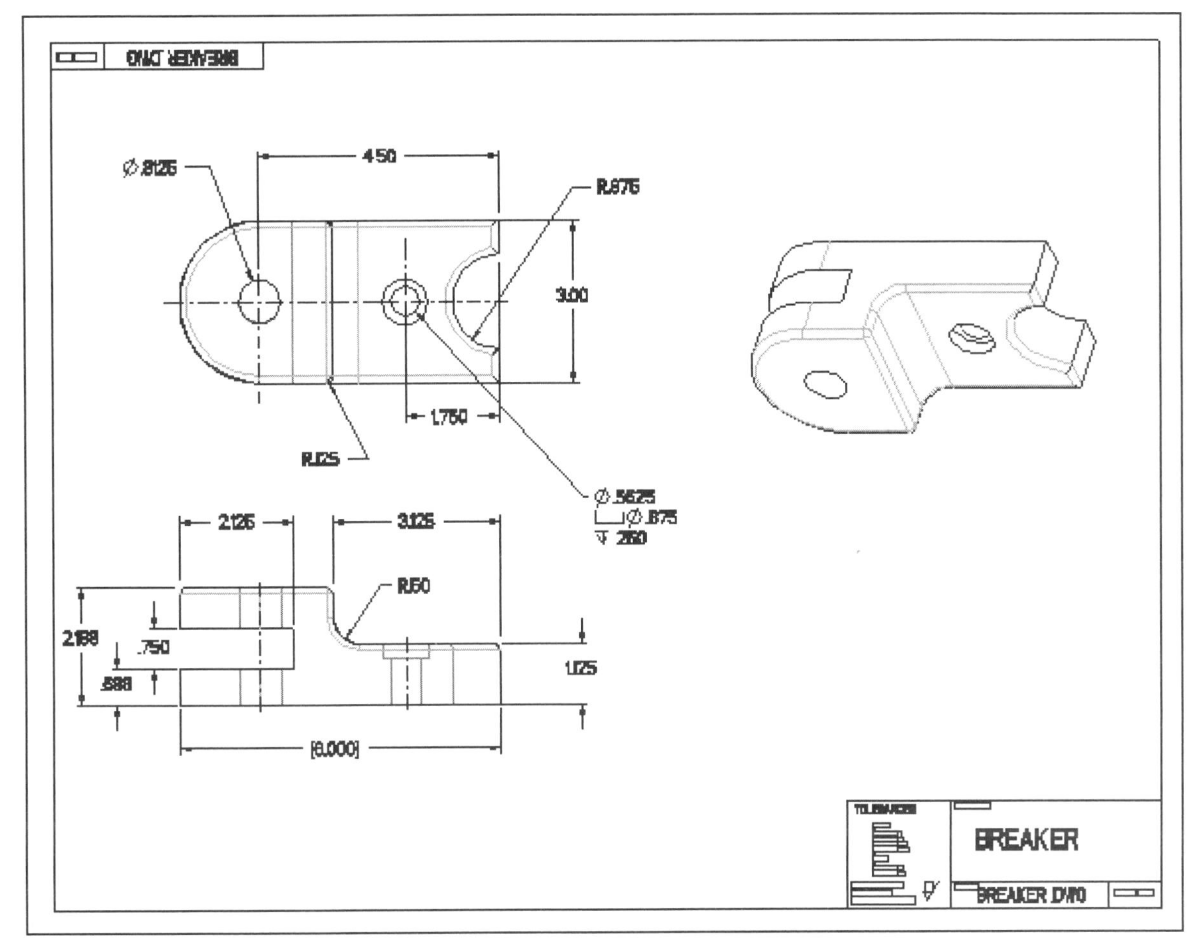

Figure 17.1
Breaker Drawing

EGD REFERENCE
Fundamentals of Engineering Graphics and Design
by L. Lamit and K. Kitto
Read Chapters 15 and 16.
See pages 321, 364, 556-559, and 622-626.

COAch™ for Pro/ENGINEER

If you have **COAch for Pro/ENGINEER** on your system, go to SEARCH and do the Segment shown in Figures 17.3, 17.7, and 17.10.

OBJECTIVES

1. **Use ASME Y14.5 1994 standards to detail drawings**
2. **Dimension a part**
3. **Create and save .dtl files**
4. **Add geometric tolerancing information to a drawing**
5. **Use Pro/MARKUP to see checker changes**
6. **Move and modify dimensions**

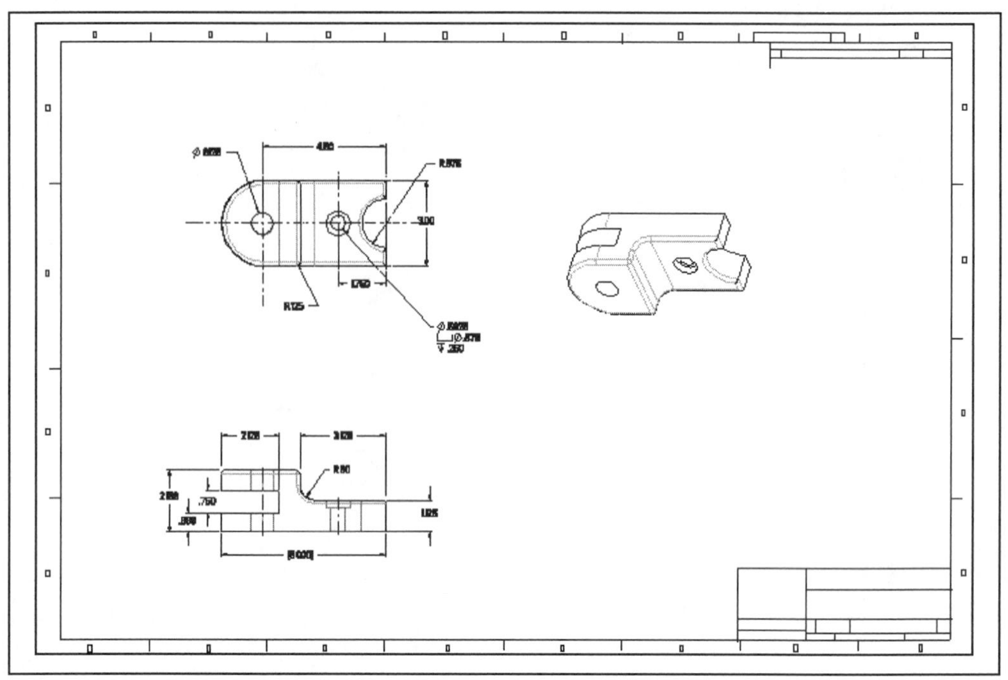

Figure 17.2
Breaker Drawing with ANSI Standard "**D**" Size Format

NOTE
Many of the concepts presented here (for parts) also apply to assemblies.

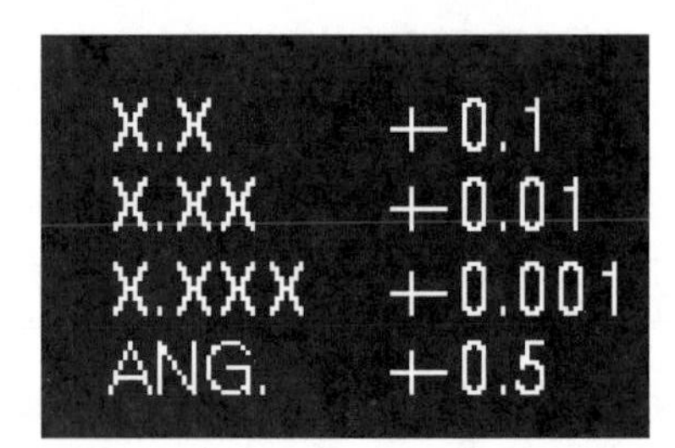

X.X	+0.1
X.XX	+0.01
X.XXX	+0.001
ANG.	+0.5

DETAILING

The purpose of an engineering drawing is to convey information so that the part can be manufactured correctly. Engineering drawings use dimensions and notes to convey this information (Fig. 17.1). Knowledge of the methods and practices of dimensioning and tolerancing is essential to the engineer or designer. The multiview projections of a model (part or assembly) provide a graphical representation of its shape (*shape description*). However, the drawing must also contain information that specifies size and other requirements.

Drawings are *annotated* with dimensions and notes. Dimensions must be provided between points, lines, or surfaces that are functionally related or to control relationships of other parts. With Pro/E, the **design intent** used in the original sequence of feature creation and the selection of dimensions used on the feature's sketch will determine the dimensions shown on the drawing. You need not create dimensions (unless desired), because the dimensions are already established during the modeling of the part. Dimensions created in Drawing mode, to describe drafting features, will not be *driving dimensions* (they will be *driven dimensions* that cannot be modified). *Only dimensions used in modeling the part drive the part feature database.*

Each dimension on a drawing has a **tolerance**, implied or specified. The general tolerance, given in the title block, is called a **general** or **sheet tolerance.** Specific tolerances are provided with each appropriate dimension. Together, the views, dimensions, and notes give the complete shape and size description of the part. Uniform practices for stating and interpreting dimensioning and tolerancing requirements are maintained in **ASME Y14.5 1994.**

Dimensioning

Figure 17.3
COAch for Pro/E, Drawings Detailing (Show Dimensions)

Views of a model (part or assembly) may be dimensioned in Drawing mode (Fig. 17.2). Pro/E displays dimensions in a view based on the way the part (or the assembly) was modeled. The dimension type is selected from options before showing the dimensions on the drawing (Figs. 17.3 and 17.4). Linear dimensions and ordinate dimensions are two of these options.

Over the next several Segments, you are going to detail the provided drawing to appear approximately as shown below. In this Demonstration, you will Show All of the feature dimensions.

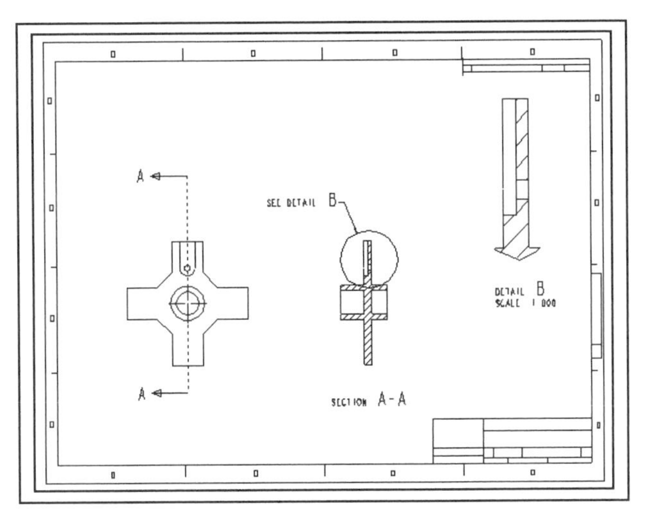

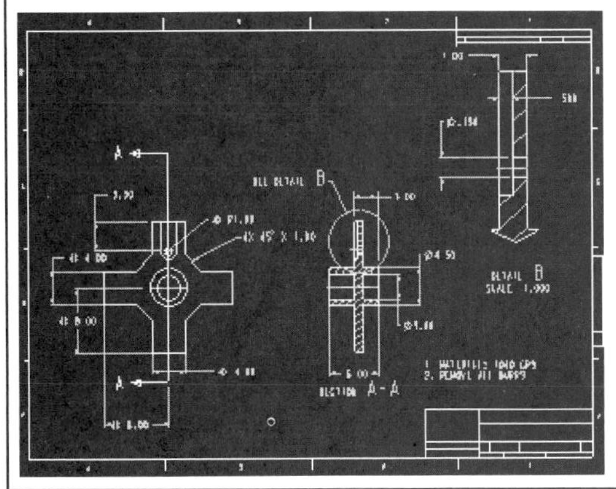

After a part's features are sketched, aligned, and dimensioned, you modify the dimension values to the exact sizes required for the design. The dimensioning scheme, the controlling features, the parent-child relationships, and the datums used to define and control the part features are determined as you design and model with Pro/E.

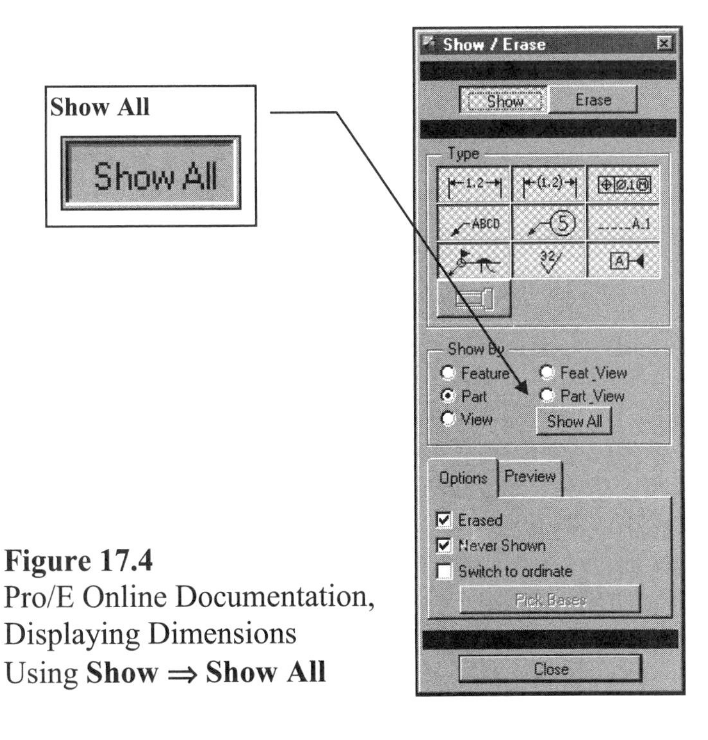

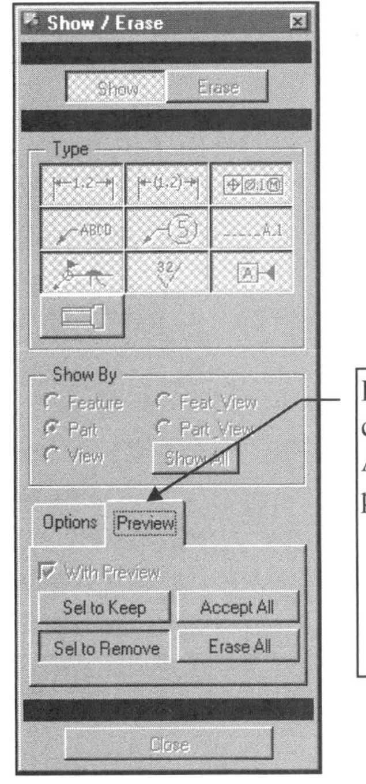

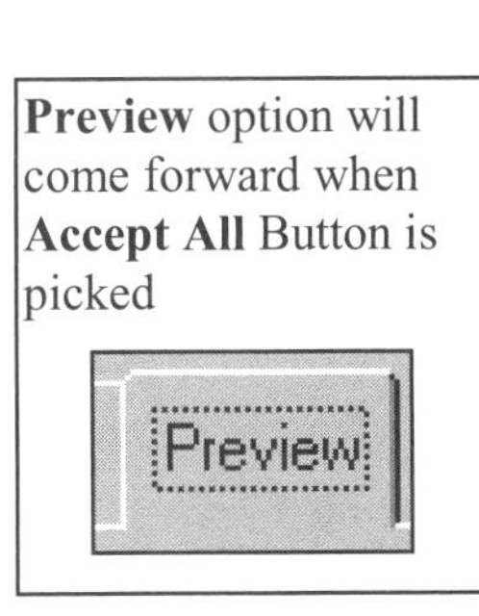

Figure 17.4
Pro/E Online Documentation, Displaying Dimensions Using **Show ⇒ Show All**

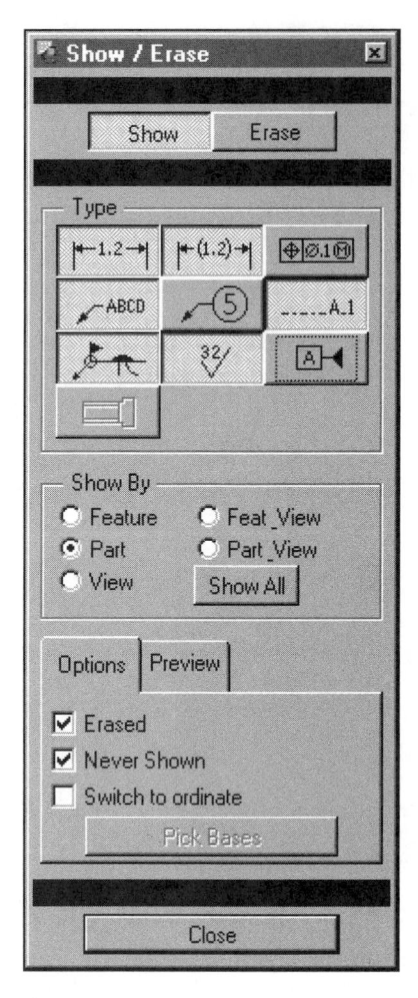

When detailing, you simply choose Pro/E commands to display views needed to describe the part and then display the dimensions used to model the part. These are the same dimensions used in the part's design. You cannot under-dimension or over-dimension, because Pro/E displays exactly what is required to model the part. Pro/E will not duplicate driving dimensions on a drawing. If a dimension is shown in one view, it will not be shown in another view. The dimension, however, can be switched to another view via detailing options.

To display dimensions on a drawing:

1. Choose **Show/Erase** from the DETAIL menu. The **Show/Erase** dialog box is displayed. Select the dimension icon in the Type area. (To specify the dimensions as ordinate, choose the **Switch to ordinate** check box at the bottom of the dialog box, note that the **Pick Bases** button highlights, and then pick a *baseline*. In this case, the baseline must already exist.)

2. Choose one of the following radio buttons:

   Feature Show all the dimensions associated with a particular feature in the appropriate views. Select a feature to be dimensioned.
   Feat_View Show all dimensions for a single feature in a single view. Select a feature and the view where the dimensions are to be displayed.
   Part Show dimensions associated with a part.
   Part_View Show dimensions associated with a part and view.
   View Show all the dimensions associated with a selected view. Select the view(s) in which you would like dimensions shown.

To show all the dimensions for the current model, choose the **Show All** command button (Fig. 17.4). To *preview* what the drawing will look like when you make the changes, choose the **Preview** button. To specify the items that you wish to show on the drawing, choose the **Preview** tab and then select **With Preview** and one of the following command buttons: **Accept All, Erase All, Sel to Keep,** or **Sel to Remove.** Figure 17.5 shows all dimensions.

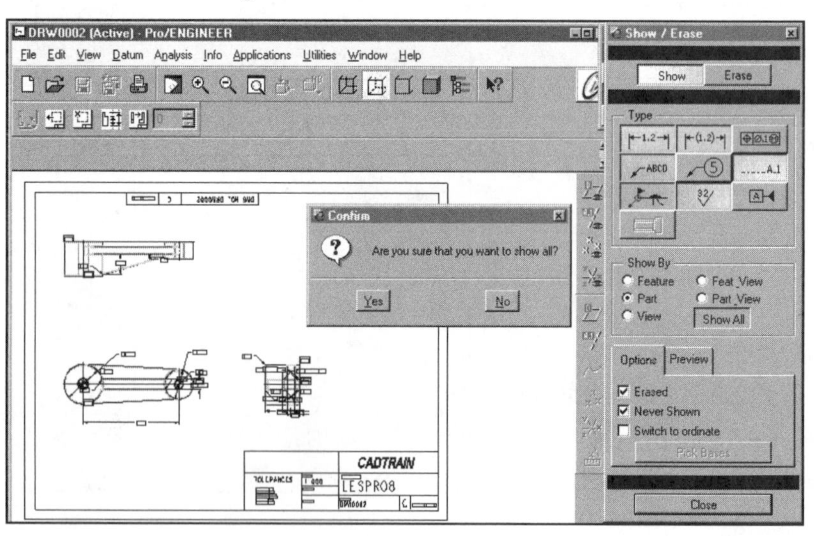

Figure 17.5
Drawing with Axes, Dimensions, and Datums Displayed (Uncleaned, Original Position)

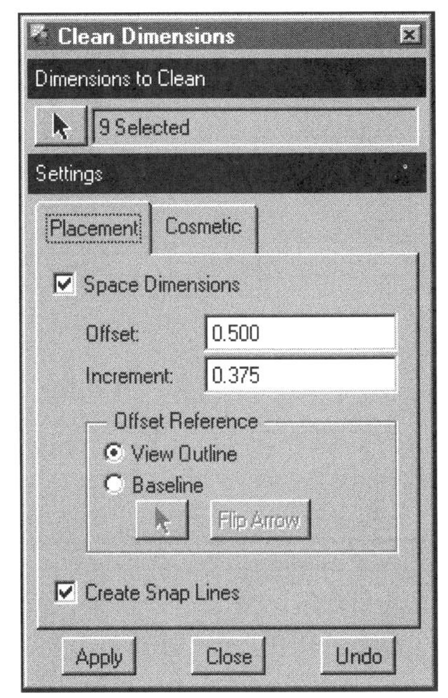

The **Clean Dims** option allows you to distribute standard and ordinate dimensions with equidistant spacing along witness lines, displaying them in a more orderly and readable fashion (Fig. 17.6).

To clean up the dimension display:

1. Select **Tools ⇒ Clean Dims.**
2. Select the view(s) or dimensions to be cleaned up.
3. In the Clean Dimensions dialog box, enter the Offset value for the first dimension line (the one that is closest to the model).
4. Enter the distances between all other dimensions (Increment).
5. Choose **Apply.**
6. Dimensions pertaining to the selected view are displayed with the specified spacing. In the event that the new display of dimensions is unsatisfactory, choose **Undo** to restore. Choose **Apply** to accept the dimension placement. Choose **Close** to close the dialog box.

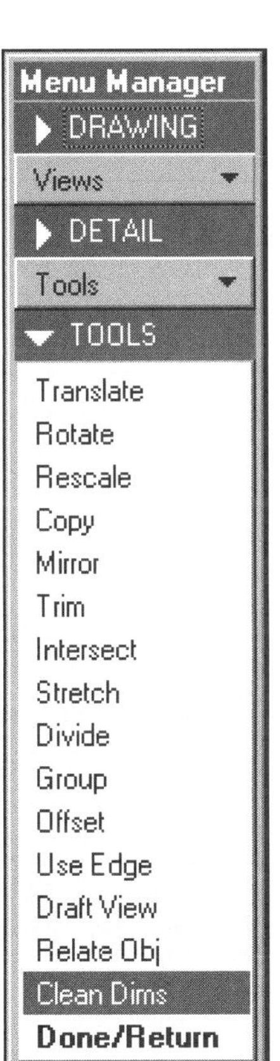

Clean Dims does not reposition radius or diameter dimensions.

The cleaned dimensions are usually not in the best positions for each dimension and note. After cleanup, the next step is to move and reposition the dimensions to create an ASME standard drawing.

Dimensions can be removed from the display by erasing them **(Show/Erase).** Erasing dimensions does not delete them from the model (driving dimensions, "true part" dimensions, *cannot be deleted*, but reference dimensions and driven dimensions can). Dimensions that have been erased can be redisplayed via the **Show/Erase** dialog box.

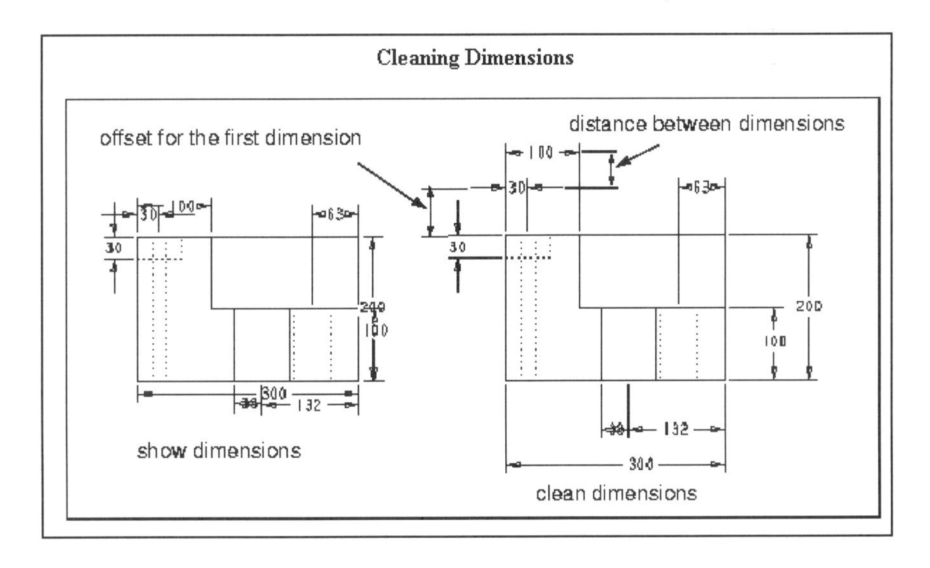

Figure 17.6
Pro/E Online Documentation, Cleaning Dimensions

Text and Notes on Drawings

> *COAch™ for Pro/ENGINEER*
>
> The **CADTRAIN CD** supplied with this text has a segment on **TEXT.**

Notes (Fig. 17.7) can be part of a dimension, attached to one or many edges on the model or "free." You can add notes by typing from the keyboard or by reading them in from a text file. Notes are created with the default values (height, font, etc.) specified in the **Draw Setup (.dtl)** file.

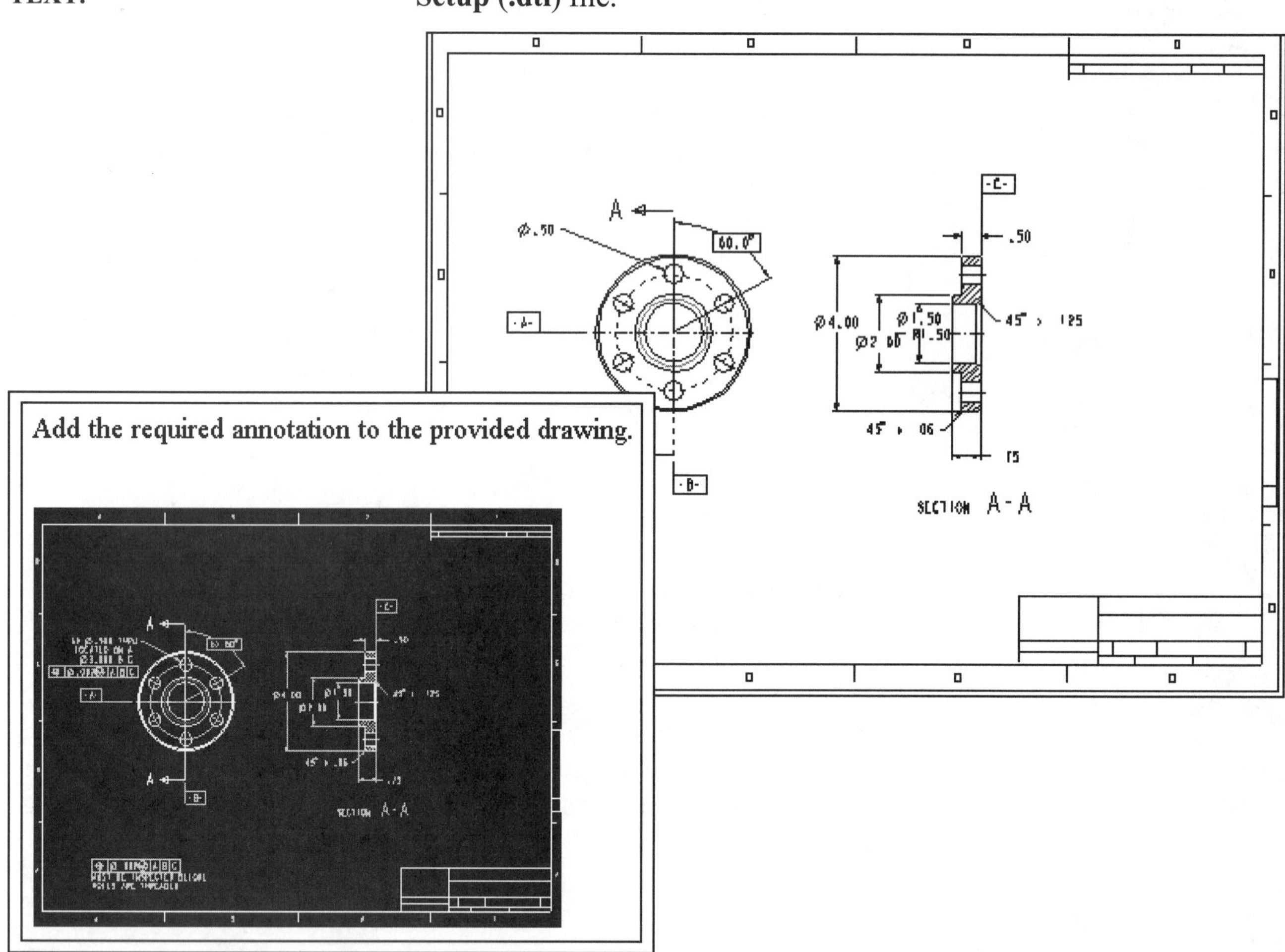

Figure 17.7
COAch for Pro/E, Drawings, Text (Parameters and Notes)

▾ NOTE TYPES
- No Leader
- Leader
- ISO Leader
- On Item
- Offset
- Enter
- File
- Horizontal
- Vertical
- Angular
- Standard
- Normal Ldr
- Tangent Ldr
- Left
- Center
- Right
- Default
- Style Lib
- Cur Style
- **Make Note**
- Done/Return

To add notes to the drawing, choose **Create ⇒ Note ⇒ Make Note**. Notes can include text and/or symbols (Fig. 17.8) and can be added with or without leaders, and the style of text can be modified via a dialog box (Fig. 17.9). The following options are available:

No Leader/Leader/On Item Create a note, with or without a leader, attached to an entity or not.
ISO Leader Create a note with ISO standard leader line.
Enter/File Enter the note from the keyboard, or read the note in from a text file.
Horizontal/Vertical/Angular Create a horizontal or vertical note, or enter an angular value between **0°** and **359°.**
Standard Create a note with multiple leaders.
Normal Ldr Create a note with a leader normal to an entity.
Tangent Ldr Create a note that is tangent to an entity.
Left/Center/Right/Default Create the note text as left-justified, center-justified, or right-justified. **Default** is left-justified.

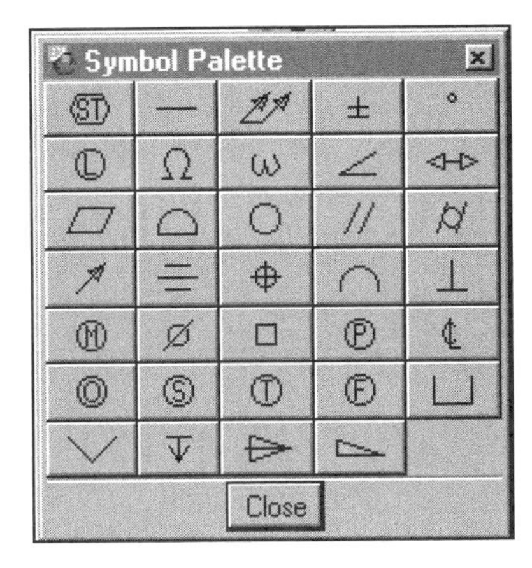

Figure 17.8
Special Symbols

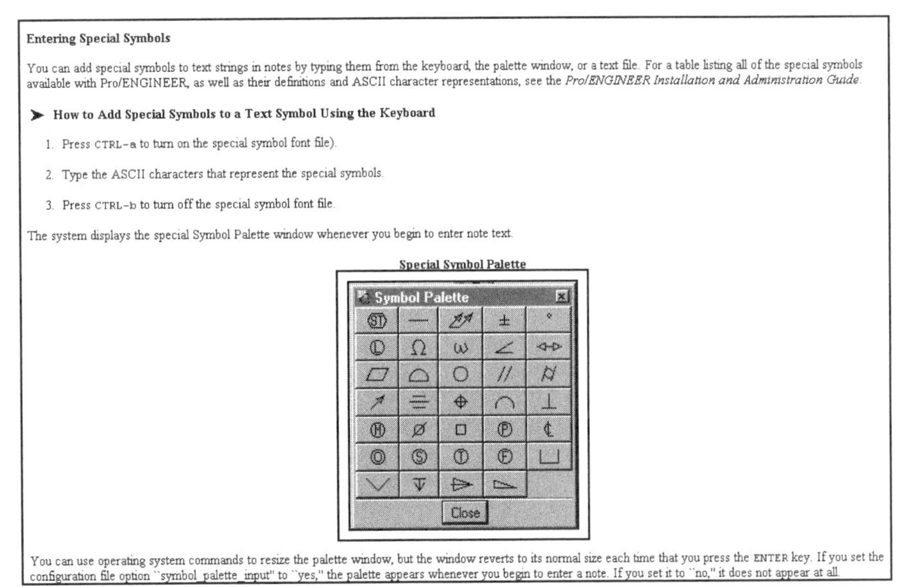

Entering Special Symbols

You can add special symbols to text strings in notes by typing them from the keyboard, the palette window, or a text file. For a table listing all of the special symbols available with Pro/ENGINEER, as well as their definitions and ASCII character representations, see the *Pro/ENGINEER Installation and Administration Guide*.

➤ **How to Add Special Symbols to a Text Symbol Using the Keyboard**

1. Press CTRL-a to turn on the special symbol font file).
2. Type the ASCII characters that represent the special symbols.
3. Press CTRL-b to turn off the special symbol font file.

The system displays the special Symbol Palette window whenever you begin to enter note text.

Special Symbol Palette

You can use operating system commands to resize the palette window, but the window reverts to its normal size each time that you press the ENTER key. If you set the configuration file option "symbol_palette_input" to "yes," the palette appears whenever you begin to enter a note. If you set it to "no," it does not appear at all.

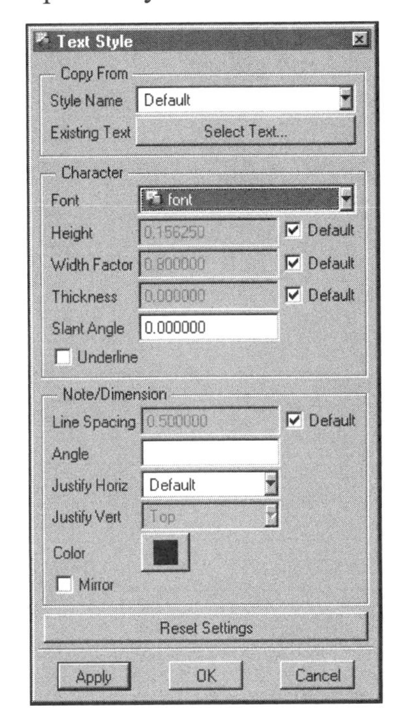

Figure 17.9
Text Style Dialog Box

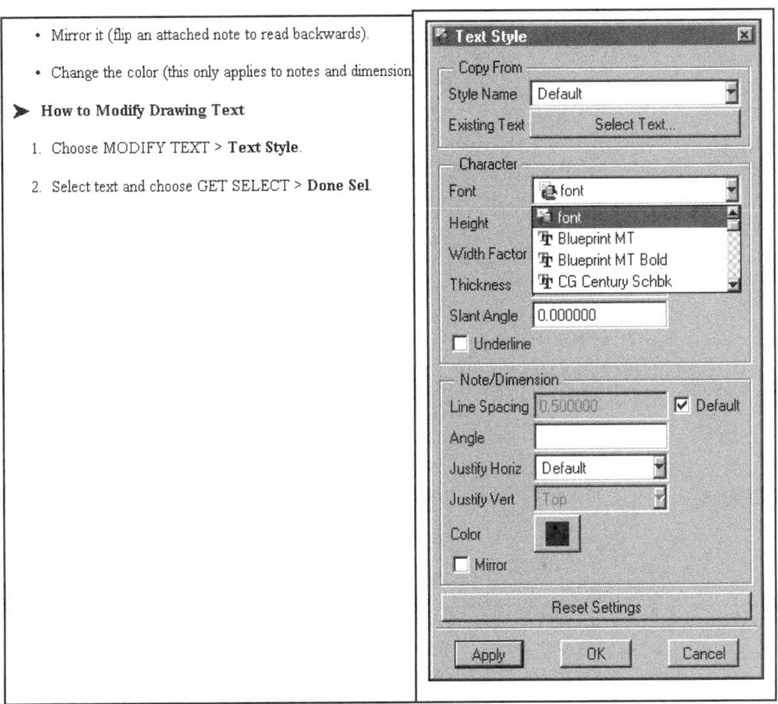

- Mirror it (flip an attached note to read backwards).
- Change the color (this only applies to notes and dimension

➤ **How to Modify Drawing Text**

1. Choose MODIFY TEXT > **Text Style**.
2. Select text and choose GET SELECT > **Done Sel**.

Adding Text to a Dimension

The **Text** option from the MODIFY DRAW menu allows you to add text to a dimension value (for example, **DIAMETER**, **REF**, and **TYP**), as well as special symbols. You can also define your own special fonts and symbols. To add text:

1. Choose **Text** from the MODIFY DRAW menu.
2. Choose **Text Line** to modify one line of the note or dimension, or choose **Full Note** to modify the whole note or dimension.
3. Pick the dimension to which you will add the text.
4. Enter the line or lines of text. Each line must end with a carriage return (**enter**).
5. To complete the text string, follow the line with an **enter**, or exit the editor when you are finished with the **Full Note**.

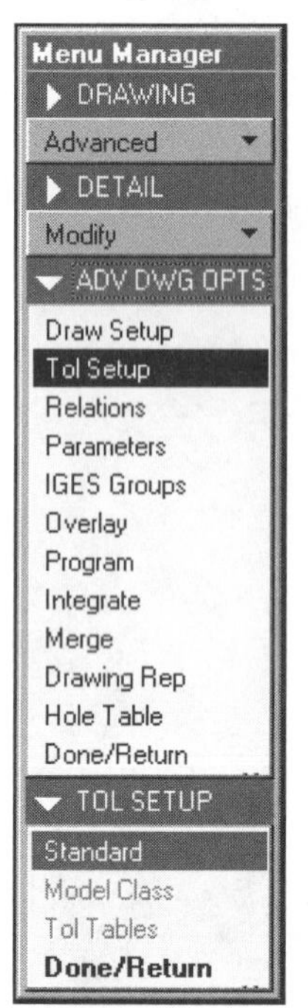

Geometric Dimensioning and Tolerancing

The manufacturing of parts and assemblies uses a degree of precision determined by **tolerances**. A typical parametric design system supports three types of tolerances:

Dimensional Specifies allowable variation of size.
Geometric Controls form, profile, orientation, and runout (Fig. 17.10).
Surface finish Controls the deviation of a part surface from its nominal value.

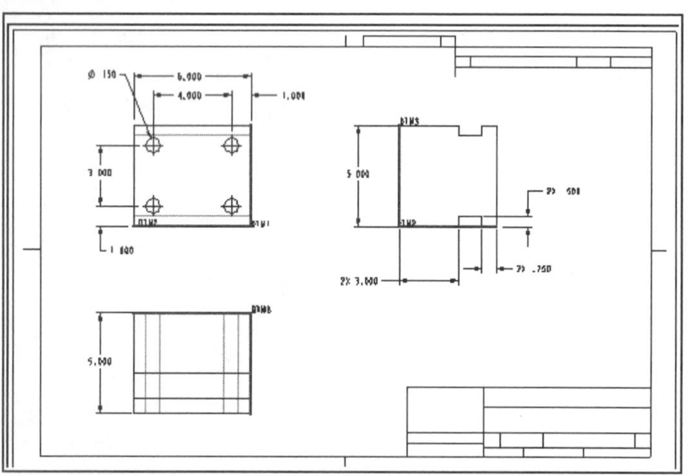

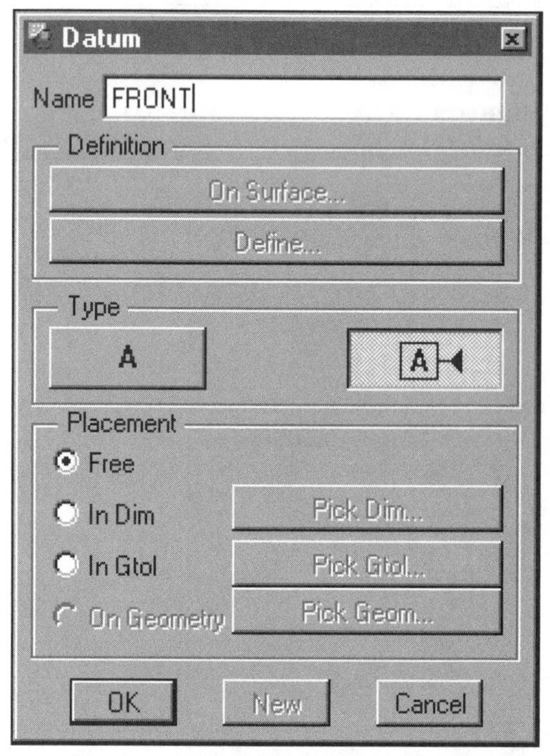

Figure 17.10
COAch for Pro/E, Drawings, GD&T (Setting Datums)

When you design a part, you specify dimensional tolerance, which means the ***allowable variations in size***. All dimensions are controlled by tolerances, except "**basic**" dimensions, which for the purpose of reference are considered to be *exact*.

Dimensional tolerances on a drawing can be expressed in two forms:

* As **general tolerances** presented in a tolerance table. These apply to those dimensions that are displayed in nominal format--that is, without tolerances.
* As **individual tolerances** specified for individual dimensions.

You can use general tolerances given as defaults in a table or set individual tolerances by modifying default values of selected dimensions. Default tolerance values are used at the moment you start to create a model; therefore, *default tolerances must be set prior to creating geometry.*

When you start to create a part, the table at the bottom of the window will display the current defaults for tolerances. If you have not specified tolerances, Pro/E defaults are assumed, and the table will look as follows:

* x.x	±0.1
* x.xx	±0.01
* x.xxx	±0.001
* ANG.	±0.5

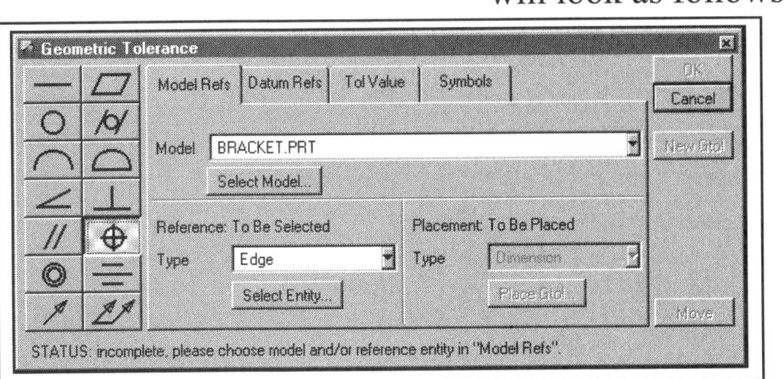

You have a choice of displaying or blanking tolerances. If tolerances are not displayed, Pro/E still stores dimensions with their default tolerances, as shown in Figure 17.11. You can specify geometric tolerances, create "basic" dimensions, and set selected datums as reference datums for geometric tolerancing. ISO tolerances are generated from a table.

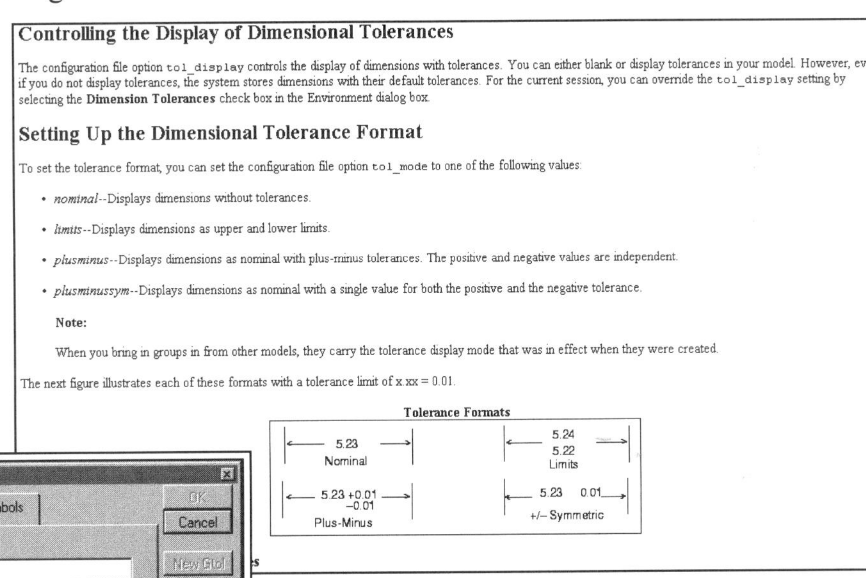

Controlling the Display of Dimensional Tolerances

The configuration file option `tol_display` controls the display of dimensions with tolerances. You can either blank or display tolerances in your model. However, even if you do not display tolerances, the system stores dimensions with their default tolerances. For the current session, you can override the `tol_display` setting by selecting the **Dimension Tolerances** check box in the Environment dialog box.

Setting Up the Dimensional Tolerance Format

To set the tolerance format, you can set the configuration file option `tol_mode` to one of the following values:

- *nominal*--Displays dimensions without tolerances.
- *limits*--Displays dimensions as upper and lower limits.
- *plusminus*--Displays dimensions as nominal with plus-minus tolerances. The positive and negative values are independent.
- *plusminussym*--Displays dimensions as nominal with a single value for both the positive and the negative tolerance.

Note:

When you bring in groups in from other models, they carry the tolerance display mode that was in effect when they were created.

The next figure illustrates each of these formats with a tolerance limit of x.xx = 0.01.

Tolerance Formats

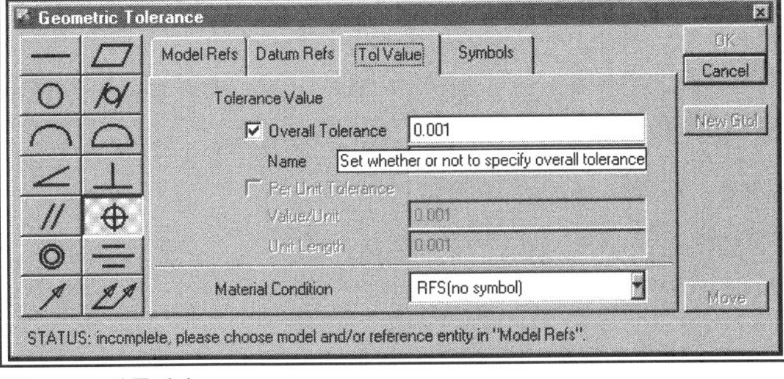

Figure 17.11
Tolerance Format

The available tolerance formats are:

Nominal Dimensions displayed without tolerances.
Limits Tolerances displayed as upper and lower limits.
Plus-Minus Tolerances displayed as nominal with plus-minus tolerance. The positive and negative values are independent.
±Symmetric Tolerances displayed as nominal with a single value for both the positive and the negative tolerance.
As Is Tolerances as is.

COAch™ for Pro/ENGINEER

The **CADTRAIN CD** supplied with this text has a segment on **GD&T.**

Pro/DETAIL enables you to add geometric tolerances to the model from Drawing mode. Note that geometric tolerances can be added in Part or Drawing mode but are reflected in all other modes. Geometric tolerances are treated by Pro/E as *annotations*, and they are always associated with the model. Unlike dimensional tolerances, *geometric tolerances do not have any effect on part geometry.*

Geometric tolerances (Fig. 17.12) provide a method for controlling the location, form, profile, orientation, and runout of features. You add geometric tolerances to the model from Part mode or Drawing mode. The geometric tolerances are treated by Pro/E as annotations, and they are always associated with the model. *Unlike dimensional tolerances, geometric tolerances do not have any effect on part geometry.*

When adding a geometric tolerance to the model, you can attach it to existing dimensions, edges, and existing geometric tolerances, or you can display it as a note without a leader.

Before you can reference a datum in a geometric tolerance, you must first indicate your intention by **setting** the datum. Once a datum is set, hyphens are added before and after the datum name, and it is enclosed in a rectangle (using the old standards). The ASME Y14.5M 1994 standards display the datum name, as in Figures 17.13 and 17.14. You can change the name of a datum, either before or after it has been set, by using the **Name** option in the **Set Up** menu from Part or Assembly mode.

Figure 17.12
Geometric Tolerances

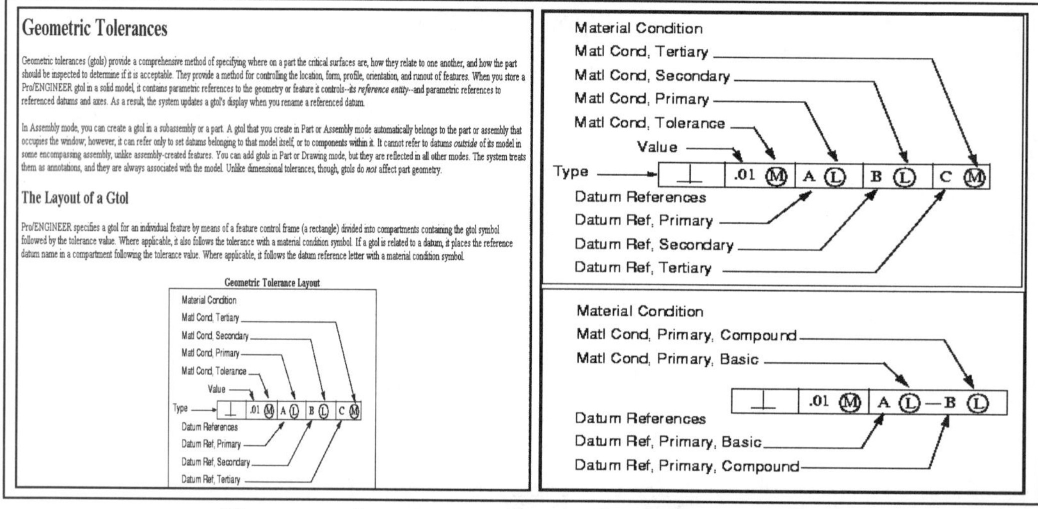

Geometric Tolerances

Geometric tolerances (gtols) provide a comprehensive method of specifying where on a part the critical surfaces are, how they relate to one another, and how the part should be inspected to determine if it is acceptable. They provide a method for controlling the location, form, profile, orientation, and runout of features. When you store a Pro/ENGINEER gtol in a solid model, it contains parametric references to the geometry or feature it controls--its *reference entity*--and parametric references to referenced datums and axes. As a result, the system updates a gtol's display when you rename a referenced datum.

In Assembly mode, you can create a gtol in a subassembly or a part. A gtol that you create in Part or Assembly mode automatically belongs to the part or assembly that occupies the window; however, it can refer only to set datums belonging to that model itself, or to components within it. It cannot refer to datums *outside* of its model in some encompassing assembly, unlike assembly-created features. You can add gtols in Part or Drawing mode, but they are reflected in all other modes. The system treats them as annotations, and they are always associated with the model. Unlike dimensional tolerances, though, gtols do *not* affect part geometry.

The Layout of a Gtol

Pro/ENGINEER specifies a gtol for an individual feature by means of a feature control frame (a rectangle) divided into compartments containing the gtol symbol followed by the tolerance value. Where applicable, it also follows the tolerance with a material condition symbol. If a gtol is related to a datum, it places the reference datum name in a compartment following the tolerance value. Where applicable, it follows the datum reference letter with a material condition symbol.

Geometric Tolerance Layout

You can choose any datum feature as a reference datum for a geometric tolerance. To set a reference datum in Drawing mode, choose **Create ⇒ Geom Tol:**

1. Choose **Set Datum** from the GEOM TOL menu.
2. Select the datum plane or axis to be set.
3. Change the name if desired. Change the type of feature control frame; then choose **OK**.
4. The datum is enclosed in a feature control frame.

A geometric tolerance for individual features is specified by means of a **feature control frame** (a rectangle) divided into compartments containing the geometric tolerance symbol followed by the tolerance value. Where applicable, the tolerance is followed by a **material condition symbol**. Where a geometric tolerance is related to a datum, the reference datum name is placed in a compartment following the tolerance value. Where applicable, the reference datum name is followed by a material condition symbol.

HINT

To set datums displayed with the correct ASME Y14.5 1994 symbology (Fig. 17.14), change your **.dtl** file:

Advanced ⇒ Draw Setup ⇒ ***gtol_datums std_ANSI*** change to ***std_ASME*** **⇒ Add/Change ⇒ Apply ⇒ Close**

Figure 17.13
Setting Up the Standard

For each class of tolerance, the types of tolerances available and the appropriate types of entities can be referenced. The available material condition symbols are shown in the dialog box (Fig. 17.11).

You are guided in the building of a geometric tolerance by Pro/E requests for each piece of required information. You respond by making menu choices, entering a tolerance value, and selecting entities and datums. As the tolerance is built, the choices are limited to those items that make sense in the context of the information you have already provided. For example, if the geometric characteristic is one that does not require a datum reference, you will not be prompted for one. Other checks are made to help prevent mistakes in the selection of entities and datums.

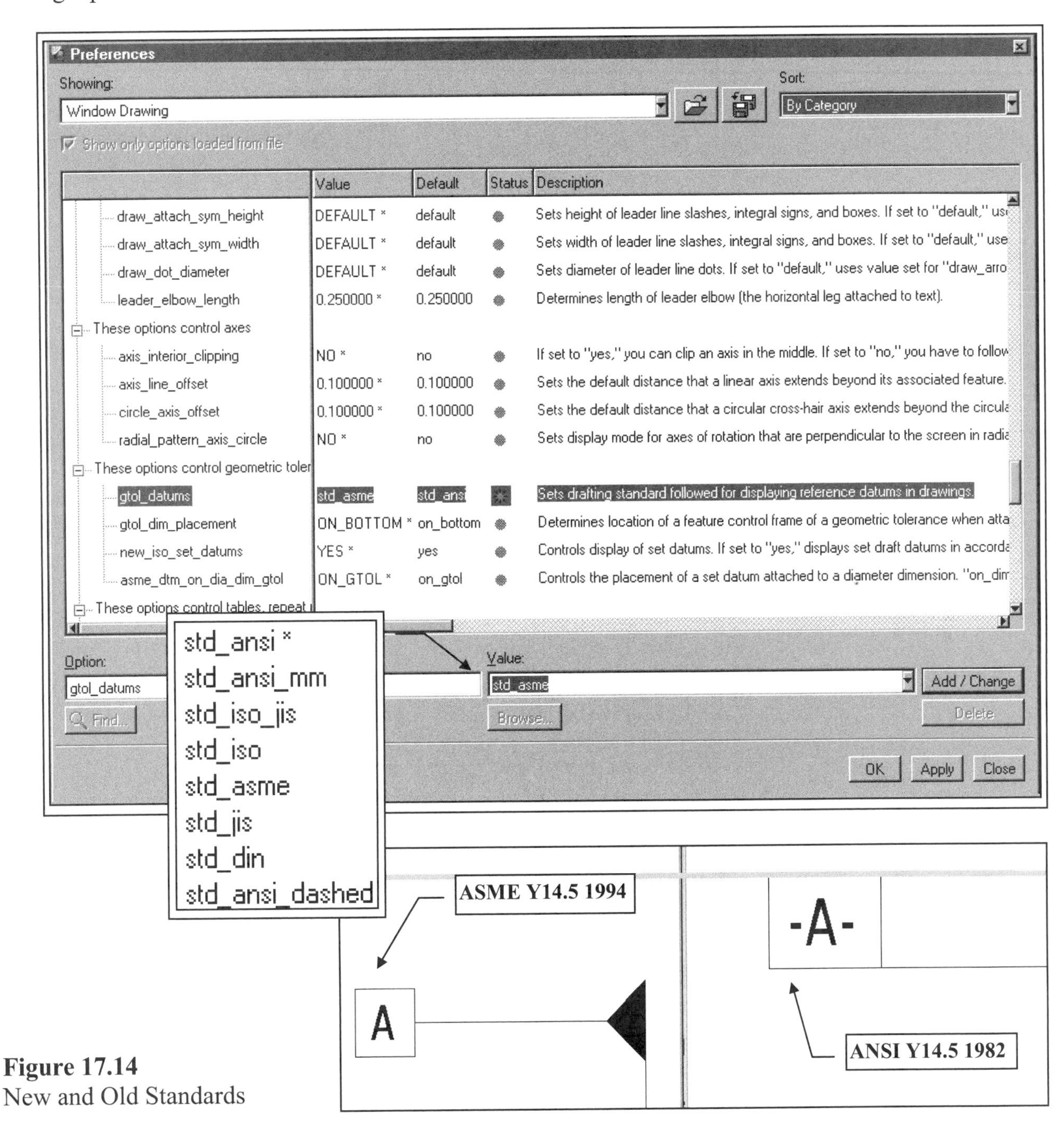

Figure 17.14
New and Old Standards

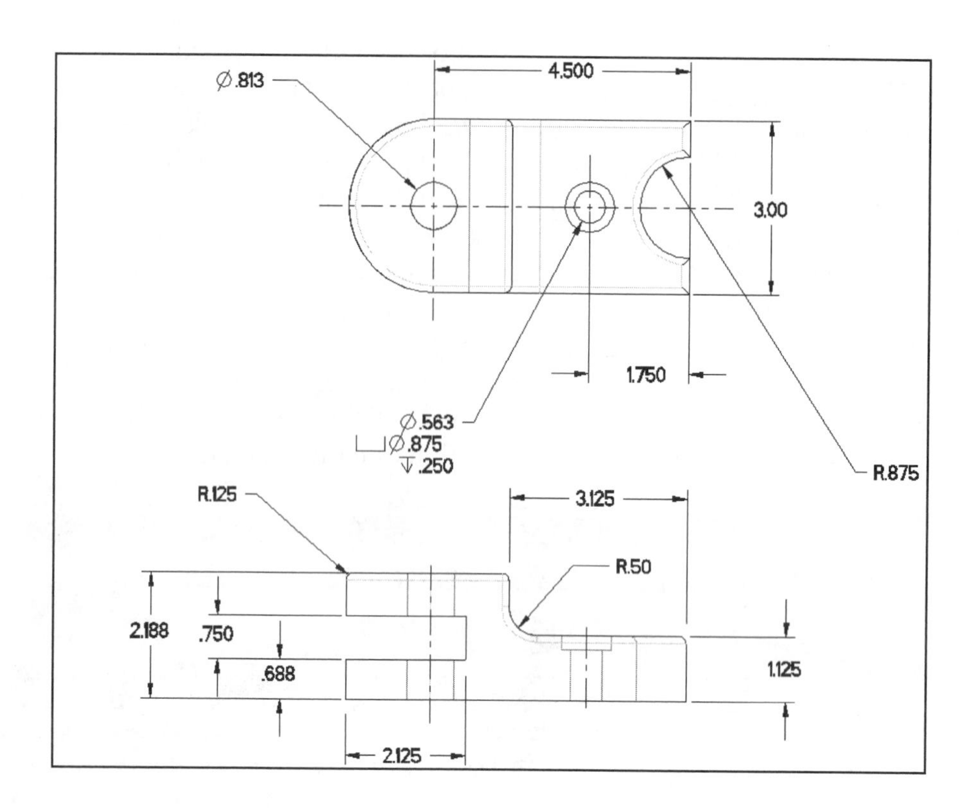

Figure 17.15
Breaker Drawing with Dimensions, Notes, and Centerlines

Breaker Drawing

Lesson procedures & commands
START HERE ➡ ➡ ➡

The Breaker (Fig. 17.15) from Lesson 3 will be detailed in this lesson. Dimensioning, centerlines (axes), annotation, and other drawing requirements will be used to complete the detailing of the model.

Pro/E allows you to dimension a drawing automatically based on the design intent dimensioning scheme used to create the part. Using this method, you are virtually assured that the part will be fully dimensioned without being overdimensioned. In addition to automatic dimensioning, Pro/E can automatically display centerlines based on datum axes and radial patterns that were used to model the part.

After the dimensions and axes are displayed, the only remaining detailing work involves cosmetically cleaning up their display and adding other annotations, such as general notes, title block information, and required tolerancing information.

Pro/E also allows you to *create associative dimensions* that are based on existing feature constraints. These dimensions are created using the same techniques you use to create sketch dimensions. These dimensions are referred to as *driven* dimensions, because their display is "driven" by changes to the model. *Driven dimensions* cannot, however, be used to make parametric changes to the model.

In this lesson, you will learn how to *display dimensions automatically* using the original constraints, *move dimensions* to another view, move dimensions to more appropriate locations, *display centerlines* on your drawing, *erase dimensions*, *modify extension lines* to show the proper gap to the appropriate edges, *add annotation* to a drawing, *alter decimal places*, and *add text* to dimension text.

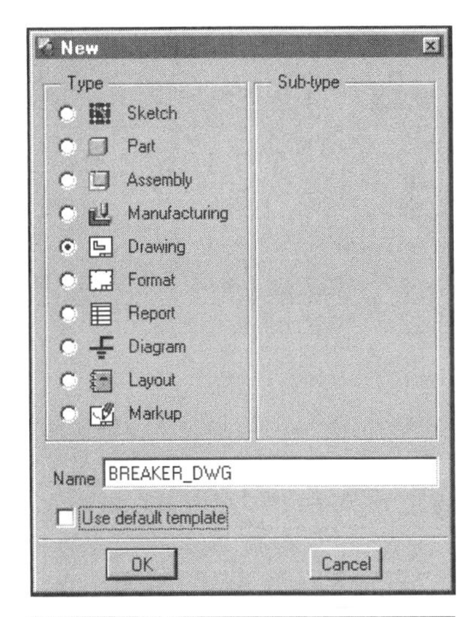

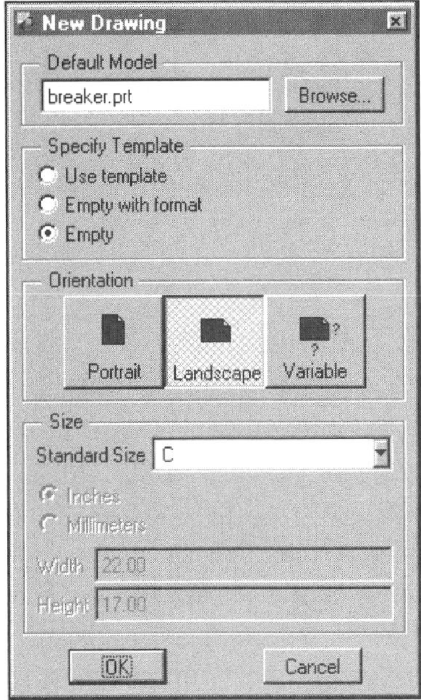

Figure 17.16
First View of the Drawing

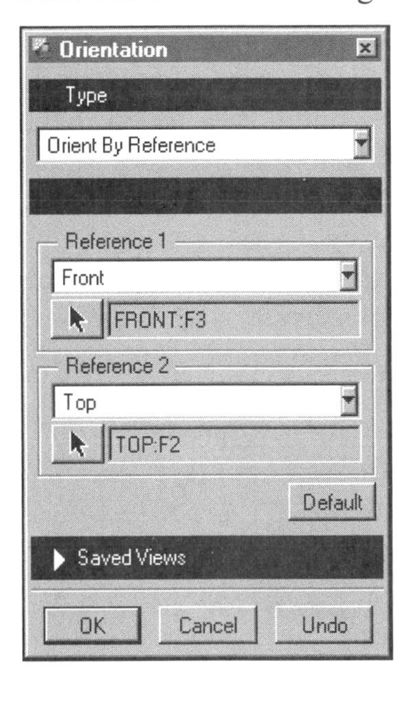

Upon entering the Drawing mode, a format is generally required prior to selecting the model to be detailed. The format can be a blank format, a Pro/E-provided ANSI standard format (with or without a default template), or a user-defined format created with parameters that will automatically display title block information and sheet callouts, as in Lesson 16.

Instead of using a format with a default template, use an empty drawing and later add a format. Use the following commands:

File ⇒ **New** ⇒ Type ●**Drawing** ⇒ Name (**BREAKER_DWG** or some other logical name) ⇒ ❑ **Use default template** ⇒ **OK** ⇒ **Browse** ⇒ Default Model [choose **Breaker** (or name you used for Lesson 3 part) from the directory list] ⇒ **Open** ⇒ Specify Template ●**Empty** ⇒ Orientation **Landscape** ⇒ Standard Size **C** ⇒ **OK** ⇒ **Views** ⇒ **Add View** ⇒ **General** ⇒ **Full View** ⇒ **No Xsec** ⇒ **No Scale** ⇒ **Done** (to accept defaults) ⇒ Select CENTER POINT for drawing view [pick a place on the drawing for the first view (Figure 17.16)] ⇒ **Front** ⇒ (pick **FRONT** as **Reference 1**) ⇒ **Top** ⇒ (pick **TOP** as **Reference 2**) ⇒ **OK** ⇒ **Add View** ⇒ **Projection** ⇒ **Full View** ⇒ **No Xsec** ⇒ **No Scale** ⇒ **Done** (to accept defaults) ⇒ Select CENTER POINT for drawing view [pick in the area to place the front view (Fig. 17.17)] ⇒ **Done/Return** ⇒ **File** ⇒ **Save** ⇒ ✔

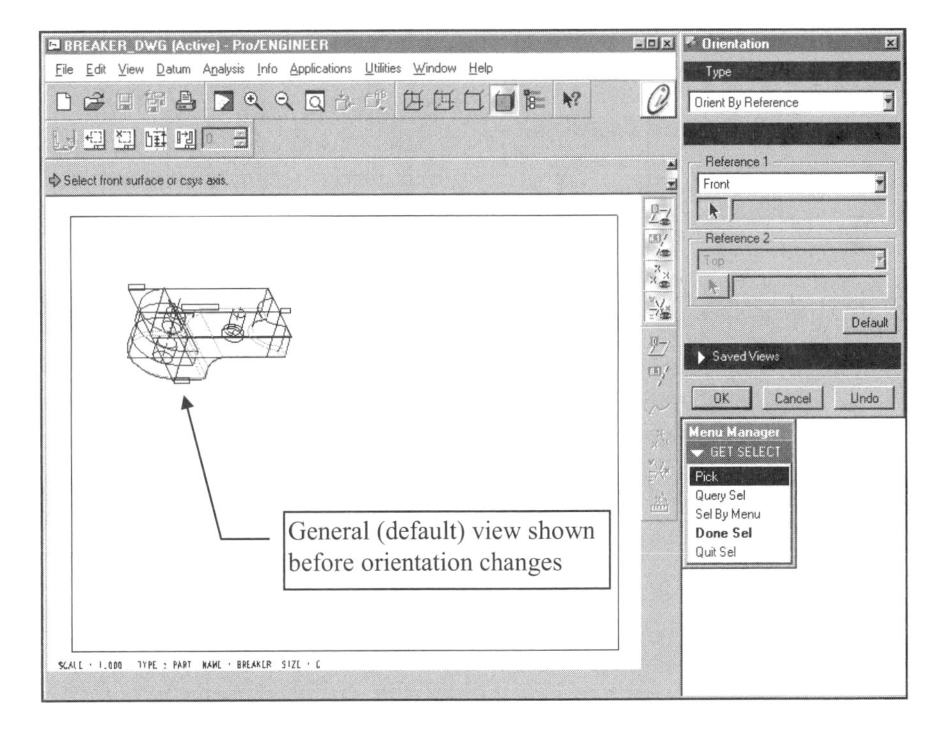

After you have the two views on your drawing, use **Move View** to relocate them to better positions for detailing. Remember, you can reposition the views at any time in the detailing process, even after the dimensions are placed.

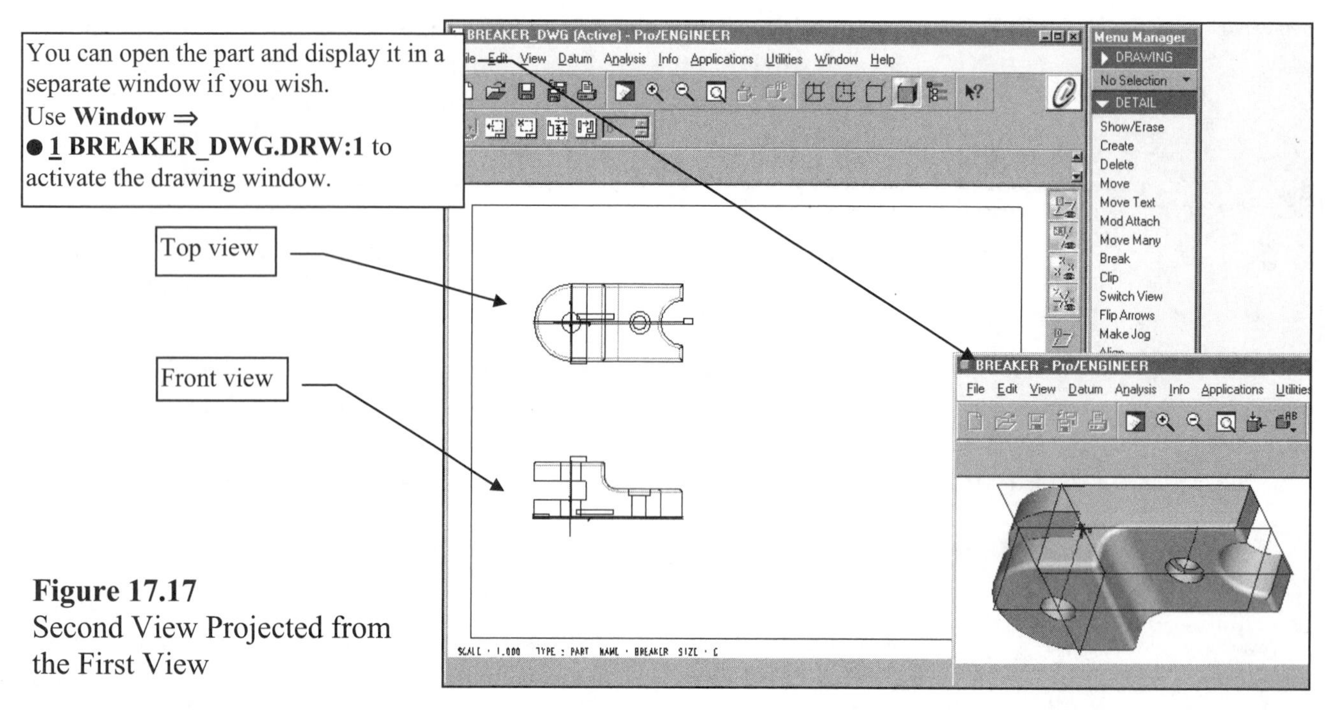

Figure 17.17
Second View Projected from the First View

Showing Axes

Before dimensions are added, Pro/E can automatically display centerlines at each location where there is a **Datum Axis**. The axis must be either parallel or perpendicular to the screen to be displayed.

If the axis is perpendicular to the screen, Pro/E displays a "crosshairs" centerline. If the axis is parallel to the screen, Pro/E displays a linear centerline. If the **Environment** option **Datum Axes** is off, the name of the axis is not displayed on the drawing. A crosshairs axis actually consists of four centerline segments. Each segment makes up one "leg" of the crosshairs. A linear axis actually consists of two centerline segments. Each segment makes up one "half" of the centerline. This distinction means very little, unless you need to trim the centerline segments because of interference with other drafting objects or edges.

There are several **Draw Setup (.dtl)** file parameters that control the display of centerlines. The parameter *circle_axis_offset* controls the distance that centerlines extend past the circle they lie on. The parameter *axis_line_offset* controls the distance that centerlines extend past the end of a cylindrical face. The parameter *axis_interior_clipping* controls whether or not you can trim the interior segments of a centerline. If this parameter is set to **no**, you can shorten or lengthen the ends of only linear and crosshairs centerlines. If the parameter is set to **yes**, you can shorten or lengthen any end of any centerline segment. The parameter *radial_pattern_axis_circle* will display a bolt circle centerline.

Turn off the datum planes, axes, points, and coordinate system before continuing.

NOTE

The **format** will have a **.dtl** file associated with it, and the **drawing** will have a different **.dtl** file associated with it. *They are separate **.dtl** files and have nothing to do with each other.* When you activate a drawing and then add a format, the **.dtl** for the format controls the font etc., for the format only. The drawing **.dtl** file still needs to be established.

If you changed the **.dtl** file in the format to have a filled font, the format will show as filled font when used on a drawing. The drawing dimensions and text will *not* show as filled because it has a separate **.dtl** file!

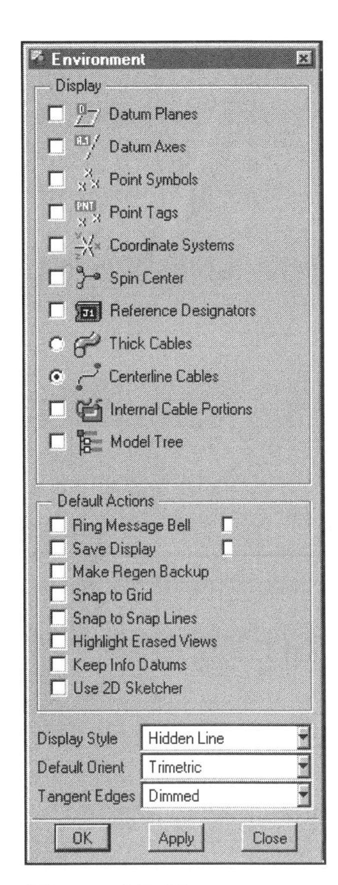

Figure 17.18
Displaying Axes

Turn off most of the **Environment** options so that they do not clutter the drawing views (left-hand column).

To show the centerlines/axes for the Breaker, choose:

Show/Erase ⇒ **Show** (button) ⇒ **A_1** (axis radio button on) ⇒ **Show All** ⇒ **Yes** (Fig. 17.18) ⇒ **Accept All** ⇒ **Close**

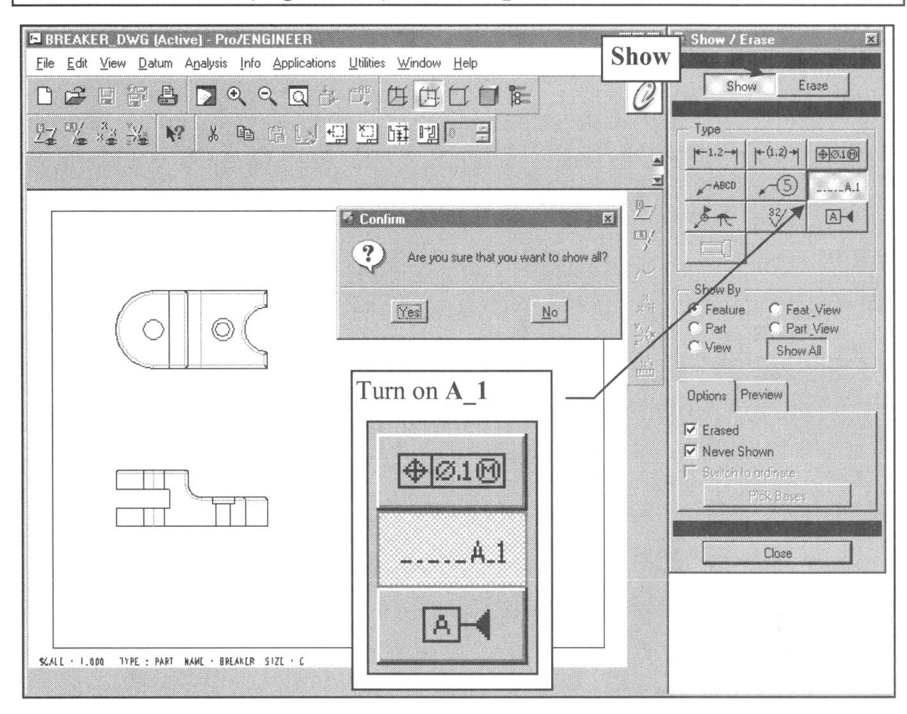

You can edit with a pop-up menu by *clicking the right mouse button while the cursor is in the main window.* Select **Modify Item** from the pop-up button; then select (in this case) the axis to modify. The **Move/Activate** option is used to move the end of the axis (Fig. 17.19). Modify the axes on both views to their required positions.

Click and hold down the right mouse button while the cursor is in the Main Window. The **Modify Item** button will appear. Pick it and then select the item you wish to edit (in this case, the centerline of the large arc).

Modify Item

The EDIT ACTIONS menu items will appear next to the mouse cursor.

Pick
Query Sel
Finish Modify
Move/Activate
Mod Attach
Redefine
Erase

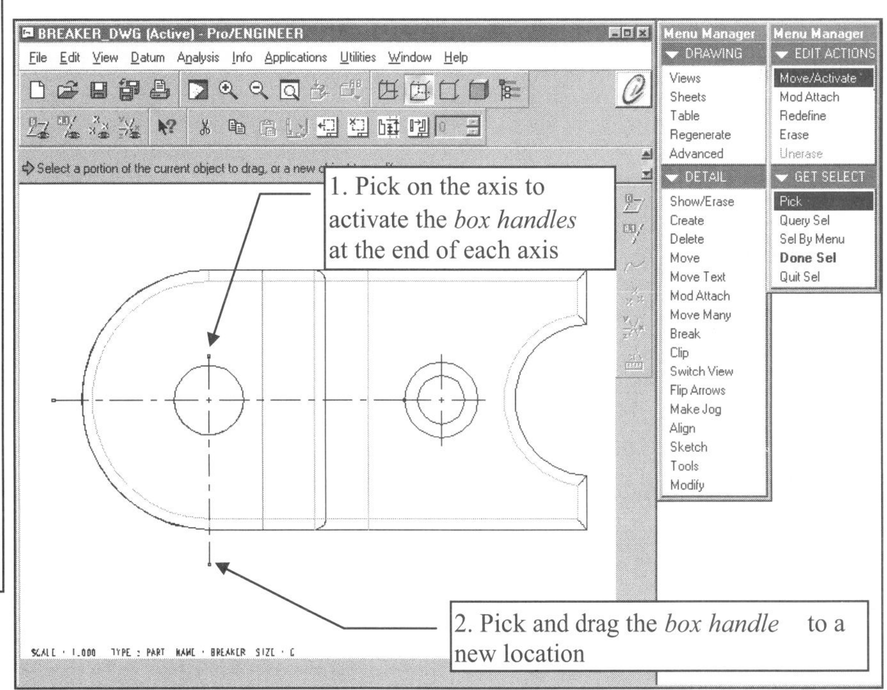

Figure 17.19
Modifying Axes

Show Dimensions

The **Show** option allows you to display dimensions based on the dimensional references and sketch dimensions that were used to create the model. You can show the dimensions of a feature, all the dimensions of all features in a particular view, or all the dimensions of all features on the model.

If you show dimensions in more than one view, and a feature can be dimensioned in more than one view, Pro/E attempts to decide which view is most appropriate to show the dimensions in.

Diameter dimensions are displayed differently based on the view where they are displayed. If a dimension is shown in a view where the cylinder axis is normal to the screen, the arrows are drawn to the circular edge of the cylinder. If the dimension is shown in a view where the cylinder axis is parallel to the screen, extension lines are added along the silhouette of the cylinder.

Changing the length of an axis can also be accomplished with the **Move** option. The **Move** option works on either end of the axis, depending on which end is selected.

You can show dimensions, axes, datums, and so on at the same time, but for a complex part, the views quickly become cluttered. Showing axes first, then the dimensions, and then the cosmetic features gives you an opportunity to modify each drawing entity type separately to see the view requirements more clearly. Figure 17.20 illustrates the dimensions displayed on the drawing using the following commands (your drawing may differ slightly):

Show/Erase ⇒ Show ⇒ (Dimension radio button on- all others off) ⇒ **Show All ⇒ Yes** (Fig. 17.20) ⇒ **Accept All ⇒ Close**

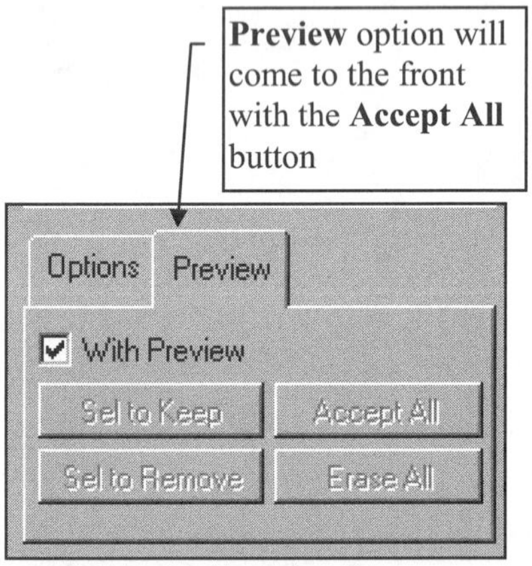

Preview option will come to the front with the **Accept All** button

Options Preview
With Preview
Sel to Keep Accept All
Sel to Remove Erase All

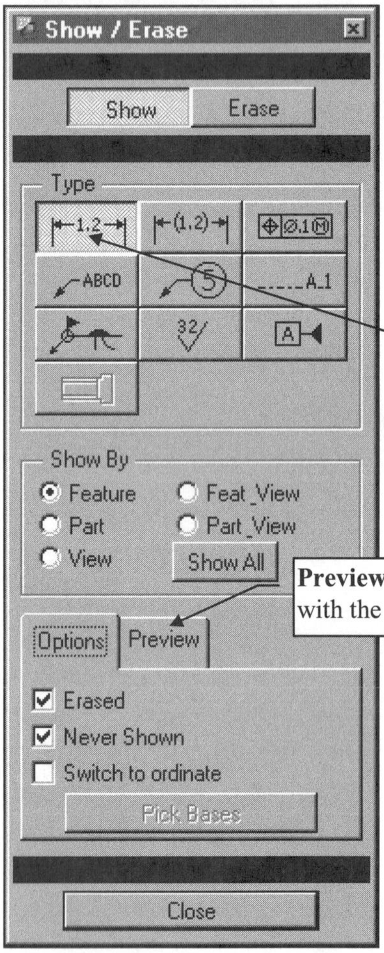

Show / Erase
Show Erase
Type
Show By
Feature Feat_View
Part Part_View
View Show All
Options Preview
Erased
Never Shown
Switch to ordinate
Pick Bases
Close

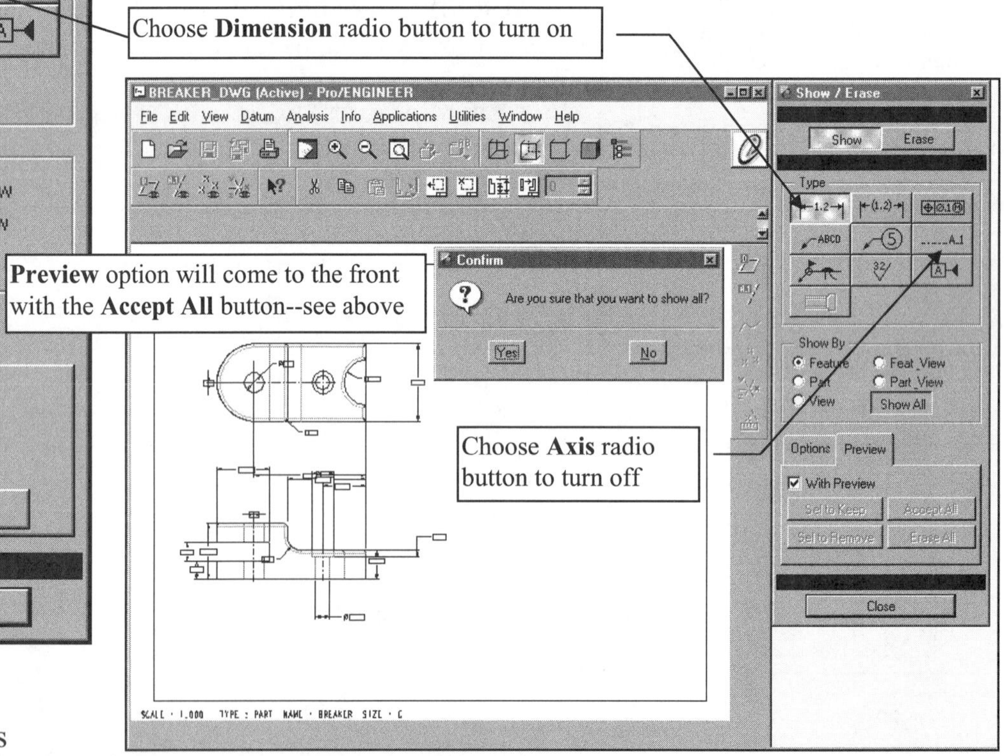

Choose **Dimension** radio button to turn on

Preview option will come to the front with the **Accept All** button--see above

BREAKER_DWG (Active) - Pro/ENGINEER
Confirm
Are you sure that you want to show all?
Yes No

Choose **Axis** radio button to turn off

Figure 17.20
Displaying Dimensions

Before modifying the dimension locations and views, change the settings in the **.dtl** file to set ***default_font*** to ***filled*** and ***draw_arrow_style*** to ***filled*** and ***gtol_datums*** to ***std_asme:***

Advanced ⇒ **Draw Setup** ⇒ Sort **By Category** ⇒ [make the changes (Fig. 17.21)] ⇒ **Add/Change** ⇒ **Apply** ⇒ **Close** ⇒ **View** ⇒ **Repaint** ⇒ **Done/Return** (Fig. 17.22)

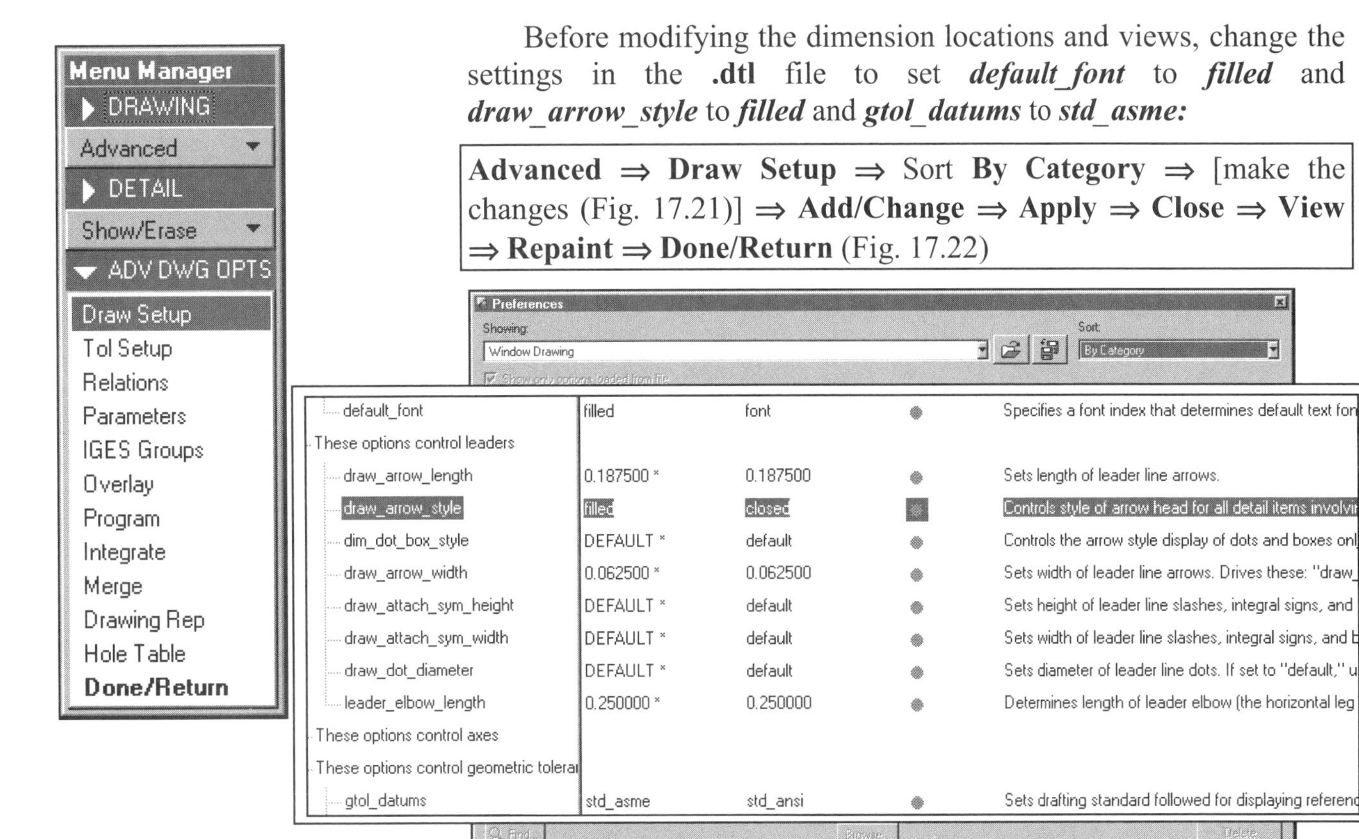

Figure 17.21
Advanced ⇒ Draw Setup

Modifying Dimensions and Drawing Entities

In Drawing mode, you can access a pop-up menu item by clicking and holding down the *right* mouse button while the cursor is in the main window. From the pop-up window, select **Modify Item**, and then select the item you wish to modify (Fig. 17.22). Another menu displays the actions that can be performed on this type of item. The **Move/Activate** option is the default. To move an item, simply select it and you will be in dynamic move.

The list of additional actions you can perform on the item is also available in the pop-up menu (see left). To modify another item, simply select it, and the menu will display the actions appropriate to it. This eliminates the need to traverse the screen to reach the menus, because the pop-up window is available at all times. For example, while the **Show/Erase** dialog box is displayed, you can use the pop-up menu option to move dimensions into the correct locations without having to close the dialog box.

Move, **Move Text**, **Clip**, and **Skew** functionality are all available using **Move/Activate** (Fig. 17.23). The selection of a particular box handle determines how items are moved. For example, if you select the dimension text, you can move the text of the dimension and the leader line. Selection of the leader line allows you to clip that leader line. Picking the extension (witness) line allows you to clip that extension line, and picking the other end of the extension line allows you to skew the dimension.

HINT

Again, after choosing **Modify Item** and selecting any entity, click and hold down the *right* mouse button while the cursor is in the Main Window. The EDIT ACTIONS menu will appear next to the mouse cursor.

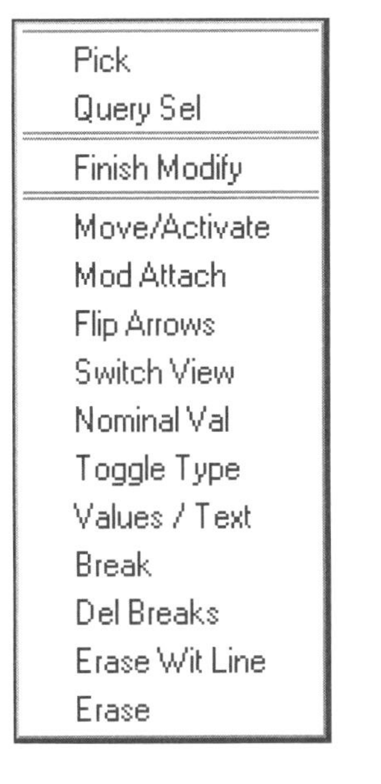

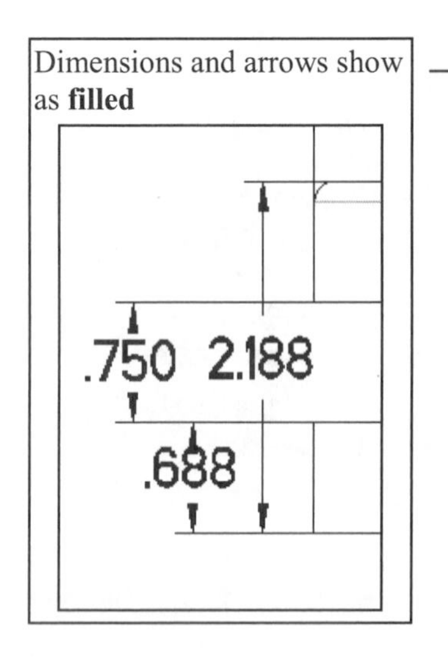

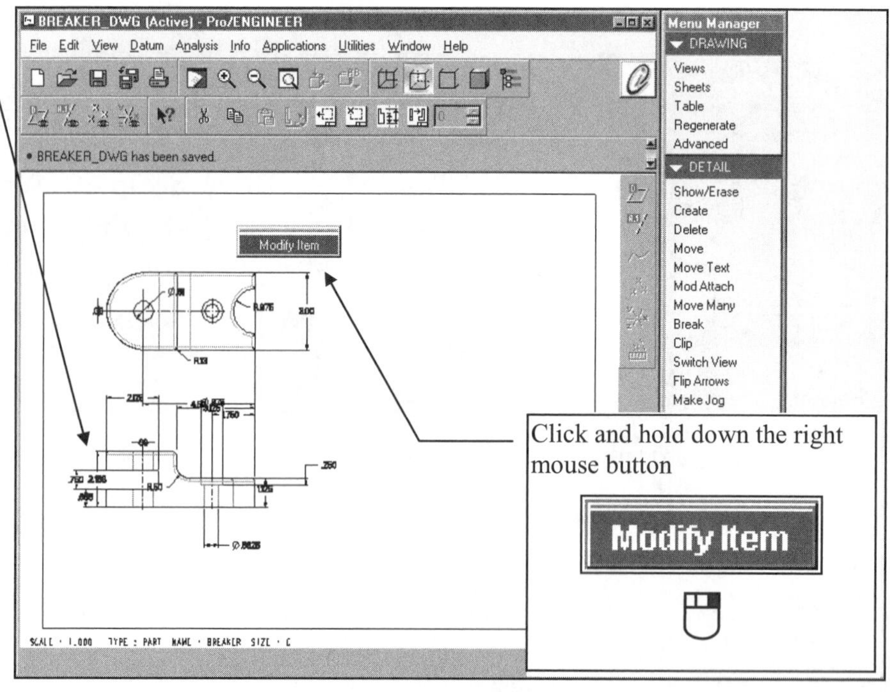

Figure 17.22
Modify Using Pop-up Window

There are four move menu options that allow you to change the position of a dimension. **Move, Move Text, Mod Attach,** and **Move Many** allow you to alter different aspects of a dimension. The most commonly used option is **Move**. When you **Move** a dimension, Pro/E drags its display along with the cursor in all directions. When you use **Move Text**, Pro/E moves the text only and does not move the extension lines. When you use **Mod Attach**, select a note, geometric tolerance, symbol, or surface finish to modify. Select the desired reference options, and then select the leader line you want to move and then the new attach point. Before modifying dimensions, make your grid size very small or turn the grid off as well as **Snap to Grid.** Reposition the dimensions and clean up the drawing. Do not concern yourself about dimensions in the wrong view at this time.

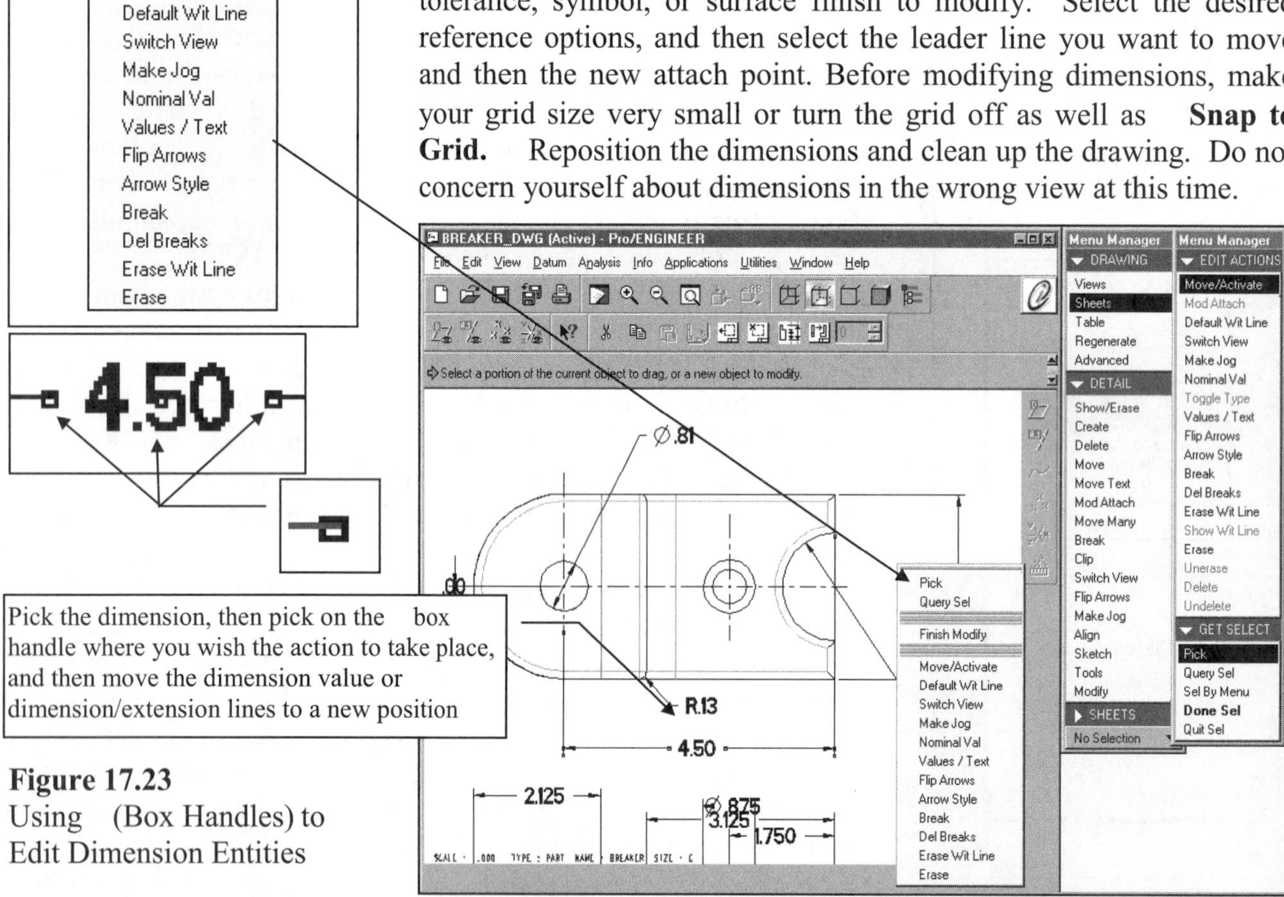

Figure 17.23
Using (Box Handles) to Edit Dimension Entities

Clipping Drafting Objects

File ⇒ Save ⇒ ✔

The **Clip** option allows you to move the endpoints of many drafting object components, such as extension lines and centerlines. This option allows you to "pick up" the end of an extension line and drag it to a new length (Fig. 17.24). Of course, clipping can also be done with the pop-up option in the **Modify Item** menu.

One of the most common applications for **Clip** is to "regap" extension lines. When you **Show** feature dimensions, the extension lines are often gapped to a point on the part that is not appropriate for the view in which the dimension is displayed.

If you are clipping *multiple dimensions*, Pro/E "grabs" the extension line in each dimension that is closest to the cursor when you click the *middle* mouse button.

Clip allows you to drag the extension line end to a location that is visually more pleasing (and consistent with drafting standards). When you **Clip** the extension line, Pro/E remembers the clip distance and continues to apply it even if the model changes or if you **Erase** and redisplay the dimension.

When you **Clip** extension lines, Pro/E prompts you to pick dimensions to clip. This allows you to clip several dimensions' extension lines together.

To stop picking dimensions, click the *middle* mouse button. Next, indicate which extension line(s) you want to clip, by moving the cursor near the desired extension line end. Pro/E then begins to drag the extension line(s). To **Clip** the line(s) after you move the cursor to the desired location, click the *left* mouse button. If you want to abort the **Clip**, click the *middle* mouse button.

Figure 17.24
Using **Clip**

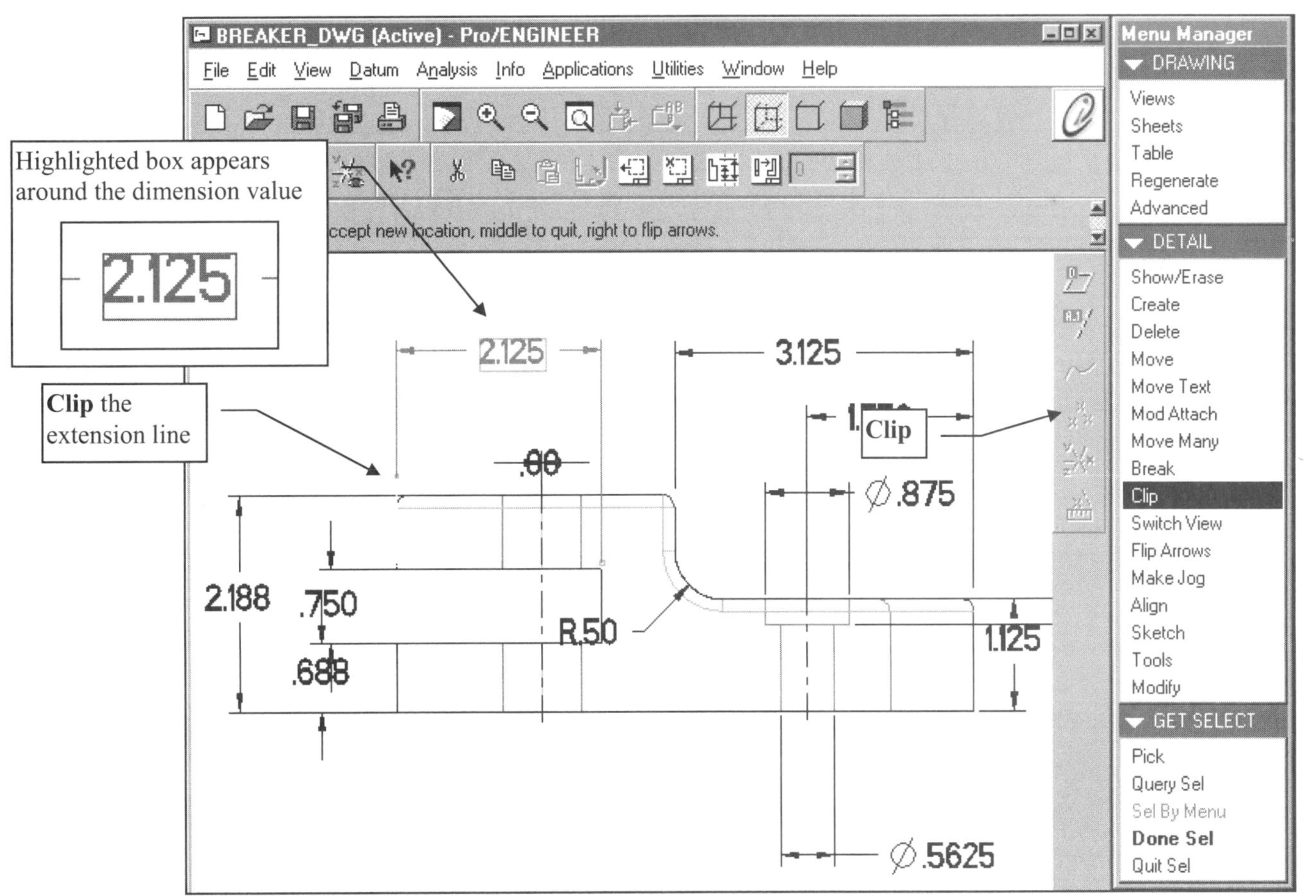

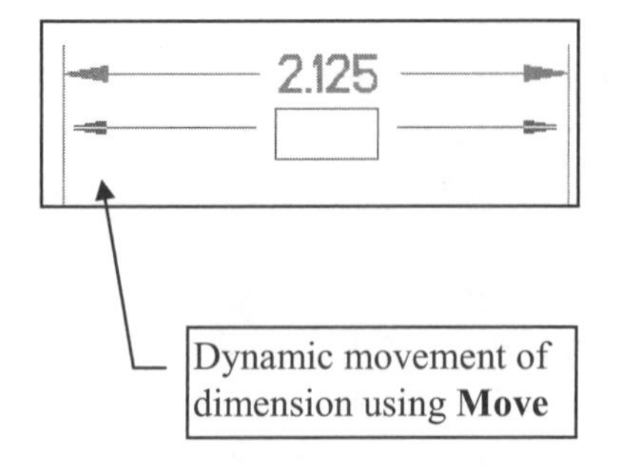

When you use **Move** to move a linear dimension, Pro/E adjusts the leader, arrows, and extension lines in proper relation to the text (which is what you are actually dragging).

If you place the text of a dimension between the extension lines and it really does not fit (because the extension lines are too close together), you may find the display of the arrows to be unacceptable. Move the origin of the text from one side of the extension lines to the other; Pro/E automatically flips the leader to the other side of the text.

When you move a radial dimension using **Move**, you can move the text only along the leader and the location where the arrow touches the arc and thus alter the angle of the leader; however, the short line (stub) between the leader and the text stays at its original length. When you use **Move Text**, you can change only the length of the stub.

Besides moving dimensions, you may also need to switch the view that a dimension appears in after choosing **Show All.** In Figure 17.25, **Switch View** was chosen from the DETAIL menu (also available after choosing a dimension when the **Modify Item** option is selected from the pop-up window menu).

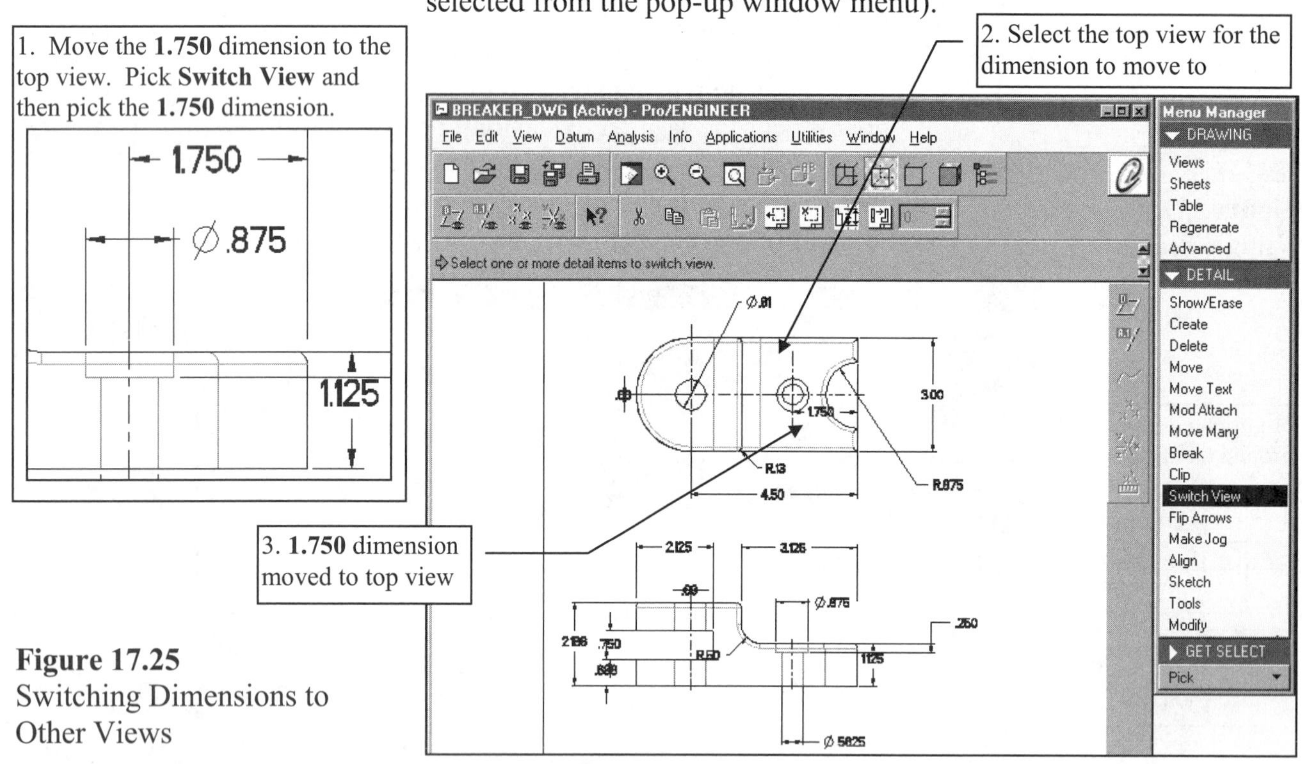

Figure 17.25
Switching Dimensions to Other Views

Erasing Feature Dimensions

Feature dimensions can either be shown on a drawing (using **Show**) or erased (using the **Erase** option). You cannot **Delete** feature ("true part-driving") dimensions from a drawing.

When you first add views to a drawing, the feature dimensions are also present; they are simply in the **Erased** state. When you choose **Show/Erase** and the desired options, you are unerasing them.

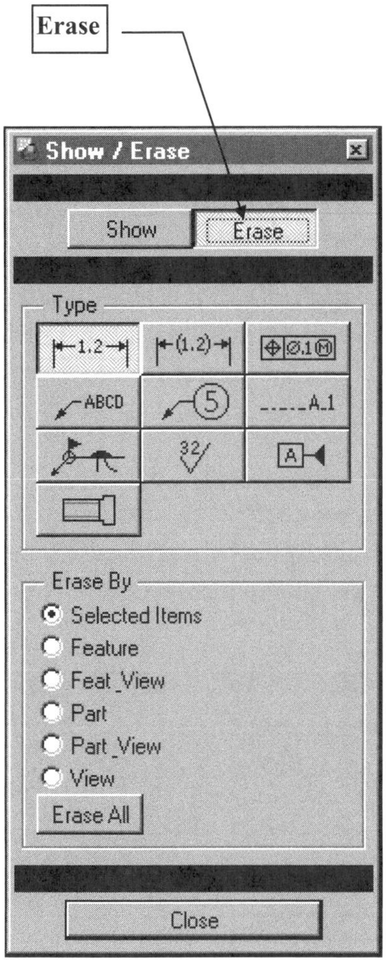

Erase

Many times, when you **Show** feature dimensions, there are dimensions you do not need. This is especially true of dimensions of thickness and pattern number callouts. You may find that there are other dimensions you do not want shown.

If you want a dimension, but it appears in the wrong view, you do *not* want to erase it. The **Switch View** option allows you to move that dimension to the view where you want it displayed.

If, however, there are actually dimensions that you do not need, you can **Erase** them. When you choose **Erase** and the desired options, Pro/E prompts you to select the dimensions to erase. You use the *middle* mouse button to finish the selecting of the desired dimensions to erase.

If you accidentally erase desired dimensions, you can redisplay them using the **Show** option. Unfortunately, the **Show** option does not know exactly which dimensions you want to redisplay (from the set of dimensions that you have erased). Therefore, you may have to redisplay several dimensions and re-erase those you really do not need. Using the **Feat & View** option with **Show** (instead of **Show All**) will minimize the number of dimensions you need to redisplay.

The **Erase** menu gives you a great deal of control over how you select the items to erase. You can choose to select a type of drafting object individually or erase all objects of a type, in the drawing, in a view, and/or by feature.

If you choose to **Erase All** of an item, Pro/E prompts you to confirm the erasure. If you pick **No**, the erasure operation is aborted. If you pick **Yes**, the items are all erased.

The simplest way to erase drawing items is to use the **Show/Erase** dialog box, choose the **Type**, and pick the **Selected Items**. You may now choose items to erase by using **Query Sel**.

Using the right mouse button, activating **Modify Item**, and picking the item to modify (as in Fig. 17.26, where the **.00** dimension was chosen) will provide a choice of several actions, including **Erase**. Picking **Erase** will make the dimension *gray out*. You can continue picking items to erase. When the selections are complete, choose **Done Sel** to remove the items from the screen. This method, although faster, provides fewer choices.

When you locate a hole from a datum at a distance of **.00** you will create what we call "zero" dimensions. These dimensions need erasing in **Drawing** mode. At this time, erase all **.00** dimensions (Figure 17.26).

Move the dimensions off the faces of the views in accordance with ASME standards.

Another aspect to change on the drawing is the number of digits displayed for individual dimensions (Fig. 17.27). In most cases, the number of digits is dependent on the tolerance for the dimension.

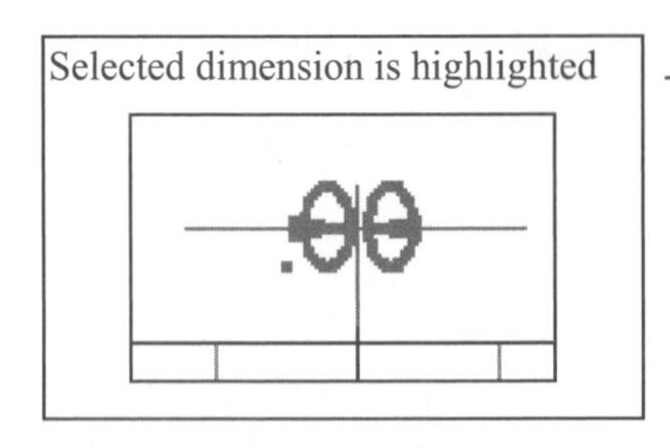

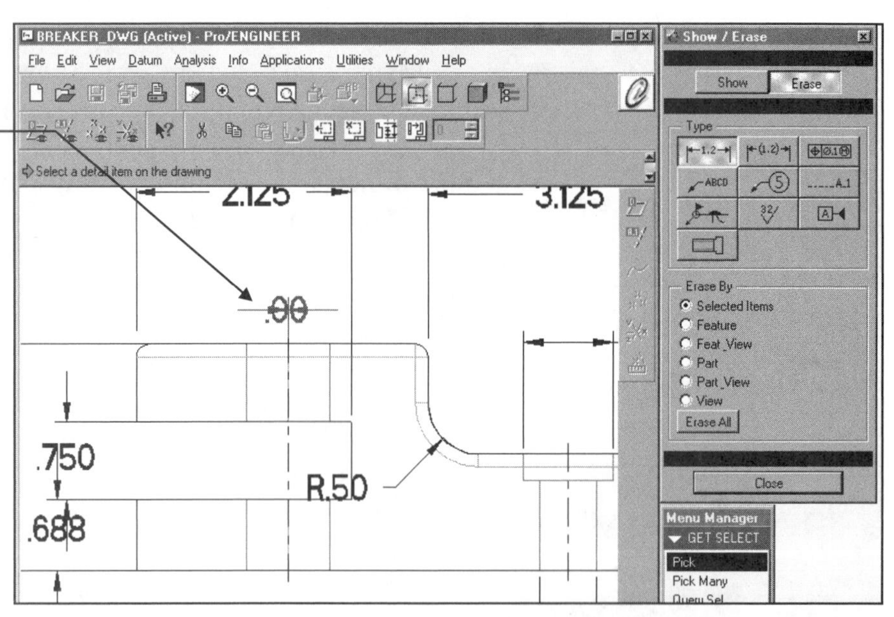

Figure 17.26
Using **Modify Item** to **Erase** a Dimension

Change the number of digits for the dimension **(R.13)** displayed with two decimal places to three **(R.125)** (Fig. 17.27) by choosing the following commands. *(Your model may be different. Practice changing the number of digits displayed on the appropiate dimensions as per your model, etc.)*

Modify ⇒ **Num Digits** ⇒ ("Number of decimal places for value"-- type **3)** ⇒ ✔ ⇒ (pick the dimension(s) to alter) ⇒ **Done Sel** ⇒ **Done Sel** ⇒ **Done/Return** ⇒ **File** ⇒ **Save** ⇒ ✔ ⇒ **File** ⇒ **Delete** ⇒ **Old Versions** ⇒ ✔

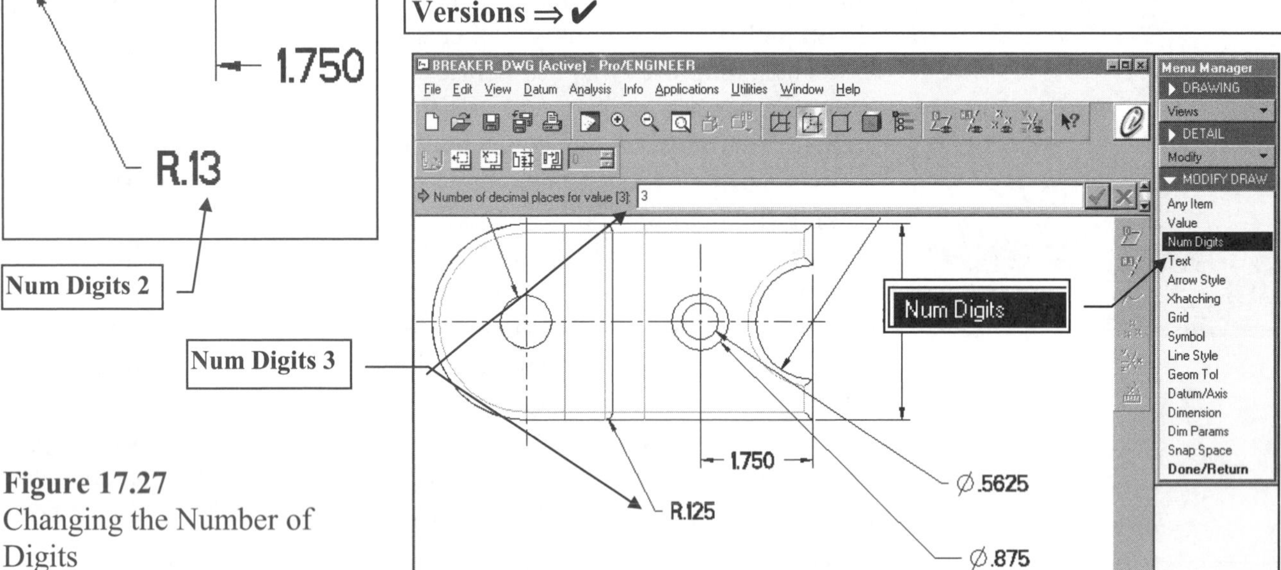

Figure 17.27
Changing the Number of Digits

HINT

When you change the number of digits of a driving dimension, the dimension color will change from *yellow* to *white* and will print bold (object line thickness). Regenerate the part and drawing to get them to return to *yellow*.

The **Flip Arrows** option is used to change the dimension arrows from inside to outside arrows or from outside to inside arrows. Small dimension values will create arrows on top of the dimension value, and radial dimensions sometimes point to the wrong side of the arc, as in Figure 17.28, where after the dimension was moved to the proper position, the arc radius is dimensioned to the side of the arc that does not exist. You may have to pick the dimension more than once to obtain the required result. Also, flip the arrows for all holes.

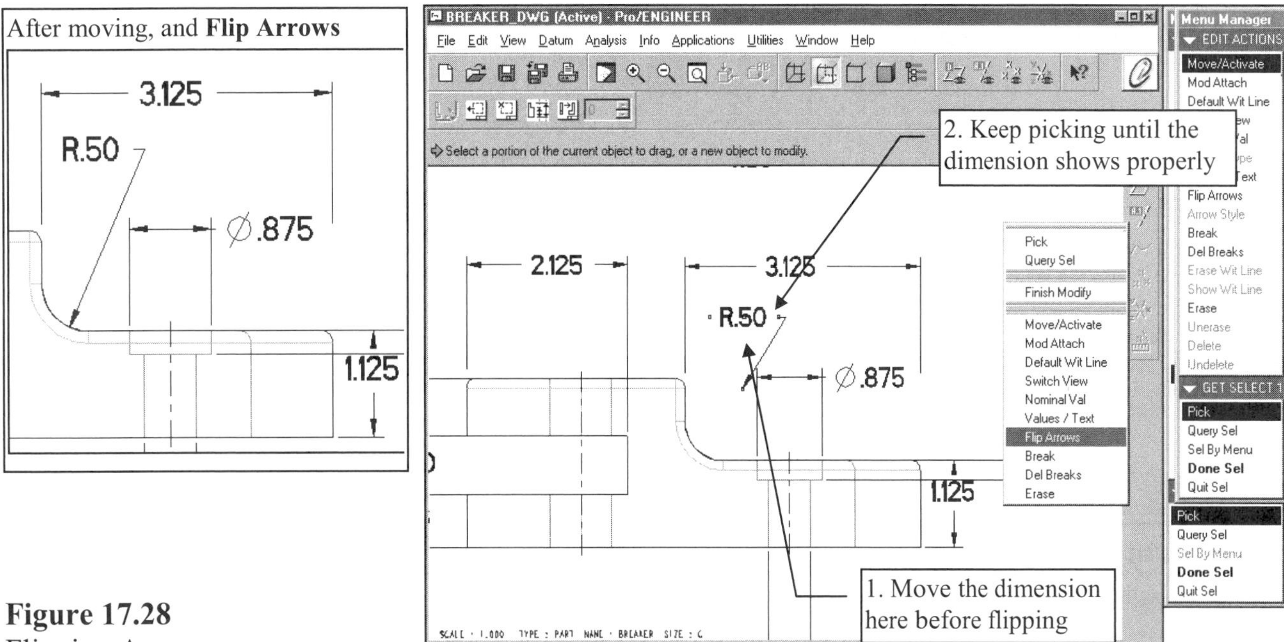

Figure 17.28
Flipping Arrows

To have diameter dimensions point to the outside of the circle with one arrowhead, instead of across the diameter using two arrowheads, flip the arrows (after moving the dimensions to the top view. The dimensions are complete (Fig. 17.29), but you still need to make a number of changes to create a correctly dimensioned detail of the part. In some cases, dimensions need to be combined into notes, and reference dimensions need to be added to the drawing.

Figure 17.29
Dimensioned Drawing

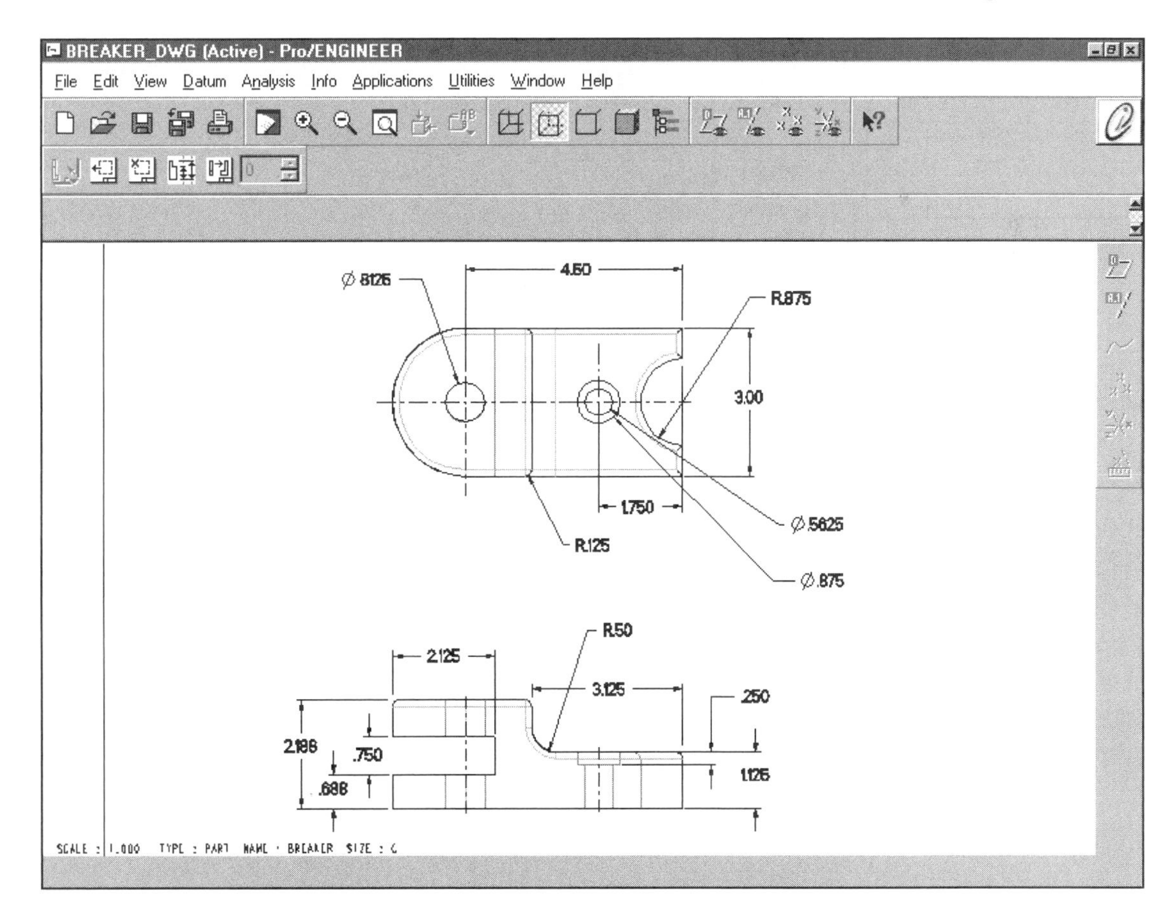

There are times when simply showing the dimensions is not enough to annotate a part completely. You can add to a drawing additional dimensions, labels, and notes that were not a part of its original definition. As an example, the counterbore depth, the thru hole, and the counterbore diameter need be combined into one note using the thru hole as the dimension to modify. The counterbore diameter and depth dimensions are then erased from the drawing.

In Figure 17.28, the diameter dimension for the counterbore was switched to the top view. After the dimension is switched, click and hold down the *right* mouse button anywhere on the screen to show **Modify Item**. Pick the **.5625** diameter dimension and choose **Values/Text** (from the EDIT ACTIONS menu or press your right mouse button to access the pop-up menu options), and the Modify Dimension dialog box will display, as shown in Figure 17.30. Select the **Dim Text** tab from the dialog box (Fig. 17.31). You will need to add the diameter symbol to the value as a prefix and to add the **.875** counterbore diameter and the **.250** depth to the note. We created this feature with a **Revolved Cut** instead of a **Hole** command for the purpose of demonstrating the following capabilities.

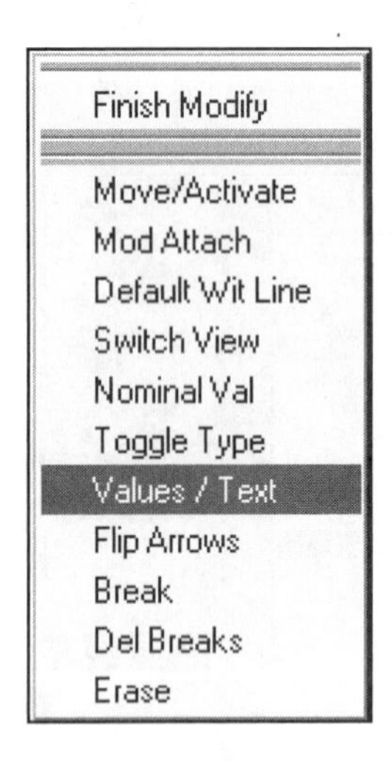

Figure 17.30
Modify Dimension Dialog Box

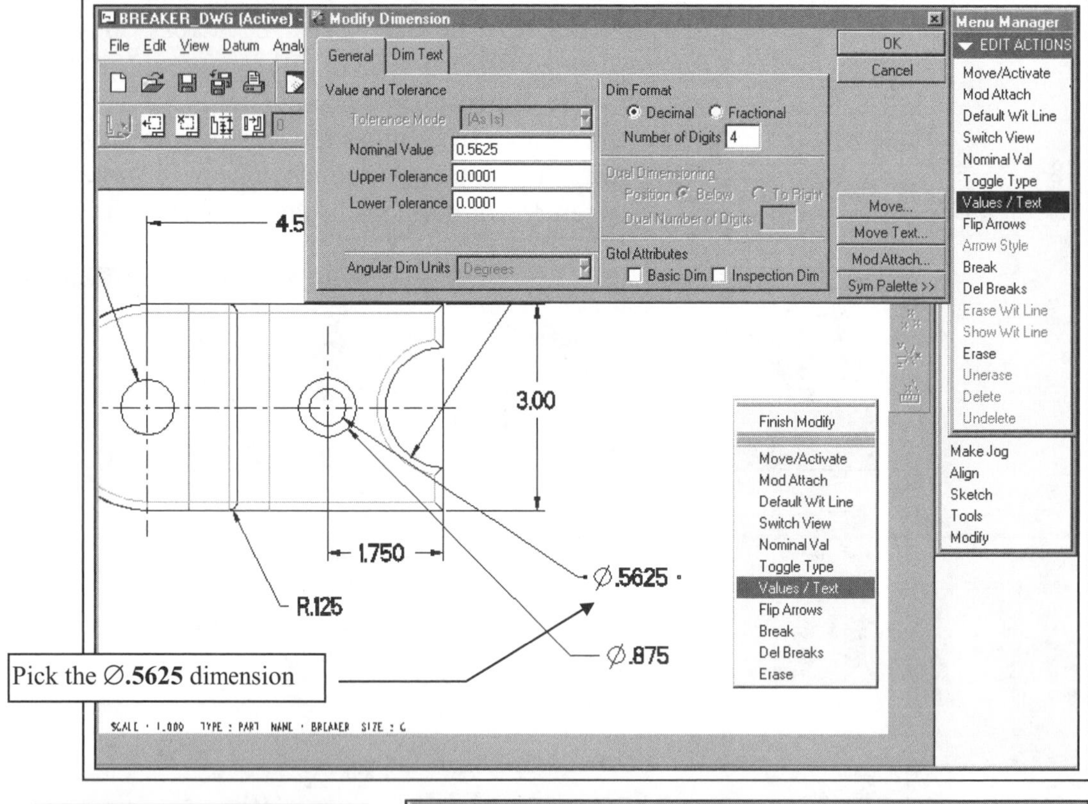

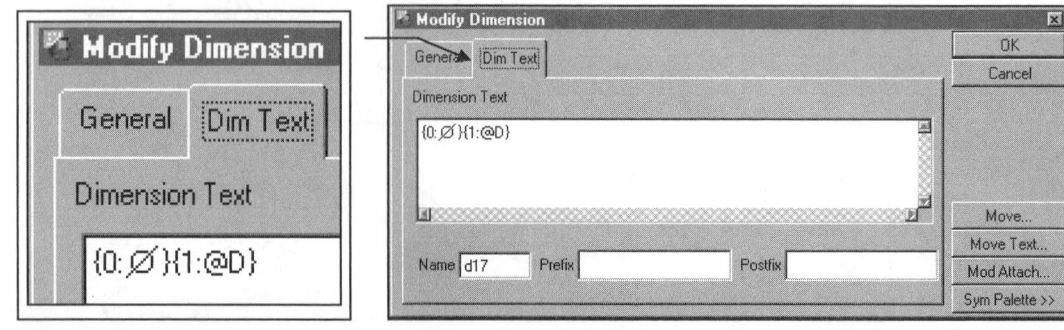

Figure 17.31
Dim Text Tab

NOTE

To make the note parametric, enter **&** followed by the dim symbols instead of the dimension values (e.g., **d14** and **d16**); the dimensions are then parameters that will reflect any modifications and changes. To see the dim symbols in Drawing mode, choose **Info** ⇒ **Switch Dims.** As an example:

Choose the **Sym Palette** button (Fig. 17.32); then place your cursor at the end of the text line in the dialog box window and press your enter key. Add a *counterbore symbol* on a new text line, along with the **Ø.875** dimension. To complete the note, add a third line with a depth symbol and the **.250** dimension. Erase the **Ø.875** and **.250** dimensions from the drawing (Fig. 17.33).

Ø.5625
⌴Ø.875
↧.250
Nonparametric note

{0:Ø}{1:@D}
⌴Ø&d14
↧&d16
Parametric note

Dimension Text
{0:Ø}{1:@D}
⌴Ø.875
↧.250

Erase the **Ø.875** dimension

Pick symbols from the palette

Figure 17.32
Modifying Dimensions

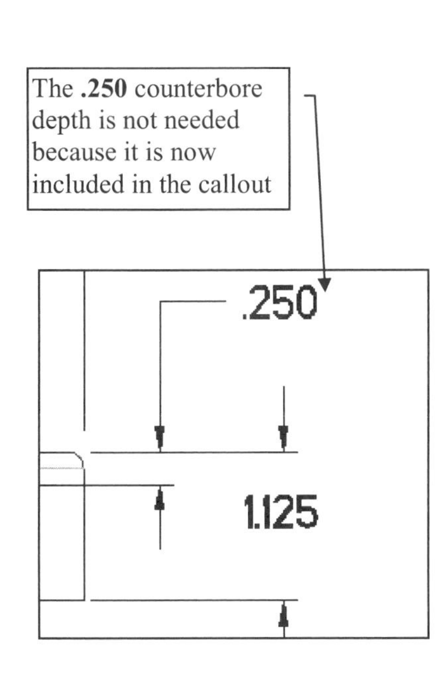

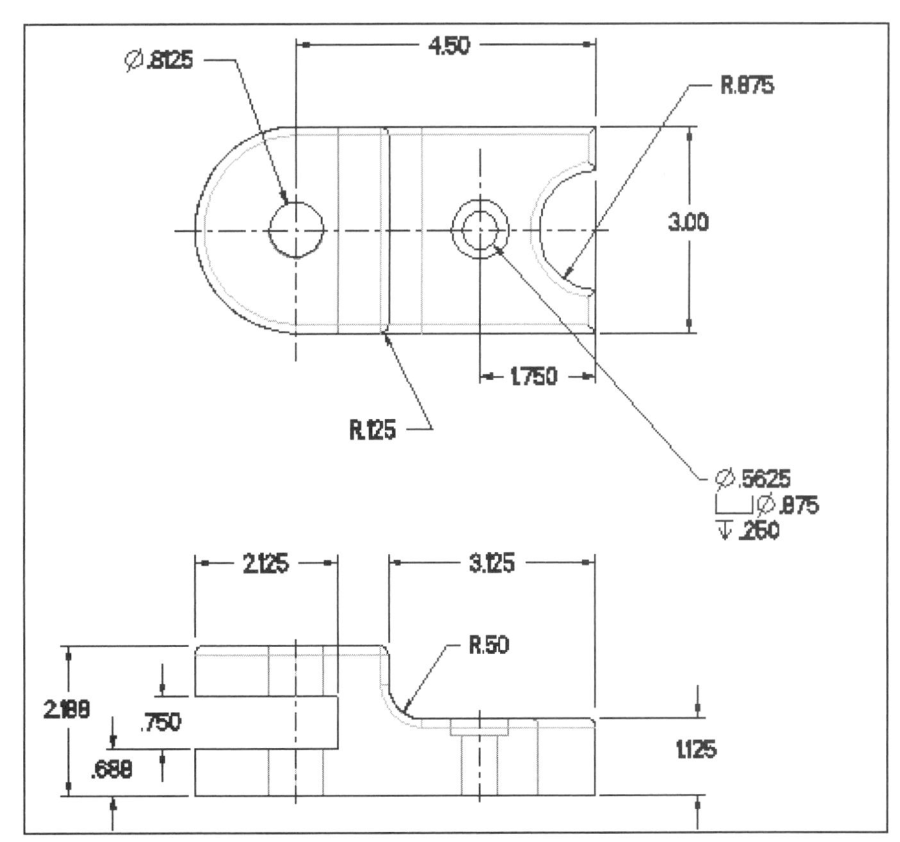

Figure 17.33
Erasing Unneeded Dimensions

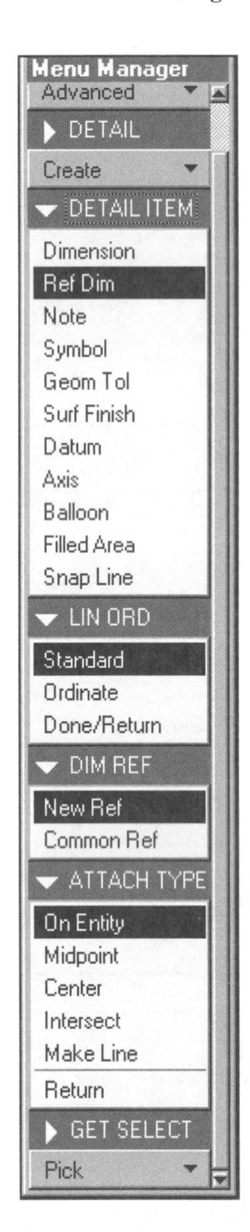

Dimensions that you add "manually" are called ***driven dimensions***. Driven dimensions change when the model changes, but they cannot be used to make changes to the model. Only dimensions that you place on the drawing using the **Show** option(s) can ***drive*** changes to the model (and they are referred to as ***driving dimensions***).

Driven dimensions can be erased in the same manner as driving dimensions. In addition, however, driven dimensions can be deleted; this permanently removes them from the drawing database.

You can also create notes and labels to add to the annotation on your drawing. A label is basically a note that has a leader. Leaders can be attached to edges or drawn to positions in space. It is better practice to modify an existing diameter dimension than it is to create a note with a leader.

You create driven dimensions using the same techniques that are available in sketching. For example, you create parallel dimensions by selecting a linear edge (with the *left* mouse button) and indicating a placement location (with the *middle* mouse button). You create diameter dimensions by double-clicking on a circular edge and indicating a placement location.

Reference dimensions can also be added to a drawing. Reference dimensions are driven dimensions. Create a reference dimension to show the total width of the part using the following commands:

Create ⇒ Ref Dim ⇒ (pick both ends of the part in the front view and place the dimension below the view, as shown in Fig. 17.34) **⇒ Done Sel ⇒ Done/Return**

NOTE

ASME standards use parentheses around the reference dimension. The use of **REF** is not standard, but it is used by some companies.

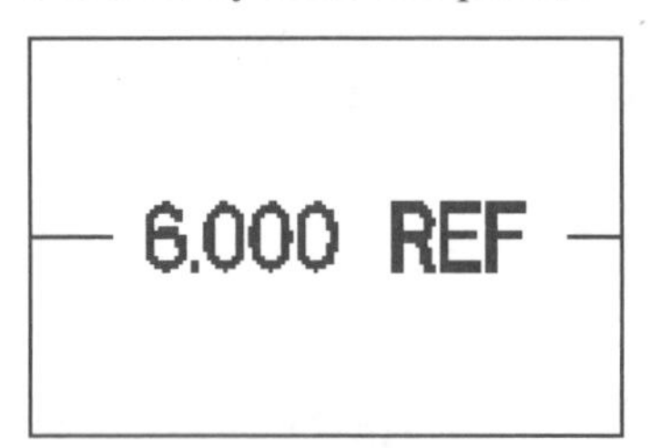

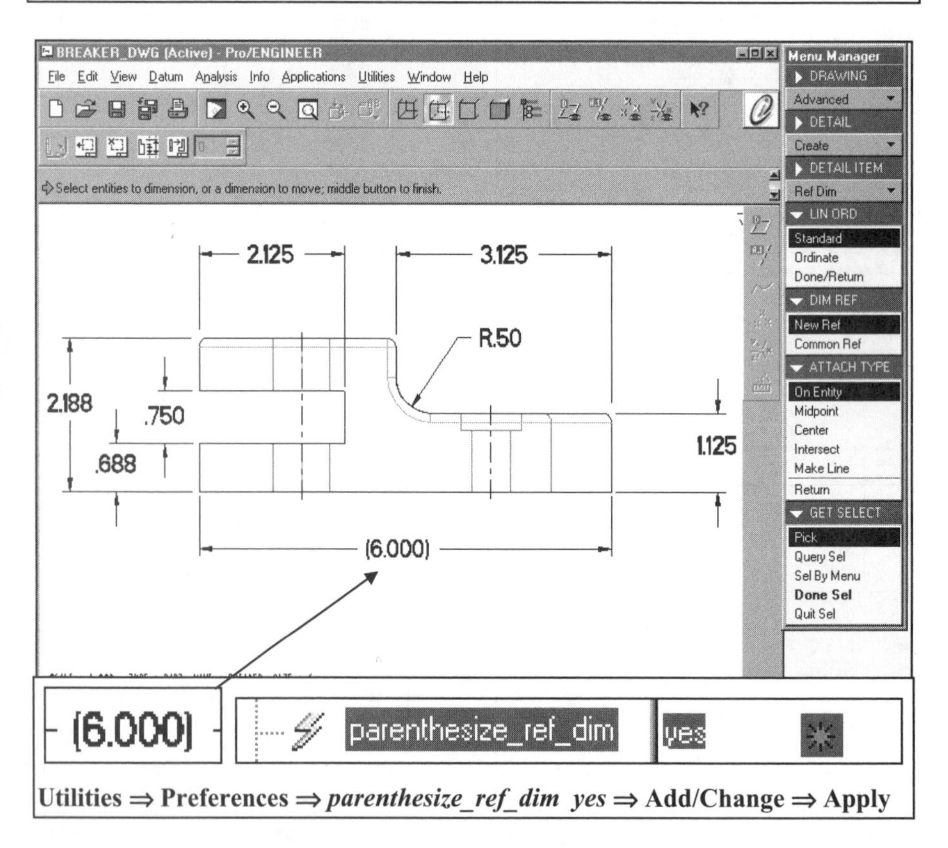

Utilities ⇒ Preferences ⇒ ***parenthesize_ref_dim yes*** **⇒ Add/Change ⇒ Apply**

Figure 17.34
Creating a Reference Dimension

HINT
If you set your **Environment** to **Isometric**, your pictorial will be isometric instead of **Trimetric**.

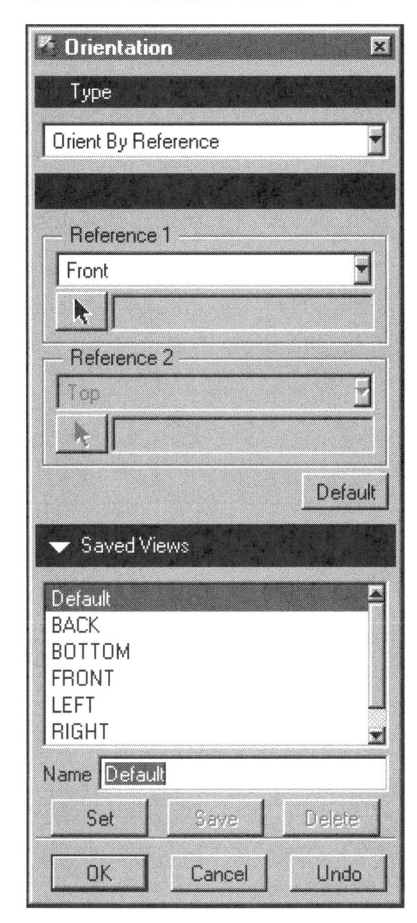

Alphabetic characters are created in a font and sized such that they comply with standards for **Geometric Dimensioning and Tolerancing (GD&T)**. These characters can be used for any purpose, but they are designed to be properly displayed in **GD&T** frames.

You may have noticed that up to this point, most of the text you enter is automatically capitalized by Pro/E. For example, you name a section by entering "**a**" but Pro/E displays it as "**A**." When you enter text in labels, for notes, or as text appended to a dimension, Pro/E displays exactly the case you enter. If you enter "**material,**" Pro/E displays "**material**" on the drawing. If you want "**MATERIAL,**" you need to type "**MATERIAL.**"

The drawing is almost complete, so let us add another general view (pictorial view) in the upper corner of the drawing using:

Views ⇒ **General** ⇒ **Done** ⇒ (pick in the upper right of the drawing, as shown in Fig. 17.35) ⇒ **Saved Views** ⇒ **Default** ⇒ **Set** ⇒ **OK** ⇒ **Done/Return**

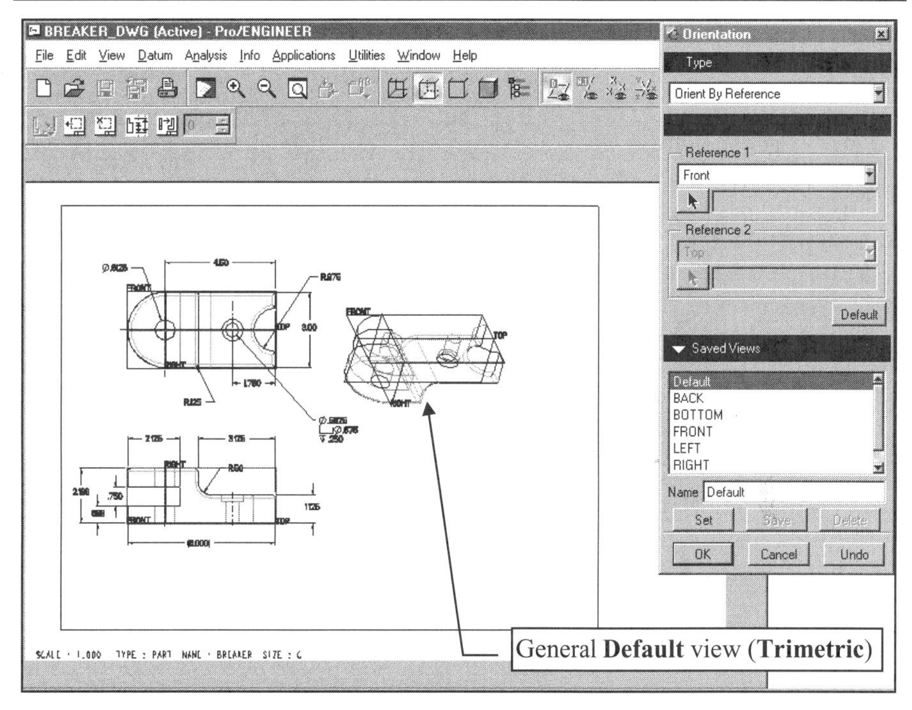

Figure 17.35
Adding a General View

Because you used a blank format to start the drawing, replace the format with a standard ANSI "**C**" size drawing format using the following commands (Fig. 17.36) :

Sheets ⇒ **Format** ⇒ **Add/Replace** ⇒ **System Formats** ⇒ (pick **c.frm**) ⇒ **Open** ⇒ **Done/Return**

Also, try to use your personal format (Fig. 17.37):

Sheets ⇒ **Format** ⇒ **Add/Replace** ⇒ ▼ (navigate to the directory where your formats are stored) ⇒ (pick ***your_format_name*.frm**) ⇒ **Open** ⇒ **Done/Return**

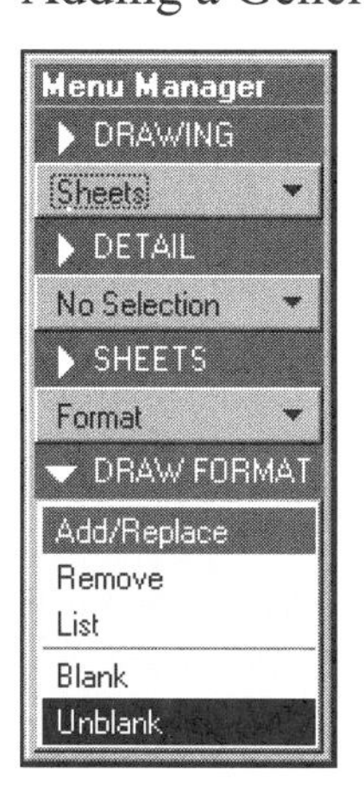

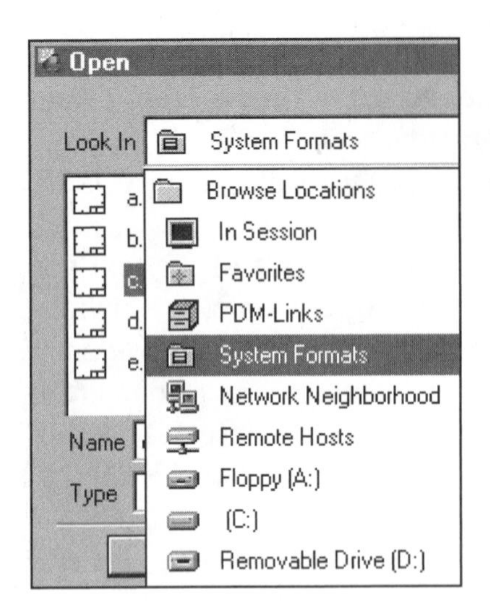

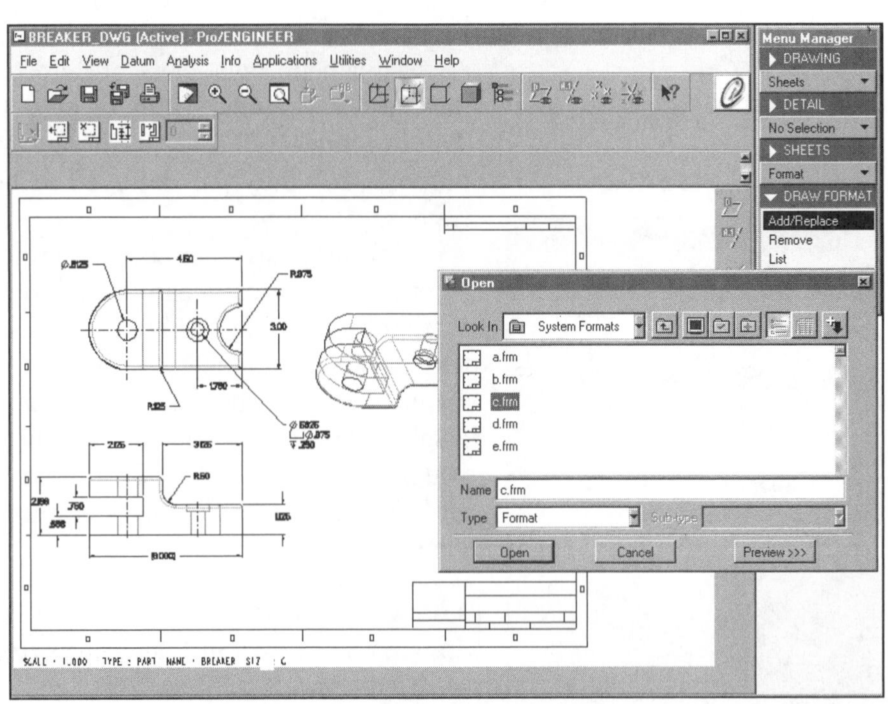

Figure 17.36
ANSI Standard "**C**" Size Format

File ⇒ Save ⇒ ✔
File ⇒ Delete ⇒
Old Versions ⇒ ✔

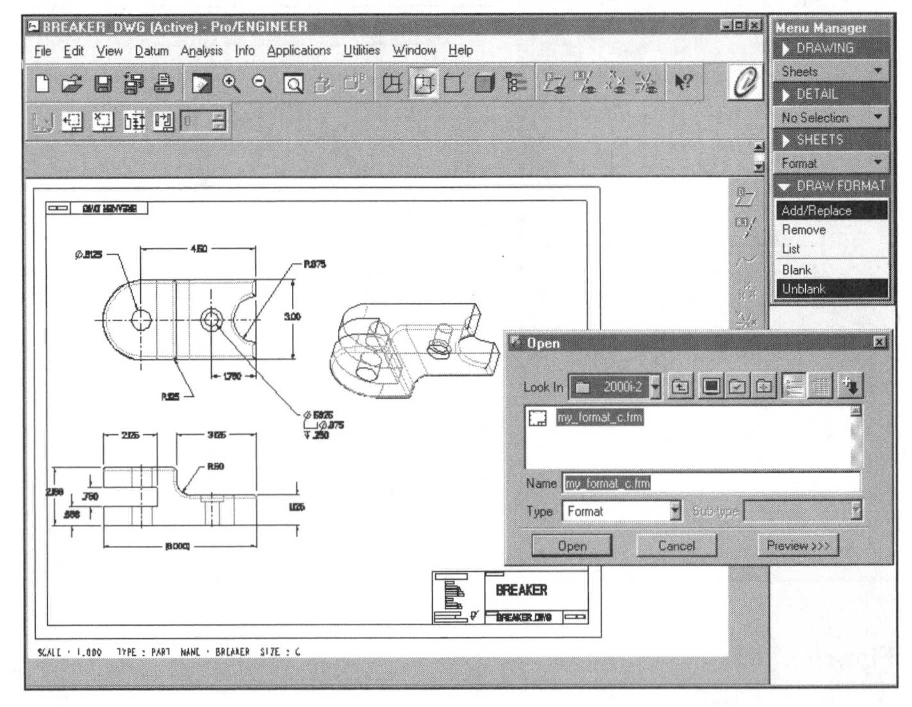

Figure 17.37
Your Personal Format

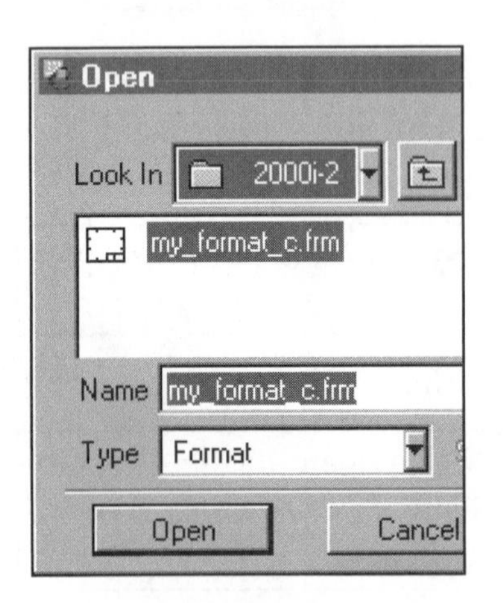

Each view of your model can have a different line display style: **Wireframe**, **Hidden Line**, **No Hidden**, or **Default**. The tangent display status can also be established independent of the environment default. **Tan Solid**, **No Disp Tan**, **Tan Ctrln**, **Tan Phantom**, **Tan Dimmed**, and **Tan Default** are available. Use the following commands to set the display of each view, starting with the general pictorial view:

Views ⇒ Disp Mode ⇒ View Disp ⇒ (pick the general view, as shown in Fig. 17.38) **⇒ Done Sel ⇒ No Hidden ⇒ Tan Solid ⇒ Done ⇒ Done Sel ⇒ Done/Return**

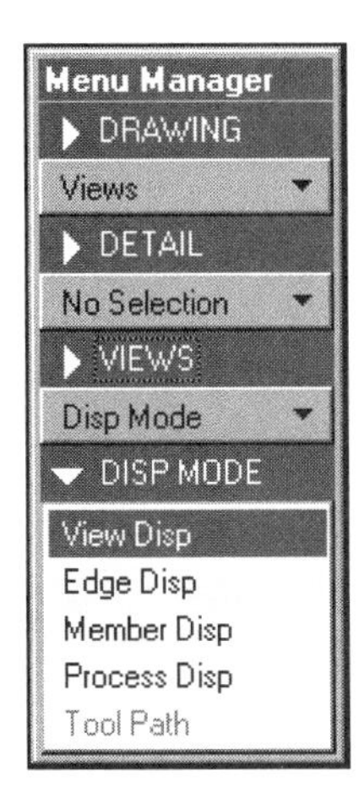

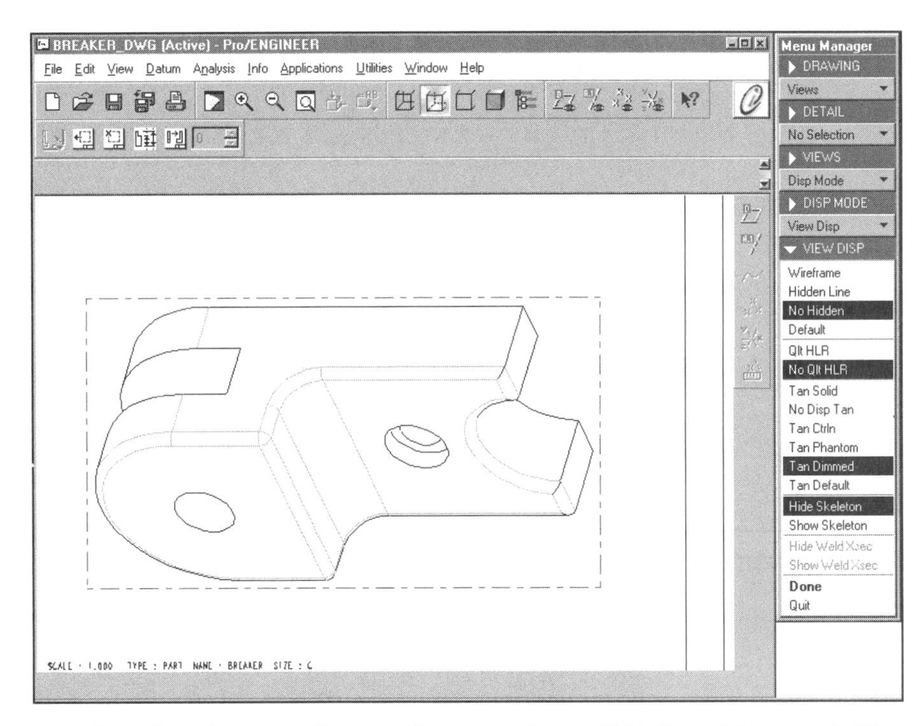

Figure 17.38
Setting the Display Status for a General View: **No Hidden** and **Tan Dimmed**

Set the front and top views to have **Hidden Line** and **Tan Dimmed** using the following commands:

Views ⇒ **Disp Mode** ⇒ **View Disp** ⇒ (pick the top and front views) ⇒ **Done Sel** ⇒ **Hidden Line** ⇒ **Tan Dimmed** ⇒ **Done** ⇒ **Done Sel Done/Return** (Fig. 17.39) ⇒ **File** ⇒ **Save** ⇒ ✔ ⇒ **File** ⇒ **Delete** ⇒ **Old Versions** ⇒ ✔

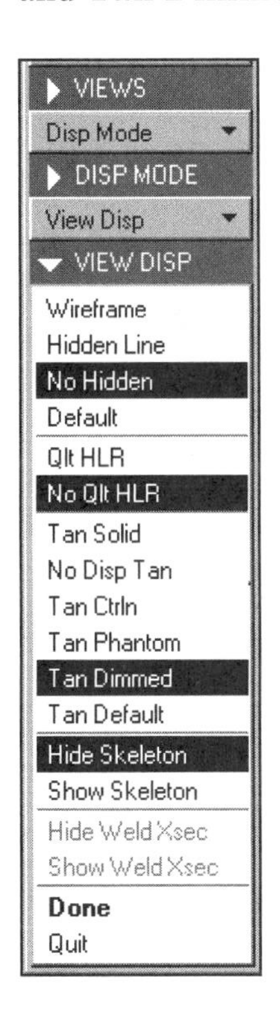

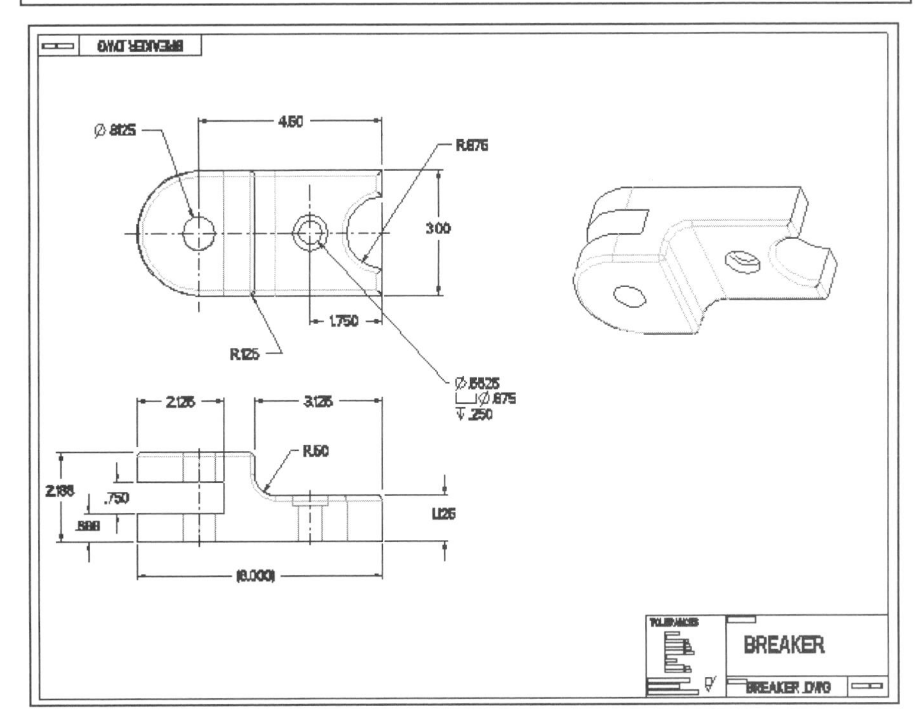

Figure 17.39
Setting the Display Status for the Top and Front Views: **Hidden Line** and **Tan Dimmed**

You believe the project is complete, so print or plot out a copy and submit it to the checker (teacher, boss, design checker). The checker will use **Markup** mode to check and mark up your drawing.

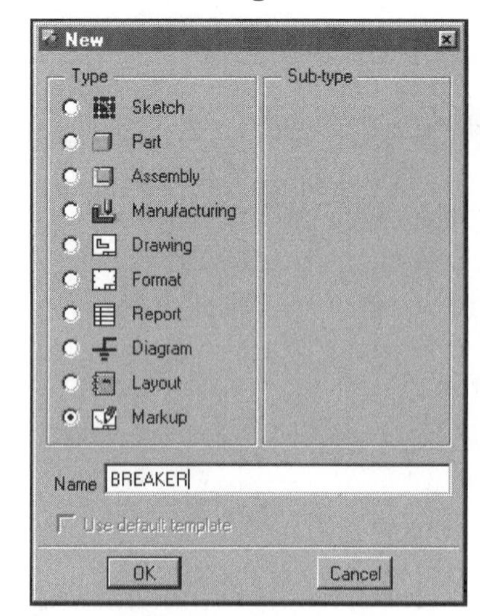

The checker will choose the following commands to enter Markup mode (Fig. 17.40) and show the **checker changes** he or she feels are necessary. The checker may also make some design changes at this time (Fig. 17.41). To enter Markup mode, choose:

File ⇒ **New** ⇒ **●Markup** ⇒ Name **BREAKER** ⇒ **OK** ⇒ Type **Drawing** ⇒ (select the object you want to markup--here select the drawing name you are presently working on, **breaker_dwg)** ⇒ **Open** ⇒ [a new window will open (Fig. 17.40)]

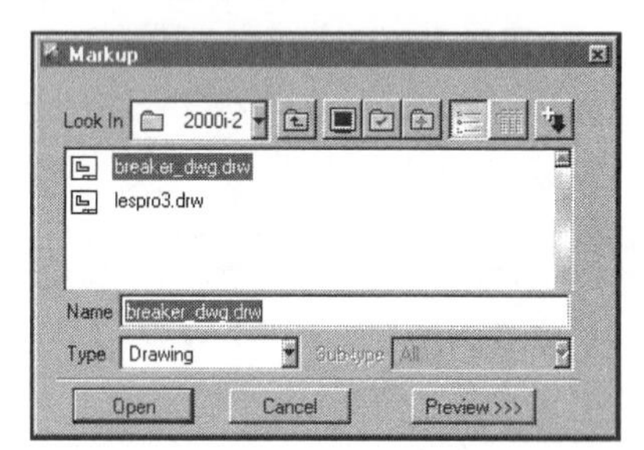

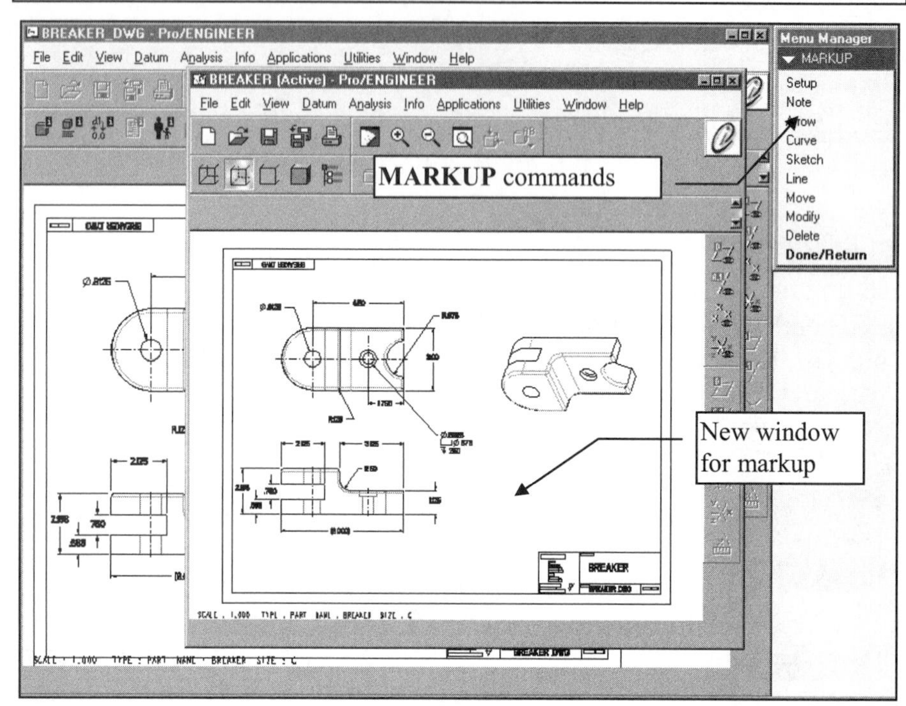

Figure 17.40
Pro/MARKUP

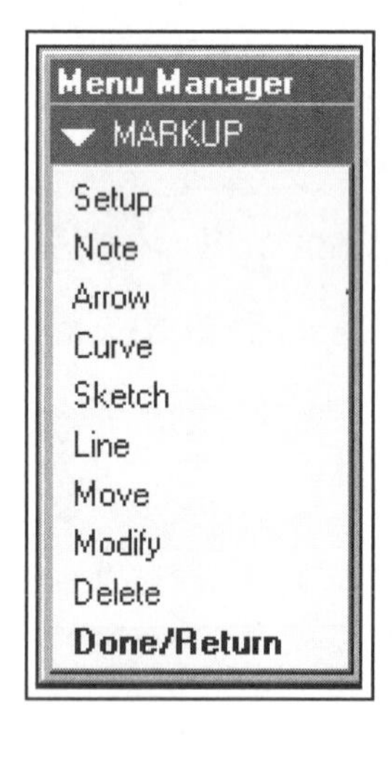

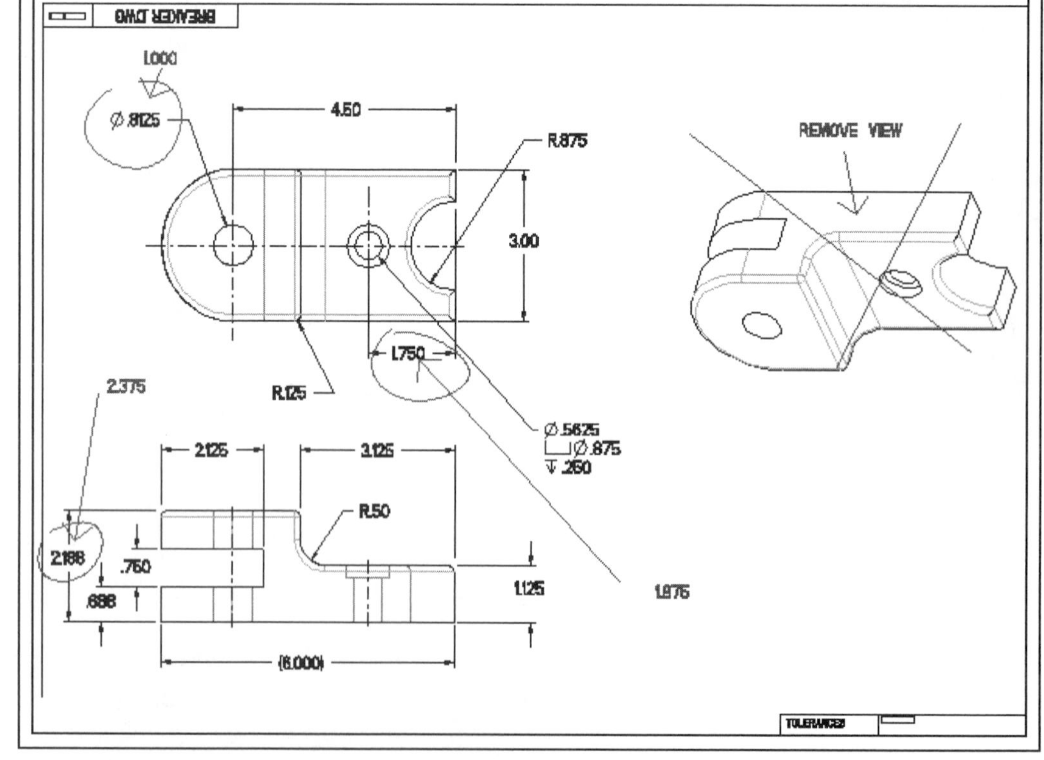

Figure 17.41
Checker Changes

After the checker saves the changes, you can bring up the markup drawing and your Breaker detail drawing. By keeping the two files in session, you can work in your detail window and still see the markup drawing. Because you will also be making some design changes, it is a good idea to keep the part window active to see the changes propagated throughout the model, including the part, the drawing, and the assembly (if used in an assembly). All files will be updated with the new design changes after regeneration. *If you are starting a new session,* use the following commands (Fig. 17.42):

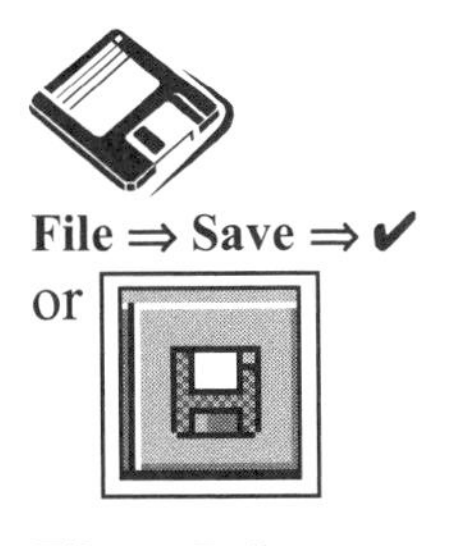

File ⇒ Save ⇒ ✔
or

File ⇒ Delete ⇒ Old Versions ⇒ ✔

1. **File ⇒ Open ⇒** Type **Drawing ⇒ breaker_dwg.drw ⇒ Open**
2. **File ⇒ Open ⇒** Type **Part ⇒ breaker.prt ⇒ Open**
3. **File ⇒ Open ⇒** Type **Markup ⇒ breaker(breaker_dwg.drw).mrk ⇒ Open**

Move the windows of each mode for easier viewing and work, as shown in Figure 17.42. Change the active window so that it is the drawing, not the part or the markup. Make the checker changes to the drawing (of course, you could also make the Part mode the active window and modify and regenerate the model there). Regardless of the mode or window you are working in, remember to **Regenerate** both the drawing *and* the part. The completed design can now be saved and plotted (Fig. 17.43). Finally, you can extract some important information from the drawing using: **Info ⇒ Drawing ⇒ Highlight by Attributes ⇒** (choose options) **Highlight** (Fig. 17.44).

Figure 17.42
Part, Drawing, and Markup

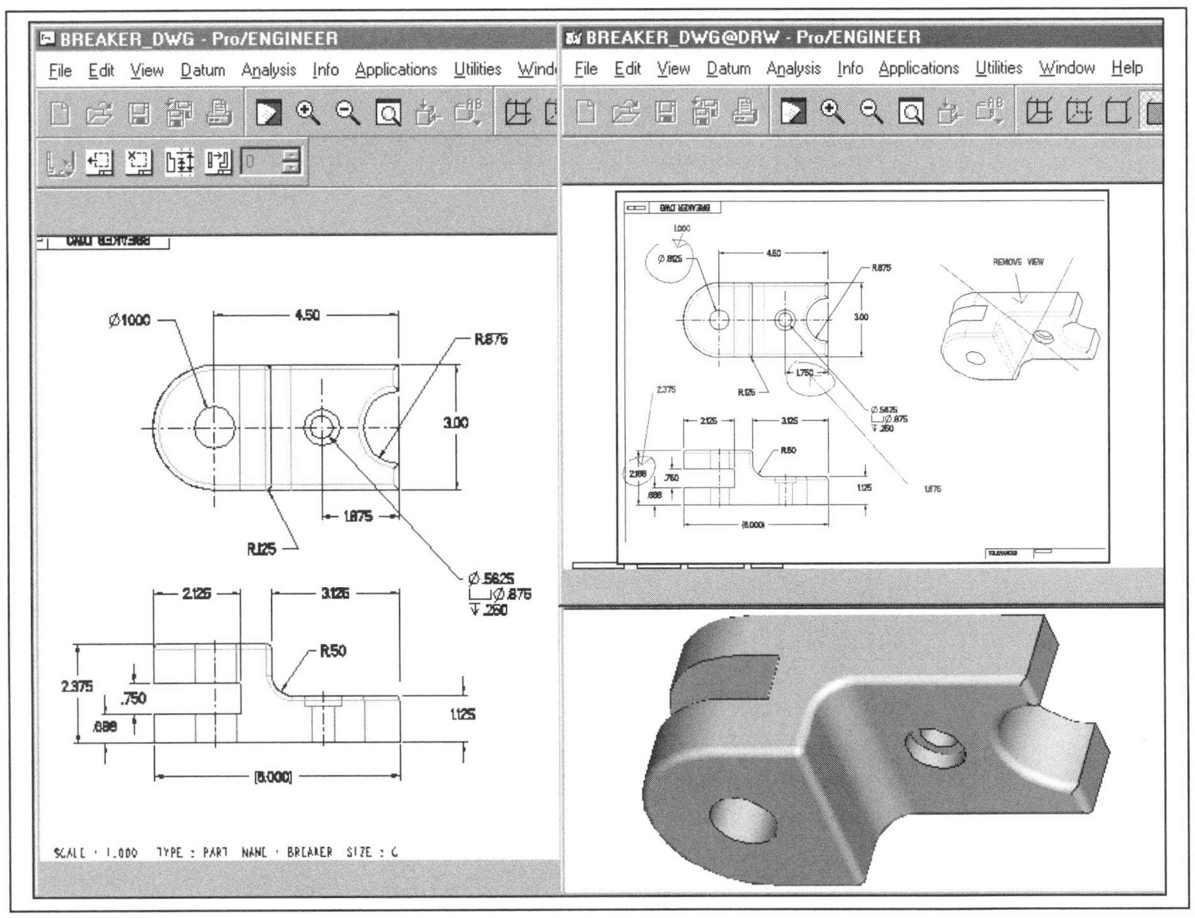

Figure 17.43
Regenerated Drawing and Part Incorporating Checker Changes

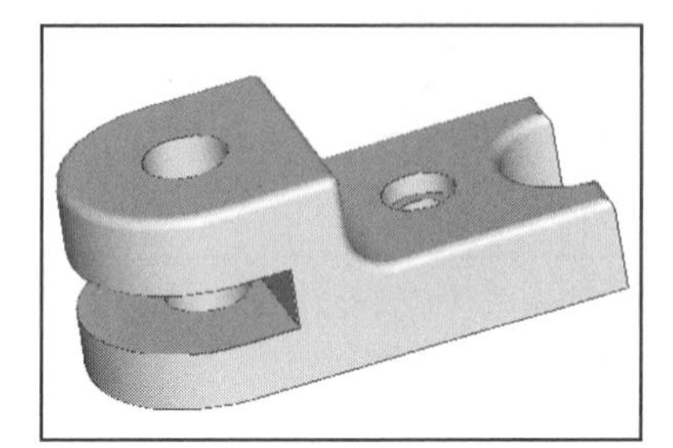

Figure 17.44
Info ⇒ Drawing ⇒ Highlight by Attributes ⇒ Highlight

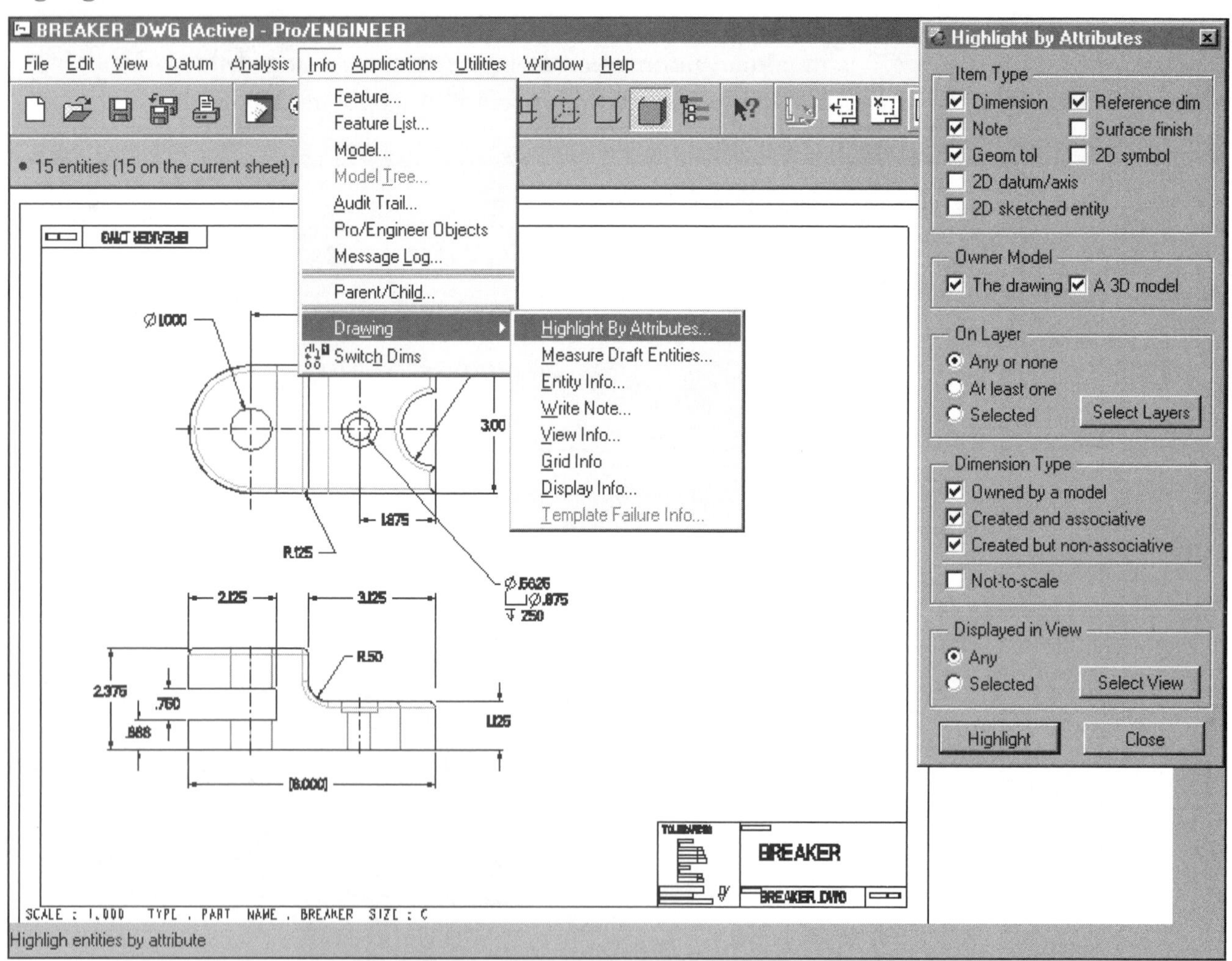

Lesson 17 Project

Cylinder Rod Drawing

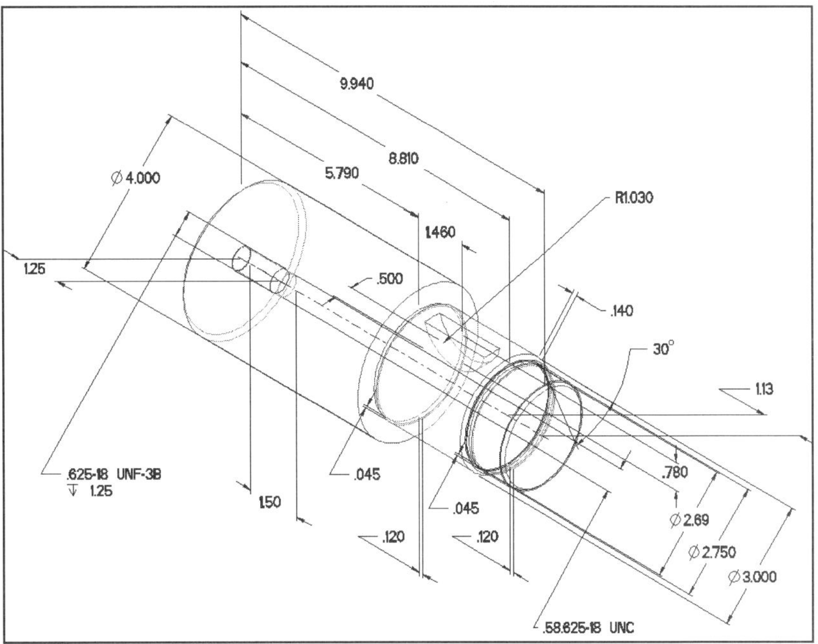

Figure 17.45
Cylinder Rod Pictorial Drawing

Cylinder Rod Drawing

The second **lesson project** for Drawing mode will use the part modeled in Lesson 6. The drawing for the Cylinder Rod (Fig. 17.45) is just one of many lesson parts and lesson projects that can be brought into Drawing mode and detailed.

Analyze the part and plan out the sheet size and the drawing views required to display the model's features for detailing (Figs. 17.46 and 17.47). Use the formats created in Lesson 16. Detail the part according to **ASME Y14.5 1994.**

EGD REFERENCE
Fundamentals of Engineering Graphics and Design
by L. Lamit and K. Kitto
See pages 487 and 674-677.

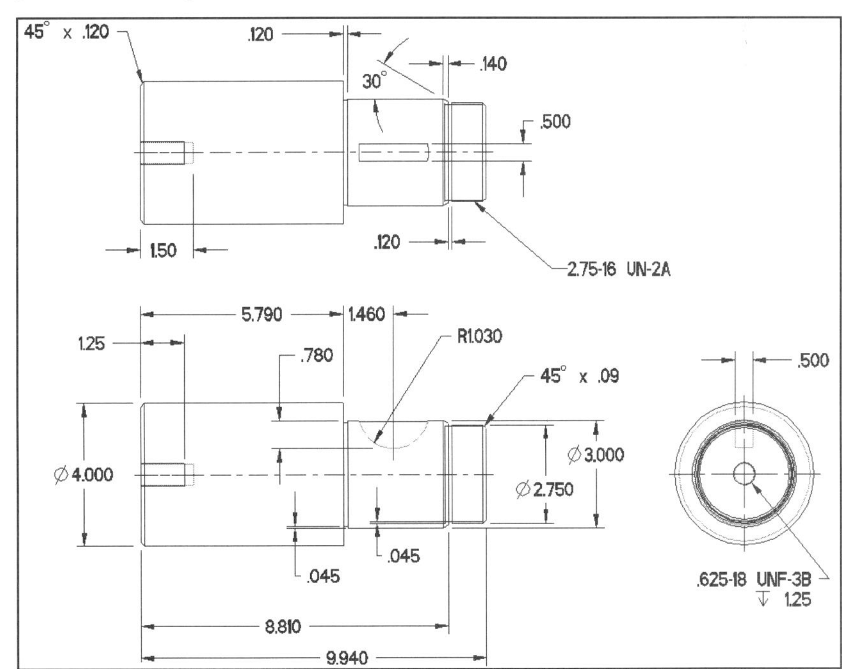

Figure 17.46
Cylinder Rod Drawing Without Format

Figure 17.47
Cylinder Rod Drawing with Format (Before Dimensioning)

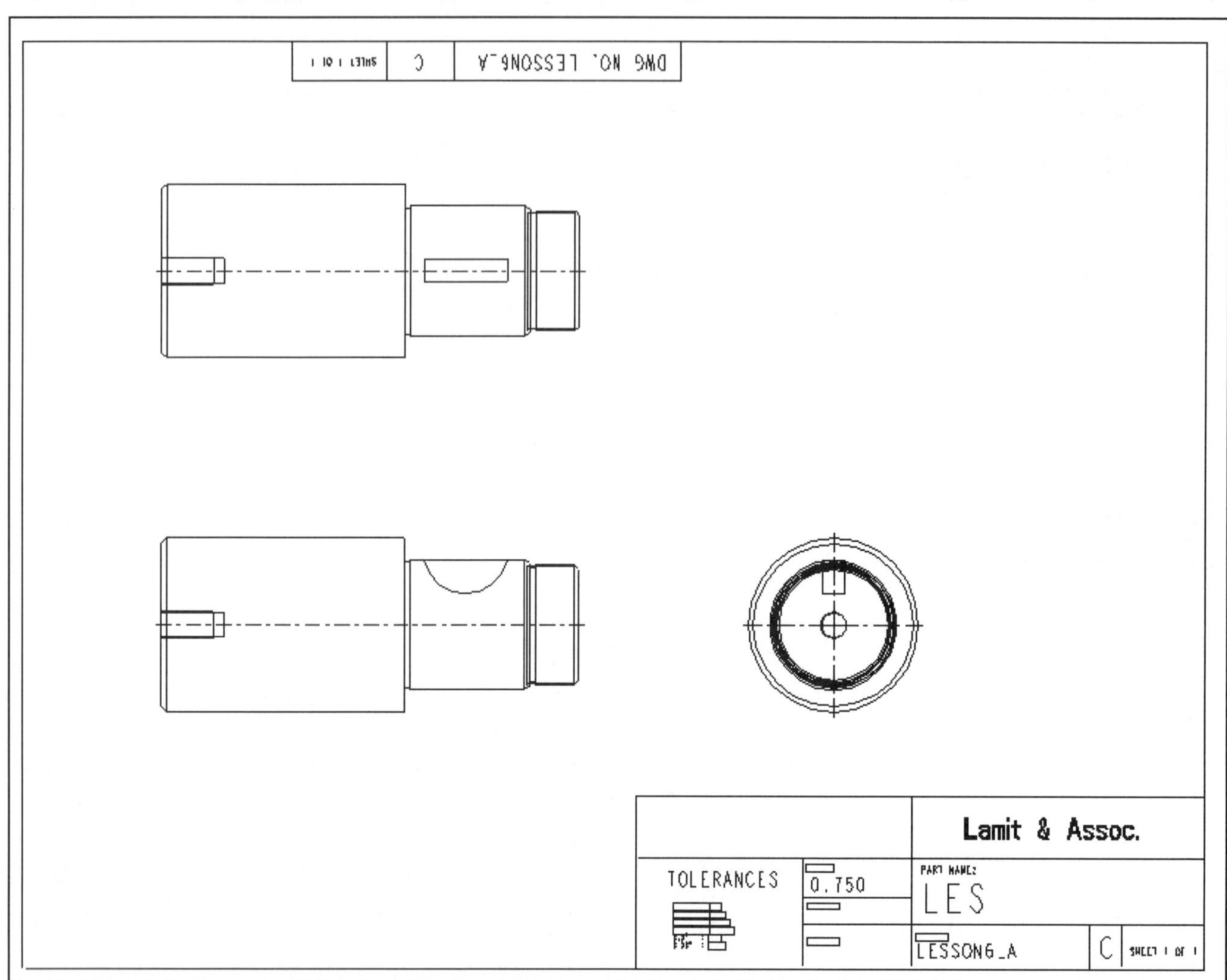

Lesson 18

Sections and Auxiliary Views

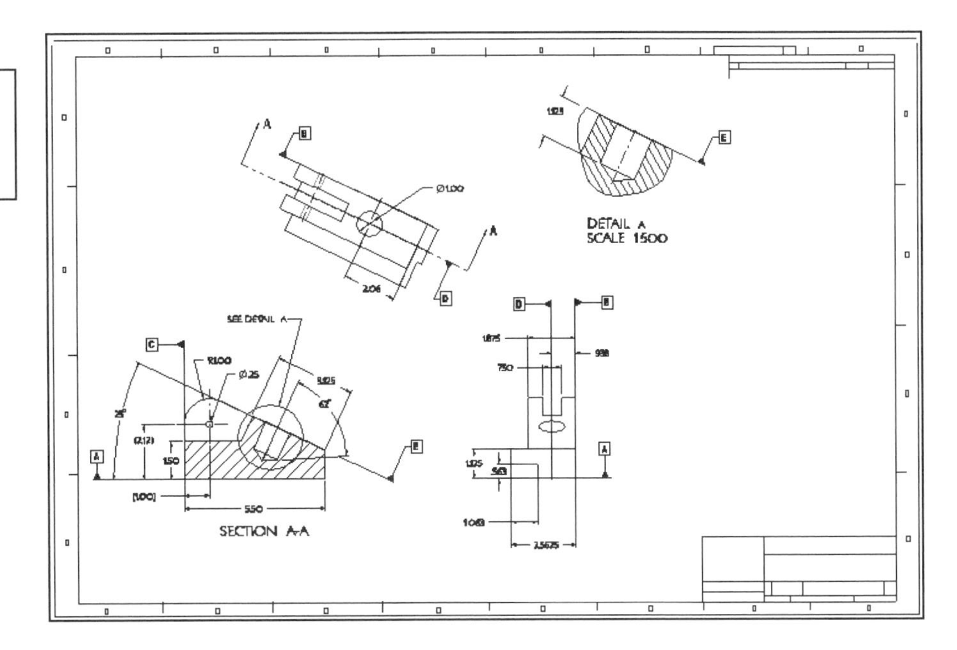

Figure 18.1
Anchor Drawing

OBJECTIVES

1. **Identify the need for sectional views to clarify interior features of a part**

2. **Establish a .dtl file to use when detailing and creating section drawings**

3. **Identify cutting planes and the resulting views**

4. **Create sections using datum planes**

5. **Detail section views**

6. **Develop the ability to produce auxiliary views**

7. **Create detail views**

8. **Create scaled detail views of complicated feature geometry**

9. **Apply standard drafting conventions and linetypes to illustrate interior features**

EGD REFERENCE
Fundamentals of Engineering Graphics and Design
by L. Lamit and K. Kitto
Read Chapter 12.
See pages 387-416 and 549.

COAch™ for Pro/ENGINEER

If you have **COAch for Pro/ENGINEER** on your system, go to SEARCH and do the Segment shown in Figures 18.3 through 18.6. A variety of CADTRAIN illustrations are also incorporated into this lesson.

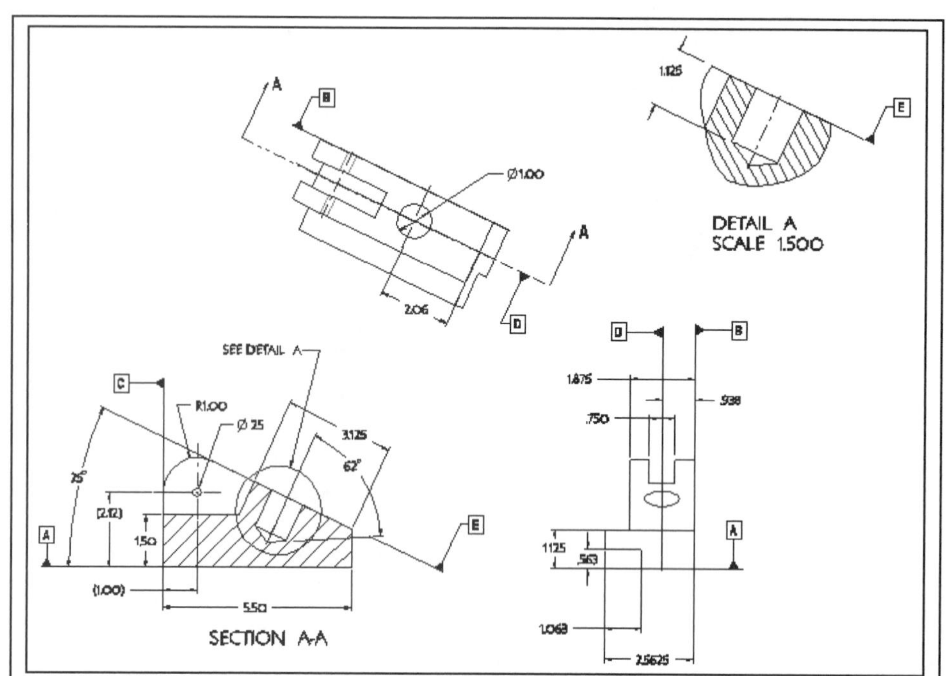

Figure 18.2
Anchor Drawing with Dimensions

SECTIONS AND AUXILIARY VIEWS

Designers and drafters use **sectional views**, also called **sections**, to clarify and dimension the internal construction of a part. Sections are needed for interior features that cannot be clearly described by hidden lines in conventional views (Figs. 18.1 and 18.2).

Auxiliary views are used to show the true shape/size of a feature or the relationship of part features that are not parallel to any of the principal planes of projection. Many parts have inclined surfaces and features that cannot be adequately displayed and described by using principal views alone. To provide a clearer description of these features, it is necessary to draw a view that will show the *true shape/size*. Besides showing a feature's true size, auxiliary views are used to dimension features that are distorted in principal views and to solve a variety of engineering problems graphically (Fig. 18.3).

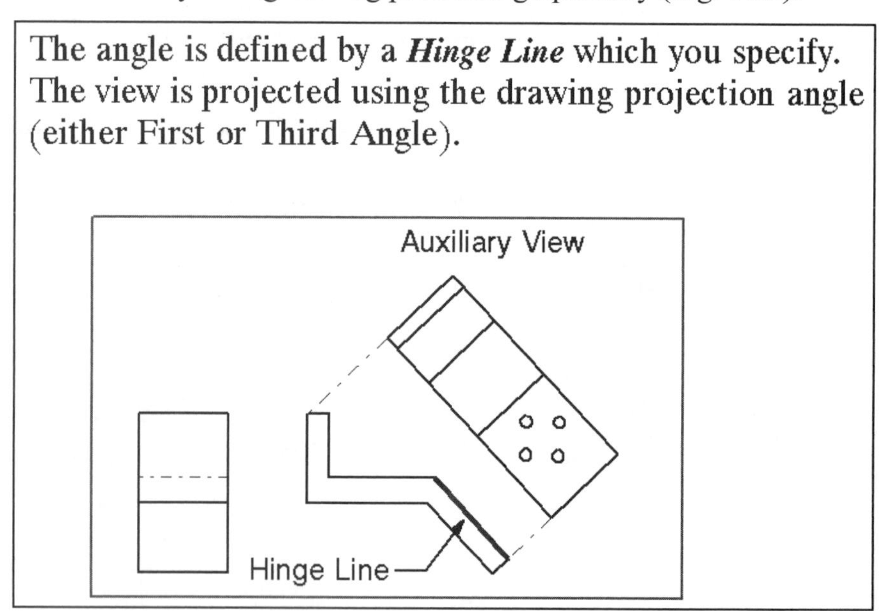

Figure 18.3
COAch for Pro/E, Drawings (Auxiliary Views)

Sections

A sectional view (Fig. 18.4) is obtained by passing an imaginary **cutting plane** through the part, perpendicular to the **line of sight**. The line of sight is the direction in which the part is viewed. The portion of the part between the cutting plane and the observer is "removed." The part's exposed solid surfaces are indicated by section lines. **Section lines** are uniformly spaced, angular lines drawn in proportion to the size of the drawing. There are many different types of section views.

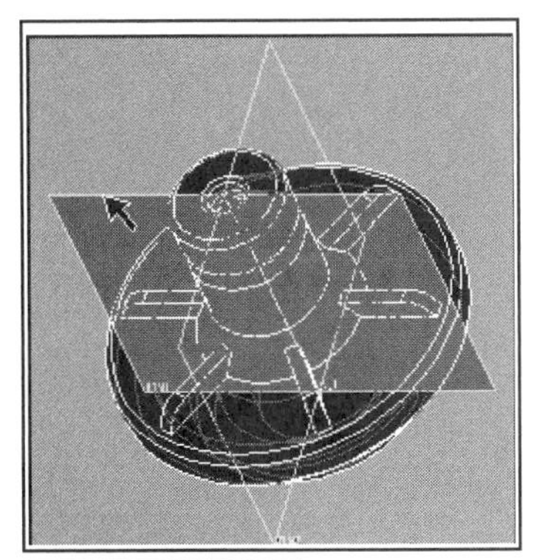

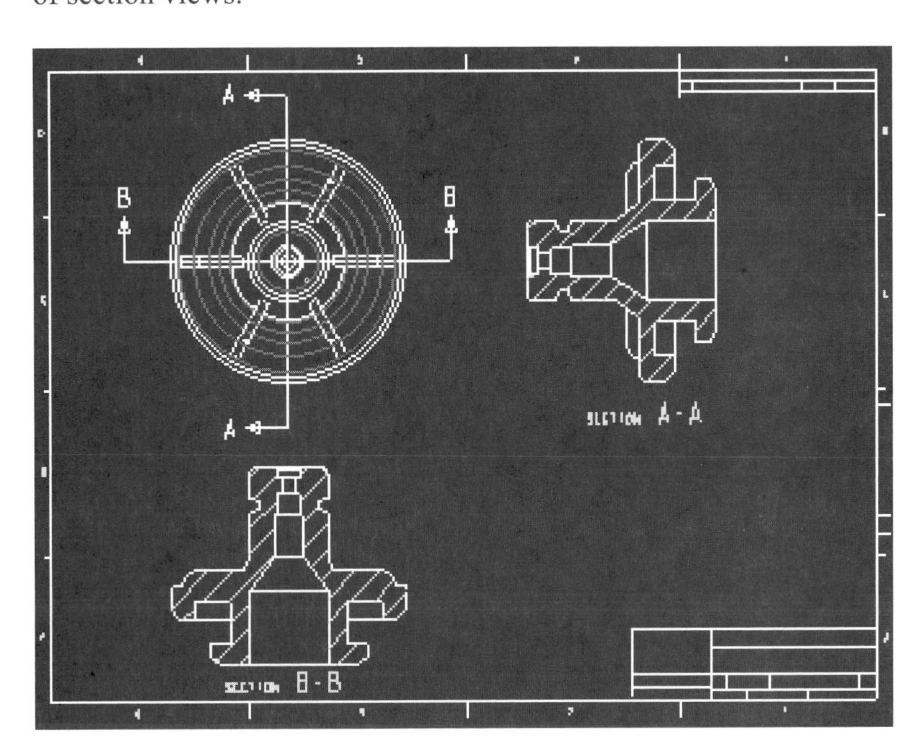

Figure 18.4
COAch for Pro/E, Sections (Planar Sections)

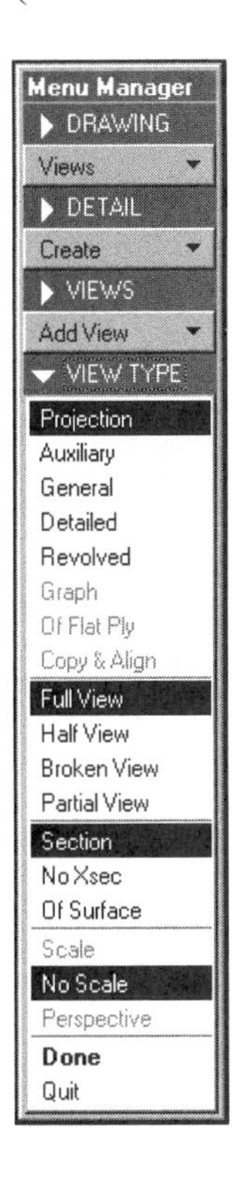

Sectional views are slices through a part or assembly and are valuable for opening up the part or assembly for displaying features and detailing in Drawing mode. Part sectional views can also be used to calculate sectional view mass properties. Each sectional view has its own unique name within the part or assembly, allowing any number of sectional views to be created and then retrieved for use in a drawing. A variety of standard (ANSI) section lining symbols (cross-hatch patterns) representing the type of material can be generated and displayed. You can create a variety of sectional view types:

* Standard planar sections of models (parts or assemblies)
* Offset sections of models (parts or assemblies)

Planar Sections

Planar sectional views are created along a datum plane. The datum may be established during the creation of the sectional view using the **Make Datum** option, or an existing plane may be selected.

To create a planar sectional view of a part:

1. Choose **X-section** from the PART menu, and then **Create** from the CROSS SEC menu.
2. Choose **Planar** from the XSEC CREATE menu, then **Done**.
3. Enter a name for the sectional view, then select (datum or surface), or make, the datum along which the section is to be generated.

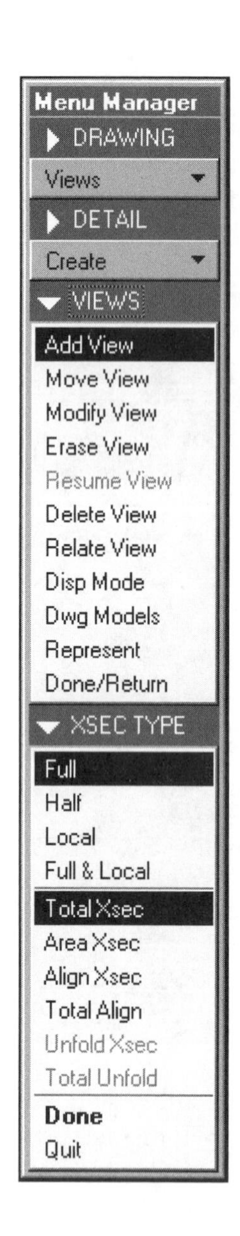

To create a planar sectional view of an assembly:

1. Choose **Set Up** from the ASSEMBLY menu and **X-Section** from the ASSEM SETUP menu, then **Create** from the CROSS SEC menu.
2. Choose **Planar** from the XSEC OPTS menu, then **Done**.
3. Enter a name for the sectional view.
4. Select (datum or surface), or create, the **assembly** datum along which the section is to be generated.

Offset Sections

An **offset** sectional view (Fig. 18.5) is created by extruding a 2D section perpendicular to the sketching plane, just like creating an extruded cut but without removing any material. This type of sectional view is valuable for opening up the part to display several features with a single section.

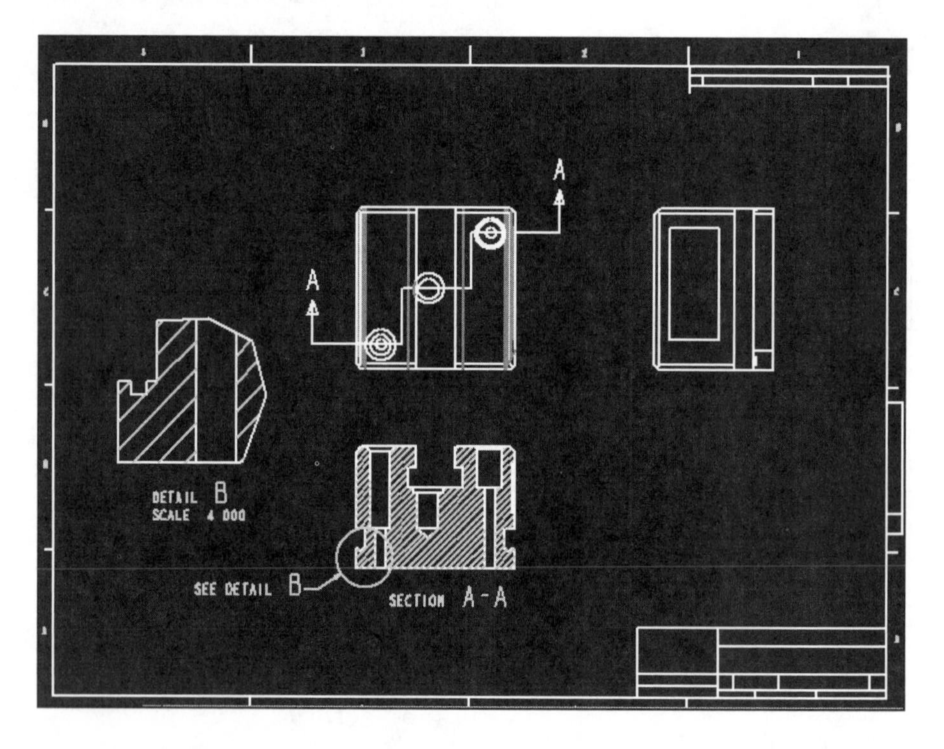

Figure 18.5
COAch for Pro/E, Sections (Stepped Sections)

The sketched section must be an *open section*. The first and last segments of the open section must be straight lines.

To create an offset section:

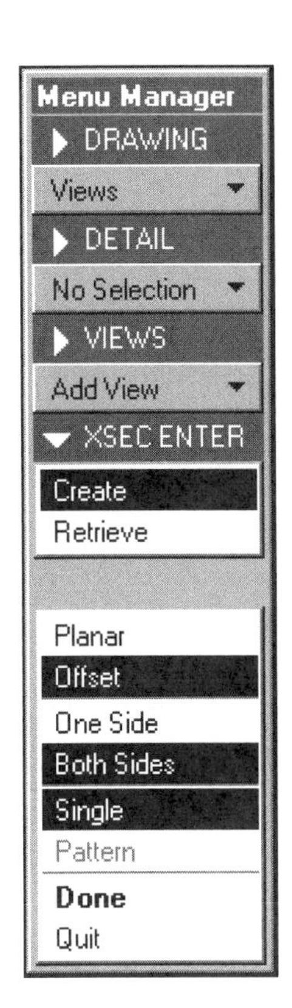

1. Choose **X-section** from the PART or ASSEM SETUP menu, then **Create** from the CROSS SEC menu.
2. Choose **Offset** and **One Side** or **Both Sides** from the XSEC CREATE or XSEC OPTS menu, then **Done**.
3. Enter a name for the sectional view.
4. Answer the prompts for entering Sketcher. The sketching plane can be created using the **Make Datum** option.
5. **Sketch** the sectional view and **Dimension** it to the model. Choose **Done** when the section has been regenerated successfully.

Auxiliary Views

The proper selection of views, view orientation, and view alignment is determined by a part's features and its natural or assembled position. Normally, the front view is the primary view and the top view is obvious, based on the position of the part in space or when assembled. The choice of additional views is determined by the part's features (Fig. 18.6) and the minimum number of views necessary to describe the part and show its dimensions.

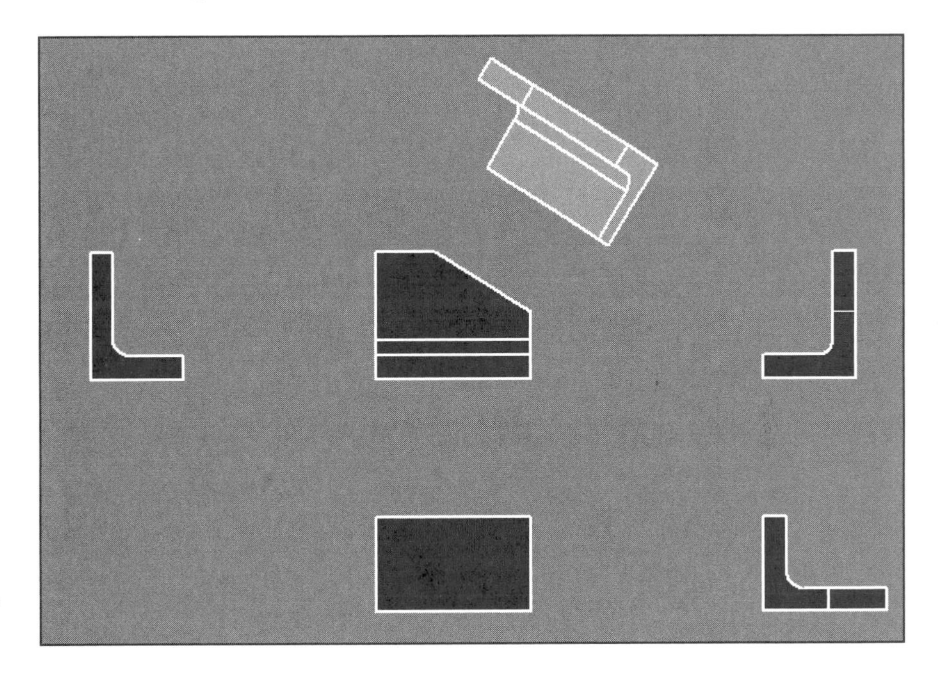

Figure 18.6
COAch for Pro/E, Auxiliary Views (Creating an Auxiliary View)

Auxiliary views are created by making a projection of the model perpendicular to a selected edge. They are normally used to discern the true size and shape of a planar surface on a part. An auxiliary view can be created from any other type of view. Auxiliary views can have arrows created for them that point back at the view(s) from which they were created.

To add an auxiliary view to the drawing, use the following command options:

1. Choose **Auxiliary** and other available options from the VIEW TYPE menu.
2. Choose **Done** to accept the options, or **Quit** to quit the creation of a new view.
3. Pick the location of the new view on the drawing (Fig. 18.7).
4. Pick an edge of, an axis through, or a datum plane of the surface of the model in the view from which the auxiliary view will be developed. If the edge selected is from a view that has a pictorial (isometric-trimetric) orientation, the new view will be oriented as the base feature section was. Otherwise, the view will be oriented with the selected surface parallel to the plane of the drawing.

Detail views will also be discussed in this lesson (Fig. 18.8).

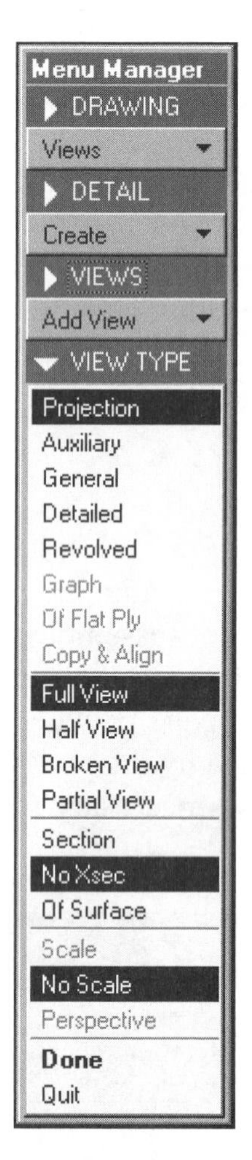

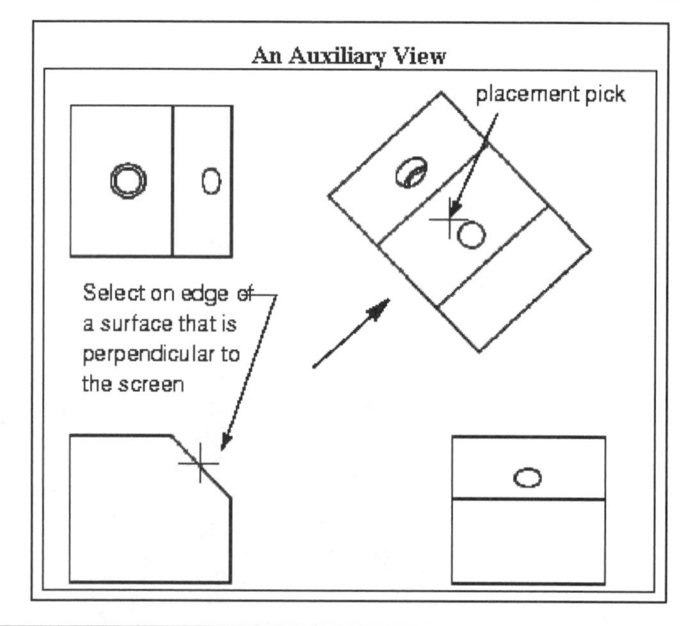

Figure 18.7
Auxiliary Views

Detailed Views

A detailed view displays a portion of the model shown in another view. To create one, you define a spline boundary around the portion of an existing view that you want to show in detail. As shown in the following figure, the system displays the view name and scale value below it, circles the portion of the parent view that is detailed, and attaches a note identifying the detailed view.

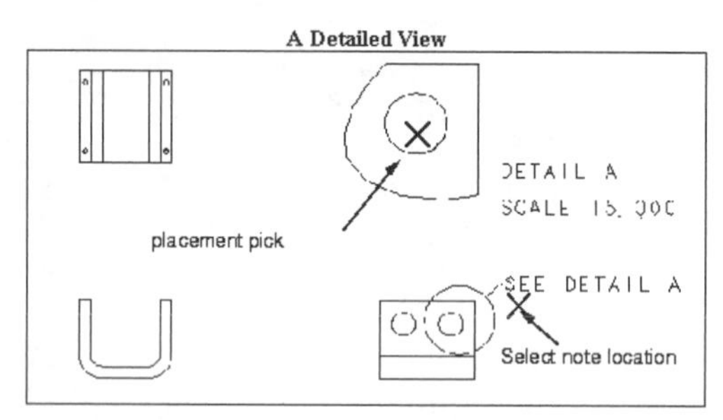

The display of a detailed view follows that of the view from which you created it. For example, if the parent view displays hidden lines in the area that is detailed, the detailed view (which is simply a portion of its parent) also displays those hidden lines. Also, if you erase a feature from the parent view, the system also erases it from the detailed view. Because of this dependency, you can modify such characteristics as cross-hatching and hidden line display in a detailed view *only if you modify the parent view as well.* However, you can make a detailed view independent of its parent, as described in Step 12 of the following procedure.

➤ **How to Add a Detailed View to a Drawing**

1. Choose **Detailed** and other available commands from the VIEW TYPE menu.
2. Choose **Done** to accept the commands, or **Quit** to quit the creation of a new view.
3. Select the location of the new view on the drawing.
4. Type the scale value for the view.

Figure 18.8
Detailed Views

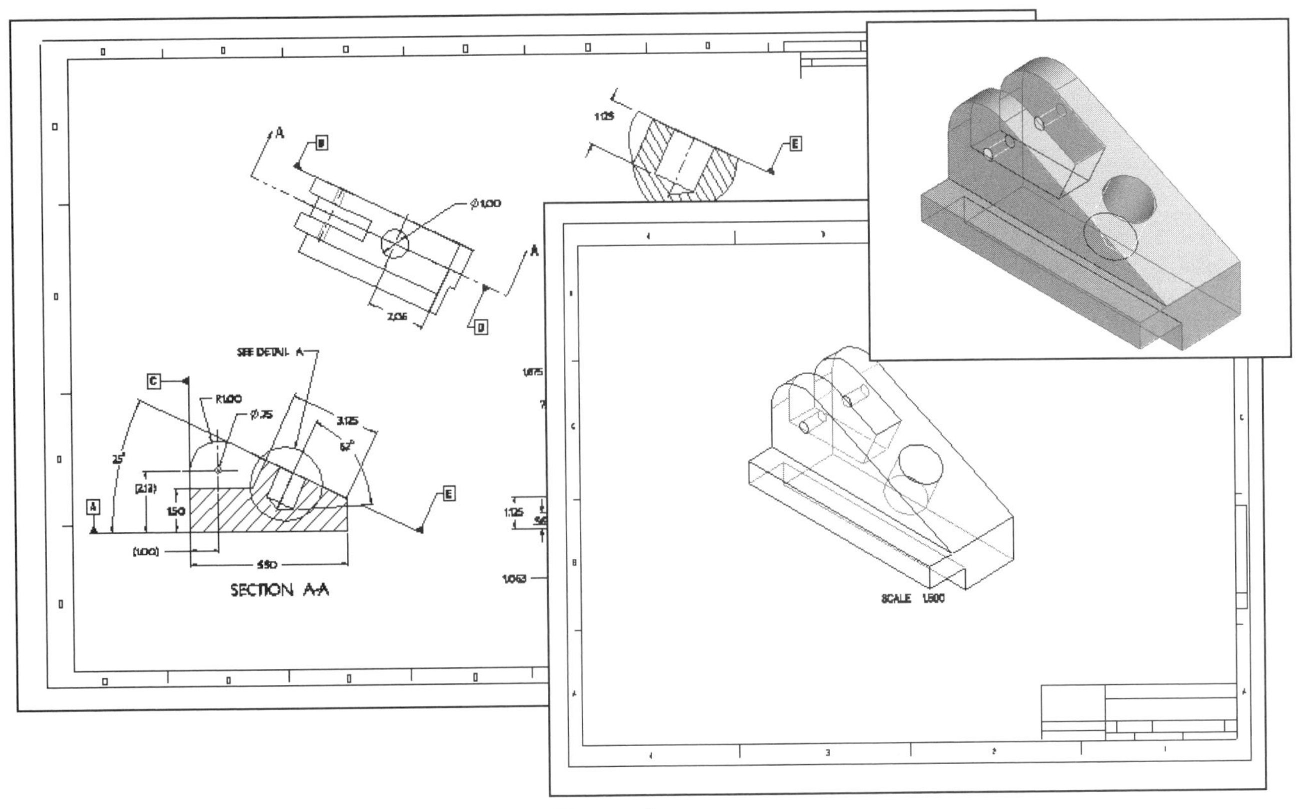

Figure 18.9
Anchor

ENVIRONMENT
Utilities (from Toolbar) ⇒
Environment ⇒
- ❑ **Datum Planes**
- ❑ **Point Symbols**
- ❑ **Datum Axes**
- ❑ **Coordinate System**
- ❑ **Spin Center**
- ❑ **Snap to Grid**
- ❑ **Snap to Snap Lines**
- **Hidden Line**
- **Isometric**
- **No Display**

OK

Anchor Drawing

Auxiliary views are created, using Pro/E, by making a projection of the model perpendicular to a selected edge or along an axis. They are normally used to discern the true size and shape of a planar surface on a part, such as the Anchor, which has an angled surface with a hole machined perpendicular to that surface.

Pro/E is able to create and display fully associative *section views* of solid models (Fig. 18.9). You can create a section view by using an existing view on a drawing in Drawing mode or by creating a section in Part mode for retrieval on a drawing. As you were completing some of the lessons in this text, you were instructed to create sections in a number of lesson parts and lesson projects.

You create a section view while looking at the drawing by sketching (with the help of Pro/E) a section line (*Offset*) or by selecting an existing **Datum Plane** for the section line (*Planar*) to pass through.

The display of the *section line symbol* and the *section view text* is controlled by the **Section Line Display Parameters** in the Drawing DTL SETUP file (**.dtl**).

The sectioning parameters apply to the two basic types of standards: the ANSI (ASME) Standards and the ISO/JIS/DIN Standards. Section line display and the manner in which the view titles (parameters) are created vary based on the standard you choose to use. For a listing of section line and view parameters, see Appendix A in the **Pro/ENGINEER Drawing User's Guide.** The pertinent parameters are shown in Figure 18.10. Figure 18.11 shows the **.dtl** file used by CADTRAIN for its sectioned drawings.

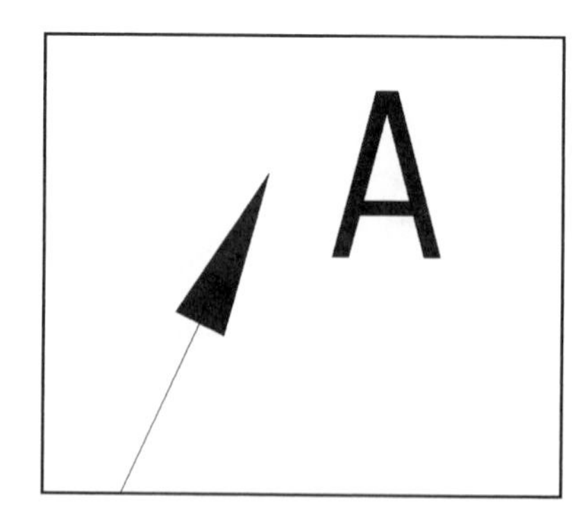

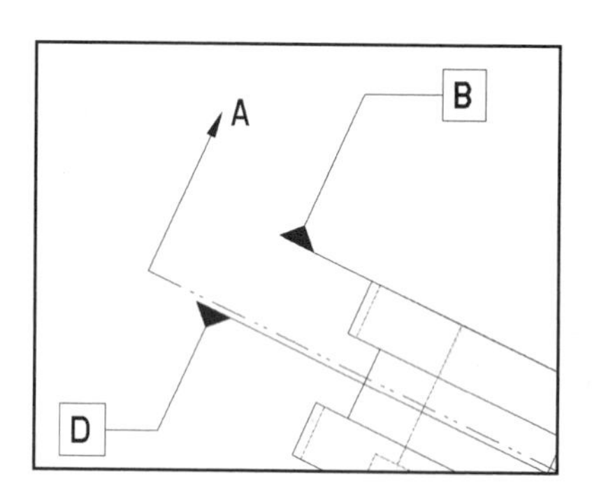

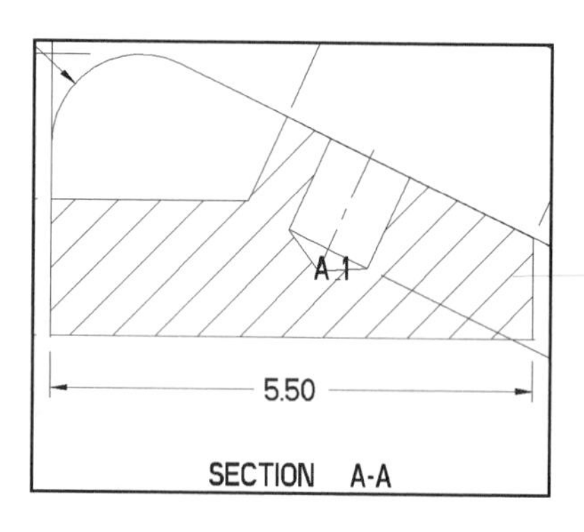

Figure 18.10
COAch for Pro/E, Section Line Symbol and the Section View Text Parameters

Parameter	Default	Hint
crossec_arrow_length	.187	
crossec_arrow_width	.0625	
crossec_arrow_style	tail_online	Tail Head
crossec_text_place	after_head	A after_head A before_tail A above_tail A above_line
cutting_line	std_ansi	std_ansi std_din std_jis std_iso std_ansi_dashed std_jis_alternate
cutting_line_segment	0	0 value
def_view_text_height	0	A
view_note	std_ansi	std_ansi std_din std_jis std_iso

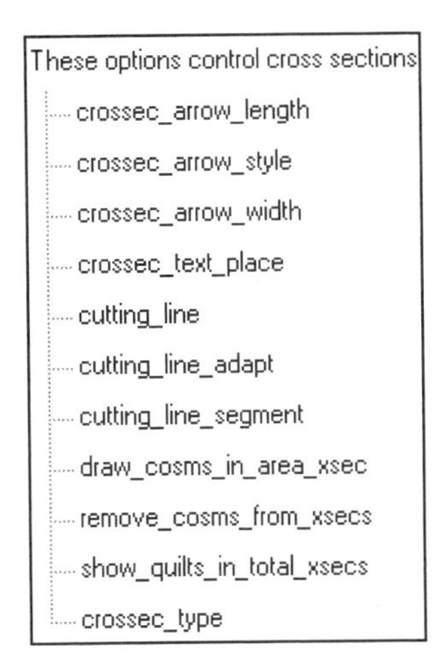

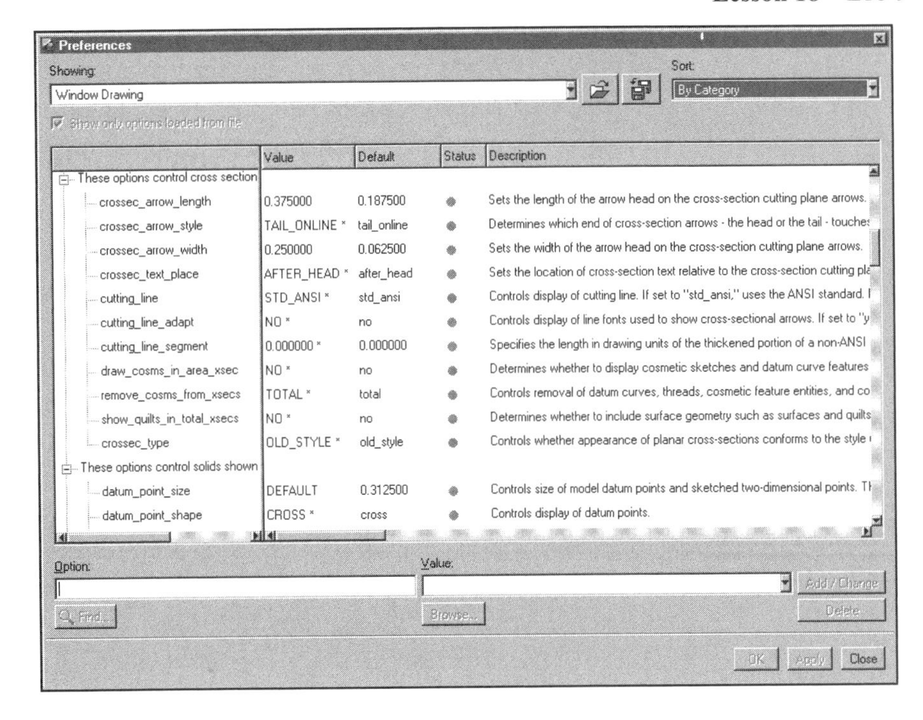

	Value	Default	Status	Description
These options control cross section				
crossec_arrow_length	0.375000	0.187500	●	Sets the length of the arrow head on the cross-section cutting plane arrows.
crossec_arrow_style	TAIL_ONLINE *	tail_online	●	Determines which end of cross-section arrows - the head or the tail - touches
crossec_arrow_width	0.250000	0.062500	●	Sets the width of the arrow head on the cross-section cutting plane arrows.
crossec_text_place	AFTER_HEAD *	after_head	●	Sets the location of cross-section text relative to the cross-section cutting pl
cutting_line	STD_ANSI *	std_ansi	●	Controls display of cutting line. If set to "std_ansi," uses the ANSI standard. I
cutting_line_adapt	NO *	no	●	Controls display of line fonts used to show cross-sectional arrows. If set to "y
cutting_line_segment	0.000000 *	0.000000	●	Specifies the length in drawing units of the thickened portion of a non-ANSI
draw_cosms_in_area_xsec	NO *	no	●	Determines whether to display cosmetic sketches and datum curve features
remove_cosms_from_xsecs	TOTAL *	total	●	Controls removal of datum curves, threads, cosmetic feature entities, and co
show_quilts_in_total_xsecs	NO *	no	●	Determines whether to include surface geometry such as surfaces and quilts
crossec_type	OLD_STYLE *	old_style	●	Controls whether appearance of planar cross-sections conforms to the style
These options control solids shown				
datum_point_size	DEFAULT	0.312500	●	Controls size of model datum points and sketched two-dimensional points. Th
datum_point_shape	CROSS *	cross	●	Controls display of datum points.

Figure 18.11
.dtl File for Drawing Parameters Showing Options That Control Cross Sections and Their Arrows

HINT

If you did not create a section when modeling the Anchor, use the following commands in Part mode:

X-section ⇒ Create ⇒ Done ⇒ (type **A** for the section name) ⇒ ✔ ⇒ **Make Datum ⇒ Through ⇒** (pick axis of the hole) ⇒ **Parallel** ⇒ [pick **FRONT** (or **B**)] ⇒ **Done ⇒ Done/Return**

If you are going to create sections while you are in Drawing mode, you may require **Datum Planes**. Because most views on a drawing are orthographic, and because they cannot be changed with **Spin**, you may want to prepare the model with the necessary **Datum Planes** *before* you enter Drawing mode. You were asked to create a section of the Anchor in Lesson 4. If you did not, you now need to create a section that passes vertically, lengthwise, through the model in Part mode (Fig. 18.12) or to create a section in Drawing mode, as described next.

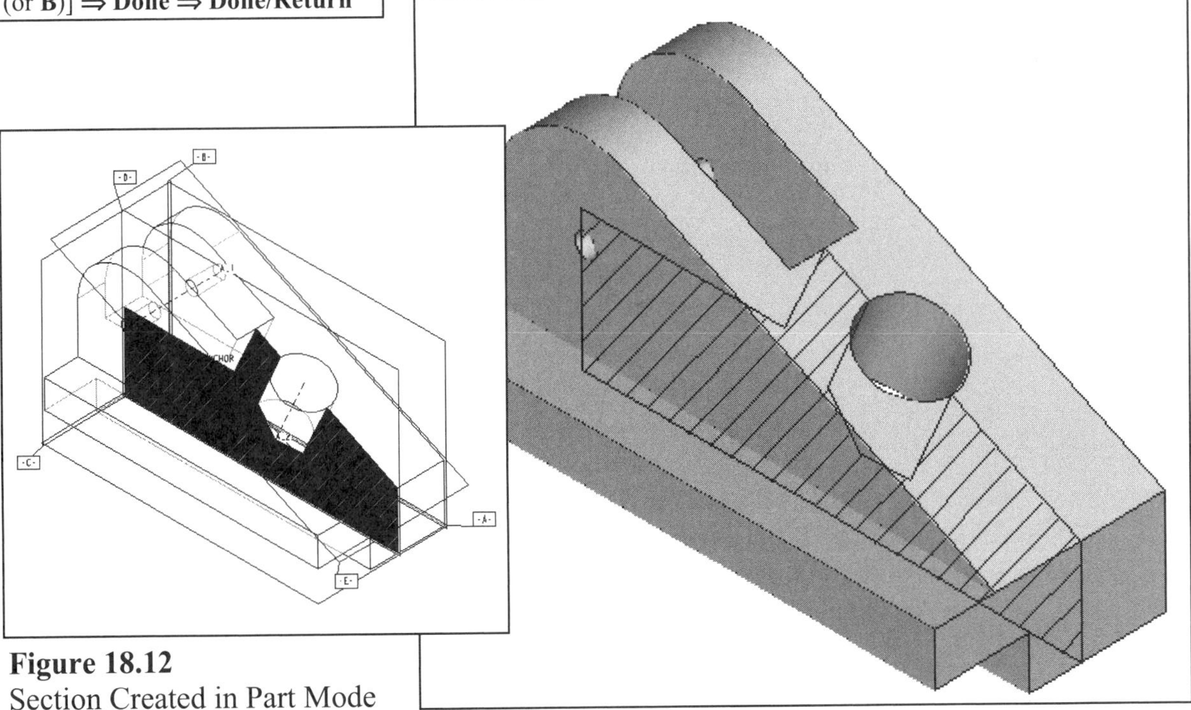

Figure 18.12
Section Created in Part Mode

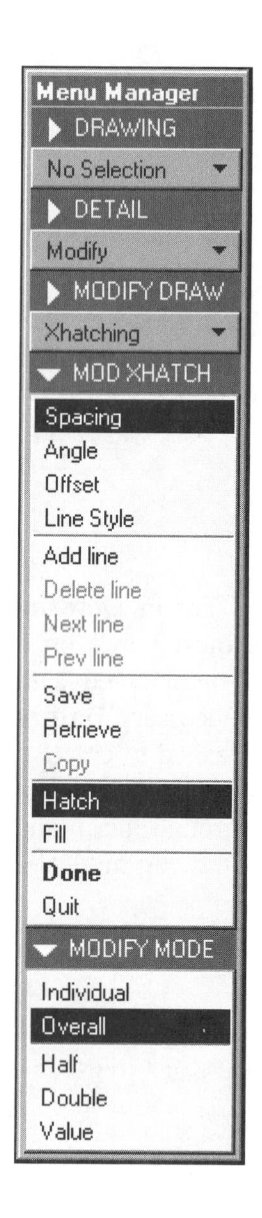

In Drawing mode, the **Planar** section option allows you to create a section that passes straight through a part without any "jogs" (steps) in the section line. **Planar** sections can be defined by retrieving a section, creating a datum plane, or selecting an existing datum plane to define the cut position.

When you want to create the section while in Drawing mode, you can select either a planar face or a **Datum Plane** as the "plane" the section cut passes through. Even though you can pick a planar face, this is generally not good practice. The section cut takes place *at* the plane you select. If you pick a face, this causes the section to be tangent to it. Not only is this poor drafting practice, but in some cases it also causes Pro/E to generate incomplete sections.

When you create a **Planar** section on the drawing, Pro/E first asks you to define the location of the center of the section view. It then displays the orthographic projection to which the section edges (created by the cut) and the crosshatching will be added. Choose **Create** ⇒ **Planar** ⇒ **Done**. Pro/E prompts you to enter a **name** for the section. This "name" is actually the section letter you wish to use. If you are creating your first section and you want it to have the letter **A** displayed at each arrow end and have the title **SECTION A-A**, you should enter **A** as the name of the section. It is a common mistake to type **AA**, thinking that you are establishing **SECTION A-A,** when in reality you are getting **SECTION AA-AA**. Type one letter only. You must then select a **Datum Plane** to define the section cutting plane. Pro/E then prompts you to pick a view in which the cutting plane is perpendicular to the screen. This is the view in which it will draw the section line and the view where the cut will actually take place. Pro/E displays the section line, with its arrows pointing in the direction it "thinks" you want to view the cut.

After the section has been created, use the right mouse button to get the pop-up menu item **Modify Item** ⇒ (pick the section view) ⇒ **Flip X-sec** (to flip the section identification arrows). The arrow direction does not affect the cross-sectioned area, which was defined when you indicated the location of the view. It does affect what you see in the "background," behind the cutting plane.

You can change the name of a section after it is created by going to **Part** mode and retrieving the part used to make the drawing. If you choose the option **X-section** ⇒ **Modify** ⇒ (pick the name to be changed) ⇒ **Name** ⇒ (enter a new name), the moment you change back to the drawing window, the section name will be updated.

In the following pages, you will be creating a detail drawing of the Anchor. The front view will be a full section. A right side view and an auxiliary view are required to detail the part. Views will be displayed according to visibility requirements per ANSI standards, such as no hidden lines in sections. The part is to be dimensioned according to ASME Y14.5M 1994. You will add the format created in Lesson 16. Detailed views of other parts will be introduced to show the wide variety of view capabilities of Pro/E's Drawing mode.

Lesson procedures & commands
START HERE ➡ ➡ ➡

Using the Anchor model, start a drawing that will include a front (sectioned) view, a right side view, an auxiliary view, and a detail view. You will not be using the default template.

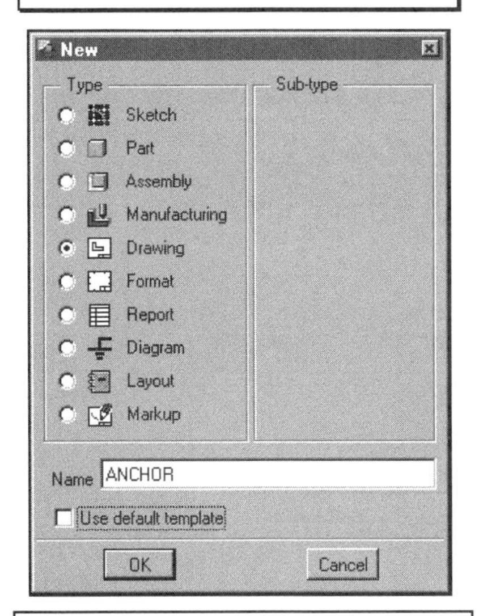

File ⇒ **New** ⇒ ●**Drawing** ⇒ (type **ANCHOR**) ⇒ ❑ **Use default template** ⇒ **OK** ⇒ **Browse** (from New Drawing dialog box) ⇒ (select the part name- **ANCHOR**) ⇒ **Open** ⇒ Standard Size **D** ⇒ **OK** ⇒ **Sheets** ⇒ **Format** ⇒ **Add/Replace** ⇒ **Format** ⇒ System Formats ⇒ **d.frm** (Fig. 18.13) ⇒ **Open** ⇒ **Done/Return** ⇒ **Advanced** ⇒ **Draw Setup** ⇒ (Open a configuration file and then pick your **.dtl** from your directory list, and/or create a new **.dtl** file using the values provided in the left-hand column under ***HINT***) ⇒ **Add/Change** ⇒ **Apply** ⇒ **Close** ⇒ **Done/Return** ⇒ **Views** (keep all defaults) ⇒ **Done** ⇒ [pick the approximate center of the view (Fig. 18.14)] ⇒ **Front** (Reference 1 from the Orientation Dialog box) ⇒ (pick datum **B**) ⇒ **Top** ⇒ (pick datum **A**) ⇒ **OK** (Fig. 18.15)

HINT

Advanced ⇒ **Draw Setup** ⇒
Option: (type the first few letters of the option you wish to change) ⇒
Value: (type the number or select ▼ to get the required value) ⇒
Add/Change ⇒ **Apply** ⇒ **Close**

draw_text_height	*.25*
default_font	*filled*
draw_arrow_style	*filled*
gtol_datums	*STD_ASME*
allow_3d_dimensions	*YES*

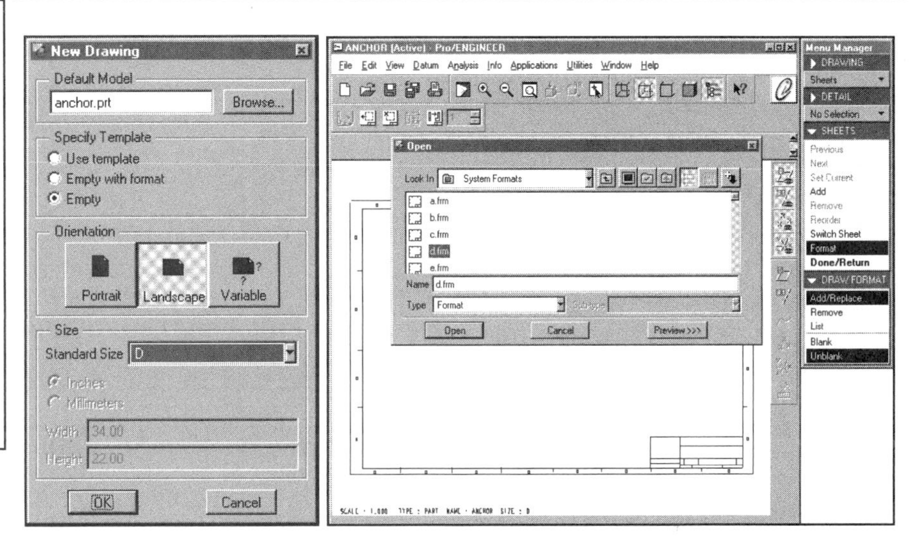

Figure 18.13
Starting a Drawing

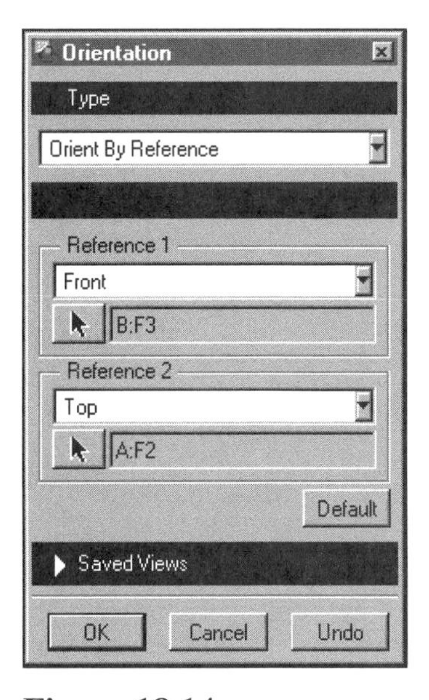

Figure 18.14
Orienting the Front View

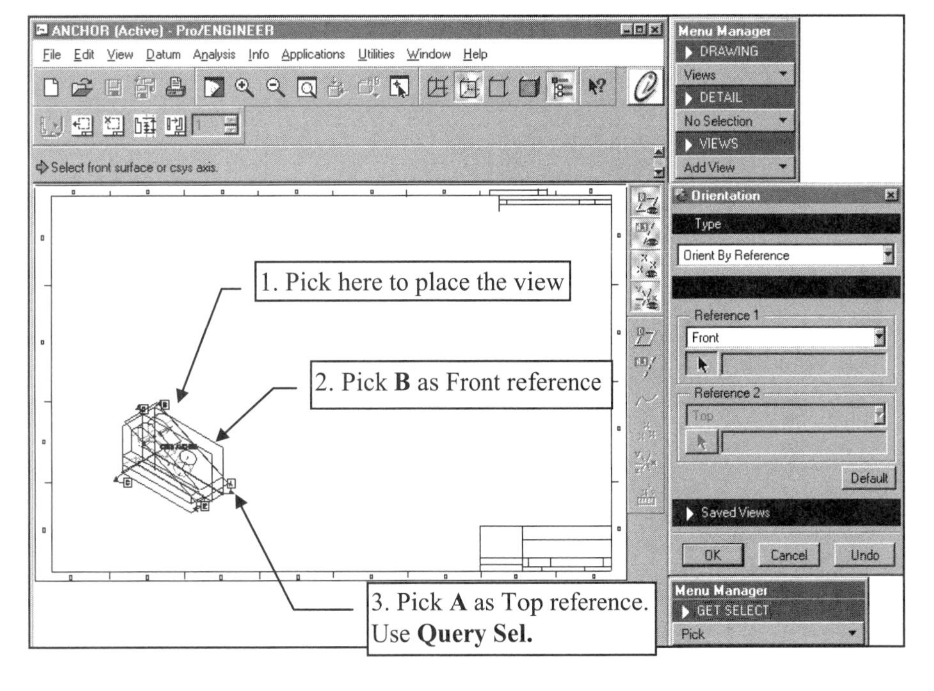

NOTE

Recall from Lesson 4 that Gtol datums were set basic (**Set Up ⇒ Geom Tol ⇒ Set Datum**)

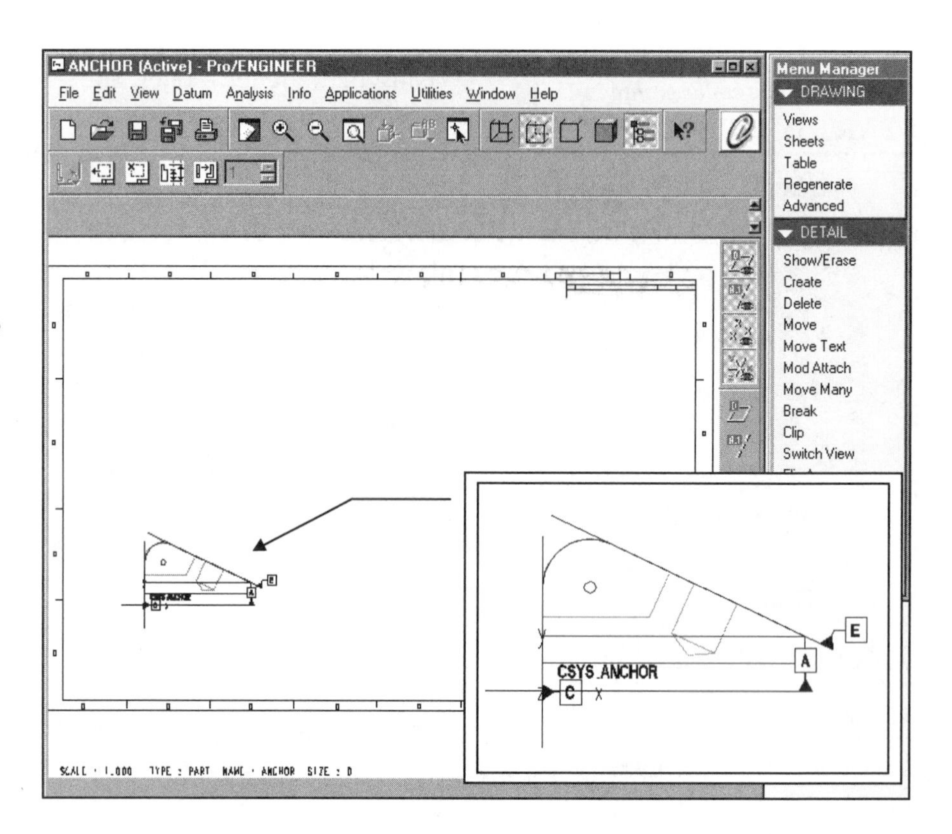

Figure 18.15
Front View Placed

Add the right side view, as shown in Figure 18.16. Choose the following commands:

HINT

Before detailing a drawing, always change the display of nonbasic datums and other Environment datum features by toggling the buttons in your Toolbar.

Done/Return ⇒ File ⇒ Save ⇒ ✔ ⇒ Views ⇒ Add View ⇒ Projection ⇒ Full View ⇒ No Xsec ⇒ No Scale ⇒ Done ⇒ (pick a position for the right side view as shown in Figure 18.16) ⇒

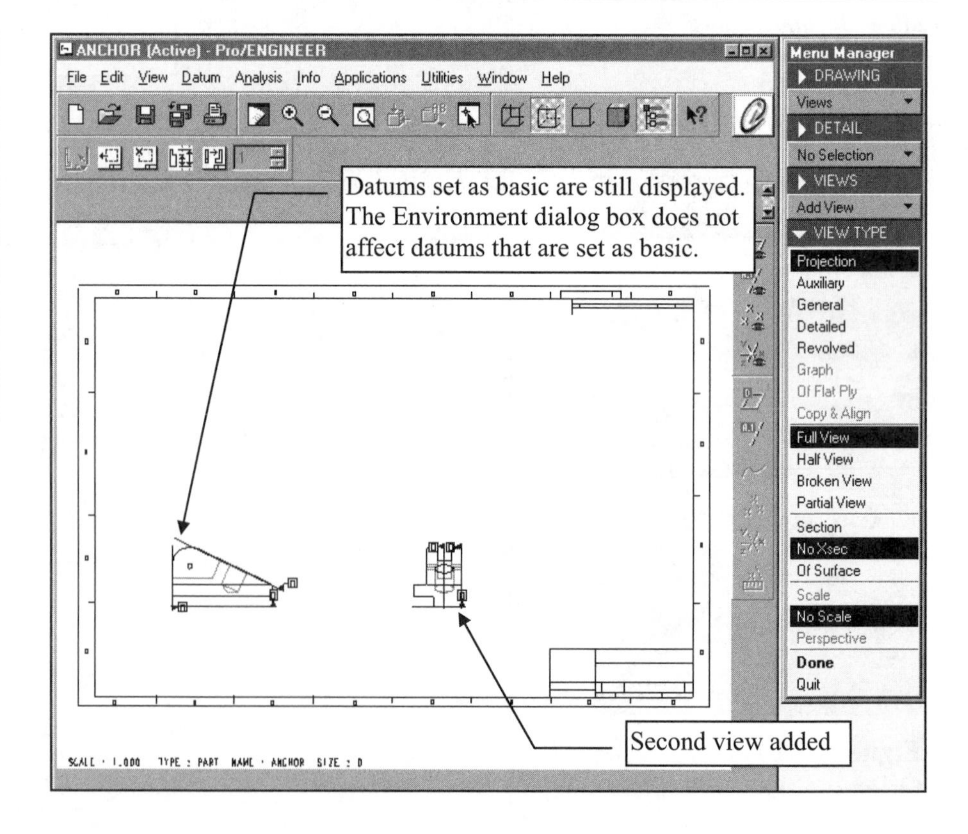

Figure 18.16
Right Side View Placed

Use the following commands to add the auxiliary view:

Add View ⇒ Auxiliary ⇒ Done ⇒ [pick a center point for view) ⇒ (pick angled edge in front view (Fig. 18.17)] ⇒ **Done/Return**

2. Pick the angled edge

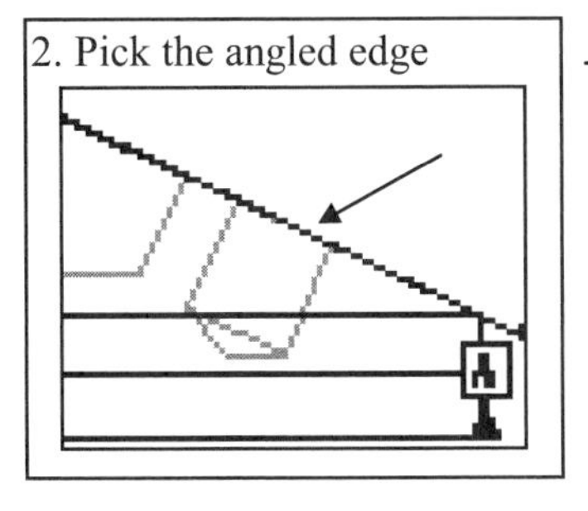

1. Pick here to locate the center of the auxiliary view

Figure 18.17
Auxiliary View

Now change the front view into a section view. The section **A** was created in Part mode. Use these commands (Fig. 18.18):

Views ⇒ Modify View ⇒ View Type ⇒ (pick the front view as the view to be modified) **⇒ Section ⇒ Done ⇒ Full** (default) **⇒ Total Xsec** (default) **⇒ Done ⇒** (pick section name from XSEC NAMES menu: **A**) ⇒ [pick a view (the auxiliary view) for the arrows where the section cutting plane is *perpendicular* (Fig. 18.18)] ⇒ **Done Sel**

Section cutting plane lines, arrows, and section identification letters

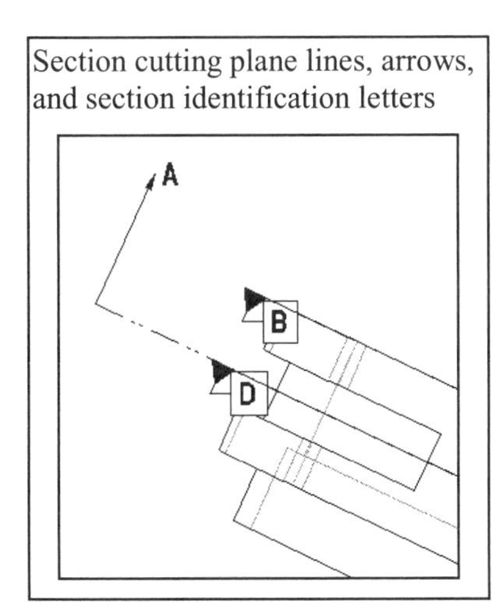

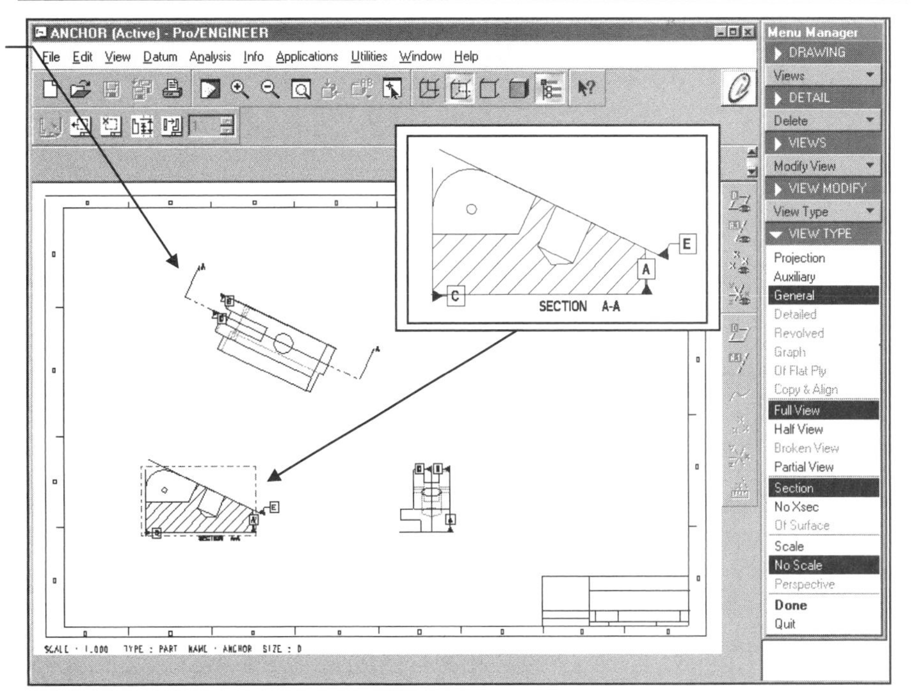

Figure 18.18
Front View Modified to Be a Front Section View

Modify the visibility of the views to remove all hidden lines:

Disp Mode ⇒ **View Disp** ⇒ (pick all three views) ⇒ **Done Sel** ⇒ **No Hidden** ⇒ **No Disp Tan** ⇒ **Done** (Fig. 18.19) ⇒ **Done Sel** ⇒ **Done/Return**

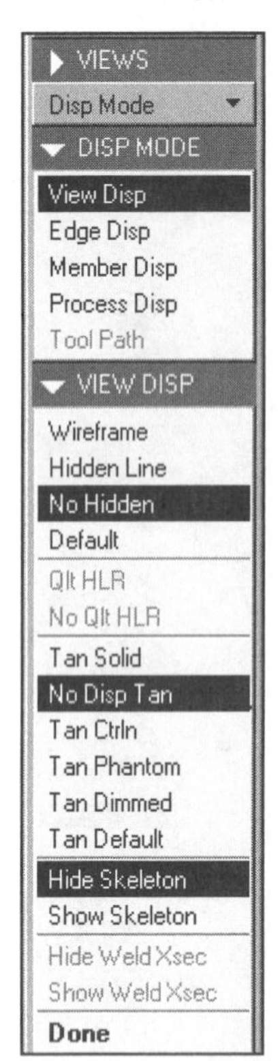

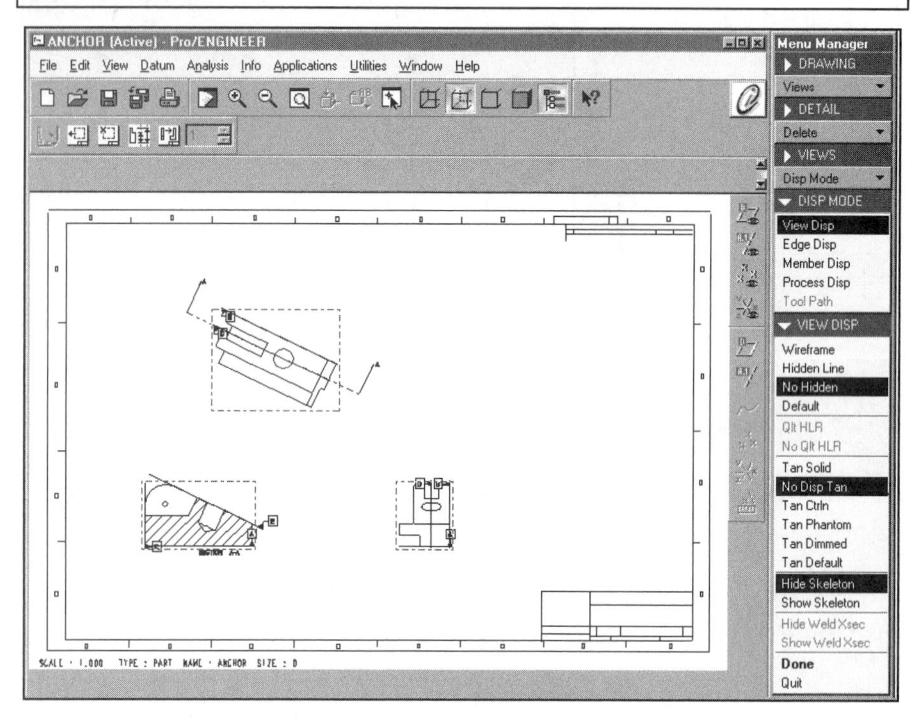

Figure 18.19
No Hidden, No Disp Tan

Show all dimensions and axis-centerlines:

Show/Erase ⇒ **Show** ⇒ (pick radio buttons for *dimensions* and *axes)* ⇒ **Show All** ⇒ **Yes** ⇒ **Accept All** (Fig. 18.20) ⇒ **Close** ⇒ **Tools** ⇒ **Clean Dims** ⇒ **Pick Many** (pick all dimensions by enclosing the views in a pick box) ⇒ **Done Sel** ⇒ ❑ **Create Snap Lines** ⇒ **Apply** ⇒ **Close** ⇒ **Done/Return**

Figure 18.20
Show All

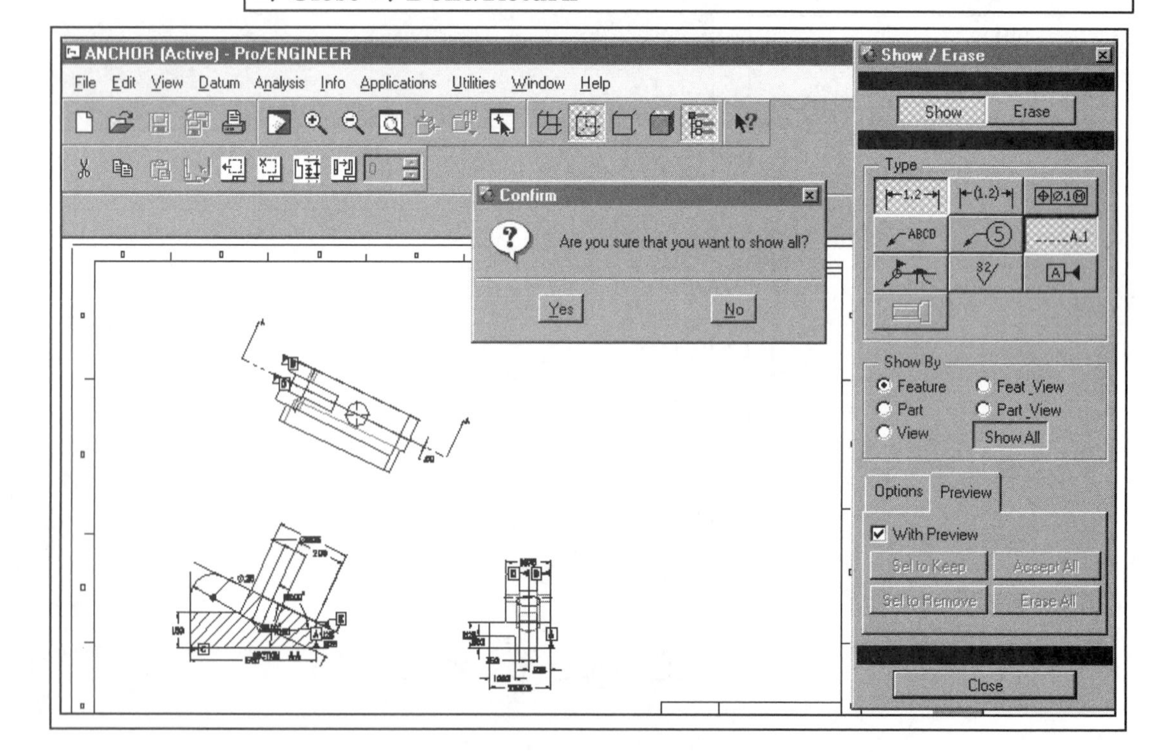

Use your *right* mouse button (**Modify Item**) to edit the dimensions and axes and to rotate the centerline of the large hole so that it aligns with the view correctly:

HINT
Change the height of the section lettering to **.375**, either in the **.dtl** file (as the new default) or use **Modify ⇒ Text ⇒ Text Height**

Modify Item ⇒ (pick the centerline (Fig. 18.21 left) **⇒ Mod Attach ⇒ Parallel ⇒** (pick the edge of the part (Fig. 18.21 right) **⇒ Done/Return ⇒ Done Sel**

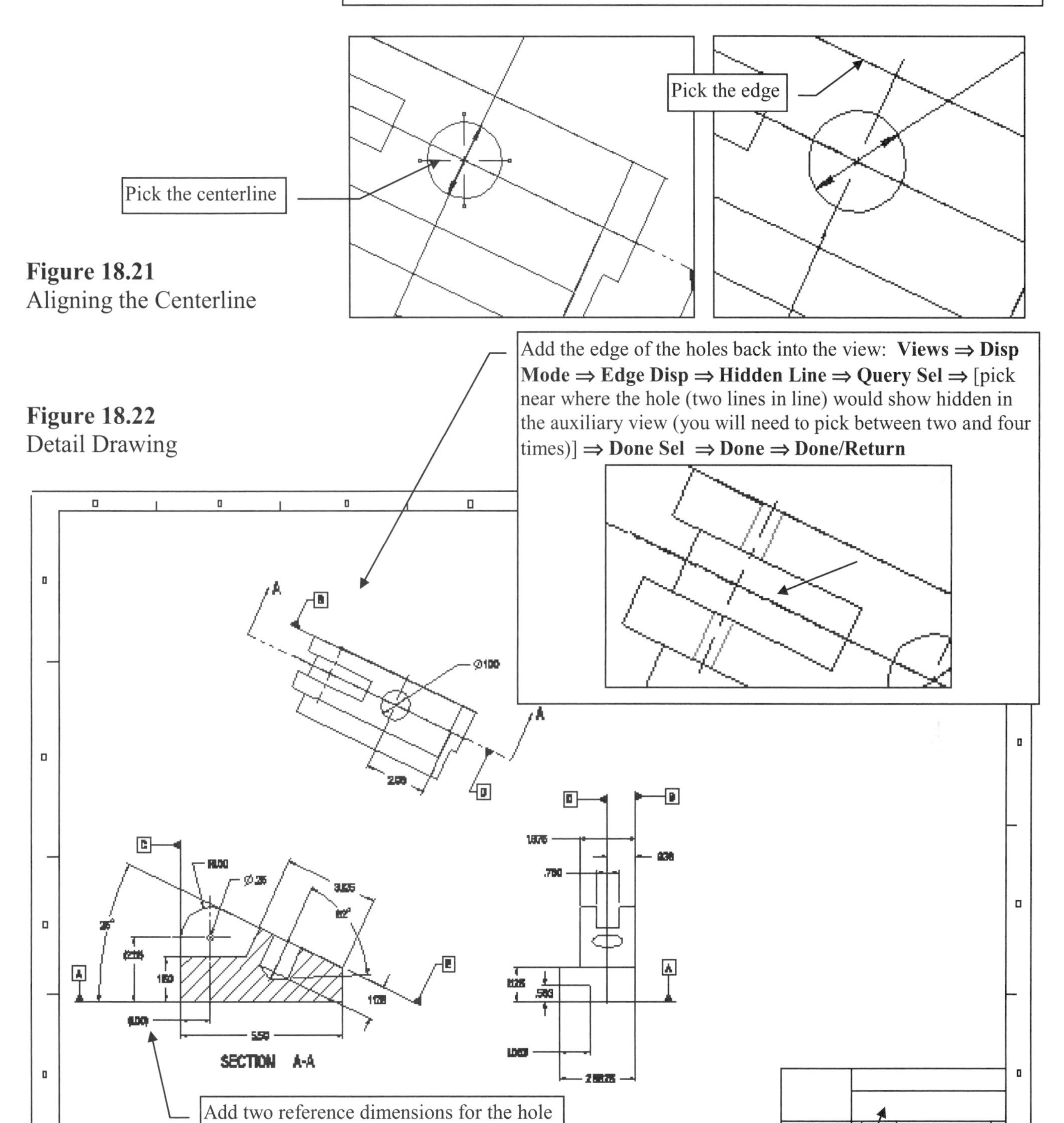

Figure 18.21
Aligning the Centerline

Figure 18.22
Detail Drawing

HINT

When Pro/E creates the section cutting plane line, it draws across the entire part. You can move the section line arrow to shorten it, based on your drafting standard.

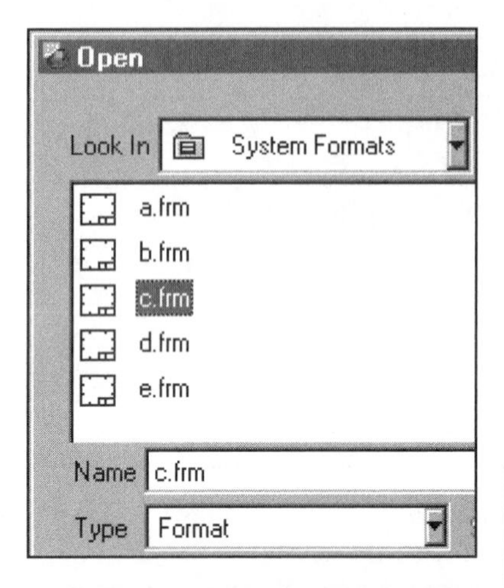

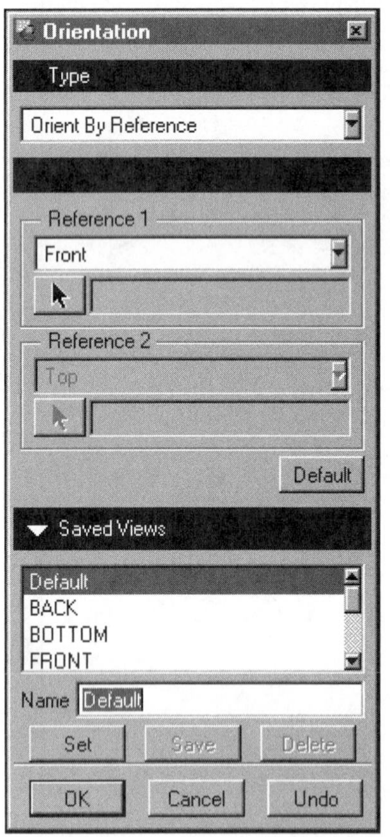

To increase the clarity of this drawing, you will need to master a number of capabilities before completing the lesson, lesson project and other, more advanced projects. Partial views, detail views, using multiple sheets, and modifying section lining (crosshatch lines) are just a few of the many options available in Drawing mode within Pro/E.

As an example, when you want to break away a small portion of a part to see internal features in a local area, you can use the **Local** section option. You define the area of the **Local** breakout by drawing a spline. The area enclosed by the spline is then removed to the depth of a selected planar face or **Datum Plane**. You draw the spline in the same manner as in the **Detailed** view option. You must define a center point along an existing edge. The spline must be closed and must contain the center point. This is one of many detailing options available in Drawing mode. (Use Pro/HELP for more information.)

Create a *second sheet* with an isometric view of the Anchor using the following commands:

Utilities ⇒ **Environment** ⇒ Default Orient **Isometric** ⇒ **OK** ⇒ **Sheets** ⇒ **Add** ⇒ **Format** ⇒ **Add/Replace** ⇒ [▼ and pick the "**C**" size format you created in Lesson 16 or pick **c.frm** (System Formats)] ⇒ **Open** ⇒ **Done/Return** ⇒ **Views** ⇒ **Scale** (leave other defaults) ⇒ **Done** ⇒ (pick a center point for the view) ⇒ (type **1.5** as the scale) ⇒ ✔ ⇒ ▼ Saved Views ⇒ **Default** ⇒ **Set** ⇒ **OK** ⇒ **View** ⇒ **Repaint** (Fig. 18.23) ⇒ **Done/Return** ⇒ **Sheets** ⇒ **Previous** (to return to **SHEET 1 OF 2**) ⇒ **Done/Return**

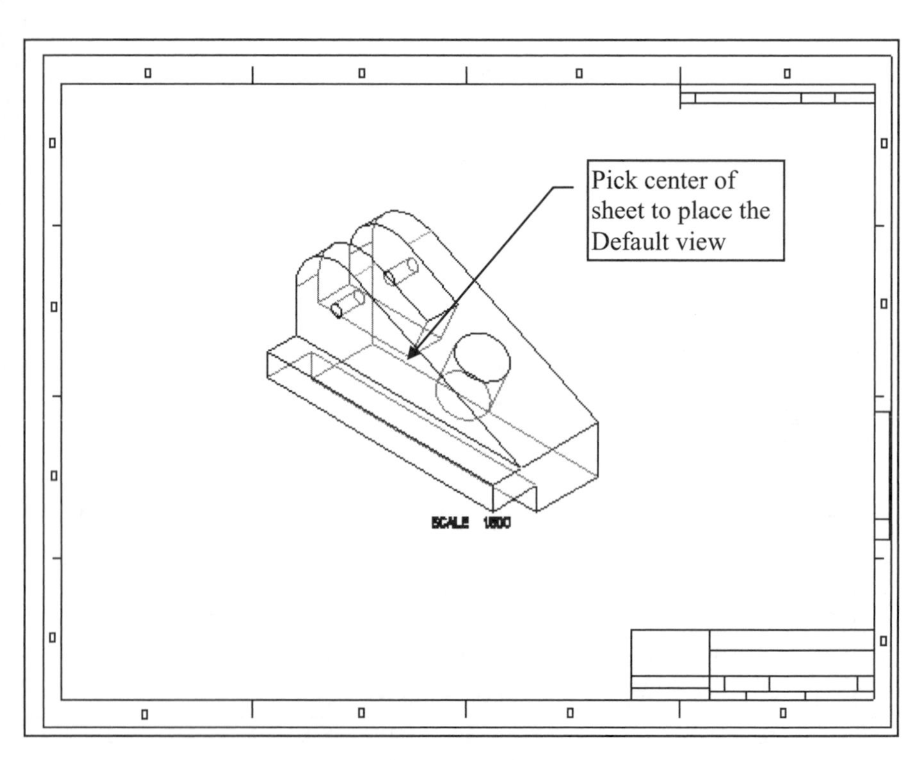

Figure 18.23
SHEET 2 OF 2 with an Isometric View of the Anchor

Create a detailed view with the following commands:

Views ⇒ Add View ⇒ Detailed ⇒ Done ⇒ (pick an open space on the drawing to position the new view, as in Figure 18.24) ⇒ (type **1.5** for the view scale) ⇒ ✔ ⇒ (select the center point for the detail on an existing view; pick the top edge of the hole in the front section view) ⇒ (sketch a spline around the hole; use the left mouse button to sketch the spline and the middle button to end it) ⇒ (enter a name for the detail view: **A**) ⇒ ✔ ⇒ (pick **Circle** from the BOUNDARY TYPE menu) ⇒ (select the location of the note as shown in Figure 18.25) ⇒ **Done/Return**

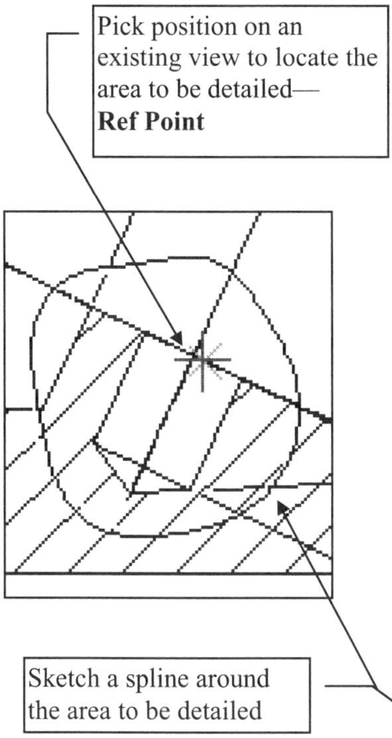

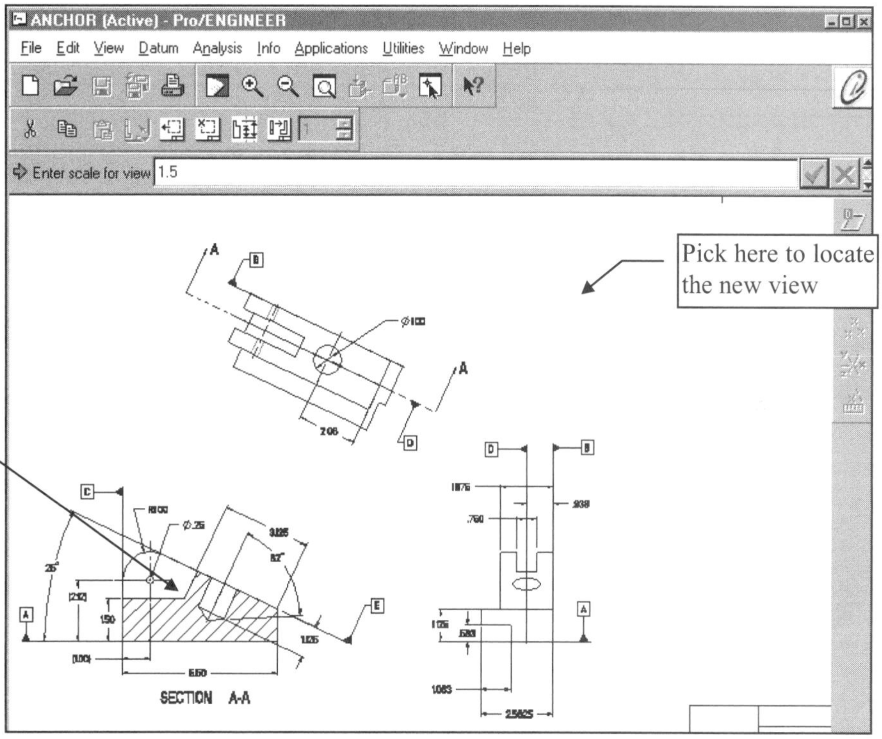

Pick position on an existing view to locate the area to be detailed—**Ref Point**

Sketch a spline around the area to be detailed

Pick here to locate the new view

Figure 18.24
Creating a Spline Around the Area to Be Detailed in a View

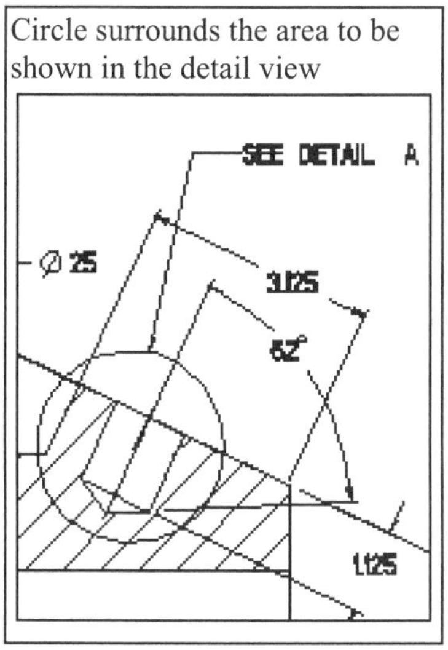

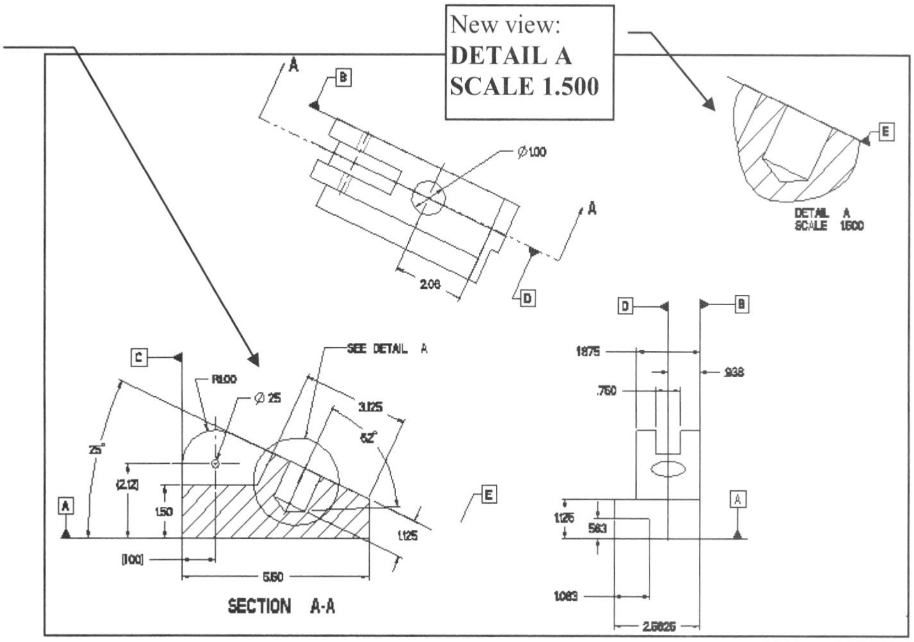

Circle surrounds the area to be shown in the detail view

New view: **DETAIL A SCALE 1.500**

Figure 18.25
Detailed View

HINT
Modify the placement of the view and view name by using your *right* mouse button: **Modify Item.**

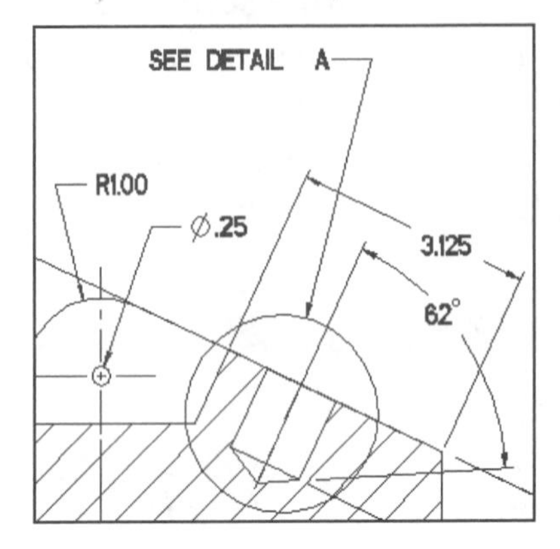

Figure 18.26
DETAIL A

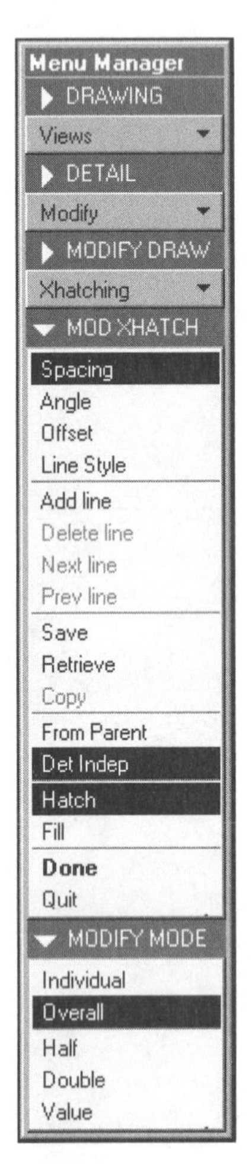

Figure 18.27
Modifying the Section Lining for the Detail View

Add an axis to the detail of the hole. Modify the text height of the detail note, name, and scale so it is **.375** (Fig. 18.26):

Show/Erase ⇒ **Show** ⇒ Axis icon ⇒ **Show All** ⇒ **Yes** ⇒ **Accept All** ⇒ **Close** ⇒ **File** ⇒ **Save** ⇒ ✔

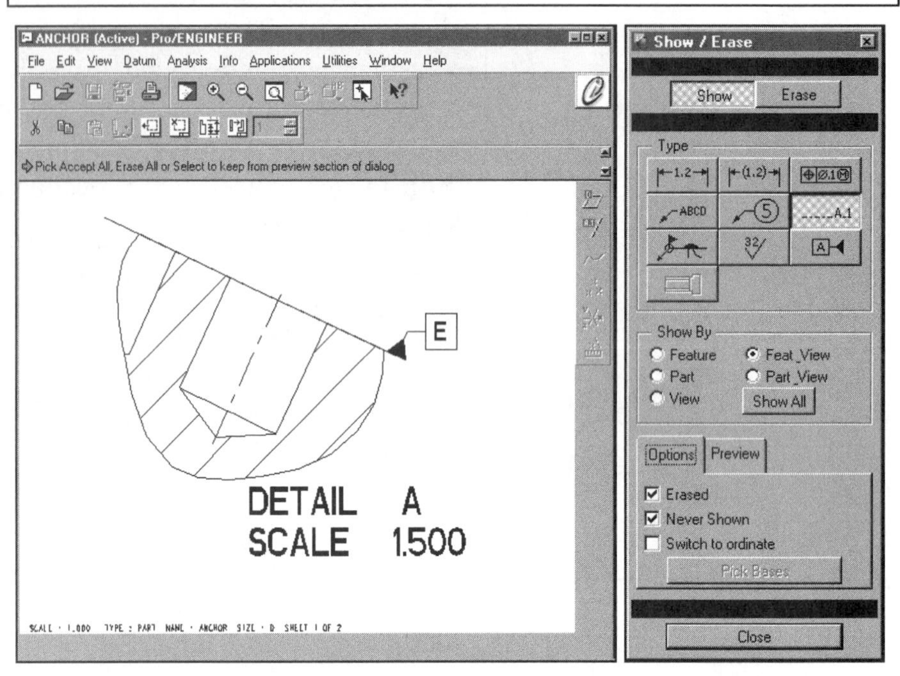

The section lining in the detail view should be modified. Change the spacing and the angle of the lining:

Modify ⇒ **Xhatching** ⇒ (pick the section lining in **DETAIL A)** ⇒ **Done Sel** ⇒ **Det Indep** (breaks the relationship to the parent view hatching, making it independent) ⇒ **Spacing** ⇒ **Hatch** ⇒ **Half** ⇒ **Angle** ⇒ **120** ⇒ **Done** (Fig. 18.27) ⇒ **Done Sel** ⇒ **Done/Return**

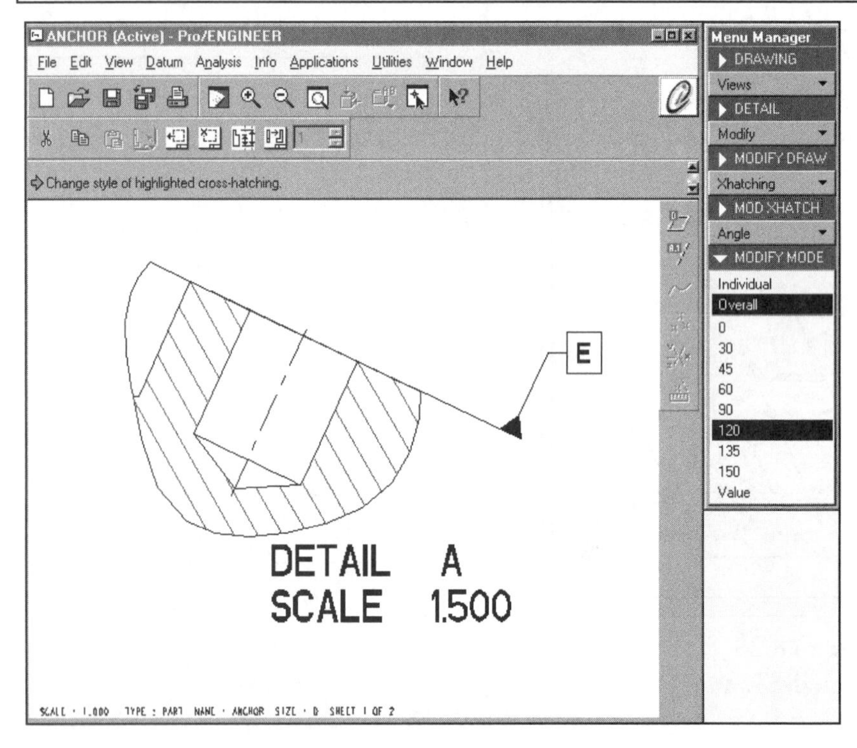

Switch the hole depth dimension from the front section view to the detail view and then modify the text style:

Modify ⇒ **Text** ⇒ **Text Style** ⇒ **Pick Many** ⇒ (enclose all of the drawing text with a pick box) ⇒ **Done Sel** ⇒ Font **Blueprint MT** (Figures 18.28 and 18.29) ⇒ **Apply** ⇒ **OK** ⇒ **File** ⇒ **Save** ⇒ ✔

Figure 18.28
Hole Depth Dimension Shown in **DETAIL A**

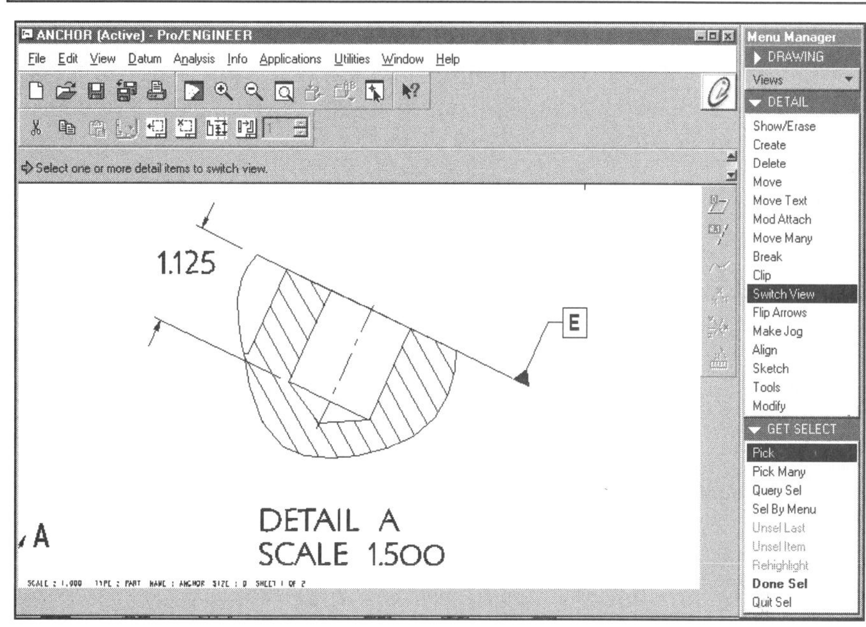

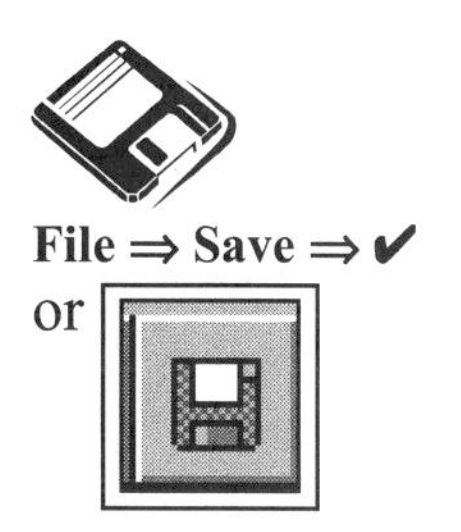

File ⇒ **Save** ⇒ ✔
or

File ⇒ **Delete** ⇒
Old Versions ⇒ ✔

Figure 18.29
Completed Detail Drawing

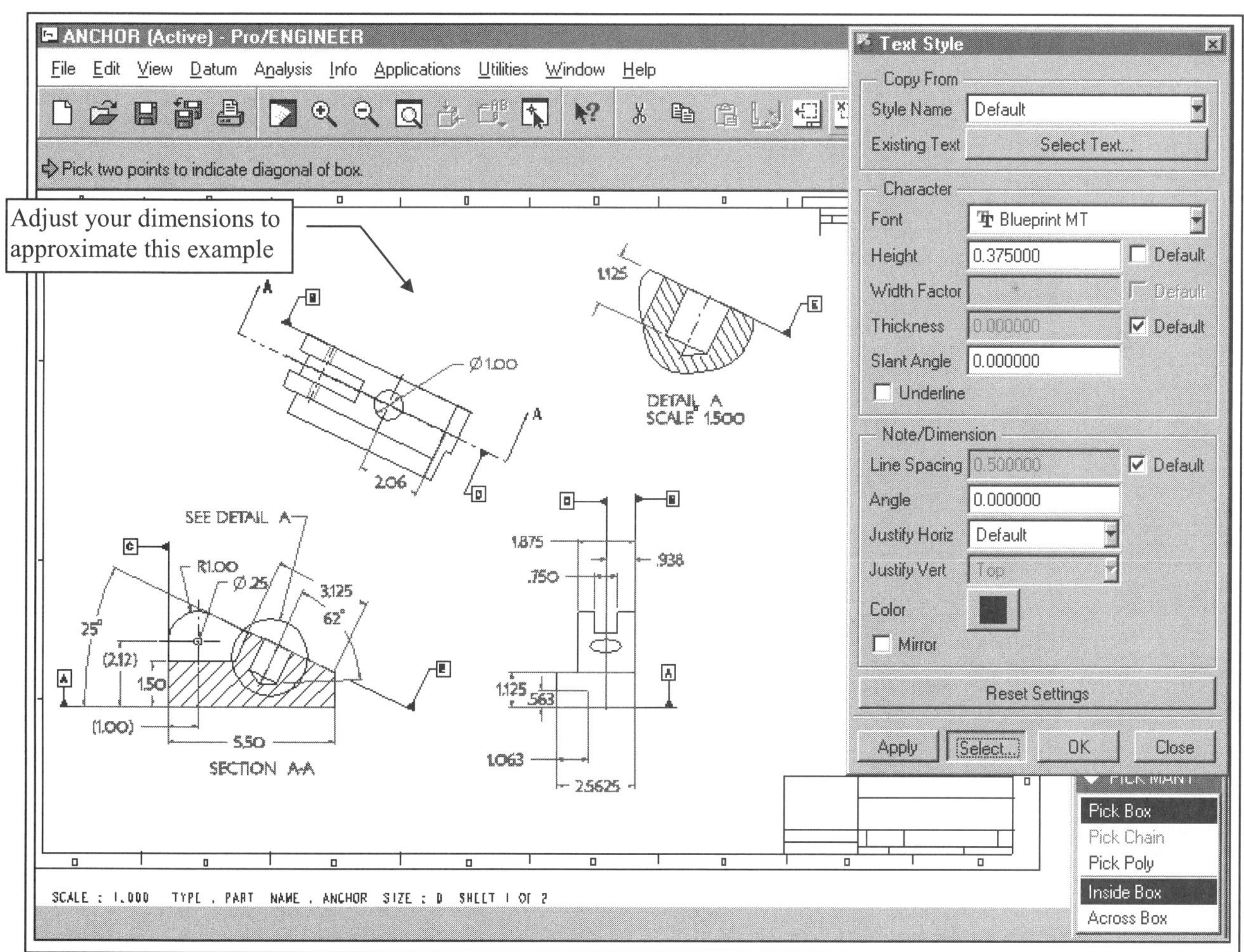

Lesson 18 Project

Cover Plate Drawing

Figure 18.30
Cover Plate Drawing, Sheet One

NOTE

You may detail any of the parts created in Lessons 1-13 at this time.

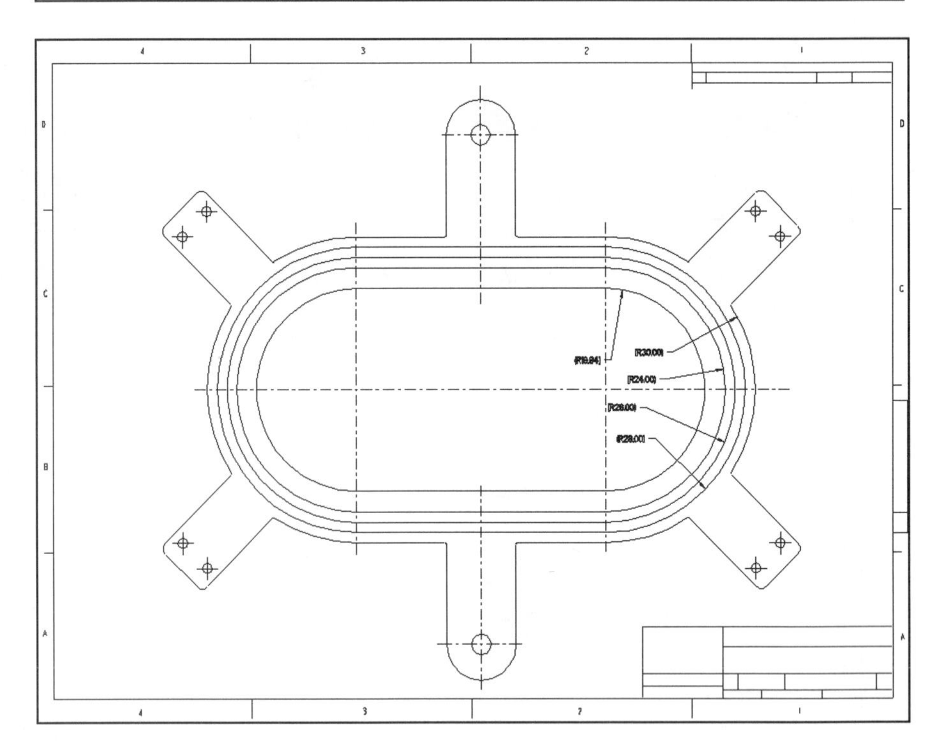

Figure 18.31
Cover Plate Drawing, Sheet Two

Cover Plate

Detail the **Cover Plate** (Figs. 18.30 and 18.31) created for the Lesson 11 Project. Use your own format and title block, created in Lesson 16. Analyze the Cover Plate for sheet size and view requirements. Create the required sections in Part mode to be used on the Drawing mode views. See the complete set of drawings and views provided in the Lesson 11 Project.

Lesson 19

Assembly Drawings and BOM

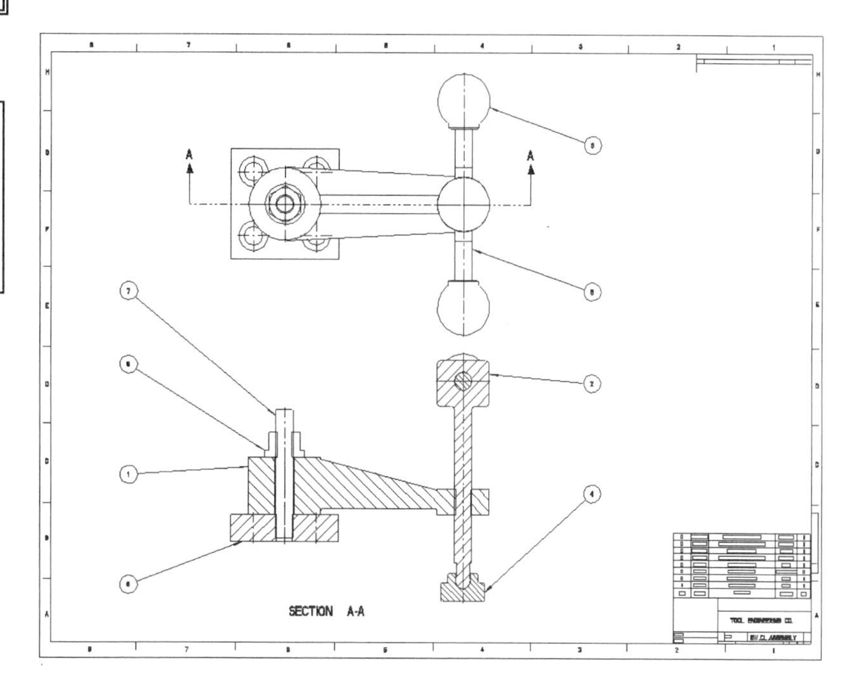

Figure 19.1
Swing Clamp Assembly Drawing

OBJECTIVES

1. **Create an assembly drawing**
2. **Generate a parts list from a bill of materials (BOM)**
3. **Balloon an assembly drawing**
4. **Create a section assembly view and change component visibility**
5. **Add parameters to parts**
6. **Create a table to generate a parts list automatically**

EGD REFERENCE
Fundamentals of Engineering Graphics and Design
by L. Lamit and K. Kitto
Read Chapters 11, 17, and 23.
See pages 358-368, 662-671, and 810-846.

COAch™ for Pro/ENGINEER

If you have **COAch for Pro/ENGINEER** on your system, go to SEARCH and do the Segments shown in Figures 19.4, and 19.5.

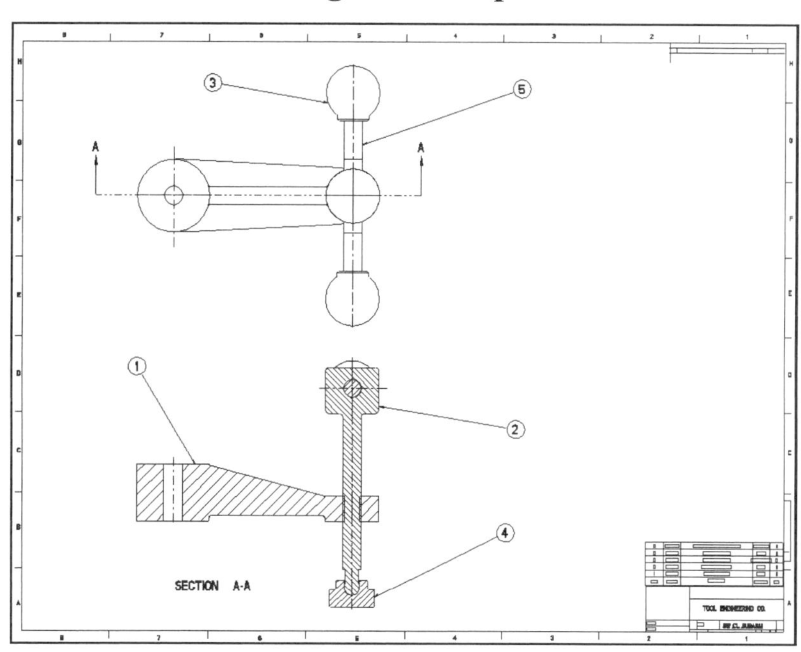

Figure 19.2
Swing Clamp Subassembly Drawing

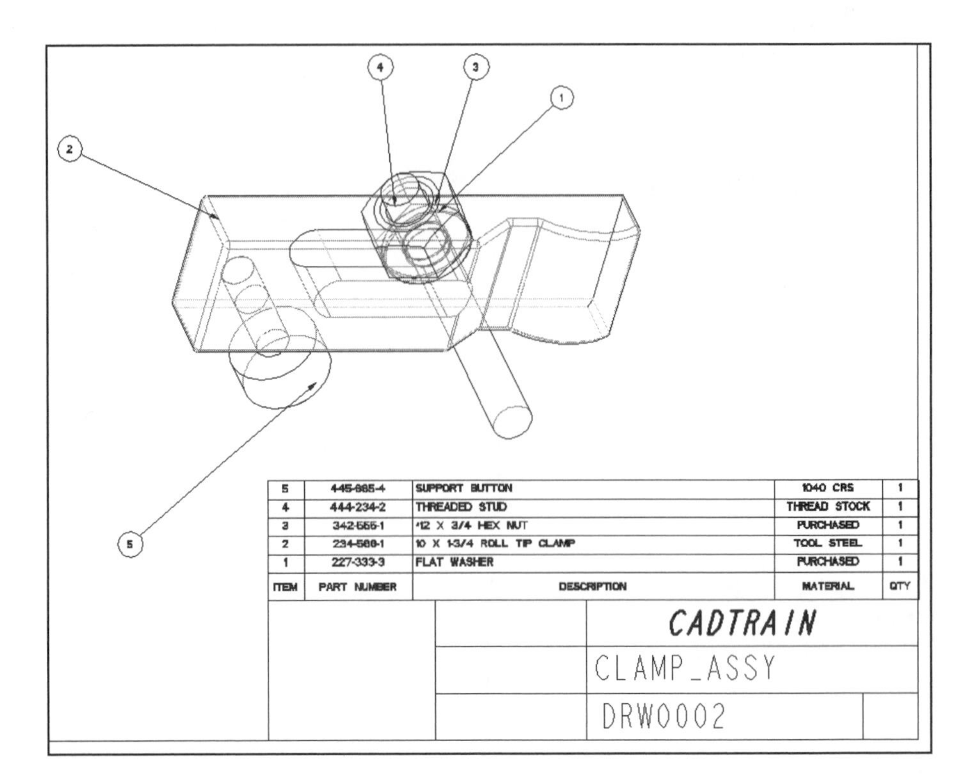

ITEM	PART NUMBER	DESCRIPTION	MATERIAL	QTY
5	445-885-4	SUPPORT BUTTON	1040 CRS	1
4	444-234-2	THREADED STUD	THREAD STOCK	1
3	342-555-1	#12 X 3/4 HEX NUT	PURCHASED	1
2	234-588-1	10 X 1-3/4 ROLL TIP CLAMP	TOOL STEEL	1
1	227-333-3	FLAT WASHER	PURCHASED	1

Figure 19.3
Strap Clamp with Parts List from CADTRAIN

ASSEMBLY DRAWINGS AND BOM

Pro/E incorporates a great deal of functionality into drawings of assemblies (Figs. 19.1 and 19.2). You can assign parameters to parts in the assembly that can be displayed on a *parts list* in an assembly drawing (Fig. 19.3). Pro/E can also generate the item balloons for each component (Fig. 19.4).

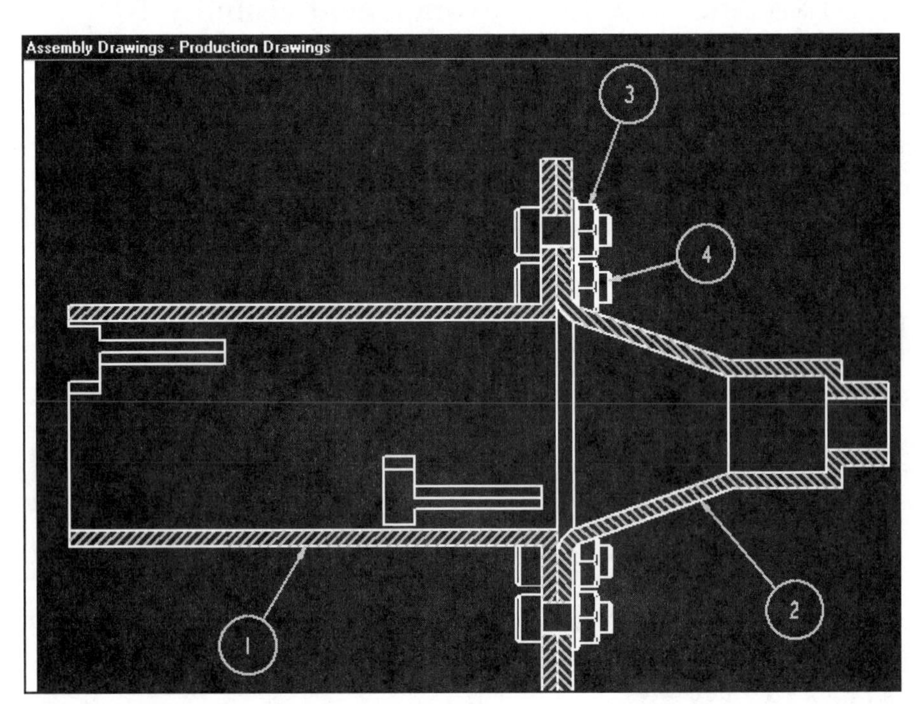

Figure 19.4
COAch for Pro/E, Assembly Drawings (Production Drawings)

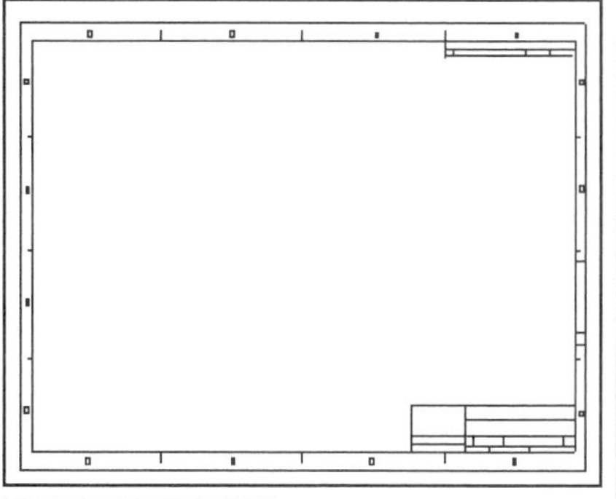

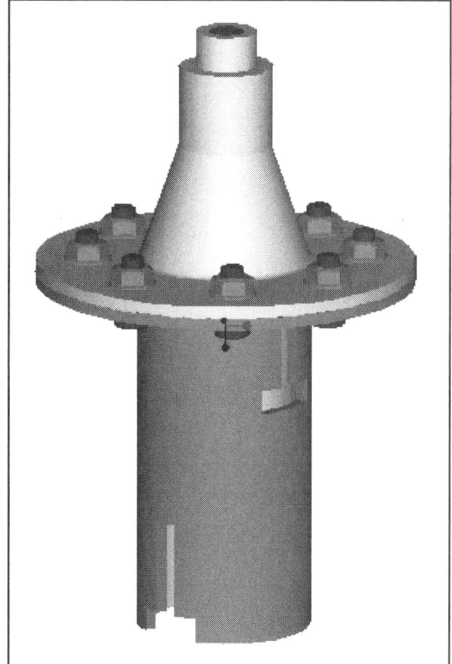

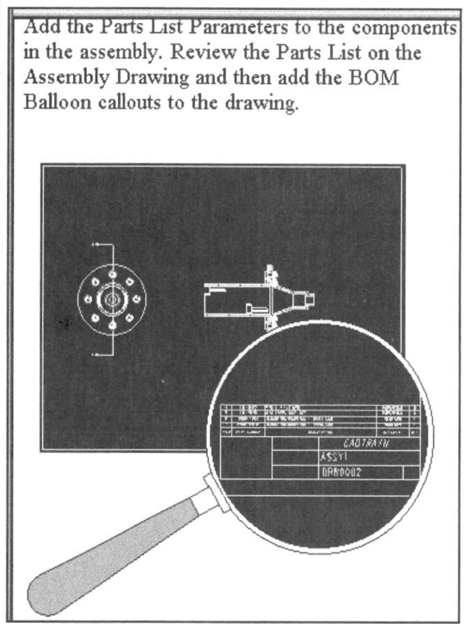

Figure 19.5
COAch for Pro/E, Assembly Drawings (Adding Parts List Data)

In addition, a variety of specialized capabilities allow you to alter the manner in which individual components are displayed in views and in sections (Fig. 19.5). The **format** for an assembly is usually (slightly) different from the format used for detail drawings. The most significant difference is the presence of a *Parts List.*

The assembly format provided has been adapted from CADTRAIN's COAch for Pro/ENGINEER. As part of this lesson, you will create a set of assembly formats and place your standard parts list on them.

A parts list is actually a *Drawing Table object* that is formatted to represent a bill of materials on a drawing. By defining *parameters* in the parts in your assembly that agree with the specific format of the parts list, you make it possible for Pro/E to add pertinent data to the parts list automatically as components are added to the assembly. After the parts list and parameters have been added, Pro/E can balloon the assembly automatically (Fig. 19.6).

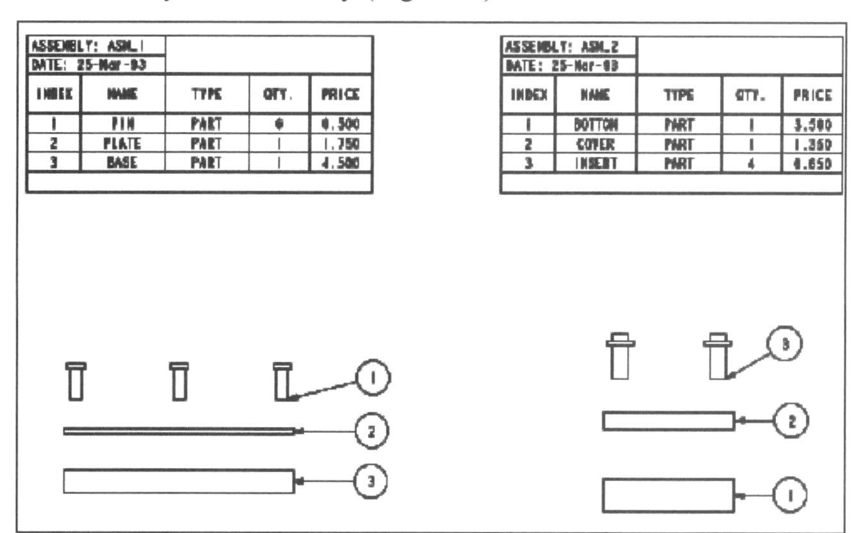

Figure 19.6
BOM Balloons

Info Applications Utilities
BOM...
Feature...
Feature List...
Model...
Component...
Model Tree...
Audit Trail...
Pro/Engineer Objects
Message Log...
Parent/Child...
Drawing
Switch Dims

BOM

An assembly drawing is created after the assembly is complete. With Pro/REPORT, you can then generate a bill of materials or other tabular data as required for the project. **Pro/REPORT** introduces a formatting environment where text, graphics, tables, and data can be combined to create a dynamic report (Fig. 19.7). Specific tools enable you to generate customized **bills of materials** (BOMs), family tables, and other associative reports:

* Dynamic, customized reports with drawing views and graphics can be created (Fig. 19.7).
* User-defined or predefined model data can be listed on reports, drawing tables, or layout tables. These reported data can be sorted by any requested data-type display.
* Regions in drawing tables, report tables, and layout tables can be defined to expand and shrink automatically with the amount of model information you have asked to have displayed.
* Filters can be added to eliminate the display of specific types of data from reports, drawing tables, or layout tables.
* Recursive or top-level assembly data can be searched for display.
* Duplicate occurrences of model data can be listed individually or as a group in a report, drawing table, or layout table.
* Assembly component balloons can be linked directly to a customized BOM and automatically updated when assembly modifications are made.

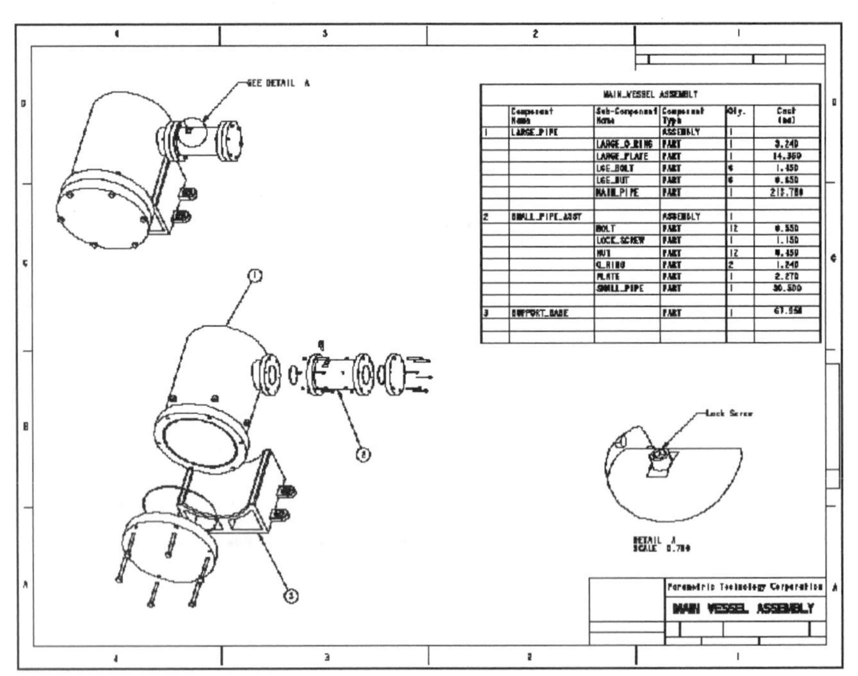

Figure 19.7
Report with Assembly Views

In **Report** mode, data can be displayed in a tabular form on reports, just as they are in drawing tables. The data reported on the tables are taken directly from a selected model and update automatically when the model is modified or changed. A common example of a report is a bill of materials report or a generic part table.

Including a Bill of Materials in a Drawing

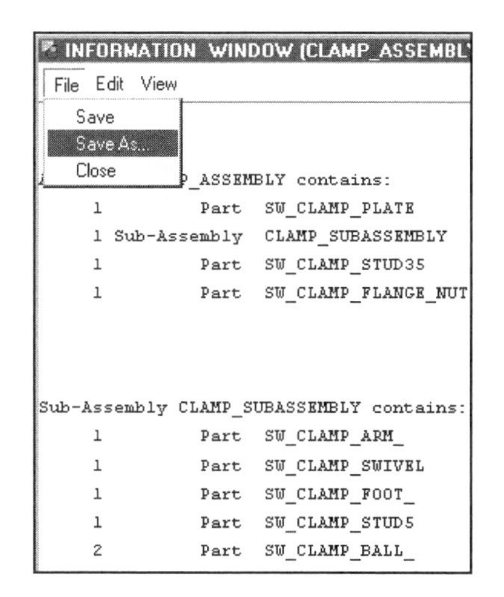

If you do not have Pro/REPORT (Fig. 19.8) and want to add a BOM, create a BOM file in Assembly mode using the **Info** option in the Menu Bar. Choose **Info** ⇒ **BOM** ⇒ **Top Level** ⇒ **OK** (see left) ⇒ **File** ⇒ **Save As** ⇒ (type a name for the .bom file). Add the BOM file to the drawing as a note entered from the file. To format or arrange the information in the BOM, you must use a text editor.

ITEM	PT NUM	DESCRIPTION	MATERIAL	QTY
8	SW100-22FLN	.500-13 X HEX FLANGE NUT	PURCHASED	1
7	SW100-21ST	.500-13 X 3.50 DOUBLE END STUD	PURCHASED	1
6	SW100-20PL	SWING CLAMP PLATE	1020 CRS	1
5	SW101-85TL	.500-13 X 5.00 DOUBLE END STUD	PURCHASED	1
4	SW101-8FT	SWING CLAMP FOOT	NYLON	1
3	SW101-7BA	SWING CLAMP BALL	BLACK PLASTIC	2
2	SW101-6SW	SWING CLAMP SWIVEL	STEEL	1
1	SW101-5AR	SWING CLAMP ARM	STEEL	1

Figure 19.8
Parts List on Assembly Drawings

A BOM that is added to a drawing as a note is not connected with the BOM file that was used to create the note. If the composition of the assembly changes, you must create a new BOM file and add it to the drawing as a new note. You can fully edit the BOM displayed on drawings as a note without affecting the original BOM file.

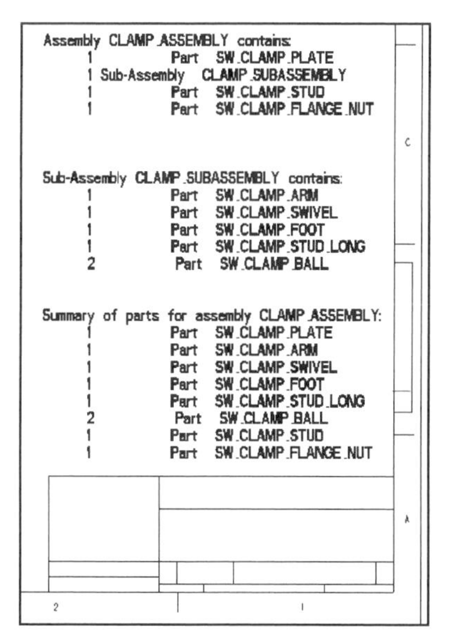

To add a BOM to a drawing as a note (see left):

1. Choose **Create** ⇒ **Note** ⇒ **File** from the NOTE TYPES menu. When adding a BOM to a drawing as a note, justify the note using **Default** or **Left.** If you use **Center** or **Right**, the BOM may be formatted incorrectly on the drawing.
2. Choose **Make Note** when you have finished choosing options.
3. Pick a location for the note (BOM) to appear.
4. Enter **Name** and type the file name, including the **.bom** full extension. Then pick **Open.**
5. The BOM is displayed on the drawing.

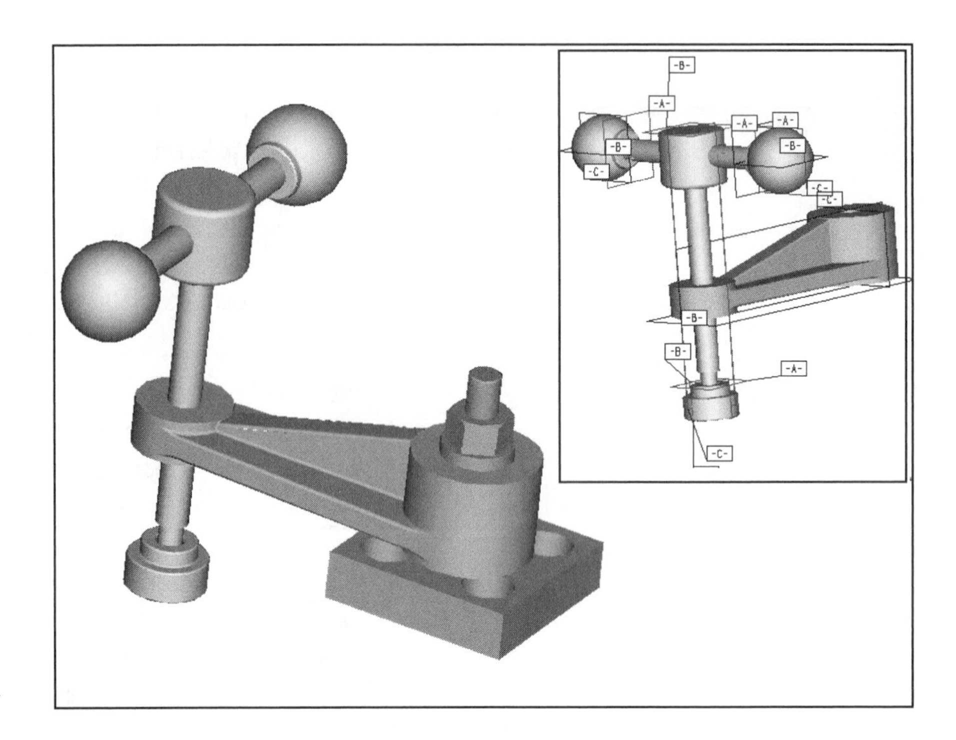

Figure 19.9
Swing Clamp Subassembly and Swing Clamp Assembly

Lesson procedures & commands
START HERE ➡ ➡ ➡

Swing Clamp Assembly Drawing

The format for an assembly is usually different from the format used for detail drawings. The most significant difference is the presence of a parts list. We will create a standard "**E**" size format and place a standard parts list on it. You should create a set of assembly drawing formats on "**B**," "**C**," and "**D**" size sheets at your convenience.

A **parts list** is actually a *Drawing Table object* that is formatted to represent a bill of material (BOM) on a drawing. By defining parameters for the parts in your assembly that agree with the specific format of the parts list, you make it possible for Pro/E to add pertinent data to the parts list as components are added to the assembly.

After you create an "**E**" size format sheet with a parts list table, you will create two new drawings (each with two views) using your new assembly format. The Swing Clamp subassembly (Fig. 19.9, top right) will be used for the first drawing. The second drawing will use the Swing Clamp assembly (Fig. 19.9, left). Both drawings use the "**E**" size format created in the first section of this lesson. The format will have a parameter-driven title block (as in Lesson 16) and an integral parts list.

Using steps similar to those outlined in Lesson 16, where a "**C**" size format was created and saved, create an "**E**" size format using the following commands:

NOTE
Use a format file name that will identify the format as an assembly format, such as **ASM_FORMAT_E**.

File ⇒ Working Directory (select your working directory) **⇒ OK ⇒ File ⇒ Open ⇒ ▼ ⇒ System Formats ⇒ e.frm ⇒ Open ⇒ File ⇒ Rename ⇒** (type a unique name for your format: ***your_format_name***) **⇒ OK ⇒ OK**

> ☑ ***NOTE***
> The **format** will have a **.dtl** associated with it, and the **drawing** will have a different **.dtl** file associated with it. *They are separate* ***.dtl*** *files and have nothing to do with each other.* When you activate a drawing and then add a format, the **.dtl** for the format controls the font etc. for the format only. The drawing **.dtl** file that will control the items on the drawing still needs to be established. Set the following in your ***format*** **.dtl** file:
>
> **Advanced ⇒ Draw Setup**
>
> | | |
> |---|---|
> | **default_font** | **filled** |
> | **draw_arrow_style** | **filled** |
> | **drawing_text_height** | **.25** |
> | **draw_arrow_length** | **.25** |
> | **draw_arrow_width** | **.08** |
>
> **Add/Change ⇒ Apply ⇒ Close**

Figure 19.10 shows the standard "**E**" size format available from Pro/E using the format directory.

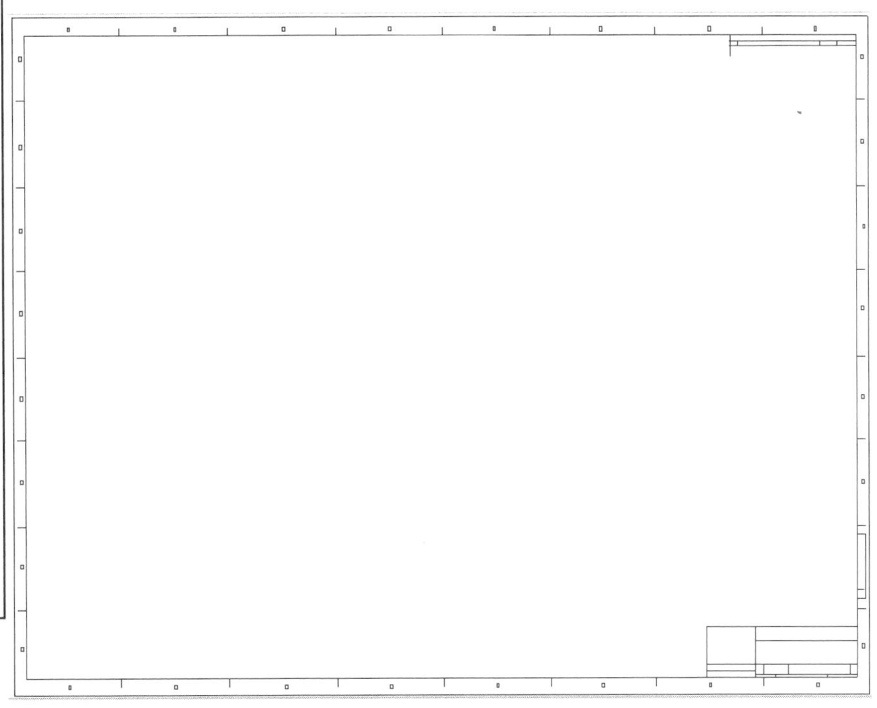

Figure 19.10
Standard "**E**" Size Format

Modify your **.dtl** file as shown in Figure 19.11:

Figure 19.11
.dtl Format File

> **Advanced ⇒ Draw Setup ⇒** (edit the file as required) **⇒ Add/Change ⇒ Apply ⇒ Close ⇒ File ⇒ Save ⇒ ✔**

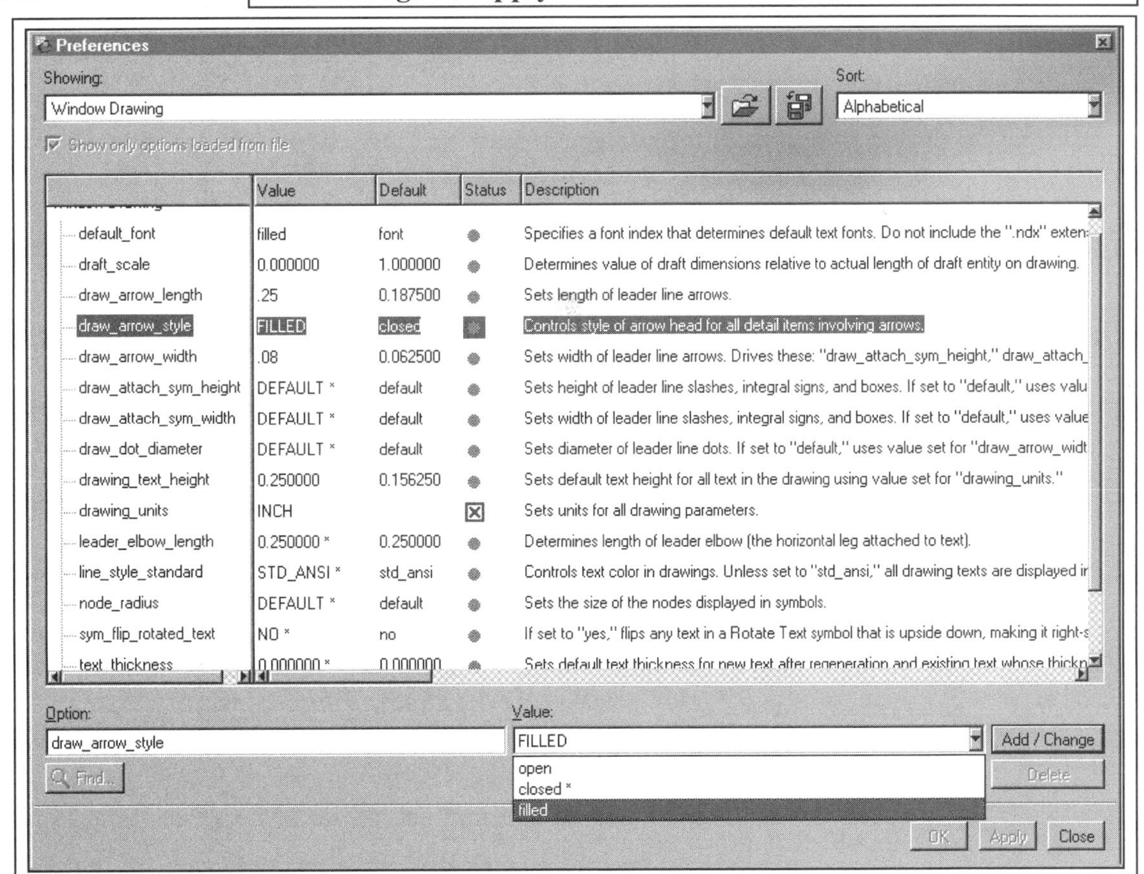

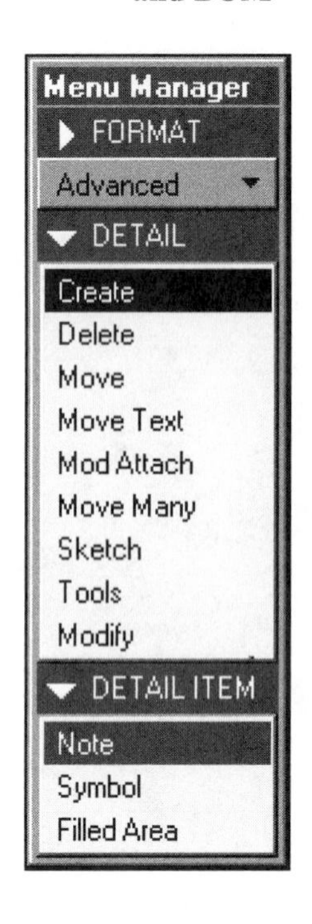

Zoom into the title block region, and fill in the titles and parameters required to display the proper information. Choose the following commands (you may wish to have **Snap to Grid** on):

> **Modify** ⇒ **Grid** ⇒ **Grid Params** ⇒ **X&Y Spacing** ⇒ (type **.2**) ⇒ ✔ ⇒ **Done/Return** ⇒ **Grid On** ⇒ **Done/Return** ⇒ **Done/Return** ⇒ **Create** ⇒ **Note** (make nonparametric notes) ⇒ **Make Note** ⇒ (pick point for note) ⇒ (type **TOOL ENGINEERING CO.**) ⇒ ✔ ⇒ ✔ ⇒ **Make Note** ⇒ [create notes for **DRAWN** and **ISSUED**; place them in the title block (Fig. 19.12)]
>
> Make parametric notes: **Make Note** ⇒ (pick point for note) ⇒ (type **&dwg_name**) ⇒ **enter** ⇒ **enter** ⇒ [create parametric notes for **&scale** and **SHEET ¤t_sheet OF &total_sheets**; place them in the title block (Fig. 19.12)] ⇒ **Done/Return**

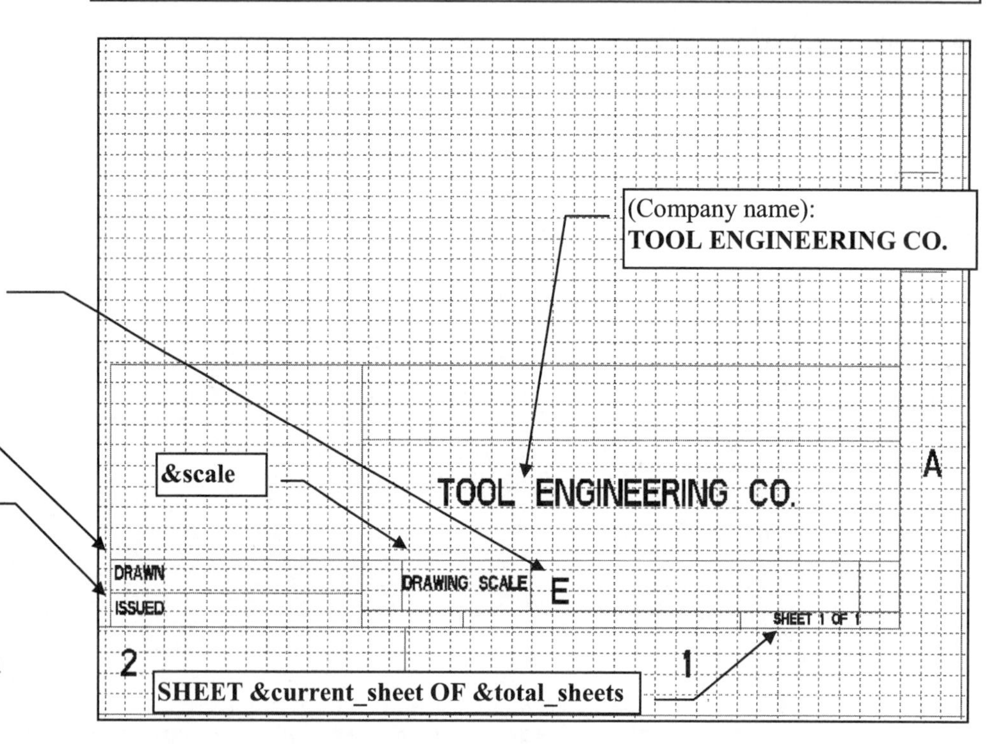

Figure 19.12
Parameters and Labels in the Title Block

Turn off the **Snap to Grid**. **Modify** the text height and the placement of the notes so that they are placed similarly to those in Figure 19.12.

The parts list table can now be created and saved with this drawing format. You can add and replace formats and still keep the table associated with the drawing. Start the parts list by creating a table using the following commands:

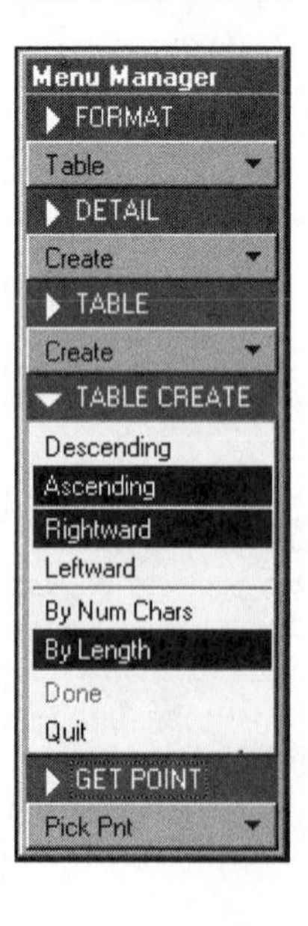

> **Modify** ⇒ **Grid** ⇒ **Grid Off** ⇒ **Done/Return** ⇒ **Done/Return** ⇒ **Table** ⇒ **Create** ⇒ **Ascending** ⇒ **Rightward** ⇒ **By Length** ⇒ (pick a point above the title block as shown in Fig. 19.13) ⇒ (enter the width of the first column in drawing units) ⇒ **1** ⇒ ✔ ⇒ **1** ⇒ ✔ ⇒ **4** ⇒ ✔ ⇒ **1** ⇒ ✔ ⇒ **.75** ⇒ ✔ ⇒ ✔ ⇒ (enter the height of the first row in drawing units) ⇒ **.5** ⇒ ✔ ⇒ **.375** ⇒ ✔ ⇒ ✔

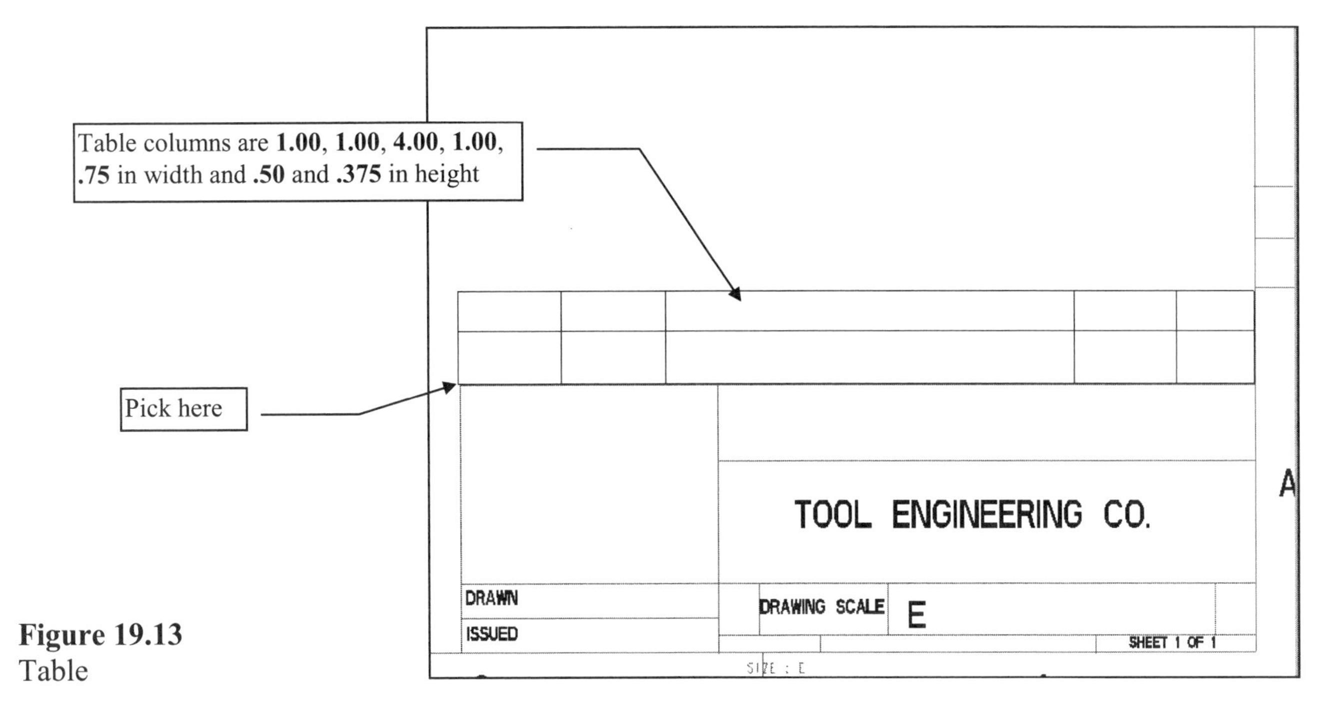

Figure 19.13
Table

Next we will add a **Repeat Region** to the table by continuing with these commands:

Repeat Region ⇒ **Add** ⇒ (locate the corners of the region as shown in Fig. 19.14) ⇒ **Attributes** ⇒ (select the Repeat Region you just created) ⇒ **No Duplicates** ⇒ **Recursive** ⇒ **Done/Return** ⇒ **Done/Return**

Menu Manager
FORMAT
Table
DETAIL
Create
TABLE
Repeat Region
TBL REGIONS
Attributes
REGION ATTR
Duplicates
No Duplicates
No Dup/Level
Recursive
Flat
Min Repeats
Start Index
No Start Idx
Bln By Part
Bln By Comp
Done/Return

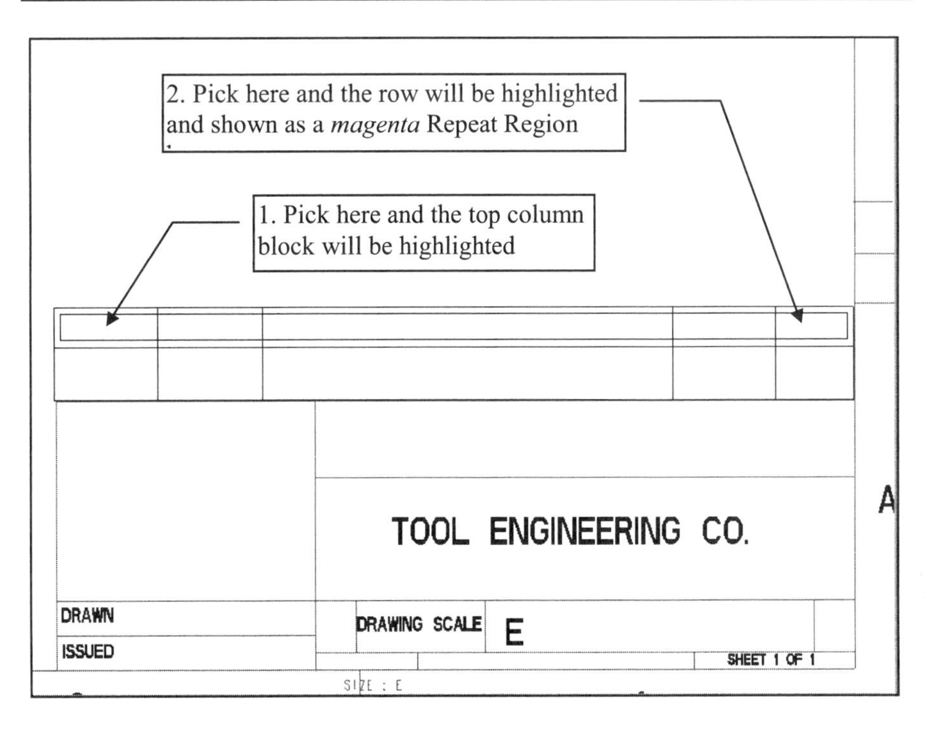

Figure 19.14
Repeat Region

The table must have parameters set in each appropriate block. The column headings should be inserted first, using plain text. Choose the following commands:

Table ⇒ **Mod Rows/Cols** ⇒ **Justify** ⇒ **Column** ⇒ **Center** ⇒ **Middle** ⇒ (pick all five columns; they will be outlined in red as they are selected) ⇒ **Enter Text** ⇒ **Keyboard** ⇒ (pick the table cell where the text is to be placed; choose the **4.00** width column in the lower row) ⇒ (type **DESCRIPTION**) ⇒ **enter** ⇒ **enter**

Continue adding the titles **MATERIAL**, **QTY**, **ITEM**, and **PT NUM,** as shown in Figure 19.15. From the DETAIL menu, **Modify** the text height to **.125**.

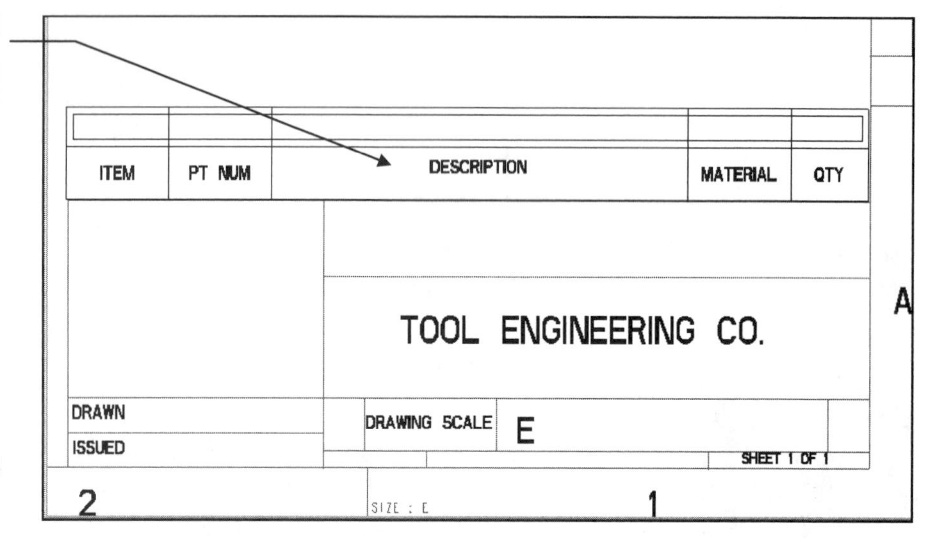

Figure 19.15
Entering Plain Text Headings for a Parts List

The **Repeat Region** now needs to have some of its headings correspond to the parameters created in Part mode for each component model. The **ITEM** and quantity (**QTY**) columns will have the **rpt.index** and **rpt.qty** parameters. Choose the following commands from the TABLE menu (Fig. 19.16):

Table ⇒ **Enter Text** ⇒ **Report Sym** ⇒ (pick the first table cell of the Repeat Region) ⇒ **rpt...** ⇒ **index**

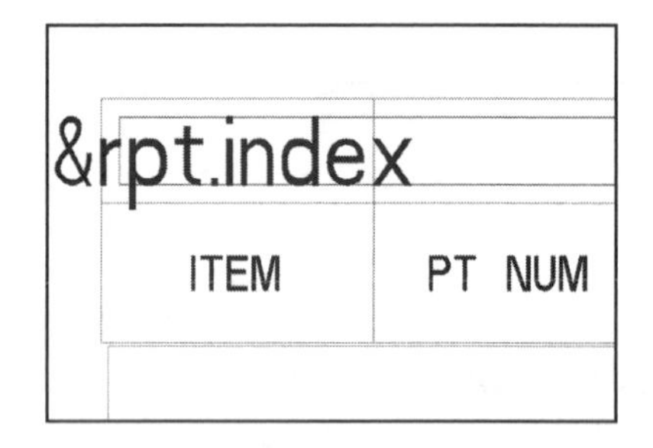

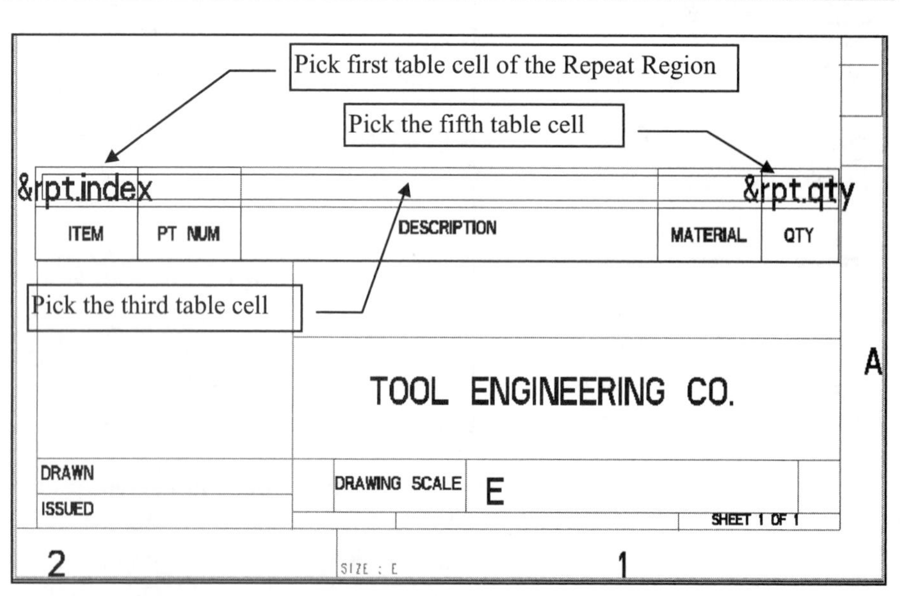

Figure 19.16
Entering Report Symbols

Enter Text ⇒ **Report Sym** ⇒ (pick the fifth table cell of the **Repeat Region**) ⇒ **rpt...** ⇒ **qty** (Fig. 19.16) ⇒ Create the parametric **User Defined** text [pick the third table cell where the next text is to be placed (Fig. 19.16)] ⇒ **asm...** ⇒ **mbr...** ⇒ **User Defined** ⇒ (enter symbol text, type **DSC**) ⇒ ✔ ⇒ [pick the fourth table cell (Fig. 19.17)] ⇒ **asm...** ⇒ **mbr...** ⇒ **User Defined** ⇒ (enter symbol text, type **MAT**) ⇒ ✔ ⇒ (pick the second table cell) ⇒ **asm...** ⇒ **mbr...** ⇒ **User Defined** ⇒ (enter symbol text, type **PRTNO**) ⇒ ✔ (Fig. 19.18) ⇒ **File** ⇒ **Save** ⇒ ✔

HINT

You will need to change the text height of the Report Symbols in the table cells of the **Repeat Region** to **.125.**

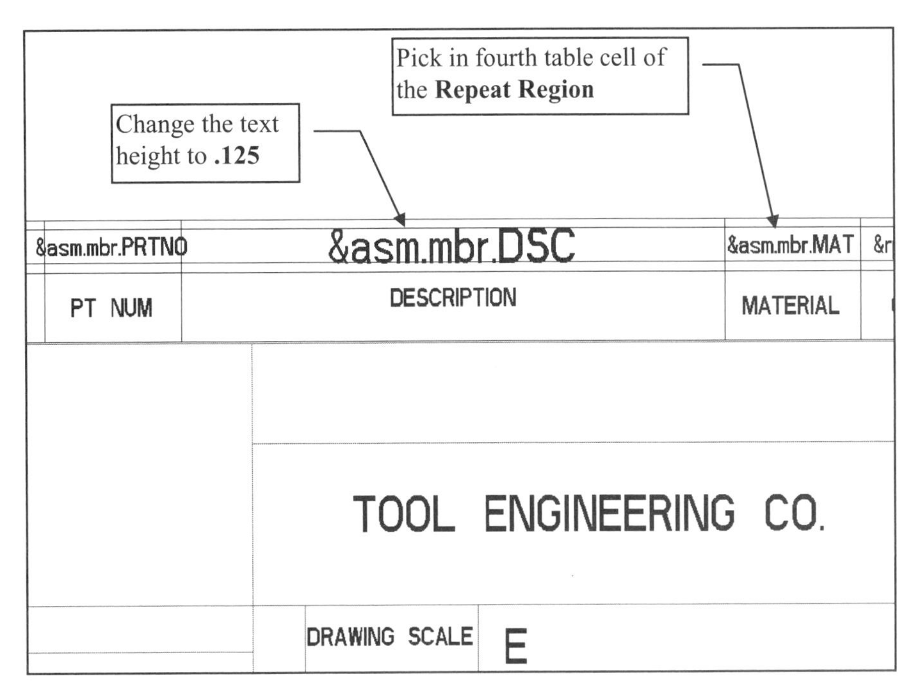

Figure 19.17
Entering Repeat Region Parameters

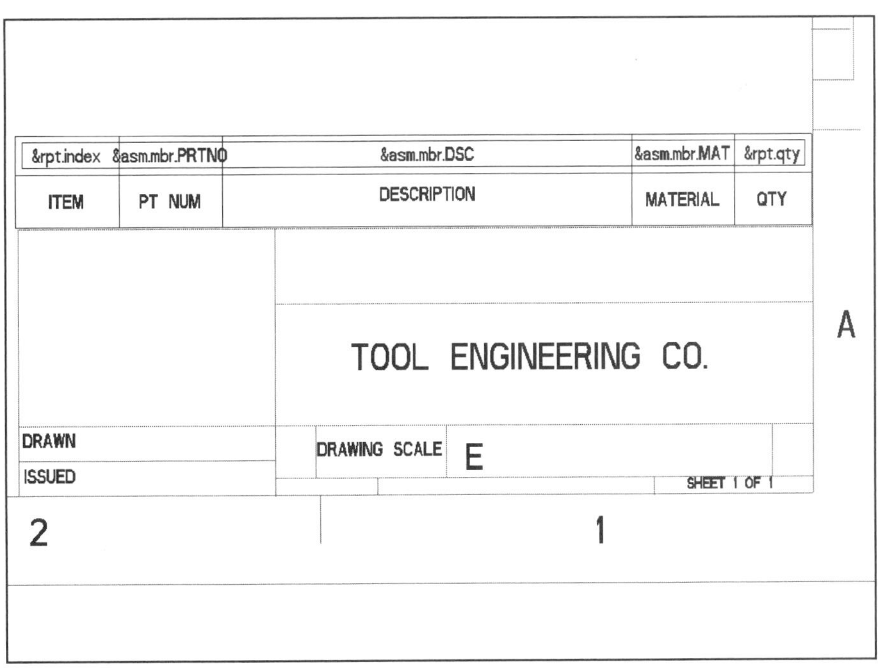

Figure 19.18
Completed Repeat Region Parameters

Adding Parts List Data

When you save your standard assembly format, the Drawing Table that represents your standard parts list is now included. You must be aware of the titles of the parameters under which these data are stored so that you can add them properly to your parts.

As you add components to the assembly, Pro/E reads the parameters from them and updates the parts list. You can also see the same effect by adding these parameters after the drawing has been created.

Pro/E also creates **Item Balloons** on the first view that was placed on the drawing. To improve their appearance, you can move these balloons to other views and alter the locations where they attach.

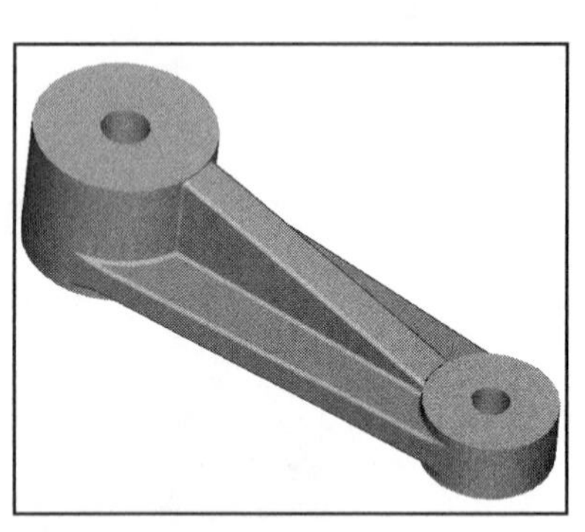

In Part mode, add the three user-defined parameters to each of the components used in the assembly using the following commands:

Retrieve the clamp arm: **File** ⇒ **Open** ⇒ **Part** ⇒ [choose **sw_clamp_arm.prt** (or the name you gave the part) from the directory list] ⇒ **Open**

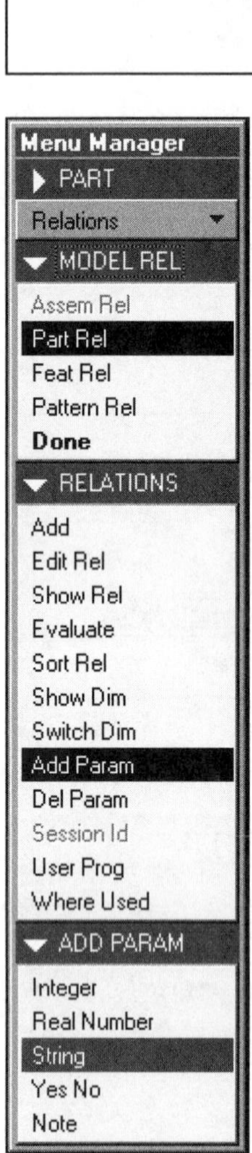

Add the parts list data:

Choose **Relations**

Choose **Add Param**

Choose **String**

Enter the *exact* title of the parameter that you previously established in the assembly drawing format table parts list:

Type **PRTNO** ⇒ ✔

Enter the part number:

Type **SW101-5AR** ⇒ ✔

Choose **String**

Enter the *exact* title of the parameter:

Type **DSC** ⇒ ✔

Enter the component description:

Type **SWING CLAMP ARM** ⇒ ✔

Choose **String**

Enter the *exact* title of the parameter:

Type **MAT** ⇒ ✔

Enter the material:

Type **STEEL** ⇒ ✔

Save and erase the current object:

File ⇒ **Save** ⇒ ✔ ⇒ **File** ⇒ **Erase** ⇒ **Current** ⇒ **Yes**

Retrieve the clamp swivel: **File** ⇒ **Open** ⇒ **Part** ⇒[choose **sw_clamp_swivel.prt** (or the name you gave the part)] ⇒ **Open**

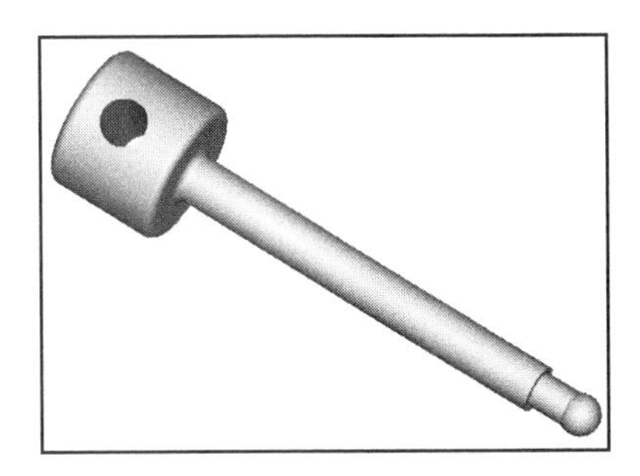

Add the parts list data:
    Choose **Relations**
    Choose **Add Param**
    Choose **String**
Enter the *exact* title of the parameter:
    Type **PRTNO** ⇒ ✔
Enter the part number:
    Type **SW101-6SW** ⇒ ✔
    Choose **String**
Enter the *exact* title of the parameter:
    Type **DSC** ⇒ ✔
Enter the component description:
    Type **SWING CLAMP SWIVEL** ⇒ ✔
    Choose **String**
Enter the *exact* title of the parameter:
    Type **MAT** ⇒ ✔
Enter the material:
    Type **STEEL** ⇒ ✔
Save and erase the current object:
    File ⇒ **Save** ⇒ ✔ ⇒ **File** ⇒ **Erase** ⇒ **Current** ⇒ **Yes**

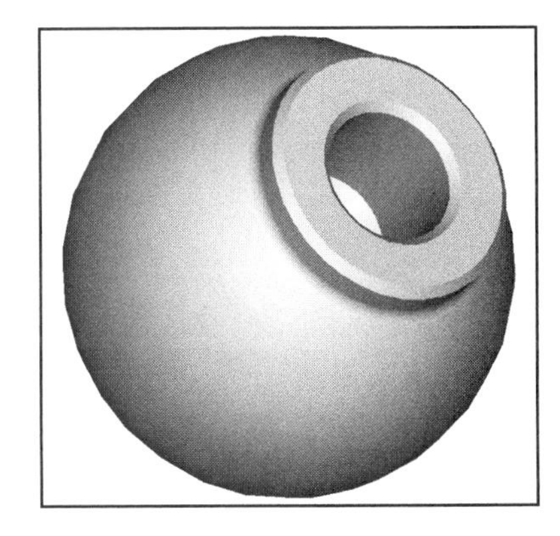

Retrieve the clamp ball: **File** ⇒ **Open** ⇒ **Part** ⇒ [choose **sw_clamp_ball.prt** (or the name you gave the part)] ⇒ **Open**

Add the parts list data:
    Choose **Relations**
    Choose **Add Param**
    Choose **String**
Enter the *exact* title of the parameter:
    Type **PRTNO** ⇒ ✔
Enter the part number:
    Type **SW101-7BA** ⇒ ✔
    Choose **String**
Enter the *exact* title of the parameter:
    Type **DSC** ⇒ ✔
Enter the component description:
    Type **SWING CLAMP BALL** ⇒ ✔
    Choose **String**
Enter the *exact* title of the parameter:
    Type **MAT** ⇒ ✔
Enter the material:
    Type **BLACK PLASTIC** ⇒ ✔
Save and erase the current object:
    File ⇒ **Save** ⇒ ✔ ⇒ **File** ⇒ **Erase** ⇒ **Current** ⇒ **Yes**

Use the following information to add parameters both to purchased components (standard parts) and to the remaining parts required for the subassembly and the assembly (remember to use the *exact* parameter names).

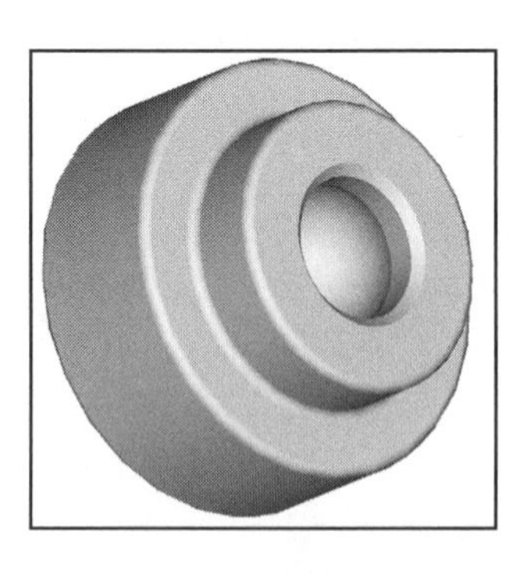

Component	**sw_clamp_foot**
Part Number	**SW101-8FT**
Description	**SWING CLAMP FOOT**
Material	**NYLON**

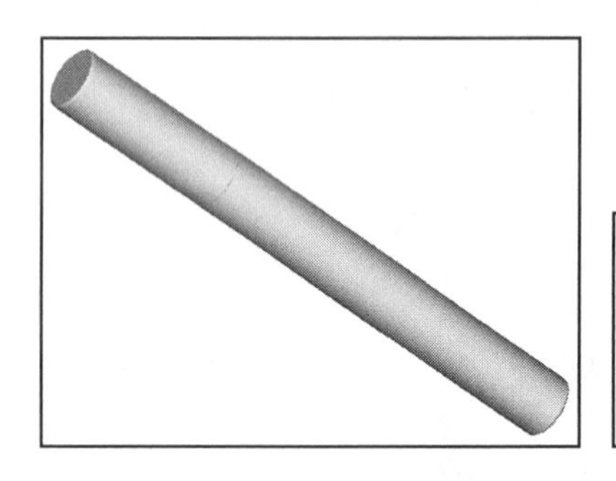

Component	**sw_clamp_stud_long**	**(STUD5)**
Part Number	**SW101-9STL**	
Description	**.500-13 X 5.00 DOUBLE END STUD**	
Material	**PURCHASED**	

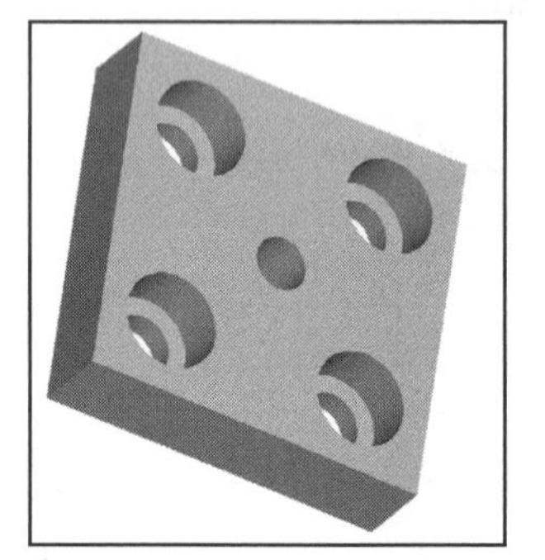

Component	**sw_clamp_plate**
Part Number	**SW100-20PL**
Description	**SWING CLAMP PLATE**
Material	**1020 CRS**

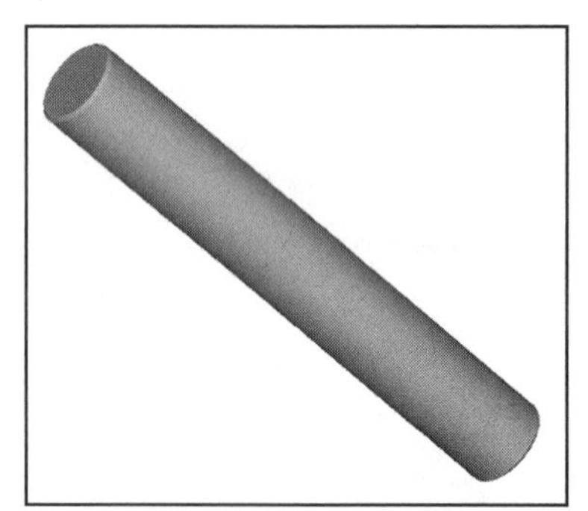

Component	**sw_clamp_stud**	**(STUD35)**
Part Number	**SW100-21ST**	
Description	**.500-13 X 3.50 DOUBLE END STUD**	
Material	**PURCHASED**	

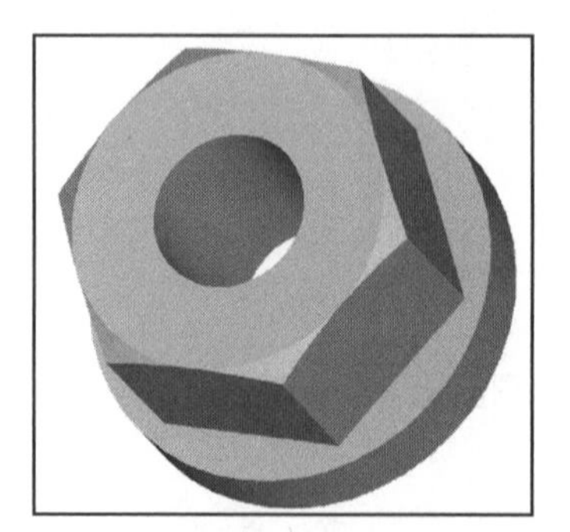

Component	**sw_clamp_flange_nut**	**(FLNGNUT)**
Part Number	**SW100-22FLN**	
Description	**.500-13 HEX FLANGE NUT**	
Material	**PURCHASED**	

NOTE

Parameters can be added, deleted, and modified in Part mode, Drawing mode, or Assembly mode. In Drawing mode, choose **Advanced ⇒ Parameters**; in Part and Assembly mode, choose **Set Up ⇒ Parameters.** You can also add *parameter columns* to the Model Tree and edit the parameter value by highlighting it in the tree and typing a new value.

As you add the required relations to each component, you can see the parameters and check to see if they were input correctly by choosing **Show Rel** from the RELATIONS menu, as shown in Figure 19.19.

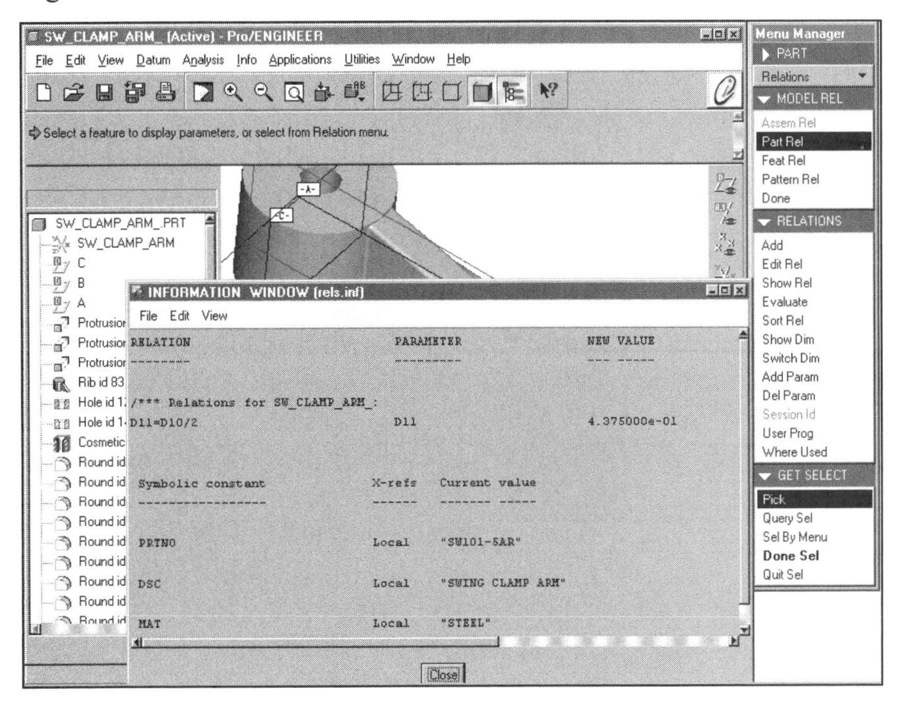

Figure 19.19
Relations for Clamp Arm

The parameters (and their values) have been established in each part. The assembly format with related parameters in a parts list table has been created and saved in the *read only* format directory. You can now create a drawing of the assembly, where the parts list will be generated automatically.

Create a new drawing with the following commands (Fig. 19.20):

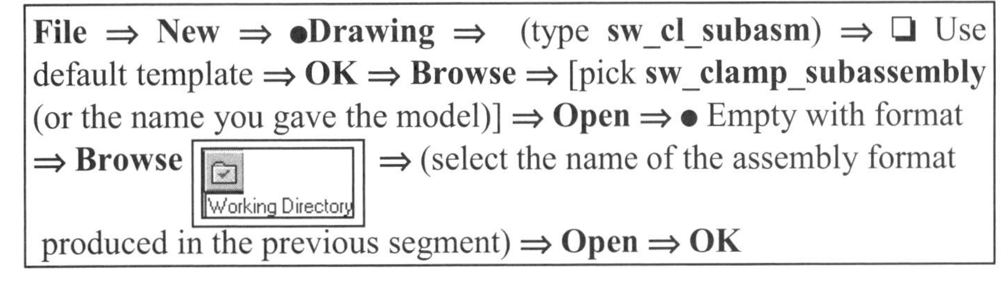

File ⇒ New ⇒ ●Drawing ⇒ (type **sw_cl_subasm**) ⇒ ❑ Use default template ⇒ **OK ⇒ Browse ⇒** [pick **sw_clamp_subassembly** (or the name you gave the model)] ⇒ **Open ⇒** ● Empty with format ⇒ **Browse** [Working Directory] ⇒ (select the name of the assembly format produced in the previous segment) ⇒ **Open ⇒ OK**

NOTE

Set the following in your *Drawing* **.dtl** file:

Advanced ⇒ Draw Setup ⇒

default_font	***filled***
draw_arrow_style	***filled***
drawing_text_height	***.50***
draw_arrow_length	***.50***
draw_arrow_width	***.17***
dim_leader_length	***1.00***
max_balloon_radius	***.50***
min_balloon_radius	***.50***

Add/Change ⇒ Apply ⇒ Close

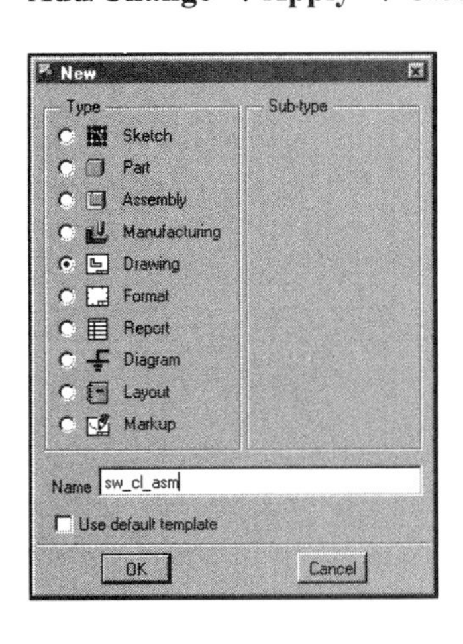

To get the drawing to appear with the correct style, you must modify the values of the **.dtl** Drawing Set Up file using **Advanced ⇒ Draw Setup** (see left-hand column).

Use the **Save** option [Save a copy of the currently displayed configuration file] to save the settings to a file name so you can recall and use the file on another drawing. Add the top view as the first view. Choose the following commands:

Done/Return ⇒ Views ⇒ Done (to accept all of the defaults) ⇒ (indicate the location for the center of the top view)

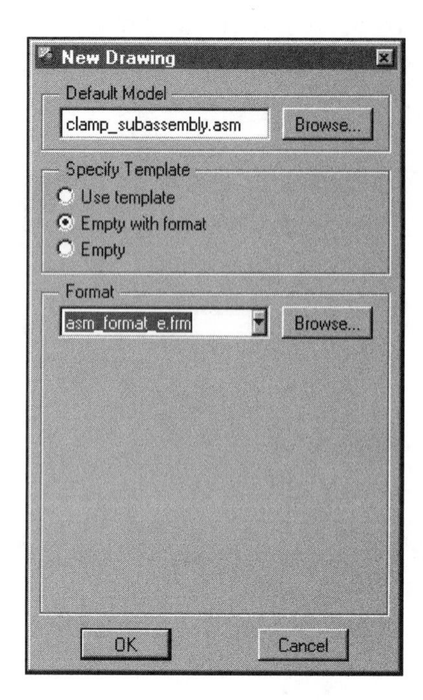

You will now use **datum planes** to establish the view orientation. *You must select assembly datums!* This would be very difficult to accomplish by picking on the screen, so you will use **Sel By Menu**:

Choose **Top** (Reference 1)
Choose **Sel By Menu ⇒ Sel By Menu**
Choose **sw_clamp_subassembly** from the list
Choose **Datum**
Choose **Name**
Choose **ASM_TOP** from the list
Choose **Front** (Reference 2)
Choose **Sel By Menu ⇒ Sel By Menu**
Choose **sw_clamp_subassembly** from the list
Choose **Datum**
Choose **Name**
Choose **ASM_FRONT** from the list
Choose **OK**
Choose **Done/Return**

Modify the scale of the drawing so that it is **2.00**:

Choose **Modify**

Select the drawing scale value (**1.00**) that is just below the drawing in the lower left hand corner:

Type **2.00** ⇒ ✔ (Fig. 19.20) ⇒ **Done Sel** ⇒ **Done/Return**

Notice that the title block and parts list were filled in before the view was created, also note that in order to get the exact view show in Figure 19.20 you have to use **Show/Erase** ⇒ **Erase** ⇒ Datums Tab ⇒ **Erase All** ⇒ **Yes** ⇒ **Close.**

HINT
Orient the view as shown in Figure 19.20. Your selections will be different if you have different datum plane names. If you do not get the correct view orientation, **Delete View** and **Add View** again (or pick **Default** and redo the orientation).

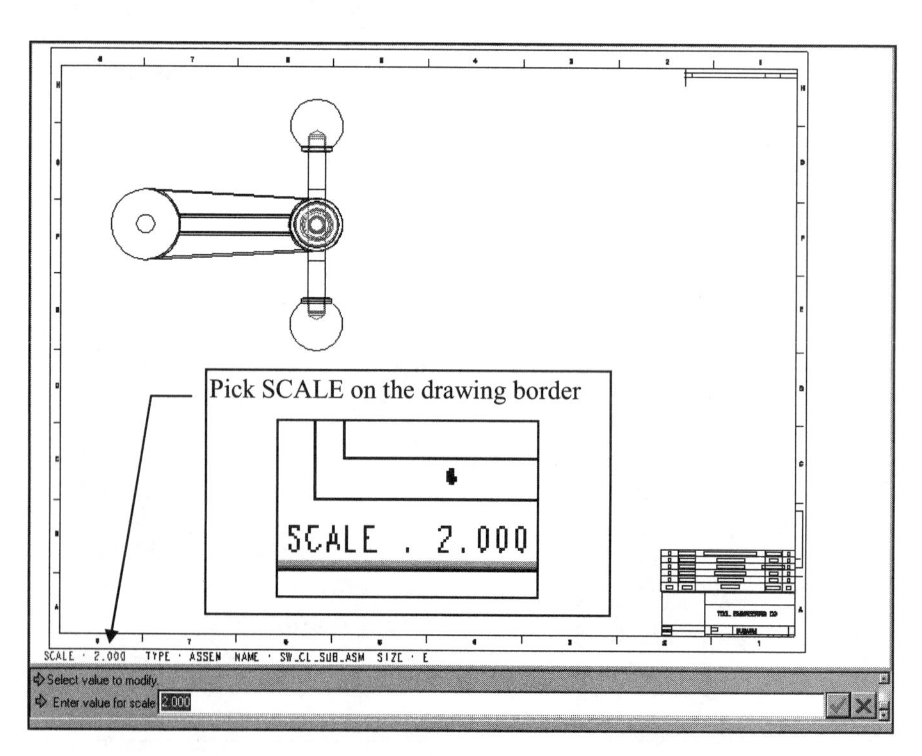

5	SW101-9STL	.500-13 X 5.00 DOUBLE END STUD	PURCHASED	1
4	SW101-8FT	SWING CLAMP FOOT	NYLON	1
3	SW101-7BA	SWING CLAMP BALL	BLACK PLASTIC	2
2	SW101-6SW	SWING CLAMP SWIVEL	STEEL	1
1	SW101-5AR	SWING CLAMP ARM	STEEL	1
ITEM	PT NUM	DESCRIPTION	MATERIAL	QTY

TOOL ENGINEERING CO.
DRAWN
ISSUED
2.000 SW CL SUBASM
SHEET 1 OF 1
2 1

Figure 19.20
Swing Clamp Subassembly Drawing with Top View

HINT

If you haven't already done so, turn off the datum planes, points, coordinate system, spin center, axes, and snap to grid in the Environment window.

To erase the set datums, use the following commands:

Show/Erase ⇒ **Erase** ⇒ (pick the datum radio button) ⇒ **Erase All** ⇒ **Yes** ⇒ **Close**

You will need to do this after each view is placed on the drawing or after all views have been established.

The only other view needed to show the assembly is the front view. We will be making a front section view using the following commands (Fig. 19.21):

Choose **Views**
Choose **Section**
Choose **Done**
Choose **Done** (because **Full** is the default)

Indicate the center of the view below the top view:

Choose **Create**
Choose **Done** (because **Planar** is the default)
Type **A** ⇒ ✔
Choose **Sel By Menu**
Choose **Sel By Menu** (again)
Choose **sw_clamp_subassembly** from the list
Choose **Datum**
Choose **Name**
Choose **ASM_TOP** from the list

Now select any location in the *top* view to define it as the view where the section line, cutting plane line, arrows, and section identification lettering will be placed:

Choose **Done/Return**

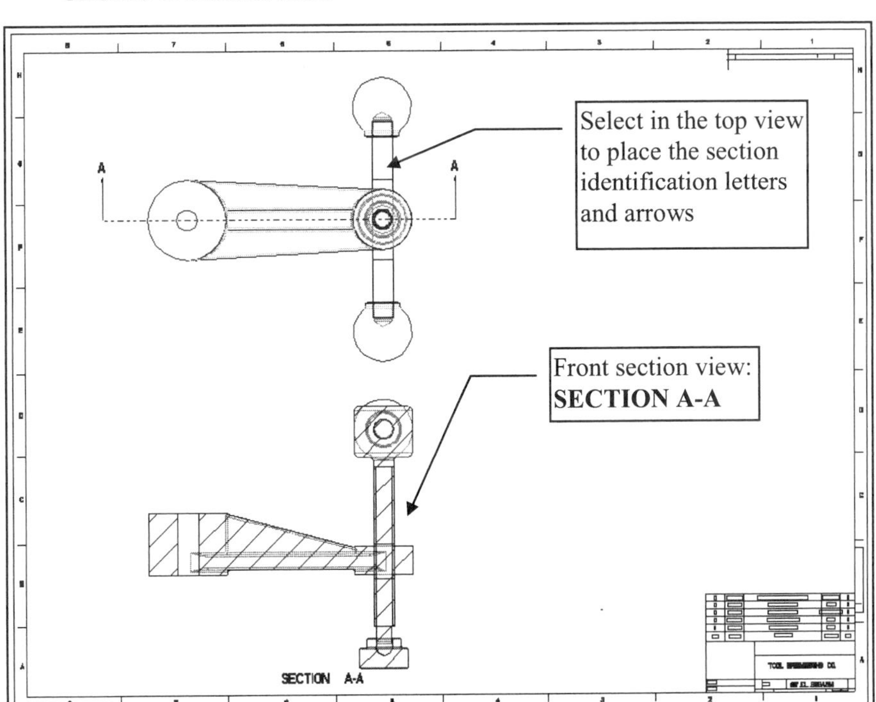

Figure 19.21
Swing Clamp Subassembly Drawing with Top View and Front Section View

Pro/E allows you to alter the display of the section view so an assembly makes more sense and to comply with industry standard practices.

Most companies require that the crosshatching on parts in section views of assemblies be "clocked" such that parts that meet do not use the same section lining (crosshatching) spacing and angle (Fig. 19.22). This makes the separation between parts more distinct. Clean up the section view to comply with industry practices. Change the spacing and the angle on the **Swivel** component (Fig. 19.22):

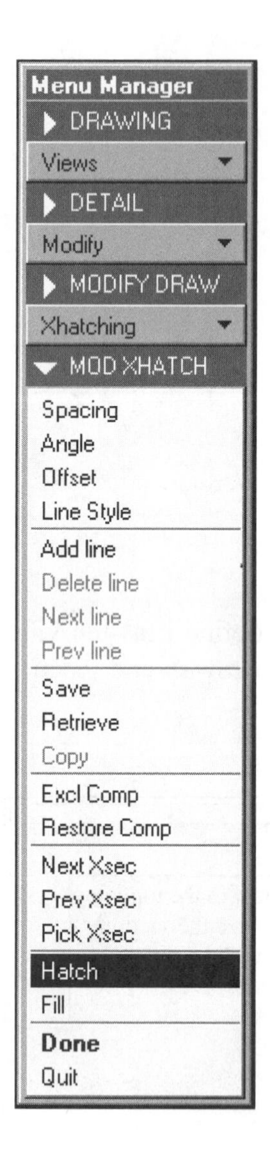

Choose **Modify**
Choose **Xhatching**

Select the crosshatching in the section view:

Choose **Done Sel**

Pro/E selects all the crosshatching in the view as a single object. You can cycle through the portion that lies on each component using the **Next Xsec**, **Prev Xsec**, and **Pick Xsec** options:

Choose **Next Xsec** until the Swivel (**sw_clamp_swivel**) is selected (highlighted).
Choose **Spacing**
Choose **Half** (pick twice)
Choose **Angle**
Choose **135**
Choose **Done**

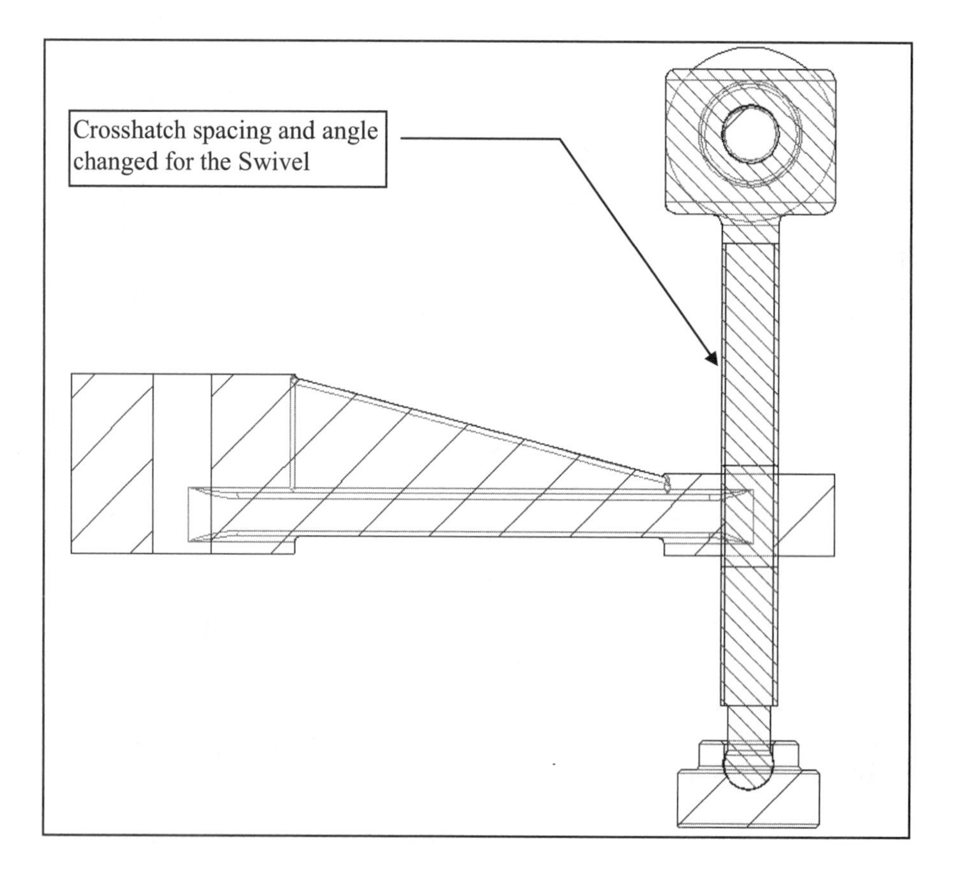

Figure 19.22
Changing the Section Lining Angle and Spacing on Assembly Components

Change the spacing on each component so that it is similar to that shown in Figure 19.23. Erase the *cosmetic threads* from both views (Fig. 19.24).

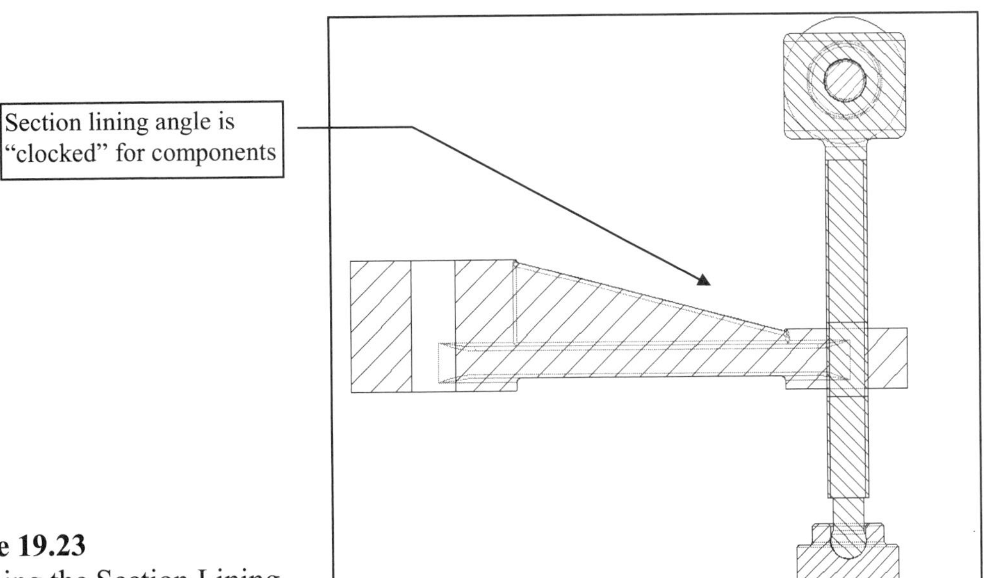

Figure 19.23
Changing the Section Lining

Because hidden lines are usually not shown on a section view, change the display of both views to remove hidden lines and tangent edges:

Choose ▼ DRAWING
Choose **Views**
Choose **Disp Mode**
Choose **View Disp**

Select both views:

Choose **Done Sel**
Choose **No Hidden**
Choose **No Disp Tan**
Choose **Done**
Choose **Done Sel**
Choose **Done/Return**

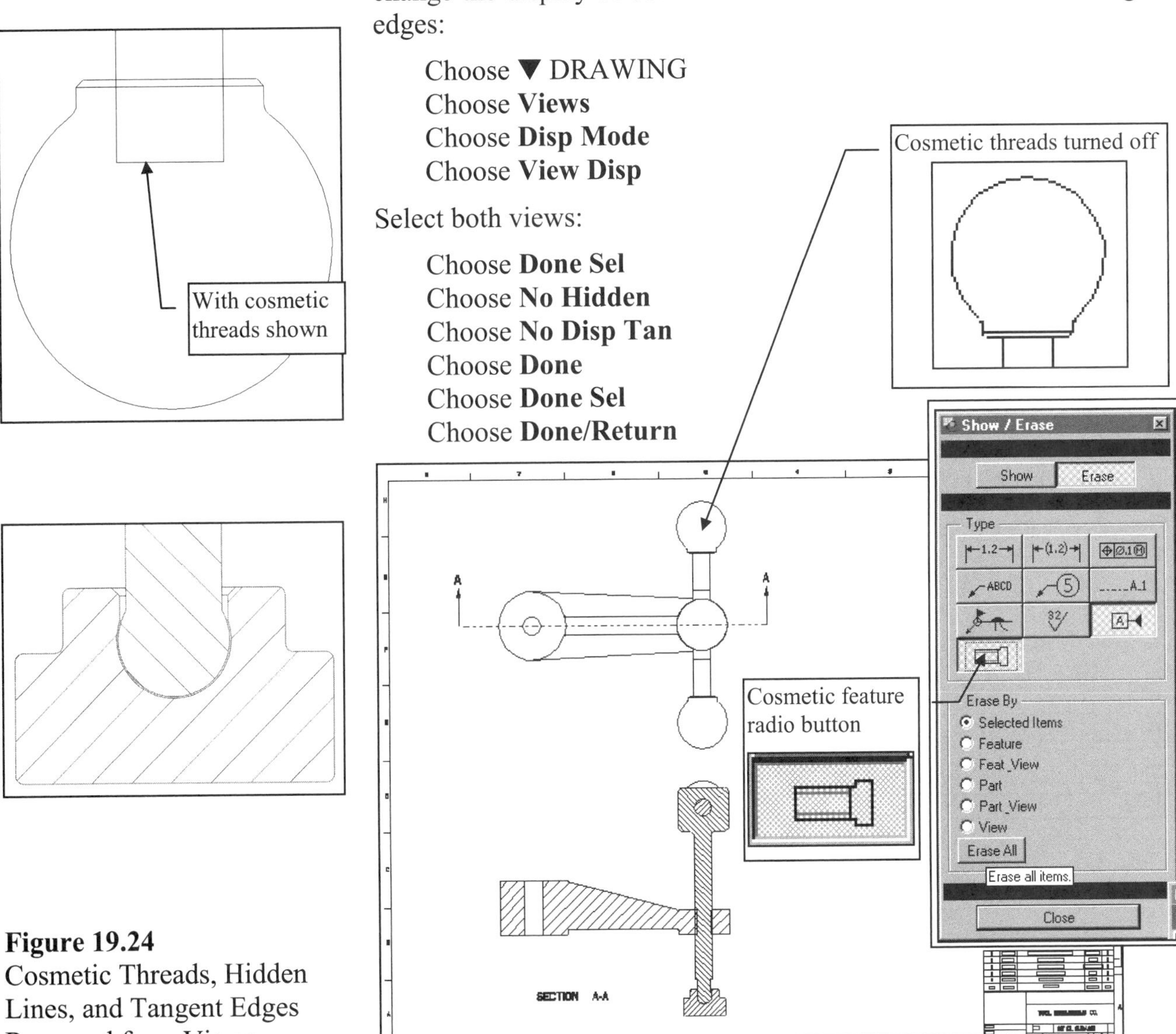

Figure 19.24
Cosmetic Threads, Hidden Lines, and Tangent Edges Removed from Views

To complete the views, show the axes (Fig. 19.25).

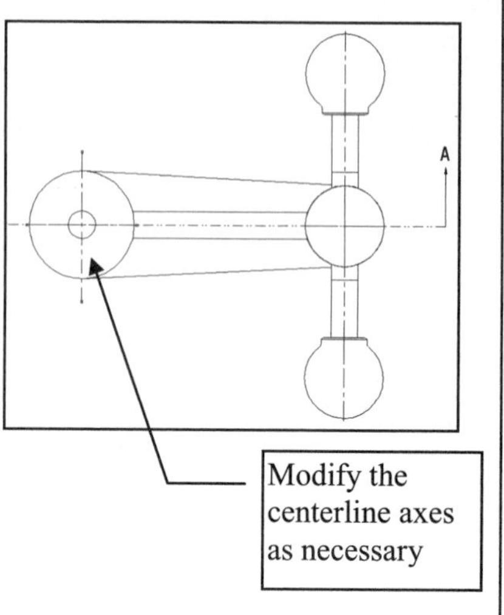

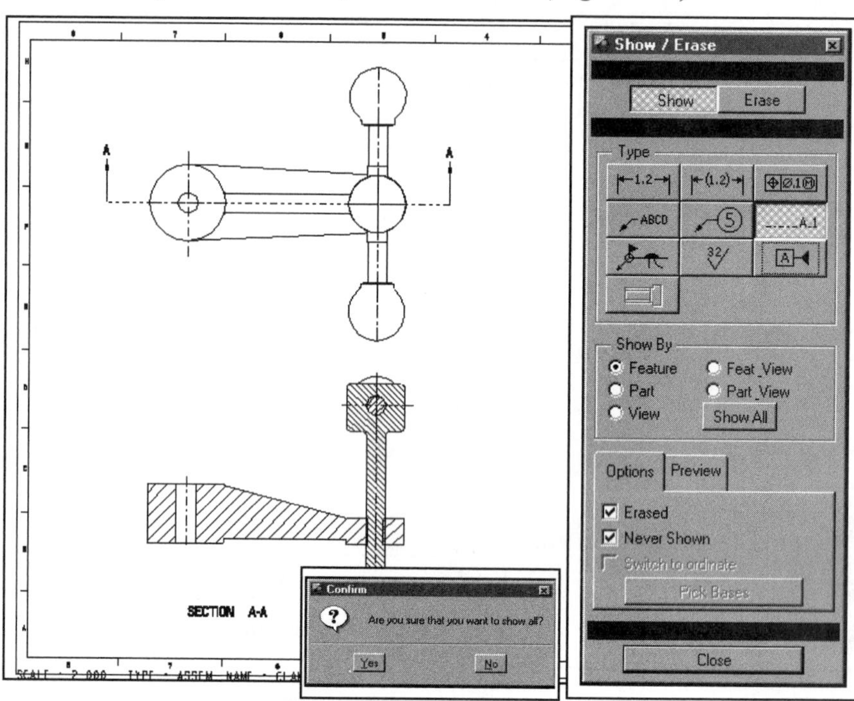

Figure 19.25
Centerline Axes Shown on Assembly Drawing

To finish the assembly drawing, you must display the balloons for each component. Show the **Item Balloons** on the drawing using the following commands (Fig. 19.26):

Choose **Table**
Choose **BOM Balloon**
Choose **Set Region**

Select the parts list on the drawing:

Choose **Show**
Choose **Show All**
Choose **Done/Return**

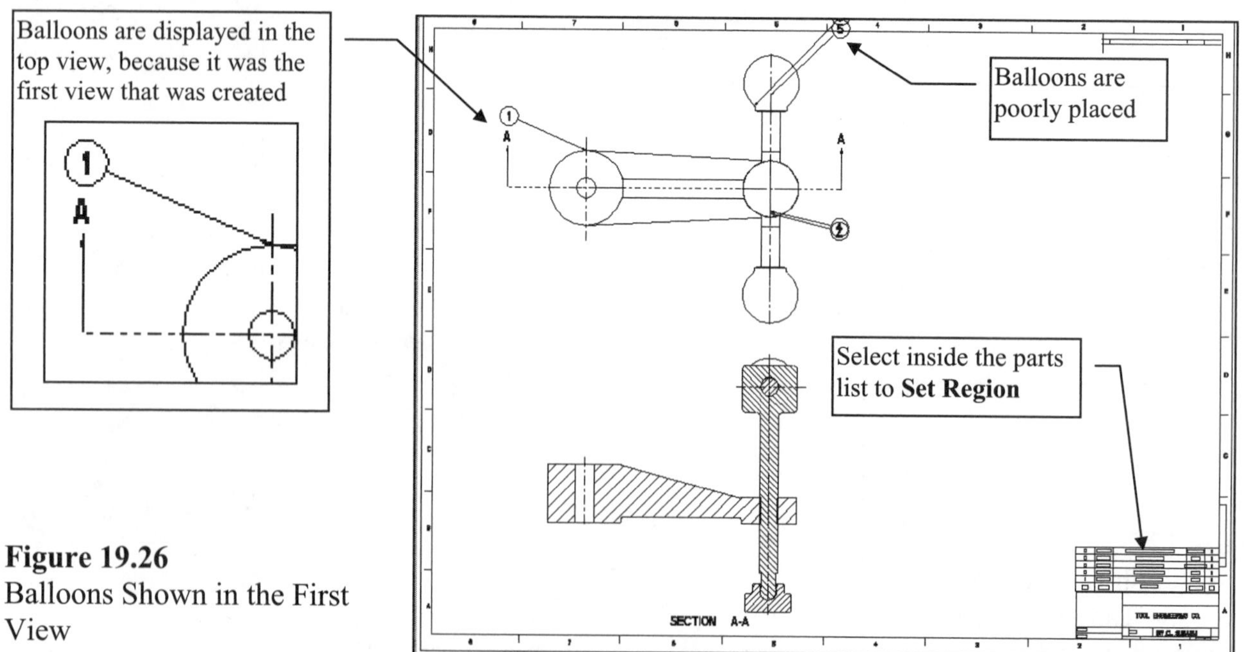

Figure 19.26
Balloons Shown in the First View

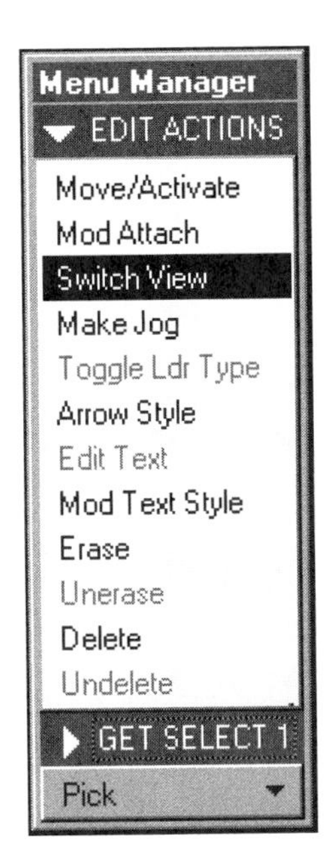

Switch the display for the Arm, Swivel, and Foot balloons to the front section view (Fig. 19.27). This is similar to switching the view of a dimension:

Choose **Switch View**
Select a balloon ⇒
Select a balloon ⇒
Select a balloon ⇒ **Done Sel**
Select the section view
Choose **View** ⇒ **Repaint**

Balloons are poorly placed in their default position

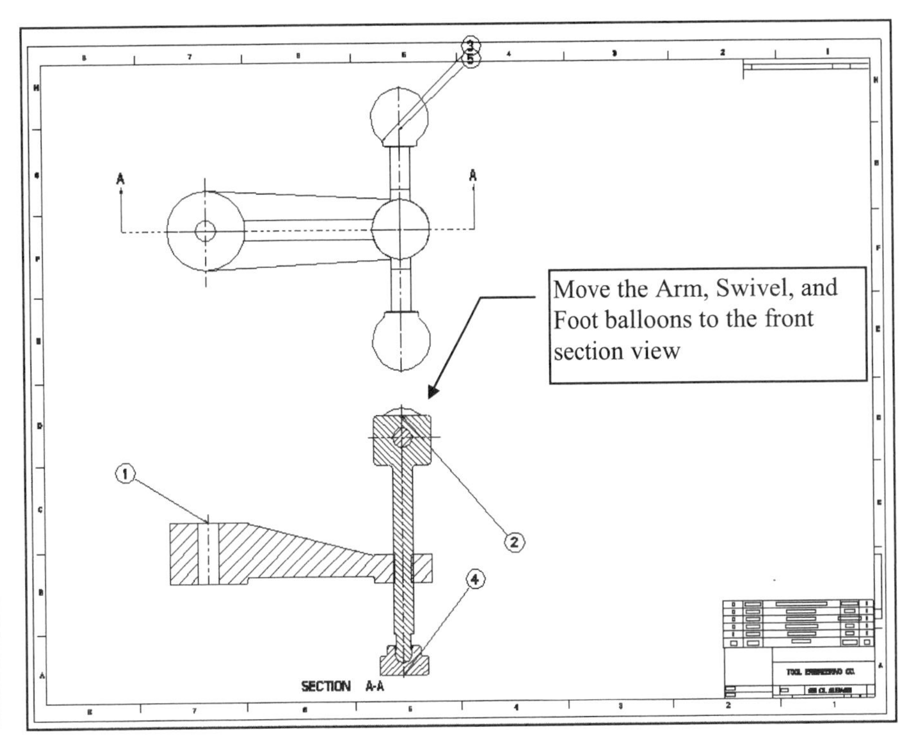

Figure 19.27
Switch Three Balloons to the Front View

Use the **Move** option to move the balloons away from each other (a bit). Move the attachment of the balloons (Figs. 19.28 and 19.29):

Choose **Mod Attach**

Select balloon item **1**:

Choose **Change Ref**

Change Ref allows you to pick a new edge to place the arrowhead on. Select the edge of the section as the new attachment:

Choose **Done Sel**
Choose **Done/Return**

Repeat the process until all the balloons are reattached and placed approximately, as shown in Figure 19.29.

File ⇒ Save ⇒ ✔

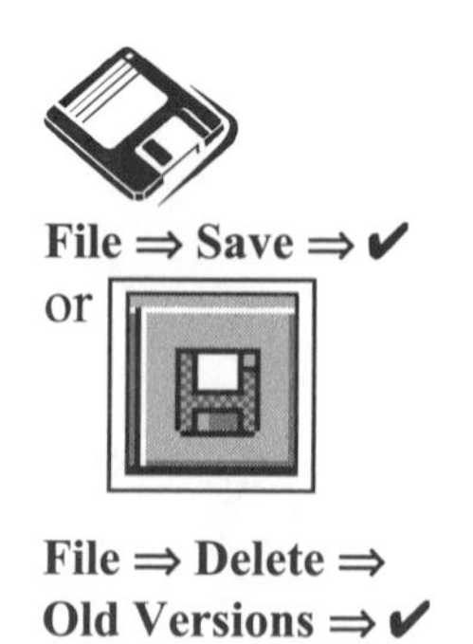

File ⇒ Save ⇒ ✔
or

File ⇒ Delete ⇒ Old Versions ⇒ ✔

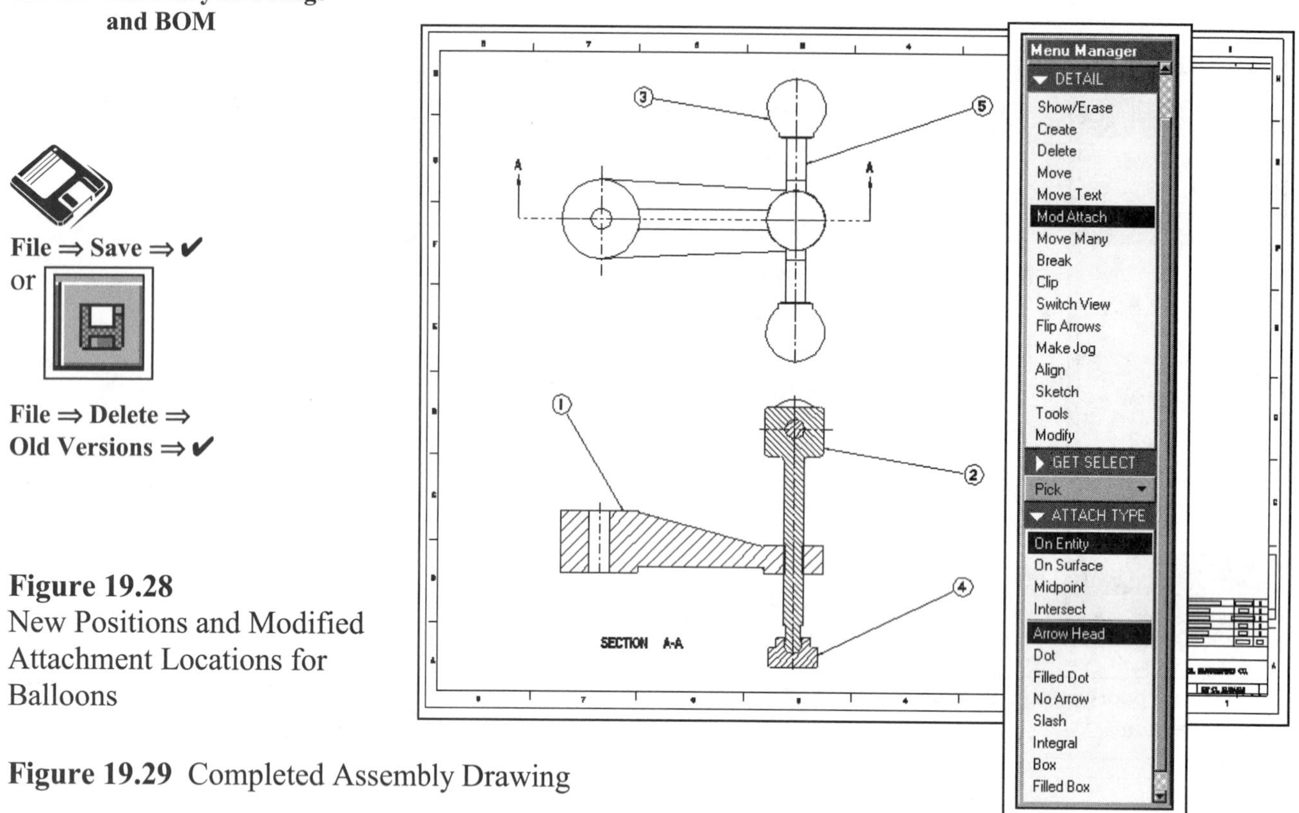

Figure 19.28
New Positions and Modified Attachment Locations for Balloons

Figure 19.29 Completed Assembly Drawing

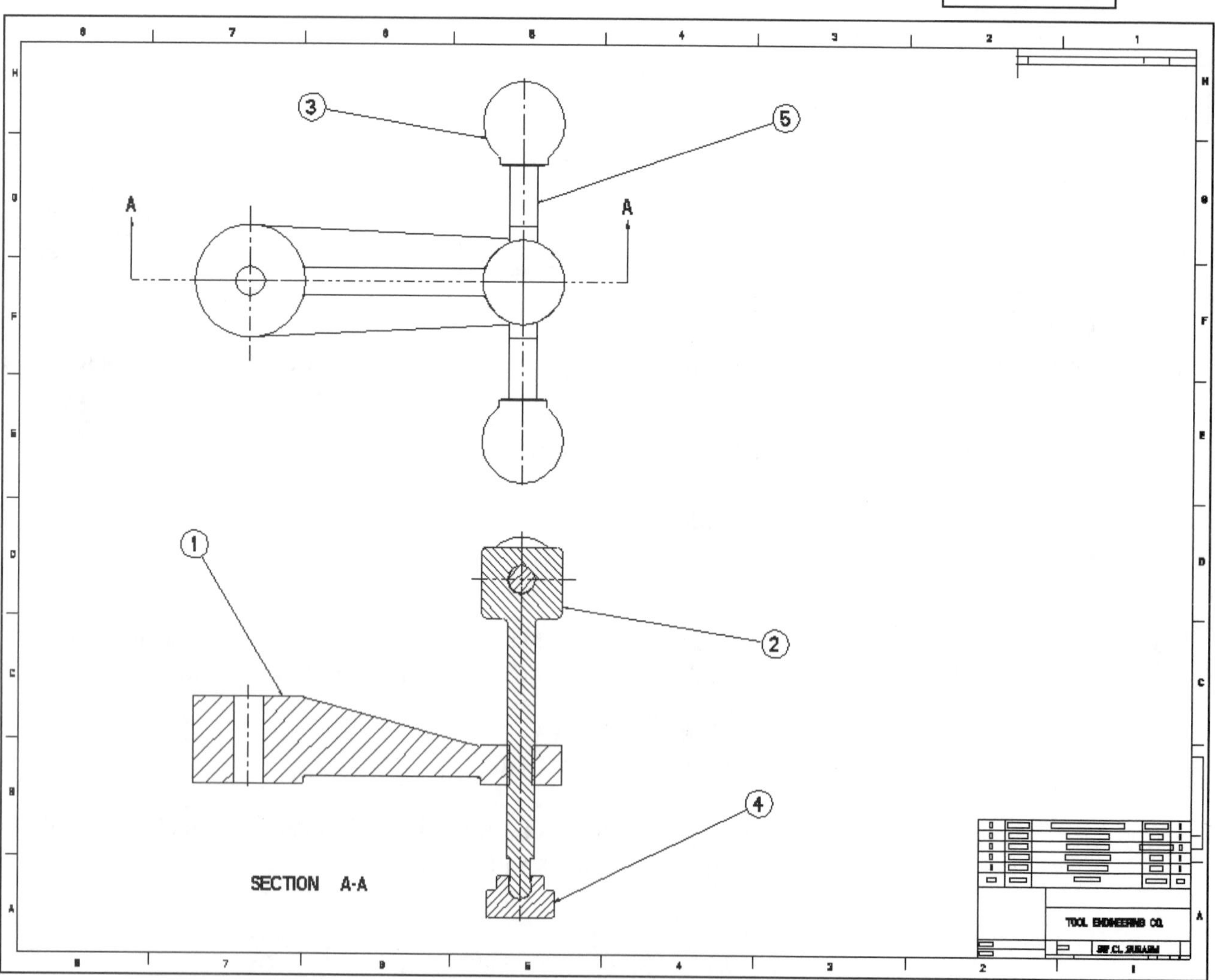

The next drawing we will create is the assembly drawing for the complete Swing Clamp. This assembly is composed of the subassembly in the previous drawing, the **Plate**, the short **Stud**, and the **Flange Nut**. The drawing will use the same format created for the subassembly. Formats are read-only files that can be used as many times as you want. Create the drawing using the following commands:

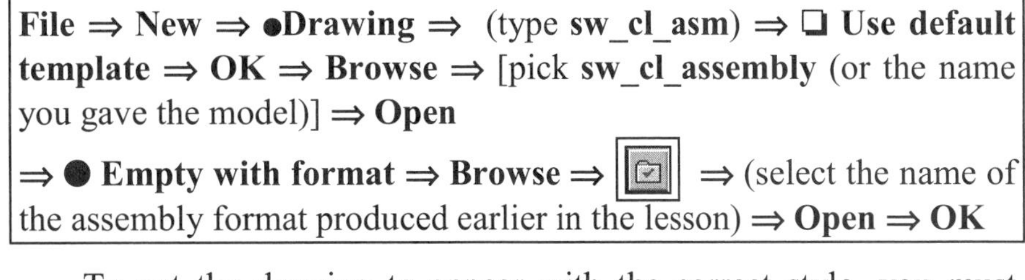

File ⇒ **New** ⇒ ●**Drawing** ⇒ (type **sw_cl_asm**) ⇒ ❑ **Use default template** ⇒ **OK** ⇒ **Browse** ⇒ [pick **sw_cl_assembly** (or the name you gave the model)] ⇒ **Open**

⇒ ● **Empty with format** ⇒ **Browse** ⇒ ⇒ (select the name of the assembly format produced earlier in the lesson) ⇒ **Open** ⇒ **OK**

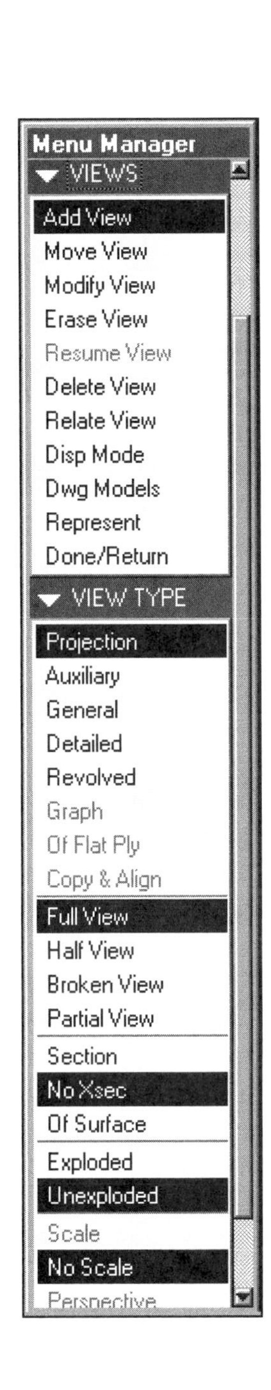

To get the drawing to appear with the correct style, you must modify the values of the **.dtl** Drawing Set Up file using **Advanced** ⇒ **Draw Setup** and modify the values as before. Another way would be to retrieve and load a previously used **.dtl** file if you had used the **Save** icon inside of the Preferences dialog box.

Now associate the complete assembly model to this Assembly drawing, and add two views. Add the top view as the first view. Choose the following commands:

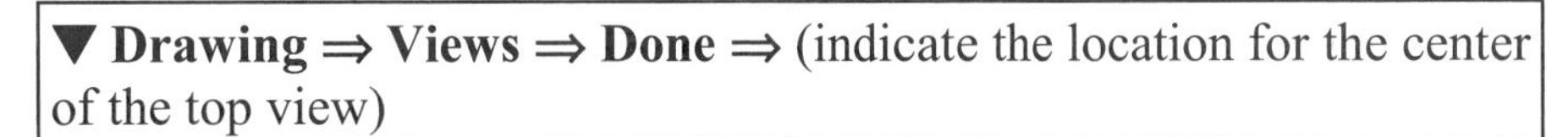

▼ **Drawing** ⇒ **Views** ⇒ **Done** ⇒ (indicate the location for the center of the top view)

Orient the top view, as shown in Figure 19.30, and then add a front section view as was done for the subassembly.

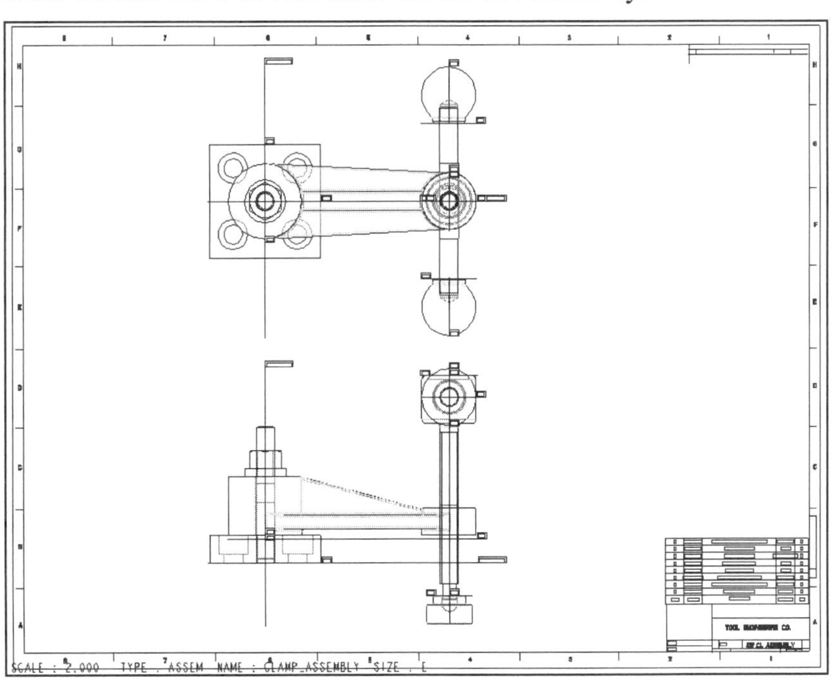

Figure 19.30
Assembly Drawing with Two Views

If you added the front view without choosing **Section** as one of the options, as was done in Figure 19.30, you must modify the view using the following commands (Fig. 19.31):

Views ⇒ **Modify View** ⇒ **View Type** ⇒ (pick the front view) ⇒ **Section** ⇒ **Done** ⇒ **Done** ⇒ **Create** ⇒ **Done** ⇒ (type **A**) ⇒ ✔ ⇒ [pick a datum plane that passes laterally through the assembly top view (Fig. 19.31)] ⇒ (pick the top view for the cutting plane line)

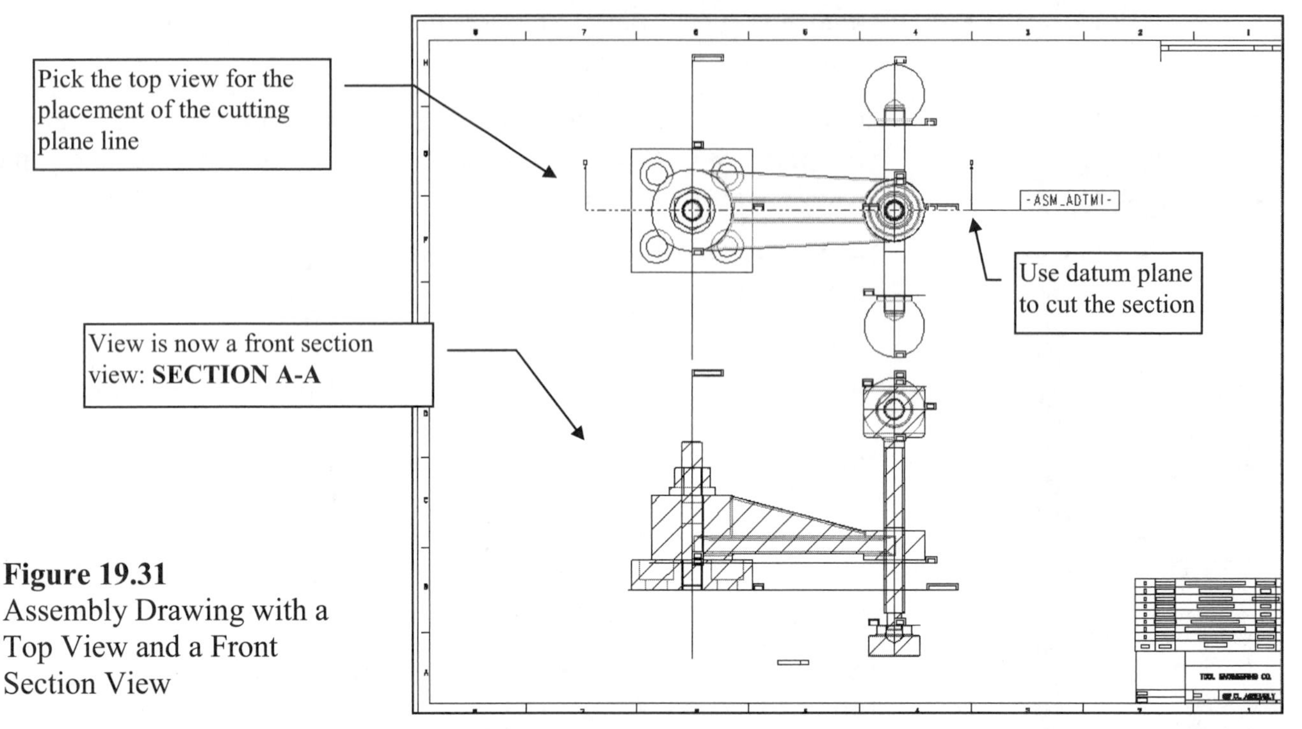

Figure 19.31
Assembly Drawing with a Top View and a Front Section View

Clean up the drawing by removing the cosmetic threads, datum axes, datum planes, and hidden lines. Display the tangent edges as you did for the subassembly. Show the centerline axes and modify them so that they are visually correct. Change the *scale* to **2.00.**

Most companies (and as per drafting standards) require that round purchased items, such as nuts, bolts, studs, springs, and die pins be excluded from sectioning even when the cutting plane passes through them.

NOTE
Modify the text height of the section identification lettering--(**SECTION A_A**) if it is too small.

Remove the section lining (crosshatching) from the short **Stud** (**3.50** length) and the **Flange Nut** in the front section view:

Choose **Modify**
Choose **Xhatching**

Select the crosshatching in the section view:

Choose **Done Sel**

Pro/E selects all the crosshatching in the view as a single object. You can cycle through the portion that lies on each component using the **Next Xsec**, **Prev Xsec**, and **Pick Xsec** options. Choose the next commands (Fig. 19.32):

Choose **Excl Comp** (this eliminates Xsec of flange nut)
Choose **Next Xsec** (the short stud happens to be next)
Choose **Excl Comp** (this eliminates Xsec of short stud)

*(See **HINT**)*

HINT
If you wish to eliminate the section lines from the long stud (end view), use the following commands after completing the command sequence to the right:

Choose **Next Xsec**
(the long stud happens to be next)
Choose **Excl Comp**
(this eliminates Xsec of long stud)

Or you can use **Query Sel** to pick the components:

Choose **Pick Xsec**
Choose **Query Sel**
Select the short **Stud** ⇒ **Accept**
Choose **Excl Comp**

and then the **Flange Nut**:

Choose **Pick Xsec**
Choose **Query Sel**
Select the **Flange Nut** ⇒ **Accept**
Choose **Excl Comp**

Choose **Done** ⇒ **Done Sel** ⇒ **Done/Return**

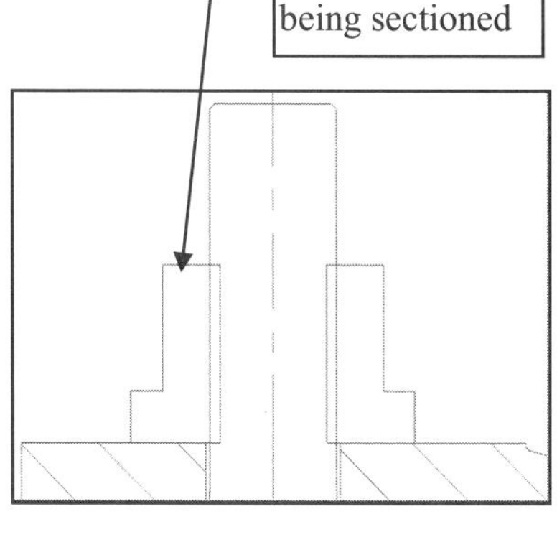

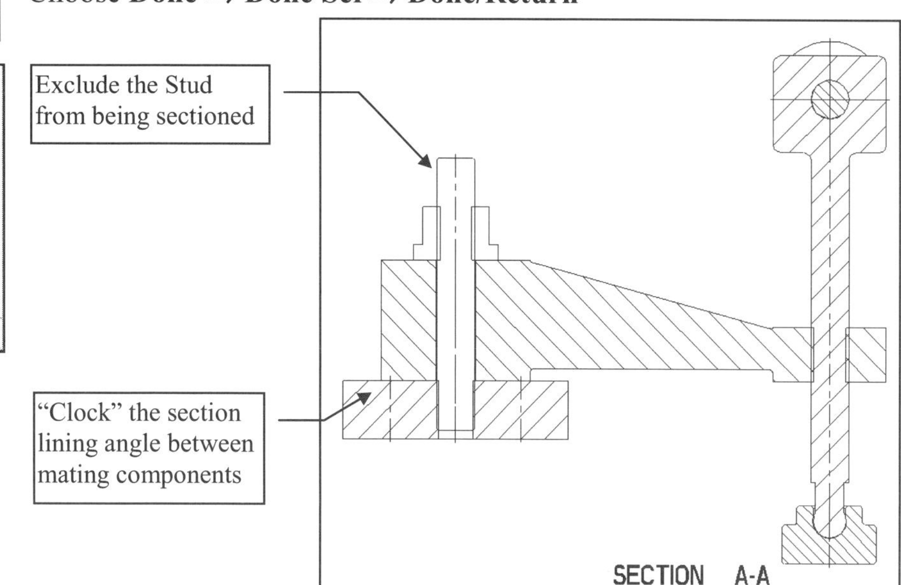

Figure 19.32
Assembly Drawing with Stud and Flange Nut Section Lining Excluded from the Views

The next step in completing the assembly drawing is to show the balloons. Choose the following commands (Fig. 19.33):

Table ⇒ **BOM Balloon** ⇒ **Set Region** ⇒ (select inside the parts list on the drawing) ⇒ **Show** ⇒ **Show All** ⇒ **Done/Return**

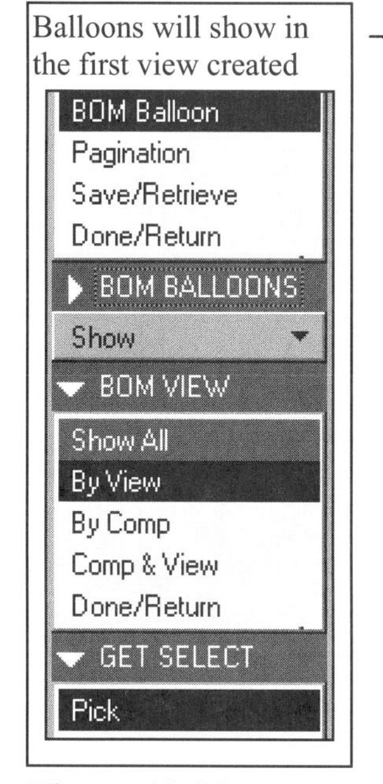

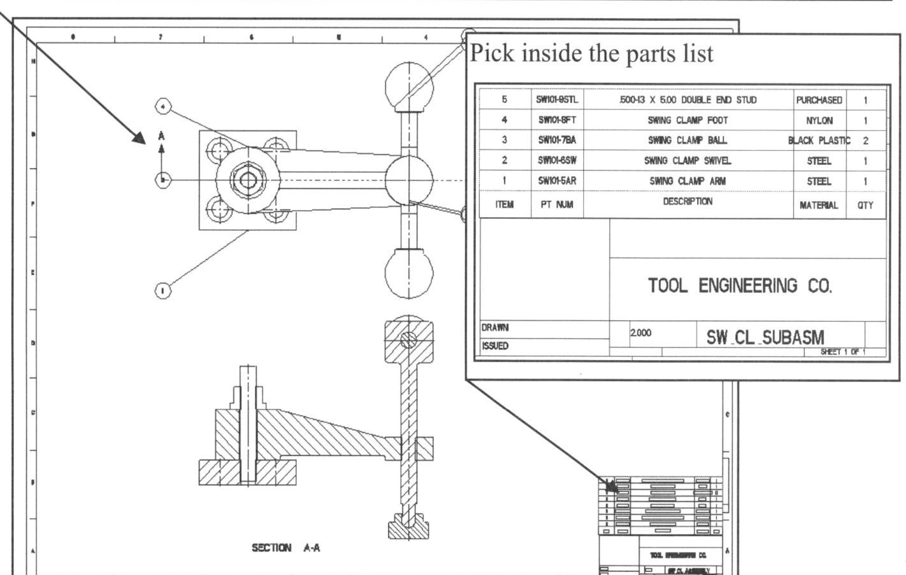

Figure 19.33
Showing the Balloons

Switch some of the balloons to the front view and reposition the others to make a more clear and balanced ballooning scheme (Fig. 19.34). You can line up the balloons by turning on **Snap to Grid** and then moving the balloons to new positions.

File ⇒ Save ⇒ ✔
File ⇒ Delete ⇒
Old Versions ⇒ ✔

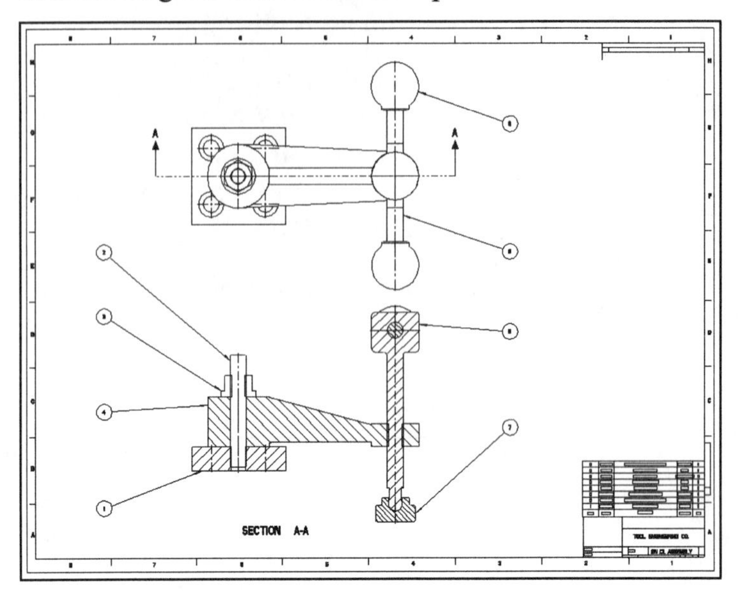

Figure 19.34
Showing the Balloons

The numbering of the components on the assembly may need to be different from the default setting. To change the ballooning, you must use **Fix Index** from the TABLE menu. Use the following commands to change the numbering scheme (Fig. 19.35):

Table ⇒ Repeat Region ⇒ Fix Index ⇒ (select in the parts list region) **⇒** (select a record in the repeat region; select the *Arm*, which is defaulted to item **4**) **⇒** (type **1**) **⇒ ✔ ⇒** (continue selecting and changing the numbering to that of the *subassembly* sequence: make the *Plate* **6**, the short *Stud* **7**, and the *Flange Nut* **8**) **⇒ Done**

When you are finished, the parts list will sequence itself (Fig. 19.36), and the ballooning will update automatically (Fig. 19.37).

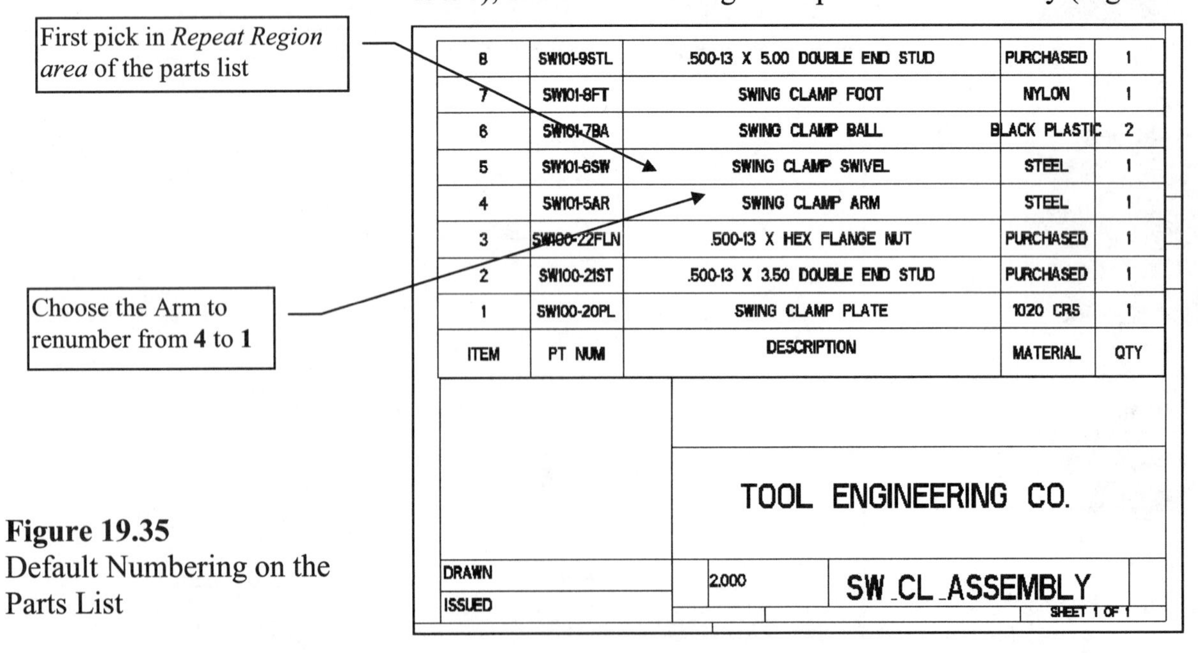

8	SW101-9STL	.500-13 X 5.00 DOUBLE END STUD	PURCHASED	1
7	SW101-8FT	SWING CLAMP FOOT	NYLON	1
6	SW101-7BA	SWING CLAMP BALL	BLACK PLASTIC	2
5	SW101-6SW	SWING CLAMP SWIVEL	STEEL	1
4	SW101-5AR	SWING CLAMP ARM	STEEL	1
3	SW100-22FLN	.500-13 X HEX FLANGE NUT	PURCHASED	1
2	SW100-21ST	.500-13 X 3.50 DOUBLE END STUD	PURCHASED	1
1	SW100-20PL	SWING CLAMP PLATE	1020 CRS	1
ITEM	PT NUM	DESCRIPTION	MATERIAL	QTY

Figure 19.35
Default Numbering on the Parts List

Arm is now number **1**, as in the subassembly

8	SW100-22FLN	.500-13 X HEX FLANGE NUT	PURCHASED	1
7	SW100-21ST	.500-13 X 3.50 DOUBLE END STUD	PURCHASED	1
6	SW100-20PL	SWING CLAMP PLATE	1020 CRS	1
5	SW101-9STL	.500-13 X 5.00 DOUBLE END STUD	PURCHASED	1
4	SW101-8FT	SWING CLAMP FOOT	NYLON	1
3	SW101-7BA	SWING CLAMP BALL	BLACK PLASTIC	2
2	SW101-6SW	SWING CLAMP SWIVEL	STEEL	1
1	SW101-5AR	SWING CLAMP ARM	STEEL	1
ITEM	PT NUM	DESCRIPTION	MATERIAL	QTY

TOOL ENGINEERING CO.

DRAWN

ISSUED

2.000

SW_CL_ASSEMBLY

SHEET 1 OF 1

Figure 19.36
Fix the Index Numbering on the Parts List

File ⇒ Save ⇒ ✔

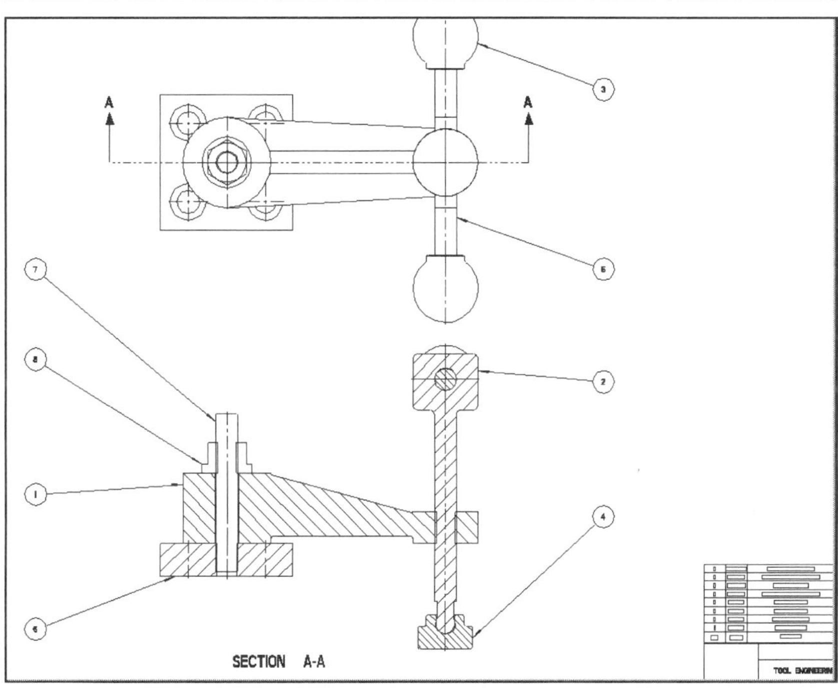

Figure 19.37
Renumbered Balloons

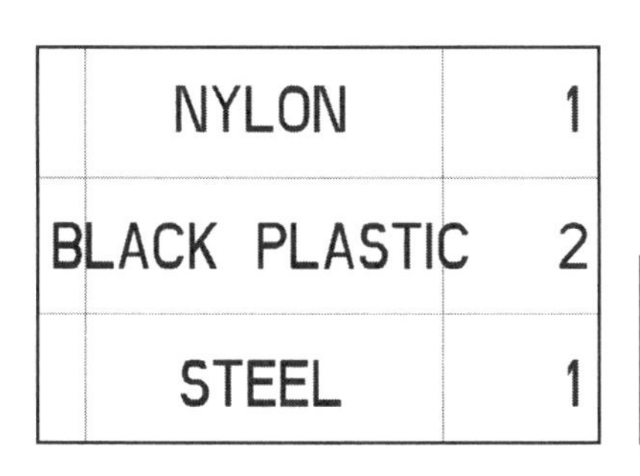

NYLON	1
BLACK PLASTIC	2
STEEL	1

To change the numbering after it has been fixed, you must **Unfix** first and then **Fix Index** again.

Figure 19.36 shows that the MATERIAL cell column is not wide enough to accommodate the length of the BLACK PLASTIC material value. You can change the size of the column using **Mod Rows/Cols** from the TABLE menu.

Choose the following commands (Fig. 19.38):

Mod Rows/Cols ⇒ Change Size ⇒ Column ⇒ By Length ⇒ (pick the MATERIAL column) ⇒ (type **1.25**) ⇒ **✔** ⇒ (pick the DESCRIPTION column) ⇒ (type **3.75**) ⇒ **✔** (Fig. 19.39)

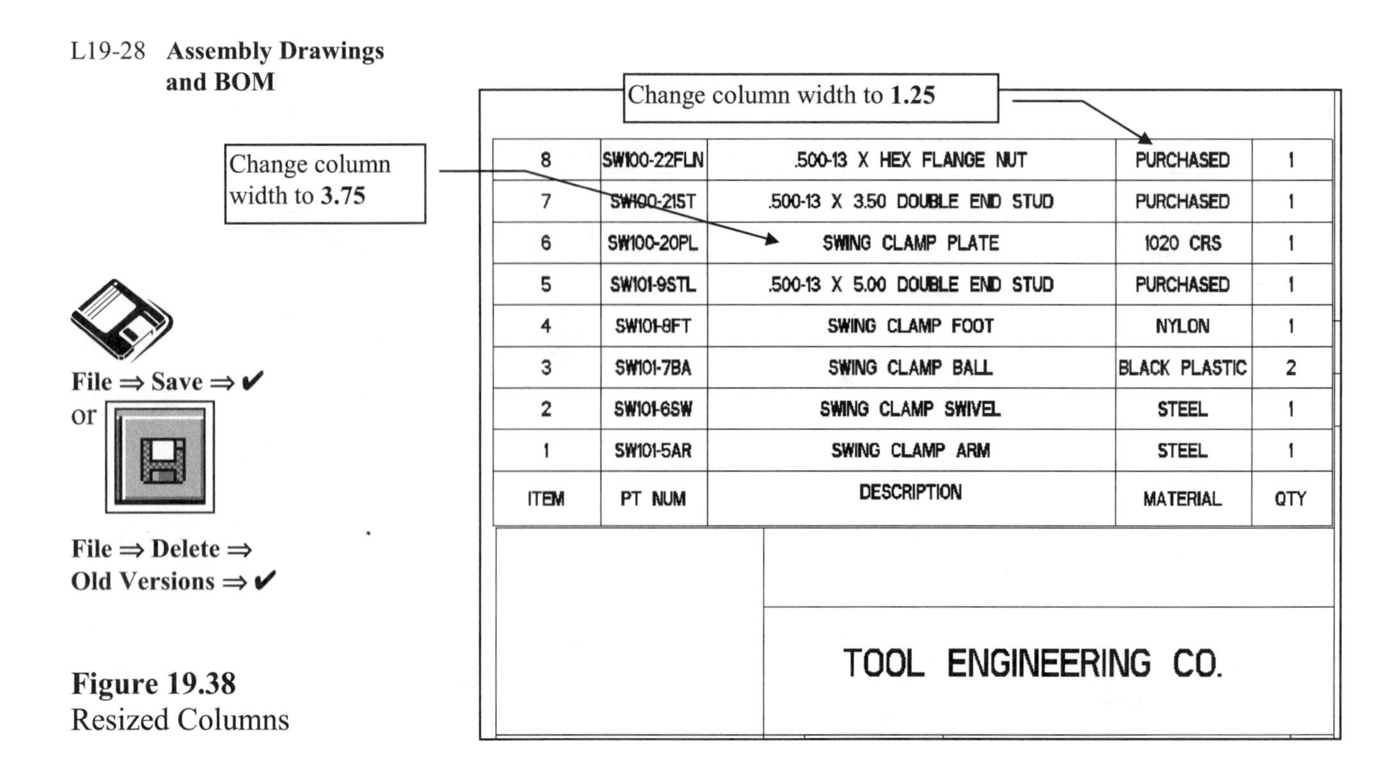

8	SW100-22FLN	.500-13 X HEX FLANGE NUT	PURCHASED	1
7	SW100-215T	.500-13 X 3.50 DOUBLE END STUD	PURCHASED	1
6	SW100-20PL	SWING CLAMP PLATE	1020 CRS	1
5	SW101-9STL	.500-13 X 5.00 DOUBLE END STUD	PURCHASED	1
4	SW101-8FT	SWING CLAMP FOOT	NYLON	1
3	SW101-7BA	SWING CLAMP BALL	BLACK PLASTIC	2
2	SW101-6SW	SWING CLAMP SWIVEL	STEEL	1
1	SW101-5AR	SWING CLAMP ARM	STEEL	1
ITEM	PT NUM	DESCRIPTION	MATERIAL	QTY

File ⇒ Save ⇒ ✔
or

File ⇒ Delete ⇒ Old Versions ⇒ ✔

Figure 19.38
Resized Columns

Figure 19.39 Assembly Drawing

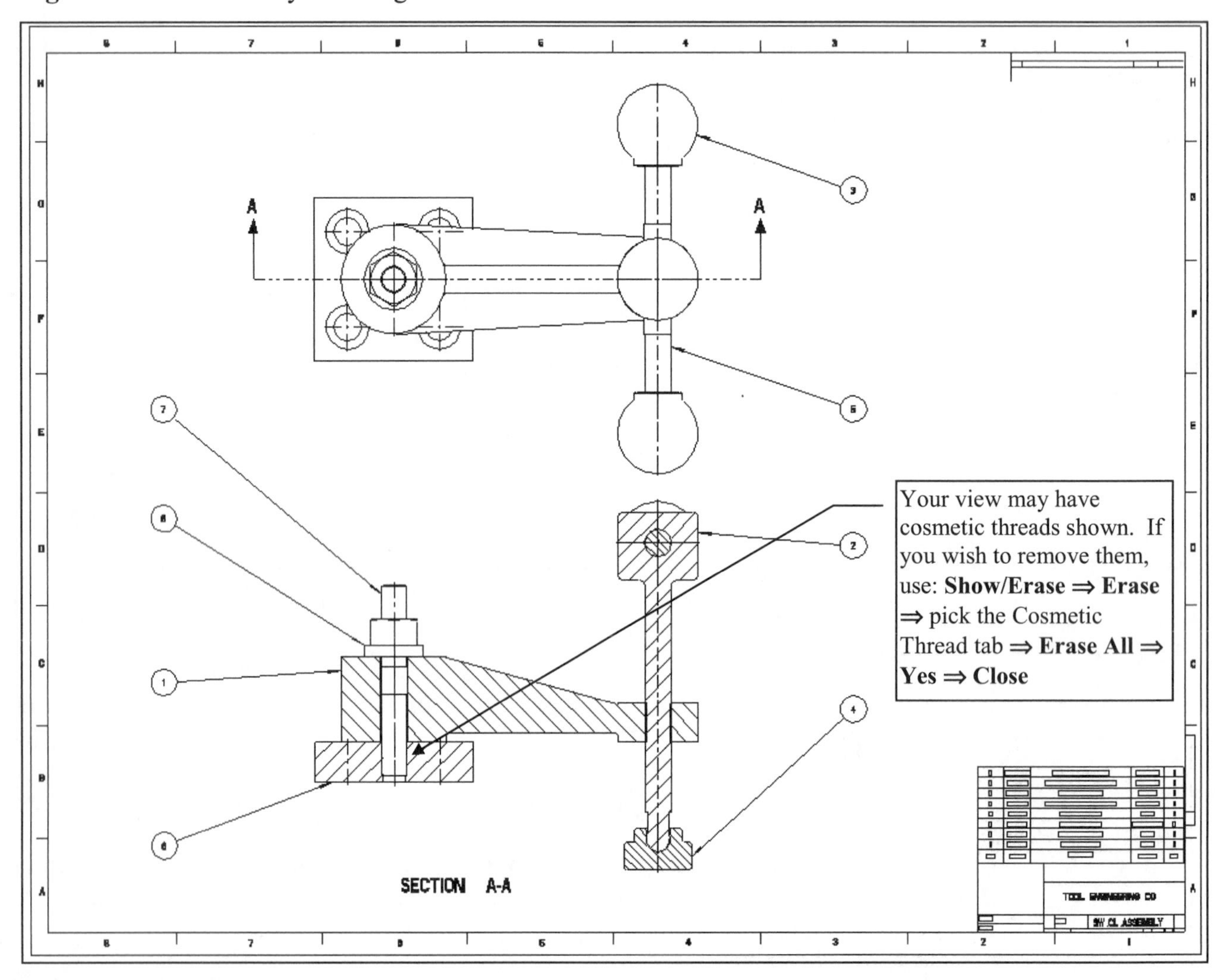

Lesson 19 Project

Coupling Assembly Drawing

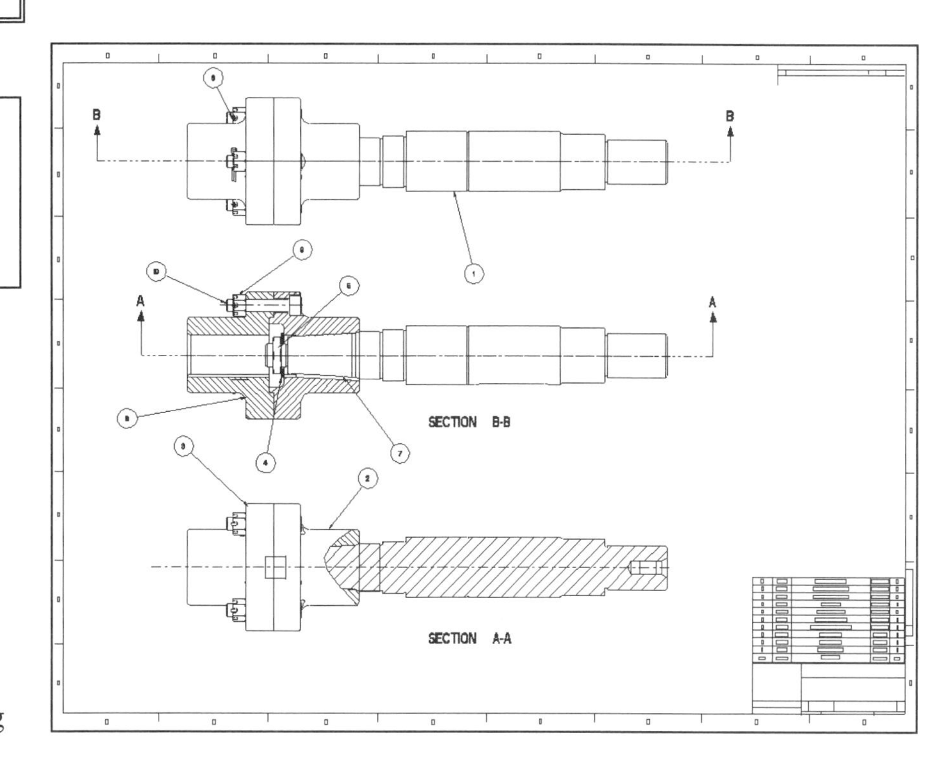

Figure 19.40
Coupling Assembly Drawing

Coupling Assembly Drawing

Create an assembly drawing of the Coupling Assembly using the format made in this lesson. Use Figures 19.40 through 19.44 as examples for this **lesson project.** The ballooned assembly drawing will have three views and a parts list.

Assign parameters to the parts in the assembly so that they can be displayed on a parts list in the assembly drawing, and generate item balloons for each component.

EGD REFERENCE
Fundamentals of Engineering Graphics and Design
by L. Lamit and K. Kitto
Read Chapter 23.
See pages 865-866.

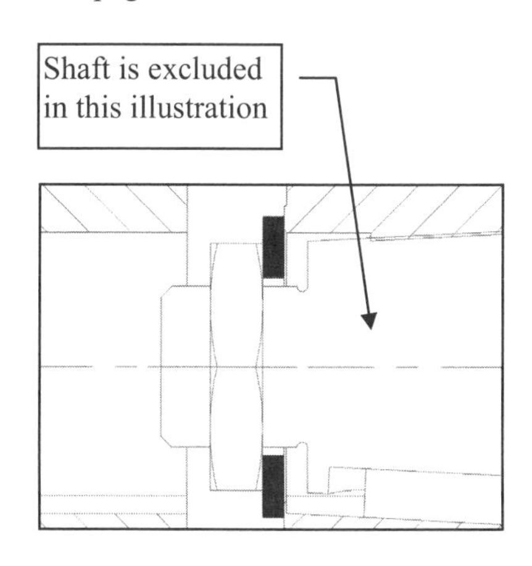

ITEM	PT NUM	DESCRIPTION	MATERIAL	QTY
10	110-2CS	SOC HD CAP SCREW	PURCHASED	3
9	109-2SN	HEX SLOT NUT 16 X 2	PURCHASED	3
8	108-2CP	COTTER PIN .150 X 1.25	PURCHASED	3
7	107-2KY	KEY 14 X 61	PURCHASED	1
6	106-2DW	DOWEL 12OD X 70	PURCHASED	2
5	105-2HN	HEX NUT M30 X 3.5	PURCHASED	1
4	104-2WA	WASHER 33ID X 50OD X 4	PURCHASED	1
3	103-2CP2	COUPLING TWO	1040 CRS	1
2	102-2CP1	COUPLING ONE	1040 CRS	1
1	101-2SH	COUPLING SHAFT	1020 CRS	1

Figure 19.41
Coupling Assembly Drawing Parts List

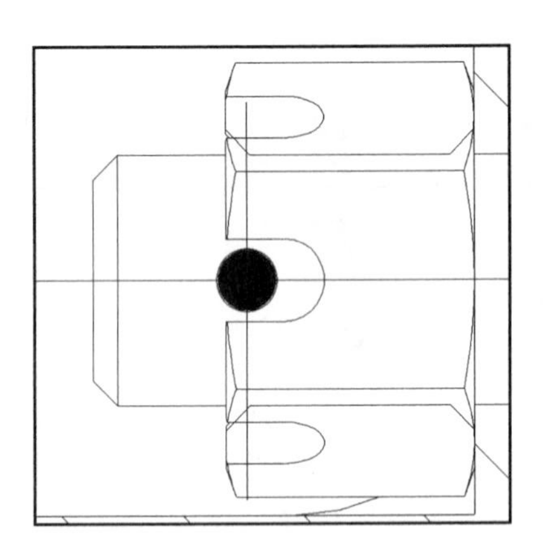

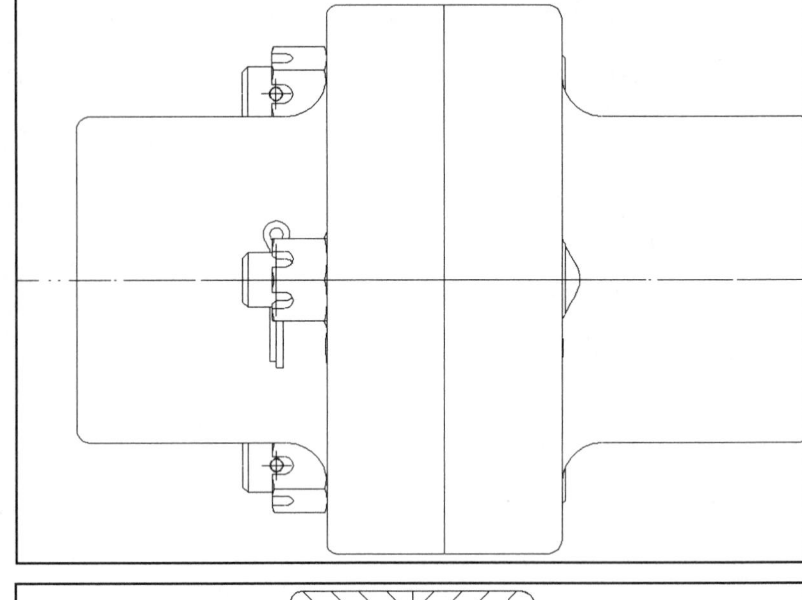

Figure 19.42
Coupling Assembly Drawing, Slotted Hex Nut

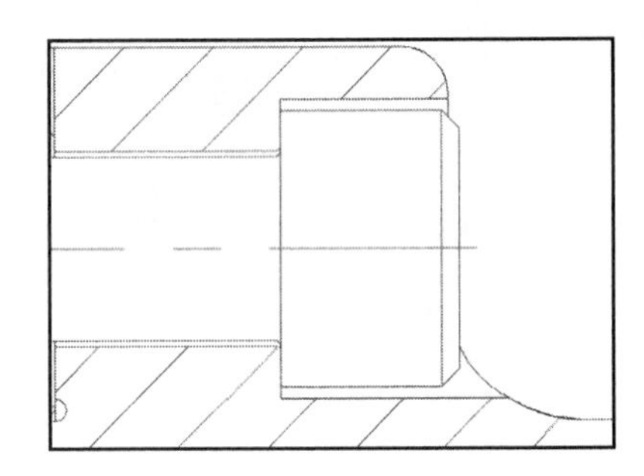

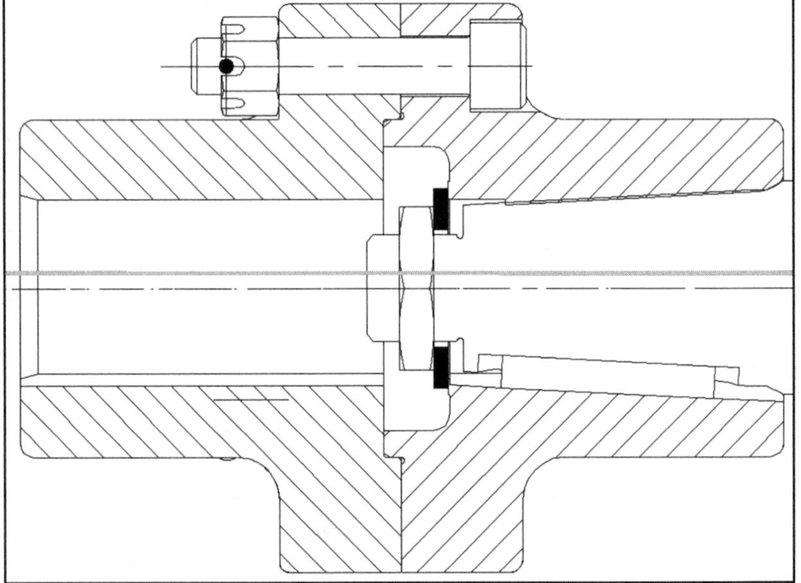

Figure 19.43
Coupling Assembly Drawing, Section Close-up

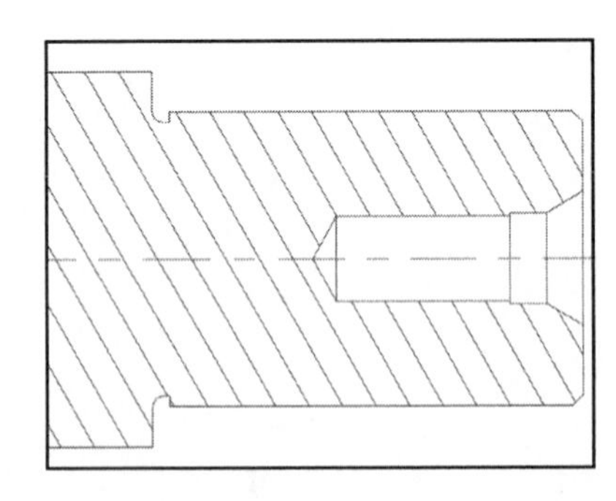

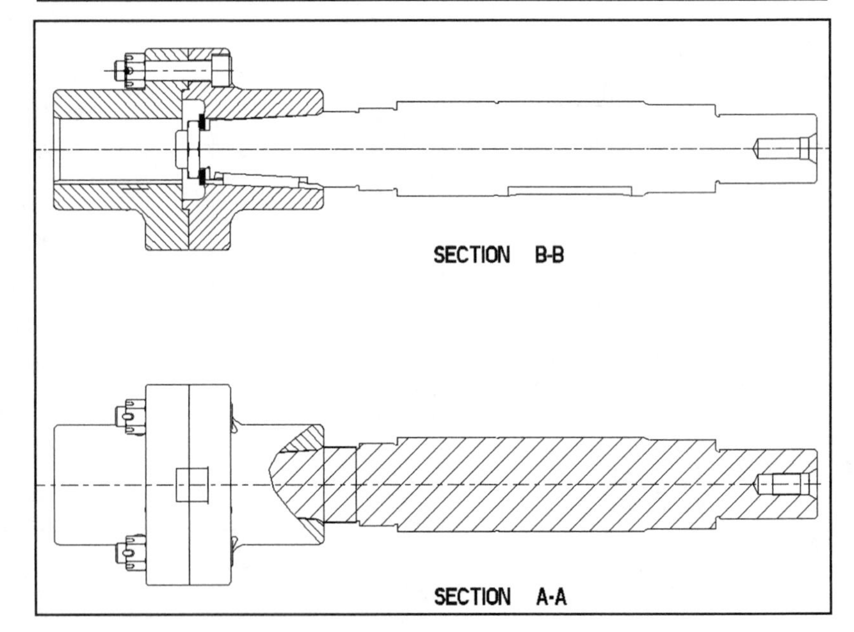

Figure 19.44
Coupling Assembly Drawing, **SECTION A-A** and **SECTION B-B**

Lesson 20

Exploded Assembly Drawings

Figure 20.1
Sheet 1 Exploded Swing Clamp Drawing with Standard Format and Balloons

☑ *EGD REFERENCE*
Fundamentals of Engineering Graphics and Design
by L. Lamit and K. Kitto
Read Chapters 13 and 23.
See pages 447, 823, 836-837, and 841.

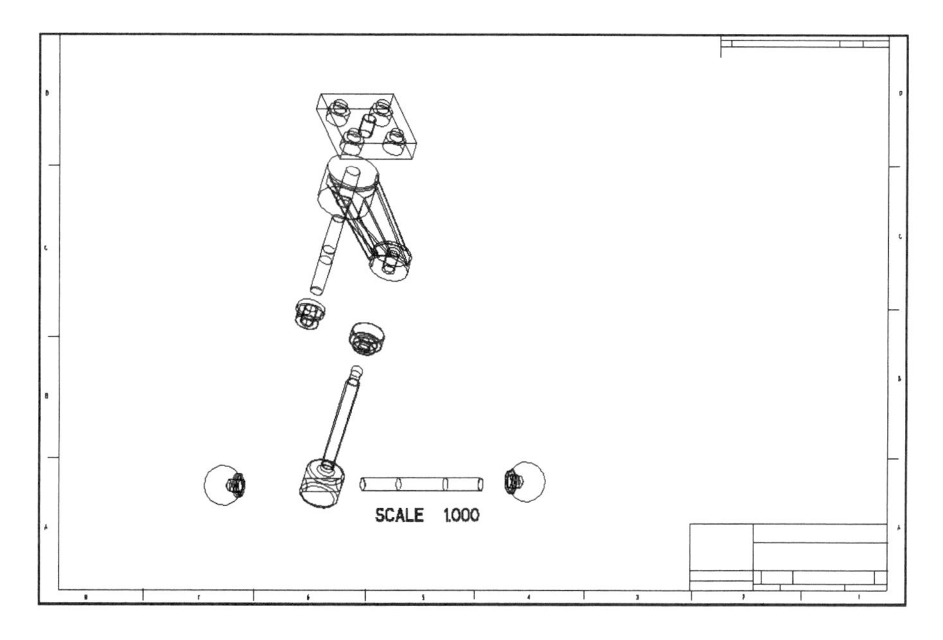

Figure 20.1 (continued)
Sheet 2 Exploded Trimetric View

OBJECTIVES

1. **Create drawings with exploded views**

2. **Use multiple sheets**

3. **Make assembly drawing sheets with multiple models**

4. **Create balloons on exploded assemblies**

COAch™ for Pro/ENGINEER

If you have **COAch for Pro/ENGINEER** on your system, go to SEARCH and do the Segments relating to Assemblies, Exploded Assemblies.

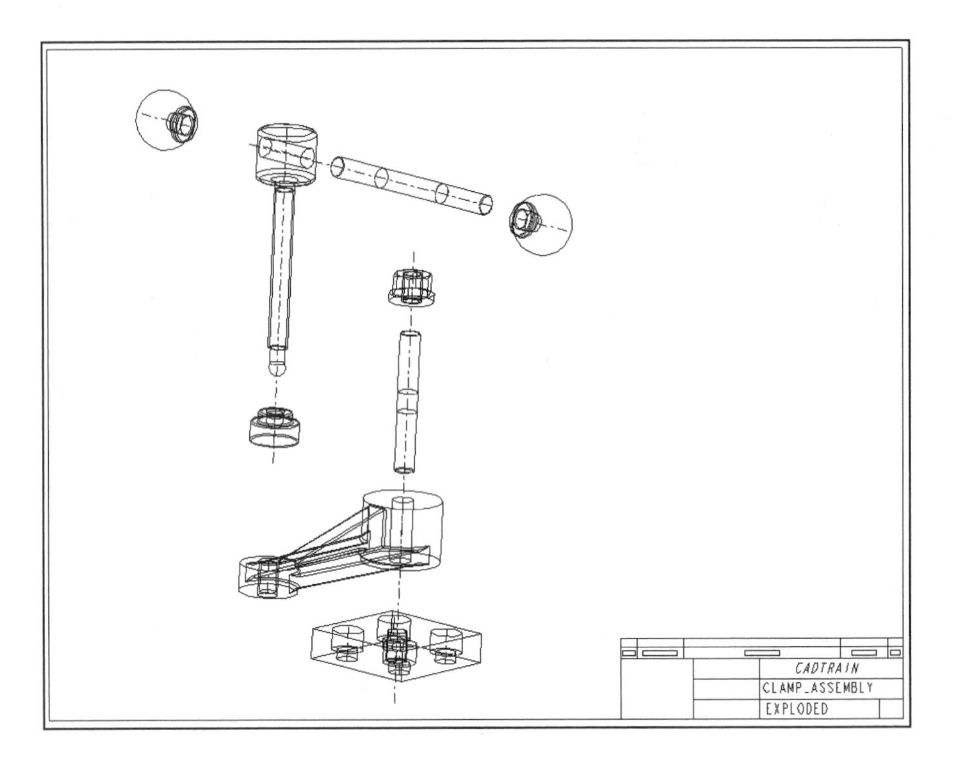

Figure 20.2
Exploded Swing Clamp Drawing

EXPLODED ASSEMBLY DRAWINGS

As we explained in Lesson 15, you can create an exploded view of an assembly. The *exploded views* created and saved with the model in Lesson 15 will be used in this lesson. The **Swing Clamp Assembly** shown in Figures 20.1 and 20.2 is used for the lesson model, and the **Coupling Assembly** is used for the Lesson 20 Project.

Exploding an assembly affects only the display of the assembly; it does not alter actual distances between components (Fig. 20.3). *Exploded states* are created and saved to allow a clear visualization and understanding of the positional relationships of all of the components in an assembly. For each explode state, you can toggle the explode status of components, change the explode locations of components, and create explode offset lines. You can define multiple explode states for each assembly and then explode the assembly using any of these explode states at any time. You can also set an exploded state for each drawing view of an assembly.

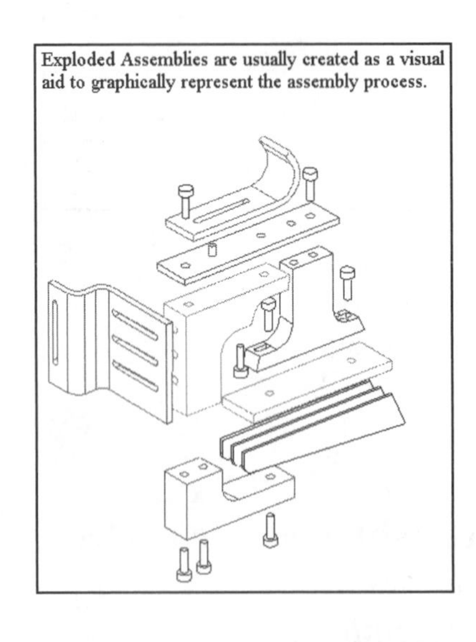
Exploded Assemblies are usually created as a visual aid to graphically represent the assembly process.

If none of the exploded views you have created is exactly the exploded position you wish to use in this lesson, bring up the model (assembly), create new exploded views, and save them for use in this lesson and lesson project.

You are required to create a complete documentation package for the two assemblies. *A documentation package contains all models and drawings needed to manufacture the parts and assemble the components.* Your instructor may change the requirements, but in general, create and plot the following:

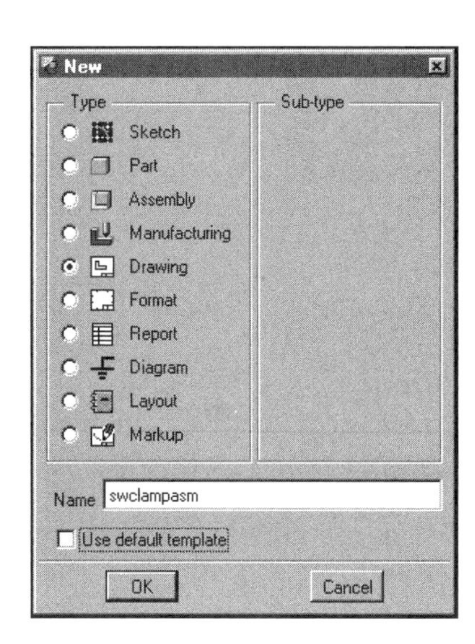

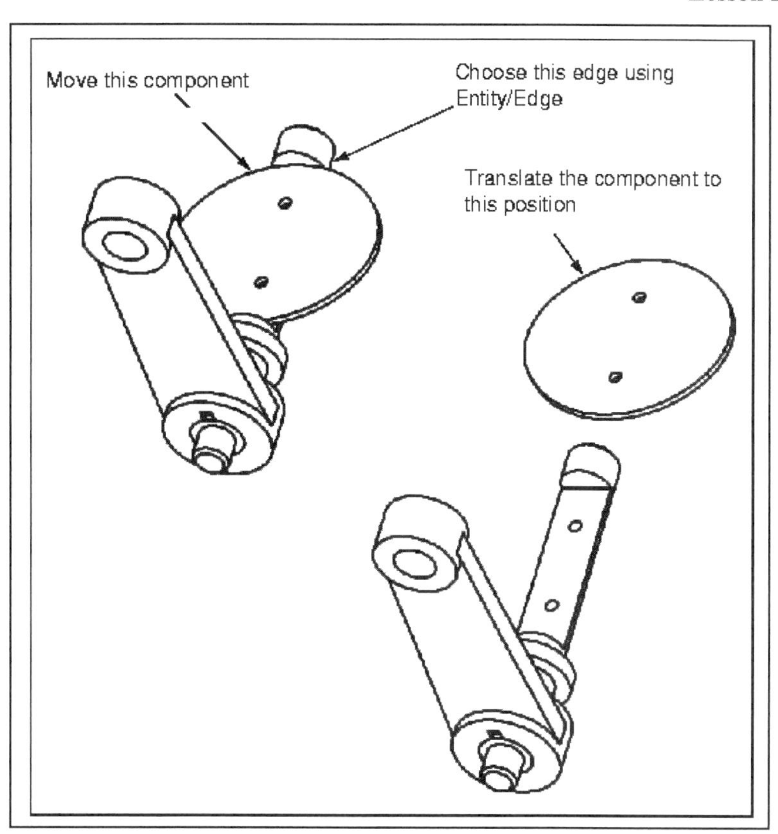

Figure 20.3
Exploded Valve Assembly

Lesson procedures & commands
START HERE ➡ ➡ ➡

Part Models for all swing clamp components
Detail Drawings for each nonstandard component, such as the Clamp Arm, Clamp Swivel, Clamp Foot, and Clamp Ball
Detail Drawings for standard components that have been altered; show only the dimensions needed to alter the component--for example, the socket head cap screw in the coupling assembly
Assembly Drawing using standard orthographic ballooned views
Exploded Assembly Drawing of the ballooned assembly
Exploded Subassembly Drawing of the ballooned subassembly

Exploded Swing Clamp Assembly Drawings

The process required to place an exploded view on a drawing is similar to that provided in Lesson 19 for adding assembly orthographic views to a drawing. Choose the following commands:

File ⇒ New ⇒ ●Drawing ⇒ [enter a name- **swclampasm** (or one of your choosing)] ⇒ **❑ Use default template ⇒ OK ⇒ Browse ⇒** [select the assembly model (Fig. 20.4) **clamp_assembly.asm** (or the name you used)] ⇒ **Open ⇒ ●Empty ⇒ Landscape ⇒ C ⇒ OK ⇒ Sheets ⇒ Format ⇒ List** [to see the formats available in the current working directory and the standard system formats (Fig. 20.5)] ⇒ **Close**

Continue to complete the process by adding a format from the standard system format directory provided by Pro/E.

File ⇒ New ⇒ ●Drawing ⇒
❑ Use default template ⇒ OK ⇒
Landscape ⇒ Browse

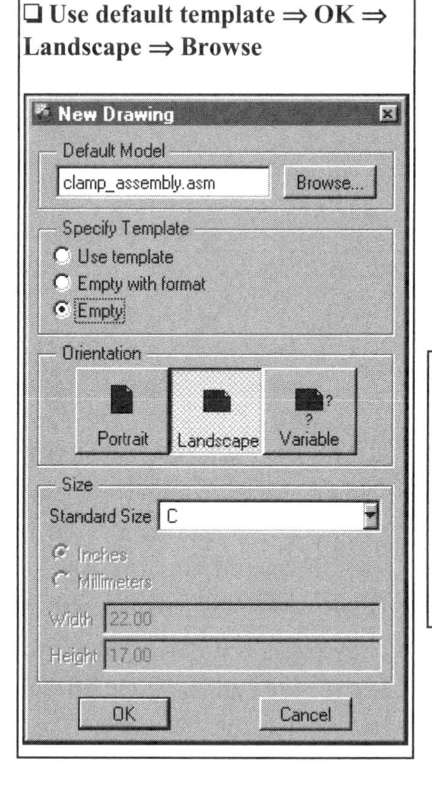

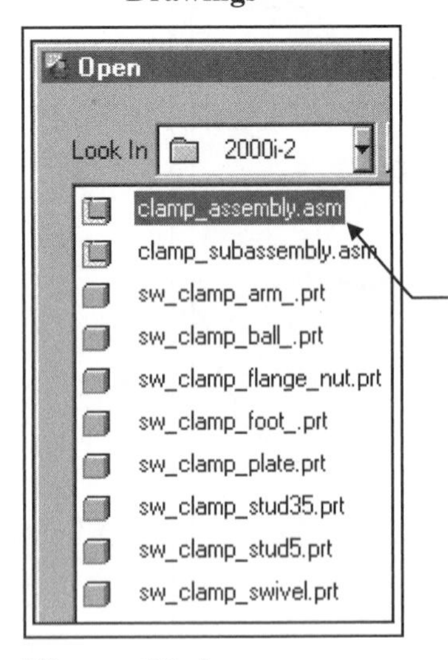

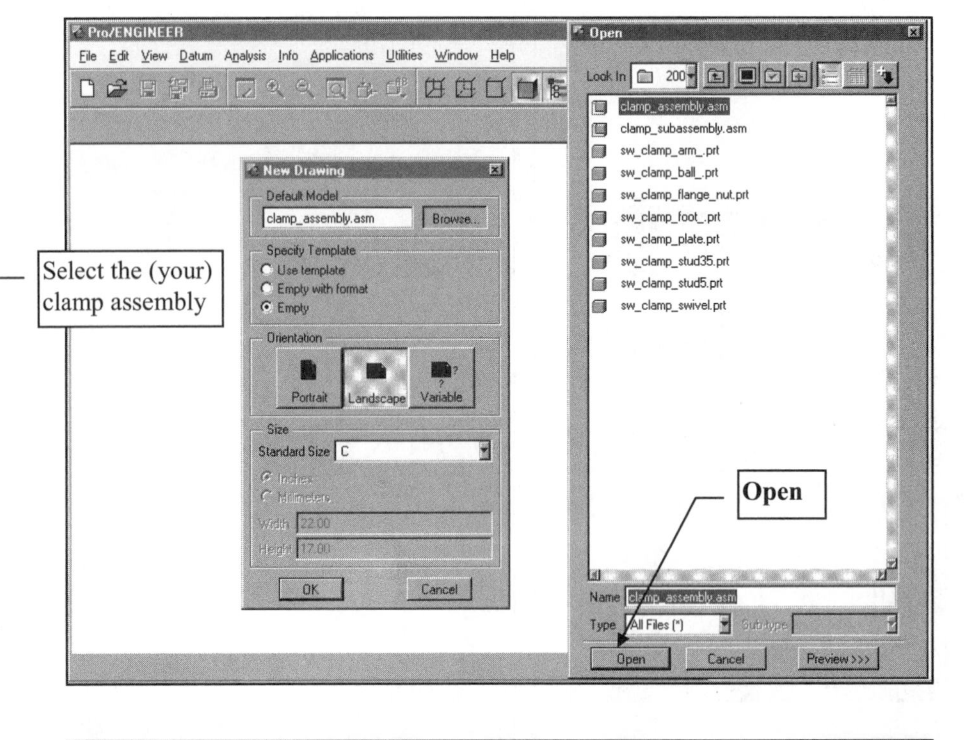

Figure 20.4
Select the Assembly Model for the Drawing

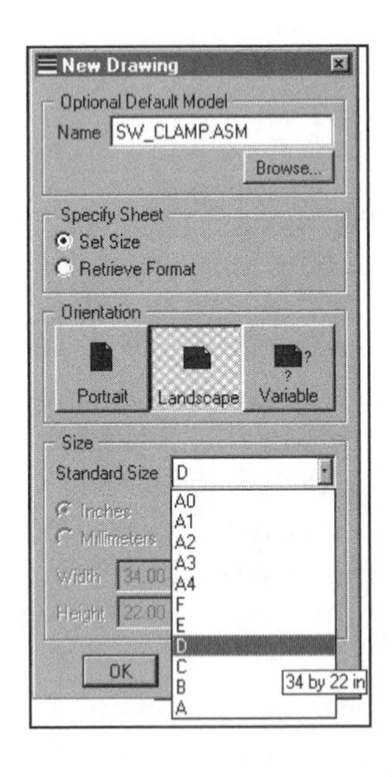

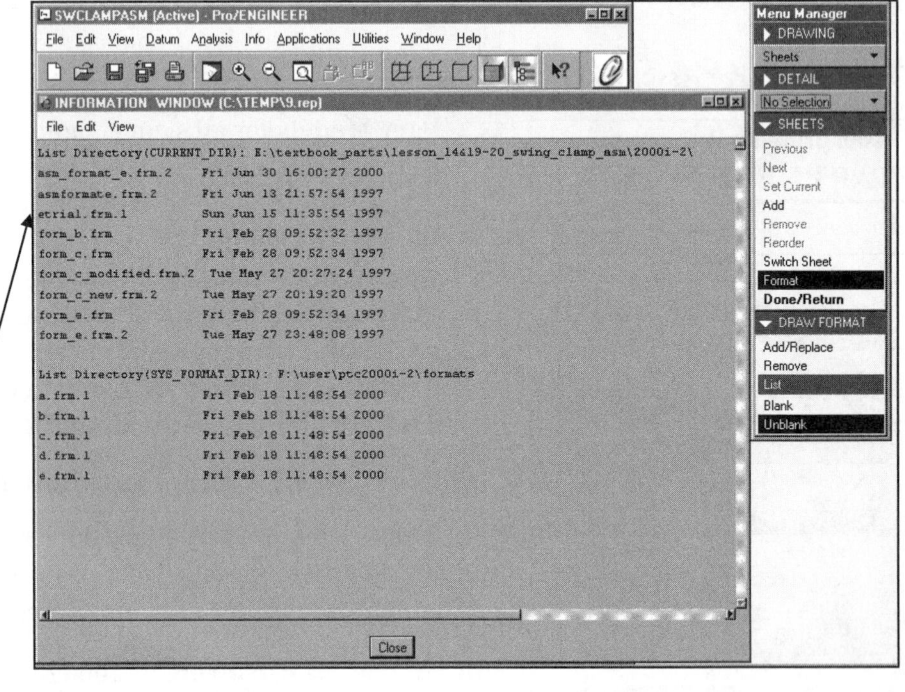

Figure 20.5
Format List

```
List Directory(CURRENT
asm_format_e.frm.2
asmformate.frm.2
etrial.frm.1
form_b.frm
form_c.frm
form_c_modified.frm.2
form_c_new.frm.2
form_e.frm
form_e.frm.2
```

Add/Replace ⇒ ▼ ⇒ **System Formats** (Fig. 20.6) ⇒ [select **d.frm** from list (Fig. 20.7)] ⇒ **Open** (Fig. 20.8) ⇒ **Done/Return** ⇒ **Utilities** ⇒ **Environment** ⇒ **Isometric** ⇒ **OK** ⇒ **Views** ⇒ **Add View** ⇒ **General** ⇒ **Full View** ⇒ **No Xsec** ⇒ **Exploded** ⇒ **Scale** ⇒ **Done** ⇒ [pick the center of the drawing (Fig. 20.8)] ⇒ **✔Default** ⇒ **Done** ⇒ (type **.75**) ⇒ **✔** ⇒ (select next to **Saved Views** to open a list of saved views) ⇒ [select **EXPLODED 1** from the list (Fig. 20.9)] ⇒ **Set** ⇒ **OK** ⇒ **Modify View** ⇒ **Change Scale** ⇒ [pick the view (Fig. 20.10)] ⇒ (type **1** as the new scale) ⇒ **✔** ⇒ **Done Sel** ⇒ **Done/Return** (Figure 20.11) ⇒ **File** ⇒ **Save** ⇒ **✔**

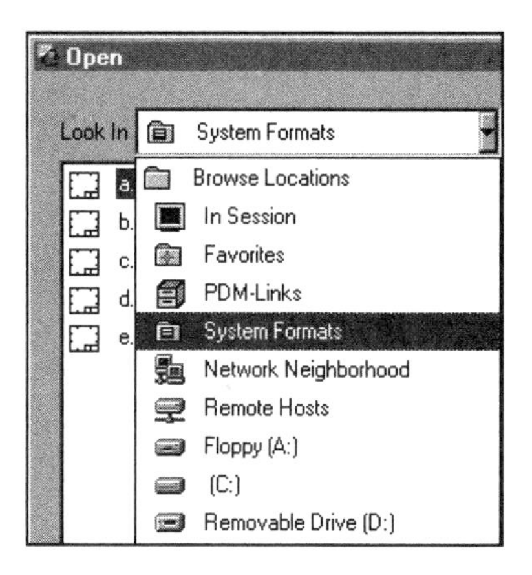

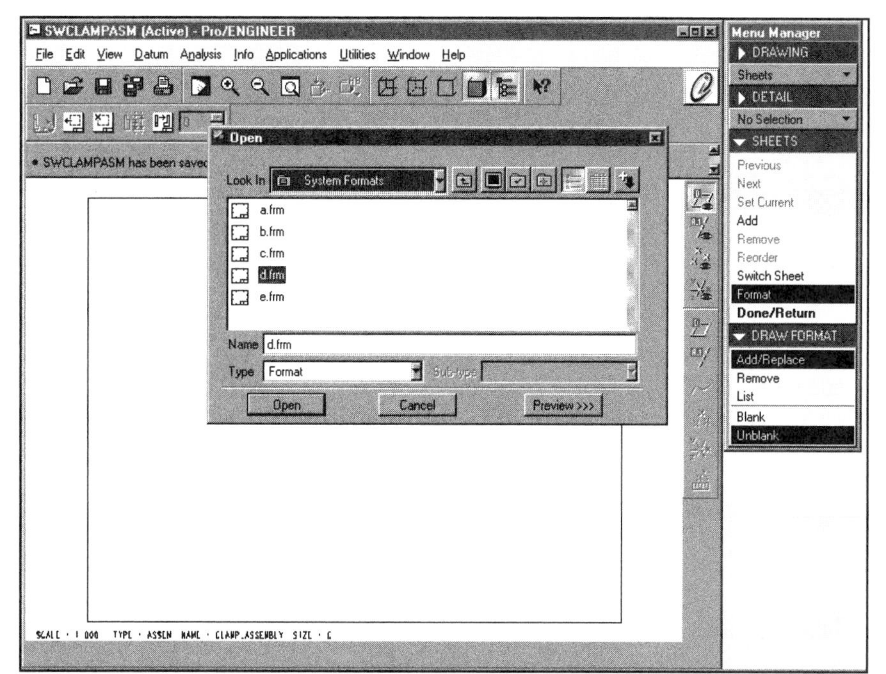

Figure 20.6
System Formats

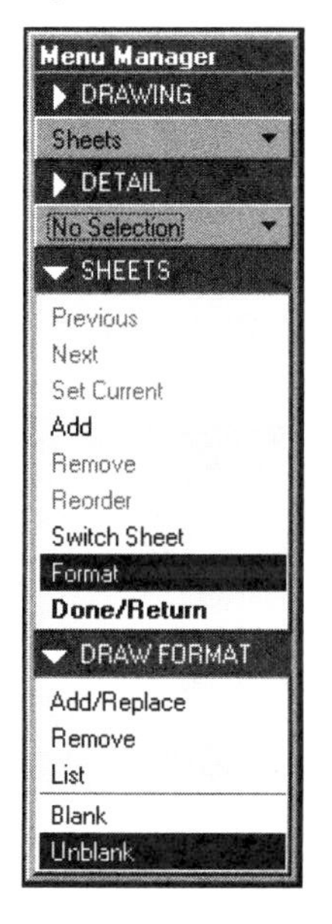

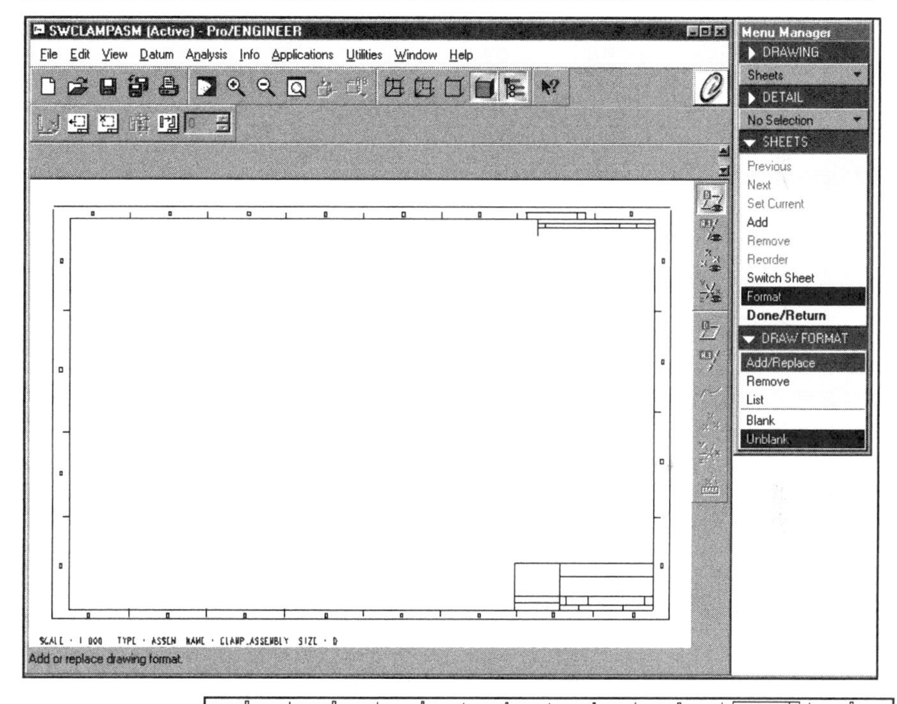

Figure 20.7
Format **d.frm**

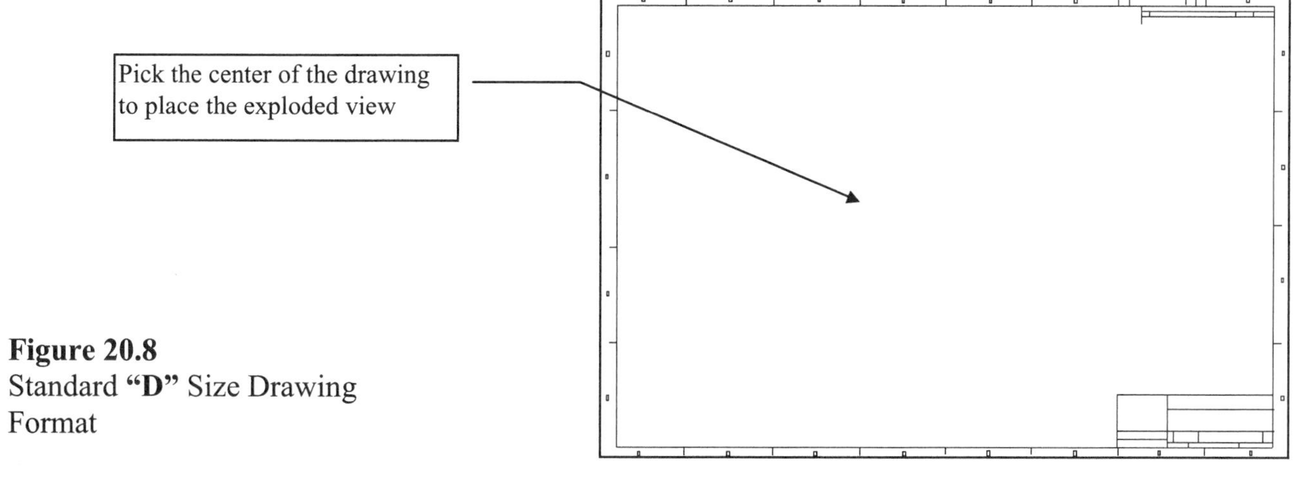

Figure 20.8
Standard **"D"** Size Drawing Format

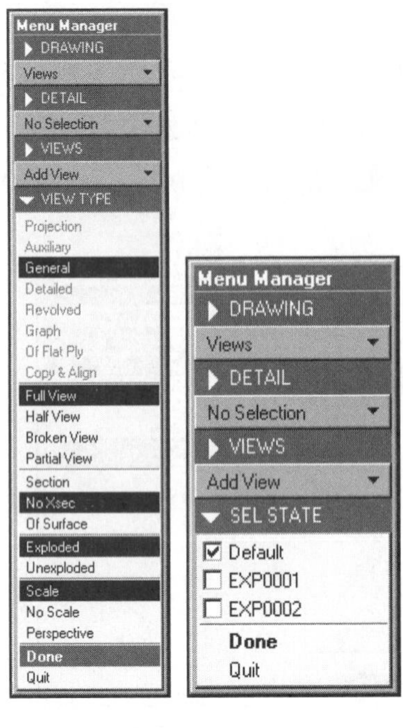

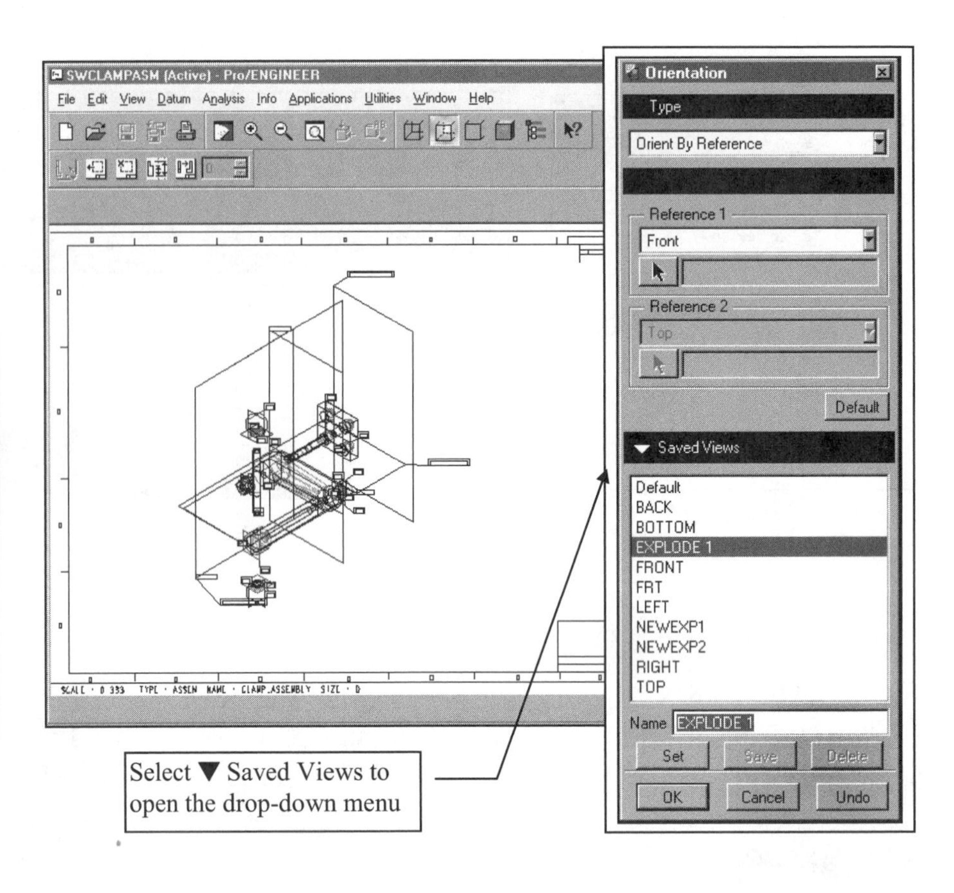

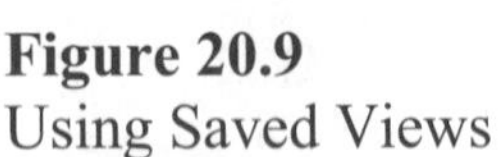

Figure 20.9
Using Saved Views

Figure 20.10
Modifying the View Scale

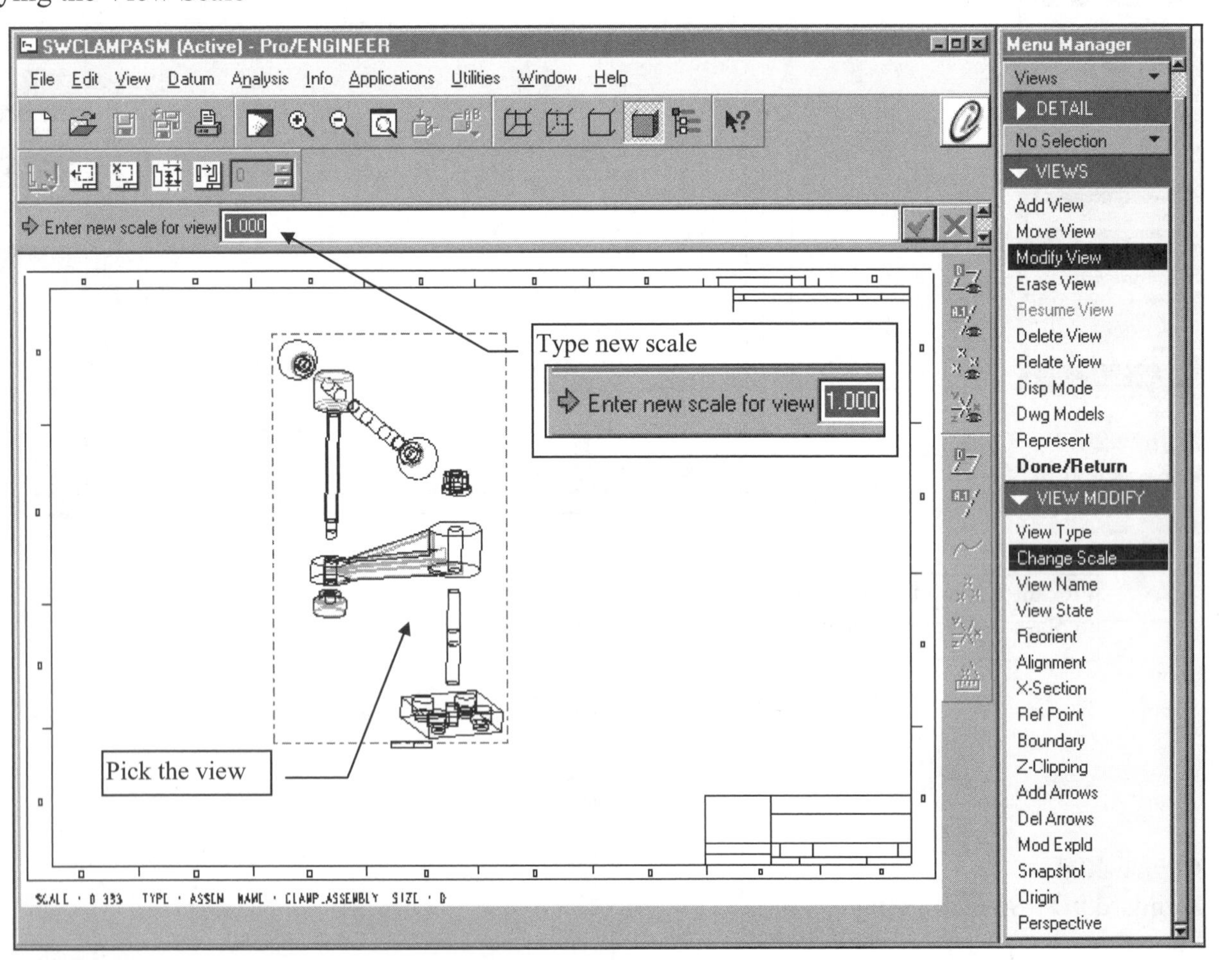

This drawing sheet will have only one view (Fig. 20.11). Add a *second sheet* with a different exploded view orientation (Fig. 20.12):

File ⇒ **Delete** ⇒ **Old Versions** ⇒ ✔ ⇒ **Sheets** ⇒ **Add** ("Sheet number **2** has been added to the drawing") ⇒ **Done/Return** ⇒ **Views** ⇒ **Add View** ⇒ **General** ⇒ **Full View** ⇒ **No Xsec** ⇒ **Exploded** ⇒ **Scale** ⇒ **Done** ⇒ (pick the center of the drawing) ⇒ **✔Default** ⇒ **Done** ⇒ [type **1** as the scale **enter** (or pick the green ✔)] ⇒ **Default** (from the Saved Views list in the Orientation dialog box) ⇒ **Set** [to keep the default orientation **(Default)**] ⇒ **OK** ⇒ **Done/Return**

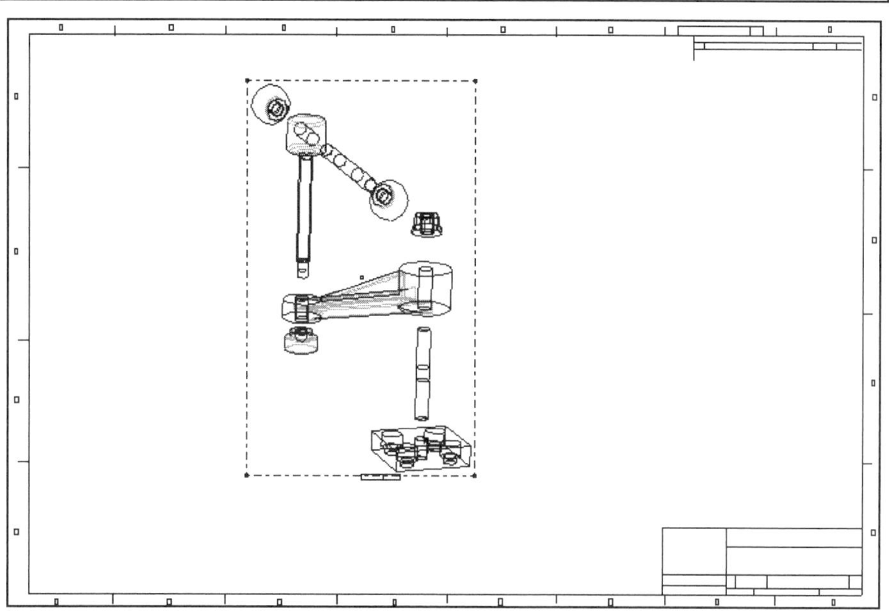

Figure 20.11
One Exploded View on Drawing

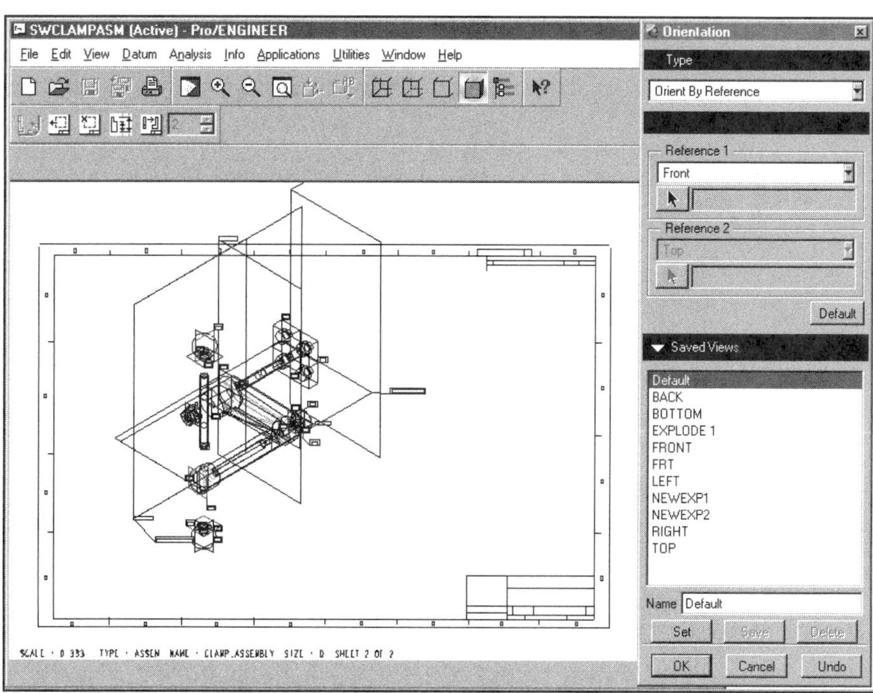

Figure 20.12
Adding a Second Sheet and a New Exploded View

Next, add another sheet and a different model. Use the **Swing Clamp subassembly** (Fig. 20.13). Modify the scale of the drawing after the view is placed. Choose the following commands:

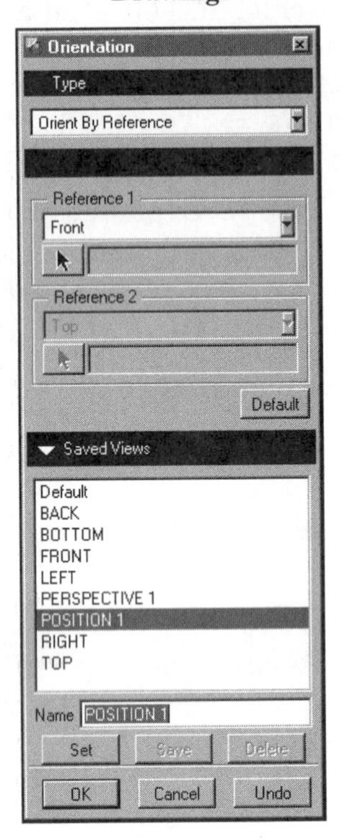

Figure 20.13
Sheet 3 with Exploded View of Subassembly

File ⇒ **Save** ⇒ ✔ ⇒ **Sheets** ⇒ **Add** ("Sheet number **3** has been added to the drawing") ⇒ **Format** ⇒ **Add/Replace** ⇒ ▼ ⇒ **System Formats** ⇒ **c.frm** ⇒ **Open** ⇒ **Done/Return** ⇒ **Views** ⇒ **Dwg Models** ⇒ **Add Model** ⇒ [pick your subassembly **(clamp_subassembly.asm)** from the directory list] ⇒ **Open** ⇒ **Done/Return** ⇒ **Views** ⇒ **Exploded** ⇒ **Scale** ⇒ **Done** ⇒ (pick the center of the new drawing sheet) ⇒ ✔**Default** ⇒ **Done** ⇒ (type **.75** as the scale) ⇒ ✔ ⇒ [pick a saved view name from the list (**Position 1** was used here)] ⇒ **Set** ⇒ **OK** ⇒ **Done/Return** ⇒ **Views** ⇒ **Modify View** ⇒ **Change Scale** ⇒ (pick the view) ⇒ (type **1.125** as the scale) ⇒ ✔ ⇒ **Done Sel** ⇒ **Done/Return** ⇒ **View** ⇒ **Repaint**

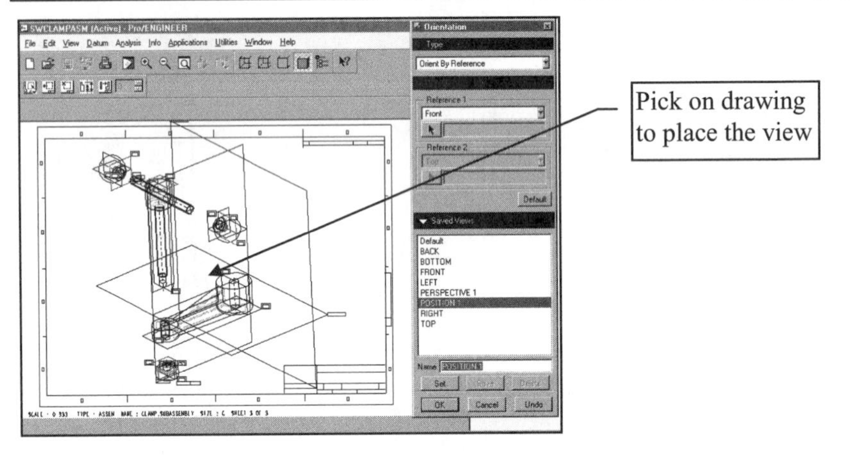

Show all the axes for the components on the first sheet. First, switch between sheets. Choose the following commands (Fig. 20.14):

Sheets ⇒ **Next** (SHEET 1 OF 3) ⇒ **Done/Return** ⇒ **Show/Erase** ⇒ **Show** (radio button) ⇒ **----A_1** (axis radio button) ⇒ **Show All** (radio button) ⇒ **Yes** (radio button) ⇒ (you may need to choose **Accept All**) ⇒ **Close** (radio button) ⇒ (press and hold down the right mouse button anywhere on the drawing, then **Modify Item)** ⇒ (pick an axis to be modified) ⇒ [modify each centerline (axis) to stretch between components that are in line] ⇒ **Done Sel** ⇒ **Repaint** (icon)

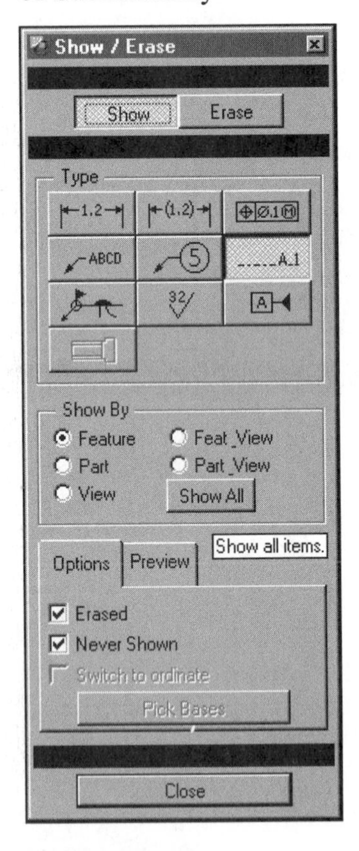

Figure 20.14
Showing and Modifying Axes

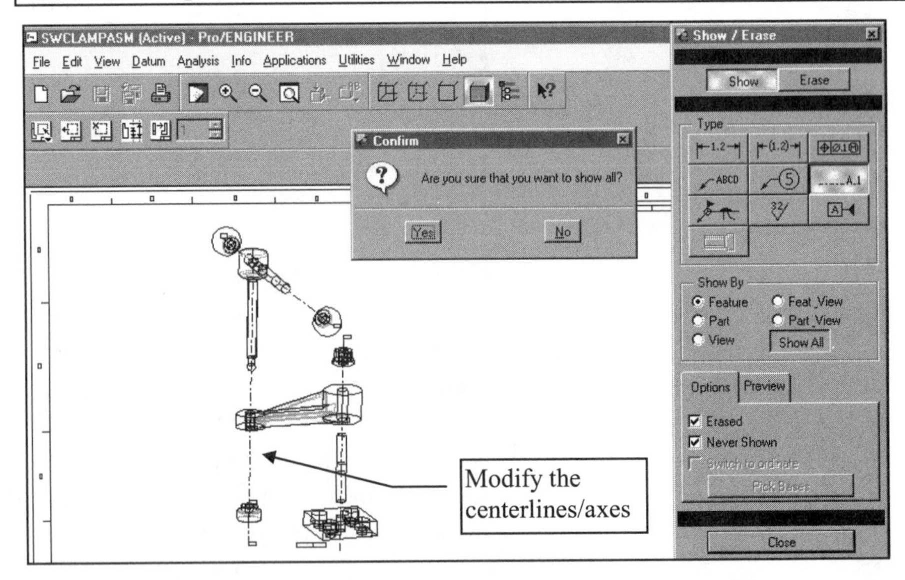

NOTE

To change the defaults on balloons:

Advanced ⇒ Draw Setup ⇒

drawing_text_height ***.50***
max_balloon_radius ***.50***
min_balloon_radius ***.50***

Add/Change ⇒ Apply ⇒ Close

To add balloons to one of the sheets not using a parametric title block, you need to create each balloon separately. Before creating balloons, change the **.dtl** file so that it has **.500** radius balloons and **.500** height for lettering (Fig. 20.15). Create balloons for the components on the first sheet. The balloons added must correspond to the component balloons on the assembly drawing completed in Lesson 19. Choose the following commands:

Create ⇒ Balloon ⇒ Leader ⇒ Make Note ⇒ (pick Swing Clamp Arm) **⇒ Done Sel ⇒ Done ⇒** (pick location for note) **⇒** (type **1**) **⇒ ✔ ⇒ ✔ ⇒ Make Note ⇒** [continue until all *seven* components are ballooned (Figures 20.15 and 20.16)] **⇒ Done/Return**

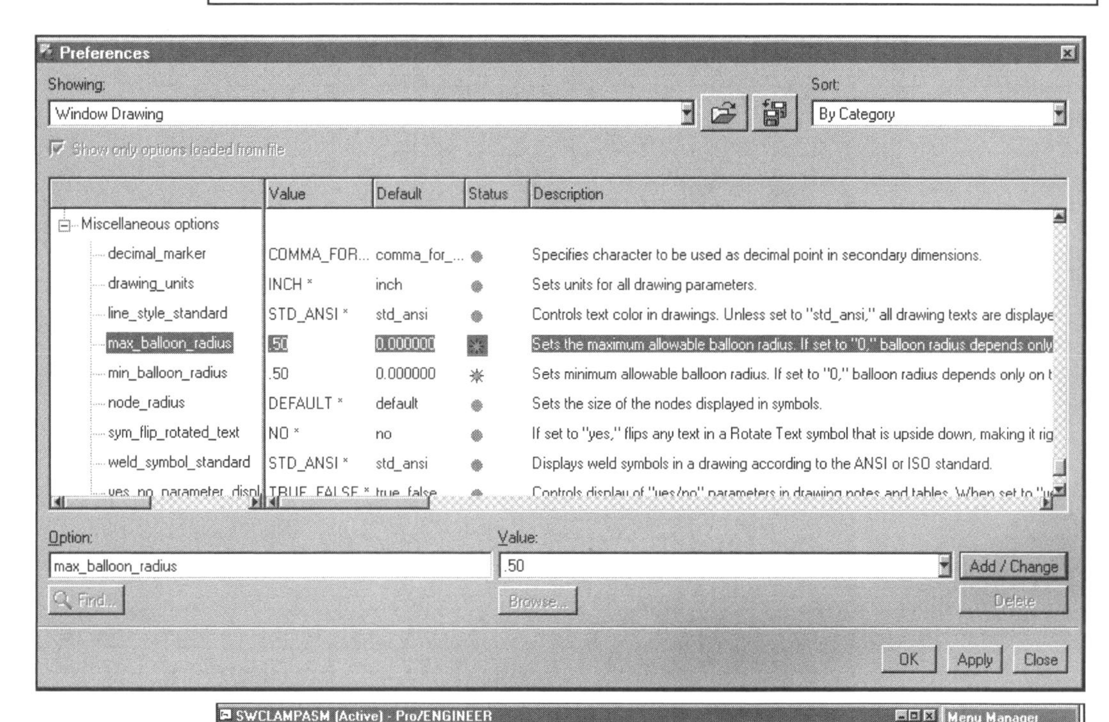

Figure 20.15
Draw Setup

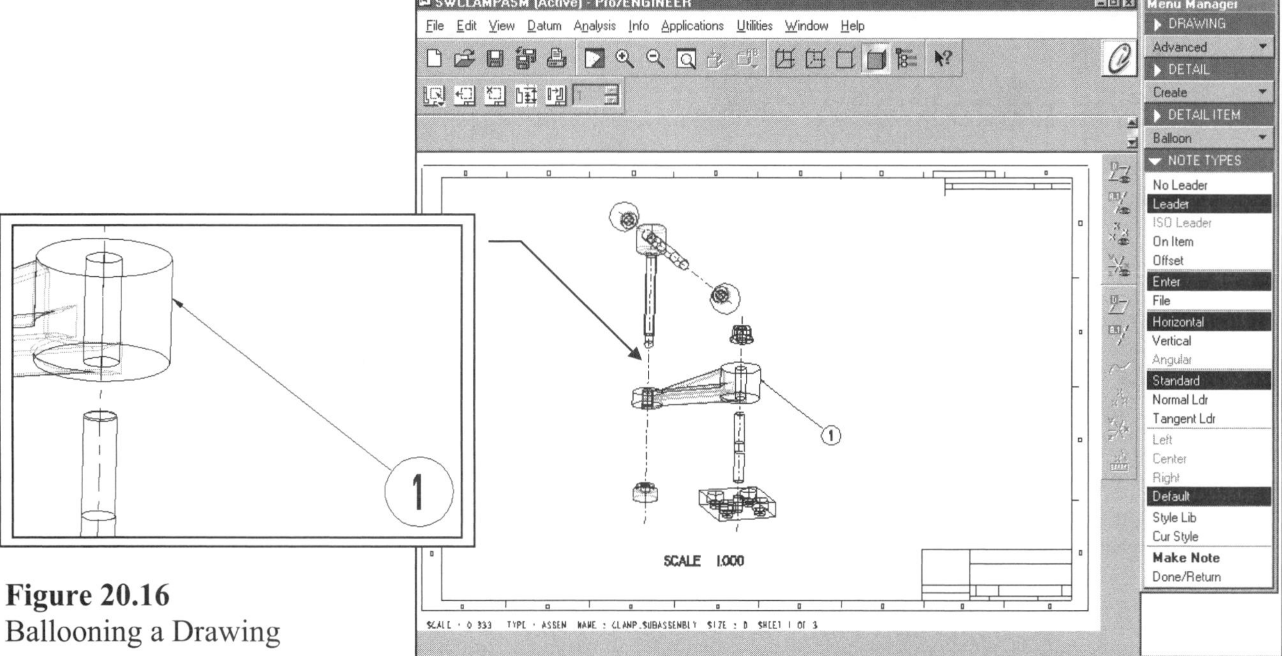

Figure 20.16
Ballooning a Drawing

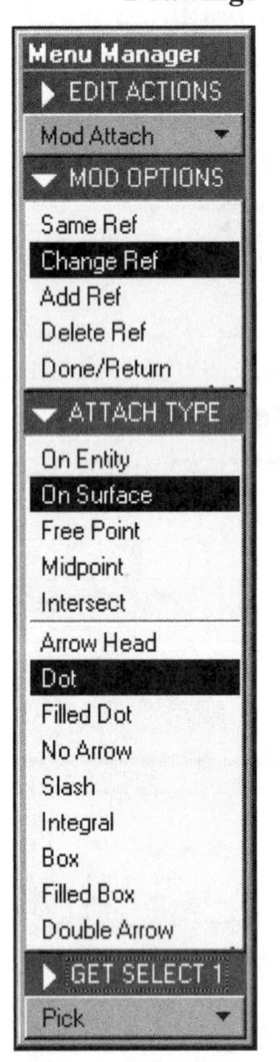

Figure 20.17
Modifying Balloons

After the ballooning is complete, move the balloons and their attachment points to clean up the drawing. Choose the following commands to modify the attachment point (from edge to surface) and change the arrow to a dot (Fig. 20.17):

Modify Item (right mouse button) ⇒ (pick balloon **3**) ⇒ **Mod Attach** ⇒ **On Surface** ⇒ **Dot** ⇒ (pick a place on the ball's surface) ⇒ **Done/Return** ⇒ **Done Sel**

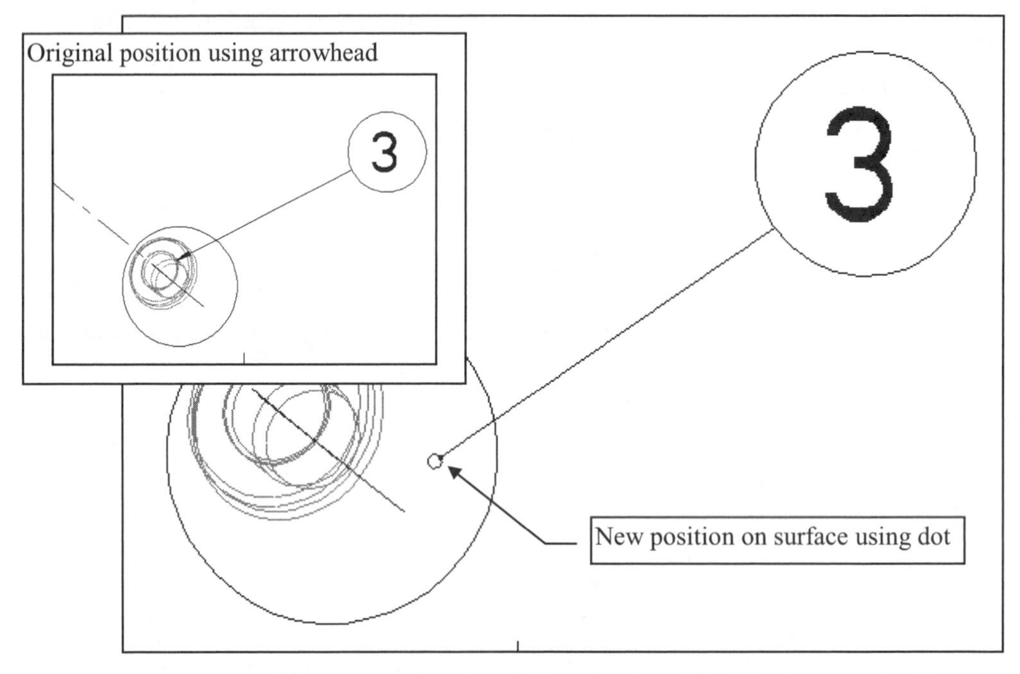

Complete the drawing by filling in the title block or switching to the parametric format you created in Lesson 19 (Fig. 20.18). Change the view display to show as:

Views ⇒ **Disp Mode** ⇒ **View Disp** ⇒ (pick the view) ⇒ **Done Sel Hidden Line** ⇒ **Tan Dimmed** ⇒ **Done** ⇒ **Done Sel** ⇒ **Done/Return** ⇒ **File** ⇒ **Save** ⇒ ✔

Always remember to turn off the datum planes, axes, points, and the coordinate system when in the Drawing Mode

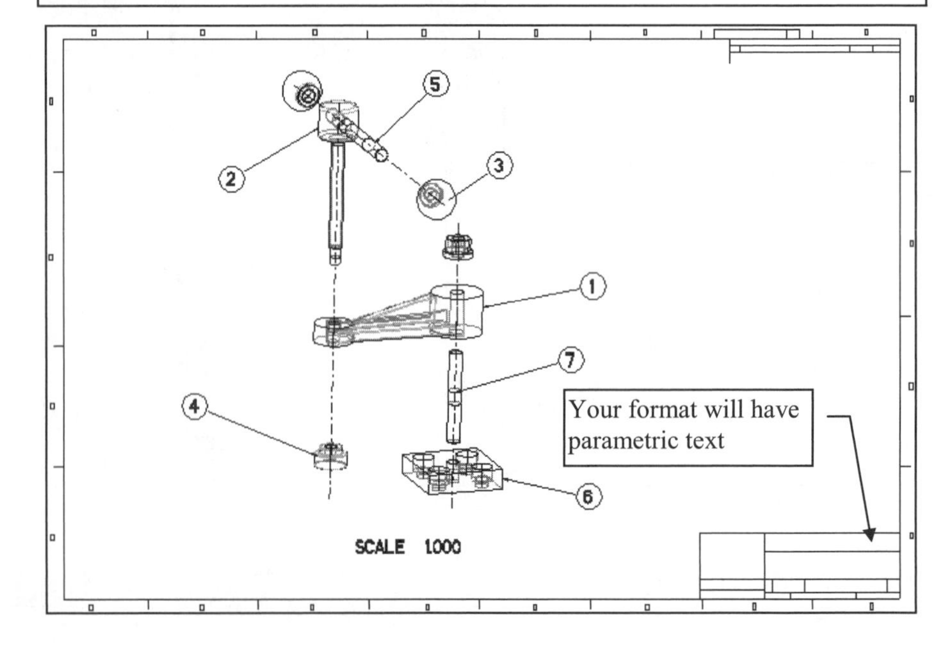

Figure 20.18
Ballooned Drawing

You can modify the positions of the exploded components for this drawing without going back to the Assembly mode and editing the exploded model (Figs. 20.19 and 20.20).

Choose the following commands to modify the drawing:

Views ⇒ **Modify View** ⇒ **Mod Expld** ⇒ (pick the view) ⇒ **Redefine** ⇒ **Position** ⇒ Motion Reference **Entity/Edge** ⇒ (select a *vertical* axis) ⇒ (pick the Arm) ⇒ (slide the Arm to a new position) ⇒ (pick to place) ⇒ [Continue moving the components until they appear as in Fig. 20.19. To move the Balls and long Stud, you will need to choose **Entity/Edge** again, ↖, and a different axis (in this case a *horizontal* axis).] ⇒ **Done Sel** ⇒ **OK** ⇒ **Done/Return** ⇒ **Done/Return** ⇒ **Done/Return** ⇒ (clean up balloon placement) ⇒ (add the missing balloon for the Nut) ⇒ **File** ⇒ **Save** ⇒ ✔ ⇒ **File** ⇒ **Delete** ⇒ **Old Versions** ⇒ ✔

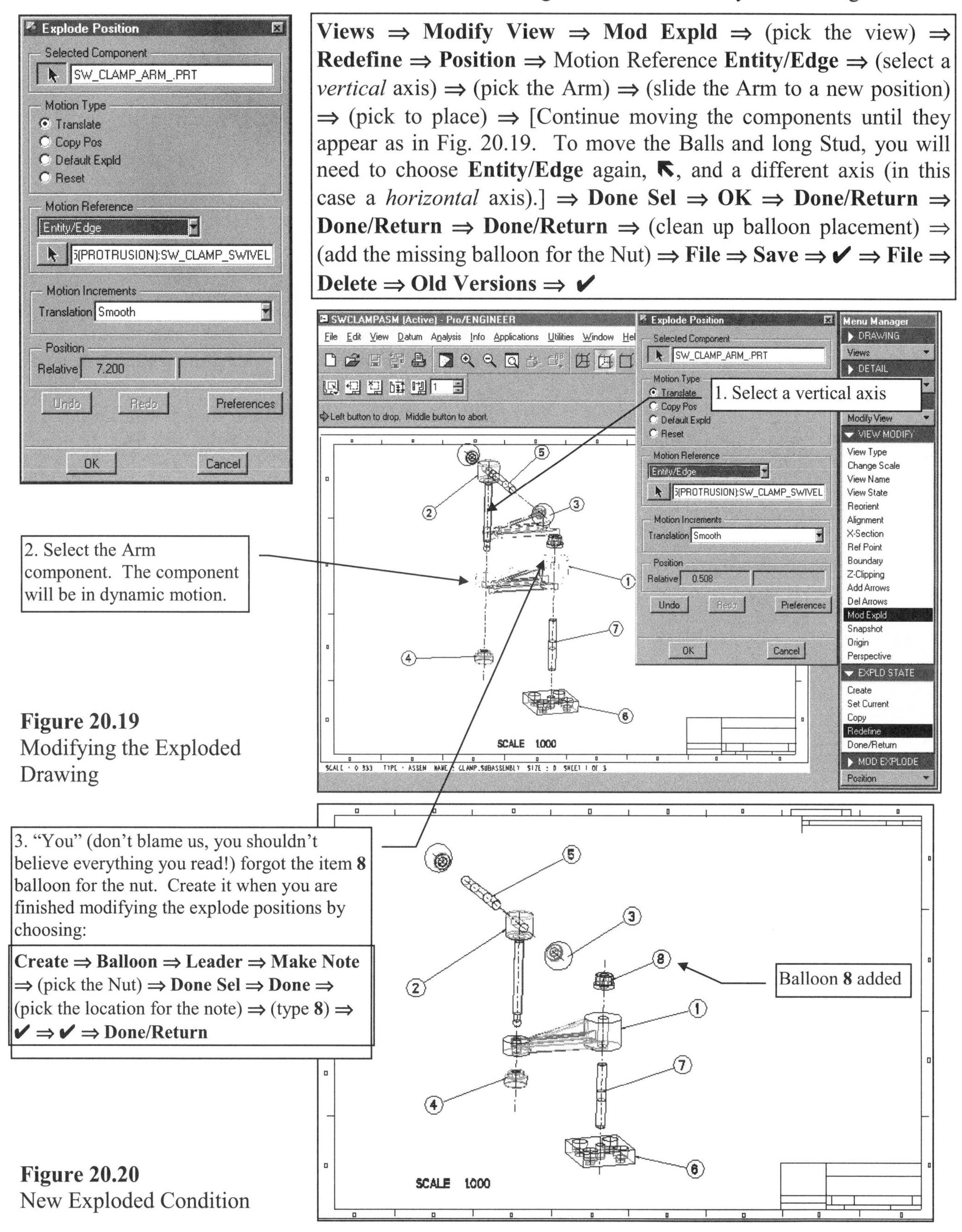

Figure 20.19
Modifying the Exploded Drawing

3. "You" (don't blame us, you shouldn't believe everything you read!) forgot the item **8** balloon for the nut. Create it when you are finished modifying the explode positions by choosing:

Create ⇒ **Balloon** ⇒ **Leader** ⇒ **Make Note** ⇒ (pick the Nut) ⇒ **Done Sel** ⇒ **Done** ⇒ (pick the location for the note) ⇒ (type **8**) ⇒ ✔ ⇒ ✔ ⇒ **Done/Return**

Figure 20.20
New Exploded Condition

Lesson 20 Project

Exploded Coupling Assembly Drawing

Exploded Coupling Assembly Drawing

Create a "complete documentation package" for the Coupling Assembly (Figs. 20.21 through 20.24). A documentation package includes all the models and drawings required to manufacture the parts and assemble the components. Some of the items listed here have been created in other lessons. Create or extract existing models and drawings, and plot the following:

Part Models for all coupling assembly components
Detail Drawings for each nonstandard component--for example, the Coupling Shaft
Detail Drawings for standard components that have been altered; show only the dimensions required to alter the component--for example, the socket head cap screw in the coupling assembly
Assembly Drawing and **Parts List (BOM)** using standard orthographic ballooned views
Exploded Assembly Drawing of the ballooned assembly
Exploded Subassembly Drawing of the ballooned subassembly

Figure 20.21
Exploded Coupling Assembly Drawing

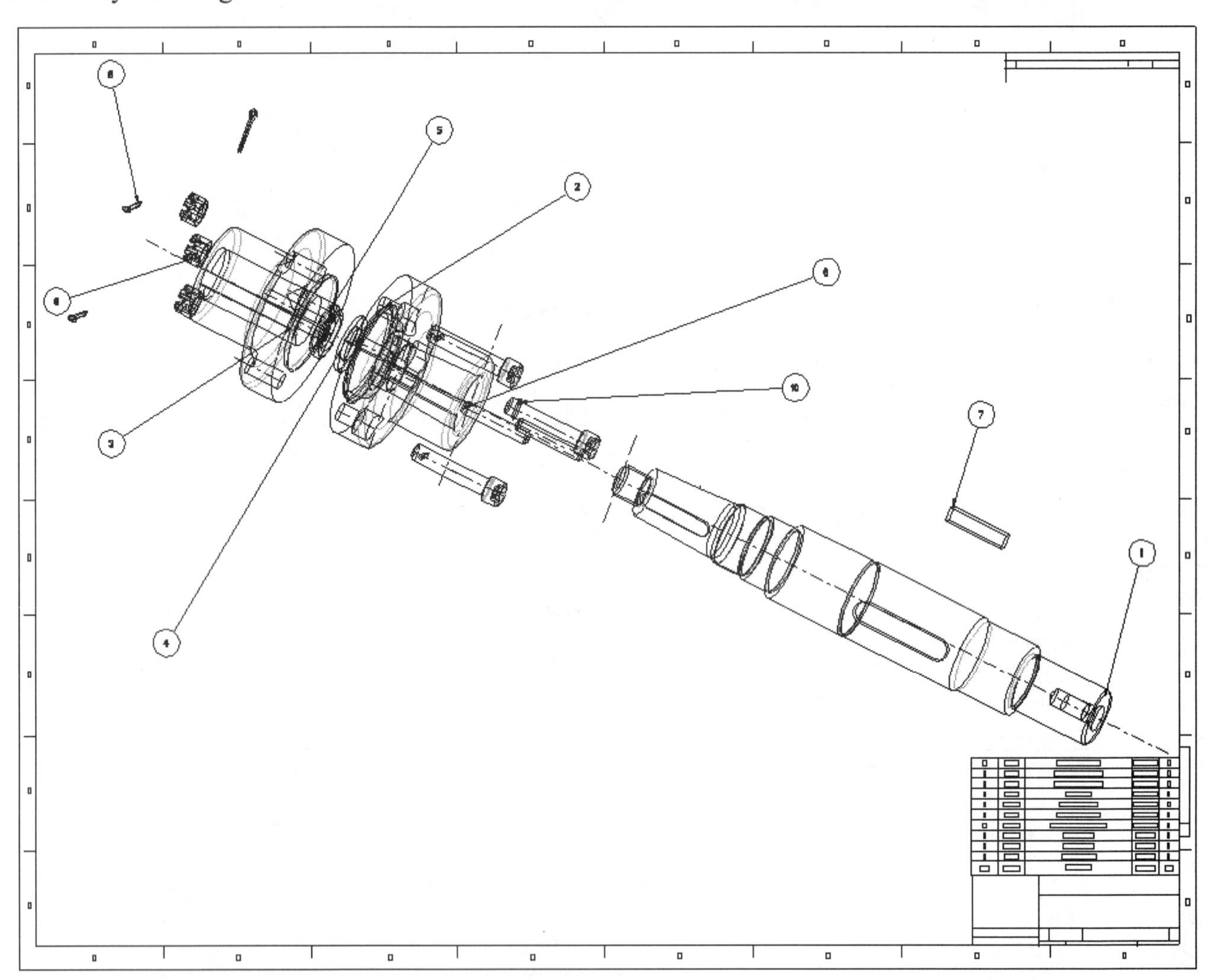

Figure 20.22 Coupling Assembly BOM

ITEM	PT NUM	DESCRIPTION	MATERIAL	QTY
10	110-2CS	SOC HD CAP SCREW	PURCHASED	3
9	109-2SN	HEX SLOT NUT 16 X 2	PURCHASED	3
8	108-2CP	COTTER PIN .150 X 1.25	PURCHASED	3
7	107-2KY	KEY 14 X 61	PURCHASED	1
6	106-2DW	DOWEL 12OD X 70	PURCHASED	2
5	105-2HN	HEX NUT M30 X 3.5	PURCHASED	1
4	104-2WA	WASHER 33ID X 50OD X 4	PURCHASED	1
3	103-2CP2	COUPLING TWO	1040 CRS	1
2	102-2CP1	COUPLING ONE	1040 CRS	1
1	101-2SH	COUPLING SHAFT	1020 CRS	1

Figure 20.23 Coupling Assembly, Close-up

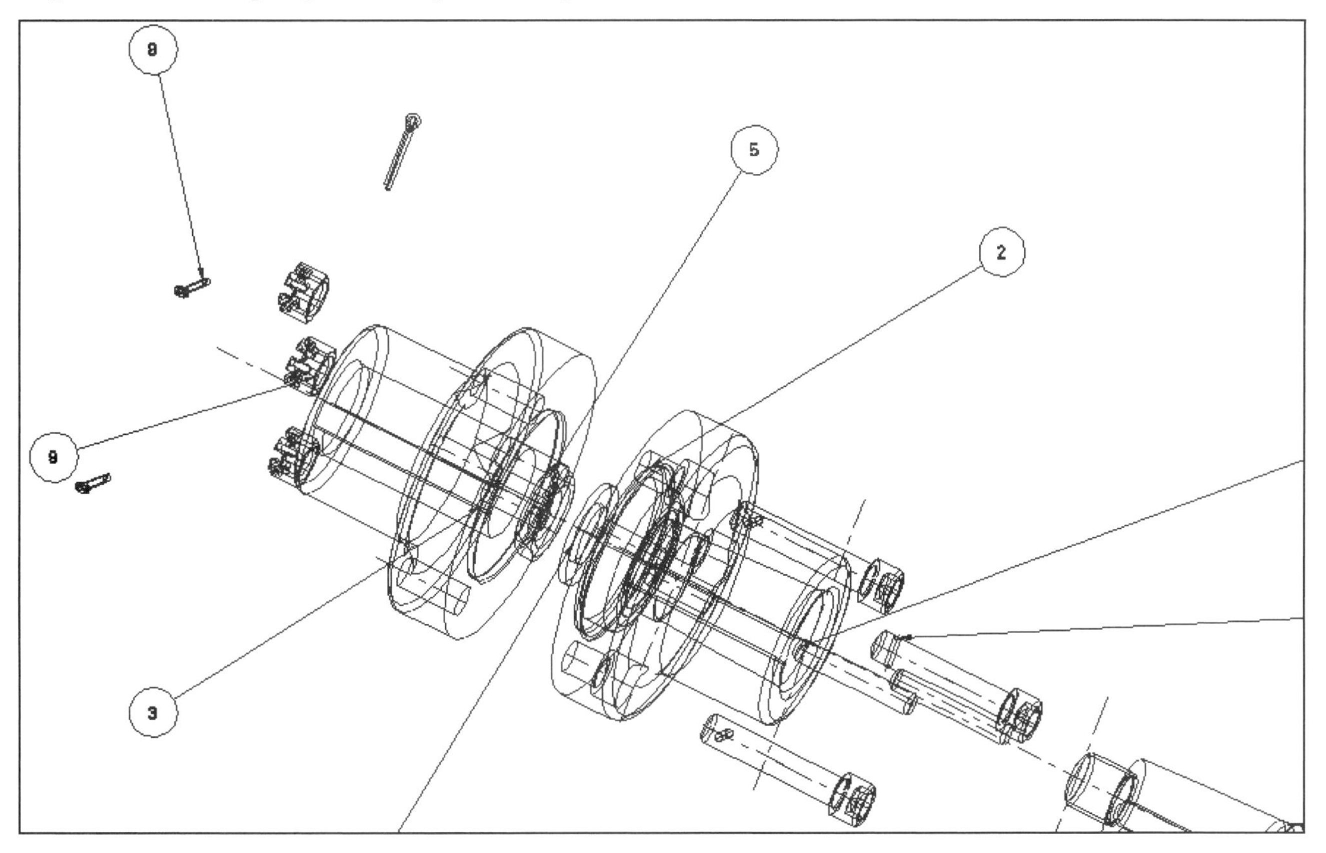

L20-14 **Exploded Assembly Drawings**

Figure 20.24 Coupling Assembly

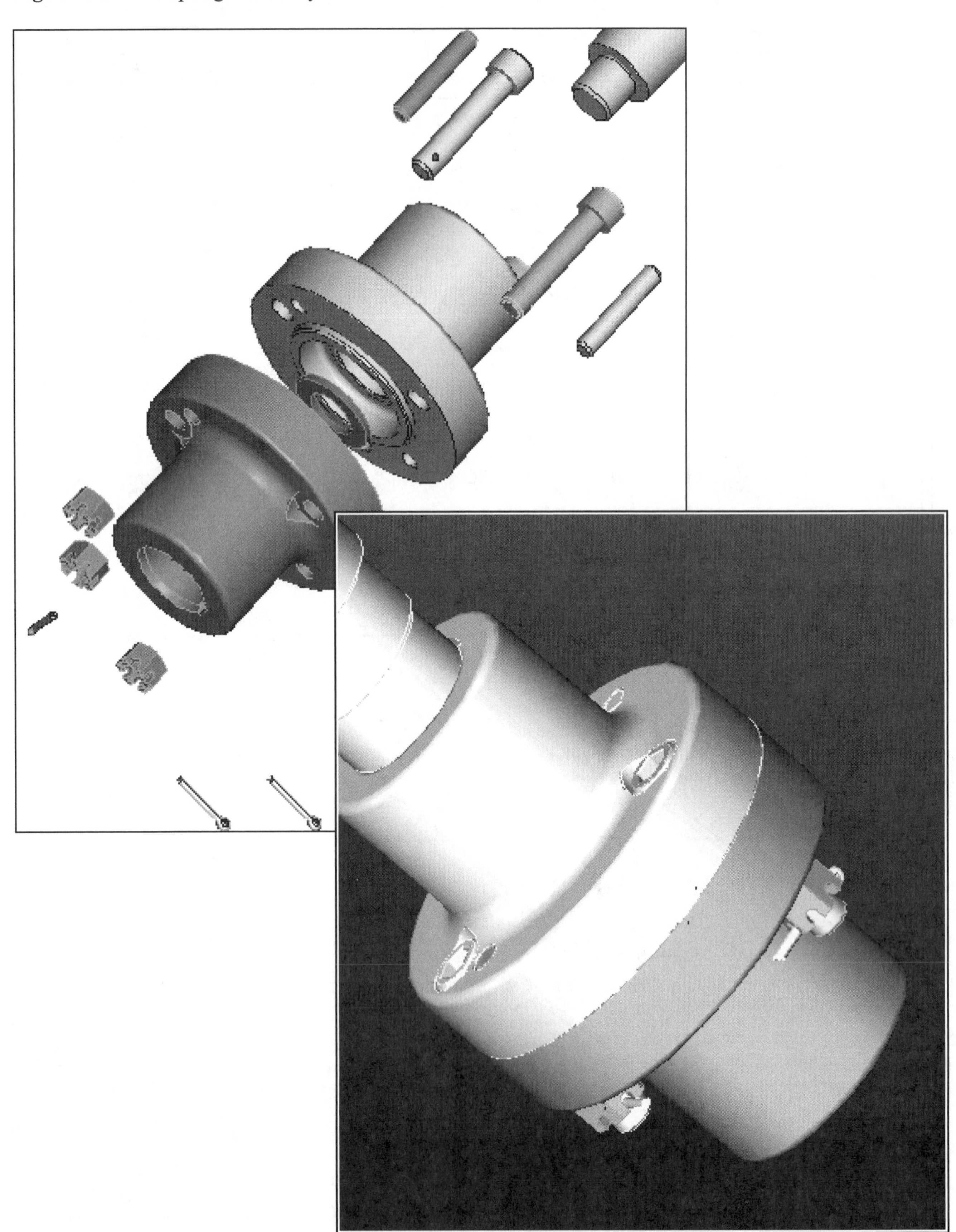

Part Four

Advanced Capabilities

Lesson 21 User-Defined Features and Family Tables
Lesson 22 Pro/NC
Lesson 23 Pro/SHEETMETAL

ADVANCED CAPABILITIES

User-Defined Features and **Family Tables, Pro/NC,** and **Pro/SHEETMETAL** are covered in Part Four. With each new change of software from PTC and subsequent revision of the text, we hope to add one lesson.

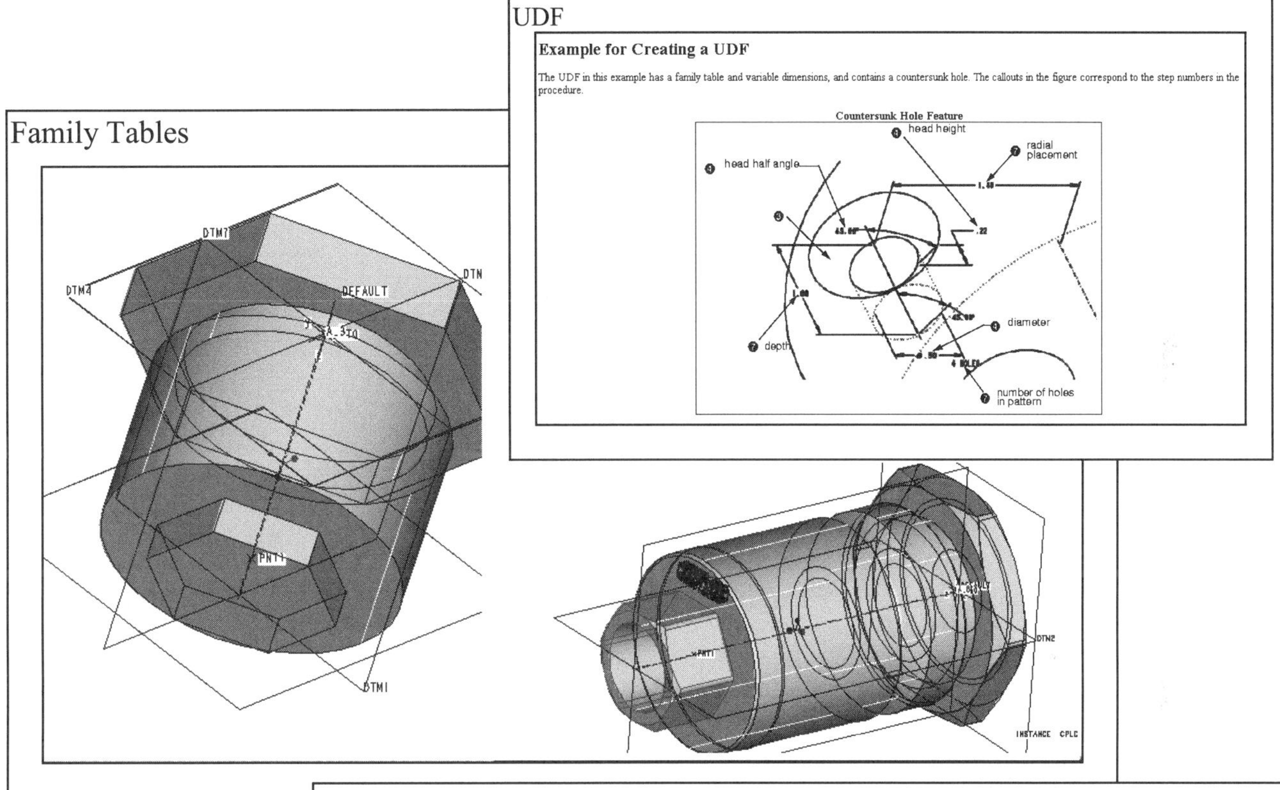

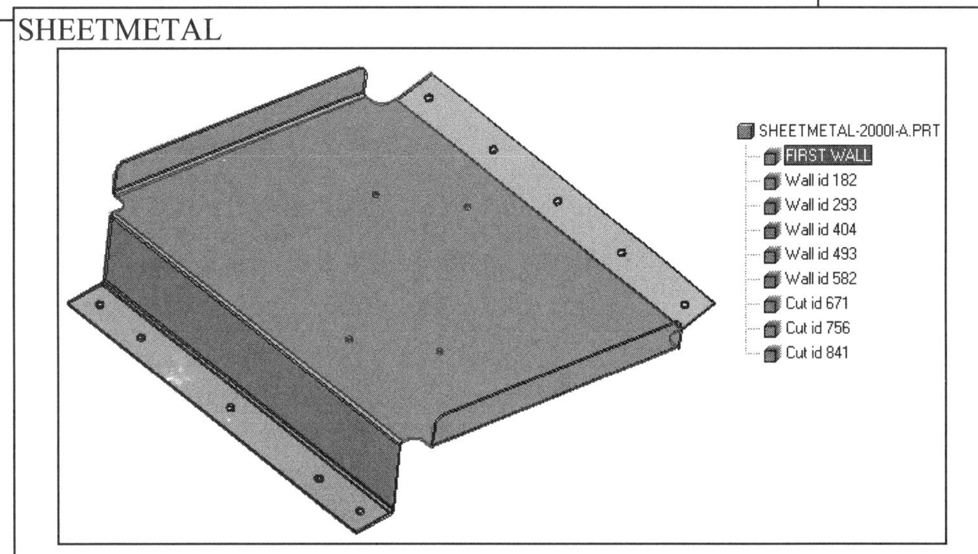

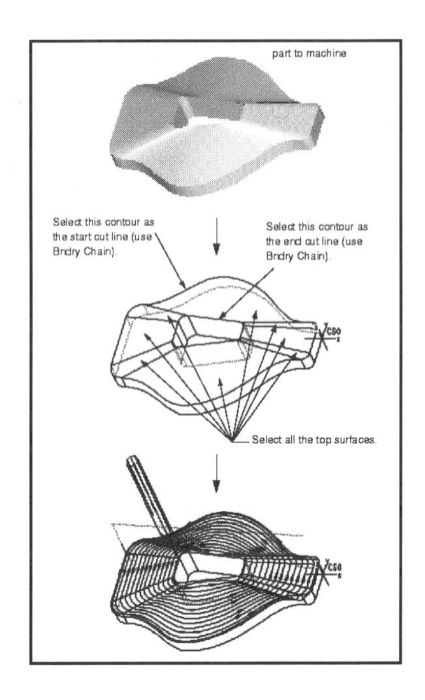

Pro/NC

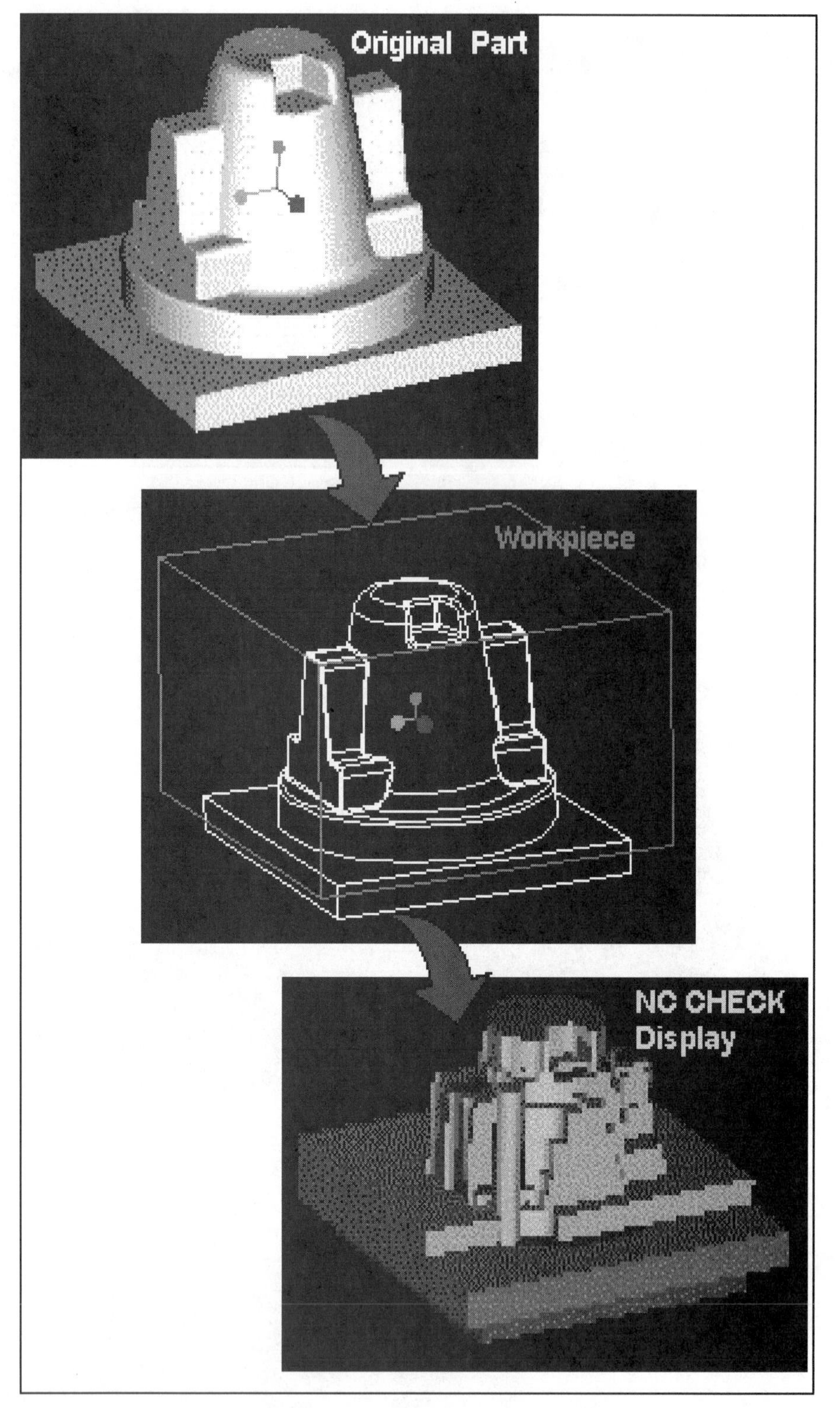

COAch for Pro/ENGINEER
Pro/NC

Lesson 21

User-Defined Features and Family Table

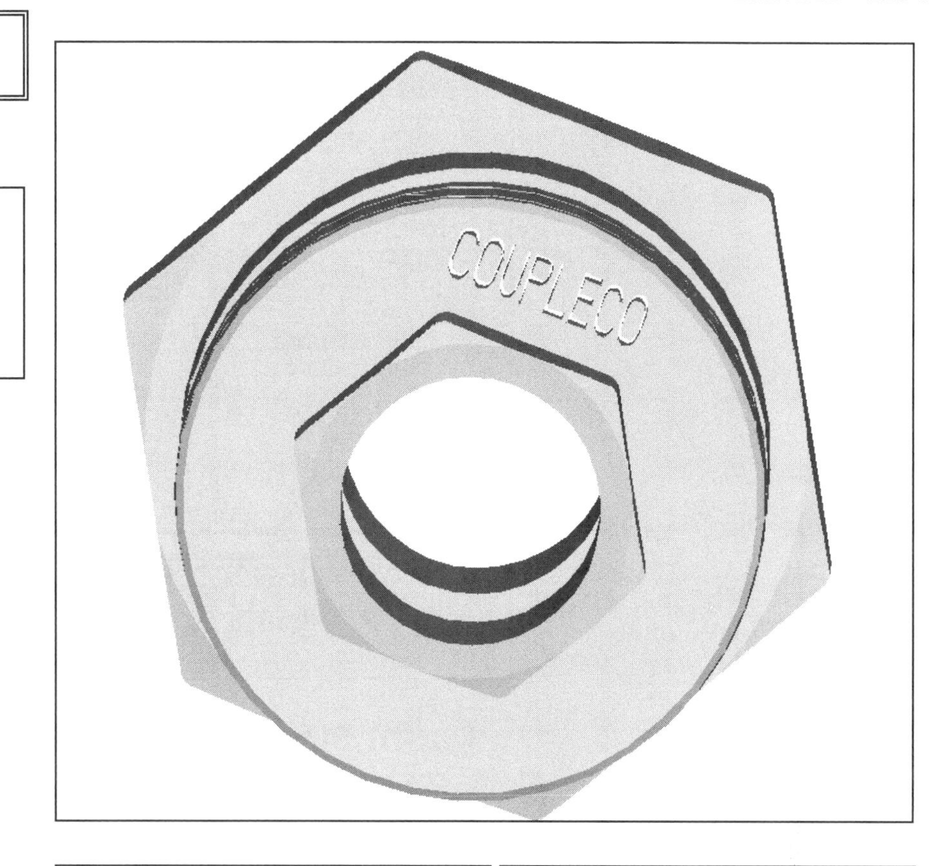

Figure 21.1
Coupling Fitting

☑ *EGD REFERENCE*
Fundamentals of Engineering Graphics and Design
by L. Lamit and K. Kitto
Read Chapter 17.
See pages 675-678.

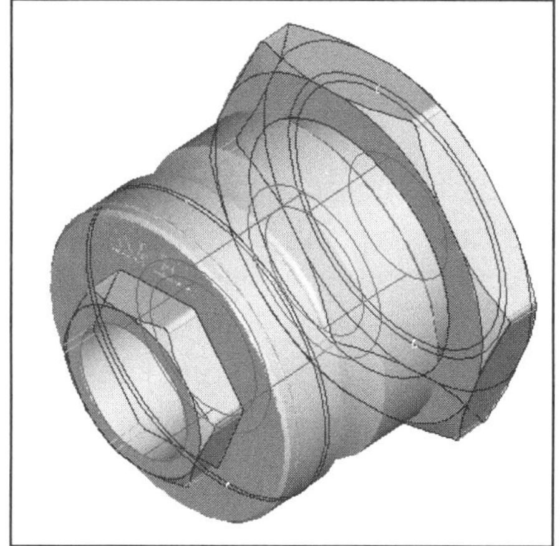

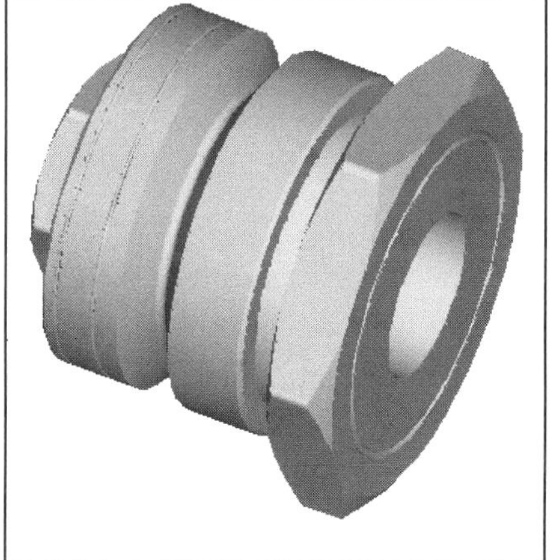

Figure 21.2
COUPLECO Coupling Fitting

COAch™ for Pro/ENGINEER

If you have **COAch for Pro/ENGINEER** on your system, go to SEARCH and do the Segments shown in Figures 21.6, 21.7, and 21. 9.

OBJECTIVES

1. **Understand User-Defined Features (UDFs)**
2. **Use User-Defined Features from Pro/LIBRARY**
3. **Comprehend the power of Family Tables, Generic Parts, and Instances**
4. **Create a Family Table for a part**

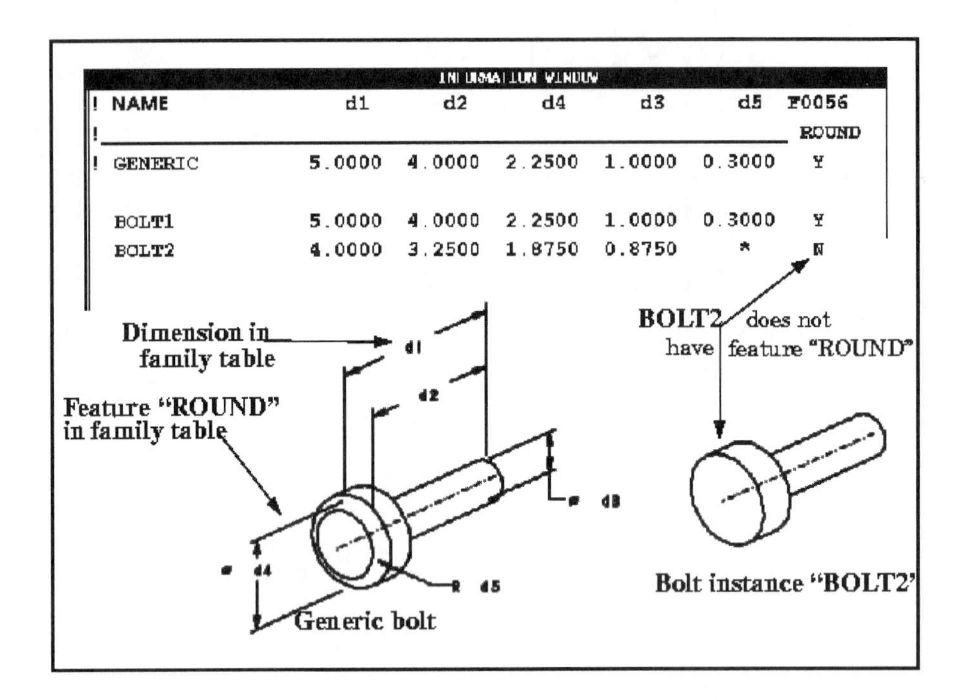

! NAME	d1	d2	d4	d3	d5	F0056 ROUND
! GENERIC	5.0000	4.0000	2.2500	1.0000	0.3000	Y
BOLT1	5.0000	4.0000	2.2500	1.0000	0.3000	Y
BOLT2	4.0000	3.2500	1.8750	0.8750	*	N

Figure 21.3
Family Table and Library Part

USER-DEFINED FEATURES AND FAMILY TABLES

User-Defined Features (UDFs) allow you to store a set of standard features, parts, or assemblies in a library for later use (Figs. 21.1 and 21.2). UDFs can then be added to existing parts (if they are features) or to existing assemblies (if they are parts or subassemblies).

A **Family Table** (Fig. 21.3) is a powerful tool for capturing engineering intelligence and promoting data reuse. Family Tables provide an automated and simple way for engineers to multiply a model into logical instances that reference the generic. This process delivers many models, which capture all the intelligence of the original, update as the generic updates, and use minimal disk space. Family Tables can be edited using Microsoft Excel (Fig. 21.4).

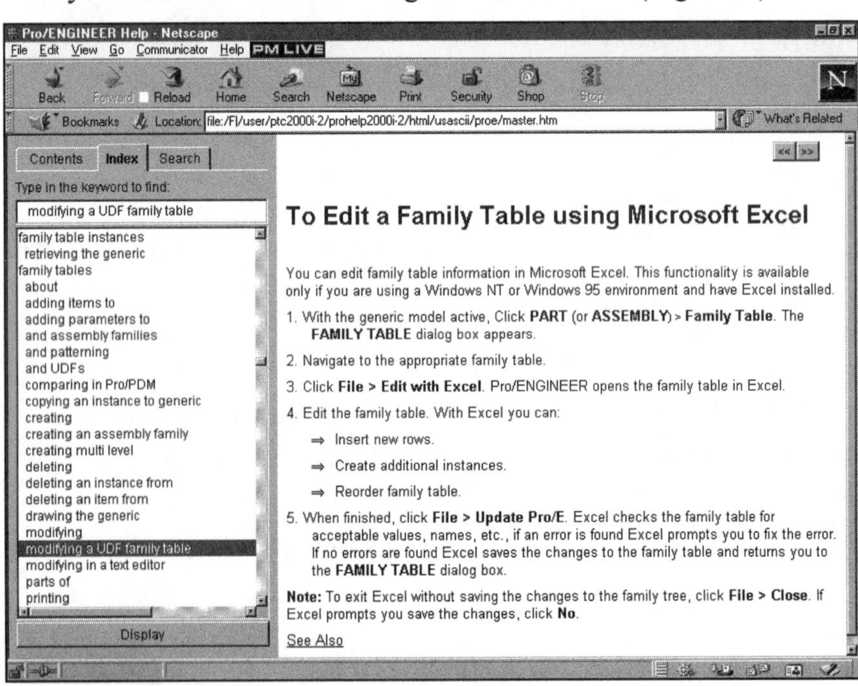

Figure 21.4
Editing Family Tables Using Microsoft Excel

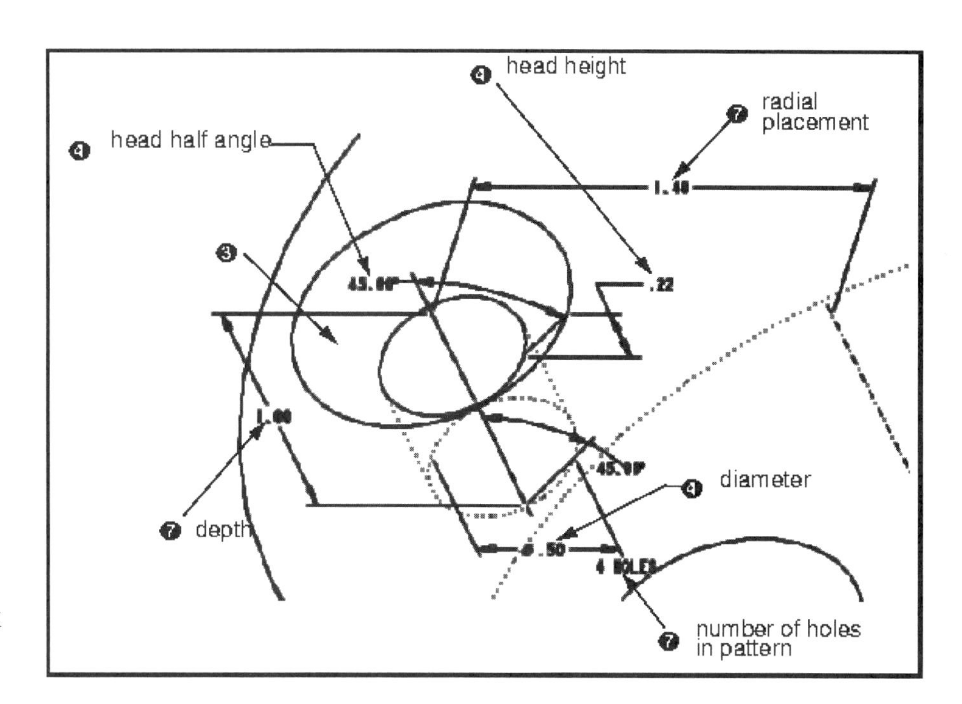

Figure 21.5
User-Defined Feature (Pro/LIBRARY Countersunk Hole/Feature)

User-Defined Features

When you create a **UDF**, you must define the placement references for its use. When you place a UDF in your part, Pro/E prompts you to use these references in the same manner as Pro/E does when you place a **Cut** or a **Slot** or a **Hole**.

A variety of UDF features are available in Pro/LIBRARY, such as the countersink shown in Figure 21.5.

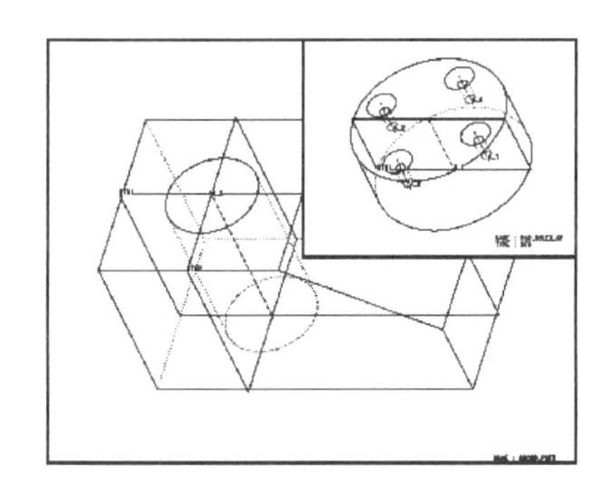

It is always advisable to plan out the references in a UDF before you create it so that, when placed, the prompts and interaction mimic Pro/E as much as possible.

If you create a UDF that represents a uniquely shaped feature (Figs. 21.6 and 21.7), define it in such a way that when it is placed, the user must select a placement plane and two reference edges to dimension from and then enter the distances of these dimensions and, if applicable, the size of the feature.

If you want the ability to place a UDF in several different ways, you are required to create a separate UDF for each method you want to use. In addition, you will have to create a separate model for each, because the manner in which you build the original defines the references and therefore controls how it can be placed.

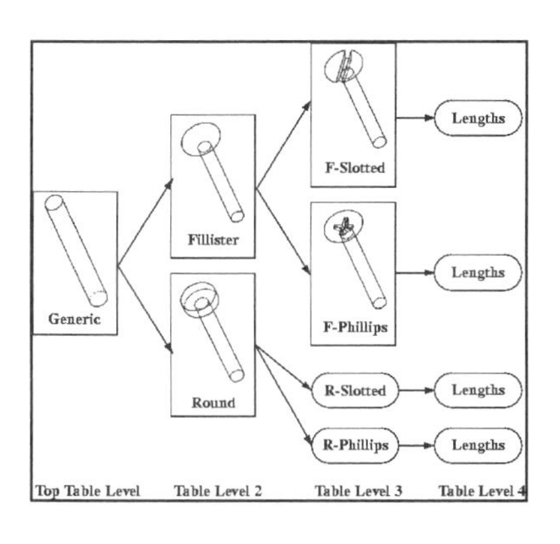

For example, if you want to be able to place a hole using either **Linear** or **Coaxial** references, you must create a hole using these methods and then create a UDF of each hole.

When you create a UDF, Pro/E creates a file, in the current directory, with the extension **.gph**. This is the UDF itself. If you also tell Pro/E to save a reference part, Pro/E also creates a part file with the text **_gp** added to the end of its name.

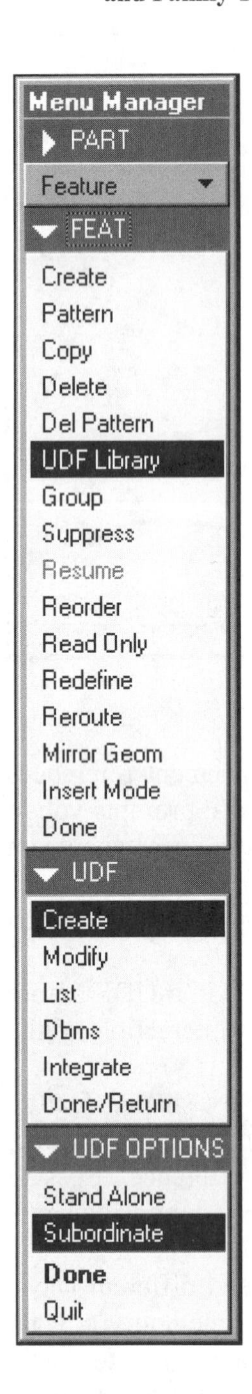

Creating a UDF

1. Choose **UDF Library** from the FEAT menu; then choose **Create** from the UDF menu. The menu lists the following options:

   Create Add a new UDF to the UDF library.
   Modify Modify an existing UDF. If there is a reference part, Pro/E displays the UDF in a separate part window.
   List List all the UDF files in the current directory.
   Dbms Perform database management functions for the current UDF.
   Integrate Resolve the differences between the source and the target UDFs.

2. Enter a name for the UDF.
3. Choose one of the options in the UDF OPTIONS menu, followed by **Done:**

   Stand Alone Pro/E copies all the required information to the UDF. Respond to the prompt to include a reference part.
   Subordinate Pro/E copies the information from the original part at run time.

4. Pro/E displays the UDF feature creation dialog box, which lists the required and optional elements.
5. Choose the **Features** element and **Define** from the dialog box.

When you create a UDF, it is not kept in RAM but, rather, stored in your current directory. Only when you use it in a model (part or assembly) does it appear in RAM.

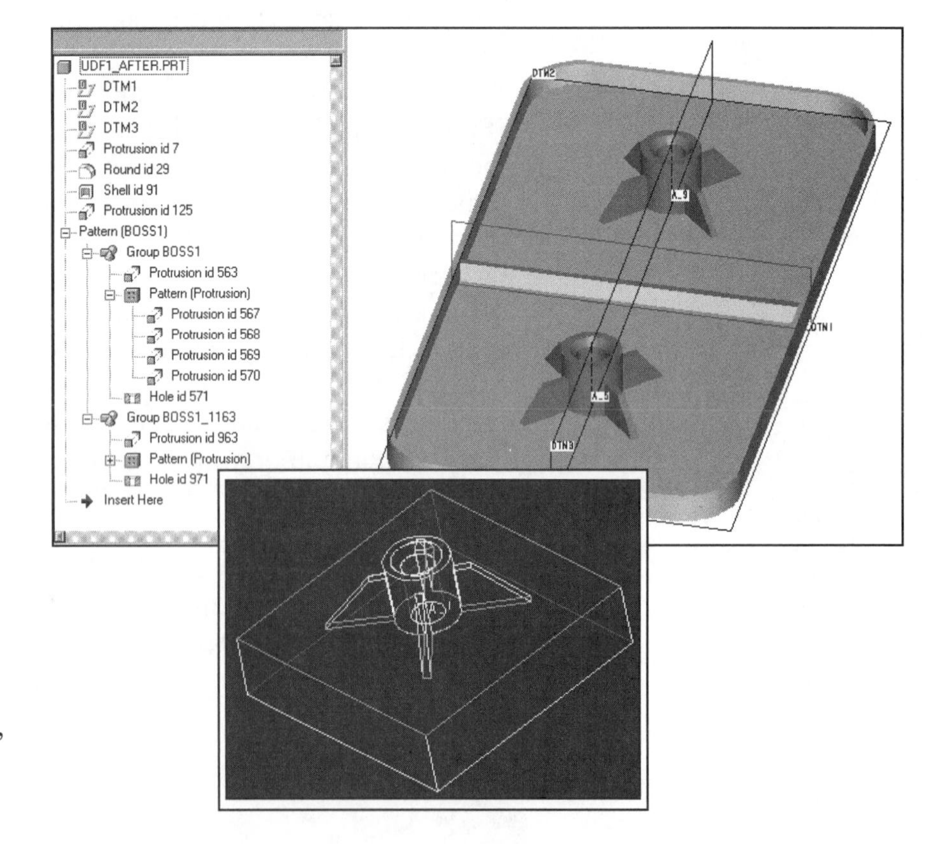

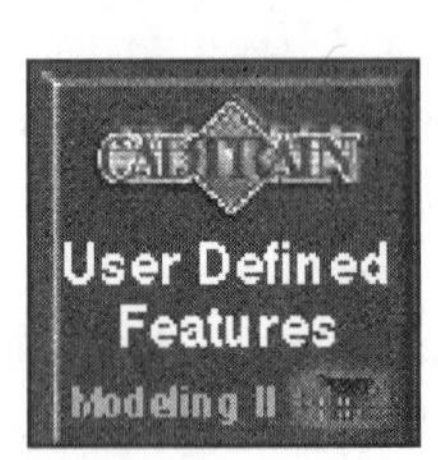

Figure 21.6
COAch for Pro/ENGINEER, Modeling II (User-Defined Features)

6. Pro/E displays the UDF FEATS menu, which lists the following options:

   Add Add a feature to the UDF.
   Remove Remove a feature from the UDF.
   Show Highlight all the features in the UDF.
   Info List all the features in the UDF.

7. Choose **Add** from the UDF FEATS menu.
8. Using the SELECT FEAT menu, select the features to add to the UDF. Note that you can include any feature except base features and shells. When you have finished, choose **Done** from the SELECT FEAT menu and **Done/Return** from the UDF FEATS menu.
9. Enter the prompts for the references used by the selected features. Pro/E highlights each reference and asks you to enter the prompt.

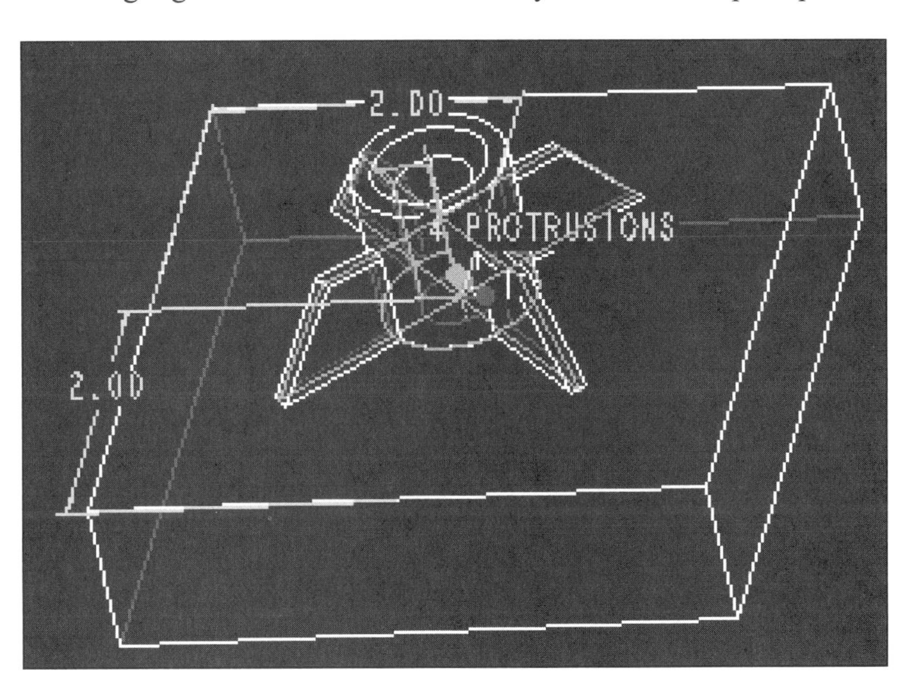

Figure 21.7
COAch for Pro/ENGINEER, Modeling II (UDF)

When you specify a prompt for a placement reference that is used by more than one feature in the UDF, Pro/E lets you specify either single or multiple prompts for this reference. Choose the desired option from the PROMPTS menu and then select **Done:**

Single Specify a single prompt for the reference used in several features. When the UDF (Fig. 21.8) is placed, the prompt appears only once, but the reference you select for this prompt applies to all features in the group that use the same reference.
Multiple Specify an individual prompt for each feature that uses this reference. If you select **Multiple**, Pro/E highlights each feature that uses this reference, so you can enter a different prompt for each of them.

10. After you have entered all the prompts, Pro/E displays the MOD PRMPT and SET PROMPT menus so you can change any prompt as follows:

Use **Next** and **Previous** from the MOD PRMPT menu to select the prompt you want to change, and enter the new prompt instead.

To change a single prompt (specified for the placement reference used in several features) into multiple prompts, find a prompt that you want to change, choose **Multiple**, and enter an individual prompt for each feature, as prompted by Pro/E.

11. If you are satisfied with the prompts, choose **Done/Return** from the SET PROMPT menu.
12. The required elements in the UDF dialog box have been defined. You can complete the creation of the UDF by selecting the **OK** button in the UDF dialog box, or you can define other optional elements.

To define variable elements in the UDF:

1. Select the **Var Elements** element in the dialog box and click on the **Define** button.
2. Select a feature that belongs to the UDF for which you want to specify variable elements.
3. Pro/E displays the SEL ELEMENT menu, which lists the elements of the selected feature. Place a check mark in front of the elements you want to define as variable and then choose **Done.**
4. When you finish selecting variable elements, choose **Done Sel** from the GET SELECT menu.
5. Choose **Done/Return** from the UDF menu.

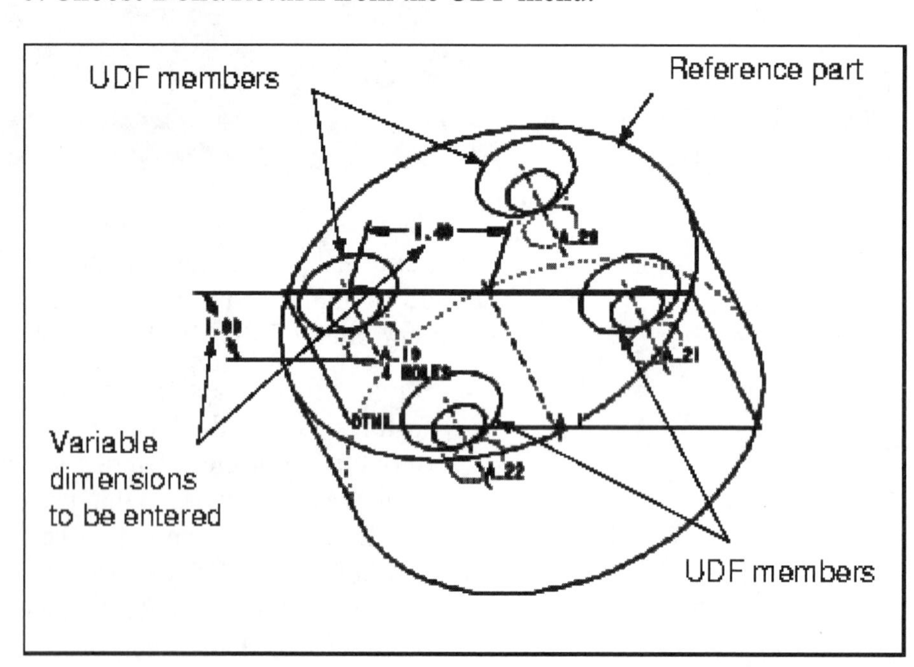

Figure 21.8
UDF Patterns and Placement

Family Tables

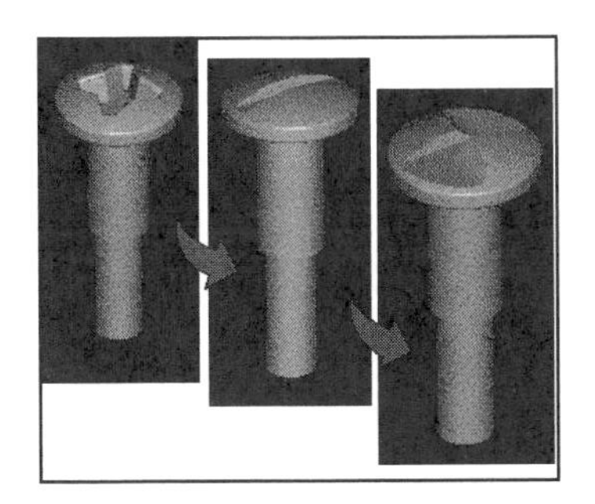

Family Tables are effective for two main reasons: they provide a beneficial tool, and they are easy to use. You need to understand the functionality of Family Tables, and you must understand when a Family Table is required and what circumstances should promote its use.

Family Tables are used any time a part or assembly (Fig. 21.9) has several unique iterations developed from the original model. The iterations must be considered separate parts, not just iterations of the same model.

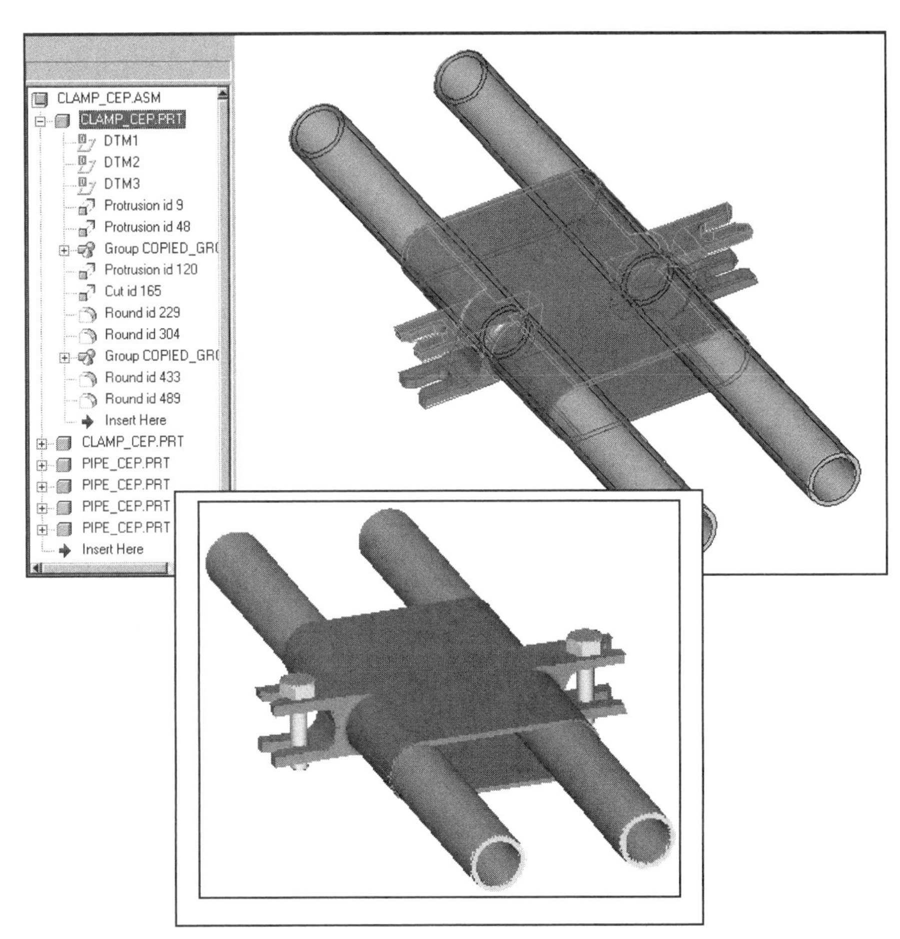

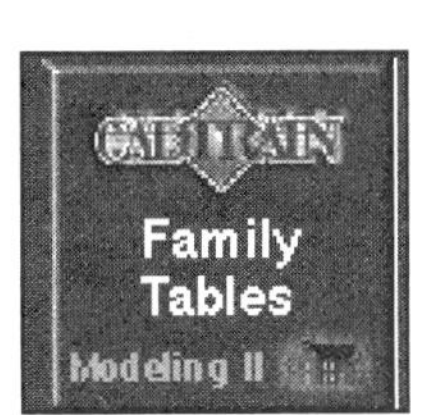

Figure 21.9
COAch for Pro/ENGINEER, Modeling II
(Assembly Family Tables)

To determine whether a model is a candidate for a Family Table, establish whether the original and the variation would ever have to co-exist at the same time (both in the same assembly, both shown in the same drawing, both with an independent Bill of Materials) and whether they should be tied together (most of the same dimensions, features, and parameters). If so, they are a candidate for the creation of a Family Table. Otherwise, the model may be a candidate for copying to an independent model or for a **Pro/PROGRAM.**

When the decision is made to create a Family Table (Fig. 21.10), the layout of the table should be considered next. The three main considerations are *what is in the table, how deep the table should be,* and *how to organize the table.*

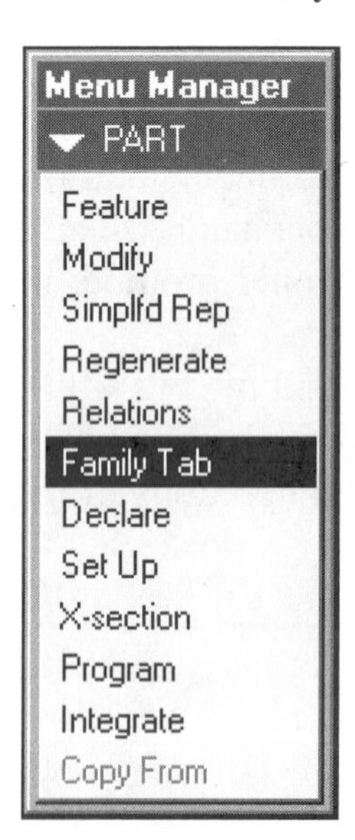

Figure 21.10
Starting a Family Table

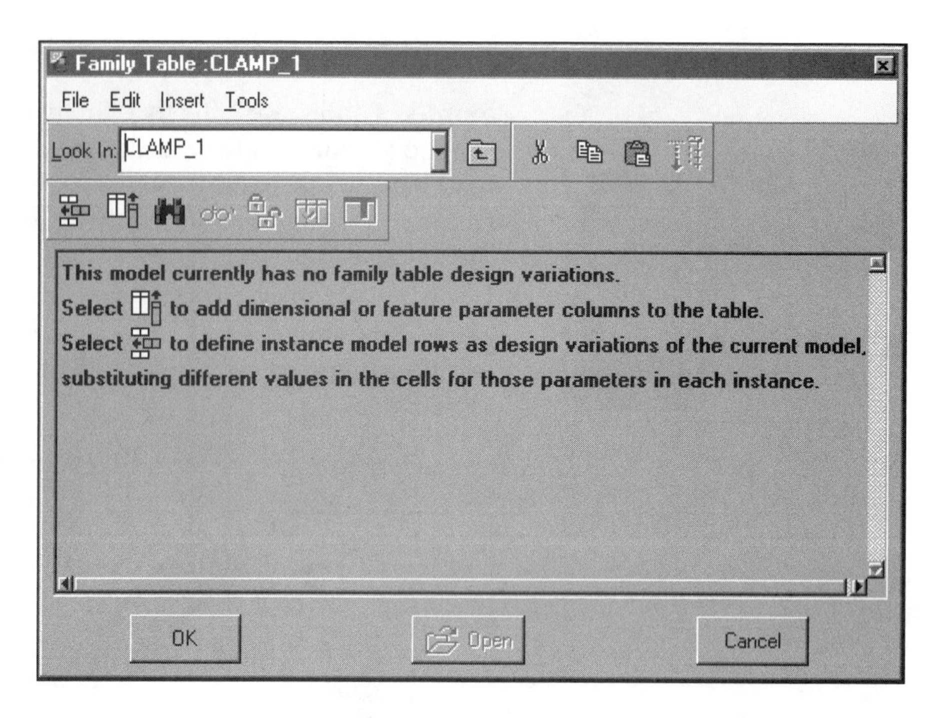

The *first consideration*, ***what is in the table,*** is defined by how the instances vary from the original. A table can include dimensions, features, components, parameters, groups, reference models, pattern tables, and system parameters. Each of these should be considered as a variation possibility, although most tables vary in their dimensions, parameters, and features.

The *second consideration* is ***how deep the table should be.*** The model can have a single-level table (e.g., different-sized Phillips flat-head screws) or a multiple-level table (e.g., screws to flat-head screws to Phillips-head screws to individual sizes). This is dependent on a trade-off between complexity (a single-level table may be unwieldy if it becomes too large) and ease of use (a multiple-level table can make instance retrieval more cumbersome).

The *third consideration* is ***how to organize the table.*** Pro/ENGINEER will prompt users for information to identify an instance by the order in which it appears in the table (Fig. 21.11). Therefore, if length is the most important differentiating aspect of the instance, it should be the first item in the table.

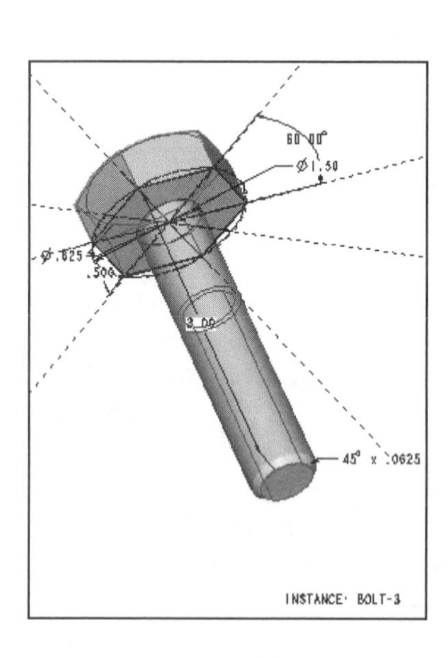

Figure 21.11
Family Table for Bolt

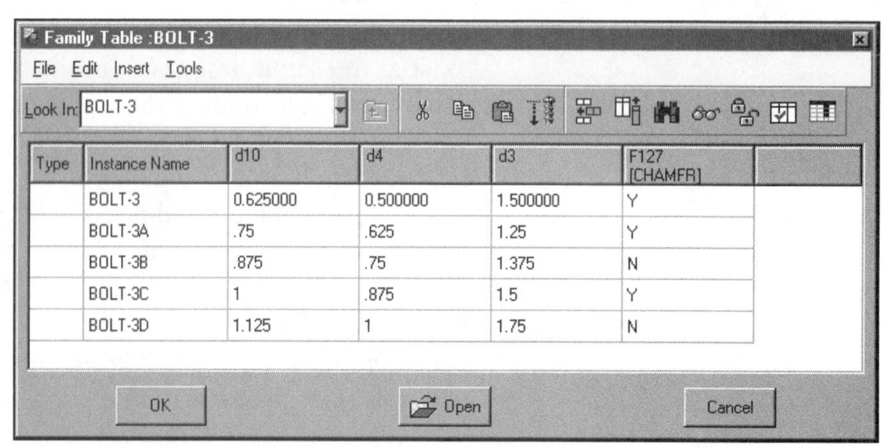
Family Table :BOLT-3
File Edit Insert Tools
Look In: BOLT-3

Type	Instance Name	d10	d4	d3	F127 [CHAMFR]
	BOLT-3	0.625000	0.500000	1.500000	Y
	BOLT-3A	.75	.625	1.25	Y
	BOLT-3B	.875	.75	1.375	N
	BOLT-3C	1	.875	1.5	Y
	BOLT-3D	1.125	1	1.75	N

OK Open Cancel

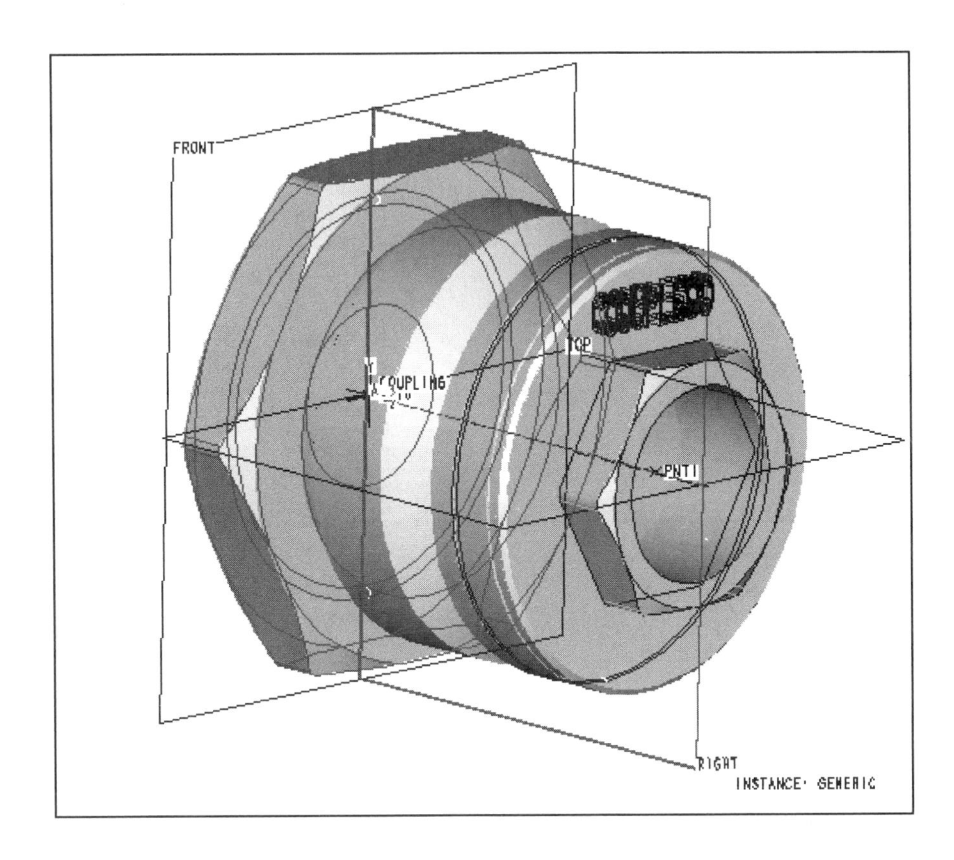

Figure 21.12
Coupling Fitting

Coupling Fitting

The **Coupling Fitting** is designed so that three versions are available to the user. Detail drawings for the COUPLING are provided in Figures 21.12 through 21.25. A Family Table is used to establish three instances of the coupling. The generic version is the smallest, and the three instances vary in size and features. The coupling uses two hexagonal features taken from Pro/LIBRARY. If your system does not have access to Pro/LIBRARY, create a hexagonal UDF and store it in your own User-Defined feature library according to the information provided in the first part of this lesson. ***Commands start on page L21-15.***

Set up the **COUPLING FITTING**:

- Material = Stainless Steel
- Units = Inch lbm Second

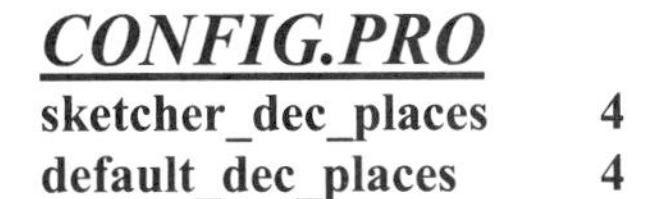

sketcher_dec_places 4
default_dec_places 4

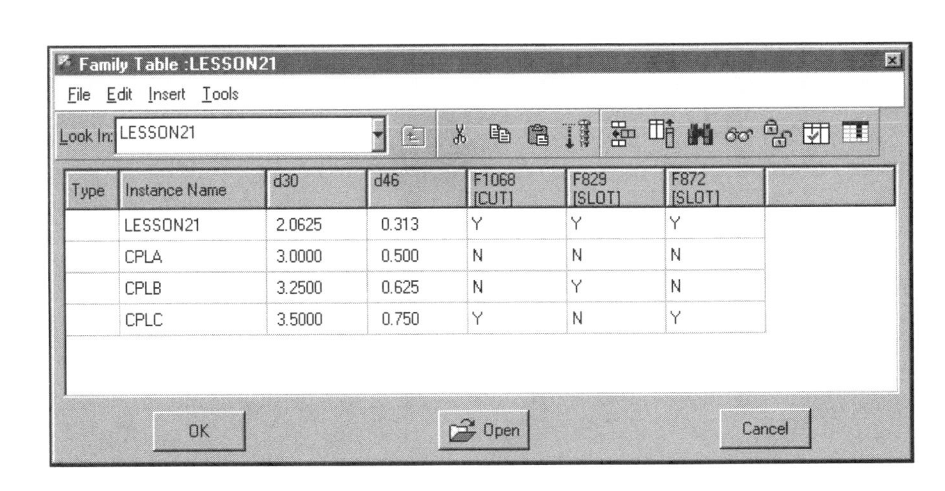

Family Table :LESSON21

File Edit Insert Tools

Look In: LESSON21

Type	Instance Name	d30	d46	F1068 [CUT]	F829 [SLOT]	F872 [SLOT]
	LESSON21	2.0625	0.313	Y	Y	Y
	CPLA	3.0000	0.500	N	N	N
	CPLB	3.2500	0.625	N	Y	N
	CPLC	3.5000	0.750	Y	N	Y

OK Open Cancel

Figure 21.13
Coupling Fitting Family Table

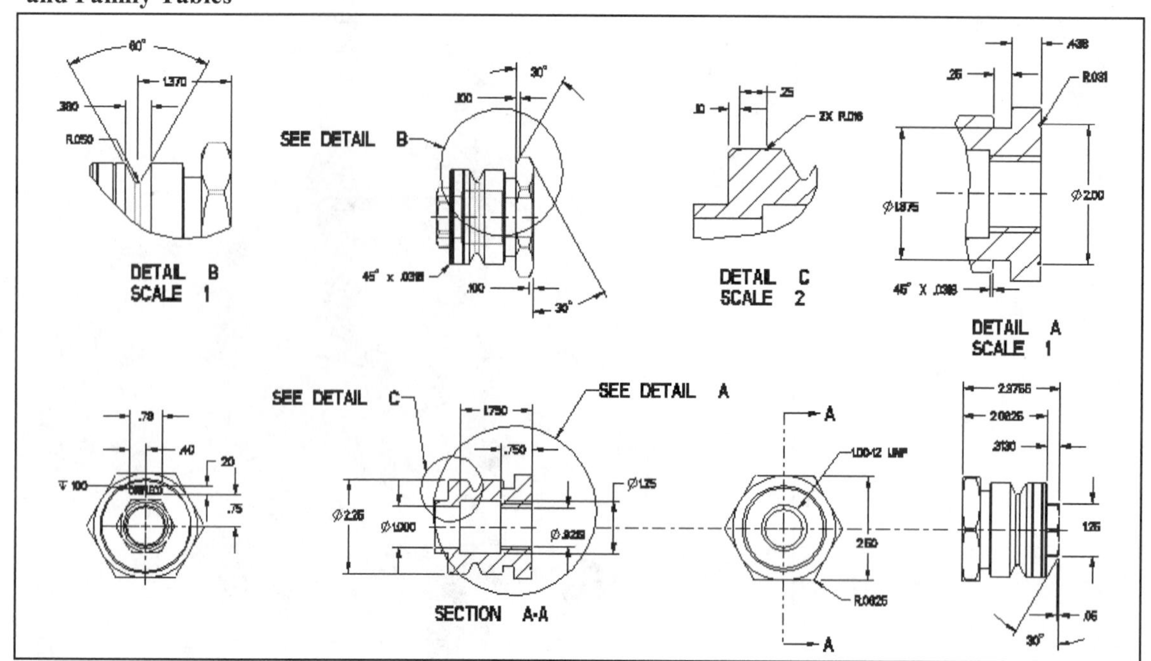

Figure 21.14
Coupling Fitting, Detail Drawing

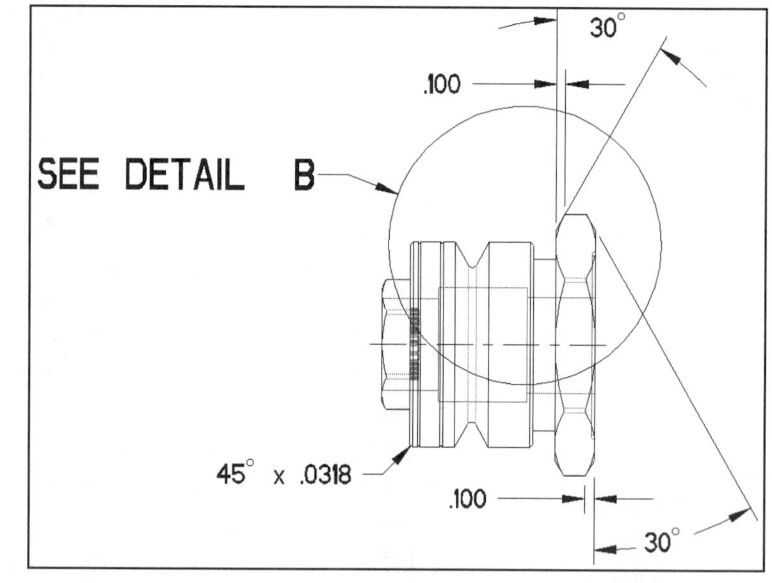

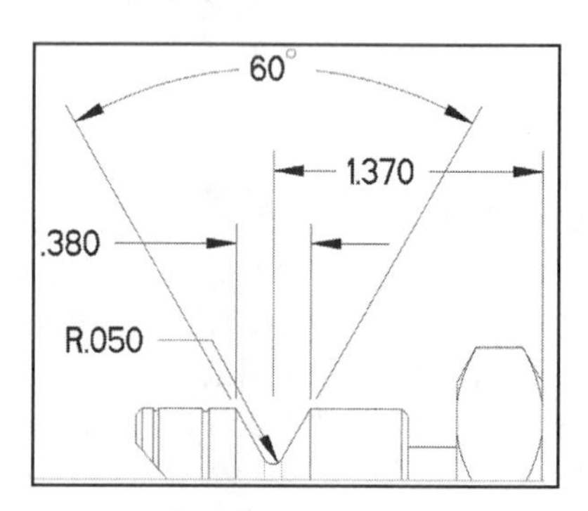

Figure 21.15
Coupling Fitting, Top View

See **EGD** or your **Machinery's Handbook** for the tap drill size (diameter)

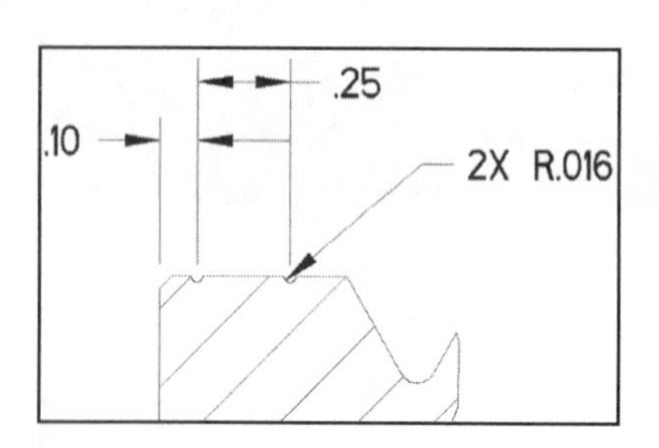

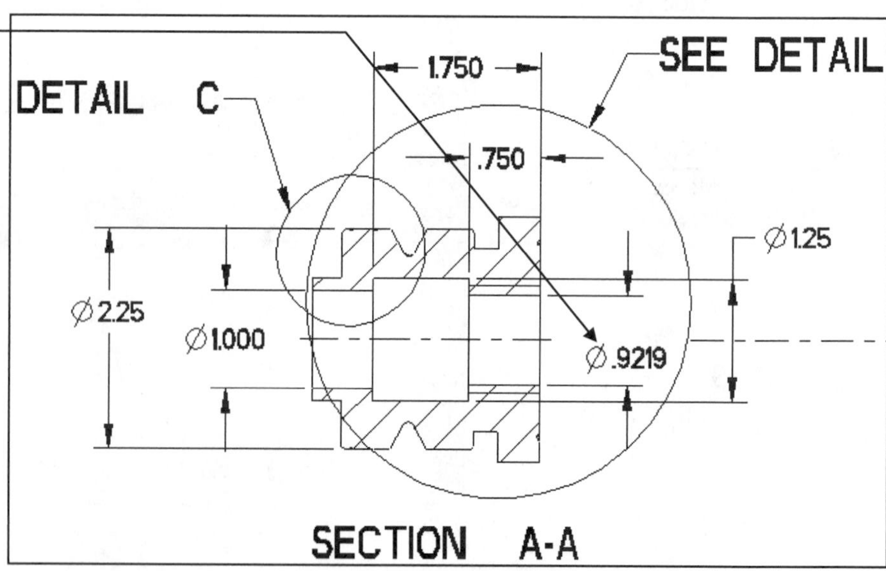

Figure 21.16
Coupling Fitting, **SECTION A-A**

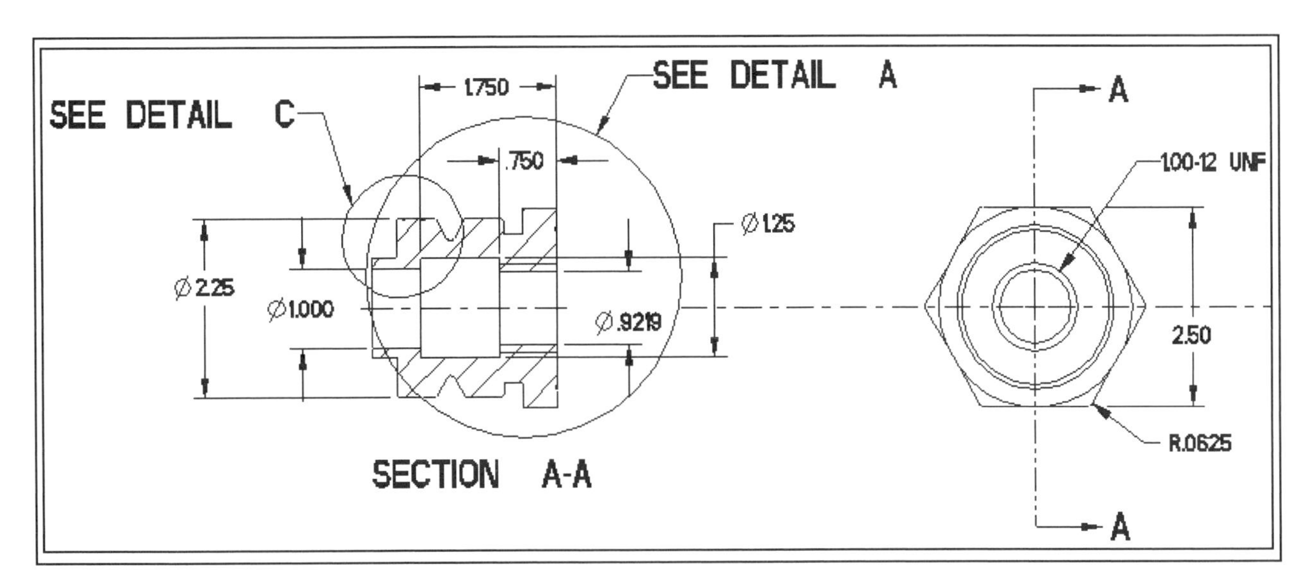

Figure 21.17
Coupling Fitting, Front View and Right Side View

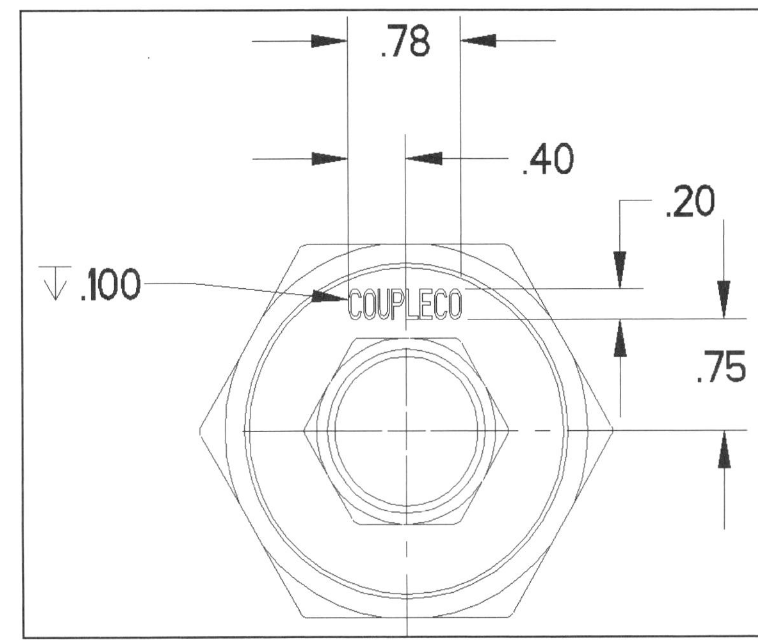

Figure 21.18
Coupling Fitting, Engraved Company Name **(COUPLECO)**

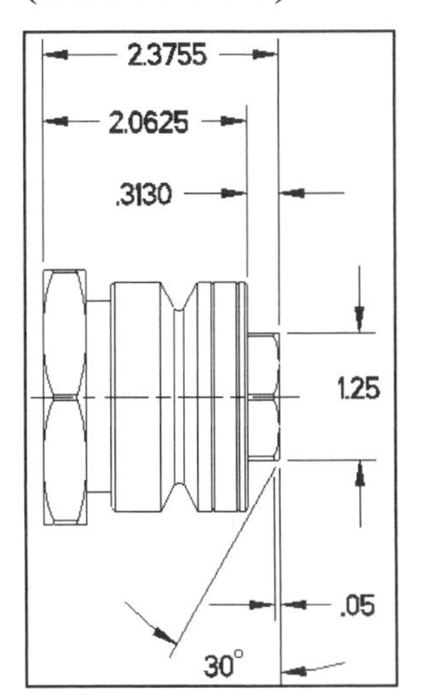

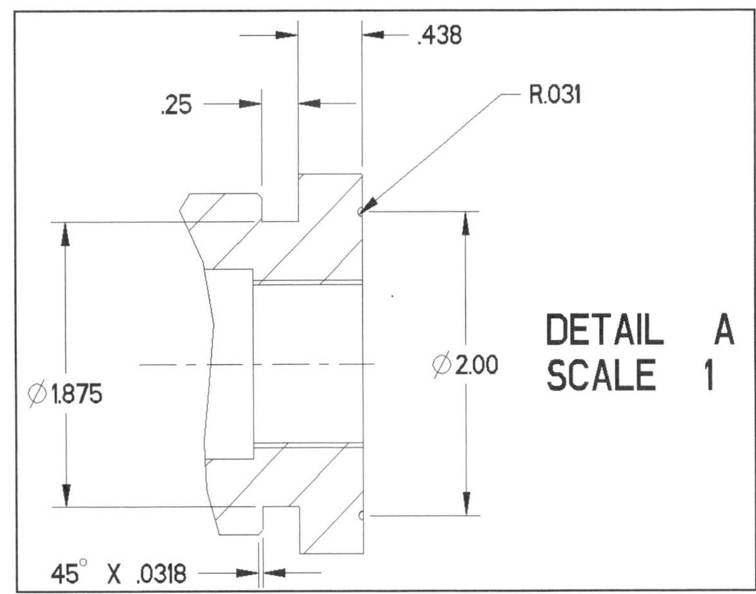

Figure 21.19
Coupling Fitting, **DETAIL A**

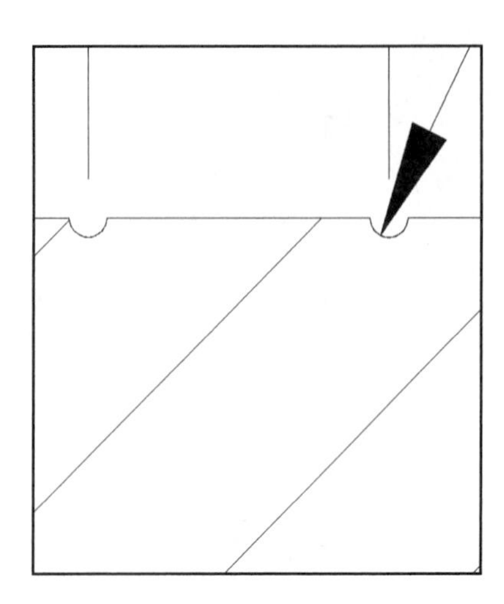

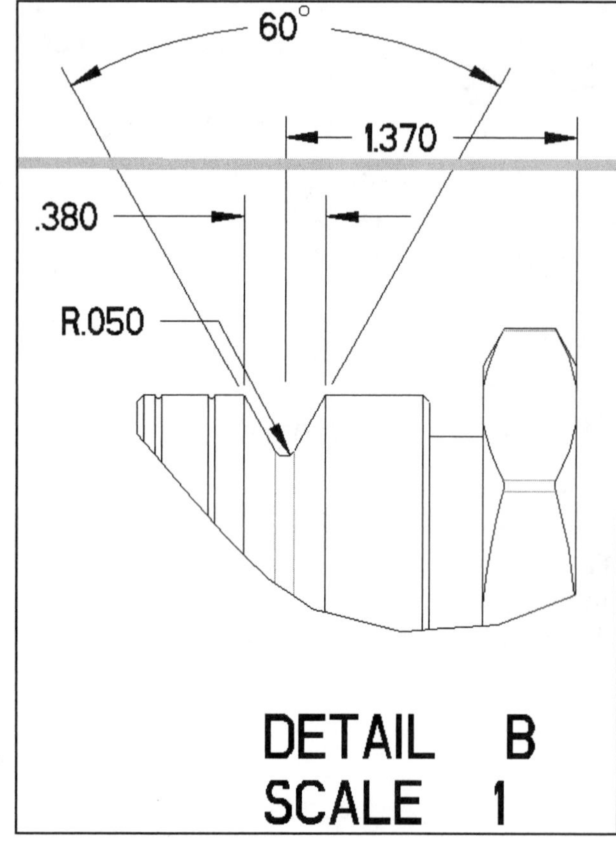

Figure 21.20
Coupling Fitting, **DETAIL B**

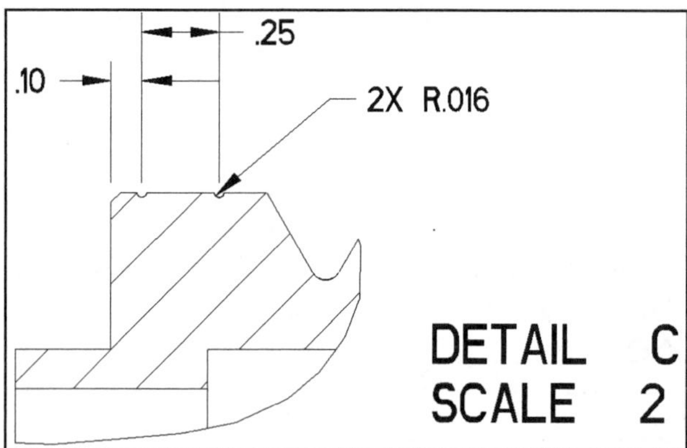

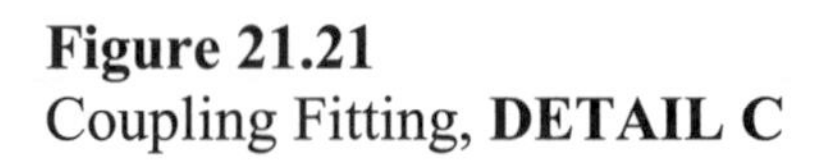

Figure 21.21
Coupling Fitting, **DETAIL C**

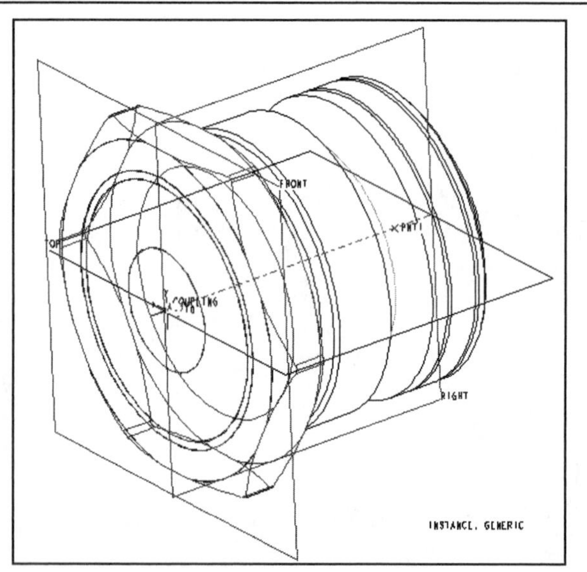

Figure 21.22
Coupling Fitting Showing Hexagonal UDF

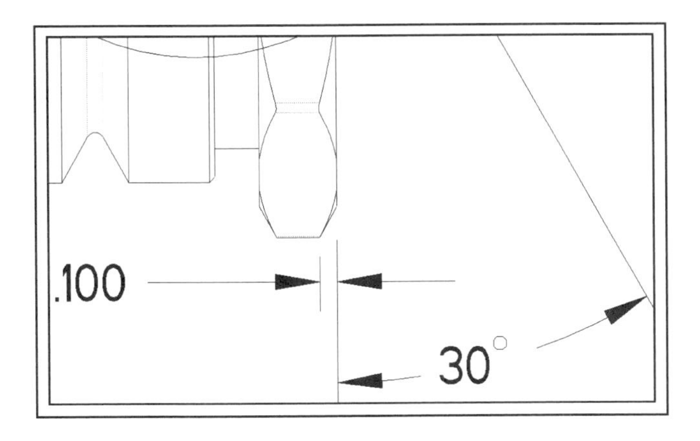

Figure 21.23
Coupling Fitting, Engraved End

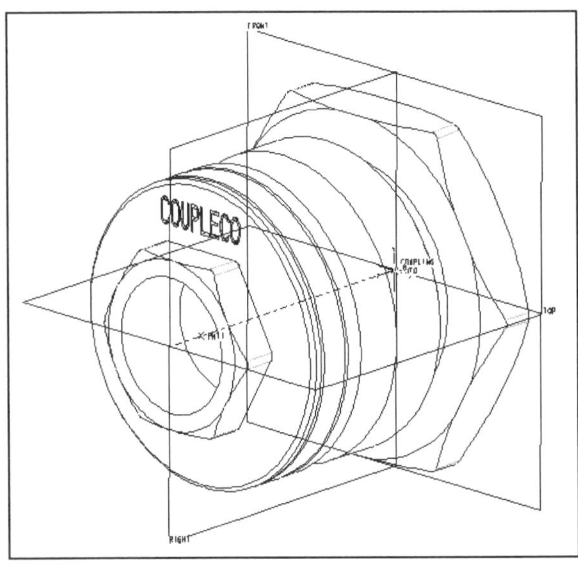

Figure 21.24
Coupling Fitting, Shaded Model (Generic)

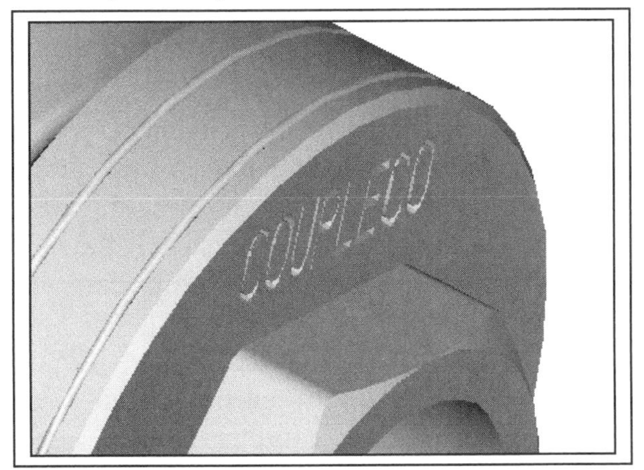

Figure 21.25
Coupling Fitting Showing Full Section

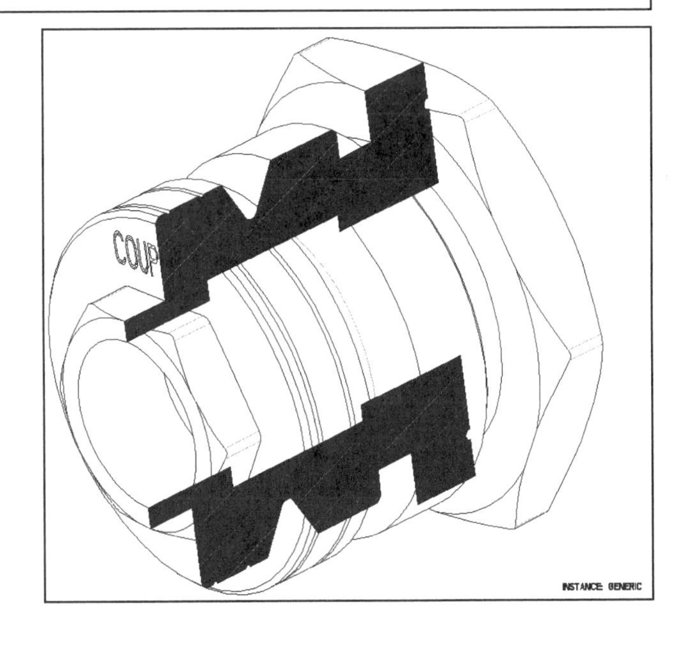

Using a User-Defined Feature

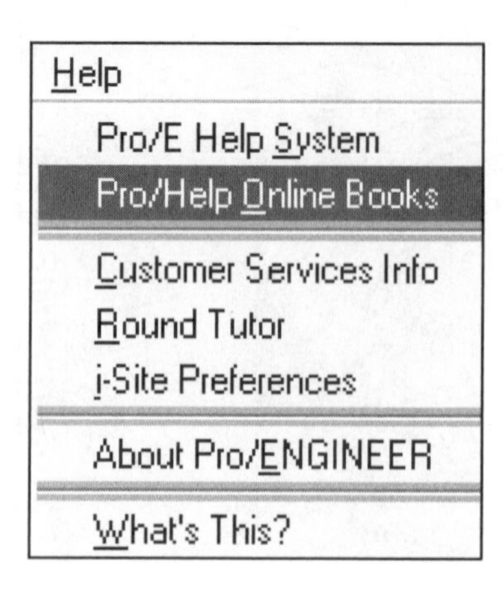

Most companies create a library of frequently used standard features for each designer to use. Pro/ENGINEER has a wide range of features available in Pro/LIBRARY. You can access these features by choosing **Feature ⇒ Create ⇒ User Defined ⇒** (navigate through your directories to **Pro/Library ⇒ objlib ⇒ featurelib**).

When you choose a **Feature**, Pro/E asks you if you want to display a ***Reference Part***. If you respond **yes**, Pro/E opens a new window to show you what the UDF looks like. This can be extremely helpful if you have not used a particular UDF before. In general, it is good practice always to use the reference part. Another practice worth mentioning is the use of online documentation to see what UDF library features are available and how to insert and use the features. Choose the following commands to see the online documentation for UDFs (Fig. 21.26):

Help ⇒ Pro/Help Online Books ⇒ Pro/ENGINEER® Libraries ⇒ BASIC™ LIBRARY Catalog ⇒ Geometric Features ⇒ Protrusions

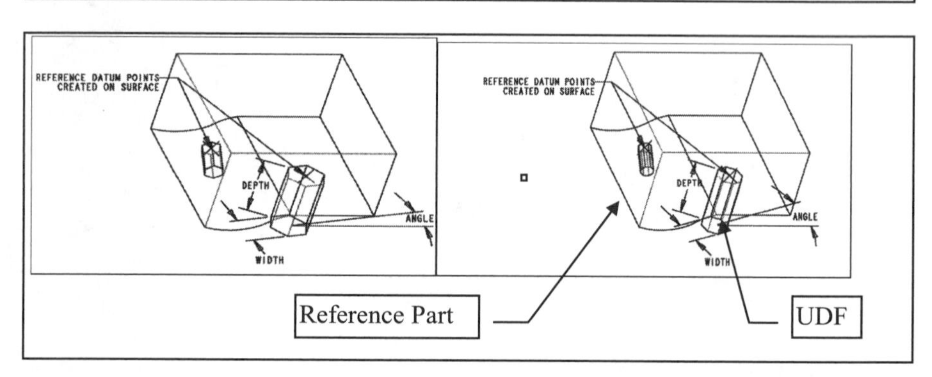

Figure 21.26
Pro/HELP for UDF Features Library Showing Hexagonal and Octagonal Protrusions Available as UDFs

Most UDFs are created on a **base feature** that is not actually part of the UDF. This feature is usually a flat rectangular solid. Even though this feature is displayed in the reference part, *it is not added to your part when you place the feature.*

The UDF is always stored when it is added to a part and must be placed using the same references used to create the original. You must plan your construction method and references before you create the original model you use to store the UDF. When you place the UDF, Pro/E prompts you through the defining of the references. If the UDF was stored with variable dimensions (e.g., height, width, diameter), Pro/E also prompts you to enter those values.

After all the data are defined, the UDF is placed. If the UDF is a dependent one, a reference is established between the placed object and the stored UDF. Then whenever the original library version changes, the object in your part will also change (at the next regeneration or retrieval of your part).

Lesson procedures & commands START HERE ➡ ➡ ➡

File ⇒ New ⇒ ●Part ⇒ (**COUPLING**) **⇒ ✔Use default template ⇒ OK**

Datum Analysis
Plane...
Axis...
Curve...
Point...
Csys...
Graph...
Analysis...
Reference...
Offset Planes...

Leave the **Environment** setting on **Trimetric** during the creation of this part. Because the reference part and the UDF are displayed in **Trimetric,** it is easier to answer the prompts for placing and sizing the UDF.

Create a new feature using a **UDF** from Pro/Library. The first part feature is a hexagonal protrusion. Online documentation states: "When you use a protrusion in your part, you must have already defined a *datum point* on the surface of the part for the center of the group, and a *reference plane* for the direction." Create a datum point on datum **FRONT** using the following commands:

Datum (from Menu Bar) **⇒ Point ⇒ On Surface ⇒** (pick **FRONT) ⇒** [then pick somewhere *on* **FRONT** (Fig. 21.27)] **⇒ Done Sel ⇒** (*Pro/E responds:* **Select two placement PLANES/EDGES for the dimensions.** Select **RIGHT** and **TOP) ⇒** (type the distance from each reference) **⇒ 0 ⇒ ✔ ⇒ 0 ⇒ ✔ ⇒ Done** [**PNT0** is located *on* datum **FRONT** at the intersection of the three datum planes (Fig. 21.28)] **⇒ Done**

COAch™ for Pro/ENGINEER

If your system has **COAch™** or your text comes with a **CADTRAIN** disk, review Advanced Modeling, User-Defined Features before completing this lesson.

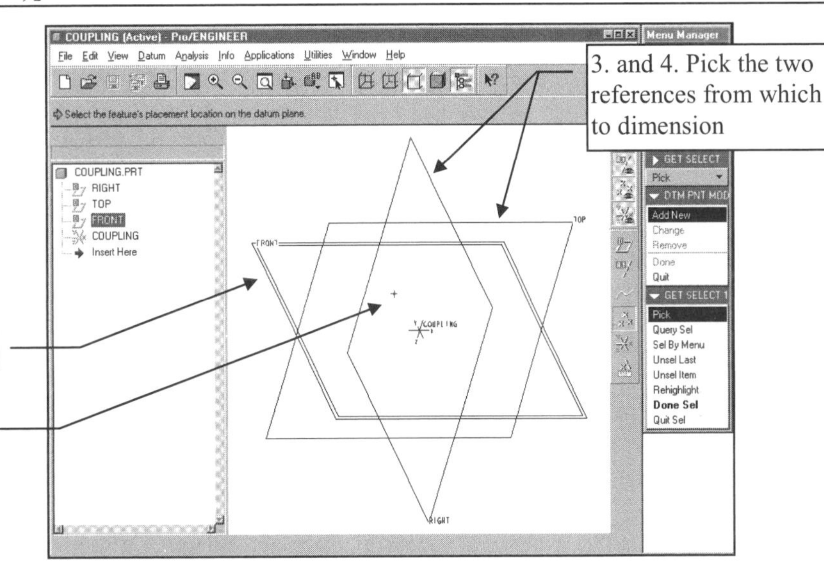

Figure 21.27
Placing the Datum Point

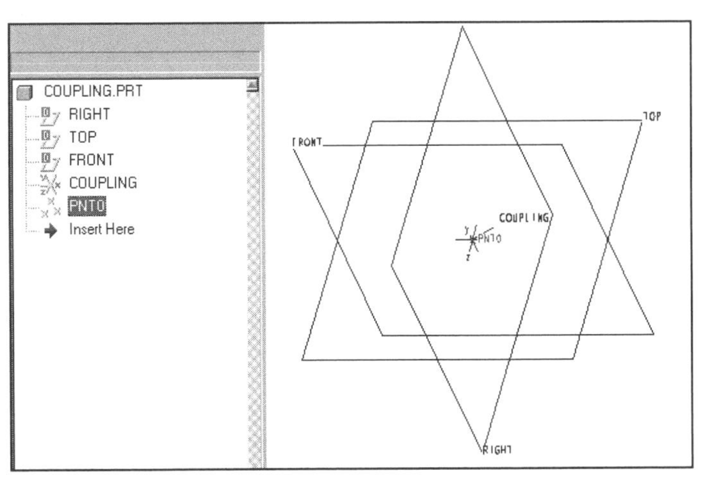

Figure 21.28
Datum Point **PNT0**

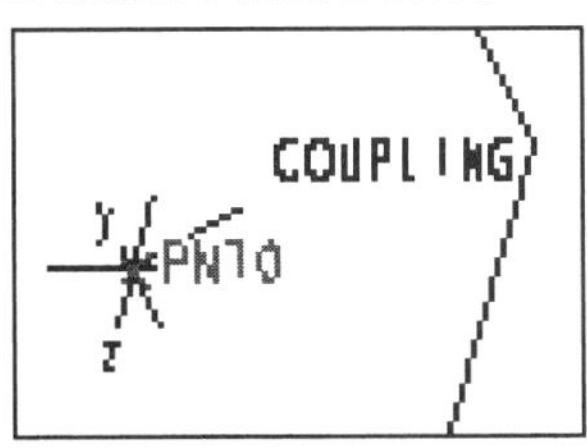

The first protrusion is created using a Pro/LIBRARY UDF. Choose the following commands:

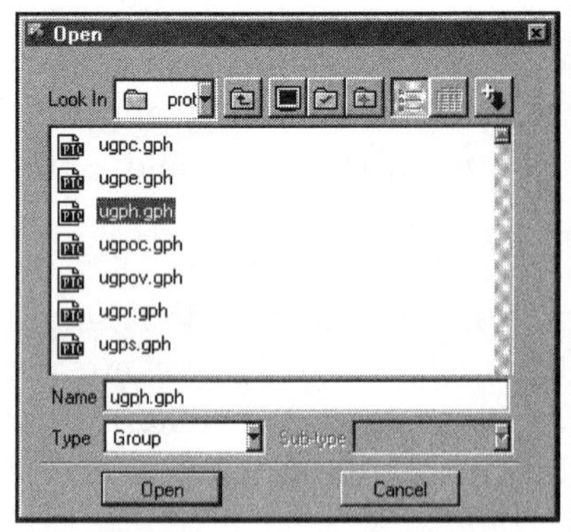

Feature ⇒ Create ⇒ User Defined ⇒ ▼ (navigate through your directories to **Pro/Library ⇒ objlib ⇒ Open ⇒ featurelib ⇒ Open ⇒ geometry_udf ⇒ Open ⇒ protrusion_udf ⇒ Open ⇒ ugph.gph ⇒ Open)**

Your directory may be different. See your system manager for the required path to Pro/LIBRARY UDFs.

⇒ Retrieve reference part UGPH_GP? **Yes ⇒ Independent ⇒ Done** (Fig. 21.29) ⇒ **Same Dims ⇒ Done** (to accept the current scale--the default scale is **1**) ⇒ (*Pro/E prompts with:* **Enter depth: 4.0000** *type* **.438**) (**.438** is the *thickness* of the larger of your coupling's two hexagonal shapes) ⇒ ✔ ⇒ (*Pro/E prompts with:* **Enter angle from reference: 10.0000** *type* **0**) ⇒ ✔ ⇒ (*Pro/E prompts with:* **Enter width: 2.0000** *type* **2.50**) (**2.50** is the distance *across the flats* of the hexagon) ⇒ ✔

HINT

As each prompt comes up, the dimensions in question are highlighted in the UDF reference part window.

The dimensions that control the placement of the UDF are displayed in the reference part, as shown in Figure 21.29.

Pro/E now wants to know how it should add the feature's dimensions to the part. The default is **Normal,** which means that the UDF's dimensions will be treated the same as any other feature's dimensions. The **Read Only** option allows the dimensions to be displayed in the part, but they cannot be changed.

(Choose) **Normal ⇒ Done**

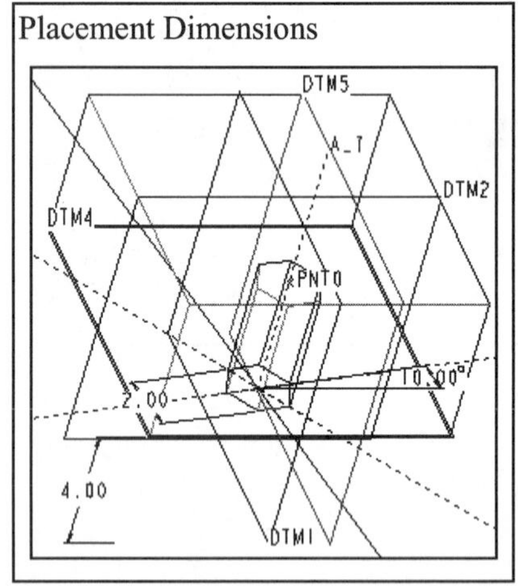

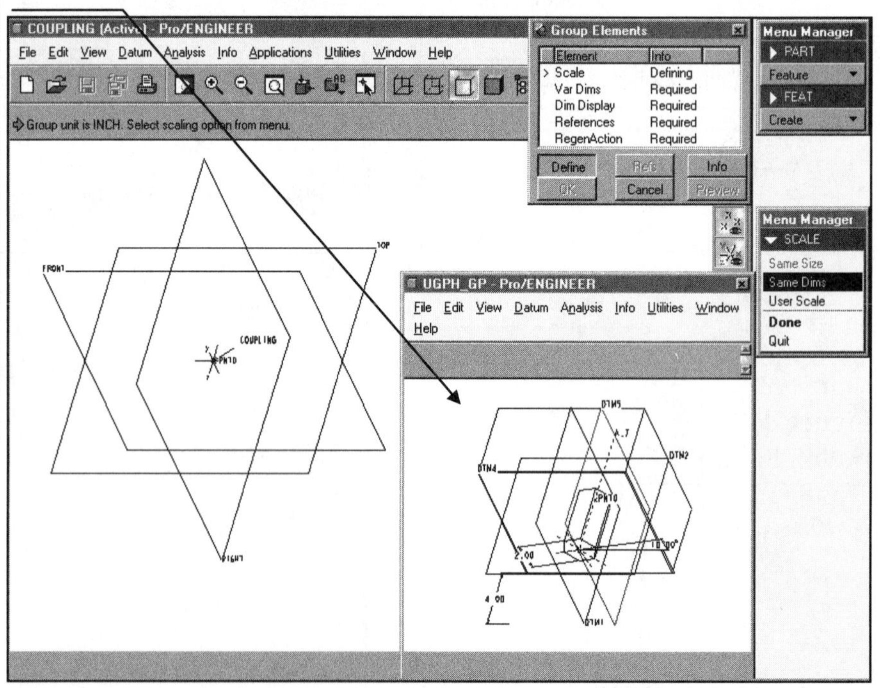

Figure 21.29
UDF with Reference Part Is Displayed with Placement Dimensions

Pro/E starts to display the placement references defined for the UDF and prompts you to select the appropriate surfaces on your part. The references requested are highlighted in the reference part as shown in Figure 21.30.

(*Pro/E prompts with:* **Select point to locate center of feature.**) *Select the datum point you created--***PNT0** ⇒ (*Pro/E prompts with:* **Select plane to specify orientation.**) Select **TOP,** which corresponds to the highlighted plane in the UDF window.

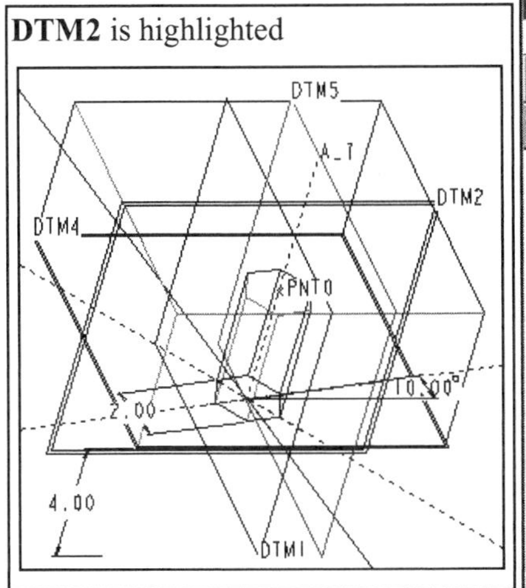

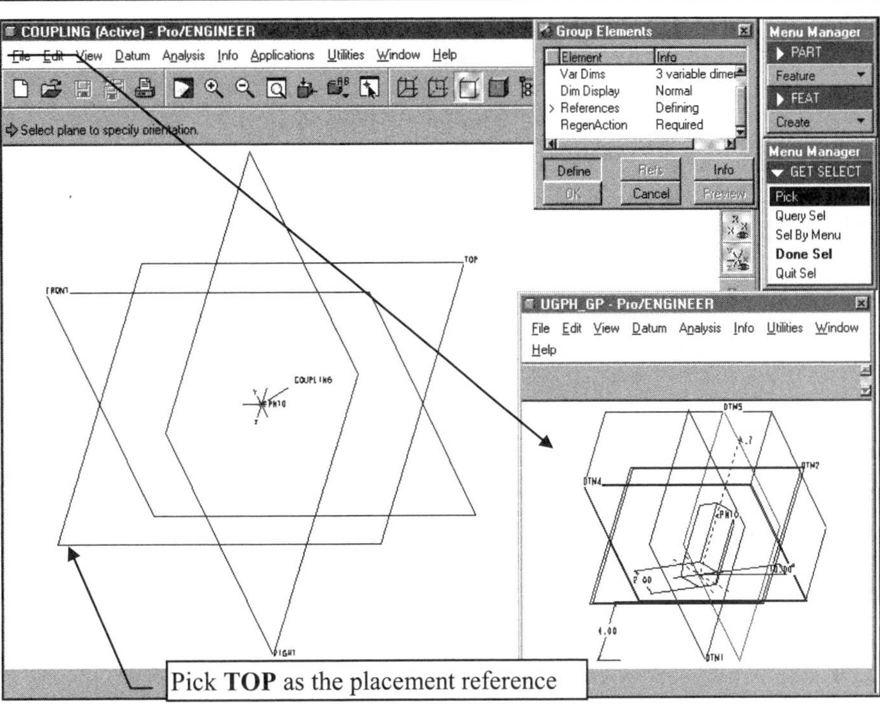

Figure 21.30
Picking the Reference Plane for UDF Orientation

Pro/E displays an arrow to show the direction in which it assumes that the positive axis points (Fig. 21.31). This direction affects not only the orientation of the feature but also the direction in which the distances you previously entered are measured.

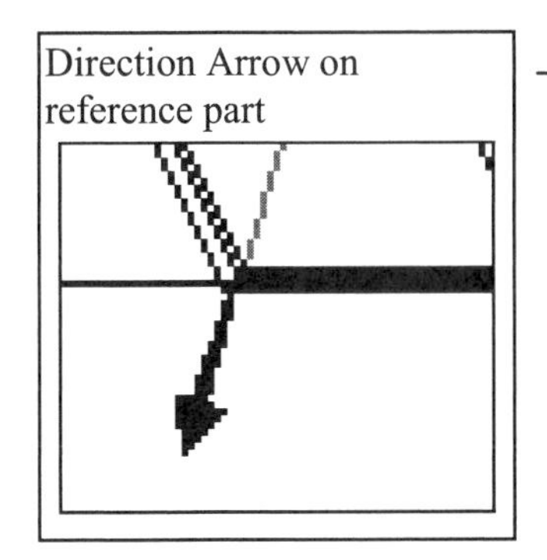

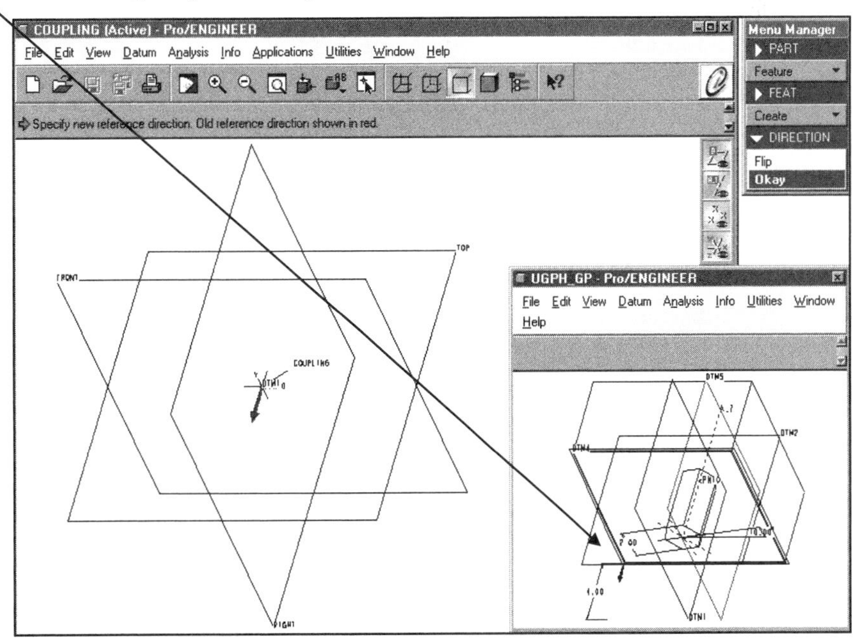

Figure 21.31
Direction Arrow for Feature Creation

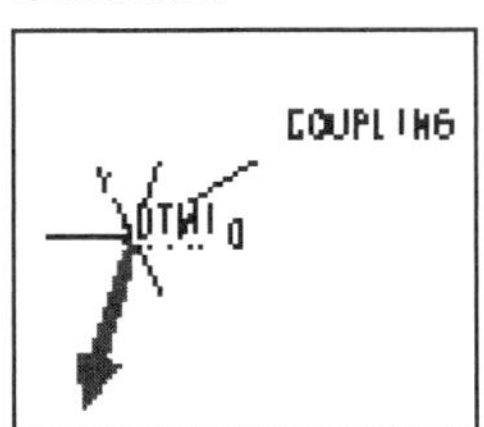

Save 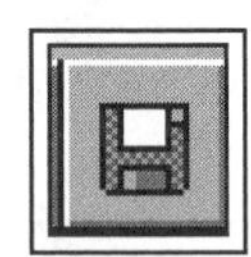

Choose **Okay** ⇒ (*Pro/E prompts with:* **Specify new reference direction. Old reference direction shown in red.**) ⇒ **Okay** ⇒ **Done** ⇒ [*Pro/E responds with:* **Group created successfully. Select Done to accept** (see Figure 21.32)] ⇒ **Done** ⇒ **View** ⇒ **Shade**

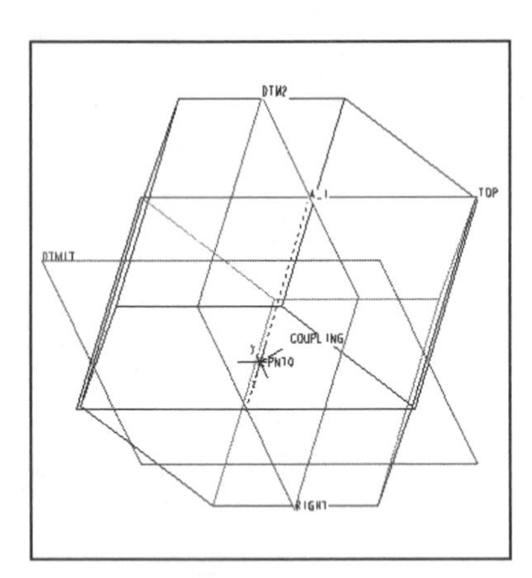

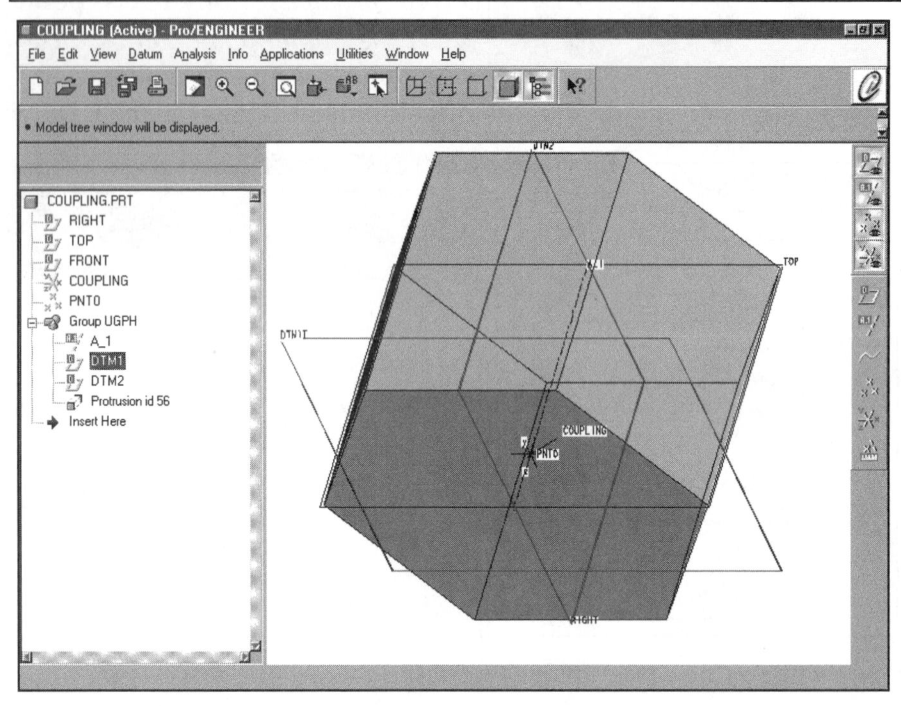

Figure 21.32
Group Created Successfully

To complete the feature, **Redefine** the UDF and change the depth to **Blind** (Fig. 21.33):

Feature ⇒ **Redefine** ⇒ (pick the UDF from the screen) ⇒ (Select **Depth** from the dialog box) ⇒ **Define** ⇒ **Blind** ⇒ **Done** ⇒ (*Pro/E prompts:* **Enter depth [1.4337]:**) *type* **.438** ⇒ ✔ ⇒ **OK** (Fig. 21.34)

NOTE

When creating the **Tool Body** in the **Lesson Project**, use the following commands instead of what is shown to the right:

Redefine ⇒ (pick the UDF) ⇒ (Select **Depth** from the dialog box) ⇒ **Define** ⇒ **UpTo Surface** (pick **DTM4**) ⇒ **Done** ⇒ **OK**

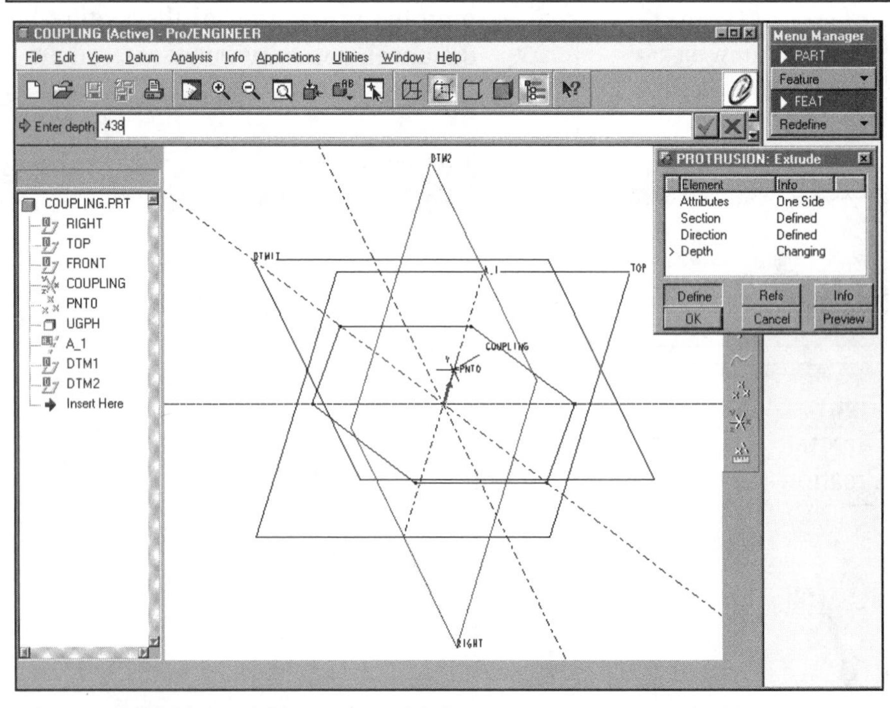

Figure 21.33
Redefining the UDF Depth

File ⇒ Save ⇒ ✔
File ⇒ Delete ⇒
Old Versions ⇒ ✔

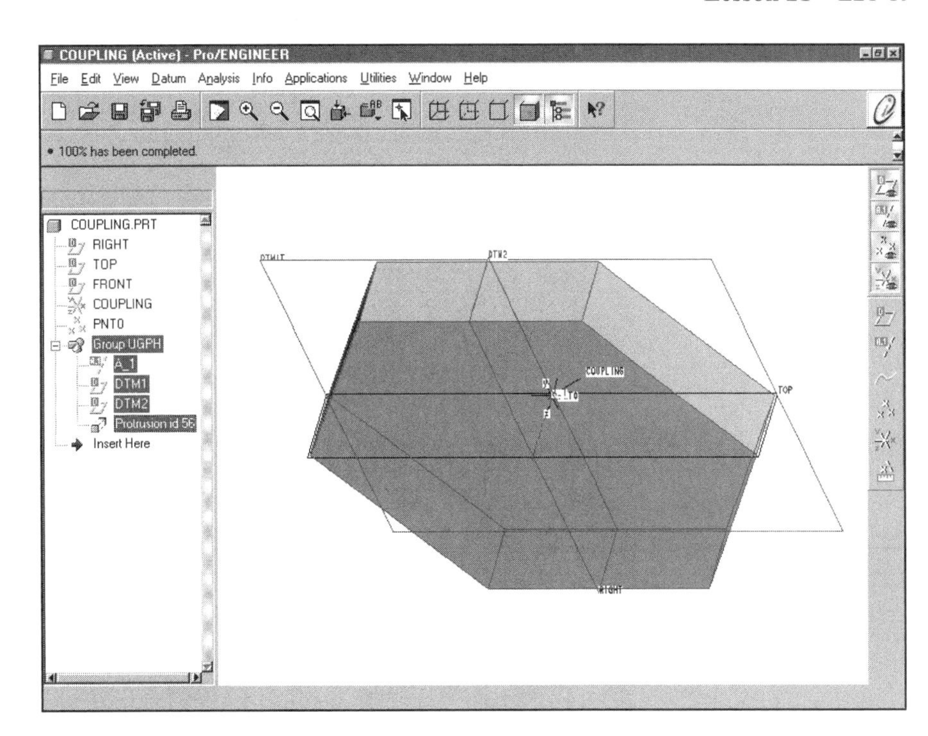

Figure 21.34
Completed UDF Feature

Seems like a lot of effort just to get a premade hexagonal protrusion. Remember that in general, UDFs can be created for almost any part or feature that is used frequently. This can be a great time saver in industry.

The next feature will be a revolved protrusion representing the body of the fitting (Fig. 21.35). You may include the **V** cut in this feature or add it later. It may be better design intent to make each cut separately so as to allow changes to the features individually and when suppressing in a Family Table. Sketch the revolved protrusions and cuts on **RIGHT.**

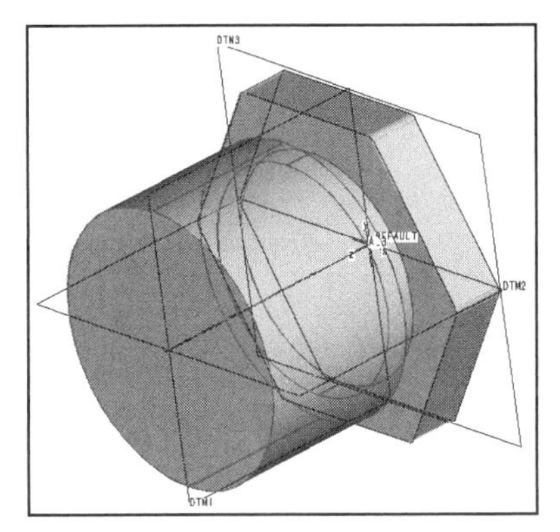

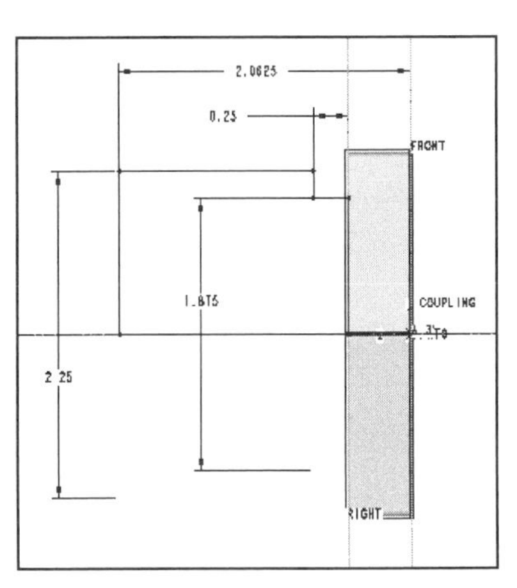

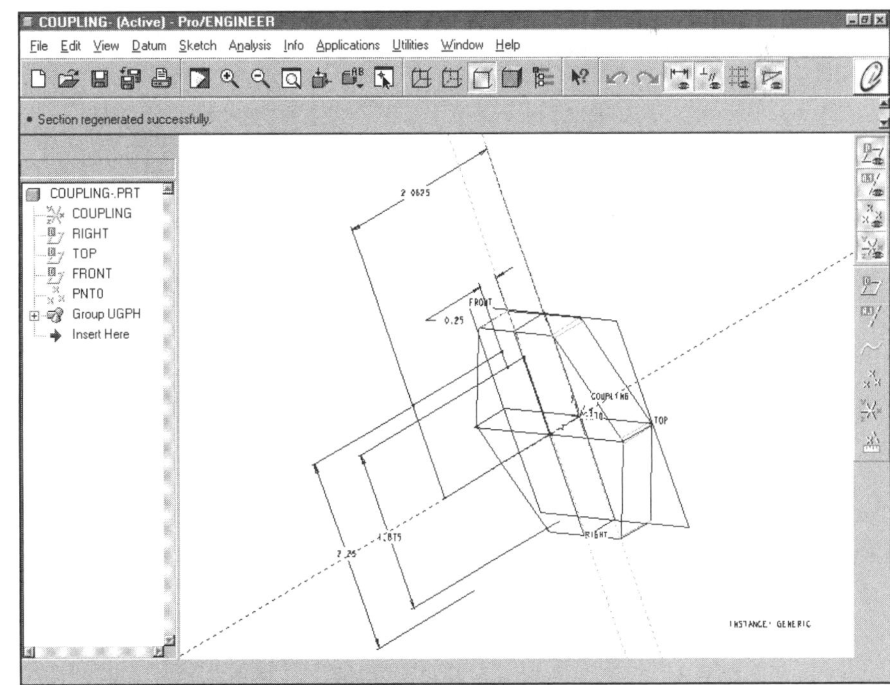

Figure 21.35
Section Sketch for Revolved Protrusion

Create a datum point on the planar end of the revolved protrusion to use for placement of a second hexagonal UDF (Fig. 21.36). Be sure to create the datum point as described previously.

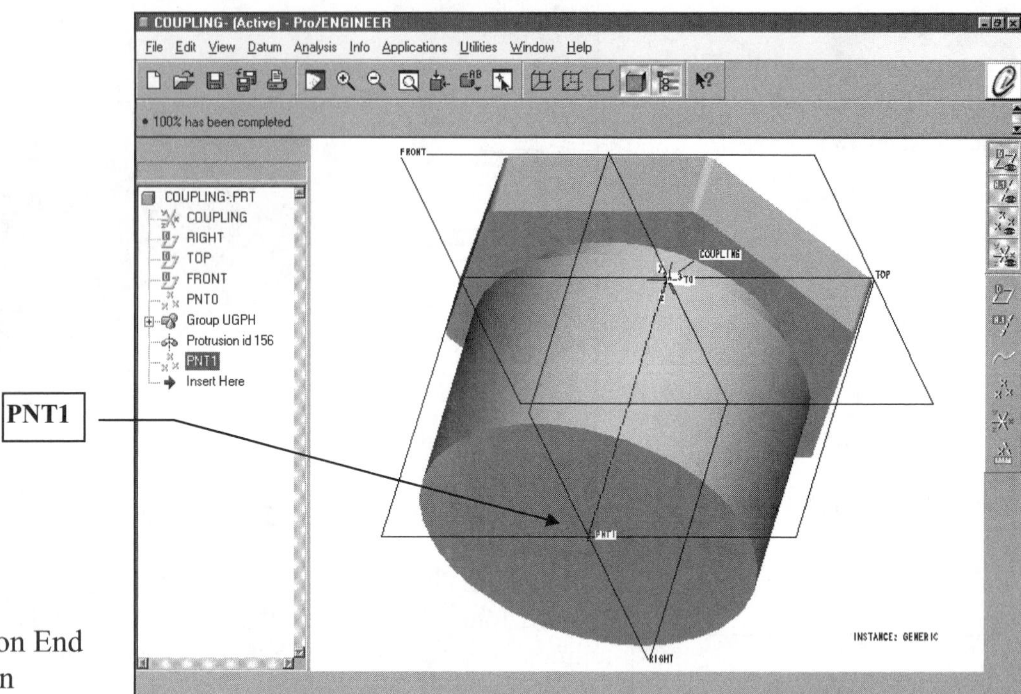

Figure 21.36
Datum Point Created on End of Revolved Protrusion

Use the same steps that were required to complete the first UDF. You should not need to redefine the UDF, because its depth of **.313** will establish the proper size of the feature (Fig. 21.37).

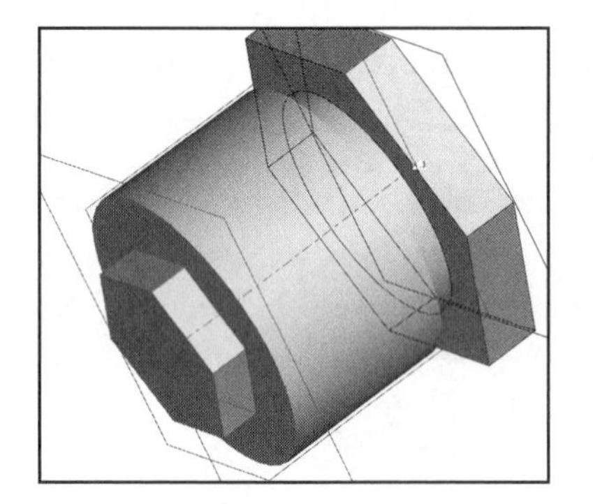

Figure 21.37
Second UDF Added

Complete the external features using the dimensions provided at the beginning of this lesson and Figure 21.38.

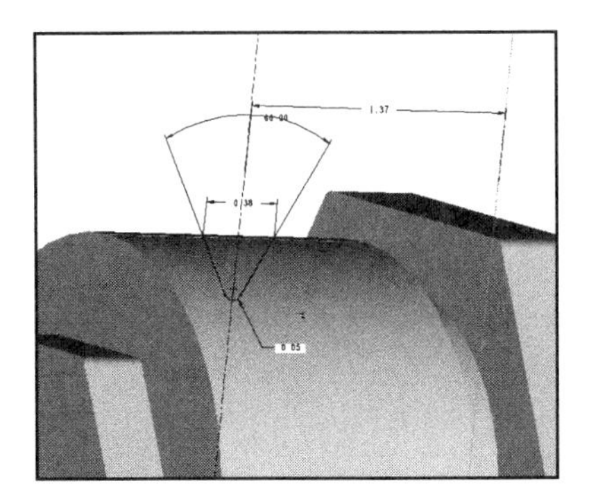

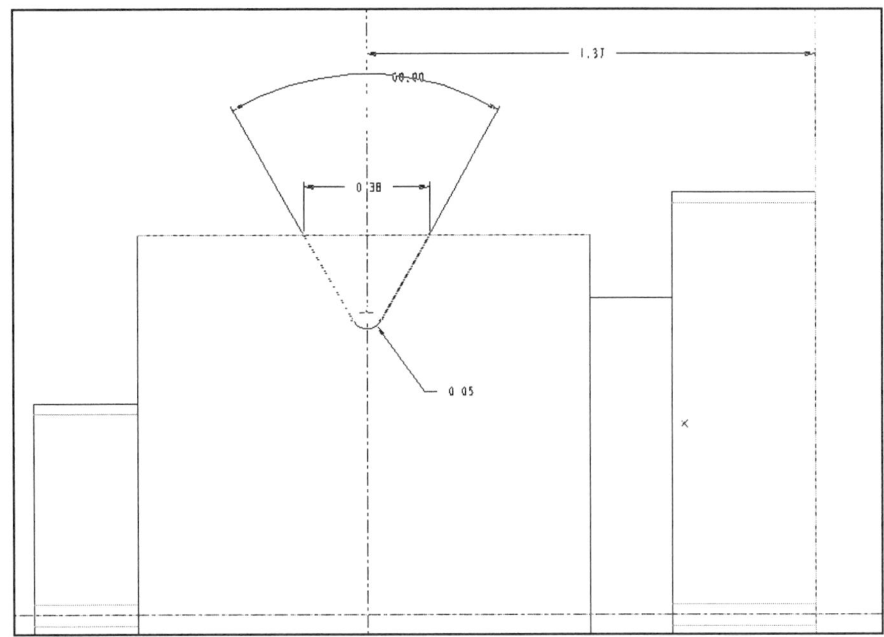

Figure 21.38
V Cut

Add the internal cuts last. The internal feature should be a revolved cut (Fig. 21.39) with both ends of the feature aligned to the outside of the part. If you align the *revolved cut* in this manner, the cut will remain *through* the part regardless of the changes different instances will have when a Family Table is created (Fig. 21.39). Do not use a *sketched hole,* because that would require a *relation* to establish the hole as *through* the part at all times.

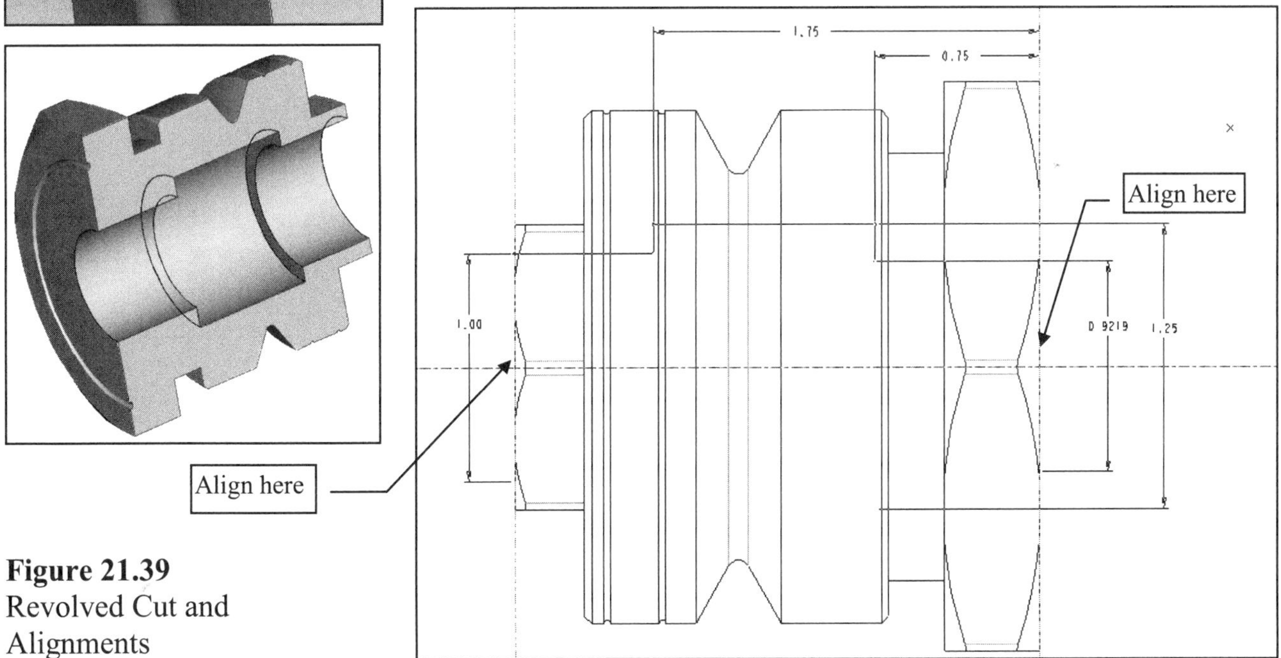

Figure 21.39
Revolved Cut and Alignments

Complete the generic part by adding cosmetic threads and cutting the **COUPLECO** company name into the part as shown in the figures provided.

Family Tables

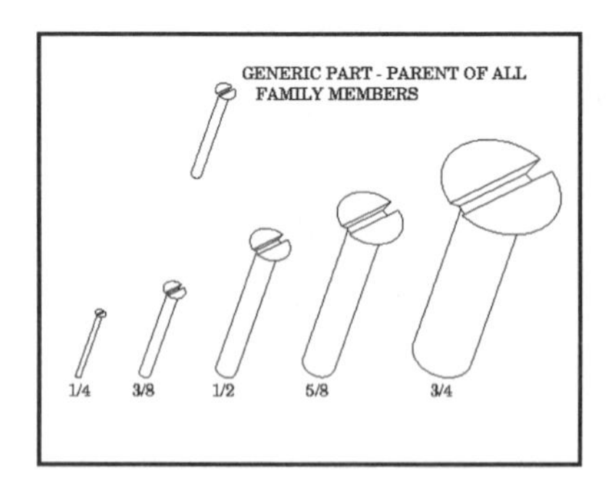

Family Tables are collections of parts (or assemblies or features) that are essentially similar but deviate slightly in one or two aspects, such as size or detail features. For example, wood screws come in various sizes, but they all look alike and perform the same function. Thus, it is useful to think of them as a family of parts. Parts in family tables are also known as table-driven parts. Using family tables, you can:

* Create and store large numbers of objects simply and compactly
* Save time and effort by standardizing part generation
* Generate variations of a part from one part file without having to re-create and generate each one
* Create slight variations in parts without having to use relations to change the model
* Create a table of parts that can be saved to a print file and included in part catalogs

COAch™ for Pro/E

If your system has **COAch™ for Pro/E** installed, review Advanced Modeling, Family Table

Family tables promote the use of standardized components. They let you represent your actual part inventory in Pro/E. Moreover, families make it easy to interchange parts and subassemblies in an assembly, because instances from the same family are automatically interchangeable with each other.

Family Table Structure

Family tables are essentially spreadsheets, consisting of columns and rows. You use the Family Table dialog box to create and modify family tables. Family tables include:

* The base object (generic object or *generic*) on which all members of the family are based
* Dimensions and parameters, feature numbers, user-defined feature names, and assembly member names that are selected to be table-driven (*items*)
* Names of all family members (*instances*) created by the table and the corresponding values for each of the table-driven items

Rows contain instances and their corresponding values; ***columns*** are used for items. The column headings include the *instance name* and the names of all of the *dimensions, parameters, features, members,* and *groups* that were selected for the table. **Dimensions** are listed by name (for example, **d9**) with the associated symbol name (if any) on the line below it (for example, depth). **Parameters** are listed by name (dim symbol). **Features** are listed by feature number (for example, F107) with the associated feature type (for example, [cut]) or feature name on the line below it. The generic model is the first row in the table. The table entries belonging to the generic can be changed only by modifying the actual part, suppressing, or resuming features; *you cannot change the generic model by editing its entries in the family table.*

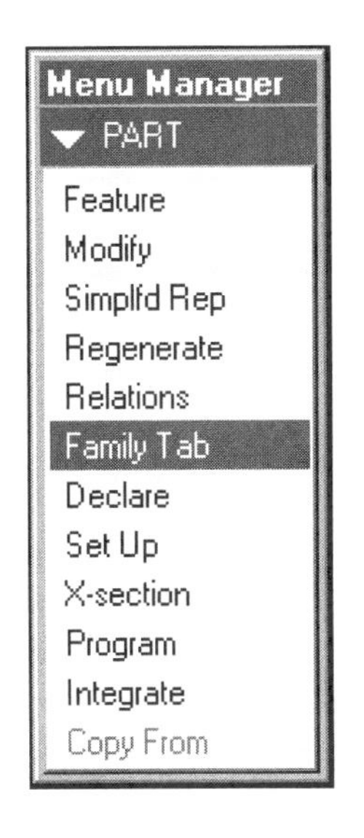

Family table names are not case-sensitive. For each instance, you can define whether a feature, parameter, or assembly name is used in the instance either by indicating whether it is present in the instance (**Y** or **N**) or by providing a numeric value (in the case of a dimension). All dimension cells must have a value, either a number or an asterisk (*) to use the generic's value. All aspects of the generic model that are not included in the family table automatically occur in each instance. As an example, if the generic model has a parameter called Material with a value Aluminum, all instances will have the same parameter and value.

To Add a Dimension to a Family Table

1. From the PART menu, pick **Family Tab.** The Family Table dialog box opens (Fig. 21.40).
2. Pick the **Add/Del** Item icon. The Family Items dialog box appears.
3. Pick **Dimension** from the Add Item group.
4. Select the dimension you want to add to the family table. You can add regular dimensions, pattern dimensions, and geometric tolerances.
5. Pick **OK**.

To Add a Feature to a Family Table

If you give the feature a name, the name appears at the top of that feature's column in the family table. When a feature is omitted, all its dimensions and children are omitted or ignored. If a feature is present, all its parents are present. In cases of conflict, omitted features supersede present features. When dealing with patterns, the system automatically records the parent feature completely.

1. From the PART or ASSEMBLY menu, pick **Family Tab.** The Family Table dialog box opens (Fig. 21.40).
2. Pick the **Add/Delete** Table Columns icon. The Family Items dialog box appears (see left-hand column).
3. Pick **Feature** from the Add Item group. The GET SELECT menu appears.
4. Select the feature on the screen or pick **Sel By Menu** and enter the feature name, number, or ID.
5. Pick **OK.**

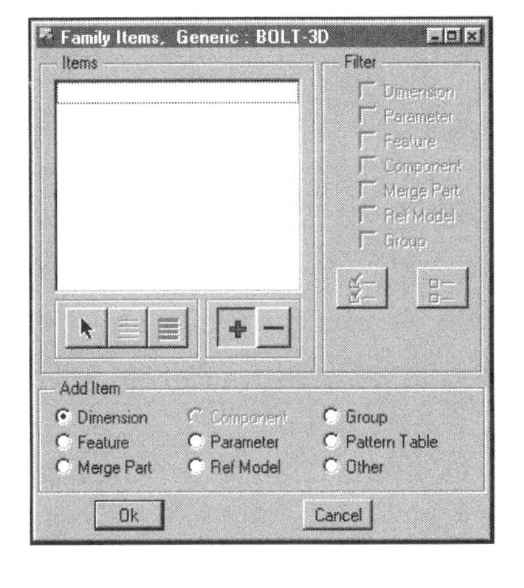

Figure 21.40
Family Table Dialog Box

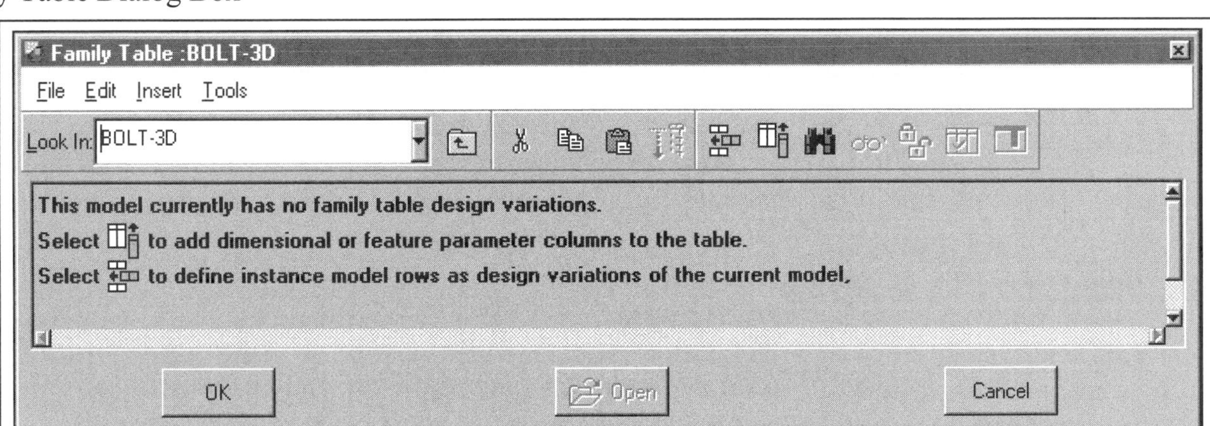

Creating a Family Table

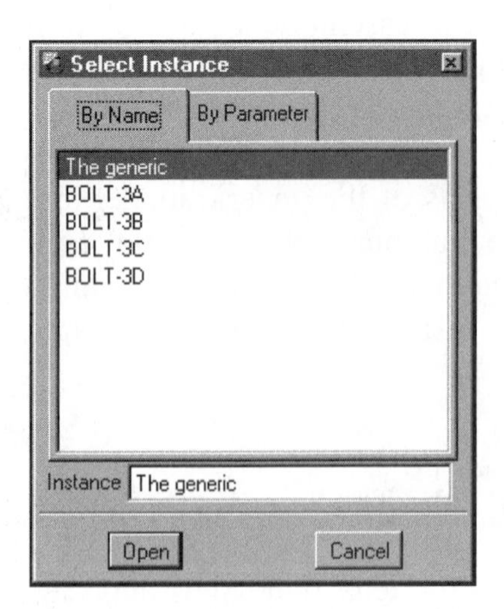

You will be creating a Family Table from the existing model, Coupling Fitting. The model is the **Generic.** Each variation is referred to as an **Instance.**

When you create a Family Table, Pro/E allows you to *select dimensions,* which can vary between instances. You can also *select features* to add to the Family Table. Features can vary by being suppressed or resumed in an instance (Fig. 21.41).

When you are finished selecting items (e.g., dimensions, features, and parameters), the Family Table is automatically generated.

When you complete the Family Table, Pro/E evaluates the Family Table entries and creates instances for each new name. The values in the table for each instance name are used to vary the dimension and parameter names that head each column in the Family Table. Instances of the generic model can also have family tables defining further the variations of the original part. Geometric tolerances, pattern dimensions, and dimensional tolerance values can be added as dimensions.

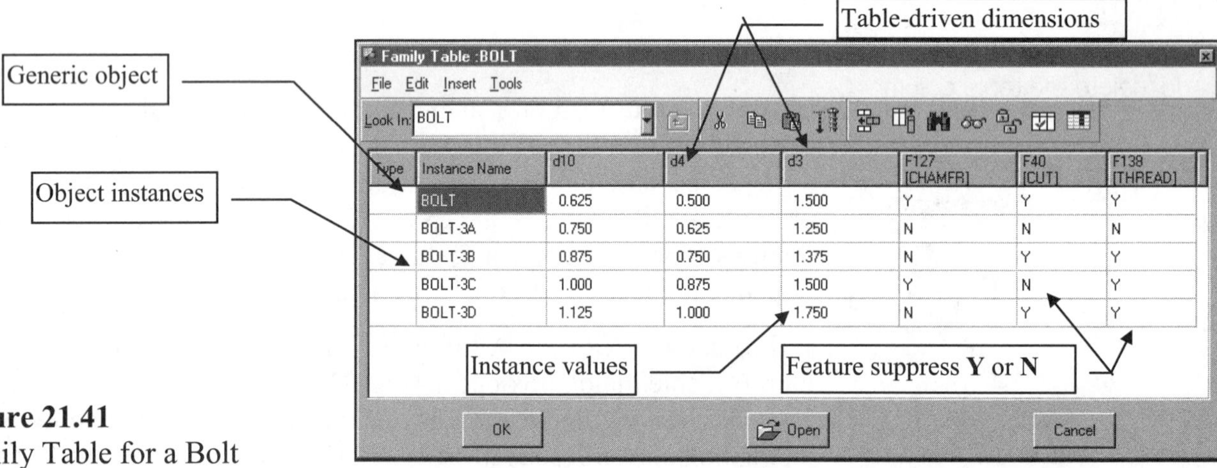

Type	Instance Name	d10	d4	d3	F127 [CHAMFR]	F40 [CUT]	F138 [THREAD]
	BOLT	0.625	0.500	1.500	Y	Y	Y
	BOLT-3A	0.750	0.625	1.250	N	N	N
	BOLT-3B	0.875	0.750	1.375	N	Y	Y
	BOLT-3C	1.000	0.875	1.500	Y	N	Y
	BOLT-3D	1.125	1.000	1.750	N	Y	Y

Figure 21.41
Family Table for a Bolt

When adding features and components, enter an **N** to suppress the feature or component, a **Y** to add the feature or component, or the *name* of a component, which will automatically **Replace** the original component in the assembly.

Each instance must have a unique name. Consider each instance a separate model, as in Figure 21.41, where the bolts are named **BOLT-3A**, **BOLT-3B**, **BOLT-3C** and **BOLT-3D.**

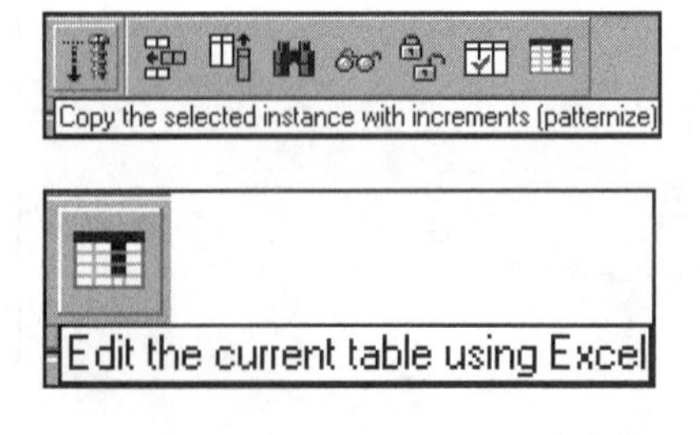

If the variation can be described by a pattern, the **Patternize** option can automatically proliferate the table.

You can edit family table information in Microsoft Excel. This functionality is available only if you are using a Windows NT or Windows 95 (or higher) environment and have Excel installed. With Excel you can insert new rows, create additional instances, and reorder family tables. You can delete a family table by picking Family Table (with the generic model active). When the FAMILY TABLE dialog box appears, pick **Edit ⇒ Delete Entire Table** and then confirm the deletion (see the next page, left-hand column).

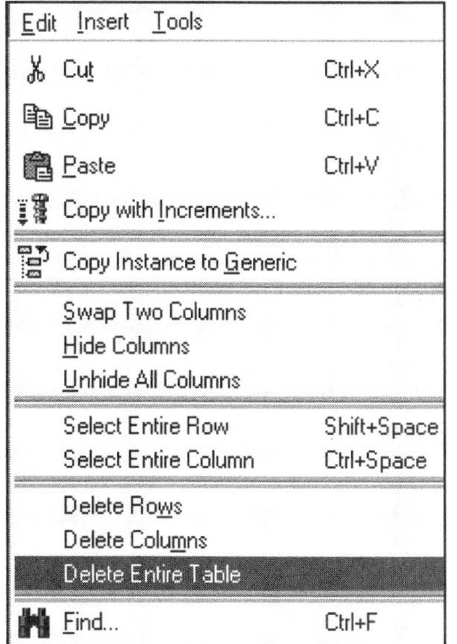
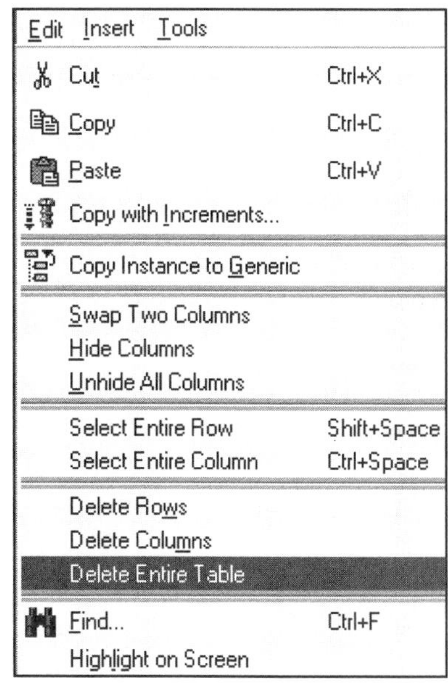

Complete the Coupling Fitting by creating three instances of the generic model. The instances will have varying dimensions for the length of the body and the length of the small hexagonal feature (Figure 21.42).

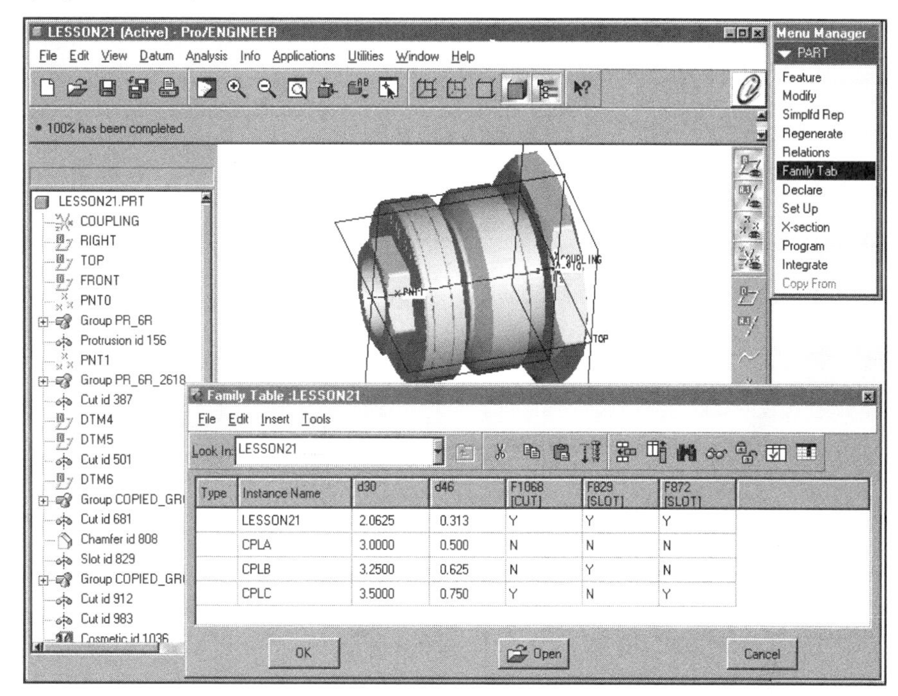

Type	Instance Name	d30	d46	F1068 [CUT]	F829 [SLOT]	F872 [SLOT]
	LESSON21	2.0625	0.313	Y	Y	Y
	CPLA	3.0000	0.500	N	N	N
	CPLB	3.2500	0.625	N	Y	N
	CPLC	3.5000	0.750	Y	N	Y

Figure 21.42
Family Table for Coupling

Use the following commands:

Family Tab [the Family Table dialog box opens (Fig. 21.43)] ⇒ (pick the **Add/Del** Item icon ⇒ [the Family Items dialog box appears (see left-hand column)] ⇒ ●**Dimension** (from the Add Item group) ⇒ [pick on the large cylindrical feature and the small hexagonal feature to display the dimensions (Fig. 21.44)] ⇒

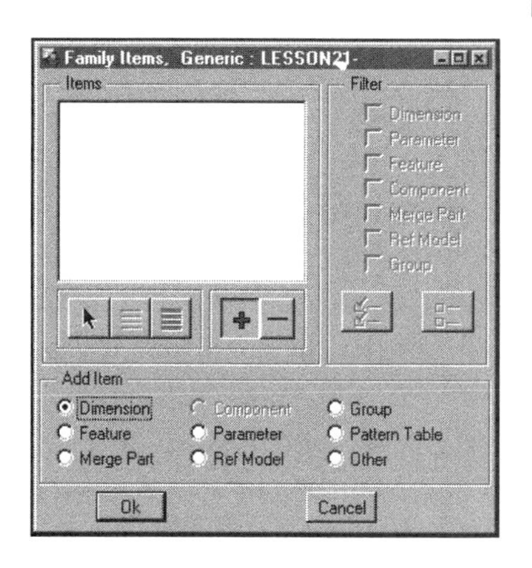

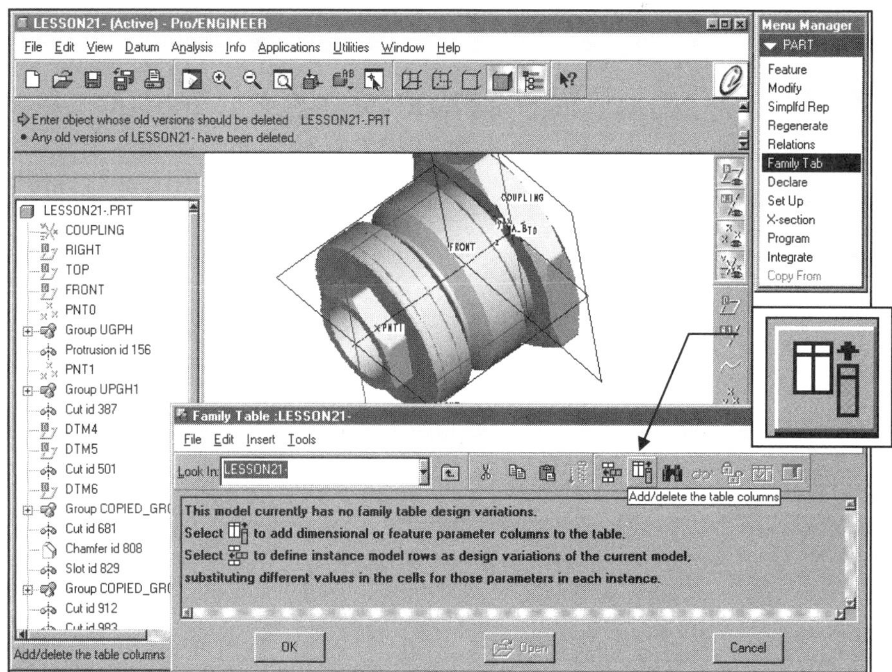

Figure 21.43
Family Table Dialog Box

[pick the **2.0625** dimension first and then the **.3130** dimension (Fig. 21.44)] ⇒ •**Feature** ⇒ (pick the cut that created the text **COUPLECO)** ⇒ (pick the first slot-groove) ⇒ (pick the second slot-groove) ⇒ **Done Sel** ⇒ **Done** (Fig. 21.45) ⇒ **OK** (ending the selection of varying dimensions and features on the Family Table) ⇒

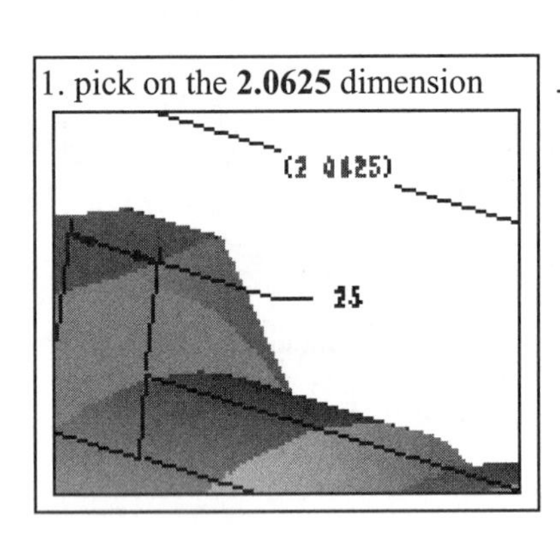

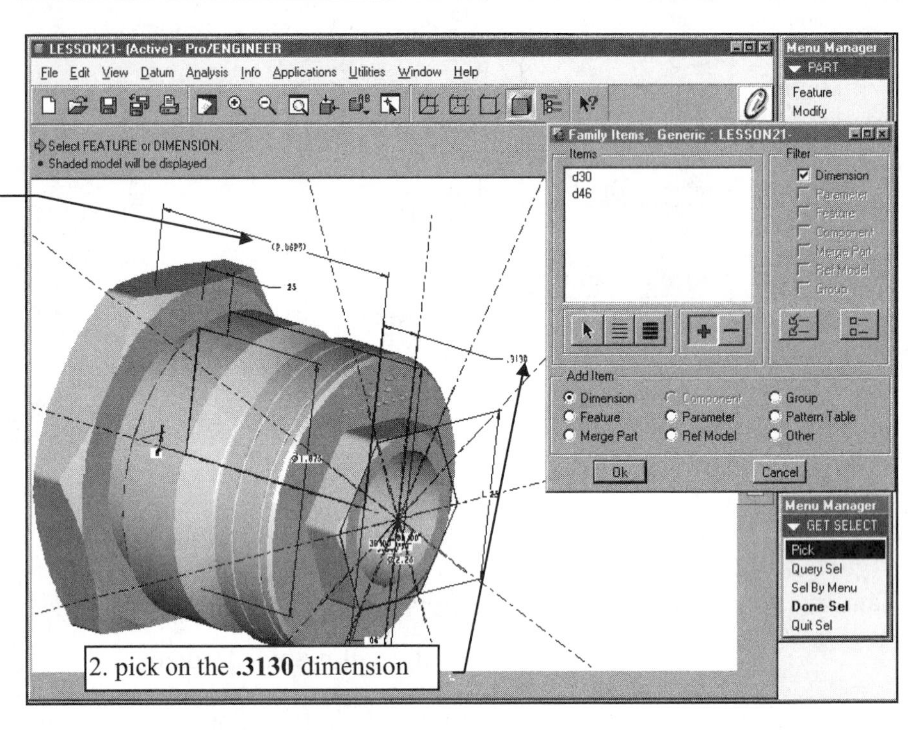

Figure 21.44
Selecting Varying Dimensions and Features

Besides dimensions, you just included *features* in Family Tables to control their presence in the design instance. Features in a table can be either present or suppressed. This allows you to create multiple versions of a design from the same basic part (using the same generic model).

When you add features to a table, they are also listed in columns to the right of the **Generic** name. As with dimensions, each feature has its own column. If the feature is displayed (and they all can be in the **Generic**), the column for that feature contains a **Y** for Yes.

You add features in the same manner as you did for dimensions. In each column for the instance, you must enter a **Y**, an **N**, or an asterisk (*). The **N** causes Pro/E to suppress that feature in the instance. An asterisk (*) causes Pro/E to use the **Generic** value or feature. Because your choice is normally a **Y**, the feature will be displayed. Add three different features to the Family Table.

Because **Y** is the entry in all the columns in the **GENERIC**, it is also the default value for any other instance in the Family Table. Therefore, you could enter an asterisk (*) instead of a **Y** if you wanted. An implication of this would be that if anyone replaced the **GENERIC** later on, the asterisk would cause that change to be reflected in any instances with asterisks.

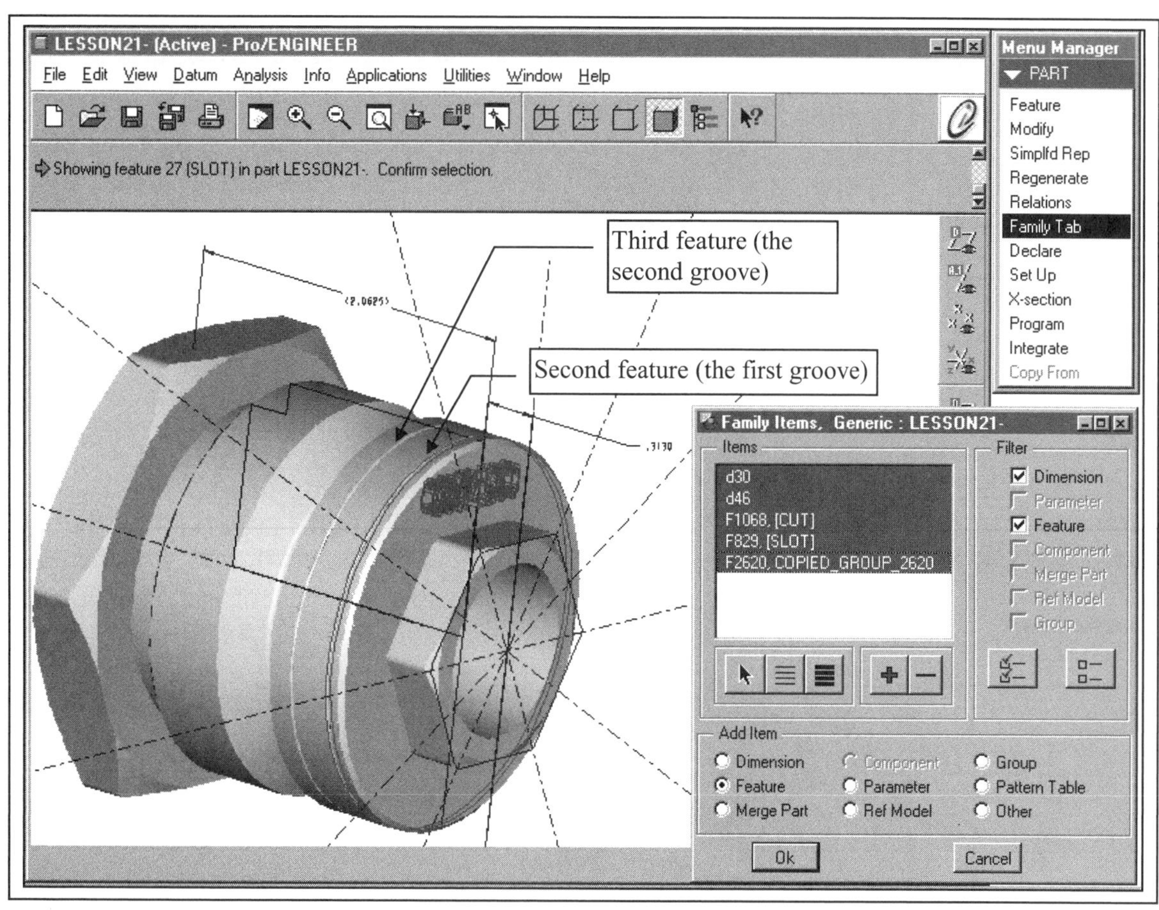

Figure 21.45
Family Items Completed

Add three instances using the following commands:

First feature (the text cut **COUPLECO**) added to the Family Table (shown highlighted)

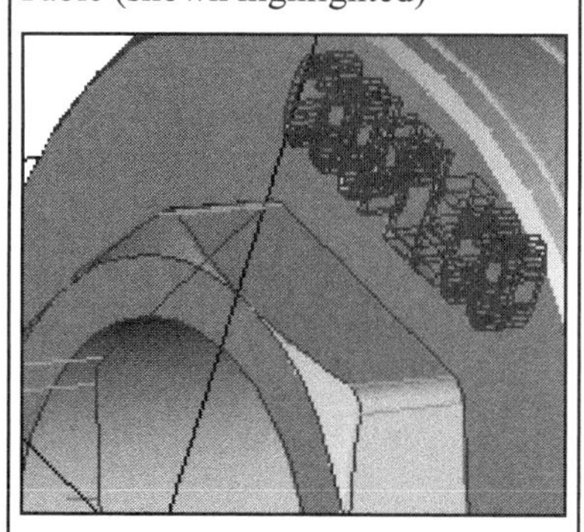

[select [icon] three times to add instances (Fig. 21.46] ⇒ (click on the name of the first instance **NEW_INSTANCE** and type **CPLA** ⇒ (click on the name of the second instance **NEW_INST_1,** and type **CPLB**) ⇒ [click on the name of the third instance **NEW_INST_2,** and type **CPLC** as the third instance (Fig. 21.47)] ⇒ (change the values for the instance dimensions)

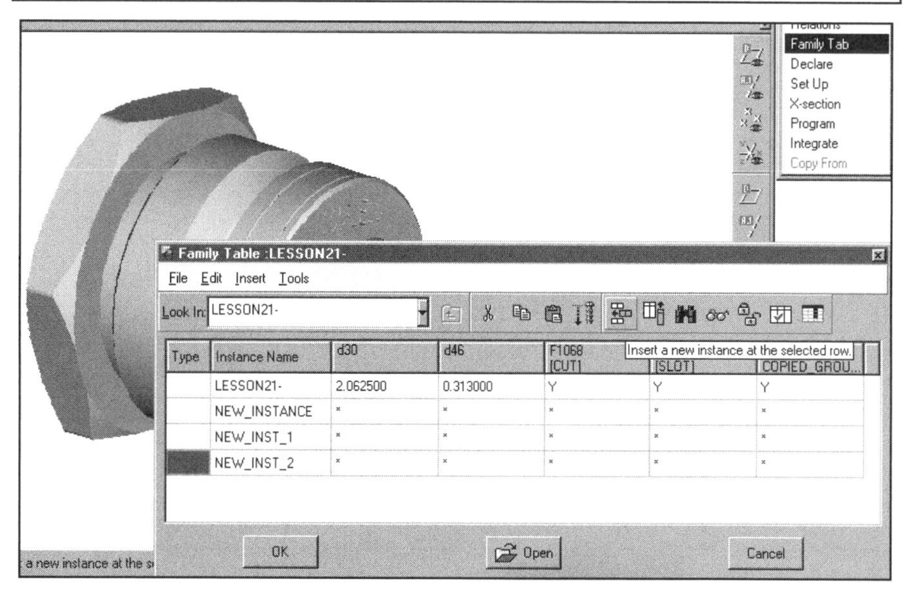

Figure 21.46
Three Instances Created

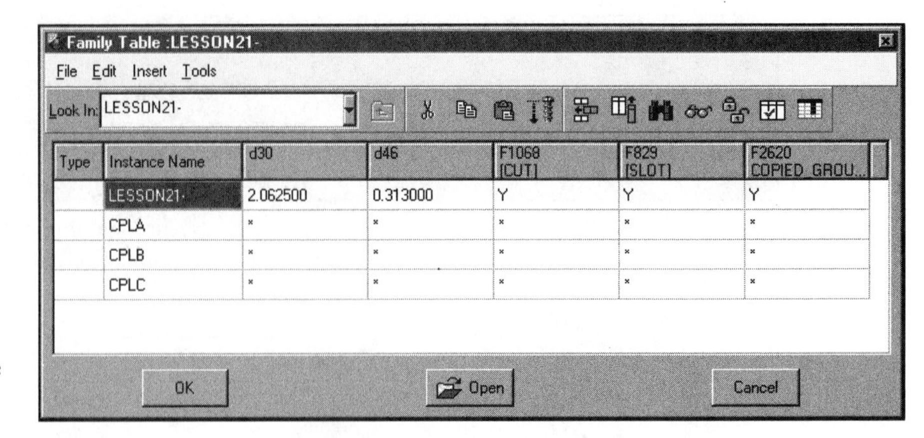

Family Table :LESSON21-

Type	Instance Name	d30	d46	F1068 [CUT]	F829 [SLOT]	F2620 COPIED_GROU...
	LESSON21-	2.062500	0.313000	Y	Y	Y
	CPLA	*	*	*	*	*
	CPLB	*	*	*	*	*
	CPLC	*	*	*	*	*

Figure 21.47
Changed Names of the Three Instances

To change the value for the first instance, click in the field just below the text **2.0625** (Fig. 21.45). Type **3.00** as the width of this instance. Click in the field just below the **.3130** value and type **.500.** This completes the dimensional values for instance **CPLA.** Continue filling in the table as shown in Figure 21.48.

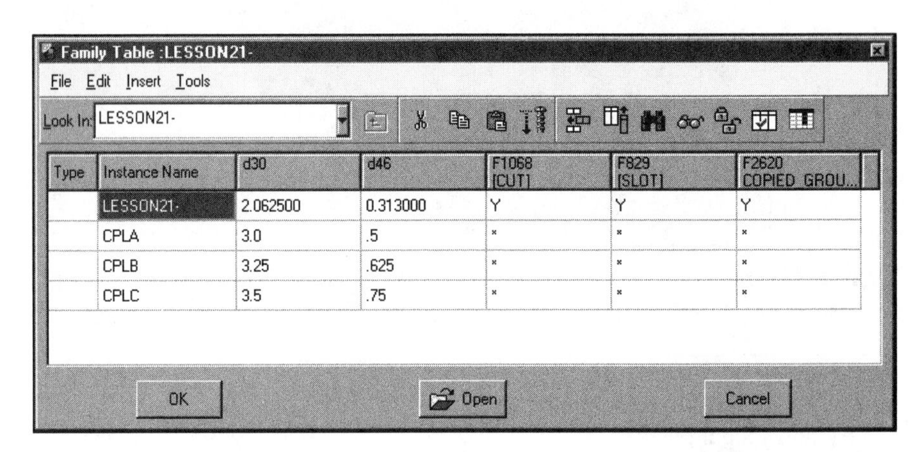

Family Table :LESSON21-

Type	Instance Name	d30	d46	F1068 [CUT]	F829 [SLOT]	F2620 COPIED_GROU...
	LESSON21-	2.062500	0.313000	Y	Y	Y
	CPLA	3.0	.5	*	*	*
	CPLB	3.25	.625	*	*	*
	CPLC	3.5	.75	*	*	*

Figure 21.48
Changed Dimensional Values of the Three Instances

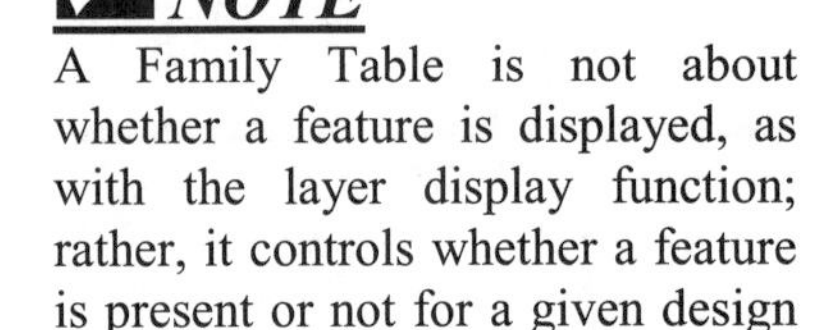

NOTE
A Family Table is not about whether a feature is displayed, as with the layer display function; rather, it controls whether a feature is present or not for a given design instance.

Next you will establish what features are suppressed in each instance. Complete the Family Table as shown in Figure 21.49(a). Pick in the field below each feature listing, and type a **Y** or an **N,** depending on the required value (or use **▼**).

Family Table :LESSON21-

Type	Instance Name	d30	d46	F1068 [CUT]	F829 [SLOT]	F2620 COPIED_GROU...
	LESSON21-	2.062500	0.313000	Y	Y	Y
	CPLA	3.0	.5	N	N	N
	CPLB	3.25	.625	N	Y	N
	CPLC	3.5	.75	Y	N	Y

Figure 21.49(a)
Changed Feature Values

Edit the current table using Excel

You can edit the table using Excel as shown in Figure 21.49(b).

? Pro/HELP

Remember to use the help available on Pro/E by highlighting a command and pressing the right mouse button.

Family Tab

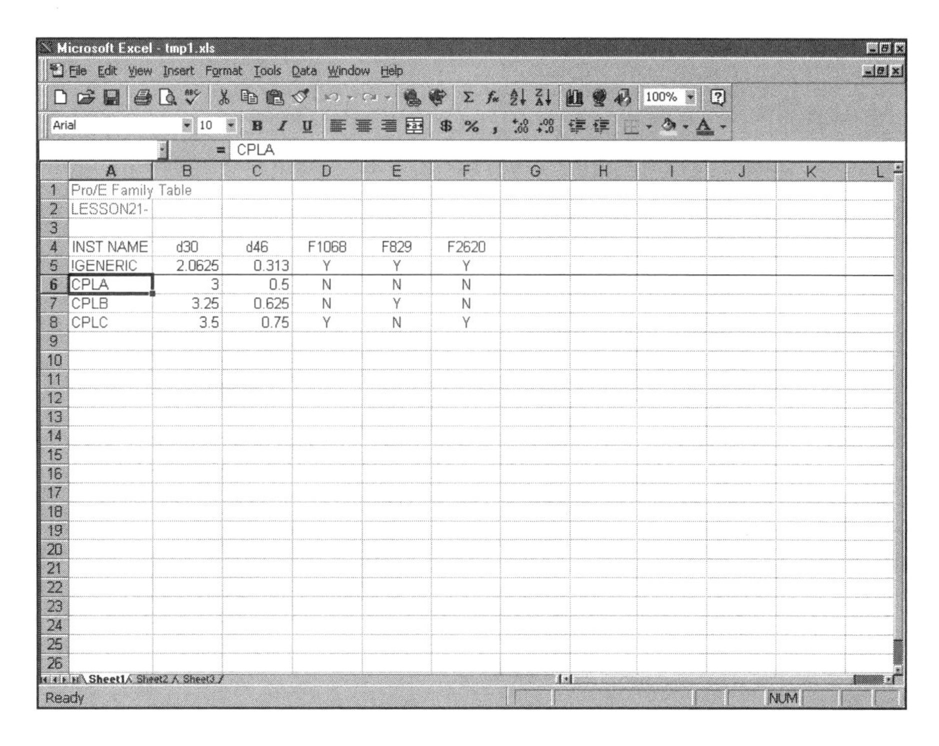

	A	B	C	D	E	F
1	Pro/E Family Table					
2	LESSON21-					
3						
4	INST NAME	d30	d46	F1068	F829	F2620
5	IGENERIC	2.0625	0.313	Y	Y	Y
6	CPLA	3	0.5	N	N	N
7	CPLB	3.25	0.625	N	Y	N
8	CPLC	3.5	0.75	Y	N	Y

Figure 21.49(b)
Editing the Table with Excel

After completing the table, retrieve the **CPLA** instance you defined in the Family Table. Notice that only the features added to the Family Table that have an entry of **Y** appear on the instance.

Use the following commands:

[pick on **CPLA** (from the Family Table dialog box) (Figure 21.50)] ⇒ **Open** ⇒ (adjust your windows to view both the generic model and instance **CPLA**) ⇒ **Window** (from the menu bar on the generic part) ⇒ **Activate** ⇒ **Family Tab** ⇒ [pick on **CPLB** (from the Family Table dialog box) ⇒ **Open** ⇒ [adjust your windows (Fig. 21.51)]

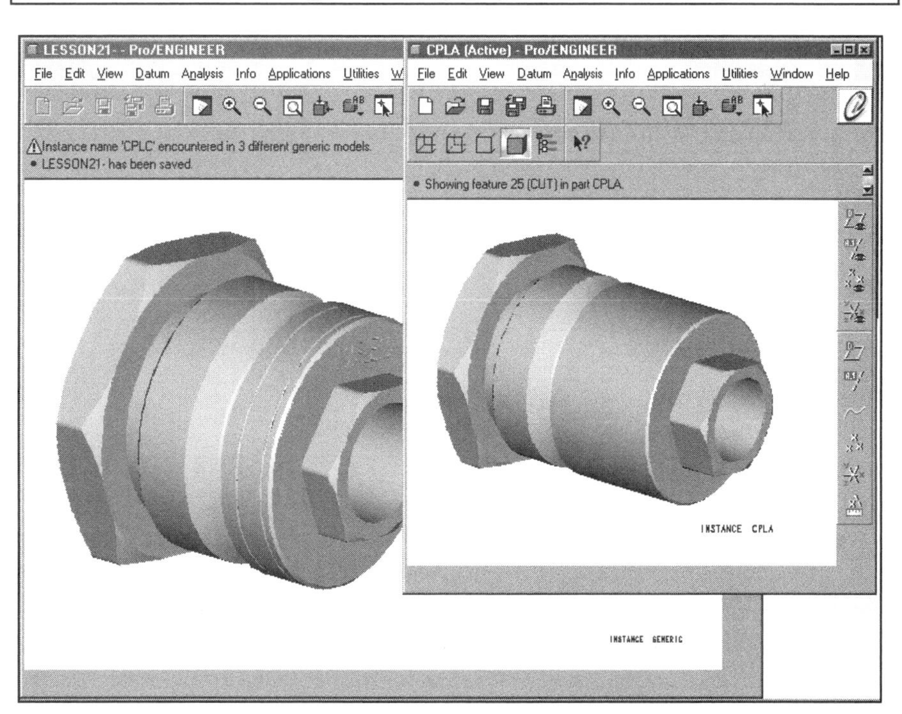

Figure 21.50
Body of instance **CPLA** is longer and has no grooves. Note that the text protrusion **COUPLECO** is suppressed

File ⇒ Save ⇒ ✔
File ⇒ Delete ⇒
Old Versions ⇒ ✔

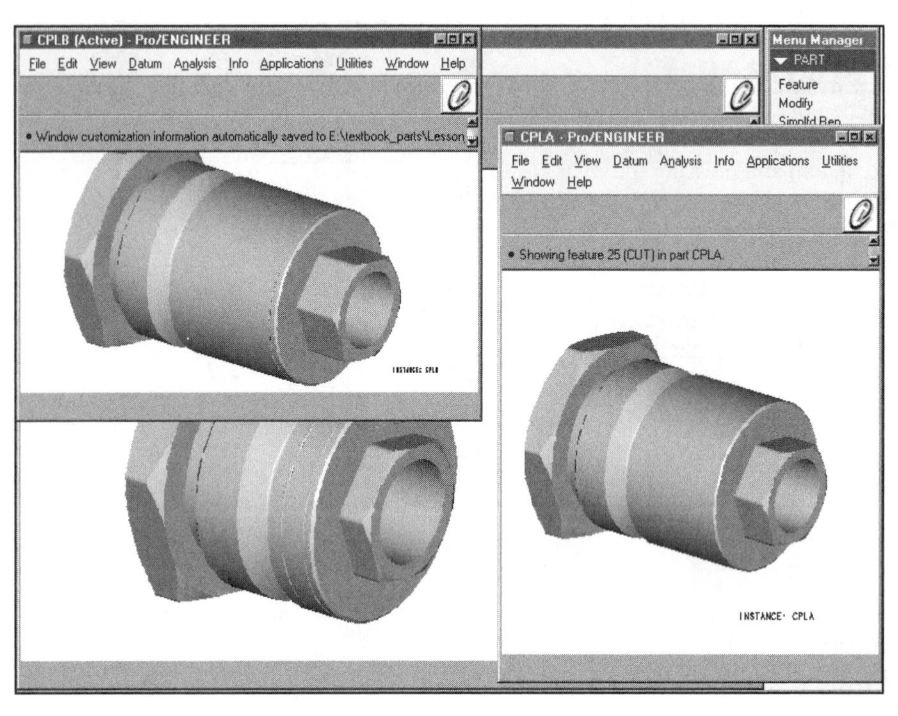

Figure 21.51
Displaying Instance **CPLA**

You now have three *instances* and a *generic* of the Coupling Fitting. When you open this part in **Part** mode or add it to an assembly or drawing, you will be prompted to select the **Generic** or **Instance** from the **Select Instance** dialog box (Fig. 21.52).

Figure 21.52 File ⇒ Open ⇒ (select the generic part from the File Open dialog box) ⇒ (pick the generic or an instance of the model from the Select Instance dialog box)

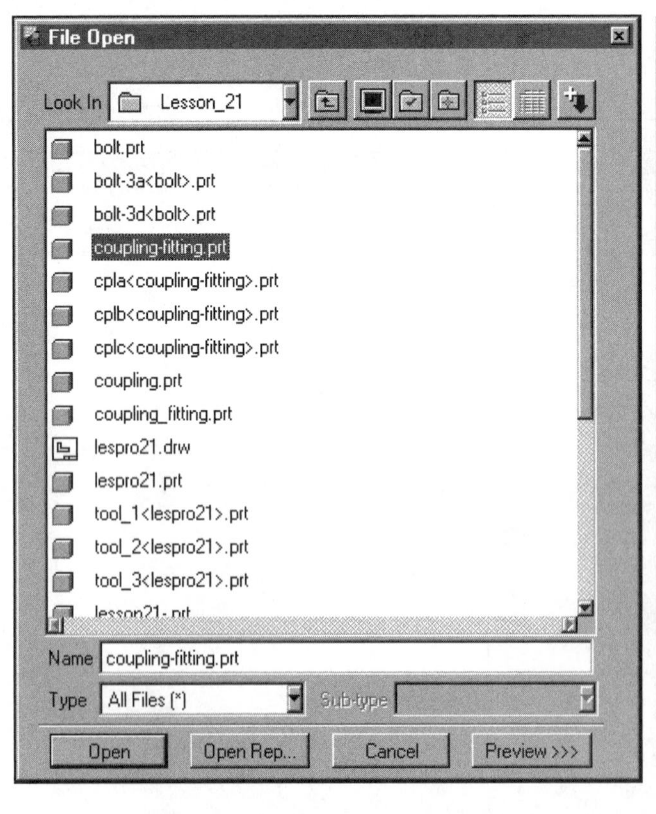

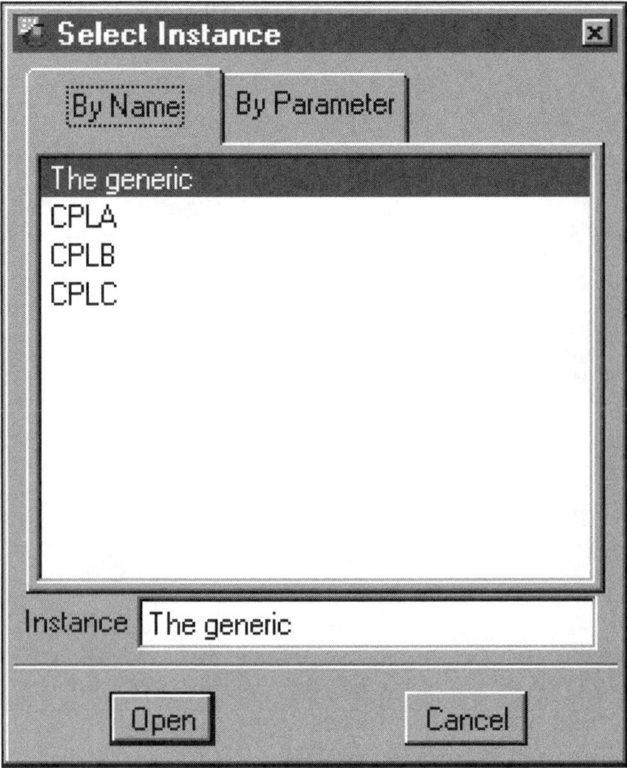

Lesson 21 Project

Tool Body

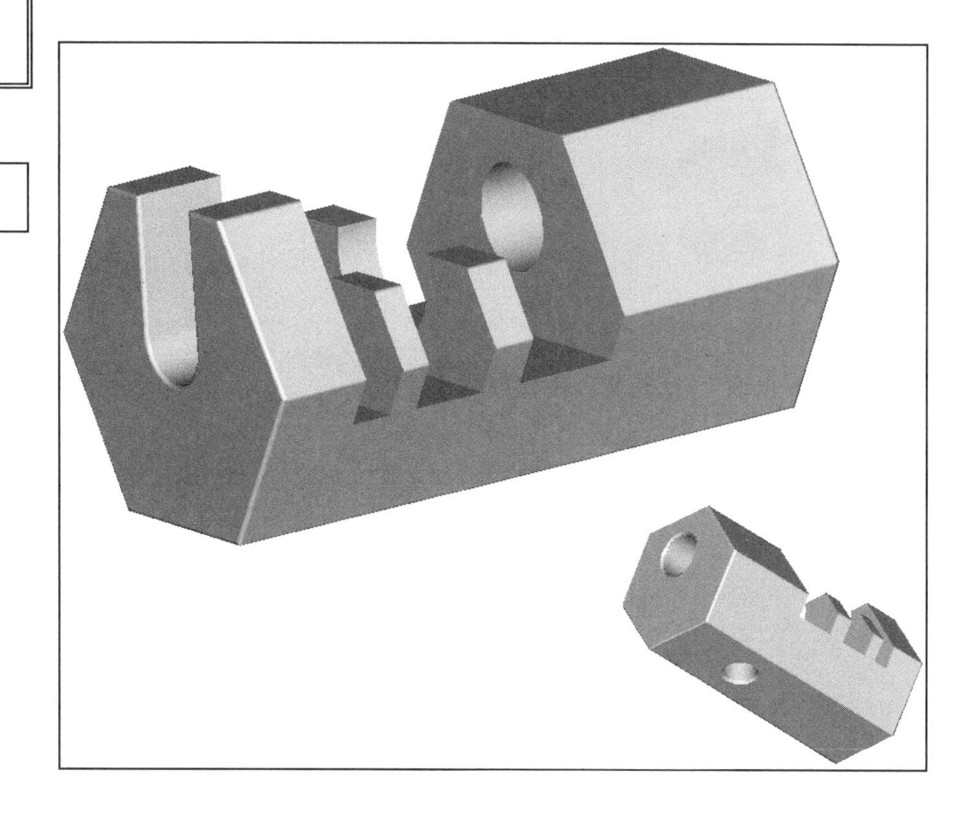

Figure 21.53
Tool Body (**Generic**)

EGD REFERENCE
Fundamentals of Engineering Graphics and Design
by L. Lamit and K. Kitto
Read Chapter 21.
See page 864.

Tool Body

This advanced **lesson project** is a **Tool Body,** which uses commands similar to those for creating the **Coupling Fitting**. Create the part shown in Figures 21.51 through 21.56. Use the **DIPS** in Appendix D to plan out the feature creation sequence and the parent-child relationships for the part. The Tool Body is made of *steel.*

Use a hexagonal UDF from your library for the first protrusion. If you do not have Pro/LIBRARY installed on your system, create a hexagonal UDF and save it in your own library. Create a Family Table with three instances. Add the following dimensions and features to the Family Table:

	Length	Hexagon	Hole (.1875)
Generic	1.500	.750	Y
(Instances)			
Tool_1	1.625	.875	N
Tool_2	1.750	1.000	Y
Tool_3	1.875	1.125	N

NOTE
When creating the **Tool Body** in the **Lesson Project**, use the following when redefining the UDF:

Redefine ⇒ (pick the UDF) ⇒ (Select **Depth** from the dialog box) ⇒ **Define** ⇒ **UpTo Surface** (pick **DTM4**) ⇒ **Done** ⇒ **OK**

Figure 21.54 Tool Body (**Generic** and **Instances**)

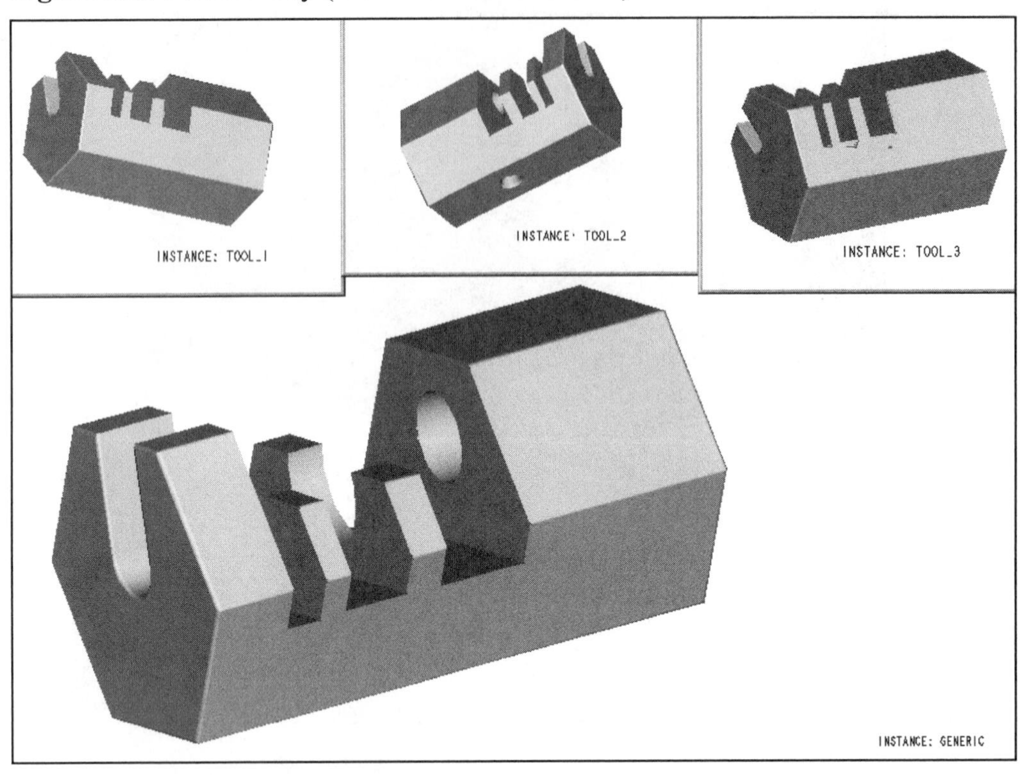

Figure 21.55 Tool Body Family Table

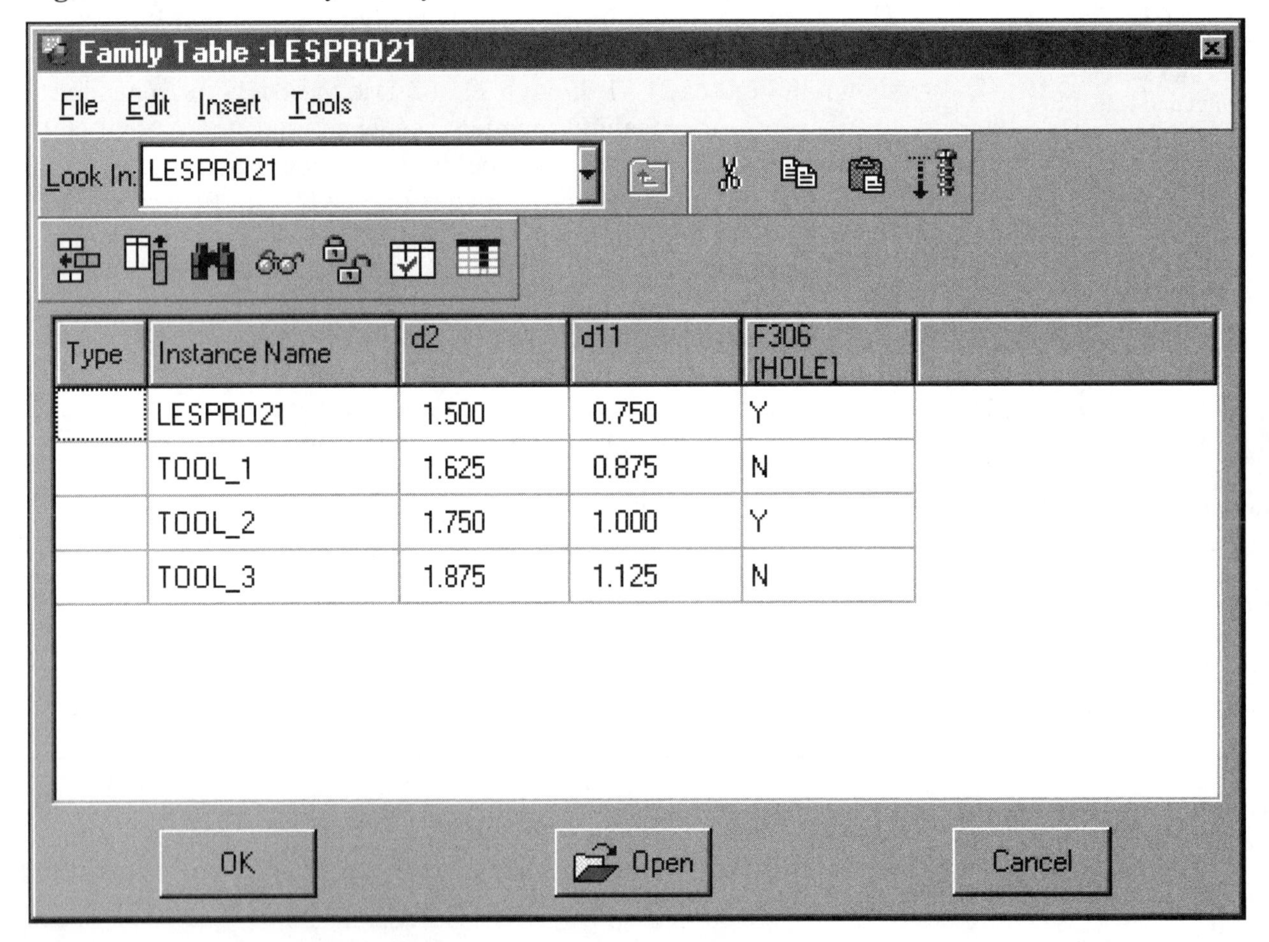

Type	Instance Name	d2	d11	F306 [HOLE]
	LESPRO21	1.500	0.750	Y
	TOOL_1	1.625	0.875	N
	TOOL_2	1.750	1.000	Y
	TOOL_3	1.875	1.125	N

Figure 21.56 Tool Body (**Generic**) Detail Drawing

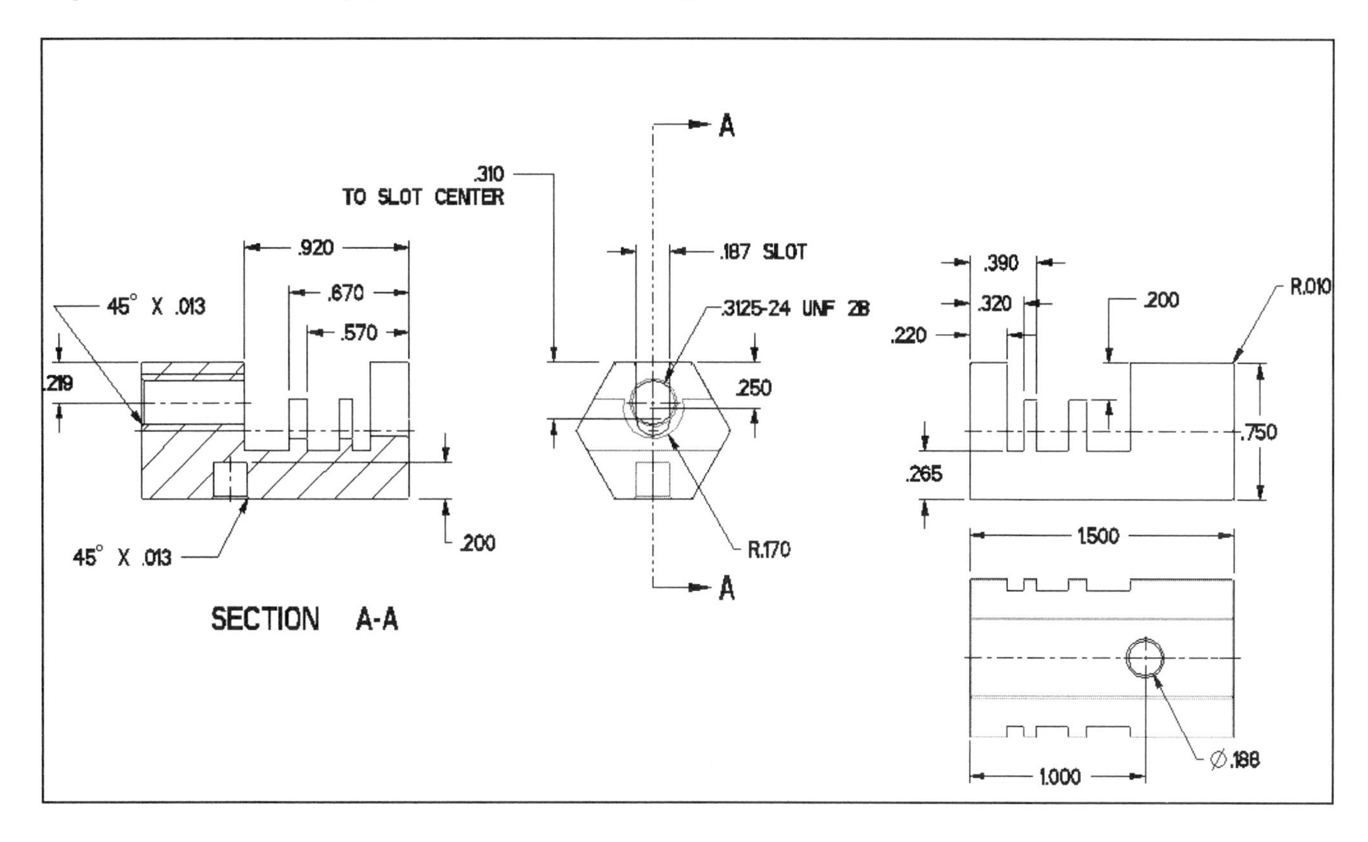

Figure 21.57 Tool Body Detail Drawing (**SECTION A-A**)

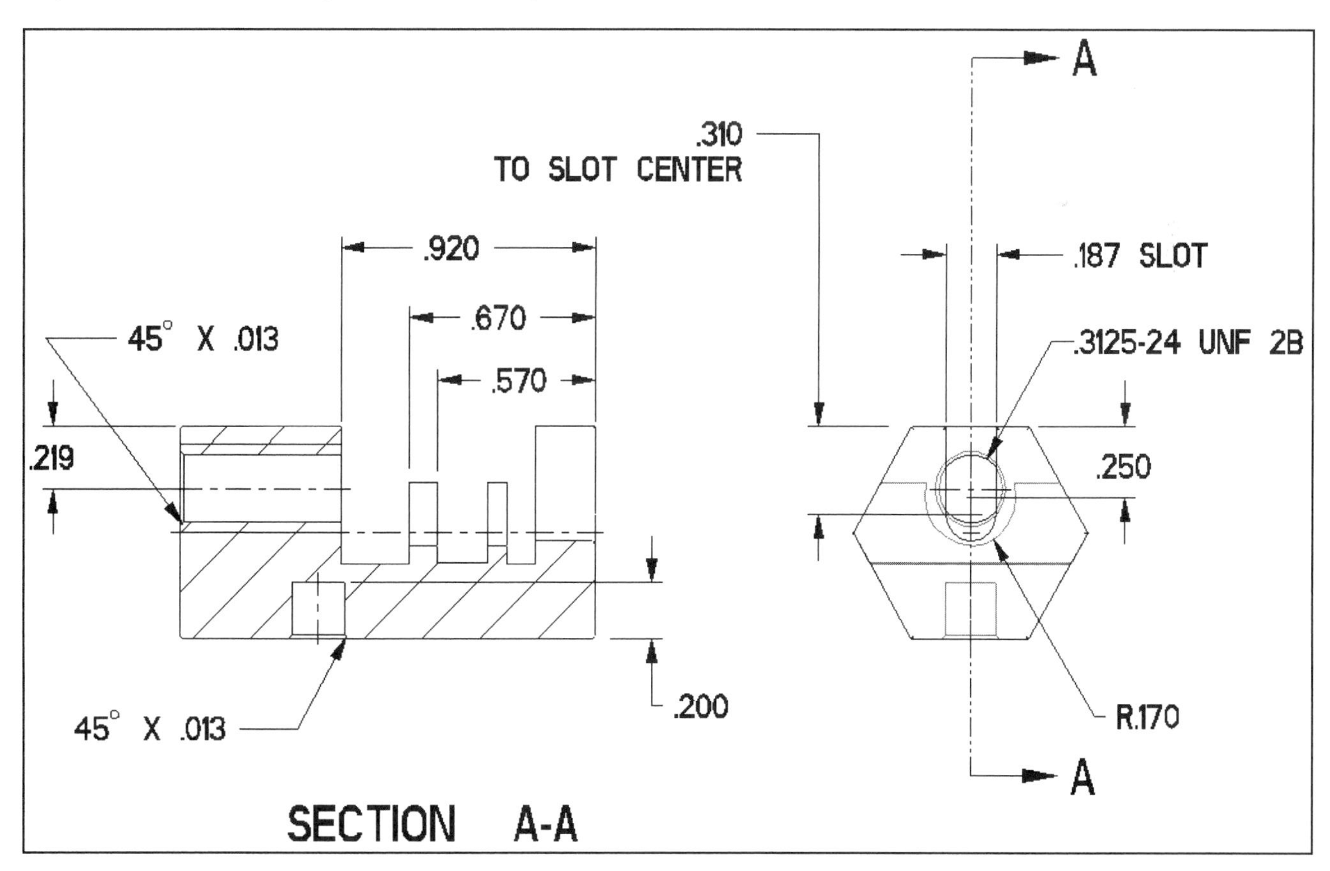

Figure 21.58 Generic Tool Body

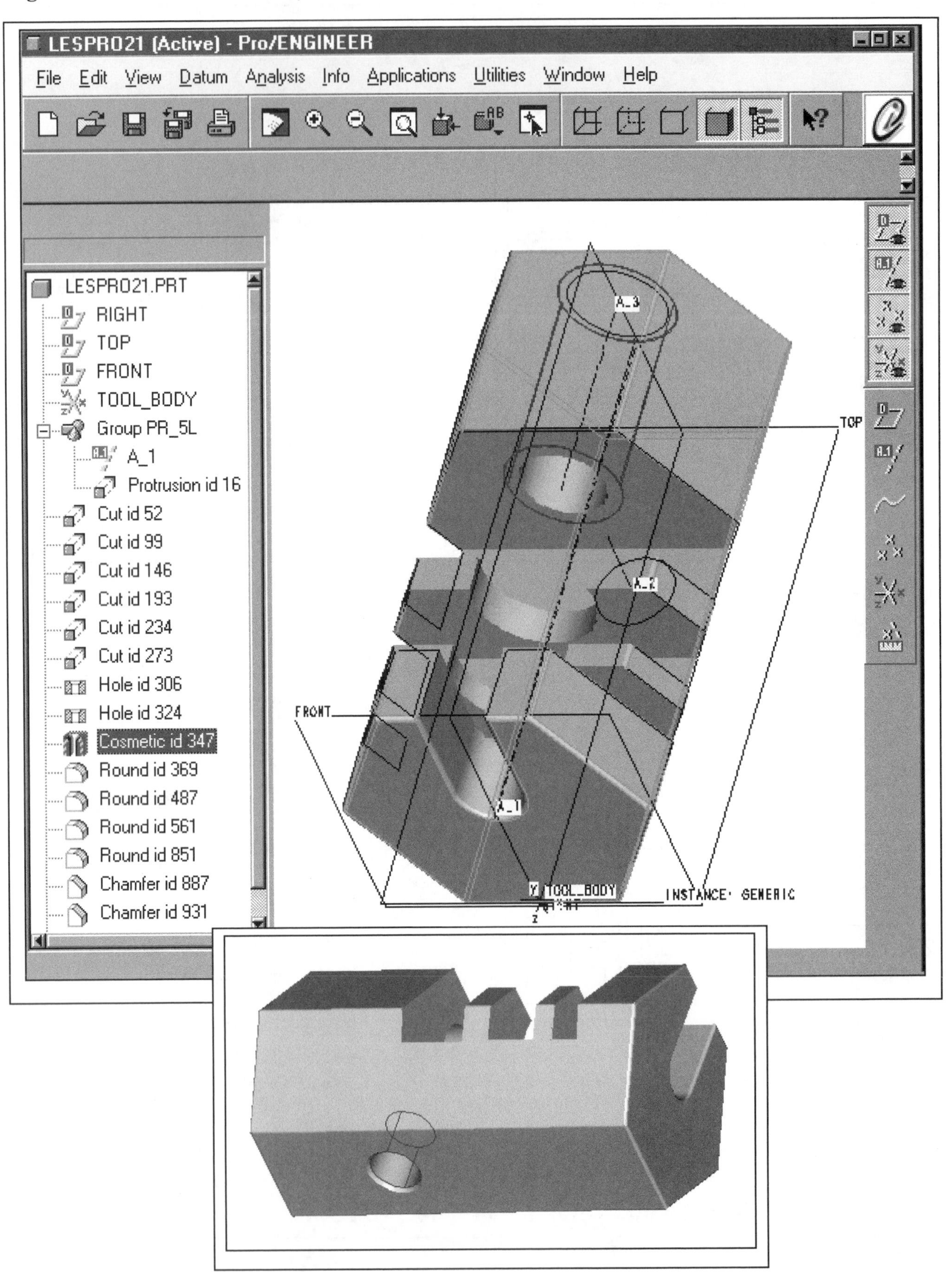

Lesson 22

Pro/NC

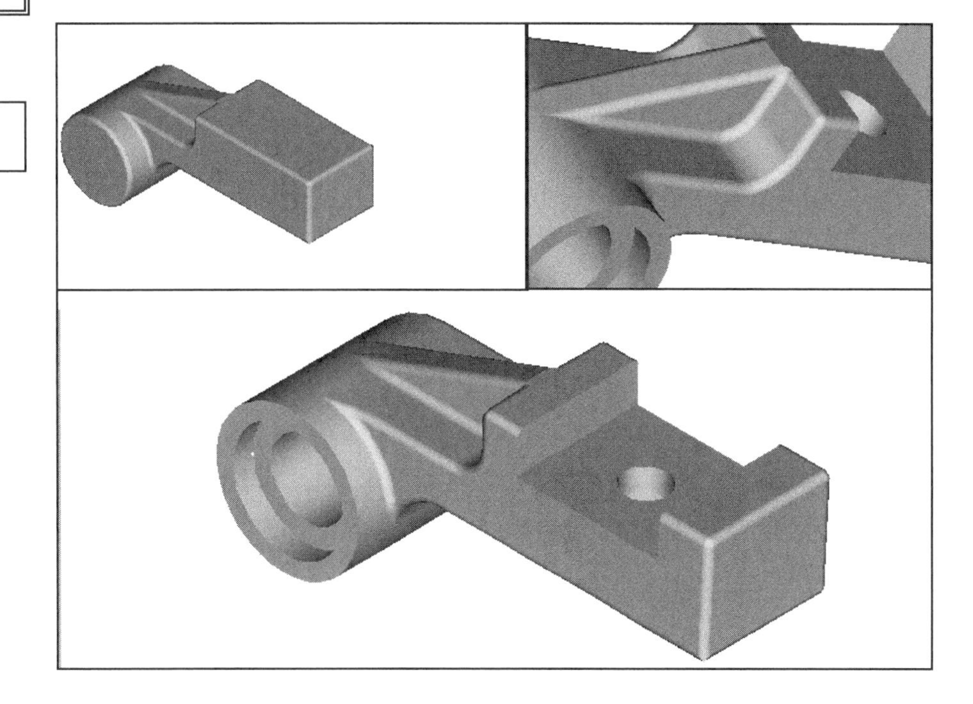

Figure 22.1
Adjustable Guide,
Casting, and Machine Part

OBJECTIVES

1. **Understand the different models used in manufacturing**

2. **Merge a design part and a workpiece to create a manufacturing model for machining**

3. **Implement machining setup elements**

4. **Understand and set manufacturing parameters**

5. **Set a tool's parameters**

6. **Create tool paths**

7. **Run simulations of tool paths**

8. **Utilize NC Check to analyze NC sequences**

EGD REFERENCE
Fundamentals of Engineering Graphics and Design
by L. Lamit and K. Kitto
Read Chapter 14.
See pages 498 and 852.

COAch™ for Pro/E

If you have **COAch for Pro/ENGINEER** on your system, go to SEARCH and do the Segment shown in Figure 22.3.

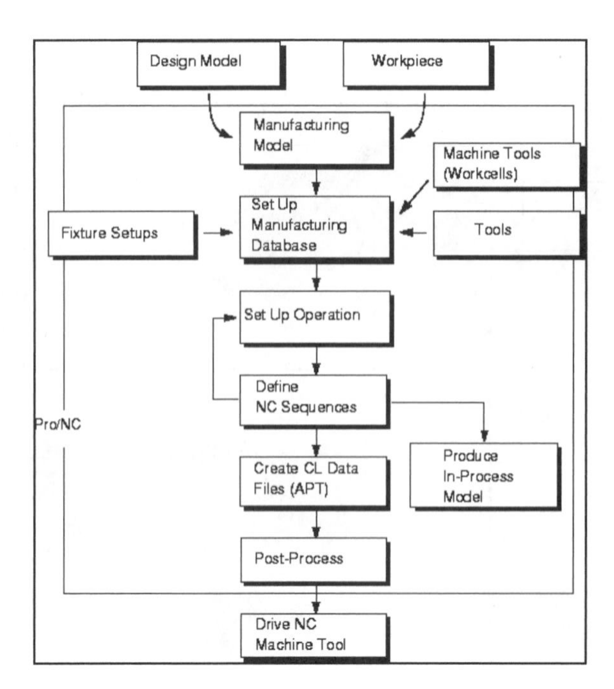

Figure 22.2
Pro/NC

Pro/NC

Pro/NC (Fig. 22.1 and 22.2) will create the data necessary to drive an NC machine tool to machine a Pro/E part. It does this by providing the tools to let the manufacturing engineer follow a logical sequence of steps to progress from a design model to ASCII CL data files that can be post-processed into NC machine data.

The following is an overview of terms that will be used throughout this lesson. These are basic Pro/NC concepts that are vital to understanding the module. Pro/NC (Fig. 22.3) is designed to map your manufacturing intent, as captured by the NC sequence parameters, onto geometry selected from the engineering part (design model). Throughout this lesson, the appropriate settings will be described and the techniques displayed to make this mapping clear.

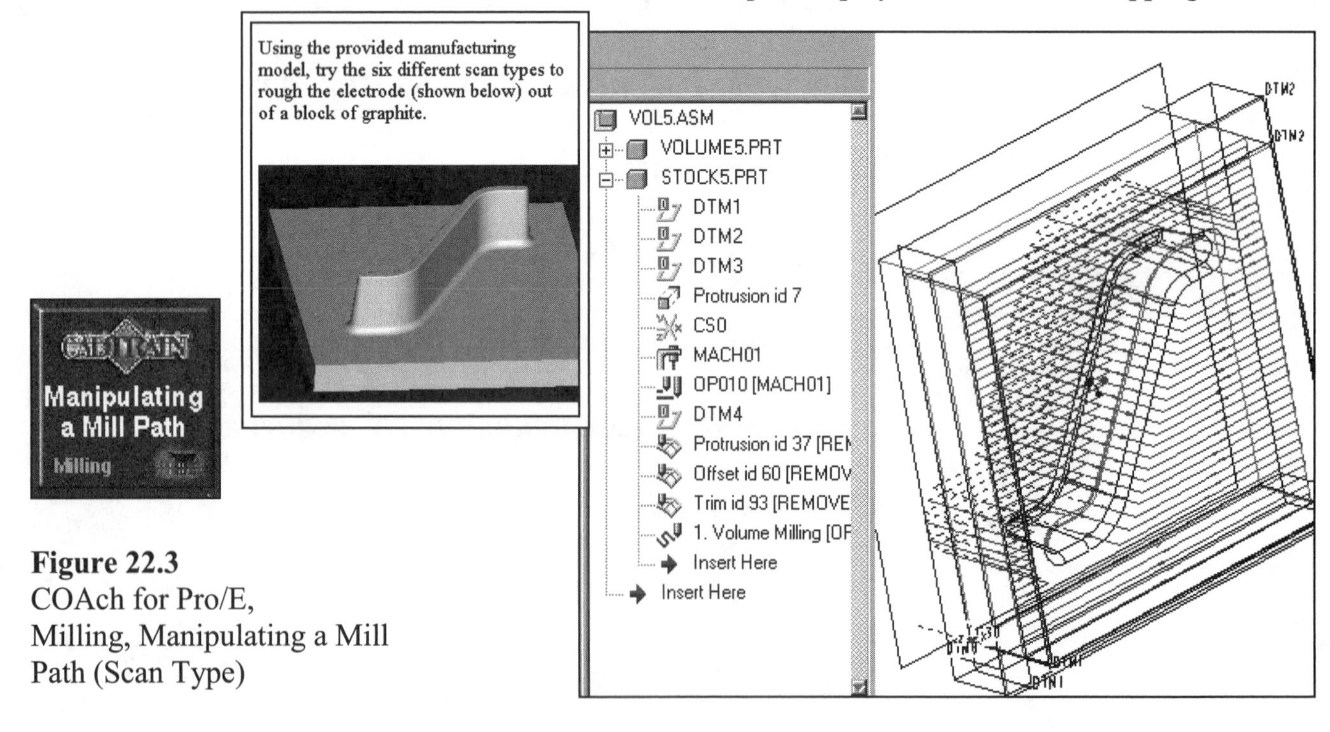

Figure 22.3
COAch for Pro/E, Milling, Manipulating a Mill Path (Scan Type)

Design Model

The Pro/E design model, representing the finished product, is used as the basis for all manufacturing operations. Features, surfaces, and edges are selected on the design model as references for each tool path. Referencing the geometry of the design model (Fig. 22.4 and Fig. 22.5) sets up an associative link between the design model and the workpiece. Because of this link, when the design model is changed, all associated manufacturing operations are updated to reflect the change. Parts, assemblies, and sheetmetal parts may be used as design models.

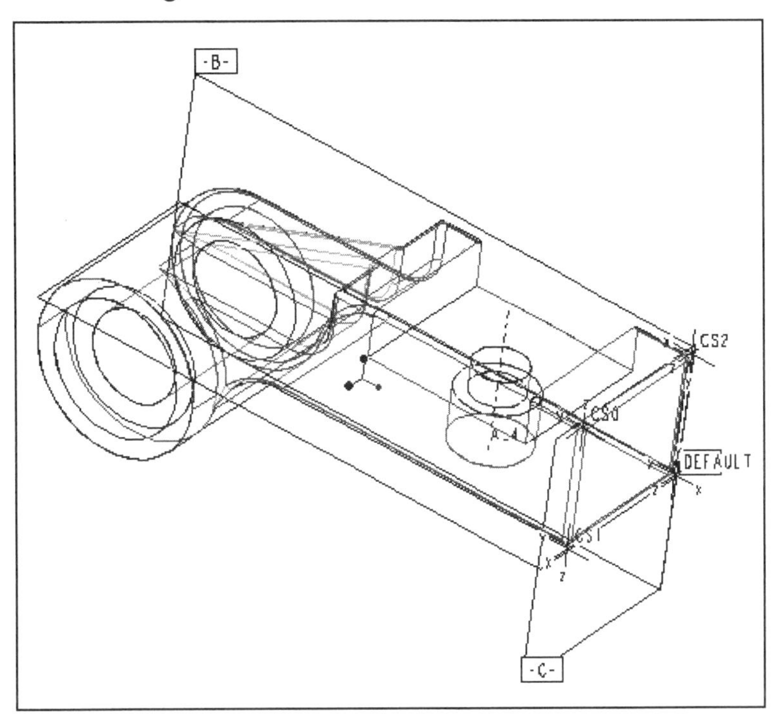

Figure 22.4
Adjustable Guide Design Part Shown in Hidden Line

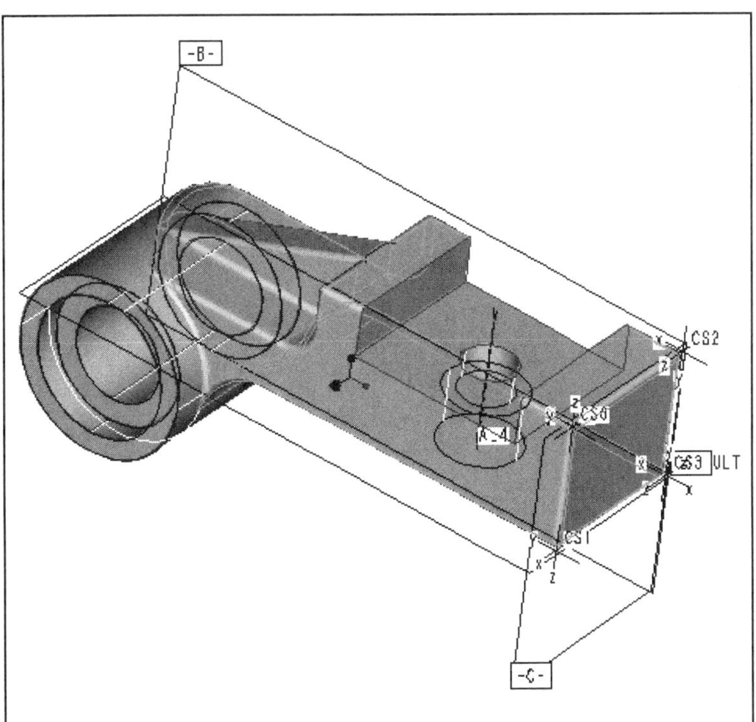

Figure 22.5
Adjustable Guide Design Part Shown in Shading

Workpiece

The **workpiece** represents the raw stock that is going to be machined by the manufacturing operations. Its use is optional in Pro/NC. The benefits of using a workpiece include:

* Automatic definition of extents of machining when creating NC sequences.
* Dynamic material removal simulation and gouge checking (available with Pro/NC-CHECK).
* In-process documentation by capturing removed material.

The workpiece can represent any form of raw stock: bar stock, casting (Fig. 22.6 and 22.7), etc. It may easily be created by copying the design model and modifying the dimensions or deleting/suppressing features to represent the real workpiece.

As a Pro/E part, the workpiece can be manipulated as any other part: it can exist as an instance of a part family table; and it can be modified, redefined, etc.

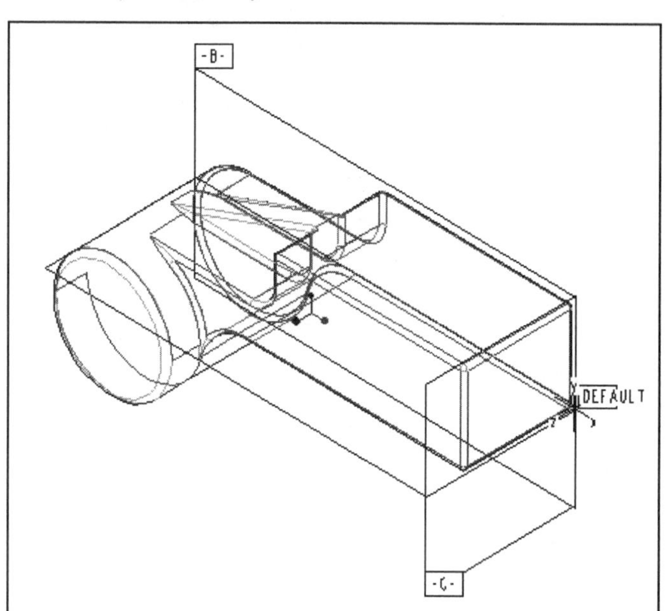

Figure 22.6
Adjustable Guide Workpiece Shown in Hidden Line

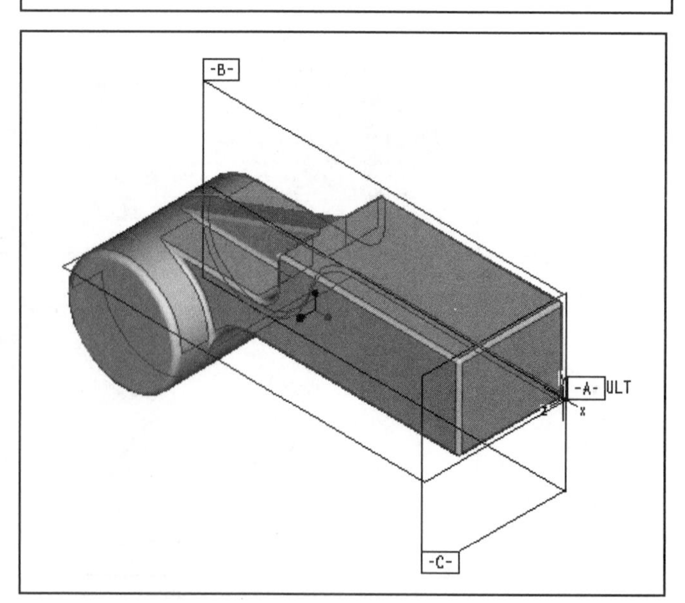

Figure 22.7
Adjustable Guide Workpiece Shown in Shading

Manufacturing Model

A regular manufacturing model consists of a design model (also called a "reference part" because it is used as a reference for creating NC sequences) and a workpiece assembled together (Fig. 22.8 and Fig. 22.9). As the manufacturing process is developed, the material removal simulation can be performed on the workpiece. Generally, at the end of the manufacturing process, the workpiece geometry should be coincident with the geometry of the design model. However, material removal is an optional step.

When a manufacturing model is created, it generally consists of four separate files:

* The design model (machine detail info) *filename.prt*
* The workpiece (stock or casting detail info) *filename.prt*
* The manufacturing model *assembly-manufacturename.asm*
* The manufacturing process *file-manufacturename.mfg*

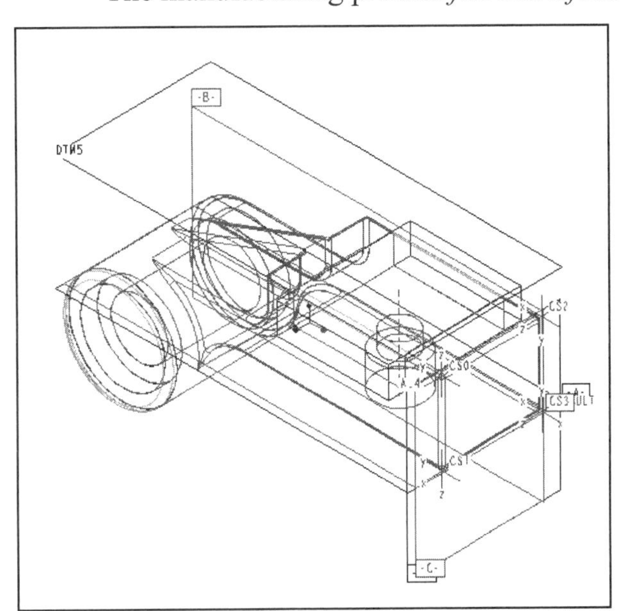

Figure 22.8
Adjustable Guide Manufacturing Model Shown in Hidden Line

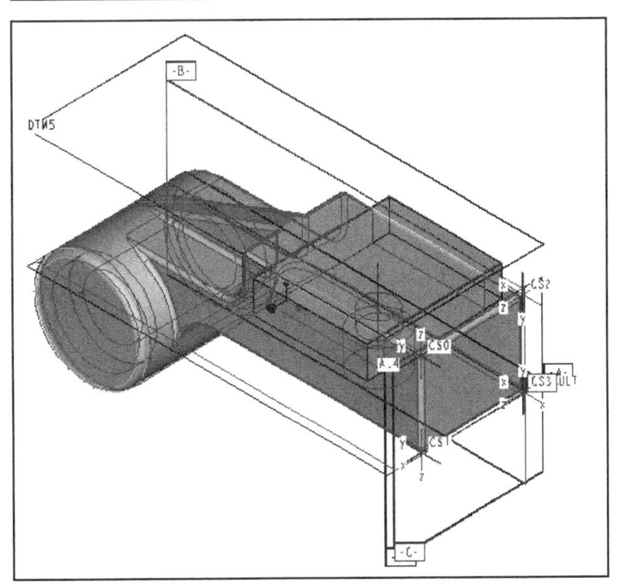

Figure 22.9
Adjustable Guide Manufacturing Model Shown in Shading

Part and Assembly Machining

There are two separate types of Pro/NC:

* *Part machining* Acts on the assumption that the manufacturing model contains one reference part and one workpiece (also a part). Multi-part allows you to assemble multiple design models or reference models, but they are automatically merged upon assembly so that the manufacturing model still consists of one reference part and one workpiece (Fig. 22.10).
* *Assembly machining* No assumptions are made by Pro/E as to the manufacturing model configuration. The manufacturing model can be an assembly of any level of complexity (with subassemblies, etc.), and it can contain any number of independent reference models and/or workpieces. It can also contain other components that may be part of the manufacturing assembly but have no direct effect on the actual material removal process (e.g., the turntable, clamps, etc.).

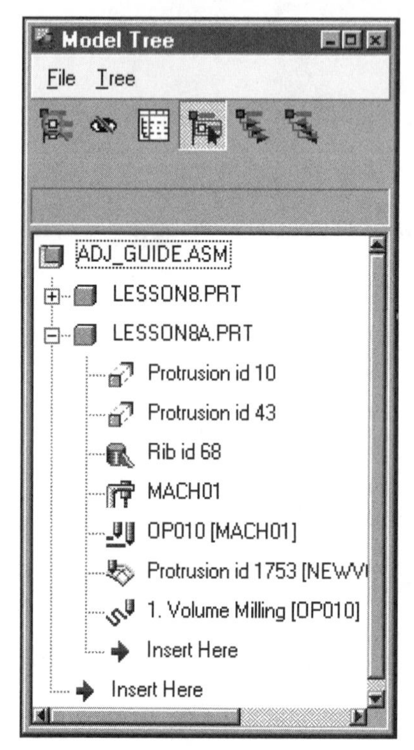

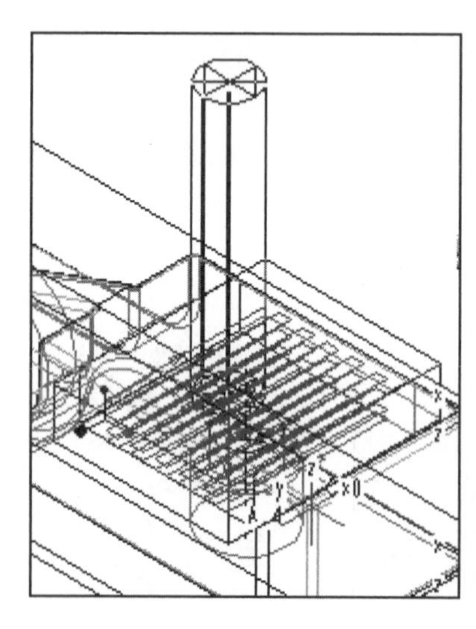

Figure 22.10
Adjustable Guide Manufacturing Model CL Data

Once the manufacturing model is created, Part and Assembly machining use similar techniques to develop the manufacturing process. If there are specific techniques for defining an NC sequence, they will be described where appropriate. Keep in mind that in Part machining, Pro/E automatically determines some of the machining aspects based on the workpiece geometry; therefore, although Assembly machining gives you more flexibility in building the manufacturing model, it may also require extra steps when creating the NC sequences.

The major difference between Part and Assembly machining is that in Part machining all the components of the manufacturing process (operations, workcells, NC sequences, etc.) are *part features that belong to the workpiece,* whereas in Assembly machining these are *assembly features that belong to the manufacturing assembly.*

For this lesson, we will be concerned with part manufacturing. Because this is an advanced project, your instructor may require you to create a jig and fixture to hold the design part while machining. The workpiece will then be added during the Pro/NC session, and you will be required to do assembly machining.

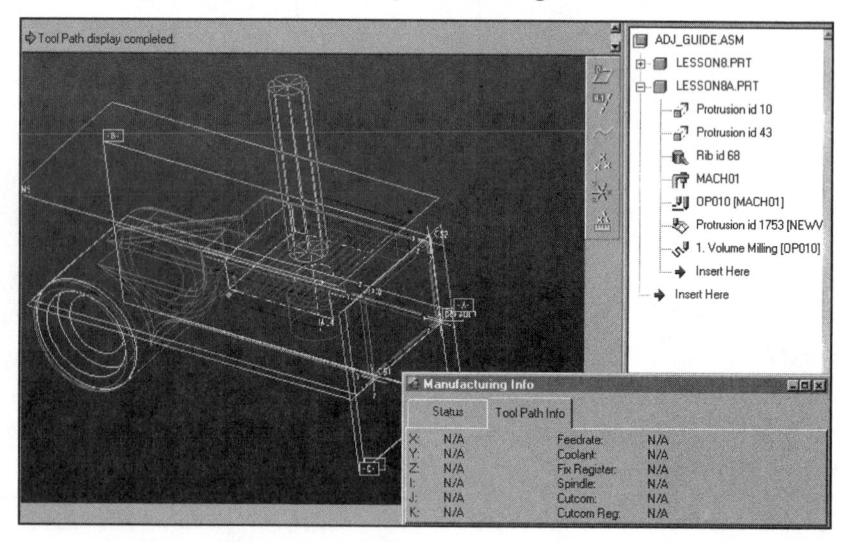

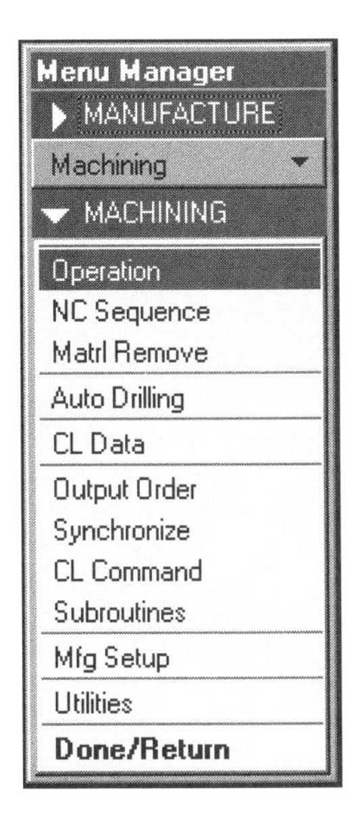

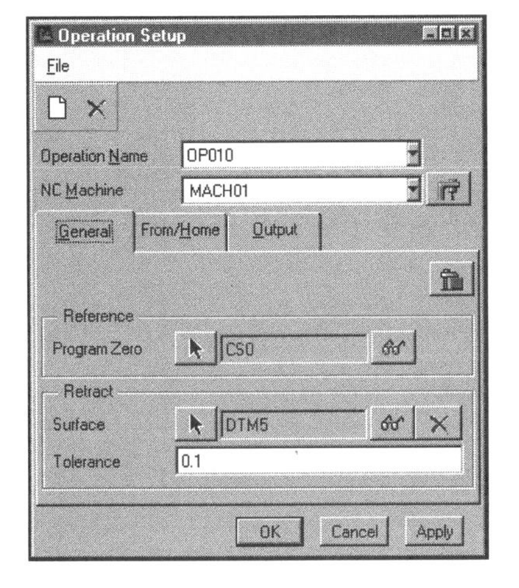

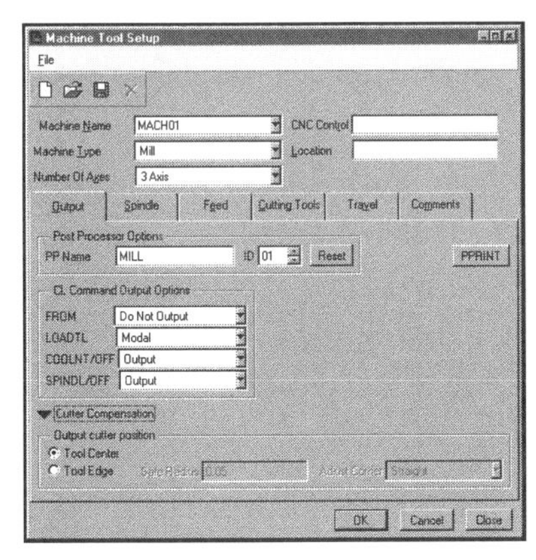

A Pro/NC Session

The Pro/NC process consists of the following basic steps:

1. Set up the manufacturing database. It may contain such items as workcells (machine tools) available, tooling, fixture configurations, site parameters, or tool tables. This step is optional. If you do not want to set up all your database up front, you can go directly into the machining process and later define any of the items above when you actually need them.
2. Define an operation. An operation setup may contain the following elements:

 * Operation name
 * Workcell (machine tool)
 * Machine coordinate system (MSYS) for CL output
 * Operation comments
 * Operation parameters
 * FROM and HOME points

You have to define a workcell and a machine coordinate system before you can start creating NC sequences. Other setup elements are optional.

3. Create NC sequences for the specified operation. Each NC sequence is a series of tool motions with the addition of specific post-processor words that are not motion-related but are required for the correct NC output.

The tool path (Fig. 22.11) is automatically generated by Pro/E based on the NC sequence type (such as **Volume Milling**, **Outside Turning**), cut geometry, and manufacturing parameters. You can apply more "low-level" control by:

 * Defining your own tool motions--that is, approach, exit, and connect motions. Tool motions include Automatic Cut motions.
 * Inserting nonmotion CL commands.

4. For each completed NC sequence, you can create a material removal feature, either by making Pro/E automatically remove material (where applicable) or by manually constructing a regular Pro/E feature on the workpiece (such as **Slot** or **Hole**).

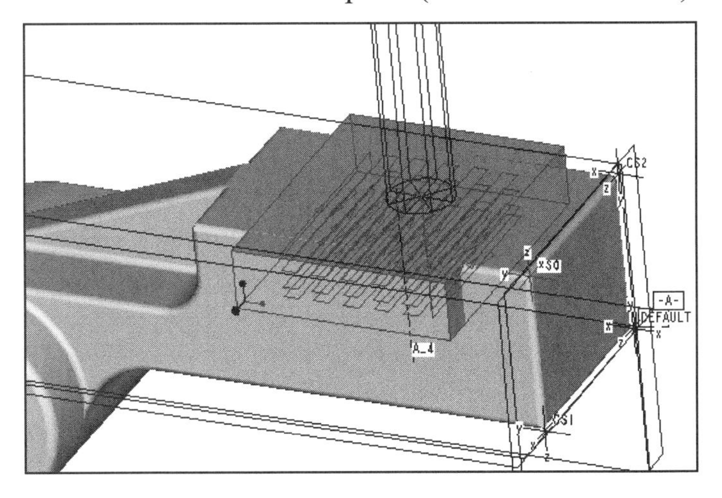

Figure 22.11
Adjustable Guide Manufacturing Model Showing Tool Path

Modal Settings

Most of the machining setup elements (Fig. 22.12) are modal; that is, all subsequent NC sequences will use this setting until you explicitly change it:

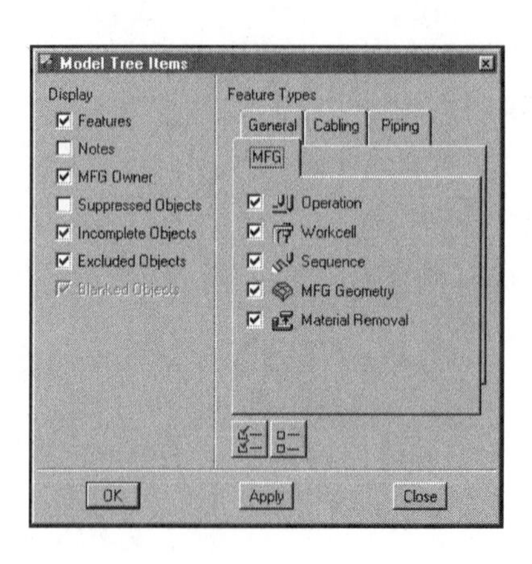

* Operation setup [including the workcell and Machine coordinate system (MSYS)]
* Fixture setup
* Tool (Fig. 22.13) (provided that the tool type is compatible with the NC sequence type).
* Manufacturing parameters of an activated site
* NC sequence coordinate system (for the first NC sequence, the Machine coordinate system specified for the operation will be implicitly used as the NC sequence coordinate system as well, unless you explicitly specify another one)
* Retract surface

Figure 22.12
Manufacturing Info

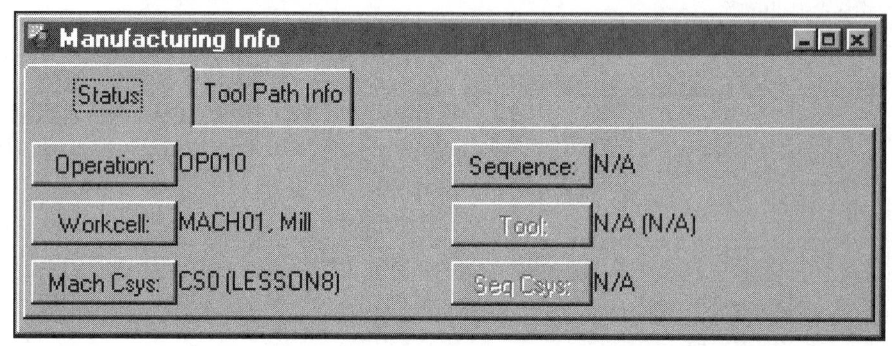

Figure 22.13
Tool Setup and Param Tree

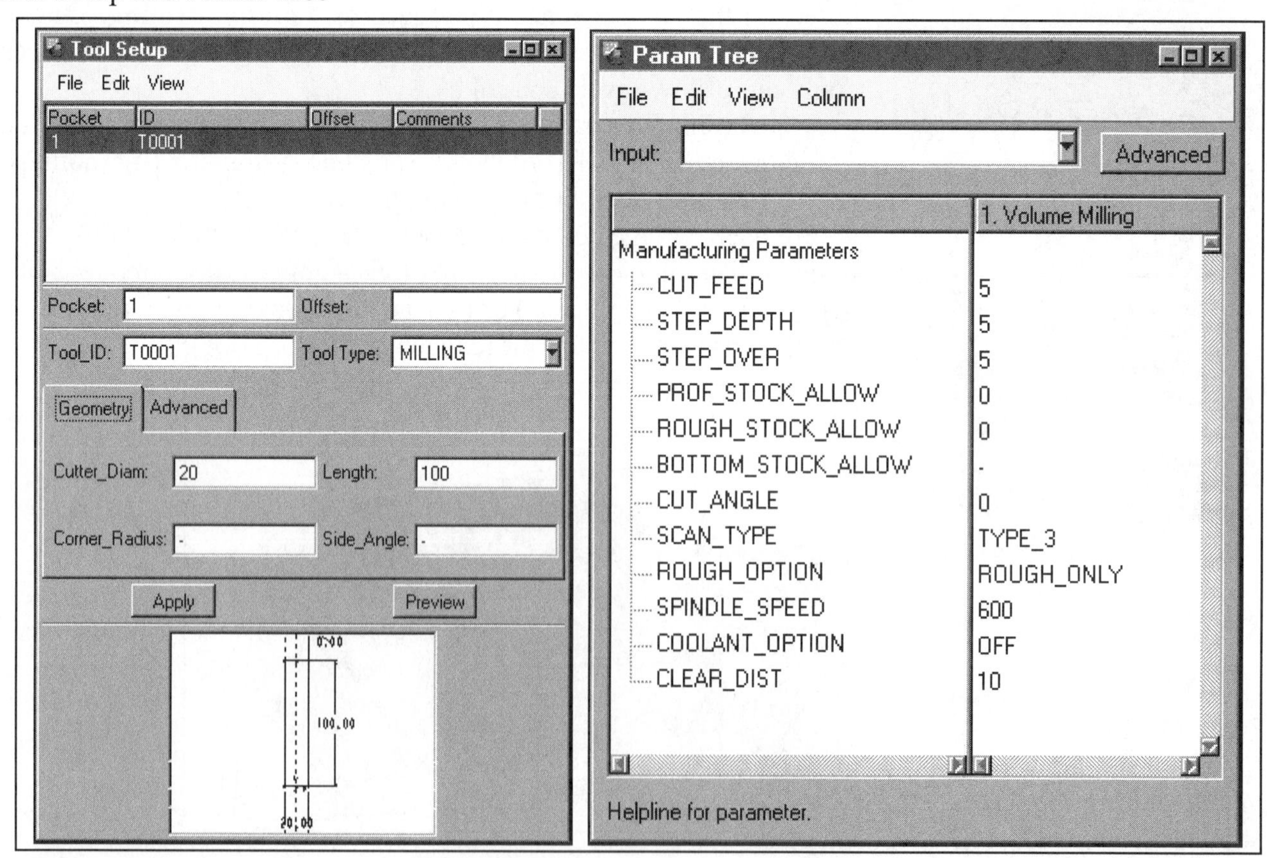

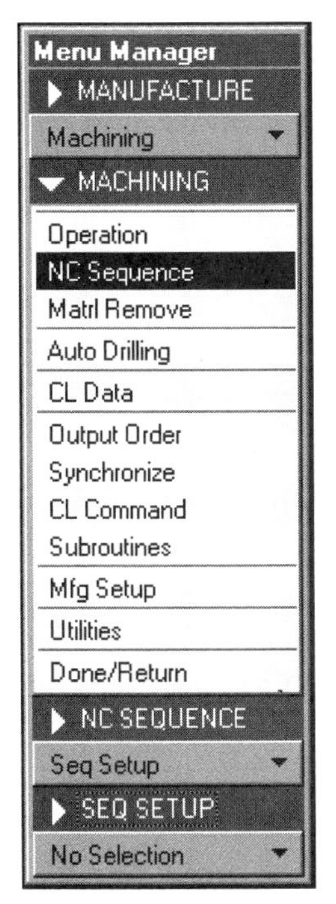

Walk-Through Menus

Most of Pro/NC menus are designed to "walk" you through the process development. These menus use check marks ✔ to select an option; more than one option may be selected at a time. When you choose **Done** from such a menu, Pro/E will invoke the appropriate user interface for each selected option in turn.

If some selection is necessary at a particular point, the check mark will be automatically turned on (checked). For example, when you first set up an operation, only the **Workcell** and **Mach Csys** (MSYS) options will have check marks next to them. You can also turn other check marks on (for example, to specify the FROM and HOME points), but you are not required to do so. Similarly, when you start defining the first NC sequence (Fig. 22.14), you will have a check mark at the **✔Tool** option. For the next NC sequence, however, the Tool option will not have a check mark next to it (provided that the previous tool is applicable). Turn it on (check mark) only if you want to specify another tool.

Another aspect of the "walk-through" functionality in the process development user interface is that if you omit a step, Pro/E will prompt you for the required information. For example, if you select the NC sequence option while the operation has not been set up, Pro/E will bring up the OPERATION menu first (as though you had selected **Operation**) and then invoke the user interface for creating an NC sequence.

This allows you to reduce the time, and minimize the number of menu selections, involved in defining an NC sequence.

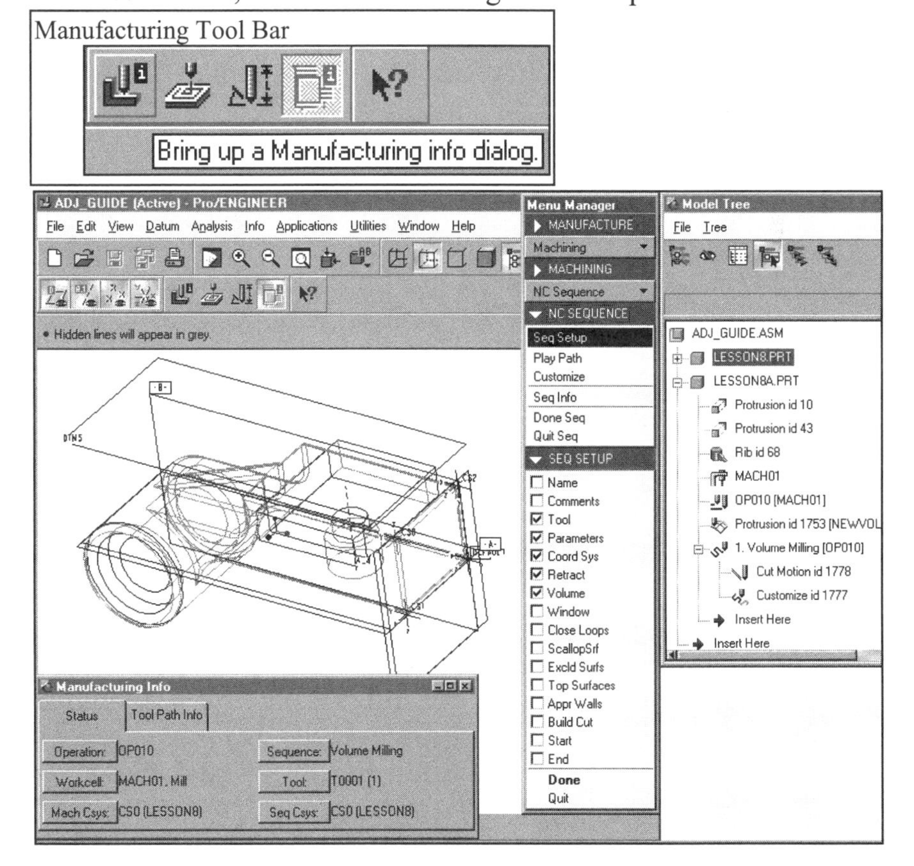

Figure 22.14
Manufacturing Model NC Sequence

Entering Manufacturing Mode

To work in manufacturing mode, you have to retrieve an existing manufacturing model or create a new one. You will want a design model (and probably a workpiece model) available. Create these in Part mode before entering Manufacturing.

To create the manufacturing model:

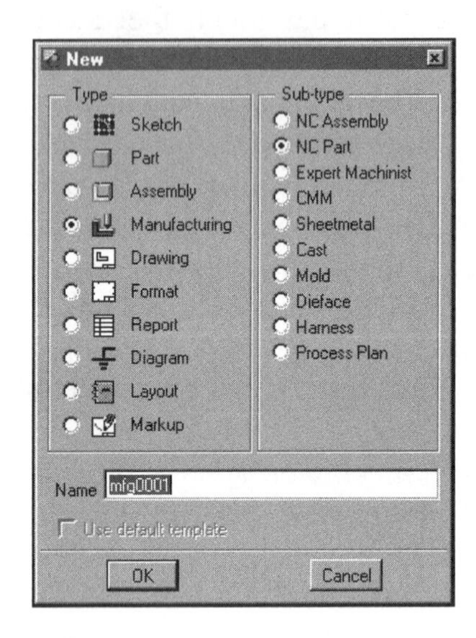

1. From the Pro/E menu bar, choose **File ⇒ New** (or pick the corresponding icon). Pro/E displays the New dialog box.
2. Choose the Manufacturing option button under Type.
3. Specify the type of the model by selecting an option button under Sub-type:

    * If you are machining a single part with only one workpiece, choose **NC Part**.
    * If you are machining an assembly of reference parts with no, one, or many workpieces, or if you do not have permission to make changes to the workpiece model, choose **NC Assembly**.

4. Type a name for the new manufacturing model in the Name text box, unless you want to accept the default.
5. Pick **OK**.
6. If you have chosen NC Part as the sub-type, Pro/E displays the browser window, listing all the part files in the current directory. Select the name of the **design model** (reference part).
7. Pro/E displays the MANUFACTURE menu, the model tree, and, in the case of Part machining, the reference part. The following commands are available in the MANUFACTURE menu:

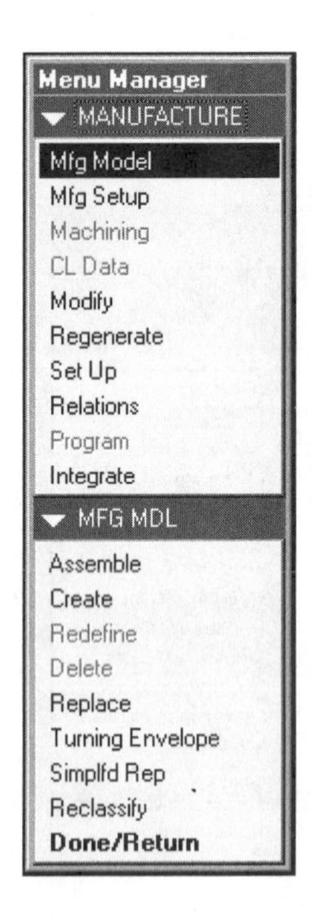

Mfg Model Define or modify the manufacturing model; i.e., assemble or create the workpiece and create simplified representations.
Mfg Setup Set up the manufacturing database: workcells, fixture setups, tools, default parameter sets, mill volumes, and surfaces.
Machining Develop the manufacturing process; i.e., perform manufacturing operations and generate all the tool motions necessary to machine the part.
CL Data Modify or generate cutter location (CL) paths for display or output to a file for a specific operation.
Modify Modify the design model, the workpiece, a manufacturing operation, or the manufacturing assembly by changing dimensions or manipulating features of the model.
Regenerate Regenerate the manufacturing assembly. Recalculates the workpiece based on changes made to the design model or manufacturing operations.
Set Up Set up model properties, units, etc.
Relations Work with relations in the manufacturing model.
Program Allows you to access Pro/PROGRAM functionality.
Integrate Allows you to compare two different versions of the same manufacturing model and to integrate the differences if necessary.

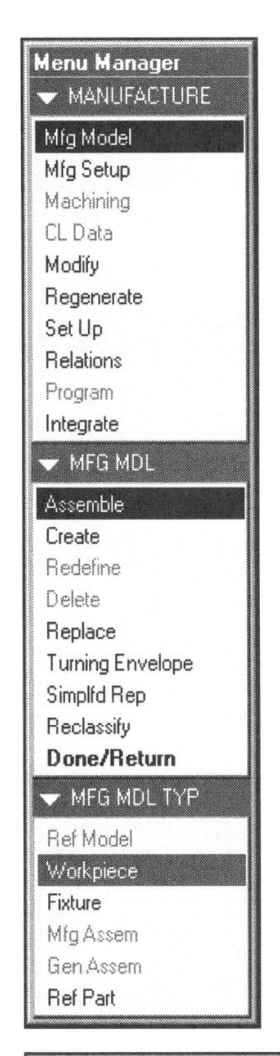

Defining a Manufacturing Model

Techniques for defining a manufacturing model are different in Part and Assembly machining.

Part Machining

In part machining, you have to assemble or create a workpiece before the manufacturing process can start (Fig. 22.15). Manufacturing features created during tool path development will be placed into the workpiece model.

Placing a Workpiece

When you create a new model for part manufacturing, you are immediately prompted to enter the name of the design model (reference part). This is the base component of the manufacturing assembly. To continue with the manufacturing assembly:

1. Choose **Mfg Model** from the MANUFACTURE menu.
2. Choose one of the options:

   Assemble To assemble the workpiece to the design model. Choose **Workpiece** and enter the name of the workpiece. The workpiece will be retrieved in a sub-window for assembling with the design model. Assemble the workpiece using the PLACE menu options. Choose **OK** when the proper placement constraints have been applied.
   Create To create the workpiece directly in Manufacturing mode. Choose **Workpiece**, enter the name of the workpiece, and create the first feature of the workpiece referencing geometry of the design model as necessary. To create more features on the workpiece, use the **Mod Work** option in the MFG MODIFY menu.

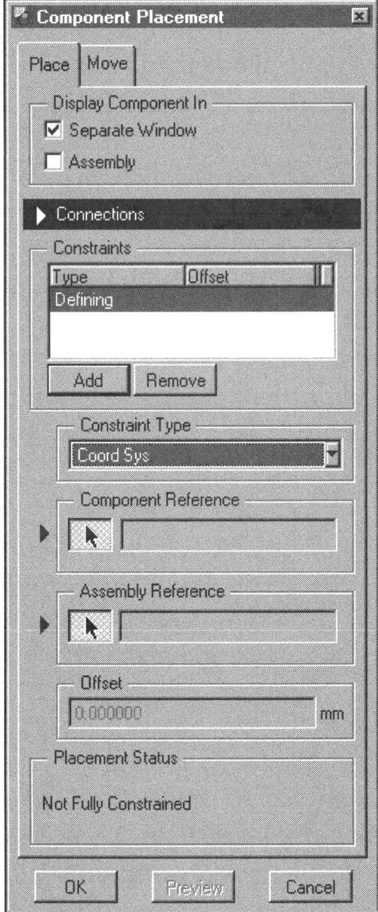

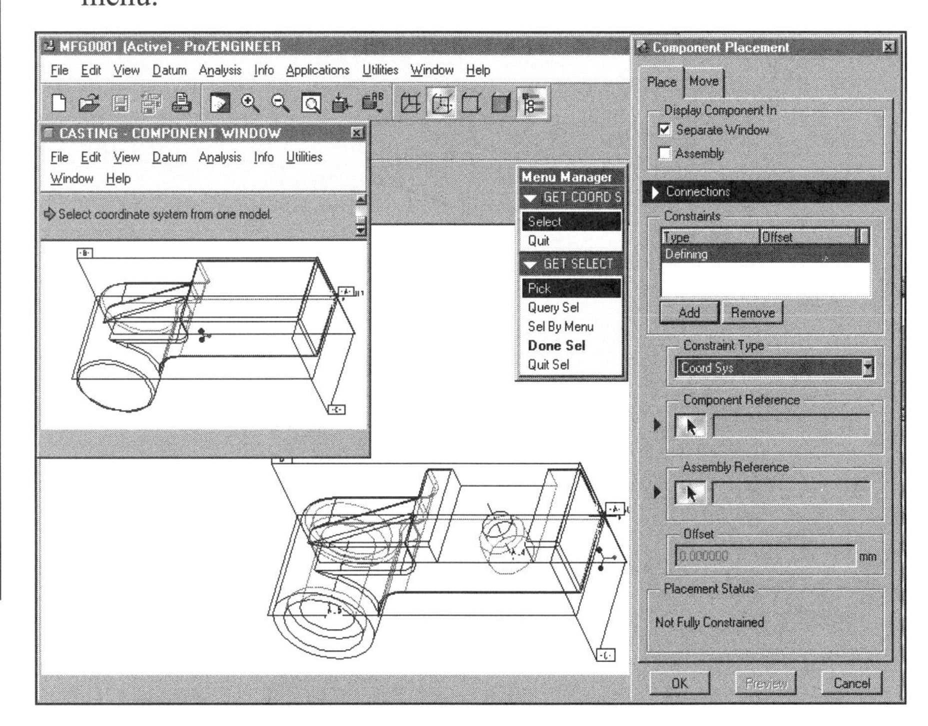

Figure 22.15
Assembling the Manufacturing Model

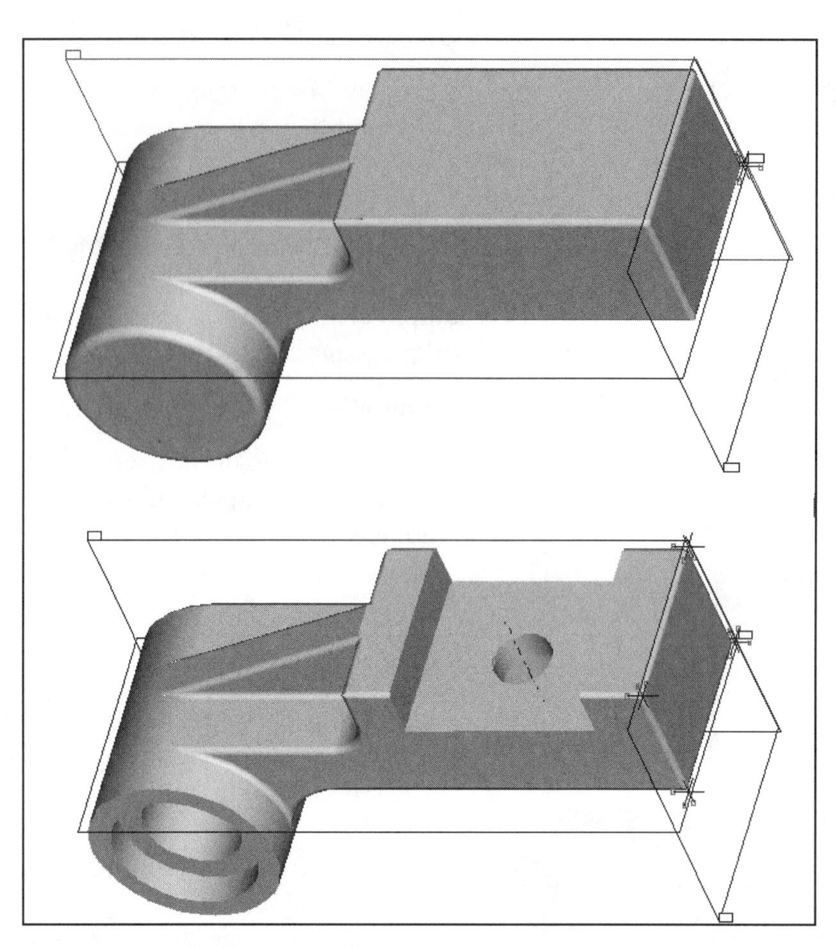

Figure 22.16
Adjustable Guide; Casting and Machined Part

Adjustable Guide

Lesson procedures & commands
START HERE ➡ ➡ ➡

You have already created and saved the workpiece and design part (Fig. 22.16) from Lesson 8. The workpiece is provided again as a detail drawing in Figure 22.19 (**ADJ_GUIDE_CAST),** and the design part is detailed in Figure 22.20 (**ADJ_GUIDE_MACHINE)**.

If you have not completed Lesson 8, do so before attempting this lesson. Model the casting first (Fig. 22.16, top) and save it when it is completed. After saving the casting under a different name, use the part model to create the machined **Adjustable Guide** (Fig. 22.16, bottom). By having a *casting part* (which is called a **workpiece** in **Pro/NC**) and a separate but almost identical *machined part* (which is called a **design part** in **Pro/NC**), you can create an operation for machining and an NC sequence. During the Pro/NC manufacturing process, you merge the workpiece into the design part and create a **manufacturing model** (Fig. 22.17). The difference between the two objects is the difference between the volume of the casting part and the volume of the machined part. The removed volume can be seen as *material removal* when you are doing an **NC Check** operation on the manufacturing model. The manufacturing model is *green;* the cuts completed in **NC Check** show as *yellow* until the machine tool reaches the design size, and then they show as *magenta.* If the machining process gouges the part, the gouge will be displayed as *cyan.* The cutter location can also be displayed as an animated machining process, as shown in Figure 22.18.

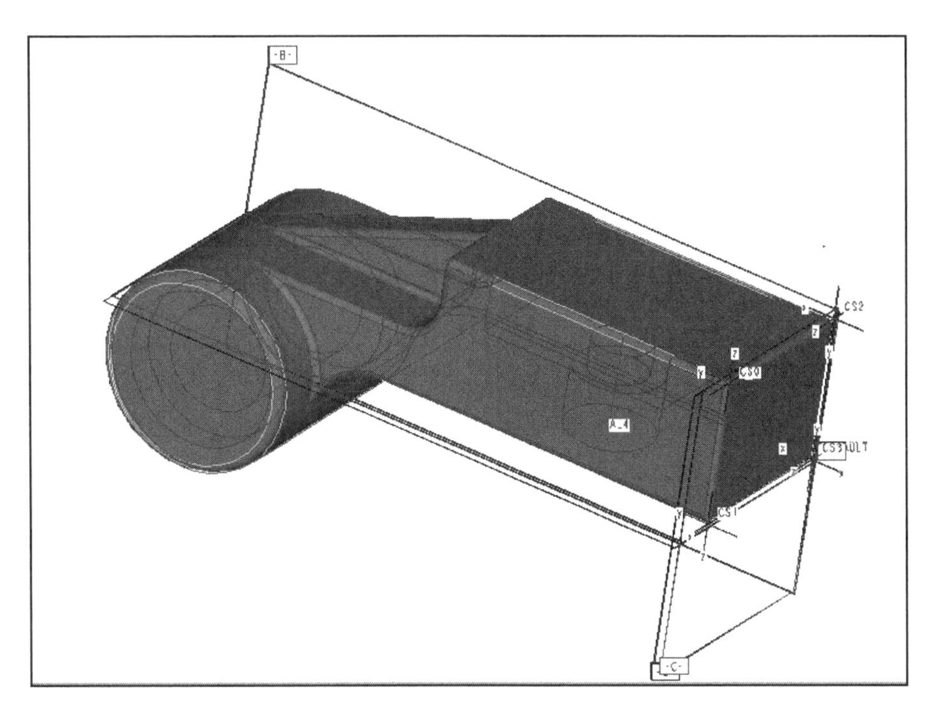

Figure 22.17
Manufacturing Model

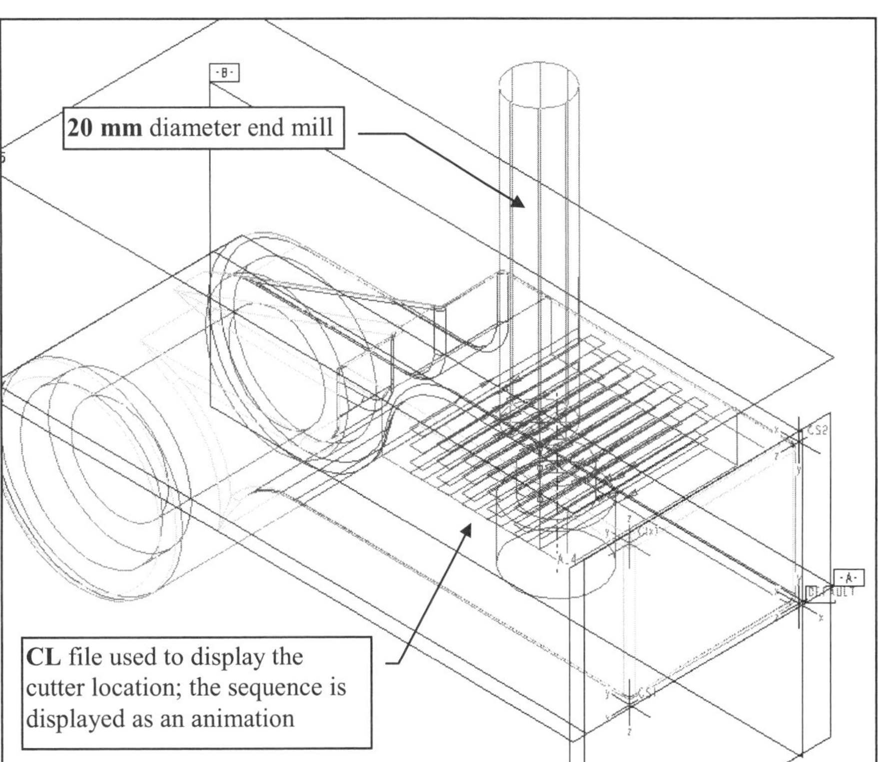

Figure 22.18
Pro/NC NC Sequence and a **CL** File

Machining

When you first enter the Manufacturing Module, you must create a Manufacturing Model. This is a file that contains all of the data associated with the resulting NC Program. The **Manufacturing Model** references the *Design Model*, the *Stock Material* (*workpiece*), the *Machining Parameters*, and the *CL File*, which contains the *NC Programming* (*GOTO*) statements.

NOTE

An alternative way of completing the workpiece and the design part is to model the design part completely and then create a Family Table where the instance has all holes and machined cuts, and surfaces suppressed.

Create the manufacturing model, assemble the design part, and then assemble the instance of the design part with the suppressed features.

If at any time in manufacturing stage an ECO is generated changing the part, both the design part and the workpiece will be updated since one is the instance of the other. The database for the is the workpiece is the same as that for the design part. Change the design part and the workpiece also changes. The casting-workpiece is an instance of the machined-design part.

Manufacturing models are created with the file extension *.mfg*. Many users give the manufacturing model the same name as the design model as a way to show which file goes with which part.

Once you create the **Manufacturing Model**, you need to assemble the **Design Model** into it. Pro/E places the Design Model at a default location and orientation when it is assembled. If the Manufacturing Model contains features such as datum planes, you must use the **Assemble** ⇒ **Component** interaction to place the Design Model.

You can create an NC Program in several different "modes" in Pro/NC. The two most common are Part and Assembly.

In the Part mode, you are required either to create or to assemble a **Workpiece** into the Manufacturing Model. *In the Assembly mode, the Workpiece is optional.* If you are going to drive the tool along a series of faces that represent the finished part, and you do not wish to remove large areas of material, you may find that you do not want to bother with a workpiece. Simple **Profiling** and **Pocketing** often fall within this category. For this lesson we will use a workpiece.

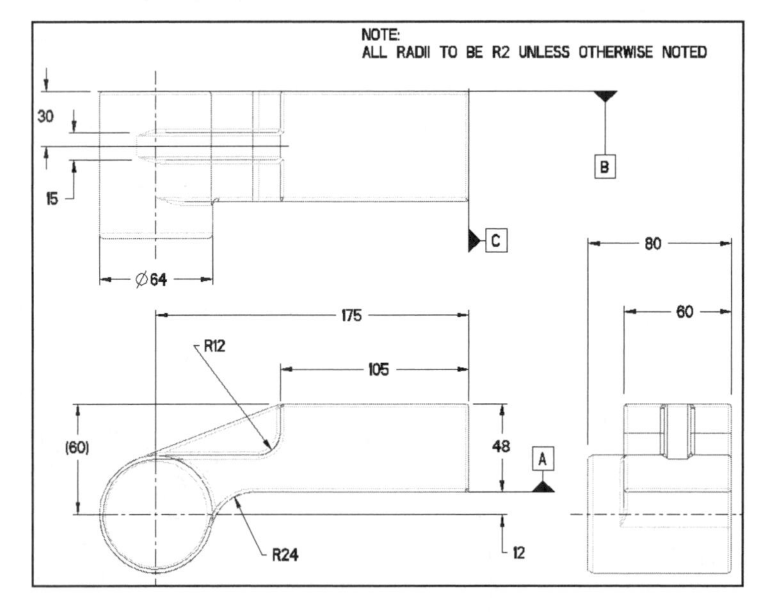

Figure 22.19
Adjustable Guide, Casting Detail

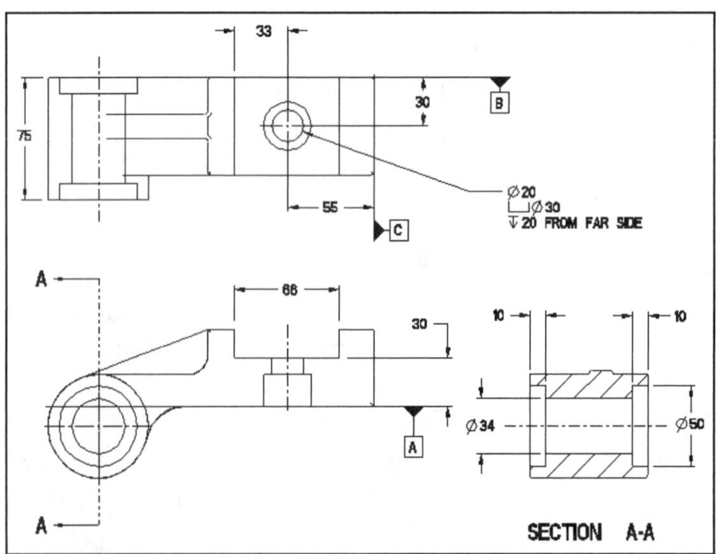

Figure 22.20
Adjustable Guide, Machining Detail

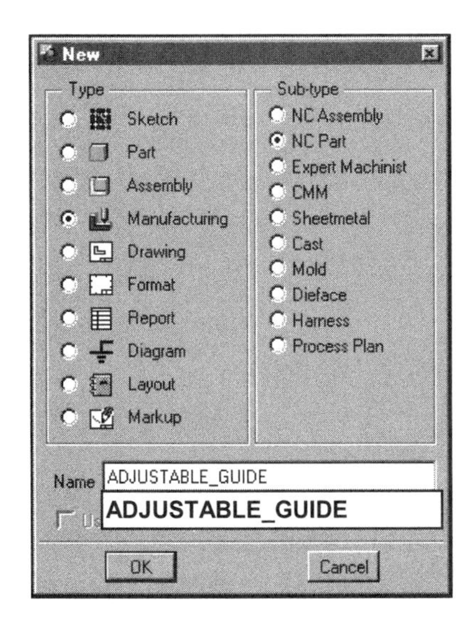

In this lesson you will create a Manufacturing Model and then add the **Design Model**. The Design Model represents the finished part (after all of the machining takes place). You are actually going to machine the casting of this part and remove material from the slot and face one of the cylindrical ends. Holes are also required.

To create a Manufacturing Model, choose the following commands:

File (from the Menu bar) ⇒ **New** ⇒ **Manufacturing** (from the Type list) ⇒ **●NC Part** (from the Sub-type list) ⇒ (Type **ADJUSTABLE_GUIDE)** ⇒ **OK** ⇒ (pick **adj_guide_machine** from the Open dialog box) ⇒ **Open** ⇒ **Mfg Model** (to Assemble the workpiece) ⇒ **Assemble** ⇒ **Workpiece** ⇒ (pick **adj_guide_cast** from the Open dialog box) ⇒ **Open** (Fig. 22.21) ⇒ **✔Separate Window** ⇒ **❑ Assembly** ⇒ Constraint Type **Coord Sys** ⇒ (pick the coordinate system on the workpiece (casting) and then the coordinate system on the design part) ⇒ **OK** ⇒ **Done/Return**

NOTE

Your file names may be different

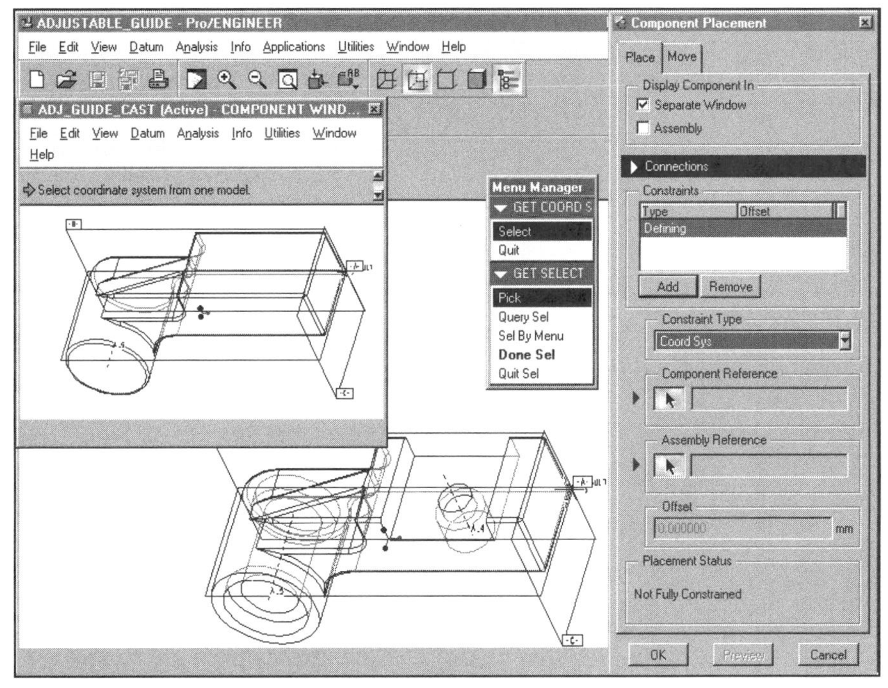

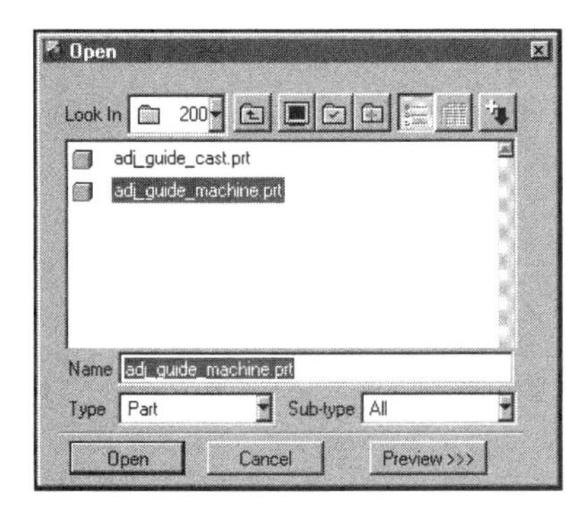

Figure 22.21
adj_guide_cast
adj_guide_machine

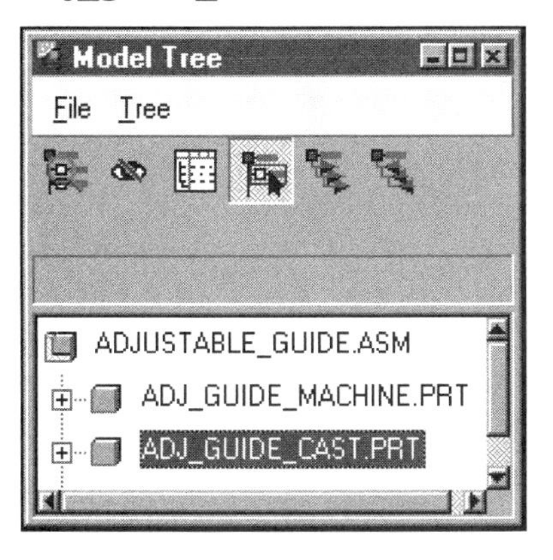

Operation Creation

You have to create an operation before you can start defining machining features. When creating the operation, the required elements are the *machine tool name* and the *Program Zero coordinate system.*

The Operation Setup dialog box opens when you select **Operation** or **Machining**. It contains the default settings for the operation name and output parameters. To change the default name, type the new name in the Operation Name text box.

The red arrows next to an item in the dialog box means you must select or specify that item before closing and completing that step.

If you start by selecting **Machining**, Pro/E will prompt you for the Operation Setup, the same as you would have been prompted for if you had selected **Operation.**

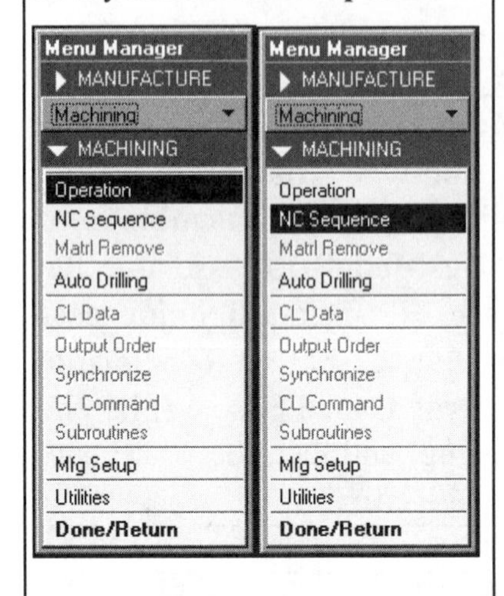

Machining ⇒ **NC Sequence** [NC Sequence is the default and Pro/E opens and displays the Operation Setup dialog box (Fig. 22.22)] ⇒ (pick the NC Machine icon)

NC Machine

(investigate the various options by selecting the Tabs) ⇒ **OK** [to accept the NC Machine defaults (Fig. 22.23)] ⇒ [pick the arrow to select the Reference Program Zero arrow (Fig. 22.24)]

Reference
Program Zero

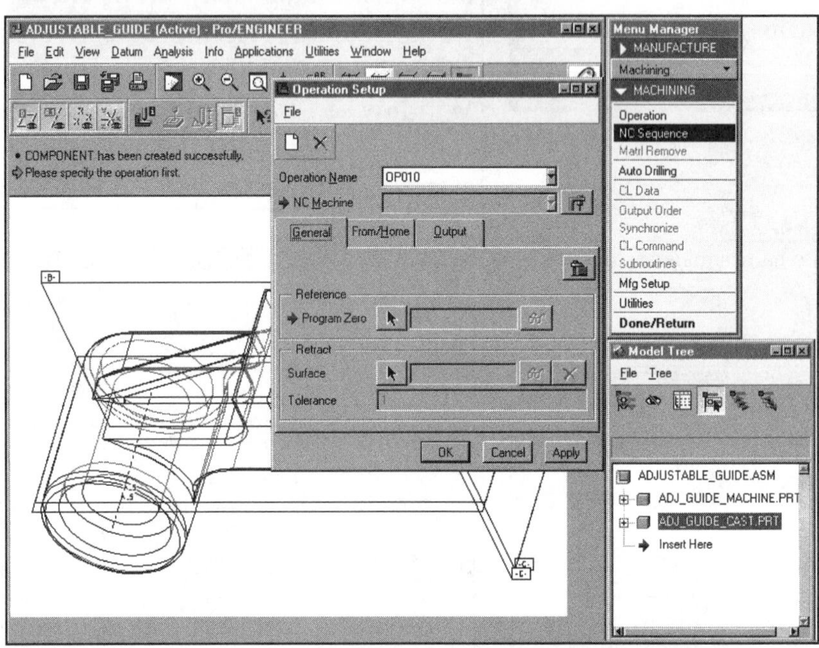

Figure 22.22
Operation Setup

Select or create a machine tool. If you have set up some machine tools prior to creating the operation, their names appear in the NC Machine drop-down list. To create a machine tool, click the Machine button. If you pick **OK,** you will be accepting the defaults for MACH01.

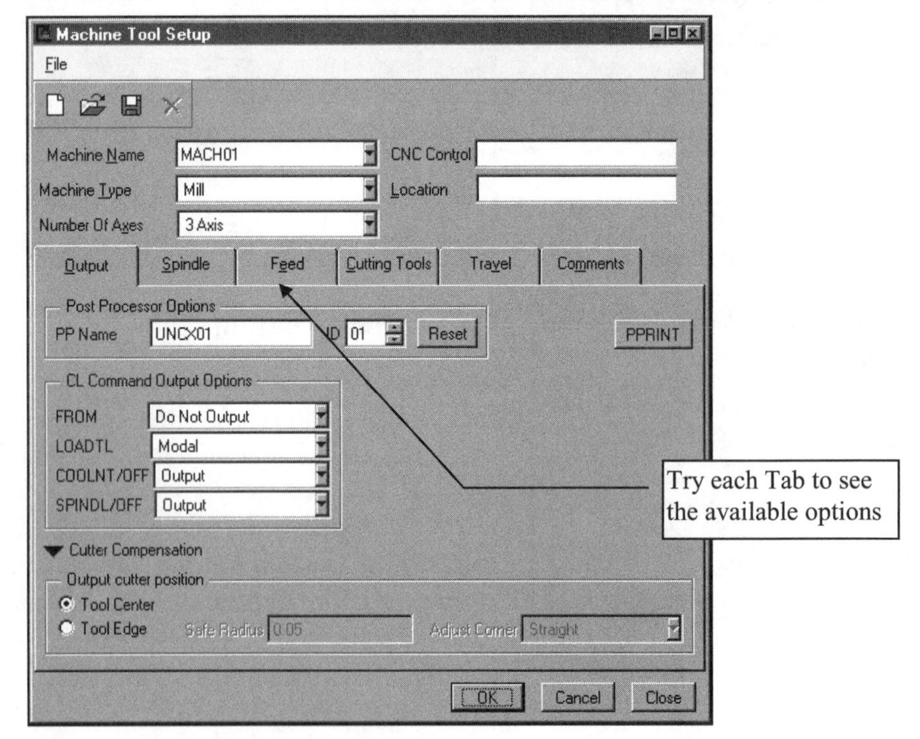

Figure 22.23
Machine Tool Setup

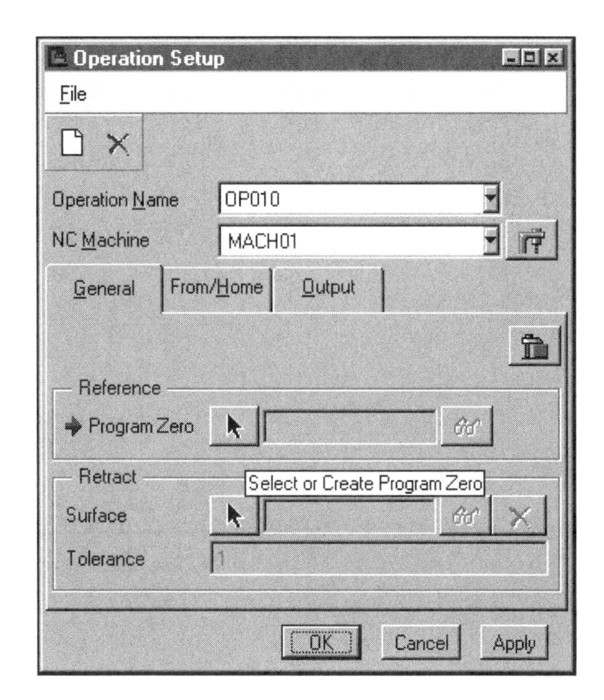

Figure 22.24
Operation Setup and Program Zero Selection

The material presented here is information only. The required command picks will resume later, in a command box.

Program Zero

To define **Program Zero**, you have to create or select a coordinate system, which will define the orientation of the stock on the machine and act as the origin (**0, 0, 0**) for CL data generation. The Program Zero coordinate system can belong to the reference model, stock, or the NC Model assembly; it can be created in Part or Assembly mode, outside of Expert Machinist, or directly at the time of defining Program Zero. Program Zero for an operation or a machining feature is specified in a similar way, as described in the following procedure.

1. To define Program Zero at the operation level, pick Program Zero in the Operation Setup dialog box. To define Program Zero at the feature level, click Program Zero in the appropriate machining feature dialog box (for example, Pocket Feature).
2. The MACH CSYS or the SEQ CSYS menu, respectively, appears with the following commands:

> **Create** Select which model the coordinate system will belong to, then create the coordinate system.
> **Select** Select an existing coordinate system, either by selecting on the screen or by using the **Sel By Menu** command.
> **Use Prev** Lets you select a coordinate system used for an earlier operation or machining feature.

After picking **Done**, if you pick **Show**, the operation Program Zero is highlighted in *red*; if you specify a different Program Zero at the feature level, it is highlighted in *magenta*.

Once the Program Zero is defined, the name of the coordinate system appears in the Program Zero text box, and clicking the **Show** button next to it will highlight the coordinate system on the screen.

Setup and Clearance

The setup orientation and location of the MSYS are ordinarily specified to match the design of the actual setup on the machine tool (Fig. 22.25). The MSYS orientation must match the fixed axis orientation of the machine tool. The origin of the MSYS is normally specified to match the zero set point on the setup, but only as a point of convenience.

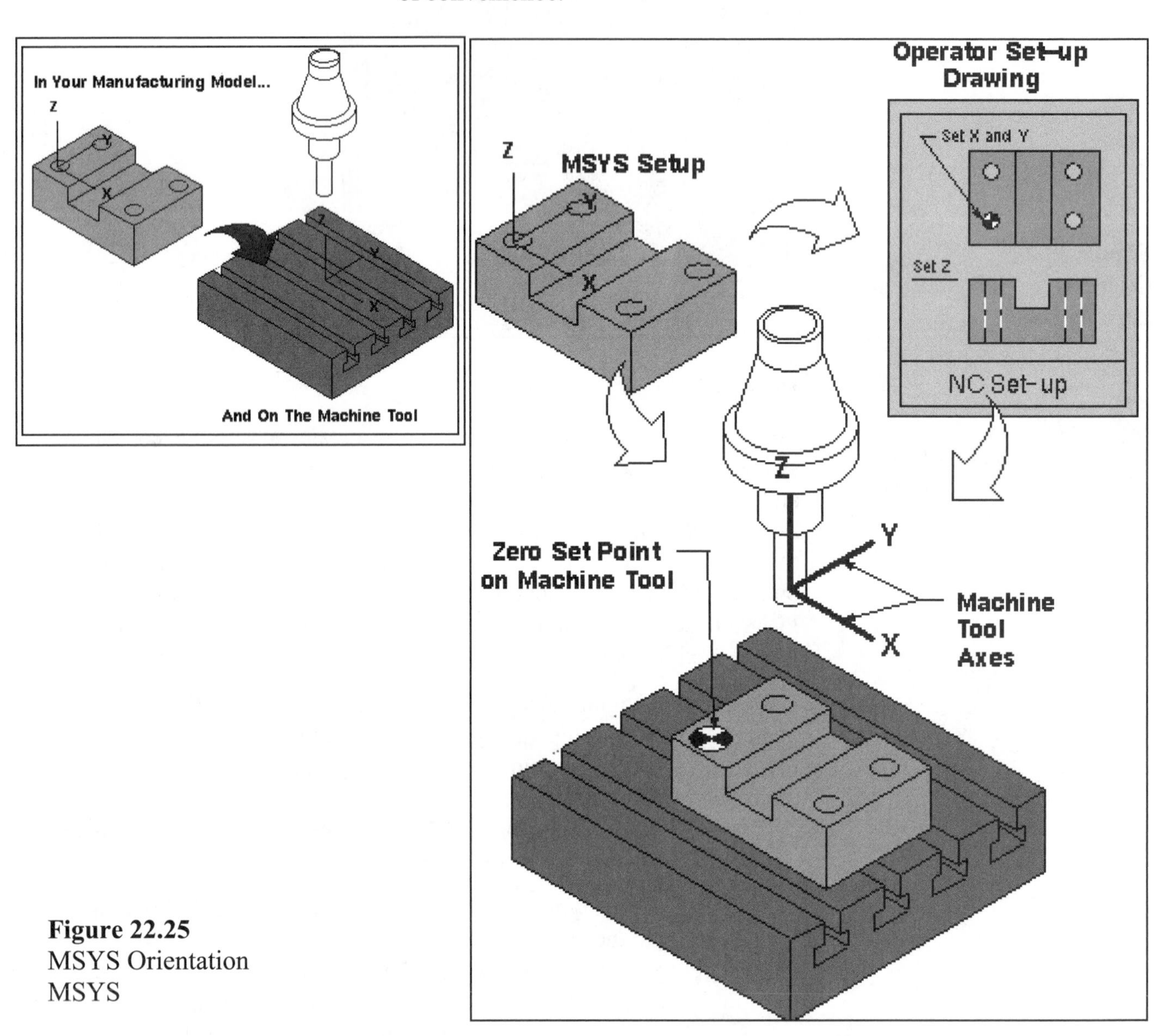

Figure 22.25
MSYS Orientation
MSYS

The orientation of the MSYS in terms of the axis orientations (**X**, **Y**, and **Z**) ***must*** represent the fixed axes of the Machine Tool.

The MSYS establishes the orientation of all other aspects of the tool path. The tool axis is always defined relative to the MSYS. In **3 Axis** machining, the tool axis is always parallel to the positive **Z** axis. **Entry** and **Exit** motions are defined relative to the MSYS. The **Retract Plane** is always parallel to the **X-Y** plane of the MSYS. **From** and **Return Points** are defined in MSYS coordinates.

Next, you will create a machining coordinate system (MSYS). You must always make sure that the **Z Axis** is pointing "up" towards the spindle on the NC machine. Remember, the tool axis is going to be parallel to the **Z Axis**. Choose the following commands:

Create [from the MACH CSYS menu (Fig. 22.26)] ⇒ [select a model on the screen. This associates the MSYS with the model. (There are three models on the screen. You may select either the *machine part*, the *workpiece*, or the *manufacturing mode.* Here we have selected the workpiece from the Model Tree.)] ⇒ **3 Planes** ⇒ **Done** ⇒ [select the front three planes of the workpiece (Fig. 22.27)]

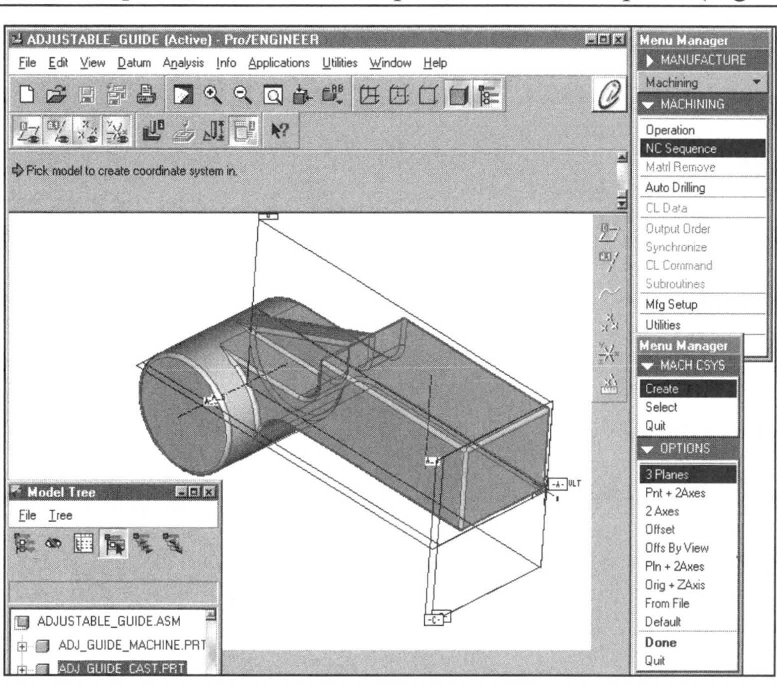

Figure 22.26
MSYS

Though it is not really necessary, pick the top surface first, the front surface second, and the right side surface third, so that your model will resemble Figure 22.27. Pro/E now displays the "chicken foot" to show you the orientation of the axes. You must define the direction and the name of two of the three axes in the chicken foot.

The red axis is the one that will be affected by choices you make on the menu. You can either affect the red axis or switch to another one using the **Next** and **Previous** options.

Choose **Z-Axis** [for the first arrow establishes the *tool axis* parallel to the **Z-Axis** (Pro/E highlights another axis)] ⇒ **Reverse** ⇒ **X-Axis** (the **X-Axis** determines the **Cut Angle** direction, the direction of the cutter. You want to cut from the front of the part towards the back, not left to right) ⇒ **OK** (Fig. 22.28) ⇒ **Machining** ⇒ **Volume** ⇒ **Done** ⇒ **Seq Setup** [the SEQ SETUP menu displays, showing you the required items (Fig. 22.29)] ⇒ **Done** (left-hand column)

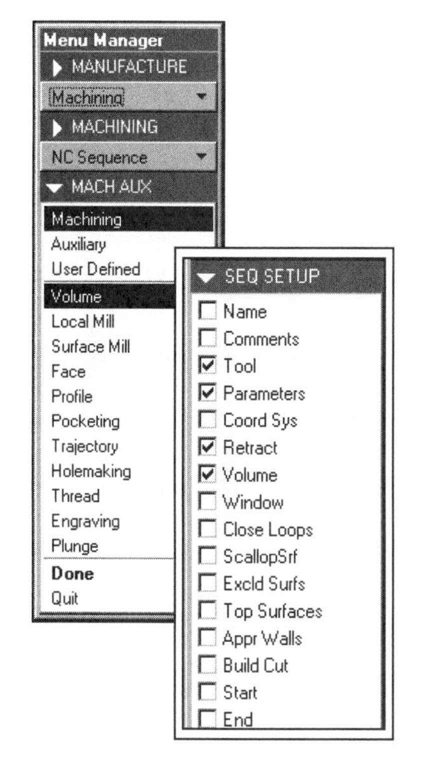

Pro/E begins to offer you the menus to specify the required information ✔**Tool,** ✔**Parameters,** ✔**Retract,** and ✔**Volume** (Fig. 22.29). The Tool Setup dialog box is displayed (Fig. 22.30).

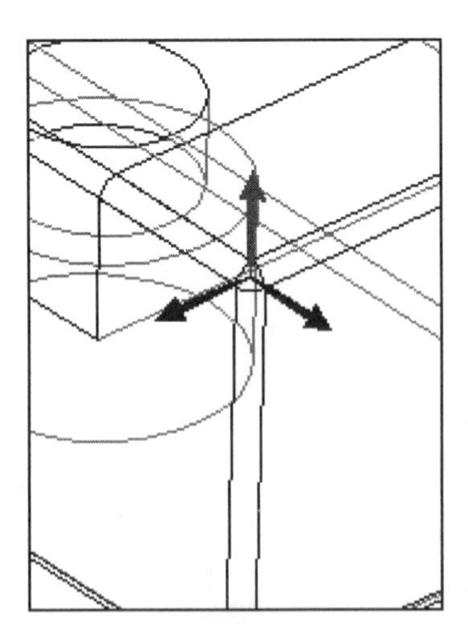

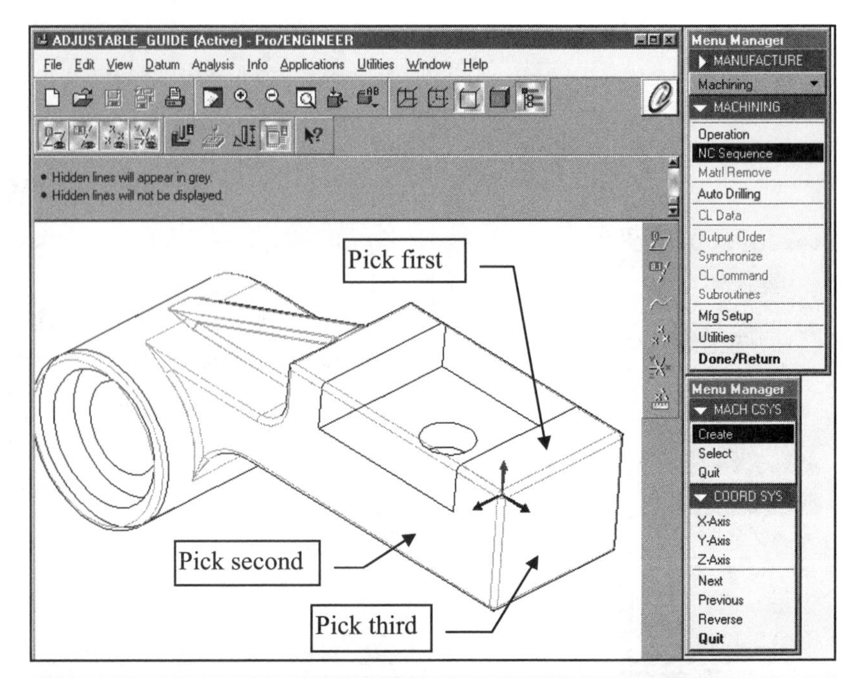

Figure 22.27
Surfaces to Establish the Three Planes for the MSYS

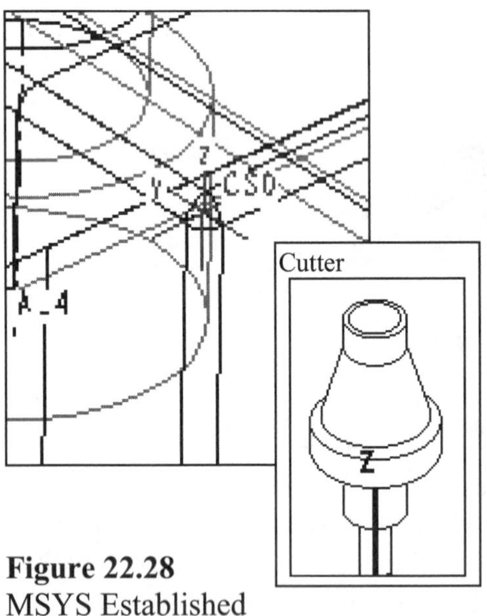

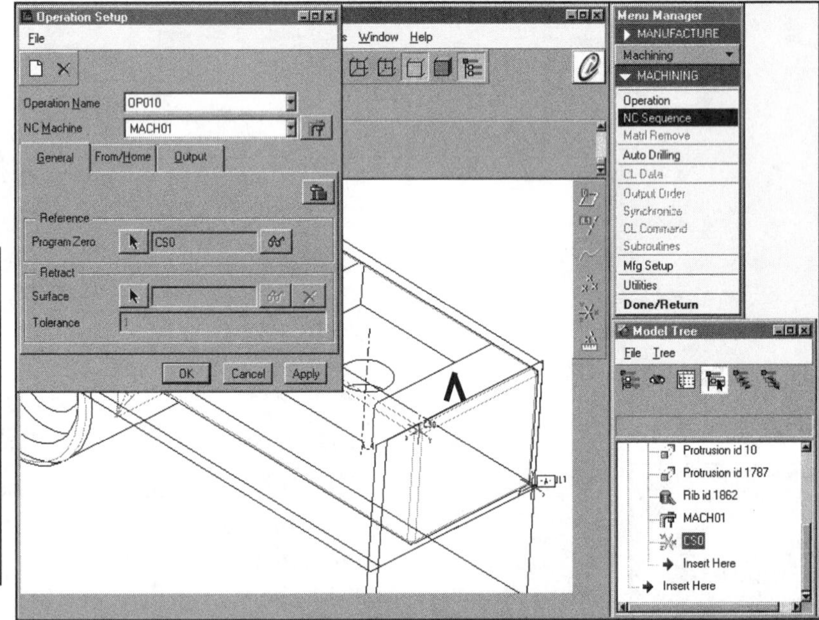

Figure 22.28
MSYS Established

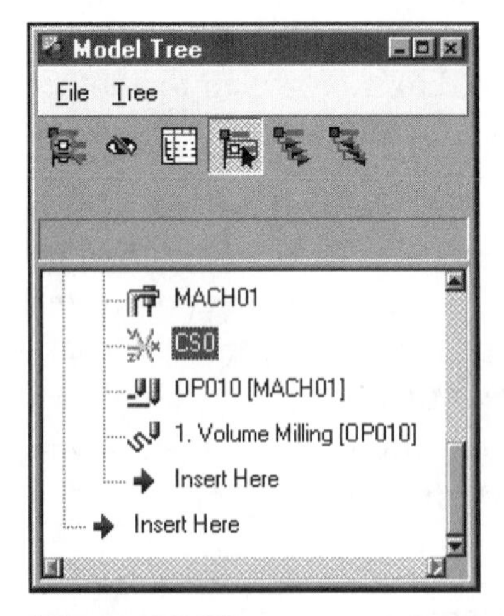

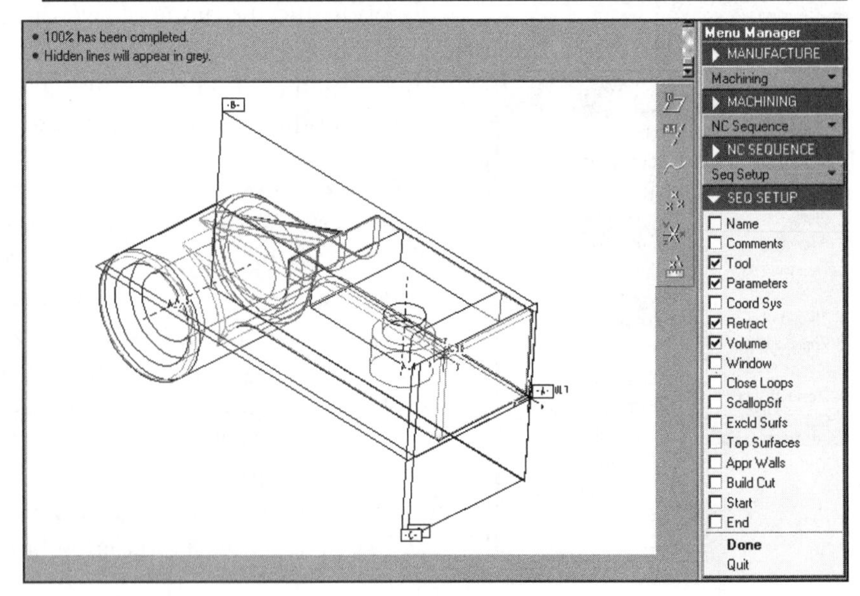

Figure 22.29
Seq Setup

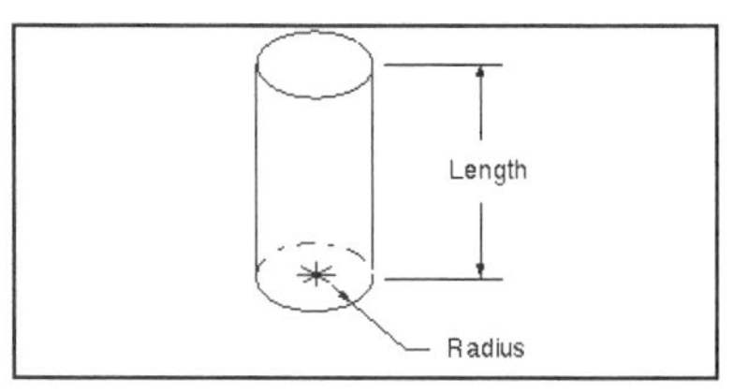

Tool Definition

Each NC Sequence requires you to specify an existing tool or to create a new tool. Standard end mills are represented as cylinders (left top) or tapered cylinders (left bottom). You define a tool using the Tool Setup dialog box (Fig. 22.30)

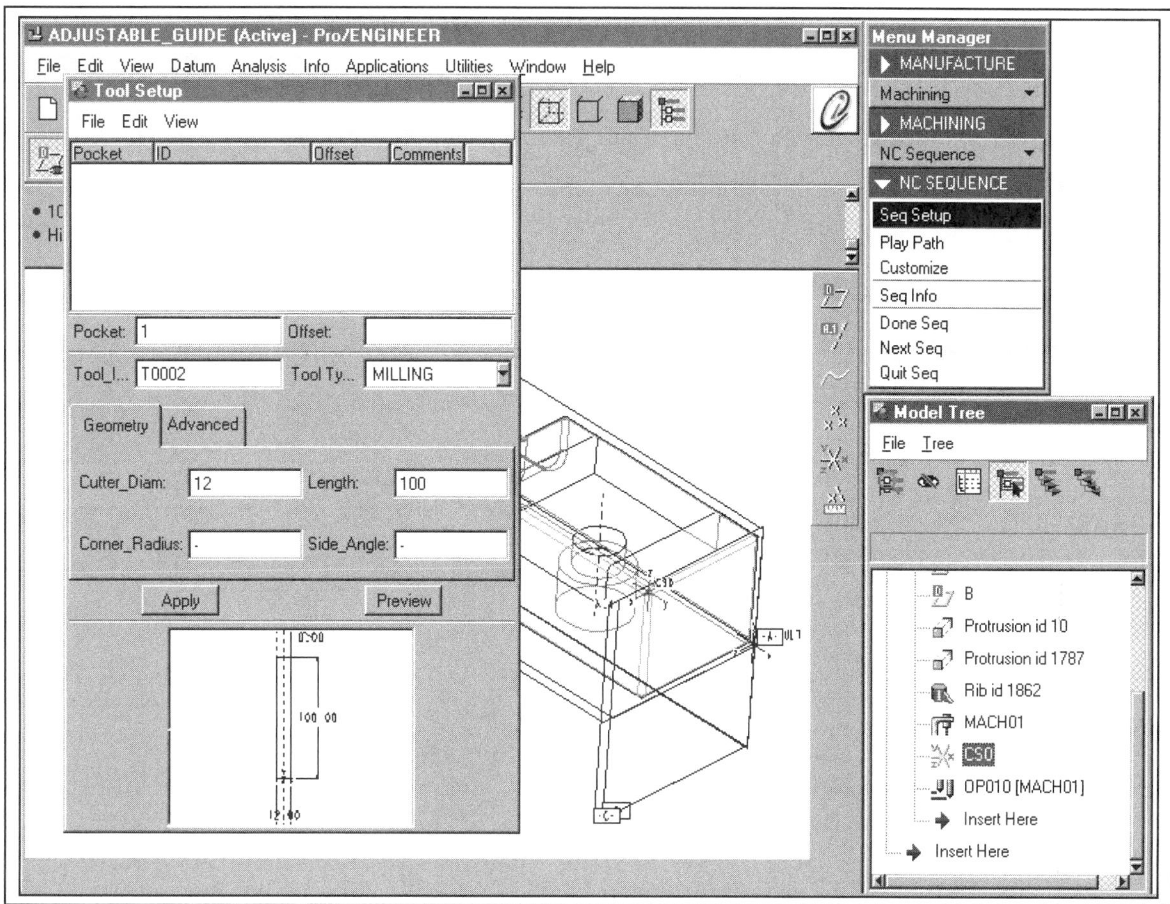

Figure 22.30
Tool Setup Dialog Box

Required inputs are cutter diameter and length. The diameter is required because it is used to calculate the tool position. The length is required because it is used to display the tool on the screen. The length does not affect the tool position calculation.

Optional parameters include **Corner Radius** (R), **Side Angle** (ANG), **Number of Teeth**, **Tool Material**, etc. All of this data is passed on to the CL File in the Tool Description Postprocessor Statement. The Corner Radius and Side Angle are also reflected in the display of the tool on the screen. These two parameters also affect the tool position calculation; however, they are not required because they have default values. The default for Corner Radius is zero (or none), and the default for the Side Angle is none, which defines a straight tool.

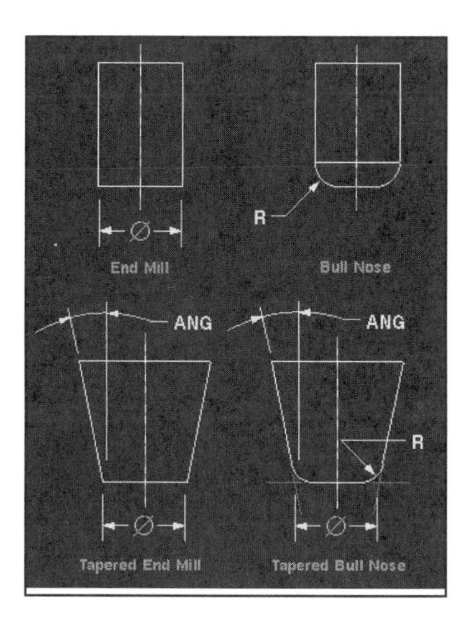

The positions written to the CL File represent the center of the tip of the tool. Pro/E performs the calculation required to derive this position, based on the machining surfaces and tool description and on any active parameters for floor or side stock. The tool path centerline display represents the continuous motion of the tool at the center of the tool tip.

The tool is positioned along a machining surface (Fig. 22.31) based on the type of machining you are doing (e.g., **Profiling, Volume Mill, Face Mill,** etc.). In basic profiling and pocketing, the bottom of the tool is placed on the surface being machined. If the surface being machined is perpendicular to the bottom of the tool, the side of the tool is placed on the surface.

HINT

It is important that each tool you define have a different name. Pro/NC does not automatically assign new names or warn you if two tools have the same name. Always be sure to specify a new name for each new tool you specify. The tool name is defined in the parameter: TOOL_ID in the tool specification table (in Pro/TABLE or in the Tool Setup dialog box).

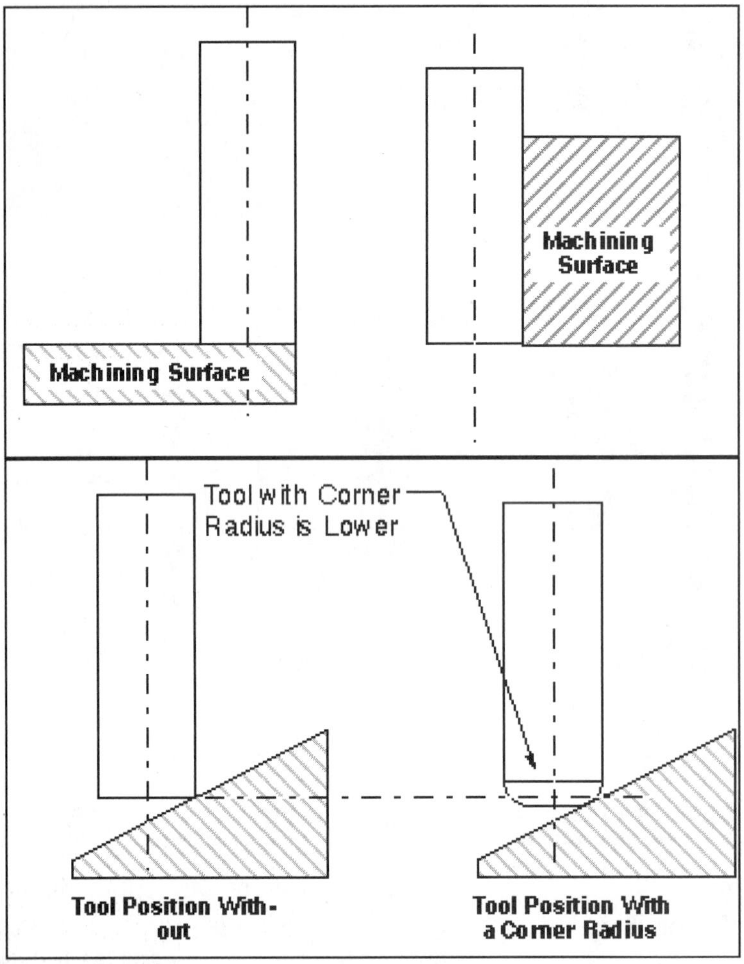

Figure 22.31
Tool Position with and Without Corner Radius

If the surface being machined is parallel to the bottom of the tool, the entire tool bottom will contact the surface. This means that the GOTO location output will be equal to the **Z** height of the surface. If however, the machining surface is at an angle, the corner radius, not the tool tip, comes in contact with it, thus affecting the **Z** value calculations.

Define the tool by specifying a series of tool parameters by choosing the following commands:

NOTE

The tool diameter and length will be in the units of the manufacturing model. The design part was created using millimeters, so the cutter will use the same units.

(Click with the mouse cursor in the field to the right of the text **Cutter_Diam** and type **20** for the cutter diameter) ⇒ (click with the mouse cursor in the field to the right of the text **Length** and type **100** for the length) ⇒ **Apply** [Pro/E displays the tool in the small graphics window within the Tool Setup dialog box (Fig. 22.32)] ⇒ **File** ⇒ **Save tool** ⇒ **File** ⇒ **Done** [to exit the Tool Setup dialog box and display the Manufacturing Info window (Fig. 22.33)]

? Pro/HELP

Pro/NC is extremely complex and rich in capabilities. Only a few aspects of the program are provided in the text. Consult Pro/HELP for each new concept presented in this lesson to get a deeper understanding of the software.

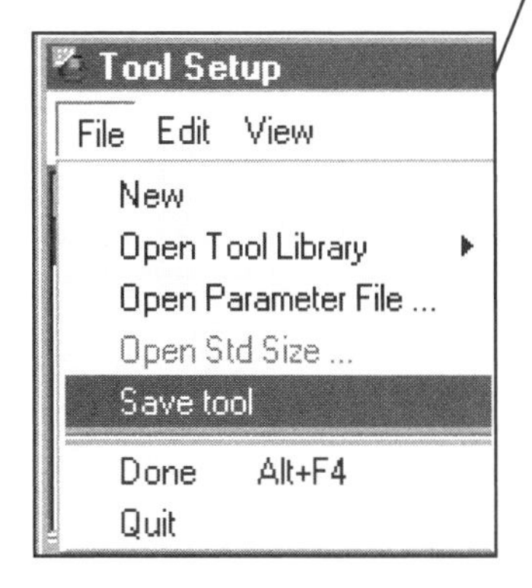

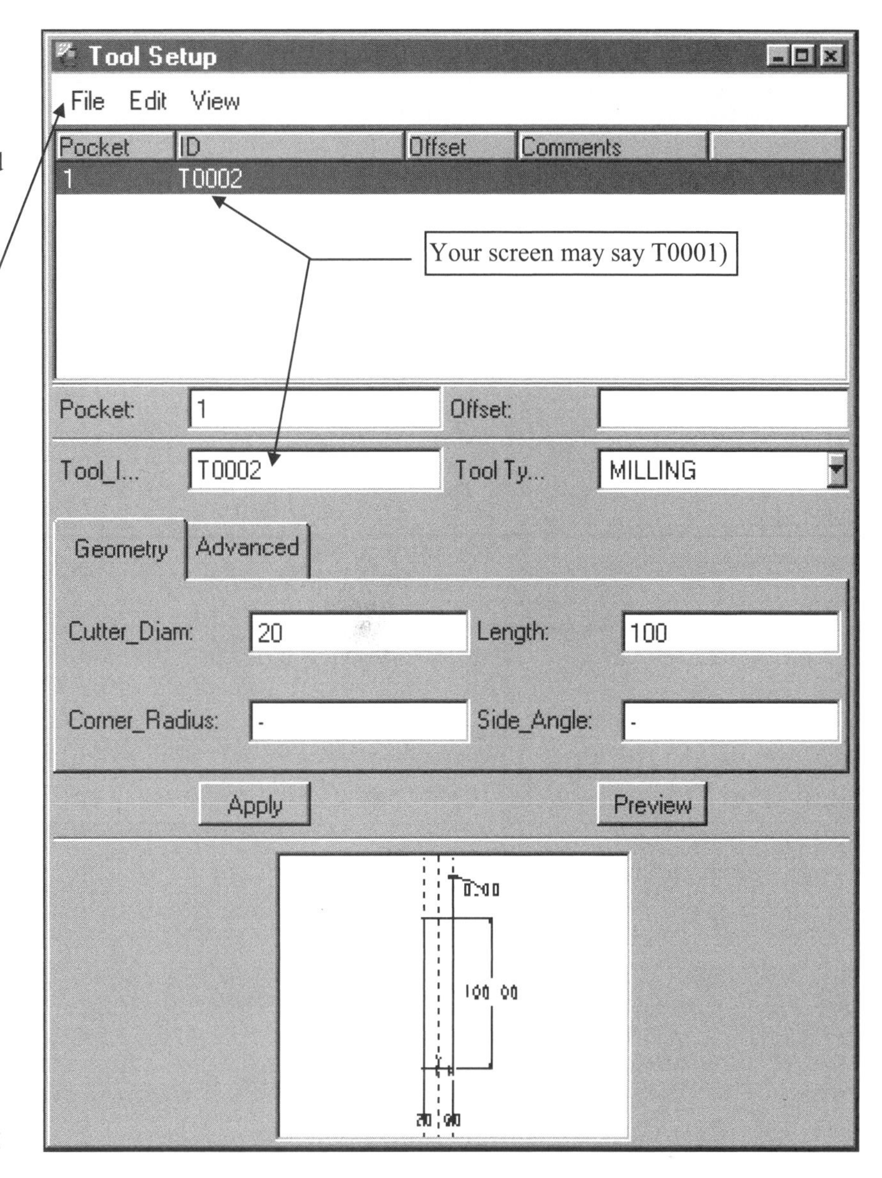

Figure 22.32
Tool Length and Tool Diameter for Volume Milling

NOTE

Notice that the Manufacturing Info window displays at the top of the screen. This window constantly displays the **Status** information and the **Tool Path** Information and allows you to access the machining parameters.

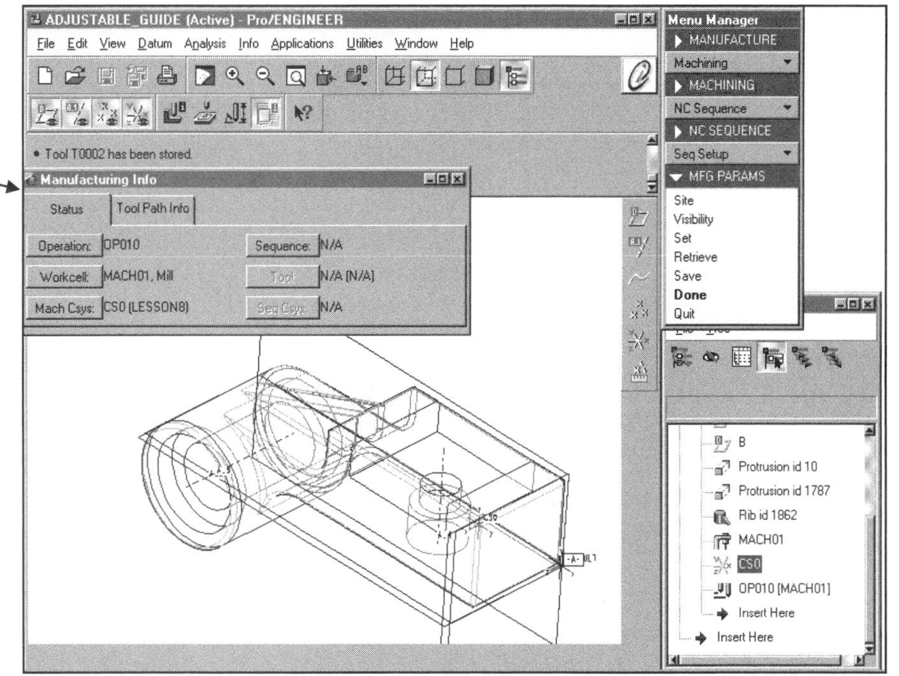

Figure 22.33
Manufacturing Info Window

Specifying Parameters

Pro/E provides many parameters to allow you to control how the tool path is calculated. You specify these parameters in the same manner as the tool, using a series of spreadsheets. Parameters vary depending on the type of **NC Sequence** you are using.

If you attempt to use a set of parameters that was defined for an **NC Sequence** type that is different from the one you are currently using, Pro/E will display a series of warnings. These warnings simply state that a parameter is not appropriate for the current **NC Sequence** and will be ignored (i.e., not included in the parameters you are currently using).

You access the machining parameters from the MFG PARAMS menu. You can name the set of parameters when you **Save** them. This allows you to **Retrieve** them for later use.

The **Set** option allows you to set up the parameters using the Param(eter) Tree dialog box (Fig. 22.34). The default display shows (basically) only the required parameters.

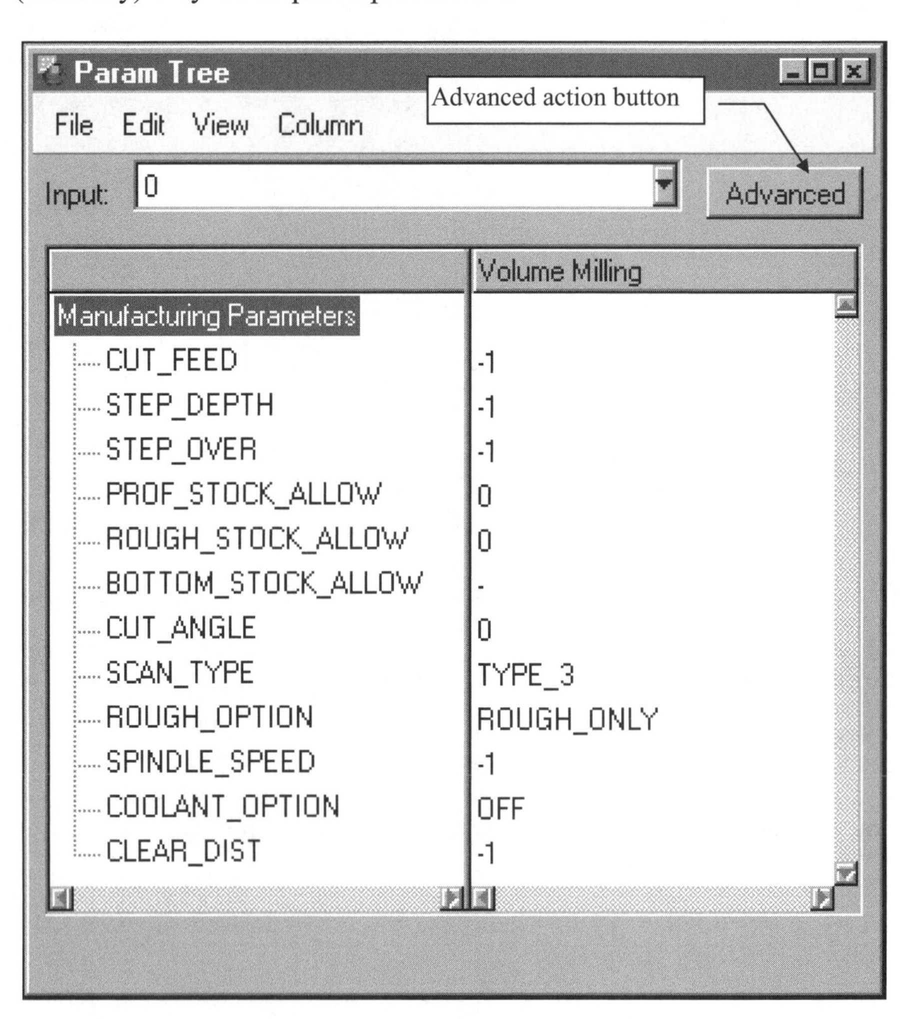

Figure 22.34
Parameter Tree

If you are defining all the parameters at the same time, you may find the **Advanced** action button useful (Fig. 22.35). This expands the display to contain every parameter available.

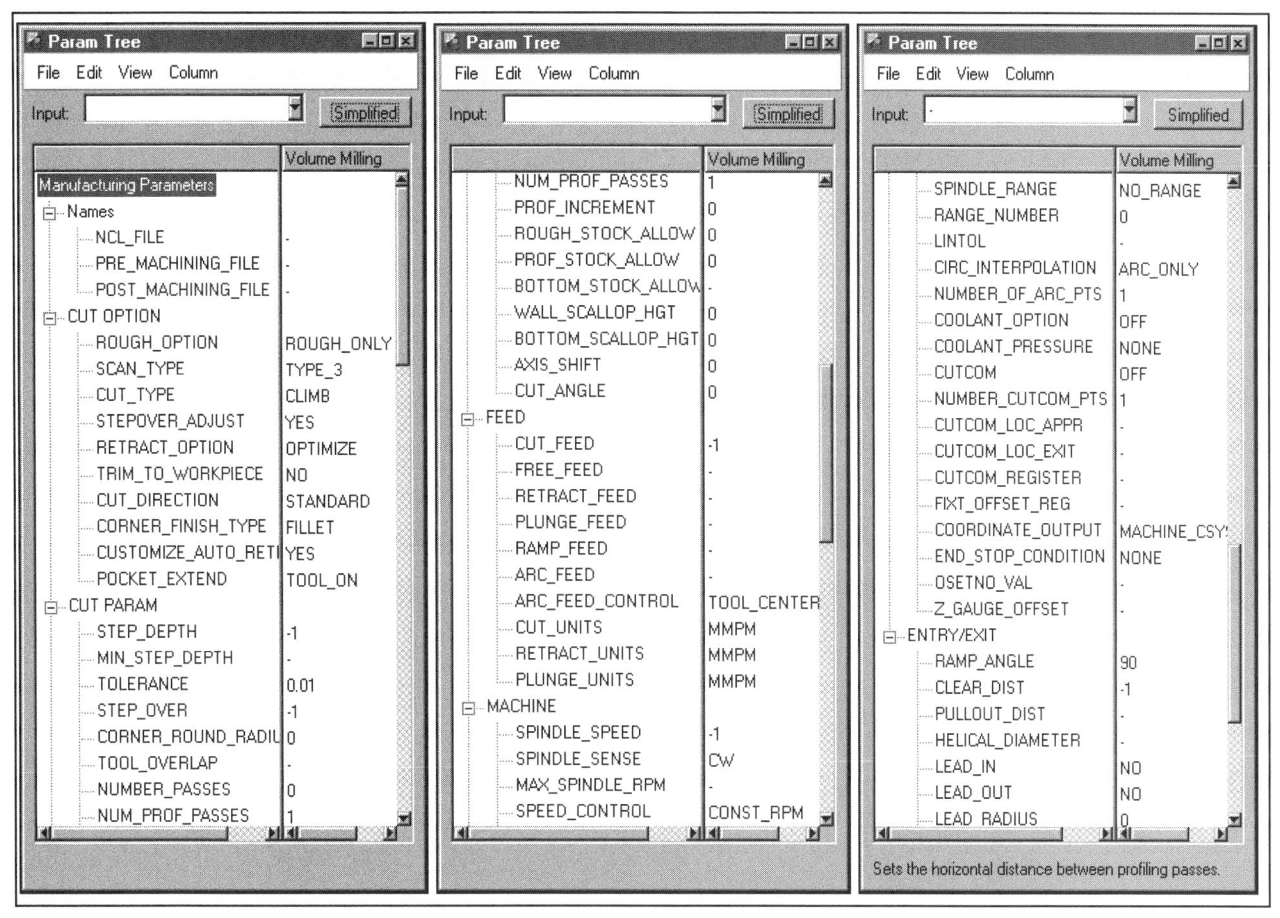

Figure 22.35
Advanced Selections

The parameters available will change based on the type of Workcell you are using (e.g., **3 Axis Mill, 5 Axis Mill, 2 Axis Lathe,** etc.) and the type of machining you are doing (e.g., **Profiling, Volume Milling, Holemaking,** etc).

There are *numeric fields* and *option choice fields* for the parameters in the Param(eter) Tree. In a numeric field, you will find either a default value, a zero (**0**), a dash (**-**), or a negative one (**-1**). If there is a default value, you may choose to leave it alone or enter a new value. If there is a zero, a numeric value must be present (you cannot enter a dash). If there is a dash (**-**), the parameter has no current setting and will be ignored. *If a negative one (**-1**) is present, that parameter must be given a value before you can calculate a tool path.*

The material presented here is information only. The required command picks will resume later, in a command box.

Choice fields allow you to choose from a specific set of options. To choose an option, you simply highlight the field. Its current value will display on the Input field near the top of the dialog box. If the field is an option field, Pro/E allows you to choose from the drop-down menu. To finish modifying the parameters, choose **File ⇒ Exit.** Choose **Save** from the MFG PARAMS menu if you want to save this file to disk (you will have to save the parameters if you want to retrieve them for another NC sequence), and enter a file name. The file will be saved with a file extension corresponding to the type of NC sequence you are creating.

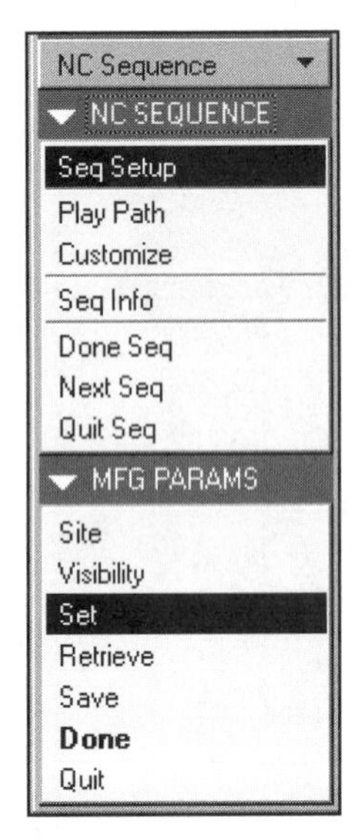

Specify the minimum required manufacturing parameters to generate a tool path for the volume milling:

Set (Pro/E now displays the Param Tree dialog box containing the minimum set of the parameters required for the type of NC Sequence you are using; for this lesson you will need to set only the items that have **-1** values) ⇒ (click with the mouse cursor in the field to the right of the text **CUT_FEED** and type **5)** ⇒ (click with the mouse cursor in the field to the right of the text **STEP_DEPTH** and type **5)** ⇒ (click in the field to the right of the text **STEP_OVER** and type **5)** ⇒ (click in the field to the right of the text **SPINDLE_SPEED** and type **500)** ⇒ [click in the field to the right of the text **CLEAR_DIST** and type **5** (Fig. 22.36)] ⇒ **File** ⇒ **Exit** ⇒ **Done**

Changed Parameter Values

Param Tree

File Edit View Column

Input: 5 Advanced

Manufacturing Parameters	Volume Milling
CUT_FEED	5
STEP_DEPTH	5
STEP_OVER	5
PROF_STOCK_ALLOW	0
ROUGH_STOCK_ALLOW	0
BOTTOM_STOCK_ALLOW	-
CUT_ANGLE	0
SCAN_TYPE	TYPE_3
ROUGH_OPTION	ROUGH_ONLY
SPINDLE_SPEED	500
COOLANT_OPTION	OFF
CLEAR_DIST	5

Distance needed to completely clear the workpiece prior t...

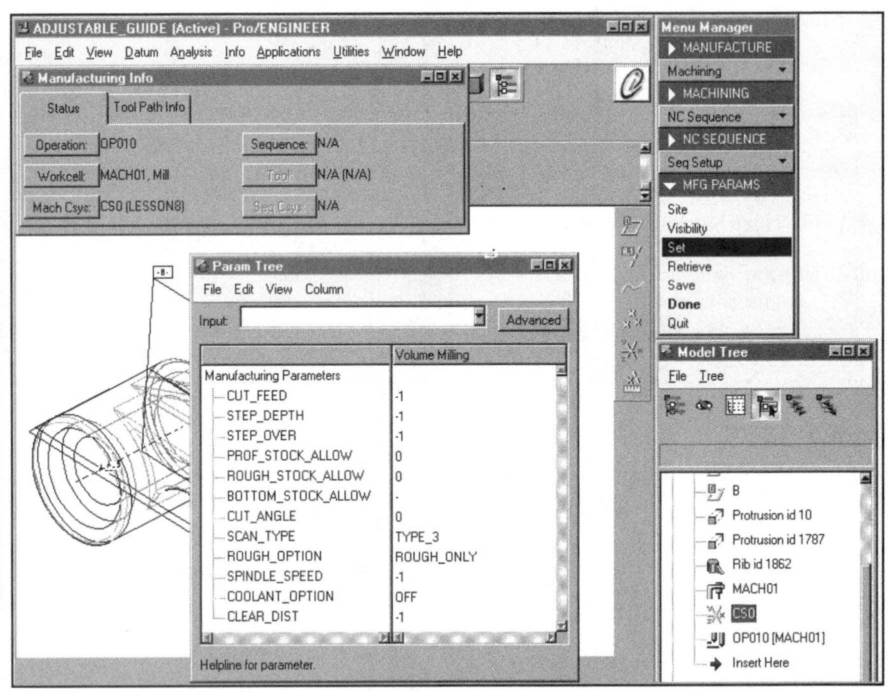

Figure 22.36
Param(eter) Tree Values

You have finished defining the manufacturing parameters and are now ready to specify the **Retract Plane** (Fig. 22.37).

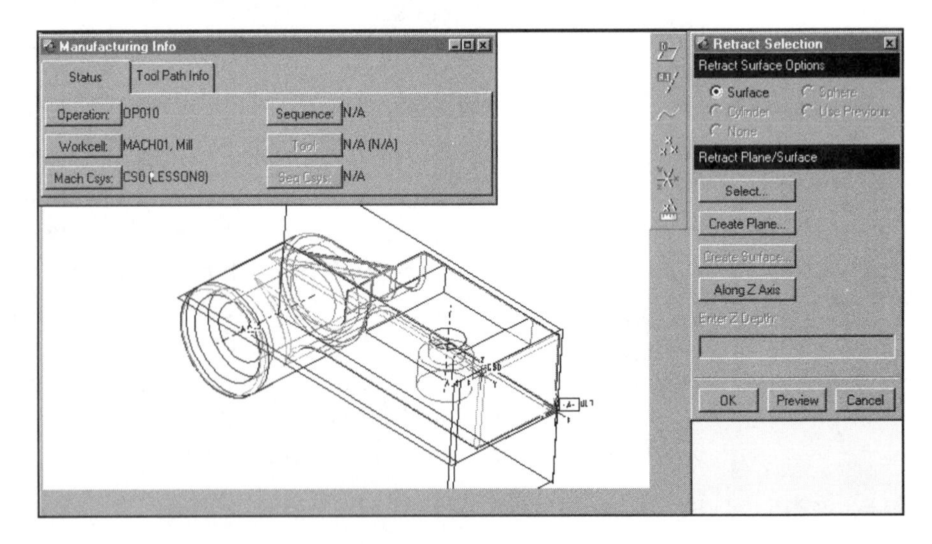

Figure 22.37
Retract Selection

NOTE
If you want to move down to the next field in the Param Tree, you can use the DOWN arrow key on the keyboard instead of clicking in the field.

Along Z Axis (from Retract Selection dialog box) ⇒ Enter **Z** Depth: **30** (Fig. 22.38) ⇒ **Preview** ⇒ **OK** (Fig. 22.39)

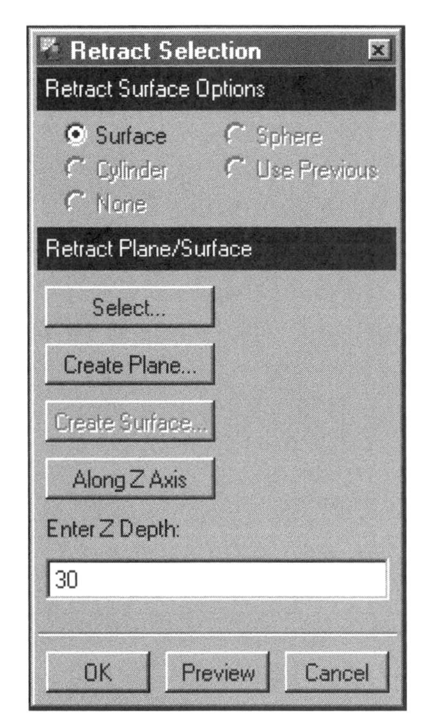

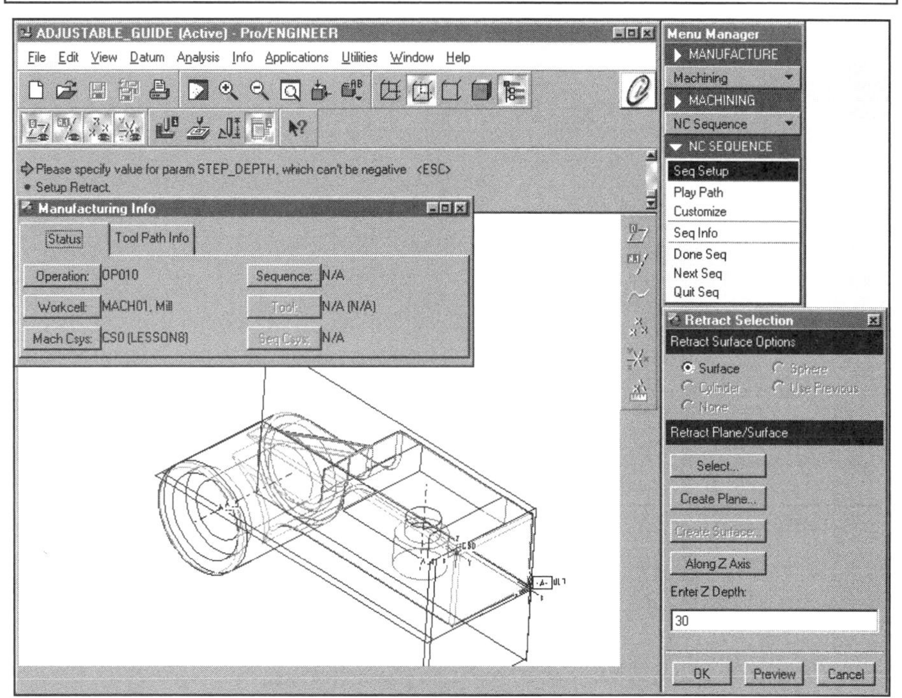

Figure 22.38
Retract Plane

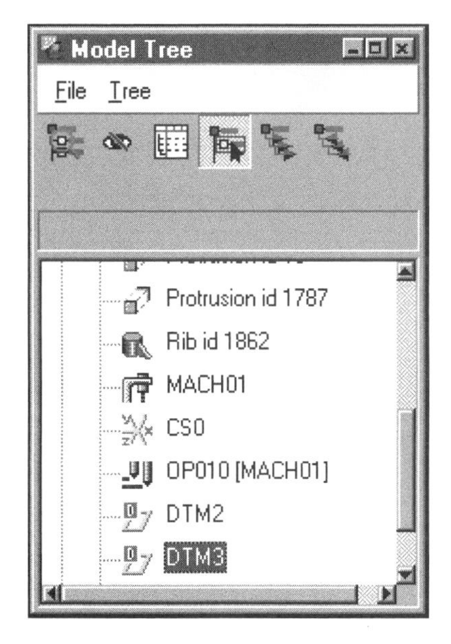

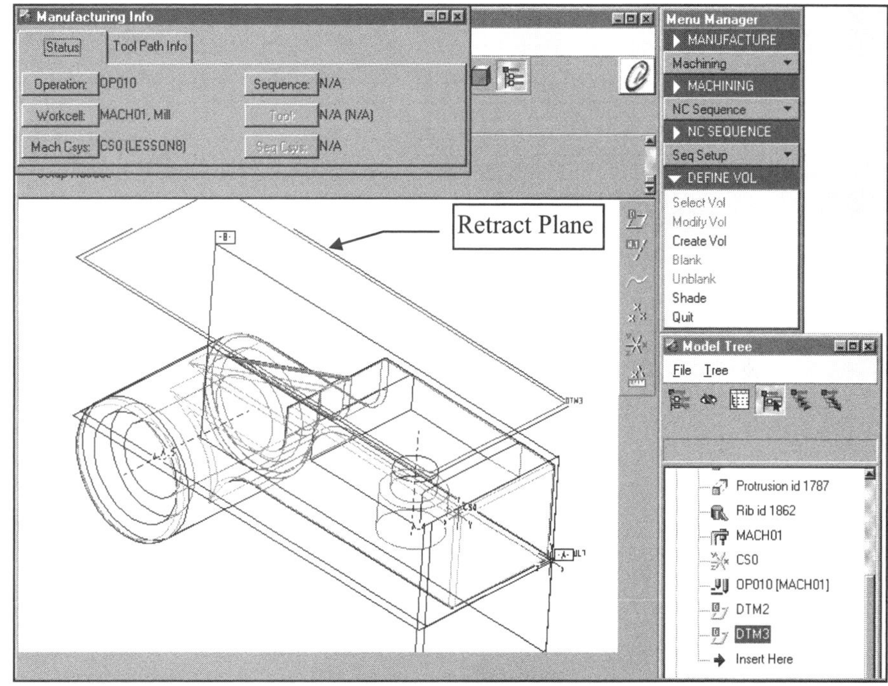

Figure 22.39
Retract Plane Created

NOTE

If you want to see manufacturing information, pick on any of the buttons shown in Figure 22.40.

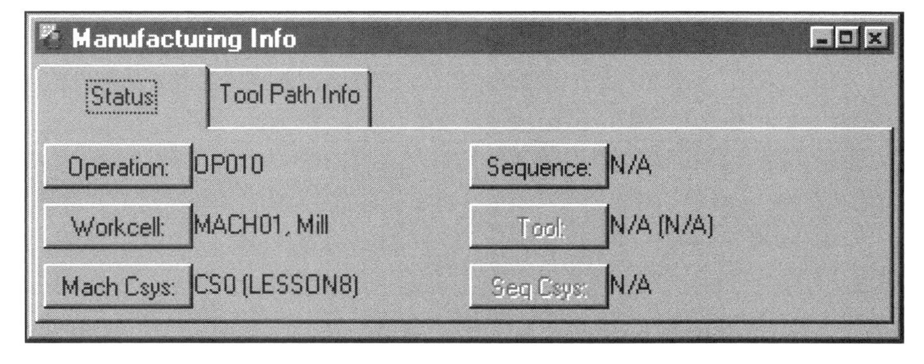

Figure 22.40
Manufacturing Info

Machining Milling Volume

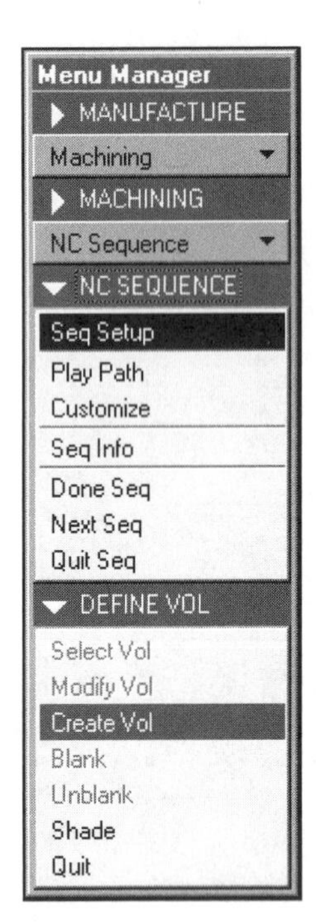

The features to be machined are defined next. You will be machining the slot first. To specify the milling surface, we will create a mill volume using the commands shown below (Fig. 22.41).

The cutter is placed in contact with the surfaces based on the order in which you select them. The orientation of the surfaces with respect to the MSYS also affects the cutter placement. If the surfaces are parallel to the axis of the tool, the side of the cutter contacts them. If the surfaces are not parallel to the tool, the bottom of the tool contacts them.

The cutter moves along a surface until its side reaches the edge of the surface. This keeps the tool contained within the boundaries of the surface (Fig. 22.42).

The placement of the tool is highly dependent on the type of machining you are doing. This description is intended only to give you an idea of how the tool is related to the surfaces you machine.

Create Vol ⇒
(Enter a name for the milling volume [Quit] **adj_slot) ⇒ ✔**

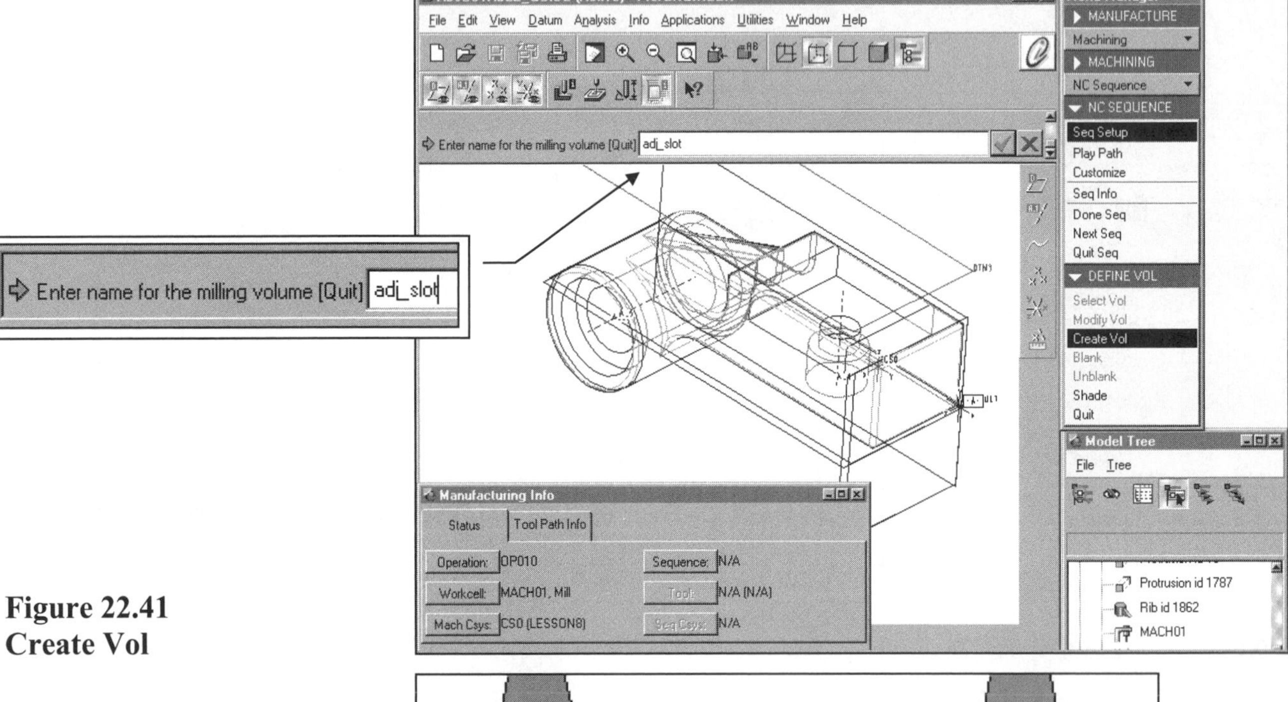

Figure 22.41
Create Vol

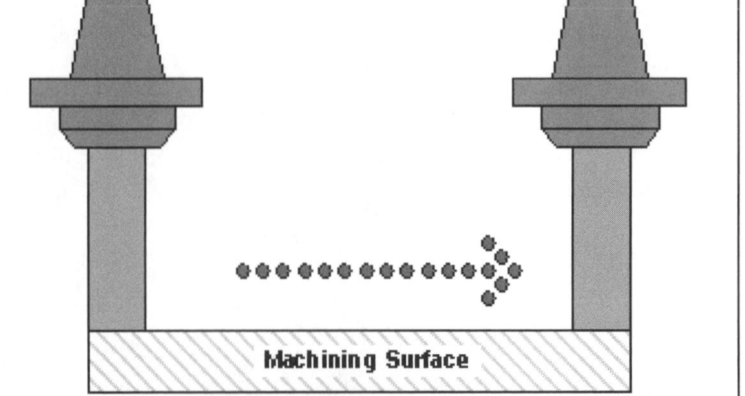

Figure 22.42
Tool Placement

Sketch ⇒ **Extrude** ⇒ **Solid** (Fig. 22.43) ⇒ **Done** ⇒ **Both Sides** ⇒ **Done** ⇒ **Make Datum** ⇒ **Offset** ⇒ **Plane** ⇒ (select the front planar surface) ⇒ **Enter Value** (Fig. 22.44) ⇒ [type **-30** (half of the adjustable guide's width of **60** mm)] ⇒ ✔ ⇒ **Done** ⇒ **Okay** [to accept the default direction of viewing (Fig. 22.45)] ⇒ **Top** [pick the top surface of the model (Fig. 22.46)]

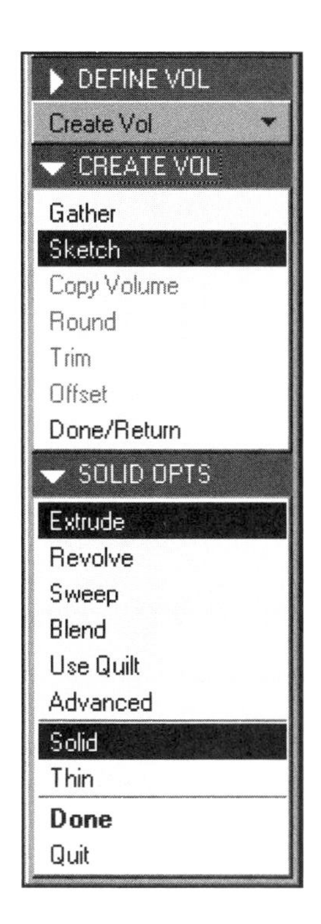

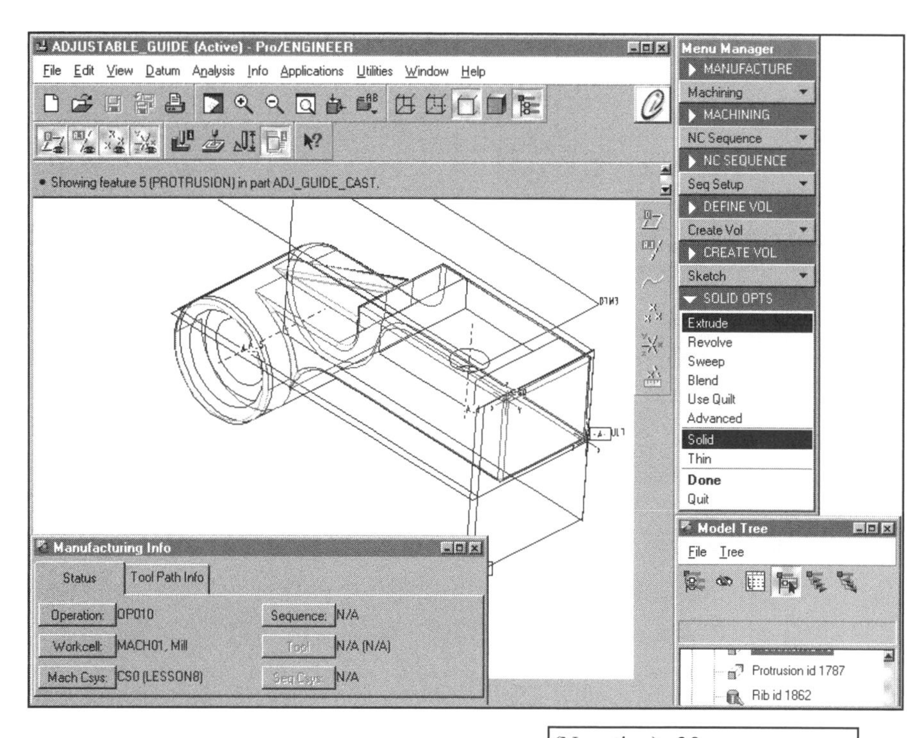

Figure 22.43
Sketch ⇒ Extrude

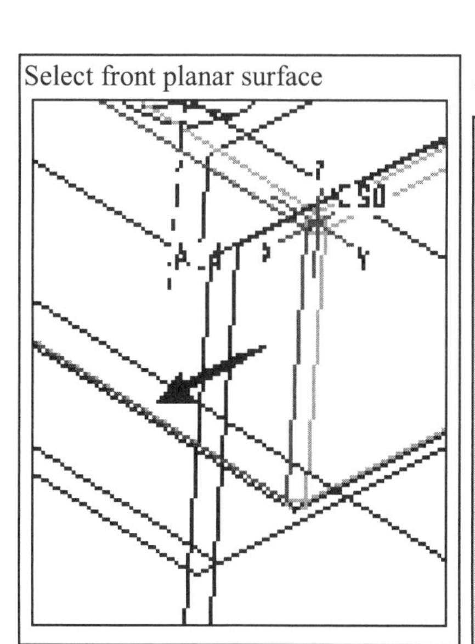

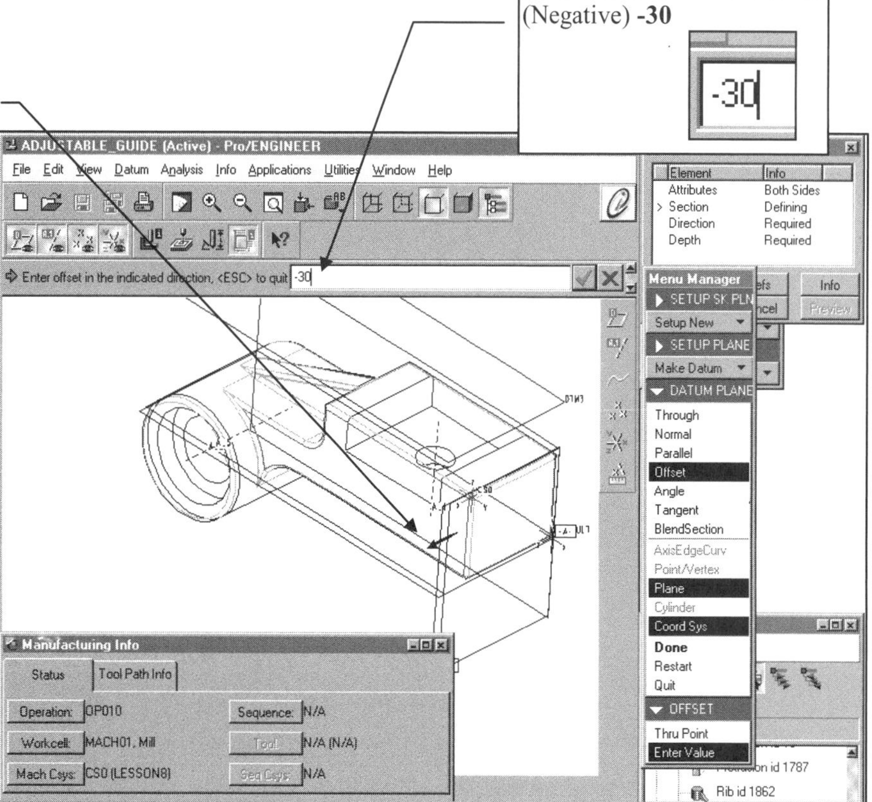

Figure 22.44
Make Datum

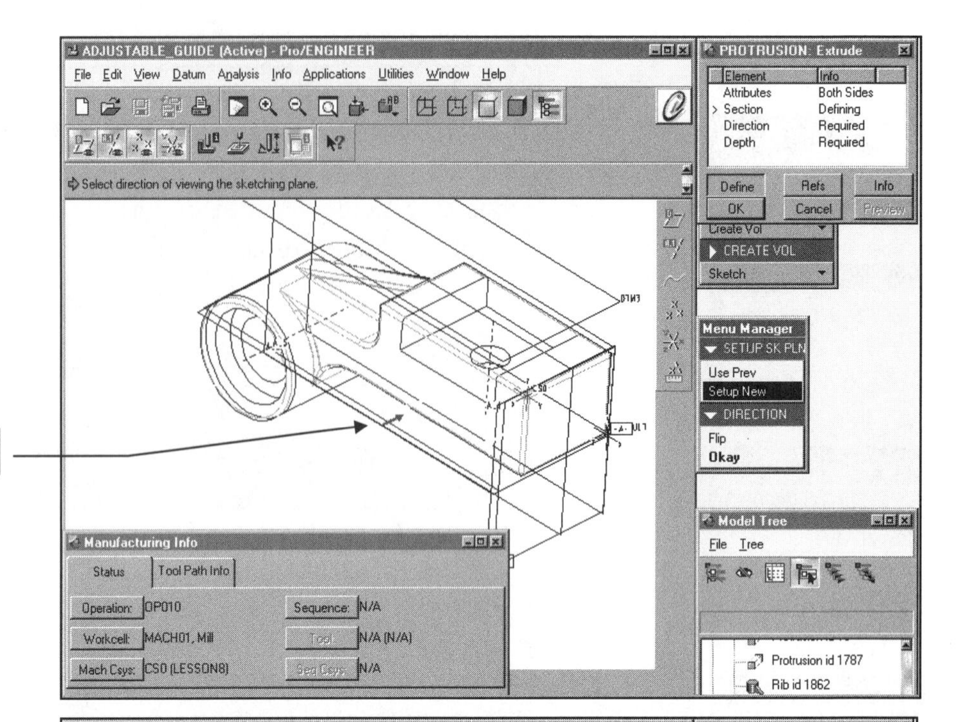

Viewing direction arrow

Figure 22.45
Viewing Direction

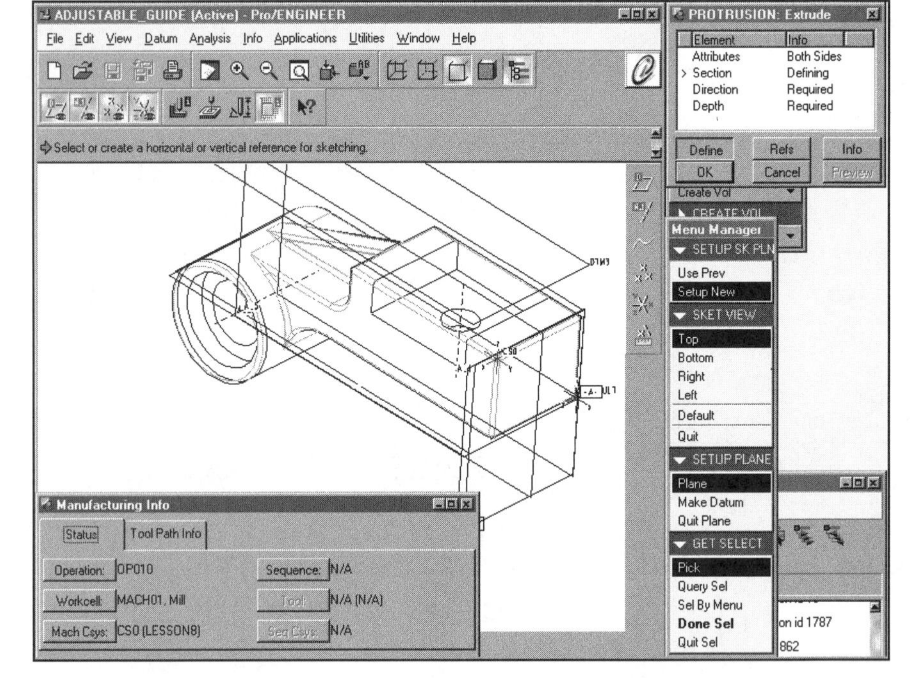

NOTE
Turn off the grid snap before sketching.

❑ **Snap to Grid**

Figure 22.46
Reference Plane **Top**

NOTE
If the cut's dimensions are modified with an ECO, the **Mill Volume** will also be updated, and the machining will still remove the required material.

HINT
Always choose **Done Seq** when you complete an NC sequence.

Rectangle (Fig. 22.47) ⇒ [sketch a rectangle approximating the cut in the design part (Fig. 22.48)] ⇒ **Sketch** ⇒ ❑ **Intent Manager** ⇒ **Alignment** [align the four lines of the rectangle to the four sides of the cut (Fig. 22.49)] ⇒ **Delete** (delete the dimensions) ⇒ **Regenerate** (the rectangle is determined by the cut, so no dimensions are needed) ⇒ **Done** ⇒ **View** ⇒ **Default** (Fig. 22.50) ⇒ **Blind** ⇒ **Done** ⇒ Enter Depth **100** ⇒ ✔ ⇒ **Preview** ⇒ **OK** (Fig. 22.51) ⇒ **Done/Return** ⇒ **Done Seq** (*always do this when you complete an NC Sequence*) ⇒ (from the **Model Tree)** ⇒ **Tree** ⇒ **Item Display** ⇒ MFG tab ⇒ (✔check all of the boxes) ⇒ **OK** (from Model Tree Items dialog box as shown in Figures 22.52 and 22.53) (Figure 22.54 shows the completed mill volume)

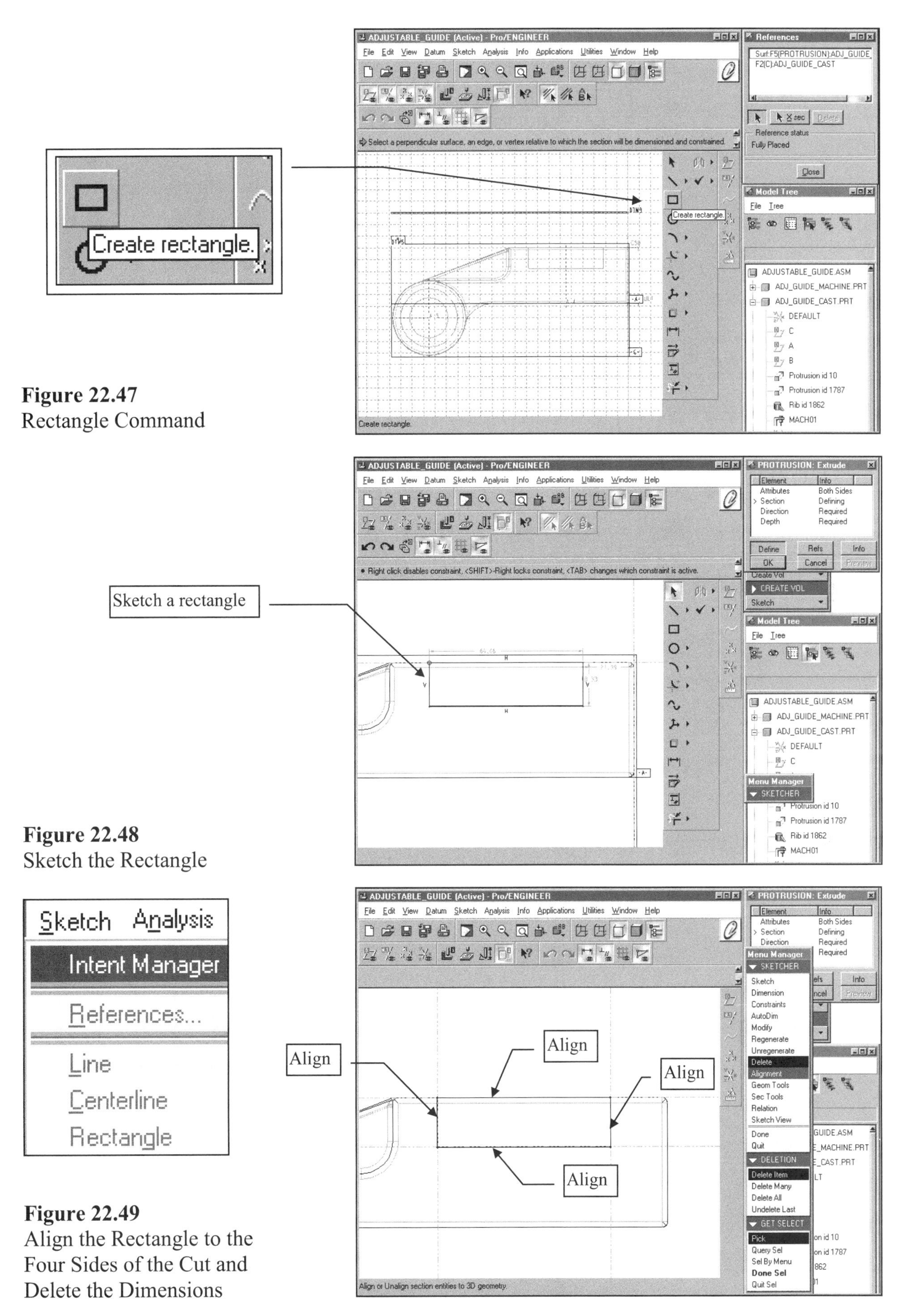

Figure 22.47
Rectangle Command

Figure 22.48
Sketch the Rectangle

Figure 22.49
Align the Rectangle to the Four Sides of the Cut and Delete the Dimensions

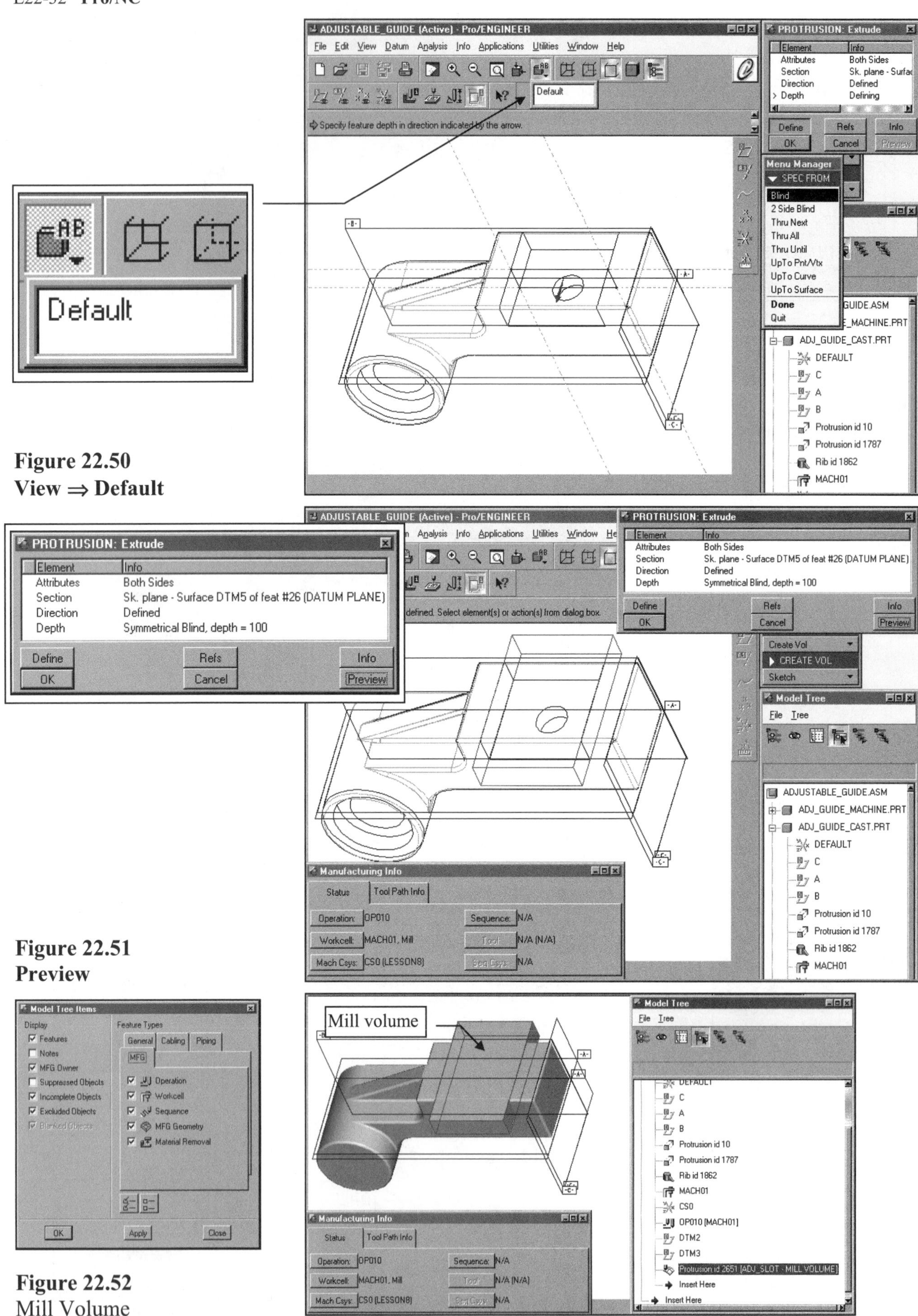

Figure 22.50
View ⇒ Default

Figure 22.51
Preview

Figure 22.52
Mill Volume

Figure 22.53
Model Tree Showing MFG Features

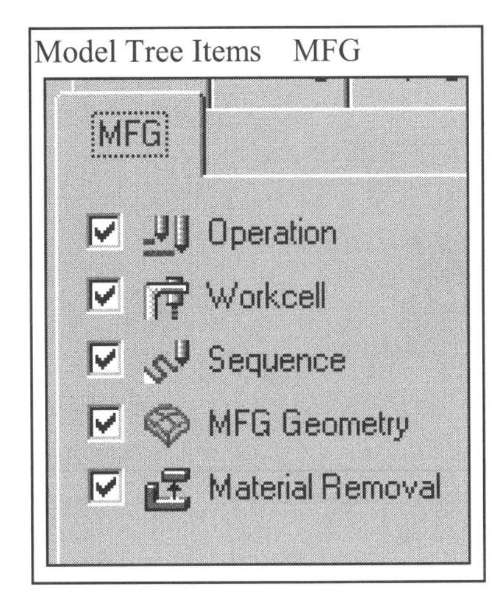

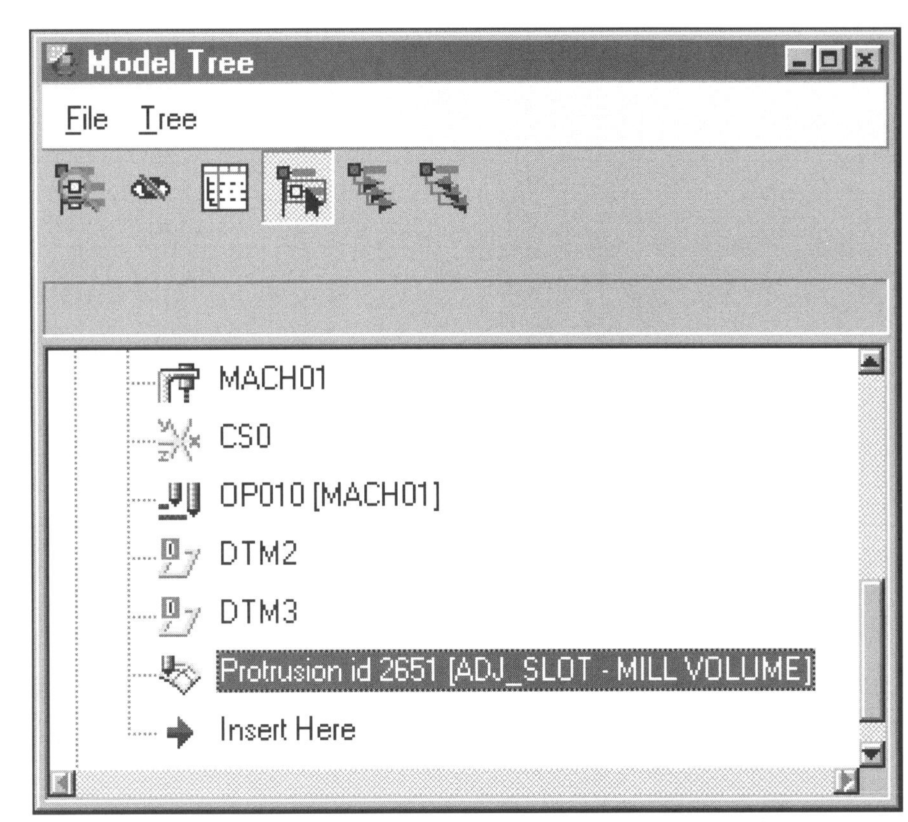

Figure 22.54
Save Your Work
File ⇒ Save ⇒ ✔

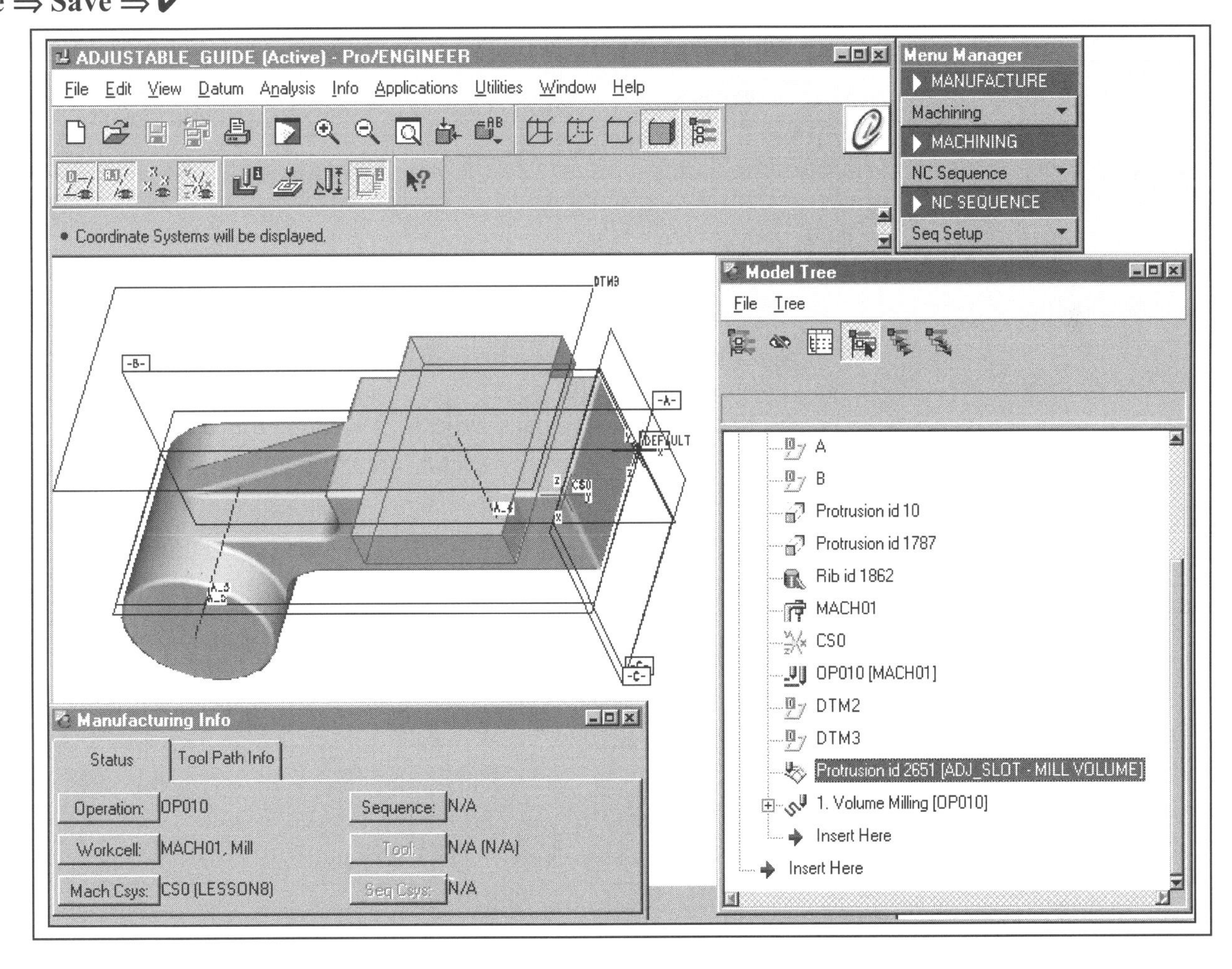

Tool Path Simulation

You can display the cutter path and a simulation of the tool, prior to completing the NC sequence, to verify the tool path and make a visual check for interference with fixtures and model features. All simulated tool dimensions represent the parameters defined during tool setup (refer to Tooling for more information). All tools except turning appear three-dimensional in isometric or trimetric views.

When you choose **Play Path** from the NC SEQUENCE menu, the PLAY PATH menu appears with the following options:

Screen Play Display the tool path on the screen.
Play Steps Appears only if the tool path has been customized. Allows you to display the tool path step by step. Use **Continue** to display the next step.
Quit To quit the display process.
Show File View the contents of the CL data file in an Information Window.
NC Check Access the NC Check functionality (refer to Pro/NC-CHECK). See the ***HINT*** in the left-hand column.
Gouge Check Access the Gouge Check functionality (see Gouge Checking). Available for Milling NC sequences only.

HINT
NC Check can be performed:
* At the time of creating an NC sequence (from the PLAY PATH menu) to check the current tool path.
* From the CL DATA menu after the NC sequence or operation is created. You will be prompted for a CL file name. At this point, you can either select an existing file or create a new one.
* When editing CL data.
You can control which **NC Check** simulation module to use by setting the configuration option ncCheck_type.

The values are:
* ***vericut*** (default) Use Vericut(tm) provided by CGTech.

* ***nccheck*** Use Pro/NC-CHECK.

Change your *config.pro* file to the following:

nccheck_type ***ncchec***

If you choose **Screen Play,** the DISPLAY CL menu appears with the following options:

Tool Depending on whether the check mark is on or off, display or do not display the tool.
Disp Cycles Appears only for Holemaking NC sequences and Thread Turning. If the check mark is on, all the tool motions included in the CYCLE command or in the thread cycle will be displayed. If the check mark is off, a simplified display will be used.
Status Box When you display the tool path, additional information appears in the **Info Box**, such as the feed rate, the spindle speed, and the current **XYZ** coordinates of the tool as shown in Figure 22.55.
StopAtStart Makes the tool stop at the beginning of the tool path, to allow you to check the cutter location coordinates in the Info Box.

The next two options represent two ways to control the frequency of consecutive tool displays on the tool path.

Time Increment Puts the tool display in the real-time mode. Enter a value for the time increment (in seconds) between two consecutive tool displays.
Cutter Step Displays the tool at uniform distances along the tool path. Enter a value for step size (in the units of the workpiece). If you enter a large value for step size, the tool will be displayed at the GOTO locations only.

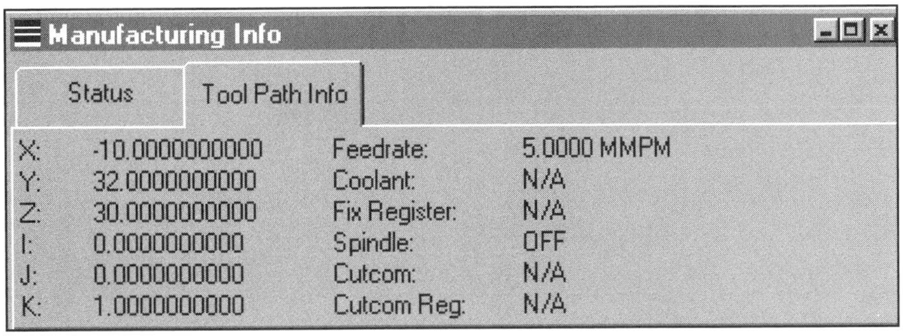

Manufacturing Info

Status | Tool Path Info

X:	-10.0000000000	Feedrate:	5.0000 MMPM
Y:	32.0000000000	Coolant:	N/A
Z:	30.0000000000	Fix Register:	N/A
I:	0.0000000000	Spindle:	OFF
J:	0.0000000000	Cutcom:	N/A
K:	1.0000000000	Cutcom Reg:	N/A

Figure 22.55
Tool Path Info

Once you display a tool path, Pro/NC will remember the options used and select them as defaults when next displaying a tool path within the manufacturing session.

Once you have set up the CL display environment, choose **Done CL**. The tool path is displayed according to the specified options. Then the CL CONTROL menu appears with the following options:

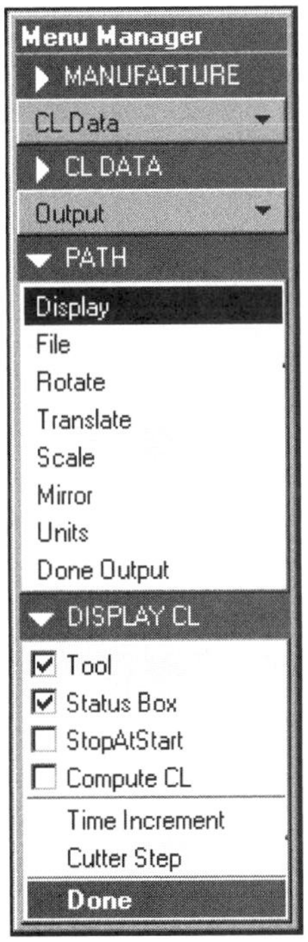

Position Select a point along the tool path. The tool will be positioned at this point.
Next The tool is displayed at the location corresponding to the next GOTO command.
Prev The tool is displayed at the location corresponding to the previous GOTO command.
CL Measure Access the Pro/ENGINEER Measure functionality to compute tool interference or clearance.
Time Increment and **Cutter Step** The same as in the DISPLAY CL menu.
Continue Proceed with the tool path display from the current position of the tool.
Done Display the tool path for the next NC sequence, or, if only a single NC sequence is present, exit the CL CONTROL menu.
Quit Exit the CL CONTROL menu. This command will appear only if multiple NC sequences are present.

Menu Manager
MANUFACTURE
CL Data
CL DATA
Output
PATH
Display
File
Rotate
Translate
Scale
Mirror
Units
Done Output
DISPLAY CL
☑ Tool
☑ Status Box
☐ StopAtStart
☐ Compute CL
Time Increment
Cutter Step
Done

The tool will always be displayed while you move it using the CL CONTROL menu options, even if the Tool check mark in the DISPLAY CL menu was turned off. Once you choose **Continue** from the CL CONTROL menu, the tool display will again be controlled by the CL display environment.

If you choose **NC Check**, the dynamic material removal simulation will be activated.

Because we have completed the NC sequence and chosen **Done Seq,** a slightly different set of menus will appear. Choose the following commands to view the CL data and cutter path (Fig. 22.56):

CL Data ⇒ **Operation** ⇒ (select **OP010** from the list) ⇒ **Display** ⇒ **✔Tool ✔Status Box** ⇒ **Done** ⇒ **Done** ⇒ **Done Output** ⇒ **NC Check** ⇒ **Display** ⇒ **Run** ⇒ **Create** ⇒ **Operation** ⇒ (select **OP010** from the list) ⇒ **OK** (Fig. 22.57) ⇒ **Done/Return** ⇒ **Done/Return**

HINT
You can interrupt the cutter path display at any time by clicking on the STOP sign at the right end of the Message Window.

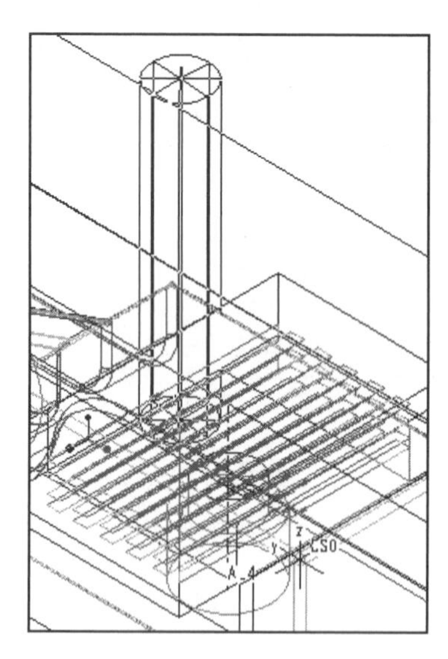

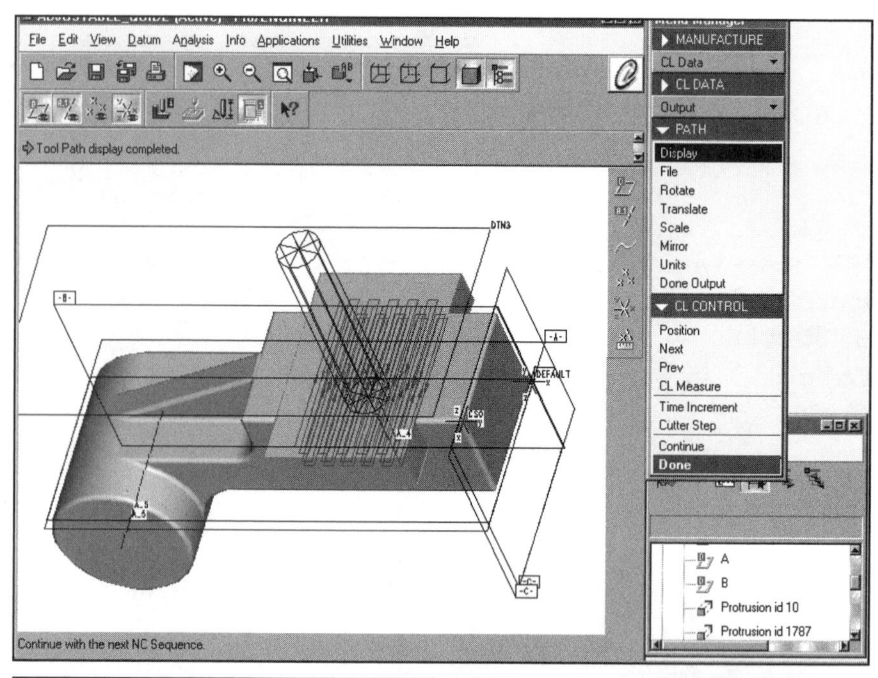

Figure 22.56
Displaying the Cutter Path

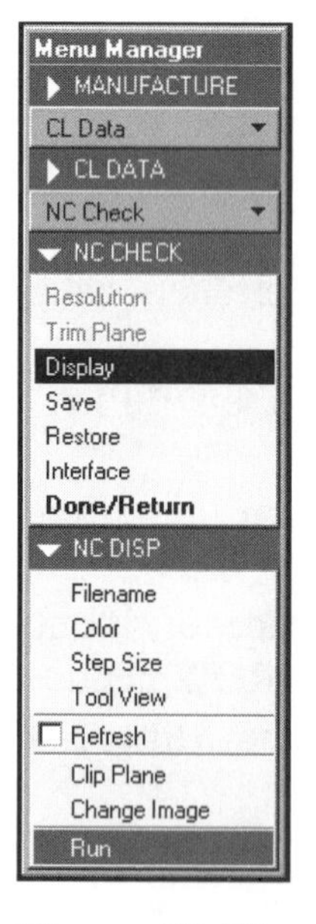

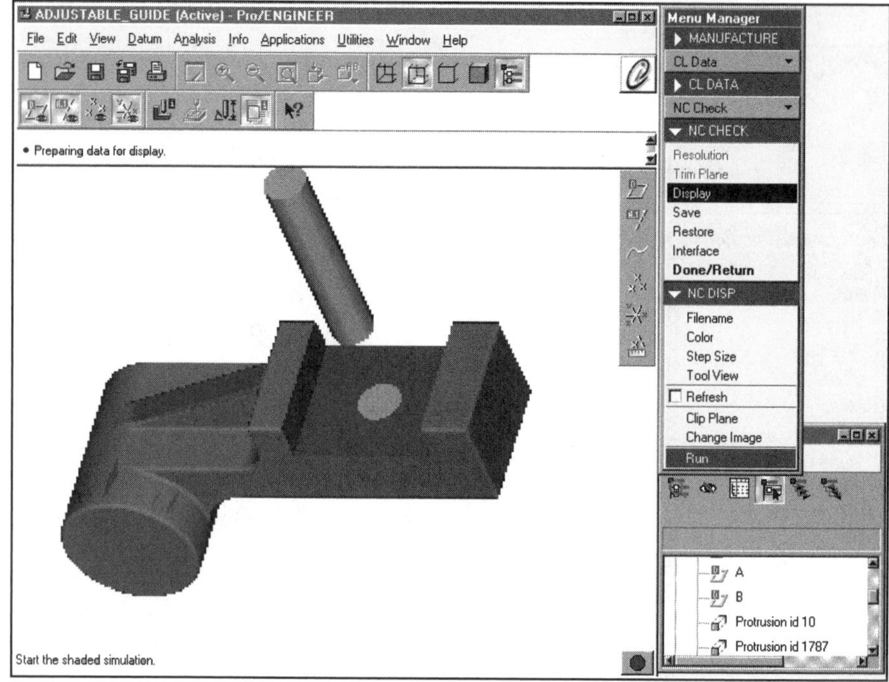

Figure 22.57
NC Check

Creating a Holemaking NC Sequence

A **Holemaking** NC sequence is created by selecting the cycle type and specifying the holes to drill by defining the **Hole Set**(s). The order of machining the holes is defined by the SCAN_TYPE parameter value; you can also build the traversal path between the selected holes either by sketching or by connecting the hole axes.

A **Hole Set** includes one or more holes to be drilled; each **Hole Set** has a drilling depth specification or countersink diameter value associated with it. You can include more than one **Hole Set** in a single **Holemaking** NC sequence; this allows drilling of holes with differing depths and multiple countersink diameter values.

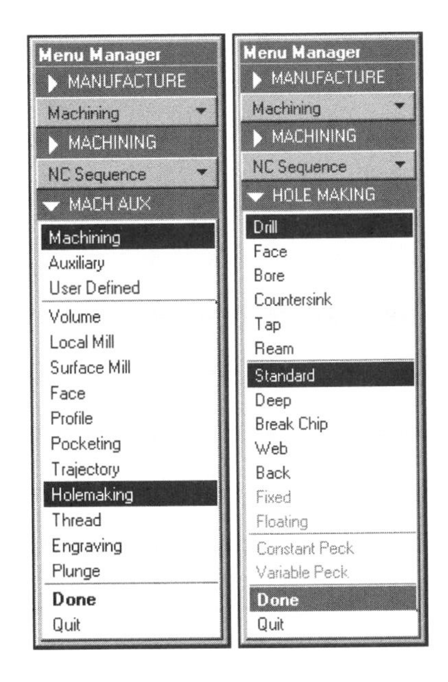

There are various methods of selecting the holes to be included in a **Hole Set:**

- * Selecting individual hole axes
- * Including all holes on a specified surface
- * Including all holes of a specified diameter
- * Including all holes with a certain value of a feature parameter
- * Including all holes with chamfers machinable by the current tool (for countersinking)
- * Selecting individual datum points to mark the drill locations
- * Including all datum points on a specified surface
- * Reading in a file containing the datum points' coordinates with respect to a specified coordinate system

If you need to perform a series of **Holemaking** NC sequences on the same group of holes, you can define a **Drill Group** using the techniques above and then reference this **Drill Group** when defining **Hole Set(s)**. This simplifies the selection process; you can also parametrically update all the NC sequences by modifying the **Drill Group**. To create a **Holemaking** NC sequence for the **∅20** hole on the part, choose the following commands:

View ⇒ **Repaint** ⇒ **Machining** ⇒ **NC Sequence** (from the MACHINING menu) ⇒ **New Sequence** ⇒ **Holemaking** ⇒ **Done** ⇒ **Drill** ⇒ **Standard** ⇒ **Done** ⇒ **Seq Setup** ⇒ **✔Tool** (Fig. 22.58) ⇒ **Done** ⇒ (the Tool Setup dialog box comes up as shown in Figure 22.59) ⇒ Tool Type **DRILLING** ⇒ Tool_ID: **T0003** ⇒ Cutter_Diam: **20** ⇒ Length: **120** ⇒ **Preview** (Fig. 22.59) ⇒ **Apply** ⇒ **File** ⇒ **Save tool** ⇒ **File** ⇒ **Done** ⇒

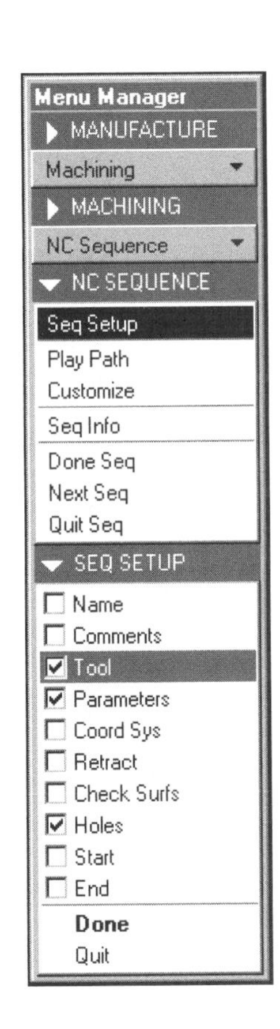

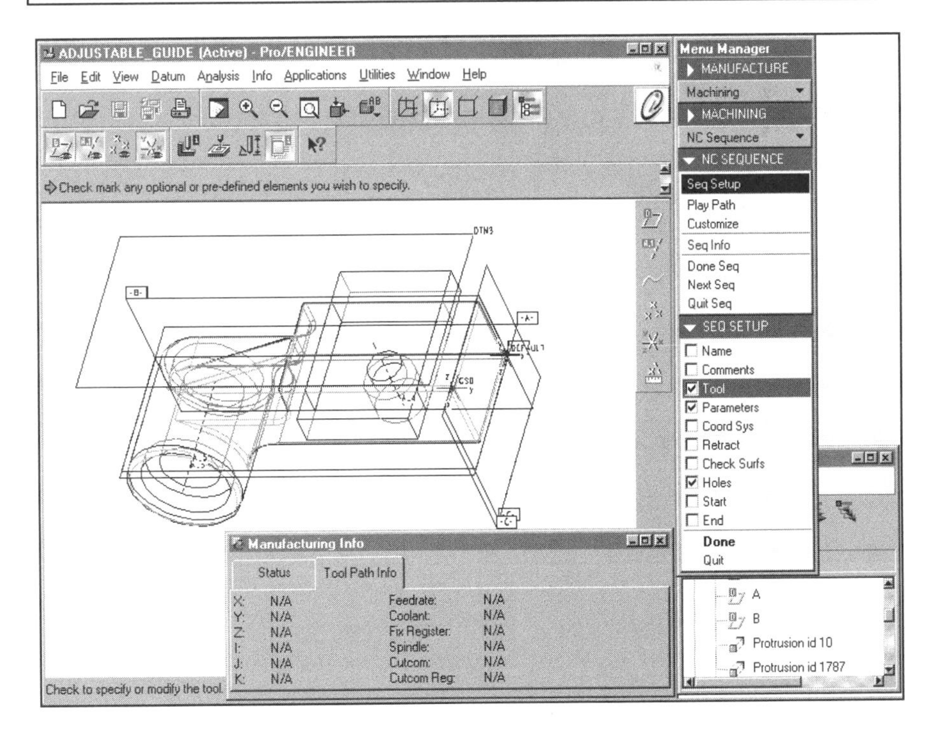

Figure 22.58
Creating a Holemaking NC Sequence

Pick to show drop-down menu

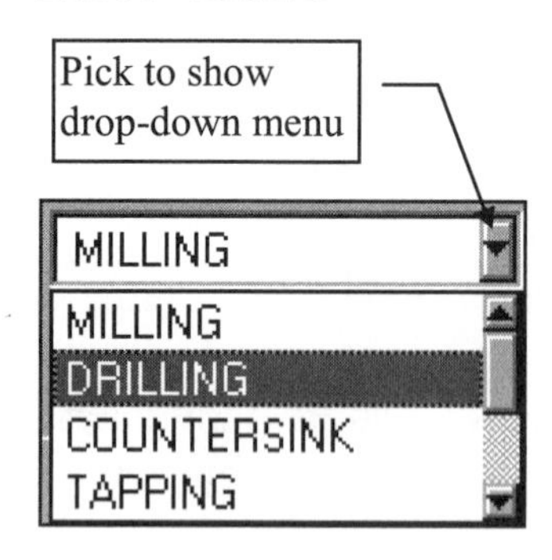

Figure 22.59
Creating a Holemaking NC Sequence

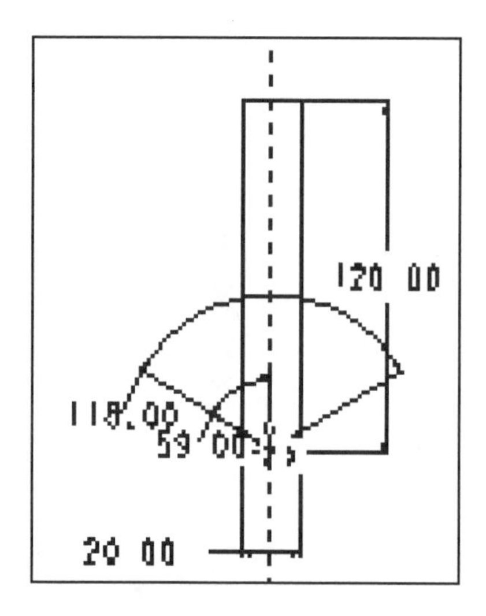

Set [from the MFG PARAMS menu (Fig. 22.60)] ⇒ (in the Param Tree dialog box, input the following values: CUT_FEED **5** SPINDLE_SPEED **500** CLEAR_DIST **5)** ⇒ **File** ⇒ **Save** ⇒ **OK** ⇒ **File** ⇒ **Exit** ⇒ **Done** (the Hole Set dialog box displays) ⇒ Select the **Diameters** tab ⇒ **Add** (Fig. 22.61) ⇒ (select the hole to drill **20.000000**) ⇒ **OK** ⇒ **Depth** ⇒ **●Auto** (Fig. 22.62) ⇒ **OK** ⇒ **OK** ⇒ **Done/Return** ⇒ **Done Seq** ⇒ **Done/Return**

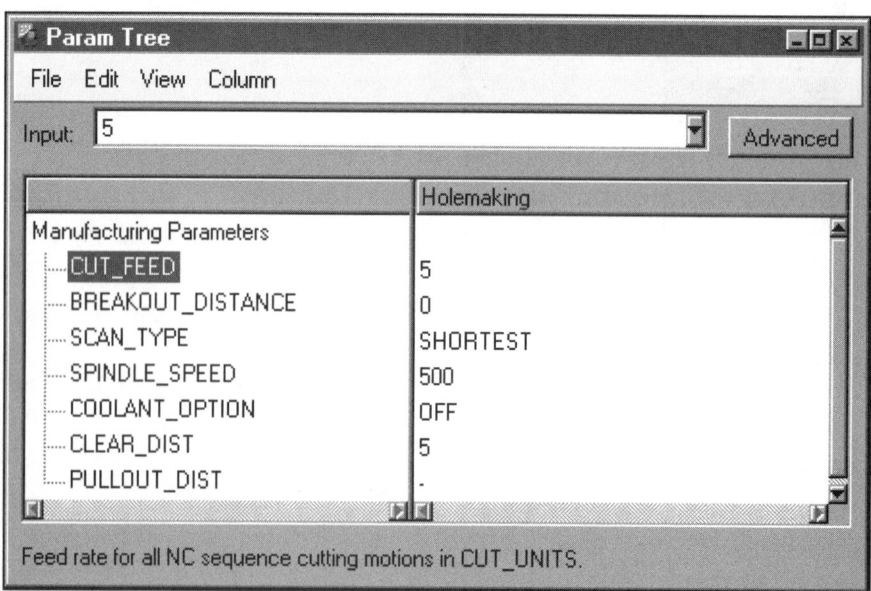

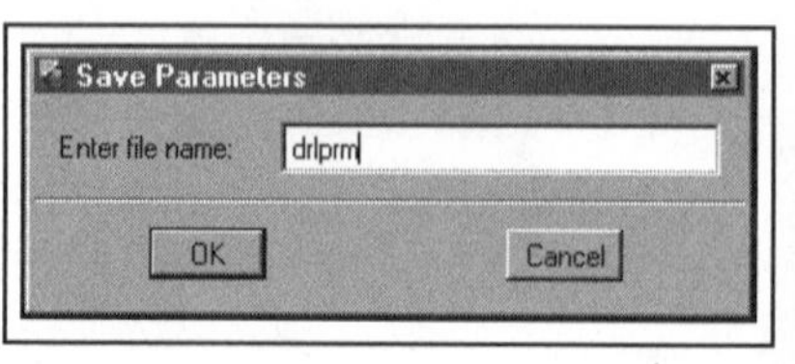

Figure 22.60
Param Tree Dialog Box

Figure 22.61
Hole Selection

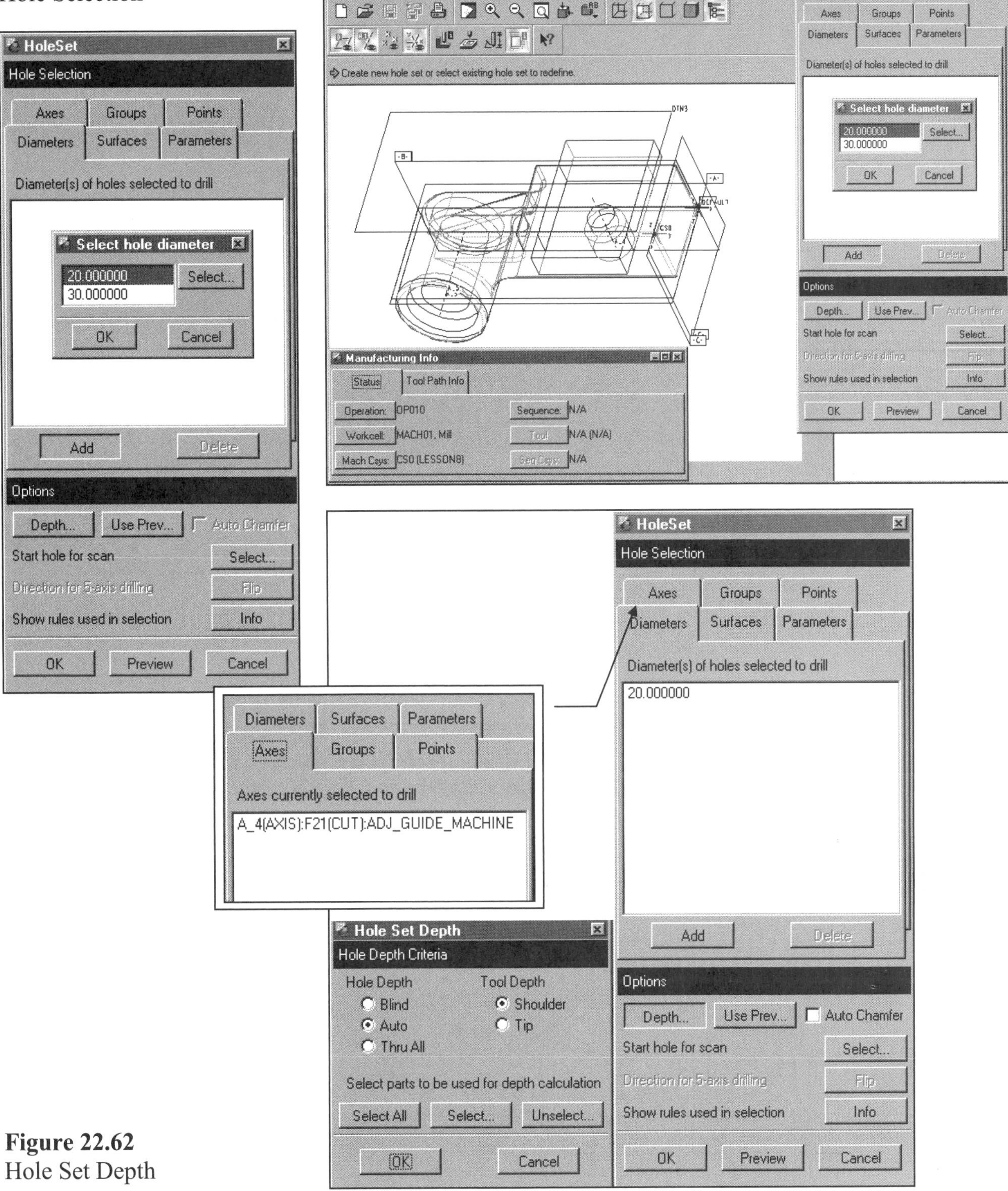

Figure 22.62
Hole Set Depth

CL Data ⇒ **NC Sequence** ⇒ (select **Holemaking** from the list) ⇒ **Display** ⇒ ✔**Tool** ✔**Status Box** ⇒ **Done** (Fig. 22.63) ⇒ **Done** ⇒ **Done Output** ⇒ **NC Check** ⇒ **Display** ⇒ **Run** ⇒ **Create** ⇒ **NC Sequence** ⇒ (select **Holemaking** from the list) ⇒ **OK** ⇒ **Done/Return** ⇒ **Done/Return** ⇒ **Done/Return**

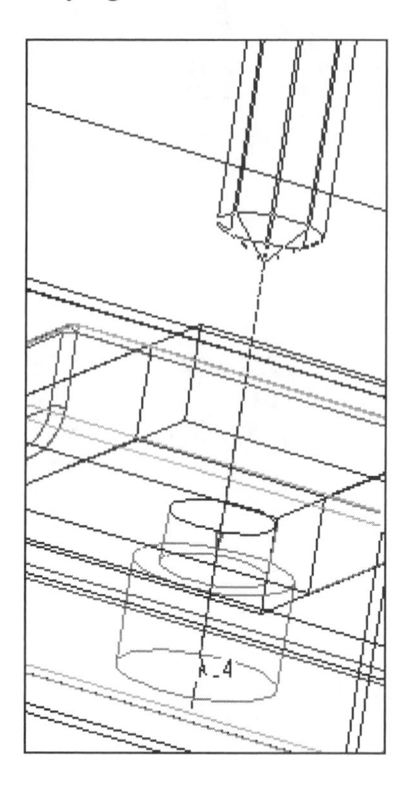

Figure 22.63
Playing Tool Path of the Drill

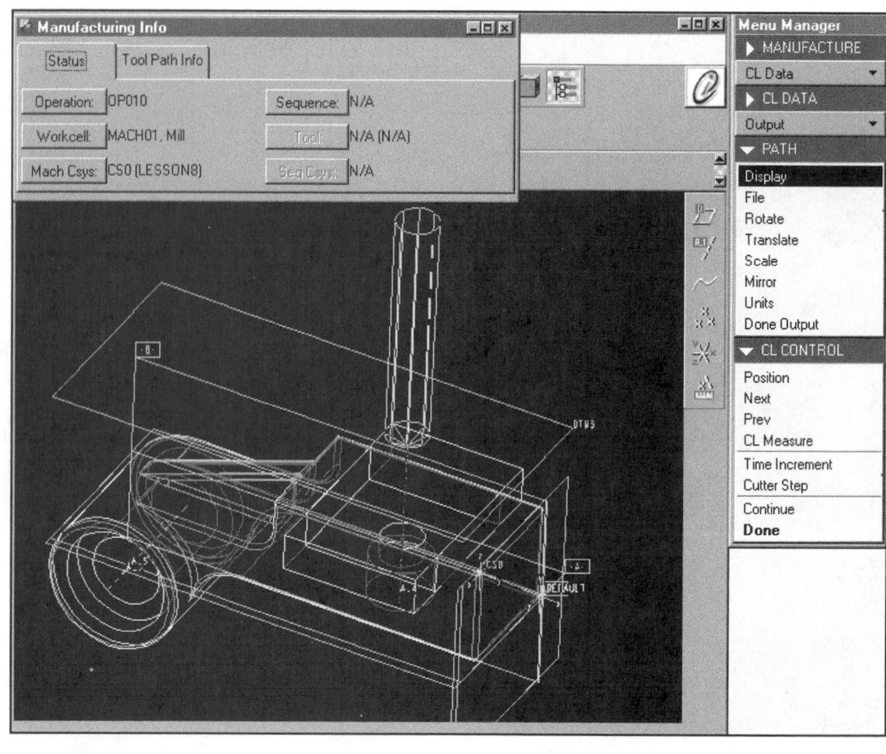

The tool path is created automatically, depending on the SCAN_TYPE parameter value. If you are not satisfied, you can modify the parameters.

You can view the two NC sequences together (as an **Operation**) by choosing the following commands:

CL Data ⇒ **Operation** ⇒ (select **OP010** from the list) ⇒ **Display** ⇒ ✔**Tool** ✔**Status Box** ⇒ **Done** (Fig. 22.64) ⇒ **Done** ⇒ **Done Output** ⇒ **NC Check** ⇒ **Display** ⇒ **Run** ⇒ **Create** ⇒ **Operation** ⇒ (select **OP010**) ⇒ **OK** (Fig. 22.65) ⇒ **Done/Return** ⇒ **Done/Return**

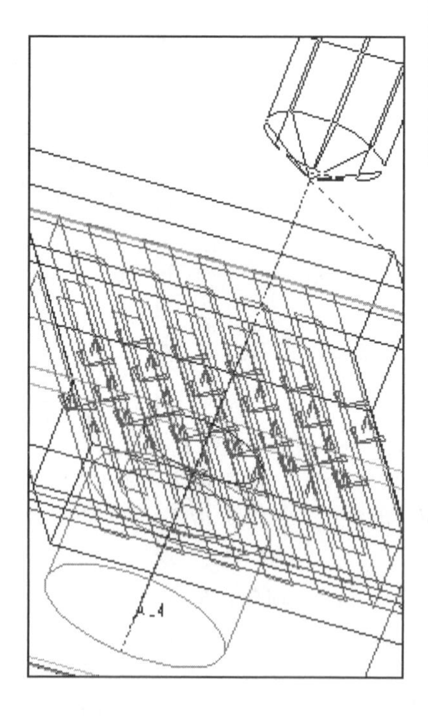

Figure 22.64
Playing **CL Data** for Operation **OP010**

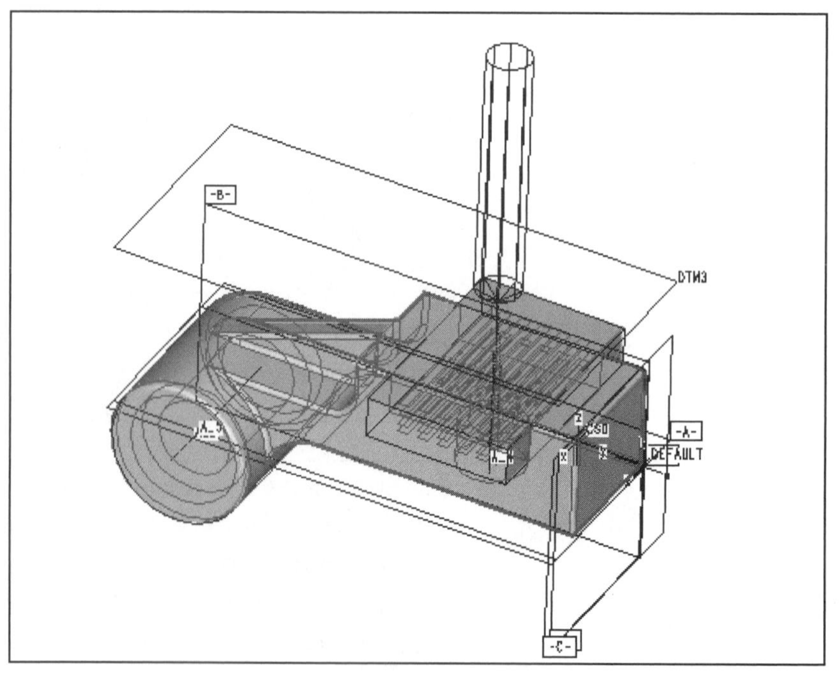

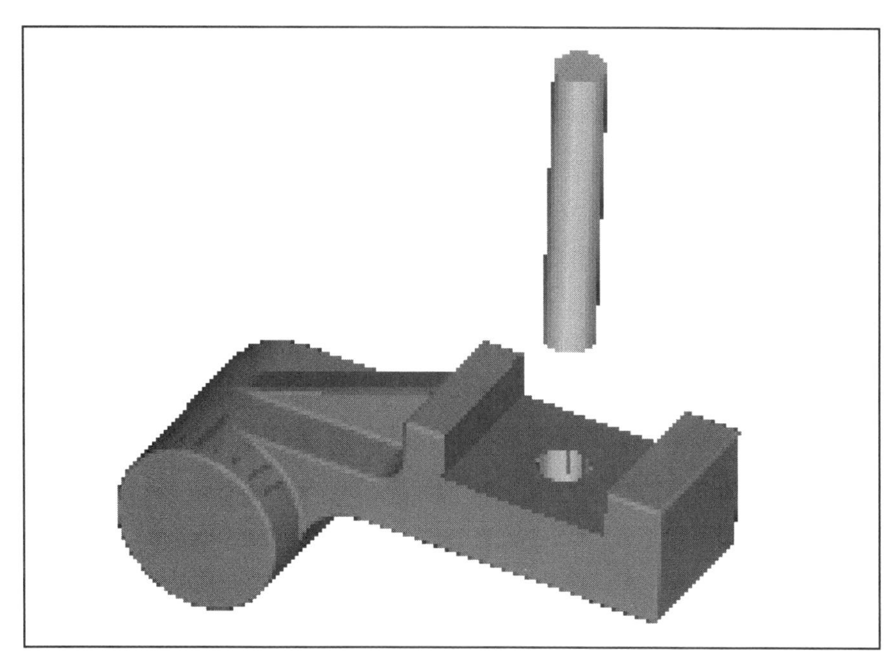

Figure 22.65
NC Check of Operation
OP010

The next NC sequence will be a facing operation (**Face**) on the front cylindrical feature of the part. You will be creating a second operation. Operation **OP020** will contain the **Face** NC sequence. If you have time, and with your instructor's permission, you may add a **Holemaking** NC sequence to the Operation to **Ream** and **Bore** the hole from this direction. If you wish to complete the part machining, you must have an **Operation** (and coordinate system) for each direction. Four operations (from four directions) would be required to complete the machining for the part.

Choose the following commands:

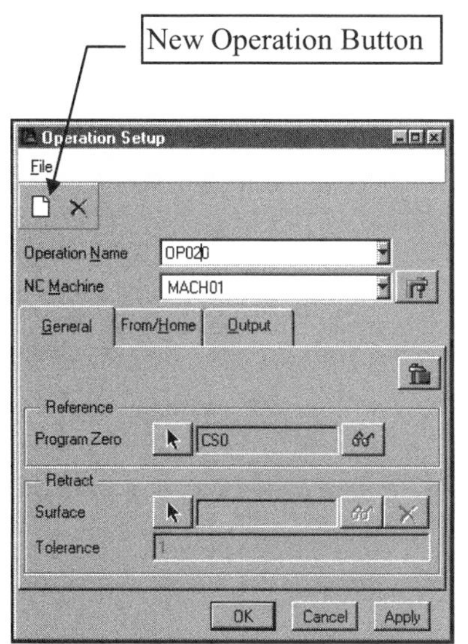

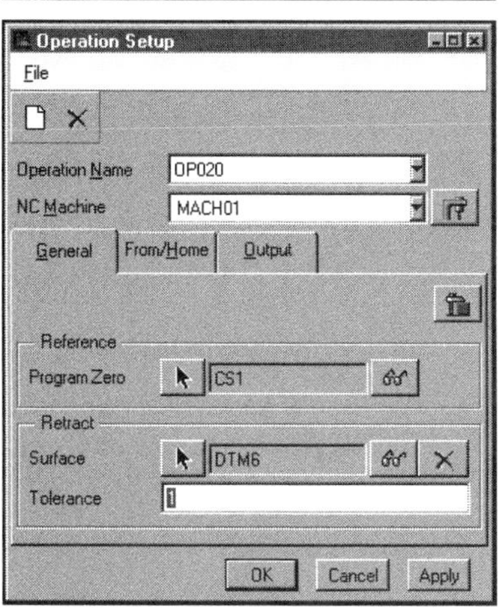

Machining ⇒ **Operation** ⇒ New Operation button ⇒ Operation Name **OP020** ⇒ (leave the same NC Machine selected **MACH01**) ⇒ Program Zero (pick the arrow button) ⇒ **Create** [from the MACH CSYS menu ⇒ (select the workpiece from the Model Tree) ⇒ **3 Planes** ⇒ **Done** ⇒ [select the three planes of the workpiece (Fig. 22.66)] ⇒ [use **Next** and **Reverse** to locate the coordinate system's (Fig. 22.66)] **Z-Axis** (this establishes the *tool axis* parallel to the **Z-Axis)** ⇒ **Done/Return** ⇒ Retract Surface (pick the arrow button) ⇒ (Retract Plane dialog box displays) ⇒ **Along Z-Axis** (from Retract Selection dialog box) ⇒ Enter Z Depth: **40)** ⇒ **Preview** (Fig. 22.67) ⇒ **OK** ⇒ **Apply** ⇒ **OK** ⇒ **NC Sequence** ⇒ **New Sequence** ⇒ **Machining** ⇒ **Face** ⇒ **Done** ⇒ (the SEQ SETUP menu displays, showing you the required items) ⇒ **Done** ⇒ **File** ⇒ **New** (adds a new tool **T0004)** ⇒ fill in the Tool Setup dialog box by inputting a cutter diameter of **30** and a length of **90**) ⇒ **Apply** ⇒ **File** ⇒ **Save tool** ⇒ **File** ⇒ **Done** ⇒ **Set** ⇒ [CUT_FEED **5**, STEP_DEPTH **2**, STEP_OVER **15**, SPINDLE_SPEED **500**, CLEAR_DIST **5** (Fig. 22.68)] ⇒ **File** ⇒ **Exit** ⇒ **Model** ⇒ **Done** ⇒ **Query Sel** (pick the machine surface as shown in Figure 22.69) ⇒ **Accept** ⇒ **Done Sel** ⇒ **Done/Return** ⇒ **Done Seq**

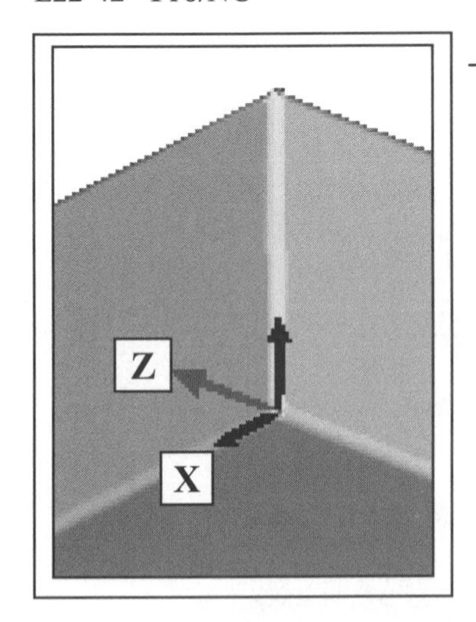

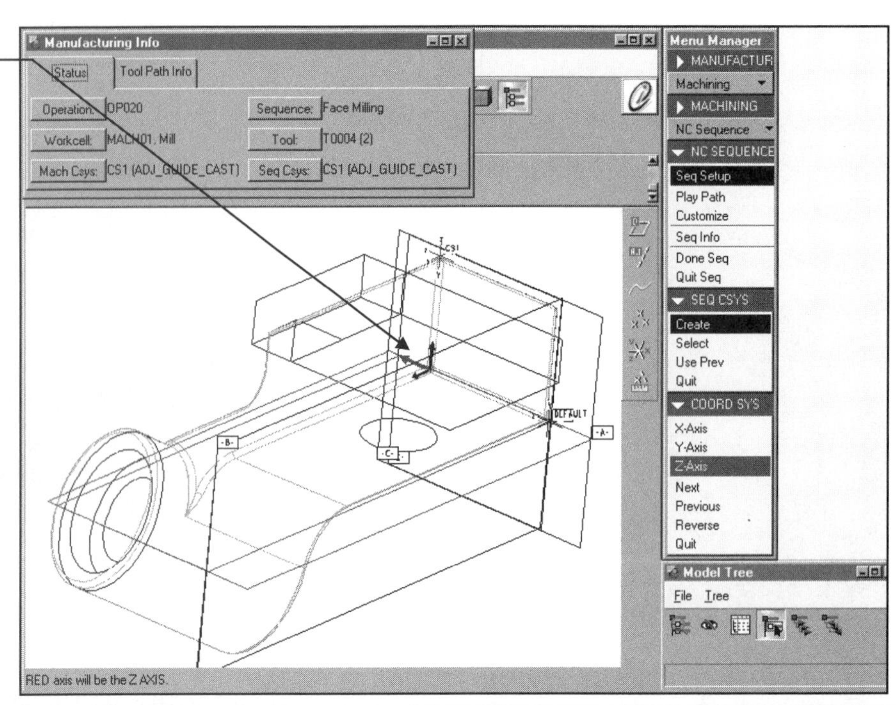

Figure 22.66
Coordinate System with the **Z Axis** Pointing Out from the Front of the Model

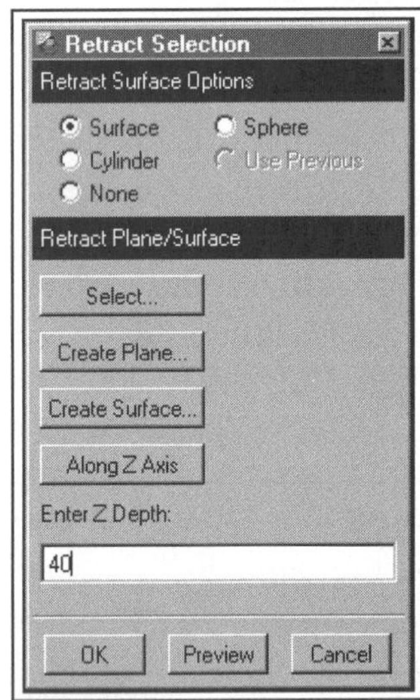

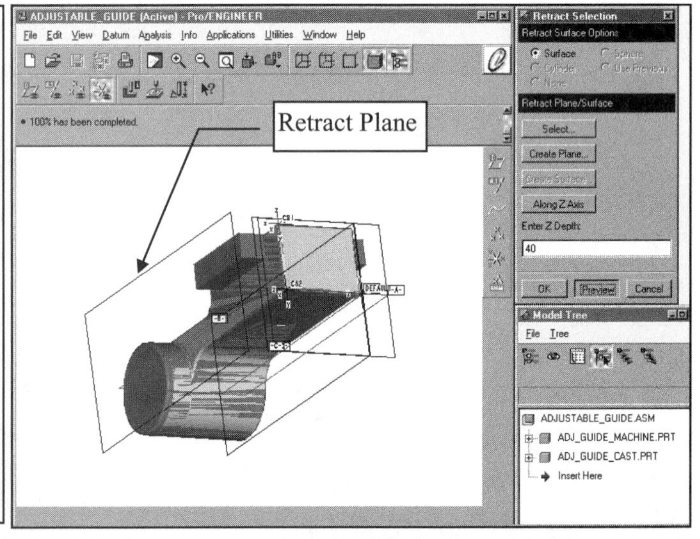

Figure 22.67
Setting Retract Plane

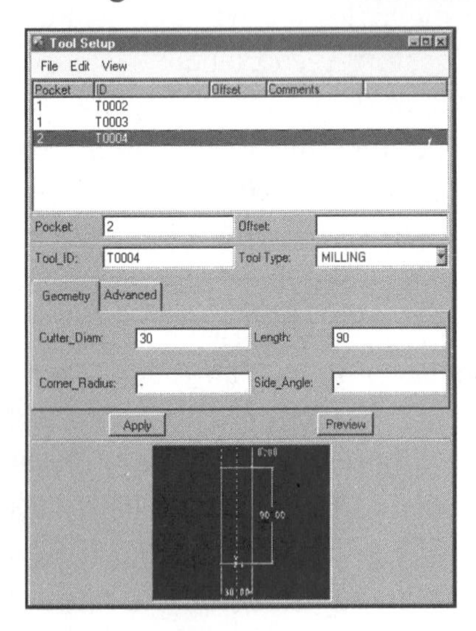

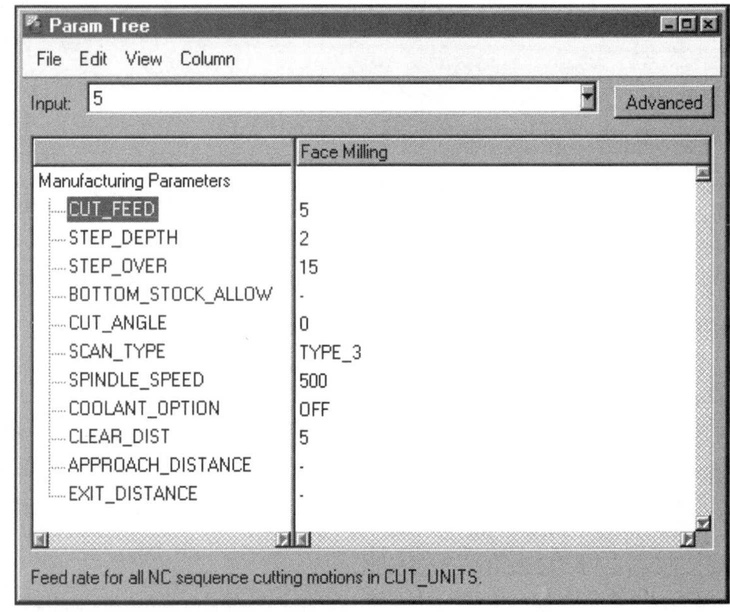

Figure 22.68
Setting Parameters

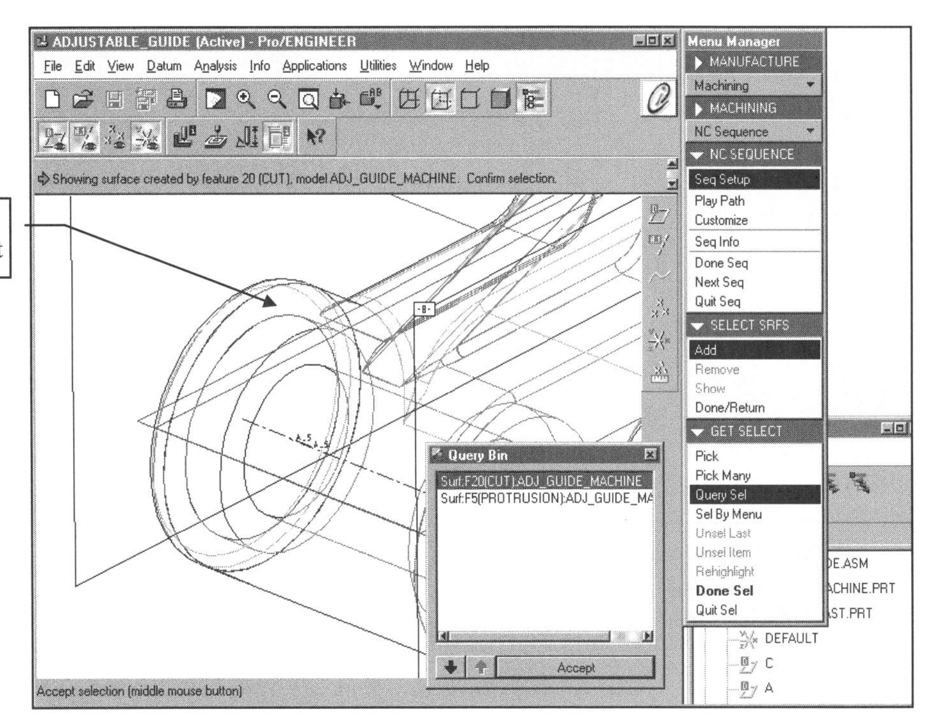

Query Sel until you highlight this surface of the design (machine) part

Figure 22.69
Selecting the Design Part's Surface to Face Mill

CL Data ⇒ **Operation** ⇒ (select **OP020** from the list) ⇒ **Display** ⇒ ✔**Tool** ✔**Status Box** ⇒ **Done** (Fig. 22.70) ⇒ **Done** ⇒ **Done Output** ⇒ **NC Check** ⇒ **Display** ⇒ **Run** ⇒ **Create** ⇒ **Operation** ⇒ (select **OP020** from the list) ⇒ **OK** (Fig. 22.71) ⇒ **Done/Return** ⇒ **Done/Return** ⇒ **View** ⇒ **Repaint** ⇒ **Machining** ⇒ **NC Sequence** ⇒ (pick the sequence from the list) ⇒ **Seq Info** (for information about the sequence) ⇒ **Done Sel** (Fig. 22.72)] ⇒ **Close** ⇒ **File** ⇒ **Save** ⇒ ✔ ⇒ **File** ⇒ **Delete** ⇒ **Old Versions** ⇒ ✔

Figure 22.70
CL Data for **OP020**

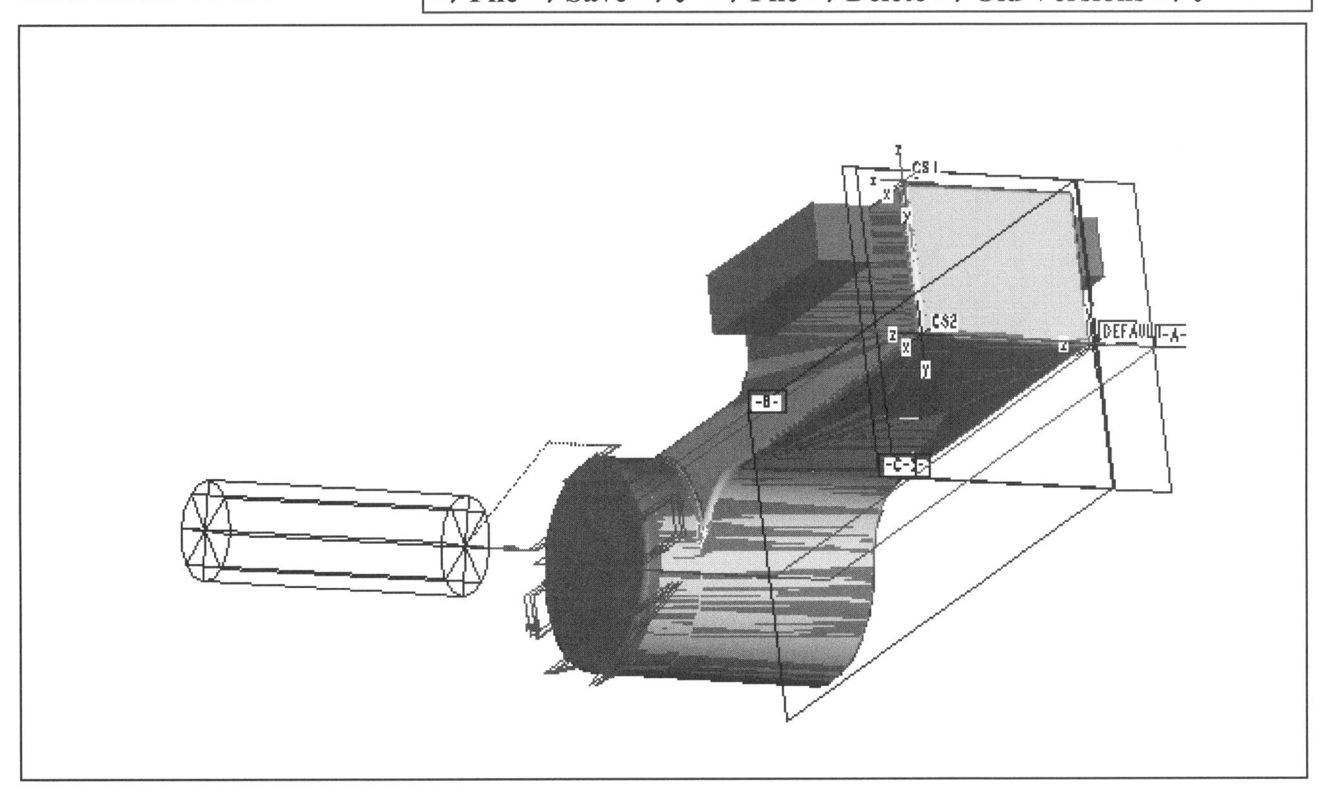

Figure 22.71
NC Check of Operation OP020

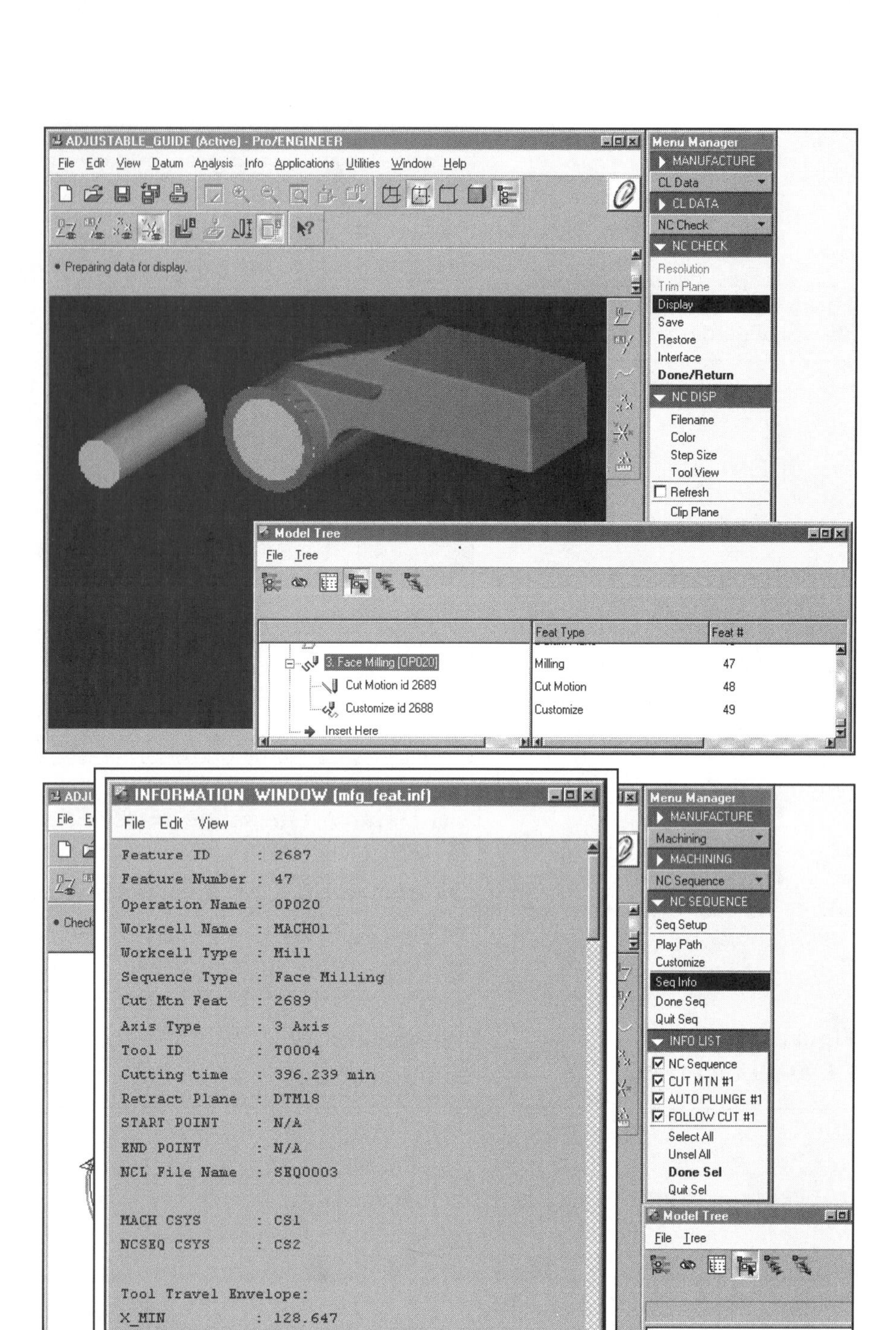

NC Sequence ⇒ Seq Info to get information on **OP020**

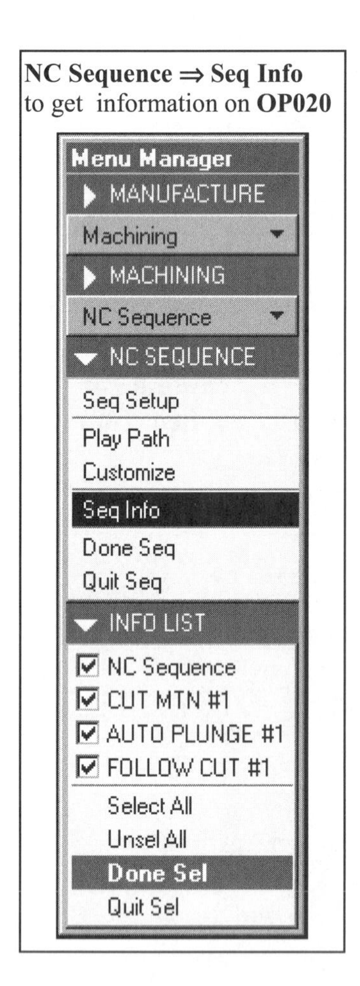

Figure 22.72
Sequence Information

You may choose to create NC sequences for the counterbore and the thru hole for this operation. The remaining machining is done from two different directions; therefore two more operations will be required (bottom small counterbore hole, and the back side for the other matching large counterbore hole).

Lesson 22 Project

Clamp Arm Machining

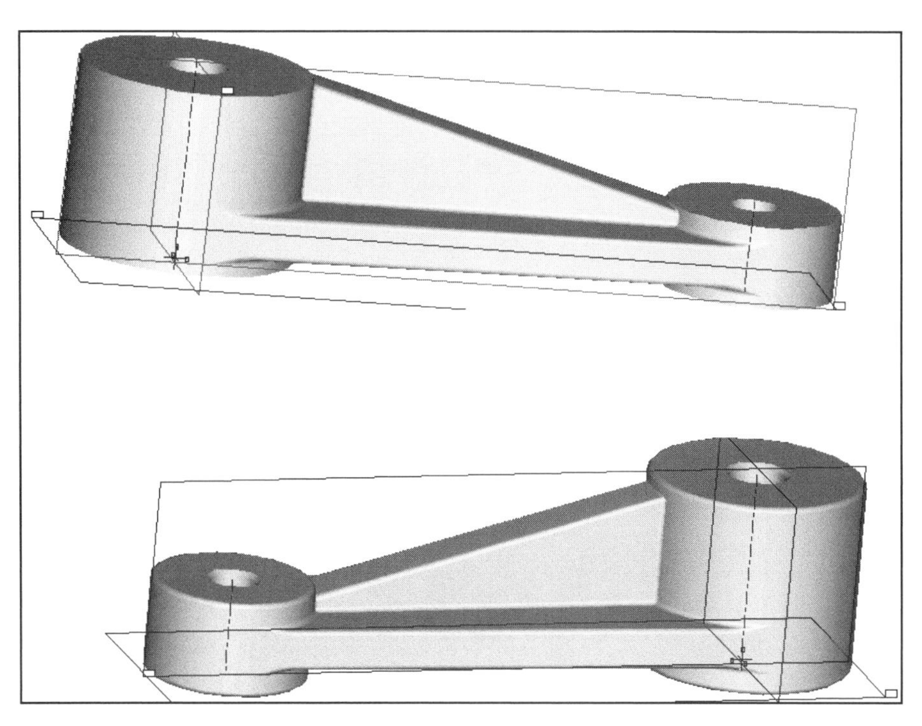

Figure 22.73
Clamp Arm Machined Part (Design Part) Top
Clamp Arm Casting Part (Workpiece) Bottom

Clamp Arm Machining

In the eighth **lesson project** you modeled a cast part (**workpiece**) and an identical version (with machined features added) called a **design part**. Here, you will use the two versions of the Clamp Arm for the **manufacturing model** and complete the *operations* and *NC sequences* required to machine the part (Fig. 22.73).

Use a *facing* NC sequence to machine the top and bottom of the Clamp Arm. The two holes will require a *holemaking* NC sequence.

As an alternative project, you may select one of the parts (Fig. 22.74 and 22.75) provided in Appendix A. Remember, you must model the casting *and* the design part.

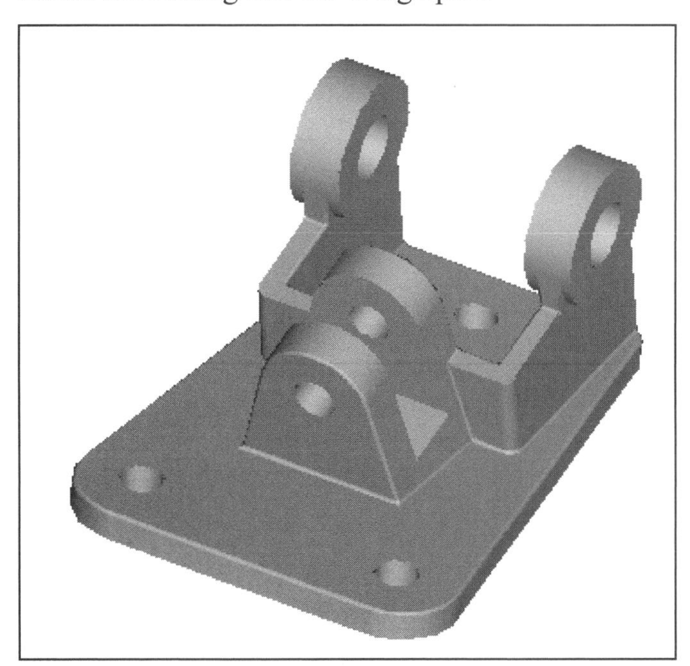

Figure 22.74
Appendix A Casting

Figure 22.75
Appendix A Cover

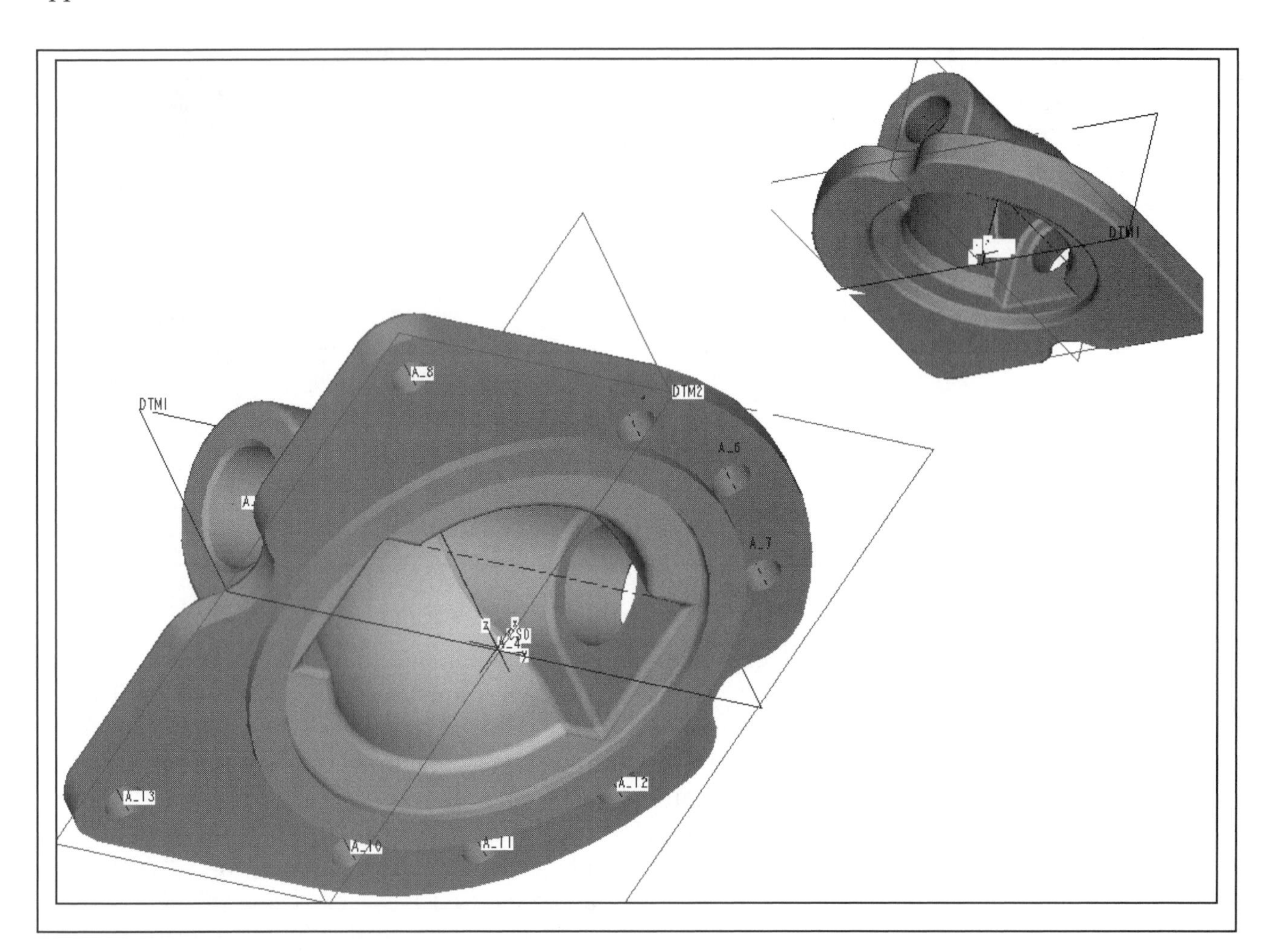

Lesson 23

Pro/SHEETMETAL

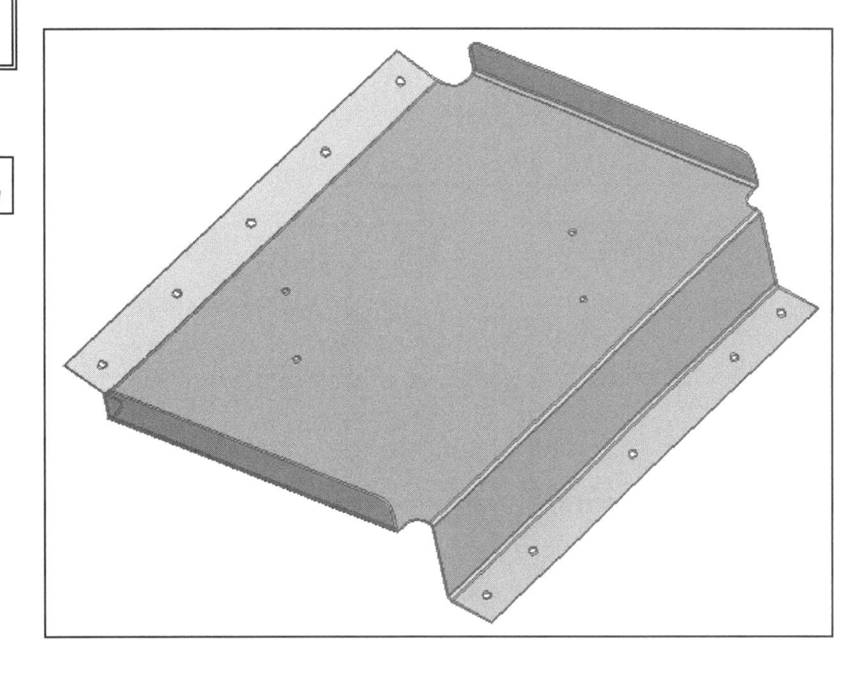

Figure 23.1
Bracket

OBJECTIVES

1. **Model sheet metal parts and create sheet metal drawings**

2. **Create walls, reliefs, and bends**

3. **Understand and create Bend Order Tables**

4. **Produce Flat Pattern Instances of sheet metal parts**

5. **Create form features and rips**

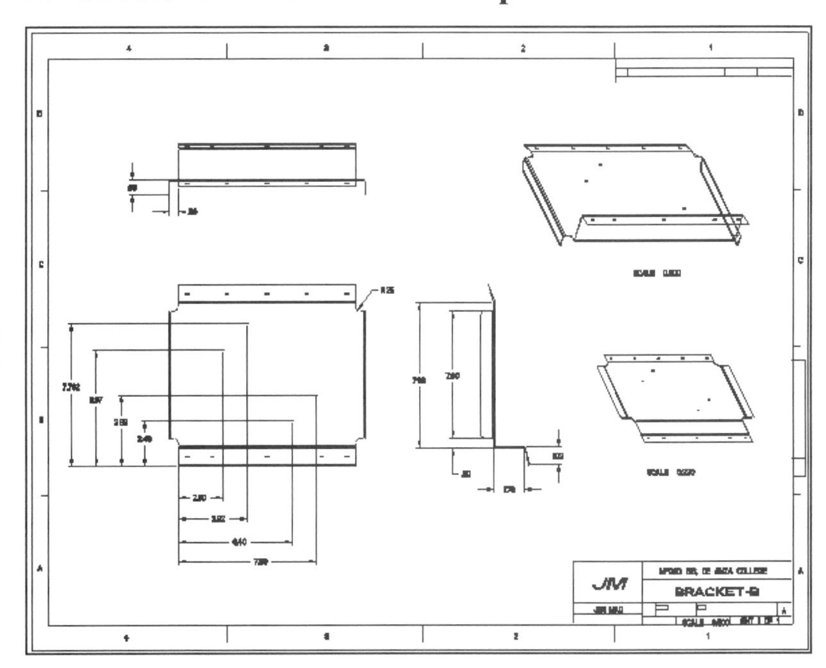

Figure 23.2
Bracket Detail Drawing

COAch™ for Pro/E

If you have **COAch for Pro/E** on your system, go to Sheet Metal Design and complete the five lessons.

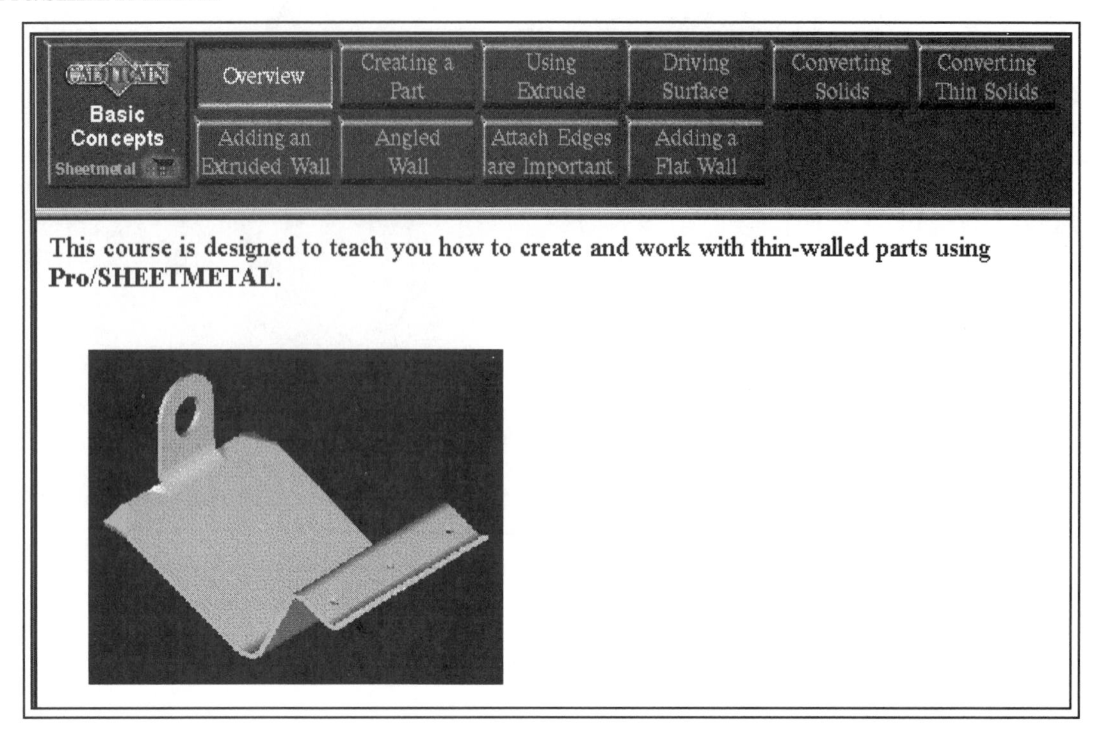

Figure 23.3
COAch for Pro/E, Pro/SHEETMETAL, Basic Concepts, Overview

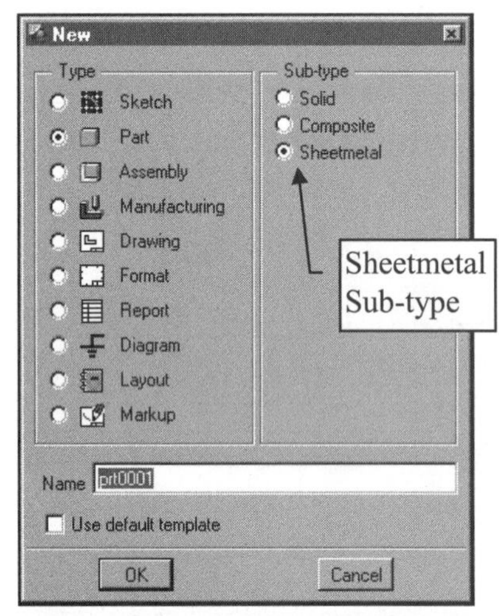

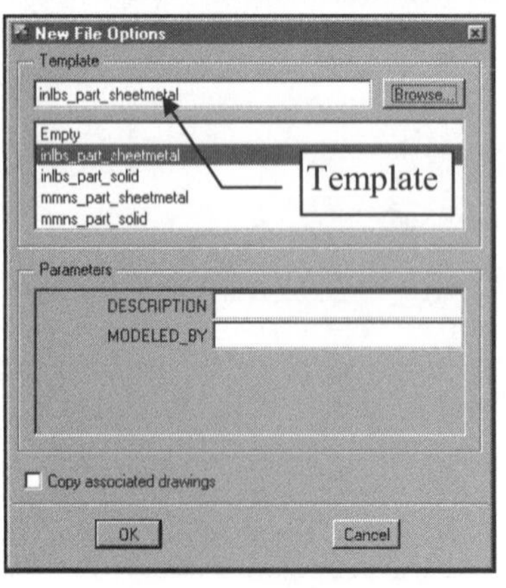

Pro/SHEETMETAL

This Lesson introduces how to create and work with thin-walled parts using **Pro/SHEETMETAL** (Fig. 23.1 and Fig. 23.2).

Pro/SHEETMETAL parts are similar to Pro/E parts (even though their database is actually a bit different). Pro/SHEETMETAL parts are a Sub-type of parts. When you create a **New** file, if **Part** is the selected **Type**, you can choose **Sheetmetal** as the Sub-type. From the New File Options dialog box you then choose the Template.

Sheet metal parts can also be used in assemblies, along with other Pro/E and/or Pro/SHEETMETAL parts. In addition, sheet metal parts can contain normal Pro/E features, such as solid protrusions, surfaces, and datum entities.

Using Pro/SHEETMETAL, you can create parts in either their formed (bent) shape (Fig. 23.3) or the flattened shape. Once created, these parts can then be formed or flattened, as required. You can also use **Family Tables** to maintain the part in both states.

You can also create a Pro/SHEETMETAL model by converting it from a Pro/E solid. The solid must be a constant thickness.

As sheet metal parts are flattened (or formed if you start with a flat state), Pro/SHEETMETAL utilizes a "**Y factor**" to allow for the fact that metal is elastic and stretches as it bends. Pro/SHEETMETAL will subtract from the actual length of the material to compensate for the elasticity. You can substitute a **Bend Table** for the **Y** factor if you so desire. Several default bend tables are provided with Pro/SHEETMETAL; they are based on the tables from the Machinery Handbook.

This lesson provides an overview of the basic concepts required to create simple parts in Pro/SHEETMETAL.

The actual Pro/E part database is different when you create parts using Pro/SHEETMETAL. All sheet metal parts are by definition thin-walled constant-thickness parts. Because of this, sheet metal parts have some unique properties, that other Pro/E parts do no have. You may convert a solid part into a sheet metal part, but not a sheet metal part into a solid part.

All sheet metal parts are constructed of features known as **Walls.** A sheet metal **Wall** feature (Fig. 23.4) is analogous to a **Solid Protrusion**. When you create Wall features, Pro/E is actually doing a great deal of work behind the scenes--it creates many internal datum planes and automatic alignment points that never appear on the screen. Another important sheet metal feature is a **Bend**. Bends are often generated automatically during the creation of Wall features.

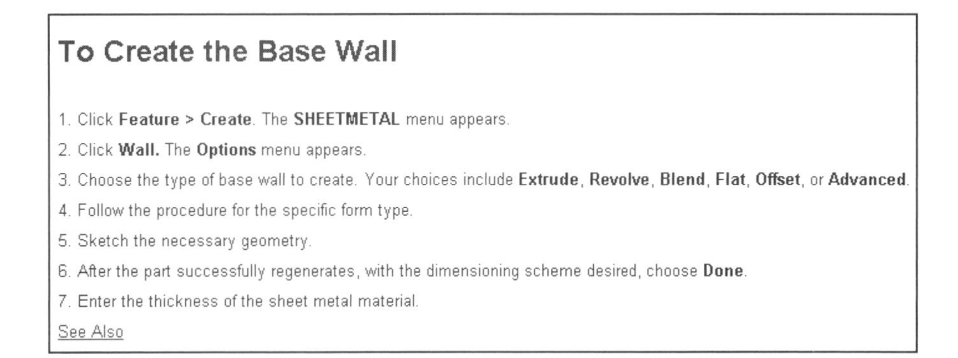

To Create the Base Wall

1. Click **Feature > Create**. The **SHEETMETAL** menu appears.
2. Click **Wall.** The **Options** menu appears.
3. Choose the type of base wall to create. Your choices include **Extrude**, **Revolve**, **Blend**, **Flat**, **Offset**, or **Advanced**.
4. Follow the procedure for the specific form type.
5. Sketch the necessary geometry.
6. After the part successfully regenerates, with the dimensioning scheme desired, choose **Done**.
7. Enter the thickness of the sheet metal material.

See Also

Figure 23.4
Online Documentation, Sheet Metal, Walls

Sheet metal drawings usually contain views both of the *fully formed part* and of the **flat pattern,** as in Figure 23.5. You can place views of the part in both states on the same drawing by creating a multi-model drawing.

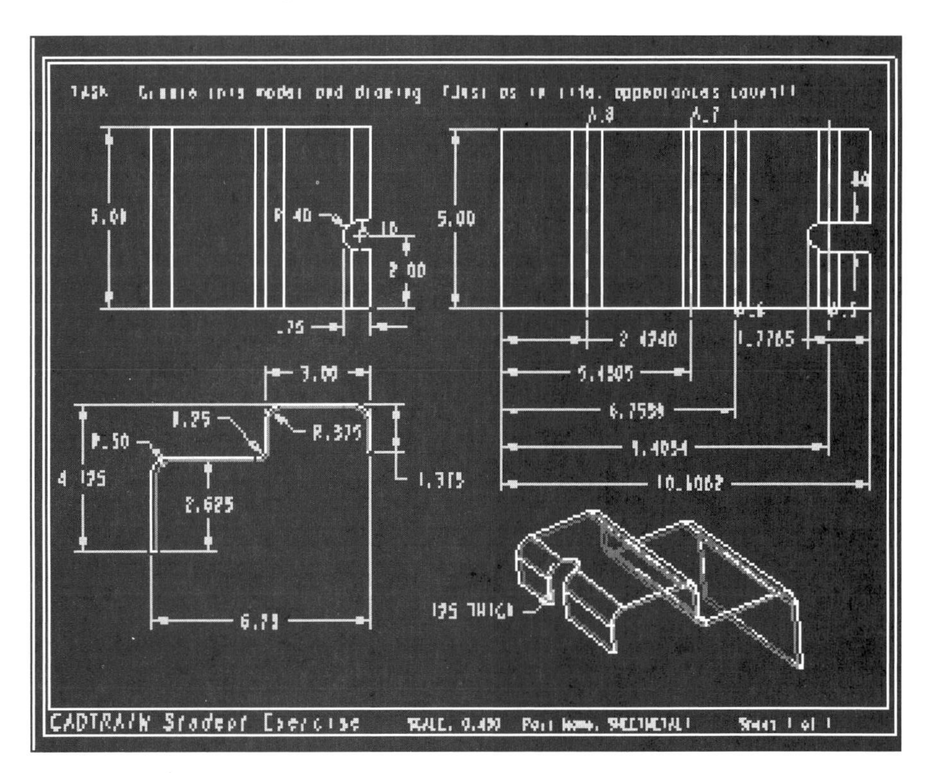

Figure 23.5
COAch for Pro/ENGINEER, Pro/SHEETMETAL, Drawings, Flat State

There are many similarities between creating solid features in Pro/E and in Pro/SHEETMETAL. However, Pro/SHEETMETAL offers some additional special features.

To create a sheet metal part in Pro/E, choose **Sheetmetal** instead of selecting **Solid** from the **New** (file) dialog. The FEAT CLASS menu structure appears differently in Pro/SHEETMETAL mode (see left column, top).

The first feature in any sheet metal part must always be a **Wall** feature, which is analogous to a **Solid** protrusion. After constructing the first Wall feature, you may create surface or solid features, as well as datum or sheet metal features.

The OPTIONS menu (see left column, bottom) is used only when constructing the first **Wall** feature. Subsequent sheet metal features are created from a different menu.

Most of these options should look familiar; their functionality is identical to their **Solid** feature creation counterparts.

There are two new options available when creating the first **Wall** feature in Sheet metal mode: **Flat** and **Offset**. The **Offset** option creates a wall that is offset from an existing surface. This is most useful for creating feature geometry in Assembly mode. With the **Flat** option, you can sketch the outline of a planar surface (Fig. 23.6).

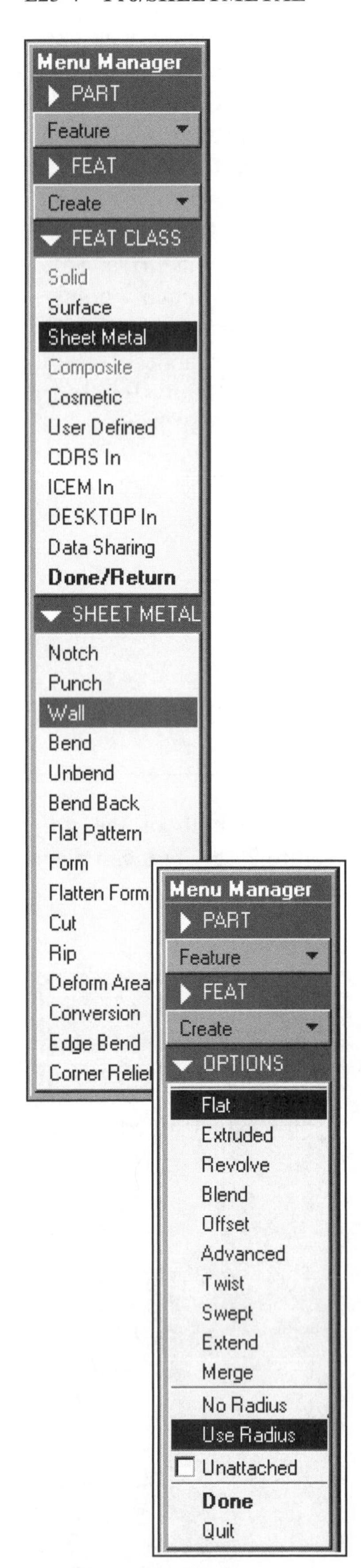

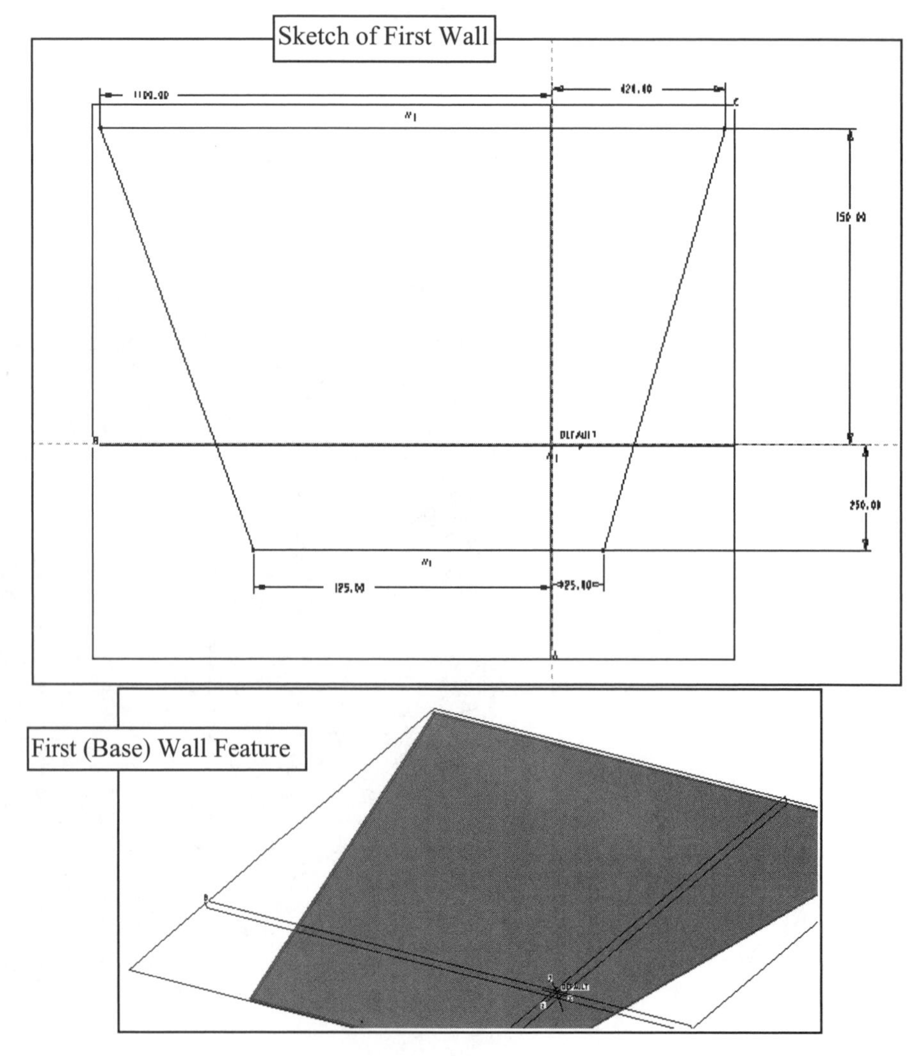

Figure 23.6
Modeling the Flat Wall

Sheet Metal Design Using Pro/SHEETMETAL

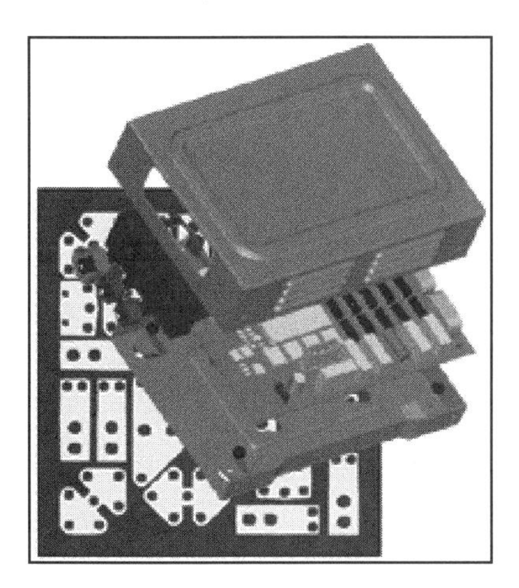

In general, it is most effective to create sheet metal designs in their bent state and then unbend them for flat pattern geometry. Create your design as per your design intent. It is not necessary to create the design in the order of manufacture. Sheet metal parts can be created in the Part mode or the Assembly mode. You can use the top-down approach to design when you use the Assembly mode.

Typical sheet metal structures that can be designed with Pro/SHEETMETAL are cabinets and supporting structures for electrical and mechanical equipment. In these instances, you should design the cabinet and supporting structures around the internal components. As with regular Pro/E parts created in Assembly mode, a sheet metal part can be dimensioned to the component parts that it is supporting.

Before starting to create sheet metal features, remember to design the part in the *"as designed"* condition, i.e., not as a flat pattern, unless you know all flat pattern details and dimensions. (Problems can arise when you modify a part that was designed as a flat pattern). Add as many bends to the part as possible before adding other features. Creating cuts at an angle or through bend areas might require larger dimensions for proper clearance.

Feature Order and References

The proper feature creation order (Fig. 23.7) and sketch references (Fig. 23.8) are essential when modifying the part and presenting the part in a drawing.

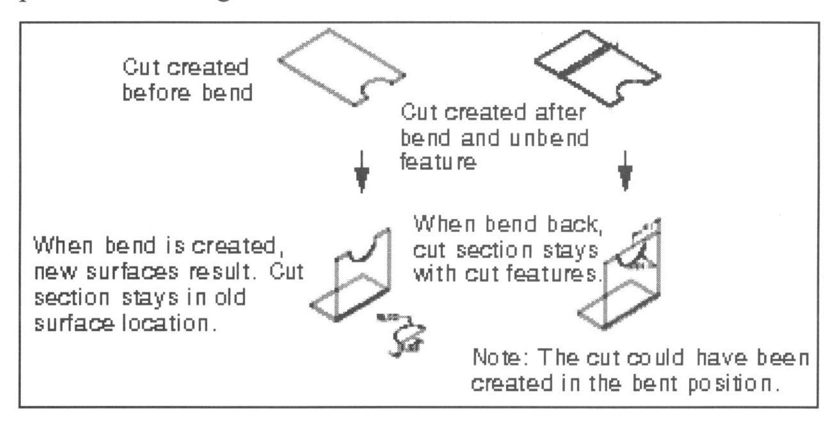

Figure 23.7
Order of Feature Creation

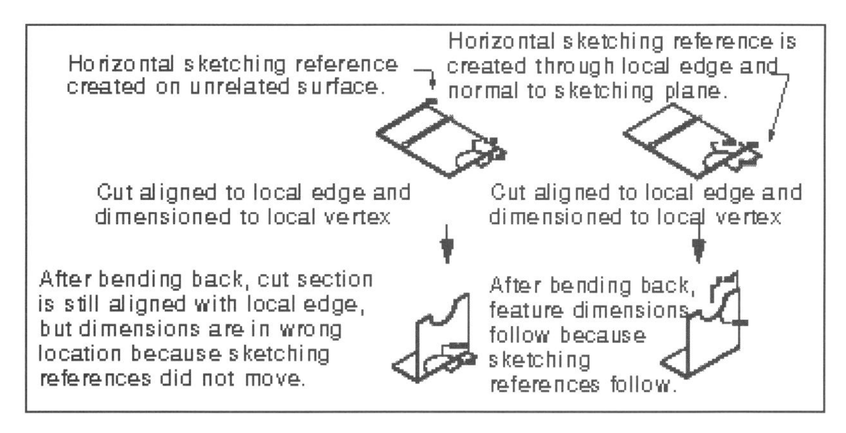

Figure 23.8
References for Feature Creation

Walls

If you choose to create a sheet metal part in Sheet Metal mode, the first feature (other than datums) must be a wall. There are two types of walls: **Base** and **Secondary.** Pro/E *grays out* all the other feature types in the SHEET METAL menu until you have created the first (base) wall. There are several options for creating the base wall that are not available when you are adding (secondary) walls to the part.

The feature forms available for the base wall include:

***Extrude** Sketch the side section of the wall and extrude it a specified depth (Fig. 23.9).

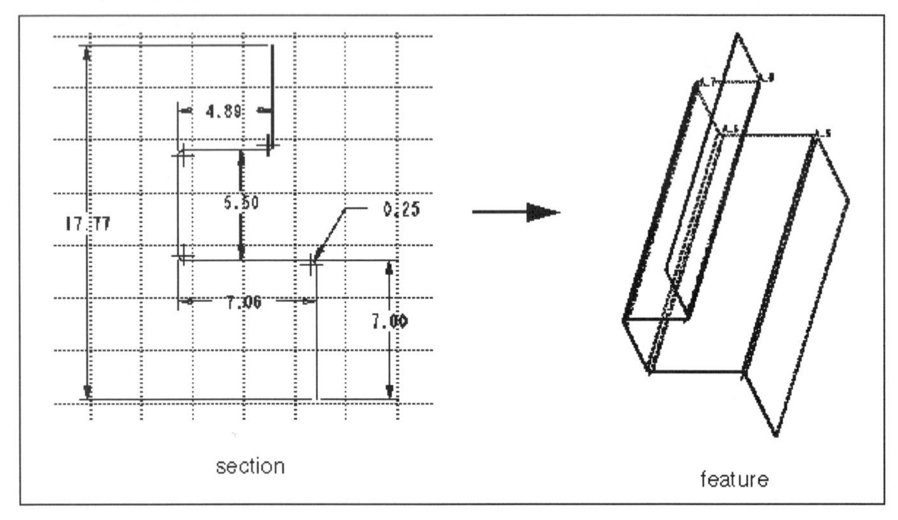

Figure 23.9
Base Feature, Extruded Wall

***Revolve** Sketch the side section of the wall and revolve about an axis (Fig. 23.10).

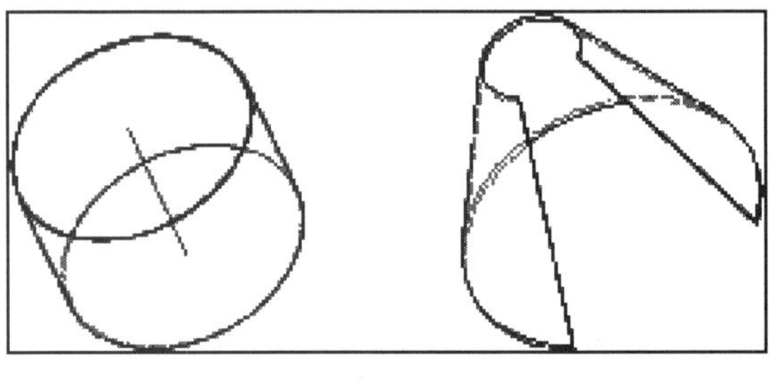

Figure 23.10
Base Feature, Revolved Base

***Blend** Create a sheet metal wall by blending several sections sketched in parallel planes (Fig. 23.11).

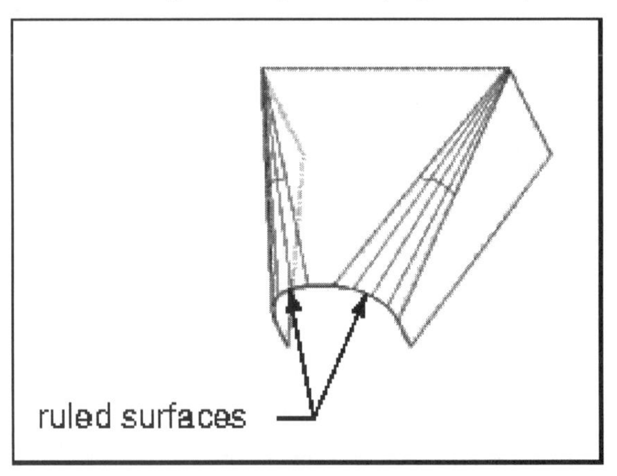

Figure 23.11
Base Feature, Blended Wall

* **Flat** Sketch the boundaries of the wall (Fig. 23.12).

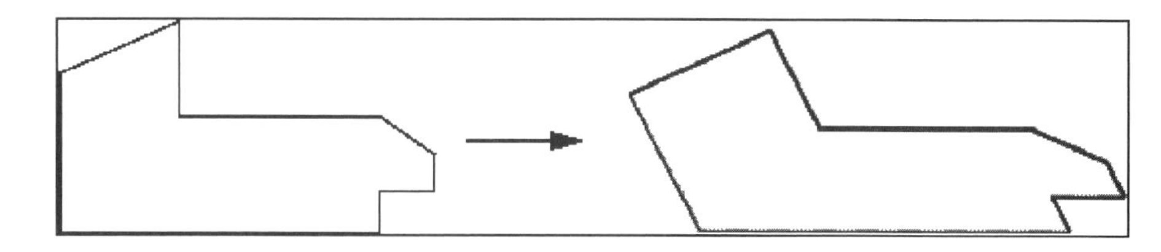

Figure 23.12
Feature, Flat Wall

* **Offset** Create a wall that is offset from a surface (Fig. 23.13).

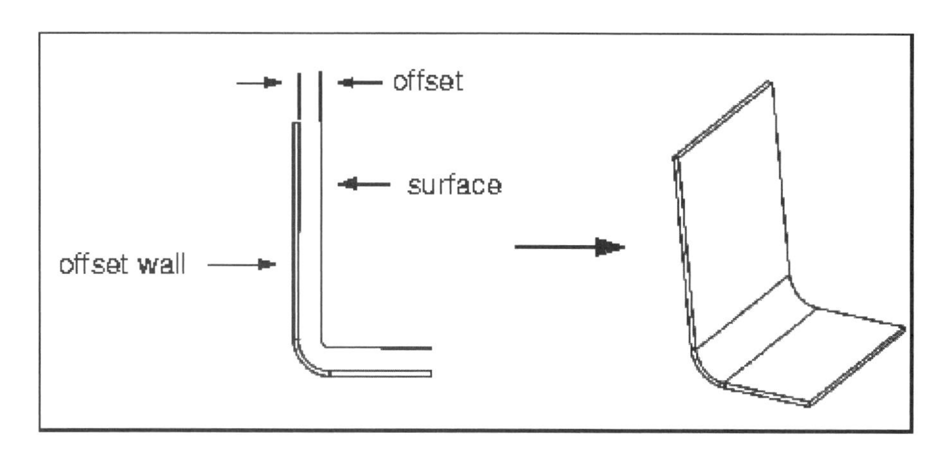

Figure 23.13
Base Feature, Offset Wall

* **Advanced** Create a sheet metal wall using datum curves, multiple trajectories, etc. Because you are in Sheet Metal mode, any features that you created through this menu are thin ones.

Creating the Base Wall

1. Pick **Feature ⇒ Create**. The SHEETMETAL menu appears.
2. Pick **Wall**. The Options menu appears.
3. Choose the type of base wall to create: **Extrude**, **Revolve**, **Blend**, **Flat**, **Offset**, or **Advanced**.
4. Follow the procedure for the specific form type.
5. Sketch the necessary geometry.
6. After the part successfully regenerates, with the dimensioning scheme desired, choose **Done**.
7. Enter the thickness of the sheet metal material.

Viewing a Sheet Metal Part

When manipulating *thin-walled* models (like sheet metal or die-cast parts) it is often difficult to select the thin side surfaces when you wish to orient the view. Because of the thinness of a sheet metal part, using edges for orienting the part is more convenient than using the side surfaces. To orient the part:

1. Choose a view command (**Front**, for example); then select a surface (wall).
2. Select the corresponding view command (**Top**, for example) and select an edge of a wall.

Secondary Walls

You can add flat walls, extruded walls, swept walls, extends, twists, and merged walls. They are attached to the edges on the part. Except for the extends, these walls can be attached either to a whole edge or to a portion of it (partial walls, sweeps, and twists). Extends extrude a wall along the full length of the specified edge. You can also add unattached walls, independent of the base wall, by checking **Unattached** in the Wall menu.

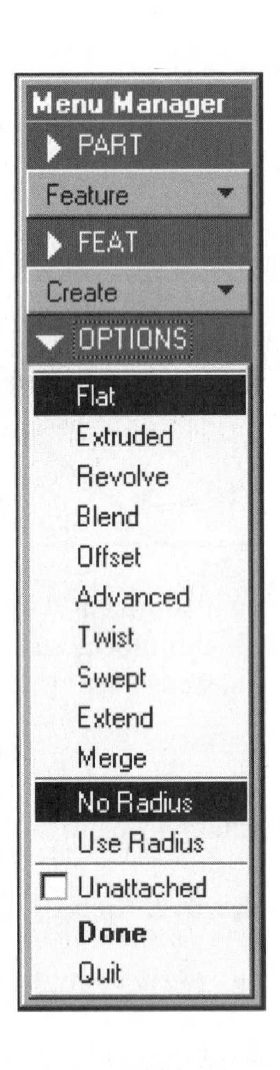

The choice of walls to add to the base wall will depend on your design intent. **Flat Walls** and extruded walls must be attached to planar surfaces. Their attachment edges must be straight lines. Flat walls can have a section of any shape, but their profile is always flat. **Extruded,** walls can be sketched with a more complex profile. Extruded walls are always created as rectangular because they are extruded to a uniform depth; however, you can afterwards add **Cut** features to modify the wall shape.

Swept walls can be attached to virtually any surface. Their attachment edge need only be a tangent chain.

The three major geometry types for additional walls are **Flat**, **Extruded** and **Swept** (**Partial** walls are a subset of **Extruded** walls). With two additional menu options, **No Radius** and **Use Radius**, possible wall features are as follows:

Flat Wall, **No Radius** Sketch the boundaries of the wall attached to the selected edge. The adjacent wall must be either planar or a twist. The attachment edge must be a straight line. The new wall is automatically created coplanar with the adjacent planar wall or tangential with the end of the twist (Fig. 23.14).

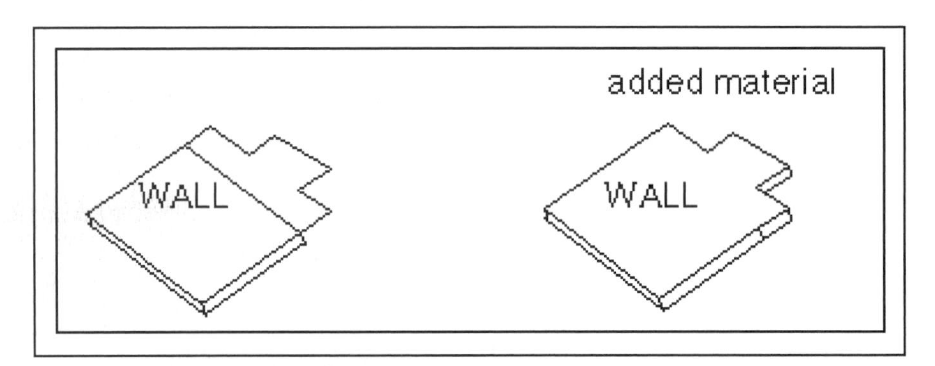

Figure 23.14
Flat Wall-No Radius

Extruded or **Partial** wall, **No Radius** Sketch the side section of the wall that will be extruded along the attachment edge.

You determine whether a bend is to be created when you are sketching the section. For a partial wall, the adjacent wall must be planar. For an extruded wall, the adjacent wall must be either planar or a twist. In all cases, the attachment edge must be a straight line. (Fig. 23.15).

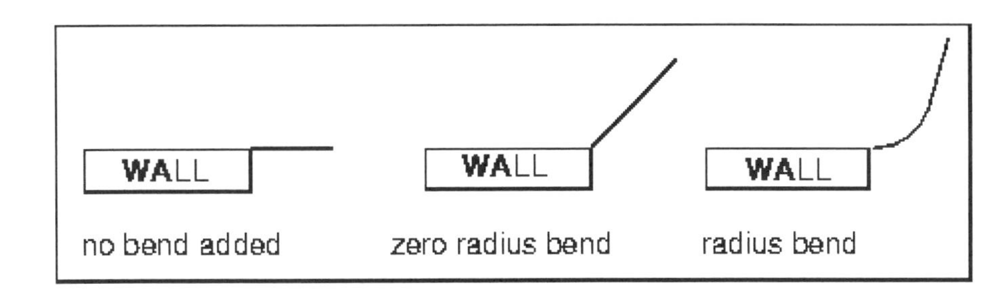

Figure 23.15
Extruded or Partial Wall-No Radius

Swept Wall, No Radius Use this to attach a wall to almost any type of surface. The sweep trajectory is the attachment edge of the adjacent wall. It must be a tangent chain but it can be three-dimensional. Sketch the side section of the wall. Determine whether a bend is to be created when you are creating the sketch.
Flat Wall, Use Radius Sketch the boundaries of the wall attached to the selected edge. The sketching plane is set up at a specified angle, and the radius is added after you create the wall. Note that material is removed from the wall-wall intersection (as shown in Figure 23.16, **Flat Wall, Use Radius**). If the wall is not attached to the edge vertices, you have to use reliefs to create the wall geometry properly. The adjacent wall must be planar, and the attachment edge must be a straight line.

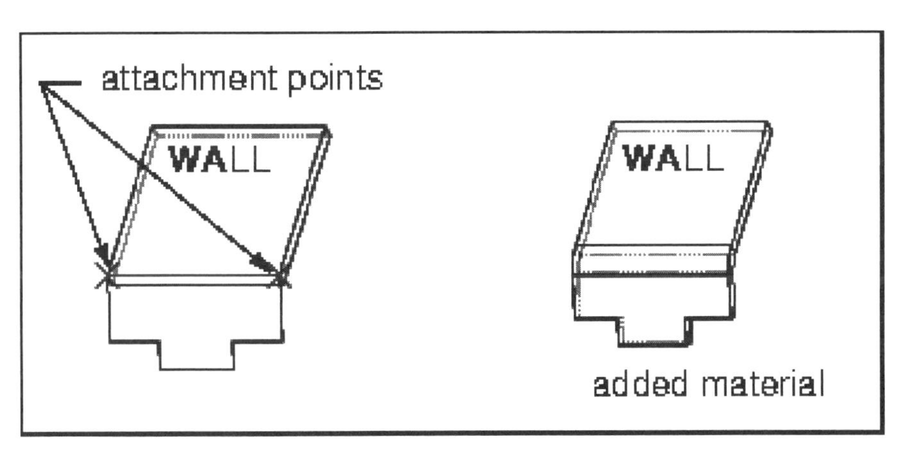

Figure 23.16
Flat Wall, Use Radius

Extruded or **Partial** wall, **Use Radius** Sketch the side section of the wall that will be extruded along the attachment edge. The bend radius for the wall is added after you create the wall. Note that material is removed from the wall at the bend location. For a **Partial** wall (Fig. 23.17), you have to use reliefs to create the wall geometry properly. The adjacent wall must be planar, and the attachment edge must be a straight line.

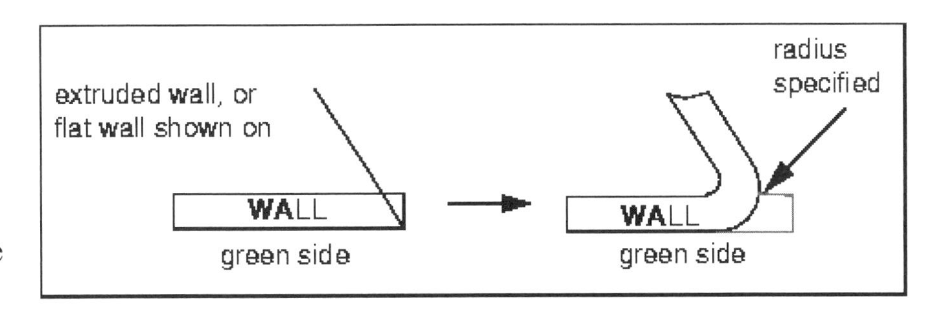

Figure 23.17
Extruded or Partial Wall, Use Radius

Swept Wall, **Use Radius** Use this to attach a wall to almost any type of surface. The sweep trajectory is the attachment edge of the adjacent wall. It must be a tangent chain, but it can be three-dimensional. Sketch the side section of the wall. The limitation is that the line of intersection between the sketching plane and the adjacent surface must be a straight line.
Twist Use this wall to change the plane of a part. Its adjacent wall must be planar, and the attachment edge must be a straight line (Fig. 23.18).

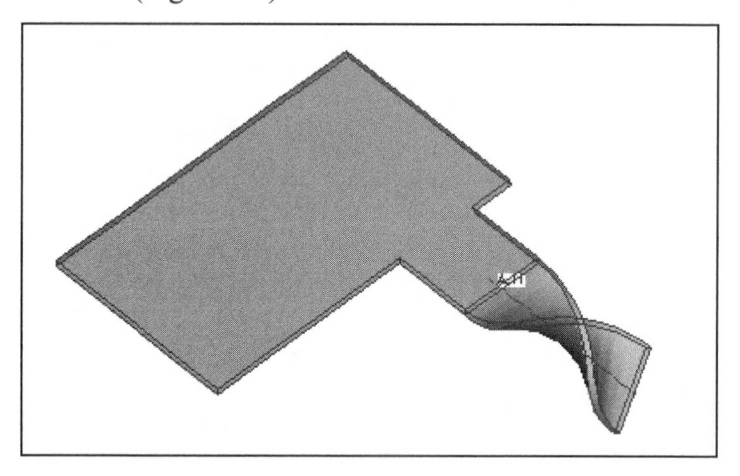

Figure 23.18
Twist Wall

Merge Use this to integrate unattached walls into one part (Figure 23.19).

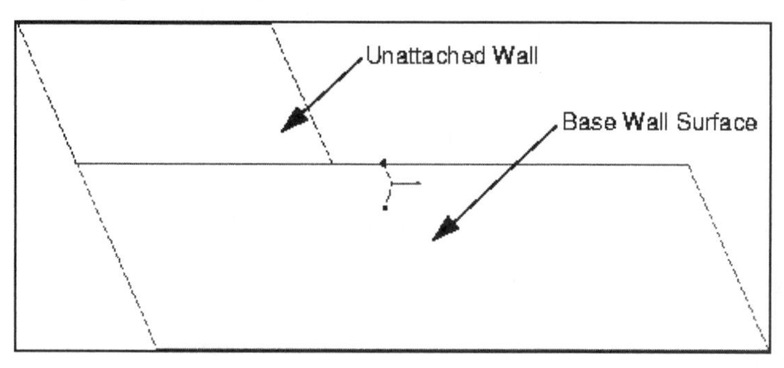

Figure 23.19
Merge

Unattached Walls You can create flat, extruded, revolved, blend, offset, and advanced unattached walls.

Automatic Reliefs

For any type of wall other than the base wall, you can specify the kind of relief that you want.

After you have sketched the new wall and the sketch has been regenerated, the Relief menu appears with the following choices:

No Relief Attach the wall without reliefs.
No Relief The bend is created without relief.
w/Relief Brings up the RELIEF TYPE menu. Pro/E highlights each attachment point in turn and prompts you to select the kind of relief option:

StrtchRelief If you choose this option, Pro/E prompts you to enter the *relief width* and *relief angle*.
Rip Relief Creates a cut for the relief.

In Figure 23.20, a simple flat wall is attached at an angle to an existing flat wall. The new wall does not run the full length of the attachment. Therefore, if it is to be attached at an angle, it must have relief. If it is to be attached without relief, then it cannot be at an angle. In that case, you can add a bend later.

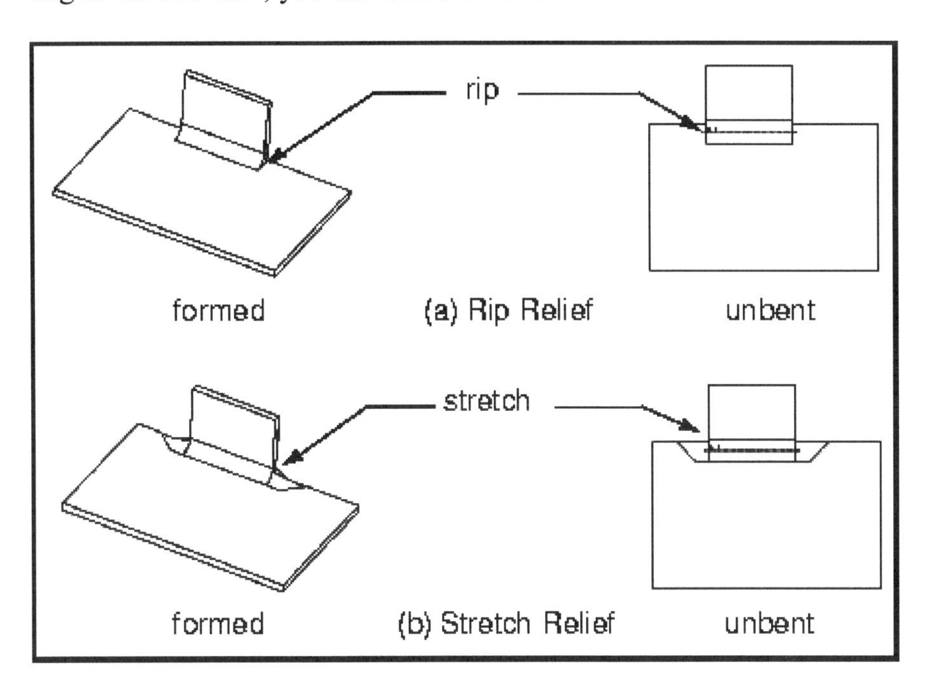

Figure 23.20
Reliefs

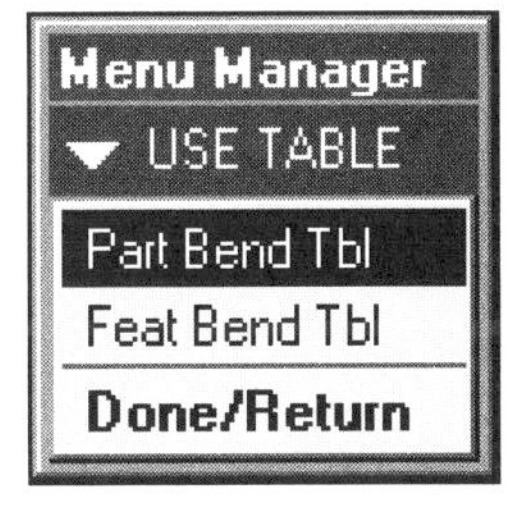

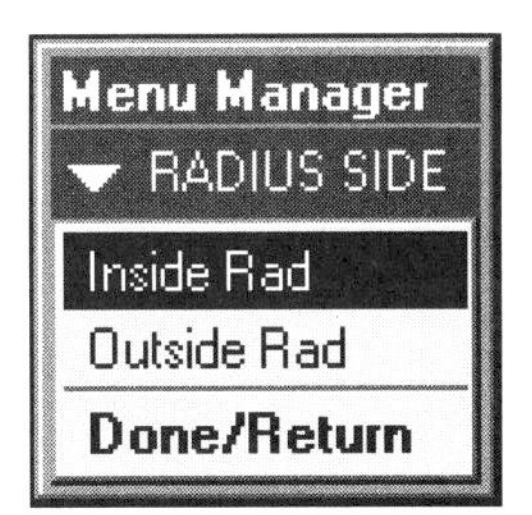

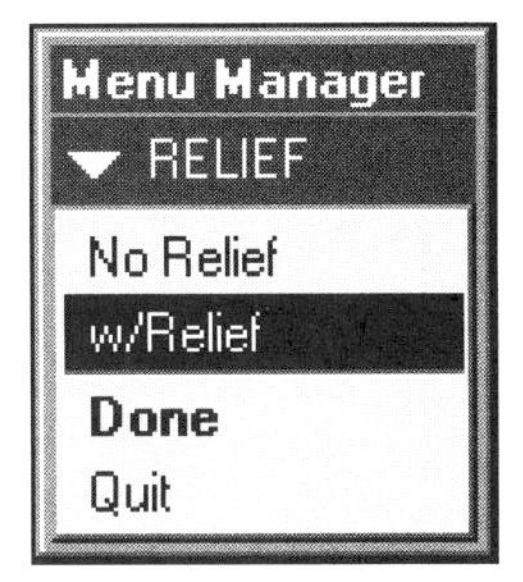

Menus Used to Create Flat, Extruded, Partial, and Swept Walls

When you create a flat, extruded, or partial wall, you must specify a bend table. If you use the **Use Radius** option when creating a flat, extruded, partial, or swept wall, you must also specify the relief and radius. Dialog boxes are used to define these types of walls. In the main dialog box for each of these types of walls, you define the bend table, relief, and radius elements. When you select one of these elements, the appropriate menu appears to aid in your definition.

Use the USE TABLE menu to specify the type of bend table to assign to the bend:

Part Bend Tbl Use the default part bend table.
Feat Bend Tbl Assign a specific bend table to this feature. The Data Files menu appears with a list of possible bend tables.

If you select the **Use Radius** option to create a flat, extruded, partial, or swept wall, use the RELIEF menu to specify the following:

No Relief Create the wall without relief.
w/Relief Create the wall with relief. The RELIEF TYPE menu appears with the following options:
No Relief Create wall without relief.
StrtchRelief Create wall with stretch relief.
Rip Relief Create wall with rip relief.

If you select the **Use Radius** option to create a flat, extruded, partial, or swept wall, use the SELECT RADIUS menu to specify the following:

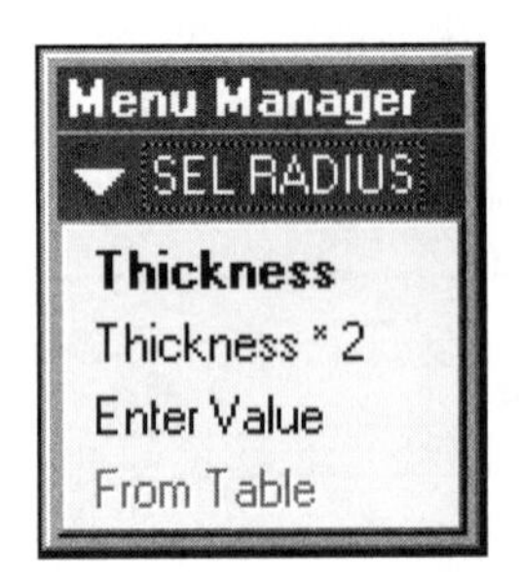

Thickness Sheet metal thickness.
Enter Value Enter a value at the keyboard.
Default Use the defined **Default Radius**. (This option appears only if the **Default Radius** has already been defined.)
From Table If a **Bend Table** has been assigned and you select this option, the FROM TABLE menu appears with a listing of all the radii in the table. You can then choose one of the values.

Flat Wall

A flat wall (Fig. 23.21) is attached to a selected edge, and you sketch the boundaries of the wall in a plane at the bend angle of the finished wall. The surface adjacent to the edge must be planar.

When you create a flat wall, you must sketch the Flat Wall as an ***open loop*** *and attach the Flat Wall to either the green or the white side of the Attach Edge, with its attachment points inside (or aligned with) the* ***highlighted vertices****.*

If the Flat Wall has a radius and rip relief, and you dimension it to the inner surfaces of adjacent walls then; you can set the offset distance to zero.

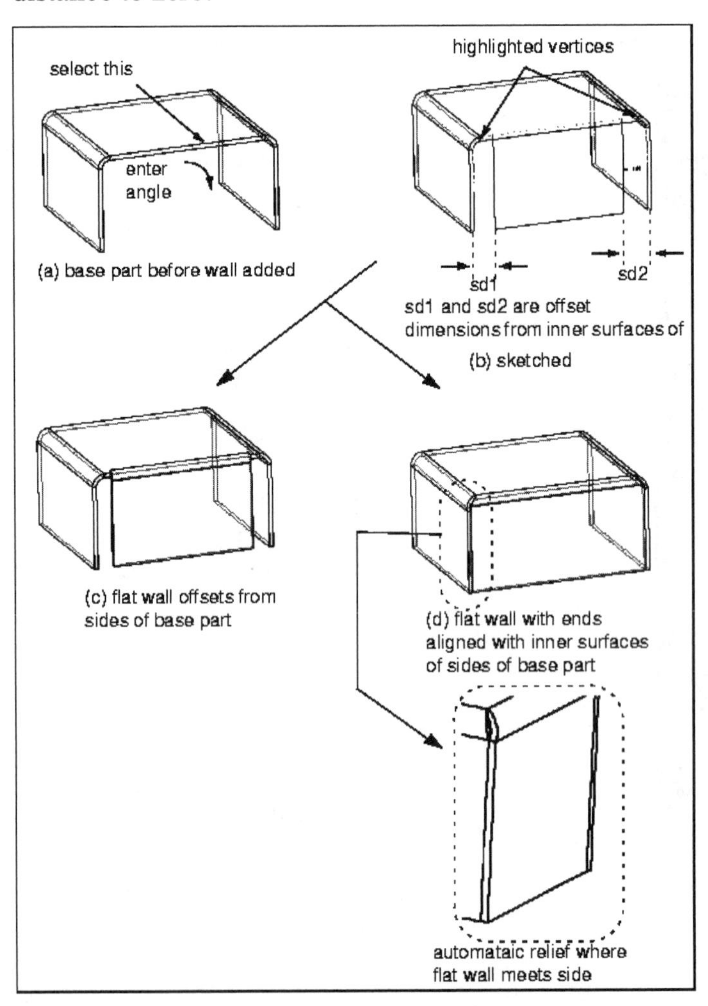

Figure 23.21
Creating a Flat Wall

Pro/E relieves the Flat Wall as shown in Figure 23.21(d). This is functionally equivalent to the effect that you would get if you created the Flat Wall with real offset dimensions, as in Figure 23.21(c), and then added Extend features at the ends.

The dialog box for a flat wall appears when you select the option Flat and the option **No Radius** or **Use Radius** from the Options menu. The dialog box lists the following elements:

Bend Table (required) Displays the USE TABLE menu. Use the default part bend table or assign a specific bend table to the feature.
Radius Type, **Inside Radius,** or **Outside Radius** Required only if **Use Radius** is selected.
Attach Edge (required) A straight edge on an existing planar wall where the new wall is attached.
Angle The angle at which the new wall is attached to the existing wall. Required only if **Use Radius** is selected.
Sketch (required) Sketched flat wall. It must be attached to an existing wall.
Relief Displays the RELIEF menu. The type of relief to be used. Required only if **Use Radius** is selected.
Radius Bend radius. Required only if **Use Radius** is selected.

Creating Flat Walls

1. Choose **Wall** from the SHEET METAL menu.
2. Choose wall option **Flat** and the option **No Radius** or **Use Radius** from the OPTIONS menu. Then choose **Done**. The WALL Options: FLAT dialog box appears.
3. Choose the **Bend Table** element, then the **Define** button. The USE TABLE menu appears. Choose the type of bend table that you want, then **Done**.
4. If you chose the **Use Radius** option in step 2, the RADIUS SIDE menu appears. Choose **Inside Rad** or **Outside Rad**, then **Done/Return**.
5. Choose the **Attach Edge** element, then the **Define** button. Pick an edge on an existing plane sheet metal wall.
6. If you chose the option in step 2, the DEF BEND ANGLE menu appears. Choose one of the standard listed values or choose **Enter Value** and then enter the exact value (in degrees). Choose **Done**.
7. Choose the **Attach Edge** element, then the **Define** button. A red arrow appears on the sheet metal, growing from one end of the **Attach Edge**. This indicates the direction of viewing the sketching plane. Choose **Flip** to reverse the direction and/or **Okay** to accept it.

Sketch the wall section. It must be an open section with the endpoints aligned to the attachment edge. If the endpoints lie at the vertices highlighted with an **X**, you do not have to explicitly align them. When the wall section successfully regenerates, choose **Done**.

8. If you chose the **Use Radius** option in step 2, the RELIEF menu appears. Choose one of the relief options, then **Done**.
9. If you chose the **Use Radius** option in step 2, the SEL RADIUS menu appears. Choose or enter the bend radius.
10. The feature is now fully defined. Choose **OK** from the dialog box. Pro/E creates the flat wall with the same thickness as that of the base feature.

Extruded Wall

An **Extruded Wall** is attached to an edge and always extrudes the complete length of the edge. The surface adjacent to the edge must be planar.

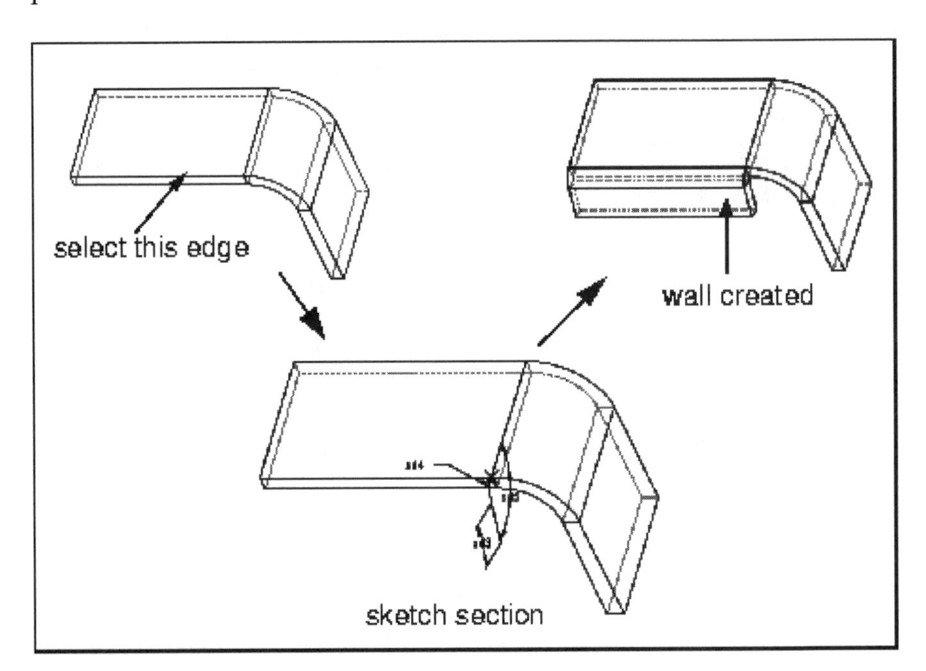

Figure 23.22
Extruded Partial Wall, with the option No Radius

To Create an **Extruded Wall,** use the following commands:

1. Choose **Wall** from the SHEET METAL menu.
2. Choose Extruded, then **No Radius** or **Use Radius** from the OPTIONS menu. Then choose **Done.** The WALL Options: Extruded dialog box appears.
3. Choose the **Bend Table** element, then the **Define** button. The USE TABLE menu appears. Choose an option, then **Done.**
4. If you chose the **Use Radius** option in step 2, the RADIUS SIDE menu appears. Choose **Inside Rad** or **Outside Rad**, then **Done/Return.**

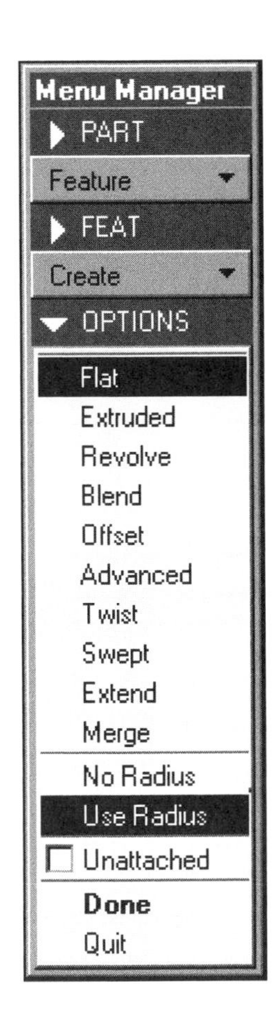

5. Choose the **Attach Edge** element, then the **Define** button. Pick an edge on an existing plane sheet metal wall.
6. The SETUP SK PLN menu appears. Select or create a sketching plane. The DIRECTION menu also appears. A red arrow appears on the sheet metal, growing from one end of the **Attach Edge**. This indicates the direction of viewing the sketching plane. Choose **Flip** to reverse the direction. Choose **Okay** to accept it.

Sketch the wall section. The sketch must be attached to the vertex *highlighted* with an **X.** When the sketch successfully regenerates, choose **Done**.
7. If you chose the **Use Radius** option in step 2, the RELIEF menu appears. Choose one of the relief options, then **Done**.
8. If you chose the **Use Radius** option in step 2, the SEL RADIUS menu appears. Choose or **Enter** the **Bend Radius**.
9. The feature is now fully defined. Choose **OK** from the dialog box. Pro/E creates the extruded wall.

Tangent Entities for an Extruded Wall

If an extruded wall section is to be tangent to the adjacent surface (Fig. 23.23), you have to make sure your entity at the attachment point is tangent.

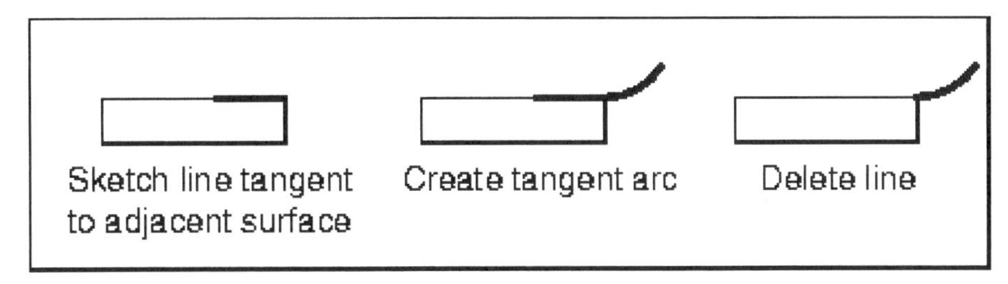

Figure 23.23
Creating Tangent Walls

To create a **Tangent Arc** or **Spline,** do the following :

1. Create a small straight line segment tangent to the existing wall and opposite to the direction in which you want the wall created.
2. Create a tangent arc or spline to this sketched entity.
3. Delete the first line.

Material Thickness for an Extruded Wall

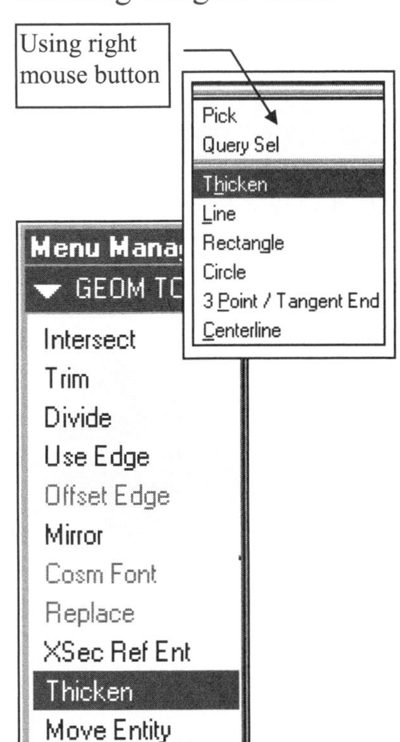

Use the **Thicken** command to dimension to both sheet metal surfaces while creating an extruded wall section (Fig. 23.24). This enables you to dimension the inside bend radii on opposite sides of the section, or to create dimensions needed for proper sizing or clearance, and so on. This way, you do not have to add material thickness to your dimension values. To thicken a wall, do the following:

1. Create the section sketch.
2. Chose **Geom Tools** [or press the right mouse button (see left-hand column)].
3. Choose **Thicken** from the GEOM TOOLS menu. Offset edges are created automatically.

4. Dimension the section. You might have to add centerlines or extra dimensions so the sketcher can solve it.
5. **Regenerate** the sketch.

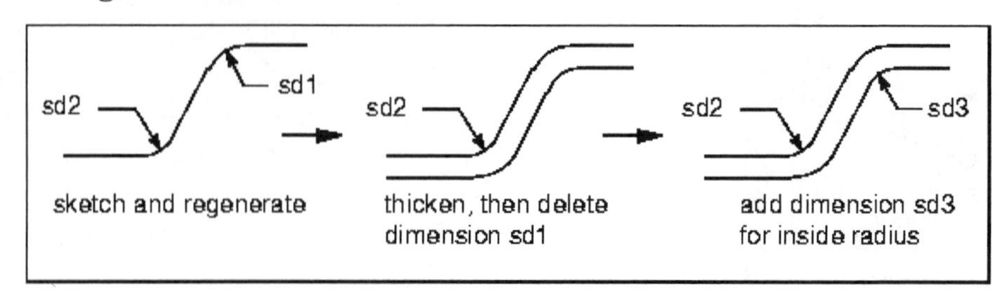

Figure 23.24
Thickening a Wall Section

Partial Walls

There are two ways to create partial walls in sheet metal parts: One way is to sketch a Flat wall not connected to the endpoints of the attachment edge (Fig. 23.25).

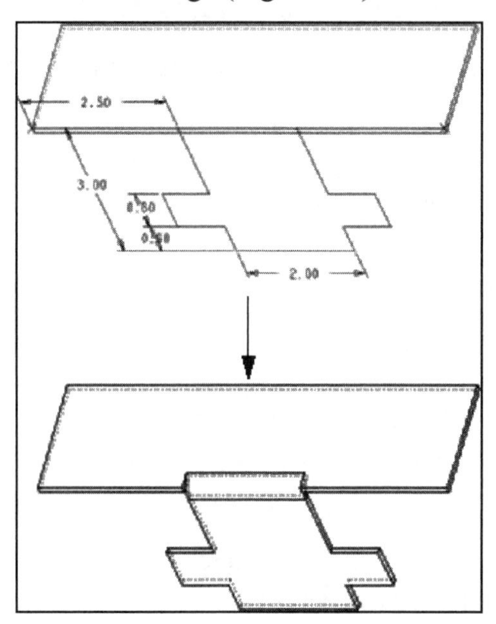

Figure 23.25
Flat Partial Wall with Reliefs (Rip Relief)

Or, you can use the **Extruded** option to create a partial wall by sketching its profile. This wall starts from a datum point (which you specify) and runs for a specific depth (which you also specify).

Partial walls can be created on an edge where they cross an existing bend feature (Fig. 23.26).

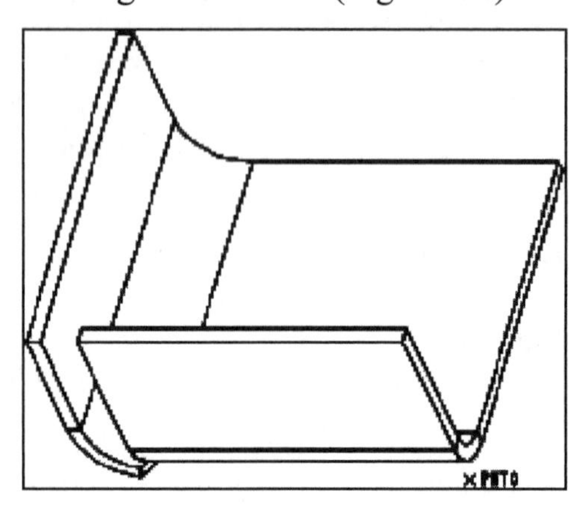

Figure 23.26
A Partial Wall Crossing a Bend

You must use reliefs when you create any type of partial wall with **Use Radius** (Fig. 23.27).

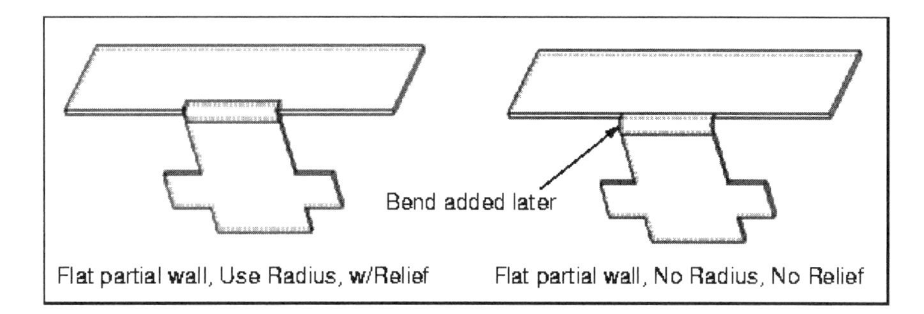

Figure 23.27
Using Reliefs for Partial Walls

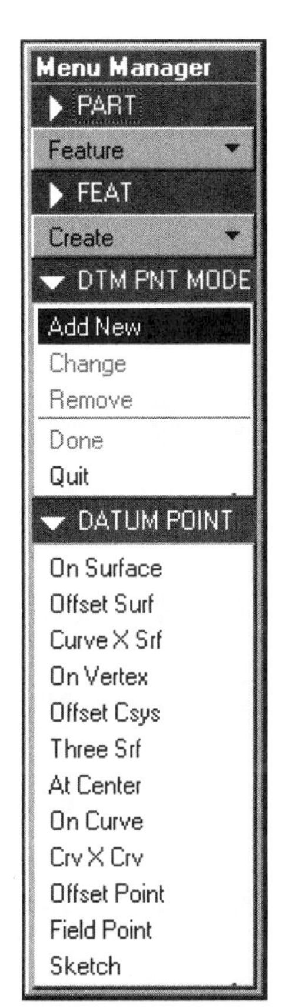

To create an **Extruded** partial wall, do the following:

1. Create a datum point on the edge (the **Attach Edge**) to which the wall is to be attached (can also be made while creating the wall).
2. Choose **Wall** from the SHEET METAL menu.
3. Choose the **Wall** option **Extruded** and the option **No Radius** or **Use Radius** from the OPTIONS menu. Then choose **Done**.
4. Choose the **Bend Table** element, then the **Define** button. The USE TABLE menu appears. Choose an option, then **Done**.
5. If you chose the **Use Radius** option in step 3, the RADIUS SIDE menu appears. Choose **Inside Rad** or **Outside Rad** and then **Done/Return**.
6. Choose **One Side** or **Both Sides.**
7. Choose the **Attach Edge** element, then the **Define** button. Pick an edge on an existing plane sheet metal wall.
8. The SETUP SK PLN menu appears. Pick **By Point.** Select or create a datum point (**Sel Point**, **Create Point** from the WALL START PNT menu).
9. The DIRECTION menu also appears. A red arrow appears on the sheet metal, growing from one end of the **Attach Edge**. This indicates the direction of viewing the sketching plane. Choose **Flip** to reverse the direction. Choose **Okay** to accept it.
10. Sketch the wall section. The sketch must be attached to the vertex *highlighted* with an **X**. When the sketch successfully regenerates, choose **Done**.
11. If you chose the **Use Radius** option in step 3, the RELIEF menu appears. Choose one of the relief options (**No Relief** or **w/Relief**), then **Done**. Select the Relief type: **No Relief**, **StrtchRelief** or **Rip Relief.** Repeat for both sides.
12. If you chose the **Use Radius** option in step 3, the SEL RADIUS menu appears. Choose **Thickness**, **2*Thickness**, or **Enter Value.**
13. Choose the **Depth** element and pick the **Define** button. Enter a value for the extrusion depth.
14. The feature is now fully defined. Choose **OK** from the dialog box. Pro/E creates the extruded partial wall (Fig. 23.28).

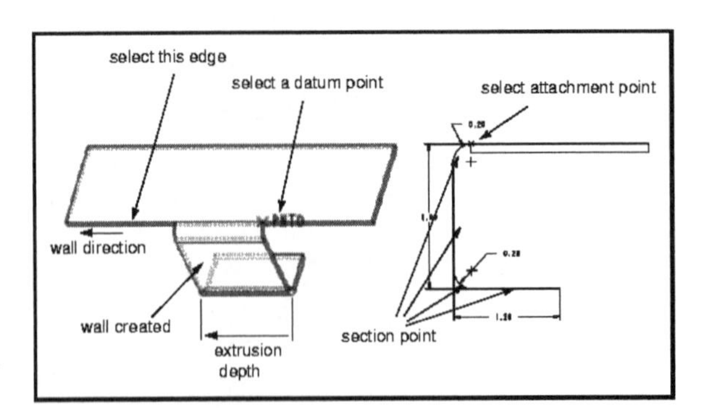

Figure 23.28
Partial Extruded Wall with **No Radius**

Extend Wall

The **Extend Wall** (Fig. 23.29) feature enables you to model various overlap conditions, mainly at corners. An **Extend Wall** feature extends out from a straight edge of an existing planar surface. It can either extend for a specified length or extend up to a specified plane (may be either a datum plane or another planar surface).

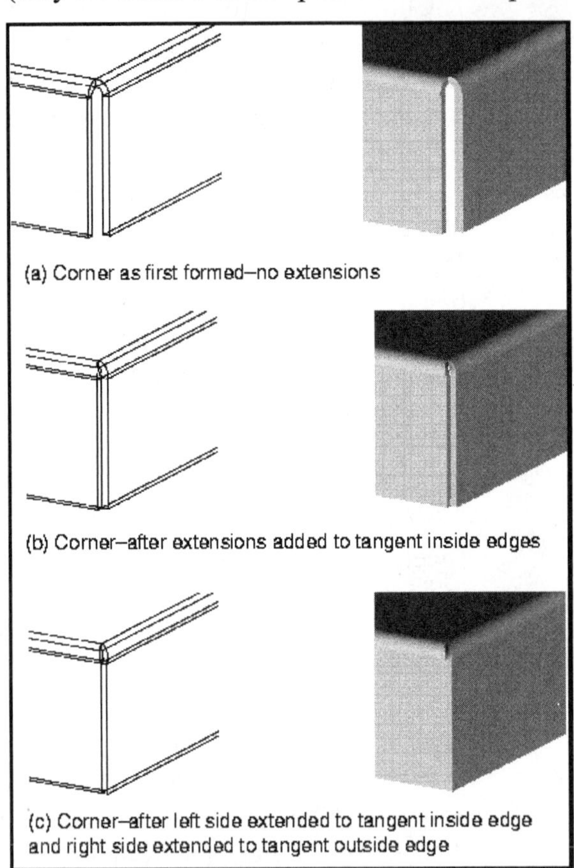

Figure 23.29
Sheet Metal Corner, with Various Degrees of Overlap

1. Choose **Sheet Metal** from the FEAT CLASS menu.
2. Choose **Wall** from the SHEET METAL menu.
3. Choose **Extend** from the OPTIONS menu and then **Done**. The WALL Options: Extend dialog box appears.
4. Choose the **Edge** element, then the **Define** button. Pick an edge on an existing plane wall. The EXT DIST menu appears.
5. Choose an option from the EXT DIST menu.
6. The feature is now fully defined. Choose **OK** from the dialog box. Pro/E creates the **Extend** feature.

Twists

A **Twist** is a special wall that can be used to change the plane of a sheet metal part. Twists serve as transitions between two areas of sheet metal, and the twist angle is relatively small (Fig. 23.30).

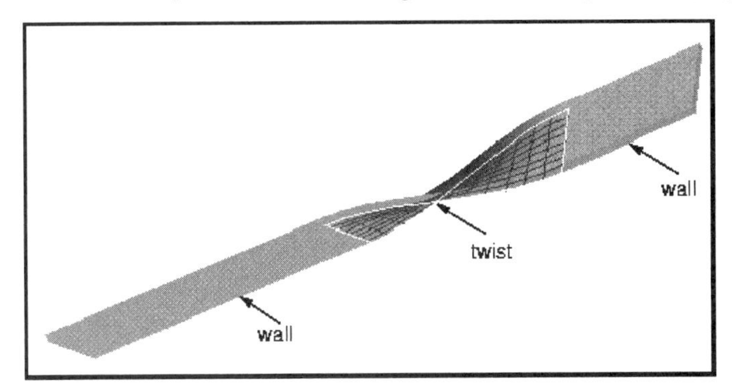

Figure 23.30
Twist

It is attached to a straight edge (the attach edge) on an existing plane wall (Fig. 23.31). It can be rectangular or trapezoidal. It has an axis (the twist axis) running through its center, perpendicular to the attach edge, and it can be twisted around that axis by a specified angle.

When creating a twist wall, you can add a flat or an extruded wall at the end of a twist only if the wall was created with no radius and tangent to the twist. You can unbend a twist using the **Unbend**, Regular menu options. Give the following commands:

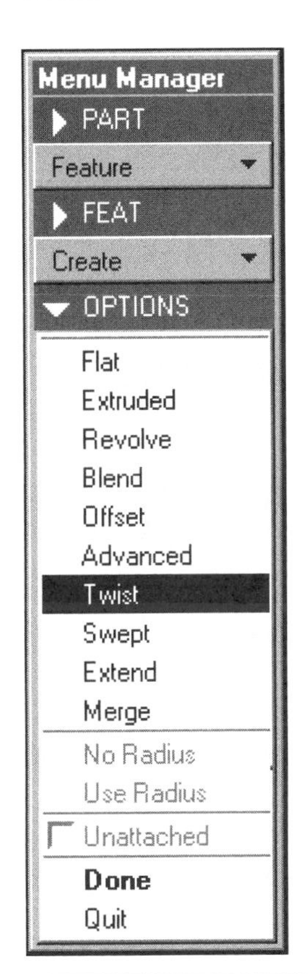

1. Choose **Sheet Metal** from the FEAT CLASS menu.
2. Choose **Twist** from the SHEET METAL menu. The TWIST dialog box appears.
3. Choose the **Attach Edge** element, then the **Define** button. Pick an edge on an existing plane sheet metal wall. The TWIST AXIS PNT menu appears. Choose one of the options.
4. Enter a value for the **Start Width** element.
5. Enter a value for the **End Width** element.
6. Enter a value for the **Length** element.
7. Enter a value for the **Twist Angle** element.
8. Enter a value for the **Develop Length** element.
9. The feature is now fully defined. Choose **OK** from the dialog box. Pro/E creates the Twist.

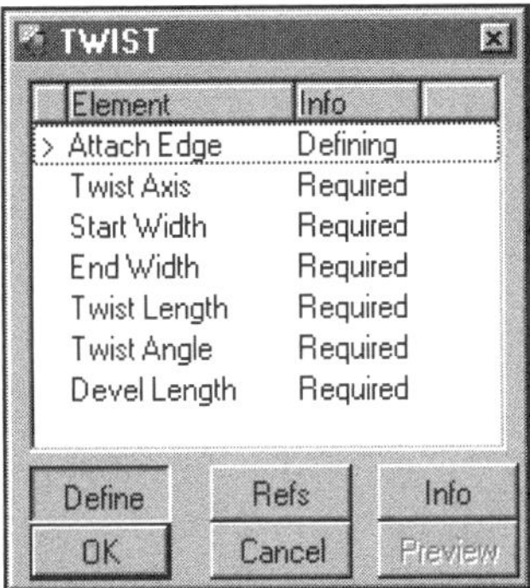

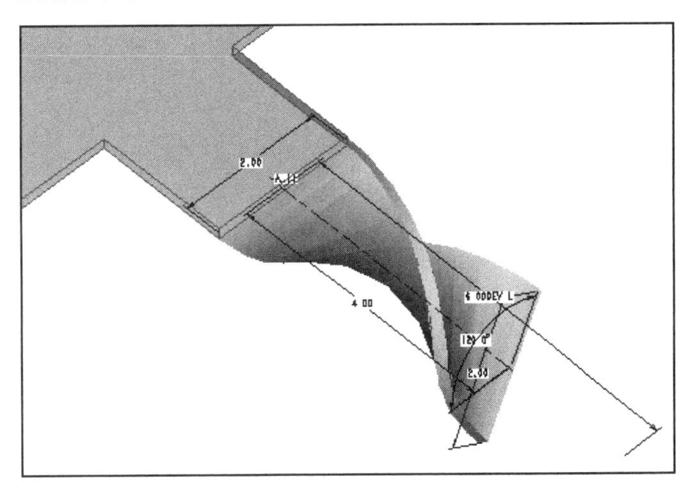

Figure 23.31
Flat Wall with Attached Twist

Cuts

The **Cut** feature can be used to create a **Thru-All** or **Thru-Next** cut or to create the UDFs used for notches and punches. The techniques used to create a cut in Sheet Metal mode are similar to those used in Part mode.

Sheet metal cuts are always made on the green or white surfaces and never on the edges. The white surface is offset from the green surface by the sheet thickness. Sheet metal cuts are sketched on a plane and then projected onto the sheet metal. The cuts are made normal to the sheet metal surface and can be driven from either the green side or the white side. See Figure 23.32 for an example.

If you want the **Cut** feature to cut the material at a defined angle, then you should use a *Solid-Class* cut rather than a *Sheet Metal-Class* cut.

Solid-Class Cut

Some *Solid-Class features* (e.g. holes, rounds, chamfers, slots, cuts) are available for use in sheet metal parts. They can be added in Sheet Metal mode. Solid-class features can be placed on the green and white surfaces and on the edges.

When you make a *Sheet Metal-Class cut*, you must select whether the cut is going to be **Solid** or **Thin**. If you choose **Solid**, then you must sketch the cut with all its dimensions. If you choose **Thin**, then you need sketch only a simple line sketch for a cut that will have uniform thickness (width). Figure 23.32 shows a Sheet Metal-Class cut in a curved sheet metal part. The cut was defined first as a solid cut and then as a thin cut. The cut could have been created as either a solid cut or a thin cut.

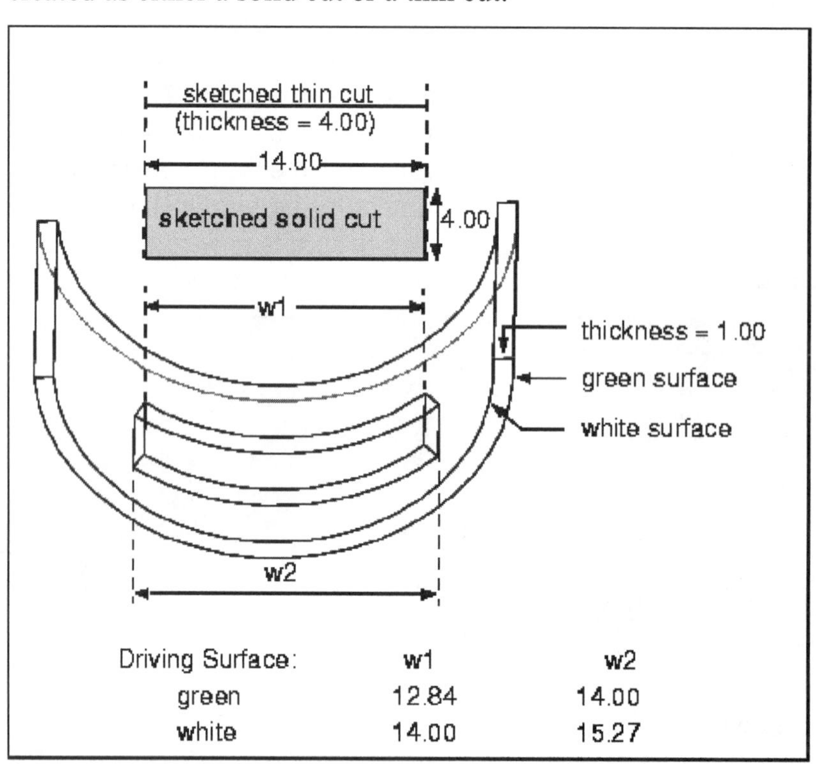

Driving Surface:	w1	w2
green	12.84	14.00
white	14.00	15.27

Figure 23.32
Sheet Metal Cuts

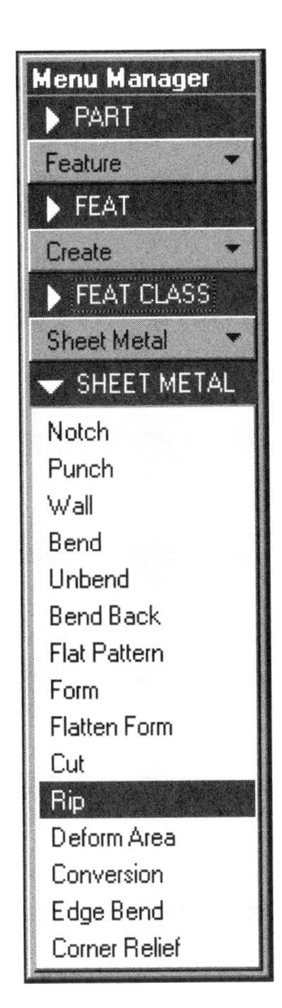

Rips

In general, *a rip is a zero-volume cut that is created on a sheet metal part.* If your part is designed as a continuous piece of material, it cannot be unbent without ripping the sheet. Create a **Rip** feature before unbending. You can choose among a number of different rips, including:

Regular Rip Select the surface on which to create the rip and then sketch the rip section. Pro/E creates a *sawcut* along the sketched *rip line*. (Fig. 23.33).
Surface Rip Select a surface and rip the geometry out. This option effectively cuts out the selected surface and so actually removes volume. (Fig. 23.34).
Edge Rip Select an edge of a surface. Pro/E creates a *sawcut* along the edge. (Fig. 23.35).

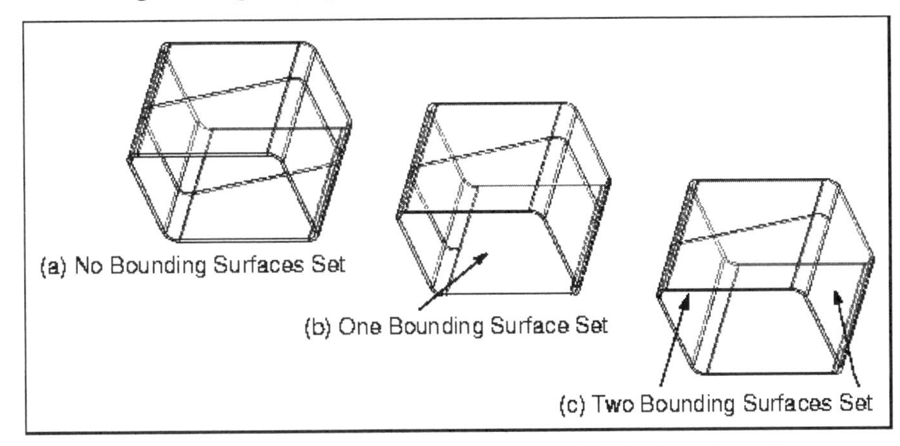

Figure 23.33
Regular Rip

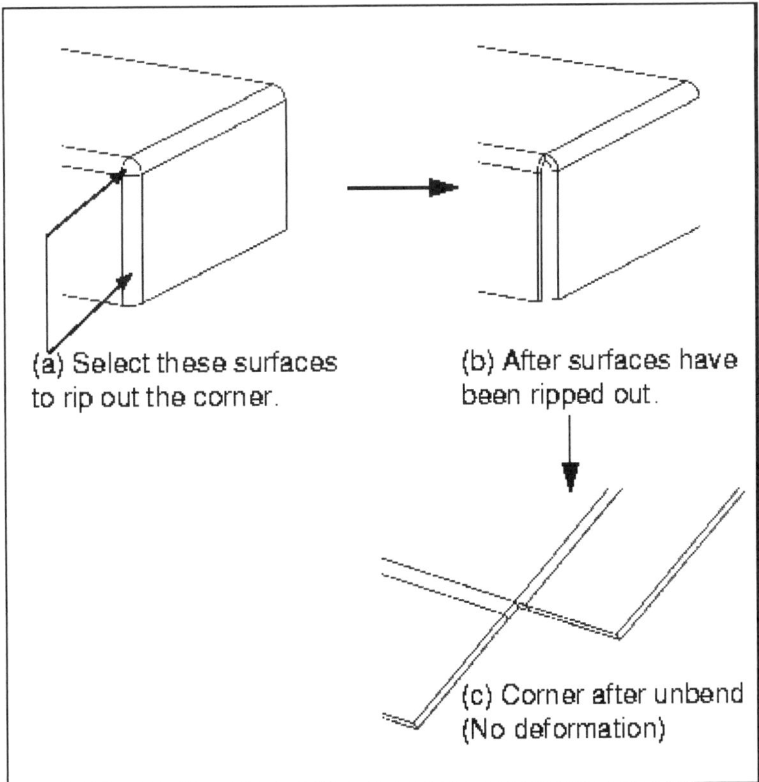

Figure 23.34
Surface Rip and Unbend

Figure 23.35
Edge Rip and Unbend

To create a **Regular Rip,** choose the following commands:

1. Choose **Rip** from the SHEET METAL menu.
2. Choose **Regular Rip** from the OPTIONS menu, then **Done.** The RIP (Regular Type) dialog box appears.
3. Choose the **Sketch** element, then the **Define** button. The SETUP SK PLN (Set up Sketching Plane) menu appears with **Setup New** preselected. The SETUP PLANE menu appears.
4. Sketch the rip section. The sketch can contain more than one entity, but all the entities must form one continuous open chain with the endpoints aligned to surface edges or silhouettes.
5. Dimension and regenerate the sketch, and then choose **Done.**
6. Select the **Bound Surf** element, then the **Define** button. The BOUND SURF and FEATURE REFS menus appear.
7. Choose **Done Refs**, then **Done/Return.**
8. The feature is fully defined. Choose **OK** on the dialog box. Pro/E creates the rip. It displays the rip as a sawcut on the sheet metal surface.

When you unbend a part with a **Regular Rip,** the material breaks along the rip section as shown in Figure 23.36.

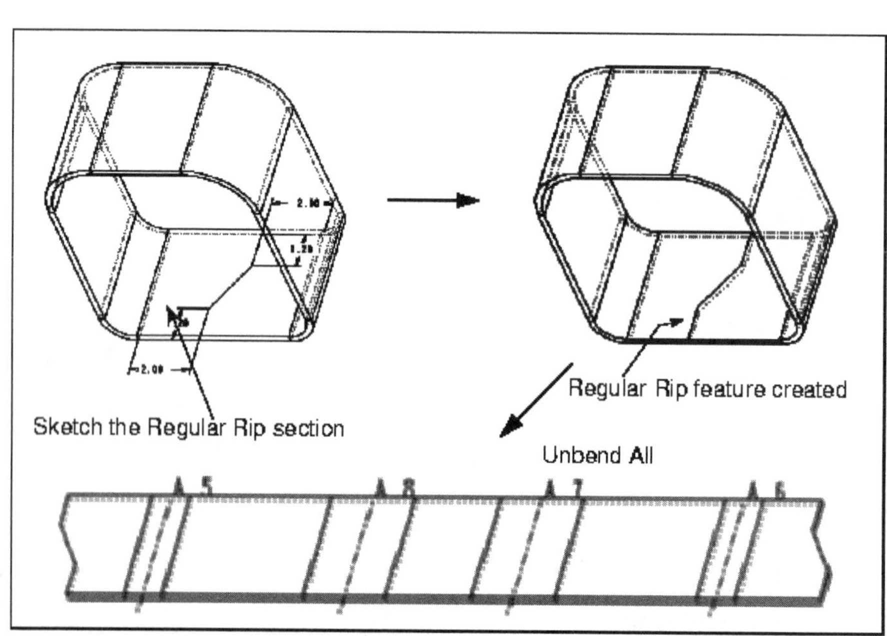

Figure 23.36
Regular Rip and Unbend

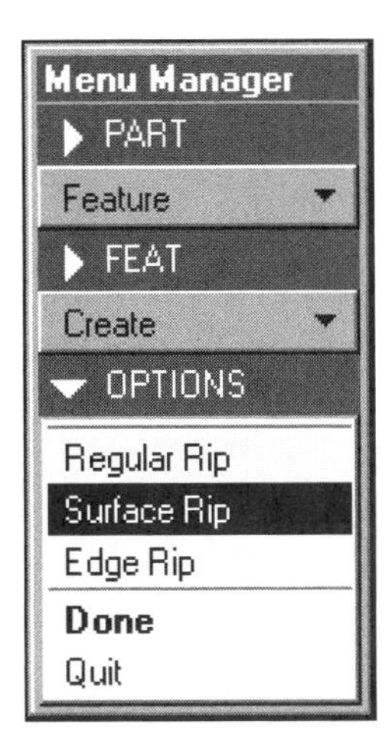

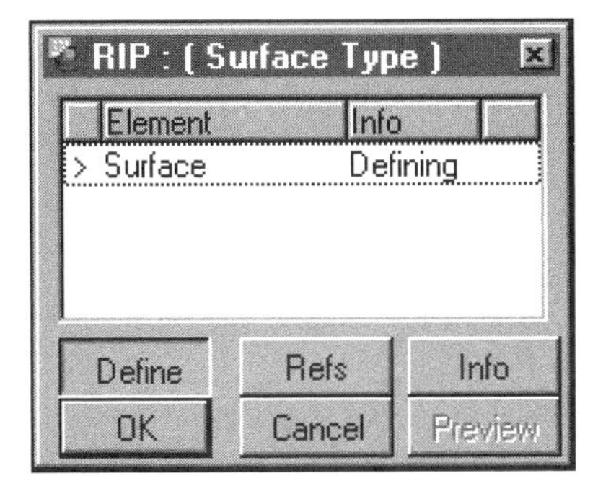

Surface Rip

Use the **Surface Rip** option to select a surface and rip the geometry out. To create a Surface Rip (Fig. 23.34), give the following commands:

1. Choose **Rip** from the SHEET METAL menu.
2. Choose **Surface Rip** from the OPTIONS menu, then **Done**.
3. The RIP (Surface Type) dialog box appears.
4. Choose the **Surface** element, then the **Define** button. The FEATURE REFS menu appears.
5. Pick the surfaces that you want to rip out, and then choose **Done Refs.** Pro/E highlights the selected surfaces in section color (*cyan*).
6. The feature is fully defined. Choose **OK** on the dialog box. Pro/E creates the **Surface Rip** by removing the selected surfaces.

When you unbend that area of the model, the material breaks along the rip section (Fig. 23.34).

Edge Rip

Use the **Edge Rip** option to create a rip along the selected edge. To create an Edge Rip, give the following commands:

1. Choose **Rip** from the SHEET METAL menu.
2. Choose **Edge Rip** from the OPTIONS menu, then **Done**. The RIP (Edge Type) dialog box appears.
3. Select the **Edge** element, then the **Define** button. The RIP PIECES menu appears (only the **Add** and **Done Sets** options are enabled initially).
4. Pick the edges that you want to rip, and then choose **Done Sets**. Pro/E highlights the selected edges in section color (*cyan*).
5. Choose **OK** on the dialog box. Pro/E creates the edge rip(s) by making a sawcut along the selected edges. When you unbend that area of the model, the material breaks along the rip sections.

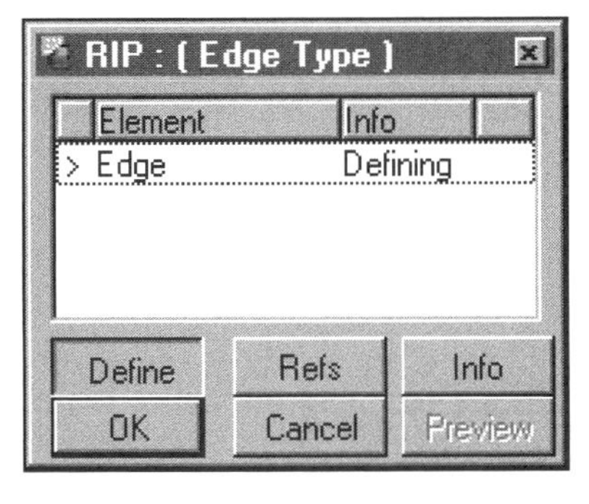

Bend, Unbend, Bend Back

This section describes the bend, unbend, and bend back features. You can also make a projected datum curve follow the unbending and bending back of sheet metal surfaces.

The *bend feature* consists of **Regular bends, Transition bends, Planar bends,** and **Edge bends.**

The *unbend feature* consists of **Regular unbends, Transition unbends,** and **Cross-section-driven unbends.**

The *bend back feature* enables you to return a bend, that has been unbent, to its original condition. Proper use of unbend and bend back features is very important for sheet metal design.

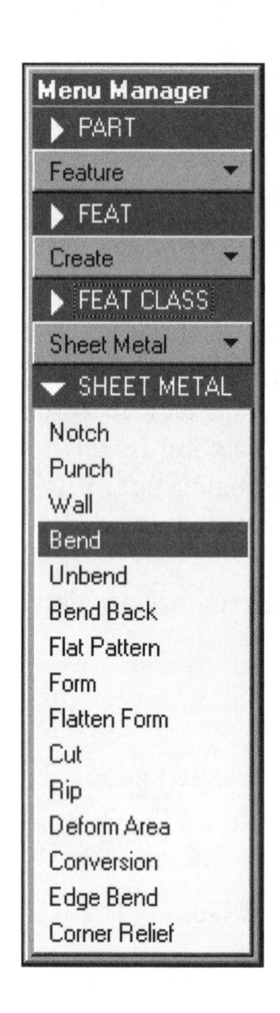

Adding a bend back feature is not the same as deleting an unbend. The suggested practice as follows:

* If you add an unbend (or bend back) feature just to see how the part looks in a flattened (unbent) condition, delete it before proceeding. Do not add unnecessary pairs of unbend/bend back features; they inflate the part size and might cause problems at regeneration.
* If you specifically want to create some features in a flattened state, add an **Unbend** feature, create the features you want, and then add a **Bend Back** feature. Do not delete the **Unbend** feature in this case; features that reference it might fail regeneration.

Bends

Not only are bends added to the part when a wall feature is created, but the sheet metal part can have bends added at any time via the **Bend** command. The part can be bent and unbent around these bends again and again by means of the **Unbend** and **Bend Back** features as shown in Figure 23.37.

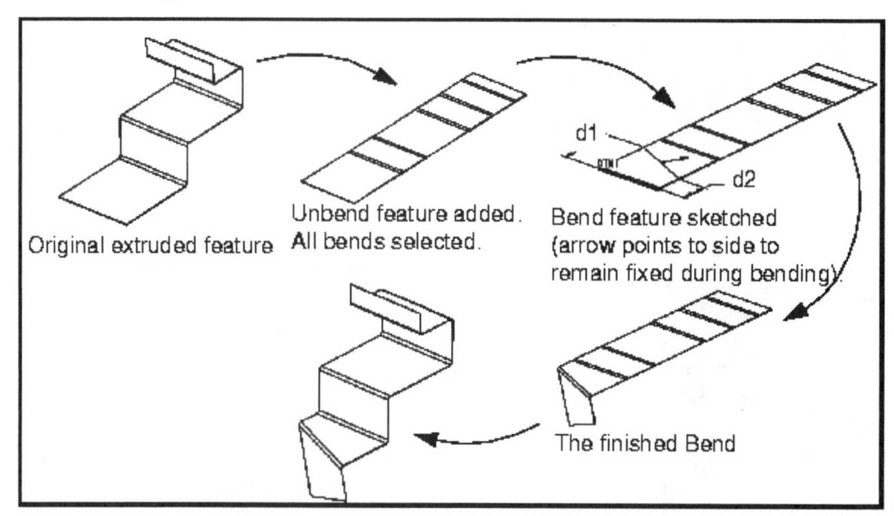

Figure 23.37
Examples of Bends

The **Bend** feature is used to add a bend to a flat section of the part. You cannot add a bend where it crosses another bend, but bends can be added across form features. When adding bends, consider the following:

* You cannot copy a Bend feature by mirroring it.
* Bends with transition areas do not accept bend relief.
* You can generally unbend zero-radius bends; however, you cannot unbend those that have slanted cuts across them.

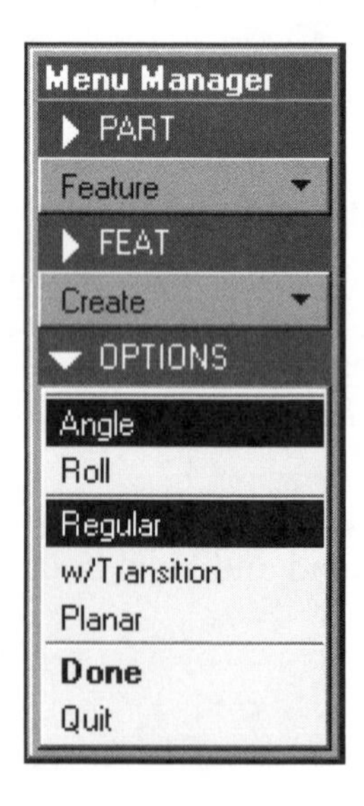

For all regular bends, you must specify the relief to be used:

No Relief Create the bend without reliefs.
Stretch Relief Use material stretching to provide bend relief where the bend crosses an existing edge of the fixed material.
Rip Relief At each end selected, cut the fixed material normal to the bend line, to provide relief.

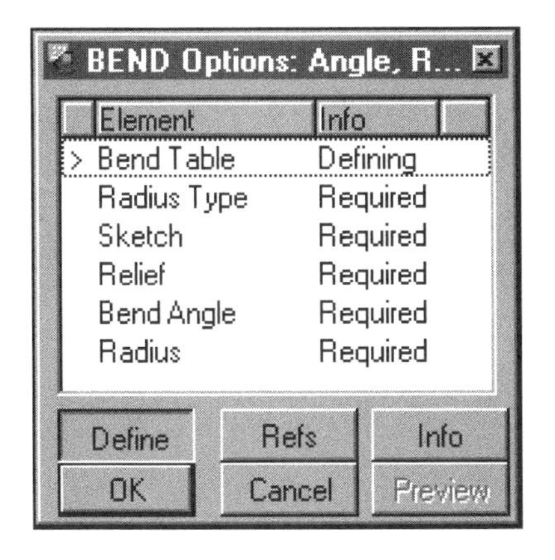

After you have sketched the bend and the sketch has been regenerated, the RELIEF menu appears.

In Figure 23.38, a simple *L-shaped* sheet is to be bent on a line that is a short distance above the imaginary line where the upper left part meets the main lower part. In the three relief examples, the large lower area was selected as the fixed area for the bend. In addition, the bend was created on the fixed side of the bend line.

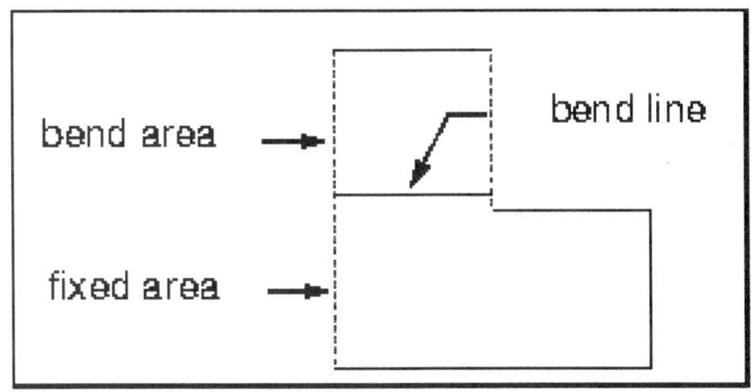

Figure 23.38
L-shaped Sheet

The results can be summarized as follows:

Bend with No Relief The upper part of the fixed area was also bent, along its full width (Fig. 23.39).

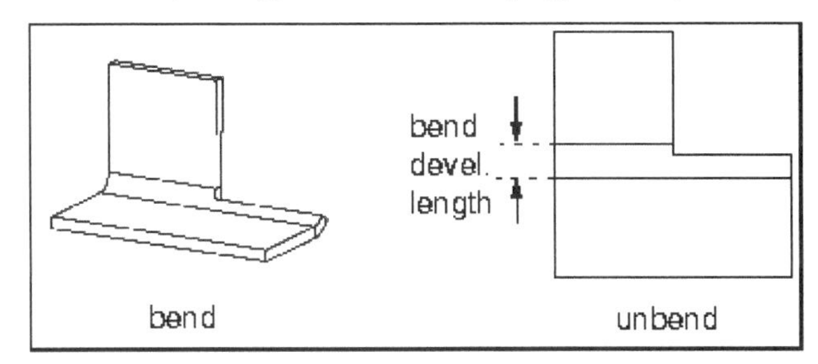

Figure 23.39
Bend with No Relief

Bend with Rip Relief The material in the fixed section was cut normal to the bend line, to provide the bend relief (Fig. 23.40).

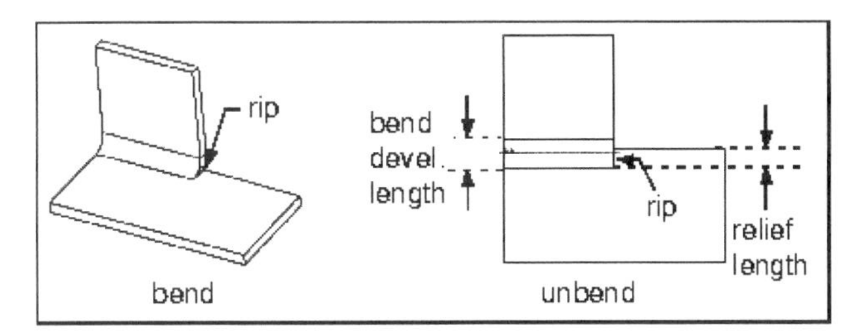

Figure 23.40
Bend with Rip Relief

Bend with Stretch Relief The material in the corner between the fixed and bent sections was stretched to provide the bend relief. Note, in the unbent view, the relief width and relief angle (Fig. 23.41).

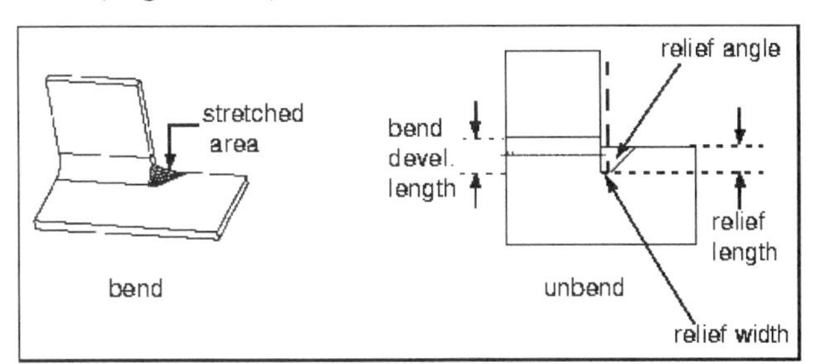

Figure 23.41
Bend with Stretch Relief

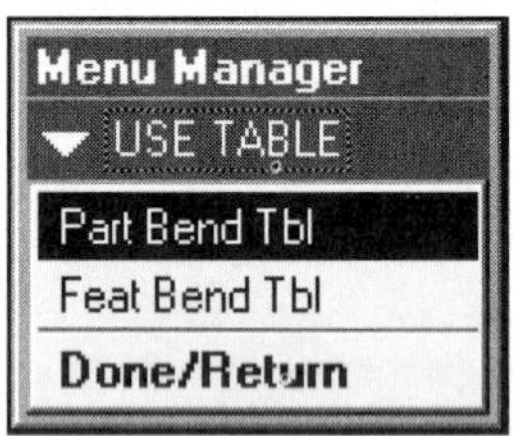

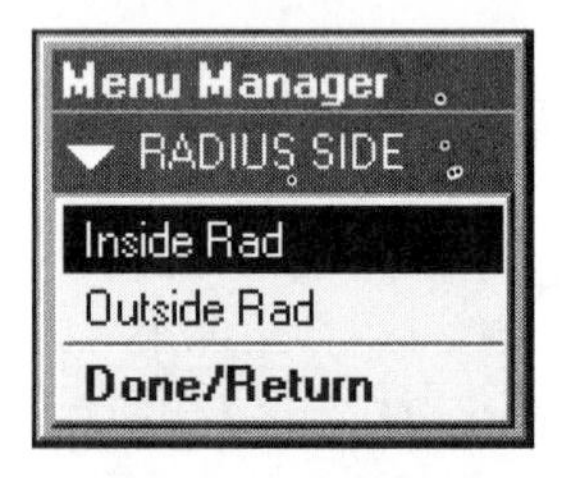

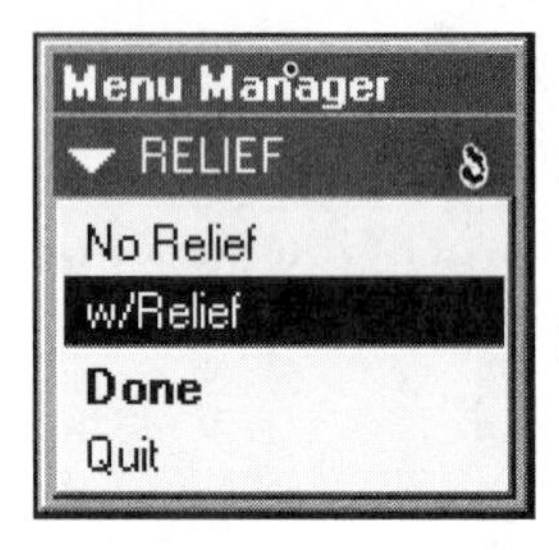

Types of Bends

Bends can be a combination of attributes--one from each of the following groups:

> **Angle** (default) Create a bend with a specified radius and angle
>
> *or*
>
> **Roll** Create a bend with a specified radius. The resulting angle is determined by both the radius and the amount of flat material to be bent;

then select

> **Regular** (default) Create a normal bend with no transition surfaces
>
> *or*
>
> **w/Transition** Deform the surface between the bend and an area you want to remain flat
>
> *or*
>
> **Planar** Create a bend around an axis that is perpendicular to the green surface

then select

> **Part Bend Table** (default) Use the bend table assigned to the part to calculate the developed length of the bend. (If no bend table has been assigned to the part, use the Y-factor.)
>
> *or*
>
> **Feature Bend Table** Use a specific bend table for this bend only

then select

> **Inside Rad** (default) Apply the radius to the inside of the bend
>
> or
>
> **Outside Rad** Apply the radius to the outside of the bend

and finally, select (for **Regular Bends** only)

> **No Relief** (default) Create a bend without reliefs
>
> or
>
> **w/ Relief** Create automatic bend reliefs.

How to Add a Regular Angle Bend

1. Choose **Bend** from the SHEET METAL menu.
2. Choose type **Angle** and the attribute **Regular** from the OPTIONS menu. Then choose **Done**. The BEND Options: Angle, Regular dialog box appears.
3. Choose the **Bend Table** element, then the **Define** button. The USE TABLE menu appears. Choose the type of bend table that you want, then **Done**.
4. To define the radius type, choose **Inside Rad** or **Outside Rad** from the RADIUS SIDE menu, then **Done/Return**.
5. Choose the **Sketch** element, then the **Define** button. Select the surface on which you want to create the bend. Create your sketching references. Try to create references to local geometry.

Sketch the bend line. This must be a line, and there can be only one entity. You must align it to an outside edge at each end. **Dimension** and **Regenerate** the bend sketch. After successful regeneration, choose **Done**.

The BEND SIDE menu appears. Select the side of the bend line on which you want to create the feature. Pointing the arrow to one side or the other keeps the bend totally to that side, regardless of the bend radius. Selecting Both creates the bend area equally on both sides of the sketched line.

The DIRECTION menu appears. Use the **Flip/Okay** options to select the surface to remain fixed.

6. Choose the **Relief** element, then the **Define** button. The RELIEF menu appears. Choose one of the options, then **Done.**
7. Choose the **Bend Angle** element, then the **Define** button. The DEF BEND ANGLE menu appears. Choose one of the standard listed values or choose **Enter Value** and enter the exact value (in degrees).
8. To change the direction of the angle, choose **Flip.** Then choose **Done**.
9. Choose the **Radius** element, then the **Define** button. Choose one of the options from the SEL RADIUS menu or enter the bend radius.
10. The feature is now fully defined. Choose **OK** on the dialog box. Pro/E creates the **Regular Angled Bend.**

Figure 23.42 is an example of a typical regular angle bend.

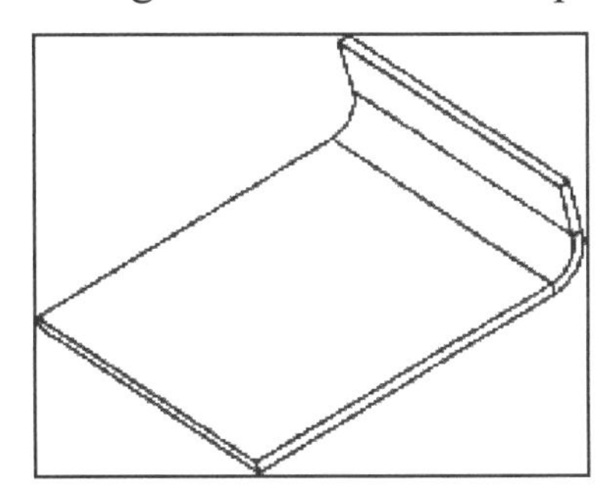

Figure 23.42
Regular Angle Bend

A **Roll** bend (Fig. 23.43) is sketched like that for an angle bend; the direction of bending, however, is determined when you select the direction of viewing the sketching plane. The bend is created in the direction you are viewing--that is, away from you.

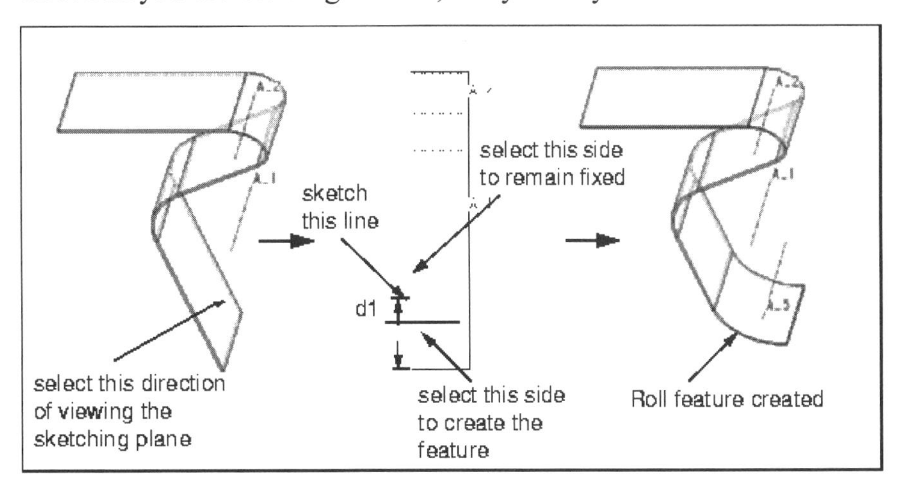

Figure 23.43
Roll Bend

Bend Order Tables

You can create bend order tables and display them in sheet metal drawings to document the order and dimensions of bend features. Bend order tables can be updated by reviewing the bend sequence. When a bend order table is stored, the file name is *<modelname>.bot.* A bend order table cannot be created or edited on a completely unfolded model. This bend order table shown in Figure 23.44 is based on the completed part that is shown in Figure 23.45. The bend sequences for the part are shown in Figure 23.46.

Figure 23.44
Bend Order Table

BEND SEQ	#BENDS	BEND#	BEND DIRECTION	BEND ANGLE	BEND RADIUS	BEND LENGTH
1	2	1	OUT	90.000	0.500	0.518
		2	OUT	90.000	0.500	0.510
2	2	1	IN	90.000	0.250	0.518
		2	OUT	90.000	0.500	0.518
3	1	1	IN	90.000	0.250	0.518
4	1	1	IN	90.000	0.250	0.518

green side
sequence 3, bend 1 (in)
sequence 1, bend 1 (out)
sequence 1, bend 2 (out)
sequence 2, bend 1 (in)
sequence 2, bend 2 (out)
sequence 4, bend 1 (in)

Figure 23.45
Completed Sheet Metal Part

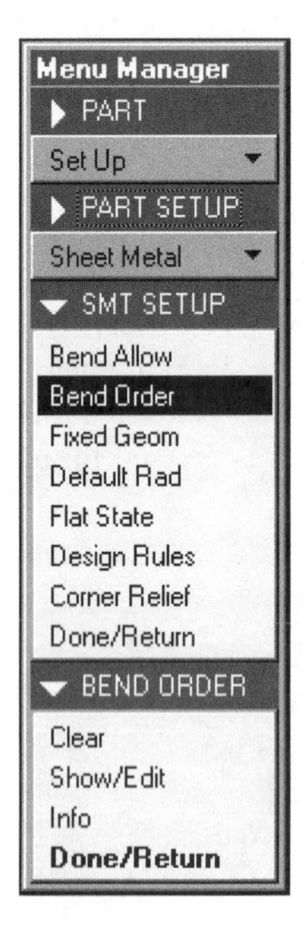

Show/Edit–Unbend all bends
First sequence highlighted

Next–Bend back first sequence
Second sequence highlighted

Next–Bend back second sequence
Third sequence highlighted

Next–Bend back third sequence
Fourth sequence highlighted

Next–Bend back fourth and last sequence

Figure 23.46
Bend Sequences

Creating a Bend Order Table

1. With a model in the bent condition, choose **Sheet Metal** from the PART SETUP menu.
2. Choose **Bend Order** from the SMT SETUP menu.
3. Choose **Show/Edit** from the BEND ORDER menu. The GET SELECT menu appears.
4. In response to Pro/E prompt, select a plane or edge to remain fixed while the model is completely unbent. The model is completely unfolded.
5. The **Add Bend** option in the SHOW/EDIT menu is enabled. The GET SELECT menu appears, and you are prompted to select the bends for the first sequence. You can select any bend, and any number of bends in any order.
6. When you have finished adding bends to the current sequence, and you want to start another sequence, choose **Next** from the SHOW/EDIT menu. The bends in the current sequence are now highlighted in magenta.
7. In response to the prompt, select a plane that is to remain fixed. The highlighted bends are then bent back.
8. The **Add Bend** option in the SHOW/EDIT menu is re-enabled and the GET SELECT menu reappears. You are now starting a new sequence.
9. Repeat Steps 6 through 8 until the whole part is bent back.
10. Choose **Done** from the SHOW/EDIT menu. Pro/E creates the bend order table.

Displaying a Bend Order Table

To write the bend order table to a file and display it on the screen, choose **Info** from the BEND ORDER menu. In the bend order table, a bend's direction is considered to be in when it is less than **180°** (i.e., acute or obtuse) on the *green* side. It is considered to be out when it is greater than **180°** (i.e., oblique) on the *green* side.

Unbends

Use the **Unbend** feature to unbend any curved surface on the sheet metal part, whether it is a bend feature or a curved wall. Use the UNBEND OPT menu to do the following:

> **Regular** (default) Pick an existing bend or wall feature to unbend. You can use this option to unbend most bends in a part. After the bend, the surfaces that were tangential to the bend are in the same plane. If you select all bends, you create a flat pattern of your part.

Transition Select stationary surfaces and transitional surfaces to create an unbend feature. This can be used on undevelopable surfaces, such as blended walls.
Xsec Driven Select stationary surfaces and specify a cross-sectional curve to determine the shape of the unbend feature. This can be used on undevelopable surfaces, such as hems and flanges.
Done Continue with the unbending process.
Quit Abort the unbending process.

Most bends in a part can be unbent with the regular option as shown in Figure 23.47 and Figure 23.48. Use the regular option to unbend undevelopable (deformed) surfaces and twists.

The dialog box to unbend a regular bend appears when you select **Regular** from the UNBEND OPTS menu. The dialog box lists the following elements:

Fixed Geom (required) The region of the model that is to remain fixed during the unbend.
Unbend Geom (required) The bend features to be unbent.
Deformation (not applicable to a regular unbend).

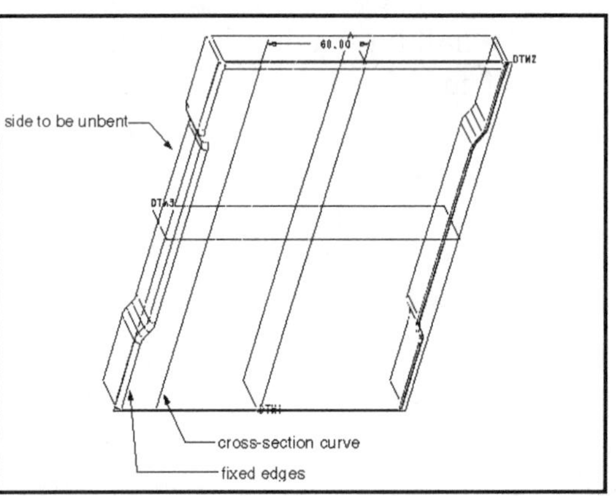

Figure 23.47
Bend Sequences

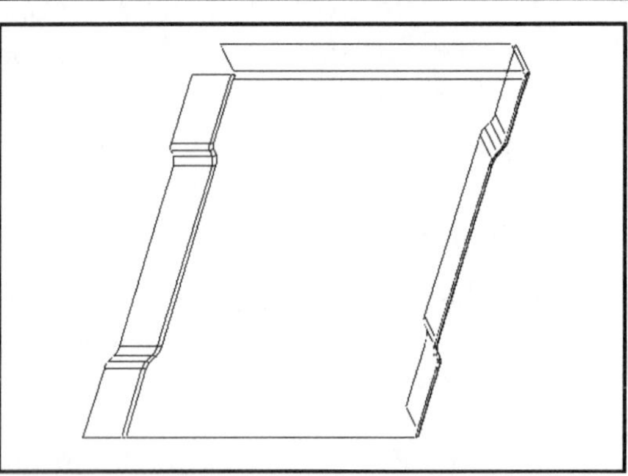

Figure 23.48
Unbent Bend

Bending Back

When a bend has been unbent, you can use the **Bend Back** command to return it to its original condition. If you partially bend back a regular unbend that contains a deform area, you might not be able to complete the bend back to the fully formed condition. As a rule, you should bend back only a fully unbent area. When creating a **Bend Back** feature, you can specify contours to remain unbent (Fig. 23.49).

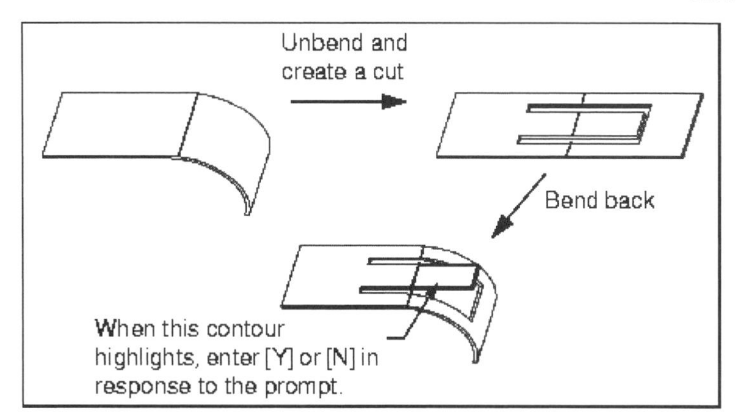

Figure 23.49
Bending Back with Fixed Surfaces

The dialog box for a regular roll bend appears when you select **Roll** and **Regular** from the OPTIONS menu. The dialog box lists the following elements:

Fixed Geom (required) The region of the model that is to remain fixed during the bend back.
BendBack Geom (required) The unbent feature to be bent back.

Use the BENDBACKSEL menu after you select the **BendBack** Geom element to do the following:

BendBack Sel The GET SELECT menu appears. Select one or more Unbends to bend back, and then choose **Done**.
BendBack All All Unbends are bent back automatically.

To bend back a sheet metal part, give the following commands:

1. Choose **Bend Back** from the SHEET METAL menu.
2. The BEND BACK dialog box appears.
3. Choose the **Fixed Geom** element, then the **Define** button. Select the surface that is to remain fixed.
4. Choose the **BendBack** Geom element, then the **Define** button. The BENDBACKSEL menu appears. Choose an option and then **Done**.
5. The feature is now fully defined. Choose **OK** on the dialog box. Pro/E bends back the selected unbends.

As Pro/E bends back each selected unbend, it examines its contours. If it finds that a contour partially intersects the bend area, it highlights that part of the contour and prompts you as to whether you want this contour to remain flat (Fig. 23.50).

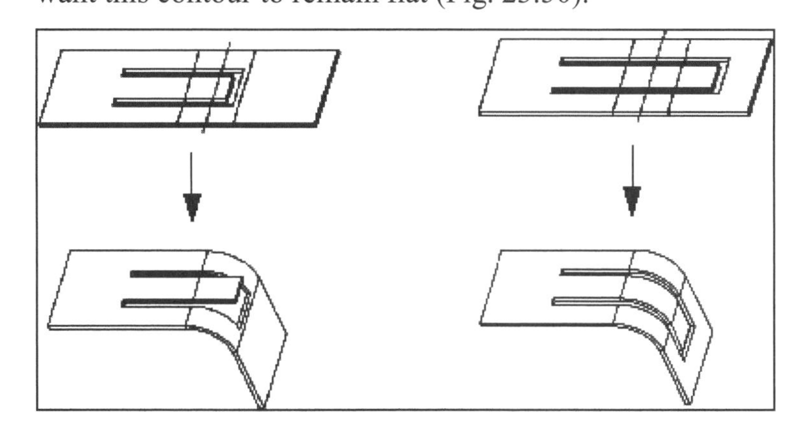

Figure 23.50
Bend Backs with Contours Intersecting Bend Areas

Flat Pattern

Flat Pattern is equivalent to an **Unbend All** feature, but it is always positioned last in the part feature sequence. You can create a Flat Pattern feature early in the design, to get started on drawings and manufacturing. If new features are later added to the part, they are automatically reordered before the Flat Pattern feature.

After you create the Flat Pattern, the part always displays in the flattened state; however, once you start creating a new feature, the Flat Pattern is temporarily suppressed, and then it is automatically resumed and reordered when the new feature is completed. If you do not want the part to be constantly flipping back and forth, you can suppress the Flat Pattern feature and resume it only when you want to see or use the flat pattern of the part. To create a Flat Pattern, give the following commands:

1. Choose **Feature** ⇒ **Create**⇒ **Sheet Metal** from the SHEET METAL menu.
2. Select a plane or edge to remain fixed.
3. All bent surfaces and edges are unbent. There can be only one Flat Pattern feature per part; after you create it, the **Flat Pattern** option becomes unavailable.

Flat State Instances

In order to simplify and streamline the creation of Flat Pattern representations for manufacturing, a FLAT STATE menu (called FLAT STAT) is provided. It is accessed through the SHEET METAL SETUP menu (see left column). It can be used to create instances directly, without having to edit the part's family table. The instances thus created are initially completely unbent (Fig. 23.51). Features created in **Flat State** instances behave just like those created in regular family table members.

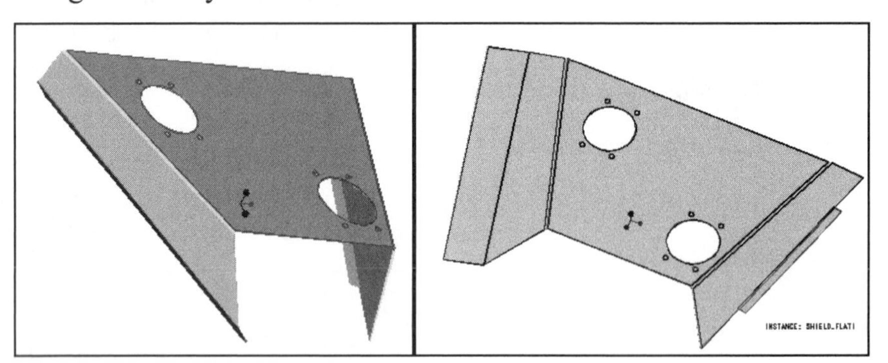

Figure 23.51
Fully Formed Part and Flat State Instance

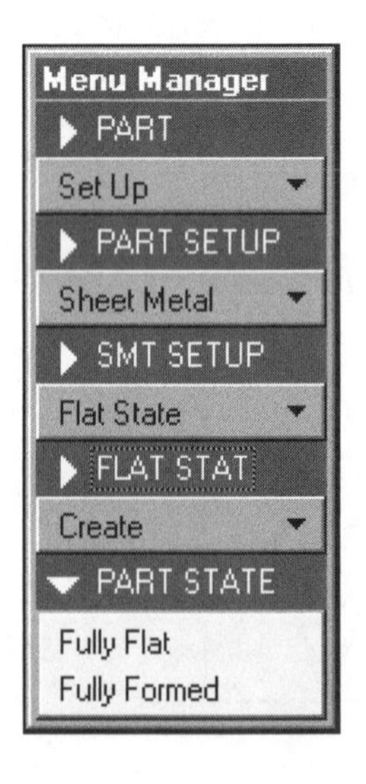

If you subsequently choose the **Update** option from the FLAT STAT menu, then all the features that were added/enabled in any and all of the *Flat State instances* are enabled (resumed) in the generic and in all the instances, except in those instances in which they were specifically suppressed. The system enables you to clean up the family table by removing from it all the features that are now enabled in the generic and all the Flat State instances.

You can create the first flat state instance from a generic that is either fully formed or fully flat. If the generic is (fully) formed, then the system brings up the Unbend feature editor and prompts you for the Unbends necessary to make it fully flat. After you choose **Done**, the system creates the instance. To create Flat State Instances:

1. Choose **Set Up** from the PART menu. The PART SETUP menu appears with **Sheet Metal** preselected, which brings up the SMT SETUP menu.
2. Choose **Flat State** from the SMT SETUP menu.
3. Choose **Create** from the FLAT STAT menu.
4. In response to the prompt, enter the name of the flat state instance. If this is the first flat state instance, the PART STATE menu appears, with the options **Fully Flat** and **Fully Formed**.
5. Choose **Fully Formed** from the PART STATE menu. The system brings up the UNBEND OPT menu and prompts you to unbend (flatten) the part. When you choose **Done** from the FEATURE EDIT menu, the system creates the flat state instance and momentarily displays it on the screen.
6. The family table for the generic now contains a new item the **Unbend** that you just created (Fig. 23.52); it is suppressed for the generic (because the generic is formed), and it is enabled (resumed) for the instance.

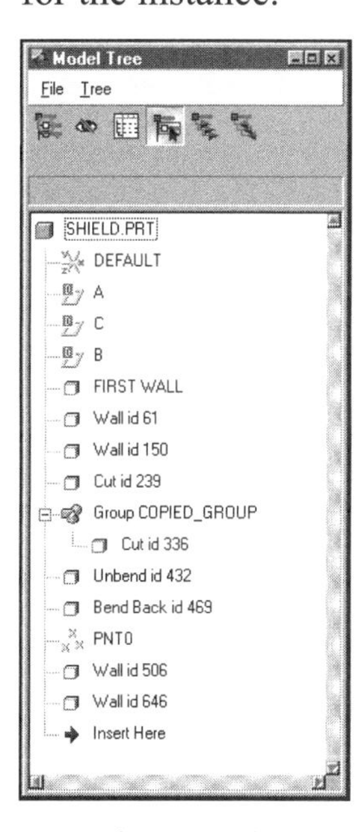

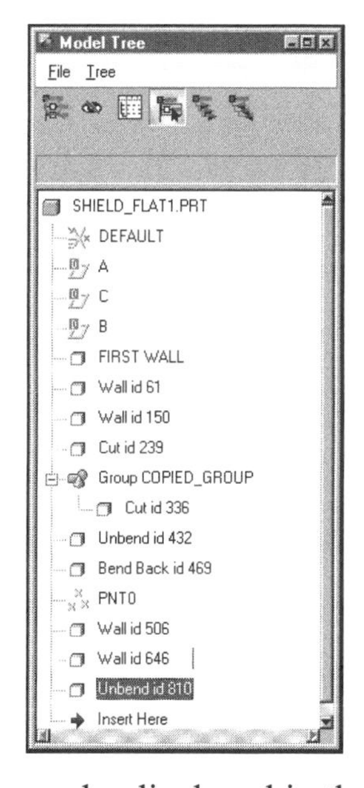

Figure 23.52
Model Tree Before and After Creation of Flat State Instance

Flat state instances can be displayed in three ways:

* If the generic is in the active window, choose **Instance** from the FAMILY TABLE menu, and then choose the particular instance from the Instances namelist menu.
* If the generic is in the active window, choose **Show** from the FLAT STAT menu; then choose the particular instance from the FLAT MODELS NAMELIST menu.
* If you are retrieving the part through the ENTERPART menu, you have the choice of retrieving the generic or one of the instances.

NOTE
You can display and dimension a *flat state instance* on the same drawing with the detailed sheet metal part and its dimensioned views.

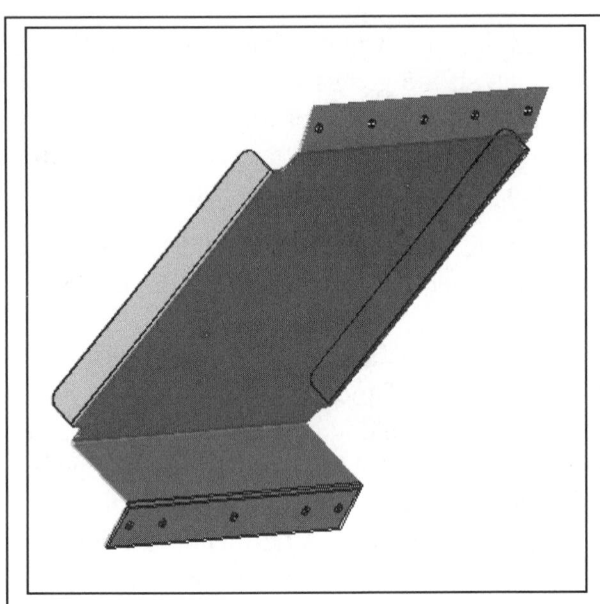

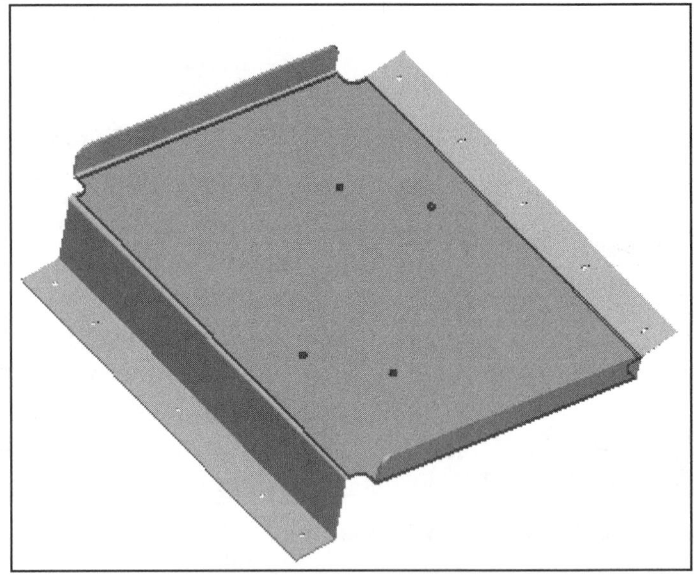

Figure 23.53
Sheet Metal Design

SHEET METAL DESIGN

You have already created a wide variety of parts, assemblies, and drawings using this text. By now you should be so proficient using Pro/E that a step-by-step cookbook recipe of a sheet metal part should not be unnecessary. Therefore, in this version of the text this new lesson introducing Pro/SHEETMETAL will utilize a different format. We will introduce a wide range of part design concepts, commands, and capabilities as they apply to sheet metal (Fig. 23.53). At the end of the lesson, you will apply this material by creating a sheet metal part and documenting it with a detail drawing. The instructional portion of the lesson will require that you create a number of very simple parts that will be used to demonstrate Pro/SHEETMETAL capabilities. The lesson projects will include a variety of parts from which to choose.

Design Approach

Here is one possible design approach to follow for creating sheet metal parts:

1. Create the basic sheet metal parts in Sheet Metal mode. Because many of the components will be held in place with screws or bent tabs, you might want to leave the creation of these features for later when the components are assembled.
2. Create the assembly by assembling all major internal components relative to each other (Fig. 23.54). Include simple supporting structures, or sheet metal parts that are not completely defined at this time, to place the components.
3. Create or modify the sheet metal parts using the internal components as references. This aids you in adding supporting walls, form features for stiffening panels, and punches and notches for fastening the components.

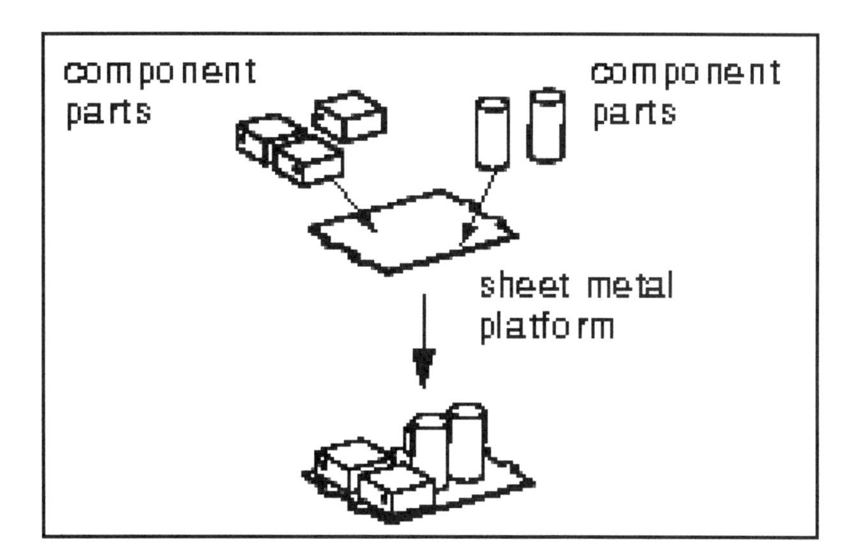

Figure 23.54
Components and Sheet Metal Platform, Before Assembly

4. After the cabinet and supporting structures are defined relative to the internal components and each other, add any remaining components, sheet metal, or assembly features (Fig. 23.55).
5. Create and/or select a bend table to provide material allowances when unbending the part. This could also be done as the first step in the design.
6. For each part, create a bend order table to define the bending sequences.

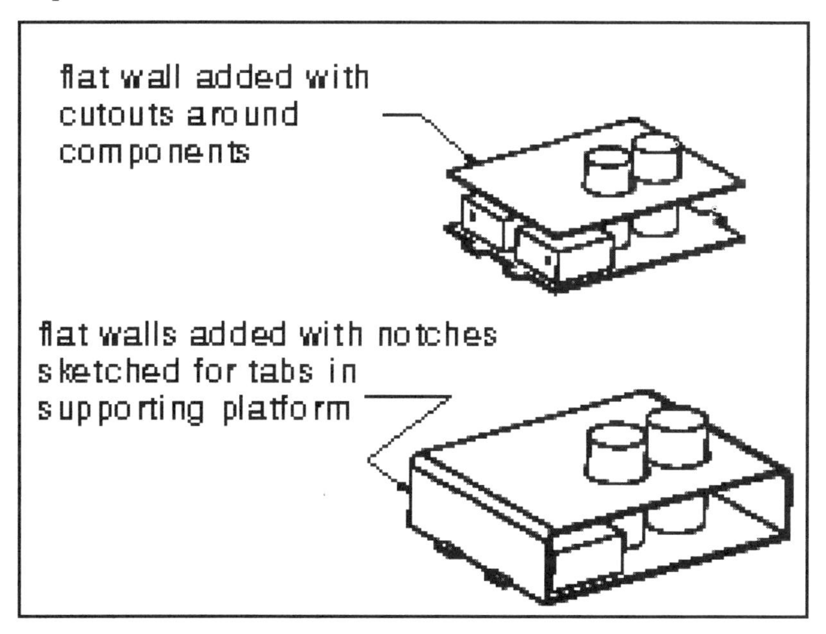

Figure 23.55
Components and Sheet Metal Enclosure, After Assembly

7. Add a **Flat State** instance. This creates your flat pattern for drawing and manufacturing. The bend table data ensures that the flat pattern's geometry is accurate.
8. Document the parts by creating drawings; you can include both the generic (as designed) part and the **Flat State** instance (multi-model drawing). Show the dimensions for the as-designed part, and show/create dimensions for the flat pattern part. Add the bend order table as a note.

Sheet Metal Parts

A sheet metal part can be created in Sheet Metal mode or in Assembly mode as a sheet metal component, or it can be a regular part that has a constant thickness that is converted to a sheet metal part. (You cannot merge sheet metal parts with other assembly components.)

Part Surfaces

A sheet metal part appears with ***green and white surfaces***, with side surfaces in between. ***The green surface is the driving surface.*** This helps in visualizing the part and geometry selection, because sheet metal parts are comparatively thin. *Sheet metal parts have a constant thickness; Pro/E creates the white surface by offsetting the material thickness from the green.* The side surfaces are not added until the part is fully regenerated.

Sheet Metal Parameters

You should set up certain parameters and conditions prior to adding features to your sheet metal part, including:

* *Determining* the method to calculate the developed length of flat sheet metal required to make a bend of a specific radius and angle.
* *Generating* a table to document the order and dimensions of bend sequences.
* *Setting* the default surface or edge that remains fixed when you bend or bend back the part.
* *Defining* the default radius used when you create a bend or add walls.
* *Creating*, *displaying,* and *updating* flat state instances of a current part.

Bend Allowance

Bend allowance is a method to calculate the length of flat sheet metal required to make a bend of a specific radius and angle. Bend tables are used for the accurate calculation of this length. Use one of the following three methods to calculate the bend allowance:

* The system default equation
* One of the provided Pro/SHEETMETAL bend tables
* Pro/TABLE to create your own

Bend tables are used to calculate the bend allowance. When Pro/E looks for a value and does not find it in the bend table data, it uses the formula equation.

Pro/SHEETMETAL Bend Tables

Bend tables are used to calculate accurately the length of flat material (developed length) required to make a bend of a specific radius and angle. For the same bend, the developed length is different for different materials and material thicknesses.

Bend tables can be read in at any time. Note, however, that once a part is associated with a bend table, its geometry depends on the table data. *The bend table information is not stored with the part.*

Every time the part is regenerated, it looks up the associated table for appropriate length values. If you modify a bend table, all parts associated with it are updated upon regeneration.

Pro/SHEETMETAL provides the following three bend tables:

Table Name	Material	Y-Factor	K-Factor
TABLE1	Soft Brass and Copper	0.55	0.35
TABLE2	Hard Brass and Copper Soft Steel and Aluminum	0.64	0.41
TABLE3	Hard Copper Bronze Cold Rolled Steel Spring Steel	0.71	0.45

(The bend tables provided with Pro/SHEETMETAL are used with permission from Machinery's Handbook, 23rd Edition)

You can make a particular **Wall** or **Bend** feature reference a different bend table from the bend table or **Y-factor** associated with the overall part. At the time of creating a **Wall** or **Bend** feature, you have two options:

* **Part Bend Tbl** The bend table associated with the part is used for developed length calculations. If no table is currently assigned to the part, the Y-factor formula is used.
* **Feat Bend Tbl** Specify a separate bend table to be used by the feature. Select one of the three supplied bend tables from the Data Files namelist menu, or choose **Names** and enter the name of the bend table file (including the path, if necessary).

Features

Pro/SHEETMETAL uses a set of features unique to sheet metal parts. Sheet metal features can be added to a sheet metal part when the part is completely unbent, when it is completely bent in its design condition, or at any stage of bend/unbend in between. Sheet metal parts cannot be retrieved in Part mode. In Sheet Metal mode, you can create datum, cosmetic, and some Solid-class features, such as holes, slots, chamfers, and (flat) protrusions.

Lesson procedures and commands
START HERE ➡ ➡ ➡

Construct the First Wall Feature Using the Flat Option

The first part we will create is a simple sheet metal part using the **Flat** option for the first wall (Fig. 23.56)

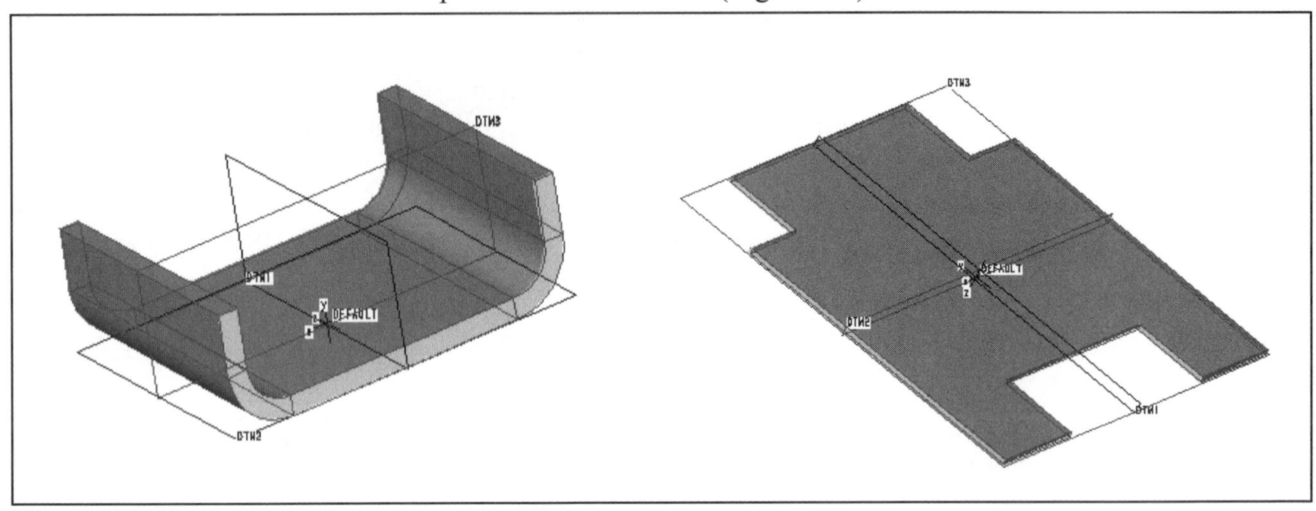

Figure 23.56
Simple Sheet Metal Parts

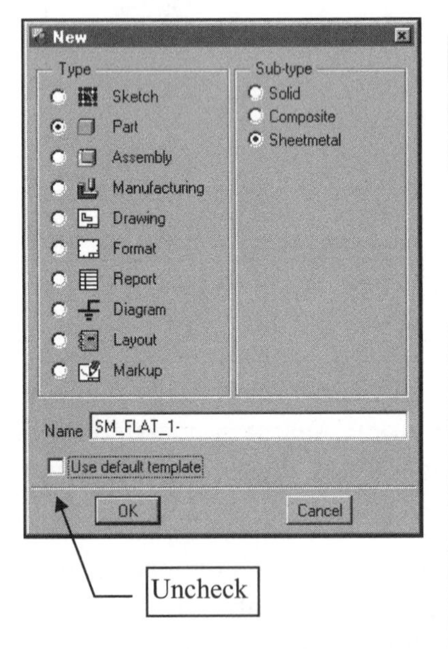

File ⇒ New ⇒ Part ⇒ Sub-type ● **Sheetmetal ⇒** (type the part name **SM_FLAT_1**) **⇒ ❑ Use default template ⇒ OK ⇒ inslb_part_sheetmetal ⇒** (fill in the Parameters) **⇒ OK**

Set up the environment:

SETUP AND ENVIRONMENT

Set Up ⇒ Material ⇒ Define ⇒ (type **STEEL**) **⇒ ✔ ⇒** (table of material properties, change or add information) **File ⇒ Save ⇒ File ⇒ Exit ⇒ Assign ⇒** (pick **STEEL**) **⇒ Accept ⇒ Done**

Rename the datums and coordinate system: **Set Up ⇒ Name** etc. [coordinate system (**PRT_CSYS_DEF** to **DEFAULT**) and the datum planes (**FRONT** to **DTM3**, **TOP** to **DTM2**, and **RIGHT** to **DTM1**)]

Utilities ⇒ Environment ⇒ ✔ Snap to Grid
Display Style **Hidden Line** Tangent Edges **Dimmed**

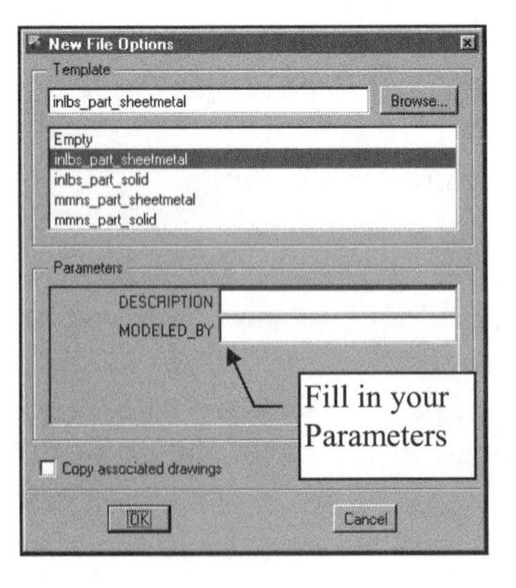

The first *sheet metal-class* feature will be a **Flat** wall feature:

Feature ⇒ Create ⇒ Wall ⇒ Flat ⇒ Done ⇒ (pick **DTM3** as the *sketching* plane) **⇒ Okay ⇒ Top ⇒** (pick **DTM2** as *orientation* plane) **⇒ Sketch ⇒ ✔Intent Manager ⇒ Close** (to accept **DTM1** and **DTM2** as the References) **⇒ Sketch ⇒ Line ⇒** (sketch, and then modify the sketch to match Figure 23.57) **⇒ Done ⇒** (enter the thickness of **.125) ⇒ ✔ ⇒ OK ⇒ File ⇒ Save ⇒ ✔** (Fig. 23.58) **⇒ Window ⇒ Close Window**

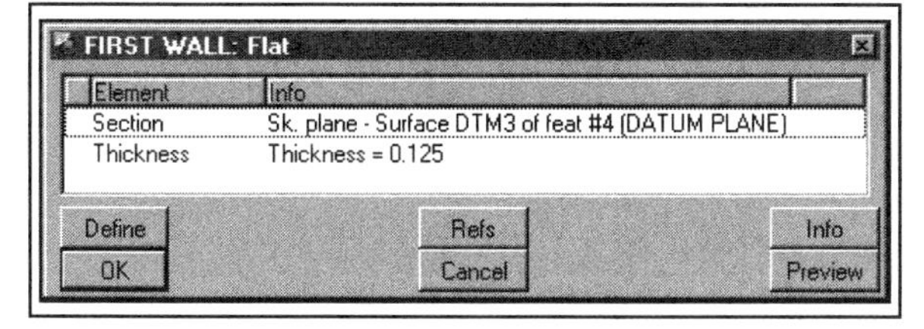

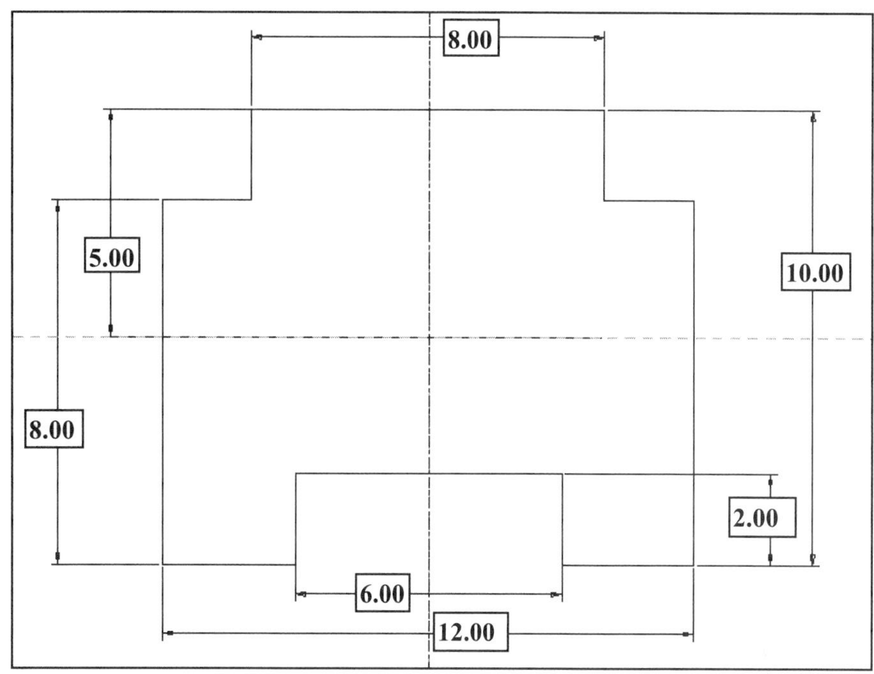

Figure 23.57
Sketch for Flat Wall

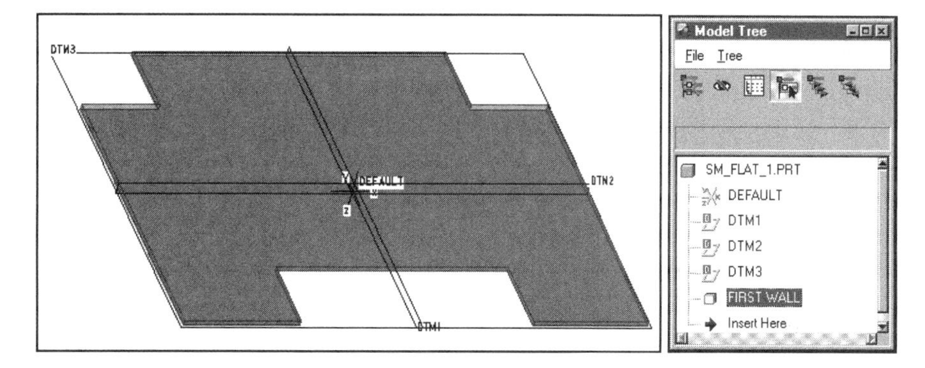

Figure 23.58
Completed Flat First Wall

Don't forget to rename your datum's and coordinate system for every sheet metal part you create in this lesson. You may wish to create a template part to reuse for each lesson exercise and project.

Rename the datums and coordinate system: **Set Up** ⇒ **Name** etc. [coordinate system (**PRT_CSYS_DEF** to **DEFAULT**) and the datum planes (**FRONT** to **DTM3**, **TOP** to **DTM2**, and **RIGHT** to **DTM1**)]

You have now created your first sheet metal part. You will use this part later to practice creating other sheet metal features.

Using the Extrude Option to Create the First Wall Feature

The next first wall will be created with the **Extrude** option. This is very similar to creating a **Solid** feature with the **Thin** option. Use the same setup and environment as with the first part, with the exception of using metric units (mm). *Don't forget to rename your datums and coordinate system.*

File ⇒ **New** ⇒ **Part** ⇒ Sub-type ● **Sheetmetal** ⇒ (type the part name **SM_EXTRUDE_2**) ⇒ ❑ **Use default template** ⇒ **OK** ⇒ **mmns_part_sheetmetal** ⇒ (fill in the Parameters) ⇒ **OK**

The sketch will create a simple **Wall** feature that conforms to the cross section. To create this feature, the Pro/E will prompt you for the *"thicken" direction,* the *thickness value*, and the *depth dimension*.

Feature ⇒ **Create** ⇒ **Wall** ⇒ **Extruded** ⇒ **Done** ⇒ **Both Sides** ⇒ **Done** ⇒ (pick **DTM3** as the *sketching* plane) ⇒ **Okay** ⇒ **Top** ⇒ (pick **DTM2** as *orientation* plane) ⇒ **Sketch** ⇒ **✔Intent Manager** ⇒ References (dialog box) Reference Status--Fully Placed (accept **DTM1** and **DTM2** as the default references) ⇒ **Close** ⇒ (sketch, dimension, and modify the sketch (Fig. 23.59) ⇒ **Sketch** ⇒ **Done** ⇒ (The *direction arrow* appears, indicating the side of the section on which the material thickness will be added. Make sure the arrow points toward the *inside* of the section.) ⇒ **Flip** (if necessary) ⇒ **Okay** (Fig. 23.60) ⇒ (enter thickness **10**) ⇒ **✔** ⇒ **Blind** ⇒ **Done** Enter Depth **130** ⇒ **✔** (Fig. 23.61)

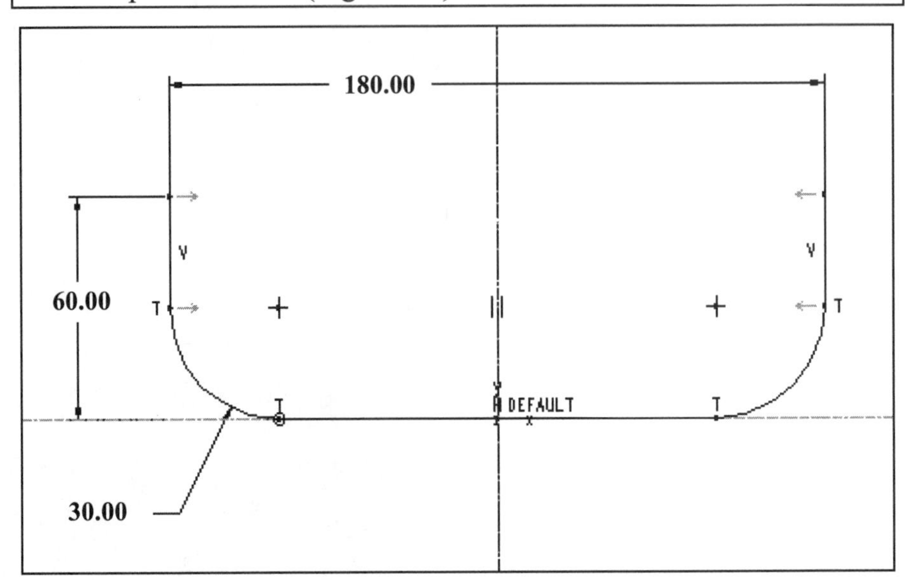

Figure 23.59
Sketch for Extruded Wall

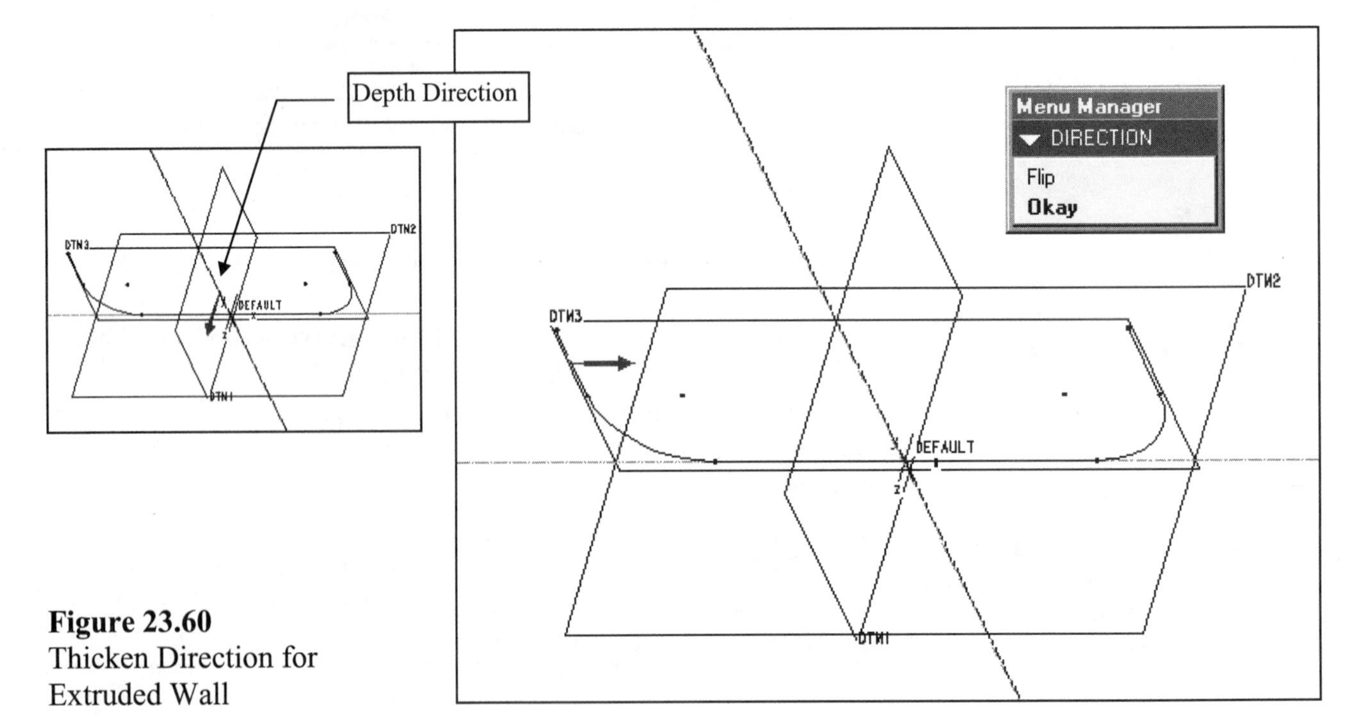

Figure 23.60
Thicken Direction for Extruded Wall

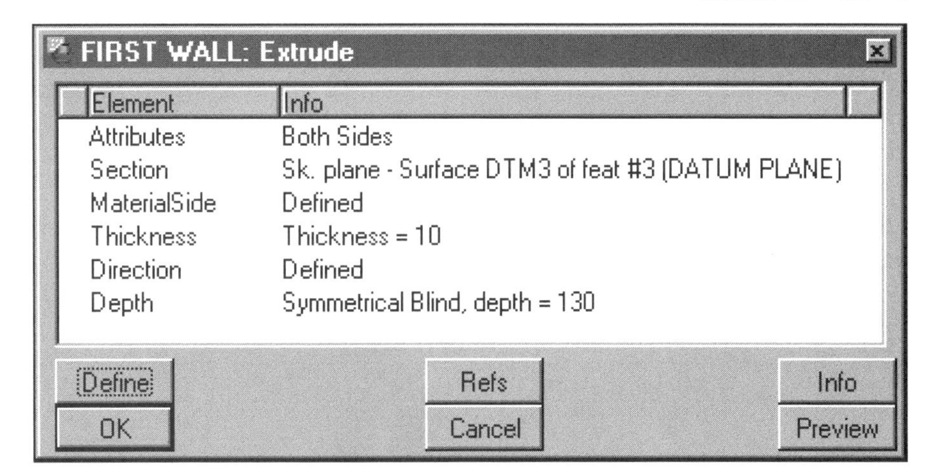

Figure 23.61
Dialog Box
FIRST WALL: Extrude

OK ⇒ File ⇒ Save ⇒ ✔ (Fig. 23.62)

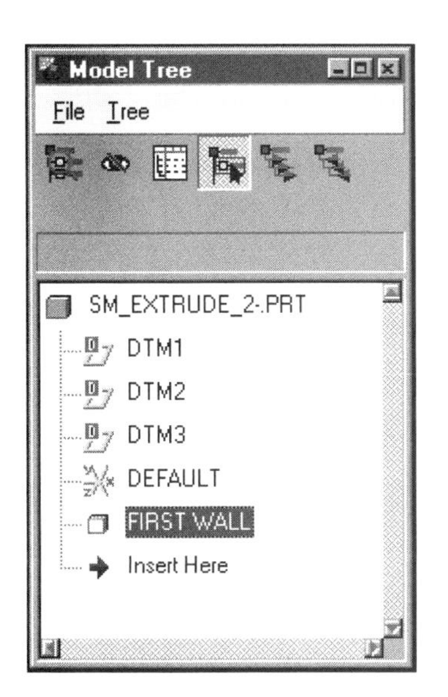

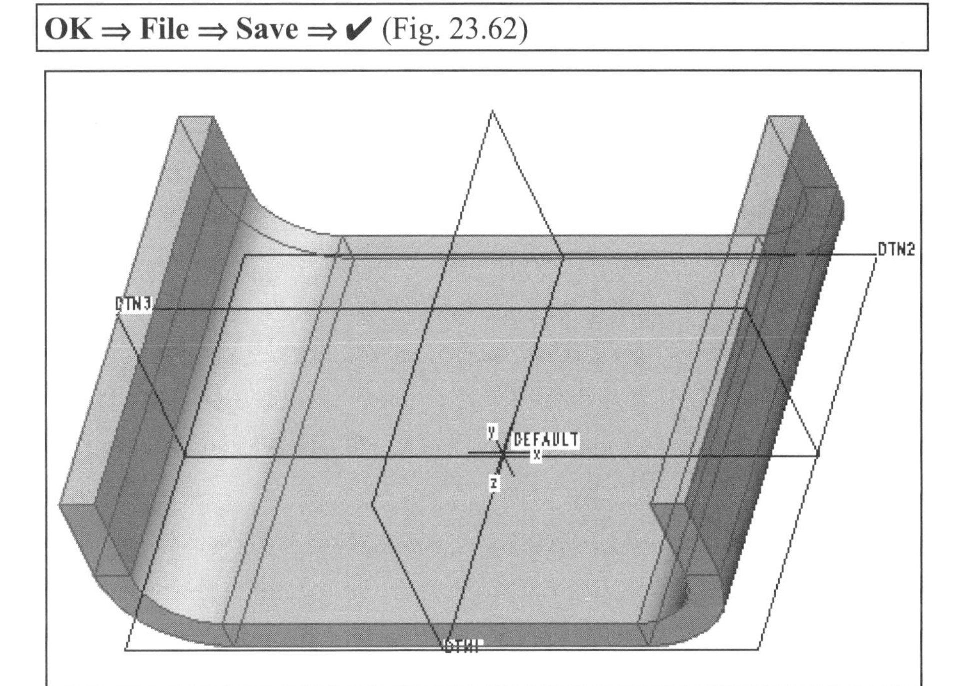

Figure 23.62
Extruded First Wall

Driving Side

Observe the relationship between the sketch and the resultant Sheet Metal Wall feature. Currently, the *green* surface is on the outside of the part. To observe the relationship between the feature and the sketch that created it, simply **Modify** the feature. **Modify** ⇒ (select the wall feature). Notice the sketch corresponds to the *green* surface--this is because the "*thicken*" direction was to the inside (Fig. 23.63). The sketch creates the *green* (driving) surface for sheet metal parts and offsets the *white* (driven) surface from it.

Redefine the "thicken" direction and observe the relationship between the sketch and the wall feature.

Feature ⇒ **Redefine** (select the wall feature) ⇒ **MaterialSide** ⇒ **Define** ⇒ **Flip** [so the arrow is pointed outward (Fig. 23.64)] ⇒ **Okay** ⇒ **OK** [button in the Feature dialog box (Fig. 23.65)]

The feature is redefined. Now the *green* surface is on the inside, and the *white* surface is on the outside.

Measure the distance between the outsides of the two "ears." It will measure **200** not **180** because the sketch is now on the inside and the wall was created with the material side pointing outward, as shown in Figure 23.66. ***Do not save the part in this condition.*** The part was saved previously; just **(File ⇒ Erase ⇒ Current ⇒ Yes).**

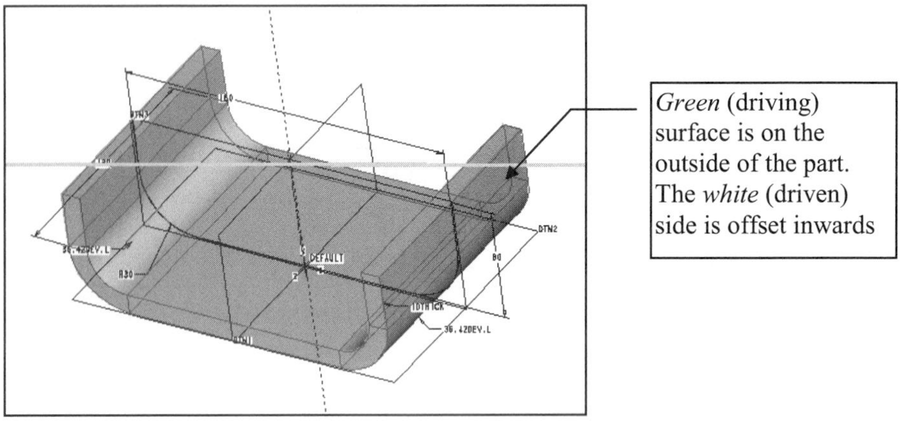

Green (driving) surface is on the outside of the part. The *white* (driven) side is offset inwards

Figure 23.63
The *Green* Driving Surface Is on the Outside of the Part. The Distance Between the Outsides of the Ears Is **180.**

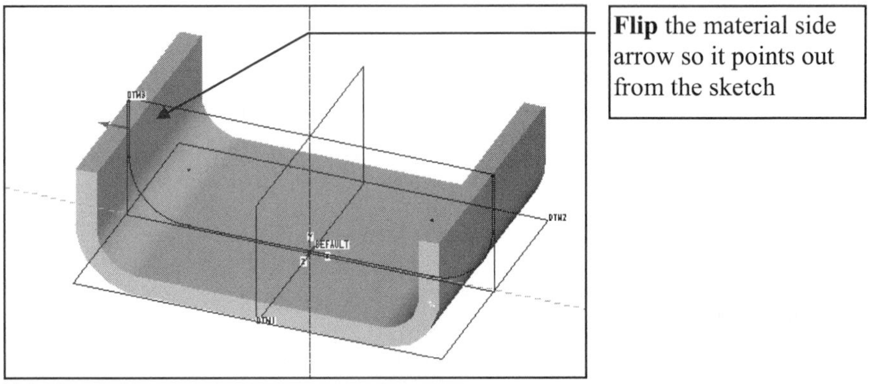

Flip the material side arrow so it points out from the sketch

Figure 23.64
Flip the Material Side So That the Material Is Added to the Outside

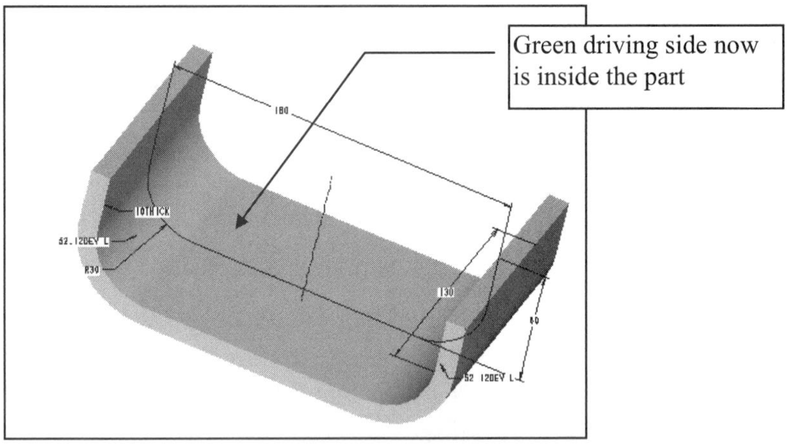

Green driving side now is inside the part

Figure 23.65
The Sketch Is Now on the Inside (as Well as the Green Driving Side)

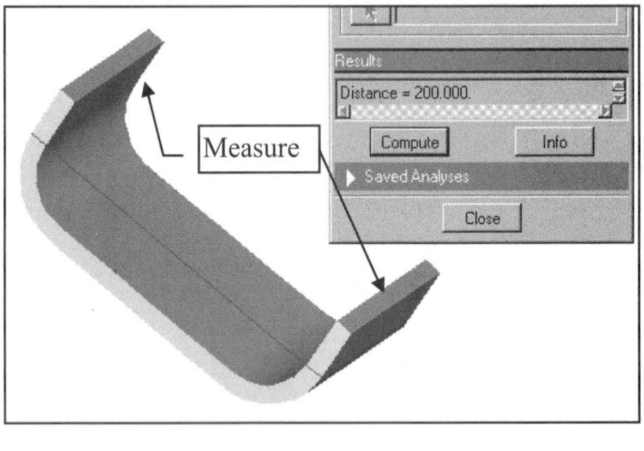

Measure

Figure 23.66
Measure the Distance Between the Outsides of the Two "Ears" (**200**)

Part Conversion

To convert a part that was created in "solid" mode into a sheet metal part, you simply retrieve it in sheet metal mode. As a part of the conversion process, Pro/E turns the part geometry into a constant-thickness configuration, based on user input.

There are two options available in Pro/SHEETMETAL to accomplish this **Driving Srf** and **Shell**. The **Shell** option behaves exactly as it does in solid mode--the outside of the geometry is offset inwards, and the part becomes hollowed out. The outside of the part becomes the *green* surface. The **Driving Srf** option assumes that the part is already constant-thickness; you are required to specify a planar surface, which then defines the *green* surface.

Create the simple solid Pro/E part **(SM_CONV_SOLID_3)** (not a sheet metal part) shown in Figure 23.67. **Save** and then **Erase** the part from memory. You will use this part for this practice session and for later sessions in this lesson.

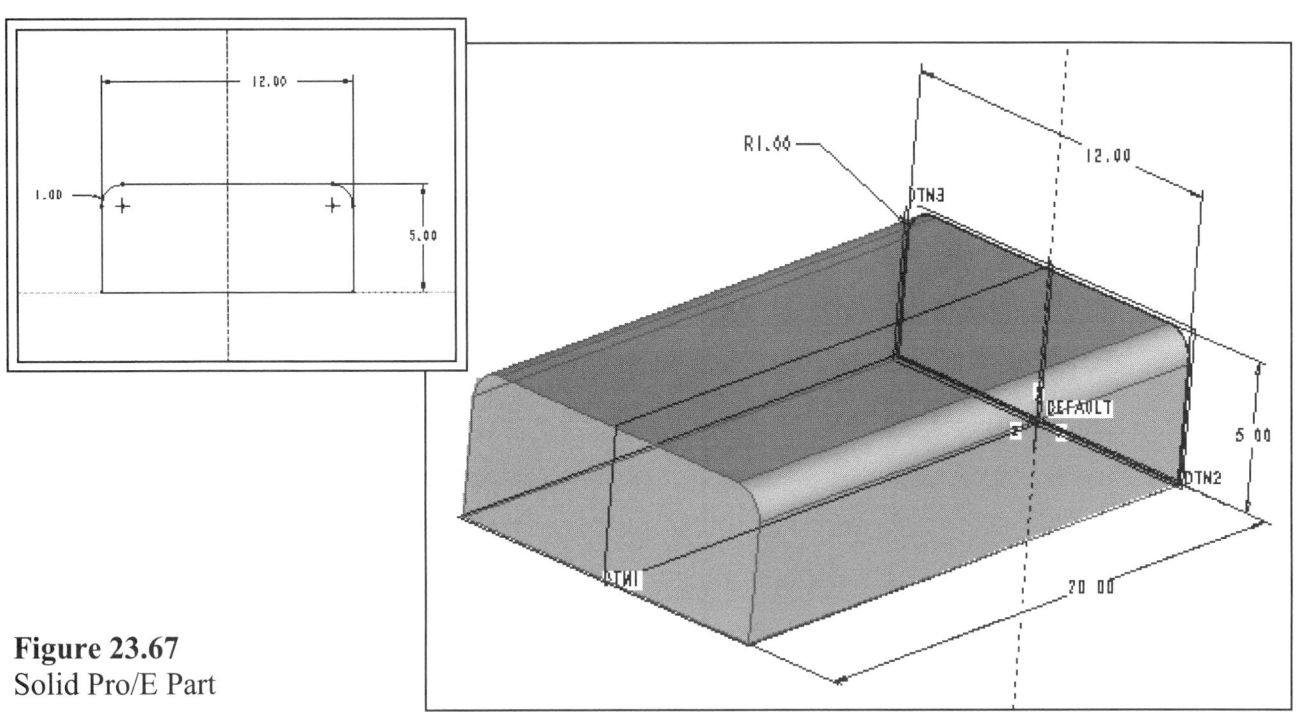

Figure 23.67
Solid Pro/E Part

You will be converting the existing Solid part into the Sheet Metal part, using the **Shell** option. This is a part you created using Solid features, not Sheet Metal features.

File ⇒ **Open** (select the solid part you just created) ⇒ **Applications** (from the Menu Bar) ⇒ ●**Sheetmetal** ⇒ **Shell** ⇒ [select the three surfaces to remove as shown in (Fig. 23.68)] ⇒ **Done Sel** ⇒ **Done Refs** ⇒ Enter Thickness **.125** ⇒ ✔ (Fig. 23.69) ⇒ **File** ⇒ **Save**

The part is shelled out and converted to a sheet metal part. Notice that the *green* surface is on the outside of the part.

You can also convert a thin-walled solid into a sheet metal part by defining a driving surface. Pro/E then converts the data base of the part and saves the driving-surface and driven-surface information.

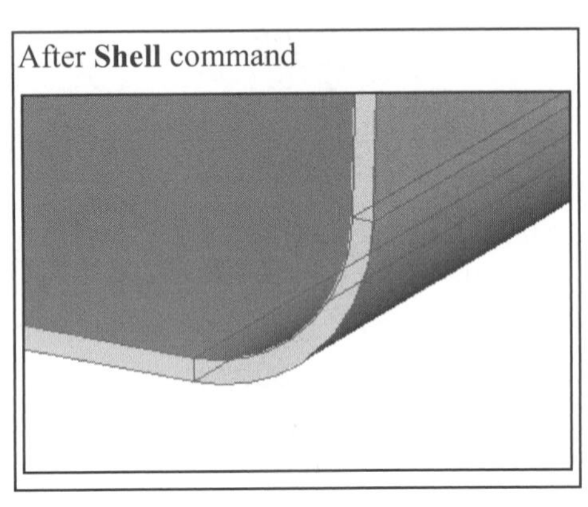

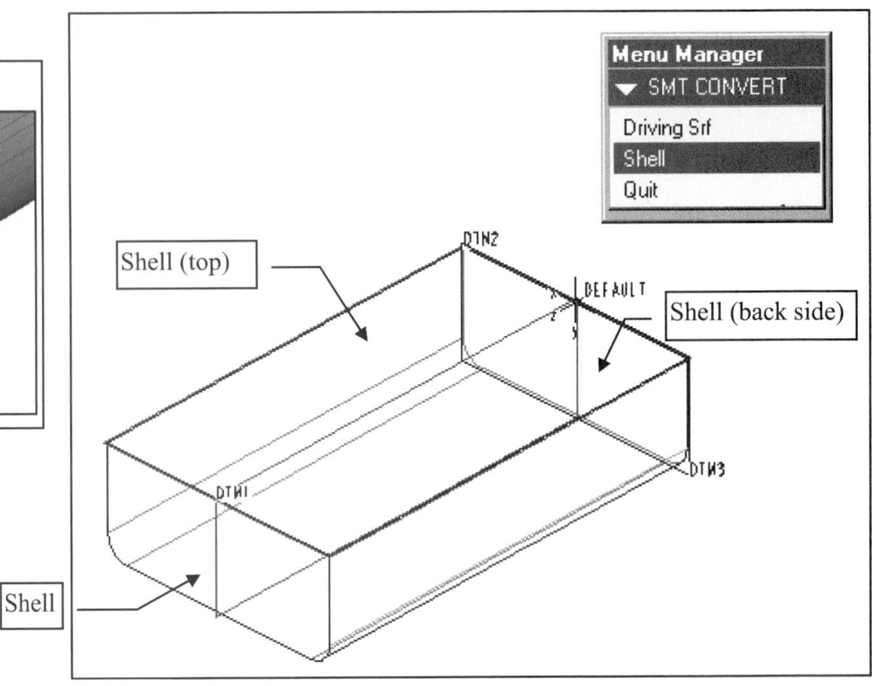

Figure 23.68
Converting the Solid Part to a Sheet Metal Part Using **Shell**

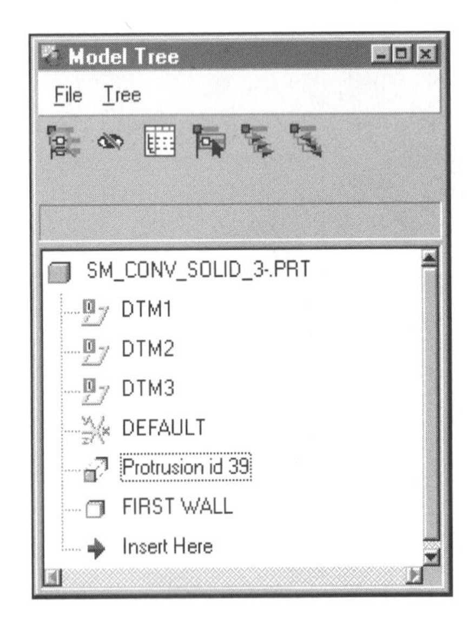

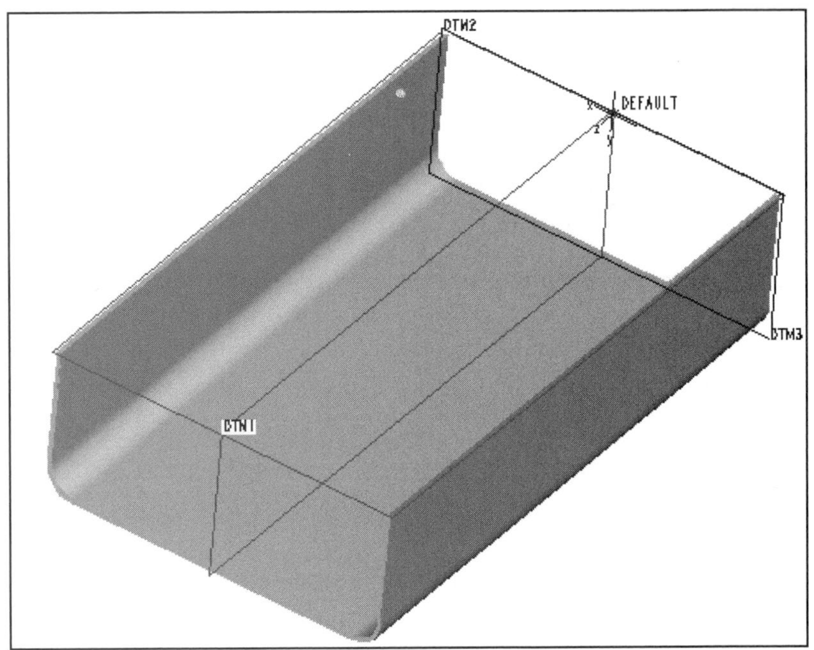

Figure 23.69
Converted Part

You must begin with a constant-wall-thickness part to use this option. The most common example of a solid with a constant thickness is one that was created with the **Thin** protrusion type or a solid which contains a **Shell** feature.

When you choose a face that lies on one "side" of the constant-thickness wall, Pro/E selects all adjacent faces and converts these surfaces to the driving surface. The surfaces that lie on the other side of the constant thickness from the driving surfaces become the driven surface. Obviously, there are some limitations to this technique. The part must truly be of a constant thickness and there must not be other features protruding from either the driving or the driven surface. Try converting a simple solid part (all one thickness) into a sheet metal part.

Extruded Walls

The **Extruded** option allows you to create a **Wall** feature by extruding a cross section sketch along a linear edge of a sheet metal part. The edge along which the cross-section of the wall is extruded is known as the **Attach Edge.** The sketch should be constructed as if it were for a **Thin Solid** feature (in Part Mode)--the cross section is automatically thickened by Pro/SHEETMETAL during feature creation.

There are six pieces of information that must be defined in order to create an **Extruded Wall** feature: *Attach Edge, Section, Radius Type, Radius Value, Bend Relief*, and *Bend Table.*

The first item necessary for any **Extruded Wall** is the **Attach Edge**. The Attach Edge may be any linear edge of the part. Next, you must choose either the **Use Radius** or the **No Radius** option for the **Wall**. Choosing the Use Radius option causes Pro/ENGINEER to place a **Bend** into the Wall feature automatically (Fig. 23.70). The **No Radius** option is used for specialized circumstances.

Figure 23.70
Extruded Walls:
Use Radius or **No Radius**, **Inside Radius**, **Outside Radius**

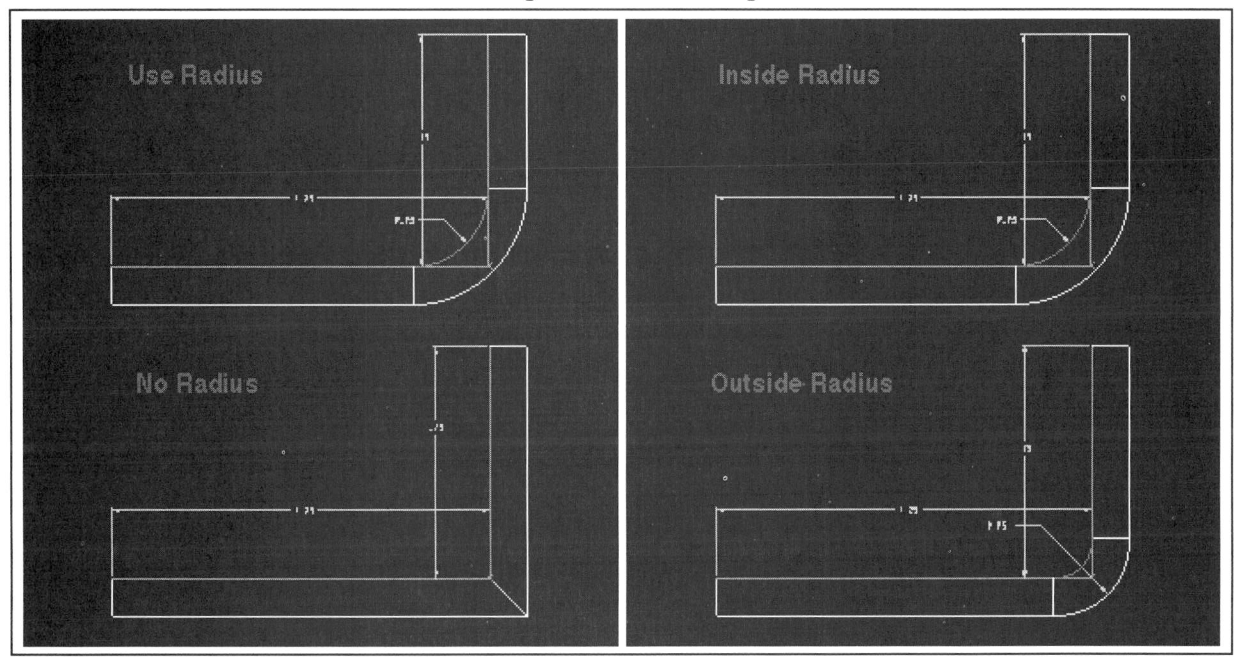

If you choose the **Use Radius** option, you must then define several attributes that relate to the **Bend** geometry. The **Radius Type** setting controls whether you want to specify the **Outer** or the **Inner** bend radius (Fig. 23.70). This setting will depend upon the particular design and manufacturing constraints for the Wall feature.

The **Radius Value** setting may be set to **Enter Value**, which allows you to enter a Radius value, or to **Material Thickness**, which automatically sets the Radius value equal to the part's material thickness parameter. The final two pieces of information necessary for every Extruded Wall feature are the **Bend Table** and the **Bend Relief** specification. Bend Tables are used to define material deformation when creating Bend features. Bend Relief must be defined whenever a Wall feature cannot be added to the part without deforming or ripping the existing material.

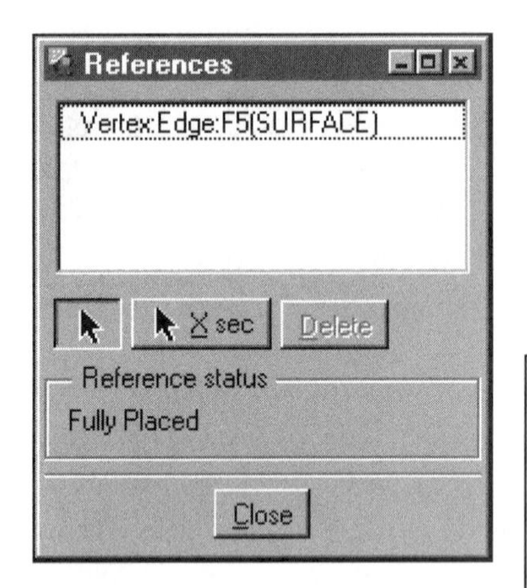

Create the **Extruded Wall** feature on both sides of the part. Explore the several ways available to specify the Bend radius. For practice, define an outer bend radius when you create a similar wall on the part's opposite side. Retrieve the sheet metal part **SM_FLAT_1.prt.**

The Extruded Wall is **1.50** inches high, with an inside Bend radius equal to the material thickness. Give the following commands:

Feature ⇒ **Create** ⇒ **Wall** ⇒ **Extruded** ⇒ **Use Radius** ⇒ **Done** ⇒ **Part Bend Tbl** ⇒ **Done/Return** ⇒ **Inside Rad** ⇒ **Done/Return** ⇒ **One Side** ⇒ **Done** ⇒ (select the **Attach Edge**) ⇒ **Default** (SETUP SK PLN menu) ⇒ **Okay** ⇒ **Close** (References dialog) ⇒ (create the sketch)

The **Attach Edge** must be a linear part edge. Select the Attach Edge as shown in Figure 23.71. Pro/E will display two Datum Planes (**DTM4** and **DTM5**) and the *red Direction Arrow* (Fig. 23.72).

Two internal datum planes are always created when any Extruded Wall feature is defined. These two datum planes represent the *Sketch plane* and the *Top reference plane* for the sketch.

Figure 23.71
Attach Edge

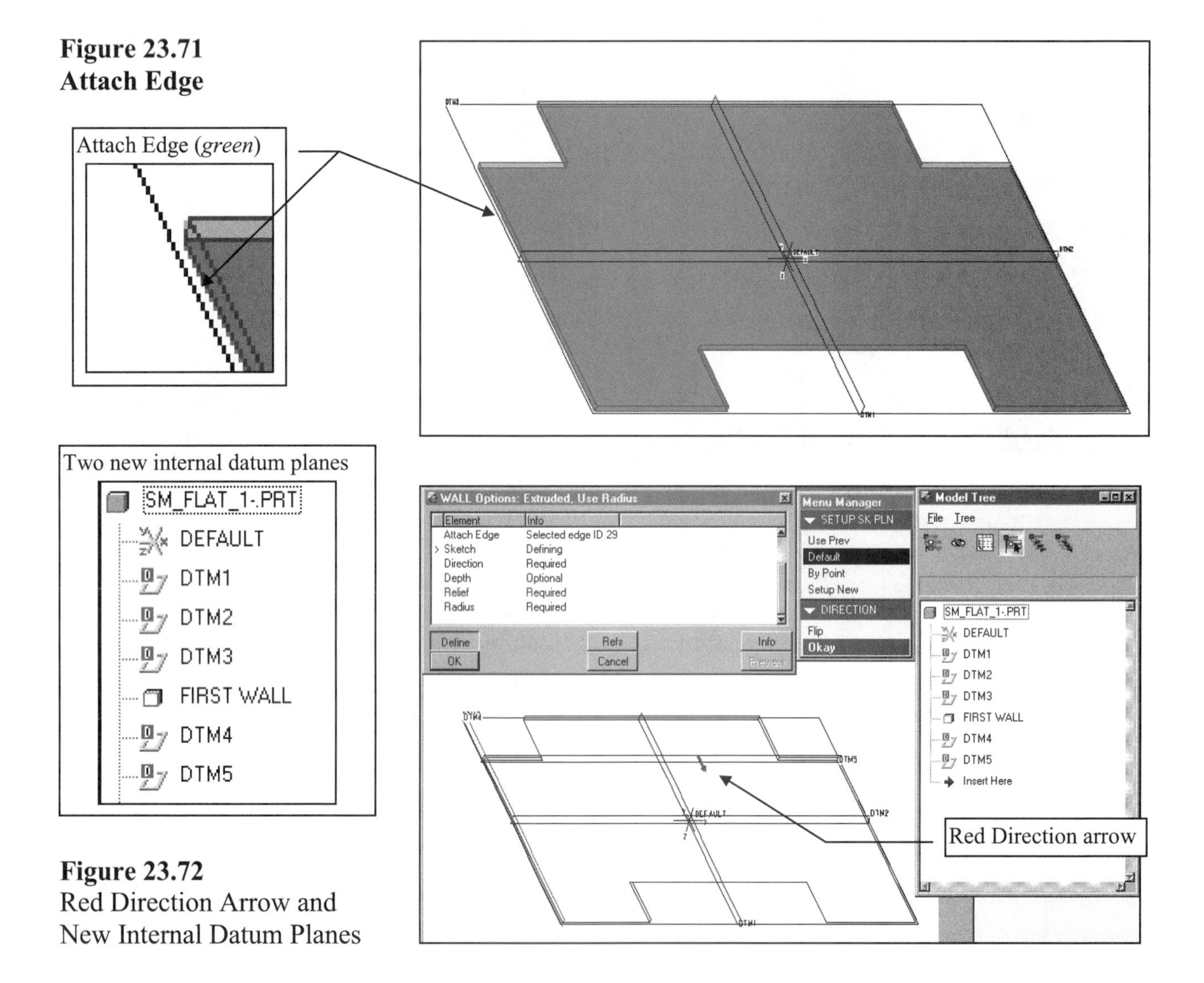

Figure 23.72
Red Direction Arrow and
New Internal Datum Planes

Notice the sketch alignment "magnet" point at the top of the part (Fig. 23.73). Any sketch entities which are located near to this point will be automatically aligned to it; ***you do not have to align the entities manually.*** Create and modify the sketch (Fig. 23.74).

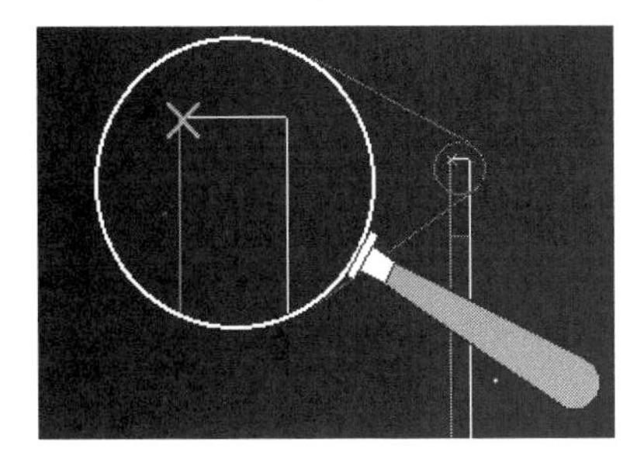

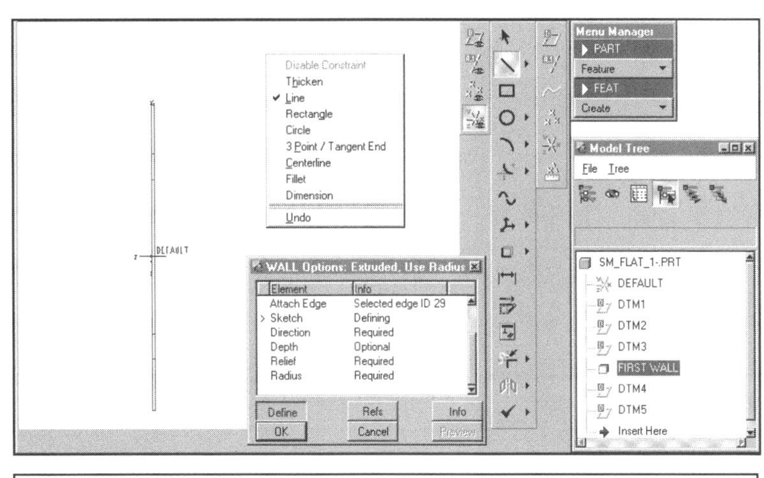

Figure 23.73
Sketch the Section

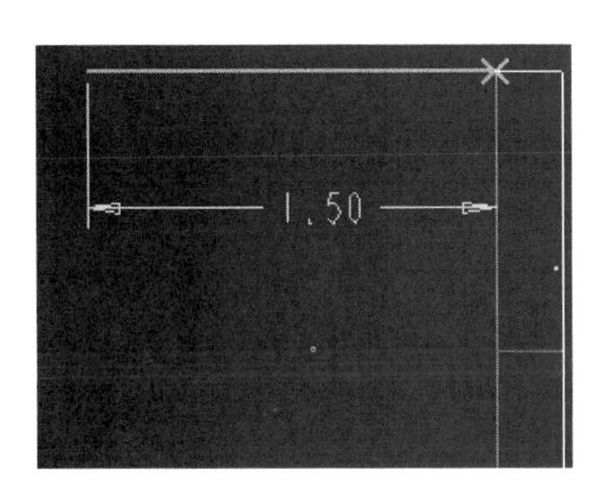

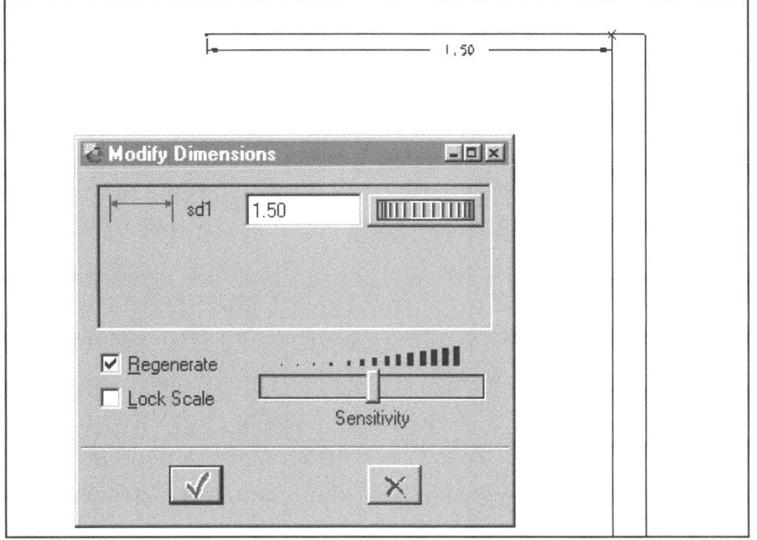

Figure 23.74
Sketch

Sketch ⇒ **Done** ⇒ **No Relief** ⇒ **Done** ⇒ (Pro/E prompts you to specify the bend radius. You may either specify a value or use the material thickness) **Thickness** ⇒ **Preview** (Fig. 23.75) ⇒ **OK**

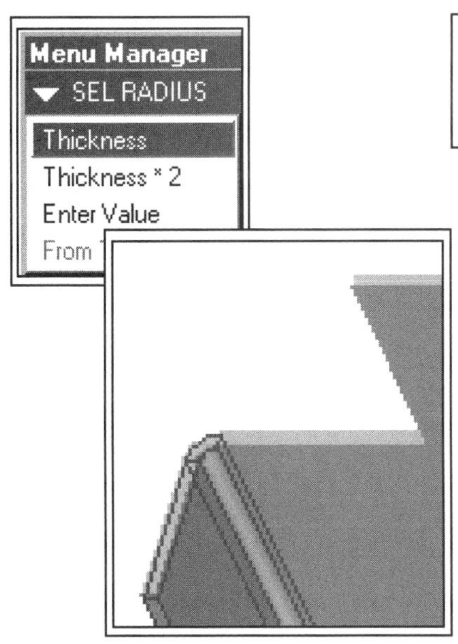

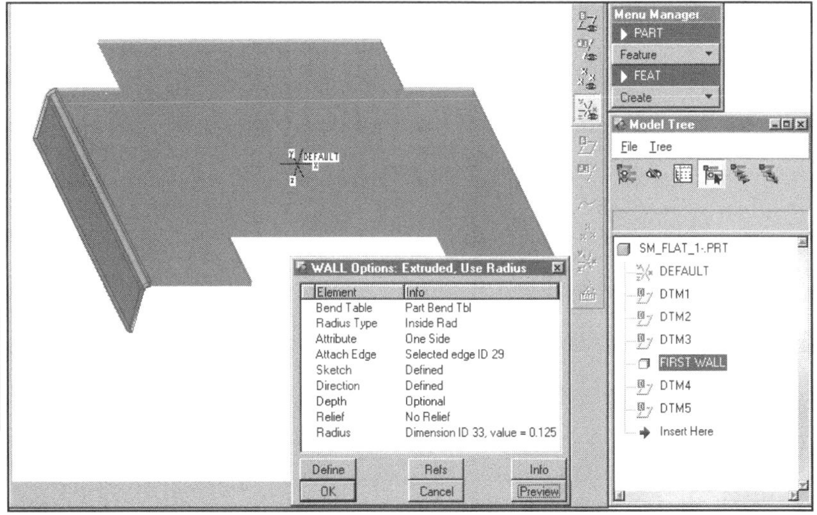

Figure 23.75
Preview Extruded Wall

Create an Extruded Wall on the opposite of the part as shown in Figure 23.76. Use **Outside Rad**, a Wall length of **2.00,** and a radius value of **.50.** **Save** the part.

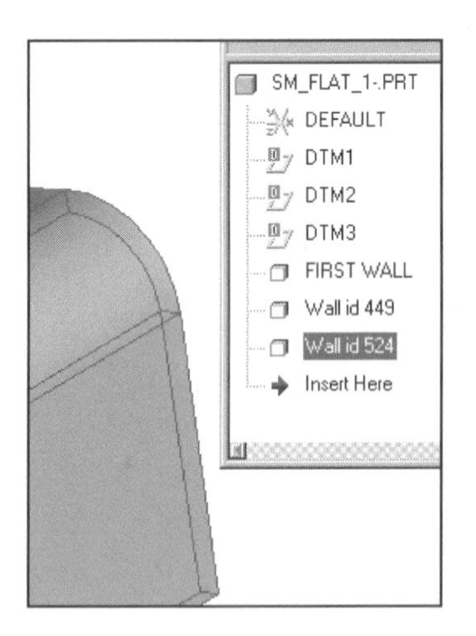

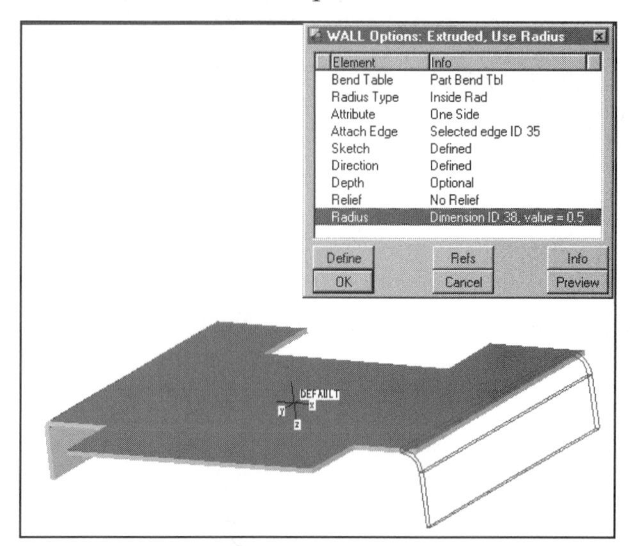

Figure 23.76
Extended Wall on Opposite Side of Part

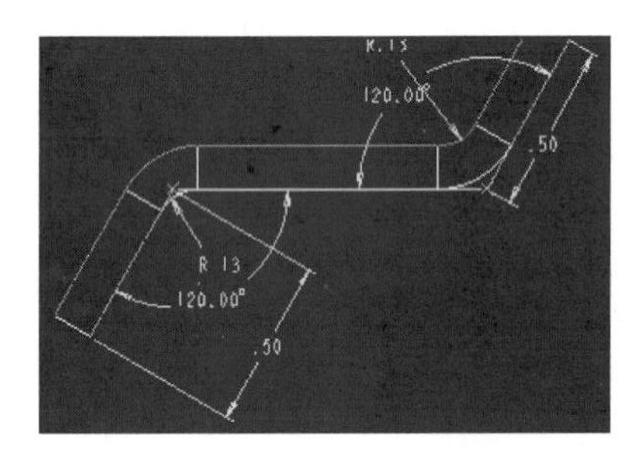

Angled wall features can be created using the Extruded option by sketching the cross section with an angular dimension.

It is good design practice to create even **90° Extruded Wall** features with an angle dimension unless you are certain that they will always remain perpendicular. By following this practice, you will build additional flexibility into your models, because you can then easily modify the angle dimension to whatever value is necessary.

To create a right-angle bend, with the flexibility to be changed to another angle, you sketch the line that represents the wall at an angle that is clearly not **90°.** If you sketch it near **90°,** the system will assume that it is so, and assign a constraint that you would need to disable. Add an angle dimension and then change the angle dimension value to **90°,** thus allowing it to be easily modified.

Using the same part, create an angled wall. Use the same commands as for the first Extruded wall you created, but select a different **Attach Edge** and draw the wall line at an angle. **Use Radius**, **Inside Rad**, with dimensions of **2.50** long and at an angle of **45°** (Fig. 23.77). Use **.50** for the radius.

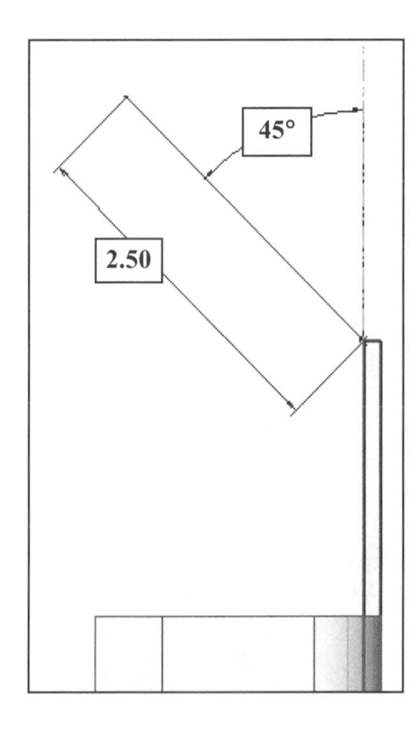

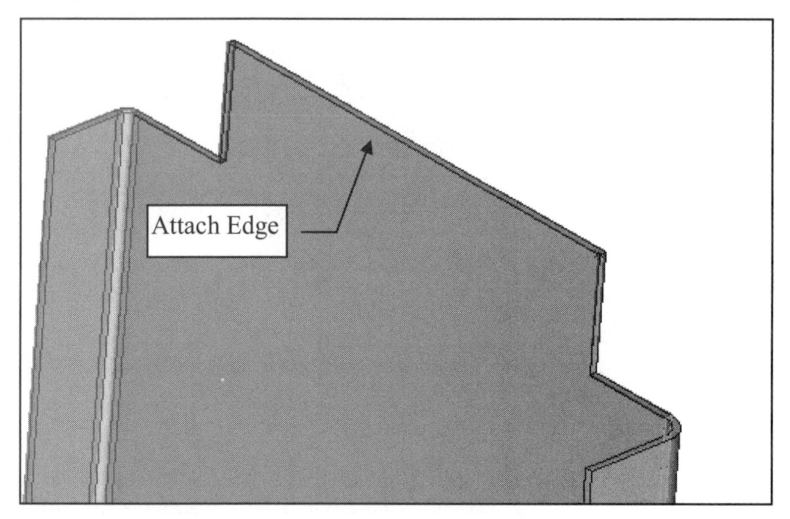

Figure 23.77
Attach Edge

Notice that the Wall's length dimension refers to the tangent point on the *green* surface of the bend, the **Attach Edge** (Fig. 23.78).

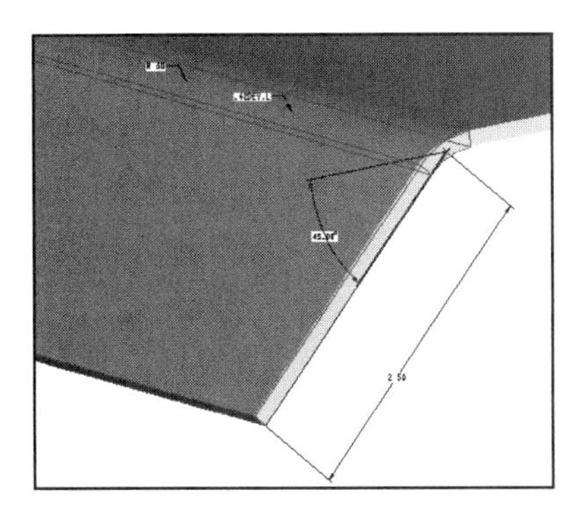

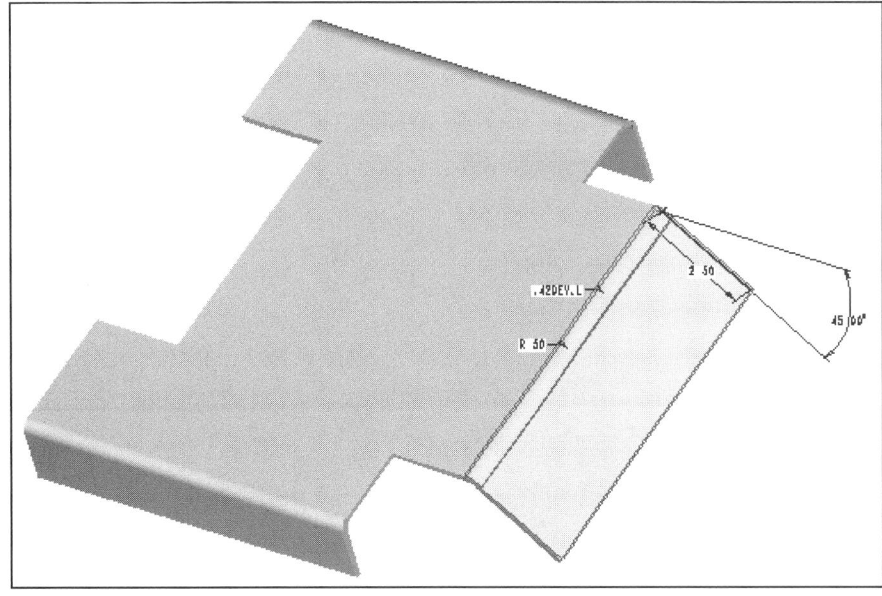

Figure 23.78
Angled Wall

Modify the angle to **135°** (Fig. 23.79). **Save** and **Erase** the part.

File ⇒ Save ⇒ ✔

Figure 23.79
Modified Angled Wall

When creating **Extruded Wall** features, it is important to select the correct **Attach Edge** in order to preserve the design intent of the part. The edge that is selected as the Attach Edge for an Extruded Wall controls the feature's dimensioning scheme. In the example [Fig. 23.80 (left)], the top edge of the part (the *green* surface) is selected to create a Wall that extends above the part. Notice the resultant dimensioning scheme--the wall grows out away from the inside edge of the part, and the **1.25** dimension is maintained on the "inside" face of the part.

A different result is obtained if the Wall feature is created in an identical fashion except that the Attach Edge is at the bottom [*white* surface (Fig. 23.80 right)].

In this case, the wall grows in towards the part, and the resultant dimensioning scheme has the **1.25** dimension maintained to the outside face of the part. As you may guess, selecting the proper Attach Edge for a Wall feature will help you to maintain the design intent for the part. Try redefining some of the Extruded Walls you have created. Select different Attach Edges and see how the part and its dimensions change.

Figure 23.80
Attach Edge Selection

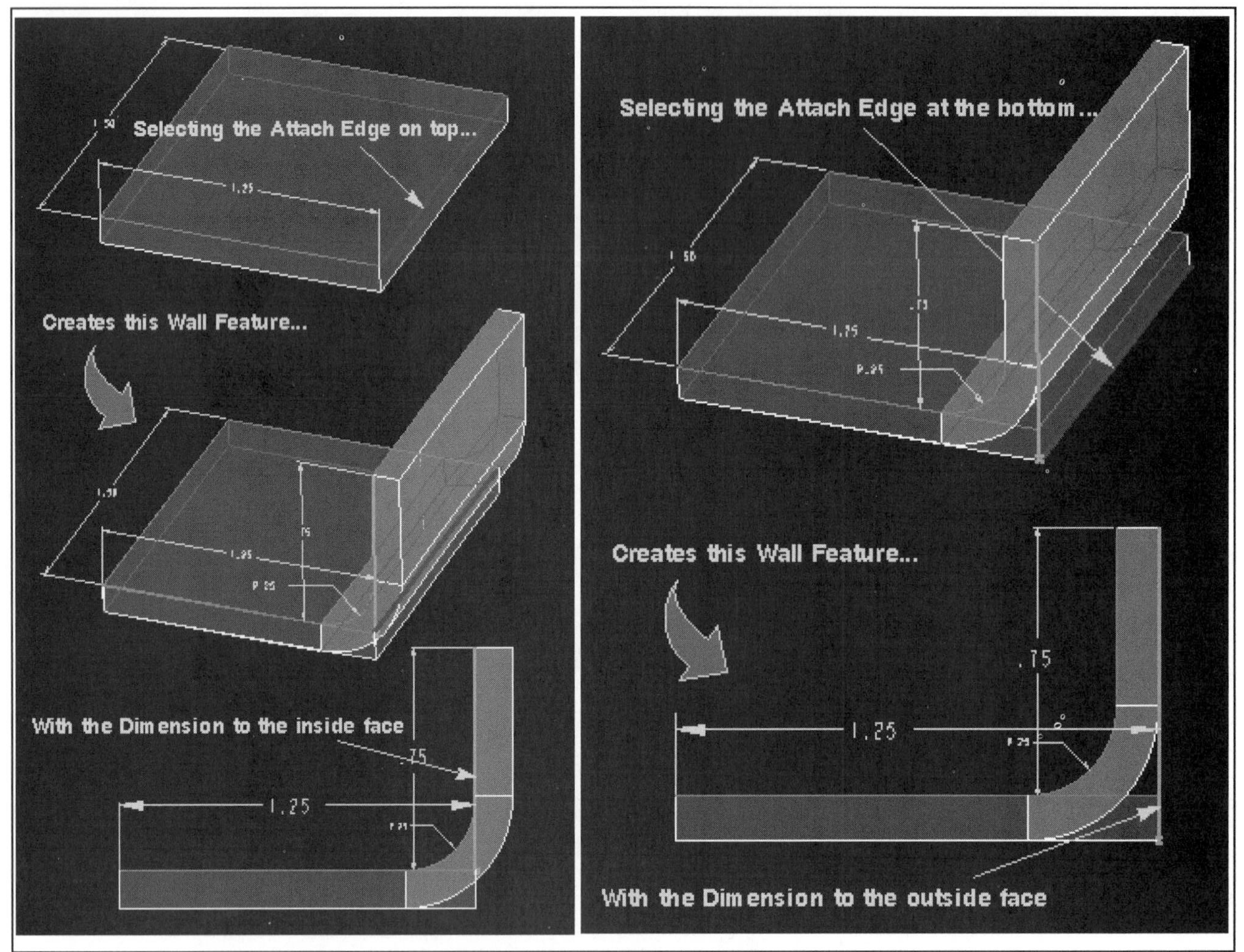

Sketching Flat Wall Outlines

Using the **Flat** option, you can create sheet metal features by sketching the outline of a Wall, rather than sketching its cross section. This is useful when you must create Walls that have a complicated profile. The information required to define a **Flat Wall** feature is the same as for Extruded Walls, with the addition of an **Angle** dimension that is not part of the sketched section. The **Attach Edge** must be specified correctly in order to obtain the desired dimensioning scheme for Flat Walls.

As with Extruded Wall features, you must define the **Attach Edge, Section, Radius Type, Radius Value, Bend Relief,** and **Bend Table** for the feature. For **Flat Wall** features you must also supply a **Bend Angle.**

Create a Wall feature using the **Flat** option. Retrieve the second part you created for this lesson **(SM_EXTRUDE_2.prt).**

Feature ⇒ Create ⇒ Wall ⇒ Flat ⇒ Use Radius ⇒ Done ⇒ Part Bend Tbl ⇒ Done/Return ⇒ Inside Rad ⇒ Done/Return ⇒ [select the **Attach Edge** (Fig. 23.81)] **⇒ 30.00 ⇒ Done ⇒ Okay**

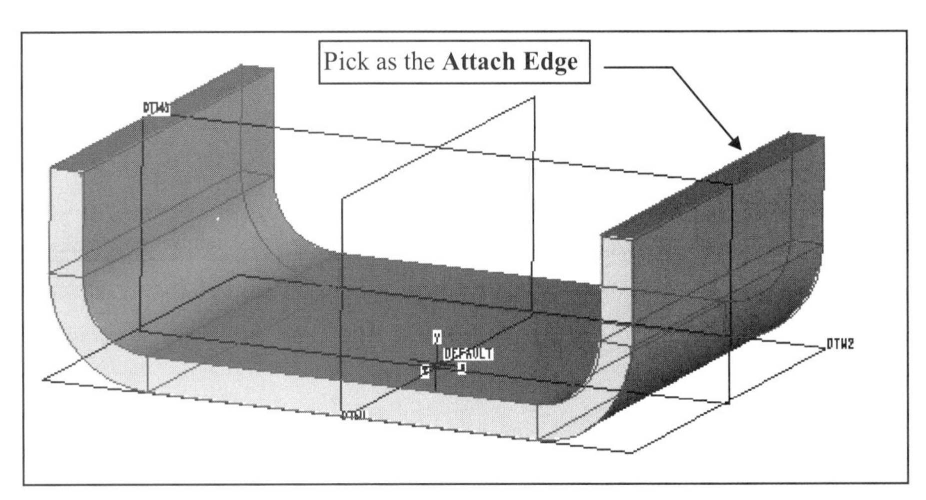

Figure 23.81
Attach Edge

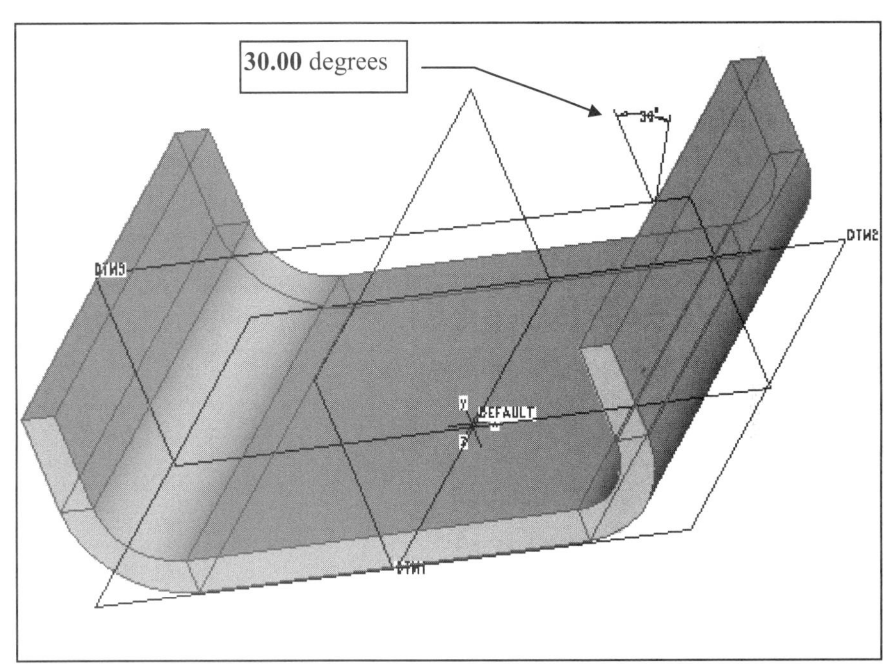

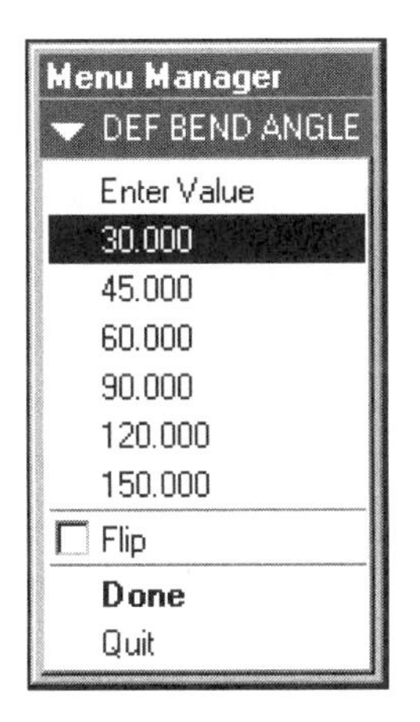

Figure 23.82
Wall Angle

To define a **Flat Wall**, you must also specify the bend angle, **30°** (Fig. 23.82). The **Flip** option enables you to reverse the specified angle *(do not flip the angle).* The system displays the sketch plane (**DTM4**) and the TOP horizontal reference plane (**DTM5**) as shown in Figure 23.83.

By default, the system will flip into the sketch view (Fig. 23.84). *Notice that there are two sketch alignment points displayed, one for each end of the **Attach Edge**.*

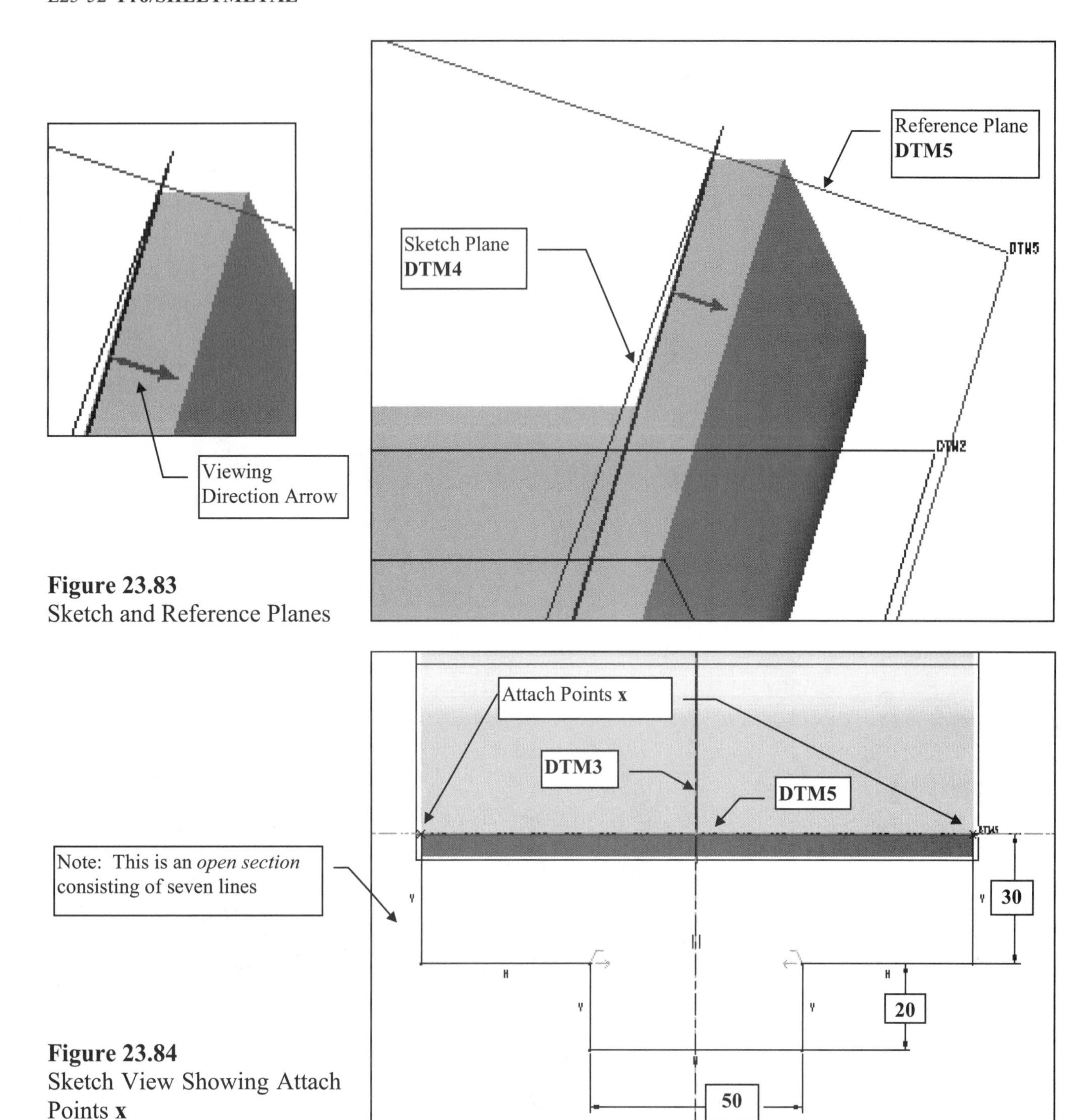

Figure 23.83
Sketch and Reference Planes

Figure 23.84
Sketch View Showing Attach Points **x**

Close (to accept the references) ⇒ (sketch and modify as shown in Figure 23.84) ⇒ **Sketch** ⇒ **Done** ⇒ **No Relief** ⇒ **Done** ⇒ **Thickness** ⇒ **OK** (Fig. 23.85) ⇒ **File** ⇒ **Save** ⇒ ✔ ⇒ **File** ⇒ **Delete** ⇒ **Old Versions** ⇒ ✔ ⇒ **File** ⇒ **Erase** ⇒ **Current** ⇒ **Yes**

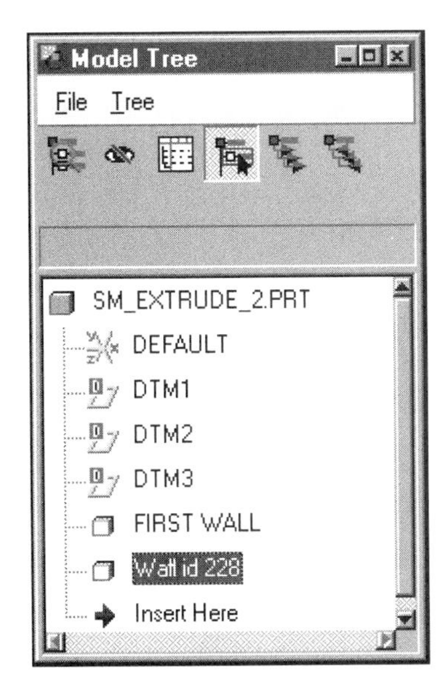

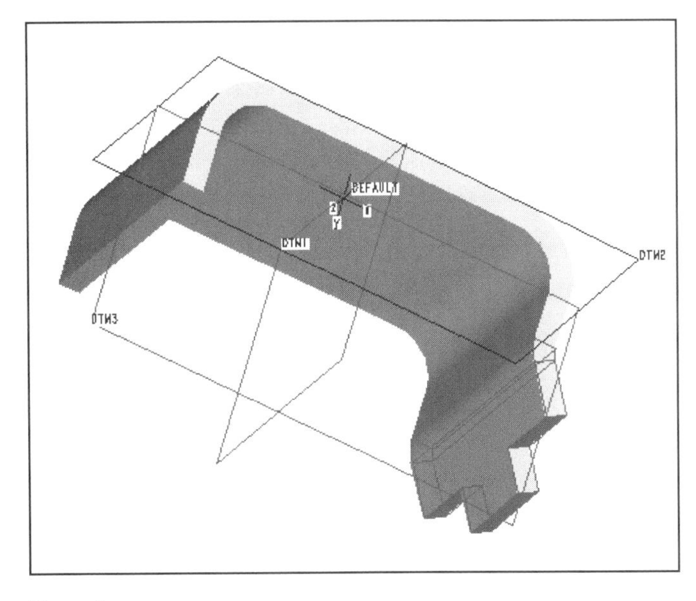

Figure 23.85
Sketched Flat Wall

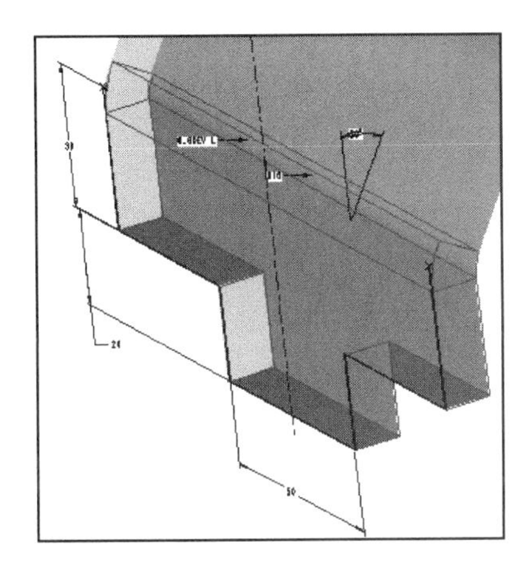

Bends

You have already seen that Pro/SHEETMETAL automatically creates a **Bend** feature when you select the **Use Radius** option while creating **Wall** features. You can also manually create a **Bend** feature in a sheet metal part, as shown in Figure 23.86.

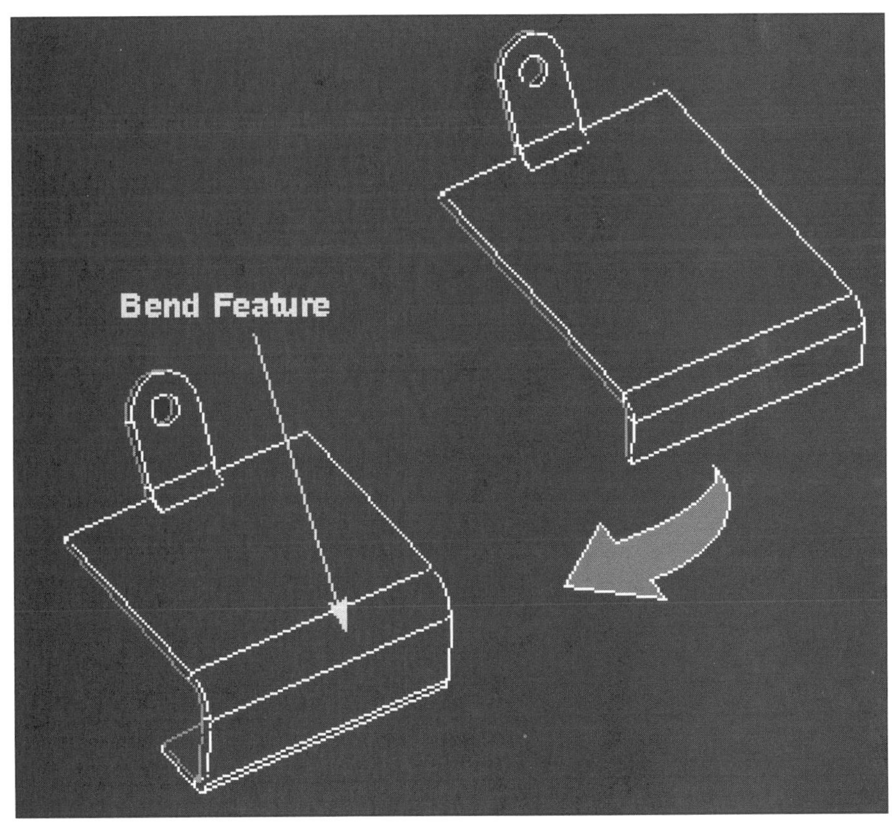

Figure 23.86
Bend Feature

The **Bend** feature is also useful whenever you wish to design in **Flat Pattern mode**--that is, beginning the design with a flattened outline shape and developing the final bent-up part from the flat pattern. To create a Bend feature in Pro/SHEETMETAL, choose **Bend** from the SHEET METAL menu.

To define any **Bend** feature in Pro/SHEETMETAL, you must sketch a **Bend Line**, supply all of the required Bend parameters (**Radius**, **Angle**, **Bend Type**), and specify the **Bend Relief.** A new concept when creating Bend features is sketching the Bend Line. The Bend Line defines the axis about which the Bend feature is created. To be valid, it must be a single-line entity, and must also be aligned to two outside edges (Fig. 23.87).

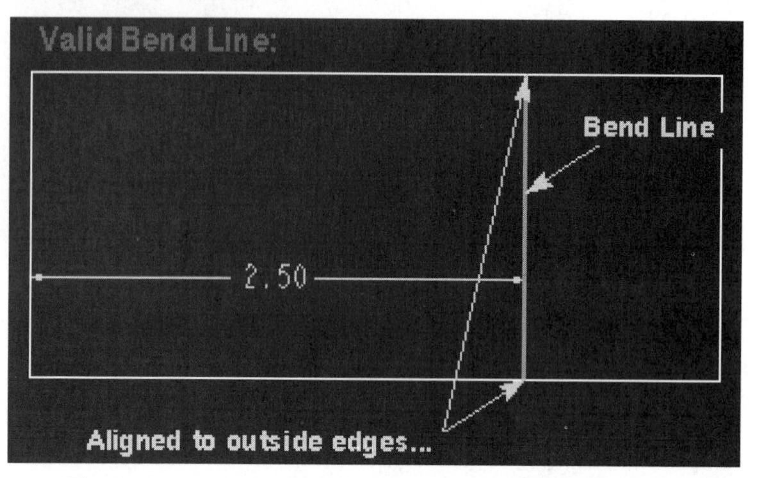

Figure 23.87
Bend Line

There are two basic types of Bend features available in Pro/SHEETMETAL: **Angle** and **Roll.** The most common type is Angle; a typical **Angle Bend** feature is shown in Figure 23.88.

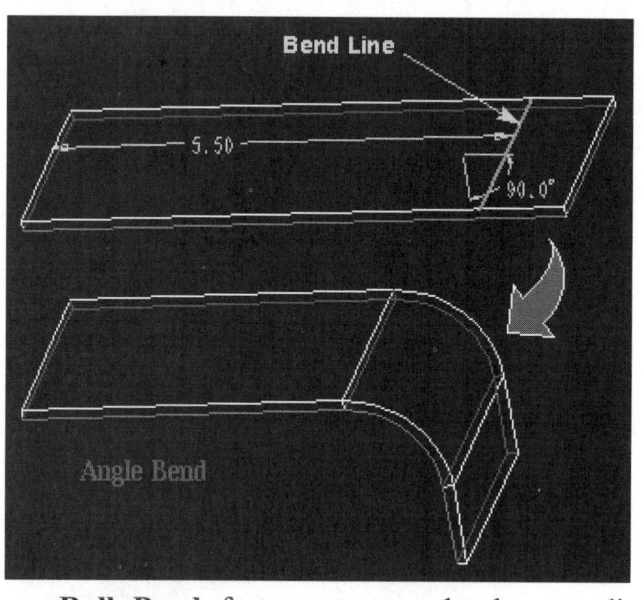

Figure 23.88
Angle Bend

Roll Bend features are used where a dimensioning scheme similar to the one shown in the left column is required.

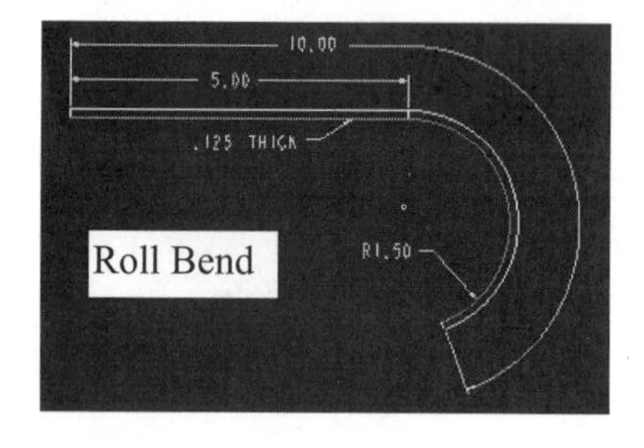

Notice that the **Bend Radius** and the distance to the start point of the bend are specified. The Roll Bend feature begins tangent and runs out along for the entire remaining length of the part.

After sketching a Bend Line, you must define on which side of the line the Bend will be created. Using the *red* **Direction Arrow**, you specify both a **Fixed Side** and a **Bend Side** for all Bend features. The Fixed Side should be the portion of the part you want to remain stationary during the Bend operation. The Bend Side determines on which side of the Bend Line the feature will be created (Fig. 23.89).

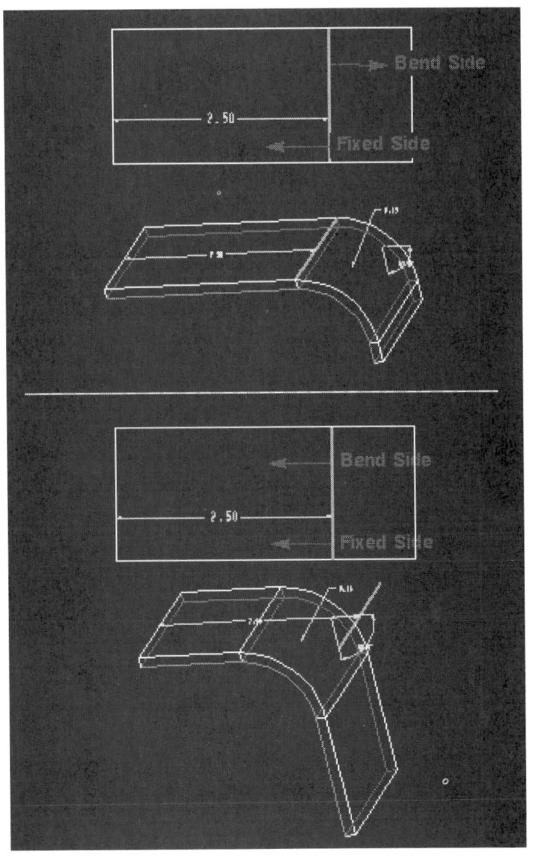

Figure 23.89
Bend Side

The examples shown in Figure 23.89 illustrate the behavior of changing the **Bend Side** of a feature while keeping the **Fixed Side** the same.

You may create *angled* Bend features by simply including an angular dimension in the Bend Line sketch (Fig. 23.90).

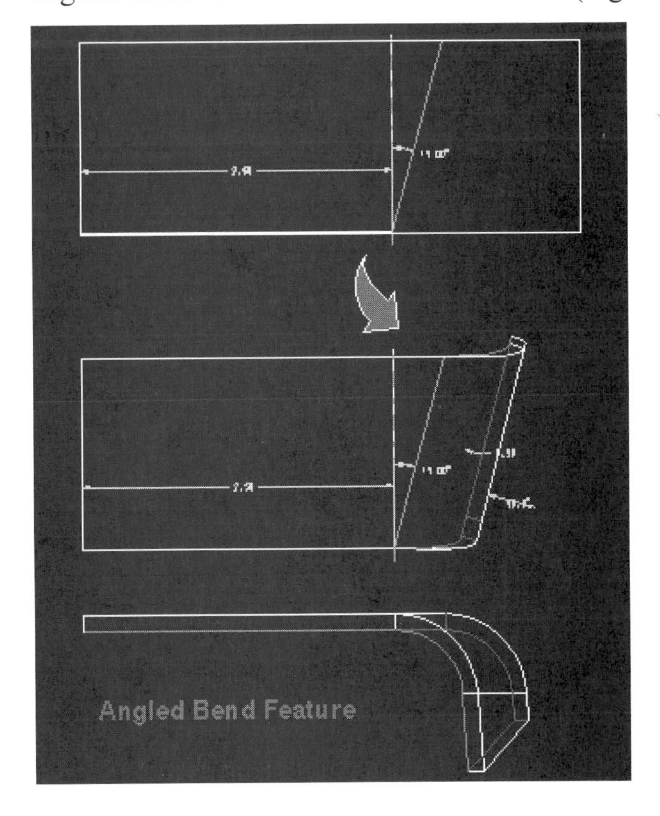

Figure 23.90
Angled Bend Feature

Creating Bends

For a practice part, create the simple, flat sheet metal part **(SM_BEND_4.prt)** shown in Figure 23.91. The sheet metal part is **10.00 X 24.00** with a thickness of **.20.**

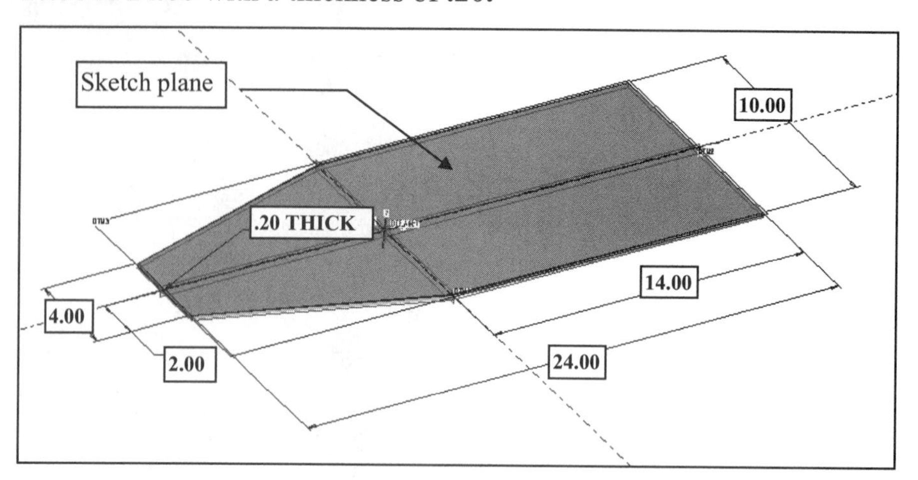

Figure 23.91
SM_BEND_4.prt

Using the following commands, add two sheet metal **Bend** features to complete the part:

Feature ⇒ Create ⇒ Bend ⇒ Angle ⇒ Regular ⇒ Done ⇒ Part Bend Tbl ⇒ Done/Return ⇒ Inside Rad ⇒ Done/Return ⇒ (select the top face (Fig. 23.91) as the Sketch plane) **⇒ Okay ⇒ Default ⇒**

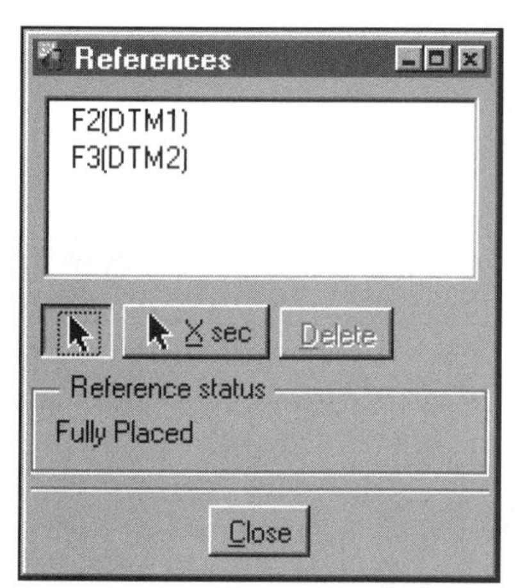

Pro/E automatically switches into Sketch mode. You are now ready to sketch the **Bend Line.**

Close (to accept references **DTM1** and **DTM2,** see left column) ⇒ [sketch the line (Fig. 23.92)] ⇒ (align to the outer edge *surfaces*) ⇒ **Modify** (change the dimension value to **6.00**) ⇒ **Sketch ⇒ Done ⇒ Okay** (for the side of bend creation) ⇒ **Flip** (for fixed side of bend) ⇒ **Okay ⇒ No Relief ⇒ Done ⇒ 60.000 ⇒ Done ⇒ Thickness ⇒ Preview** (Fig. 23.93) ⇒ **OK**

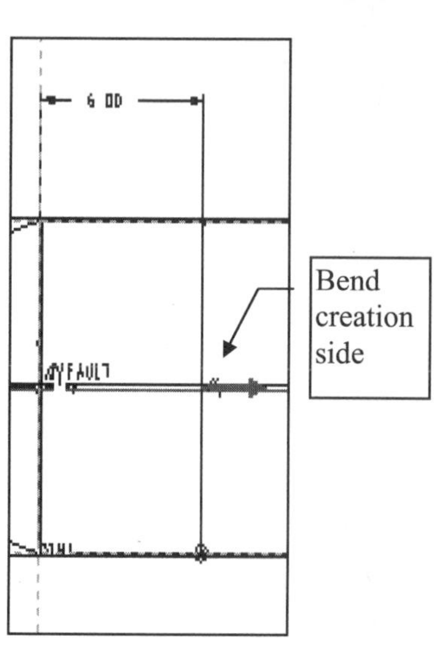

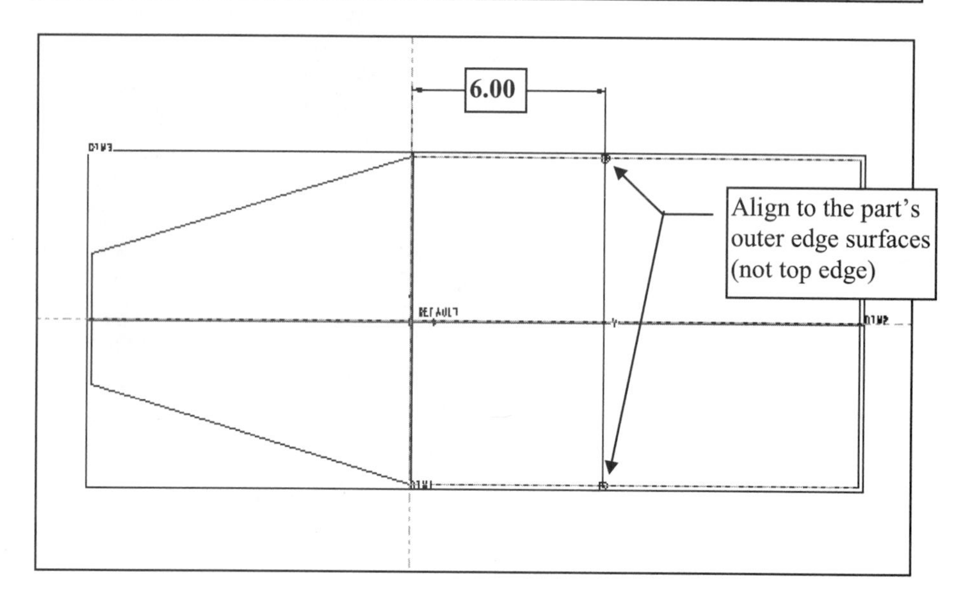

Figure 23.92
Sketch Bend Line

Element	Info
Bend Table	Part Bend Tbl
Radius Type	Inside Rad
Sketch	Defined
Relief	No Relief
Bend Angle	Dimension ID 10, angle = 60
Radius	Dimension ID 11, value = 0.2

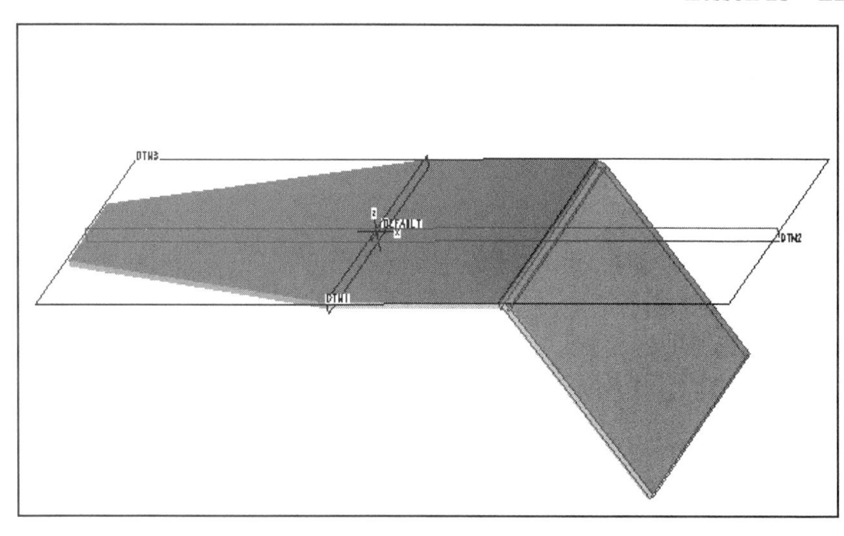

Figure 23.93
Preview

Now create a second bend on the *tab* side of the part:

Feature ⇒ Create ⇒ Bend ⇒ Angle ⇒ Regular ⇒ Done ⇒ Part Bend Tbl ⇒ Done/Return ⇒ Inside Rad ⇒ Done/Return ⇒ Use Prev ⇒ Okay ⇒ Close ⇒ (sketch the line) ⇒ (align to outer slanted edge *surfaces*) ⇒ (modify the dimension value to **3.00**) ⇒ **Sketch ⇒ Done** (Fig. 23.94) ⇒ **Done ⇒ Flip ⇒ Okay ⇒ Okay ⇒ No Relief ⇒ Done ⇒ 30.000 ⇒ Done ⇒ Thickness*2 ⇒ Preview** (Fig. 23.95) **⇒ OK ⇒ File ⇒ Save**

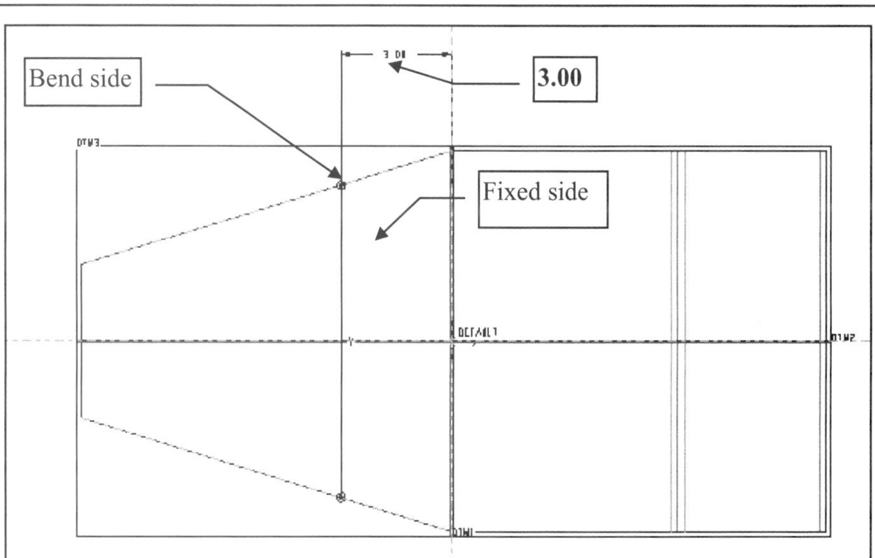

Figure 23.94
Second Bend Sketch

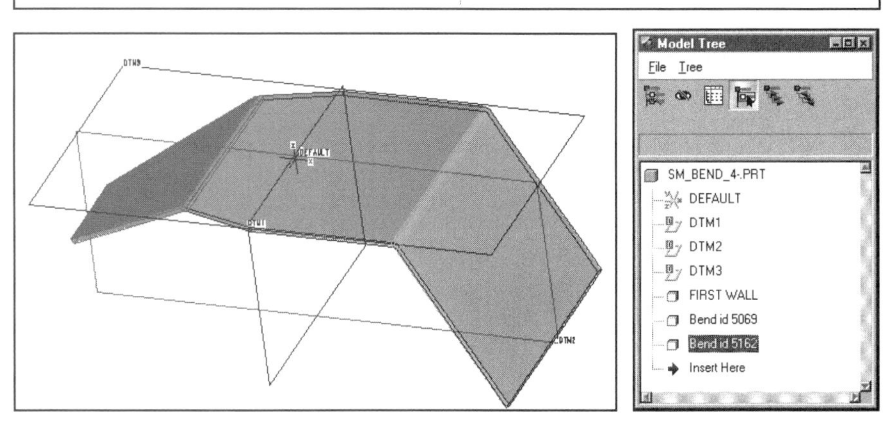

Figure 23.95
Tab Bend

Bend Reliefs

Bend Relief is a method of relieving the material in the region near the radius of a Bend feature. It is intended to model the behavior of the bent material when the part is manufactured (Fig. 23.96).

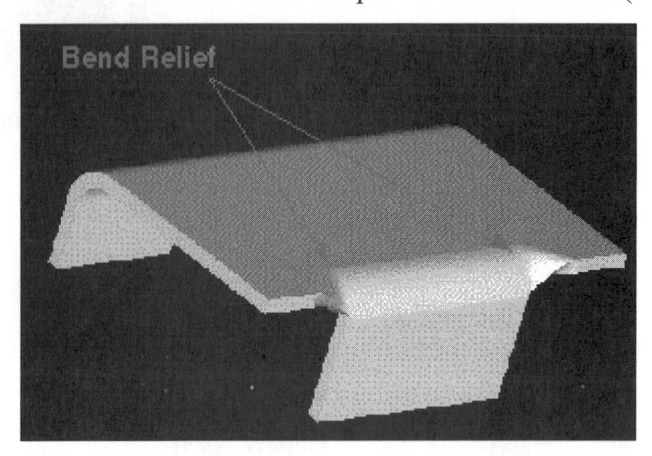

Figure 23.96
Bend Relief

NOTE

In general, the bend relief must be specified when creating partial Wall features, or whenever two Wall features intersect. Without specifying **Bend Relief**, you cannot create this type of Wall. Pro/E will give you a message which is slightly misleading - it says that it is unable to create a non-zero radius bend on the edge.

In Pro/SHEETMETAL, you may specify **Bend Relief** during the creation of either **Wall** or **Bend** features. If you choose to define Bend Relief, you also must specify the type at each "attachment point" (end) of the Bend.

An automatic relief is used in sheet metal fabrication to allow for control of high-deformation areas. For example, stretching may be appropriate for relatively little deformation; however, if a rip is used for larger deforming areas, unpredictable material behavior (unwanted ripping) may result if left unrelieved. After you have sketched the new wall and the sketch has been regenerated, the RELIEF menu appears with the following choices:

No Relief The wall is created without relief.
w/Relief The RELIEF TYPE menu appears. Pro/E highlights each attachment point in turn and asks you to select the kind of relief you want to use there. Choose one of the following:
No Relief Attach the wall without reliefs.
StrtchRelief Use material stretching to provide bend relief at the wall attachment points (Fig. 23.97).
Rip Relief At each attachment point, rip the existing material normal to the edge and back to the tangent line (Fig. 23.97).
RectRelief To create a rectangular relief.
ObrndRelief To create an obround relief.

The SEL WIDTH menu appears when you choose the rectangular or round reliefs. You then type the relief dimensions and determine whether you want to create the relief up to bend or tan to bend.

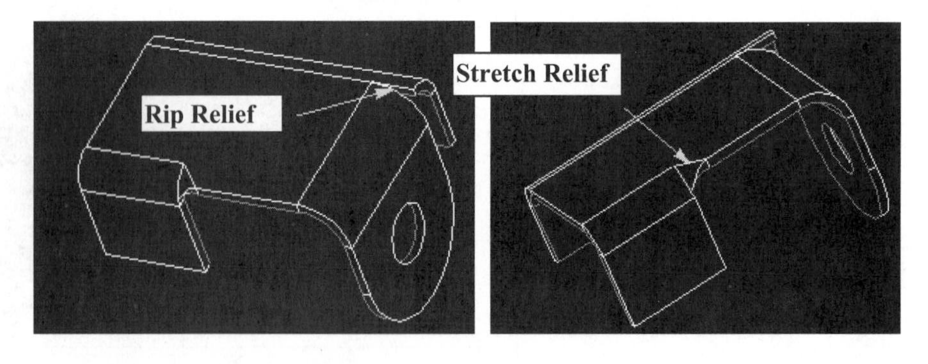

Figure 23.97
Rip Relief and **Stretch Relief**

Create the **Flat Wall** feature on SM_BEND_4. Because it does not extend along the entire length of the edge, you will have to specify **Bend Relief** in order to create the Wall.

Feature ⇒ Create ⇒ Wall ⇒ Flat ⇒ Use Radius ⇒ Done ⇒ Part Bend Tbl ⇒ Done/Return ⇒ Inside Rad ⇒ Done/Return [select the **Attach Edge** (Fig. 23.98)] ⇒ **45.000 ⇒ Done ⇒ Okay ⇒ Close**

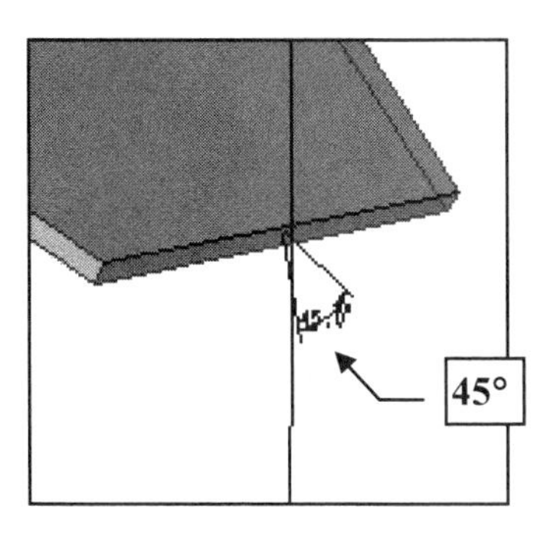

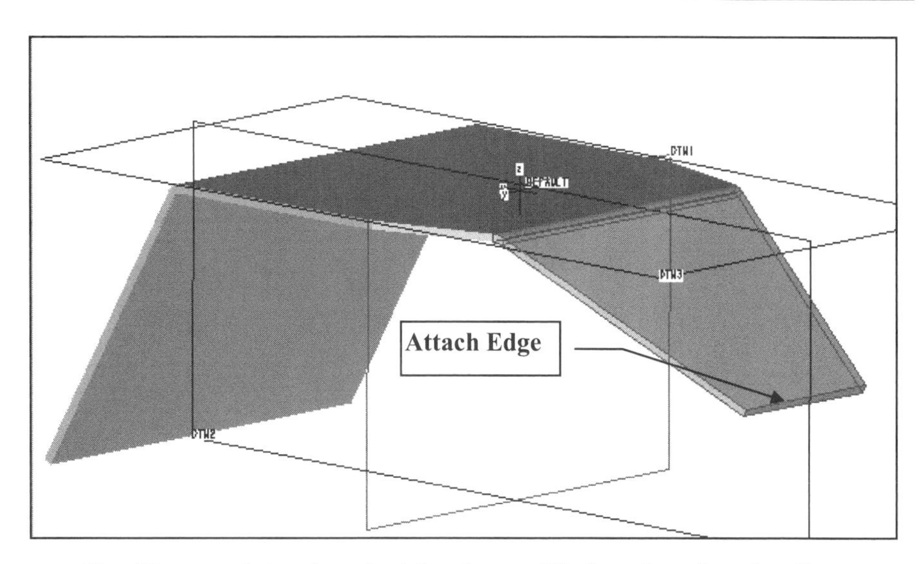

Figure 23.98
Attach Edge for New Wall

Pro/E pops into the sketch view. Notice the sketch alignment points at each end of the **Attach Edge**. Create the sketch shown in Figure 23.99. *Sketch a horizontal centerline that passes through the sketch alignment points* and a vertical centerline passing through the part. Modify the dimensions and **Sketch ⇒ Done.**

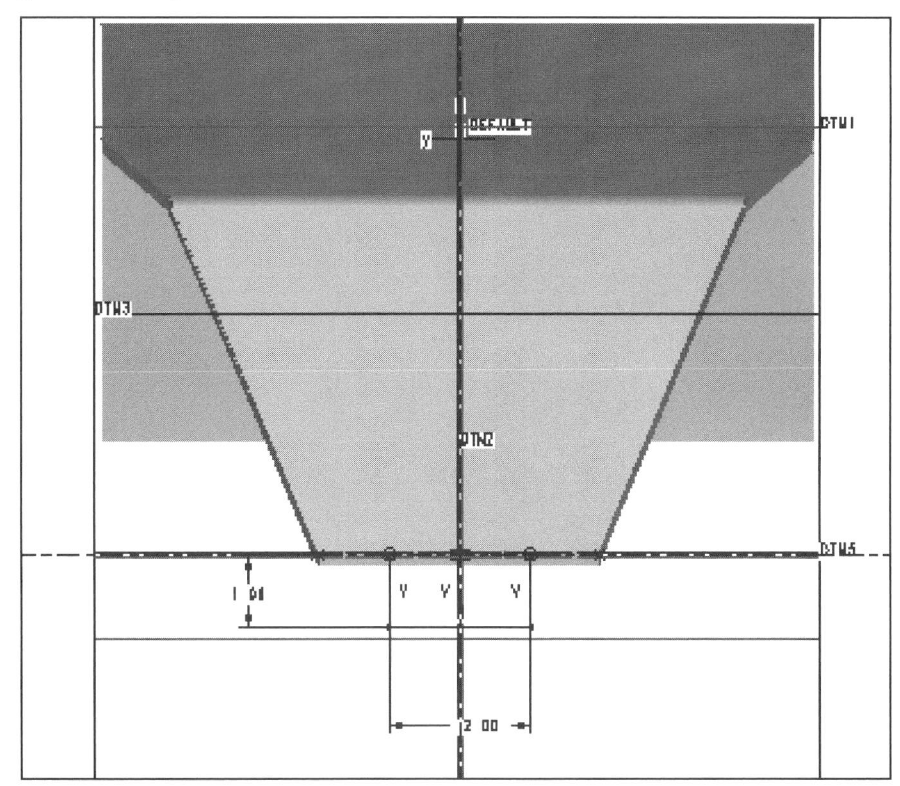

Figure 23.99
Sketch the **2.00** (width) by **1.00** Wall

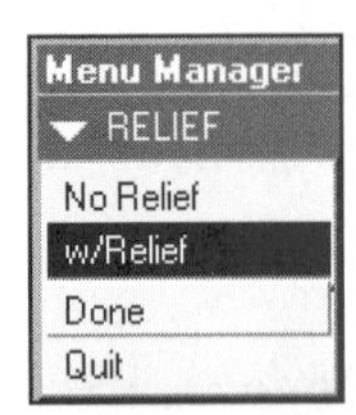

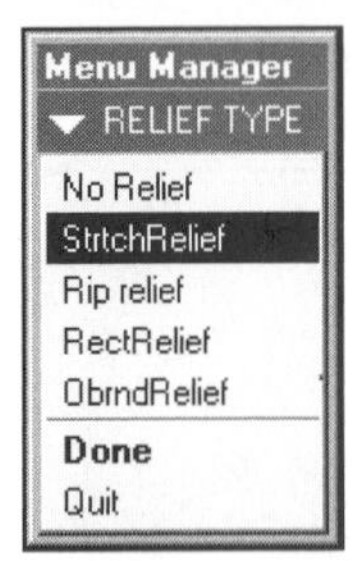

(Pro/E displays the RELIEF menu. The default is not to specify Bend Relief) ⇒ **w/Relief** ⇒ **Done** ⇒ **StrtchRelief** ⇒ **Done** (You are now defining the Bend Relief for the first end of the Wall feature. The end is *highlighted* with a **X**) ⇒ Define relief's width ⇒ **Enter Value** ⇒ **.02** ⇒ ✔⇒ Define relief's angle ⇒ **45.0** ⇒ ✔ ⇒ (now repeat this sequence for the Relief at the other end of the Wall) ⇒ **Enter Value** ⇒ (type **.80** for the bend radius value) ⇒ ✔ ⇒ **Preview** (Fig. 23.100)

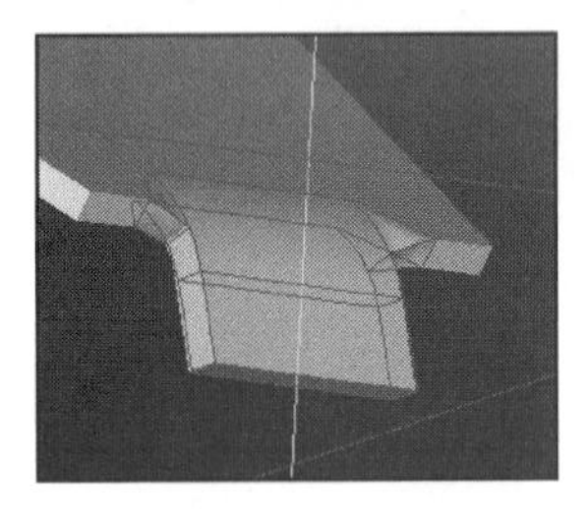

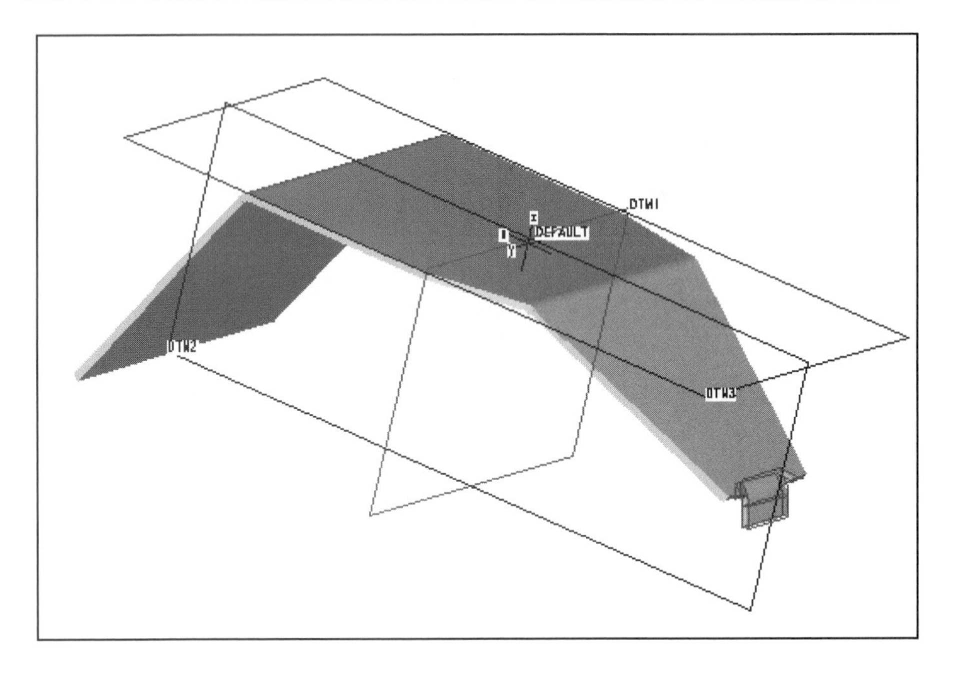

Figure 23.100
New Wall with Stretch Reliefs

Pick on **Relief** and then **Define** in the dialog box to change the two relief types (see left column). Follow the prompts and select **Rip Relief** for both ends (Fig. 23.101).

Relief (dialog box) ⇒ **Define (**dialog box button) ⇒ **w/Relief** ⇒ **Done** ⇒ **Rip Relief** ⇒ **Done** ⇒ **Rip Relief** ⇒ **Done** ⇒ **OK** ⇒ **File** ⇒ **Save** ⇒ ✔

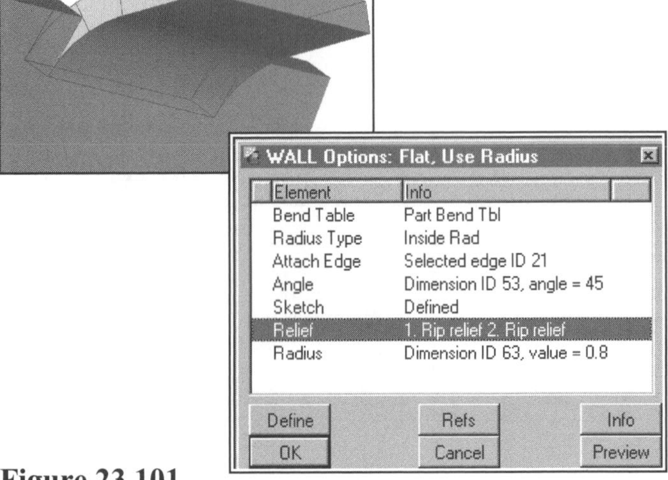

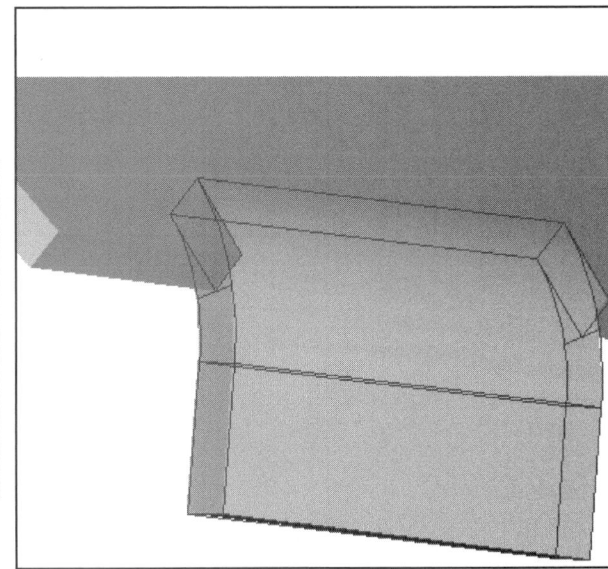

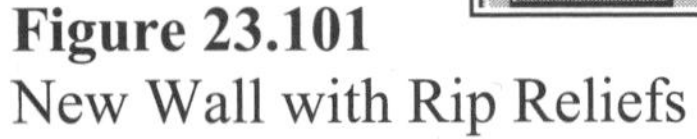
Figure 23.101
New Wall with Rip Reliefs

Sheet Metal Cut Features

The **Sheet Metal Cut** feature is used when you must remove material from a sheet metal part (Fig. 23.102).

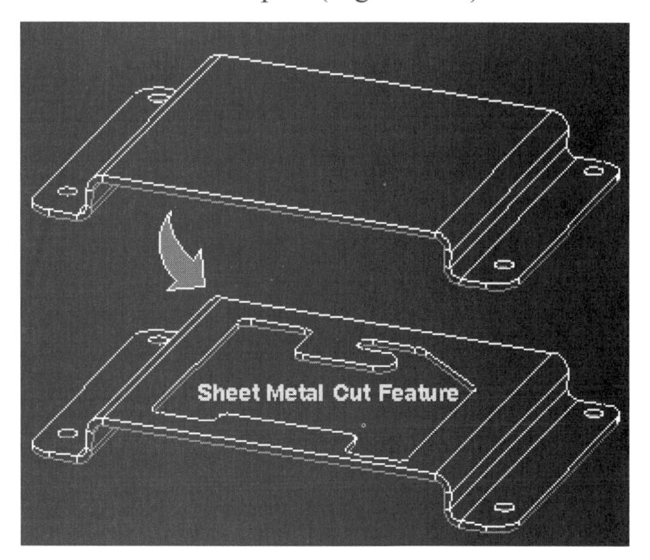

Figure 23.102
Sheet Metal Cut Feature

Sheet Metal Cut features are very similar to **Solid Cut** features. There is one very important difference to understand, however. Whenever you create a Sheet Metal Cut feature, Pro/E first projects the sketch section onto the *green* surface, then creates the cut. The result of this construction technique actually affects the resultant geometry. Because the *white* surface is developed by offsetting from the surface, the side faces that are in the bend region will be "distorted." However, the feature does appear correctly when displayed in the flat pattern drawing (Fig. 23.103).

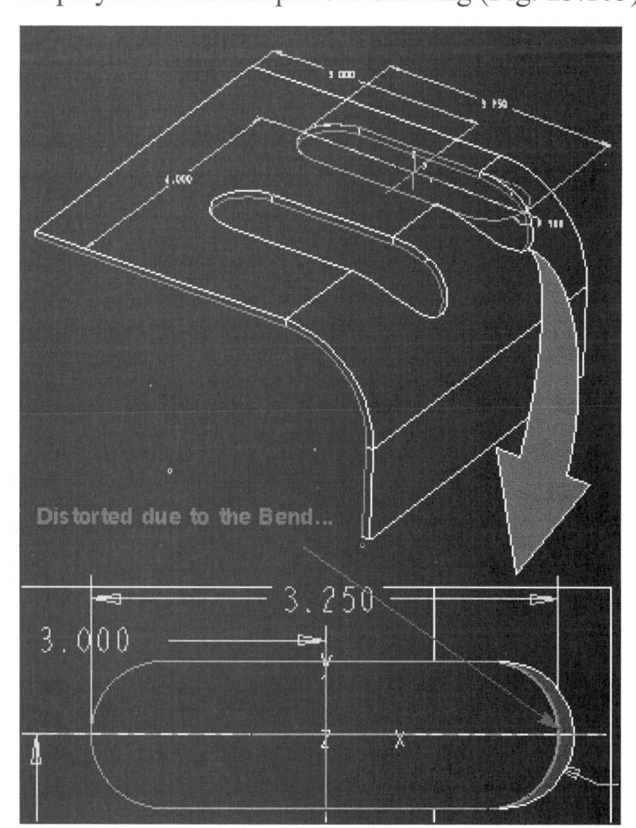

Figure 23.103
Sheet Metal Cut Feature and Flat Pattern Drawing

In order to see the difference between a **Sheet Metal Cut** and a **Solid Cut,** we will have you create both types on the same part (**SM_BEND_4**) with the exact same dimensions. The first Cut will be a Solid Cut. Use the dimensions provided in Figure 23.104 for both types. The Solid Cut will be as shown in this figure. The Sheet Metal Cut will be placed below. Create the **Cut** (**Create ⇒ Solid ⇒ Cut** etc.) using the sketch dimensions provided. Save the section (rename it to **CUT_SEC**) so that you can reuse it for the second cut.

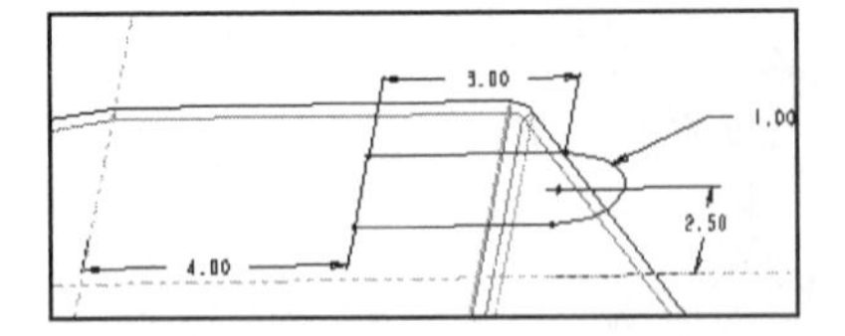

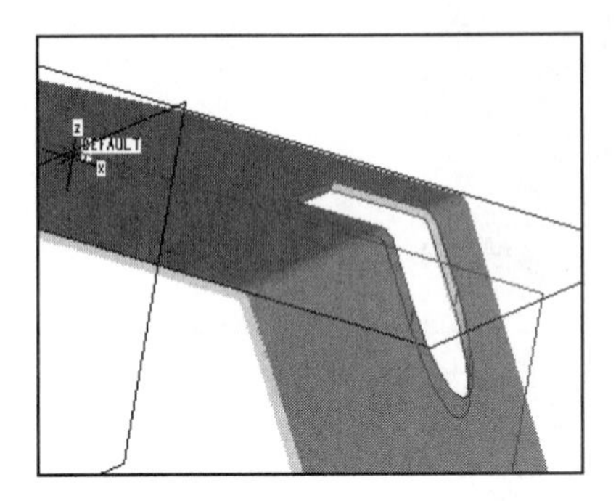

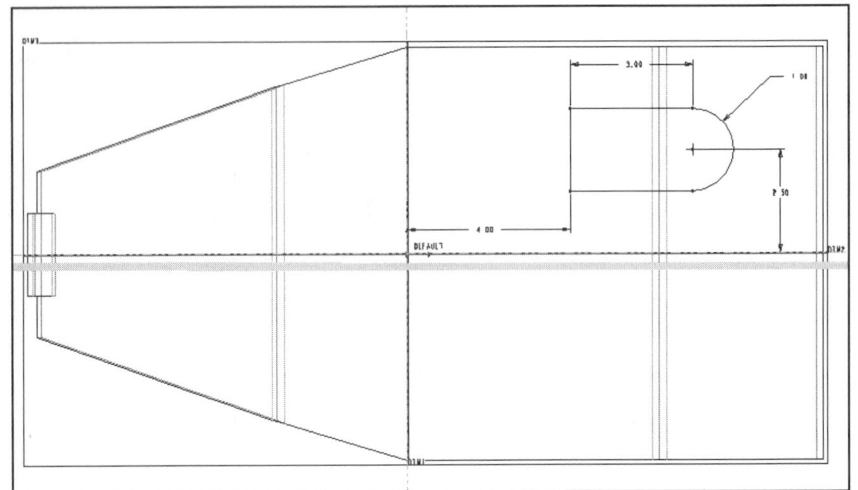

Figure 23.104
Solid Cut

Next, create the **Sheet Metal Cut**:

 ***NOTE***

If you leave the Intent Manager on you can import the section using:

✔Intent Manager ⇒ File ⇒ Import ⇒ Append to Model

(place the section using ⊗)

Create ⇒ Sheet Metal ⇒ Cut ⇒ Extrude ⇒ Solid ⇒ Done ⇒ (use **DTM3** as the sketch plane) **⇒ Okay ⇒ Use Prev ⇒ Okay ⇒** (accept **DTM1** and **DTM2** as the References) **⇒ Close ⇒** (reuse the section from the first cut) **⇒ ❑ Intent Manager ⇒** [**Sec Tools ⇒ Place Section** (**CUT_SEC**), etc. (Fig. 23.105)] **⇒** (use the same placement dimensions as the **Solid Cut**) **⇒ Regenerate ⇒ Done ⇒ Okay ⇒ Thru All ⇒ Done ⇒ Green ⇒ Done ⇒ OK** (Fig. 23.106)

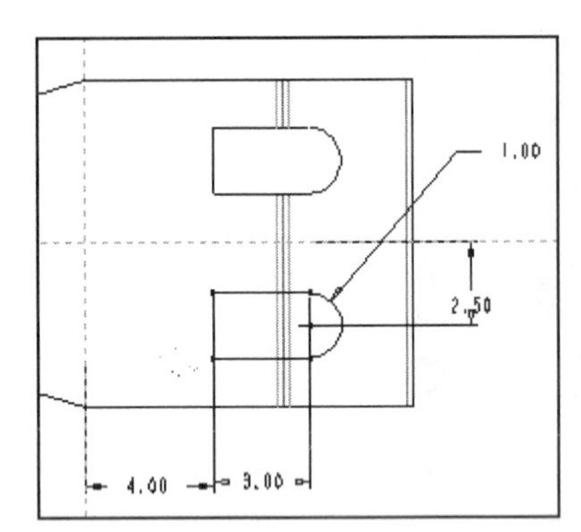

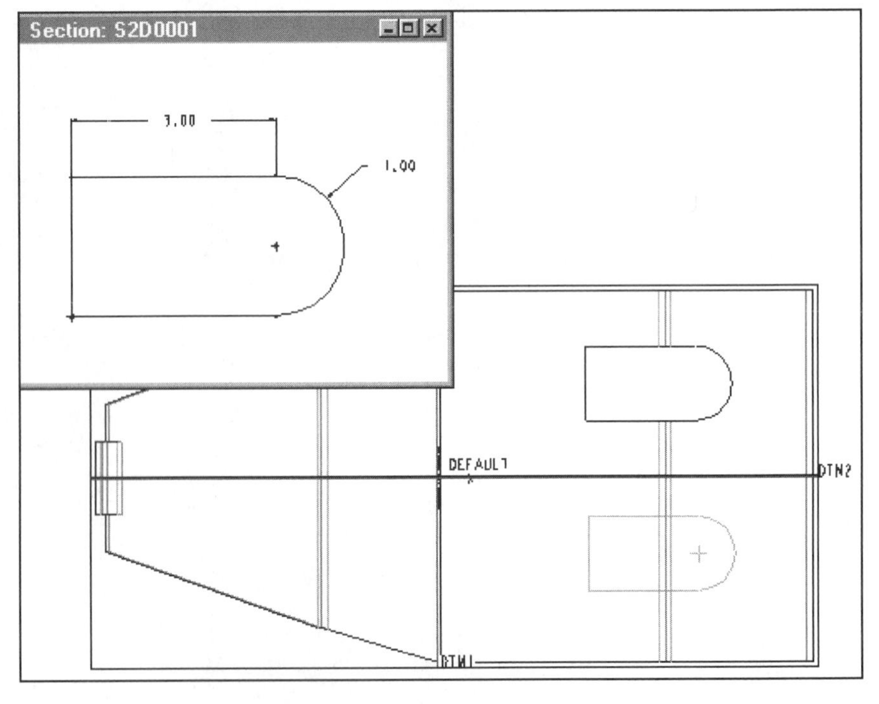

Figure 23.105
Sheet Metal Cut Sketch
Using Place Section

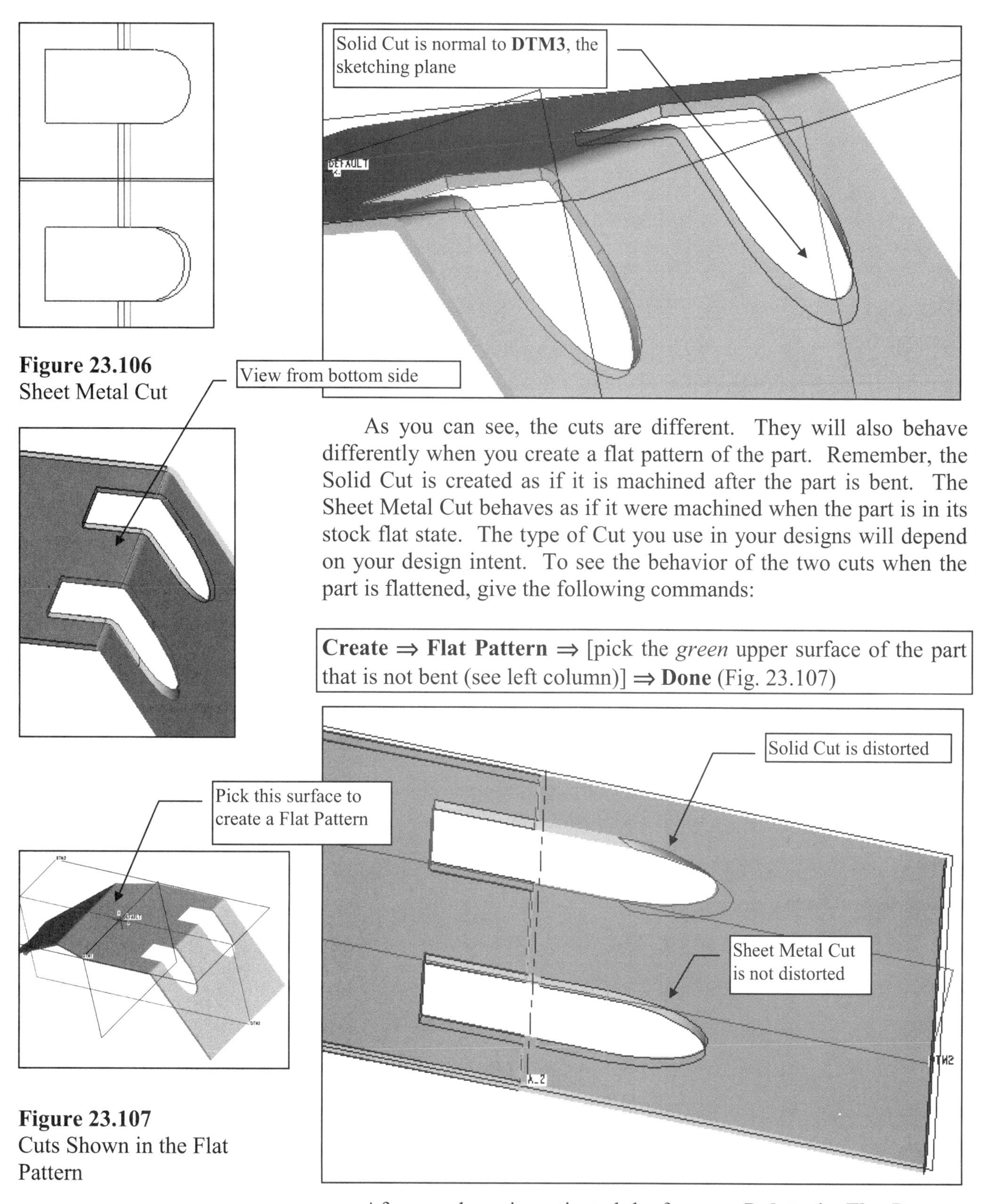

Figure 23.106
Sheet Metal Cut

As you can see, the cuts are different. They will also behave differently when you create a flat pattern of the part. Remember, the Solid Cut is created as if it is machined after the part is bent. The Sheet Metal Cut behaves as if it were machined when the part is in its stock flat state. The type of Cut you use in your designs will depend on your design intent. To see the behavior of the two cuts when the part is flattened, give the following commands:

Create ⇒ **Flat Pattern** ⇒ [pick the *green* upper surface of the part that is not bent (see left column)] ⇒ **Done** (Fig. 23.107)

Figure 23.107
Cuts Shown in the Flat Pattern

After you have investigated the features, **Delete** the **Flat Pattern** feature before saving and closing your part. You do not want to leave the Flat Pattern feature on your part at this time. Later, you will learn how to create a **Flat Pattern Instance** in a **Family Table** and use the instance on a drawing for detailing the sheet metal part in its before-bent state.

Form Features

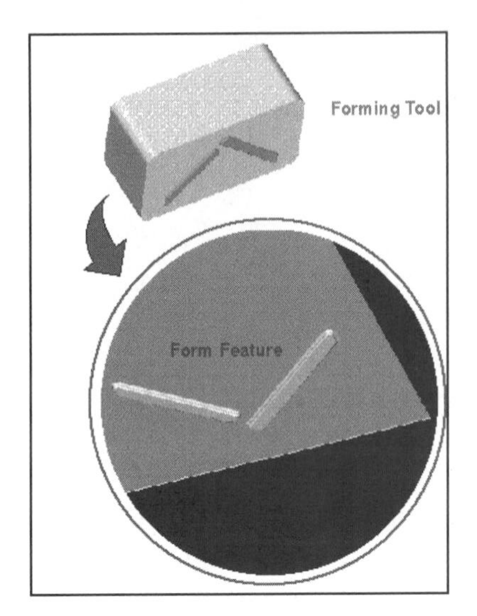

Form features are used to create shapes that are stamped into sheet metal. You can control the shape of the Form as either a **Punch** or a **Die**. The **Flatten Form** feature is used to flatten the Form feature for display in the part's Flat State.

A Form feature is typically used to create shapes that deform the surfaces of a sheet metal part, such as an extruded hole or a stamped pocket. Form features get their shape definition from the surfaces of an existing part, known as the **Reference Part**. There are two types of Form features available in Pro/SHEETMETAL: Punch and Die. Which one you use depends upon whether your reference geometry is in the form of the punch (the "positive" shape) or the die (the "negative" shape).

In the example shown to the left, the **Forming Tool** was modeled as a Solid part. The Form feature was created, using the Forming Tool model as the Reference Part.

When creating Form features, you must choose between the **Copy** and **Reference** options. These options control the behavior of the **Form** geometry. If you select Copy, the geometry is copied into the part from the Reference Part. The Form feature will then behave just like any other feature in Pro/E.

If you choose Reference, the feature is actually tied to the reference part. When the Reference Part changes, so does the Form feature, because they are associated. This functionality is extremely useful if you have a complicated punch tool with a design that is dynamic; whenever you modify the punch (the Reference Part), the resultant Form feature in the sheet metal part is updated correspondingly. If you use this option, you must be certain not to delete the Reference Part (and also to ensure that it is available to Pro/E through the search directory path).

Form features are not affected by Unbend features. To unbend a Form feature for display in the flat state, you must use the Flatten Form feature (Fig. 23.108).

In order to have a part that corresponds *exactly* to the project in the next sample exercise **(SM_FORM_5)** , sketch on **DTM3** (**FRONT**) and project the flat wall protrusion outward

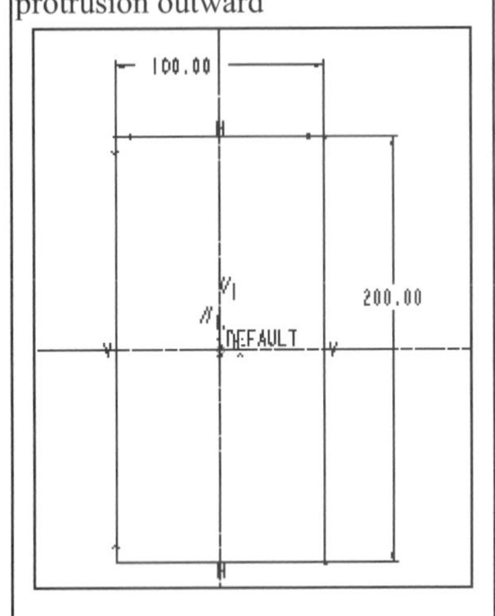

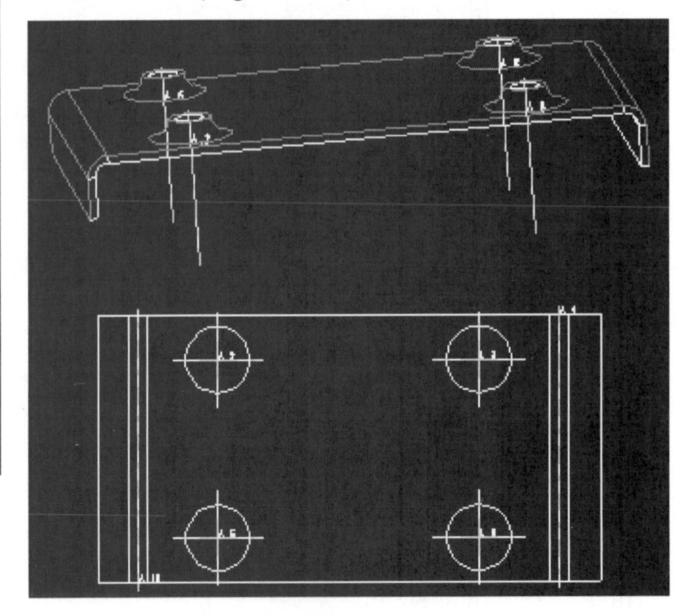

Figure 23.108
Form Feature and **Flatten Form Feature**

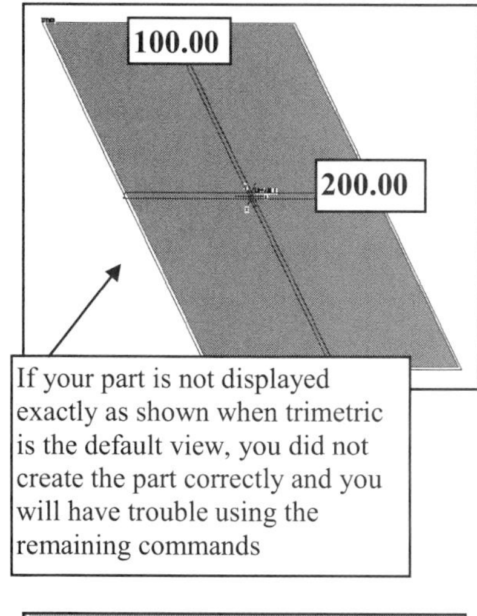

Create the **Flat Wall Sheet Metal** part (**SM_FORM_5**) with the following dimensions **100.00** (width) **X 200.00** (height) **X .20** (thick). The Form feature comes directly from Pro/LIBRARY. The Form Louver Die (**open_flat_louver.prt**) will be used.

Create ⇒ **Form** ⇒ **Die** ⇒ **Reference** ⇒ **Done** ⇒ (double-click on each directory option: **user** ⇒ **prolibrary** ⇒ **objlib** ⇒ **featurelib** ⇒ **stm_udf** ⇒ **louvers** ⇒ **open_flat_louver.prt** ⇒ **Open)** ⇒ **Mate** ⇒ [pick the part surface (Fig. 23.109)] ⇒ (pick the mating surface on the form part) ⇒ **Align Offset** ⇒ (pick the top horizontal surface on both the part and the form) ⇒ Offset in indicated direction: **30** ⇒ ✔ ⇒ **Align Offset** ⇒ (pick the right side surface on both the part and the form) ⇒ Offset in indicated direction: **45** ⇒ ✔ ⇒ **Done** ⇒ [Select the boundary plane from the reference part (Fig. 23.110)] ⇒ [Select the seed surface from the reference part (Fig. 23.110)]

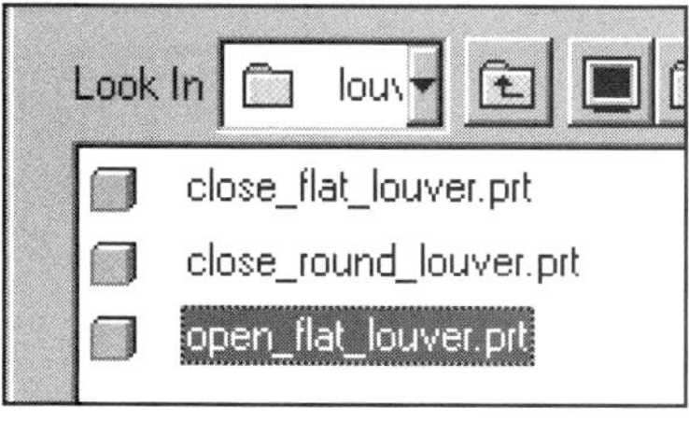

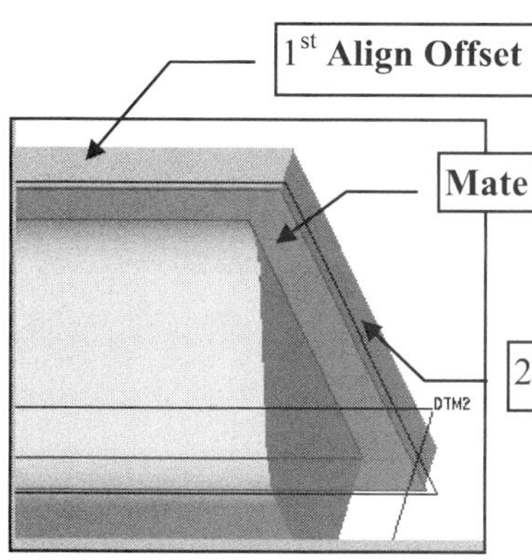

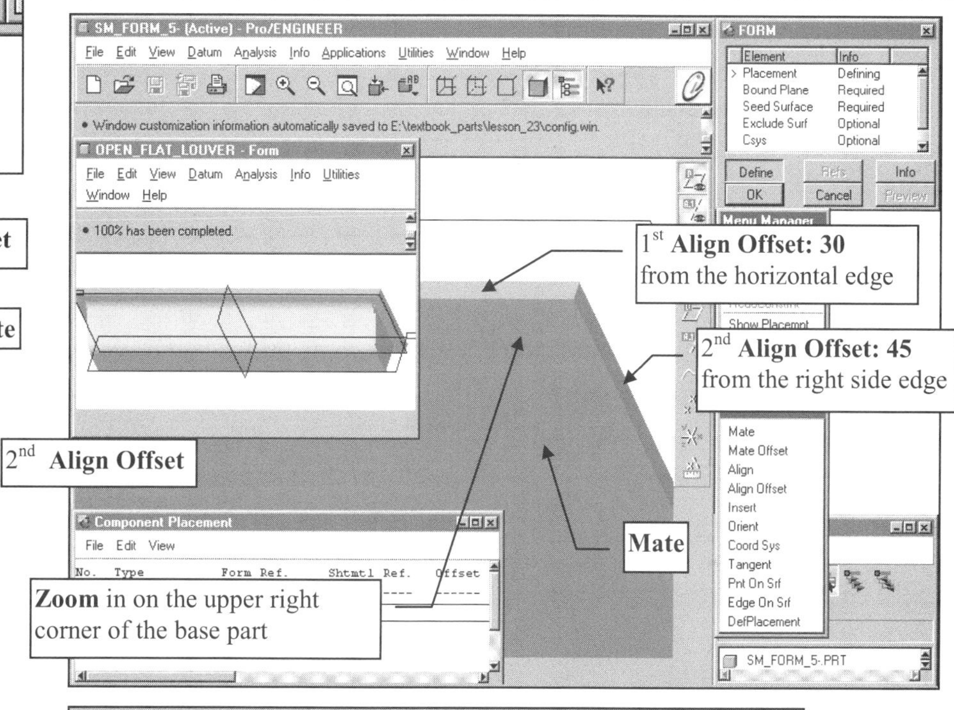

Figure 23.109
Placing the Form Feature

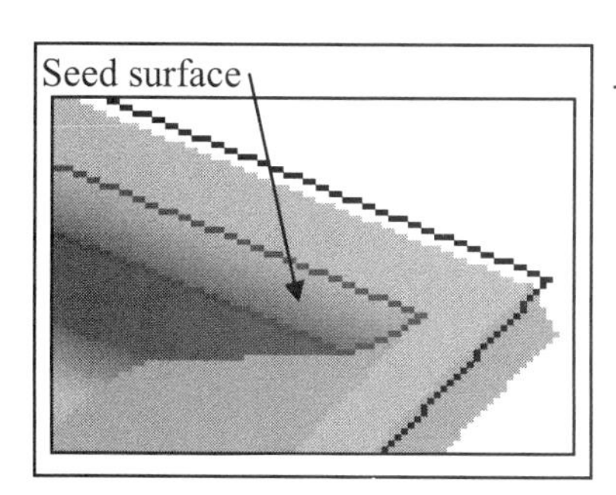

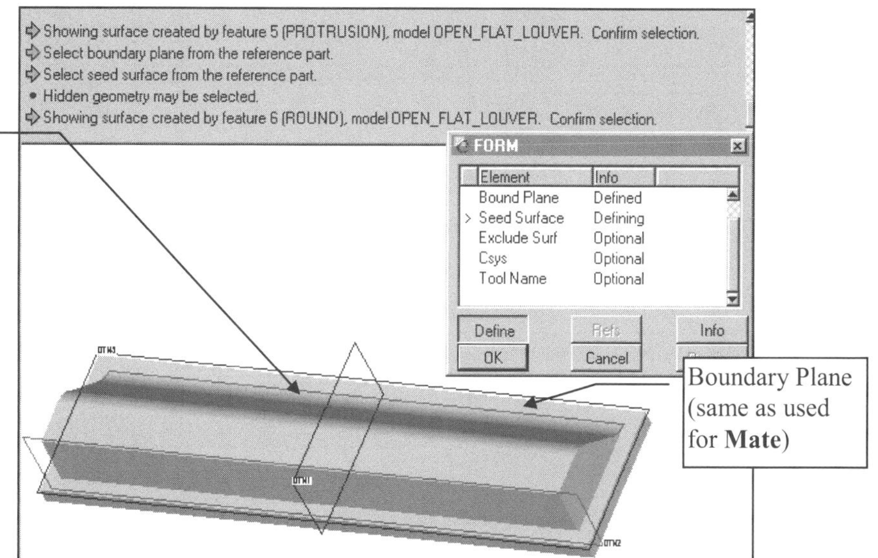

Figure 23.110
Boundary Plane (same as used for **Mate**).
Seed Surface (see above).

Exclude Surf (Fig. 23.111) ⇒ **Define** ⇒ (select the three surfaces to remove) ⇒ **Done Sel** ⇒ **Done Refs** ⇒ **OK** (Fig. 23.112)

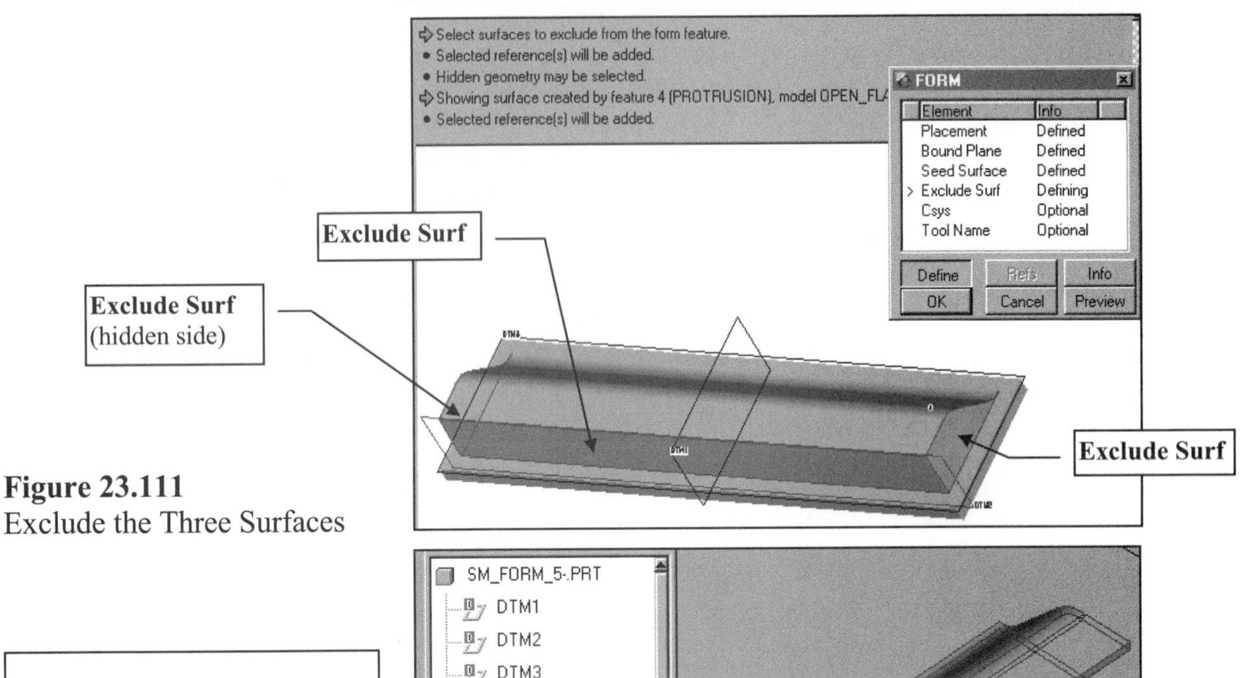

Figure 23.111
Exclude the Three Surfaces

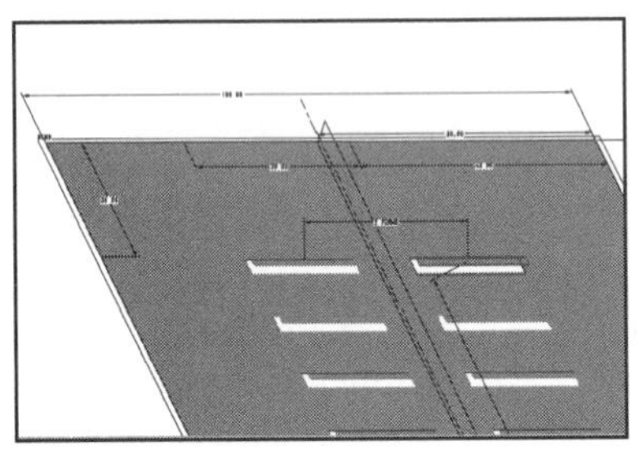

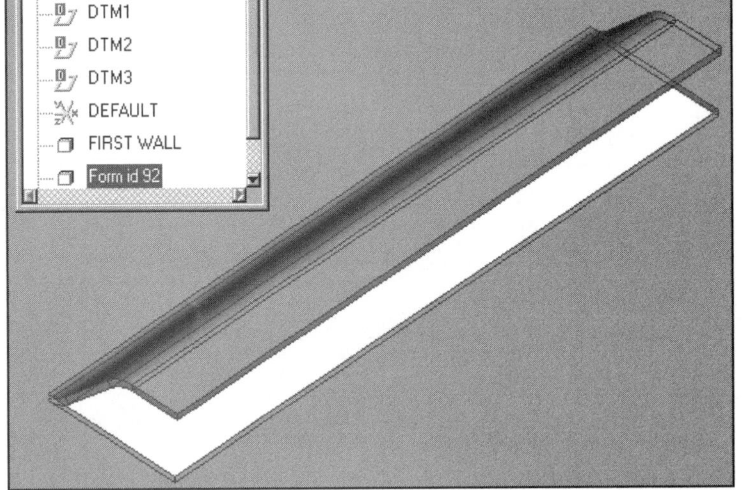

Figure 23.112
Completed Die Form

File ⇒ Save ⇒ ✔

Complete the part by patterning the Form feature (Fig. 23.113). Two features spaced at **30.00** in the short direction and ten spaced at **15.00** in the long direction (twenty form features). **Save** the part and **Close** the window.

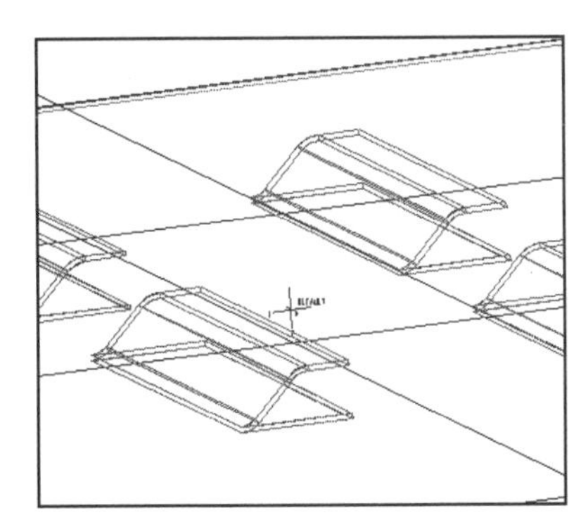

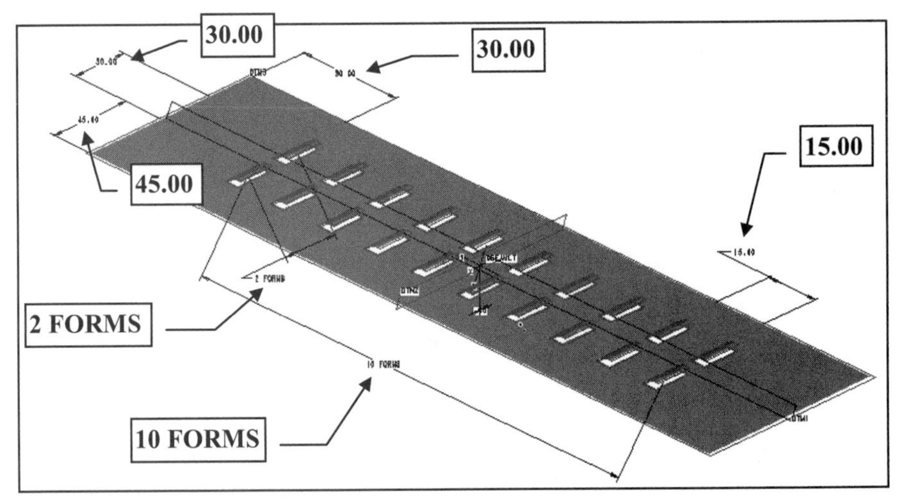

Figure 23.113
Completed Die Form

Unbend and Bend Back Features

> **NOTE**
> For **Flatten Form Features**
>
> Form features will not be unbent when you **Unbend All** the Bend geometry, but you can create a **Flatten Form** feature that will unbend all of the Form features:
>
> **Create** ⇒ **Flatten Form** ⇒ (choose the **Form** element) ⇒ **Define** ⇒ (select the Form features) ⇒ ✔ ⇒ **Done Sel** ⇒ **Done Refs** ⇒ **OK** (see below)

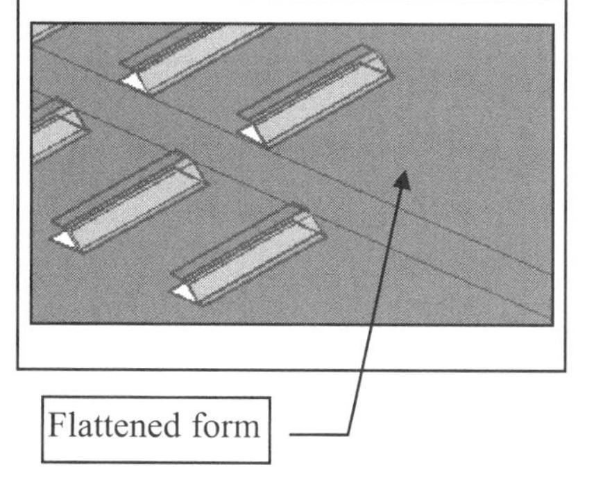

Flattened form

If you want to add a feature to a sheet metal part in its flattened state, you normally create an **Unbend** feature (as opposed to creating a flat state instance). This feature becomes a part of the original part (not an instance) and is affected whenever the part regenerates.

A **Bend Back** feature is used to "re-bend" a part after an Unbend feature has been created.

You create an Unbend feature by specifying a fixed face on the part. You can then **Unbend All** or select bends to Unbend. Any features added to the part in this state are automatically bent when the part is bent (if they lie on a face that moves when the part is bent).

When you create a Bend Back feature, you select a fixed plane. You can then **Bend Back All** or select bends to Bend Back. Any features added to the part in this state are automatically bent when the part is bent (if they lie on a face that moves when the part is bent).

This combination of Unbend and Bend Back features is often used to create cuts that are difficult to model with the part in the formed position, such as the *tab* feature (see left column). In addition, you may find that you can create some shapes with this technique that cannot be modeled in any other way. For example, if you want to create a straight "tang" in the middle of a bend, you first create the bent part, then flatten it with an Unbend feature; then add the cut in the form of the tang outline, and finally "re-bend" the part using a Bend Back feature. When you "re-bend" the part, Pro/E permits you to leave the tang piece unbent, even though the material was originally bent as shown in Figure 23.114.

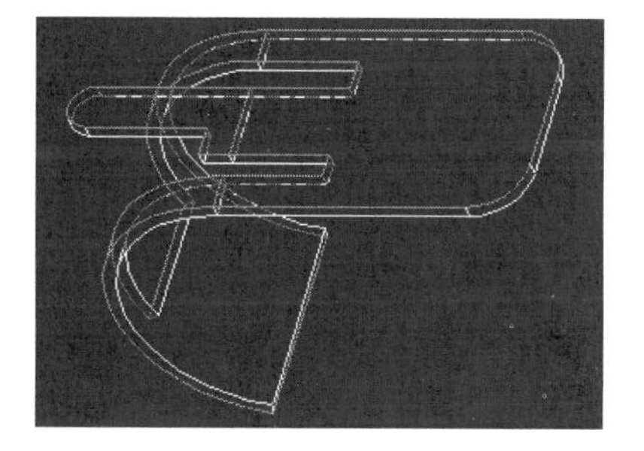

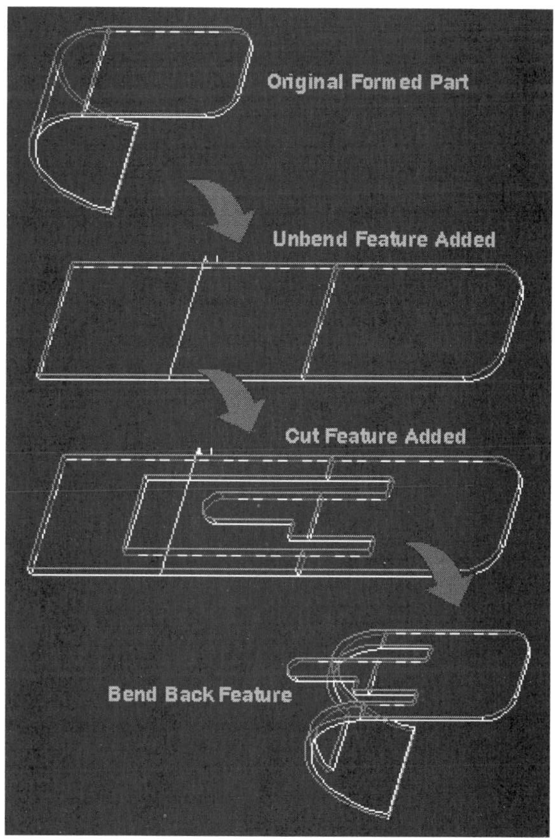

Figure 23.114
Unbend and **Bend Back**

Create the sheet metal part shown in Figure 23.115. The first wall is an extruded wall. All dimensions are in millimeters.

Pick this plane surface to remain fixed while unbending

140

5 THICK

R190

325

Figure 23.115
SM_BENDBACK_6.prt
First Wall, Extruded Wall

Next, you will add an **Unbend** feature:

Feature ⇒ Create ⇒ Sheet Metal ⇒ Unbend ⇒ Regular ⇒ Done ⇒ [select the plane to remain fixed (Fig. 23.116)] ⇒ **Unbend All ⇒ Done ⇒ Preview ⇒ OK**

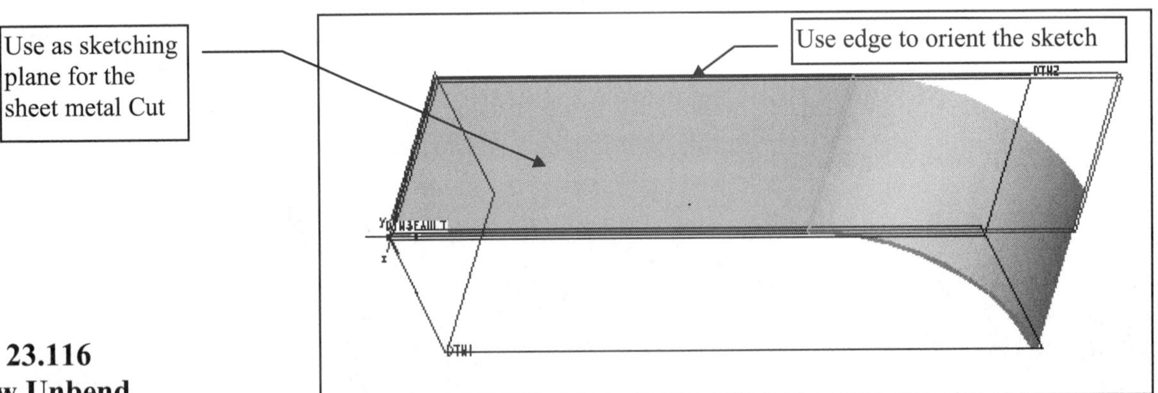

Figure 23.116
Preview Unbend

The next feature will be a sheet metal cut:

Create ⇒ Sheet Metal ⇒ Cut ⇒ Extrude ⇒ Solid ⇒ Done ⇒ (select the top of the part as the sketching plane) ⇒ **Okay ⇒ Top ⇒** (select the upper edge of the part) ⇒ etc. (Fig. 23.117)

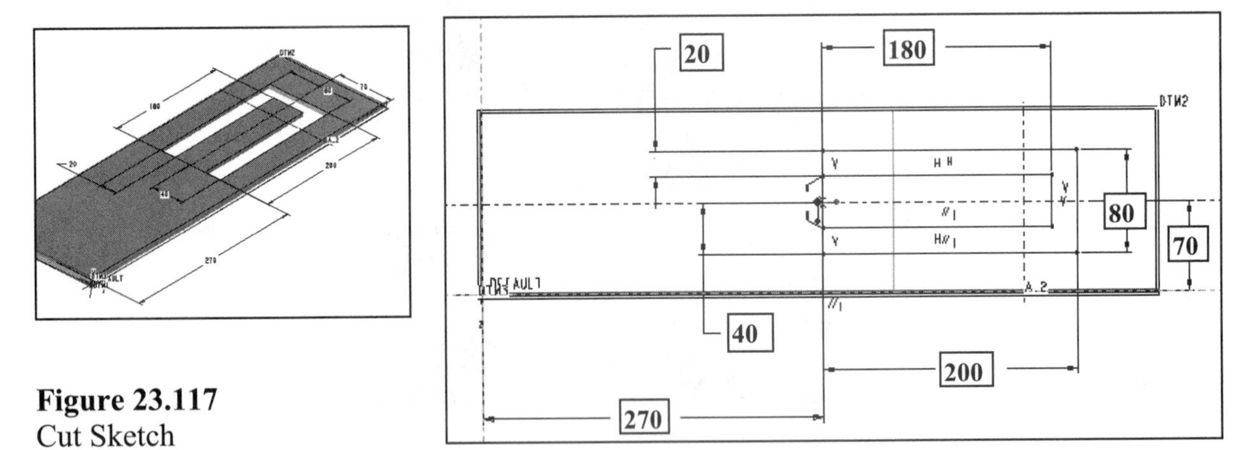

Figure 23.117
Cut Sketch

A **Bend Back** feature is now added to the model so that you can select the tab to remain unbent while returning the first bend to its original curved position.

Feature ⇒ **Create** ⇒ **Sheet Metal** ⇒ **Bend Back** ⇒ [select the plane to remain *fixed* (Fig. 23.118)] ⇒ **Bendback All** ⇒ **Done** ⇒ [Select all contours in unbent areas to remain flat (Fig. 23.118)]-*(select next until the contour to remain flat is highlighted in blue)* ⇒ **Accept** (the *blue* highlighted tab) ⇒ **Done** ⇒ **Preview** (Fig. 23.119) ⇒ **OK** ⇒ **Save** ⇒ ✔(Fig. 23.120)

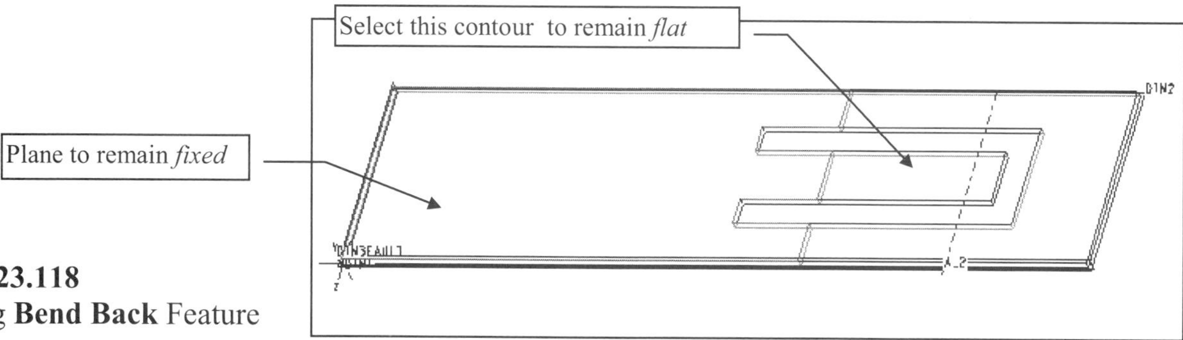

Figure 23.118
Creating **Bend Back** Feature

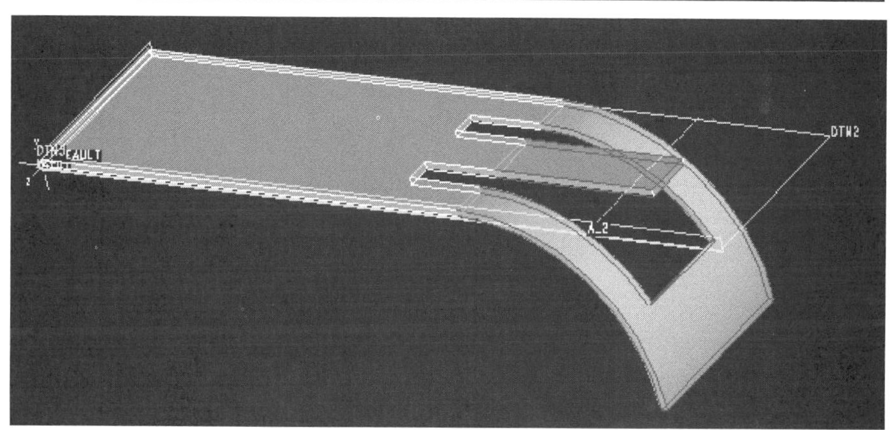

Figure 23.119
Preview Bend Back

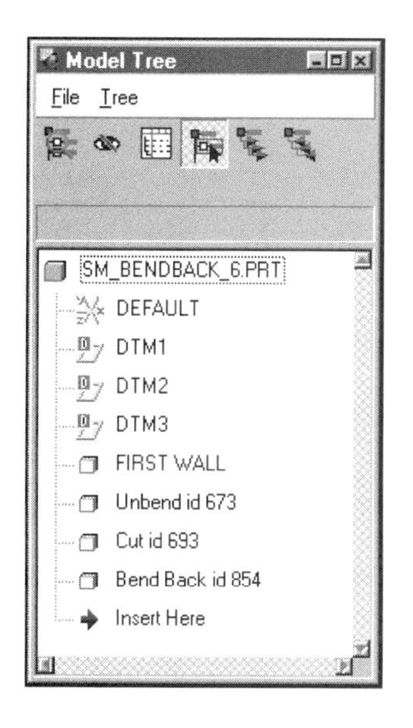

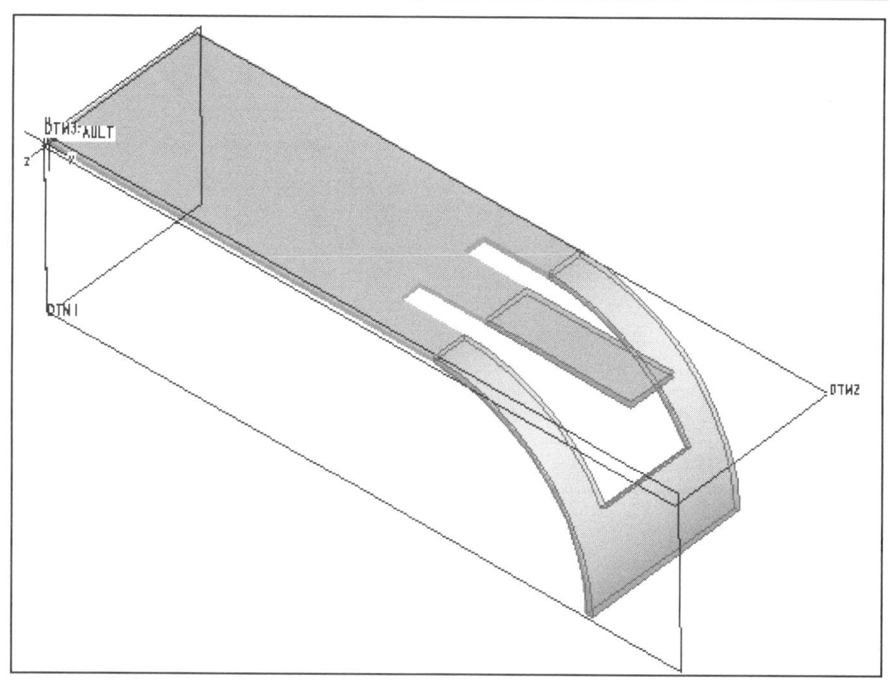

Figure 23.120
Completed Part

Rips

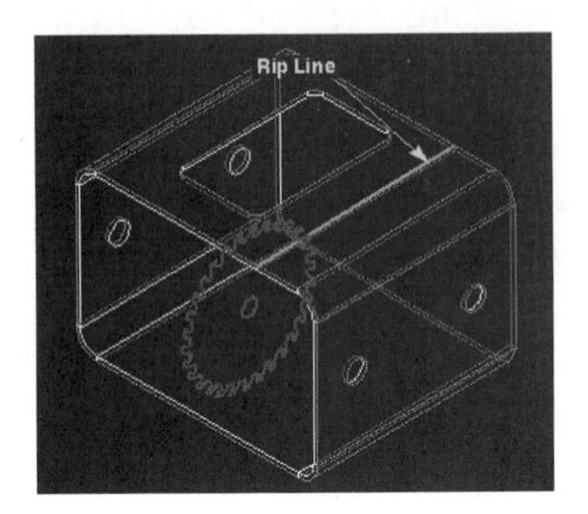

A Rip feature is a cut that is used to "saw" a part along a specified line or edge. The most commonly used **Rip** feature is a **Regular** Rip. Essentially, it saws the part in two along a sketched line, as shown in the left-hand column. To define a Regular Rip, you simply sketch a **Rip Line** that represents the cut. The Rip Line may be made up of multiple entities of any type, as long as they form a continuous chain. The Rip Line sketch must be aligned with the outside edges of a part. *A Regular Rip removes no volume from the part.* A common use for a Regular Rip feature is to sever a part that has no parting line so that it can be unbent into a flat pattern (Fig. 23.121).

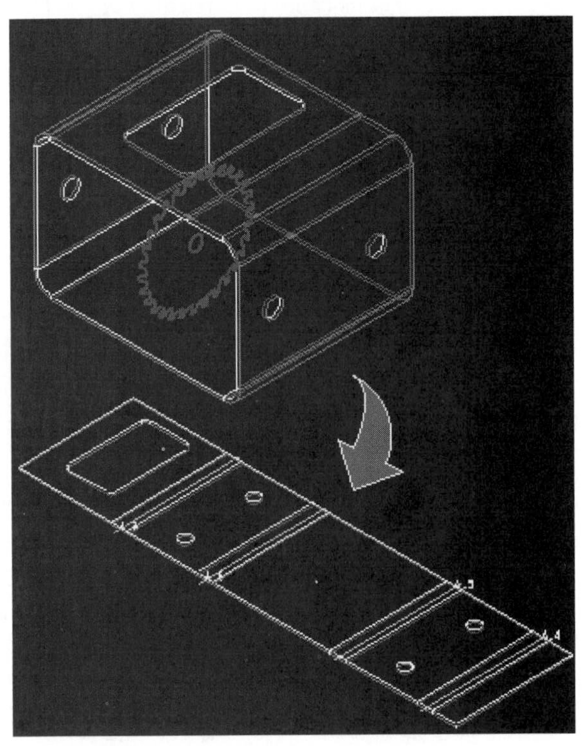

Figure 23.121
Rip

Create a new sheet metal part called **SM_ERIP_7.prt.** The first wall will be an Extruded Wall using the dimensions (Fig. 23.122).

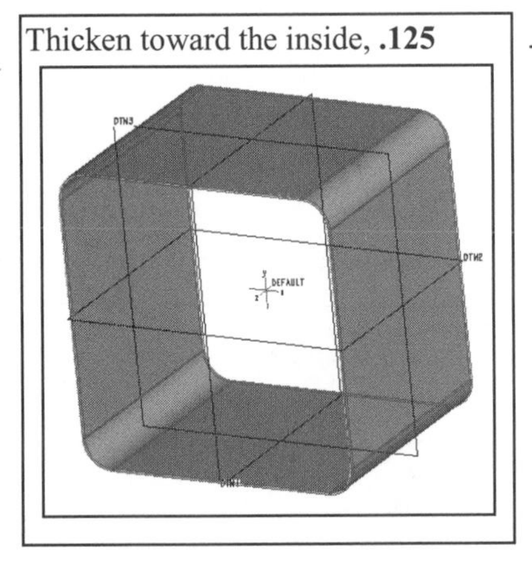

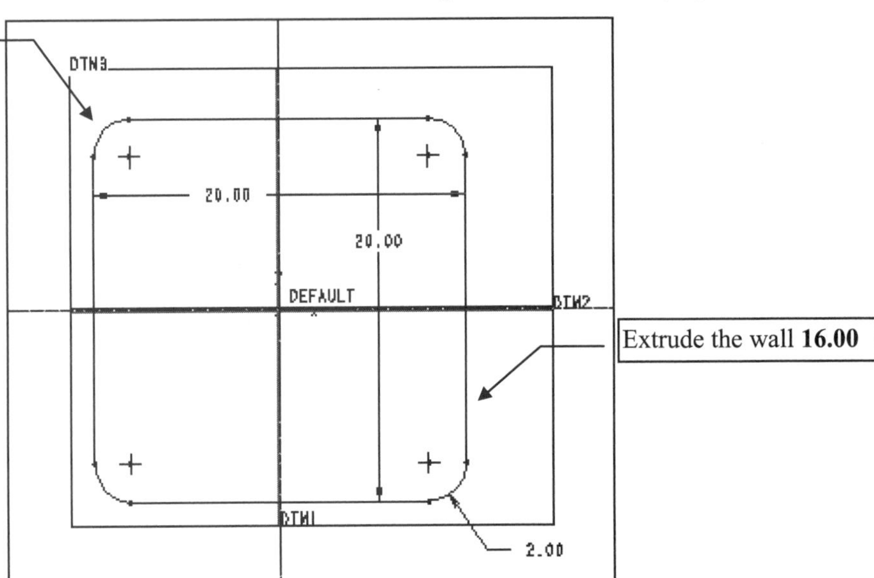

Figure 23.122
SM_ERIP_7.prt Sketch

Add a **Regular Rip** so the part can be unbent using **Flat Pattern:**

Sketching Plane

Reference edge

Feature ⇒ Create ⇒ Sheet Metal ⇒ Rip ⇒ Regular Rip ⇒ Done ⇒ [select the top surface as the sketching plane (see left column] ⇒ **Top** ⇒ (select the back edge to orient the sketch) ⇒ (create the sketch as shown in Figure 23.123) ⇒ [align to the outside edge *surfaces* (Fig. 23.123)] ⇒ [follow the prompts and complete the **Rip** (Fig. 23.124)]

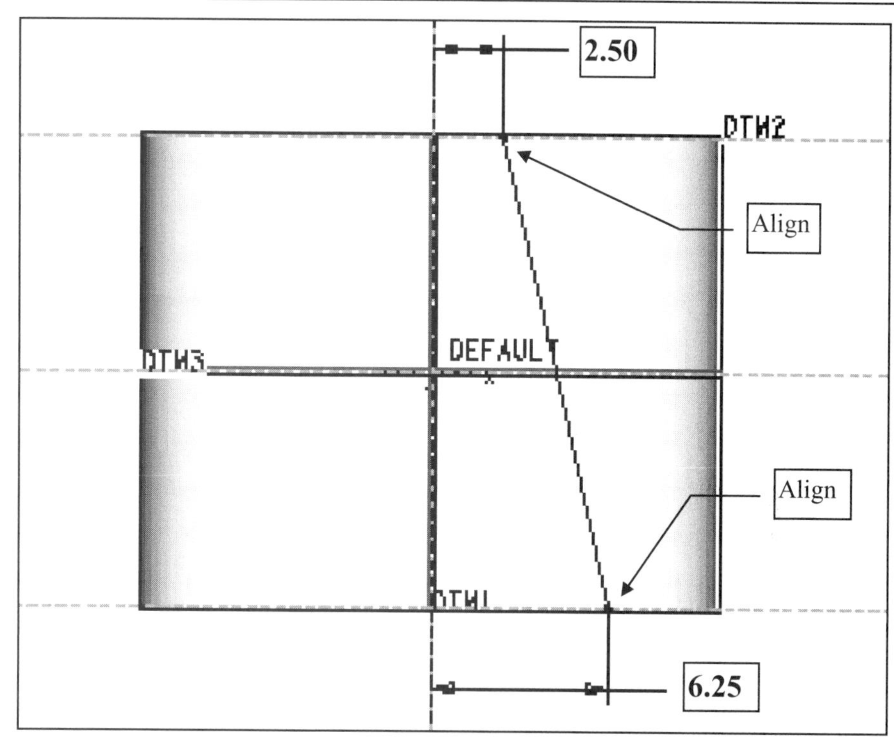

Figure 23.123
SM_ERIP_7.prt Sketch

Create ⇒ Flat Pattern ⇒ [pick the top surface, the part is now shown in its flat condition (Fig. 23.125)] ⇒ **Delete ⇒** (delete the Flat Pattern feature) ⇒ **File ⇒ Save ⇒ File ⇒ Close Window**

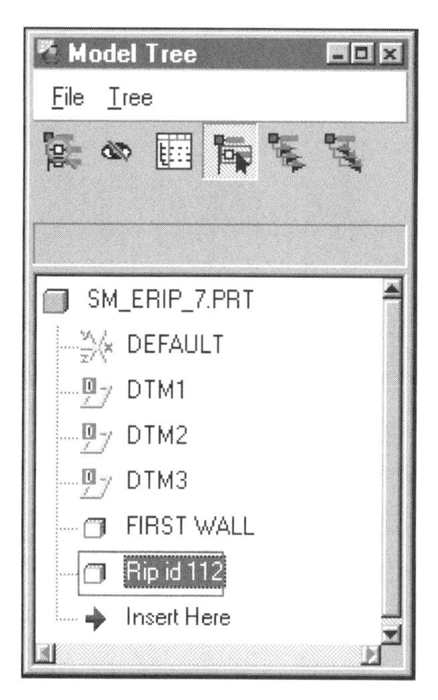

Figure 23.124
Completed Rip

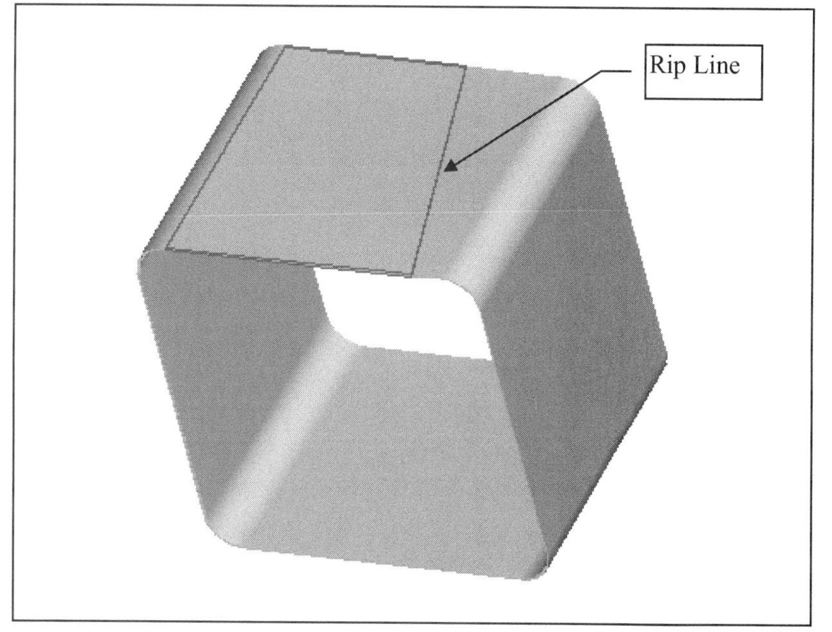

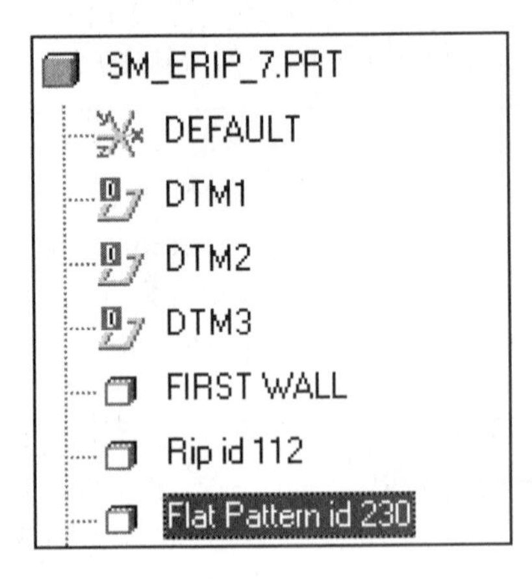

Figure 23.125
Flat Pattern

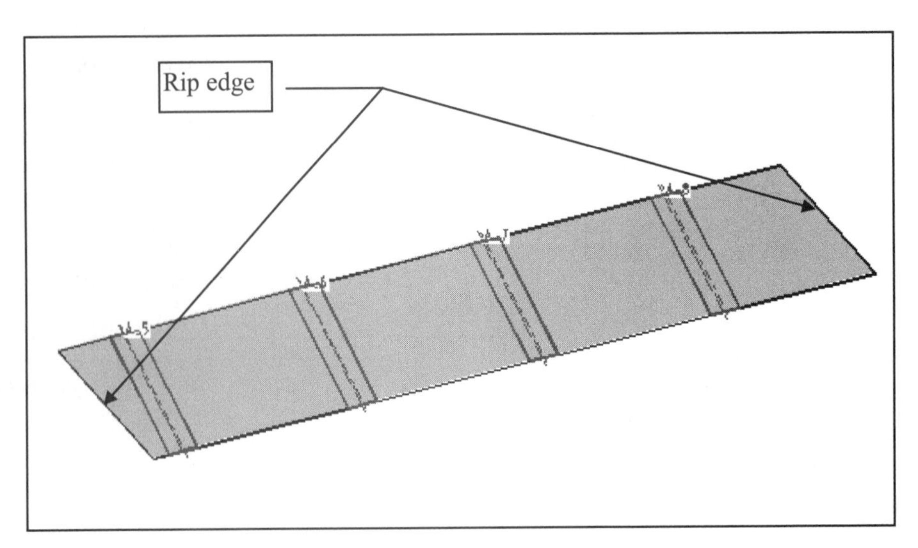

You can also create **Rip** features that allow you to unbend geometry that cannot otherwise be unbent. There are two types of features available in Pro/SHEETMETAL for this purpose: **Edge Rip** and **Surface Rip**.

Edge Rip features create a *"zero-volume"* cut on an edge of a part; it is similar to a Regular Rip, except that you specify a part edge instead of a sketched Rip Line. Surface Rip features are used to "rip out" the designated surface from the model.

When an Edge Rip feature is created. The appearance of the part does not change, because it is a "zero-volume" cut, and the location coincides with an edge. What has actually occurred is that the part has been sliced (Fig. 23.126). This will become apparent when the **Flat Pattern** is displayed.

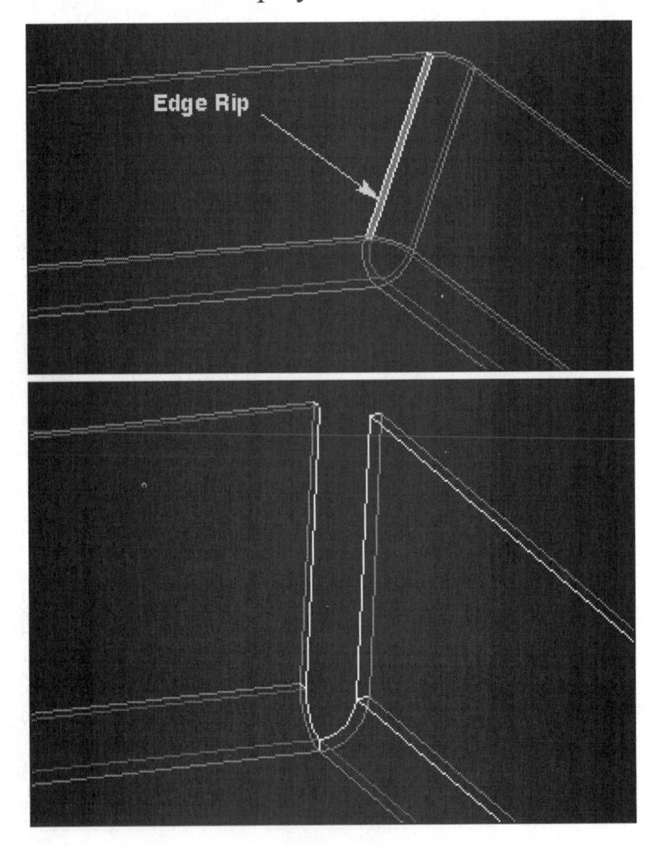

Figure 23.126
Edge Rip (top)
Surface Rip (bottom)

Create a metric part (Solid Part, not a sheet metal part) using the dimensions shown in Figure 23.127.

Solid Part **is 150** deep **X 350** wide **X 250** height. All rounds are **R20.**

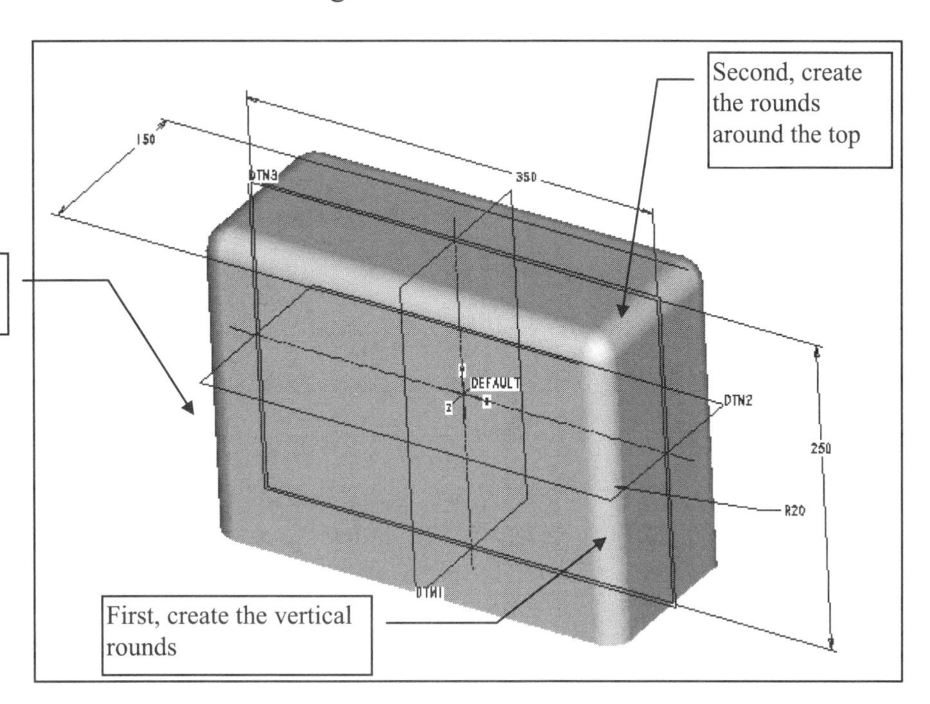

Figure 23.127
Solid Part (**SM_SRIP_8.prt**)

Convert the part into a sheet metal part. Use the **Shell** option.

Applications
- Standard
- Sheetmetal
- Legacy
- Scantools
- Mechanica
- Plastic Advisor
- Mold/Casting

Applications ⇒ **Sheetmetal** ⇒ **Shell** ⇒ [select the bottom surface to remove (Fig. 23.128)] ⇒ **Done Sel** ⇒ **Done Refs** ⇒ **5** ⇒ ✔

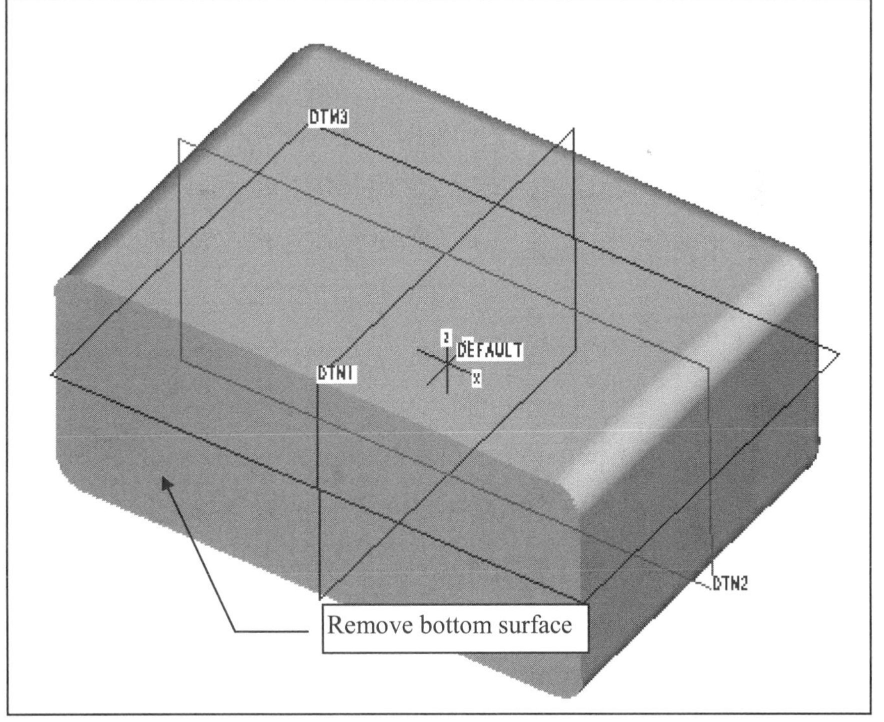

Figure 23.128
Convert the Solid Part to a Sheet Metal Part Using the Shell Option

The part is now shelled and is a sheet metal part. The **Shell** feature becomes the **First Wall** of the part (Fig. 23.129).

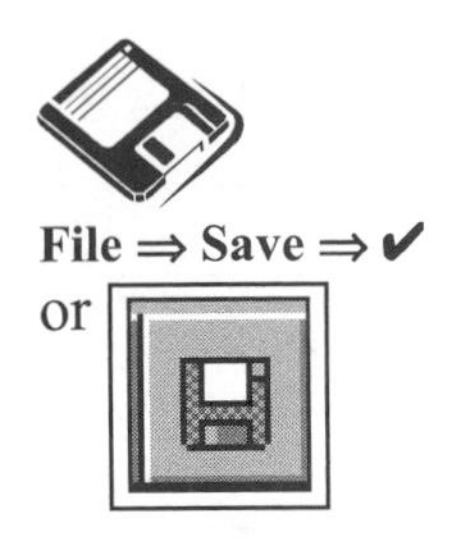

File ⇒ Save ⇒ ✔
or

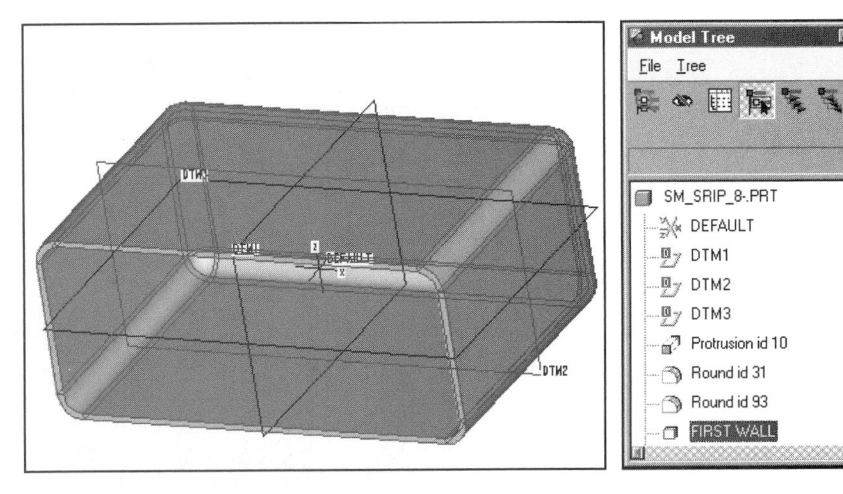

Figure 23.129
Sheet Metal Part with the First Wall as the Shell Feature

Next, create an **Edge Rip.**

Feature ⇒ Create ⇒ Rip ⇒ Edge Rip ⇒ Done ⇒ (select the two edges shown in Figure 23.130) **⇒ Done Sel ⇒ Done Sets ⇒ OK**

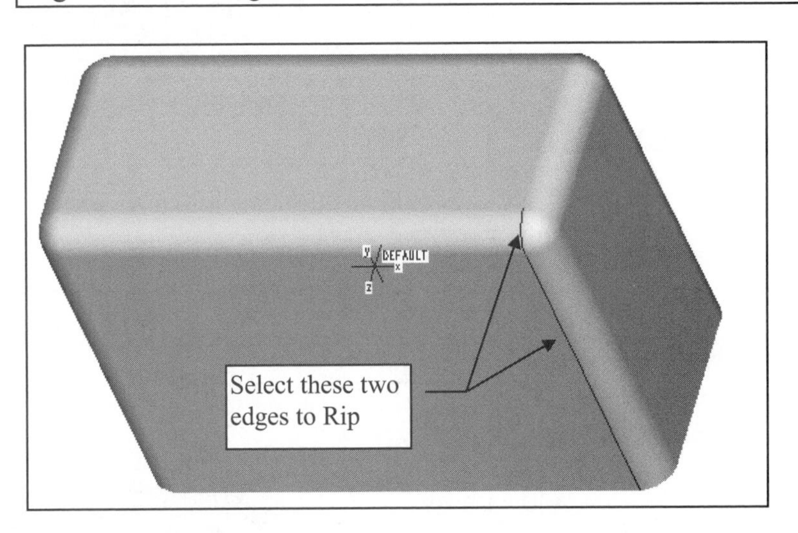

Figure 23.130
Edge Rips

Feature ⇒ Create ⇒ Rip ⇒ Surface Rip ⇒ Done ⇒ [select the two surfaces (Fig. 23.131)] **⇒ Done Sel ⇒ Done Refs ⇒ OK**

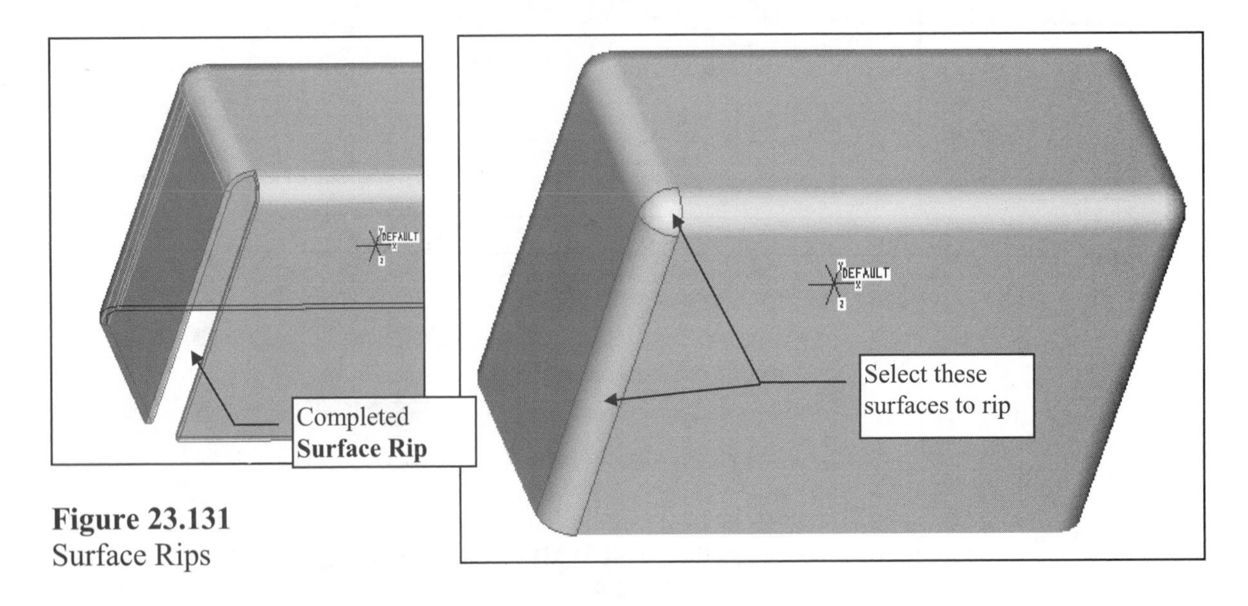

Figure 23.131
Surface Rips

Now, rip the back edges and surfaces. Use a combination of edge and surface rips on each. For the right rear corner use an **Edge Rip** for the straight edge and **Surface Rip** for the curved corner. For the left rear corner use a **Surface Rip** for the long straight surface and an **Edge Rip** for the corner curved section (Fig. 23.132). After you have completed the rips, try and create a **Flat Pattern** and investigate the differences that the rip type makes. If you get a failure, investigate the problem and solve it by using more Edge Rips and Surface Rips at the problem edges or surfaces. You should need one of each to solve the problem (Fig. 23.133). Delete the Flat Pattern feature before saving.

File ⇒ Save ⇒ ✔

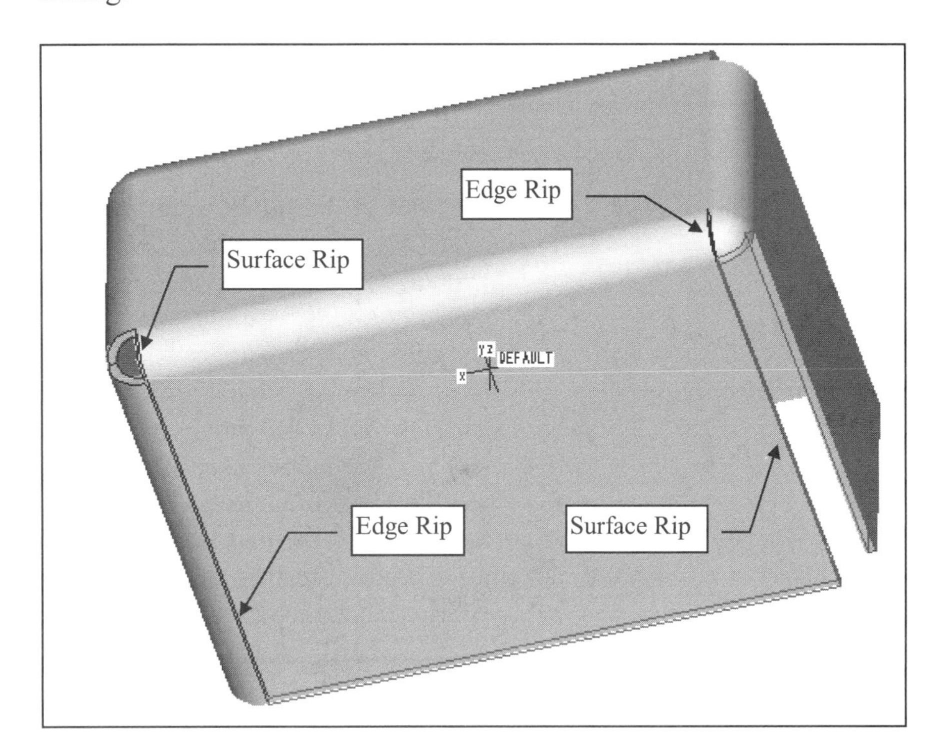

Figure 23.132
Surface and Edge Rip Combinations

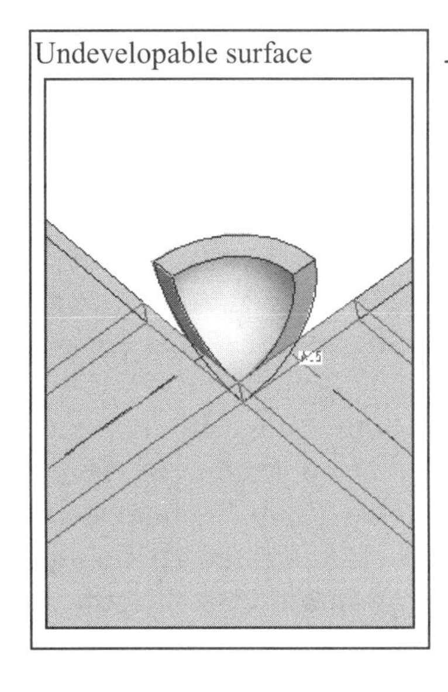

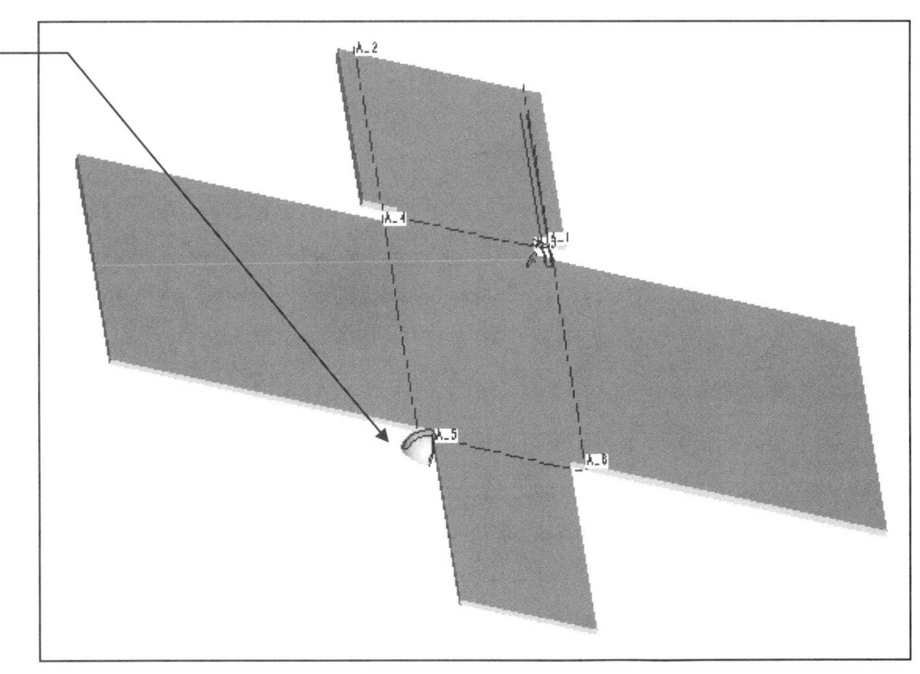

Figure 23.133
Flat Pattern

Bend Table

Pro/SHEETMETAL allows you to specify a **Bend Table** for a Sheet Metal part. You may also define a different Bend Table for each feature, if required. You can specify which Bend Table you want to use when you create any **Bend** feature (as you may have noticed).

The default Bend Tables that are supplied with Pro/SHEETMETAL are copied from (and used with the permission of) the *Machinery Handbook*. The tables are based on the hardness and *"bendability"* of the material being used:

TABLE1	soft brass and copper
TABLE2	hard brass and copper, soft steel and aluminum
TABLE3	hard copper, bronze, cold rolled steel (CRS) and spring steel

Bend Tables apply only to constant-radius bends. The flat pattern size of variable-radius bends is determined by the Y-factor formula. If you have some other method of developing **Flat Pattern** lengths of bends, you must create user-defined Bend Tables.

When you create any sheet metal Bend feature, you can specify that it reference either the **Part Bend Table** or else a **Feature Bend Table.** Most of the time, you should select Part Bend Table, since the behavior of bends is usually the same for all Bend features in the part. Possible exceptions to this rule are multiple-material models or when you have user-defined Bend Tables that take the grain of the material into account. In this case, you may have one table for across-grain features, and another table for along-grain features. Note that the default Bend Tables supplied with Pro/SHEETMETAL do not account for the material grain direction.

The Part Bend Table is specified with the **Set Up** option. Each bend can be made to reference it by selecting the **Part Bend Tbl** option from the USE TABLE menu.

If you choose the **Feat Bend Tbl** option, Pro/E displays menus that allow you to select from the tables located in the Bend Table directory. The default directory is *proXX/text/bend_tables.*

Creating Sheet Metal Drawings

Sheet metal drawings usually contain views both of the fully formed part and of the flat pattern (Fig. 23.133).

You can place views of the part in both states on the same drawing by creating a multi-model drawing. *The best way to create a Flat Pattern model of a part is to use its Family Table.* A function in the **Setup** options for Pro/SHEETMETAL is designed to do this automatically. The Flat State option stores an instance of the part, in its flat state, in the Family Table. The original model remains in the bent position (Fig. 23.134).

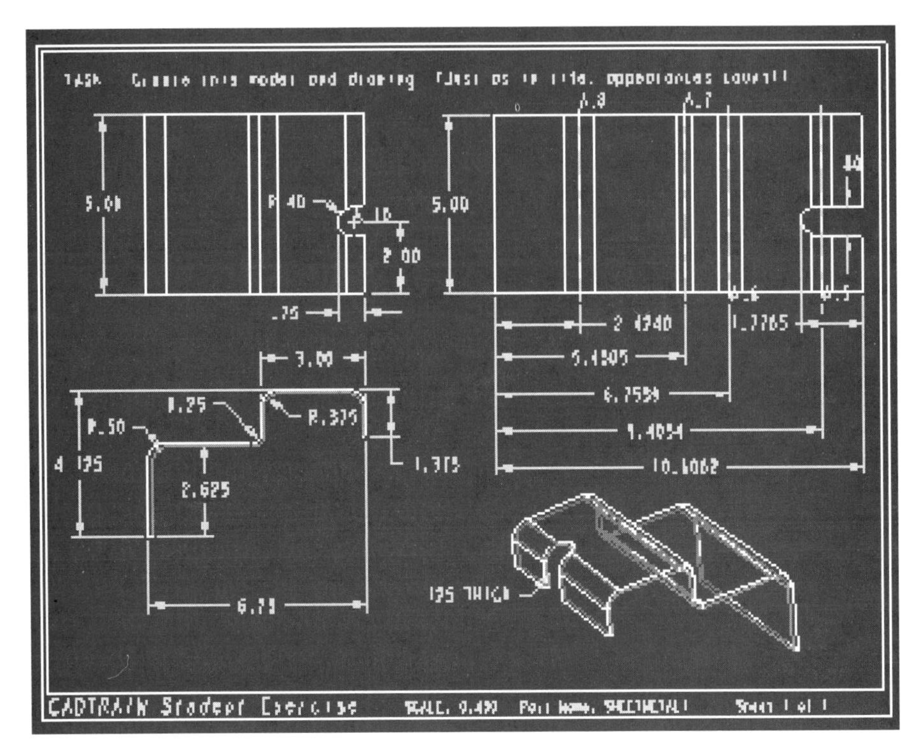

Figure 23.134
Sheet Metal Detail Drawing

After creating the Flat State, the original part model will become the **Generic** in the Family Table; the flat pattern is defined as an instance of the generic part. By default, this instance will be called ***partname*****_FLAT1.** You may then add both of these models to the sheet metal drawing.

A Family Table instance (Fig. 23.135) is not a separate part file and therefore does not show up in your parts directory listing when you are using the operating system. The instance does appear in the Pro/E listing of parts when you use the **File ⇒ Open** option (see left-hand column). When you create a flat state instance of a sheet metal part, Pro/E creates a file called *sheetmetal.idx* in your current directory. If this file already exists, Pro/E edits the file. You must have write access to your current directory to be able to create a flat state instance.

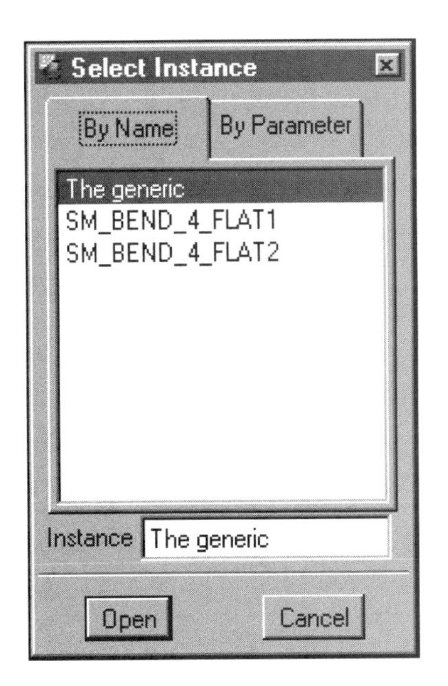

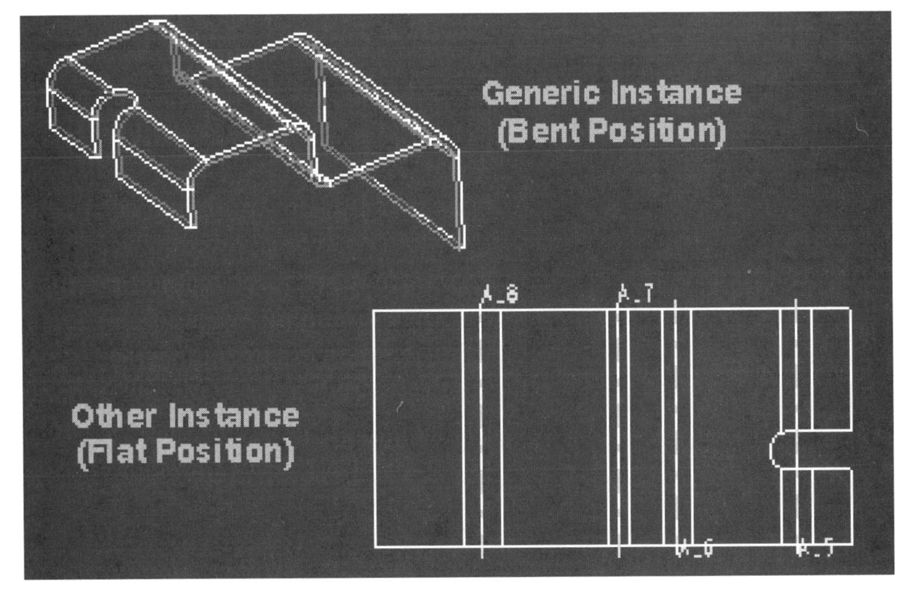

Figure 23.135
Generic and Flat Pattern Instance

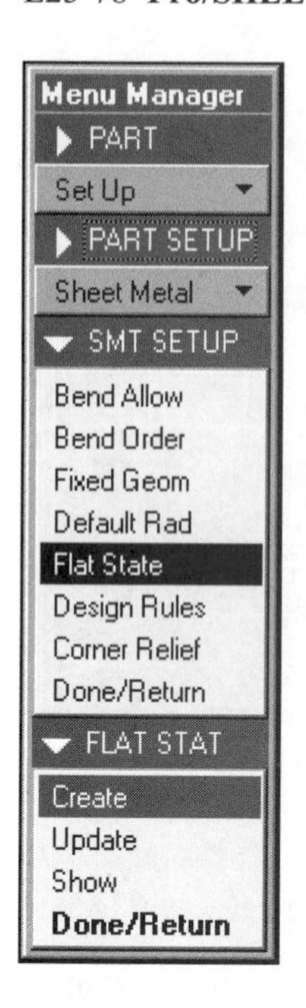

Create a Family Table containing both the bent and flat states for the provided part. Retrieve the part: **SM_BEND_4.prt**. Give the following commands to set up the flat state instance for the part.

Set Up ⇒ **Flat State** ⇒ **Create** ⇒ ✔(to accept the default name for the instance) ⇒ (Select the *current* state of the part. Because it is currently bent, pick **Fully Formed** from the PART STATE menu) ⇒

Pro/E displays the Feature dialog box and prompts you to select the Fixed Geometry for the part. The fixed geometry is a plane that remains stationary during the Bending operations.

[select the top face (Fig. 23.136) as the **Fixed Geometry**] ⇒ **OK** ⇒ **Show** (to show the Flat State instance) ⇒ (Select **SM_BEND_4_FLAT1.prt** from the list) ⇒ ✔ (Fig. 23.137)

Close the **SM_BEND_4_FLAT1.prt** window, but do not erase it. This closes the flat state window but keeps the flat state in memory for use later. **Save** the part (this also saves the flat pattern instance).

Figure 23.136
SM_BEND_4.prt

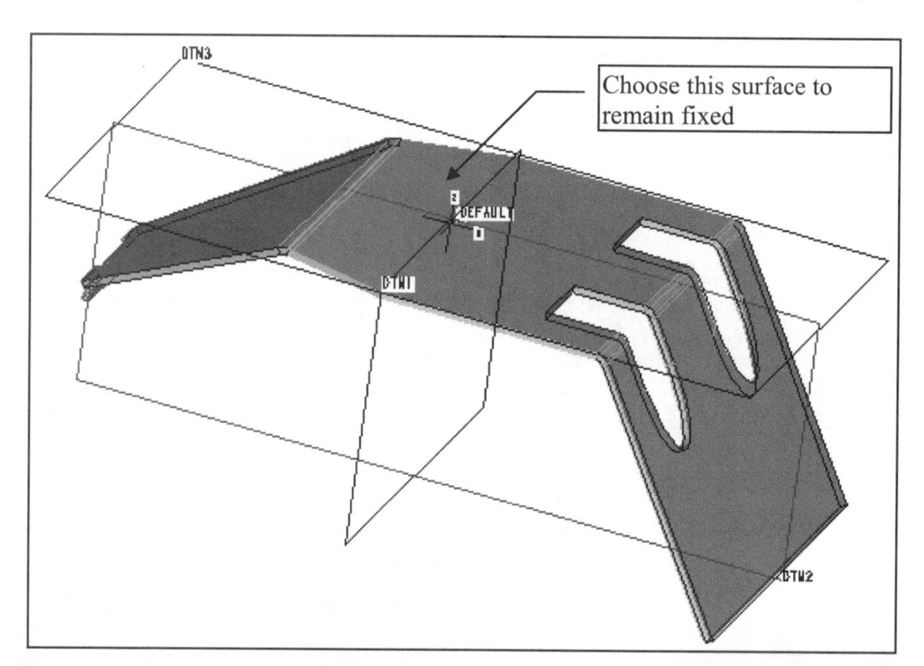

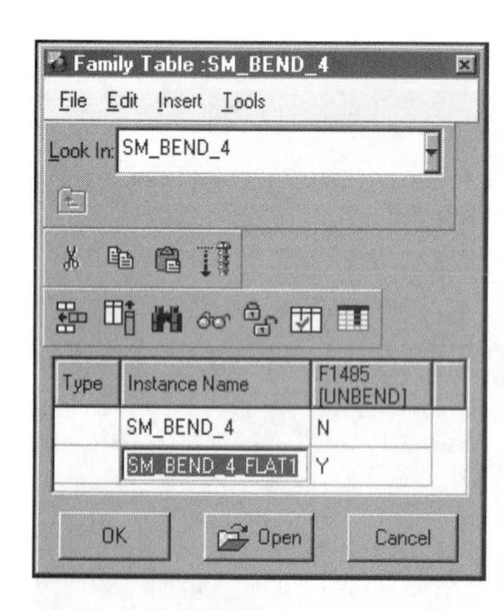

Figure 23.137
Flat State Instance
SM_BEND_4_FLAT1.prt

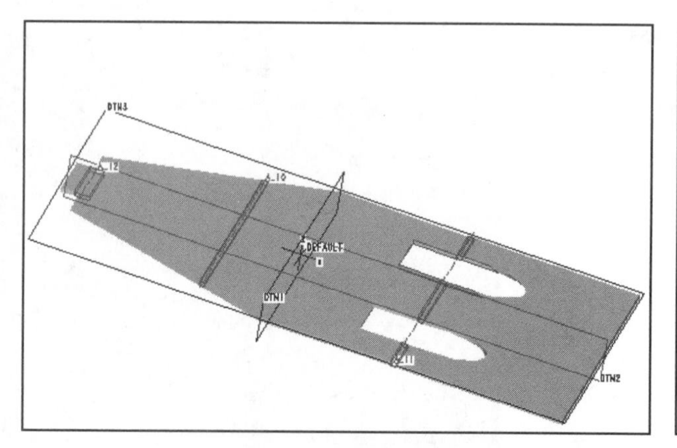

SM_BEND_4_FLAT1.PRT
- DEFAULT
- DTM1
- DTM2
- DTM3
- FIRST WALL
- Bend id 83
- Bend id 176
- Wall id 485
- Cut id 632
- Cut id 681
- Unbend id 1485

Bend Order Table

Pro/SHEETMETAL has the ability to generate automatically a **Bend Order Table** that can be placed as a note on your sheet metal drawing. *Do not confuse the **Bend Order Table** with the **Bend Table**.* The Bend Order Table (shown below) is a list of **Bend Sequences**, with their individual Bends. *The Bend Order Table is text only--it is not associated to the model once it is placed on the drawing:*

BEND SEQ	#BENDS	BEND#	BEND DIRECTION	BEND ANGLE	BEND RADIUS	BEND LENGTH
1	2	1	OUT	90.000	2.000	3.008
		2	OUT	90.000	1.000	1.437
2	1	1	IN	90.000	1.000	1.633
3	1	1	IN	90.000	2.000	3.204

When you generate a Bend Order Table, Pro/E writes a text file that is automatically named using the name of the part. The file has a ***.bot*** extension. The Bend Order Table file is re-created whenever the Bend Order of the part is modified.

You can place the contents of the *.bot* file on a sheet metal drawing by creating a note from a saved text file. If the Bend Order of a part is modified, this note must be deleted and re-created in order to contain the correct Bend Sequences.

Next, you will create a Bend Order Table for **SM_BEND_4.prt** (Fig. 23.138). Finally, you will create a detail drawing and add both the formed part (with required views) and the Flat State instance. A note that contains the Bend Order Table text is also required.

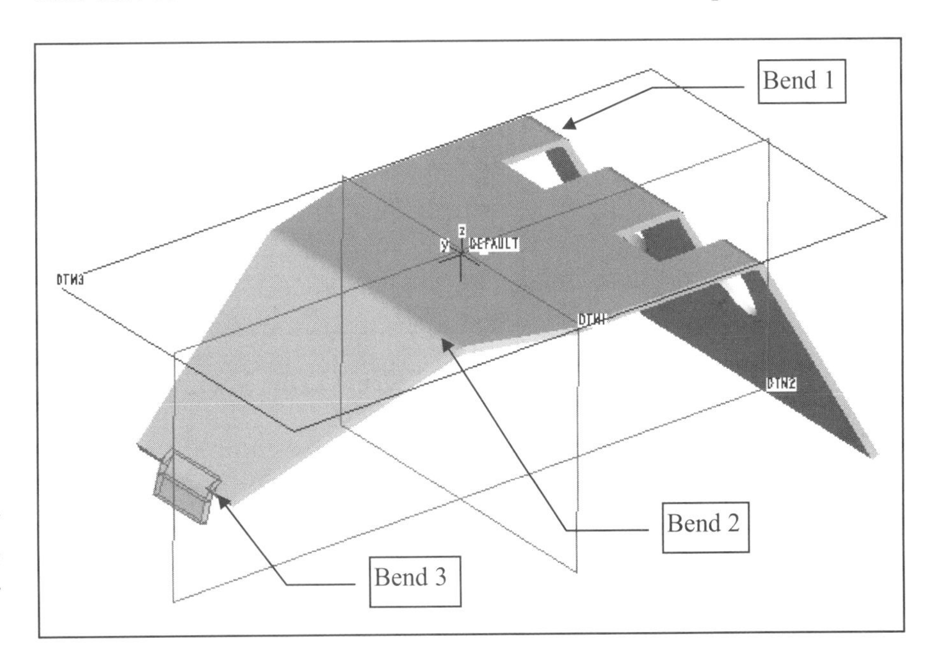

Figure 23.138
SM_BEND_4.prt
Showing the Order Number for the Bend Sequence to Be Selected for the Bend Order Table

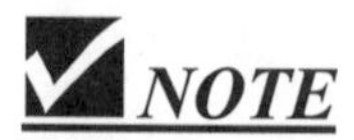

NOTE

When a Bend Order Table already exists for a part, the system already knows which is the Fixed.

SM_BEND_4.prt should still be active on your screen. If not, retrieve it now. Create the Bend Order Table using the following commands:

Set Up ⇒ **Bend Order** ⇒ **Show/Edit** ⇒ (select the face as shown in left column as the stationary face; use the same one that you selected when you created the Flat State instance) ⇒ (select the three bends (surfaces) in the order shown in Figure. 23.139) ⇒ **Done Sel** ⇒ **Done** ⇒ (select the face to remain fixed while bending back)

Pick this face again to establish the plane to remain fixed while bending back

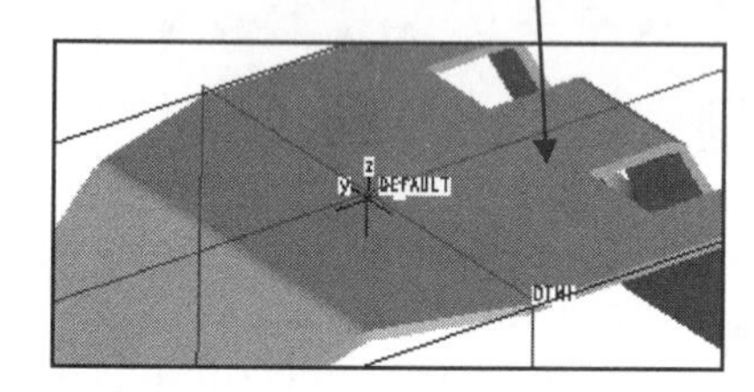

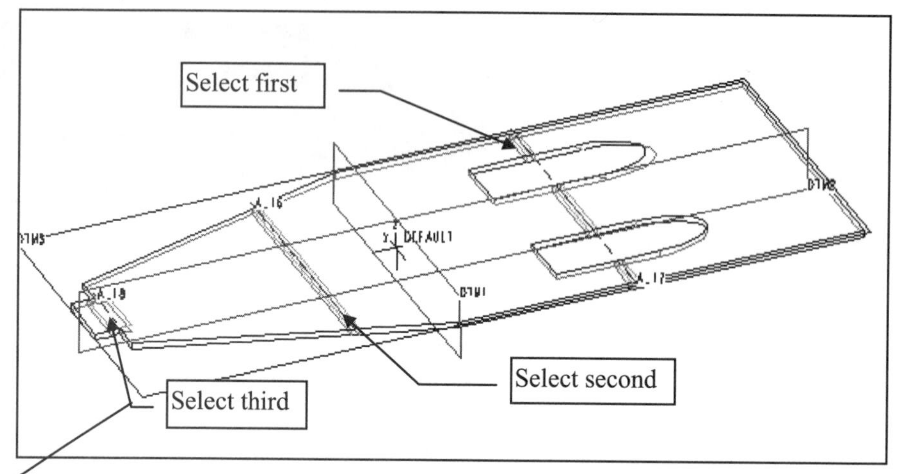

Figure 23.139
Select the Bends to Create the Bend Order Table

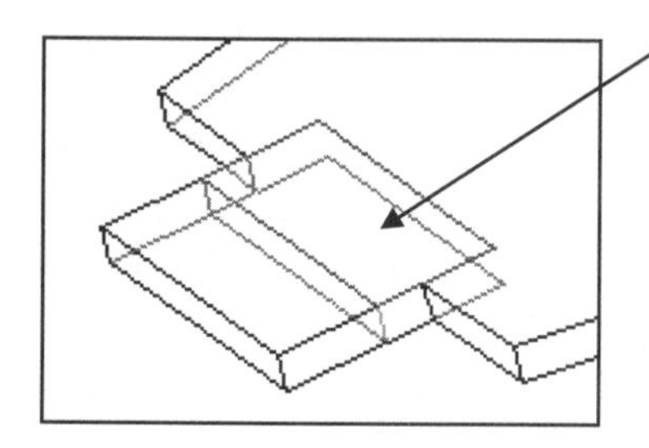

You have created a Bend Order Table with one Bend Sequence, containing three Bends. Display the Bend Order Table on the screen (Fig. 23.140):

Info (from the BEND ORDER menu) ⇒ **Close** (to exit the Information Window) ⇒ **Done/Return** ⇒ **Done/Return** ⇒ ▼ ⇒ **Done**

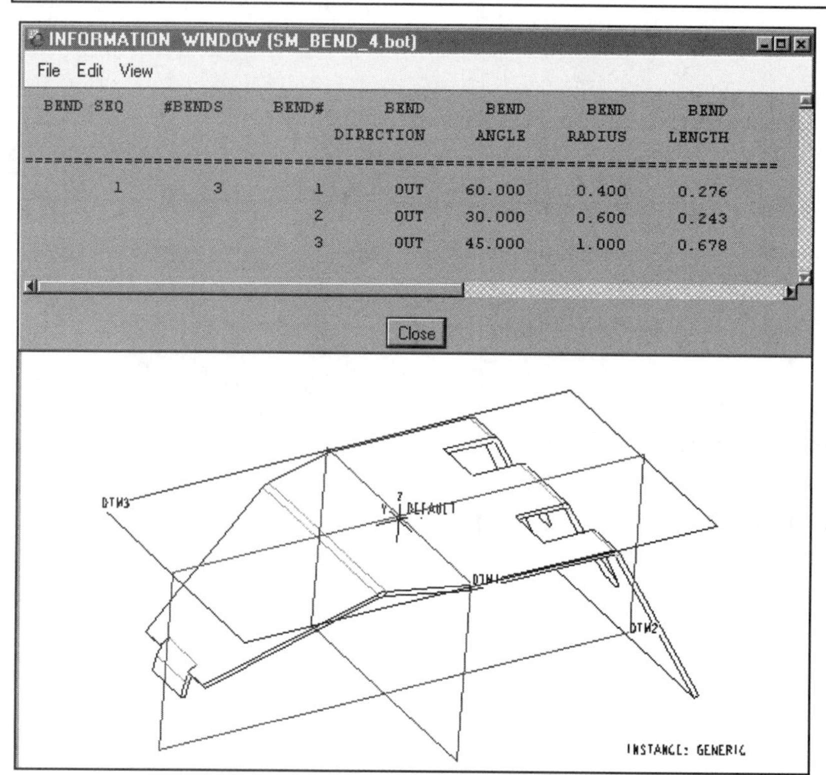

INFORMATION WINDOW (SM_BEND_4.bot)

File Edit View

BEND SEQ	#BENDS	BEND#	BEND DIRECTION	BEND ANGLE	BEND RADIUS	BEND LENGTH
1	3	1	OUT	60.000	0.400	0.276
		2	OUT	30.000	0.600	0.243
		3	OUT	45.000	1.000	0.678

Close

File ⇒ **Save**

Figure 23.140
Select the Bends to Create the Bend Order Table

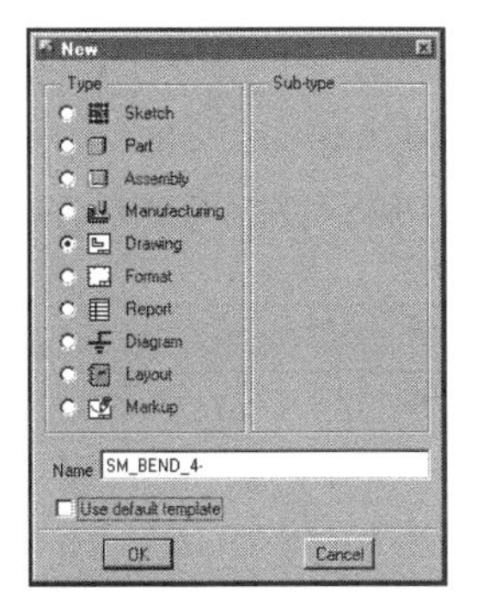

Close the **SM_BEND_4.prt** part window, but do not erase it from memory. Start a new drawing:

File ⇒ New ⇒ Drawing ⇒ SM_BEND_4 ⇒ ❑ Default template ⇒ OK ⇒ ● Empty with Format ⇒ Browse ⇒ d.frm ⇒ Open ⇒ OK ⇒ The generic (from the Select Instance dialog box) **⇒ Open ⇒** (place the first view of the bent model)

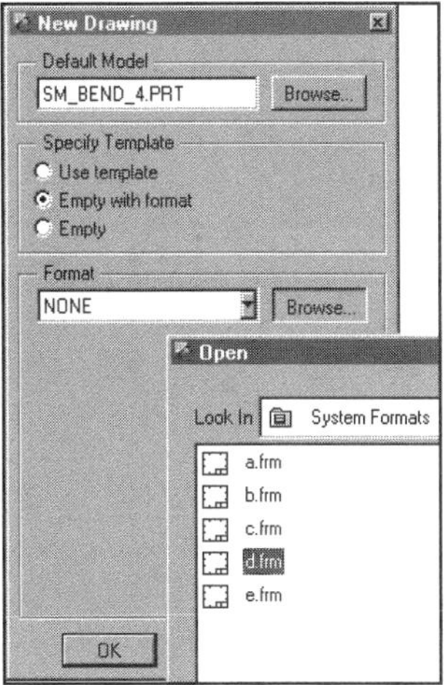

Add the required views as shown in Figure 23.141. Adjust the views and select the scale that best fits your needs.

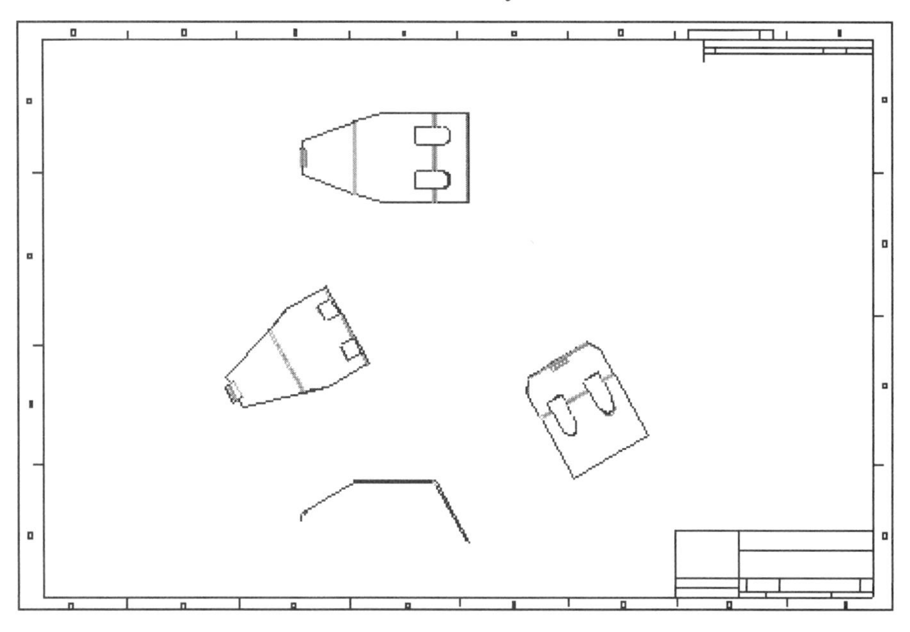

Figure 23.141
Select the Bends to Create the Bend Order

Add a view of the Flat Pattern (Fig. 23.142):

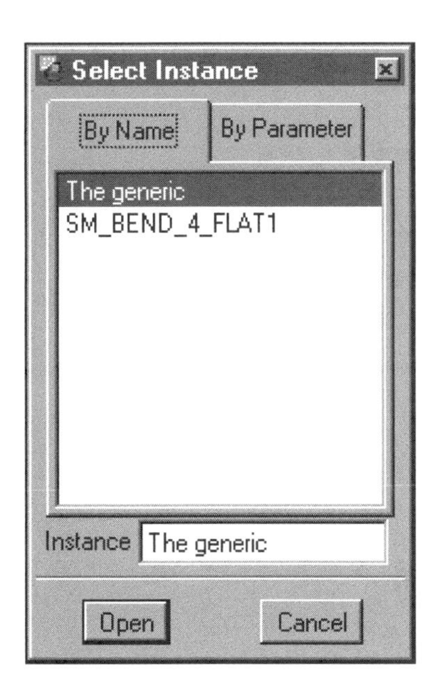

Views ⇒ Dwg Models ⇒ Add Model ⇒ SM_BEND_4_FLAT1.prt (from the list) **⇒ Open ⇒ Add View ⇒** (add a view of the Flat Instance)

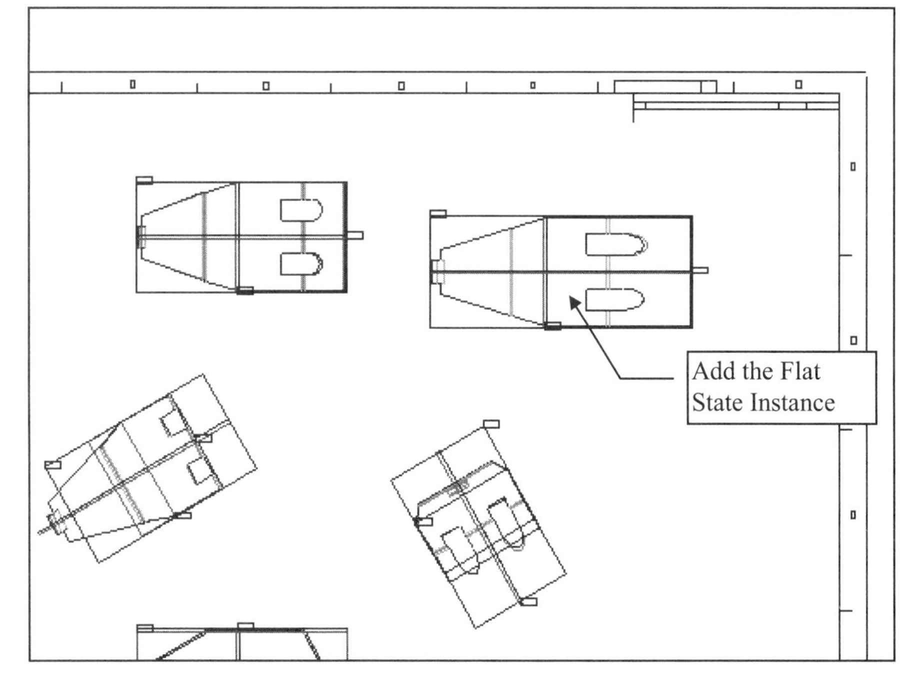

Figure 23.142 Adding a View of the Flat Instance

Add the Bend Order Table to the drawing (Fig. 23.143):

Create ⇒ Note ⇒ File ⇒ Right ⇒ Make Note ⇒ (indicate the location of the note) ⇒ Name **SM_BEND_4.bot ⇒ Open ⇒ Done/Return**

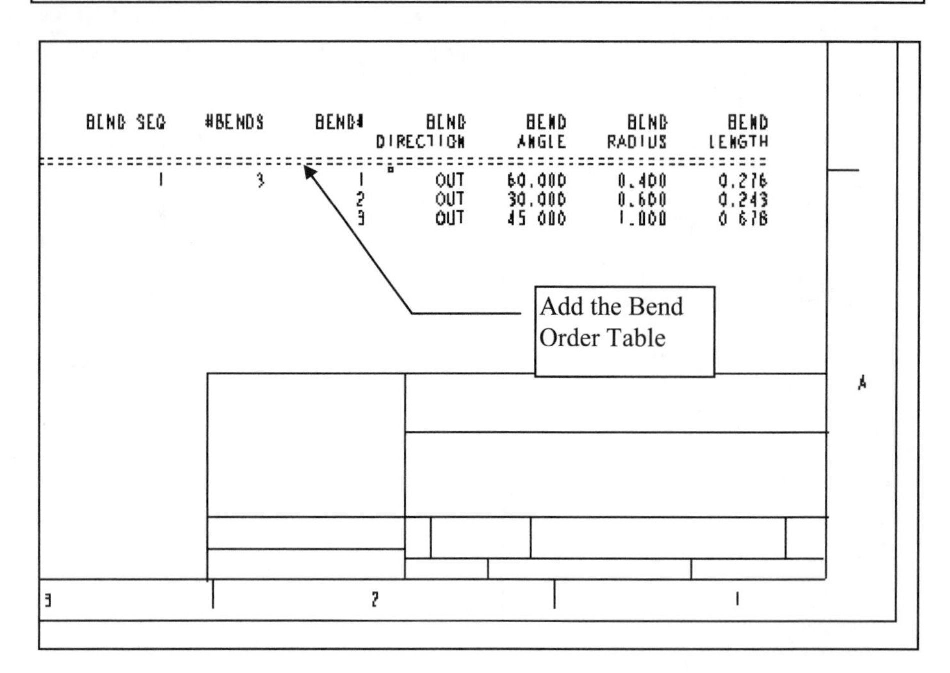

Figure 23.143
Adding a Bend Order Table to the Drawing

Show all dimensions and centerlines (Fig. 23.144). Clean up the view and note placement. Complete the drawing by showing the centerlines, dimensions, etc. Follow ASME detailing standards.

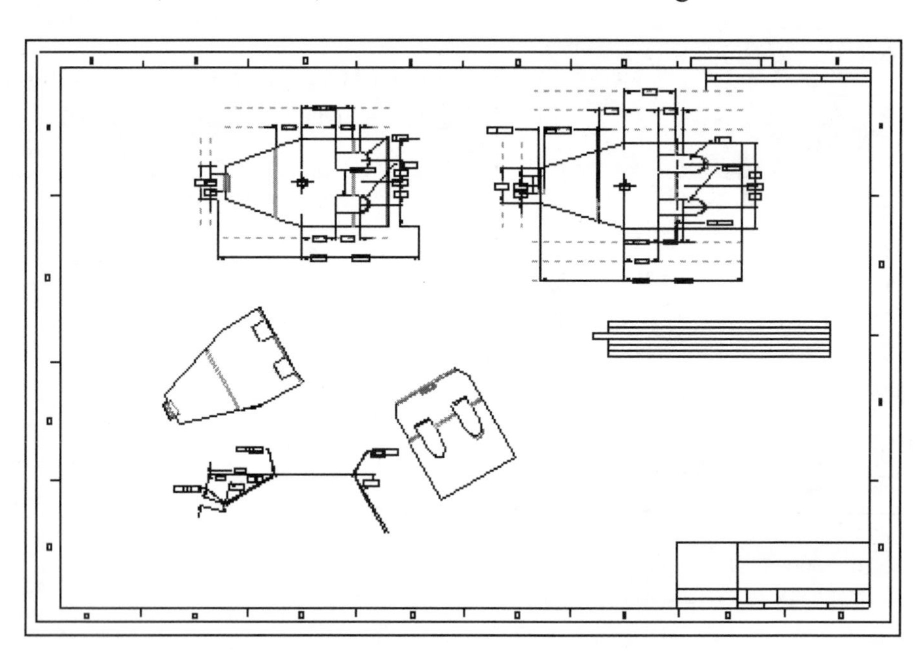

Figure 23.144
Drawing Before Finalizing to ASME Standards

You are now ready to complete the lesson projects, utilizing the commands and tools you have mastered in this lesson.

Lesson 23 Projects

Sheet Metal Parts

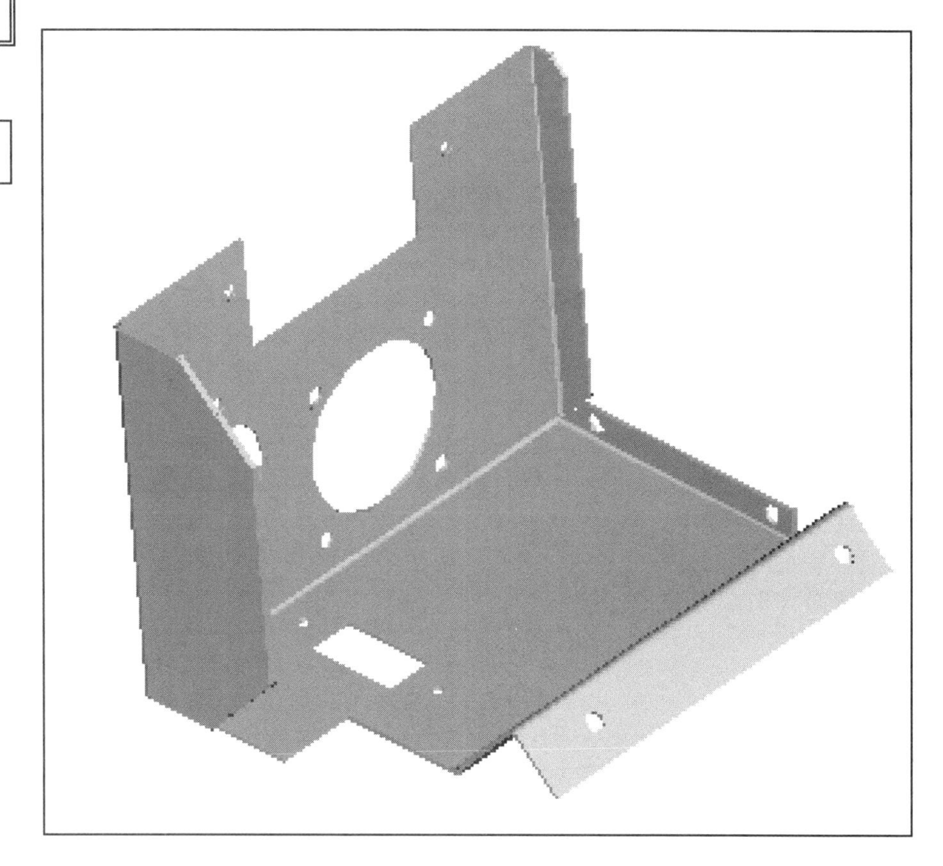

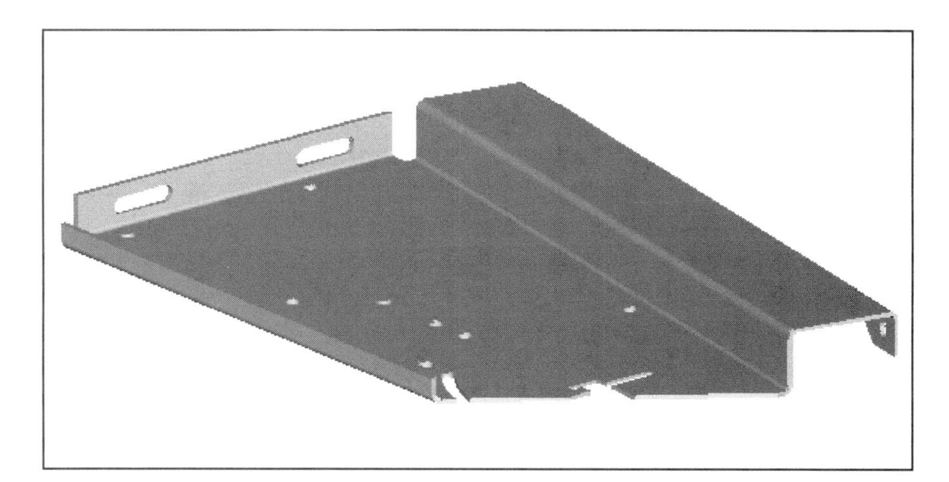

Figure 23.145
Bracket_1
Bracket_2

Sheetmetal Parts

Model one or more of the projects provided (Fig. 23.145 to Fig. 23.170). After the sheet metal part is modeled, create a drawing. On the detail drawing, include an appropiate set of dimensioned orthographic views of the fully formed as-designed part, a flat state dimensioned view, a bend order table, and a pictorial view. Transfer the flat pattern to a sheet of thin cardboard, cut out the outline, and bend as required to make a physical model.

Bracket_1

Figure 23.146
Bracket_1, Drawing

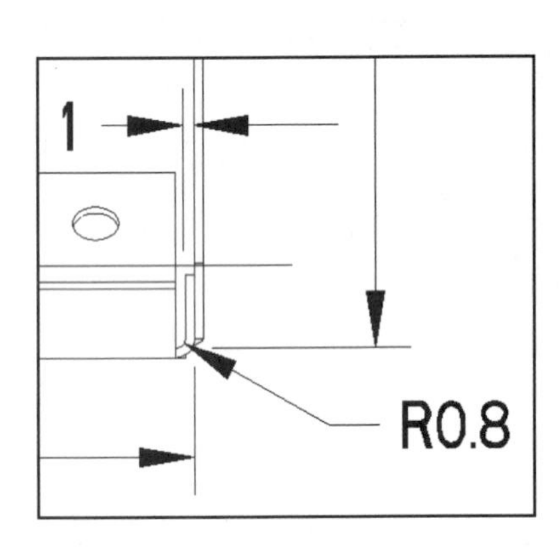

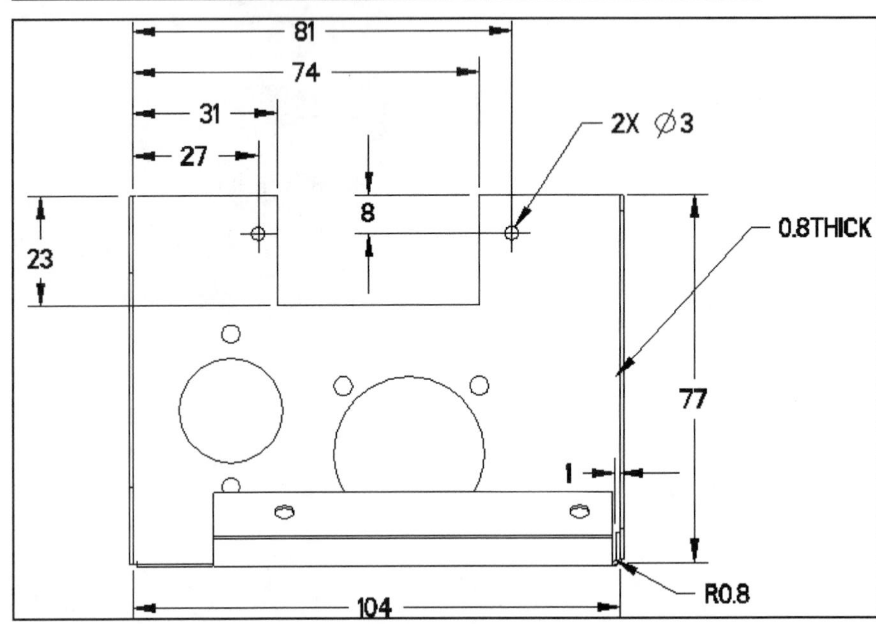

Figure 23.147
Bracket_1, Front View

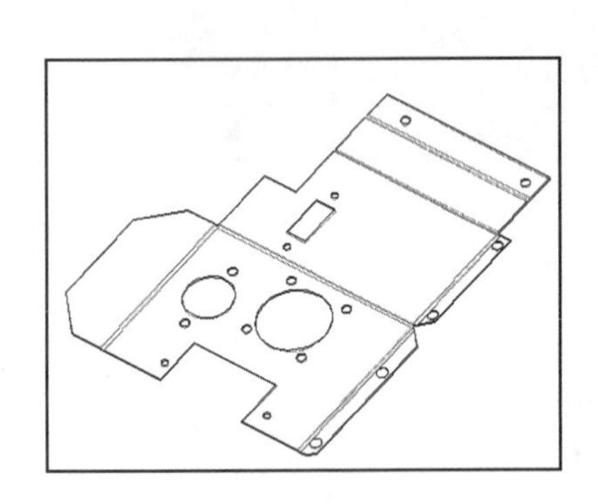

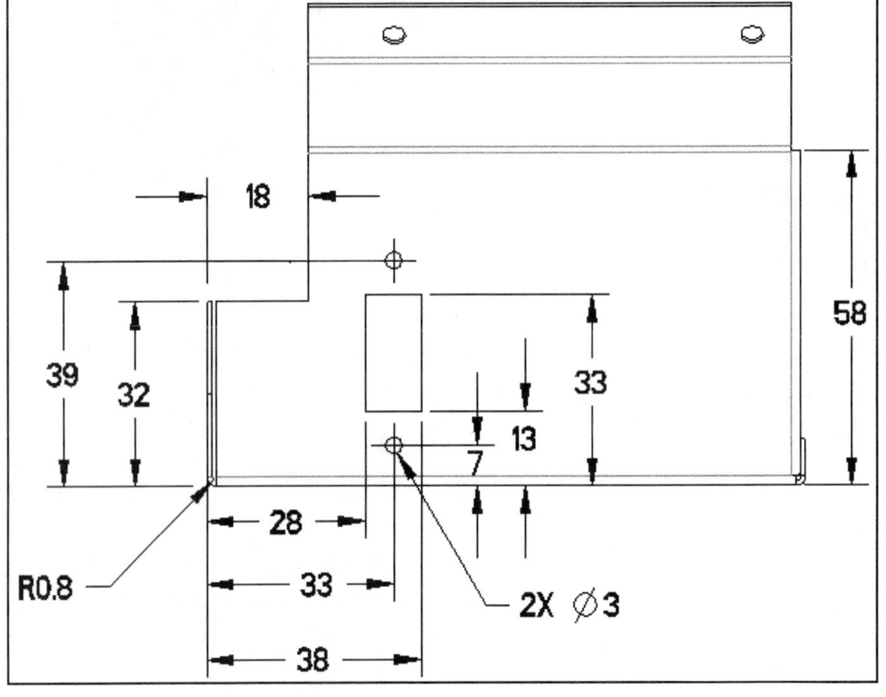

Figure 23.148
Bracket_1, Bottom View

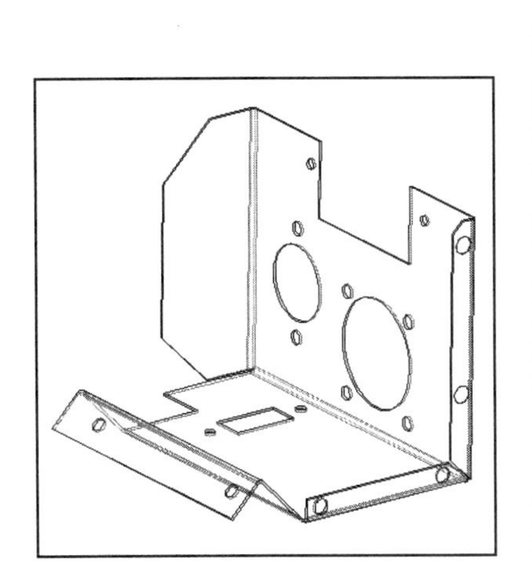

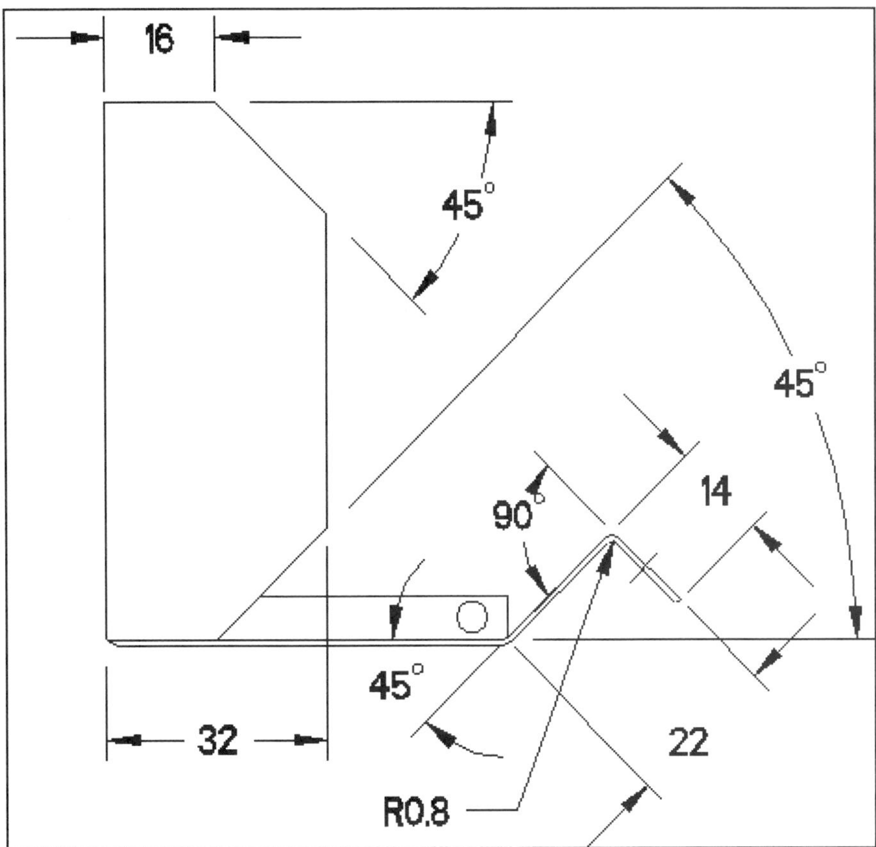

Figure 23.149
Bracket_1, Left Side View

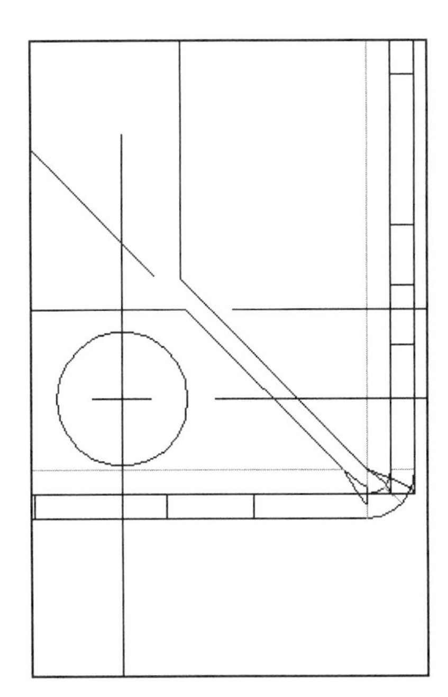

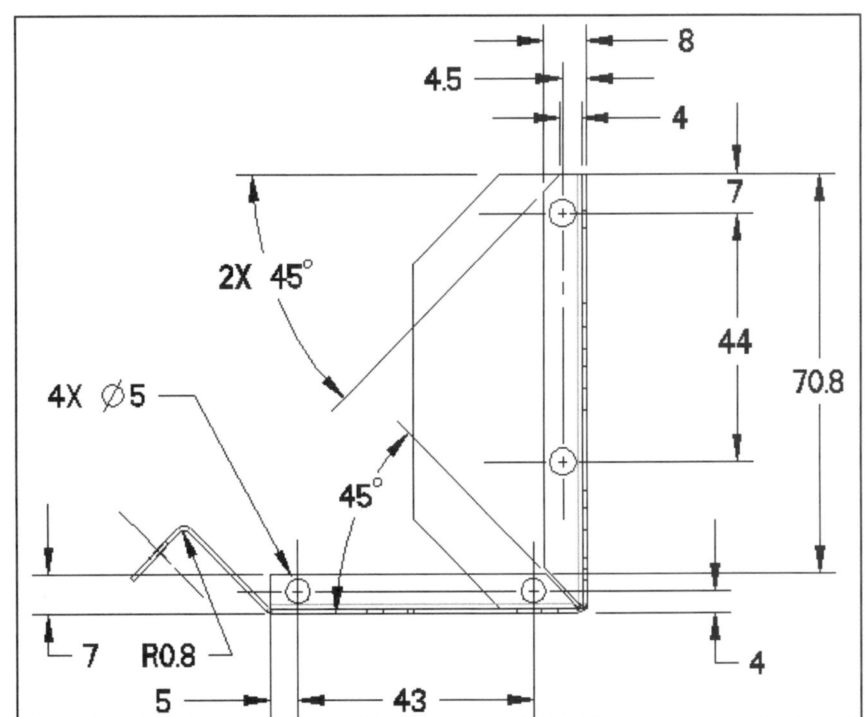

Figure 23.150
Bracket_1, Right Side View

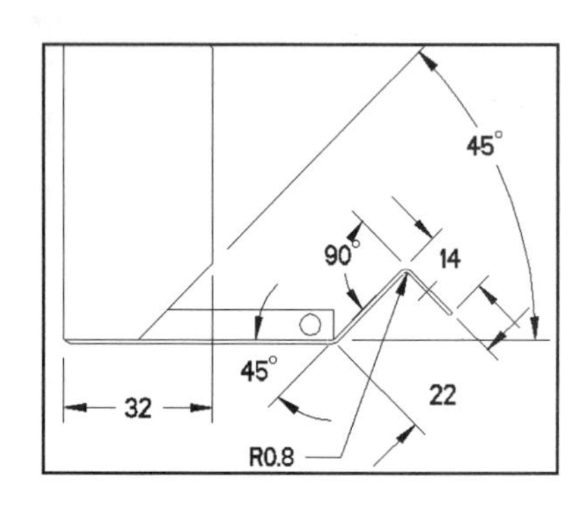

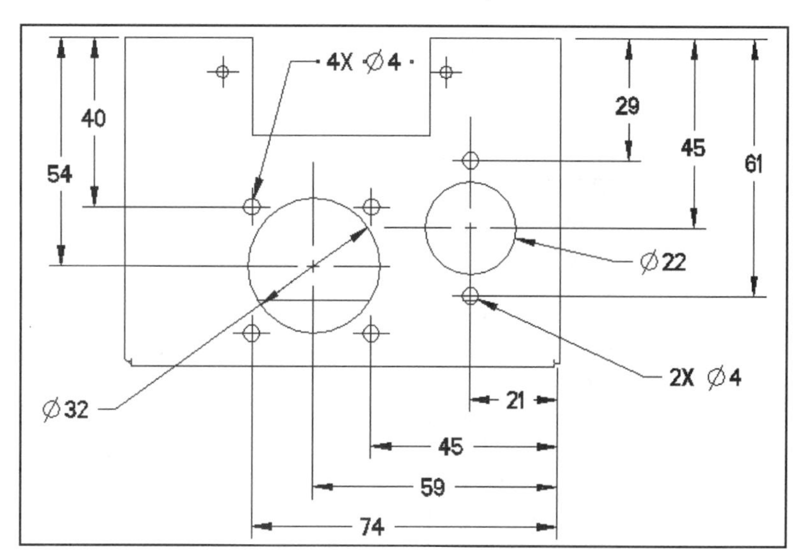

Figure 23.151
Bracket_1, Back View

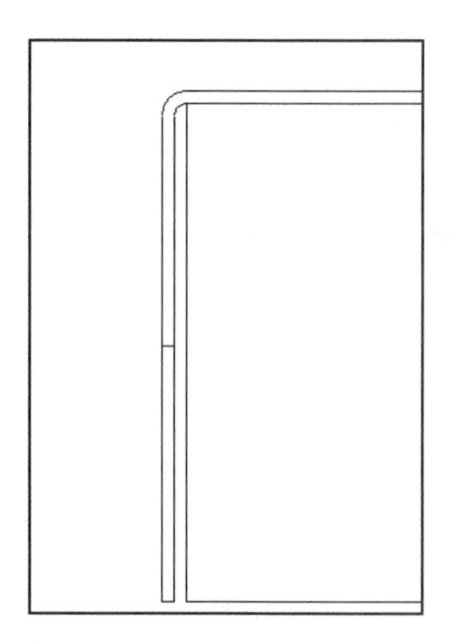

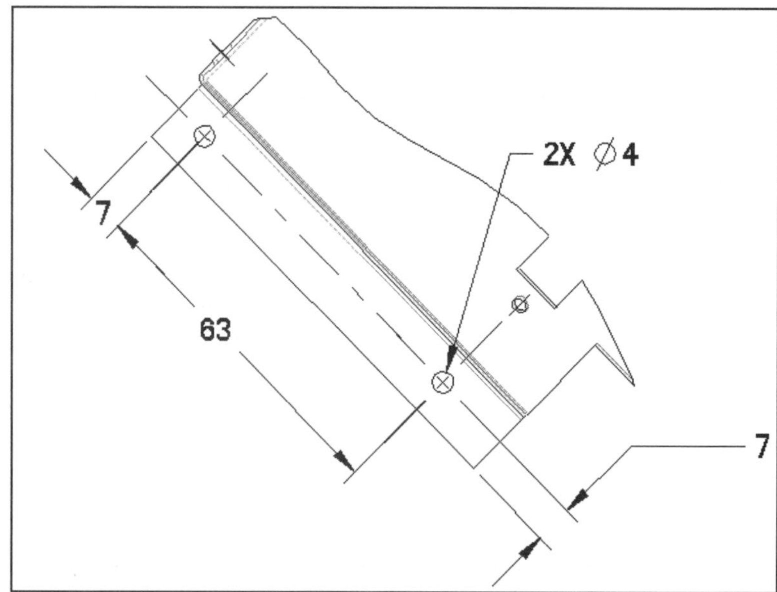

Figure 23.152
Bracket_1, Auxiliary View

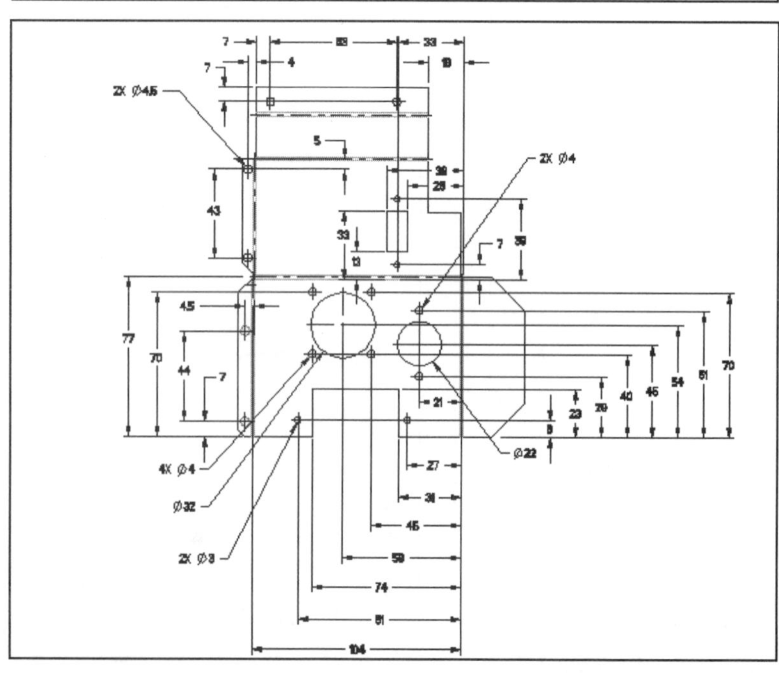

Figure 23.153
Bracket_1, Flat Pattern

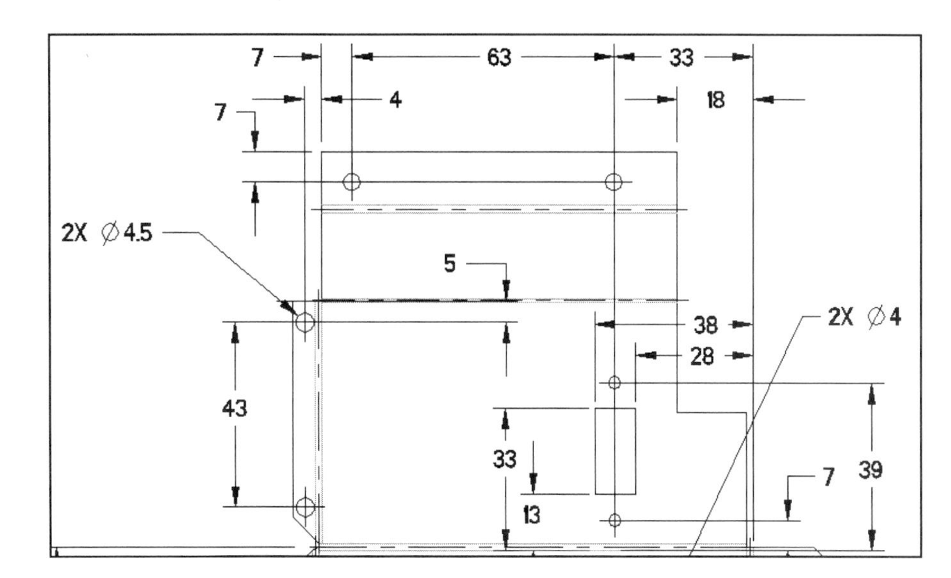

Figure 23.154
Bracket_1, Flat Pattern, Top Close-up

Figure 23.155
Bracket_1, Flat Pattern, Bottom Close-up

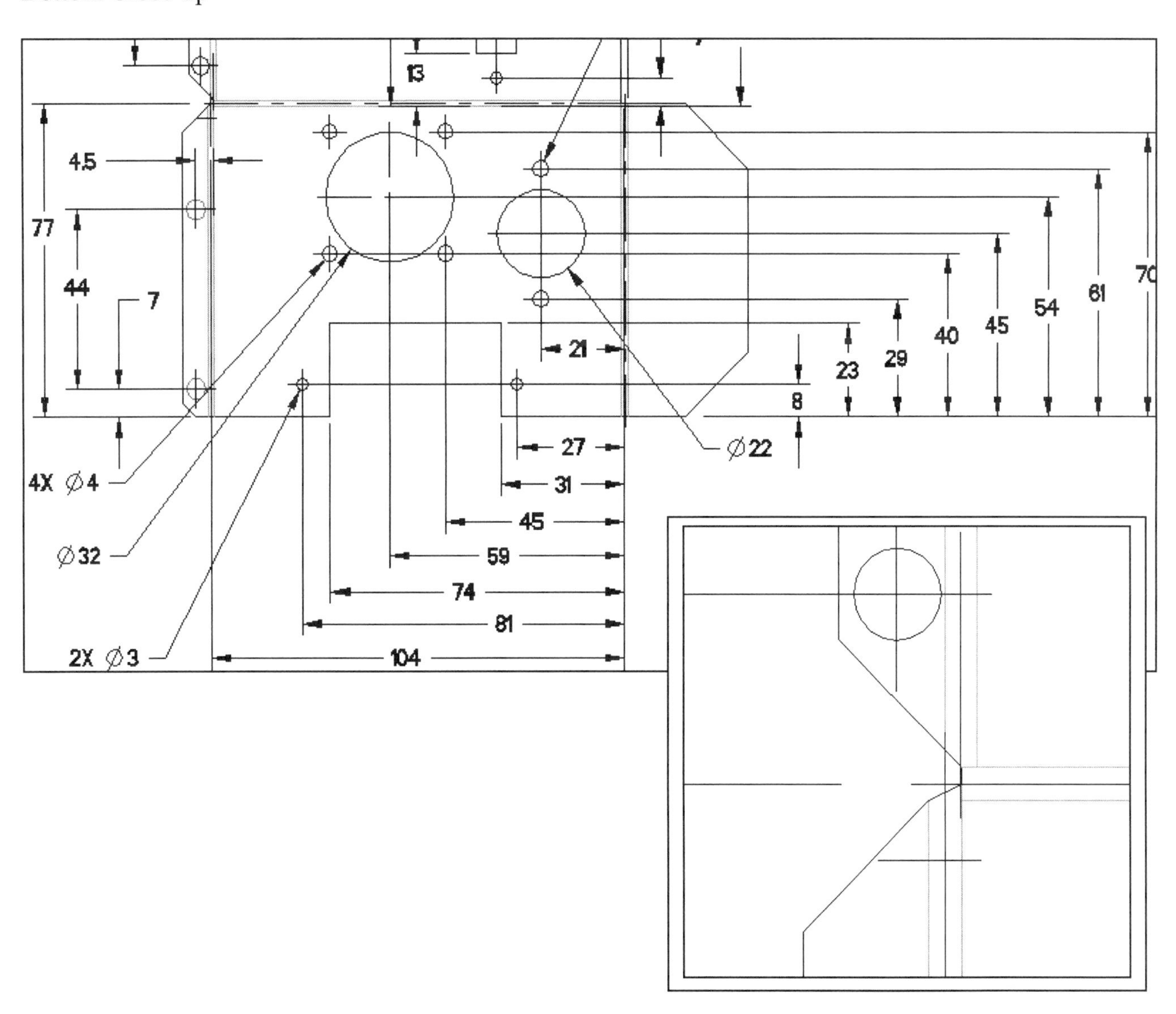

Bracket_2

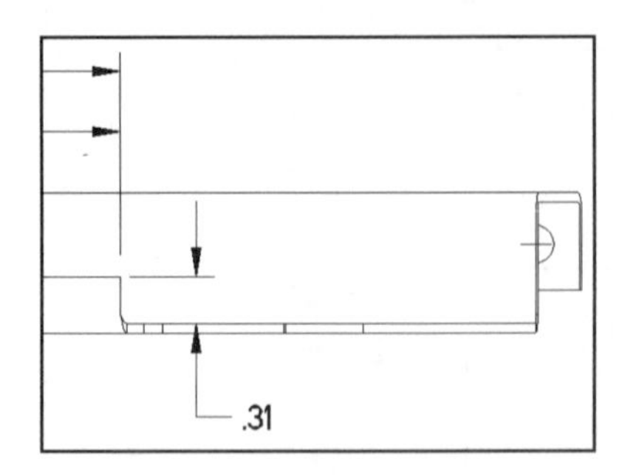

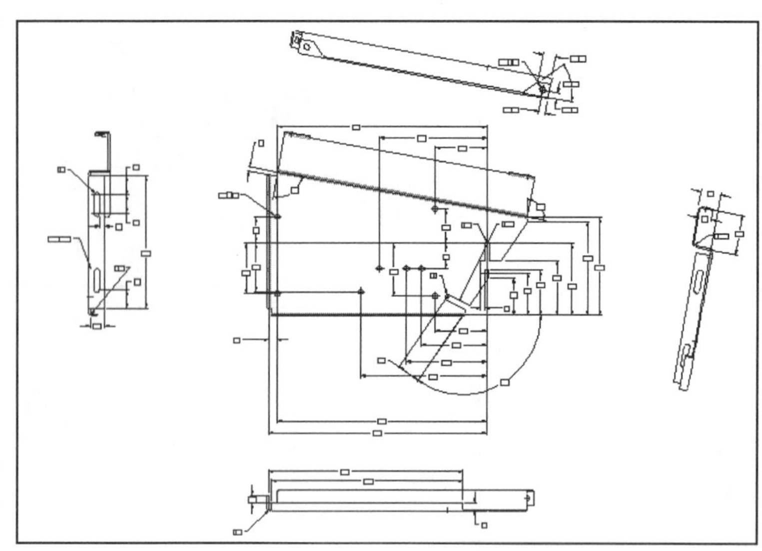

Figure 23.156
Bracket_2, Drawing

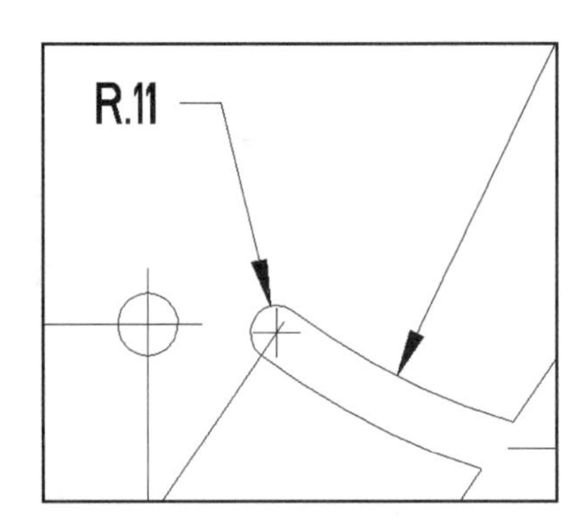

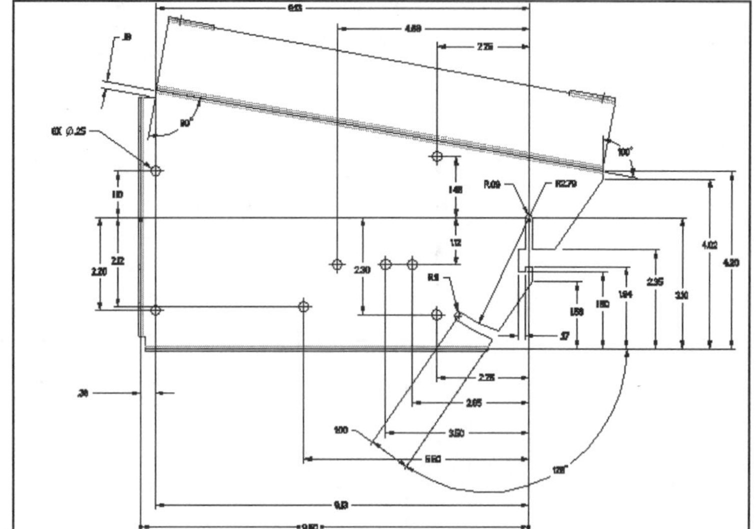

Figure 23.157
Bracket_2, Front View

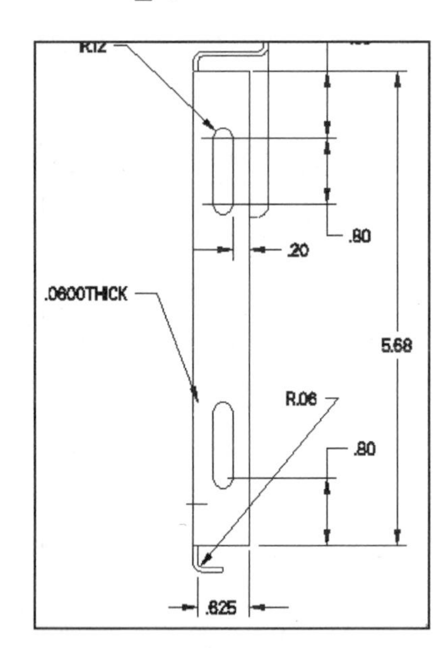

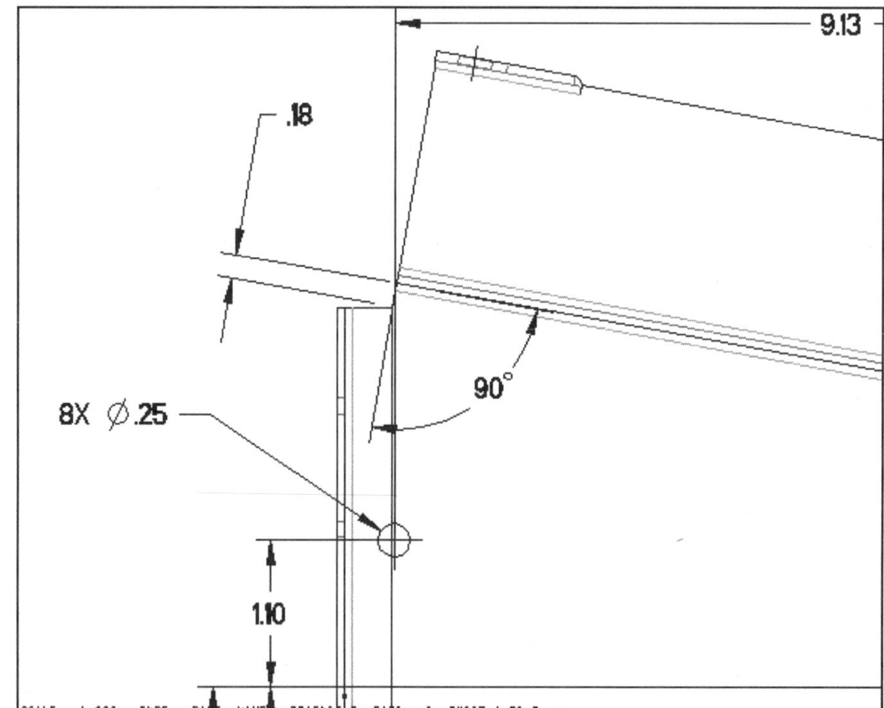

Figure 23.158
Bracket_2, Front View
Upper Left

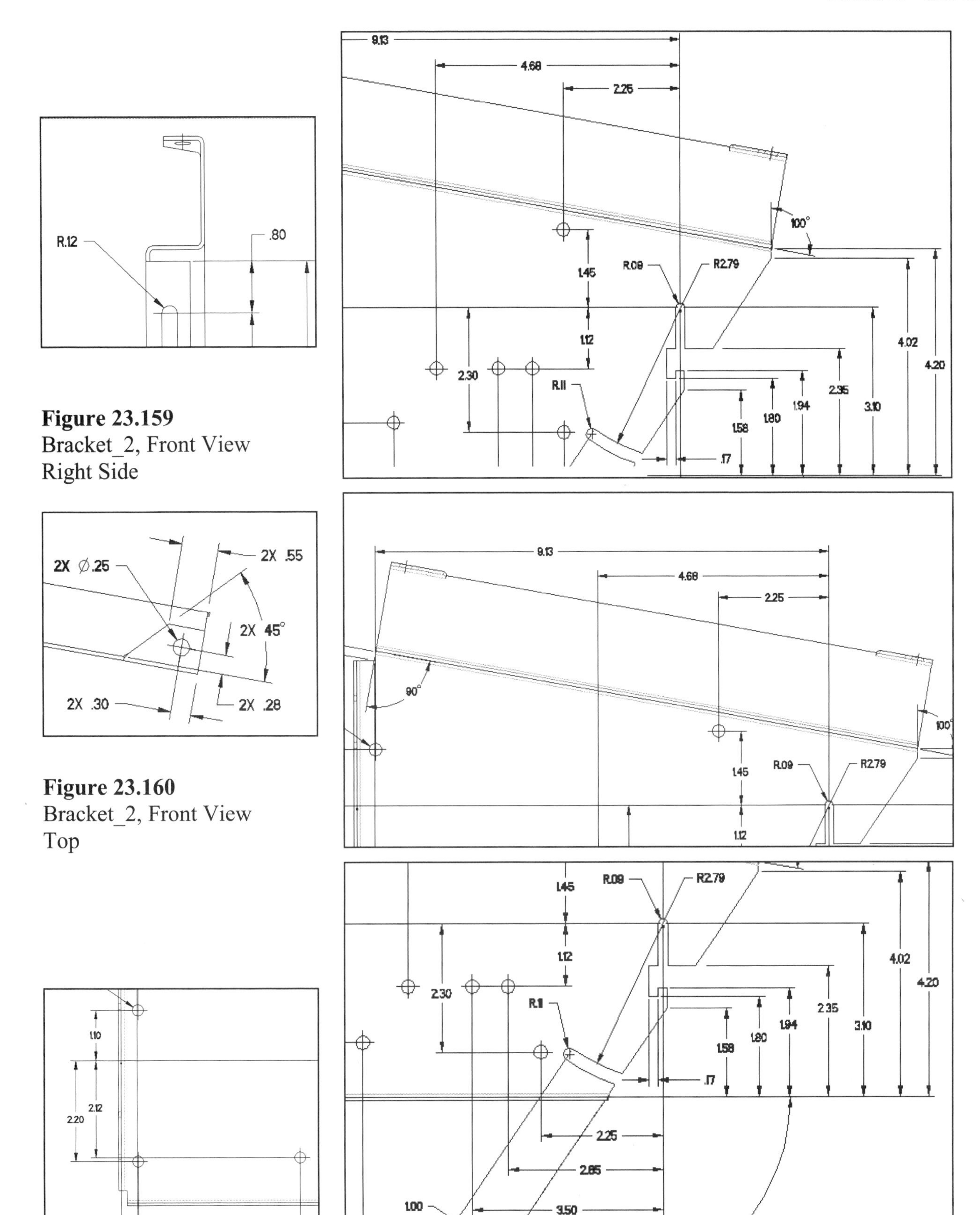

Figure 23.159
Bracket_2, Front View
Right Side

Figure 23.160
Bracket_2, Front View
Top

Figure 23.161
Bracket_2, Front View
Lower Right

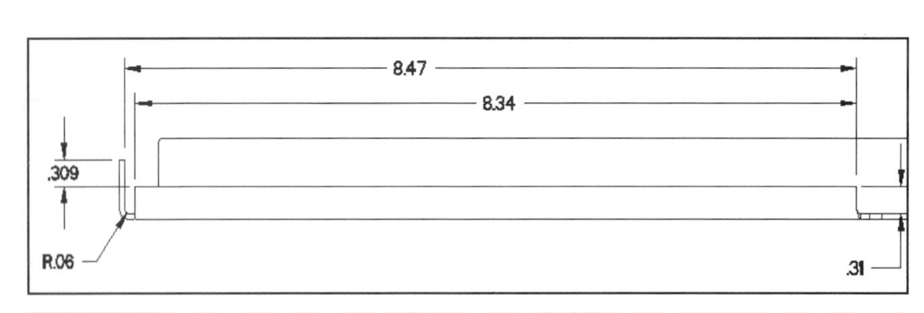

Figure 23.162
Bracket_2, Bottom View

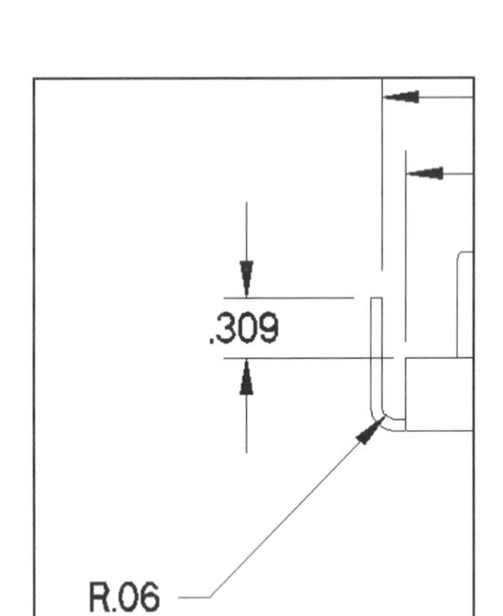

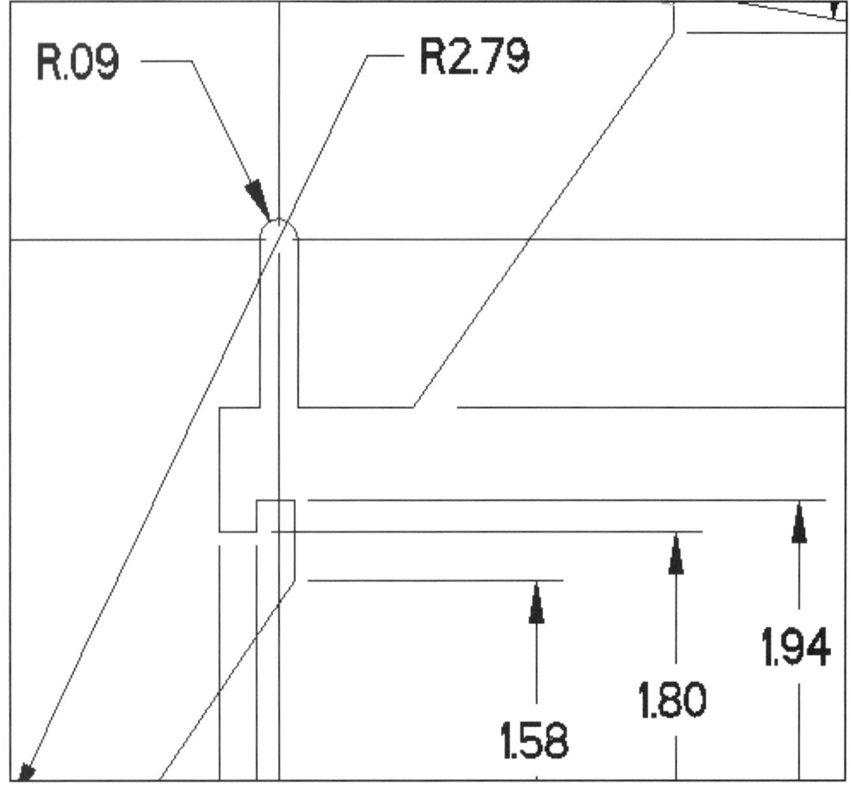

Figure 23.163
Bracket_2, Front View
Slot

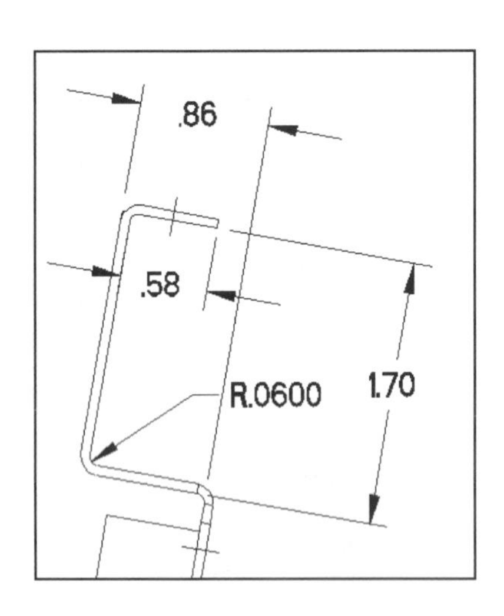

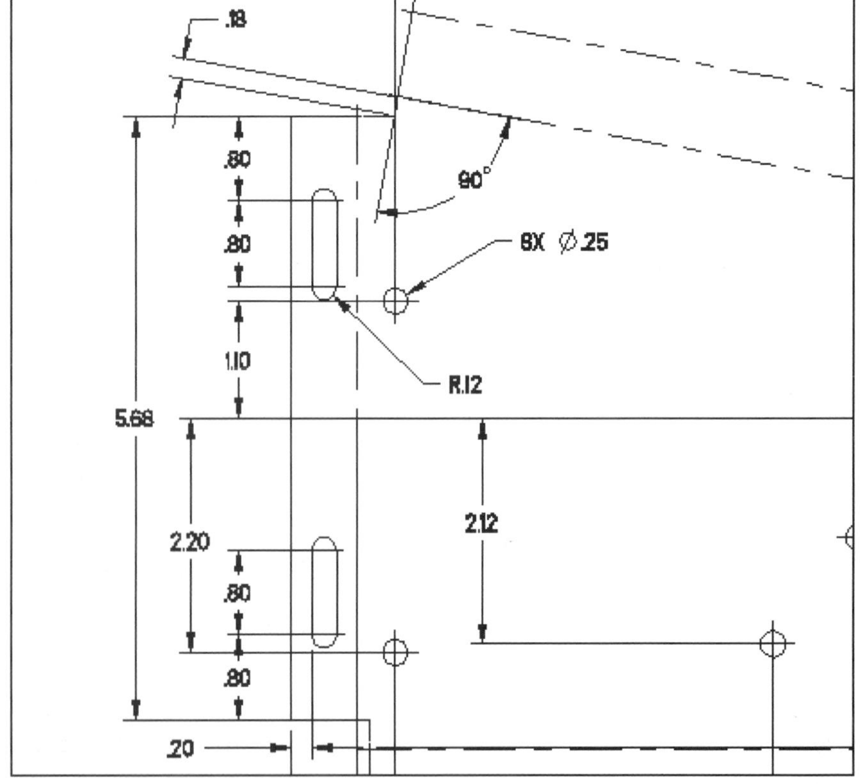

Figure 23.164
Bracket_2, Flat Pattern

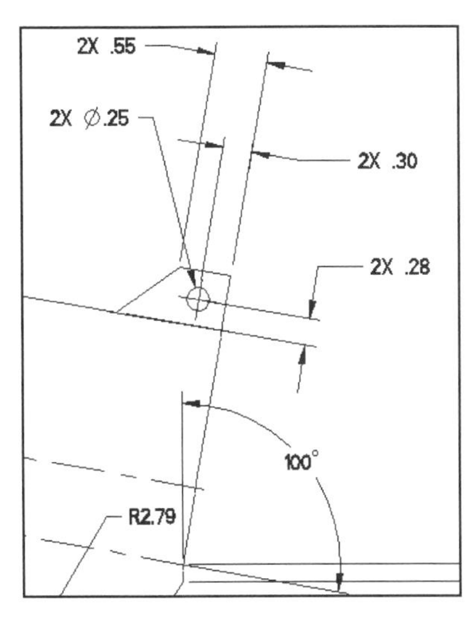

Figure 23.165
Bracket_2, Flat Pattern,
Right Side

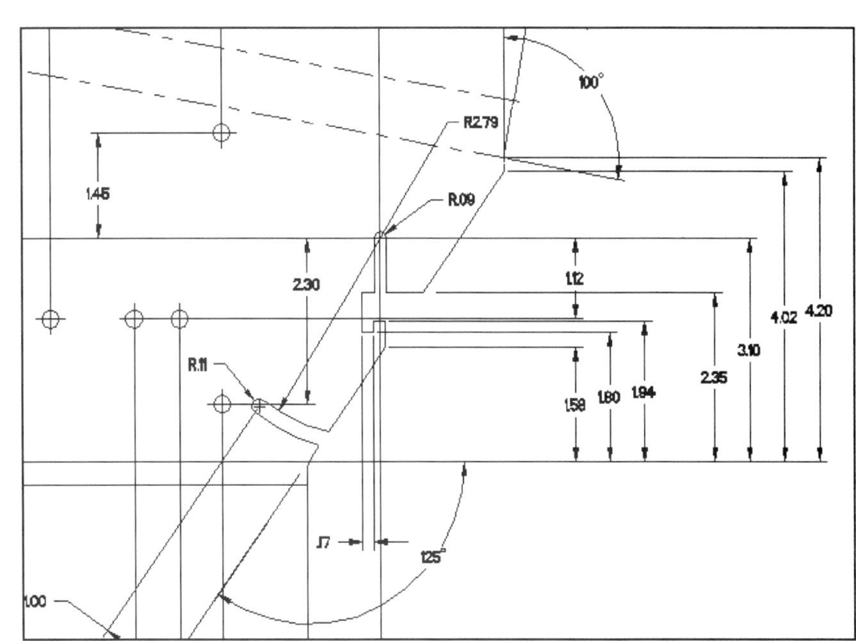

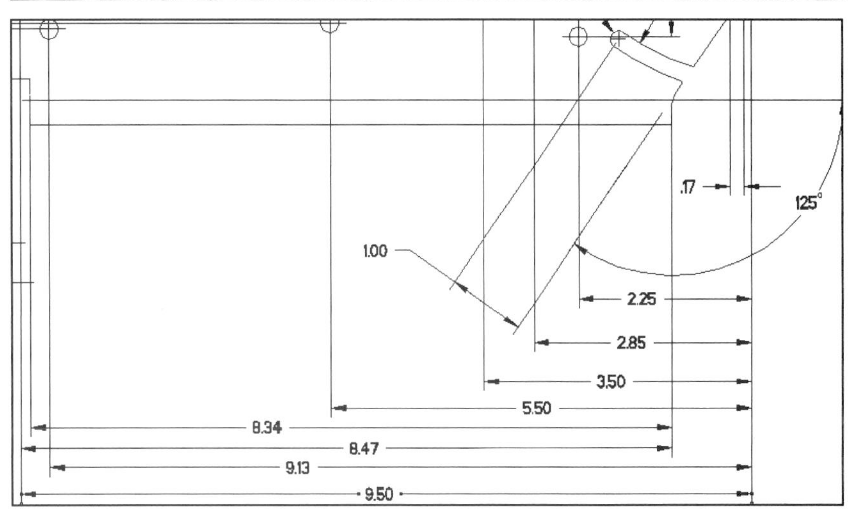

Figure 23.166
Bracket_2, Flat Pattern,
Lower Right Side

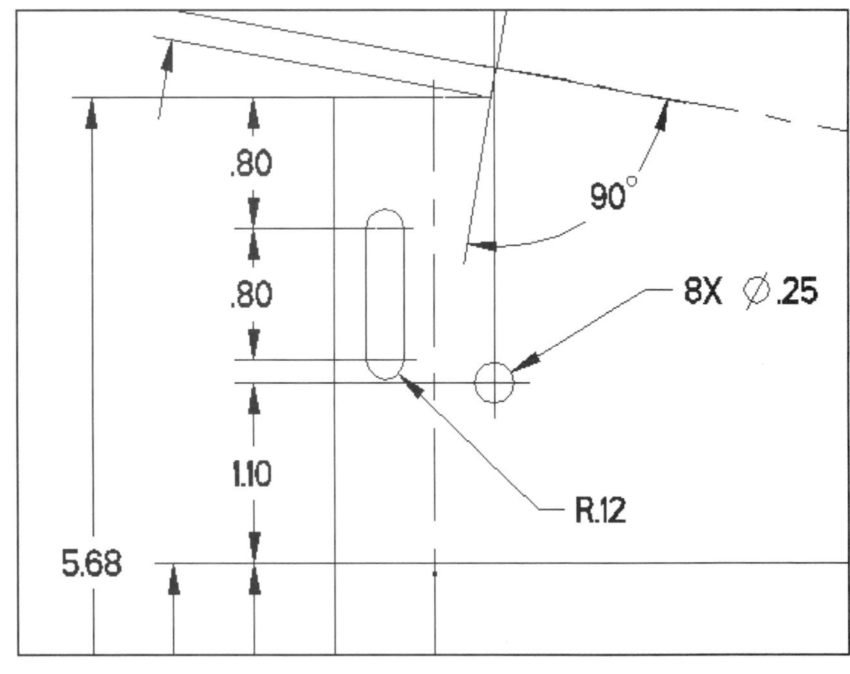

Figure 23.167
Bracket_2, Flat Pattern,
Slot

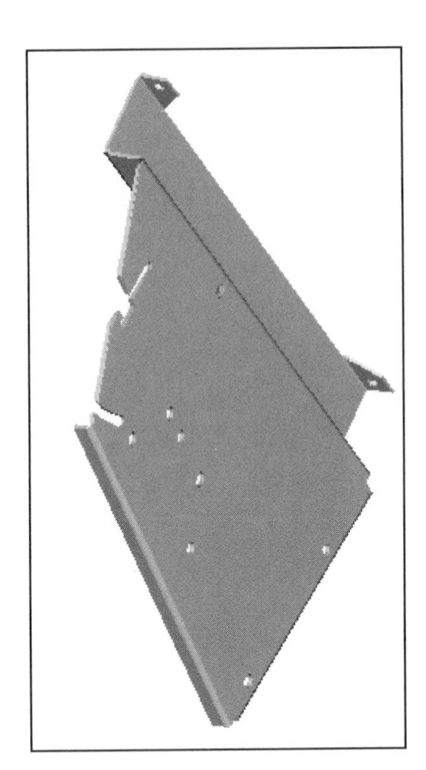

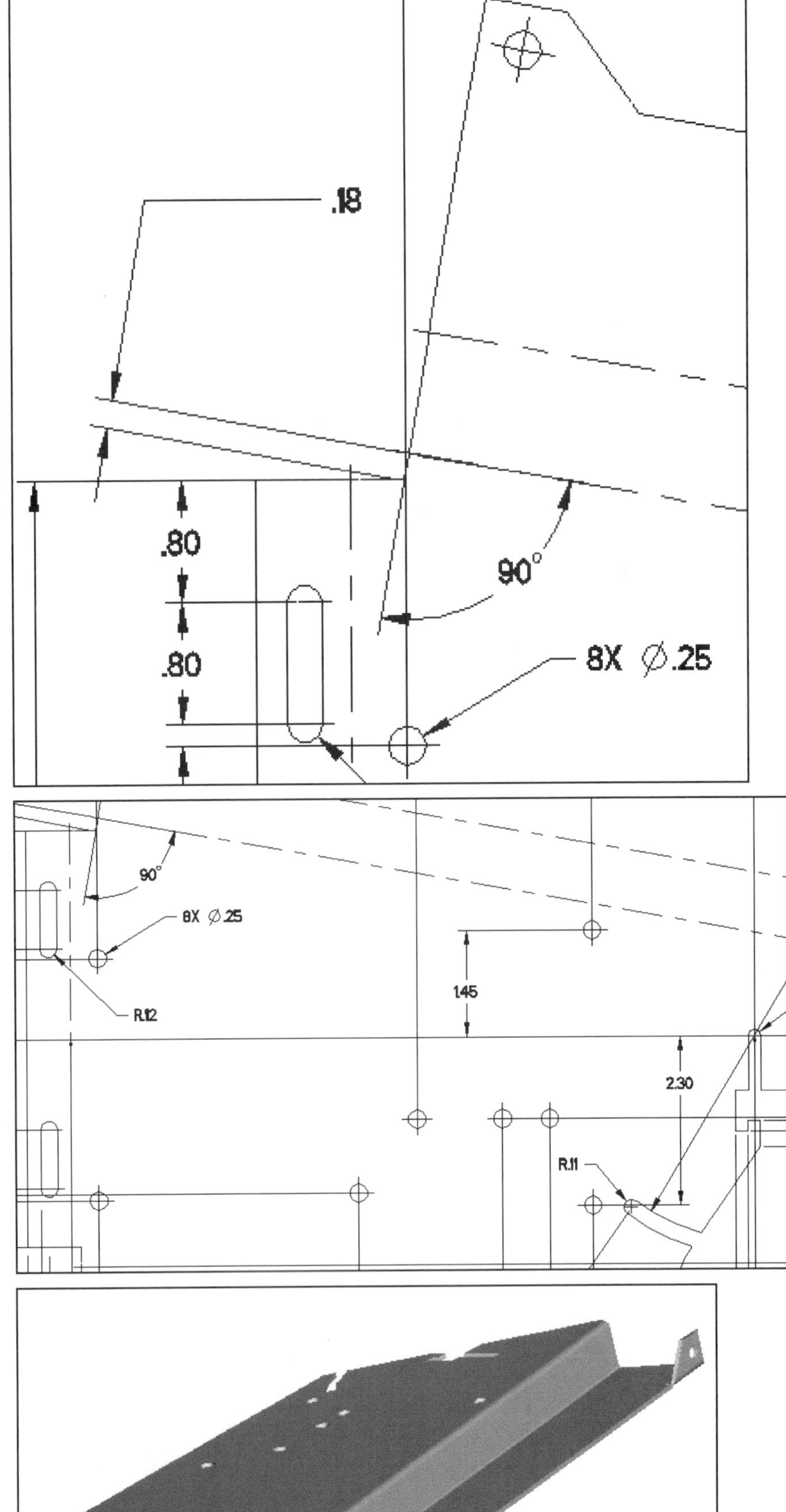

Figure 23.168
Bracket_2, Flat Pattern, Upper Left

Figure 23.169
Bracket_2, Flat Pattern, Center

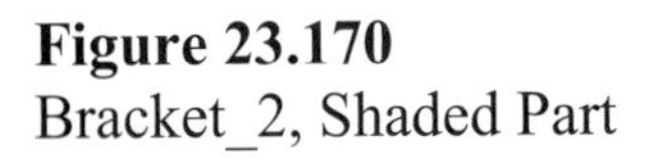

Figure 23.170
Bracket_2, Shaded Part

Appendices

Advanced Project

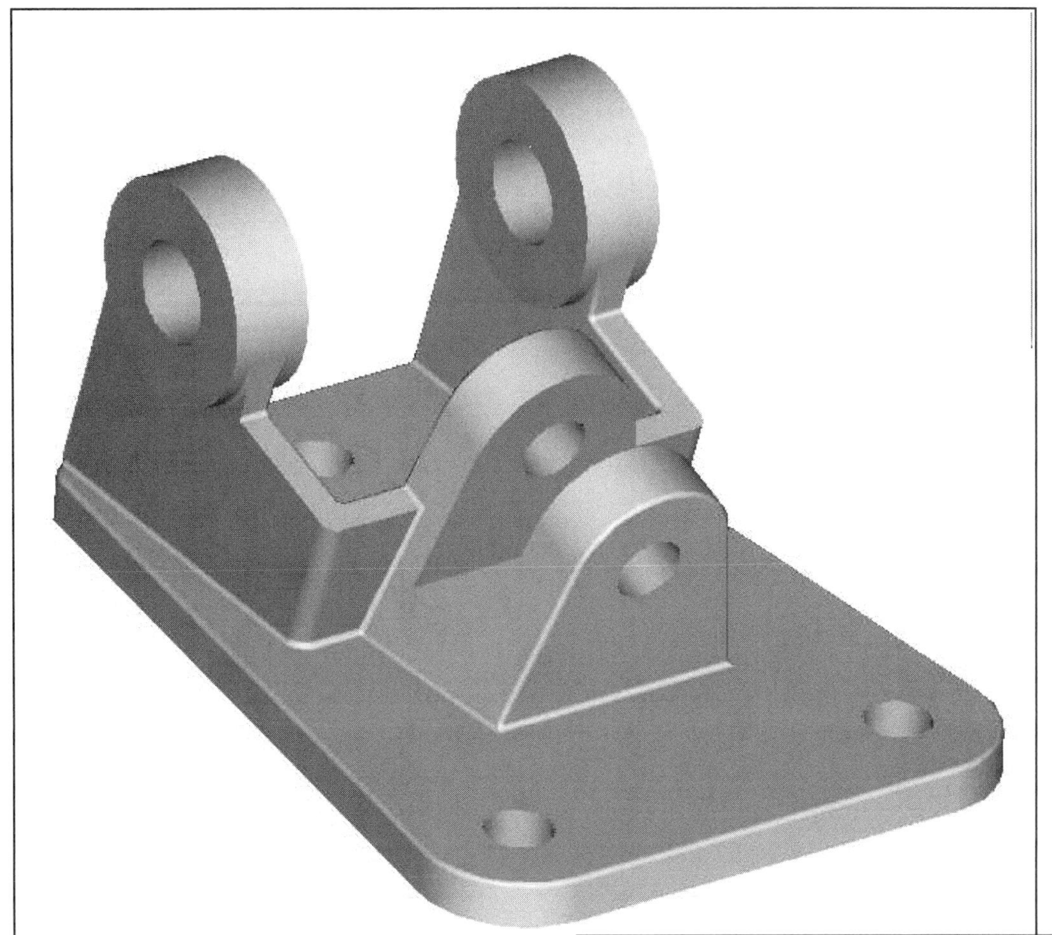

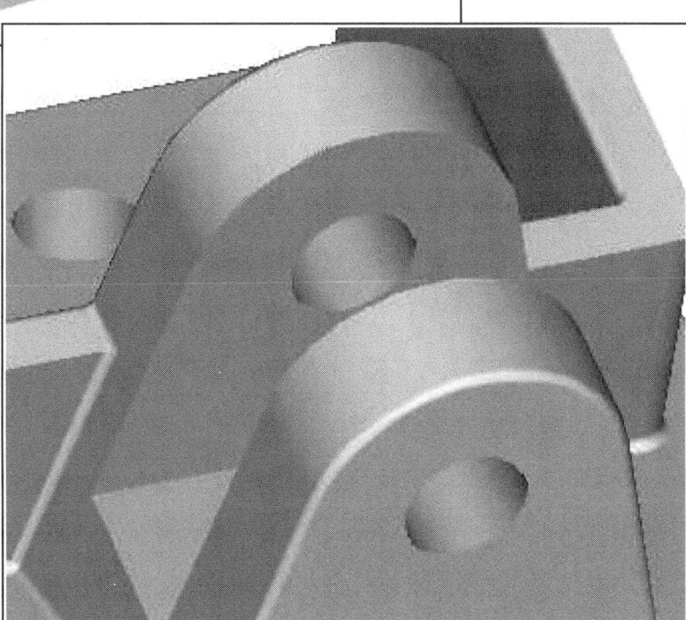

Config.pro File

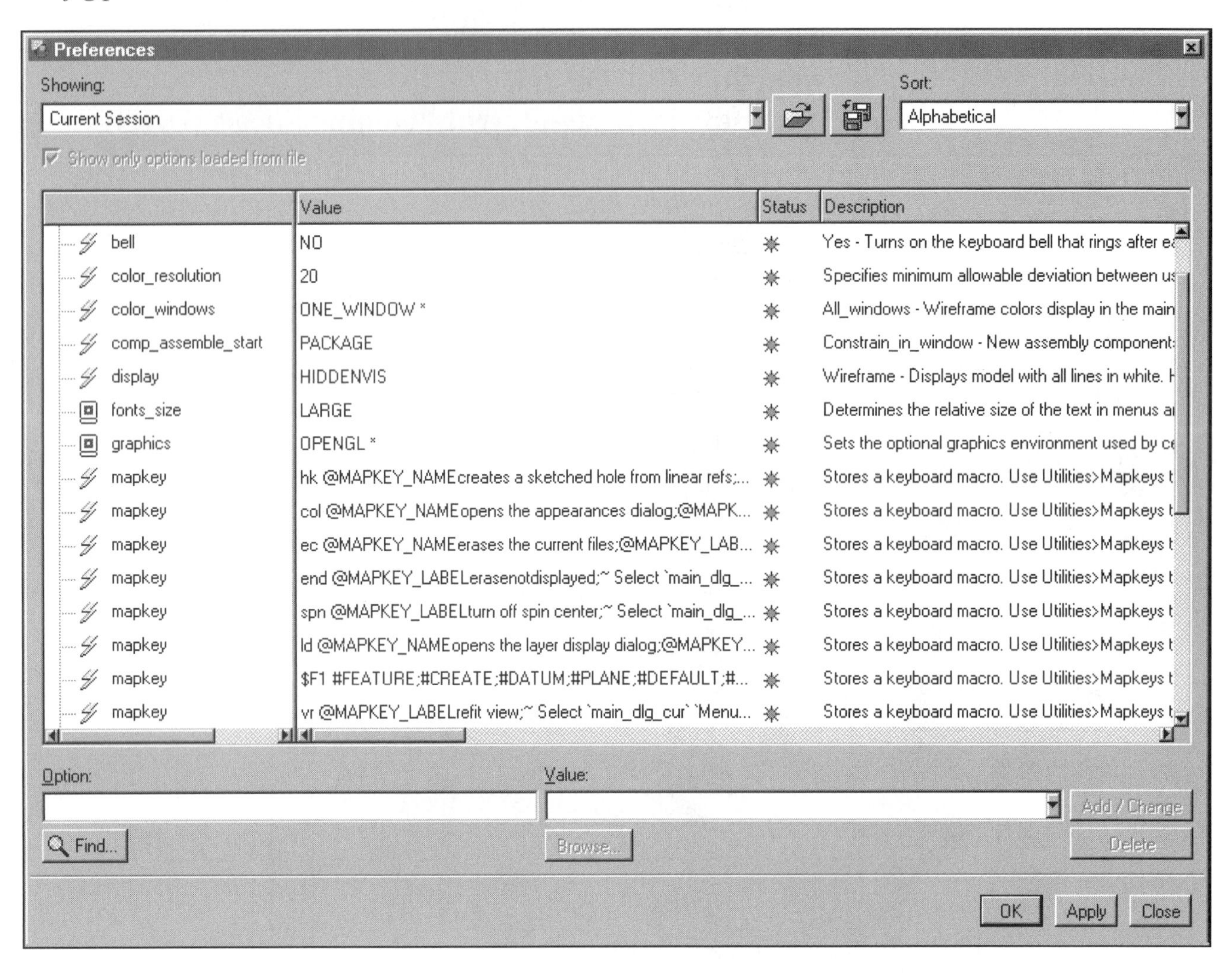

Appendix A

Advanced Projects

Casting
Cover

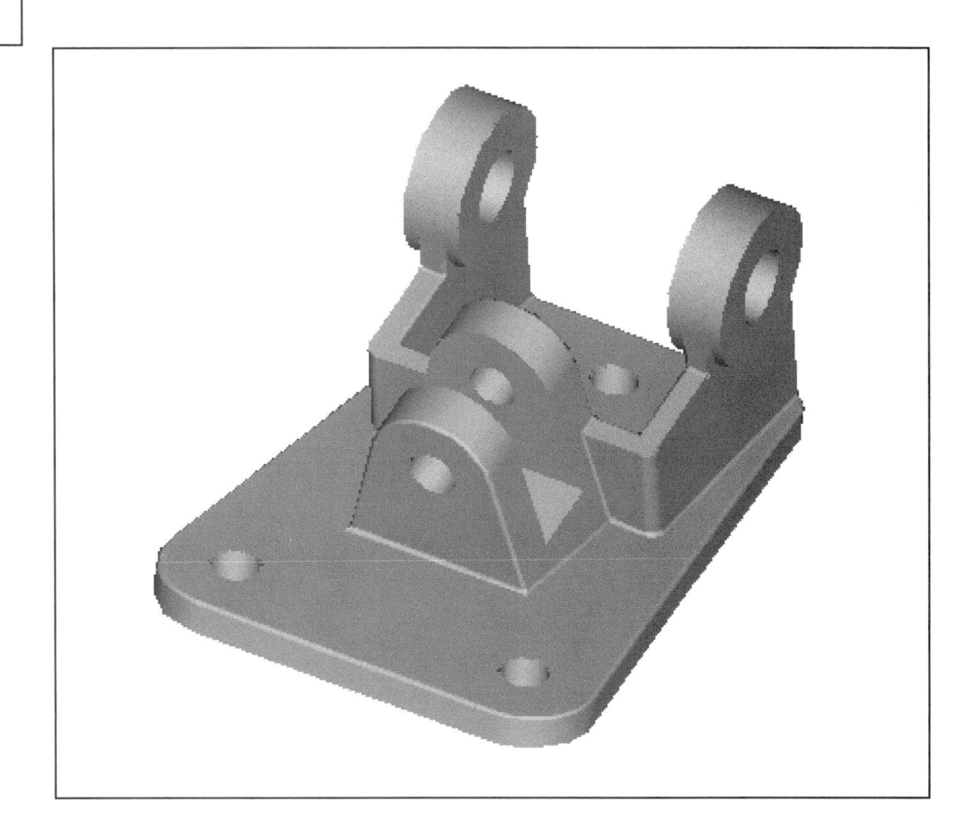

Figure A.1
Casting

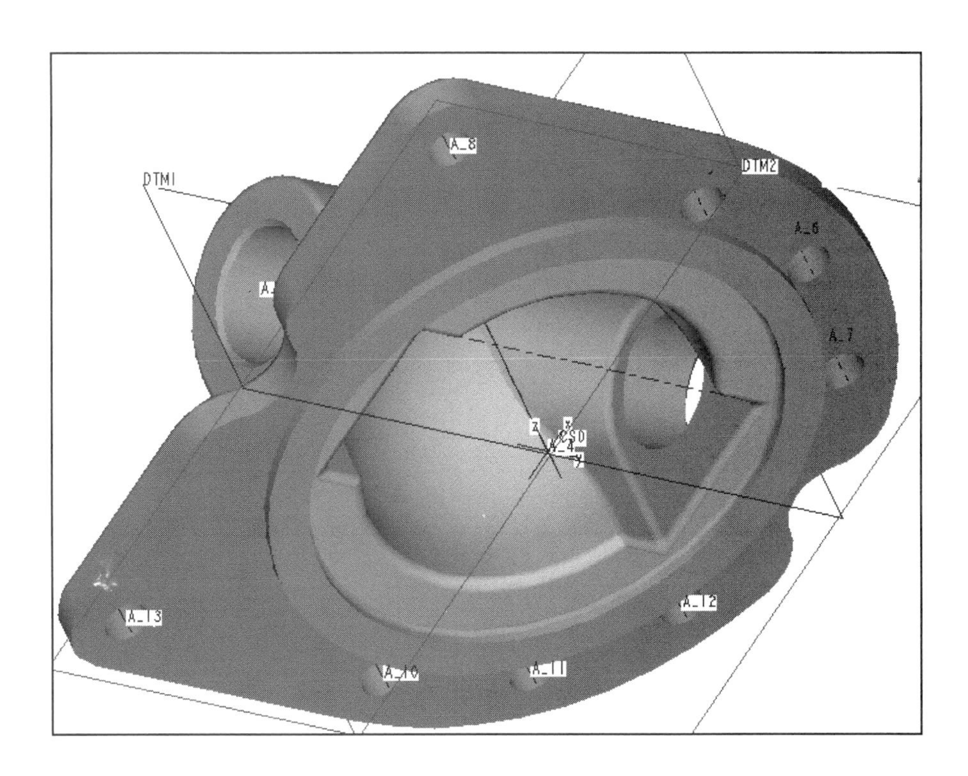

Figure A.2
Cover

Casting

Figure A.3
Casting Drawing

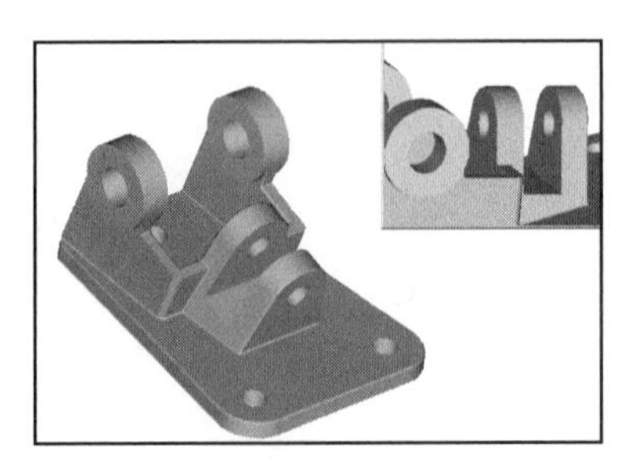

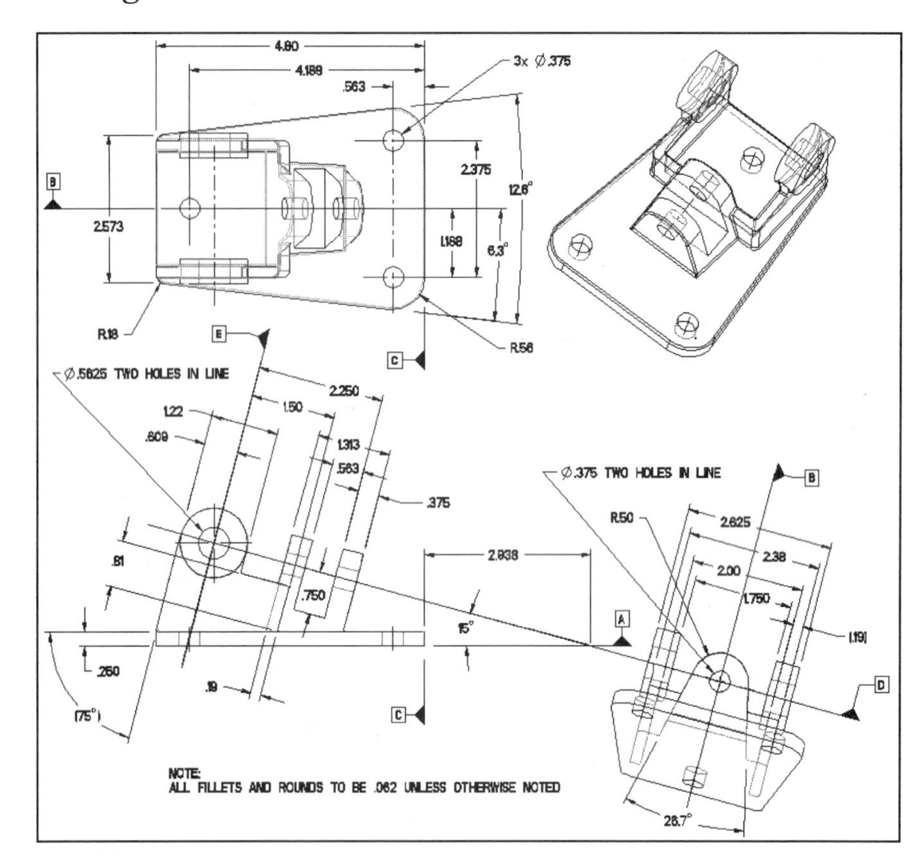

☑ *EGD REFERENCE*
Fundamentals of Engineering Graphics and Design
by L. Lamit and K. Kitto
See page 411, Problem 12.13.

Figure A.4
Casting Drawing, Top View

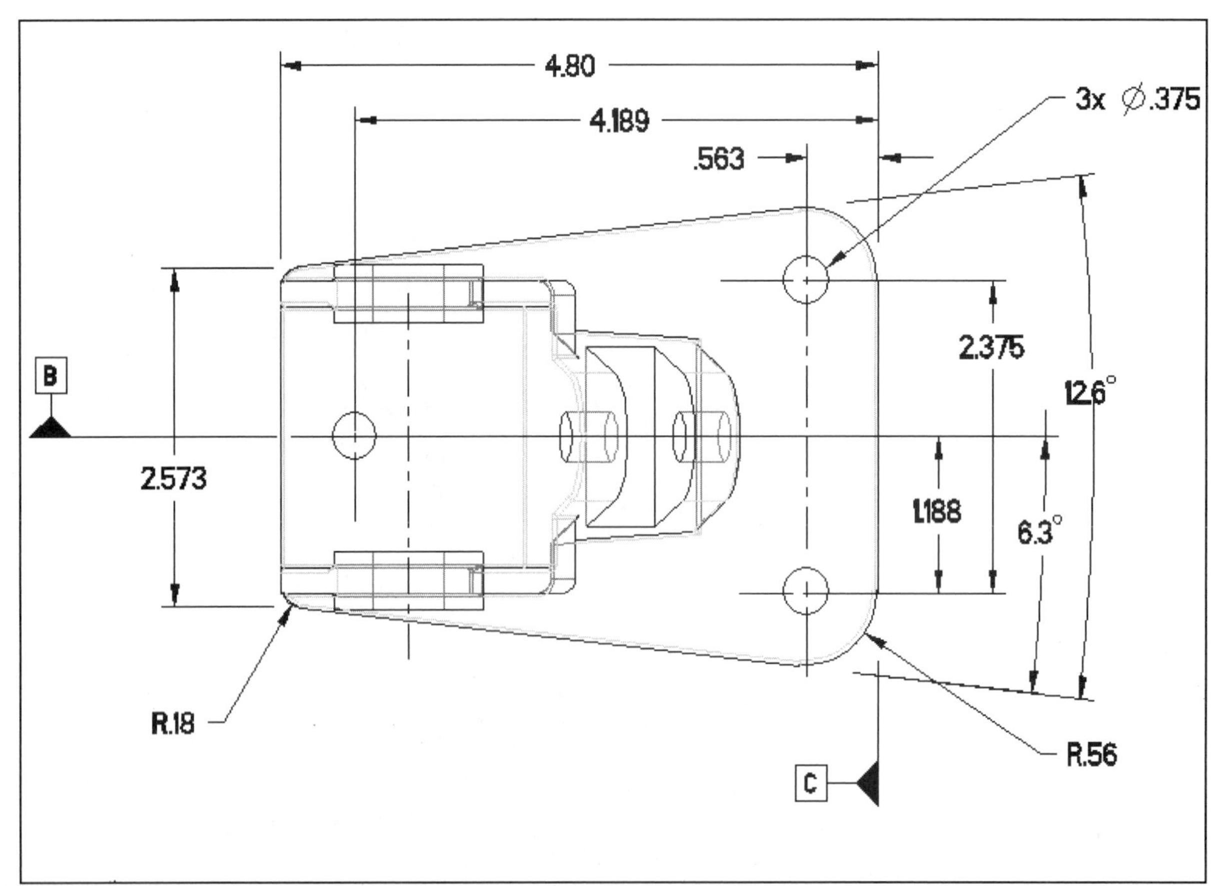

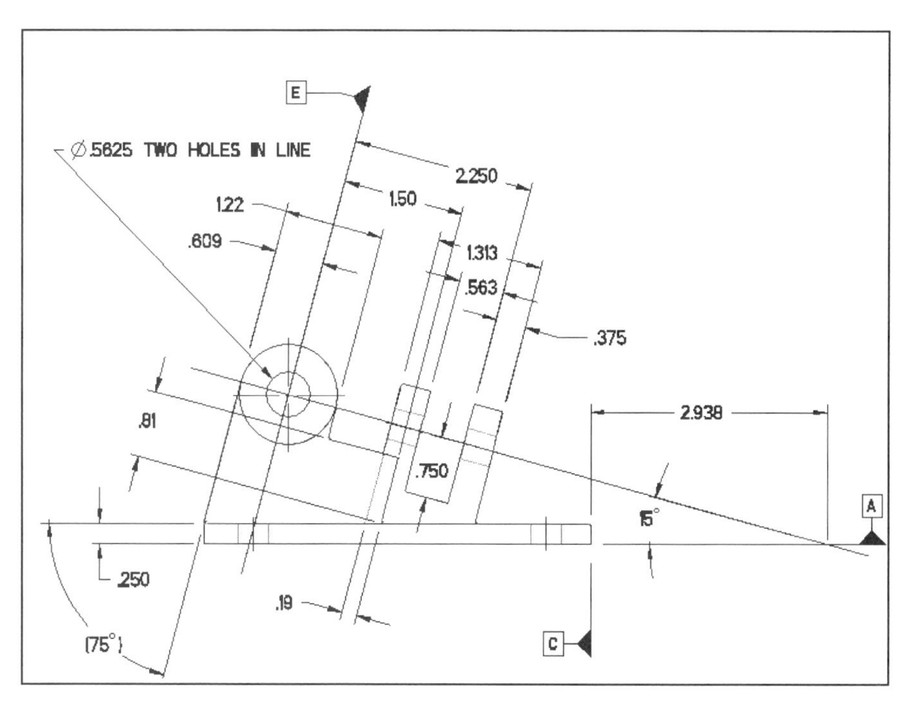

Figure A.5
Casting Drawing, Front View

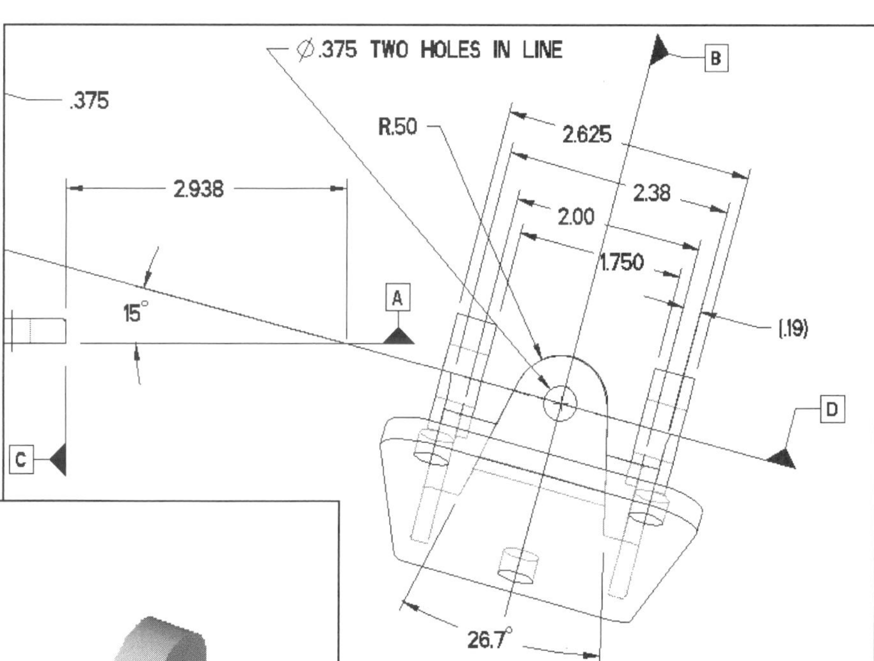

Figure A.6
Casting Drawing,
Auxiliary View

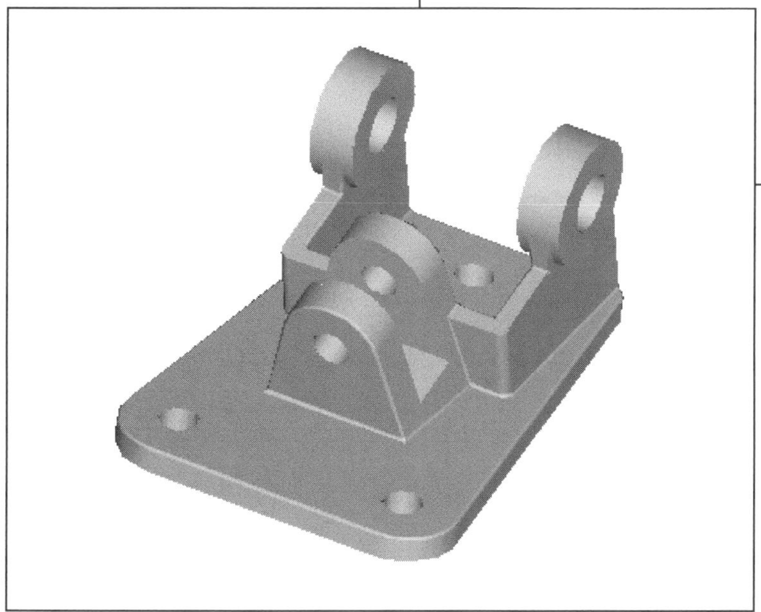

Cover

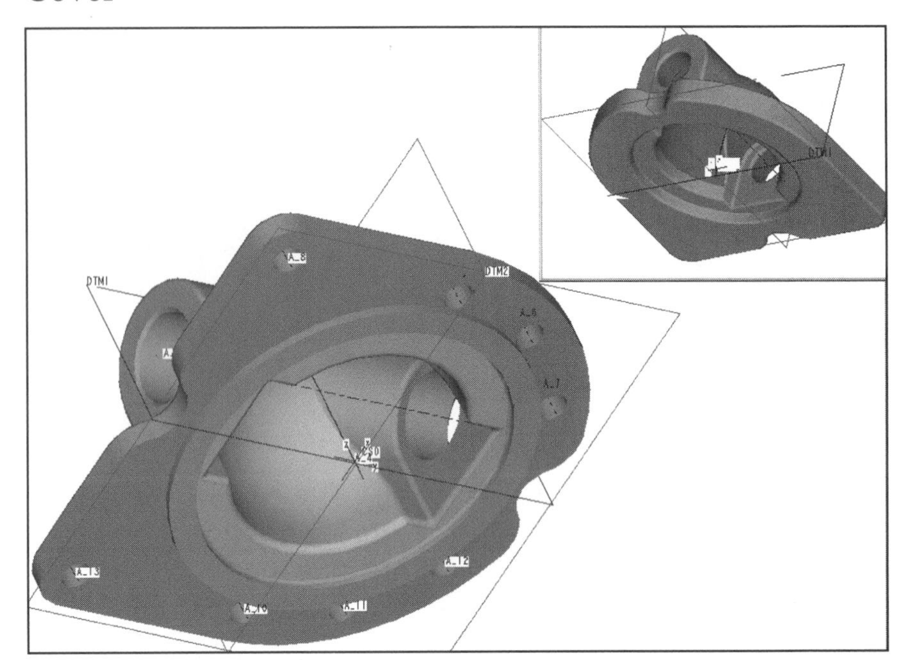

☑ *NOTE*

Model the casting, save it under a new name, and continue modeling the machine part.

When it is complete, you will have two separate parts, a casting (**workpiece**) and a machined part (**design part**), that can be used to create the *manufacturing model* for use in Pro/NC.

Figure A.7
Cover Machined Part and Cover Casting (inset)

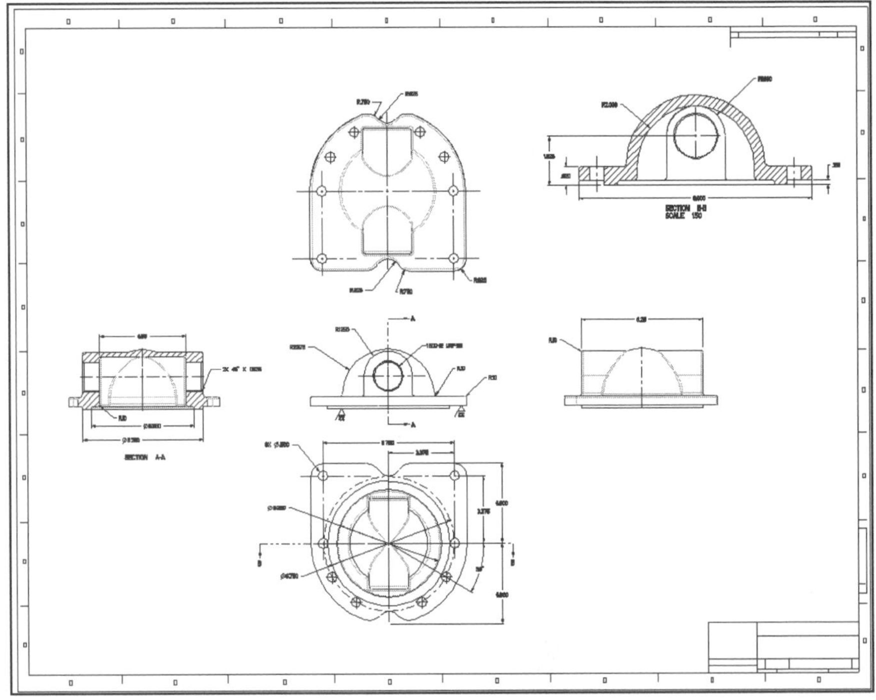

☑ *EGD REFERENCE*

Fundamentals of Engineering Graphics and Design
by L. Lamit and K. Kitto
See page 338.
Problem 10.33.

Figure A.8
Cover Drawing

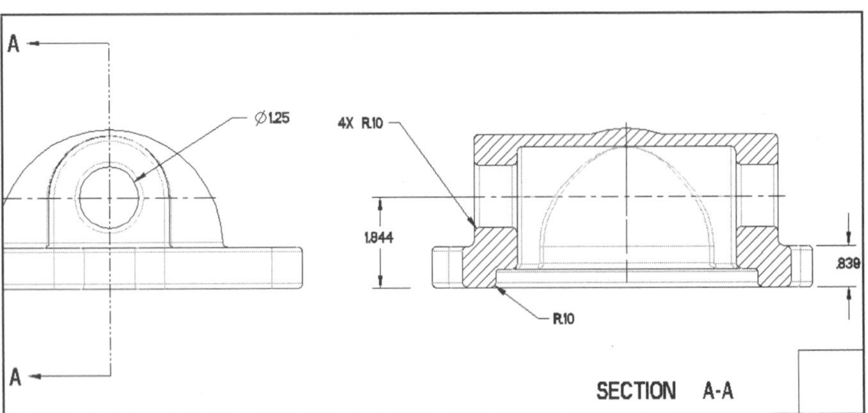

Fig
Cover C

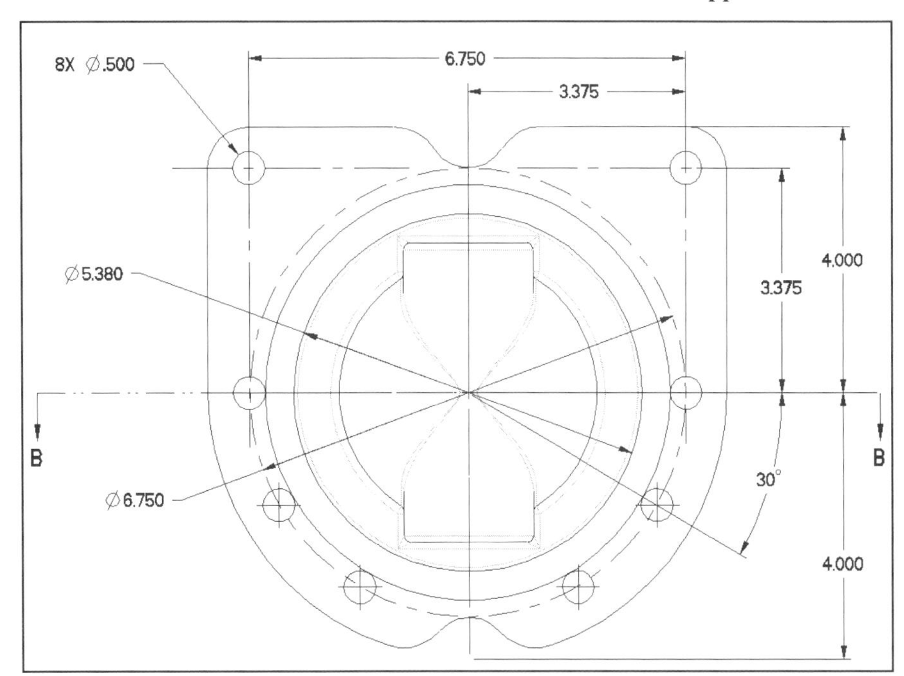

Figure A.10
Cover Drawing, Bottom View

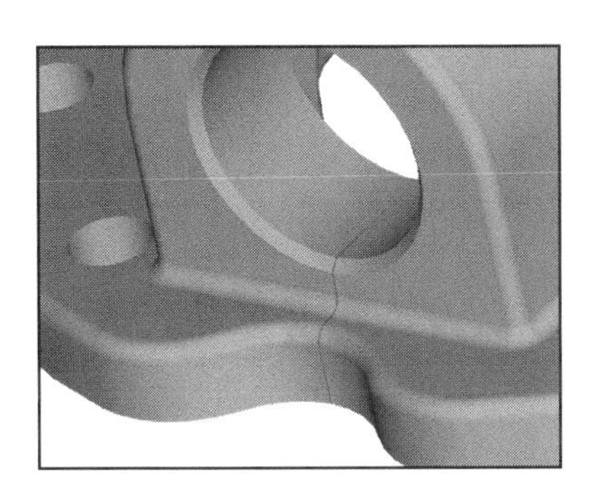

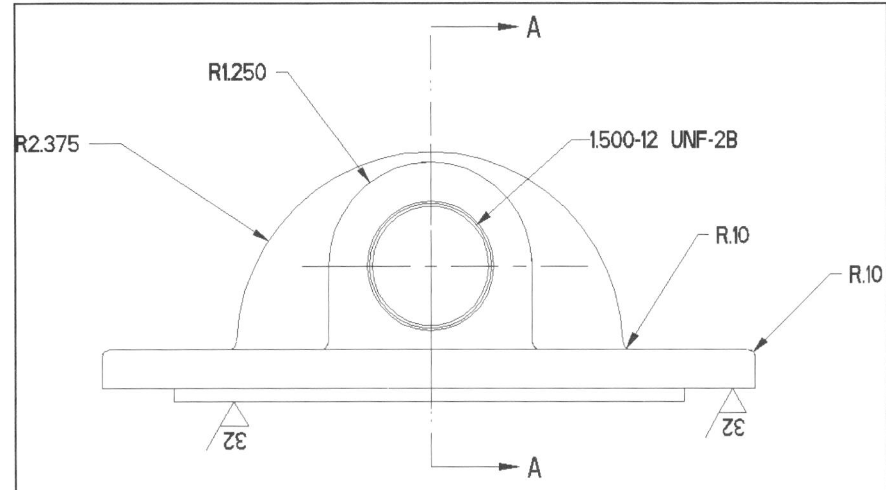

Figure A.11
Cover Drawing, Front View

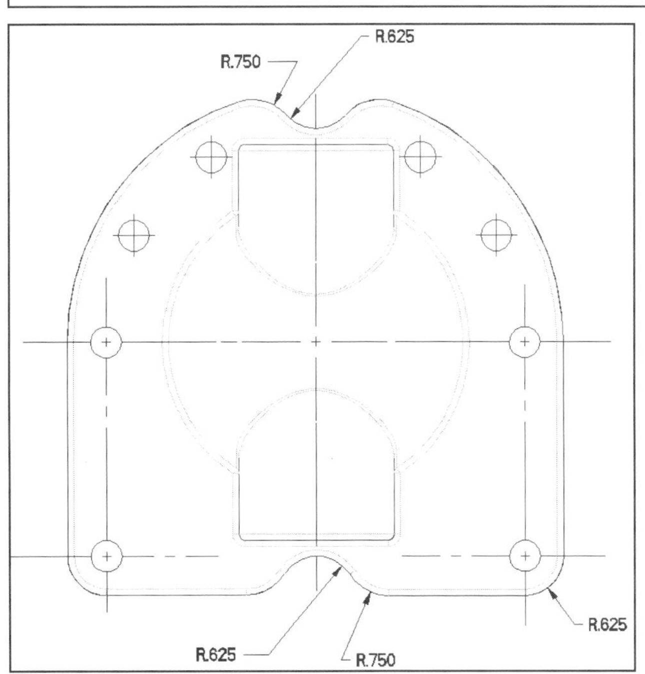

Figure A.12
Cover Drawing, Top View

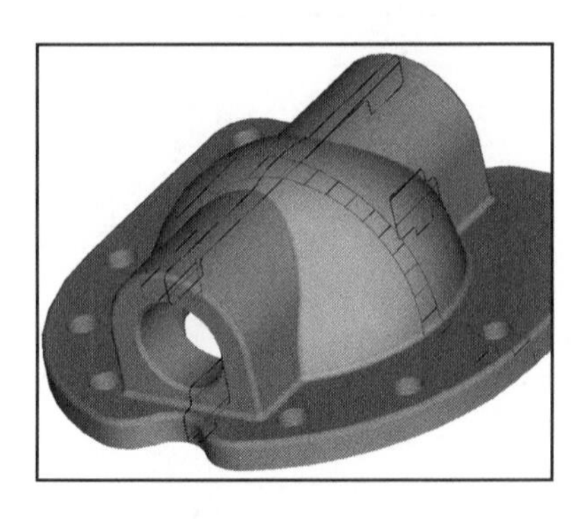

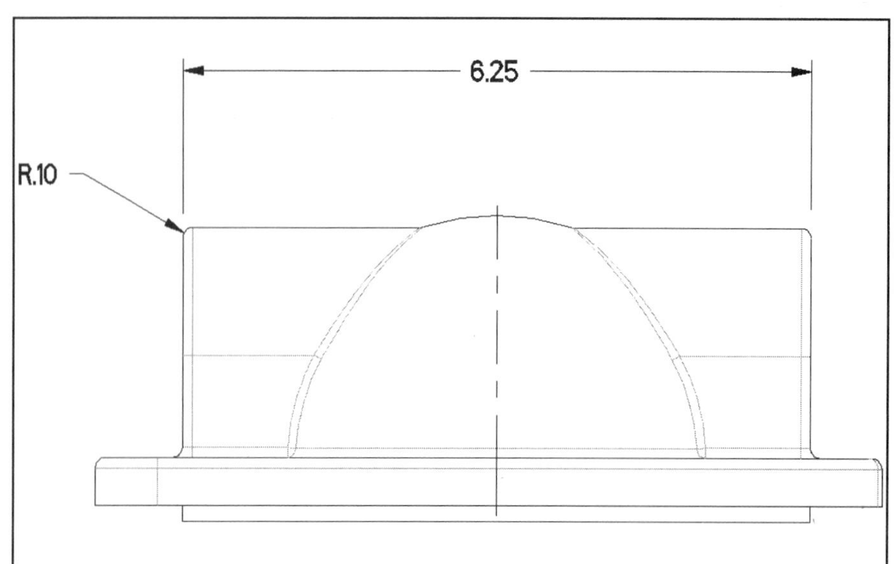

Figure A.13
Cover Drawing,
Right Side View

2X 45° X .0625

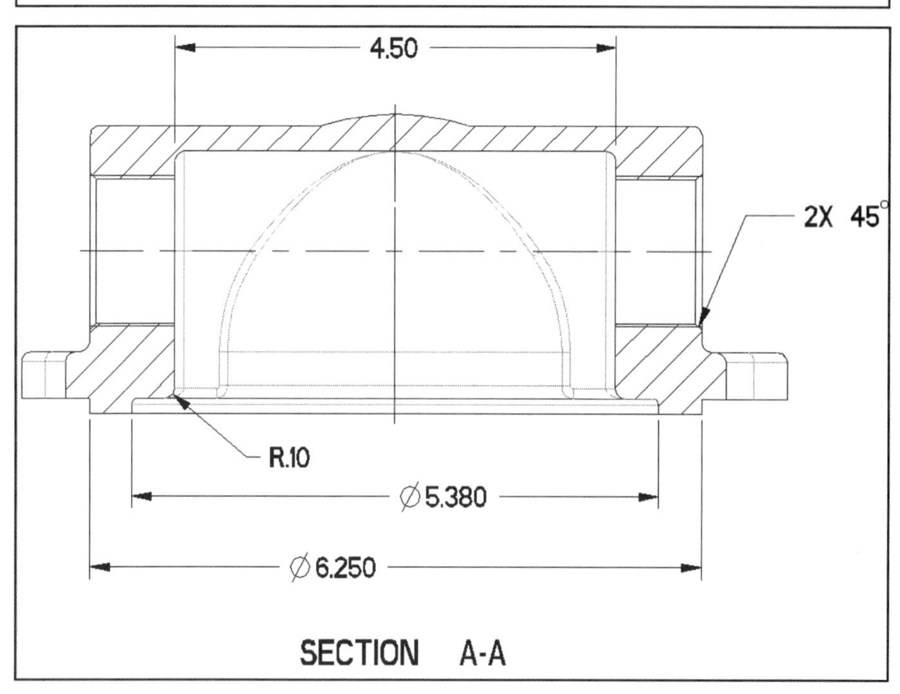

Figure A.14
Cover Drawing,
SECTION A-A

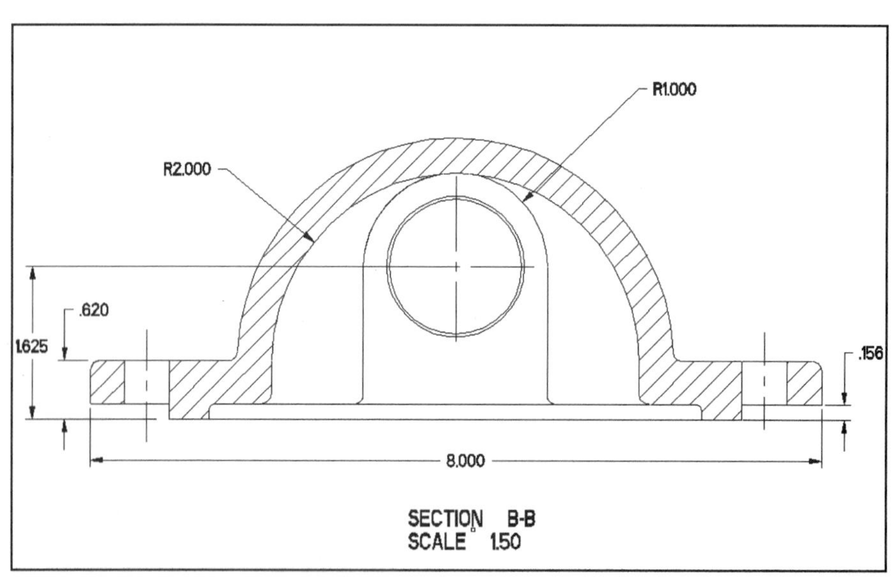

Figure A.15
Cover Drawing,
SECTION B-B

Appendix B

Config.pro Files and Mapkeys

> ***NOTE:***
>
> A CD with all book Pro/E files is available from the author for instructors who adopt the text:
>
> **lglamit@yahoo.com**
> or
> **www.netcom.com/~llamit/**
>
> This disk also contains a list of common ***config.pro*** options and a short description of each.

Config.pro Files

The following is an example of various entries in a *config.pro* file.

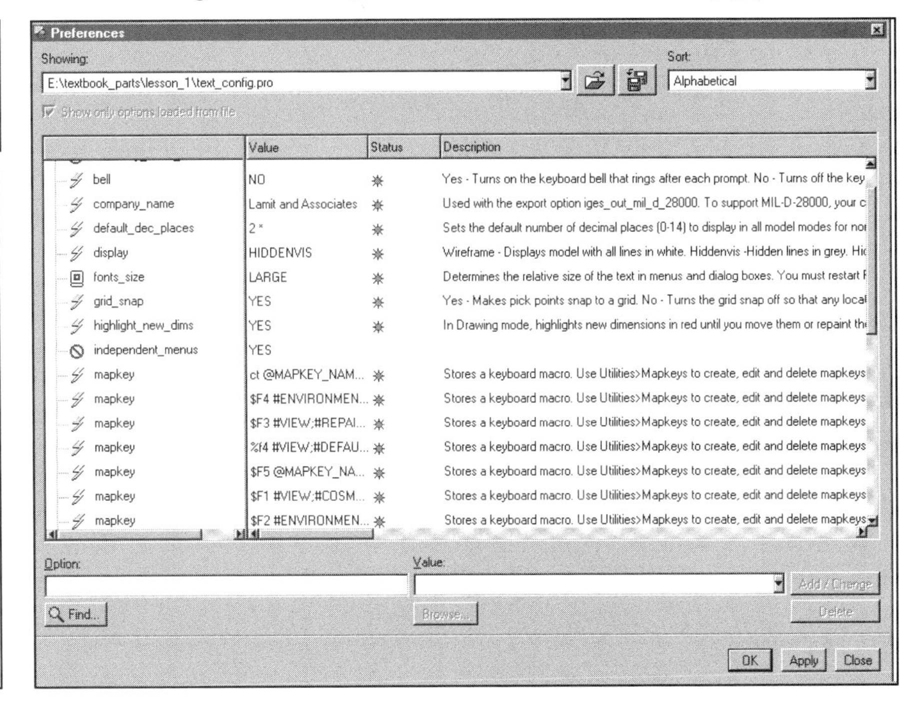

Mapkeys

Mapkey entries can be added to a configuration file (*config.pro* file) at any time in the design process. These keyboard macros help streamline the design process. Feel free to experiment with your own macro creation. If you find yourself using a set of commands for a project over and over, it may be to your advantage to create a mapkey macro that will automate the process.

You may also wish to make icons for a number of frequently used mapkeys and place them on a **Toolbar.**

Creating mapkeys is important for efficient use of Pro/E in an industrial setting. However, if you cannot remember all the keys because you created too many, this does not increase efficiency. If your desktop **Toolbar** becomes cluttered with icons, and the amount of usable graphic window space diminishes substantially (to a point where you have to move windows, move menus, or zoom in and out constantly to do your modeling), then you have exceeded the limit of usefulness in mapkey and icon creation. A few examples follow:

MAPKEY SD #VIEW; #SHADE

MAPKEY SW #RELATIONS; #SWITCH DIM

MAPKEY ZR #VIEW; #SPIN/PAN/ZOOM; #REFIT

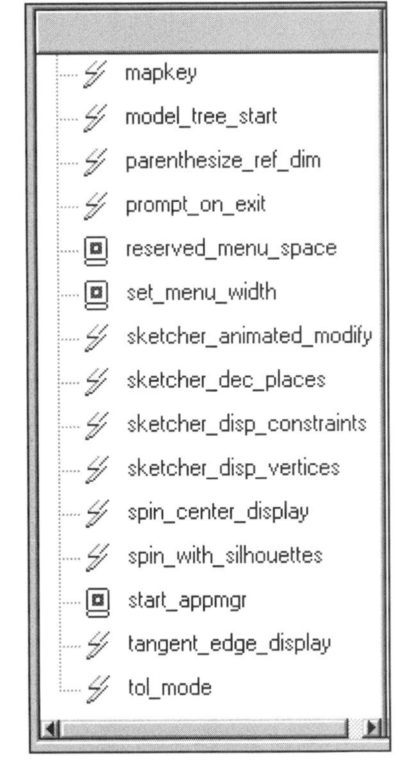

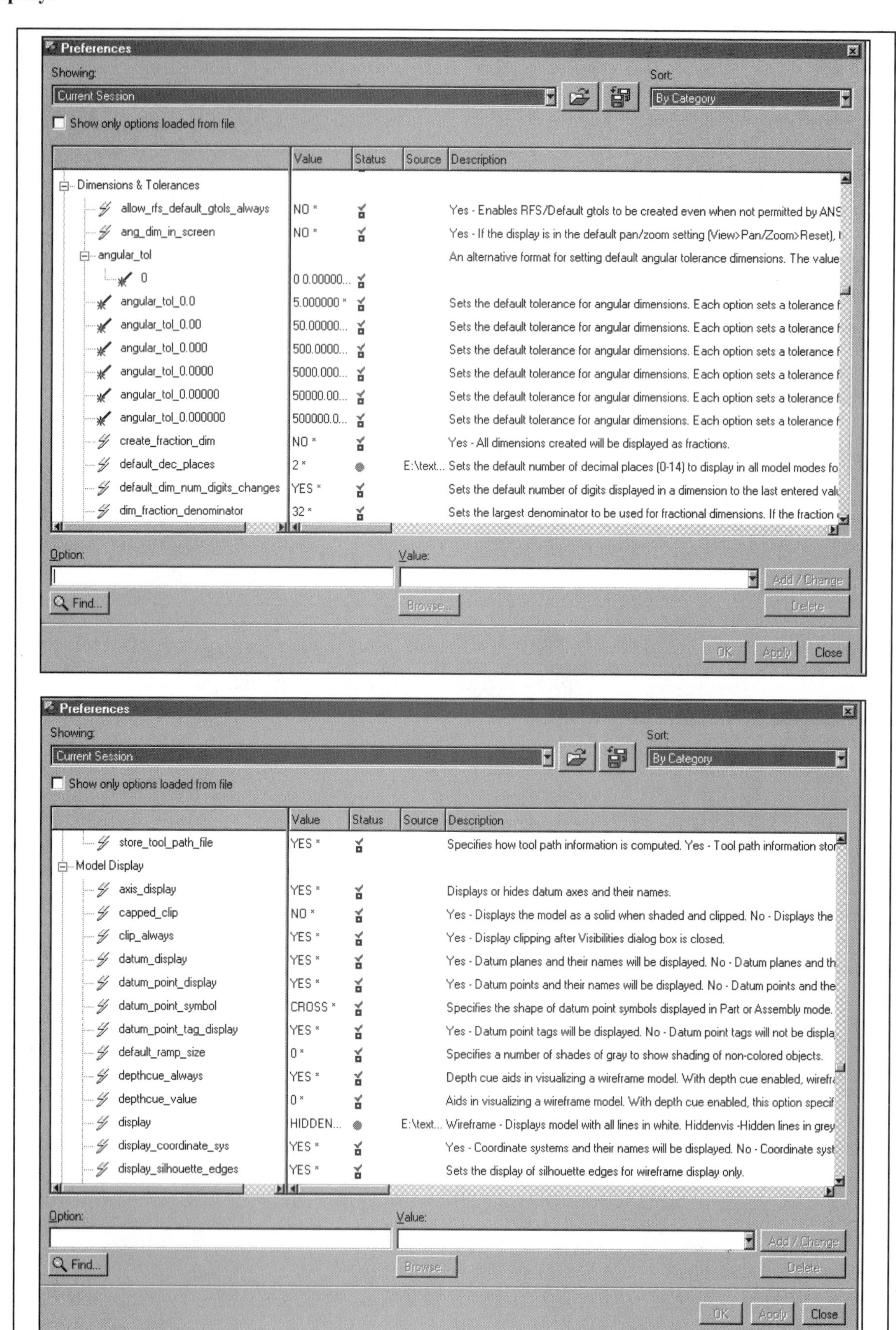
Preferences
Showing:
Current Session
Sort:
By Category
Show only options loaded from file
Value
Status
Source
Description
Dimensions & Tolerances
allow_rfs_default_gtols_always
NO *
Yes - Enables RFS/Default gtols to be created even when not permitted by ANS
ang_dim_in_screen
NO *
Yes - If the display is in the default pan/zoom setting (View>Pan/Zoom>Reset),
angular_tol
An alternative format for setting default angular tolerance dimensions. The value
0
0 0.00000...
angular_tol_0.0
5.000000 *
Sets the default tolerance for angular dimensions. Each option sets a tolerance f
angular_tol_0.00
50.00000...
angular_tol_0.000
500.0000...
angular_tol_0.0000
5000.000...
angular_tol_0.00000
50000.00...
angular_tol_0.000000
500000.0...
create_fraction_dim
NO *
Yes - All dimensions created will be displayed as fractions.
default_dec_places
2 *
E:\text...
Sets the default number of decimal places (0-14) to display in all model modes fo
default_dim_num_digits_changes
YES *
Sets the default number of digits displayed in a dimension to the last entered val
dim_fraction_denominator
32 *
Sets the largest denominator to be used for fractional dimensions. If the fraction
Option:
Value:
Add / Change
Find...
Browse...
Delete
OK
Apply
Close
Preferences
Showing:
Current Session
Sort:
By Category
Show only options loaded from file
Value
Status
Source
Description
store_tool_path_file
YES *
Specifies how tool path information is computed. Yes - Tool path information sto
Model Display
axis_display
YES *
Displays or hides datum axes and their names.
capped_clip
NO *
Yes - Displays the model as a solid when shaded and clipped. No - Displays the
clip_always
YES *
Yes - Display clipping after Visibilities dialog box is closed.
datum_display
YES *
Yes - Datum planes and their names will be displayed. No - Datum planes and th
datum_point_display
YES *
Yes - Datum points and their names will be displayed. No - Datum points and the
datum_point_symbol
CROSS *
Specifies the shape of datum point symbols displayed in Part or Assembly mode.
datum_point_tag_display
YES *
Yes - Datum point tags will be displayed. No - Datum point tags will not be displa
default_ramp_size
0 *
Specifies a number of shades of gray to show shading of non-colored objects.
depthcue_always
YES *
Depth cue aids in visualizing a wireframe model. With depth cue enabled, wirefr
depthcue_value
0 *
Aids in visualizing a wireframe model. With depth cue enabled, this option specif
display
HIDDEN...
E:\text...
Wireframe - Displays model with all lines in white. Hiddenvis -Hidden lines in grey
display_coordinate_sys
YES *
Yes - Coordinate systems and their names will be displayed. No - Coordinate syst
display_silhouette_edges
YES *
Sets the display of silhouette edges for wireframe display only.
Option:
Value:
Add / Change
Find...
Browse...
Delete
OK
Apply
Close

Appendix C

Glossary

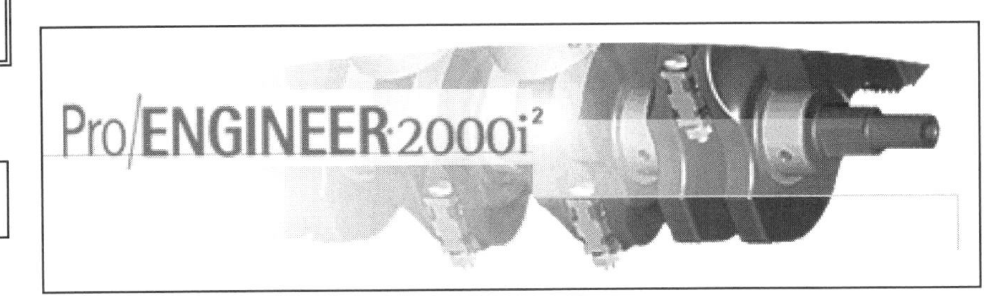

Active Window The window to which the displayed menus apply. To change the active window, choose **Window** from the menu bar and pick **Activate.**
Align Aligns a sketched entity with a part or assembly edge and is used as an assumption when you are solving the section.
Attributes (feature) The various characteristics of a feature. For example, a blind hole will have the following attributes: *position,* how the hole is placed (linearly, radially, coaxially); *section,* the cross section that defines the shape of the blind hole; *intersection,* simple or complex.
Attributes (user) The attributes that are added by a user to supply a description of the object beyond the geometric definition. For example, stock number, price, and cost per unit are all user attributes.
Base Feature The very first feature created for a part.
Base Member The very first component of an assembly.
Child An item, such as a view, part, or feature, that is dependent on another item for its existence. *See also* **Parent.**
Collinear Two or more objects that occur along the same axis.
Configuration File A special text file that contains default settings for many Pro/ENGINEER functions. Default environment, units, files, directories, and so on are set when Pro/ENGINEER reads this file when it is started. A configuration file can reside in the startup directory to set the values for your working session only, or it can reside at the *load point directory* to set values for all users running a given version of Pro/ENGINEER. Also known as the *config.pro* file.
Dependent Parameter A parameter (dimension or user-defined) in a relation that is defined as a function of other parameters and values. A relation has one and only one dependent parameter, and it *always* appears on the left side of the equals sign.
Dimmed Option An option that appears grayed out when a menu is displayed. You cannot choose menu items when they are dimmed.
Flip Arrow An arrow that appears on surfaces and edges of features to specify in which direction an operation should grow or project. If the arrow is pointing in the proper direction, choose **Okay**. If not, choose **Flip** and the arrow will be flipped **180°**. Then choose **Okay**.
Independent Parameter A parameter (dimension, value, relation) in a relation that is used to specify the value of the dependent parameter. Independent parameters *always* appear on the right side of the equals sign.

Information Window A Pro/ENGINEER window that displays information: object lists, mass properties, BOMs, and so on.
Macro Keys The keyboard function keys or key sequences for which you predefine a menu option or sequence of menu options. This allows you to pick these keys to perform frequently used menu sequences.
Main View The first view added to a drawing.
Main Window The large primary window (main graphics window) created by Pro/ENGINEER.
Menu A list of options presented by Pro/ENGINEER that you select using the mouse or predefined macro keys.
Mode An environment in which Pro/ENGINEER allows you to perform closely related functions (e.g., Drawing, Sketcher).
Model A part or an assembly.
Object A Pro/ENGINEER object can be a drawing, a part, an assembly, a layout, and so on.
Parent An item that has other items dependent on it for their existence. For example, the base feature will have all other features dependent upon it. If a parent is deleted, all dependent items (**children**) will be deleted.
Part Type A part can be either standard (standard off-the-shelf component from the library) or nonstandard. The part type affects the number of times a part can be assembled.
Pattern A method of feature creation in which one construction feature is used to create several related features.
Pick To select an item onscreen by clicking the left mouse button.
Placement Plane The plane on which a construction feature is located for placement.
Regenerate A menu option that recreates a model, incorporating any changes that have been made since the last time the model was stored or regenerated.
Startup Directory The directory from which you started Pro/ENGINEER.
System Editor The text set of editing functions available within your operating system.
Toggle A menu option that lets you switch between two settings; for example, **Flip Arrow.**
Trail File A record of all the menu picks, screen picks, keyboard entries, and so on that occur during a Pro/ENGINEER session. The trail file can be run to re-create a work session.
Working Directory The working directory is the directory where you store and retrieve your Pro/ENGINEER files.
Work Session The period between your starting and stopping Pro/ENGINEER.
View A particular display of the model. View parameters include view orientation matrix, center, and scale.
Zoom Magnify or reduce your view of an object currently on the screen.

Appendix D

Design Intent Planning Sheets (DIPS)

This appendix provides a variety of sketching formats for planning your design. The **design intent** of a feature, a part, or an assembly (or even a drawing) should be established before any work is started with Pro/ENGINEER. Here we have provided a number of different formats in which to sketch and plan your feature, part, assembly, or drawing. Copy the sheets so that you have a number available for each lesson in the text. ***Inch, metric, and isometric engineering grid paper can be substituted for the DIPS.***

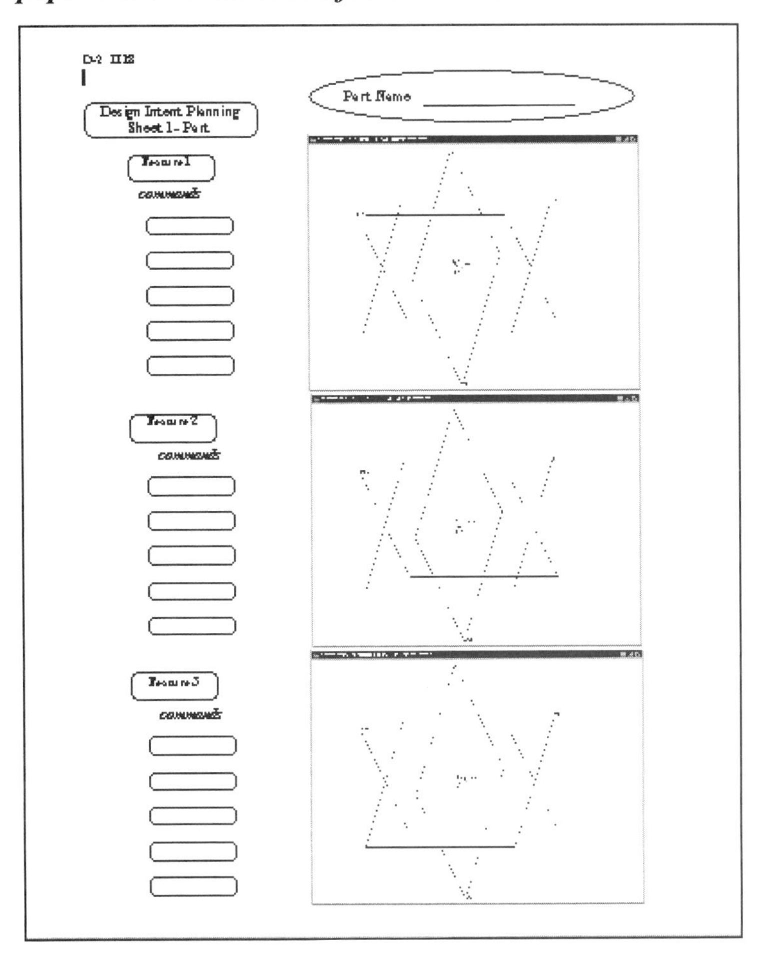

This appendix contains the following sheets:

DIPS1 Part Planning Sheet (trimetric)

DIPS2 Part Planning Sheet (pictorial with axes)

DIPS3 Assembly Planning Sheet (trimetric)

DIPS4 Assembly Planning Sheet (pictorial with axes)

DIPS5 Drawing Planning Sheet (no format)

DIPS6 Drawing Planning Sheet (format)

DIPS7 Feature/Sketch Planning Sheet (three sketches)

DIPS8 Feature/Sketch Planning Sheet (two sketches)

DIPS9 Part Model Trees & Assembly Model Trees

Part Name ______________________

Design Intent Planning Sheet 1 for Parts

Feature 1

commands

Feature 2

commands

Feature 3

commands

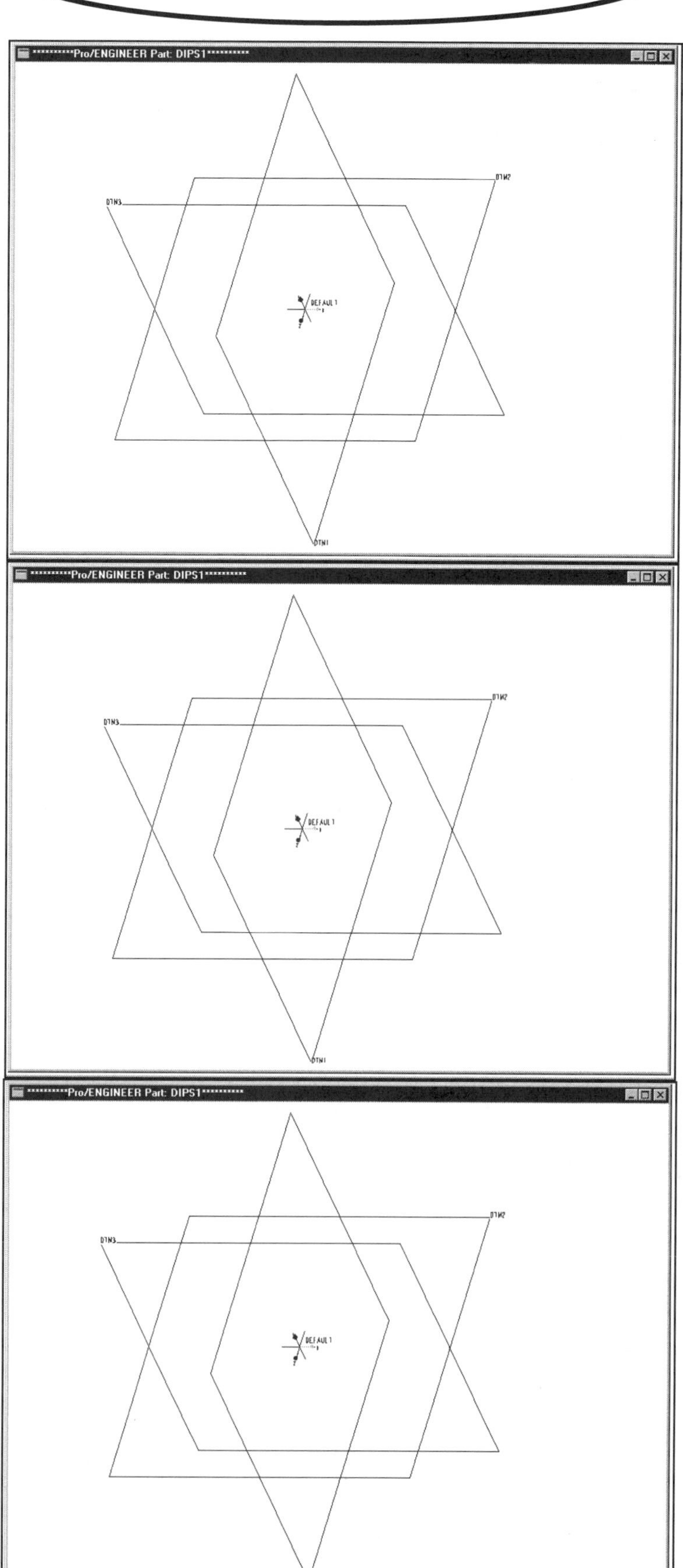

Part Name ____________________

Design Intent Planning Sheet 2 for Parts

Feature 1

commands

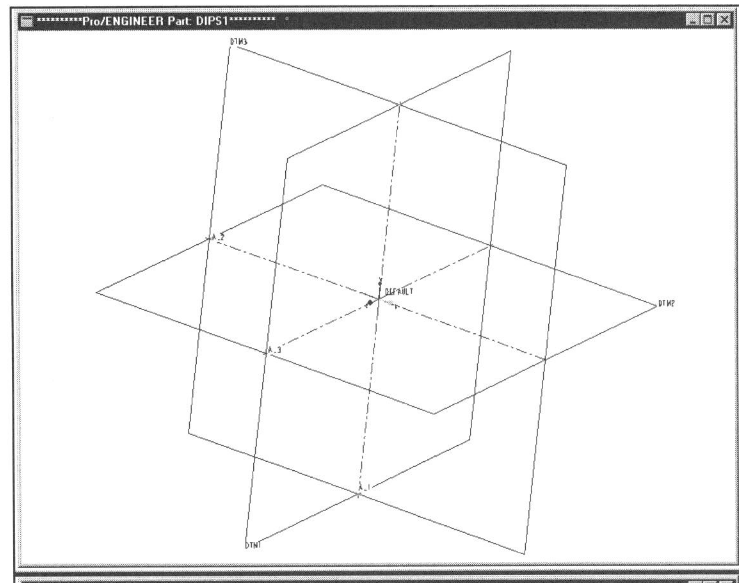

Feature 2

commands

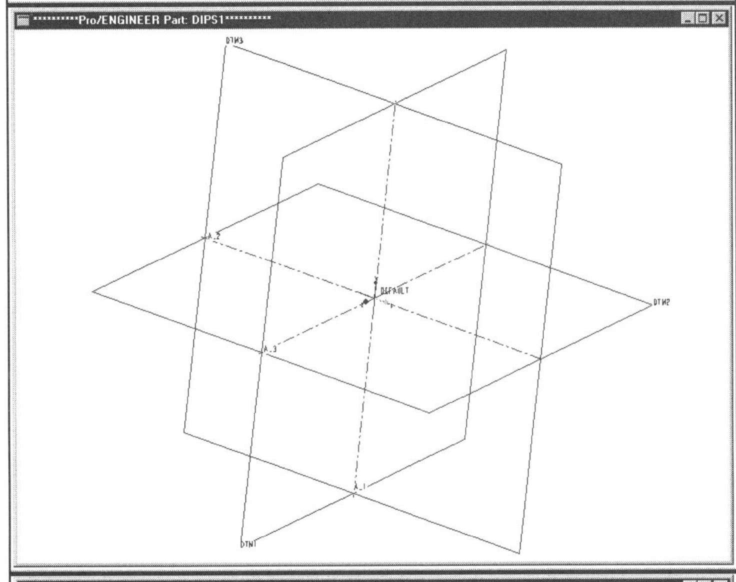

Feature 3

commands

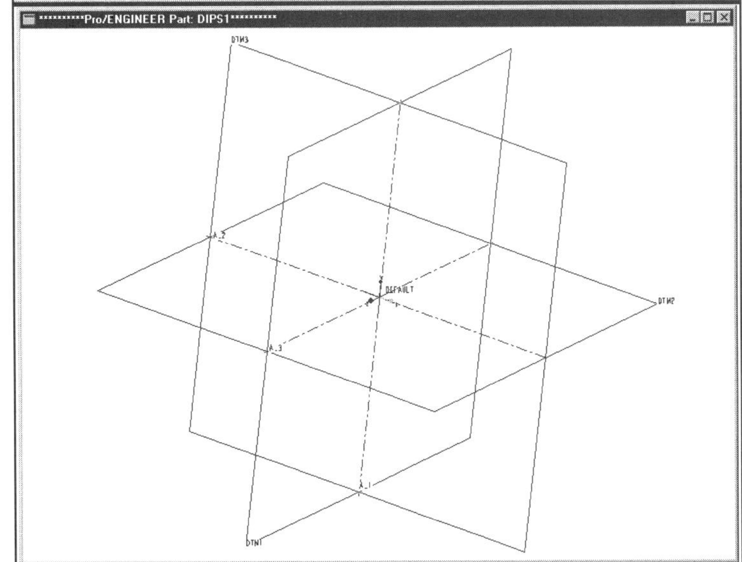

Design Intent Planning Sheet 3 for Assemblies

Assembly Name ____________________

Component

constraints

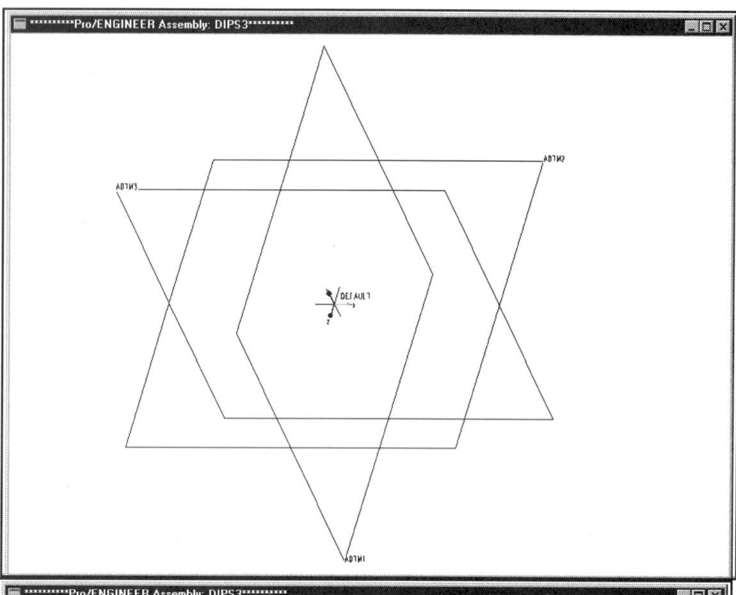

Component

constraints

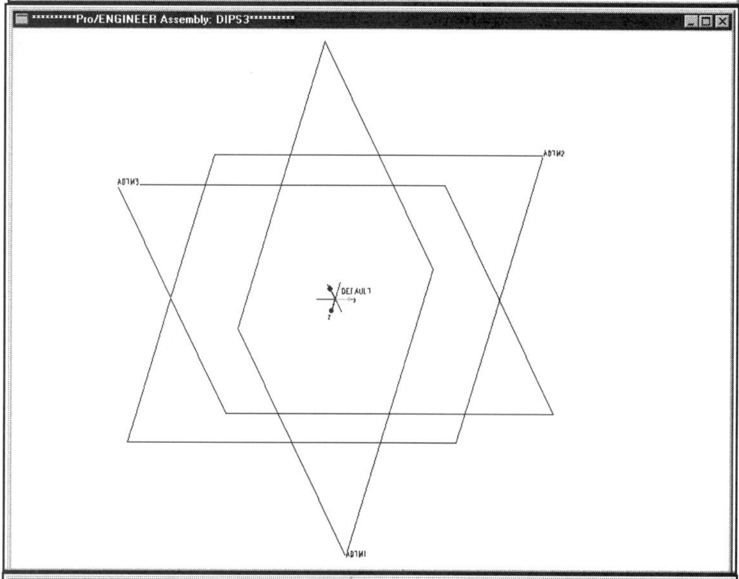

Component

constraints

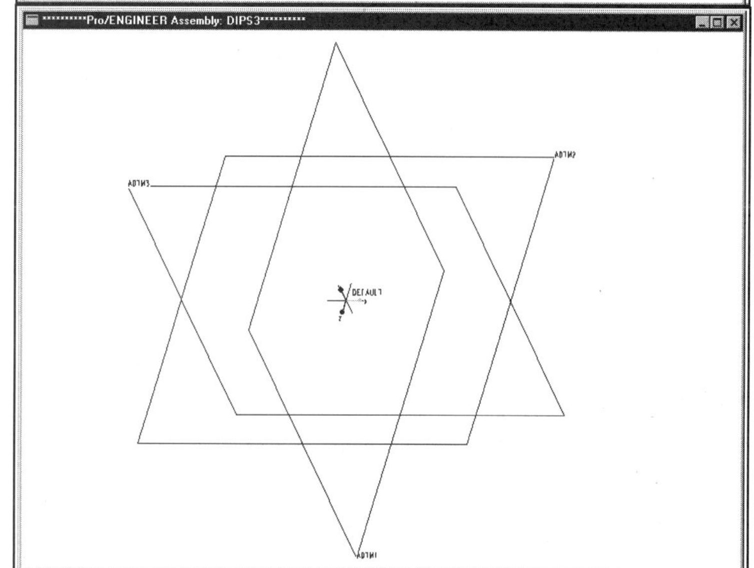

Design Intent Planning Sheet 4 for Assemblies

Assembly Name ____________________

Component

constraints

Component

constraints

Component

constraints

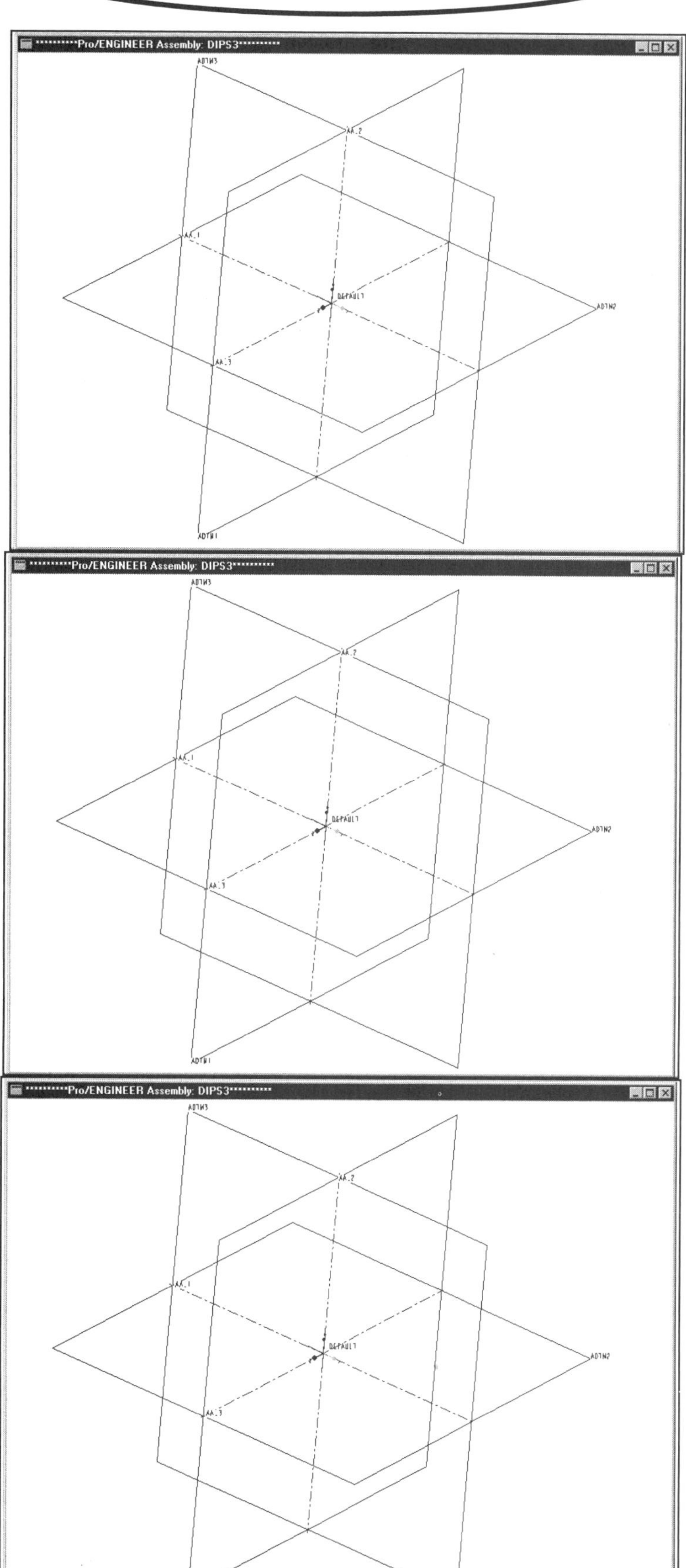

Design Intent Planning Sheet 5 for Drawings

Drawing Name ____________________

Part/Assembly

Specifications

Pro/ENGINEER Drawing: DIPS5

TYPE · DRAFT NAME · NONE SIZE · C

NOTES:

Pro/ENGINEER Drawing: DIPS5

TYPE · DRAFT NAME · NONE SIZE · C

Design Intent Planning Sheet 6 for Drawings

Drawing Name ____________________

Part/Assembly

Specifications

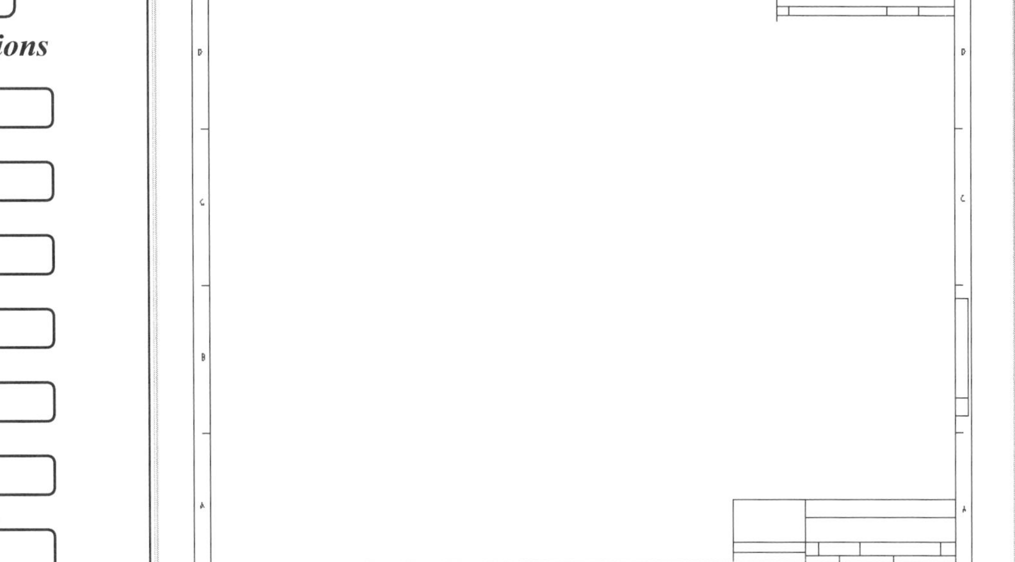

NOTES:

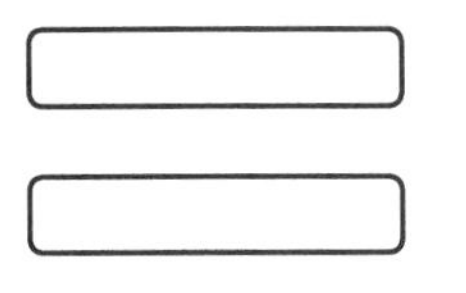

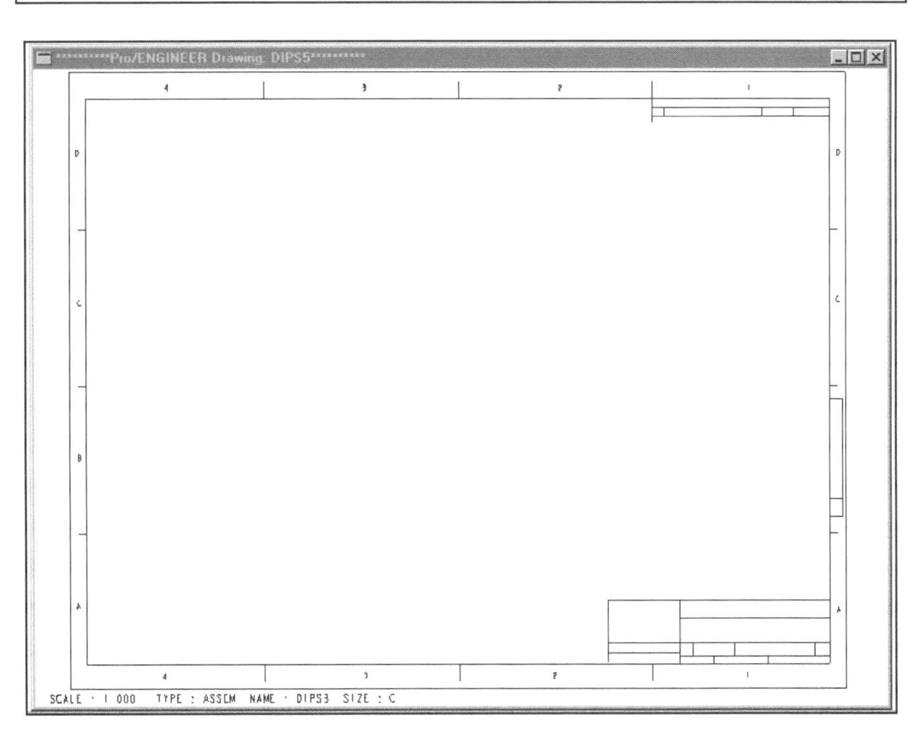

Design Intent Planning Sheet 7 for Features/Sketches

Part/Section Name ____________________

Sketch/Section 1

Sketch/Section 2

Sketch/Section 3

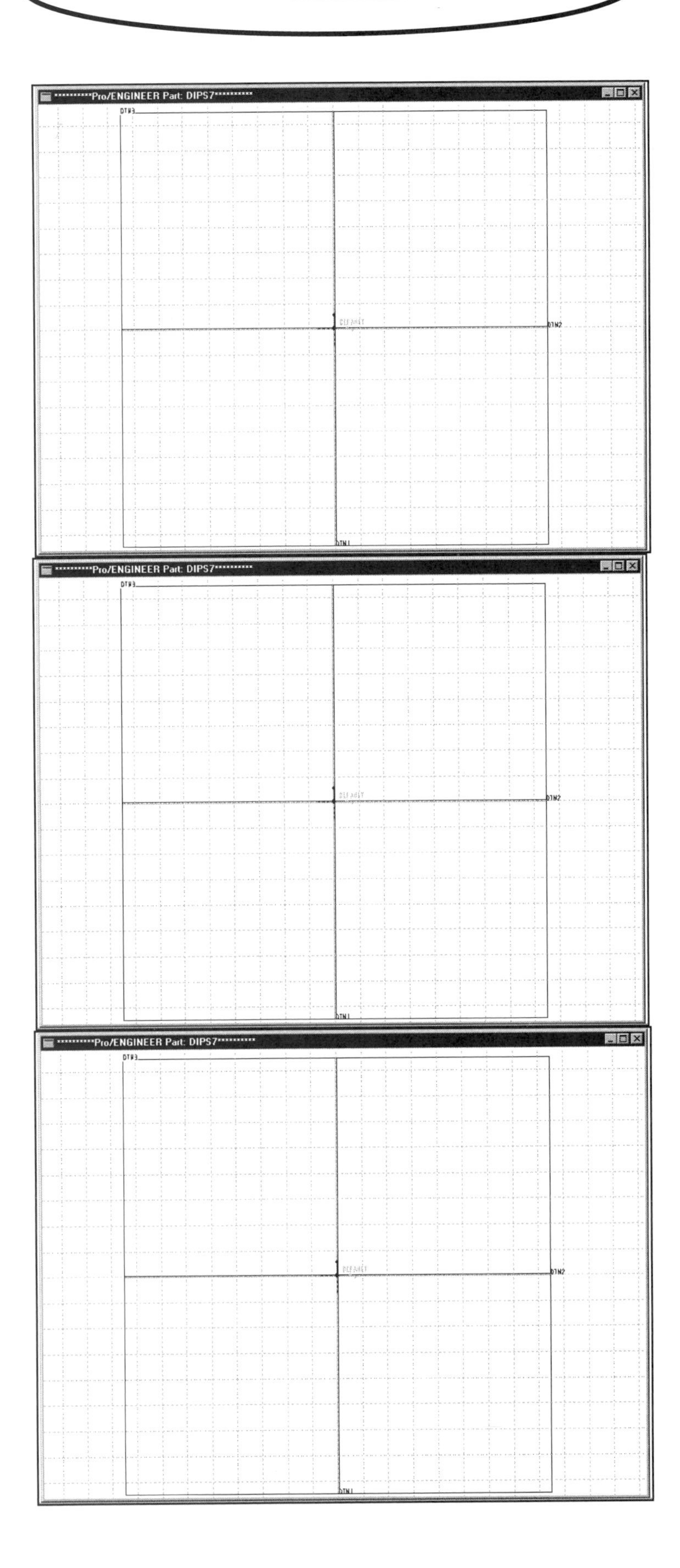

Part/Section Name____________________

Design Intent Planning Sheet 8 for Features/Sketches

Sketch/Section 1

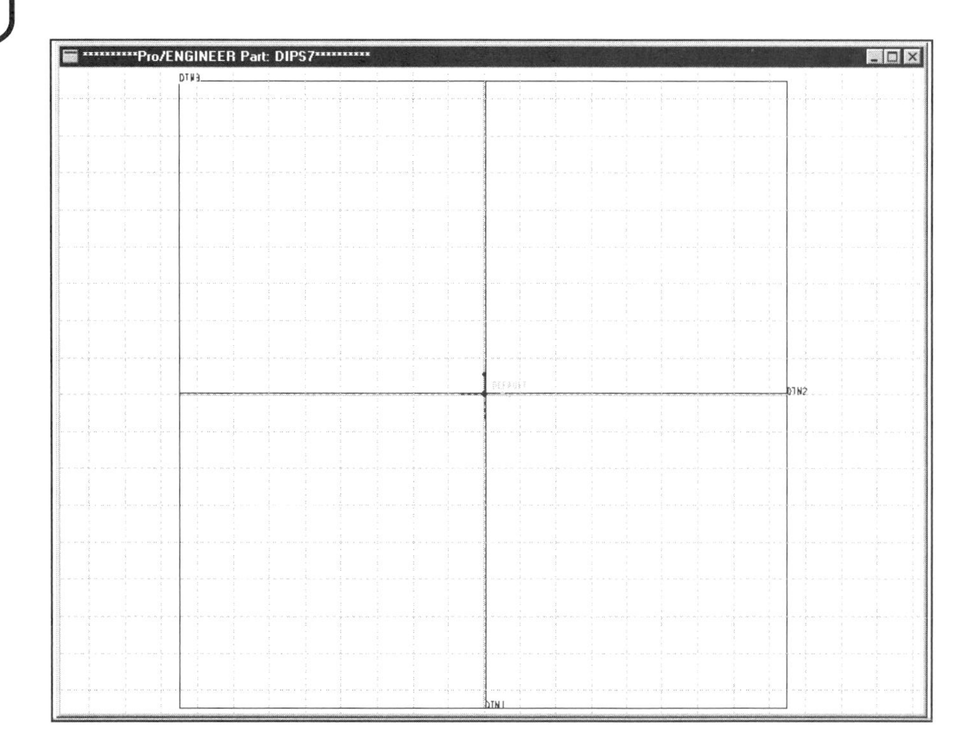

NOTES:

Sketch/Section 2

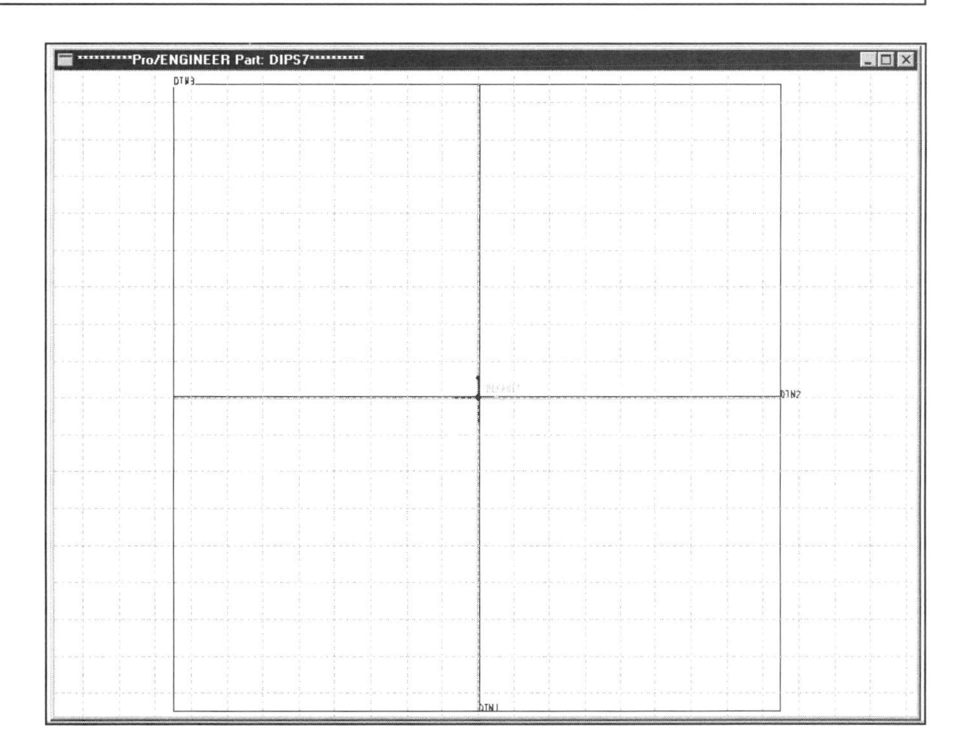

Design Intent Planning Sheet 9
for
Part Model Tree &
Assembly Model Tree

Part Name = ____________________
Assembly Name = ____________________
Subassembly Name = ____________________

Part Model Tree

Part Features

 # Type **Name**

1. default prt datum (TOP) = ____________________
2. default prt datum (FRONT) = ____________________
3. default prt datum (RIGHT) = ____________________
4. default csys (PRT CSYS DEF) = ____________________
5. first protrusion = ____________________
6. ____________________
7. ____________________
8. ____________________
9. ____________________
10. ____________________
11. ____________________
12. ____________________
13. ____________________
14. ____________________
15. ____________________
16. ____________________
17. ____________________
18. ____________________
19. ____________________
20. ____________________
21. ____________________
22. ____________________
23. ____________________
24. ____________________
25. ____________________
26. ____________________
27. ____________________
27. ____________________
29. ____________________
30. ____________________
31. ____________________
32. ____________________
33. ____________________
34. ____________________
35. ____________________
36. ____________________
37. ____________________

Assembly Model Tree

Assembly Features

 # Type **Name**

1. default asm dtm (TOP) = ____________________
2. default asm dtm (FRONT) = ____________________
3. default asm dtm (RIGHT) = ____________________
4. default csys (ASM CSYS DEF) = ____________________
5. ____________________
6. ____________________
7. ____________________

Assembly Components

1. ____________________
2. ____________________
3. ____________________
4. ____________________
5. ____________________
6. ____________________
7. ____________________

Subassembly Features

 # Type **Name**

1. default subasm dtm (TOP) = ____________________
2. default subasm dtm (FRONT) = ____________________
3. default subasm dtm (RIGHT) = ____________________
4. default subasm csys (ASM CSYS DEF) = ____________________
5. ____________________
6. ____________________
7. ____________________

Subassembly Components

1. ____________________
2. ____________________
3. ____________________
4. ____________________
5. ____________________
6. ____________________

Index

D

E

F

N

O

P

Q

R

S

T

U

V

W

X

Y

Z

Config.pro and *.dtl* references